Daniel Bland

Touch + Dic

Human Anatomy and Physiology

FOURTH EDITION

FOURTH EDITION

Human Anatomy and Physiology

Alexander P. Spence, Ph.D.
STATE UNIVERSITY OF NEW YORK COLLEGE AT CORTLAND

Elliott B. Mason, Ph.D.
STATE UNIVERSITY OF NEW YORK COLLEGE AT CORTLAND

WEST PUBLISHING COMPANY
Saint Paul New York Los Angeles San Francisco

To my wife Marion
and our children Mark, Carol, and Cindy
ALEXANDER P. SPENCE

To my wife Marsha
and our children Jennifer, Julie, and Jessica
ELLIOTT B. MASON

Their love, patience, encouragement, and understanding made this book possible

PRODUCTION CREDITS

Composition: Parkwood Composition
Copyediting: Patricia Lewis
Indexing: E. Virginia Hobbs
Illustrations: Rolin Graphics
Interior design: K. M. Weber
Cover design: K. M. Weber
Cover photograph: Myron, Discobolos, Firenze, Pal. Vecchio. Scala/Art Resource, New York
Appendix 1 photographs: Reproduced by permission from *Photographic Atlas of the Human Body* by Drs. B. Vidić and F. R. Suarez. Copyright © 1984 by Mosby Year Book, Inc.

◆

Printed in the United States of America
99 98 97 96 95 94 93 92 8 7 6 5 4 3 2 1 0

Library of Congress Cataloging-in-Publication Data

Spence, Alexander, P., 1929—
Human anatomy and physiology/Alexander P. Spence, Elliott B. Mason.—4th ed.
p. cm.
Includes index.
ISBN 0-314-87693-6 (hard)
1. Human physiology. 2. Human anatomy. I. Mason, Elliott B., 1943- . II. Title.
[DNLM: 1. Anatomy. 2. Physiology. O$ 4 S744h]
QP34.5.S72 1992
612-dc20
DNLM/DLC 91-21622
for Library of Congress CIP

CONTENTS IN BRIEF

CONTENTS

◆ CHAPTER 8 ◆
The Muscular System: General Structure and Physiology 242

◆ CHAPTER 9 ◆
The Muscular System: Gross Anatomy 278

CHAPTER 19
The Cardiovascular System: The Heart 592

CHAPTER 20
The Cardiovascular System: Blood Vessels 626

◆ CHAPTER 21 ◆
The Lymphatic System 674

◆ CHAPTER 22 ◆
Defense Mechanisms of the Body 686

◆ CHAPTER 23 ◆
The Respiratory System 716

◆ CHAPTER 24 ◆
The Digestive System 754

CHAPTER 25
Metabolism, Nutrition, and Temperature Regulation 802

CHAPTER 26
The Urinary System 832

PREFACE

Working on a new edition of a textbook is always a rewarding experience. A new edition provides the opportunity to update information and revise the manner in which it is presented. This fourth edition of *Human Anatomy and Physiology* is especially exciting because we are working with a new publisher. West Publishing Company has committed themselves to a major upgrading of our book and stimulated a renewed vigor for the project.

Our major goal for this edition remains the same as in previous editions: To provide a textbook that is written at the introductory level and yet is comprehensive enough to meet the needs of diverse groups of human anatomy and physiology students. We have found that courses in human anatomy and physiology vary considerably in length, depth, and objectives. However, we believe our book provides enough information, without being overwhelming, to make it suitable for most courses.

Our book is written primarily for students who are preparing for careers in physical education, nursing, or various health-related professions. The book is also well suited for students majoring in the liberal arts. Because students taking anatomy and physiology courses have such diverse backgrounds and goals, we have assumed that the students using our text will have only a general familiarity with science. All the information needed to understand anatomy and physiology at the introductory level is presented in the text. In some instances we have placed certain topics in special boxes (entitled "A Closer Look") and examined them in greater detail than is generally done in anatomy and physiology texts. We consider this in-depth information to be optional, but believe it will be of interest to the more motivated students.

As the book's title indicates, it is designed for courses that teach anatomy and physiology at the same time. Consequently, we have made a concerted effort to integrate structure and function throughout the text. In addition, we emphasize that structure and function complement one another. It is important for students to realize that the structures of the various organs and other components of the body are specifically suited to contribute to their optimal functioning.

New in This Edition

New Art Program

This edition contains an entirely new art program. Every figure has been redrawn, and most photomicrographs are new. The art program from the previous three editions has been carefully modified so that each illustration depicts precisely what it is designed to show. In many cases, the new figures closely follow the detail shown in the figures in previous editions, which were generally well received. Retaining the detail of these proven figures while presenting them in an entirely new style has been a rare opportunity.

Increased Use of Color

The art program has been expanded to full color throughout the book. We have chosen pastel colors to avoid the rather garish appearance of some recent textbooks. We have also worked closely with the artists to ensure that color is used in a functional manner that serves as a learning aid—that is, to focus attention, differentiate related structures, or guide the viewer through a complex illustration—rather than simply for its own sake.

The use of color is consistent throughout the book. The color used to illustrate a particular structure in early chapters is used to illustrate the same structure in later chapters.

Updated Physiology

The sections of the text dealing with physiology have been extensively rewritten and updated. Many of the sections have been modified to make them more easily understood by students in an introductory course.

Aspects of Exercise Physiology

Many students enrolled in human anatomy and physiology courses are preparing for careers in physical education. In addition, the general public is increasingly aware of the value of exercise for the maintenance of health. To provide some interesting and pertinent information about the effects of exercise on specific body systems, we have included *Aspects of Exercise Physiology* boxes in most chapters. Each box discusses an effect of exercise that is probably familiar to most people and explains why the body responds to the exercise in the manner that it does.

Special Features

This edition retains several special features that were well received in the previous editions. These features enhance the text's usefulness as a teaching tool and increase student interest.

Integration of Embryology

We have found that structural and functional relationships within the body are understood better when students have some knowledge of embryonic development. For this reason, the discussion of each body system begins with a brief consideration of the embryonic development of the system. These discussions are self-contained and, if the instructor chooses, may be omitted without detracting from the remainder of the text.

Conditions of Clinical Significance

The emphasis throughout the text is on normal human anatomy and physiology. However, brief discussions of diseases, dysfunctions, and aging are included when they enhance and reinforce an understanding of normal structure and function. These discussions appear in separate *Conditions of Clinical Significance* boxes. This special treatment allows instructors to emphasize or deemphasize these discussions according to the objectives of their particular courses.

Clinical Correlations

In several chapters, we have included *Clinical Correlation* boxes that are of special interest to students majoring in nursing, premedicine, and the allied health sciences. Each box presents an actual case history of a patient suffering from a condition that illustrates basic physiological principles.

Regional Anatomy Appendix

Regional relationships are a significant aspect of human anatomy. Consequently, a regional anatomy appendix composed of full-color, labeled photographs of dissected cadavers appears at the end of the book (Appendix 1, page A1). This appendix will be valuable to anyone wishing to study regional anatomy, especially those with access to a cadaver in the laboratory.

In-Text Learning Aids

Different students learn in different ways. We have, therefore, provided various pedagogical aids to assist the diverse groups of students who take courses in human anatomy and physiology.

Learning Objectives

Each chapter begins with a list of learning objectives that direct the student toward significant aspects of the chapter.

Chapter Contents

Chapter contents placed at the beginning of each chapter list the major topics included in the chapter.

In-Text Pronunciations

When a scientific or technical term is first used in the text, it is accompanied by a phonetic pronunciation.

Study Outline

A study outline at the end of each chapter provides a summary of the chapter. Page references are included in the outline to help students locate information they may want to review.

Self-Quiz

A list of questions in the form of a self-quiz is included at the end of each chapter to help students determine how well they have learned the information in the chapter. The answers to the questions are listed in Appendix 2 (page A26).

Glossary

A glossary, which provides phonetic pronunciations and definitions for over 1,500 terms, is located on page G1.

Metric Appendix

A metric appendix, on page A34, provides metric/English conversion constants.

Supplements

Instructor's Manual/Test Bank

This supplement includes an extensive testing program to accompany *Human Anatomy and Physiology,* fourth edition. Each chapter includes 75 multiple choice, fill-in-the-blank, or true-false questions, some with anatomical line art, and at least five essay/thought questions, all arranged by level of difficulty. Also included is an extensive and newly updated guide to resources for teaching the course and course related media. This supplement is prepared by Professor Beth Howard of Rutgers University.

Computerized Testing

WESTEST is a full-featured microcomputer implementation of the test bank, which is available to qualified adopters. Available on both the Macintosh and IBM formats, WESTEST for Spence and Mason, fourth edition is fully editable, and is accompanied by clip art of line drawings based on anatomical illustrations in the text.

Study Guide

Newly developed for this edition by Alease Bruce of the University of Massachusetts at Lowell, this book is an interactive ancillary for sale to students, providing them with skills for self-study, drill exercises with answers, and "coloring exercises" to reinforce comprehension of the textual material.

Transparencies

This edition includes 200 full-color acetates based on art and tables from the text, with type reset for effective in-class use. It is available to qualified adopters.

Laboratory Manuals

Two versions of the laboratory manuals are available. Designed to parallel the text, they can be used independently with *any* book. Each of the laboratory manuals is accompanied by an instructor's manual outlining procedures, equipment, and suggested answers to exercises.

Laboratory Manual for Anatomy and Physiology with Dissection Guide for the Cat, by Katherine Malone and Jane Schneider, is a comprehensive manual designed to cover all aspects of the course. Qualified adopters can also receive a 20-minute videotape of a cat dissection, produced by Carolina Biological Supply Company.

Laboratory Manual for Anatomy and Physiology with Dissection Guide for the Fetal Pig, by Katherine Malone, Jane Schneider, and Eileen Walsh, is a comprehensive manual designed to cover all aspects of the course. Qualified adopters can also receive a 20-minute videotape of a pig dissection, produced by Carolina Biological Supply Company.

Video Library for Class Use

Qualified adopters have access to the extensive library of the film series: "The Human Body," developed by Films for the Humanities and Sciences.

Videodisc

Developed specifically for the fourth edition of Spence and Mason, this videodisc contains extensive visual resources to allow instructors to draw together all aspects of this course into coherent and integrated presentations. Included on the disc are all the images from the parent text, human cadaver dissection photos, all the images from the lab manuals, images from "The Human Body" produced by Films for the Humanities and Sciences, Mayo Clinic images utilizing the ANALYZE system, highly animated sequences of anatomical and physiological processes, histology slides, and more. To facilitate ease of use, the videodisc is accompanied by a guide to its use and bar code labels. It is available to qualified adopters.

Acknowledgments

We believe that one reason our text has been so successful for the past twelve years is that each edition has been extensively reviewed, and we have taken the reviewers' suggestions seriously. More than 45 people read all or part of the manuscript for this fourth edition, and their suggestions have been extensively incorporated into it. We are very grateful to the following for their thorough reviews and their many helpful suggestions;

Donna Alder
Roberts Wesleyan College

Barry Anderson
University of Scranton

Tom Baldus
North Dakota State College of Science

William Belzer
Clarion University

Jeffrey H. Black
East Central University

Robert J. Boettcher
Lane Community College

James A. Bridger
Prince George's Community College

Alan H. Brush
University of Connecticut

Ray D. Burkett
Shelby State Community College

Jerry Button
Portland Community College

Cynthia Carey
University of Colorado-Boulder

A. Carey Carpenter
Mt. Hood Community College

Anthony Chee
Houston Community College

Jean Cons
College of San Mateo

Philip Cooper
Suffolk Community College

Irene M. Cotton
Lorain County Community College

Darrell Davies
Kalamazoo Valley Community College

Gerald Dotson
Front Range Community College

William E. Dunscombe
Union County College

Douglas B. Fonner
Ferris State University

George Fortunato
Nassau Community College

John L. Frehn
Illinois State University

Greg Garman
Centralia College

Norman Goldstein
California State University-Hayward

Judy Goodenough
University of Massachusetts-Amherst

Donald W. Green
New Mexico Junior College

James E. Hall
Central Piedmont Community College

Ann Harmer
Orange Coast College

H. Kendrick Holden
Northern Essex Community College

Reinhold Hutz
University of Wisconsin-Milwaukee

R. Bruce Judd
Edison Community College

Jerry T. Justus
Arizona State University

Kenneth Kaloustian
Quinnipiac College

Joseph R. Koke
Southwest Texas State University

Linda Kollett
Massasoit Community College

Alan S. Kolok
University of Colorado-Boulder

Charles Leavell
Fullerton College

Harvey Liftin
Broward Community College

Linda L. MacGregor
Bucks County Community College

Kathryn Malone
Westchester Community College

Joseph W. McDaniel
Vermont College

Randall M. McKee
University of Wisconsin-Parkside

A. Kenneth Moore
Seattle Pacific University

Robert Nabors
Tarrant County Junior College

W. Brian O'Connor
University of Massachusetts-Amherst

Ann Marie Olson
Bunker Hill Community College

Patricia O'Mahoney-Damon
University of Southern Maine

Steven J. Person
Lake Superior State University

Michael Postula
Parkland College

Ralph E. Reiner
College of the Redwoods

James W. Russell
Georgia Southwestern College

Mary Schwanke
University of Maine-Farmington

David S. Smith
San Antonio College

Carol Spaulding
University of Maryland

Eugene Volz
Sacramento City College

Elizabeth Walker
West Virginia University

Edith Wallace
William Paterson College

Edward P. Wallen
University of Wisconsin-Parkside

James F. Waters
Humboldt State University

Richard E. Welton
Southern Oregon State College

Barry J. Wicklow
St. Anselm College

Clarence C. Wolfe
Northern Virginia Community College

One improvement in this edition that will be noticed immediately by anyone familiar with our previous editions is the beautiful full-color art program. We are delighted with the work done by Rolin Graphics, and we appreciate the patience shown by their artists as we made, what must have seemed to them, endless revisions in the artwork.

We also want to express our appreciation to Dr. Rachel Yeater, professor, Sports Exercise Program, and director, Human Performance Laboratory, School of Physical Education, West Virginia University for providing the Aspects of Exercise Physiology boxes used in the text. They indeed add a new dimension to the study of anatomy and physiology.

It has been our good fortune to work with West Publishing Company, a truly outstanding publisher. The people at West are enthusiastic, energetic, and full of ideas. We are particularly indebted to Ron Pullins, acquiring editor, for professionally and expertly overseeing all aspects of the preparation of this fourth edition. We thank Denise Bayko, developmental editor, for organizing the most effective reviewing process we have ever encountered, and Pat Lewis, freelance copyeditor, for her efficient editing of the manuscript. We are grateful to Laura Nelson, who conducted the photo research, for finding just the right photos to complement the text, and to Kristen Weber, freelance interior designer, for giving the book such a pleasing and effective appearance. We extend our very special thanks to the production editor, Deanna Quinn. We stand in awe of her ability to handle the many details and difficulties involved in the preparation of this book and still keep the project on schedule. To all these people, it seems inadequate to merely say thanks. Each one has contributed in a significant way to this text.

Alexander P. Spence
Elliott B. Mason

Department of Biological Sciences
State University of New York College at Cortland

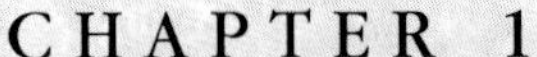

CHAPTER 1

Introduction to Anatomy and Physiology

CHAPTER CONTENTS

LEARNING OBJECTIVES

After completing this chapter, you should be able to:

1. Distinguish between anatomy and physiology.
2. Cite several examples of the interrelationship of structure and function in the body.
3. Name four types of tissues produced by the three embryonic cell layers.
4. Name the major organ systems in the human body.
5. Describe common directional and regional terms.
6. Name the planes and cavities of the body.
7. Distinguish between the parietal and the visceral membranes of the ventral body cavities.
8. Describe the mesenteries of the abdominopelvic cavity.
9. Describe what is meant by homeostasis.
10. Explain positive and negative feedback mechanisms.

CHAPTER 1

One of the most exciting and meaningful accomplishments that a person can strive for is to gain an understanding of his or her body. The main goals of this book are to develop in the reader an understanding of how the human body is constructed and how it functions, and to correlate structure with function. Along with this factual knowledge, we hope to instill in the reader an appreciation of what a marvelous organism the human body is. With the rapid advances that have been made in scientific knowledge in recent years—particularly in the health sciences—and the prominent coverage given these advances in newspapers and popular magazines, it has become increasingly important for everyone to know more about the human body.

Anatomy is the study of the *structure* of an organism and the relationship among its parts. The term *anatomy* is derived from the Greek words meaning "apart" and "to cut." As this derivation indicates, anatomy is based largely on dissection of the body. In some of the newer fields of anatomy, however, the use of electron microscopes and other instruments provides valuable supplements to dissection. **Physiology** is the study of the *functions* of a living organism. It attempts to explain in physical and chemical terms the factors and processes involved in these functions.

Fields of Anatomy

The study of anatomy involves examination of the general structures of the body **(gross anatomy)** as well as those structures that can be seen only with the aid of a microscope **(microscopic anatomy).** Gross anatomy can be studied by regions, such as the head, neck, thorax, abdomen, pelvis, or limbs. This approach, referred to as **regional anatomy,** is often used in dissection, in which all structures in a region are studied simultaneously. For our purposes, however, the most helpful approach is **systemic anatomy,** which is organized around organ systems that perform common functions. This book uses that approach. It is often useful to study anatomy by identifying underlying structures or regions whose contours are visible at the surface of the body. This approach is called **surface anatomy.** Microscopic anatomy includes the study of cells *(cytology)* and the study of tissues *(histology).* When anatomy is studied under the extremely high magnifications possible with the electron microscope, it is referred to as *fine structure* or *ultrastructure.* **Developmental anatomy,** another subdivision of anatomy, focuses on the development of the body from the fertilized egg to the adult form. Developmental anatomy includes *embryology,* which is limited to prenatal development.

Fields of Physiology

Physiology is a vast field, with many subdivisions. The subdivisions are often concerned with different levels of organization—from the subcellular to the multicellular. The subdivisions include specialties such as *viral physiology, bacterial physiology, cellular physiology, plant physiology,* and *animal physiology.* In this text we are primarily interested in physiology at the organ level in humans. To study *human physiology,* however, even at the organ level, it is necessary to study certain aspects of cellular physiology.

Scientists whose work is primarily concerned with physiology are referred to as *physiologists.* Human physiologists tend to concentrate their studies on one or another of the body systems. Thus, there are *renal physiologists, reproductive physiologists, neurophysiologists,* and so on. We follow that same approach in this book; the physiology—as well as the anatomy—of each organ system is described as the particular system is studied.

Medical Imaging Methods

The ability to visualize internal body structures is particularly valuable in the diagnosis of disorders and injuries. For many years however, the X-ray and fluoroscope machines were the only instruments available for this purpose. Both instruments produce *X rays (roentgen rays),* which pass through the structure under examination and either expose an X-ray film or illuminate a fluroescent screen. When an exposed film is developed, the result is a photographic image called a *roentgenogram,* commonly known as an X ray. Although X rays are diagnostically useful, they are somewhat limited because the three-dimensional relationships of the body's parts are lost on the film, where the body image is flat. Moreover, small differences in the densities of tissues are not always detectable on an X ray.

Tomography

In the 1970s, technological advances made an imaging technique called **tomography** possible. With tomography, structures at a specific level in the body are clearly depicted, whereas structures above and below this level are blurred. This effect is achieved by rotating an X-ray tube and film around the selected level while repeatedly

◆ **FIGURE 1.1 CT scanner**

(a) Patient inside a CT scanner. The table on which the patient lies slides back and forth as scans are shot in 10–20-mm cross sections. In this instance an intravenous fluid is being used to disperse opaque iodine contrast medium, which will aid in highlighting hard-to-see areas. (b) CT scan from the midtrunk region. Density is indicated by the degree of lightness of an area. White areas are the most dense, followed by the grey, and then the black. CT scans are performed primarily as an accurate and rapid means of searching for tumors or other space-occupying lesions.

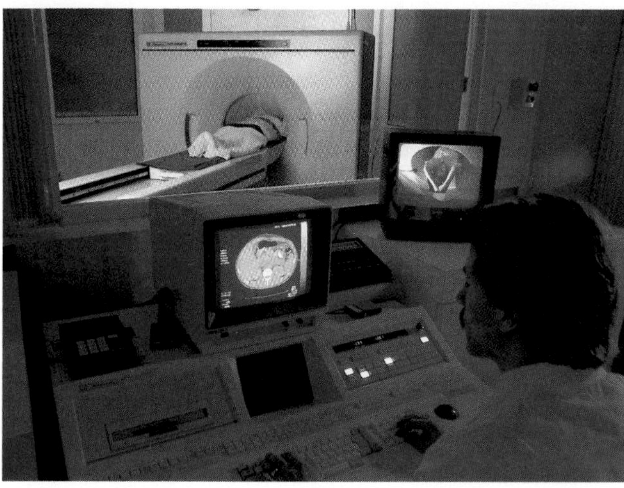

(a)

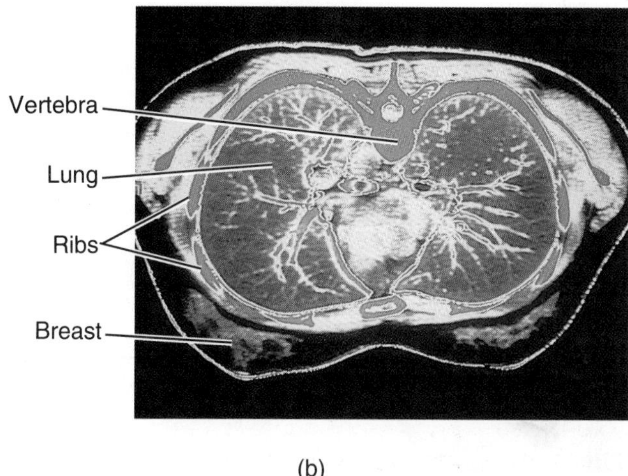

(b)

exposing this region of the body to X rays. The technique, in effect, produces a cross section of the body at the selected level (Figure 1.1). Since the rotation of the X-ray tube and film are coordinated by computer, the technique is referred to as *computed tomography (CT scan)* or *computed axial tomography (CAT scan).* A CT scan shows the three-dimensional relationships of the body parts and differentiates among tissues more clearly than X rays can. In fact, images produced by a CT scan are so detailed that they can detect even subtle changes within the brain, bulges of blood vessel walls, very small tumors, and the like. The use of CT scans has greatly reduced the need to do exploratory surgery to diagnose the cause of vague symptoms. Some CT scanners are capable of examining the entire body, while smaller models are used to examine only the head.

Dynamic Spatial Reconstructor

A complex X-ray machine called the *dynamic spatial reconstructor (DSR)* is another device used for the visual imaging of body structures. The DSR uses multiple X-ray tubes that rapidly revolve around the patient, producing thousands of cross sections within a few seconds. The DSR produces moving three-dimensional images of an organ that can be rotated and tipped in any direction, allowing all sides of an organ to be viewed (Figure 1.2). A CT scan, in contrast, produces only cross-sectional slices of the body. In addition, the image produced by the DSR can be "sliced open" on the screen so that the interior of an organ can be viewed.

Magnetic Resonance Imaging

Magnetic resonance imaging (MRI) is an imaging technique that has the advantage of using non-ionizing radiation, which is less damaging to cells than are X rays. In MRI the patient is placed in a body-size chamber within a large magnet. The magnetic field causes the nuclei of hydrogen atoms, as well as certain other nuclei, throughout the body to become aligned. By adjusting the magnetic energy generated by the magnet, it is possible to detect the amount of energy absorbed by the various nuclei. This information, when fed into a computer, can be used to plot the distribution of the nuclei and thus provide images of body organs. Because MRI records the behavior of hydrogen nuclei dissolved in water, it permits visualization of soft tissues of the body but not the more dense structures, such as bones. Thus, the skeleton does not interfere with viewing of the underlying soft tissues. Since MRI uses a powerful magnetic field, it cannot be used on patients with metal objects in their bodies, such as pacemakers or artificial joints. It has not been clearly established that MRI is completely free of risk, and therefore it is not used on pregnant women.

◆ **FIGURE 1.2 Dynamic spatial reconstructor (DSR)**
(a) The instrument and control unit. Reproduction with permission from the JIRA. (b) Three-dimensional DSR images of the upper abdominal organs.

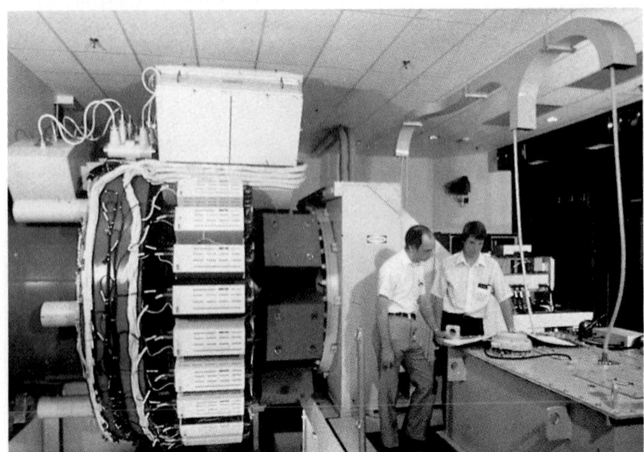

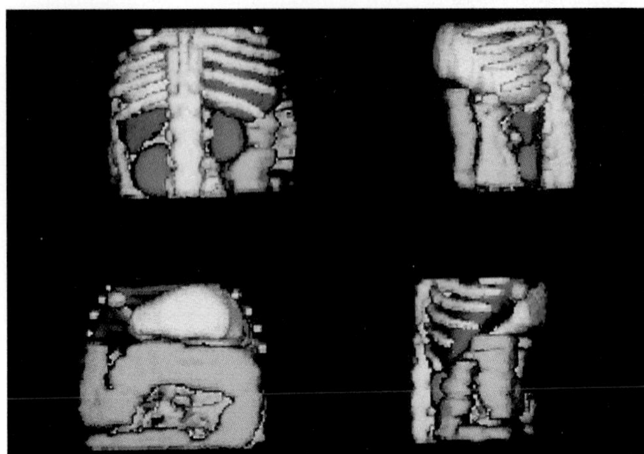

Ultrasonography

The use of *ultrasonography* or *ultrasound imaging* provides another means of obtaining images of body organs. In this procedure, high-frequency sound waves are sent into the body, and the echoes they make when they strike various tissues are used to define the boundaries of the organs. The sound waves appear not to have any significant harmful side effects on cells; therefore ultrasonography can be used safely to produce images of unborn fetuses (Figure 1.3). It can also produce images of moving objects, such as the blood flowing within a vessel. However, because sound waves are not able to penetrate the body very well and are quickly dissipated in air, ultrasound imaging is not very useful for examining the lungs or the brain and spinal cord.

Positron-emission Tomography

A branch of medicine called *nuclear medicine* uses radioisotopes to diagnose and treat various diseases. *Positron-emission tomography (PET)* is a specialized nuclear medicine imaging technique that provides information about organ functioning. Whereas CT scans produce images that show what an organ looks like, PET produces images that give a measure of the activity of the cells within the organ. During a PET examination, short-lived radioisotopes that have been attached to a molecule used by the body (such as glucose) are injected into the patient's bloodstream, and the patient is placed in a PET scanner. As the injected solution circulates and is metabolized within various body tissues, the radioisotopes cause high-energy gamma rays to be produced. The scanner detects the gamma rays, and its computer constructs a multicolored image showing the rates at which the injected solution is being metabolized in various regions and tissues (Figure 1.4). The greatest disadvantage of PET is that the specialized equipment is very expensive and thus is not available at every hospital.

Interrelationship of Structure and Function

When you study anatomy and physiology at the same time, as in this text, you will benefit by keeping in mind the interrelation of structure and function. By describing how a particular body structure is suited to its function, the study of anatomy helps make the physiological processes of the body more meaningful.

Just as the structures, shapes, and organization of the parts of a machine—such as an automobile—are appropriate to their functions, so the structures, shapes, and organization of the parts of the body are intimately associated with their functions. This interrelation of structure and function is evident at all levels of body organization. At the whole-body level, for example, the structure of the joints in the human hand makes possible an opposable thumb that is essential for efficiently grasping and manipulating objects. At the cellular level, one example of the interrelation of structure and function is the nerve cell, which has long, thin processes extending from the cell body. These processes are well suited to the cell's function of transmitting information

◆ **FIGURE 1.3 Ultrasound image of a fetus**

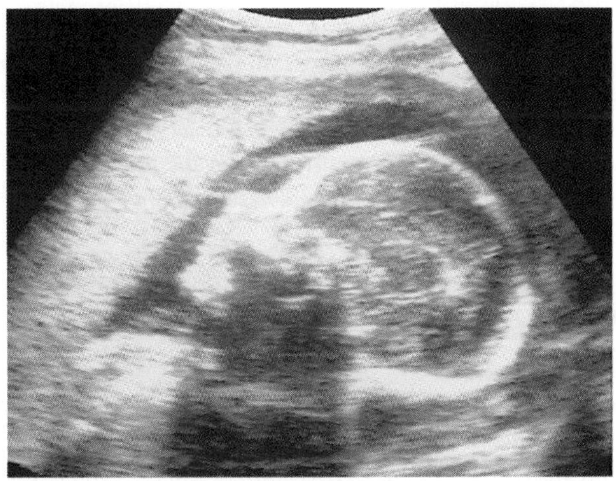

in the form of nerve impulses from one body region to another. Even at the molecular level, structure is critical to function. Enzymes, for example, which are molecules that speed up the rates at which chemical reactions occur in the body, can act only if they have shapes that specifically "fit" the shapes of the reacting molecules.

Levels of Structural Organization

The human body contains several levels of structural organization. Here we will briefly consider the chemical, cellular, tissue, organ, and organ system levels.

Chemical Level

The simplest level of organization is the **chemical level.** At this level the body is composed of nonliving chemical elements such as carbon, hydrogen, oxygen, and nitrogen. Atoms of chemical elements combine with one another to form molecules, such as carbohydrates, proteins, and fats, which are also nonliving. The chemical level of organization is examined in more detail in Chapter 2.

Cellular Level

Although the body is composed of nonliving chemical elements, the basic living units of the body are the **cells.** The body's tissues, organs, and organ systems are composed of cells, and it has been estimated that there are more than 75 trillion cells in the adult human body. Therefore, in order to comprehend the workings of the body's tissues, organs, and organ systems, we must understand the structure and function of cells. We consider cells in Chapter 3.

Tissue Level

All cells share many common characteristics and, in general, perform the basic functions required for their survival. However, the structure and function of various cells differ greatly, and the body's cells are specialized for particular purposes. In the body, groups of similar cells join together to form **tissues.** The presence of tissues allows the body to perform more complex physiological activities than is possible for individual cells.

In the early embryo, where tissue formation first occurs, similar cells group together into three layers: the **ectoderm,** which forms both the outer covering of the body and the nervous tissue; the **endoderm,** which forms the inner lining of the digestive tube and its associated structures; and the **mesoderm,** the layer located between the ectoderm and endoderm tissues, which forms the skeleton and the muscles of the body (Figure 1.5). These three embryonic cell layers—the ectoderm, endoderm, and mesoderm—give rise to four types of tissues, which are briefly discussed here and examined in detail in Chapter 4.

Epithelial Tissues

Epithelial tissues cover the surface of the body and the line the various body cavities, ducts, and vessels. The epithelial tissue that forms the outer protective layer of the body, the epidermis of the skin, is derived from

◆ **FIGURE 1.4 Positron-emission tomography image**

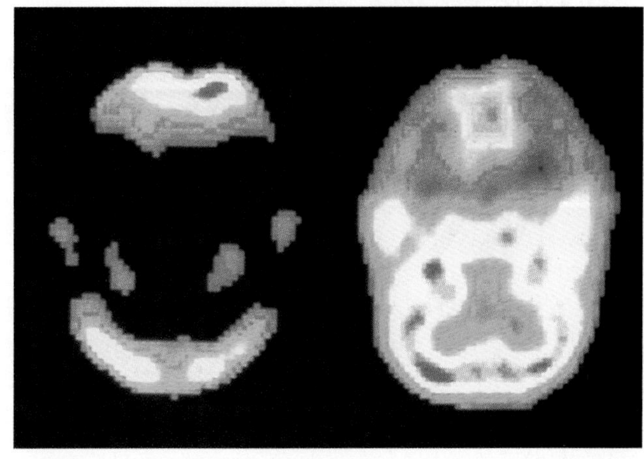

embryonic ectoderm. The rest of the epithelial tissues originate from either the mesoderm or the endoderm of the embryo.

◆ **FIGURE 1.5 Schematic cross section of an embryo showing the location of ectoderm, endoderm, and mesoderm**

The dotted line on the embryo indicates the site of the cross section.

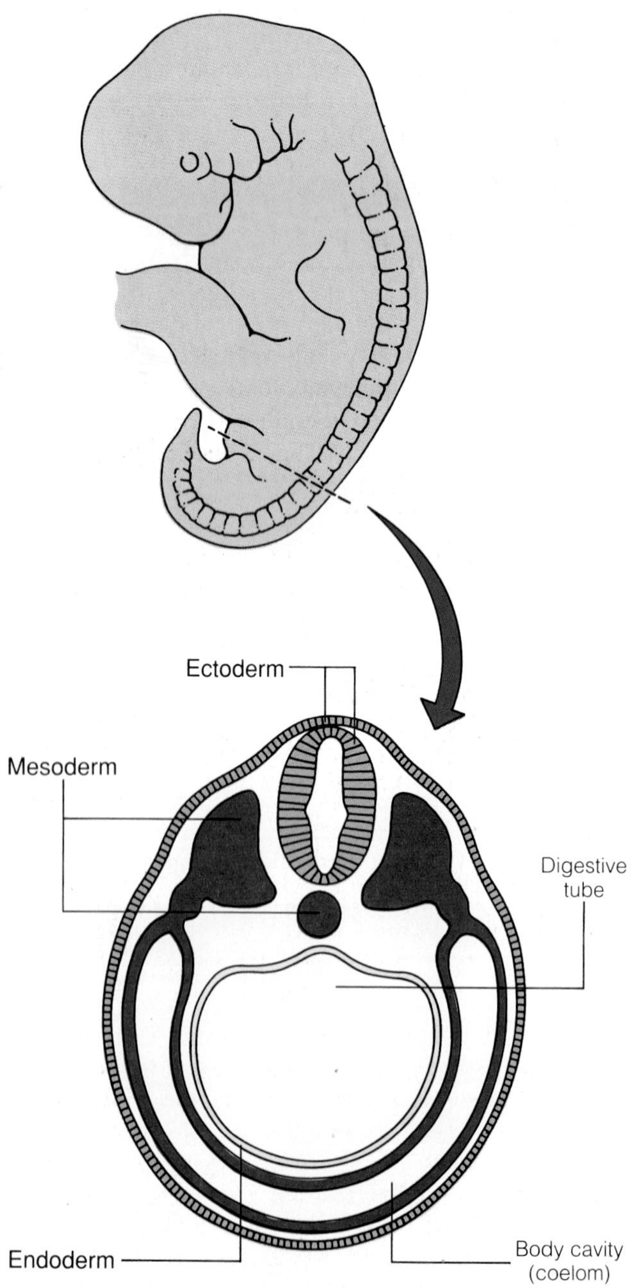

Muscular Tissues

Muscular tissues are composed of specialized cells that are capable of contracting and thereby decreasing in length. These tissues move the skeleton, propel the blood throughout the body, and aid in digestion by moving food through the digestive tract. As we will explain in Chapter 8, there are three types of muscle tissues: *skeletal, cardiac,* and *smooth.* Each type is derived from embryonic mesoderm.

Nervous Tissues

Nervous tissues, which form the brain, spinal cord, and nerves, consist largely of cells with long protoplasmic extensions. These nerve cells (neurons) transmit messages throughout the body. They originate from the ectodermal layer of the embryo.

Connective Tissues

Several types of cells are involved in the formation of connective tissues. These tissues, most of which are derived from mesoderm, are used for support (bones and cartilage), for the attachment of other tissues (tendons, ligaments, and fasciae), or for other specialized functions (such as blood).

Organ Level

Physiologic processes that are even more complex than those occurring in individual tissues are made possible when two or more tissues combine to form an **organ.** The stomach, for instance, is an organ that is lined with epithelial tissue, and its walls are formed by muscular tissue. These tissues are held together as a discrete structure by various connective tissues and are innervated by nervous tissue. Each type of tissue contributes in a specific manner to the functioning of the stomach. Without such tissue combinations, it would not be possible to process large particles of complex foods. This same principle holds true for all organs: each organ is a specialized physiologic center for the body.

Organ System Level

The ability of organs to function for the general well-being of the body is enhanced by the fact that groups of organs commonly work together, with each organ performing a specific part of a general body function. For example, one shared function is obtaining energy from food for use throughout the body. Food is prepared and partially digested in the mouth; it is trans-

◆ **TABLE 1.1 Organ Systems of the Body**

SYSTEM	MAJOR COMPONENTS	REPRESENTATIVE FUNCTIONS
Integumentary	Skin and associated structures such as hair and nails	Protects internal body structures against injury and foreign substances; prevents fluid loss (dehydration); important in temperature regulation
Skeletal	Bones	Supports and protects soft tissues and organs
Muscular	Skeletal muscles	Moves body and its parts
Nervous	Brain, spinal cord, nerves, special sense organs	Controls and integrates body activities; responsible for "higher functions" such as thought and abstract reasoning
Endocrine	Hormone-secreting glands such as the pituitary, thyroid, parathyroid, adrenals, pancreas, and gonads	Controls and integrates body activities; function closely allied with that of the nervous system
Cardiovascular	Heart, blood vessels, and blood	Links internal and external environments of the body; transports materials between different cells and tissues
Lymphatic	Lymphatic capillaries, collecting vessels, lymph nodes, and various lymphoid organs	Returns interstitial fluid to the blood
Respiratory	Nose, trachea, lungs	Transfers oxygen from the atmosphere to the blood and carbon dioxide from the blood to the atmosphere
Digestive	Mouth, esophagus, stomach, small intestine, large intestine; accessory structures include salivary glands, pancreas, liver, gallbladder	Supplies body with substances (food materials) from which energy for activity is derived and from which components for synthesis of required substances are obtained.
Urinary	Kidneys, ureters, urinary bladder, urethra	Eliminates variety of metabolic end products such as urea; conserves or excretes water and other substances as required
Reproductive	Male: seminal vesicles, testes, prostate gland, bulbourethral glands, penis, associated ducts	Produces male gametes (sperm); provides method for introducing sperm into the female
	Female: ovaries, uterine tubes, uterus, vagina, mammary glands	Produces female gametes (ova); provides proper environment for development of fertilized ovum

ported by the esophagus to the stomach, where it is further prepared and digested; it is absorbed into the blood vessels through the walls of the intestines; and finally, the residue that is not absorbed is eliminated from the body through the anus.

Organs that function cooperatively to accomplish a common purpose (such as the digestion and absorption of food) are said to be part of an **organ system.** There are eleven major organ systems in the human body: *integumentary (in-teg-u-men´-tar-ee), skeletal, muscular, nervous, endocrine, cardiovascular, lymphatic, respiratory, digestive, urinary,* and *reproductive.* The structure and function of each of these systems are listed in Table 1.1 and discussed in greater detail in later chapters.

Anatomical and Physiological Terminology

Every branch of science has its own special terminology, and anatomy and physiology are no exceptions. Although you will find many terms in this text that are new to you, the terms themselves are not new. A sizable number of them originated centuries ago and have Greek or Latin origins. The terms may appear formidable, but they are quite descriptive if you understand their roots. For example, *ilio* refers to the hip bone (ilium), and *costal* refers to the ribs. Therefore, the *iliocostalis* muscle is clearly a muscle that passes from the ilium to the rib cage.

A knowledge of prefixes and suffixes is also helpful in understanding anatomical and physiological terms. For example, the prefix *endo* means "within" and is used in many scientific terms, including the following:

◆ **endocardium** (*en-do-kar´-di-um; cardium* refers to the heart) the inner lining of the heart

◆ **endocarditis** (*en-do-kar-di´-tis; itis* refers to an inflammation) an inflammation of the inner lining of the heart

◆ **endogenous** (*en-doj´-en-us; genous* refers to producing) produced within the organism

◆ **endometrium** (*en-do-me´-tree-um; metra* refers to the uterus) the inner lining of the uterus

The meanings of new prefixes, suffixes, and roots will be explained as they are introduced in the text. A list that will be helpful in understanding new words appears in Appendix 3 at the back of the book.

Body Positions

While studying the detailed description of each body structure, you must also understand the positional relationships among body structures. Thus, you must become familiar with the terms used to describe these relationships.

If the body is lying horizontally with the face downward, the body is in the *prone position.* If the body is on its back, with the face upward, it is in the *supine (soo´-pine) position.* The relationships of the various body structures to each other differ in these positions. To communicate effectively concerning human anatomy, therefore, we always refer to the body as if it were in a standard position so that the structural relationships are clear and consistent. This standard position is referred to as the **anatomical position** (Figure 1.6a). In this position, the body is erect, with the feet together. The upper limbs hang at the side, with the palms of the hands facing forward, the fingers extended, and the thumbs pointing away from the body. With the hands in this position, the bones of the hands and fingers are exposed, and their relationships, therefore, are easily described. Moreover, when the palms are facing forward, the bones of the forearm are uncrossed. *Unless stated otherwise, all anatomical descriptions refer to a body in the anatomical position.*

Directional Terms

The terms used to denote direction come in pairs, each indicating an opposite direction (Figures 1.6b and c). **Anterior** (or **ventral**) refers to the front, whereas its opposite, **posterior** (or **dorsal**) refers to the back. **Superior** (or **cranial**) means "toward the head"; **inferior** (or **caudal;** *kaw´-dal*) means "away from the head." It should be noted that the direction indicated by these terms differs according to whether they refer to humans, who stand upright, or to four-legged animals. In this text we are interested only in their meaning as they relate to humans. The directional pairs are listed in Table 1.2.

Regional Terms

In addition to directional terms, several frequently used terms refer only to special areas of the body (Figure 1.6a):

◆ **cervical** refers to the neck

◆ **thoracic** the portion of the body between the neck and the abdomen that is commonly referred to as the chest (thorax)

◆ **lumbar** the portion of the back between the thorax and the pelvis

◆ **sacral** the lower portion of the back, just superior to the buttocks

◆ **plantar** the sole of the foot; the top of the foot is the *dorsal* surface

◆ **palmar** the anterior surface of the hand; the posterior surface of the hand is the *dorsal* surface

◆ **axilla (armpit)** the depression on the inferior surface of the attachment of the upper limb and the body trunk

◆ **groin (inguinal region)** the junction of the thigh wall with the abdominal wall

◆ **arm** the portion of the upper limb between the shoulder and the elbow

◆ **forearm** the portion of the upper limb between the elbow and the wrist

◆ **thigh** the portion of the lower limb between the hip and the knee

◆ **leg** the portion of the lower limb between the knee and the ankle

To make it easier to describe the location of the abdominal organs, this cavity is divided into nine regions using four imaginary lines: two vertical lines that bisect the clavicles, or collarbones; and two horizontal lines, one along the lower edge of the rib cage and another across the upper edges of the hip bones (iliac crests) (Figure 1.7a). These abdominal regions are:

◆ **umbilical** located centrally, surrounding the *umbilicus* (navel)

◆ **lumbar** the regions to the right and left of the umbilical region

◆ **epigastric** (*epi* means "on or above"; *gastric* refers to the stomach) the midline region superior to the

◆ **FIGURE 1.6 Anatomical terms**
(a) Anatomical position and regions of the body. (b,c) Directional terms.

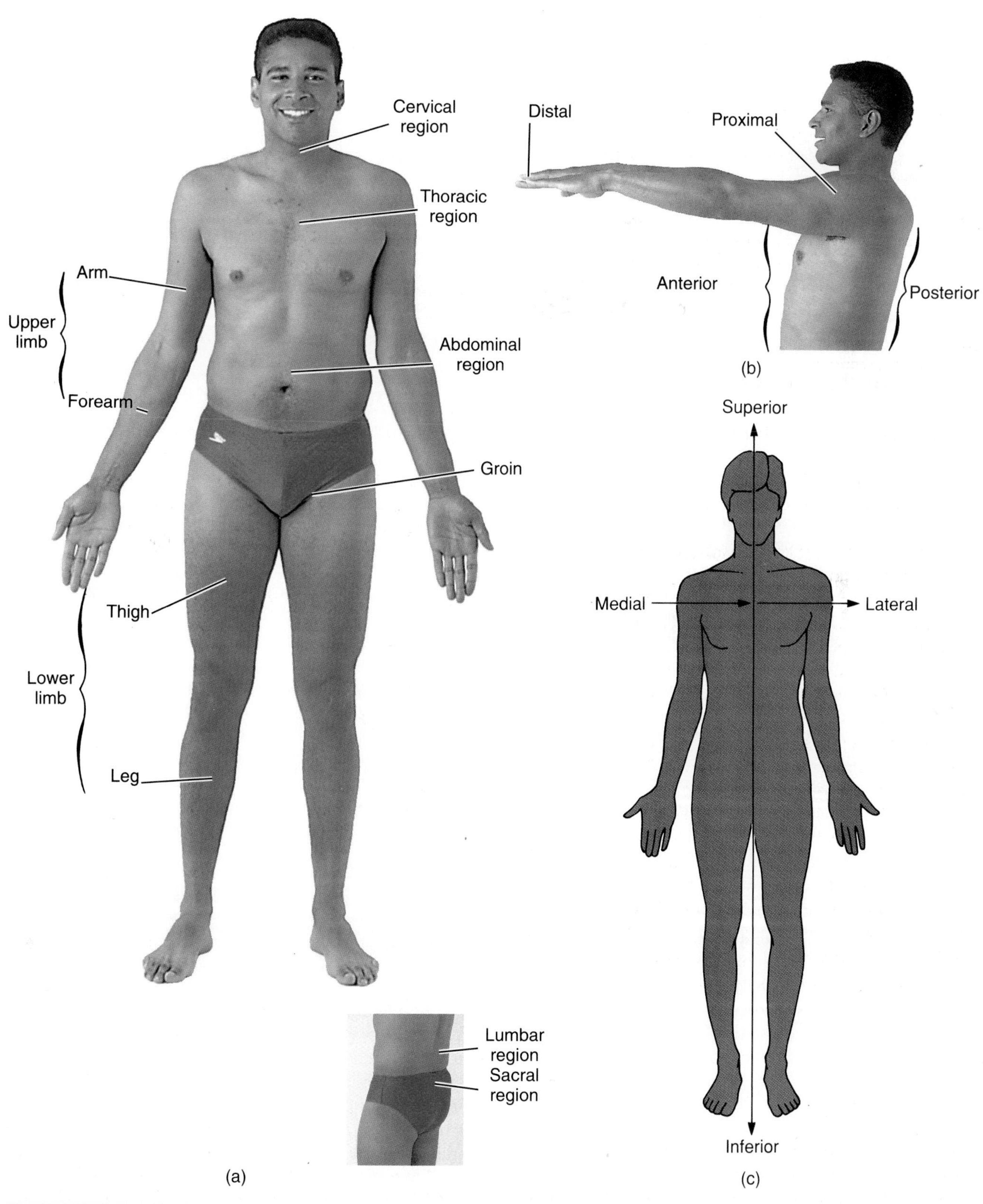

◆ **TABLE 1.2 Directional Terms**

TERM	DEFINITION	EXAMPLE
Anterior (ventral)	Situated in front of; the front of the body	The chest is on the anterior surface of the body.
Posterior (dorsal)	Situated in back of; the back of the body	The buttocks are on the posterior surface of the body.
Superior (cranial)	Toward the head; relatively higher in position	The eyebrows are superior to the eyes.
Inferior (caudal)	Away from the head; relatively lower in position	The mouth is inferior to the nose.
Medial	Toward the midline of the body	The breast is medial to the armpit.
Lateral	Away from the midline of the body	The hip is on the lateral surface of the body.
Proximal	Closer to any point of reference, such as the attached end of a limb, the origin of a structure, or the center of the body	The arm is proximal to the forearm.
Distal	Farther from any point of reference, such as the attached end of a limb, the origin of a structure, or the center of the body	The hand is distal to the wrist.
Superficial (external)	Located close to or on the body surface	The skin is superficial to the muscles.
Deep (internal)	Located further beneath the body surface than superficial structures	The muscles are deep to the skin.

umbilical region. As the name implies, most of the stomach is located in this region.

◆ **hypochondriac** (*hypo* means "beneath or under"; *chondral* refers to cartilage) the regions to the right and left of the epigastric region. The name indicates that the hypochondriac regions are located beneath the cartilage of the rib cage.

◆ **hypogastric** the midline region directly inferior to the umbilical region

◆ **iliac** the regions on either side of the hypogastric region. The name is derived from the iliac (hip) bones that form the lateral boundaries of the regions. These areas are also referred to as the *inguinal regions* because their lower margins end at the inguinal ligament, which follows the fold of the groin.

In practice, it is more common to divide the abdominopelvic cavity into four quadrants by means of an imaginary horizontal plane that passes through the umbilicus and a vertical midsagittal plane (Figure 1.7b). These two intersecting planes divide the abdominopelvic cavity into a **right upper (superior) quadrant;** a **right lower (inferior) quadrant;** a **left upper (superior) quadrant;** and a **left lower (inferior) quadrant.**

Body Planes

In the study of anatomy, it is useful to visualize the body as cut or sectioned through various planes of reference (Figure 1.8). A **sagittal plane** *(saj´-i-tal)* is a longitudinal section that divides the body or any of its parts into right and left portions. If the section passes through the midline of the body, it is referred to as a **median sagittal (midsagittal) section.** Such a section divides the body into *equal* right and left halves. Sagittal sections other than the median sagittal section are often referred to as **parasagittal sections.** Parasagittal sections divide the body into *unequal* right and left portions. The **frontal (coronal) plane** is also a longitudinal section, but it runs at right angles to the sagittal plane, dividing the body into anterior and posterior portions. A **transverse plane (cross section** or **horizontal section)** divides the body or any of its parts into superior and inferior portions.

Body Cavities

The body contains two main cavities: the **dorsal (posterior) cavity** and the **ventral (anterior) cavity** (Figure 1.9). Each of these cavities is lined with membranes and contains a small amount of fluid surrounding the organs that fill the cavities. The dorsal cavity has two subdivisions: the **cranial cavity,** which houses the brain, and the **spinal (vertebral) cavity,** which contains the spinal cord. The spinal cavity communicates with the cranial cavity through the *foramen magnum,* a large opening in the base of the skull. The membranes

◆ **FIGURE 1.7 Abdominal wall**

(a) Abdominal regions. The top horizontal line passes along the lower edge of the rib cage. The lower horizontal line passes across the upper margins of the hip bones. The vertical lines pass through the midpoints of the clavicles and the inguinal ligaments. (b) The abdominal wall and abdominopelvic cavity subdivided into four quadrants.

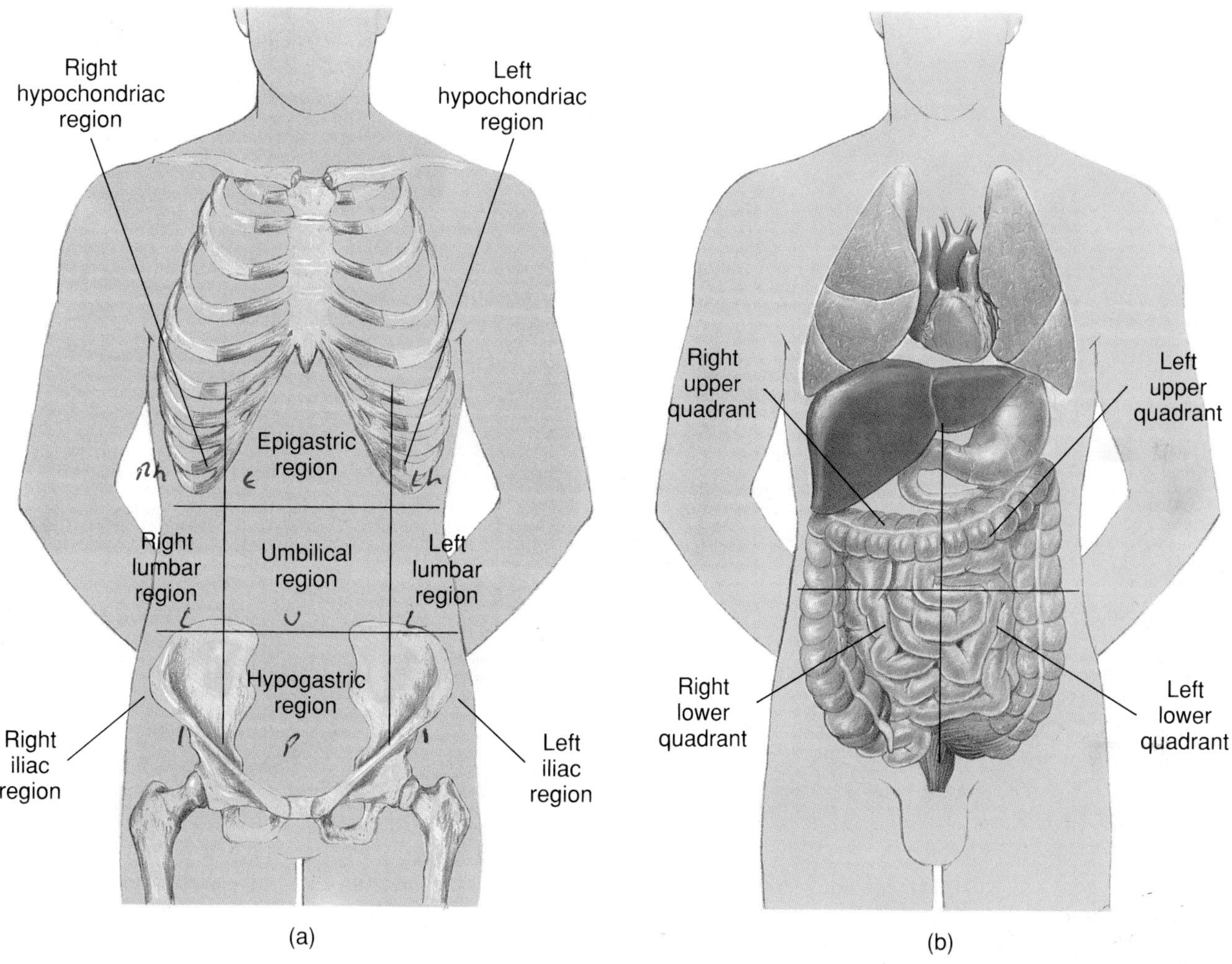

associated with these dorsal cavities are examined in greater detail in Chapter 12. For now, it is sufficient to call the membranes that cover the brain and the spinal cord the *meninges (men-in´-jeez).* The fluid found in these dorsal cavities is the *cerebrospinal fluid.* It too is considered in more detail with the nervous system.

The ventral cavity of the body also has two subdivisions. It is divided by a muscle called the *diaphragm* into an upper **thoracic cavity** and a lower **abdominopelvic (peritoneal) cavity.** Each of these cavities is, in turn, further subdivided. The thoracic cavity is divided into a **pericardial cavity,** which surrounds the heart, and right and left **pleural cavities,** each of which encompasses a lung. The portion of the thoracic cavity between the two pleural cavities is called the **mediastinum** *(mee-dee-as-tigh´num).* The pericardial cavity and the heart are located in the mediastinum. The trachea, esophagus, thymus gland, and several major blood vessels are also located within, or pass through, the mediastinum.

The abdominopelvic cavity, the lower subdivision of the ventral body cavity, is divided for descriptive purposes into a superior **abdominal cavity** and an inferior **pelvic cavity (true pelvis)** by an imaginary oblique plane that passes from the superior margin of the *symphysis pubis* anteriorly to the *sacral promontory* posteriorly (Figure 1.10, page 14). The circumference of this plane is called the **pelvic brim.** The pelvic cavity is completely surrounded by the bones of the pelvis. In contrast, the lower portion of the abdominal cavity is bounded posteriorly by the flat hip bones, but its anterior wall is formed by the abdominal wall. This region, which is located just above the pelvic brim, is the **false pelvis.** The abdominal cavity contains the stomach,

◆ **FIGURE 1.8 Body planes**

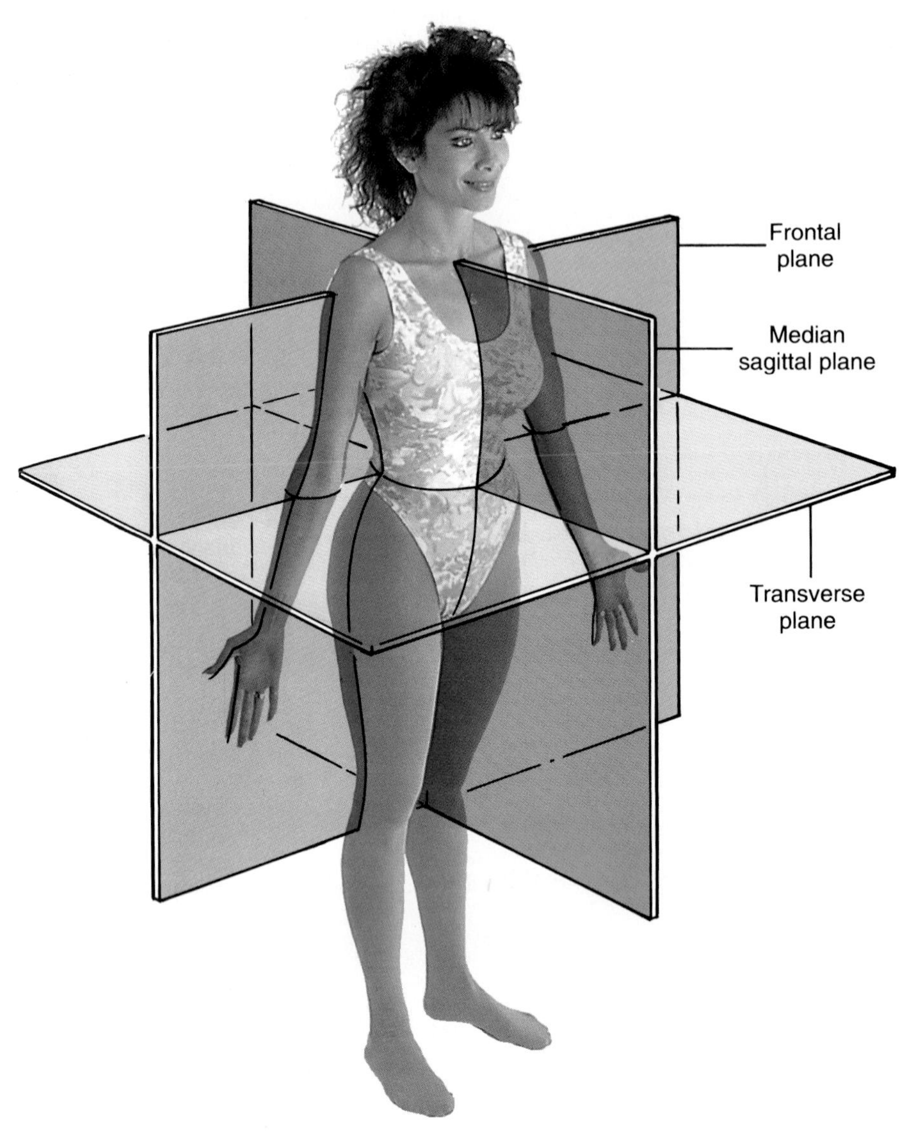

spleen, liver, gallbladder, pancreas, and the small and large intestines. The pelvic cavity contains the lower part of the digestive system (rectum), the urinary bladder, and, in the female, the internal reproductive organs.

Membranes of the Ventral Body Cavities

To understand the membranes associated with the ventral body cavities, imagine your fist being thrust into an inflated balloon, pushing in one side, as in Figure 1.11a. Notice that the inner wall of the balloon (that is, the wall that has been pushed in) lies close against the fist, whereas the outer wall is separated from the inner wall by the air in the balloon. Now suppose that your fist is an organ. The membranes of the ventral body cavities have the same relationship with the organs in the cavities, except that they are kept apart by fluid rather than air.

In the pericardial cavity, the heart (rather than a fist) pushes in one side of a membranous sac (rather than a balloon) (Figure 1.11b). The membrane that lies against the heart (the inner wall of the balloon) is the **visceral pericardium.** The membrane that covers both the

◆ **FIGURE 1.9 Body cavities**

The body has two major cavities, dorsal and ventral. Each of these is subdivided into smaller cavities. (a) Sagittal view. (b) Frontal view showing subdivisions of the thoracic cavity.

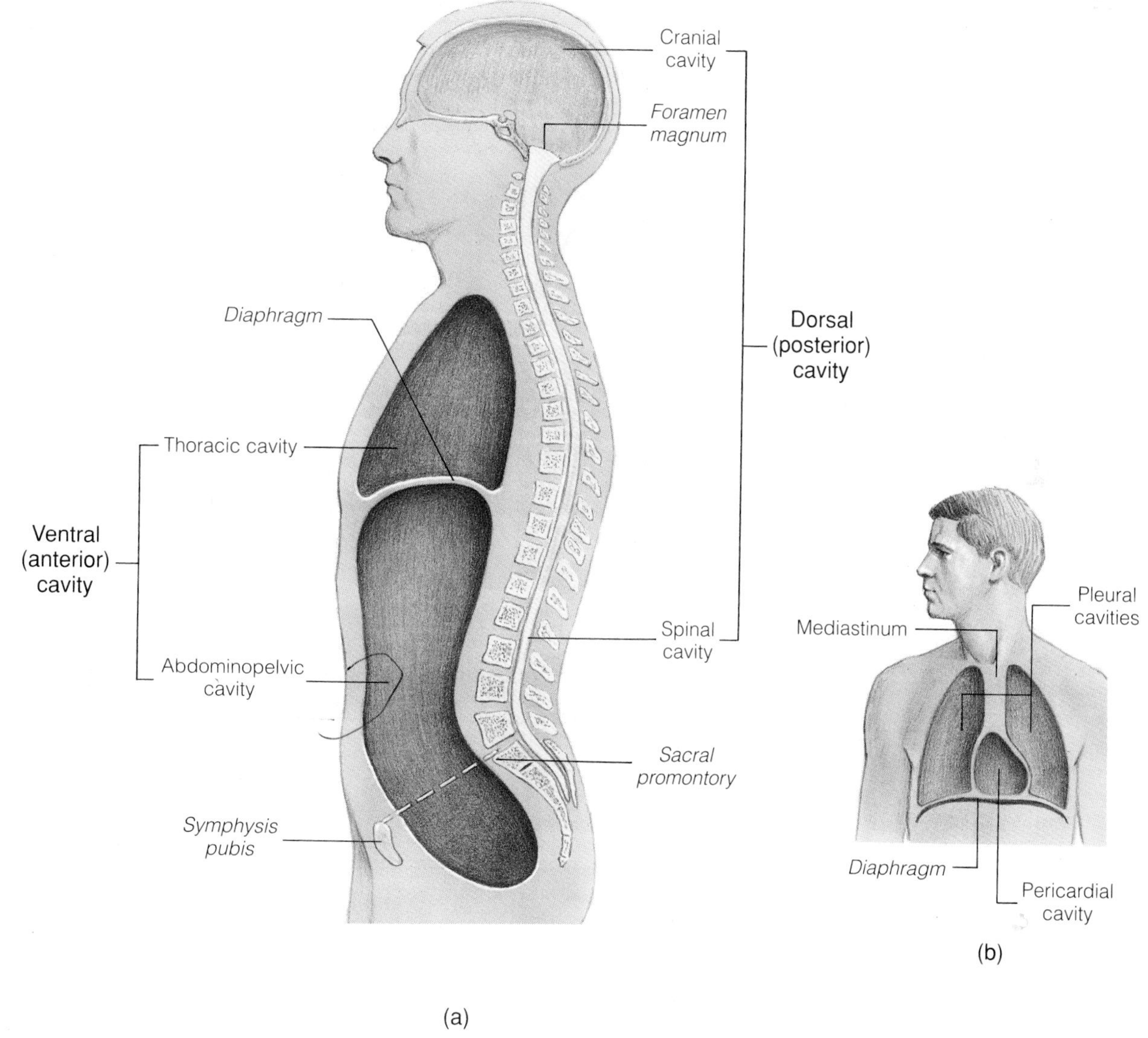

heart and the visceral pericardium (the outer wall of the balloon) is the **parietal pericardium** (*parietal* refers to the walls of a body cavity). The parietal pericardium is separated from the visceral pericardium by **pericardial fluid,** which is secreted by the cells of the pericardial membranes. In other words, the outer wall of the pericardial cavity is *lined* with parietal pericardium, and the heart is *covered* by visceral pericardium. However, these are both simply different regions of the same membrane. The amount of pericardial fluid that separates these two membranes is very small—just enough to reduce friction and maintain the tissues in a healthy state. If the membranes of the heart become inflamed, the condition is known as *pericarditis.*

The membrane relationships of the pleural cavities are similar to those of the pericardial cavity. The membrane that lies tightly against the surface of the lungs is the **visceral pleura.** The outer walls of the pleural cavities are lined by **parietal pleura.** These two membranes are separated by the **pleural fluid** they secrete. An inflammation of these membranes may result in the secretion of excessive amounts of pleural fluid into the pleural cavity. Prolonged inflammation may cause the visceral and parietal layers of the pleura to adhere to each other. This condition, which makes breathing painful, is called *pleurisy.*

The membrane relationships within the abdominopelvic cavity are similar to those of the thoracic cavity,

◆ **FIGURE 1.10 Sagittal section of the body showing membrane relationships of the abdominopelvic (peritoneal) cavity**

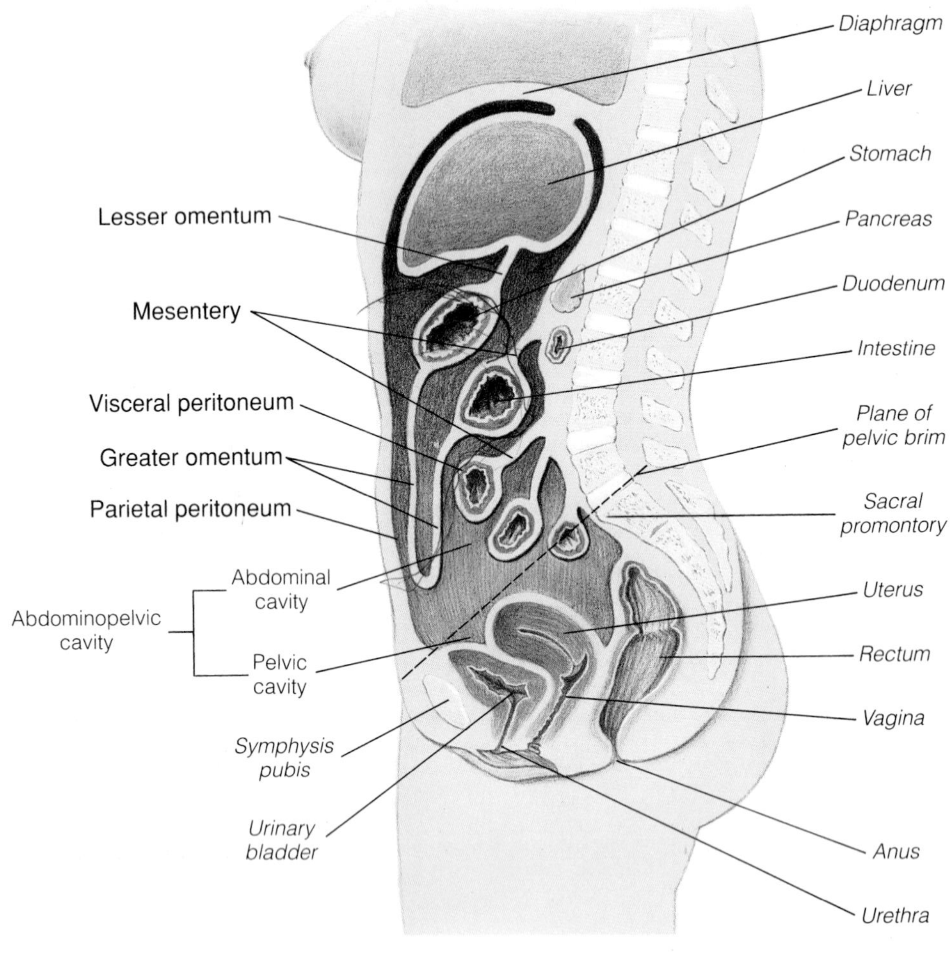

◆ **FIGURE 1.11 Membrane relationships in the ventral body cavities**

(a) Schematic representation using the analogy of a fist thrust into a balloon. (b) Membranes surrounding the heart.

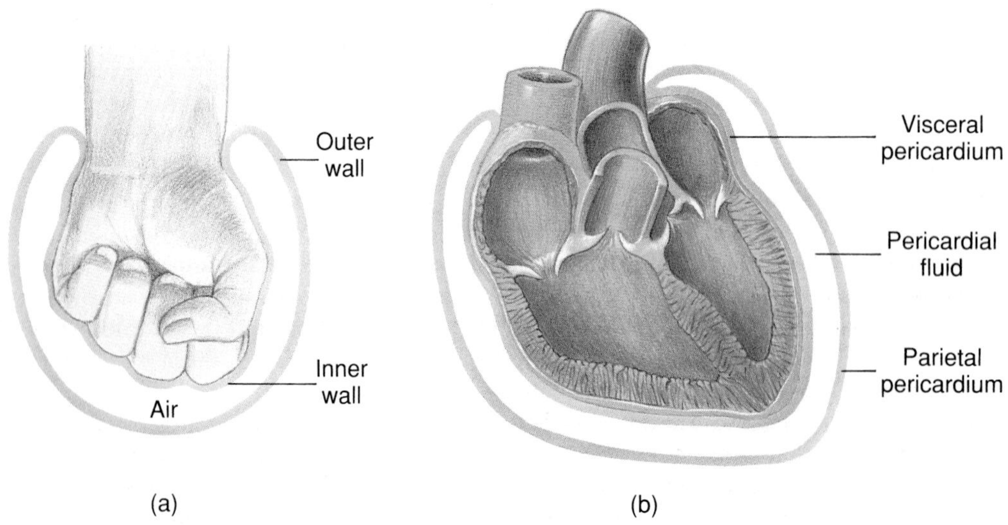

Outer wall

Inner wall

Air

(a)

Visceral pericardium

Pericardial fluid

Parietal pericardium

(b)

◆ **FIGURE 1.12 Transverse body section showing the retroperitoneal position of the kidneys**

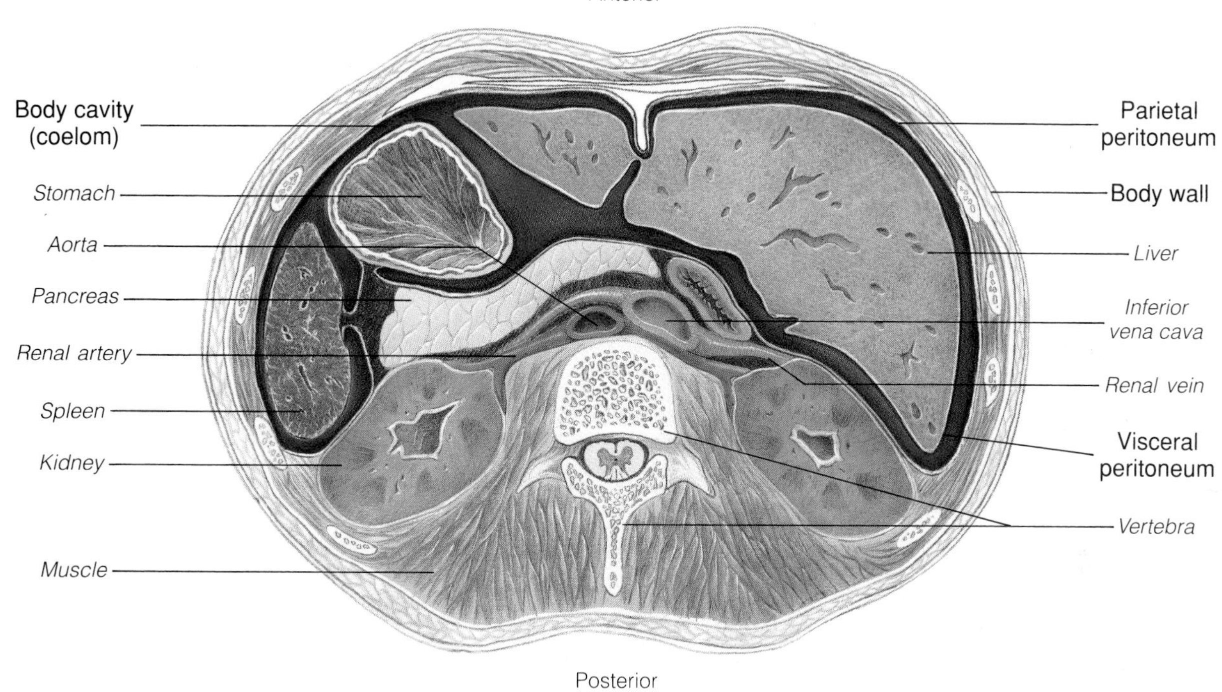

but here the membrane is called the **peritoneum.** The organs of this cavity are covered with **visceral peritoneum,** and the outer walls of the cavity are lined with **parietal peritoneum.** The space between these two membranes is filled with **peritoneal fluid,** which is secreted by the peritoneum. The peritoneum can become inflamed, causing a very serious condition called *peritonitis.*

Most of the organs in the abdominopelvic cavity are suspended from the posterior wall of the cavity by a double-layered membrane of peritoneum (Figure 1.10). These membrane supports are called **mesenteries.** The mesenteries that support particular organs or structures are given specific names such as mesocolon (mesentery of the large intestine), mesoappendix (mesentery of the appendix), mesovarium (mesentery of the ovary), and so forth. The mesenteries not only hold the organs in position, but they also provide a pathway through which blood vessels, lymphatic vessels, and nerves can reach the organs. As the membranes that form the mesenteries continue over the organ that they suspend, they become visceral membranes. Some structures, such as the kidneys, do not hang into the cavity on mesenteries. Instead, they are located outside the cavity, between the body wall and the parietal peritoneum (Figure 1.12). These structures are *retroperitoneal*—that is, they are located *behind* the peritoneum.

 Homeostasis

The body's cells can survive and function efficiently only under relatively constant conditions of temperature, pressure, acidity, and so forth. However, the body as a whole is surrounded by an *external environment* in which temperature, humidity, and other factors fluctuate rather widely. If a person is to survive, therefore, his or her cells must be protected from the variability and extremes of this environment. The cells are protected from the variability of the external environment because they exist within the body in an aqueous **internal environment.** Consequently, the cells are generally not exposed directly to the external environment.

The body contains two main categories of fluid: the **intracellular fluid,** or fluid within the cells, and the **extracellular fluid,** or fluid outside the cells. The internal environment is made up of the extracellular fluid. The principal components of the extracellular fluid are the fluid portion of the blood, which is called the **blood plasma,** and the fluid that surrounds and bathes the cells, which is called the **interstitial fluid** (Figure 1.13).

The body maintains relatively constant chemical and physical conditions within the internal environment, and the existence of a relatively constant internal environment is referred to as **homeostasis.** The fact that

◆ **FIGURE 1.13 Organ systems and homeostasis**
The body's cells exist in an aqueous internal environment, which is made up of the fluid portion of the blood and the interstitial fluid that continually bathes the cells. Essentially all the organ systems of the body help maintain relatively constant conditions within this environment. The existence of a relatively constant internal environment is referred to as *homeostasis.*

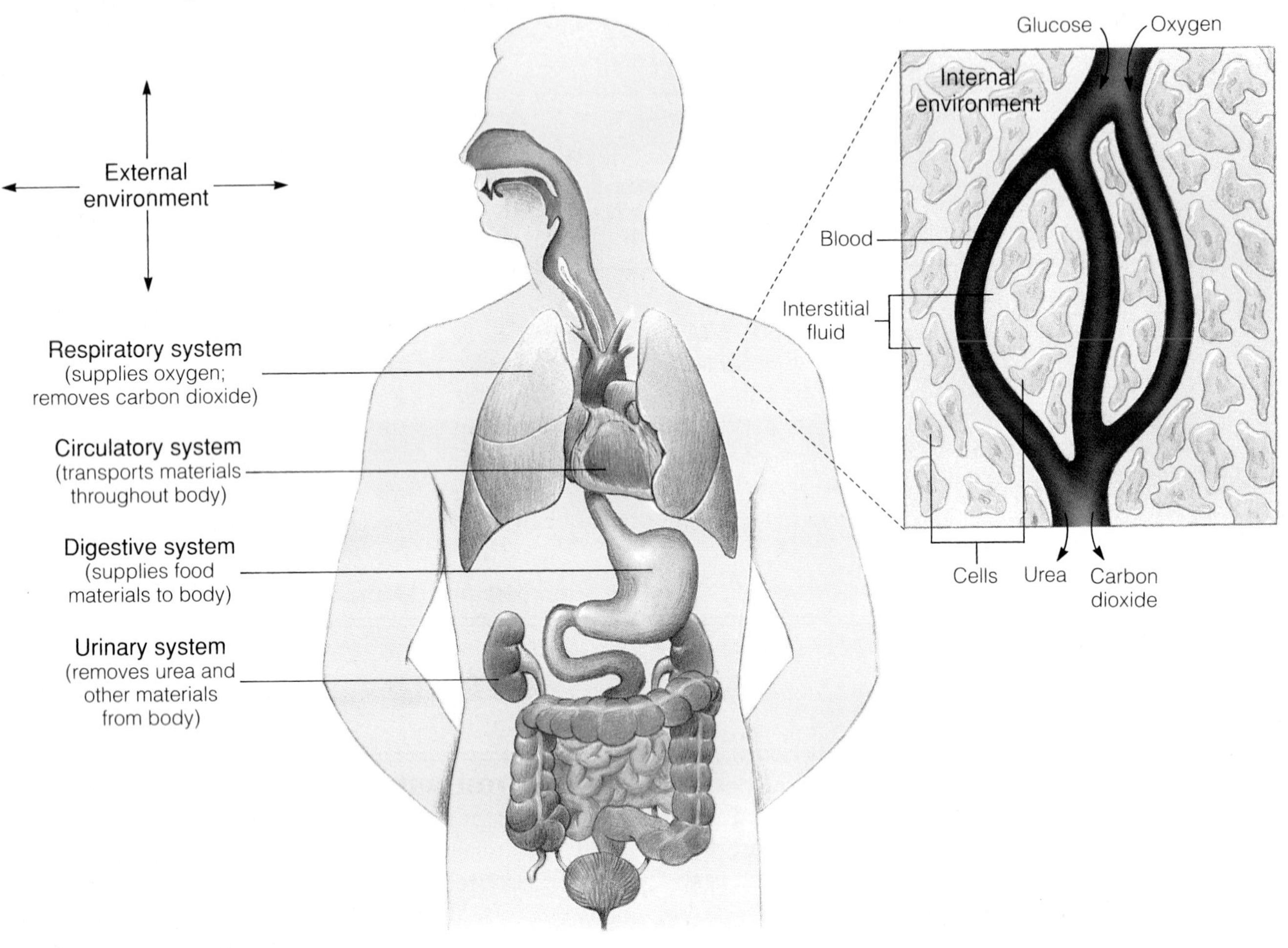

conditions within the internal environment are relatively constant, however, does not mean that this environment is static or unchanging. Rather, a variety of occurrences continually tend to cause changes in the internal environment. For example, the activities of the body's cells remove materials such as glucose and oxygen from the internal environment and add materials such as urea and carbon dioxide. As a result, homeostasis can be maintained only if materials are added to the internal environment as rapidly as they are removed or removed as quickly as they are added. The internal environment, therefore, is not static, but rather exists in a dynamic steady state in which the input and output of materials are balanced.

Essentially all the organ systems of the body contribute to the maintenance of homeostasis. For example, as the cells remove glucose and oxygen from the internal environment, the digestive and respiratory systems replace them. In addition, certain materials produced by the cells and added to the internal environment are removed by the urinary system. The circulatory system transports needed materials from areas such as the gastrointestinal tract or the lungs to the cells, and it carries materials produced by the cells to organs such as the kidneys for removal from the body.

As long as the various organ systems operate properly, the relative constancy of the internal environment is maintained, and the cells can survive and function efficiently. If the organ systems do not operate properly, however, the composition of the internal environment can change to such a degree that cellular function is impaired, resulting in disease or even death.

ASPECTS OF EXERCISE PHYSIOLOGY

What Is Exercise Physiology?

Exercise physiology is the study of both the functional changes that occur in response to a single bout of exercise and the adaptations that occur as a result of regular, repeated bouts of exercise. Exercise disrupts homeostasis. The changes that occur in response to exercise are the body's attempt to reduce the stress that has been placed on the entire organism.

Changes that are normal during exercise would be considered abnormal if they occurred in a nonexercising individual. For example, the level to which the body temperature rises during exercise would be considered a fever if the person were not exercising.

Heart rate is one of the easiest factors to monitor that shows both an immediate response to exercise and long-term adaptation to a regular exercise program. When a person begins to exercise, the active muscle cells use more oxygen to support their increased energy demands. Heart rate increases to deliver more oxygenated blood to the exercising muscles. The heart adapts to regular exercise of sufficient intensity and duration by increasing its strength and efficiency so that it pumps more blood per beat. Because of increased pumping ability, the heart does not have to beat as rapidly during exercise in order to pump a given quantity of blood to perform a particular activity as it did before physical training.

Exercise physiologists study the mechanisms responsible for the changes that occur as a result of exercise. Much of the knowledge gained from the study of exercise is used to develop appropriate exercise programs to increase the functional capacities of people, ranging from athletes to the infirm.

Because many students who enroll in anatomy and physiology courses are interested in the physiological changes associated with exercise, boxes that discuss some aspects of exercise physiology are included in most chapters.

Homeostatic Mechanisms

To maintain homeostasis, the body must be able to sense changes in the internal environment, and it must be able to compensate for those changes. Thus, the body must be able to control the organ systems concerned with maintaining the composition of the internal environment. The nervous and endocrine systems are the body's principal sensing and controlling systems.

Negative Feedback

The body uses a regulatory principle known as negative feedback to maintain relatively constant, or stable, conditions in the internal environment.

Negative feedback mechanisms have several components (Figure 1.14a). One component, the *controlled system,* is a system whose activity is regulated to maintain the appropriate level of a particular variable—for example, temperature or oxygen. A second component, the *set point,* is a reference that calls for or indicates the level at which the variable is to be maintained. A third component, the *receptor,* monitors the actual level of the variable and transmits information—referred to as feedback—to a fourth component, the *processing center.* The processing center compares and integrates information from the receptor about the actual level of the variable with information from the set point about the level of the variable called for. If necessary, the processing center increases or decreases the activity of the controlled system in order to bring the actual level of the variable to the level called for by the set point.

A common example of a negative feedback mechanism is the operation of a thermostatically controlled heater, which keeps the temperature in a room comfortable and relatively constant when the temperature outside the room is low (Figure 1.14b). The heater is the controlled system whose activity—heat production—is regulated in order to maintain the appropriate level of the variable, the room temperature. The set point level of the variable is the temperature called for by setting the thermostat at a particular value. The thermostat contains a receptor that provides information about the actual room temperature and a processing center that compares this information with infor-

◆ **FIGURE 1.14 Negative feedback**

(a) Schematic representation of the components of a negative feedback mechanism. (b) A thermostatically controlled heater is an example of a negative feedback mechanism.

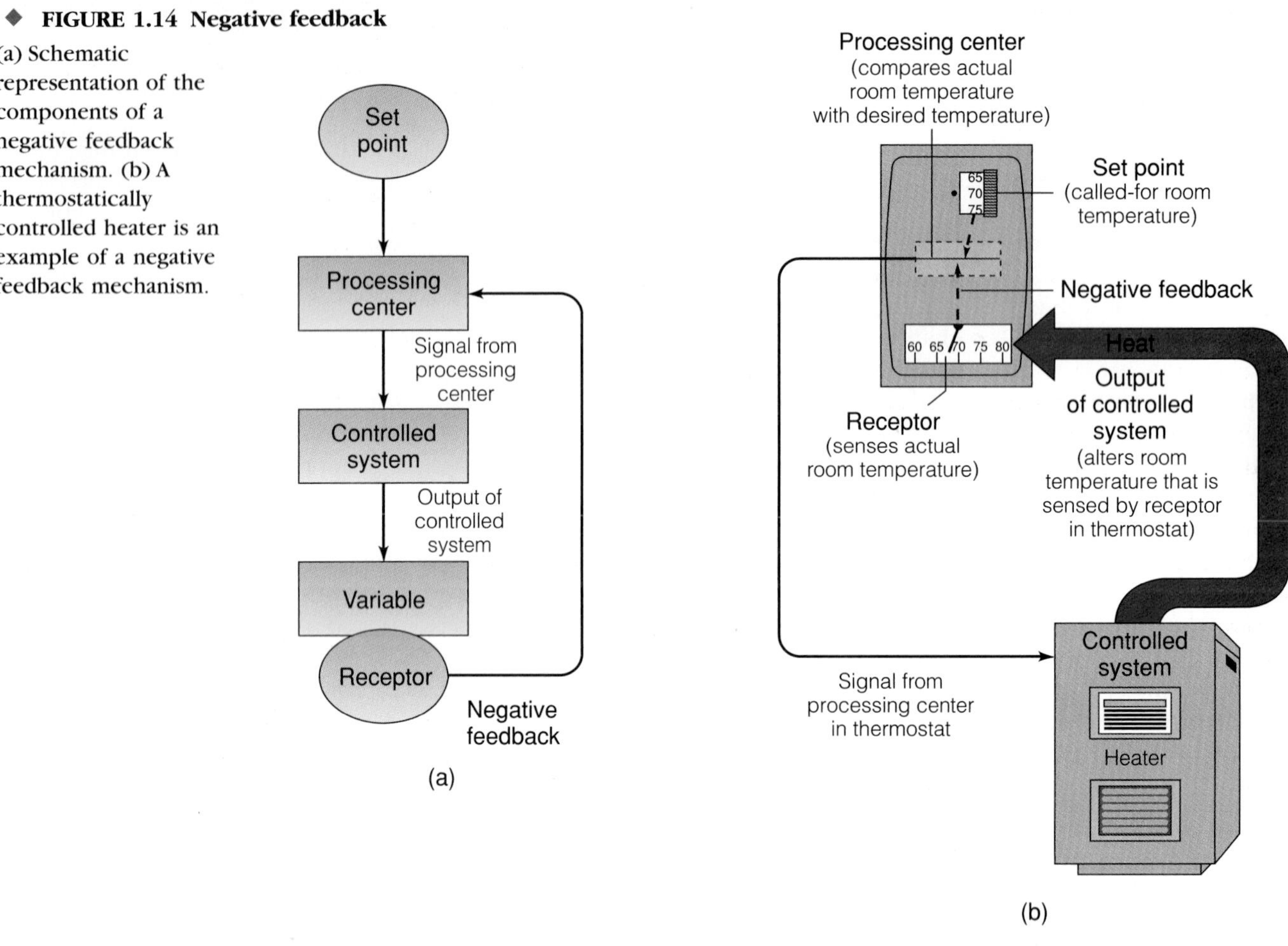

mation about the set point temperature. If the actual temperature differs from the set point temperature, the thermostat turns the heater on or off as necessary to bring the actual room temperature to the set point temperature.

Negative feedback mechanisms are called negative because the feedback tends to cause the level of a variable to change in a direction opposite to that of an initial change. In the case of a thermostatically controlled heater, for example, an increase in room temperature above the set point level leads to increased feedback, which acts in an inhibitory fashion to turn the heater off and thereby allow the room temperature to decrease toward the set point level. Conversely, when the room temperature falls below the set point level, the inhibitory feedback decreases, the heater turns on, and the heat produced raises the room temperature toward the set point level. Thus, negative feedback mechanisms minimize the difference between the actual level and the set point level of a variable and, consequently, tend to maintain relatively constant and stable conditions under circumstances that would otherwise cause the conditions to change.

Many negative feedback mechanisms are present in the body. For example, the body temperature control mechanism is believed to include a set point that calls for the maintenance of a particular body temperature. The body contains receptors that monitor the actual body temperature and transmit feedback information to the brain, which contains a processing center. The processing center compares and integrates the information from the receptors about the actual body temperature with information from the set point about the called-for temperature. If necessary, the processing center initiates appropriate action to bring the actual body temperature to the called-for temperature. For example, if the actual body temperature is below the set point temperature, shivering may be initiated. The muscular contractions of shivering produce heat that tends to raise the body temperature to the set point level. Conversely, if the body temperature is above the set point temperature, sweating may be stimulated. The evaporation of the water in the sweat cools the body surface and tends to lower the body temperature to the set point level.

Positive Feedback

In contrast to negative feedback mechanisms, positive feedback mechanisms maximize, rather than minimize, the difference between the actual and the set point level of a variable. They are called positive because the feedback tends to cause the level of the variable to change in the same direction as an initial change. With a positive feedback mechanism, an increase in the level of a particular variable above the set point level leads to increased feedback, which acts in a stimulatory fashion to increase the system's activity even more and thereby further increase the level of the variable. This leads to still greater feedback, which in turn leads to a further increase in the activity of the system, and so on and so on. Thus, positive feedback produces a cyclical effect in which a change in the level of a variable leads to further change in the same direction. In a positive feedback situation, however, the level of a variable does not necessarily continue to change indefinitely. For example, the level to which a variable can rise may ultimately be limited by the controlled system's maximum level of activity or by the amount of energy or raw materials available to the system.

Positive feedback does not lead to the maintenance of stable, homeostatic conditions, and consequently it does not occur in the body as frequently as negative feedback. However, positive feedback responses are occasionally evident. For example, during the birth of a baby, the pressure of the baby's head against the area around the opening of the mother's uterus stimulates the contraction of the uterine muscles (Figure 1.15). The contractions, in turn, increase the pressure of the head against the area around the uterine opening, which further stimulates the contraction of the uterine muscles. In this instance, a positive feedback type of response is clearly useful in promoting the expulsion of the baby from the uterus.

◆ **FIGURE 1.15 Positive feedback**

During birth, a positive feedback type of response stimulates increased uterine contractions.

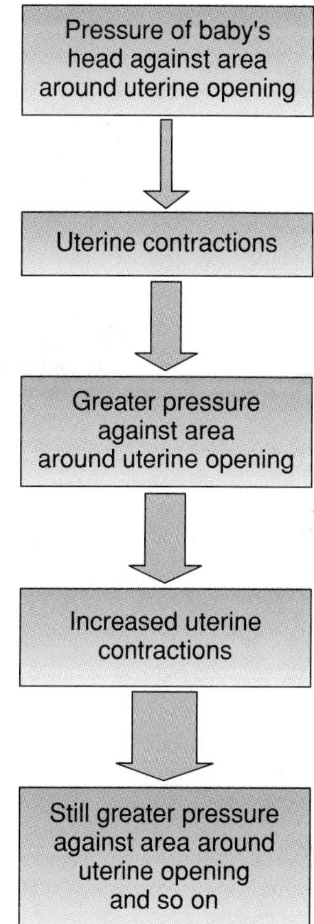

This chapter has provided a general background in the anatomy of the human body and an introduction to the mechanisms by which the internal environment of the body is kept relatively constant, even under fluctuating external conditions. In the forthcoming chapters, we will expand on these topics as we examine the body in greater detail.

Study Outline

◆ **FIELDS OF ANATOMY** p. 2

Subdivided into gross anatomy and microscopic anatomy. Gross anatomy can be studied by regions or by systems; microscopic anatomy can involve the study of cells or tissues.

◆ **FIELDS OF PHYSIOLOGY** p. 2

Physiology is a vast field with many subdivisions, often concerned with different levels of organization.

◆ **MEDICAL IMAGING METHODS** pp. 2–4

Provide ability to visualize internal body structures and provide information about organ functioning. Methods include X ray, fluoroscopy, computed tomography, dynamic spatial reconstructor, magnetic resonance imaging, ultrasonography, and positron-emission tomography.

◆ **INTERRELATIONSHIP OF STRUCTURE AND FUNCTION** pp. 4–5

Structures, shapes, and organization of the parts of the body are closely associated with their functions.

◆ **LEVELS OF STRUCTURAL ORGANIZATION** pp. 5–7
Include chemical, cellular, tissue, organ, and organ system levels.

Chemical Level. At the simplest level, the body is composed of nonliving chemical elements.

Cellular Level. The basic living units of the body.

Tissue Level. Three embryonic cell layers—the ectoderm, endoderm, and mesoderm—give rise to four tissue types:

EPITHELIAL TISSUES. Cover the body surface and line body cavities, ducts, and vessels; they develop from embryonic ectoderm, mesoderm, and endoderm.

MUSCULAR TISSUES. Move the skeleton, pump blood, and move food through the digestive tract; they develop from embryonic mesoderm.

NERVOUS TISSUES. Form the brain, spinal cord, and nerves; they develop from embryonic ectoderm.

CONNECTIVE TISSUES. Are used for support and the attachment of other tissues; they develop from embryonic mesoderm.

Organ Level. Tissues combine to form organs, such as the stomach.

Organ System Level. Eleven major organ systems make up the body: integumentary, skeletal, muscular, nervous, endocrine, cardiovascular, lymphatic, respiratory, digestive, urinary, reproductive.

◆ **ANATOMICAL AND PHYSIOLOGICAL TERMINOLOGY** pp. 7–8
Understanding word roots, prefixes, and suffixes is helpful in understanding anatomical terms; for example, the prefix *endo* means "within," as in *endocardium,* the inner lining of the heart.

◆ **BODY POSITIONS** p. 8
Anatomical position is achieved when body stands erect, feet together, upper limbs hanging at sides, palms forward, fingers extended, thumbs pointing away from body. The prone position is lying face down; the supine position, lying face up.

◆ **DIRECTIONAL TERMS** p. 8
Denote direction in pairs of opposites, such as medial (toward midline of body) and lateral (away from midline of body).

◆ **REGIONAL TERMS** pp. 8–10
Refer to special areas, such as the cervical (neck), thoracic (chest), and plantar (sole of foot).

Abdominal Cavity. Divided into nine regions by two vertical lines and two horizontal lines; it is also divided into four quadrants by a horizontal plane and a vertical plane.

◆ **BODY PLANES** p. 10

Sagittal Plane. The longitudinal section that divides the body, or its parts, into right and left parts; may be median sagittal or parasagittal.

Frontal Plane. A longitudinal section that runs at right angles to the sagittal plane, dividing the body into anterior and posterior parts.

Transverse Plane. Divides the body or its parts into superior and inferior parts.

◆ **BODY CAVITIES** pp. 10–12
The body contains two main cavities.

Dorsal (Posterior) Cavity. Has two subdivisions:

CRANIAL CAVITY. Houses the brain.

SPINAL (VERTEBRAL) CAVITY. Contains the spinal cord.

Ventral (Anterior) Cavity. Has two subdivisions:

THORACIC CAVITY. Divided into:

PERICARDIAL CAVITY.

PLEURAL CAVITIES. (right and left, around lungs)

ABDOMINOPELVIC CAVITY. Divided into:

ABDOMINAL CAVITY.

PELVIC CAVITY.

◆ **MEMBRANES OF THE VENTRAL BODY CAVITIES** pp. 12–15
Both thoracic and abdominopelvic cavities have visceral membranes that cover organs and parietal membranes that line the outer walls of the cavities. Space between two membrane regions is filled with fluid.

◆ **HOMEOSTASIS** pp. 15–16
Refers to the relatively constant chemical and physical conditions that are maintained within the internal environment of the body.

◆ **HOMEOSTATIC MECHANISMS** pp. 17–19
Means by which the body senses and compensates for changes in the internal environment. The nervous and endocrine systems are the body's principal sensing and controlling systems.

Negative Feedback. Used to maintain homeostasis. It involves several components: a controlled system, set point, receptor, and processing center.

◆ **POSITIVE FEEDBACK** p. 20
Maximizes, rather than minimizes, the difference between the actual and the set point level of a variable. It is called positive because the feedback tends to cause the level of a variable to change in the same direction as an initial change. Not used by body to maintain homeostasis, but is used in particular circumstances, such as childbirth.

Self-Quiz

1. When studying anatomy and physiology at the same time, one should keep in mind the interrelation of: (a) nerves and muscles; (b) nerves and the skeleton; (c) structure and function.
2. The basic living units of the body are: (a) chemical elements; (b) cells; (c) tissues.
3. Nervous tissue is derived from the embryonic: (a) ectoderm; (b) endoderm; (c) mesoderm.
4. The various body cavities are lined with: (a) muscular tissue; (b) epithelial tissue; (c) connective tissue.
5. The component of the word *endocarditis* that signifies "inflamation" is: (a) endo; (b) card; (c) itis.
6. Match the following types of tissues with the appropriate lettered descriptions:

Epithelial tissues	(a) Derived from embryonic mesoderm
Muscular tissues	(b) Derived for the most part from embryonic mesoderm
Nervous tissues	(c) Derived from embryonic ectoderm, mesoderm, and endoderm
Connective tissues	(d) Form the brain, spinal cord, and nerves
	(e) Cover the surface of the body and line the body cavities
	(f) Derived from the embryonic ectoderm
	(g) Aid in digestion by moving food through the digestive tract
	(h) Used for support or attachment

7. Match the following body positions and directional terms with the appropriate description:

Prone position	(a) Toward the head
Anterior (ventral)	(b) Away from the midline of the body
Posterior (dorsal)	(c) Lying face up
Anatomical position	(d) Away from the head
Superior (cranial)	(e) Toward the attached end of a limb
Medial	(f) Sole of the foot
Supine position	(g) Front
Inferior (caudal)	(h) Standing erect with feet together, arms at side, palms forward
Lateral	(i) Away from the attached end of a limb
Proximal	(j) The part of the back between the thorax and pelvis
Cervical	(k) Lying face down
Plantar	(l) Back
Distal	(m) Neck
Lumbar	(n) Toward the midline of the body
Palmar	(o) Anterior surface of hands

8. Which of the following directional terms are paired correctly? (a) superficial and deep; (b) medial and distal; (c) proximal and lateral.
9. The midline region superior to the umbilical region is called the epigastric region, and it contains most of the stomach. True or False?
10. The midline region directly inferior to the umbilical region is the: (a) hypochondriac; (b) lumbar; (c) hypogastric.
11. All sagittal sections other than the median sagittal section are: (a) intersagittal; (b) parasagittal; (c) intrasagittal.
12. The fluid associated with the body's dorsal cavities is the cerebrospinal fluid. True or False?
13. The portion of the thoracic cavity between the two pleural cavities is called the: (a) mediastinum; (b) meninges; (c) foramen magnum.
14. If the membranes of the heart become inflamed, the condition is known as: (a) pericarditis; (b) peritonitis; (c) pleuritis.
15. Which of the following does the abdominal cavity *not* contain? (a) spleen; (b) liver; (c) lungs.
16. Which of the following does the pelvic cavity *not* contain? (a) urinary bladder; (b) liver; (c) female internal reproductive organs.
17. Most of the organs within the abdominopelvic cavity are suspended from the anterior wall of the cavity by a single-layered membrane of peritoneum. True or False?
18. The mesenteries; (a) support the kidneys; (b) are mostly associated with the dorsal cavities; (c) are doubled-layered membranes.
19. The existence of a relatively constant internal environment around the cells of the body is referred to as *homeostasis.* True or False?
20. The homeostasis of the body is generally maintained by means of: (a) positive feedback mechanisms; (b) negative feedback mechanisms; (c) a combination of a and b.

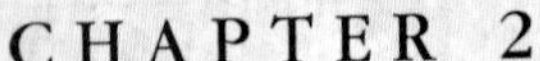

CHAPTER 2

The Chemical and Physical Basis of Life

CHAPTER CONTENTS

LEARNING OBJECTIVES

After completing this chapter, you should be able to:

1. Define the term *chemical element,* and describe the three principal particles that make up an atom.
2. Distinguish between polar and nonpolar covalent bonds, and give an example of each.
3. Describe the composition of carbohydrates, lipids, proteins, and nucleic acids, and give an example of each.
4. Describe how enzymes catalyze metabolic reactions, and cite two examples of the regulation of enzymatic activity.
5. Distinguish among solutions, colloids, and suspensions.
6. Explain three different ways in which the concentrations of solutions are expressed.
7. Define an acid and a base.
8. Explain the use of pH units in the measurement of the hydrogen ion concentration of a solution.
9. Distinguish between diffusion and osmosis, and cite an example of each.
10. Explain the process of dialysis, and describe its application in the artificial kidney.
11. Distinguish between filtration and bulk flow, and cite an example of each.

Although the organs and tissues of the human body differ from one another in both form and function, they are composed of the same basic materials. The body is made up of chemical elements that interact with one another to form the anatomical structures and carry out the physiological processes characteristic of a living organism.

Chemical Elements and Atoms

A **chemical element** is a substance that cannot be broken down into simpler material by chemical means, and a **compound** is a substance composed of two or more different elements in a fixed ratio. Each chemical element has its own name and a one- or two-letter symbol. For example, the symbol for the element oxygen is O, and the symbol for the element sodium is Na. At the present time, approximately 109 chemical elements are recognized, but only about 24 of these are normally found in the body (Table 2.1).

A chemical element is made up of extremely small units of matter called **atoms,** which are themselves composed of even smaller particles (Figure 2.1). Positively charged particles called **protons** are located in the central area, or nucleus, of an atom. Uncharged particles called **neutrons,** if present, are also located in the nucleus. Negatively charged particles called **electrons** are in constant motion around the nucleus.

The number of protons and the number of electrons in an atom—and, therefore, the number of positive charges and the number of negative charges—are equal.

◆ **FIGURE 2.1 An atom of the element helium**

Two protons (plus signs) and two neutrons (black spheres) are located in the nucleus, and two electrons (minus signs) are in constant motion around the nucleus. The electrons do not follow fixed pathways as they move around the nucleus, but the light colored area indicates the region where they are located most of the time.

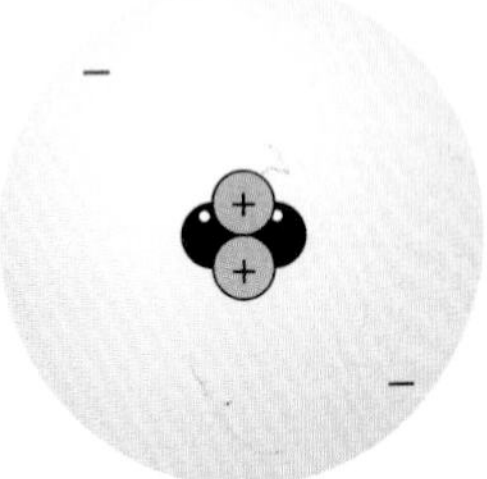

◆ **FIGURE 2.2 Diagrammatic representation of the first four electron energy levels ($n = 1, n = 2, n = 3,$ and $n = 4$)**

The sphere in the center is the nucleus. The maximum number of electrons that can occupy each energy level is given.

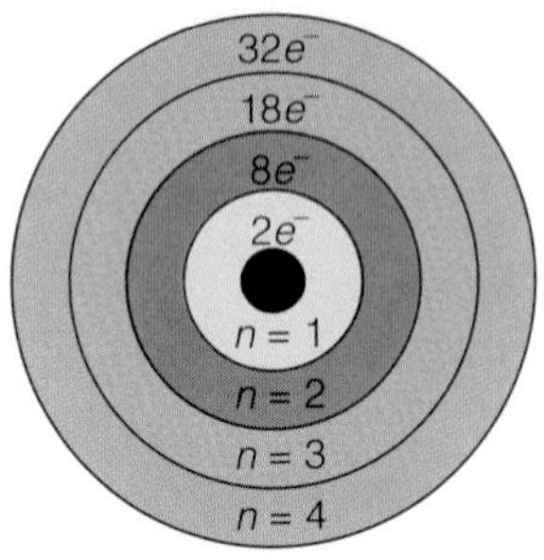

Consequently, an atom has no overall electrical charge and is electrically neutral.

Protons, neutrons, and electrons, like the atoms they compose, are matter, and, like all matter, they occupy space and possess mass. Each proton or neutron, however, has over 1800 times the mass of an electron. Thus, most of the mass of an atom is concentrated in the nucleus.

Electron Energy Levels

Energy is the capacity to do work, and electrons possess different amounts of energy. In fact, the present view of atomic structure suggests that the electrons of an atom should be assigned to particular energy levels, which are numbered, starting with the lowest as one. The numbers are called **principal quantum numbers** (n). The lowest energy level ($n = 1$) can contain a maximum of 2 electrons, the second energy level ($n = 2$) can contain 8 electrons, the third energy level ($n = 3$) can contain 18 electrons, and the fourth energy level ($n = 4$) can contain 32 electrons (Figure 2.2). Above the fourth energy level are still higher energy levels. (The electron energy levels are sometimes referred to as shells, with the K shell being equivalent to the $n = 1$ energy level, the L shell to the $n = 2$ energy level, the M shell to the $n = 3$ level, and so on.)

Electrons do not follow fixed pathways at they move around the nucleus of an atom, and it is impossible to determine the precise position of a specific electron at any one moment. However, electrons occupying higher energy levels are generally located farther from the nucleus than electrons occupying lower energy levels.

 Atomic Number, Mass Number, and Atomic Weight

An atom's **atomic number** indicates the number of protons in the atom. The combined number of protons and neutrons is given by the atom's **mass number.** The number of neutrons in an atom is equal to the difference between the atom's atomic number and mass number. The atomic number of an atom is often indicated by a subscript preceding the symbol of the atom's chemical element, and the mass number is indicated by a superscript preceding the symbol. For example, an atom of oxygen with an atomic number of 8 and a mass number of 16 is written as ${}^{16}_{8}O$.

The total mass of an atom is called its **atomic weight.** The atomic weight of an atom is almost but not exactly equal to the sum of the masses of its constituent

◆ **TABLE 2.1 Chemical Elements of the Human Body**

ELEMENT	SYMBOL	REPRESENTATIVE FUNCTIONS
Carbon	C	A primary constituent of organic molecules, such as carbohydrates, lipids, and proteins.
Hydrogen	H	A component of organic molecules and water. As an ion (H^+), it affects the pH of body fluids.
Nitrogen	N	A component of amino acids, proteins, and nucleic acids.
Oxygen	O	A component of many molecules, including water. As a gas, it is important in cellular respiration.
Calcium	Ca	A component of bones and teeth. It is required for proper muscle activity and blood clotting.
Chlorine	Cl	Ionic chlorine (Cl^-) is one of the major anions of the body.
Iodine	I	A constituent of the thyroid hormones thyroxine and triiodothyronine.
Iron	Fe	A constituent of the hemoglobin molecule. It is also a component of a number of respiratory enzymes.
Magnesium	Mg	Found in bone. It is also an important coenzyme in a number of reactions.
Phosphorus	P	A component of bones, teeth, many proteins, and nucleic acids. It is also a constituent of energy compounds such as adenosine triphosphate (ATP) and creatine phosphate (CP).
Potassium	K	As an ion, potassium (K^+) is the major intracellular cation. Potassium is important in the conduction of nerve impulses and in muscle contraction.
Sodium	Na	As an ion, sodium (Na^+) is the major extracellular cation. It is important in water balance and in the conduction of nerve impulses.
Sulfur	S	A component of many proteins, particularly the contractile protein of muscle.
Chromium	Cr	These substances are required by the body in very small amounts. They are referred to as *trace elements.*
Cobalt	Co	
Copper	Cu	
Fluorine	F	
Manganese	Mn	
Molybdenum	Mo	
Selenium	Se	
Silicon	Si	
Tin	Sn	
Vanadium	V	
Zinc	Zn	

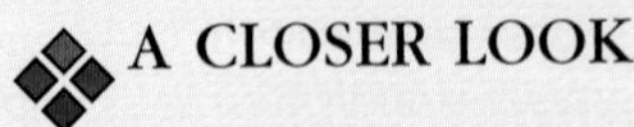
A CLOSER LOOK

Isotopes and Radiation

All the atoms of a particular chemical element have the same number of protons. However, they can have different numbers of neutrons. These different forms of a given element are called **isotopes.**

The isotopes of a given element have the same atomic number (number of protons) but different mass numbers (due to different numbers of neutrons). For example, the most common isotope of the element oxygen has eight protons and eight neutrons in its nucleus ($^{16}_{8}O$). However, less common isotopes of oxygen containing eight protons and nine neutrons ($^{17}_{8}O$) and eight protons and ten neutrons ($^{18}_{8}O$) also exist. In nature, oxygen occurs as a mixture of these isotopes (99.76%, $^{16}_{8}O$; 0.039%, $^{17}_{8}O$; and 0.20%, $^{18}_{8}O$).

A number of isotopes are radioactive; that is, they emit various kinds of radiation. Radioactive emissions can be either particlelike, or they can take the form of electromagnetic rays.

Particulate Radiation

Alpha Radiation. An alpha particle is essentially a helium nucleus (composed of two protons and two neutrons) that is positively charged. Alpha particles are emitted by some radioactive isotopes, such as uranium ($^{238}_{92}U$).

Beta Radiation. A beta particle has essentially the same mass as an electron, but it can be either positively or negatively charged. Positive beta particles, which are called positrons, are emitted by such isotopes as $^{30}_{15}P$. Negative beta particles, which are called negatrons, are emitted by $^{14}_{6}C$ and $^{136}_{53}I$.

Electromagnetic Radiation: Gamma Radiation

A gamma ray can be regarded as a bundle of energy that is similar to a high-energy X ray. It has no detectable mass or electrical charge. Gamma ray emissions frequently accompany positive or negative beta emissions.

Transmutation

When a radioactive element emits radiation, it is itself altered. The radioactive element disappears and a different element appears. Thus, atoms of one element spontaneously change into atoms of another element. This change of one element into another is called **transmutation.** For example, the emission of an alpha particle ($^{4}_{2}He$) by uranium ($^{238}_{92}U$) forms thorium ($^{234}_{90}Th$) as follows:

$$^{238}_{92}U \rightarrow \, ^{234}_{90}Th + \, ^{4}_{2}He$$

Biological Uses of Radioactive Isotopes

Radioactive isotopes have proven very useful in studies of living organisms and the chemical reactions that occur within them. The radioactive emissions of different isotopes can be measured by a variety of means. Consequently, radioactive carbon and other radioactive isotopes that can be introduced into the body by ingestion, inhalation, or injection are used as tracers and followed through a variety of physiological processes. For example, the thyroid gland normally removes iodine from the blood and uses it in the formation of thyroid hormones. Thus, radioactive iodine, which is also taken up by the thyroid gland, can be used to study thyroid function.

Since radioactive emissions can damage tissues, high levels of radiation are used to treat certain disease conditions. In some forms of cancer, radiation treatments are used to destroy actively dividing cancer cells.

protons, neutrons, and electrons. The discrepancy is due to the fact that when protons, neutrons, and electrons combine to form an atom, some of their mass is converted to energy and is given off. The atomic numbers, mass numbers, and atomic weights of atoms of the chemical elements present in the body are given in Table 2.2.

Chemical Bonds

When atoms are close enough to one another, the outer electrons of one atom may interact with those of others. As a result, attractive forces can develop between atoms that are strong enough to hold the atoms together. These attractive forces are called **chemical bonds.**

In general, many atoms in the human body appear to be particularly stable when their highest electron energy levels are either filled or contain eight electrons. Therefore, much chemical bonding results in atoms whose highest electron energy levels are either filled or contain eight electrons.

Covalent Bonds and Molecules

In many cases, chemical bonds result from atoms sharing electrons. Bonds based on electron sharing are called **covalent bonds,** and two or more atoms held together as a unit by covalent bonds are known as a **molecule.** Many covalent bonds are *single covalent bonds,* in which two atoms share one pair of electrons. However, *double covalent bonds,* in which two atoms share two pairs of electrons, and *triple covalent bonds,* in which two atoms share three pairs of electrons, also occur.

Nonpolar Covalent Bonds

A covalent bond in which two atoms share electrons equally, and one atom does not attract the shared electrons more strongly than the other atom, is called a **nonpolar covalent bond.** For example, the hydrogen atom

◆ TABLE 2.2 Atomic Numbers, Mass Numbers, and Atomic Weights of Atoms of Body Elements*

ELEMENT	SYMBOL	ATOMIC NUMBER	MASS NUMBER	ATOMIC WEIGHT
Calcium	Ca	20	40	40.08
Carbon	C	6	12	12.011
Chlorine	Cl	17	35	35.453
Chromium	Cr	24	52	51.996
Cobalt	Co	27	59	58.933
Copper	Cu	29	63	63.546
Fluorine	F	9	19	18.998
Hydrogen	H	1	1	1.008
Iodine	I	53	127	126.905
Iron	Fe	26	56	55.847
Magnesium	Mg	12	24	24.305
Manganese	Mn	25	55	54.938
Molybdenum	Mo	42	98	95.94
Nitrogen	N	7	14	14.007
Oxygen	O	8	16	15.999
Phosphorus	P	15	31	30.974
Potassium	K	19	39	39.098
Selenium	Se	34	80	78.96
Silicon	Si	14	28	28.086
Sodium	Na	11	23	22.99
Sulfur	S	16	32	32.064
Tin	Sn	50	120	118.69
Vanadium	V	23	51	50.942
Zinc	Zn	30	64	65.38

*The mass numbers are those of the most common isotope of the element. The atomic weights (expressed in atomic mass units, also called daltons) are weighted averages of the atomic weights of the naturally occurring isotopes of the element.

◆ **FIGURE 2.3 A hydrogen molecule**

(a) Covalent bonding of two hydrogen atoms to form a hydrogen molecule. In this situation, the electrons (minus signs) are shared equally between the two nuclei (plus signs), resulting in a nonpolar covalent bond and a nonpolar hydrogen molecule. (b, c) Alternative methods of representing the composition of a molecule (in this case, a hydrogen molecule) that will be used in later illustrations. In molecular diagrams (b), a shared pair of electrons (a single covalent bond) is represented by a straight line connecting the two bonded atoms. In molecular formulas (c), the number of atoms of each element that are present in a molecule is indicated, but there is little or no information as to how the atoms are connected.

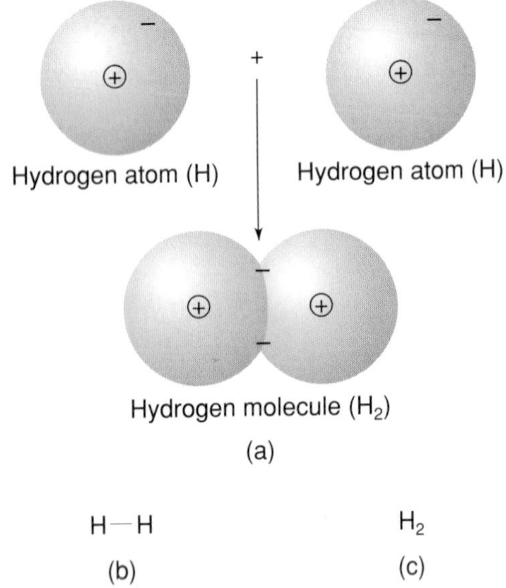

possesses one proton in its nucleus and one electron in its $n = 1$ energy level. When two hydrogen atoms combine to form a hydrogen molecule, their single electrons are shared equally between the two nuclei so that each electron spends equal time in the vicinity of each nucleus (Figure 2.3). Although neither atom gains complete possession of the other's electron, this sharing allows both to fill their highest ($n = 1$) electron energy levels.

Because the two atoms of a hydrogen molecule share electrons equally, the molecule's center of positive charge and center of negative charge are at the same location. Such a molecule is called a **nonpolar molecule.** Thus, the hydrogen molecule is an example of a nonpolar molecule.

Polar Covalent Bonds

A covalent bond in which there is an unequal sharing of electrons between two atoms, with one atom attracting the shared electrons more strongly than the other atom, is called a **polar covalent bond.** The degree of polarity of such bonds can vary, depending on how strongly one atom is able to attract shared electrons from another atom.

The unequal electron sharing that occurs in polar covalent bonds can give rise to polar molecules. A **polar molecule** has its center of positive charge at a different location than its center of negative charge, and it has both positive and negative areas. For example, a water molecule is a polar molecule composed of two hydrogen atoms, each of which is linked by a polar covalent bond to a single oxygen atom (Figure 2.4). In these bonds, the shared electrons are more strongly attracted to the oxygen of the water molecule than to the hydrogens. As a result, the shared electrons spend more time in the vicinity of the oxygen nucleus than in the vicinities of the hydrogen nuclei. Since electrons are negatively charged, the oxygen portion of the molecule becomes somewhat negative, and the hydrogen portions, with the positively charged protons of their nuclei less balanced by the presence of electrons, become somewhat positive.

Ionic Bonds and Ions

Often, the attraction of one atom for the electrons of another is so strong that electrons are not shared but are actually transferred from one atom to another; that is, they spend essentially all of their time in the vicinity of one nucleus and none in the vicinity of the other. This leaves one atom negatively charged (the one that gains electrons) and one atom positively charged (the one that loses electrons). Such charged atoms (or aggre-

◆ **FIGURE 2.4 A water molecule**

(a) Two hydrogen atoms (H) bonded covalently to an oxygen atom (O) to form a molecule of water. The shared electrons in this situation are attracted more strongly to the oxygen nucleus than to the hydrogen nuclei, resulting in polar covalent bonds and giving rise to a polar water molecule. (b) A molecular diagram of a water molecule. (c) The molecular formula of a water molecule.

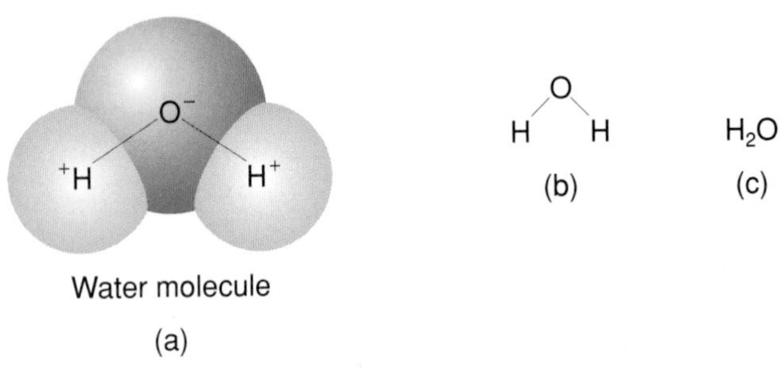

◆ **FIGURE 2.5 Ions and ionic compounds**
(a) In the reaction between sodium and chlorine, each sodium atom loses the single electron (black dot) in its highest ($n = 3$) electron energy level to a chlorine atom. This produces positively charged sodium ions, which have their highest ($n = 2$) electron energy levels filled, and negatively charged chloride ions, which have a stable eight electrons in their highest ($n = 3$) electron energy levels. The number of protons (P) and neutrons (N) in the different nuclei is indicated. (b) Positively charged sodium ions and negatively charged chloride ions attract one another, forming crystals of solid sodium chloride (table salt). In a sodium chloride crystal, each sodium ion is surrounded by six chloride ions, and each chloride ion is surrounded by six sodium ions.

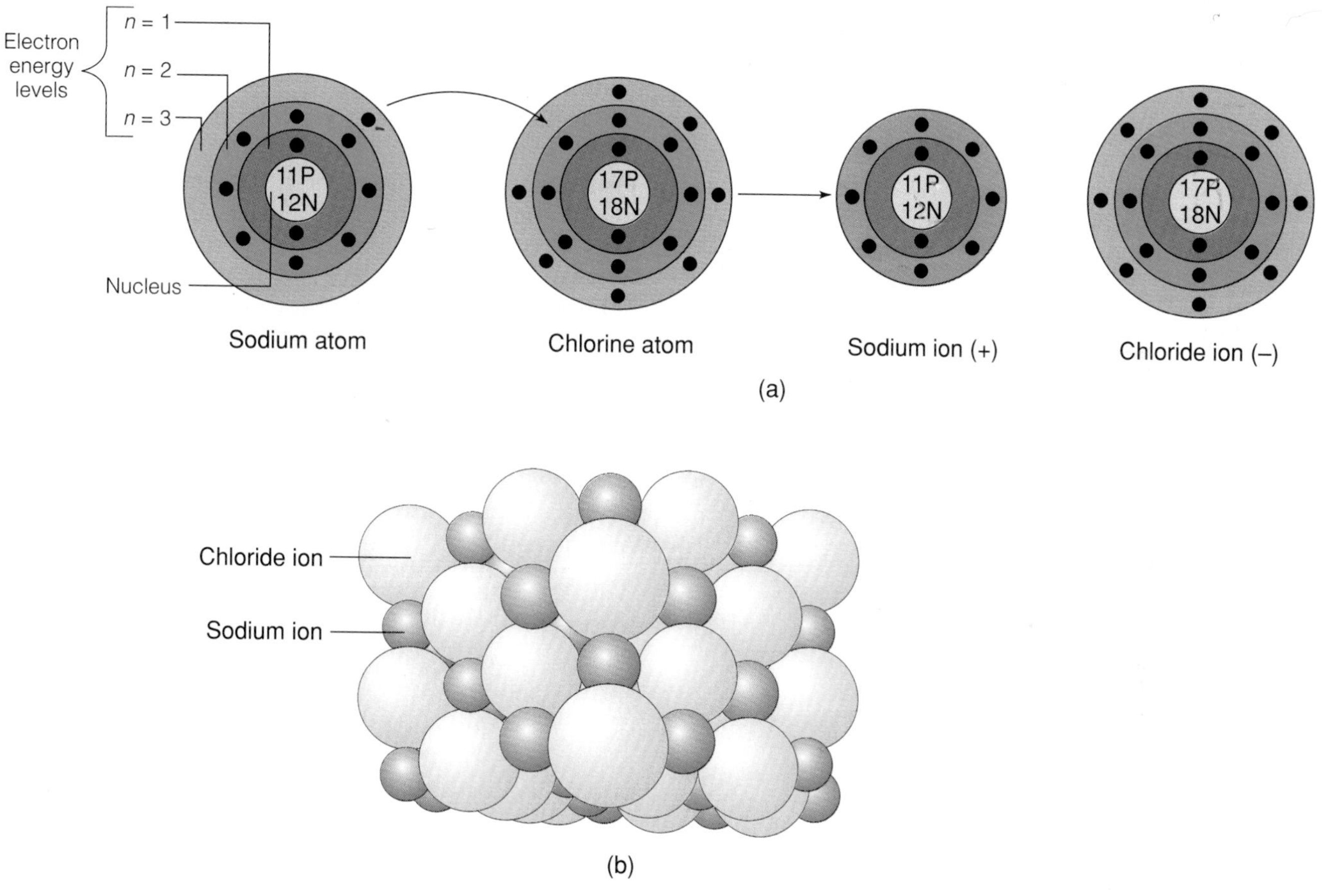

gates of atoms) are called **ions.** Positively charged ions are called **cations,** and negatively charged ions are called **anions.**

Opposite charges attract one another, and oppositely charged ions can be held together by this attraction to form electrically neutral ionic compounds. Such attractions are called **ionic attractions** or **ionic bonds.** For example, in the reaction between sodium and chlorine, each sodium atom loses the single electron in its highest ($n = 3$) electron energy level to a chlorine atom (Figure 2.5a). This produces positively charged sodium ions, which have their highest ($n = 2$) electron energy levels filled, and negatively charged chloride ions, which have a stable eight electrons in their highest ($n = 3$) electron energy levels. The positively charged sodium ions and negatively charged chloride ions attract one another, forming crystals of solid sodium chloride. In a sodium chloride crystal, sodium ions and chloride ions are packed into a three-dimensional lattice in such a way that each positive sodium ion is surrounded on four sides and top and bottom by negative chloride ions, and each chloride ion is similarly surrounded by six sodium ions (Figure 2.5b). This arrangement is a particularly stable one of positive and negative charges. Strictly speaking, there are no molecules in ionic compounds; there are only ordered arrays of ions in which no one positively charged ion belongs to any one negatively charged ion.

Crystals of sodium chloride are commonly called table salt. In chemical terms, however, a **salt** is generally considered to be an ionic compound that is composed of cations other than hydrogen ions (H^+) and anions other than hydroxide ions (OH^-) or oxide ions (O^{-2}). Thus, there are many salts in addition to sodium chloride.

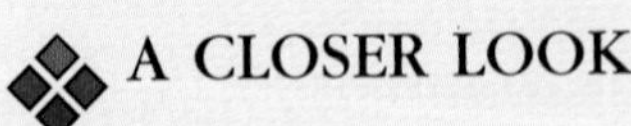

A CLOSER LOOK

Electronegativity

When considering the chemical bonds that occur between atoms, covalent bonds and ionic bonds can be thought of as parts of a single continuous pattern. At one end of the pattern are nonpolar covalent bonds in which the bonded atoms each attract electrons equally. Consequently, the atoms share electrons equally between them. In the middle are polar covalent bonds in which one of the bonded atoms attracts electrons more strongly than the other atom. Although the atoms still share electrons, the sharing is unequal. At the other end are ions and ionic bonds, which are the result of one atom attracting electrons so much more strongly than another that electrons are transferred from one atom to another. Thus, ions and ionic bonds represent an extreme degree of polarity, but because the pattern is continuous, there is no distinct line between covalent bonds and ionic bonds. In fact, some chemical compounds spend part of their time in a polar covalent state and part of their time being ionic.

The **electronegativity** of an atom is a measure of the attraction that the atom has for electrons in a bond it has formed with another atom. An atom with a high electronegativity attracts electrons more strongly than an atom with a lower electronegativity. Nonpolar covalent bonds commonly occur between atoms that have similar electronegativities, and ions and ionic bonds frequently result when atoms that have greatly different electronegativities interact. Polar covalent bonds often occur between atoms that have different, but not greatly different, electronegativities (Table C2.1).

◆ **TABLE C2.1 Electronegativity and Bond Type**

ATOMS AND ELECTRONEGATIVITIES	ELECTRONEGATIVITY DIFFERENCE	BOND TYPE
H (2.1) H (2.1)	0.0	Nonpolar covalent
N (3.0) H (2.1)	0.9	Polar covalent
Na (0.9) Cl (3.0)	2.1	Ionic

If the electronegativity difference between two atoms is greater than about 2.0, the bond between the two atoms is usually ionic.

Hydrogen Bonds

Oppositely charged regions of polar molecules can attract one another. An attraction of this sort that involves hydrogen and certain other atoms such as oxygen or nitrogen is called a **hydrogen bond.** For example, as noted previously, in the polar water molecule, the hydrogen portions of the molecule are somewhat positive, and the oxygen portion is somewhat negative. Consequently, the hydrogen portions of a water molecule can attract the oxygen portion of nearby water molecules, forming hydrogen bonds (Figure 2.6). Hydrogen bonds also occur in proteins and other large molecules found in the body.

◆ **FIGURE 2.6 Hydrogen bonds (dotted lines) between polar water molecules**

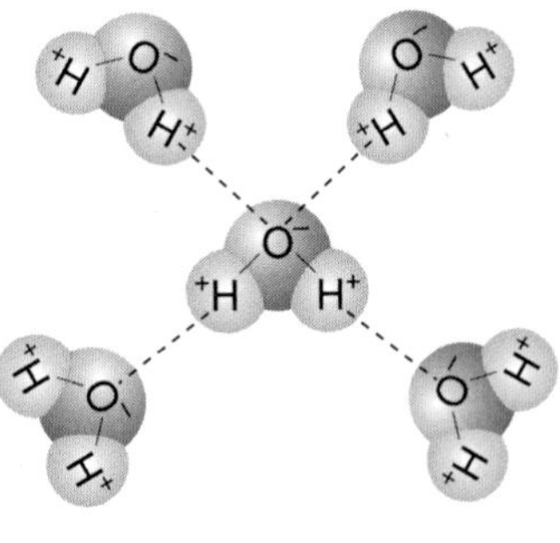

◆ **FIGURE 2.7 A carbon atom bonded covalently to four hydrogen atoms to form a molecule of methane (CH_4), which is an example of a hydrocarbon**

```
    H
    |
H — C — H
    |
    H
```

Carbon Chemistry

A tremendous number of molecules, especially those present in living organisms, contain the element carbon. A carbon atom has four electrons in its highest ($n = 2$) electron energy level, which it can share with other atoms. Carbon atoms, therefore, can each form four covalent bonds, and they commonly bond with other carbon atoms or with hydrogen, oxygen, or nitrogen atoms (Figure 2.7).

With a few exceptions, chemicals that contain carbon are classified as *organic chemicals;* all other chemicals are classified as *inorganic chemicals.* The exceptions include pure carbon in the form of diamond and graphite, carbon dioxide (CO_2), carbon monoxide (CO), carbonates such as limestone ($CaCO_3$), bicarbonates such as baking soda ($NaHCO_3$), and cyanides such as sodium cyanide (NaCN). Even though these chemicals contain carbon, they are considered to be inorganic chemicals. (Organic chemicals are so named because at one time they were thought to come only from living or once-living sources.)

Although both organic and inorganic chemicals are essential to life, much of the body's chemistry is organic in nature. Several major groups of organic chemicals are of particular importance to the body.

Carbohydrates

Carbohydrates, which include sugars and starch, are a major energy source for the body. In general, carbohydrates are composed of carbon, hydrogen, and oxygen, and the hydrogen and oxygen are frequently present in the same 2:1 ratio as they are in water. An important group of carbohydrates is the **monosaccharides,** or single-unit sugars, such as glucose ($C_6H_{12}O_6$) (Figure 2.8). Monosaccharide molecules can be linked together into larger molecules by synthetic reactions that generally involve the removal of a molecule of water (dehydration) at each linkage. The combination of two monosaccharide units produces a molecule called a **disaccharide.** Sucrose, or table sugar, is a disaccharide formed by bonding a glucose molecule to another monosaccharide called fructose. Much larger carbohydrate molecules can be formed by linking together many monosaccharide units. These molecules are called **polysaccharides.** Glycogen, a storage form of body carbohydrates, is a polysaccharide formed from thousands of glucose units bonded together into a single large molecule.

Lipids

Lipids are stored by the body as energy reserves and are utilized as structural components. Lipids vary greatly in molecular structure, and substances are classified as lipids primarily on the basis of their solubility. Lipids are almost insoluble in water, but they are very soluble in organic solvents such as benzene, ether, and chloroform. The elements carbon, hydrogen, and oxygen are present in lipids, but the proportion of oxygen is much lower than it is in carbohydrates.

Many lipids incorporate fatty acids into their molecular structures. A **fatty acid** consists of a nonpolar chain of carbon atoms that has an acid carboxyl group (COOH) at one end of the chain (Figure 2.9). The most important of the fatty acids are stearic ($C_{17}H_{35}COOH$); palmitic ($C_{15}H_{31}COOH$); oleic ($C_{17}H_{33}COOH$); linolenic ($C_{17}H_{29}COOH$); linoleic ($C_{17}H_{31}COOH$): and arachidonic ($C_{19}H_{31}COOH$).

Fatty acids often combine with the alcohol glycerol (Figure 2.10). The combination occurs at the carboxyl-group end of the fatty acid by the removal of a molecule of water, which commonly occurs when the organic molecules of the body join with one another. If a single fatty acid molecule attaches to the glycerol molecule, the product is called a **monoglyceride (monoacylglycerol).** If two fatty acids attach to the glycerol molecule, the product is a **diglyceride (diacylglycerol),** and if three fatty acids attach, a **triglyceride (triacylglycerol)** is formed. Triglycerides are simple lipids that are classified as **fats** or **oils** on the basis of their melting points. Fats are solids at room temperature whereas oils are liquids.

Another category of lipids is the **phospholipids.** Phospholipids have a phosphate group (derived from phosphoric acid) incorporated into their molecular structures (Figure 2.11a and b, page 34). In addition, the phospholipids present in the body usually have an alcohol such as choline attached to the phosphate, and they possess two chains of carbon atoms.

The phosphate portion of a phospholipid molecule is polar, and the alcohol attached to the phosphate is generally polar as well. The two chains of carbon atoms are nonpolar. Consequently, a phospholipid has a polar region in the area of the molecule where the phosphate

◆ **FIGURE 2.8 Carbohydrates**

(a) The monosaccharide glucose. (b) Monosaccharides are joined by bonds in a reaction that generally involves the removal of a molecule of water (a dehydration synthesis reaction, as shown here by the color areas). (c) A portion of the polysaccharide glycogen; the ring structures represent glucose molecules.

Glucose

(a)

Monosaccharides:

Glucose
Fructose
Galactose

Glucose + Fructose → Sucrose + H_2O

(b)

Disaccharides:

Sucrose
(Glucose & Fructose)
Lactose
(Glucose & Galactose)
Maltose
(Glucose & Glucose)

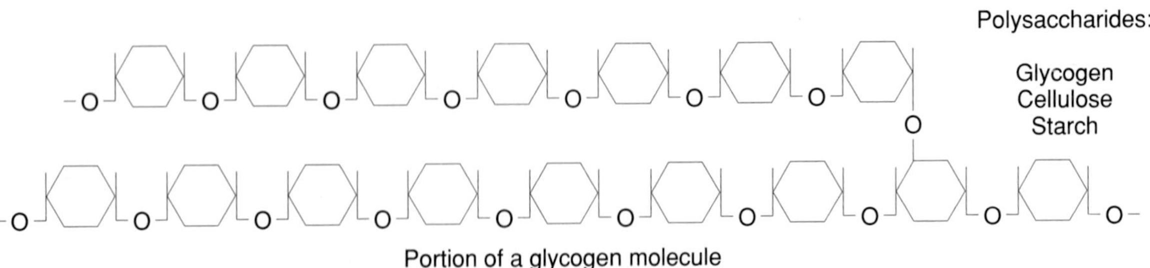

Portion of a glycogen molecule

(c)

◆ **FIGURE 2.9 Fatty acids**

Saturated fatty acids contain only single covalent bonds between carbon atoms. Unsaturated fatty acids contain at least one double covalent bond between carbon atoms (a double covalent bond is one in which two pairs of electrons are shared between atoms; it is indicated by two straight lines connecting the bonded atoms).

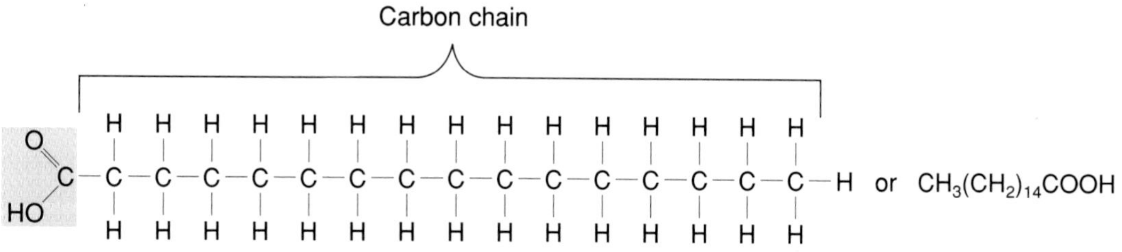

Carbon chain

$CH_3(CH_2)_{14}COOH$

Palmitic acid, a typical saturated fatty acid

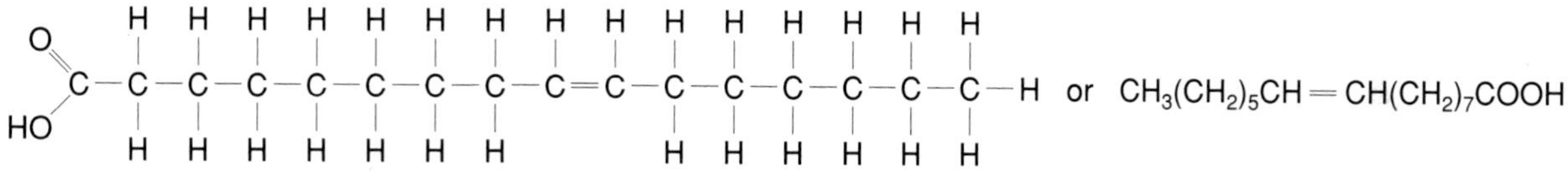

$CH_3(CH_2)_5CH{=}CH(CH_2)_7COOH$

Oleic acid, a typical unsaturated fatty acid

◆ **FIGURE 2.10 Triglycerides**

(a) Glycerol and three fatty acids combine to form a triglyceride.
(b) Diagrammatic representation of the structure of a triglyceride.

Glycerol + 3 Fatty acids → Triglyceride + 3 H_2O

(a)

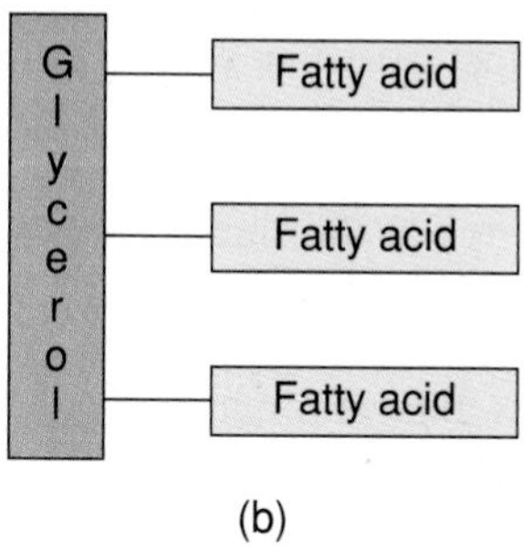

(b)

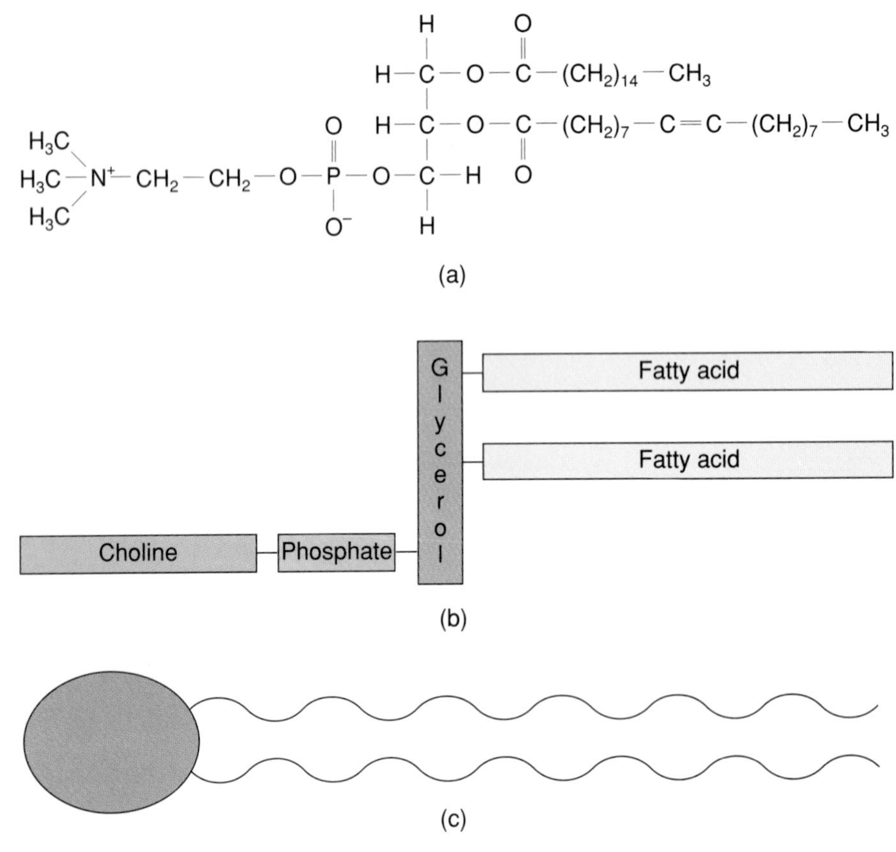

◆ **FIGURE 2.11 Phospholipids**
(a) An example of a phospholipid. This phospholipid is called phosphatidyl choline.
(b) Diagrammatic representation of the components of the phospholipid depicted in part (a).
(c) Diagrammatic representation of the polar and nonpolar regions of the phospholipid depicted in part (a). The choline and phosphate portions of the chemical structure contribute to the polar head of the molecule. The carbon chains of the fatty acid portions contribute to the nonpolar tails.

and attached alcohol are located (Figure 2.11c). It also has a nonpolar region in the area of the molecule where the two nonpolar chains of carbon atoms are located.

A molecule such as a phospholipid molecule that has both polar and nonpolar regions is called **amphipathic.** In an aqueous (water) environment, many amphipathic molecules form spherical clusters known as **micelles** (Figure 2.12). The polar regions of the molecules are located at the surface of the micelle, where they associate (in part by hydrogen bonds) with water molecules, and the nonpolar regions are oriented toward the center of the micelle.

The two nonpolar chains of carbon atoms of a phospholipid molecule are too bulky to permit phospholipids to form micelles easily. Instead, in an aqueous environment, phospholipid molecules form *lipid bilayers* (also called *bimolecular sheets*), in which the polar regions of the molecules are located at the surfaces of the bilayer, where they associate with water molecules, and the nonpolar regions are oriented toward the interior of the bilayer (Figure 2.13). This bilayer organization of phospholipid molecules forms the basic structure of cell membranes.

An additional category of lipids is the steroids. **Steroids** basically consist of four interconnected rings of carbon atoms that have few polar groups attached (Figure 2.14). Cholesterol and some hormones—for example, the sex hormones—are steroids that are important in the body.

Proteins

Proteins are components of many body structures. In addition, a large number of proteins function as enzymes, which play critical roles in the chemical reactions that occur within the body. **Proteins** are large, complex molecules that are formed from smaller molecules called **amino acids.** Generally, amino acids have a central or alpha carbon to which is attached a hydrogen atom (H), an acid carboxyl group (COOH), an amino group (NH_2), and a fourth group that differs from one amino acid to another and is often indicated by the letter R (Figure 2.15). Thus, amino acids, and consequently proteins, contain nitrogen (from the amino group) in addition to carbon, hydrogen, and oxygen. They may also contain other elements, such as sulfur, depending on the constitution of the individual R groups. Approximately 20 different amino acids (different because they possess different R groups) are com-

◆ **FIGURE 2.12 A micelle**

In an aqueous environment, amphipathic molecules become organized into micelles.

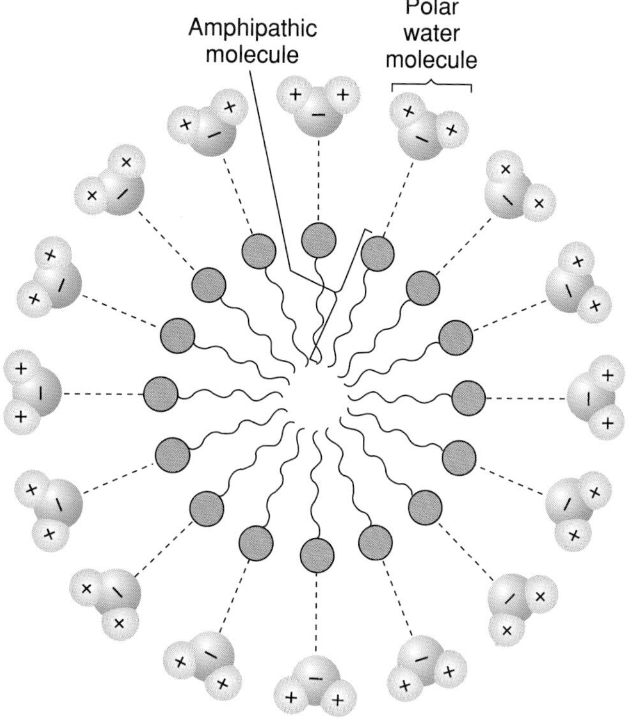

monly found in the proteins of the human body (Table 2.3). However, no one protein necessarily has all these different amino acids in its structure.

Proteins are formed from amino acids by reactions that bond the amino group of one amino acid to the acid carboxyl group of another, with the simultaneous loss of a molecule of water (Figure 2.16). This bond is called a **peptide bond.** Two amino acids joined together by a peptide bond form a **dipeptide.** Approximately ten or more amino acids linked into a chain by peptide bonds form a **polypeptide.** A **protein** is a chain of approximately 100 or more amino acids linked by peptide bonds.

The sequence of amino acids in a polypeptide chain or protein constitutes what is called the **primary structure** of the molecule. Hydrogen bonds that occur principally between the constituents of the different peptide bonds of the linked amino acids provide the polypeptide or protein with a **secondary structure.** For example, the hydrogen bonds cause some amino acid chains to form a coiled, helical structure called the alpha helix (Figure 2.17, page 38). Interactions between atoms of the R groups of different amino acids of an amino acid chain also occur. These interactions cause the amino acid chain (which may be in a helical configuration) to fold into a particular three-dimensional configuration. This folding provides the polypeptide or protein with a **tertiary structure** (Figure 2.18, page 38). Protein molecules can consist of a single amino acid chain, or several chains may link together through R-group interactions to form a multichain protein molecule. These interactions between the different amino acid chains of

◆ **FIGURE 2.13 A lipid bilayer**

In an aqueous environment, phospholipid molecules form lipid bilayers.

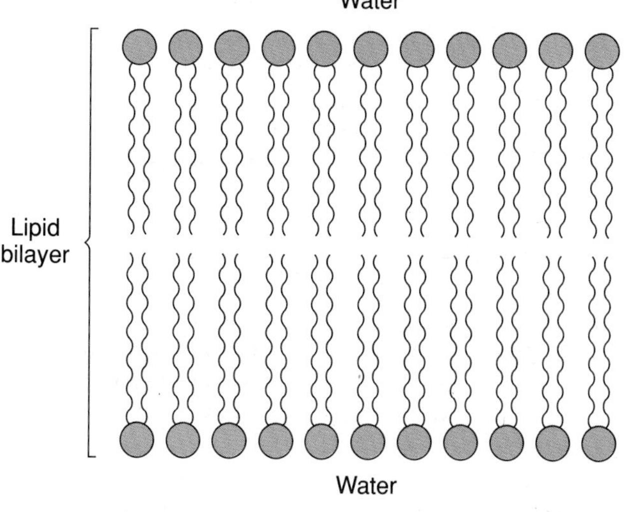

◆ **FIGURE 2.14 Steroids**

Testosterone is a male sex hormone. Estradiol and progesterone are female sex hormones. As can be seen from the rings of carbon atoms, carbon atoms and the hydrogen atoms bound to them are often not specifically indicated by the letters C or H, respectively, in the representation of the structure of a carbon-containing molecule. This method of representing molecular structure will be encountered in later figures.

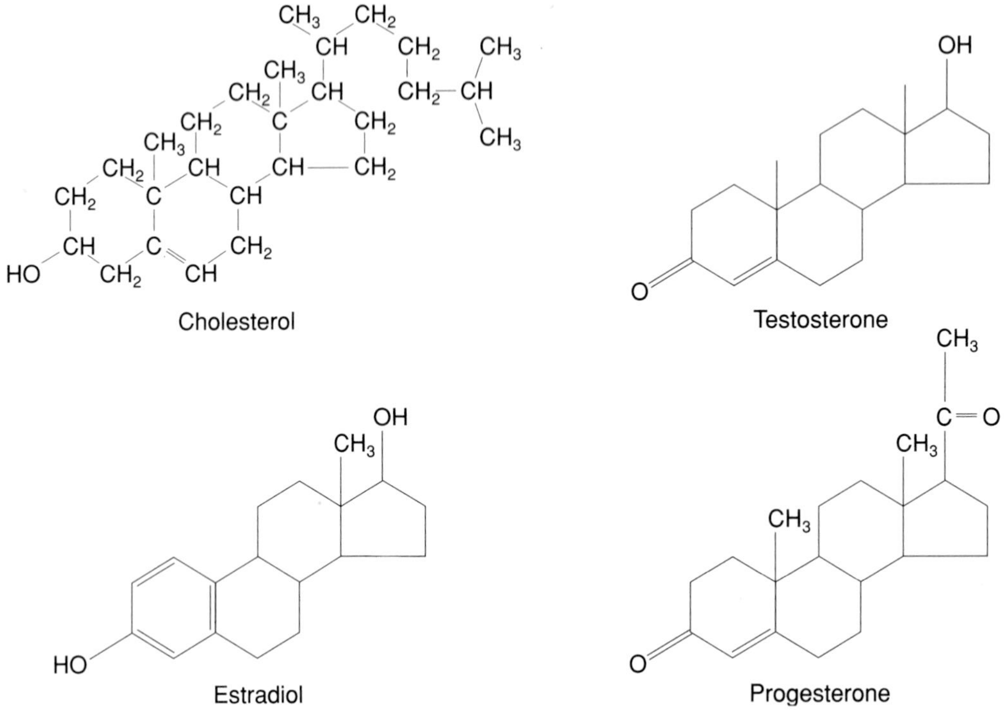

Cholesterol

Testosterone

Estradiol

Progesterone

a multichain protein molecule provide still another level of structural organization to protein molecules—the **quaternary structure** (Figure 2.19).

A protein that exists in its normal three-dimensional configuration is described as being in its *native state.* However, if the interactions that hold a protein in its native state configuration are disrupted, the protein assumes a more random, disorganized orientation and is said to be *denatured.* Denaturation can result when proteins are subjected to heating, variations in pH, or treatment with specific chemicals such as urea, alcohol, or heavy metal ions. In some cases, denaturation is reversible, and under the appropriate conditions, a denatured protein can return to its native state (that is, the protein can renature). In other cases, denaturation is irreversible, as exemplified by the changes that occur in the white of an egg as it is cooked.

Some proteins have carbohydrate attached to them, forming *glycoproteins.* Glycoproteins are present in cell membranes, and several hormones are glycoproteins. Some proteins associate with and play a role in the transport of lipids in the body, particularly in the blood. These protein-lipid associations are referred to as *lipoproteins.*

◆ **TABLE 2.3 The Twenty Amino Acids Found in Proteins**

NAME	THREE-LETTER ABBREVIATION	ONE-LETTER ABBREVIATION
Alanine	Ala	A
Argnine	Arg	R
Asparagine	Asn	N
Aspartic acid	Asp	D
Cysteine	Cys	C
Glutamic acid	Glu	E
Glutamine	Gln	Q
Glycine	Gly	G
Histidine	His	H
Isoleucine	Ile	I
Leucine	Leu	L
Lysine	Lys	K
Methionine	Met	M
Phenylalanine	Phe	F
Proline	Pro	P
Serine	Ser	S
Threonine	Thr	T
Tryptophan	Trp	W
Tyrosine	Tyr	Y
Valine	Val	V

◆ **FIGURE 2.15 General structure of amino acids and specific representative amino acids**

Different amino acids have different R groups, shown here as the color portions of the structures.

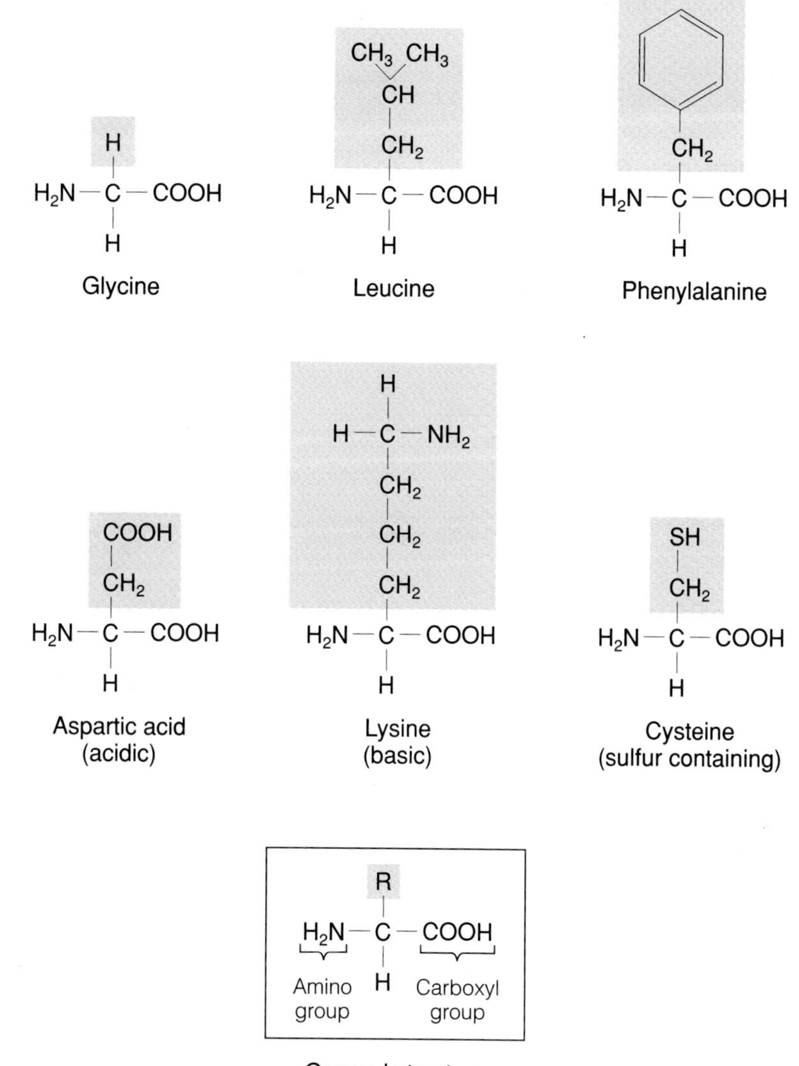

◆ **FIGURE 2.16 Peptide bonds**

Peptide bonds link the amino group of one amino acid to the acid carboxyl group of another.

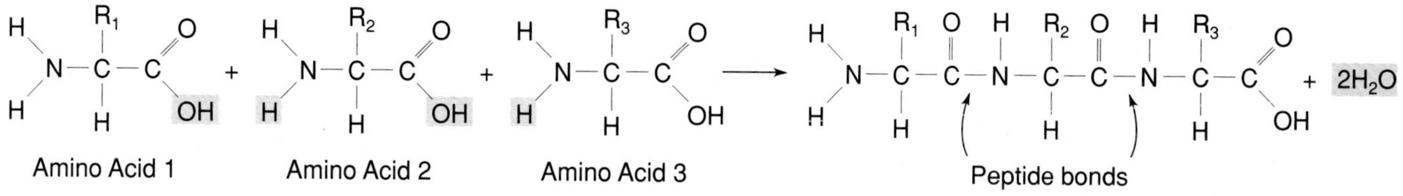

Amino Acid 1 + Amino Acid 2 + Amino Acid 3 → + $2H_2O$

Peptide bonds

Nucleic Acids

Nucleic acids store and transmit information that is needed to synthesize the particular polypeptides and proteins present in the body's cells. Nucleic acids are complex molecules composed of structures known as purine and pyrimidine bases, five-carbon sugars (pentoses), and phosphate groups (which contain phosphorus and oxygen). A single base-sugar-phosphate unit is called a **nucleotide** (Figure 2.20a). Individual nucleotides are linked together into a polynucleotide chain by bonds between the phosphate group of one nucleotide and the sugar of the next (Figure 2.20b). If the nucleo-

◆ FIGURE 2.17 Secondary alpha helical structure of proteins

Dotted lines indicate hydrogen bonds.

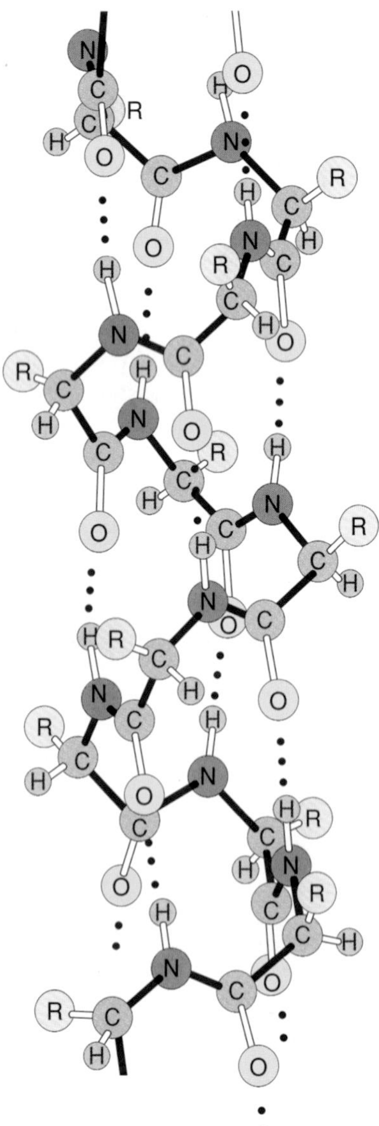

◆ FIGURE 2.18 Tertiary structure of a molecule of the protein myoglobin

Areas of helical secondary structure are also evident. The blue tube that appears to enclose the helical secondary structure has been drawn here to make it easier to visualize the molecule's tertiary structure.

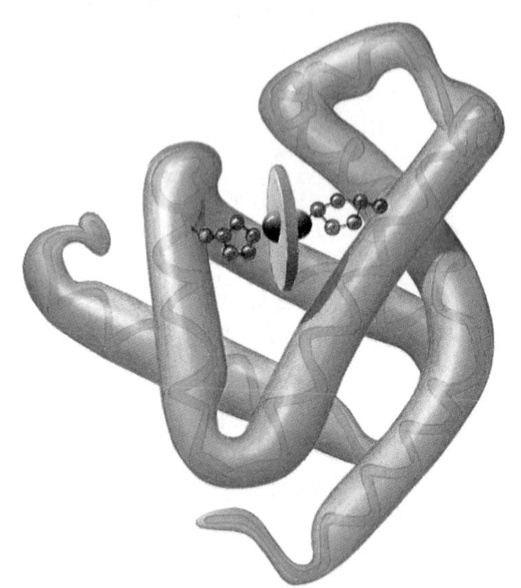

◆ FIGURE 2.19 Quaternary structure of the protein hemoglobin

A single hemoglobin molecule is composed of four polypeptide chains linked to one another. The particular configuration of the linked chains is called the quaternary structure of the protein.

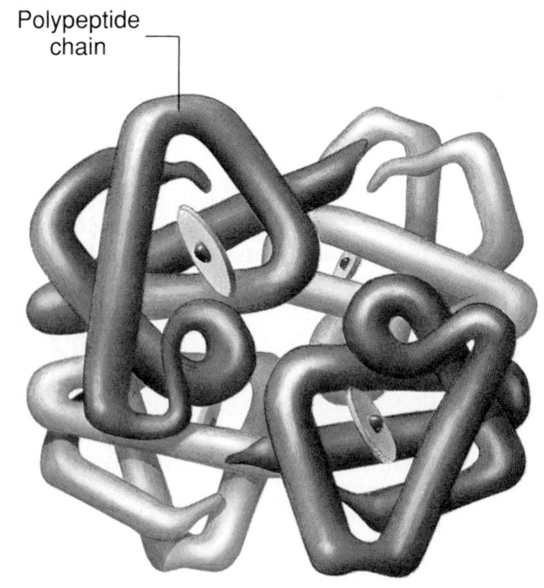

◆ **FIGURE 2.20 A nucleotide and a polynucleotide chain**

(a) The nucleotide pictured here contains the base cytosine and the sugar deoxyribose. The sugar ribose has the same structure, with the exception that the hydrogen indicated by the arrow is replaced by a hydroxyl (OH) group. (b) Four nucleotides linked in a polynucleotide chain. Adenine and guanine are purine bases, and cytosine and thymine are pyrimidine bases.

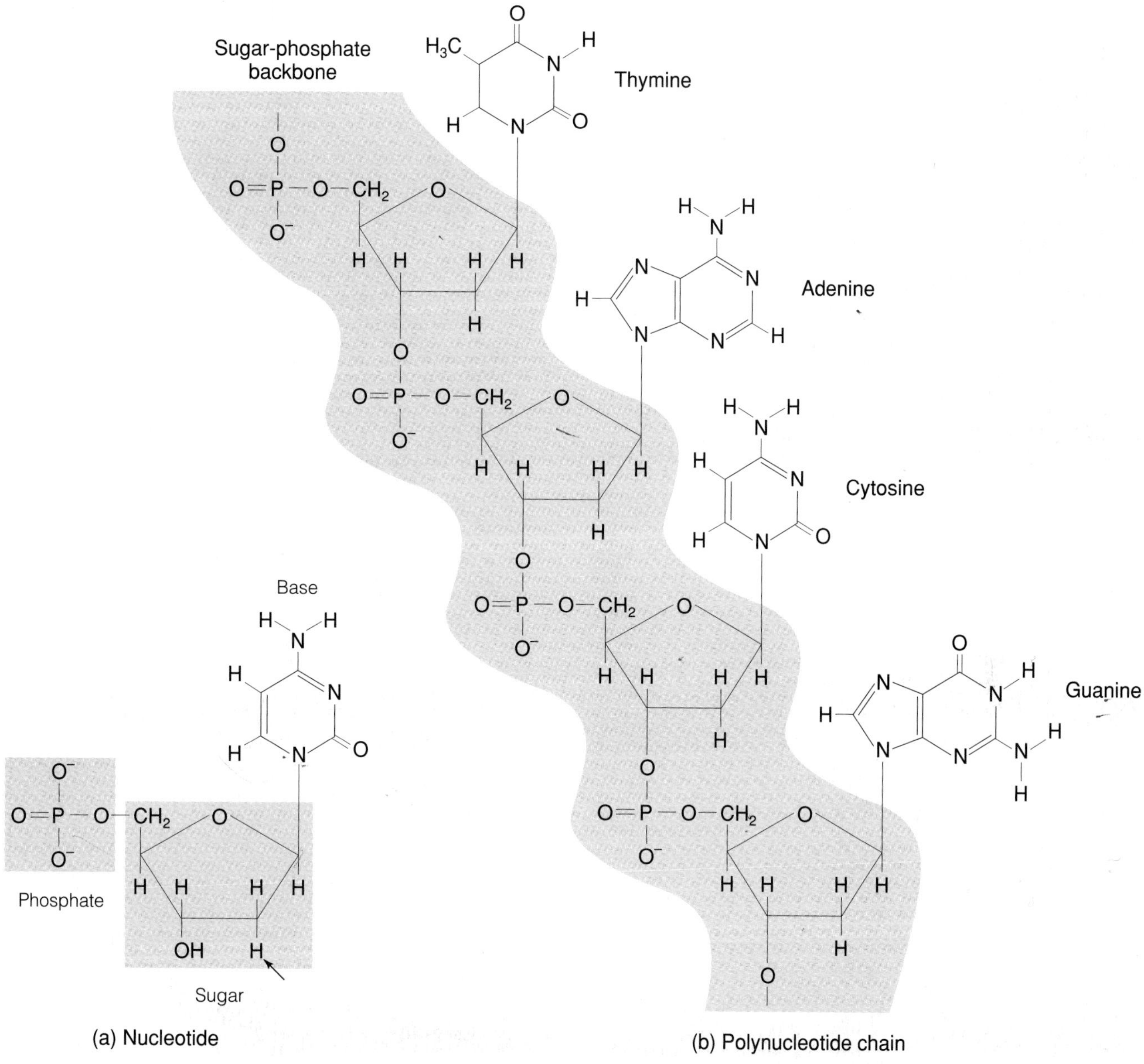

tides in the polynucleotide chain contain the sugar ribose, the chain is called **ribonucleic acid,** or **RNA.** If the sugar is deoxyribose, the chain constitutes one portion of the two-chain molecule **deoxyribonucleic acid** *(dee-ox-i-rye-bo-nu-klee´-ik),* or **DNA.** A complete DNA molecule consists of two polynucleotide chains that run in opposite directions to one another. The purine and pyrimidine bases that are opposite one another in each polynucleotide chain link together by hydrogen bonds, and the two linked chains form a double spiral coil known as a *double helix* (Figure 2.21). The complete, two-chain DNA structure is commonly called a DNA molecule even though the two polynucleotide chains are held together by hydrogen bonds rather than by covalent bonds.

The purine bases of DNA are adenine and guanine, and the pyrimidine bases are cytosine and thymine. Because of structural and bonding considerations, when

◆ **FIGURE 2.21 DNA**

(a) The ribbons of the model on the left and the strings of colored atoms in the space-filling model on the right represent the sugar-phosphate "backbones" of the two polynucleotide chains. The bases are stacked in the center of the molecule between the two backbones. The bases are 0.34 nm apart (nm = nanometer; 1 nm = 10^{-9} m). (b) A portion of a two-chain DNA molecule. Complementary base pairs (A–T; G–C) are held together by hydrogen bonds (dotted lines). The sugar-phosphate "backbone" of each chain is in the colored region. The phosphate groups are shown in ionized form. Note that the two chains run in opposite directions—that is, they are antiparallel.

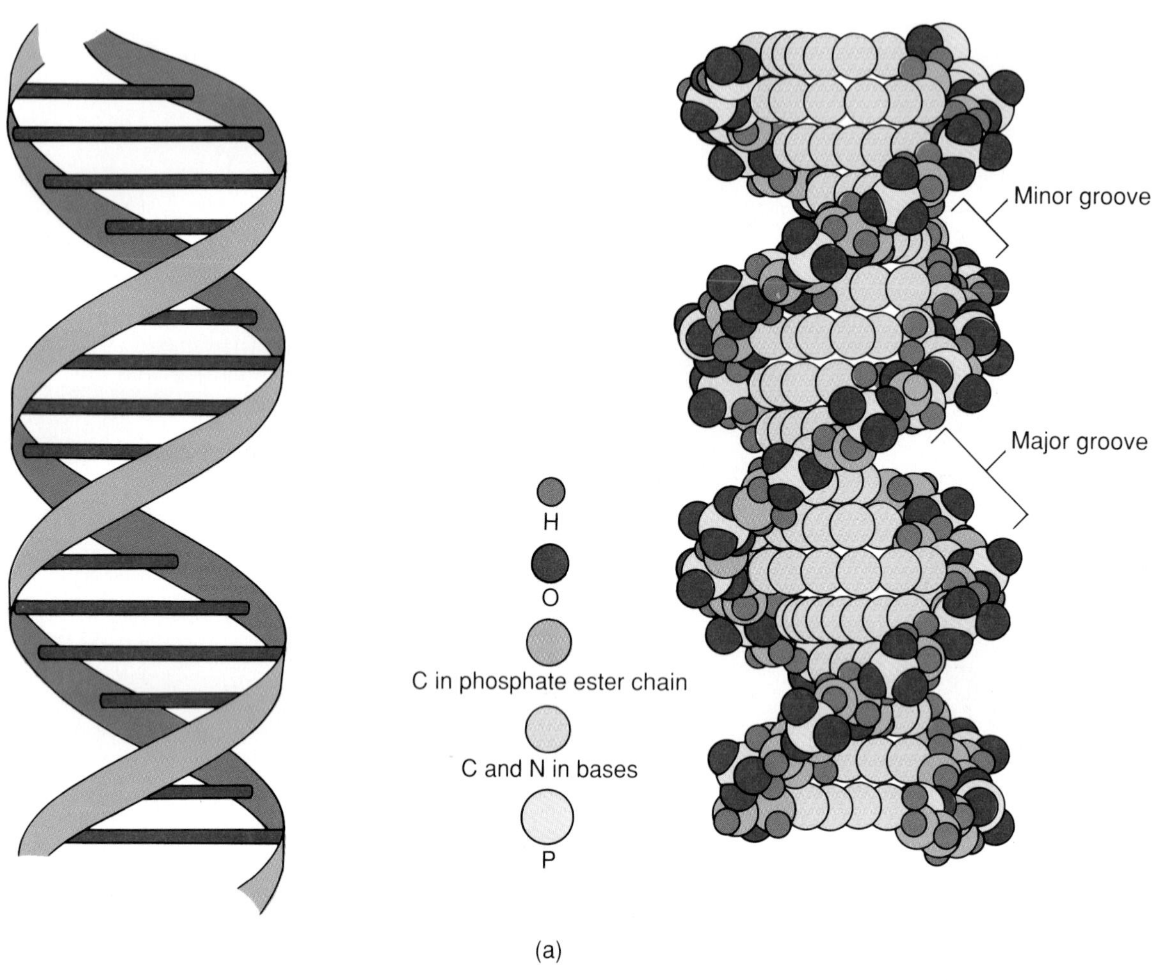

the two polynucleotide chains of DNA link with one another, an adenine of one chain bonds with a thymine of the other chain and vice versa, and a cytosine bonds with a guanine and vice versa (Figure 2.21). This is called **complementary base pairing.** As a result of complementary base pairing, if the base sequence of one chain is known, the base sequence of the other can be predicted. The same bases are also found in RNA, with the exception that the base uracil substitutes for thymine.

DNA is the genetic material of the cell, and it makes up a major portion of structures called chromosomes. DNA contains coded information within its base sequence that specifies the order in which amino acids are to be linked together to form particular polypeptides or proteins. RNA is involved in the transmission of the information of DNA to the areas of the cell where polypeptides and proteins are synthesized. DNA and RNA molecules are compared in Table 2.4 and are discussed further in Chapter 3.

Adenosine Triphosphate

A substance called **adenosine triphosphate (ATP;** *a-den´-o-sene tri-fos´-fate*) is the immediate source of energy for bodily activity—for example, muscle contraction. ATP is composed of the nitrogenous base adenine,

◆ **FIGURE 2.21 continued**

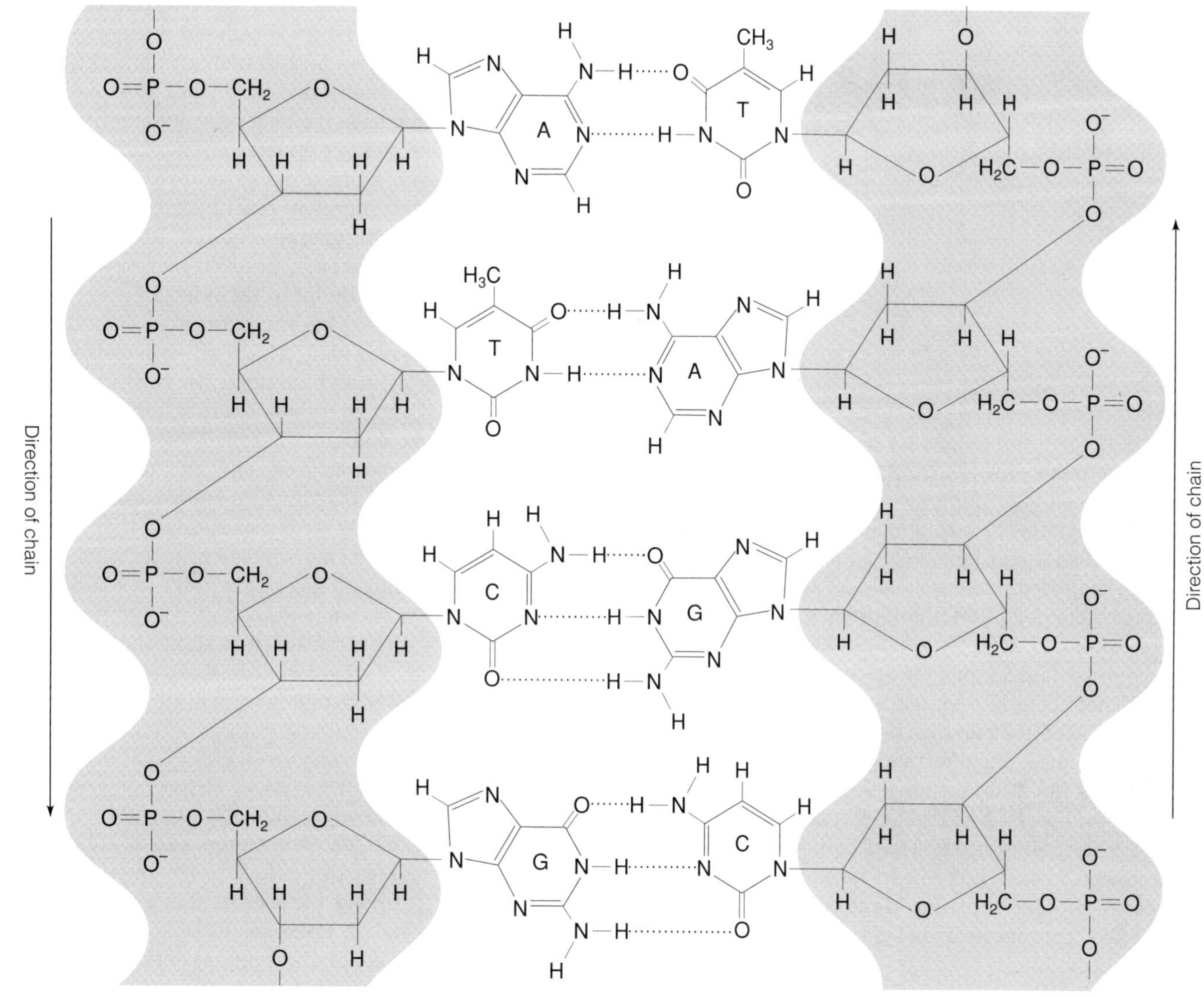

(b)

the five-carbon sugar ribose, and three phosphate groups (Figure 2.22). The phosphate groups are linked by high-energy chemical bonds that, when broken, provide energy to support the activities of the body. For example, when the terminal phosphate group is split away from a molecule of ATP, a molecule of adenosine diphosphate (ADP) is produced and energy is released.

$$\text{ATP} \rightleftharpoons \text{ADP} + \text{Phosphate} + \text{Energy}$$

Once ADP has been formed, it can be resynthesized into ATP, provided energy is available. As will be discussed later (see Chapter 25), the breakdown of various food materials by chemical reactions that occur in the body releases energy that is utilized in ATP synthesis. In this way, energy contained within the chemical bonds of food materials is made available to the body in a usable form as ATP.

Enzymes and Metabolic Reactions

The chemical reactions that constantly occur within the body are lumped together under the classification of **metabolism.** Metabolic reactions, in turn, are subdivided into anabolic, or synthesis, reactions that build up

◆ **TABLE 2.4 Comparison of DNA and RNA Molecules**

COMPONENT OR CHARACTERISTIC	DNA	RNA
Purine bases	Adenine Guanine	Adenine Guanine
Pyramidine bases	Cytosine Thymine	Cytosine Uracil
Sugar	Deoxyribose	Ribose
Number of strands	Double stranded	Single stranded

body structure, and catabolic, or decomposition, reactions that break down materials for various purposes such as the supply of energy. Metabolic reactions produce heat, and this heat can be of value because humans must maintain a constant body temperature. When humans are exposed to cold, metabolic rates increase, and the heat generated is important in maintaining body temperature.

At normal body temperature, most metabolic reactions do not occur fast enough to benefit the body. Consequently, special catalysts—that is, substances that accelerate chemical reactions without undergoing any net chemical change during the reactions—are utilized to increase the rates of metabolic reactions to levels that can meet the body's needs. These biological catalysts are collectively termed **enzymes.** As a general rule, enzymes are protein in nature. However, in at least some cases, RNA can act as a biological catalyst for certain chemical reactions.

Action of Enzymes

Enzymes increase reaction rates by lowering the **activation energy** required for metabolic reactions to occur (Figure 2.23). Under normal body conditions, few of the atoms and molecules that participate in a particular metabolic reaction have the necessary amount of energy to react with one another. In the presence of the proper enzyme, however, more of these atoms and molecules have the necessary energy to react because the enzyme lowers the amount of energy required. As a result, reactions that would otherwise proceed very slowly occur rapidly enough to be useful to the body.

Enzymes act by forming a temporary union with the reacting molecules, which are called **substrates.** This union is called an **enzyme-substrate (ES) complex.** The particular portion of an enzyme molecule with which a substrate combines is called the **active site** of the enzyme. Enzymes are very specific, and each catalyzes only individual reactions or limited classes of reactions. This specificity is due to the fact that a given enzyme has particular characteristics (such as its three-dimensional structure, or shape, and its electrical charge), and only certain substrates have the necessary complementary characteristics that allow them to unite with the enzyme. Therefore, only these substrates can form enzyme-substrate complexes with the enzyme and react (Figure 2.24).

Since the specific characteristics, such as three-dimensional shape and electrical charge, of a protein that acts as a enzyme are essential to its ability to form an enzyme-substrate complex, factors that disrupt these characteristics can inactivate the enzyme and destroy its catalytic ability. Among the factors that can inactivate enzymatic proteins are variations in such internal environmental conditions of the body as temperature and

◆ **FIGURE 2.22 Adenosine triphosphate (ATP)**
High-energy bonds are indicated by ~.

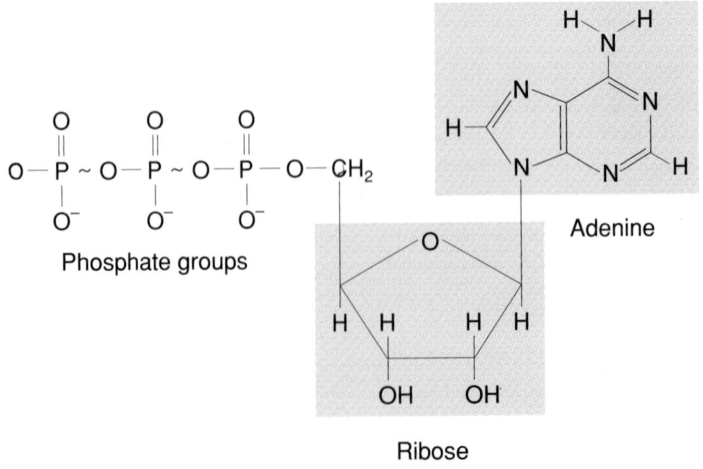

◆ **FIGURE 2.23 Diagrammatic representation of activation energy**

(a) Just as the balls must have sufficient energy to roll up the small slope before they can roll down the large one, the atoms and molecules of the body must have sufficient energy before they can react with one another. Under normal body conditions, very few atoms and molecules have the energy required to react. (b) In the presence of an enzyme, the amount of energy required to react is lowered (the small uphill slope gets smaller); more atoms and molecules will have this lower amount of energy required to react with one another.

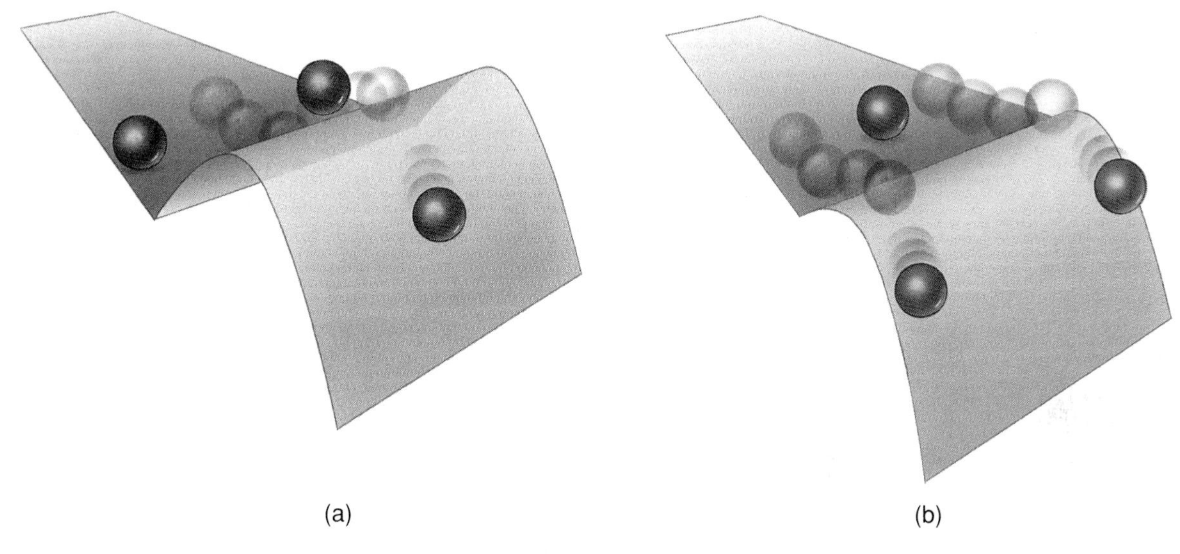

◆ **FIGURE 2.24 Diagrammatic representation of enzyme action**

(a) Only certain substrates are able to unite with the active site of a given enzyme (E_A), and only these substrates (S_A) will form enzyme-substrate complexes and react. (b) Other materials (S_B) will not be able to form an enzyme-substrate complex with the particular enzyme illustrated.

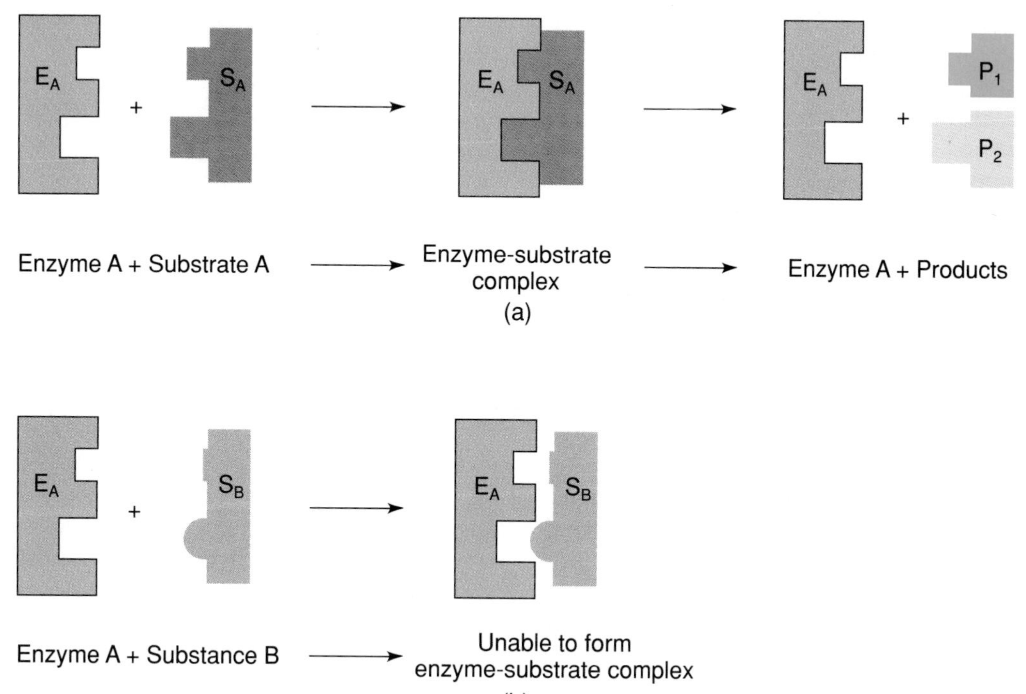

he internal environment of the body, therefore, must be relatively constant—that is, homeostasis must be maintained—if the chemical reactions required for survival are to proceed in a stable fashion. Generally, enzymes are not destroyed during the course of the reactions they catalyze, and they appear at the conclusion of the reactions in the same states as when they entered. Many of the body's enzymes are initially produced in inactive forms (precursors) that must be activated before they will be effective catalysts.

Regulation of Enzymatic Activity

Because enzymes increase the rates of metabolic reactions, the amounts and types of different enzymes present in the body at any given moment play an important role in determining how rapidly different reactions can occur. Therefore, the regulation of enzymatic activity provides a method of regulating metabolic reactions.

Control of Enzyme Production and Destruction

One means of regulating enzymatic activity is to control the rates of production and destruction of particular enzymes, thereby controlling the total amount of those enzymes present. If a particular enzyme is present in relatively large amounts, the reaction it catalyzes may proceed at a rapid rate, and much product may form. If only relatively small amounts of the enzyme are present, the reaction may proceed very slowly, and only little product may form.

Control of Enzyme Activity

A second means of controlling enzymatic activity is to inhibit or enhance the activities of particular enzymes. In one type of inhibition, called **competitive inhibition,** an inhibitor molecule rather than a substrate molecule attaches reversibly to the active site of an enzyme molecule. Although the inhibitor molecule is generally similar in structure to the substrate molecule, it does not react to form a product as the substrate molecule does. Because both the inhibitor molecule and the substrate molecule can combine reversibly with the active site of the enzyme molecule, they compete for the active site. If a great number of inhibitor molecules are present, many of them occupy the active sites of the enzyme molecules, and relatively few substrate molecules are able to form enzyme-substrate complexes and react.

Another type of enzyme inhibition, called **noncompetitive inhibition,** can take several forms. In some cases, a noncompetitive inhibitor molecule combines irreversibly with the active site of an enzyme molecule. As a result, substrate molecules cannot combine with the active site. In other cases, a noncompetitive inhibitor molecule combines with an enzyme molecule at a site other than the active site. As a result of this combination, the structure of the enzyme molecule is altered so that it is less able to form an effective enzyme-substrate complex with its substrate.

Enzymes subject to regulation by small molecules have special binding sites called allosteric effector sites to which regulatory molecules attach by weak bonds. The combination of a regulatory molecule with an allosteric site alters the structure of the enzyme molecule and either activates or inhibits the enzyme. Often, in what is essentially a negative feedback response, the final product of a series of enzymatically catalyzed reactions allosterically inhibits the first enzyme in the series. Thus, as the amount of product increases, the activity of the system that produces the product declines. Allosteric inhibition is a type of noncompetitive inhibition.

Some enzyme molecules are activated or inhibited by the chemical addition or removal of a phosphate (or other) group. The regulation of enzyme activity by phosphorylation or dephosphorylation differs from allosteric regulation in that it involves changes in covalent bonds in enzyme molecules, whereas allosteric regulation involves only patterns of weak bonds. The addition or removal of covalently bound groups requires the intervention of still other enzymes; kinase enzymes add phosphate groups, whereas phosphatase enzymes remove them. Since kinase and phosphatase enzymes are themselves subject to regulation—frequently by feedback mechanisms—enzyme regulation can involve a series of interacting events.

Cofactors

Often, enzymes require the presence of nonprotein structures called **cofactors** to actively catalyze reactions. A cofactor may be either a metal ion or a complex organic molecule called a **coenzyme.** Many vitamins, for example, act as coenzymes.

Solutions

Water is the medium in which all living processes occur, and life as we know it would be inconceivable in the absence of this molecule. In fact, the chemical reactions that occur continuously within the body involve, for the most part, reactants that are in aqueous (water) solutions. A **solution** is a homogeneous mixture of two or more components that can be gases, solids, or liquids. The components of a true solution cannot be distinguished in the mixture, and they do not settle out at an appreciable rate. If a beam of light is passed through a true solution, the light path will not be visible. The particles dispersed within a true solution are very small—generally in the atomic- and molecular-size range. For example, sodium chloride dissolved in water forms a true solution. When dealing with solutions, the material present in the greatest amount is generally called the **solvent,** whereas substances present in smaller amounts are generally called **solutes.**

Properties of Water

Water is the body's principal solvent, and the solutions that occur most commonly in the body result from dissolving gases, liquids, or solids in water. Water has a number of properties that make it particularly well suited for its role in the body:

1. Many of the body's chemical components are either polar molecules or ionic compounds, and water is a polar liquid whose chemical properties are such that many different polar molecules and ionic compounds can dissolve in it. For example, many salts dissolve easily in water because the positive and negative charges on the polar water molecules can substitute for the positive and negative charges on the ions that compose the crystal lattices of the salts. When a salt crystal dissolves in water, each positively charged ion becomes surrounded by water molecules that have their negative oxygen portions turned toward the ion, and each negatively charged ion becomes surrounded by water molecules whose positively charged hydrogen portions are closest (Figure 2.25). In this condition the ions from the salt crystal are said to be **hydrated.** Thus, when a salt crystal dissolves in water, it does not simply come apart into ions, but is taken apart by the water molecules. Some physiologically important salts and the ions into which they separate or dissociate when they dissolve in water are listed in Table 2.5.

◆ **FIGURE 2.25 Breakup of a salt crystal by water molecules, with hydration of ions**

Each salt ion in solution is surrounded by polar water molecules with the opposite charge to that of the ion turned toward it.

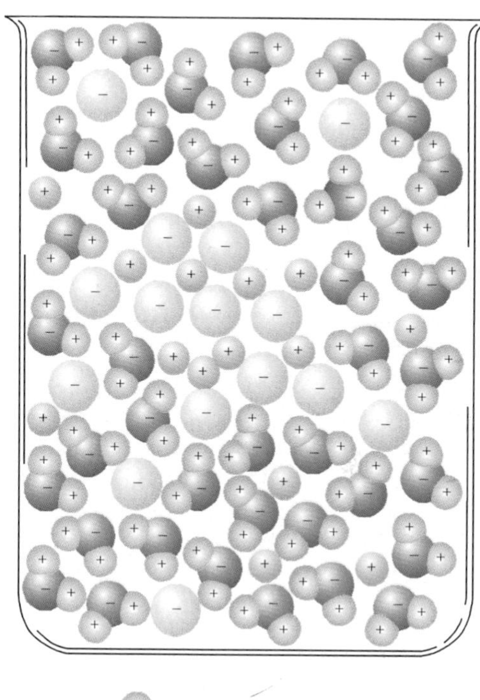

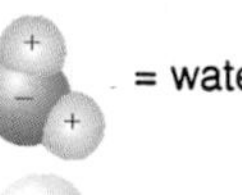

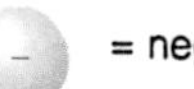

◆ **TABLE 2.5 Some Salts of Importance to the Body***

SALT		IONS
Sodium chloride (NaCl)	→	Sodium (Na^+) and chloride (Cl^-)
Potassium chloride (KCl)	→	Potassium (K^+) and chloride (Cl^-)
Calcium chloride ($CaCl_2$)	→	Calcium (Ca^{+2}) and chloride (2 Cl^-)
Magnesium chloride ($MgCl_2$)	→	Magnesium (Mg^{+2}) and chloride (2 Cl^-)
Calcium carbonate ($CaCO_3$)	→	Calcium (Ca^{+2}) and carbonate (CO_3^{-2})
Calcium phosphate ($Ca_3[PO_4]_2$)	→	Calcium (3 Ca^{+2}) and phosphate (2 PO_4^{-3})
Sodium sulfate (Na_2SO_4)	→	Sodium (2 Na^+) and sulfate (SO_4^{-2})

*The arrows indicate the ions into which these salts dissociate when they are dissolved in water.

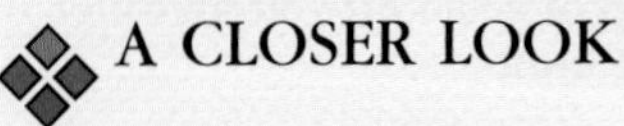

Expressing the Concentration of a Solution

It is often necessary to know the concentration of a solution, and concentrations are expressed in a number of ways.

One way is simply to indicate the percentage of solute in a solution by weight (wt./wt.), by volume (v./v.), or by a combination of the two (wt./v.). For example, a 10% solution by weight can be made by dissolving 10 grams of a solute (such as glucose) in enough solvent (such as water) to make 100 grams of solution. Similarly, a 10% solution by volume can be made by mixing 10 milliliters of a solute (such as ethyl alcohol) with enough solvent (water, for example) to make 100 milliliters of solution. Alternatively, a 10% solution by weight/volume can be made by dissolving 10 grams of a solute (such as glucose) in enough solvent (such as water) to make 100 milliliters of solution. When the concentration of a solution is expressed in percent, the measurements employed—wt./wt., v./v., or wt./v.—should always be indicated.

A second way of expressing the concentration of a solution is in terms of its **molarity.** In this method, the amount of solute is not expressed in terms of weight or volume but in terms of moles. A **mole** of a substance is the amount of the substance in grams that is equal to the molecular weight of the substance. The **molecular weight** of a substance can be determined by adding together the atomic weights of the atoms that make up a molecule of the substance (as indicated by the chemical formula for a molecule of the substance). For example, a molecule of glucose, which has a chemical formula of $C_6H_{12}O_6$, is composed of 6 carbon atoms, 12 hydrogen atoms, and 6 oxygen atoms. The atomic weight of carbon is 12.011, the atomic weight of hydrogen is 1.008, and the atomic weight of oxygen is 15.999. A mole of glucose would therefore weigh:

$$\begin{aligned} 6 \times 12.011 &= 72.066 \text{ (for carbon)} \\ 12 \times 1.008 &= 12.096 \text{ (for hydrogen)} \\ 6 \times 15.999 &= \underline{95.994} \text{ (for oxygen)} \\ & \quad 180.156 \text{ grams} \end{aligned}$$

A mole of any substance contains the same number of molecules—6.022×10^{23}—as a mole of any other substance. This number is called *Avogadro's number.*

Expressing the concentration of a solution in terms of molarity indicates the number of moles of solute in a liter of solution. For example, a one molar (mol/L) solution can be made by adding one mole of a solute to enough solvent to make one liter (1000 milliliters) of solution. If the molecules of the solute remain intact in the solution, the solution will contain 6.022×10^{23} molecules of the solute.

The term *mole* can be applied to atoms and ions as well as to molecules. For example, a mole of potassium is equal to the atomic weight of potassium in grams (39.098 grams). A mole (39.098 grams) of potassium contains 6.022×10^{23} potassium atoms. In the case of salts, which are made up of ions, the number of grams in a mole of a salt can be calculated by adding together the atomic weights of the atoms in the chemical formula of the salt. Sodium chloride, for example, has the chemical formula NaCl, and a mole of sodium chloride is 58.443 grams of sodium chloride because the atomic weight of sodium (22.99) plus the atomic weight of chlorine (35.453) totals 58.443. A mole of sodium chloride contains 6.022×10^{23} sodium ions (one mole of sodium ions) and 6.022×10^{23} chloride ions (one mole of chloride ions). Similarly, the salt calcium chloride has the chemical formula $CaCl_2$, and a mole of calcium chloride is 110.986 grams of calcium chloride because the atomic weight of calcium (40.08) plus two times the atomic weight of chlorine ($2 \times 35.453 = 70.906$)

totals 110.986. A mole of calcium chloride contains 6.022×10^{23} calcium ions (one mole of calcium ions) and $2 \times 6.022 \times 10^{23} = 12.044 \times 10^{23}$ chloride ions (two moles of chloride ions).

Frequently, the amounts of ionized inorganic constituents in the body fluids are expressed in terms of **equivalents (Eq)** per liter of solution rather than in terms of moles per liter of solution. For present purposes, the number of grams in one equivalent of a particular ion can be said to be equal to the sum of the atomic weights of the atoms in the chemical formula of the ion, divided by the charge of the ion, without regard for the sign (+ or −) of the charge (Table C2.2). For example, one equivalent of phosphate ions (PO_4^{-3}) is equal to:

$$\begin{aligned} 1 \times 30.974 &= 30.974 \text{ (for phosphorus)} \\ 4 \times 15.999 &= \underline{63.996} \text{ (for oxygen)} \\ & \quad 94.970 \div 3 = 31.657 \text{ grams of phosphate ions} \end{aligned}$$

Similarly, one equivalent of calcium ions (Ca^{+2}) is equal to 40.08/2 = 20.04 grams of calcium ions.

The solutions encountered in the body generally have low concentrations, and the amounts of solute present are often expressed in terms of millimoles (1 millimole = 1/1000th of a mole) or milliequivalents (1 milliequivalent = 1/1000th of an equivalent), rather than in terms of moles or equivalents. The concentrations of solutions considered in this manner are then expressed as millimoles per liter (millimolar; mmol/L) or milliequivalents per liter (mEq/L).

◆ **TABLE C2.2 Charges of Common Body Ions**

ION	CHARGE
Bicarbonate (HCO_3^-)	−1
Calcium (Ca^{+2})	+2
Chloride (Cl^-)	−1
Hydrogen (H^+)	+1
Magnesium (Mg^{+2})	+2
Phosphate (PO_4^{-3})	−3
Potassium (K^+)	+1
Sodium (Na^+)	+1
Sulfate (SO_4^{-2})	−2

2. Water is not a good solvent for nonpolar molecules (or for the nonpolar portions of large molecules). Nonpolar molecules do not interact well with water molecules, and they interfere with and disrupt hydrogen bonds between water molecules. Because water molecules have a tendency to form hydrogen bonds, they tend to exclude nonpolar molecules. Consequently, in an aqueous environment, nonpolar molecules tend to cluster together, associating with one another instead of with water. Such associations of nonpolar molecules in an aqueous environment are called **hydrophobic interactions.** These interactions are important in determining the tertiary structure of proteins, in the organization of cell membranes, and in the assembly of cellular structures.

3. Water has a high **specific heat,** which means that, compared with other liquids, water requires a good deal of heat to raise its temperature. Thus, the heat produced by metabolism does not affect body temperature as much as if some other solvent were present.

4. Water has a high **latent heat of vaporization,** which means that, compared with other liquids, water requires a good deal of heat to change it from the liquid to the vapor state. Thus, the evaporation of water from body surfaces carries away large amounts of heat and provides the body with an effective cooling mechanism.

Suspensions

Other types of mixtures besides true solutions are possible. Among these are **suspensions.** In a suspension, the dispersed particles are so large that they can be kept dispersed only by constant agitation. The components of the suspension remain distinct from one another, thus creating a heterogeneous mixture. If left to stand, the dispersed particles settle. A mixture of sand in water, for example, is an obvious suspension; in the body, blood cells are suspended in the fluid portion of the blood.

Colloids

The transition from the homogeneity of true solutions to the heterogeneity of obvious suspensions is not a sudden one, and there are many gradations in between. **Colloids** are intermediate between true solutions and obvious suspensions. Colloids consist of particles that are dispersed in a medium, much like the composition of obvious suspensions. However, the particles are small enough that they do not readily settle out if left to stand. Colloidal systems can be distinguished from true solutions by passing a beam of light through them. If the system is a colloid, the beam will be scattered by the dispersed particles, and the light path will be visible when viewed from the side. In nature, colloids are very common. Milk is a colloid, and the protoplasm of living cells is considered to have colloidal properties.

Acids, Bases and pH

Water molecules exist mostly in an undissociated state, with the two hydrogens chemically bonded to the oxygen of the molecule. However, a very small percentage of water molecules (0.0000002%) dissociate into hydrogen ions (H^+) and hydroxide ions (OH^-). Although only about 1 in 500 million water molecules actually dissociates, this small amount of dissociation is one of the most important properties of water, and many metabolic reactions are critically dependent on the hydrogen ion concentration of the solution in which they occur.

Substances that alter the hydrogen ion concentration can be added to water. Substances that increase the hydrogen ion concentration are called **acids,** and substances that decrease the hydrogen ion concentration are called **bases.** Alternatively, acids may be defined as *proton donors* (a hydrogen ion is equivalent to a proton), whereas bases are *proton acceptors.* For example, hydrochloric acid (HCl) is an acid because it can dissociate into hydrogen ions (H^+) and chloride ions (Cl^-) and thereby serve as a hydrogen ion (proton) donor that can increase the hydrogen ion concentration of a solution:

$$HCl \rightleftharpoons H^+ + Cl^-$$

Conversely, ammonia (NH_3) is a base because it can accept hydrogen ions (protons) to form ammonium ions (NH_4^+) and thereby decrease the hydrogen ion concentration of a solution:

$$NH_3 + H^+ \rightleftharpoons NH_4^+$$

The hydrogen ion concentration of the body fluids must be maintained within narrow limits, and it is often necessary to know the hydrogen ion concentration of a solution. In pure water the small dissociation of water molecules results in a hydrogen ion concentration of 0.0000001 (1×10^{-7}) moles per liter. When expressed in this manner, the hydrogen ion concentration is such a small number that it is difficult to work with. Consequently, the hydrogen ion concentration of a solution is commonly expressed as a logarithmic value called the **pH,** which is determined according to the following relationship:

$$pH = \log_{10} \frac{1}{[H^+]}$$

where $[H^+]$ is the molar concentration of hydrogen ions in the solution. The pH of pure water is 7, and a solution with a pH of 7 is considered to be a neutral solution. As the hydrogen ion concentration of a solution increases, the pH value drops. Each drop of one unit on the pH scale indicates a tenfold increase in the hydrogen

◆ **FIGURE 2.26 The pH scale**
pH values below 7 indicate acidic conditions, and values above 7 indicate basic, or alkaline, conditions.

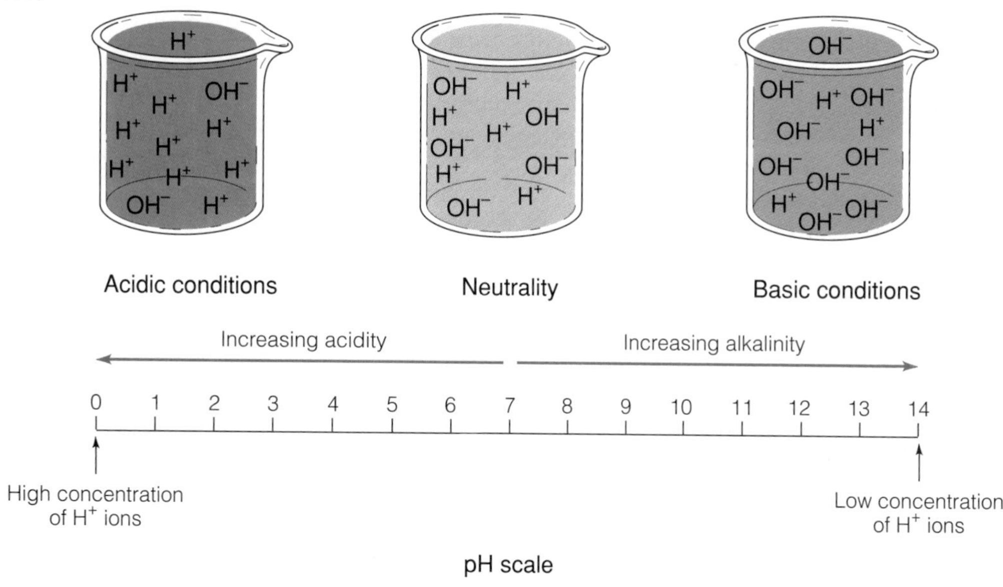

ion concentration. A solution with a pH of 6, therefore, has ten times the hydrogen ion concentration of a solution with a pH of 7, and a solution with a pH of 5 has $10 \times 10 = 100$ times the hydrogen ion concentration of a solution with a pH of 7. The lower the pH, therefore, the more acidic the solution (Figure 2.26). Similarly, pH values above 7 indicate progressively more basic, less acidic solutions. Typical pH values for several body fluids as well as for a number of other aqueous solutions are listed in Table 2.6.

Gradients

In the body, the level of a particular variable, such as pressure, temperature, degree and kind of electrical charge, or the concentration of a substance, often changes in a more or less continual fashion between one region and another. Such a change in the level of a variable over distance is referred to as a **gradient.** For example, in the case of blood flowing along a blood vessel, the blood pressure is relatively high at the end of the vessel where the blood enters. However, the pressure continually falls as the blood flows along the vessel, and it is relatively low at the end of the vessel where the blood leaves. Thus, a pressure gradient exists along the vessel between one end and the other.

The movement of a substance such as blood along a pressure gradient from a region of higher pressure to a region of lower pressure is described as movement *down* the gradient (or *with* the gradient). Conversely, the movement of a substance from a region of lower pressure to a region of higher pressure is referred to as movement *up* the gradient (or *against* the gradient). Similarly, movement along a temperature gradient from a region of higher temperature to a region of lower temperature is movement down the gradient, and movement from a region of lower temperature to a region of higher temperature is movement up the gradient. In like manner, the movement of a substance along a concentration gradient (also called a chemical gradient) from a region of higher concentration of the substance to a region of lower concentration is movement down the gradient, and movement of the substance from a region of lower concentration to a region of higher concentration is movement up the gradient.

In the case of an electrical gradient, the movement of a positively charged substance toward a region of increasing negative charge or the movement of a negatively charged substance toward a region of increasing positive charge is movement down the gradient. The movement of a positively charged substance toward a region of increasing positive charge or the movement of a negatively charged substance toward a region of increasing negative charge is movement up the gradient. A charged substance (for example, an ion) is often described as moving down or up an electrochemical gradient, which takes into account both the differing concentrations of the substance and the differing electrical conditions that exist between two regions.

Just as energy does not have to be supplied for an automobile to coast down a hill (it can coast down even with the engine turned off), so energy does not have to be supplied for a substance to move down a gradient. However, energy must be supplied to move an automobile up a hill, and likewise, it must be supplied to move a substance up a gradient. In the body, substances frequently move down gradients, but substances are also moved up gradients. The energy required to move substances up gradients is supplied by energy-yielding chemical reactions that occur within the body's cells.

Diffusion

Atoms, molecules, and ions are constantly in motion. Thus, they possess kinetic energy, or energy of motion. The velocity at which an atom, molecule, or ion moves is a function of temperature—the higher the temperature, the greater the velocity of movement.

Diffusion is the movement of atoms, molecules, or ions from one location to another as a consequence of their thermal motion. For example, if a cube of sugar is placed in a beaker of water, the sugar will dissolve, and the sugar molecules will eventually diffuse throughout the water (Figure 2.27). In the diffusion process, individual sugar molecules move in a random fashion. However, since initially there are more sugar molecules in the area that surrounds the sugar cube (where the sugar molecules are entering solution) than in areas of the water farther from the cube, it is probable that more sugar molecules will move away from the area around the sugar cube than will move toward the area. Thus, although individual sugar molecules move at random, the net movement of sugar molecules by diffusion (that is, the net diffusion) is from regions of high concentrations of sugar molecules to regions of low concentrations of sugar molecules (provided the temperature and pressure throughout the system are constant). **Net diffusion,** then, is the movement of a substance from a region of higher to a region of lower concentration as a consequence of the thermal motion of the atoms, molecules, or ions of the substance when the temperature and pressure throughout the system are constant. Note that the net movement of a substance by diffusion—that is, its net diffusion—is down a gradient.

In the preceding example, the movement of sugar molecules by diffusion will eventually result in a uniform distribution of sugar molecules throughout the water, and no differences in concentration will exist. When this occurs, the system is said to be in equilibrium. A state of equilibrium, however, does not imply a state where there is no longer any movement. Rather, it means that as many atoms, molecules, or ions of a substance—in this case, sugar molecules—enter a particular area at any one time as leave it. Thus, although the same atoms, molecules, or ions of a substance may not be in a given area, the same number of atoms, molecules, or ions of the substance will be. As a result, there is no net change in concentration.

◆ TABLE 2.6 Typical pH Values of Body Fluids and Common Aqueous Solutions

SUBSTANCE	TYPICAL pH
1 molar hydrochloric acid	0
0.1 molar hydrochloric acid	1
Gastric juice	1.4–1.8
Lemon juice	2.1–2.3
Vinegar	2.4–3.4
Orange juice	2.8
Soda water	3.8
Tomato juice	4.1–4.2
Black coffee	5.0
Milk	6.3–6.9
Urine	4.8–7.4
Saliva	6.0–7.0
Pure water	7.0
Intestinal juice	6.5–7.5
Venous blood	7.35
Arterial blood	7.4
Bile	7.8–8.6
Pancreatic juice	8.0
Seawater	8.4
Milk of magnesia	10.5
Household ammonia	11.5–11.9
0.1 molar sodium hydroxide	13.0
1 molar sodium hydroxide	14.0

Water molecules can also diffuse, and they can exhibit net diffusion from regions of higher to regions of lower concentration. But how can there be different concentrations of water? Consider the following: If one beaker is filled with pure water, there will be 100% water in the beaker, and the entire volume of the beaker will be occupied by water molecules. If an identical second beaker is filled with sugar solution, some of the volume will be occupied by water molecules and some by sugar molecules. Therefore, there will not be 100% water molecules in this second beaker, and the concentration of

◆ **FIGURE 2.27 Diffusion**

(a) Initially, all the sugar molecules are within the sugar cube. (b) As the cube dissolves, sugar molecules disperse in the water. Note that their concentration is highest near the sugar cube, where they are entering solution. (c) When all of the sugar has dissolved and the system is at equilibrium, sugar molecules are dispersed evenly throughout the water as a result of their random movement.

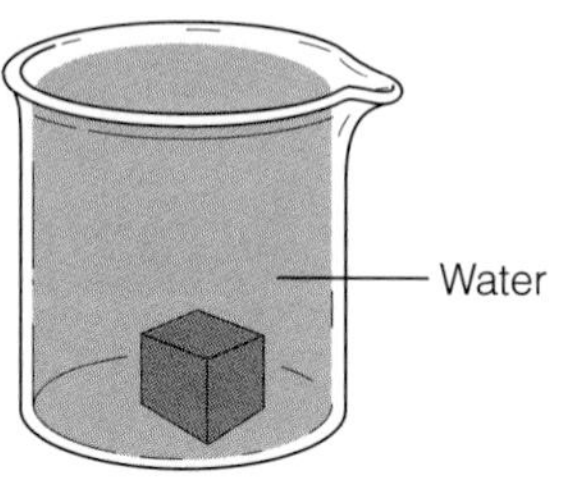

Sugar cube intact

(a)

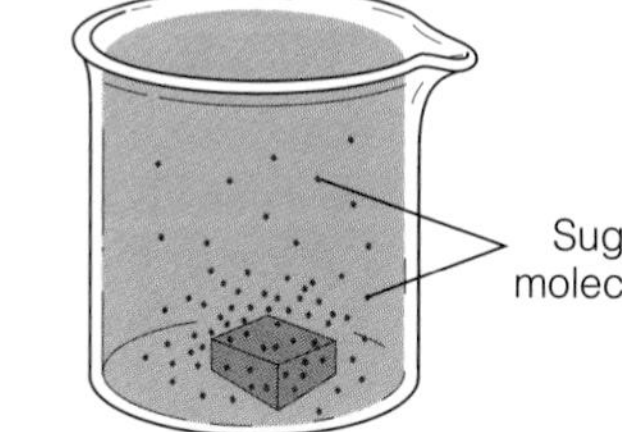

Sugar cube partially dissolved

(b)

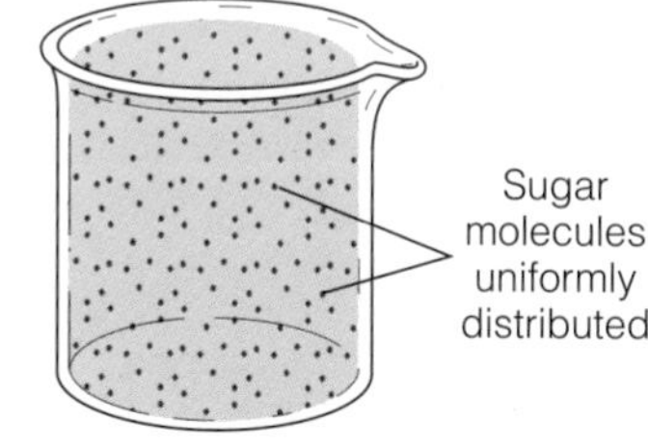

Sugar cube dissolved, showing state of equilibrium

(c)

water in this beaker will be lower than the concentration of water in the first beaker (Figure 2.28).

If, instead of keeping the pure water and the sugar solution in separate beakers, they are placed on separate sides of a single container that has a removable barrier between the sides, the following will occur when the barrier is removed (Figure 2.29): The sugar molecules will show a net diffusion from their region of higher concentration (the sugar solution side) to their region of lower concentration (the pure water side). Likewise, water molecules will exhibit a net diffusion from their region of higher concentration (the pure water side) to their region of lower concentration (the sugar solution side). Both processes will continue until equilibrium is reached—that is, until both sugar molecules and water molecules are uniformly distributed throughout the system. At this point, no regions of different concentration exist, and thus no further net diffusion of either sugar molecules or water molecules will occur.

Osmosis

Consider another situation. Suppose pure water placed on one side of a container is separated by a membrane from a sugar solution placed on the other side of the container. Suppose further that the membrane is a **semipermeable membrane** that allows the passage of solvent (water molecules) but not solutes (sugar molecules) (Figure 2.30). In this situation, water will exhibit a net movement from its region of higher concentration through the membrane to its region of lower concentration. Sugar molecules, however, will not be able to

◆ **FIGURE 2.28 Different concentrations of water**

When solute molecules (in this case, sugar) are added to water, the concentration of water molecules in a given volume decreases. There can be different concentrations of water (solvent), just as there can be different concentrations of solute in a solution.

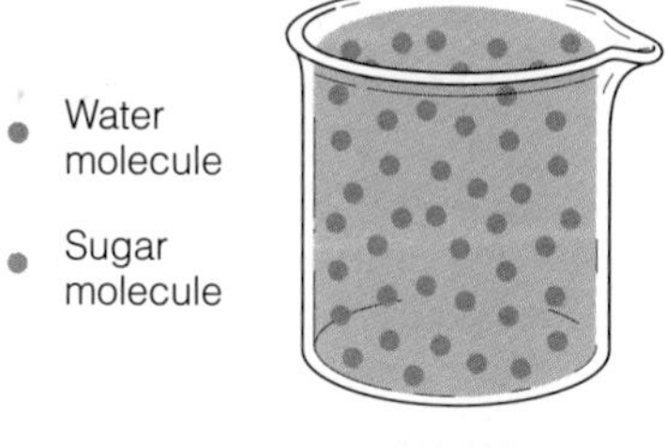

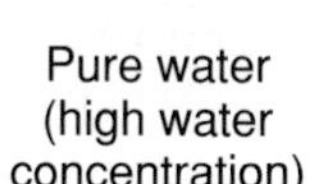

Pure water (high water concentration)

Sugar solution (lower water concentration)

◆ FIGURE 2.29 Redistribution of water and sugar molecules when a barrier to molecular movement is removed

Sugar molecules (red) show a net diffusion from their region of high concentration to their region of low concentration. Likewise, water molecules (blue) show a net diffusion from their region of high concentration to their region of low concentration. At equilibrium, both sugar and water molecules are uniformly distributed throughout the container.

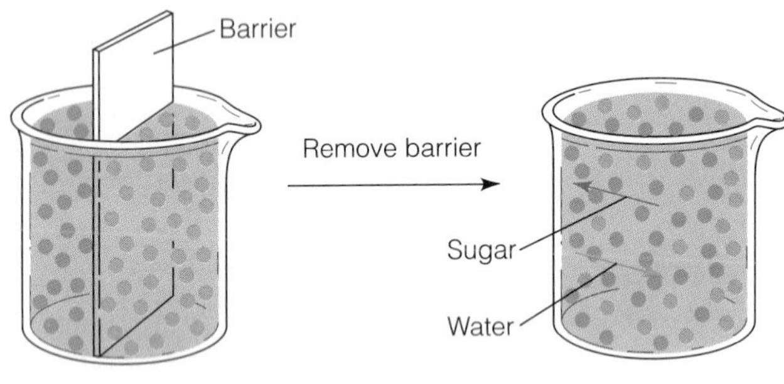

move through the membrane. As a result, there will be a net movement of water into the sugar solution, but no movement of sugar into the pure water. The movement of water that takes place across the semipermeable membrane in this instance is an example of osmosis. More generally, **osmosis** is the movement of solvent through any membrane in response to a concentration difference (a concentration gradient) across the membrane.

In the preceding example, the occurrence of osmosis will cause the volume of pure water on the one side of the membrane to decrease gradually, and the volume of sugar solution on the other side to increase gradually. However, there will always be a higher concentration of water on the pure water side of the membrane than on the sugar solution side. Nevertheless, the net movement of water into the sugar solution does not go on indefinitely. Eventually, the pressure of the additional volume of fluid on the sugar solution side (the **hydrostatic pressure**) rises to a point at which it is able to balance the force tending to move water into the sugar solution by osmosis, and no further net movement of water into the sugar solution occurs. At this point, the hydrostatic pressure that is exerted against the membrane on the sugar solution side is great enough to force water molecules across the membrane from the sugar solution into the pure water as fast as they move from the pure water into the sugar solution. The result is an equilibrium in which the respective concentrations of sugar molecules and water molecules are not equal in all parts of the system, but in which opposing forces prevent the net movement of water molecules and a membrane prevents any movement of sugar molecules.

The pressure required to prevent the net movement of pure water into a solution when the water is separated from the solution by a semipermeable membrane is a measure of the solution's **osmotic pressure.** The osmotic pressure of a solution, which is expressed in units called osmoles or milliosmoles, indicates the tendency of water to move by osmosis into the solution. The osmotic pressure of a solution depends basically on the number of solute particles present and not on their nature. The greater the number of solute particles in a given volume of solution, the greater the osmotic pressure of the solution. If two solutions have the same osmotic pressure, they are said to be **isosmotic** (*iso* = same). If two solutions have different osmotic pressures, the one with the higher osmotic pressure is said to be **hyperosmotic** to the one with the lower osmotic pressure, and the solution with the lower osmotic pressure is said to be **hypoosmotic** to the solution with the higher osmotic pressure (*hyper* = above; *hypo* = below).

The membrane that surrounds living cells is not a simple semipermeable membrane but behaves like a **selectively permeable membrane.** A selectively permeable membrane is a membrane that does not permit the free, unhampered movement of all solutes present, but maintains a differential concentration (a concentration gradient) of at least one solute across itself. Thus, a selectively permeable membrane is not equally permeable to all solute particles present. Osmosis can occur across a selectively permeable membrane in response to different concentrations of water on either side of the membrane.

Dialysis

Some membranes are permeable to water molecules and small particles—for example, sodium ions and chloride ions—that may be present in the water, but they are not permeable to large particles such as protein molecules. If such a membrane is placed between a sodium chloride-protein solution on one side and pure water on the other, then water molecules, sodium ions, and chloride ions will be able to pass through the membrane but protein molecules will not. As a result, there will be a net diffusion of sodium ions and chloride ions from the solution into the pure water, and water will exhibit a net movement into the sodium chloride-protein solution. If the pure water side is constantly drained away and replenished so that no equilibrium is established, the sodium chloride can be removed from the sodium chloride-protein solution. This process of selectively separating substances in a liquid by taking advantage of their differing diffusibilities through porous membranes is called **dialysis.**

◆ **FIGURE 2.30 The process of osmosis**
Set the text for a detailed discussion. Sugar molecules are red and water molecules are blue.

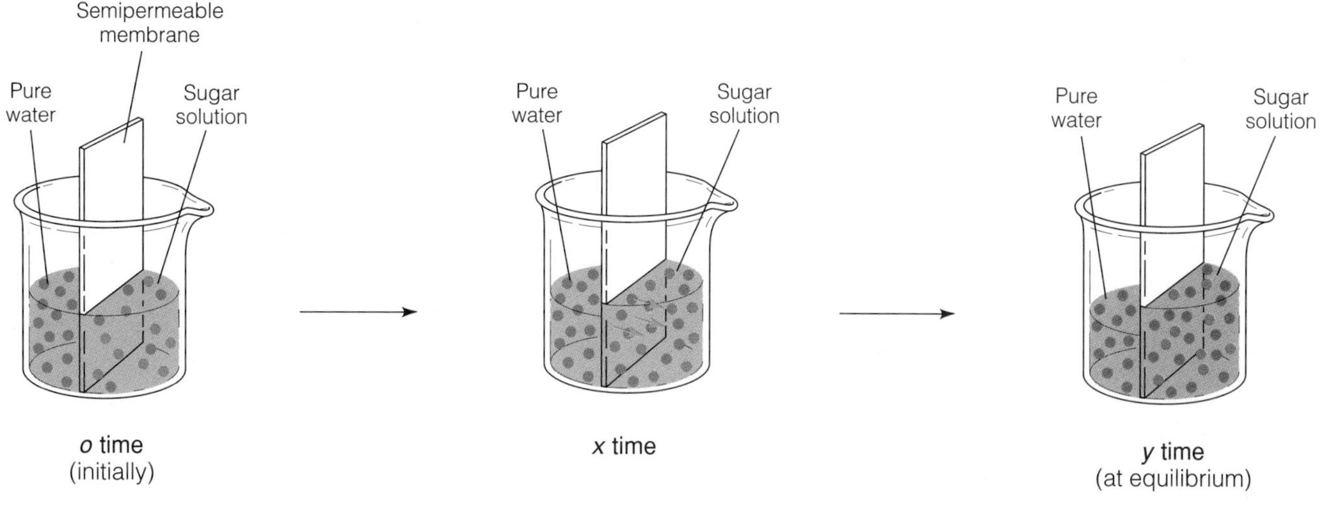

Dialysis achieves its most dramatic application in the artificial kidney. In this device, blood is passed through a membranous tube immersed in a bathing medium of known composition. The membranous tube is permeable to water molecules and other small particles, but it is not permeable to large particles such as protein molecules. Since the atoms, molecules, or ions of most wastes are small particles, these can be removed from the blood while vital protein molecules are retained. Substances required by the body that are also composed of small particles can be retained by having them present in the bathing fluid in the same concentration as they are in the blood. Thus, there will be no concentration difference of these substances on either side of the membrane and, therefore, no net diffusion.

Bulk Flow

Bulk flow is the movement of atoms, molecules, ions, or other particles as a unit in one direction as the result of forces that push them from one point to another. For example, when a sodium chloride solution is pumped through a pipe, the ions and molecules of the solution travel in one direction as a unit.

The bulk flow of a liquid or gas depends on an inequality of pressure that acts on the liquid or gas. If different regions of a liquid or gas are subjected to unequal pressures, the liquid or gas will flow from the region of higher pressure to the region of lower pressure. The flow of blood within the blood vessels and the movement of air into and out of the lungs during respiration are examples of bulk flow that occur in the body.

Filtration

Filtration is a process that separates one or more components of a mixture from the other components. In this process, the mixture is forced through a porous filter or membrane by mechanical forces such as hydrostatic pressure. The direction of movement is from a region of higher pressure to a region of lower pressure, and large particles that are present in the mixture may not be able to pass through the filter or membrane. For example, consider a glucose solution to which sand has been added. If this mixture is forced through a membrane containing pores that are too small to pass the sand, water molecules and glucose molecules will move through the membrane, but sand particles will remain behind.

Within the body, a filtration process that allows the passage of only small particles commonly occurs. The blood pressure within blood vessels called capillaries acts to force the fluid portion of the blood, including small dissolved particles such as glucose molecules or the ions of salts, across the capillary walls and out of the circulatory system. However, blood cells and protein molecules that are too large to leave the vessels remain behind in the blood.

Study Outline

◆ CHEMICAL ELEMENTS AND ATOMS p. 24

Chemical Element. A substance that cannot be broken down into simpler material by chemical means.

Atoms. Make up chemical elements and are composed of protons (in nucleus; have positive charge); neutrons (in nucleus; electrically neutral); electrons (move around nucleus; have negative charge).

◆ ELECTRON ENERGY LEVELS p. 24

Present view of atomic structure suggests that electrons should be assigned to particular energy levels. The statistical distribution of electrons is such that electrons occupying higher energy levels will generally be found farther from the nucleus of an atom than electrons occupying lower energy levels.

◆ ATOMIC NUMBER, MASS NUMBER, AND ATOMIC WEIGHT pp. 25–27

Atomic Number. Number of protons in an atom.

Mass Number. Combined number of protons and neutrons in an atom.

Atomic Weight. Total mass of an atom.

A Closer Look: Isotopes and Radiation.

◆ CHEMICAL BONDS pp. 27–30

Attractive forces between atoms that are strong enough to hold atoms together.

Covalent Bonds and Molecules. Bonds based on electron sharing are called covalent bonds, and two or more atoms held together by covalent bonds are known as a molecule.

NONPOLAR COVALENT BONDS. Involve equal sharing of electrons between atoms.

POLAR COVALENT BONDS. Involve unequal sharing of electrons between atoms; condition can give rise to polar molecules that have both positive and negative areas.

Ionic Bonds and Ions. Transfer of electrons from one atom to another creates charged atoms (or aggregates of atoms) called ions; ionic bonds are attractions between oppositely charged ions that can hold ions together to form electrically neutral ionic compounds.

A Closer Look: Electronegativity.

Hydrogen Bonds. Attractions between oppositely charged regions of polar molecules that involve hydrogen and certain other atoms such as oxygen or nitrogen.

◆ CARBON CHEMISTRY pp. 31–41

With a few exceptions, chemicals that contain carbon are organic chemicals; all other chemicals are classified as inorganic chemicals.

Carbohydrates. Are composed of carbon, hydrogen, oxygen, with hydrogen and oxygen frequently in a 2:1 ratio.

MONOSACCHARIDES. Single-unit sugars, such as glucose.

DISACCHARIDES. Two monosaccharide units, for example, sucrose.

POLYSACCHARIDES. Many monosaccharide units, for example, glycogen.

Lipids. Almost insoluble in water but very soluble in organic solvents like benzene, ether, chloroform.

TRIGLYCERIDES. Union of fatty acids with glycerol.

PHOSPHOLIPIDS. Contain phosphate; important components of cellular membranes.

STEROIDS. Basically consist of four interconnected rings of carbon atoms that have few polar groups attached; examples are cholesterol and sex hormones.

Proteins. Large, complex molecules formed from amino acids. About 20 different amino acids form human proteins. Proteins are formed from amino acids by linking the amino acids with peptide bonds.

Nucleic Acids. Composed of purine and pyrimidine bases, five-carbon sugars, and phosphate groups; single base-sugar-phosphate unit called a nucleotide.

RNA. Formed when the sugar in a polynucleotide chain is ribose.

DNA. Formed when the sugar in two polynucleotide chains is deoxyribose; makes up major portion of chromosomes.

Adenosine Triphosphate. Composed of adenine, ribose, and three phosphate groups; serves as immediate source of energy for bodily activity.

◆ ENZYMES AND METABOLIC REACTIONS pp. 41–44

Sum total of chemical reactions in body is called metabolism. Metabolic reactions are subdivided into anabolic, or synthesis, reactions and catabolic, or decomposition, reactions. Enzymes are biological catalysts that speed up metabolic chemical reactions by lowering activation energies.

Action of Enzymes. Enzyme forms enzyme-substrate complex, with substrate at enzyme active site.

Regulation of Enzymatic Activity. Provides a method of regulating metabolic reactions.

CONTROL OF ENZYME PRODUCTION AND DESTRUCTION. Controls amounts of particular enzymes present.

1. Competitive inhibition: inhibitor molecule binds reversibly to active site of enzyme.
2. Noncompetitive inhibition: inhibitor molecule binds irreversibly to active site or to site other than active site of enzyme.

Cofactors. Substances often needed by enzymes to catalyze reactions.

◆ **SOLUTIONS** pp. 44–48
Homogeneous mixtures of two or more components that can be gases, solids, liquids; material present in a solution in greatest amount is generally called the solvent; materials present in lesser amounts are generally called solutes.

Properties of Water. Water is body's principal solvent.

1. Many different polar molecules and ionic compounds dissolve in water.
2. Water is not a good solvent for nonpolar molecules.
3. Water has a high specific heat.
4. Water has a high latent heat of vaporization.

A Closer Look: Expressing the Concentration of a Solution.

◆ **SUSPENSIONS** p. 48
Mixtures in which the dispersed particles are so large that they tend to settle out.

◆ **COLLOIDS** p. 48
Between true solutions and obvious suspensions; dispersed particles are small enough that they do not readily settle out.

◆ **ACIDS, BASES, AND pH** pp. 48–49
In pure water, there is a balance between hydrogen ions and hydroxide ions, since dissociation of a water molecule produces one ion of each type.

Acids. Substances that increase hydrogen ion concentration of an aqueous solution (proton donors).

Bases. Substances that decrease the hydrogen ion concentration of an aqueous solution (proton acceptors).

pH Scale. A logarithmic mathematical scale used to indicate the hydrogen ion concentration of a solution.

◆ **GRADIENTS** pp. 49–50
The level of a particular variable such as pressure, temperature, degree and kind of electrical charge, or the concentration of a substance often changes in a more or less continual fashion between one region and another. Such a change in the level of a variable over distance is referred to as a gradient.

◆ **DIFFUSION** pp. 50–51
The movement of atoms, molecules, or ions from one location to another as a consequence of their thermal motion. Net diffusion is the movement of a substance from a region of higher concentration to a region of lower concentration as a consequence of the thermal motion of the atoms, molecules, or ions of the substance.

◆ **OSMOSIS** pp. 51–52
The movement of solvent through any membrane in response to a concentration difference across the membrane.

Osmotic Pressure. The tendency of water molecules to enter a solution in response to a concentration gradient.

Isosmotic. Describes two solutions having the same osmotic concentration.

Hyperosmotic and Hypoosmotic. Describe two solutions of different osmotic concentrations where the more concentrated one is hyperosmotic to the less concentrated, and the less concentrated is hypoosmotic to the more concentrated.

◆ **DIALYSIS** pp. 52–53
A process that selectively separates substances in a liquid by taking advantage of their differing diffusibilities through porous membranes.

Application in Artificial Kidney. Blood is passed through a membranous tube immersed in bathing medium of known composition; wastes that are small particles are removed; vital protein molecules are retained; essential small particles are also retained by having them at same concentration in bathing medium as in blood, hence no net diffusion across membrane.

◆ **BULK FLOW** p. 53
The movement of atoms, molecules, ions, or other particles as a unit in one direction as the result of forces that push them from one point to another.

◆ **FILTRATION** p. 53
A process that separates one or more components of a mixture from the others. In this process, the mixture is forced through a porous filter or membrane by mechanical forces such as hydrostatic pressure.

Self-Quiz

1. At the present time, approximately 109 elements are recognized, but only about 24 of these are normally found in the body. True or False?
2. Most of the mass of an atom is concentrated in the atom's: (a) outer electron energy levels; (b) ionic bonds; (c) nucleus.
3. A neutral atom contains the same number of electrons as it does: (a) protons; (b) neutrons; (c) energy levels.
4. The number of neutrons contained in $^{17}_{8}O$ is: (a) 8; (b) 9; (c) 17.
5. Salts are made up of positive and negative ions that form nonpolar covalent bonds with one another. True or False?
6. Organic molecules: (a) do not contain carbon atoms; (b) do not contain polar covalent bonds; (c) are always soluble in water; (d) none of these.
7. Which of the following is not present in a triglyceride? (a) fatty acids; (b) glycerol; (c) glucose.
8. Proteins are formed from amino acids by linking the amino group of one amino acid to the acid carboxyl group of another with: (a) an ionic bond; (b) a peptide bond; (c) a disulfide bond.
9. Match the terms with the appropriate lettered descriptions:

Proteins	(a) The sequence of amino acids in a polypeptide chain.
Amino acids	(b) The means by which an amino group of one amino acid is joined to the acid carboxyl group of another amino acid.
Peptide bond	(c) Composed of 100 or more amino acids linked into a chain.
Dipeptide	(d) Result of R-group interactions of different amino acid chains of multichain protein molecules.
Polypeptide	(e) Two amino acids bonded chemically.
Primary structure	(f) Result of hydrogen bonds that occur principally between the constituents of a protein's peptide bonds.
Secondary structure	(g) These compounds are the basic units of proteins.
Tertiary structure	(h) The folding of an amino acid chain into a particular three-dimensional configuration.
Quaternary structure	(i) Ten amino acids linked into a chain.

10. RNA makes up a major portion of chromosomes, which contain hereditary genetic information for directing the activities of the body's cells. True or False?
11. Enzymes are very specific, and each will catalyze only individual reactions or limited classes of reactions. True or False?
12. Generally, enzymes are destroyed during the course of the reactions they catalyze. True or False?
13. The type of enzyme inhibition in which an inhibitor molecule rather than a substrate molecule combines reversibly with an enzyme molecule at the active site of the enzyme is called: (a) noncompetitive inhibition; (b) competitive inhibition; (c) allosteric inhibition.
14. Match the terms with the appropriate lettered descriptions:

Solvent	(a) In a solution, the substance present in the smaller amount is generally this substance.
Solution	(b) The dispersed particles tend to settle out in this mixture.
Solute	(c) In the body, water is this substance.
Suspension	(d) A homogeneous mixture of two or more components: gases, solids, or liquids.
Colloid	(e) In this mixture, dispersed particles tend not to settle out, and a beam of light passing through the mixture is scattered by the particles.

15. Substances that increase the hydrogen ion concentration of a solution are called: (a) salts; (b) bases; (c) acids.
16. Compared to a solution with a pH of 7, a solution with a pH of 5 has a hydrogen ion concentration that is: (a) one hundred times greater; (b) two times greater; (c) two times lesser.
17. Net diffusion can be viewed as the movement of a substance from a region of lower concentration to a region of higher concentration. True or False?
18. The movement of solvent through any membrane in response to a concentration difference across the membrane is termed: (a) diffusion; (b) osmosis; (c) dialysis.
19. If two solutions have the same osmotic concentration, they are said to be: (a) isosmotic; (b) hyperosmotic; (c) hypoosmotic.
20. The movement of atoms, molecules, ions, or other particles as a unit in one direction as the result of forces that push them from one point to another is called: (a) osmosis; (b) diffusion; (c) bulk flow.

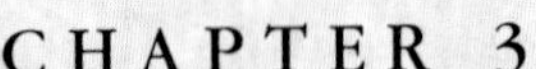

CHAPTER 3

The Cell

CHAPTER CONTENTS

LEARNING OBJECTIVES

After completing this chapter, you should be able to:

1. Describe the structure and function of the plasma membrane.
2. Distinguish in detail between mediated and nonmediated transport.
3. Describe the ways in which a cell takes in and expels substances by means of membrane-bounded vesicles.
4. Describe three types of plasma-membrane receptors to which signaling molecules bind.
5. Cite and explain the main functions of DNA.
6. Distinguish among the functions of mRNA, tRNA, and rRNA.
7. Describe the functions of ribosomes, the endoplasmic reticulum, and the Golgi apparatus, indicating the relationship among these structures.
8. Describe the functions of cilia, flagella, basal bodies, and centrioles, indicating the relationship among these structures.
9. Distinguish between mitosis and meiosis.
10. Cite two occurrences during meiosis that lead to genetic diversity, and explain how each occurs.

PTER

3

The cell is the basic structural and functional unit of the human body, and the cellular nature of the body's tissues, organs, and systems is evident upon microscopic examination. Consequently, a knowledge of cellular function is essential to understanding human anatomy and physiology.

Human cells are specialized both anatomically and physiologically. Muscle cells, for example, have a well-developed property of **contractility** (the ability to move or contract), whereas nerve cells are specialized for **conductivity** (the ability to transmit impulses). Other cells exhibit highly developed properties of **metabolism** (the ability to process foods, obtain energy, and synthesize products), **irritability** (the capacity to respond to stimuli), or **reproduction** (the ability to duplicate themselves). It is important to remember, however, that these properties are present to some degree in all cells and as such may be regarded as general cell characteristics.

Cell Components

Cells are highly organized units made up of many different components (Figure 3.1; Table 3.1). They contain a variety of structures, collectively called **organelles** (little organs), that are responsible for specific cellular functions. Cells also contain chemical substances, such as glycogen granules or lipid droplets, that are collectively called **inclusion bodies.**

A cellular component called the **nucleus** is the control center of the cell. The nucleus is surrounded by a region known as the **cytoplasm.** A membrane called the **plasma membrane** (or **cell membrane**) sur-

◆ **FIGURE 3.1 A "typical" cell showing subcellular organelles**
There is probably no actual cell that can be considered "typical" in all respects.

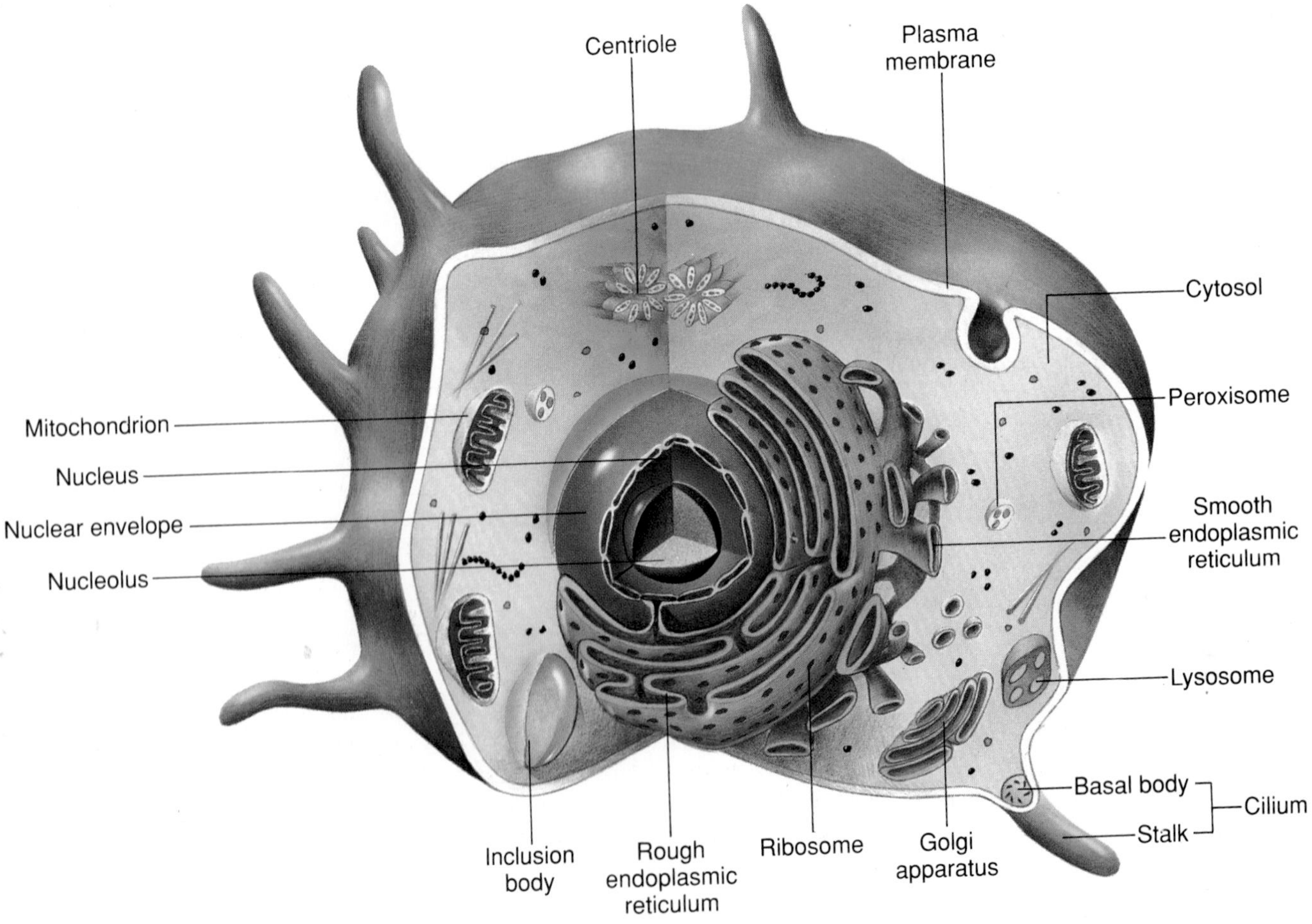

TABLE 3.1 Cell Components

COMPONENT	STRUCTURAL ELEMENTS	REPRESENTATIVE FUNCTIONS
Nucleus	DNA and specialized proteins enclosed by a nuclear envelope	Control center of the cell containing genetic material; provision of specifications for synthesis of polypeptides and proteins that determine the specific nature of each cell
Plasma membrane	Lipid bilayer containing proteins and carbohydrates	Selective barrier between cell contents and extracellular fluid
Cytoplasm components		
Endoplasmic reticulum	Network of interconnected, membrane-bounded tubules and flattened sacs	Production of lipids for new cell membranes; involved in manufacture of products for secretion
Golgi apparatus	Flattened, membrane-bounded sacs	Processing, modification, and sorting of polypeptides and proteins
Lysosomes	Membrane-bounded vesicles containing digestive enzymes	Destruction of material such as foreign substances and cellular debris
Peroxisomes	Membrane-bounded sacs containing enzymes including catalase	Detoxification of potentially harmful substances
Mitochondria	Membrane-bounded structures surrounded by two membranes; inner membrane forms partitions called cristae	Major site of ATP production
Ribosomes	Particles composed of RNA and proteins	Assembly of amino acids into polypeptides and proteins
Secretory vesicles	Membrane-bounded packages of secretory products	Storage of secretory products
Inclusion bodies	Chemical substances in the form of particles or droplets	Storage forms of carbohydrate and triglycerides
Cytoskeleton	Microtubules, microfilaments, and intermediate filaments	Provide complex and dynamic structural network for the cell; involved in variety of movement processes

rounds the cytoplasm and forms the limiting boundary of the cell. The cytoplasm contains a variety of organelles that are surrounded by their own membranes. The portion of the cytoplasm that includes everything other than the membrane-bounded organelles is called the **cytosol.**

Membranes are selectively permeable barriers that regulate the movement of different materials into and out of cells, as well as into and out of different organelles. They also provide points of attachment for various cell components. For example, the contractile elements of muscle cells are attached to the plasma membrane, and the enzymes that mediate certain chemical reactions are bound to the membranes of particular organelles. Although some differences are evident, the basic organization and fundamental properties of the various cell membranes are believed to be similar to those of the plasma membrane.

Plasma Membrane

The **plasma membrane** is composed of a bimolecular layer of lipid (particularly phospholipid and cholesterol), and various proteins are associated with the lipid (Figure 3.2). Proteins called *integral proteins* are embedded in the lipid, and proteins called *peripheral proteins* are loosely bound to the membrane surface, primarily to integral proteins on the inner surface of the membrane. Carbohydrate molecules are often covalently linked to lipid and protein molecules at the extracellular surface of the membrane. (Proteins that have carbohydrates attached to them are called *glycoproteins,* and lipids that have carbohydrates attached are called *glycolipids.*)

In some cases, integral proteins extend completely through the membrane, forming water-filled channels or pores that connect the interior of the cell with the

◆ **FIGURE 3.2 The structure of the plasma membrane**

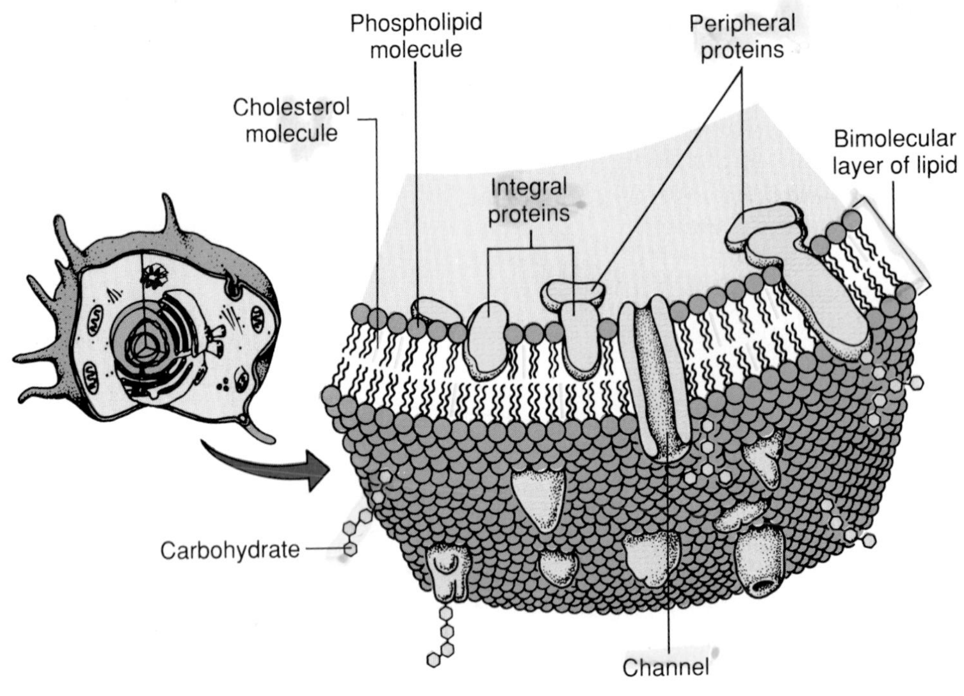

extracellular fluid. Moreover, some membrane proteins and glycoproteins serve as **receptors.** These proteins and glycoproteins possess specific *binding sites* to which particular molecules or ions can bind by noncovalent interactions. A molecule or ion that binds noncovalently to the surface of a protein is called a **ligand.** Thus, the molecules and ions that bind noncovalently to the binding sites of membrane proteins and glycoproteins that serve as receptors are ligands.

Movement of Materials across the Plasma Membrane

All materials that enter or leave a cell must pass either through the plasma membrane or through channels in the membrane. Thus, the properties of the membrane (as well as the properties of the penetrating materials) are important in determining the ease with which substances enter or leave cells.

Diffusion through the Lipid Portion of the Membrane. Many nonpolar substances that are soluble in lipids move easily through the plasma membrane by simple diffusion. However, water-soluble polar substances that are not very lipid-soluble diffuse through the membrane only with difficulty, if at all. Thus, the lipid portion of the membrane is a selectivity permeable barrier to substances entering or leaving cells.

Movement through Membrane Channels. The membrane channels provide an alternative route through the membrane for polar molecules and ions. However, this route is limited to substances small enough to fit through the channels, which are estimated to be about 0.8 nanometer (nm) in diameter (1 nm = 1×10^{-6} mm).

The movement of ions through the channels is highly specific, and certain ions pass most easily through certain channels. For example, sodium ions pass easily through channels called *sodium channels,* and potassium ions pass easily through *potassium channels.* This specificity is believed to be due to specific arrangements of charge on the portions of the proteins that form the interior surfaces of the channel walls. The net movement of ions through membrane channels is down a gradient.

Various signals can open or close membrane channels and thereby alter membrane permeability. The signals, which are frequently chemical or electrical in nature, are believed to open or close channels by causing changes in the conformations (shapes) of the proteins that form the channels.

Mediated Transport. Many polar molecules larger than 0.8 nm in diameter enter cells readily, even though they are quite insoluble in lipids and are too large to fit through membrane channels. Various forms of mediated transport account for the ability of these substances to cross the plasma membrane. **Mediated transport** makes use of molecules called *carrier molecules,* which are part of the membrane itself. A number of integral membrane proteins serve as carrier molecules.

Substances that move across the membrane by mediated transport attach to specific binding sites on the carrier molecules in a manner similar to that by which a substrate attaches to an enzyme. This attachment enables the transported substances to cross the membrane and, depending on the direction of transport, either enter or leave the cell.

The exact mechanisms by which carrier molecules enable substances to pass through the membrane are not known. One theory, however, proposes that a carrier molecule possesses a binding site oriented in such a manner that a substance to be transported that is located on one side of the membrane can attach reversibly to the site (Figure 3.3). Once the attachment occurs, the carrier molecule undergoes a conformational change that exposes the binding site and the attached substance on the other side of the membrane, where the substance is released.

◆ **FIGURE 3.3 A proposed mechanism of mediated transport**

(a) A carrier molecule possesses a binding site oriented in such a manner that a substance to be transported that is located on one side of the membrane can attach reversibly to the site. (b) Once the attachment occurs, the carrier undergoes a conformational change (a change in shape). (c) The conformational change exposes the binding site and the attached substance on the other side of the membrane, where the substance is released.

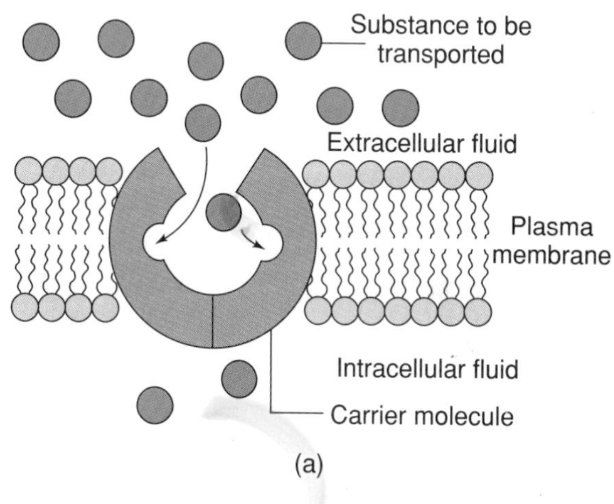

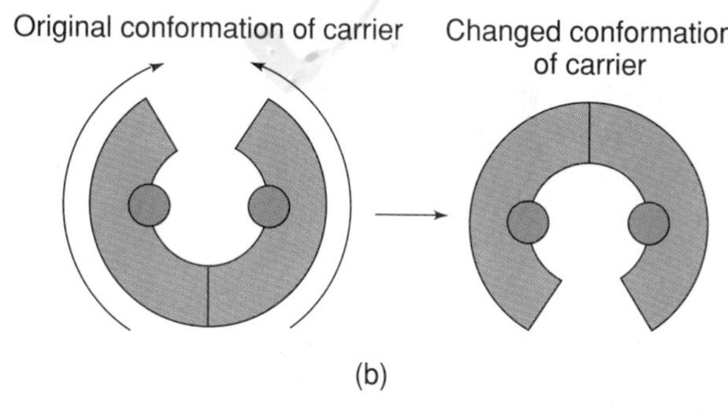

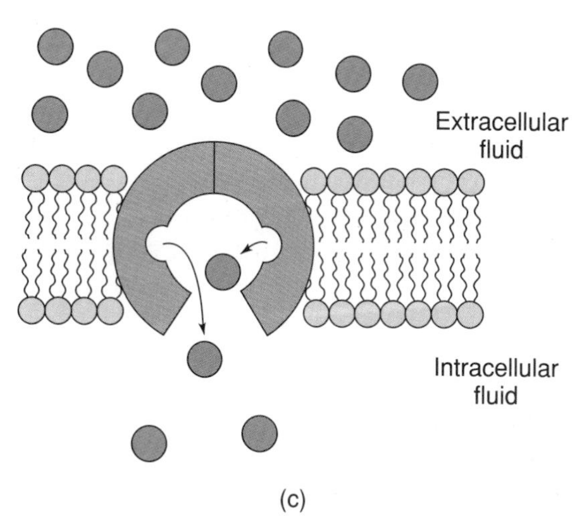

Characteristics of Mediated-Transport Systems. Mediated-transport systems exhibit the characteristics of specificity, saturation, and competition:

1. **Specificity** means that only certain substances can move across the membrane by mediated transport (Figure 3.4a). Each carrier molecule binds only with a select group of substances, and only substances that can attach to the various carrier molecules can cross the membrane by mediated transport.
2. **Saturation** means there is a limit to the amount of a substance that can cross the membrane by mediated transport in a given time (Figure 3.4b). When all of the carrier molecules for a given substance are being utilized, the addition of more of the substance will not increase its rate of transport.
3. **Competition** means that different substances can compete for the services of the same carrier (Figure 3.4c). Suppose, for example, that substance A and substance B are both transported across the membrane by attaching to the same binding site on carrier X. If a substantial amount of substance A but no substance B is present, then all of the carrier X molecules will be utilized in transporting substance A. If substance B is added, it will compete with substance A for the services of the carrier X molecules. In this situation, some substance A and some substance B will be transported,

◆ **FIGURE 3.4 Characteristics of mediated-transport systems**
Mediated-transport systems exhibit characteristics of (a) specificity, (b) saturation, and (c) competition. See the text for details.

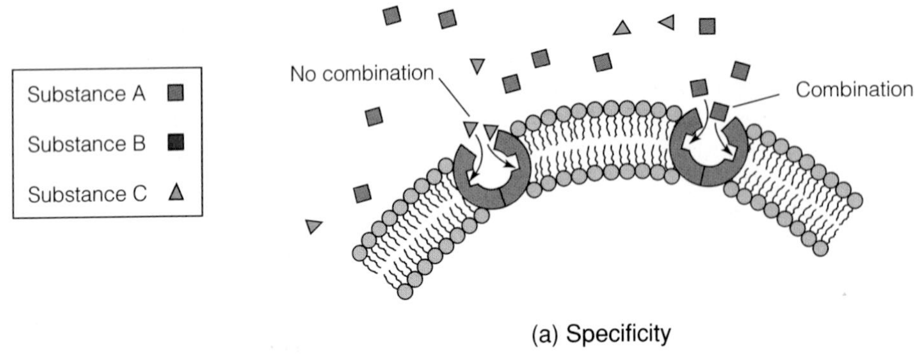

(a) Specificity

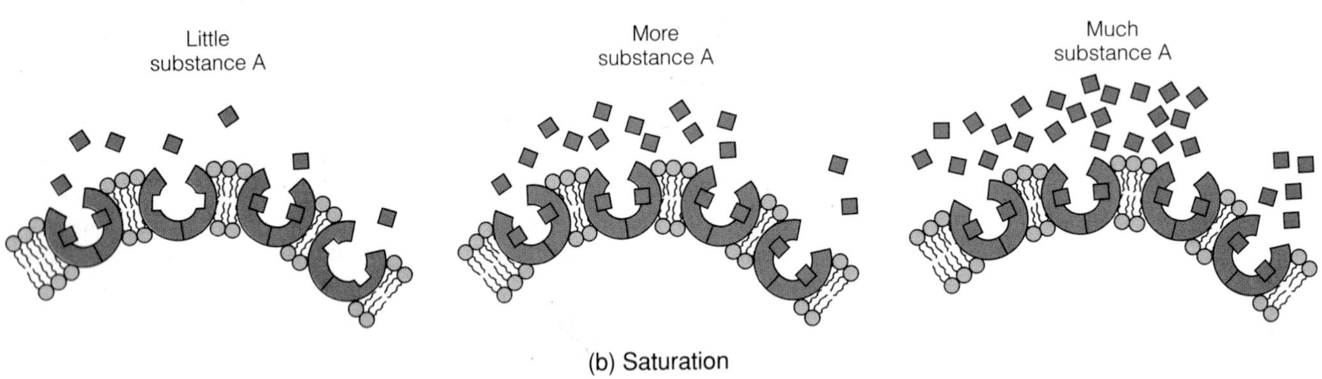

(b) Saturation

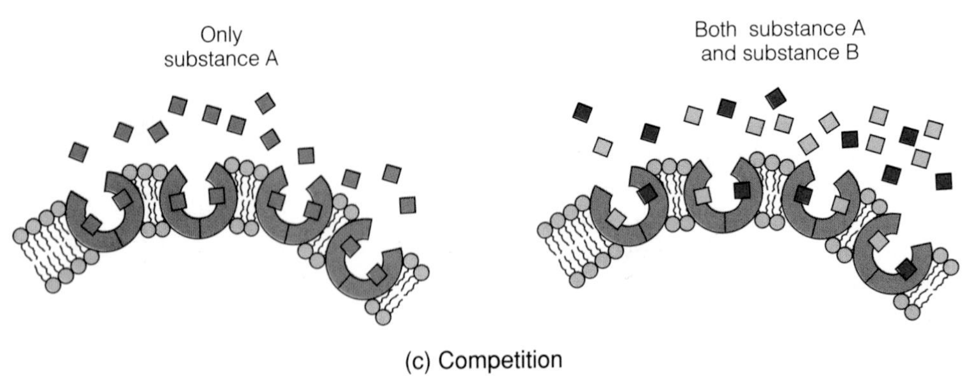

(c) Competition

and the transport rate of substance A will decrease compared to the rate when it was the only substance present.

Facilitated Diffusion. One type of mediated transport is called **facilitated diffusion.** In this type of transport, substances that cannot readily cross the membrane by themselves (because they are not very lipid soluble, for example) attach to carrier molecules. This attachment enables the substances to move across the membrane.

Carrier molecules that function in facilitated diffusion permit the transfer of substances across the membrane equally well in either direction—into or out of the cell. However, the net movement of a substance by facilitated diffusion is down a gradient. For example, glucose enters many cells by facilitated diffusion. In this process, carrier molecules for glucose permit glucose transfer across the membrane in either direction. However, more glucose molecules cross the membrane from their region of higher concentration outside the cell to their region of lower concentration inside the cell than move in the opposite direction, resulting in a net movement of glucose into the cell.

Facilitated diffusion is considered to be a passive process that requires no expenditure of cellular energy. In fact, facilitated diffusion is generally thought of as simply providing a mechanism for a movement that would occur anyway if it were not for the inability of a substance to cross the membrane readily by itself.

Active Transport. Another type of mediated transport is called **active transport.** In this type of transport, carrier molecules move substances across the membrane and up gradients (for example, from a region of lower concentration of a substance to a region of higher concentration). Consequently, active-transport systems can accumulate concentrations of substances on one side of the membrane that are many times the concentrations of the substances on the opposite side of the membrane.

Energy must be expended to move substances up gradients, and active transport depends on energy-yielding chemical reactions that occur within cells to supply the needed energy. If a cell is unable to supply the required amount of energy, active transport cannot occur.

Active-transport systems have inherent directionality. A system that actively transports a substance across the membrane in one direction will not actively transport the substance in the opposite direction. In the absence of an energy supply, however, some carriers that normally function in the active transport of a substance can facilitate the passive transport of the same substance. In such a case, the carrier can facilitate the transfer of the substance across the membrane in either direction, but the net movement of the substance is down a gradient (for example, from a region of higher concentration of the substance to a region of lower concentration). A carrier of this sort is not itself inherently directional. Rather, when the carrier is functioning in active transport, directionality is imposed upon it by the energy-supplying mechanism to which it is coupled.

Endocytosis. Cells are able to form membrane-bounded intracellular vesicles that contain extracellular material (Figure 3.5). In this process, a cell surrounds some extracellular material with a section of its plasma membrane. This section of the membrane then separates from the plasma membrane, forming a membrane-bounded vesicle within the cell. The general term for this process is **endocytosis.** If the vesicle is small and contains materials in solution or very small particles in suspension, the process of vesicle formation is called **pinocytosis** (*pi-no-sigh-to´-sis;* cell drinking). If the vesicle is large and contains particles such as microorganisms, cellular debris, or even whole cells, the process is called **phagocytosis** (*fag-o-sigh-to´-sis;* cell eating). Endocytosis is an active process that requires the expenditure of cellular energy.

Pinocytosis. Virtually all body cells continually form endocytic (pinocytic) vesicles that contain small amounts of extracellular fluid. The process of taking in extracellular fluid by means of vesicle formation is called **fluid-phase endocytosis.** In most cases, vesicle formation occurs at specialized depressed regions of the plasma membrane known as *coated pits.* At a coated pit, a protein called *clathrin* is present at the inner surface of the membrane, coating this surface. The coated pit invaginates and separates from the membrane, forming a coated vesicle that contains extracellular fluid.

◆ **FIGURE 3.5 The process of endocytosis**

A cell surrounds some extracellular material with a section of its plasma membrane. This section of the membrane then separates from the plasma membrane, forming a membrane-bounded vesicle within the cell.

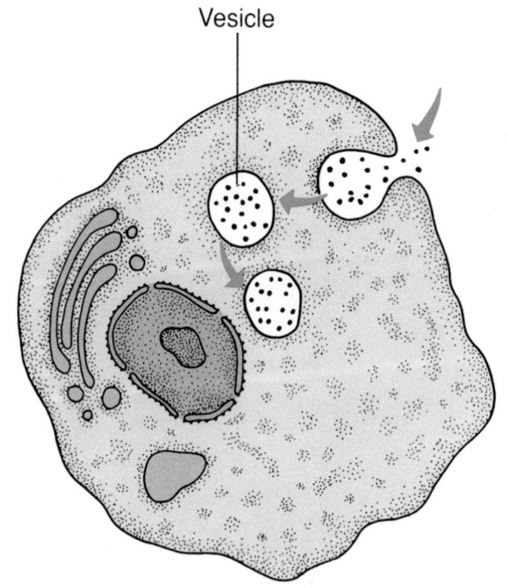

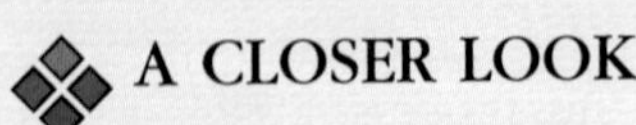

A CLOSER LOOK

Active Transport and the Sodium-Potassium Pump

Although the precise details by which active-transport systems operate are not known, some information about their operation is available. For example, the body's cells maintain a relatively low concentration of sodium ions and a high concentration of potassium ions in their intracellular fluid compared to the concentrations of these ions in the extracellular fluid. To maintain these concentrations, they continually transport sodium ions outward and potassium ions inward (that is, up their respective electrochemical gradients). In fact, the energy expended by cells in the active transport of sodium and potassium ions accounts for about one-third of the total energy expenditure of a resting individual.

The active-transport mechanisms responsible for the transport of sodium and potassium ions are called sodium-potassium pumps. One of the most intensely studied sodium-potassium pumps is located in the plasma membranes of red blood cells. This pump consists of an integral membrane protein that extends from one side of the membrane to the other. The protein possesses enzymatic activity, and it can split ATP present in the cell into ADP and inorganic phosphate, releasing energy that drives the active-transport process. In addition, the protein can interconvert between two different conformations, designated E_1 and E_2.

In the E_1 conformation, sites that have a high affinity for sodium ions are exposed to the intracellular fluid (Figure C3.1a). The binding of sodium ions to these sites triggers the splitting of ATP by the protein. The phosphate released when ATP is split attaches to the protein, and upon phosphorylation the protein assumes its E_2 conformation.

In the E_2 conformation, the sodium binding sites are exposed to the extracellular fluid, and they have only a low affinity for sodium ions (Figure C3.1b). Consequently, sodium ions leave the sites and enter the extracellular fluid. Also, when the protein is in its E_2 conformation, sites that have a high affinity for potassium ions are exposed to the extracellular fluid. The binding of potassium ions to these sites triggers the release of the phosphate from the protein, and upon dephosphorylation the protein reverts to its E_1 conformation.

In the E_1 conformation, the potassium binding sites are exposed to the intracellular fluid, and they have only a low affinity for potassium ions (Figure C3.1c). Consequently, potassium ions leave the sites and enter the intracellular fluid. With the protein in its E_1 conformation, sodium ions can once again bind, the protein can be phosphorylated, and the cycle repeated. For each ATP molecule used by this pump, three sodium ions are transported from the intracellular fluid to the extracellular fluid, and two potassium ions are transported in the opposite direction.

◆ **FIGURE C3.1 The sodium-potassium pump**

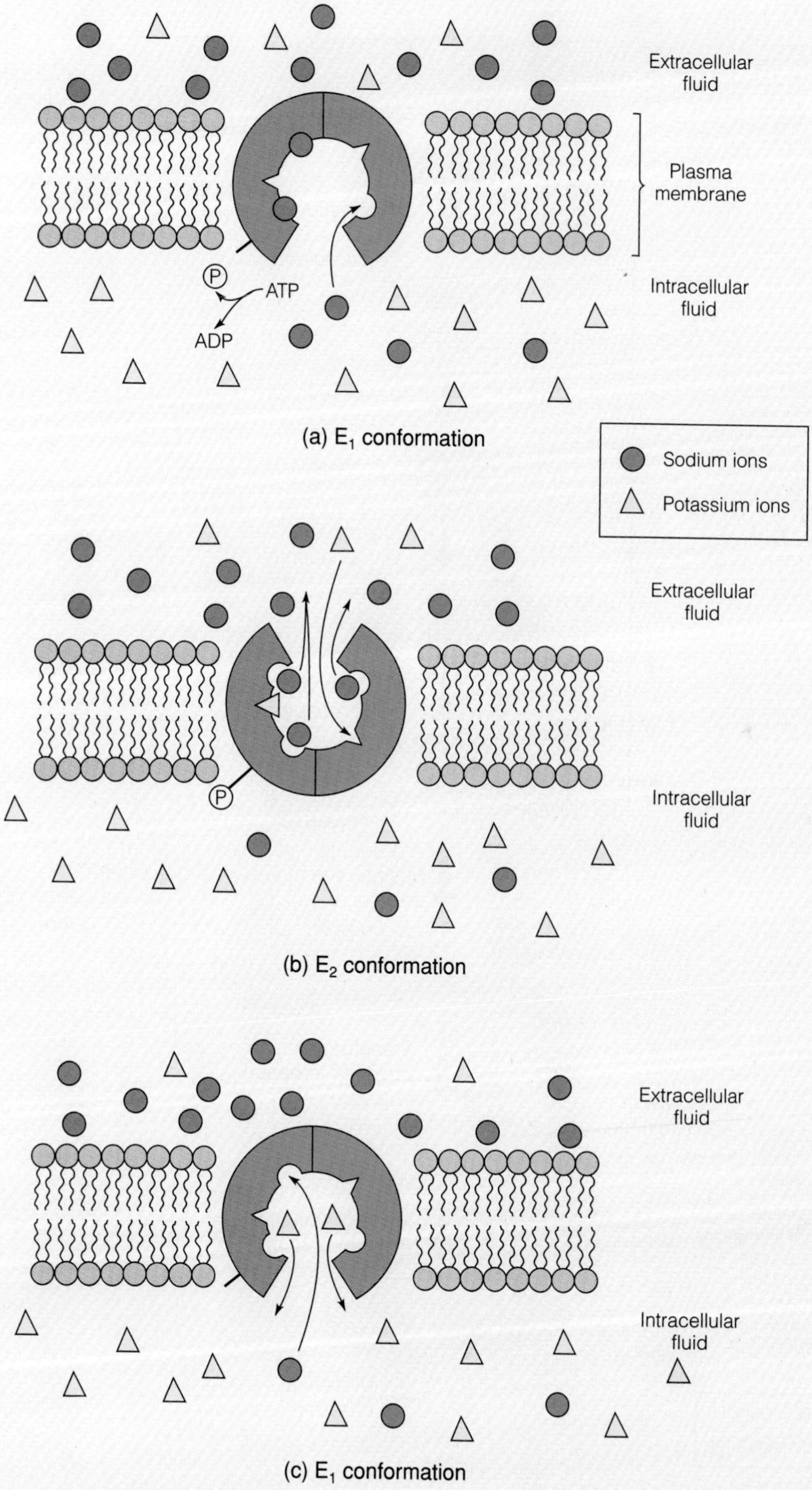

Extracellular fluid

Plasma membrane

Intracellular fluid

P

ATP

ADP

(a) E_1 conformation

Sodium ions

Potassium ions

Extracellular fluid

Intracellular fluid

P

(b) E_2 conformation

Extracellular fluid

Intracellular fluid

(c) E_1 conformation

◆ **FIGURE 3.6 Receptor-mediated endocytosis**
See the text for details.

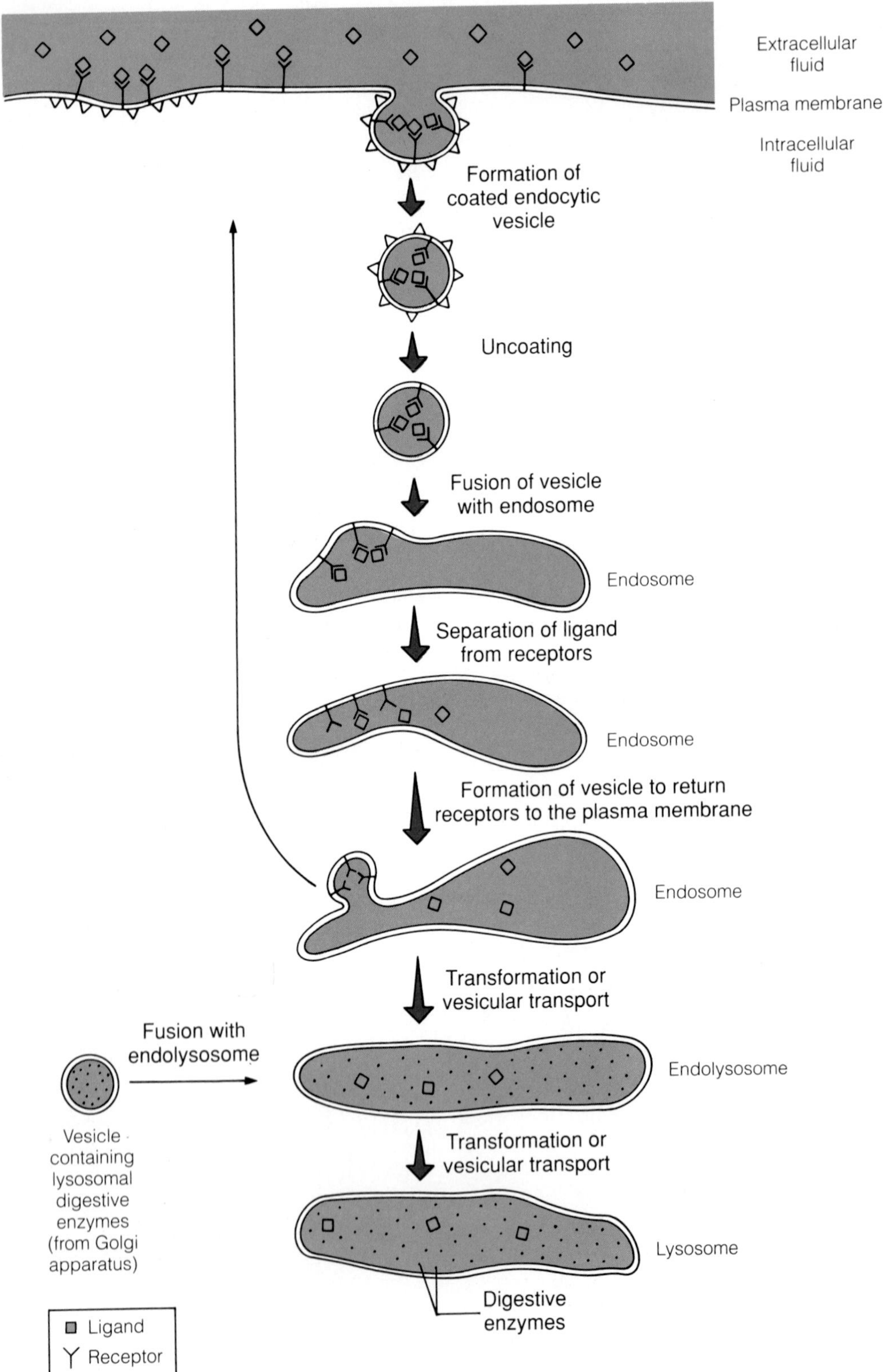

The formation of coated pits and vesicles also provides cells with a means of selectively taking in proteins and other large molecules. This selective process is called **receptor-mediated endocytosis.** A cell has specific receptors on the outer surface of the plasma membrane for each substance to be selectively taken in. The receptors are generally proteins or glycoproteins, and the particular substances that bind to them are ligands (page 62).

In receptor-mediated endocytosis, a particular ligand binds to its receptors, which are either located at a coated pit or move laterally in the membrane to a coated pit after the ligand binds (Figure 3.6). When the coated pit invaginates and separates from the membrane, forming a coated vesicle, the vesicle contains the ligand-receptor complexes as well as extracellular fluid.

Within the cell, a coated vesicle loses all or part of its clathrin coat, and the resulting vesicle commonly fuses with membrane-bounded vesicular or tubular structures called **endosomes.** (The clathrin can return to the plasma membrane.)

Unless it is specifically retrieved, most of the material in endosomes appears next in membrane-bounded **endolysosomes,** and ultimately in vesicles called **lysosomes,** which contain enzymes that break down the material. In many cases, however, material is retrieved from endosomes. For example, within endosomes, ligands often separate from their receptors. Transport vesicles containing the receptors arise from the endosomes and return the receptors to the plasma membrane.

It is uncertain how endocytosed material moves from endosomes to endolysosomes to lysosomes. Some researchers suggest that vesicles arising from endosomes carry material to endolysosomes, and vesicles arising from endolysosomes carry material to lysosomes. Other investigators propose that endosomes themselves are transformed into endolysosomes and that endolysosomes are converted into lysosomes.

In most cells, the formation of clathrin-coated pits and vesicles is the principal mechanism for the pinocytic uptake of both extracellular fluid and receptor-bound ligands. Although pinocytic mechanisms involving other types of vesicles also exist, little is known about their relative importance in the uptake of materials by cells.

Phagocytosis. Essentially all body cells carry on pinocytosis. However, only certain cells, such as some white blood cells and cells called *macrophages,* engage in phagocytosis. In this process, a particle to be taken into a cell binds to receptors on the plasma membrane. This binding induces a portion of the membrane to extend around the bound particle and enclose it. This portion of the membrane separates from the plasma membrane, forming an intracellular phagocytic vesicle called a *phagosome,* which contains the particle.

In general, phagosomes fuse with lysosomes (or endolysosomes), and the lysosomal enzymes break down the phagocytized material. (In some cases, material can be retrieved by transport vesicle formation and returned to the plasma membrane.)

Exocytosis. Substances contained within certain membrane-bounded intracellular vesicles leave cells by **exocytosis** (Figure 3.7). In this process, a membrane-bounded vesicle within a cell fuses with the plasma membrane, and the contents of the vesicle are released from the cell.

All cells constantly release some substances by exocytosis. These substances are contained within transport vesicles that continually fuse with the plasma membrane. In addition, certain cells (for example, cells of the pancreas that produce digestive enzymes) secrete particular products by exocytosis when they receive appropriate signals. In these cells, which are called *secretory cells,* the products are stored in special *secretory vesicles.* When a secretory cell receives an appropriate signal, secretory vesicles fuse with the plasma membrane and release their contents. The signal is often a chemical messenger (for example, a hormone) that binds to receptors located on the cell's plasma membrane and induces a transient increase in the concentration of free calcium ions within the cell. The increased

◆ **FIGURE 3.7 The process of exocytosis**
A membrane-bounded vesicle fuses with the plasma membrane, and the contents of the vesicle are released from the cell.

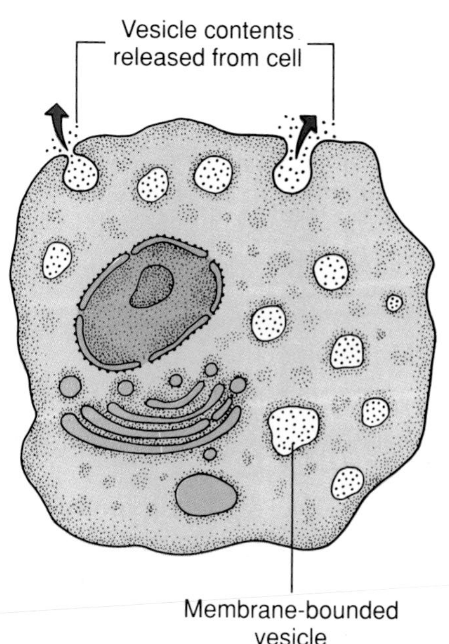

concentration of calcium ions triggers the fusion of secretory vesicles with the plasma membrane.

Additional membrane is added to the plasma membrane when vesicles fuse with it during exocytosis. In most cells, however, as much membrane is removed by endocytosis as is added by exocytosis, and the overall area of the plasma membrane and cell volume remain remarkably constant. Nevertheless, exocytosis does provide a means by which new components (for example, particular receptors) can be added to the plasma membrane, and endocytosis provides a means by which components can be removed from the membrane. Like endocytosis, exocytosis is an active process that requires the expenditure of cellular energy.

When considering the different ways in which materials enter or leave cells, it is important to remember that a given substance may enter or leave a cell by several different routes. For example, sodium ions and potassium ions can move through membrane channels, and they are also actively transported. The different ways in which materials enter or leave cells are summarized in Table 3.2.

Tonicity

The plasma membrane is permeable to water. When a cell is immersed in an aqueous solution, the movement of water across the plasma membrane can influence the state of tension or tone of the cell. The cell may swell due to water entry or shrink due to water loss. Consequently, a solution can be described in terms of its **tonicity;** that is, in terms of its ability to influence the state of tension or tone of cells immersed in it as a result of the movement of water into or out of the cells.

An **isotonic** solution is a solution in which cells maintain their normal tone and no net movement of water into or out of the cells occurs. A **hypotonic** solution produces a change in the tone of cells immersed in it as a result of the net movement of water from the solution into the cells. A **hypertonic** solution produces a change in the tone of cells immersed in it as a result of the net movement of water out of the cells and into the solution.

Signaling Molecules and Plasma-Membrane Receptors

The body's cells need to communicate with one another in order to coordinate their activities. Cells frequently communicate by way of **signaling molecules** that are produced by one cell and act on another. Signaling molecules include hormones and chemicals called neurotransmitters, which are released by nerve cells.

In many cases, signaling molecules produced by one cell attach to receptors that are part of the plasma membrane of another cell. When a signaling molecule (a ligand) binds to a specific membrane receptor, the receptor undergoes a conformational change, and a series of events is triggered that influences such cellular activities as the rates of certain chemical reactions or the permeability of the plasma membrane. Signaling molecules bind to three main types of plasma-membrane receptors: G-protein–linked receptors, catalytic receptors, and channel-linked receptors.

G-Protein–Linked Receptors. Many signaling molecules bind to receptors on plasma membranes called **G-protein–linked receptors.** In general, the binding of signaling molecules to these receptors activates peripheral membrane proteins called **G proteins,** which bind the substance guanosine triphosphate (GTP) (Figure 3.8). In most cases, activated G proteins participate in a series of events that leads to a change in the intracellular concentration of one or more substances known as **intracellular mediators.**

One intracellular mediator is the substance *3′, 5′ cyclic adenosine monophosphate (cyclic AMP).* Depending on cell type, increases or decreases in the intracellular concentration of cyclic AMP can affect cel-

◆ **FIGURE 3.8 A mechanism by which the binding of signaling molecules to G-protein–linked receptors influences cell function**

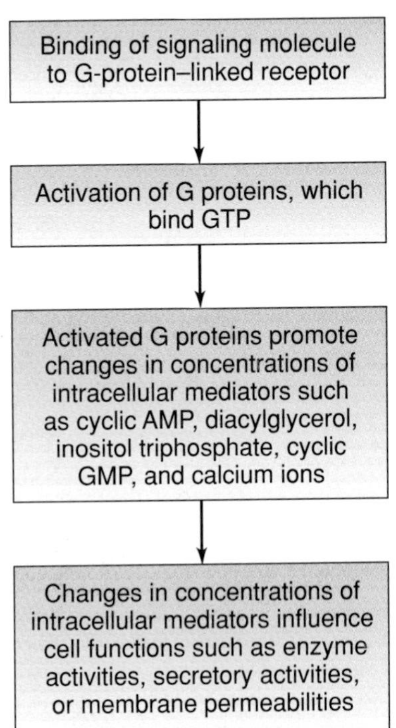

◆ **TABLE 3.2 Methods of Membrane Transport**

METHODS OF TRANSPORT	REPRESENTATIVE SUBSTANCES INVOLVED	ENERGY REQUIREMENTS AND FORCE PRODUCING MOVEMENT	COMMENTS
DIFFUSION			
Through lipid bilayer	Large or small nonpolar molecules (e.g., O_2, CO_2, fatty acids)	Passive; concentration gradient (from high to low concentration)	Continues until the gradient is abolished (state of equilibrium with no net diffusion)
Through membrane channels	Specific small ions (e.g., Na^+, K^+, Ca^{+2}, Cl^-)	Passive; electrochemical gradient (from high to low concentration and attraction of ion to area of opposite charge)	Continues until there is no net movement and a state of equilibrium is established
Special case of osmosis	Water only	Passive; water concentration gradient (water moves to area of lower water concentration)	Continues until concentration difference is abolished or until stopped by an opposing hydrostatic pressure or until cell is destroyed
MEDIATED TRANSPORT			
Facilitated diffusion	Specific polar molecules for which a carrier is available (e.g., glucose)	Passive; concentration gradient (from high to low concentration)	Carrier can become saturated
Active Transport	Specific ions for which carriers are available (e.g., Na^+, K^+, H^+); specific polar molecules for which carriers are available (e.g., glucose, amino acids)	Active; ATP required; moves up gradient	Carrier can become saturated
VESICULAR TRANSPORT			
Endocytosis			
Pinocytosis	Small volume of extracellular fluid, perhaps with specific bound substances (e.g., proteins); also important in membrane recycling	Active; ATP required	In instances of transport of a specific substance, binding occurs to receptors on membrane surface
Phagocytosis	Multimolecular particles (e.g., bacteria and cellular debris)	Active; ATP required	Binding occurs to receptors on membrane surface
Exocytosis	Particular secretory products (e.g., hormones and enzymes); also important in membrane recycling	Active; ATP required; increase in cytosolic Ca^{+2} triggers fusion of secretory vesicles with plasma membrane	Secretion of particular products triggered by appropriate signals

lular functions such as enzyme activities, secretory activities, or membrane permeabilities. For example, when the hormone glucagon, which is produced by the pancreas, binds to G-protein–linked receptors on liver cells, an elevated cyclic AMP concentration promotes the breakdown of glycogen into glucose. When thyroid-stimulating hormone from the pituitary gland binds to G-protein–linked receptors on cells of the thyroid gland, an increased cyclic AMP concentration leads to the release of the thyroid hormones.

In some cases, the binding of signaling molecules to G-protein–linked receptors influences cell function by way of intracellular mediators other than cyclic AMP. These mediators include *diacylglycerol, inositol triphosphate, cyclic guanosine monophosphate (cyclic GMP),* and *calcium ions.*

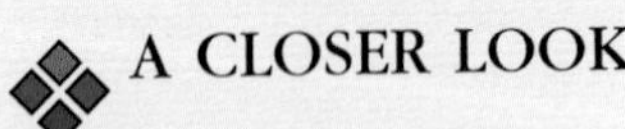

A CLOSER LOOK

G-Protein-Linked Receptors and Intracellular Mediators

In most cases, signaling molecules that bind to G-protein-linked receptors influence cell function by influencing the concentration of intracellular mediators such as cyclic AMP, diacylglycerol, inositol triphosphate, cyclic GMP, or calcium ions.

Cyclic AMP

Cyclic AMP is formed from adenosine triphosphate (ATP) by the enzyme *adenylate cyclase,* which is an integral membrane protein that acts at the inner surface of the plasma membrane. Once cyclic AMP is formed, it is rapidly and continuously broken down by enzymes called *cyclic AMP phosphodiesterases.* Consequently, the concentration of cyclic AMP within a cell is determined in large measure by the activity of adenylate cyclase. (In general, the greater the number of active adenylate cyclase molecules, the higher will be the concentration of cyclic AMP within a cell.)

Signaling molecules that bind to certain G-protein-linked receptors influence the activity of adenylate cyclase—and thus the concentration of cyclic AMP within a cell (Figure C3.2). In some cases, the binding of signaling molecules to G-protein-linked receptors leads to the activation of G proteins known as stimulatory G proteins (G_s), which bind GTP. The stimulatory G proteins with bound GTP activate adenylate cyclase molecules, thereby increasing the concentration of cyclic AMP in the cell.

In other cases, the binding of signaling molecules to G-protein-linked receptors leads to the activation of inhibitory G proteins (G_i), which also bind GTP. The inhibitory G proteins with bound GTP inhibit adenylate cyclase molecules either directly or by preventing the stimulatory G proteins from activating them. As a result, the concentration of cyclic AMP within the cell decreases as it is broken down by cyclic AMP phosphodiesterases.

One of the major functions of cyclic AMP is to activate a protein kinase enzyme called *cyclic AMP-dependent protein kinase.* Protein kinase enzymes attach phosphate groups to particular proteins, which are often enzymes themselves. This phosphorylation can activate or inhibit enzymatic proteins or alter the function of other regulatory molecules, leading to the stimulation or inhibition of various cellular reactions or processes.

A given protein kinase phosphorylates specific proteins, and different types of cells generally contain different combinations of these proteins. Thus, different types of cells may respond differently to a change in cyclic AMP concentration because they contain different proteins that are phosphorylated by cyclic AMP-dependent protein kinase.

Diacylglycerol, Inositol Triphosphate, and Cyclic GMP

In some cases, signaling molecules bind to G-protein-linked receptors on plasma membranes, and the binding activates certain G proteins (tentatively called G_p), which bind GTP (Figure C3.3). These G proteins with bound GTP activate the enzyme *phosphoinositide-specific phospholipase C,* which is bound to the inner surface of the plasma membrane. This enzyme catalyzes the splitting of a membrane phospholipid called *phosphatidylinositol 4,5 biphosphate (PIP_2)* into *diacylglycerol* and *inositol triphosphate.* Often, the activation of this pathway is accompanied by the conversion of GTP into *cyclic guanosine monophosphate (cyclic GMP)* by the enzyme *guanylate cyclase,* indicating that the pathway may also promote cyclic GMP formation.

◆ **FIGURE C3.2 G-protein–linked receptors and cyclic AMP**

Some signaling molecules that bind to certain G-protein–linked receptors at plasma membranes influence cell function by way of the intracellular mediator cyclic AMP.

One of the major functions of cyclic AMP is to activate a protein kinase enzyme called cyclic AMP–dependent protein kinase.

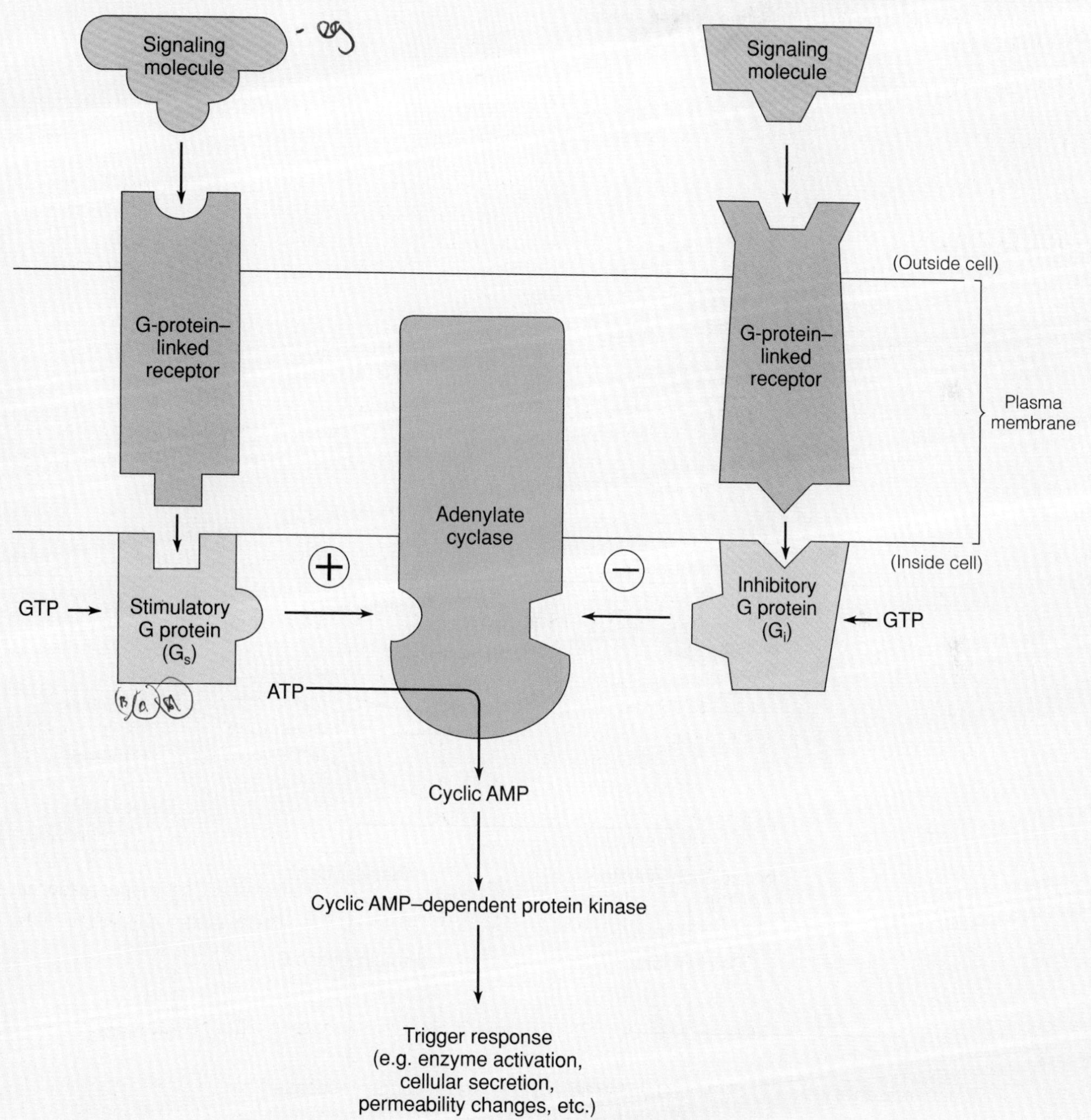

Diacylglycerol activates a protein kinase enzyme called *protein kinase C.* Cyclic GMP activates a protein kinase known as *cyclic GMP-dependent protein kinase* and in some cases acts directly on regulatory molecules in cells to influence their functions. As is the case for cyclic AMP-dependent protein kinase,

continued on next page

◆ **FIGURE C3.3 G-protein–linked receptors and inositol triphosphate, diacylglycerol, cyclic GTP, and calcium ions**

Some signaling molecules that bind to certain G-protein-linked receptors at plasma membranes influence cell function by way of intracellular mediators such as inositol triphosphate, diacylglycerol, cyclic GTP, and/or calcium ions.

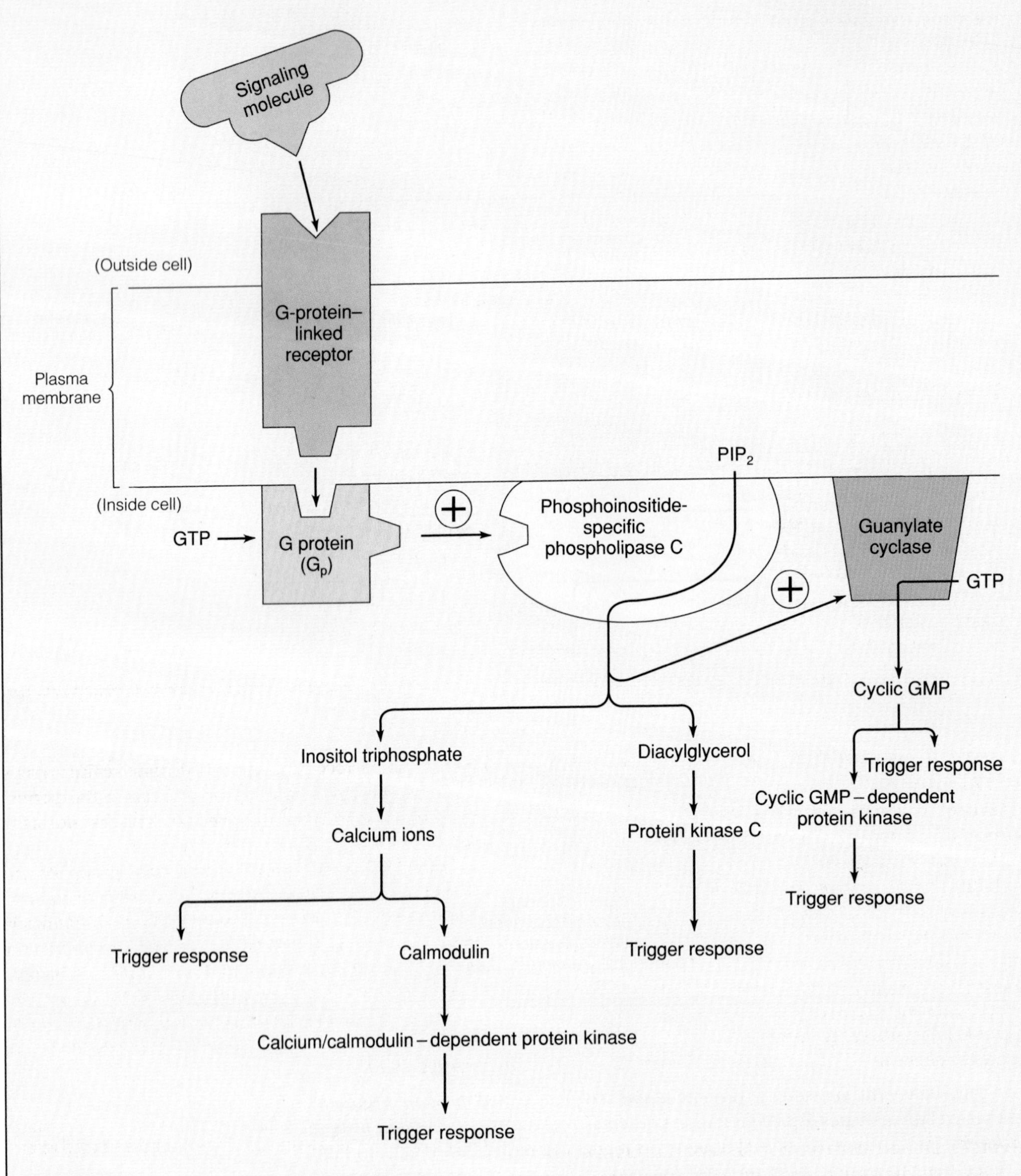

the responses of a given cell type to activated protein kinase C and cyclic GMP-dependent protein kinase are determined in large part by the specific proteins present that are phosphorylated by these enzymes.

Calcium Ions

Signaling molecules often influence the concentration of calcium ions within cells, and inositol triphosphate (as well as cyclic AMP and perhaps cyclic GMP) can modulate intracellular calcium ion concentrations. For example, inositol triphosphate stimulates the release of calcium ions into the cytosol from a membrane-bounded intracellular calcium-sequestering compartment. The ability of signaling molecules—perhaps acting by way of inositol triphosphate and/or cyclic AMP—to influence the concentration of calcium ions within cells is important because calcium ions are themselves a widely used intracellular mediator. In some cases, calcium ions bind directly to proteins, thereby altering their functions. In other cases, calcium ions bind to a protein called *calmodulin,* and the calmodulin with bound calcium regulates the activities of specific proteins, including certain membrane transport proteins and protein kinase enzymes called ***calcium/calmodulin-dependent protein kinases.***

Termination of Activity

In many cases, a signaling molecule remains bound to a receptor for only a short time, and in general, the series of events that occurs when a signaling molecule binds to a G-protein-linked receptor is turned off when the signaling molecule is no longer bound. G proteins have enzymatic activity and slowly break down their bound GTP into guanosine diphosphate (GDP) and inorganic phosphate. Without bound GTP, the G proteins are inactive. In addition, enzymatic reactions rapidly break down diacylglycerol and inositol triphosphate and dephosphorylate phosphorylated proteins. Moreover, calcium ions are rapidly transported out of the cytosol by active-transport mechanisms.

Catalytic Receptors. **Catalytic receptors** act directly as enzymes, and the binding of a signaling molecule to a catalytic receptor activates the receptor. The catalytic portion of the receptor is exposed at the inner surface of the plasma membrane.

Many catalytic receptors function as enzymes called *tyrosine-specific protein kinases,* which attach phosphate groups to tyrosine amino acid residues of specific proteins. Interestingly, catalytic receptors that function as tyrosine-specific protein kinases phosphorylate themselves when they are activated. Receptors for the hormone insulin, which is produced by the pancreas, are catalytic receptors.

Channel-Linked Receptors. **Channel-linked receptors** are actually ion channels in the plasma membrane that possess receptor regions to which certain signaling molecules bind. When a signaling molecule binds to a channel-linked receptor, the ion channel either opens or closes, thereby increasing or decreasing the permeability of the membrane to the ion or ions that normally pass through the channel.

Channel-linked receptors belong to a group of ion channels called **chemically gated channels** or **ligand-gated channels.** The opening or closing of chemically gated channels is regulated by various chemical substances (ligands) that bind to the channels. Channel-linked receptors and chemically gated channels are particularly important in the transmission of information by the nervous system, and they are considered further in Chapter 11.

Electrical Conditions at the Plasma Membrane

Ions are electrically charged particles, and the distribution and movements of ions across the plasma membrane have electrical consequences for cells. In order to understand these consequences, however, it is necessary to understand the basic principles of electricity.

Basic Electrical Concepts. There are two types of electrical charge: positive and negative. The total amount of positive charge in the universe is believed to equal the total amount of negative charge; consequently, the universe is electrically neutral. However, limited areas within the universe—and within the human body as well—can have more positive than negative charge, or vice versa. These areas are said to be either positively or negatively charged, depending on which charge predominates.

Like charges repel one another; that is, positive charge repels positive charge, and negative charge repels negative charge. However, an electrical force occurs between opposite—that is, positive and negative—charges that attracts them to one another and tends to draw them together. This force increases as the amount of charge increases and as the distance between the charges decreases.

Because of the attractive force between positive and negative charges, energy must be expended and work must be done to separate them. Conversely, if positive and negative charges are allowed to come together, energy is liberated, and this energy can be used to perform work. Thus, when positive and negative charges are separated from one another, they have the *potential* to perform work. The measure of this potential is known as **voltage,** and the units of measurement are volts or millivolts (1 millivolt = 0.001 volt). Voltage is always measured between two points in a system and is often referred to as the electrical potential, or potential difference, between the points.

The actual movement, or flow, of electric charge from one point to another is known as **current.** The amount of charge that moves between two points depends on the voltage and on the hindrance to the movement of charge offered by the material between the points. This hindrance is called **resistance.**

The relation between voltage *(E),* current *(I),* and resistance *(R)* is expressed by Ohm's law:

$$I = E/R$$

Thus, when the resistance is constant, the current flow between two points increases when the voltage between the points increases. When the voltage is constant, the current flow decreases when the resistance increases.

As previously indicated, ions are electrically charged particles. If positively charged ions are separated from negatively charged ions, a voltage develops. Moreover, the movement of ions from one point to another produces a current.

◆ **TABLE 3.3 Representative Concentrations of Selected Ions in the Extracellular Fluid and the Intracellular Fluid of a Typical Cell**

ION	EXTRACELLULAR FLUID	INTRACELLULAR FLUID
Potassium (K^+)	4 mM	139 mM
Sodium (Na^+)	145 mM	12 mM
Chloride (Cl^-)	116 mM	4 mM
Calcium (Ca^{+2})	1.8 mM	<1 μM

mM = millimoles/liter; μM = micromoles/liter.

Membrane Potential. The ionic composition of a cell's intracellular fluid and the ionic composition of the extracellular fluid differ in several ways (Table 3.3). These differences are due to the characteristics and function of the plasma membrane, including the following:

1. The plasma membrane of a cell contains energy-requiring, active-transport mechanisms that move certain ions into the cell and other ions out of the cell. For example, the membrane contains a *sodium-potassium pump* that transports sodium ions outward and potassium ions inward. As a result, the intracellular fluid of a cell contains a higher concentration of potassium ions and a lower concentration of sodium ions than the extracellular fluid.

2. The plasma membrane is not equally permeable to all ions. For example, the plasma membrane contains a number of permanently open channels called *potassium leak channels* that are quite permeable to potassium ions but relatively impermeable to sodium ions. Consequently, the plasma membrane is 50 to 100 times more permeable to potassium ions than to sodium ions.

As a consequence of the differences in ionic composition between a cell's intracellular fluid and the extracellular fluid, a very slight excess of positively charged ions accumulates in the extracellular fluid immediately outside the plasma membrane, and an equal number of negatively charged ions accumulates in the intracellular fluid immediately inside the membrane (Figure 3.9). This separation of positive and negative ions across the membrane involves only a minute quantity of the total ions present, and it does not significantly affect the overall electrical neutrality of the intracellular or extracellular fluids. However, as a result of this separation of positive and negative ions, an electrical potential, or voltage, exists across the membrane between the intracellular fluid immediately inside the membrane and the extracellular fluid immediately outside. This voltage is known as the **membrane potential.**

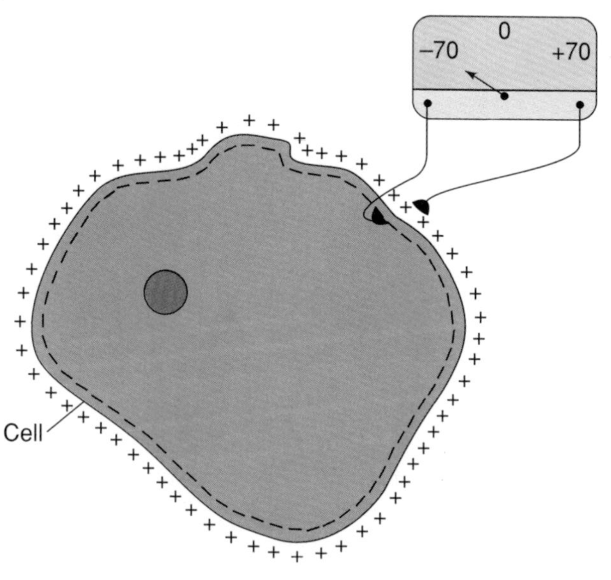

◆ **FIGURE 3.9 The membrane potential**
A slight excess of positively charged ions accumulates in the extracellular fluid immediately outside the plasma membrane of a cell, and an equal number of negatively charged ions accumulates in the intracellular fluid immediately inside the membrane. This separation of positive and negative ions across the membrane gives rise to the membrane potential.

Depending on the type of cell, the value of the membrane potential is usually between −50 and −100 millivolts. (The sign of the potential—plus or minus—indicates the electrical condition that exists in the intracellular fluid immediately inside the membrane relative to the extracellular fluid immediately outside.)

Changes in the Membrane Potential. Alterations in the ionic permeability of the plasma membrane that result in a net movement of particular ions across the membrane can change the membrane potential. A number of cells, including nerve cells and muscle cells, can transiently alter the ionic permeabilities of their plasma membranes, thereby bringing about changes in their membrane potentials.

For example, in an unstimulated nerve cell, as in other cells, sodium ions are present in lower concentration in the intracellular fluid than in the extracellular fluid. Also, as a result of the separation of charges that occurs across the plasma membrane, the intracellular fluid immediately inside the membrane is negative relative to the extracellular fluid immediately outside, and an electrical force attracts positively charged ions like sodium ions into the cell. Thus, an electrochemical gradient exists for sodium ions that favors a net movement of sodium ions into the cell. Nevertheless, in an unstimulated nerve cell, there is no net movement of sodium ions into the cell because the plasma membrane is relatively impermeable to sodium ions, and the sodium-potassium pump within the membrane actively transports enough sodium ions out of the cell to compensate for any sodium ions that enter.

When a nerve cell is sufficiently stimulated, however, the plasma membrane briefly becomes much more permeable to sodium ions, and there is a rapid net movement of sodium ions into the cell (down their electrochemical gradient). The net movement of positively charged sodium ions into the cell from the extracellular fluid alters the membrane potential, causing the intracellular fluid immediately inside the membrane to become less negative in relation to the extracellular fluid immediately outside the membrane; that is, the electrical polarity that exists across the membrane is reduced, and the membrane becomes **depolarized.** In fact, when enough sodium ions enter the cell, the membrane potential reverses, and the intracellular fluid immediately inside the membrane becomes positive relative to the extracellular fluid immediately outside. This rapid depolarization and polarity reversal is called an **action potential,** and it is unique to nerve and muscle cells.

Changes in the ionic permeability of the plasma membrane that result in the occurrence of action potentials are responsible for producing nerve impulses in nerve cells and triggering contraction in muscle cells. The events responsible for these changes are discussed in Chapter 11.

The Nucleus

The **nucleus** is a large organelle that is separated from the rest of the cell by a structure called the **nuclear envelope** (Figure 3.10). Most cells contain a single nucleus. However, some cells contain two or more nuclei, and mature red blood cells contain no nucleus.

The nuclear envelope, which is about 40 nm thick, consists of two parallel membranes separated by a *perinuclear space.* Numerous *nuclear pore complexes* are present in the nuclear envelope. Each complex consists of a series of large protein granules arranged in an octagonal pattern with a water-filled channel in the center. The nuclear pore complexes are believed to be the routes by which certain proteins manufactured in the cytosol are actively transported into the nucleus and by which some materials produced in the nucleus are actively transported into the cytosol.

DNA is the genetic material of a cell. It is present in the nucleus in the form of *chromatin,* which is composed of DNA combined with protein. Two of the main types of DNA-binding proteins are histones and gene

◆ **FIGURE 3.10 An electron micrograph of the nucleus of a pancreatic cell**

Note the nuclear pore complexes (indicated by arrows) in the nuclear envelope that surrounds the nucleus. The dark area within the nucleus is the nucleolus.

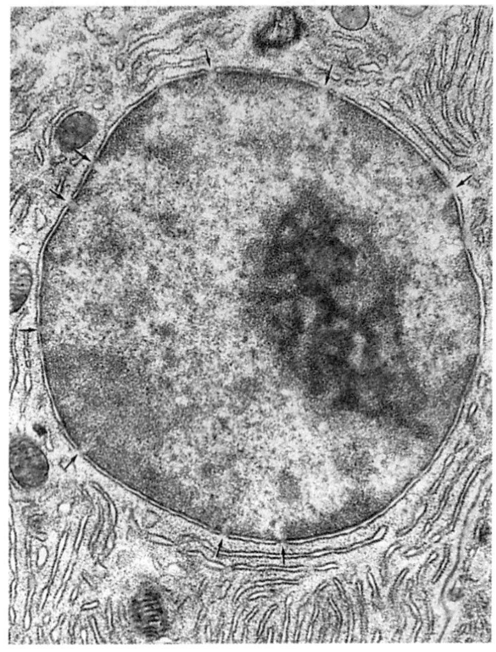

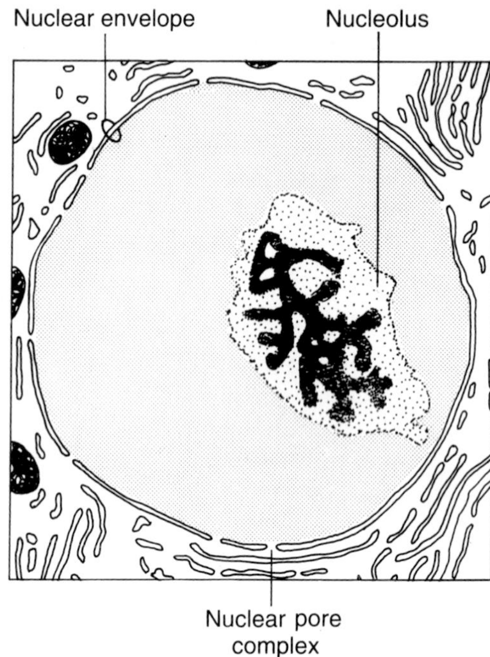

regulatory proteins. Histones are involved in the structural organization and packaging of DNA in the nucleus. Gene regulatory proteins are important in controlling the interactions of DNA with other molecules.

DNA contains coded information within its base sequence that specifies the order in which amino acids are to be linked together to form particular polypeptides or proteins. Once the polypeptides or proteins are synthesized, they act as enzymes, as structural proteins, or in other ways to accomplish the work of the cell.

All cells except the reproductive cells contain a person's full genetic complement of DNA. However, all the possible polypeptides and proteins that DNA can specify are not present at all times in all cells. Various control mechanisms (including those involving gene regulatory proteins) regulate the production of polypeptides and proteins by cells. Consequently, a particular polypeptide or protein may be produced by a given cell at one time but not at another, or it may be produced by one type of cell but not by another.

The ability of a cell to regulate which polypeptides or proteins it produces and when it produces them is important because a cell's functions depend in large measure on the polypeptides and proteins it produces. For example, if a cell is to carry out a specific sequence of metabolic reactions, it must produce the particular enzymatic proteins required for the reactions. Thus, thyroid gland cells must produce the enzymes needed to synthesize thyroid hormones, and adrenal gland cells must produce the enzymes required to manufacture adrenal hormones. Because DNA contains the information needed to synthesize particular polypeptides or proteins, it plays a central role in determining cell function.

RNA Synthesis (Transcription)

DNA in the nucleus is unable to pass through the nuclear envelope into the cytoplasm; yet it is in the cytoplasm that polypeptides and proteins are synthesized. Because of this situation, DNA must transfer its specifications for synthesizing particular polypeptides or proteins to molecules that carry the specifications from the nucleus to the cytoplasm. It does this by a process called **transcription,** in which DNA serves as a template for the assembly of molecules of RNA.

As discussed in Chapter 2, both DNA and RNA are made up of base-sugar-phosphate units called *nucleotides,* and a DNA molecule is composed of two polynucleotide chains. The two chains are joined by hydrogen bonds between the bases of the different chains, and the bases pair with one another in a predictable, complementary fashion. Adenine links with thymine and vice versa, whereas cytosine links with guanine and vice versa.

When DNA serves as a template for RNA synthesis, a multisubunit enzyme called *RNA polymerase* attaches to a DNA molecule at a region called a *promoter.* A promoter is a specific sequence of DNA bases that signals where RNA synthesis should begin. (The attachment of

RNA polymerase to a promoter requires the presence of proteins called transcription factors, which are present on DNA.)

After attaching to a promoter, RNA polymerase moves along the DNA molecule. As RNA polymerase moves, it separates the two linked chains of DNA from one another, exposing part of the base sequence of the DNA, and it links together RNA nucleotides in a stepwise fashion to form an RNA chain (Figure 3.11).

One of the two DNA chains serves as the template chain for RNA synthesis. The RNA nucleotides that are linked together by RNA polymerase are selected by their ability to pair according to the complementary base-pairing pattern with the bases exposed along the DNA template chain. (Recall from Chapter 2 that in RNA the base uracil substitutes for thymine, and the sugar unit of RNA nucleotides is ribose rather than deoxyribose as in DNA.)

According to the complementary base-pairing pattern, adenine-containing RNA nucleotides pair with exposed thymines of the DNA template chain, guanine-containing RNA nucleotides pair with exposed cytosines of the DNA template, uracil-containing RNA nucleotides pair with exposed adenines, and cytosine-containing RNA nucleotides pair with exposed guanines. This complementary base pairing determines the order in which RNA nucleotides are linked together by RNA polymerase to form an RNA chain. Consequently, the base sequence of the RNA chain is specified by and is complementary to the base sequence of the DNA template chain.

As RNA polymerase moves along DNA, it rejoins the two DNA chains it initially separated, thereby displacing newly formed portions of the RNA chain from the DNA template. Ultimately, the RNA chain is released completely as an independent RNA molecule.

RNA polymerase moves along DNA, linking together RNA nucleotides until it reaches a sequence of DNA bases that serves as a *termination (stop) signal.* Upon reaching a termination signal, RNA polymerase separates from the DNA.

Only one of the two chains of any given segment of DNA serves as a template for RNA synthesis. However, one DNA chain may serve as the template at some sites, and the other DNA chain may be the template at other sites. When RNA polymerase attaches to a promoter at a particular site, the promoter orients the polymerase in such a way that it uses one DNA chain and not the other as a template.

Once RNA molecules have been synthesized, they undergo various types of processing. For example, groups of nucleotides can be removed from the center of an RNA molecule, and the remaining portions of the molecule can be spliced together to produce a shorter RNA molecule. Ultimately, three major types of RNA molecules—*messenger RNA, transfer RNA,* and *ribosomal RNA*—are produced.

◆ **FIGURE 3.11 RNA synthesis**

The enzyme RNA polymerase catalyzes the addition of nucleotides to the growing RNA chain in the sequence specified by the order of bases in the DNA template.

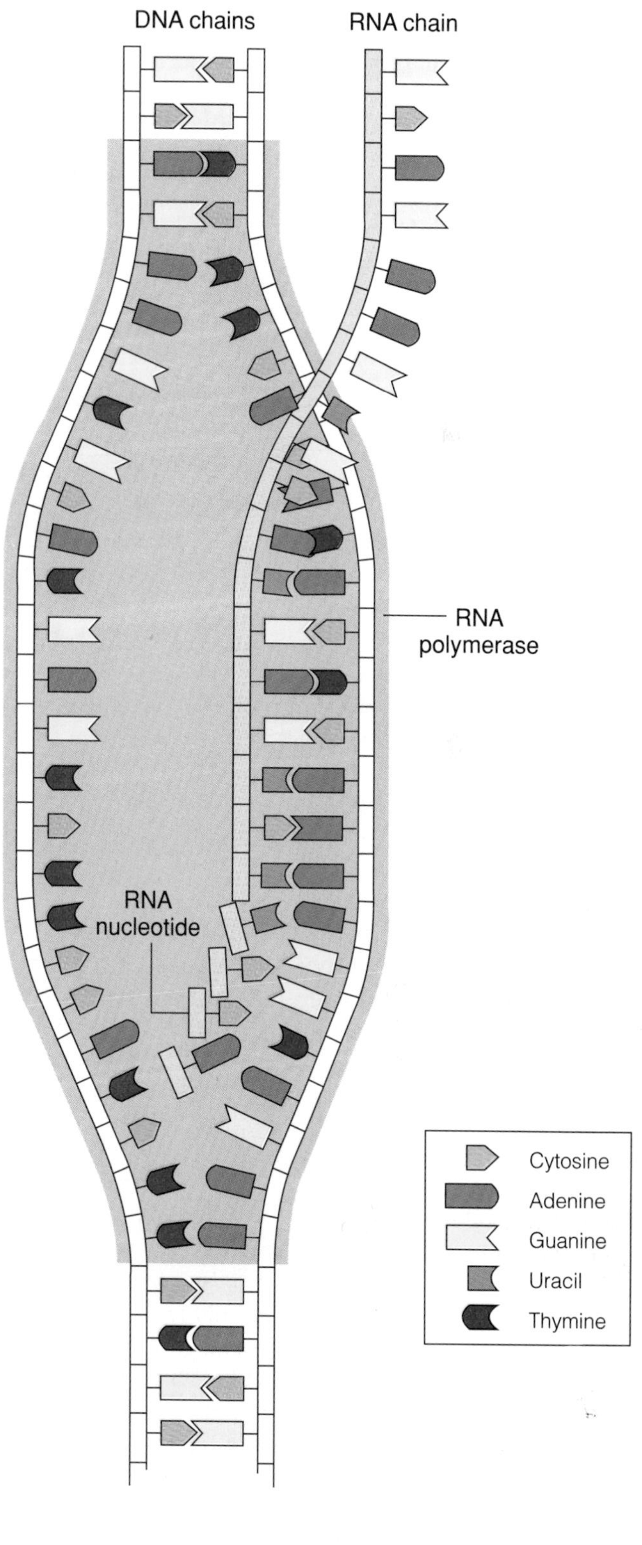

Messenger RNA

Messenger RNA (mRNA) molecules carry information from DNA in the nucleus to structures called *ribosomes* in the cytoplasm, where polypeptides and proteins are synthesized. Each mRNA molecule contains a coded message that specifies the order in which different amino acids are to be linked together to form a particular polypeptide or protein. The message is contained within the base sequence of the mRNA, and this base sequence is ultimately determined by the base sequence of the DNA that serves as a template for RNA synthesis.

Each sequence of three bases along an mRNA molecule constitutes a **codon** that specifies a particular amino acid of a polypeptide or protein. For example, the three-base sequence guanine-cytosine-uracil (GCU) is a codon that specifies the amino acid alanine.

The four principal bases of mRNA—adenine (A), cytosine (C), guanine (G), and uracil (U)—can form 64 different three-base codons. Since only about 20 different amino acids are present in polypeptides and proteins, there are more than enough codons to specify each amino acid. In fact, several different codons can specify the same amino acid (Table 3.4). For example, the codons GCU, GCC, GCA, and GCG all specify the amino acid alanine. As noted in Table 3.4, three codons do not specify amino acids but instead serve as *stop codons* that indicate the end of an mRNA message.

Transfer RNA

Like messenger RNA molecules, **transfer RNA (tRNA)** molecules play an important role in polypeptide and

◆ **TABLE 3.4 The Genetic Code***

First Position	Second Position: U		Second Position: C		Second Position: A		Second Position: G		Third Position
U	UUU	Phe	UCU	Ser	UAU	Tyr	UGU	Cys	U
	UUC		UCC		UAC		UGC		C
	UUA	Leu	UCA		UAA	End	UGA	End	A
	UUG		UCG		UAG	End	UGG	Trp	G
C	CUU	Leu	CGU	Pro	CAU	His	CGU	Arg	U
	CUC		CCC		CAC		CGC		C
	CUA		CCA		CAA	Gln	CGA		A
	CUG		CCG		CAG		CGG		G
A	AUU	Ile	ACU	Thr	AAU	Asn	AGU	Ser	U
	AUC		ACC		AAC		AGC		C
	AUA		ACA		AAA	Lys	AGA	Arg	A
	AUG	Met	ACG		AAG		AGG		G
G	GUU	Val	GCU	Ala	GAU	Asp	GGU	Gly	U
	GUC		GCC		GAC		GGC		C
	GUA		GCA		GAA	Glu	GGA		A
	GUG		GCG		GAG		GGG		G

*The three bases in an mRNA codon are designated, respectively, as the first position, second position, and third position of the codon. See Table 2.3 for amino acid abbreviations. (Three-base sequences designated "End" indicate termination points for reading the mRNA message.)

◆ **FIGURE 3.12 A transfer RNA molecule**

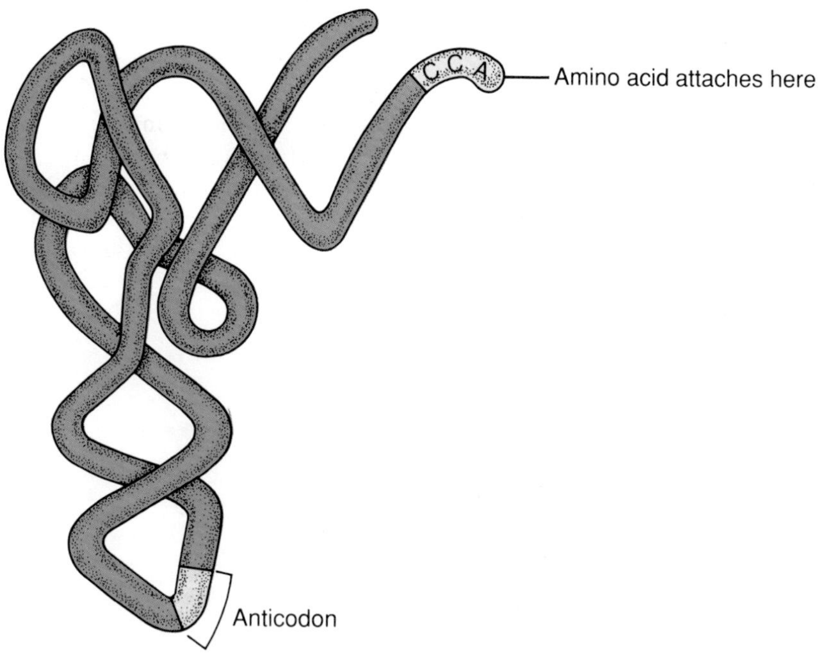

protein synthesis. In the cytoplasm, amino acids combine with tRNA molecules. The tRNA molecules carry the amino acids to ribosomes, where the amino acids are linked together in the order specified by mRNA to form particular polypeptides and proteins.

There are different types of tRNA molecules, and each type carries a specific amino acid. The different types of tRNA have the same general three-dimensional structure, which resembles a letter L (Figure 3.12). The amino acid that a tRNA molecule carries is located at the end of the short arm of the L. A three-base sequence called the **anticodon** is located at the end of the long arm of the L. Different types of tRNA have different anticodons.

Ribosomal RNA

Ribosomal RNA (rRNA) molecules are present in the cytoplasm in the form of rRNA-protein complexes that become organized into ribosomes. Precursors of these rRNA-protein complexes originate in a region of the nucleus called the **nucleolus** (Figure 3.10). The nucleolus is the site of synthesis (on DNA templates) of precursor RNA molecules that are ultimately processed into rRNA molecules, and the precursor RNA combines with protein in the nucleolar region.

Ribosomes

Ribosomes are small structures in the cytoplasm that contain rRNA and protein. At a ribosome, amino acids are linked together in the order specified by mRNA to form a polypeptide or protein.

A ribosome has enzymelike activity. It catalyzes the formation of peptide bonds, which link amino acids to one another.

Basic Features of Polypeptide and Protein Synthesis (Translation)

The process of synthesizing a polypeptide or protein according to the specifications provided by mRNA is called **translation.** During translation, a ribosome becomes attached to an mRNA molecule at a special *start codon* near one end of the mRNA. From the start codon, the mRNA is moved through the ribosome in steps, one codon at a time (Figure 3.13). As the mRNA moves, the ribosome links together amino acids in the order specified by the mRNA, thereby forming a growing polypeptide chain.

The amino acids that are incorporated into the growing polypeptide chain are brought to the ribosome by tRNA molecules. During translation, the bases of the

◆ **FIGURE 3.13 Polypeptide and protein synthesis**

At a ribosome, amino acids are linked together in the order specified by mRNA to form a polypeptide or protein chain of amino acids. Molecules of tRNA bring amino acids to the ribosome for incorporation into the growing chain.

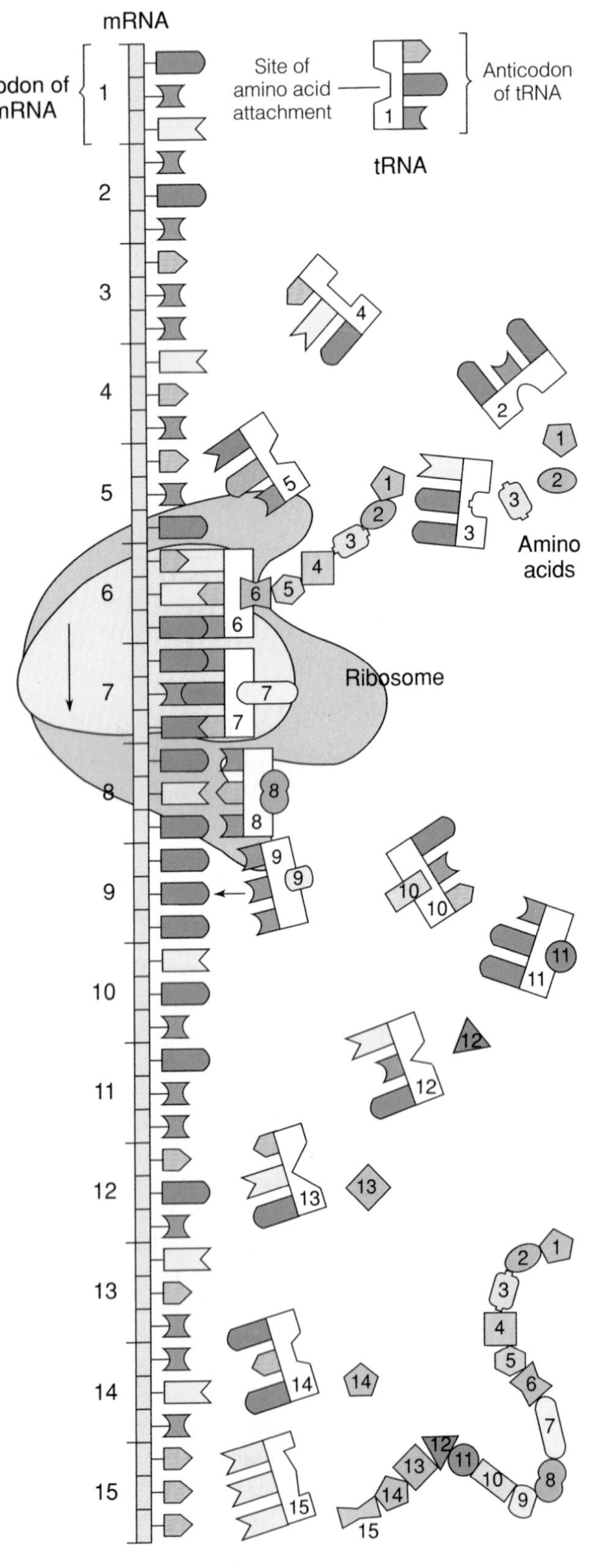

◆ **FIGURE 3.14 A polyribosome**

A polyribosome is formed by a number of ribosomes simultaneously translating an mRNA molecule.

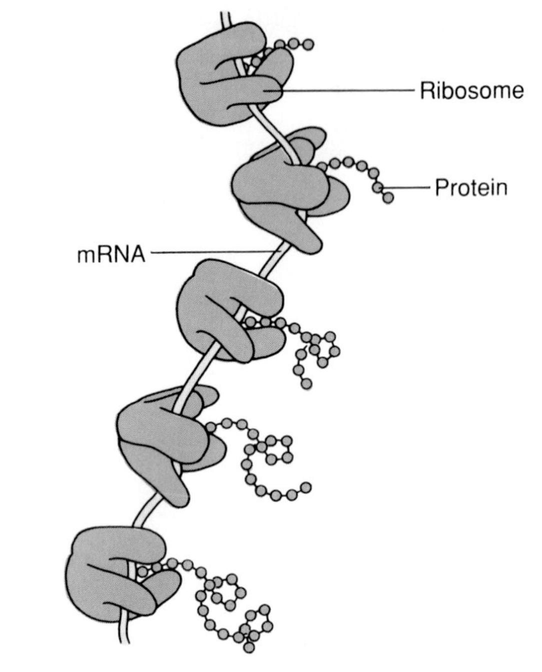

anticodons of particular tRNA molecules pair with complementary bases of specific mRNA codons. This pairing allows the amino acids carried by the tRNAs to be linked to the growing polypeptide chain at the positions specified by the mRNA. For example, the mRNA codon that specifies the amino acid tryptophan is UGG, and the anticodon of the tRNA that carries tryptophan is CCA. When a UGG codon reaches the ribosome, base pairing occurs between this codon and the CCA anticodon of a tryptophan-carrying tRNA. The tryptophan-carrying tRNA becomes attached to the ribosome, and the catalytic activity of the ribosome joins the tryptophan to the growing polypeptide. (By convention, the bases of a tRNA anticodon are written in the reverse of the order in which they actually pair with the bases of an mRNA codon.)

The ribosome continues adding amino acids one at a time to the growing polypeptide chain until an mRNA stop codon reaches it. When a stop codon reaches the ribosome, the newly formed polypeptide or protein is released, and the ribosome separates from the mRNA.

Frequently, a number of ribosomes are attached to the same mRNA molecule, forming a *polyribosome,* or *polysome* (Figure 3.14). Each ribosome assembles a polypeptide or protein according to the specifications of the mRNA.

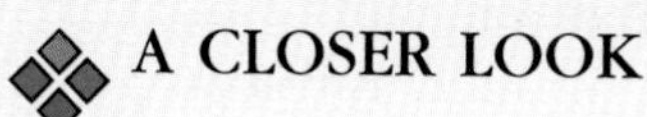

Ribosomes and Protein Synthesis

The basic features of polypeptide and protein synthesis were described on page 81. Here, polypeptide and protein synthesis are examined more completely.

Ribosomes

A single ribosome consists of two subunits: a smaller 40S subunit and a larger 60S subunit (40S and 60S refer to the relative rates of sedimentation of the subunits in a centrifugal field). Each subunit is composed of rRNA and protein.

A ribosome has two sites, an A site and a P site, to which tRNA molecules can attach. It also has a third site that can bind mRNA.

Polypeptide and Protein Synthesis

The synthesis of a polypeptide or protein at a ribosome takes place in the following manner. A special initiator tRNA molecule binds to a 40S ribosomal subunit. The initiator tRNA has an anticodon that can pair with a specific AUG *start codon* of an mRNA molecule. The initiator tRNA also carries the amino acid methionine, which becomes the first amino acid of an amino acid chain.

The 40S ribosomal subunit, with its bound initiator tRNA, attaches to an mRNA molecule at the end of the mRNA called the 5′ end. The 40S subunit moves along the mRNA until it reaches the AUG start codon. When the 40S subunit reaches the start codon, the anticodon of the initiator tRNA pairs with the start codon. A 60S ribosomal subunit then binds with the 40S subunit, forming a functional ribosome that can link together amino acids in the order specified by the mRNA (Figure C3.4a).

When a functional ribosome forms, the initiator tRNA is located at the P site of the ribosome, and its anticodon is paired with the start codon of the mRNA. The A site of the ribosome is located at the next codon farther along the mRNA.

A tRNA molecule whose anticodon can pair with the codon at the A site becomes attached to the site. The catalytic activity of the ribosome removes the amino acid from the tRNA occupying the P site (in this case, methionine is removed from the initiator tRNA), and the amino acid is linked to the amino acid attached to the tRNA occupying the A site (Figure C3.4b).

The tRNA occupying the P site of the ribosome (in this case, the initiator tRNA) is then released, and the tRNA occupying the A site (which has the growing amino acid chain attached to it) is translocated to the P site. During this translocation, the mRNA is moved exactly one codon farther through the ribosome. At the completion of the translocation, the A site is unoccupied, and it is located one codon farther along the mRNA (Figure C3.4c).

A tRNA molecule whose anticodon can pair with the mRNA codon now at the A site becomes attached to the site (Figure C3.4d). The catalytic activity of the ribosome removes the growing amino acid chain from the tRNA now at the P site and links it to the amino acid attached to the tRNA occupying the A site.

The tRNA occupying the P site of the ribosome is then released. The tRNA occupying the A site is translocated to the P site, and the mRNA is moved one codon farther through the ribosome. Another amino acid is then added to the growing amino acid chain. In this manner, amino acids are added one at a time to the growing amino acid chain in the sequence specified by the mRNA.

When a stop codon reaches the A site of the ribosome, a protein called a *release factor* binds to the codon, and the newly formed polypeptide or protein is released. The ribosome then separates from the mRNA and dissociates into

continued on next page

its 40S and 60S subunits, which can be reused. The mRNA can be used by other ribosomes until it is degraded.

An mRNA molecule always moves through a ribosome in the same direction (from the 5′ end of the mRNA toward the end called the 3′ end), and a polypeptide or protein is synthesized from the end of the amino acid chain with the free amino group to the end with the free acid carboxyl group. Proteins called *initiation factors* participate in the initiation of translation. Proteins called *elongation factors* are invcived in the process by which an amino acid is added to a growing amino acid chain.

◆ **FIGURE C3.4 Polypeptide and protein synthesis at a ribosome**

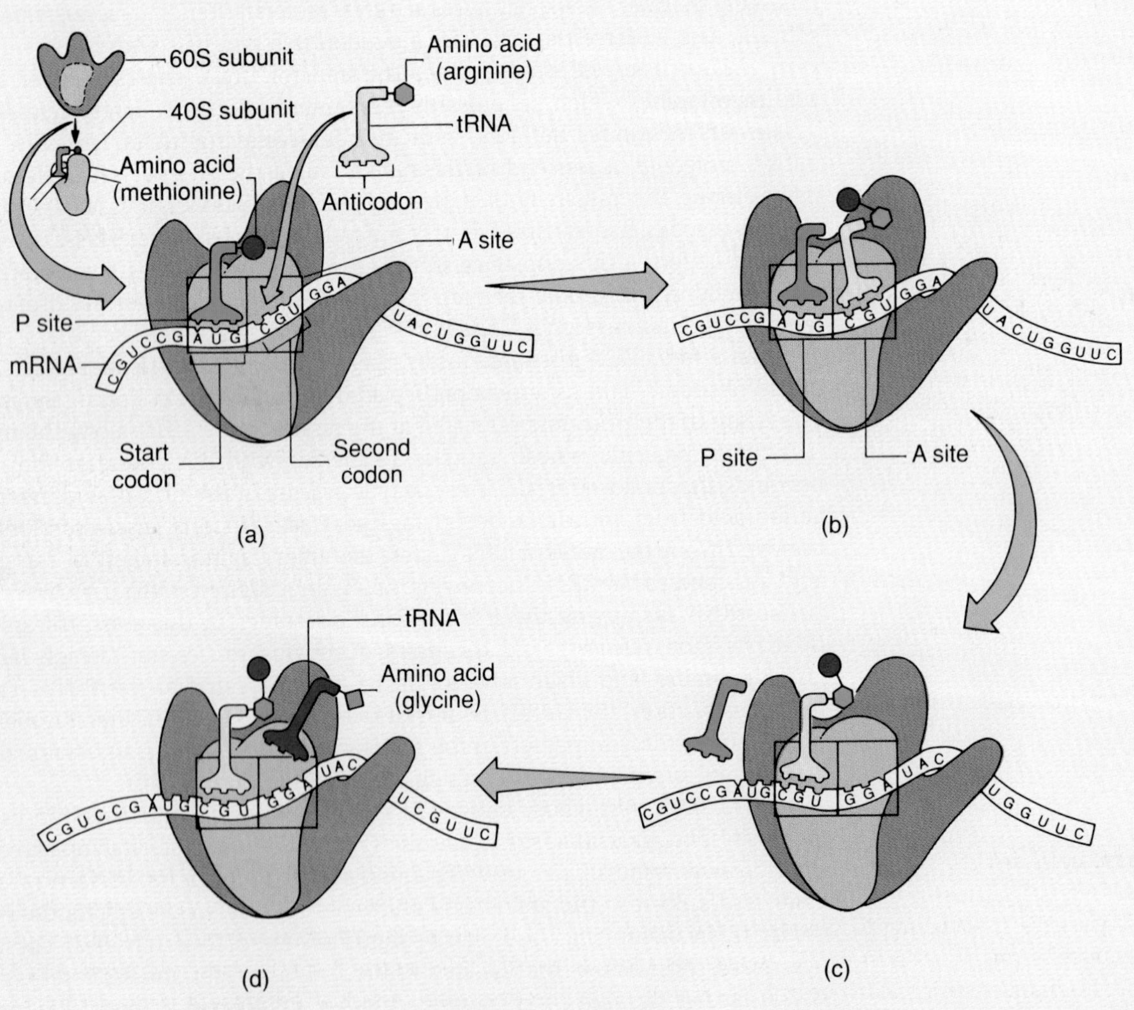

◆ **FIGURE 3.15 The endoplasmic reticulum**
(a) A drawing of the endoplasmic reticulum. (b) An electron micrograph of rough endoplasmic reticulum. The granules attached to the reticulum are ribosomes.

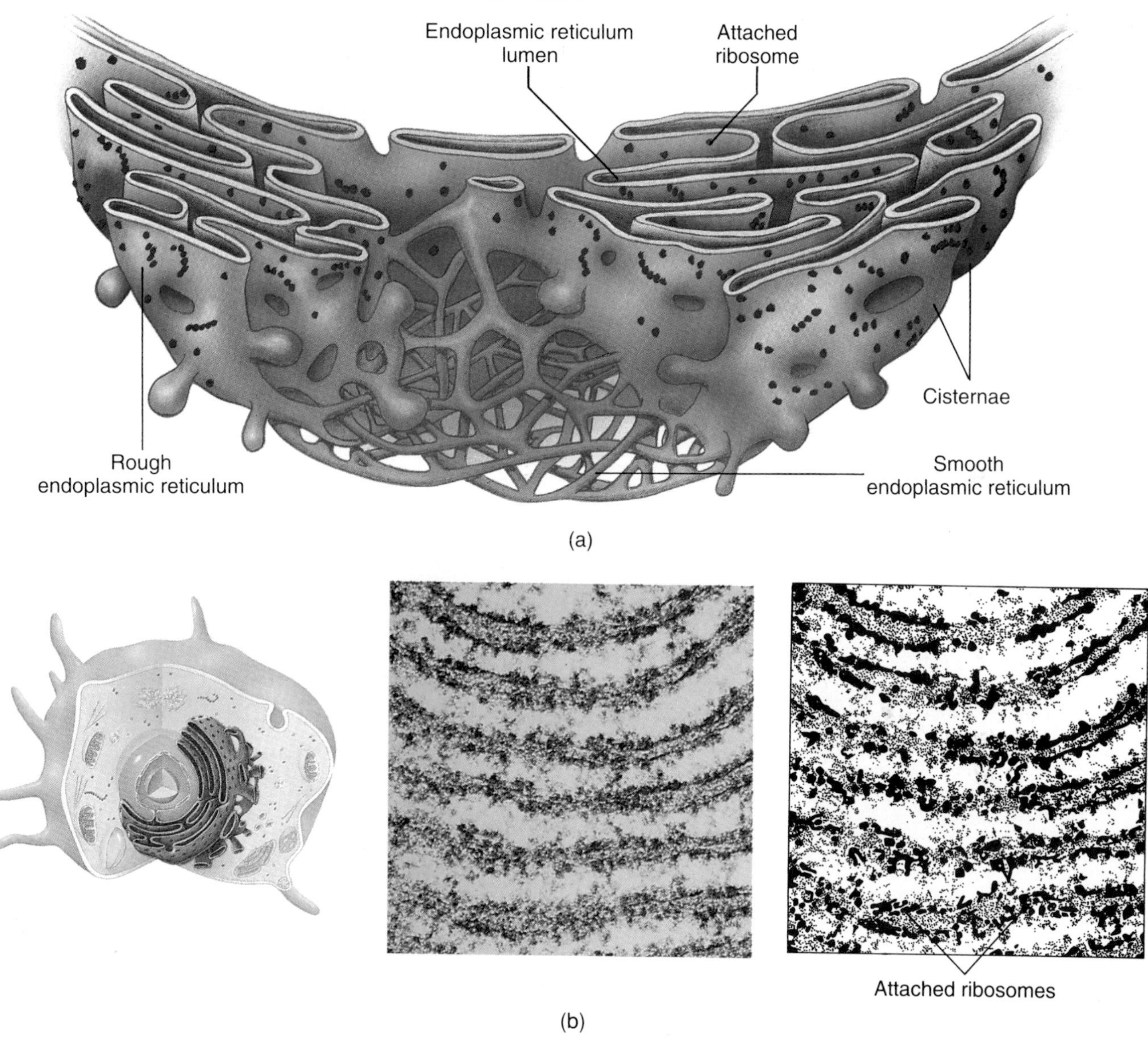

Endoplasmic Reticulum

The **endoplasmic reticulum** *(en-do-plaz´-mik re-tik´-u-lum)* consists of a network of interconnected, membrane-bounded tubules and flattened sacs that extends throughout the cytoplasm of a cell (Figure 3.15). The space within the tubules and sacs is called the *lumen* of the endoplasmic reticulum, and it is filled with fluid. The endoplasmic reticulum membrane is continuous with the outer membrane of the nuclear envelope, and the lumen is continuous with the perinuclear space.

Ribosomes are often attached to the cytosolic surface of the endoplasmic reticulum membrane (as well as to the cytosolic surface of the outer membrane of the nuclear envelope). Endoplasmic reticulum that has ribosomes attached is called **rough** endoplasmic reticulum, and it commonly exists in the form of flattened sacs called *cisternae.* Endoplasmic reticulum that does not have ribosomes attached is called **smooth** endoplasmic reticulum. It frequently occurs as a network of tubules.

The endoplasmic reticulum membrane contains enzymes that synthesize many different lipids, including

◆ **FIGURE 3.16 The Golgi apparatus**
(a) A drawing of the Golgi apparatus. (b) An electron micrograph of the Golgi apparatus.

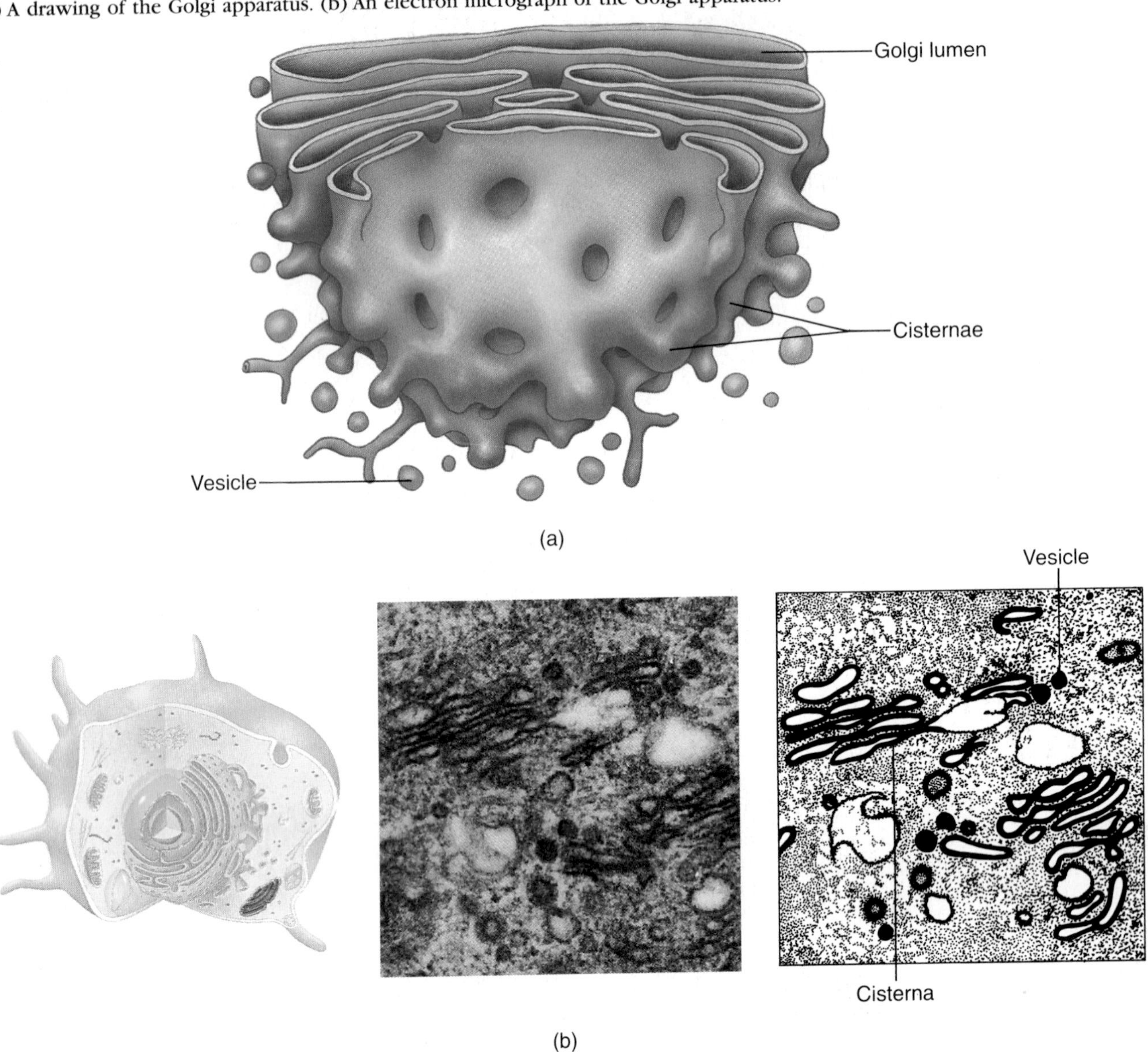

phospholipids, cholesterol, and steroids. The endoplasmic reticulum membrane produces almost all the lipids required for the manufacture of new cell membranes, and an extensive endoplasmic reticulum is present in cells that secrete steroid hormones.

Some polypeptides and proteins synthesized at ribosomes attached to the endoplasmic reticulum are inserted into the endoplasmic reticulum membrane and become membrane components. Others are transported into the lumen of the endoplasmic reticulum.

Within the endoplasmic reticulum, enzyme-catalyzed reactions add carbohydrate to many polypeptides and proteins, and the added carbohydrate is modified. Ultimately, sections of membrane separate from the endoplasmic reticulum membrane, giving rise to membrane-bounded transport vesicles that contain polypeptides and proteins, most of which have carbohydrate attached.

Golgi Apparatus

The **Golgi apparatus** *(gol´-jee)* is a cytoplasmic organelle that consists of a series of flattened, membrane-bounded sacs called *cisternae,* which are frequently located near the nucleus of a cell (Figure 3.16). Vesicles from the endoplasmic reticulum, which contain polypeptides and proteins, fuse with the membranes of the Golgi apparatus. Within the Golgi apparatus, the polypeptides and proteins are processed and modified. For example, enzyme-catalyzed reactions remove portions of the carbohydrate that was added to many polypeptides and proteins in the endoplasmic reticulum, and additional carbohydrate is attached. Also within the Golgi apparatus, the polypeptides and proteins are sorted for delivery to different destinations. Ultimately, membrane-bounded vesicles separate from the Golgi membranes and carry the polypeptides and proteins to their destinations.

In all cells, transport vesicles that originate from the Golgi apparatus continually release their contents from the cells by exocytosis. In secretory cells, secretory vesicles that originate from the Golgi apparatus release their contents by exocytosis when the cells receive appropriate signals. Both of these processes were discussed in the section on exocytosis (page 69).

Endosomes, Endolysosomes, and Lysosomes

Endosomes, endolysosomes, and **lysosomes** are related groups of cytoplasmic membrane-bounded vesicles or tubules. Lysosomes contain strong digestive enzymes that are capable of breaking down proteins, lipids, certain carbohydrates, DNA, and RNA (Figure 3.17). These enzymes, which are called *acid hydrolases,* function best under acidic conditions (pH optimum about 5.0). The lysosome membrane contains an active-transport mechanism that moves hydrogen ions into the lysosomes, thereby maintaining acidic conditions within them. Vesicles that contain the digestive enzymes present in lysosomes originate from the Golgi apparatus and fuse with endolysosomes (see Figure 3.6).

As described in the section on endocytosis (page 65), endocytic vesicles formed at coated pits fuse with endosomes. Although some material is specifically retrieved, most of the material in endosomes appears next in endolysosomes and ultimately in lysosomes, where the digestive enzymes break down the material. (The digestive processes probably begin in endolysosomes.) Many of the products of the digestive processes (for example, sugars, amino acids, and nucleotides) leave the lysosomes to be utilized by the cell. Lysosomes that contain undigested material after the digestive processes are completed are called *residual bodies.*

In phagocytic cells, phagosomes fuse with lysosomes (or endolysosomes), and phagocytized material that is not specifically retrieved is broken down. Lysosomes are particularly abundant in certain white blood cells whose principal activity is the phagocytosis of foreign materials in the body.

Parts of a cell itself are sometimes broken down by a process called *autophagy.* In this process, a cell forms membrane-bounded *autophagosomes,* which enclose portions of the cell. For example, the average life of a mitochondrion in a liver cell is about ten days, and membrane believed to be derived from endoplasmic reticulum membrane surrounds aged or damaged mitochondria, forming autophagosomes. Autophagosomes fuse with lysosomes (or endolysosomes), and the contents of the autophagosomes are broken down by lysosomal enzymes. During starvation, cells break down parts of their

◆ **FIGURE 3.17 An electron micrograph of lysosomes in a white blood cell**
The large structure is the nucleus.

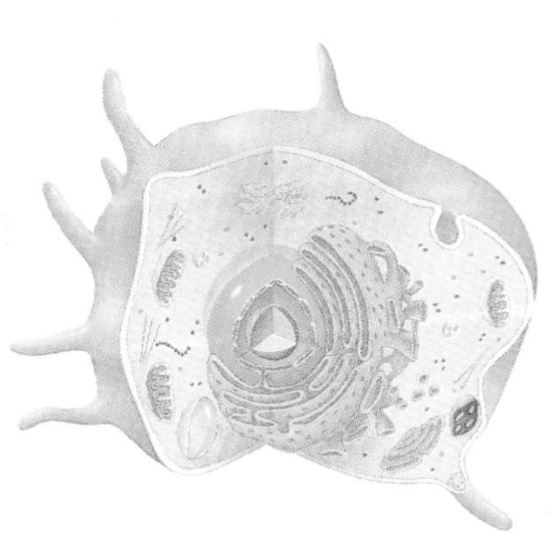

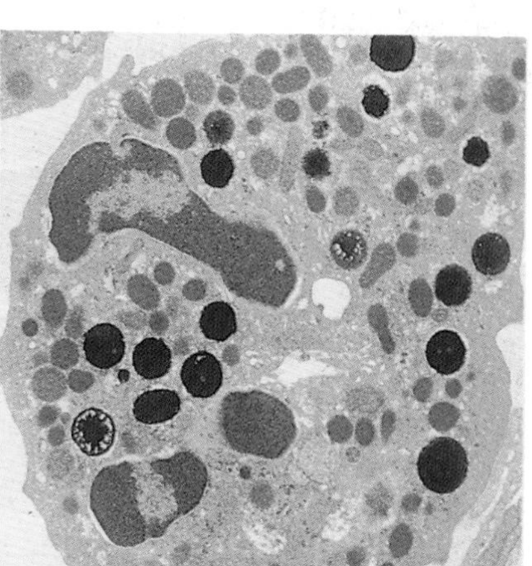

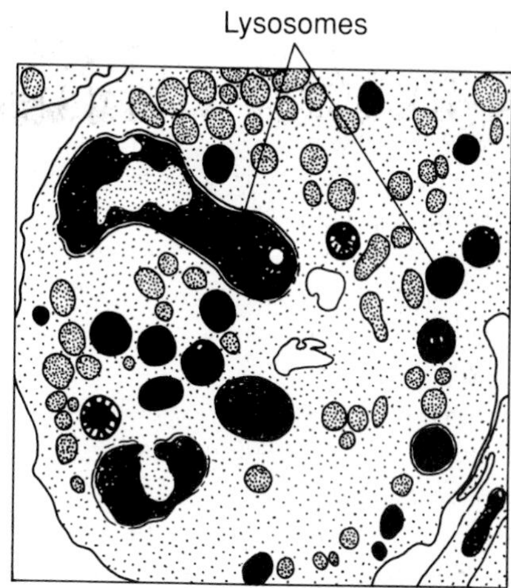

own substance by autophagy and utilize the products that result as energy sources.

In a process called *autolysis,* a controlled destruction of cells by lysosomal enzymes occurs. In this process, lysosomal enzymes are apparently released into a cell, and reactions catalyzed by the enzymes destroy the cell. Autolysis plays an important role in normal embryonic development and in the regression of the mother's mammary glands when her infant is no longer nursing. In both situations, there is an excess of cells that must be eliminated. The destruction of unneeded cells by lysosomal enzymes seems to perform a necessary role in such cases of cell death.

Peroxisomes

Peroxisomes (*per-ox´-i-somes;* microbodies) are small membrane-bounded cytoplasmic organelles. In most cells, peroxisomes are between 0.15 and 0.25 μm in diameter, but in some cells, such as liver cells, they are considerably larger (about 0.5 μm in diameter).

Peroxisomes contain a variety of enzymes that catalyze reactions believed to be important in detoxifying potentially harmful substances. In a number of reactions occurring within peroxisomes, hydrogen atoms are removed from specific organic substrates and combined with oxygen to form hydrogen peroxide (Figure 3.18a). The hydrogen peroxide, in turn,, can participate in reactions catalyzed by the enzyme *catalase,* which is abundant in peroxisomes.

Catalase has two modes of functioning: peroxidatic and catalatic. In some cases, catalase functions in its peroxidatic mode to catalyze reactions in which hydrogen ions removed from specific substrates combine with hydrogen peroxide to form water. For example, about 50% of the ethyl alcohol consumed by an individual is converted to acetaldehyde by this type of reaction (Figure 3.18b). Usually, however, catalase functions in its catalatic mode to remove excess hydrogen peroxide, which is itself a potentially harmful substance. When functioning in this mode, catalase facilitates a reaction that converts hydrogen peroxide to water and oxygen (Figure 3.18c).

Mitochondria

Mitochondria are cytoplasmic organelles that are bounded by two membranes (Figure 3.19). The outer membrane is smooth and surrounds the mitochondrion itself. The inner membrane folds at intervals into the central portion of the mitochondrion, forming partitions known as *cristae.* The compartment between the two mem-

◆ **FIGURE 3.18 Reactions occurring within peroxisomes**

(a) In some reactions, hydrogen ions are removed from organic substances and combined with oxygen to form hydrogen peroxide. (b) Reaction exemplifying the peroxidatic functioning of catalase. (c) Catalase functioning in its catalatic mode to convert hydrogen peroxide to water and oxygen.

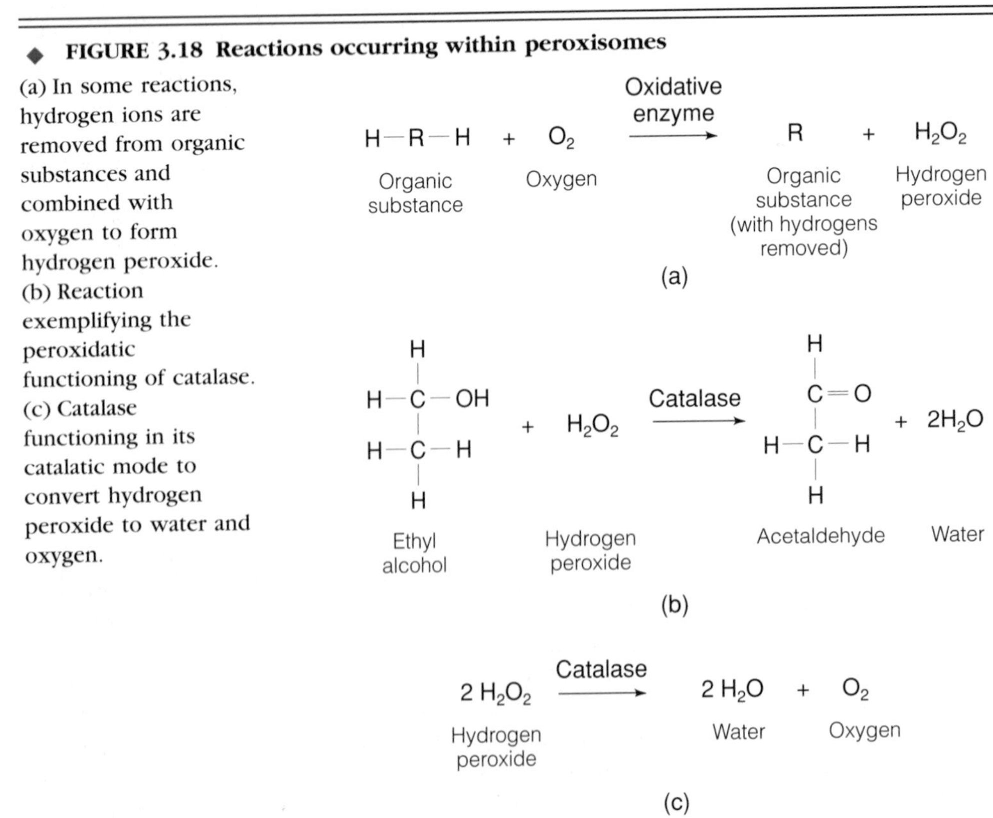

◆ **FIGURE 3.19 A mitochondrion**
(a) A drawing of a mitochondrion. (b) An electron micrograph of a mitochondrion.

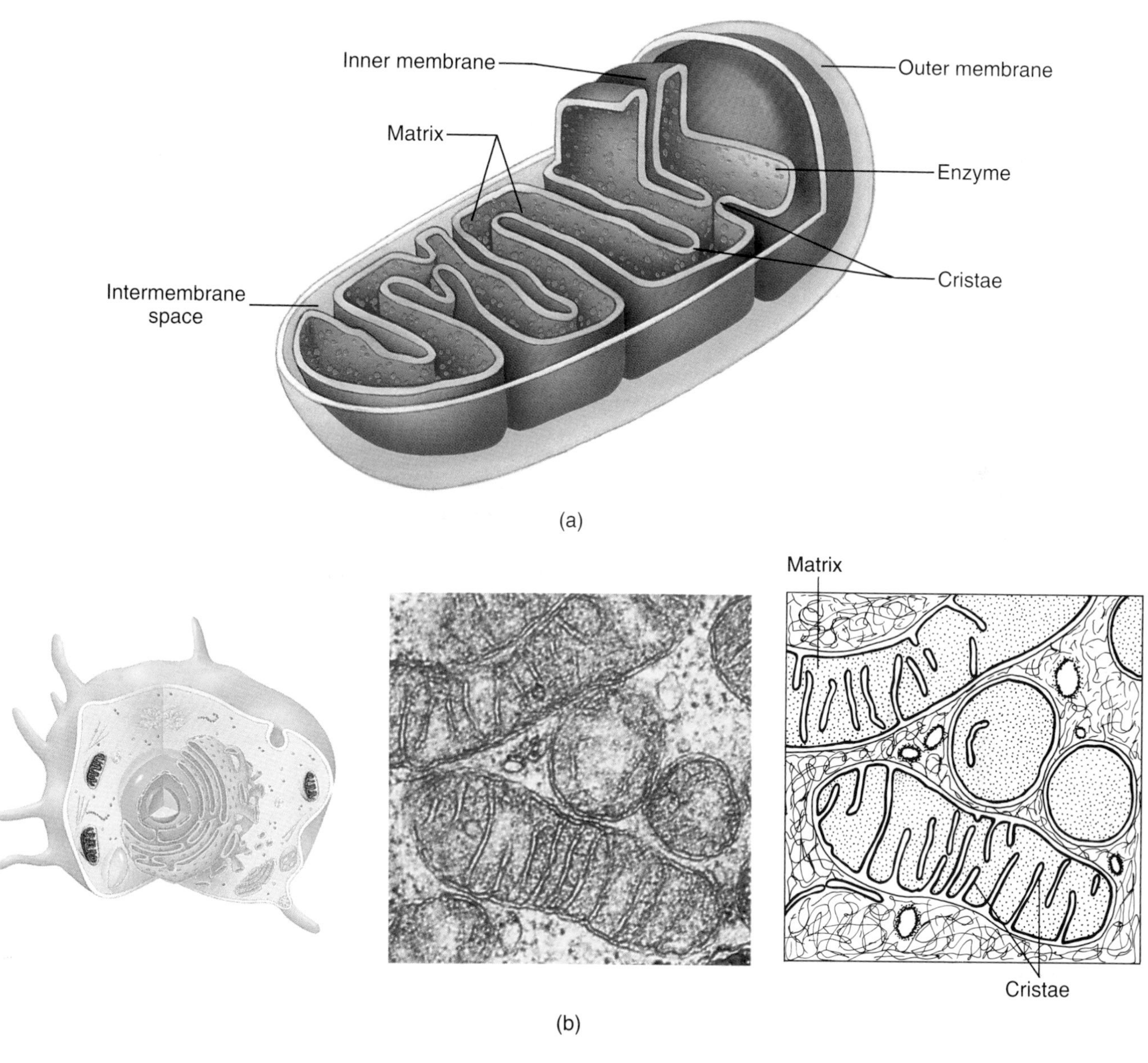

branes is called the *intermembrane space,* and the material within it contains essentially the same types of small molecules as the cytosol. The inner compartment of the mitochondrion is known as the *matrix space,* and the material within it is called the *matrix.*

The matrix contains hundreds of different enzymes, and additional enzymes are embedded in the inner membrane. Many of the chemical reactions that generate ATP (which is used to provide energy for cellular activities) occur within mitochondria. In fact, in some cells such as heart muscle cells, mitochondria are localized near sites of high-energy use. The particular chemical reactions that take place within mitochondria are discussed in Chapter 25.

Mitochondria contain ribosomes and DNA, and they are capable of self-duplication. For example, when certain skeletal muscle cells are repeatedly stimulated to contract for prolonged periods, the number of mitochondria within the cells can increase to as much as five to ten times their original number.

Cytoskeleton

The cytoplasm of a cell contains a **cytoskeleton** (cell skeleton) that consists of an intricate, three-dimensional network of tubular or filamentous structures (Figure 3.20; Table 3.5). These structures maintain the shape and organization of the cell. They are also involved in a variety of cellular activities, including cell movement, the movement of organelles within the cell, and cell division. Prominent among the structures that make up the cytoskeleton are *microtubules, intermediate filaments,* and *microfilaments.*

Microtubules

Microtubules are small, hollow, cylindrical, unbranched tubules about 25 nm in diameter (Figure 3.21). They are composed primarily of subunits of a protein called *tubulin,* which link with one another to form microtubules.

Microtubules play a structural role in the development and maintenance of cell shape. They are also in-

◆ **FIGURE 3.20 Photomicrograph of the cytoskeleton of a cell**

Microtubules appear green; microfilaments appear red. Most of the other filaments are intermediate filaments.

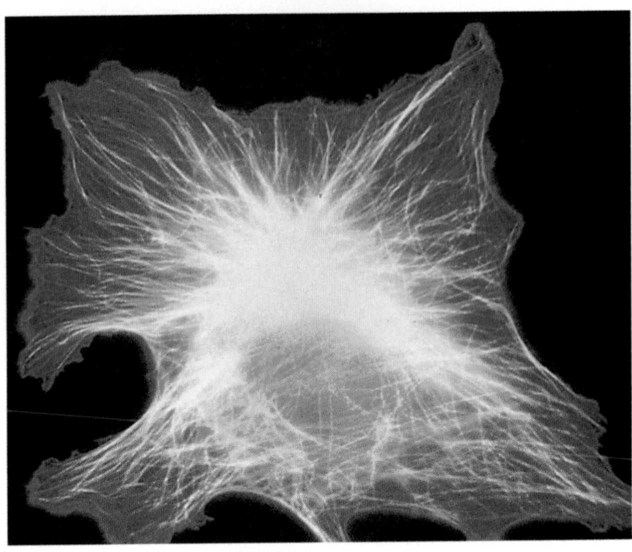

◆ **TABLE 3.5 Functions of the Cytoskeleton**

GENERAL FUNCTIONS
Responsible for the shape, rigidity, and spatial geometry of each type of cell
Responsible for directing many aspects of intracellular transport
Responsible for many cellular movements
FUNCTIONS OF INDIVIDUAL ELEMENTS
MICROTUBULES
Play a structural role in the development and maintenance of cell shape
Involved in cell movement processes
Facilitate transport of certain secretory vesicles in nerve cells
Serve as a component of cilia and flagella
Form mitotic spindle during cell division
INTERMEDIATE FILAMENTS
Provide structural support for a cell and its nucleus
MICROFILAMENTS
Provide support for the plasma membrane
Serve as a component of microvilli
Microvilli increase the surface area available for absorption in the intestines and kidneys
Specialized microvilli detect sound and positional changes in the ears.
Play a significant role in the contractile activities that enable a cell to change shape or move
Play a dominant role in muscle contraction
Form nonmuscle contractile assemblies, such as the contractile ring that divides a cell in half during cell division
Important in movements of motile cells

◆ **FIGURE 3.21 Microtubules**

(a) Structure of a microtubule. (b) An electron micrograph of microtubules.

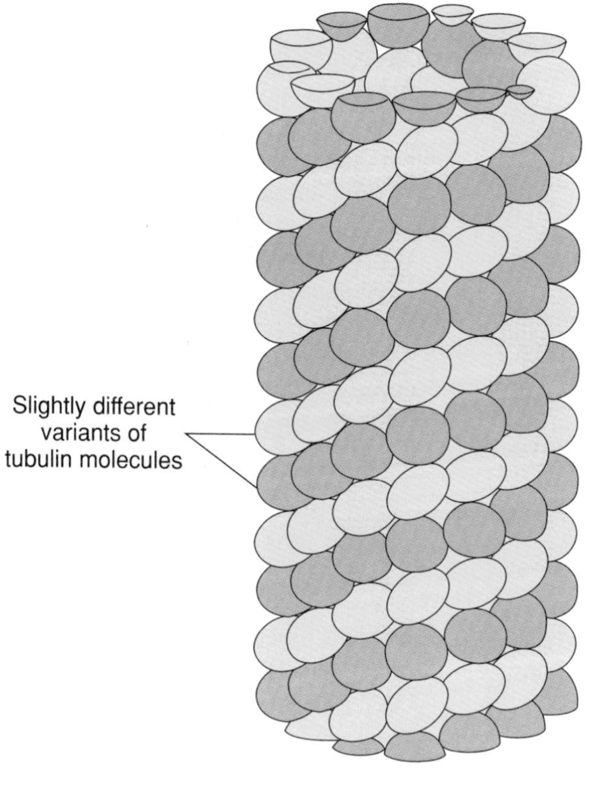

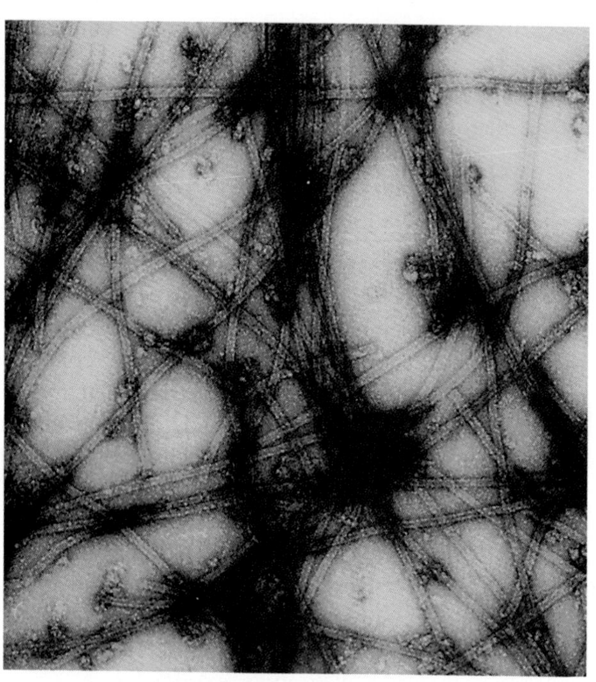

volved in cell movement processes such as the translocation of organelles from one place to another.

Intermediate Filaments

Intermediate filaments, which are about 10 nm in diameter, differ in composition from one cell to another. Many epithelial cells contain intermediate filaments composed of subunits of fibrous proteins called keratins. Nerve cells contain *neurofilaments,* which are intermediate filaments composed of neurofilament proteins.

The nucleus of a cell is surrounded by a network of intermediate filaments, and intermediate filaments extend outward from this network throughout the cytoplasm. Within the nucleus, intermediate filaments composed of proteins called nuclear lamins form a latticework known as the *nuclear lamina,* which is associated with the nuclear envelope.

Intermediate filaments provide structural support for the cell and its nucleus. They are particularly prominent in areas where cells are exposed to mechanical stress.

Microfilaments

Microfilaments are very small fibrils about 7 nm in diameter (Figure 3.22). They are composed primarily of a protein called *actin* and can occur in bundles or other groupings rather than singly. Although microfilaments are present throughout the cytoplasm, an extensive network of microfilaments is attached to the inner surface of the plasma membrane.

Microfilaments provide support for the plasma membrane and play a role in determining cell shape. For example, the plasma membrane at the free surfaces of certain epithelial cells is folded into numerous, tiny, fingerlike projections called *microvilli,* and the core of a microvillus consists of a rigid bundle of 20 to 30 parallel actin filaments. Microfilaments play a significant role in the contractile activities that enable a cell to change shape or move. Muscle cells possess an extensive array of microfilaments.

Cilia, Flagella, and Basal Bodies

Many cells have one or more thin, cylindrical structures that are motile (capable of movement) projecting from their surfaces. These structures can move substances over the cell surface, or they can move an entire cell through a liquid medium. If the structures are short and numerous, they are called **cilia.** If they are longer and fewer, they are called **flagella** *(fla-jel´-ah).* (In the body, the only cell with a flagellum is the male reproductive cell, which is called a *spermatozoon.*)

Both cilia and flagella have the same basic organizational pattern, and both originate from cytoplasmic

◆ **FIGURE 3.22 An electron micrograph of microfilaments**

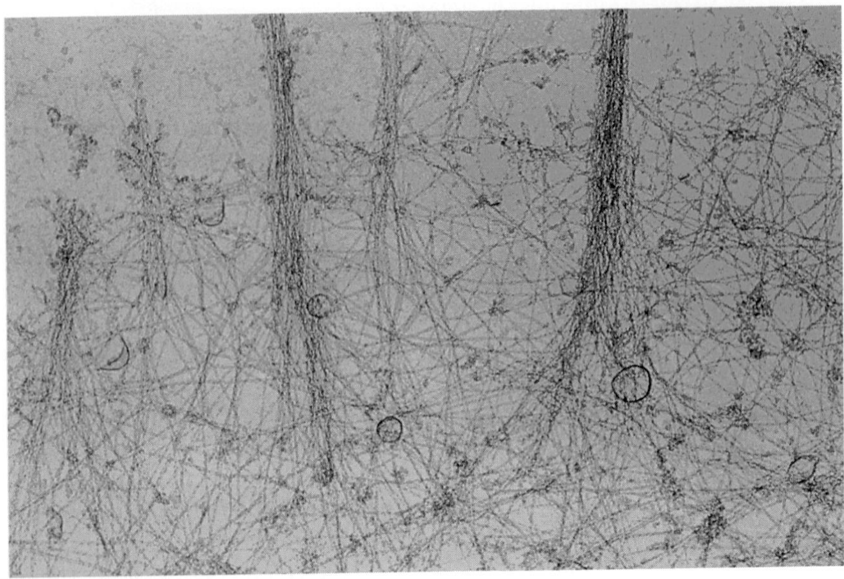

structures called **basal bodies.** Cilia and flagella consist of outward extensions of the plasma membrane that enclose a series of microtubules. The microtubules are arranged in a characteristic circular pattern of nine groups of tubules, with two tubules per group (Figure 3.23a). Two additional microtubules are located in the center of this pattern. Basal bodies also have a characteristic circular pattern of groups of microtubules. Basal bodies, however, have three tubules per group, and they do not have the two central tubules (Figure 3.23b).

A protein complex called *dynein* is associated with the circularly arranged groups of microtubules of cilia and flagella. The dynein attached to one group of microtubules interacts with an adjacent group of microtubules, generating a force that causes the microtubules to bend. This bending produces movement of the cilium or flagellum.

Centrosome and Centrioles

The **centrosome** (cell center) is a region of amorphous material located near the nucleus of a cell. The centrosome serves as a *microtubule-organizing center,* and numerous microtubules extend from it into the cytoplasm.

The centrosome usually contains a pair of cylindrical structures called **centrioles,** with each member of the pair oriented at right angles to the other. A centriole contains a series of microtubules that are arranged in the same pattern as the microtubules of basal bodies (Figure 3.23b). In fact, basal bodies are centrioles.

Inclusion Bodies

The cytosol of many cells contains chemical substances in the form of particles or droplets, which are collectively called **inclusion bodies.** For example, some cells—particularly liver and muscle cells—contain granules of the polysaccharide glycogen, which is a storage form of carbohydrate. Adipose tissue cells contain droplets of almost pure triglycerides.

Extracellular Materials

Many body substances are located outside cells rather than within them. These substances are collectively called **extracellular materials.** Extracellular materials include the extracellular body fluids and the extracellular framework in which many cells are embedded. Many extracellular materials are products of the cells themselves. Among these materials are chondroitin sulfate, which is a jellylike substance present in bone, cartilage, and heart valves; and hyaluronic acid, which is a viscous, fluidlike substance present in a number of tissues. A variety of fibrous materials, such as the proteins collagen and elastin, also occur extracellularly. Connective tissue is particularly rich in extracellular materials.

◆ **FIGURE 3.23 Cilia, flagella, basal bodies, and centrioles**

(a) Cross-sectional view of a typical 9 + 2 arrangement of microtubules in cilia and flagella. (b) Cross-sectional view of a typical arrangement of tubules in basal bodies and centrioles. Note that there are three tubules per group and no tubules in the center.

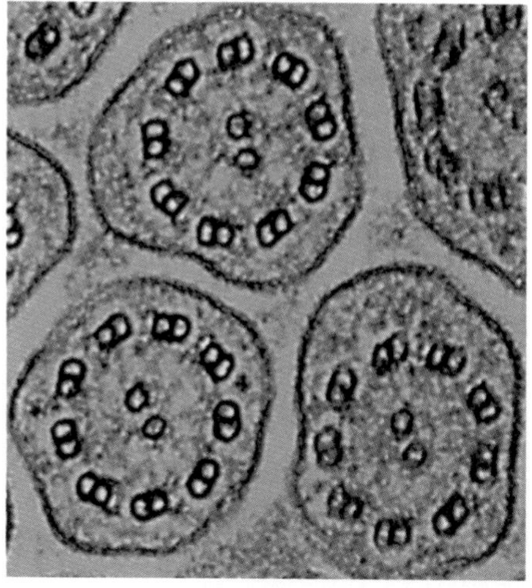

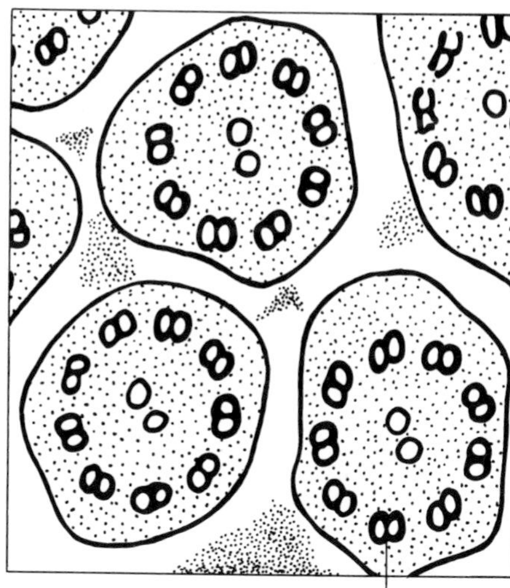

(a) Cilia or flagella

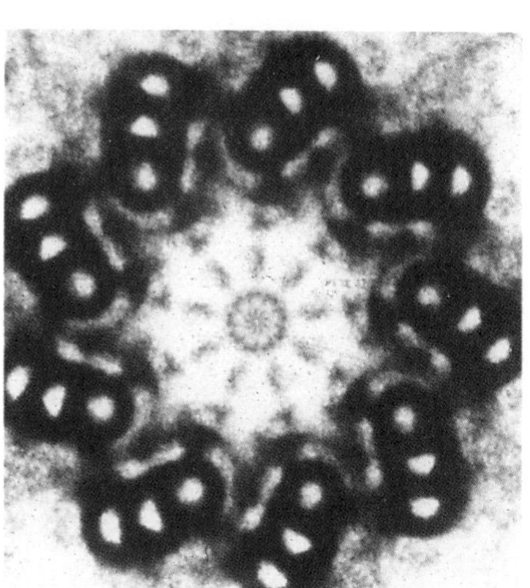

(b) A centriole or basal body

Cell Division

Many of the body's cells (for example, liver, intestinal, bone marrow, and epidermal cells) are able to divide and reproduce themselves. The processes by which such cells divide encompass several basic events. One event is the replication of the DNA within the nucleus of a cell. A second event is the redistribution of the DNA into two new nuclei. A third event is the division of the cell's cytoplasm into two new cells (called *daughter cells*), each with its own nucleus.

The process of redistributing DNA into two new nuclei, each of which contains the same genetic information as the original nucleus, is called **mitosis** *(my-toe´-sis).* The process by which a cell's cytoplasm is divided into two new cells, each with its own nucleus, is called **cytokinesis.** Usually, mitosis and cytokinesis occur simultaneously.

Interphase

The period of time between one mitosis and the next is called **interphase.** Interphase is generally divided into three separate phases: G_1 (first gap), S (synthesis), and G_2 (second gap).

1. The G_1 phase follows the completion of mitosis. During the G_1 phase, an active synthesis of RNA and protein occurs, and the nucleus and cytoplasm enlarge.
2. The S phase follows the G_1 phase. During the S phase, DNA molecules are replicated.
3. The G_2 phase follows the S phase. During the G_2 phase, the metabolic activities of the cell decrease as changes occur in preparation for mitosis. (Mitosis is called the M phase.)

Two events that occur during interphase are particularly significant in terms of cell division. These events are centriole replication and DNA replication.

Centriole Replication

During the G_1 phase of interphase, the two centrioles that are located within the centrosome separate slightly

◆ **FIGURE 3.24 Levels of organization of DNA**
(a) DNA molecule. (b) DNA molecule wound around histones, forming a "beads-on-a-string" structure typical of chromatin. (c) Further folding and supercoiling of the DNA-histone complex. (d) Chromosomes—the most condensed form of DNA. Chromosomes are visible in a cell's nucleus during cell division.

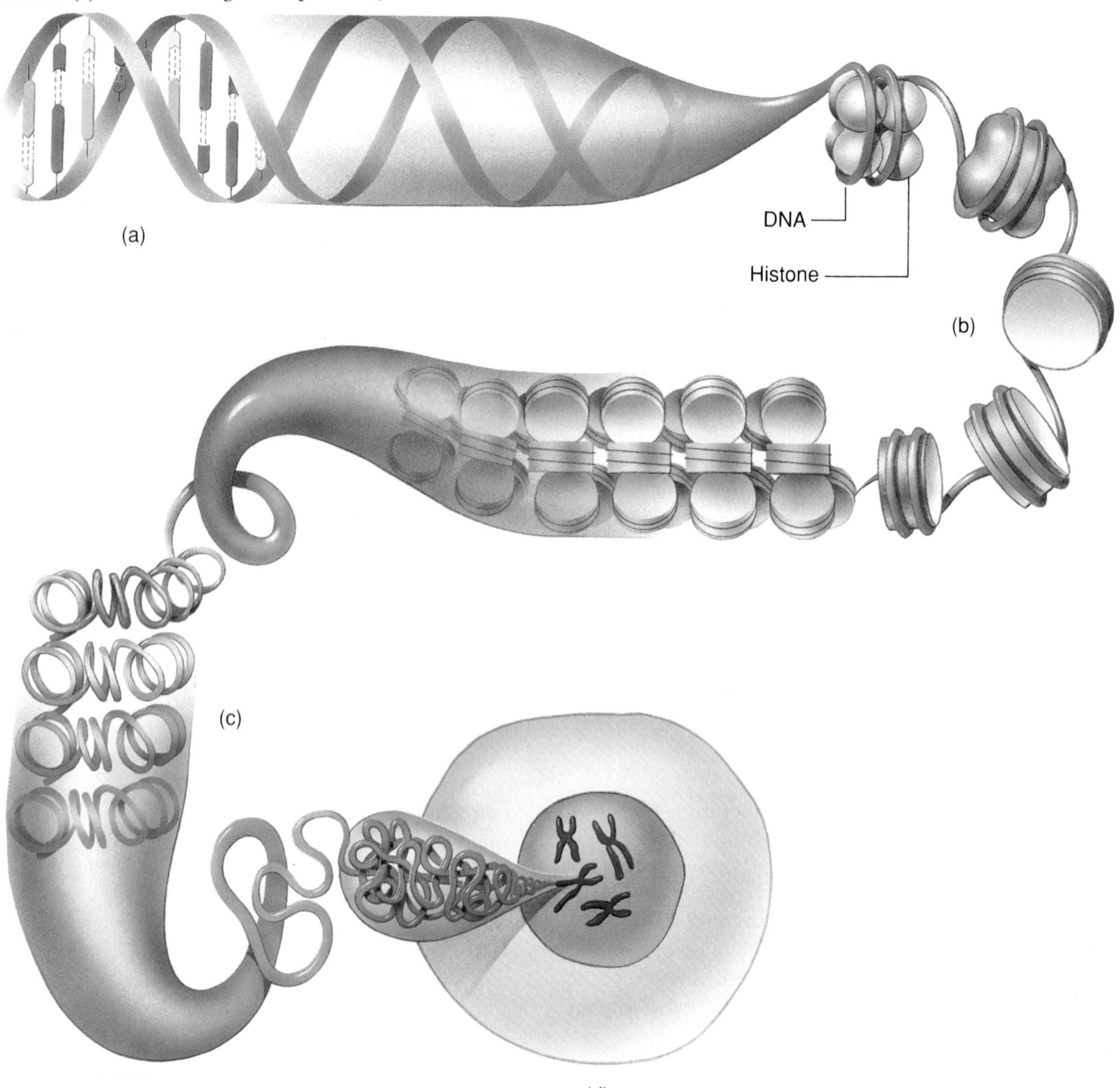

from one another. During the S phase, a new centriole forms at right angles to each of these centrioles. Thus, when centriole replication is completed, the centrosome contains two pairs of centrioles.

DNA Replication

During interphase, DNA molecules and the proteins (principally histones) associated with them are present within the nucleus as indistinct chromatin threads (Figure 3.24). During the S phase of interphase, DNA molecules serve as templates for the replication of additional DNA molecules.

During DNA replication, enzymes separate the two linked chains of a DNA molecule from one another for some distance, and each chain serves as a template that specifies the order in which individual DNA nucleotides are to be incorporated into a new DNA chain (Figure 3.25). DNA polymerase and other enzymes link the nucleotides into the new DNA chain according to the complementary base-pairing pattern. That is, an adenine-containing nucleotide is incorporated into the new DNA chain wherever the template chain has thymine; a guanine-containing nucleotide is incorporated wherever the DNA template has cytosine; a thymine-containing nucleotide, wherever the template has adenine; and a cytosine-containing nucleotide, wherever the template has guanine. The end result is a new DNA chain that is the complement of the original DNA template chain.

The new DNA chain remains attached to the template chain, thereby forming a new, two-chain DNA molecule that is exactly like the original DNA molecule. (Note that the new DNA molecule contains one chain from the original DNA, which served as a template, and one newly synthesized chain.) Because each chain of a DNA molecule serves as a template during DNA replication, two identical DNA molecules are produced from each original DNA molecule. By the end of interphase, therefore, the cell contains twice the amount of DNA as when it entered interphase. During DNA replication, newly synthesized proteins (histones) bind with DNA, thereby providing the additional proteins necessary to establish the DNA-protein associations of chromatin.

◆ **FIGURE 3.25 DNA replication**
Enzymes separate the two chains of a DNA molecule. Each chain then serves as a template for the assembly of a new complementary chain. This process results in two DNA molecules, each exactly like the original molecule.

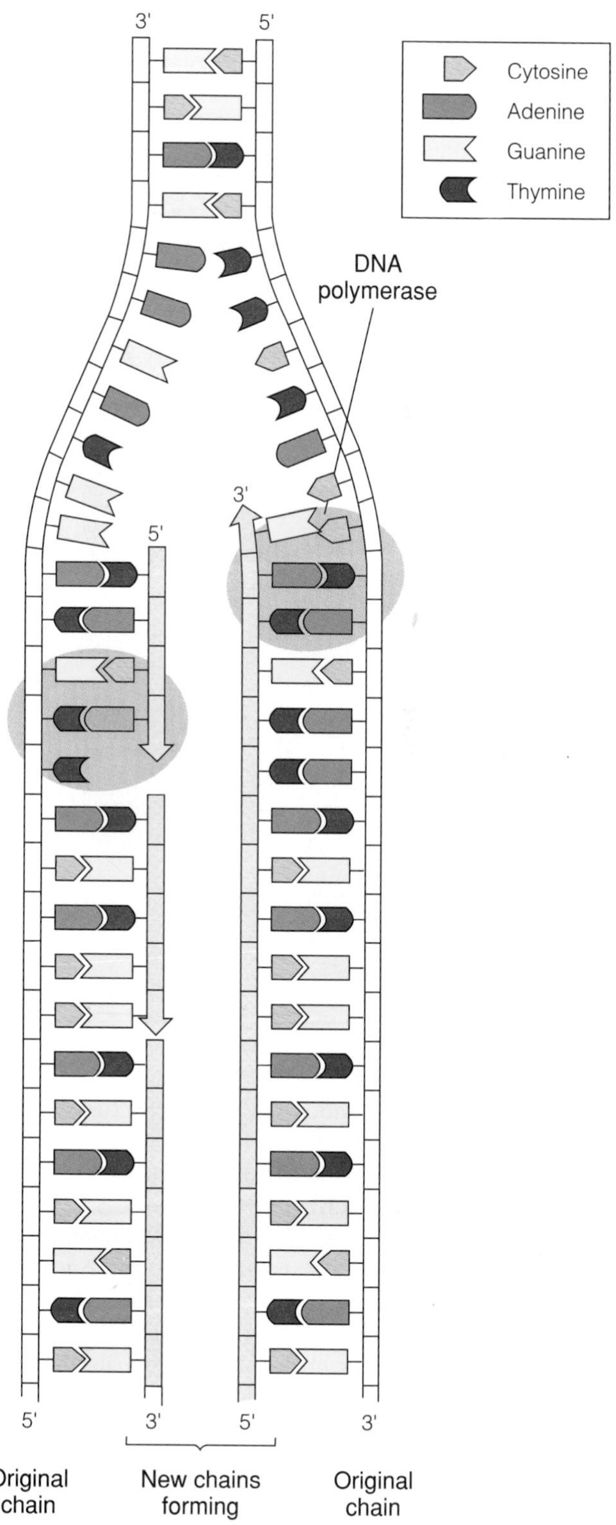

Mitosis

Several phases in the overall process of mitosis are recognized (Figure 3.26). However, it must be emphasized that mitosis is a *continuous* event and not a series of discrete steps.

Prophase

The initial phase of mitosis is called **prophase** (Figure 3.26b). During prophase, the nucleolus disappears, the chromatin threads of DNA and protein become

◆ **FIGURE 3.26 Interphase and the phases of mitosis**

See the text for a detailed discussion.

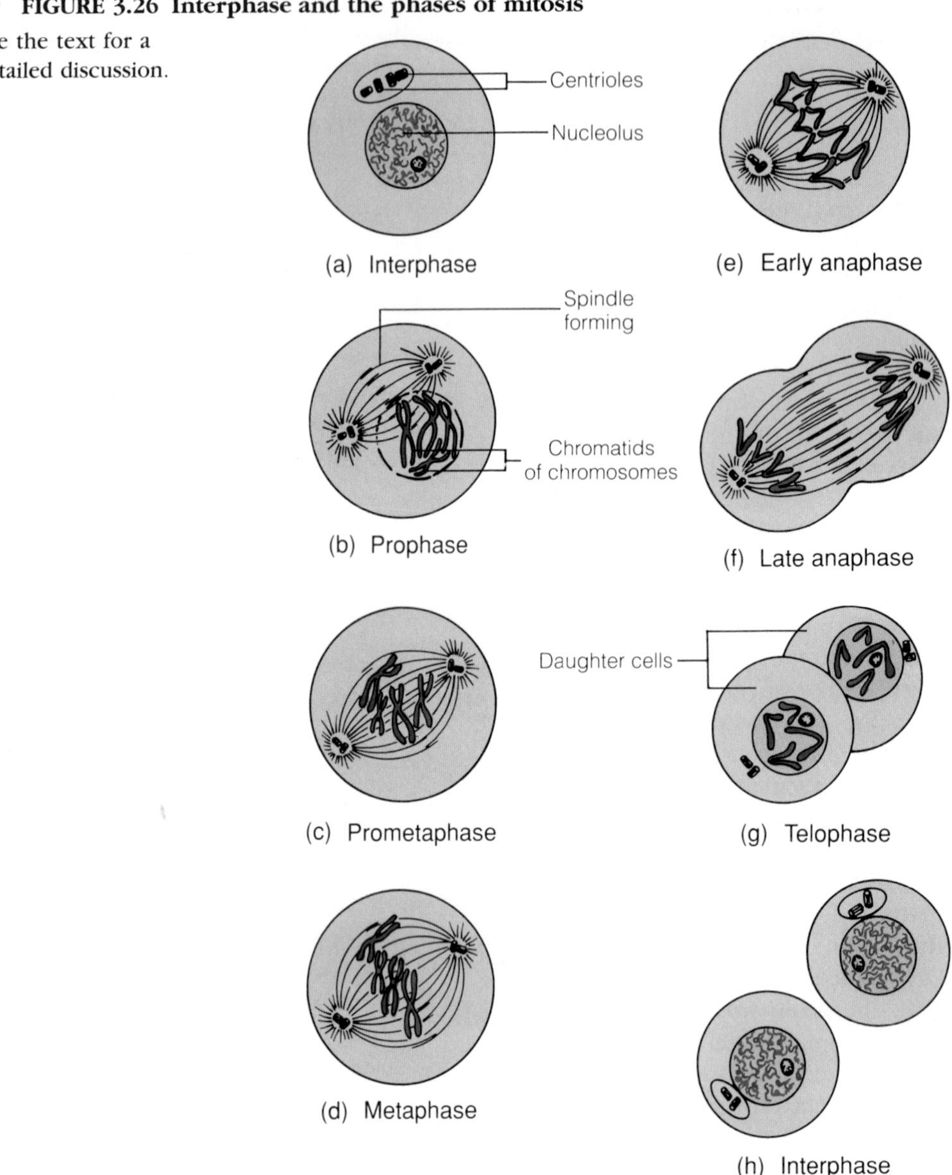

tightly coiled and condensed, and structures called **chromosomes** become clearly visible.

A chromosome is made up of two condensed chromatin threads called **chromatids** (Figure 3.27). One chromatid includes a new, two-chain DNA molecule that was produced during DNA replication using one chain of an original DNA molecule as a template. The second chromatid includes the new DNA molecule that was formed using the other chain of the original DNA as a template. At one point along its length, each chromatid has a special region called a *centromere,* which is a specific DNA sequence element. The two chromatids of a chromosome are attached to one another near their centromeres. During late prophase, a multiprotein complex called a *kinetochore* matures at the centromere of each chromatid.

During prophase, the centrosome splits into two centrosomes, each of which contains one pair of centrioles. Microtubules extend outward from each centrosome. Some microtubules from one centrosome overlap with microtubules from the other centrosome. The overlapping microtubules, which are called *polar microtubules,* interact with one another, and the two centrosomes move apart toward opposite ends of the cell.

Prometaphase

Prometaphase begins when the nuclear envelope disintegrates (Figure 3.26c). During prometaphase, microtubules interact with the chromosomes.

◆ **FIGURE 3.27 Diagrammatic representation of a chromosome as it appears during late prophase**

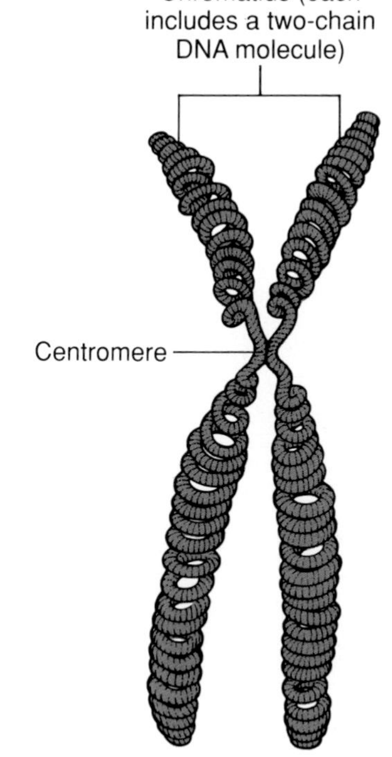

As the centrosomes approach opposite ends of the cell, microtubules extending from one centrosome attach to the kinetochore of one chromatid of each chromosome, and microtubules extending from the other centrosome attach to the kinetochore of the other chromatid of each chromosome. When these microtubules, which are called *kinetochore microtubules,* attach to the chromosomes, the chromosomes begin to move toward a position halfway between the two centrosomes.

Together, the polar microtubules and the kinetochore microtubules comprise the *mitotic spindle.* The centrosomes constitute the *spindle poles.* In addition to polar and kinetochore microtubules, microtubules called *astral microtubules* extend from each spindle pole.

Metaphase

At **metaphase,** the chromosomes become aligned at a location called the *metaphase plate* of the spindle, which is halfway between the two spindle poles (Figure 3.26d).

Anaphase

At **anaphase,** the two chromatids of each chromosome separate (Figure 3.26e and f). One chromatid of each chromosome moves toward one spindle pole, and the other chromatid moves toward the opposite spindle pole. As these movements occur, the kinetochore microtubules shorten. Once the chromatids of a chromosome separate, each chromatid is considered to be an independent chromosome. Also during anaphase, the polar microtubules elongate, and the spindle poles move farther apart.

Telophase

Telophase is the final phase of mitosis. During telophase, the chromosomes reach the spindle poles, and the spindle disappears (Figure 3.26g). A nuclear envelope forms around each of the two groups of chromosomes, and nucleoli appear. The chromosomes decondense and become less distinct, and they gradually assume their interphase appearance of chromatin threads.

Cytokinesis

When mitosis and cytokinesis occur together, the beginning of cytokinesis is generally evident during anaphase as an inward movement of the plasma membrane, usually near the middle of the cell. This movement forms a cleavage furrow that continues to deepen during telophase. Ultimately, the cytoplasm is completely divided into two separate cells, each with its own nucleus.

The inward movement of the plasma membrane is due to actin filaments that become attached to the inner surface of the membrane in a contractile ring (Figure 3.28). These filaments interact with a protein called myosin in a process that narrows the contractile ring and pulls the plasma membrane inward.

Meiosis

There are 46 chromosomes in each of the human somatic cells (all cells except the reproductive cells). Two of these are **sex chromosomes** (two X chromosomes in females, and one X and one Y chromosome in males).

◆ **FIGURE 3.28 The contractile ring that forms during cytokinesis**

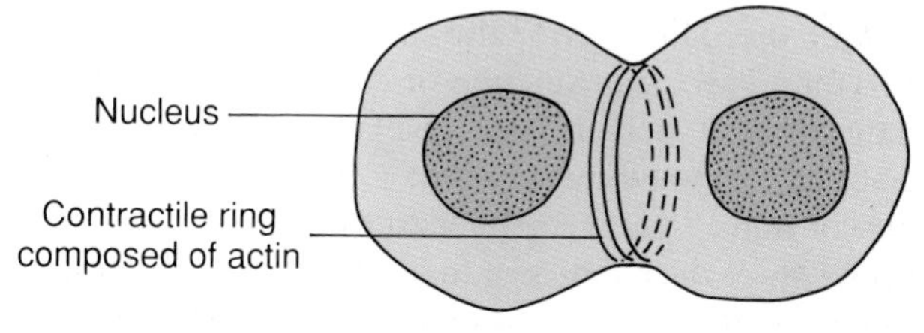

◆ **FIGURE 3.29 Human chromosomes**

(a) Chromosomes of a female, with X sex chromosomes indicated. (b) Chromosomes of a male, with homologous chromosomes arranged in pairs. Note X and Y chromosomes—the sex chromosomes—in the fourth row.

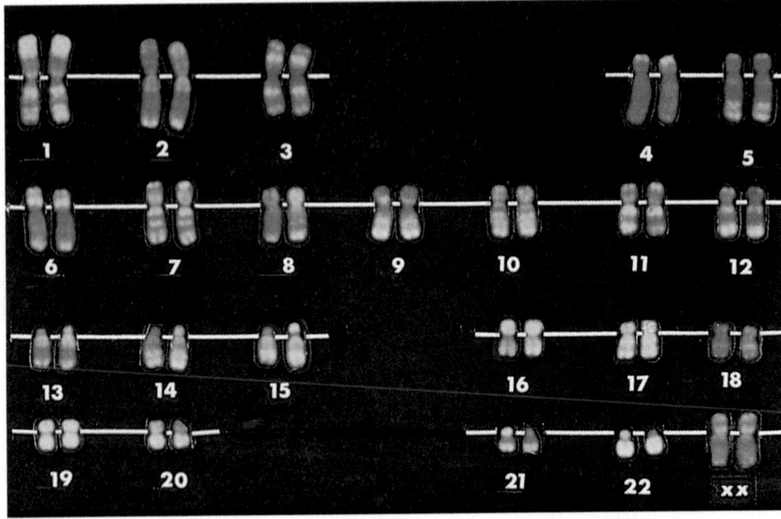

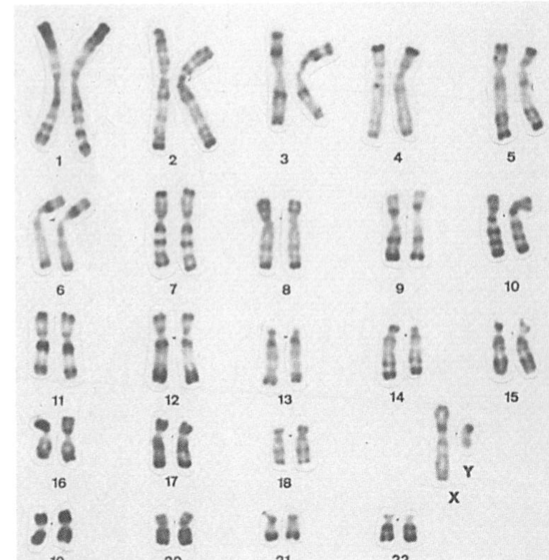

The remaining 44 chromosomes are called **autosomes.** The 44 autosomes consist of 22 pairs of similar-appearing chromosomes (Figure 3.29). One member of each pair contains genetic information derived from the person's father. The other member of each pair contains information derived from the person's mother. Each pair makes up a set of **homologous chromosomes** *(ho-mol´-o-gus).* The two sex chromosomes of the female (XX) are also homologous, but the two sex chromosomes of the male (X and Y) are not.

Homologous chromosomes each possess genetic information that controls the same functions or characteristics. Often the genetic information for a particular function or characteristic on one chromosome of a homologous pair takes precedence over the corresponding information on the other chromosome of the pair. If the genetic information derived from the person's father takes precedence over the genetic information derived from the mother, the paternal function or characteristic will be **dominant,** and the person will display the paternal function or characteristic. In such a case, the maternal function or characteristic is said to be **recessive.** If the maternal genetic information takes precedence over the corresponding paternal information, the person will display the maternal function or characteristic. If neither the paternal nor the maternal genetic information for a particular function or characteristic takes precedence, the person may display some intermediate function or characteristic. (For a more detailed discussion of human genetics, refer to Chapter 29.)

The 46 chromosomes of human somatic cells actually consist of two 23-chromosome sets (22 autosomes and 1 sex chromosome per set), with one set having been derived from the person's father and one from the person's mother. Thus the **gametes** *(gam´-eets),* or reproductive cells, of the male (the spermatozoa from the testes) and female (the ova from the ovaries) each contain only 23 chromosomes. When a sperm cell fertilizes an ovum, each contributes its 23 chromosomes, thereby establishing the full 46 chromosomes of the new individual.

Cells with two complete sets of chromosomes (46 chromosomes) are known as **diploid cells.** The formation of gametes, however, must result in the formation of cells that have only one set of chromosomes (23 chromosomes rather than 46 chromosomes). Such cells are called **haploid cells,** and the normal processes of mitosis do not produce such cells. A second type of cell division, known as **meiosis** *(my-o´-sis)* is responsible for the production of haploid reproductive cells.

Phases of Meiosis

Two successive division sequences—division I and division II—occur during meiosis (Figure 3.30). During prophase of division I, homologous chromosomes pair with one another (Figure 3.30a). (The nonhomologous sex chromosomes of the male also pair, and during meiosis, they behave basically in the same manner as homologous chromosomes.)

Like the chromosomes present during prophase of mitosis, each chromosome present during prophase of division I is made up of two attached chromatids. However, the kinetochores of the two chromatids appear to

◆ **FIGURE 3.30 The phases of meiosis**
See the text for a detailed discussion.

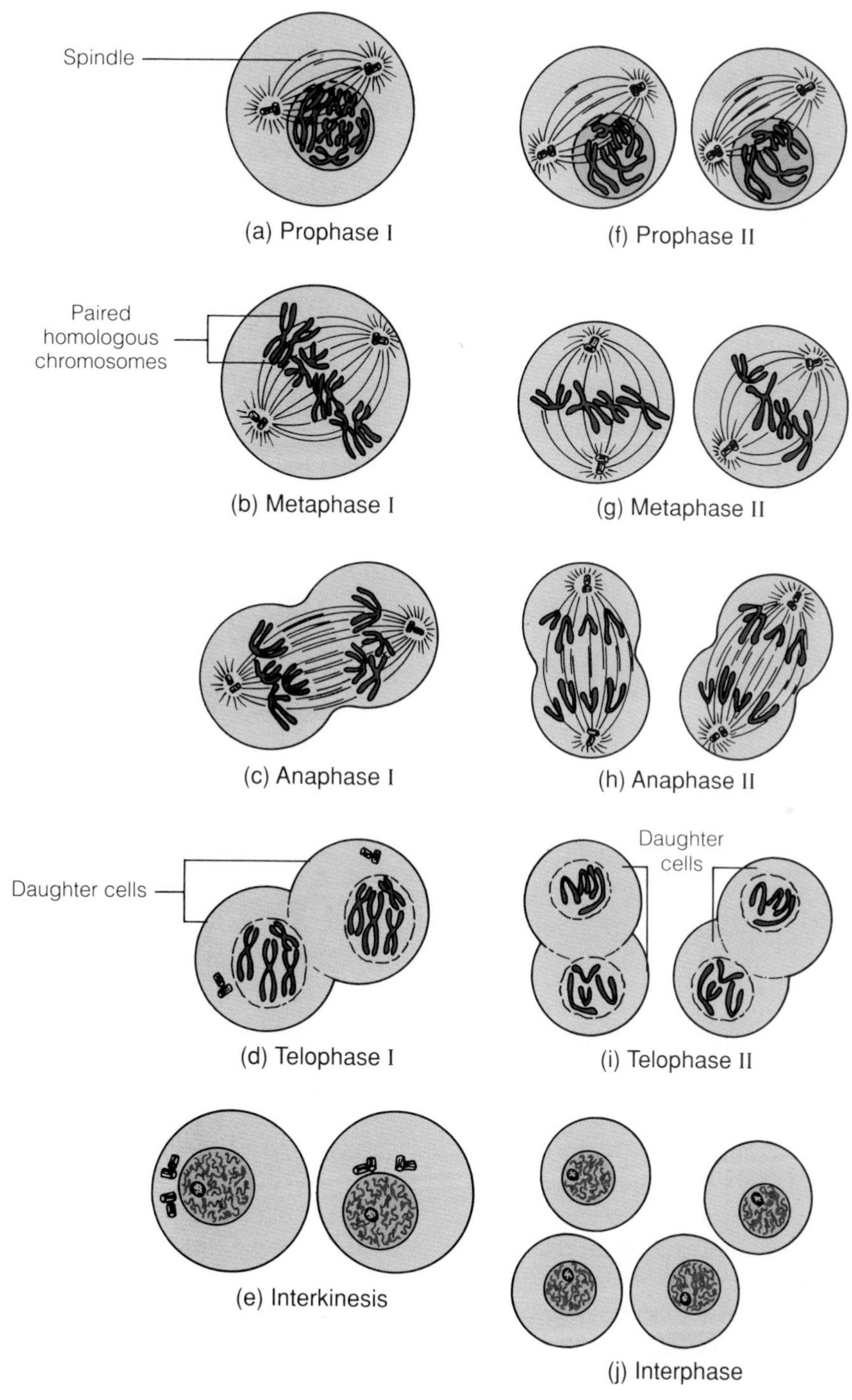

be fused together. As division I proceeds, kinetochore microtubules extending from one spindle pole attach to the fused kinetochores of one chromosome of a homologous pair, and kinetochore microtubules extending from the other spindle pole attach to the fused kinetochores of the other chromosome of the pair.

At metaphase of division I, the chromosomes become aligned at the metaphase plate of the spindle as homologous pairs (Figure 3.30b). At anaphase, one chromosome of each homologous pair moves toward one spindle pole, and the other chromosome of the pair moves toward the other spindle pole (Figure 3.30c). Thus, 23

◆ **FIGURE 3.31 Redistribution of chromosomes during meiosis**

Chromosomes that contain genetic information derived from the male parent (blue) do not necessarily remain together; nor do chromosomes containing genetic information derived from the female parent (red) necessarily stay together. The resulting cells have a mixture of chromosomes that contain genetic information derived from both parents.

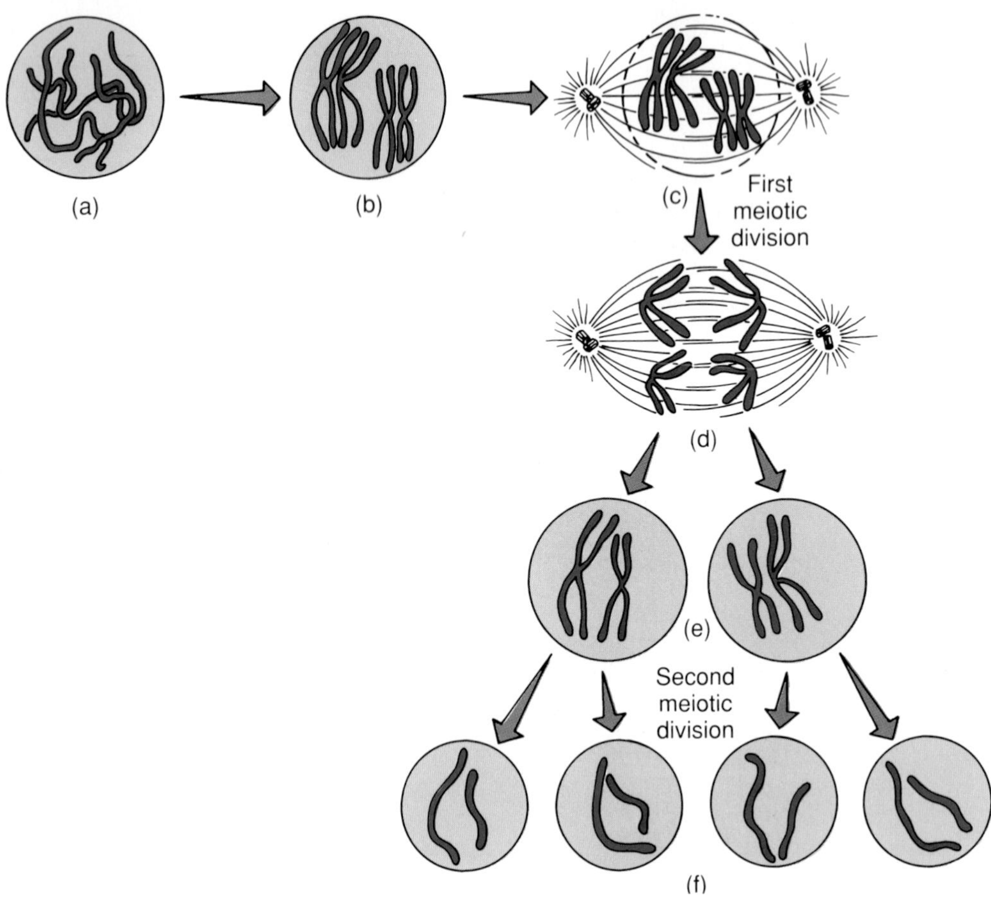

chromosomes move to each spindle pole during division I of meiosis. Each chromosome is made up of two attached chromatids.

During telophase of division I, a nuclear envelope forms around each of the two groups of chromosomes, and the chromosomes decondense to some degree (Figure 3.30d). Cytokinesis takes place in conjunction with division I, producing two cells that each contain 23 chromosomes.

A period called **interkinesis** follows division I (Figure 3.30e). Interkinesis is somewhat similar to the interphase period between mitotic divisions. However, DNA replication does not occur during interkinesis. Therefore, when interkinesis ends and the two cells produced during division I go through prophase of division II of meiosis, each cell contains 23 chromosomes, and each chromosome is made up of two attached chromatids (Figure 3.30f).

Division II of meiosis is essentially a mitotic division (Figure 3.30f through i). (The two chromatids of each prophase chromosome separate at anaphase.) Cytokinesis occurs in conjunction with division II, producing four cells from the two that began division II (Figure 3.30j). Each of the four is a haploid cell containing 23 chromosomes.

Genetic Diversity

Meiosis allows a great deal of genetic diversity in the makeup of reproductive cells (spermatozoa and ova). In fact, the gametes a person produces commonly differ from one another genetically. The differences are due in large measure to the manner in which the person's genetic material (DNA) is redistributed during meiosis.

The chromosomes that contain genetic information derived from a person's male parent do not necessarily all move toward one spindle pole, nor do the chromosomes that contain genetic information derived from the person's female parent all move toward the other spindle pole during division I of meiosis. Rather, some chromosomes containing genetic information derived from the person's male parent move toward one spindle pole, and some move toward the other spindle pole (Figure 3.31). The same is true for chromosomes containing genetic information derived from the person's female parent. Consequently, the cells produced during division I of meiosis each receive, in random assortment, some chromosomes that contain genetic information derived from the person's male parent and some that contain genetic information derived from the female parent.

Further genetic diversity results from the process of **crossing over,** which takes place between paired homologous chromosomes during prophase of division I of meiosis. In this process, enzyme-catalyzed reactions exchange a portion of the DNA molecule of a chromatid of one chromosome of a homologous pair with a portion of the DNA molecule of a chromatid of the other chromosome of the pair (Figure 3.32). When the paired homologous chromosomes move apart at anaphase of division I and cytokinesis ultimately takes place, the genetic compositions of the resulting cells can be different than if crossing over did not occur.

◆ **FIGURE 3.32 The process of crossing over**

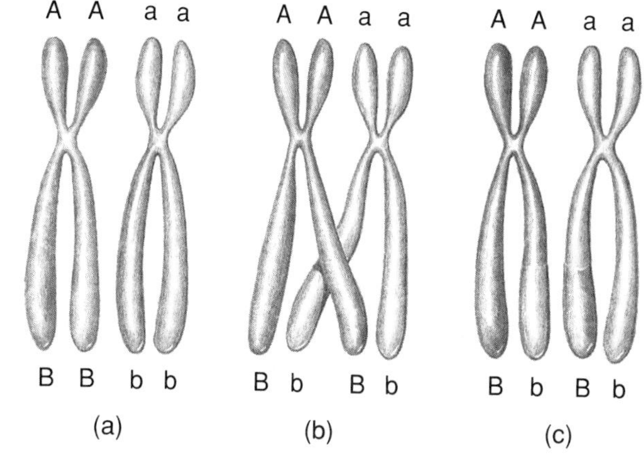

CONDITIONS OF CLINICAL SIGNIFICANCE

Genetic Disorders

The Cell

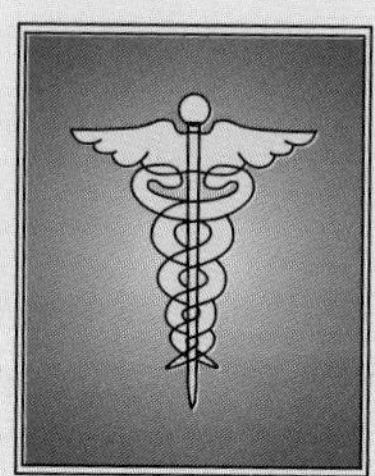

Many diseases and abnormalities can be traced to genetic defects. Occasionally, incorrect bases are incorporated into DNA chains, and because of complementary base pairing, RNA molecules produced using these DNA chains as templates will also contain incorrect bases. In mRNA, three-base codons containing incorrect bases may specify amino acids that are different from those that would normally have been specified. When ribosomes use these erroneous mRNA molecules during translation, incorrect amino acids are inserted into the proteins being formed. These incorrect proteins may not function properly, and disease can be one result.

Such is the case in the genetic disease sickle-cell anemia, which occurs when an incorrect amino acid becomes incorporated into one of the polypeptide chains of the oxygen-transporting molecule hemoglobin. Red blood cells that contain the abnormal hemoglobin molecules are fragile and misshapen. They rupture within and block blood vessels, leading to an impaired delivery of oxygen to the body's cells.

Also, since many proteins function as enzymes that catalyze cellular chemical reactions, errors in their construction can make it impossible for certain reactions to occur. The genetic disease phenylketonuria results from the lack of a functional enzyme required to convert the amino acid phenylalanine to tyrosine. The inability of the cells to carry out this reaction efficiently can ultimately produce mental retardation.

Genetic disorders range from those in which an incorrect polypeptide or protein is produced to those in which a polypeptide or protein is either not produced at all or is produced in excess. Changes in the DNA of a cell that lead to such occurrences are called

continued on next page

CONDITIONS OF CLINICAL SIGNIFICANCE

mutations. In most instances, mutations are harmful or, at best, neutral to cell function. Occasionally, however, they may lead to an improved condition as far as the organism is concerned. When a cell that contains a mutation divides, the mutation can be passed to the cell's progeny, and a mutation that occurs in a reproductive cell can be passed from parent to offspring. Genetic disorders are discussed further in Chapter 29.

Cancer

In some cases, the mechanisms that regulate cell division become disrupted, and an excessive division of cells occurs that gives rise to an abnormal growth called a *tumor.* In many cases, a tumor remains localized, and tumor cells do not spread beyond the region where the tumor occurs. This type of tumor is called a *benign tumor,* and it can usually be removed surgically. In other cases, tumor cells have the ability to invade surrounding tissues. This type of tumor is called a *malignant tumor.* Usually, cells of a malignant tumor are able to travel in the blood or lymph to areas of the body removed from the site of the initial tumor, which is called the *primary tumor.* In these areas, the tumor cells form *secondary tumors,* or *metastases.*

In its essence, a **cancer** is a malignant tumor. Once cancer cells spread, or metastasize, from the site of the primary tumor to areas of the body where they form secondary tumors, the cancer generally cannot be removed surgically. In such a case, other means of treatment must be employed. Radiation therapy is often used because rapidly dividing cells are very susceptible to radiation. Various anticancer drugs and chemotherapeutic agents that attack rapidly dividing cells are also employed. However, radiation and most anticancer drugs and chemotherapeutic agents can damage normal body cells as well as cancer cells, and they must be used with care.

It is difficult to define a gene precisely. For present purposes, however, a **gene** can be defined as a region of DNA that becomes transcribed into a discrete RNA molecule. In a case in which an mRNA molecule is ultimately formed, the polypeptide or protein produced when the mRNA is translated at ribosomes is commonly called the *gene product.* In a case in which an RNA molecule not translated at ribosomes is ultimately formed (for example, a tRNA or rRNA molecule), the RNA molecule formed is frequently considered to be the gene product.

Most cancers are believed to originate from a single abnormal cell in which mutations affect genes that regulate cell division and invasiveness. As a result, the cell and its progeny become able to proliferate and spread to other areas of the body. Two types of genes and their gene products are particularly important in regulating cell proliferation. The products of genes called **proto-oncogenes** help stimulate cell proliferation, and the products of genes called **tumor-suppressor genes** help inhibit cell proliferation. Mutations that cause proto-oncogenes to become hypereffective—that is, to function as **oncogenes**—or mutations that cause tumor-suppressor genes to become ineffective may ultimately lead to cancer.

It is unlikely that a single mutation will cause a normal cell to become a cancer cell. Rather, researchers believe that a number of mutations must occur to generate a cancer cell, and the mutations probably involve both proto-oncogenes and tumor-suppressor genes. In fact, one of the reasons that cancer is much more common in older people than in younger people may be that it takes time for sufficient mutations to accumulate to produce a cancer cell.

A number of substances called *chemical carcinogens* increase the likelihood of cancer development. Some chemical carcinogens act directly on cells, but others act only after they have been changed to more reactive forms by cellular reactions. Most chemical carcinogens cause mutations. In many cases, the effects of exposure to a chemical carcinogen are not immediately evident, and years may elapse before cancer develops. Researchers believe that, during this time, mutations accumulate until a cancer cell is finally produced.

Aging and Cells

A number of theories have been proposed to explain what causes the body to age. Some of these theories suggest that aging occurs at the organismic level. For example, one organismic theory proposes that aging is programmed by a region of the brain called the hypothalamus. The hypothalamus influences the production of hormones by the pituitary, thyroid, and adrenal glands, as well as by the testes and ovaries. According to this theory, with the passage of time, the influence exerted by the hypothalamus

CONDITIONS OF CLINICAL SIGNIFICANCE

on hormone production changes, and consequently, changes occur in hormone production. The changes in hormone production have wide-ranging effects on the body that contribute to aging.

Other theories suggest that aging is due mainly to changes occurring at the cellular level. Some of these theories propose that aging changes are programmed into cells. Experiments have shown that when connective tissue cells called fibroblasts are grown in culture (outside the body), they will divide only until a certain number of doublings of the cell population occur. The cells then stop dividing and die. Fibroblasts taken from an older person generally stop dividing and die after fewer population doublings than fibroblasts taken from a younger person. For example, fibroblasts taken from an 80-year-old person stop dividing after about 30 population doublings, but fibroblasts taken from a 40-year-old person continue to divide until about 40 population doublings occur.

Another cellular theory of aging proposes that, with the passage of time, cross-links form between cellular proteins, thereby altering their structure and function. The cellular proteins believed to be most often affected by cross-linking are enzymes, and impaired enzyme function could contribute to impaired cell function and cell death.

Still another cellular theory suggests that, over time, chemical substances called free radicals gradually accumulate within cells. (Free radicals are oxygen molecules that contain unpaired electrons.) Free radicals react with lipids and other substances within a cell. When they accumulate, they are thought to alter the plasma membrane and the membranes surrounding mitochondria and lysosomes. They may even damage DNA.

Whatever the cause, changes in cell structure and function occur with age. One of the more documented changes is an increased incidence of damaged DNA, perhaps due in part to a decline in the effectiveness of processes that normally repair damaged DNA. There is also some evidence that, in older cells, a substance that inhibits DNA formation is produced in the cytoplasm and moves into the nucleus.

A common cytoplasmic change in older cells is the accumulation of deposits of a substance called lipofuscin. The presence of lipofuscin in older cells is so consistent that it is referred to as the "age pigment." Many older cells also have a decreased number of mitochondria, and there may be a reduced number of folds of the inner membrane of the mitochondria (that is, a reduced number of cristae). Lysosomes are also thought to undergo changes in older cells. The accumulation of lipofuscin and the changes that occur in mitochondria and lysosomes may impair cell function and contribute to aging at the cellular level.

Study Outline

◆ **CELL COMPONENTS** pp. 60–92

Nucleus surrounded by cytoplasm, which contains numerous structures called organelles. Limiting boundary of cell is plasma membrane.

Plasma Membrane. A highly selective barrier that affects the movement of substances into or out of cells. Membrane is lipid bilayer with protein interspersed; channels in membrane connect cell interior with extracellular fluid. Materials that enter or leave cell must pass through membrane or channels.

MOVEMENT OF MATERIALS ACROSS THE PLASMA MEMBRANE. All materials that enter or leave a cell must pass either through the plasma membrane or through channels in the membrane.

DIFFUSION THROUGH THE LIPID PORTION OF THE MEMBRANE. Many nonpolar, lipid-soluble substances move easily through plasma membrane by simple diffusion.

MOVEMENT THROUGH MEMBRANE CHANNELS. Small polar molecules and ions move through channels.

MEDIATED TRANSPORT. Makes use of carrier molecules, which are part of membrane. Mediated-transport systems exhibit characteristics of specificity, saturation, and competition.

Facilitated Diffusion. Substances move across membrane with aid of carrier molecules. Substances can move in either direction. Net movement is down a gradient; does not require cellular energy.

Active Transport. Carrier molecules move substances across the membrane; substances are moved up gradients; requires cellular energy.

ENDOCYTOSIS. Cells form membrane-bounded intracellular vesicles that contain extracellular material.

Pinocytosis. Endocytic process in which vesicle formed is small and contains materials in solution or very small particles in suspension.

A Closer Look: Active Transport and the Sodium-Potassium Pump.

Phagocytosis. Endocytic process in which vesicle formed is large and contains particles such as microorganisms, cellular debris, or even whole cells.

EXOCYTOSIS. Substances leave cell by the fusion of membrane-bounded vesicles with plasma membrane.

TONICITY. Refers to the ability of a solution to influence the state of tension, or tone, of cells placed in it due to water movement into or out of the cells.

SIGNALING MOLECULES AND PLASMA-MEMBRANE RECEPTORS. In many cases, signaling molecules attach to receptors that are part of the plasma membrane of a cell.

G-PROTEIN–LINKED RECEPTORS. Binding of signaling molecule to these receptors influences cell function by way of G proteins, which can bind GTP.

A Closer Look: G-Protein–Linked Receptors and Intracellular Mediators.

CATALYTIC RECEPTORS. Act directly as enzymes.

CHANNEL-LINKED RECEPTORS. Ion channels in membrane that possess receptor regions to which certain signaling molecules bind. Binding either opens or closes channel.

ELECTRICAL CONDITIONS AT THE PLASMA MEMBRANE. Distribution and movements of ions across the plasma membrane have electrical consequences for cells.

BASIC ELECTRICAL CONCEPTS.

1. Two types of electrical charge: positive and negative.
2. Like charges repel one another; unlike charges attract one another.
3. Voltage: A measure of the potential of separated positive and negative charges to perform work.
4. Current: Actual movement of electric charge from one point to another.
5. Resistance: Hindrance to the movement of charge.

MEMBRANE POTENTIAL. All body cells are electrically polarized; voltage between inside and outside of cell is the membrane potential. Membrane potential is result of differences in ionic composition of intracellular and extracellular fluids due to characteristics and function of plasma membrane.

CHANGES IN THE MEMBRANE POTENTIAL. Alterations in the ionic permeability of the plasma membrane that result in a net movement of particular ions across the membrane can change the membrane potential. A rapid depolarization and polarity reversal called an action potential occurs in nerve and muscle cells.

The Nucleus.

1. Contains DNA in the form of chromatin, which is composed of DNA combined with protein.
2. DNA contains coded information within its base sequence that specifies the order in which amino acids are to be linked together to form particular polypeptides or proteins.
3. Polypeptides or proteins synthesized according to DNA specifications act as enzymes, as structural proteins, or in other ways to accomplish the work of the cell.

RNA SYNTHESIS (TRANSCRIPTION). DNA serves as template for RNA synthesis.

MESSENGER RNA.

1. mRNA contains coded information, derived ultimately from DNA, that is used to direct protein synthesis.
2. The principal bases of mRNA—adenine, cytosine, guanine, uracil—can be arranged to form 64 different three-base codons.
3. Codon sequence specifies the amino acid sequence of a polypeptide or protein.

TRANSFER RNA. Carries amino acids to ribosomes, where the amino acids are incorporated into polypeptides or proteins.

RIBOSOMAL RNA.

1. Nucleoli are present in the nucleus and are associated with rRNA production.
2. rRNA becomes part of ribosomes.

Ribosomes. Cytoplasmic organelles that consist of rRNA and protein.

BASIC FEATURES OF PROTEIN SYNTHESIS (TRANSLATION). At ribosomes, amino acids are linked together in the sequence specified by mRNA to form polypeptides and proteins.

A Closer Look: Ribosomes and Protein Synthesis.

Endoplasmic Reticulum. Network of interconnected, membrane-bounded tubules and flattened sacs; membrane contains enzymes that synthesize many different lipids. Some proteins synthesized at ribosomes enter endoplasmic reticulum, and protein-containing vesicles can break away.

Golgi Apparatus. Flattened membranous sacs with which vesicles from endoplasmic reticulum can fuse. Proteins are processed, modified, and sorted in Golgi apparatus; vesicles formed from Golgi membranes carry proteins to their destinations.

Endosomes, Endolysosomes, and Lysosomes. Related groups of cytoplasmic membrane-bounded vesicles or tubules. Lysosomes contain digestive enzymes that break down proteins, lipids, certain carbohydrates, DNA, and RNA.

Perioxisomes. Contain enzymes that catalyze reactions believed to be important in detoxifying potentially harmful substances. The enzyme catalase is abundant in peroxisomes.

Mitochondria. Many of the chemical reactions that generate ATP occur in mitochondria.

Cytoskeleton. Consists of an intricate network of tubular or filamentous structures.

MICROTUBULES. Small, hollow, cylindrical, unbranched tubules; involved in development and maintenance of cell shape and cell movement processes.

INTERMEDIATE FILAMENTS. Provide structural support for the cell and its nucleus.

MICROFILAMENTS. Composed primarily of actin; can occur in bundles or other groupings; provide support for the plasma membrane and play a role in determining cell shape; play significant role in contractile activities that enable cell to change shape or move.

Cilia, Flagella, and Basal Bodies. Cilia and flagella are motile projections of the plasma membrane that can move substances over cell surfaces or move entire cells about. Both arise from basal bodies. Cilia, flagella, and basal bodies contain microtubules.

Centrosome and Centrioles. Centrosome serves as microtubule-organizing center. Centrioles are similar in structure to basal bodies; in fact, basal bodies are centrioles.

Inclusion Bodies. Chemical substances in cytoplasm in form of particles or droplets.

◆ EXTRACELLULAR MATERIALS p. 92

Include extracellular body fluids and extracellular framework in which many cells are embedded. Connective tissue is particularly rich in extracellular materials.

◆ CELL DIVISION pp. 93–101

Several basic events are involved, including replication of DNA in nucleus, redistribution of DNA into two new nuclei, and the division of cytoplasm into two new cells, each with its own nucleus.

Interphase. Period between active cell divisions.

CENTRIOLE REPLICATION. New centriole forms at right angles to each of the two centrioles in the centrosome.

DNA REPLICATION. Two chains of a DNA molecule separate, and each chain serves as a template for the assembly of a new DNA chain; result is two DNA molecules like original molecule.

Mitosis. A continuous event, although it is divided into several phases.

PROPHASE. Chromosomes become clearly visible; kinetochore matures at centromere of each chromatid; centrosome splits into two centrosomes; centrosomes move toward opposite ends of cell.

PROMETAPHASE. Begins when nuclear envelope disintegrates; kinetochore microtubules attach to chromosomes; chromosomes begin to move toward a position halfway between the two centrosomes.

METAPHASE. Chromatids become aligned along metaphase plate, halfway between the two spindle poles.

ANAPHASE. Chromatids separate; one chromatid of each chromosome moves toward one spindle pole, and the other chromatid moves toward the other spindle pole. Once chromatids separate, each is considered to be an independent chromosome.

TELOPHASE. Chromosomes reach spindle poles, spindle disappears, and chromosomes assume interphase appearance of chromatin threads.

Cytokinesis. When occurring with mitosis, beginning of cytokinesis is generally evident during anaphase as inward movement of plasma membrane. Ultimately, cytoplasm is completely divided into two separate cells, each with its own nucleus.

Meiosis. Cell division process that produces haploid cells.

PHASES OF MEIOSIS. Two successive division sequences—division I and division II. During division I, chromosomes become aligned at metaphase plate as homologous pairs, and one member of each pair moves toward each spindle pole. Division II is essentially a mitotic division.

GENETIC DIVERSITY. Gametes a person produces commonly differ from one another genetically; differences are due primarily to the manner in which the person's DNA is redistributed during meiosis.

◆ CONDITIONS OF CLINICAL SIGNIFICANCE: THE CELL pp. 101–103

Genetic Disorders. Many diseases and abnormalities are traceable to genetic defects. Genetic disorders can cause incorrect protein synthesis; erroneous protein might then lead to disease. Alterations in the genetic material of the cell are called mutations; in most cases mutations are harmful.

Cancer. In its essence, a cancer is a malignant tumor. Radiation therapy and chemotherapy are forms of treatment if cancer cells metastasize.

Aging and Cells. Many theories including some that suggest aging changes may be programmed into cells.

Self-Quiz

1. Many nonpolar substances that are soluble in lipids move relatively easily through the plasma membrane by: (a) exocytosis; (b) active transport; (c) simple diffusion.
2. Match the following terms associated with the plasma membrane with the appropriate lettered descriptions:

Specificity	(a) Substances move across membrane with aid of carrier molecules; substances move in either direction; movement is down a gradient
Saturation	(b) Each carrier molecule binds with only a select group of substances
Competition	(c) Carrier molecules move materials across the plasma membrane up a gradient
Facilitated diffusion	(d) The amount of a substance that can cross the plasma membrane by mediated transport in a given time is limited
Active transport	(e) A particular binding site on a carrier molecule can bind with more than one substance

3. The term for the process by which a cell takes in extracellular material by means of membrane-bounded vesicles is: (a) exocytosis; (b) endocytosis; (c) cytokinesis.
4. A solution that produces a change in the tone of cells immersed in it as a result of the net movement of water from the solution into the cells is a(an): (a) isotonic solution; (b) hypotonic solution; (c) hypertonic solution.
5. The intracellular fluid: (a) contains a lower concentration of sodium ions than the extracellular fluid; (b) contains a higher concentration of sodium ions than the extracellular fluid; (c) contains the same concentration of sodium ions as the extracellular fluid.
6. The nuclear envelope consists of two parallel membranes separated by a perinuclear space. True or False?
7. tRNA contains coded messages that specify the order in which amino acids are to be linked together to form particular polypeptides or proteins. True or False?
8. A three-base sequence in mRNA that specifies a particular amino acid of a polypeptide or protein is called a(an): (a) codicil; (b) codon; (c) anticodon.
9. Amino acids are linked together to form polypeptides or proteins at: (a) lysosomes; (b) centrosomes; (c) ribosomes.
10. Enzymes that synthesize many different lipids are associated with the membrane of the: (a) endoplasmic reticulum; (b) peroxisome; (c) lysosome.
11. Enzymes capable of digesting proteins, lipids, DNA, and RNA are present within: (a) nucleoli; (b) microtubules; (c) lysosomes.
12. Organelles involved in the production of ATP are: (a) peroxisomes; (b) mitochondria; (c) lysosomes.
13. Thin, short, and numerous projections that can move substances over the surface of a cell are known as: (a) cilia; (b) flagella; (c) basal bodies.
14. Which structures are composed primarily of actin? (a) microtubules; (b) intermediate filaments; (c) microfilaments.
15. Cilia, flagella, basal bodies, and centrioles all contain: (a) mitochondria; (b) microtubules; (c) peroxisomes.
16. Match the following cytoplasmic organelles with their related functions:

Centrosome	(a) Organelles that contain the enzyme catalase
Golgi apparatus	(b) Structures from which cilia and flagella arise
Peroxisomes	(c) Structures that are associated with contractile activities involved in cell movement
Basal bodies	(d) Structure that serves as a microtubule-organizing center
Microfilaments	(e) A membranous organelle that gives rise to secretory vesicles

17. Mitosis and cytokinesis always occur simultaneously. True or False?
18. The DNA molecules that comprise the genetic material of the cell appear only as indistinct chromatin threads within the nucleus during: (a) metaphase; (b) anaphase; (c) interphase.
19. During both RNA production and DNA replication, nucleotides attach to an exposed template DNA chain according to a complementary base-pairing pattern. True or False?
20. Match the following items associated with meiosis with the appropriate lettered descriptions:

Autosomes	(a) XX
XY	(b) The period between division I and division II of meiosis
Homologous chromosomes	(c) Cells with two complete sets of chromosomes are said to be this type
46	(d) This process contributes to genetic diversity
Diploid	(e) This cell type is characteristic of gametes
23	(f) The total number of chromosomes in a human somatic cell
Haploid	(g) Male sex chromosomes
Interkinesis	(h) Chromosome classification that does not include sex chromosomes
Crossing over	(i) The number of chromosomes contained in each of the human male and female gametes

CHAPTER 4

Tissues

CHAPTER CONTENTS

LEARNING OBJECTIVES

After completing this chapter, you should be able to:

1. Name the four primary tissues, and cite one example of each.
2. Describe the specializations by which adjacent cells may be attached to one another.
3. List three means of classifying epithelial tissue, and cite at least one example of each tissue.
4. Describe the shape that characterizes each of these cell types: squamous, cuboidal, and columnar.
5. Distinguish between simple epithelium and stratified epithelium.
6. Distinguish between exocrine glands and endocrine glands.
7. Classify three types of glands by mode of secretion, and describe how each type functions.
8. List the types of connective tissue, and state at least one function of each.
9. Cite several structural differences between bone and cartilage.
10. Name the three main types of muscle tissue, and describe the appearance of the cells of each type.
11. Cite two kinds of tissues that cannot undergo regeneration in a mature person, and two kinds that can.

CHAPTER 4

Chapter 3 described the structure of a typical cell and pointed out that each cell contains various organelles, which carry on a number of physiological processes. It is important to realize, however, that cells more commonly work together as a group for the benefit of the organism as a whole than as discrete individuals simply attending to their own needs. Groups of cells that are similar in structure, function, and embryonic origin and that are bound together with varying amounts of intercellular material are referred to as **tissues.** There are four primary tissues in the body: *epithelial, connective, muscular,* and *nervous.* Since these four tissues join together to form the organs of the body, an understanding of the structure and function of each tissue type contributes to our understanding of the organ systems.

Epithelial Tissues

Epithelial tissues (epithelia; singular, **epithelium)** are formed of closely joined cells with only a minimum of intercellular material between them. Epithelial cells are always underlain by connective tissue, to which they are attached by a thin layer called the **basement membrane.** The basement membrane consists of two layers: one layer, called the *basal lamina,* is composed of collagen and glycoproteins and is a product of the epithelial cells; the deeper layer is called the *reticular lamina* because it is composed of reticular fibers that develop from the connective tissue. Epithelia may develop from either the ectoderm, endoderm, or mesoderm of the embryo.

Epithelia are, by definition, sheets of cells that cover body surfaces and line body cavities. In general, they cover most of the free surfaces of the body, both internal and external. For instance, they form the outer layer of the skin, the lining of the digestive tube, the linings of the ventral body cavities, the lining of the blood vessels, and those glandular ducts and tubules that develop from epithelial linings or coverings. In addition, some epithelial tissues are incorporated into the various glands, where they serve as the functional part of the gland. With such a variety of locations, it is not surprising that epithelial tissues have diverse functions. The epidermis of the skin, for instance, is a *protective* layer that forms a barrier between the organism and its external environment, whereas the linings of the internal body organs are involved in the *absorption* of materials into the body, the *excretion* of waste products, and the *secretion* of special products into the cavities. Regardless of their functions, none of the epithelial tissues contain blood vessels. Blood vessels within the connective tissue underlying the epithelial tissues supply nutrients to, and remove wastes from, the epithelial cells.

Specializations of Epithelial Cell Surfaces

The portion of epithelial cells that forms the surface of the body or lines the cavities and lumina (interior space) of the various tubes in the body is referred to as the **free surface.** The free surfaces of the epithelial cells that line the blood vessels are smooth. The electron microscope shows that some other epithelial cells have their free surfaces folded into tiny protoplasmic projections called **microvilli** (Figure 4.1). Because microvilli greatly increase the area of the free surface, they are especially abundant in locations where absorption is the main activity, as in the lining of the digestive tract. Before the electron microscope made it possible to view their structure clearly, these dense groups of microvilli were identified as *striated borders* or *brush borders.* Microvilli line some surfaces that are not absorptive, and their function is less well understood there. They may serve to anchor mucus to the cell surface. Microvilli are not motile. In some locations, the free surfaces of epithelial cells are modified by the presence of **cilia.** Most cilia are motile and move rhythmically, thus serving to propel materials along the epithelial surface.

Specializations for Cell Attachments

We have noted that one characteristic of epithelial tissues is that their cells are situated close together, with little intercellular material between them. In fact, adjacent epithelial cells are generally joined together into coherent sheets. It is these cellular junctions that make epithelial tissues so well suited to cover the body surfaces and to line the body cavities.

Functionally, there are three groups of cell junctions: (1) *occluding junctions,* which join the plasma membranes of adjacent cells tightly together; (2) *anchoring junctions,* which physically connect adjacent cells and their cytoskeletons, but leave a space separating the plasma membranes; and (3) *communicating junctions,* which permit the passage of chemical and electrical signals between the joined cells. Such specialized cell junctions are found in many tissues throughout the body, but are especially abundant in epithelial tissues where some of them are organized into groups called **junctional complexes.**

Junctional Complexes

A junctional complex generally has three distinct components—a *tight junction,* an *intermediate junction,* and a *desmosome*—all of which are associated with the plasma membranes of adjacent cells (Figure 4.1).

◆ **FIGURE 4.1 Typical epithelial cell as seen with an electron microscope**

The drawing shows a highly magnified junctional complex consisting of a tight junction, an intermediate junction, and a desmosome. (a) Photomicrograph of a junctional complex. (b) Photomicrograph of a gap junction.

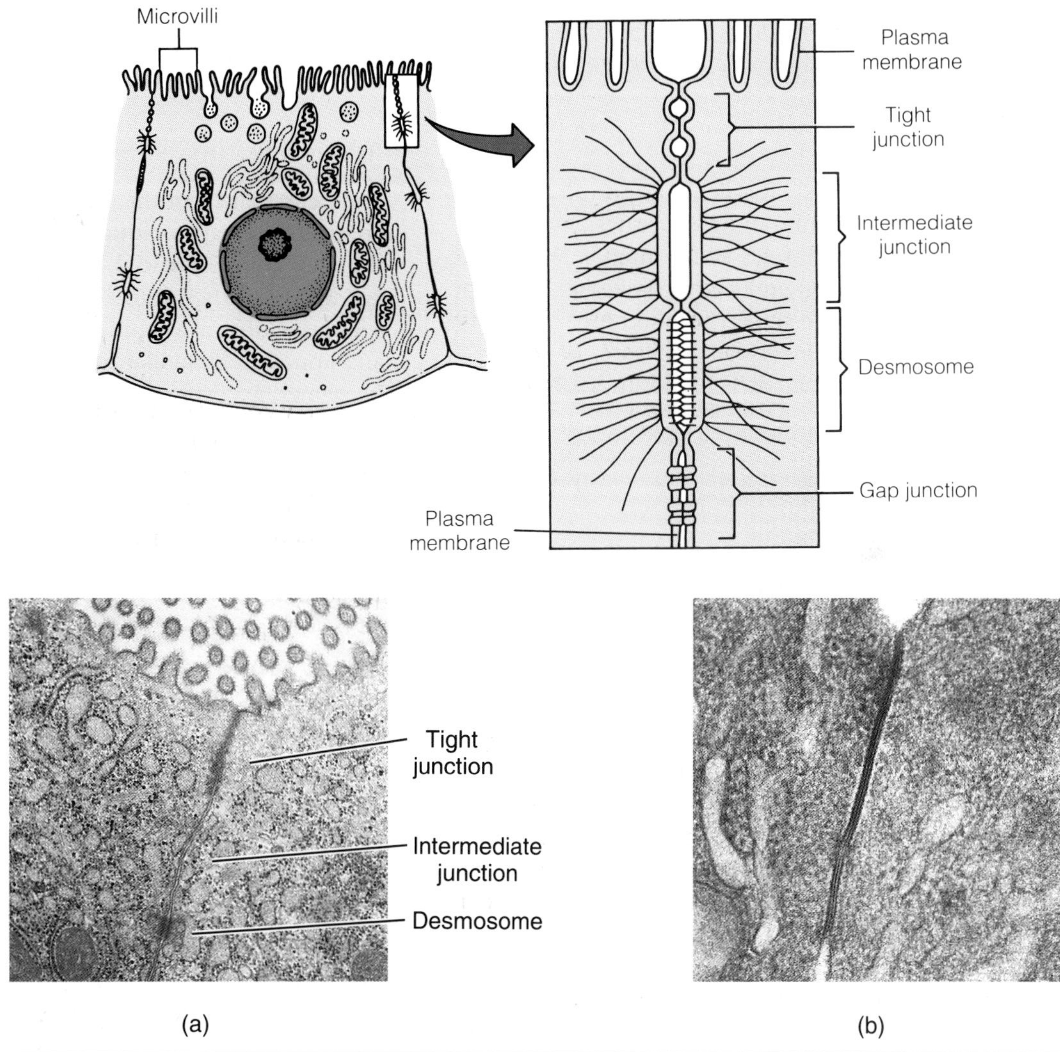

Tight Junctions. Tight junctions (zonula occludens) are occluding junctions located just below the free surfaces of adjacent cells. In this region, a network of protein strands extends across the plasma membranes of adjacent cells and causes the outer surfaces of the membranes to fuse in several places. Intercellular separations remain between the sites of membrane fusion. Tight junctions extend around the outer margin of the cell like a belt. These junctions not only connect adjacent cells, but they also restrict the movement of substances through an epithelial membrane by way of cellular spaces between adjacent cells.

Intermediate Junctions. Intermediate junctions (zonula adherens) are anchoring junctions located just below tight junctions. In the intermediate junction, the plasma membranes of adjacent cells are separated by a gap of about 200 Å. Bundles of actin fibers lie against the inside surface of the plasma membrane and run parallel with it. The actin fibers form a continuous *adhesion belt* that extends around the circumference of the cell, as in the tight junction. The adhesion belts of adjacent epithelial cells are directly apposed and are held together by glycoproteins of the cytoskeleton that extend across the plasma membrane and into the extracellular space. In this manner, intermediate junctions connect the cytoskeletal elements of adjacent cells.

Desmosomes. The third component of a typical junctional complex is another anchoring junction, the **desmosome** *(des´-mo-some)* or **macula adherens** *(mak´-u-la ad-hear´-uns)*. Each desmosome is an individual point of cell attachment rather than a beltlike zone. The cell membranes remain about 200 Å apart in a desmosome, and the inner layer of each cell membrane is thickened by a dense cytoplasmic plaque composed of various intracellular proteins. Cytoplasmic filaments that form part of the cytoskeleton of the cells attach to the proteins of the plaque. Indistinct filaments of glycoprotein run across the intercellular space between adjacent desmosomes. It is thought that these filaments are related to the cell-to-cell binding at these points. Desmosomes can occur anywhere around the periphery of an epithelial cell.

While desmosomes are particularly prevalent in epithelia, they are also found in other tissues, such as cardiac muscle and the muscles of the uterus. Where the cell membrane contacts connective tissue, as along the basal lamina, half desmosomes (**hemidesmosomes**) are sometimes found. Hemidesmosomes are common in stratified epithelium, such as is found in the outer layer of the skin.

Gap Junctions

Separate from the components of the junctional complex is another type of intercellular specialization called the **gap junction** or **communicating junction.** In this junction, the plasma membranes of adjacent cells are separate but very close together—only approximately 20Å apart. This extremely narrow gap is bridged by small tubular channels formed by proteins that pass through the plasma membrane and extend into the extracellular space, where they join with similar channels from the adjacent cell. These channels, which are called *connexons,* directly link the cytoplasm of adjacent cells and allow ions and small water-soluble molecules to pass from one cell to another. Thus, gap junctions play an important role in coupling adjacent cells both metabolically and electrically. Unlike junctional complexes, gap junctions are not restricted to epithelia; they are also found connecting the cells of some smooth muscles, cardiac muscle, some nerve cells, and the bone-forming cells located within compact bone.

Classification of Epithelia

Epithelial tissues are generally classified on the basis of the *number and arrangement of cell layers* within the tissue and the *shape of the cells at the free surface* of the tissue.

According to Cell Layers

If an epithelium is formed by a single layer of cells, all of which are in contact with the basal lamina, it is called **simple epithelium.** If it has two or more layers of cells, and only the deepest layer is in contact with the basal lamina, it is termed **stratified epithelium.** If the tissue appears to consist of several layers but is actually formed of a single layer with all cells touching the basal lamina, it is **pseudostratified epithelium** (*soo-dō-strat´-a-fide; pseudo* = false). This false impression of stratification occurs because some cells are shorter than others and the taller cells overlap the short ones, preventing them from reaching the free surface of the tissue.

According to Cell Shape

The cells that form the free surface of epithelial tissues are of three shapes. **Squamous cells** *(skway´-mus)* are flat and thin. **Cuboidal cells** are about as tall as they are wide and therefore appear almost square in vertical section. **Columnar cells** are taller than they are wide and appear rectangular in vertical section. Epithelia can be named according to which of these cell types forms their free surfaces.

General Classification

The general classification of epithelial tissues takes into consideration both the shape of the cells that form the free surface and the number of layers of cells in the tissue.

Simple Squamous Epithelium. Simple squamous epithelium is formed of a single layer of squamous cells (Figure 4.2). Since this thin sheet does not form a very effective barrier, substances can move across it easily. In addition, the flat cells do not contain enough cytoplasmic inclusions to aid secretion or absorption.

In general, simple squamous epithelium is found in regions where diffusion and filtration occur. Specifically, it lines the heart and the blood vessels and is the only barrier that separates the blood in capillaries from the tissue fluid. The simple squamous lining of the vascular system is called **endothelium.** Similarly, it lines the air sacs (alveoli) of the lungs, where it separates the air from the tissue fluid, and lines the surface of body cavities. The simple squamous lining of the body cavities is called **mesothelium.** Simple squamous epithelium also forms the glomerular capsules of the kidneys, the sites where substances are filtered from the blood to form urine.

Stratified Squamous Epithelium. As the name implies, **stratified squamous epithelium** (Figure 4.3) is composed of many layers, the precise number of which varies in different locations. The deeper cells, close to the basal lamina, tend to be cuboidal, but those against the surface are typical squamous cells. The deeper cells undergo mitosis and thus increase in number. These newly formed cells are pushed toward the surface, where they replace the older surface cells that are being continually sloughed off. Because of the ability of strat-

◆ **FIGURE 4.2 Simple squamous epithelium (surface view)**

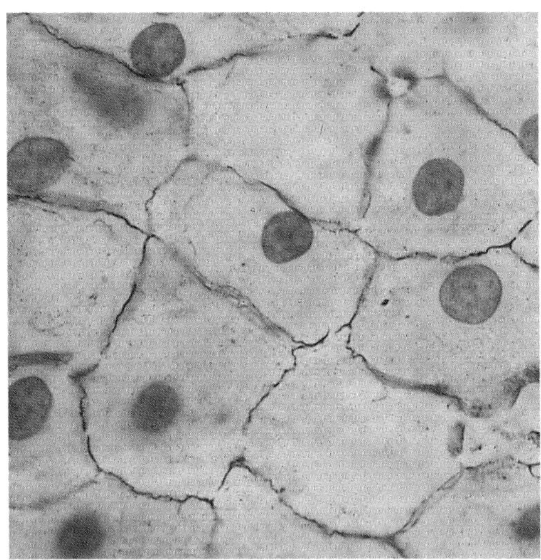

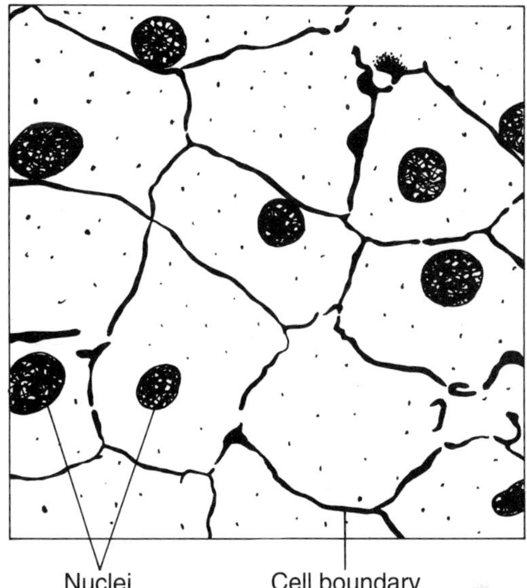

ified squamous epithelium to replace the cells of the superficial layers, this tissue is capable of compensating for the loss of cells due to such actions as abrasion. Thus, stratified squamous epithelium forms a protective layer on the body surface, as the epidermis of the skin, and also in areas that are subjected to friction, such as the linings of the mouth, pharynx, esophagus, anus, and vagina.

Transitional Epithelium. Transitional epithelium is a specialized stratified tissue that lines the urinary bladder and a few other hollow organs (Figure 4.4). The surface cells of transitional epithelium vary between cuboidal and squamous, depending upon whether the bladder is empty or expanded. The deepest cells are columnar and are overlaid by many layers of cells when the bladder is empty. When the bladder is full and its

◆ **FIGURE 4.3 Stratified squamous epithelium**

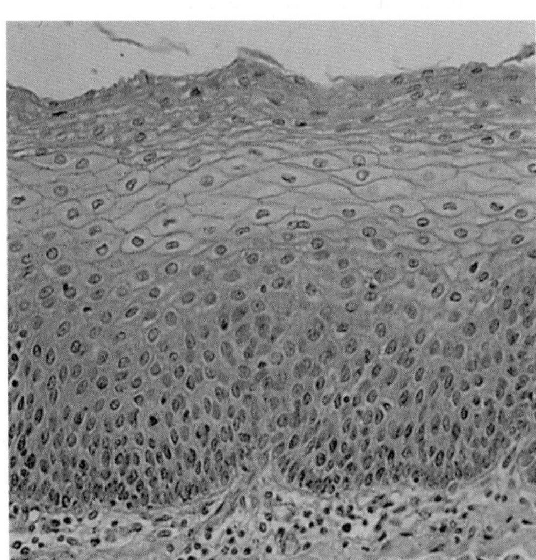

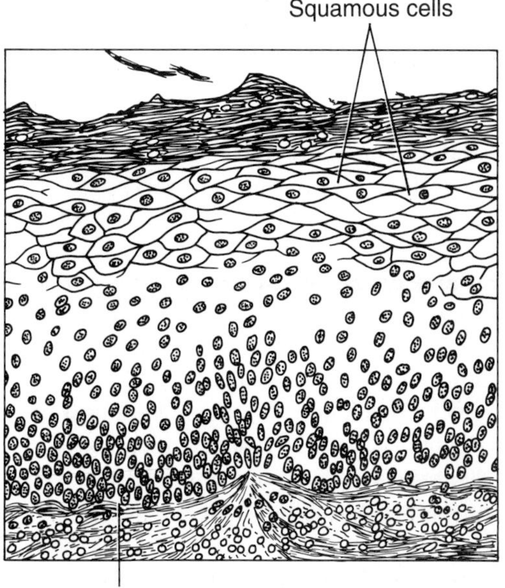

◆ **FIGURE 4.4 Transitional epithelium**

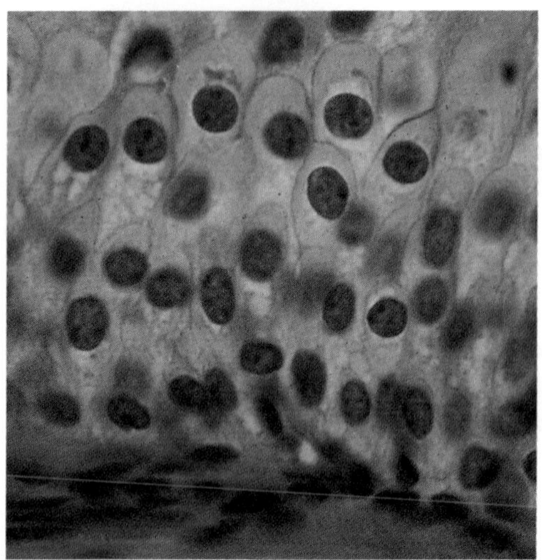

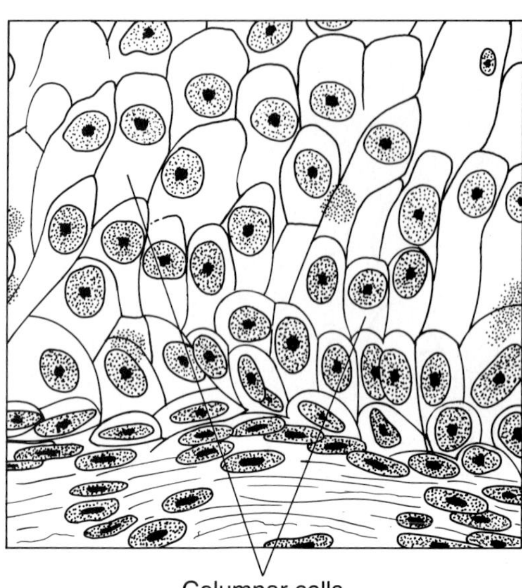

walls are stretched, the cells flatten and slide over each other, leaving only three or four strata (layers) between the deepest layer and the free surface. In this condition, the surface cells are flattened, like squamous cells. This specialized tissue allows an organ to expand with only minimal resistance from the tissue, thus lessening the chance of the organ rupturing and reducing the discomfort that occurs as the organ becomes full.

Simple Cuboidal Epithelium. Cuboidal cells are generally found as a single layer of cells that form **simple cuboidal epithelium** (Figure 4.5). Only rarely are cuboidal cells found in layers (stratified). Some cuboidal cells are capable of forming secretions and consequently are found in glands such as the thyroid, the sweat glands, and the salivary glands. Cuboidal cells also form the ducts of glands and parts of the kidney tubules as well as the outer cell layer covering the ovary.

Simple Columnar Epithelium. Columnar cells, with their large amounts of cytoplasm and cytoplasmic organelles, can perform rather complex chemical reactions and are therefore found in regions where secretion and absorption occur. The nuclei of columnar cells tend to be elongated and are usually located close to the bases of the cells. **Simple columnar epithelium** (Figure 4.6) lines the digestive tube from the stomach to the anal canal. It also forms the ducts of many glands. Cilia are present on the free surfaces of the columnar cells that form the membranes lining the bronchi of the lungs, the nasal cavity, the oviducts, and scattered regions of the uterus. The presence of cilia causes these tissues to be classified as **simple columnar ciliated epithelium.**

Stratified Columnar Epithelium. There are only a few locations where columnar tissue is truly stratified. In these tissues, the cells next to the basal lamina are small and rounded, and the columnar cells that form the free surface of the tissue do not contact the basal lamina (Figure 4.7). **Stratified columnar epithelium** appears on the epiglottis, in parts of the pharynx and anal canal, and in the male urethra.

Pseudostratified Columnar Epithelium. In **pseudostratified columnar epithelium,** which is much more common than the truly stratified columnar epithelium, all the cells contact the basal lamina, but some are shorter than others and do not reach the free surface. The nuclei are also found at different levels within the cells, which adds to the impression of stratification. This epithelium is found in the large ducts of some glands, such as the parotid gland, and in regions of the male urethra. The free surface of pseudostratified tissue often has cilia, and the tissue is then called **pseudostratified columnar ciliated epithelium** (Figure 4.8). These ciliated tissues line the mucous membranes of the respiratory passageways and the auditory (eustachian) tubes.

A summary of the classification of epithelial tissues according to cell layers and shape, as well as a typical location for each type of tissue, appears in Figure 4.9.

◆ **FIGURE 4.5 Simple cuboidal epithelium**

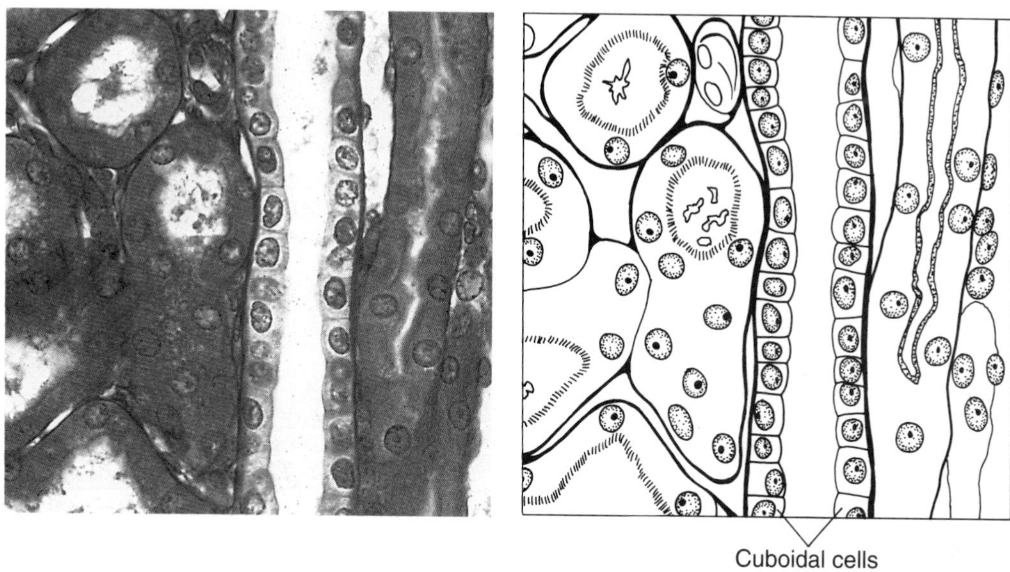

Epithelial Membranes

There are two important types of body membranes—*mucous membranes* and *serous membranes*—that, while not actually epithelial tissues, are composed of an epithelial layer on their free surface and an underlying connective tissue layer. For this reason they are often referred to as *epithelial membranes;* we will consider them here with the epithelial tissues.

Mucous Membranes. Mucous membranes or **mucosae** are moist epithelial membranes forming the linings of the digestive, respiratory, urinary, and reproductive tracts—all of which open to the exterior of the body. Their free surfaces vary in type, but they are usually either stratified squamous epithelium (mouth and esophagus) or simple columnar epithelium (stomach and intestine).

◆ **FIGURE 4.6 Simple columnar epithelium**

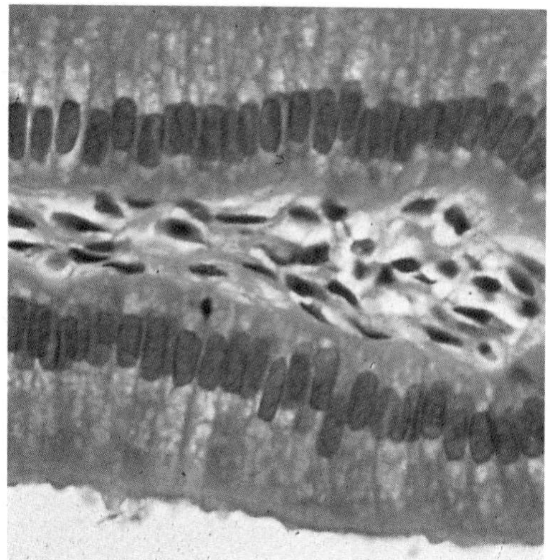

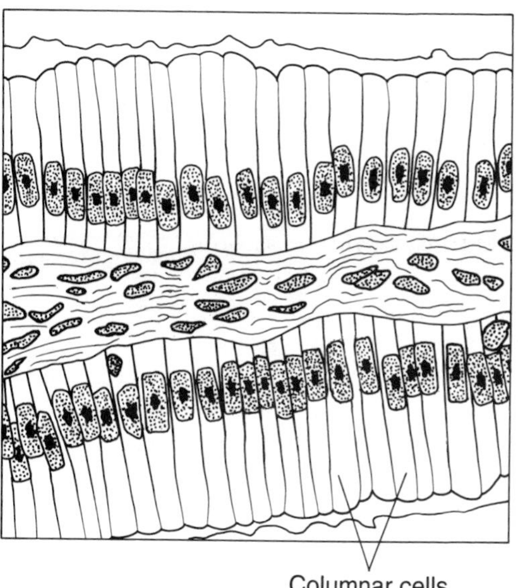

◆ **FIGURE 4.7 Stratified columnar epithelium**

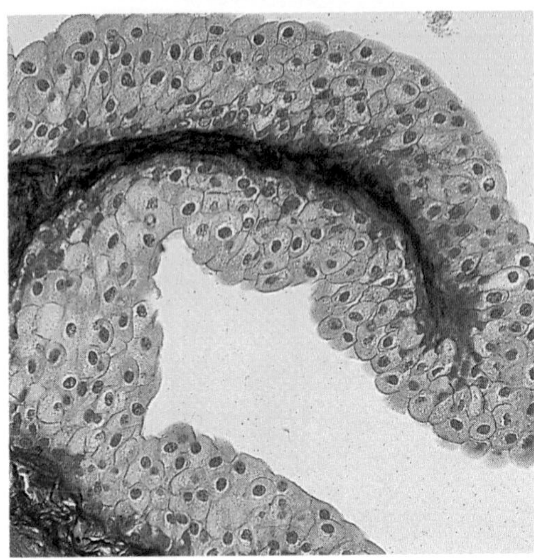

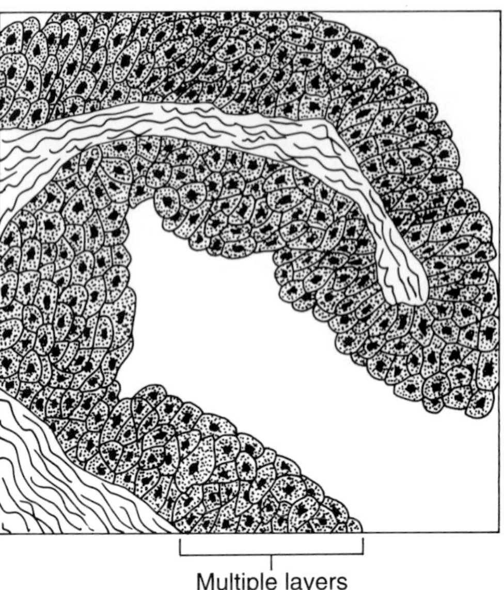

Mucous membranes are absorptive and secretory, making them particularly well suited to function in digestion and respiration. For example, food substances are absorbed into the body through the mucous membranes of the digestive tract. Most mucous membranes secrete **mucus,** a viscous fluid that protects and lubricates the membrane.

Most mucous membranes have a layer of loose connective tissue—the **lamina propria**—located deep to the innermost epithelial layer. Beneath the lamina propria, there is often a thin layer of smooth muscle called the **muscularis mucosae.** Because mucous membranes are composed of several types of tissues, they are often considered to be simple organs.

◆ **FIGURE 4.8 Pseudostratified columnar ciliated epithelium**

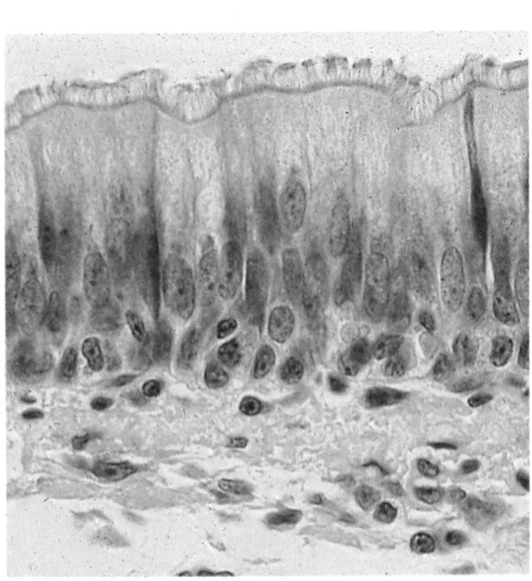

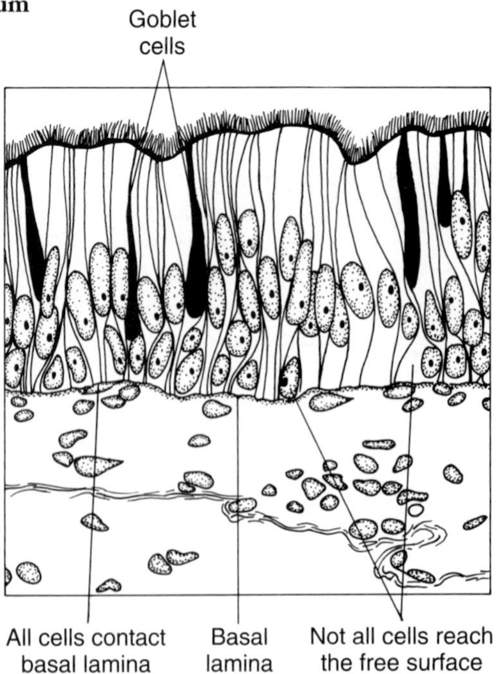

◆ **FIGURE 4.9 Classification of epithelial tissues according to cell layers and shape**

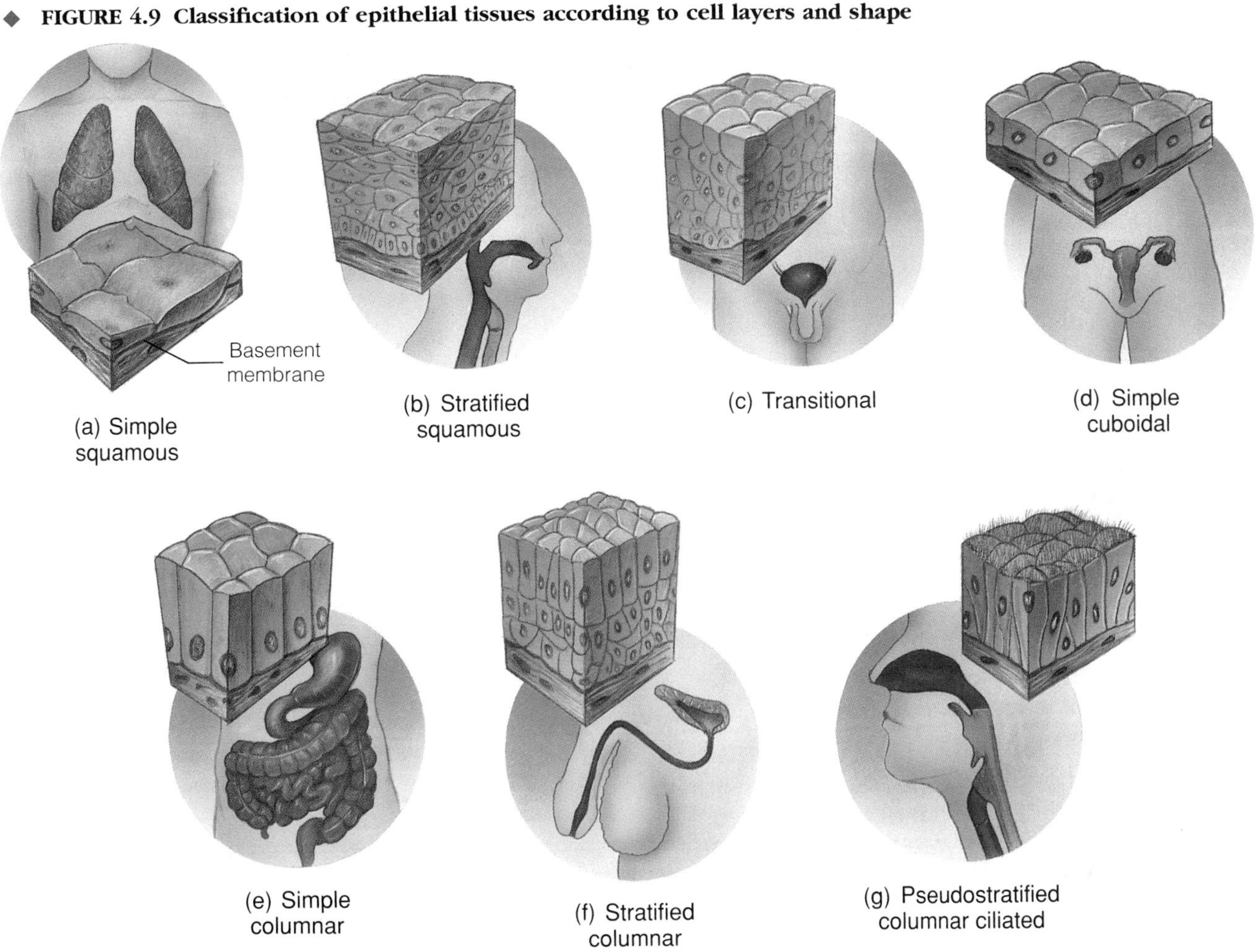

Serous Membranes. The ventral body cavities, which do not open to the exterior of the body, are lined with **serous membranes** or **serosae.** Serous membranes form both the parietal and the visceral portions of the pleura, the pericardium, and the peritoneum (see pages 12–15). The serous membranes consist of a thin layer of loose connective tissue covered by a surface layer of simple squamous epithelium called **mesothelium.** Mesothelium is derived from embryonic mesenchyme (mesoderm). The cells of the mesothelium secrete a clear, watery fluid, called **serous fluid,** that keeps the membranes moist. Serous membranes, like mucous membranes, are composed of more than one type of tissue and may therefore be considered simple organs.

Glandular Epithelium

Most glands of the body are composed of epithelial cells that produce a specific secretion (such as sweat, milk, a hormone, or an enzyme) or excrete certain waste products (such as bile pigments). The mucus-secreting **goblet cells** of the respiratory and digestive tracts (see Figure 4.8) are examples of *individual cells* that function as glands. Most glands, however, are *multicellular*—that is, they are formed of clusters of cells.

Embryologically, all glands originate from an epithelium. And most glands retain their connection with the epithelium—a connection that serves as a duct through which the secretions of the gland are carried to a particular site. Such glands are called **exocrine glands.** Some glands, however, lose their connection with the epithelium and empty their secretions directly into the blood. These are the **endocrine glands** and their secretions are **hormones.**

The multicellular exocrine glands are classified according to (1) their structure and (2) the manner in which they produce their secretions. Structurally, the ducts of the glands may be *unbranched* or *branched.* Glands whose ducts do not branch are called *simple glands;* glands whose ducts branch repeatedly are called *compound glands.* Simple and compound glands can be further subdivided according to whether their secreting

◆ **FIGURE 4.10 Classification of exocrine glands according to structure**

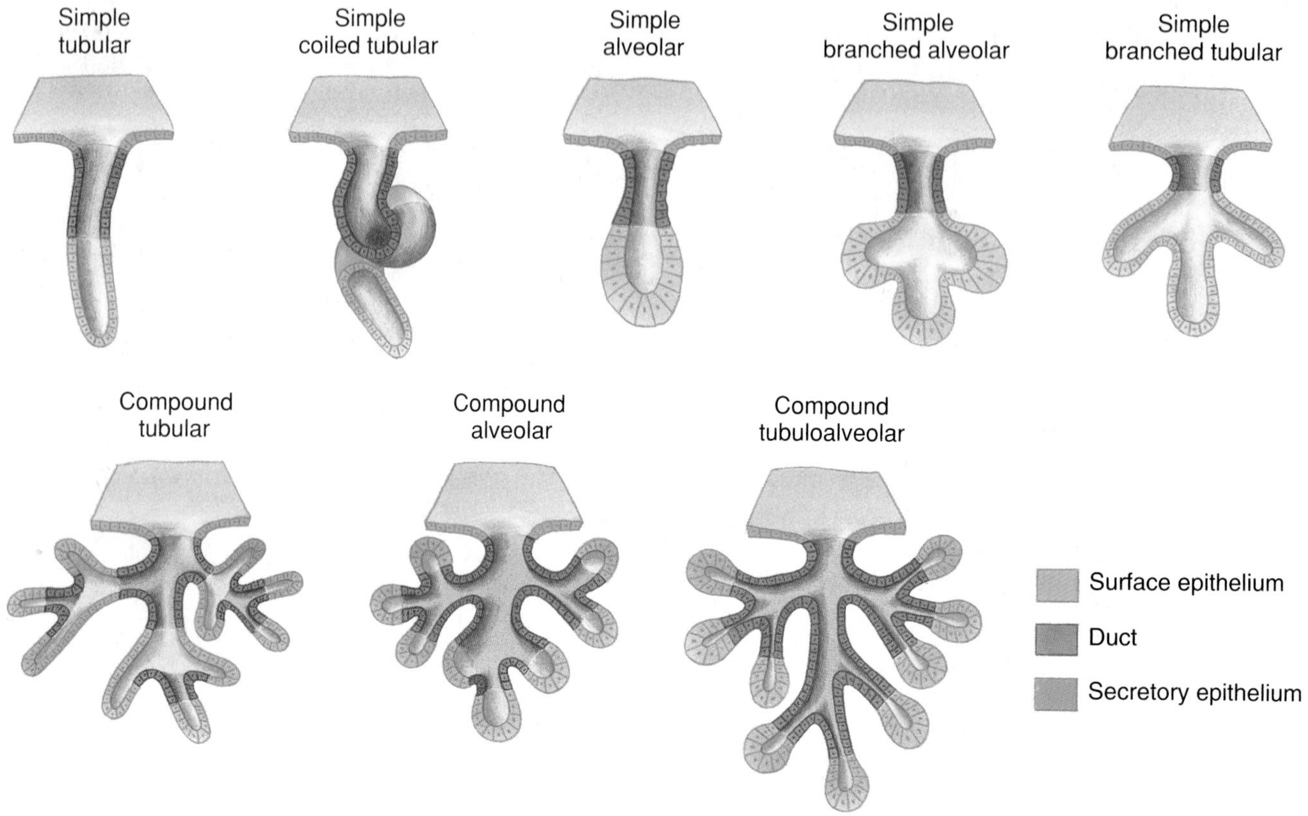

Simple tubular
Simple coiled tubular
Simple alveolar
Simple branched alveolar
Simple branched tubular
Compound tubular
Compound alveolar
Compound tubuloalveolar
Surface epithelium
Duct
Secretory epithelium

portions are (1) *tubular,* (2) composed of small sacs called *alveoli (al-ve´-o-lie)* or *acini,* or (3) a combination of blindly ending tubules and alveoli *(tubuloalveolar)* (Figure 4.10). Simple glands whose secretory portions are branched are referred to as *simple branched tubular* or *simple branched alveolar,* depending on the form of their secretory portions. Compound tubuloalveolar glands are the most common type of exocrine gland, being present in the pancreas, prostate, salivary glands, and mammary glands.

When glands are classified according to their *mode of secretion,* there are three different types: merocrine, holocrine, and apocrine (Figure 4.11).

1. **Merocrine glands** produce secretions that do not accumulate significantly in the gland cells. Rather, the secretions pass through the plasma membrane within membranous vesicles by the process of exocytosis. In merocrine glands, no destruction of glandular cells occurs during secretion. The pancreas, the salivary glands, and most sweat glands are merocrine glands. In fact, most exocrine glands are of this type.
2. **Holocrine glands** accumulate their secretions in their cells and discharge them only when the cells rupture and die. New cells then form to replace those that have died. The sebaceous (oil) glands of the skin are the only true example of holocrine glands.
3. **Apocrine glands** produce secretions that accumulate toward the outer ends of the gland cells. The secretions are released and a small amount of cytoplasm is lost when these ends pinch off. But rather than dying, as happens in holocrine glands, the cell is only slightly damaged, and it repeats the accumulation of secretory products. Some sweat glands are apocrine glands, and the mammary glands are generally considered to be apocrine glands, although they are actually mixed glands as some of their secretion is of a merocrine type.

Connective Tissues

Connective tissues vary considerably in form as well as function. Some serve as the framework upon which epithelial cells cluster to form organs; others bind various tissues and organs together, supporting them in their proper locations; some contain the media (tissue fluid) through which nutrients and wastes pass while traveling between blood and body cells; others serve as storage

◆ **FIGURE 4.11 Classification of glands according to mode of secretion**
(a) Merocrine.
(b) Holocrine.
(c) Apocrine.

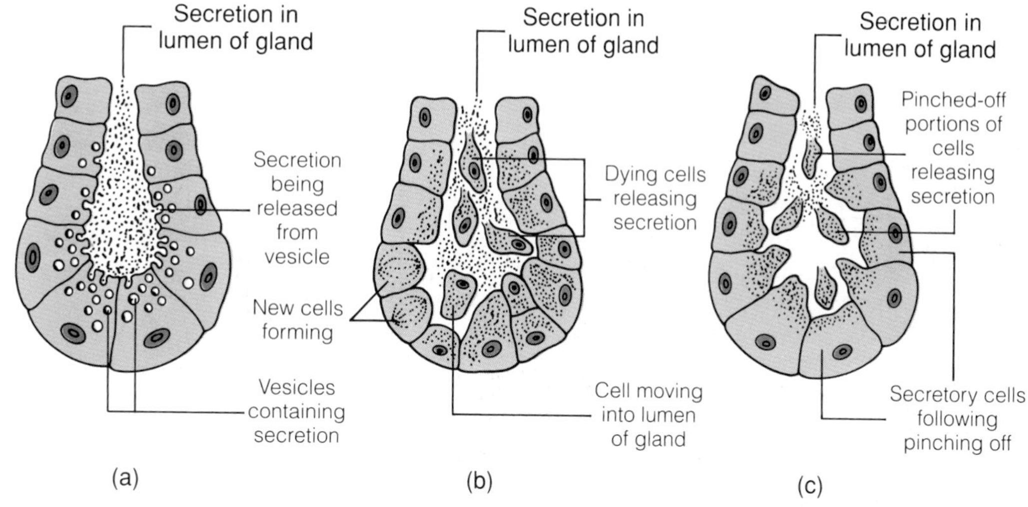

sites for excess food materials in the form of fat; and still others form the rigid skeletal framework of the body.

All adult connective tissues develop from undifferentiated embryonic connective tissue called **mesenchyme.** Thus, mesenchymal cells have the potential of developing into bone cells, cartilage cells, and blood cells, as well as the cells that produce the fibers of connective tissue. Mesenchyme develops from the mesodermal layer of the embryo; therefore all adult connective tissues are of mesodermal origin.

As you have seen, epithelial tissues are formed of closely packed cells that have very little material *(intercellular matrix)* between adjoining cells. In contrast, connective tissues are characterized by abundant intercellular matrix surrounding relatively few cells.

Several types of cells are associated with the connective tissues, but fibroblasts and macrophages are the most common. **Fibroblasts** are spindle-shaped cells that form the various fibers characteristic of connective tissues. Fibroblasts that are in a resting or less active phase are called **fibrocytes. Macrophages** *(mak-ro-faj-ez)*, which are generally not as abundant as fibroblasts, are active **phagocytes.** The macrophages may be *fixed* (attached to the connective tissue fibers), or they may remain *free* and be capable of moving through the matrix of the connective tissue. Both fixed and free macrophages engulf foreign matter and dead or dying cells; they are particularly active at sites of infection. Free macrophages have the advantage of being able to move to an infected area. However, in response to an infection, fixed macrophages can detach and become free cells.

The activities of macrophages are so important in protecting the body against invasion by microorganisms that they are often referred to collectively as the **macrophage system** (formerly called the *reticuloendothelial system*), even though they do not form a discrete system and are distributed widely and rather randomly throughout the body. For instance, macrophages are found in many tissues throughout the body, including the loose connective tissue, lymphatic tissues, mesenteries of the digestive tract, bone marrow, spleen, adrenal gland, and pituitary gland. In some locations, macrophages are given special names—such as the *Kupffer cells* that line the blood sinusoids of the liver, the *dust cells* of the lungs, the *histiocytes* of loose connective tissues, and the *microglia* of the central nervous system.

Whereas all these cells have a similar function—phagocytosis—macrophages in specific tissues or organs are often selective as to what they ingest. The macrophages of the spleen and the liver, for example, are particularly active in breaking down aging red blood cells. Macrophages are also involved in the activities of the body's immune system. These selective processes, in addition to their phagocytic role, make the cells of the macrophage system a major defense mechanism of the body. Because macrophages develop from a type of blood cell called a monocyte, which has a single unlobed nucleus, they are also referred to as the **mononuclear phagocytic system.**

Intercellular Matrix

The **intercellular matrix** of connective tissue is formed of *ground substance* and *fibers.* The ground substance is a homogeneous product of the connective tissue cells that it surrounds. It is composed of tissue fluid and *proteoglycans,* which consist of a small amount of proteins linked to long chains of polysaccharides called *glycosaminoglycans (GAGs).* The polysaccharides comprise about 95% of the proteoglycan molecule. Hyaluronic acid and chondroitin sulfate are GAGs molecules that are commonly found in connective tissues. Because of the ability of the proteoglycans to bind with water, the ground substance of connective tissue varies in consistency from fluid to a semisolid gel. In order for dissolved substances to pass between cells and blood capillaries, they must diffuse through the ground substance.

The fibers, which are also produced by the connective tissue cells, are found in varying amounts within the ground substance. There are three types of fibers: *collagenous, elastic,* and *reticular.*

Collagenous Fibers

Collagenous fibers, the most abundant type, appear as wavy bands under the microscope. Each fiber is made up of bundles of smaller fibrils. Collagenous fibers are very strong and inelastic and are composed primarily of the protein *collagen.* Collagenous fibers that are closely packed have a white color and are sometimes referred to as white fibers.

Elastic Fibers

Elastic fibers are long, threadlike branching fibers that often form interwoven networks. Their main protein, called *elastin,* gives the fibers the capacity of returning to their original lengths after being stretched. They therefore function to give resilience to connective tissue. Collagenous fibers are always found in the same tissue with elastic fibers. The collagneous fibers are capable of only a limited amount of stretch, after which they prevent further stretching. When the tension on the tissue is lessened, elastic fibers return the connective tissue to its normal length. Consequently, elastic fibers are found in abundance in tissues that routinely undergo stretching, such as the walls of blood vessels and the lungs. Large masses of elastic fibers have a slightly yellow color.

Reticular Fibers

Reticular fibers are short and very thin. They branch freely, forming a tight network called a **reticulum.** These fibers often form a gland's internal framework *(stroma),* to which the epithelial cells that make up the bulk of the gland are attached. Reticular fibers also join connective tissues to other types of tissues. Reticular fibers are inelastic and composed primarily of a type of collagen called *reticulin.* In fact, reticular fibers are believed to be quite similar to collagenous fibers, but are considerably thinner.

Types of Connective Tissue

Connective tissues are classified according to the nature of the ground substance and the types and organization of the fibers in the ground substance.

◆ **FIGURE 4.12 Loose connective tissue**

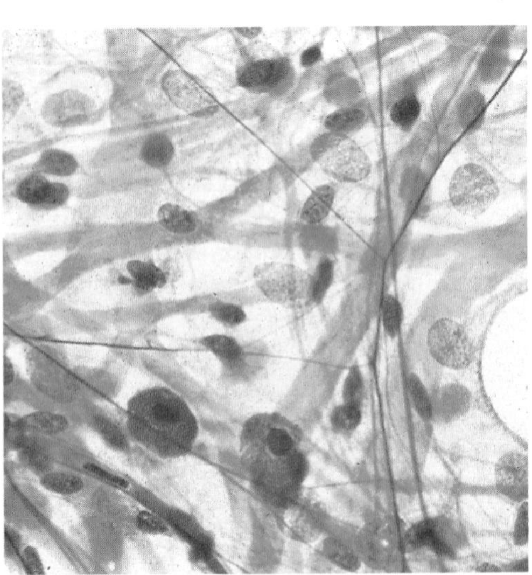

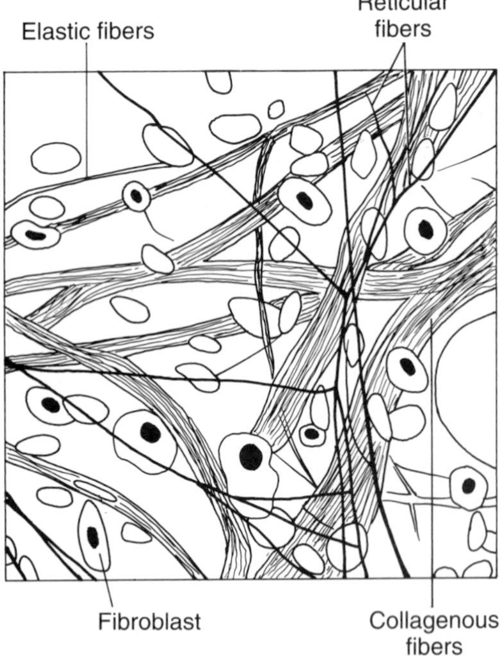

◆ **FIGURE 4.13 Adipose tissue**

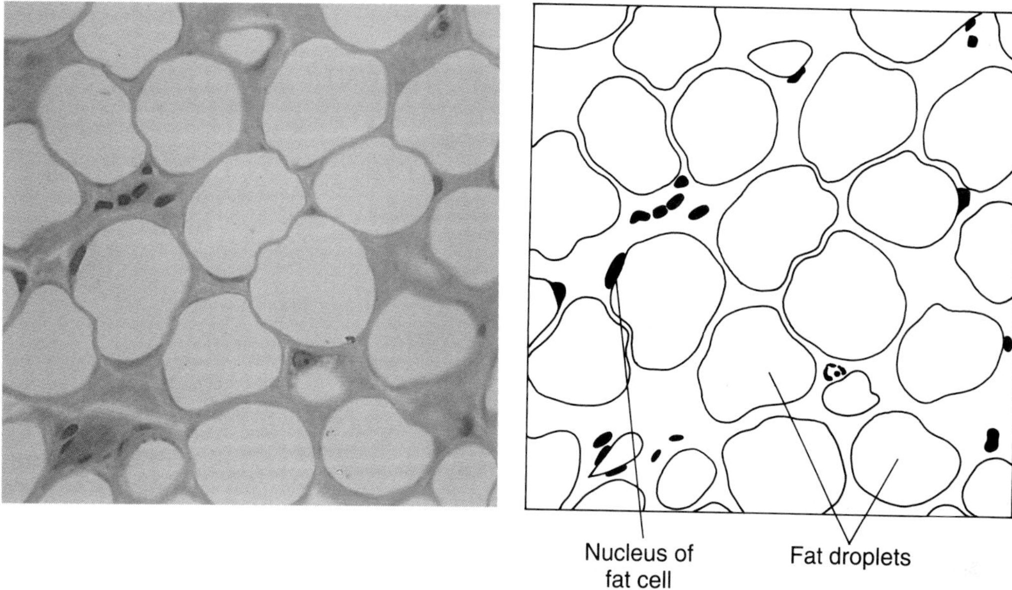

Loose Connective Tissue

Because the unorganized arrangement of the fibers in **loose connective tissue** leaves many spaces between them, it is also called **areolar connective tissue** (*areolar* = space). The ground substance of loose connective tissue is semifluid and is composed primarily of hyaluronic acid. Most of the fibers in loose connective tissue are collagenous, but elastic and reticular fibers are also present (Figure 4.12). Loose connective tissue contains several different types of cells, but fibroblasts and macrophages are the most common. It is soft and pliable and is the most widespread connective tissue of the body, being used to (1) attach the skin to the underlying tissue *(subcutaneous tissue)*, (2) fill the spaces between the various organs and thus hold them in place, and (3) surround and support the blood vessels. Because of the large spaces between cells and fibers, loose connective tissue contains a large amount of intercellular fluid (tissue fluid), which enters the spaces from capillaries. Tissue fluid is used to carry nutrients to, and waste products away from, the cells. If excessive fluid accumulates in these spaces, the affected area becomes swollen—a condition called *edema (e-dee´-mah)*. The presence of macrophages in loose connective tissue provides the body with a widespread defense against microorganisms.

Adipose Tissue

Adipose tissue *(ad´-i-pos)* is essentially composed of fat cells *(adipocytes)* dispersed in loose connective tissue (Figure 4.13). Each cell contains a large droplet of fat that squeezes and flattens the nucleus and forces the cytoplasm into a thin ring around the cell's periphery. Fat cells become larger or smaller as they take up or release fat. Adipose tissue is well supplied with blood vessels, indicating a high metabolic activity.

Adipose tissue serves as a storage site for fats and may develop anywhere loose connective tissue is found. Its most common site is in the subcutaneous tissue beneath the skin where, since fat is a poor conductor of heat, it serves as an insulating layer. Deposits of adipose tissue are also usually found around the kidneys and the heart, behind the eyeballs, and in the marrow cavities of long bones.

Reticular Connective Tissue

Reticular connective tissue resembles loose connective tissue in that it contains considerable ground substance located between a network of interlacing fibers (Figure 4.14). However, the fibers in this tissue are primarily reticular fibers, which are thinner than the collagenous fibers that predominate in loose connective tissue.

Although reticular fibers are commonly present in other types of connective tissues, the distribution of reticular connective tissue is rather limited. The tissue forms a delicate internal framework *(stroma)* in the liver, spleen, lymph nodes, and other organs. The glandular cells of the organs are anchored to the reticular framework. Reticular connective tissue also binds some muscle cells together and is present in the basal lamina of epithelial tissues.

◆ **FIGURE 4.14 Reticular connective tissue**

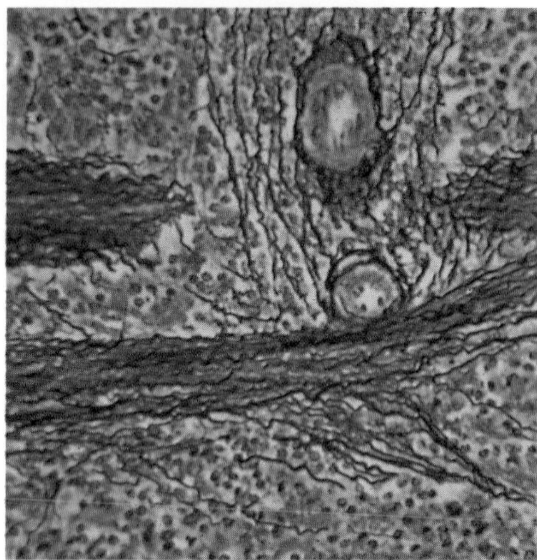

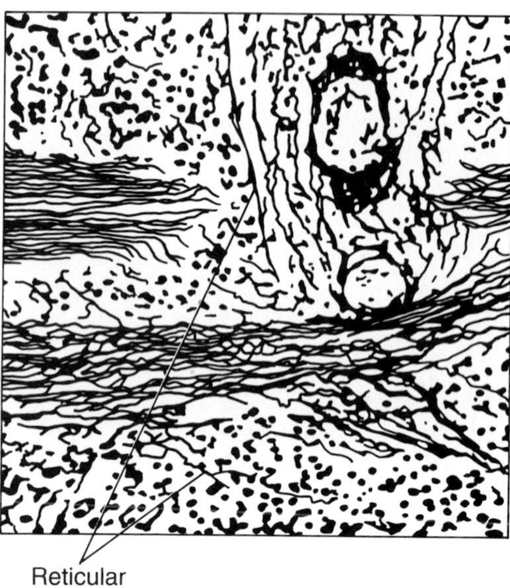

Dense Irregular Connective Tissue

The dense connective tissues are distinguished by an abundance of collagenous fibers, which provide the capacity to resist exceptional degrees of tension. **Dense irregular connective tissue** (Figure 4.15) contains all the same elements as loose connective tissue but has fewer cells and more numerous collagenous fibers. The fibers are closely interwoven, forming a compact tissue with fewer spaces. Because these tissues need to resist tensions that come from all directions, the fibers are oriented randomly.

The dermis layer of the skin, the fibrous coverings of cartilage (perichondrium), bone (periosteum), and nerves (perineurium), and the strong fibrous capsules that surround the liver, kidneys, spleen, and some other organs are all formed of dense irregular connective tissue.

Dense Regular Connective Tissue

Dense regular connective tissue (Figure 4.16) is characterized by a predominance of collagenous fibers that are tightly packed in parallel bundles. In this tissue, the tensions to be resisted come from a single direction, parallel to the orientation of the fibers. Because of the prevalence of collagenous fibers, this tissue is sometimes referred to as *white fibrous connective tissue.* The only cells present are fibroblasts, which are located between the fiber bundles. The abundance of fibers gives this tissue great strength. It forms the tendons of muscles, the ligaments of joints (which also contain some elastic fibers), and various fibrous membranes, such as fascia and aponeuroses. **Fascia** *(fash´-ee-ah)* surrounds the organs and the muscles; **aponeuroses** *(ap-o-noo-ro´-sees)* are broad sheets that function as thin tendons, attaching muscles to other structures.

Elastic Connective Tissue

In contrast to dense connective tissue, **elastic connective tissue** (Figure 4.17) contains more elastic fibers than collagenous fibers. Although the collagenous fibers cause this tissue to be quite strong, the predominance of elastic fibers allows it to stretch and return to its usual length. Elastic tissue is found in the walls of arteries, in the trachea and bronchi, and in the vocal cords, as well as in the walls of some hollow organs.

Cartilage

Cartilage is a specialized fibrous connective tissue that has a firm matrix containing numerous collagenous fibers, with the ground substance formed of various proteoglycans, chondroitin sulfate, and hyaluronic acid. The connective tissues that we have studied up to this point all have fluid or, at most, semisolid matrices. Because of its firm matrix, cartilage is able to function as a structural support. At the same time, the presence of fibers in the matrix imparts a certain amount of flexibility to cartilage.

The matrix of cartilage is formed by cells called **chondroblasts** *(kon´-dro-blasts).* Each chondroblast be-

◆ **FIGURE 4.15 Dense irregular connective tissue**

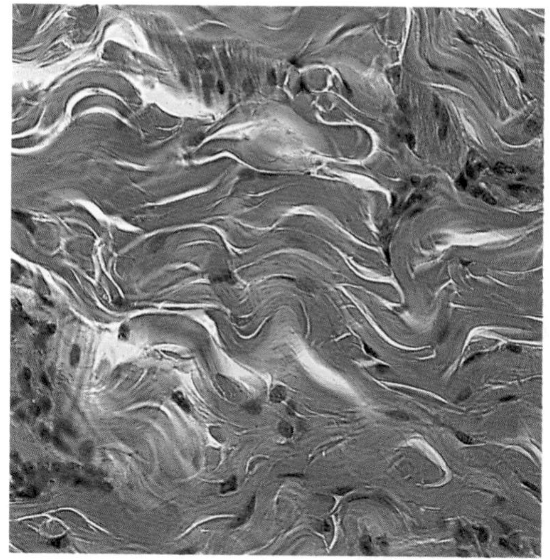

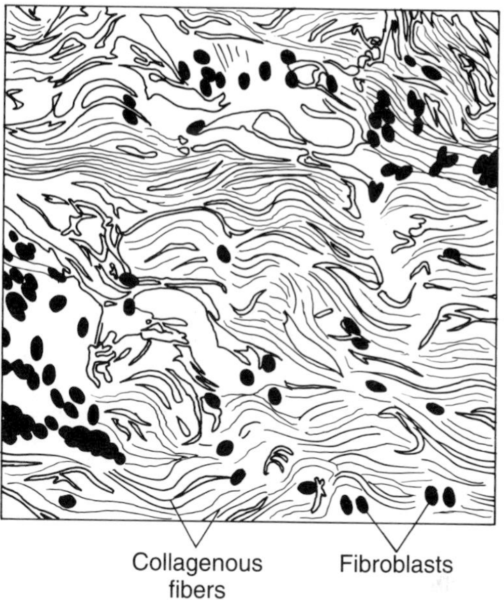

comes surrounded by matrix that it produces. As a result, the cartilage-forming cells eventually occupy small spaces called **lacunae** *(la-ku´-ni)*. When cartilage formation is complete, the chondroblast produces only enough matrix to maintain the cartilage. These mature cells are then called **chondrocytes.** The matrix of cartilage is avascular—that is, it contains no blood vessels. The only blood supply to cartilage is provided by blood vessels that enter the cartilage through the **perichondrium** *(per-i-kon´-dree-um)*. The perichondrium is a membrane of dense irregular connective tissue that covers the external surfaces of all cartilaginous structures (with the exception of the articular cartilages of joints) and is vitally important in the growth of cartilage. Because there is no direct blood supply to the matrix of cartilage, the nourishment of chondrocytes depends on

◆ **FIGURE 4.16 Dense regular connective tissue**

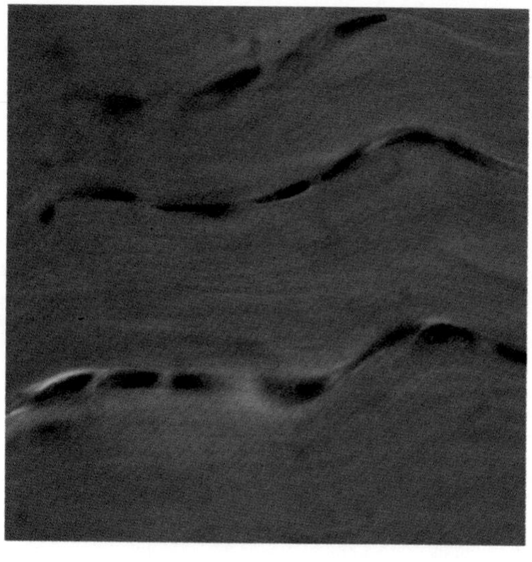

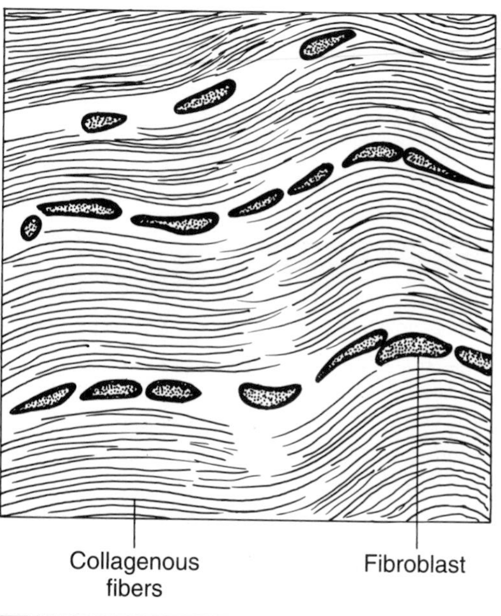

◆ **FIGURE 4.17 Elastic connective tissue**

the diffusion of nutrients through the matrix from capillaries located around the cartilage or from the synovial fluid of joint cavities. Similarly, waste materials must diffuse from the chondrocytes to the surrounding blood vessels.

Cartilage is especially prevalent in the embryo, but it also forms many adult structures. It is divided into three types according to variations in its fibrous structure: *hyaline, elastic,* and *fibrocartilage.*

Hyaline Cartilage. Cartilage that contains many closely packed collagenous fibers dispersed throughout the matrix is called **hyaline cartilage** (*high´-a-lin*) (Fig-

◆ **FIGURE 4.18 Hyaline cartilage**

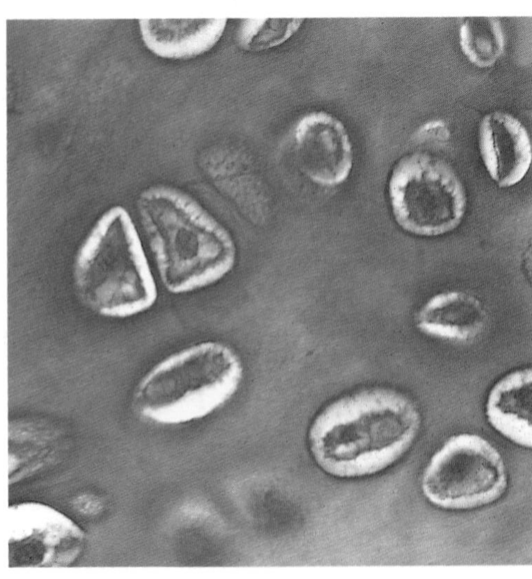

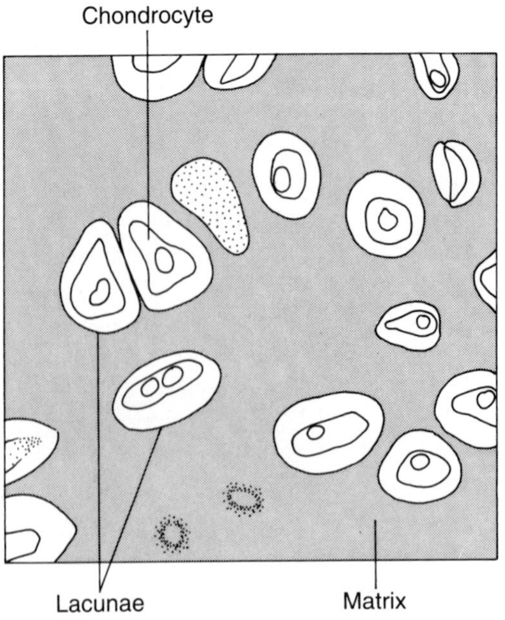

◆ **FIGURE 4.19 Elastic cartilage**

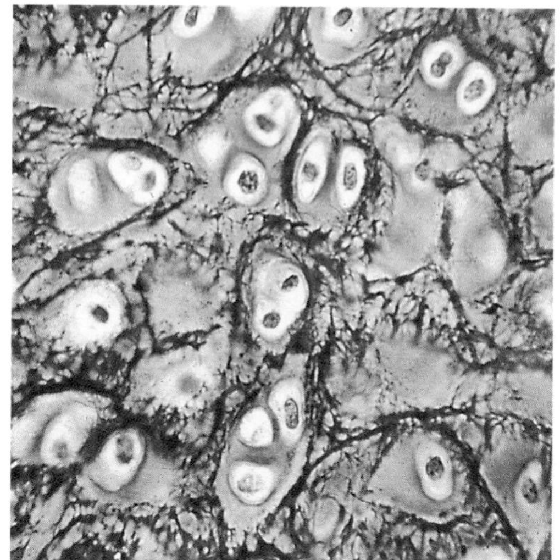

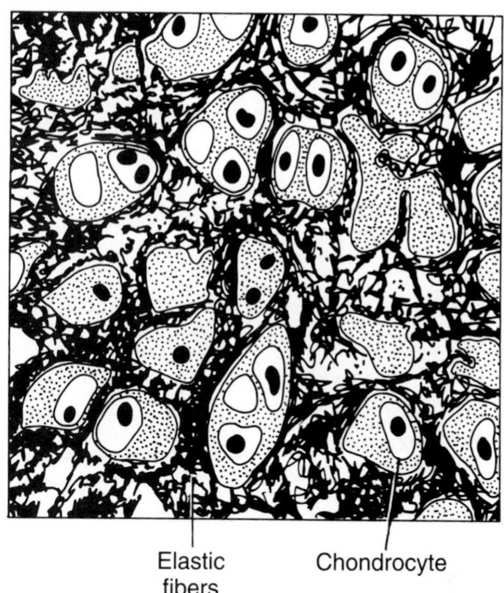

ure 4.18). Since the fibers have the same refractive index and staining properties as the ground substance, they are not distinguishable by ordinary microscopic examination. Hyaline cartilage is semitransparent and is smooth and firm, but flexible.

This type of cartilage, the most abundant in the body, is found primarily in places where strong support is needed but some flexibility is desirable. Hyaline cartilage forms most of the embryonic skeleton, which is gradually replaced by bone. It also makes up the costal cartilages, which attach the ribs to the sternum (breastbone) and allow the thorax to expand during respiration. The cartilage rings of the trachea are also hyaline cartilage, as are the articular cartilages on the ends of bones where two bones meet to form a movable joint. These joint cartilages provide smooth, moist surfaces that permit body movement with minimal friction.

Elastic Cartilage. Some body structures must furnish firm but elastic support, and this is the function of **elastic cartilage** (Figure 4.19). Elastic cartilage contains collagenous fibers like hyaline cartilage, but the fibers are not so closely packed. Moreover, elastic cartilage contains a generous network of elastic fibers, which are stainable and therefore show up under the microscope. This cartilage forms the external ear, the epiglottis, and the auditory tubes.

Fibrocartilage. The construction of **fibrocartilage** (Figure 4.20) differs from that of hyaline cartilage in that its collagenous fibers are arranged in thick, parallel bundles that give the matrix a coarse appearance. Actually, fibrocartilage resembles dense irregular connective tissue. Because the fibers are not compacted as much as those in hyaline cartilage, fibrocartilage is slightly compressible—which makes it beneficial in regions that support the body weight or that must withstand heavy pressure. It occurs in the intevertebral discs, which provide cushions between the vertebrae; the articular discs, which are located in the knee joint; and the pad of the pubic symphysis, which creates a partially movable joint between the two sides of the pelvis.

Bone

Because it has become mineralized, the matrix of **bone** or **osseous tissue** (Figure 4.21) is even harder than that of cartilage. Like cartilage, bone contains collagenous fibers, but its rigidity and strength are greatly increased by the presence of inorganic salts among the fibers. There are two other structural differences between bone and cartilage: (1) bone is well supplied by blood vessels throughout its matrix, and (2) its **lacunae** are interconnected by very small canals called **canaliculi** *(kan-al-ik´-u-lie)*. Bone forms the major portion of the adult skeleton. Its structure is considered in greater detail in Chapter 6, where the skeletal system is discussed.

Blood

Because the cells of blood (called *formed elements*) are interspersed in abundant matrix, blood is often considered to be a type of connective tissue. The matrix of blood, which surrounds the blood cells, is a fluid called **plasma.** Both the cells and the fluid matrix of blood are discussed in Chapter 17.

◆ **FIGURE 4.20 Fibrocartilage**

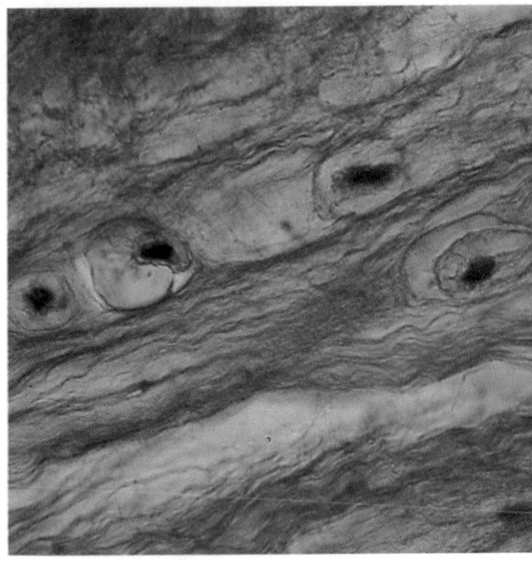

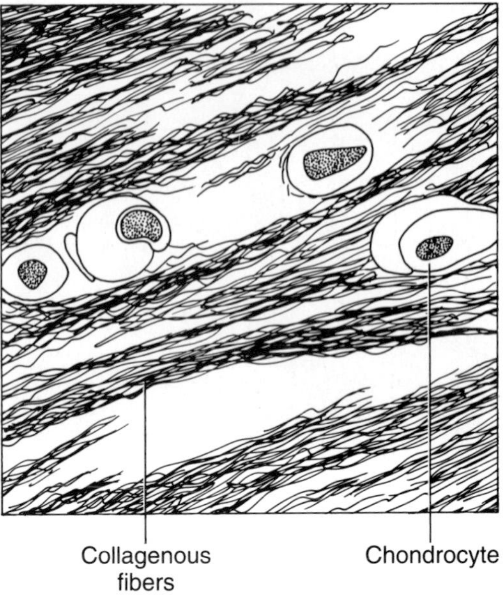

Muscle Tissue

The long, thin cells of **muscle tissue** are called **fibers.** It is important to realize that muscle cells are living cells and are in no way similar to the fibers of connective tissue. Muscle tissue is well supplied with blood vessels, and muscle fibers are highly contractile. There are three structurally different types of muscle tissue: *skeletal, cardiac,* and *smooth.*

Skeletal Muscle

Skeletal muscle (Figure 4.22) is attached to various bones of the skeleton. The cells of skeletal muscle are

◆ **FIGURE 4.21 Microscopic structure of bone showing lacunae and canaliculi**

◆ **FIGURE 4.22 Skeletal (straited) muscle tissue**

long and cylindrical—in fact, some skeletal muscle cells are thought to extend the entire length of the muscle. Running longitudinally throughout the skeletal muscle cells are regularly ordered threadlike arrays of proteins called **myofibrils.** Transverse light and dark bands that alternate along the myofibrils give skeletal muscle cells a characteristic *striated* (striped) appearance. Each skeletal muscle cell is multinucleate—that is, it has more than one nucleus. These nuclei are located on the periphery of the cell, just inside the plasma membrane.

Cardiac Muscle

Cardiac muscle (Figure 4.23) forms the wall of the heart. The cells of cardiac muscle, unlike those of skeletal muscle, form branching networks throughout the tissue. Where adjoining cells meet end to end, their junctions form structures called *intercalated discs* that are visible under the microscope and are unique to cardiac muscle. Cardiac muscle cells contain myofibrils that are arranged in a pattern similar to that of skeletal muscle and give cardiac muscle cells a similar *striated* appearance. In contrast to skeletal muscle, each cardiac muscle cell has a single nucleus that is centrally located.

Smooth Muscle

Smooth muscle (Figure 4.24) is so named because its cells do not have the striated appearance of skeletal and cardiac muscle cells. Smooth muscle is also called *visceral muscle (vis´-ser-al)* because it is located in the walls of hollow internal structures, such as ducts, blood vessels, and the digestive tract, as well as in numerous other locations. Each cell of smooth muscle is shaped like a long spindle, with each end of the cell tapering to a point. A smooth muscle cell contains a single, centrally located nucleus.

Because contraction of skeletal muscle is under our conscious control, it is considered to be a *voluntary muscle.* In contrast, smooth and cardiac muscles cannot ordinarily be consciously controlled and are referred to as *involuntary muscles.* Skeletal and smooth muscles are considered in more detail in Chapter 8; cardiac muscle is discussed in Chapter 18.

Nervous Tissue

Nervous tissue is composed of **neurons**—highly specialized cells capable of generating and transmitting impulses very rapidly—plus supportive cells, including **neuroglia** and **Schwann cells.** The structure of the neuron is adaptive to its function. Each neuron consists of a cell body with two or more thin cytoplasmic extensions. Because of the anatomical arrangement of nervous tissue, certain of these cytoplasmic extensions transmit impulses toward the cell body of the neuron while others carry impulses away from the cell body—either to another neuron or to a specific structure. Nervous tissue is studied in greater detail with the nervous system in Chapter 10.

◆ **FIGURE 4.23 Cardiac muscle tissue**

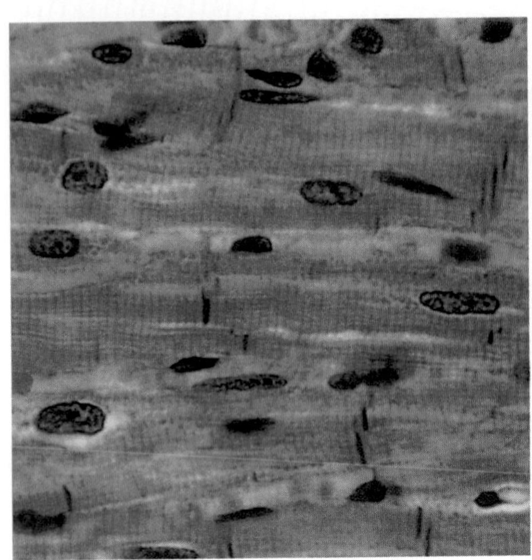

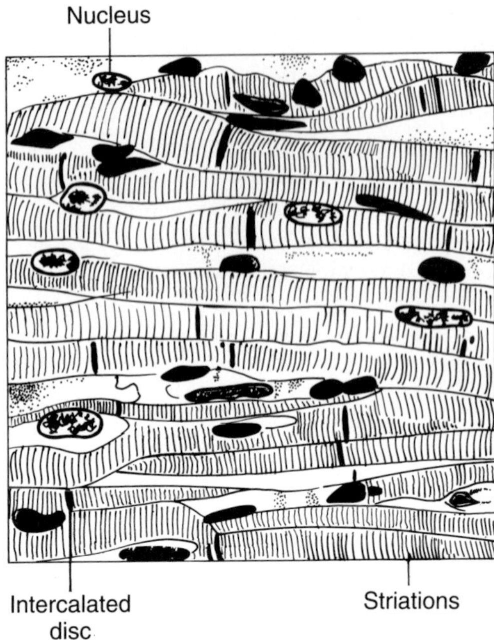

 Tissue Repair

In the embryo, the cells of all tissues are capable of dividing by mitosis, thus enabling the tissues to grow and to repair damage. As the body continues to develop following birth, however, the ability of the cells of certain tissues to divide is greatly reduced or lost completely. Thus, the capability of postembryonic tissues to undergo growth and repair depends on the tissue involved. Nerve cells and skeletal muscle cells generally become mitotically inactive once the tissues have completed their development. In contrast, cells of the epithelial tissues—including those of the skin, digestive

◆ **FIGURE 4.24 Smooth (visceral) muscle tissue**

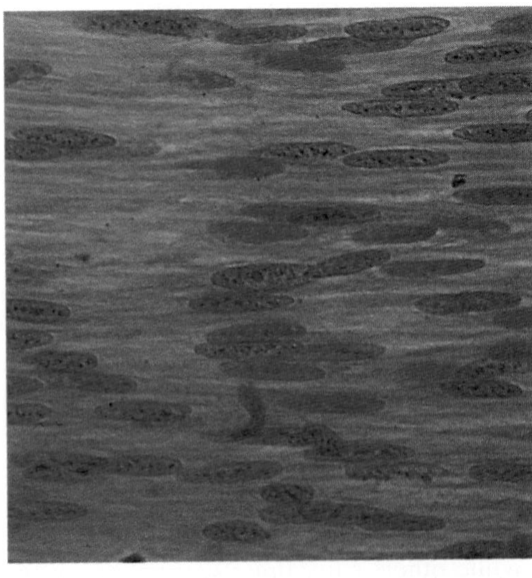

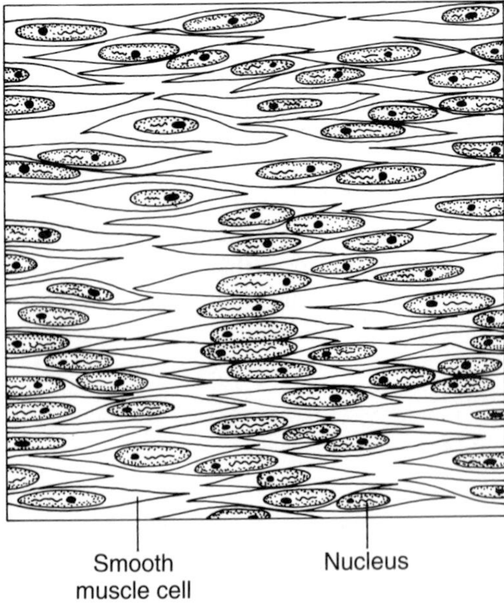

tract, respiratory tract, urogenital tract, and various glands and organs—remain mitotically active and are thus capable of undergoing repair. Fibroblasts also retain the capacity to divide; therefore, like epithelial tissue, connective tissue is able to undergo repair.

Tissue repair is achieved either by **regeneration** of new cells of the same kind as the destroyed cells or by **replacement** of the destroyed cells, no matter what type they were, with fibrous connective tissue. Most tissues contain two groups of cells: (1) connective tissue cells that form the *stroma* (structural framework) of the tissue and (2) the cells that form the *parenchyma* (functional portion) of the tissue.

The success of tissue repair is greatly dependent upon whether the parenchyma undergoes regeneration or replacement. For example, if destroyed parenchymal cells are regenerated, new functional cells are produced, and the repair of the damaged area may be almost indistinguishable from undamaged tissue. Furthermore, the functioning of the tissue may be returned to normal. On the other hand, if destroyed parenchymal cells are replaced with fibrous connective tissue—a process referred to as **fibrosis—scar tissue** is formed, and, because of the permanent loss of parenchymal cells, the functioning of the damaged tissue may never be completely normal.

Study Outline

◆ EPITHELIAL TISSUES pp. 110–118

Occur as coverings for most free surfaces of the body—internal and external; all are avascular.

Specializations of Epithelial Cell Surfaces.

MICROVILLI. Tiny protoplasmic projections that increase the area of the free surface.

CILIA. Motile processes that move rhythmically to propel materials along the epithelial surface.

Specializations for Cell Attachments.

JUNCTIONAL COMPLEXES. Attach adjacent cells in columnar and some cuboidal epithelia. Each junctional complex consists of a *tight junction (zonula occludens)*, an *intermediate junction (zonula adherens)*, and a *desmosome (macula adherens)*.

GAP JUNCTIONS. Hold adjacent cells together in some smooth muscle and nerve tissue as well as in epithelia; form channels that directly link the cytoplasm of adjacent cells.

Classification of Epithelia. Epithelia can be classified according to (1) the number and arrangement of cell layers and (2) the shape of cells on the free surface.

ACCORDING TO CELL LAYERS.

SIMPLE EPITHELIUM. Has a single layer of cells, all contacting basal lamina.

STRATIFIED EPITHELIUM. Has two or more layers, with only deepest layer contacting basal lamina.

PSEUDOSTRATIFIED EPITHELIUM. Appears multilayered, but is actually only a single layer with all cells touching basal lamina.

ACCORDING TO CELL SHAPE.

SQUAMOUS EPITHELIUM. Cells are flat and thin.

CUBOIDAL EPITHELIUM. Cells are about as tall as they are wide; appear almost square in vertical section.

COLUMNAR EPITHELIUM. Cells are taller than they are wide; appear rectangular in vertical section.

GENERAL CLASSIFICATION. Based both on the shape of cells that form free surface and on the number of layers.

SIMPLE SQUAMOUS EPITHELIUM. A single layer of squamous cells; found in regions where diffusion and filtration occur; includes endothelium that lines the blood vessels and mesothelium that lines the body cavities.

STRATIFIED SQUAMOUS EPITHELIUM. Consists of multiple layers; forms protective layer on body surface (epidermis) and other sites of abrasion.

TRANSITIONAL EPITHELIUM. Specialized stratified tissue lining urinary bladder and certain other hollow organs; permits expansion.

SIMPLE CUBOIDAL EPITHELIUM. Generally a single layer of cuboidal cells; occurs in many glands, such as the salivary glands.

SIMPLE COLUMNAR EPITHELIUM. Single layer of columnar cells; capable of performing complex reactions, such as secretion and absorption; lines digestive tube.

STRATIFIED COLUMNAR EPITHELIUM. Truly stratified columnar tissue; found on epiglottis and parts of pharynx, anus, and urethra.

PSEUDOSTRATIFIED COLUMNAR EPITHELIUM. Columnar tissue in which all cells contact basal lamina but not all reach the free surface; found in large ducts of some glands, such as parotid; is often ciliated, as in respiratory tract.

EPITHELIAL MEMBRANES. Composed of a surface layer of epithelial cells underlain with loose connective tissue.

MUCOUS MEMBRANES. Absorptive and secretory tissue that lines digestive, respiratory, urinary, and reproductive tracts; surface layer of stratified squamous or simple columnar epithelium.

SEROUS MEMBRANES. Line central body cavities; surface layer of simple squamous epithelium called mesothelium.

GLANDULAR EPITHELIUM. Consists of cells that secrete various substances or excrete wastes such as bile pigments.

GLAND TYPES.

1. *Merocrine glands.* Produce secretions that do not accumulate in the gland cell; secrete with no glandular destruction; most common type.
2. *Holocrine glands.* Accumulate secretions in their cells, discharging the secretions only when cells rupture and die.
3. *Apocrine glands.* Accumulate secretions in outer ends of gland cells; released when end of cell pinches off.

◆ **CONNECTIVE TISSUES** pp. 118–126

Vary considerably in form and function; all have abundant intercellular matrix; serve as internal framework of organs; bind tissues and organs together for support; some provide media through which nutrients and wastes pass between blood and body cells; serve as food storage sites; form rigid skeletal framework of body; contain fixed and free macrophages which serve as phagocytes and comprise the macrophage system.

Intercellular Matrix. Formed of ground substance and fibers.

GROUND SUBSTANCE. Varies from fluid to gel; produced by connective tissue cells; composed of tissue fluid and proteoglycans.

FIBROBLASTS. Form various fibers characteristic of connective tissues.

COLLAGENOUS FIBERS. Inelastic fibers composed of bundles of strong fibrils.

ELASTIC FIBERS. Long, threadlike, branching fibers that often form interwoven networks.

RETICULAR FIBERS. Short, thin fibers that branch freely, forming tight, inelastic networks.

Types of Connective Tissue.

LOOSE CONNECTIVE TISSUE. Has unorganized fiber arrangement; is most widespread type. Used to attach skin to underlying tissue; fill spaces between organs, holding them in place; surround and support blood vessels.

ADIPOSE TISSUE. Consists of fat cells dispersed in loose connective tissue; serves as storage site for fats and also pads certain body regions.

RETICULAR CONNECTIVE TISSUE. Resembles loose connective tissue but contains primarily reticular fibers; forms internal framework of many organs.

DENSE IRREGULAR CONNECTIVE TISSUE. Contains more randomly oriented collagenous fibers than loose tissue does; forms dermis layer of skin.

DENSE REGULAR CONNECTIVE TISSUE. Very strong because of numerous parallel collagenous fibers in tightly packed bundles; forms tendons of muscles, ligaments of joints.

ELASTIC CONNECTIVE TISSUE. Allows some stretching; found, for instance, in walls of arteries and trachea.

CARTILAGE. Has firm avascular matrix of various proteoglycans, chondroitin sulfate, and hyaluronic acid; functions as structural support but is somewhat flexible; divided into three types:

HYALINE CARTILAGE. Most abundant type; contains tightly packed collagenous fibers; provides strong support, as in costal cartilage and rings of trachea.

ELASTIC CARTILAGE. Contains elastic fibers; furnishes firm but elastic support, as in outer ear and epiglottis.

FIBROCARTILAGE. Tough and slightly compressible; found in intervertebral discs of vertebral column and articular discs of knee joint.

BONE. Has great strength and rigidity provided by inorganic salts; osteoblasts located in lacunae; forms major part of adult skeleton.

BLOOD. The matrix that surrounds blood cells is fluid plasma.

◆ **MUSCLE TISSUE** pp. 126–127

Living cells of muscle tissue are called fibers.

Skeletal Muscle. Attached to skeleton and moves it.

Cardiac Muscle. Forms walls of heart.

Smooth Muscle. Located in walls of hollow internal structures (organs, ducts, and blood vessels).

◆ **NERVOUS TISSUE** p. 127

Composed of neurons—highly specialized cells capable of generating and transmitting impulses; also contains supportive cells called neuroglia and Schwann cells.

◆ **TISSUE REPAIR** pp. 128–129

Embryonic cells of all tissues are capable of mitosis. Some postembryonic tissues are also able to grow and repair damage. With continued development, cells of certain tissues are rendered incapable of mitosis, or mitotic activity is greatly reduced. Tissue repair is achieved either by regeneration of destroyed cells or their replacement by connective tissue cells.

Self-Quiz

1. In which one of the following are the cell membranes of adjacent cells closest together? (a) desmosomes; (b) zonula occludens; (c) zonula adherens.
2. Which cell junction seems to allow for the passage of ions and small molecules? (a) gap junction; (b) zonula occludens; (c) zonula adherens.
3. Epithelial tissues are generally classified on the basis of the number and arrangement of cell layers within the tissue, the shape of the cells on the free surface, and/or the location or function of certain epithelia. True or False?
4. Epithelia with two or more layers of cells, with only the deepest layer in contact with the basal lamina, are: (a) stratified epithelia; (b) simple epithelia; (c) pseudostratified epithelia.
5. Match the following terms associated with epithelial cells with the appropriate lettered description:

Squamous cells	(a) Cells taller than they are wide
Stratified epithelium	(b) A single layer of squamous cells
Simple columnar	(c) Cells that are flat and thin
Columnar cells	(d) Cells about as tall as they are wide
Cuboidal cells	(e) A single layer of epithelial cells all in contact with the basal lamina
Simple squamous epithelium	(f) Epithelium with two or more layers of cells, only the deepest layer in contact with the basal lamina

6. Specialized tissue that allows for the expansion of an organ with only minimal resistance from the tissue is composed of: (a) simple cuboidal epithelium; (b) transitional epithelium; (c) stratified squamous epithelium.
7. Which one of these epithelial tissue types has ciliated cells that line the mucous membranes of the respiratory passageways? (a) simple columnar; (b) stratified columnar; (c) pseudostratified columnar.
8. Mucous membranes are particularly well suited to function in digestion and respiration. True or False?
9. Match the following items with the appropriate lettered description.

Mucous membranes	(a) Tissue lining the wall of the heart
Lamina propria	(b) Tissue whose cells secrete serous fluid
Serous membranes	(c) Loose connective tissue that is part of the mucous membranes
Mucus	(d) Absorptive and secretory tissue lining the respiratory tract
Mesothelium	(e) Function as glands and secrete mucus
Endothelium	(f) Tissue lining the ventral body cavity
Goblet cells	(g) Viscous fluid that lubricates mucous membranes

10. These glands accumulate their secretions and release them only when the individual cells that store the secretions rupture and die: (a) merocrine; (b) holocrine; (c) apocrine.
11. Match the following terms associated with connective tissues with the appropriate lettered description.

Fibroblasts	(a) Fiber cells in a resting phase
Macrophages	(b) Fibers that form the stroma of glands
Ground substance	(c) Filamentous structures that occur in the ground substance
Fibrocytes	(d) Homogeneous product of the connective tissue cells that it surrounds
Fibers	(e) Long, threadlike, branching fibers yellow in color
Collagenous	(f) Active phagocytes of loose connective tissue
Elastic	(g) Cells that form various fibers of connective tissues
Reticular	(h) Most common type of fiber; composed of bundles of strong fibrils

12. Loose connective tissue is the most common type and forms the dermis layer of the skin. True or False?
13. This connective tissue forms the tendons of muscles and the ligaments of joints: (a) dense irregular; (b) areolar; (c) dense regular.
14. This connective tissue contains numerous collagenous fibers embedded in a firm ground substance: (a) cartilage; (b) elastic; (c) adipose.
15. The matrix of cartilage is formed by: (a) chondrocytes; (b) osteoblasts; (c) chondroblasts.
16. The cartilage that forms the intervertebral discs is termed: (a) hyaline; (b) elastic; (c) fibrocartilage.
17. Although all the cells of the macrophage system have a similar function (phagocytosis), macrophages in specific tissues or organs are often selective as to what they ingest. True or False?
18. Intercalated discs are found in cardiac, skeletal, and smooth muscle tissues alike. True or False?
19. With continued development, the ability of the cells of certain tissues to divide is greatly reduced or is completely lost. True or False?
20. Which one of the following contains cells that remain mitotically active and thus capable of undergoing repair during a human's life span? (a) muscle tissue; (b) nervous tissue; (c) epithelial tissue.

CHAPTER 5

The Integumentary System

CHAPTER CONTENTS

LEARNING OBJECTIVES

After completing this chapter, you should be able to:

1. Name the four body structures that are a part of the integumentary system.
2. Name the two layers that form the skin, and describe the composition of each.
3. Describe the process by which a cell becomes cornified.
4. Describe the factors responsible for skin color.
5. Name and describe the two layers that compose the dermis.
6. Distinguish between eccrine sweat glands and apocrine sweat glands, and cite one example of the latter.
7. List five functions of the integument.
8. Cite three ways in which the skin plays a major role in homeostasis.
9. Describe the "rule of nines."

CHAPTER 5

In Chapter 4, we discussed the organization of individual cells into tissues. It was noted that when two or more tissues join together, as occurs in the formation of serous membranes and mucous membranes, an organ is formed. This chapter describes another combination of tissues into a simple organ—the **skin.** Although the skin is not often viewed as an organ, it is, in fact, one of the larger organs of the body in terms of surface area (approximately two square yards) and weight. The skin and its accessory structures—hair, nails, and glands—comprise the **integumentary system.**

The skin forms the entire external covering of the body. It is continuous with, but differs structurally from, the mucous membranes lining the external openings of the respiratory, digestive, and urogenital systems (mouth, nose, anus, urethra, and vagina). The skin is composed of two main layers: (1) the *epidermis,* a surface layer of closely packed epithelial cells, and (2) the *dermis,* a deeper layer of dense irregular connective tissue. The dermis is connected to the underlying fascia of the muscles by a layer of loose connective tissue called the *hypodermis.* In many areas, fat is deposited in the loose connective tissue, thus forming adipose tissue. The hypodermis connects the skin and the underlying fascia of the muscles only loosely, thus allowing the muscles to contract without pulling on the skin. In some areas, where muscles do not lie beneath the skin, only a small amount of hypodermis is present, and the integument is more tightly attached. For example, on the shins the skin is connected directly to the membrane (periosteum) that covers the bone.

Epidermis

The outer portion of the skin is called the **epidermis** (Figure 5.1). The epidermis develops from the single layer of surface ectoderm of the embryo. By the time of birth, it consists of several layers of squamous cells that form a stratified squamous epithelium. The epidermis is generally quite thin, not exceeding 0.12 mm over most of the body. However, it is considerably thicker in areas that are subjected to constant pressure or friction, such as the soles of the feet and the palms of the hands. Continued pressure at a particular location causes the

◆ **FIGURE 5.1 Photomicrograph of the epidermis of thick skin**

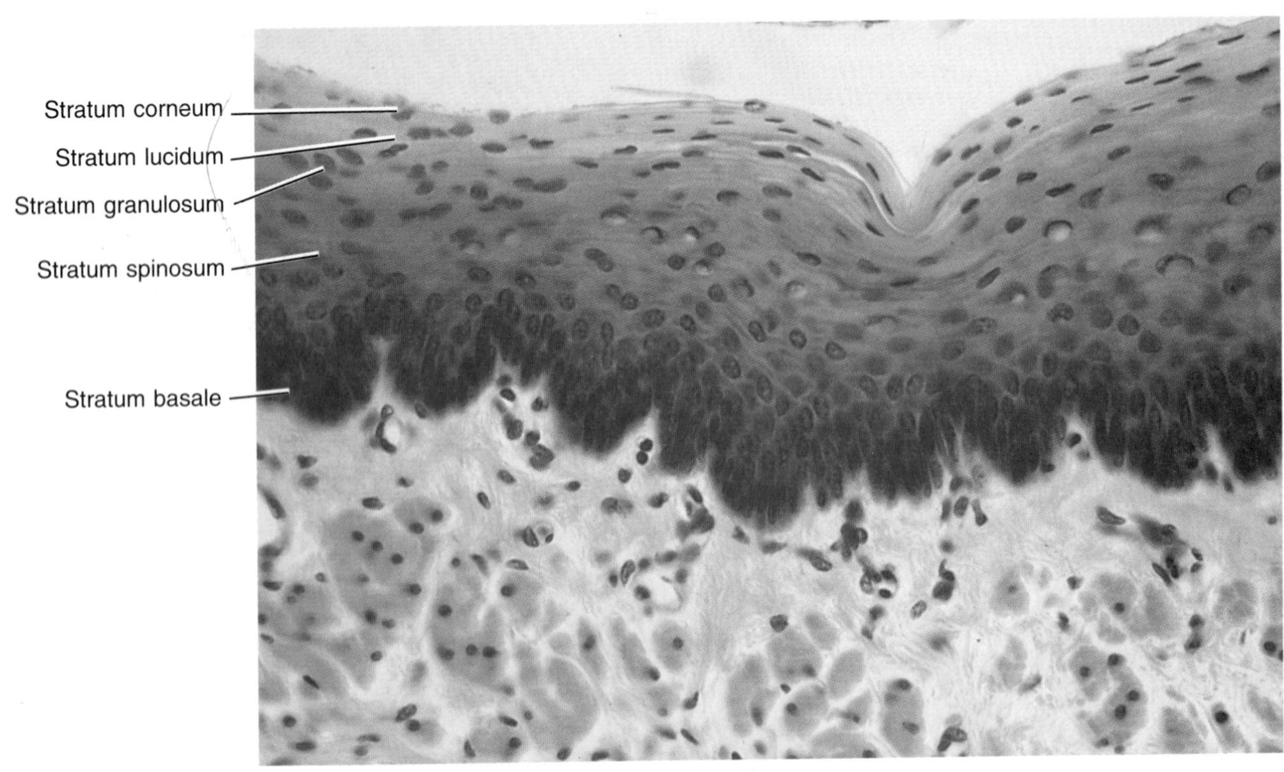

epidermis to thicken, forming calluses and corns. The cells comprising the epidermis only live for about one month; therefore, the epidermis is constantly regenerating itself.

The most common cell type in the epidermis is called the **keratinocyte** because it produces a fibrous protein called **keratin.** The keratinocytes, which are connected to one another by desmosomes, are pushed toward the surface of the skin by the production of new cells in the deeper layers of the epidermis. Keratin accumulates within the cytoplasm, and by the time the keratinocytes have reached the free surface of the skin, they have died and their cytoplasm has been almost completely replaced by keratin. Thus, the superficial layers of the epidermis serve as a thin protective shield composed of keratinized cells.

Epidermal Layers

Where the epidermis is thick, it is possible to identify five layers, or **strata**, of keratinocytes. The innermost layer is the *stratum basale.* This layer is overlaid by the *stratum spinosum,* the *stratum granulosum,* the *stratum lucidum,* and the *stratum corneum*—in that order. In regions where the epidermis is thin, the stratum lucidum is often absent.

Stratum Basale

The **stratum basale** is the deepest layer of the epidermis, lying directly on the dermis (Figures 5.1 and 5.2). It is within this layer that mitosis occurs, furnishing cells

◆ **FIGURE 5.2 Skin structure**

(a) Structure of the epidermis and dermis layers of the skin, and the hypodermis. (b) Vertical section through a hair follicle showing the structure of a hair and the relationship of a sebaceous gland and arrector pili muscle to the follicle.

Hair
Sebaceous gland
Sweat pore
Arrector pili muscle
Duct of sweat gland
Epidermis
Dermis
Hypodermis
Blood vessel
Nerve fiber
Hair follicle
Sweat gland
Adipose tissue
Connective tissue
(a)

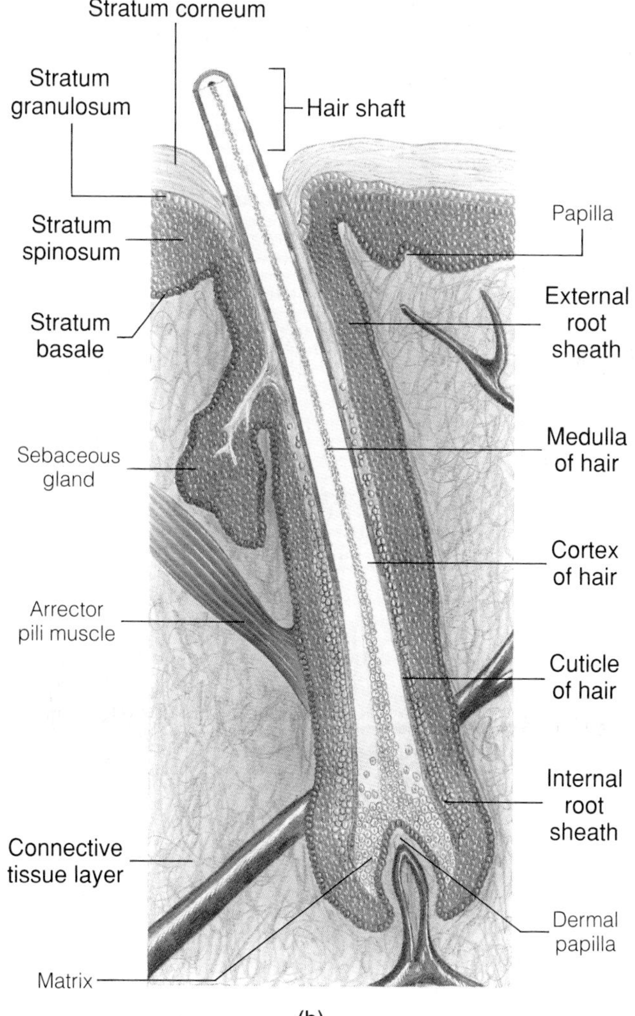

to replace those lost from the more superficial strata of the epidermis. For this reason, this layer was formerly referred to as the stratum germinativum. The cells of the stratum basale are attached to one another by desmosomes and contain bundles of microfibrils called *tonofibrils* in their cytoplasm.

The cells in the deepest layer of the stratum basale, which contacts the dermis, are columnar cells. It is within this deep layer that most mitoses take place. Specialized cells called **melanocytes** are located in the stratum basale. The melanocytes produce a pigment called **melanin,** which provides protection against ultraviolet radiation. A few **Merkel cells** are also distributed throughout the stratum basale. Merkel cells are found close to nerve endings and are thought to be involved in sensory reception.

Stratum Spinosum

Superficial to the basal layer, the cells become somewhat flattened and polyhedral in shape. Under the microscope there appear to be cytoplasmic extensions that connect adjacent cells. Because of these extensions, the layers of cells superficial to the stratum basale are referred to as the **stratum spinosum** (*spine* = projection).

A type of cell called **Langerhans cells,** which are distinct from the keratinocytes, is fairly prevalent in the upper layers of the stratum spinosum. The actual function of Langerhans cells is not certain, but they have been shown to migrate from the skin to lymph nodes in the region. For this reason, they are thought to serve as macrophages and may be involved in the body's immune responses.

Stratum Granulosum

The cells of the **stratum granulosum** (*granulosum* = granular) are flattened and are arranged in about three layers just superficial to the stratum spinosum (Figures 5.1 and 5.2). This stratum derives its name from the presence of granules of **keratohyalin** within the cytoplasm of its cells. As the granules increase in size, the nucleus disintegrates, which results in the outermost cells of the stratum granulosum dying.

Stratum Lucidum

The **stratum lucidum** (*lucid* = clear) is a clear band superficial to the stratum granulosum (Figure 5.1). It consists of several layers of flattened, closely packed cells, most of which have only indistinct outer boundaries and have lost all of their cytoplasmic inclusions except for keratin fibrils and some droplets of a substance called **eleidin.** Eleidin is transformed into keratin as the cells of the stratum lucidum become part of the outer stratum corneum. The stratum lucidum is most prominent in areas of thick skin and is absent in some locations.

Stratum Corneum

The **stratum corneum** (*cornu* = horn) is the most superficial layer of the epidermis. It is formed of varying numbers of flat, closely packed cells (Figures 5.1 and 5.2). These cells are dead, their cytoplasm having been largely replaced by keratin. Such cells are referred to as being *cornified,* or *horny.* The cornified cells form a covering over the entire body surface that not only protects the body against invasion by substances in the external environment, but also helps to restrict the loss of body water. The cells of the stratum corneum are held together by modified desmosomes that include dense extracellular material. The outermost layers of the stratum corneum are constantly being lost as the result of abrasion—caused by, for example, friction with clothing. However, the lost cells are constantly being replaced by cells from the deeper layers of the epidermis. It takes from two weeks to one month for a living cell from the stratum basale to die and become incorporated into the stratum corneum. It has been estimated that all of the cells of the epidermis are replaced about every 28 days. If subjected to repeated episodes of trauma, the stratum corneum thickens and may form *calluses* if the trauma is long lasting.

On the outer surface of the epidermis, a series of ridges and grooves is apparent. On the palms of the hands and the soles of the feet, these **epidermal ridges** form a pattern of loops and whorls that is unique in each person. The epidermal ridges are actually the result of the epidermis conforming to the contours of the underlying dermis. The epidermal ridges increase the friction of the epidermal surface and enhance the gripping ability of the fingers and feet. Because the ducts of sweat glands open on the epidermal ridges, fingerprints that display the pattern of the ridges are left when smooth objects are touched.

Nourishment of the Epidermis

As is typical of all epithelia, there are no blood vessels within the epidermis, although the underlying dermis is well vascularized. As a result of this situation, the only method by which cells of the epidermis can obtain nourishment is by diffusion from the capillary beds of the dermis. This method is sufficient for those cells closest to the dermis, but as cells divide and some are forced toward the body surface—and thus farther away from the source of nourishment—they die. As was indicated earlier, their cytoplasm is gradually replaced with keratin, thus forming the structures typical of the outer layers of the epidermis.

Dermis

Lying deep to the stratum basale is a layer of dense irregular connective tissue called the **dermis** (Figure 5.2). In contrast to the epidermis, the dermis develops from the mesoderm of the embryo, as do the muscles and the skeleton of the body. The dermis contains some elastic and reticular fibers, but is composed primarily of collagenous fibers. It is well supplied with blood vessels, lymph vessels, and nerves. It also contains specialized glands and sense organs. The thickness of the dermis varies in different locations, but it averages about 2 mm. The dermis is composed of two indistinctly separated layers, the *papillary layer* and the *reticular layer*.

Papillary Layer

The outer **papillary layer** *(pa´-pil-lar-ee)* fits closely against the basal layer of the epidermis. This layer is so named because it has many **dermal papillae** (projections) that protrude into the epidermal region. As was indicated in the discussion of epidermal ridges, on the palms and the soles, these papillae are in the form of curving parallel ridges that cause the overlying epidermis to form the characteristic fingerprint and footprint patterns. The patterns of the dermal papillae are determined genetically and thus are unique for each person. Many papillae contain capillary loops, whereas others contain specialized sensory receptors that react to external stimuli such as temperature and pressure changes. These receptors are described in Chapter 10.

Reticular Layer

The deeper **reticular layer** of the dermis consists of dense bundles of collagenous fibers that run in various directions (thus forming a reticulum). Elastic fibers are scattered in networks among the bundles of collagenous fibers; thus, the dermis can stretch as needed and return to its usual form. Stretching of the dermis occurs most dramatically during a substantial weight gain or on the abdomen and breasts during pregnancy. Such extreme stretching may exceed the ability of the elastic fibers to extend, causing small tears in the dermis that remain as white scars called *striae* or "stretch marks." A loss of elasticity by the elastic fibers in the dermis due to age, hormones, excessive ultraviolet radiation, and so on may affect the contractility of the dermis and cause the skin to sag and form wrinkles.

Separations between the bundles of collagenous fibers in the reticular layer form **lines of cleavage** *(tension lines)* in the skin (Figure 5.3). The cleavage lines form a characteristic pattern in each part of the body. An incision at right angles tends to gap more than an incision parallel to the cleavage lines. Therefore, surgeons tend to follow the pattern of the cleavage lines when cutting through the skin in order to enhance healing and reduce scar tissue formation.

The fibers of the reticular layer of the dermis are continuous with the fibers of the underlying hypodermis.

Hypodermis

The **hypodermis** (*hypo* = beneath) is not a part of the skin, but it is important because it attaches the skin to the underlying structures. As was mentioned earlier, the hypodermis is composed of loose connective tissue,

◆ **FIGURE 5.3 Cleavage lines of the skin**

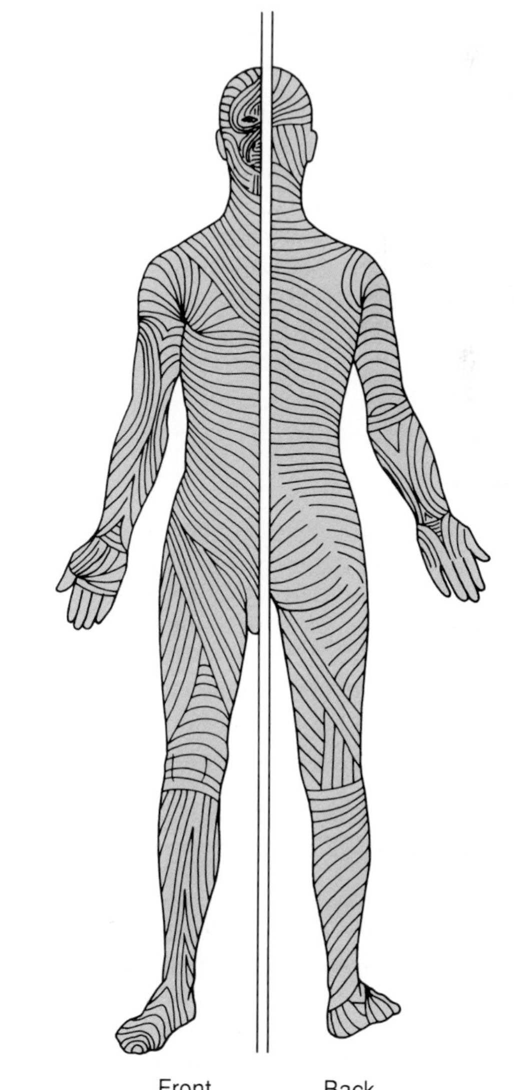

ASPECTS OF EXERCISE PHYSIOLOGY

What the Scales Don't Tell You

Body composition refers to the percentage of body weight that is composed of lean tissue and of adipose tissue. The assessment of body composition is an important component in evaluating the health status of individuals. The age-height-weight tables used by insurance companies can be misleading for determining healthy body weight. Many athletes, for example, would be considered overweight by these charts. A football player may be 6′5″ tall and weigh 300 lbs, but have only 12% body fat. This player's extra weight is muscle, not fat, and therefore is not a detriment to his health. A sedentary person, on the other hand, may be normal on the height-weight charts but have 30% fat. This person should maintain body weight while increasing muscle mass and decreasing fat. Ideally, men should have 15% fat or less, and women should have 20% fat or less.

The most accurate method for assessing body composition is underwater weighing. This technique is based on the fact that lean tissue is denser than water and fat tissue is less dense than water. You can readily demonstrate this for yourself by dropping a piece of lean meat and a piece of fat in a glass of water; the lean meat will sink and the fat will float.

The most common method of underwater weighing requires the person to expel as much air as possible from his or her lungs and then completely submerge in a tank of water while sitting in a swing that is attached to a scale. Because a person learns to expel more air from the lungs with a little practice, this procedure is repeated 10 to 12 times. When the results from trial to trial become consistent, the last few trials are averaged and used to determine body density by means of equations that take into consideration the density of water, the difference of the person's weight in air and underwater, and the volume of air remaining in the lungs. Because of the difference in density between lean and fat tissue, people who have more fat have a lower density and weigh relatively less underwater than in air compared to their lean counterparts. Percentage fat is determined by an equation developed by Dr. William Siri:

$$\text{Percentage fat} = 495/\text{Body density} - 450$$

Another common way to assess body composition if laboratory facilities for underwater weighing are not available is skinfold thickness. Because approximately half of the body's total fat content is located just beneath the skin, total body fat can be estimated from measurements of skinfold thickness taken at various sites on the body. Skinfold thickness is determined by pinching up a fold of skin at one of the designated sites and measuring its thickness by means of a caliper, a hinged instrument that fits over the fold and is calibrated to measure thickness. Mathematical equations specific for the person's age and sex can be used to predict the percentage of fat from the skinfold-thickness scores. If the investigator is skilled and a valid equation is used to predict percentage of fat, skinfold determinations are a valuable tool for assessing the body composition of large numbers of people. A major criticism of skinfold assessments is that accuracy depends on the investigator's skill.

Exercise physiologists often assess body composition as an aid in prescribing and evaluating exercise programs. Exercise generally reduces the percentage of body fat and, by increasing muscle mass, increases the percentage of lean tissue.

often having fat cells deposited among its fibers. This tissue is also referred to as *subcutaneous tissue* or the *superficial fascia.* In some regions, such as over the abdomen and the buttocks, the accumulation of fat within the subcutaneous tissue can become quite extensive. The hypodermis is well supplied by blood vessels and nerve endings.

Skin Color

Skin color is determined by the presence and distribution of two pigments in the skin and the hemoglobin in the blood capillaries of the dermis. One pigment, **melanin,** ranges in color from yellow to dark brown or black. Melanin is produced by cells called *melanocytes,* which migrate into the epidermis and transfer the pigment to cells of the stratum basale and the underlying dermis. There are no great differences in the number of melanocytes found in the skin of the various human races, but the melanocytes of dark-skinned people produce copious amounts of darker melanin that is distributed in all layers of the epidermis. In light-skinned people, relatively little melanin is distributed among the layers of the epidermis except in heavily pigmented areas such as the nipples of the breasts. Exposure of the skin to sunlight increases the activity of melanocytes, causing them to produce more melanin. This darkens the skin, thereby protecting its deeper cells from excessive ultraviolet radiation.

Carotene is another pigment that affects skin color. It is a yellowish pigment found in the stratum corneum and in the fatty layer beneath the skin. In combination with melanin, carotene produces the yellowish hue that is typical of Oriental people.

Skin color is also influenced by a reddish hue that results from the blood vessels of the dermis being visible through the epidermis. Whereas the amount of pigment in the skin is relatively constant for each person, a change in the amount of blood in the capillaries of the dermis—or in the amount of oxygen carried within the blood of these capillaries—can cause intense temporary changes in skin color. For example, *blushing* is caused by an expansion of the capillaries, whereas the bluish skin characteristic of *cyanosis* results from a decrease in the amount of oxygenated hemoglobin in the blood within these capillaries.

Glands of the Skin

Two types of glands have a widespread distribution in the skin: the sweat glands and the sebaceous glands. In addition, the ceruminous (wax) glands of the external ear canal, the ciliary and meibomian glands of the upper eyelids, and the mammary glands are specialized skin glands. The skin glands begin their embryonic development as solid downgrowths from the ectoderm. The downgrowing cords become hollow tubes as they continue to develop and extend into the dermis, forming the skin glands and their associated ducts.

Sweat Glands

The **sweat glands,** which are also called **sudoriferous glands** (*sudor* = sweat), are distributed over most of the body surface (Figure 5.2a). In only a few places, such as the lips, the nipples, and portions of the skin of the genital organs, are they absent. Typical sweat glands—the *eccrine sweat glands*—are merocrine glands, each in the form of a simple tubule that becomes coiled within the dermis. Stimulation of the sympathetic nerves to these glands causes them to secrete a watery solution of sodium chloride, with traces of urea, sulfates, and phosphates. The amount of sweat secreted depends on such factors as environmental temperature and humidity, amount of muscular activity, and various conditions that cause stress.

Sweat glands that are located in the axilla, around the anus, on the scrotum, and on the labia majora of the female external genitalia are unusually large and extend into the subcutaneous tissue. Glands in these locations often empty into a hair follicle rather than directly onto the surface of the skin. These large glands are *apocrine sweat glands*—that is, part of the cytoplasm of the secreting cells is included within the secretion, which is thicker and more complex than true sweat. Although the secretion of apocrine sweat glands is itself odorless, bacteria that thrive in the rather protected regions where the glands are located can cause odors. The bacteria act on the secretions of the sweat glands and produce by-products that contribute to "body odor." In the female, apocrine glands periodically become enlarged and hyperactive in conjunction with the menstrual cycle. The **ceruminous glands,** which produce "wax" (**cerumen**; *se-roo´-men*) in the ear canal, are also apocrine glands that are considered to be modified sweat glands.

Sebaceous Glands

Most **sebaceous glands** *(se-bay´-shus)* develop from, and empty their secretions into, hair follicles (Figure 5.2a). Their secretion (**sebum**), an oily substance rich in lipids, travels along the shaft of the hair to the surface of the skin. Sebum not only serves to oil the skin and the hair, preventing them from drying, but also contains substances that are toxic to certain bacteria. The sebaceous glands, which are known to be stimulated by the presence of sex hormones (mainly testosterone), are particularly active during adolescence. If their secretion accumulates within the duct of the gland, it forms a white pimple. This blocked sebum may become oxidized, darken, and form a "blackhead." Most hairless regions of the body, such as the palms of the hands and the soles of the feet, lack sebaceous glands. However, some areas that lack hair, such as the lips, the glans penis, and the labia minora, do have sebaceous glands. In these regions, the glands empty their secretions directly onto the surface of the epidermis.

Structurally, sebaceous glands are typically of the simple alveolar type, although some are compound alveolar (for instance, the *meibomian glands* of the upper eyelids). Functionally, all sebaceous glands are holocrine glands.

Hair

Although **hair** is most obvious on the head and in the axillary and pubic regions, it is also present—though much less conspicuous—over most of the body. The only hairless skin is on the lips, the palms of the hands, the soles of the feet, the nipples, and parts of the external genitalia. Hair grows as the result of the mitotic activity of epidermal cells at the bottom of a **hair follicle.** The follicles extend from the epidermis into the dermis (Figure 5.2b). The outermost layer of the follicle, the **external root sheath,** is a downgrowth of the epidermis. From the bottom of the follicle up to the level of the sebaceous glands, the follicles are lined by the **internal root sheath,** which consists of several layers of keratinized cells. Covering the follicle is an outer layer of connective tissue that develops from the dermis. A portion of the dermis protrudes into the bottom of each follicle, forming a **papilla.** Papillae contain blood capillaries that nourish the overlying follicle cells and permit them to undergo repeated mitosis. Emptying into each hair follicle is one or more sebaceous glands, whose secretions help to soften the hair.

Each hair is essentially a column of keratinized cells. The mitotically active cells that cover the papilla are called the **matrix.** The **shaft** of the hair develops from the cells of the matrix; the free end of the shaft extends beyond the surface of the skin (Figure 5.4). The part of the hair that is below the surface of the skin, within the hair follicle, is the **root.** The **medulla,** the central core of the hair shaft, consists of loosely connected horny cells with air spaces between them. The **cortex,** which surrounds the medulla, is formed of tightly compressed keratinized cells. Outside the cortex is a **cuticle** of very hard keratinized cells. Straight hairs are cylindrical or oval; curly hairs are somewhat flattened.

Hair follicles exhibit cyclic activity, having active periods alternating with periods of inactivity. During the active periods, cells in the matrix of a follicle undergo mitosis, pushing the older cells upward and causing the hair to elongate. The cells die and become keratinized as they are pushed farther from the nourishment provided by the blood vessels in the papillae. The keratinized cells are incorporated into the hair. During inactive periods, when the matrix cells are not undergoing mitosis, the shaft of a hair becomes detached from the matrix, and the hair gradually moves up the follicle. The detached hair may be pulled from the follicle by brushing or combing, or it may remain within the follicle until the next active period, when the new hair produced by matrix cells pushes it out.

Hair follicles in different parts of the body follow different patterns of cyclic activity. For instance, in the scalp, individual follicles can remain active and cause their hairs to elongate continuously for several years before becoming inactive for a period of months. In other

◆ **FIGURE 5.4 Scanning electron micrograph of hair shafts extending from hair follicles**

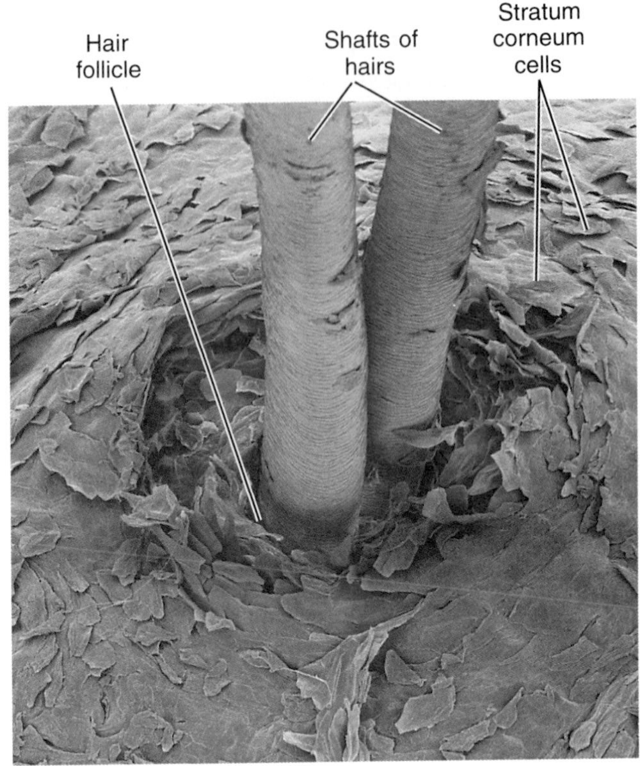

regions of the body, the follicles may be active for only a few months before entering an inactive phase. Cutting or shaving hair does not affect the cyclic activity of the follicles, and therefore has no effect on the growth of hair.

Because of the repeated formation of new hairs during the active periods, normal hair loss does not generally lead to baldness. Baldness is a genetic trait that requires the presence of the male hormones—the androgens—for the hereditary tendency to be expressed. For this reason, baldness is more common in males than in females, who may have inherited the trait but lack the androgens necessary to activate it. A further indication of the relationship between male hormones and baldness is the observation that hair follicles stop growing hair because of the increased presence of a simple protein molecule within the follicle cells. The protein molecule carries the male hormone testosterone into the nuclei of the hair follicle cells, causing the disruption of the hair growth process. Hair follicle cells that are vigorously growing hairs contain only small amounts of this simple protein molecule, and thus the amount of testosterone entering their nuclei is limited.

The color of hair is determined primarily by the pigment melanin. Melanin, which ranges from black to brown to yellow, is formed by melanocytes in the follicle and becomes located in the cortex and medulla of each hair. As people age, their hair gradually tends to become gray. This process is due to a decrease in the amount of pigments present, possibly as a result of a decline in the level of a specific enzyme necessary for the production of melanin. In the complete absence of pigments, the hair appears white.

The hair follicles are generally at an oblique angle to the surface of the skin, as are the hairs themselves. Running diagonally from the connective tissue covering of each follicle to the papillary layer of the dermis is a smooth muscle: the **arrector pili** (*ar-rek´-tor pih´-lee;* Figure 5.2). Contraction of this muscle pulls the follicle and causes the hair to "stand up"—that is, to be perpendicular to the skin surface—and causes the skin to bulge in front of the follicle, producing the "goose pimples" that form in response to cold or to frightening situations. In animals whose bodies are heavily covered with hairs, the erection of hairs traps air between them and the body surface, thus producing an insulating effect and reducing the loss of body heat. This response is probably of little importance in humans, on whom body hair is generally quite sparse.

Nails

On the dorsal surfaces of the distal portions of the fingers and toes, the outer two epidermal layers—the strata corneum and lucidum—are heavily cornified, forming the **nails** (Figure 5.5). The **nail bed,** upon which the nail rests, is formed by the stratum basale and the stratum spinosum. These strata are thickened under the proximal end of the nail, forming a whitish area called the **lunula** (*luna* = moon), which is visible through the nail. The region of thickened strata is called the **nail matrix.** It is within the nail matrix that mitosis occurs, pushing forward the previously formed cells that have cornified and thus causing the nail to grow. At the proximal end of the nail, a narrow fold of epidermis extends onto the free surface, forming the **eponychium** *(ep-o-neech´-ee-um)* or cuticle. Under the free edge of the nail, the stratum corneum is thickened and is called the **hyponychium** *(high-po-neech´-ee-um).* Nails generally have a pink coloration because the capillary network beneath them is visible through the cornified cells.

◆ **FIGURE 5.5 Structure of a nail**

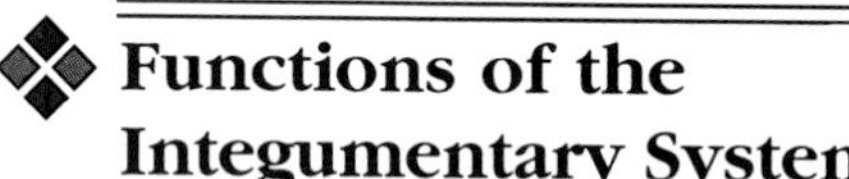

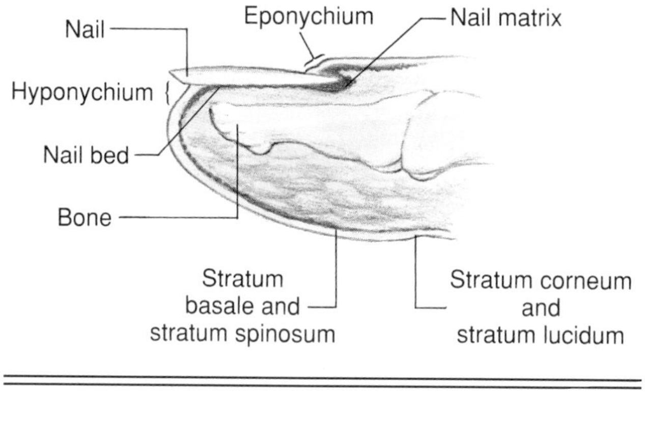

Functions of the Integumentary System

Having an understanding of the structure of the skin and its associated organs should make it easier to understand the various functions the skin performs. These functions can be grouped into several categories: protection, regulation of body temperature, excretion, sensation, and production of vitamin D.

Protection

The skin serves as a barrier that prevents microorganisms and other substances from invading the body. The multiple layers of keratinized cells in the epidermis form a physical barrier that blocks the diffusion of water and water-soluble substances either into or out of the body. The barrier is not absolute, however, and some substances—such as those that are soluble in lipids—are able to pass through the skin. This selective permeability makes it possible to give a number of medications in the form of Band-Aid-like patches applied to the skin rather than orally or by injection. The medication passes continuously through the skin from the patch in such minute quantities that even some drugs that would be toxic when given by other means can be administered by skin patches. Drugs commonly delivered by skin patches include female hormones, scopalamine for motion sickness, and nitroglycerin for heart patients. A barrier of another type, a pigment barrier, is formed by melanin. Because of this pigment barrier, the cells in the deeper layers of the skin are protected against ultraviolet damage.

Another form of protection is provided by the acidity of the secretions of the skin glands. These secretions cause the surface of the skin to be coated with a thin liquid film that tends to be acidic (pH 4 to pH 6.9). This acid film acts as an antiseptic layer and retards the growth of microorganisms that are always present in large numbers on the surface of the skin.

Yet another form of protection afforded by the skin results from the presence of phagocytic cells called *macrophages* in the dermis and macrophage-like cells called *Langerhans cells* in the epidermis. Both these specialized cells are thought to activate the immune system against foreign substances. The precise role that the skin plays in immune activity is not yet clear, but recent evidence suggests that the skin is an integral component of the immune system. For instance, it has been shown that a very large dose of ultraviolet radiation to the skin can affect the immune capacity of the spleen. In addition, Langerhans cells send signals that cause immune cells called helper T-cells to move to the area and assist them in destroying bacteria. It also appears that keratinocytes, which are best known for producing keratin, also produce a hormone that enhances the growth and development of T-cells. Thus, the protective value of the skin is a combination of its ability to provide a physical barrier against invasion from the external environment and its ability to activate the immune system when the physical barrier is breached.

Body Temperature Regulation

Even under conditions of high environmental temperature or during exercise, the body temperature remains almost normal, in part because considerble heat is lost through the skin. As body temperature begins to rise, the arterioles in the dermis dilate, bringing a greater volume of blood to the body surface and thus allowing more internal heat to be lost to the environment. At the same time, the body surface may become wet because of increased secretory activity by the sweat glands. The evaporation of this sweat further facilitates the loss of body heat. In a similar manner, under cold conditions, body heat can be conserved by the constriction of the dermal arterioles. This constriction reduces the amount of blood that flows to the body surfaces so that less heat will be lost to the external environment. The role of the skin in regulating body temperature is discussed further in Chapter 25.

Excretion

In addition to its cooling effect, the secretion of sweat functions, to a limited extent, as a means of excretion. Small amounts of nitrogenous waste products and sodium chloride leave the body via the sweat. Both the volume and the composition of sweat vary according to the changing needs of the body.

Sensation

Because of the presence within it of nerve endings and specialized receptors (see page 346), the skin provides the body with much information concerning the external environment. Events such as temperature change, light touch, pressure, or painful trauma all stimulate integumentary receptors. These receptors, in turn, alert the central nervous system to the particular event, thus enabling appropriate action to be taken. This action might be simple and automatic, such as withdrawing the hand from a harmful situation; or it might require a more complicated act, such as deciding that a warmer coat should be worn.

Vitamin D Production

The skin is also involved in the production of vitamin D. In the presence of sunlight or ultraviolet radiation, one of the sterols (7-dehydrocholesterol) found within the skin is altered in such a way that it forms vitamin D_3 (cholecalciferol). After being metabolically transformed, vitamin D_3 assists in the absorption of calcium and phosphate from ingested food. Vitamin D_3, therefore, is important in maintaining the calcium and phosphate levels of the body at optimum levels, thus facilitating the normal growth of bones and their repair following a fracture.

Because of its role in such activities as protection of the deeper-lying structures of the body, regulation of body temperature, and prevention of excessive loss of water to the environment, the skin plays a major role in maintaining the internal homeostasis of the body and in ensuring the continued normal activity of individual cells.

CONDITIONS OF CLINICAL SIGNIFICANCE

The Integumentary System

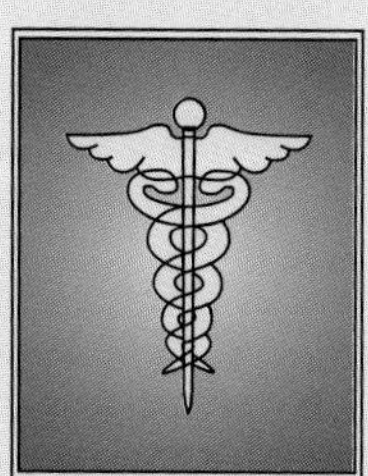

The importance of the skin in preventing the invasion of the body by microorganisms is apparent from the abundance of such organisms that are normally present, even on healthy skin, and yet cause no bodily harm unless the epidermis is damaged, thus allowing them to enter the body. The secretions of the sweat glands and the sebaceous glands provide ample nutrients as well as a favorable environment in which these microorganisms can thrive. The **fungi** that cause *athlete's foot* are often present on the soles of the feet and between the toes without causing harm. Then, due to some change in their environment, the fungi rapidly proliferate and cause the disease condition. Some **yeasts** also live harmlessly on healthy skin. By far the most abundant microorganisms on the skin are **bacteria.** These include the rod-shaped forms and the spherical cocci. Most cocci are harmless, but one species, *Staphylococcus aureus,* can cause pimples, boils, and other more serious infections. However, even these powerful disease-producing microorganisms, which are normally present on certain areas of the skin, do not cause skin diseases unless the epidermis is penetrated.

The number of bacteria that are present on the skin varies in different regions of the body as well as from person to person. The largest bacterial populations are found on the face and neck and in the axillae and the groin. Reported population densities range from 2.41 million bacteria per square centimeter of epidermis in the male axillae to 314 bacteria per square centimeter on the back.

There are many recognized diseases of the integumentary system, but we will consider only a few of the more common pathologies.

Acne

Acne is an inflammatory disease caused by a rod-shaped bacterium *(Corynebacterium acnes)*. These microorganisms provoke excessive secretion by the sebaceous glands, which, in turn, causes the formation of pimples, blackheads, and dandruff. Acne is most prevalent during puberty because of the hormonal changes that occur during that period. After several years, the skin usually becomes adapted to the higher levels of sex hormones, and the condition disappears.

Warts

The common *wart* is the result of a viral invasion of the skin. This condition is most common in adolescents and young adults. Warts are often found in groups because they are capable of spreading to adjacent areas. Warts that occur on the sole of the foot—*plantar warts*—are particularly painful because they are almost constantly subjected to pressure.

Dermatitis and Eczema

Dermatitis and *eczema* are general terms that refer to many inflammatory skin conditions. Also included within this category are nonspecific allergic responses of the skin to many different substances.

Psoriasis

Psoriasis is a fairly common condition that is characterized by small reddish brown elevations and patches that are covered by layers of silvery scales. When the patches are scraped away, bleeding occurs from minute points that correspond to the tops of the papillae of the dermis. Tiny abscesses form under the stratum corneum, producing an exudate.

Psoriasis is caused by an abnormally high rate of mitosis in epidermal cells. Attacks may be triggered by infection, hormonal changes, trauma, and stress.

Impetigo

Impetigo is a highly contagious skin infection that is most common in children. It results from the invasion of the epidermis by various strains of *Staphylococci* and *Streptococci* bacteria. Pus-filled sacs *(pustules)* form beneath the stratum corneum, causing inflammation and swelling. The pustules rupture and form a crust.

continued on next page

CONDITIONS OF CLINICAL SIGNIFICANCE

Moles

Moles, which are elevations of the skin that are generally pigmented, are very common. Almost everyone has at least one mole, and the average person has about 20 on various locations of the body. Moles are considered to be congenital, although they often do not appear until adulthood. It has been suggested that their eruption may be stimulated by steroid hormones. Most moles are *benign*—that is, they do not develop into tumors. They grow slowly over a period of time, remain stable for a long period, and then gradually diminish in size (atrophy). A few, however, may become *malignant* or capable of spreading to other parts of the body. This change is indicated by an increase in size and pigmentation, a reddening of the skin around the mole, and itching.

Herpes Simplex

Herpes simplex infection is commonly called a *fever blister* or a *cold sore.* It occurs when a particular virus, having been dormant in a spinal nerve, travels along the processes of the nerve cells and becomes active on the skin and mucous membranes. The active virus causes clusters of watery blisters to form. The blisters generally occur on the lips or the external genitalia. They are often associated with any disease that causes an elevated body temperature.

Shingles

Like a cold sore, *shingles* results from the invasion of the body by a virus that at first remains dormant in the spinal nerves, generally in the thoracic region. Once the virus becomes active, it affects the sensory nerves of that region, causing an aching pain that follows the paths of the nerves. Groups of small vesicles develop in the skin that overlies the nerve paths. The same virus *(Herpes zoster)* that causes the vesicles of shingles is also responsible for the skin vesicles of chicken pox.

Cancers

Numerous types of tumors arise within the skin. Some originate in the various layers of the epidermis, some in the dermis, and others in the sweat glands and the sebaceous glands. Most of these tumors are benign and do not spread to other parts of the body. Warts are an example of benign tumors. Other tumors are malignant and have the capability of spreading *(metastasizing)* to other regions of the body. These latter tumors are generally called *cancers.*

The three most common forms of skin cancer are basal cell carcinoma, squamous cell carcinoma, and malignant melanoma. *Basal cell carcinomas* account for over 75% of all skin cancers, but are the least likely to metastasize widely. In this condition, cells of the stratum basale proliferate, producing growths that may reach the surface of the epidermis, where they often form a crust. If not removed surgically, the growths may invade and destroy the underlying dermis and hypodermis. *Squamous cell carcinomas* begin their development in the stratum spinosum. The tumors are often in the form of reddish nodules—somewhat similar in appearance to warts. Squamous cell carcinomas often metastasize. *Malignant melanomas* are relatively rare, but frequently metastasize widely. They develop in melanocytes, often within a preexisting epidermal mole. However, melanomas can develop anywhere there are melanocytes, and any brown or black patch that develops on the skin should be examined by a physician.

The cause of most skin tumors is not known. However, prolonged overexposure to the ultraviolet rays of sunlight appears to be directly related to the development of many of them. There is a greater incidence of skin tumors in farmers and others whose occupations require that they work outdoors over a period of years. A higher incidence is also noted in the southern United States compared to the northern regions. Skin cancers are seldom found in dark races, where the skin is heavily pigmented.

Burns

While burns cannot be considered to be pathological conditions of the skin, they disrupt the homeostasis of the body so drastically that we consider them here.

The seriousness of burns results from the destruction of the skin and clearly demonstrates the skin's importance to the other body systems. When the skin is destroyed, there is a large loss of body water (**tissue fluid**) and blood plasma. Plasma proteins and

CONDITIONS OF CLINICAL SIGNIFICANCE

mineral salts leave the body with these fluids. The loss of plasma proteins upsets the osmotic equilibrium of the body, and the loss of salts produces an electrolyte imbalance (discussed in Chapter 27). The results are dehydration, kidney malfunction, and shock. In addition, with the protection of the skin gone, it is very easy for infectious agents to invade the body.

Burns are classified according to their severity. In *first-degree burns,* only the epidermal layers of the skin are damaged. Symptoms include localized pain, redness, and swelling. Sunburn is usually a first-degree burn. In *second-degree burns,* both the epidermis and the dermis are damaged. However, the damage is not severe enough to prevent the skin from regenerating quickly. In *third-degree burns,* both the epidermis and the dermis are so severely damaged that they can regenerate only from the edges of the wound. If the burned area is extensive, this regeneration can be a slow process, during which body fluids are constantly being lost from the damaged area and the possibility of infection is high. In addition, such wounds can result in extensive scar tissue formation, which is not only disfiguring but can also restrict the movement of the damaged part. To hasten healing (and thus reduce the loss of body fluids) and to minimize scar formation, large burn areas are often covered with *skin grafts* taken from other regions of the body (Figure C5.1).

In one method of skin grafting—called *split skin grafting*—skin taken from one part of a person's body is used to cover another part. Split skin grafting is useful because it eventually increases the total amount of skin on the body surface. With this method of skin grafting, the outer portion of the skin (the epidermis and perhaps half the dermis) is removed from an undamaged area and placed over a region where serious skin damage has occurred (Figure C5.2a). The cells of the graft are nourished by interstitial fluid from the damaged surface. Gradually, connective tissue cells form new intercellular substance that attaches the graft in place, and eventually the graft becomes vascularized.

The exposed, nonepidermal surface in the region from which the graft was taken becomes covered with new epidermis, which grows over the surface from the remaining hair follicles and the ducts of sweat glands (which originate embryonically from the same tissue that gives rise to the epidermis) (Figure C5.2b).

◆ **FIGURE C5.1 Skin graft**

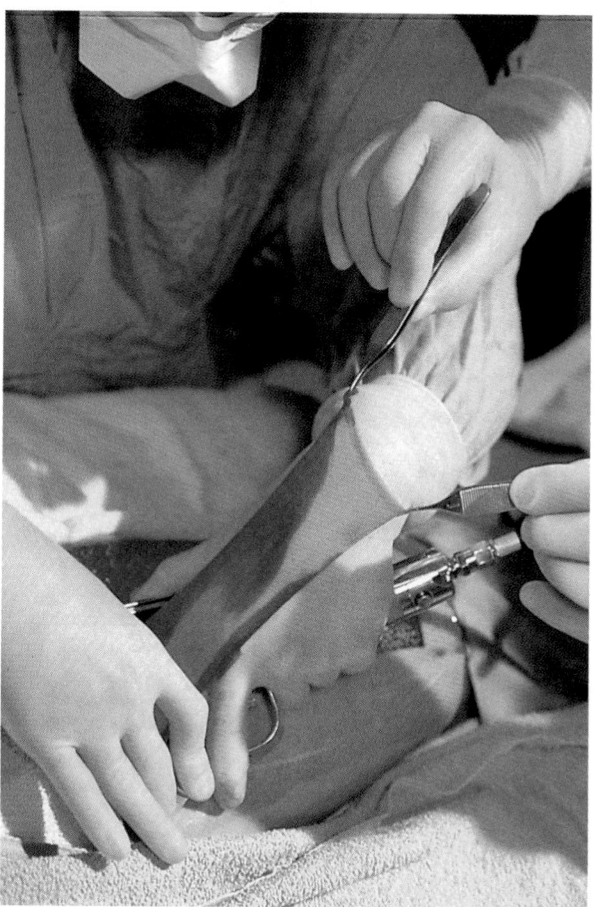

If a patient has suffered extensive burns, removing skin to perform a skin graft such as that described above is ill-advised. Instead, artificial skin made of silicone, collagen, and a polysaccharide is sometimes used in place of real skin.

Because the treatment of burns depends to some degree on the amount of body surface area that has been damaged, it is useful to be able to estimate quickly the extent of a burn. There are methods by which rather precise estimates can be obtained, but a less exact method is commonly used because it is so easy to apply. This method is called the **rule of nines** (Figure C5.3). In this estimation, the body surface area is divided in the following way: each upper

continued on next page

CONDITIONS OF CLINICAL SIGNIFICANCE

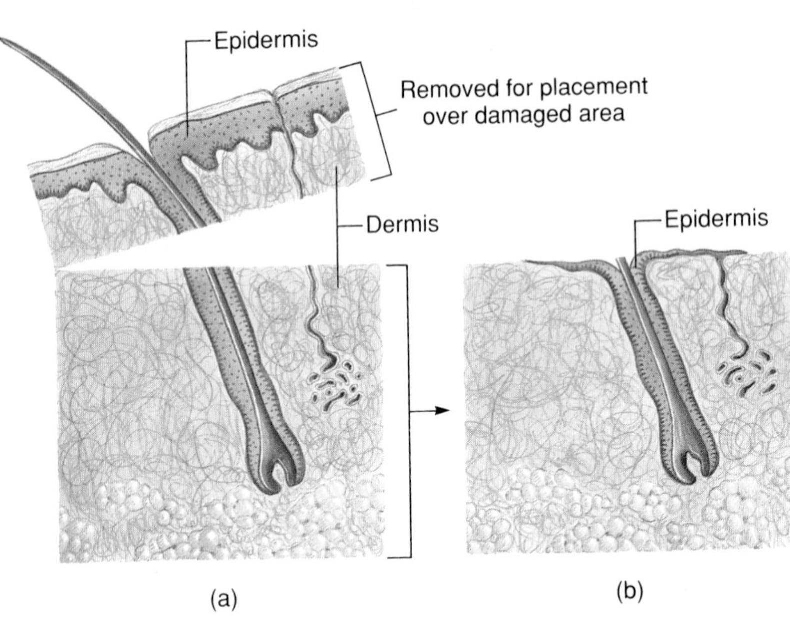

◆ **FIGURE C5.2 Split skin grafting**
(a) Epidermis (and perhaps half of the dermis) of undamaged skin is removed for placement over a damaged region. (b) New epidermis grows over the donor site.

limb is considered to have 9% of the body surface area; each lower limb 18%; the anterior and posterior trunk regions also each have 18%; the head and neck together have 9%; and the perineum has the remaining 1%.

Wound Healing in Skin

When the skin is cut or otherwise damaged, the underlying tissues are immediately exposed to the danger of infection, and the processes involved in wound healing begin. If the damage involves only the epidermis, and the dermis remains intact, repair is a relatively simple matter. In such cases, cells in the stratum basale located along the edge of the damaged region lose their contacts with the dermis and begin to migrate across the damaged region (Figure C5.4). The cells migrate as a sheet, remaining attached to the adjacent cells and pulling them behind them. When the migrating cells meet the cells that are migrating from the other edges of the wound, further cellular movement is inhibited. Thus, the damaged area is bridged over with basal epidermal cells. As this bridging is occurring, the epidermal cells begin to divide fairly rapidly, causing the formation of new strata and thickening the epidermis. A minor epidermal wound may be repaired in this manner within one to two days.

When the damaged area goes deeper than the epidermis and extends into the dermis, blood vessels are cut and the repair process is more complex. One of the first responses of the body to such a wound is the triggering of an inflammatory response in which phagocytes dispose of bacteria and other foreign material, as well as dead or damaged tissue cells. Thus, even though the protection afforded by the epidermis is breached, the chances of infection are minimized. At the same time the inflammatory response is activated, a blood clot forms in the wound, preventing additional blood loss (Figure C5.5a). Strands of fibrin develop within the clot, connecting the edges of the wound, and a **scab** forms on the surface. Epithelial cells and fibroblasts migrate along the fibrin strands, bridging the wound and forming scar tissue. The damaged blood vessels begin to redevelop. At this stage of repair, the tissue filling the damaged region is called **granulation tissue** (Figure C5.5b).

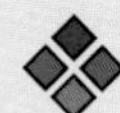

CONDITIONS OF CLINICAL SIGNIFICANCE

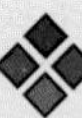

◆ **FIGURE C5.3 Estimating the extent of burns on the body surface area by using the rule of nines**

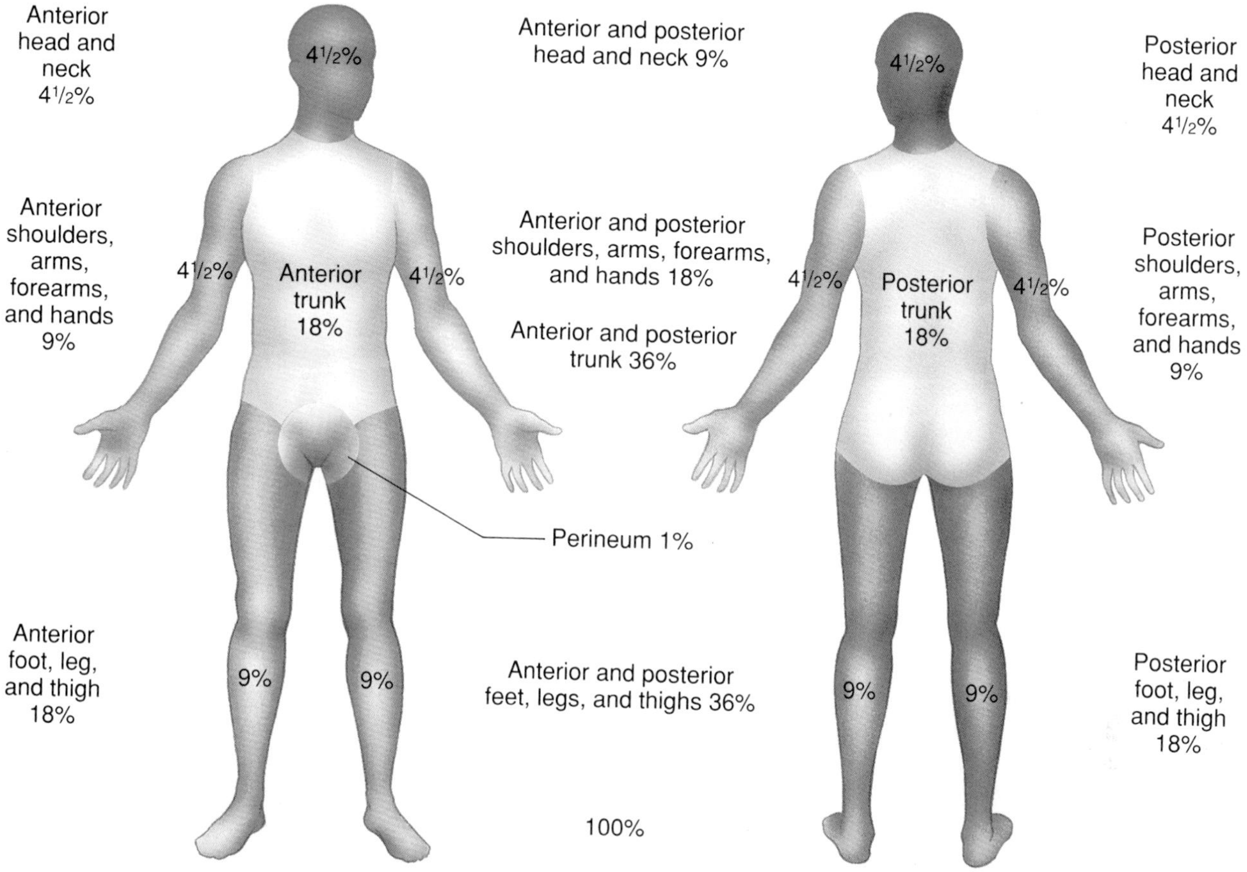

Anterior head and neck 4½%
4½%
Anterior and posterior head and neck 9%
4½%
Posterior head and neck 4½%
Anterior shoulders, arms, forearms, and hands 9%
4½%
Anterior trunk 18%
4½%
Anterior and posterior shoulders, arms, forearms, and hands 18%
Anterior and posterior trunk 36%
4½%
Posterior trunk 18%
4½%
Posterior shoulders, arms, forearms, and hands 9%
Perineum 1%
Anterior foot, leg, and thigh 18%
9%
9%
Anterior and posterior feet, legs, and thighs 36%
9%
9%
Posterior foot, leg, and thigh 18%
100%

◆ **FIGURE C5.4 Healing of an epidermal wound**

(a) Basal cells migrate across the damaged region; (b) complete bridging over of the damaged area.

Detached, enlarged basal epithelial cells migrating across wound
Dividing basal epithelial cell
Epidermis
Basement membrane
Dermis
Resurfacing of wound
(a)
(b)

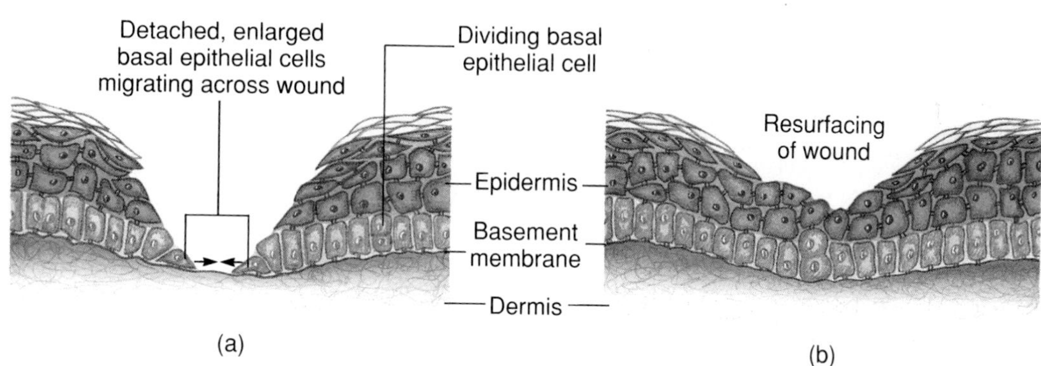

continued on next page

CONDITIONS OF CLINICAL SIGNIFICANCE

◆ FIGURE C5.5 Stages of tissue repair (wound healing)
(a) Damage to skin also damages blood vessels and a blood clot forms in the wound. (b) Fibrin strands form within the clot, and the epidermis extends down into the dermis. (c) The epidermis extends across the bottom of the wound; fibroblasts and capillaries form a ridge of new tissue. (d) New tissue fills in the wound.

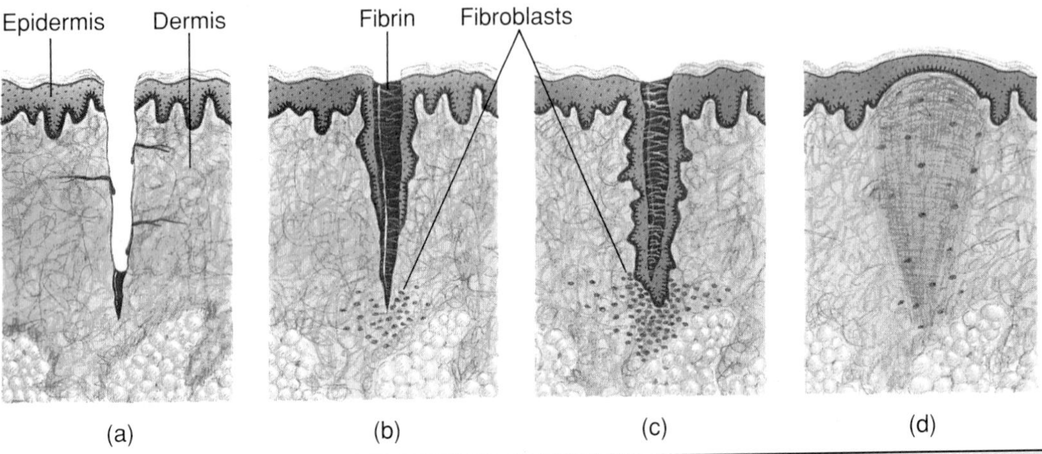

Gradually, epidermis from the surface grows down along the edges of the wound. After about one week, the epidermis extends down into the dermis. The epidermis continues its downward growth, adhering to the healthy dermis. Epidermal continuity is restored after about two weeks when the epidermis growing down one side of the wound meets the epidermis growing down the opposite side near the bottom of the wound (Figure C5.5c).

While the epidermis is growing down the sides of the wound, fibroblasts from the subcutaneous tissue continue to move into the granulation tissue and capillary buds invade the clot. These occurrences hasten the repair of the connective tissue of the dermis. As the connective tissue is repaired, the blood clot is pushed upward by the granulation tissue and eventually occupies only a small depression on the surface of the skin (Figure C5.5d). After the scab is shed, a shallow depression may remain at the site of injury, but with continued cellular division the epidermal surface eventually becomes level.

Effects of Aging

With aging, the skin tends to become thinned, somewhat wrinkled, dry, and occasionally scaly. The thinning of the epidermis is due in part to an increased scaling off of its cells and a declining rate of division. As the epidermis becomes thinner, it may become more permeable, allowing substances to pass through it more readily. Moreover, with aging, the collagen fibers in the dermis become thicker, the elastic fibers less elastic, and the underlying subcutaneous fat gradually decreases. These changes contribute to the formation of wrinkles and sags that are common in older persons.

Hair follicles, sweat glands, and sebaceous glands also decrease in number and activity with aging. Consequently, aging is often accompanied by a loss of hair, reduced sweating, and decreased oil (sebum) production. The hair also tends to become gray due to reduced amounts of pigments present. Melanocytes in the skin tend to atrophy with age. However, the melanocytes that remain tend to be larger and group together, forming dark pigment plaques called *aging spots* that are typical of older persons.

Skin that has been exposed to sunlight over a lifetime will show changes that are more severe than those due to aging alone. Such skin shows more marked wrinkling and furrowing and may develop nodules of an abnormal type of collagen. Moreover, aging skin that has been exposed to large amounts of sunlight tends to develop more cutaneous cancers than skin that has had less exposure.

Study Outline

◆ EPIDERMIS pp. 134–136

Outermost layer of skin; develops from embryonic ectoderm; lacks blood vessels; skin color is primarily determined by a dark pigment called melanin, but it is also influenced by the yellow pigment carotene and by dermal blood vessels; generally thin, but can thicken as calluses; five epidermal layers.

Epidermal Layers.

STRATUM BASALE. Deepest layer; where mitosis occurs; supplies epidermis with new cells.

STRATUM SPINOSUM. Composed of flattened cells with cytoplasmic extensions.

STRATUM GRANULOSUM. Composed of cells that contain granules of keratohyalin within cytoplasm; as granules expand, cell nucleus dies, so outermost cells of this layer are dead.

STRATUM LUCIDUM. Clear band superficial to the stratum granulosum; cells of this tissue continuously become part of stratum corneum through the presence of eleidin, which is transformed into keratin.

STRATUM CORNEUM. Outermost layer; composed of closely packed dead cells filled with fibrous protein, keratin.

Nourishment of the Epidermis. Epidermis obtains nourishment by diffusion from capillary beds of dermis.

◆ DERMIS p. 137

Lies deep to the stratum basale; second main layer of skin; well supplied with blood vessels, lymph vessels, nerves, glands, sense organs; has two indistinctly separated layers.

Papillary Layer. Next to basal layer of epidermis; contains specialized sensory receptors and capillary loops.

Reticular Layer. Deep layer consisting of bundles of collagenous fibers; continuous with the deeper hypodermis layer.

◆ HYPODERMIS p. 137

Not part of skin, but important because it attaches skin to underlying structures; composed of loose connective tissue.

◆ SKIN COLOR p. 138

Determined by presence and distribution of melanin, carotene, and hemoglobin.

◆ GLANDS OF THE SKIN p. 139

Sweat Glands. Also called sudoriferous glands; distributed over most of body surface.

ECCRINE SWEAT GLANDS. Coiled tubules within dermis; secrete a watery solution of salt, with traces of urea, sulfates, and phosphates.

APOCRINE SWEAT GLANDS. Secrete part of their cell contents, so secretion is more complex than true sweat.

CERUMINOUS GLANDS. Produce "wax" in ears; modified sweat glands.

Sebaceous Glands. Empty their secretion (sebum) into hair follicles; serve to oil skin and hair. Especially active in adolescence. In regions of skin lacking hair, glands empty secretions onto epidermis surface.

◆ HAIR pp. 140–141

Covers almost entire body; its growth is due to mitotic activity of epidermal cells at bottom of hair follicles.

Hair Follicles. Extend from epidermis into dermis; composed of two layers: (1) inner layer gives rise to hair; (2) outer layer of connective tissue develops from dermis.

PAPILLAE. At bottom of hair follicles; contain blood capillaries for nourishment and mitosis.

ARRECTOR PILI MUSCLE. Pulls on follicle; causes hair to "stand up."

Single Hair. Consists of root (part within follicle), shaft (part above skin surface); shaft has central core (medulla) of loose horny cells, cortex of tightly compressed keratinized cells that surround medulla, and outside cuticle of hard keratinized cells. Hair color primarily due to melanin.

◆ NAILS p. 141

Heavily cornified layers of strata corneum and lucidum. Each nail rests on nail bed of stratum basale and stratum spinosum. Mitosis, which produces nail growth, occurs in thickened matrix under proximal end of nail.

◆ FUNCTIONS OF THE INTEGUMENTARY SYSTEM pp. 141–142

Protection. Skin forms physical barrier against invasion of body by foreign substances; reduces water loss; melanin protects against ultraviolet radiation; acid fluid film acts as antiseptic layer; macrophages and Langerhans cells activate immune system.

Body Temperature Regulation.

OVERHEATING OF BODY. Prevented as capillaries in dermis dilate and bring greater volume of blood to body surface to lose heat to the environment; body surface also becomes wet, providing additional cooling by evaporation.

HEAT CONSERVATION. Accomplished during cold by constriction of dermal capillaries.

Excretion. Some nitrogenous wastes and salt leave the body via sweat.

Sensation. Nerve endings and specialized receptors in skin provide body with much information, such as temperature change, increased pressure.

Vitamin D Production. Occurs in skin in presence of sunlight, ultraviolet radiation; helps maintain optimum levels of calcium and phosphate.

◆ CONDITIONS OF CLINICAL SIGNIFICANCE: THE INTEGUMENTARY SYSTEM pp. 143–148

Fungi, yeasts, and bacteria live on body skin, yet cause no harm unless the epidermis is damaged, allowing them to enter the body.

Acne. Inflammatory disease caused by bacteria that provoke excessive secretion by sebaceous glands; results in pimples, blackheads, dandruff; most frequent during puberty.

Warts. Caused by viral invasion of skin; most common in adolescents and young adults.

Dermatitis and Eczema. General terms for numerous inflammatory skin conditions.

Psoriasis. Common condition of small reddish brown elevations and patches covered by layers of silvery scales; accompanied by bleeding and tiny abscesses; cause unknown.

Impetigo. Highly contagious bacterial infection common in children; pustules that form beneath stratum corneum cause inflammation and swelling.

Moles. Common pigmented elevations of skin; considered to be congenital; may become malignant, but most grow, stabilize, and finally atrophy.

Herpes Simplex. Fever blister, or cold sore, caused by viral activity.

Shingles. Caused by viral activity in spinal nerves; most often affects sensory nerves of thoracic region; causes vesicle formation and pain.

Cancers. Numerous types of tumors in epidermis, dermis, and skin glands; most are nonspreading (benign), but some do spread (malignant); cause of most skin tumors unknown, but some may be caused by prolonged overexposure to ultraviolet radiation.

Burns. The seriousness of burns results from destruction of skin, which can drastically disrupt the homeostasis of the body. Classified according to severity;

FIRST-DEGREE. Only epidermal layers of skin are damaged.

SECOND-DEGREE. Both epidermis and dermis are damaged, but skin quickly regenerated.

THIRD-DEGREE. Both epidermis and dermis are damaged so extensively that skin can regenerate only from the edges of the wound.

Wound Healing in Skin. Cut skin is healed by new connective tissue derived chiefly from subcutaneous tissue.

Effects of Aging. Skin tends to become thinned, wrinkled, dry, and sometimes scaly. Collagen and elastic fibers change, and subcutaneous fat decreases. Number and activity of hair follicles, sweat glands, and sebaceous glands decrease. Melanocytes tend to form aging spots. Ultraviolet radiation increases aging changes in the skin.

Self-Quiz

1. The subcutaneous connective tissue is called the: (a) dermis; (b) hypodermis; (c) stratum corneum.
2. In regions where the epidermis is thin, the stratum lucidum is often absent. True or False?
3. Match the following terms associated with the epidermis to the appropriate lettered description:

Stratum basale	(a) The layer that helps restrict loss of body water
Keratohyalin	(b) The substance that is transformed into keratin
Melanin	(c) A yellow pigment
Stratum corneum	(d) A clear band of several layers of flattened, closely packed cells
Eleidin	(e) The deepest layer of the epidermis, where mitosis occurs
Stratum granulosum	(f) A fibrous protein
Carotene	(g) Granules associated with the disintegration of the nucleus
Stratum lucidum	(h) The stratum deriving its name from the presence of keratohyalin
Keratin	(i) A dark pigment

4. Skin color is determined primarily by the presence and distribution of: (a) carotene; (b) melanin; (c) hemoglobin.
5. The dermis develops from embryonic mesoderm as do the muscles and the skeleton of the body. True or False?
6. Those glands that empty their secretions into hair follicles are termed: (a) sebaceous; (b) ceruminous; (c) sudoriferous.
7. Hair grows due to the mitotic activity of epidermal cells at the bottom of the arrector pili. True or False?
8. Nail growth occurs at the: (a) nail tips; (b) nail bed; (c) matrix.
9. The skin tends to be slightly basic, with a pH of 6.8. True or False?
10. Body heat is *not* conserved when the dermal capillaries: (a) dilate; (b) constrict.
11. Vitamin D_3 assists in maintaining the calcium and phosphate levels of the body at optimum levels. True or False?

12. The most abundant organisms on the skin are: (a) yeasts; (b) bacteria; (c) fungi.
13. Acne is an inflammatory disease caused by: (a) bacteria; (b) virus; (c) fungi.
14. The common wart is the result of a viral invasion of the skin and is most common in adolescents and young adults. True or False?
15. The cause of which is these disorders is unknown? (a) warts; (b) eczema; (c) psoriasis.
16. Which one of the the following conditions is congenital? (a) herpes simplex; (b) warts; (c) moles.
17. Match the terms associated with common pathologies of the integumentary system with the appropriate lettered description:

Acne	(a) Caused when a dormant virus becomes activated on the skin and mucous membranes
Warts	(b) The name for any number of inflammatory skin conditions
Psoriasis	(c) Pigmented elevations of the skin that are congenital
Impetigo	(d) The formation of pimples, blackheads, and dandruff due to bacteria
Moles	(e) Metastasizing tumors
Herpes simplex	(f) Small reddish brown patches covered by layers of silvery scales
Shingles	(g) Caused by a viral invasion of the skin
Cancers	(h) A highly contagious skin infection common in children
Eczema	(i) Aching pain and formation of vesicles along paths of thoracic sensory nerves

18. Which one of the following is caused by a viral infection of sensory spinal nerves? (a) shingles; (b) warts; (c) impetigo.
19. A burn victim who has experienced damage to both the epidermis and dermis but whose lost tissue will quickly be regenerated is suffering what degree burn? (a) first; (b) second; (c) third.
20. A burn of the anterior trunk region involves about the same amount of body surface area as does a burn of the anterior surfaces of both lower limbs. True or False?

CHAPTER 6

The Skeletal System

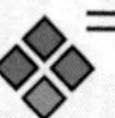

CHAPTER CONTENTS

LEARNING OBJECTIVES

After completing this chapter, you should be able to:

1. List the functions of the skeleton.
2. Cite the main components of bone that provide it with its strength.
3. List and define the major kinds of bones in the human skeleton.
4. Describe the two methods by which bone develops in the embryo.
5. Distinguish between the axial skeleton and the appendicular skeleton, and name the components of each.
6. Name the bones that form the various regions of the skull.
7. Name the components of the skeleton of the thorax.
8. Distinguish between the functions of the pectoral and pelvic girdles, and name the major components of each.
9. Describe the differences between the various types of vertebrae.
10. Describe the differences between the pelvises of males and females.

CHAPTER 6

The human skeleton is an endoskeleton—that is, it lies within the soft tissues of the body. It is a living structure capable of growth, adaptation, and repair. An endoskeleton differs greatly from the exoskeleton of arthopods like beetles and crayfish. Because an exoskeleton is a nonliving structure located on the outside of the body, an animal that has an exoskeleton must shed its exterior skeletal structure and form a new, larger one if it is to continue to grow. As you know from your own growth, your skeleton, in contrast, has grown at the same time as the rest of your body structures.

Functions of the Skeleton

The skeleton performs several important functions.

Support

The skeleton acts as the framework of the body, giving support to the soft tissues and providing points of attachment for most of the body muscles.

Movement

Because many of the body muscles attach to the skeleton, and many of the bones meet *(articulate)* in movable joints, the skeleton plays an important role in determining the kind and extent of movement of which the body is capable.

Protection

Many of the vital internal organs are protected from injury by the skeleton. The brain is encased within the cranial cavity of the skull, the spinal cord is within the canal formed by the vertebrae, the thoracic organs are protected by the rib cage, and the urinary bladder and internal reproductive organs are protected by the bony pelvis.

Mineral Reservoir

Calcium and phosphorus are the main minerals that are stored within the bones of the skeleton, but smaller amounts of sodium, potassium, and magnesium also accumulate within the matrix of bone. These minerals can be mobilized and distributed by the blood vascular system to other regions as they are required by the body. During pregnancy, for instance, calcium is removed from the mother's skeleton and used in the development of the baby's bones if the mother's diet does not include enough calcium. Because of the large mineral content of bones, they can remain intact for many years after death.

Blood-Cell Formation (Hemopoiesis)

Following birth, the red marrow within certain bones produces the blood cells that are found within the cardiovascular system.

Classification of Bones

Bones can be classified according to their shape as long, short, flat, or irregular.

Long Bones

Most of the bones of the upper and lower limbs have a long axis; that is, they are longer than they are wide. These are classified as long bones (humerus, radius, ulna, femur, tibia, fibula, phalanges).

Short Bones

Bones that do not have a long axis, such as those of the wrist (carpals) and ankle (tarsals) are called short bones.

Flat Bones

The rather thin bones that form the roof of the cranial cavity, the ribs, and the sternum (the breastbone) are flat bones.

Irregular Bones

Bones of various shapes that do not fit any of the other categories are classed as irregular bones. Some skull bones, the vertebrae, and the bones of the pectoral and pelvic girdles are examples of irregular bones.

◆ **FIGURE 6.1 Structure of a long bone**
Longitudinal section and insets of higher magnification.

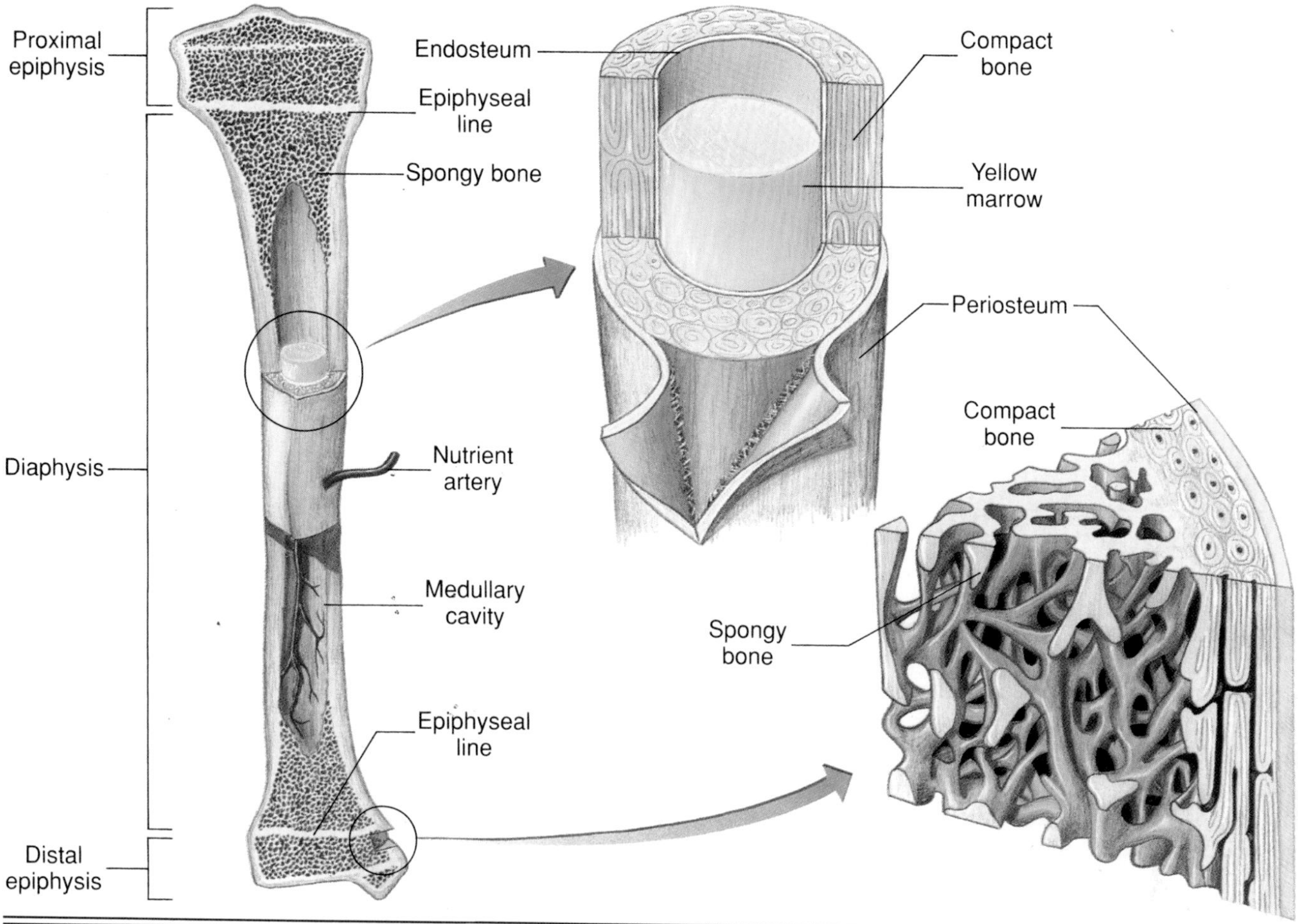

Proximal epiphysis
Endosteum
Epiphyseal line
Spongy bone
Compact bone
Yellow marrow
Periosteum
Compact bone
Diaphysis
Nutrient artery
Medullary cavity
Spongy bone
Epiphyseal line
Distal epiphysis

Structure of Bone

It is instructive to study the structure of bone at two different levels: at the gross level, where no microscope is used, and at the microscopic level.

Gross Anatomy

A typical long bone has a shaft, called a **diaphysis,** and two ends, called proximal and distal **epiphyses** (Figure 6.1). The diaphysis is composed of a hollow cylinder of **compact bone** that surrounds a **medullary cavity.** The medullary cavity, which is used as a fat storage site, is also called the **yellow bone marrow cavity.** It is lined by a thin membrane called the **endosteum.** The endosteum contains both bone-forming cells and bone-destroying cells. The outer surfaces of the epiphyses are also formed of compact bone. However, their central regions are filled with interconnecting plates of **spongy (cancellous) bone.** The cavities between the bony plates of spongy bone are lined with endosteum. The spongy bone in the proximal epiphyses of the bones of the arm (humerus) and thigh (femur) contains **red bone marrow.** In children and young adults, the diaphysis and epiphysis are separated by an **epiphyseal cartilage** or **plate** that provides the means for the bone to increase in length. In the adult, when skeletal growth has been completed, the epiphyseal cartilage is replaced by bone, firmly uniting the epiphysis with the rest of the bone. This bony junction is called the **epiphyseal line.**

There is no medullary cavity in a flat bone. This kind of bone is formed of spongy bone called **diploe,** which is sandwiched between two surface layers of compact bone (Figure 6.2). The spongy bone contains red marrow.

Bones are covered with a double-layered membrane called the **periosteum.** The outer layer is composed of dense irregular connective tissue and thus is referred to as the *fibrous layer.* The inner layer of the periosteum is called the *osteogenic layer,* because it is composed

◆ **FIGURE 6.2 Structure of a flat bone**
(a) Cross section. (b) Photomicrograph of spongy bone.

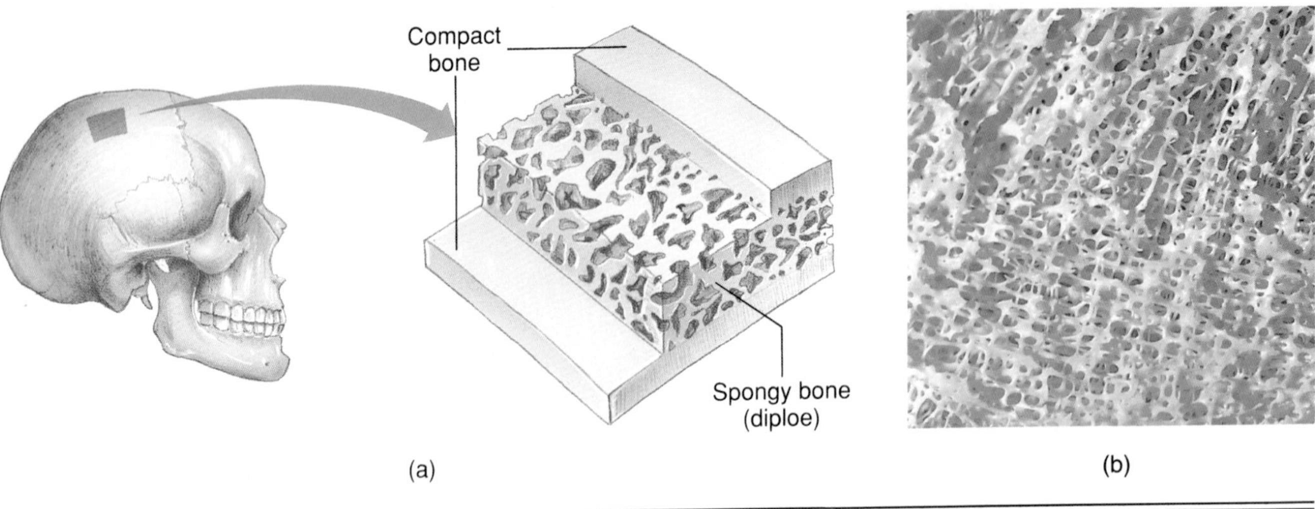

primarily of bone-forming cells (osteoblasts). The periosteum is well supplied with blood vessels, lymphatic vessels, and nerve fibers, some of which enter the bone through nutrient canals. The inner layer of the periosteum is anchored to the bone by collagenous fibers **(Sharpey's fibers)** that extend from the fibrous layer. The periosteum is absent in those areas of a joint where the bone is covered with articular cartilage.

Microscopic Anatomy

When examined under the microscope, compact bone is seen to be composed of many organized systems of interconnecting canals (Figures 6.3 and 6.4). The unit of structure of adult compact bone is the **osteon** or **Haversian system.** Each osteon has a **central (Haversian) canal** that is surrounded by concentrically arranged **lamellae** (layers) of bone. Because the osteons generally run parallel to the long axis of a bone, in longitudinal sections the central canals appear as long tubes. This orientation of osteons contributes to the capacity of bone to resist compressive forces. Located between adjacent lamellae in an osteon are small cavities called **lacunae.** Each lacuna contains a cell called an **osteocyte** (mature bone cell). All of the lacunae within each osteon are interconnected by tiny canals called **canaliculi.** The osteocytes have cytoplasmic processes that extend into the canaliculi and make contact via gap junctions with the cytoplasmic processes of neighboring osteocytes (Figure 6.5).

Each central canal contains at least one blood capillary, which provides a source of nutrients and a means of waste removal for those osteocytes that are embedded within the lacunae. The nutrients and wastes need to diffuse only a short distance through the tissue fluid within the lacunae and canaliculi from or to the central canal. The blood vessels reach the canals from larger vessels that are located either on the surface of the bone (that is, in the periosteum) or within the marrow cavity. Blood vessels, as well as lymph vessels and nerves, enter and leave the marrow cavity by means of **nutrient canals** that penetrate the bone from the surface and communicate with the marrow cavity. Blood vessels from either of these sources reach the central canals through **perforating canals (Volkmann's canals),** which run at right angles to the central canals. At the external surface of a bone, just beneath the periosteum, there may be several **circumferential lamellae,** which follow the circumference of the shaft rather than surrounding a central canal.

Spongy bone does not show the organization that is characteristic of compact bone. The osteocytes are embedded within lacunae, and the lacunae intercommunicate via canaliculi, as in compact bone. However, the lamellae are not arranged in concentric layers. Rather, they are arranged in various directions that correspond with the lines of maximum pressure or tension. Blood capillaries reach the vicinity of the osteocytes by passing within the bone marrow spaces between the plates of bone that are formed by the lamellae.

From this consideration of the microscopic anatomy of bone, it is apparent that the skeletal system is a living system, well supplied with blood vessels and nerves. As such, it is capable of performing the dynamic functions of hemopoiesis and of serving as a mineral reservoir, in addition to the static functions of support and protection.

◆ **FIGURE 6.3 Diagram of magnified osteons as seen in compact bone tissue**
The periosteum has been pulled back to show a blood vessel entering the osteons through a perforating canal. The inset is a highly magnified sketch showing osteocytes within lacunae. Note that the lacunae are interconnected by canaliculi.

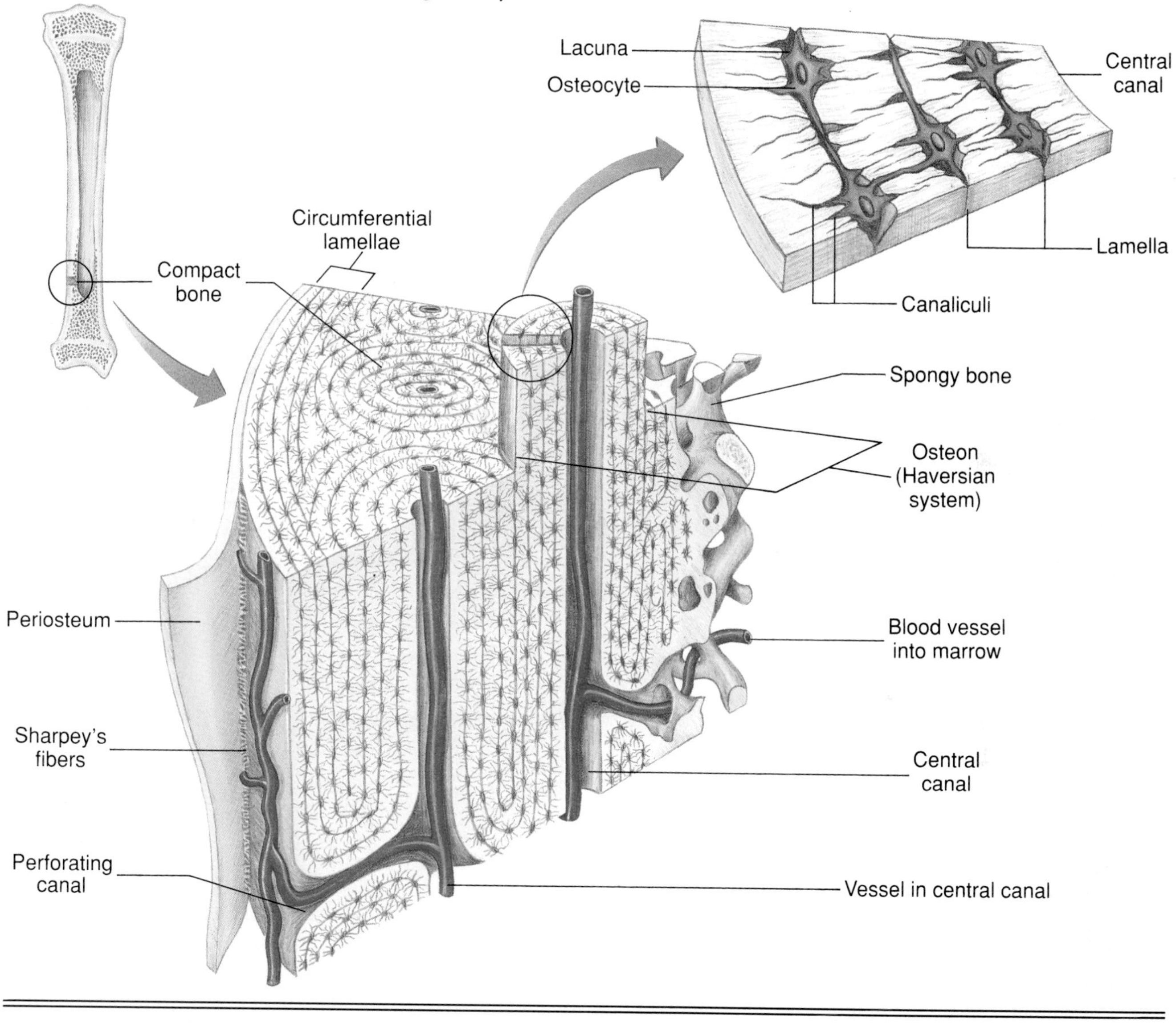

Composition and Formation of Bone

Bone is a type of connective tissue, and its intercellular matrix contains both organic components and inorganic salts. The organic components include glycosaminoglycans present in the ground substance as well as numerous collagen fibers. The inorganic salts consist primarily of calcium phosphate [$Ca_3(PO_4)_2$], which is present in the form of highly insoluble crystals of hydroxyapatite [$3Ca_3(PO_4)_2 \cdot Ca(OH)_2$].

The collagen fibers are capable of resisting stretching and twisting, and they provide bone with great tensile strength. The inorganic salts allow bone to withstand compression. This combination of collagen fibers and inorganic salts makes bone exceptionally strong without being brittle. The same principle is used in reinforced concrete, where steel rods provide tensile strength, and cement, sand, and gravel give compressional strength.

Many of the chemical events of bone formation are still incompletely understood. In general, cells called **osteoblasts** secrete the organic components of the intercellular matrix of bone, including glycosaminoglycans

◆ **FIGURE 6.4 Scanning electron micrograph of an osteon**

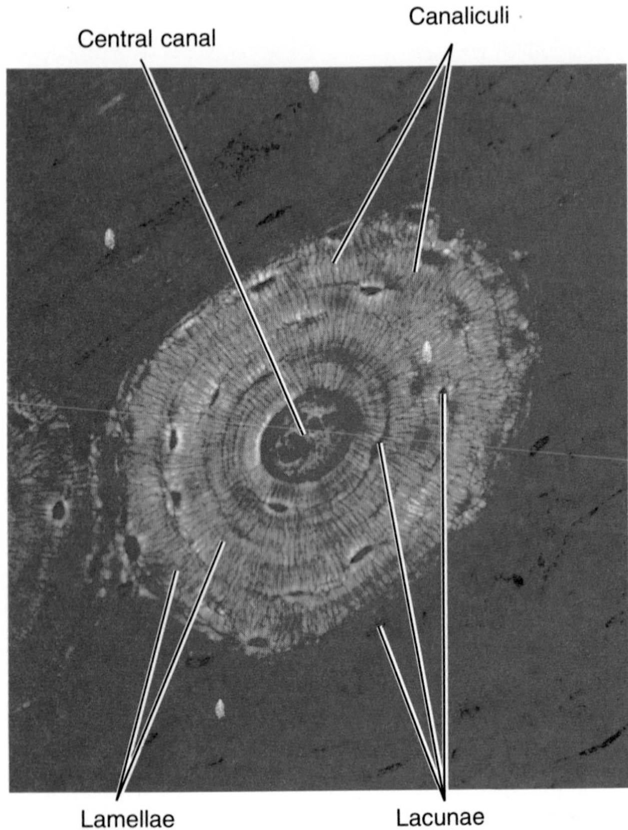

and collagen. Within a period of days, the intercellular matrix becomes calcified by the deposition of insoluble calcium salts within it. Normally, calcium phosphate salts are in solution in the extracellular fluid, but in an area where bone is forming, these salts precipitate and are progressively transformed into crystals of hydroxyapatite. (Newly formed matrix that is not yet calcified is called *osteoid.*)

Some investigators believe that membrane-bounded vesicles called *matrix vesicles* act as calcification initiators. In many cases, matrix vesicles are present in the intercellular matrix at sites where calcification is occurring, and some of the first-formed crystals of hydroxyapatite seem to develop in close association with these vesicles. Matrix vesicles are believed to arise from osteoblasts as outpocketings of the plasma membrane, which separate from the membrane. An enzyme called *alkaline phosphatase* is present in matrix vesicles, and this enzyme is believed to promote the deposition of calcium salts within the intercellular matrix.

The intercellular matrix contains certain sulfated glycosaminoglycans and a calcium-binding protein called osteocalcin, which may also be involved in calcification. The sulfated glycosaminoglycans may provide nucleation sites for crystal formation, and osteocalcin may augment the extracellular calcium concentration of the matrix, thereby promoting calcification.

Development of Bone

Early Development of Bone

An embryonic tissue called **mesenchyme** forms during the early development of an embryo. Mesenchyme gives rise to all types of connective tissue, including bone.

Mesenchymal cells give rise to cells called **osteogenic cells** *(osteoprogenitor cells).* In areas that are well supplied by blood vessels, osteogenic cells differentiate into osteoblasts. In areas that are not well supplied by blood

◆ **FIGURE 6.5 Scanning electron micrograph of an osteocyte within a lacuna**

Note the protoplasmic extensions from the surface of the cell entering canaliculi.

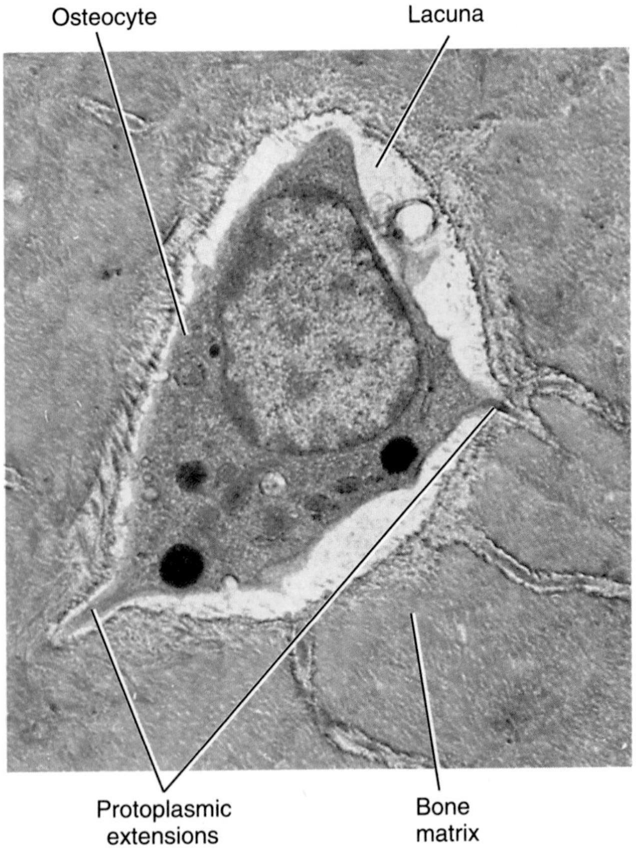

vessels, osteogenic cells differentiate into *chondroblasts,* which can form cartilage.

The flat bones of the roof of the skull, some facial bones, and the clavicles (collarbones) form by a process called *intramembranous ossification.* Long bones and short bones form by a process called *endochondral ossification.* (The bone produced by either intramembranous ossification or endochondral ossification has the same composition.)

Intramembranous Ossification

Intramembranous ossification is a process of bone formation that takes place within mesenchyme. Intramembranous ossification begins near the end of the second month of gestation. At a site where bone will form, blood capillaries grow into mesenchyme, and an *ossification center* begins to develop (Figure 6.6). Mesenchymal cells within the center give rise to osteogenic cells, which differentiate into osteoblasts, and bone formation begins. As bone formation continues, the osteoblasts become enclosed within lacunae that are surrounded by bone matrix. After the osteoblasts are enclosed within lacunae, their activity declines, and they become mature bone cells, or osteocytes.

As bone formation occurs, a layer of mesenchyme containing an extensive supply of blood vessels condenses at the external surfaces of the bone and develops into a periosteum. Mesenchymal cells near the bone surfaces give rise to osteogenic cells, which differentiate into osteoblasts. Mesenchymal cells farther away from the bone surfaces give rise to fibroblasts, which form the fibrous outer layer of the periosteum.

During intramembranous ossification, bone is initially present in the form of small slivers called *spicules,* but as more bone forms, the spicules become thicker. Well-developed spicules of bone are known as *trabeculae* ("little beams"). The trabeculae radiate in all directions, uniting with one another to form a network of spongy bone. In areas that will eventually form compact bone, the trabeculae continue to thicken as additional bone is deposited. Gradually, the spaces between the trabeculae are narrowed, and bone replaces the spaces, forming compact bone.

As additional bone forms during intramembranous ossification, some previously formed bone is resorbed (removed) in areas where it is no longer needed. Bone is resorbed by cells called **osteoclasts.** Osteoclasts are large, multinucleated cells that arise from precursor cells traveling within the bloodstream. The precursor cells collect at sites where bone resorption is occurring, and fuse to form osteoclasts. A *ruffled border,* which consists of branching, fingerlike processes, develops on the part of the osteoclast that is in contact with bone (Figure 6.7). Enzymes and hydrogen ions are secreted along this ruffled border. The enzymes break down the organic components of the intercellular matrix of bone, and the hydrogen ions create an acidic environment in which the inorganic salts of the matrix dissolve.

◆ **FIGURE 6.6 Intramembranous ossification**
See the text for details.

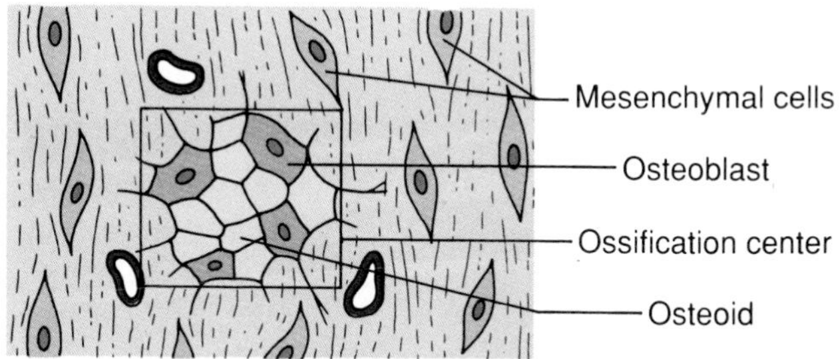

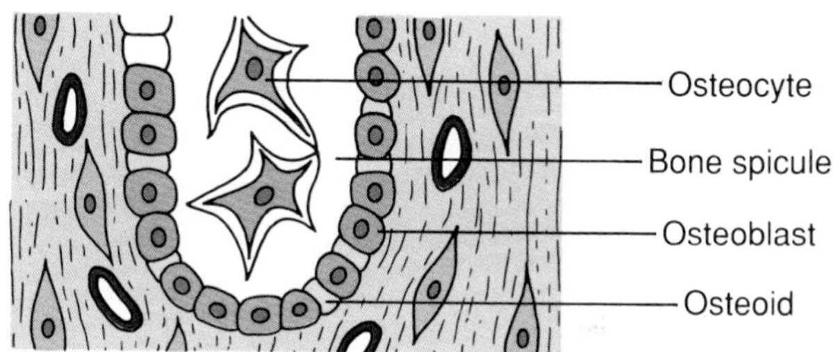

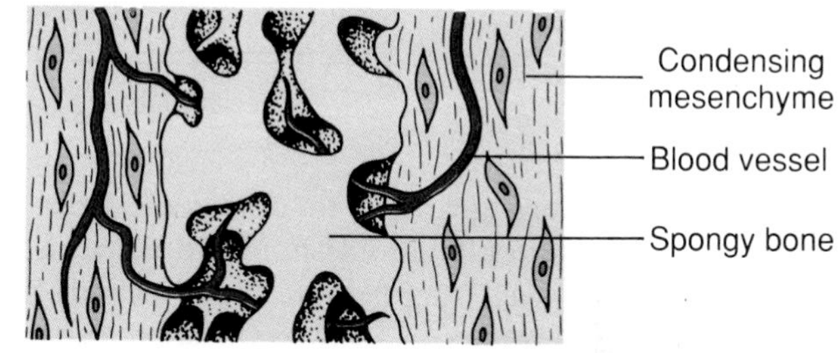

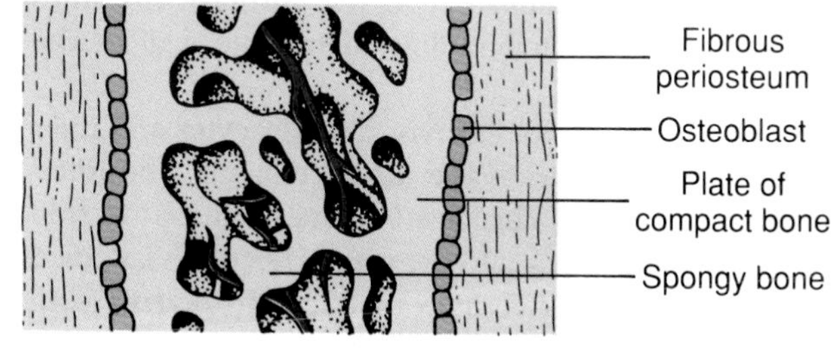

Endochondral Ossification

Endochondral ossification is a process of bone formation that takes place within hyaline cartilage. The hyaline cartilage develops from mesenchyme. At a site where bone will form, mesenchymal cells aggregate in a pattern that resembles the shape of the future bone.

◆ **FIGURE 6.7 An osteoclast**
(a) Scanning electron micrograph. (b) Diagram.

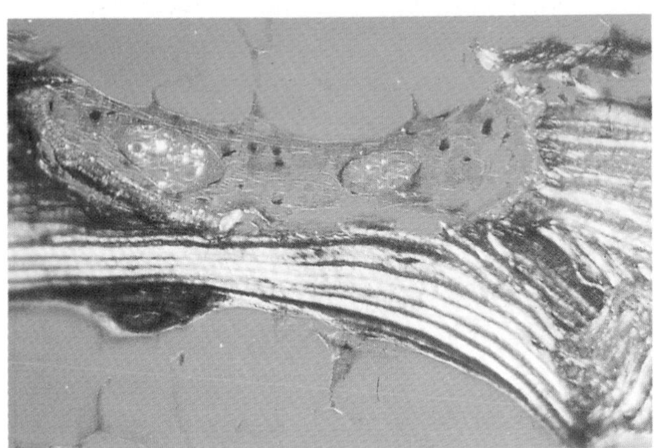

(a)

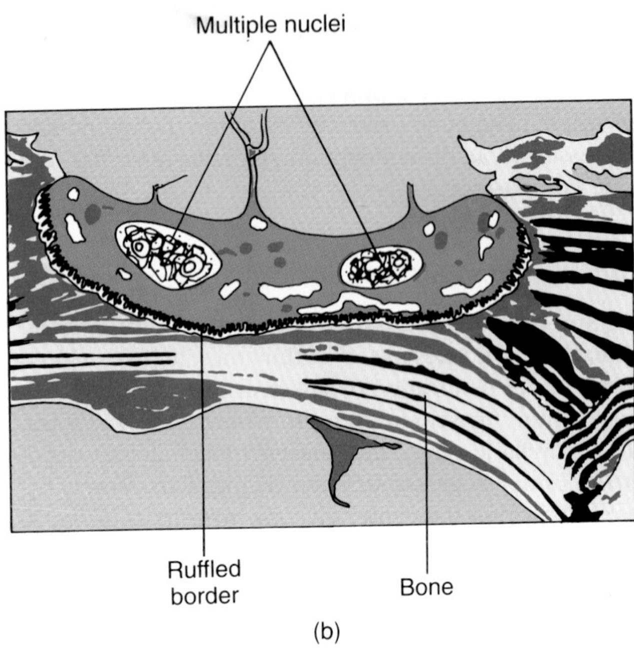

(b)

The mesenchymal cells give rise to osteogenic cells, which differentiate into chondroblasts. (Note that this mesenchyme is not well supplied by blood vessels.) The chondroblasts produce a growing hyaline cartilage model of the future bone (Figure 6.8a). Mesenchyme immediately adjacent to the cartilage becomes organized into a surrounding membrane called the **perichondrium.**

As the cartilage model grows, chondrocytes and their lacunae in the middle of the model enlarge. The intercellular matrix of the cartilage becomes reduced to thin partitions and spicules, and the matrix becomes calcified (Figure 6.8b). (Matrix vesicles that appear to arise from chondrocytes are present in calcifying cartilage, and the matrix contains sulfated glycosaminoglycans and a calcium-binding protein called chondrocalcin.) When examined with a microscope, many of the lacunae in heavily calcified cartilage appear to be empty, and the intercellular matrix appears to be breaking down either by disintegration or erosion, leaving spaces within it.

As the cartilage in the middle of the model calcifies, blood vessels infiltrate the perichondrium around the midsection of the model. Osteogenic cells at the inner surface of the perichondrium differentiate into osteoblasts, and a collar of compact bone forms around the cartilage (Figure 6.8b). When osteoblasts develop at the inner surface of the perichondrium and bone formation occurs, the perichondrium becomes a periosteum because it surrounds bone.

Blood vessels from the periosteum grow into the calcified cartilage, and osteogenic cells, osteoblasts, and osteoclasts invade the area, forming a *primary ossification center* (Figure 6.8c). Within the primary ossification center, trabeculae of spongy bone form on what is left of the calcified cartilage. As bone formation continues, the primary ossification center enlarges, and in a developing long bone, some of the newly formed spongy bone is broken down by osteoclasts, forming the medullary cavity (Figure 6.8d).

In most long bones, the primary ossification center develops near the end of the second month of gestation. At birth, most long bones have a diaphysis of compact bone surrounding remnants of spongy bone and the medullary cavity. The epiphyses are still hyaline cartilage.

Sometime after birth (or just prior to birth in a few cases), chondrocytes within the epiphyses of a long bone enlarge, the cartilage becomes calcified, and *secondary ossification centers* develop within the epiphyses in essentially the same manner as the primary ossification center develops within the diaphysis (Figure 6.8e). Spongy bone forms within the secondary ossification centers, but no medullary cavity is formed. The secondary ossification centers increase in size, and spongy bone forms throughout the interiors of the epiphyses (Figure 6.8f). Eventually, the only cartilage that remains is a thin surface layer at the end of each epiphysis, which becomes the articular cartilage, and a

◆ **FIGURE 6.8 Endochondral ossification as it occurs in a long bone**
See the text for details.

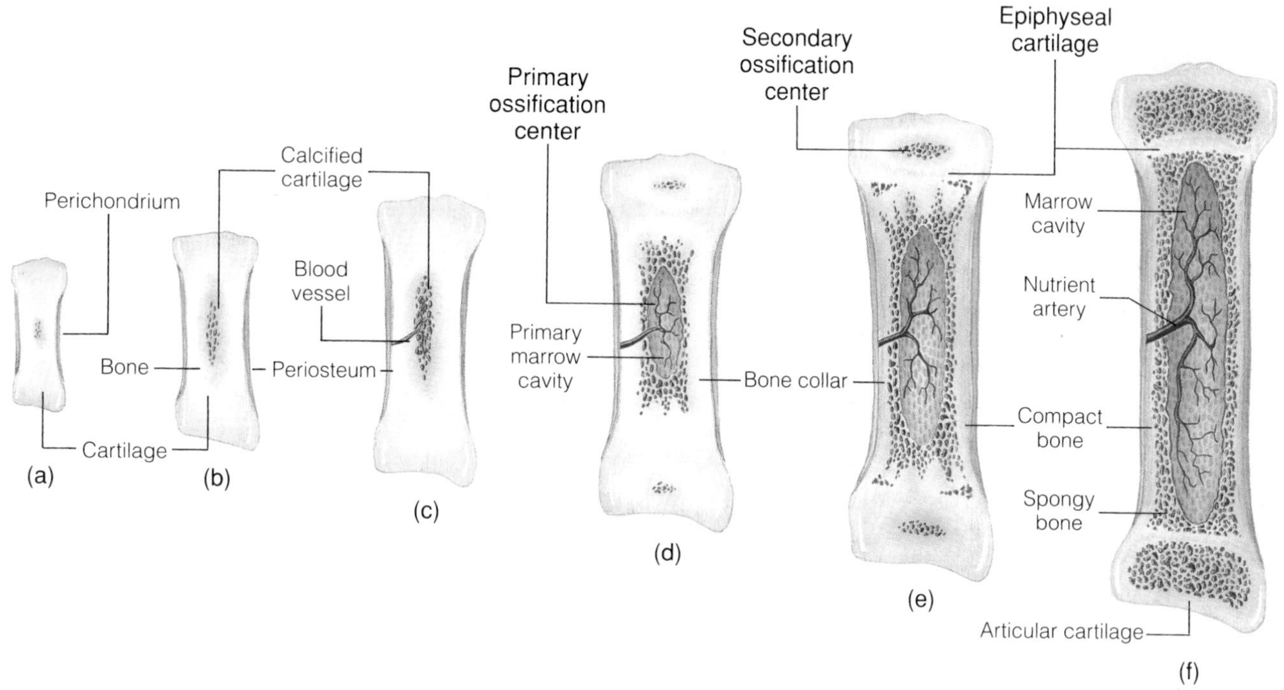

thicker layer that separates each epiphysis from the bony diaphysis. This thicker layer is called the **epiphyseal cartilage,** or **epiphyseal plate.**

Although a secondary ossification center develops in each epiphysis of most long bones, in a few long bones a secondary ossification center develops at only one end. Only a primary ossification center develops in most short bones.

Increase in Long Bone Length and Diameter

As long as the epiphyseal plates of a long bone remain present, it is possible for the bone to increase in length. It does so in the following manner. Cartilage cells near the sides of the epiphyseal plates toward the epiphyses undergo mitosis and form new cartilage that tends to increase the thickness of the plates (Figure 6.9). Simultaneously, cartilage cells near the sides of the plates toward the diaphysis enlarge, and the cartilage becomes calcified. Osteoblasts infiltrate the calcified cartilage, and bone forms within it. Under normal conditions of growth, the formation of new cartilage near the epiphyseal sides of the epiphyseal plates and the formation of bone at the diaphyseal sides of the plates balance one another, and the diaphysis increases in length while the thickness of the epiphyseal plates remains constant.

During the late teen years, the rate of cartilage formation near the epiphyseal sides of the epiphyseal plates slows, but bone formation continues at the diaphyseal sides of the plates. As a result, the plates gradually become thinner. By about age 21 in males and age 18 in females, the epiphyseal plates are completely replaced by bone, leaving only **epiphyseal lines** to mark their previous locations. Once the epiphyseal plates are replaced by bone, the bone is no longer able to increase in length. It does, however, retain the ability to increase in diameter.

In an area where a long bone is increasing in diameter, osteoblasts are present at the inner surface of the periosteum. In this area, new bone forms around the diaphysis, just beneath the periosteum. As the diameter of the diaphysis is increased by the addition of this new bone, other bone is usually resorbed from within—beneath the endosteum. This resorption, which increases the volume of the marrow cavity, is generally less extensive than bone formation. Consequently, the overall thickness of the bone increases, but the increase is not as great as it would be if resorption did not occur.

Remodeling of Bone

Once a bone is formed, it undergoes continual remodeling. Remodeling is accomplished by the resorption of

◆ **FIGURE 6.9 Events occurring at an epiphyseal plate during the growth in length of a long bone**

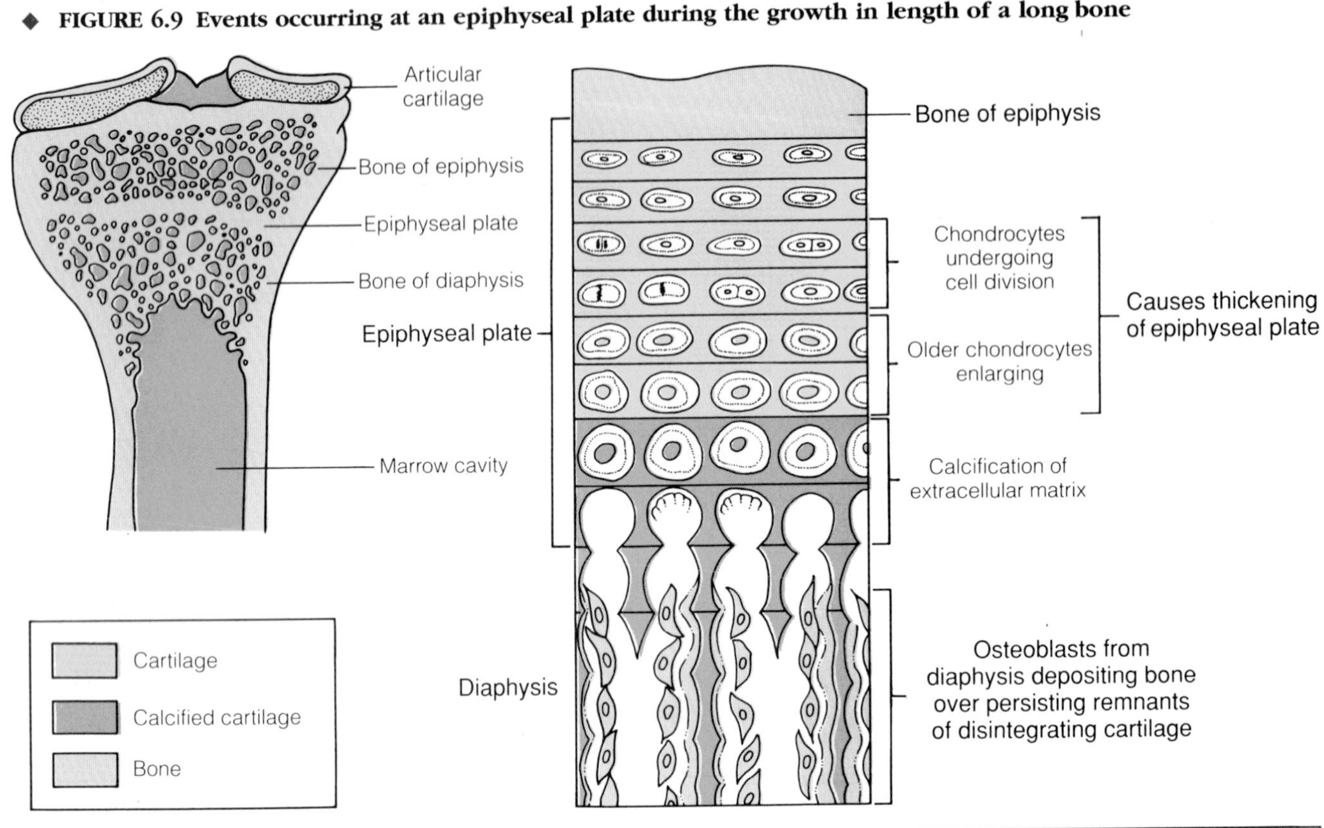

previously formed bone and the formation of new bone. Osteoclasts resorb previously formed bone, and osteoblasts secrete the organic components of the intercellular matrix of new bone, which becomes calcified.

In compact bone, previously formed osteons are replaced by new ones. Small groups of osteoclasts resorb the bone of previously formed osteons, producing a channel in the bone (Figure 6.10a). The channel is invaded by cells, and a small blood vessel grows into the channel from an existing vessel. The channel wall becomes lined by a layer of osteoblasts, and concentric layers of new bone form around the wall. The continued formation of concentric layers of new bone progressively narrows the channel until only the small central canal remains around the blood vessel. As the concentric layers of new bone form, many of the osteoblasts become enclosed within lacunae. The activity of these osteoblasts declines, and they become osteocytes of the new osteon (Figure 6.10b).

Remodeling enables a bone to adapt to the changing dimensions of a growing body or to new mechanical stresses placed on it. For example, as a long bone increases in length, remodeling maintains the correct proportional relationship between the sizes of the epiphyses and the size of the diaphysis.

Factors That Affect Bone Development

A number of factors influence bone development. Among the more important factors are mechanical stress and various hormones.

Mechanical Stress

Bone is a living tissue that is capable of adjusting its strength in proportion to the degree of mechanical stress to which it is subjected. Increased amounts of collagen fibers and inorganic salts can be deposited in a bone in response to prolonged heavy loads. Conversely, if a bone is not subjected to mechanical stress, salts are withdrawn from the bone.

As new stress patterns occur, remodeling may change the orientation of the collagen fibers in a bone so that the fibers are aligned in such a manner as to provide maximal tensile strength to withstand the new stress patterns. When a bone is compressed or bent, bone is formed at the area of compression.

Bones are normally subjected to two major mechanical stresses: *gravitational forces,* such as those that result from supporting the weight of the body, and *func-*

◆ **FIGURE 6.10 Compact bone**

(a) Formation of a new osteon in compact bone. (b) Photomicrograph of a transverse section through compact bone showing a newly forming osteon and a previously formed osteon.

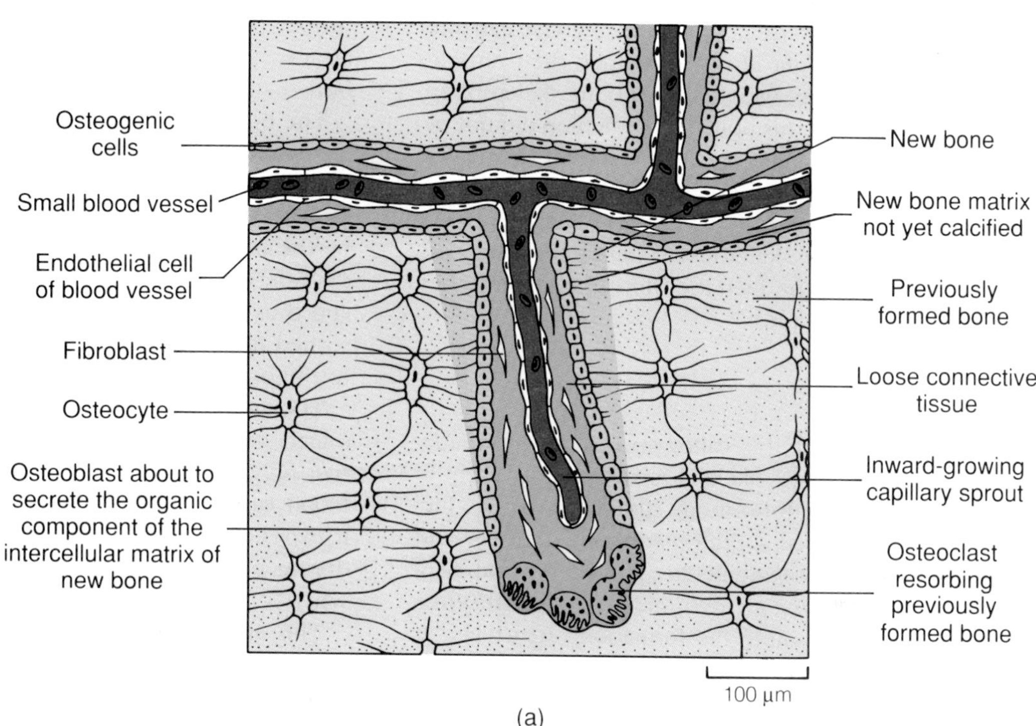

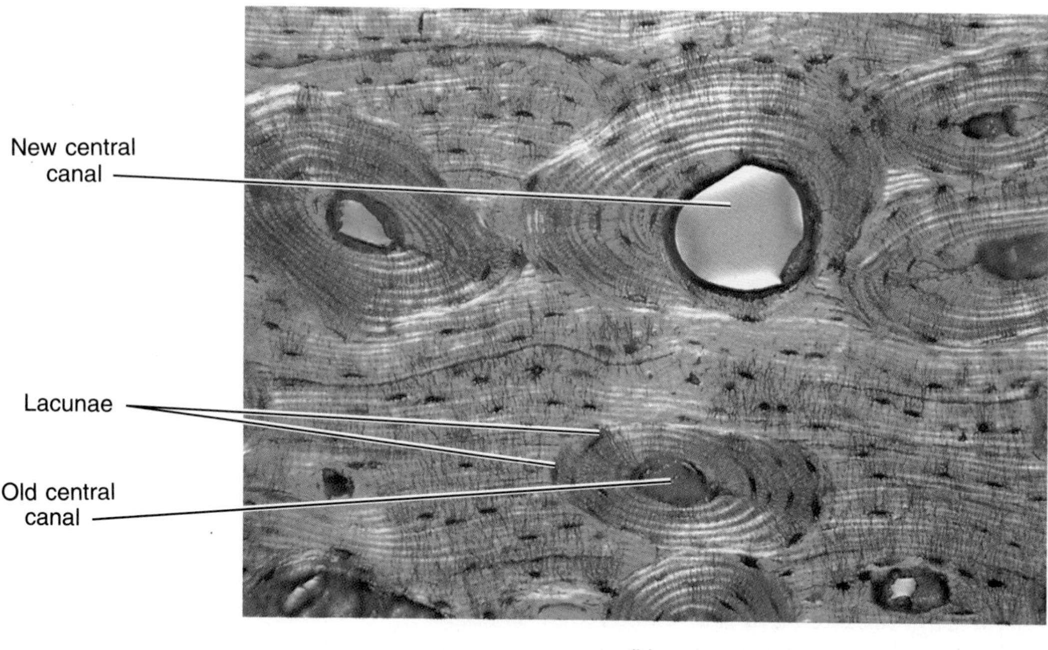

tional forces, such as those that result from the pull exerted on a bone by contracting muscles. A bone does not develop normally in the absence of either of these forces. When gravitational forces are removed for extended periods, such as under the weightless conditions of space travel, or when functional forces are greatly reduced, such as occurs when a limb is paralyzed or immobilized by a cast, the bones in the affected areas do not grow and may actually atrophy (that is, degenerate).

There is evidence that bone growth is promoted by intermittently applied forces like those applied to bones during muscular exercise, and regular exercise can alter the skeleton. As weight lifters strengthen their muscles so they can lift heavier weights, their bony skeletons are also strengthened. If this bone strengthening did not occur, greatly strengthened muscles could damage the bones to which they are attached.

Continuously applied forces can also alter the skeleton. For example, the skeletons of obese people become heavier because of the increased forces to which they are constantly subjected.

Hormones

A pituitary-gland hormone called *growth hormone (somatotropin)* influences skeletal growth, particularly in young people. Growth hormone stimulates the production of peptides called *somatomedins* by the liver and other tissues. The somatomedins, which include peptides called *insulinlike growth factor I* and *insulinlike growth factor II,* promote the formation of cartilage at the epiphyseal plates of long bones and thereby promote the growth in length of the bones. *Thyroid hormones* from the thyroid gland are necessary for the proper production and action of growth hormone, and an absence of thyroid hormones leads to the arrest of bone elongation and to retarded bone maturation.

The bones contain large amounts of calcium that can be utilized to maintain the proper level of calcium in the blood. An increased level of a hormone called *parathyroid hormone (parathormone),* which is produced by the parathyroid glands, leads to an increase in both the number and the activity of osteoclasts within bones and thus to an increase in bone resorption. Calcium resorbed from bones can enter the blood and contribute to the blood calcium level.

The hormone *calcitonin,* which is produced by C (clear) cells of the thyroid gland, decreases the resorptive activity of osteoclasts. Calcitonin tends to lower the blood calcium level, particularly in young people, and it may help protect bones from excessive calcium loss during periods of high calcium demand (for example, when a woman is pregnant or breast feeding).

The activities of these hormones are discussed in greater detail in Chapter 17.

CONDITIONS OF CLINICAL SIGNIFICANCE

The Skeletal System

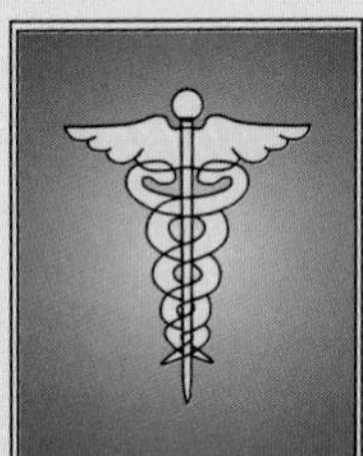

Fractures

Even though the composition of bones is such that they are well suited to withstand twisting and compressional forces, it is possible to break them. Broken bones are referred to as *fractures.* Most fractures can be restored to normal alignment by manipulation. This procedure, which does not require surgery, is called a **closed reduction.** In more severe fractures, the break must be exposed surgically in order to position the bone fragments properly. This surgical procedure is referred to as an **open reduction.**

Types of Fractures

Fractures are named according to various conditions at the site of the break (Figure C6.1). The following are the more common types of fractures:

◆ *Simple (closed) fracture.* The broken ends of the bone do not penetrate through the skin.

◆ *Compound (open) fracture.* The broken ends of the bone protrude through the skin, leading to the possibility of a severe bone infection.

◆ *Comminuted fracture.* Rather than being broken in a single plane, the bone is splintered into many fragments at the site of the break.

◆ *Depressed fracture.* The broken region is pushed inward, as often occurs in fractures of the flat skull bones that form the roof over the brain.

◆ *Impacted (compression) fracture.* The broken ends of the bone are driven into each other. Such fractures occur in falls in which the person lands on the ends of the bone.

◆ *Spiral fracture.* The bone is fractured by excessive twisting forces.

◆ *Greenstick fracture.* Only one side of the bone breaks. The other side resists the force causing the break by bending. This type of fracture is common

CONDITIONS OF CLINICAL SIGNIFICANCE

◆ **FIGURE C6.1 Types of fractures**

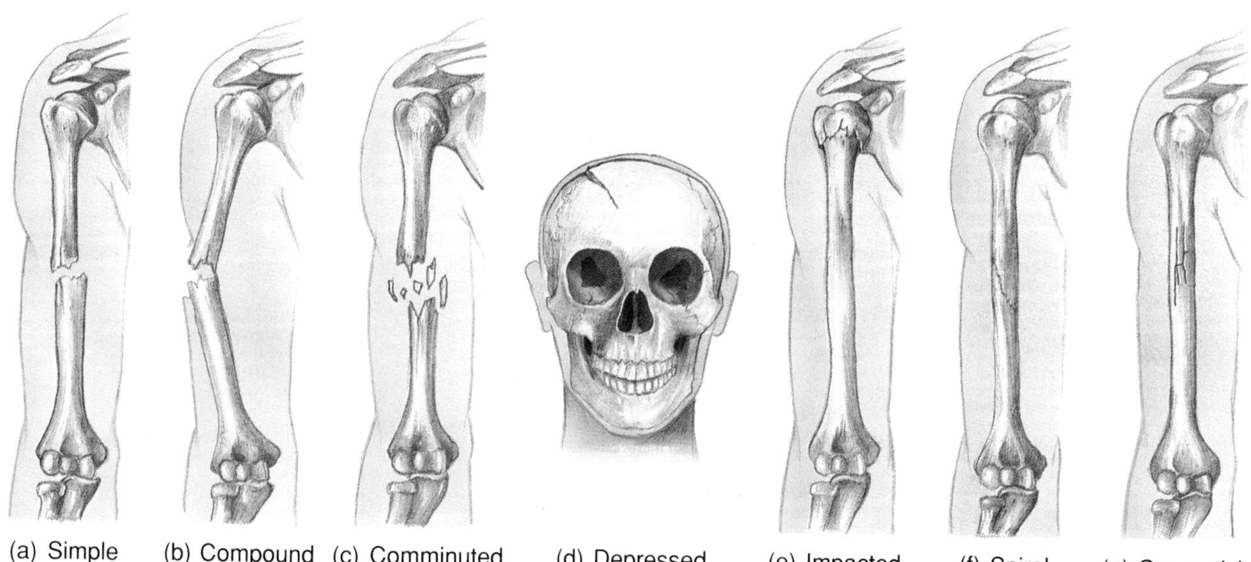

in children, whose bones are more flexible than those of older persons.

Healing of Fractures

Fractures undergo a series of progressive changes during the healing process (Figure C6.2).

1. *Formation of a hematoma.* When a bone is broken, bleeding occurs from the vessels of the osteons, the marrow cavity, and the periosteum. This bleeding produces a swelling and forms a blood clot in and around the site of the fracture that is called a **hematoma.**

2. *Formation of a fibrocartilaginous callus.* Capillaries grow into the fracture hematoma within a few days, and phagocytes, fibroblasts, and osteoblasts invade the fracture site. The phagocytes clean up the area of the hematoma by removing dead cells and other debris. The fibroblasts produce collagenous fibers that form a bridge across the break, temporarily reconnecting the broken ends. Some fibroblasts differentiate into chondroblasts and secrete a cartilage matrix around the fibers. Osteoblasts then deposit regions of spongy bone within the matrix, forming a firm **fibrocartilaginous callus** that reunites the broken bones. The part of the callus that extends beyond the usual surface of the bone is the **external callus.** The part of the callus between the broken ends of the bone, including the medullary cavity, is called the **internal callus.**

3. *Formation of a bony callus.* Gradually, osteoblasts derived from the inner layer of the periosteum and the endosteum form a **bony callus** that knits the ends of the bone firmly together.

4. *Remodeling.* At first, the bony callus is spongy bone, but it is slowly remodeled to form compact bone. The formation of compact bone is partially under the influence of stress as the repaired bone is again used for body support and movement.

During the months that follow the formation of a bony callus, osteoclasts resorb the bone of the external callus and the part of the internal callus that blocks the medullary cavity so that eventually a slight external enlargement is all that is left to mark the location of the break.

continued on next page

CONDITIONS OF CLINICAL SIGNIFICANCE

◆ **FIGURE C6.2 Healing of a fracture**

The initial repair begins with the formation of a blood clot called a *hematoma*. Connective tissue invades the hematoma, replacing it with a fibrocartilaginous callus that contains a latticework of spongy bone. This fibrous callus is eventually replaced by compact bone that develops from cells of the periosteum, and the healed fracture is remodeled by clearing the medullary cavity and reducing the external callus.

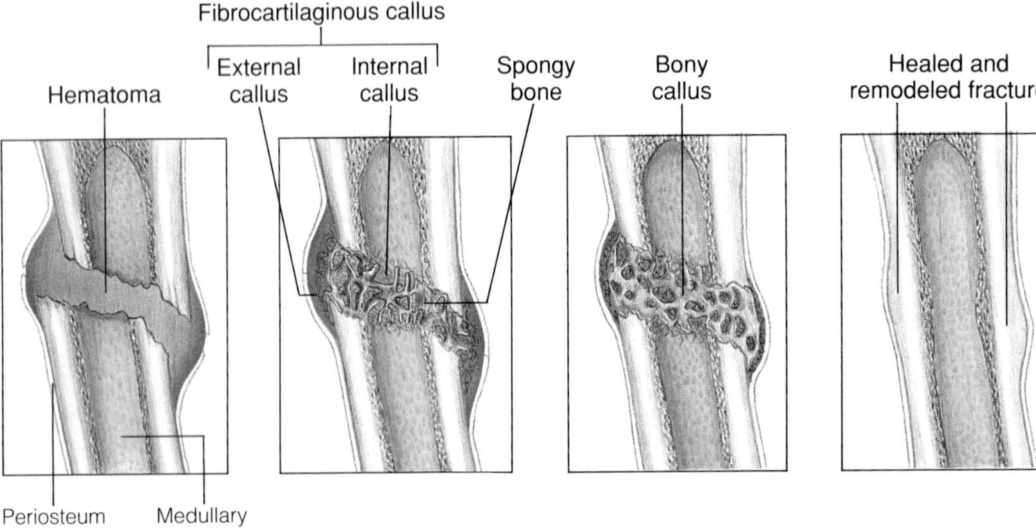

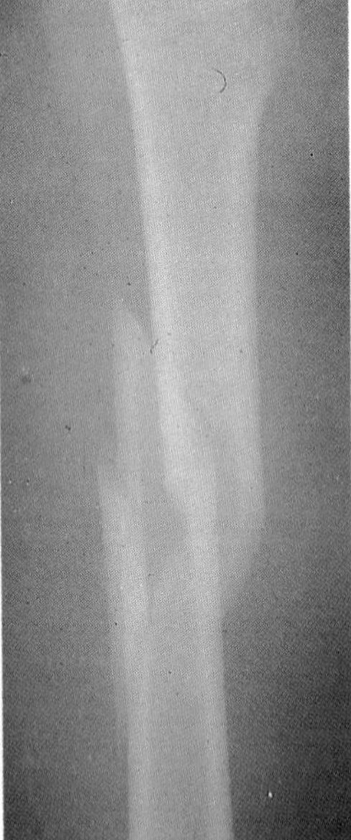

Fracture

Healing fracture after 3 weeks

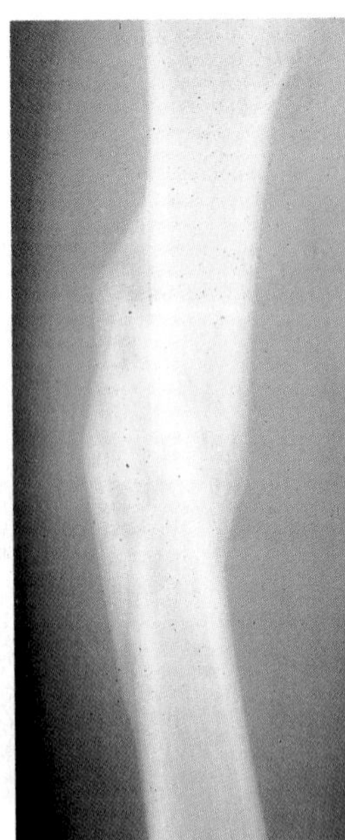

Healing fracture after 7 weeks

CONDITIONS OF CLINICAL SIGNIFICANCE

Because of the importance of stress in strengthening the bone, prolonged immobilization of a broken bone may be detrimental to the healing process. For this same reason, pins are sometimes used to hold the ends of a broken bone together, thus allowing the bone to support weight almost immediately. The stresses of use increase the activity of osteoblasts and facilitate healing.

Metastatic Calcification

The deposition of calcium in tissues that do not usually become calcified is called *metastatic calcification.* It often results if blood calcium levels become too high, as occurs during some decalcifying diseases of bone or when the amount of parathyroid hormone is increased. The kidney is the most common site of metastatic calcification (which results in kidney stones), but it occurs in many other tissues as well.

Spina Bifida

Occasionally, during embryonic development, the posterior portions of the vertebrae of the spinal column fail to form a complete bony arch around the spinal cord. This condition, which is called *spina bifida,* is most common in the lumbar and sacral regions of the spinal column but may occur in a vertebra of any region. If the defect is large enough, the coverings of the spinal cord and the wall of the spinal cord itself may protrude.

Osteoporosis

Osteoporosis, or "porous bone," is a common condition in older persons—especially women. However, it can occur at any age within the bones of paralyzed or immobilized limbs. Osteoporosis is a consequence of bone resorption occurring at a faster rate than bone deposition. This results in a decrease in bone mass, but the composition of the intercellular matrix of the remaining bone is normal. The loss of bone mass weakens the bones, and they break relatively easily. The most common fractures occurring in persons suffering from osteoporosis are of the hip, spinal vertebrae, and forearm.

Because some degree of osteoporosis occurs in most women over age 50 but in relatively few older men, reduced secretions of estrogens associated with menopause are thought to be a major factor in its development. There appear to be at least two stages of osteoporosis: (1) In women between 50 and 60 years of age, bone resorption increases, and the rate of bone deposition cannot keep up. (2) In women older than 60, the problem appears to be a decline in the rate of bone formation, rather than an increase in the rate of bone resorption. In both cases, reduced estrogen levels characteristic of menopause are thought to be involved in the activity changes.

Osteoporosis is a complex condition, and not all of the interactions associated with its development are understood. A certain amount of demineralization often occurs in the disease, and there is evidence that the removal of calcium and other minerals from bone may be slowed by increasing the level of estrogen in the blood. For this reason, estrogen therapy has been used to treat osteoporosis in some postmenopausal women. Treatment also may involve supplemental calcium and vitamin D, along with a program of weight-bearing exercises.

Osteomyelitis

In *osteomyelitis,* the periosteum, the contents of the marrow cavity, and the bone tissue become infected. Since the causative agent is usually *Staphylococcus aureus,* a bacterium that enters the body through a boil or some other break in the skin, osteomyelitis may follow trauma. The initial damage to the bone has the appearance of an abscess. The abscess spreads throughout the bone, converting the fatty tissue of the marrow cavity into pus and destroying the bony tissue. The infection can pass from the shaft of the bone, where it generally appears first, to the epiphyses, where it perforates the articular cartilage and enters the joint cavity.

Before the discovery of antibiotics, osteomyelitis had a high mortality rate. Because the disease responds well to antibiotic therapy, death from it is now rare.

continued on next page

CONDITIONS OF CLINICAL SIGNIFICANCE

Tuberculosis of Bone

Tuberculosis of bone is a type of osteomyelitis that is caused by another bacterium, *Mycobacterium tuberculosis.* The bacterium is generally carried by the bloodstream from an infection in the lung or lymph nodes. It is characterized by excessive bone destruction.

Rickets and Osteomalacia

Both *rickets* and *osteomalacia* result from the demineralization and subsequent softening of bone. Rickets occurs in children, while osteomalacia refers to the softening of adult bones.

In rickets, the epiphyseal cartilages continue to grow and are not replaced by bone; thus, the epiphyseal plates become wider than normal. At the same time, the osteoid produced in the shaft of the bone fails to ossify, causing the bones to remain comparatively soft and weak. In fact, the bones are so weak they bend as they support the weight of the body as the child walks. The bones of adults suffering from osteomalacia undergo an excessive loss of calcium and phosphorus, causing them to become soft and bend. The loss of minerals in adults is especially high in the bones of the spine, pelvis, and legs.

Rickets and osteomalacia are due to deficiencies of calcium, phosphorus, or vitamin D, or to a lack of sunlight. Ultraviolet rays in the sunlight convert sterols in the skin into vitamin D_3. Metabolically transformed vitamin D_3 facilitates the absorption of calcium and phosphate from the intestine into the bloodstream, making them available for bone formation.

Tumors of Bone

Many types of tumors, both benign and malignant, have been identified in bone. Some malignant bone tumors have traveled via the bloodstream from the lungs, breast, prostate, and other structures. Tumors weaken the bone by destroying the tissue. Their presence may be detected for the first time when a patient is x-rayed following a fracture.

Abnormal Growth Patterns

The amount of growth hormone secreted by the anterior pituitary gland can have a dramatic effect on bone development. The presence of excessive growth hormone delays the ossification of the epiphyseal cartilage. As a consequence, bone development continues for a longer period than normal, producing a *pituitary giant.* Conversely, a deficiency of growth hormone, below the level needed to maintain active epiphyseal cartilages, results in the early replacement of those cartilages by bone. This closure of the epiphyseal plates causes a halt in the development of most bones early in life, producing a midget referred to as a *pituitary dwarf.*

If excessive growth hormone is secreted after the epiphyseal cartilages have been replaced by bone, an abnormal pattern of bone growth occurs, particularly in the hands, face, and feet. This condition is called *acromegaly.*

Sometimes, under the influence of hereditary factors as well as hormone levels, the epiphyseal cartilages of the bones of the limbs function for only a short time, while the bones of the rest of the body continue developing fairly normally. This condition, called *achondroplasia,* results in a dwarf with short arms and legs but normal trunk and head. Dachshunds are achondroplastic dogs that have been selectively bred.

Effects of Aging

The loss of calcium from the bones, which is the major effect of aging on the skeletal system, is more severe in women than in men. In women, the amount of calcium in the bones steadily decreases after the age of 40, so that by the age of 70 as much as 28% of the calcium in the skeletal system has been lost. Men basically have higher bone calcium levels to start with, and generally do not begin to lose calcium until after the age of 60. The cause of the calcium loss is not known, and there are no certain methods of preventing the loss. However, it has been shown that the loss of calcium from bone is more prevalent in people who are immobilized for prolonged periods. And there is strong evidence that continued weight-bearing activities such as walking can increase the deposition of calcium into bone. Therefore, it is very

 CONDITIONS OF CLINICAL SIGNIFICANCE 

important for elderly persons, particularly women, to exercise regularly.

Normally, the matrix of bone is constantly being broken down and replaced by new matrix. In elderly people, however, along with the calcium loss, the rate of protein formation may be so slow that the organic portion of the matrix is not replaced as rapidly as it is broken down. As a consequence, the matrix of bone gradually comes to contain a greater proportion of inorganic salts. This may cause the bones of elderly people to become brittle and fracture rather easily.

Individual Bones of the Skeleton

The human skeleton consists of 206 bones (Figure 6.11). The bones can be grouped into the axial skeleton and the appendicular skeleton (Table 6.1). Your study of the skeletal system will be easier if you first familiarize yourself with some of the more common terms used to describe the structural features of bones. These terms are listed in Table 6.2. Throughout the chapter, each bone is discussed in a general way, and the reader is usually provided with both a figure that illustrates the bone and a table that gives more specific information about it.

Axial Skeleton

The axial skeleton consists of the bones that form the skull, the vertebral column, and the thorax. This portion of the skeleton provides the main axial support for the body and protects the central nervous system and the organs of the thorax.

Skull

The skull is formed of 29 bones, 11 of which are paired. With the exception of the mandible (lower jaw) and three small bones (ossicles) within each middle-ear cavity, all of the bones of the adult skull are joined together in immovable joints called **sutures.** At birth and for some years afterwards, most of these sutures are held together by fibrous connective tissue rather than by bone and therefore are capable of undergoing some movement. This permits the skull to narrow somewhat during birth by allowing the bones of the **calvarium** (the roof of the skull) to override one another when subjected to pressures within the birth canal. The presence of connective tissue sutures also allows for further growth of the skull to accommodate the normal development of the brain. At some points of junction of two or more sutures, there are fibrous membrane areas that remain prominent for up to 18 months after birth. These "soft spots" of the skull are called **fontanels** *(fon-tan-els´)* (Figure 6.12).

Calvarium

The calvarium is formed of the *frontal, parietal,* and *occipital* bones.

Frontal Bone. The **frontal** is a single bone that forms the anterior superior region of the skull (Figures 6.13 and 6.14). This bone begins its development as two separate bones that meet in a midline **metopic suture.** This suture is generally not distinguishable in the adult. The frontal bone forms the forehead and the roof of the orbital cavities. Inside of the bone, just above its junction with the nasal bones, are the **frontal sinuses.** These, like the other sinuses of the skull, are air spaces that are

◆ **TABLE 6.1 Divisions of the Skeleton**

CATEGORY	NUMBER OF BONES	
Axial skeleton		80
Skull	29	
Vertebral column	26	
Thorax (ribs and sternum)	25	
Appendicular skeleton		126
Pectoral girdle	4	
Upper limbs	60	
Pelvic girdle	2	
Lower Limbs	60	
Total	206	

◆ **FIGURE 6.11 The human skeleton, anterior and posterior views**

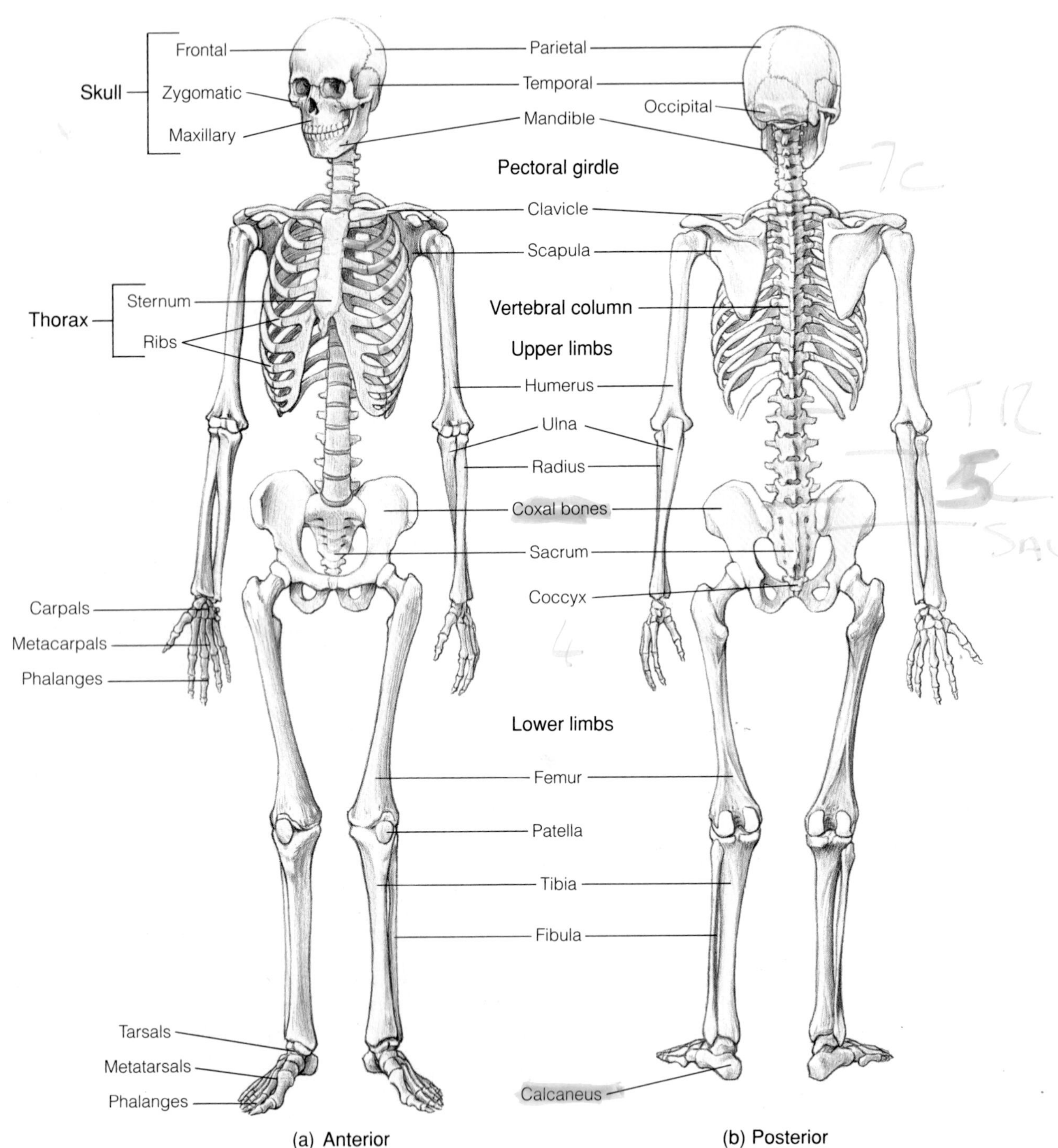

◆ **TABLE 6.2 Common Skeletal Structure Terms**

CREST
A sharp, prominent bony ridge.

CONDYLE
A rounded prominence that articulates with another bone.

EPICONDYLE
A small projection located on or above a condyle.

FACET
A smooth, nearly flat articular surface.

FISSURE
A narrow cleftlike passage.

FORAMEN
A hole.

FOSSA
A depression; often used as an articular surface.

FOVEA
A pit; generally used for attachment rather than for articulation.

HEAD
Generally, the larger end of a long bone; often set off from the shaft of the bone by a constricted neck.

LINE
A slight bony ridge.

MEATUS
A canal.

PROCESS
A prominence or projection.

RAMUS
A projecting part or elongated process.

SPINE
A slender pointed projection.

SULCUS
A groove.

TROCHANTER
A large, somewhat blunt process.

TUBERCLE
A nodule or small, rounded process.

TUBEROSITY
A broad process, larger than a tubercle.

◆ **FIGURE 6.12 Fetal skull, showing the fontanels**
(a) Superior view. (b) Lateral view.

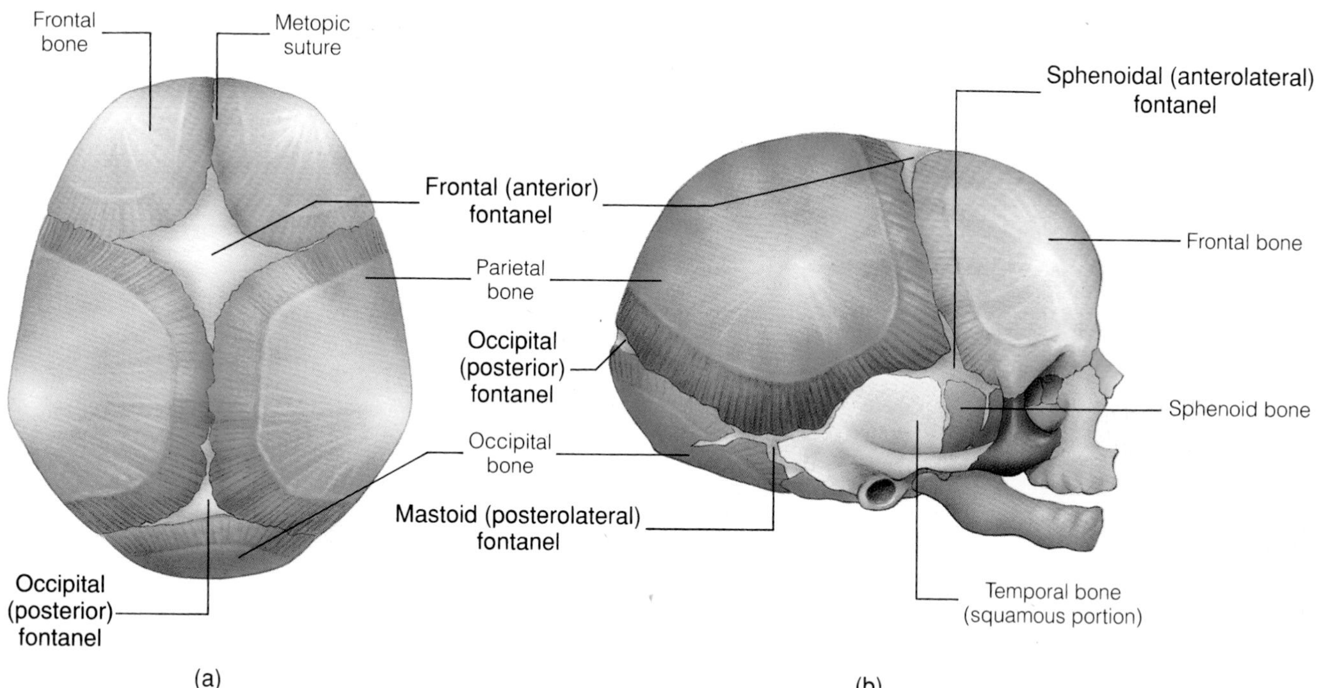

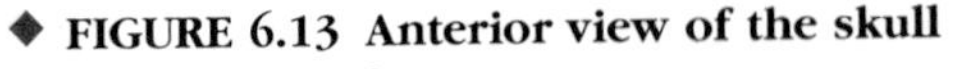

◆ **FIGURE 6.13 Anterior view of the skull**

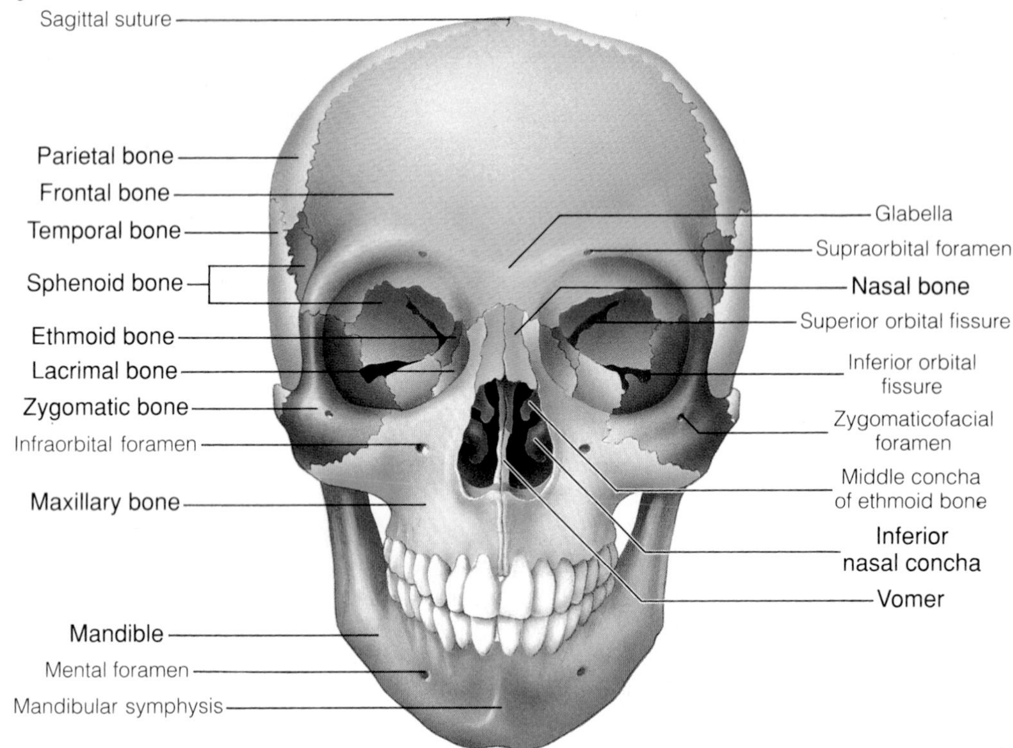

◆ **FIGURE 6.14 Lateral view of the skull**

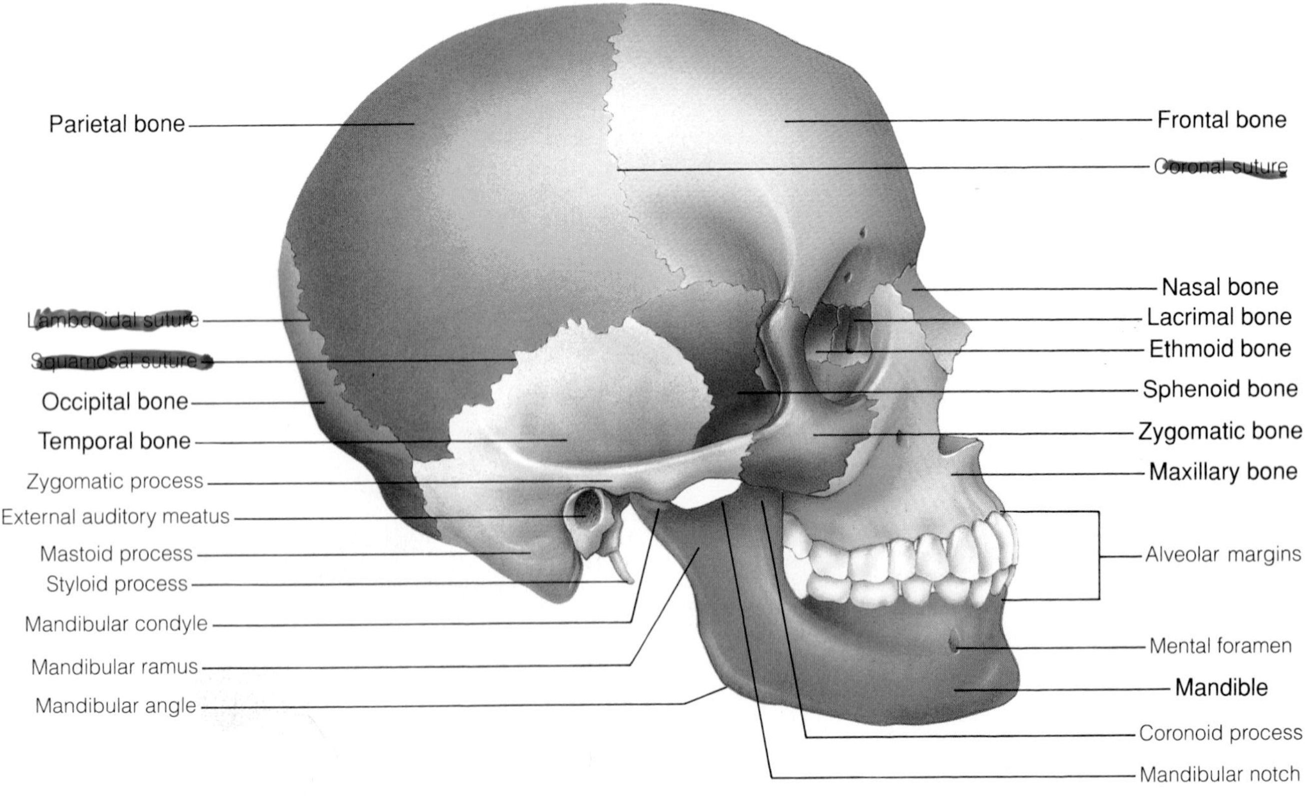

lined with mucous membrane. Posteriorly, the frontal bone joins with the two parietal bones, forming the **coronal suture.** Table 6.3 lists the features associated with the frontal bone.

Parietal Bones. Posterior to the frontal bone are the two **parietal bones,** which form most of the calvarium (Figures 6.13 and 6.14). The parietals meet in the midline, forming the **sagittal suture.** The parietal bones form the **coronal suture** across the top of the skull, where they meet with the frontal bone; the **lambdoidal suture** at the back of the skull, where they meet with the occipital bone; and the **squamosal sutures** along the lower sides of the skull where they meet with the temporal bones. In a young child's skull, the fibrous fontanels occupy the points of junction of these sutures (see Figure 6.12). The **frontal (anterior) fontanel** is located at the junction of the sagittal and coronal sutures. The **occipital (posterior) fontanel** is located at the junction of the sagittal and lambdoidal sutures. There are right and left **sphenoidal (anterolateral) fontanels** at the junction of the coronal and squamosal sutures, and **mastoid (posterolateral) fontanels** where the squamosal sutures meet the lambdoidal sutures.

Occipital Bone. The single **occipital bone** forms the lower posterior wall of the roof of the skull, as well as the posterior portion of the floor of the cranial cavity (Figures 6.13 and 6.14). Its most obvious landmark is the large **foramen magnum,** by which the cranial cavity communicates with the vertebral canal. The medulla oblongata of the brain passes through this foramen. Anterior and lateral to the foramen magnum are two **occipital condyles,** one on each side. These condyles articulate with the superior surface of the first cervical vertebra and provide the only connection between the skull and the vertebral column. In addition to articulating with the parietal bones (lambdoidal suture), the occipital bone is also supported against the temporal and sphenoid bones. Table 6.3 lists the features of the occipital bone.

Bones That Form the Floor of the Cranial Cavity

The floor of the cranial cavity, upon which the brain rests, is formed by six bones: the midline *frontal, ethmoid, sphenoid* and *occipital* and the paired *temporals* (Figures 6.15, 6.16, and 6.17). We have already discussed the frontal and occipital bones. The frontal bone

◆ **FIGURE 6.15 Inferior view of the skull**

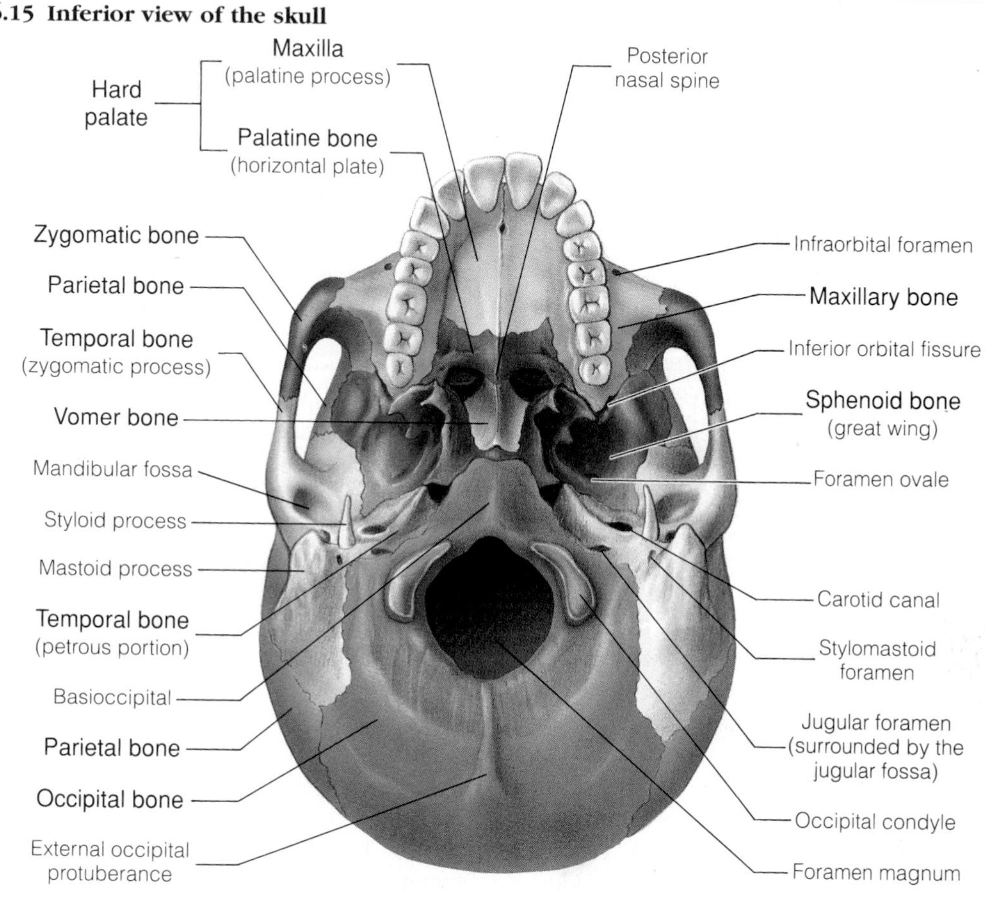

◆ **FIGURE 6.16 Superior view of the skull with the calvarium removed, showing the floor of the cranial cavity**

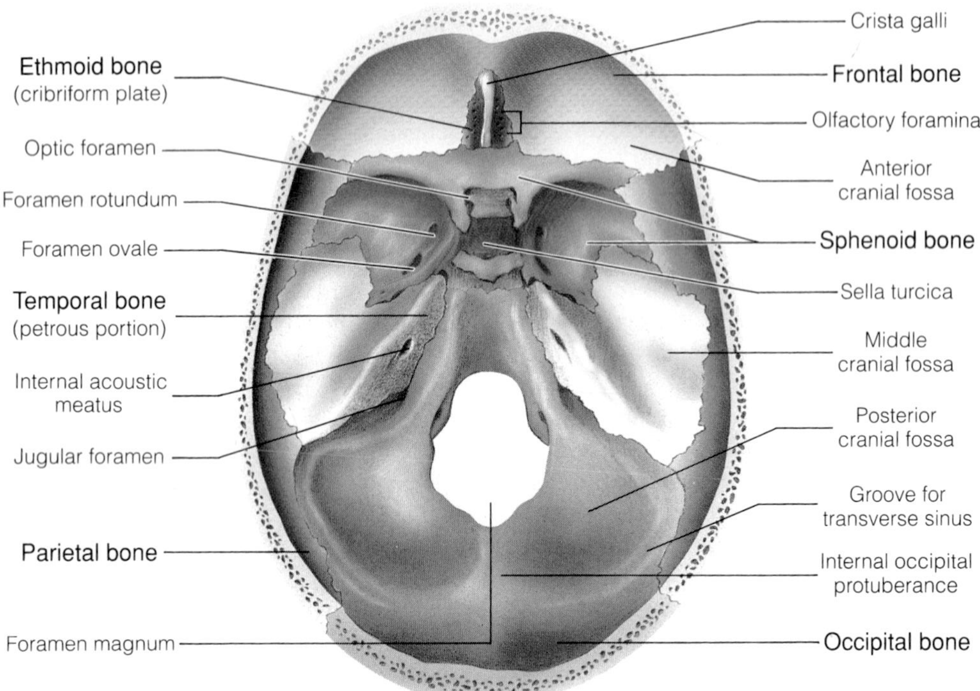

(together with the ethmoid) forms the **anterior cranial fossa;** the occipital bone forms the **posterior cranial fossa.** We will now consider the ethmoid, the sphenoid, and the temporal bones. The sphenoid and temporal bones form the **middle cranial fossa.**

Ethmoid Bone. The **ethmoid bone** is situated in the middle of the floor of the anterior cranial fossa, where it forms most of the walls of the upper portion of the nasal cavity (Figures 6.16, 6.17, and 6.18). The ethmoid bone is lightweight and delicate, containing many air sinuses. It has four parts: the **horizontal (cribriform) plate,** with the midline **perpendicular plate,** and two **lateral masses** that project downward from the horizontal plate.

The horizontal plate joins with the frontal bone to form the floor of the anterior cranial fossa. This plate is called the cribriform plate (*cribriform* = sievelike) because it is perforated by many tiny **olfactory foramina.** The olfactory nerves pass through these foramina in traveling between the mucous membranes of the nasal cavity and the olfactory bulbs of the brain. The perpendicular plate forms the major part of the nasal septum, which divides the nose into right and left nasal cavities. On the medial surfaces of the lateral masses are projections called **superior** and **middle conchae (turbinates),** which form the sidewalls of the nasal cavities; the smooth lateral surfaces of the lateral masses are referred to as the **lamina orbitalis,** because they form part of the medial walls of the orbital cavities. Table 6.3 lists additional information concerning the ethmoid bone.

Sphenoid Bone. The **sphenoid bone** extends completely across the floor of the middle cranial fossa (Figures 6.15, 6.16, and 6.19). The sphenoid is surrounded on all sides by other bones, articulating posteriorly with the basioccipital portion of the occipital bone, laterally with the temporal and parietal bones, and anteriorly with the frontal and ethmoid bones.

The sphenoid has a complex shape, with a central **body,** from which pairs of **small (lesser) wings, great wings,** and **pterygoid processes** *(ter´-a-goid)* project. The anterior surfaces of the great wings form most of the posterior walls of the orbital cavities. The **optic foramina,** located in the bases of the small wings, provide for the passage of the optic nerves from the eyes to the base of the brain. The superior surface of the body of the sphenoid contains a deep depression called the **sella turcica** (Turk's saddle). The sella turcica houses the pituitary gland. The sella turcica is bounded posteriorly by a ridge of bone called the **dorsum sellae.** Table 6.3 lists additional information concerning the sphenoid bone.

◆ **FIGURE 6.17 The skull**
(a) Sagittal view. (b) Photo of sagittal view.

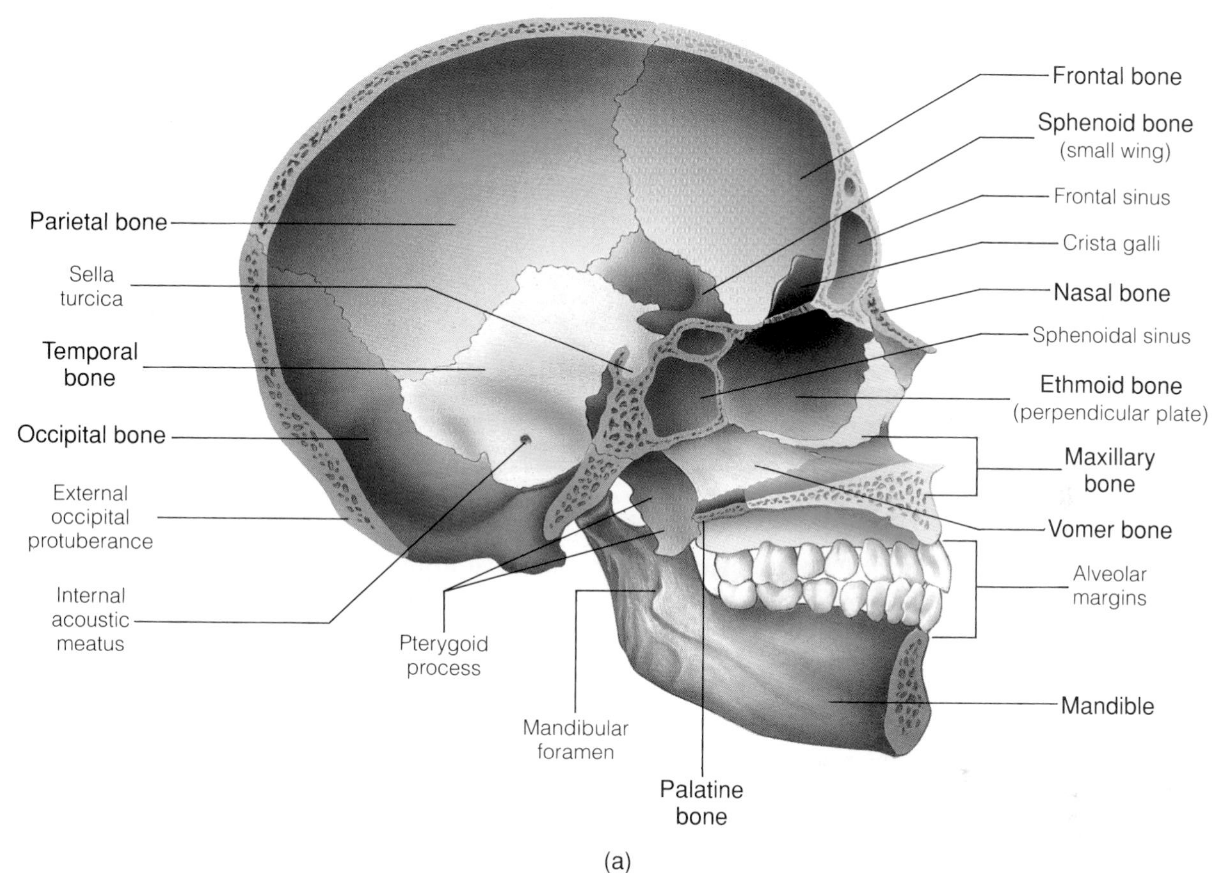

(a)

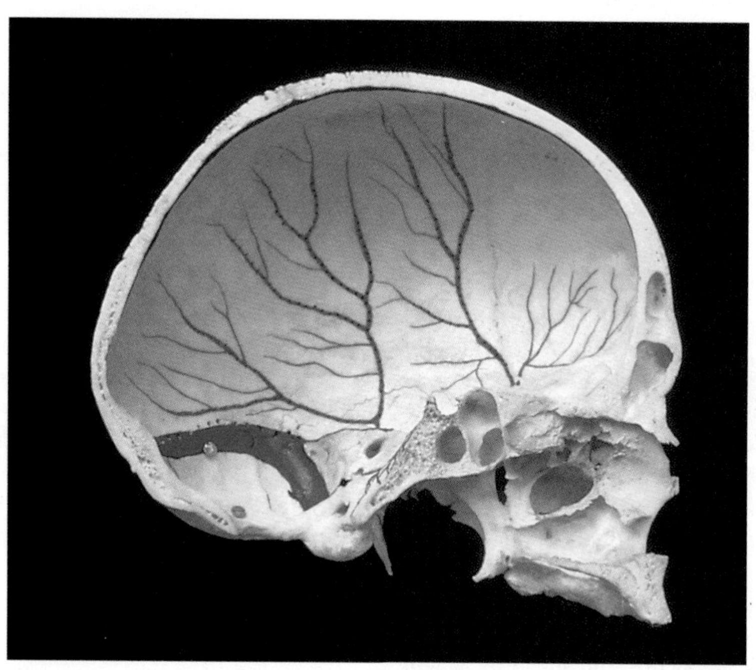

(b)

◆ **FIGURE 6.18 Anterior view of the ethmoid bone**

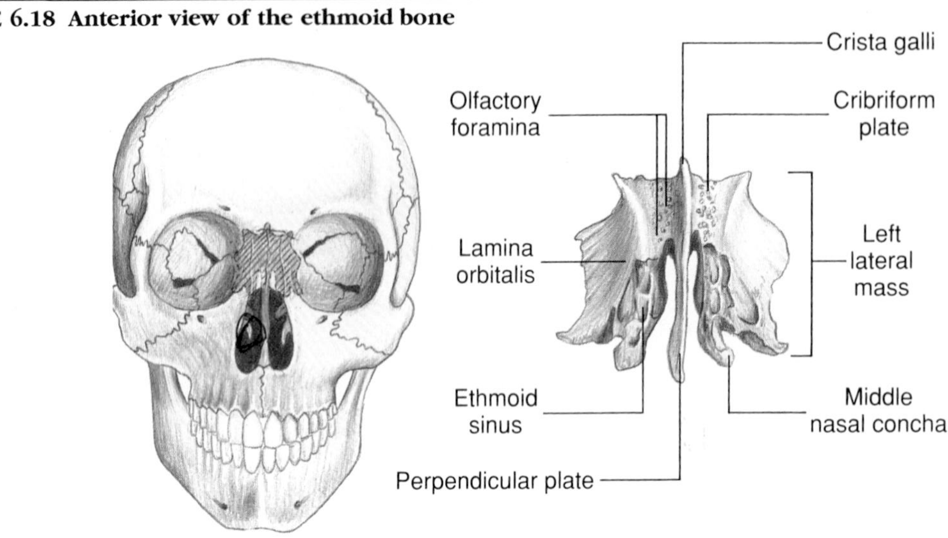

Temporal Bones. The two **temporal bones,** together with the sphenoid bone, form the *middle cranial fossa* (Figures 6.14, 6.15, 6.16, and 6.17). Each bone consists of four regions (Figure 6.20):

1. The thin **squamous portion** projects upward, articulating with the parietal bone in the squamous suture.
2. The **tympanic portion** forms the walls of the external auditory meatus and the region of the bone that closely surrounds the meatus.
3. The **petrous portion** (Figures 6.15 and 6.16) projects medially between the sphenoid and occipital bones. It contains the middle-ear and inner-ear cavities.
4. The **mastoid portion** is located posterior to the external auditory meatus.

Table 6.3 lists additional information concerning the temporal bones.

Facial Skeleton

Ten bones form most of the facial skeleton (Figure 6.13): the unpaired *frontal* and *mandible* bones and the paired *maxillary, zygomatic, nasal,* and *lacrimal* bones. We have previously described the frontal bone.

Maxillary Bones. The **maxillary bones (maxillae)** form the central part of the facial skeleton (Figure 6.21). With the exception of the mandible, all of the facial bones articulate directly with the maxillae. The two maxillary bones join in the midline to form the upper jaw. In addition, each assists in forming the roof of the mouth, the floor and lateral wall of the nasal cavity, and the floor of the orbit. The teeth of the upper jaw are embedded in the lower margin of the maxillae. Large maxillary sinuses are within the body of the bone. Table 6.3 lists additional information concerning the maxillary bones.

Zygomatic Bones. The **zygomatic (malar) bones** articulate with the maxillary and temporal bones to form the prominences of the cheek (Figures 6.13 and 6.14). They also articulate with the frontal and sphenoid (great wing) to form part of the floor and lateral wall of the orbit.

Nasal Bones. The **nasal bones** (Figure 6.21) are two small oblong bones that meet at the midline of the face to form the bridge of the nose. In addition, they articulate with the frontal, ethmoid (perpendicular plate), and maxillary bones (frontal process).

Lacrimal Bones. The right and left **lacrimal bones** (Figure 6.21) are small delicate bones that help to form the medial surface of the orbital cavity. They articulate above with the frontal bone, behind with the ethmoid (orbital surfaces of the lateral masses), and in front with the maxillary bones (frontal process). Near their anterior edges, each lacrimal bone has a **lacrimal sulcus** for the lacrimal sac and the nasolacrimal duct. The lacrimal sac and the nasolacrimal duct transport tear fluid from the surface of the eye to the nasal cavity.

Mandible. The facial skeleton is completed by the **mandible** (Figure 6.14), which forms the lower jaw.

◆ **FIGURE 6.19 Sphenoid bone**

(a) Superior view. (b) Posterior view.

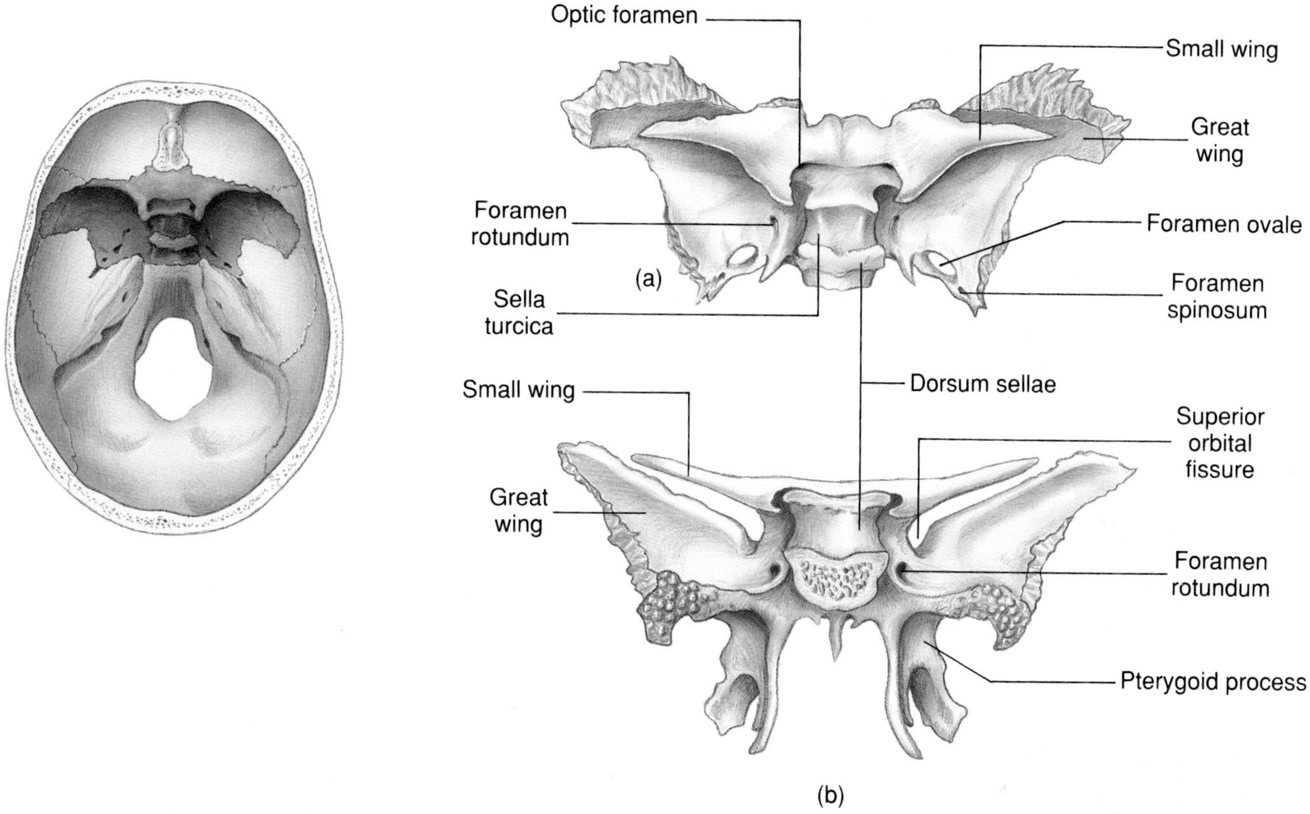

Optic foramen
Small wing
Great wing
Foramen rotundum
Foramen ovale
Foramen spinosum
(a)
Sella turcica
Dorsum sellae
Small wing
Superior orbital fissure
Great wing
Foramen rotundum
Pterygoid process
(b)

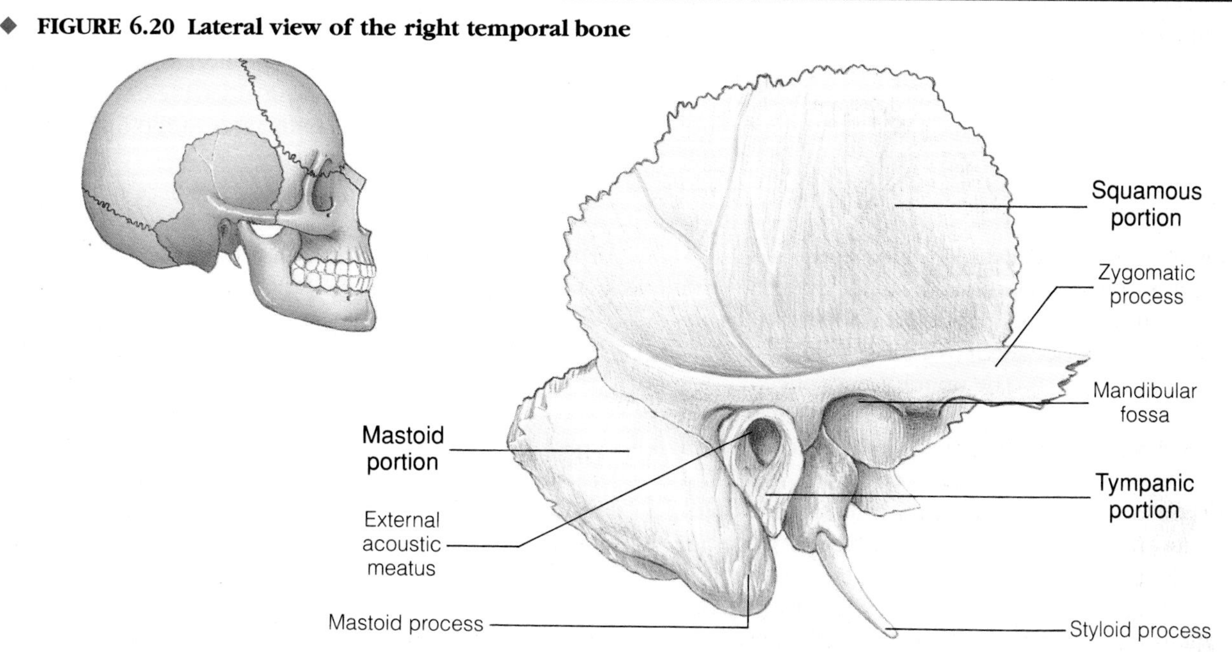

◆ **FIGURE 6.20 Lateral view of the right temporal bone**

Squamous portion
Zygomatic process
Mandibular fossa
Mastoid portion
Tympanic portion
External acoustic meatus
Mastoid process
Styloid process

◆ **FIGURE 6.21 Lateral view of the right maxillary, nasal, and lacrimal bones**

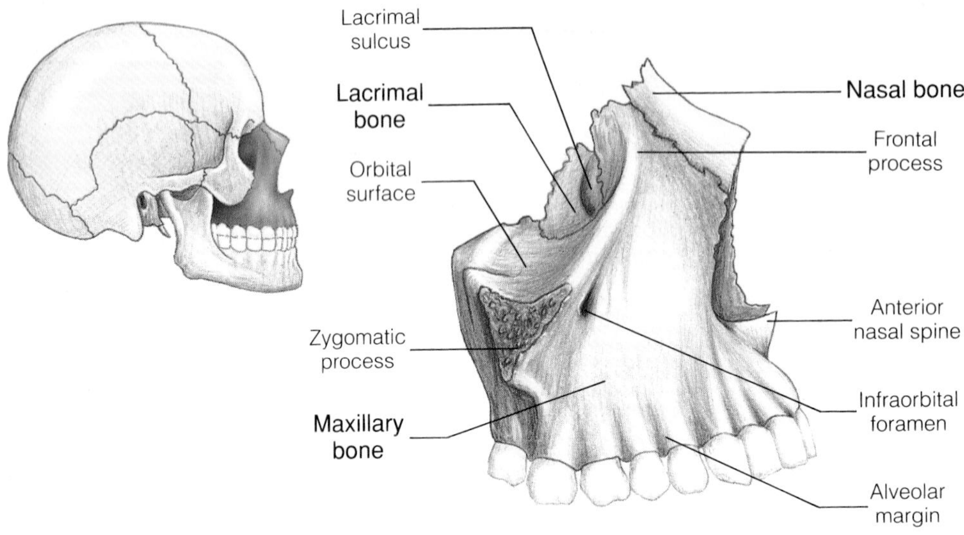

The mandible consists of a horizontal horseshoe-shaped **body** and two perpendicular **rami.** The teeth of the lower jaw are embedded in the upper margin of the body. Condyloid processes located on the superior margins of the rami form movable joints with the mandibular fossae of the temporal bones. Table 6.3 lists additional information concerning the mandible.

Bones That Form the Nasal Cavity

We have already described how most of the nasal septum is formed by the perpendicular plate of the ethmoid. The superior and middle conchae of the ethmoid bone form much of the lateral walls of the nasal cavity (see Table 6.3). Three additional bones also contribute to the formation of the nasal cavity: the *vomer* and the two *inferior nasal conchae.*

The **vomer** (Figure 6.17) is a thin quadrangular bone that forms the posterior inferior portion of the nasal septum. Its superior border articulates with the sphenoid, between the pterygoid processes; its inferior border articulates with the upper surface of the hard palate (maxillae and palatine bones). The upper part of the vomer's anterior border articulates with the perpendicular plate of the ethmoid, and the lower part is in contact with the cartilage portion of the nasal septum.

The paired **inferior nasal conchae** (Figure 6.22) form elongated shelves that protrude medially from the lateral walls of the nasal cavity. They are located just below the middle concha of the ethmoid bone. The inferior nasal conchae articulate with the maxillae, lacrimals, ethmoid, and palatine bones.

Bones That Form the Hard Palate

The hard palate forms the roof of the mouth (Figures 6.15 and 6.22). The anterior portion of the hard palate is formed by the **palatine processes** of the **maxillary bones.** The posterior portion is formed by the **horizontal plates** of the two **palatine bones.**

Each palatine bone is L-shaped, with a horizontal and a vertical portion (Figure 6.23). On the posterior edge of the horizontal plates, at their midline point of junction, is a sharp projection called the **posterior nasal spine.** This spine serves for the attachment of the uvula, a small fleshy mass that hangs from the soft palate. The vertical portion of each palatine bone forms part of the posterior lateral wall of the nasal cavities. A small portion of the vertical plate contributes to the formation of the orbital cavity.

Bones That Form the Orbital Cavity

We have already described the contribution of the various bones to the formation of the orbit (Figure 6.24). These are summarized in Table 6.4 on page 184.

Paranasal Sinuses

Located within the *frontal, ethmoid, maxillary,* and *sphenoid* bones are a series of air spaces called the **paranasal sinuses** (Figure 6.25). The paranasal sinuses, which drain into the nasal cavity, are lined with a mucous membrane that is continuous with the mucous membrane of the nasal cavity. The sinuses lighten the

◆ **FIGURE 6.22 Bones that form the left lateral wall of the nasal cavity**

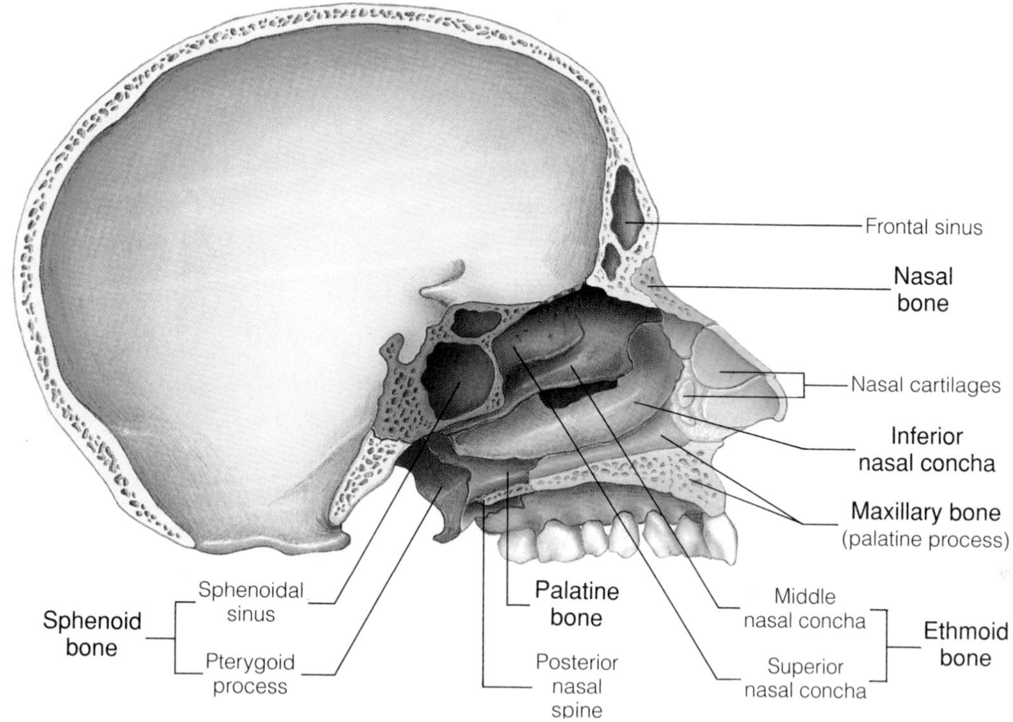

skull and enhance the resonance of the voice. The passages connecting the sinuses with the nasal cavity allow air to enter the sinuses from the nasal cavity. The passages also allow mucus formed by the mucous membranes lining the sinuses to drain into the nasal cavity. The **frontal sinuses** (Figures 6.17 and 6.22) are found above the medial ends of the orbits in the region of the glabella. The **ethmoid sinuses** (Figure 6.18) are a series of small spaces located in the lateral masses of the bone. The **maxillary sinuses,** the largest of the paranasal sinuses, occupy much of the bone from the orbit to the alveolar processes. The **sphenoidal sinuses** (Figures 6.17 and 6.22) are contained within the body of the bone.

Auditory Ossicles

Three tiny bones are called **auditory ossicles** are located in the *middle-ear (tympanic) cavities* (Figure 6.26). The cavities are inside the petrous portion of each temporal bone. The ossicles form a bridge across the tympanic cavity from the eardrum (tympanic membrane) to a membrane-covered opening (oval window) that separates the middle ear from the inner ear.

The **malleus** (hammer) is attached to the inside surface of the tympanic membrane. The **stapes** (stirrup) fits against the oval window. The **incus** (anvil) forms a connection between the malleus and the stapes. The role played by these bones in transmitting sound is discussed in Chapter 16.

Hyoid Bone

The **hyoid bone** (Figure 6.27) is a U-shaped bone suspended by ligaments from the styloid processes of the temporal bones. It is located just above the larynx. The hyoid consists of a central **body** and pairs of **greater**

◆ **FIGURE 6.23 Left palatine bone (posterior view)**

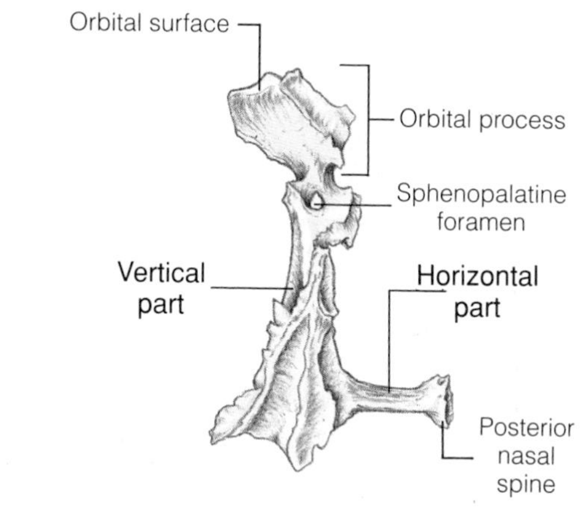

◆ **FIGURE 6.24 Frontal view of the right orbit, showing the bones that form the orbital cavity**

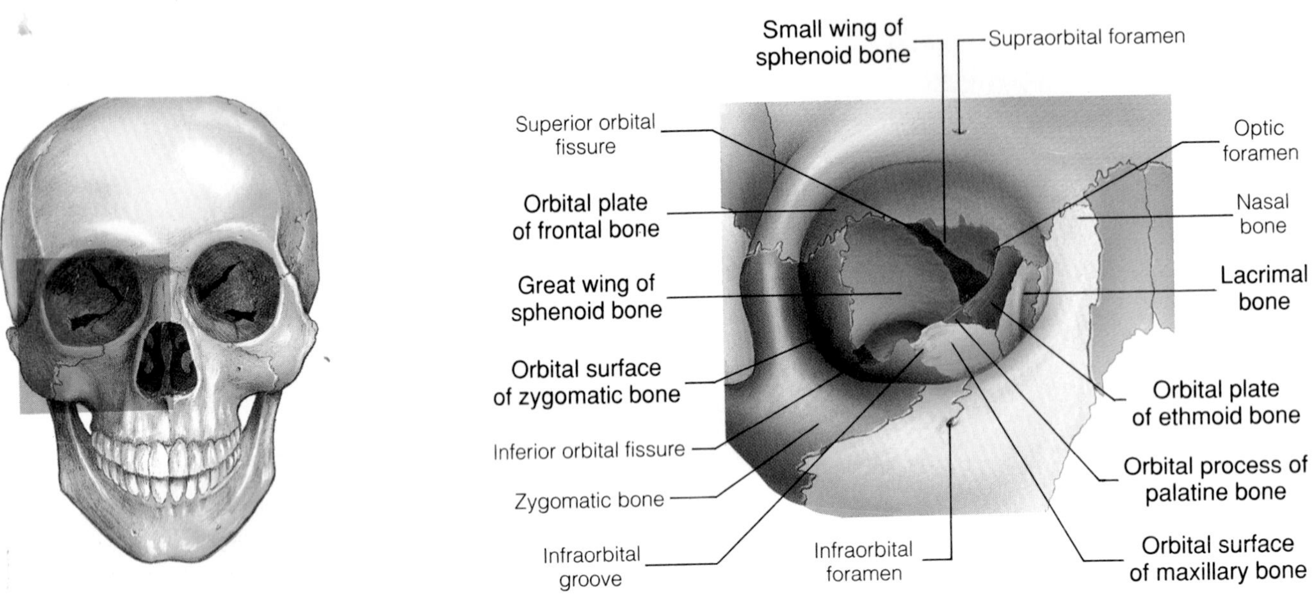

◆ **FIGURE 6.25 X ray of the skull, showing several paranasal sinuses**

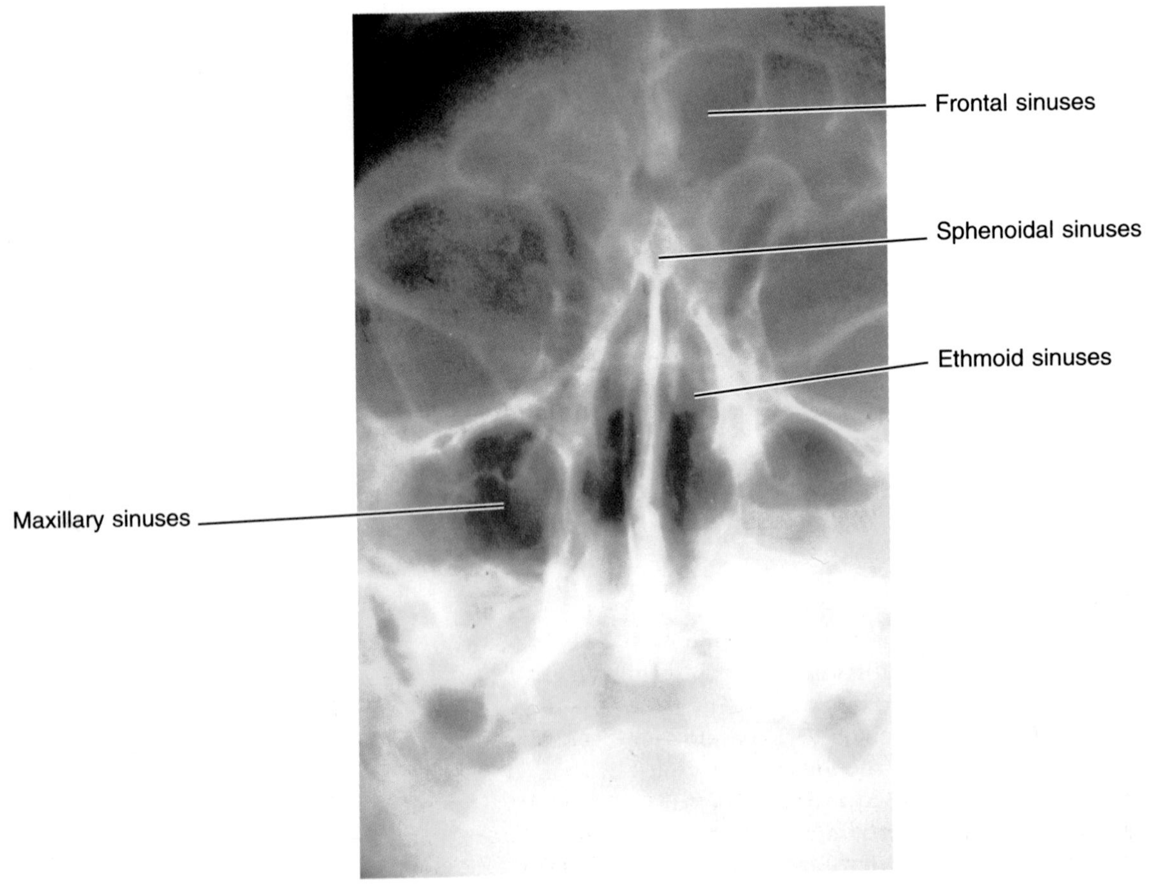

◆ **FIGURE 6.26 The middle-ear cavity and the auditory ossicles**

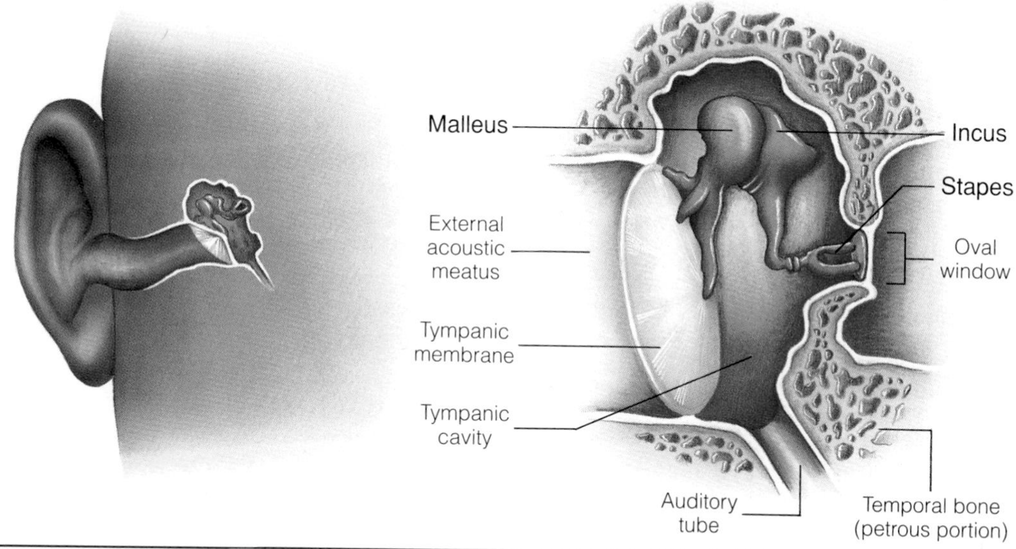

◆ **FIGURE 6.27 Anterior view of the hyoid bone**

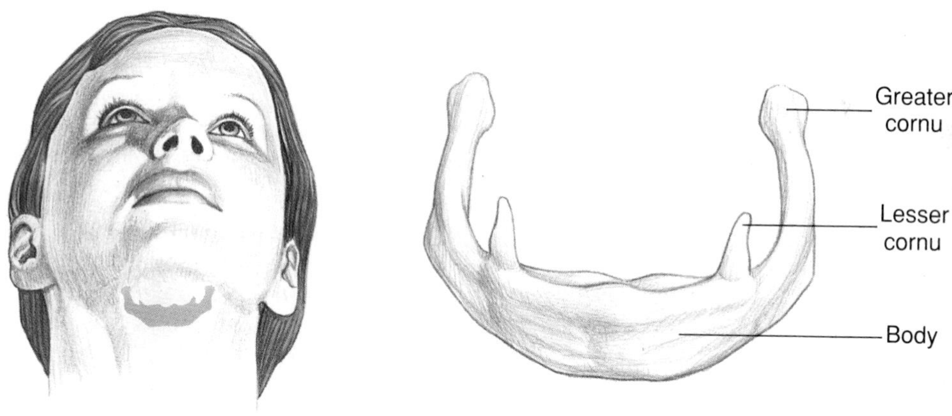

and **lesser cornua (horns).** It serves as points of attachment for muscles of the tongue and throat.

Vertebral Column

The embryonic vertebral column consists of 33 vertebrae. The vertebrae are separated into five different types, depending on the regions of the body in which they are located. The upper 7 are **cervical** vertebrae; these are followed by 12 **thoracic, 5 lumbar, 5 sacral,** and 4 **coccygeal** *(kok-sij´-ee-al)* vertebrae. In the adult, the sacral vertebrae fuse into a single **sacrum,** and the coccygeal vertebrae fuse to form a **coccyx.** Therefore, the adult vertebral column has 26 separate bones (Figure 6.28, page 185).

Curvatures of the Vertebral Column

When viewed from the lateral aspect, the vertebral column of a newborn infant has a single curve, called a *primary curve,* which is convex posteriorly. As the child begins to raise its head, a **cervical curve** that is convex anteriorly develops in the neck region. In a similar manner, an anterior **lumbar curve** develops in the small of the back as the child begins to walk. Because the cervical and lumbar curves are modifications of the primary curve, they are referred to as *secondary curves.* In the thoracic and sacral regions, remnants of the primary curve form a **thoracic curve** and a **sacral curve,** respectively, that are convex posteriorly.

If the anterior lumbar curve is excessive, it is called a **lordosis** (swayback). If the posterior thoracic curve is excessive, it is called a **kyphosis** (hunchback). The vertebral column is normally straight, without lateral curvatures. If a lateral curve does exist, it is called a **scoliosis.**

Functions of the Vertebral Column

The vertebral column is the main axial support for the body, providing attachments for the skull, the thorax,

◆ **TABLE 6.3 Summary of Specific Features of Individual Skull Bones**

FRONTAL BONES [*Figs. 6.13; 6.14; 6.16; 6.17; 6.24*]	
METOPIC SUTURE The line of junction between the two separate embryonic ossification centers. Generally not present in the adult skull. **FRONTAL SINUSES** [*Fig. 6.17*] Mucuous-membrane–lined air cavities located within the bone, close to the orbital cavities.	**SUPRAORBITAL FORAMINA OR NOTCHES** [*Fig. 6.13*] Openings for blood vessels and nerves located just above the orbital cavities. They may appear as holes (foramina) or notches. **GLABELLA** [*Fig. 6.13*] The smooth area located between the two orbital cavities just above the nose.
OCCIPITAL BONE [*Figs. 6.14; 6.15; 6.16; 6.17*]	
FORAMEN MAGNUM [*Figs. 6.15; 6.16*] The opening through which the medulla oblongata of the brain stem leaves the skull to become continuous with the spinal cord. **CONDYLES** [*Fig. 6.15*] Smooth convex external projections on either side of the foramen magnum. They articulate with the first cervical vertebra. **BASIOCCIPITAL** [*Fig. 6.15*] A narrow portion that extends anteriorly from the foramen magnum. It articulates with the sphenoid bone. **EXTERNAL OCCIPITAL PROTUBERANCE** [*Figs. 6.15; 6.17*] A midline prominence on the outer surface, a short distance above the foramen magnum.	**NUCHAL LINES*** Slight ridges on the external surface. *Medial nuchal line.* Runs vertically between the external occipital protuberance and the foramen magnum. *Superior nuchal line.* Extends laterally from external occipital protuberance. *Inferior nuchal line.* Extends laterally from the medial nuchal line at about its midpoint. **INTERNAL OCCIPITAL PROTUBERANCE** [*Fig. 6.16*] A prominence on the inner surface of the bone. This marks the confluence of grooves for the sagittal, transverse, and occipital venous blood sinuses of the brain.
ETHMOID BONE [*Figs. 6.16; 6.17; 6.18; 6.22; 6.24*]	
HORIZONTAL (CRIBRIFORM) PLATE [*Figs. 6.16; 6.18*] The transverse portion that forms the roof of the nasal cavity and floor of the anterior cranial cavity. The plate is perforated by the olfactory foramina to allow for the passage of the olfactory nerves (first cranial nerve). **CRISTA GALLI** [*Figs. 6.16; 6.17; 6.18*] A midline projection from the horizontal plate into the cranial cavity. It serves as the anterior point of attachment for the **falx cerebri,** a midline connective tissue septum that anchors the brain within the anterior cranial fossa. **PERPENDICULAR PLATE** [*Figs. 6.17; 6.18*] A downward projection from the midline of the undersurface of the horizontal plate. It forms the upper portion of the nasal septum. The remainder of the septum is formed by the vomer bone and hyaline cartilage.	**LATERAL MASSES** [*Fig. 6.18*] Thin-walled processes that extend downward from the lateral margins of the horizontal plate. They contain the **ethmoid sinuses,** which are mucous-membrane–lined air cavities. The smooth lateral surfaces *(lamina orbitalis)* of the lateral masses form the medial walls of the orbital cavities. **SUPERIOR AND MIDDLE CONCHAE (TURBINATES)** [*Fig. 6.22*] Thin plates of bone that form the medial surfaces of the lateral masses. They also form part of the lateral walls of the nasal cavity. Recesses called **superior, middle,** and **inferior meatuses** are located beneath the shelves of the conchae.

◆ **TABLE 6.3 Summary of Specific Features of Individual Skull Bones (continued)**

SPHENOID BONE [*Figs. 6.13; 6.14; 6.15; 6.16; 6.17; 6.19*]

BODY [*Fig. 6.19*]
The central portion of the bone. It contains a large mucous-membrane-lined air sinus.

SELLA TURCICA [*Figs. 6.16; 6.19*]
A saddle-shaped depression on the superior surface of the body, bounded posteriorly by the **dorsum sellae.** It serves as the protective cavity for the pituitary gland.

SMALL WINGS [*Figs. 6.17; 6.19*]
Sharp lateral projections from the superior portion of the body of the sphenoid. They form part of the posterior walls of the orbital cavities.

OPTIC FORAMINA [*Figs. 6.16; 6.19*]
Openings through the bases of each small wing for the passage of the optic nerves (second cranial nerves) into the orbital cavities.

GREAT WINGS [*Figs. 6.15; 6.19*]
Large lateral projections from the body of the sphenoid. They form most of the posterior wall of the orbital cavity.

SUPERIOR ORBITAL FISSURES [*Figs. 6.13; 6.19; 6.24*]
Slitlike openings between the great and small wings. They allow for the passage of the third, fourth, part of the fifth (opthalmic division), and the sixth cranial nerves from the brain into the orbital cavity.

FORAMEN ROTUNDUM [*Figs. 6.16; 6.19*]
The opening through the base of each of the great wings for the passage of the maxillary division of the fith cranial nerves.

FORAMEN OVALE [*Figs. 6.15; 6.16; 6.19*]
The opening through the base of each of the great wings for the passage of the mandibular division of the fifth cranial nerves.

PTERYGOID PROCESSES [*Figs. 6.17; 6.22*]
Two downward projections from the region where the great wings unite with the body. Each process consists of **medial** and **lateral plates.** The processes articulate anteriorly with the palatine bones.

TEMPORAL BONE [*Figs. 6.13; 6.14; 6.15; 6.16; 6.17; 6.20*]

SQUAMOUS PORTION [*Fig. 6.20*]
The thin vertical projection that forms the anterior and superior portion of the bone. It meets with a parietal bone to form the squamous suture.

ZYGOMATIC PROCESS [*Fig. 6.20*]
The anterior projection from the squamous portion. It articulates with the zygomatic (malar) bone to form the cheek (zygomatic arch).

MANDIBULAR FOSSA [*Fig.6.20*]
An oval depression on the inferior surface of the base of the zygomatic process. It articulates with the condyle of the mandible to form the temporomandibular joint.

TYMPANIC PORTION [*Fig. 6.20*]
Forms and surrounds the external acoustic meatus.

EXTERNAL ACOUSTIC MEATUS [*Fig. 6.20*]
The opening that leads into the middle-ear cavity from the exterior of the skull.

PETROUS PORTION [*Figs. 6.15; 6.16*]
A medial wedge of bone that forms the floor of the middle cranial fossa between the sphenoid and the occipital bones. It houses the middle- and inner-ear structures.

INTERNAL ACOUSTIC MEATUS [*Figs. 6.16; 6.17*]
The opening on the posterior surface of the petrous portion. It transmits the seventh cranial nerve as it travels to the facial structures, and the eighth cranial nerve as it travels to the inner ear.

STYLOID PROCESS [*Figs. 6.14; 6.15; 6.20*]
A sharp spine that projects from the inferior lateral surface of the petrous portion. It serves as a point of attachment for the hyoid bone and for several ligaments and muscles of the pharynx and tongue.

CAROTID CANAL [*Fig. 6.15*]
The passageway for the internal carotid artery as it travels through the petrous portion.

JUGULAR FOSSA [*Fig. 6.15*]
The depression for the internal jugular vein on the inferior surface of the petrous portion.

STYLOMASTOID FORAMEN [*Fig. 6.15*]
The opening between the styloid process and the mastoid process through which the seventh cranial nerve leaves the skull. (The nerve enters through the internal acoustic meatus.)

JUGULAR FORAMEN [*Figs. 6.15; 6.16*]
The large opening that allows for passage of several blood vessels, and the ninth, tenth, and eleventh cranial nerves. It is located at the junction of the petrous portion with the occipital bone.

MASTOID PROCESS [*Figs. 6.14; 6.15; 6.20*]
A prominent downward projection from the mastoid portion, just posterior to the external acoustic meatus.

MASTOID SINUSES*
Mucous-membrane-lined air spaces within the mastoid process. These sinuses, which communicate with the middle-ear cavity, are the only cranial sinuses that do not drain into the nasal cavity.

continued on next page

◆ **TABLE 6.3 Summary of Specific Features of Individual Skull Bones (continued)**

MAXILLARY BONE (MAXILLA) [*Figs. 6.13; 6.14; 6.15; 6.17; 6.21*]

MAXILLARY SINUS [*Fig. 6.25*]
A large mucous-membrane-lined cavity within the bone.

FRONTAL PROCESS [*Fig. 6.21*]
The vertical process that forms part of the bridge of the nose. It articulates above with the frontal bone, anteriorly with the nasal, and posteriorly with the lacrimal.

ZYGOMATIC PROCESS [*Fig. 6.21*]
A rough triangular eminence that articulates with the zygomatic (malar) bone.

ALVEOLAR MARGIN [*Figs. 6.14; 6.21*]
The inferior border that holds the teeth. When the two maxillae are articulated with each other, their alveolar processes together form the alveolar arch.

PALATINE PROCESS [*Figs. 6.15; 6.22*]
The medial horizontal shelf that runs from the inner surface of the alveolar process. It joins with the palatine process of the other maxillary bone to form most of the hard palate.

ANTERIOR NASAL SPINE [*Fig. 6.21*]
A pointed process just below the nasal cavity. It joins with the nasal spine of the other maxillary bone to form a point of attachment for the cartilage portion of the nasal septum.

INFRAORBITAL FORAMEN [*Figs. 6.13; 6.21; 6.24*]
The opening just below the margin of the orbit. It transmits blood vessels and nerves.

ORBITAL SURFACE [*Fig. 6.24*]
The smooth, flat surface that forms the floor of the orbit.

MANDIBLE [*Figs. 6.13; 6.14; 6.17*]

BODY [*Fig. 6.14*]
The curved, horizontal portion that forms the chin.

RAMI [*Fig. 6.14*]
Two perpendicular projections that join the posterior lateral margins of the body at approximately right angles.

MANDIBULAR SYMPHYSIS [*Fig. 6.13*]
The vertical midline fusion between the two embryonic ossification centers that form the body.

ALVEOLAR MARGIN [*Figs. 6.14; 6.17*]
The superior edge of the body that contains the sockets for the teeth.

MENTAL FORAMINA [*Figs. 6.13; 6.14*]
Two openings on the external surface of the body that allow for the passage of blood vessels and nerves.

ANGLE [*Fig. 6.14*]
A sharp curve on the posterior inferior portion of the ramus.

MANDIBULAR FORAMINA [*Fig. 6.17*]
Openings on the inner surfaces of each ramus for the passage of blood vessels and nerves.

CORONOID PROCESSES [*Fig. 6.14*]
Thin upward projection on the anterior surface of each ramus. They provide attachment for the temporalis muscle.

MANDIBULAR CONDYLES [*Fig. 6.14*]
Smooth convex surface on the superior borders of each ramus. They articulate with the mandibular fossae of the temporal bones.

MANDIBULAR NOTCHES [*Fig. 6.14*]
Deep depression between the coronoid process and the condyle of each ramus.

*Not illustrated.

◆ **TABLE 6.4 Bones That Form the Orbit**

Roof of orbit
- Frontal
- Sphenoid (small wing)

Medial wall of orbit
- Maxilla (frontal process)
- Lacrimal
- Ethmoid

Lateral wall of orbit
- Zygomatic
- Sphenoid (great wing)

Floor of orbit
- Maxilla
- Palatine

and the pelvic girdle. Although it is a major support structure, its construction is such that it permits the trunk of the body to have appreciable flexibility, and the presence of the four anterior-posterior curves allows it to absorb the shocks associated with walking or running. In addition, the vertebral column protects the spinal cord while providing openings between adjacent vertebrae for the passage of spinal nerves.

Characteristics of a Typical Vertebra

Although there are differences among the vertebrae of the various regions of the spinal column, enough similarities exist so that it is possible to describe a typical vertebra (Figure 6.29).

◆ **FIGURE 6.28 The vertebral column**
(a) Lateral view of the vertebral column. (b) Lateral X ray of the cervical vertebrae.

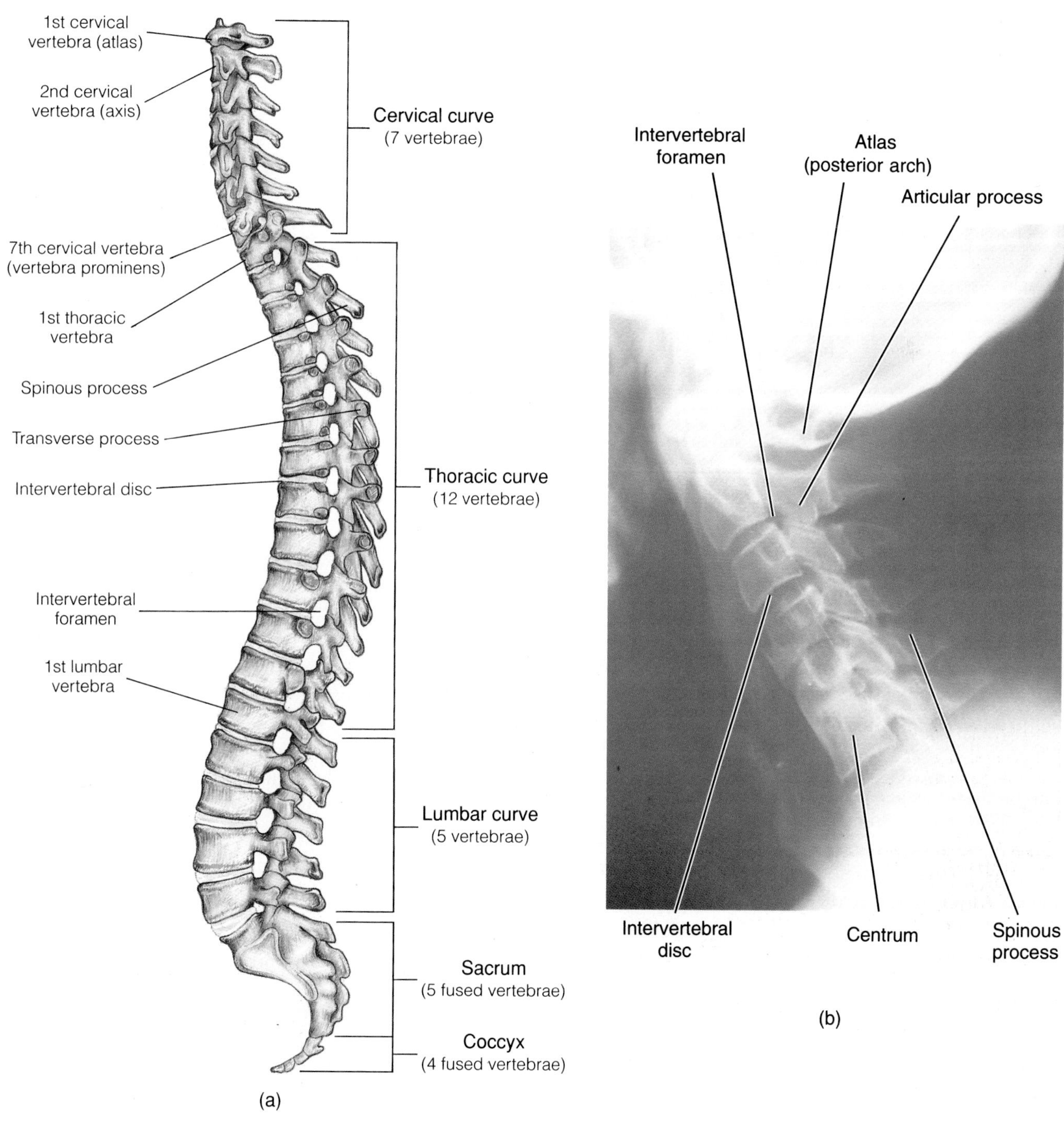

◆ **FIGURE 6.29 Structural features of a typical vertebra**
(a) Superior view. (b) Lateral view.

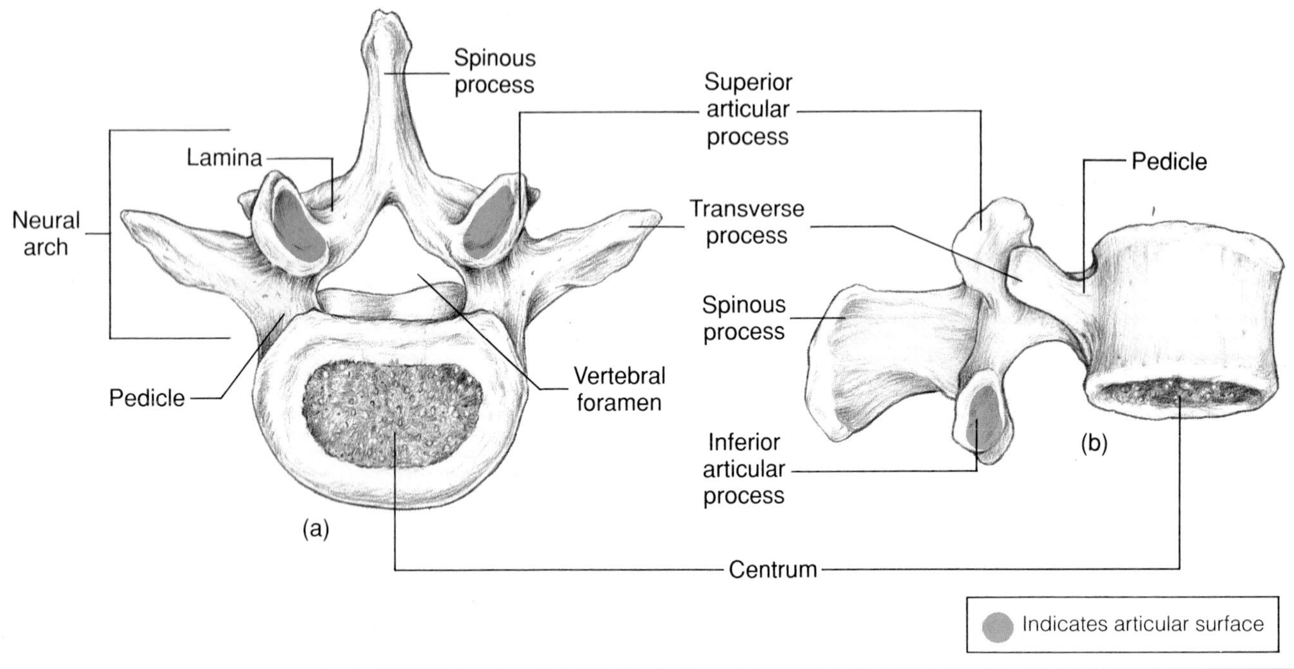

A typical vertebra has a thick anterior **centrum** (body), with a **neural (vertebral) arch** that arises from the posterior surface of the centrum. The centra of adjacent vertebrae are separated by **intervertebral discs.** Each intervertebral disc is composed of a central gelatinous *nucleus pulposus* surrounded by a strong ring of fibrocartilage called the *annulus fibrosus.* Because of their construction, the discs serve as cushions between the bodies of adjacent vertebrae. Each neural arch combines with the posterior surface of the centrum and encloses a **vertebral foramen.** The vertebral foramina of adjacent vertebrae are aligned to form a **vertebral canal** through which the spinal cord passes. **Transverse processes** extend laterally from each neural arch. Projecting from the posterior region of the neural arch is the midline **spinous process.** The spinous and transverse processes allow for the attachment of muscles and ligaments. That portion of the neural arch between the centrum and the transverse process is the **pedicle.** The portion between the transverse process and the spinous process is the **lamina.** Projecting upward from each side of the neural arch is a pair of **superior articulating processes;** their articular surfaces face posteriorly. Projecting downward is a pair of **inferior articulating** processes; their articular surfaces face anteriorly. The smooth articular surface of each process meets with the process of the vertebra above or below it, thereby increasing the rigidity of the vertebral column. The **intervertebral foramina,** through which the spinal nerves pass, are located between the pedicles of adjacent vertebrae (Figure 6.28).

Regional Differences in Vertebrae

The vertebrae of each region have specific characteristics that vary from the "typical" vertebra and enable them to be easily identified. These variations are illustrated in Figures 6.30 through 6.34 and are discussed in Table 6.5.

Thorax

The skeleton of the thorax (Figure 6.35) is formed by the *sternum,* the *ribs,* and the *costal cartilages.* The thoracic vertebrae form its posteriormost portion.

Sternum

The **sternum** is an elongated flat bone that forms the midline portion of the anterior wall of the thorax (Figure 6.35). It is composed of three parts: the **manubrium,** the **body,** and the **xiphoid process** *(zif´-oid).* The superior portion of the manubrium articulates with the medial end of each clavicle (collarbone). Between these articulations is a shallow depression called the **jugular (suprasternal) notch.** The lateral margins of the

◆ **FIGURE 6.30 Superior view of a typical cervical vertebra**

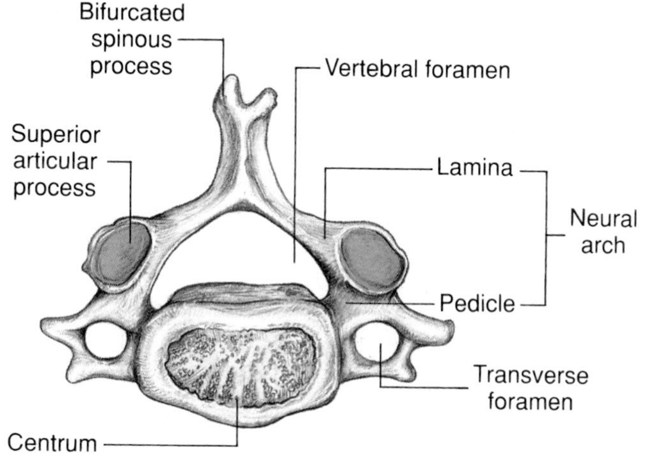

◆ **FIGURE 6.31 The first and second cervical vertebrae**

(a) Superior view of the first cervical (the atlas).
(b) Lateral view of the second cervical (the axis).
(c) Superior lateral view of articulated first and second cervical vertebrae.

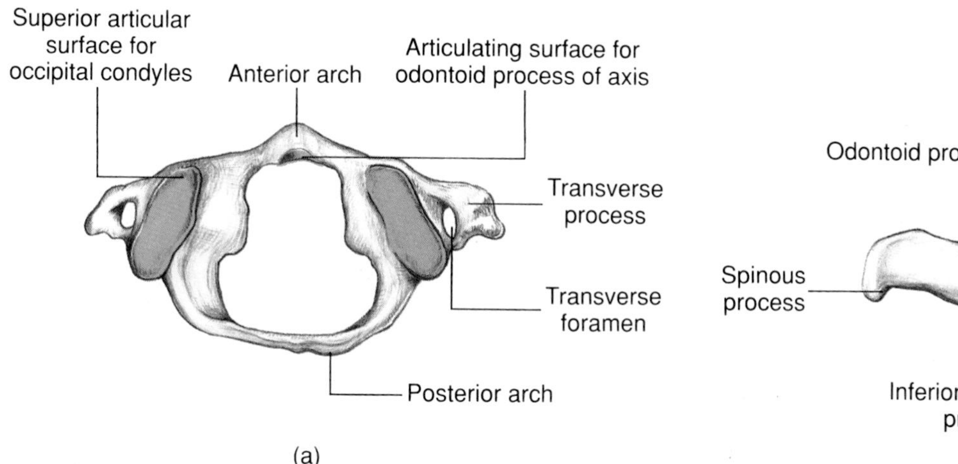

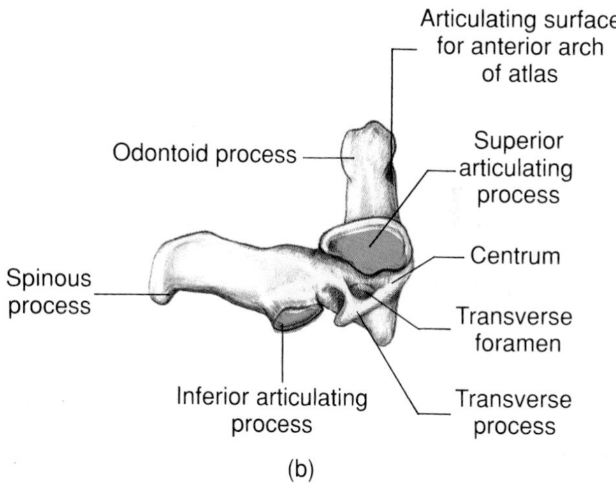

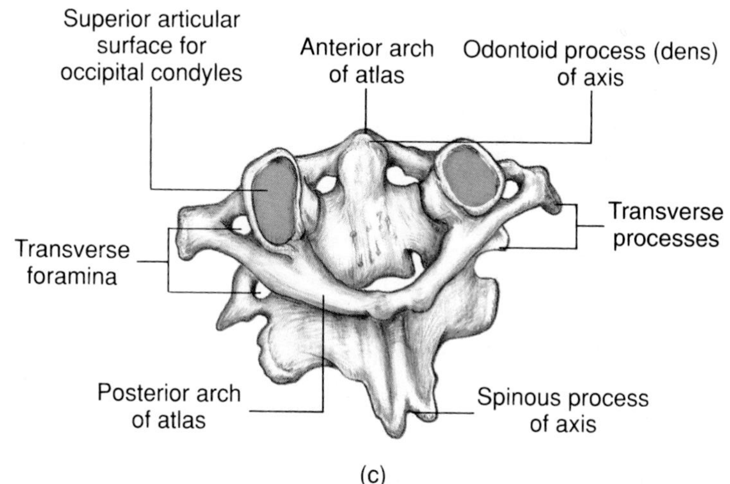

◆ **FIGURE 6.32 Two typical thoracic vertebrae**

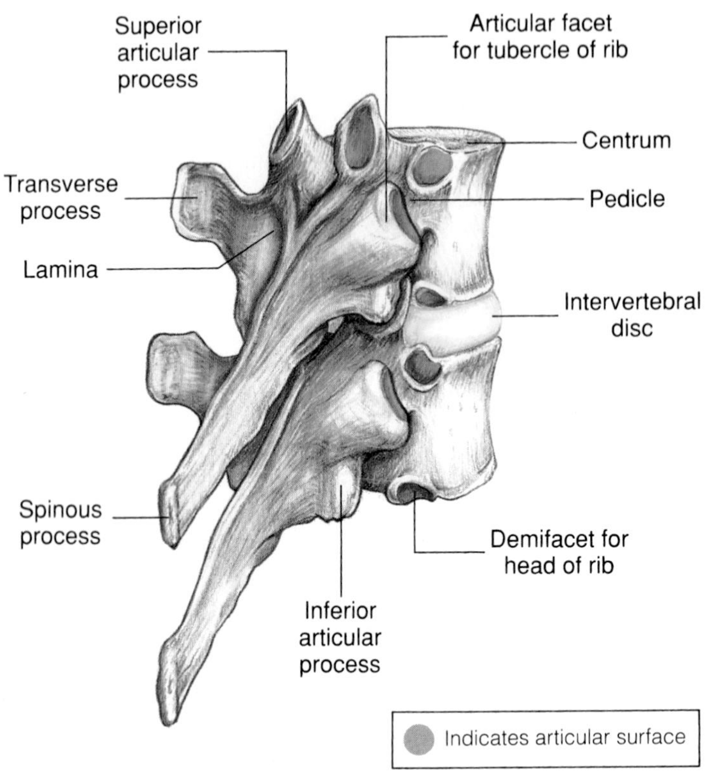

◆ **FIGURE 6.33 Two typical lumbar vertebrae**

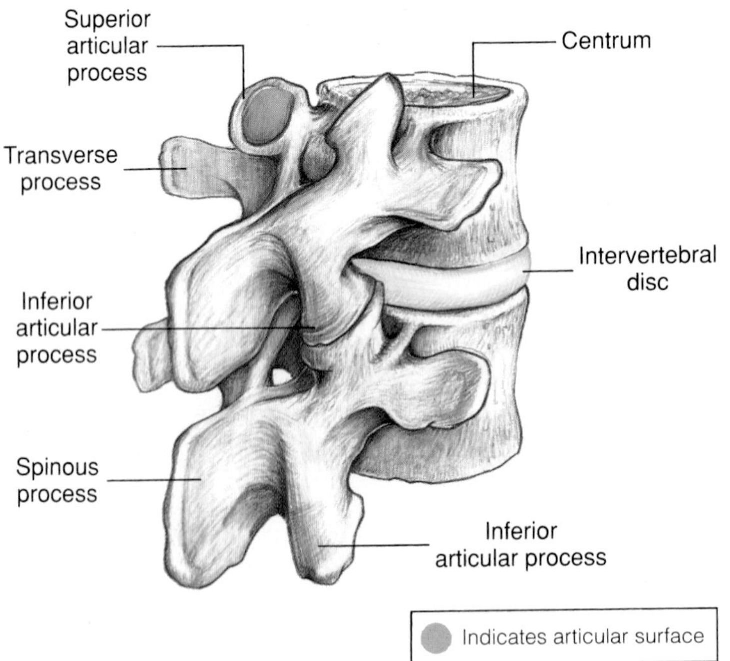

◆ **FIGURE 6.34 The sacrum and the coccyx**
(a) Anterior view. (b) Posterior view.

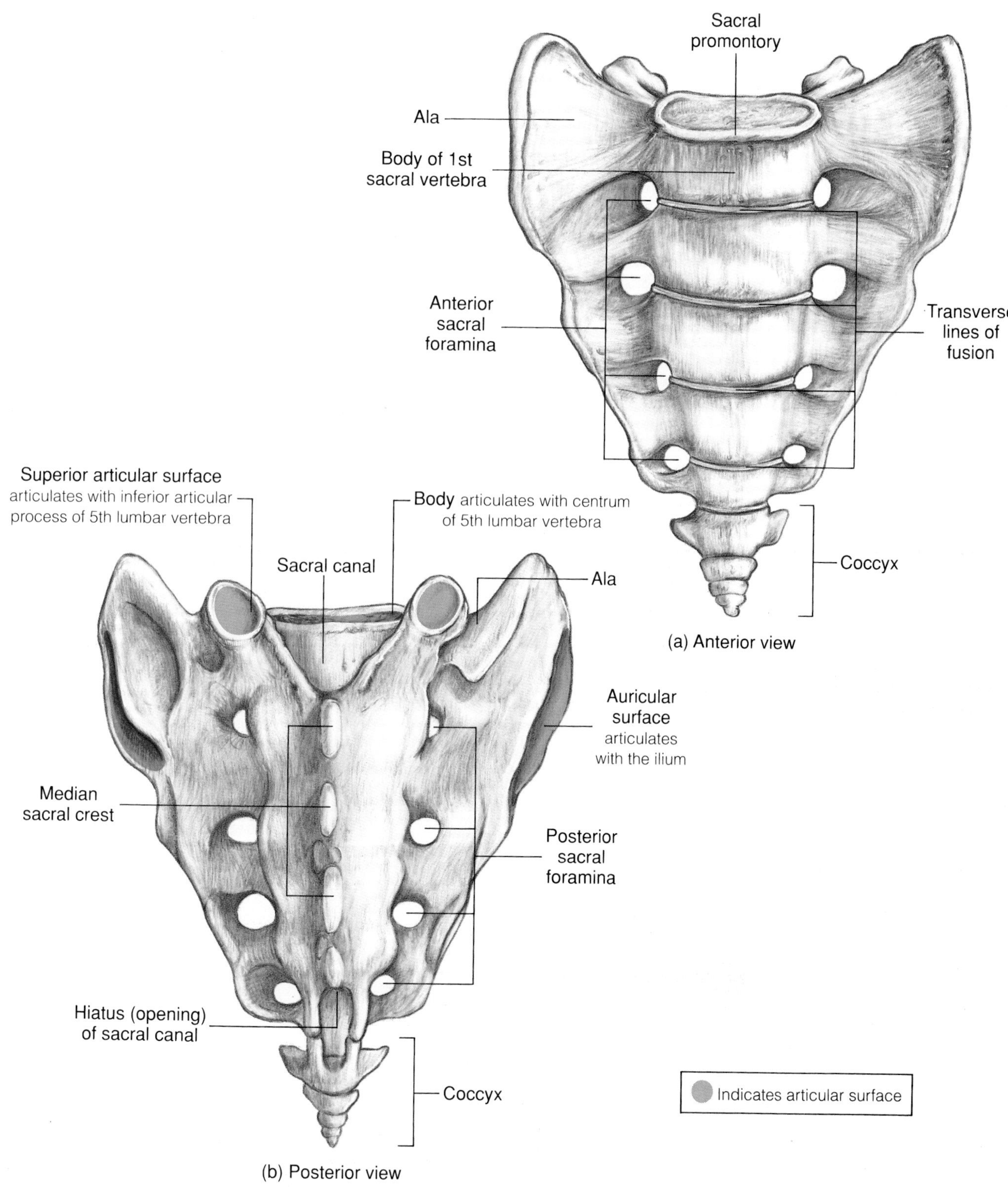

◆ **TABLE 6.5 Identifying Features of Specific Vertebrae**

CERVICAL VERTEBRAE [*Figs. 6.28; 6.30; 6.31*]	
TRANSVERSE FORAMINA [*Figs. 6.30; 6.31*] The openings in the transverse processes of each cervical vertebra. They allow for the passage of vertebral arteries and veins to and from the brain. **BIFURCATED SPINOUS PROCESSES** [*Figs. 6.30; 6.31*] The spinous processes of the cervical vertebrae have a double tip (with the exception of the first and the seventh). **VERTEBRA PROMINENS** [*Fig. 6.28*] The seventh cervical vertebra, so named because its long, prominent spinous process protrudes beyond those of the other cervical vertebrae, making it useful as a landmark in counting the other spinous processes.	**ATLAS** [*Figs. 6.28; 6.31*] The first cervical vertebra, which articulates with the occipital condyles of the skull, has no centrum or spinous process. It is ringlike, consisting of anterior and posterior arches. **AXIS** [*Figs. 6.28; 6.31*] The second cervical vertebra has a vertical projection called the **odontoid process** or **dens** that arises from the superior surface of its centrum. This process provides a pivot around which the atlas rotates.
THORACIC VERTEBRAE [*Figs. 6.28; 6.32*]	
SPINOUS PROCESSES [*Fig. 6.32*] Long and slender protruberances that project sharply downward. This is not as noticeable in the lower thoracic vertebrae.	**FACETS AND DEMIFACETS** [*Fig. 6.32*] Articular surfaces for the ribs on the transverse processes and the bodies of all thoracic vertebrae. (The eleventh and twelfth vertebrae are exceptions because they do not have articular facets on their transverse processes.)
LUMBAR VERTEBRAE [*Figs. 6.28; 6.33*]	
CENTRA [*Fig. 6.33*] Larger and heavier than the centra in other regions. **SPINOUS PROCESSES** [*Fig. 6.33*] Short and blunt as compared to the spinous processes in other regions.	**ARTICULAR PROCESSES** [*Fig. 6.33*] The superior articular processes face inward rather than posteriorly; the inferior articular processes face outward rather than anteriorly. This positioning locks the vertebrae together by preventing rotation.
SACRAL VERTEBRAE [*Figs. 6.28; 6.34*]	
In the adult, the 5 sacral vertebrae are fused into a single triangular **sacrum.** The transverse lines of fusion are visible on its anterior surface. The spinous processes form the **median sacral crest** on its posterior surface. The fused transverse process form the **alae** (wings), which articulate with the pelvic bones. The **sacral foramina** represent the intervertebral foramina. The superior edge of the ventral border of the first sacral vertebra forms a projection called the **sacral promontory.**	
COCCYX [*Figs. 6.28; 6.34*]	
The fused coccygeal vertebrae. It articulates with the apex of the sacrum.	

manubrium articulate with the costal cartilages of the first ribs and part of the second ribs. The manubrium articulates inferiorly with the body of the sternum in a joint that projects slightly anterior, forming the **sternal angle.** The body of the sternum articulates at its lateral margins with the costal cartilages of the second through the seventh ribs (the second ribs also articulate in part with the manubrium). The lower border of the body of the sternum fuses with the xiphoid process. The xiphoid process is composed of hyaline cartilage in young persons, but is often changed into bone in adults. The xiphoid process does not articulate with the ribs. Rather, it serves as a point of attachment for several ligaments and muscles, including the diaphragm and the rectus abdominis muscle. The linea alba, a longitudinal fibrous band that marks the midline of the abdomen, is also attached to it.

Ribs

There are twelve pairs of **ribs** (Figure 6.35). All of the ribs articulate posteriorly with the thoracic vertebrae, but only the first seven pairs articulate directly with the sternum through the costal cartilages. For this reason, the first seven pairs are referred to as **true** or **vertebrosternal ribs.** The remaining five pairs are called **false ribs.**

◆ **FIGURE 6.35 Anterior view of the thorax**

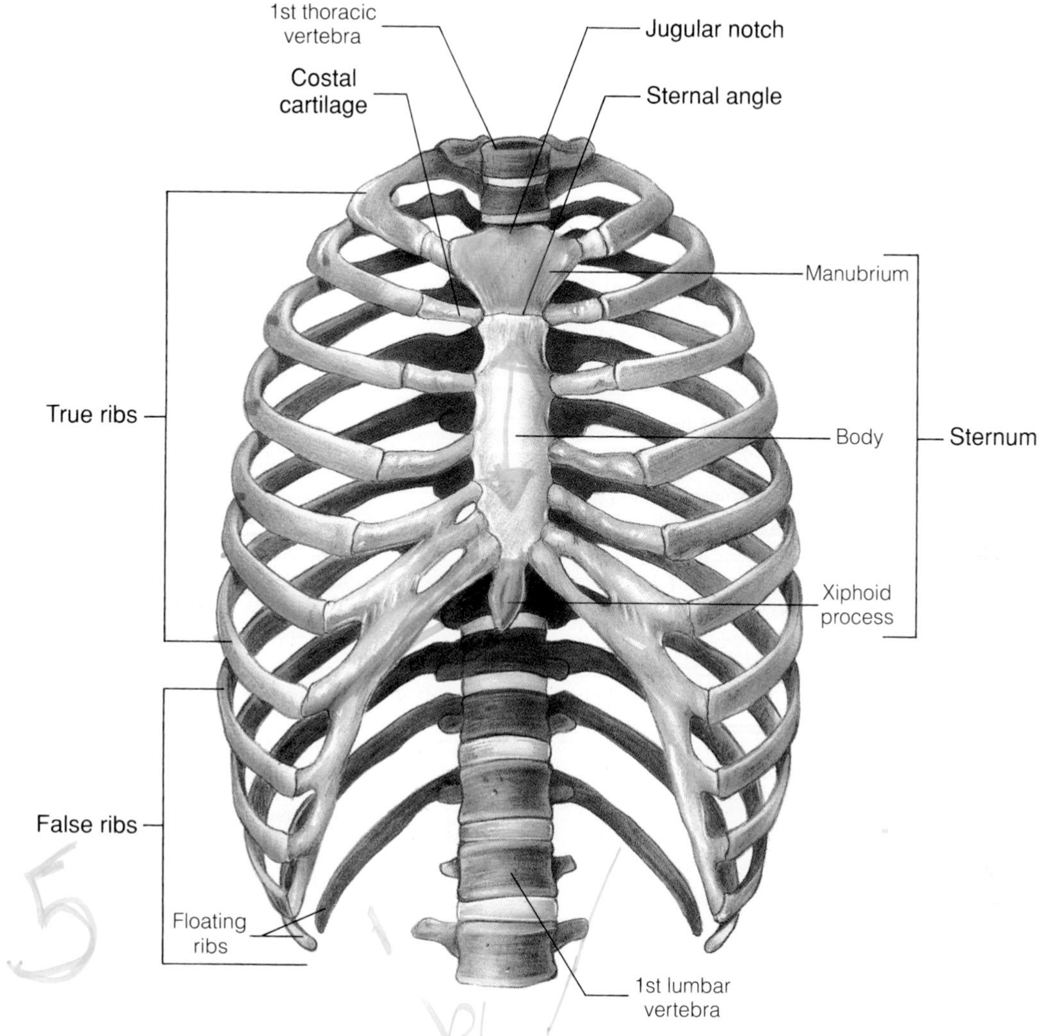

The first three pairs of false ribs (that is, the eighth, ninth, and tenth ribs) have their costal cartilages attached to the cartilages of the rib above, rather than directly to the sternum. These are called **vertebrochondral ribs.** The costal cartilages of the eleventh and twelfth ribs are short and have no anterior articulation, their costal cartilages being embedded in the muscles of the body wall. For this reason they are called **floating** or **vertebral ribs.**

The head of a typical rib articulates with the demifacets of two adjacent thoracic vertebrae (Figure 6.36). However, the heads of the first, tenth, eleventh, and twelfth ribs each articulate entirely on the facets of one vertebra. A short distance from the head is a **tubercle** *(too´-bur-kul),* which articulates with the transverse process of a thoracic vertebra. Between the head and the tubercle is a constricted **neck.** Curving anteriorly from the neck is the **shaft** or body of the rib.

Costal Cartilages

The costal cartilages are composed of hyaline cartilage. They strengthen the thorax by serving as the anterior anchors for most of the ribs (Figure 6.35). At the same time, because they are cartilage, they provide flexibility that allows the rib cage to expand during respiration.

A summary of the skeletal features of the thorax appears in Table 6.6. Table 6.7 summarizes the bones of the axial skeleton.

Appendicular Skeleton

The appendicular skeleton includes the bones of the upper and lower limbs, and those bones by which these

◆ **FIGURE 6.36 Superior view of a thoracic vertebra showing its articulations with the head and tubercle of a rib**

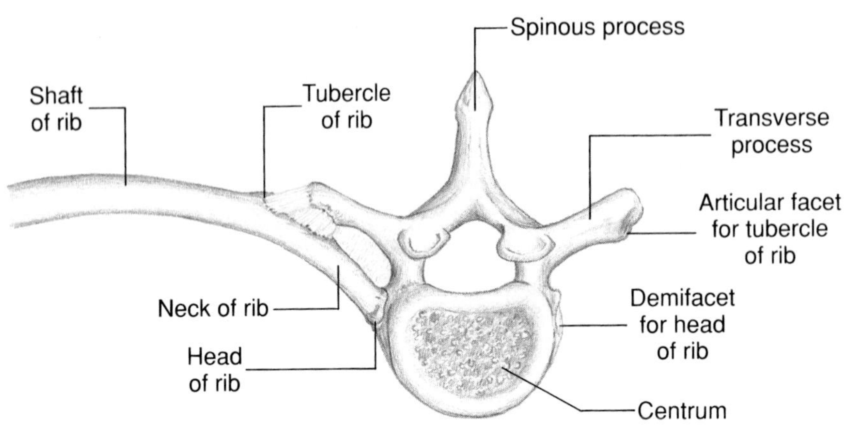

◆ **TABLE 6.6 Summary of Specific Features of the Thoracic Skeleton**

STERNUM [*Fig. 6.35*]
MANUBRIUM The broad upper segment that articulates with the medial ends of each clavicle, the costal cartilages of the first pair of ribs, and part of the second pair of ribs. Has a small depression called the ***jugular notch*** on its superior border. **BODY** The elongated middle segment to which the costal cartilages of the second through the seventh ribs attach. Forms the ***sternal angle*** at its junction with the manubrium. **XIPHOID PROCESS** A small inferior cartilage that serves for the attachment of several ligaments and muscles.
RIBS [*Fig. 6.35*]
HEAD The posterior, medial end that articulates with the bodies of the thoracic vertebrae. **NECK** The constricted portion just lateral to the head. **TUBERCLE** A small projection just beyond the neck that articulates with the transverse process of a thoracic vertebra. It is not present in the tenth, eleventh, and twelfth ribs.
TRUE RIBS
VERTEBROSTERNAL RIBS The first through seventh pairs. They attach directly to the sternum.
FALSE RIBS
VERTEBROCHONDRAL RIBS The eighth, ninth, and tenth pairs. They attach to the costal cartilages of the rib above. **VERTEBRAL (FLOATING) RIBS** The eleventh and twelfth pairs. Embedded anteriorly in muscles of the body wall.

◆ **TABLE 6.7 Summary of Bones That Form the Axial Skeleton**

	NUMBER OF BONES		NUMBER OF BONES
SKULL	29	*AUDITORY OSSICLES*	6
CRANIUM (calvarium and floor of cranial cavity)	8	*Malleus* 2	
		Incus 2	
		Stapes 2	
Parietal 2			
Temporal 2		*HYOID*	1
Frontal 1			
Occipital 1		VERTEBRAL COLUMN	26
Ethmoid 1			
Sphenoid 1		*Cervical* 7	
		Thoracic 12	
*FACE AND NASAL CAVITY**	14	*Lumbar* 5	
		Sacrum 5 fused to form 1	
Maxillary 2		*Coccyx* 4 fused to form 1	
Zygomatic 2			
Lacrimal 2		THORAX	25
Nasal 2			
Inferior nasal concha 2		*Sternum* 1	
Palatine 2		*Ribs* 24	
Mandible 1			
Vomer 1		TOTAL AXIAL SKELETON BONES	80

*The frontal and ethmoid bones also contribute to the face but are counted under the cranium.

limbs articulate with the axial skeleton—that is, the pectoral girdle and the pelvic girdle.

The pectoral girdle, which is attached to the axial skeleton only at the sternum, does not provide very firm support. This support is sufficient, however, because the upper limbs are not involved in supporting the body weight. The pectoral girdle does allow for a wide range of movements at the shoulder.

The pelvic girdle, in contrast, does support the body weight. To accomplish this, it not only has more extensive attachments to the axial skeleton through its articulation with the sacrum, but the two sides of the girdle also attach to each other at the pubic symphysis. In addition, the pelvic girdle is aligned with the bones of the lower limbs in such a manner that it transfers much of the weight it supports to the skeleton of the lower limbs.

Upper Limbs

The 64 bones that comprise the upper limbs are listed in Table 6.8.

Pectoral Girdle

The **pectoral girdle** (Figure 6.37) straddles the upper part of the thorax. Its only joint with the axial skeleton is where the medial end of each clavicle articulates with the manubrium of the sternum. The lateral end of each clavicle articulates with the acromion process of a scapula. The scapulae are attached to the posterior thorax

◆ **TABLE 6.8 Bones of the Upper Limbs**

NUMBER OF BONES IN EACH LIMB	NUMBER OF BONES IN BOTH LIMBS
Pectoral girdle	4
Clavicle 1	
Scapula 1	
Arm	2
Humerus 1	
Forearm	4
Ulna 1	
Radius 1	
Hand	54
Carpals 8	
Metacarpals 5	
Phalanges 14	
Total upper limb bones	64

◆ **FIGURE 6.37 The scapula**

The inset is an anterior view of the pectoral girdle with the rib cage removed. (a) Anterior view of the left scapula. (b) Posterior view. (c) Lateral view.

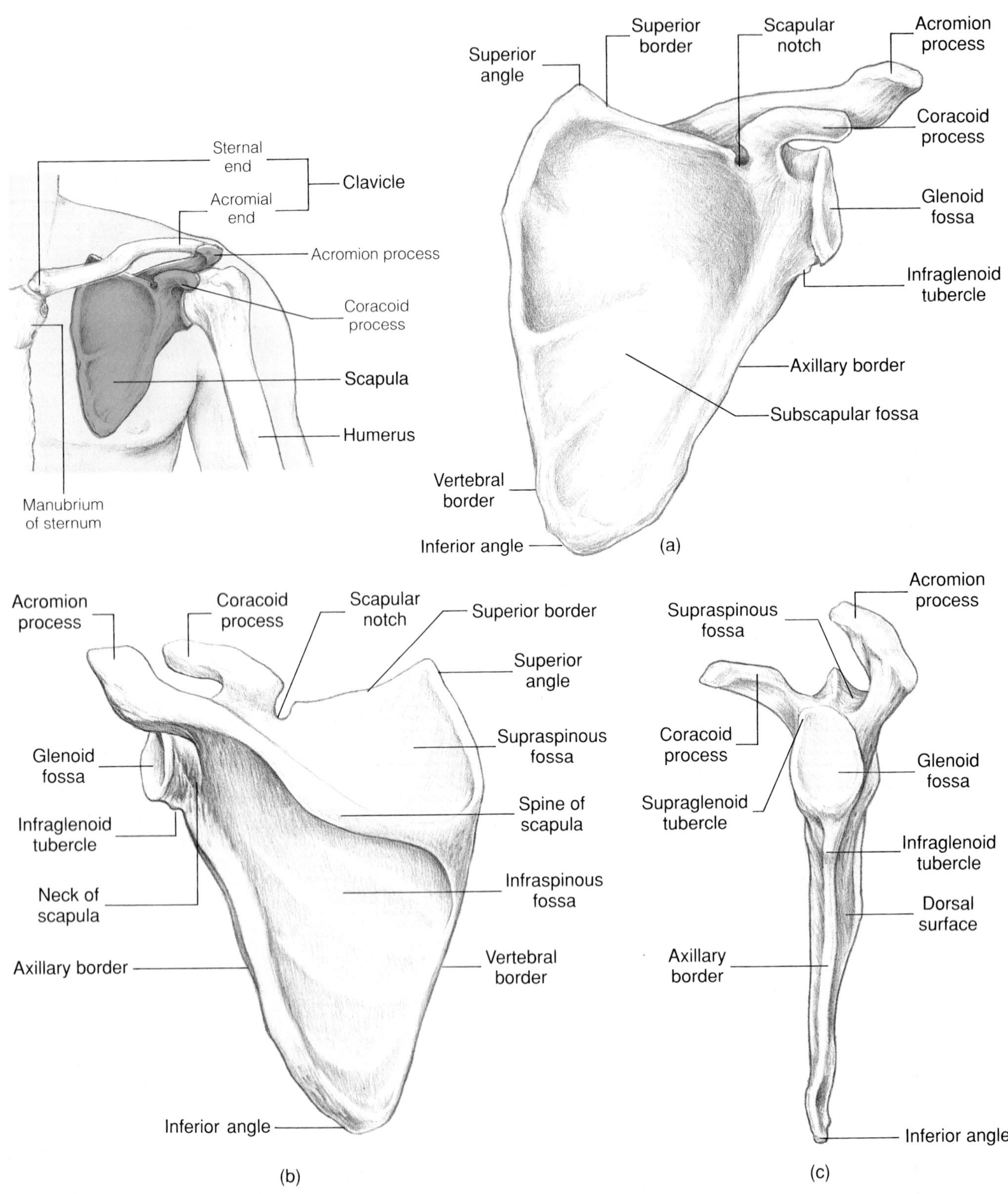

ASPECTS OF EXERCISE PHYSIOLOGY

Osteoporosis: The Bane of Brittle Bones

Osteoporosis, a decrease in bone density resulting from reduced deposition of the bone's organic matrix, is a major health problem in the United States. It is responsible for the greater incidence of bone fractures among women over the age of 50 than among the population at large. Because bone mass is reduced, the bones are more susceptible to fracture in response to a fall, blow, or lifting action that would not normally strain stronger bones. Osteoporosis is the underlying cause for approximately 1.2 million fractures each year, of which 530,000 are vertebral fractures and 227,000 are hip fractures. The cost of rehabilitation is in excess of $6 billion per year. The cost in pain and suffering is not measurable. One-half of all American women have spinal pain and deformity by age 75.

There appear to be two types of osteoporosis, which are caused by different mechanisms. Type I osteoporosis affects women soon after menopause and is characterized by vertebral crush fractures or fractures of the arm just above the wrist. It is hypothesized that these fractures occur as a result of the reduction in bone density that accompanies the estrogen deficiency of menopause. Type II osteoporosis occurs in men as well as women, although it affects females twice as often as males. It is characterized by hip fractures as well as fractures at other sites. Because Type II osteoporosis occurs later in life, the decreased ability to absorb calcium associated with advancing age may play a key role in the development of the condition, although estrogen deficiency probably also contributes, accounting for the higher incidence in women.

Estrogen replacement therapy, calcium supplementation, and a regular weight-bearing exercise program are among the therapeutic approaches used to minimize or reverse bone loss. Because treatment of osteoporosis is difficult and often less than satisfactory, prevention is by far the best approach to managing this disease. Developing strong bones before menopause through a calcium-rich diet and adequate exercise appears to be the best preventive measure. A large reservoir of bone at midlife may delay the clinical manifestations of osteoporosis in later life. Continued physical activity throughout life appears to retard or prevent bone loss, even in the elderly.

It is well documented that osteoporosis can result from disuse—that is, from reduced mechanical loading of the skeleton. Space travel has clearly shown that lack of gravity results in a decrease in bone density. Studies of athletes, on the other hand, demonstrate that physical activity increases bone density. Within groups of athletes, bone density correlates directly with the load that the bone must bear. If one looks at athletes' femurs (thigh bones), the greatest bone density is found in weight lifters, followed in order by throwers, runners, soccer players, and finally swimmers. In fact, the bone density of swimmers does not differ from that of nonathletic controls. The bone density in the playing arm of male tennis players has been found to be as much as 35% greater than in their other arm; female tennis players have been found to have 28% greater density in their playing arm than in their other arm. One study found that very mild activity in nursing-home patients, whose average age was 82 years, not only slowed bone loss but even resulted in bone buildup over a 36-month period. Thus, exercise is a good defense against osteoporosis.

The exact mechanism responsible for an increase in bone mass as a result of exercise is unknown. According to one proposal, exercise places strain on bone, which causes changes in electrical potential that induce bone formation.

by muscles, but they do not contact the ribs directly, being separated from them by other muscles.

Clavicle. The **clavicle** (Figure 6.37) is an S-shaped bone that serves as a brace for the scapula. Its medial **sternal end,** which articulates with the manubrium of the sternum, is enlarged and blunt. Its lateral **acromial end,** which articulates with the acromion process of the scapula, is flattened. Through its articulations with the sternum and the scapula, the clavicle holds the shoulder away from the rib cage, thus allowing the arms to swing freely, without first necessitating the lifting of the arm away from the body wall. When a clavicle is broken, the entire shoulder collapses. Table 6.9 lists the terms associated with the clavicle.

◆ **FIGURE 6.38 The right humerus**
(a) Anterior view. (b) Posterior view. (c) Anterior photograph.

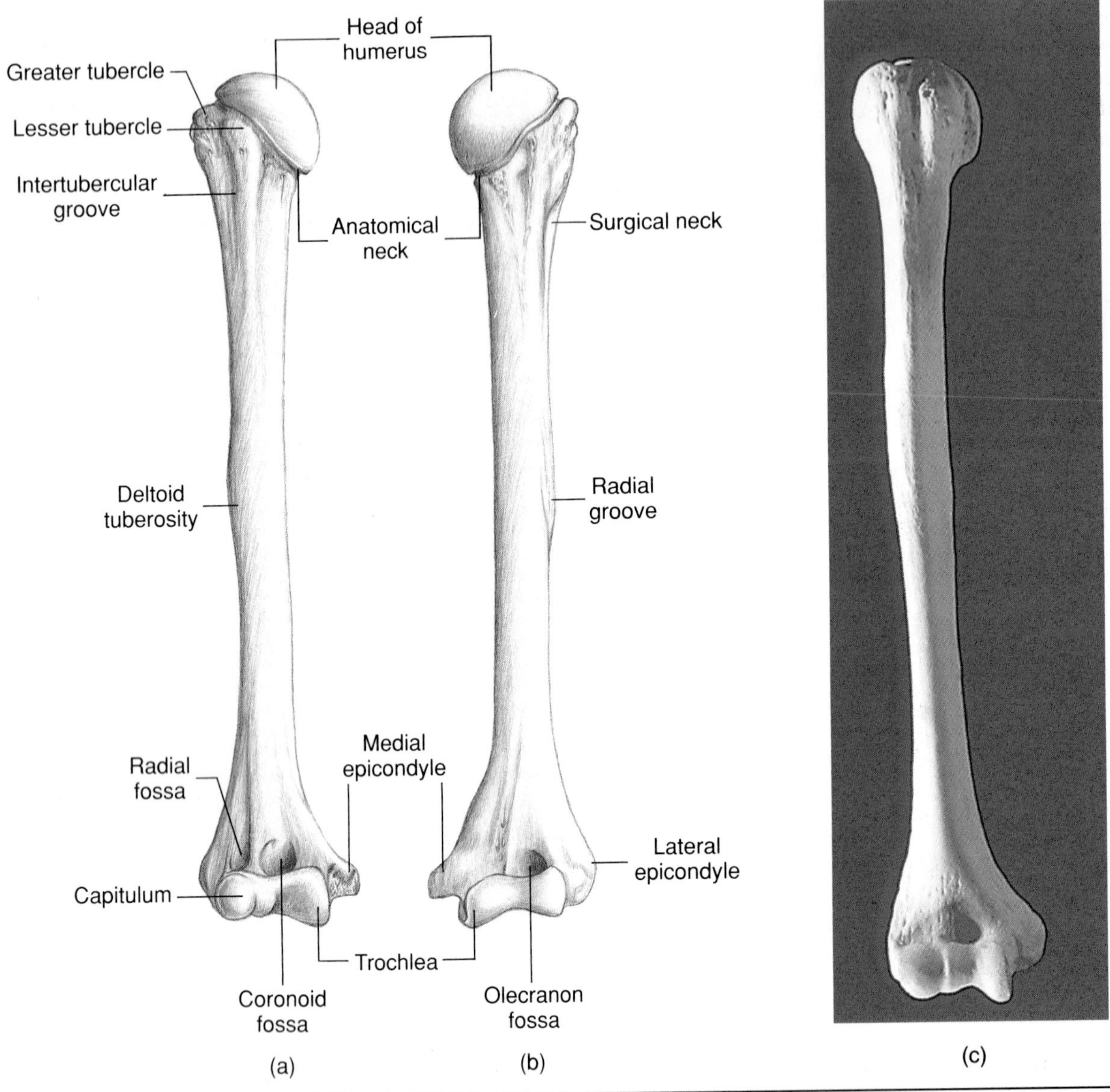

Scapula. The **scapula** (Figure 6.37) is a thin, flat triangular bone that lies over the posterior surfaces of the second to the seventh ribs. The **superior border** forms the base of the triangle, with the **vertebral (medial) border** and the **axillary (lateral) border** joining at the **inferior angle.** The **lateral** angle forms the **glenoid fossa,** which articulates with the humerus. On the superior border, just medial to the glenoid fossa, is the hooked **coracoid process** (*coracoid* = beaklike). The dorsal surface is divided into upper and lower regions by a ridge called the **spine** of the scapula. The spine terminates laterally in a flat **acromion process,** which articulates with the lateral end of the clavicle. The features of the scapula are summarized in Table 6.9.

Arm

The **humerus** is the only bone in the arm (Figure 6.38). Its proximal epiphysis, which is called the **head,** is smooth and round. The head articulates with the glenoid fossa of the scapula. The **anatomical neck** is a slight constriction below the head. There are two projections just distal to the anatomical neck: the lateral

greater tubercle and the anterior **lesser tubercle.** The tubercles are separated from each other by an **intertubercular (bicipital) groove.** The distal end of the humerus is flattened, with prominent **medial** and **lateral epicondyles.** Between the epicondyles are two articular surfaces—the lateral rounded **capitulum,** which articulates with the radius, and the medial, deeply grooved **trochlea,** which articulates with the ulna. The features of the humerus are summarized in Table 6.9.

Forearm

There are two parallel bones in the forearm. In the anatomical position, the *ulna* is medial and the *radius* is lateral (Figure 6.39). The features of the radius and ulna are summarized in Table 6.9.

Ulna. The proximal end of the ulna has two prominent processes: the large posterior **olecranon process** and the smaller anterior **coronoid process** (Figure 6.39). A smooth, concave surface, the **trochlear (semilunar) notch,** lies on the anterior surface of the olecranon process and extends onto the superior surface of the coronoid process. The trochlear notch articulates with the trochlea of the humerus. The **radial notch** is a smooth surface on the lateral side of the coronoid process. It articulates with the edge of the head of the radius.

◆ **FIGURE 6.39 Bones of the forearm**
(a) Anterior view of the bones of the right forearm. (b) Anterior photograph of the right radius and ulna.

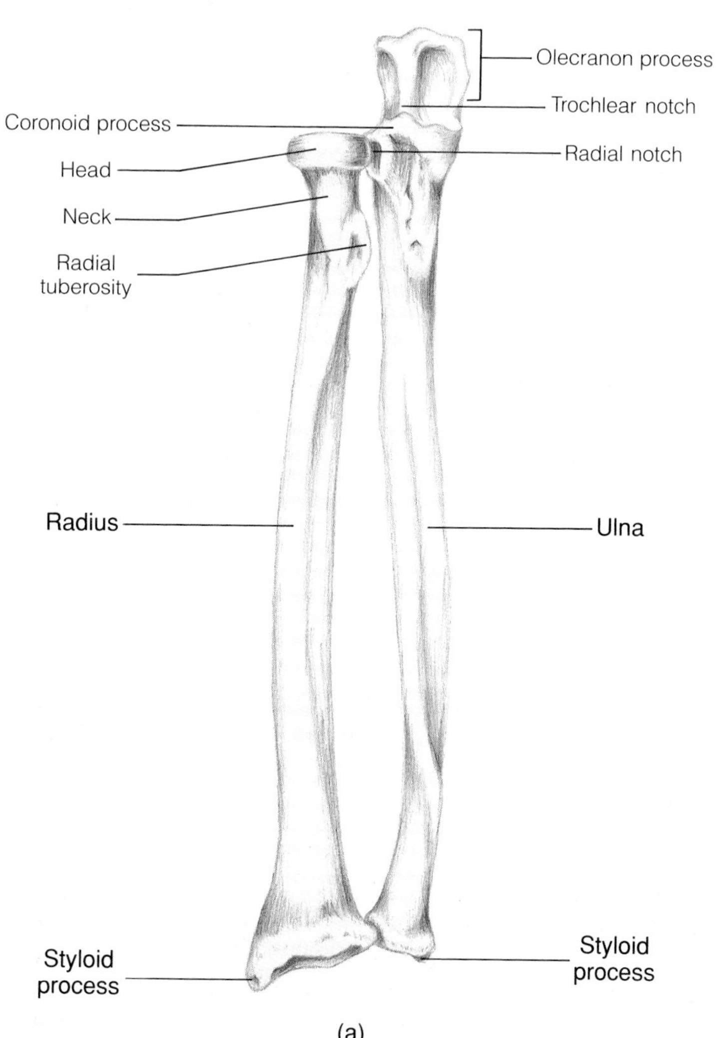

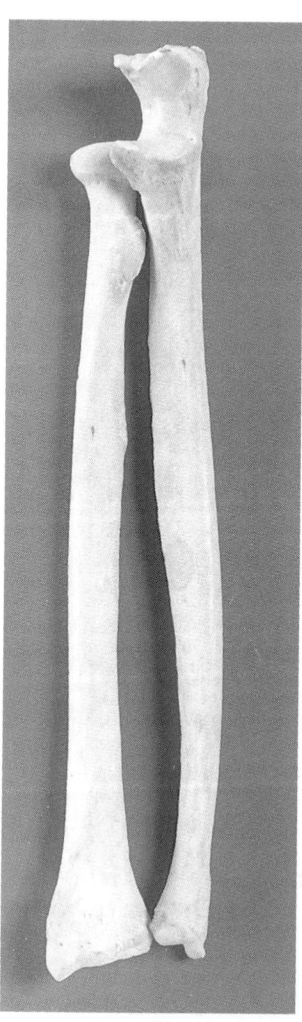

◆ **TABLE 6.9 Summary of Specific Features of the Bones of the Upper Limbs**

CLAVICLE [*Fig. 6.37*]

STERNAL END
The blunt medial end that articulates with the manubrium of the sternum.

ACROMIAL END
The flattened lateral end that articulates with the acromion process of the scapula.

SCAPULA [*Fig. 6.37*]

SUPERIOR BORDER
The upper horizontal margin.

VERTEBRAL (MEDIAL) BORDER
The vertical margin just lateral to the vertebral column.

AXILLARY (LATERAL) BORDER
The thicker oblique lateral margin.

INFERIOR ANGLE
The most inferior point of the bone. It marks the junction of the vertebral and axillary borders.

SUPERIOR ANGLE
The junction of the vertebral and superior borders.

LATERAL ANGLE
The junction of the superior and axillary borders. It contains the glenoid fossa.

GLENOID FOSSA
A shallow depression on the lateral angle. It articulates with the humerus.

SUPRAGLENOID TUBERCLE
A slight elevation just above the glenoid fossa. This is the point of attachment of the long head of the biceps brachii muscle.

INFRAGLENOID TUBERCLE
The roughened area just below the glenoid fossa. The long head of the triceps brachii muscle originates here.

CORACOID PROCESS
The projection that hooks anteriorly from the superior border. It provides for the attachment of ligaments and muscles.

SCAPULAR NOTCH
A deep notch in the superior border at the base of the coracoid process. It allows for the passage of the suprascapular nerve.

SPINE
A prominent ridge that runs horizontally across the posterior surface.

ACROMION PROCESS
The flattened lateral end of the spine. It articulates with the clavicle, thus bracing the scapula.

SUPRASPINOUS FOSSA
The dorsal surface above the spine.

INFRASPINOUS FOSSA
The dorsal surface below the spine.

SUBSCAPULAR FOSSA (COSTAL SURFACE)
The slightly concave ventral surface.

HUMERUS [*Fig. 6.38*]

HEAD
The rounded proximal epiphysis. It articulates with the glenoid fossa of the scapula to form the shoulder joint.

ANATOMICAL NECK
A shallow constriction that circles the bone just below the head.

GREATER TUBERCLE
The rounded projection from the lateral margin of the bone just distal to the anatomical neck.

LESSER TUBERCLE
The rounded projection from the anterior surface of the bone just distal to the anatomical neck.

INTERTUBERCULAR (BICIPITAL) GROOVE
A deep groove between the greater and lesser tubercles. The tendon of the long head of the biceps brachii muscle passes through the groove to reach the supraglenoid tubercle of the scapula.

SURGICAL NECK
A slightly constricted region just inferior to the tubercles. This is frequently the site of fracture.

DELTOID TUBEROSITY
A triangular roughened area on the anterior lateral surface near the middle of the shaft. The deltoid muscle inserts here.

RADIAL GROOVE (SULCUS)
An oblique groove on the posterior surface just below the deltoid tuberosity. It marks the path of the radial nerve.

EPICONDYLES (MEDIAL AND LATERAL)
The projections from the margins of the distal epiphysis.

CAPITULUM
The lateral convex portion of the distal condyles. It articulates with the head of the radius.

RADIAL FOSSA
A slight depression on the anterior surface above the capitulum. It receives the margin of the head of the radius when the elbow is flexed (bent).

◆ **TABLE 6.9 Summary of Specific Features of the Bones of the Upper Limbs (continued)**

HUMERUS *[Fig. 6.38] continued*	
TROCHLEA The medial concave portion of the distal condyles. It articulates with the semilunar notch of the ulna. **CORONOID FOSSA** A small depression on the anterior surface above the trochlea. It receives the coronoid process of the ulna when the elbow is flexed (bent).	**OLECRANON FOSSA** A deep depression on the posterior surface above the trochlea. It receives the olecranon process of the ulna when the elbow is extended (straightened).
ULNA *[Fig. 6.39]*	
OLECRANON PROCESS The thick posterior projection from the proximal end that forms the point of the elbow. It is received by the olecranon fossa of the humerus when the elbow is extended (straightened). **CORONOID PROCESS** The anterior projection from the proximal end. It is received by the coronoid fossa of the humerus when the elbow is flexed (bent). **TROCHLEAR (SEMILUNAR) NOTCH** A curved depression formed by the olecranon and coronoid processes. It articulates with the trochlea of the humerus.	**RADIAL NOTCH** A small depression on the lateral side of the coronoid process. It articulates with the margins of the head of the radius, allowing the forearm to rotate, turning the palm down (pronate). **HEAD** The small distal end that articulates with the fibrocartilaginous disc of the wrist joint. **STYLOID PROCESS** The posterior medial projection from the distal end. It serves as the point of attachment for the ulnar collateral ligament of the wrist.
RADIUS *[Fig. 6.39]*	
HEAD The proximal disc-shaped end. The superior surface articulates with the capitulum of the humerus; the edges articulate with the radial notch of the ulna. **NECK** The constriction just distal to the head. **RADIAL TUBEROSITY** A flat projection on the medial side, distal to the neck. The biceps brachii muscle inserts here.	**STYLOID PROCESS** The lateral downward projection from the distal end. It is the point of attachment for the radial collateral ligament and the brachioradialis muscle. **ULNAR NOTCH** A depression on the medial margin of the distal end. It articulates with the ulna.

Distally, the ulna has a small rounded **head** and a posterior medial **styloid process.**

Radius. In the anatomical position, the radius lies lateral to the ulna (Figure 6.39). Its small cylindrical proximal epiphysis is the **head.** The head articulates with the capitulum of the humerus and medially with the radial notch of the ulna. On the medial surface, a short distance below the head, is the **radial tuberosity** *(too-bur-os´-i-tee).* The distal end of the radius is broad, and it articulates medially with the ulna and distally with two carpal bones of the wrist. The distal end has a conical **styloid process** projecting from its lateral margin. The space between the radius and ulna is occupied by a strong interosseus membrane.

Hand

The skeleton of the hand consists of *carpal bones, metacarpal bones,* and *phalanges.* The proximal portion of the hand, toward the wrist, is composed of eight **carpal bones** arranged in two transverse rows of four bones each (Figure 6.40). The bones of the proximal row, from lateral to medial, are the **scaphoid, lunate, triquetral,** and **pisiform.** Those of the distal row, from lateral to medial are **trapezium, trapezoid, capitate,** and **hamate.** The scaphoid, lunate and triquetral articulate with the distal end of the radius to form the wrist joint.

Five **metacarpal bones** form the skeleton of the palm of the hand (Figure 6.40). Rather than being named, they are numbered from the lateral (thumb) to medial. Proximally, the metacarpals articulate with the

◆ **FIGURE 6.40 Bones of the hand**

(a) Ventral view of the bones of the right hand. (b) X rays of the hand of a two-year-old (top left), a three-year-old (bottom left), a fourteen-year-old (middle), and a sixty-year-old (right). The spaces between bones of the two-year-old are the cartilaginous epiphyses of the various bones. In the hand of the sixty-year-old, the epiphyses have ossified, leaving thin epiphyseal plates.

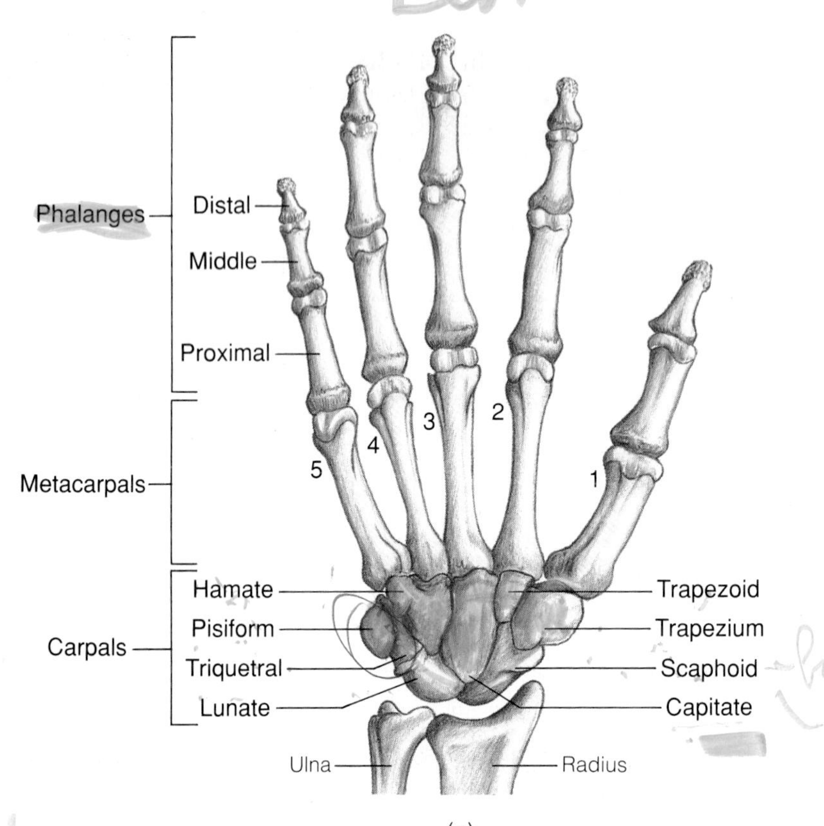

(a)

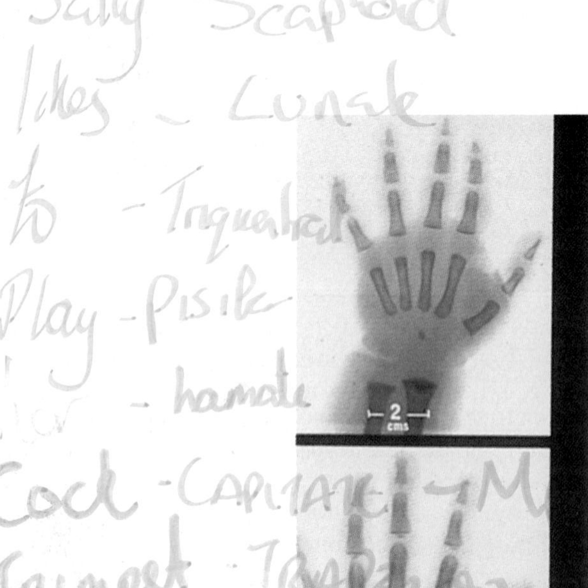

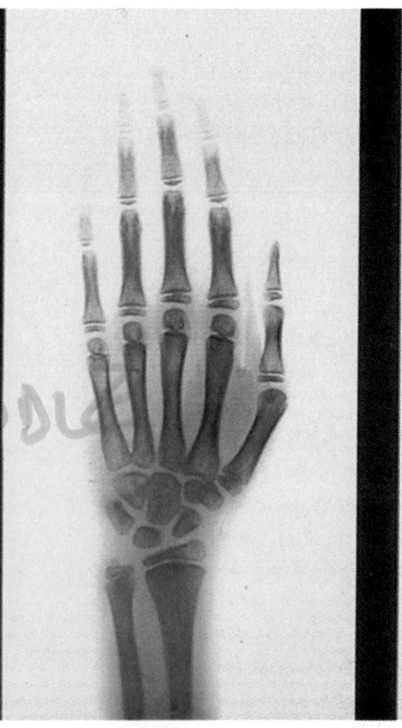

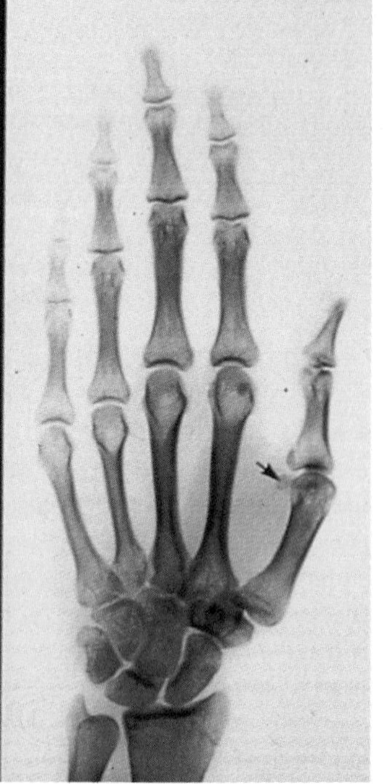

(b)

distal row of the carpal bones and with each other. Distally, each articulates with the proximal end of a phalanx.

The skeleton of the fingers is formed by 14 **phalanges** *(fay-lan´-jeez).* They are numbered from lateral to medial, in the same manner as the metacarpals. Each finger contains three phalanges, named **proximal, middle,** and **distal.** An exception is the first digit (thumb), which has only proximal and distal phalanges.

Lower Limbs

The 62 bones that comprise the lower limbs are listed in Table 6.10.

Pelvic Girdle

The pelvic girdle is formed by a pair of **coxal bones,** or **ossa coxae** (*ossa* = bones; *coxa* = hip; Figure 6.41). These bones are also commonly referred to as pelvic bones or innominate bones. The two coxal bones are firmly braced through posterior articulations with the sacrum (the sacroiliac joint) and an anterior articulation with each other (the symphysis pubis), forming a basinlike structure called the **pelvis.** The coccyx, which articulates with the inferior margin of the sacrum, is also part of the pelvis.

Each coxal bone is a single bone formed by the fusion of three separate embryonic bones: the *ilium, ischium,* and *pubis.* In the adult bone, these individual names are retained for their respective parts. On the lateral surface of the coxal bone, where the ilium, ischium, and pubis bones meet, is a deep cup, the **acetabulum.** The head of the femur articulates with the acetabulum. Below the acetabulum is a large **obturator foramen.** The features of the coxal bones are summarized in Table 6.12.

Ilium. The **ilium** is a broad, expanded portion of the coxal bone that extends upward from the acetabulum (Figure 6.41). Its superior border is called the **iliac crest.** This crest ends anteriorly at the **anterior superior iliac spine.** A short distance below this spine is the **anterior inferior iliac spine.** The iliac crest ends posteriorly at the **posterior superior iliac spine.** A short distance below this spine is the **posterior inferior iliac spine.** Below the posterior inferior iliac spine is the deep **greater sciatic notch.** The **iliac fossa** is the smooth, slightly concave internal surface. Behind the fossa is a roughened area called the **auricular surface** that articulates with the sacrum, forming the **sacroiliac joint.** Running diagonally downward and forward from this articular surface, and demarking the lower boundary of the iliac fossa, is the **arcuate line.**

◆ **TABLE 6.10 Bones of the Lower Limbs**

NUMBER OF BONES IN EACH LIMB		NUMBER OF BONES IN BOTH LIMBS
Pelvic girdle		2
Coxal bone		
Ilium	fused to form 1	
Ischium		
Pubis		
Thigh		2
Femur 1		
Leg		6
Tibia 1		
Fibula 1		
Patella 1		
Foot		52
Tarsals 7		
Metatarsals 5		
Phalanges 14		
Total lower limb bones		62

Ischium. The **ischium** forms the posterior inferior portion of the coxal bone and part of the acetabulum (Figure 6.41). On the posterior margin of the ischium, below the greater sciatic notch, is the **ischial spine.** The **lesser sciatic notch** is below the spine. The notch is bounded inferiorly by a prominent **ischial tuberosity,** which supports the body weight in the sitting position. The **ischial ramus** is an anterior projection from the tuberosity. It joins with the inferior ramus of the pubis to form the lower border of the obturator foramen.

Pubis. The **pubis** is the anterior part of the coxal bone. It forms the anterior inferior portion of the acetabulum (Figure 6.41). Its **superior ramus** is supported against the superior ramus of the opposite side, forming the **symphysis pubis.** A projection called the **pubic tubercle** is located on the superior ramus, close to the symphysis pubis. A ridge called the **pecten** extends along the superior ramus, from the pubic tubercle to the arcuate line of the ilium. A short distance lateral to the symphysis pubis, the **inferior ramus** of the pubis extends downward and posteriorly to join with the ischial ramus. The junction of the two inferior rami at the symphysis pubis forms the **pubic arch.**

Pelvic Cavities

The cavity of the pelvis is divided into two parts by a horizontal plane that passes from the sacral promontory

◆ **FIGURE 6.41 Right coxal bone**

(a) Internal surface. (b) External surface. The dotted lines indicate the approximate junctions between the ilium, ischium, and pubis.

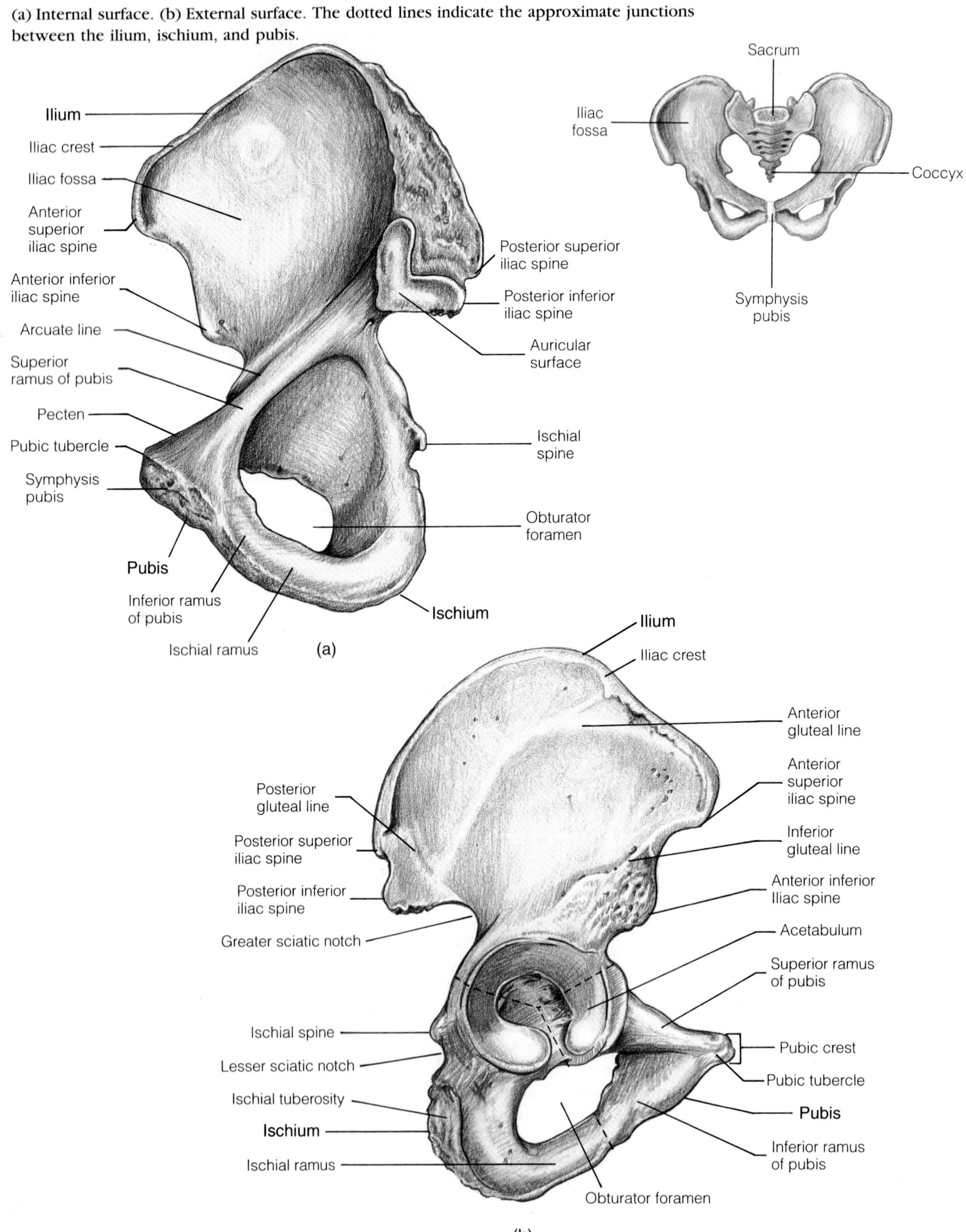

(anterior: superior border of first sacral vertebra) to the upper margin of the symphysis pubis, following the arcuate lines on the inner surface of the ilium. The circumference of this plane is the **pelvic brim** (Figure 6.42a). The expanded cavity above the pelvic brim is the **greater,** or **false, pelvis.** The **lesser,** or **true, pelvis** is the cavity below the pelvic brim. The false pelvis is bounded posteriorly by the iliac fossa. Anteriorly, it is bounded by the abdominal wall and is therefore capable of expansion. In contrast, the cavity of the true pelvis is much more restricted, being surrounded on all sides by bone (ilium, ischium, pubis, sacrum, and coccyx). The superior circumference of the true pelvis is called the **pelvic inlet** because it marks the superior entrance to the true pelvis. The margin of the pelvic inlet coincides with the pelvic brim. The **pelvic outlet** (Figure 6.42b) is the lower circumference of the true pelvis. It is bounded posteriorly by the coccyx and the two ischial spines and tuberosities, and anteriorly by the lower margin of the symphysis pubis. The features of the pelvic cavities are summarized in Table 6.12.

◆ **FIGURE 6.42 Pelvic cavities**

(a) Superior view showing the expanded false pelvis separated from the true pelvis by the pelvic brim.
(b) Inferior view showing the pelvic outlet.

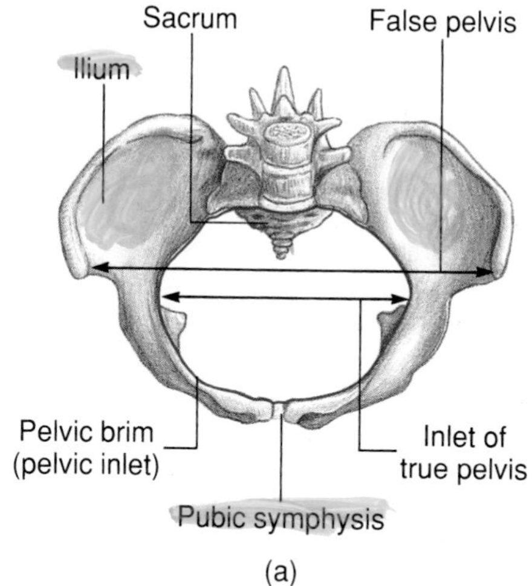

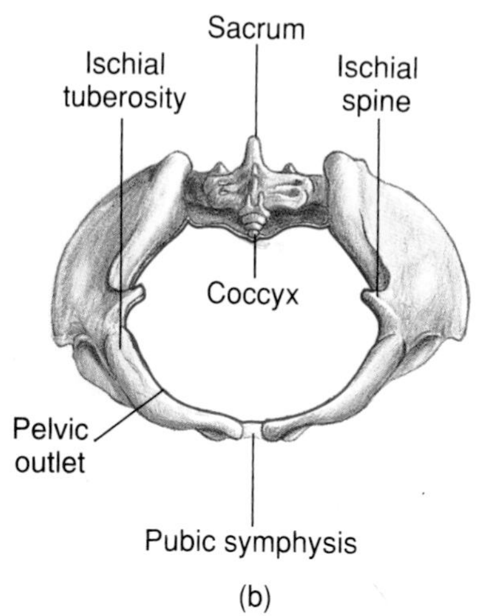

Sexual Differences in the Pelvis. There are several structural differences between the male and the female pelves, most of which are related to childbearing. Because during birth the fetus must pass from the false pelvis through the true pelvis, the measurements of the pelvic inlet and outlet are of particular importance in the female. Some sexual differences in the pelvis are listed in Table 6.11 and illustrated in Figure 6.43.

Thigh

The **femur,** which forms the skeleton of the thigh, is the longest bone of the body (Figure 6.44). The **head** of the femur is the spherical proximal epiphysis, which is directed medially and is received into the acetabulum of the coxal bone. The surface of the head is smooth, except for a central pit, the **fovea capitis.** A strong ligament *(ligamentum teres)* is attached to the fovea and to the acetabulum, helping to maintain the integrity of the hip joint. A constricted **neck** joins the head with the shaft of the bone. The neck forms an angle of about 125° with the shaft. At the point where the neck joins with the shaft, there are two large processes that serve as sites of muscle attachment. The larger lateral process is the **greater trochanter;** the smaller process, which projects medially and posteriorly, is the **lesser trochanter.** On the posterior surface of the shaft is a distinct longitudinal ridge, the **linea aspera.** At its distal end, the linea aspera divides into **medial** and **lateral supracondylar ridges,** which enclose a flat triangular **popliteal surface** between them.

The distal end of the femur is enlarged into **medial** and **lateral condyles.** The smooth surfaces of the condyles articulate with the tibia. Between the condyles, on the posterior surface, is a deep **intercondylar fossa.** Anteriorly, there is a smooth **patellar surface** between the condyles. This surface articulates with the patella when the leg is extended. The features of the femur are summarized in Table 6.12.

Leg

The skeleton of the leg consists of a strong medial **tibia** and a slender lateral **fibula.** Protecting the knee joint, between the thigh and the leg, is the **patella** (kneecap). The features of the tibia and fibula are summarized in Table 6.12.

◆ **FIGURE 6.43 The pelvis**

Anterior view of (a) the male pelvis and (b) the female pelvis. (c) Anteroposterior photo of the female pelvis.

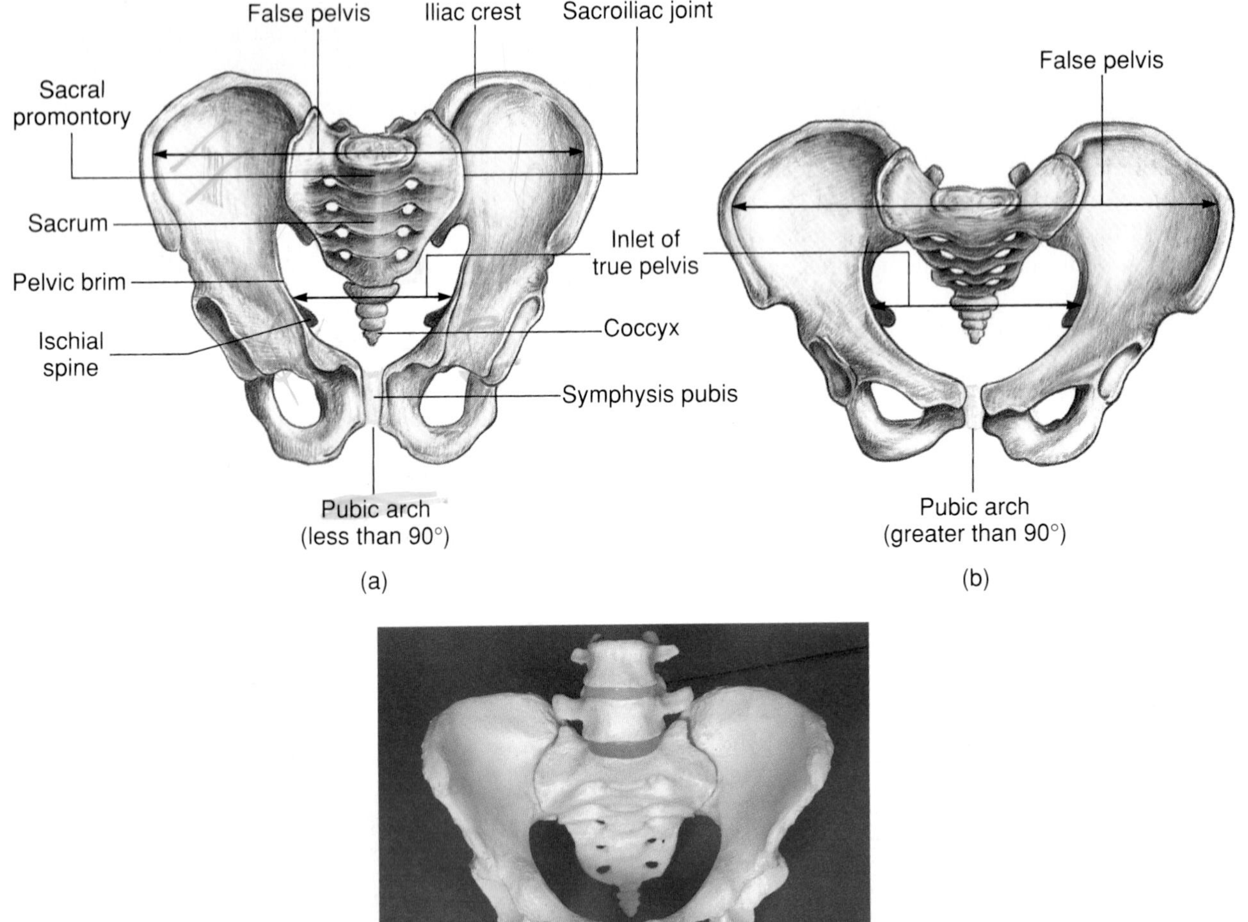

(c)

◆ **TABLE 6.11 Some Structural Differences between the Male Pelvis and the Female Pelvis**

CHARACTERISTIC	MALE PELVIS	FEMALE PELVIS
General structure	More massive; narrow	More delicate; broad
Anterior iliac spines	Less widely separated	More widely separated
Pelvic inlet	Heart-shaped	Larger; circular
Pelvic outlet	Narrower	Wider; ischial tuberosities farther apart
Pubic arch	Acute (less than 90°)	Obtuse (greater than 90°)
Obturator foramen	Oval	Triangular
Acetabulum	Faces laterally	Faces slightly more anteriorly
Ischial spines	Point medially	Point posteriorly
Sacrum	Long; narrow; more curved	Broad; short; less curved

◆ **FIGURE 6.44 Right femur**
(a) Anterior view. (b) Posterior view. (c) Anterior (left) and posterior (right) photographs.

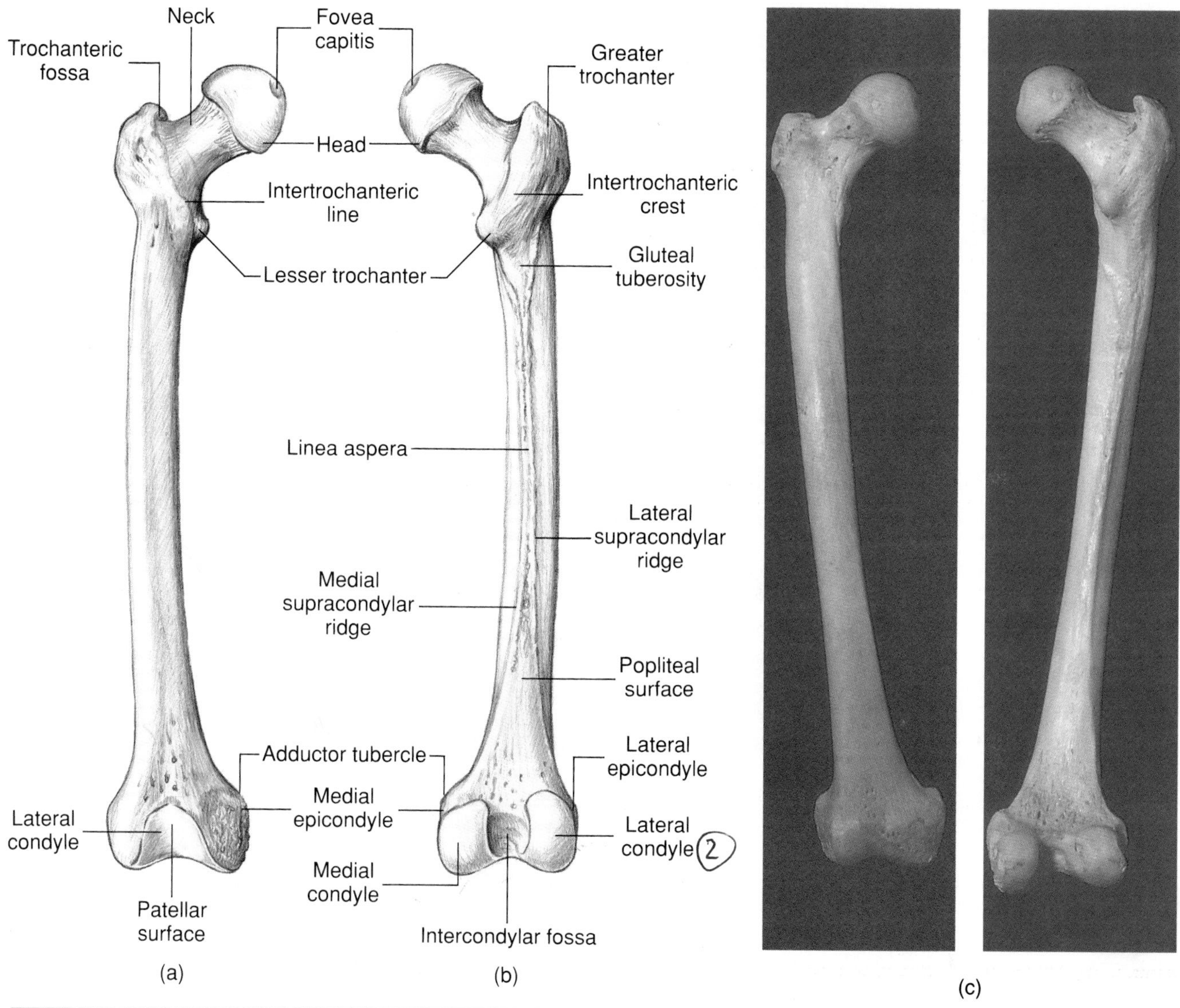

Patella. The **patella** (Figure 6.45) is a *sesamoid bone,* which means that it forms within the tendons of muscles and is not firmly anchored to the skeleton. It is located in front of the knee joint. In addition to protecting the knee joint, the patella improves the leverage of the quadriceps femoris muscle group, in whose tendon it is embedded. The posterior surface of the patella has a smooth articular surface for contact with the patellar surface of the femur.

Tibia. The **tibia** (Figure 6.46), which is the medial bone of the leg, supports the body weight transmitted to it from the femur. Its proximal end is expanded into **medial** and **lateral condyles.** The superior surfaces of the condyles are smooth and flattened. They articulate with the condyles of the femur. Between these articular surfaces is a prominence called the **intercondylar eminence** or **spine.** The **tibial tuberosity** is a prominent elevation on the anterior surface of the bone, just below the condyles. It provides an anchoring point for the tendons of the muscles of the anterior thigh—that is, those that ensheath the patella and form the *ligamentum patellae.* A sharp **anterior crest** extends almost the entire length of the shaft. The distal end of the tibia has a downward projection on its inner side called the **medial malleolus.** This is the prominent lump that can be felt on the medial side of the ankle. The distal surface of the tibia is flattened for articulation with the talus, which is a tarsal bone.

◆ **FIGURE 6.45 Lateral X ray of the knee**

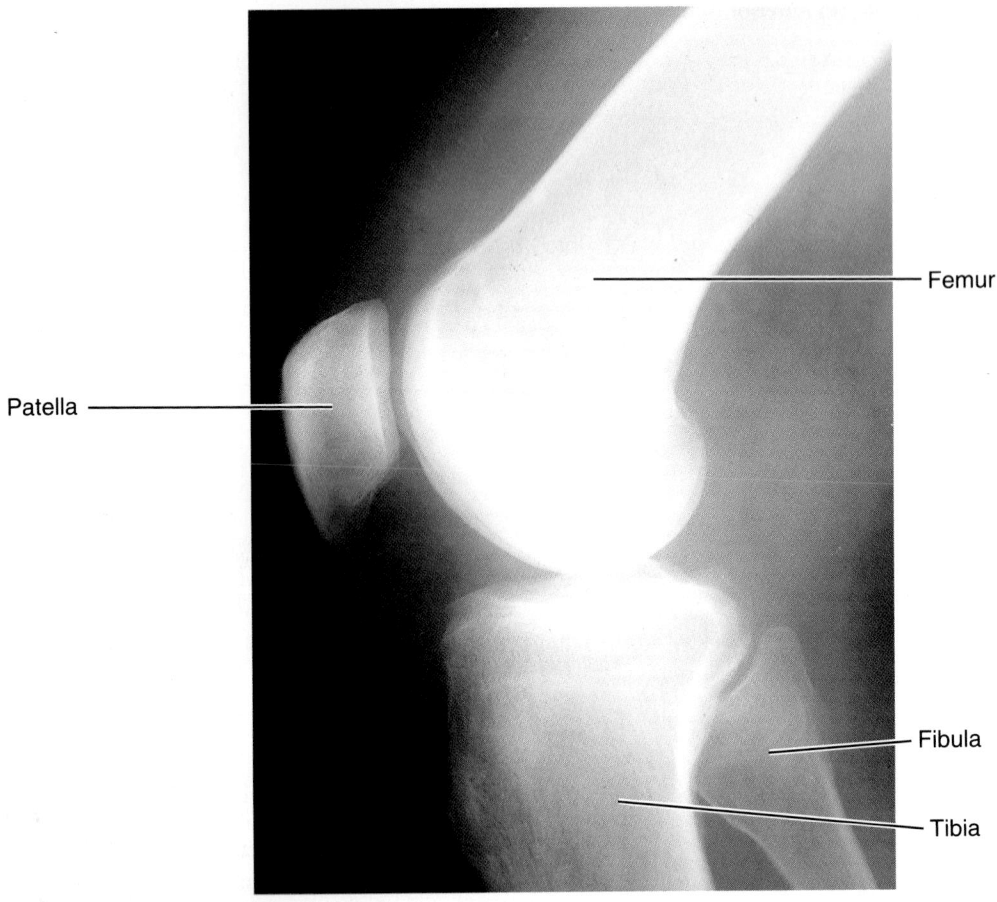

Fibula. The **fibula** is a slender bone that is lateral to the tibia (Figure 6.46). Its proximal end is expanded into a **head.** It articulates with the lateral condyle of the tibia. The distal end of the fibula is flattened, forming the **lateral malleolus,** which can be felt on the lateral side of the ankle. The lateral malleolus articulates with the talus and, together with the medial malleolus, forms the ankle joint. Just above the lateral malleolus, the fibula articulates with the distal end of the tibia. The tibia and the fibula are tightly bound together by an interosseous membrane.

Foot

The skeleton of the foot consists of *tarsal bones, metatarsal bones,* and *phalanges.* The proximal portion of the foot, toward the ankle, is composed of seven **tarsal bones** (Figure 6.47): the **talus, calcaneus, navicular, cuboid, medial cuneiform, intermediate cuneiform,** and **lateral cuneiform.**

The calcaneus forms the prominent heel bone, which provides attachment for several muscles of the calf. Resting on the superior surface of the calcaneus is the talus. The talus is located between the malleoli, where it articulates with the tibia and fibula. In this location, the talus receives the entire weight of the body, which it distributes to the other tarsal bones. The three cuneiforms and the cuboid articulate with the proximal ends of the metatarsal bones.

Five **metatarsal bones** form the skeleton of the intermediate region of the foot (Figure 6.47). They are numbered from medial to lateral. Proximally, they articulate with the tarsal bones and with each other. Distally, each articulates with the proximal end of a phalanx.

The skeleton of the toes is similar to that of the fingers. It is formed of 14 **phalanges,** with three *(proximal, middle* and *distal)* in each digit, except for the great toe, where there is no middle phalanx. The phalanges of the toes are shorter than those of the fingers.

Arches of the Foot. The tarsal and metatarsal bones are joined in such a manner that they form three arches.

◆ **FIGURE 6.46 Bones of the right leg**
(a) Anterior view. (b) Posterior view. (c) Anterior photograph.

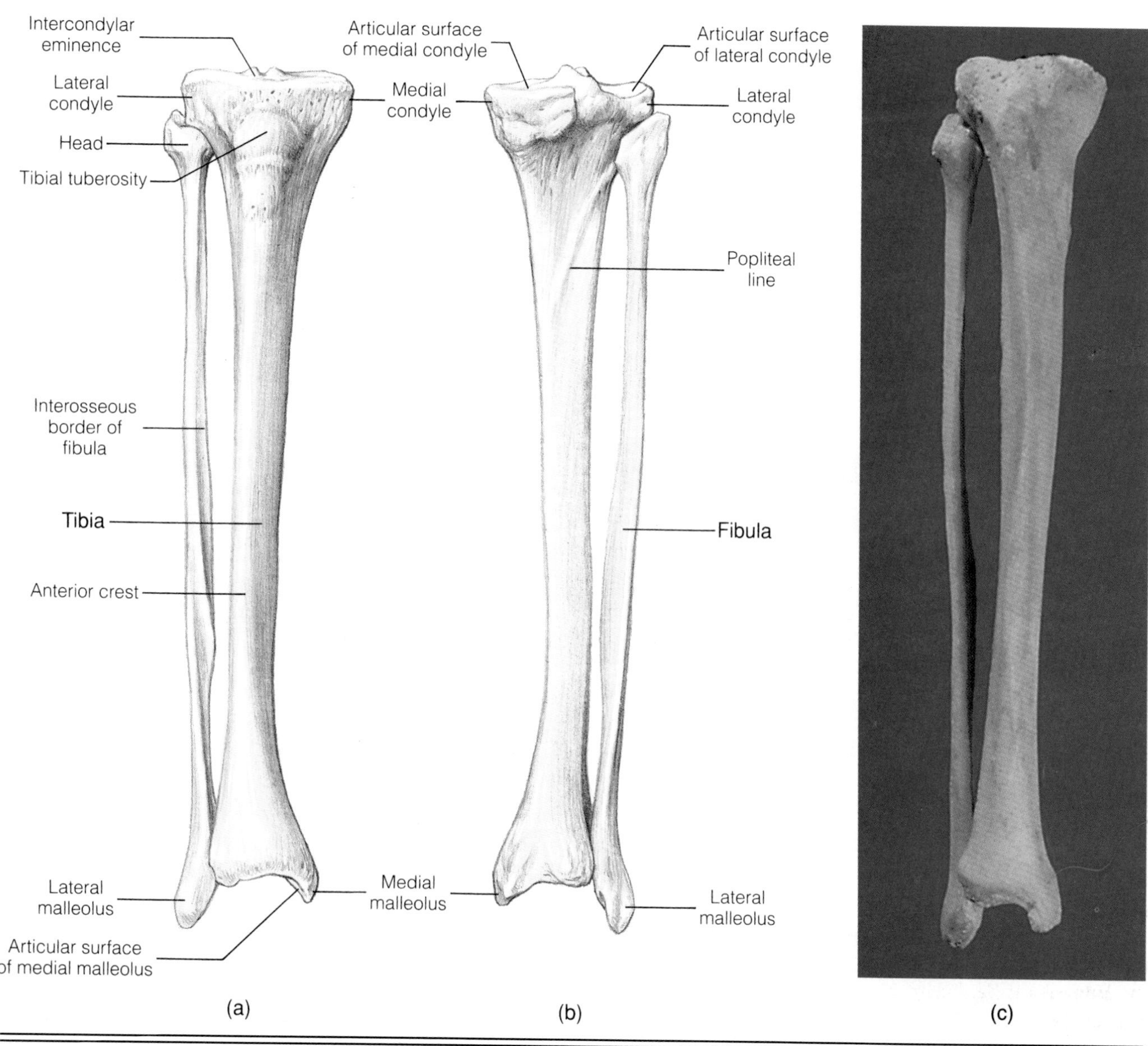

There are **medial** and **lateral longitudinal arches,** which have the calcaneus as their posterior pillar and the metatarsals and other tarsals as their anterior pillar. The **transverse (metatarsal) arch** is formed mostly by the bases of the metatarsal bones and passes obliquely from one side of the foot to the other.

The arches are supported primarily by the shapes of the bones of the feet and by ligaments, with some support gained from muscles and their tendons. They function to distribute the weight of the body fairly evenly between the heel and the metatarsals. People whose longitudinal arches have collapsed suffer from *flat feet* and often find that their feet tire easily. They may also experience low-back discomfort because the arches normally absorb many of the shocks that occur during walking.

Sesamoid Bones

In certain tendons that are subjected to compression or to unusual tensile stress, small bones called **sesamoid bones** form (Figure 6.47). Sesamoid bones are generally located around a joint. The patella is the largest sesamoid bone in the body. Although the locations of sesamoid bones vary, the most common locations include the areas around the metacarpophalangeal and the metatarsophalangeal joints.

◆ **FIGURE 6.47 Bones of the ankle and foot**
(a) Lateral view of the foot. (b) Superior view of the bones of the right ankle and foot. (c) Medial X ray of the foot.

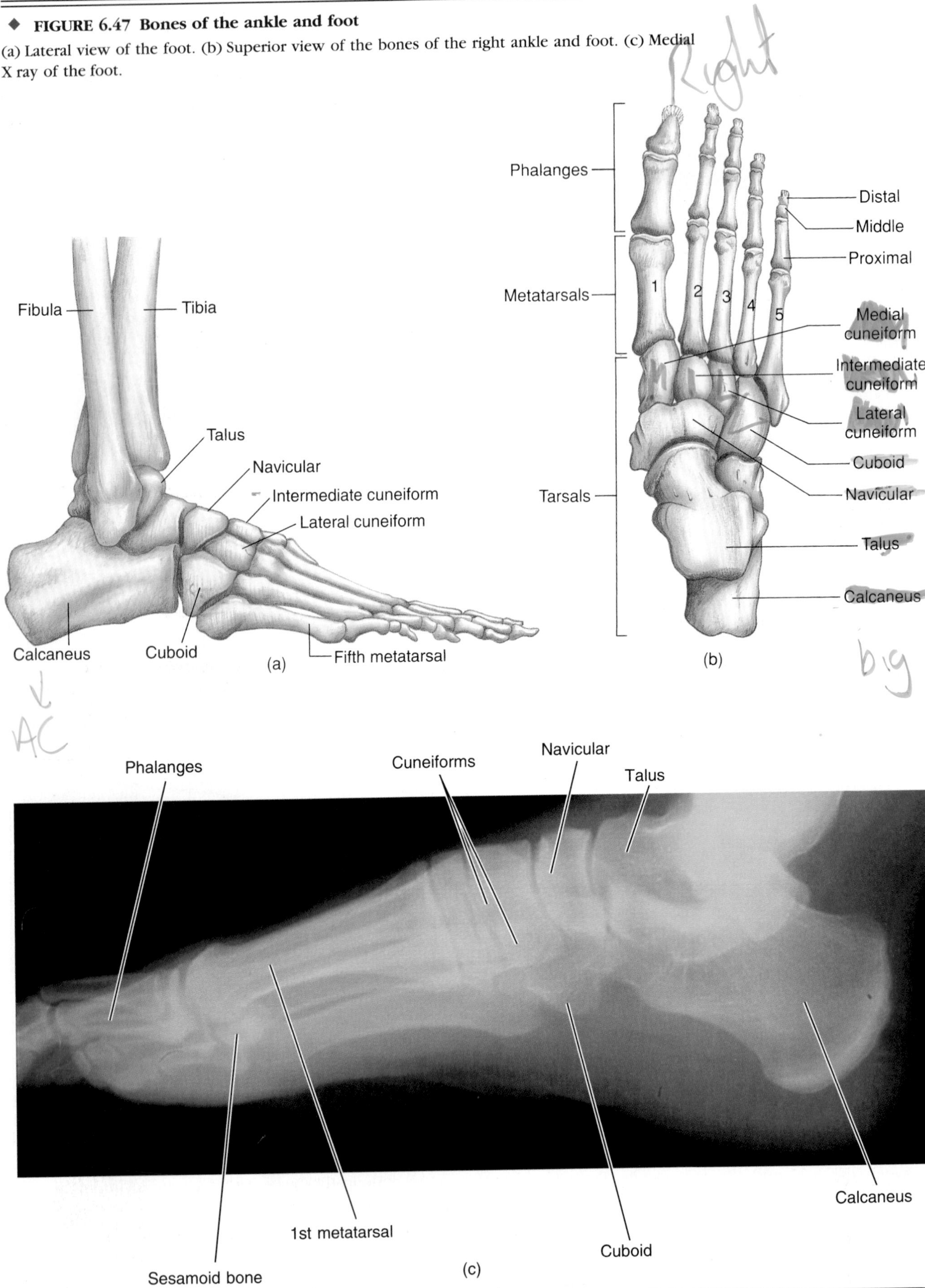

◆ **TABLE 6.12 Summary of Specific Features of the Bones of the Lower Limbs**

COXAL BONE (os coxae) [*Fig. 6.41*]

ILIUM [*Fig. 6.41*] The upper, flat expanded portion of the coxal bone.

ILIAC CREST
The superior margin of the ilium.

ILIAC SPINES
Anterior superior A projection at the anterior end of the iliac crest.

Anterior inferior A rounded projection just below the anterior spine.

Posterior superior A projection at the posterior end of the iliac crest.

Posterior inferior A projection just below the posterior superior spine.

GREATER SCIATIC NOTCH
A deep notch just below the posterior inferior spine. It provides passage for the sciatic nerve, as well as other nerves, vessels, and the piriformis muscle.

GLUTEAL LINES
Three arched lines—posterior, anterior, and inferior—on the outer surface of the ilium. The origins of the three gluteus muscles are located between these lines.

AURICULAR SURFACE
Surface that articulates with the sacrum.

ARCUATE LINE
A slight ridge on the internal surface of the ilium. It extends forward and downward from the top of the sacrum to the pecten of the pubis. The line forms the internal margins of the pelvic inlet.

ILIAC FOSSA
The smooth, concave internal surface of the ilium, above the arcuate line.

ISCHIUM [*Fig. 6.41*] The posterior portion of the coxal bone.

ISCHIAL SPINE
A triangular projection from the posterior border behind the acetabulum.

ISCHIAL TUBEROSITY
A roughened enlargement, on the posterior inferior margin.

LESSER SCIATIC NOTCH
A small indentation that separates the ischial spine and tuberosity. It transmits the tendon of the obturator internus muscle, as well as nerves and blood vessels.

ISCHIAL RAMUS
The flattened anterior projection that arises from the tuberosity. It joins with the inferior ramus of the pubis to form the lower border of the obturator foramen.

PUBIS [*Figs. 6.41; 6.42*] The anterior inferior portion of the coxal bone.

SUPERIOR RAMUS
Extends anteriorly from the acetabulum to form the upper border of the obturator foramen.

SYMPHYSIS PUBIS
The midline joint between the superior rami of both pubic bones.

PUBIC TUBERCLE
The projection from the superior ramus just lateral to the symphysis pubis.

PECTEN
A ridge that extends upward and laterally along the superior ramus. It runs from the pubic tubercle to the arcuate line.

PUBIC CREST
A ridge that extends medially from the pubic tubercle to the symphysis pubis.

INFERIOR RAMUS
The projection that passes downward and backward from the symphysis pubis. It joins with the ischium to form the lower border of the obturator foramen.

PUBIC ARCH
Formed by the convergence of the inferior rami.

ACETABULUM
A cup-shaped depression formed by the junction of the ilium ischium, and pubis. It receives the head of the femur to form the hip joint.

OBTURATOR FORAMEN
The large opening in the inferior region of coxal bone. It allows for the passage of the obturator nerves and blood vessels.

PELVIC BRIM (INLET)
The boundary of the opening that leads into the true pelvis. It is formed by the sacral promontory, arcuate lines, and the superior margin of the symphysis pubis.

FALSE PELVIS
The expanded space above the pelvic inlet. It is actually a portion of the abdominopelvic cavity.

TRUE PELVIS
The smaller cavity below the pelvic inlet, bounded by the coxal bones, sacrum, and coccyx.

PELVIC OUTLET
The lower margin of the true pelvis. It is bounded by the coccyx and the lower parts of the coxal bones.

continued on next page

◆ **TABLE 6.12 Summary of Specific Features of the Bones of the Lower Limbs (continued)**

FEMUR [*Fig. 6.44*]

HEAD
The rounded proximal end that fits into the acetabulum.

FOVEA CAPITIS
The pit on the head where the ligamentum teres is attached, which in turn is attached to the acetabulum.

NECK
The constriction that connects the head with the shaft.

GREATER TROCHANTER
The large lateral process just below the neck.

LESSER TROCHANTER
The smaller medial process just below the neck.

INTERTROCHANTERIC CREST
A ridge on the posterior surface that connects the two trochanters.

INTERTROCHANTERIC LINE
A slight line on the anterior surface that connects the two trochanters.

GLUTEAL TUBEROSITY
A small projection just below the greater trochanter.

TROCHANTERIC FOSSA
A depression on the medial surface of the greater trochanter.

LINEA ASPERA
A longitudinal ridge that extends along the middle third of the posterior surface of the shaft.

SUPRACONDYLAR RIDGES
The medial and lateral ridges that are formed by the divergence of the linea aspera at its distal end.

POPLITEAL SURFACE
A smooth triangular surface on the posterior surface. It is bounded above by the medial and lateral supracondylar ridges.

CONDYLES
Rounded medial and lateral enlargements at the distal end of the shaft. They have smooth surfaces for articulation with the tibia.

EPICONDYLES
Roughened prominences on the lateral surfaces of the condyles. They provide points of attachment for the medial and lateral collateral ligaments of the knee joint.

ADDUCTOR TUBERCLE
A small projection just above the medial condyle, at the termination of the medial supracondylar ridge.

INTERCONDYLAR FOSSA
The deep notch that separates the condyles posteriorly. It provides points of attachment for the anterior and posterior cruciate ligaments of the knee joint.

PATELLAR SURFACE
The smooth anterior surface above the condyles. It articulates with the patella when the leg is extended (straightened).

TIBIA [*Fig. 6.46*]

CONDYLES
Flattened enlargements at the proximal end of the shaft. The upper surfaces are smooth for articulation with the condyles of the femur.

INTERCONDYLAR EMINENCE OR SPINE
The vertical projection from the superior surface, between the articular surfaces of the condyles.

TIBIAL TUBEROSITY
The midline projection from the anterior surface just below the condyles. It serves as the point of attachment for the ligamentum patellae.

ANTERIOR CREST
A sharp longitudinal ridge on the anterior surface.

POPLITEAL LINE
A ridge on the posterior surface that extends downward and medially from the lateral condyle. It marks the junction between the insertion of the popliteus muscle and the origin of the soleus muscle.

MEDIAL MALLEOLUS
A downward projection from the medial side of the distal end of the tibia. It articulates with the talus.

FIBULA [*Fig. 6.46*]

HEAD
The proximal, expanded portion of the bone. It articulates with the lateral condyle of the tibia.

STYLOID PROCESS A rough vertical prominence on the head of the fibula.

LATERAL MALLEOLUS
A triangular expansion at the distal end that articulates medially with the tibia and with the talus.

Study Outline

◆ FUNCTIONS OF THE SKELETON p. 154

Support. Body framework.

Movement. Muscle attachment to skeleton; movable joints.

Protection. Of vital internal organs.

Mineral Reservoir. Storage of calcium phosphorus, sodium, potassium.

Blood-Cell Formation. Red marrow of certain bones produces the blood cells of adult.

◆ CLASSIFICATION OF BONES p. 154

Long Bones. Have long axis; most bones of upper and lower limbs.

Short Bones. Lack a long axis; carpals, tarsals.

Flat Bones. Thin bons; ribs.

Irregular Bones. Vertebrae.

◆ STRUCTURE OF BONE pp. 155–157

Studied at gross and microscopic levels.

Gross Anatomy.

DIAPHYSIS. Bone shaft.

MEDULLARY CAVITY. Fat-storage site.

EPIPHYSIS. End of long bone.

EPIPHYSEAL CARTILAGE OR PLATE. Separates diaphysis and epiphysis in children; permits increase in bone length.

PERIOSTEUM. Double-layered connective tissue that covers bones.

Microscopic Anatomy.

COMPACT BONE.

1. Unit of structure is the osteon or Haversian system.
2. Central canals surrounded by concentric lamellae.
3. Osteocytes embedded within lacunae.
4. Lacunae interconnected by canaliculi.

SPONGY BONE. Osteocytes embedded within lacunae; lamellae not arranged in concentric layers but according to lines of tension.

◆ COMPOSITION AND FORMATION OF BONE pp. 157–158

1. Intercellular matrix of bone contains both organic components and inorganic salts.
2. Osteoblasts secrete the organic components of the intercellular matrix of bone.
3. Within a period of days, the intercellular matrix becomes calcified by the deposition of insoluble calcium salts within it.

◆ DEVELOPMENT OF BONE pp. 158–164

Early Development of Bone. Mesenchymal cells give rise to osteogenic cells, which differentiate into osteoblasts in areas well supplied by blood vessels.

INTRAMEMBRANOUS OSSIFICATION.

1. Bone formation takes place within mesenchyme.
2. Mesenchymal cells give rise to osteogenic cells, which differentiate into osteoblasts, and bone formation begins.
3. A layer of mesenchyme containing an extensive supply of blood vessels condenses at the external surface of the bone and develops into a periosteum.
4. As additional bone forms, some previously formed bone is resorbed by osteoclasts.

ENDOCHONDRAL OSSIFICATION.

1. Bone formation takes place within hyaline cartilage.
2. Hyaline cartilage model of future bone develops from mesenchyme, and mesenchyme adjacent to the cartilage becomes organized into a perichondrium.
3. Chondrocytes in middle of the cartilage model enlarge, and the intercellular matrix of the cartilage becomes calcified.
4. Blood vessels infiltrate the perichondrium around the midsection of the model, osteogenic cells at the inner surface of the perichondrium differentiate into osteoblasts, and a collar of compact bone forms around the cartilage. The perichondrium becomes a periosteum.
5. Primary ossification center develops in calcified cartilage in the center of the model, and trabeculae of spongy bone form. In long bones, osteoclasts break down some newly formed bone, forming medullary cavity.
6. Secondary ossification centers develop in long bones.

Increase in Long Bone Length and Diameter.

1. Length: cartilage forms near the sides of the epiphyseal plates toward the epiphyses; cartilage becomes calcified; bone forms at diaphyseal sides of epiphyseal plates.
2. Diameter: new bone forms around the diaphysis, just beneath the periosteum; other bone is usually resorbed from within, beneath the endosteum.

Remodeling of Bone. Osteoclasts resorb previously formed bone, and osteoblasts secrete the organic components of the intercellular matrix of new bone, which becomes calcified.

Factors That Affect Bone Development.

MECHANICAL STRESS. A bone can adjust its strength in proportion to the degree of stress to which it is subjected. Bones are normally subjected to gravitational and functional forces.

HORMONES. Growth hormone influences skeletal growth,

particularly in young people; thyroid hormones are necessary for the proper production and action of growth hormone; parathyroid hormone increases the number and the activity of osteoclasts within bones; calcitonin decreases the resorptive activity of osteoclasts.

◆ CONDITIONS OF CLINICAL SIGNIFICANCE: THE SKELETAL SYSTEM pp. 164–169

Fractures.

TYPES OF FRACTURES.

SIMPLE. Bone ends do not penetrate skin.

COMPOUND. Broken bone ends penetrate skin.

COMMINUTED. Bone splintered at site of break.

DEPRESSED. Broken region pushed inward.

IMPACTED. Broken ends of bone driven into each other.

SPIRAL. Result of excessive twisting.

GREENSTICK. Break does not extend clear through bone.

HEALING OF FRACTURES.

FORMATION OF A HEMATOMA; FORMATION OF A FIBROCARTILAGINOUS CALLUS; FORMATION OF A BONY CALLUS.

Metastatic Calcification. Calcium deposits in tissues that are not normally calcified; kidney is a common site.

Spina Bifida. Failure of posterior portions of vertebrae to form bony arch around spinal cord; most common in lumbosacral area.

Osteoporosis. Reduced bone formation rate, normal bone absorption rate.

Osteomyelitis. Infection of marrow-cavity contents and bone tissue, usually caused by *Staphylococcus aureus.*

Tuberculosis of Bone. Infection of marrow-cavity contents and bone tissue caused by *Mycobacterium tuberculosis.*

Rickets and Osteomalacia. Both due to deficiencies of calcium, phosphorus, vitamin D, or lack of sunlight.

RICKETS. Demineralization and bone softening in children.

OSTEOMALACIA. Softening of adult bones.

Tumors of Bone. Benign or malignant tumors.

Abnormal Growth Patterns.

PITUITARY GIANT. Excessive pituitary growth hormone delays ossification of epiphyseal cartilage.

PITUITARY DWARF. Growth hormone deficiency resulting in early replacement of epiphyseal cartilages by bone.

ACROMEGALY. Excess growth hormone secreted after epiphyseal cartilages replaced.

ACHONDROPLASIA. Epiphyseal cartilages function for short time only, resulting in shortened arms and legs.

Effects of Aging. Gradual loss of calcium from bone; more severe in women. Matrix not replaced as rapidly as it is broken down.

◆ INDIVIDUAL BONES OF THE SKELETON p. 169

◆ AXIAL SKELETON pp. 169–191

Skull.

CALVARIUM.

FRONTAL BONE. Anterior superior region of skull.

TWO PARIETAL BONES. Form most of calvarium.

OCCIPITAL BONE. Posterior floor of cranial cavity.

BONES THAT FORM THE FLOOR OF THE CRANIAL CAVITY.

FRONTAL, OCCIPITAL, ETHMOID, AND SPHENOID BONES. Form midline.

TWO TEMPORAL BONES. Form side of floor.

SPHENOID BONE AND PAIRED TEMPORALS. Form middle cranial cavity.

FACIAL SKELETON.

FRONTAL. Forehead.

PAIRED MAXILLARY. Upper jaw.

PAIRED ZYGOMATIC. Cheek prominences.

PAIRED NASAL BONES. Bridge of nose.

PAIRED LACRIMAL. Medial orbital cavity.

MANDIBLE. Lower jaw.

BONES THAT FORM THE NASAL CAVITY.

ETHMOID. Perpendicular plate forms part of septum; superior and middle conchae form lateral walls of nasal cavity.

VOMER. Posterior inferior nasal septum.

PAIRED INFERIOR NASAL CONCHAE. Form elongated shelves from lateral walls of nasal cavity.

BONES THAT FORM THE HARD PALATE. Palatine processes of maxillary bones; horizontal processes of palatine bones.

BONES THAT FORM THE ORBITAL CAVITY. Frontal, sphenoid, maxilla, lacrimal, ethmoid, zygomatic, palatine.

PARANASAL SINUSES. Mucous-membrane–lined air spaces in the frontal, ethmoid, maxillary, and sphenoid bones.

AUDITORY OSSICLES.

MALLEUS (HAMMER), STAPES (STIRRUP), INCUS (ANVIL).

HYOID BONE. U-shaped; attachment point of tongue and throat muscles.

Vertebral Column.

CURVATURES OF THE VERTEBRAL COLUMN.

INFANT. Single primary curve, convex posteriorly.

CERVICAL CURVE. Convex anteriorly; develops with head raising.

THORACIC CURVE. Convex posteriorly; remains from primary curve of newborn.

LUMBAR CURVE. Convex anteriorly; develops as child walks.

SACRAL CURVE. Convex posteriorly; remains from primary curve of newborn.

ABNORMALITIES.

Lordosis (Swayback). Excessive anterior lumbar curve.

Kyphosis (Hunchback). Excessive posterior thoracic curve.

Scoliosis. Lateral curvature of spine.

FUNCTIONS OF THE VERTEBRAL COLUMN.

SUPPORT AND FLEXIBILITY.

SPINAL CORD PROTECTION. Intervertebral foramina provide passage for spinal nerves.

CHARACTERISTICS OF A TYPICAL VERTEBRA.

1. Thick anterior centrum with neural arch that arises posteriorly.
2. Spinous process extends from neural arch.
3. Vertebral foramen—passageway for spinal cord.
4. Transverse processes.
5. Superior and inferior articulating processes.

REGIONAL DIFFERENCES IN VERTEBRAE. Summarized in Table 6.5.

Thorax.

STERNUM. Elongated flat bone; midline anterior thorax wall; rib articulation.

RIBS. 12 pairs. All articulate posteriorly with thoracic vertebrae.

TRUE RIBS: VERTEBROSTERNAL RIBS. Pairs 1 through 7; articulate anteriorly with sternum through costal cartilage.

FALSE RIBS: VERTEBROCHONDRAL RIBS. Pairs 8 through 12; costal cartilages of pairs 8–10 attach to cartilage of rib above.

VERTEBRAL (FLOATING) RIBS. Pairs 11 and 12; short costal cartilages embedded anteriorly in muscles of body wall.

COASTAL CARTILAGES. Hyaline cartilage; strengthen thorax and provide flexibility.

◆ **APPENDICULAR SKELETON** pp. 191–210

Upper Limbs.

PECTORAL GIRDLE. Straddles upper thorax.

CLAVICLE. S-shaped bone bracing scapula; holds shoulder away from rib cage.

SCAPULA. Thin, flat, triangular bone over posterior surfaces of rib pairs 2 through 7.

ARM. ***Humerus*** is the only bone.

FOREARM. 2 bones.

ULNA. Medial to radius in anatomical position.

RADIUS. Lateral to ulna in anatomical position.

HAND. 8 carpal bones in 2 transverse rows of 4 each toward wrist.

PROXIMAL ROW. Lateral to medial—scaphoid, lunate, triquetral, pisiform.

DISTAL ROW. Lateral to medial—trapezium, trapezoid, capitate, hamate.

PALM OF HAND. 5 ***metacarpal bones.***

FINGERS. 14 ***phalanges.***

Lower Limbs.

PELVIC GIRDLE. Formed by a pair of coxal bones.

ILIUM, ISCHIUM, PUBIS. Individual bones during embryonic development; fuse to form each adult coxal bone.

PELVIC CAVITIES.

GREATER OR FALSE PELVIS. Above pelvic brim; expansible.

LESSER OR TRUE PELVIS. Cavity below pelvic brim; restricted.

SEXUAL DIFFERENCES IN PELVIS. Most are related to childbearing.

THIGH. ***Femur*** forms thigh skeleton.

LEG. ***Tibia*** Strong, medial. ***Fibula*** Slender, lateral. ***Patella*** Protects knee joint between thigh and leg.

FOOT. 7 ***tarsal bones*** toward ankle: talus, calcaneus, navicular, cuboid, medial cuneiform, intermediate cuneiform, and lateral cuneiform.
5 ***metatarsal bones.***

TOES. 14 ***phalanges.***

ARCHES OF FOOT. 3 arches formed: 2 longitudinal, 1 transverse; supported by shapes of bones of the feet and by ligaments as well as muscles and their tendons; distribute weight evenly between heel and metatarsals.

Sesamoid Bones. Form in tendons subjected to compression; generally around a joint; e.g., patella.

Self-Quiz

1. Bones may act as a storehouse for: (a) calcium; (b) phosphorus; (c) both calcium and phosphorus.
2. Flat bones lack: (a) periosteum; (b) a medullary cavity; (c) diploe.
3. The diaphysis is: (a) the shaft of a long bone; (b) the spongy bone of a flat bone; (c) the membrane that covers the surfaces of short bones.
4. A lacuna of an osteon: (a) is lined with periosteum; (b) contains blood vessels; (c) is connected to other lacunae by canaliculi.
5. During intramembranous ossification, bone formation takes place within hyaline cartilage. True or False?
6. Bone is resorbed by: (a) osteogenic cells; (b) osteoblasts; (c) osteoclasts.
7. The epiphyseal plates of the long bones of a four-year-old child are composed of: (a) mesenchyme; (b) cartilage; (c) compact bone.
8. The liver produces: (a) calcitonin; (b) growth hormone; (c) somatomedins.
9. Parathyroid hormone: (a) is produced by cells of the endosteum; (b) increases the number and activity of osteoclasts within bones; (c) prevents the formation of a medullary cavity within the secondary ossification centers of long bones.
10. The single occipital bone forms the lower posterior wall of the calvarium as well as the anterior portion of the floor of the cranial cavity. True or False?
11. An abnormal lateral curve that occurs in the vertebral column is called a: (a) scoliosis; (b) lordosis; (c) kyphosis.
12. Match the common skeletal structure terms with the appropriate lettered description:

Process
Trochanter
Tubercle
Condyle
Sulcus
Crest
Facet
Fossa
Fovea
Foramen
Meatus

(a) A smooth, nearly flat articular surface
(b) A pit; generally used for attachment rather than for articulation
(c) A hole
(d) A rounded prominence that articulates with another bone
(e) A canal
(f) A large, somewhat blunt process
(g) A depression; often used as an articular surface
(h) A sharp, prominent bony ridge
(i) A prominence or projection
(j) A nodule or small rounded process
(k) A groove

13. Match the structures with their appropriate lettered description:

Fontanels
Frontal bone
Sinuses
Parietal bones
Occipital bone
Ethmoid bone
Temporal bones
Maxillary bones
Zygomatic bones
Lacrimal bones
Mandible
Vomer bone
An auditory ossicle
Hyoid bone
Cervical vertebrae
Thoracic vertebrae
Sacral vertebrae

(a) Form most of the calvarium
(b) One of the bones of the floor of the cranial cavity
(c) Form the central part of the facial skeleton
(d) Air spaces in bone lined with mucous membranes
(e) The lower jaw
(f) "Soft spots" of the skull
(g) Forms the posterior inferior portion of the nasal septum
(h) Its most obvious landmark is the large foramen magnum
(i) Form the prominences of the cheeks
(j) Single bone that forms the anterior superior region of the skull
(k) Help form the medial surface of the orbital cavity
(l) Form part of the middle cranial cavity
(m) Form the spinal column in the neck
(n) Serves for attachment of muscles of the tongue and throat
(o) Are fused and form a wedge between the coxal bones
(p) Malleus
(q) There are 12 of these

14. The xiphoid process is a component of the: (a) sternum; (b) ribs; (c) thoracic vertebrae.
15. Those ribs that have their costal cartilages attached to the cartilages of the rib above are called: (a) true ribs; (b) vertebrosternal ribs; (c) false ribs.
16. The first seven pairs of ribs articulate anteriorly with the sternum through the: (a) costal cartilages; (b) tubercles; (c) xiphoid process.
17. Which of the following are part of the appendicular skeleton? (a) pectoral girdle; (b) pelvic girdle; (c) mandible; (d) both a and b.
18. the scaphoid, lunate, triquetral, and pisiform are bones of the: (a) foot, (b) hand; (c) tibia-fibula complex.
19. Which of the following forms the junction of the superior and axillary borders of the scapula and contains the glenoid fossa? (a) inferior angle; (b) superior angle; (c) lateral angle.
20. The coxal bone includes which of the following? (a) ischium; (b) fibula; (c) patella.
21. The lesser or true pelvis is the cavity: (a) above the pelvic brim; (b) anterior to the pelvic brim; (c) below the pelvic brim.
22. Compared with the male pelvis, the female pelvis: (a) is more massive; (b) is narrower at the pelvic outlet; (c) has an obtuse pubic angle.

23. Match the following terms with the appropriate lettered description.

Ilium	(a) Forms anterior portion of each coxal bone
Ischium	(b) Expanded space above the pelvic inlet, actually part of abdominopelvic cavity
Pubis	(c) Rounded medial and lateral enlargements at distal end of femur shaft
Acetabulum	(d) The medial bone of the leg
False pelvis	(e) Bone that forms the thigh
Pelvic outlet	(f) Sharp longitudinal ridge on anterior surface of the tibia
Femur	(g) Small portion just above medial condyle of femur
Greater trochanter	(h) Upper flat expanded portion of each coxal bone
Condyles	(i) Large lateral process just below the neck of the femur
Adductor tubercle	(j) An opening bounded by coccyx and lower parts of coxal bones
Tibia	(k) Posterior inferior portion of each coxal bone
Anterior crest	(l) A slender bone lateral to the tibia
Fibula	(m) Cup-shaped depression formed by junction of ilium, ischium, and pubis

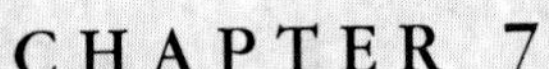

CHAPTER 7

Articulations

CHAPTER CONTENTS

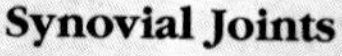

LEARNING OBJECTIVES

After completing this chapter, you should be able to:

1. State two especially useful criteria for classifying body joints.
2. Name and distinguish between the two types of fibrous joints, and give an example of each.
3. Name and distinguish between the two types of cartilaginous joints, and give an example of each.
4. Describe the distinguishing features of synovial joints.
5. Name the four types of synovial joints, and give examples of each.
6. Name the ligaments of the shoulder, hip, and knee joints.
7. Describe several common joint disorders.

CHAPTER 7

In the previous chapter, the support function of the skeleton was discussed. This chapter will examine how the individual bones of the skeleton join, or **articulate.** While some joints are rather rigid and permit little, if any, movement, most joints allow the bones to move in relation to each other. Thus, while some joints function to secure adjacent bones together, the most obvious joints are those that make it possible for our bony skeleton to be mobile and permit us to undertake many different activities that require changing the positions of the body parts.

Classification of Joints

Joints may be classified on the basis of either their function or their structure. The *functional* classification of joints is based on the degree of movement permitted by each joint. On this basis, there are **synarthroses,** *(sin″-ar-thro′-sēz)* which are immovable joints; **amphiarthroses,** *(am″-fee-ar-thro′-sēz)* which are joints that permit slight movement; and **diarthroses,** *(di″-ar-thro′-sēz)* which are freely movable joints. The restricted movement allowed by synarthroses and amphiarthroses provides firm support and protection, and these joints are largely restricted to the axial skeleton. In contrast, diarthroses make possible the free movements typical of the limbs.

The *structural* classification of joints is based on the type of connective tissue binding the bones together and the presence or absence of a synovial (joint) cavity between articulating bones. Using these bases, joints may be classified as **fibrous, cartilaginous,** or **synovial.**

Of course, these two methods of classification overlap somewhat. Fibrous joints are generally immovable, and synovial joints are freely movable. However, while most cartilaginous joints allow slight movement and may be considered to be amphiarthroses, some do not. In the following discussion, we will use the structural classification and relate it to the functional properties of the joint where appropriate.

Fibrous Joints

The **fibrous joints** include all those articulations in which the bones are held tightly together by fibrous connective tissue and no joint cavity is present. Because of its role in strengthening the joint, the connective tissue is referred to as the *sutural ligament.* Very little material separates the ends of the bones, and no appreciable movement is allowed. For this reason, most fibrous joints are synarthroses, although some fibrous joints do, in fact, permit slight movement. There are three types of fibrous joints, based on the length and type of connective tissue fibers that hold the bones together: *sutures, syndesmoses,* and *gomphoses.*

Sutures

In **sutures,** the edges of the bones have interdigitations, or grooves, that fit closely and firmly together (Figure 7.1). Consequently, the connecting fibers spanning the small gap between the bones are very short. This type of joint is found only between the flat bones of the skull. In early adulthood, the fibers of the suture begin to be replaced by bone. Eventually, if the fibers are com-

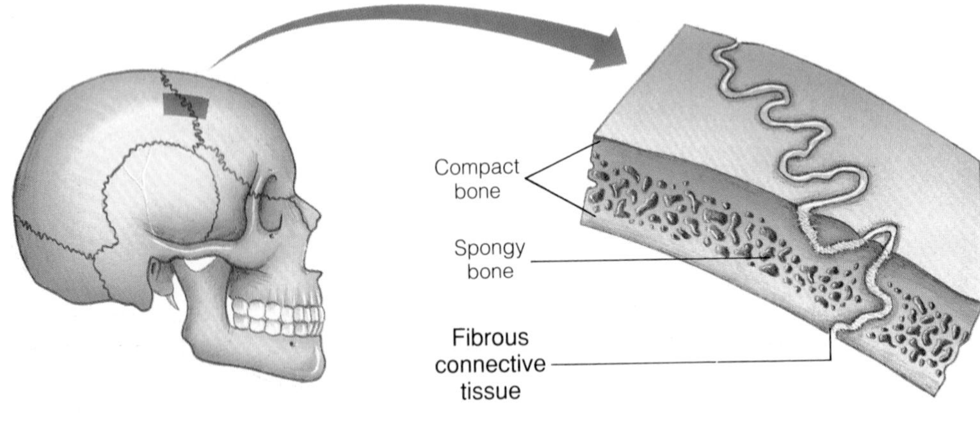

◆ **FIGURE 7.1 Suture**
Note the short connective tissue fibers joining the two bones.

pletely replaced, the bones on either side of the suture become firmly fused together. This condition is called *synostosis* (that is, held together by bone).

Syndesmoses

Like a suture, a **syndesmosis** *(sin″-des-mō′-sēz)* joint is held together by fibrous connective tissue. However, the ends of the bones are farther apart in a syndesmosis than in a suture. Consequently, the fibers joining the bones are longer and are referred to as a *ligament* or, if in the form of a sheet, as an *interosseous membrane.* The bones joined by syndesmoses are not held as firmly as those joined by sutures. And although syndesmoses can permit a kind of movement best described as "give," they do not allow any true movement. Therefore, most of these joints are classified functionally as synarthroses. The joint between the distal ends of the tibia and fibula allows very little movement and is an example of such a syndesmosis (Figure 7.2). In contrast, the radius and ulna are connected along their length by a longer interosseous membrane, and thus form a syndesmosis, but this joint permits greater movements in the form of supination and pronation.

◆ **FIGURE 7.2 Syndesmosis**
The distal ends of the tibia and fibula are held together by relatively long connective tissue fibers.

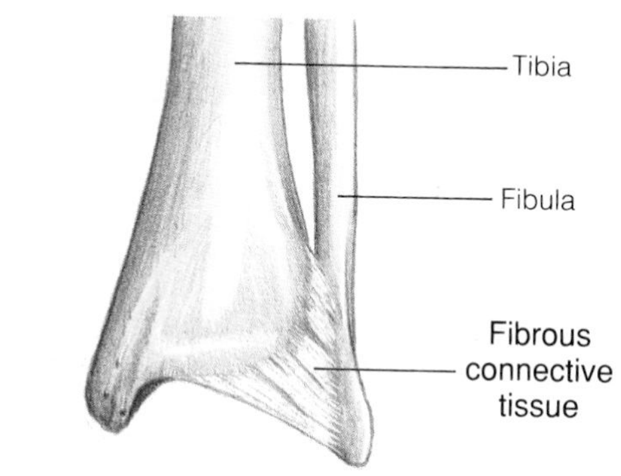

Gomphoses

A **gomphosis** *(gom-fō′-sis)* is a fibrous joint specialized for the articulation of the roots of the teeth with the alveolar sockets in the mandible and maxillae. In a gomphosis, the teeth are anchored to the walls of the alveoli by a thin fibrous membrane called the *periodontal ligament.* Gomphoses are functionally classified as synarthroses.

 Cartilaginous Joints

In **cartilaginous joints,** the bones are united by cartilage. Like fibrous joints, they lack a joint cavity. There are two types of cartilaginous joints: *synchondroses* and *symphyses.*

Synchondroses

The bones in a **synchondrosis** *(sin″-kon-drō′-sis)* joint are held together by hyaline cartilage (Figure 7.3). Most synchondroses are immovable and are functionally synarthroses. However, some allow slight movement and are functionally amphiarthroses. Many synchondroses are temporary joints, with the cartilage eventually being replaced by bone. Such joints then become synostoses. An example of a cartilaginous joint that ossifies is the articulation between the epiphyses and the diaphysis of growing bones, where the epiphyseal cartilages are replaced by bone in the adult. These joints are functionally synarthroses, as no movement is permitted. The joint between the first rib and the manubrium of the sternum is also a synchondrosis that eventually ossifies. In contrast, the joints formed between ribs 2–10 and their costal cartilages are permanent synchondroses. These joints allow some movement and are functionally amphiarthroses.

Symphyses

The articular surfaces of bones that are joined by a **symphysis** *(sim′-fi-sis)* are covered with a thin layer of hyaline cartilage. The hyaline cartilages of adjacent bones are separated by, and fused to, a fibrocartilaginous pad (Figure 7.4). These pads are compressible, allowing the symphyses to serve as shock absorbers, and they permit the joints to move slightly. For this reason, symphyses are functionally amphiarthroses. The junction of the two pubic bones and the junctions between the centra of adjacent vertebrae are examples of symphyses. The fibrocartilaginous pads between the vertebrae are called *intervertebral discs.* During development, the two halves of the mandible are joined by a midline symphysis; however, this joint becomes completely ossified by adulthood.

 Synovial Joints

Most joints of the body are **synovial joints,** *(sa-no′-vee-al)* which are characterized by the presence of a

◆ **FIGURE 7.3 Synchondroses**

Hyaline cartilage attaches the ribs to the sternum. The attachments are referred to as *costal cartilages.*

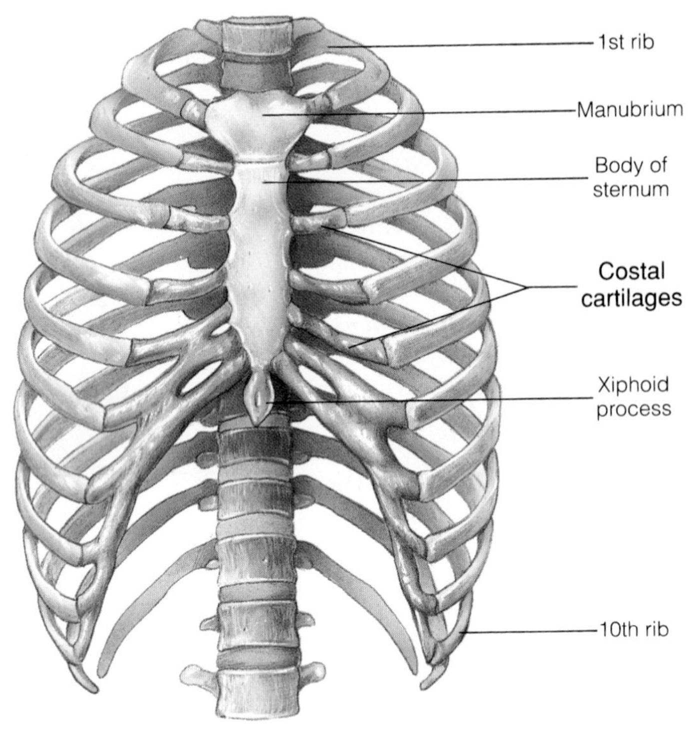

fluid-filled joint cavity. Synovial joints are freely movable, with the movement limited only by ligaments, muscles, tendons, or adjoining bones. Because of this freedom, synovial joints are classified functionally as diarthroses.

Synovial joints have five distinguishing features (Figure 7.5a):

1. **Articular cartilage.** A thin layer of hyaline cartilage that covers the smooth articular surfaces of the bones.
2. **Synovial cavity.** The synovial cavity (joint cavity) is a very small space that separates the ends of the articulating bones. The cavity is filled with synovial fluid.
3. **Articular capsule.** The synovial cavity is surrounded by a double-layered articular capsule that en-

◆ **FIGURE 7.4 Symphyses**

A pad of fibrocartilage separates the bodies of the vertebrae.

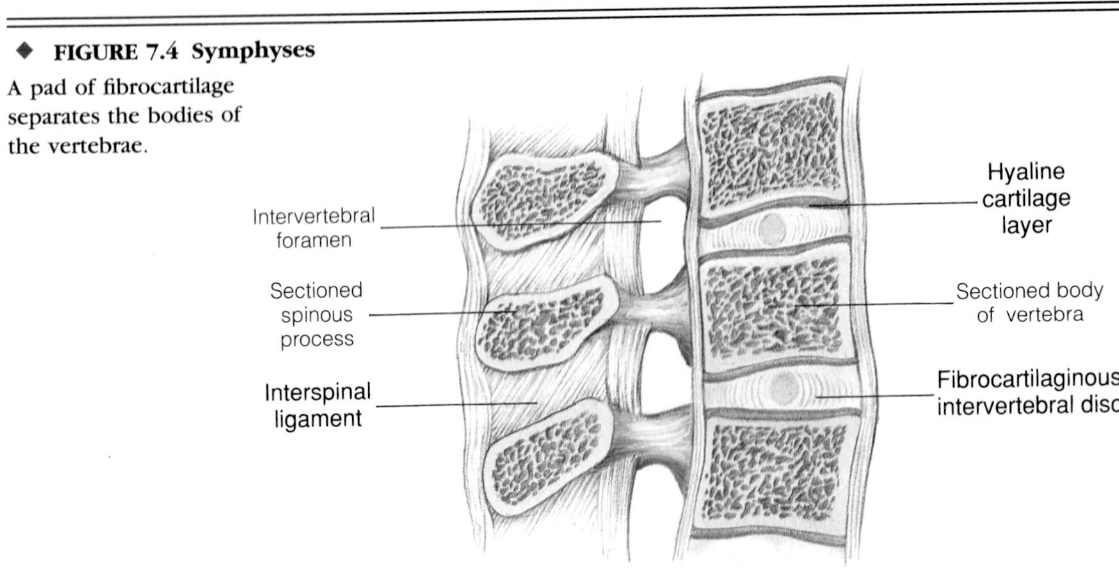

◆ **FIGURE 7.5 Structure of synovial joints**

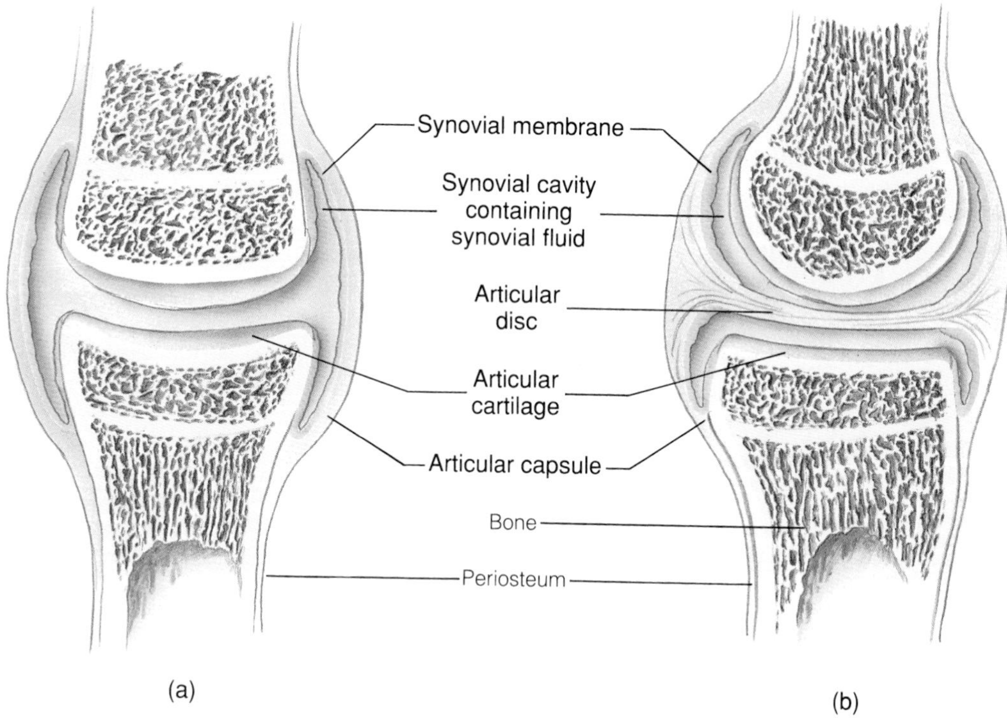

closes the joint. The outer layer is a strong **fibrous capsule** whose fibers are firmly joined to the periosteum of the bones. The inner layer of the articular capsule is the **synovial membrane.** The synovial membrane consists of loose connective tissue that is well supplied with capillaries. This membrane, which is often thrown into folds that project into the joint cavity, lines the entire joint cavity. However, it does not cover the surfaces of the articular cartilages or, if present, the articular disc. The articular capsule is well supplied with nerve fibers, which not only make the perception of pain possible but also provide constant information concerning movement and position of the joint.

4. Synovial fluid. The synovial membrane produces a thick fluid called **synovial fluid.** Synovial fluid provides nourishment to the articular cartilages while it lubricates the joint surfaces. In fact, synovial fluid serves as a weight-bearing element in the joint as it keeps the articular cartilages from contacting one another, thus preventing their erosion. Normally, only enough synovial fluid is secreted to form a thin film over the surfaces within the joint. However, fluid production may be stimulated in a joint that is injured or becomes inflamed, and enough synovial fluid may accumulate to cause swelling and discomfort.

5. Supporting ligaments. The bones forming synovial joints are strengthened and maintained in normal position largely by supporting ligaments. Some ligaments, called *intrinsic* or *capsular* ligaments, consist of parallel bundles of fibers within the outer fibrous layer of the capsule. Larger *extracapsular* ligaments are located outside the capsule and extend from bone to bone. Some joints are further strengthened internally with *intracapsular* ligaments located within the synovial cavity itself. We will name various examples of these ligaments when considering specific joints later in the chapter.

In addition to these five features, some synovial joints have **articular discs** (Figure 7.5b), or in the knee, **menisci,** of fibrocartilage, which extend inward from the articular capsule. Articular discs divide the synovial cavity into two separate cavities. In this form of joint, the synovial membrane lining the cavities extends only a short distance onto the surfaces of the disc. The jaw, the sternoclavicular joint, and the distal radioulnar joint contain articular discs.

In addition to the strengthening provided by ligaments, various muscles and their tendons, which cross the joints, serve to stabilize synovial joints while still permitting them to move freely.

Bursae and Tendon Sheaths

Synovial membranes form two other structures that, while not actually part of the synovial joints, are often associated with them. These are *bursae* and *tendon sheaths.* Both of these structures contain synovial fluid

◆ **FIGURE 7.6 Bursae and tendon sheaths**
(a) Subcutaneous bursa of the elbow. (b) Structure of a tendon sheath. (c) Longitudinal section showing the position of a tendon sheath.

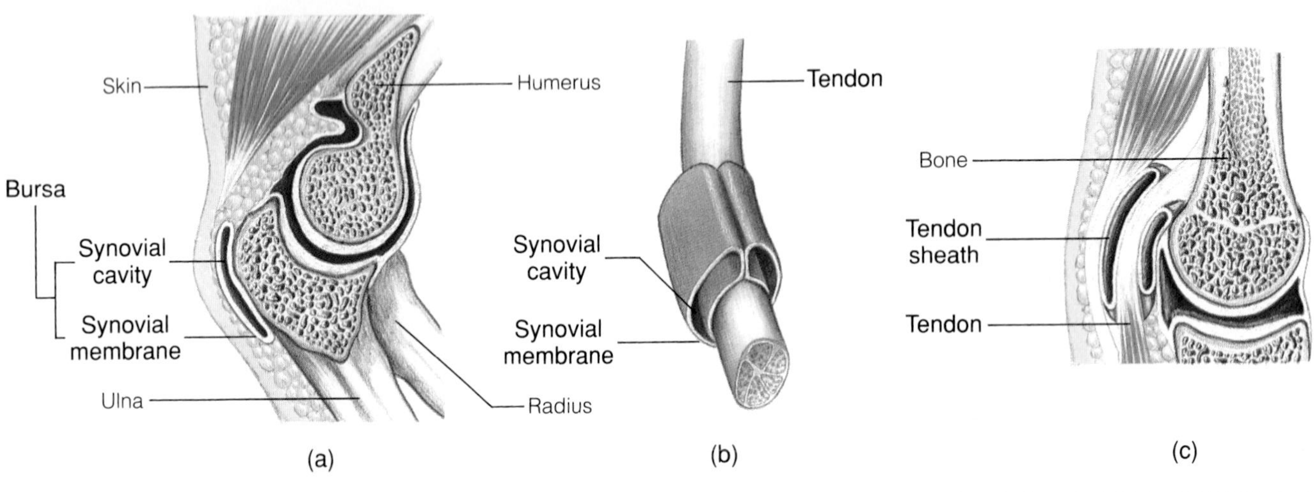

and function to reduce the friction that would occur during movement between a structure—such as skin, muscle, tendons, or ligaments—and the bone.

Bursae *(ber´-sē)* are small sacs lined with synovial membranes. Because they are filled with synovial fluid, they act as cushions between the structures they separate. Many bursae are distributed throughout the body. Some are subcutaneous, lying between the bone and the skin, such as the bursa separating the olecranon process of the elbow from the skin (Figure 7.6a). Most bursae, however, are located between the tendons and the bone. In some cases, bursae may be continuous with a joint cavity through small membrane tunnels.

Tendon sheaths (Figure 7.6b and c) are found where tendons cross joints, as in the wrist and fingers, and, without the sheaths, would be subjected to constant friction against the bones. The sheaths are cylindrical synovial sacs similar to bursae. They wrap around the tendons, forming fluid-filled, double-walled cushions for the tendons to slide through.

Movements of Synovial Joints

Synovial joints, unlike fibrous and cartilaginous joints, are not classified according to the material that connects the bones. Rather, they are named on the basis of the movements they permit. The shapes of the bony structures that surround a joint, and often the articular surface itself, generally limit the movements that are possible by any one joint. Many joints have *axes of rotation* that allow bones to move in various planes. A particular plane of movement is generally perpendicular to the axis. For instance, in the movement of the elbow, the axis is a horizontal line that passes through the joint from side to side. The bones rotate around this fulcrum (pivot point) in a vertical plane.

Joints that have only one axis of motion and can therefore move in only one plane, such as the elbow and knee, are called **uniaxial joints** (Figure 7.7a). Some joints have two axes, thus allowing movement in two planes that are at right angles to each other. Such joints are called **biaxial joints** (Figure 7.7b). Still other joints have more than two axes and permit movement in three planes. These are called **triaxial joints** (Figure 7.7c). The movement in many of the smaller joints is not restricted by the shapes of the articular surfaces, and slight movement is possible in any direction. Because their movements do not follow particular axes, these joints are referred to as **nonaxial.**

The general movements allowed in synovial joints can be placed into four groups: *gliding, angular, circumduction,* and *rotation.* In addition, several synovial joints allow movements unique to those particular joints. These unique movements are considered in the section on "Special Movements."

Gliding

The simplest and most common type of motion that can occur in a synovial joint is **gliding.** In this motion, the surfaces of adjoining bones move back and forth in relation to one another. In many cases, the articulating surfaces are flat or slightly curved, but gliding can occur between any two adjoining surfaces regardless of their forms. The joints between the heads of the ribs and the bodies of the vertebrae and between the tubercles of

◆ **FIGURE 7.7 Movements typical of synovial joints**

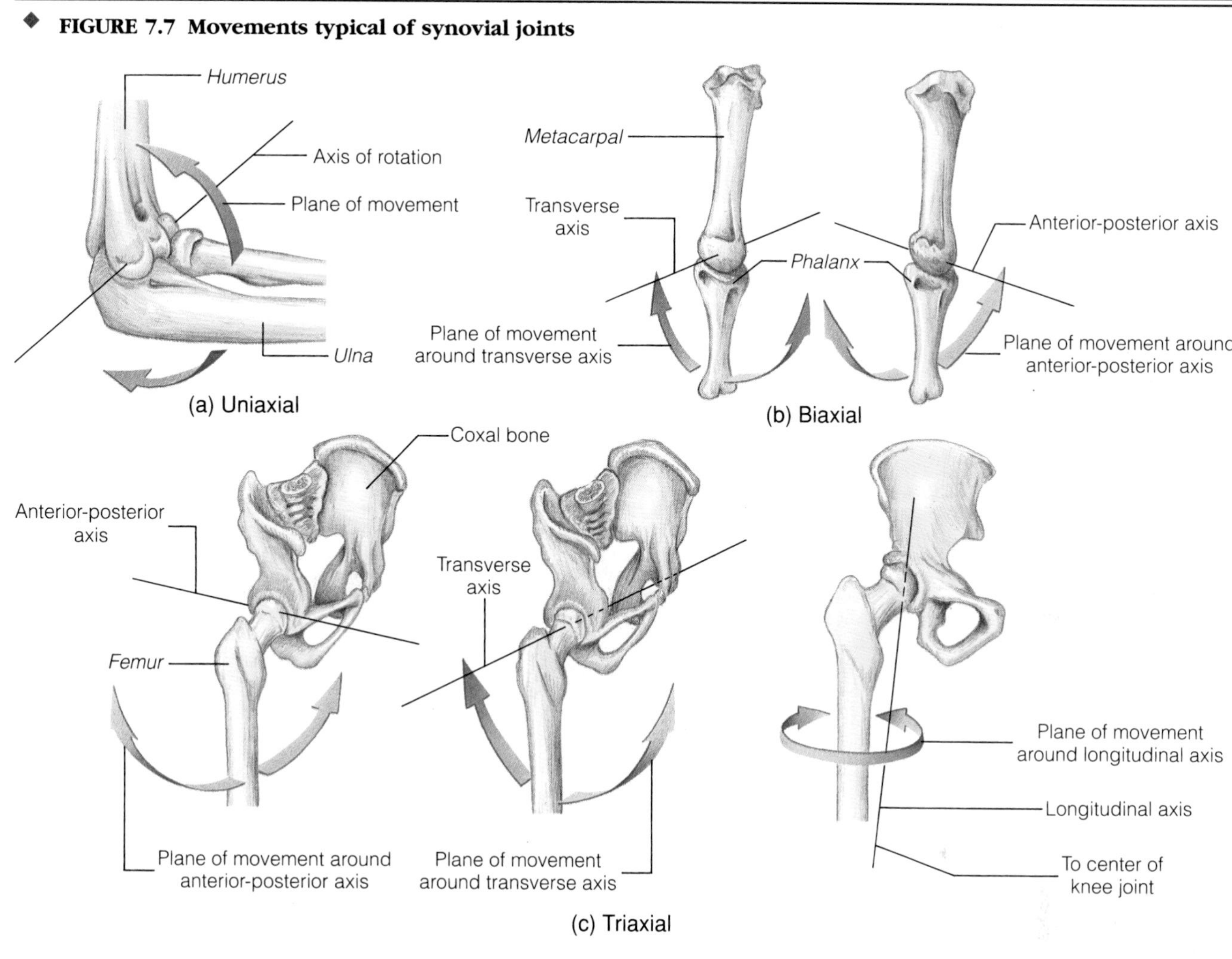

the ribs and the transverse processes of the vertebrae allow gliding movement, as do numerous other joints.

Angular Movements

Angular movements increase or decrease the angle between two adjoining bones by moving in a single plane. Four kinds of angular movements may occur in various synovial joints: *flexion, extension, abduction,* and *adduction* (Figure 7.8).

Flexion. When a bone is moved in an anterior-posterior plane in such a manner as to *decrease* the angle between it and the adjoining bone, **flexion** *(flek´-shun)* occurs. Examples include bending the elbow, bringing the thigh up toward the abdomen, and bringing the calf of the leg up toward the back of the thigh. Flexion of the ankle by raising the toe region toward the shin is called **dorsiflexion.**

Extension. Extension is the opposite of flexion. It causes the angle between adjoining bones to *increase.* Extension occurs when a flexed joint is moved back to the anatomical position, such as straightening the arm, thigh, or knee. **Hyperextension** occurs when the part is moved beyond the straight position, such as arching the back or bringing the limbs posteriorly beyond the plane of the body. Extension of the ankle by lowering the toe region is referred to as **plantar flexion.**

Abduction. When a part such as a limb is moved laterally, away from the midline of the body, **abduction** occurs. In the case of the fingers, abduction involves moving them away from the midline of the hand (third digit). Abduction of the toes is accomplished by moving them away from the longitudinal axis of the second toe.

Adduction. Adduction, the opposite of abduction, involves the movement of a part toward the midline of

◆ **FIGURE 7.8 Angular and circular movements of synovial joints**

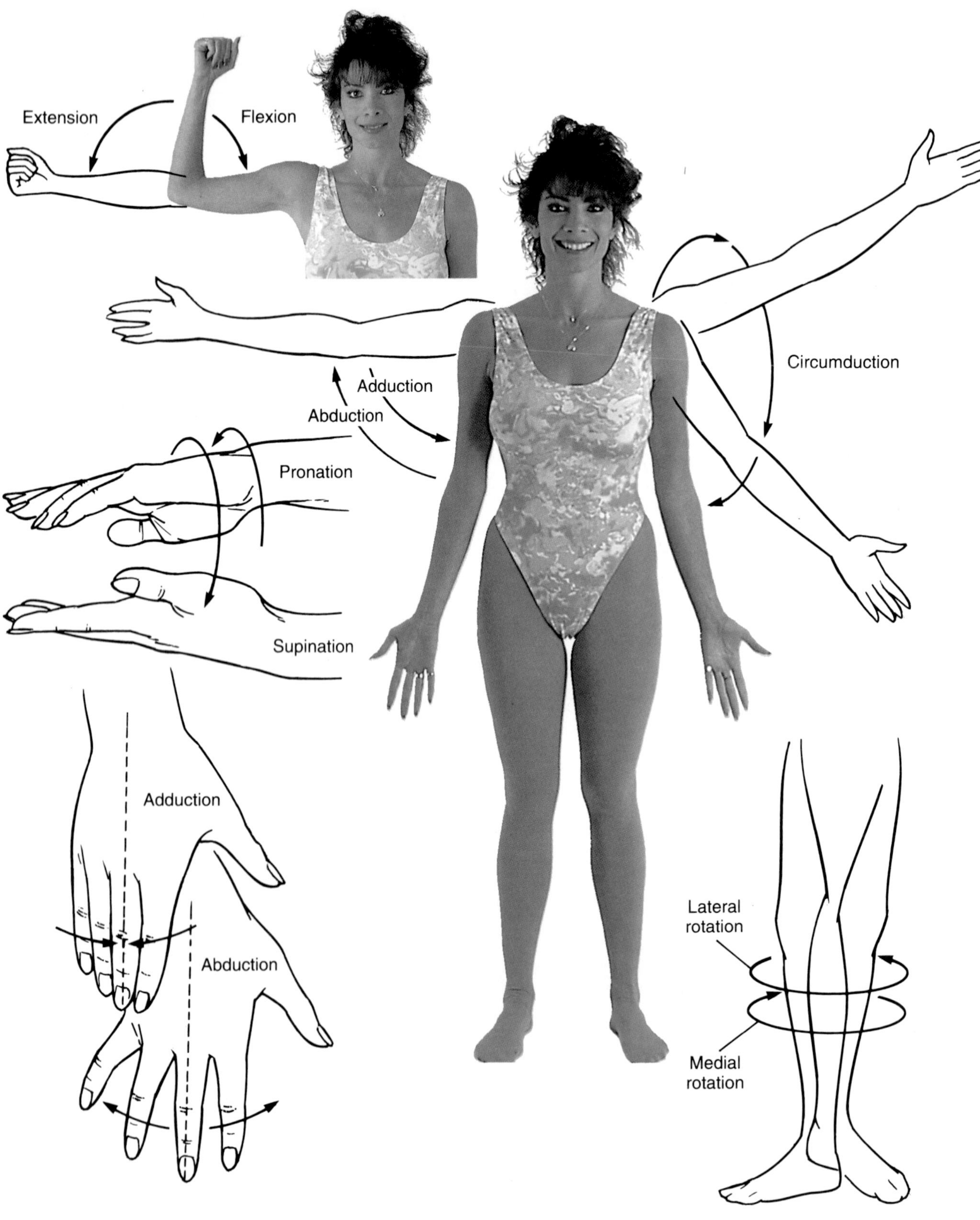

the body, back toward the anatomical position. In the case of the fingers, the movement is toward the midline of the hand (third digit). Adduction of the toes is accomplished by moving them toward the longitudinal axis of the second toe.

Circumduction

The joint motion known as **circumduction** delineates a cone (Figure 7.8). The base of the bone is outlined by the movement of the distal end of the bone, with the apex of the cone lying in the articular cavity. The movement is actually a sequential combination of flexion, abduction, extension, and adduction. Circumduction is common at the hip and shoulder joints and is also possible in other joints.

Rotation

The motion of a bone around a central axis without any displacement of that axis is **rotation** (Figure 7.8). If the anterior surface of a bone such as the humerus or femur moves inward, it is called *inward (medial) rotation.* When the anterior surface turns outward it is *outward (lateral) rotation.*

Supination. The term used to describe the outward rotation of the forearm, causing the palm to face upward or forward and the radius and the ulna to be parallel, is **supination** *(soo-pa-nay´-shun)*. The forearms are supinated in the anatomical position.

Pronation. The term used to describe the inward rotation of the forearm, causing the distal end of the radius to cross diagonally over the ulna and the palm to face downward or backward, is **pronation** *(prō-nā´-shun)*.

Special Movements

Some synovial joints allow special movements that cannot be described by any of the previously mentioned movements. These movements are *elevation, depression, inversion, eversion, protraction,* and *retraction* (Figure 7.9).

◆ **FIGURE 7.9 Special body movements**

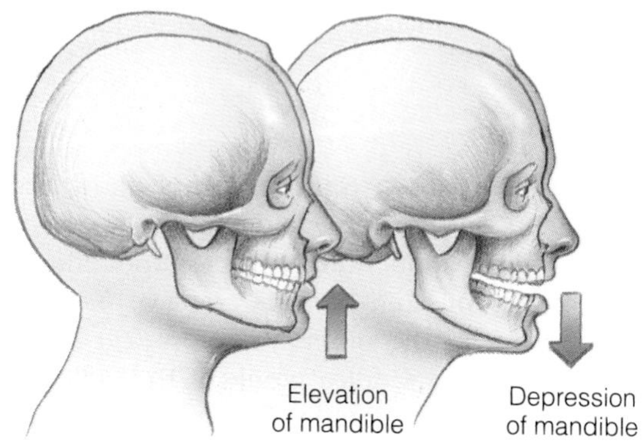

(a) Elevation and depression

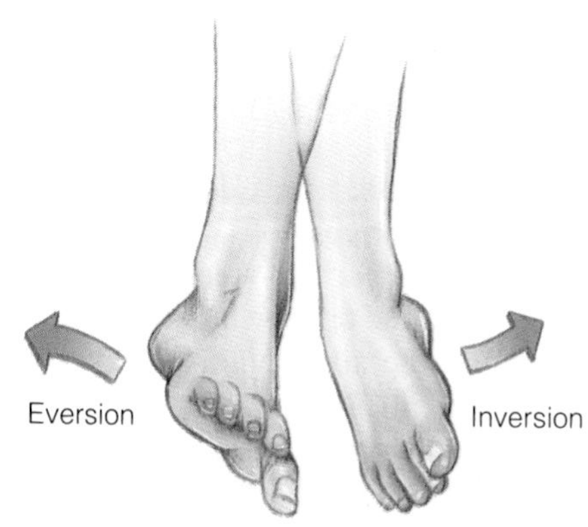

(b) Eversion and inversion

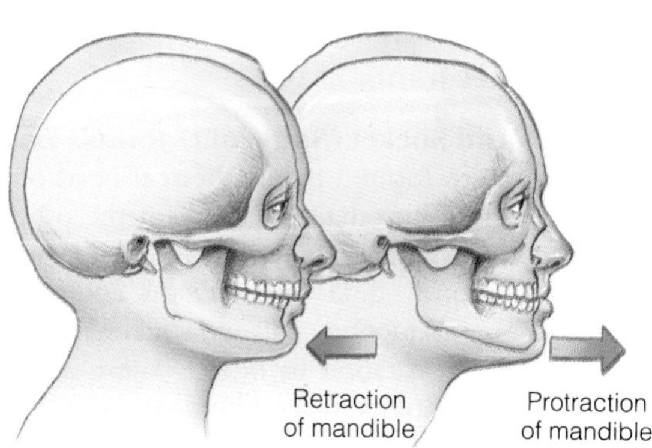

(c) Protraction and retraction

Elevation. The motion that raises a part is **elevation.** This term is most commonly used to refer to the raising of the scapula, as when shrugging the shoulders, or raising the mandible, as when closing the mouth.

Depression. The motion that lowers a part is **depression.** This term is often used to refer to the lowering of the scapula or the mandible.

Inversion. The twisting of the foot so that the sole faces inward with its inner margin raised is **inversion.**

Eversion. The twisting of the foot so that the sole faces outward with its outer margin raised is **eversion.**

Protraction. The motion that moves a part, such as the mandible, forward is **protraction.**

Retraction. The motion that returns a protracted part to its usual position is **retraction.**

Types of Synovial Joints

On the basis of the movements allowed and the shapes of the articular surfaces involved, it is possible to separate the synovial joints into six types. These types can be grouped according to whether they are *nonaxial, uniaxial, biaxial,* or *triaxial.*

Nonaxial Joints

Gliding (Arthrodial) Joints. The **gliding joints** are formed primarily by the apposition of flat, or only slightly curved, articular surfaces. Movement is allowed in any direction, being limited only by ligaments or bony processes that surround the articulation. Gliding joints are found between the articular process of vertebrae and between most carpal and tarsal bones (Figure 7.10a).

Uniaxial Joints

Hinge (Ginglymus) Joints. In **hinge joints,** the articular surfaces are shaped such that the only movements possible are flexion and extension. The elbow, the knee, and the joints between the phalanges of the fingers and toes (interphalangeal joints) are examples of hinge joints (Figure 7.10b).

Pivot (Trochoid) Joints. The only movement allowed in a **pivot joint** is rotation around the longitudinal axis of the bone. Examples are the rotation of the first cervical vertebra (atlas) around the odontoid process of the second cervical vertebra (axis), and the proximal articulations between the radius and the ulna. In the atlas/axis joint, the odontoid process is held against the inside surface of the anterior arch of the atlas by a transverse ligament that passes behind the process while connecting the two sides of the anterior arch. In the radial/ulnar joint, the head of the radius is held firmly against the radial notch of the ulna by a strong annular ligament that encircles its head (Figure 7.10c). The radius rotates within the annular ligament, allowing the forearm to pronate and supinate.

Biaxial Joints

Condyloid (Ellipsoid) Joints. Condyloid joints have one articular surface slightly concave and the other slightly convex; thus, movement is allowed in two planes that are at right angles to each other. Flexion, extension, abduction, and adduction can occur in condyloid joints. Circumduction is possible also, but axial rotation is not. The articulations between the radius and the carpals, the occipital condyles of the skull on the first cervical vertebra, the metacarpophalangeal, and the metatarsophalangeal joints are condyloid joints (Figure 7.10d).

Saddle Joints. Saddle joints allow the same movements as the condyloid joints—flexion, extension, abduction, adduction, and circumduction. The articular surface of each bone is concave in one direction and convex in the other; therefore, the bones fit together just as two saddles would if the riding surface of one saddle were rotated 90° in relation to the other and the two surfaces were placed on top of one another. The only true saddle joint in the body is the carpometacarpal joint of the thumb (Figure 7.10e).

Triaxial Joints

Ball-And-Socket (Spheroid) Joints. Ball-and-socket joints are formed by a spherical head of one bone fitting into a cup-shaped cavity on the other. Such joints allow movement around an indefinite number of axes. In addition to flexion, extension, abduction, adduction, and circumduction, ball-and-socket joints allow medial and lateral rotation to occur. There are only two ball-and-socket joints in the body—the shoulder and the hip (Figure 7.10f).

Table 7.1 summarizes the main joints of the body.

◆ **FIGURE 7.10 Types of synovial joints**
(a) Gliding joint.
(b) Hinge joint.
(c) Pivot joint.
(d) Condyloid joint.
(e) Saddle joint.
(f) Ball-and-socket joint.

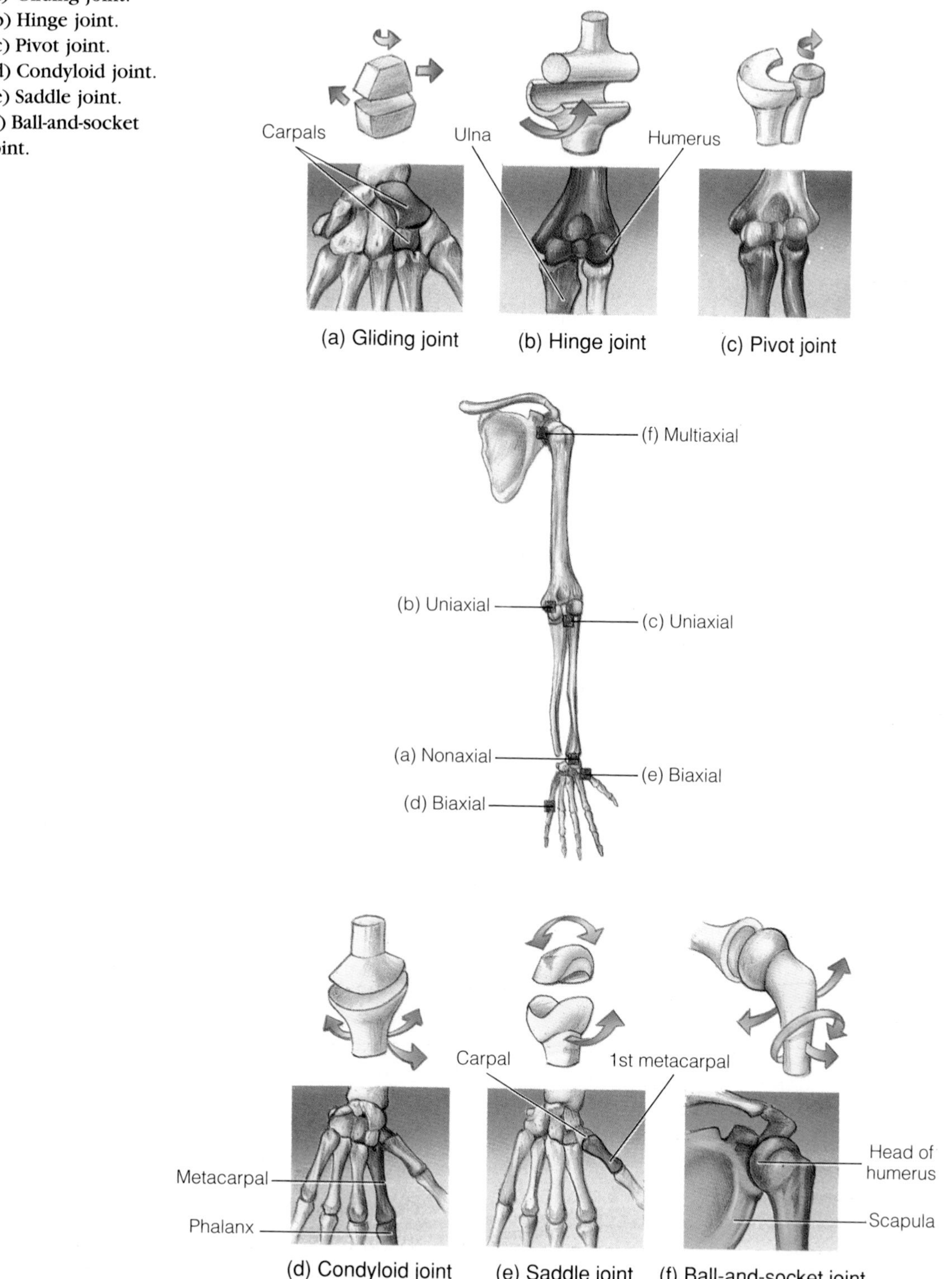

Ligaments of Selected Synovial Joints

Ligaments play an important role in maintaining the proper positioning of bones that articulate in synovial joints while at the same time allowing relatively free movement of the joints. Therefore, we will consider the ligaments of four of the more important joints: the shoulder, elbow, hip, and knee.

◆ **TABLE 7.1 Summary of the Main Joints of the Body**

JOINT	TYPE	MOVEMENT
Between the cranial bones	Fibrous (suture)	No appreciable movement
Between the distal tibia and the fibula	Fibrous (syndesmosis)	Slight movement ("give")
Between the mid-radius and the ulna (interosseous membrane)	Fibrous (syndesmosis)	Slight movement
Between the ribs and the sternum (sternocostal)	Cartilaginous (synchondrosis)	Slight movement
Between the two pubic bones	Cartilaginous (symphysis)	Slight movement
Between the bodies of the vertebrae	Cartilaginous (symphysis)	Slight movement
Between the sacrum and the ilium	Partly cartilaginous (synchondrosis) and partly synovial (gliding)	Generally no movement, but slight gliding movement is possible. In older people, fibers may hold the two bones firmly together.
Between the articular processes of the vertebrae	Synovial (gliding)	Nonaxial; gliding
Between the head of the rib and the body of the vertebra	Synovial (gliding)	Nonaxial; gliding
Between the occipital and the atlas	Synovial (condyloid)	Biaxial; flexion, extension, abduction, adduction, circumduction
Between the atlas and the odontoid process of the axis	Synovial (pivot)	Uniaxial; pivoting around the odontoid process
Between the sternum and the clavicle	Synovial (gliding)	Nonaxial; gliding
Between the acromion process of the scapula and the clavicle	Synovial (gliding)	Nonaxial; gliding and rotation of the scapula upon the clavicle
Between the humerus and the scapula	Synovial (ball-and-socket)	Triaxial: flexion, extension, abduction, adduction, circumduction, medial and lateral rotation
Between the ulna and the humerus	Synovial (hinge)	Uniaxial; flexion and extension
Between the head of the radius and the ulna	Synovial (pivot)	Uniaxial; pivoting longitudinal axis, as in pronation and supination

Ligaments of the Shoulder Joint

In the shoulder joint, the head of the humerus is received by the shallow glenoid fossa of the scapula. The joint is loosely constructed, which permits extremely free movement but also allows it to become frequently dislocated. It is protected above by the coracoid process and the acromion process of the scapula. Therefore, it is most often dislocated in the inferior direction.

Like most synovial joints, the shoulder joint is enclosed by an **articular capsule** (Figure 7.11). This capsule attaches to the rim of the glenoid fossa and extends outward to the anatomical neck of the humerus. The capsule is strengthened anteriorly by two ligaments: the **coracohumeral ligament,** *(kor″-ah-ko-hu′-mer-al)* which extends from the coracoid process of the scapula to the greater tubercle of the humerus; and the **glenohumeral ligaments,** *(gle″-no-hu′-mer-al)* which are several thickenings of the lower portion of the capsule itself. The **glenoid labrum** (*labrum* = lip), a rim of fibrocartilage that surrounds the glenoid fossa, adds somewhat to the stability of the joint by deepening the fossa.

◆ **TABLE 7.1 Summary of the Main Joints of the Body (continued)**

JOINT	TYPE	MOVEMENT
Between the radius and the carpals (scaphoid, lunate)	Synovial (condyloid)	Biaxial; flexion, extension, abduction, adduction, circumduction
Between the carpals	Synovial (gliding)	Nonaxial; gliding
Between the first metacarpal and carpal (trapezium)	Synovial (saddle)	Biaxial; flexion, extension, abduction, adduction, circumduction
Between the second through fifth metacarpals and the carpals	Synovial (gliding)	Nonaxial; gliding
Between the second through fifth metacarpals and the phalanges	Synovial (condyloid)	Biaxial; flexion, extension, abduction, adduction, circumduction
Between the phalanges (hand and foot)	Synovial (hinge)	Uniaxial; flexion, extension
Between the femur and the coxal bone	Synovial (ball-and-socket)	Triaxial; flexion, extension, abduction, adduction, circumduction, medial and lateral rotation
Between the tibia and the femur	Synovial (hinge)	Uniaxial; flexion, extension (some rotation)
Between the proximal end of the fibula and the tibia	Synovial (gliding)	Nonaxial; gliding
Between the distal ends of the fibula and the tibia and the talus	Synovial (hinge)	Uniaxial; flexion, extension
Between the tarsals	Synovial (gliding)	Nonaxial; gliding
Between the tarsals and the metatarsals	Synovial (gliding)	Nonaxial; gliding
Between the metatarsals and the phalanges	Synovial (condyloid)	Biaxial; flexion, extension, abduction, adduction, circumduction

In addition to these ligaments, the shoulder joint depends greatly on the surrounding muscles for strength. The biceps brachii muscle has a unique arrangement: the tendon of its long head, which arises from the superior border of the glenoid fossa, passes inside the joint capsule of the shoulder and through the intertubercular groove of the humerus. Thus, in effect, it helps to hold the head of the humerus against the scapula.

The tendons of the supraspinatus, infraspinatus, subscapularis, and teres minor muscles also strengthen the shoulder joint by almost completely encircling the joint as they pass along the sides of the articular capsule. Because the muscles of these tendons are important in rotation movements of the arm, the tendons are collectively referred to as the *rotator cuff.* Severe rotation or circumduction of the arm may stretch these tendons. Thus, they are commonly injured by baseball pitchers.

Ligaments of the Elbow Joint

The elbow joint is a hinge joint where the trochlea of the humerus is received into the trochlear notch of the

◆ **FIGURE 7.11 Ligaments of the right shoulder**
(a) Lateral view, with the humerus removed. (b) Anterior view.

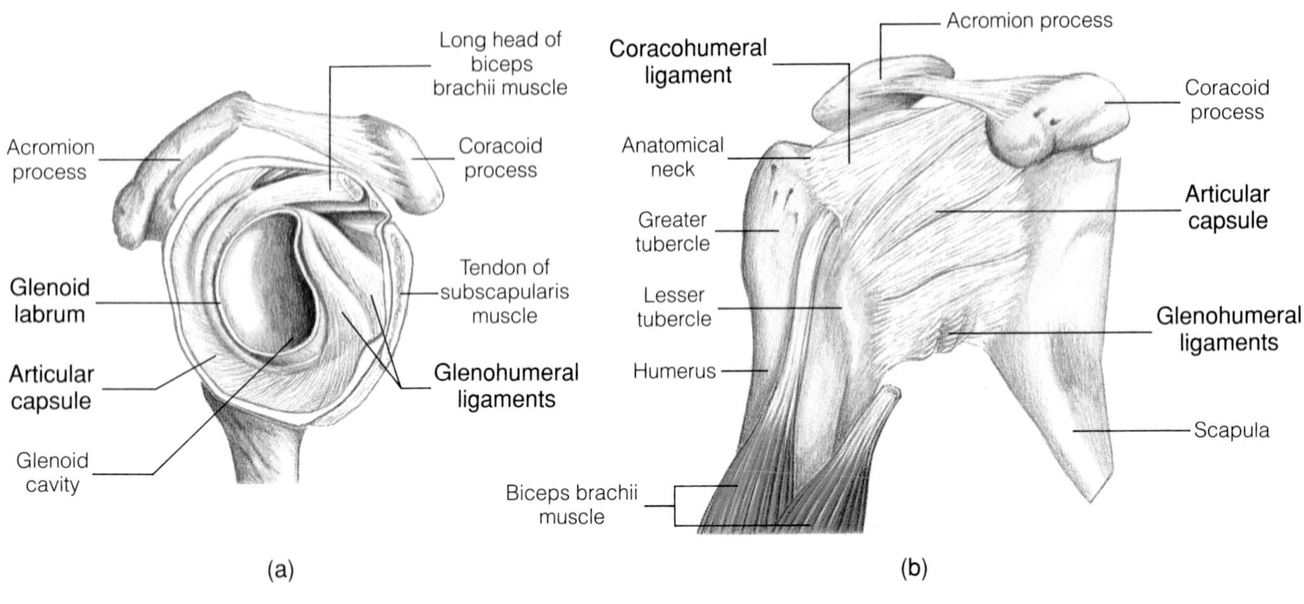

ulna. Closely associated with the elbow joint is a gliding joint between the capitulum of the humerus and the superior surface of the head of the radius. These two articulations share a common joint cavity and are enclosed by a single fibrous **articular capsule** (Figure 7.12).

Medial and lateral thickenings of the articular capsule serve to stabilize the elbow joint. The medial thickening, the **ulnar collateral ligament,** passes from the medial epicondyle of the humerus to the medial surface of the ulna between the olecranon process and the coronoid process. The lateral thickening, the **radial collateral ligament,** passes from the lateral epicondyle of the humerus to the annular ligament and to the lateral surface of the ulna. The **annular ligament** *(an´-u-lar)* is a strong band of fibers in the distal margin of the articular capsule of the elbow joint. The annular ligament does not strengthen the elbow joint to any great degree. Instead, it encircles the head and the upper part of the neck of the radius as it passes between the anterior and

◆ **FIGURE 7.12 Ligaments of the right elbow joint**
(a) Medial view. (b) Lateral view.

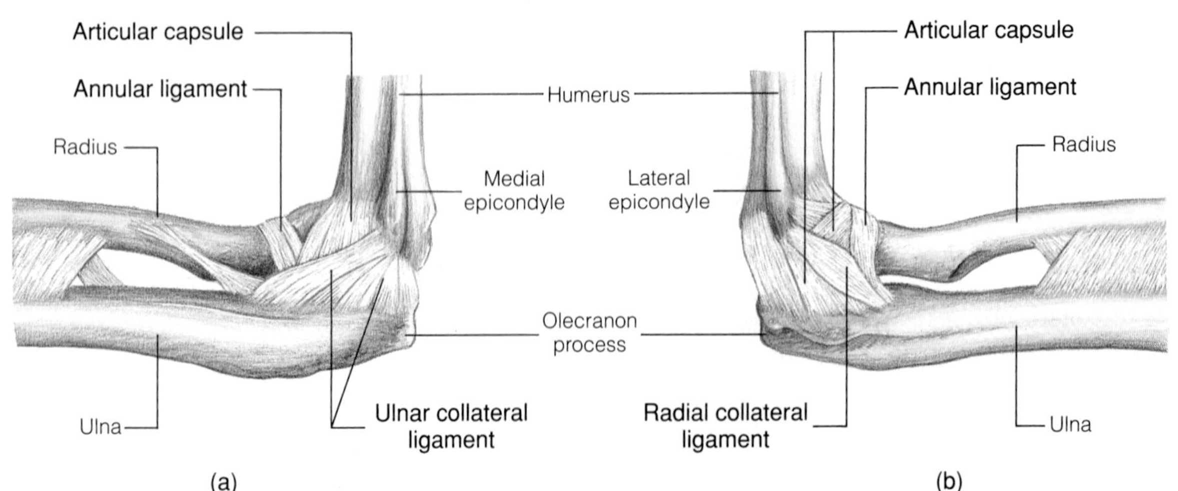

ASPECTS OF EXERCISE PHYSIOLOGY 

Warm-Up: An Effective Protection for Joints

Exercise professionals recommend that participants gradually warm up or engage in some type of preparatory exercise before starting an activity. The reasons given for warm-up usually include (1) stretching muscles to reduce the chance of injury; (2) gradually increasing heart rate and blood flow to active muscles; and (3) increasing the temperature in both nerve and muscle tissue to produce faster reaction times.

Although it is not often mentioned, warm-up is also beneficial for joints. Studies have shown that articular cartilage absorbs fluid after a brief period of exercise. In fact, after about ten minutes of exercise, the cartilage in a normal joint swells about 10%. This swelling helps protect and cushion the joint. (In older people, the cartilage swells less than it does in younger people, and in an osteoarthritic joint, the cartilage swells less than it does in a normal joint.)

The viscosity of synovial fluid also changes during exercise. When there is little or no movement of a joint, the synovial fluid is relatively viscous. If the viscosity of the fluid remained unchanged as joint movement increased, the joint would feel stiff and be difficult to move. However, synovial fluid becomes less viscous as joint movement increases.

A gradual warm-up, then, benefits joints by allowing articular cartilage to absorb fluid, which has a protective and cushioning effect. A gradual warm-up also enhances performance by initiating a decrease in the viscosity of synovial fluid that makes joints easier to move.

posterior margins of the trochlear notch of the ulna. In this manner, the radius is held tightly against the ulna and yet is allowed to rotate freely, as occurs during pronation and supination.

Ligaments of the Hip Joint

The head of the femur fits into the deep acetabulum of the coxal bone (Figure 7.13), making the hip joint a more stable joint than the shoulder.

The **articular capsule,** which extends from the margin of the acetabulum to the anatomical neck of the femur, completely encloses the joint. The capsule is strengthened anteriorly by the **iliofemoral** *(il″-ē-ō-fem′-ō-ral)* and the **pubofemoral ligaments** *(pu″-bō-fem′-ō-ral)*. On its posterior surface, the capsule is strengthened by the **ischiofemoral ligament** *(is-kee-o-fem′-ō-ral)*.

The acetabulum is surrounded by a fibrocartilaginous rim called the *acetabular labrum.* The acetabular labrum is incomplete at its inferior margin, which gives it a horseshoe shape. Like the glenoid labrum of the shoulder joint, this labrum deepens the joint cavity. A unique ligament called the **ligamentum teres** (or *capitate ligament*) extends through the joint cavity from the fovea on the head of the femur to the gap at the lower portion of the acetabular labrum. Because the ligamentum teres is slack during most movements of the hip, it is believed not to contribute significantly to the strength of the joint. It does, however, help prevent the head of the femur from slipping upward.

Ligaments of the Knee Joint

The knee joint (Figure 7.14) is a complicated joint that is vulnerable to injury. It actually consists of three joints: medial and lateral joints between the distal end of the femur and the proximal end of the tibia, and a more central joint between the posterior surface of the patella and the lower anterior surface of the femur.

The knee is classified as a hinge joint because its movements are restricted by the surrounding ligaments to, for the most part, flexion and extension. It has the structure, however, of a condyloid joint, with the condyles of the femur articulating with the slightly concave condyles of the tibia. The articular surface on the medial condyle of the femur is somewhat longer from front to back and is less curved than the articular surface on the lateral condyle. As a consequence of these structural differences, the final phase of complete extension of the knee joint primarily involves movement of the medial condyle of the femur on the tibia. This causes the tibia to undergo some lateral rotation (or the femur to undergo some medial rotation). Similarly, the extended

◆ **FIGURE 7.13 Ligaments of the right hip joint**
(a) Frontal section through the hip joint. (b) Anterior view.

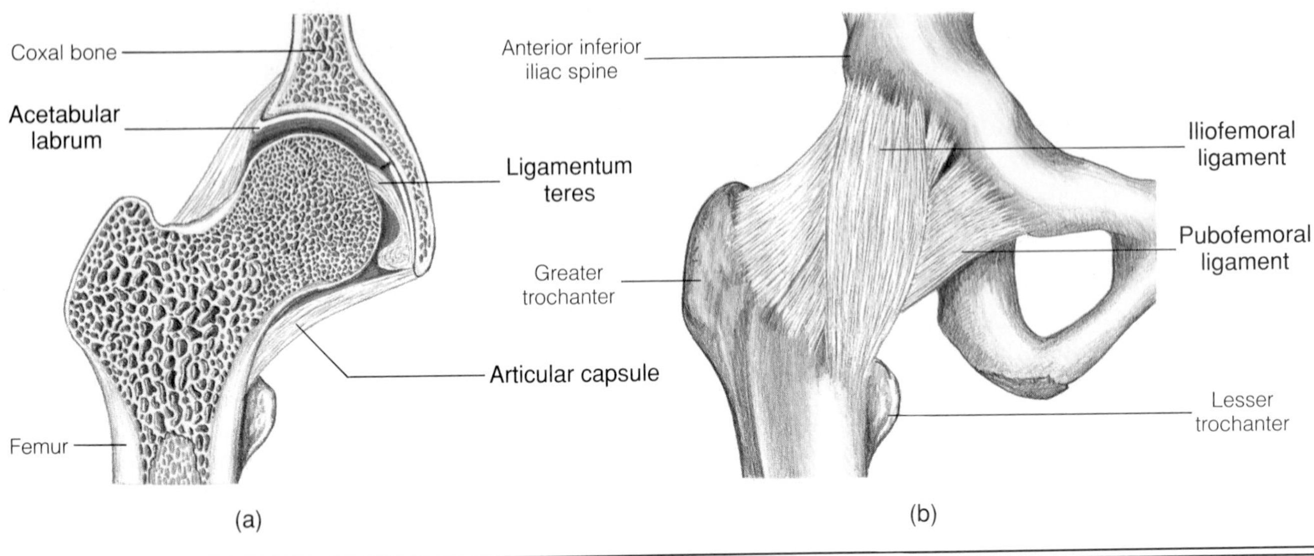

◆ **FIGURE 7.14 The right knee joint**
(a) Ligaments of the knee joint. The femur is flexed slightly to allow the ligaments to be seen. The articular capsule and the patella have been removed. (b) Anteroposterior X ray of the right knee.

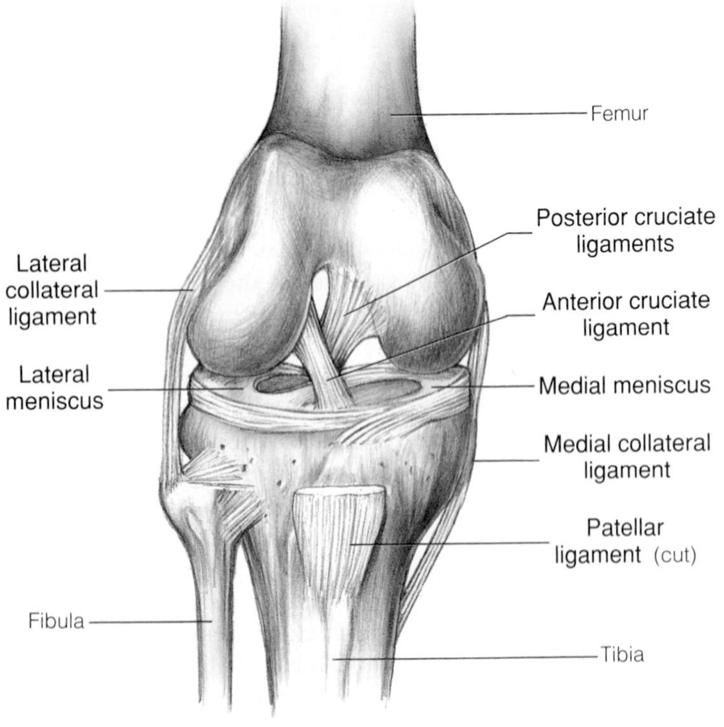

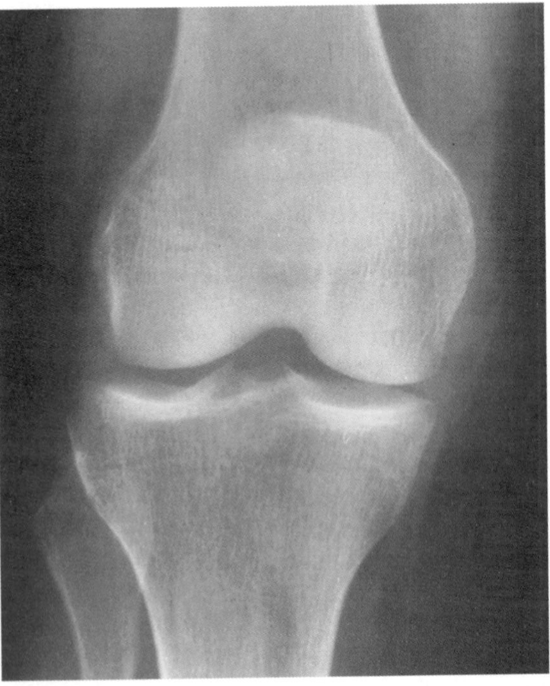

◆ **FIGURE 7.15 Midsagittal section of knee showing several bursae**

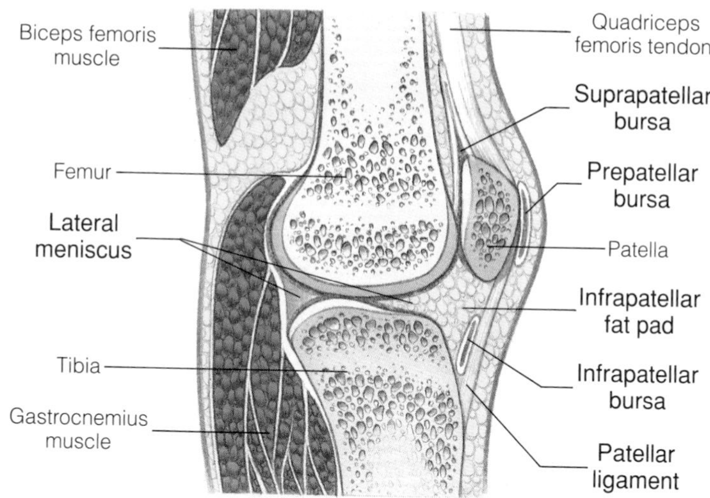

joint must be unlocked by a slight medial rotation of the tibia before flexion of the knee can occur. When the knee joint is in a partially flexed position, it is possible for it to undergo even more rotation. The joint between the patella and the femur is a gliding joint, and during knee movements the patella glides across the patellar surface of the femur.

The **articular capsule** of the knee is different from the capsule of most joints in that it does not completely enclose the joint. Anteriorly, the fibrous layer of the capsule is lacking above the patella. Thus, the synovial membrane protrudes upward in this region, forming a *suprapatellar bursa* deep to the superior margin of the patella and the tendon of the quadriceps femoris muscle of the anterior thigh (Figure 7.15). The complexity of the knee joint is such that the synovial cavity is subdivided into three separate cavities, and some 13 bursae are associated with the joint.

The knee joint is strengthened anteriorly by the **patellar ligament,** which extends from the patella to the tibial tuberosity. This ligament is a continuation of the tendon of the quadriceps femoris muscle. The posterior region of the capsule is quite thin, but is strengthened by the **oblique popliteal ligament** *(pop-la-tee´-al)* and the **arcuate popliteal ligament** (Figure 7.16). The oblique popliteal ligament is a broad, flat band attached proximally to the posterior surface of the femur just above the articular surface of the lateral condyle. From here it extends downward and medially to attach to the posterior surface of the head of the tibia. The arcuate popliteal ligament passes from the posterior surface of the lateral condyle of the femur to the styloid process of the head of the fibula. The knee joint is stabilized medially and laterally by very strong **medial** and **lateral collateral ligaments,** which extend from the condyles of the femur to the tibia or the fibula. The collateral ligaments limit the amount of rotation that is possible by the knee joint and help prevent hyperextension of the joint.

The flat superior surface of the condyles of the tibia, which are the largest weight-bearing surfaces in the body, are deepened by crescent-shaped cartilages called **medial** and **lateral menisci.** The menisci serve as shock absorbers and reduce side-to-side movement of the femur on the tibia. The menisci are attached only at their outer margins, and they frequently become damaged or torn loose in athletic injuries.

Additional stability is added to the knee joint by the presence within the joint cavity of **anterior** and **posterior cruciate ligaments** (*kroo´-she-āt; cruciate* = cross-shaped). These ligaments extend diagonally from the superior surface of the tibia, between the condyles, to the distal end of the femur. They are called cruciate because their paths cross each other. Because of their unique structural arrangements, the cruciate ligaments perform very specialized functions. When the knee is extended, the anterior cruciate ligament is taut, thus guarding against hyperextension of the joint by preventing anterior movement of the tibia (Figure 7.17a). When the knee is flexed, the posterior cruciate ligament becomes taut, preventing the tibia from slipping posteriorly (Figure 7.17b).

◆ **FIGURE 7.16 Posterior ligaments of the right knee joint**

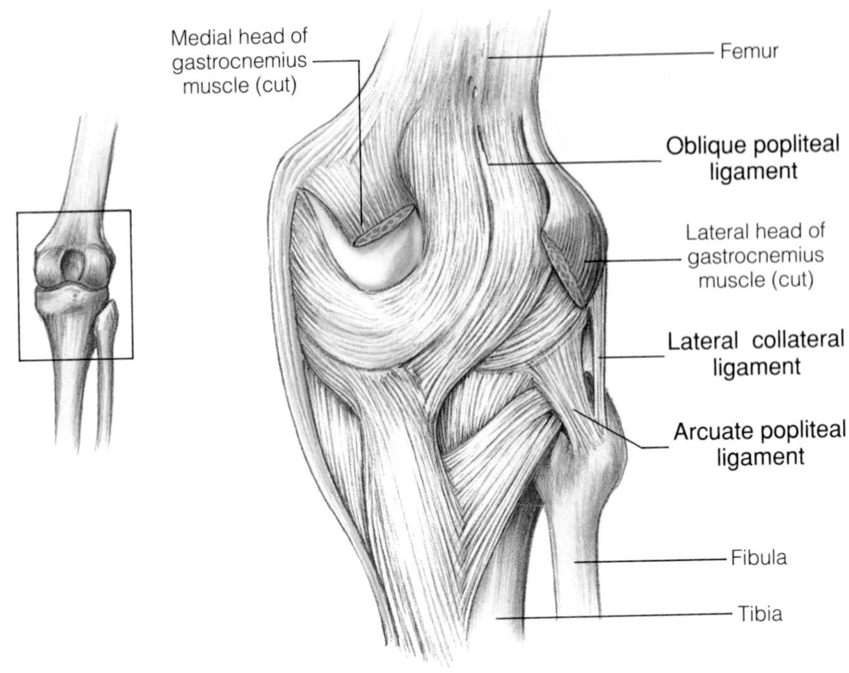

◆ **FIGURE 7.17 Functions of the cruciate ligaments**

(a) Knee extended: taut anterior cruciate ligament prevents the tibia from moving anteriorly, thereby hindering overextension of the joint.
(b) Knee flexed: taut posterior cruciate ligament prevents the tibia from slipping posteriorly. (The medial condyle of the femur has been removed to expose the cruciate ligaments.)

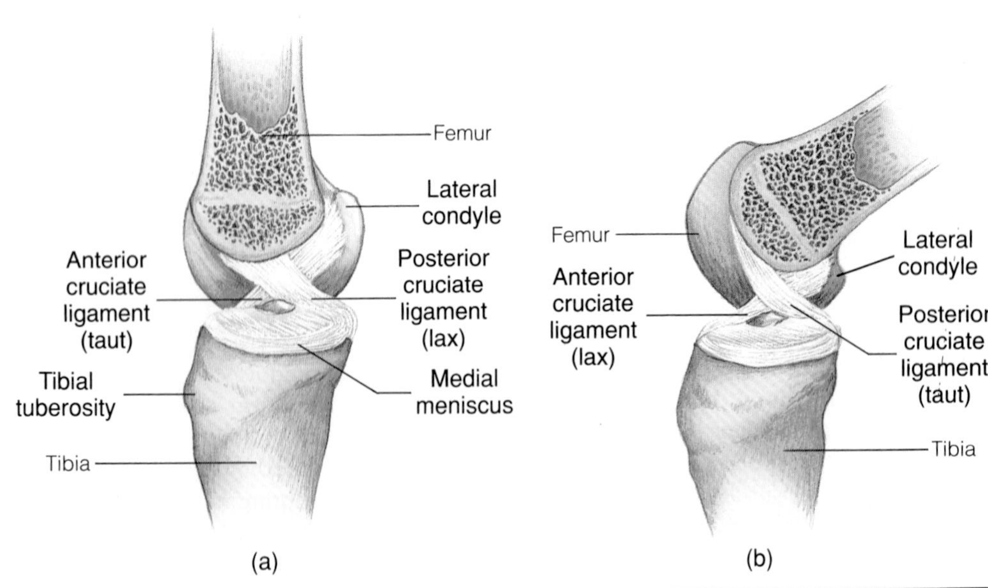

CONDITIONS OF CLINICAL SIGNIFICANCE

Articulations

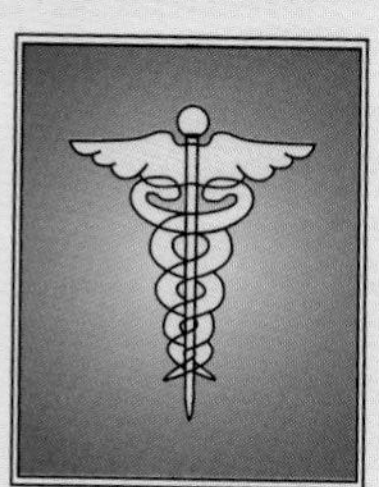

Sprains

Sprains result when the twisting or overstretching of a joint causes a ligament to tear or to separate from its bony attachment. Because ligaments are not well supplied with blood vessels, sprains heal quite slowly. In fact, ligaments that are completely torn must be repaired surgically because the healing process is so slow that the inflammatory response to the joint injury may cause the damaged ligament to break down. Ligaments that are too severely damaged to repair may be surgically removed and replaced with strips of tendon or certain artificial substances that can serve as substitute ligaments.

Dislocations

A *dislocation* or a *luxation* occurs when the bones forming a joint are forcibly displaced from their normal positions. When a dislocation is severe, the bones, as well as the surrounding tendons and ligaments, may be damaged, causing intense pain and inflammation. The joints of the shoulder, thumb, and fingers are most commonly dislocated. Dislocations are treated by manipulating the bones back to their normal positions and immobilizing the joint while the ligaments and tendons surrounding the joint heal.

Bursitis

Bursitis is the inflammation of one or more of the bursae surrounding a joint. The inflammation can result from injury, heavy exercise, or bacterial infection. The affected bursae fill with excessive synovial fluid, causing redness, local swelling, and pain. As a result, movement of the joint becomes limited. The knee is one of the more common sites of bursitis, where it is referred to as *water on the knee.*

Tendinitis

Tendinitis is the inflammation of tendon sheaths around a joint. The condition is generally characterized by local tenderness at the point of inflammation and severe pain upon movement of the affected joint. Tendinitis can result from trauma to, or excessive use of, a joint. The wrist, elbow (where it is referred to as "tennis elbow"), and shoulder joints are most often affected.

Herniated (Slipped) Disc

Among the more common discomforts that people endure are those associated with the back. There are numerous causes of back pain; one cause that involves joints is a *herniated* or *slipped disc.* In this condition, which is most common in the lower back, the relatively soft nucleus pulposus within an intervertebral disc is squeezed to one side of the disc (Figure C7.1). This can result from trauma or from improper distribution of weight along the vertebral column resulting from poor posture or deformity of the vertebrae. The displacement of the nucleus pulposus causes the relatively firm outer portion of the disc called the anulus fibrosus to rupture. If the disc protrudes into the vertebral canal, it can compress spinal nerves or, in the thoracic or cervical regions, the spinal cord itself. The pressure can cause severe pain along the paths of the nerves as well as numbness of the regions supplied by the nerves. If the pressure on the spinal nerves continues, actual nerve damage can result. The nerve damage, in turn, can cause weakness and degeneration of muscles supplied by the damaged nerves.

Torn Menisci

One of the most common injuries to the knee involves the menisci, which separate the tibia and the femur. The fibrocartilage menisci serve to somewhat adapt the surfaces of the tibia to the shape of the femoral condyles. The menisci are not firmly attached, and during rotation they move slightly on the tibia. Because of this movement, sudden changes of direction while bearing the body weight can cause

continued on next page

CONDITIONS OF CLINICAL SIGNIFICANCE

◆ **FIGURE C7.1 Superior view of a herniated (slipped) disc**

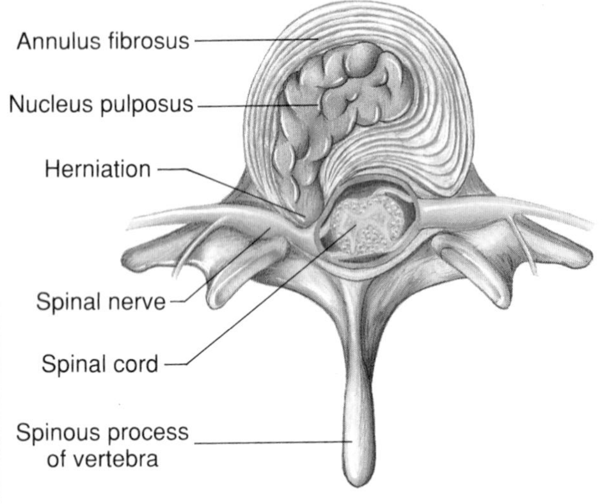

the menisci to tear loose. Severe pain and swelling of the joint can result.

An innovative surgical technique called *arthroscopic surgery* has greatly increased the success of operations performed to repair damaged menisci as well as other joint injuries. In this technique, a needlelike viewing instrument called an *arthroscope* is inserted into the joint through a tiny incision (Figure C7.2). The arthroscope contains a fiber-optic light source that makes it possible to examine the interior of the joint visually and to observe the positions of cutting instruments inserted through other tiny incisions. This procedure can be performed under local anesthesia, and the patient can often return home from the hospital the same day.

Arthritis

Many different types of inflammation or degeneration of the joints fall under the general term *arthritis.* In these conditions, there may be pathological changes in the joint membranes, cartilage, and bone that cause swelling and pain. The causes of arthritis are unknown, but trauma to a joint, bacterial infection (staphylococci, streptococci, and gonococci), and metabolic disorders have been implicated. At least one form of arthritis (rheumatoid) is thought to be caused by an immune response to inflammation of the synovial membrane of the affected joint cavity. There is evidence suggesting that arthritis may be genetically inherited, since some families show a predisposition toward the condition.

Osteoarthritis

Osteoarthritis, the most common form of arthritis, is a chronic inflammation that causes the articular cartilage in the affected joint to gradually degenerate. The inflammation is accompanied by pain, swelling, and stiffness. As the articular cartilage

◆ **FIGURE C7.2 Arthroscopic surgery**

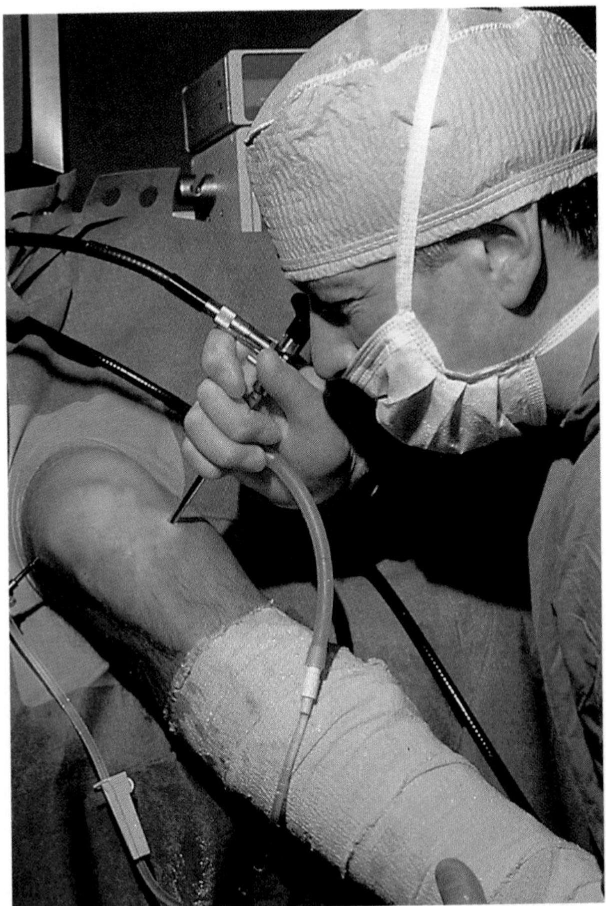

CONDITIONS OF CLINICAL SIGNIFICANCE

degenerates, bony spurs develop from the exposed ends of the bones that form the joint. These spurs tend to restrict the movement of the joint and often cause discomfort.

The cause of osteoarthritis is not definitely known. Some of the changes associated with it are thought to be related to various changes that occur during normal aging. These include age-related changes in the collagen and elastic fibers of bone and cartilage and a reduced blood supply to the articular cartilages. Because osteoarthritis is most common in older people, it is considered to be due in part to the "wear and tear" that joints are subjected to throughout one's lifetime. Thus, except in joints that have been injured or have experienced excessive prolonged stress, osteoarthritis generally does not become apparent until after middle age.

The joints most often affected by osteoarthritis include the intervertebral joints, the phalangeal joints, the knees, and the hips. Osteoarthritis of the spine may be associated with chronic backache. Often, bony spurs develop on the vertebrae and eventually fuse several adjacent vertebrae together. The fusion of vertebrae stabilizes the spinal column, but limits the movement allowed by the column.

Rheumatoid Arthritis

Rheumatoid arthritis is a severely damaging form of the disease that tends to affect certain joints of the body, especially those in the hands, feet, knees, ankles, elbows, and wrists (Figure C7.3). It affects women more frequently than men. The condition begins with inflammation of the synovial membrane of the joint, causing swelling and pain. The inflamed synovial membrane produces an abnormal tissue known as a *pannus,* which grows over the surface of the articular cartilage. As the disease progresses, the articular cartilage beneath the pannus, and in some cases the bone itself, is gradually destroyed by enzymes released by the pannus. Eventually, the pannus fills the joint space and becomes invaded by fibrous tissue, thereby restricting joint movement. In severe cases, calcification of the pannus may ankylose (fuse) the joint, making it immovable.

Rheumatoid arthritis can begin at any age, but the first symptoms usually appear before a person is 50 years old. The cause of the disease is not known, but it has been suggested that it may be initiated by a bacterial or viral infection. Some persons also seem to have a genetic tendency to develop the disease. Once the disease process has been initiated, rheumatoid arthritis often appears to be an immune

◆ FIGURE C7.3 X ray of rheumatoid arthritis in a hand

The enlarged joints are a product of an inflamed synovial membrane. In most of the serious cases, the articular cartilage completely erodes, allowing the bones to fuse together.

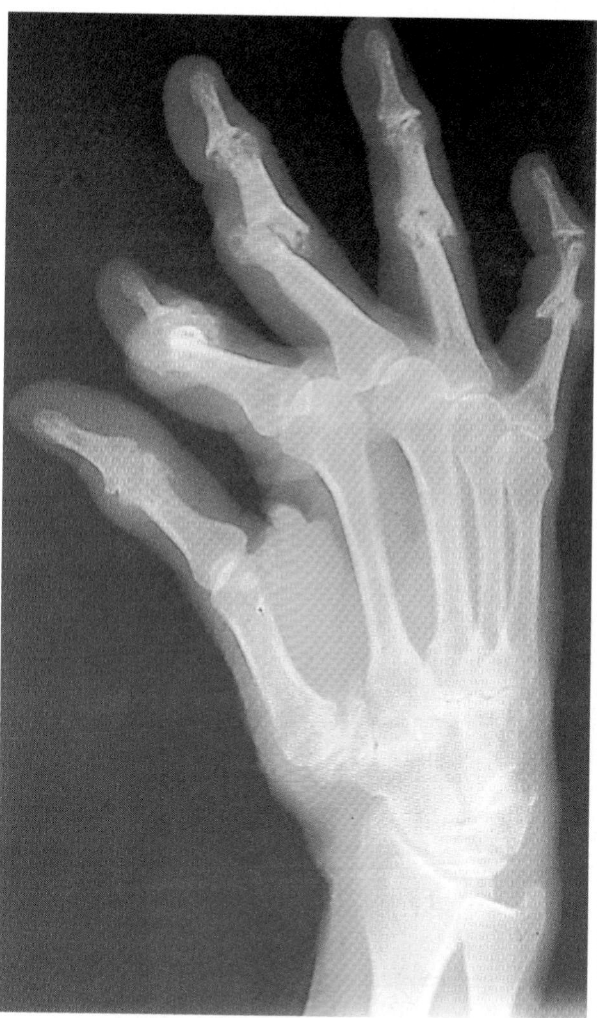

continued on next page

CONDITIONS OF CLINICAL SIGNIFICANCE

disease in which a foreign substance (an antigen) becomes located in the synovial membrane of the joint and causes an inflammatory response by the body's immune system. Some researchers suggest that, rather than requiring invasion by a foreign antigen, rheumatoid arthritis may be an *auto*immune disease, in which the body's immune system produces antibodies that act against one's own tissues.

Rheumatoid arthritis can become so severe that surgery may be necessary to repair the affected joints or to replace them with artificial joints. In some cases, the knee joint is surgically fused, causing it to be permanently immovable but making it stable enough that it can support the weight of the body and thus be used for walking.

Gouty Arthritis

Gout is a condition characterized by sudden severe pain and swelling of the joints. It affects primarily the toes, insteps, ankles, heels, knees, and wrists. Gout is more common in males. It is due to an inherited genetic defect that causes either an increased production of uric acid or a reduction in the ability of the kidneys to excrete uric acid. In either case, the result is an increase in the level of uric acid in the blood (*hyperuricemia*). Uric acid is an end product of purine metabolism. The excessive level of uric acid in the blood causes the body fluids to become supersaturated and eventually results in the formation of sodium urate crystals in the soft tissues of the body as well as in the joints.

When the joints are affected, the condition is called *gouty arthritis.* As crystals accumulate in the joints and the soft tissues surrounding the joints, they cause an inflammation that eventually may erode the articular cartilage and the underlying bone, causing intense pain and immobility of the joint.

Effects of Aging

With aging, there is a progressive loss of the cartilaginous surface of the joints, which may be accompanied by the appearance of bony, spurlike overgrowths in the joint. This often produces a condition known as *degenerative osteoarthropathy.*

The effects of aging on the joints—particularly of degenerative osteoarthropathy—vary greatly from individual to individual and are influenced both by genetic factors and by environmental factors such as the heavy use of certain joints. Degenerative osteoarthropathy increases progressively starting as early as the age of 20. By the age of 80, virtually everyone has some degenerative osteoarthropathy of the knee and the elbow joints, with somewhat lower frequencies for the hip and the shoulder joints. Women often develop bony swellings (called Heberden's nodes) in the terminal phalanges. In men, the spine is the most common site of degenerative osteoarthropathy. Here the intervertebral discs degenerate, which results in new bone formation between the bodies of the vertebrae. This degeneration can produce a narrowing of the intervertebral foramina, which causes chronic and painful pressure on the nerve roots.

Study Outline

◆ **CLASSIFICATION OF JOINTS** p. 218

Joints may be classified on the basis of function or structure.

Functional classification

1. Synarthroses—immovable
2. Amphiarthroses—slightly movable
3. Diarthroses—freely movable

Structural classification

1. Fibrous
2. Cartilaginous
3. Synovial

◆ **FIBROUS JOINTS** pp. 218–219

Articulations in which bones are held together tightly by fibrous connective tissue; three types classed by length and type of fibers that hold bones together.

Sutures. Grooves on edges of bones fit closely together; short connecting fibers; found only between flat bones of skull. In adulthood, fibers replaced by bone and sutures may fuse together to form *synostoses.*

Syndesmoses. Joint fibers longer, called ligaments; allow some "give" movement; joint between distal ends of tibia and fibula is example.

Gomphoses. Periodontal ligament holds teeth into alveolar sockets.

◆ **CARTILAGINOUS JOINTS** p. 219
Bones united by cartilage.

Synchondroses. Bones held together by hyaline cartilage; some synchondroses are temporary, with cartilage replaced by bone; others are permanent.

Symphyses. Articular surfaces of bones covered with thin layer of hyaline cartilage; fibrocartilage pad separates bones within joint; shock-absorbing joint. Junction of pubic bones and junctions between adjacent vertebrae are examples.

◆ **SYNOVIAL JOINTS** pp. 219–235
Freely movable, with movement limited only by joint surfaces, ligaments, muscles, or tendons. Have five distinguishing features:

1. *Articular cartilage.* Thin layer of hyaline cartilage that covers smooth articular bone surfaces.
2. *Synovial cavity.* Small fluid-filled space separating ends of joining bones.
3. *Articular capsule.* Double-layered; outer layer of fibrous connective tissue encloses joint; inner layer of vascular synovial membrane.
4. *Synovial fluid.* Clear, thick secretion of synovial membrane; lubricates and nourishes.
5. *Supporting ligaments.* Capsular, extracapsular, and intracapsular ligaments maintain normal position of bones.

Articular Discs. Present in some synovial joints; fibrocartilage that extends inward from capsule, dividing synovial cavity.

Bursae and Tendon Sheaths. Formed by synovial membranes; reduce friction.

BURSAE. Sacs lined with synovial membranes, filled with synovial fluid; subcutaneous or between tendon and bone.

TENDON SHEATHS. Cylindrical synovial sacs around tendons; found where tendons cross joints.

Movements of Synovial Joints.

GLIDING. Bones move back and forth upon one another.

ANGULAR MOVEMENTS. Increase or decrease angle between two adjoining bones by moving in a single plane.

FLEXION. Bone moved in anterior-posterior plane to decrease angle between it and adjoining bone.

EXTENSION. Increases angle between adjoining bones.

ABDUCTION. Body part moved away from midline.

ADDUCTION. Body part moved toward midline.

CIRCUMDUCTION. Delineates a cone; base outlined by movement of distal end of bone and apex in articular cavity; hip and shoulder joints are examples.

ROTATION. Motion of bone around a central axis.

SUPINATION. Outward rotation of forearm; palm anterior.

PRONATION. Inward rotation of forearm; palm posterior.

SPECIAL MOVEMENTS.

ELEVATION. Raising a body part.

DEPRESSION. Lowering a body part.

INVERSION. Twisting foot so that sole faces inward.

EVERSION. Twisting foot so that sole faces outward.

PROTRACTION. Moving a part forward.

RETRACTION. Returning protracted body part to usual position.

Types of Synovial Joints. Classed by movements allowed and shapes of articular surfaces.

NONAXIAL JOINTS.

GLIDING (ARTHRODIAL) JOINTS.

Articular Surface. Formed by apposition of flat or slightly curved surfaces.

Movement. Any direction.

Example. Found between vertebrae, carpal, and tarsal bones.

UNIAXIAL JOINTS.

HINGE (GINGLYMUS) JOINTS.

Movement. Flexion and extension.

Example. Articulation of humerus with ulna.

PIVOT (TROCHOID) JOINTS.

Movement. Rotation around longitudinal axis of bone.

Example. Proximal articulations of radius and ulna.

BIAXIAL JOINTS.

CONDYLOID (ELLIPSOID) JOINTS.

Articular Surface. One slightly concave; other slightly convex.

Movement. In two perpendicular planes (flexion, extension, abduction, adduction, circumduction).

Example. Radiocarpal articulations.

SADDLE JOINTS.

Articular Surface. For each bone, is concave in one direction and convex in the other.

Movement. Same as condyloid joints.

Example. Carpometacarpal joint of thumb.

TRIAXIAL JOINTS.

BALL-AND-SOCKET (SPHEROID) JOINTS.

Articular Surface. Spherical head of one bone fits in cup-shaped socket on second bone.

Movement. Allows medial and lateral rotation plus all condyloid movements.

Example. Shoulder and hip joints.

Ligaments of Selected Synovial Joints.

LIGAMENTS OF THE SHOULDER JOINT.

ARTICULAR CAPSULE. Encloses joint.

CORACOHUMERAL LIGAMENT. Strengthens joint anteriorly; extends from coracoid process to greater tubercle.

GLENOHUMERAL LIGAMENTS. Several thickenings of lower portion of capsule; glenoid labrum deepens fossa.

BICEPS BRACHII MUSCLE. Tendon enhances joint strength.

LIGAMENTS OF THE ELBOW JOINT.

ARTICULAR CAPSULE. Encloses joint.

ULNAR COLLATERAL LIGAMENT. From medical epicondyle to ulna.

RADIAL COLLATERAL LIGAMENT. From lateral epicondyle to annular ligament and ulna.

LIGAMENTS OF THE HIP JOINT.

ARTICULAR CAPSULE. Encloses joint.

ILIOFEMORAL AND PUBOFEMORAL LIGAMENTS. Provide anterior joint strength.

ISCHIOFEMORAL LIGAMENT. Strengthens posteriorly.

ACETABULAR LABRUM. Deepens joint cavity.

LIGAMENTUM TERES. Extends through joint cavity; no significant contribution to hip-joint strength.

LIGAMENTS OF THE KNEE JOINT.

ARTICULAR CAPSULE. Does not completely enclose joint.

OBLIQUE AND ARCUATE POPLITEAL LIGAMENTS. Strengthen capsule posteriorly.

MEDIAL COLLATERAL LIGAMENT. Stabilizes joint medially.

LATERAL COLLATERAL LIGAMENT. Stabilizes joint laterally.

PATELLAR LIGAMENT. Strengthens joint anteriorly.

MEDIAL AND LATERAL MENISCI. Fibrocartilages that deepen condyles of tibia.

ANTERIOR AND POSTERIOR CRUCIATE LIGAMENTS. Prevent hyperextension and posterior tibial slip.

◆ CONDITIONS OF CLINCIAL SIGNIFICANCE: ARTICULATIONS pp. 235–238

Strains. Joint being overstretched or twisted results in ligament tearing or separation.

Dislocations. Articular surfaces of bones forcibly displaced.

Bursitis. Inflamed bursa resulting from injury, exercise, or infection.

Tendinitis. Inflammation of tendon sheath.

Herniated Disc. Nucleus pulposus squeezed to cause protrusion from intervertebral disc; causes pressure on spinal nerves.

Torn Menisci. Menisci torn loose; generally occurs during sudden changes of direction.

Arthritis. Joint inflammation caused by trauma or bacterial infection, metabolic disorders, or other unknown causes; may be inherited.

OSTEOARTHRITIS. Most common form; gradual degeneration of articular cartilage and development of bony spurs.

RHEUMATOID ARTHRITIS. Severely damaging form; pannus develops on surfaces of articular cartilage; articular cartilage and bone beneath it often destroyed; joint may fuse.

GOUTY ARTHRITIS. Sudden severe pain and swelling; due to excessive production of uric acid or inability to excrete uric acid; sodium urate crystals form in joints and soft tissues; articular cartilage may be eroded.

Effects of Aging. Progressive loss of articular cartilage and growth of bony spurs. Often results in degenerative osteoarthropathy.

Self-Quiz

1. The humerus/ulna joint is an example of a fibrous joint. True or False?
2. The joint found between the flat bones of the skull is classed functionally as: (a) syndesmosis; (b) suture; (c) amphiarthrosis.
3. Most joints of the body are: (a) synchondroses; (b) symphyses; (c) synovial.
4. Which of the following exemplifies a symphysis? (a) junction of the two pubic bones; (b) junctions between the costal cartilages; (c) the epiphyses of a long bone to the diaphysis.
5. The term *diarthrosis* refers to a synovial joint that is kept apart by a fluid-filled cavity. True or False?
6. Match the terms with the appropriate lettered description:

Synarthroses	(a) A joint in which the bones on either side of a suture become firmly fused
Sutures	(b) The fibers in this synarthrosis joint are relatively long
Syndesmoses	(c) These joints usually serve as shock absorbers
Synostosis	(d) Most fibrous joints
Amphiarthroses	(e) Most joints of the body are of this type
Synchondroses	(f) Articulations that permit slight movement
Symphyses	(g) The edges of the bones that form these joints have grooves that fit very firmly and closely together
Synovial	(h) Temporary joints in which the original cartilage is usually replaced by bone

7. The inability to produce the fluid that keeps most joints moist would likely be due to a disorder in the: (a) bursae; (b) synovial membrane; (c) articular cartilage.
8. Bursae and tendon sheaths are associated with synovial joints. True or False?
9. Synovial joints, like fibrous and cartilaginous joints, are classified according to the material that connects the bones. True or False?
10. The elbow is an example of a: (a) uniaxial joint; (b) biaxial joint; (c) triaxial joint.
11. Match the angular movement terms with the appropriate lettered descriptions. There may be more than one answer per term.

Flexion	(a) When you move a toe toward the midline of your foot
Extension	(b) Arching the back
Abduction	(c) When you move a limb away from the midline of your body
Adduction	(d) A bone moved in an anterior-posterior plane, decreasing the angle between it and adjoining bone
Plantar flexion	(e) Lowering the toe region of the foot
Dorsiflexion	(f) The opposite of flexion
	(g) Raising the toe region of the foot toward the shin

12. *Supination* is the term used to describe the inward rotation of the forearm, which causes the radius to cross diagonally over the ulna. True or False?
13. This movement is characteristic of the hip and shoulder joints: (a) pronation; (b) supination; (c) circumduction.
14. Twisting the foot so the sole faces outward, with its outer margin raised, is called: (a) inversion; (b) eversion; (c) protraction.
15. The joints found between the articular processes of vertebrae and between most carpal and tarsal bones are termed: (a) hinge; (b) gliding; (c) condyloid.
16. Match the joint types listed with the appropriate lettered descriptions. There may be more than one answer per joint type.

Nonaxial	(a) Condyloid (ellipsoid)
Gliding	(b) Carpometacarpal joint of the thumb
Uniaxial	(c) The shoulder and hip joints
Hinge	(d) Gliding (arthrodial)
Pivot	(e) The only movement here is rotation around the longitudinal axis of the bone.
Biaxial	(f) Hinge (ginglymus)
Condyloid	(g) Found between most carpal bones
Saddle	(h) Articulation between the radius and the carpals
Triaxial	(i) The elbow joint
Ball-and-socket	

17. The first metacarpal/carpal joint is which of the following? (a) saddle; (b) condyloid; (c) suture.
18. Both the shoulder joint and the hip joint contain: (a) an articular capsule; (b) an iliofemoral ligament; (c) a ligamentum teres.
19. The ligament that prevents the tibia from slipping posteriorly is the: (a) medial meniscus; (b) posterior cruciate; (c) ligamentum patellae.
20. Match the joint types listed with the appropriate lettered descriptions. There may be more than one answer per joint type.

Distal tibia/fibula	(a) Fibrous (syndesmosis)
Pubic/pubic	(b) Cartilaginous (symphysis)
Ulna/humerus	(c) Synovial (condyloid)
Sternum/clavicle	(d) Synovial (gliding)
Radius/carpals	(e) Nonaxial
Tarsal/tarsal	(f) Uniaxial; flexion; extension
Occipital bone/atlas	(g) This joint is capable of slight movement ("give")

21. Arthritis refers to many different types of inflammation of the joints. True or False?
22. A painful condition due to an inherited genetic defect that causes an increased level of uric acid in the blood is called: (a) bursitis; (b) arthritis; (c) gouty arthritis.

CHAPTER 8

The Muscular System: General Structure and Physiology

CHAPTER CONTENTS

LEARNING OBJECTIVES

After completing this chapter, you should be able to:

1. Describe three types of muscle.
2. Describe the subcellular structure of a skeletal muscle fiber.
3. Discuss the events involved in excitation-contraction coupling.
4. Explain the mechanism of skeletal muscle contraction.
5. Cite three ways that adenosine triphosphate (ATP) is supplied to a muscle during activity.
6. Describe the influence of length on the development of tension by a skeletal muscle.
7. Explain what is meant when muscles are called prime movers, antagonists, synergists, or fixators.
8. Describe three different types of skeletal muscle fibers.
9. Describe the events that lead to the contraction of a smooth muscle cell when the cell is stimulated.
10. Describe three functional differences between smooth muscle and skeletal muscle.

CHAPTER 8

Muscle is composed of contractile cells that actively develop tension and shorten. As a result, muscle is important in the movement of body parts, the alteration of the diameters of tubes within the body, the propulsion of materials within the body, and the excretion of substances from the body. In addition, muscle contractions produce significant amounts of heat that can be used to maintain normal body temperature. Because of its many functions, muscle tissue contributes significantly to the maintenance of homeostasis.

Muscle Types

The body contains three types of muscle: skeletal muscle, smooth muscle, and cardiac muscle.

Skeletal Muscle

As the name implies, most **skeletal muscle** attaches to the bones of the skeleton. The contractions of skeletal muscle exert force on the bones, thereby moving them. Consequently, skeletal muscle is responsible for activities such as walking and manipulating objects in the external environment.

Skeletal muscle is *voluntary* muscle—that is, its contractions are normally under the conscious control of the individual. Under many conditions, however, skeletal muscle contractions require no conscious thought. For example, a person does not usually have to think about contracting the skeletal muscles involved in maintaining posture. Skeletal muscle contractions are regulated by signals transmitted to the muscle by a portion of the nervous system known as the somatic nervous system.

When viewed microscopically, skeletal muscle cells exhibit alternating transverse light and dark bands that give the cells a striped, or *striated,* appearance.

Smooth Muscle

Smooth muscle is so named because its cells lack the striations evident in skeletal muscle cells. Smooth muscle is found in the walls of hollow organs and tubes such as the stomach, intestines, and blood vessels, and its contractions govern the movement of materials through these structures.

Smooth muscle is *involuntary* muscle—that is, its contractions are not normally under the conscious control of the individual. However, under appropriate circumstances (see Chapter 14), a person can gain some voluntary control over smooth muscle. Smooth muscle contractions are regulated by factors intrinsic to the muscle itself, by hormones, and by signals transmitted to the muscle by a portion of the nervous system known as the autonomic nervous system.

Cardiac Muscle

Cardiac muscle is a specialized type of muscle that forms the wall of the heart. Cardiac muscle is *involuntary* muscle, and its contractions are regulated by intrinsic factors, hormones, and the autonomic nervous system. Cardiac muscle cells are *striated,* much like skeletal muscle cells. The anatomy and the physiology of cardiac muscle are considered in Chapter 19.

Embryonic Development of Muscle

Skeletal Muscle

With the exception of the muscles of the limbs and some of the muscles of the head, skeletal muscles develop embryonically from **somites,** which are masses of mesodermal cells. The somites are located dorsally, along both sides of the axial skeleton of an embryo. Only cells from one portion of a somite, called the **myotome** *(mi´-ō-tōme),* differentiate into muscle cells. The mesodermal cells from the myotomes spread downward between the skin and the body cavity, until the right and left sides meet ventrally in the midline. The skeletal muscles of the trunk develop from this sheet of mesoderm. Typical somites do not form in the head of the embryo; consequently, most of the muscles of the head develop from the general mesoderm of that region. The muscles of the limbs begin development from condensations of mesoderm within the embryonic limb buds. Some of the mesodermal cells that form the limb muscles probably migrate to the area from the myotomes.

Individual muscle cells are called **muscle fibers,** and the immature cells that give rise to muscle fibers are called *myoblasts.* To form a skeletal muscle fiber, individual mononucleated myoblasts become aligned with one another. The plasma membranes of adjacent myoblasts then fuse, producing an elongated cell that contains all of the nuclei from the fused myoblasts. Therefore, each mature skeletal muscle fiber contains several nuclei. With further maturation, contractile proteins and an extensive internal membrane system develop within the elongated cell, forming a skeletal muscle fiber.

Mature skeletal muscle fibers are generally incapable of mitosis. However, scattered among the muscle fibers are *satellite cells,* which are considered to be inactive myoblasts—and thus are potentially capable of dividing. Satellite cells are prevalent in the muscles of young children, but as a muscle matures, the number of satellite cells present decreases; in a mature muscle, they represent less than 1% of the tissue. Nevertheless, their presence would seem to represent the potential for formation of new muscle fibers, even within a fully developed muscle.

Smooth Muscle

As the digestive tube and the body organs form within the embryo, mesodermal cells migrate to them and form a thin layer around them. These mesodermal cells develop into the smooth muscles of the body.

Cardiac Muscle

The muscle of the heart forms in a manner similar to that of the smooth muscles. Mesodermal cells migrate to and surround the early heart while it is still in the form of a tubule. Cardiac muscle begins contracting very early in the development of the embryo, even before the peripheral blood vessels have completely formed.

Gross Anatomy of Skeletal Muscle

When examining skeletal muscles without the aid of a microscope, as during gross dissection, it is possible to note various distinguishing features of individual muscles. Among the most obvious features are the connective tissue coverings, the attachments, and the shape of each muscle.

Connective-Tissue Coverings

Each skeletal muscle is composed of many individual muscle fibers (muscle cells) that are held together by thin sheets of fibrous connective tissue called *fascia.* The fascia that invests an entire muscle is called the **epimysium** *(ep˝-ĭ-mis´-ē-um)* (Figure 8.1). Fascia also penetrates the muscle, separating the muscle fibers into bundles called **fasciculi.** This fascia is called the **perimysium** *(per˝-ĭ-mis´-ē-um).* Very thin extensions of the fascia, called the **endomysium** *(en˝-dō-mis´-ē-um),* envelop each individual muscle fiber. Blood vessels and nerves pass into the muscle with the fascial sheaths to reach the individual muscle fibers. Beds of capillaries are located between the muscle fibers, and each fiber is supplied by a branch of a nerve cell.

◆ **FIGURE 8.1 Connective tissue of a muscle**

(a) Entire muscle, with the belly sectioned. (b) Enlargement of a cross section of the belly.

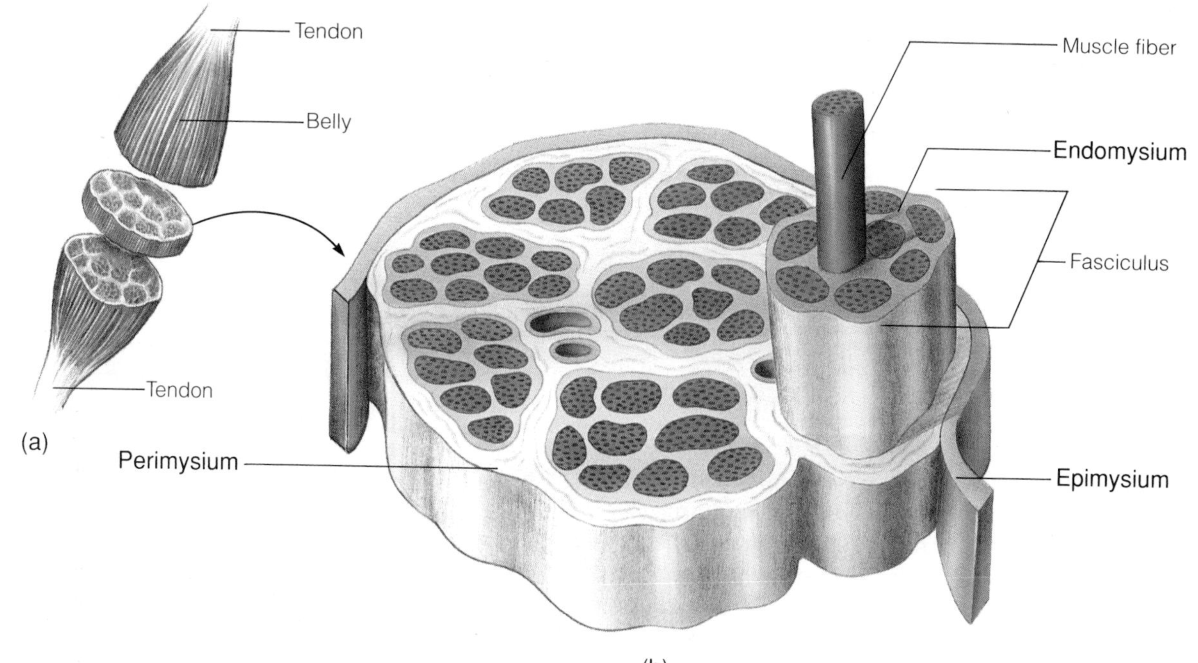

Skeletal Muscle Attachments

Skeletal muscles are anchored to the skeleton by extensions of the endomysium, perimysium, and epimysium. These connective tissues continue beyond the end of a muscle and either attach directly to the periosteum of a bone or attach to the perichondrium of a cartilage. In many cases, the connective tissues blend into a strong fibrous connection called a **tendon,** which then becomes continuous with the periosteum of a bone or with the perichondrium of a cartilage. Some tendons are quite short, whereas others are more than a foot in length. Tendons that take the form of broad thin sheets are called **aponeuroses** *(ap″-o-nu-ro′-seez).*

The attachments of both ends of a skeletal muscle are given specific names. The **origin** is the less movable end and is generally proximal. The **insertion** is the more movable end and is generally distal. The widest portion of a muscle, between the origin and the insertion, is known as the **belly** of the muscle. A muscle is described as *arising* from the origin and *inserting* at the insertion. The origin may be rather broad, arising from several different places on a bone, or even from several different bones. The insertion, in contrast, tends to be much more restricted. With the exception of sphincter muscles that surround body openings and some of the facial muscles that insert on the skin, joints are located between the origins and the insertions of skeletal muscles. When a skeletal muscle contracts, it shortens, using the joint as a pivot point to pull the insertion closer to the origin.

The origins and insertions that are described in Chapter 9 are the most common. However, you should keep in mind that for some muscles it is functionally possible to reverse the origin and the insertion—that is, the more fixed end, which is normally called the origin, can be used in some actions as the more movable end. For instance, certain of the superficial muscles of the chest are described as having their origins on the thorax and their insertions on the humerus. Clearly, the humerus is generally more movable than the thorax. However, if an individual is doing pull-ups, the thorax is moved towards the humerus and is therefore serving as an insertion.

Skeletal Muscle Shapes

The arrangement of the bundles of muscle fibers (fasciculi) varies within the different skeletal muscles (Figure 8.2). In some muscles, the fasciculi are oriented **parallel** to the long axis of the muscle, forming straplike muscles. The contraction of muscles that possess such longitudinally arranged fasciculi produces considerable movement; however, such muscles do not have much power. Less movement, though greater power, is produced by muscles that have a central tendon extending their entire length, with the fasciculi inserting diagonally into this tendon. In some muscles of this type, all of the fasciculi insert onto one side of the tendon. This arrangement is called **unipennate** *(yoo-na-pen′-it).* **Bipennate** muscles have fasciculi inserting obliquely on both sides of the tendon. The fasciculi of certain muscles have a complex arrangement in which the tendon branches within the muscle and the fasciculi are arranged around these branches. These are **multipennate** muscles. In a few muscles, the fasciculi converge from a broad origin into a single narrow tendon, an arrangement known as **convergent.** In some muscles, the fasciculi form a **circular** pattern. Such muscles are found surrounding the external openings of the body and are called sphincters.

Microscopic Anatomy of Skeletal Muscle

When a skeletal muscle is examined with the aid of a microscope, it is apparent that the muscle fibers have a regular subcellular structure. Skeletal muscle fibers are

◆ **FIGURE 8.2 Variation in muscle shapes**

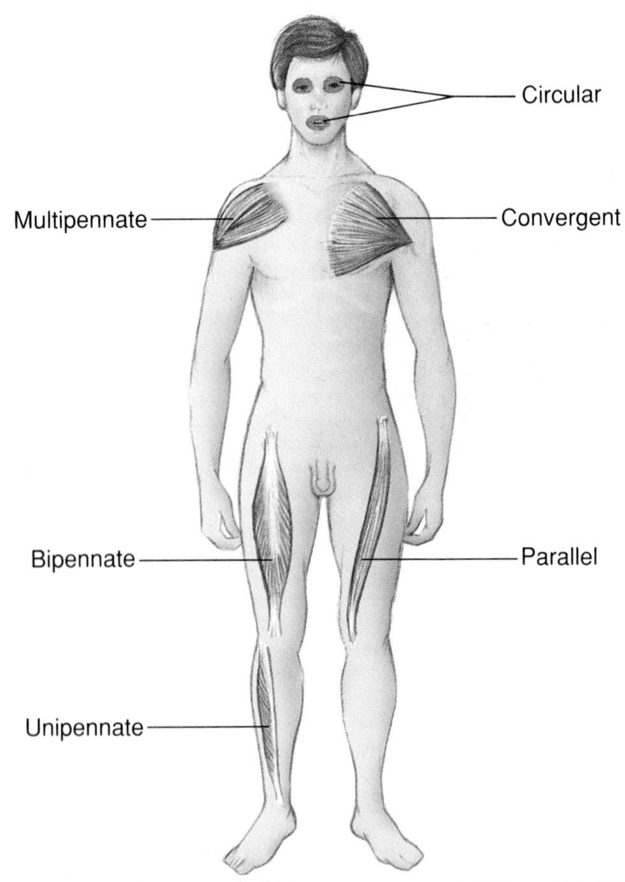

◆ **FIGURE 8.3 Microscopic anatomy of an individual skeletal muscle fiber (cell)**
Note the striated (striped) appearance of the muscle fiber and the myofibrils.

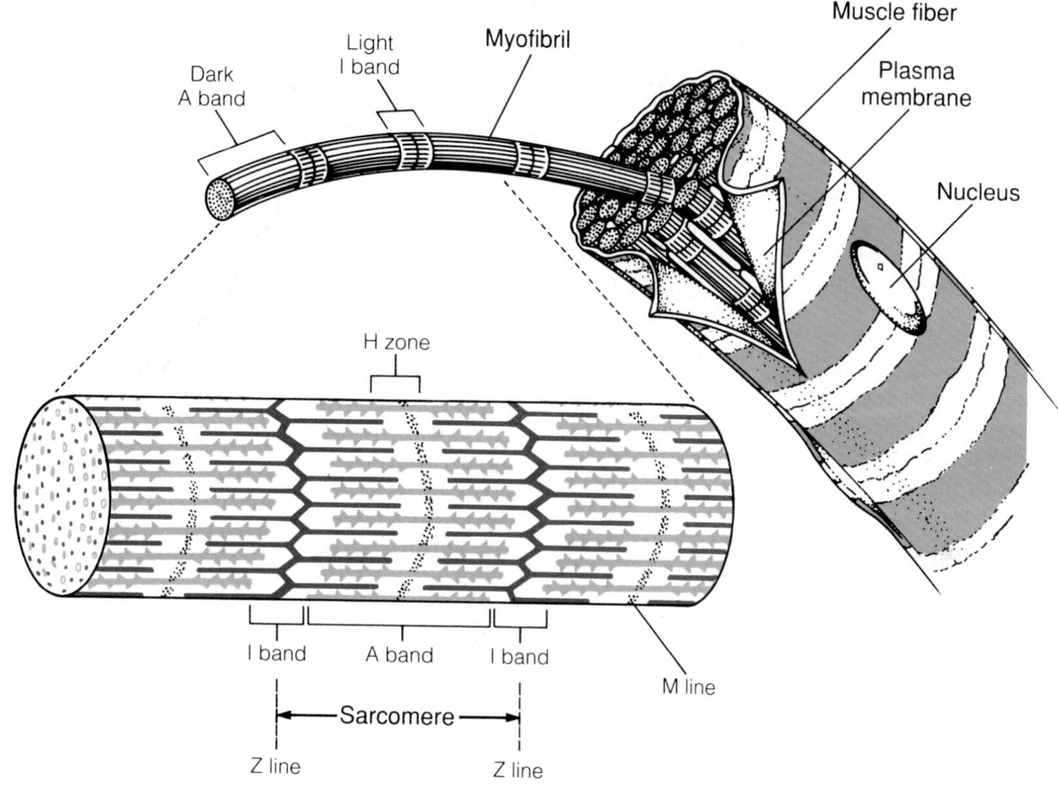

multinucleate cells approximately 10 to 100 microns in diameter and frequently many centimeters long. Each fiber contains several hundred to several thousand regularly ordered, threadlike **myofibrils** (Figure 8.3). The myofibrils extend lengthwise throughout the cell and are connected to the plasma membrane by intermediate filaments. When highly magnified, a myofibril exhibits alternating transverse light and dark bands, which are responsible for the striated appearance of the cell. The light bands are called **isotropic bands,** or **I bands,** and the dark bands are named **anisotropic bands,** or **A bands.** Crossing the center of each I band is a dense, fibrous **Z line,** or **Z disc,** which contains a protein called alpha-actinin. The Z lines divide the myofibrils into a series of repeating segments known as **sarcomeres.** In the center of a sarcomere and, therefore, in the center of an A band, is a somewhat less dense region, the **H zone.** A thin, dark **M line** crosses the center of the H zone.

A sarcomere contains two distinct types of longitudinally oriented **myofilaments:** thick filaments and thin filaments (Figure 8.4). **Thick filaments** occur in the A band where they overlap at either end with thin filaments. The H zone contains only thick filaments. The M line is formed by fine, filamentous structures that interconnect the thick filaments, holding them in a parallel arrangement. These filamentous structures contain a protein called *myomesin.* **Thin filaments** occupy the I band and part of the A band. The thin filaments attach to the Z lines. In the region of the A band where thick and thin filaments overlap, there is a hexagonal arrangement of thin filaments around each thick filament (Figure 8.5).

Composition of the Myofilaments

The thick filaments consist mainly of the protein **myosin.** A myosin molecule is made up of six polypeptide chains—two identical heavy chains and four light chains. Each heavy chain is shaped something like a golf club with a globular head and a long tail. The tail portions of the two heavy chains are tightly coiled around each other, forming a structure that has two rather bulbous heads protruding from one end of an extended shaft. Two light polypeptide chains are bound to each head.

A thick filament contains approximately 200 myosin molecules arranged in such a way that the shafts of the molecules are bundled together with the heads of the

◆ **FIGURE 8.4 Longitudinal view of the structure of sarcomeres**

The I band consists only of thin filaments and is divided by the Z line. The A band consists of thick filaments that overlap at either end with the thin filaments. The region where only thick filaments occur is the H zone. A single sarcomere extends from one Z line to the next. Thus, half of an I band is associated with one sarcomere and half with the neighboring sarcomere.

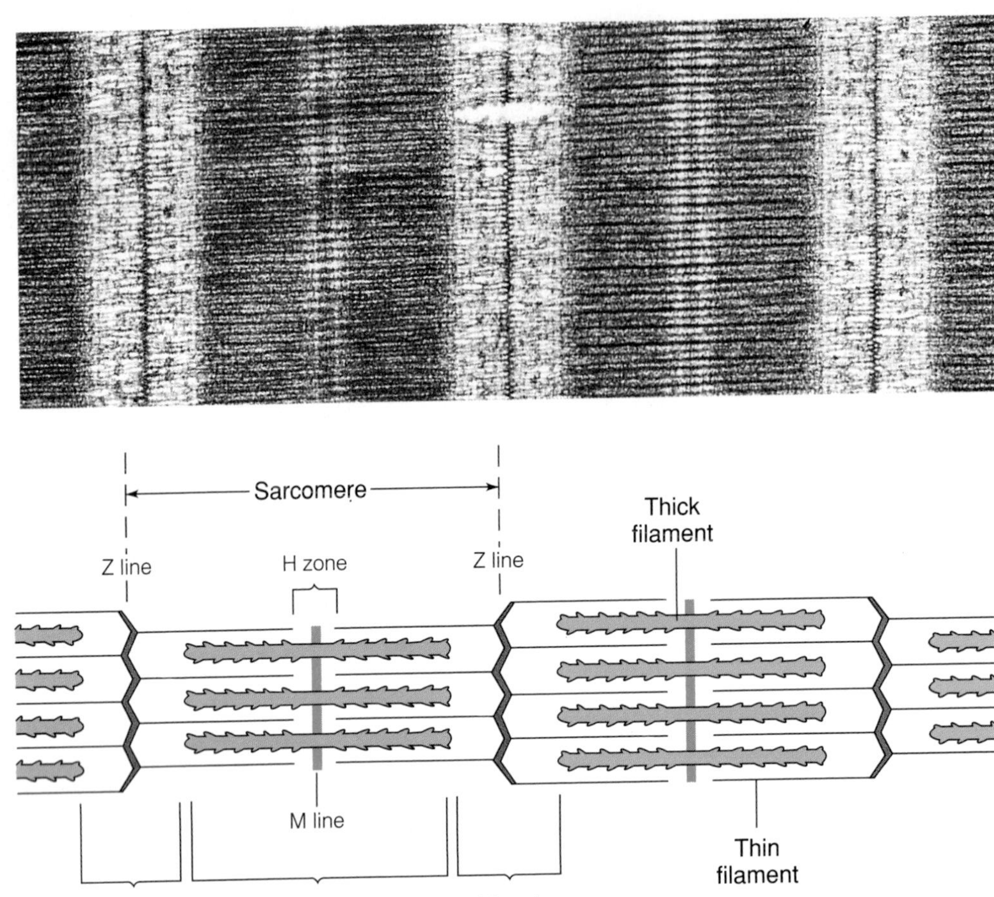

molecules (called **cross bridges**) facing outward (Figure 8.6a). The myosin molecules face in opposite directions on either side of the center of a thick filament, with the shafts of the molecules directed toward the center. Because of this arrangement, the central area of the filament contains the shaft portions of myosin molecules but no myosin heads (Figure 8.6c).

The thin filaments consist mainly of the proteins **actin, tropomyosin,** and **troponin** (Figure 8.6b). The actin portion of a thin filament is made up of subunits of globular (G) actin, each of which possesses two globular (or essentially spherical) regions. The G-actin subunits are linked together into a filamentous form of actin called F-actin, which resembles two strings of beads twisted around one another in a spiral. Although the G-actin subunits are globular in nature, they have a definite polarity and link to one another from front to back. In a thin filament, threadlike molecules of tropomyosin lie end to end along the surface of the actin, with each tropomyosin molecule extending along approximately seven G-actin subunits. Attached to each tropomyosin molecule and also to actin is a smaller molecule of the protein troponin, which can bind calcium ions. The arrangement of thick and thin filaments in a sarcomere is illustrated in Figure 8.6c.

Transverse Tubules and Sarcoplasmic Reticulum

Tubular invaginations of the plasma membrane known as **transverse tubules (t tubules)** extend deep into a skeletal muscle fiber from the membrane (Figure 8.7). In addition, a membranous network, the **sarcoplasmic reticulum,** extends throughout the fiber and surrounds each myofibril. The sarcoplasmic reticulum is in some respects similar to the smooth endoplasmic reticulum of other cells. Elements of the sarcoplasmic reticulum called **terminal cisternae** (lateral sacs) lie close to t tubules at the junctions of the A and I bands of the sarcomeres. At these locations, structures consisting of three membranous channels (terminal cisterna–t tubule–

◆ **FIGURE 8.5 Cross sections through different regions of a sarcomere of a myofibril**

Note that in the region of the A band where thick and thin filaments overlap, each thick filament is surrounded by six thin filaments, and each thin filament is surrounded by three thick filaments.

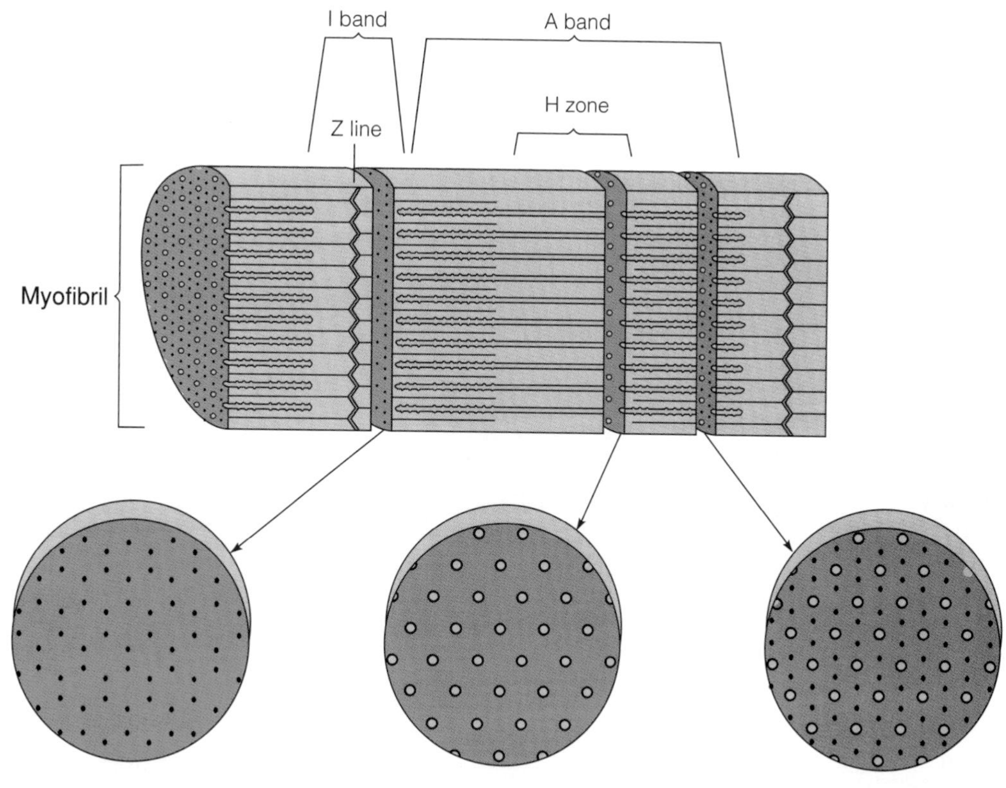

terminal cisterna) are formed. These structures are called triads.

The sarcoplasmic reticulum contains a high concentration of calcium ions, which are loosely bound to a protein called calsequestrin. When a skeletal muscle fiber is stimulated to contract, calcium ions are released from the sarcoplasmic reticulum into the cytosol. The significance of this activity will be explained in the following sections.

Contraction of Skeletal Muscle

Experimentally, two basic types of muscle contraction—isometric contraction and isotonic contraction—are frequently employed to study muscle function. An **isometric** (*iso* = equal; *metric* = measure) contraction is a contraction during which the length of a muscle remains constant (Figure 8.8a). In this type of contraction, the muscle actively develops tension and exerts force on an object but does not shorten. In the body, isometric types of muscle contractions occur when a person supports an object in a fixed position or attempts to lift an object that is too heavy to move.

An **isotonic** (*iso* = equal; *tonic* = tension) contraction is a contraction during which a muscle shortens while under a constant load (Figure 8.8b). In this type of contraction, the force against which the muscle shortens remains constant even though the length of the muscle changes considerably. In the body, where most skeletal muscles attach to bones and more than one muscle is usually involved in a movement, pure isotonic contractions seldom occur because the load on a muscle frequently changes as the muscle shortens. Nevertheless, contractions during which a muscle shortens, such as the contractions that move the legs during walking or the arms when lifting an object, are often referred to as isotonic contractions. Regardless of whether the load on a muscle changes or not, before a muscle can shorten, it must first develop sufficient tension to equal and overcome the resistance of the load against which it contracts—only then can shortening occur (Figure 8.9 on page 252).

Contraction of a Skeletal Muscle Fiber

At the cellular level, the same basic events occur during both isometric and isotonic skeletal muscle contractions. A skeletal muscle contracts when stimuli from the

◆ **FIGURE 8.6 Composition of myofilaments**

(a) A thick filament consists mainly of myosin molecules bundled together so the heads of the molecules project from the filament in a spiral pattern. (b) In a thin filament, G-actin subunits link together, forming F-actin. Threadlike molecules of tropomyosin lie end to end along the surface of actin, with each tropomyosin molecule extending along approximately seven G-actin subunits. Each tropomyosin molecule is attached to a molecule of troponin. Troponin is also attached to actin. (c) Longitudinal view of thick and thin filaments as arranged in a sarcomere. Note that the myosin molecules project in opposite directions on either side of the bare zone.

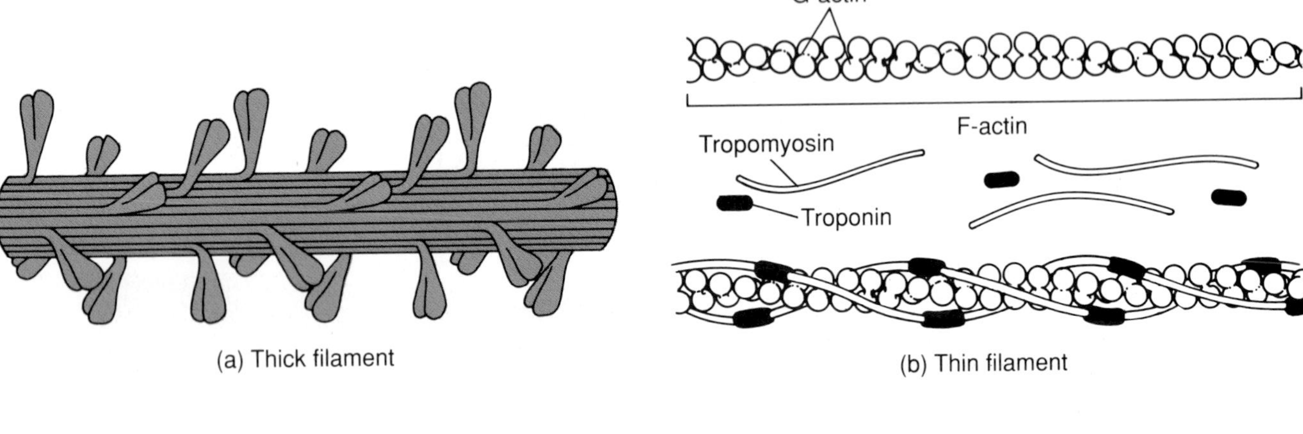

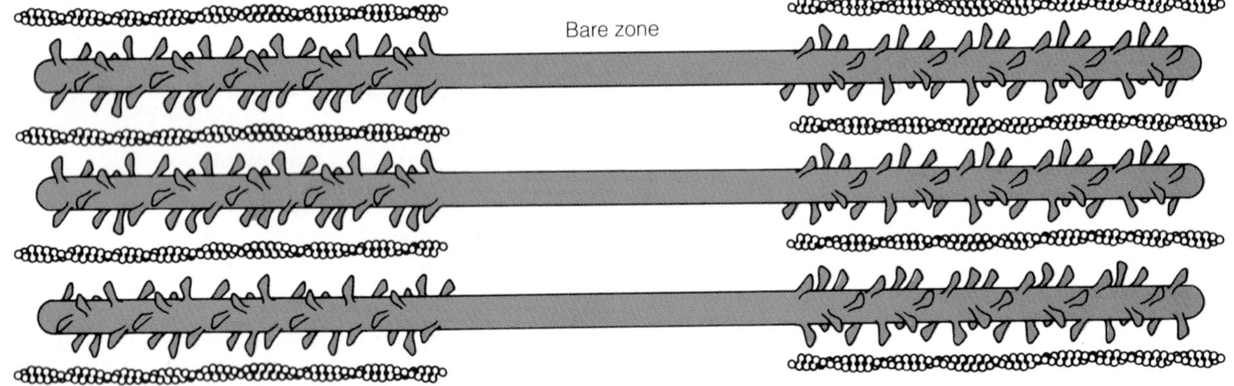

(c) Longitudinal section of filaments

nervous system excite the individual muscle fibers. This initiates a series of events that lead to interactions between the thick and thin filaments of the sarcomeres of the fibers. These interactions are responsible for the development of tension and the shortening of the fibers.

The Neuromuscular Junction

Nerve cells known as **motor neurons** supply the neural stimulation that skeletal muscle fibers require in order to contract. These neurons form specialized junctions, called **neuromuscular junctions,** with skeletal muscle fibers (Figure 8.10a). At a neuromuscular junction, an ending of a motor neuron closely approaches a specialized point along the plasma membrane of a skeletal muscle fiber, but it does not contact the membrane directly. Instead, a small gap separates the motor neuron ending from the muscle fiber membrane. Most skeletal muscle fibers have only one neuromuscular junction.

Excitation of a Skeletal Muscle Fiber

Motor neurons transmit brief, intermittent electrical signals called **nerve impulses** toward skeletal muscle fibers. The events that occur when a nerve impulse arrives at a neuromuscular junction are discussed in detail in Chapter 11. Here it is sufficient to state that a nerve impulse does not directly stimulate a skeletal muscle fiber because it cannot cross the gap that separates a motor neuron ending from the muscle fiber plasma membrane. Instead, a nerve impulse stimulates a skeletal muscle fiber indirectly in the following way.

◆ **FIGURE 8.7 Schematic representation of transverse tubules and the sarcoplasmic reticulum of a skeletal muscle fiber**

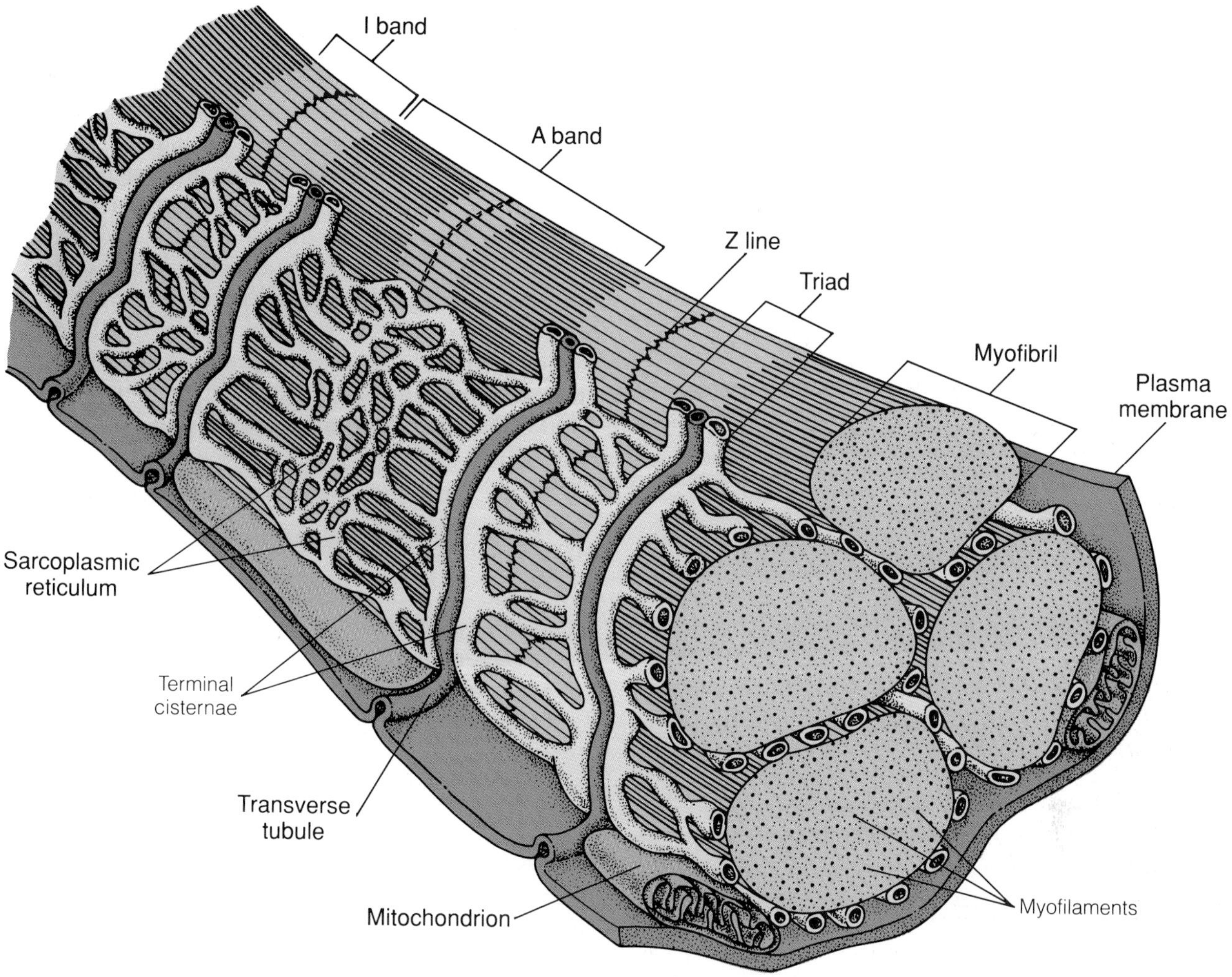

◆ **FIGURE 8.8 Isometric and isotonic muscle contractions**

(a) In an isometric contraction, the muscle develops tension but does not shorten. (b) In an isotonic contraction, the muscle shortens while under a constant load.

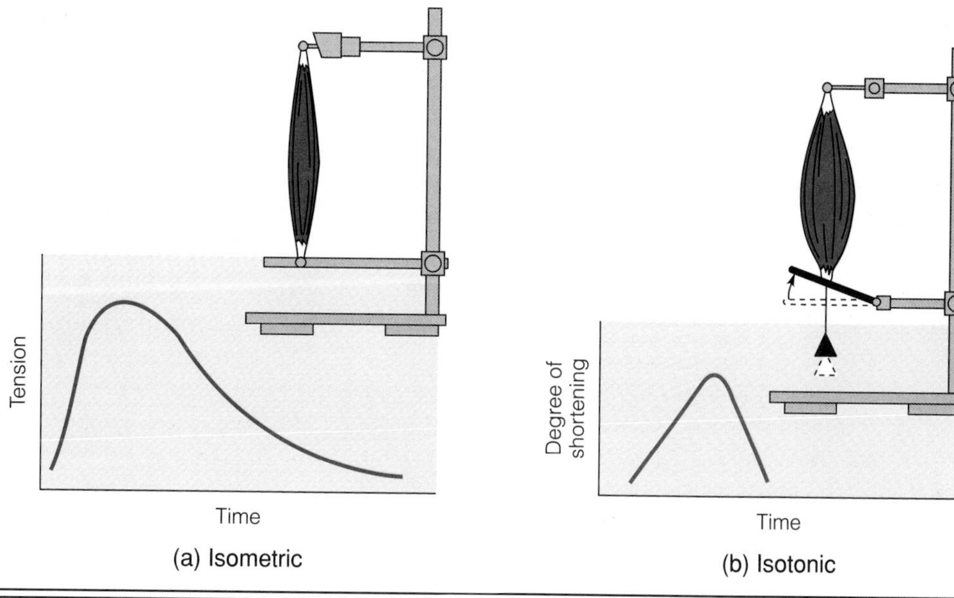

◆ FIGURE 8.9 Record of a contraction during which a muscle shortens and lifts a 10-gram load

Immediately following its stimulation, there is a period (green) during which the muscle develops sufficient tension to equal and overcome the resistance of the load but does not shorten. Following this, the muscle shortens and lifts the load.

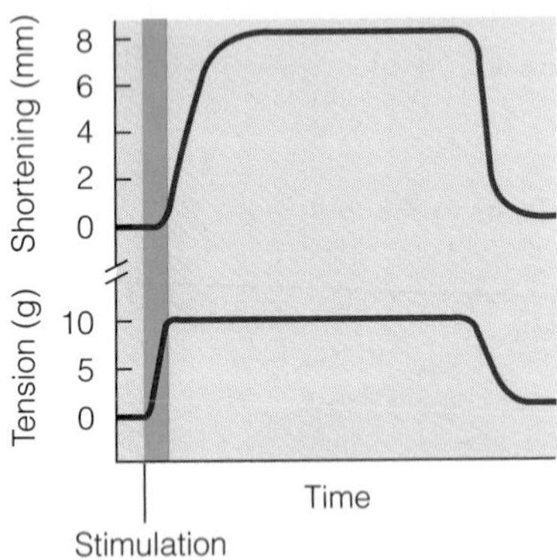

When a nerve impulse reaches a neuromuscular junction, molecules of the chemical substance *acetylcholine* are released from the motor neuron ending (Figure 8.10b). The acetylcholine molecules diffuse to the muscle fiber plasma membrane, where they bind to acetylcholine receptors.

The muscle fiber plasma membrane, like the plasma membranes of other cells, is electrically polarized, and a voltage called the *membrane potential* exists across the membrane. When acetylcholine binds to acetylcholine receptors, membrane channels that are permeable to small, positively charged ions, including sodium ions, open for a brief period. The ionic movements that take place through the open channels depolarize the muscle fiber membrane in the region of the neuromuscular junction. The depolarization triggers a stimulatory electrical impulse called an *action potential,* which is propagated (travels) along the membrane. The propagated action potential initiates a series of intracellular events that culminate in interactions between the thick and thin filaments of the sarcomeres and the contraction of the fiber.

Excitation-Contraction Coupling

The series of events by which a propagated action potential in the plasma membrane of a skeletal muscle fiber causes interactions between the thick and thin filaments of the sarcomeres and the contraction of the fiber are grouped together under the heading of **excitation-**

◆ FIGURE 8.10 A neuromuscular junction

(a) Diagrammatic representation of the general structure of a neuromuscular junction. (b) A more highly magnified section of the junctional area illustrating the events that occur at a neuromuscular junction. When a nerve impulse arrives at a terminal ending of a motor neuron, acetylcholine is released. The acetylcholine binds with receptors on the muscle fiber plasma membrane at the junction. This binding leads to a change in the membrane's permeability to small, positively charged ions (including sodium ions), and it triggers a propagated action potential that travels along the membrane.

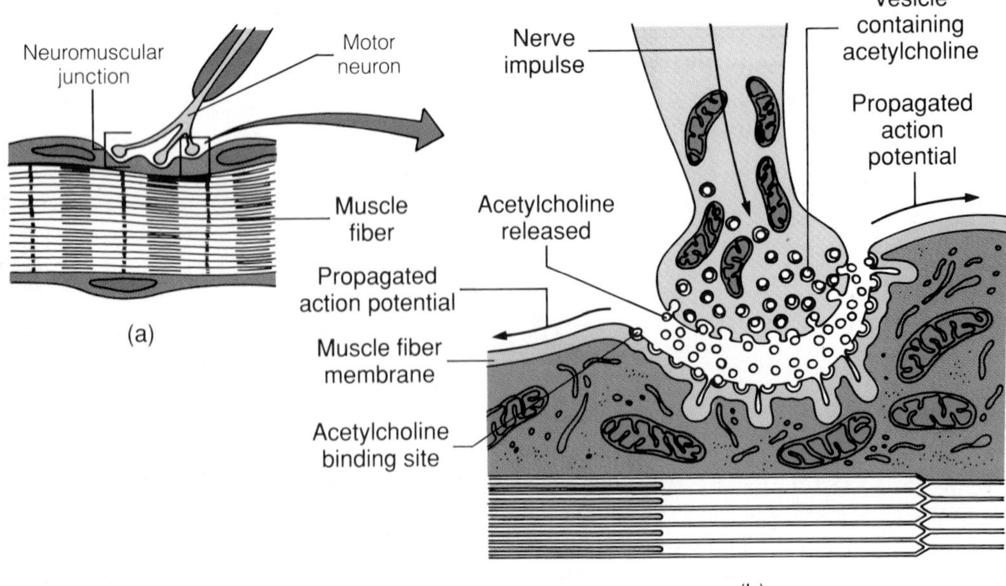

contraction coupling. From the plasma membrane, the propagated action potential passes along the t tubules into the central areas of the fiber. As it moves along the t tubules, it triggers the opening of large calcium ion release channels in the sarcoplasmic reticulum membrane, and calcium ions enter the cytosol from the sarcoplasmic reticulum. The calcium ions bind to troponin molecules of the thin filaments, leading—as is explained in the following section—to interactions between the thick and thin filaments of the sarcomeres. These interactions are directly responsible for muscle contraction.

Mechanism of Contraction

Muscle contraction requires energy, which is supplied by adenosine triphosphate (ATP). In a muscle fiber, an ATP molecule occupies a binding site on a globular head of a myosin molecule of a thick filament (Figure 8.11). The myosin molecule possesses ATPase enzymatic activity, and it splits the ATP into adenosine diphosphate (ADP) and phosphate, which remain bound to the myosin. This reaction releases energy, which is transferred to the myosin, producing a high-energy form of myosin.

In addition to an ATP binding site, a myosin head contains a binding site that can attach to a complementary site on an actin subunit of a thin filament. A high-energy form of myosin has a tendency to bind to actin. In a relaxed, unstimulated muscle fiber, this binding is prevented by tropomyosin, which lies along the surface of the actin and physically blocks interactions between high-energy myosins and actin subunits. However, when a muscle fiber is stimulated and calcium ions are released from the sarcoplasmic reticulum, the calcium ions bind to troponin molecules, which are linked to both actin and tropomyosin. The binding of calcium ions to troponin weakens the linkage between troponin and actin. The weakening of this linkage allows tropomyosin to move away from its blocking position. With the tropomyosin out of the way, high-energy myosins can bind to G-actin subunits.

The initial binding of a high-energy myosin to an actin subunit is relatively weak, but this binding triggers the release of the inorganic phosphate, which is bound to the myosin. With the release of the inorganic phosphate, the myosin binds tightly to the actin. The energy stored within the myosin is discharged, with the resultant production of a force that causes the myosin head (cross bridge) to move. In essence, the myosin head swivels toward the center of the sarcomere, pulling on the actin-containing thin filament. The ADP bound to the myosin is then released, but the myosin remains attached to the actin. When another ATP molecule binds to the myosin head, the myosin releases from the actin. The myosin splits the ATP into ADP and phosphate, producing a high-energy form of myosin that attaches to an actin subunit, and the cycle is repeated. At any instant during the contraction of a skeletal muscle fiber, approximately 50% of the myosin heads are attached to actin subunits, and the rest are at intermediate stages of the activity cycle. (Each of the two heads of a myosin molecule is believed to cycle independently of the other.)

The force of the myosin heads of the thick filaments pulling on the actin-containing thin filaments is transmitted to the plasma membrane of the muscle fiber and ultimately to the load. If enough force is developed by the fiber—and by other fibers involved in the muscle contraction—to overcome the resistance of the load, the repeated cycling of the myosin heads pulls the thin filaments past the thick filaments toward the centers of the sarcomeres (Figure 8.12). This draws the Z lines closer together, and the muscle fiber shortens.

Regulation of the Contractile Process

Once a nerve impulse stimulates a skeletal muscle fiber and calcium ions are released from the sarcoplasmic reticulum, why doesn't the formation of attachments between thick and thin filaments and, therefore, the contractile process continue indefinitely? In fact, the calcium ions released in response to a single nerve impulse are free for only a short time. Following their release, an active transport mechanism quickly pumps the calcium ions back into the sarcoplasmic reticulum. With the return of the calcium ions to the sarcoplasmic reticulum, troponin strengthens its connection with actin, pulling the tropomyosin back into its blocking position. When this occurs, no further interactions between high-energy myosins and actin subunits are possible. Consequently, the contractile process ceases, and the muscle fiber relaxes. When another nerve impulse stimulates the muscle fiber and calcium ions are again released from the sarcoplasmic reticulum, the contractile process occurs once more. If many nerve impulses arrive at a skeletal muscle fiber in rapid succession so that calcium ions continue to be available to bind with troponin, the fiber does not relax between successive impulses, and the contractile process continues until the impulses cease and the calcium ions are returned to the sarcoplasmic reticulum.

The sequence of events involved in the excitation and contraction of a skeletal muscle fiber is outlined in Table 8.1.

Sources of ATP for Muscle Contraction

ATP is the immediate source of energy for muscle contraction. However, the amount of ATP in skeletal muscle fibers is sufficient to support muscle contraction during strenuous exercise for only a few seconds. If muscular

◆ **FIGURE 8.11 Mechanism of muscle contraction**

A myosin head of a thick filament has ATP bound to it. The ATP is split into ADP and inorganic phosphate, producing a high-energy form of myosin. The high-energy myosin binds weakly to an actin subunit of a thin filament, and the inorganic phosphate is released. With the release of the inorganic phosphate, the myosin binds tightly to the actin. The energy of the high-energy myosin is discharged, and the myosin head swivels, pulling on the actin-containing thin filament. The ADP bound to the myosin is then released, but the myosin remains bound to the actin until another ATP binds to the myosin's head. When this binding occurs, the myosin releases from the actin, the ATP is split, and the cycle is repeated. For this contractile cycle to occur, troponin must bind calcium. This binding results in the removal of tropomyosin from a position blocking the myosin-actin interaction.

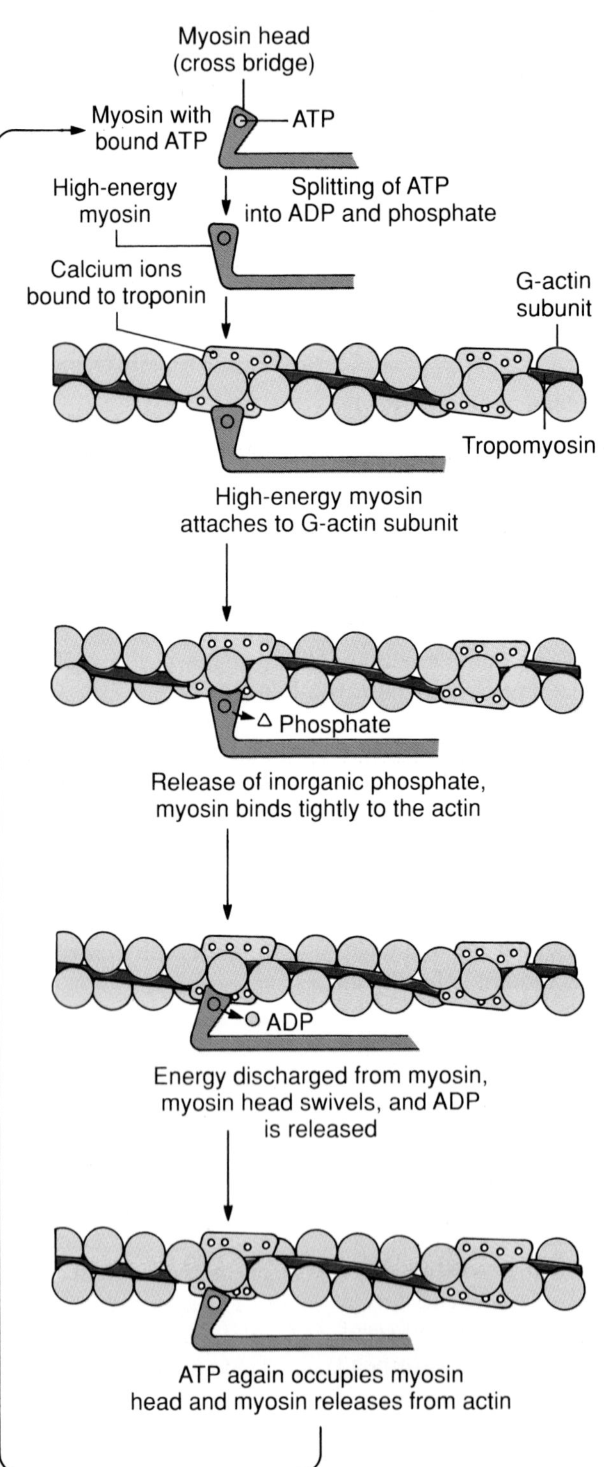

◆ **FIGURE 8.12 Filament movement and muscle fiber shortening**

A muscle fiber shortens when the thin filaments move past the thick filaments toward the centers of the sarcomeres, and the Z lines are drawn closer together.

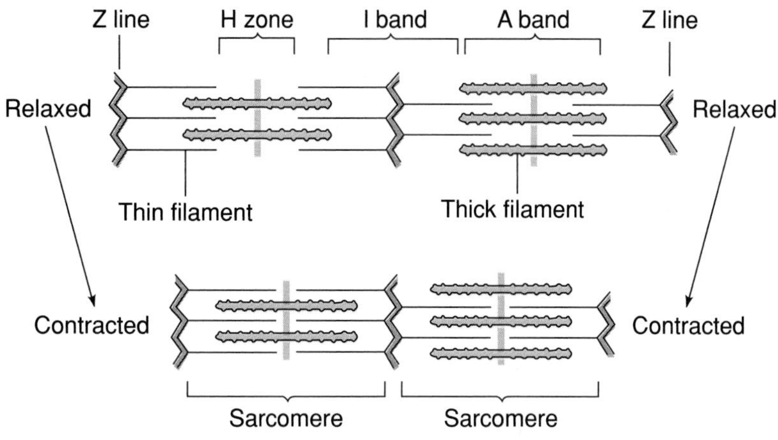

activity is to continue for longer periods, additional ATP must be produced.

Creatine Phosphate

Skeletal muscle fibers contain a substance called **creatine phosphate** that provides them with a means of forming ATP rapidly. Creatine phosphate contains phosphate and energy that can be transferred to ADP to produce ATP in a reversible reaction catalyzed by the enzyme *creatine kinase:*

$$\text{Creatine Phosphate} + \text{ADP} \underset{}{\overset{\textit{Creatine Kinase}}{\rightleftharpoons}} \text{Creatine} + \text{ATP}$$

◆ **TABLE 8.1 Sequence of Events Involved in the Excitation and Contraction of a Skeletal Muscle Fiber**

1. A nerve impulse arrives at a neuromuscular junction. Acetylcholine is released from the motor neuron and binds to receptors on the muscle fiber plasma membrane.
2. When acetylcholine binds to its receptors, membrane channels that are permeable to small, positively charged ions open for a brief period, and the muscle fiber membrane depolarizes in the region of the neuromuscular junction.
3. The depolarization triggers an action potential, which is propagated (travels) along the membrane and the t tubules.
4. The propagated action potential triggers the release of calcium ions from the sarcoplasmic reticulum.
5. Calcium ions bind to troponin.
6. Tropomyosin moves away from its blocking position, permitting actin and myosin to interact.
7. High-energy myosins (which are formed when the ATPase activity of myosin splits ATP into ADP and inorganic phosphate) bind weakly to actin subunits of the thin filaments. Inorganic phosphate is released from the myosins, and the myosins bind tightly to the actin subunits.
8. Energy stored in the high-energy myosins is discharged, and the myosin heads swivel, pulling on the thin filaments. ADP is then released from the myosins, but the myosins remain bound to the actin subunits.
9. ATP binds with the myosin heads, which release from the actin subunits.
10. ATP is split into ADP and inorganic phosphate, again producing high-energy myosins.
11. Steps 7, 8, 9, and 10 are repeated as long as calcium ions are bound to troponin and ATP is available.
12. When calcium ions are returned to the sarcoplasmic reticulum, tropomyosin moves back into its blocking position, preventing further interaction between high-energy myosins and actin subunits.
13. Contraction ceases, and the muscle fiber relaxes.

During periods of exercise, when ATP is being utilized to provide energy for muscle contraction, this reaction can form additional ATP.

Although skeletal muscle fibers contain more creatine phosphate than they do ATP, the utilization of creatine phosphate provides only enough ATP to support muscle contraction during strenuous exercise for a few additional seconds. Nevertheless, creatine phosphate is an important source of ATP during the period immediately following the initiation of muscle contraction when other means of producing ATP are not yet operating at high levels (Figure 8.13). In fact, early in contraction, muscle ATP levels may decline relatively little, but creatine phosphate levels drop substantially as creatine phosphate is used to form additional ATP. Activities that require short bursts of intense muscle contraction, such as high jumps or sprints, are supported primarily by ATP derived from creatine phosphate.

◆ FIGURE 8.13 Energy sources for muscle contraction during strenuous exercise (running at 18 km/hr on a treadmill inclined upward at an angle of 15°)

Energy from "phosphagen" (ATP and creatine phosphate) that is present within muscles at the beginning of exercise is the initial and major energy source during superexertion. Energy from aerobic metabolism increases exponentially from the onset of exercise, but the mechanism is sluggish and accounts for only a small amount of the energy used during the first few seconds. The remaining energy is derived from anaerobic glycolysis.

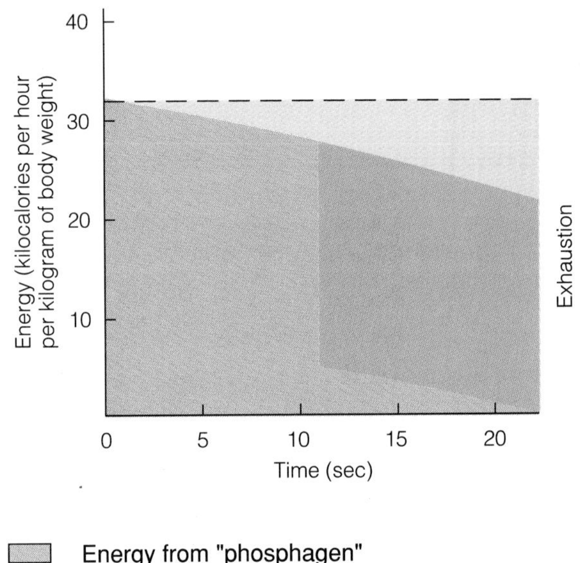

Nutrients

Ultimately, the metabolic breakdown of nutrient substances such as glucose, glycogen, and fatty acids by skeletal muscle fibers provides the ATP required to support continued muscular activity. In resting and slightly active muscles, fatty acids are the principal nutrient molecules used to produce ATP. As muscular activity increases, muscles use more and more glycogen and glucose for ATP production. Muscle fibers store some glycogen. Glucose and fatty acids are brought to the muscles by the blood.

Aerobic Metabolism. If sufficient oxygen is available to a muscle, ATP is produced by **aerobic** (that is, oxygen-utilizing) metabolic processes that break down glucose, glycogen, and fatty acids to carbon dioxide and water. The oxygen required by these aerobic processes is delivered from the lungs to the muscles by the blood. During exercise, the blood flow to active skeletal muscles increases, as do the rate and depth of breathing. These adjustments increase the delivery of oxygen to the muscles.

Aerobic metabolism produces a considerable amount of ATP for each nutrient molecule processed. For example, the aerobic breakdown of one molecule of glucose can generate up to 38 molecules of ATP. On the other hand, the breakdown process has many steps and proceeds relatively slowly. Moreover, aerobic metabolism requires a continual and adequate supply of oxygen.

Under conditions of light to moderate exercise, such as walking or jogging, muscle contraction is supported primarily by ATP derived from aerobic metabolism. This type of exercise, which normally can be continued for a considerable period of time, is called *aerobic exercise* or *endurance-type exercise.*

Anaerobic Metabolism of Glucose and Glycogen. During periods of intense muscular activity, oxygen cannot be supplied to many muscle fibers fast enough, and aerobic metabolism cannot produce all the ATP required for muscle contraction. During such periods, additional ATP is produced by **anaerobic** (that is, non-oxygen-utilizing) metabolic processes that break down glucose and stored glycogen into lactic acid.

Both the aerobic and the anaerobic breakdown of glucose begin with a series of anaerobic chemical reactions collectively termed **glycolysis** (Figure 8.14). The first reaction of glycolysis converts a glucose molecule into a molecule called glucose-6-phosphate. In addition, stored glycogen can be broken down anaerobically into molecules of glucose-6-phosphate, which are processed by the glycolytic pathway.

◆ **FIGURE 8.14 The breakdown of glucose and stored glycogen in skeletal muscle fibers**

The chemical reactions that break down glucose to pyruvic acid (blue box) are collectively termed glycolysis. Glycolysis is an anaerobic process, as is the breakdown of glycogen into molecules of glucose-6-phosphate. The anaerobic breakdown of glucose into lactic acid generates 2 molecules of ATP for each glucose molecule processed. The aerobic breakdown of glucose into carbon dioxide and water generates up to 38 molecules of ATP for each glucose molecule.

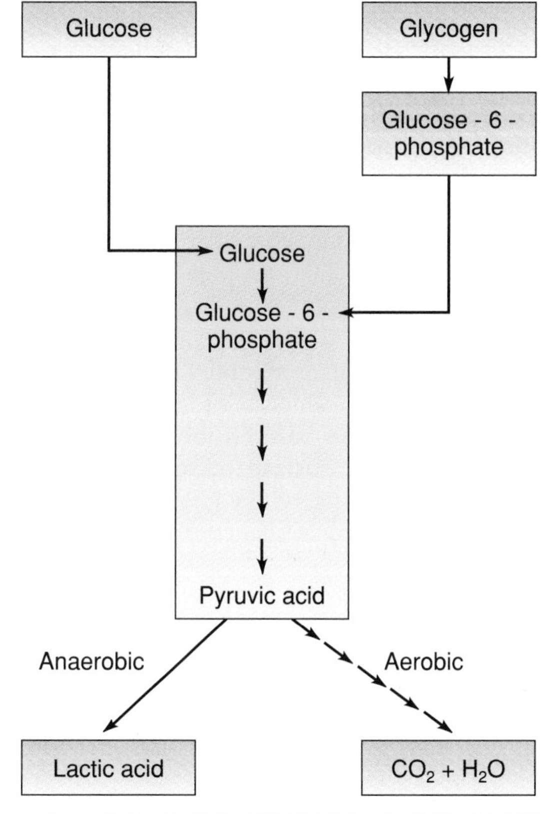

The glycolytic reactions ultimately produce two molecules of pyruvic acid from each molecule of glucose. During aerobic metabolism, the pyruvic acid molecules are broken down into carbon dioxide and water by oxygen-utilizing reactions. During anaerobic metabolism, the pyruvic acid molecules are converted into lactic acid by a reaction that does not utilize oxygen. Thus, the final products of the aerobic breakdown of glucose (and stored glycogen) are carbon dioxide and water, and the final product of the anaerobic breakdown of glucose (and stored glycogen) is lactic acid.

The anaerobic breakdown of glucose into lactic acid generates 2 molecules of ATP for each glucose molecule processed, and as previously indicated, the aerobic breakdown of glucose generates up to 38 molecules of ATP for each glucose molecule. Thus, the anaerobic processing of glucose produces substantially less ATP per glucose molecule than does the aerobic processing of glucose. On the other hand, the anaerobic breakdown of glucose proceeds more rapidly than aerobic metabolism, and over a limited period of time, it can produce more ATP than aerobic metabolism. To do so, however, the anaerobic processes must use large amounts of glucose (or glycogen). Moreover, hydrogen ions are generated by glycolysis, and a buildup of hydrogen ions may contribute to the onset of muscle fatigue. Nevertheless, the anaerobic breakdown of glucose and stored glycogen to provide ATP for muscle contraction allows muscles to maintain a high level of activity for a longer period of time than would be possible if creatine phosphate and aerobic metabolism were the only sources of ATP.

Muscle Fatigue

Intense skeletal muscle activity cannot continue indefinitely, and muscles eventually become fatigued. **Muscle fatigue** can be defined as the inability of a muscle to maintain a particular strength of contraction or tension (that is, a particular power output) over time.

The immediate events that cause muscle fatigue are not well understood and are thought to differ with different types of exercise (for example, running a marathon as opposed to lifting heavy weights). However, it is widely believed that a major factor underlying the occurrence of muscle fatigue is the inability of a muscle to generate energy at a rate sufficient to meet its energy requirements. This inability may be due to a depletion of metabolic reserves of nutrient substances such as muscle glycogen, or it may be due to a buildup of substances produced during contraction, such as hydrogen ions. A buildup of hydrogen ions, which increases muscle acidity, is believed to inhibit the activation of certain enzymes involved in cross-bridge cycling and to reduce the activity of certain enzymes involved in energy production.

Often, "psychological fatigue" can cause a person to stop exercising before any of the factors just discussed create an actual muscle fatigue. For example, a person may discontinue a strenuous exercise because of discomfort associated with the exercise or discontinue a repetitive exercise due to boredom or the monotonous nature of the exercise.

Oxygen Debt

When muscular activity ceases, the creatine phosphate level within the muscle fibers must be returned to normal (as must the ATP level, if it has dropped), and any

ASPECTS OF EXERCISE PHYSIOLOGY

Uptake of Glucose Increases in Exercising Muscles

During exercise, muscle cells use more glucose and other nutrient fuels than usual to power their increased contractile activity. The rate of glucose transport into exercising muscle may increase more than tenfold during moderate or intense physical activity. The mechanisms responsible for increased glucose uptake by exercising muscle are still unclear. In many cells, including resting muscle cells, facilitated diffusion of glucose into the cells depends on the hormone insulin. Because plasma-insulin levels fall during exercise, however, insulin is probably not responsible for the increased transport of glucose into exercising muscles. One possibility currently being studied is that exercise increases the availability of glucose carriers in the plasma membrane of muscle cells. This has been demonstrated in rats that have undergone physical training.

Exercise influences glucose transport into cells in yet another way. Regular aerobic exercise (see p. 809) has been shown to increase both the affinity (degree of attraction) and the number of plasma-membrane receptor sites that bind specifically with insulin. This adaptation results in an increase in insulin sensitivity; that is, the cells are more responsive than normal to a given level of circulating insulin.

Because insulin enhances the facilitated diffusion of glucose into most cells, an exercise-induced increase in insulin sensitivity is one of the factors that makes exercise a beneficial therapy for controlling diabetes mellitus, a disorder characterized by insulin deficiency (see Chapter 17). As a result of inadequate insulin action, glucose entry into most cells is impaired. Plasma levels of glucose become elevated because glucose remains in the plasma instead of being transported into the cells. In the type II form of the disease, insulin is being produced, but not in sufficient quantities to meet the body's need for glucose uptake. By increasing the cells' responsiveness to the limited amount of insulin available, regular aerobic exercise helps drive glucose into the cells, where it can be used for energy production, instead of remaining in the plasma, where it leads to detrimental consequences for the body.

lactic acid produced by the anaerobic breakdown of glucose and stored glycogen must be metabolized. These activities are accomplished by aerobic metabolic processes. These oxygen-utilizing processes provide ATP for the resynthesis of creatine phosphate (which occurs by a reversal of the creatine phosphate breakdown reaction described earlier) and, if necessary, for the replenishment of ATP. In addition, aerobic metabolic processes convert a portion of any lactic acid back into glucose or glycogen. (Lactic acid produced in skeletal muscle fibers can leave the fibers and enter the blood, and this conversion occurs to a large extent in the liver.) Aerobic metabolic processes also utilize some of the lactic acid as an energy source for ATP production.

To supply the oxygen required by these processes, increased breathing continues for some time after strenuous muscular activity ceases. In essence, an **oxygen debt** is built up during periods of muscular activity, when nonoxidative sources of ATP are used to support muscle contraction. This debt is paid back by the increased breathing that continues after the end of the activity. The increased breathing provides the oxygen required by the aerobic processes that replenish creatine phosphate and ATP supplies and metabolize lactic acid.

Following exercise, it is not only necessary to restore creatine phosphate and ATP levels to normal, but it is also necessary to replace any muscle glycogen that was used for ATP production. The rate at which muscle glycogen stores are replenished is strongly influenced by diet. After exhaustive exercise that severely depletes muscle glycogen stores, a person on a high-carbohydrate diet can replenish them in about two days. If the person is on a high-fat, high-protein diet, however, glycogen stores will not be fully replenished even five days after the exercise.

◆ **FIGURE 8.15 Two motor units of a skeletal muscle**

Each motor unit consists of a motor neuron and all the muscle fibers supplied by the neuron.

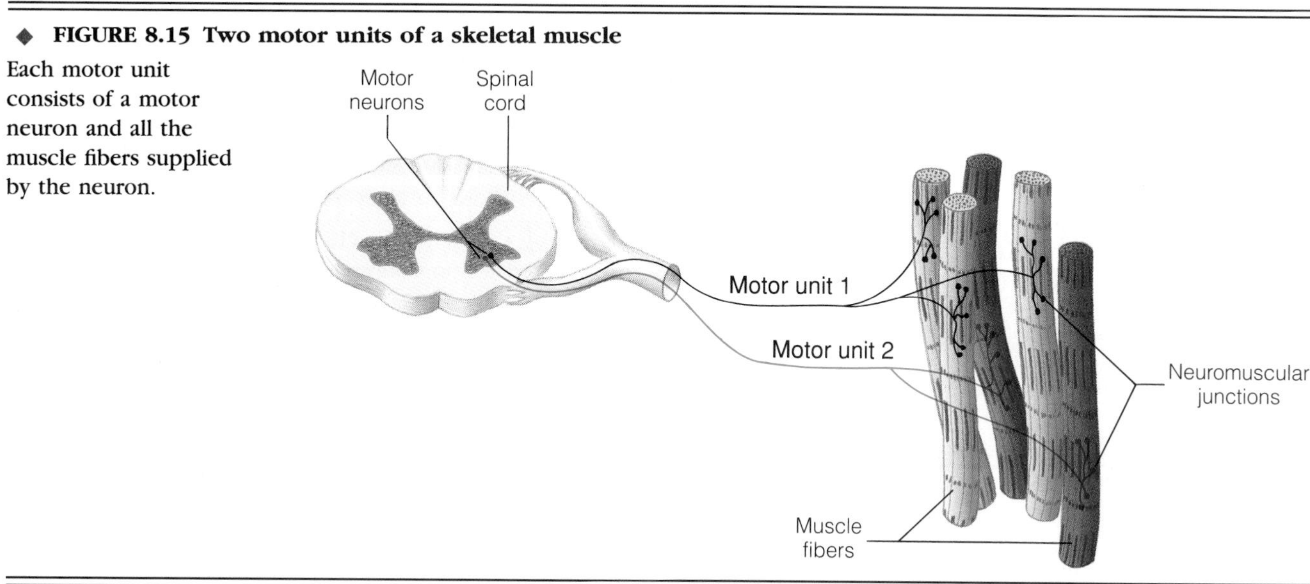

The Motor Unit

There are more muscle fibers in a skeletal muscle than there are neurons to supply the muscle. Consequently, each neuron must branch to supply several muscle fibers, and a nerve impulse transmitted along a particular neuron will reach all the fibers supplied by the neuron.

A single neuron and all the muscle fibers it supplies make up a **motor unit** (Figure 8.15). In the body, the motor unit and not the individual muscle fiber can be thought of as the functional unit of muscle activity because a nerve impulse in a single neuron stimulates all the muscle fibers supplied by the neuron. Muscles used for fine movements over which great control is exercised, such as the muscles of the hand, generally contain motor units in which each neuron supplies a relatively small number of muscle fibers—perhaps only a dozen or so. Muscles whose contractions are less precisely controlled, such as the muscles of the back or calf, have motor units in which a single neuron may supply several hundred muscle fibers.

Responses of Skeletal Muscle

The contractions of the muscle fibers of the motor units of a muscle combine to produce contractions of the whole muscle that vary in both duration and strength.

Muscle Twitch

If the fibers of a skeletal muscle are stimulated with a single brief stimulus, the muscle will contract once rapidly and then relax. This response, which is called a **muscle twitch,** exhibits three distinct phases (Figure 8.16). Immediately following the arrival of the stimulus at the muscle, there is a short *latent period* during which no response is seen. During this period, the processes associated with excitation-contraction coupling occur. Following the latent period, a *period of contraction* occurs. During this period, the muscle actively develops tension, and if enough tension develops to overcome the resistance of the load, the muscle shortens. The final phase of the muscle twitch is the *period of relaxation.* During this period, the tension that was actively developed by the muscle diminishes, and if the muscle had shortened, it returns to its original, unstimulated length.

Graded Muscular Contractions

When a skeletal muscle fiber is stimulated and a propagated action potential travels over the plasma membrane and along the t tubules, enough calcium ions are released from the sarcoplasmic reticulum to activate the fiber completely. Consequently, when a nerve impulse triggers a propagated action potential in the membrane of a skeletal muscle fiber, the fiber contracts to the maximum extent possible for the existing conditions. (Note that this does not mean the contraction of a skeletal muscle fiber is exactly the same at all times. The condition of a fiber at the time it is stimulated affects its contraction, and different conditions can exist at different times. For example, a fiber may or may not have contracted previously, and it may contain greater or lesser amounts of hydrogen ions, which may contribute to fatigue.)

◆ **FIGURE 8.16 A muscle twitch**

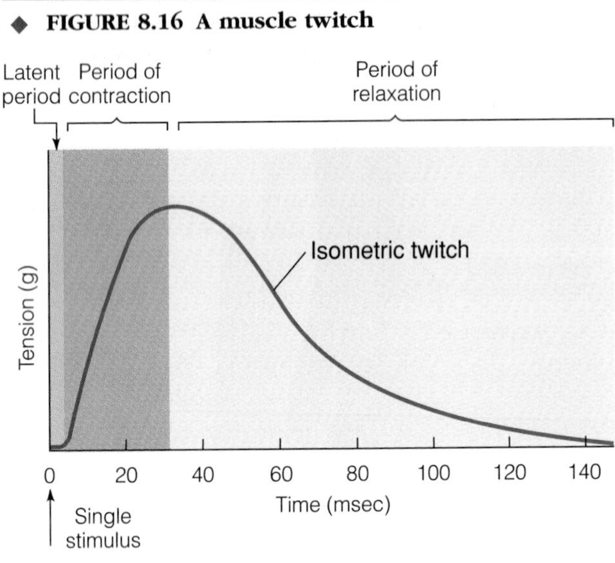

Even though individual skeletal muscle fibers contract to the maximum extent possible for the existing conditions, a muscle as a whole responds in a graded fashion to meet the demands of the task at hand. For example, the biceps brachii muscle of the arm does not contract to the same extent when a person lifts a feather as when the person lifts a bowling ball. A smooth, graded response by a muscle depends on such factors as the number of motor units activated at any particular time, the frequency of nerve impulses, and the asynchronous activation of different motor units.

Multiple Motor Unit Summation. The term **multiple motor unit summation** (spatial summation) refers to the ability of the individual motor units of a muscle to combine their simultaneous activities to influence the degree of contraction of the entire muscle. If only a few motor units are active at any one time, the muscle contraction will be relatively weak. If many motor units are active, the muscle contraction will be relatively strong.

Wave Summation and Tetanus. Following the stimulation of a skeletal muscle fiber and the triggering of a propagated action potential that leads to the contraction of the fiber, there is a brief period of time during which a second stimulus will not produce another propagated action potential. This period is called the *refractory period.* The refractory period usually lasts only 1 or 2 milliseconds and ends well before the actual contraction of the muscle fiber (which can take 20 to 100 milliseconds) is completed. Consequently, a skeletal muscle fiber can be stimulated to contract a second time before it has relaxed from an initial contraction, a third time before it has relaxed from the second contraction, and so on.

In the body, the neurons that supply the muscle fibers of the motor units of a skeletal muscle normally do not transmit just a single stimulatory impulse that would produce only a twitch-type response. Instead, the neurons transmit volleys of impulses to the fibers, with one impulse closely following another. Consequently, the muscle fibers are stimulated a second time before they have completely relaxed from their initial contractions, a third time before they have relaxed from their second contractions, and so on. This results in a summation of individual contractions, or twitches, called **wave summation** (temporal summation) that can create a state of more-or-less sustained contraction called **tetanus** (Figure 8.17). If the successive stimuli arrive far enough apart in time that the muscle is allowed to relax partially between stimuli, a condition of *incomplete tetanus* is seen. If the stimuli arrive so rapidly that no relaxation of the muscle occurs between stimuli, a condition of *complete tetanus* exists.

◆ **FIGURE 8.17 Record of response of an isolated skeletal muscle stimulated with increasing frequencies of stimuli of sufficient intensity to produce a maximal response from the muscle**

(a) Single muscle twitches. (b) Partial or incomplete tetanus (summation of twitches). (c) Complete tetanus.

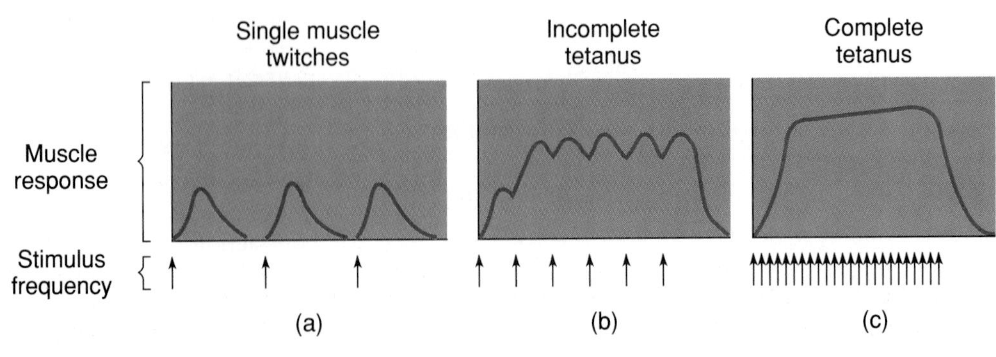

◆ **FIGURE 8.18 Asynchronous motor unit summation**
The maintenance of a nearly constant tension in an entire muscle is a result of the asynchronous activity of individual motor units of the muscle.

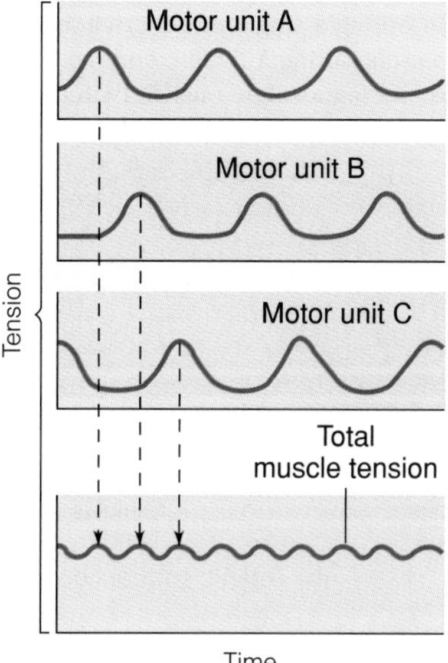

Asynchronous Motor Unit Summation. A single muscle, such as the biceps brachii of the arm, consists of many motor units, and a sustained, coordinated contraction of the muscle is due in part to the asynchronous activation of different motor units (Figure 8.18). Some motor units are activated initially; then they relax while other motor units are activated; then these relax and still other motor units become active. The result is a smooth, sustained contraction of the muscle as a whole.

Factors Influencing the Development of Muscle Tension

The amount of force, or tension, developed by a skeletal muscle is influenced by the composition of the muscle and by the length of the muscle at the time it contracts.

Contractile and Series Elastic Elements

A muscle contains both contractile elements and series elastic elements. The **contractile elements** are those structures actively involved in contraction, such as the thick and thin filaments of the muscle fibers. The **series elastic elements** are structures that resist stretching (but can be stretched) that are located between the contractile elements and the load. They include connective tissue and some parts of the muscle fibers themselves. When a muscle contracts, the force generated by the contractile elements stretches the series elastic elements, which, in turn, exert force on the load.

Influence of Series Elastic Elements on Muscle Tension. Because of the presence of series elastic elements, the tension developed by the contractile elements during a muscle contraction—that is, the internal tension or active state of the muscle fibers—is not always equivalent to the external tension exerted on the load. When a muscle is stimulated, the formation of attachments between the thick and thin filaments and the development of internal tension by the contractile elements of the muscle occur rather rapidly. However, following the formation of attachments between the thick and thin filaments and the development of internal tension, some time and effort are required to stretch or take up any slack in the series elastic elements of the muscle. During a single muscle twitch, the internal tension reaches its peak and decreases before the series elastic elements are stretched to a tension equal to the maximum internal tension. As a result, less than the full internal tension is transmitted to the load (Figure 8.19a). During a tetanic contraction, however, the repeated stimulation of the muscle maintains the internal tension long enough for the series elastic elements to be stretched to a tension similar to the internal tension (Figure 8.19b). As a result, more of the internal tension is transmitted to the load during a tetanic contraction than during a single muscle twitch, and the response of a muscle during a tetanic contraction is greater than that during a single muscle twitch (see also Figure 8.17).

Active and Passive Tension. Because a muscle contains both contractile and elastic elements, it can develop two types of tension: active tension and passive tension. **Active tension** is the tension due to the activity of the contractile elements when a muscle is stimulated and contracts. **Passive tension,** in contrast, is a consequence of a muscle's elasticity, and a muscle does not have to contract to exert passive tension. When a skeletal muscle is stretched so that it lengthens, it behaves much like a spring or rubber band. Within limits, the more it is stretched, the greater the passive tension it exerts.

Influence of Length on the Development of Muscle Tension

The amount of active tension developed by a skeletal muscle fiber when it contracts varies with the length of the fiber at the time of contraction (Figure 8.20). This relation between active tension and fiber length is related to the fact that the fiber contracts as the result of

◆ **FIGURE 8.19 Influence of series elastic elements on muscle tension**

(a) During a single muscle twitch, the internal tension developed by the contractile elements (the active state of the muscle fibers) reaches its peak and decreases before the series elastic elements are stretched to a tension equal to the maximum internal tension. As a result, less than the full internal tension is transmitted to the load.
(b) During a tetanic contraction, the repeated stimulation of the muscle maintains the internal tension long enough for the series elastic elements to be stretched to a tension similar to the internal tension. As a result, more of the internal tension is transmitted to the load during a tetanic contraction than during a single muscle twitch, and the response of a muscle during a tetanic contraction is greater than that during a single muscle twitch.

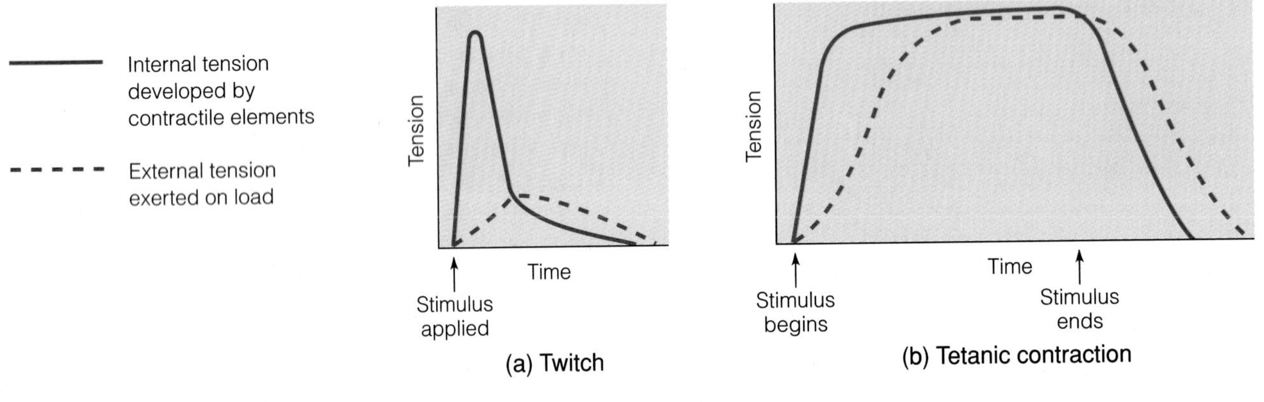

◆ **FIGURE 8.20 Active tension developed by a skeletal muscle fiber at different initial fiber lengths, expressed as a percent of sarcomere length**

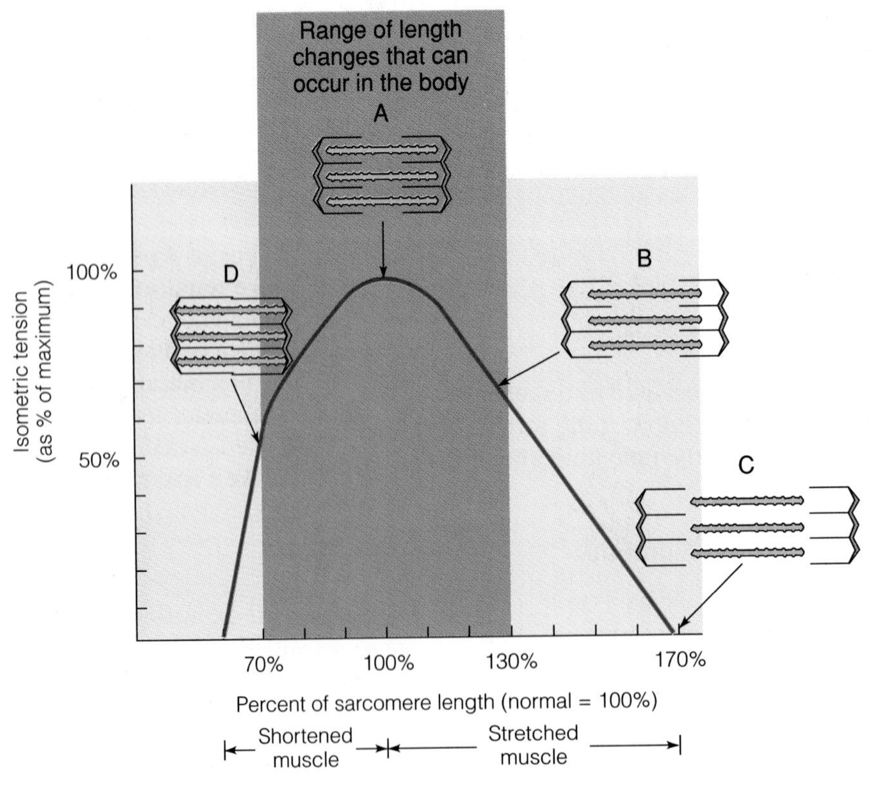

the formation of attachments between the thick and thin filaments of its sarcomeres. When the thin filaments completely overlap the portions of the thick filaments that possess cross bridges, more attachments can form than when the muscle fiber is stretched so there is less overlap. If a muscle fiber is stretched to the point where no overlap occurs, no attachments between thick and thin filaments can form, and the fiber cannot contract. On the other hand, when a muscle fiber is compressed, the thin filaments overlap and interfere with one another, and the thick filaments are forced against the Z lines of the sarcomeres. As a result, the active tension the muscle fiber can develop when the fiber contracts is less than maximal.

A whole skeletal muscle exhibits a similar relationship between its length and the amount of active tension it develops when it contracts (Figure 8.21). Each skeletal muscle has an optimal length from which it can develop the maximum active tension. In addition, as a muscle is lengthened—that is stretched—it exerts increasing amounts of passive tension, and the total tension exerted by a muscle when it contracts is equal to the sum of its active and passive tensions. In the body, the lengths of most relaxed, unstimulated muscles are such that they allow the development of maximal active tension when the muscles are stimulated.

Relation of Load to Velocity of Shortening

The greater the load on a muscle, the slower the velocity of shortening (Figure 8.22). With only a small load, the velocity of shortening is relatively fast. As the load increases, the velocity of shortening decreases until a point is reached at which the load is too great for the muscle to move. At this point, the velocity of shortening is zero, and the contraction of the muscle is isometric.

◆ FIGURE 8.21 Graph indicating the influence of muscle length on the tension developed by a skeletal muscle

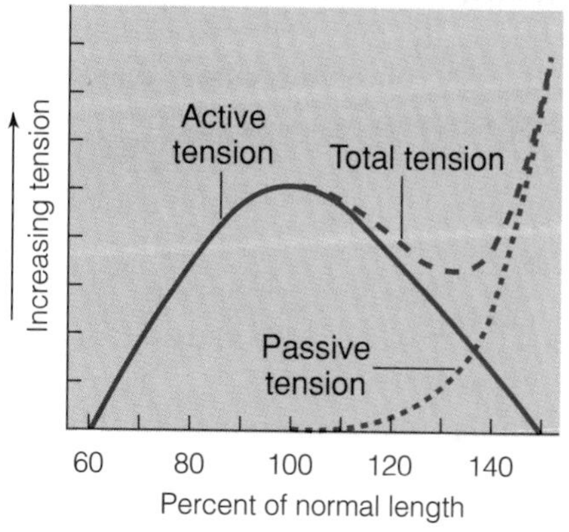

◆ FIGURE 8.22 Relation of load to velocity of shortening of a skeletal muscle

As the load increases, the velocity of shortening decreases.

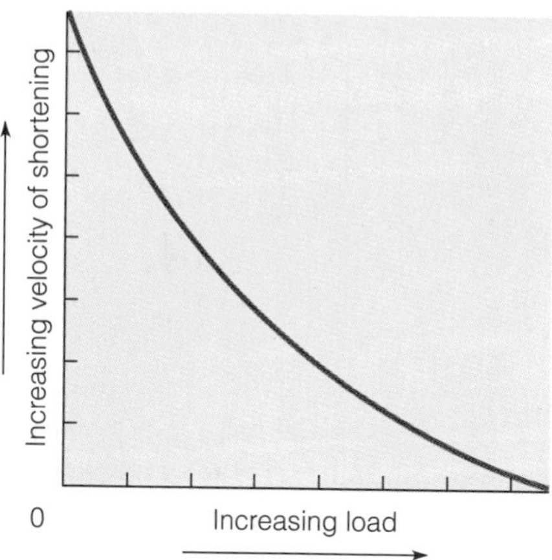

Muscle Actions

The contractions of skeletal muscles cause the various movements at the different joints, as described in Chapter 7. Some muscles pass in front of a joint and thus flex the bone to which they are attached; others pass behind a joint and extend the bone to which they are attached. Some muscles move a part away from the midline of the body, or abduct it; others move the part back toward the midline, or adduct it. Some muscles rotate the bones that form certain joints, and so forth.

In order to bring about these movements, muscles usually work in groups rather than individually. Those muscles whose contractions are primarily responsible for a particular movement are called **prime movers** or **agonists.** In any movement, there are always some muscles, generally situated on the opposite side of the joint, whose actions oppose the particular movement. These muscles, whose contraction offers resistance to the movement, are called **antagonists.** When prime movers contract and produce a movement, the antagonists are stretched. It is important to realize that a particular muscle is not always either a prime mover or an antagonist; rather, its role changes, depending on the movement that is being produced. For instance, when the forearm is flexed, the anterior muscles of the arm are the prime movers, and the posterior arm muscles are the antagonists. When the forearm is brought back to the anatomical position—or extended—the posterior muscles of the arm become the prime movers, and the anterior arm muscles are the antagonists.

In addition to prime movers and antagonists, most joint movements involve muscles that act as **synergists** *(sin´-er-jists).* Synergists are muscles that either directly assist the prime mover in a particular movement or prevent unwanted movements that often occur when a prime mover contracts. For example, flexion may be accompanied by an undesired rotation in the joint. In this case, the contraction of synergistic muscles may assist the prime-mover muscles by opposing the rotation. In a similar manner, if synergistic muscles did not act to immobilize the wrist and thus keep it from moving, the wrist would flex every time a person made a fist. This action would occur because the muscles that flex the fingers also pass anteriorly across the wrist. Therefore, some synergistic muscles may assist a movement by acting as antagonists.

When a synergist acts to immobilize a joint or an individual bone, it is referred to as a **fixator** *(fiks´-ay-tor)*. The muscles described in the previous paragraph as immobilizing the wrist were functioning as fixators. Many of the muscles that attach the scapula to the axial skeleton have important actions as fixators. The scapula is freely movable, and in order for it to serve as a firm origin for those muscles that move the arm, it must be held steady when the muscles contract. The contractions of fixators hold the scapula firmly against the thorax so that contractions of the arm muscles can move only their insertions, which are on the bones of the arm and forearm.

Although the prime movers cause the actual movements, the contractions of antagonistic and synergistic muscles are necessary to produce the smooth coordinated movements that are typical of a normal person. The strength of an antagonist's contraction affects the strength and the speed of the prime mover's contraction. For instance, if the extensor muscles of the forearm remain partially contracted—and therefore act as antagonists—while the flexor muscles of the forearm are causing the elbow to bend—and therefore are serving as prime movers—the flexor muscles have to contract harder to overcome the opposition, and the joint movement will be slower than it might otherwise be. The actions of antagonists and synergists make very fine and precise movements possible.

Relationship between Levers and Muscle Actions

The movements brought about by the actions of most skeletal muscles involve the use of levers. A **lever** is a rigid structure capable of moving around a pivot point, called a **fulcrum,** when a force is applied. In the body, the bones of the skeleton function as levers, the joints serve as fulcrums, and skeletal muscles provide the force to move the bones. Depending on the location of the fulcrum, a lever can make it possible to move heavier loads than could otherwise be moved, or to alter both the rate of movement and the distance over which a load can be moved.

Classes of Levers

There are three *classes of levers:* Class I, Class II, and Class III.

Class I Levers

In a **Class I lever,** the fulcrum is located between the point at which the force is applied and the load that is to be moved (Figure 8.23a). A seesaw is a common example of a Class I lever. In the body, this type of lever is used when the head is tipped back to raise the face: The occipital condyles on the atlas serve as the fulcrum, the facial portion of the skull is the load, and the force (pull) is applied to the back of the skull by the posterior muscles of the neck.

Class II Levers

In a **Class II lever,** the load to be moved is between the fulcrum and the point of force (Figure 8.23b). A wheelbarrow involves this type of lever. There are not many Class II levers in the body. The best example is raising the body on the toes. In this case, the base of the toes serves as the fulcrum, the toes support the load (body weight) and the contraction of the posterior muscles of the calf causes a force (pull) to be exerted on the calcaneous bone.

Class III Levers

In a **Class III lever,** the load is at one end, the fulcrum is at the other, and the force is applied between them (Figure 8.23c). Lifting a shovel utilizes this type of leverage. There are many examples of Class III levers in the body, since it is the most common lever system used. One example is flexion of the forearm: The load is at the wrist, the fulcrum is the elbow joint, and the force (pull) is exerted by the contraction of flexor muscles on the anterior of the arm that insert on the radius or the ulna, between the fulcrum and the load.

Effects of Levers on Movements

The portion of a lever located between the fulcrum and the point where a force is applied is called the **power arm;** the portion between the fulcrum and the load is the **load arm.** When the load arm is long in relation to

◆ **FIGURE 8.23 Classes of levers**

(a) Class I lever. The fulcrum (F) is located between the load (L), or resistance, and the force (P), or pull. Arrows indicate the direction of movement. (b) Class II lever. The load (L), or resistance, is located between the fulcrum (F) and the point of force (P), or pull. Arrows indicate the direction of movement. (c) Class III lever. The force (P), or pull, is applied between the fulcrum (F) and the load (L), or resistance. Arrows indicate the direction of movement.

(a) Class I lever

(b) Class II lever

(c) Class III lever

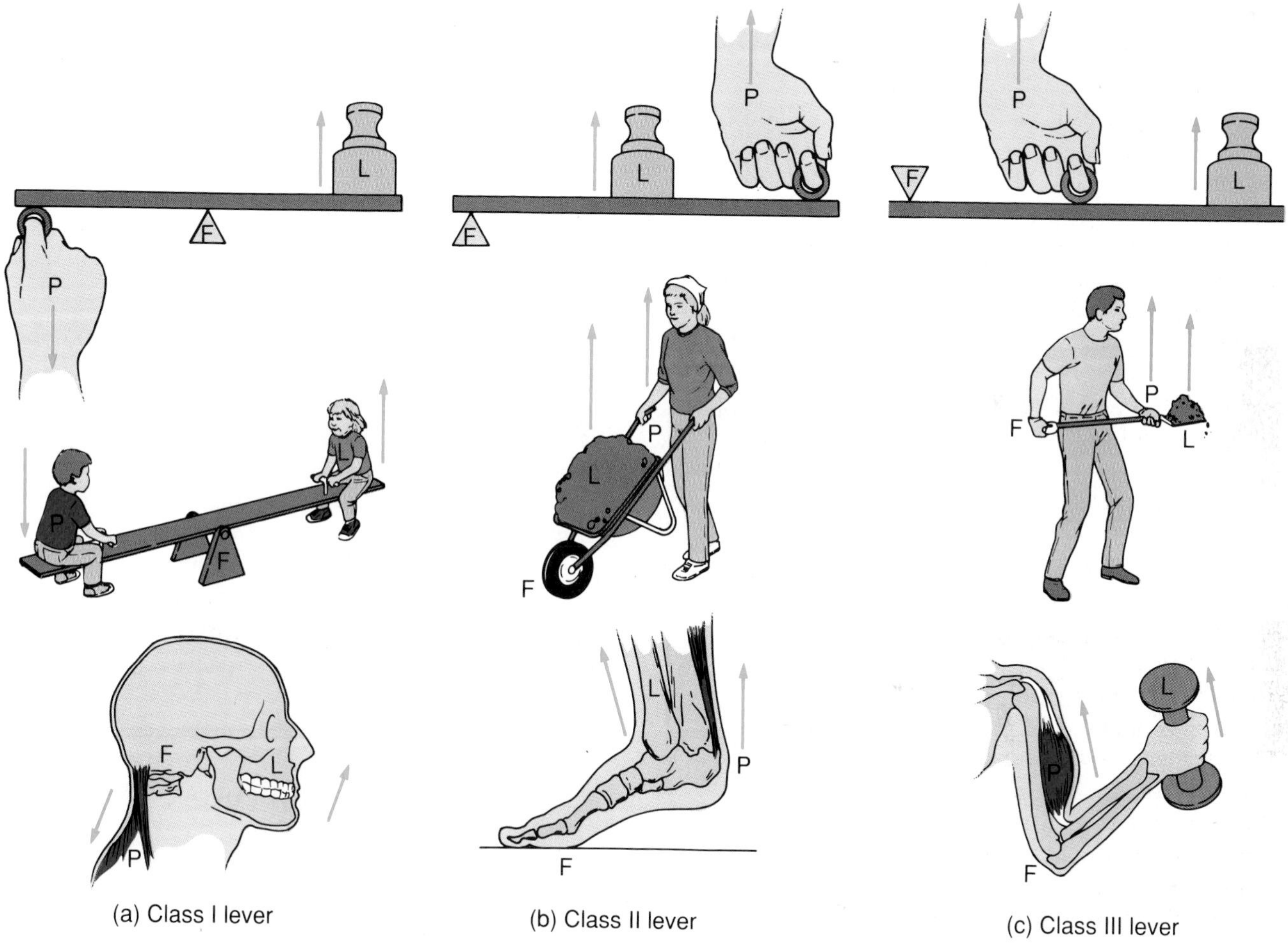

the power arm, a load can be moved rapidly over a considerable distance, but a strong force is required. Conversely, when the load arm is short in relation to the power arm, the same load can be moved with less force, but both the speed of movement and the range of movement are reduced. Thus, depending on the arrangement of the particular muscles and bones, the levers of the body may enable muscles to move loads faster over greater distance than would otherwise be possible, or they may enable muscles to move heavier loads than they otherwise could. For example, in the lever system involved in flexing the forearm, the muscles are inserted close to the fulcrum. Thus, this particular lever system has a short power arm and a long load arm, providing for fast and extensive movements characteristic of the forearm. In contrast, the lever system involved in raising the body on the toes has a power arm that is longer than the load arm, making it possible for the calf muscles to lift the entire body weight easily.

Types of Skeletal Muscle Fibers

Not all skeletal muscle fibers are identical. For example, the myosin molecules of some fibers split ATP more rapidly than the myosin molecules of other fibers. Because of this, cross-bridge cycling occurs more rapidly in some fibers than others, and some fibers contract faster than others. In addition, the metabolic processes by which muscle fibers produce the ATP necessary for contraction differ from one fiber to another, and the ability of a fiber to produce ATP influences its resistance to fatigue.

◆ **TABLE 8.2 Characteristics of Skeletal Muscle Fibers**

CHARACTERISTIC	SLOW TWITCH FATIGUE-RESISTANT (TYPE I)	FAST TWITCH FATIGUE-RESISTANT (TYPE IIa)	FAST TWITCH FATIGABLE (TYPE IIb)
Myosin-ATPase activity	Low	High	High
Speed of contraction	Slow	Fast	Fast
Resistance to fatigue	High	Intermediate	Low
Aerobic metabolic capacity	High	High	Low
Enzymes for anaerobic glycolysis	Low	Intermediate	High
Mitochondria	Many	Many	Few
Capillaries	Many	Many	Few
Myoglobin content	High	High	Low
Glycogen content	Low	Intermediate	High
Fiber diameter	Small	Intermediate	Large
Intensity of contraction	Low	Intermediate	High

Based on their contraction speed and fatigue resistance, three types of skeletal muscle fibers can be distinguished (Table 8.2).

Slow Twitch, Fatigue-Resistant Fibers

Slow twitch, fatigue-resistant fibers (type I fibers) contain myosin molecules that split ATP at a slow rate. Consequently, cross-bridge cycling occurs slowly, and the fibers contract at a slow speed. Slow twitch, fatigue-resistant fibers contain many mitochondria, and they have a highly developed capacity for aerobic (oxidative) metabolism. (The oxygen-utilizing reactions of aerobic metabolism take place within mitochondria.) The fibers are surrounded by many blood vessels, and they contain large amounts of an oxygen-binding protein known as **myoglobin** *(my-o-glo´-bin)*. Myoglobin increases the rate of oxygen diffusion into the fibers, and it provides them with small stores of oxygen. Slow twitch, fatigue-resistant fibers can meet their needs for ATP almost entirely through aerobic processes, and they are extremely resistant to fatigue.

Fast Twitch, Fatigue-Resistant Fibers

Fast twitch, fatigue-resistant fibers (type IIa fibers) also contain many mitochondria and have a highly developed capacity for aerobic metabolism. The fibers are well supplied by blood vessels, and they contain large amounts of myoglobin. However, they contain myosin molecules that split ATP at a rapid rate. Consequently, cross-bridge cycling occurs rapidly, and they contract at a fast speed. The highly developed oxidative processes of these fibers can supply most of their needs for ATP, and they are quite resistant to fatigue (although less so than slow twitch, fatigue-resistant fibers).

Fast Twitch, Fatigable Fibers

Fast twitch, fatigable fibers (type IIb fibers) contain myosin molecules that split ATP rapidly, and they contract at a fast speed. However, these fibers contain comparatively few mitochondria. The fibers are not well supplied by blood vessels, and they do not contain large amounts of myoglobin. Instead, they contain large stores of glycogen, and they are especially geared to the utilization of anaerobic metabolic processes. These processes are unable to supply the fibers continually with the amount of ATP they require, and they fatigue easily.

Utilization of Different Fiber Types

In some skeletal muscles, one type of skeletal muscle fiber predominates. However, most skeletal muscles contain all three fiber types, in proportions that vary from muscle to muscle. Consequently, muscles show a range of contraction speeds and fatigue resistances.

The presence of varying proportions of the three fiber types in different skeletal muscles is consistent with the following facts.

1. Not all skeletal muscles perform the same functions, and different functions often require different types of muscle contraction. For example, the postural muscles of the back and legs, which support the body against the force of gravity, must be able to undergo sustained contractions without fatigue, and the muscles that move the arms must be able to develop large amounts of tension quickly to allow the rapid lifting of heavy objects.

2. The same skeletal muscle often performs different functions at different times. For example, certain muscles of the legs may be involved in supporting the body against the force of gravity at one time, and they may participate in moving the legs during walking or running at another.

CLINICAL CORRELATION

Skeletal Muscle Contracture During Sustained Activity

Case Report

THE PATIENT: A 25-year-old male.

PRINCIPAL COMPLAINT: Extreme muscle stiffness during exercise.

HISTORY: The patient's birth and development were normal. His earliest recollection of a disorder of movement was at five years of age, when he fell during a foot face because his muscles stiffened. Thereafter, he was aware that muscle stiffness occurred during vigorous exercise or sudden, rapid movements. Strenuous exercise caused painless stiffness within seconds, and the stiffness was of such magnitude that the exercise could not be continued. However, the stiffness disappeared after only a few seconds of rest. Despite his problem, the patient was able to carry out daily activities, provided he paced his movements. If he was careful, he could even play tennis and swim. He managed to qualify for the college wrestling team by using "brute strength," but when speed was required, he lost competitive matches. Cold temperatures led to the rapid occurrence of stiffness, and the patient often noted clumsiness in his hands under these conditions. He had been accepted into military service, but subsequently was discharged because he could not perform "double time" marches. The patient denied weakness and had no history of myoglobinuria (myoglobin in the urine). He had no siblings, and no one in his known ancestry had a similar condition.

CLINICAL EXAMINATION: At examination, the patient appeared healthy and well developed. Abnormal findings were limited to the skeletal muscles. Under gross examination, the muscles were normal in contour, tone, and strength, but a progressive lengthening of the relaxation time was observed during vigorous exercise. After 10 to 15 seconds of repetitive contraction with maximal effort, the arm muscles became paralyzed in a contracted state, which was recognizable grossly. After 5 to 15 seconds of rest, the muscles relaxed again, and exercise could be resumed. The momentary contracture* was painless unless the patient continued his efforts to contract the shortened muscles. This phenomenon was identified in muscles of the limbs, the face, and the jaws. There were no symptoms of uncoordination or disturbance of gait, and passive movements did not induce the persistent shortening in any muscles. Electrical studies using external electrodes showed that the velocities of nerve impulse transmission were normal and that the muscle responded normally to either single or repetitive excitation. During the contracture induced by exercise, no electrical activity could be detected in the muscle.

A hollow needle was used to obtain samples of muscle tissue for study. Examination of samples by both light and electron microscopy disclosed no abnormal morphology, and biochemical analysis revealed a normal concentration of ATP. A sample of muscle was homogenized, and a cell fraction containing sarcoplasmic reticulum was obtained from the sample by ultracentrifugation. The uptake of calcium by the fraction was measured using radioactive calcium (^{45}Ca). The calcium uptake was found to be significantly less than the calcium uptake of control tissues.

COMMENT: The studies that were done suggest a defect in the uptake of calcium ions by the sarcoplasmic reticulum. Apparently, calcium ions accumulate in the cytoplasm of the muscle cells during repeated contractions until contracture is produced. The condition is made worse by cold, which further decreases the rate of uptake of calcium ions by the sarcoplasmic reticulum. At present, the cause of the abnormal function of the sarcoplasmic reticulum is not known.

OUTCOME: The patient's condition has not progressed, and despite the lack of any useful therapy, he lives a reasonably normal life. He accepts his limitations and avoids muscular activity that is rapid or prolonged enough to induce contracture.

*Contracture: a retarded relaxation of a muscle due to maintained force in the contractile apparatus of the muscle cells that is independent of any electrical activity in the plasma membranes of the cells.

CONDITIONS OF CLINICAL SIGNIFICANCE

Skeletal Muscle

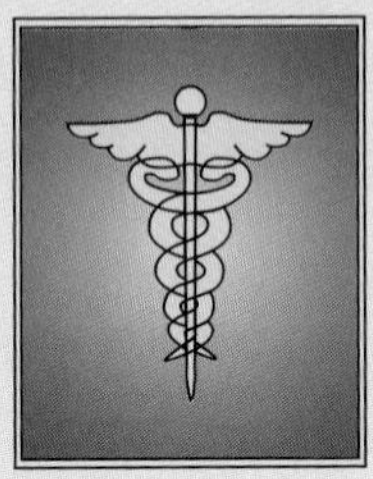

Muscle Atrophy

The shrinkage and death of muscle cells—*muscle atrophy*—cause a reduction in the size of the affected muscles. Atrophy of muscles can be caused by prolonged disuse or by a number of disorders, most of which reduce the blood supply or interfere with the nerve supply to the muscle. Muscle atrophy can be widespread or localized. Localized atrophy may not reduce the size of the muscle, however, since the unaffected cells undergo compensatory enlargement (hypertrophy).

Cramps

Cramps are painful, involuntary muscle contractions that are slow to relax. They can occur during exercise or at rest, and their precise cause is not known. Cramps may be caused by conditions within the muscle itself—for example, a low oxygen supply—or by stimulation from the nervous system. There is some evidence that cramps that occur during heavy exercise are caused by low blood levels of sodium and chloride ions, resulting from the loss of those ions by sweating. However, it is not clear whether the depletion of sodium chloride acts on the muscles or on the nervous system.

Muscular Dystrophy

The term *muscular dystrophy* refers to a group of diseases characterized by progressive muscular weakness. The weakness is the result of the degeneration of muscle fibers, an increase in connective tissue within the muscles, and, in some forms, the replacement of muscle fibers by fatty tissue. The muscular dystrophies are genetically transmitted. Some forms of muscular dystrophy are fatal, but other forms have a more favorable outlook.

Myasthenia Gravis

Myasthenia gravis is a chronic condition characterized by extreme muscle weakness. It is believed to be caused by an abnormal response of the body's immune system that disrupts acetylcholine receptors on the muscle cell membranes at the neuromuscular junctions. This decreases the responsiveness of the

Even though most skeletal muscles contain all three fiber types, all the fibers of any one motor unit are of the same type. Moreover, the different fiber types tend to be utilized in characteristic fashions when a muscle contracts. For example, during activities of short duration, if only a weak contraction is required from a muscle, just the slow twitch, fatigue-resistant fibers are activated. If a stronger contraction is required, the fast twitch, fatigue-resistant fibers are also activated, and if a still stronger contraction is necessary, the fast twitch, fatigable fibers are added.

Effects of Exercise on Skeletal Muscle

Regular exercise can produce increases in muscle size, strength, and endurance. However, different types of exercise cause different kinds of changes in the fibers of a skeletal muscle.

Aerobic or endurance-type exercise, such as distance running, leads to an increase in the number of capillaries around muscle fibers. The muscle fibers synthesize more myoglobin, and the number of mitochondria within the fibers increases. However, only relatively small increases in the diameters or strength of the fibers occur. Although all types of muscle fibers are affected to some degree, fast twitch, fatigue-resistant fibers show the greatest response.

Aerobic exercise produces cardiovascular and respiratory changes that enable muscles to be better supplied with substances such as oxygen and carbohydrates. Thus, the improved muscular performance that results from aerobic exercise is not due just to muscular changes, but to changes in other systems as well.

High-intensity, short-duration exercise, such as weight lifting, causes muscle fibers to hypertrophy (increase in size). This response is most evident in fast twitch, fatigable fibers. The fibers increase in diameter, and there is an increased synthesis of actin and myosin filaments that results in a large increase in the strength of the fibers.

CONDITIONS OF CLINICAL SIGNIFICANCE

muscle fibers to acetylcholine released from motor neuron endings. About 10% of the people who have this disease die from it. However, if an afflicted individual survives the first three years, there is a good chance that the condition will stabilize, with some degree of recovery.

Effects of Aging

One of the most obvious age-related changes in skeletal muscle is a slow but continuous reduction in skeletal muscle mass. The loss of mass is thought to be due to atrophy of muscle fibers. Because muscle fibers are postmitotic, new fibers are not formed to replace those lost. In fact, many atrophied muscle fibers are replaced by fat tissue, and islands of fat often develop between muscle cells in elderly persons. Up to 30% of the fibers in skeletal muscles may be lost by age 80. The loss of muscle fibers reduces the size of individual motor units in skeletal muscle. As a consequence, it becomes necessary to activate more motor units in order to move a particular weight, and thus moving the weight seems to require increased effort in older persons.

Accompanying the reduction in skeletal muscle mass is an age-related reduction in muscle strength. In most persons, the reduction in strength is minimal until about 70 years of age. The best defense against muscle atrophy and loss of strength is exercise. Regular moderate exercise slows the rate of muscular atrophy and can actually increase the strength of the muscles—even in quite elderly persons.

Muscle strength is generally associated with the quick and powerful contractions of fast twitch muscle fibers, and these fibers seem to atrophy earlier than do slow twitch fibers. Slow twitch fibers contract relatively slowly and maintain contractions for longer periods. These slow, somewhat prolonged contractions are used for such things as maintaining posture—which is not significantly, if at all, affected until very late in life.

The age-related decline in skeletal muscle mass and strength is partially associated with age-related changes that occur in the nervous system. For instance, a progressive loss of motor neurons occurs with aging, and there appears to be a reduction in the synthesis of acetylcholine by neurons. The loss of motor neuron innervation contributes to muscle fiber atrophy, and lowered levels of acetylcholine reduce the efficiency of muscle stimulation.

The amount of connective tissue between the fibers also increases, contributing to increased muscle size.

The increases in strength that occur with exercise do not always seem to be accounted for simply by increased muscle size. It has been suggested that in addition to increased muscle size, more nerve pathways that activate more motor units may be utilized. Normally, all the motor units of a muscle are not activated simultaneously. If they were, the muscle's tendons could be torn from their attachments. As exercise develops a muscle, however, perhaps more motor units can be activated simultaneously, thereby increasing strength.

Smooth Muscle

Smooth muscle fibers are uninucleate, spindle-shaped cells that are considerably smaller than skeletal muscle fibers. Smooth muscle fibers have no t tubules, and their sarcoplasmic reticulum is poorly developed. Smooth muscle fibers possess thick filaments that contain myosin and thin filaments that contain actin (but not troponin). In contrast to skeletal muscle fibers, which have 2 thin filaments for each thick filament, smooth muscle fibers have 10 to 15 thin filaments for each thick filament. Smooth muscle fibers also possess intermediate filaments, which are believed to serve as part of their cytoskeletal framework.

The thick and thin filaments of smooth muscle fibers are not organized into regularly ordered sarcomeres, and smooth muscle fibers are not striated. Structures called **dense bodies** are present within smooth muscle fibers. The dense bodies contain the protein alpha-actinin, which is also present in the Z lines of skeletal muscle fibers. The thin filaments are anchored either to the dense bodies or to the internal surface of the plasma membrane.

Smooth Muscle Arrangements

Two basic arrangements of smooth muscle—single unit and multiunit—occur in the body. Some smooth muscles display characteristics that are typical of one of the arrangements. Other smooth muscles exhibit characteristics that fall somewhere between the two arrangements.

Single-Unit Smooth Muscle

Single-unit smooth muscle, which is also called **visceral smooth muscle,** is the most common smooth muscle arrangement. Single-unit smooth muscle is present in small arteries and veins, the intestines, the uterus, and other structures. The cells of single-unit smooth muscle are connected by gap junctions at which electrical signals can spread from cell to cell. As a result, many cells respond as a unit to stimulation.

Single-unit smooth muscle is self-excitable and contracts without external stimulation. Occasionally, one cell of the unit becomes spontaneously stimulated, and this stimulus passes to other cells, causing the entire unit to contract.

The membrane potential of a cell that becomes spontaneously stimulated is not constant, but instead fluctuates even in the absence of external stimulation. These inherent fluctuations in membrane potential generally take the form of either pacemaker activity or slow-wave potentials.

Pacemaker activity is the result of automatic changes in the permeability of ion channels in the plasma membrane. In response to these permeability changes, alterations occur in the passive movements of ions across the membrane, and the altered ionic movements depolarize the membrane (Figure 8.24a). When the depolarization (the pacemaker potential) reaches a level called *threshold,* an action potential occurs.

Slow-wave potentials are the result of automatic, cyclical changes in the rate at which sodium ions are actively transported across the plasma membrane. These changes in sodium transport produce cyclical changes in the membrane potential (Figure 8.24b). If a cyclical depolarization reaches threshold, a burst of action potentials occurs. However, threshold is not always reached, and action potentials do not always occur with each slow-wave oscillation.

Multiunit Smooth Muscle

Multiunit smooth muscle, which is less common than single-unit smooth muscle, is present in large arteries, the large airways to the lungs, and some other structures. There are very few gap junctions in multiunit smooth muscle, and the individual muscle cells are usually sufficiently separated from one another so that a stimulus to one cell is not transferred to neighboring cells. Thus, each cell (or small group of cells) responds independently to stimulation.

Multiunit smooth muscle is generally not self-excitable and requires external stimulation to initiate a contraction. When muscle cell membranes are depolarized, a contractile response takes place. The degree of depolarization that occurs in response to appropriate stimulation varies with the intensity of the stimulation, but action potentials do not occur in most multiunit smooth muscles.

◆ **FIGURE 8.24 Self-generated electrical activity in smooth muscle**
(a) Pacemaker activity. (b) Slow-wave potentials.

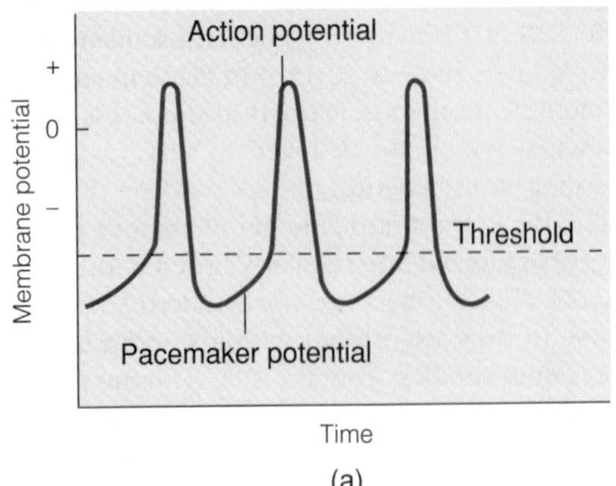

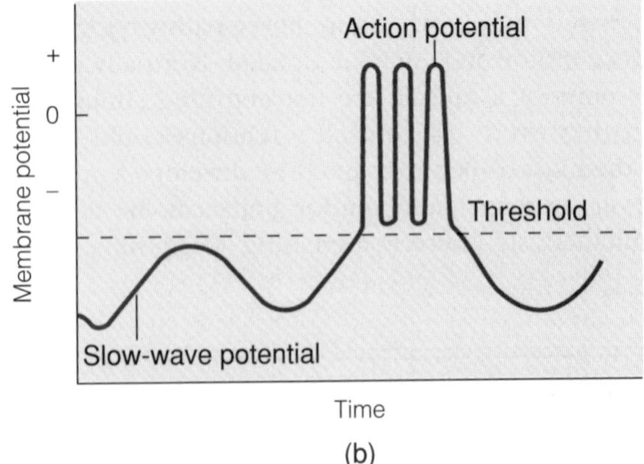

Influence of External Factors on Smooth Muscle Contraction

Smooth muscle contraction is influenced by external factors that include neural activity and chemical substances such as hormones. As just mentioned, external factors are generally required to initiate the contraction of multiunit smooth muscle. Moreover, external factors modulate—that is, enhance or inhibit—the frequency and intensity of the spontaneous contractions of single-unit smooth muscle.

In general, factors that influence smooth muscle contraction cause changes in the concentration of calcium ions in the cytosol of smooth muscle fibers. In most cases, changes in cytosolic calcium concentration are brought about by mechanisms that produce changes in the membrane potential of the muscle fibers. In some cases, however, changes in cytosolic calcium concentration are brought about by mechanisms that do not produce changes in membrane potential.

The responses of different smooth muscles to particular factors vary greatly, and a factor that stimulates one smooth muscle may inhibit another. As particular smooth muscles are encountered in later portions of the text, their responses to specific substances will be considered.

Smooth Muscle Contraction

Smooth muscle uses cross-bridge movements between myosin and actin to generate force, and it is generally believed that smooth muscle contraction occurs by a sliding filament mechanism similar to that in skeletal muscle. Smooth muscle contraction is triggered by calcium ions obtained from two sources—the sarcoplasmic reticulum and the extracellular fluid—and most smooth muscle fibers rely on both sources to some extent. When a smooth muscle fiber is stimulated, calcium channels in the sarcoplasmic reticulum and the plasma membrane open, and calcium ions enter the cytosol from the sarcoplasmic reticulum and the extracellular fluid. The calcium ions trigger smooth muscle contraction by binding to a protein called **calmodulin,** which then activates an enzyme called **myosin light chain kinase.** The activated myosin light chain kinase phosphorylates (adds phosphate groups to) light chains of myosin molecules of the thick filaments. This phosphorylation leads to the interaction of the thick myosin-containing filaments and the thin actin-containing filaments, and the muscle fiber contracts.

Smooth muscle relaxation occurs when the calcium ions are actively transported back into the sarcoplasmic reticulum or across the plasma membrane and out of the fiber. The removal of calcium ions from the cytosol leads to the inactivation of the myosin light chain kinase, with the result that myosin light chains cease being phosphorylated. Moreover, another enzyme within the muscle fiber, called **myosin light chain phosphatase,** removes phosphate groups from myosin light chains. This dephosphorylation prevents the interaction of the thick myosin-containing filaments and the thin actin-containing filaments, and the muscle fiber relaxes.

Overall, the contraction and relaxation of smooth muscle fibers depend on the relative activities of the enzymes myosin light chain kinase and myosin light chain phosphatase. When a smooth muscle fiber is stimulated and calcium ions are present in the cytosol in relatively high concentrations, the phosphorylating activity of myosin light chain kinase is greater than the dephosphorylating activity of myosin light chain phosphatase. Consequently, myosin light chains tend to have phosphate groups attached, and the muscle fiber contracts. Following stimulation, when calcium ions are present in lower concentrations and the myosin light chain kinase is inactive, the dephosphorylating activity of myosin light chain phosphatase is dominant. As a result, myosin light chains lose their phosphate groups, and the muscle fiber relaxes.

Smooth Muscle Contraction Speed and Energy Supply

Per unit of cross-sectional area, smooth muscle can generate the same amount of contractile tension as skeletal muscle. However, smooth muscle consumes ATP less rapidly than skeletal muscle, and it contracts more slowly than skeletal muscle. The myosin molecules of smooth muscle split ATP very slowly, and cross-bridge cycling and filament sliding occur very slowly.

Smooth muscle does not contain a rapidly available energy reserve such as creatine phosphate, but such a reserve is not required because smooth muscle does not consume ATP rapidly when it contracts. Smooth muscle generally receives enough nutrients and oxygen to enable it to meet its relatively modest needs for ATP by aerobic metabolic processes. If oxygen delivery is inadequate, and aerobic metabolism cannot supply enough ATP, smooth muscle can produce additional ATP by anaerobic processes. In fact, some smooth muscles can produce sufficient ATP for contraction anaerobically, and even when enough oxygen is available, the production of lactic acid is still high.

Calcium ions are actively transported out of the smooth muscle cytosol more slowly than calcium ions are actively transported out of skeletal muscle cytosol. Consequently, smooth muscle relaxes more slowly than skeletal muscle, and a cycle of smooth muscle contraction and relaxation can last 3000 milliseconds (3 seconds) compared to a skeletal muscle twitch, which generally lasts less than 100 milliseconds.

Stress-Relaxation Response

Smooth muscle can be stretched much more than can skeletal muscle before exhibiting any marked changes in tension. When a smooth muscle is suddenly stretched, it initially displays an increased tension. This tension begins to decrease almost immediately, however, and within a few minutes it has returned to its original level. This phenomenon, which is called **stress-relaxation,** is important because many smooth muscles are located in the walls of hollow structures such as the stomach, intestines, or urinary bladder. As these structures become filled, they enlarge, stretching their walls and the smooth muscles associated with them. Within limits, however, the stress-relaxation response of smooth muscle allows these hollow structures to enlarge with little appreciable change in the pressures exerted on their contents.

Ability to Contract When Stretched

Smooth muscle is able to contract and actively generate tension under greater stretching than skeletal muscle. This ability is at least partly explained by the fact that the thick and thin filaments of smooth muscle fibers are not organized into a regular pattern of sarcomeres but rather exist in a less ordered array. As a result, the problem of stretching smooth muscle fibers to the point where thick and thin filaments no longer overlap one another is less serious than in skeletal muscle cells. At many degrees of stretch, some thick and thin filaments still overlap one another, and the fibers can still contract.

The ability to contract when stretched is important to smooth muscles that are associated with hollow structures. During the filling and distension of these structures, the smooth muscles associated with them

◆ **TABLE 8.3 Comparison of Muscle Types**

CHARACTERISTIC	SKELETAL	SINGLE-UNIT SMOOTH	MULTIUNIT SMOOTH	CARDIAC
Location	Attached to skeleton (most)	Walls of hollow organs in digestive, reproductive, and urinary tracts and in small blood vessels	Large blood vessels, eye, and hair follicles	Heart only
Function	Movement of body in relation to external environment	Movement of contents within hollow organs	Varies with structure involved	Pumps blood out of heart
Innervation	Somatic nervous system	Autonomic nervous system	Autonomic nervous system	Autonomic nervous system
Level of control	Under voluntary control; also subject to subconscious regulation	Under involuntary control	Under involuntary control	Under involuntary control
Role of nervous stimulation	Initiates contraction; accomplishes gradation	Modifies contraction; can excite or inhibit; contributes to gradation	Initiates contraction; contributes to gradation	Modifies contraction; can excite or inhibit; contributes to gradation
Modifying effect of hormones	No	Yes	Yes	Yes
Presence of thick myosin and thin actin filaments	Yes	Yes	Yes	Yes
Striated due to orderly arrangement of filaments	Yes	No	No	Yes
Presence of troponin	Yes	No	No	Yes
Presence of t tubules	Yes	No	No	Yes (large)

undergo considerable stretching. Despite this stretching, the muscles maintain the ability to contract and actively generate tension.

Degree of Shortening During Contraction

When smooth muscles contract, they can shorten far more as a percentage of their length than skeletal muscles can. The useful distance of contraction for a skeletal muscle is about 25% to 35% of the length of the muscle, whereas a smooth muscle can contract from twice its normal length to one-half its normal length. This allows structures such as the stomach, intestines, and urinary bladder to vary the diameters of their lumens (cavities) from considerably large values to pratically zero.

Smooth Muscle Tone

In many single-unit smooth muscle cells, enough calcium ions are present in the cytosol to maintain a low level of muscle tension, even in the absence of action potentials. When action potentials occur, additional calcium ions enter the cytosol, and the slow contraction and relaxation response typical of smooth muscle is superimposed on the low level of tension.

The sustained low level of tension is called **smooth muscle tone,** and it is important in the cardiovascular system, where the smooth muscles of blood vessels called arterioles must maintain some degree of contraction for long periods. Also, the tonic contraction of smooth muscles of the intestine exerts a steady pressure on the intestinal contents.

Table 8.3 compares and summarizes the characteristics of skeletal, smooth, and cardiac muscle.

◆ **TABLE 8.3 Comparison of Muscle Types (continued)**

CHARACTERISTIC	SKELETAL	SINGLE-UNIT SMOOTH	MULTIUNIT SMOOTH	CARDIAC
Level of development of sarcoplasmic reticulum	Well developed	Poorly developed	Poorly developed	Moderately developed
Cross bridges turned on by Ca^{+2}	Yes	Yes	Yes	Yes
Source of increased cytosolic Ca^{+2}	Sarcoplasmic reticulum	Extracellular fluid and sarcoplasmic reticulum	Extracellular fluid and sarcoplasmic reticulum	Extracellular fluid and sarcoplasmic reticulum
Site of Ca^{+2} regulation	Troponin in thin filaments	Myosin in thick filaments	Myosin in thick filaments	Troponin in thin filaments
Mechanism of Ca^{+2} action	Physically repositions troponin-tropomyosin complex to uncover actin cross-bridge binding sites	Chemically brings about phosphorylation of myosin cross bridges so they can bind with actin	Chemically brings about phosphorylation of myosin cross bridges so they can bind with actin	Physically repositions troponin-tropomyosin complex
Gap junctions present	No	Yes	Yes (very few)	Yes
ATP used directly by contractile apparatus	Yes	Yes	Yes	Yes
Myosin-ATPase activity; speed of contraction	Fast or slow, depending on type of fiber	Very slow	Very slow	Slow
Means by which gradation accomplished	Varying number of motor units contracting (multiple motor unit summation) and frequency at which they are stimulated (wave summation)	Varying cytosolic Ca^{+2} concentration by inherent, spontaneous activity and via influences by autonomic nervous system, hormones, mechanical stretch, and local metabolites	Varying number of muscle fibers contracting and varying cytosolic Ca^{+2} concentration in each fiber by autonomic and hormonal influences	Varying lengths of fibers (depending on extent of heart filling) and varying cytosolic Ca^{+2} concentration through autonomic, hormonal, and local metabolite influence

Study Outline

◆ MUSCLE TYPES p. 244

Skeletal Muscle. Most attached to skeleton; striated cells; voluntary control; regulated by somatic nervous system.

Smooth Muscle. In walls of hollow organs and tubes; cells lack striations; involuntary control; regulated by factors intrinsic to muscle itself, hormones, and autonomic nervous system.

Cardiac Muscle. Forms wall of heart; striated cells; involuntary control; regulated by factors intrinsic to muscle itself, hormones, and autonomic nervous system.

◆ EMBRYONIC DEVELOPMENT OF MUSCLE p. 244

Skeletal Muscle.

1. Except for muscles of head and limbs, skeletal muscles develop from embryonic masses of mesoderm called somites—located dorsally along axial skeleton.
2. Myotome—that portion of a somite that differentiates into muscle cells.
3. Head musculature develops from general mesoderm of that region.
4. Limb musculature develops from mesodermal condensations within embryonic limb buds.
5. Individual muscle cells are called muscle fibers.
6. Myoblasts give rise to skeletal muscle fibers.

Smooth Muscle. Mesodermal cells migrate to embryonic digestive tube and body organs and surround them in a thin layer.

Cardiac Muscle. Mesodermal cells migrate to and surround early tubular heart.

◆ GROSS ANATOMY OF SKELETAL MUSCLES pp. 245–246

Connective-Tissue Coverings. Skeletal muscle is composed of muscle fibers held together by thin sheets of fibrous connective tissue called fascia.

EPIMYSIUM. Fascia that encases entire muscle.

PERIMYSIUM. Penetrates muscle; separates fibers into bundles called fasciculi.

ENDOMYSIUM. Thin extensions of fascia; envelop plasma membrane of each muscle fiber.

Skeletal Muscle Attachments. Extensions of endomysium, perimysium, and epimysium may directly attach to bone or may blend into strong fibrous connection called a tendon.

TENDONS. Extensions of connective tissue beyond end of muscle; length varies; continuous with periosteum.

APONEUROSES. Broad, thin tendon sheets.

ORIGIN. Less movable end of muscle; usually proximal.

INSERTION. More movable end of muscle; usually distal.

Skeletal Muscle Shapes. Fasciculi may run **parallel** to long axis of muscle, producing considerable movement but little strength; or fasciculi may insert **diagonally** into a tendon running length of muscle, producing less movement but greater power.

UNIPENNATE. All fasciculi insert on one side of tendon.

BIPENNATE. Fasciculi insert on both sides of tendon.

MULTIPENNATE. Convergence of several tendons.

CONVERGENT. Fasciculi converge from broad origin to single narrow tendon.

External openings are surrounded by **circular** muscles.

◆ MICROSCOPIC ANATOMY OF SKELETAL MUSCLE pp. 246–249

Skeletal muscle fibers (cells) contain:

Myofibrils. Longitudinal threadlike arrays of proteins; cross-striated because of alternating light and dark bands.

ANISOTROPIC (A, DARK) BANDS. Have less dense H zone in center; H zone is crossed by M line.

ISOTROPIC (I, LIGHT) BANDS. Have dense Z line crossing center; Z lines divide myofibrils into sarcomeres.

Sarcomeres. Repeating units of myofibrils; contain filamentous structures called myofilaments.

THICK FILAMENTS. Only in A band; H zone has only thick filaments; each thick filament surrounded by six thin filaments.

THIN FILAMENTS. I band; part of A band; attach to Z lines.

Composition of the Myofilaments.

1. *Thick filaments.* Composed mainly of protein myosin.
2. *Thin filaments.* Composed of proteins actin, tropomyosin, and troponin.

Transverse Tubules and Sarcoplasmic Reticulum.

TRANSVERSE TUBULES (t TUBULES). Invaginations of plasma membrane that extend deep into muscle cell.

SARCOPLASMIC RETICULUM. Membranous network that surrounds myofibrils.

◆ CONTRACTION OF SKELETAL MUSCLE pp. 249–263

1. *Isometric.* Muscle actively develops tension but does not shorten.
2. *Isotonic.* Muscle actively develops tension and shortens.

Contraction of a Skeletal Muscle Fiber. At cellular level, same basic events occur during both isometric and isotonic skeletal muscle contractions.

THE NEUROMUSCULAR JUNCTION. Ending of a motor neuron approaches specialized point along plasma membrane of skeletal muscle fiber, forming neuromuscular junction.

EXCITATION OF A SKELETAL MUSCLE FIBER. Nerve impulse reaches neuromuscular junction; acetylcholine is released; acetylcholine binds to receptors on muscle fiber membrane; binding changes permeability of muscle fiber plasma membrane at junction; propagated action potential produced.

EXCITATION-CONTRACTION COUPLING. Propagated action potential moves along t tubules to fiber interior; calcium ions released from sarcoplasmic reticulum; calcium ions bind to troponin molecules of thin filaments of sarcomeres; binding leads to interactions between thick and thin filaments and to muscle contraction.

MECHANISM OF CONTRACTION.

1. High-energy myosin formed when ATP is split.
2. When calcium ions bind to troponin molecules, tropomyosin molecules move aside.
3. High-energy myosin links with G-actin subunit.
4. Energy discharged from high-energy myosin; myosin head swivels and pulls on actin-containing filament.
5. When ATP again occupies myosin head, cycle repeats.

REGULATION OF THE CONTRACTILE PROCESS. Active transport mechanism returns calcium ions to sarcoplasmic reticulum; when calcium ions return to sarcoplasmic reticulum, tropomyosin returns to blocking position.

Sources of ATP for Muscle Contraction. Creatine phosphate can rapidly produce ATP. Aerobic metabolism and anaerobic breakdown of glucose and glycogen also produce ATP.

Muscle Fatigue. Can be defined as inability of muscle to maintain particular strength of contraction or tension (that is, particular power output) over time.

Oxygen Debt. Built up during periods of muscular activity when anaerobic sources of ATP are used to support muscle contraction; oxygen debt repaid by continued increased breathing after activity.

The Motor Unit. Single neuron and all muscle cells it supplies.

Responses of Skeletal Muscle. Contractions of muscle fibers of motor units combine to produce contractions of entire muscle.

MUSCLE TWITCH.

1. *Latent period.* Following arrival of stimulus at muscle.
2. *Period of contraction.* Active development of tension and, possibly, shortening.
3. *Period of relaxation.*

GRADED MUSCULAR CONTRACTIONS. When action potential occurs in membrane of skeletal muscle fiber, fiber contracts to maximum extent possible for existing conditions; however, entire muscle responds in graded fashion to meet demands of particular tasks.

Factors Influencing the Development of Muscle Tension.

1. *Contractile elements.* Those structures actively involved in contraction, such as thick and thin filaments of muscle fibers.
2. *Series elastic elements.* Structures that resist stretch (but can be stretched) located between contractile elements and load.

INFLUENCE OF LENGTH ON THE DEVELOPMENT OF MUSCLE TENSION. Greatest active tension develops at optimal muscle length, decreases when a muscle is stretched or compressed. Within limits, passive tension increases with stretch of the muscle.

Relation of Load to Velocity of Shortening. The greater the load on a muscle, the slower the velocity of shortening.

◆ MUSCLE ACTIONS pp. 263–264

Prime Movers. Those muscles whose contractions are primarily responsible for a particular movement.

Antagonists. Those muscles whose contraction offers resistance to the movement; generally on opposite side of joint.

Synergists. Those muscles that indirectly aid a particular movement by preventing unwanted movements; for example, steadying a joint.

FIXATOR. A synergist acting to immobilize.

◆ RELATIONSHIP BETWEEN LEVERS AND MUSCLE ACTIONS pp. 264–265

Lever. Rigid structure capable of moving around pivot point (fulcrum) when force is applied.

Classes of Levers.

CLASS I LEVERS. Fulcrum is between point of force and load to be moved; for example, a seesaw, or head tipping back to raise face.

CLASS II LEVERS. Load to be moved is between fulcrum and point of force; for example, a wheelbarrow, or raising body on toes.

CLASS III LEVERS. Load at one end, fulcrum at other, point of force between them; for example, shovel lifting, or flexion of forearm.

Effects of Levers on Movements.

POWER ARM. Portion of lever between fulcrum and point of force.

LOAD ARM. Portion of lever between fulcrum and weight.

LONG LOAD ARM. Rapid movement, but strong force required.

LONG POWER ARM. Slower movement, but less force required.

◆ TYPES OF SKELETAL MUSCLE FIBERS pp. 265–266

Three types, based on contraction speed and fatigue resistance.

Slow Twitch, Fatigue-Resistant Fibers. Myosin splits ATP at slow rate; slow speed of contraction; many blood vessels, much myoglobin; well-developed aerobic metabolic processes; very fatigue-resistant.

Fast Twitch, Fatigue-Resistant Fibers. Myosin splits ATP at rapid rate; fast speed of contraction; many blood vessels; much myoglobin; well-developed aerobic metabolic processes; quite fatigue-resistant.

Fast Twitch, Fatigable Fibers. Myosin splits ATP at rapid rate; fast speed of contraction; few blood vessels; little myoglobin; geared to anaerobic metabolic processes; easily fatigued.

◆ UTILIZATION OF DIFFERENT FIBER TYPES pp. 266–268

Most skeletal muscles contain all three fiber types; each type tends to contribute in characteristic fashion to a muscle contraction. All fibers of any one motor unit are of same type.

◆ CONDITIONS OF CLINICAL SIGNIFICANCE: SKELETAL MUSCLE pp. 268–269

Muscle Atrophy. Reduction in size of muscles due to shrinkage and death of muscle cells; may be widespread or localized.

Cramps. Involuntary, painful muscle contractions, slow to relax; may be caused by low oxygen supply in muscles or by nervous system stimulation; those that occur during heavy exercise may be due to low levels of sodium and chloride ions in blood.

Muscular Dystrophy. Genetically transmitted progressive muscular weakness resulting from muscle cell degeneration; increase in connective tissue or replacement of muscle cells by fatty tissue.

Myasthenia Gravis. Rare chronic condition of extreme muscle weakness; related to inability of muscle fiber plasma membrane at neuromuscular junction to respond to acetylcholine.

Effects of Aging. Progressive and continuous loss of muscle mass and strength.

◆ EFFECTS OF EXERCISE ON SKELETAL MUSCLE pp. 268–269

1. Muscle size and strength can increase with exercise.
2. Aerobic exercise leads to increased number of capillaries around muscle fibers; more myoglobin and mitochondria within fibers; and cardiovascular and respiratory changes.
3. High-intensity, short-duration-exercise causes muscle fiber hypertrophy, especially of fast twitch, fatigable fibers. Fibers increase in diameter, and there is increased synthesis of actin and myosin filaments.

◆ SMOOTH MUSCLE pp. 269–273

1. Differs from skeletal muscle.
2. Smaller cells; no regular cross striations.
3. No regularly ordered myofibrils, but does possess thin filaments that contain actin (but not troponin), and thick filaments that contain myosin.

Smooth Muscle Arrangements.

SINGLE-UNIT SMOOTH MUSCLE.

1. Cells connected by gap junctions; impulses spread from cell to cell.
2. Self-excitable; spontaneous contractions.
3. Present in small arteries, veins, intestines, uterus.

MULTIUNIT SMOOTH MUSCLE.

1. Very few gap junctions; stimulus to one cell usually not transferred to others; cells generally require external stimulation.
2. Present in large arteries; large airways to lungs.

Influence of External Factors on Smooth Muscle Contraction. Neural activity and chemical substances such as hormones can influence contraction.

Smooth Muscle Contraction.

1. Calcium source extracellular as well as from sarcoplasmic reticulum; calcium enters cytosol, is then pumped out.
2. Calcium ions bind to calmodulin, which then activates myosin light chain kinase.
3. Myosin light chain kinase phosphorylates myosin light chains of thick filaments, and muscle fiber contracts.
4. Relaxation occurs when calcium ions are actively transported out of cytosol; myosin light chain kinase becomes inactive, and myosin light chain phosphatase removes phosphate groups from myosin light chains.

Smooth Muscle Contraction Speed and Energy Supply.

1. Smooth muscle consumes ATP less rapidly than skeletal muscle and contracts more slowly than skeletal muscle.
2. Myosin molecules of smooth muscle split ATP very slowly, and cross-bridge cycling and filament sliding occur very slowly.
3. Smooth muscle does not contain a rapidly available energy reserve such as creatine phosphate.
4. Smooth muscle generally can meet its ATP needs by aerobic metabolic processes, and some smooth muscles can produce sufficient ATP anaerobically.

Stress-Relaxation Response. Does not increase tension when greatly stretched.

Ability to Contract When Stretched. Can contract and actively develop tension when stretched.

Degree of Shortening During Contraction. Can shorten by relatively large amounts.

Smooth Muscle Tone. Sustained low level of tension in many single-unit smooth muscles, even in absence of action potentials.

Self-Quiz

1. Skeletal muscles are voluntary muscles. True or False?
2. Skeletal muscle fibers: (a) exhibit cross striations; (b) are innervated by the autonomic nervous system; (c) are uninucleate.
3. Match the following terms with the appropriate lettered descriptions:

Epimysium	(a) The less movable end of a skeletal muscle
Perimysium	(b) When all the fasciculi insert onto one side of the tendon
Endomysium	(c) The fascia that envelops an entire muscle
Aponeuroses	(d) Tendons that take the form of long, thin sheets
Origin	(e) When the fasciculi of certain muscles have a complex arrangement involving the convergence of several tendons
Insertion	(f) Thin extensions of the fascia that envelop the plasma membrane of each muscle fiber
Unipennate	(g) The more movable end of a skeletal muscle
Bipennate	(h) Fascia that penetrates a muscle, separating the fibers into bundles
Multipennate	(i) Muscles that have fasciculi inserting obliquely on both sides of the tendon

4. The thick filament of a sarcomere of a skeletal muscle cell contains: (a) actin; (b) troponin; (c) myosin.
5. The threadlike molecules that lie along the surface of actin are: (a) myosin; (b) tropomyosin; (c) troponin.
6. Invaginations of the plasma membrane that extend into the interior of a skeletal muscle fiber form the: (a) t tubule network; (b) sarcoplasmic reticulum; (c) myofibers.
7. The membrane system of skeletal muscle fibers that contains calcium ions necessary for contraction is the: (a) sarcomere; (b) sarcoplasmic reticulum; (c) myofibril.
8. When acetylcholine binds to acetylcholine receptors on a skeletal muscle fiber plasma membrane at a neuromuscular junction, membrane channels that are permeable to small, positively charged ions open for a brief period. True or False?
9. During a skeletal muscle contraction, ATP occupies: (a) actin; (b) myosin; (c) troponin.
10. Tropomyosin acts as an enzyme to split ATP into ADP and phosphate. True or False?
11. In an unstimulated skeletal muscle fiber, the interaction of actin and myosin is believed to be directly blocked by: (a) calcium ions; (b) t tubules; (c) tropomyosin.
12. During a skeletal muscle contraction, calcium ions bind with: (a) tropomyosin; (b) myosin; (c) troponin.
13. During a contraction in which a skeletal muscle fiber shortens: (a) troponin shortens; (b) the Z lines are drawn closer together; (c) calcium binds with myosin.
14. During sustained, intense, rapid exercise, a skeletal muscle produces: (a) acetylcholine; (b) glycogen; (c) lactic acid.
15. Which substance provides skeletal muscle fibers with a means of rapidly forming ATP immediately following the initiation of muscular activity? (a) fatty acids; (b) creatine phosphate; (c) troponin.
16. Within a skeletal muscle: (a) there are more neurons than muscle fibers; (b) there are no neuromuscular junctions between nerve endings and muscle fibers; (c) each neuron branches to supply a number of muscle fibers.
17. An individual neuron, together with all of the skeletal muscle fibers it supplies, makes up a: (a) sarcomere; (b) motor unit; (c) myofibril.
18. When a nerve impulse triggers a propagated action potential in the membrane of a skeletal muscle fiber, the fiber contracts to the maximum extent possible for the existing conditions. True or False?
19. Fast twitch, fatigable skeletal muscle fibers are: (a) well supplied by blood vessels; (b) myoglobin rich; (c) geared for the utilization of anaerobic metabolic processes.
20. Smooth muscle fibers are: (a) multinucleate; (b) present in the walls of internal or visceral organs; (c) principally under voluntary control.
21. Which type of muscle is most likely to become spontaneously stimulated? (a) single-unit smooth muscle; (b) skeletal muscle; (c) multiunit smooth muscle.

CHAPTER 9

The Muscular System: Gross Anatomy

CHAPTER CONTENTS

LEARNING OBJECTIVES

After completing this chapter, you should be able to:

1. Cite several criteria used to name muscles, and give an example of each.
2. Distinguish between intrinsic and extrinsic muscles, citing several examples of each.
3. Name the major muscles of the head and neck, describing the origin, insertion, action and innervation of each.
4. Contrast the general functions of the neck muscles that are found in the posterior and anterior triangles.
5. Name the major muscles of the trunk, describing the origin, insertion, action, and innervation of each.
6. Name the major muscles that move the vertebral column.
7. Name the major muscles of the upper limbs, describing the origin, insertion, action and innervation of each.
8. Name the major muscles of the lower limbs, describing the origin, insertion, action, and innervation of each.

There are over 600 muscles within the human body (Figures 9.1 and 9.2). Only the more commonly studied muscles are included in this chapter. Several criteria are used to name muscles; each describes a particular characteristic of the muscle being named, such as its shape, action, or location. You will find it quite useful in your study of the muscles to familiarize yourself with these criteria:

◆ **Shape** The names of some muscles include references to their shape. For example, the trapezius muscles are shaped like trapezoids and the rhomboideus muscles resemble rhomboids.

◆ **Action** Various muscle names include references to the actions of the muscle by using the terms flexor, ex-

◆ **FIGURE 9.1 Anterior view of the muscles of the body**

The left external oblique muscle has been removed.

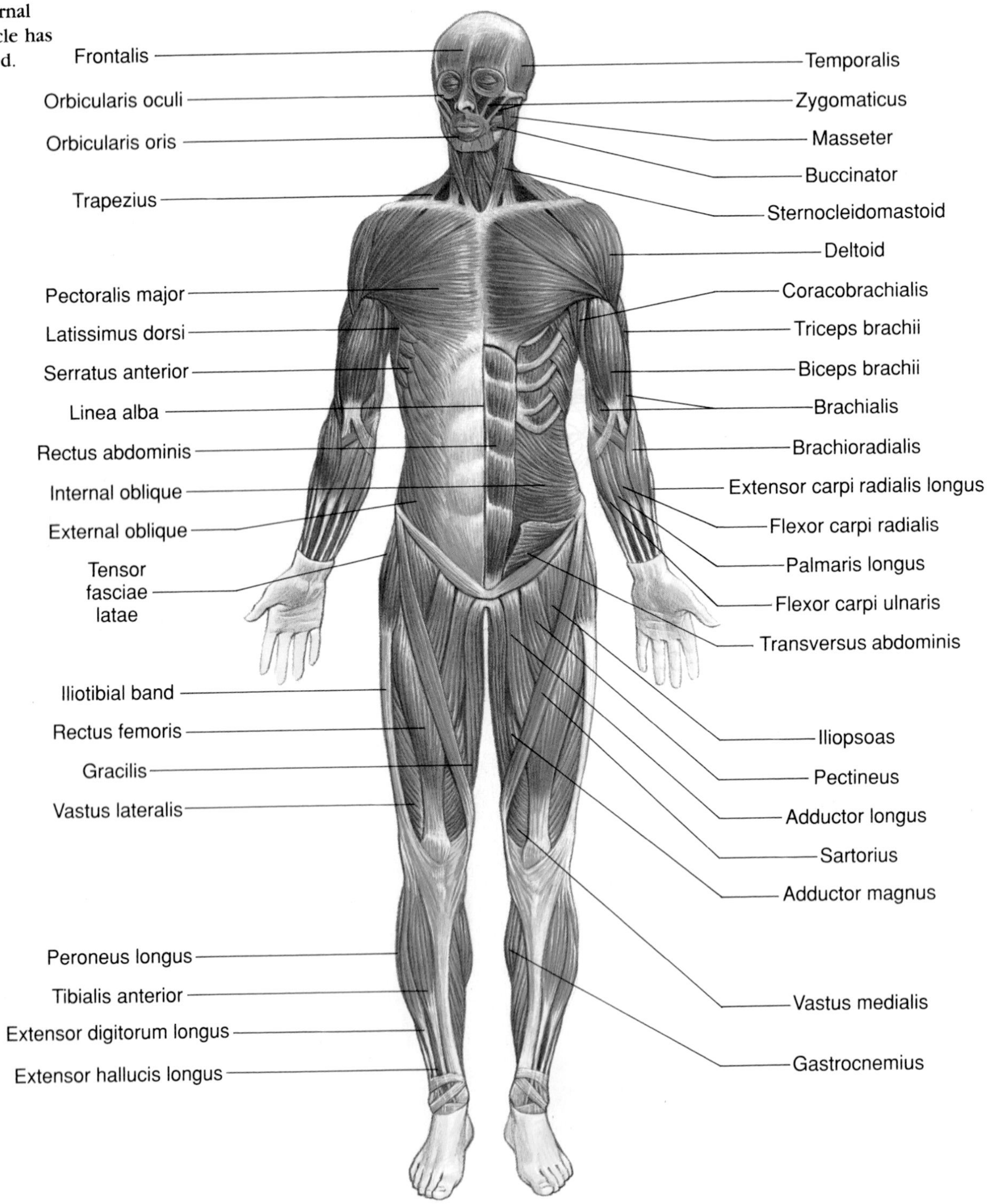

Frontalis
Orbicularis oculi
Orbicularis oris
Trapezius
Pectoralis major
Latissimus dorsi
Serratus anterior
Linea alba
Rectus abdominis
Internal oblique
External oblique
Tensor fasciae latae
Iliotibial band
Rectus femoris
Gracilis
Vastus lateralis
Peroneus longus
Tibialis anterior
Extensor digitorum longus
Extensor hallucis longus
Temporalis
Zygomaticus
Masseter
Buccinator
Sternocleidomastoid
Deltoid
Coracobrachialis
Triceps brachii
Biceps brachii
Brachialis
Brachioradialis
Extensor carpi radialis longus
Flexor carpi radialis
Palmaris longus
Flexor carpi ulnaris
Transversus abdominis
Iliopsoas
Pectineus
Adductor longus
Sartorius
Adductor magnus
Vastus medialis
Gastrocnemius

tensor, adductor, or pronator. For example, the flexor carpi radialis muscles flex the hands, and the extensor digitorum longus muscles extend the toes.

◆ **Location** It is possible to locate certain muscles by their names. For example, the intercostal muscles (*inter* = between; *costal* = rib) are located between the ribs, and the tibialis anterior muscles lie alongside the anterior margin of each tibia.

◆ **Attachments** The attachments of a muscle to the skeleton are included in some names. For example, the sternocleidomastoid muscles have origins on the sternum and clavicles and insert on the mastoid processes of the temporal bones; the coracobrachialis muscles have their origins on the coracoid processes of the scapulae and insert on each brachium—which refers to the arm (humerus).

◆ **Number of Divisions** Some muscles are separated into two, three, or four divisions, and this is indicated

◆ **FIGURE 9.2 Posterior view of the muscles of the body**

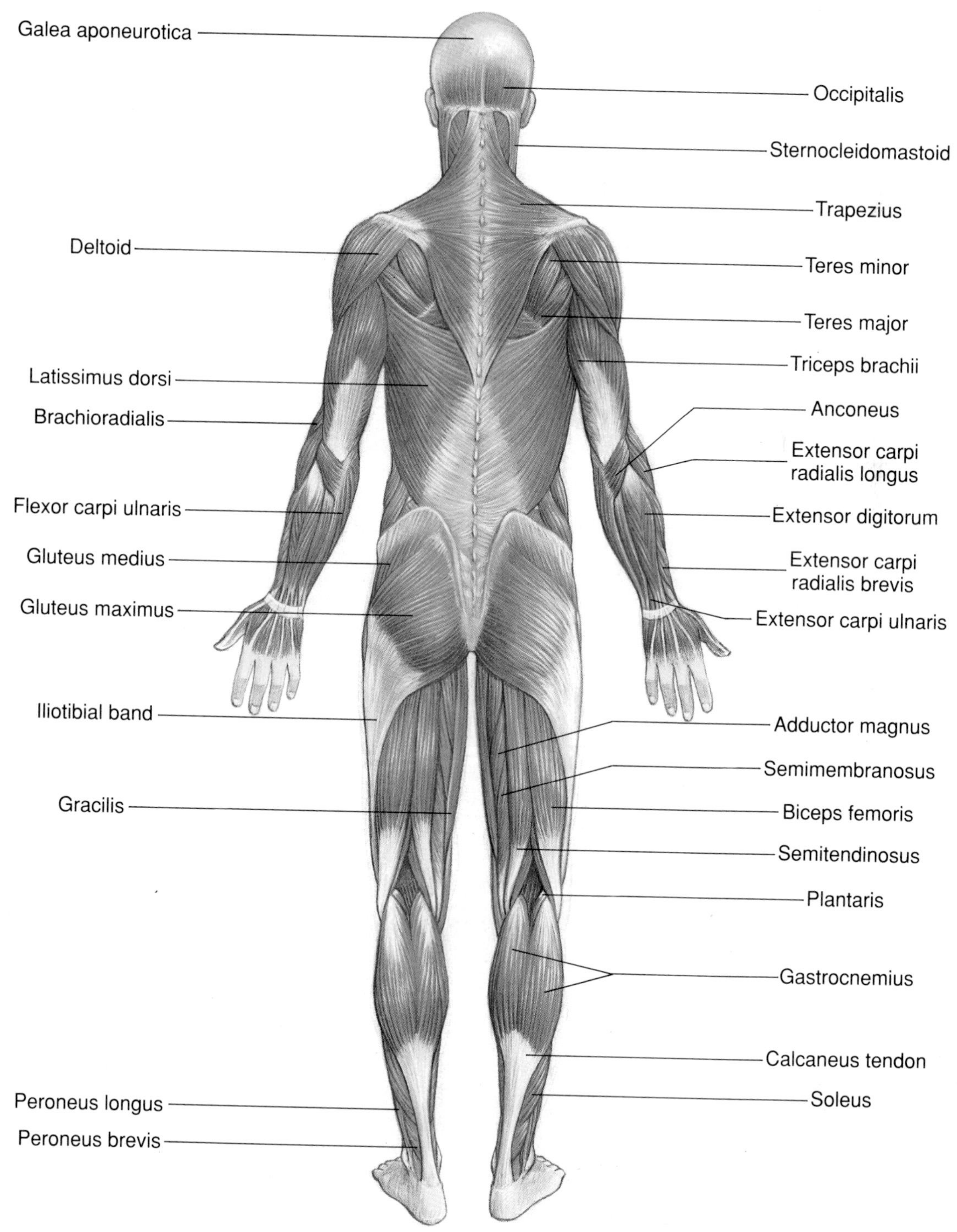

in their names. For example, the biceps brachii muscles have two divisions, the triceps brachii have three, and the quadriceps femoris muscles have four.

◆ **Size Relationships** Terms referring to size are often included in muscle names—for example, the gluteus maximus and gluteus minimus muscles of the buttocks are large and small, respectively, and the peroneus longus and peroneus brevis muscles of the leg are long and short, respectively.

In many cases, muscle names include more than one of these criteria. For example, the name of the flexor digitorum longus muscle indicates the muscle's action (flexion), its insertion (digits), and its size relationship (long, in comparison to the flexor digitorum brevis muscle).

To organize their study, we will consider the muscles in various groups (but note that some muscles belong to more than one group). Each group of muscles is discussed in a general way, and the reader is provided with figures that illustrate the muscles as well as tables that give a more detailed description of each muscle, including information on each muscle's origin, insertion, principal actions, and nerve innervation.

Muscles of the Head and Neck

Muscles of the Face

Because of their actions, the muscles of the face (Figure 9.3; Table 9.1) are also referred to as the muscles of facial expression. Whereas some facial muscles arise from the bones of the skull, others arise from the superficial fascia of the face. Most of them insert into the skin of the region, and therefore serve to move the skin

◆ **FIGURE 9.3 Muscles of the face and neck**

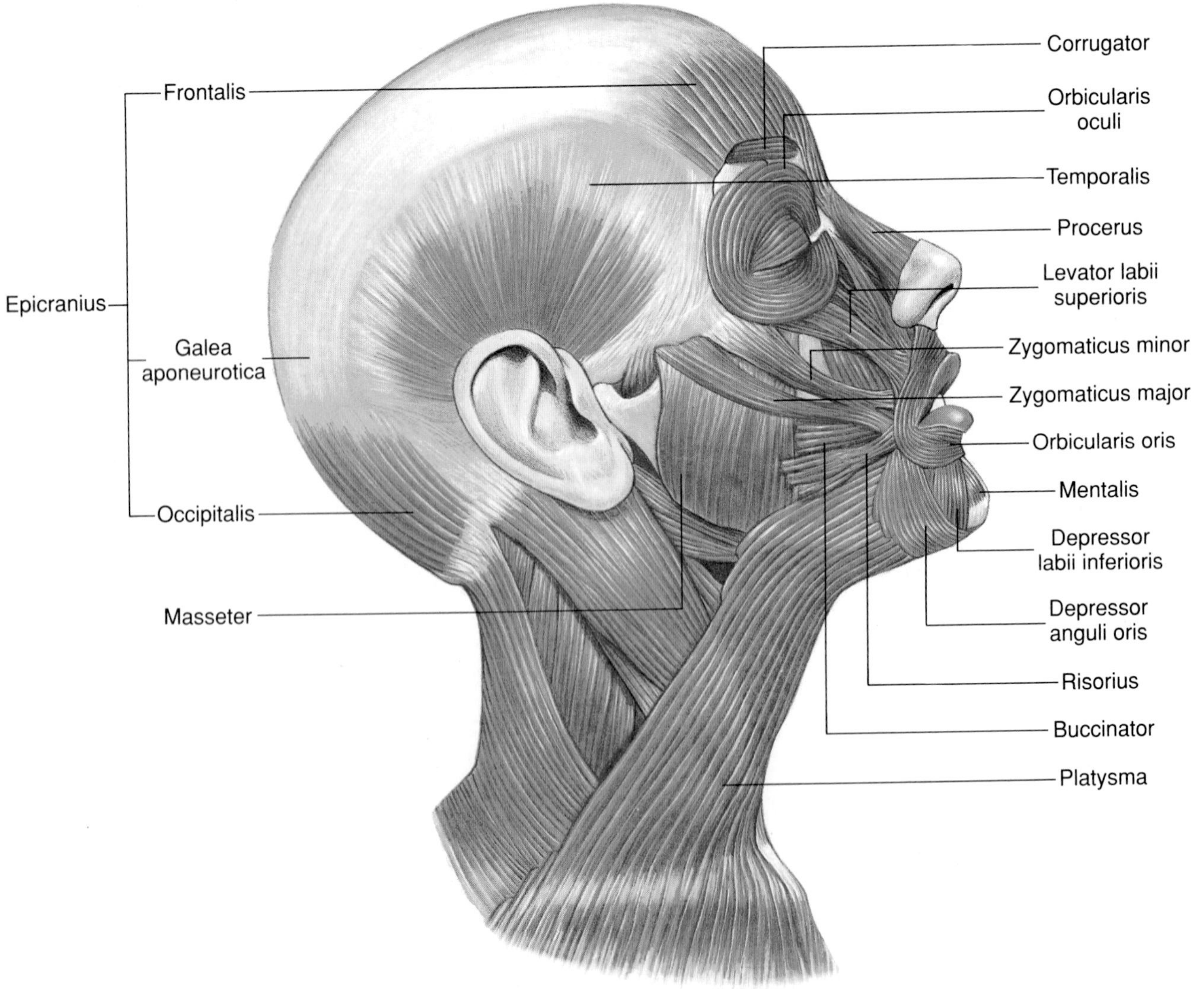

◆ **TABLE 9.1 Muscles of the Face [*Fig. 9.3*]**

MUSCLE	ORIGIN	INSERTION	ACTION	INNERVATION
Buccinator *(buk´-sĭ-na˝-tor)*	Alveolar process of the mandible and the maxillary bone	Orbicularis oris and skin at the angle of the mouth	Compresses cheek; pulls corner of the mouth laterally	Facial (cranial nerve VII)
Corrugator *(cor´-u-ga˝-tor)*	Frontal bone, lateral to the glabella	Skin of the eyebrows	Draws the eyebrows together, as in frowning	Facial
Depressor anguli oris *(de-pres´-or ang´-gu-li or´-is)*	Body of the mandible, below the mental foramen	Skin and muscles at the angle of the mouth	Pulls the angle of the mouth downward	Facial
Depressor labii inferioris *(de-pres´-or la´-bē-i in-fer˝-ē-or´-is)*	Body of the mandible between the symphysis and the mental foramen	Skin and muscle of the lower lip	Pulls the lower lip downward	Facial
Epicranius *(ep˝-ĭ-kra´-nē-us)*				
Frontalis *(frun-ta´-lis)*	Galea aponeurotica	Skin and muscles of the forehead	Raises the eyebrows; wrinkles the skin of the forehead	Facial
Occipitalis *(ok-sip˝-ĭ-ta´-lis)*	Occipital bone (superior nuchal line)	Galea aponeurotica	Draws the scalp posteriorly	Facial
Levator labii superioris *(le-vā´-tor la´-bē-i su-pēr˝-ē-or´-is)*	Lower margin of orbit (maxillary and zygomatic bones)	Skin and muscles of the upper lip, and wing of the nose	Raises the upper lip; dilates the nares (nostrils)	Facial
Mentalis *(men-ta´-lis)*	Mandible, near the symphysis	Skin of the chin	Raises and protrudes the lower lip	Facial
Orbicularis oculi *(or-bik˝-u-lar´-is ok´-u-li)*	Frontal and maxillary bones; medial palpebral ligament	Circles the orbit and extends within the eyelids	Closes the eyelids; tightens the skin of the forehead	Facial
Orbicularis oris *(or-bik˝-u-lar´-is or´-is)*	Muscles surrounding the mouth	Skin surrounding the mouth	Closes and protrudes the lips	Facial
Platysma *(plah-tiz´-mah)*	Fascia over the pectoralis major and the deltoid muscles	Lower border of the mandible, and the skin of the chin and cheek	Depresses the mandible; draws the angle of the mouth downward; tightens and wrinkles the skin of the neck	Facial
Procerus *(pro-se´-rus)*	Lower portion of the nasal bone; upper part of the lateral nasal cartilage	Skin between the eyebrows	Wrinkles the skin between the eyebrows	Facial
Risorius *(rih-zor´-ē-us)*	Fascia of the masseter muscle	Skin at the angle of the mouth	Pulls the angle of the mouth backward	Facial

continued on next page

◆ **TABLE 9.1 Muscles of the Face** [*Fig. 9.3*] **(continued)**

MUSCLE	ORIGIN	INSERTION	ACTION	INNERVATION
Zygomaticus (*zi″-go-mat′-ik-us*) major and zygomaticus minor	Zygomatic bone	Skin and muscles above the angle of the mouth	Raise the angle of the mouth	Facial

rather than a joint. Among the unusual types of muscles in this group are the *sphincters (sfink′-ters),* which are ring-shaped muscles that surround body openings. They can enlarge or close the opening by relaxing or contracting. The **orbicularis oculi** is a sphincter used in closing the eye, winking and squinting. Contraction of the sphincter muscle named **orbicularis oris** closes the mouth and purses the lips. Several of the other facial muscles insert onto the fascia that covers the orbicularis oris.

Another unusual facial muscle is the **epicranius.** This muscle has two parts: the anterior **frontalis** and the posterior **occipitalis.** These two muscular portions are connected by a broad, flat connective-tissue sheet, the **galea aponeurotica,** which lies tight against the top of the skull. Contraction of one or the other of the muscular portions pulls the scalp forward or backward.

While the **platysma** is not actually a facial muscle, we will consider it here because its main actions are on the mandible and the skin around the mouth. It is a superficial sheetlike muscle that covers the ventral surface of the upper thorax and the neck and extends over the chin to the region of the mouth. Contraction of the platysma lowers the mandible, the lower lip, and the corners of the mouth, as well as tightening the skin of the neck.

Muscles of Mastication

Four pairs of muscles are involved in biting and chewing. The large fan-shaped **temporalis,** which passes deep to the zygomatic arch of the cheek, and the quadrilateral-shaped **masseter,** which arises from the zygomatic arch, both serve to raise the mandible (Figure 9.4; Table 9.2). These muscles can be felt when the teeth are forcibly clenched. The other two pairs of mus-

◆ **FIGURE 9.4 Muscles of mastication**
(a) The temporalis and masseter muscles are the strongest masticatory muscles. (b) The temporalis and masseter muscles have been removed, and the zygomatic arch and mandible have been sectioned to reveal the pterygoid muscles.

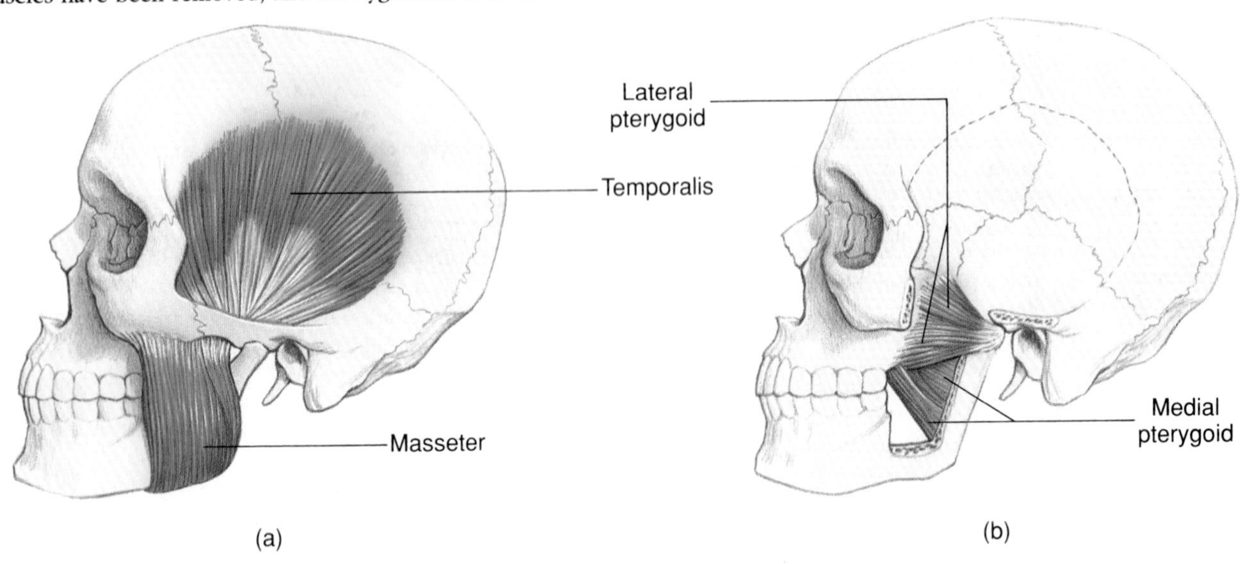

◆ **TABLE 9.2 Muscles of Mastication [*Fig. 9.4*]**

MUSCLE	ORIGIN	INSERTION	ACTION	INNERVATION
Temporalis *(tem-po-ra´-lis)*	Temporal fossa	Coronoid process and ramus of the mandible	Raises the mandible, closing the jaws; retracts the mandible	Trigeminal (cranial nerve V)
Masseter *(mas´-se-ter)*	Zygomatic arch	Angle and ramus of the mandible	Raises the mandible, closing the jaws	Trigeminal
Medial pterygoid *(ter´-ĭ-goid)*	Medial surface of the lateral pterygoid plate of the sphenoid bone, and the tuberosity of the maxillary bone	Inner surface of the mandible, at the angle	Closes the jaws; together with the lateral pterygoid, it aids in sideways movement of the jaws	Trigeminal
Lateral pterygoid	Lateral surface of the lateral pterygoid plate and the great wing of the sphenoid bone	Mandible just below the condyle	Opens and protrudes the mandible; moves the mandible from side to side	Trigeminal

cles involved in mastication are the medial and lateral **pterygoid** muscles, which move the mandible sideways in grinding movements, as well as assist in opening and closing the mouth.

Muscles of the Tongue

The tongue is a muscular organ covered with mucous membrane. Some of the muscles lie entirely within the tongue. These are called **intrinsic muscles.** The fibers of the intrinsic muscles are arranged in longitudinal, vertical, and transverse planes; consequently, when they contract, they squeeze, fold, and curl the tongue. These actions are particularly useful in speaking and manipulating food within the mouth.

The **extrinsic muscles** (Figure 9.5; Table 9.3) anchor the tongue to the skeleton (hyoid, mandible, and temporal bones) and control the protrusion, retraction, and sideward movement of the tongue.

Muscles of the Neck

The muscles of the neck (Figure 9.6; Table 9.4) are often described as being located within one of two triangles. Those within the anterior triangle are separated from those within the posterior triangle by the **sternocleidomastoid** muscle. This muscle runs diagonally across the lateral margins of the neck, from the mastoid process of the temporal bone to the sternum and the clavicle. It is beyond the scope of this text to describe all

◆ **TABLE 9.3 Extrinsic Muscles of the Tongue [*Fig. 9.5*]**

MUSCLE	ORIGIN	INSERTION	ACTION	INNERVATION
Genioglossus *(jē˝-nē-ō-glos´-us)*	Internal surface of the mandible, near the symphysis	Undersurface of the tongue; body of the hyoid	Protracts, retracts, and depresses the tongue	Hypoglossal (cranial nerve XII)
Hyoglossus *(hī˝-ō-glos´-us)*	Body and greater cornu of the hyoid bone	Side of the tongue	Depresses the tongue; draws its sides down	Hypoglossal
Styloglossus *(sti˝-lo-glos´-us)*	Styloid process of the temporal bone	Side of the tongue	Retracts and elevates the tongue	Hypoglossal

◆ **FIGURE 9.5 Extrinsic muscles of the tongue and the suprahyoid muscles of the neck**

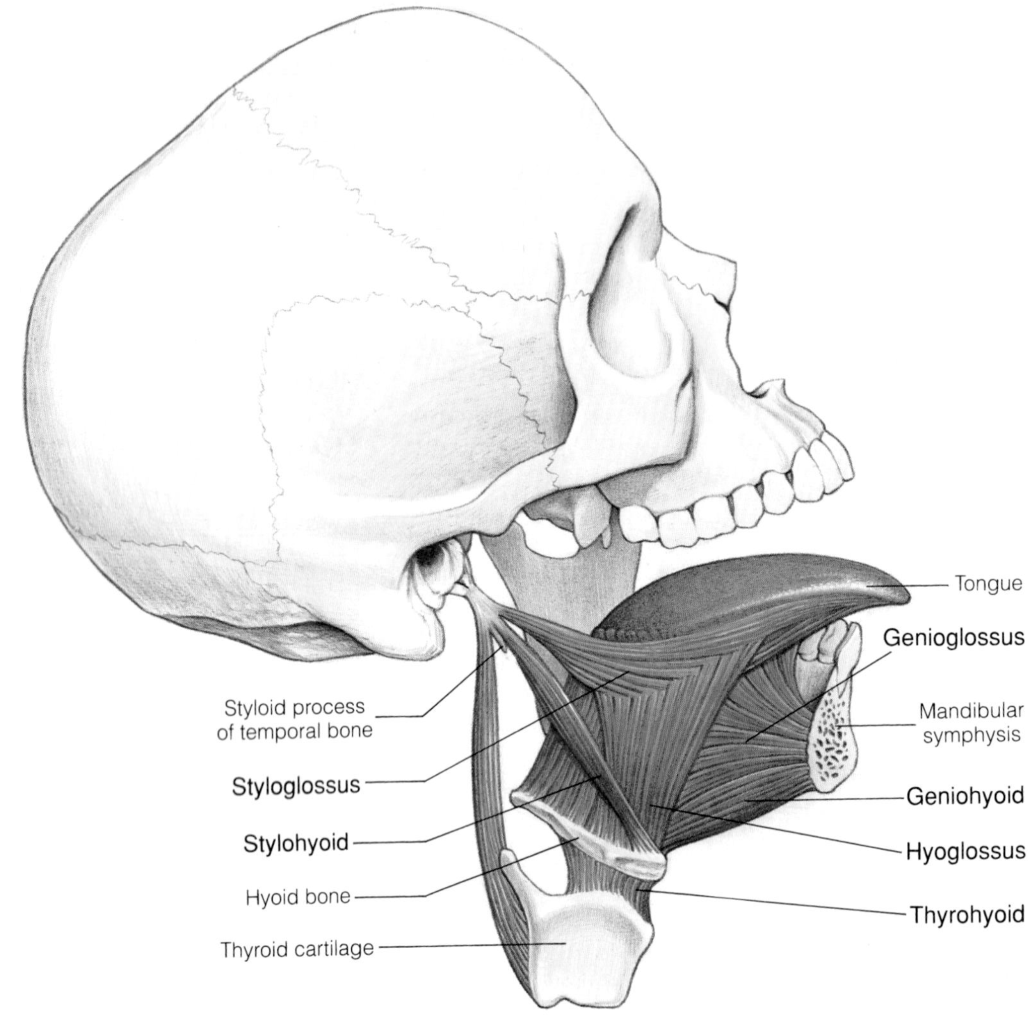

of the muscles within these two triangles. Rather we will consider at this time only selected muscles of the anterior triangle—in particular, the muscles of the throat. The muscles of the posterior triangle are included with those muscles that move the vertebral column and the head (Table 9.5).

Muscles of the Throat

The muscles of the throat are the deep muscles of the anterior triangle (Figure 9.5; Table 9.4). They help to form the floor of the oral cavity and are attached to the hyoid bone. Because the tongue is also attached to the hyoid bone, these muscles are involved with movements of the tongue. In addition, some of the throat muscles are attached to the larynx and therefore aid in swallowing. These muscles are often divided into two groups, depending upon whether they are located above or below the hyoid bone:

1. The *suprahyoid muscles:* **digastric, stylohyoid, mylohyoid,** and **geniohyoid.** As a group, these muscles raise the hyoid bone during swallowing and lower the jaw when the hyoid bone is fixed.
2. The *infrahyoid muscles:* **sternohyoid, sternothyroid, thyrohyoid,** and **omohyoid.** These muscles pull down on the larynx and hyoid, returning them to their normal positions after swallowing.

 Muscles of the Trunk

The muscles of the trunk include those that are associated with the vertebral column, the back, the thorax,

◆ **FIGURE 9.6 The suprahyoid and infrahyoid muscles of the neck**
The sternocleidomastoid and digastric muscles have been removed on the right.

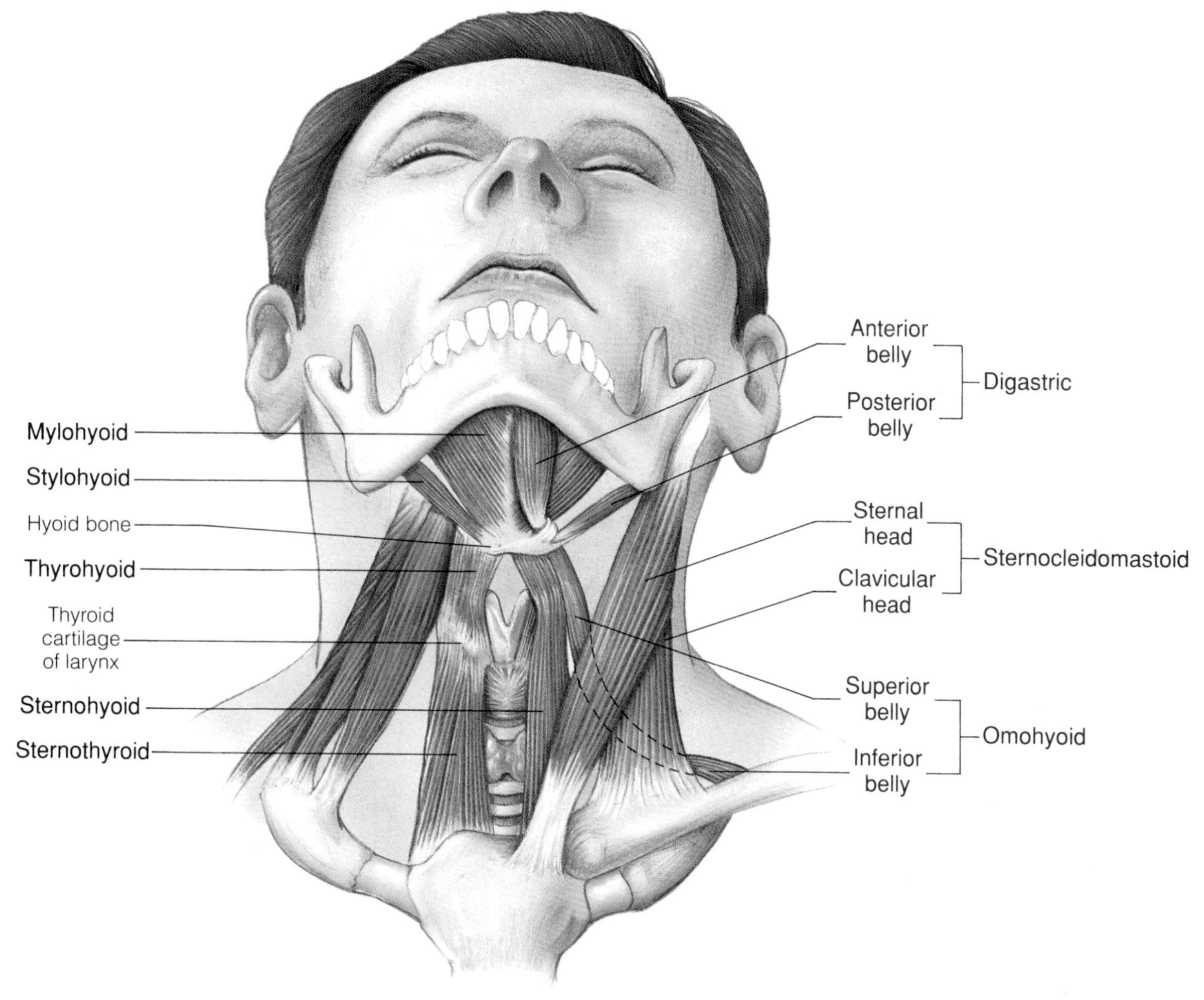

the floor of the pelvic cavity, and the wall of the abdomen. Trunk muscles have various actions, depending on their locations. Some move the vertebral column, others move the head; some are involved in respiratory movements, other function to move the upper limbs; and so forth. We will study them in groups according to their actions.

Muscles of the Vertebral Column

Most of the muscles that move the vertebral column are located on the posterior surface of the spine (Table 9.5). A few, such as the *splenius,* have fibers that insert onto the skull and therefore move the head as well as the vertebral column. The deepest of these muscles are located medially and travel only a few segments superiorly before inserting onto the transverse processes or spinous processes of the vertebrae (Figure 9.7). These include the **multifidus, rotatores, interspinales,** and **intertransversarii,** and the **semispinalis thoracis, cervicis,** and **capitis.** The **scalenes** (anterior, middle, and posterior) pass from the transverse processes of the cervical vertebrae to the upper two ribs (Figure 9.8).

Lateral to these muscles, located in the depression between the spinous processes and the transverse processes and the ribs, is a longitudinal muscle mass that extends from the sacrum to the skull. This is the **erector spinae (sacrospinalis)** muscle (Figure 9.9). The erector spinae has three subdivisions in the form of columns. The **iliocostalis,** which is the most lateral column, inserts on the ribs; the medial column is the **spinalis,** the fibers of which insert on the vertebrae; the **longissimus** subdivision is located between the other two columns. Each of these columns is further separated into **lumborum, thoracis, cervicis,** and/or **capitis**

◆ **TABLE 9.4 Muscles of the Anterior Triangle of the Neck** *[Fig. 9.6]*

MUSCLE	ORIGIN	INSERTION	ACTION	INNERVATION
Sternocleidomastoid *(ster″-no-kli″-do-mas′-toid)*	By two heads: the manubrium of the sternum, and the medial portion of the clavicle	Mastoid process of the temporal bone	Both muscles acting together flex the cervical vertebral column; acting singly, each rotates head to the opposite side	Accessory (cranial nerve XI) and upper cervical spinal nerves
SUPRAHYOID MUSCLES				
Digastric *(dī″-gas′-trik)*	*Anterior belly:* inner surface of the mandibular symphysis *Posterior belly:* mastoid process of the temporal bone	Hyoid bone, via the intermediate tendon	Raises the hyoid and assists in lowering the jaw	Trigeminal (anterior belly); facial (posterior belly)
Stylohyoid *(stī′-lō-hi″-oid)*	Styloid process of the temporal bone	Hyoid bone	Raises the hyoid and pulls it backward	Facial
Mylohyoid *(mī′-lō-hi′′-oid)*	Inner surface of the mandible, from the symphysis to the angle	Hyoid bone	Raises the hyoid and the floor of the mouth	Trigeminal
Geniohyoid *(jē′-nē-ō-hi″-oid)* [*Fig. 9.5*]	Inner surface of the mandibular symphysis	Hyoid bone	Pulls the hyoid anteriorly	1st cervical spinal nerve (through hypoglossal)
INFRAHYOID MUSCLES				
Sternohyoid *(ster′-nō-hī″-oid)*	Manubrium and the medial end of the clavicle	Hyoid bone	Pulls the hyoid inferiorly	1st–3rd cervical spinal nerves (through ansa cervicalis—see p. 437)
Sternothyroid *(ster′-nō-thī″-roid)*	Manubrium	Thyroid cartilage of the larynx	Pulls the larynx inferiorly	1st–3rd cervical spinal nerves (through ansa cervicalis)
Thyrohyoid *(thī′-rō-hi″-oid)*	Thyroid cartilage of the larynx	Hyoid bone	Pulls the hyoid inferiorly and raises the larynx	1st cervical spinal nerve (through hypoglossal)
Omohyoid *(ō′-mō-hi′′-oid)*	Superior border of the scapula	Hyoid bone	Pulls the hyoid inferiorly	1st–3rd cervical spinal nerves (through ansa cervicalis)

parts, which are named according to their points of insertion. In Table 9.5 the origins, insertions, and actions of these parts have been combined for each subdivision of the erector spinae.

All the muscles that insert on the vertebral column act to extend the vertebral column, and when acting on one side only, they bend the vertebral column to that side and may assist in its rotation. These muscles are not the only ones that move the vertebral column. Some of the muscles of the abdominal wall, such as the **rectus abdominis** and the **quadratus lumborum** (Table 9.7, page 294), also act on the vertebral column. In addition, the psoas major (Table 9.17, page 315), which acts on the hip joint, can cause the vertebral column to flex if the thighs are fixed.

Deep Muscles of the Thorax

Most of the deep muscles of the thorax insert on the ribs and assist in breathing by drawing the ribs together or by elevating or depressing the rib cage (Table 9.6).

◆ **TABLE 9.5 Muscles That Move the Vertebral Column**

MUSCLE	ORIGIN	INSERTION	ACTION	INNERVATION
Semispinalis *(sem″-ē-spī-na′-lis)* [*Fig. 9.9*] thoracis cervicis capitis	Transverse processes of the thoracic and the seventh cervical vertebrae	Spinous processes of the second cervical through the fourth thoracic vertebrae, and the occipital bone	Extend the vertebral column and the head (capitis); rotate them to the opposite side	Branches of the spinal nerves
Multifidus *(mul″-tif′-ĭ-dus)* [*Fig. 9.7*]	Posterior surface of the sacrum and the ilium, and the transverse processes of the lumbar, thoracic, and lower cervical vertebrae	Spinous processes of the lumbar, thoracic, and cervical vertebrae	Extend the vertebral column; rotate it toward the opposite side	Branches of the spinal nerves
Rotatores *(ro″-tah-tō′-rēs;* long and short) [*Fig. 9.7*]	Transverse processes of all the vertebrae	Base of the spinous process of the vertebra above the vertebra of origin (short) or the second vertebra above (long)	Extend the vertebral column; rotate it toward the opposite side	Branches of the spinal nerves
Interspinales *(in″-ter-spī′-nal-es)* [*Fig. 9.7*]	Superior surface of all the spinous processes	Inferior surface of the spinous process of the vertebra above the vertebra of origin	Extend the vertebral column	Branches of the spinal nerves
Scalenes *(skā′-lēnz)* [*Fig. 9.8*]	Transverse process of cervical vertebrae	Upper two ribs	Flex and rotate the neck; assist in inspiration	Branches of the lower cervical nerves
Intertransversarii *(in″-ter-trans-ver-sar′-ri-i)* [*Fig. 9.7*]	Transverse processes of all the vertebrae	Transverse processes of the vertebra above the vertebra of origin	Bend the vertebral column laterally	Branches of the spinal nerves
Splenius *(splē′-nē-us)* [*Fig. 9.9*] capitis cervicis	Spinous processes of the upper thoracic and the seventh cervical vertebrae, and from the ligamentum nuchae	Occipital bone, mastoid process of the temporal bone, and the transverse processes of the upper three cervical vertebrae	Acting together, they extend the head and the neck; acting singly, they abduct and rotate the head toward the same side	Branches of the spinal nerves
ERECTOR SPINAE *(ē-rek′-ter spī′-nē;* SACROSPINALIS) [*Fig. 9.9*]				
Iliocostalis *(il″-ē-o-kos-ta′-lis)* lumborum thoracis cervicis	Crest of the sacrum; spinous processes of the lumbar and lower thoracic vertebrae; iliac crests; angles of the ribs	Angles of the ribs; transverse processes of the cervical vertebrae	Extend the vertebral column and bend it laterally	Branches of the spinal nerves
Longissimus *(lon-jis′-i-mus)* thoracis cervicis capitis	Transverse processes of the lumbar, thoracic, and lower cervical vertebrae	Transverse processes of the vertebra above the vertebra of origin, and the mastoid process of the temporal bone (capitis)	Extend the vertebral column and head; rotate the head toward the same side	Branches of the spinal nerves
Spinalis *(spī-na′-lis)* thoracis cervicis	Spinous process of the upper lumbar, lower thoracic, and seventh cervical vertebrae	Spinous processes of the upper thoracic and the cervical vertebrae	Extend the vertebral column	Branches of the spinal nerves

◆ **FIGURE 9.7 Muscles of the vertebral column**

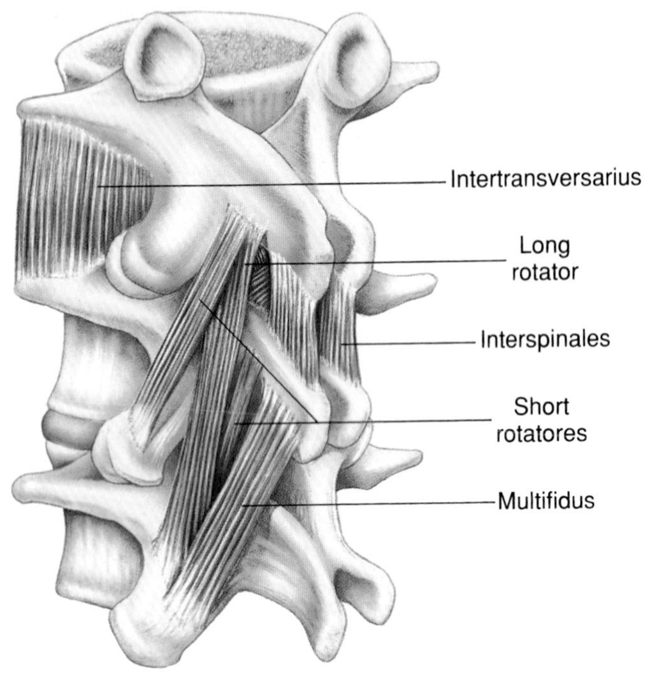

◆ **FIGURE 9.8 The scalene muscles viewed from the front**

The right anterior scalene muscle has been removed.

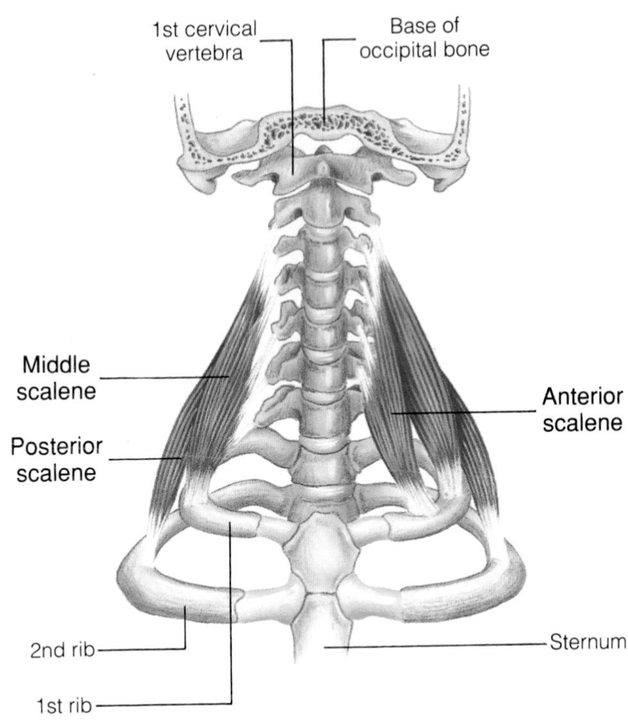

◆ **TABLE 9.6 Deep Muscles of the Thorax (Respiratory Muscles)**

MUSCLE	ORIGIN	INSERTION	ACTION	INNERVATION
Diaphragm *(dī´-ah-fram)* [*Fig. 9.11*]	The xiphoid process; inner surfaces of lower six ribs; and the lumbar vertebrae	Central tendon of the diaphragm	Pulls central tendon downward, increasing the size of the thoracic cavity and therefore causing inspiration	Phrenic
External intercostals *(in″-ter-kos´-tals)* [*Fig. 9.10*]	Inferior border of the ribs and the costal cartilages	Superior border of the rib below the rib of origin	Draw ribs together, aiding respiration	Intercostal
Internal intercostals [*Fig. 9.10*]	Inner surface of the ribs and the costal cartilages	Superior border of the rib below the rib of origin	Draw ribs together, aiding respiration	Intercostal
Subcostals* *(sub-kos´-tals)*	Inner surface of the ribs, near their angles	Inner surface of the second or third rib below the rib of origin	Draw ribs together, aiding expiration	Intercostal
Transversus thoracis* *(trans″-vers´-us tho″-ras´us)*	Inner surface of the sternum and the xiphoid process	Inner surface of the costal cartilages	Draw anterior portion of the rib cage downward, aiding expiration	Intercostal

*Not illustrated

◆ **FIGURE 9.9 Some muscles that move the vertebral column**
(a) The erector spinae muscles showing the longissimus, iliocostalis, and spinalis portions.
(b) The splenius capitis and semispinalis capitis muscles.

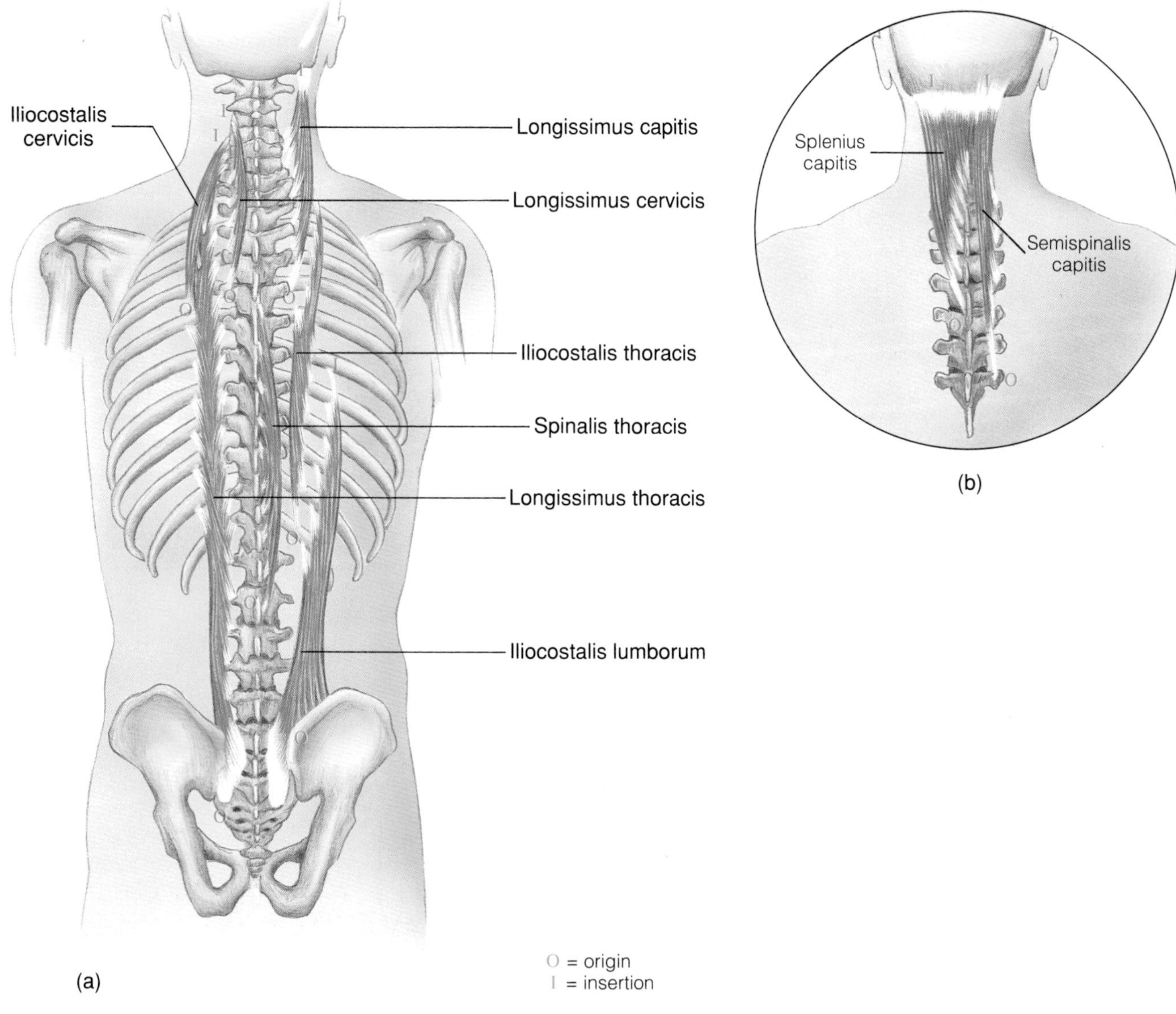

The ribs slope downward as they pass forward; therefore, any muscle that elevates them increases the volume of the thoracic cavity, causing inspiration. Conversely, muscles that depress the ribs back to their usual positions decrease the volume of the thoracic cavity, forcing air from the lungs in expiration. The muscles described in Table 9.6 are those involved in normal, quiet respirations. In forced breathing, when overexpansion of the thoracic cage is beneficial, additional muscles may be involved, such as the **scalenes** (Table 9.5), the **sternocleidomastoid** (Table 9.4), and the **quadratus lumborum** (Table 9.7). Because these muscles more commonly perform other actions, they are listed within other groups, as their table references indicate.

The spaces between adjacent ribs are reinforced primarily by **external** and **internal intercostal** muscles (Figure 9.10). In addition, small **innermost intercostal** muscles lie deep to the internal intercostals. The fibers of the external intercostal muscles run at right angles to the fibers of the internal and innermost muscles, as in a bias-ply automobile tire, thus providing a strong muscular wall between the ribs, without requiring heavy musculature.

The **diaphragm** is the muscle chiefly responsible for quiet breathing. Dome-shaped, it separates the thoracic

◆ **FIGURE 9.10 The external and internal intercostal muscles**

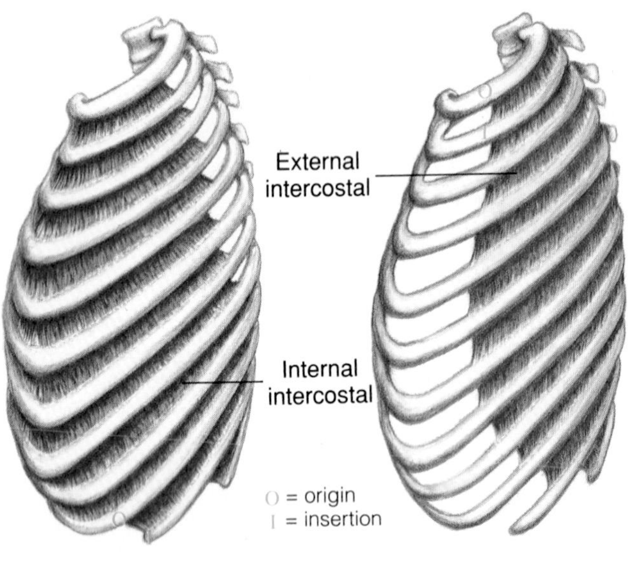

cavity from the abdominopelvic cavity. The upper surface of the diaphragm is in contact with the heart and lungs; the lower surface contacts the liver, the stomach, the spleen, and the pad of fat that surrounds the suprarenal glands and kidneys.

The muscle fibers of the diaphragm are grouped into sternal, costal, and lumbar portions (Figure 9.11). The small *sternal* portion arises from the inner surface of the xiphoid process; the *costal* fibers arise from the inner surfaces of the seventh, eighth, and ninth ribs and the distal ends of the last three ribs; the *lumbar* portion arises from the front of the lumbar vertebrae by two tendinous bands called crura (singular *crus*). The muscle fibers from these three portions insert on a common *central tendon,* which they surround. When the diaphragm contracts, its dome is pulled downward, flattening the muscle and increasing the volume of the thoracic cavity.

The diaphragm is pierced by a number of openings that permit the passage of structures between the thorax and abdomen. The largest openings are for the aorta, the inferior vena cava, and the esophagus.

◆ **FIGURE 9.11 The abdominal surface of the diaphragm**

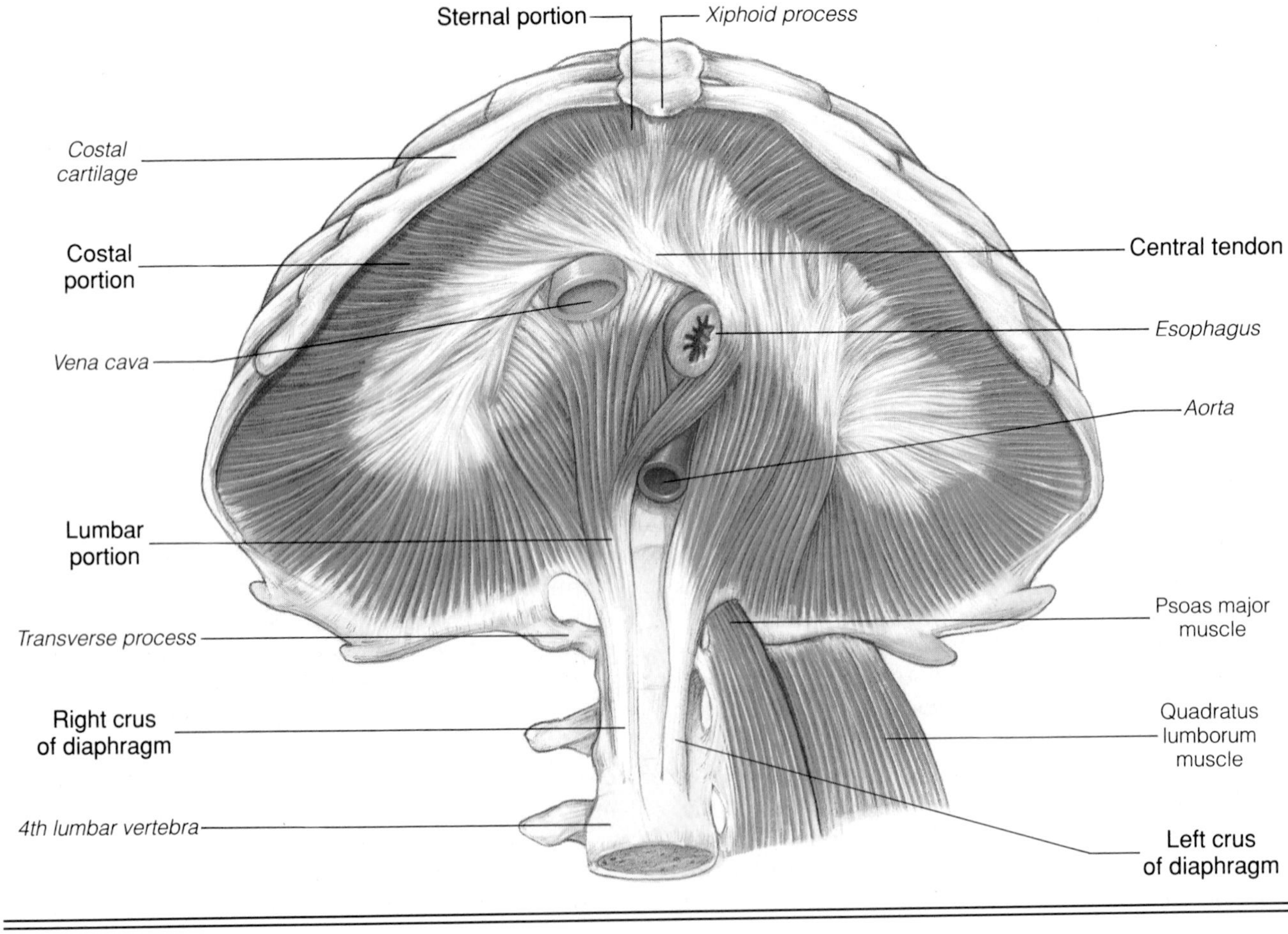

◆ **FIGURE 9.12 Muscles of the abdominal wall**

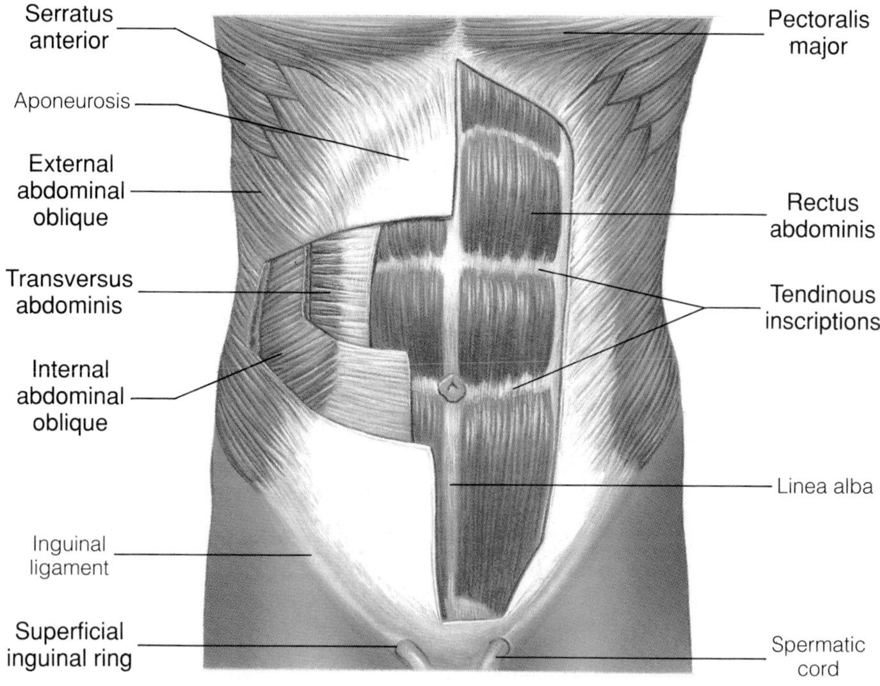

The mechanics of respiration are considered in greater detail in Chapter 23.

Muscles of the Abdominal Wall

Because there are no skeletal supports within the walls of the abdominal cavity, the abdominal cavity derives its strength entirely from muscles. There are three layers of muscles in the wall (Figure 9.12; Table 9.7). The fibers of each of the muscles run in different directions, providing additional strength. The outermost layer is the **external abdominal oblique** muscle, whose fibers pass medially and downward as a continuation of the external intercostal muscles. The **internal abdominal oblique** lies just beneath the external oblique. Its fibers run upward and medially, becoming continuous over the ribs with the internal intercostal muscles. Deep to both of the oblique muscles is a thin muscle whose fibers run horizontally, encircling the abdominal cavity. This is the **transversus abdominis** muscle. The tendons of the three abdominal-wall muscles pass medially in the form of broad aponeuroses that insert on a midline **linea alba** ("white line"). The linea alba is a fibrous band that extends from the xiphoid process of the sternum to the symphysis pubis.

At the lower margin of the external abdominal oblique muscle, the fascia of the muscle forms a tendinous border called the **inguinal ligament** *(ing´-gwi-nal)*. This ligament runs between the pubic tubercle and the anterior superior iliac spine. It marks the separation between the body wall and the thigh. At one point, there is an opening between the muscle fascia and the inguinal ligament. This opening, which is called the *superficial inguinal ring,* is the external opening of the **inguinal canal.** The canal passes laterally, above and parallel with the inguinal ligament. About midway between the anterior superior iliac spine and the pubic tubercle, the canal opens into the abdominal cavity through an aperture in the fascia of the transversus abdominis muscle. This opening is called the *deep inguinal ring.* In the male, the spermatic cord passes through the canal. In the female, the canal provides for passage of the round ligament of the uterus. If the deep inguinal ring is weak, increased abdominal pressure can force some of the abdominal contents into the inguinal canal. This condition is called an *inguinal hernia,* or *rupture.* Inguinal hernias are more common in males because during embryonic development their inguinal canals are expanded and weakened as a result of the passage of the testes through them into the scrotal sacs.

The **rectus abdominis** is a narrow, flat muscle on the ventral aspect of the abdominal wall. Its fibers run vertically from the pubis to the rib cage, alongside the linea alba. Each rectus abdominis is completely ensheathed by the fascia of the oblique and transversus abdominal muscles. The fasciae separate in various combinations to pass superficially and deep to the rectus abdominis muscle. Each rectus abdominis is crossed by

◆ **TABLE 9.7 Muscles of the Abdominal Wall** [*Fig. 9.12*]

MUSCLE	ORIGIN	INSERTION	ACTION	INNERVATION
External abdominal oblique *(ō-blēk´)*	External surface of the lower eight ribs	Linea alba and the anterior half of the iliac crest	Compresses the abdominopelvic cavity; assists in flexing and rotating the vertebral column	Intercostal, iliohypogastric, and ilioinguinal
Internal abdominal oblique	Inguinal ligament, the iliac crest, and the lumbodorsal fascia	Linea alba, the pubic crest, and the lower four ribs	Compresses the abdominopelvic cavity; assisting in flexing and rotating the vertebral column	Intercostals, iliohypogastric, and ilioinguinal
Transversus abdominis *(trans´´-vers´-us ab-dom´-ĭ-nis)*	Inguinal ligament, the iliac crest, the lumbodorsal fascia, and the costal cartilages of the last six ribs	Linea alba and the pubic crest	Compresses the abdominopelvic cavity	Intercostals, iliohypogastric, and ilioinguinal
Rectus abdominis *(rek´-tus ab-dom´-ĭ-nis)*	Pubic crest	Xiphoid process and the costal cartilages of the fifth through the seventh ribs	Compresses the abdominopelvic cavity; flexes the vertebral column	Intercostals
Quadratus lumborum *(kwod-ra´-tus lum-bōr´-um)* [*Fig. 9.28*]	Iliac crest, and the iliolumbar ligament	Lower border of the twelfth rib; the transverse processes of the upper lumbar vertebrae	Pulls the thoracic cage toward the pelvis; bends the vertebral column laterally toward the side that is being contracted	Twelfth thoracic and first lumbar

three transverse fibrous bands called the **tendinous inscriptions.** In a person who has developed his or her rectus abdominis muscle through exercise, the portions of the muscle between these inscriptions enlarge, causing the inscriptions to appear through the skin as horizontal depressions.

The three oblique muscles and the rectus abdominis compress the abdominal cavity, assisting in such actions as forced expiration, defecation, and urination.

Most of the posterior portion of the abdominal wall is formed by the **quadratus lumborum** muscle (Figure 9.28). This is a broad quadrilateral muscle that runs from the posterior region of the iliac crest to the twelfth rib. The psoas major muscle also forms part of the posterior wall, but it acts primarily on the femur and so is described with that group of muscles (Table 9.17). A small psoas minor muscle that acts on the lumbar vertebral column is sometimes present ventral to the psoas major.

Muscles That Form the Floor of the Abdominopelvic Cavity

The viscera of the abdominopelvic cavity are supported by a muscular floor called the **pelvic diaphragm** (Figure 9.13; Table 9.8). Two muscles, the **levator ani** and the **coccygeus,** form the pelvic diaphragm. The levator ani is composed of several parts, of which the **pubococcygeus** muscle and the **iliococcygeus** muscle are the most prominent. There are openings through these muscles for the anal canal, the urethra, and, in the female, the vagina.

Muscles of the Perineum

The **perineum** is the lower end of the trunk between the thighs (Figure 9.14; Table 9.9). It is bounded anteriorly by the pubic arch, posteriorly by the coccyx, and laterally by the ischiopubic rami and the sacrotuberous ligaments, which run between the lateral margins of the sacrum and coccyx and the ischial tuberosities.

The perineal muscles are located inferior to the pelvic diaphragm. They consist of superficial muscles associated with the external genital organs and deeper muscles that form the **urogenital diaphragm.** The urogenital diaphragm is located just below the anterior portion of the pelvic diaphragm, to which it adds support. The urogenital diaphragm is composed primarily of the **deep transverse perinei** and **sphincter urethrae** muscles and various fascial sheets. The superficial perineal mus-

◆ **FIGURE 9.13 Pelvic diaphragm of the female, viewed from the inside of the pelvic cavity**

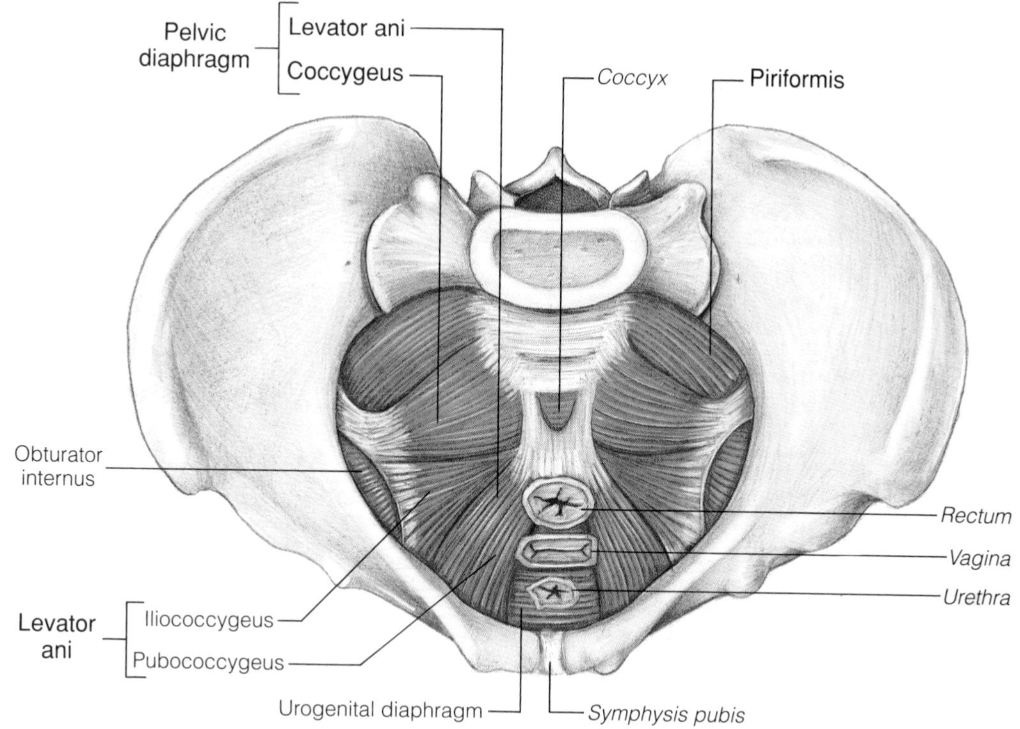

cles include the **ischiocavernosus, bulbospongiosus (bulbocavernosus),** and **superficial transverse perinei** muscles. An **external anal sphincter** muscle surrounds the intestine after it passes through the perineum.

Muscles of the Upper Limbs

Muscles That Act on the Scapula

Included in the muscles of the upper limbs are those muscles of the pectoral girdle that anchor the scapula (and to a lesser extent, those muscles that anchor the clavicle) to the axial skeleton (Table 9.10). Although it is possible for the scapula to move over the ribs, these muscles act primarily as fixators of the scapula. When it is immobilized by these muscles, the scapula serves as a stable point of origin for most of the muscles that move the arm.

Four posterior muscles anchor the scapula (Figure 9.15, page 298): the **trapezius,** the **rhomboideus major** and **minor,** and the **levator scapulae.** The trapezius and rhomboids are named according to their shapes, the levator according to its action.

Because of the shape of the *trapezius,* its fibers pull in various directions, and its actions depend on which portion of the muscle contracts. If the upper portion

◆ **TABLE 9.8 Muscles That Form the Floor of the Abdominopelvic Cavity** [*Fig. 9.13*]

MUSCLE	ORIGIN	INSERTION	ACTION	INNERVATION
PELVIC DIAPHRAGM				
Levator ani (*le-va´-tor ā´-nī*)	Inner surface of the superior ramus of the pubic bone, the lateral pelvic wall, and the spine of the ischium	Inner surface of the coccyx	Supports the pelvic viscera	Third through fifth sacral
Coccygeus (*kok-sij´-e-us*)	Spine of the ischium and the sacrospinous ligament	Sides of the coccyx and the sacrum	Supports the pelvic viscera	Fourth and fifth sacral

◆ **TABLE 9.9 Muscles of the Perineum [*Fig. 9.14*]**

MUSCLE	ORIGIN	INSERTION	ACTION	INNERVATION
Ischiocavernosus *(is´-kē-ō-ka˝-ver-no´-sis)*	Tuberosity and rami of ischium	Crus of penis or clitoris	Retards return of blood through veins, thereby maintaining erection of penis or clitoris	Pudendal (second through fifth sacral)
Bulbospongiosus *(bul˝-bō-spun˝-je-ō-sus;* bulbocavernosus)	*Male:* from ventral median raphe on base of penis	Encircles base of penis and joins with fibers from opposite side on dorsum of penis	Empties urethral canal; assists in erection of penis	Pudendal
	Female: central tendinous point of perineum near anus	Pass on either side of vagina to insert on base of clitoris	Assists in erection of clitoris	Pudendal

◆ **FIGURE 9.14 Muscles of the perineum**

(a) Superficial muscles. (b) Deep muscles forming the urogenital diaphragm.

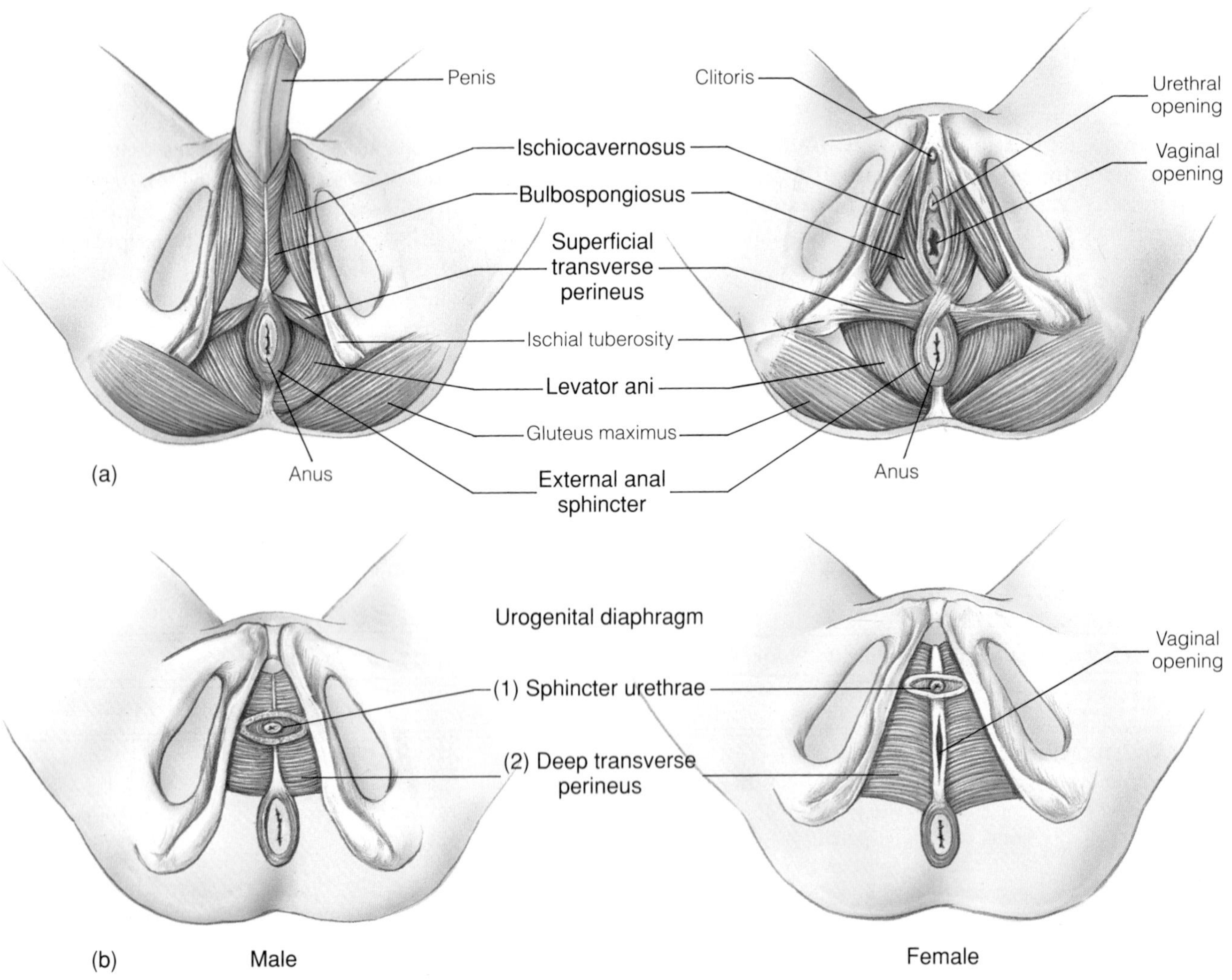

◆ **TABLE 9.9 Muscles of the Perineum** [*Fig. 9.14*] (continued)

MUSCLE	ORIGIN	INSERTION	ACTION	INNERVATION
Superficial transverse perineus *(per″-i-ne′-us)*	Tuberosity of ischium	Central tendinous point of perineum	Fixes (tightens) central tendinous point	Pudendal
Urogenital diaphragm *(u″-rō-jen′-i-tal dī′-ah-fram)*				
Deep transverse perineus	Inferior rami of ischium	Median raphe where it joins with corresponding muscle from opposite side	Both muscles of urogenital diaphragm act as constrictors of urethra	Pudendal
Sphincter urethrae	Encircle membranous portion of urethra			

◆ **TABLE 9.10 Muscles That Act on the Scapula** [*Figs. 9.15; 9.16*]

MUSCLE	ORIGIN	INSERTION	ACTION	INNERVATION
Trapezius *(trah-pē′-ze-us)*	Occipital bone, the ligamentum nuchae, and the spinous processes of the seventh cervical and all of the thoracic vertebrae	Lateral third of the clavicle, the acromion process, and the spine of the scapula	Elevates (upper portion) or depresses (lower portion), rotates, adducts, and stabilizes the scapula	Accessory (cranial nerve XI)
Rhomboideus major *(rom″-boid′-e-us)*	Spinous processes of the second through the fifth thoracic vertebrae	Vertebral border of the scapula, below the spine of the scapula	Adduct, stabilize, and rotate the scapula, lowering its lateral angle	Dorsal scapular (fifth cervical)
Rhomboideus minor	Spinous processes of the seventh cervical and first thoracic vertebrae	Vertebral border of the scapula, at the base of the spine of the scapula		
Levator scapulae *(le-va′-tor skap′-u-lē)*	Transverse processes of the upper four cervical vertebrae	Vertebral border of the scapula, above the spine of the scapula	Elevates scapula and bends the neck laterally when the scapula is fixed	Dorsal scapular
Pectoralis minor *(pek″-to-ra′-lis)*	Anterior surface of the third through the fifth ribs	Coracoid process of the scapula	Depresses the scapula and pulls it anteriorly	Medial pectoral (eighth cervical and first thoracic)
Serratus anterior *(ser-ā′-tis)*	Outer surface of the first nine ribs	Entire length of the ventral surface of the vertebral border of the scapula	Stabilizes, abducts, and rotates the scapula upward	Long thoracic (fifth through seventh cervical)
Subclavius *(sub-klā′-vē-us)*	Outer surface of the first rib	Inferior surface of the lateral portion of the clavicle	Stabilizes and depresses the pectoral girdle	Fifth and sixth cervical

◆ **FIGURE 9.15 Posterior muscles, including those that anchor the scapula to the axial skeleton**
Note that the superficial muscles have been removed on the left side.

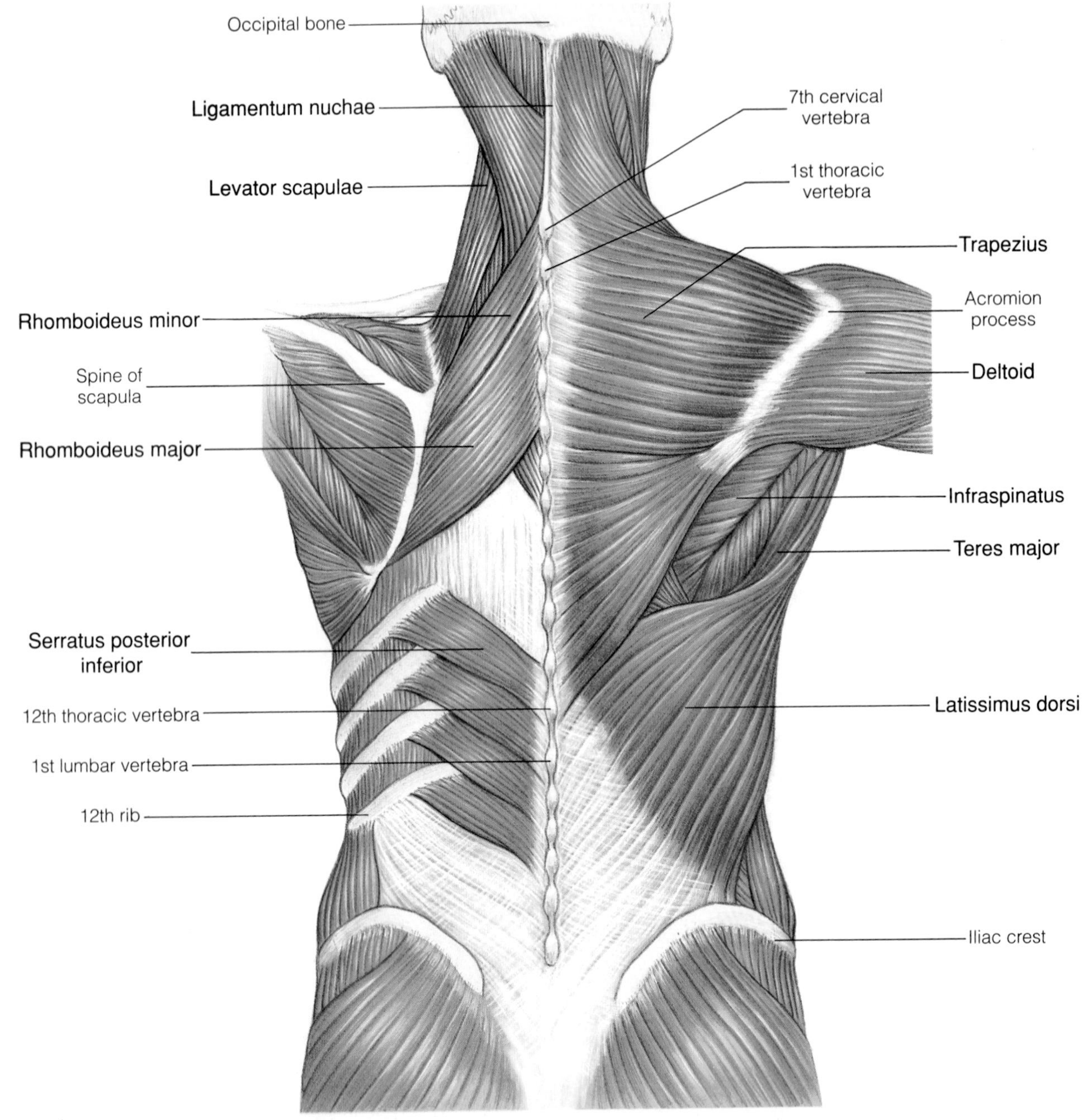

contracts, the scapula is elevated, as in shrugging the shoulders. If the lower portion contracts, the scapula is depressed. If the entire muscle contracts, the scapula is pulled toward the vertebral column—that is, it is adducted. If the scapula is fixed, the trapezius muscle assists in moving the head posteriorly.

The *rhomboids* insert on the vertebral border of the scapula and act to pull the scapula medially as well as rotating it, thus lowering the lateral angle. The **levator scapulae,** which runs from the upper cervical vertebrae to the superior angle of the scapula, elevates the scapula, acting as a synergist to the upper portion of the trapezius muscle.

Two anterior muscles anchor the scapula to the thorax (Figure 9.16). The **pectoralis minor** pulls the scapula anteriorly and downward, lowering the lateral angle. In this manner, it acts antagonistically to the trapezius, rhomboid, and levator scapulae muscles. The

◆ **FIGURE 9.16 Deep anterior muscles, including those that anchor the scapula and the clavicle to the thorax**

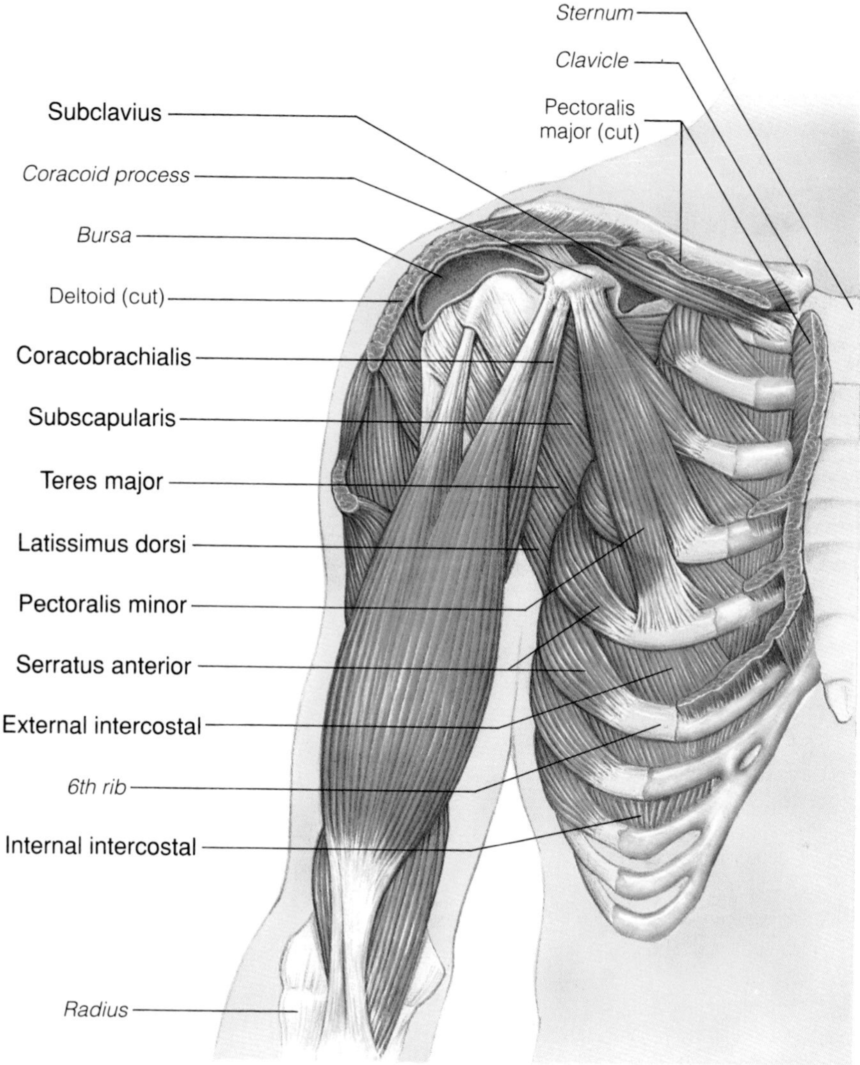

serratus anterior derives its name from its scalloped origin on the anterior surfaces of the ribs. From this origin it passes posteriorly, between the dorsal surface of the ribs and the subscapular fossa of the scapula, to insert on the vertebral border of the scapula. The serratus anterior pulls the scapula laterally; therefore, it acts antagonistically to the rhomboid muscles.

In addition to these muscles, the **subclavius** muscle anchors the pectoral girdle to the thoracic cage through its insertion onto the clavicle. For this reason it is included in Table 9.10, even though it does not act directly on the scapula.

Muscles That Act on the Arm (Humerus)

Nine muscles cross the shoulder joint and insert on the humerus (Table 9.11, page 302). Seven of these muscles arise from the scapula, indicating how important it is for the muscles discussed in Table 9.10 to fix the scapula. The other two muscles arise from the axial skeleton and have no attachments to the scapula. These are the *pectoralis major* and the *latissimus dorsi* muscles. The actions of all these muscles are summarized in Table 9.12, page 304.

The **pectoralis major** (Figure 9.17) is a large fan-shaped chest muscle that completely covers the smaller pectoralis minor. The pectoralis major passes from the thoracic cage to the humerus, forming the anterior border of the axilla. It acts to adduct, to flex, and—because it passes anterior to the shoulder joint—to rotate the humerus medially.

The **latissimus dorsi** (Figure 9.15) arises from the vertebrae of the lower back and from the pelvis. It twists upon itself as it passes between the humerus and the scapula to insert on the anterior surface of the hu-

◆ **FIGURE 9.17 Superficial muscles of the chest, shoulder, and anterior arm**

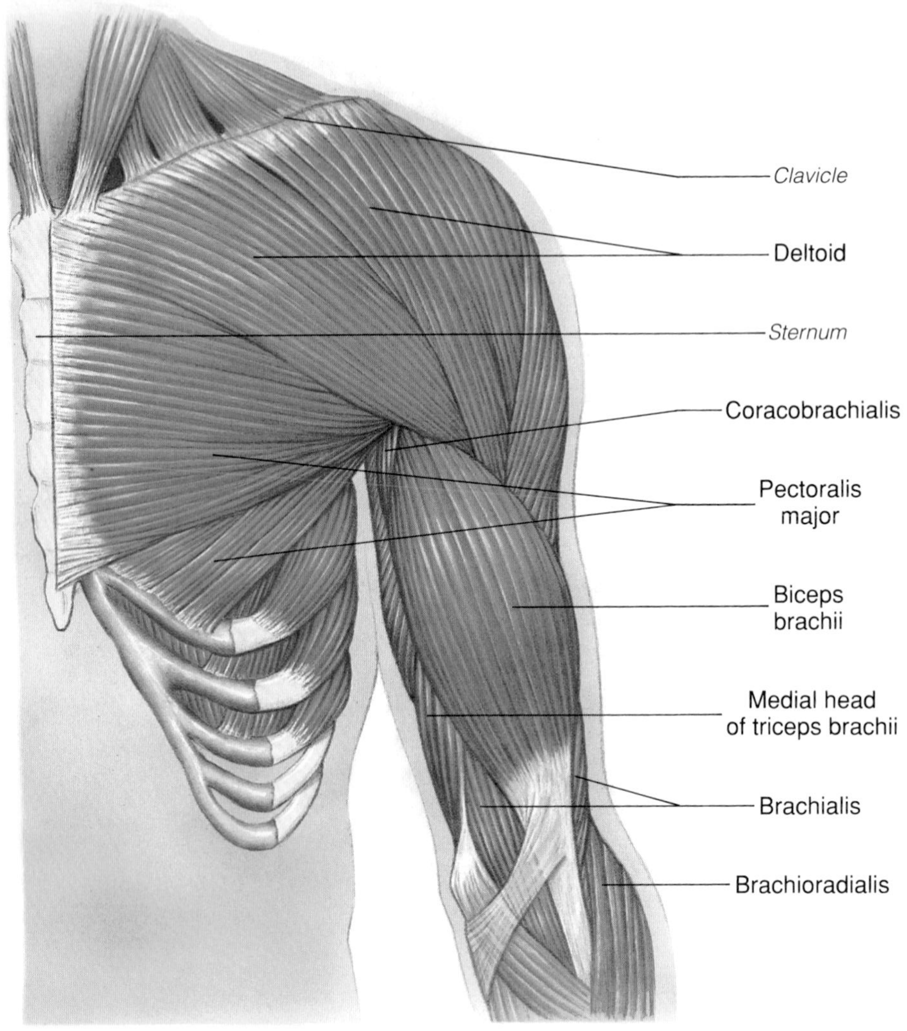

merus. In doing so, it forms the posterior border of the axilla. The latissimus dorsi extends the arm, pulling it downward or backward, thereby acting antagonistically to the pectoralis major. At the same time, it acts synergistically to the pectoralis major in adducting and—because of how it wraps around the humerus to insert on the anterior surface—in rotating the arm medially.

The **deltoid** (Figures 9.15 and 9.17) is a large muscle that forms a cap over the shoulder. It arises anteriorly from the clavicle and posteriorly from the scapula. Some of its fibers, therefore, pass in front of the shoulder joint; some pass behind; and some pass directly over the lateral surface of the joint. Because of the different positions of their fibers, the anterior and posterior parts of the muscle have actions that are antagonistic to each other. If the entire muscle contracts, it abducts the arm (antagonistic to the pectoralis major and the latissimus dorsi). The anterior fibers act to flex and medially rotate the arm (synergistic to the pectoralis major). The posterior fibers extend and laterally rotate the arm (antagonistic to the action of the anterior fibers).

The origins of the **supraspinatus, infraspinatus,** and **subscapularis** muscles cover most of the ventral and dorsal surfaces of the scapula (Figures 9.16 and 9.18). They derive their names from the scapular fossae from which they arise. The *supraspinatus* crosses the upper part of the shoulder joint, allowing it to serve as an abductor of the arm. The *infraspinatus* passes posteriorly to the shoulder joint, acting as a lateral rotator of the arm. The *subscapularis,* whose origin is on the ventral surface of the scapula, separates the scapula from the serratus anterior muscle. It inserts on the anterior surface of the humerus; consequently, it acts as a medial rotator of the arm. In addition to their prime actions, all these deep muscles that cross the shoulder joint assist in strengthening and stabilizing the joint.

◆ **FIGURE 9.18 Deep posterior muscles that attach the arm to the scapula**

The deltoid muscle has been removed.

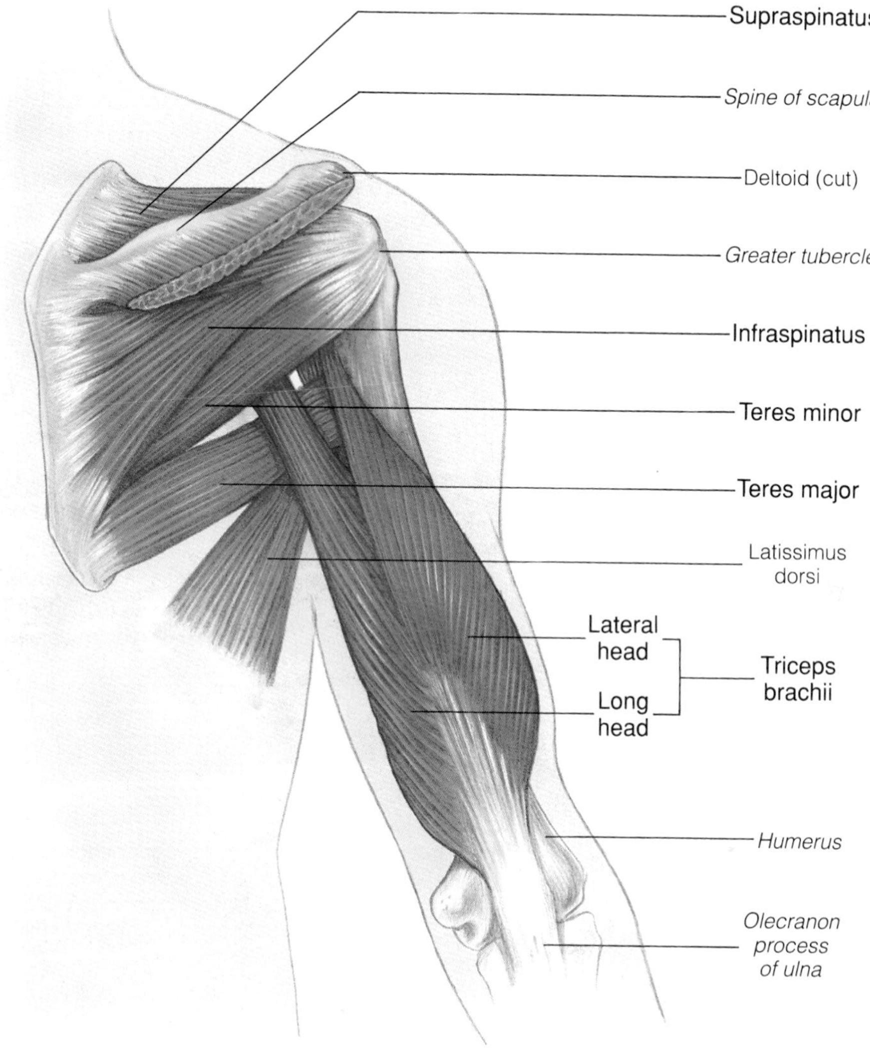

The **teres major** and **teres minor** muscles both originate from the axillary border of the scapula (Figure 9.18). The teres major is the longer, its origin being from the inferior angle. The two muscles bracket the humerus, the teres major inserting on the anterior surface, the teres minor inserting on the posterior surface. Because of these insertions, the major is a medial rotator of the arm, and the minor is a lateral rotator.

The **coracobrachialis** (Figures 9.16 and 9.20) is the remaining muscle of scapular origin. As the name indicates, it arises from the coracoid process of the scapula and inserts onto the humerus. It is an anterior muscle that assists in flexion and adduction of the arm.

Two additional muscles, the *biceps brachii* and the *triceps brachii,* have origins on the scapula and pass into the arm. However, because their major actions are on the elbow joint, they are discussed with that group of muscles (Table 9.13, page 304).

Muscles That Act on the Forearm (Radius and Ulna)

The more powerful of the muscles that move the elbow and/or the proximal radioulnar joint (producing supination and pronation) are located in the arm (Table 9.13). They are assisted, however, by several muscles whose bellies lie in the forearm. Most of the latter muscles have their prime actions on the hand and are described with that group (Table 9.14).

The two anterior muscles of the arm, the *biceps brachii* and the *brachialis,* flex the forearm. A third anterior arm muscle, the *coracobrachialis,* has its prime action on the shoulder joint; it was described earlier with that group of muscles (Table 9.11).

As the name indicates, the **biceps brachii** (Figures 9.17 and 9.19) has two heads, both of which have their origins on the scapula. The tendon of the *long*

◆ **TABLE 9.11 Muscles That Act on the Arm (Humerus)**

MUSCLE	ORIGIN	INSERTION	ACTION	INNERVATION
ORIGIN ON AXIAL SKELETON				
Pectoralis major *(pek″-to-ra′-lis)* [*Fig. 9.17*]	Medial half of the clavicle, the sternum, the costal cartilages of the upper six ribs, and the aponeurosis of the external oblique muscle	Greater tubercle of the humerus	Flexes, adducts, and medially rotates the arm	Medial and lateral pectoral
Latissimus dorsi *(lah-tis′-ĭ-mus dor′-sī)* [*Fig. 9.15*]	Spinous processes of the lower six thoracic and the lumbar vertebrae, the sacrum, the posterior iliac crest—all via the lumbodorsal fascia	Medial margin of the intertubercular groove of the humerus	Extends, adducts, and medially rotates the arm	Thoracodorsal
ORIGIN ON SCAPULA				
Deltoid *(del′-toid)* [*Fig. 9.15*]	Lateral third of the clavicle, the acromion process, and the spine of the scapula	Deltoid tuberosity of the humerus	Abducts arm; anterior fibers flex and medially rotate the arm; posterior fibers extend and laterally rotate the arm	Axillary
Supraspinatus *(su″-prah-spī-nah′-tus)* [*Fig. 9.18*]	Supraspinous fossa of the scapula	Greater tubercle of the humerus	Abducts the arm; slight lateral rotation	Suprascapular
Infraspinatus *(in′-frah-spī-nah′-tus)* [*Fig. 9.18*]	Infraspinous fossa of the scapula	Greater tubercle of the humerus (posterior to the supraspinatus)	Rotates the arm laterally; slight adduction	Suprascapular
Subscapularis *(sub-skap″-u-layr′-us)* [*Fig. 9.16*]	Subscapular fossa of the scapula	Lesser tubercle of the humerus	Rotates the arm medially	Subscapular
Teres major *(tayr′-ēz)* [*Fig. 9.18*]	Dorsal surface of the inferior angle of the scapula	Lesser tubercle of the humerus	Adducts, extends, and medially rotates the arm	Subscapular
Teres minor [*Fig. 9.18*]	Axillary border of the scapula	Greater tubercle of the humerus (posterior to the infraspinatus)	Rotates the arm laterally; weakly adducts and extends the arm	Axillary
Coracobrachialis *(kor″-ah-kō-brā′-kē-al′-us)* [*Fig. 9.20*]	Coracoid process of the scapula	Middle of the humerus, medial surface	Flexes and adducts the arm	Musculocutaneous

head passes over the top of the shoulder joint and travels through the intertubercular groove on the humerus. The two heads blend into the thick belly of the muscle. The main insertion of the biceps brachii is on the tuberosity of the radius. When the forearm is supinated, the biceps flexes it. However, when the forearm is pronated with the tuberosity of the radius rotated toward the ulna, the biceps acts to supinate the forearm before flexing it. In addition, it strengthens the shoulder joint and assists in flexion of the joint.

The **brachialis** (Figures 9.17 and 9.20), which is deep to the biceps, is also a strong flexor of the forearm. Because it arises on the humerus and inserts on the ulna, it acts only on the elbow joint.

There are two muscles that act to extend the forearm. The first, the **triceps brachii** (Figure 9.18), is the only

◆ **FIGURE 9.19 The biceps brachii muscle**

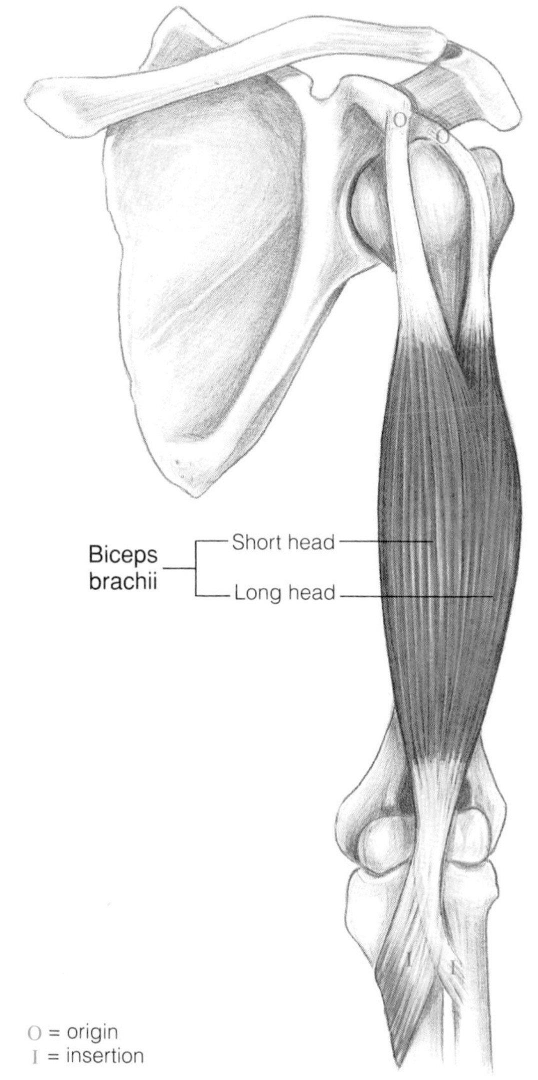

There is one additional muscle that is located in the forearm but acts to flex the forearm. This is the **brachioradialis** muscle (Figure 9.21). It runs from the distal end of the humerus to the distal end of the radius, forming the lateral surface of the forearm.

Muscles That Act on the Hand and Fingers

Most of the muscles that form the bulge of the proximal end of the forearm have at least a part of their origins on the distal end of the humerus. As a result, they cross the elbow joint as well as the wrist. However, their

◆ **FIGURE 9.20 The brachialis and coracobrachialis muscles**

The more superficial biceps brachii muscle has been removed.

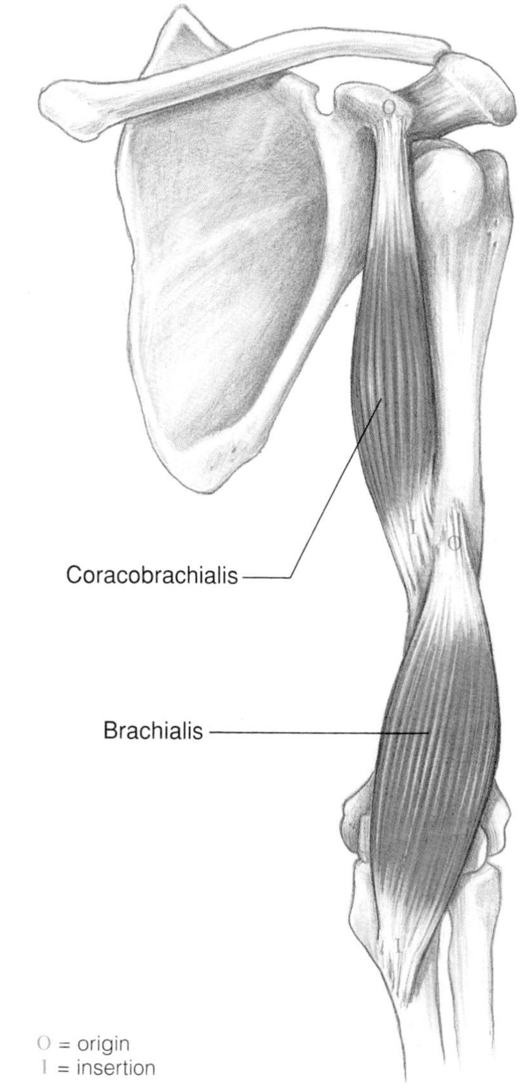

muscle in the posterior compartment of the arm. One of its heads—the *long head*—arises from the scapula; the other two—the *lateral* and *medial heads*—arise from the humerus. All three heads insert by a common tendon onto the olecranon process of the ulna. Because the long head, which passes between the teres major and minor muscles, crosses the shoulder joint, it acts as a weak synergist to the latissimus dorsi muscle to extend and adduct the arm. Its main action, however, is extension of the forearm.

The second extensor of the forearm is the **anconeus** (Figures 9.24 and 9.25), a small muscle that appears to be a lateral continuation of the triceps brachii. It runs from the lateral epicondyle of a humerus to the lateral side of the olecranon process. Although it is considered to be a muscle of the forearm, the anconeus does not act on the wrist; its primary action is to assist the triceps in extending the forearm.

◆ **TABLE 9.12 Summary of Muscle Actions on the Arm**

FLEXION	EXTENSION	ADDUCTION	ABDUCTION	MEDIAL ROTATION	LATERAL ROTATION
Deltoid	Deltoid	Pectoralis major	Deltoid	Pectoralis major	Deltoid
Pectoralis major	Latissimus dorsi	Latissimus dorsi	Supraspinatus	Latissimus dorsi	Supraspinatus
Coracobrachialis	Teres major	Teres major Coracobrachialis Infraspinatus		Deltoid Subscapularis Teres major	Infraspinatus Teres minor

◆ **TABLE 9.13 Muscles That Act on the Forearm (Radius and Ulna)**

MUSCLE	ORIGIN	INSERTION	ACTION	INNERVATION
Biceps brachii *(bī´-seps brā´-kē-ī)* [*Fig. 9.19*]	*Long head:* Supraglenoid tubercle of the scapula *Short head:* Coracoid process of the scapula	Tuberosity of the radius	Flexes the forearm and the arm; supinates the forearm	Musculocutaneous
Brachialis *(brā˝-kē-al´-us)* [*Fig. 9.20*]	Anterior surface of the distal half of the humerus	Coronoid process of the ulna	Flexes the forearm	Musculocutaneous
Triceps brachii *(trī´-seps brā´-kē-ī)* [*Fig. 9.18*]	*Long head:* Infraglenoid tubercle of the scapula *Lateral head:* Posterior surface of the humerus above the radial groove *Medial head:* Posterior surface of the humerus below the radial groove	Olecranon process of the ulna	Extends the forearm; long head also extends the arm	Radial
MUSCLES OF FOREARM				
Anconeus *(an-kō´-nē-us)* [*Fig. 9.24*]	Lateral epicondyle of the humerus	Lateral surface of the olecranon process of the ulna	Extends the forearm	Radial
Brachioradialis *(brā˝-kē-ō-rā-dē-a´-lus)* [*Fig. 9.21*]	Lateral supracondylar ridge of the humerus	Styloid process of the radius	Flexes the forearm	Radial
Pronator teres *(prō´-nā˝-tor tayr-ēz)* [*Fig. 9.21*]	Medial epicondyle of the humerus and the coronoid process of the ulna	Middle of the lateral surface of the shaft of the radius	Pronates and weakly flexes the forearm	Median
Pronator quadratus *(prō´-nā˝-tor kwod-ra´-tus)* [*Figs. 9.21; 9.22; 9.23*]	Distal ventral surface of the ulna	Distal ventral surface of the radius	Pronates the forearm	Median
Supinator *(su´-pĭ-nā-ter)* [*Figs. 9.22; 9.23; 9.25*]	Lateral epicondyle of the humerus, annular ligament, and proximal end of ulna	Proximal end of the lateral surface of the shaft of the radius	Supinates the forearm	Radial

◆ **FIGURE 9.21 Superficial anterior muscles of the right forearm and hand**

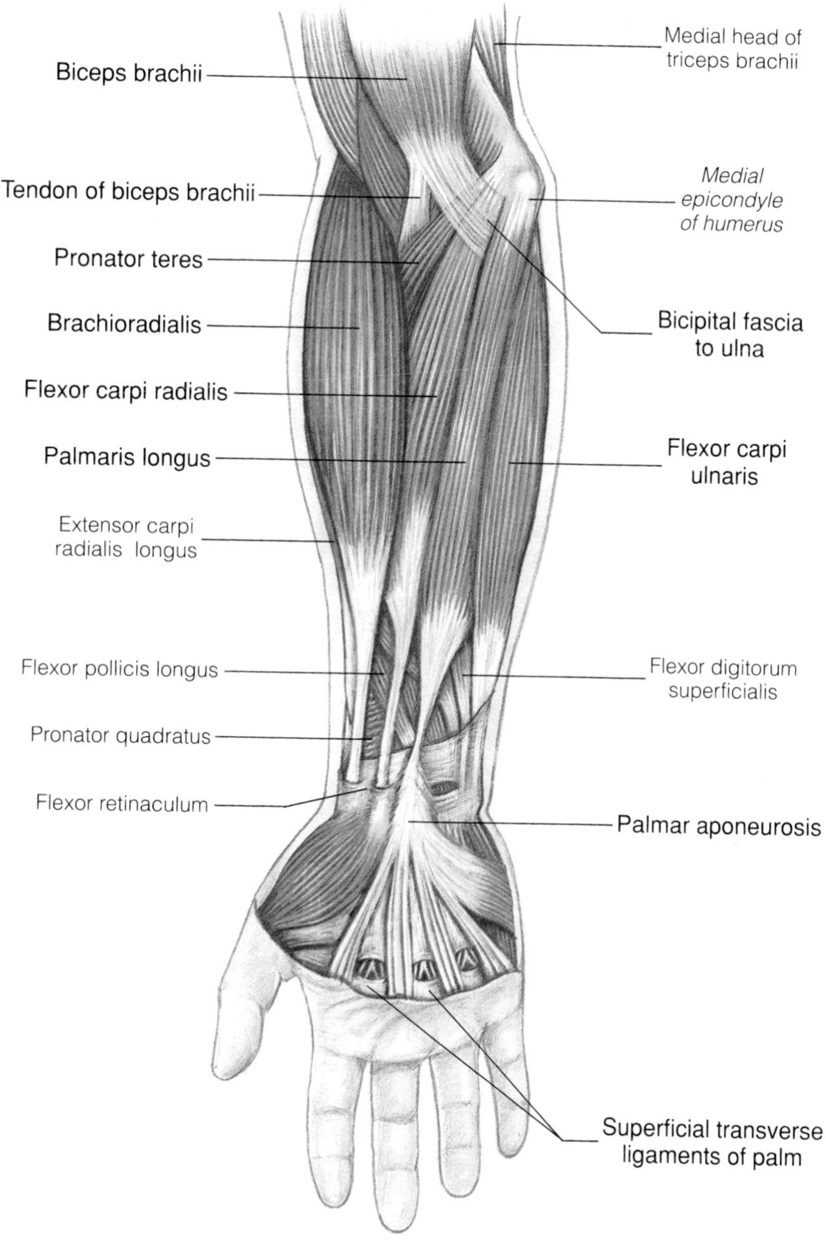

actions on the elbow joint are very slight. Their prime actions are on the hand and fingers.

Although this is a complex group of muscles with formidable-sounding names, most of the names indicate the muscle's action, origin, or insertion. These muscles can be divided into two groups on the basis of location and function. The muscles of the *anterior group* serve as flexors or pronators. Most of these anterior muscles have their origins on the medial epicondyle of the humerus. A few of them insert on the radius, but most insert on the carpals, metacarpals, or phalanges. The *posterior group* of forearm muscles serve as extensors and supinators. Most of these muscles have their origins on the lateral epicondyle of the humerus and insert on the metacarpals or phalanges. Both the anterior and posterior groups can be divided further into superficial and deep muscles. The tendons of these forearm muscles are held down at the wrist by heavy thickenings of the fascia called **flexor** and **extensor retinacula** (*retinaculum* = halter). If these transverse bands of fascia were not present, the tendons would protrude when the hand is flexed or extended.

Although the muscles that act on the hand and fingers are not described here individually, they are illustrated

◆ **FIGURE 9.22 Second layer of anterior muscles of the right forearm**
The superficial muscles have been removed.

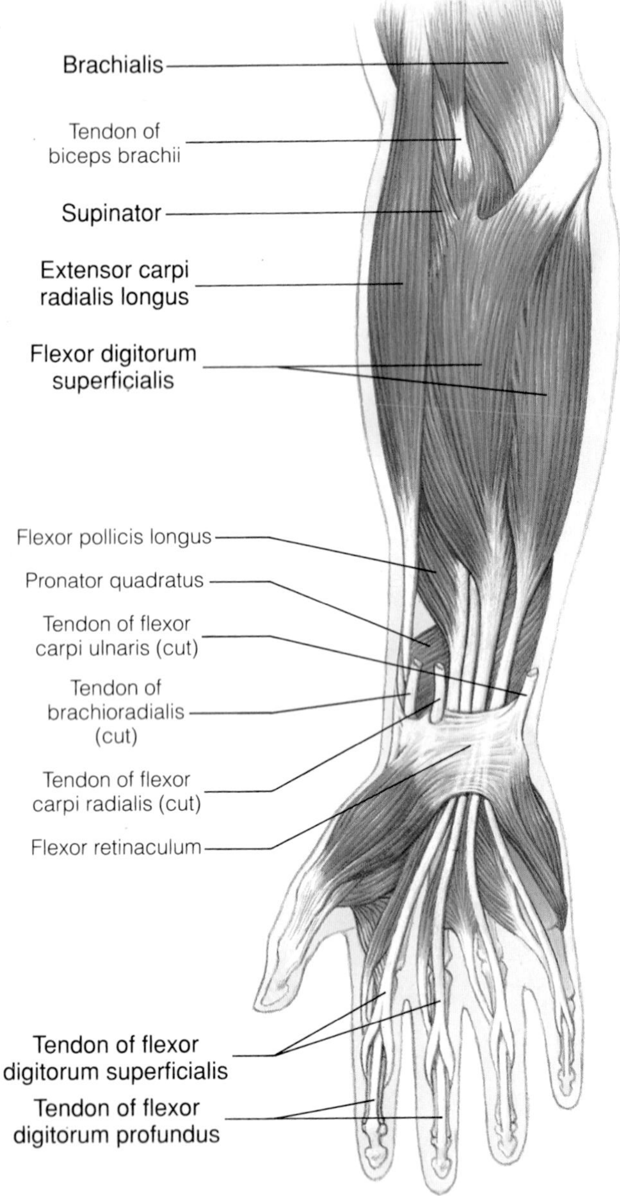

◆ **FIGURE 9.23 Deep anterior muscles of the right forearm**
The more superficial muscles that are illustrated in Figures 9.21 and 9.22 have been removed.

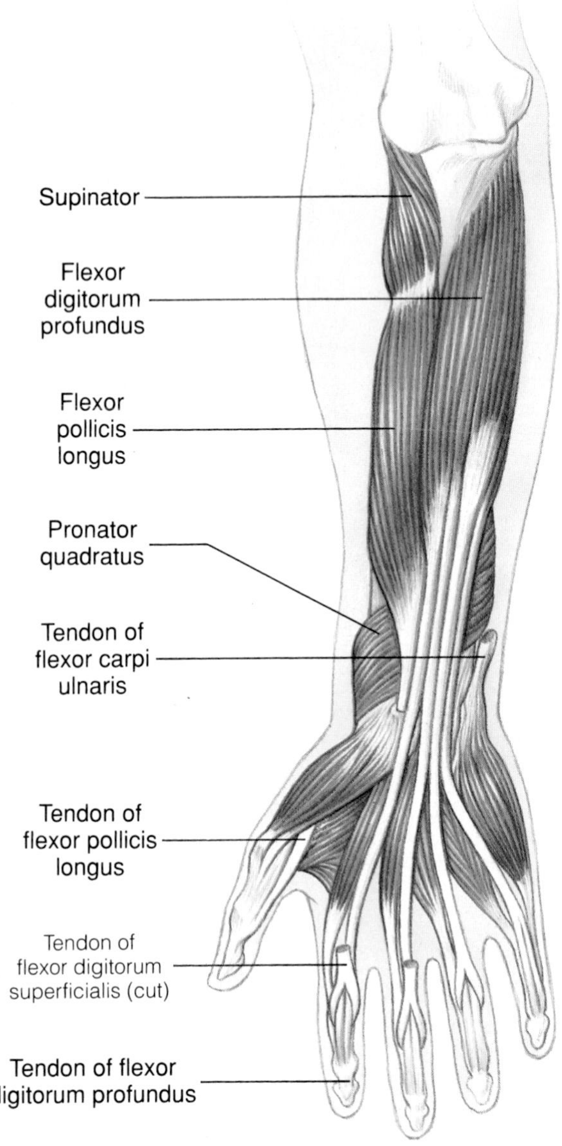

in Figures 9.21 through 9.25, and their precise locations and actions are listed in Table 9.14. In addition, Table 9.15 summarizes their actions and makes it possible to easily identify synergists and antagonists. Notice that while the three muscles that extend the wrist are antagonistic to the flexors of the forearm, the **extensor carpi radialis longus** acts synergistically with the **flexor carpi radialis** to abduct the hand. In a similar manner, the **extensor carpi ulnaris** acts synergistically with the **flexor carpi ulnaris** to adduct the hand.

Intrinsic Muscles of the Hand

We have seen that several of the muscles of the forearm have long tendons that reach the phalanges and serve to move the fingers. Apart from these forearm muscles,

◆ **FIGURE 9.24 Superficial posterior muscles of the right forearm and hand**

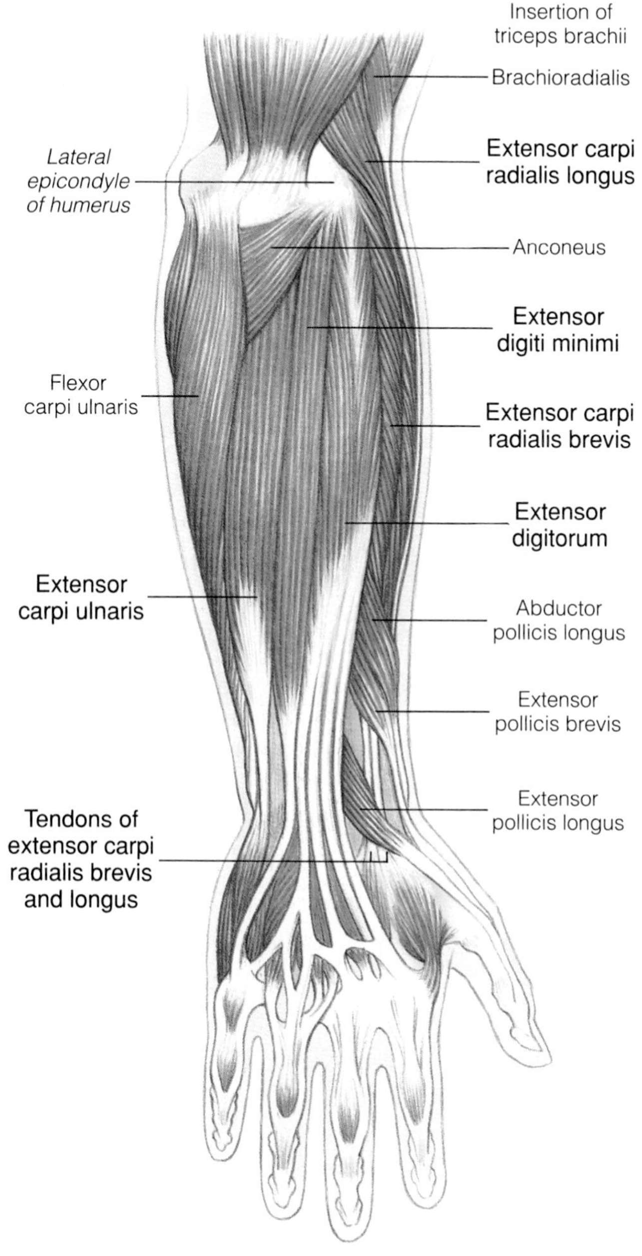

◆ **FIGURE 9.25 Deep posterior muscles of the right forearm and hand**

The superficial muscles have been removed.

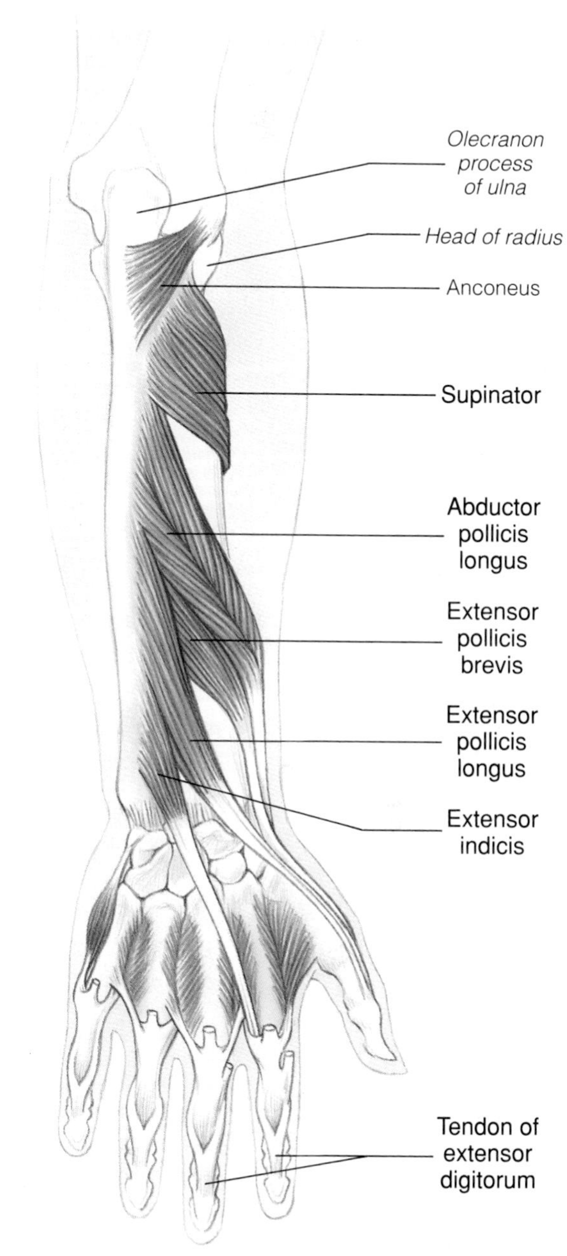

there are several groups of small muscles whose origin and insertion are both in the hand. As we learned earlier, such muscles are called *intrinsic* muscles. The intrinsic muscles make possible the fine and precise movements that are typical of the fingers.

The intrinsic muscles of the hand (Figure 9.26, page 311) are divided into three groups. Those that act on the thumb form the **thenar eminence** at the base of the thumb. Those that act on the little finger form the **hypothenar eminence** on the medial side of the hand. The intermediate, or **midpalmar,** muscles act on all the phalanges except the thumb. These intrinsic muscles of the hand are described in Table 9.16. Notice that there are no intrinsic muscles on the dorsum of the hand, since the dorsal interossei are located between the metacarpal bones.

◆ **TABLE 9.14 Muscles That Act on the Hand and Fingers**

MUSCLE	ORIGIN	INSERTION	ACTION	INNERVATION
ANTERIOR GROUP				
Superficial Muscles (Listed From Lateral to Medial)				
Flexor carpi radialis *(flek´-sor kar´-pē rā˝-dē-a´-lis)* [*Fig. 9.21*]	Medial epicondyle of the humerus	Ventral surface of the second and third metacarpals	Flexes and abducts the hand; aids in flexion and pronation of the forearm	Median
Palmaris longus *(pal-ma´-ris lon´-gus)* [*Fig. 9.21*]	Medial epicondyle of the humerus	Palmar aponeurosis	Flexes the hand	Median
Flexor carpi ulnaris *(flek´-sor kar´-pē ul-na´-ris)* [*Figs. 9.21; 9.24*]	Medial epicondyle of the humerus, olecranon process, and the proximal two-thirds of the posterior surface of the ulna	Pisiform, hamate, and fifth metacarpal	Flexes and adducts the hand	Ulnar
Flexor digitorum superficialis *(flek´-sor dij˝-ĭ-tor´-um su˝-per-fish˝-ē-a´-lis;* beneath the other superficial muscles) [*Figs. 9.21; 9.22*]	Medial epicondyle of the humerus, coronoid process of the ulna, and the anterior surface of the radius	Ventral surface of the middle phalanges of the second through the fifth fingers	Flexes the phalanges and the hand	Median
Deep Muscles				
Flexor digitorum profundus *(flek´-sor dij˝-ĭ-tor´-um prō-fun´-dus; profundus* = deep) [*Fig. 9.23*]	Upper one-half of anterior and medial surfaces of the ulna, the coronoid process, and the interosseous membrane	Ventral surface base of the distal phalanges of the second through the fifth fingers	Flexes the phalanges and the hand	Median and ulnar
Flexor pollicis longus *(flek´-sor pol´-ĭ-kis lon´-gus; pollex* = thumb) [*Figs. 9.21; 9.22; 9.23*]	Ventral surface of the radius and the interosseus membrane	Ventral surface base of the distal phalanx of the thumb	Flexes the thumb; aids in flexing the hand	Median
POSTERIOR GROUP				
Superficial* Muscles (Listed From Lateral to Medial)				
Extensor carpi radialis longus *(ek-sten´-sor kar´-pē rā˝-dē-a´-lis lon´gus)* [*Figs. 9.21; 9.22; 9.24*]	Lateral supracondylar ridge of the humerus	Dorsal surface of the base of the second metacarpal	Extends and abducts the hand	Radial
Extensor carpi radialis brevis *(ek-sten-sor kar´-pē rā˝-dē-a´-lis brev´-is; brevis* = short) [*Fig. 9.24*]	Lateral epicondyle of the humerus	Dorsal surface of the base of the third metacarpal	Extends the hand	Radial
Extensor digitorum *(ek-sten´-sor dij´-ĭ-tor´-um)* [*Fig. 9.24*]	Lateral epicondyle of the humerus	Dorsal surface of the phalanges of the second through the fifth fingers	Extends the fingers and the hand	Radial
Extensor digiti minimi *(ek-sten´-sor dij´-ĭ-tī min´-ĭ-mī;* = little finger) [*Fig. 9.24*]	Tendon of the extensor digitorum	Tendon of the extensor digitorum on the dorsum of the little finger	Extends the little finger	Radial

◆ **TABLE 9.14 Muscles That Act on the Hand and Fingers (continued)**

MUSCLE	ORIGIN	INSERTION	ACTION	INNERVATION
POSTERIOR GROUP				
Superficial* Posterior Muscles, continued				
Extensor carpi ulnaris *(ek-sten´-sor kar´-pē ul-na´-ris) [Fig. 9.24]*	Lateral epicondyle of the humerus	Base of the fifth metacarpal	Extends and adducts the hand	Radial
POSTERIOR GROUP				
Deep Muscles (Listed Lateral to Medial)				
Abductor pollicis longus *(ab-duk´-ter pol´-ĭ-kis lon´-gus) [Figs. 9.24; 9.25]*	Posterior surface of the middle of the radius and ulna, and the interosseus membrane	Base of the first metacarpal	Extends the thumb and abducts the hand	Radial
Extensor pollicis brevis *(ek-sten´-sor pol´-ĭ-kis brev´-is) [Figs. 9.24; 9.25]*	Posterior surface of the middle of the radius, and the interosseus membrane	Base of the first phalanx of the thumb	Extends the thumb and abducts the hand	Radial
Extensor pollicis longus *(ek-sten´-sor pol´-ĭ-kis lon´-gus) [Figs. 9.24; 9.25]*	Posterior surface of the middle of the ulna, and the interosseus membrane	Base of the last phalanx of the thumb	Extends the thumb and abducts the hand	Radial
Extensor indicis *(ek-sten´-sor in´-dih-kis) [Fig. 9.25]*	Posterior surface of the distal end of the ulna, and the interosseus membrane	Tendon of the extensor digitorum to the index finger	Extends the index finger	Radial

*The brachioradialis is located with the posterior superficial muscles but was described in Table 9.13 because it is a flexor of the forearm.

◆ **TABLE 9.15 Summary of Muscle Actions on the Hand**

FLEXION	EXTENSION	ADDUCTION	ABDUCTION
Flexor carpi radialis	Extensor carpi radialis longus	Flexor carpi ulnaris	Flexor carpi radialis
Palmaris longus	Extensor carpi radialis brevis	Extensor carpi ulnaris	Extensor carpi radialis longus
Flexor carpi ulnaris	Extensor carpi ulnaris		Abductor pollicis longus
Flexor digitorum superficialis	Extensor digitorum		Extensor pollicis brevis
Flexor digitorum profundus			Extensor pollicis longus
Flexor pollicis longus			

◆ **TABLE 9.16 Intrinsic Muscles of the Hand** [*Fig. 9.26*]

MUSCLE	ORIGIN	INSERTION	ACTION	INNERVATION
Abductor pollicis brevis *(ab-duk´-ter pol´-ĭ-kis brev´-is)*	Flexor retinaculum, scaphoid, and trapezium	Proximal phalanx of the thumb	Abducts the thumb	Median
Opponens pollicis *(o-po´-nenz pol´-ĭ-kis; opponens = one that opposes)*	Flexor retinaculum and trapezium	Lateral border of the metacarpal of the thumb	Pulls the thumb in front of the palm to meet the little finger	Median
Flexor pollicis brevis *(flek´-sor pol´-ĭ-kis brev´-is)*	Flexor retinaculum, trapezium, and first metacarpal	Base of the proximal phalanx of the thumb	Flexes and adducts the thumb	Median and ulnar
Adductor pollicis *(ad-duk´-tor pol´-ĭ-kis)*	Capitate, and second and third metacarpals	Proximal phalanx of the thumb	Adducts the thumb	Ulnar
HYPOTHENAR MUSCLES				
Palmaris brevis *(pal-ma´-ris brev´-is)*	Flexor retinaculum	Skin on the ulnar border of the hand	Pulls the skin toward the middle of the palm	Ulnar
Abductor digiti minimi *(ab-duk´-ter dij´-ĭ-tī min´-ĭ-mī)*	Pisiform, and the tendon of the flexor carpi ulnaris	Base of the proximal phalanx of the little finger	Abducts the little finger	Ulnar
Flexor digiti minimi brevis *(flek´-sor dij´-ĭ-tī min´-ĭ-mī brev´-is)*	Flexor retinaculum and hamate	Base of the proximal phalanx of the little finger	Flexes the little finger	Ulnar
Opponens digiti minimi *(o-po´-nenz dij´-ĭ-tī min´-ĭ-mī)*	Flexor retinaculum and hamate	Metacarpal of the little finger	Brings the little finger out to meet the thumb	Ulnar
MIDPALMAR MUSCLES				
Lumbricales *(lum´-brĭ-kals)*	Tendons of the flexor digitorum profundus	Tendons of the extensor digitorum	Flex the proximal phalanx and extend the middle and distal phalanges of the second through fifth fingers	Median and ulnar
Dorsal interossei (4) *(in″-ter-os´-ē-ī)*	Adjacent sides of all of the metacarpals	Proximal phalanx of second, third, and fourth fingers	Abduct the fingers from the middle finger; flex the proximal phalanx	Ulnar
Palmar interossei (3)	Medial side of the second metacarpal, and lateral side of the fourth and fifth metacarpals	Proximal phalanx of the same finger	Adduct the fingers toward the middle finger; flex the proximal phalanx	Ulnar

◆ **FIGURE 9.26 Palmar view of the intrinsic muscles of the hand**
(a) Superficial muscles. (b) Interossei. The superficial muscles have been removed.

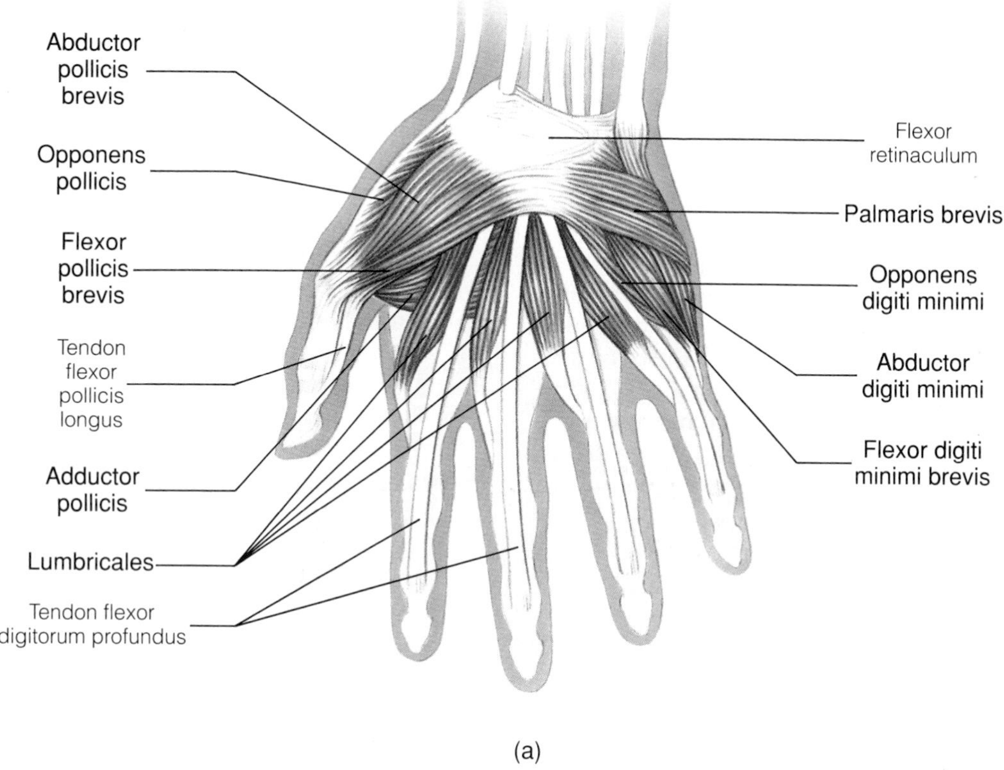

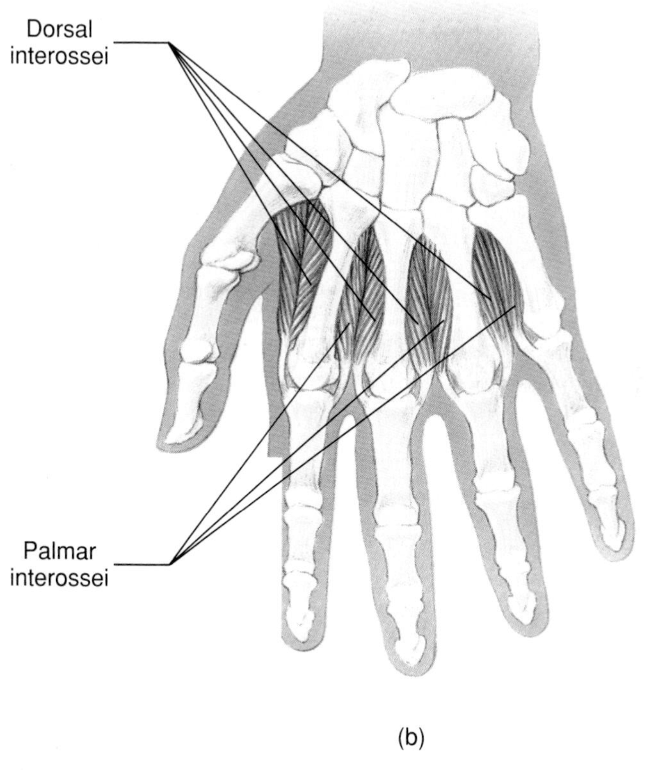

 Muscles of the Lower Limbs

When compared to the muscles of the upper limbs, those of the lower limbs tend to be bulkier and more powerful. The versatile movements that are characteristic of the upper limbs are somewhat sacrificed in the lower limbs in favor of strength, stability, and locomotion. Many of the muscles of the lower limbs are used in maintaining upright posture and therefore must constantly resist the pull of gravity. Unlike the pectoral girdle, the pelvis does not rely entirely on muscles to stabilize or fix it. The only movement possible in the bony pelvis is a slight gliding between the sacrum and the ilium. Many of the muscles of the lower limbs cross two joints—either the hip and the knee, or the knee and the ankle—and may act equally strongly on both joints. These double actions are listed in Tables 9.17–9.19.

Muscles That Act on the Thigh (Femur)

Most of the muscles that act on the femur arise from the pelvic girdle. One muscle, the **psoas major** (Figures 9.27a and 9.28), arises from the lumbar vertebrae. Because the psoas major muscle and the **iliacus** muscle (Figures 9.27b and 9.28), share a common insertion and are synergistic to each other, these two muscles are often referred to as the **iliopsoas** muscle. The tendon of the iliopsoas passes beneath the inguinal ligament to reach the femur, which it flexes. When the lower limbs are fixed, the psoas muscles flex the vertebral column, as when bending over.

Three large gluteal muscles (Figures 9.29 and 9.30) give shape to the buttocks and serve as powerful mobilizers of the hip joint. The largest and most superficial is the **gluteus maximus.** The gluteus maximus covers the posterior third of the smaller **gluteus medius;** deep to the gluteus medius is the still smaller **gluteus minimus.** The broad tendon of the gluteus maximus passes behind the hip joint, causing it to extend and laterally rotate the femur. In contrast, the tendon of the gluteus medius passes above, and that of the minimus passes in front of, the hip joint, causing them to abduct and medially rotate the femur. In the rotation of the thigh, therefore, the gluteus maximus is an antagonist to the two smaller gluteal muscles. Acting synergistically with the gluteus maximus in rotating the thigh laterally are six deep muscles that extend from the sacrum and the coxal bone to the posterior surface of the proximal end of the femur. These lateral rotators of the thigh are included in Table 9.17.

The **tensor fasciae latae** (Figure 9.28) is a lateral hip muscle that inserts on a strong band of connective tissue called the **iliotibial tract** of the **fascia lata** ("broad fascia"). The fascia lata encloses all of the muscles of the thigh, but it is especially thick laterally—thus forming the iliotibial tract. The tensor serves primarily to stabilize the knee by tightening the fascia lata, but it also assists in flexing the thigh.

◆ **FIGURE 9.27 Two muscles that act on the thigh**
(a) The psoas major muscle. (b) The iliacus muscle.

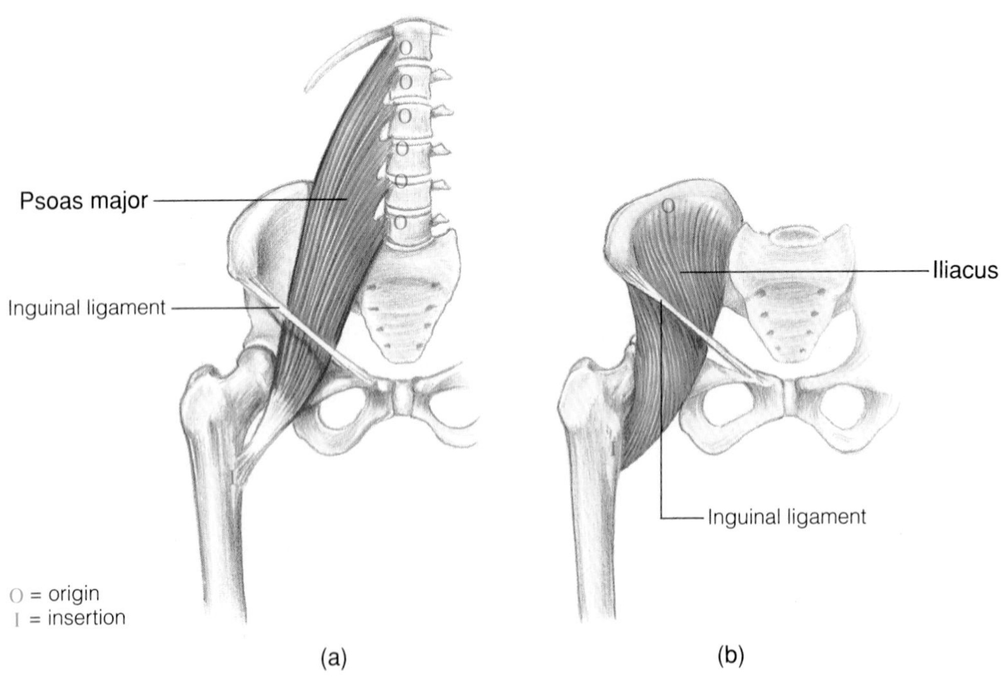

◆ **FIGURE 9.28 Anterior view of the muscles that attach the femur to the pelvis and the lumbar vertebrae**

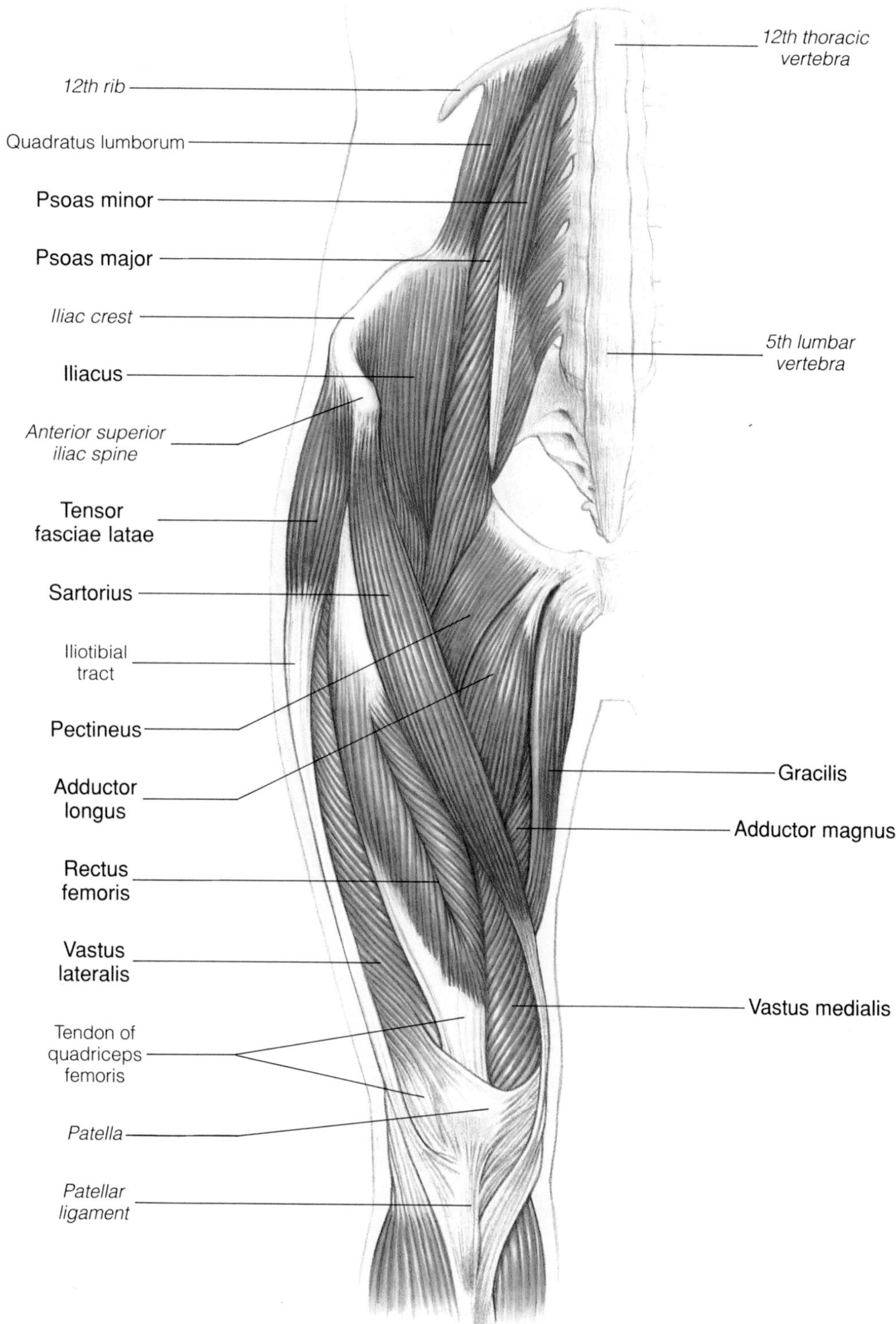

◆ **FIGURE 9.29 Superficial muscles of the posterior hip and thigh**

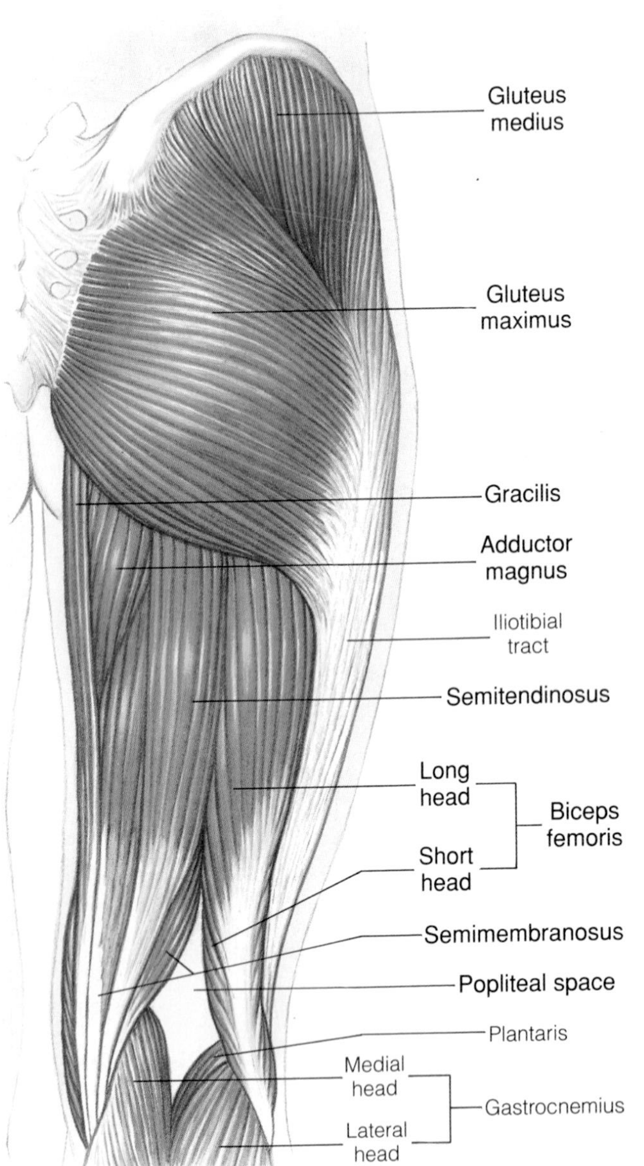

◆ **FIGURE 9.30 Deep muscles of the posterior hip**
The gluteus maximus and gluteus medius have been cut to expose the deep muscles.

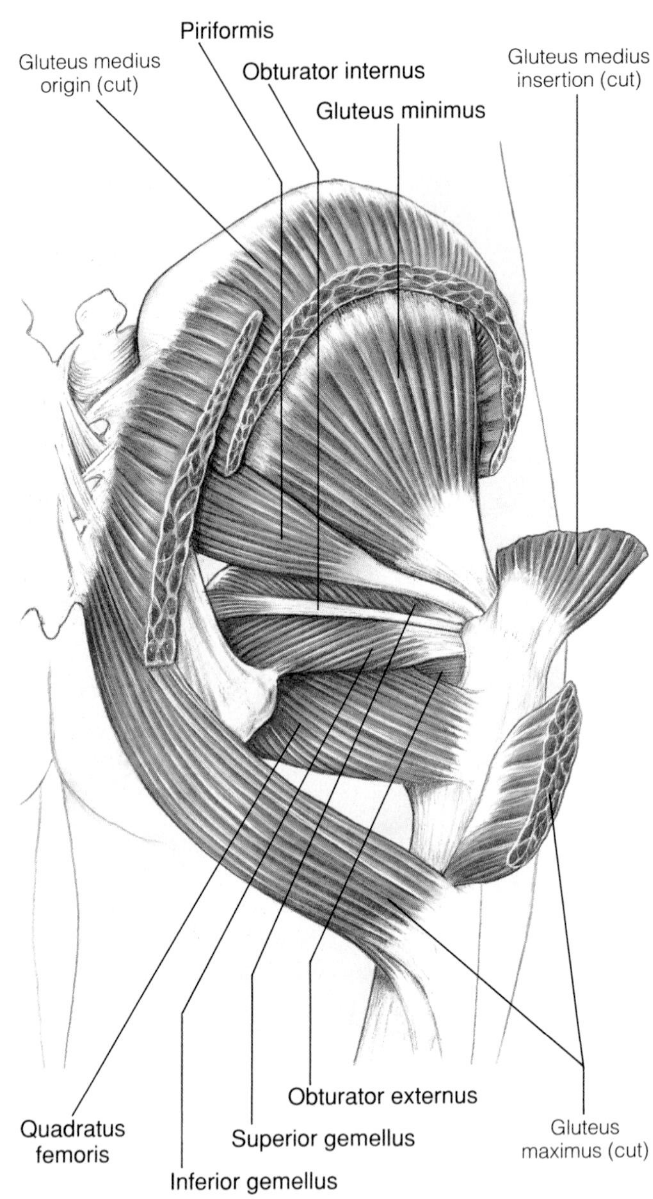

◆ **TABLE 9.18 Muscles That Act on the Leg (Muscles of the Thigh)**

MUSCLE	ORIGIN	INSERTION	ACTION	INNERVATION
MEDIAL COMPARTMENT				
Adductor magnus Adductor longus Adductor brevis Pectineus	These muscles act only on the femur. They are illustrated in Figures 9.28 and 9.31 and described in Table 9.17.		Adduct and laterally rotate the thigh	Obturator
Gracilis *(grub-sil´-us)* [*Fig. 9.33*]	Symphysis pubis and the pubic arch	Medial surface of the tibia just below the condyle	Adduct the thigh; flex the leg	Obturator
ANTERIOR COMPARTMENT				
Sartorius *(sar-tor´-ē-us)* [*Fig. 9.35*]	Anterior superior iliac spine	Proximal medial surface of the tibia, below the tuberosity	Flex the thigh and the leg; laterally rotate the thigh	Femoral
Quadriceps femoris *(kwod´-rĭ-ceps fem´-ō-rus)* [*Fig. 9.34*]				
Rectus femoris *(rek´-tus fem´-ō-rus)*	Anterior inferior iliac spine and just above the acetabulum of the coxal bone	Tibial tuberosity, via the patella and the patellar ligament	Extend the leg; the rectus femoris also flexes the thigh	Femoral
Vastus lateralis *(vas´-tis la-ter-a´-lis)*	Greater trochanter and lateral lip of the linea aspera of the femur			
Vastus medialis *(vas´-tis mē-dē-a´-lis)*	Medial lip of the linea aspera of the femur			
Vastus intermedius *(vas´-tis in˝-ter-mē´-dē-us)*	Anterior surface of the shaft of the femur			
POSTERIOR COMPARTMENT				
Hamstrings [*Fig. 9.29*]				
Biceps femoris *(bī´-seps fem´-ō-rus)*	*Long head:* ischial tuberosity *Short head:* lateral lip of the linea aspera	Lateral surface of the head of the fibula, and the lateral condyle of the tibia	Flexes the leg; long head extends the thigh	Sciatic
Semitendinosus *(sem˝-ē-ten˝-di-nō´-sus)*	Ischial tuberosity	Medial surface of the proximal end of the tibia	Flexes the leg; extends the thigh	Tibial
Semimembranosus *(sem˝-ē-mem˝-brab-nō´-sus)*	Ischial tuberosity	Medial surface of the proximal end of the tibia	Flexes the leg; extends the thigh	Tibial

continued on next page

◆ **TABLE 9.19 Summary of Muscle Actions on the Thigh (Including Muscles of the Thigh)**

FLEXION	EXTENSION	ADDUCTION	ABDUCTION	MEDIAL ROTATION	LATERAL ROTATION
Iliopsoas	Gluteus maximus	Adductor magnus	Gluteus medius	Gluteus medius	Gluteus maximus
Sartorius	Biceps femoris	Adductor longus	Gluteus minimus	Gluteus minimus	Piriformis
Rectus femoris	Semitendinosus	Adductor brevis	Tensor fasciae latae	Tensor fasciae latae	Obturator internus
Pectineus	Semimembranosus	Pectineus	Piriformis		Obturator externus
Adductor longus	Piriformis	Gracilis			Superior and inferior gemelli
Tensor fasciae latae	Adductor magnus				Quadratus femoris Adductor magnus Adductor longus Adductor brevis Pectineus

◆ **FIGURE 9.34 The individual muscles that form the quadriceps femoris muscle**

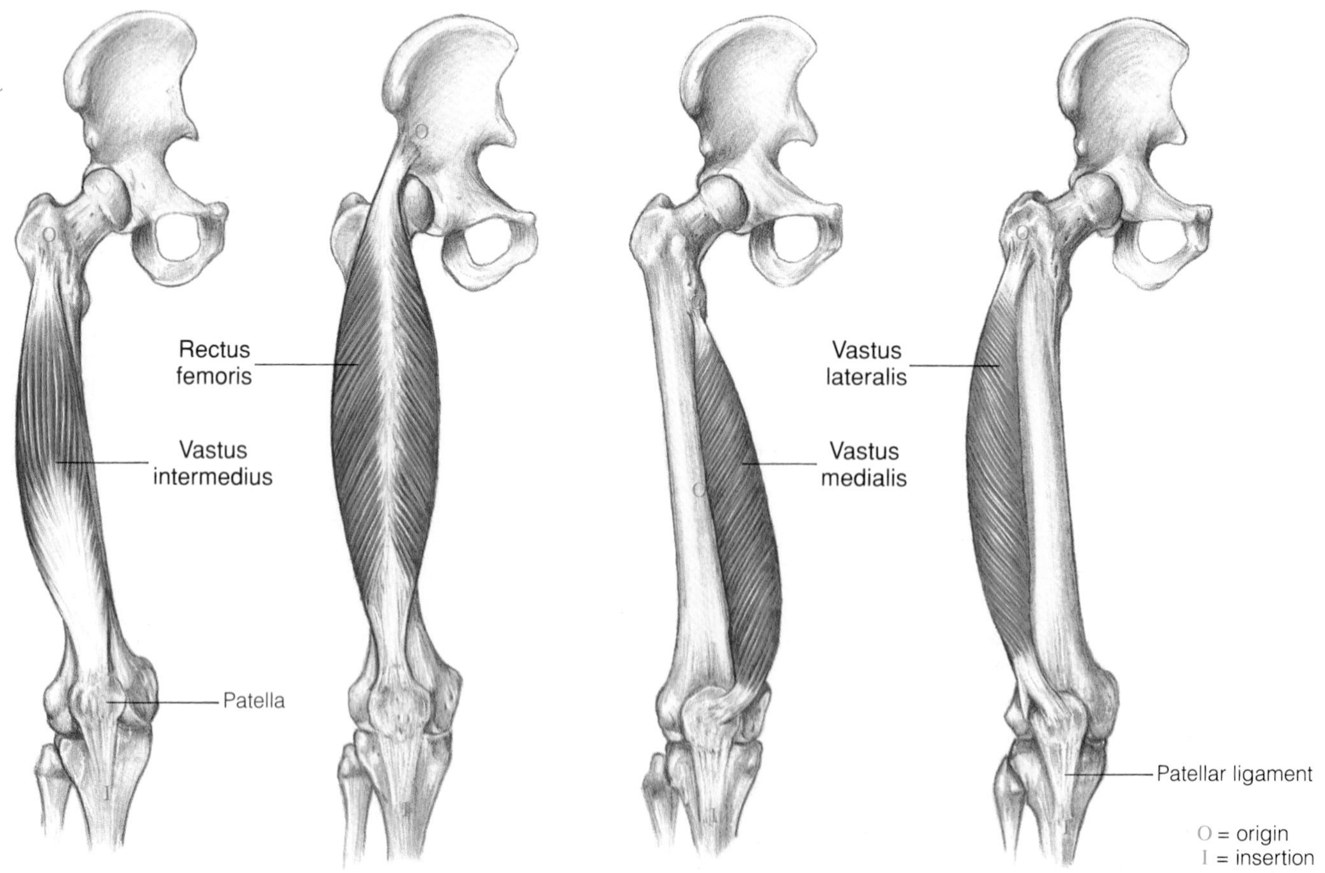

◆ **FIGURE 9.35 The sartorius muscle**

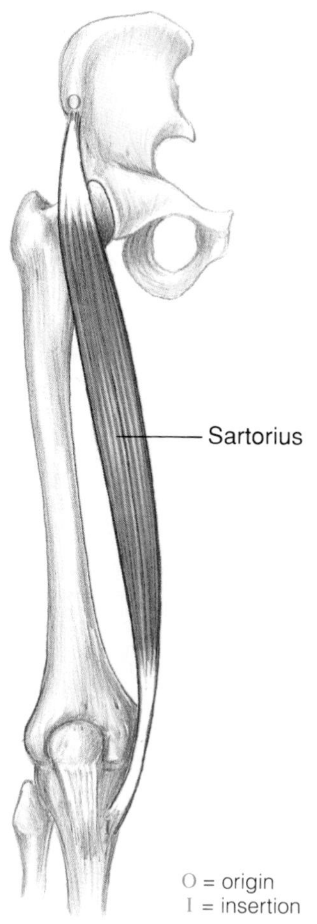

◆ **FIGURE 9.36 Cross section of the right leg showing the three compartments and general actions of the muscles in them**

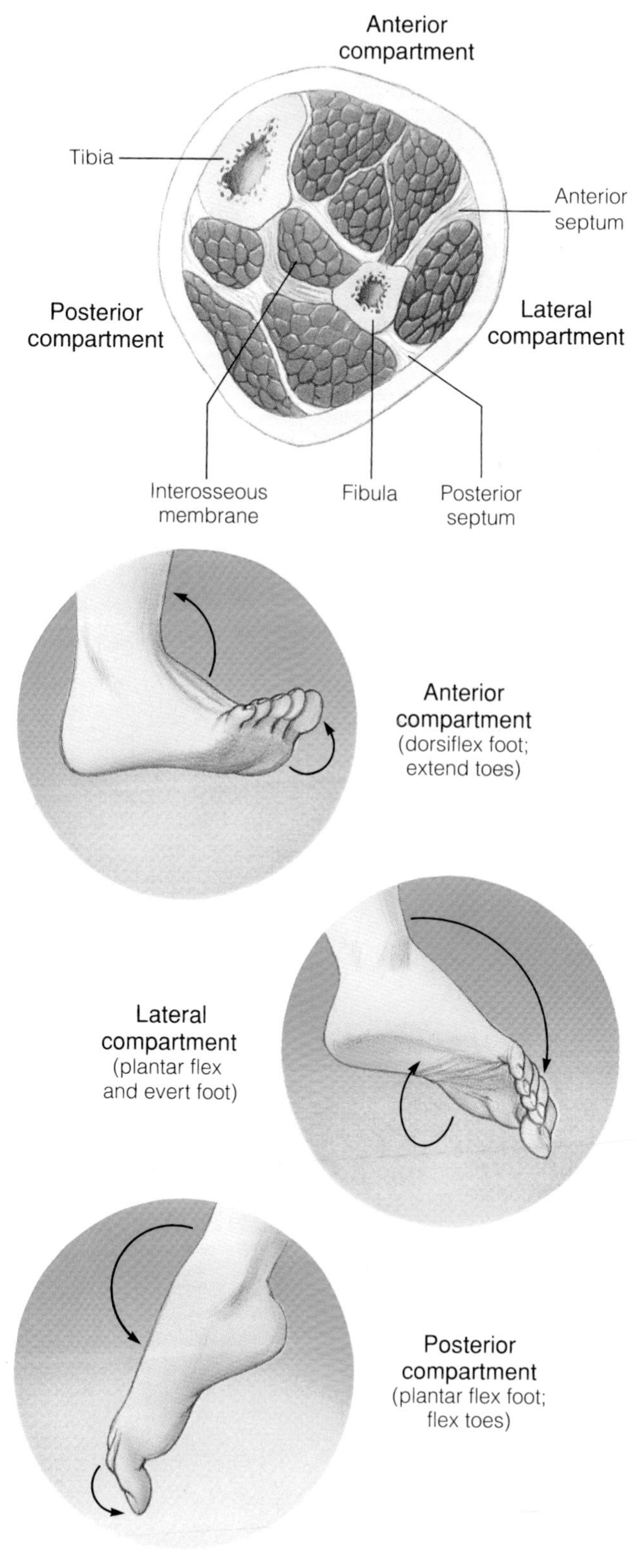

Muscles That Act on the Foot and Toes

The muscles of the leg, like the muscles of the thigh, are grouped into three compartments (Figure 9.36; Table 9.20). The interosseous membrane between the tibia and fibula, and connective-tissue septa that extend anteriorly and posteriorly from the fibula, separate the muscles into *anterior, posterior,* and *lateral compartments.*

The muscles of the *anterior compartment* (Figures 9.37 and 9.38) act to extend the toes and/or dorsiflex the foot. The tendons of these anterior muscles are held firmly to the ankle by **superior** and **inferior extensor retinacula,** in a manner similar to the tendons at the wrist.

The *lateral* or *peroneal compartment* (Figures 9.39 and 9.40) contains two of the three peroneal muscles. The **peroneus tertius** is fused to the extensor digitorum longus muscle and is included in the anterior compartment. The tendons of the **peroneus longus** and **brevis** pass behind the lateral malleolus of the fibula to insert on the plantar and lateral surfaces of the metatarsals. The peroneus longus tendon crosses the sole of the foot obliquely, from the lateral edge to the medial tarsal

◆ FIGURE 9.37 Muscles of the anterior compartment of the right leg

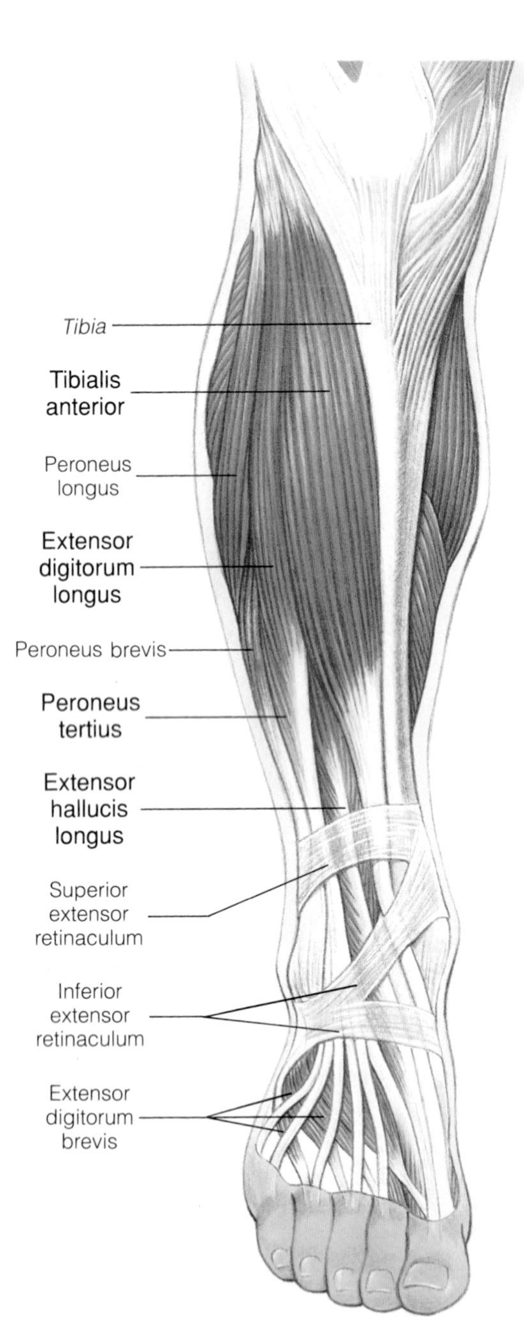

◆ FIGURE 9.38 The individual muscles of the anterior compartment of the right leg

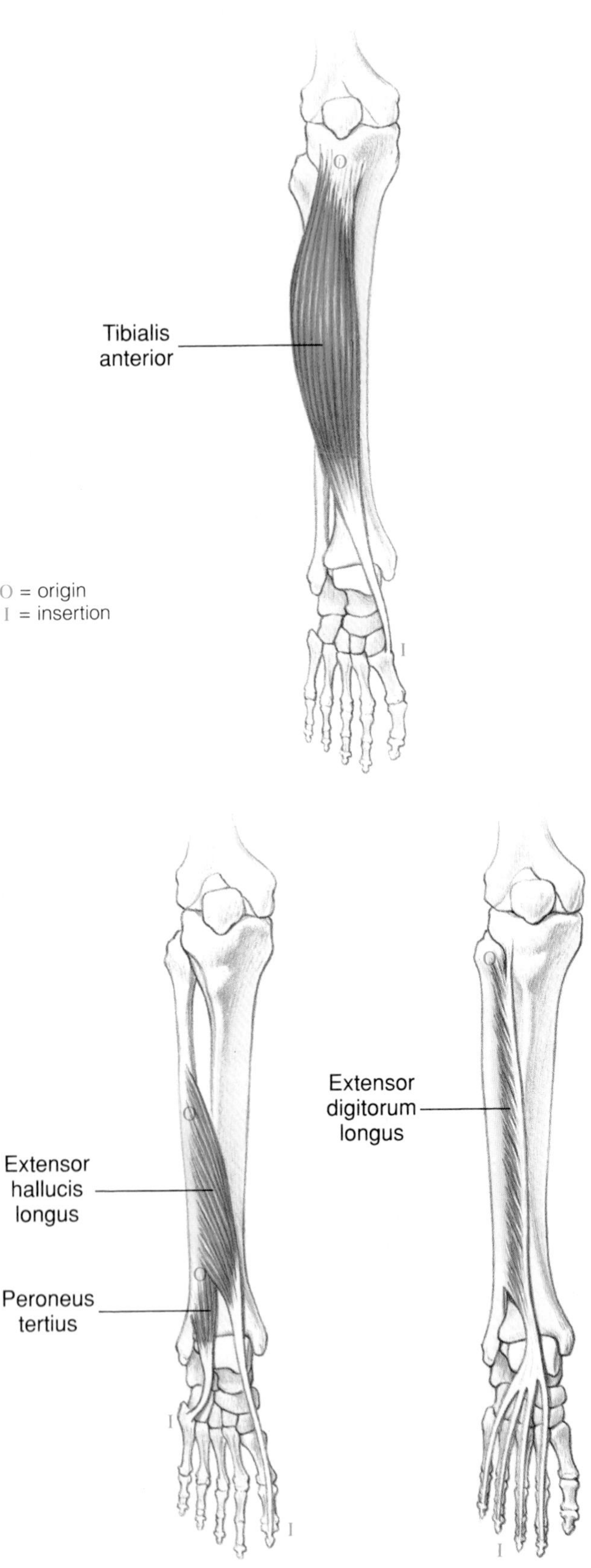

◆ **FIGURE 9.39 Muscles of the lateral compartment of the right leg**

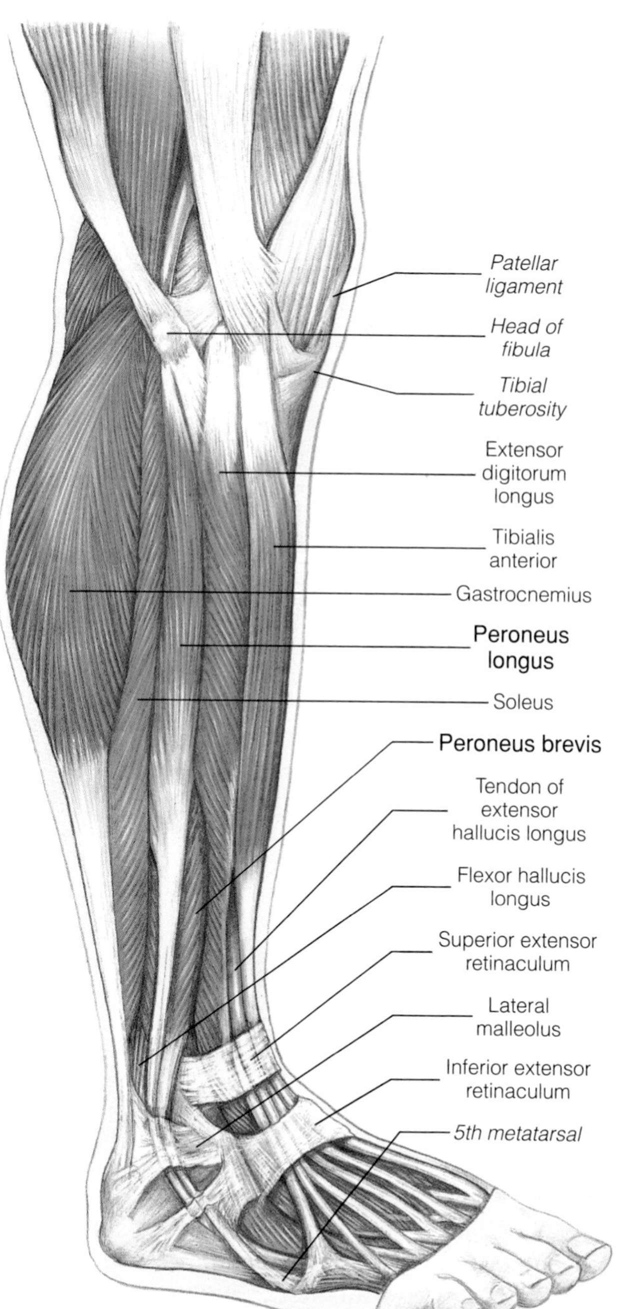

◆ **FIGURE 9.40 The individual muscles of the lateral compartment of the right leg**

The inset is a plantar view of the right foot, showing the insertion of the peroneus longus muscle.

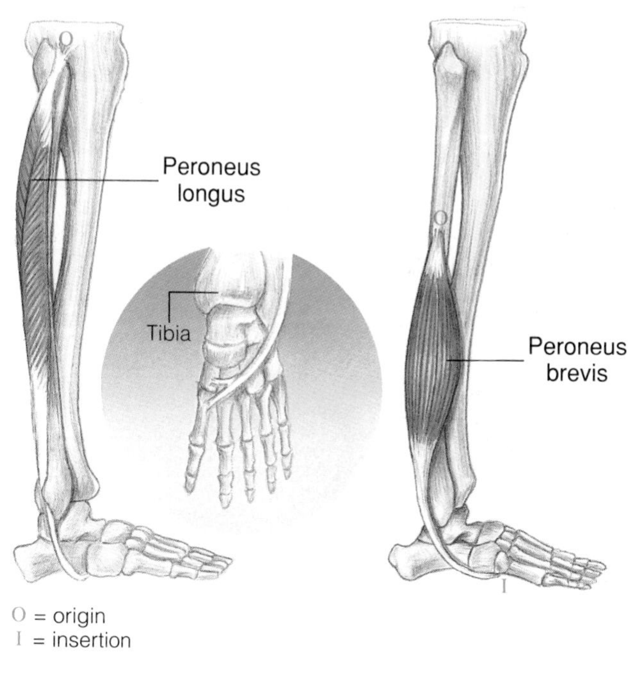

◆ **FIGURE 9.41 Superficial muscles of the posterior compartment of the right leg**
(a) Superficial layer. (b) The origin of the gastrocnemius has been removed to reveal the underlying muscles.

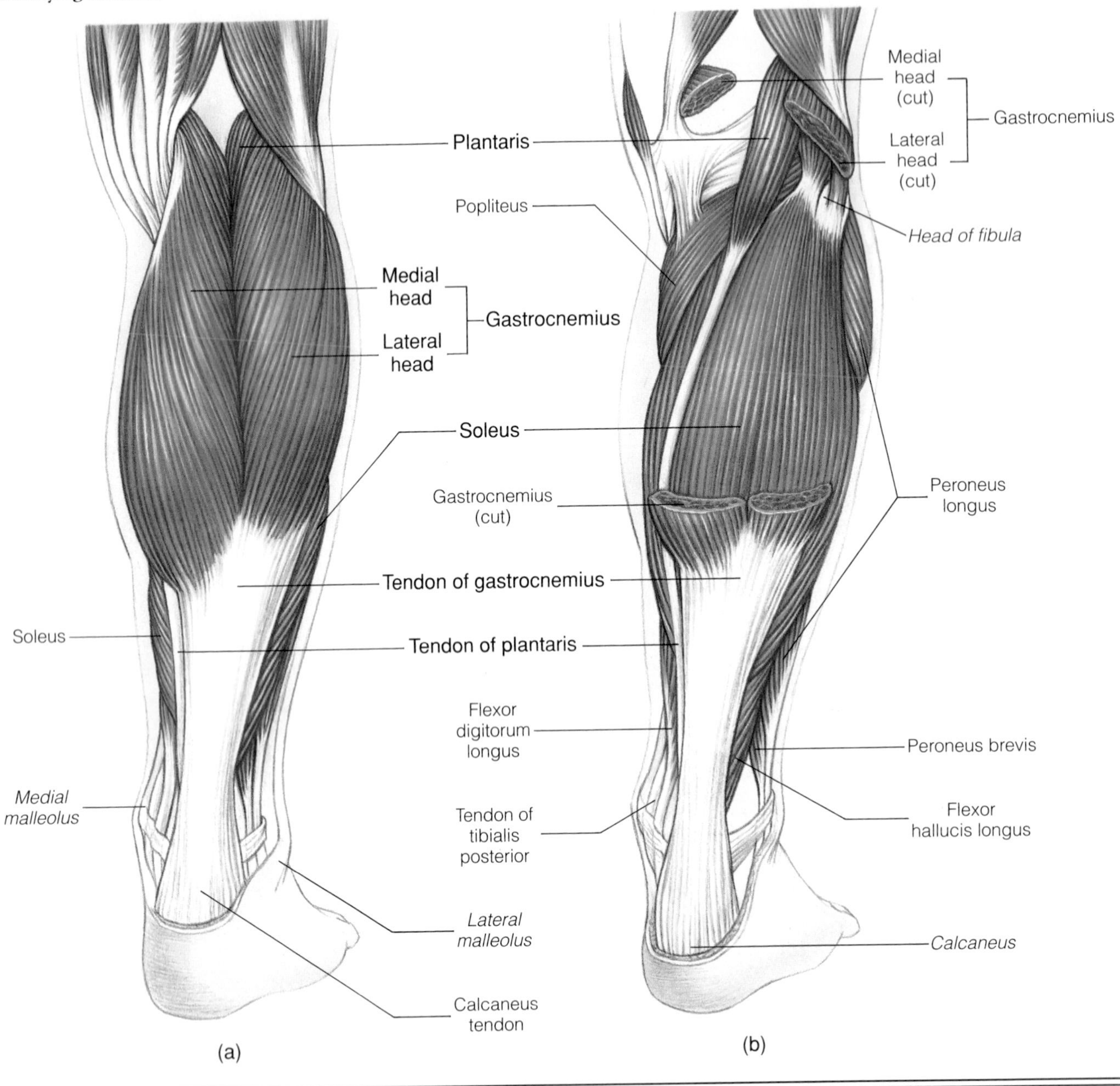

and metatarsal bones. Using the lateral malleolus as a pulley, the peroneal muscles of the lateral compartment act to plantar flex and evert the foot.

The muscles of the *posterior compartment* are grouped into three superficial posterior muscles (Figure 9.41)—the **gastrocnemius, soleus,** and **plantaris**—and four deep posterior muscles (Figures 9.42 and 9.43)—the **popliteus, flexor hallucis longus, flexor digitorum longus,** and the **tibialis posterior.** The tendons of the latter three muscles pass behind the medial malleolus of the tibia and use it as a pulley. The gastrocnemius and the underlying soleus form the characteristic bulge of the calf. They share a common tendon, the **calcaneus (Achilles) tendon,** which inserts onto the calcaneus. With the exception of the popliteus, which flexes and medially rotates the leg, the muscles of the posterior compartment plantar flex the foot, and two of them also flex the toes.

As is the case in the thigh, the muscles within each compartment of the leg are innervated by branches from the same nerve (Table 9.20). The muscles in the anterior compartment are innervated by the deep peroneal

◆ **FIGURE 9.42 Deep muscles of the posterior compartment of the right leg**

The gastrocnemius and the soleus have been removed.

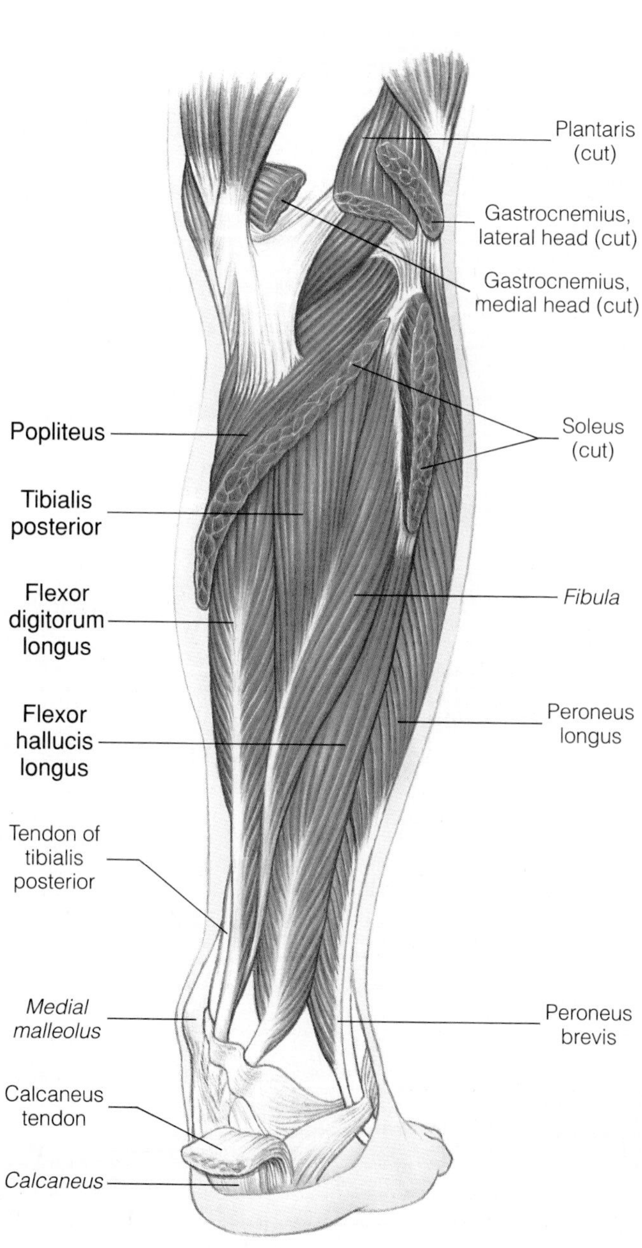

◆ **FIGURE 9.43 The individual deep muscles of the posterior compartment of the right leg**

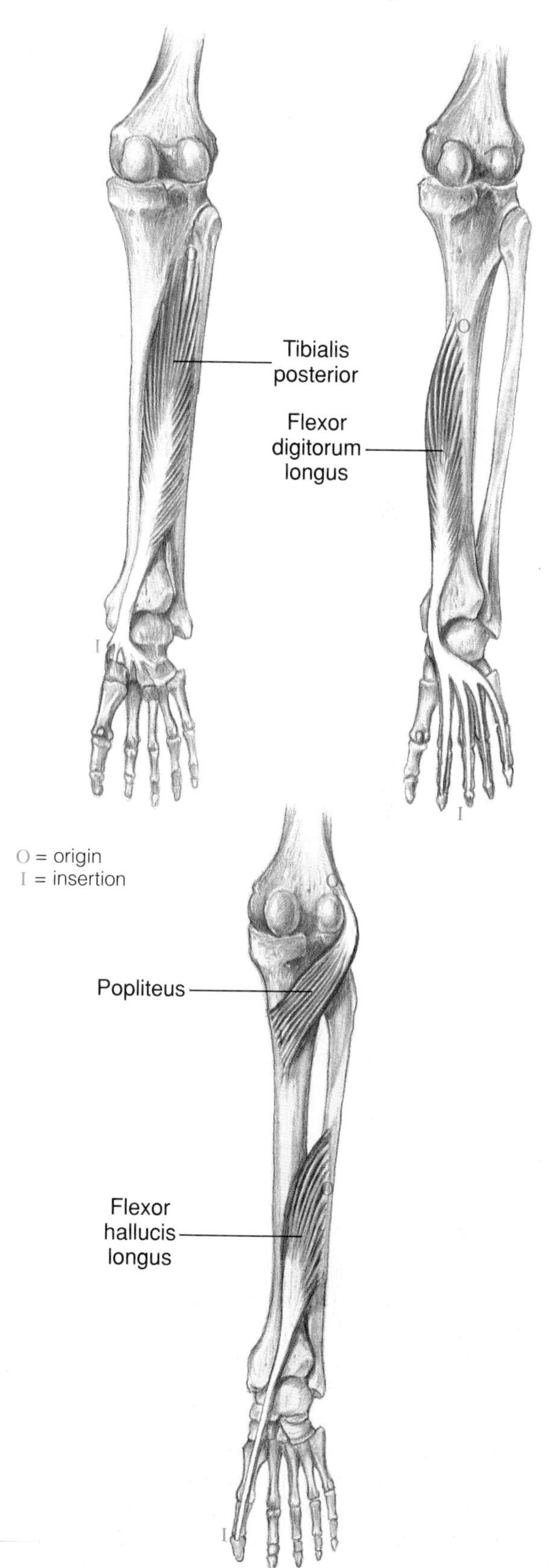

◆ **TABLE 9.20 Muscles That Act on the Foot and Toes (Muscles of the Leg)**

MUSCLE	ORIGIN	INSERTION	ACTION	INNERVATION
ANTERIOR COMPARTMENT				
Tibialis anterior *(tib´-ē-al-us)* [*Figs. 9.37; 9.38*]	Lateral condyle and proximal two-thirds of the shaft of the tibia, and the interosseous membrane	Medial surface of first cuneiform and first metatarsal	Dorsiflexes and inverts foot	Deep peroneal
Extensor hallucis longus *(ek-sten´-sor hal´-u-kis lon´-gus; hallux* = great toe) [*Fig. 9.38*]	Anterior surface of the middle of the fibula, and the interosseous membrane	Dorsal surface of the distal phalanx of the great toe	Dorsiflexes and inverts foot; extends the great toe	Deep peroneal
Extensor digitorum longus *(ek-sten´-sor dij˝-ĭ-tor´-um lon´-gus)* [*Fig. 9.38*]	Lateral condyle of the tibia, proximal three-fourths of the anterior surface of the fibula, and the interosseous membrane	Dorsal surface of the phalanges of the second through fifth toes	Dorsiflexes and everts the foot; extends the toes	Deep peroneal
Peroneus tertius *(payr˝-ō-nē´-us ter´-shus; tertius* = third) [*Fig. 9.38*]	Distal third of the anterior surface of the fibula, and the interosseous membrane	Dorsal surface of the fifth metatarsal	Dorsiflexes and everts the foot	Deep peroneal
LATERAL COMPARTMENT				
Peroneus longus *(payr˝-ō-nē´-us lon´-gus)* [*Figs. 9.39; 9.40*]	Proximal two-thirds of the lateral surface of the fibula	Ventral surface of the first metatarsal and the medial cuneiform	Plantar flexes and everts the foot	Superficial peroneal
Peroneus brevis *(payr˝-ō-nē´-us brev´-is)* [*Fig. 9.39*]	Distal two-thirds of the fibula	Lateral side of the fifth metatarsal	Plantar flexes and everts the foot	Superficial peroneal
POSTERIOR COMPARTMENT				
Superficial Muscles				
Gastrocnemius *(gas´´-trok-ne´-mē-us)* [*Figs. 9.39; 9.41a*]	Medial and lateral condyles of the femur	Calcaneus, via the calcaneus tendon	Flexes the leg; plantar flexes the foot	Tibial
Soleus *(sō´-lē-us)* [*Figs. 9.39; 9.41*]	Posterior surface of the proximal third of the fibula, and the middle third of the tibia	Calcaneus, via the calcaneus tendon	Plantar flexes the foot	Tibial
Plantaris *(plan-tah´-ris)* [*Fig. 9.41b*]	Posterior surface of the femur above the lateral condyle	Calcaneus	Flexes the leg; plantar flexes the foot	Tibial
Deep Muscles				
Popliteus *(pop˝-lĭ-tē´-us)* [*Figs. 9.41b; 9.42*]	Lateral condyle of the femur	Proximal portion of the tibia	Flexes the leg and rotates it medially	Tibial
Flexor hallucis longus *(flek´-sor hal´-u-kis lon´-gus)* [*Figs. 9.41b; 9.42*]	Lower two-thirds of the fibula	Distal phalanx of the great toe	Flexes the great toe; plantar flexes and inverts the foot	Tibial

◆ **TABLE 9.20 Muscles That Act on the Foot and Toes (Muscles of the Leg) (continued)**

MUSCLE	ORIGIN	INSERTION	ACTION	INNERVATION
Deep Posterior Muscles (continued)				
Flexor digitorum longus *(flek´-sor dij´´-ĭ-tor´-um lon´-gus)* [*Figs. 9.41b; 9.42*]	Posterior surface of the tibia	Distal phalanx of the second through fifth toes	Flexes the toes; plantar flexes and inverts the foot	Tibial
Tibialis posterior *(tib´-ē-al-us)* [*Fig. 9.42*]	Posterior surface of the interosseous membrane, the tibia, and the fibula	Navicular, cuneiforms, cuboid; second through fourth metatarsals	Plantar flexes and inverts the foot	Tibial

◆ **TABLE 9.21 Summary of Muscle Actions on the Foot**

DORSIFLEXION	PLANTAR FLEXION	ADDUCTION PLUS INVERSION	ABDUCTION PLUS EVERSION
Tibialis anterior	Peroneus longus	Tibialis anterior	Peroneus tertius
Extensor hallucis longus	Peroneus brevis	Tibialis posterior	Peroneus longus
Extensor digitorum longus	Gastrocnemius	Extensor hallucis longus	Peroneus brevis
Peroneus tertius	Soleus Plantaris Tibialis posterior Flexor hallucis longus Flexor digitorum longus	Flexor hallucis longus Flexor digitorum longus	Extensor digitorum longus

nerve; those in the lateral compartment by the superficial peroneal nerve; and those in the posterior compartment by the tibial nerve.

Table 9.21 summarizes the muscle actions on the foot.

Intrinsic Muscles of the Foot

The skeletal structure of the foot is very similar to that of the hand; however, it performs quite different functions. Whereas the hand has great versatility, being capable of grasping and other fine movements, the foot is adapted for support and locomotion. Consequently, the muscles of the foot tend to be heavier than those of the hand, thus making it possible for them to support the arches of the foot. The intrinsic (and extrinsic) muscles are aided in this support by a tough fibrous **plantar aponeurosis** that extends from the calcaneus to the phalanges. Another difference in the musculature of the foot as compared to that of the hand is the presence of one intrinsic muscle, the **extensor digitorum brevis,** on the dorsum of the foot. Recall that there are no intrinsic muscles on the dorsum of the hand.

The remaining intrinsic muscles are on the plantar surface. They are illustrated in Figure 9.44 and described in Table 9.22, page 330. These muscles are separated into four layers (as illustrated), and they can be divided functionally into those that act on the big toe, those that move only the small toe, and those that move all the toes except the big toe.

◆ **FIGURE 9.44 Plantar view of the intrinsic muscles of the right foot, showing successively deeper layers of muscles**

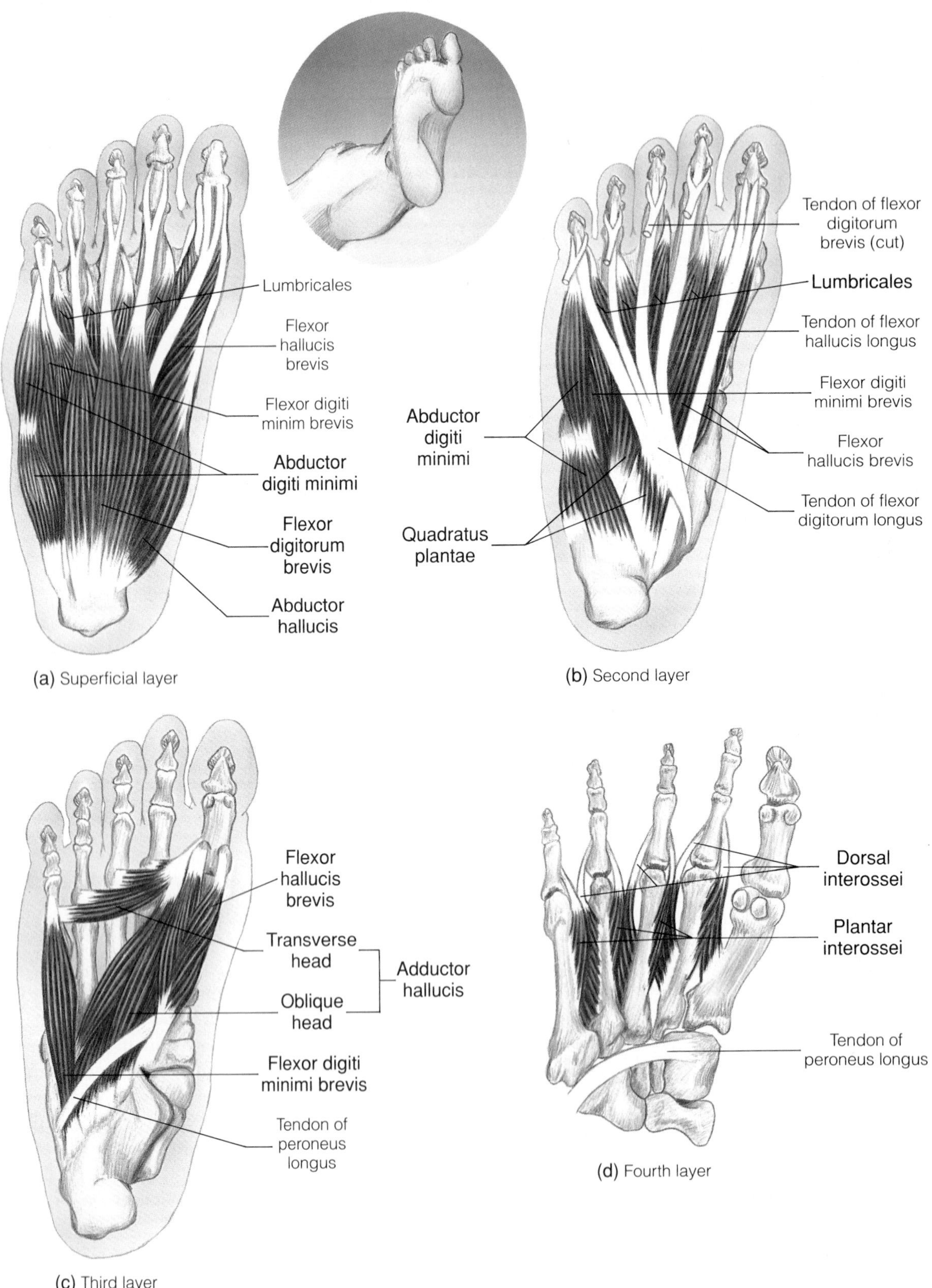

 ASPECTS OF EXERCISE PHYSIOLOGY 

Adverse Effects of Steroid Use

◆

The testing of athletes for drugs and the exclusion from competition of those found to be using substances outlawed by sports federations have stirred considerable controversy and have been much publicized. One such group of drugs are **anabolic androgenic steroids** (*anabolic* = build up of tissues; *androgenic* = male producing; *steroids* are a class of hormone). These agents are closely related to testosterone, the natural male sex hormone, which is responsible for promoting the increased muscle mass characteristic of males.

Before their use was outlawed, these agents were taken by many athletes who specialized in power events, such as weight lifting and sprinting, in an attempt to gain a competitive edge by increasing muscle mass and, accordingly, increasing muscle strength. Both male and female athletes have resorted to using these substances. The adverse effects of these drugs, however, outweigh any benefits derived. In males, who already have an abundance of their own naturally produced testosterone, sought-after increases in muscle mass, strength, and body weight are not well demonstrated. Small, variable gains occur only in conjunction with a regular muscle-building exercise program and are not experienced by all who take the agents. In fact, some suggest that the changes that do occur are largely due to the exercise alone. In females, who lack potent androgenic hormones, the effect of promoting "male-type" muscle mass and strength is readily demonstrable when these agents are used. However, these drugs also "masculinize" females in other ways, such as inducing growth of facial and overall body hair and lowering the voice.

These are but a few of the side effects of using anabolic steroids. In both males and females, these agents adversely affect the liver and the reproductive and cardiovascular systems.

ADVERSE EFFECTS ON THE REPRODUCTIVE SYSTEM

In males, testosterone secretion and sperm production by the testes are normally controlled by hormones released from the anterior pituitary gland. In negative-feedback fashion, testosterone inhibits secretion of these controlling hormones so that a constant circulating level of testosterone is maintained. The anterior pituitary is similarly inhibited by androgenic steroids taken as a drug. As a result, because the testes do not receive their normal stimulatory input from the anterior pituitary, sperm production decreases, the testes atrophy, and testosterone secretion declines.

In females, inhibition of the anterior pituitary by the androgenic drugs results in repression of the hormonal output that controls ovarian function. The result is failure to ovulate, menstrual irregularities, and decreased secretion of female sex hormones. Since the latter are "feminizing hormones," their decline results in diminution in breast size and other female characteristics.

ADVERSE EFFECTS ON THE LIVER

Liver dysfunction is not uncommon with high steroid intake, perhaps because the liver, which normally inactivates steroid hormones and prepares them for urinary excretion, is overloaded by the excess steroid intake.

ADVERSE EFFECTS ON THE CARDIOVASCULAR SYSTEM

Use of anabolic steroids induces several changes in the cardiovascular system that increase the risk of developing atherosclerosis, which in turn is associated with an increased incidence of heart attacks and strokes. Among these adverse cardiovascular effects are an elevation in blood pressure. In addition, damage to the heart muscle itself has been demonstrated in animal studies.

ADVERSE EFFECTS ON BEHAVIOR

Although still controversial, anabolic steroid use appears to promote aggressive, even hostile behavior.

Thus, for health reasons, not even taking into account the ethical issues involved, athletes should not use anabolic steroids in an attempt to gain a competitive advantage. Despite these potential dangers, however, some athletes continue to use these drugs illegally and seek measures to mask their use from testing procedures.

◆ **TABLE 9.22 Intrinsic Muscles of the Foot**

MUSCLE	ORIGIN	INSERTION	ACTION	INNERVATION
DORSAL MUSCLE				
Extensor digitorum brevis *(ek-sten´-sor dij″-ĭ-tor´-um brev´-is)* [*Fig. 9.37*]	Lateral surface of the calcaneus	Proximal phalanx of the great toe and the tendons of the extensor digitorum longus	Extends the first through fourth toes	Deep peroneal
PLANTAR MUSCLES				
***Superficial Layer* [*Fig. 9.44a*]**				
Abductor hallucis *(ab-duk´-ter hal´-u-kis)*	Calcaneus	Proximal phalanx of the great toe (with the tendon of the flexor hallucis brevis)	Abducts the great toe	Medial plantar
Flexor digitorum brevis *(flek´-sor dij″-ĭ-tor´-um brev´-is)*	Calcaneus and plantar aponeurosis	Middle phalanx of the second through the fifth toes	Flexes the second through fifth toes	Medial plantar
Abductor digiti minimi *(ab-duk´-ter dij-ĭ-tī min´-ĭ-mī)*	Calcaneus and plantar aponeurosis	Proximal phalanx of the small toe	Abducts the small toe	Lateral plantar
***Second Layer* [*Fig. 9.44b*]**				
Quadratus plantae *(kwod-rā´-tus plan´-tē)*	Calcaneus	Into tendons of the flexor digitorum longus	Aids in flexing the second through fifth toes by straightening the pull of the flexor digitorum longus	Lateral plantar
Lumbricales *(lum´-brĭ-kals)* (4)	From tendons of the flexor digitorum longus	Into tendons of the extensor digitorum longus	Flex the proximal phalanx; extend the middle and distal phalanges of the second through fifth toes	medial and lateral plantar
***Third Layer* [*Fig. 9.44c*]**				
Flexor hallucis brevis *(flek´-sor hal´-u-kis brev´-is)*	Cuboid and lateral cuneiform	Proximal phalanx of the great toe	Flexes the great toe	Medial plantar
Adductor hallucis *(ad-duk´-tor hal´-u-kis)*	*Oblique head:* second, third, and fourth metatarsals; *Transverse head:* ligaments of the metatarsophalangeal joints	Proximal phalanx of the great toe	Adducts the great toe	Lateral plantar
Flexor digiti minimi brevis *(flek´-sor dij″-ĭ-tī min´-ĭ-mī brev´-is)*	Fifth metatarsal	Proximal phalanx of the small toe	Flexes the small toe	Lateral plantar

◆ **TABLE 9.22 Intrinsic Muscles of the Foot (continued)**

MUSCLE	ORIGIN	INSERTION	ACTION	INNERVATION
PLANTAR MUSCLES (continued)				
Fourth Layer [Fig. 9.44d]				
Plantar interossei (3) *(in″-ter-os′-ē-ī)*	Third, fourth, and fifth metatarsals	Proximal phalanx of the same toe	Adduct the toes toward the second toe	Lateral plantar
Dorsal interossei (4)	Bases of the adjacent metatarsals	Proximal phalanges; both sides of the second toe; lateral side of the third and fourth toes	Abduct the toes from the second toe; move the second toe medially and laterally	Lateral plantar

Study Outline

◆ CRITERIA USED TO NAME MUSCLES p. 280

Muscle Shape. For example: trapezius, rhomboid.

Muscle Action. For example: flexor, extensor, adductor, pronator.

Muscle Location. For example: intercostal, tibialis anterior.

Muscle Attachments. For example: sternocleidomastoid, coracobrachialis.

Muscle Divisions. For example: biceps brachii, triceps brachii, quadriceps femoris.

Muscle Size Relationships. For example: gluteus maximus, gluteus minimus, peroneus longus, peroneus brevis.

◆ MUSCLES OF THE HEAD AND NECK pp. 282–286

Muscles of the Face. (Table 9.1) Also called muscles of facial expression. Some arise from skull bones, others from superficial fascia of the face. Most insert into skin of the region and thus serve to move the skin rather than joints. Facial muscles control mouth and lips; compress cheeks; control eyebrow, eyelid, and scalp movement; dilate nares; tighten and wrinkle neck and forehead skin; and depress mandible.

Muscles of Mastication. (Table 9.2) Four pairs of muscles involved in biting and chewing. Two pairs raise mandible. Two pairs move mandible sideways and assist in opening and closing mouth.

Muscles of the Tongue. Intrinsic muscles of tongue squeeze, fold, and curl tongue; extrinsic muscles of tongue (Table 9.3) anchor it to skeleton and control tongue's protrusion, retraction, and sideways movement.

Muscles of the Neck. Include muscles grouped within one of two triangles—anterior (Table 9.4) or posterior—that are separated by sternocleidomastoid muscle.

Muscles of the Throat. Consist of deep muscles of anterior triangle; help form floor of oral cavity and are attached to hyoid bone; aid in movements of tongue and in swallowing.

◆ MUSCLES OF THE TRUNK pp. 286–295

Include muscles of vertebral column, the back, thorax, abdominal wall, and muscles that form floor of abdominopelvic cavity.

Muscles of the Vertebral Column. (Table 9.5) Most are located on posterior surface of spine. Some have fibers that insert onto skull and thus also move the head. Others (some that comprise abdominal wall and those that act on hip) also act on vertebral column.

Deep Muscles of the Thorax. (Table 9.6) Most insert on ribs and assist in breathing by elevating or depressing rib cage; spaces between ribs are reinforced by double layer of intercostal muscles; diaphragm is muscle chiefly responsible for quiet breathing.

Muscles of the Abdominal Wall. (Table 9.7) Wall of abdominal cavity lacks skeletal support and thus derives strength entirely from muscles, which are composed of three layers:

1. *External abdominal oblique.* Outermost layer, with fibers that pass medially and downward.
2. *Internal abdominal oblique.* Layer immediately beneath external oblique; has fibers that run upward and medially.
3. *Transversus abdominis muscle.* Deep to both oblique muscles; a thin muscle whose fibers run horizontally and encircle abdominal cavity.

RECTUS ABDOMINIS MUSCLE. Ensheathed by fasciae of oblique and transversus muscles; passes vertically from pubic symphysis to rib cage.

QUADRATUS LUMBORUM MUSCLE. Forms most of posterior abdominal wall.

Muscles That Form the Floor of the Abdominopelvic Cavity. (Table 9.8) Viscera of abdominopelvic cavity are supported by muscular floor called pelvic diaphragm, which is composed of two principal muscles: levator ani and coccygeus.

Muscles of the Perineum. (Table 9.9) Five muscles in two layers; located just below pelvic diaphragm; urogenital diaphragm formed by two of the muscles.

◆ MUSCLES OF THE UPPER LIMBS pp. 295–311

Include muscles that act on scapula; clavicle; arm and forearm; hand and fingers.

Muscles That Act on the Scapula. (Table 9.10) Six muscles—four posterior and two anterior—act directly on scapula and anchor it to thorax.

Muscles That Act on the Arm (Humerus). (Tables 9.11 and 9.12) Nine muscles cross shoulder joint and insert on humerus; seven of these arise from scapula.

Muscles That Act on the Forearm (Radius and Ulna). (Table 9.13) Elbow joint and/or proximal radial/ulnar joints are moved by powerful muscles in arm, but are assisted by muscles in forearm.

Muscles That Act on the Hand and Fingers. (Tables 9.14 and 9.15) Moved by muscles that form bulge of proximal end of forearm; anterior group of muscles serves as flexors and/or pronators; the posterior group serves as extensors and/or supinators.

Intrinsic Muscles of the Hand. (Table 9.16) Several muscles of forearm move fingers; intrinsic finger muscles produce fine, precise finger movement.

◆ MUSCLES OF THE LOWER LIMBS pp. 311–331

Include muscles that act on thigh (femur), leg (tibia and fibula), foot and toes, and intrinsic muscles of foot. Lower limb muscles tend to be bulkier and more powerful than those of upper limbs because the former are concerned with support and locomotion. Many muscles of lower limbs cross and act on two joints, either hip and knee or knee and ankle.

Muscles That Act on the Thigh (Femur). (Tables 9.17 and 9.19) Most arise from pelvis.

Muscles That Act on the Leg (Tibia and Fibula). (Table 9.18) Located in thigh; their tendons cross knee; grouped into anterior (extensor), posterior (flexor), and medial (adductor) compartments.

Muscles That Act on the Foot and Toes. (Tables 9.20 and 9.21) Most located in leg; grouped into anterior (dorsiflex ankle and/or extend toes), posterior (plantar flex foot and/or flex toes), and lateral (plantar flex and evert foot) compartments.

Intrinsic Muscles of the Foot. (Table 9.22) Four layers on plantar surface and one intrinsic muscle on dorsum of foot; tend to be heavier than hand muscles because foot is adapted for support and locomotion.

Self-Quiz

1. Match each muscle with the correct lettered criterion or criteria on which the name is based:

Intercostalis	(a) Shape
Rhomboideus	(b) Size relationship
Pronator	(c) Attachment
Coracobrachialis	(d) Location
Gluteus maximus	(e) Number of divisions
Trapezius	(f) Action
Rectus abdominis	
Pectoralis minor	
Flexor carpi radialis	
Biceps brachii	
Quadriceps femoris	

2. Most of these muscles insert into the skin of their particular region: (a) intrinsic muscles of the hand; (b) facial muscles; (c) intrinsic muscles of the foot.

3. Match each facial muscle with the correct lettered statement about its action:

Depressor labii inferioris	(a) Pulls angle of mouth backward
Epicranius occipitalis	(b) Draws eyebrows together in frowning
Corrugator	(c) Draws scalp posteriorly
Orbicularis oculi	(d) Pulls lower lip down
Orbicularis oris	(e) Wrinkles skin between eyebrows
Procerus	(f) Closes eyelids; tightens forehead skin
Risorius	(g) Closes and protrudes lips

4. Which muscle retracts and elevates the tongue? (a) styloglossus; (b) hypoglossus; (c) genioglossus.

5. The muscles of the throat are the deep muscles of the anterior triangle. True or False?
6. Raising the hyoid, pulling it backward and anteriorly, and raising the floor of the mouth are actions performed by the: (a) infrahyoid muscles; (b) suprahyoid muscles; (c) sternothyroid muscle.
7. The diaphragm is the muscle chiefly responsible for quiet breathing. True or False?
8. Match each vertebral column muscle with its correct lettered statement:

Multifidus	(a) *Origin:* spinous processes of upper lumbar, lower thoracic, and 7th cervical vertebrae
Interspinales	(b) *Insertion:* occipital bone, mastoid process of temporal bone, and transverse processes of upper three cervical vertebrae
Splenius capitis and cervicis	(c) *Action:* extends vertebral column; rotates it toward opposite side
Spinalis thoracis and cervicis	(d) *Origin:* superior surface of all spinous processes.

9. Select the correct statement about the temporalis muscle. (a) It originates in the zygomatic arch. (b) It opens and protrudes the mandible. (c) It is innervated by the trigeminal nerve.
10. The internal intercostal muscles, the external intercostal muscles, and the transversus muscles are all involved in respiration. True or False?
11. Inguinal hernias are more common in females than in males. True or False?
12. Which of these muscles of the abdominal wall pulls the thoracic cage toward the pelvis and bends the vertebral column toward the side being contracted? (a) quadratus lumborum; (b) internal abdominal oblique; (c) rectus abdominis.
13. A muscle that forms the floor of the abdominopelvic cavity is the: (a) transversus abdominis; (b) levator ani; (c) internal abdominal oblique.
14. Match the following muscles that act on the scapula with the correct lettered statement about each muscle's action:

Subclavius	(a) Depresses the scapula and pulls it anteriorly
Rhomboideus major	(b) Stabilizes and depresses the pectoral girdle
Pectoralis minor	(c) Adducts, stabilizes, and rotates the scapula, lowering its lateral angle

15. Match the muscle action with the appropriate muscles. (*Note:* A given muscle may have more than one action associated with it.)

Deltoid	(a) Flexion
Latissimus dorsi	(b) Extension
Teres minor	(c) Adduction
Teres major	(d) Abduction
Coracobrachialis	(e) Medial rotation
Infraspinatus	(f) Lateral rotation

16. Which of the following muscles *flex* the forearm? (a) biceps brachii; (b) brachialis; (c) triceps brachii; (d) anconeus; (e) brachioradialis.
17. Match the muscle action with the appropriate hand or finger muscle:

Flexor carpi radialis	(a) Flexes the hand and phalanges
Extensor indicis	(b) Extends and abducts the hand
Extensor carpi ulnaris	(c) Extends the little finger
Extensor pollicis longus	(d) Extends the index finger
Extensor carpi radialis longus	(e) Extends and adducts the hand
Extensor digiti minimi	(f) Extends the thumb and abducts the hand
Flexor carpi ulnaris	(g) Flexes and adducts the hand
Flexor digitorum superficialis	(h) Flexes and abducts the hand

18. Which two of the following muscles are involved when you move your little finger in to meet your thumb, and when you move your thumb in to meet your little finger? (a) flexor digiti minimi brevis; (b) opponens digiti minimi; (c) flexor pollicis brevis; (d) palmaris brevis; (e) opponens pollicis.
19. Match the muscle action with the appropriate muscles:

Gluteus maximus	(a) Abducts and medially rotates the thigh
Adductor magnus	(b) Adducts, flexes, and laterally rotates the thigh
Pectineus	(c) Adducts and laterally rotates the thigh; assists in flexion of the thigh
Adductor longus	(d) Extends and laterally rotates the thigh
Gluteus medius	(e) Adducts and laterally rotates the thigh; assists in extension of the thigh

20. Those muscles of the thigh that flex the leg and extend the thigh are located in the: (a) medial compartment; (b) anterior compartment; (c) posterior compartment.
21. What is the action of the muscles that are located in the anterior compartment of the leg? (a) flex the leg; (b) dorsiflex foot and/or extend toes; (c) plantar flex foot and flex toes.
22. Match the muscle action with the appropriate leg muscle that acts on the foot and/or toes:

Tibialis anterior	(a) Flexes the great toe; plantar flexes and inverts the foot
Peroneous longus	(b) Plantar flexes and everts the foot
Extensor digitorum longus	(c) Dorsiflexes and inverts the foot
Flexor hallucis longus	(d) Flexes the leg and rotates it medially
Popliteus	(e) Dorsiflexes and everts the foot; extends the toes

CHAPTER 10

The Nervous System: Its Organization and Components

CHAPTER CONTENTS

LEARNING OBJECTIVES

After completing this chapter, you should be able to:

1. Distinguish between receptors and effectors
2. Name the two divisions of the nervous system according to structural location, and list the principal components of each.
3. Name the divisions of the nervous system according to function.
4. Distinguish between the somatic nervous system and the autonomic nervous system, and cite the principal function of each.
5. Differentiate between myelinated and unmyelinated nerve fibers.
6. Identify the connective-tissue sheaths that cover a nerve.
7. Distinguish between a dendrite and an axon.
8. Name the types of neuroglia, and state the principal function of each.
9. Name the types of neurons according to structural and functional classification.
10. Cite the various ways receptors may be classified.
11. Describe the specialized peripheral neuron endings.

CHAPTER 10

The survival of any multicellular organism depends on its having some means of regulating and coordinating the activities of its cells. If there were no coordination, each cell would function just as genetically programmed, without regard to the functions other cells were performing. At best, such haphazard organization would be inefficient; at worst, it could be fatal.

The human organism, which is composed of billions of cells, has two systems that serve primarily as means of *internal communication* among the cells—the nervous system and the endocrine system. Because both systems cooperate to provide the body's internal communications, they can be viewed as closely allied networks operating together to ensure proper body functioning. By monitoring and regulating how the various body functions interrelate, the nervous system and the endocrine system help maintain the homeostasis of the body. In Chapters 10 through 16, we consider the nervous system. The endocrine system is discussed in Chapter 17.

The cells of the nervous system that are specialized for the transmission of electrical signals are called **neurons.** A particularly important kind of electrical signal transmitted by neurons is called a **nerve impulse.** Nerve impulses often originate within a nerve cell as a result of the activity of sensory structures called **receptors.** Receptors are generally activated by changes in either the internal or external environments of the body. The changes that activate the receptors are called **stimuli.** When a stimulus activates a receptor, nerve impulses are initiated within **sensory neurons.** These impulses are carried by the sensory neurons to the spinal cord and brain (Figure 10.1), where other neurons may be activated. And these neurons may then conduct nerve impulses to various other locations within the spinal cord and brain. Ultimately, nerve impulses carried by **motor neurons** leave the brain and the spinal cord

◆ **FIGURE 10.1 Schematic representation of the basic organization of the nervous system**

The arrows follow a typical pathway of nerve impulses from receptor to effector.

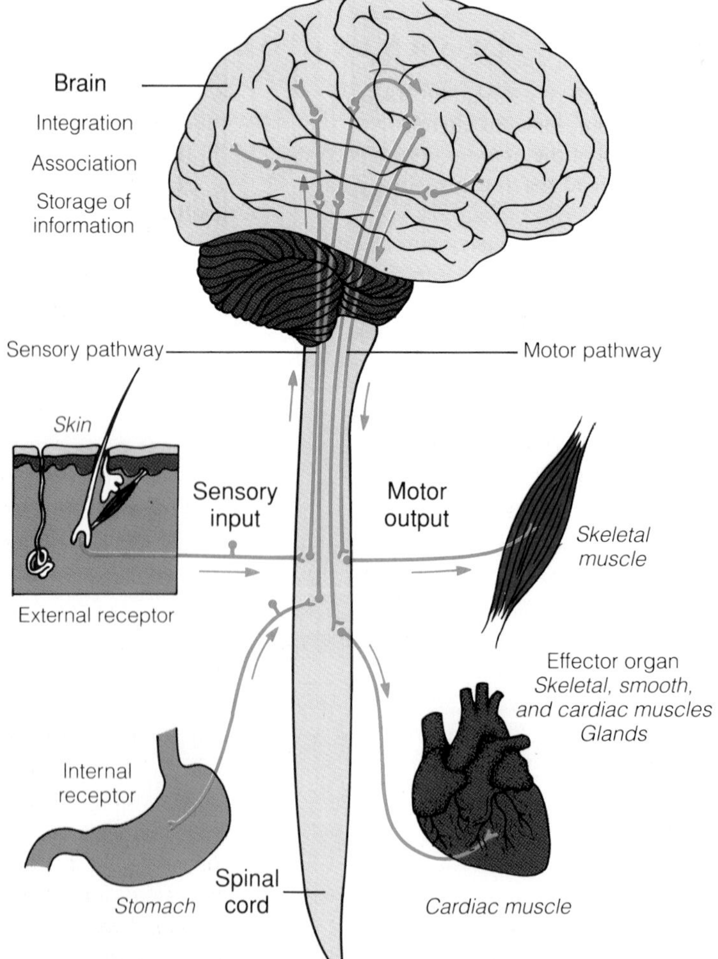

◆ **FIGURE 10.2 The central nervous system and the proximal portions of the peripheral nervous system**

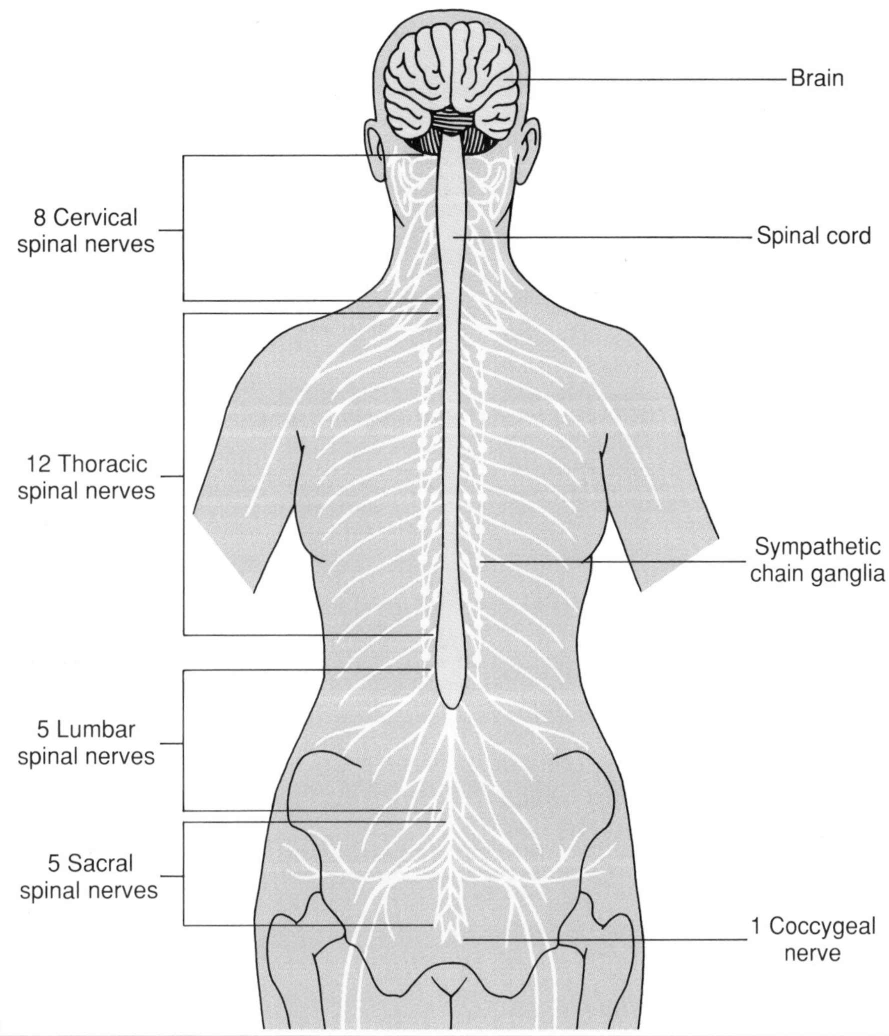

and bring about responses to environmental changes by selectively activating various **effectors.** Effectors capable of responding to nerve impulses include muscle cells and the secretory cells of the glands and organs. In addition to responding to environmental stimuli, the human nervous system has the ability to integrate and store the information that it receives, thus providing for such capacities as memory, abstract reasoning, and conceptualizing.

Organization of the Nervous System

Although there is actually only one nervous system, it can be separated conceptually into various divisions based on either structural locations or functional characteristics. Keep in mind, however, that these divisions—which are themselves called *nervous systems*—are all integral parts of a single nervous system. Structurally, the nervous system may be divided into two parts: the *central nervous system* and the *peripheral nervous system* (Figure 10.2).

Central Nervous System

The **central nervous system (CNS)** consists of the brain and the spinal cord. It is completely encased within bony structures—the brain within the cranial cavity of the skull and the spinal cord within the vertebral canal of the spinal column. The central nervous system is the *integrative and control center* of the nervous system. It receives sensory input from the peripheral nervous system and formulates responses to this input.

Peripheral Nervous System

The **peripheral nervous system (PNS;** *per-if´-er-al*) is composed of all nervous structures located outside of the central nervous system. **Nerves,** which connect effectors and receptors in the outlying regions of the body with the CNS, are the main components of the PNS. Associated with these nerves are groups of nerve cell bodies called **ganglia** *(gan´-glee-uh).*

The peripheral nervous system includes 12 pairs of **cranial nerves,** which arise from the brain and the brain stem and leave the cranial cavity through foramina in the skull, and 31 pairs of **spinal nerves,** which arise from the spinal cord and leave the vertebral canal through the intervertebral foramina. The paired spinal nerves include 8 cervical, 12 thoracic, 5 lumbar, 5 sacral, and 1 coccygeal nerve (Figure 10.2). The peripheral nervous system can be divided functionally into afferent (sensory) and efferent (motor) divisions.

Afferent Division

The **afferent division** *(af´-er-ent)* of the peripheral nervous system includes *somatic sensory* neurons, which carry impulses to the central nervous system from receptors located in the skin, in the fasciae, and around the joints, and *visceral sensory* neurons, which carry impulses from the viscera of the body to the central nervous system.

Efferent Division

The **efferent division** *(ef´-er-ent)* of the peripheral nervous system is divided into the *somatic nervous system* and the *autonomic nervous system.*

1. The **somatic nervous system** *(so-mat´-ik)* is also called the **voluntary nervous system,** because many of its motor functions may be consciously controlled. It consists of *somatic motor* neurons, which carry impulses from the central nervous system to the skeletal muscles. The impulses carried by somatic motor neurons produce contractions of the skeletal muscles. Muscle contractions brought about by the somatic nervous system may be under the conscious control of the individual, or in the case of reflex responses, they may not be consciously controlled.
2. The **autonomic nervous system** *(aw-to-nom´-ik)*—or **involuntary nervous system**—in contrast to the somatic nervous system, is composed of *visceral motor* neurons, which transmit impulses to smooth muscles, cardiac muscle, and glands. Visceral motor impulses generally cannot be consciously controlled.

The autonomic nervous system may be further subdivided functionally into **sympathetic** *(sim-pa-thet´-ik)* and **parasympathetic divisions** *(par-a-sim-pa-thet´-ik),* each of which is studied in more detail in Chapter 14. As noted earlier, the autonomic and somatic nervous "systems" are actually overlapping divisions of the body's single nervous system. Moreover, although the motor neurons of both the somatic and the autonomic divisions are generally considered to be parts of the peripheral nervous system, they are under the control of centers located in the central nervous system. The organization of the nervous system is summarized in Figure 10.3.

◆ **FIGURE 10.3 Organization of the nervous system**

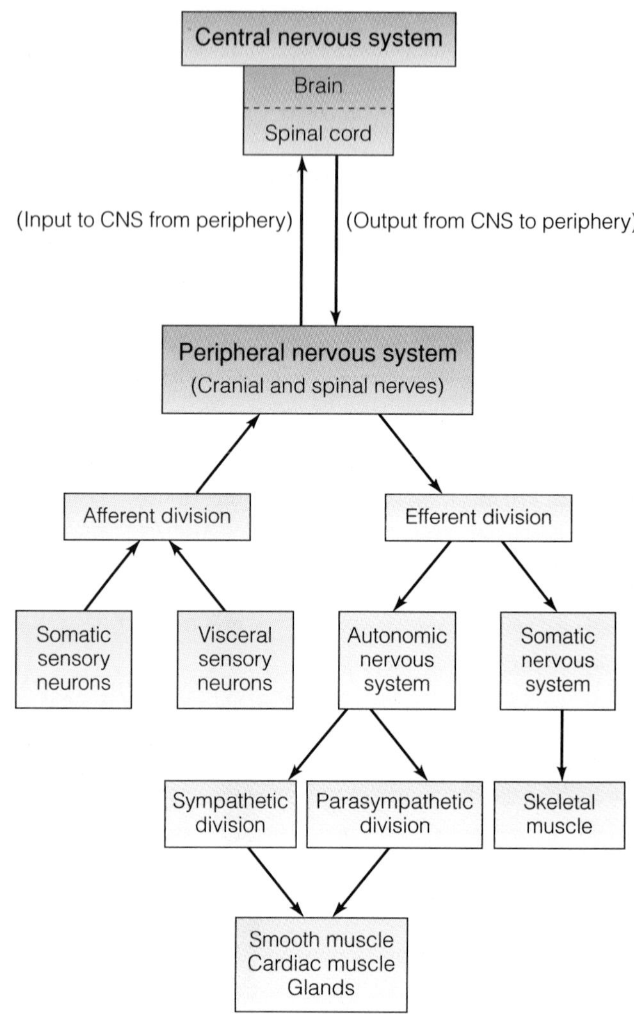

◆ Embryonic Development of the Nervous System

The nervous system develops from ectodermal cells that, by the second week of gestation, form a flat thickening on the dorsal surface of the embryo. This thickening, called the **neural plate,** gives rise to all of the nerve cells of the nervous system. In addition, neural plate cells give rise to most of the supportive cells *(neuroglial cells)* of the nervous system.

As development continues, the midline of the neural plate sinks inward. At the same time, the proliferation of cells along the margins of the plate produces elevations. The result is a **neural groove** flanked by **neural folds** that extend the entire length of the embryo (Figure 10.4).

The neural groove deepens as the neural folds increase in height. Eventually, the folds meet and fuse in the midline, converting the groove into the **neural tube** (Figure 10.5). The neural tube separates from the surface ectoderm and subsequently develops into the brain and the spinal cord. In addition, cells within the neural tube send out processes to peripheral structures. These cells and their processes form the motor neurons of both the somatic and the autonomic nervous systems.

While the neural folds are fusing, ectodermal cells from the tops of the folds move laterally to form columns of cells on each side of the neural tube. The cells within these columns are called **neural crest cells.** These neural crest cells develop connections with the central nervous system and with peripheral structures, thus forming sensory neurons within cranial nerves and spinal nerves. Some neural-crest cells migrate to other locations and develop into motor neurons of the sympathetic nervous system, into Schwann cells that wrap around neurons, or into other structures not directly associated with the nervous system.

The anterior end of the neural tube undergoes the most rapid growth and gives rise to the brain. By the fourth week of gestation, the brain is in the form of

◆ FIGURE 10.4 Early embryonic development of the central nervous system showing the neural groove flanked by neural folds

The wide region of the anterior end is the future brain.

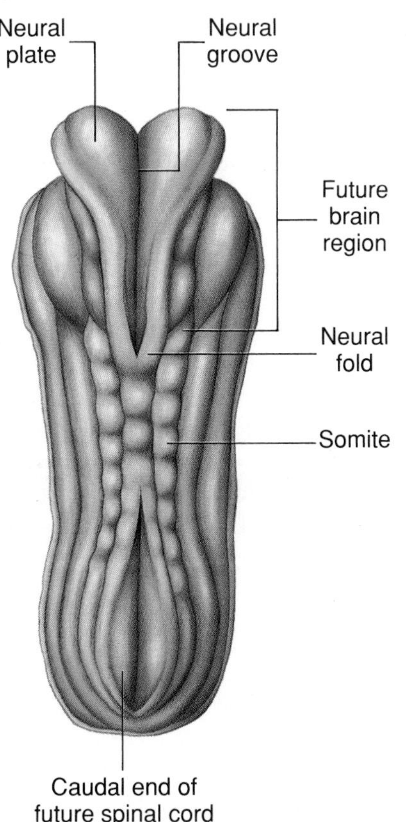

◆ FIGURE 10.5 Cross section of a neural plate showing its progressive development into a neural groove and a neural tube (future brain and spinal cord)

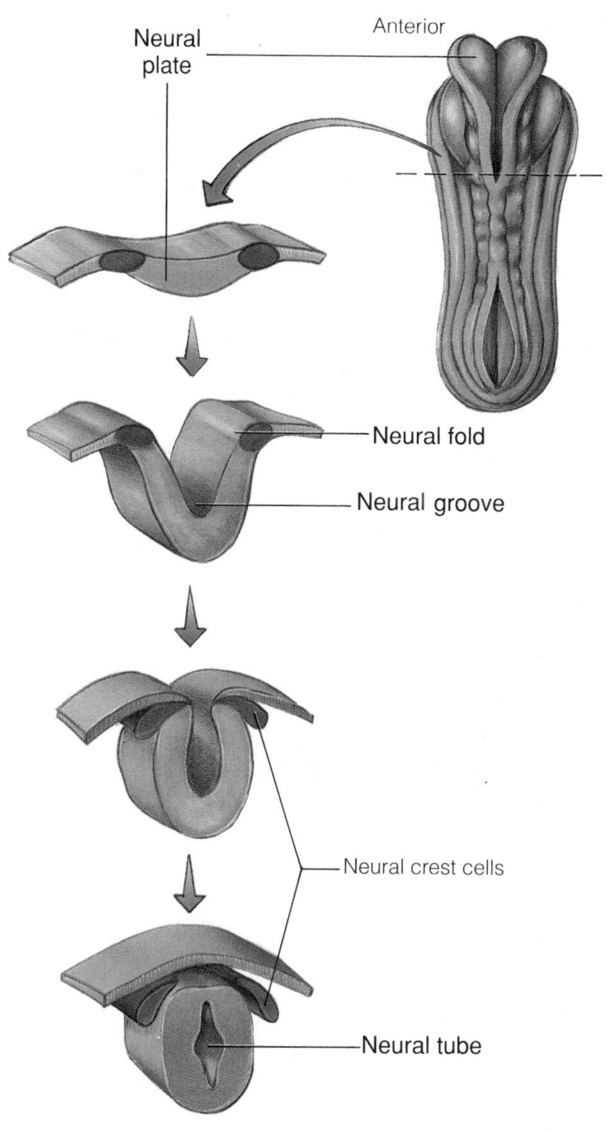

◆ **FIGURE 10.6 Division of the early embryonic brain (three to four weeks) into prosencephalon, mesencephalon, and rhombencephalon**

(a) Frontal section. (b) Lateral view.

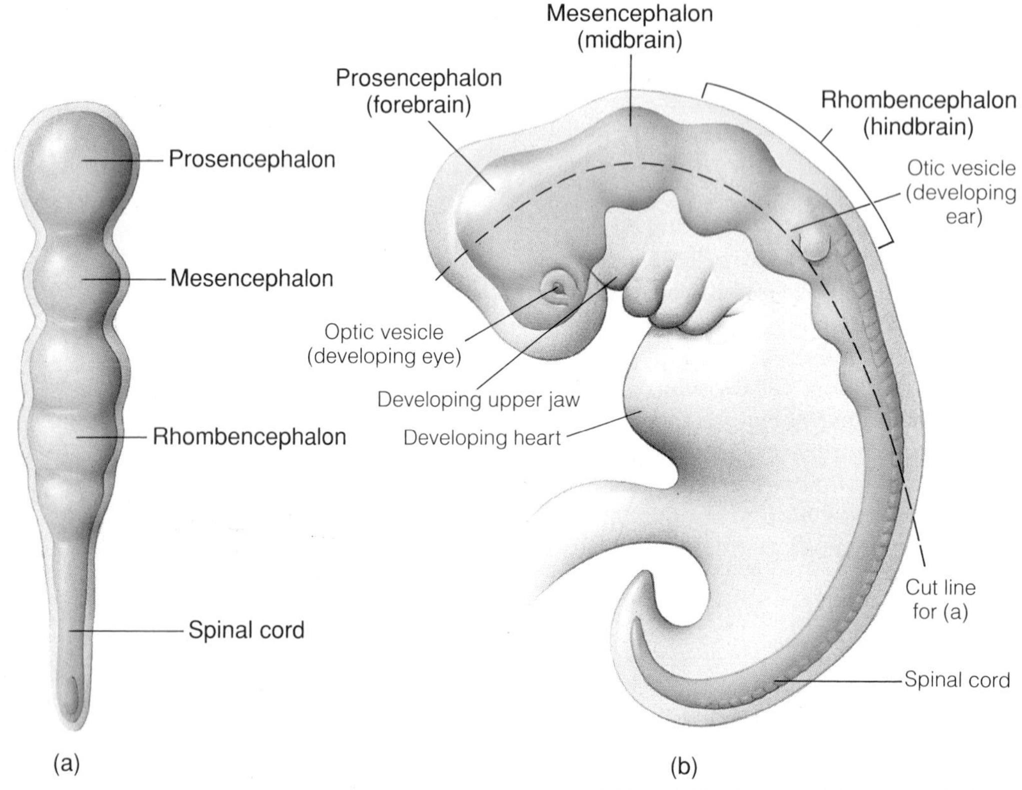

three fluid-filled enlargements, or vesicles: the **prosencephalon** *(pross-en-sef´-uh-lon)* or **forebrain,** the **mesencephalon** *(mes-en-sef´-uh-lon)* or **midbrain,** and the **rhombencephalon** *(rom´´-ben-sef´-uh-lon)* or **hindbrain** (Figure 10.6). During subsequent development, this region undergoes several bendings or flexures and the vesicles further subdivide. As a result, by the fifth week, the brain consists of five vesicles tightly curved upon themselves at the anterior end of the embryo (Figure 10.7). The prosencephalon subdivides into two vesicles: an anterior **telencephalon** *(tel-en-sef´-uh-lon)* and, just behind it, the **diencephalon** *(dye-en-sef´-uh-lon).* The mesencephalon does not further subdivide, but the rhombencephalon divides into the **metencephalon** *(met-en-sef´-uh-lon)* and the most posterior brain vesicle, the **myelencephalon** *(mile-len-sef´-uh-lon).* The cavities in these brain vesicles will become the *ventricles* of the adult brain, and the fluid in them is the **cerebrospinal fluid** *(ser-ee-bro-spy´-nul).* Brain development is discussed further in Chapter 12.

The region of the neural tube posterior to the myelencephalon gives rise to the spinal cord. The development of the spinal cord is also discussed in Chapter 12.

Components of the Nervous System

Nervous tissue is composed of two classes of cells that differ structurally and functionally: (1) **neurons (nerve cells),** which are specialized for transmitting nerve impulses; and (2) **neuroglial cells** *(nu-rog´-lee-al)* or **glial cells,** which do not generate nerve impulses, but provide structural and functional support to neurons. Neuroglial cells can be further subdivided into supporting cells of the central nervous system (astrocytes, oligodendrocytes, microglia, and ependymal cells) and supporting cells of the peripheral nervous system (Schwann cells and satellite cells). We will consider these cells in greater detail later in this chapter.

Neurons

The **neuron (nerve cell)** is the basic structural and functional component of the nervous system. It has the

◆ **FIGURE 10.7 Development of the brain (at five weeks) into telencephalon, diencephalon, mesencephalon, metencephalon, and myelencephalon**

(a) Frontal section. (b) Lateral view. Roman numerals indicate cranial nerves.

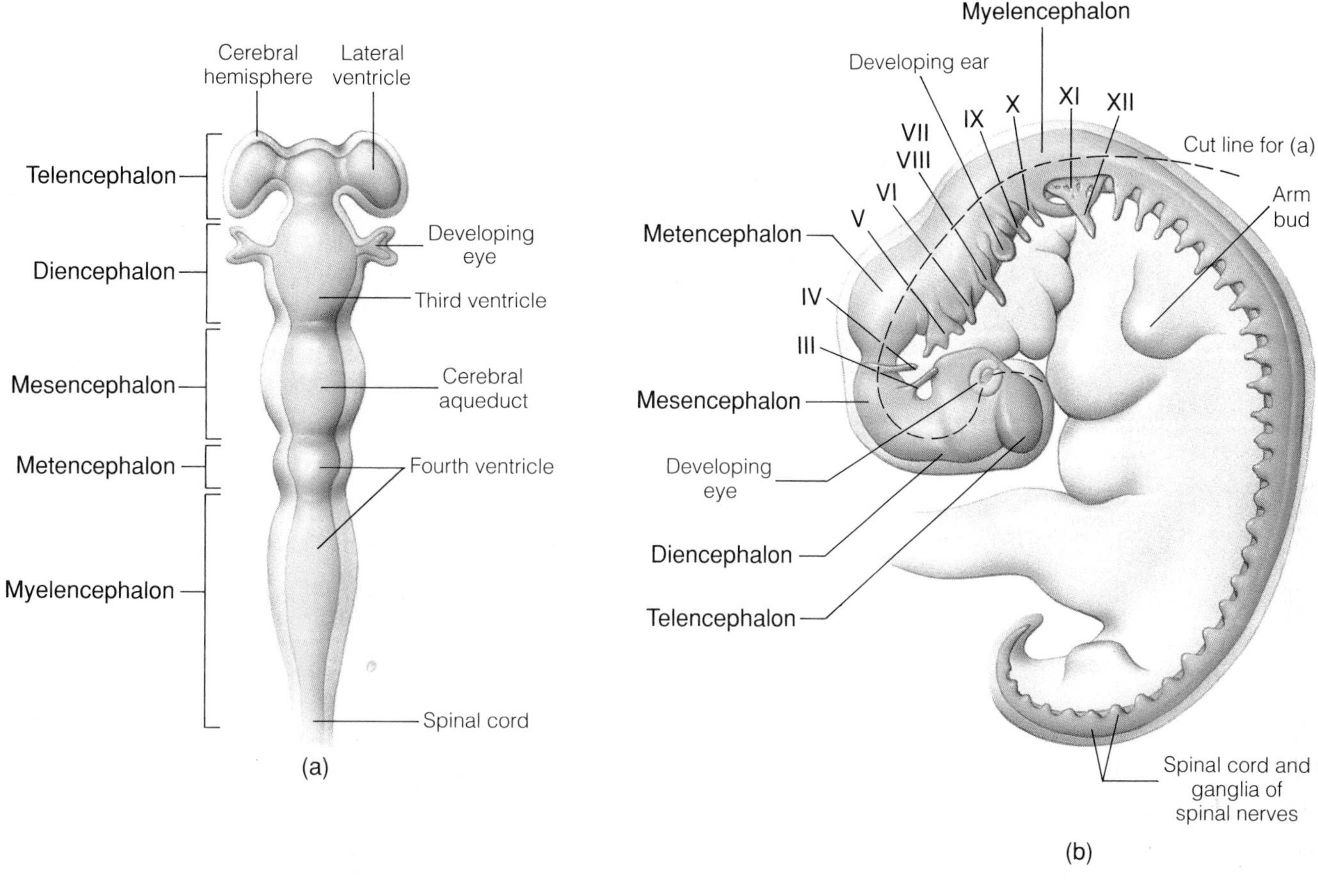

ability to respond to stimulation by initiating and conducting electrical signals. Certain other cells, such as muscle fibers, are also capable of conducting electrical signals, but the unique shape of the neuron makes it especially well suited for transmitting information from one point to another. Neurons have processes that can be quite long. For instance, a single neuron may extend from the spinal cord to the tip of a toe!

With a few specific exceptions, mature neurons are unable to undergo mitosis. For this reason, once embryonic development is complete, most neurons cannot be replaced when they die or are destroyed. Under the proper conditions, however, peripheral neuronal processes that have been damaged or severed can be repaired or regenerated, thus reestablishing the nerve supply to the affected structures.

Structure of a Neuron

Each neuron is composed of a **cell body (soma,** or **perikaryon)** and one or more processes containing cytoplasm that extend from the cell body (Figure 10.8). Within the cytoplasm of the cell body is a large nucleus containing a prominent nucleolus. As in most cells, the cytoplasm of neurons also contains mitochondria and Golgi apparatus. In addition, the cytoplasm contains a dark-staining **chromatophilic substance** *(Nissl bodies)* and slender fibrils called **neurofibrils.** Under the electron microscope, the chromatophilic substance is seen to be composed of free ribosomes and parallel layers of rough endoplasmic reticulum, and the neurofibrils are seen to be composed of microfilaments, intermediate filaments, and microtubules. It is thought that the fibrils provide support to the cell, whereas the microtubules may be involved in the transport of materials within the cell.

Locations of the Cell Bodies of Neurons

The cell bodies of most neurons are located in the central nervous system, although large numbers are located in the peripheral system. Nerve cell bodies in the CNS

◆ **FIGURE 10.8 Structure of a typical motor neuron**

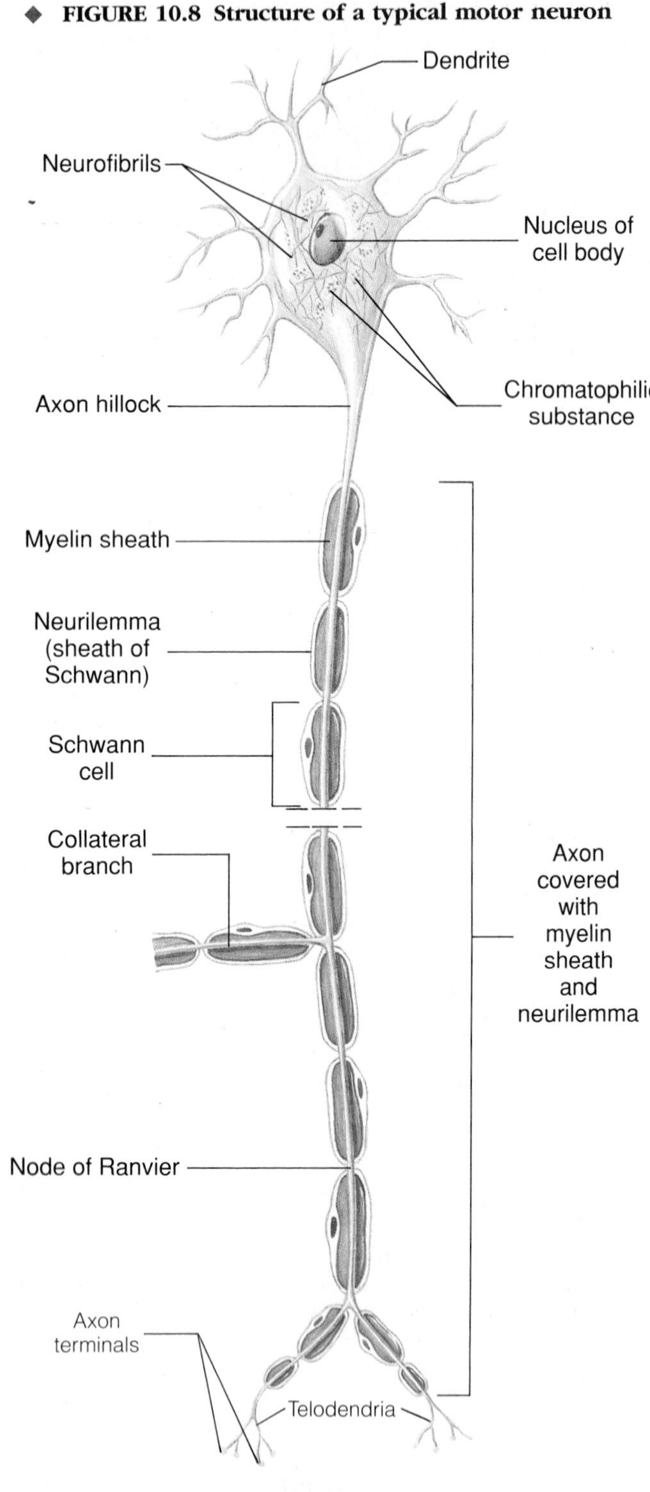

tend to be located in clusters called **nuclei.** (Note that this use of the term *nucleus* has no relation to the nucleus of a cell.) A nucleus or group of nuclei whose neurons govern a particular function is referred to as a **center.** Nerve cell bodies located outside the CNS, in the peripheral nervous system, are generally found in groups called **ganglia.**

Processes of Neurons

The processes of a neuron are very thin extensions of the cell. There are two types of processes: *dendrites* and *axons.* According to classical usage of the terms, dendrites are the neuronal process or processes that conduct electrical signals *toward* a cell body; an axon (of which there is only one per cell) is the neuronal process that conducts electrical signals *away* from a cell body. It is becoming more common, however, to differentiate dendrites and axons in the following way:

◆ **Dendrites** The **dendritic zone** is the *receptive* portion of a neuron, where electrical signals originate in response to released chemicals. The dendritic zone can include the cell body as well as branching processes, or **dendrites,** that are either direct extensions of the cell body or more remote branchings separated from the cell body by a length of axon. The number, length, and extent of branching of the dendrites vary in different types of neurons. Dendrites possess many of the structures found in the cell body, including filaments and microtubules that are oriented parallel to the long axis of the dendrite.

◆ **Axons** The **axon (nerve fiber)** is the *conductive* process of the neuron—that is, the part that transmits the electrical signals. In motor neurons, the cell body is located between the axon and the dendrites. The axon arises from a cone-shaped process on the cell body called the **axon hillock.** In sensory neurons, the cell body is located to one side of the axon, to which it is attached by a short process. The lengths of axons vary considerably. They may be quite short, traveling only a short distance in the central nervous system, or they may be long, traveling a considerable distance within the CNS. In addition, some axons extend a meter or more out into the peripheral nervous system. Each neuron has only one axon, but each axon generally has several branches called **collaterals.** Axons contain mitochondria, microtubules, and neurofibrils, but they lack chromatophilic substance. An axon and its collaterals end by separating into many fine branches called **telodendria** *(tel″-ō-den′-drē-uh).* The distal end of each telodendron expands into small bulblike structures called **axon terminals,** or **synaptic knobs.**

Formation of Myelinated Neurons

Most axons are covered with a fatty substance called **myelin** *(my′-uh-lin).* Such axons are called **myelinated fibers** *(my′-uh-li-nay-ted)*; axons not covered with myelin are called **unmyelinated fibers.** Because of the lipid content of myelin, myelinated fibers appear

◆ **FIGURE 10.9 The ultrastructure of myelinated neurons**

(a) Schematic. (b) Photomicrograph of a cross section of a nerve fiber surrounded by a myelin sheath.

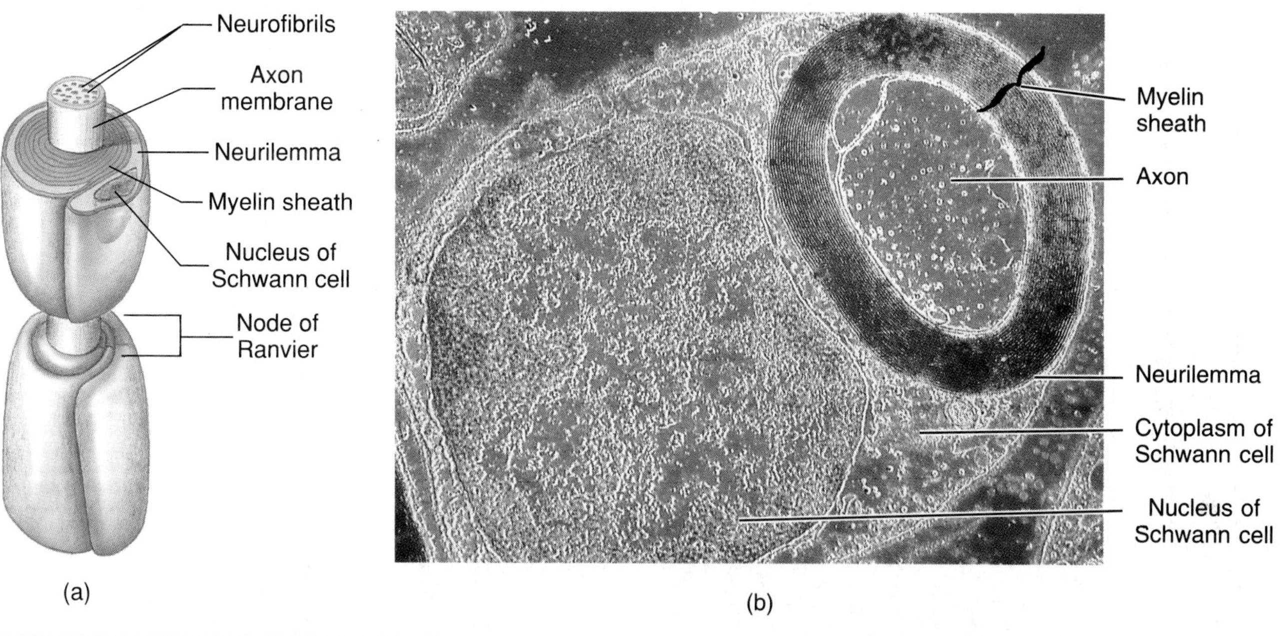

white; when traveling in groups, they form white pathways in the nervous system.

When myelinated axons of the peripheral nervous system are viewed under the light microscope, there appears to be a thin membrane between the myelin and the connective-tissue sheath (endoneurium) surrounding the axon. This membrane is called the **neurilemma,** or **sheath of Schwann** (Figure 10.9). The neurilemma and the myelin of these myelinated axons are interrupted at regular intervals along the length of the axon. Each of these points of interruption is called a **node of Ranvier** (Figures 10.8 and 10.9a).

Both myelinated and unmyelinated axons of the peripheral nervous system are surrounded by **Schwann cells,** which are arranged sequentially along the length of the axon. During embryonic development, Schwann cells, which are derived from embryonic neural crest cells, migrate along the axon and envelop it. We consider the myelinated axons first.

Under the electron microscope, it can be seen that the myelin sheath of myelinated axons is formed when each Schwann cell wraps itself around a portion of the axon several times, pushing the cytoplasm and nucleus of the Schwann cell peripherally (Figure 10.9). Consequently, the myelin sheath is composed of alternating layers of the proteins and lipids that comprise the plasma membrane of the Schwann cell. The outermost part of the Schwann cell, which contains the cytoplasm and the nucleus that were pushed peripherally as the Schwann cell wrapped around the axon, is the neurilemma (sheath of Schwann). The nodes of Ranvier are the junctions between two successive Schwann cells. It is at the nodes of Ranvier that any collateral branches of the axon occur. The nodes are also instrumental in increasing the rate of transmission of the nerve impulse, as is discussed in Chapter 11.

Not all axons are myelinated, however. As shown in Figure 10.10, it is not unusual for the axons of several neurons to become embedded within the cytoplasm of a single Schwann cell. The Schwann cell may simply envelop some of these axons without forming a myelin sheath around them. Such axons are unmyelinated. The unmyelinated axons of the peripheral nervous system are covered by a neurilemma formed by the Schwann cells. However, the unmyelinated axons lack the additional spiraled wrappings of the Schwann cells that form the myelin covering.

Many neurons in the central nervous system are myelinated, but there are no Schwann cells in the CNS. Instead, the central nervous system contains a type of neuroglial cell called an **oligodendrocyte** that sends out processes that spiral around axons, thus forming a myelin covering. However, because in this case the myelin sheath is formed by coiling of processes from the cells—rather than by coiling of the entire cell—the cytoplasm that is squeezed out of the coiled regions is forced back toward the cell body (Figure 10.11). Thus, no neurilemma is formed in the central nervous system. Myelinated axons form the white matter of the central nervous system.

◆ **FIGURE 10.10 A single Schwann cell encompassing eight unmyelinated nerve fibers**

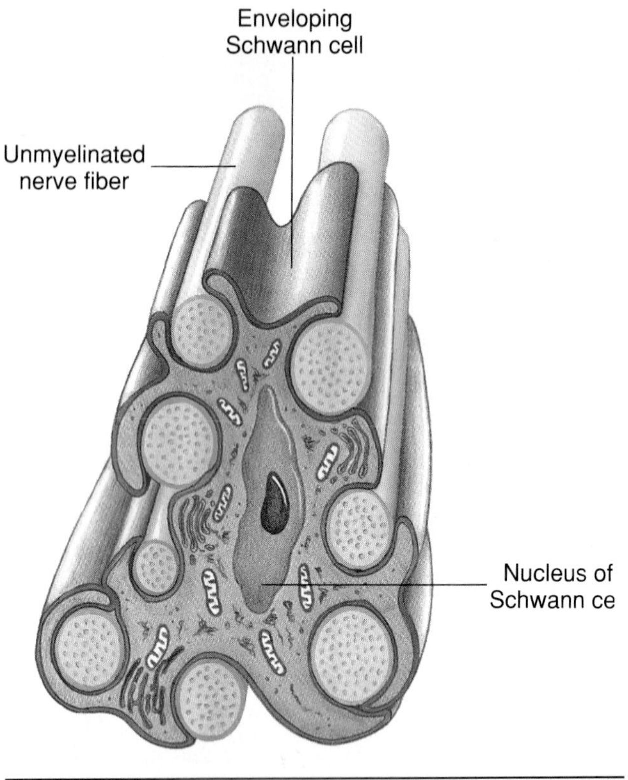

◆ **FIGURE 10.11 A single oligodendrocyte forming myelin sheaths around three axons in the central nervous system**

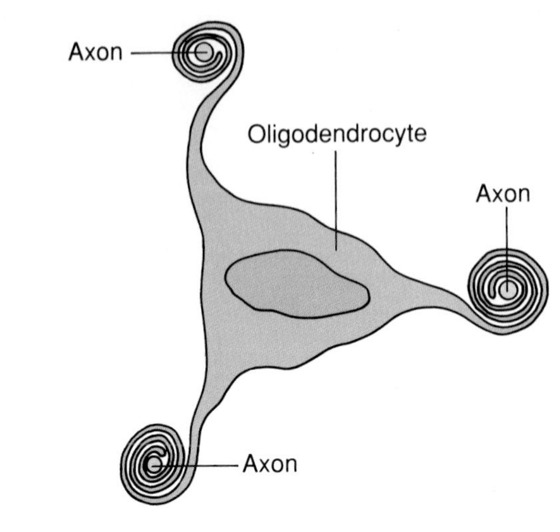

Types of Neurons

Neurons can be classified according to their form, or structure, and according to their function—that is, the role they perform in the nervous system.

◆ **FIGURE 10.12 Types of neurons**

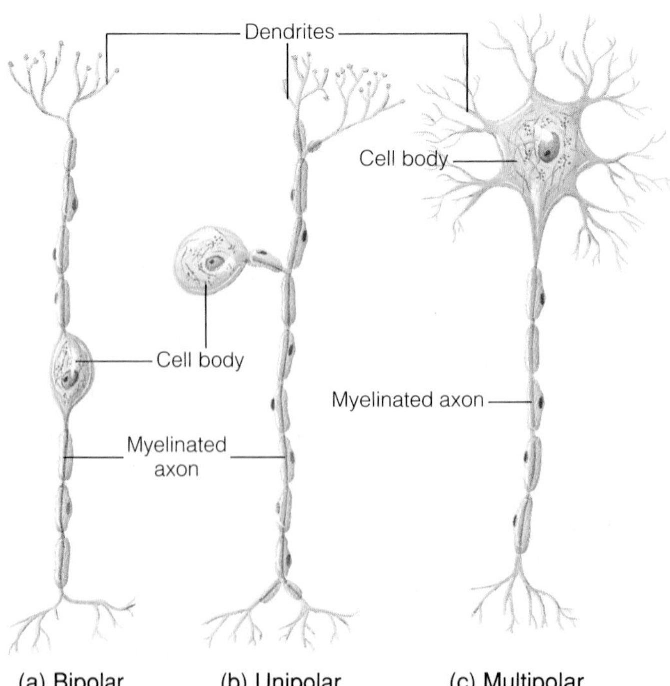

Classification According to Structure. Structurally, neurons can be classified into three types, based on the number of processes that extend from the cell body.

1. **Bipolar neurons** (Figure 10.12a) have two processes, one extending from each end of the cell body. There are only a few examples of bipolar neurons in the body.
2. **Unipolar neurons** (Figure 10.12b) are formed during embryonic development when the two processes of certain bipolar neurons fuse together so that only a single process arises from the cell body. Beyond this point of fusion, the two processes remain separate, with the central branch serving as an axon and the peripheral branch serving as a dendrite.
3. **Multipolar neurons** (Figures 10.12c and 10.13), the most common type of neuron, have one long process that arises from the cell body and functions as an axon; numerous other processes that arise from the cell body function as dendrites.

Classification According to Function. Functionally, there are also three types of neurons (Figure 10.14).

1. **Motor (efferent) neurons** transmit impulses away from the CNS to an effector, or from a higher center within the CNS to a lower center.
2. **Sensory (afferent) neurons** carry impulses from receptors to the CNS, or from a lower center within the CNS to a higher center.
3. **Interneurons (association neurons)** transmit impulses from one neuron to another. They often connect sensory neurons with motor neurons.

Interneurons, which are multipolar, are found only in the central nervous system. Motor neurons are also multipolar. Most sensory neurons are unipolar, but those that carry impulses from the retina of the eye, inner ear, and the olfactory epithelium are bipolar.

◆ **FIGURE 10.13 Photomicrograph of multipolar neurons**

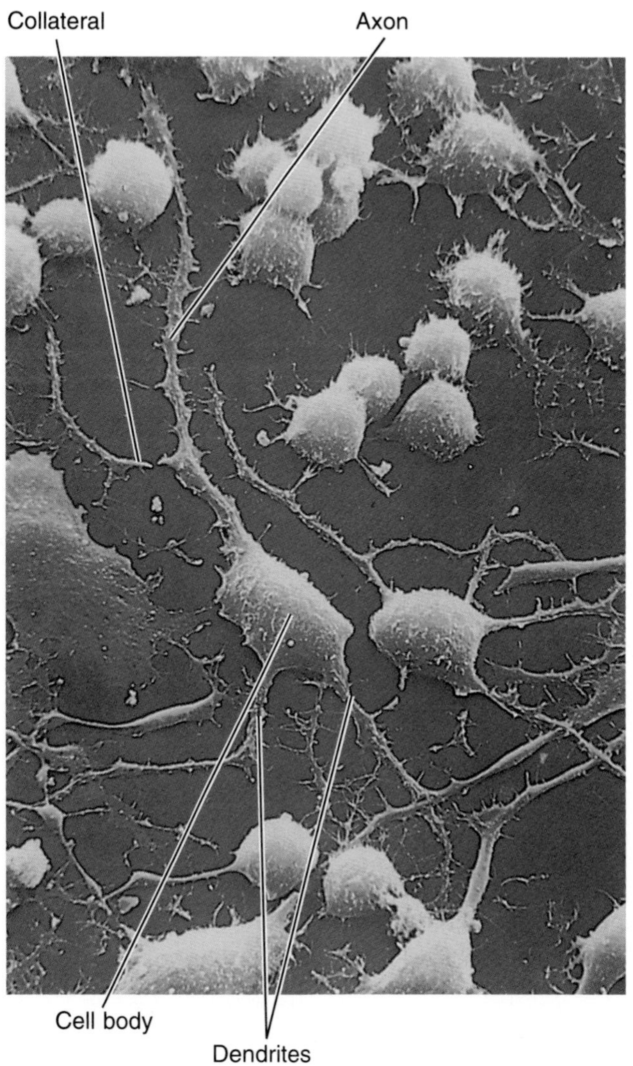

Nerves

A **nerve** is composed of the processes of many neurons held together by connective-tissue sheaths (Figure 10.15). Each nerve fiber is individually wrapped in a thin connective-tissue sheath called the **endoneurium** *(en″-do-nu′-re-um)*. The processes of the neurons are separated into groups within the nerve by another connective-tissue wrapping, the **perineurium** *(per″-ē-nu′-re-um)*. Each bundle of nerve fibers surrounded by perineurium is called a **fasciculus** *(fah-sik′-u-lus)*. Several fasciculi surrounded by a connective-tissue sheath called the **epineurium** *(ep″-ē-nu′-re-um)* constitute a single nerve. Blood vessels and lymphatic vessels travel within the connective-tissue sheaths to supply the neurons.

Nerves are found only in the peripheral nervous system, and they vary in size and composition. Nerves that contain only the processes of sensory neurons and carry impulses to the CNS are **sensory nerves.** Nerves that contain only the processes of motor neurons and transmit nerve impulses from the CNS to effectors in the peripheral nervous system are **motor nerves.** Purely sensory or purely motor nerves are very rare. Most cranial nerves and all spinal nerves are **mixed nerves**—that is, they contain the processes of both sensory and motor neurons. As a result of their composition, nerve impulses within mixed nerves travel both to and from the central nervous system.

To help you understand the terminology used for the main components of the nervous system, Table 10.1

◆ **FIGURE 10.14 Structure and location of the three types of neurons**

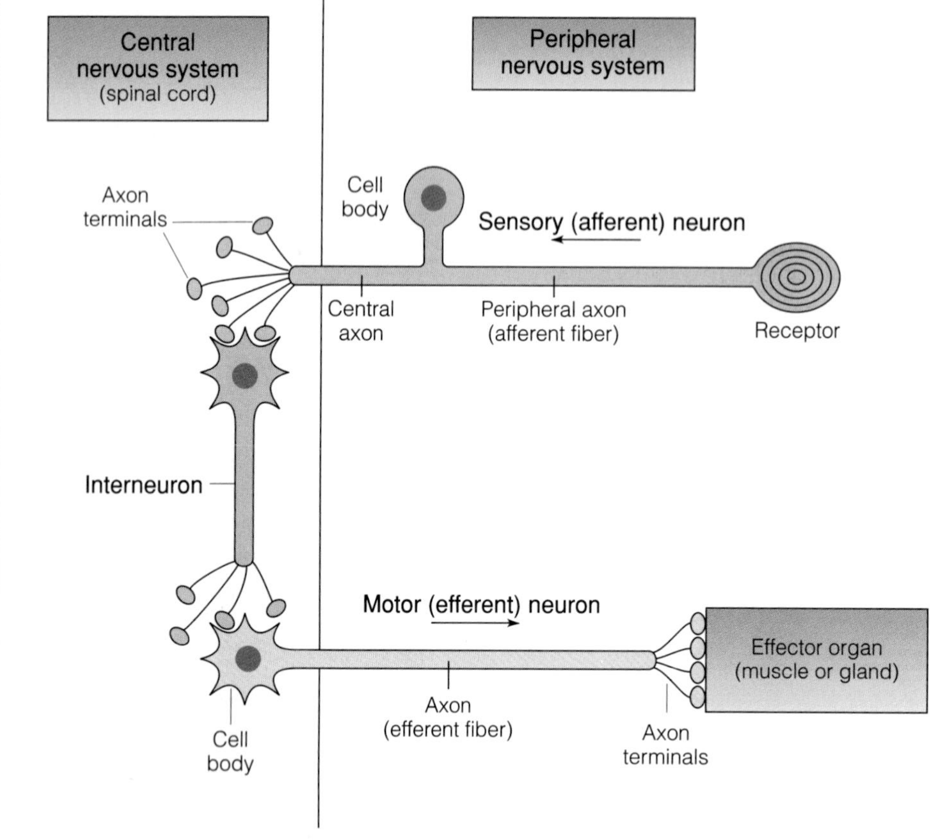

provides a reference for distinguishing between a neuron, a nerve fiber, and a nerve.

Specialized Endings of Peripheral Neurons

The endings of neuronal processes in the peripheral regions of the body are generally specialized structures. They range from simple free nerve endings to complex encapsulated structures (Figure 10.16).

Endings of Motor Neurons

Motor neurons (**somatic motor neurons,** or **efferent neurons**) form a **neuromuscular** *(nu-ro-mus´-ku-lar)* or **myoneural junction** with the skeletal muscles they supply (Figure 10.16a). The axon of the somatic motor neuron divides into many terminal branches that end in neuromuscular junctions at skeletal muscle cells. The myelin sheath of the axon does not extend to the terminal end of the process; thus, the bare axon, which is covered on its outer surface with neurilemma, is able to come very close to the membrane of the muscle cell. The process by which a nerve impulse crosses a neuromuscular junction is described in Chapter 11.

Endings of Sensory Neurons

The terminal portions of the peripheral processes of sensory neurons are dendrites. The terminal ends of the dendrites of many sensory neurons are sensitive to changes in their environment. For this reason, the dendritic endings of many sensory neurons function as **receptors.** In the skin, for example, there are modified

◆ **TABLE 10.1 Comparison of Neurons, Nerve Fibers, and Nerves**

Neuron	A nerve cell
Nerve fiber	Any long process of a neuron. The term usually refers to axons, but also includes the peripheral processes of sensory neurons.
Nerve	A collection of nerve fibers in the peripheral nervous system.

◆ **FIGURE 10.15 Structure of a nerve**

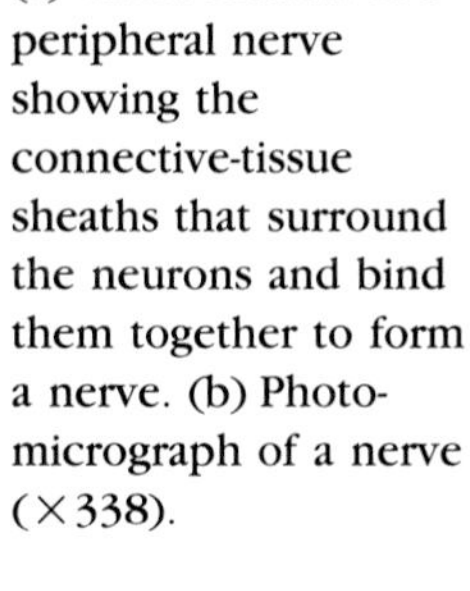

(a) Cross section of a peripheral nerve showing the connective-tissue sheaths that surround the neurons and bind them together to form a nerve. (b) Photomicrograph of a nerve (×338).

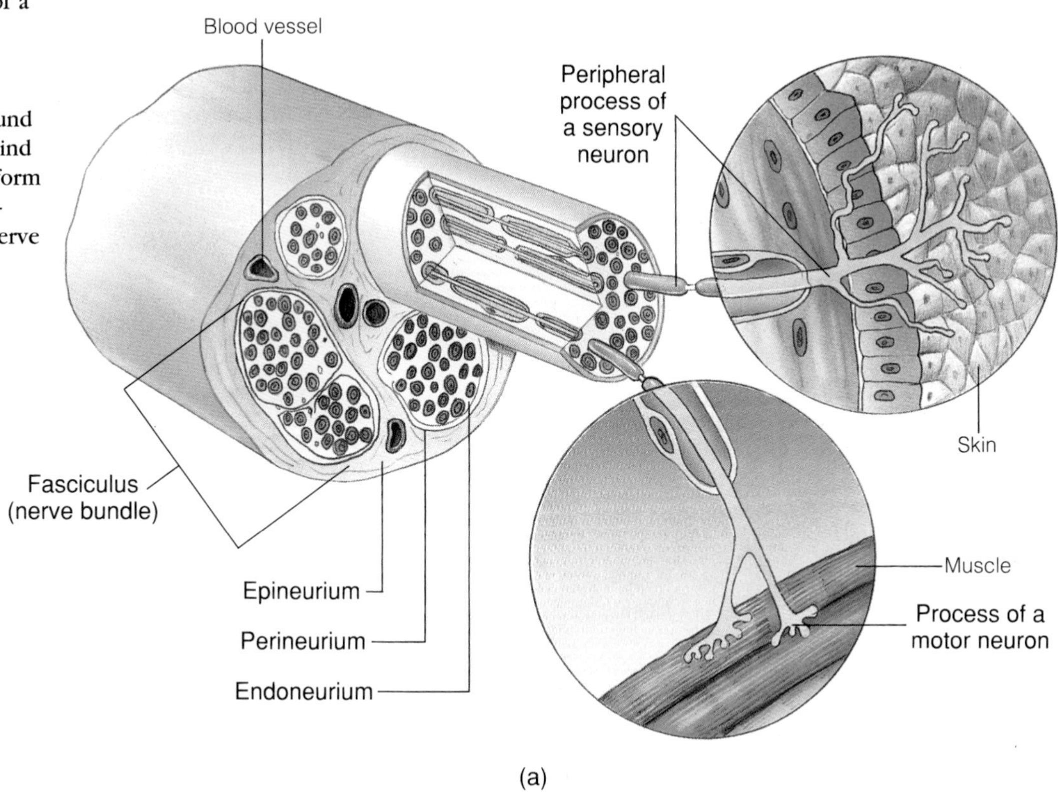

(a)

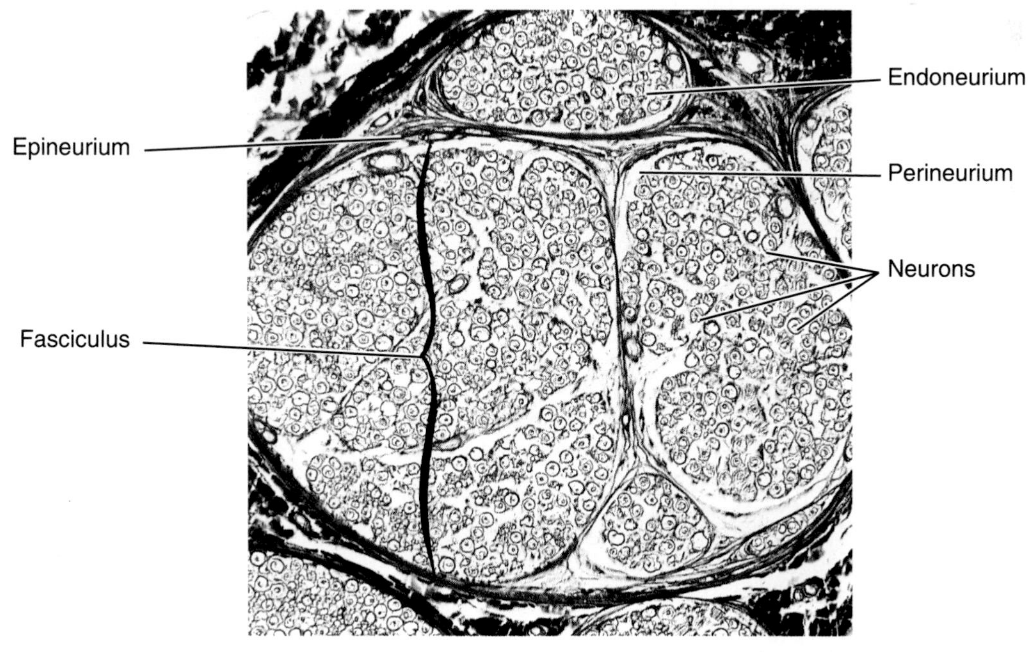

(b)

dendrites that serve as receptors associated with the various cutaneous senses.

The cutaneous senses include those of pain, touch, pressure, heat, and cold. For many years, it has been believed that the body's ability to distinguish one cutaneous sense from another is due to the presence of structurally distinct and functionally specialized receptors—a different type for the reception of each type of

◆ **FIGURE 10.16 Specialized endings of neurons**

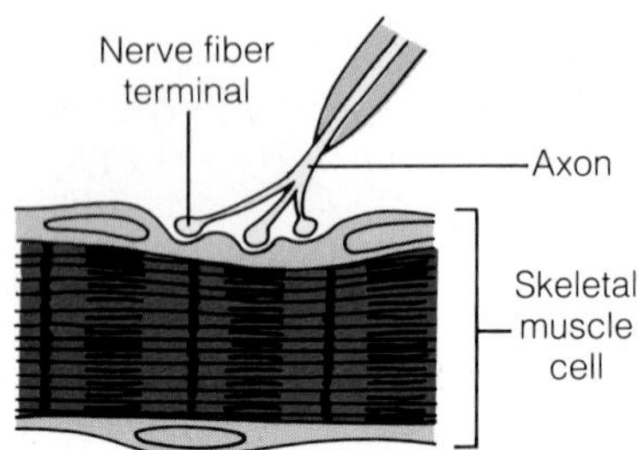

(a) Neuromuscular junction

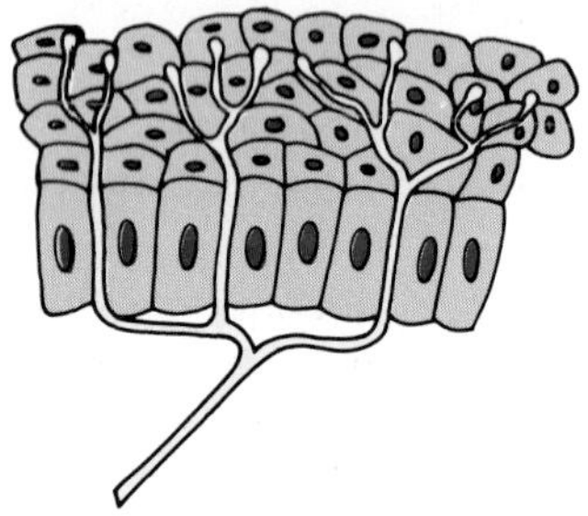

(b) Free nerve endings

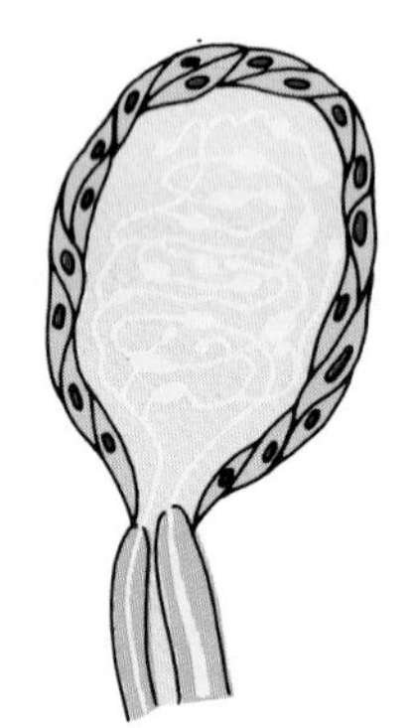

(c) Meissner's corpuscle

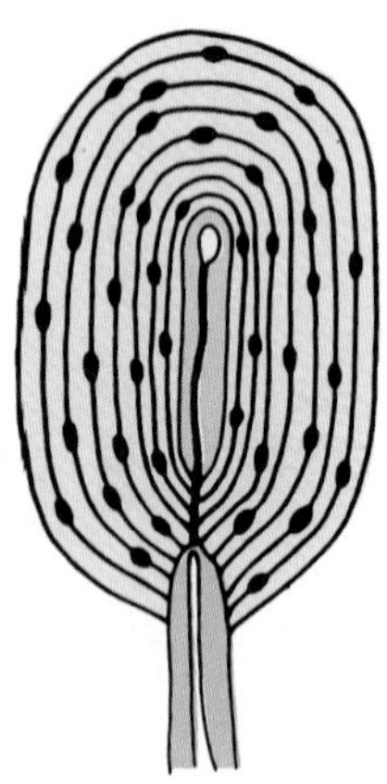

(d) Pacinian corpuscle

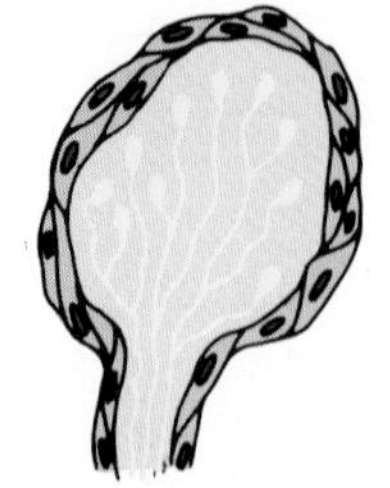

(e) End-bulb of Krause

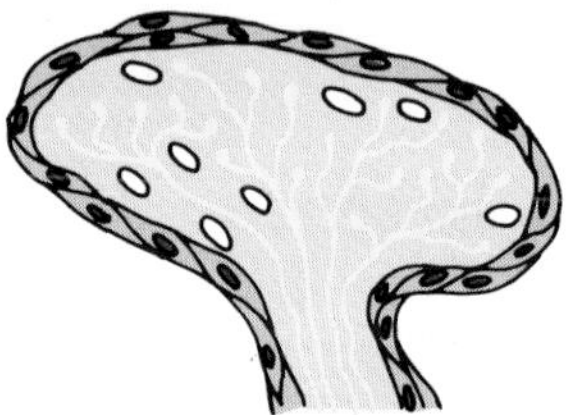

(f) Ruffini's corpuscle

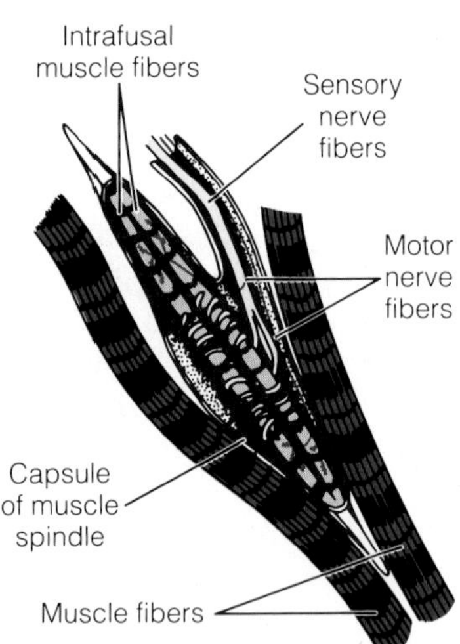

(g) Muscle spindle

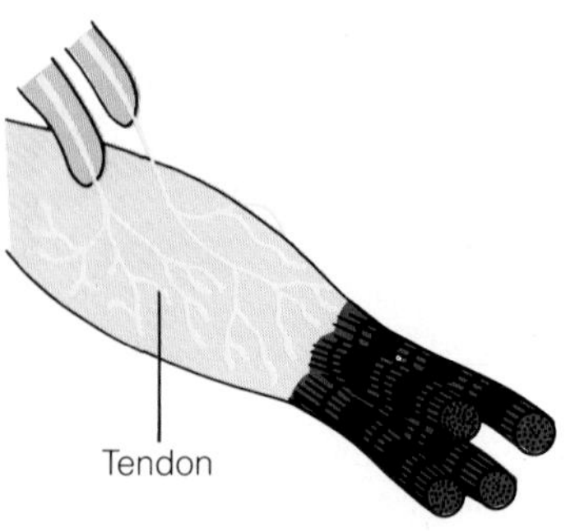

(h) Neurotendinous organ

sensation. There is now some doubt about the accuracy of this one-to-one structural-functional specificity. It may be that each anatomically distinct type of receptor in the skin is, in fact, sensitive to a variety of stimuli rather than to only one basic type of stimulus. According to this view, one type of receptor may have a somewhat different sensitivity to a variety of stimuli than another type. Although the following discussion indicates that each type of receptor may be responsible for sensing a specific kind of stimulus, bear in mind that this notion may require modification as additional information becomes available. Regardless of their precise functions, however, the presence of structurally distinct types of receptors in the skin has been clearly established.

Free Nerve Endings. Free nerve endings are the least modified receptors. They consist of bare dendrites (Figure 10.16b). The free nerve endings branch between epithelial cells, connective-tissue cells, muscle cells, cells of mucous membranes, and so forth. They are thought to serve primarily as *pain* receptors of the body, although free nerve endings that surround hair follicles are believed to be important *touch* receptors, and some free nerve endings serve as heat and cold receptors.

Encapsulated Sensory Endings. The rest of the general sensory receptors are surrounded by connective-tissue capsules that contribute to the activity of the receptors in ways that are not completely understood. These encapsulated receptors are Meissner's corpuscles, Pacinian corpuscles, end-bulbs of Krause, Ruffini's corpuscles, muscle spindles, and neurotendinous organs.

Meissner's corpuscles *(mīz´-nerz)* are small elliptical connective-tissue capsules that surround a spiraled ending of a dendrite (Figure 10.16c). These receptors are thought to be particularly sensitive to *light touch* and are abundant in the dermal papillae of the skin (especially the fingertips), in the mucous membranes of the tongue, and in other sensitive regions of the body.

Located deeper than the Meissner's corpuscles, **Pacinian corpuscles** *(pak-sin´-e-un)* (Figure 10.16d) are found in the deeper layers of the skin, in the mesenteries, and in loose connective tissue. They are composed of a dendrite and several concentric layers of connective tissue surrounding the dendrite. Because they are surrounded by a comparatively heavy capsule, Pacinian corpuscles are thought to be sensitive to heavier *pressure* but not to light touch.

End-bulbs of Krause (Figure 10.16e), which are quite common throughout the body, are thought to serve as *cold* receptors. Oval capsules called **Ruffini's corpuscles** *(ru-fin´-ēz)* (Figure 10.16f) are found primarily in subcutaneous tissue and are generally believed to function as *heat* receptors.

Muscle spindles are complex capsules found in skeletal muscles (Figure 10.16g). Within the capsules are thin skeletal muscle fibers called *intrafusal fibers.* The intrafusal fibers are supplied by sensory neurons, some of which spiral around the fibers. When the muscle is stretched, the intrafusal fibers are also stretched. This increases the number of nerve impulses carried back to the spinal cord by the neurons that surround the intrafusal muscle fibers. In response to the sensory nerve impulses produced by the stretching of the skeletal muscle, there is an increase in the frequency of motor nerve impulses to the same muscle. In this way, the stretch is reflexly resisted.

Neurotendinous organs *(Golgi tendon organs)* function in close association with the muscle spindles (Figure 10.16h). They are composed of dendrites that divide into many small branches in a tendon, near its junction with a muscle. The contraction of the muscle causes varying amounts of tension to be exerted on the tendon. The development of excessive tension by the muscle activates the neurontendinous organs. Sensory impulses from the organs are then carried back to the spinal cord, where motor neurons to the same muscle are inhibited, thus relaxing the muscle.

Types of Receptors

Receptors are structures that are generally activated by changes (stimuli) in either the internal or external environment of the body. As a result of the activity of the receptors, nerve impulses are initiated within sensory nerve cells. Receptors may be the endings of peripheral sensory neurons such as just described, or they may be specialized cells associated with such peripheral endings. Receptors can be classified in two ways: according to the *location* of stimulus or the *type* of stimulus.

Classification According to Location of Stimulus

One means of classification is according to the location of the stimuli to which the receptors respond. Thus, **exteroceptors** *(eks˝-ter-ō-sep´-tors)* respond to stimuli from the body surface, including touch, pressure, pain, temperature, light, and sound. **Interoceptors** *(in˝-ter-ō-sep´-tors)* or **visceroceptors** are sensitive to pressure, pain, and chemical changes in the internal environment of the body. Much of the information transmitted over interoceptors does not reach the level of consciousness. Therefore, a person normally is not aware of small changes in blood pressure, the amount of gases carried within the blood, or contractions of the smooth muscles of the organs. A type of interoceptor called a **proprioceptor** *(pro˝-prē-ō-sep´-tor)* provides information

 ASPECTS OF EXERCISE PHYSIOLOGY 

Back Swings and Pre-Jump Crouches: What Do They Have in Common?

Proprioception, the sense of the body's position in space, is critical to any movement and is especially important in athletic performance, whether it be a figure skater performing triple jumps on ice, a gymnast performing a difficult floor routine, or a football quarterback throwing perfectly to a spot 60 yards down field. In order to control skeletal muscle contraction to achieve the desired movement, the central nervous system (CNS) must be continuously apprised of the results of its actions by means by sensory feedback information.

A number of receptors provide proprioceptive input. Muscle proprioceptors provide feedback information on muscle tension and length. Joint proprioceptors provide feedback on joint acceleration, angle, and direction of movement. Skin proprioceptors inform the CNS of weight-bearing pressure on the skin. Proprioceptors in the inner ear, along with those in neck muscles, provide information about head and neck position so that the CNS can orient the head correctly. For example, neck reflexes facilitate essential trunk and limb movements during somersaults, and divers and tumblers use strong movements of the head to maintain spins.

The most complex and probably one of the most important proprioceptors is the muscle spindle. Muscle spindles are found throughout a muscle but tend to be concentrated in its center. Each spindle lies parallel to the muscle fibers within the muscle. The spindle is sensitive to both the muscle's rate

concerning the position of body parts, without the necessity of visually observing the parts. Thus, for example, it is possible to know the position of the fingers even with the eyes closed. The muscle spindles and the neurotendinous organs are proprioceptors. There are also sensory receptors that monitor the degree of stretch in joint capsules. Since these stretch receptors provide information concerning the position of joints, they, too, serve as proprioceptors.

Classification According to Type of Stimulus

Receptors can also be classified according to the type of stimulus to which they are sensitive. **Mechanoreceptors** are sensitive to physical deformation, such as pressure or stretch. These receptors include Pacinian corpuscles, muscle spindles, neurotendinous organs, and perhaps Meissner's corpuscles. Receptors that respond to changes in temperature are known as **thermoreceptors.** Ruffini's corpuscles and the end-bulbs of Krause, as well as some free nerve endings, are generally classified as thermoreceptors, but they are considered by some to be mechanoreceptors sensitive to touch and closely related to Meissner's corpuscles. **Chemoreceptors** are stimulated by various chemicals in food, in the air, or in the blood. Thus, the senses of taste and smell rely upon chemoreceptors. The receptors that monitor the pH and the levels of gases in the blood are also chemoreceptors. Specialized receptors that are sensitive to light energy are called **photoreceptors.** The only photoreceptors in the body are located in the retinas of the eyes. These receptors, along with the chemoreceptors associated with the senses of taste and smell, are considered to be organs of special senses; they are discussed in Chapter 16.

Neuroglial Cells

There are billions of neurons in the central nervous system, but there is an even greater number of supportive cells distributed among the neurons. These cells are called **neuroglial cells (glial cells).** Some neuroglial cells provide structural support for the neurons; others are involved in the transfer of nutrients from the blood vessels to the neurons and assist in the removal of waste products from the neurons. Still others serve as active phagocytes within the central nervous system. One type of neuroglial cell forms the myelin covering of axons. In contrast to neurons, neuroglial cells are capable of undergoing mitotic division. Included in the neuroglial cells are *astrocytes, oligodendrocytes, microglia,* and *ependymal cells.* With the exception of the microglia, all of the neuroglial cells are of ectodermal origin. Microglia originate from the mesoderm that forms connective tissues.

ASPECTS OF EXERCISE PHYSIOLOGY

of change in length and the final length achieved. If a muscle is stretched, each muscle spindle within the muscle also is stretched, and the afferernt neuron whose peripheral axon terminates on the muscle spindle is stimulated. The afferent fiber passes into the spinal cord and synapses directly on the motor neurons that supply the same muscle. Stimulation of the stretched muscle as a result of this stretch reflex causes the muscle to contract sufficiently to relieve the stretch.

Older persons or those with weak quadriceps (thigh) muscles unknowingly take advantage of the muscle spindle by pushing on the center of the thighs when they get up from a sitting position. Contraction of the quadriceps muscle extends the knee joint, thus straightening the leg. The act of pushing on the center of the thighs when getting up slightly stretches the quadriceps muscle in both limbs, stimulating their muscle spindles. The resultant stretch reflex aids in contraction of the quadriceps muscles and helps the person assume a standing position.

In sports, people use the muscle spindle to advantage all the time. To jump high, as in a basketball jumpball, an athlete starts by crouching down. This action stretches the quadriceps muscles and increases the firing rate of their spindles, thus triggering the stretch reflex that reinforces the quadriceps muscles' contractile response so that these extensor muscles of the legs gain additional power. The same is true for crouch starts in running events. The back swing in tennis, golf, and baseball similarly provides increased muscular excitation through reflex activity initiated by stretched muscle spindles.

There are also two types of cells present in the peripheral nervous system that are considered to be neuroglial cells. These are *satellite cells,* which are located in the capsules that surround the bodies of sensory neurons, and *Schwann cells,* which surround axons of peripheral nerves.

Astrocytes

Astrocytes (*as´-trō-sites*) are rather large cells with star-shaped bodies from which numerous processes radiate outward and attach to neurons and capillaries by means of expanded "endfeet" (Figure 10.17a). They are the most abundant type of neuroglial cell, and they provide most of the structural support for the CNS. Moreover, those astrocytes whose processes contact capillaries within the brain induce the endothelial cells of the capillaries to form tight junctions, sealing adjacent cells firmly together. In other regions of the body, there are usually pores between the endothelial cells forming the capillary walls that allow for free exchange of all plasma components, with the exception of large plasma proteins. The tight junctions in the capillaries of the brain interfere with the free passage of substances between capillaries and neurons, forming what is referred to as a **blood-brain barrier.** The blood-brain barrier is thought to regulate the transport of substances between capillaries and neurons in the brain because the tight junctions prevent molecules from leaving the blood by passing between adjacent capillary cells. Lipid-soluble substances such as O_2, CO_2, and steroid hormones can cross the endothelial cells directly by dissolving in the lipid of the plasma membrane. However, water-soluble substances require the assistance of selective membrane-bound carriers to transport them across the cells by facilitated diffusion. These substances include essential molecules such as glucose, amino acids, lactates, and ions.

Thus, the blood-brain barrier protects the neurons of the brain from sudden and extreme fluctuations in the composition of the tissue fluid by controlling the movement of substances out of the capillaries that supply the brain. The barrier also minimizes the possibility that potentially harmful substances present in the blood might reach the brain tissue. The blood-brain barrier makes the treatment of some brain and spinal cord disorders more difficult, however, because certain drugs that would be effective against the disorders are unable to pass through the barrier.

Oligodendrocytes

Oligodendrocytes *(ol´´-ĭ-gō-den´-dro-sites)* are smaller than astrocytes and have fewer processes (Figure 10.17b). Some of these processes wrap around axons

◆ **FIGURE 10.17 Types of neuroglial cells**

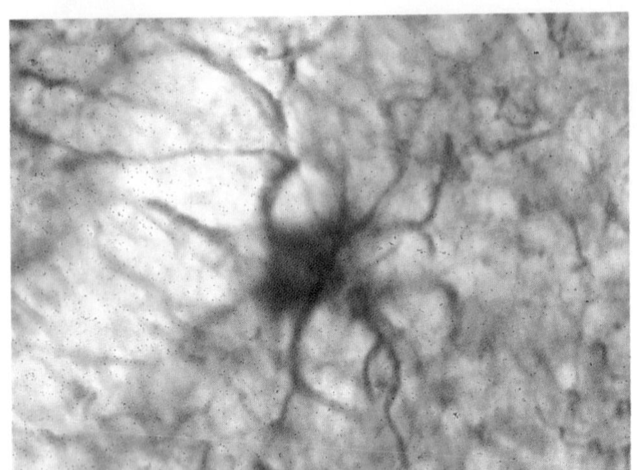
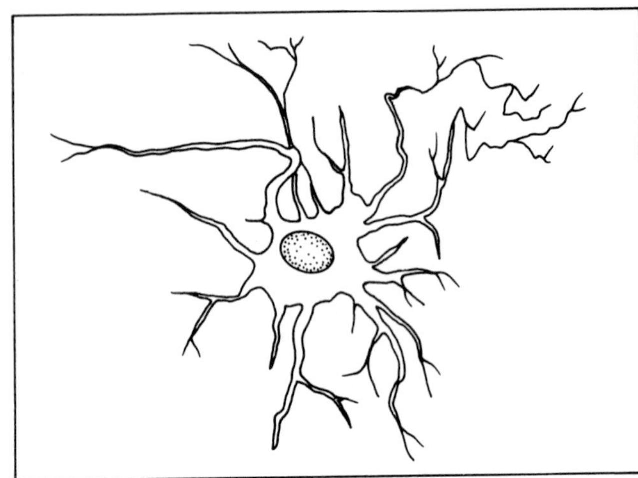

(a) Astrocyte (×527)

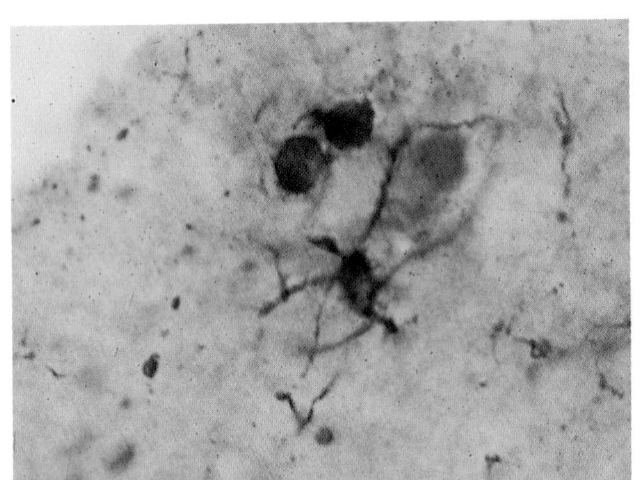
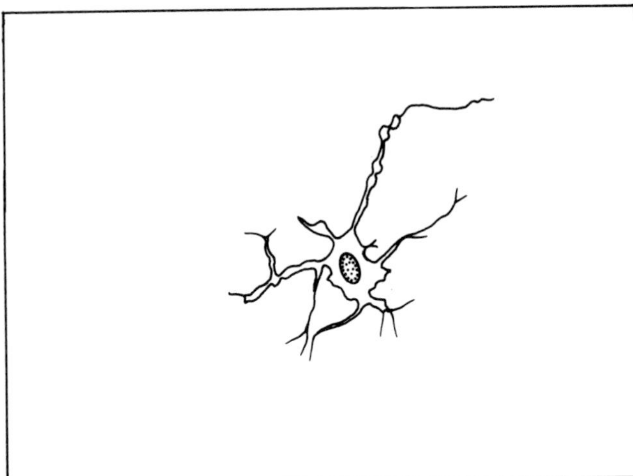

(b) Oligodendrocyte (×969)

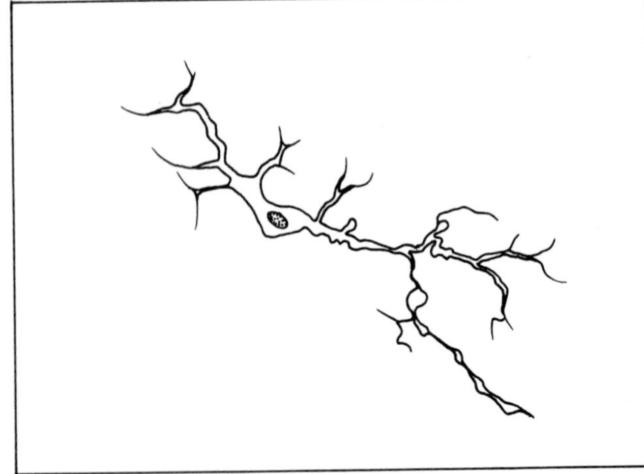

(c) Microglia (×969)

within the central nervous system, forming a myelin sheath, much as the Schwann cells do in the peripheral nervous system.

Microglia

Microglia *(mi-krog´-le-ah)* are the smallest neuroglial cells (Figure 10.17c). They are phagocytic and therefore function in the removal of dead tissue or foreign materials from the CNS. For this reason, microglia are considered to be a part of the macrophage system (see Chapter 4).

Ependymal Cells

Ependymal cells *(ē-pen´-dĭ-mul)* are neuroglial cells that line the ventricles (cavities) of the brain and the central canal of the spinal cord. Some of these cuboidal epithelial cells are ciliated. The ependymal cells in the ventricles of the brain are further modified, having microvilli extending from their free surfaces. These modified ependymal cells are closely associated with invaginations from a covering of the brain called the pia mater and with networks of capillaries, forming structures called **choroid plexuses.** The choroid plexuses produce cerebrospinal fluid, which fills the ventricles.

Effects of Aging on the Nervous System

Because most mature neurons are unable to undergo mitosis, nerve cells that die or are destroyed are generally not replaced. In nervous tissue, as in all tissues, many cells die each day in the normal course of aging. Consequently, with aging there are fewer neurons within the nervous system. How this reduction in neuron number affects the functioning of the nervous system varies among individuals, depending upon the regions of the nervous system in which the reductions occur. The loss of neurons is thought to contribute to a reduction of as much as 10% in the weight of the brain by the age of 90 years.

With aging, there seems to be some loss of myelin from the neurilemma, and the rate of conduction of impulses by neurons decreases in older persons. There is also an increase in the time required to transmit impulses from one neuron to another—that is, to cross a synapse. This increase is thought to be due to an age-related reduction in the transmitter substances released by the presynaptic neurons and a reduction in the numbers or sensitivity of the receptors on the postsynaptic neurons.

Age-related changes within the neurons themselves include a reduction in the number of Golgi apparatus and the chromatophilic substance. The chromatophilic substance contains RNA. Therefore, its reduction could indicate a decline in metabolism within the neurons.

Study Outline

◆ **ORGANIZATION OF THE NERVOUS SYSTEM** pp. 337–338

Central Nervous System (CNS). Brain and spinal cord.

Peripheral Nervous System (PNS). Spinal nerves, cranial nerves, ganglia, and receptors.

AFFERENT DIVISION. Includes somatic and visceral sensory neurons.

EFFERENT DIVISION. Includes somatic and visceral motor neurons—*somatic nervous system (voluntary), autonomic nervous system (involuntary).*

◆ **EMBRYONIC DEVELOPMENT OF THE NERVOUS SYSTEM** pp. 338–340

Ectodermal Cells Thicken. Form neural plate.

Neural Groove Becomes Neural Tube. Develops into brain and spinal cord.

Neural Crest Cells. Form sensory neurons in cranial and spinal nerves, sympathetic motor nerves, and Schwann cells.

ANTERIOR END OF NEURAL TUBE. Forms brain.

◆ **COMPONENTS OF THE NERVOUS SYSTEM** pp. 340–353

Composed of neurons and neuroglial cells.

Neurons.

STRUCTURE OF A NEURON.

1. *Cell body (soma, perikaryon).* And one or more processes containing cytoplasm.
2. *Large nucleus.* Prominent nucleolus.
3. *Chromatophilic substance.* Parallel layers of rough endoplasmic reticulum.

4. *Neurofibrils.* Fibrils for cell support and microtubules for transport within cell.

LOCATIONS OF THE CELL BODIES OF NEURONS.

1. *Nuclei.* Cell body clusters in CNS.
2. *Center.* Group of nuclei with same function.
3. *Ganglia.* Groups of nerve cell bodies in PNS.

PROCESSES OF NEURONS.

1. *Dendrites (dendritic zone).* Receptive portion of neuron where electrical signal originates.
2. *Axon.* Conducting process of neuron along which transmission of electric signal occurs; one axon per neuron; axon branches called collaterals; an axon plus its covering is called a nerve fiber; lack chromatophilic substance.

FORMATION OF MYELINATED NEURONS.

1. *Unmyelinated fibers.* Axons not covered with myelin; are covered by Schwann cells.
2. *Myelinated fibers.* Axons covered with myelin, a fatty substance that causes them to appear white; spiral wrappings of Schwann cells form myelin in PNS; oligodendrocytes form myelin in CNS.

TYPES OF NEURONS.

CLASSIFICATION ACCORDING TO STRUCTURE.

1. *Bipolar neurons.* Two processes; found mostly in the embryo.
2. *Unipolar neurons.* Single process that arises from cell body.
3. *Multipolar neurons.* One axon, several dendrites.

CLASSIFICATION ACCORDING TO FUNCTION.

1. *Motor neurons (efferent).* Transmit from CNS to effector or to lower center in CNS; multipolar.
2. *Sensory neurons (afferent).* Transmit from receptor to CNS or to higher center in CNS; most unipolar, some bipolar.
3. *Interneurons (association neurons).* Transmit from neuron to neuron; multipolar; only in CNS.

Nerves.

1. Individual neuron (nerve cell) wrapped in connective-tissue endoneurium.
2. Neuron processes in bundles wrapped in perineurium; bundle termed a fasciculus.
3. Nerve is a group of fasciculi surrounded by connective-tissue epineurium.
4. Sensory nerves contain only sensory neurons; transmit *to* CNS.
5. Motor nerves contain only motor neurons; transmit to effectors in the PNS.
6. Mixed nerves contain sensory and motor neuron processes; transmit *to* and *from* CNS.

Specialized Endings of Peripheral Neurons.

ENDINGS OF MOTOR NEURONS.

1. *Neuromuscular or myoneural junctions.* Where terminal branches of axon meet but do not touch skeletal muscles.
2. *Myelin.* Not present at end of axon.

ENDINGS OF SENSORY NEURONS. Dendrites function as receptors, of which there are histologically distinct types. Each type of receptor may or may not be responsible for sensing just one specific kind of stimulus.

1. *Free nerve endings.* Bare dendrites; primarily pain receptors; also touch and temperature.
2. *Encapsulated sensory endings.* Surrounded by connective-tissue capsules.

   Meissner's corpuscles; Pacinian corpuscles; end-bulbs of Krause; Ruffini's corpuscles; muscle spindles; neurotendinous organs

Types of Receptors.

CLASSIFICATION ACCORDING TO LOCATION OF STIMULUS.

1. *Exteroceptors.* Respond to body-surface stimuli, such as touch, light.
2. *Interoceptors (visceroceptors).* Respond to pressure, pain, and chemical changes in internal body environment.
3. *Proprioceptors.* Respond to position of body parts, for example, muscle spindles and neurotendinous organs.

CLASSIFICATION ACCORDING TO TYPE OF STIMULUS.

1. *Mechanoreceptors.* Sensitive to physical deformations such as pressure or stretch; Pacinian corpuscles, muscle spindles, neurotendinous organs.
2. *Thermoreceptors.* Sensitive to temperature changes; Ruffini's corpuscles, end-bulbs of Krause, and some free nerve endings.
3. *Chemoreceptors.* Sensitive to chemicals; taste, smell, and pH.
4. *Photoreceptors.* Sensitive to light; retinas of the eyes.

Neuroglial (Glial) Cells. Structural support for neurons; nutrient transfer; phagocytosis; insulation of electrical activity; undergo mitotic divisions; all are ectodermal in origin, except for microglia, which are mesodermal. Satellite cells and Schwann cells considered to be peripheral neuroglia.

ASTROCYTES. Large star-shaped bodies with many processes. CNS structural support.

OLIGODENDROCYTES. Form myelin sheath in CNS.

MICROGLIA. Considered part of macrophage system; remove dead tissue and foreign materials from CNS.

EPENDYMAL CELLS. Line ventricles of brain and central canal of spinal cord; some in roof of ventricles have microvilli.

◆ EFFECTS OF AGING ON THE NERVOUS SYSTEM p. 353

Reduction in number of neurons; loss of myelin slows the rate of conduction of impulses; increased time required to cross synapses.

Self-Quiz

1. There are two systems that serve primarily as means of internal communications—the nervous system and the endocrine system. True or False?
2. The central nervous system includes: (a) ganglia; (b) an autonomic division; (c) the spinal cord.
3. The peripheral nervous system includes: (a) the spinal nerves; (b) the brain; (c) the spinal cord.
4. Impulses that produce contraction of skeletal muscles are carried by: (a) somatic motor neurons; (b) visceral motor neurons; (c) visceral sensory neurons.
5. Both the somatic and the autonomic divisions are under the control of centers located in the peripheral nervous system. True or False?
6. The brain develops from embryonic: (a) mesoderm; (b) endoderm; (c) ectoderm.
7. Nerve impulses in mixed nerves travel: (a) both to and from the central nervous system; (b) only from the CNS; (c) only to the CNS.
8. Mature neurons are generally able to undergo mitosis. True or False?
9. Clusters of nerve cell bodies in the central nervous system are generally called: (a) ganglia; (b) soma; (c) nuclei.
10. Match the following nervous system components with the appropriate lettered item:

Nuclei	(a) Groups of nerve cell bodies in the peripheral nervous system
Soma	(b) The portion of the neuron in which an electrical impulse originates
Center	(c) An axon together with certain sheaths
Ganglia	(d) Clusters of cell bodies of neurons in the CNS
Dendrite	(e) A thin membrane between the myelin and the endoneurium of myelinated axons of the peripheral nervous system
Axon	(f) The name of the cell body of each neuron
Nerve fiber	(g) A fatty substance that covers most axons; formed by spiraling of Schwann cells
Sheath of Schwann	(h) A group of nuclei whose neurons all have a specific function
Myelin	(i) A neuronal process that conducts a nerve impulse

11. Axon processes in the peripheral nervous system may have which of the following associated with them? (a) neurilemma; (b) nodes of Ranvier; (c) both.
12. Schwann cells are not found in the central nervous system. True or False?
13. The most common type of neurons are those called: (a) unipolar; (b) bipolar; (c) multipolar.
14. These efferent neurons transmit impulses away from the central nervous system to an effector: (a) motor; (b) sensory; (c) interneuron.
15. The terminal portions of the peripheral processes of sensory neurons are dendrites. True or False?
16. Match the following terms related to specialized nerve endings with the appropriate lettered item:

Motor neuron endings	(a) Modified dendrites that serve as receptors associated with cutaneous senses
Sensory neuron endings	(b) These receptors are stimulated by heavy pressure
Pacinian corpuscles	(c) These are the least modified receptors, consisting of bare dendrites
End-bulbs of Krause	(d) Heat receptors
Ruffini's corpuscles	(e) Form a neuromuscular junction with the muscles
Muscle spindles	(f) Cold receptors
Neurotendinous organs	(g) These receptors are particularly sensitive to light touch
Free nerve endings	(h) Complex capsules in skeletal muscles
Meissner's corpuscles	(i) Function in close association with muscle spindles

17. These receptors provide information concerning the position of body parts without an individual having to observe the parts: (a) exteroceptors; (b) proprioceptors; (c) mechanoreceptors.
18. Ruffini's corpuscles and the end-bulbs of Krause are thought to be: (a) thermoreceptors; (b) chemoreceptors; (c) photoreceptors.
19. Neuroglial cells that act as phagocytes in the central nervous system are: (a) oligodendrocytes; (b) microglia; (c) ependymal cells.
20. In contrast to neurons, neuroglial cells are capable of undergoing mitotic divisions. True or False?
21. Which type of neuroglial cells are thought to be involved with inducing the formation of the blood-brain barrier? (a) microglia; (b) astrocytes; (c) oligodendrocytes.
22. Some neurons in the central nervous system have a myelin sheath around them that is formed by: (a) Schwann cells; (b) microglia; (c) oligodendrocytes.
23. The ventricles of the brain and the central canal of the spinal cord are lined with: (a) ependymal cells; (b) microglia; (c) astrocytes.

CHAPTER 11

Neurons, Synapses, and Receptors

CHAPTER CONTENTS

LEARNING OBJECTIVES

After completing this chapter, you should be able to:

1. Describe the movements of potassium, sodium, and chloride ions across an unstimulated neuronal membrane.
2. Describe the role of active-transport mechanisms in maintaining the resting membrane potential of a neuron.
3. Describe the movements of sodium and potassium ions during an action potential.
4. Explain the all-or-none nature of an action potential and a nerve impulse.
5. Cite two factors that influence the velocity at which an axon conducts a nerve impulse.
6. Discuss the events involved in the transfer of information from one neuron to another at a chemical synapse.
7. Distinguish between an excitatory postsynaptic potential and an inhibitory postsynaptic potential.
8. Distinguish between temporal summation and spatial summation.
9. Describe the characteristics of a generator potential.
10. Describe the process of adaptation.

Chapter 10 described the general organization of the nervous system and examined the structure of its basic functional component—the nerve cell, or neuron. This chapter focuses on the physiology of the neuron. It describes how a neuron transmits information from one point to another in the form of electrical signals, and it examines the ways in which one neuron communicates with another at specialized junctions known as *synapses.* It also considers the function of *receptors,* which are structures that convert information about conditions in the body's internal or external environments into neural signals.

Because a neuron transmits information from one point to another in the form of electrical signals, it is necessary to understand the basic principles of electricity in order to understand the function of a neuron. At this time, therefore, review the section in Chapter 3 that deals with electrical conditions at the plasma membrane (page 75).

Resting Membrane Potential

The distribution and movements of ions across the plasma membranes of body cells (including neurons) have electrical consequences for the cells, and there are a number of differences between the ionic composition of a cell's intracellular fluid and the ionic composition of the extracellular fluid (Table 11.1). These differences are due to the characteristics and function of the plasma membrane, including the following:

1. The plasma membrane of a cell contains energy-requiring, active-transport mechanisms that move certain ions into the cell and other ions out of the cell. For example, the membrane contains a *sodium-potassium pump* that transports sodium ions outward and potassium ions inward. As a result, the intracellular fluid of a cell contains a higher concentration of potassium ions and a lower concentration of sodium ions than the extracellular fluid.

2. The plasma membrane is not equally permeable to all ions. For example, the plasma membrane contains a number of permanently open channels (called *potassium leak channels*) that are quite permeable to potassium ions but relatively impermeable to sodium ions. Consequently, the plamsa membrane is 50 to 100 times more permeable to potassium ions than to sodium ions.

◆ TABLE 11.1 Representative Concentrations of Selected Ions in the Extracellular Fluid and the Intracellular Fluid of a Typical Cell

ION	EXTRACELLULAR FLUID	INTRACELLULAR FLUID
Potassium (K^+)	4 mM	139 mM
Sodium (Na^+)	145 mM	12 mM
Chloride (Cl^-)	116 mM	4 mM

mM = millimoles/liter.

◆ FIGURE 11.1 Resting membrane potential
A slight excess of positively charged ions accumulates in the extracellular fluid immediately outside the plasma membrane of a cell, and an equal number of negatively charged ions accumulates in the intracellular fluid immediately inside the membrane. As a result of this separation of positive and negative ions, a voltage exists across the membrane. In an unstimulated neuron, this voltage is called the resting membrane potential.

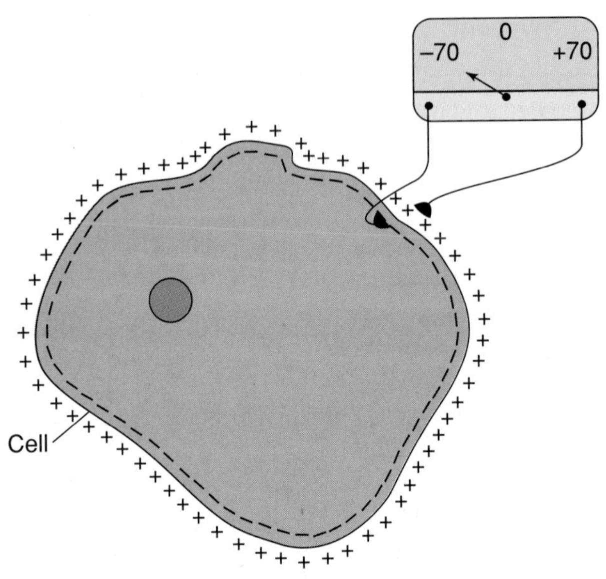

As a consequence of the differences in ionic composition between a cell's intracellular fluid and the extracellular fluid, a very slight excess of positively charged ions accumulates in the extracellular fluid immediately outside the plasma membrane, and an equal number of negatively charged ions accumulates in the intracellular fluid immediately inside the membrane (Figure 11.1). This separation of positive and negative ions across the membrane involves only a minute quantity of the total ions present, and it does not significantly affect the overall electrical neutrality of the intracellular or extracellular fluids. However, as a result of this separation of positive and negative ions, an electrical potential, or voltage, exists across the membrane between the intracellular fluid immediately inside the membrane and the extracellular fluid immediately outside. In an unstimulated neuron, this voltage is called the **resting membrane potential.** Its value is usually between −70 and −85 millivolts. (The sign of the potential—plus or minus—indicates the electrical condition that exists in the intracellular fluid immediately inside the plasma membrane relative to the extracellular fluid immediately outside.)

Development of the Resting Membrane Potential

Two facts are of particular importance in establishing the resting membrane potential: (1) the intracellular fluid of an unstimulated neuron contains a higher concentration of potassium ions and a lower concentration of sodium ions than the extracellular fluid, and (2) the plasma membrane of an unstimulated neuron is 50 to 100 times more permeable to potassium ions than to sodium ions. As a result, potassium ions can diffuse across the neuronal membrane from the intracellular fluid, where they are in high concentration, to the extracellular fluid, where they are in lower concentration, relatively easily. However, sodium ions have difficulty diffusing from the extracellular fluid, where they are in high concentration, to the intracellular fluid, where they are in lower concentration. Since both potassium ions and sodium ions are positively charged, the overall result is that, initially, a somewhat greater number of positively charged ions move out of the neuron than into it, leaving behind a slight excess of negatively charged ions (particularly, negatively charged proteins). This occurrence contributes to the separation of positive and negative charges across the neuronal membrane that gives rise to the resting membrane potential. However, like charges repel and unlike charges attract one another. Thus, as the resting membrane potential develops and the intracellular fluid immediately inside the plasma membrane of the neuron becomes negative relative to the extracellular fluid immediately outside, an electrical force begins to oppose the outward movement of positively charged ions such as potassium and tends to attract positively charged ions into the neuron. Ultimately, a steady state is reached at which the forces tending to move potassium ions—and also sodium and other ions—out of the neuron are balanced by forces tending to move the ions into the neuron. In this state, which is the normal state of unstimulated neurons in the body, the neuron is polarized at its resting membrane potential.

Movement of Ions across the Unstimulated Neuronal Membrane

The movements of potassium, sodium, and chloride ions across the plasma membrane of an unstimulated neuron that is in a steady state and polarized at its resting membrane potential are influenced by both concentration and electrical forces as well as by active-transport mechanisms.

Potassium Ions

The plasma membrane of an unstimulated neuron is relatively permeable to potassium ions, and a concentration force favors a net diffusion of potassium ions from the intracellular fluid, where they are in high concentration, to the extracellular fluid, where they are in lower concentration (Table 11.1). However, in an unstimulated neuron that is polarized at its resting membrane potential, the intracellular fluid immediately inside the plasma membrane is negative relative to the extracellular fluid immediately outside. Consequently, an electrical force opposes a net outward movement of positively charged ions such as potassium ions and tends to attract positively charged ions into the neuron. This electrical force is not sufficient by itself to prevent a net outward diffusion of potassium ions, but the concentration force tending to move potassium ions outward is balanced by a combination of this electrical force and the active transport of potassium ions from the extracellular fluid to the intracellular fluid by the sodium-potassium pump (Figure 11.2). Thus, equal numbers of potassium ions move across the membrane in each direction, and there is no net movement in either direction.

Sodium Ions

In an unstimulated neuron that is polarized at its resting membrane potential, a concentration force favors a net diffusion of sodium ions from the extracellular fluid, where they are in high concentration, to the intracellular fluid, where they are in lower concentration (Table 11.1). In addition, the electrical force resulting from the polarity of the neuronal membrane also favors a net inward movement of positively charged sodium ions (Figure 11.2). Nevertheless, there is no net movement of sodium ions across the neuronal membrane because the membrane is relatively impermeable to sodium ions, and the sodium-potassium pump within the membrane actively transports any sodium ions that enter the neuron back into the extracellular fluid.

◆ **FIGURE 11.2 Diagrammatic representation of the balanced movements of potassium and sodium ions across the stable, unstimulated neuronal membrane**

The membrane is polarized, with the intracellular fluid immediately inside the membrane negative relative to the extracellular fluid immediately outside.

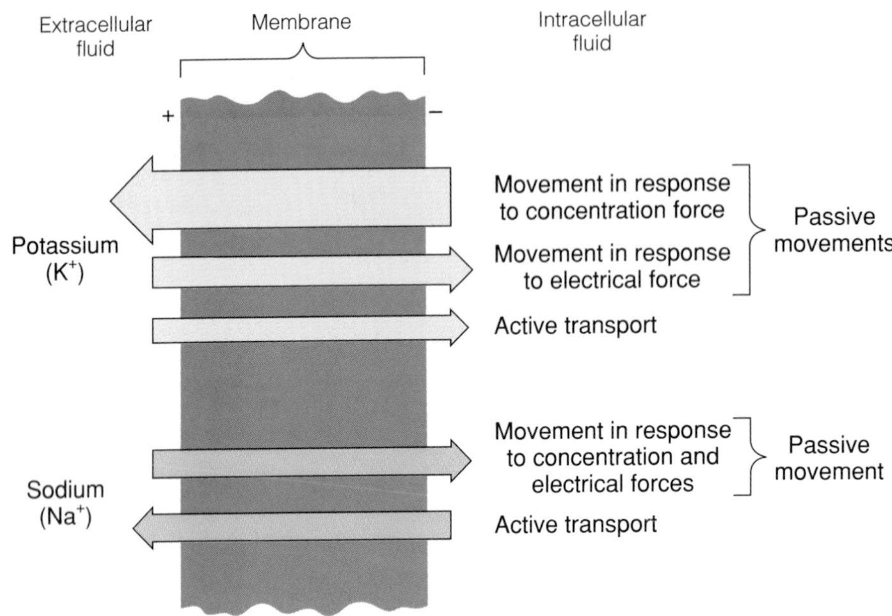

Chloride Ions

The plasma membrane of an unstimulated neuron that is polarized at its resting membrane potential is permeable to chloride ions, and a concentration force favors a net diffusion of chloride ions from the extracellular fluid, where they are in high concentration, to the intracellular fluid, where they are in lower concentration (Table 11.1). This force is opposed by the electrical force due to the polarity of the membrane, which favors a net outward movement of negatively charged ions. In some cases, the concentration and electrical forces may balance one another. In such cases, chloride ions are passively distributed—that is, in electrochemical equilibrium—across the membrane. However, in many neurons, the electrical force is not sufficient by itself to prevent a net entry of chloride ions. Consequently, an active-transport mechanism called a chloride pump, which moves chloride ions outward, is believed to be important in preventing a net entry of chloride ions into these neurons.

Role of Active-Transport Mechanisms in Maintaining the Resting Membrane Potential

As previously indicated, the resting membrane potential is a result of differences in the ionic compositions of the extracellular and intracellular fluids, and these differences are maintained by active-transport mechanisms. For example, if the sodium-potassium pump stops operating, a net movement of potassium ions out of the neuron occurs in response to a concentration force that is not balanced by an opposing electrical force, and a net movement of sodium ions into the neuron occurs in response to both concentration and electrical forces. These movements alter the ionic compositions of the extracellular and intracellular fluids, and this, in turn, alters the resting membrane potential.

Gated Ion Channels

In addition to permanently open potassium leak channels, the plasma membrane of a neuron also contains ion channels called gated ion channels. **Gated ion channels** open or close in response to various signals, thereby increasing or decreasing the permeability of the membrane to the ion or ions that normally move through the channels.

Voltage-gated channels are one class of gated ion channels. These channels open or close in response to alterations in membrane polarity. Voltage-gated ion channels are particularly important in the transmission of nerve impulses along the axons of neurons.

Chemically gated ion channels, or *channel-linked receptors* (see page 75), form a second class of gated ion channels. These channels possess receptor regions to

◆ **FIGURE 11.3 Types of changes in the resting membrane potential**

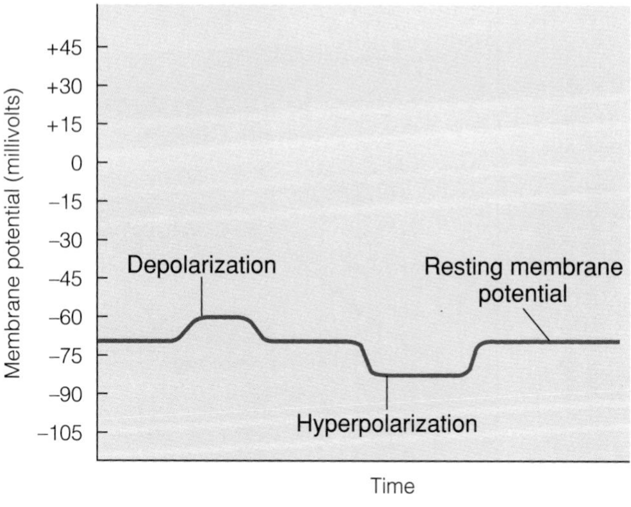

which particular chemical substances (ligands) bind, and this binding regulates the opening or closing of the channels. Chemically gated ion channels convert extracellular chemical signals into electrical signals. These channels are particularly important in the communication that occurs between one neuron and another.

Graded Potentials

Certain events can trigger a change in the membrane potential of a neuron. A change from the resting membrane potential that brings the membrane potential closer to 0 millivolts is called a *depolarization* (Figure 11.3). A change from the resting membrane potential that moves the membrane potential farther away from 0 millivolts is called a *hyperpolarization.*

A change in membrane potential that varies in magnitude with the magnitude of the event that triggers the change is called a **graded potential.** Thus, the stronger the triggering event, the larger the graded potential (Figure 11.4).

A graded potential (which can be either a depolarization or a hyperpolarization) is a local response that occurs at the membrane site where the triggering event takes place. As the distance along the membrane from the site of the triggering event increases, the magnitude of the graded potential decreases—that is, a graded potential is *decremental* (Figure 11.5). In fact, a graded potential that occurs in a neuron dies away entirely a few millimeters away from the site of the triggering event. Consequently, a graded potential can serve as a signal only for a very short distance. A neuron transmits signals for longer distances by means of a nondecremental change in its membrane potential called an action potential.

Action Potential

The plasma membrane of a neuron contains voltage-gated ion channels that are selectively permeable to sodium ions. In most neurons, these *voltage-gated sodium channels* are concentrated in the axon of the neuron. Few of them are present in the dendrites and cell body.

The voltage-gated channels can exist in one of three states:

1. Closed (but able to open)
2. Open
3. Inactivated (closed and unable to open)

In an unstimulated neuron that is polarized at its resting membrane potential, almost all the voltage-gated sodium

◆ **FIGURE 11.4 Graded potentials**

Note that the magnitude of the potential varies with the magnitude of the triggering event.

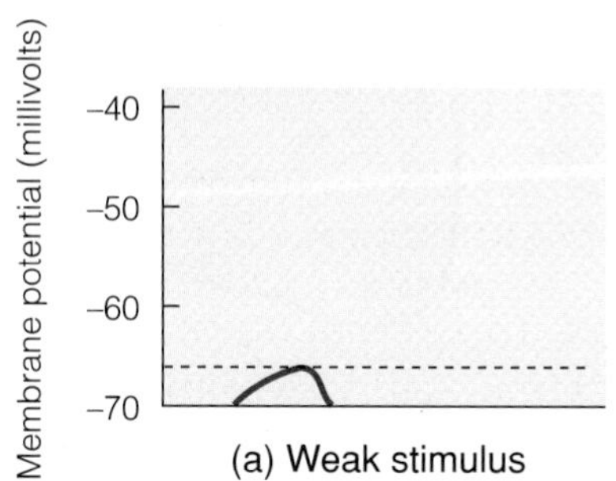

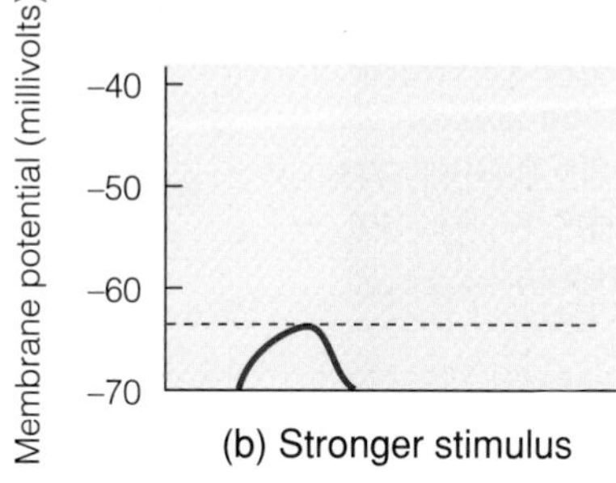

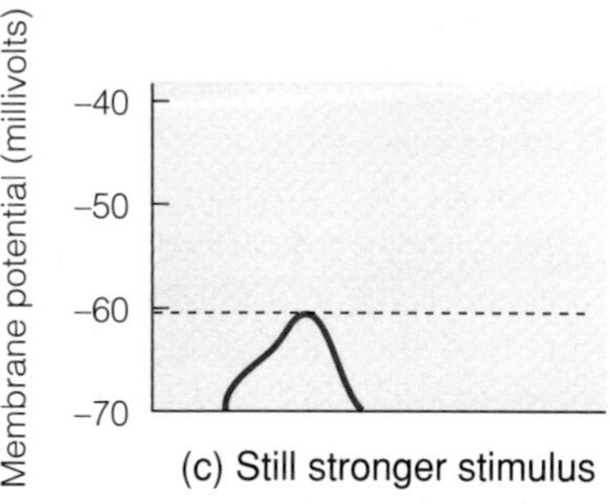

channels are in the closed state, and the plasma membrane is relatively impermeable to sodium ions.

The appropriate stimulation of a neuron can trigger a graded potential that depolarizes a section of the axonal membrane of the neuron. If the depolarization is only slight, most of the voltage-gated sodium channels in the depolarized area remain closed, and only the graded potential occurs in the neuron. However, the depolarization may be of sufficient magnitude to reach a critical level called **threshold.** (Generally, threshold is reached when a section of axonal membrane is depolarized by 15 to 20 millivolts from its resting level.)

Once threshold is reached, the axonal membrane continues to depolarize on its own, without the further involvement of the triggering event or the graded potential. The membrane potential quickly decreases to zero and then reverses—that is, the intracellular fluid immediately inside the membrane becomes positive relative to the extracellular fluid immediately outside (Figure 11.6). This rapid depolarization and polarity reversal, which is called an **action potential,** is unique to nerve and muscle cells.

The next two sections describe how an action potential develops and subsides in one area of the plasma membrane of an axon. A subsequent section explains how an action potential travels along the membrane as a nerve impulse. Table 11.2 compares the characteristics of graded potentials and action potentials, some of which are yet to be discussed.

Depolarization and Polarity Reversal

When a section of axonal membrane is depolarized to threshold, a substantial number of voltage-gated sodium channels open. Thus, in the depolarized area, the permeability of the membrane to sodium ions increases considerably, and there is a significant net movement of sodium ions across the membrane from the extracellular fluid into the intracellular fluid. (Recall that both a concentration force and an electrical force favor this movement.) Since the axonal membrane is polarized with the intracellular fluid immediately inside the membrane negative relative to the extracellular fluid immediately outside, and since sodium ions are positively charged, the net movement of sodium ions into the intracellular fluid further depolarizes the membrane. This further depolarization causes even more voltage-gated sodium channels to open. The opening of these channels further increases the permeability of the membrane to sodium ions and leads to a net movement of even more sodium ions into the intracellular fluid. The net inward movement of more sodium ions further depolarizes the axonal membrane, causing still more voltage-gated sodium channels to open, and so on.

Thus, when a section of axonal membrane is depolarized to threshold, a regenerative, positive-feedback cycle

◆ **FIGURE 11.5 Decremental spread of a graded potential**

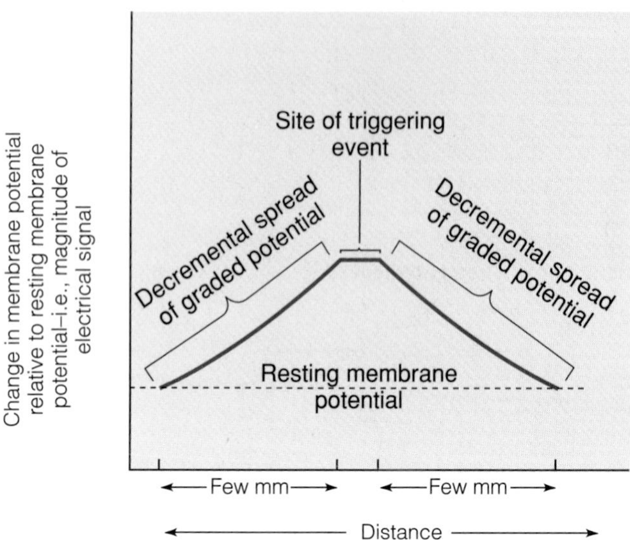

◆ **FIGURE 11.6 Record of an action potential in an axon**

When a graded potential reaches threshold, the membrane potential quickly decreases to zero and then reverses—that is, the intracellular fluid immediately inside the membrane becomes positive relative to the extracellular fluid immediately outside. Following the reversal of the membrane potential, the axonal membrane returns to its original membrane potential.

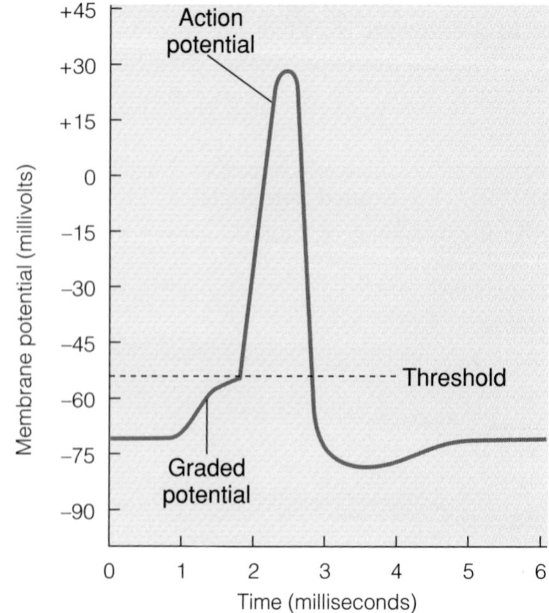

occurs in which the permeability of the membrane to sodium ions increases greatly. As a consequence of this increase in sodium permeability, there is a net movement of sodium ions into the intracellular fluid of the axon. The net inward movement of sodium ions reduces the polarity of the axonal membrane to zero and then reverses the polarity so that the intracellular fluid immediately inside the membrane becomes positive relative to the extracellular fluid immediately outside.

Return to the Original Membrane Potential

The voltage-gated sodium channels remain open for only a brief interval. Then they rapidly close and enter the inactivated state. Thus, the actual time of increased permeability to sodium ions at any one point on the axonal membrane during an action potential is quite short—on the order of a millisecond (Figure 11.7). Moreover, following the depolarization and polarity reversal due to the net movement of sodium ions into the intracellular fluid, there is a significant net movement of positively charged potassium ions from the intracellular fluid into the extracellular fluid. (Note that the axonal membrane is relatively permeable to potassium ions and that both concentration and electrical forces favor a net outward movement of potassium ions when the polarity reverses during an action potential.) The net outward movement of potassium ions causes the intracellular fluid immediately inside the axonal membrane to become less positive and then more negative compared to the extracellular fluid immediately outside, thereby reestablishing the original polarity of the membrane. Thus, following the depolarization and polarity reversal that occur during an action potential, the axonal membrane is returned to its original membrane potential by

1. a rapid decrease in membrane permeability to sodium ions that occurs when the voltage-gated sodium channels close and enter the inactivated state.
2. a net movement of potassium ions from the intracellular fluid into the extracellular fluid.

In unmyelinated axons, the net movement of potassium ions from the intracellular fluid into the extracellular fluid is facilitated by the opening of voltage-gated channels that are selectively permeable to potassium

◆ **TABLE 11.2 Comparison of Graded Potentials and Action Potentials**

GRADED POTENTIALS	ACTION POTENTIALS
Graded potential change; magnitude varies with magnitude of triggering event	All-or-none membrane response; once initiated, magnitude of response independent of triggering event; magnitude of triggering event coded in frequency rather than amplitude of action potentials
Decremental conduction; magnitude diminishes with distance from initial site	Propagated along membrane in undiminishing fashion
Passive spread to neighboring inactive areas of membrane	Self-regeneration in neighboring inactive areas of membrane
No refractory period	Refractory period
Can be summed	Summation impossible
No threshold	Threshold
Can be depolarization or hyperpolarization	Always depolarization and polarity reversal
Triggered by stimulus, by combination of neurotransmitter with receptor, or by spontaneous shifts in leak-pump cycle	Triggered by depolarization to threshold, usually through spread of graded potential
Occurs in specialized regions of membrane designed to respond to triggering event	Occurs in regions of membrane with abundance of voltage-gated sodium channels

◆ **FIGURE 11.7 Change in the permeability of the plasma membrane of an axon to sodium ions that is associated with an action potential**

Also shown is the change in the permeability of the membrane to potassium ions that occurs in unmyelinated axons.

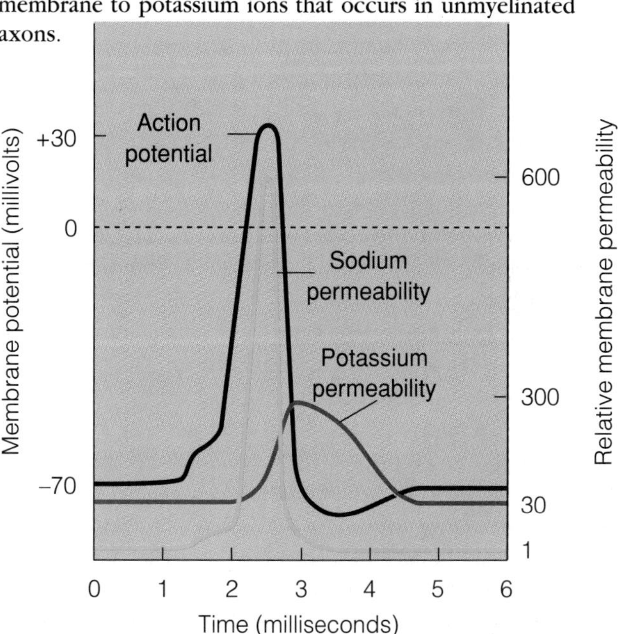

ions. These *voltage-gated potassium channels* open a short time after the voltage-gated sodium channels open, and they remain open somewhat longer than the voltage-gated sodium channels. While the voltage-gated potassium channels are open, the permeability of the axonal membrane to potassium ions is increased, and this increased permeability facilitates the net movement of potassium ions from the intracellular fluid into the extracellular fluid. In myelinated axons, voltage-gated potassium channels are present only in very small numbers, and they do not play a significant role in the movement of potassium ions from the intracellular fluid into the extracellular fluid.

A few milliseconds after the membrane potential returns to its original value, the inactivated voltage-gated sodium channels return to the closed (but able to open) state. When the voltage-gated sodium channels return to the closed state, the section of axonal membrane can again produce an action potential if a graded potential depolarizes it to threshold.

Sometimes, the factors responsible for returning the membrane potential to its original polarity overshoot the original value to some degree, producing a temporary hyperpolarization called an *after hyperpolarization.* For example, in unmyelinated axons, some voltage-gated potassium channels may not close quickly enough, and somewhat more potassium ions move out of the axon than are required to return the membrane potential to its original value.

Maintenance of Sodium and Potassium Ion Concentrations

During the occurrence of an action potential and the return of the axon to its original polarity, only relatively small numbers of sodium ions actually enter the axon, and only relatively small numbers of potassium ions actually leave. Considered alone, these ionic movements have essentially no effect on the sodium ion and potassium ion concentrations of the intracellular and extracellular fluids. However, if thousands of action potentials were to occur, the sodium ion and potassium ion concentrations of these fluids could become altered to such a degree that no additional action potentials could be generated.

Even though numerous action potentials can occur in an axon, the sodium ion and potassium ion concentrations of the intracellular and extracellular fluids are maintained at normal levels by the activity of the sodium-potassium pump. The sodium-potassium pump moves sodium ions out of the axon and into the extracellular fluid and potassium ions into the axon and out of the extracellular fluid. Thus, this active-transport system compensates for the movements of these ions that occur during action potentials and the return of the axon to its original polarity. Note, however, that the sodium-potassium pump is not directly involved in the occurrence of an action potential or in the return of the axon to its original polarity.

 The Nerve Impulse

In an unmyelinated axon, when one area of the axonal membrane reverses its polarity during an action potential, the voltage difference between that area and the immediately adjacent area of the membrane leads to a local current flow between the areas (Figure 11.8). The current is carried by ions and, by convention, the direction of current flow is the direction of movement of positively charged ions. Along the inside of the membrane, positively charged ions are repelled from the area of polarity reversal, and negatively charged ions are attracted toward the area. Along the outside of the mem-

◆ FIGURE 11.8 Current flow between an area of polarity reversal and adjacent areas of the axonal membrane of an unmyelinated axon during an action potential

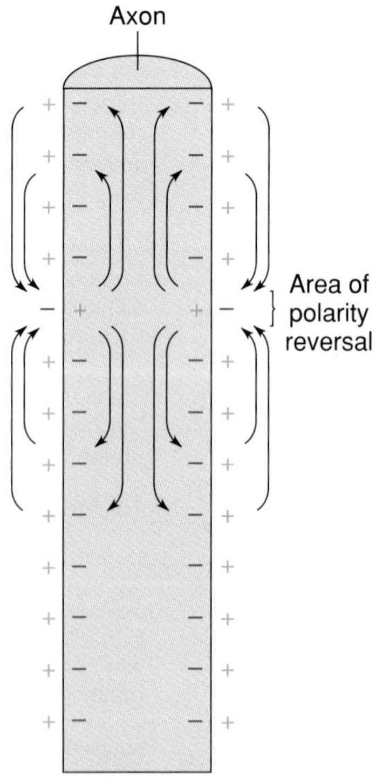

brane, positively charged ions are attracted toward the area of polarity reversal, and negatively charged ions are repelled from the area. Consequently, current flows away from the area of polarity reversal along the inside of the membrane and toward the area of polarity reversal along the outside of the membrane. The current flow depolarizes the area of the membrane immediately adjacent to the area of polarity reversal. When the depolarization of the area reaches threshold, substantial numbers of voltage-gated sodium channels open, and an action potential occurs. A local current flow between this area and the next adjacent area of the membrane depolarizes the next adjacent area to threshold, and an action potential occurs there. This activity continues along the length of the membrane, producing a **propagated action potential,** or **nerve impulse,** that moves along the axonal membrane.

In a myelinated axon, a type of nerve impulse conduction called **saltatory conduction** ("jumping" conduction) occurs. In a myelinated axon, practically all the voltage-gated sodium channels are concentrated at the nodes of Ranvier. There are almost no voltage-gated sodium channels in areas of the membrane covered by myelin. Moreover, myelin is a good insulator, with a high resistance to current flow. Therefore, in a myelinated axon, the local current that gives rise to a propagated action potential flows from one node of Ranvier, where the myelin is interrupted, to the next (Figure 11.9). Thus, action potentials do not occur all along the axon, but only at the nodes of Ranvier. As a result, a nerve impulse in a myelinated axon "jumps" quickly along the axon from node to node. Because of this saltatory (or "jumping") conduction, a nerve impulse travels faster in a myelinated axon than it would if the axon were unmyelinated.

The saltatory conduction of nerve impulses in myelinated axons is energetically more efficient than the conduction of impulses in unmyelinated axons. In a myelinated axon, where action potentials occur only at the nodes of Ranvier, fewer sodium ions enter the axon and fewer potassium ions leave during the conduction of a nerve impulse than in an unmyelinated axon, where action potentials occur all along the axon. Consequently, less metabolic energy is required to restore the resting distribution of ions in a myelinated axon than in an unmyelinated axon.

◆ **FIGURE 11.9 Local current flow in a myelinated axon**

In a myelinated axon, the local current that gives rise to a propagated action potential flows from one node of Ranvier to the next.

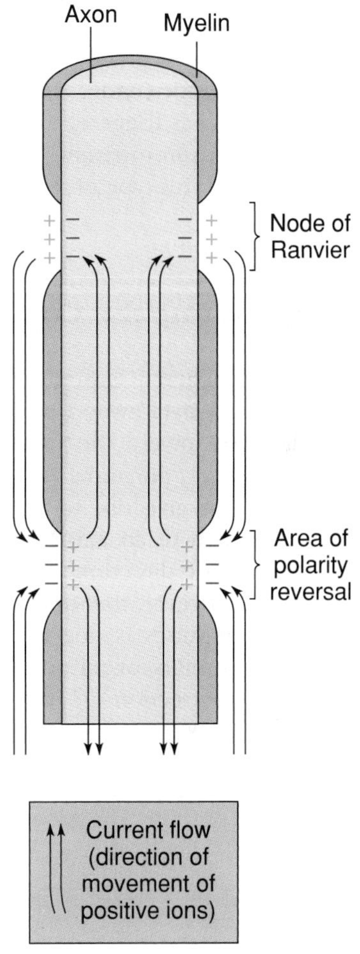

Refractory Periods

The increased permeability to sodium ions of any given section of an axonal membrane during an action potential lasts only about one millisecond. During the short period following the action potential when the voltage-gated sodium channels are in the inactivated state, no additional stimulus applied to this section of the membrane can trigger another action potential, regardless of how strong that stimulus is. This period is called the **absolute refractory period,** and its length limits the number of action potentials—and therefore the number of nerve impulses—that can be produced in a given period of time. Intact nerve cells can generally produce action potentials at frequencies between 0 and 500 per second, although some cells are capable of much higher frequencies for brief periods of time.

During a brief period following the absolute refractory period, a stimulus stronger than that normally required to reach threshold may be able to initiate another action potential. This period is known as the **relative refractory period.**

All-or-None Response

As is evident from the events involved in the generation of a propagated action potential, the conduction of a nerve impulse depends on the properties of the nerve cell membrane (for example, the membrane's ability to change its permeability to sodium ions). Once an axon is depolarized to threshold and a nerve impulse is triggered, the impulse travels along the axon with a magnitude that is characteristic of the particular neuron, and independent of the strength of the triggering event. Thus, a nerve impulse is not a graded response, but instead has the same magnitude (under similar physiological conditions) whether it is triggered by a weak stimulus that just causes the axon to reach threshold or by a much stronger stimulus. This type of response is called an *all-or-none response.*

Direction of Nerve Impulse Conduction

If nerve impulses are triggered experimentally in the middle of an axon, they travel away from their site of origin in both directions toward the two ends of the axon. In the body, however, the initial action potentials and polarity reversals that give rise to nerve impulses normally occur at one end of an axon. Consequently, nerve impulses travel in one direction toward the other end of the axon. Note, however, that this unidirectional transmission of nerve impulses is due to their site of origin and not to any inability of an axon to conduct impulses in the opposite direction.

Conduction Velocities

Both the diameter of an axon and the presence or absence of myelin influence the velocity at which the axon conducts a nerve impulse. In general, the larger the diameter of an axon, the faster it will conduct a nerve impulse. Also, as previously indicated, a nerve impulse travels faster in a myelinated axon than it would if the axon were unmyelinated.

Synapses

For information to be transmitted throughout the nervous system, not just one neuron but chains of them must be traversed. This activity requires a means of passing information from neuron to neuron as well as the ability to transmit information along the axon of a single neuron in the form of nerve impulses.

Information is transferred from one neuron to another at neuronal junctions called **synapses** *(sin´-ap-siz).* A neuron that transmits information toward a synapse is called a **presynaptic neuron,** and a neuron that transmits information away from a synapse is called a **postsynaptic neuron.** A single neuron can be a presynaptic neuron at one synapse and a postsynaptic neuron at another (Figure 11.10).

Near its end, an axon branches into numerous **axon terminals,** and most synapses occur between an axon terminal of a presynaptic neuron and a dendrite or cell body of a postsynaptic neuron. However, in some cases, synapses occur between two axons, two dendrites, or a dendrite and a cell body.

There are two types of synapses: electrical and chemical.

◆ **FIGURE 11.10 Presynaptic and postsynaptic neurons**

A single neuron can be a presynaptic neuron at one synapse and a postsynaptic neuron at another.

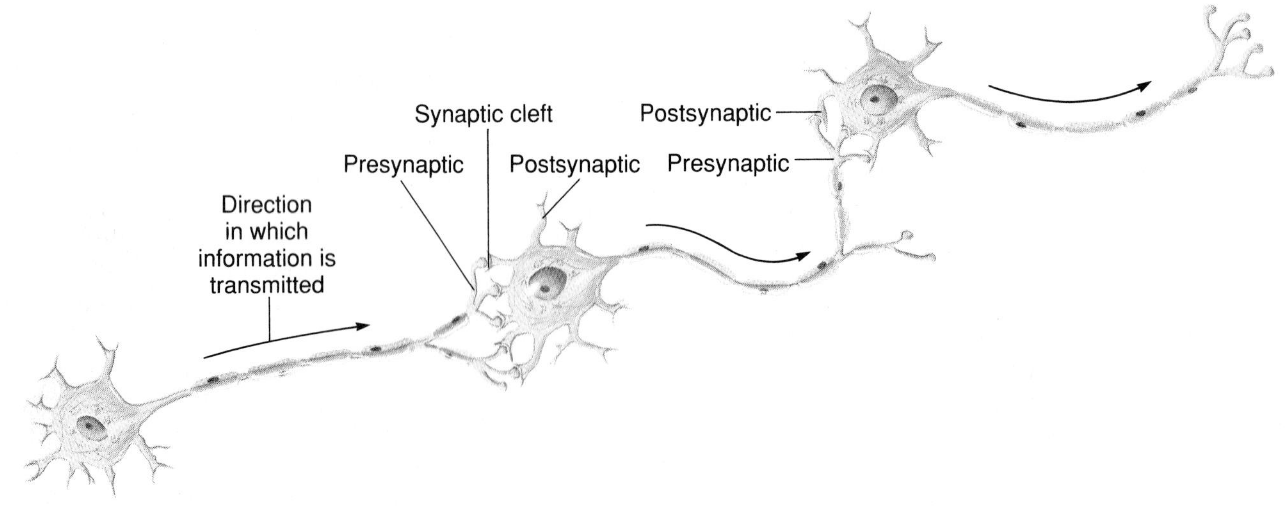

◆ **FIGURE 11.11 A chemical synapse**
When a nerve impulse arrives at an axon terminal, chemical neurotransmitter molecules are released. The molecules diffuse across the synaptic cleft and bind to receptors on the membrane of the postsynaptic neuron. This binding initiates a series of events that influence the activity of the postsynaptic neuron.

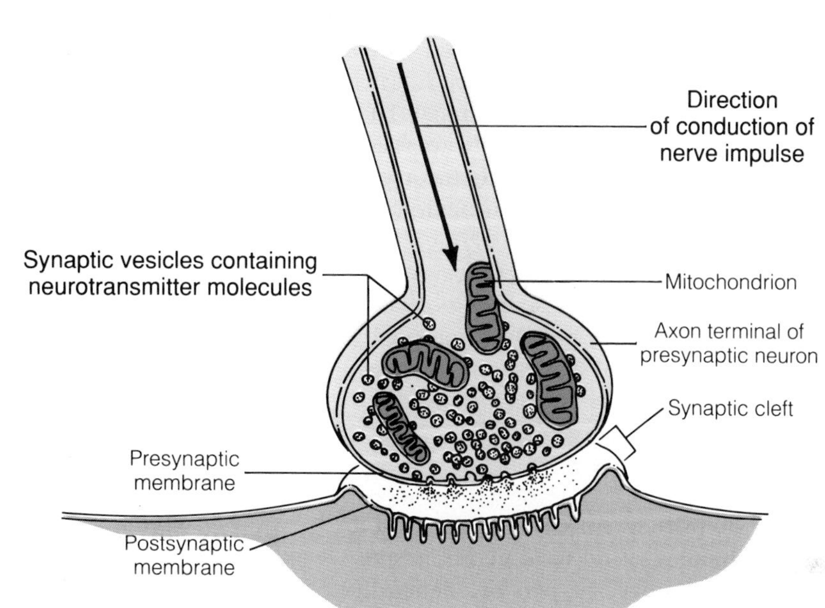

Electrical Synapses

At an **electrical synapse,** a presynaptic and postsynaptic neuron are connected by gap junctions (see page 112). This connection allows local electrical currents resulting from action potentials in the presynaptic neuron to pass directly to the postsynaptic neuron and influence its activity. In adults, electrical synapses are relatively rare, although some are present in certain regions of the brain.

Chemical Synapses

In the human nervous system, chemical synapses occur much more often than electrical synapses. At a **chemical synapse,** an axon terminal of a presynaptic neuron closely approaches a dendrite or cell body of a postsynaptic neuron (Figure 11.11). The two neurons, however, remain separated by a small **synaptic cleft,** which prevents a nerve impulse in the presynaptic neuron from crossing directly to the postsynaptic neuron. Instead, when a nerve impulse arrives at an axon terminal of a presynaptic neuron, chemical signaling molecules are released from the terminal. The signaling molecules diffuse across the synaptic cleft and bind to receptors on the membrane of the postsynaptic neuron. This binding initiates a series of events that influence the activity of the postsynaptic neuron.

A chemical signaling substance that is released by a neuron and that activates specific receptors on a second neuron (or on a muscle cell or a secretory cell) is called a **neurotransmitter.** In many cases, neurotransmitter molecules bind to receptors that are part of *chemically gated ion channels (channel-linked receptors).* In other cases, neurotransmitter molecules bind to *non-channel-linked receptors,* such as G-protein–linked receptors (page 70) or catalytic receptors (page 75). The following sections deal with the function of chemical synapses involving chemically gated ion channels. A subsequent section considers the function of chemical synapses involving non-channel-linked receptors.

Chemical Synapses Involving Chemically Gated Ion Channels

Synaptic signaling that involves chemically gated ion channels occurs rapidly, and the signal is terminated quickly. Moreover, this type of signaling is very accurate—the neurotransmitter released from one axon terminal acts only on a single postsynaptic cell.

Release of Neurotransmitters. A number of different substances serve as neurotransmitters. In general, these substances are manufactured by the neurons that release them. Within an axon terminal, neurotransmitter molecules are stored in small, membrane-bounded sacs called **synaptic vesicles.**

When a nerve impulse arrives at an axon terminal, voltage-gated channels that are selectively permeable to calcium ions open, thereby increasing the permeability of the axon-terminal membrane to calcium ions. Calcium ions are present in higher concentration in the extracellular fluid than in the intracellular fluid, and when the voltage-gated calcium channels open, there is a net movement of calcium ions into the axon terminal. The net entry of calcium ions triggers a process of exocytosis in which some synaptic vesicles fuse with the membrane of the axon terminal and release their neurotransmitter molecules.

The calcium ions that enter the axon terminal remain free to trigger the release of neurotransmitter molecules for only a brief period. They are rapidly taken up by calcium-binding proteins, calcium-sequestering vesicles, and mitochondria, and they are actively transported back into the extracellular fluid.

Effects of Neurotransmitters on Chemically Gated Ion Channels The binding of released neurotransmitter molecules to receptors that are part of chemically gated ion channels in the membrane of a postsynaptic neuron triggers the opening or closing of the channels, and thereby changes the permeability of the membrane to the ions that normally move through the channels. Depending on the response of the postsynaptic neuron to the change in permeability, particular neurotransmitter molecules and their effects are described as either excitatory or inhibitory.

Excitatory Effects. The binding of excitatory neurotransmitter molecules to chemically gated ion channels in the membrane of a postsynaptic neuron increases the likelihood that the neuron will transmit a nerve impulse. In general, the binding triggers changes in the permeability of the membrane that lead to a depolarization of the postsynaptic neuron called an **excitatory postsynaptic potential (EPSP).**

An excitatory postsynaptic potential is a graded potential that varies in magnitude with the degree of stimulation of the postsynaptic neuron by neurotransmitter molecules. If the stimulation is sufficiently intense, the axon of the postsynaptic neuron depolarizes to threshold, and a nerve impulse is triggered.

Inhibitory Effects. The binding of inhibitory neurotransmitter molecules to chemically gated ion channels in the membrane of a postsynaptic neuron decreases the likelihood that the neuron will transmit a nerve impulse. In general, the binding triggers changes in the permeability of the membrane that lead to a hyperpolarization of the postsynaptic neuron called an **inhibitory postsynaptic potential (IPSP).** In some cases, however, the permeability change leads to a stabilization of the membrane potential without a hyperpolarization. In any event, a stronger-than-normal excitatory stimulus is required to trigger a nerve impulse in the postsynaptic neuron.

An inhibitory postsynaptic potential, like an excitatory postsynaptic potential, is a graded potential that varies in magnitude with the degree of stimulation of the postsynaptic neuron by neurotransmitter molecules.

Synaptic Delay. A neuron can conduct nerve impulses quite rapidly (speeds in excess of 100 meters per second have been recorded). However, the speed of transmission at chemical synapses is much slower. The time required to cross a chemical synapses is called the **synaptic delay.** At a chemical synapse involving chemically gated ion channels, this delay—which is approximately 0.5 to 1 millisecond—is primarily due to the time required for the release of neurotransmitter molecules from the presynaptic ending upon the arrival of a nerve impulse.

Removal of Neurotransmitters. Depending on the particular neurotransmitter substance and synapse involved, neurotransmitter molecules are ultimately removed from the region of the receptors by enzymatic inactivation, by diffusion away from the region, or by being taken up by the neuron that released them, by other neurons, or by neuroglial cells.

Chemical Synapses Involving Non-Channel-Linked Receptors

Synaptic signaling that involves non-channel-linked receptors is generally slower and longer lasting than signaling that involves chemically gated ion channels. Moreover, this type of signaling is often spatially diffuse—the neurotransmitter released from one axon terminal may act on many cells in the vicinity of the terminal.

In most cases, the non-channel-linked receptors to which neurotransmitter molecules bind are G-protein–linked receptors, and the binding activates particular G proteins. The activated G proteins may open or close certain ion channels, thereby changing the permeability of the neuronal membrane to the ions that normally move through the channels. They may also influence the production and level of intracellular messengers, such as cyclic AMP, which can directly or indirectly regulate the behavior of ion channels or other aspects of cell function.

In some cases, the binding of neurotransmitter molecules to non-channel-linked receptors on neuronal membranes leads to changes in membrane permeability that produce excitatory or inhibitory effects similar to the effects produced by the binding of neurotransmitter molecules to chemically gated ion channels. In other cases, the binding of neurotransmitter molecules to non-channel-linked receptors produces effects that alter or modulate the responses of chemically gated ion channels to the neurotransmitter molecules that bind to them.

Kinds of Neurotransmitters

More than 50 different substances are known or suspected to serve as neurotransmitters. The following are some of the more common and widely accepted neurotransmitter substances.

Neurotransmitters that Bind to Chemically Gated Ion Channels. Only a relatively small number of neurotransmitters are believed to bind to chemically gated ion channels.

Acetylcholine. **Acetylcholine** *(a-see-til-ko´-leen)* is a neurotransmitter that binds to specific chemically gated ion channels, causing them to open briefly. The channels are permeable to small, positively charged ions, and the opening of these channels in a neuronal membrane increases the permeability of the membrane to potassium ions and sodium ions. The increased permeability leads to a net movement of some potassium ions out of the neuron and to a net movement of even more sodium ions into the neuron, because the total force favoring a net inward movement of sodium ions is stronger than the total force favoring a net outward movement of potassium ions. (Recall that in a neuron polarized at its resting membrane potential, a concentration force favors a net outward movement of potassium ions, but this force is opposed to some degree by the electrical force due to the polarity of the membrane. However, both a concentration force and this electrical force favor a net inward movement of sodium ions.)

The overall result of the ionic movements that take place through the chemically gated ion channels to which acetylcholine binds is a depolarization of the membrane—that is, an excitatory postsynaptic potential. Thus, when acetylcholine binds to these channels in neuronal membranes, it is an excitatory neurotransmitter with excitatory effects.

Acetylcholine is synthesized in the neurons from which it is released. Within the neurons, molecules of acetylcoenzyme A react with choline molecules to form acetylcholine. This reaction is catalyzed by the enzyme choline acetyltransferase.

$$\text{Acetylcoenzyme A} + \text{Choline} \xrightleftharpoons{\text{Choline acetyltransferase}} \text{Acetylcholine} + \text{Coenzyme A}$$

Following its release from an axon terminal of a neuron, acetylcholine is rapidly broken down into acetate and choline by the enzyme acetylcholinesterase, which is present in the synaptic cleft. The choline can be taken up by the axon terminal and used in the synthesis of new acetylcholine.

Glutamate and Aspartate. **Glutamate** (glutamic acid) is an amino acid that functions as a neurotransmitter. Glutamate appears to trigger the opening of chemically gated ion channels that have essentially the same permeabilities as the chemically gated ion channels to which acetylcholine binds. Thus, when glutamate binds to these channels in neuronal membranes, it is an excitatory neurotransmitter with excitatory effects.

Aspartate (aspartic acid) is an amino acid that may act on the same chemically gated ion channels as glutamate and thereby serve as an excitatory neurotransmitter with excitatory effects.

Gamma-Aminobutyric Acid and Glycine. **Gamma-aminobutyric acid (GABA)** is a neurotransmitter that is synthesized from glutamate, and **glycine** is an amino acid neurotransmitter. Gamma-aminobutyric acid and glycine trigger the opening of chemically gated ion channels that are permeable to small, negatively charged ions—particularly chloride ions. In some cases, the opening of chemically gated chloride channels in neuronal membranes leads to a net movement of chloride ions into the cell, which hyperpolarizes the membrane—that is, produces an inhibitory postsynaptic potential. In other cases, the opening of the channels stabilizes the membrane potential without a hyperpolarization. In any event, when gamma-aminobutyric acid and glycine bind to these channels in neuronal membranes, they are inhibitory neurotransmitters with inhibitory effects.

Neurotransmitters That Bind to Non-Channel-Linked Receptors. A substantial number of neurotransmitters bind to non-channel-linked receptors. These neurotransmitters can have varying effects on different cells, depending on the particular receptors each cell possesses and on the responses that occur when neurotransmitter molecules bind to them.

Some neurotransmitters, such as acetylcholine, bind both to chemically gated ion channels and to non-channel-linked receptors. The chemical nicotine mimics the action of acetylcholine at chemically gated ion channels. Consequently, the receptor portions of the chemically gated ion channels to which acetylcholine binds are called *nicotinic receptors.* The chemical muscarine mimics the action of acetylcholine at non-channel-linked receptors, and the non-channel-linked receptors to which acetylcholine binds are called *muscarinic receptors.*

Norepinephrine. **Norepinephrine** *(nor-ep-i-nef´-rin)* is a neurotransmitter that binds to non-channel-linked receptors. It is released by most neurons of the sympathetic division of the autonomic nervous system at their junctions with effectors and is also released by a number of neurons in the brain. Norepinephrine is thought to be a neurotransmitter in certain nerve pathways involved with the maintenance of arousal, in the brain system of reward, in dreaming sleep, and in the regulation of mood.

Norepinephrine is synthesized from the amino acid tyrosine (Figure 11.12). Once it is released, norepinephrine is removed primarily by being taken up by the axon terminals that released it. Norephinephrine can also be inactivated by the enzyme catechol-O-methyltransferase (COMT), which is associated with the membranes of postsynaptic cells. However, enzymatic inactivation is not a major way that norepinephrine is removed from synaptic junctions. Norepinephrine taken up by axon terminals can eventually be released again. Also, some of it may be metabolized in a reaction sequence involving the enzyme monoamine oxidase (MAO), which is associated with mitochondria present in the axon terminals.

◆ **FIGURE 11.12 Reactions that synthesize dopamine, norepinephrine, and epinephrine from tyrosine**

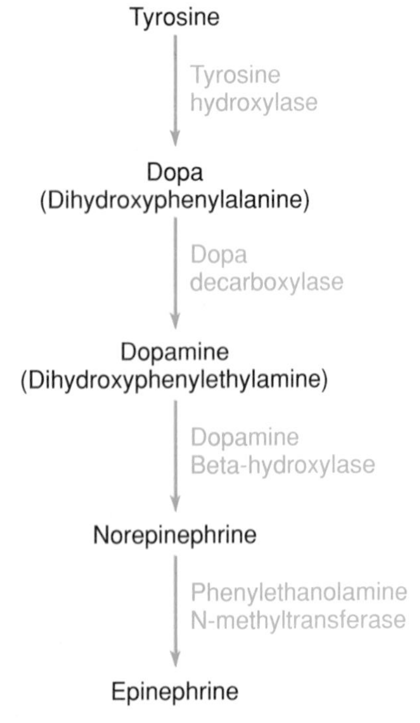

Epinephrine. The chemical substance **epinephrine** *(ep-i-nef´-rin)* is very closely related to norepinephrine. It is present in certain neurons of the brain stem and is believed to be a neurotransmitter in some nerve pathways concerned with behavior, mood, and perhaps emotions.

Like norepinephrine, epinephrine is synthesized from tyrosine (Figure 11.12). Following its release, epinephrine is removed in the same manner as norepinephrine.

Dopamine. **Dopamine** is a chemical substance present in the brain that is widely accepted as a neurotransmitter. It has been implicated in the regulation of emotional responses and in the control of complex movements.

Dopamine is similar to norepinephrine in chemical structure, and it is synthesized from tyrosine by the same series of reactions that produce norepinephrine (Figure 11.12). Dopamine is also removed from synaptic junctions in the same manner as norepinephrine. Because of their chemical structures, norepinephrine, epinephrine, and dopamine are referred to as *catecholamines (kat-i-kol´-uh-meenz).*

Serotonin. **Serotonin** (5-hydroxytryptamine) is another chemical substance present in the brain that is widely considered to be a neurotransmitter. It is believed to play a role in temperature regulation, in sensory perception, and in the onset of sleep.

Serotonin is synthesized from the amino acid tryptophan. Following its release, serotonin can be removed by being taken up by axon terminals.

Neuropeptides. A group of substances known collectively as **neuropeptides** are present in the nervous system. These substances are believed to function as neurotransmitters that bind to non-channel-linked receptors. Neuropeptides are chains of amino acids, and essentially all neuropeptides discovered to date also have nonneural effects. In fact, many of them were originally identified in nonneural tissues. Table 11.3 lists some neuropeptides and describes some of their neural effects.

Neuromodulators

Neuromodulators are chemical substances other than neurotransmitters that alter neuronal activity by altering neurons directly or by influencing the effectiveness of neurotransmitters. For example, a neuromodulator may alter the synthesis or release of a neurotransmitter from a presynaptic neuron or it may increase or decrease the sensitivity of a postsynaptic neuron to a neurotransmitter. A number of hormones are believed to function as neuromodulators, and many substances currently believed to be neuromodulators also exert nonneural effects.

Neural Integration

The nervous system integrates information from many different sources in order to produce useful, coordinated responses by the body to a wide variety of conditions. At the cellular level, this integrative ability depends on such occurrences as divergence, convergence, summation, and facilitation.

Divergence and Convergence

The axon of a presynaptic neuron may branch many times and, thus, may synapse with many postsynaptic neurons (Figure 11.13). This phenomenon, which is called **divergence,** permits a nerve impulse in a single presynaptic neuron to affect many postsynaptic neurons. Conversely, axon terminals of many different presynaptic neurons may all synapse with a single postsynaptic cell. This phenomenon is called **convergence.** Divergence and convergence allow for a wide variety of neuronal interactions and information transfers within the nervous system.

Summation

If the effects of the neurotransmitter molecules released at a chemical synapse are excitatory to the postsynaptic cell, the synapse is called an excitatory synapse. If the effects of the neurotransmitter molecules are inhibitory to the postsynaptic cell, the synapse is called an inhibitory synapse.

The postsynaptic potentials produced by the release of neurotransmitter molecules at chemical synapses can add together, or **summate,** to influence the activity of a postsynaptic neuron. In fact, summation is normally required to trigger a nerve impulse in a postsynaptic neuron because the excitatory postsynaptic potential (EPSP) produced by the neurotransmitter released in response to the arrival of a single nerve impulse at a single excitatory synapse is rarely, if ever, large enough to depolarize the axon of the postsynaptic neuron to threshold. There are two types of summation: temporal and spatial.

Temporal Summation

Temporal summation is the summation that occurs when many nerve impulses arrive at a single synapse within a short period of time. For example, the arrival of one nerve impulse at an excitatory synapse leads to the release of some neurotransmitter molecules, which produce a relatively small EPSP in the postsynaptic neuron. However, if a second nerve impulse arrives and a second EPSP is produced before the initial EPSP dies away, the two EPSPs can add together, or summate. This summation produces a greater depolarization of the postsynaptic neuron than would result from either EPSP alone. If enough nerve impulses arrive at the synapse close enough together in time, the summation of the EPSPs that are produced can depolarize the axon of the postsynaptic neuron to threshold and trigger a nerve impulse in it.

◆ **TABLE 11.3 Representative Neuropeptides**

NAME	NEURAL EFFECTS
Substance P	Stimulates spinal cord neurons that respond to painful stimuli; may be a neurotransmitter involved in the transmission of pain signals; may play a role in emotional behavior
Enkephalins	Appear to inhibit the release of sensory pain neurotransmitters such as substance P in spinal cord
Leu-enkephalin	May be involved in regulating emotional behavior
Met-enkephalin	May be involved in integrating sensory perception; may be principal mediator of analgesic effects of enkephalins
Neurotensin	May play a role in pain perception; has potent analgesic effect
Cholecystokinin	May be involved in pain-integrating functions; is powerful excitant of cerebral cell firing
Vasoactive intestinal polypeptide	May activate and synchronize neuronal activity of certain cerebral cortical cells; is a potent and rapid excitant of neurons in hippocampus of brain
Bradykinin	May be involved in blood pressure regulation and in pain perception

Spatial Summation

Spatial summation is the summation that occurs when nerve impulses arrive very close together in time at a number of synapses between different presynaptic axon terminals and the same postsynaptic neuron. For ex-

◆ **FIGURE 11.13 Divergence and convergence**
(a) Neuronal processes of a single cell diverge to a number of other cells. (b) Many neuronal processes converge on a single cell.

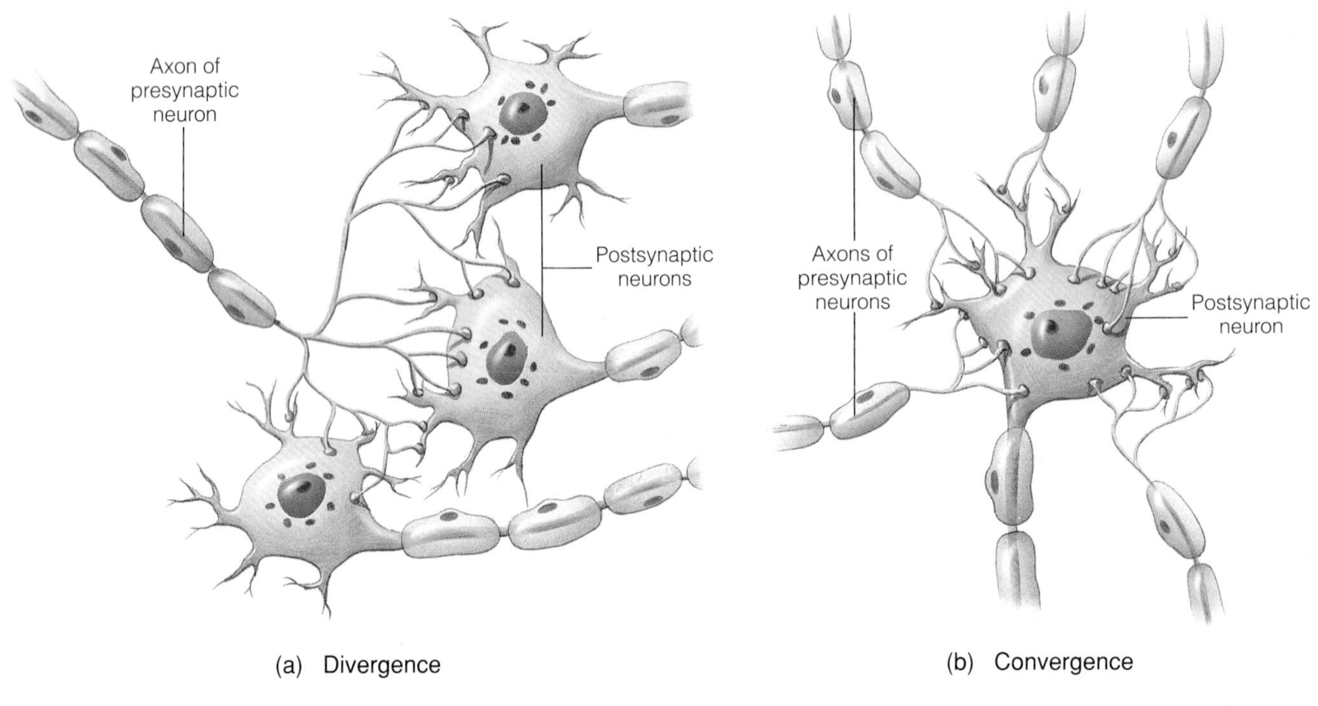

ample, as previously indicated, the EPSP produced by the neurotransmitter released in response to the arrival of a single nerve impulse at a single excitatory synapse is normally not very great. But, if nerve impulses arrive at many excitatory synapses at about the same time, the EPSPs produced at each synapse can summate and depolarize the axon of the postsynaptic neuron to threshold.

Facilitation

Quite often, as a result of spatial summation, the axon of a postsynaptic neuron is depolarized toward threshold by EPSPs produced at some excitatory synapses. Consequently, less depolarization is required for the axon to reach threshold when additional EPSPs occur. This phenomenon is called **facilitation** because the initial EPSPs that depolarize the axon toward threshold facilitate the attainment of threshold and the generation of a nerve impulse in the axon when additional EPSPs occur.

Determination of Postsynaptic Neuron Activity

As is the case for EPSPs, inhibitory postsynaptic potentials (IPSPs) produced by neurotransmitter molecules released at inhibitory synapses can also undergo temporal and spatial summation. Moreover, a postsynaptic neuron often receives thousands of presynaptic inputs, some of which may form excitatory synapses with the neuron and some of which may form inhibitory synapses.

Whether or not the axon of a postsynaptic neuron is depolarized to threshold depends on the relationship between the excitatory and inhibitory events influencing it at any moment. If excitatory events are sufficiently dominant, the axon may reach threshold and transmit a nerve impulse. If inhibitory events predominate, the axon will be unlikely to transmit a nerve impulse. Thus, the activity of a postsynaptic neuron is determined by the integrated activities of many presynaptic inputs from various sources. Integrative processes of this sort, involving many neurons in multineuronal pathways, allow the nervous system to generate appropriate responses to many different circumstances.

ASPECTS OF EXERCISE PHYSIOLOGY

Relation between Excitatory Postsynaptic Potentials and Muscle Strength

Neurons called alpha motor neurons innervate skeletal muscles. Action potentials in alpha motor neurons initiate the contractile process in the muscle on which they terminate. Both excitatory and inhibitory presynaptic inputs from a variety of sources impinge on the cell body and dendrites of the alpha motor neurons. These neurons fire, thereby initiating muscle contraction, when their membrane is brought to threshold by summation of excitatory postsynaptic potentials (EPSPs). A person can voluntarily contract a given skeletal muscle by consciously activating nerve pathways that generate EPSPs at the appropriate alpha motor neurons.

Among the inhibitory inputs to alpha motor neurons are those arising from the neurotendinous organ, a sensory device located within the tendon that attaches a muscle to the skeleton (see page 349). The neurotendinous organ is sensitive to the tension exerted on the tendon by the muscle's contraction. When substantial tension is exerted on the tendon, the neurotendinous organ activates a nerve pathway that generates inhibitory postsynaptic potentials (IPSPs) at the alpha motor neurons that innervate the contracting muscle. If the membrane potential of the involved alpha motor neurons is reduced below threshold by the inhibitory input from the neurotendinous organ, the muscle is no longer stimulated and it relaxes. This reflex relaxation is protective in nature; its purpose is to prevent the muscle from contracting hard enough to rupture the muscle or tendon or break a bone.

Weight training requires the voluntary control of this natural muscular inhibitory reflex. When persons begin to weight train, their ability to lift increases dramatically in the first two weeks even though no physiologic change has yet occurred in the muscle. This initial improvement in strength is thought to be brought about by the body learning to voluntarily produce more EPSPs, thus voluntarily overriding the inhibitory influences of the neurotendinous organs. As weight training progresses, muscles get physiologically stronger as the muscle cells enlarge, the tendons get thicker, and the neurotendinous organs decrease their sensitivity.

Power lifters sometimes yell as they lift to try to further increase the excitatory stimuli during heavy lifting. In sports such as arm wrestling, the inhibitory reflex is overridden by voluntary control to the point that participants have ruptured muscles or tendons and have even broken the bones in the arm during the excitement of competition.

Presynaptic Inhibition and Presynaptic Facilitation

In some cases, the amount of chemical neurotransmitter released when a nerve impulse arrives at an excitatory synapse is reduced by the process of **presynaptic inhibition.** In this process, an inhibitory neuronal ending makes synaptic contact with the ending of a presynaptic neuron at an excitatory synapse (Figure 11.14). When the inhibitory neuronal ending is activated, less neurotransmitter is released from the ending of the presynaptic neuron when a nerve impulse arrives than would otherwise be the case. As a consequence, the stimulation of the postsynaptic neuron is reduced.

In a similar manner, the amount of chemical neurotransmitter released when a nerve impulse arrives at an excitatory synapse can be increased by the process of **presynaptic facilitation.** In this process, a stimulatory neuronal ending makes synaptic contact with the ending of a presynaptic neuron at an excitatory synapse. When the stimulatory neuronal ending is activated, more neurotransmitter is released from the ending of the presynaptic neuron when a nerve impulse arrives than would otherwise be the case. As a consequence, the stimulation of the postsynaptic neuron is increased.

The occurrence of presynaptic inhibition and presynaptic facilitation provides the nervous system with a means of regulating the influence that nerve impulses in

◆ **FIGURE 11.14 Presynaptic inhibition**

An inhibitory neuronal ending makes synaptic contact with the ending of the presynaptic neuron at an excitatory synapse. When the inhibitory neuronal ending is activated, less neurotransmitter is released from the ending of the presynaptic neuron when a nerve impulse arrives than would otherwise be the case. As a consequence, the stimulation of the postsynaptic neuron is reduced.

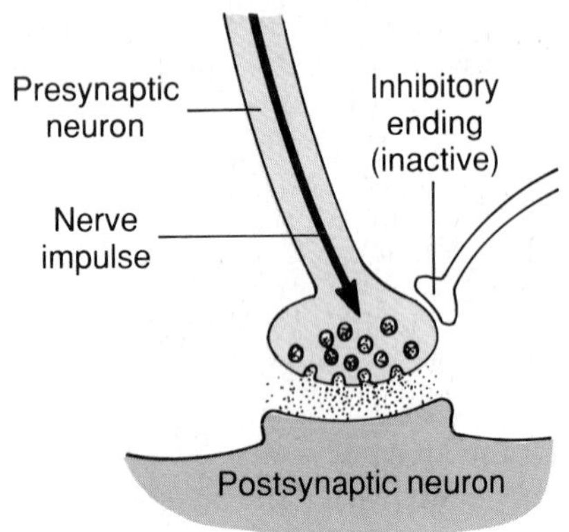

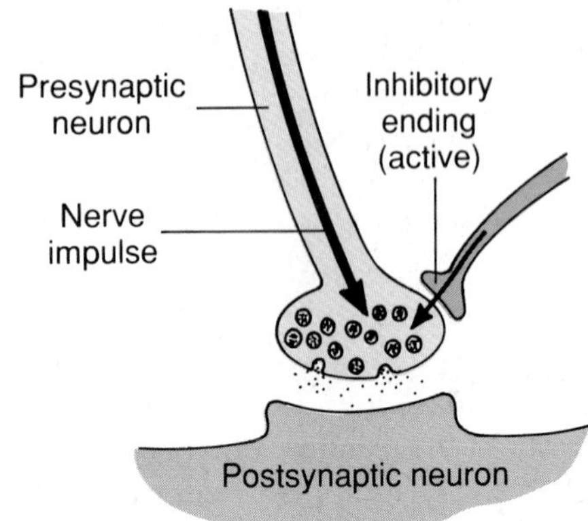

a particular presynaptic neuron will have on a postsynaptic neuron.

Receptors

Receptors are structures that convert information about conditions in the body's internal or external environments into neural signals. Some receptors, such as those in the nose for smell, are specialized endings of neurons. Others, such as those of the ears for sound, are separate cells that are synaptically connected to neurons. Regardless of their nature, however, receptors act as *transducers*—that is, they convert various forms of energy (light, heat, pressure, sound, and so on) into the ionic-electrical phenomena that ultimately result in the generation of neural signals.

The environmental factors that activate receptors are called *stimuli.* However, not all receptors respond to all stimuli. The receptors of the eyes, for example, are normally activated by light but not by sound, and the receptors of the ears are usually activated by sound, but not by light. The particular stimuli that activate a given receptor are referred to as *adequate stimuli* for that receptor.

Generator Potentials

When an adequate stimulus is applied to a receptor that is a specialized neuronal ending, the receptor responds with a general increase in permeability to all small ions that produces a depolarization called a **generator potential** (Figure 11.15). A generator potential is a graded potential. The stronger the stimulus applied to a receptor ending, the greater will be the magnitude of the generator potential. In addition, the rate of application or removal of a stimulus can affect the magnitude of a generator potential. Thus, a stimulus that rapidly builds or diminishes in intensity usually produces a greater generator potential than a stimulus that only slowly builds or diminishes in intensity. A generator potential generally lasts longer than the 1 or 2 milliseconds of an action potential, and it does not exhibit a refractory period. Consequently, if a second stimulus is applied before a generator potential resulting from an initial stimulus disappears, a second generator potential can add to the first, producing an even greater depolarization. Thus, a *summation* of generator potentials can occur. A generator potential of sufficient magnitude depolarizes the axon of the neuron to threshold and triggers a nerve impulse in the axon.

Receptor Potentials

In most cases, when an adequate stimulus is applied to a receptor that is a separate cell synaptically connected to a neuron, the receptor responds with a depolarization that is called a **receptor potential.** A receptor potential is a graded potential that exhibits characteristics similar to those of a generator potential. The receptor potential causes the release of a chemical neurotransmitter from the receptor cell, and the neurotransmitter alters the permeability and polarity of the associated neuron. If the axon of the associated neuron is depolarized to threshold, a nerve impulse is triggered.

Discrimination of Differing Stimulus Intensities

A nerve impulse is conducted in an all-or-none fashion regardless of the intensity of the stimulus that triggers it. Consequently, the magnitude of a nerve impulse does not vary with the intensity of the stimulus. How then does the nervous system transmit information about stimulus intensity? For example, how can a light touch on the hand be distinguished from a much stronger touch? One answer is that, in many cases, more neurons are activated by strong stimuli than by weak stimuli. Another, less obvious answer is that different frequencies of nerve impulses are generated in a single neuron in response to varying intensities of stimuli.

The frequency of nerve impulses in a neuron associated with a receptor is related to the magnitude of the generator or receptor potential produced by a stimulus. A strong stimulus produces a large generator or receptor potential, which, in turn, triggers a high frequency of nerve impulses in the neuron. A weak stimulus produces a smaller generator or receptor potential, which, in turn, triggers a lower frequency of nerve impulses in the neuron. In this way, nerve impulse frequency provides the nervous system with a means of distinguishing stimulus intensity.

Adaptation

When a stimulus is continuously applied, the frequency of nerve impulses in a neuron may diminish with time, even though the intensity of the applied stimulus remains the same. This phenomenon is called **adaptation.** In some cases, adaptation is due to the fact that the responsiveness of the receptor membrane diminishes with time, causing the magnitude of the generator or receptor potential evoked by the stimulus to diminish also. Consequently, a lower frequency of nerve impulses is triggered in the neuron. In some cases, adaptation may explain the observation that certain sensations—taste, for instance—diminish in intensity when stimuli are applied for long periods.

Some sensations—pain, for one—do not diminish in intensity when stimuli are applied for long periods. A pain sensation that does not diminish in intensity with time can be beneficial to a person because it may warn of a potentially harmful situation. Unfortunately, the person cannot always control the situation—or the pain.

◆ **FIGURE 11.15 Adequate stimuli applied to neuronal endings that act as receptors produce generator potentials**

Generator potentials are graded potentials—that is, the stronger the stimulus, the greater the magnitude of the generator potential.

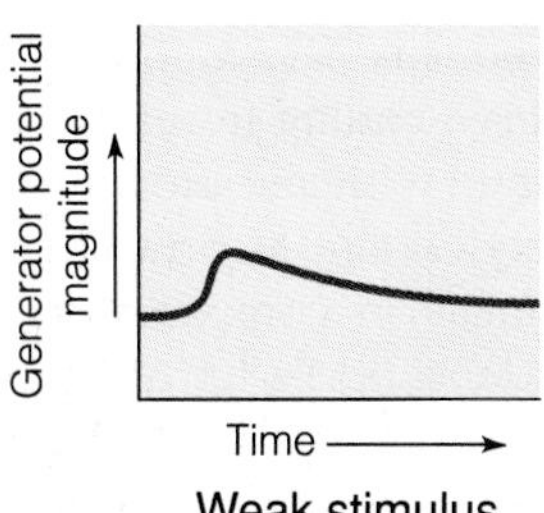

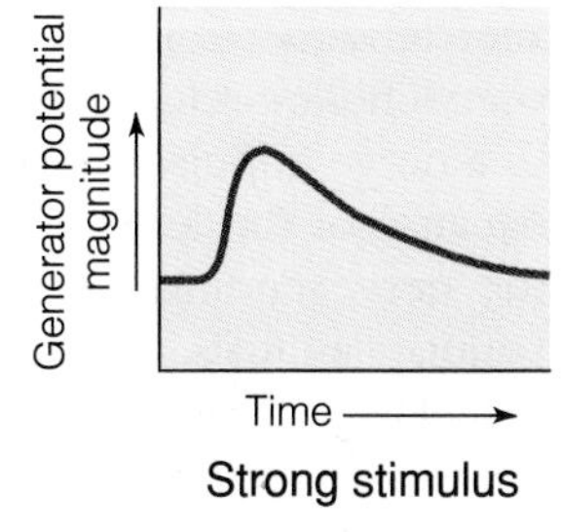

Effectors

Ultimately, neural signals are transmitted to structures called **effectors** to bring about responses to the various stimuli received by the nervous system. Effectors capable of responding to neural signals include muscle cells and the secretory cells of glands and organs. A neuron-effector junction is similar to a chemical synapse between two neurons, and information is transmitted from neurons to effectors by chemical neurotransmitters. When a nerve impulse arrives at the ending of a neuron at a neuron-effector junction, chemical neurotransmitter molecules are released. The neurotransmitter molecules diffuse to the effector cell and alter its activity. For example, at a *neuromuscular junction* between a neuron and skeletal muscle cell, an ending of a neuron closely approaches the membrane of the muscle cell (Figure 11.16). When a nerve impulse arrives at the junction, the neurotransmitter *acetylcholine* is released from the neuronal ending. The acetylcholine diffuses to the plasma membrane of the muscle cell, where it binds to the receptor portion of chemically gated ion channels like those described in the earlier section dealing with acetylcholine (page 369). This binding leads to an increase in the permeability of the membrane to both sodium and potassium ions, and the muscle cell depolarizes in the region of the junction. This depolarization, which is called an *end-plate potential,* is of sufficient magnitude to initiate a propagated action potential that travels along the muscle cell membrane. The propagated action potential triggers the intracellular events that lead to the contraction of the muscle cell.

The effective combination of acetylcholine with the receptors of the muscle cell membrane at the neuromuscular junction lasts only a few milliseconds. Following its release from a neuronal ending at a neuromuscular junction, acetylcholine is rapidly inactivated by the

CLINICAL CORRELATION

Myasthenia Gravis

Case Report

THE PATIENT: A 34-year-old woman.

PRINCIPAL COMPLAINT: Muscle weakness.

HISTORY: About two months before the current admission, the patient began to notice episodes of double vision, drooping of the eyelids, difficulty chewing and swallowing, and generalized weakness. The severity of the symptoms increased with activity, decreased with rest, and varied with time. About one month before the current admission, she experienced a particularly severe episode of muscular weakness, during which she was unable to walk. The patient was seen by a physician, who suspected that she had myasthenia gravis. She was given 0.5 mg of neostigmine methylsulfate intramuscularly, which inhibits acetylcholinesterase, thereby prolonging the action of the acetylcholine released at synapses and effector sites. Her condition temporarily improved. Another severe attack resulted in her present admission.

CLINICAL EXAMINATION: The patient was breathing with difficulty, and both eyelids were drooping. She had some loss of ability to move her eyes and turn her head, inability to close her mouth completely, and weakness of her arms and legs. Repeated attempts to open her eyes widely or to clench her teeth produced greater fatigue. Because of the fatigue of the respiratory muscles, an endotracheal tube was inserted and mechanical respiration was provided.

◆ **FIGURE 11.16 A neuromuscular junction**
When a nerve impulse arrives at the ending of the neuron, acetylcholine is released. The acetylcholine binds to the receptor portion of chemically gated ion channels in the muscle cell membrane. This binding leads to a change in the membrane's permeability to sodium and potassium ions and produces a propagated action potential that travels along the muscle cell membrane.

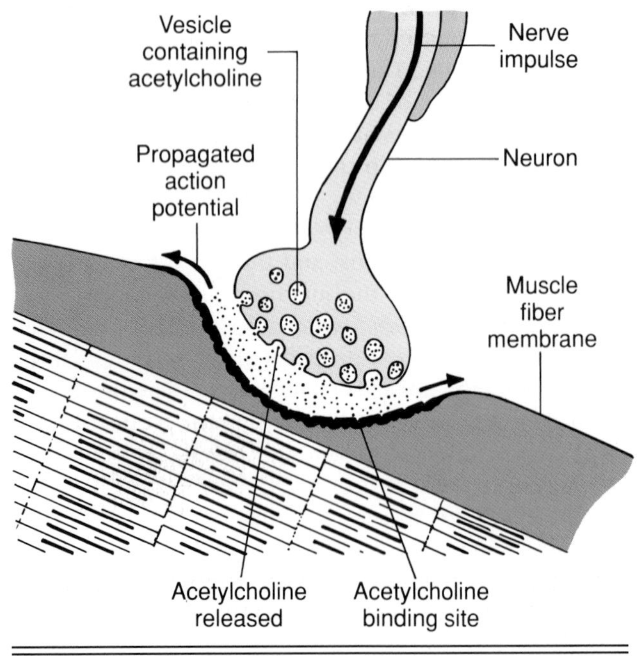

enzyme *acetylcholinesterase,* which is secreted by the muscle cell and is present in the gap between the neuronal ending and the muscle cell membrane. The inactivation of acetylcholine terminates its action on the muscle cell.

Effects of Aging on Neuromuscular Junctions

As a person ages, his or her physical strength generally declines. Recent evidence from animal experiments indicates that the reduced physical strength associated with aging may be due in part to changes that occur at neuromuscular junctions.

Accompanying the aging process is a reduction in the amount of the neurotransmitter acetylcholine present in nerve endings at neuromuscular junctions. This reduction is thought to be due to an increased leakage of acetylcholine from the nerve endings. The amount of acetylcholine released from a nerve ending in response to a nerve impulse does not change with age. However, because of the leakage, less acetylcholine is present in the nerve endings of older animals in comparison to younger animals. As a result, older animals have a tendency to run out of acetylcholine faster. Moreover, there are two known types of acetylcholine receptors on skeletal muscle cells, and many of the receptors on the muscle cells of older animals are of the less effective of the

CLINICAL CORRELATION

TREATMENT: Neostigmine methylsufate was administered intraveneously in increments of 0.125 mg/5 min until a satisfactory response was obtained (indicated by strength of handgrip). The endotracheal tube was removed, and neostigmine bromide was given orally in increments of 7.5 mg until a satisfactory response was obtained. Pyridostigmine bromide, which also inhibits acetylcholinesterase and prolongs the action of the acetylcholine released at synapses and effector sites, was used at night in sustained-release tablets (6-8 hours duration).

COMMENT: The muscular weakness of myasthenia gravis is caused by a decreased responsiveness of skeletal muscle cells to the acetylcholine released from motor neuron endings at neuromuscular junctions. The condition appears to be due to an abnormal response of the body's immune system that disrupts acetylcholine receptors on skeletal muscle cell membranes at the junctions. (Antireceptor antibodies have been identified in patients who have myasthenia gravis.) Exacerbations and remissions of the myasthenic condition occur frequently, sometimes in response to such variables as upper respiratory tract infection, loss of sleep, intake of alcohol, or menstruation.

Drugs that inhibit acetylcholinesterase benefit the patient by decreasing the rate of destruction of acetylcholine. Presumably, the inhibition of acetylcholine destruction results in levels of acetylcholine that are high enough to stimulate the skeletal muscle cells. The fluctuations in the myasthenic condition require that patients learn to modify the dosage schedule as their needs change.

OUTCOME: The patient was stabilized on neostigmine bromide by day and sustained-release pyridostigmine bromide at night. Her condition has not increased in severity, but the prognosis (probable outcome) with myasthenia gravis is uncertain.

two types. The reduction of acetylcholine and the proportional increase of less effective receptors are thought to contribute to the decline in physical strength typical of aging.

Study Outline

◆ **RESTING MEMBRANE POTENTIAL** pp. 358-359
Result of differences in ionic composition of intracellular and extracellular fluids due to characteristics and function of plasma membrane of neuron.

1. Plasma membrane contains active-transport mechanisms that move ions into and out of cells.
2. Plasma membrane is not equally permeable to all ions.
3. Differences in ionic compositions of intracellular and extracellular fluids result in accumulation of slight excess of positively charged ions immediately outside membrane and slight excess of negatively charged ions immediately inside membrane that gives rise to resting membrane potential.

Development of the Resting Membrane Potential. Of particular importance are the facts that

1. Intracellular fluid of unstimulated neuron contains higher concentration of potassium ions and lower concentration of sodium ions than extracellular fluid.
2. Unstimulated neuron is 50 to 100 times more permeable to potassium ions than to sodium ions.

◆ **MOVEMENT OF IONS ACROSS THE UNSTIMULATED NEURONAL MEMBRANE** pp. 359-360

Movements of potassium, sodium, and chloride ions across plasma membrane of an unstimulated neuron that is in steady state and polarized at resting membrane potential are influenced by both concentration and electrical forces as well as by active-transport mechanisms.

Potassium Ions. Concentration force favoring net outward movement of potassium ions balanced by electrical force favoring net inward movement of positively charged ions and by inward pumping of potassium ions.

Sodium Ions. Electrical and concentration forces favoring net inward movement of sodium ions balanced by outward pumping of sodium ions; membrane not very permeable to sodium ions.

Chloride Ions. Concentration force favoring net inward movement of chloride ions balanced by electrical force favoring net outward movement of negatively charged ions and in at least some neurons by outward pumping of chloride ions.

◆ **ROLE OF ACTIVE-TRANSPORT MECHANISMS IN MAINTAINING THE RESTING MEMBRANE POTENTIAL** p. 360

Active-transport mechanisms maintain differences in ionic compositions of extracellular and intracellular fluids that give rise to resting membrane potential.

◆ **GATED ION CHANNELS** pp. 360-361

Channels that open or close in response to various signals. Include voltage-gated channels and chemically gated channels (that is, channel-linked receptors).

◆ **GRADED POTENTIALS** p. 361

A change in membrane potential that varies in magnitude with the magnitude of the event that triggers the change. A local response that is decremental.

◆ **ACTION POTENTIAL** pp. 361-364

When axon reaches threshold, membrane potential quickly decreases to zero, then reverses so inside of axon is positive relative to outside.

Depolarization and Polarity Reversal. Due to entry of sodium ions into axon through voltage-gated channels.

Return to the Original Membrane Potential. Due to decreased entry of sodium ions into axon and movement of potassium ions out of axon.

Maintenance of Sodium and Potassium Ion Concentrations. Sodium-potassium pump restores normal distribution of ions.

◆ **THE NERVE IMPULSE** pp. 364-366

In unmyelinated axon, axonal membrane adjacent to area of initial action potential is depolarized and generates action potential. Action potentials continue to be generated along axon, producing a propagated action potential, or nerve impulse. In myelinated axon, saltatory conduction occurs.

Refractory Periods. Absolute refractory period is period following action potential when no additional stimulus can evoke another action potential. Relative refractory period is period following absolute refractory period when stimulus greater than that normally required can initiate action potential.

All-or-None Response. Under similar physiological conditions, nerve impulse has same magnitude regardless of stimulus strength.

Direction of Nerve Impulse Conduction. In the body, nerve impulse normally originates at end of axon and is conducted in one direction toward other end of the axon.

Conduction Velocities.

1. In general, the larger the diameter of an axon, the faster it conducts a nerve impulse.
2. Nerve impulse travels faster in myelinated axon than it would if the axon were unmyelinated.

◆ **SYNAPSES** pp. 366-371

Junctions between neurons at which information is transferred from one neuron to another.

Electrical Synapses. Presynaptic and postsynaptic neurons connected by gap junction. Electrical currents in presynaptic neuron can pass directly to postsynaptic neuron.

Chemical Synapses. Axon terminal of presynaptic neuron is separated from postsynaptic neuron by synaptic cleft. Axon terminal releases neurotransmitter molecules that diffuse across synaptic cleft and bind to receptors on membrane of postsynaptic neuron.

CHEMICAL SYNAPSES INVOLVING CHEMICALLY GATED ION CHANNELS. Signaling is very accurate, occurs rapidly, and is terminated quickly.

RELEASE OF NEUROTRANSMITTERS. Released by exocytosis from synaptic vesicles when nerve impulse arrives at axon terminal.

EFFECTS OF NEUROTRANSMITTERS ON CHEMICALLY GATED ION CHANNELS. Can be excitatory or inhibitory.

Excitatory Effects. The binding of excitatory neurotransmitter molecules to chemically gated ion channels in membrane of postsynaptic neuron generally triggers changes in permeability of the membrane that lead to depolarization of neuron called an *excitatory postsynaptic potential.*

Inhibitory Effects. The binding of inhibitory neurotransmitter molecules to chemically gated ion channels in membrane of postsynaptic neuron triggers changes in permeability of the membrane that usually cause an increase in resting polarity of neuron called an ***inhibitory postsynaptic potential.*** In some cases, the permeability change leads to a stabilization of the membrane potential without a hyperpolarization.

SYNAPTIC DELAY. Time required to cross chemical synapse.

REMOVAL OF NEUROTRANSMITTERS. Depending on particular neurotransmitter substance and synapse involved, neurotransmitter molecules are ultimately removed by enzymatic inactivation, by diffusion away, or by being taken up by neuron that released them, other neurons, or neuroglial cells.

CHEMICAL SYNAPSES INVOLVING NON-CHANNEL-LINKED RECEPTORS. Signaling is generally slower and longer lasting than signaling that involves chemically gated ion channels and is often spatially diffuse. Binding of neurotransmitter molecules can influence production and level of intracellular messengers, which directly or indirectly regulate behavior of ion channels or other aspects of cell function. In some cases, binding leads to changes in membrane permeability that produce excitatory or inhibitory effects similar to the effects produced by the binding of neurotransmitter molecules to chemically gated ion channels.

KINDS OF NEUROTRANSMITTERS. More than 50 different substances are known or suspected to serve as neurotransmitters.

NEUROTRANSMITTERS THAT BIND TO CHEMICALLY GATED ION CHANNELS. Only a relatively small number of neurotransmitters.

Acetylcholine.

Glutamate and Aspartate.

Gamma-Aminobutyric Acid and Glycine.

NEUROTRANSMITTERS THAT BIND TO NON-CHANNEL-LINKED RECEPTORS. Can have varying effects on different cells. Some neurotransmitters, such as acetylcholine, bind to both chemically gated ion channels and to non-channel-linked receptors.

Norepinephrine.

Epinephrine.

Dopamine.

Serotonin.

Neuropeptides.

NEUROMODULATORS. Chemical substances other than neurotransmitters that alter neuronal activity by altering neurons directly or by influencing the effectiveness of neurotransmitters.

◆ **NEURAL INTEGRATION** pp. 371-374
Nervous system integrates information from many sources to produce useful, coordinated responses by the body to wide variety of conditions.

Divergence and Convergence.

1. *Divergence.* Presynaptic neuron processes branch many times to synapse with many postsynaptic neurons.
2. *Convergence.* Processes of many presynaptic neurons synapse with single postsynaptic cell.

Summation. Postsynaptic potentials produced by release of neurotransmitter molecules at chemical synapses can add together, or summate, to influence activity of postsynaptic neuron.

TEMPORAL SUMMATION. Occurs when many nerve impulses arrive at single synapse within short period of time.

SPATIAL SUMMATION. Occurs when nerve impulses arrive close together in time at a number of synapses between different presynaptic axon terminals and same postsynaptic neuron.

Facilitation. Initial depolarization of axon of postsynaptic neuron by excitatory postsynaptic potentials produced at some excitatory synapses; facilitates attainment of threshold and generation of nerve impulse when additional excitatory postsynaptic potentials occur.

Determination of Postsynaptic Neuron Activity. Determined by integrated activities of many presynaptic inputs from various sources.

Presynaptic Inhibition and Presynaptic Facilitation. Can alter transmission of information from one neuron to another at synapses.

◆ **RECEPTORS** pp. 374-375
Act as transducers that convert various forms of environmental energy into ionic-electrical phenomena that result in generation of neural signals.

Generator Potentials. When an adequate stimulus is applied to a receptor that is a specialized neuronal ending, the receptor responds with a general increase in permeability to all small ions that produces a depolarization called a generator potential. It is a graded potential with longer duration than action potential and no refractory period.

Receptor Potentials. In most cases, when an adequate stimulus is applied to a receptor that is a separate cell synaptically connected to a neuron, the receptor responds with a depolarization that is called a receptor potential. It is similar to a generator potential, but receptor cell releases chemical neurotransmitter.

Discrimination of Differing Stimulus Intensities. Strong stimuli may activate more neurons than weak stimuli; different frequencies of nerve impulses in single neuron in response to varying intensities of stimuli.

Adaptation. A phenomenon that occurs when stimulus is applied continuously but frequency of nerve impulses diminishes with time even though intensity of applied stimulus is constant.

◆ **EFFECTORS** pp. 375-377
Muscle cells and secretory cells of glands and organs.

Self-Quiz

1. In an unstimulated neuron: (a) the intracellular fluid immediately inside the plasma membrane is negative relative to the extracellular fluid immediately outside; (b) the intracellular fluid immediately inside the plasma membrane is positive relative to the extracellular fluid immediately outside; (c) the sodium-potassium pump does not operate.

2. The resting, unstimulated neuron has: (a) a greater concentration of sodium ions in the intracellular fluid than are in the extracellular fluid; (b) a greater concentration of potassium ions in the intracellular fluid than are in the extracellular fluid; (c) the same concentration of sodium and potassium ions in the intracellular fluid as in the extracellular fluid.

3. In an unstimulated neuron, a concentration force favors the net movement of: (a) potassium ions into the neuron; (b) sodium ions out of the neuron; (c) chloride ions into the neuron.
4. In an unstimulated neuron, an electrical force favors the movement of: (a) potassium ions out of the neuron; (b) sodium ions into the neuron; (c) chloride ions into the neuron.
5. The plasma membrane of an unstimulated neuron is: (a) essentially impermeable to chloride ions; (b) quite permeable to potassium ions; (c) extremely permeable to sodium ions.
6. When the axon of a neuron reaches threshold and an action potential is generated: (a) potassium ions rapidly enter the axon; (b) the membrane permeability to both sodium and potassium ions decreases substantially; (c) sodium ions rapidly enter the axon.
7. During an action potential, the voltage-gated sodium channels remain open for only a brief interval and then rapidly close and enter the inactivated state. True or False?
8. Which of the following is an all-or-none response? (a) action potential; (b) excitatory postsynaptic potential; (c) generator potential.
9. Saltatory, or "jumping," conduction of nerve impulses occurs in: (a) large-diameter myelinated axons; (b) large-diameter unmyelinated axons; (c) small-diameter unmyelinated axons.
10. The axons of different neurons all transmit nerve impulses at the same velocity. True or False?
11. At a chemical synapse, the plasma membranes of the presynaptic and postsynaptic neurons are: (a) fused with one another; (b) separated by a synaptic cleft; (c) connected by a gap junction.
12. The binding of excitatory neurotransmitter molecules to chemically gated ion channels in the membrane of a postsynaptic neuron would most likely lead to: (a) a depolarization of the postsynaptic neuron; (b) an increase in the resting polarity of the postsynaptic neuron; (c) a stabilization of the polarity of the postsynaptic neuron at the normal resting membrane potential.
13. Synaptic signaling that involves chemically gated ion channels generally occurs slowly, and the signal persists for a long period before it is terminated. True or False?
14. Which substance is an enzyme that breaks down acetylcholine? (a) choline acetyltransferase; (b) acetylcholinesterase; (c) acetylcoenzyme A reductase.
15. Which neurotransmitter is an amino acid that binds to chemically gated ion channels? (a) norepinephrine; (b) serotonin; (c) glycine.
16. The postsynaptic potentials produced by the release of neurotransmitter molecules at chemical synapses can add together, or summate, to influence the activity of a postsynaptic neuron. True or False?
17. Generator potentials: (a) can summate; (b) have a long absolute refractory period; (c) generally are of shorter duration than action potentials.
18. Differing stimulus intensities may lead to different: (a) magnitudes of nerve impulses in a given neuron; (b) velocities of conduction of nerve impulses in a given neuron; (c) frequencies of nerve impulses in a given neuron.
19. The sensation of pain adapts very rapidly. True or False?
20. Acetylcholine is a chemical neurotransmitter substance released by neurons at their junctions with skeletal muscle cells. True or False?

CHAPTER 12

The Central Nervous System

CHAPTER CONTENTS

LEARNING OBJECTIVES

After completing this chapter, you should be able to:

1. Name the types of tracts in the brain's white matter, and state the function of each.
2. Name the major structures located in the brain, and describe their functions.
3. Locate and describe the major functional areas of the cerebral cortex.
4. Describe the ventricles of the brain and the function and flow of cerebrospinal fluid.
5. Distinguish between the white and gray matter of the spinal cord in terms of structure and function.
6. Name and describe the ascending and descending tracts of the spinal cord.
7. Name and describe the meninges.
8. Describe the formation of a typical spinal nerve.
9. Distinguish between a stretch reflex and a tendon reflex.
10. Correlate the symptoms of several dysfunctions of the central nervous system with the region affected.

The **central nervous system** (CNS) consists of the brain and spinal cord, both of which develop from the embryonic neural tube (see Chapter 10). The spinal cord is essentially an extension of the nerve tracts of the brain. Neurons in the spinal cord carry messages to and from the brain, and the spinal cord serves as a center for reflexes. For convenience, the brain and spinal cord are discussed as if they were separate entities, but at the same time their interrelationship is emphasized.

The Brain

The brain is formed by extensive development of the anterior end of the embryonic neural tube. Chapter 10 described how three enlargements of the neural tube, called the **forebrain (prosencephalon), midbrain (mesencephalon),** and **hindbrain (rhombencephalon)** are formed in the region of the future brain. We also saw how these three enlargements then subdivide into five chambers: the forebrain divides into the **telencephalon** and **diencephalon;** the midbrain remains as a single chamber; and the hindbrain divides into the **metencephalon** and **myelencephalon** (Figure 10.7).

This chapter discusses the adult structures that develop from these five divisions. Table 12.1 lists some of these structures and names the portion of the fluid-filled neural canal that is within each subdivision of the brain. Table 12.2 provides a summary of the functions of the major components of the brain. The blood supply to the brain is discussed in Chapter 20.

Forebrain

The forebrain is the largest, most complex, and most visible region of the brain. The structures that are included in the adult forebrain develop from the telencephalon and the diencephalon.

Telencephalon

In the adult, the **telencephalon** consists of right and left **cerebral hemispheres,** which together are referred to as the **cerebrum.** Because the cerebrum grows so extensively, it completely envelops the diencephalon and obscures much of the rhombencephalon. The cerebrum has an outer surface of **gray matter,** which is composed primarily of nerve cell bodies and unmyelinated fibers. This surface layer is called the **cerebral cortex** (Figure 12.1). Deep inside each cerebral hemisphere are several additional structures of gray

◆ **TABLE 12.1 Subdivisions of the Neural Tube and the Major Adult Structures Derived from Each**

PRIMARY DIVISION	SUBDIVISION	ADULT BRAIN STRUCTURES	NEURAL CANAL REGION
Prosencephalon (forebrain)	Telencephalon	Cerebral hemispheres (cerebrum)	Lateral ventricles and upper portion of the third ventricle
		Cerebral cortex	
		Basal nuclei	
		Olfactory bulbs and tracts	
	Diencephalon	Epithalamus	Most of the third ventricle
		Thalamus	
		Hypothalamus	
Mesencephalon (midbrain)	Mesencephalon	Corpora quadrigemina*	Cerebral aqueduct
		Cerebral peduncles*	
Rhombencephalon (hindbrain)	Metencephalon	Cerebellum	Fourth ventricle
		Pons*	
	Myelencephalon	Medulla oblongata*	Part of the fourth ventricle
Spinal cord	Spinal cord	Spinal cord	Central canal

*These structures comprise the brain stem.

◆ **FIGURE 12.1 Frontal section of brain**

(a) Frontal section of the cerebrum and the diencephalon (red) showing the cerebral cortex (gray matter) surrounding the white matter and, deep within the cerebrum, the basal nuclei. (b) Photograph of a frontal section of the brain.

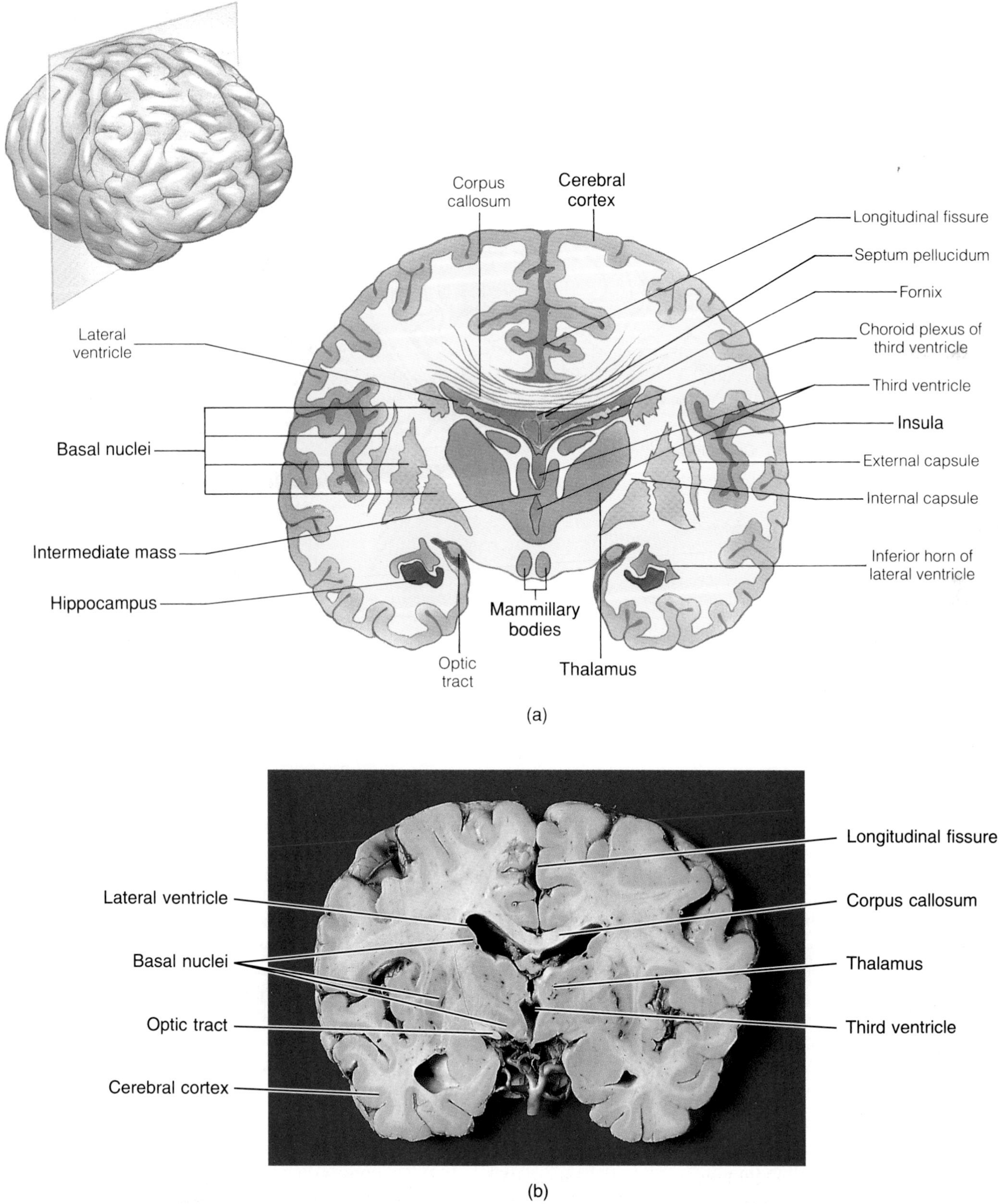

Corpus callosum
Cerebral cortex
Longitudinal fissure
Septum pellucidum
Fornix
Choroid plexus of third ventricle
Third ventricle
Insula
External capsule
Internal capsule
Inferior horn of lateral ventricle
Lateral ventricle
Basal nuclei
Intermediate mass
Hippocampus
Mammillary bodies
Optic tract
Thalamus

(a)

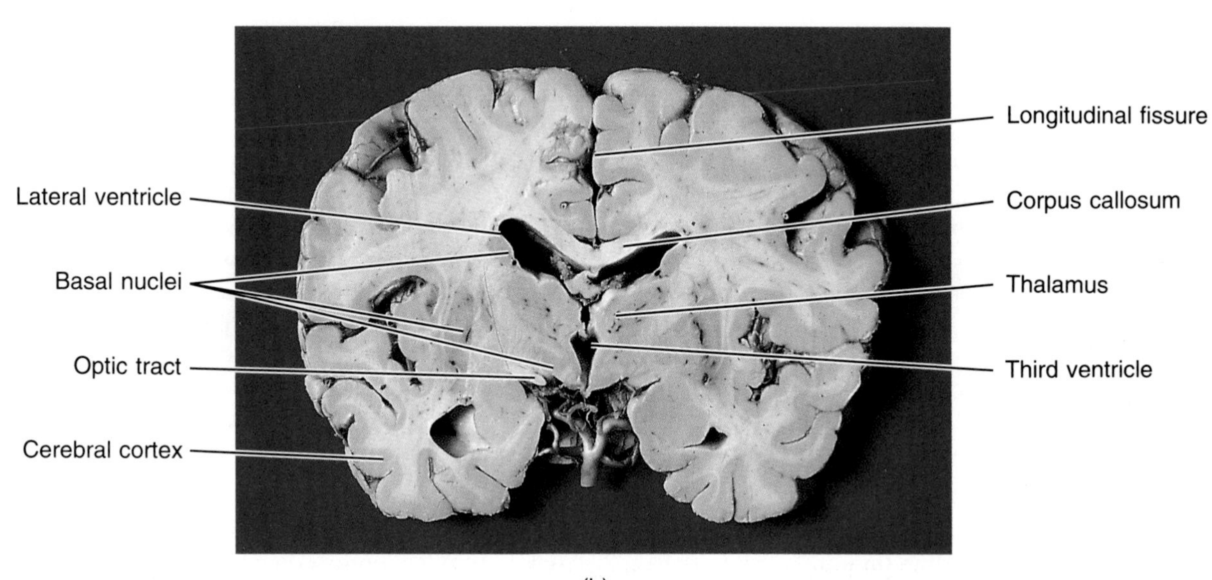

(b)

◆ **TABLE 12.2 Overview of Structures and Functions of the Major Components of the Brain**

BRAIN COMPONENT
Cerebral cortex
Basal nuclei
Thalamus
Hypothalamus
Limbic system (not entirely delineated)
Cerebellum
Brain stem (midbrain, pons and medulla)

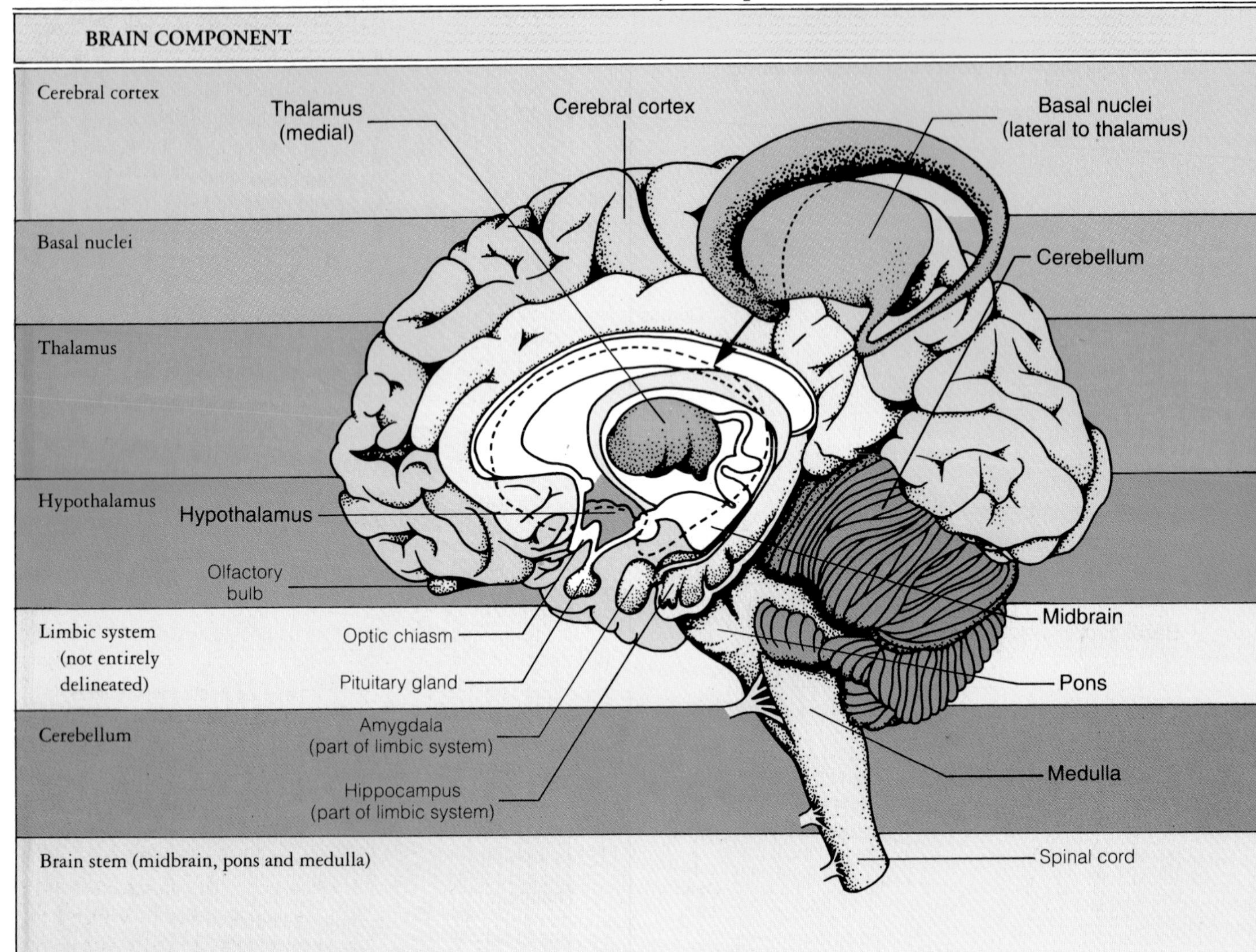

matter called the **basal (cerebral) nuclei** (or *basal ganglia*). The gray matter of the cortex is separated from the basal nuclei by **white matter,** which is composed of tracts of myelinated nerve fibers. A fluid-filled cavity called a **lateral ventricle** is located within each cerebral hemisphere.

Myelinated Nerve Fiber Tracts. In the central nervous system, bundles of nerve fibers are called **tracts.** There are three types of tracts in the white matter of the cerebrum:

1. Projection tracts are pathways formed by neurons that transmit either descending (motor) nerve impulses from the cerebral cortex to other regions of the brain and spinal cord, or ascending (sensory) impulses from the spinal cord and lower regions of the brain (such as the thalamus) to the cerebral cortex.

2. Association tracts are pathways formed by neurons that connect various areas of the cerebral cortex within the same hemisphere. Association neurons vary in length: some are quite short, whereas others extend the entire length of the hemisphere.

3. Commissural tracts are pathways formed by neurons that connect the left and right sides of the brain. Two main commissural tracts connect the cerebral hemispheres: the **anterior commissure** and the large **corpus callosum.**

Rather surprising results are obtained when the corpus callosum is severed, as is sometimes done to provide relief from epileptic convulsions. After recovery, the person appears to function normally. With special testing procedures, however, it is possible to show that fol-

MAJOR FUNCTIONS
1. Voluntary control of skeletal muscles 2. Sensory perception 3. Language ability 4. Higher intellectual activities (e.g., foresight, decision making, thinking, memory)
1. Inhibit muscle contractions 2. Coordinate slow sustained contractions 3. Suppress useless movements
1. Relay center for almost all sensory input 2. Routes sensory signals to specific regions of brain 3. Crude awareness of sensation 4. Some degree of consciousness 5. Some involvement in motor control
1. Link between autonomic nervous aystem and endocrine system via pituitary gland 2. Regulates autonomic activities, including body temperature, water balance, appetite, sexual activity, and emotions
1. Produces and modifies emotional behavior
1. Regulate skeletal muscle tone 2. Maintenance of balance 3. Helps coordinate skilled, voluntary movement 4. Involved in planning and initiating voluntary activity
1. Involved in balance and coordinated movements 2. Site of origin of cranial nerves III - XII 3. Arousal and activation of cerebral cortex 4. Site of several autonomic centers (e.g. respiration; coughing; swallowing; cardiovascular control

lowing the cutting of the commissure, a task learned with one hand cannot be performed by the other hand unless the task is relearned with this hand. A person whose corpus callosum is intact can generally perform a task with either hand, although perhaps not with equal dexterity. Therefore, it appears that the corpus callosum makes possible the transfer of information between cerebral hemispheres; and if the commissure is severed, information learned by one cerebral hemisphere is unavailable to the other.

Gyri, Fissures, Sulci, and Lobes of the Cerebrum. The surface of the cerebrum has many rounded ridges called **convolutions**, or **gyri** (singular: *gyrus; jigh´-rus*) (Figure 12.2). Separating the gyri are furrows. The deeper furrows are called **fissures;** the shallower ones are **sulci** (singular: *sulcus; sul´-kus*). The folding of the cortex that produces the gyri and sulci makes the surface area of the cerebral cortex much greater than it would be if the brain's surface were smooth. As it is, a significant percentage of the cerebral cortex is located in the fissures and sulci and is not visible on the surface.

The patterns of the gyri and fissures or sulci vary somewhat from one brain to another. Nevertheless, the locations of certain fissures and sulci are constant enough to serve as surface landmarks by which each hemisphere can be divided into *frontal, parietal, temporal,* and *occipital* lobes (Figure 12.3). Each lobe is located in the same general region as the correspondingly named skull bones.

The **longitudinal fissure** (Figure 12.1) is a deep furrow that extends down to the corpus callosum in the central region of the cerebrum. It is oriented anteriorly and posteriorly, dividing the cerebrum into right and left hemispheres. Each hemisphere is further divided into a **frontal lobe** and a **parietal lobe** by the **central sulcus** (*fissure of Rolando*) (Figure 12.3), which is oriented at right angles to the longitudinal fissure. Two gyri are parallel to the central sulcus: the one anterior to the sulcus is the **precentral gyrus,** and the one posterior to the sulcus is the **postcentral gyrus.** The functional significance of these gyri is explained in the next section.

The parietal lobe is separated posteriorly from the **occipital lobe** by an indistinct **parietooccipital sulcus.** The **temporal lobes** extend forward along the lateral sides of the cerebral hemispheres. Each temporal lobe is separated from the lower portions of the frontal and parietal lobes by a deep **lateral fissure** (*fissure of Sylvius*) (Figure 12.3). A portion of the cerebral cortex called the **insula,** which is considered to be a fifth lobe of the cerebrum, is located deep within the lateral fissure (Figure 12.1). The insula is covered by portions of the frontal, parietal, and temporal lobes. The cerebrum is completely separated posteriorly from the cerebellum by a deep **transverse fissure** (Figure 12.3).

Functional Areas of the Cerebral Cortex. On the basis of the effects of electrical stimulation of specific areas of the cerebral cortex in humans, from observing the clinical manifestations of brain disease or damage in humans, and from the results of detailed experiments on other mammals, it has been determined that certain areas of the cortex are related to specific functions. Some of these areas have been precisely mapped and numbered in a system called the *Brodmann classification,* but for our purposes it is sufficient to consider only the general locations of the major functional areas (Figure 12.4). Keep in mind, however, that Figure 12.4 represents an oversimplification of a very complex organ: no area of the brain functions alone. Because the various

◆ **FIGURE 12.2 Left cerebral hemisphere**

(a) Lateral view. (b) Photograph. The surface of each cerebral hemisphere has numerous convolutions separated by either sulci or fissures.

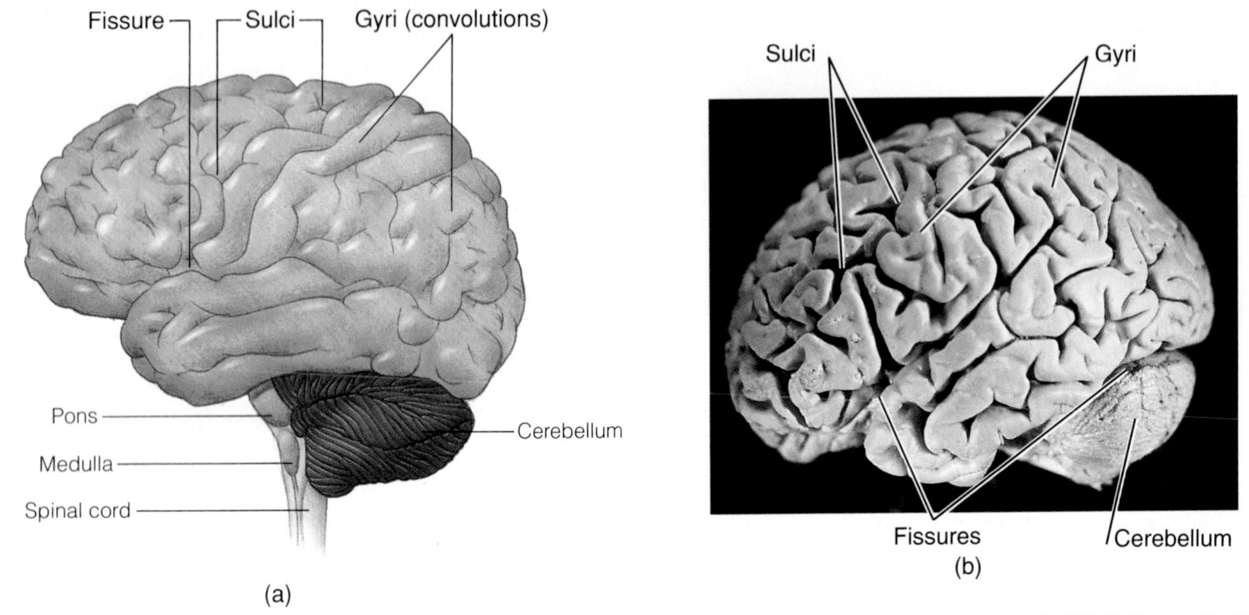

cortical areas are extensively interconnected by commissural and association neurons, any function attributed to a specific cortical area actually probably involves several cortical areas.

Primary Motor Area. The **primary motor area** of the cerebral cortex is located in the precentral gyrus of the frontal lobe, just anterior to the central sulcus. Since the neurons in this gyrus control the conscious and pre-

◆ **FIGURE 12.3 Division of a cerebral hemisphere into lobes by specific fissures and sulci**

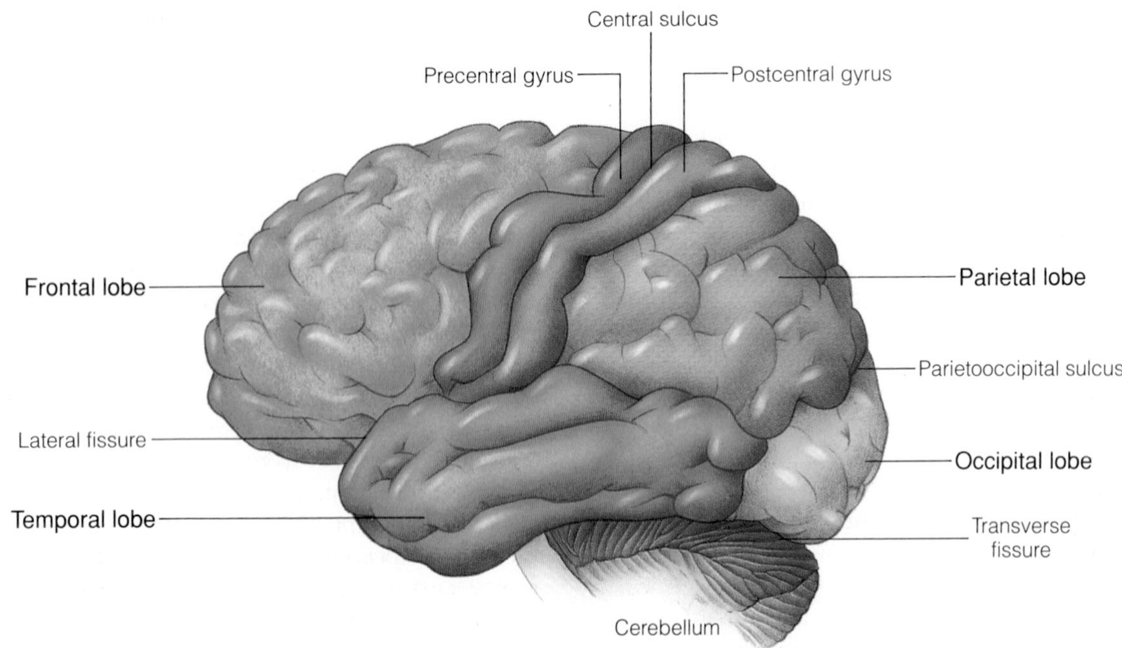

◆ **FIGURE 12.4 General functional areas of the left cerebral cortex (lateral aspect)**

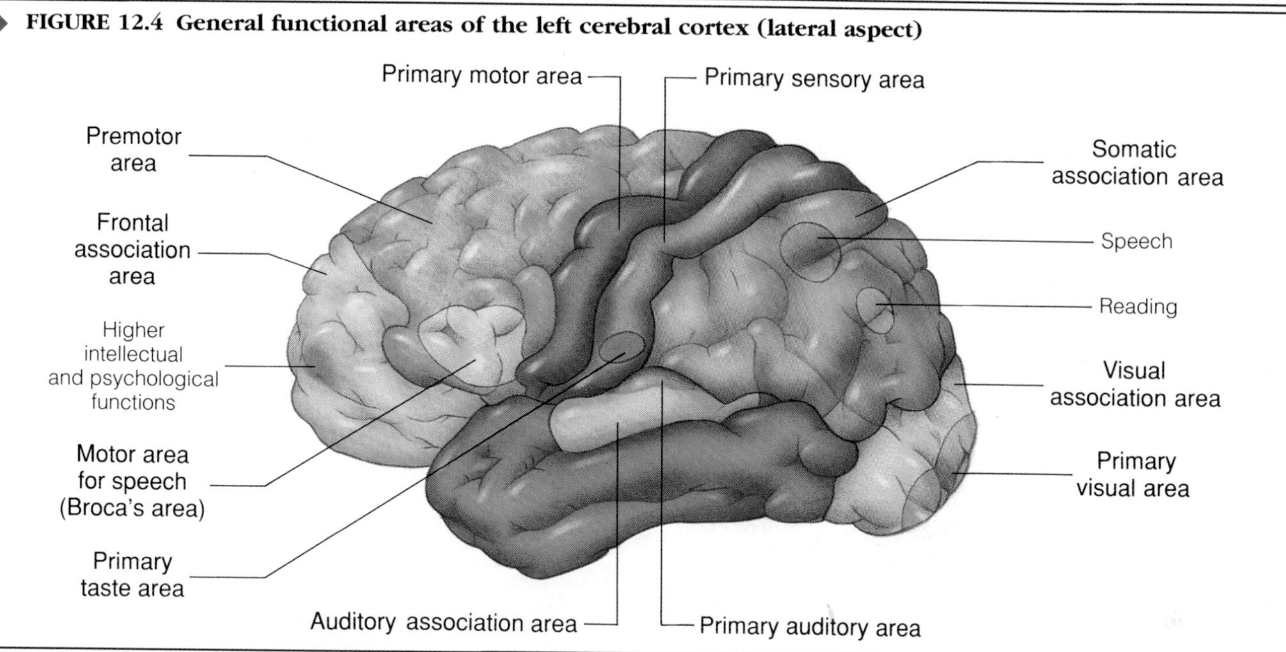

cise voluntary contractions of skeletal muscles, this area is also referred to as the **primary somatic motor area.** The neurons of the primary motor area are distributed within the precentral gyrus in an organized manner. Neurons involved in controlling toe movements are located medially, deep in the longitudinal fissure; neurons involved in controlling the movements of other body parts are located in a regular but disproportionate sequence laterally along the gyrus (Figure 12.5). Originating in the precentral gyrus are descending motor nerve tracts called **pyramidal tracts** (so named because they form pyramid-shaped structures that are visible on the ventral surface of the medulla oblongata).

Premotor Area. Located just anterior to the primary motor area is a cortical region referred to as the **premotor area.** The neurons in the premotor area cause groups of muscles to contract in a specific sequence, thereby producing stereotyped movements. These repetitive movements are involved in learned activities such as playing a musical instrument and typing. The motor neurons whose cell bodies are located in the premotor area travel within **extrapyramidal tracts.**

At the lower margin of the premotor area is a motor area associated with the ability to speak. This region, called **Broca's area,** seems to be located only in the left cerebral hemisphere in most individuals, whereas the other functional areas that are identifiable are found in both hemispheres.

Primary Sensory Area. Located just posterior to the central sulcus, in the postcentral gyrus of the parietal lobe, is the **primary sensory area (primary somatic sensory area)** of the cerebral cortex. Within this area are the terminations of pathways that transmit general sensory information concerning temperature, touch, pressure, pain, and proprioception from the body to the cerebral cortex. The pathways cross from one side of the nervous system to the other as they ascend to the cortex. Thus, the primary sensory area of the right cerebral hemisphere receives information from the left side of the body, and the primary sensory area of the left cerebral hemisphere receives information from the right side of the body. The neurons of the primary sensory area are organized sequentially within the postcentral gyrus in a manner similar to the motor neurons of the primary motor area (Figure 12.6).

Special Sense Areas. The **primary visual area** of the cerebral cortex is in the posterior portion of the occipital lobe. Located along the upper margin of the temporal lobe is the **primary auditory area,** which receives nerve impulses associated with hearing. The area concerned with the sense of smell, the **primary olfactory area,** is located on the medial surface of the temporal lobe. The **primary taste area** is located in the parietal lobe, near the bottom of the postcentral gyrus.

When dealing with sensory areas, it is important to realize that only nerve impulses, not sensations, are transmitted from receptors to the brain. In the brain, the nerve impulses are consciously interpreted as particular sensations (pain, touch, sound, taste, and so on). The brain then projects the sensations—usually with considerable accuracy—to the locations of the stimuli that are activating the receptors.

◆ **FIGURE 12.5 Map of the primary motor area of the cerebral cortex**
(a) Top view of cerebral hemispheres showing the location of the primary motor area in the precentral gyrus. (b) The distribution of motor output from the primary motor area of the cerebral cortex to different parts of the body. The distorted representation of the body parts is indicative of the relative proportion of the primary motor cortex devoted to controlling skeletal muscles in each area.

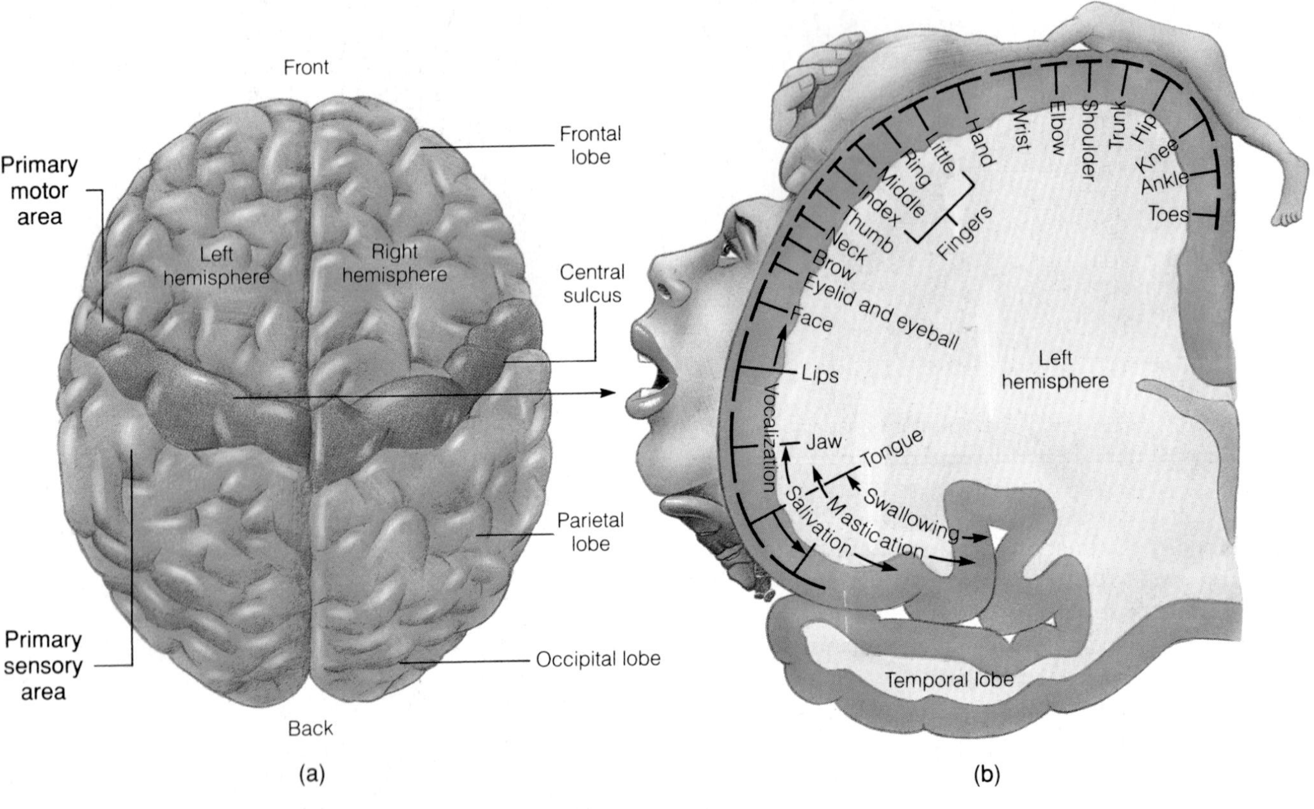

Association Areas. Several **association areas** surround the sensory and motor areas of the cerebral cortex. These areas are involved in processing, integrating, and interpreting sensory information and in formulating patterns of motor responses.

The **frontal association area,** located anterior to the premotor area, is considered to be the site of origin of the higher intellectual activities characteristic of humans. These activities include foresight, the ability to make judgments, and the capacity to select behavior for a variety of circumstances.

The **somatic association area** is located on the parietal lobe, posterior to the primary sensory area. This integration and interpretation center makes it possible to determine an object's shape and texture without viewing it and provides information about the positional relationships of body parts.

Located posterior to the somatic association area is a **visual association area,** and in the temporal lobe is an **auditory association area.** These areas contribute to the interpretation of visual and auditory experiences.

Basal Nuclei. Located deep within each cerebral hemisphere are several masses of gray matter known collectively as the **basal (cerebral) nuclei** (Figure 12.7). The nuclei, which are surrounded by white matter, are composed of groups of nerve cell bodies. Included in the basal nuclei are: the long arching **caudate nucleus;** the **amygdaloid nucleus,** which is located at the tip of the tail of the caudate nucleus; the **lentiform nucleus,** which is subdivided into the **putamen** and the **globus pallidus;** and the **claustrum,** a thin layer of gray matter just deep to the cortex of the insula (Figure 12.8). The band of white matter located between the basal nuclei and the thalamus is called the **internal capsule.** The internal capsule is composed of projection neurons of the major motor and sensory tracts as they pass to and from the cerebral cortex. Because of their appearance, the caudate nucleus, the internal capsule, and the lentiform nucleus are sometimes referred to as the **corpus striatum** ("striped body").

Most of the basal nuclei, like the neurons of the primary motor area, are involved in controlling skeletal muscle activity. Because they are located outside the precentral gyrus, the basal nuclei are part of the *extrapyramidal system.* In other words, the somatic motor activities of the body are controlled both by pyramidal-tract neurons, which originate in the cerebral cortex, and by motor neurons located elsewhere in the brain (extrapyramidal system), including the basal nuclei and the premotor area. In contrast to the pyramidal tracts, however, the neurons of the basal nuclei act in part to *inhibit* muscle contraction. This inhibition, together with the stimulatory effects of the pyramidal system, provides a means by which muscular movements can be precisely controlled.

The skeletal muscles are normally in a state of slight contraction, which is called *skeletal muscle tone,* and the basal nuclei have an inhibitory influence on skeletal muscle tone. The basal nuclei also suppress useles or unwanted patterns of movement, and they help coordinate slow, sustained muscle contractions such as those involved in maintaining posture.

Disorders of the basal nuclei result in involuntary contractions of skeletal muscles, such as the muscular rigidity and persistent tremors of the limbs associated with Parkinson's disease.

Olfactory Bulbs. On the ventral surface of each cerebral hemisphere is a small **olfactory bulb** and its associated **olfactory tract** (Figure 12.9, page 394). These structures, which are associated with the sense of smell, are located in the portion of the forebrain called the *rhinencephalon (rye-nen-sef´-a-lon).* The neurons of the **olfactory nerve (cranial nerve I)** pass from the nasal mucosa, through the cribriform plate of the ethmoid bone, and into the olfactory bulb, where they synapse with neurons of the olfactory tract. The neurons of the tracts pass to the olfactory area of the cortex on the medial surface of the temporal lobe. In addition to its olfactory function, the rhinencephalon is thought to be involved, in some obscure manner, with certain emo-

◆ **FIGURE 12.6 Map of the primary sensory area of the cerebral cortex**
(a) Top view of cerebral hemispheres showing the location of the primary sensory area in the postcentral gyrus. (b) The distribution of sensory input to the primary sensory area of the cerebral cortex from different parts of the body. The distorted representation of the body parts is indicative of the relative proportion of the primary sensory cortex devoted to reception of sensory input from each area.

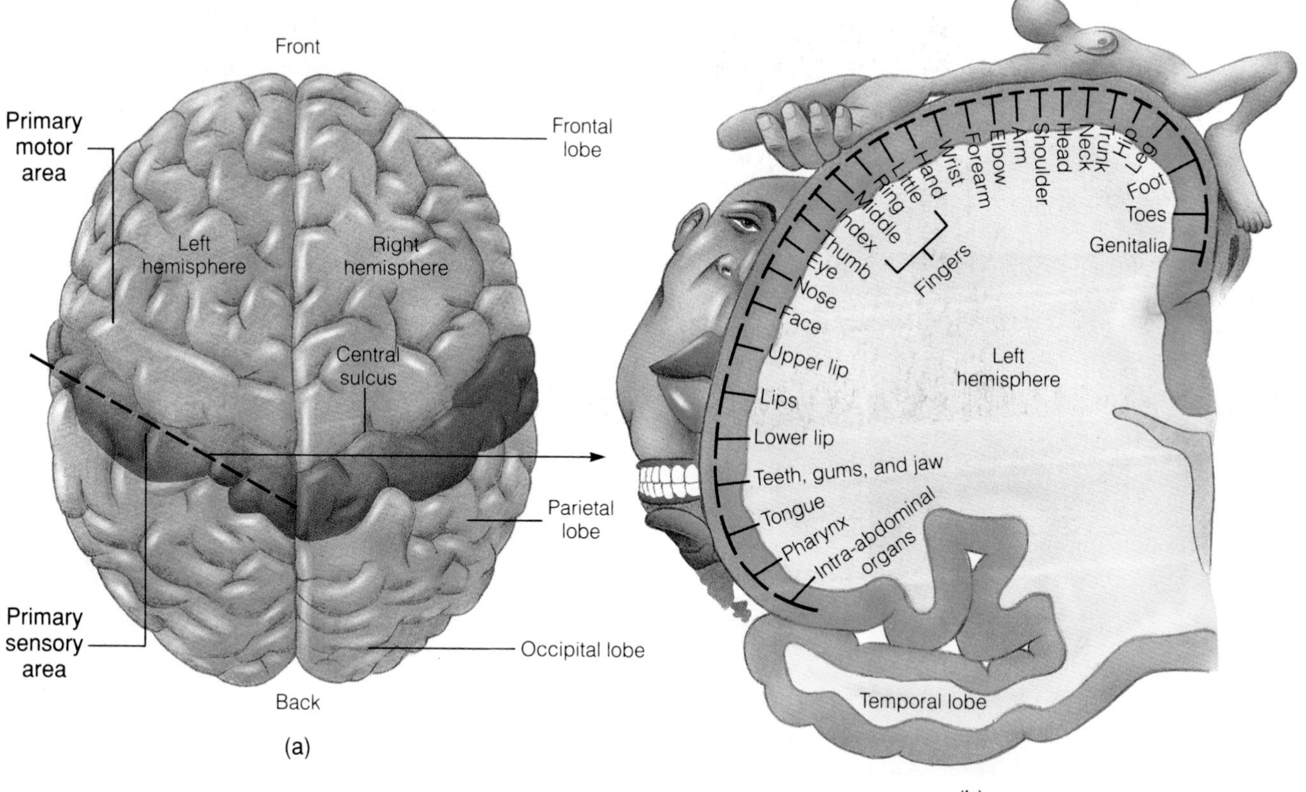

◆ **FIGURE 12.7 Sections through the brain showing basal nuclei (green) and thalamus** (a) Frontal section of the cerebrum and the diencephalon. (b) Transverse section. (c) Photograph of a transverse section.

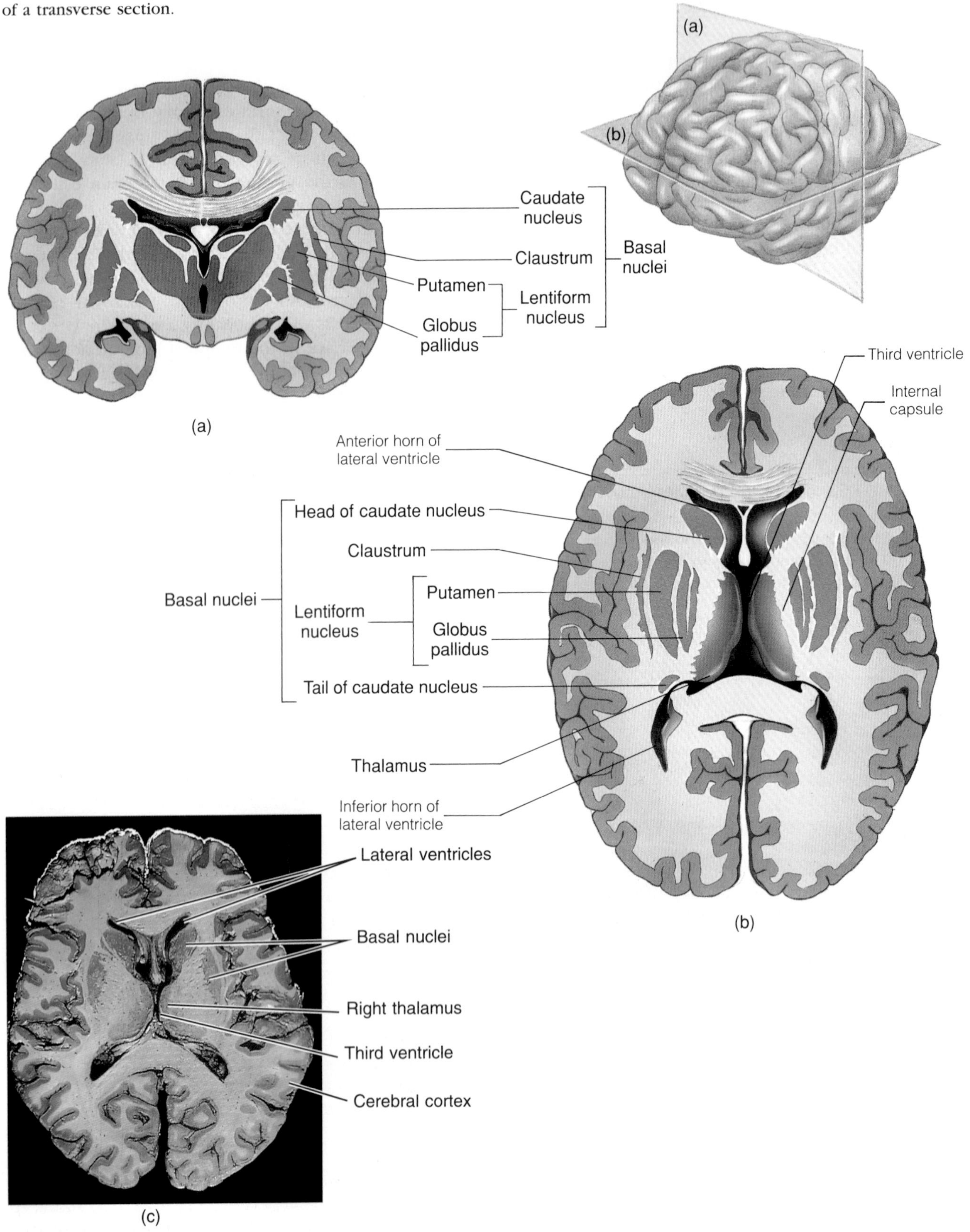

◆ **FIGURE 12.8 Three-dimensional relationships among the structures that comprise the basal nuclei**

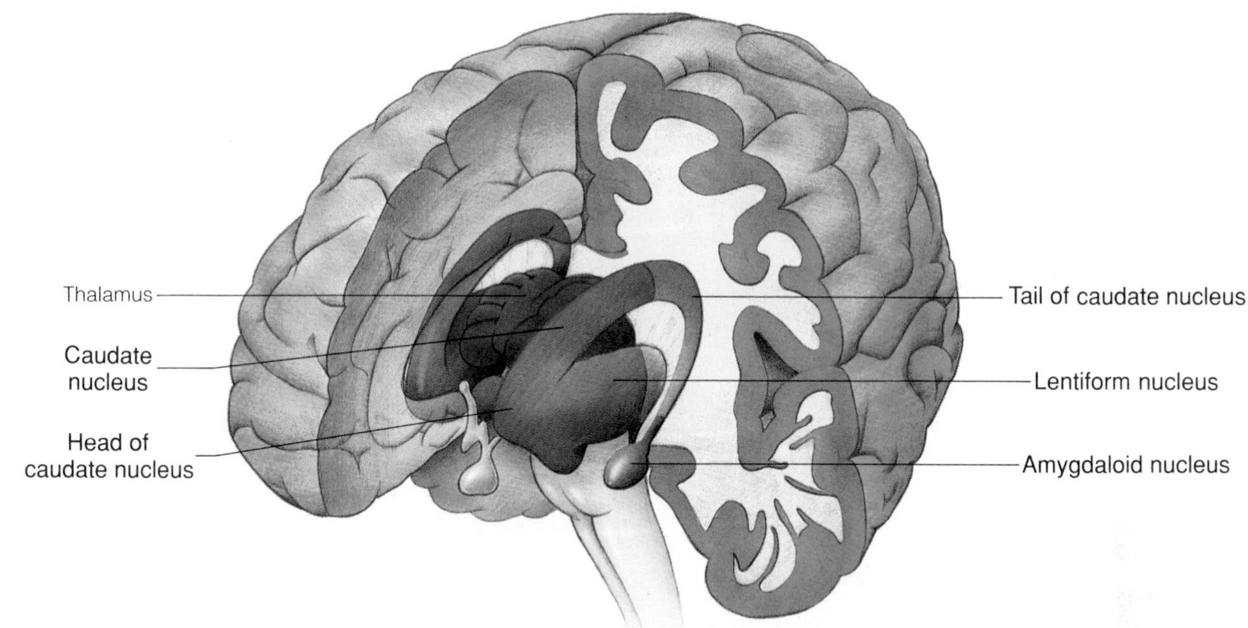

tional and behavioral responses. Those portions of the rhinencephalon that are not involved with olfaction are considered to be part of the *limbic system,* which we will consider shortly.

Hemispheric Specialization. Each cerebral hemisphere is somewhat specialized for carrying out certain kinds of mental processes. For example, language ability tends to be localized in the left cerebral hemisphere. In general, the left hemisphere is concerned with verbal and sequential processes and behaviors such as writing business letters and solving simple equations. This hemisphere excels at performing rational, linear, verbally oriented tasks, and it seems to process information in a fragmentary or analytical way. In contrast, the right cerebral hemisphere is primarily concerned with the recognition of complex visual patterns or with mentally picturing objects in three-dimensional space. The right hemisphere is also more involved in the expression and recognition of emotion and in certain musical or artistic abilities, such as identifying a theme in an unfamiliar piece of music. The information processing in the right hemisphere tends to be holistic and unitary rather than fragmentary and analytical.

The ability to recognize faces is well localized within the brain without being a function of primarily one cerebral hemisphere or the other. Damage to the medial undersides of both occipital lobes and the ventromedial surfaces of the temporal lobes produces a failure to recognize a person by sight. People with brain damage in these areas can usually correctly name objects but cannot identify faces. Such individuals may even fail to recognize their parents, spouses, or children by sight. Nevertheless, when a familiar but unrecognized person speaks, the person with brain damage may immediately recognize the other person's voice and can then name him or her.

Diencephalon

The second subdivision of the forebrain is the **diencephalon.** Since the cerebral hemispheres extend downward and almost completely surround the diencephalon, it is not visible from the exterior of the brain, except for a portion that can be seen when the brain is viewed ventrally. The **third ventricle** forms a median cavity within the diencephalon (Figure 12.10). The most important parts of the diencephalon are the *thalamus, hypothalamus,* and *epithalamus.*

Thalamus. The **thalamus** *(thal´-a-muss)* consists of two oval masses of nerve cell bodies (gray matter) that form the lateral walls of the third ventricle (Figure 12.7). A small bridge of commissural neurons called the **intermediate mass** *(massa intermedia)* passes across the third ventricle and connects the two thalamic masses. Each thalamic mass is deeply embedded in a cerebral hemisphere and is bounded laterally by the internal capsule. The thalamus contains over 20 functionally separate nuclei (aggregations of cell bodies). Functionally,

◆ **FIGURE 12.9 Ventral view of the brain**
(a) Drawing. (b) Photograph.

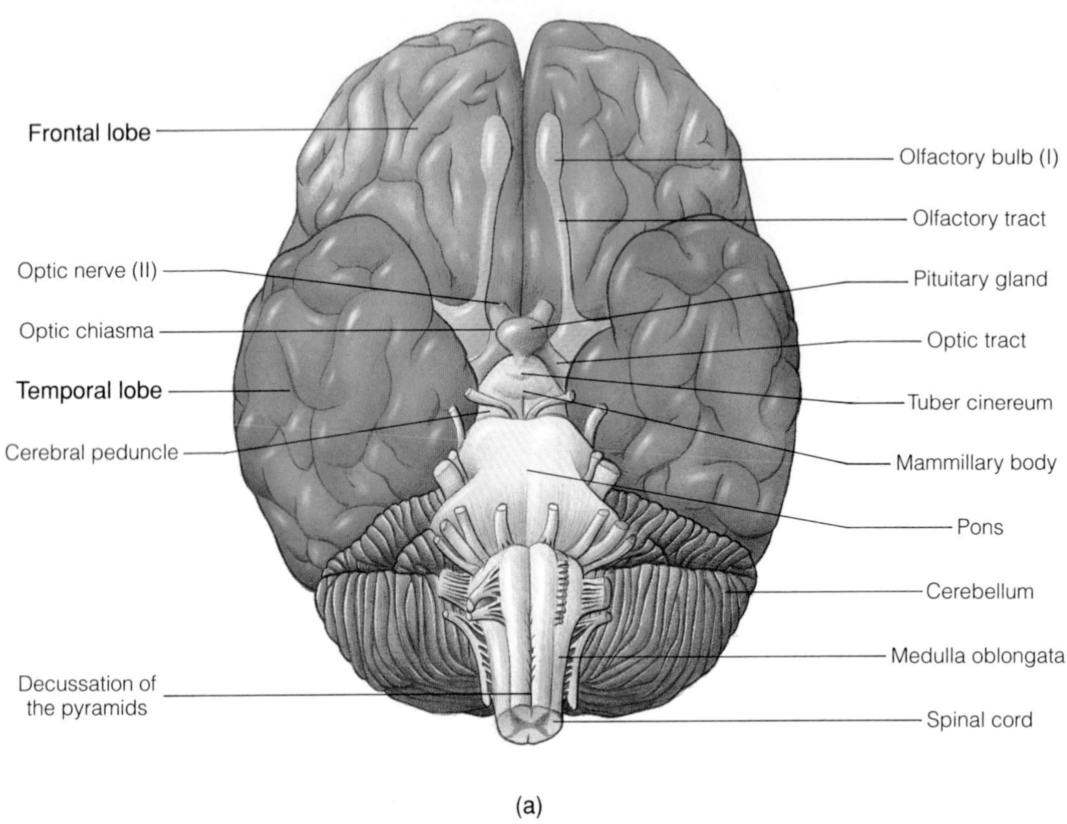

(a)

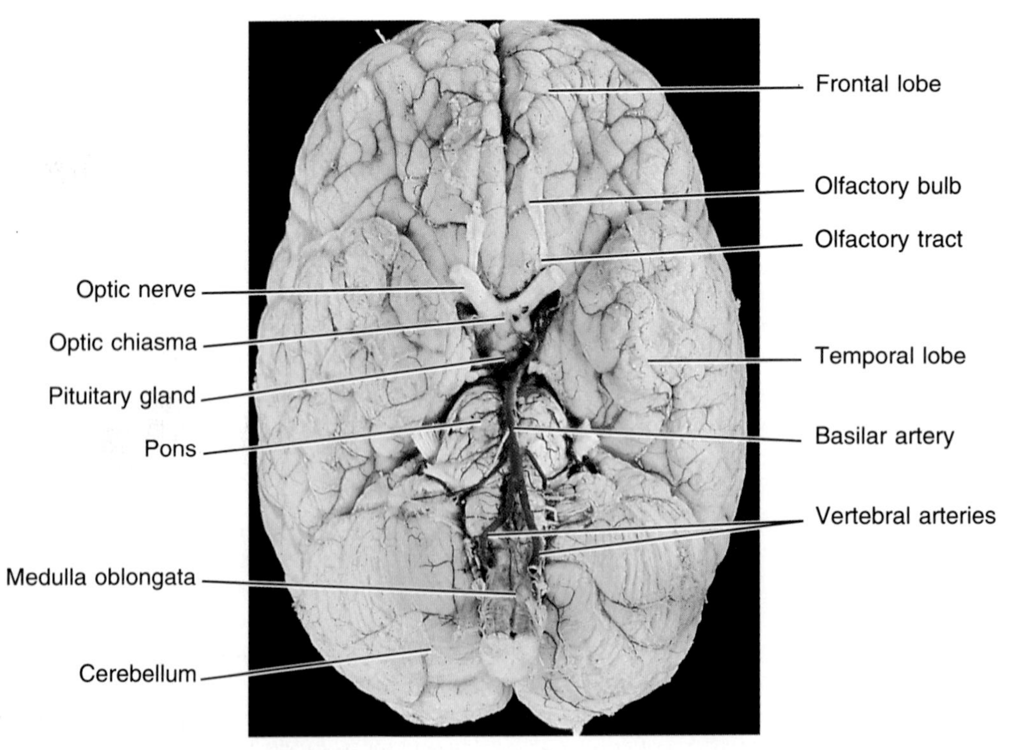

(b)

◆ **FIGURE 12.10 Midsagittal section of the brain and brain stem**
(a) Drawing. (b) Photograph.

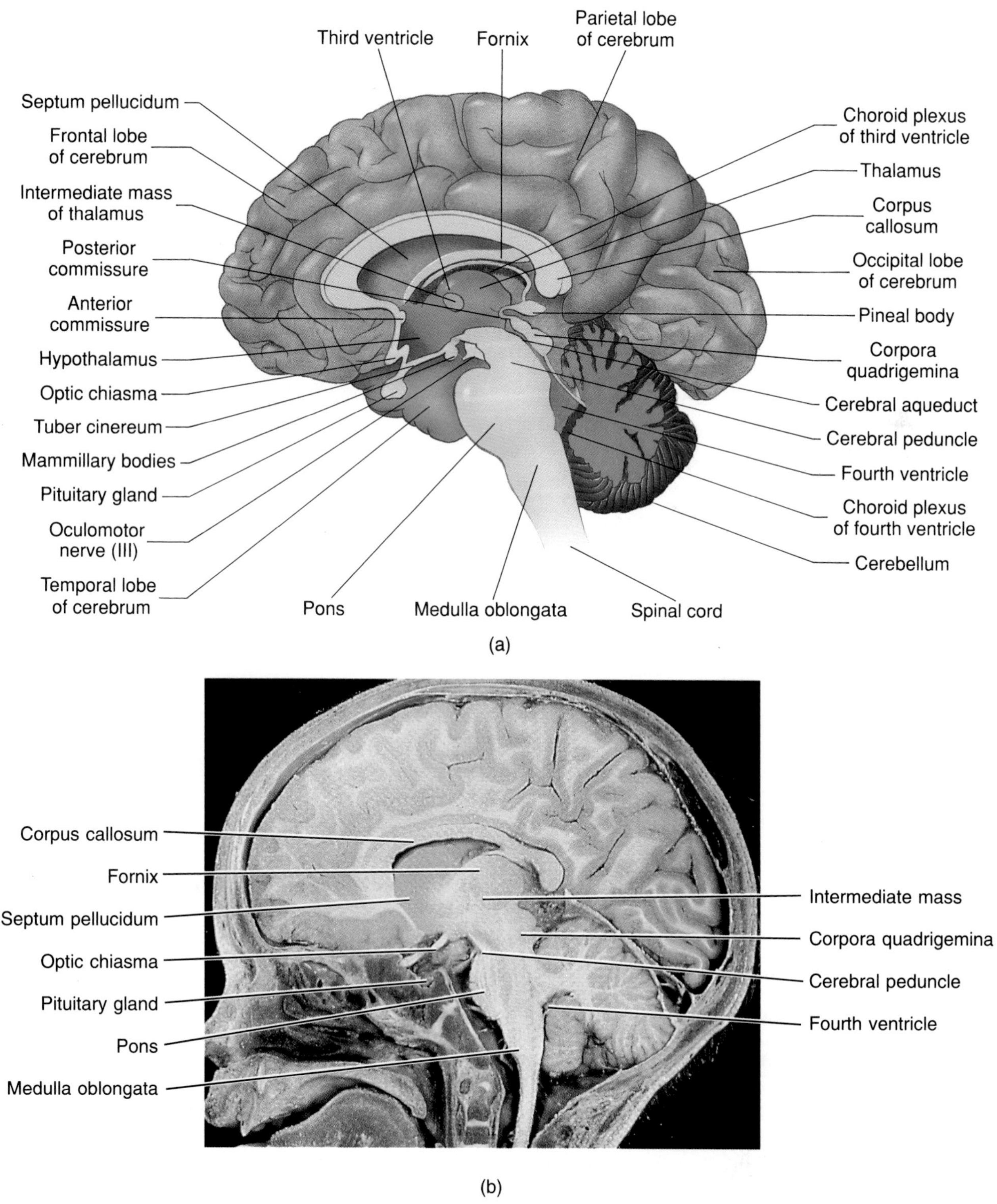

the thalamus acts as a major sensory relay and integrating center of the brain. All sensory fiber tracts that transmit nerve impulses from receptors to the cerebral cortex synapse within one of the thalamic nuclei. The thalamus screens out insignificant signals and routes the important sensory impulses to specific regions of the cerebral cortex as well as to subcortical areas such as the basal nuclei and hypothalamus. It is also capable of

◆ **FIGURE 12.11 Midsagittal section showing the nuclei of the hypothalamus**

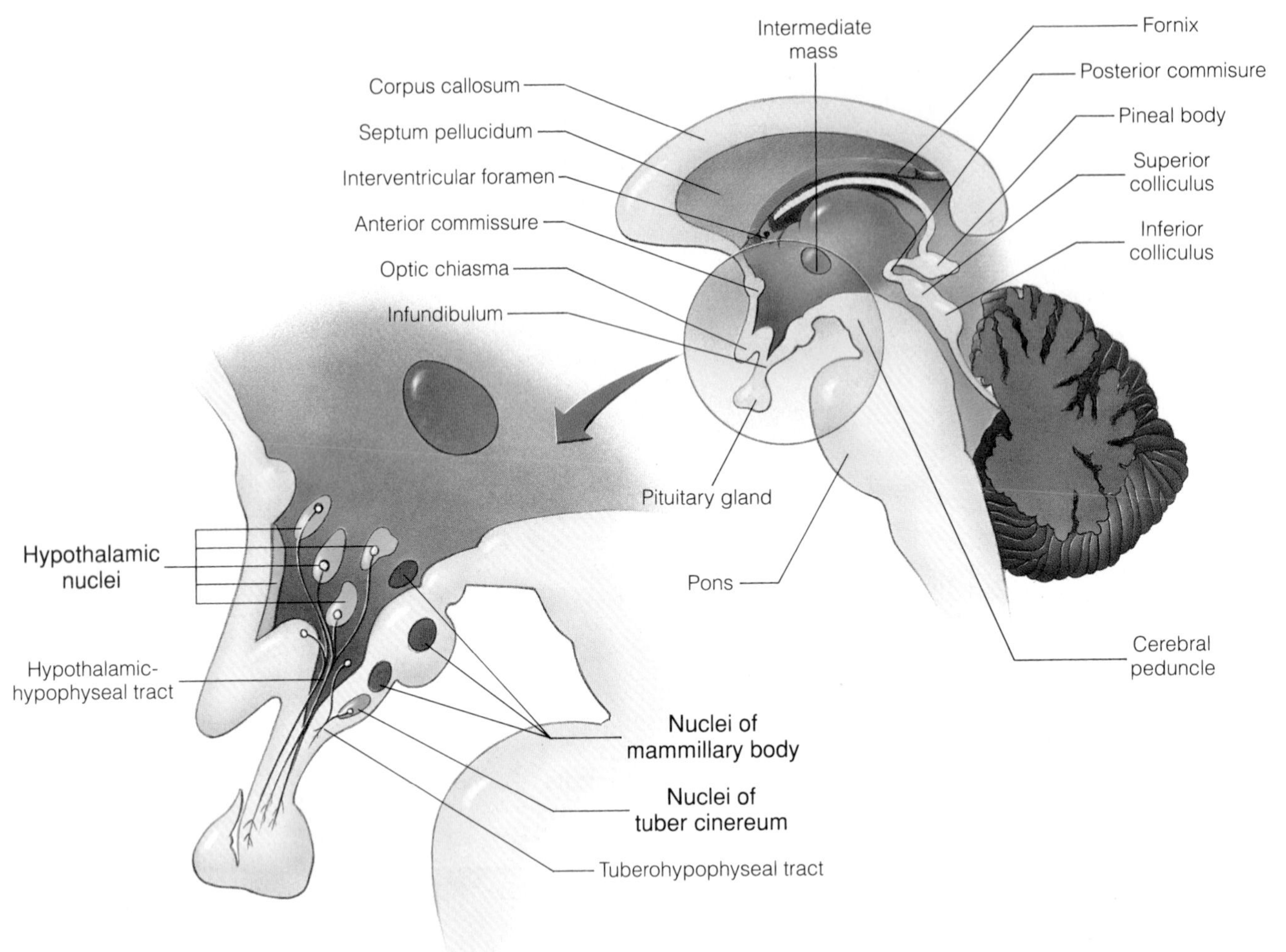

providing crude awareness of various types of sensation and contributes to some degree of consciousness. Apart from its sensory role, the thalamus is also involved with some of the motor tracts that leave the cerebral cortex.

Hypothalamus. As the name indicates, the **hypothalamus** lies below the thalamus, where it forms part of the walls and floor of the third ventricle. Like the thalamus, the hypothalamus is composed of several nuclei, each of which is involved with specific functions (Figure 12.11). Several hypothalamic structures are visible externally, including the *mammillary bodies, tuber cinereum, infundibulum,* and the *optic chiasma* (or *chiasm*).

The two **mammillary bodies** *(mam´-i-ler´´-ē)* are small, round nuclear masses that form external bulges from the undersurface of the brain posterior to the infundibulum. The mammillary bodies function primarily as relay stations for olfactory neurons and are involved in olfactory reflexes.

Located just anterior to the mammillary bodies is the **tuber cinereum** *(tu´-ber si-ne´-re-um),* which contains neurons that transport regulatory hormones (or factors) from the hypothalamus to the infundibulum. These neurons of the tuber cinereum form the *tuberohypophyseal tract.* In the infundibulum, the hormones from the neurons of the tuberohypophyseal tract enter blood vessels that transport them to the adenohypophysis of the pituitary gland (Chapter 17).

Extending downward from the tuber cinereum is the stalklike **infundibulum** *(in-fun-dib´-yoo-lum).* Nerve fibers from some of the hypothalamic nuclei pass through the infundibulum on their way to the pars nervosa of the posterior lobe of the pituitary gland. These neurons, which form the *hypothalamic-hypophyseal tract,* transport hormones (oxytocin and ADH) synthesized in hypothalamic nuclei to the posterior pituitary, from which they are released.

Anterior to the infundibulum is the **optic chiasma,** which is formed by the decussation (crossing) of some of the neurons in the optic nerves.

◆ **FIGURE 12.12 The main structures that constitute the limbic system**

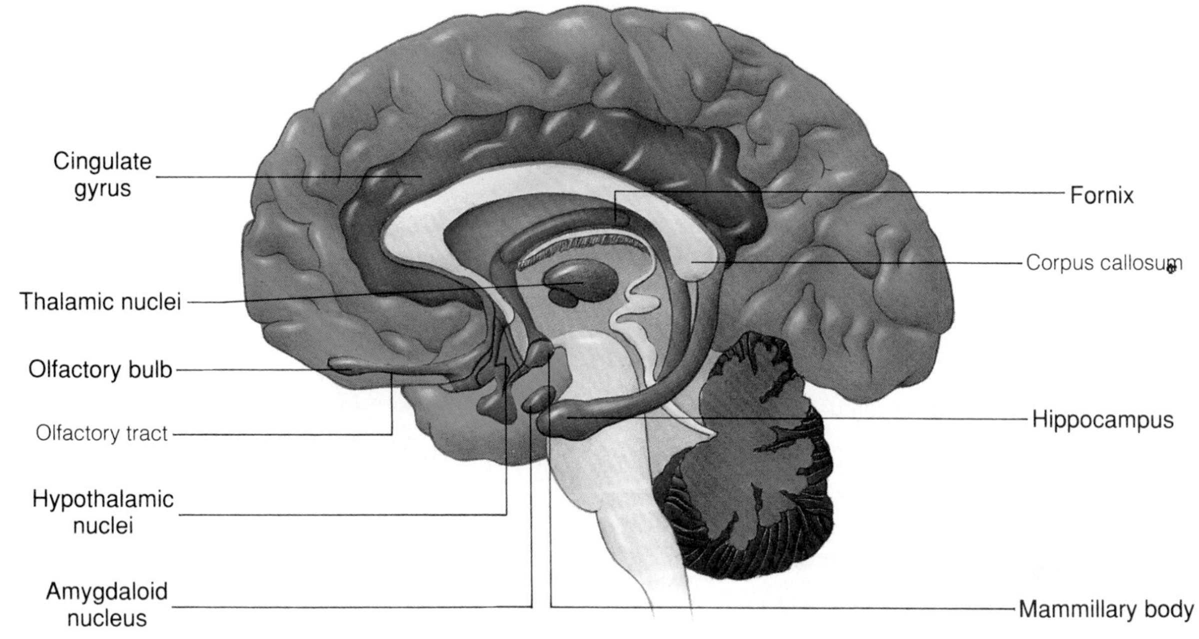

The hypothalamus controls many vital processes, most of them associated with the autonomic nervous system. Some of the hypothalamic nuclei have been shown experimentally to regulate sympathetic activity; others control parasympathetic functions. The hypothalamus is involved in regulating body temperature, water balance, appetite, gastrointestinal activity, sexual activity, and even emotions such as fear and rage. The hypothalamus also regulates the release of the hormones of the pituitary gland; and thus it greatly affects the endocrine system (see Chapter 17).

Epithalamus. The **epithalamus,** the most dorsal portion of the diencephalon, forms a thin roof over the third ventricle. The roof has a vascular choroid plexus located on its internal surface. A small mass called the **pineal body (epiphysis)** extends outward from the posterior end of the epithalamus. The possible neuroendocrine function of the pineal body is discussed in Chapter 17. The **posterior commissure** is located just ventral to the pineal body.

The Limbic System. Although emotions are influenced by the hypothalamus, emotional responses involve a complex interaction of structures in several different regions of the brain, including the cerebrum and the diencephalon. A group of structures collectively referred to as the **limbic system** are particularly important in emotional responses (Figure 12.12). The limbic system includes the **olfactory bulbs;** a band of fibers called the **fornix** *(for´-niks),* which passes from beneath the corpus callosum to the mammillary bodies of the hypothalamus; a cerebral gyrus called the **cingulate gyrus** *(sin´-gu-late),* located just above the corpus callosum; parts of the **basal nuclei,** including the **amygdaloid nucleus** *(ah-mig´-dah-loid); the* **hippocampus** *(hip˝-ō-kam´-pus),* which is a part of the cerebrum located in the floor of the lateral ventricle close to the amygdaloid nucleus; the **mammillary bodies;** and various **thalamic, hypothalamic** and **septal nuclei.**

In animal experiments it is possible to electrically stimulate various centers within the limbic system. When certain centers are stimulated, the animal responds to the stimulation as if it were pleasant ("pleasure centers"). When other centers are stimulated, the animal responds to the stimulation as if it were unpleasant ("punishment centers"). A wide variety of emotional behavior patterns can be produced by stimulating or removing specific regions of the limbic system. It has been concluded therefore that the structures of the limbic system play an important role in producing and modifying emotional behavior.

Midbrain

The **midbrain** or **mesencephalon** is a short, constricted region between the forebrain and the hindbrain. The mesencephalon is the most anterior portion of the **brain stem,** which also includes the pons (located in the metencephalon) and the medulla oblongata (myelencephalon). Within the mesencephalon is a small

◆ **FIGURE 12.13 Posterolateral view of the brain stem**
The cerebellum has been removed.

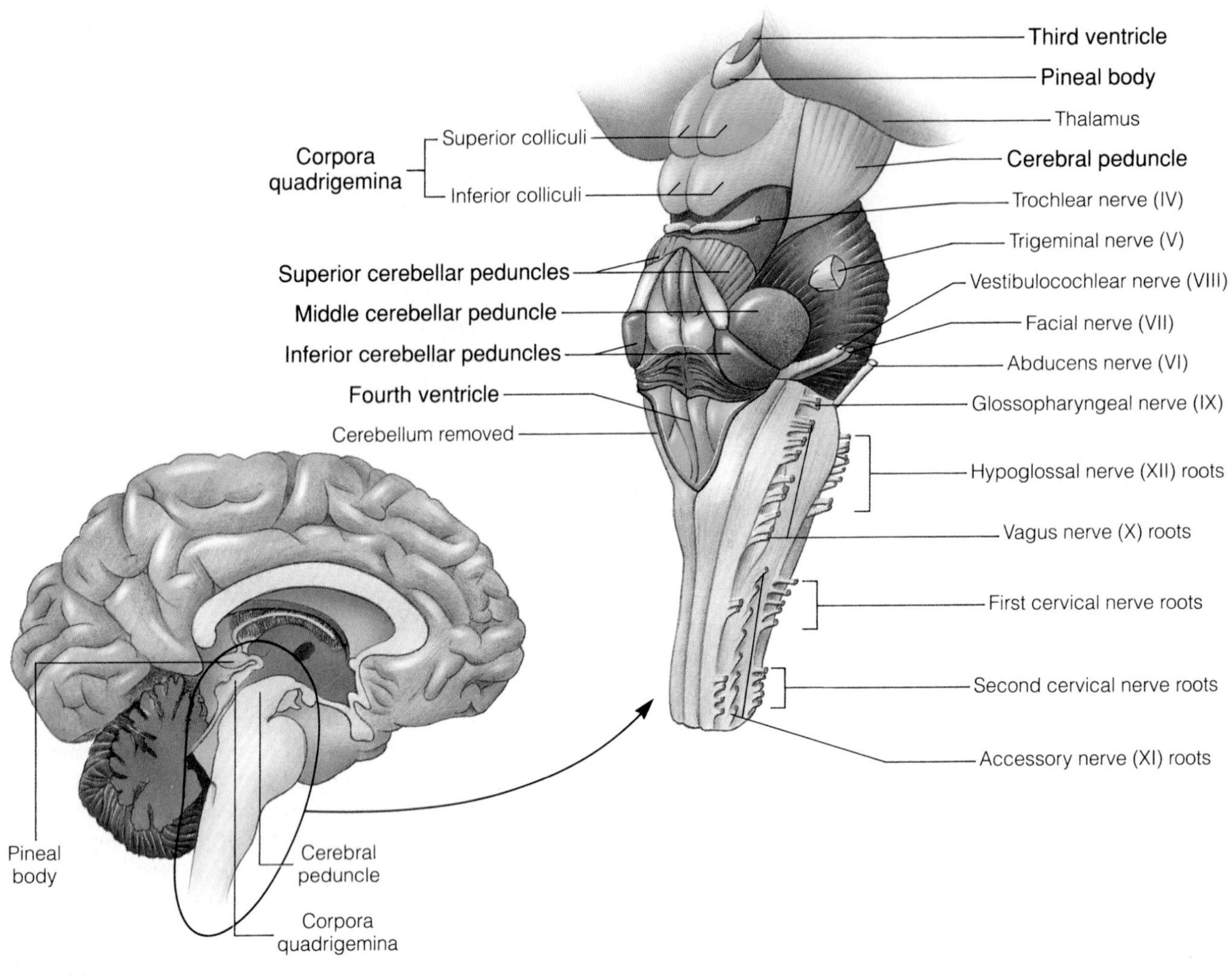

tunnel called the **cerebral aqueduct** (*aqueduct of Sylvius;* Figure 12.15), which connects the third ventricle (located in the diencephalon) with the fourth ventricle (located in the metencephalon).

Cerebral Peduncles

On the ventral surface of the mesencephalon are two cylindrical bulges called the **cerebral peduncles** (Figure 12.13). The peduncles are composed of motor nerve fibers that travel from the primary motor area of the cerebral cortex to the pons and spinal cord and sensory nerve fibers that travel from the spinal cord to the thalamus. The **oculomotor nerves (cranial nerve III)** emerge between the peduncles. Deeper within the mesencephalon, between the peduncles and cerebral aqueduct, is a small island of gray matter called the **red nucleus.** The cell bodies of neurons that compose the *rubrospinal tract* (discussed later in the chapter) are located in the red nucleus. The red nucleus serves as a relay station that coordinates impulses between the cerebellum and the cerebral hemispheres, thereby contributing to the coordination of movements and to the sense of balance.

Corpora Quadrigemina

The dorsal surface of the mesencephalon, which forms the roof of the cerebral aqueduct, consists of four rounded prominences called the **corpora quadrigemina** (*kop´-po-rah kwod´´-ri-jem´-i-nah;* "four twin bodies") (Figure 12.13). The upper pair of prominences is called the **superior colliculi** (*ko-lik´-u-li*). Some neurons of the optic tracts from the retinas of the eyes travel to the superior colliculi, where they participate in activities concerned with certain reflex responses to visual stimuli. The lower pair of prominences, the **inferior colliculi,** serve as relay stations and reflex centers for auditory stimuli. The **trochlear nerves (cranial nerve IV)** emerge from the roof of the mesencephalon just below the inferior colliculi.

Hindbrain

The hindbrain is the most caudal region of the brain. It connects the midbrain with the spinal cord. The hindbrain includes the metencephalon and the myelencephalon.

Metencephalon

The major structures of the **metencephalon** are the *cerebellum* and the *pons.* The cerebral aqueduct of the mesencephalon expands into the **fourth ventricle** in the metencephalon. The inferior portion of the fourth ventricle extends into the myelencephalon. As is true of all the ventricles of the brain, there is a vascular choroid plexus in the fourth ventricle.

Cerebellum. Projecting from the dorsal surface of the metencephalon, the **cerebellum** *(ser-a-bel´-lum)* (Figures 12.9 and 12.10) is separated from the cerebral hemispheres by a strong membrane called the *tentorium cerebelli.* The tentorium lies within the transverse fissure of the brain and supports the occipital lobes of the cerebrum, thus minimizing the pressure that the lobes exert on the cerebellum.

The cerebellum is composed of two lateral **cerebellar hemispheres** connected in the midline by a region called the **vermis.** The surface of the cerebellum consists of a thin cortex of gray matter. The cortex dips deeply below the apparent surface of the cerebellum in a manner similar to the fissures and sulci of the cerebrum, although the cerebellar fissures are more parallel, giving the appearance of a series of flattened plates called **folia.**

The cerebellum is connected to the mesencephalon by a pair of nerve tracts called the **superior cerebellar peduncles;** to the pons by a pair of **middle cerebellar peduncles;** and to the medulla oblongata by a pair of **inferior cerebellar peduncles** (Figure 12.13). The superior cerebellar peduncles are composed principally of efferent nerve fibers from the cerebellum; the middle and inferior cerebellar peduncles are composed mostly of afferent nerve fibers that transmit impulses from the pons, the medulla oblongata, and the spinal cord to the cerebellum. These extensive interconnections with other regions of the central nervous system provide the cerebellum with widespread input and output capabilities.

The cerebellum helps regulate muscle tone and is important in maintaining balance. The cerebellum helps coordinate skilled, voluntary movements, and it is believed to play a role in planning and initiating voluntary activity.

The cerebellum coordinates the activities of the skeletal muscles through sensory information carried to it from receptors for proprioception, equilibrium, and balance. Moreover, the cerebellum receives some sensory information concerning touch, vision, and sound. Further coordination occurs by way of nerve impulses sent from the cerebellum to higher brain centers. In particular, nerve impulses from the cerebellum may dictate specific movement sequences to the primary motor area of the cerebral cortex, which then initiates the nerve impulses necessary to carry out the movements. A person whose cerebellum has been damaged experiences muscular weakness, a loss of muscle tone, and uncoordinated movements. All the functions with which the cerebellum is concerned remain below the level of consciousness. Thus, the cerebellum is able to mediate certain responses without having them reach the conscious level.

Pons. The **pons** (''bridge''), which is located on the ventral surface of the metencephalon, consists of bands of nerve fiber tracts and several nuclei (Figures 12.9 and 12.10). The tracts in the pons are both transverse and longitudinal. The transverse tracts consist of neurons that enter the cerebellar hemispheres through the middle cerebellar peduncles. The longitudinal tracts are composed of neurons that travel between the brain stem and the cerebrum. The pons, therefore, functions primarily to connect the cerebellum with the cerebrum and the brain stem, thus providing connections between upper and lower levels of the central nervous system. The cerebral cortex, in particular, achieves most of its connections to the cerebellum by way of nerve fiber tracts that pass through the pons. In addition, the stimulation of nuclei within the pons affects the rate of respiration. The nuclei of the **trigeminal (V), abducens (VI), facial (VII),** and **vestibulocochlear (VIII)** cranial nerves are located in the pons.

Myelencephalon

The second portion of the hindbrain, the **myelencephalon,** is also known as the **medulla oblongata.** At its lower end, the medulla is continuous with the spinal cord. The cavity in the medulla forms the lower portion of the fourth ventricle and continues into the spinal cord as the central canal of the cord.

On the anterior surface of the medulla are two large columns of nerve fiber tracts called the **pyramids.** The pyramids contain the same motor tracts found in the cerebral peduncles. Therefore, the tracts in the pyramids carry the voluntary motor output from the primary motor area of the cerebral cortex. The tracts of the pyramids originate from cell bodies in the precentral gyrus of the cerebral cortex and continue into the spinal cord as the *corticospinal tracts* (discussed later in this chapter). Some of the nerve tracts in the pyramids cross from one pyramid to another. This crossing, called the *decussation of the pyramids,* is visible on the ventral surface of the medulla in the groove that separates the pyramids (Figure 12.9). As a consequence of the decussation of these nerve tracts, motor areas located on one side of the cerebral cortex can control muscular movements on the opposite side of the body.

The medulla oblongata also contains nuclei that give rise to the last four cranial nerves: the **glossopharyngeal (IX),** the **vagus (X),** the cranial portion of the **accessory (XI),** and the **hypoglossal (XII).**

Located in the medulla are *medullary centers,* which are groups of neurons that are involved in the control of a variety of vital functions, such as heart rate, respiration, dilation and constriction of blood vessels, coughing, swallowing, and vomiting.

◆ **FIGURE 12.14 Schematic representation of the reticular formation**

The arrows indicate input to and output from the reticular activating system.

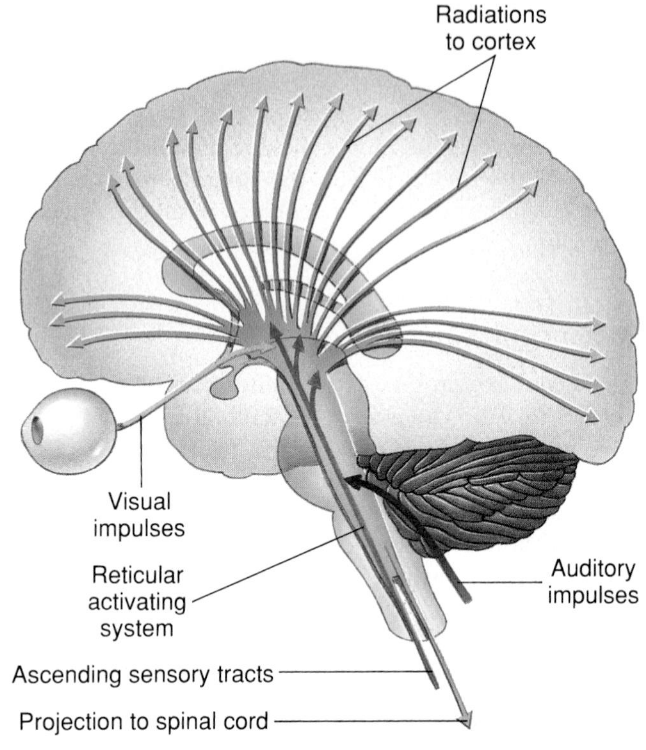

Reticular Formation. Inside the medulla oblongata and extending throughout the brain stem and up into the thalamus is a widespread network of interconnected neurons called the **reticular formation.** Neurons within the reticular formation have connections with cells in the spinal cord, medulla, cerebellum, hypothalamus, thalamus, basal nuclei, and the cerebral cortex (Figure 12.14). Thus, the reticular formation influences the activity of diverse regions of the nervous system.

Some neurons of the reticular formation synapse with neurons in the spinal cord that control contraction of skeletal muscles. These reticular neurons act together with cerebellar centers to maintain muscle tone and regulate muscle reflexes involved with equilibrium and posture.

Some of the ascending fibers originating in the reticular formation carry impulses that activate the cerebral cortex and maintain it in an aroused, conscious state. These reticular fibers are referred to as the **reticular activating system (RAS).** Input from the major sensory tracts helps keep the neurons of the RAS active. The RAS can be depressed by alcohol and various drugs. Injury or diseases that affect the RAS may produce unconsciousness *(coma).*

Ventricles of the Brain

The **ventricles** of the brain (Figure 12.15) develop from expansions of the lumen of the anterior region of the embryonic neural tube and form a continuous fluid-filled system in the brain. Their early development is discussed in Chapter 10 and illustrated in Figures 10.6 and 10.7. The roof of each ventricle is thin and contains no neurons. Each ventricle contains a **choroid plexus** *(ko´-roid),* which consists of a network of capillaries covered by modified ependymal cells. The choroid plexuses are sites of production of **cerebrospinal fluid.** The fluid fills the ventricles of the brain, the central canal of the spinal cord, and the subarachnoid space, which surrounds the brain and spinal cord (Figure 12.16).

The choroid plexuses and the arachnoid membrane (discussed on page 404) act together as a barrier between the blood and the cerebrospinal fluid (CSF). This **blood-CSF barrier** serves to keep harmful compounds from leaving the blood and entering the cerebrospinal fluid. The arachnoid membrane is generally impermeable to water-soluble substances, and its role in the blood-CSF barrier is largely passive. In contrast, the choroid plexuses are selective as to which molecules are allowed to pass across their walls, making the blood-CSF barrier a selective one. The selectivity of the blood-CSF barrier, like that of the blood-brain barrier (page 351), is the result of tight junctions that connect adjacent cells. In the case of the blood-CSF barrier, the tightly joined cells are modified ependymal cells of the choroid plexuses. The tight junctions hinder the passage of even small

◆ **FIGURE 12.15 Three-dimensional view of the ventricles of the brain (viewed as if they could be seen from the surface of the brain)**

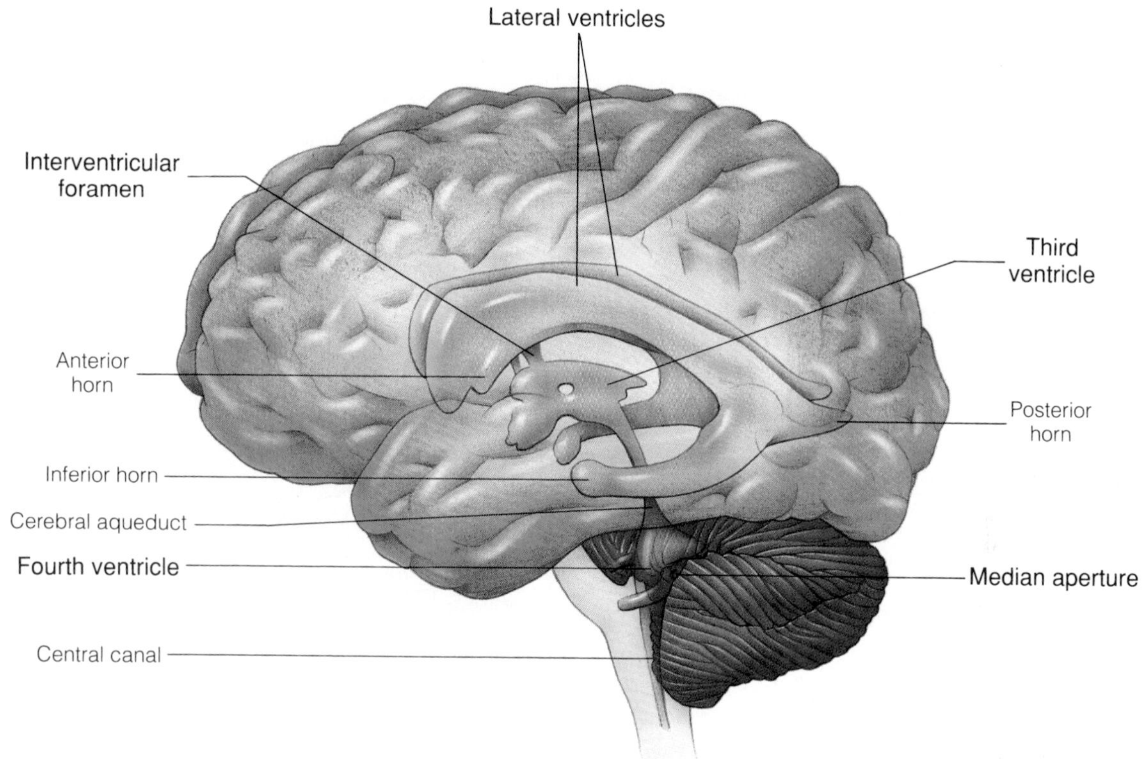

water-soluble molecules between the blood and the cerebrospinal fluid. Thus, in order for substances to pass from the blood into the cerebrospinal fluid, they must be transported across the plasma membranes of the cells of the choroid plexuses by specific carrier molecules.

Although the blood-CSF barrier and the blood-brain barrier both regulate the exchange of substances between the brain and the blood, each is specialized to transport specific molecules, and they function by somewhat different mechanisms (Figure 12.17). The blood-CSF barrier controls the transfer of substances that are needed by the brain in only small amounts from the blood within the choroid plexuses to the cerebrospinal fluid. These substances are transported by active-transport mechanisms. Consequently, transported substances can reach higher concentrations in the cerebrospinal fluid than in the blood. In contrast, the blood-brain barrier controls the transfer of substances the brain consumes rapidly and needs in large quantities (e.g., glucose, amino acids, and lactates) from the blood within the cerebral capillaries to the interstitial fluid surrounding brain cells. The movement of these substances across the barrier depends on facilitated diffusion. Therefore, the substances must be in higher concentration in the blood than they are in the interstitial fluid of the brain if there is to be a net movement of the substances into the interstitial fluid.

Lateral Ventricles

Within each cerebral hemisphere is a **lateral ventricle** that has its major portion located in the parietal lobe. Extensions from this portion protrude into the frontal lobe *(anterior horn),* the occipital lobe *(posterior horn),* and the temporal lobe *(inferior horn).* The lateral ventricles are separated from each other medially by a thin vertical partition called the **septum pellucidum** (Figure 12.10). Each lateral ventricle communicates with the third ventricle by a small opening called the **interventricular foramen** *(foramen of Monro).*

Third Ventricle

The **third ventricle** is a narrow midline chamber in the diencephalon. The right and left masses of the thalamus form most of its lateral walls. A commissure called the *intermediate mass* passes through the ventricle. The third ventricle opens into the fourth ventricle by means of the **cerebral aqueduct** of the mesencephalon.

Fourth Ventricle

The **fourth ventricle** is a pyramidal cavity located in the hindbrain just ventral to the cerebellum. There are

◆ **FIGURE 12.16 Location of the cerebrospinal fluid (blue) that surrounds the brain and spinal cord**
The arrows indicate the direction of the fluid's flow. Blood is shown in red.

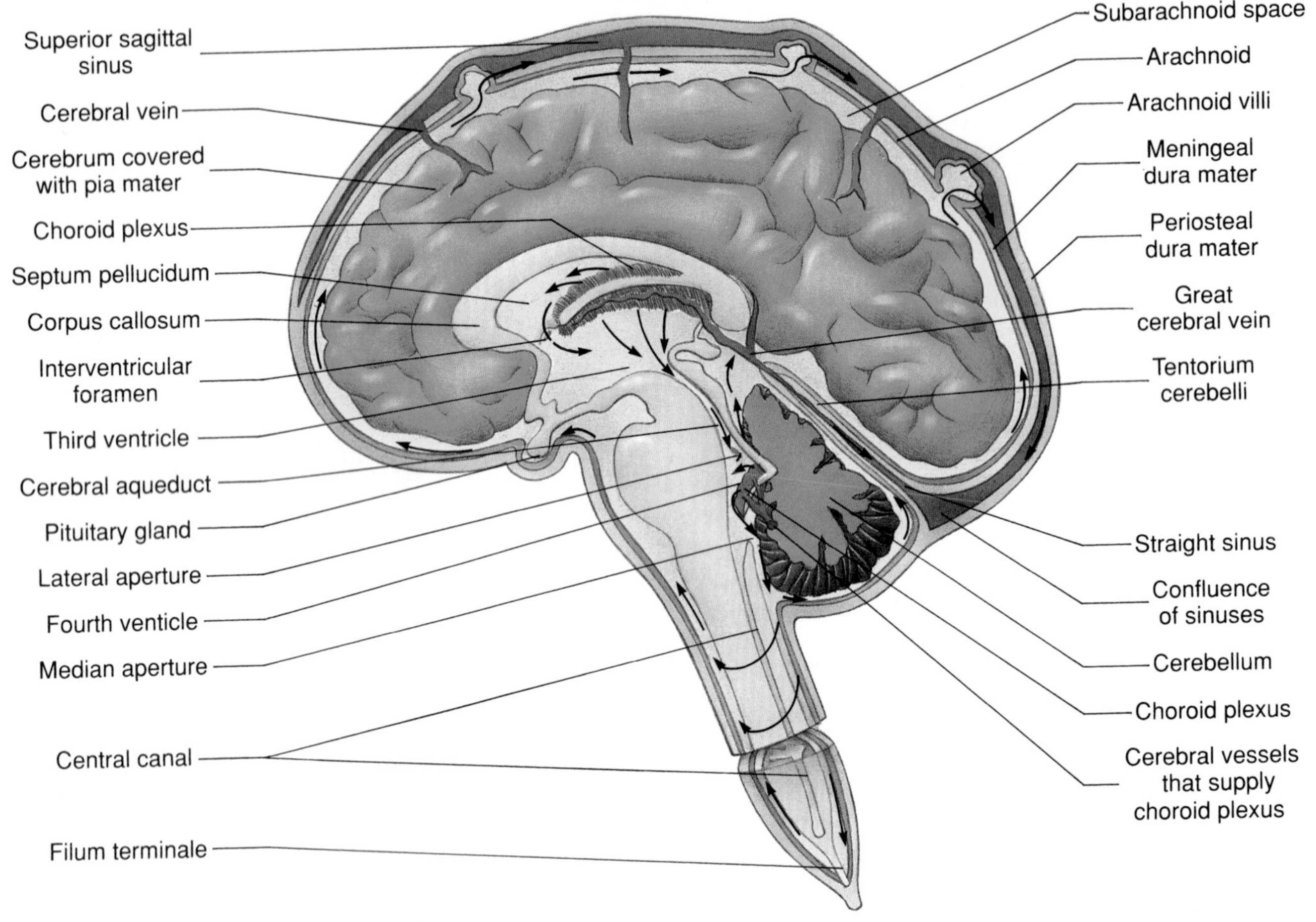

two openings in the lateral walls of the fourth ventricle called the **lateral apertures** *(foramina of Luschka)* (Figure 12.16). In the roof is a single opening, the **median aperture** *(foramen of Magendie).* The ventricles communicate through these three openings with a *subarachnoid space* that surrounds the brain and spinal cord. Inferiorly, the fourth ventricle is continuous with the narrow *central canal* that extends the length of the spinal cord.

The Meninges

The entire central nervous system is covered by three layers of connective tissue called the **meninges** (singular: *meninx*). The meninges are the *dura mater,* the *arachnoid,* and the *pia mater.*

Dura Mater

The **dura mater** is the outermost meninx. It is a strong membrane composed of fibrous connective tissue. Around the brain, the dura mater is a double-layered structure (Figures 12.16 and 12.18). The outer layer of the dura mater *(periosteal layer)* adheres closely to the bones of the skull, forming the periosteum of the cranial bones. The inner layer of the dura mater *(meningeal layer)* is continuous with the dura mater of the spinal cord. The two layers of the dura that surround the brain are fused together over most of the brain. In certain regions, however, the layers are separated, forming venous sinuses *(dural sinuses)* that carry blood to the internal jugular veins of the neck.

In several places the meningeal layer of the dura mater extends inward within fissures of the brain to form strong septa that anchor the brain to the skull. These extensions include:

1. The **tentorium cerebelli** *(ten-to´-re-um ser´´-ĕ-bel´-li),* which passes in the transverse fissure separating the cerebrum and the cerebellum (Figure 12.16). It is attached to the inner surface of the occipital bone, the petrous portions of the temporal bones, and the sphenoid bone. The tentorium helps support the cerebrum

◆ **FIGURE 12.17 Comparison of the blood-cerebrospinal fluid barrier and the blood-brain barrier**

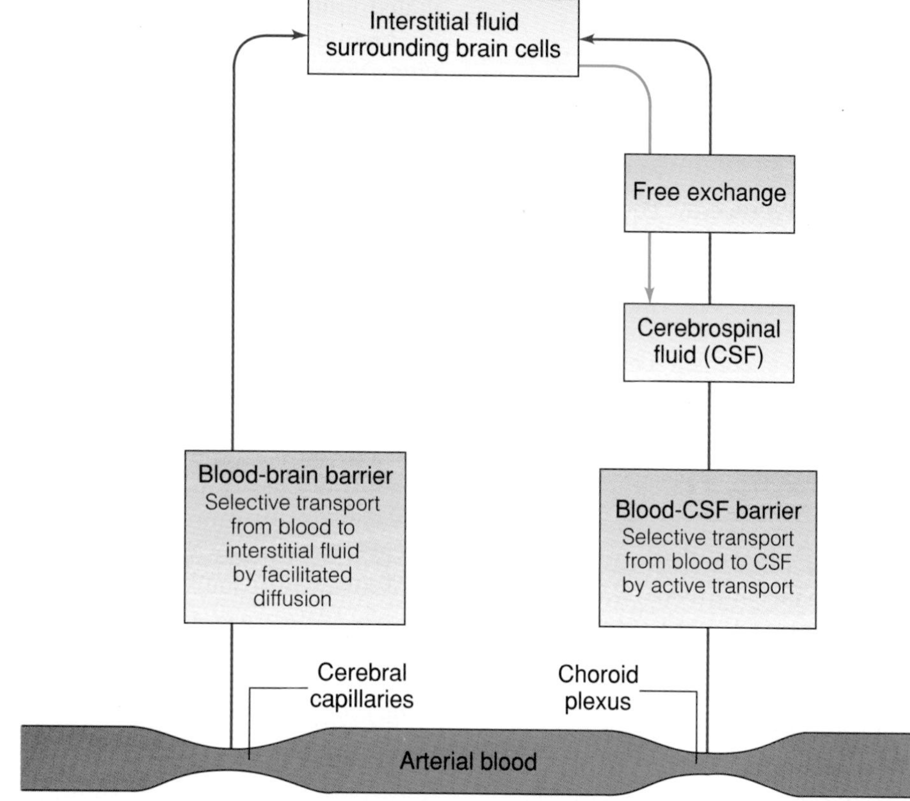

and keeps its full weight from resting on the cerebellum. **2.** The **falx cerebri** *(falks se-rēb´-ri),* which forms a midsagittal septum within the longitudinal fissure separating the cerebral hemispheres (Figure 12.18). The falx cerebri is anchored anteriorly to the crista galli of the ethmoid bone and posteriorly to the tentorium cerebelli. **3.** The **falx cerebelli** *(falks ser´´-ĕ-bel´-li),* a midsagittal septum separating the two cerebellar hemispheres.

◆ **FIGURE 12.18 Frontal section showing the relationships of the dural venous sinuses, meninges, and the subarachnoid space**

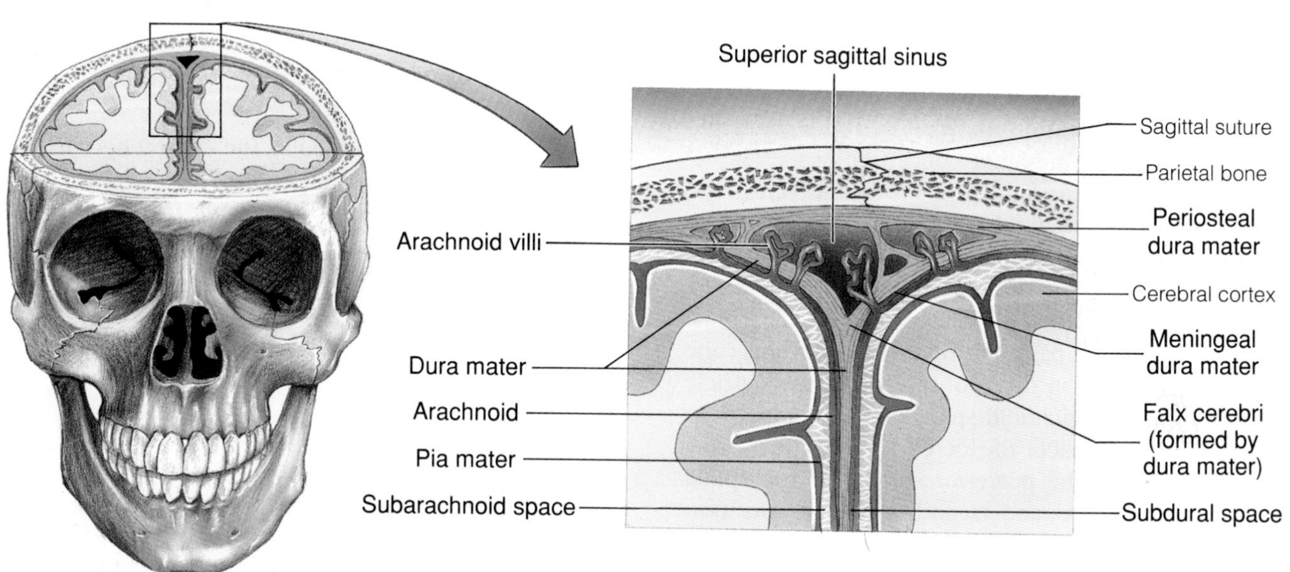

It is attached above to the tentorium cerebelli and below to the inner surface of the occipital bone.

Arachnoid

The middle of the three meninges is the **arachnoid** *(a-rak´-noid)* meninx (Figure 12.18), which is located deep to the dura mater. The arachnoid, a delicate membrane, adheres rather closely to the inner surface of the dura mater, with only a very narrow **subdural space** separating the two membranes. Between the arachnoid and the deepest meninx, the pia mater, is a wider separation called the **subarachnoid space.** The subarachnoid and the subdural spaces contain cerebrospinal fluid. The subarachnoid space is bridged by weblike strands of the arachnoid.

Pia Mater

The innermost meninx is the **pia mater.** A delicate vascular membrane of loose connective tissue, the pia mater adheres closely to the brain and spinal cord, dipping deeply into the fissures and the sulci. In the ventricles, the pia mater plus associated ependymal cells become modified and contribute to the formation of the choroid plexuses.

Cerebrospinal Fluid

Cerebrospinal fluid is a watery fluid that serves as a cushion for the entire central nervous system, protecting the soft tissue from jolts and blows. Besides filling the ventricles of the brain, the fluid surrounds the brain and spinal cord, so the central nervous system actually floats in the fluid and is effectively lightened by it. Cerebrospinal fluid is actively secreted into the ventricles by the choroid plexuses. Because it is a selective secretion, the composition of cerebrospinal fluid differs from that of the blood plasma. For example, proteins, potassium, and calcium are present in lower concentrations in the cerebrospinal fluid than in the plasma, and sodium and chloride are present in higher concentrations.

There is normally a slight pressure in the ventricles, and the cerebrospinal fluid circulates slowly from the lateral ventricles into the third and then the fourth ventricle. From the fourth ventricle, some of the cerebrospinal fluid flows into the central canal of the spinal cord, but most of its passes through the apertures (one median; two lateral) in the roof of the fourth ventricle and enters the subarachnoid space of the meninges. In the subarachnoid space, the cerebrospinal fluid circulates slowly down the posterior surface of the spinal cord, around the cord, and ascends in the anterior portion of the space to reach the brain (Figure 12.16).

If cerebrospinal fluid were allowed to accumulate, it would exert enough pressure to compress and damage the brain. If the circulation of cerebrospinal fluid is blocked during infancy, before the skull bones have united firmly, the head enlarges as the pressure within the brain increases. This condition is called *hydrocephalus.* Normally, however, cerebrospinal fluid is reabsorbed into the blood at the same rate at which it is formed. This reabsorption is accomplished through thin projections of the arachnoid meninx called **arachnoid villi,** which project into the dural venous sinuses of the brain (Figure 12.18). Cerebrospinal fluid therefore is formed from the blood, and after circulating through and around the central nervous system, it returns to the blood.

Substances such as nutrients and the end products of metabolism can move between the neural tissue and the cerebrospinal fluid, and an examination of cerebrospinal fluid provides a means of determining whether infectious organisms are present in the central nervous system. Samples of cerebrospinal fluid are withdrawn for diagnostic purposes by inserting a needle between the third and fourth lumbar vertebrae and into the subarachnoid space—a procedure called a **lumbar puncture** (Figure 12.19). Because the needle is inserted that far down the spinal column, there is little danger of damaging the spinal cord, which ends at the level of the first or second lumbar vertebra. Spinal anesthesia *(spinal block)* is sometimes administered in a similar manner. To identify damaged intervertebral discs or the presence of structures such as tumors, which may cause pressure on the spinal cord, contrast medium that appears opaque on an X ray is sometimes injected into the subarachnoid space. The X ray taken after the injection of the contrast medium is referred to as a *lumbar myelogram* (Figure 12.19c). Any obstruction of the contrast medium may be due to constriction of the subarachnoid space by a ruptured disc or by a tumor.

The Spinal Cord

Below the level of the medulla, the central nervous system continues as the **spinal cord.** The spinal cord performs two main functions: (1) it *conducts* nerve impulses to and from the brain, and (2) it *processes* sensory information in a limited manner, making it possible for the cord to initiate stereotyped reflex actions (called *spinal reflexes*) without input from higher centers in the brain.

General Structure of the Spinal Cord

The spinal cord passes through the vertebral canal of the vertebrae. It extends from the foramen magnum of

◆ **FIGURE 12.19 Lumbar puncture technique**
(a) Body position during the insertion of the needle. (b) Position of the needle within the subarachnoid space below the termination of the spinal cord. (c) Lumbar myelogram. The contrast medium that was injected into the subarachnoid space appears yellow in this colorized X ray. The region where the medium is constricted indicates a herniated intervertebral disc.

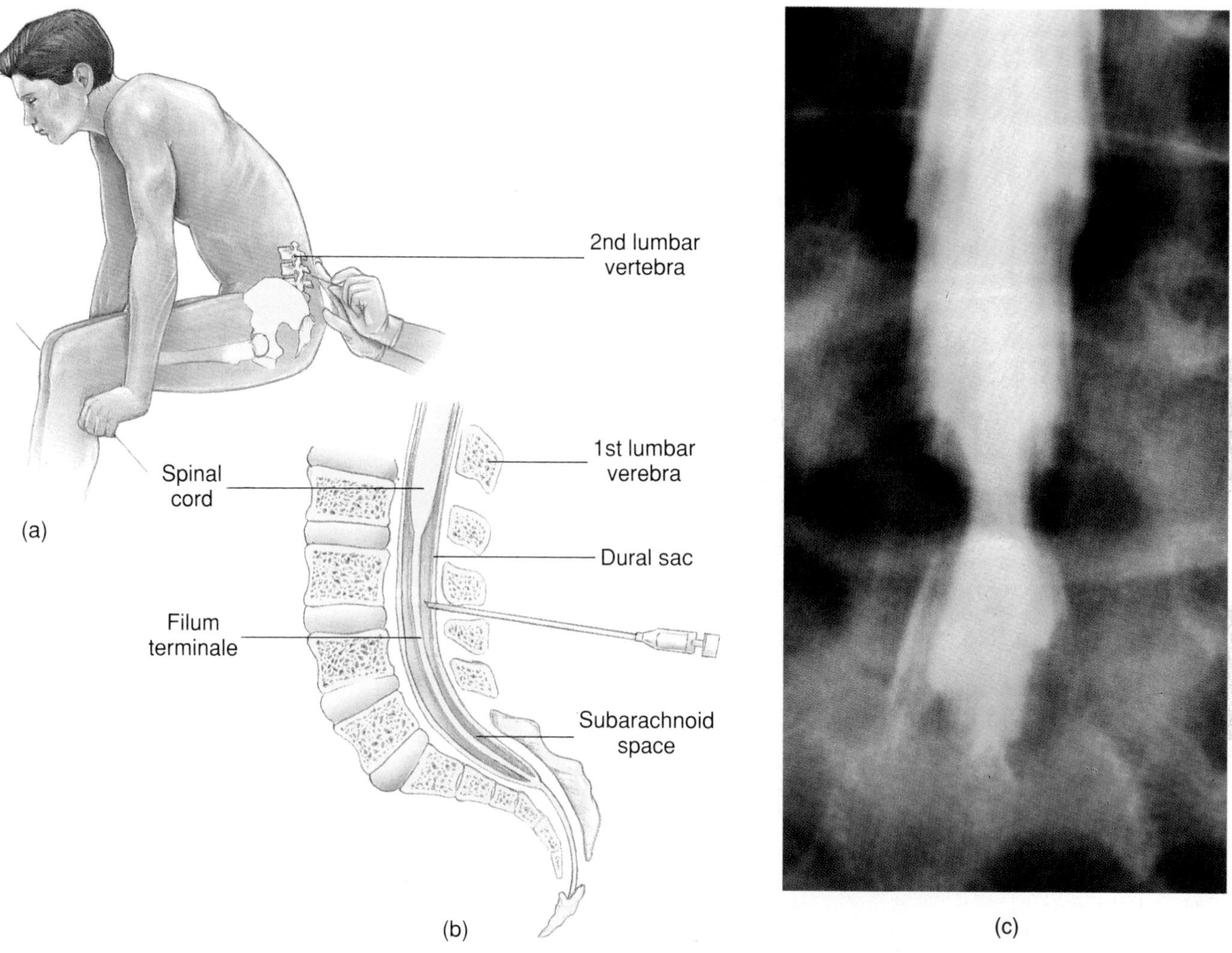

the skull to the level of the first or second lumbar vertebra (Figure 12.20a). Until the third month of fetal development, the spinal cord is as long as the vertebral column. As the embryo continues to develop, however, the vertebral column grows at a faster rate than the spinal cord. As a result, the spinal cord does not extend the entire length of the vertebral column in the adult. A thin fibrous filament of the spinal meninges called the **filum terminale** *(fī´-lum ter˝-mĭ-nal´-ē)* extends from the tip of the spinal cord **(conus medullaris)** to the coccyx.

Thirty-one pairs of spinal nerves arise from the spinal cord and pass through the intervertebral foramina between adjacent vertebrae of the vertebral column. The spinal nerves that leave the vertebral column between the cervical vertebrae are called *cervical* nerves; those that leave the vertebral column between the thoracic vertebrae are called *thoracic* nerves. Similarly, *lumbar, sacral,* and *coccygeal* nerves leave the vertebral column between the lumbar, sacral, and coccygeal vertebrae, respectively. Each portion of the cord that gives rise to a pair of spinal nerves is called a *spinal segment.* Because the vertebral column grows at a faster rate than the spinal cord, the spinal nerves are pulled downward as the column lengthens. As a result, the roots of the lower spinal nerves pass some distance inferiorly before reaching the appropriate intervertebral foramina (Figure 12.20b). At the end of the spinal cord, the mass of descending lumbar and sacral nerve roots has the appearance of a horse's tail and is therefore called the **cauda equina** *(kaw´-dah ē-kwī´-nuh).*

The spinal cord displays two prominent enlargements (Figure 12.20b). The *cervical enlargement* is located in the portion of the cord that gives rise to the spinal

◆ **FIGURE 12.20 Spinal cord and spinal nerves**
(a) The spinal cord and the proximal portions of the spinal nerves in their normal positions, viewed with the neural arches of the vertebrae removed and the dura mater that surrounds the cord cut open. (b) Schematic representation of the spinal cord and spinal nerves to illustrate the cauda equina and the cervical and lumbar enlargements. The letters indicate specific spinal nerves: C = cervical; T = thoracic; L = lumbar; S = sacral; COC = coccygeal.

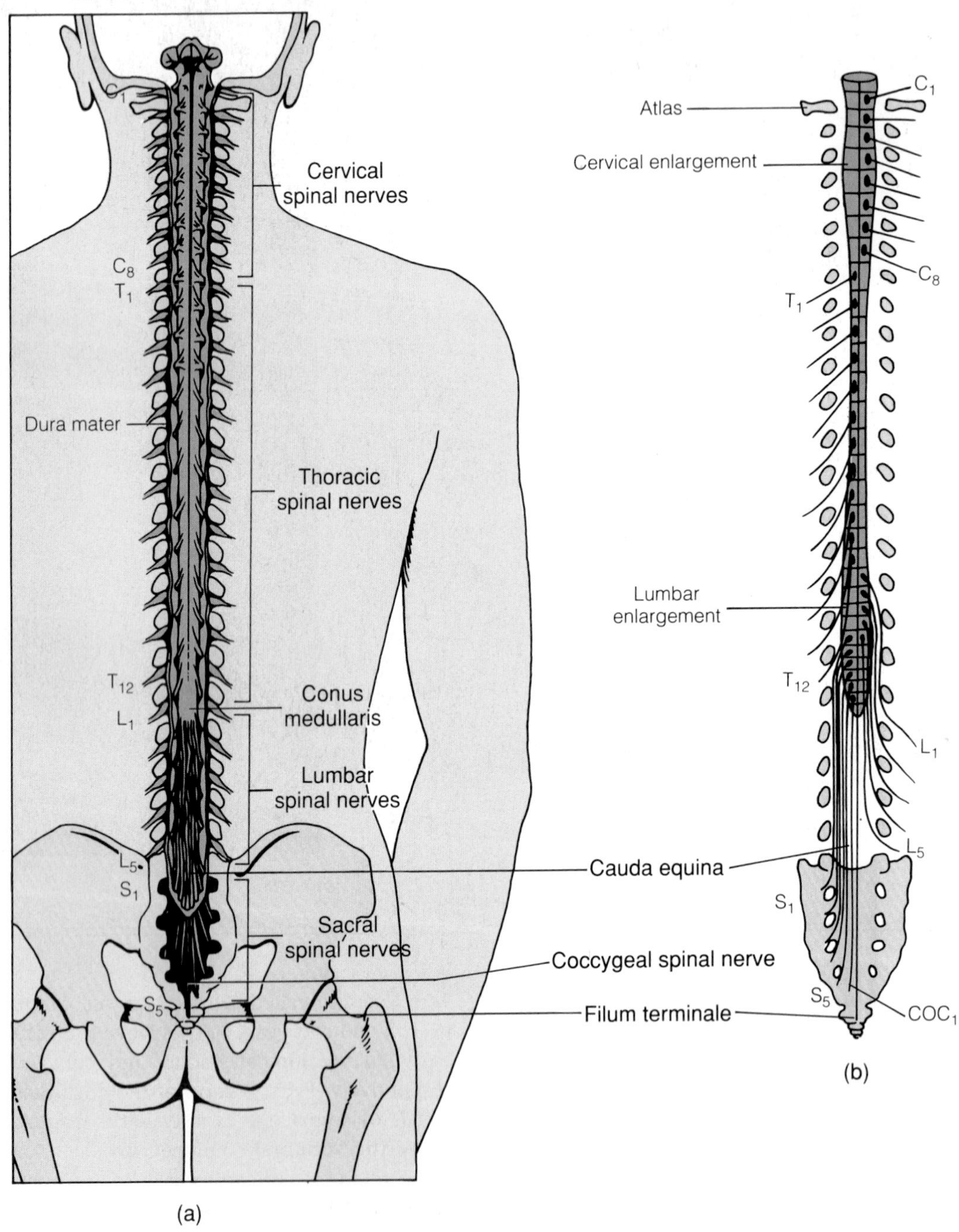

nerves that supply the upper limbs. These nerves form the brachial plexus (see Chapter 13). The *lumbar enlargement* is in the region of the cord that gives rise to the nerves that innervate the lower limbs. These nerves form the lumbosacral plexus.

In Chapter 10 we discussed how the spinal cord and brain develop embryonically from a neural groove into a neural tube. With further development, the lateral walls of the neural tube (and brain stem) show greater development than the roof or floor of the tube (Figure

◆ **FIGURE 12.21 Development of alar and basal plates within the spinal cord**
(a) Neural fold stage. (b) Neural tube stage. (c) Separation of alar and basal plates by the sulcus limitans. (d) Fully developed spinal cord with a pair of spinal nerves.

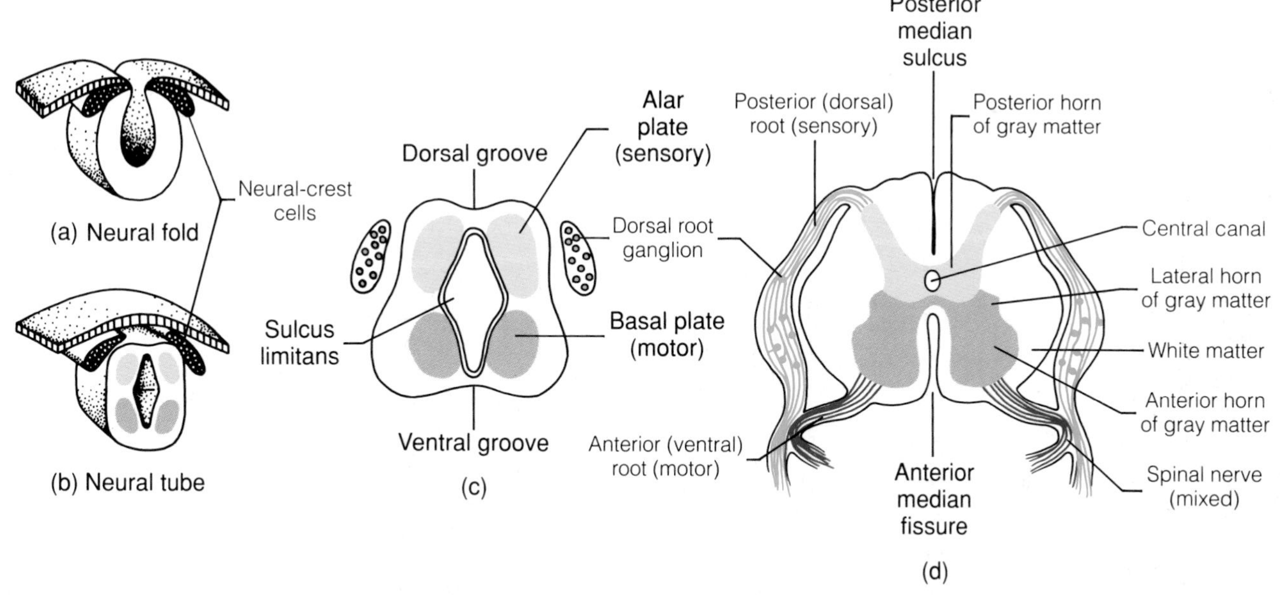

12.21). These lateral thickenings become separated into dorsal and ventral portions by a groove **(sulcus limitans)** along each wall of the central canal. The two dorsal thickenings are called **alar plates.** The neurons in the alar plates are *interneurons* that receive *sensory* and *coordinative* information from afferent neurons. The two ventral thickenings of the developing neural tube are called **basal plates.** The neurons in the basal plates develop into *motor* neurons. As the lateral regions of the neural tube develop, grooves form along the roof and floor of the tube. A fairly deep longitudinal fissure called the **anterior median fissure** forms on the ventral surface of the spinal cord (Figure 12.22). A shallow longitudinal groove called the **posterior median sulcus** is located on the posterior surface of the spinal cord.

Meninges of the Spinal Cord

The spinal cord is covered with the same three meninges that cover the brain: the dura mater, the arachnoid, and the pia mater (Figure 12.22).

The spinal **dura mater** is continuous at the foramen magnum with the inner layer of the dura mater of the brain. Unlike the situation in the skull, the spinal dura does not fuse to the bone of the surrounding vertebrae. Rather, there is a small fat-filled epidural space between the spinal dura and the vertebral column that serves to cushion the spinal cord. The spinal dura extends beyond the lower end of the cord, enclosing the cauda equina (Figure 12.20a). In the sacral region, it forms a covering around the filum terminale. The dura extends laterally to blend with the connective tissue that covers each spinal nerve.

As is the case in the coverings of the brain, the spinal **arachnoid** forms a close lining of the dura mater. The subarachnoid space, which contains cerebrospinal fluid, is largest in the region of the cauda equina.

The innermost spinal meninx is the **pia mater.** It adheres closely to the surface of the cord and the roots of the spinal nerves. The pia mater contains a rich network of blood vessels.

The spinal cord is held in a somewhat fixed position within the meninges by fibrous bridges that cross the subarachnoid space, joining the pia mater with the arachnoid and dura mater. The heaviest of the bridges are the **denticulate ligaments** (Figure 12.23), located along the lateral margins of the spinal cord.

Composition of the Spinal Cord

The spinal cord, like the brain, consists of areas of white matter and gray matter (Figure 12.24). As in the brain, the white matter is composed primarily of the myelinated processes of neurons, and the gray areas are composed primarily of nerve cell bodies and unmyelinated interneuronal nerve fibers. Neuroglia are present in both the white and gray matter. As we have seen, the gray

◆ **FIGURE 12.22 General internal structure of the spinal cord and the meninges surrounding it**

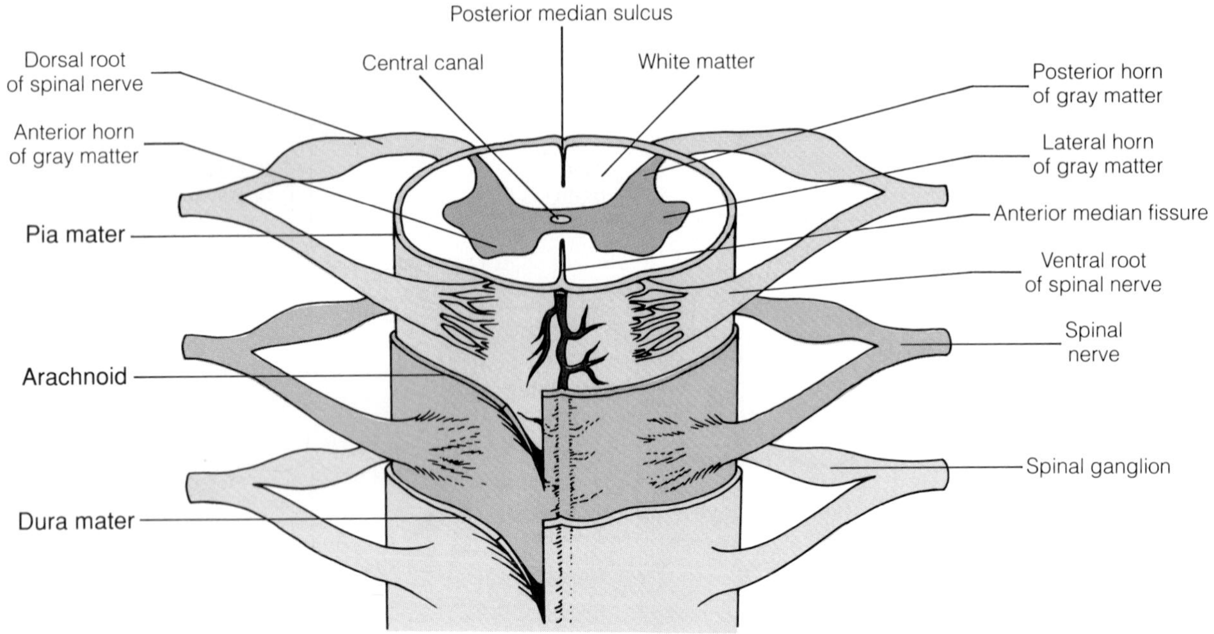

matter in the cerebrum and cerebellum of the brain forms the surface layer (cortex). By contrast, the gray matter of the spinal cord is centrally located and is surrounded by the white matter (Figures 12.22, 12.23, and 12.24).

Gray Matter of the Spinal Cord

The gray matter of the spinal cord is roughly in the form of the letter H. The transverse bar of gray matter that connects the two lateral gray areas is the **gray commissure** (Figure 12.23). Within the gray commissure is the narrow fluid-filled **central canal,** which is continuous with the fourth ventricle. The vertical bars of the gray H, on either side of the gray commissure, are separated into a pair of **posterior (dorsal) horns** and a pair of **anterior (ventral) horns.** In the thoracic and upper lumbar regions, the spinal cord also has a pair of **lateral horns** of gray matter located between the other two gray horns.

The posterior horns of gray matter develop from the alar plates (sensory) of the embryonic neural tube and are composed of axons of sensory neurons from the spinal nerves and interneurons that transmit sensory information within the central nervous system. Some of the axons of the sensory neurons enter the white matter of the posterior area of the spinal cord. These axons then travel to higher levels of the cord or to the brain. Other sensory axons enter the gray substance and synapse either directly with neurons in the anterior horns (thus forming a spinal reflex arc) or with interneurons. The interneurons, in turn, may synapse with anterior horn motor neurons at the same level, pass to higher or lower levels within the spinal cord, or travel all the way up to various regions in the brain.

The anterior and lateral horns of gray matter develop from the basal plates (motor) of the embryonic neural tube. The anterior horns contain the cell bodies of somatic motor neurons whose axons leave the cord and enter a spinal nerve. The cell bodies of visceral motor neurons are found in the lateral horns (see Chapter 14).

Dorsal and Ventral Roots of Spinal Nerves

Groups of nerve fibers called **dorsal roots** enter the spinal cord where the tips of the posterior horns of gray matter come close to the surface of the cord. Similarly, groups of nerve fibers called **ventral roots** leave the spinal cord where the tips of the anterior horns of gray matter come close to the surface of the cord. The dorsal root and ventral root on each side of each spinal segment unite to form a **spinal nerve** (Figures 12.22 and 12.23).

The dorsal roots contain only axons of sensory neurons (somatic and visceral) that pass from the spinal nerve into the posterior horn of gray matter in the spinal cord. The cell bodies of these sensory neurons are located outside the spinal cord within enlargements of the dorsal roots. These enlargements, which lie in the intervertebral foramina, are called **dorsal root ganglia,** or **spinal ganglia** (Figure 12.23).

◆ **FIGURE 12.23 Cross section showing the relationships of the spinal cord and spinal nerve roots to a vertebra**

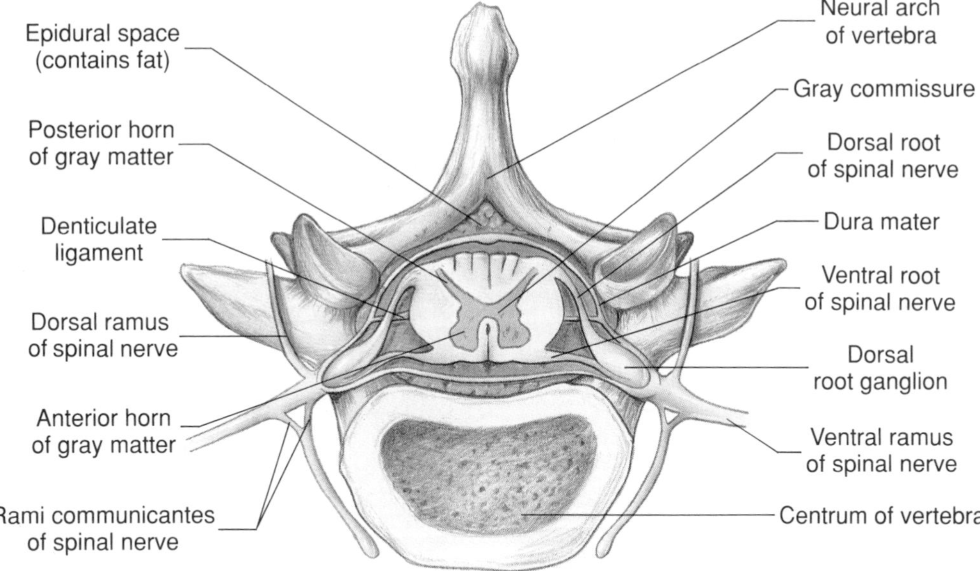

The ventral roots are formed by the axons of neurons located in the anterior and lateral horns of gray matter. The ventral roots therefore contribute processes of both somatic motor and visceral motor (autonomic) neurons to the spinal nerves.

White Matter of the Spinal Cord

The white matter of the spinal cord completely surrounds the gray matter. It is composed primarily of myelinated axons. These axons travel in three directions: (1) up the spinal cord to higher levels in the cord or brain; (2) down the spinal cord from the brain or higher levels of the cord; or (3) across the cord, transmitting impulses from one side to the other.

In each half of the spinal cord, the white matter is divided by the gray matter into three areas: the **posterior funiculus** (*fu-nik´-u-las;* column), the **lateral funiculus,** and the **anterior funiculus.** Within the funiculi are smaller bundles of nerve fibers called **tracts,** or **fasciculi** (*fah-sik´-yoo-lye;* bundles) (Figure 12.25). The tracts are composed of the processes of neurons that carry similar types of impulses to a specific destination. Some tracts are *ascending (sensory) tracts,* carrying impulses to the brain. (The impulses reach the spinal cord by way of afferent neurons of a spinal nerve.) Other tracts are *descending (motor) tracts,* carrying impulses from the brain down to the motor neurons in the anterior or lateral gray horns of the spinal cord. The tracts are not visibly discernible, but their locations have been determined by experimental methods. Most of the spinal tracts have descriptive names that indicate where they begin and where they terminate (Figure 12.25).

Ascending (Sensory) Spinal Tracts. The ascending tracts of the spinal cord carry afferent (sensory) impulses from peripheral sensory receptors to various centers in the brain. These tracts generally contain three successive neurons called first-order, second-order, and

◆ **FIGURE 12.24 Photograph of cross section of a spinal cord**

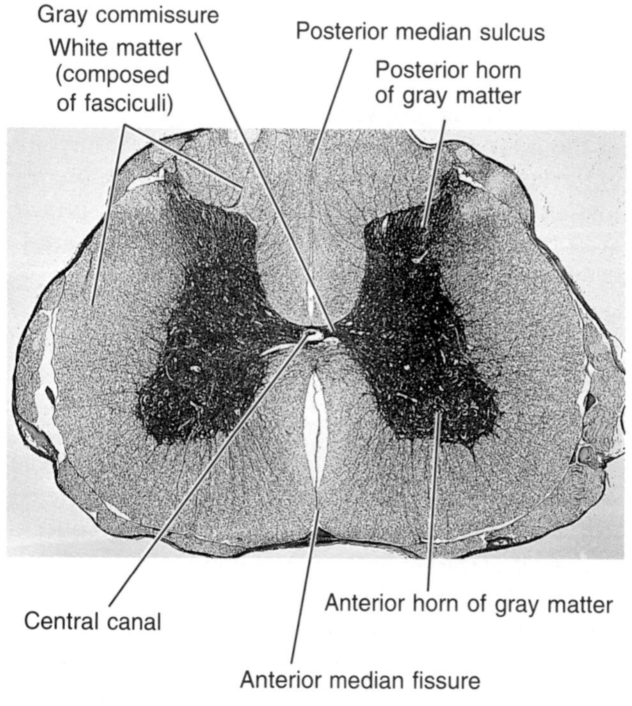

◆ **FIGURE 12.25 Main fasciculi of the spinal cord**

The ascending (sensory) tracts are shown in blue and are labeled only on the left side. The descending (motor) tracts are shown in orange and are labeled only on the right side.

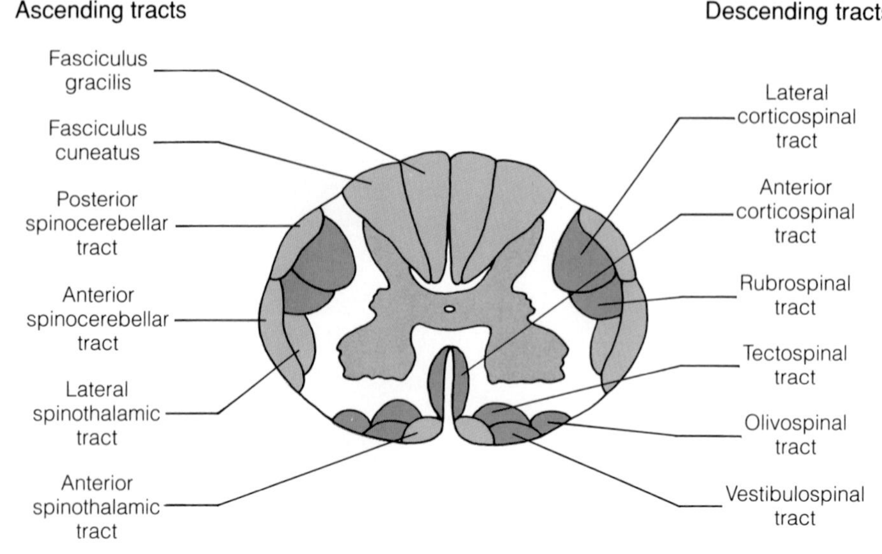

third-order neurons. A **first-order neuron** is a sensory neuron that has the peripheral portion of its nerve fiber in the spinal nerve and its cell body in a dorsal root ganglion. From the dorsal root ganglion, the central portion of the nerve fiber enters the spinal cord, where it synapses with a second-order neuron. A **second-order neuron** is a neuron whose cell body is located in the spinal cord or the medulla. A second-order neuron connects a first-order neuron to a **third-order neuron,** whose cell body is located in the thalamus. The nerve fiber of a third-order neuron extends to the cerebral cortex. All of the ascending spinal tracts cross to the other side of the central nervous system—either at the level of entry into the spinal cord or a few segments above the entry level or within the medulla. As a result, sensory information received by receptors on the right side of the body is interpreted in the primary sensory area of the left cerebral cortex, and sensory information from the left side of the body is interpreted in the primary sensory area of the right cerebral cortex. The major ascending (sensory) spinal tracts are the:

1. Fasciculus gracilis
2. Fasciculus cuneatus
3. Spinothalamic tracts
4. Spinocerebellar tracts

The **fasciculus gracilis** *(fah-sik´-u-lus)* and **fasciculus cuneatus** are two tracts in the posterior funiculus that carry similar types of sensory information concerning *proprioception* from muscles and joints and *fine-touch localization* from different parts of the body (Figure 12.26). Information carried by these tracts makes it possible to know the position of a body part without having to see it. This information also enables a person to locate where an object is touching the body and helps identify the object by shape, texture, weight, and so on. The fasciculi gracilis and cuneatus receive impulses generated by proprioceptors in the muscles and joints and mechanoreceptors in the skin. The fasciculus cuneatus, which is located lateral to the fasciculus gracilis, transmits these impulses from the upper limbs, the trunk, and the neck. The fasciculus gracilis transmits impulses that arise from receptors in the lower limbs and lower trunk.

Nerve fibers of the fasciculus gracilis synapse in a center in the medulla called the **nucleus gracilis.** Nerve fibers of the fasciculus cuneatus synapse in a medullary center called the **nucleus cuneatus.** The processes of the second-order neurons that leave the nucleus gracilis or nucleus cuneatus cross to the other side of the medulla and give rise to a tract called the *medial lemniscus.* The nerve fibers in the lemniscus synapse in the thalamus with third-order neurons that ascend to the cortex of the postcentral gyrus.

In the spinothalamic tracts, most of the second-order neurons cross within the cord, after synapsing with a sensory neuron from a spinal nerve, and ascend in the white matter of the opposite side as the lateral and anterior spinothalamic tracts. The nerve fibers of the spinothalamic tracts synapse in the thalamus. From the thalamus, third-order neurons ascend to the postcentral gyrus of the cerebral cortex. The **lateral spinothalamic tracts** (Figure 12.27) convey impulses concerned with *pain* and *temperature*. The **anterior spinothalamic tracts** carry impulses from receptors sensitive to *touch* and *pressure.*

There are four spinocerebellar tracts: a pair of **posterior** and a pair of **anterior spinocerebellar tracts.**

◆ **FIGURE 12.26 Sensory pathways for touch, pressure, and proprioception (conscious and unconscious) within the fasciculi gracilis, the fasciculi cuneatus, and the spinocerebellar tracts**

◆ **FIGURE 12.27 Sensory pathways for pain and temperature within the lateral spinothalamic tracts**

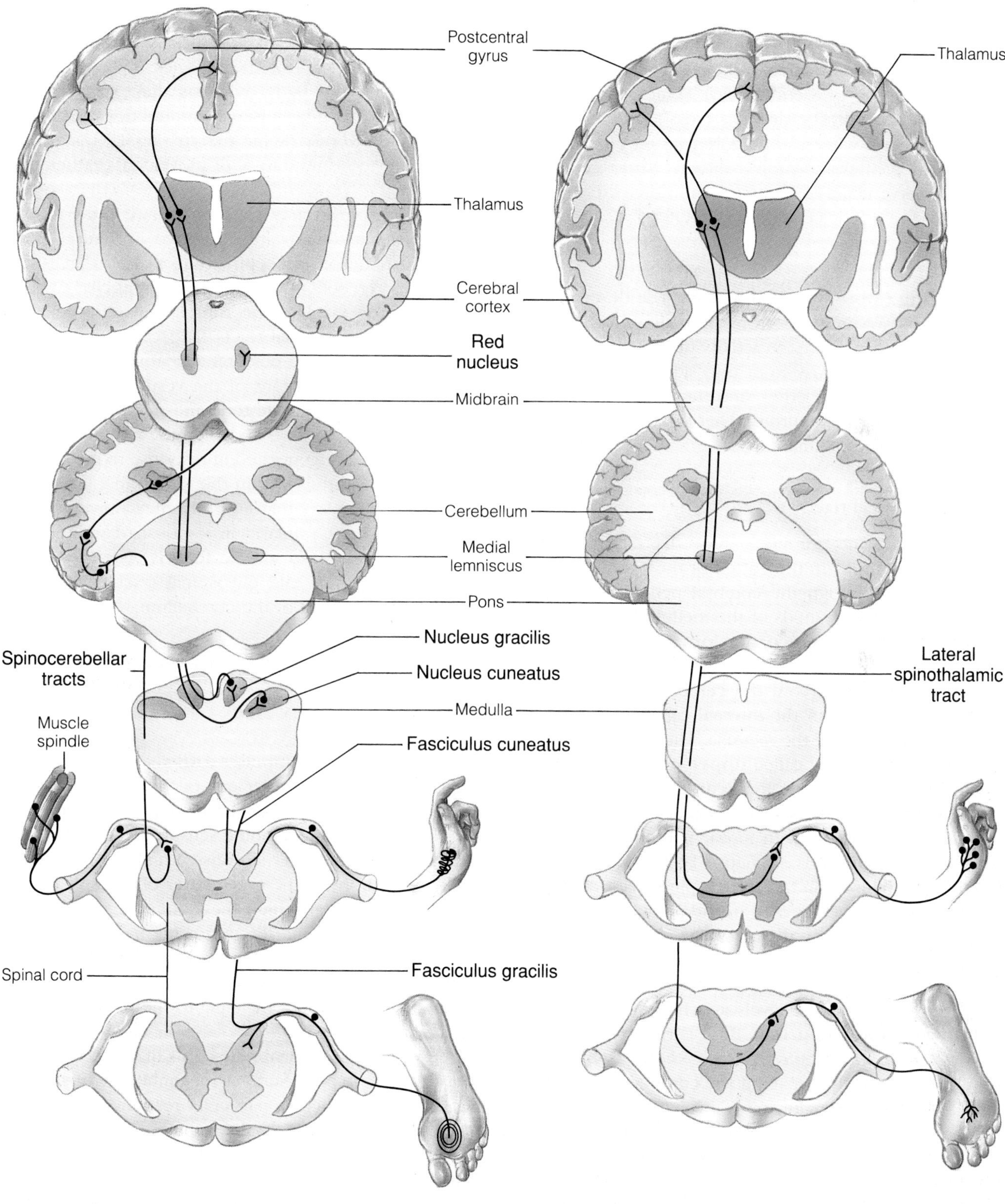

The spinocerebellar tracts carry information concerning unconscious *proprioception* from neuromuscular receptors. As the name of the tracts indicates, their nerve fibers end in the cerebellum. The nerve fibers of these tracts do not synapse with higher-order neurons that pass to the cerebral cortex, or even to the thalamus (Figure 12.28). Nerve fibers of the spinocerebellar tracts do, however, synapse with neurons in the cerebellum that can regulate contraction of skeletal muscles (by way of the red nucleus of the midbrain). Some of the nerve fibers in the spinocerebellar tracts cross to the opposite side of the cord, whereas others remain on the side of the cord they enter and project directly to the cerebellum. The nerve fibers of the tracts enter the cerebellum by way of the inferior cerebellar peduncles.

Descending (Motor) Spinal Tracts. The descending spinal tracts carry impulses from the brain to lower motor neurons that regulate the activity of skeletal muscles. All of these tracts cross from one side of the central nervous system to the other, and they all contain two or three consecutive neurons. There are two types of descending (motor) spinal tracts:

1. Pyramidal tracts
2. Extrapyramidal tracts

Pyramidal tracts are motor tracts that originate primarily from neurons in the cortex of the precentral gyrus and travel through the cerebral peduncles of the midbrain and the pyramids of the medulla. The pyramidal tracts are also called **corticospinal tracts,** a name that indicates their origin (cerebral cortex) and termination (spinal cord). From the cerebral cortex, these tracts descend through the internal capsule, midbrain, pons, and pyramids of the medulla. The nerve fibers of these tracts (which are called **upper motor neurons**) synapse primarily with motor neurons in the anterior horns of the gray matter of the spinal cord, although some of them synapse with motor neurons associated with certain cranial nerves (Figure 12.29). Most of the upper motor neurons in the corticospinal tracts cross to the opposite side of the cord in the medulla and form the **lateral corticospinal tracts.** The remaining upper motor neurons descend uncrossed as the **anterior corticospinal tracts.** Some of the nerve fibers in the anterior tracts cross at the levels at which they synapse with **lower motor neurons.** Both corticospinal tracts carry *motor stimuli* to *skeletal muscles.* The lateral corticospinal tracts, which extend the entire length of the spinal cord, are the major tracts involved in the voluntary control of skeletal muscles. The anterior corticospinal tracts generally do not extend below the thoracic level of the cord.

The remaining motor tracts, which originate from various regions of the cerebral cortex and subcortical areas, are referred to as **extrapyramidal tracts.** These tracts descend from various nuclei in the brain stem and influence muscular actions, coordination, balance, visual and auditory stimuli, and other functions. They include the **rubrospinal tracts, vestibulospinal tracts, tectospinal tracts,** and **olivospinal tracts** (Figure 12.25). Because the actions of these tracts and the corticospinal tracts seem to overlap considerably, clearly separating the effects of each tract is not possible. In general, the pyramidal tracts control the muscles involved in fine movements of the body, whereas the extrapyramidal tracts tend to modify muscular contractions related to *posture* and *balance.* Some extrapyramidal tracts have been found to be inhibitory of movements rather than excitatory.

The rubrospinal tracts, which originate in the red nucleus of the midbrain, are illustrated in Figure 12.28. Notice that the proprioceptive information that is carried to the cerebellum over the spinocerebellar tracts is ultimately conveyed to the rubrospinal tracts. In this manner, the rubrospinal tracts help coordinate the reflexes concerned with postural adjustment by affecting muscle tone. The neurons of the vestibulospinal tracts also affect skeletal muscle tone and thus help maintain posture and equilibrium. However, the impulses carried by these tracts are generated in response to movements of the head. Neurons within the tectospinal tracts conduct motor impulses that control reflex movements of the head and eyes, primarily in response to auditory and visual stimuli. Skeletal muscle tone is also affected by the neurons within the olivospinal tracts, which are largely under control of the cerebellum.

The Spinal Reflex Arc

Not all of the afferent impulses carried to the spinal cord over sensory neurons enter one of the ascending tracts of the spinal cord and thus reach a higher center of the central nervous system. Some sensory neurons synapse directly or through interneurons with motor neurons in the anterior horn of the gray matter of the spinal cord at the same level at which they enter the cord. Other neurons extend only a few segments up or down the cord before synapsing with a motor neuron. The neural pathways by which sensory impulses from receptors reach effectors without traveling to the brain are called **spinal reflex arcs.**

The presence of spinal reflex arcs makes possible automatic, stereotyped reactions to stimuli in which a particular stimulus elicits a particular response. Such reactions are called **reflexes.** Since spinal reflexes occur at the level of the spinal cord without involving the pyramidal tracts from the cortex of the brain, they are involuntary responses, even though they often involve skeletal muscles. At the same time that a reflex is occurring, however, information about the stimulus that initiated

◆ **FIGURE 12.28 Sensory and motor pathways for unconscious muscle movement within the posterior spinocerebellar and rubrospinal tracts**

◆ **FIGURE 12.29 Pathways of the pyramidal tracts (lateral and anterior corticospinal tracts) carrying motor impulses to anterior horn cells that innervate skeletal muscles**

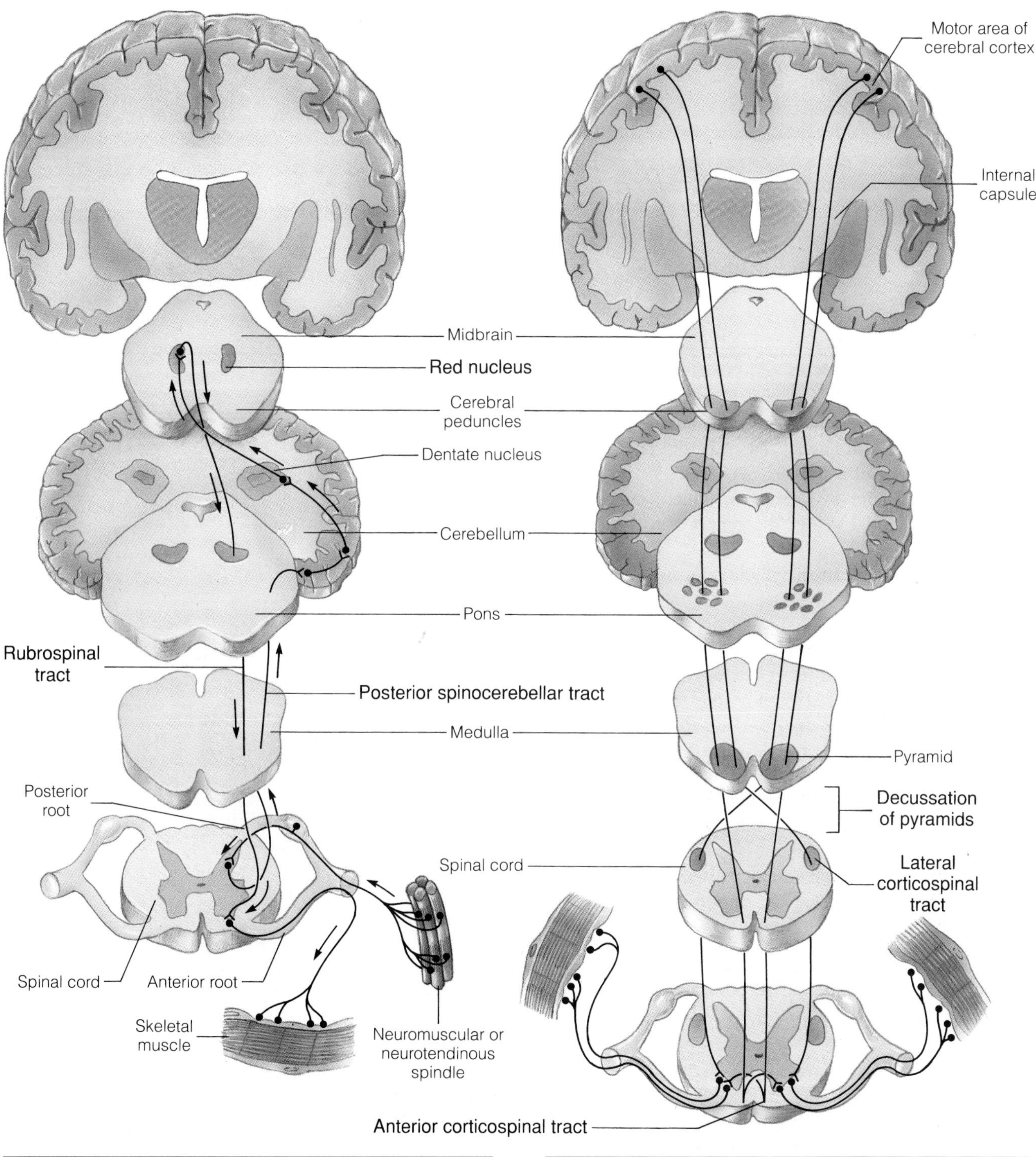

the reflex may also be transmitted to the brain, where a conscious sensation is elicited. When you touch something hot with your hand, for example, you become aware of it through impulses transmitted to the brain by way of the lateral spinothalamic tracts. By the time you feel the burning sensation, however, you have already withdrawn your hand from the hot object as a result of reflex action.

A spinal reflex arc has at least five components (Figure 12.30):

◆ **FIGURE 12.30 Components of a spinal reflex arc**

The left side of the diagram illustrates a monosynaptic reflex; the right side shows a polysynaptic reflex.

Monosynaptic reflex
Polysynaptic reflex
Synapses
Interneuron
Sensory receptor in skin
Afferent neuron
Sensory receptor (muscle spindle)
Afferent neuron
Efferent neuron
Efferent neuron
Effector organ (skeletal muscle)
Effector organ (skeletal muscle)

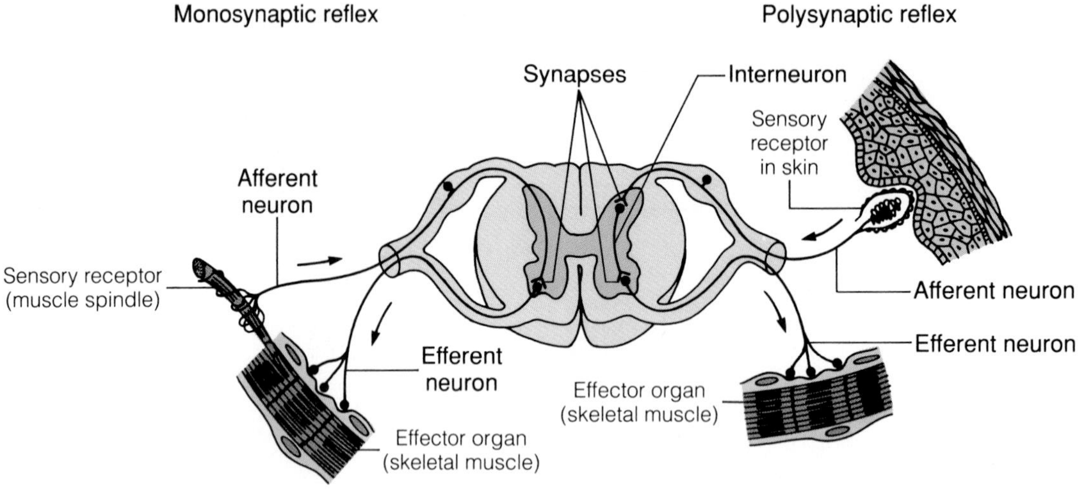

1. A **receptor,** which can be a peripheral ending of a sensory neuron or a specialized cell associated with a peripheral ending of a sensory neuron.
2. A **sensory (afferent) neuron,** which transmits afferent nerve impulses from the receptor to the spinal cord.
3. An **integrating center,** which is located within the spinal cord and basically consists of synaptic interconnections between neurons. If a sensory neuron synapses directly with a motor neuron, the reflex arc formed is known as a **monosynaptic reflex arc.** If there are one or more interneurons between a sensory and a motor neuron, requiring more than one synapse, the arc is a **polysynaptic reflex arc.** Most CNS reflexes are polysynaptic.
4. A **motor (efferent) neuron,** which transmits efferent nerve impulses from the integrating center to an effector.
5. An **effector,** which responds to the efferent nerve impulses. Muscle cells (skeletal, smooth, or cardiac) and secretory cells serve as effectors.

Two important spinal reflexes—the stretch reflex and the tendon reflex—influence the contractions of skeletal muscles.

Stretch Reflex

The muscle **stretch reflex** is initiated at skeletal muscle receptors called **muscle spindles** (Figure 12.31). The muscle spindles provide information about the lengths of skeletal muscles, and they respond to both the rate and the magnitude of a length change.

Within a muscle spindle are three to ten specialized muscle cells called *intrafusal fibers.* The intrafusal fibers

◆ **FIGURE 12.31 A muscle spindle**

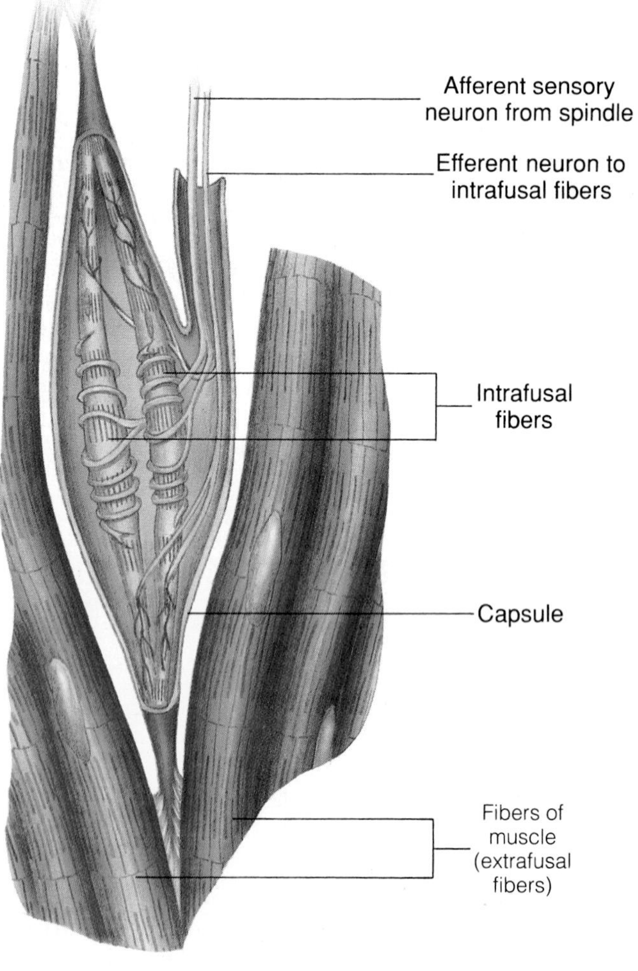

◆ **FIGURE 12.32 Pathways involved in the stretch reflex**

Excitation is indicated by a +; inhibition is indicated by a −.

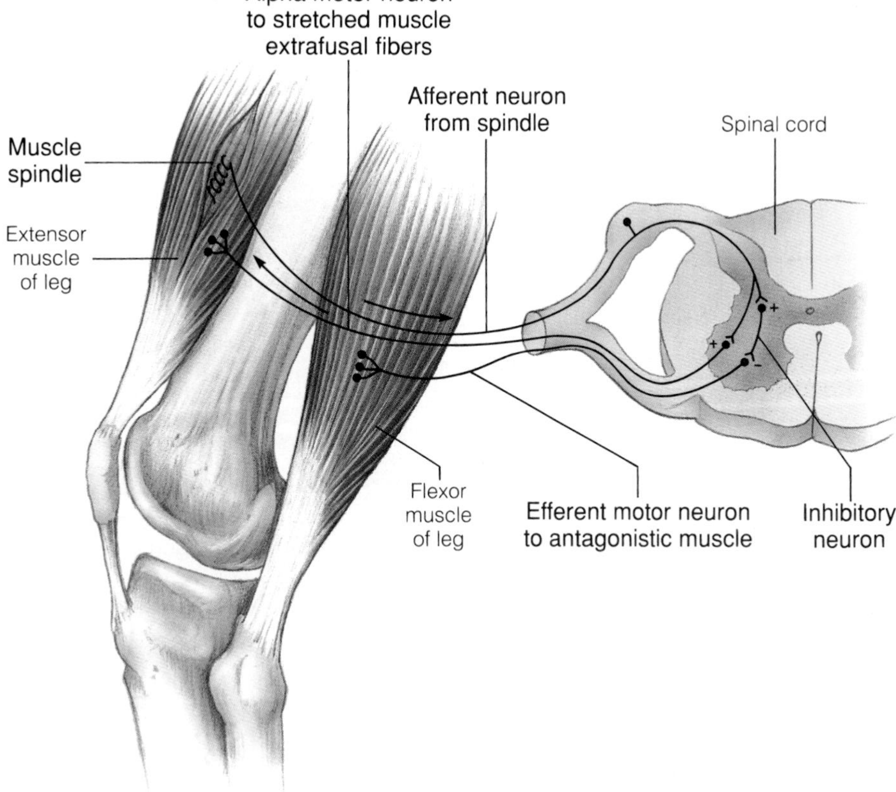

differ from the cells of the muscle as a whole, which are called *extrafusal fibers.* The intrafusal fibers are contractile only at their ends, which are supplied by efferent neurons called *gamma efferent neurons.* The noncontractile, central portions of the intrafusal fibers are supplied by afferent neurons that spiral around the intrafusal fibers. The frequency of nerve impulses in the afferent neurons increases in response to both the sudden and maintained stretch of the central areas of the intrafusal fibers and decreases when the central areas of the fibers are compressed. A muscle spindle is attached in parallel with the extrafusal fibers of a skeletal muscle so that stretching the muscle stretches the muscle spindle (and its intrafusal fibers), and contracting the muscle compresses the muscle spindle.

In the stretch reflex, the stretching of a muscle results in an increased frequency of nerve impulses in the afferent neurons associated with the muscle spindles (Figure 12.32). Within the spinal cord, the afferent neurons synapse with efferent neurons called *alpha motor neurons* that supply the extrafusal fibers of the muscle. As a result, the increased frequency of nerve impulses in the afferent neurons increases the stimulation of the alpha motor neurons, causing the muscle to contract and resist the stretch. The afferent neurons also synapse with inhibitory neurons that, in turn, synapse with efferent neurons controlling muscles whose activities oppose the contraction of the stretched muscle (that is, antagonistic muscles). Thus, when the stretch reflex stimulates the stretched muscle to contract, antagonistic muscles that oppose the contraction are inhibited. This occurrence is called *reciprocal inhibition,* and the neuronal mechanism that causes it is called *reciprocal innervation.*

The *patellar reflex* (knee jerk) is a common example of a stretch reflex. Striking the patellar tendon at the knee pulls on the tendon and thus stretches the quadriceps muscles of the anterior thigh. If all the components of the reflex arc are intact, the muscle spindles within the quadriceps muscles initiate a stretch reflex that causes the quadriceps muscles to contract and swing the leg forward.

In addition to their synapses with neurons involved in the stretch reflex, the afferent neurons from muscle spindles also synapse with neurons that relay impulses to the brain. As a result, nerve impulses from muscle spindles provide information about the state of stretch or contraction (that is, the length) of skeletal muscles that is important in maintaining body posture and in coordinating muscular activity.

◆ **FIGURE 12.33 Pathways involved in the tendon reflex**

Excitation is indicated by a +; inhibition is indicated by a −.

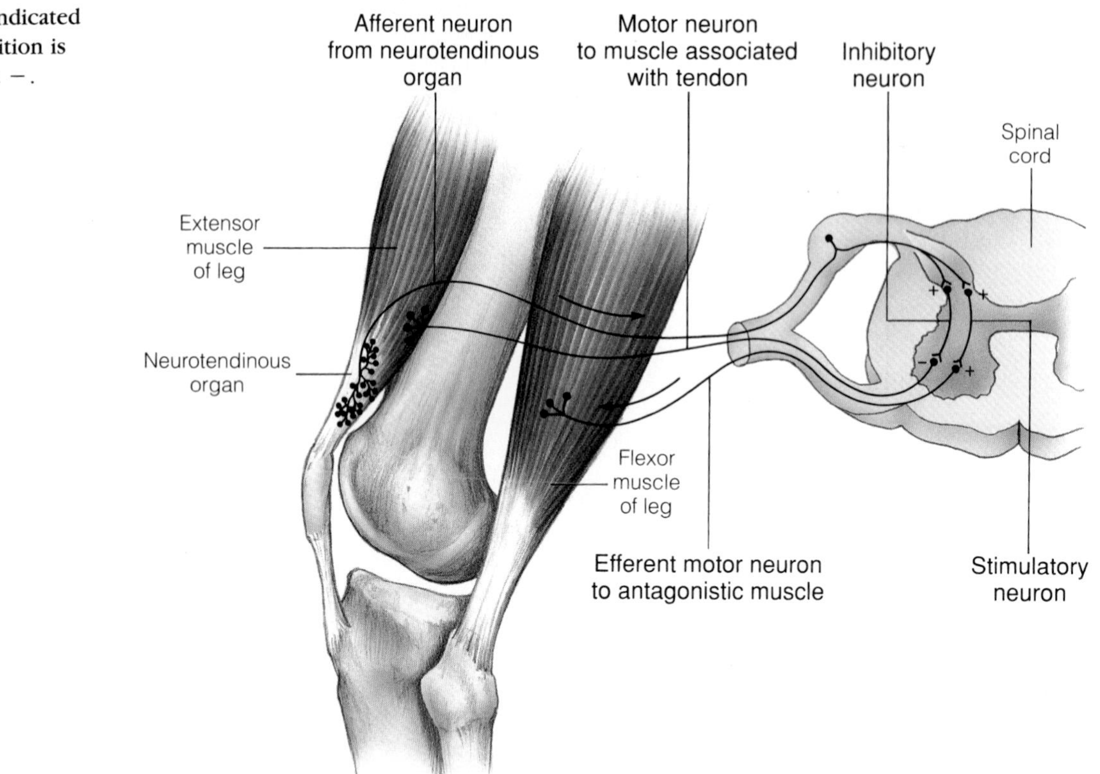

The activity and sensitivity of a muscle spindle is influenced by nerve impulses carried to the intrafusal fibers of the spindle by the gamma efferent neurons. The contraction of the ends of the intrafusal fibers due to gamma efferent stimulation stretches the central, noncontractile portions of the fibers. This stretching increases the activity of the afferent neurons and enhances the response of the spindle to further stretch. Thus, the excitability of the muscle spindle and the amount of muscle stretch necessary to initiate a stretch reflex can be altered by altering the degree of contraction of the intrafusal fibers.

Nerve impulses to skeletal muscles from higher centers often simultaneously stimulate both the extrafusal fibers of a muscle (by way of the alpha motor neurons) and the intrafusal fibers of the spindles (by way of the gamma efferent neurons). This simultaneous stimulation maintains the responsiveness of the spindles when the muscle shortens, and it may allow the basic stretch reflex mechanism to assist in attaining the proper degree of muscular contraction. For example, suppose picking up an object requires a certain degree of contraction by a muscle. If both the extrafusal fibers of the muscle and the intrafusal fibers of the muscle spindles are stimulated, and if the contraction of the muscle is equal to the contraction of the intrafusal fibers of the spindles, the degree of stimulation of the muscle spindles will not change as a result of the contraction of the muscle. However, if the object to be picked up is heavier than expected, the stimulation of the extrafusal fibers may produce a muscle contraction that is less than the contraction of the intrafusal fibers of the spindles. As a result, the central portions of the intrafusal fibers will be stretched, and the activity of the afferent neurons from the spindles will increase. This increased activity further stimulates the alpha motor neurons to the extrafusal fibers and thereby helps achieve the needed degree of muscular contraction (perhaps by activating additional motor units). In addition, nerve impulses from the muscle spindles are relayed to higher brain centers, and these centers may also alter their output to the alpha motor neurons in order to attain the proper degree of muscular contraction.

Tendon Reflex

The **tendon reflex** is initiated at receptors called **neurotendinous organs.** Neurotendinous organs are encapsulated structures located within tendons near the junction of a tendon with a muscle. Unlike muscle spindles, which are sensitive to muscle length, neurotendi-

ASPECTS OF EXERCISE PHYSIOLOGY

Swan Dive or Belly Flop: It's a Matter of CNS Control

Sport skills must be learned. Much of the time, strong basic reflexes must be overridden in order to perform the skill. Learning to dive into water, for example, is very difficult initially. Strong head-righting reflexes controlled by sensory organs in the neck and ears initiate a straightening of the neck and head before the beginning diver enters the water, causing what is commonly known as a "belly flop." In a backward dive, the head-righting reflex causes the beginner to land on his or her back or even in a sitting position.

To perform any motor skill that involves body inversions, somersaults, back flips, or other abnormal postural movements, the person must learn to consciously inhibit basic postural reflexes. This is accomplished by having the person concentrate on specific body positions during the movement. For example, in the somersault, the person must concentrate on keeping the chin tucked and grabbing the knees in order to perform the skill. After the skill is performed repeatedly, new synaptic patterns are formed in the central nervous system (CNS), and the new or conditioned response substitutes for the natural reflex responses. Sport skills must be practiced until the movement becomes automatic; then the athlete is free during competition to think about strategy or the next move to be performed in a routine.

nous organs are sensitive to tension. Within a neurotendinous organ are afferent neuron endings wrapped around small bundles of collagen fibers. When the tension applied to a tendon increases, so does the frequency of nerve impulses in the afferent neurons; when the tension decreases, so does the afferent impulse frequency. Because of their locations within tendons, the neurotendinous organs are in series with the muscles themselves, and they respond to changes in tension that result from either passive stretch or muscular contraction.

In the tendon reflex, an increase in the tension applied to a tendon—most commonly as a result of muscular contraction—increases the frequency of nerve impulses in the afferent neurons associated with the tendon organs (Figure 12.33). Within the spinal cord, the afferent neurons synapse with inhibitory neurons that, in turn, synapse with alpha motor neurons that supply the muscle associated with the tendon. As a result, the increased frequency of nerve impulses in the afferent neurons from the tendon organs ultimately diminishes the stimulation of the alpha motor neurons to the muscle and may even completely inhibit muscular contraction. Thus, the tendon reflex protects the muscle and tendon from excessive tension. The afferent neurons from the neurotendinous organs also synapse with stimulatory neurons that, in turn, synapse with motor neurons controlling antagonistic muscles. As a result, the tendon reflex is generally accompanied by a reciprocal stimulation of the antagonistic muscles.

In addition to synapsing with neurons involved in the tendon reflex, the afferent neurons from the neurotendinous organs also synapse with neurons that relay impulses to the brain. Thus, just as nerve impulses from muscle spindles provide the brain with information about the length of a skeletal muscle, nerve impulses from neurotendinous organs provide information about the tension developed by the muscle.

Neuron Pools

Within the central nervous system, neurons are organized into functional groups called **neuron pools,** and information transmitted from a receptor is often received by a particular neuron pool. Within the pool, this information is processed and integrated with information received from other sources. It is then transmitted from the pool to various destinations. Within a neuron pool, the processing of information involves integrative activities such as divergence, convergence, summation, and facilitation (see Chapter 11).

Neuron pools occur at all levels of the central nervous system, including the cerebral cortex, and there are numerous interconnections among the various pools. Although the neural pathways involved in particular activities are commonly depicted as simple neuron-to-neuron chains, the performance of even the simplest actions generally requires the participation of many neurons whose activities are coordinated in complex circuits in neuron pools. For example, the alpha motor neurons to a skeletal muscle are influenced by nerve impulses arriving from muscle spindles, neurotendinous organs, and higher brain centers. As a result, the degree of activity of the alpha motor neurons at any one moment is determined by a combination of influences.

CONDITIONS OF CLINICAL SIGNIFICANCE

The Central Nervous System

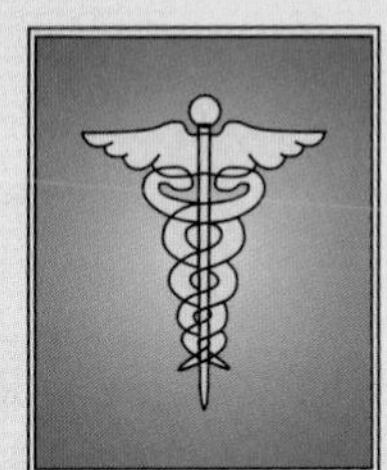

Spinal Cord Dysfunctions

Damage to tissues or organs that results in impairment or loss of function is referred to as a *lesion.* Lesions of the spinal cord, either from injury or from disease, generally involve the motor and sensory fiber tracts of which the cord is composed.

Paralysis

Lesions of the motor spinal tracts cause *paralysis* of the structures supplied by the neurons involved. Paralysis of both of the lower limbs is called *paraplegia.* The condition in which both the upper limb and the lower limb on one side of the body are paralyzed is *hemiplegia. Quadriplegia* is the paralysis of all four limbs. Paralysis may be *flaccid,* when the affected muscles lose their normal slight contraction (tonus) and reflexes are absent; or it may be *spastic,* with increased tonus and reflexes. Spastic paralysis generally results from lesions of descending motor tracts from the brain. These lesions are often in the extrapyramidal system, which eliminates the normal inhibitory effects of the higher centers on the spinal reflexes. Flaccid paralysis generally results when the pyramidal tracts or the lower motor neurons are affected.

Lesions of Sensory Tracts of the Spinal Cord

Generally, lesions of the spinal cord tracts are not so precisely localized that they involve only one tract. Damage to sensory tracts can be located by identifying the combination of sensory losses that are experienced. Damage to the tracts in the posterior funiculi (fasciculi gracilis and cuneatus) results in a loss of proprioceptive ability and touch discrimination. A loss of thermal and pain sensations from a particular region of the body indicates damage to the lateral spinothalamic tracts above the affected region. Damage to the anterior spinothalamic tracts causes a slight decrease in the sense of touch.

Specific Dysfunctions of the Spinal Cord

TABES DORSALIS. *Tabes dorsalis* is a condition caused by the progressive degeneration of the posterior funiculi of the spinal cord and the dorsal roots of the spinal nerves. It results from the invasion of the CNS by the spirochete bacteria of syphilis. Because of the loss of the proprioceptive pathways in the posterior columns and dorsal roots, tabes dorsalis causes a loss of muscular coordination. Consequently, people affected with tabes dorsalis walk unsteadily *(ataxia)* and must watch the ground in order to maintain their balance. There may be severe leg and abdominal pain in the early stages of the disease due to the irritation of the sensory neurons in the dorsal roots. As these dorsal roots degenerate, the pain diminishes along with the sense of touch.

POLIOMYELITIS. The disease *poliomyelitis* is caused by a virus that destroys primarily the motor nerve-cell bodies in the anterior horns of the spinal cord, especially those in the cervical and lumbar enlargements. This degeneration produces fever, severe headache, and a flaccid paralysis of the muscles supplied by these neurons. After several weeks of paralysis, the muscles begin to atrophy, and eventually the muscle tissue may be almost completely replaced by connective tissue and adipose tissue. Death can result from respiratory failure or heart failure if the virus invades the nerve cells in the regulatory centers of the medulla.

CONDITIONS OF CLINICAL SIGNIFICANCE

SYRINGOMYELIA. The condition *syringomyelia* occurs when small fluid-filled sacs (cysts) form in the gray matter of the spinal cord and brain stem, and the ependymal cells that line the central canal proliferate. Numerous commissural pathways are interrupted in syringomyelia, producing various sensory dysfunctions and muscular weakness accompanied by muscle atrophy. For example, the lateral spinothalamic tracts are often destroyed as they cross within the gray commissure of the cord. As a result, impulses concerned with the sensations of pain and temperature may not be transmitted to the brain from a particular region of the body.

MULTIPLE SCLEROSIS. *Multiple sclerosis* is a chronic condition that results in widespread destruction of the myelin sheaths of the nerve fibers in the spinal cord and brain. The destroyed myelin sheaths are replaced by hardened plaques that interfere with the normal transmission of nerve impulses. Multiple sclerosis is believed to be due to an autoimmune response in which the immune system attacks a person's own tissues. In multiple sclerosis, a virus may trigger the autoimmune response.

Multiple sclerosis causes a great variety of symptoms, involving both motor and sensory tracts, depending on the areas of the CNS where plaque formation occurs. Common symptoms include abnormal sensations, spastic paralysis, and exaggerated reflexes. Although multiple sclerosis is a chronic condition, it is not unusual for the symptoms of the disease to regress for years at a time. However, with each recurrence, there is permanent damage to the neurons of the CNS, thus causing progressive incapacitation.

Dysfunction of the Brain Stem

All the sensory and motor tracts of the central nervous system travel through the brain stem (medulla, pons, and mesencephalon). The brain stem also contains various centers that control the autonomic nervous system, and most cranial nerves originate there. Consequently, lesions in this region of the CNS can produce a complex mixture of motor and sensory symptoms. Moreover, a condition such as tumor, hemorrhage, or trauma can affect the reticular activating system in the brain stem, producing a *coma.*

Dysfunctions of the Cerebellum

Lesions of the cerebellum or its pathways can cause dysfunctions in the smoothly coordinated actions that normally occur among groups of muscles. These dysfunctions may be evident as disturbances of gait and posture, as the inability to perform smoothly a task that involves the movement of several joints, or as an inability to estimate the range of movement and thus to stop a movement at a particular point.

Dysfunctions of the Basal Nuclei

The basal nuclei, which are part of the extrapyramidal system, are small islands of gray matter deep within the cerebral hemispheres. The neurons in the basal nuclei regulate skeletal muscle activity, in part by inhibiting contractions. This inhibition is in contrast to the excitatory effects of the neurons in the pyramidal tracts. For this reason, lesions of the basal nuclei generally result in spastic movements that appear to be voluntary but are actually beyond the individual's control. The person's limbs may be flung wildly about, writhe continuously, or move in a rapid, jerky manner.

Parkinsonism

Parkinsonism is the most common dysfunction of the basal nuclei. Since its onset is gradual, the disease primarily affects people over age 50. A number of factors are thought to produce Parkinson's syndrome, all of which cause the degeneration of cells in the basal nuclei. The condition is characterized by useless contractions of skeletal muscles, causing tremor and rigidity of the muscles, as well as by a decrease in muscular movements, such as swinging the arms while walking and changing facial expressions related to emotions.

There is impressive evidence that parkinsonism may be related to the abnormal metabolism of the neurotransmitter *dopamine* by the cells of the basal

continued on next page

CONDITIONS OF CLINICAL SIGNIFICANCE

nuclei. In Parkinson's disease, the level of dopamine in the basal nuclei is found to be substantially reduced. It is not known how the reduced dopamine level brings about the symptoms of Parkinson's disease. Since dopamine itself is unable to cross the blood-brain barrier (see page 351) effectively, the administration of supplementary dopamine is ineffective in controlling parkinsonism. However, *L-dopa*, a precursor (forerunner) of dopamine, will pass from the blood to the brain cells and has proved effective in treating the condition.

Huntington's Chorea

Another dysfunction that results from degeneration of cells in the basal nuclei is *Huntington's chorea.* The symptoms of this progressive hereditary disorder usually do not appear until about age 40. Huntington's chorea is characterized by occasional involuntary flailing movements of the arms and legs and writhing movements of the hands, head, trunk, and feet. As the condition progresses, mental deterioration occurs, often resulting in personality changes. In contrast to parkinsonism, Huntington's disease is treated with drugs that block the action of dopamine.

Inflammatory Diseases of the Central Nervous System

Encephalitis and Myelitis

The invasion of nervous tissue by viruses (and in some instances by bacteria, fungi, and other agents) can product inflammation of the CNS. Inflammation of the brain is called *encephalitis.* Inflammation of the spinal cord is known as *myelitis.* Both conditions display a variety of possible motor and sensory symptoms, including paralysis, coma, and death.

Meningitis

Meningitis is an inflammation of the meninges that cover the brain and spinal cord. It is caused by a number of microorganisms, the most common being the meningococci, streptococci, pneumococci, and tubercle bacilli. The organisms are thought to enter the body through the nose and throat. Meningitis produces a high fever with a severe headache and stiffness of the neck. In the more serious cases, meningitis can cause coma and death.

Tumors of the Central Nervous System

Most tumors of the CNS develop from glial cells, including the oligodendrocytes that form the neurilemma in the CNS, but it is not uncommon for cells of the meninges to form tumors also. Only rarely do the neurons themselves give rise to tumors. The presence of tumors in the brain and spinal cord can produce a variety of dysfunctions, depending on their location. The symptoms of a tumor can include headache, convulsions, a change in behavior patterns, pain, and paralysis. These dysfunctions result from the destruction of nervous tissue by the tumor, increased pressure in the skull, or edema of the nervous tissue. Tumors may be treated by surgical removal or destroyed by chemical and radiation therapy.

Study Outline

◆ **THE BRAIN** pp. 384–404

Forebrain. Consists of telencephalon and diencephalon.

TELENCEPHALON. Cerebrum in adult; contains two lateral ventricles.

Cerebral cortex. Outer surface of gray matter; unmyelinated neurons.

Basal nuclei. Gray-matter structures deep inside cerebral hemispheres.

White matter. Myelinated nerve fiber tracts.

MYELINATED NERVE FIBER TRACTS.

1. *Projection tracts.* Carry motor or sensory impulses from one level of brain or spinal cord to another.
2. *Association tracts.* Connect areas in cortex of same cerebral hemisphere.
3. *Commissural tracts.* Connect left and right hemispheres.

GYRI, FISSURES, SULCI, AND LOBES OF THE CEREBRUM.

Gyri. Convolutions on surface of cerebrum.

Fissures. Deeper furrows between gyri.

Sulci. Shallower furrows between gyri.

Lobes of cerebrum. Division of fissures and sulci into frontal, parietal, temporal, and occipital lobes; insula located deep in lateral fissure.

FUNCTIONAL AREAS OF THE CEREBRAL CORTEX.

Primary Motor (Primary Somatic Motor) Area. In precentral gyrus of frontal lobe; involved in controlling voluntary contractions of skeletal muscle.

Premotor Area. Anterior to primary motor area; causes groups of muscles to contract, producing stereotyped movements.

Primary Sensory (Primary Somatic Sensory) Area. In postcentral gyrus of parietal lobe; receives sensory information concerning temperature, touch, pain, proprioception.

Special Sense Areas.

1. *Primary visual area.* Posterior occipital lobe.
2. *Primary auditory area.* Upper margin of temporal lobe.
3. *Primary olfactory area.* Medial surface of temporal lobe.
4. *Primary taste area.* Parietal lobe, near bottom of postcentral gyrus.

Association Areas. Interconnect motor and sensory areas.

1. *Frontal association area.* Site of intellectual activities.
2. *Somatic association area.* Integration and interpretation center.
3. *Visual and auditory association areas.* Contribute to interpretation of visual and auditory experiences.

BASAL NUCLEI. Involved in control of skeletal muscle activity; part of extrapyramidal system; act in part to inhibit muscle contraction.

OLFACTORY BULBS. On ventral surface of cerebral hemisphere; smell.

HEMISPHERIC SPECIALIZATION. Each cerebral hemisphere is somewhat specialized for carrying out certain kinds of mental process.

DIENCEPHALON.

THALAMUS. Two oval masses of gray matter forming walls of third ventricle; sensory relay and integrating center of brain; some motor involvement.

HYPOTHALAMUS. Located below thalamus; regulates:

1. *Sympathetic and parasympathetic activity.* (For example, body temperature, water balance, appetite, gastrointestinal activity, sexual activity).
2. *Emotions.* (Fear and rage, for example).
3. *Pituitary hormone release.*

Mammillary bodies. Olfactory reflexes.

Tuber cinereum. Contains neurons that transport regulatory hormones (or factors) from hypothalamus to pituitary gland.

Infundibulum. Hormone transport to posterior lobe of pituitary gland.

Optic chiasma. Point of crossing of some optic nerve fibers.

EPITHALAMUS. Forms roof of third ventricle; includes pineal body, which has possible neuroendocrine function.

THE LIMBIC SYSTEM. Includes structures that affect emotional responses.

Midbrain or Mesencephalon. Between forebrain and hindbrain.

CEREBRAL AQUEDUCT. Connects third and fourth ventricles.

CEREBRAL PEDUNCLES. Nerve fiber tracts from primary motor area of cerebral cortex to pons and spinal cord, and sensory nerve fibers from spinal cord to thalamus.

RED NUCLEUS. Coordinates cerebellum and cerebral hemispheres.

CORPORA QUADRIGEMINA. Visual and auditory functions.

Hindbrain. Consists of metencephalon and myelencephalon.

METENCEPHALON. Superior portion of hindbrain; contains fourth ventricle.

CEREBELLUM. Concerned with functions below conscious level; directs precise, smooth movements and maintains equilibrium.

PONS. Consists of nerve fiber tracts that connect cerebellum with brain stem; connects cerebellum with various levels of central nervous system; also affects respiration rate.

MYELENCEPHALON (MEDULLA OBLONGATA). Includes motor nerve tracts called *pyramids* on ventral surface and *medullary centers* that control vital functions such as heart rate, respiration, blood vessel dilation and constriction, coughing, swallowing, and vomiting.

RETICULAR FORMATION. Receives input from several areas of brain; sensory input from spinal tracts; portion that activates cerebral cortex and maintains wakefulness is called *reticular activating system (RAS).*

Ventricles of the Brain. Four interconnected spaces filled with cerebrospinal fluid.

CHOROID PLEXUSES. In each ventricle; form cerebrospinal fluid.

The Meninges. Three layers of connective tissue that cover entire CNS:

DURA MATER.

ARACHNOID.

PIA MATER.

Cerebrospinal Fluid. Cushions CNS: formed from blood via choroid plexuses; circulates in and around CNS; returns to blood via arachnoid villi.

◆ THE SPINAL CORD pp. 404–417

General Structure of the Spinal Cord. Gives rise to 31 pairs of spinal nerves.

CERVICAL ENLARGEMENT. Forms brachial plexus; supplies upper limbs.

LUMBAR ENLARGEMENT. Forms lumbosacral plexus; supplies lower limbs.

Meninges of the Spinal Cord. Same as brain: dura mater, arachnoid, pia mater.

Composition of the Spinal Cord. White myelinated areas surround gray, unmyelinated central area.

GRAY MATTER OF THE SPINAL CORD. H-shaped.

Posterior horns. Axons of sensory neurons from spinal nerves; develop from alar plate.

Anterior horns. Cell bodies of somatic motor neurons whose axons leave cord for a spinal nerve; develop from basal plate.

Lateral horns. Only in thoracic and upper lumbar regions; cell bodies of visceral motor neurons; develop from basal plate.

DORSAL AND VENTRAL ROOTS OF SPINAL NERVES. Dorsal and ventral roots unite on each side of each spinal segment, forming a spinal nerve.

Dorsal roots. Enter spinal cord at tips of posterior gray horns; sensory nerves; contain dorsal root ganglia.

Ventral roots. Leave spinal cord at tips of anterior gray horns; somatic and visceral motor nerves.

WHITE MATTER OF THE SPINAL CORD. Surrounds gray matter.

Funiculi. Posterior, lateral, and anterior regions of white matter in each half of spinal cord.

Tracts. Small bundles of neuron processes within funiculi:

1. *Ascending (sensory) spinal tracts.* Carry afferent sensory impulses from peripheral sensory receptors to brain centers; all cross over in CNS.
   - **a.** *Fasciculus gracilis* and *fasciculus cuneatus.* Awareness of joint sense, proprioceptive information from muscles and joints, and fine-touch localization.
   - **b.** *Spinothalamic tracts.* Pain, temperature, touch, pressure.
   - **c.** *Spinocerebellar tracts.* Unconscious proprioception.
2. *Descending (motor) spinal tracts.* Carry impulses from brain to lower motor neurons that regulate skeletal muscle.
   - **a.** *Pyramidal tracts (corticospinal tracts).* Originate from cells in cortex of precentral gyrus and travel through medulla pyramids; voluntary control of skeletal muscles, especially fine movements.
   - **b.** *Extrapyramidal tracts.* Originate from various nuclei in brain stem; modify muscular contractions for posture and balance.

The Spinal Reflex Arc. Neural pathway by which sensory impulses from receptors reach effectors without traveling to brain.

RECEPTOR → SENSORY NEURON → INTEGRATING CENTER → MOTOR NEURON → EFFECTOR.

STRETCH REFLEX. Initiated by muscle spindles that respond to stretch; stretch of spindle increases activity of afferent neurons from spindle, which increases stimulation of alpha motor neurons to muscle (for example, patellar reflex). Muscle spindles provide information about lengths of skeletal muscles that is important in coordination of muscular activity. Activity of muscle spindles may assist in attaining proper degree of muscular contraction.

TENDON REFLEX. Neurotendinous organs respond to increased tension with increased afferent neuron activity that ultimately inhibits alpha motor neurons to muscle associated with tendon. Neurotendinous organs provide information about tension developed by muscle.

◆ NEURON POOLS pp. 417–418

Functional groups of neurons within central nervous system; divergence, convergence, summation, and facilitation occur within pools; even simple actions require participation of many neurons, whose activities are coordinated in complex circuits in neuron pools.

◆ CONDITIONS OF CLINICAL SIGNIFICANCE: THE CENTRAL NERVOUS SYSTEM pp. 418–420

Spinal Cord Dysfunctions.

PARALYSIS. Caused by lesions of motor spinal tracts.

LESIONS OF SENSORY TRACTS OF THE SPINAL CORD. Various sensory losses, depending on lesion site.

SPECIFIC DYSFUNCTIONS OF THE SPINAL CORD.

TABES DORSALIS. Progressive degeneration of posterior funiculi and dorsal roots of spinal nerves caused by spirochete bacteria of syphilis.

POLIOMYELITIS. Caused by virus that destroys nerve cell bodies in anterior horns of spinal cord.

SYRINGOMYELIA. Cyst formation in gray matter of cord and brain stem, with proliferation of neuroglia of central canal.

MULTIPLE SCLEROSIS. Chronic, widespread destruction of myelin sheaths of neurons in spinal cord and brain.

Dysfunctions of the Brain Stem. Lesions, tumors, hemorrhage, or trauma can produce variety of motor and sensory symptoms; damage to reticular activating system can produce coma.

Dysfunctions of the Cerebellum. Lesions cause dysfunction in coordinated actions between muscle groups.

Dysfunctions of the Basal Nuclei. Lesions result in spastic movements.

PARKINSONISM. May be related to abnormal metabolism of dopamine by basal nuclei.

HUNTINGTON'S CHOREA. Treated with drugs that block dopamine.

Inflammatory Diseases of the Central Nervous System.

ENCEPHALITIS. Brain inflammation.

MYELITIS. Spinal cord inflammation.

MENINGITIS. Inflammation of meninges that cover brain and spinal cord.

Tumors of the Central Nervous System. Symptoms include headaches, convulsions, behavior change, pain, paralysis; tumors destroy nervous tissue, increase pressure within skull, or cause edema of nervous tissue; treatments include surgical removal or chemical and radiation therapy.

Self-Quiz

1. The neurons that connect the left and right cerebral hemispheres form the: (a) projection tracts; (b) association tracts; (c) commissural tracts.
2. The cerebrum is divided into left and right hemispheres by the: (a) longitudinal fissure; (b) occipital lobe; (c) parietal lobe.
3. Each hemisphere of the cerebrum is divided by the central sulcus into: (a) frontal and parietal lobes; (b) parietal and occipital lobes; (c) occipital and frontal lobes.
4. The taste area is located in the parietal lobe, deep in the longitudinal fissure. True or False?
5. The control of emotions is influenced largely by the: (a) thalamus; (b) hypothalamus; (c) epithalamus.
6. "Pleasure centers" and "punishment centers" of the brain are located in the: (a) mammillary bodies; (b) limbic system; (c) intermediate mass.
7. The cerebellum is part of the: (a) mesencephalon; (b) telencephalon; (c) metencephalon.
8. The part of the brain that directs precise, smooth movements and the maintenance of equilibrium is the: (a) cerebrum; (b) cerebellum; (c) medulla oblongata.
9. The myelencephalon is the division of the brain that is continuous with the spinal cord. True or False?
10. The control of heart rate, coughing, and swallowing is a function of the: (a) limbic system; (b) pons; (c) medullary centers.
11. The vagus cranial nerve arises from the: (a) mesencephalon; (b) myelencephalon; (c) telencephalon.
12. Each ventricle contains a plexus of capillaries. True or False?
13. Cushioning the CNS against jolts and blows is a function of the: (a) meninges; (b) cerebrospinal fluid; (c) ventricles of the brain.
14. The portion of the spinal cord that gives rise to the spinal nerves supplying the upper limbs is the: (a) cervical enlargement; (b) lumbar enlargement; (c) lumbosacral plexus.
15. The white matter of the brain and spinal cord is composed of: (a) meninges; (b) myelinated processes of neurons; (c) nerve cell bodies.
16. Match the terms associated with the brain with the appropriate lettered description:

Cerebrum
Projection tracts
Basal nuclei
Cerebral hemisphere
Insula
Primary motor area
Visual area
Olfactory area
Thalamus
Cerebral peduncles
Metencephalon
Pons
Hypothalamus
Ventricles
Telencephalon
Reticular formation

(a) This structure is divided into frontal, parietal, temporal, and occipital lobes
(b) Located in the posterior portion of the occipital lobe
(c) Acts as the sensory relay and integrating center of the brain
(d) Connects the cerebellum with the brain stem
(e) Carry motor and/or sensory nerve impulses from one level of the CNS to another
(f) Contains gyri, fissures, and sulci
(g) Controls many vital processes, including appetite and sexual activity
(h) A fluid-filled system of cavities in the brain
(i) The structure that envelops the diencephalon and obscures much of the rhombencephalon
(j) Network of interconnecting nerve fibers that exerts control over the cerebral cortex and other regions
(k) A portion of the cerebral cortex located deep in the lateral fissure
(l) Located on the medial surface of the temporal lobe
(m) Gray-matter structures located deep in the cerebral hemispheres
(n) Two cylindrical nerve tracts on the ventral surface of the mesencephalon
(o) Includes the cerebellum
(p) Located in the precentral gyrus

17. The ascending tracts of the spinal cord carry afferent (sensory) impulses from peripheral receptors to various centers in the brain. True or False?

18. Sensory information received by receptors on the right side of the body is transmitted to the: (a) left cerebral cortex; (b) right cerebral cortex; (c) both cerebral cortices.

19. Match the items associated with the spinal cord with the appropriate lettered description:

Cauda equina
Meninges
White matter
Gray matter
Ventral roots
Fasciculus gracilis
Spinothalamic tracts
Spinocerebellar tracts
Pyramidal tracts
Extrapyramidal tracts

(a) Contains myelinated axons that travel up or down the spinal cord to higher or lower levels in the CNS
(b) Convey impulses concerned with pain and temperature
(c) These structures contribute processes of both somatic motor neurons and visceral motor neurons to the spinal nerves
(d) Dura mater, arachnoid, pia mater
(e) Modify muscular contractions related to posture and balance
(f) Major structures involved in the voluntary control of skeletal muscles
(g) A mass of descending nerve roots below the end of the spinal cord
(h) Carry information concerning unconscious proprioception from neuromuscular receptors
(i) Enables you to locate where an object is touching your body
(j) Substance whose anterior horns contain the cell bodies of somatic motor neurons

20. Increasing the tension applied to a neurotendinous organ within a tendon: (a) decreases the activity of the afferent neurons from the tendon organ; (b) decreases the stimulation of alpha motor neurons to the muscle associated with the tendon; (c) increases the contraction of the intrafusal fibers of the tendon organ.

21. Parkinsonism is a dysfunction of the: (a) basal nuclei; (b) brain stem; (c) cerebellum.

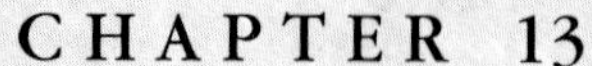

CHAPTER 13

The Peripheral Nervous System

CHAPTER CONTENTS

LEARNING OBJECTIVES

After completing this chapter, you should be able to:

1. State the function of the peripheral nervous system, and list its two divisions.
2. Name the 12 pairs of cranial nerves, and specify the functions of each.
3. Describe the distribution patterns of the spinal nerves.
4. Distinguish between dorsal rami and dorsal roots and between ventral rami and ventral roots.
5. List the three major nerve plexuses, and state the body regions they supply.
6. Name and describe the specific distribution of each nerve formed by the major nerve plexuses.
7. Distinguish between the somatic and autonomic nervous systems by describing the function of each.
8. Cite the most common disorders of the peripheral nervous system.

CHAPTER 13

In order for the brain to function effectively it must constantly receive input from the various regions of the body as well as from outside the body and be able to send messages to all parts of the body. This is the main role of the **peripheral nervous system (PNS).**

The peripheral nervous system includes all neurons except those that are restricted to the brain and spinal cord—that is, all except the neurons in the central nervous system. It consists of pathways of nerve fibers between the central nervous system and all outlying structures of the body. Included in the peripheral nervous system are 12 pairs of **cranial nerves** and 31 pairs of **spinal nerves.** In terms of function, the peripheral nervous system is divided into the:

1. Afferent (sensory) division, whose nerve fibers relay impulses from all areas of the body, including the viscera, to the central nervous system.

2. Efferent (motor) division, which is subdivided into the:

a. Somatic nervous system, whose fibers carry motor impulses between the central nervous system and the skeletal muscles.

b. Autonomic nervous system, which connects motor fibers from the central nervous system with smooth muscles, cardiac muscle, and glands.

 Cranial Nerves

Twelve pairs of cranial nerves arise from the brain. The first two pairs originate from the forebrain; the remaining ten pairs originate from the brain stem (Figure 13.1). Most of the cranial nerves are mixed nerves, being composed of both motor and sensory neurons, although a few cranial nerves carry only sensory impulses. Previ-

◆ **FIGURE 13.1 The ventral surface of the brain showing the cranial nerves**
The pituitary gland has been removed in order to expose the optic chiasma.

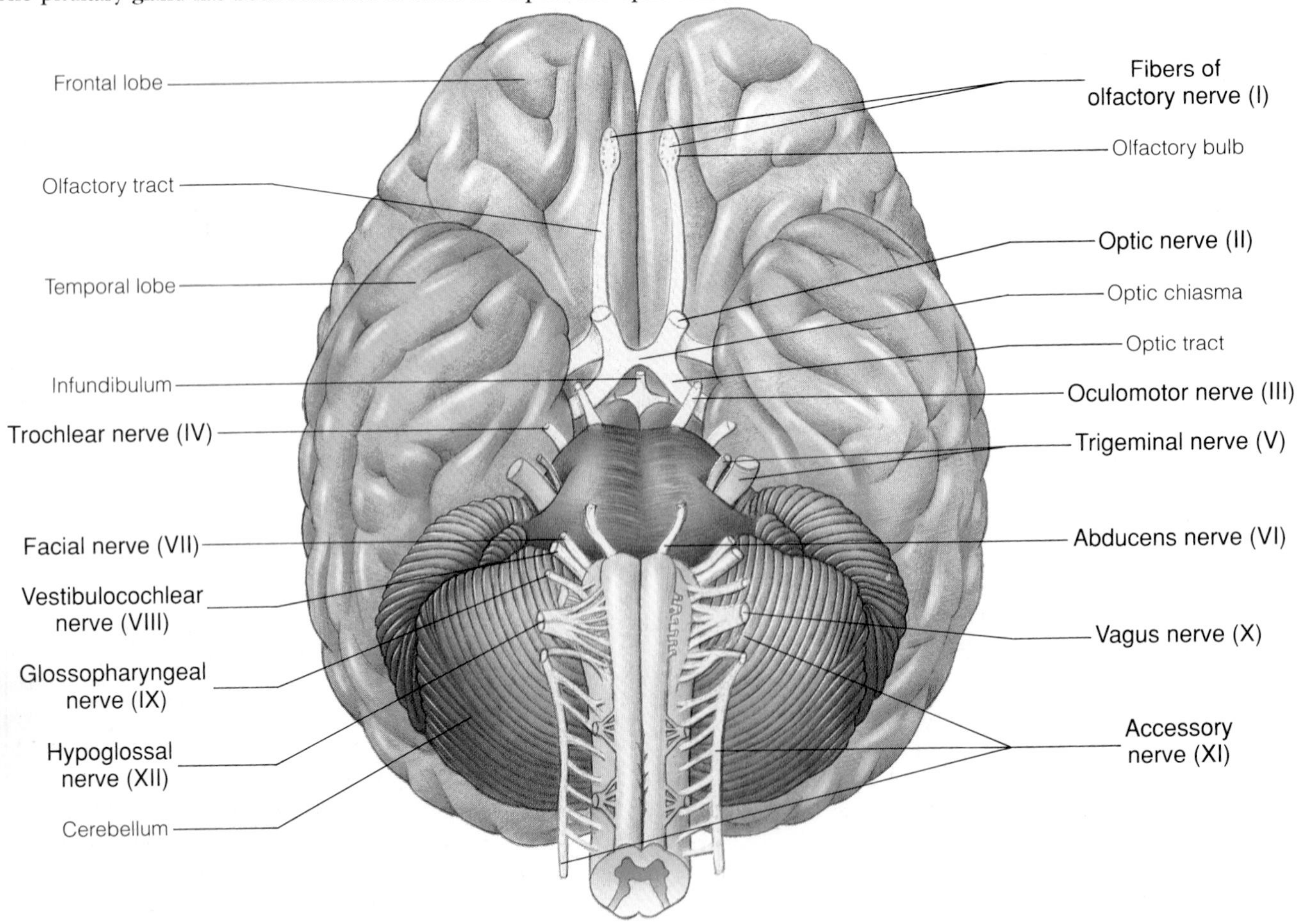

◆ **FIGURE 13.2 Olfactory nerve (cranial nerve I)**

Sagittal section of the face showing the position of the olfactory bulbs and tracts just above the cribriform plate of the ethmoid bone. The fibers of the olfactory nerve pass through the openings in the cribriform plate to enter the nasal cavity.

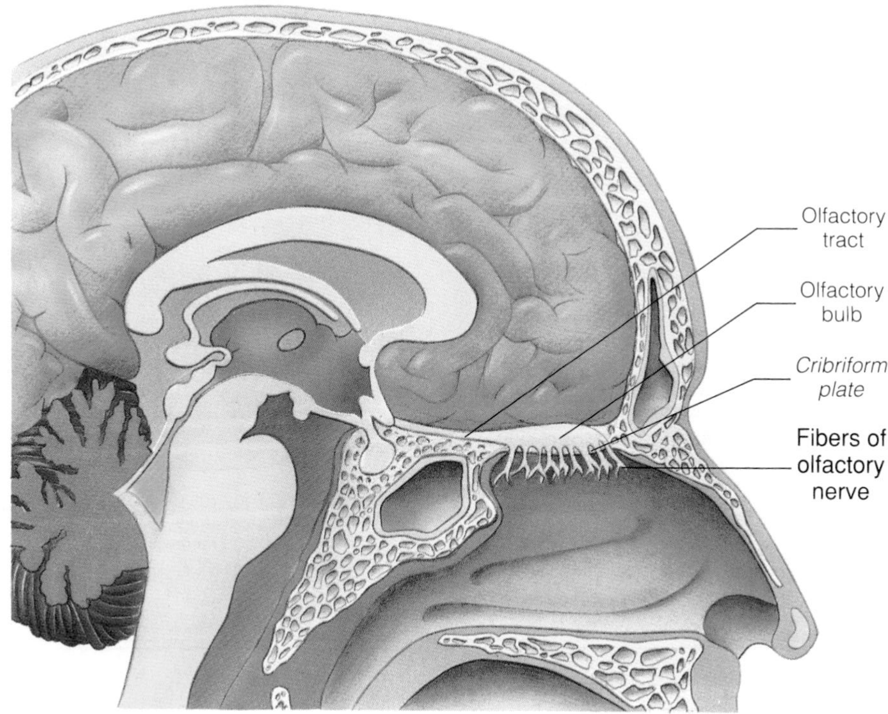

ously it was thought that some cranial nerves transmitted only motor impulses. It is now known, however, that the cranial nerves formerly considered to be entirely motor actually contain some sensory neurons from proprioceptors in the muscles they innervate. Thus, they convey information to the central nervous system concerning the lengths of specific muscles and the tension generated when the muscles contract or are stretched.

In the cranial nerves, as in the spinal nerves, the cell bodies of the sensory neurons are in ganglia located *outside* the central nervous system. The cell bodies of the motor neurons are located in nuclei *within* the central nervous system. Some motor neurons in the cranial nerves supply skeletal muscles. These neurons, which are under conscious control, are called **somatic motor neurons.** Other motor neurons in the cranial nerves are not under conscious control. These supply smooth muscles, cardiac muscle, or glands and are called **visceral motor neurons.** The visceral motor neurons in the cranial nerves are part of the parasympathetic division of the autonomic nervous system, which is discussed in Chapter 14. With the exception of the vagus nerve, the cranial nerves supply only structures in the head and neck. The cranial nerves are numbered (using Roman numerals) from anterior to posterior, in the order in which they connect to the brain. Table 13.1 (pages 438–440) summarizes the cranial nerves and their functions.

I: Olfactory Nerves

The first pair of cranial nerves, the **olfactory nerves,** arise from receptor cells in the epithelial lining of the nose—that is, the nasal mucosa (Figure 13.2; Table 13.1). Processes of these receptor cells pass through the perforations of the *cribriform plate* of the ethmoid bone and enter the olfactory bulbs of the telencephalon portion of the brain. In the olfactory bulbs, the nerve fibers synapse with neurons that pass posteriorly in the olfactory tracts. The fibers of the olfactory tracts enter the brain, and many of them travel to the cerebral cortex of the medial sides of the temporal lobes. The olfactory nerves are entirely sensory, carrying impulses associated with the sense of smell.

II. Optic Nerves

The **optic nerves** carry impulses associated with vision. Like the olfactory nerves, they are entirely sensory. The optic nerves are actually brain tracts rather than true nerves, since the are formed from outgrowths of the embryonic diencephalon.

The optic nerves originate in the retinas of the eyes, on which images are focused. After leaving the posterior surface of the eyeball, each optic nerve exits from the orbital cavity and enters the cranial cavity through an *optic foramen* in the sphenoid bone.

◆ **FIGURE 13.3 Optic nerves (cranial nerve II)**

Ventral view of the brain showing the optic nerves and the visual pathways to the cortex of the occipital lobes.

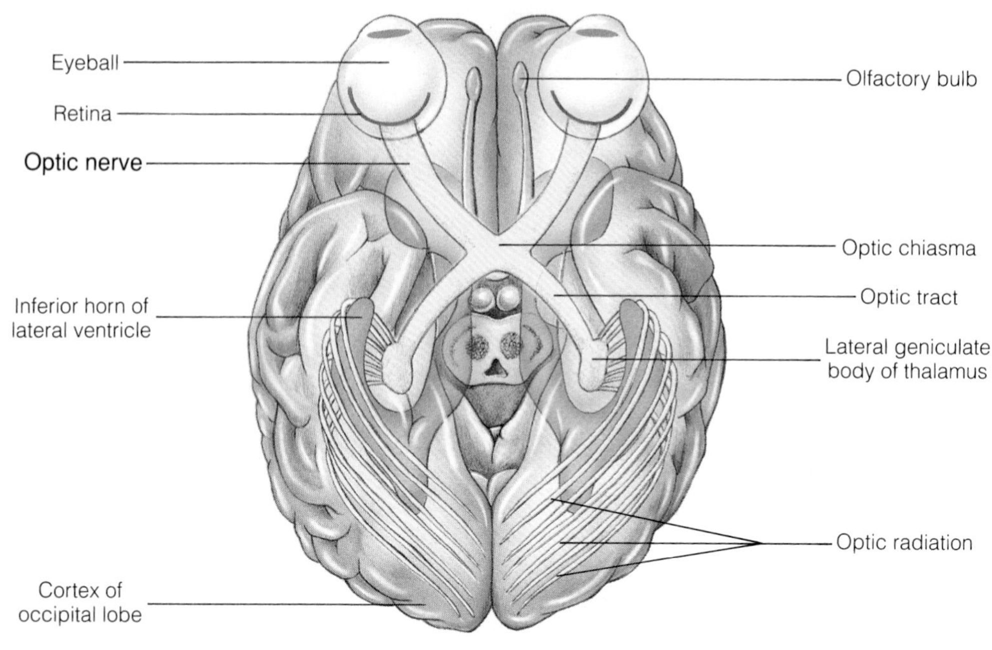

Shortly after entering the cranium, the two optic nerves meet in the **optic chiasma,** *(op´-tik kigh-az´-mah),* just anterior to the pituitary gland (Figure 13.3; Table 13.1). In the optic chiasma, nerve fibers from the medial half of each retina cross to the opposite side, and those from the lateral half of each retina remain on the same side. The fibers then continue to the brain as the **optic tracts.** Because of the crossing of nerve fibers at the optic chiasma, each optic tract consists of fibers from the retinas of both eyes.

Some nerve fibers in the optic tracts terminate in the superior colliculi of the midbrain, where they function in subconscious visual reflexes. However, most of the fibers in the optic tracts travel to the lateral geniculate bodies of the thalamus, where they synapse with neurons that form pathways called **optic radiations.** The neurons of the optic radiations, which are third-order neurons, pass through the internal capsule and terminate in the visual cortex of the occipital lobes.

III: Oculomotor Nerves

The **oculomotor nerves** *(ok´´-u-lō-mō´-ter),* emerge from the midbrain, just superior to the pons, and enter the orbits through the *superior orbital fissures,* which are located between the small wings and great wings of the sphenoid bone. The oculomotor nerves consist of somatic motor neurons traveling to, and proprioceptive sensory neurons traveling from, four of the six extrinsic muscles that move the eyeball (Figure 13.4; Table 13.1). Specifically, the oculomotor nerves innervate the superior rectus, medial rectus, inferior rectus, and inferior oblique muscles of the eyeball. In addition, the oculomotor nerves supply the levator palpebrae superioris muscles, which function to elevate the upper eyelids.

The oculomotor nerves also contain neurons of the parasympathetic division of the autonomic nervous system (Chapter 14). The efferent pathways of the autonomic nervous system have two neurons between the central nervous system and the innervated structures. The first neuron, called a **preganglionic (presynaptic) neuron** *(pree-gang-lee-on´-ik),* always travels from the central nervous system to a ganglion located outside the central nervous system. Within the ganglion, the preganglionic neuron synapses with second neurons, called **postganglionic (postsynaptic) neurons** *(post-gang-lee-on´-ik),* which leave the ganglion and travel to the innervated structures. Preganglionic parasympathetic neurons in the oculomotor nerves synapse in ganglia called **ciliary ganglia,** which are located behind the eyeballs. From a ciliary ganglion, postganglionic neurons enter the eye and innervate the intrinsic smooth muscles that regulate the size of the pupil and shape of the lens. Therefore, apart from regulating the voluntary movement of the eyeballs, the oculomotor nerves are also involved in the reflex adjustments of the eyes to varying intensities of light and in focusing the eyes for near and far vision.

IV: Trochlear Nerves

The **trochlear nerves** *(trok´-lē-ar)* are small nerves that arise below the inferior colliculi on the dorsal sur-

◆ **FIGURE 13.4 Oculomotor nerve (cranial nerve III)**

Lateral view of the right eye illustrating the oculomotor nerve. Note the somatic motor neurons to muscles that surround the right eye and parasympathetic fibers that enter the eye. (The lateral rectus muscle has been cut.)

Preganglionic parasympathetic neurons
Superior orbital fissure
Superior rectus muscle
Medial rectus muscle
Levator palpebrae superioris muscle
Colliculi
Oculomotor nerve
Superior eyelid
Pons
Inferior oblique muscle
Ciliary ganglion
Postganglionic parasympathetic neurons to eye
Inferior rectus muscle

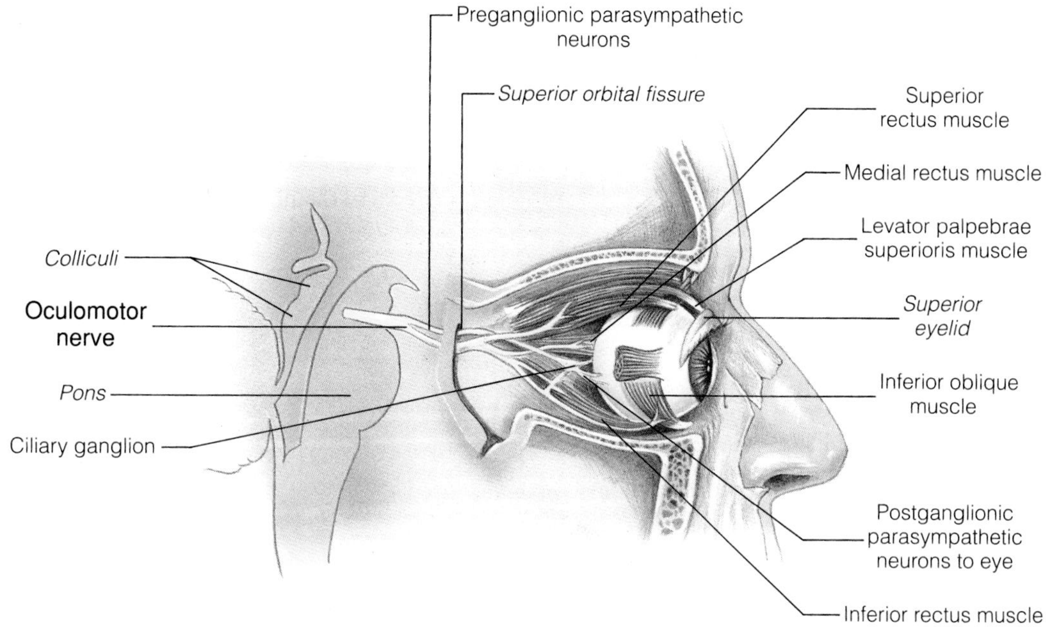

face of the midbrain. These nerves, which are the only cranial nerves to exit from the dorsal surface of the brain, curve around the lateral sides of the brain and enter the orbits through the *superior orbital fissures* along with the oculomotor nerves (Figure 13.5; Table 13.1). The trochlear nerves carry somatic motor neurons to, and proprioceptive neurons from, one of the extrinsic muscles of the eyes, the superior oblique muscles. Thus, the trochlear nerves aid in the voluntary movements of the eyeballs.

V: Trigeminal Nerves

The large **trigeminal nerves** *(tri-jem´-i-nal)* emerge from the lateral sides of the pons. As their name indicates, each trigeminal nerve has three divisions: the **ophthalmic** *(of-thal´-mik)*, **maxillary,** and **mandibular nerves** (Figure 13.6; Table 13.1). The three divisions exit the skull through openings in the sphenoid bone: the ophthalmic division leaves the skull through the *superior orbital fissure;* the maxillary division passes through the *foramen rotundum;* and the

◆ **FIGURE 13.5 Trochlear nerve (cranial nerve IV)**

Lateral view of the right eye showing the trochlear nerve to the superior oblique muscle. (The lateral rectus muscle has been cut.)

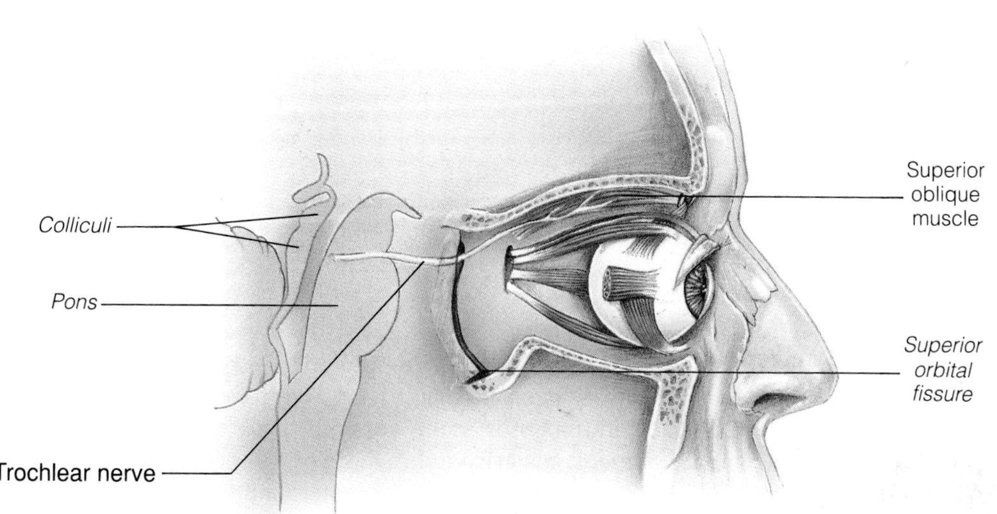

◆ FIGURE 13.6 Trigeminal nerve (cranial nerve V) showing its three divisions: ophthalmic, maxillary, and mandibular

Insert (a) shows the distribution of sensory neurons in each division; (b) shows the motor branches of the mandibular division

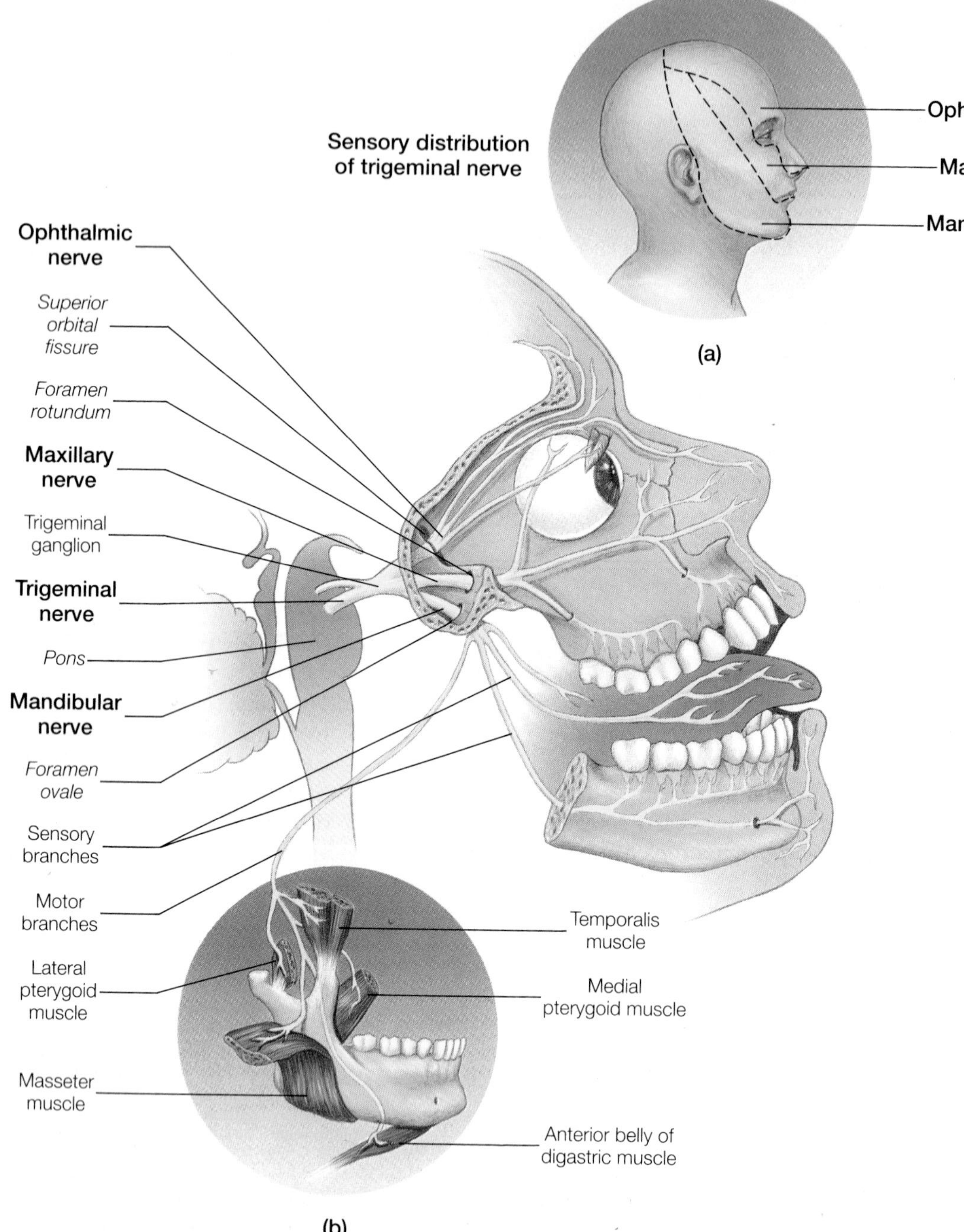

mandibular division exits the skull through the *foramen ovale.* The trigeminal nerves are the major sensory nerves of the face. They contain sensory neurons that originate from the skin of the face and anterior scalp and from the mucous membranes of the nasal cavity and mouth. The cell bodies of these sensory neurons are found in large **trigeminal (semilunar) ganglia** located at the points where the ophthalmic, maxillary, and mandibular nerves join before entering the brain.

The trigeminal nerves also contain somatic motor neurons that travel within the mandibular nerve to the muscles of mastication (medial and lateral pterygoids, masseter, and temporalis), to the mylohyoid muscle, and to the anterior belly of the digastric muscle.

◆ **FIGURE 13.7 Abducens nerve (cranial nerve VI)**
Lateral view of the right eye showing the abducens nerve to the lateral rectus muscle.

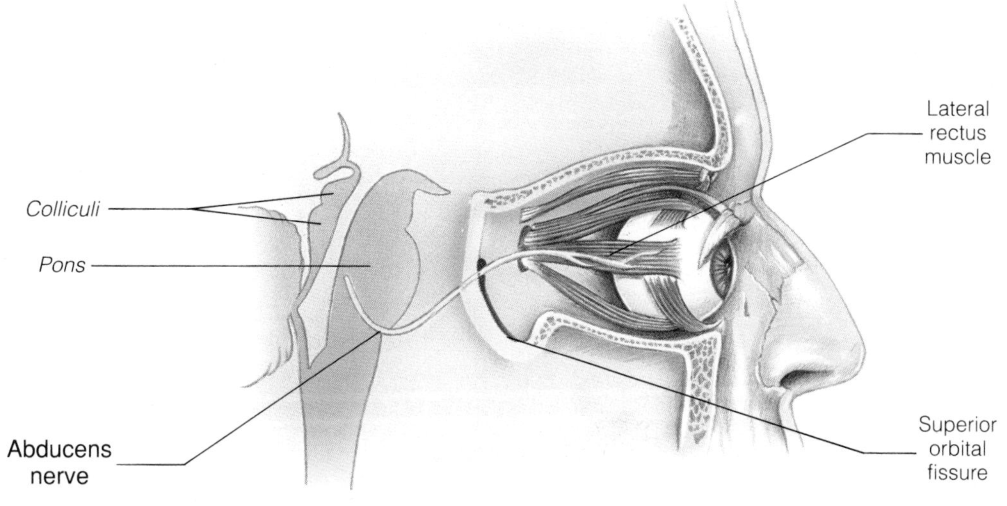

VI: Abducens Nerves

The sixth cranial nerves originate in the metencephalon and exit the brain stem just below the pons (Figure 13.7; Table 13.1). The **abducens nerves** *(ab-du´-senz)* enter the orbits through the *superior orbital fissures* along with the oculomotor (III) and trochlear (IV) cranial nerves. The abducens nerves carry somatic motor neurons to, and proprioceptive sensory neurons from, the remaining extrinsic muscle of each eye, the lateral rectus muscles.

Eye movements involve the coordinated contraction of the extrinsic muscles of both eyes. This coordination, in turn, requires a synchronization of the nerve impulses carried by the oculomotor (III), trochlear (IV), and abducens (VI) cranial nerves.

VII: Facial Nerves

The **facial nerves** leave the metencephalon at the lower border of the pons, just lateral to the abducens nerves, and enter the petrous portions of the temporal bones through the *internal auditory meatus.* After traveling through the temporal bones, the facial nerves leave the skull by way of the *stylomastoid foramina.* They then pass forward across the cheek through the parotid salivary glands and divide into numerous branches that supply somatic motor neurons to the muscles of the face and scalp (Figure 13.8; Table 13.1).

The facial nerves also contain sensory neurons that originate from the taste buds on the anterior two-thirds of the tongue. The cell bodies of these sensory neurons are located in the **geniculate ganglia,** which lie within the petrous portions of the temporal bones.

Parasympathetic neurons to the lacrimal (tear) glands, mucous glands in the nasal cavity, and the submandibular and sublingual salivary glands are also carried in the facial nerves. Preganglionic parasympathetic neurons to the lacrimal glands synapse in the **pterygopalatine ganglia** *(ter˝-ĭ-go-pal´-ah-tīn)* with postganglionic parasympathetic neurons that then pass to the lacrimal glands. The preganglionic parasympathetic neurons to the submandibular and sublingual glands synapse in the **submandibular ganglia** with postganglionic parasympathetic neurons that then pass to the glands.

VIII: Vestibulocochlear Nerves

The eighth cranial nerves have two divisions: the **cochlear** *(kōk´-lē-er)* and the **vestibular nerves** (Figure 13.9; Table 13.1). Both of these divisions are entirely sensory and originate from inner-ear receptors located in the petrous portions of the temporal bones. The two divisions join to form a common trunk, the **vestibulocochlear nerve** *(ves-tib˝-u-lō-kōk´-lē-er)* which leaves the temporal bones through the *internal auditory meatus* and enters the brain stem just below the pons.

The cochlear divisions transmit impulses related to hearing from the spiral organ located in the cochlea of the ear (see Figure 16.29, page 515). The cell bodies of the cochlear nerves lie in **spiral ganglia** located within the cochlea.

The vestibular divisions are concerned with equilibrium. Their receptors are located in the ampullae of the semicircular ducts and in the saccule and utricle of the vestibule of the ear (see Figure 16.28, page 514). The cell bodies of the vestibular nerves are located in the **vestibular ganglia.**

◆ **FIGURE 13.8 Facial nerve (cranial nerve VII)**

(a) Sensory and parasympathetic neurons. (b) Somatic motor branches.

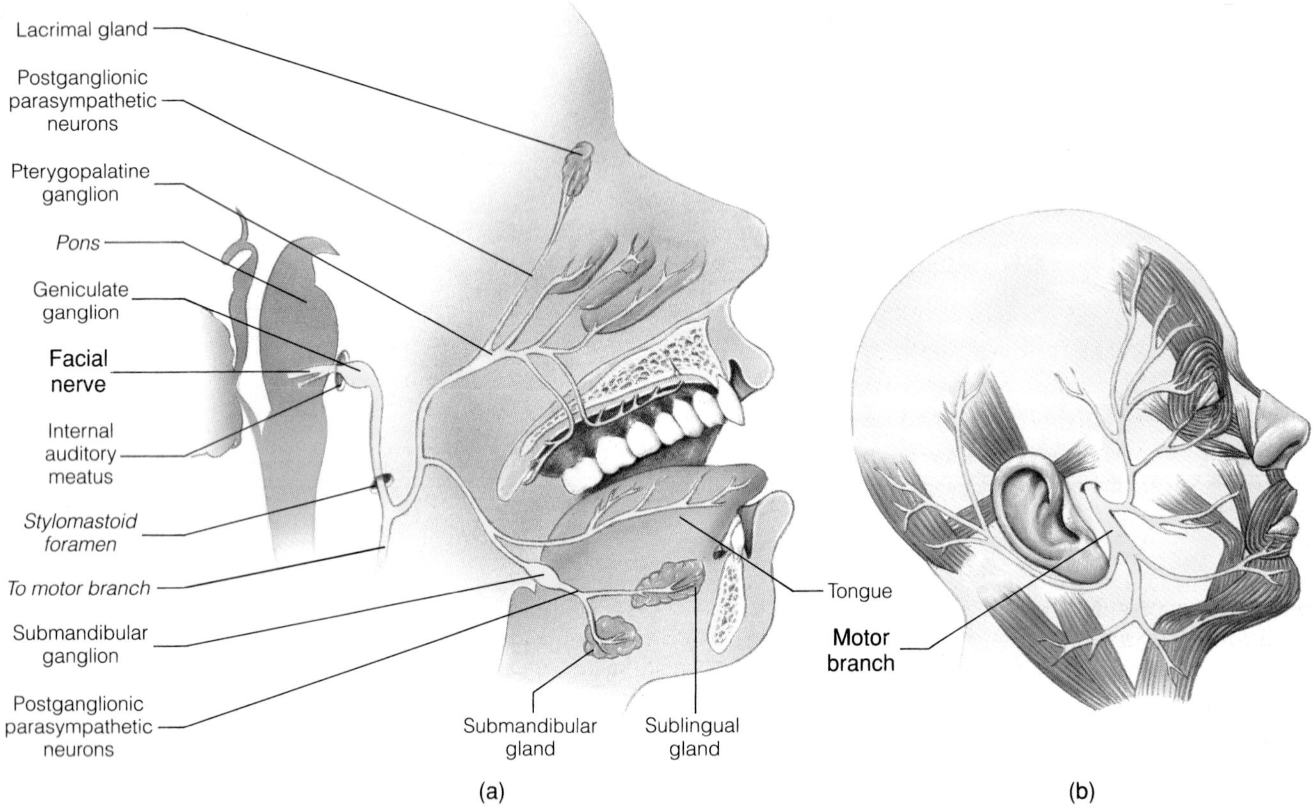

Lacrimal gland
Postganglionic parasympathetic neurons
Pterygopalatine ganglion
Pons
Geniculate ganglion
Facial nerve
Internal auditory meatus
Stylomastoid foramen
To motor branch
Submandibular ganglion
Postganglionic parasympathetic neurons
Submandibular gland
Sublingual gland
Tongue
Motor branch
(a)
(b)

◆ **FIGURE 13.9 Vestibulocochlear nerve (cranial nerve VIII), showing the vestibular nerve that supplies the vestibule and ampullae and the cochlear nerve that supplies the cochlea**

All these structures are within the inner ear.

Ampulla
Semicircular ducts
Vestibular ganglion
Internal auditory meatus
Vestibular nerve
Cochlear nerve
Vestibulocochlear nerve
Pons
Cochlea (contains spiral ganglia)
External auditory meatus
Tympanic membrane
Vestibule

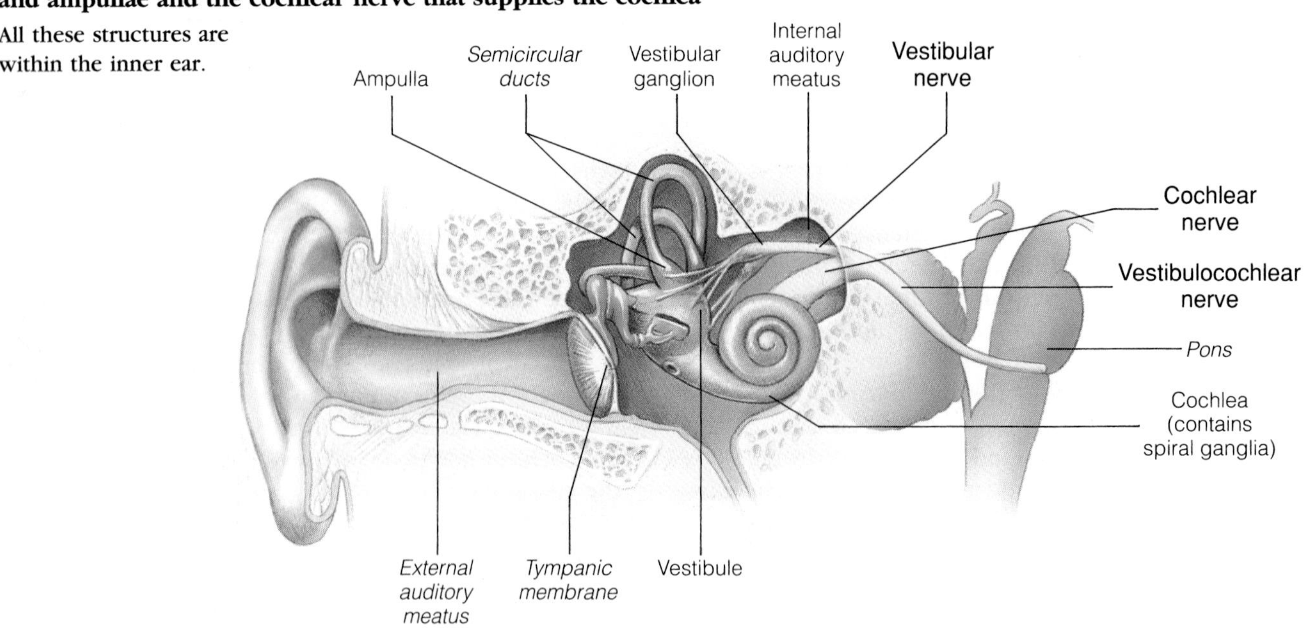

IX: Glossopharyngeal Nerves

The **glossopharyngeal nerves** *(glos″-ō-fah-rin′-jē-al)* are mixed nerves, carrying both motor and sensory impulses. As the name indicates, these nerves supply the tongue and the pharynx. They emerge from the medulla and leave the skull through the *jugular foramina* of the temporal bones (Figure 13.10; Table 13.1).

Sensory neurons in the glossopharyngeal nerves carry impulses from the taste buds of the posterior third of the tongue; from the mucous membranes of the pharynx, tonsils, and middle-ear cavities; from receptors sensitive to changes in blood levels of oxygen and carbon dioxide in the carotid bodies; and from receptors that monitor the blood pressure in the carotid sinuses (Chapter 20). The cell bodies of these sensory neurons are located in the **superior** and **inferior ganglia** of the nerves.

The glossopharyngeal nerves also contain somatic motor neurons that supply the stylopharyngeus muscles of the pharynx, which are involved in swallowing. In addition, some motor neurons of the glossopharyngeal nerves intermix with the vagus (X) and accessory (XI) cranial nerves to innervate several other pharyngeal muscles.

Preganglionic parasympathetic neurons that travel in the glossopharyngeal nerves synapse in the **otic ganglia,** from which postganglionic parasympathetic neurons travel to the parotid salivary glands.

X: Vagus Nerves

The **vagus nerves** *(vā′-gus)* are the only cranial nerves that are not restricted to the head and neck regions. They leave the sides of the medulla by several rootlets, pass through the *jugular foramina* of the temporal bones, and descend along the pharynx close to the common carotid arteries and the internal jugular veins (Figure 13.11; Table 13.1). After leaving the neck, the vagus nerves enter the thorax and abdomen.

The vagus nerves carry somatic motor impulses to the voluntary muscles of the pharynx and larynx, and sensory impulses from taste receptors located toward the base of the tongue. Moreover, they have a broad parasympathetic distribution, carrying preganglionic neurons to the involuntary muscles of the thoracic and abdominal viscera as far caudally as the transverse colon of the large intestine. Located close to, or in, the walls of the innervated structures are small terminal ganglia from which short postganglionic parasympathetic neurons supply the structures. The vagus nerves also carry sensory fibers from the viscera. The sensory input from the viscera generally does not reach the conscious level; rather, it provides information that leads to the automatic regulation of heart rate, depth of respiration, blood pressure, digestive processes, and so forth. Under certain conditions, these visceral sensations can reach conscious levels, however, as evidenced by sensations of distension and nausea.

◆ **FIGURE 13.10 Glossopharyngeal nerve (cranial nerve IX) to the tongue, throat, and parotid gland**

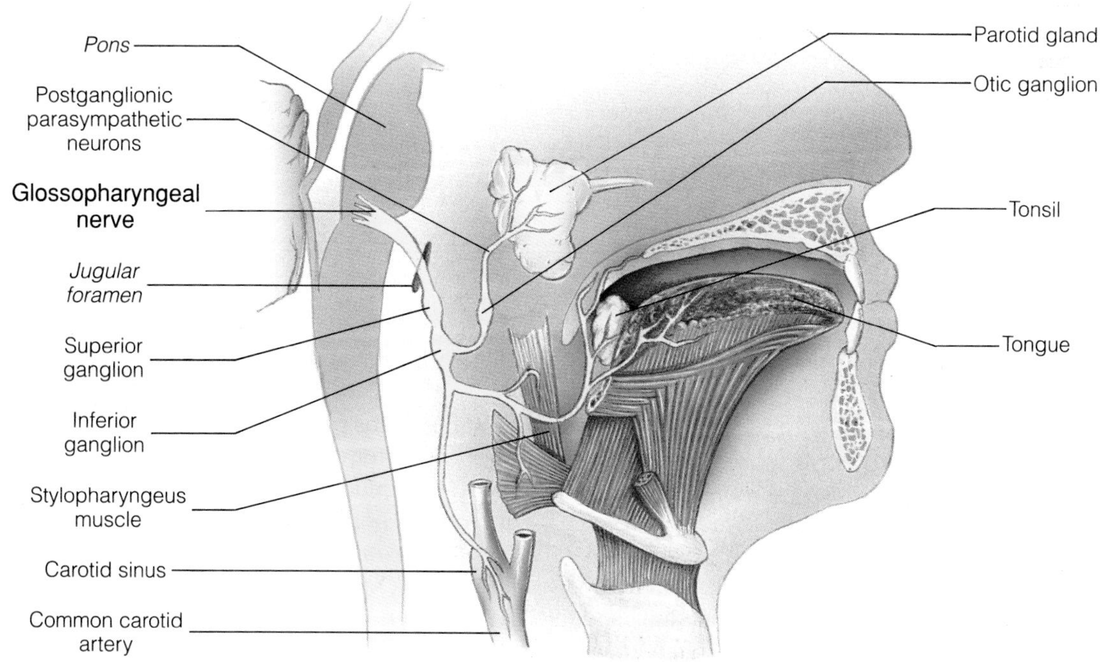

◆ **FIGURE 13.11 Vagus nerve (cranial nerve X), showing its distribution to the neck, thoracic cavity, and abdominal cavity**

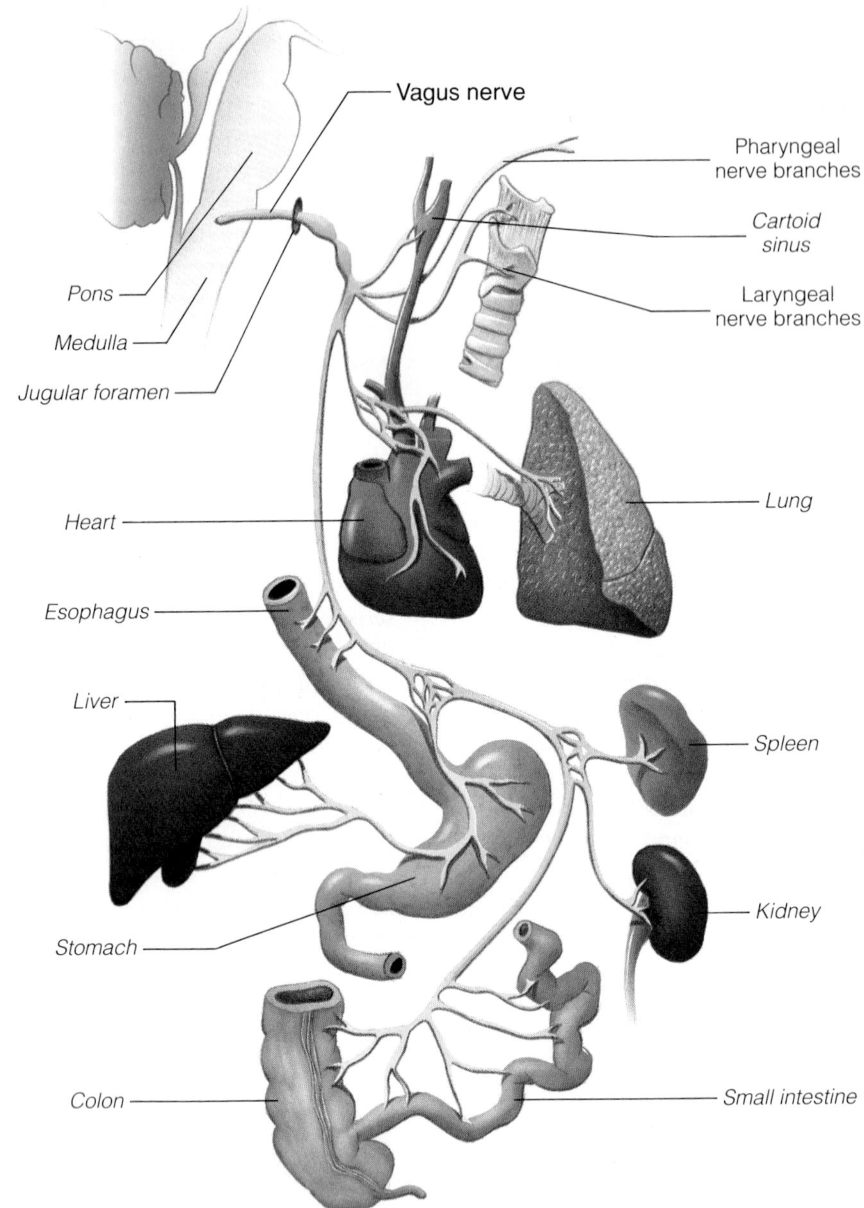

XI: Accessory Nerves

The **accessory nerves** are composed of somatic motor neurons to voluntary muscles and some sensory neurons from proprioceptors. Each accessory nerve is actually formed by two nerves. One arises from the medulla and is thus a true cranial nerve, whereas the other arises from the cervical region of the spinal cord and is actually a spinal nerve (Figure 13.12; Table 13.1). The spinal portions pass upward along the sides of the spinal cord and enter the skull through the *foramen magnum* of the occipital bone. In the cranial cavity, the spinal and cranial portions join and leave the skull through the *jugular foramina* along with the glossopharyngeal (IX) and vagus (X) cranial nerves.

The fibers of the cranial portions intermix with the vagus nerves to supply the muscles of the larynx and pharynx. The fibers of the spinal portions supply the trapezius and sternocleidomastoid muscles.

◆ **FIGURE 13.12 Accessory nerve (cranial nerve XI), with cranial and spinal portions separated**

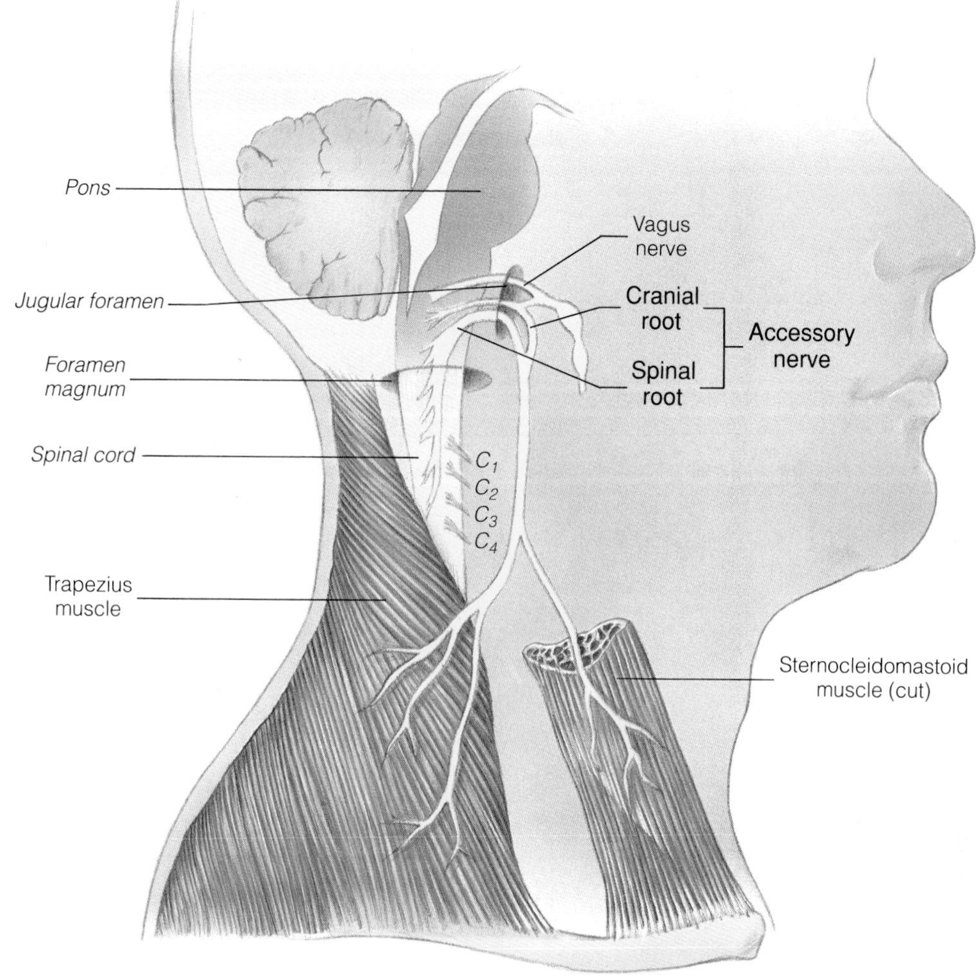

XII: Hypoglossal Nerves

The **hypoglossal nerves** *(hi″-pō-glos′-al)* leave the anterior surface of the medulla as a series of rootlets and pass out of the skull through the *hypoglossal canals* of the occipital bone. As the name indicates, they are located beneath the tongue, where they travel anteriorly in close relationship with the first cervical spinal nerve (Figure 13.13; Table 13.1). The hypoglossal nerves consist of somatic motor neurons that supply the intrinsic and extrinsic muscles of the tongue, and some sensory neurons from proprioceptors. The first three cervical nerves send motor fibers to some of the muscles of the neck through branches called the *ansa cervicalis,* which are so closely associated with the hypoglossal nerves that they appear to arise from them.

Spinal Nerves

There are 31 pairs of **spinal nerves,** including 8 cervical (designated C_1-C_8), 12 thoracic (T_1-T_{12}), 5 lumbar (L_1-L_5), 5 sacral (S_1-S_5), and 1 coccygeal. With the exception of the first pair of cervical nerves, the spinal nerves leave the vertebral canal by passing through the *intervertebral foramina.* The first pair of cervical nerves exit between the occipital bone and the atlas. The second through the seventh pairs of cervical nerves emerge *above* the vertebrae for which they are named. The eighth pair of cervical nerves emerge between the seventh cervical and first thoracic vertebrae. All remaining pairs of spinal nerves pass *below* the vertebrae for which they are named.

◆ **FIGURE 13.13 Hypoglossal nerve (cranial nerve XII), which supplies the muscles of the tongue**

Branches of the first three cervical spinal nerves interconnect with the hypoglossal nerve through the ansa cervicalis to supply certain muscles of the throat.

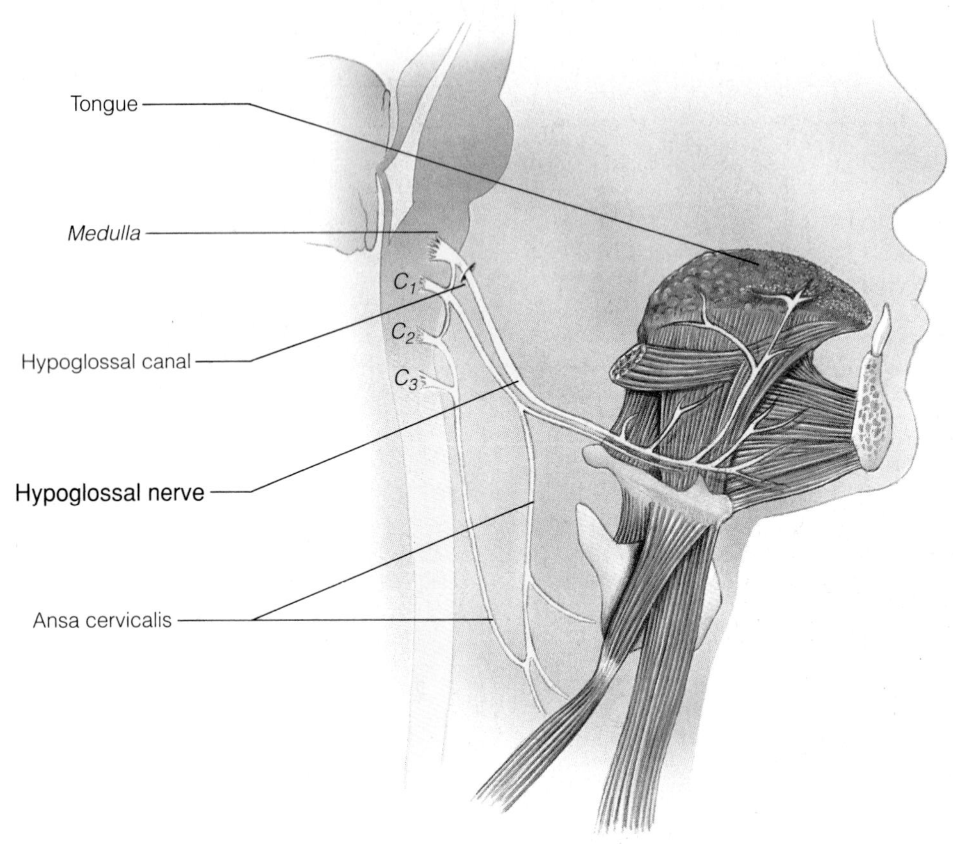

◆ **TABLE 13.1 Summary of the Cranial Nerves**

NERVES	SITE OF CONNECTION WITH BRAIN	SITE OF EXIT FROM SKULL	FUNCTIONS
I: OLFACTORY [*Fig. 13.2*] *sensory*	Telencephalon (cerebral hemisphere)	Cribriform plate of ethmoid	*Sensory* olfaction (sense of smell)
II: OPTIC [*Fig. 13.3*] *sensory*	Diencephalon	Optic foramen	*Sensory* vision
III: OCULOMOTOR [*Fig. 13.4*] *motor and proprioception*	Mesencephalon (midbrain)	Superior orbital fissure	*Somatic motor* levator palpebrae superioris and the external eye muscles, except superior oblique and lateral rectus *Proprioception* from the innervated muscles *Parasympathetic* ciliary muscle of the lens and the sphincter of the pupil
IV: TROCHLEAR [*Fig. 13.5*] *motor and proprioception*	Mesencephalon (midbrain)	Superior orbital fissure	*Somatic motor* superior oblique muscle of the eye *Proprioception* from the superior oblique muscle

◆ **TABLE 13.1 Summary of the Cranial Nerves (continued)**

NERVES	SITE OF CONNECTION WITH BRAIN	SITE OF EXIT FROM SKULL	FUNCTIONS
V: TRIGEMINAL [*Fig. 13.6*] *mixed*	Metencephalon (pons)		
Ophthalmic division		Superior orbital fissure	*Sensory* cornea; skin of nose, forehead, and scalp
Maxillary division		Foramen rotundum	*Sensory* nasal cavity, palate and upper teeth, skin of cheek, and upper lip
Mandibular division		Foramen ovale	*Sensory* tongue, lower teeth, skin of chin, lower jaw, and temporal regions *Somatic motor* muscles of mastication *Proprioception* from muscles of mastication
VI: ABDUCENS [*Fig. 13.7*] *motor and proprioception*	Metencephalon (pons)	Superior orbital fissure	*Somatic motor* lateral rectus muscle of the eye *Proprioception* from lateral rectus muscle
VII: FACIAL [*Fig. 13.8*] *mixed*	Metencephalon (pons)	Stylomastoid foramen	*Somatic motor* muscles of facial expression *Proprioception* from muscles of facial expression *Sensory* taste from the anterior two-thirds of tongue *Parasympathetic* sublingual and submandibular salivary glands; lacrimal glands; mucous glands of nasal cavity
VIII: VESTIBULOCOCHLEAR [*Fig. 13.9*] *sensory*	Metencephalon (pons)		
Vestibular division		Internal auditory meatus	*Sensory* equilibrium
Cochlear division		Internal auditory meatus	*Sensory* hearing
IX: GLOSSOPHARYNGEAL [*Fig. 13.10*] *mixed*	Myelencephalon (medulla)	Jugular foramen	*Somatic motor* stylopharyngeus muscle; other pharyngeal muscles via cranial nerves X and XI *Proprioception* from innervated muscles *Parasympathetic* parotid salivary glands *Sensory* taste and general sensation from the posterior one-third of the tongue; pharynx, middle-ear cavities, carotid sinuses
X: VAGUS [*Fig. 13.11*] *mixed*	Myelencephalon (medulla)	Jugular foramen	*Somatic motor* muscles of the pharynx and the larynx *Proprioception* from innervated muscles *Sensory* taste from the rear of the tongue; visceral sensory from the thoracic and the abdominal organs *Parasympathetic* organs of the thoracic and abdominal cavities

continued on next page

◆ **TABLE 13.1 Summary of the Cranial Nerves (continued)**

NERVES	SITE OF CONNECTION WITH BRAIN	SITE OF EXIT FROM SKULL	FUNCTIONS
XI: ACCESSORY [*Fig. 13.12*] *motor and proprioception*	Myelencephalon (medulla)	Jugular foramen	*Somatic motor* trapezius and sternocleidomastoid muscles *Proprioception* from innervated muscles
XII: HYPOGLOSSAL [*Fig. 13.13*] *motor and proprioception*	Myelencephalon (medulla)	Hypoglossal canal	*Somatic motor* intrinsic and extrinsic muscles of the tongue *Proprioception* from innervated muscles

Formation of Spinal Nerves

The spinal nerves are formed from the union of **ventral** and **dorsal roots** that leave or enter the spinal cord (Figure 13.14). The ventral roots contain axons of motor neurons that leave the anterior and lateral gray horns of the spinal cord. The cell bodies of these motor neurons are located in the spinal cord. The dorsal roots contain axons of sensory neurons that enter the posterior horns of the gray matter. The cell bodies of these sensory neurons lie outside the spinal cord in **dorsal root ganglia (spinal ganglia),** one ganglion on each dorsal root. After the roots join, all spinal nerves are mixed nerves containing the processes of motor (somatic and visceral) and sensory neurons.

Branches of Spinal Nerves

Soon after passing through an intervertebral foramen, each spinal nerve divides into two branches: a **dorsal ramus** and a **ventral ramus** (Figure 13.15). The dorsal rami pass posteriorly to supply the skin and muscles of the back. The ventral rami are longer, and their distribution varies in different body regions. Like the spinal nerves, both rami are mixed, containing motor and sensory fibers. In the thoracic region, the ventral rami of T_2 through T_{12} extend peripherally in the intercostal spaces between the ribs to supply the skin and muscles of the lateral and anterior body walls. In the cervical, lumbar, and sacral regions, the ventral rami of succes-

◆ **FIGURE 13.14 A segment of the spinal cord showing the formation of a pair of spinal nerves from dorsal and ventral roots**

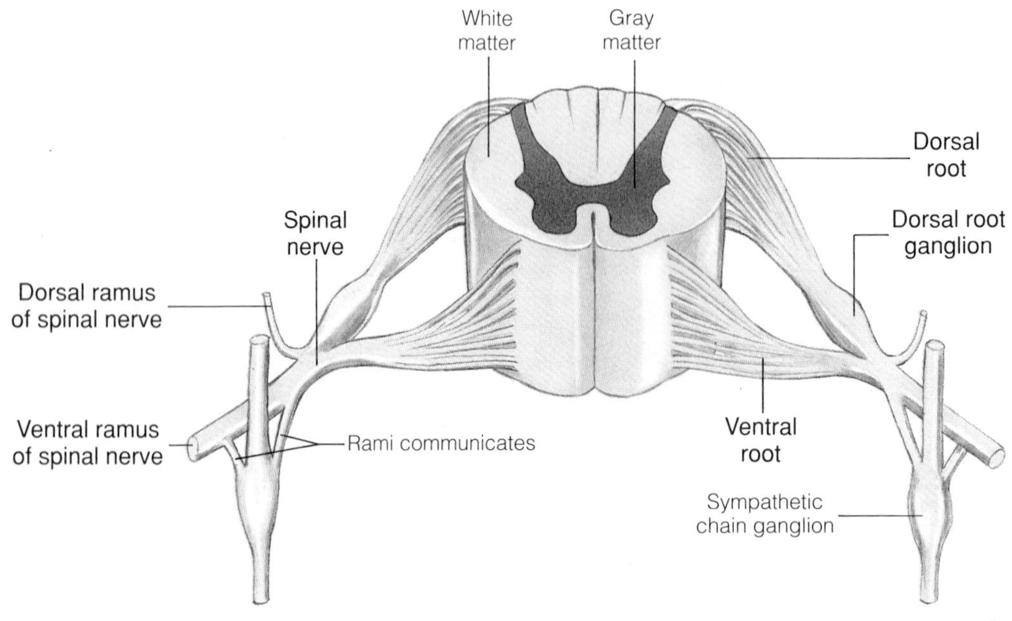

◆ **FIGURE 13.15 The pathways of the ventral and dorsal rami of a thoracic spinal nerve**

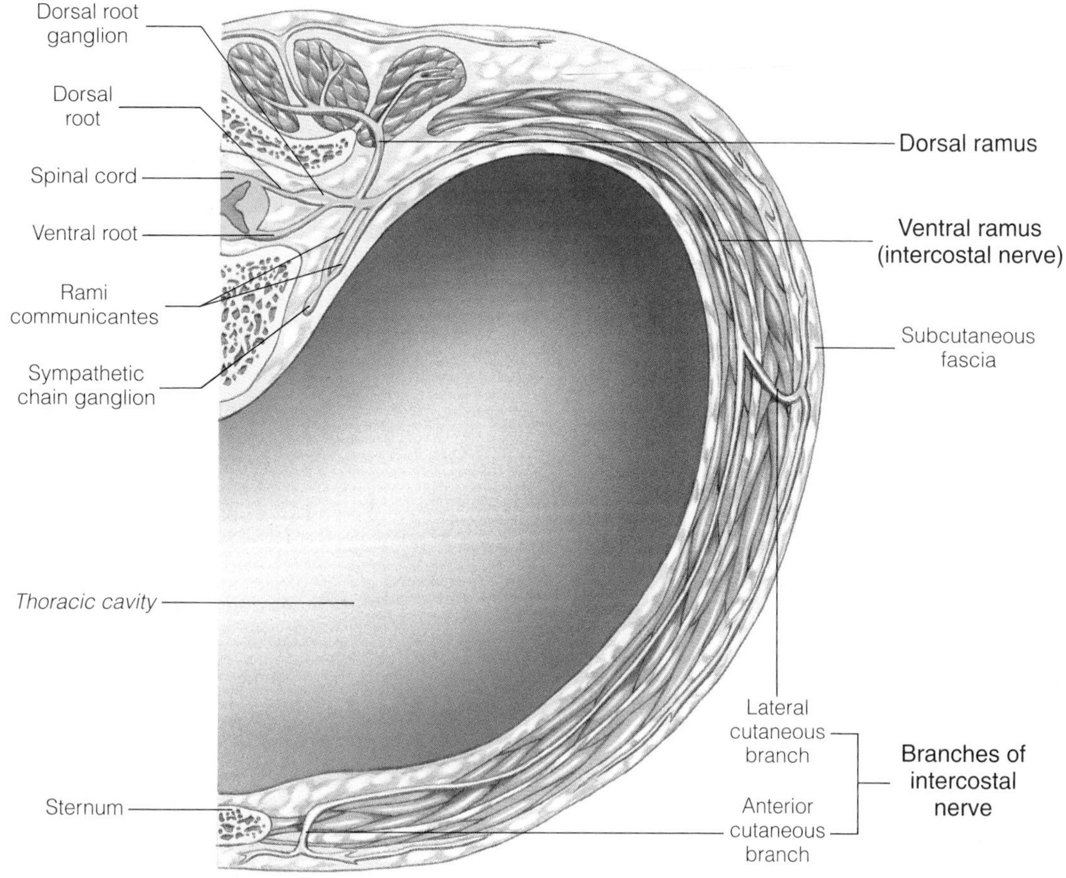

sive spinal nerves unite to form *plexuses* (networks) that give rise to nerves that supply the skin, muscles, and joints of the upper and lower limbs.

Distribution of Spinal Nerves

The rami of the spinal nerves are distributed throughout the body in a fairly systematic pattern, as evidenced by the distribution of sensory fibers to the skin in uniform regions called **dermatomes** (Figure 13.16). Each dermatome is an area of skin that is supplied by the sensory fibers in a single spinal nerve. The pattern is particularly regular in the trunk region, where the rami of each spinal nerve supply a horizontal band of skin. The regular pattern of nerve distribution is also present, although somewhat modified, in the skin of the limbs. In the upper limbs, the ventral rami of cervical nerves supply the posterior and anterolateral surfaces of the entire limb, as well as the ventral surfaces of the hands. The ventral ramus of the first thoracic nerve (T_1) supplies the anteromedial surfaces of the limbs (in the anatomical position). In the lower limbs, the ventral rami of the lumbar nerves supply the anterior surfaces, and the ventral rami of the sacral nerves supply the posterior surfaces.

Plexuses and Peripheral Nerves

The peripheral nerves that are formed by the intermixing of the ventral rami of the spinal nerves in plexuses have specific names. These nerves primarily supply the skin and the underlying muscles of the limbs. Since each of these named peripheral nerves is formed of fibers from more than one spinal nerve, their distribution patterns (Figure 13.16) do not duplicate the cutaneous distribution patterns of the spinal nerves. The main nerve plexuses, all of which are paired, are the *cervical, brachial,* and *lumbosacral* plexuses.

Cervical Plexus

Each **cervical plexus** is formed by the ventral rami of the first four cervical nerves (Figure 13.17). Branches

◆ **FIGURE 13.16 Dermatome and peripheral distribution of spinal nerve innervations**

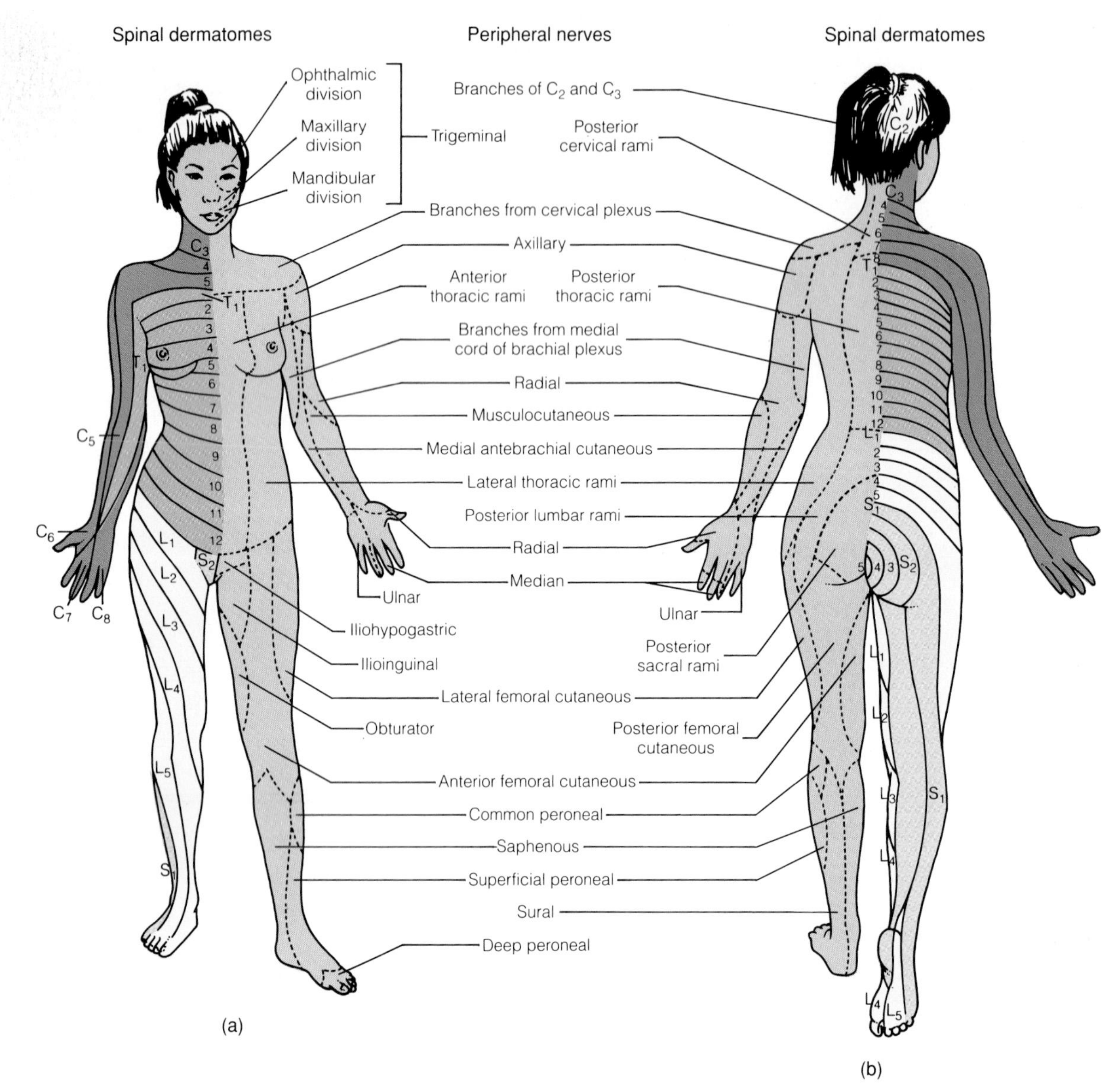

from the plexus supply the muscles and skin of the neck, shoulder, and chest. Branches of the cervical plexus also interconnect with cranial nerves X (vagus), XI (accessory), and XII (hypoglossal). An important branch from each of the cervical plexuses is the **phrenic nerve** *(fren´-ik)*. The phrenic nerve, which receives fibers from C_5 in addition to C_3 and C_4, passes through the thorax to supply the diaphragm. Although this innervation may seem unusual for a cervical nerve, during embryonic development the diaphragm arises from cervical myotomes. As the diaphragm assumes its adult position in the lower portion of the thoracic cavity, it retains its embryonic nerve supply. The major branches from the cervical plexus are summarized in Table 13.2.

Brachial Plexus

Each **brachial plexus** is formed by the ventral rami of the last four cervical nerves (C_5 to C_8) and the first thoracic nerve (T_1). The brachial plexus extends downward and laterally, passing behind the clavicle, to enter the axilla (Figure 13.18). The *roots* of the plexus, which are

ASPECTS OF EXERCISE PHYSIOLOGY

Loss of Muscle Mass: A Problem of Space Flight

It's a case of "use it or lose it" with skeletal muscles. Stimulation of skeletal muscles by motor neurons is essential not only for inducing contraction of the muscles but also for maintaining the muscles' size and strength. Muscles that are not routinely stimulated gradually atrophy, or diminish in size and strength.

When humans entered the weightlessness of space, it became apparent that the muscular system requires the stress of work or gravity to maintain its size and strength. The effort required to move the body is remarkably less in space than on earth, and there is no need for active muscular opposition to gravity. The result is what some refer to as *functional atrophy.*

The muscles most affected are those in the lower extremities, the gluteal (buttocks) muscles, the extensor muscles of the neck and back, and the muscles of the trunk. Changes include a decrease in muscle volume and mass, decrease in strength and endurance, increased breakdown of muscle protein, and loss of muscle nitrogen (an important component of muscle protein). The exact biologic mechanisms that induce muscle atrophy are unknown, but a majority of scientists believe that the lack of customary forcefulness of contraction plays a major role.

Space programs in the United States and the Soviet Union have employed intervention techniques that emphasize both diet and exercise in an attempt to prevent muscle atrophy. Faithful performance of vigorous, carefully designed physical exercise has helped reduce the severity of functional atrophy. Studies of nitrogen and mineral balances, however, suggest that muscle atrophy continues to progress during exposure to weightlessness despite efforts to prevent it.

◆ **FIGURE 13.17 Major branches of the cervical plexus**

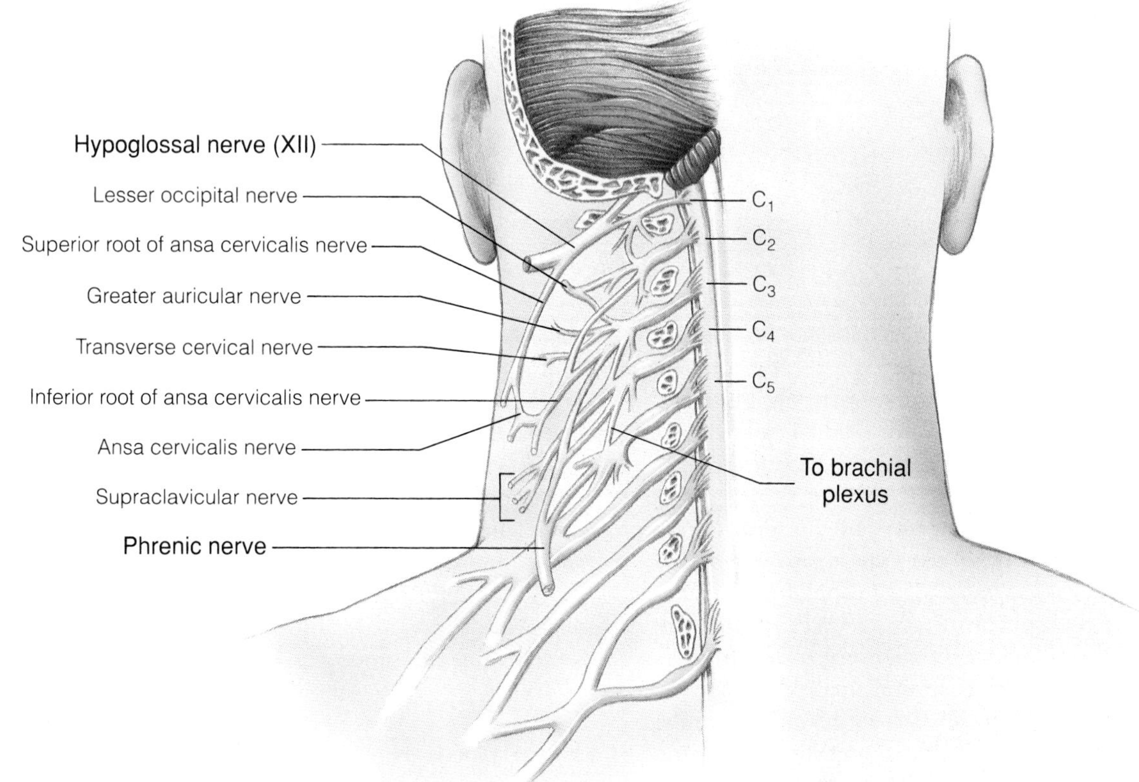

◆ **FIGURE 13.18 The brachial plexus and the nerves of the upper limb**

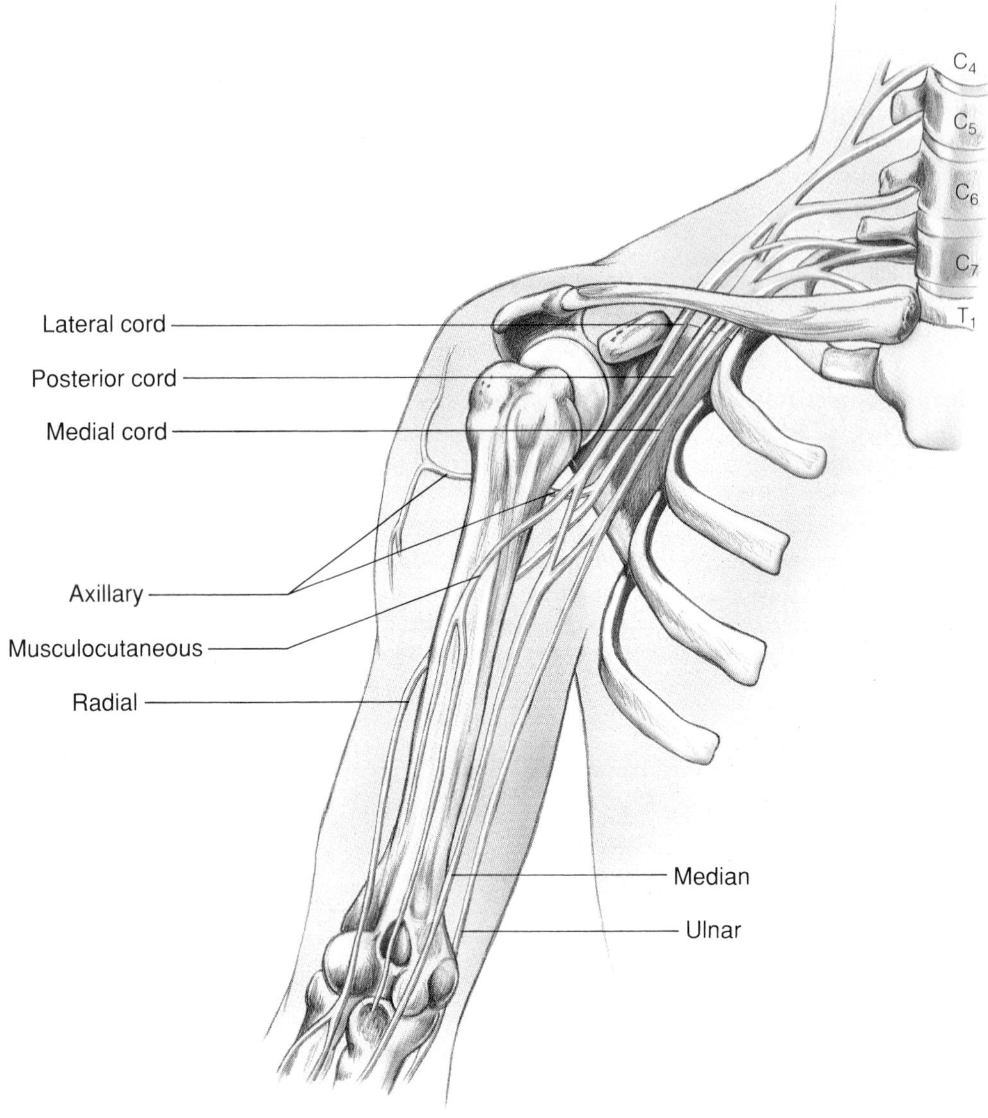

the ventral rami of the spinal nerves, unite to form a *superior trunk* (C_5–C_6), a *middle trunk* (C_7), and an *inferior trunk* (C_8–T_1). Branching off of the roots are a *dorsal scapular nerve* to the levator scapulae and rhomboid muscles, and a *long thoracic nerve* to the serratus anterior muscle.

Each trunk divides into *anterior* and *posterior divisions.* Branching off of the trunks are a *suprascapular nerve* to the supraspinatus and infraspinatus muscles and a nerve to the subclavius muscle. The trunks, in turn, separate into *posterior, lateral,* and *medial* cords. Five major nerves, as well as several smaller ones, arise from these cords and provide the entire nerve supply to the skin and muscles of the upper limbs (Figure 13.19). The major branches of the brachial plexus are summarized in Table 13.3.

Posterior Cord. The **axillary nerve** passes laterally from the posterior cord to supply the skin and muscles of the shoulder. The **radial nerve,** the main branch of the posterior cord, passes behind the humerus and curves around it (in the radial groove) to supply the skin on the posterior surface of the arm, forearm, and hand as well as the extensor muscles of the arm and forearm. Smaller branches of the posterior cord include the *superior* and *inferior subscapular nerves* to the subscapularis and teres major muscles and the *thoracodorsal nerve* to the latissimus dorsi muscle.

Lateral Cord. The **musculocutaneous nerve** from the lateral cord supplies the skin on the lateral surface of the forearm and several anterior muscles of the arm.

◆ **TABLE 13.2 The Cervical Plexus** [*Figs. 13.16 and 13.17*]

NERVE	SPINAL NERVES INVOLVED (VENTRAL RAMI)	DISTRIBUTION
SUPERFICIAL BRANCHES		
Lesser occipital	C_2	*Skin* of scalp behind and above ear
Greater auricular	C_2 through C_3	*Skin* over parotid gland, over mastoid process, and on back of ear
Transverse cutaneous	C_2 through C_3	*Skin* over side and front of neck
Supraclavicular	C_3 through C_4	*Skin* over upper region of shoulder and chest
DEEP (MUSCULAR) BRANCHES		
Ansa cervicalis (leaves the hypoglossal nerve)		
Superior root	C_1 through C_2	Deep *muscles* of neck, including geniohyoid and thyrohyoid
Inferior root	C_2 through C_3	Infrahyoid *muscles*, including sternohyoid and sternothyroid
Phrenic	C_3 through C_5	*Diaphragm*
Muscular branches	C_2 through C_4	*Muscles* sternocleidomastoid, trapezius, levator scapulae, and scalenus medius

The *lateral pectoral nerve* branches off of the lateral cord and supplies the pectoralis major muscle.

Medial Cord. The **ulnar nerve** leaves the medial cord and passes behind the medial epicondyle of the humerus to supply the skin on the medial surface of the hand and some flexor muscles of the forearm as well as several intrinsic muscles of the hand. Other branches from the medial cord include the *medial pectoral nerve* to the pectoralis major and minor muscles, the *medial brachial cutaneous nerve* to the skin of the medial side of

◆ **TABLE 13.3 Major Branches of the Brachial Plexus** [*Figs. 13.16, 13.18, and 13.19*]

NERVE	SPINAL NERVES INVOLVED (VENTRAL RAMI)	DISTRIBUTION
POSTERIOR CORD		
Axillary	C_5 and C_6	*Skin* of the shoulders *Muscles* teres minor and deltoid
Radial	C_5 through C_8; T_1	*Skin* of posteiror lateral surface of arm, forearm and hand *Muscles* triceps brachii, supinator, anconeus, brachioradialis, extensor carpi radialis brevis, extensor carpi radialis longus, extensor carpi ulnaris, and several muscles that move the fingers (see Table 9.14)
LATERAL CORD		
Musculocutaneous	C_5 through C_7	*Skin* of the lateral surface of the forearm *Muscles* brachialis, biceps brachii, coracobrachialis
MEDIAL CORD		
Ulnar	C_8 and T_1	*Skin* of the medial third of the hand *Muscles* flexor carpi ulnaris, flexor digitorum profundus (½), and several intrinsic muscles of the hand (see Table 9.16)
MEDIAN NERVE	C_5 through C_8; T_1	*Skin* of the lateral two-thirds of the hand *Muscles* pronator teres, pronator quadratus, palmaris longus, flexor carpi radialis, flexor pollicis longus, flexor digitorum superficialis, flexor digitorum profundus (½), and several intrinsic muscles of the hand (see Tables 9.14 and 9.16)

◆ **FIGURE 13.19 Major branches of the brachial plexus**

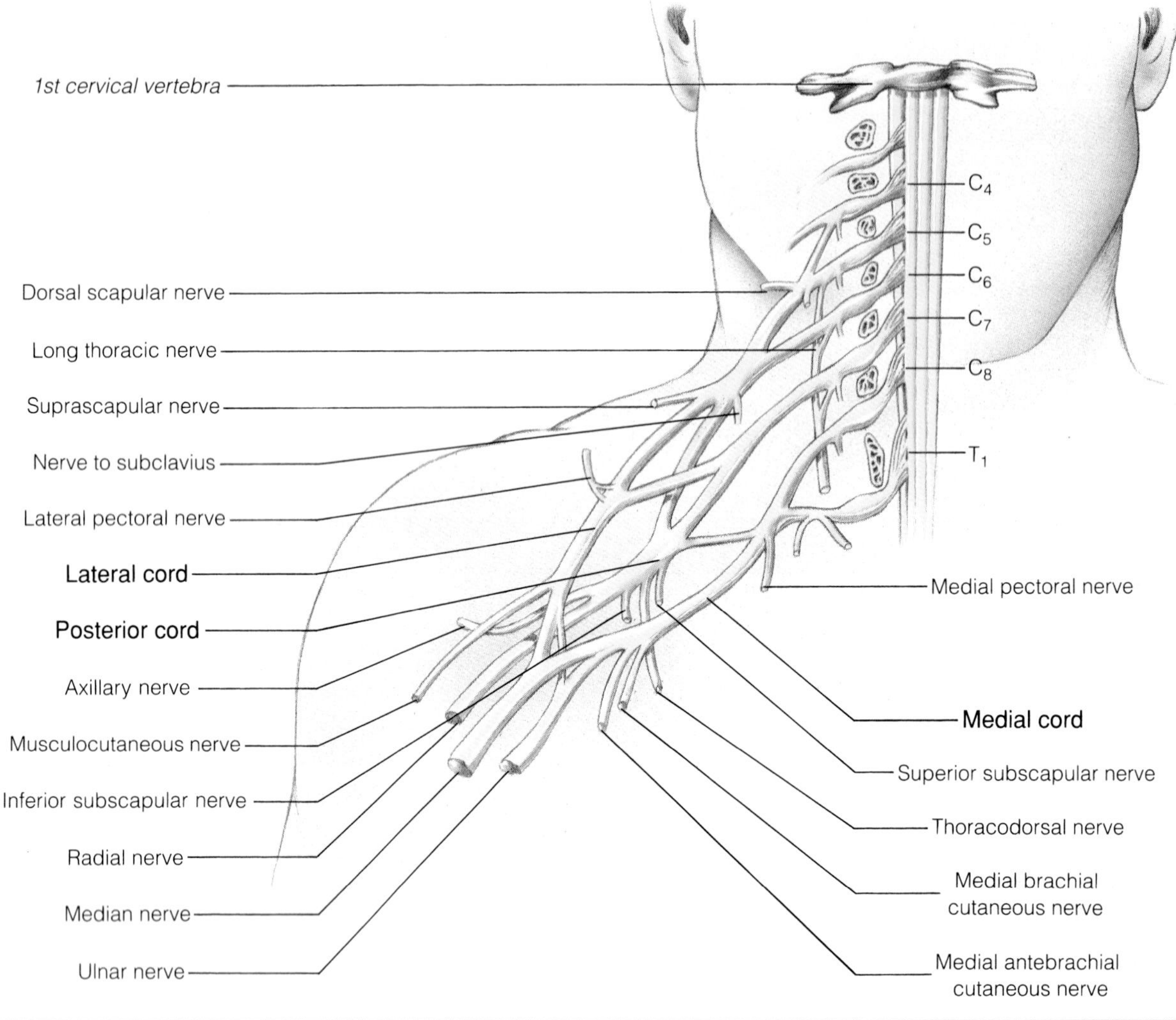

◆ **TABLE 13.4 The Lumbar Plexus [*Figs. 13.16 and 13.20*]**

NERVE	SPINAL NERVES INVOLVED (VENTRAL RAMI)	DISTRIBUTION
Iliohypogastric	T_{12} and L_1	*Skin and muscles* of lower back, hip, and lower abdomen
Ilioinguinal	L_1	*Skin* of upper medial thigh and external genitalia *Muscles* of lower abdominal wall
Genitofemoral	L_1 and L_2	*Skin* of anterior thigh, almost to knee, and external genitalia
Lateral femoral cutaneous	L_2 and L_3	*Skin* of lateral thigh
Femoral	L_2 through L_4	*Skin* of anterior, medial surface of thigh, leg, and foot through anterior femoral cutaneous and saphenous branches *Muscles* sartorius, iliopsoas, quadriceps femoris, pectineus
Obturator	L_2 through L_4	*Skin* of medial thigh *Muscles* adductor longus, adductor magnus, adductor brevis, gracilis, pectineus, obturator externus

◆ **FIGURE 13.20 The lumbosacral plexus**

(a) Major branches of the plexus. (b) Peripheral distribution of the femoral nerve and its branches to the anterior surface of the limb. (c) Peripheral distribution of the sciatic nerve and its branches to the posterior and lateral surfaces of the limb.

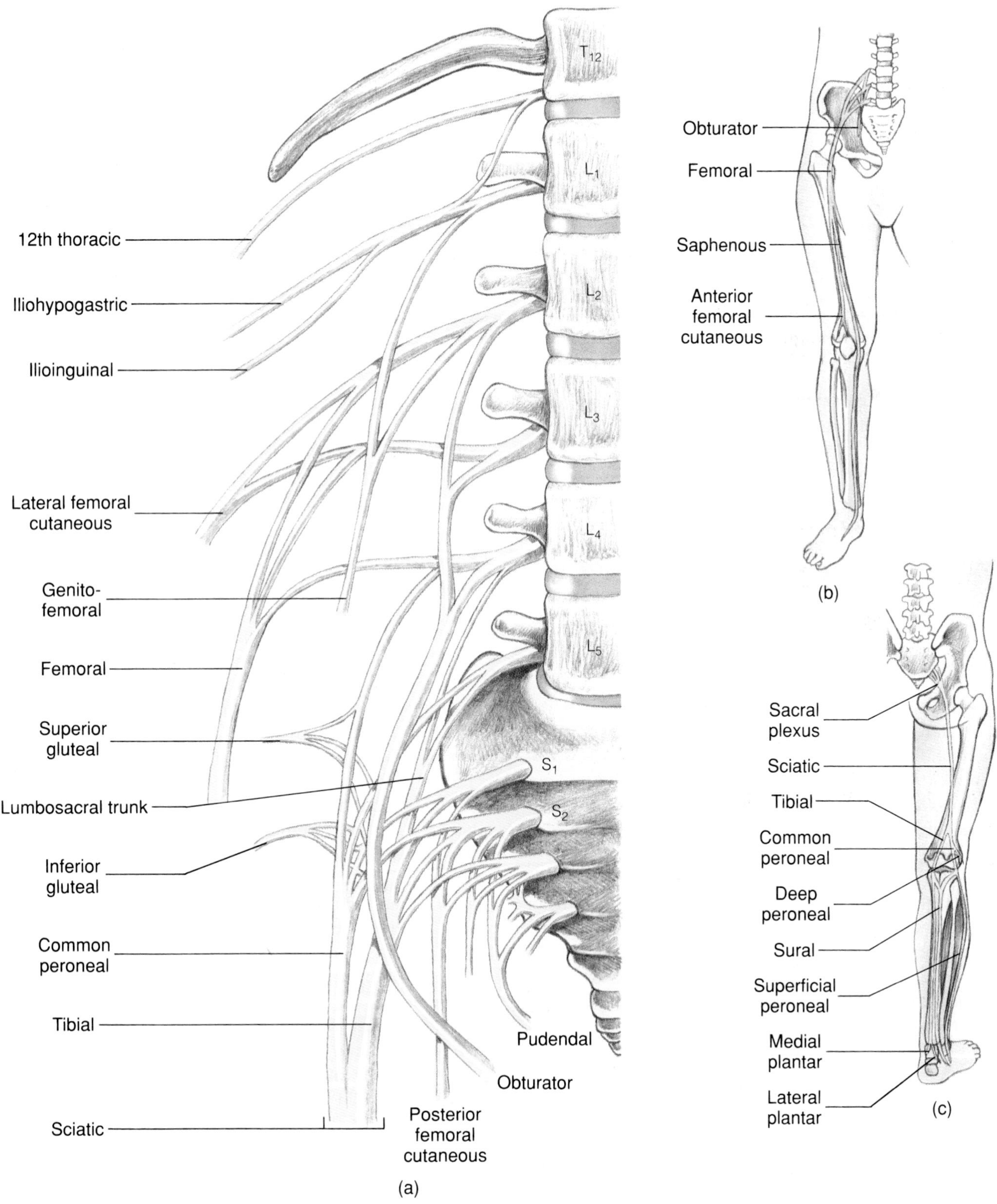

◆ **TABLE 13.5 The Sacral Plexus** [*Figs. 13.16 and 13.20*]

NERVE	SPINAL NERVES INVOLVED (VENTRAL RAMI)	DISTRIBUTION
Superior gluteal	L_4 and L_5; S_1	*Muscles* gluteus minimus, gluteus medius, and tensor fasciae latae
Inferior gluteal	L_4 and L_5; S_1	*Muscles* gluteus maximus
Posterior femoral cutaneous	S_1 through S_3	*Skin* on posterior surface of the thigh
Pudendal	S_2 through S_4	*Skin and muscles* of perineum External genitalia
Sciatic		
Tibial (Sural and medial and lateral plantar branches)	L_4 and L_5; S_1 through S_3	*Skin* of posterior surface of leg and sole of foot *Muscles* biceps femoris, semimembranosus, semitendinosus, gastrocnemius, soleus, flexor digitorum longus, flexor hallucis longus, tibialis posterior, popliteus, and intrinsic muscles of foot (see Table 9.22)
Common peroneal (Superficial and deep peroneal branches)	L_4 and L_5; S_1 and S_2	*Skin* of anterior surface of leg and dorsum of foot *Muscles* peroneus brevis, peroneus longus, peroneus tertius, tibialis anterior, extensor hallucis longus, extensor digitorum longus, extensor digitorum brevis

the arm, and the *medial antebrachial cutaneous nerve* to the skin on the medial side of the forearm.

Median Nerve. The **median nerve** is formed by branches from the medial and lateral cords of the brachial plexus. It supplies the skin and muscles of the lateral portion of the palmar surface of the hand and the flexor muscles of the forearm.

Lumbosacral Plexus

The nerves of each **lumbosacral plexus** supply the skin and muscles of the buttocks, pelvis, lower abdomen, and lower limbs (Figure 13.20). The plexus is divisible into two sections, the *lumbar plexus* and the *sacral plexus,* which are connected by a *lumbosacral trunk* of nerves.

Lumbar Plexus. Each **lumbar plexus** is formed from the ventral rami of the first four lumbar nerves and some nerve fibers from the twelfth thoracic nerve. The nerves that arise from the lumbar plexus supply the lower abdomen and the anterior and medial portions of the lower limb (Figure 13.16). The major nerves that arise from the lumbar plexus are the femoral and the obturator nerves. Smaller branches of this plexus are described in Table 13.4.

The **femoral nerve** passes underneath the inguinal ligament to supply the anterior muscles of the thigh (Figure 13.20b). Two superficial branches of the femoral nerve, the **anterior femoral cutaneous nerve** and the **saphenous nerve,** innervate the skin of the anterior medial surface of the thigh, leg, and foot.

The **obturator nerve** leaves the pelvis through the obturator foramen and supplies the skin on the medial surface of the thigh and the muscles of the medial (adductor) compartment of the thigh.

Sacral Plexus. The **sacral plexus** is formed from the ventral rami of the last two lumbar spinal nerves (L_4 and L_5) and the first four sacral spinal nerves (S_1–S_4). The fibers of the lumbar neurons reach the plexus by way of the **lumbosacral trunk.** The nerves from the sacral plexus supply the lower back, the pelvis, and the posterior surface of the thigh and leg, as well as the dorsal and ventral surfaces of the foot (Figure 13.16). The nerves that form the sacral plexus are summarized in Table 13.5.

The **sciatic nerve,** which is the largest nerve in the body, is the main branch of the sacral plexus. It leaves the pelvis through the greater sciatic notch and travels down the posterior surface of the thigh, innervating the muscles and skin of the region (Figure 13.20c). The sciatic nerve is actually two nerves wrapped in a common sheath. In the lower thigh, these two nerves separate into the common peroneal nerve and the tibial nerve.

The **common peroneal nerve** passes obliquely along the lateral side of the popliteal fossa to supply muscles in the anterior and lateral compartments of the leg and the skin on the anterior surface of the leg and the dorsum of the foot. The common peroneal nerve has branches called the **superficial** and **deep peroneal nerves.** The superficial peroneal supplies the muscles of the lateral compartment of the leg; the deep peroneal supplies the muscles of the anterior compartment of the leg.

The **tibial nerve** supplies the muscles and skin on the posterior surface of the leg and the sole of the foot. The tibial nerve ends on the sole of the foot as the **medial** and **lateral plantar nerves.** Branches from the tibial nerve and the common peroneal nerve give rise to the **sural nerve,** which supplies the posterior and lateral surfaces of the leg and foot.

CONDITIONS OF CLINICAL SIGNIFICANCE

The Peripheral Nervous System

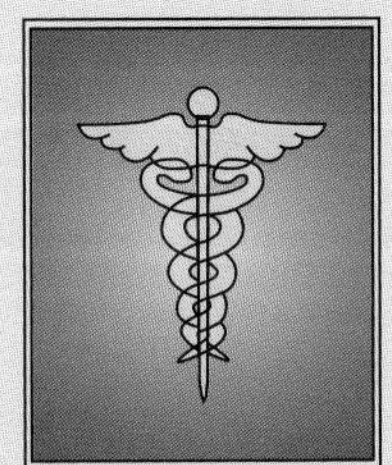

Injury and Regeneration of Peripheral Nerves

The most common disorders involving neurons of the peripheral nervous system are those associated with damage or inflammation. When a peripheral nerve is severely damaged or severed, the portion of the nerve that is distal to the injury undergoes degenerative changes (Figure C13.1). Within a few days following the injury, the nerve fibers and their myelin sheaths are broken down by macrophages from the endoneurium. Cells of the neurilemma (sheath of Schwann) seem to assist in the degeneration by exerting phagocytic action themselves.

Following the injury, the Schwann cells proliferate and form cords in the endoneurial tubes. In a matter of days, sprouts form on the stumps of the damaged nerve fibers. Some of these sprouts grow into the endoneurial tubes, and if there are no obstructions (such as scar tissue) in the tubes, the fibers may again grow out to the periphery and eventually innervate the structures that were separated from their nerve supply. As the new fibers from the proximal stump of the neuron grow along the endoneurial tubes, they are surrounded by Schwann cells. Because nerve fibers regenerate at a rate of from 1 to 4 mm per day, it is possible to estimate the length of time required for the nerve supply to return to the denervated structure.

The first indication that a peripheral nerve has reached the vicinity of a denervated structure is an improved blood supply to the area. This stage is followed by the return of sensory function to the structure. In the case of a paralyzed skeletal muscle, motor function is the last function to return.

Numerous surgical techniques have been used in an attempt to rejoin the severed portions of a damaged nerve. Unfortunately, the success rate of these operations has been only about 15%. Two new procedures have been very successful in reconnecting severed nerves in animals, and they hold great promise for use in humans.

One of these procedures uses a silicone tube filled with a collagen-carbohydrate material to bridge the gap between the ends of a severed nerve. This material provides support for new growths from the proximal portions of the severed nerve fibers. By the time the carbohydrate material degrades six weeks later, some nerve fibers have bridged the gap between the severed ends. Critical to the usefulness of any neurosurgical technique, however, is the number of nerve fibers that reinnervate the same structure they innervated prior to being severed. Whether enough functional nerve fiber reconnections occur when this procedure is used has not yet been satisfactorily determined.

The second procedure attempts to increase the number of functional nerve fiber reconnections between proximal and distal ends of a severed nerve by eliminating the gap that occurs between the nerve endings at the site of injury. After freezing the severed ends of the nerve, soaking them in a solution with the same ionic concentration as the cytoplasm of the nerve cells, and cutting the severed ends of the nerve with a vibrating blade so that they are very smooth, the nerve is held firmly together with a specially designed rubber support. The reported success rates of this procedure have been very high, and the hope is that it will prove just as successful in humans.

Neuritis

The term *neuritis* means "inflammation of a nerve," although many of the conditions referred to as neuritis are more degenerative than inflammatory. Neuritis is characterized by a range of sensations—from

continued on next page

CONDITIONS OF CLINICAL SIGNIFICANCE

◆ **FIGURE C13.1 Regeneration of a peripheral nerve**
(a) Nerve fibers degenerate distal (and a short distance proximal) to the site of the injury. (b) Schwann cells become active, forming cords within the endoneurial tube. (c) Nerve fibers then form sprouts that grow into the endoneurial tube alongside the Schwann cell cords. (d) Nerve fibers follow the Schwann cell cords. The Schwann cells then wrap around the nerve fibers, forming a myelin sheath around the new fiber sprouts.

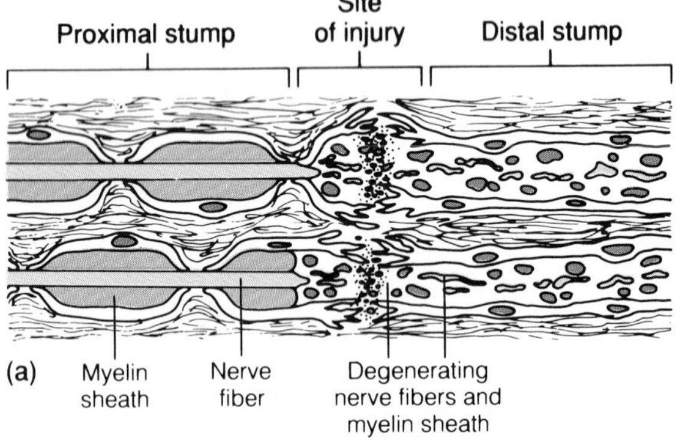

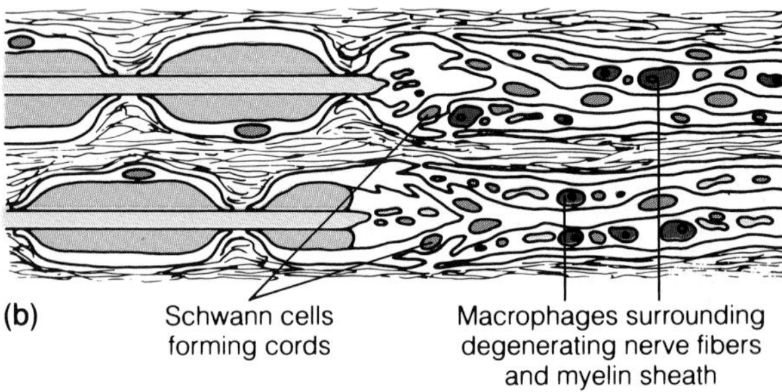

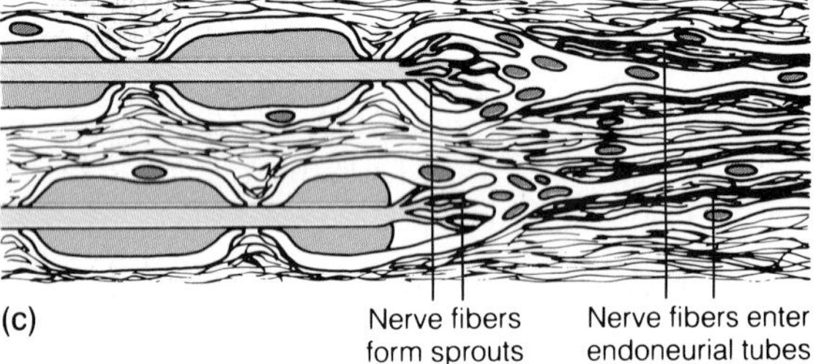

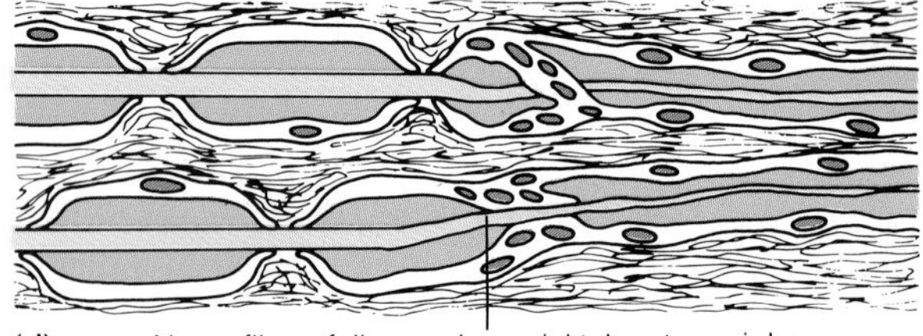

mild tingling to sharp, stabbing pains. It can result from a number of conditions, including mechanical damage to the involved nerves, prolonged pressure on the nerves, vascular disorders involving the nerves, and direct invasion of the nerves by pathological organisms.

Bell's palsy is a term used to describe peripheral inflammation of the facial nerve (cranial nerve VIII). In Bell's palsy, all the muscles of facial expression are paralyzed on the affected side, causing them to sag. This paralysis causes difficulties in speaking and eating, and it results in the lower portion of the face being pulled to the unaffected side when the person smiles.

Neuralgia

Neuralgia refers to attacks of severe pain along the path of a peripheral nerve. There are many varieties of neuralgia, most of which have unknown causes. In *trigeminal neuralgia (tic douloureux),* there are short attacks of excruciating pain along the course of the maxillary and mandibular divisions of the trigeminal nerve (cranial nerve V). The attacks of pain are short at first and may be followed by periods of several weeks without pain. With time, however, the attacks generally become more frequent and of longer duration. The cause of trigeminal neuralgia is unknown.

CONDITIONS OF CLINICAL SIGNIFICANCE

Shingles

The disease known as *shingles* is an infection of the dorsal root ganglia of the spinal nerves by the *herpes zoster* virus. The infection causes pain and produces fluid-filled vesicles of the skin along the path of the peripheral sensory neurons that are infected. (Shingles is discussed in greater detail on page 144.)

Study Outline

◆ **PERIPHERAL NERVOUS SYSTEM** p. 428
Includes 12 pairs of cranial nerves and 31 pairs of spinal nerves; divided into afferent (sensory) and efferent (motor) divisions; efferent divisions divided functionally into somatic and autonomic nervous systems.

◆ **CRANIAL NERVES** pp. 428–437
12 pairs.

I: Olfactory Nerves. Arise from receptor cells in nasal mucosa; entirely sensory; control sense of smell.

II: Optic Nerves. Entirely sensory; carry impulses concerned with vision; arise from retina of eye.

Optic chiasma. Where the two optic nerves meet and the medial fibers of each cross-over.
Optic tract. Beyond optic chiasma; each optic tract consists of fibers from retinas of both eyes.
Optic radiations. Neurons from thalamus to visual cortex of occipital lobe.

III: Oculomotor Nerves.

1. Emerge from midbrain, superior to pons.
2. Innervate four extrinsic muscles of eyeball, which assist in voluntary movements of eyeball, and also the levator palpebrae superioris muscle, which elevates upper eyelids.
3. Also contain neurons of parasympathetic nervous system that cause reflex adjustments of eyes to varying light intensities and focus eyes for near and far vision.

IV: Trochlear Nerves.

1. Arise below inferior colliculi on dorsal surface of midbrain.
2. Somatic motor neurons to, and proprioceptive neurons from, superior oblique eye muscle.
3. Assist in voluntary movements of eyeball by controlling contraction of superior oblique muscle.

V: Trigeminal Nerves. Ophthalmic, maxillary, and mandibular nerves.

1. Emerge from lateral sides of pons.
2. Major sensory nerves of face.
3. Somatic motor neurons to muscles of mastication.

VI: Abducens Nerves.

1. Originate from metencephalon below pons.
2. Somatic motor neurons to, and proprioceptive neurons from, lateral rectus eye muscle.
3. Coordinate with oculomotor and trochlear nerves to cause contraction of extrinsic muscles of both eyes.

VII: Facial Nerves.

1. Arise from metencephalon at lower border of pons.
2. Somatic motor neurons to face and scalp muscles.
3. Carry sensory neurons from taste buds on anterior two-thirds of tongue.
4. Carry parasympathetic neurons to lacrimal glands and submandibular and sublingual salivary glands.

VIII: Vestibulocochlear Nerves. Cochlear and vestibular nerves.

1. Originate from inner-ear receptors; entirely sensory.
2. Cochlear division transmits impulses related to hearing.
3. Vestibular division concerned with equilibrium.

IX: Glossopharyngeal Nerves.

1. Emerge from medulla.
2. Mixed nerves—motor and sensory.
3. Sensory—impulses from taste buds of posterior third of tongue, from membranes of pharynx and tonsils, from receptors in carotid sinus and carotid bodies.
4. Somatic motor—innervate pharyngeal muscles.
5. Parasympathetic neurons to parotid salivary glands.

X: Vagus Nerves. Only cranial nerves not restricted to head and neck regions.

1. Arise from sides of medulla.
2. Somatic motor neurons to, and sensory neurons from, pharynx and larynx.
3. Sensory input from viscera regulates heart rate, respiration depth, blood pressure, digestion, etc.
4. Parasympathetic neurons to thoracic and abdominal viscera and blood vessels.

XI: Accessory Nerves.

1. Composed of a true cranial nerve (which arises from medulla) and a cervical spinal nerve.
2. Cranial portions supply pharynx and larynx muscles.
3. Spinal portions supply trapezius and sternocleidomastoid muscles.

XII: Hypoglossal Nerves.

1. Arise from anterior surface of medulla.
2. Supply intrinsic and extrinsic tongue muscles.
3. Sensory neurons from proprioceptors.

◆ SPINAL NERVES pp. 437–449

31 pairs.

Formation, Branches, and Distribution.

1. Union of ventral and dorsal roots.
2. All are mixed (motor and sensory) nerves.
3. Each spinal nerve divides into dorsal and ventral rami.
4. Sensory nerves in rami are distributed in bands called dermatomes.

Plexuses and Peripheral Nerves.

CERVICAL PLEXUS. Formed by ventral rami of first four cervical nerves; supply muscles and skin of neck and shoulder; phrenic nerve supplies diaphragm.

BRACHIAL PLEXUS. Formed by ventral rami of last four cervical nerves and first thoracic nerve. Dorsal scapular and long thoracic nerves branch from roots; suprascapular nerve and nerve to subclavius muscle branch from the trunks of plexus.

POSTERIOR CORD.

1. *Axillary nerve.* Skin and muscles of shoulder.
2. *Radial nerve.* Skin on posterior surface of arm, forearm, and hand; extensor muscle of arm and forearm.
3. *Superior* and *inferior subscapular nerves* and *thoracodorsal nerves* branch from posterior cord.

LATERAL CORD.

1. *Musculocutaneous nerve.* Skin of lateral surface of forearm; anterior muscles of arm.
2. *Lateral pectoral nerve.* Supplies pectoralis major muscle.

MEDIAL CORD.

1. *Ulnar nerve.* Skin of medial surface of hand; forearm; intrinsic hand muscles.
2. *Medial pectoral nerve; medial brachial cutaneous nerve; medial antebrachial cutaneous nerve.*

MEDIAN NERVE. From branches of *medial* and *lateral* cords; flexor muscles of forearm; skin and muscles of lateral palmar portion of hand.

LUMBOSACRAL PLEXUS. Skin and muscles of buttocks, pelvis, lower abdomen, lower limbs.

LUMBAR PLEXUS. Ventral rami L_1–L_4

1. *Femoral nerve.* Anterior thigh muscles, skin of anterior medial thigh, leg, and foot; has anterior femoral cutaneous and saphenous branches.
2. *Obturator nerve.* Skin of medial thigh; adductor muscles of thigh.

SACRAL PLEXUS. Ventral rami L_4, L_5, and S_1–S_4.

1. *Sciatic nerve.* Muscles and skin of posterior surface of thigh.
2. *Common peroneal nerve.* Muscles of anterior and lateral compartments of leg; skin on anterior surface of leg and dorsum of foot; has superficial and deep peroneal branches.
3. *Tibial nerve.* Muscles and skin on posterior surface of leg and sole of foot; has sural as well as medial and lateral plantar branches.

◆ CONDITIONS OF CLINICAL SIGNIFICANCE: THE PERIPHERAL NERVOUS SYSTEM pp. 449–451

Injury and Regeneration of Peripheral Nerves.

1. Macrophages break down nerve fibers and myelin sheath distal to site of injury.
2. Schwann cells proliferate, forming cords.
3. Buds form on stumps of nerves.
4. New fibers develop from buds and grow toward periphery.
5. Schwann cells surround new fibers.

Neuritis.

1. Degenerative or inflammatory condition.
2. Tingling to sharp pain.
3. Results from mechanical damage, prolonged pressure, vascular disorders, pathological organism invasion.
4. Bell's palsy is neuritis of facial nerve.

Neuralgia. Attacks of severe pain along peripheral nerve path.

Trigeminal neuralgia (tic douloureux). Short attacks of severe pain along maxillary and mandibular divisions of trigeminal cranial nerve.

Shingles. Viral infection of dorsal root ganglia of spinal nerves.

Self-Quiz

1. The peripheral nervous system includes the: (a) brain; (b) spinal cord; (c) spinal nerves.
2. Which motor neurons are not under conscious control, are within the cranial nerves and supply smooth or cardiac muscle? (a) visceral motor neurons; (b) somatic motor neurons; (c) sympathetic neurons.
3. The olfactory nerves are: (a) motor and leave the mesencephalon; (b) sensory and enter the pons; (c) sensory and enter the telencephalon.
4. The optic nerves are: (a) sensory and leave the skull through the optic foramen; (b) motor and leave the telencephalon; (c) motor and leave the skull through the superior orbital fissure.
5. The oculomotor nerves: (a) include proprioceptor fibers; (b) are located in the diencephalon; (c) leave the skull through the foramen rotundum.
6. The trochlear nerves: (a) arise in the diencephalon; (b) aid in voluntary movements of the eyeball; (c) are large nerves.
7. The trigeminal nerves are the: (a) major sensory nerves of the face; (b) principal nerves that control eyeball movement; (c) nerves that control the tongue.
8. The abducens nerves: (a) originate from the telencephalon; (b) are sensory nerves; (c) leave the skull through the superior orbital fissure.
9. The facial nerves: (a) leave the skull through the foramen magnum; (b) also supply the viscera of the thorax; (c) are mixed nerves.
10. The vestibulocochlear nerves: (a) are concerned with equilibrium and hearing; (b) are motor nerves; (c) arise in the telencephalon.
11. The glossopharyngeal nerves: (a) leave the skull through the stylomastoid foramen; (b) carry motor and sensory impulses; (c) regulate contractions of the abdominal viscera.
12. The vagus nerves: (a) supply most organs of the thorax and abdomen; (b) are restricted to the head and neck region; (c) carry sensory impulses to the pharynx.
13. The accessory nerves are composed of: (a) somatic motor neurons; (b) sensory neurons from proprioceptors; (c) parasympathetic neurons; (d) both (a) and (b).
14. The hypoglossal nerves: (a) supply the intrinsic and extrinsic muscles of the pharynx; (b) arise in the telencephalon; (c) consist of motor neurons that supply the tongue.
15. The spinal nerves are formed from the union of ventral and dorsal roots that leave or enter the spinal cord. True or False?
16. After the ventral and dorsal roots join, all the spinal nerves are mixed nerves containing the processes of motor neurons (somatic and visceral) and sensory neurons. True or False?
17. The cell bodies of the motor neurons are located outside the spinal cord, whereas those of the sensory neurons lie within the spinal cord. True or False?
18. The phrenic nerve supplies the diaphragm after arising from which plexus? (a) lumbosacral; (b) brachial; (c) cervical.
19. The skin on the posterior surface of the arm, forearm, and hand, and the extensor muscles of the arm and forearm are supplied by the: (a) radial nerve; (b) phrenic nerve; (c) musculocutaneous nerve.
20. The palmaris longus and flexor carpi radials muscles are supplied by the: (a) ulnar nerve; (b) radial nerve; (c) median nerve.
21. The sciatic nerve, which is the largest nerve in the body, is the main branch of the sacral plexus. True or False?
22. It is possible for a peripheral nerve that has been severed to regeneate and innervate the same structures it supplied before it was damaged. True or False?

CHAPTER 14

The Autonomic Nervous System

CHAPTER CONTENTS

LEARNING OBJECTIVES

After completing this chapter, you should be able to:

1. State the function of the autonomic nervous system.
2. Distinguish between the sympathetic and parasympathetic divisions of the autonomic nervous system in terms of structure and function.
3. List the paths that the preganglionic sympathetic axons can follow upon entering the chain ganglia.
4. Describe the roles of nicotinic and muscarinic receptors.
5. Describe biofeedback as it relates to the autonomic nervous system.

CHAPTER 14

The **autonomic nervous system (ANS)** is a part of the efferent division of the peripheral nervous system. It is composed entirely of visceral motor (efferent) neurons that innervate and thus regulate the activity of cardiac muscle, smooth muscle, and the glands of the body. This system is normally an involuntary system, functioning below the conscious level.

We will consider the autonomic nervous system separately from the peripheral nervous system, but keep in mind that the ANS is structurally and functionally an integral part of the body's single nervous system. In fact, many nerve fibers of the ANS travel in the spinal nerves and certain cranial nerves. Even though the autonomic nervous system has only motor functions, visceral sensory fibers of the afferent division of the peripheral nervous system travel along the same pathways as the visceral motor fibers of the autonomic nervous system. The cell bodies of these visceral sensory fibers are located in the dorsal root ganglia of the spinal nerves or in one of the outlying ganglia of certain cranial nerves.

Anatomy of the Autonomic Nervous System

The autonomic nervous system pathways that pass from the central nervous system to the effectors are composed of two neurons. One of these neurons, called a **preganglionic (presynaptic) neuron,** has its cell body in the central nervous system. The axon of the preganglionic neuron extends to an **autonomic ganglion** located outside the central nervous system. In the autonomic ganglion, the preganglionic neuron synapses with other neurons called **postganglionic (postsynaptic) neurons.** Shortly after leaving an autonomic ganglion, the axons of postganglionic neurons generally enter into branching networks known as **autonomic plexuses** and then travel to the various effectors. This two-neuron chain is in contrast to the somatic nervous system, where a single motor neuron extends from the central nervous system to the structure innervated.

The autonomic nervous system can be separated both structurally and functionally into two divisions: the *sympathetic division* and the *parasympathetic division* (Figure 14.1).

Sympathetic Division

The cell bodies of the preganglionic neurons of the **sympathetic division** of the autonomic nervous system are located in the lateral horns of the gray matter of the spinal cord, from the first thoracic segment (T_1) through the second lumbar segment (L_2). For this reason, the sympathetic division is also called the **thoracolumbar division** *(thō″-rah-kō-lum′-bar)*. The axons of these visceral motor neurons leave the spinal cord through the ventral roots, along with somatic motor axons, and enter the ventral rami of the spinal nerves of the various segments. After going only a short distance in the ventral rami, all the preganglionic sympathetic nerve fibers leave the spinal nerve and enter one of a series of interconnected **sympathetic chain (paravertebral) ganglia** (Figure 14.2). The chain ganglia form longitudinal pathways, called **sympathetic trunks,** on both sides of the bodies of the vertebrae along the entire length of the vertebral column, including the cervical and sacral regions. Except in the cervical region, where several sympathetic chain ganglia fuse together to form three or four larger ganglia, the ventral ramus of each spinal nerve generally has a sympathetic chain ganglion associated with it. Since most of the preganglionic sympathetic nerve fibers are myelinated, the short pathways they form in passing from the ventral ramus of the spinal nerve to the sympathetic trunks appear white and are called **white rami communicantes** *(kō-mu″-ne-kan′-tēz)*. There are 14 pairs of white rami communicantes connecting the first thoracic through the second lumbar spinal nerves to the sympathetic chain ganglia.

Upon entering the sympathetic chain ganglia, the preganglionic sympathetic axons follow one of three courses:

1. *Preganglionic sympathetic axons may synapse with the cell bodies of postganglionic neurons in the sympathetic chain ganglion located at the same level at which the preganglionic fibers entered the chain.* The axons of the postganglionic neurons return directly to the spinal nerve and travel to the periphery in the dorsal and ventral rami of the spinal nerve. These axons innervate effectors in the skin, including smooth muscles in the walls of the blood vessels of the skin, sweat glands, and the arrector pili muscles of the hairs. Because most postganglionic axons are unmyelinated, the pathways they form as they pass from the sympathetic chain ganglia to the spinal nerves appear gray and they are called **gray rami communicantes.**
2. *Preganglionic sympathetic axons may pass up or down within the sympathetic trunk before synapsing with postganglionic neurons at a higher or lower level.* The axons of some of these postganglionic neurons enter the cervical and sacral spinal nerves through gray rami communicantes and ultimately innervate the skin (blood vessels, sweat glands, arrector pili muscles) of the regions supplied by those nerves. Notice that, although preganglionic axons enter the sympathetic chain ganglia through white rami from only the first thoracic through the second lumbar segments, every chain ganglion is connected to a spinal nerve by a gray ramus.

◆ **FIGURE 14.1 The autonomic nervous system**

The parasympathetic division is shown in red; the sympathetic division is shown in black. The solid lines indicate preganglionic nerve fibers; the dashed lines indicate postganglionic nerve fibers. Postganglionic nerve fibers that originate from the thoracic and lumbosacral chain ganglia are not shown.

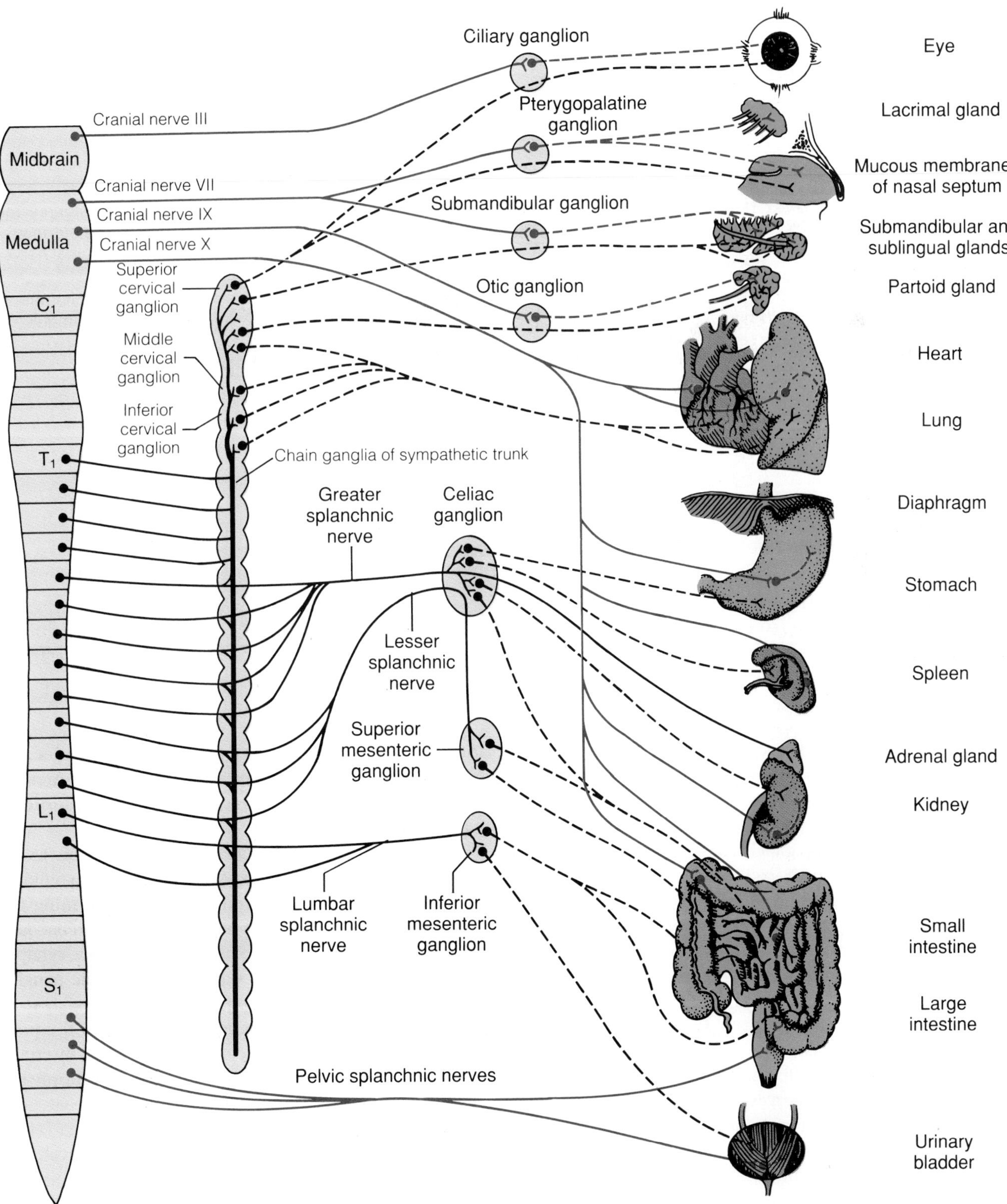

◆ **FIGURE 14.2 The sympathetic division of the autonomic nervous system**

The central connections of the sympathetic neurons are shown as seen in one segment of the thoracic spinal cord. Solid black lines indicate sympathetic preganglionic neurons; dotted black lines indicate sympathetic postganglionic neurons; blue lines indicate sensory neurons. The arrows show the direction of the nerve impulse.

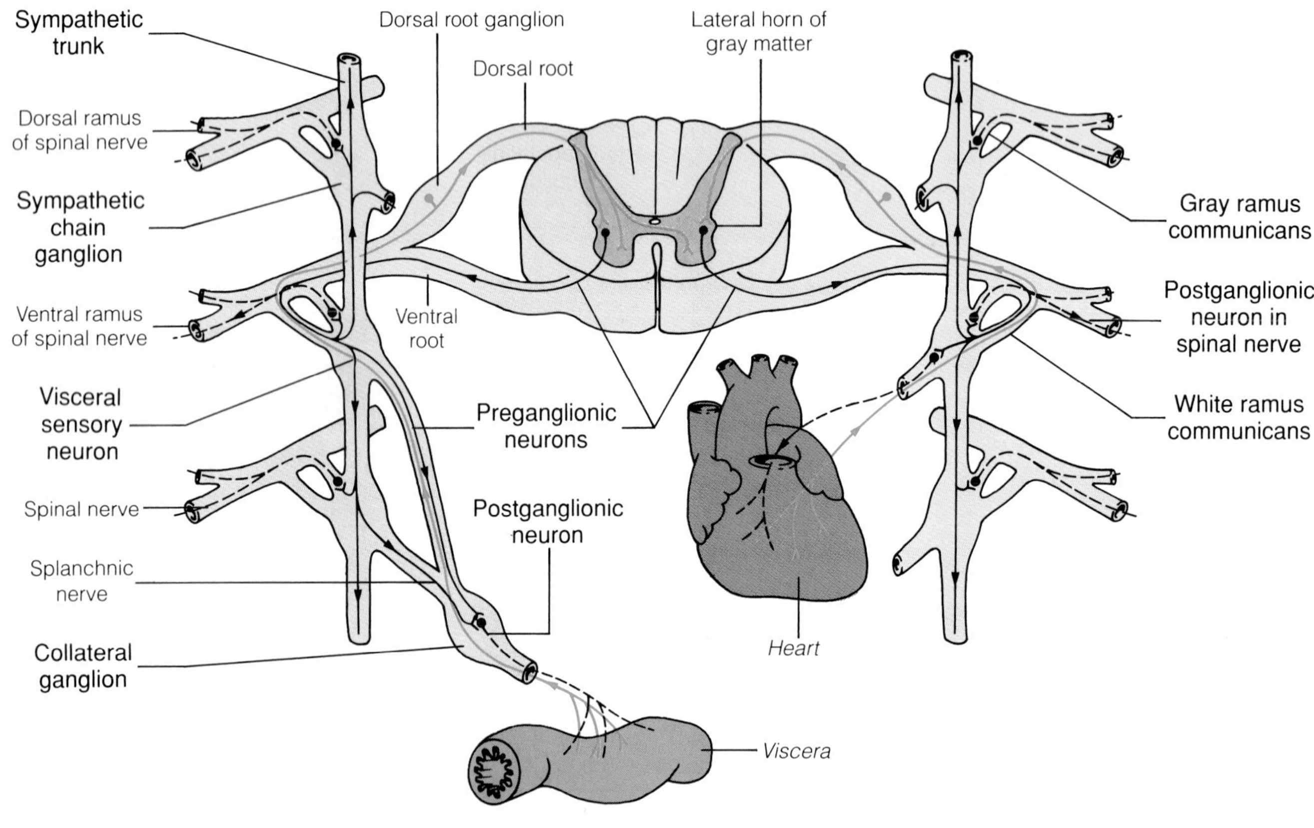

Thus, because preganglionic axons may extend to higher or lower levels in the sympathetic trunk, every spinal nerve receives axons of postganglionic sympathetic neurons. However, some postganglionic axons in the thoracic and cervical ganglia pass directly from the ganglia to innervate the thoracic viscera and structures in the head, rather than entering the spinal nerves.

3. *Preganglionic sympathetic axons may pass through the sympathetic chain ganglia without synapsing.* Some preganglionic axons from T_5 through L_2 do not synapse within the sympathetic chain ganglia; rather, they form pathways called **greater, lesser,** and **lumbar splanchnic nerves** *(splank´-nik).* The splanchnic nerves pass through the diaphragm and form an interconnecting network of nerves called the **abdominal aortic plexus** on the anterior surface of the abdominal aorta (which is the large blood vessel that carries blood from the heart and distributes it to the body). Within the aortic plexus are several **collateral (prevertebral) ganglia.** The major collateral ganglia, which are named for the vascular branches of the aorta they are located close to, are the paired **celiac** and **superior mesenteric ganglia** and the single **inferior mesenteric ganglion.**

Within the collateral ganglia, the preganglionic axons of the splanchnic nerves synapse with postganglionic neurons. Most of the preganglionic axons forming the greater splanchnic nerves synapse within the celiac ganglia. The fibers of the lesser splanchnic nerves synapse in either the celiac or the superior mesenteric ganglia. The fibers of the lumbar splanchnic nerves synapse in the inferior mesenteric ganglion. The axons of the postganglionic neurons leave the collateral ganglia, interconnect to form several autonomic plexuses, and supply the viscera of the abdominopelvic cavity.

Preganglionic sympathetic neurons that innervate the adrenal medulla travel in the splanchnic nerves and do not synapse before reaching the gland. Thus, there are no postganglionic sympathetic neurons innervating the

◆ **FIGURE 14.3 Neurotransmitters of the efferent division of the peripheral nervous system**

(a) Preganglionic neurons (solid line) of the sympathetic division of the autonomic nervous system release acetylcholine at their synapses with postganglionic neurons (dashed line). Although exceptions occur, the postganglionic neurons release mainly norepinephrine at their junctions with effectors. (b) Preganglionic neurons (solid line) of the parasympathetic division of the autonomic nervous system release acetylcholine at their synapses with postganglionic neurons (dashed line), and the postganglionic neurons also release acetylcholine at their junctions with effectors. (c) The adrenal medulla is supplied by preganglionic sympathetic neurons that release acetylcholine. The adrenal medulla releases epinephrine and norepinephrine into the bloodstream, which carries the secretions to effectors. (d) Somatic efferent neurons release acetylcholine at their junctions with skeletal muscle cells.

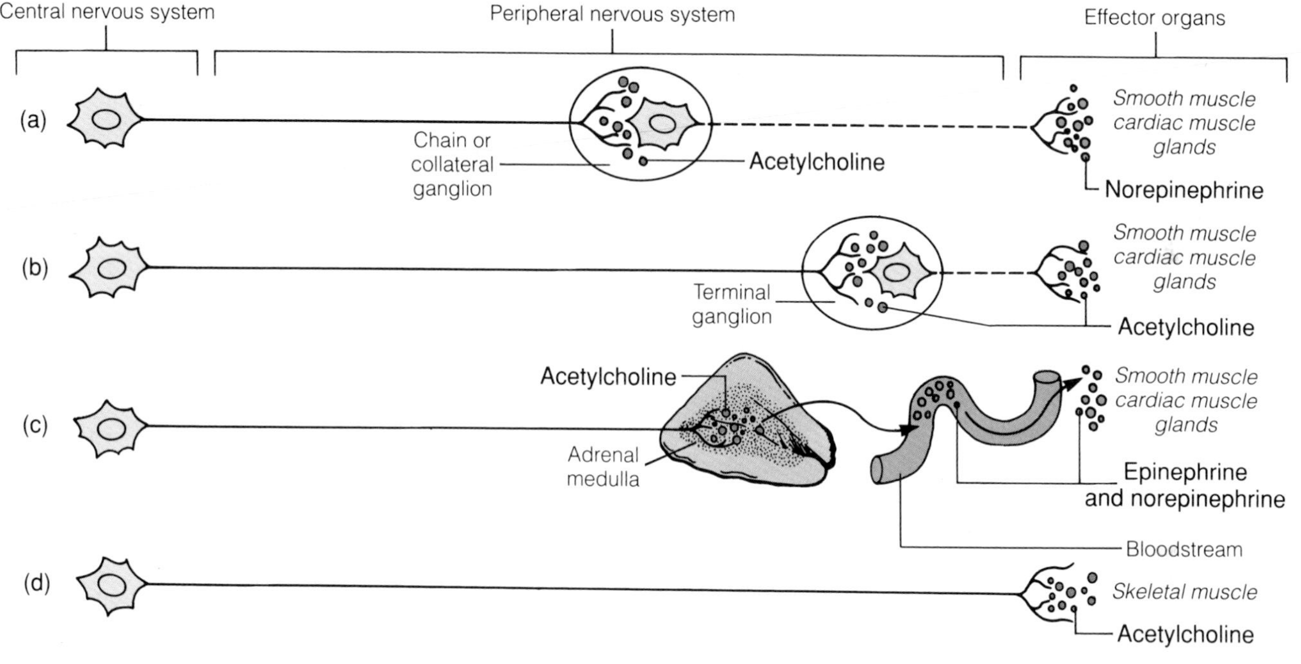

adrenal medulla—the only exception to the usual two-neuron chain in autonomic efferent pathways. The reason for this exception is explained later in the chapter (page 461) and it is illustrated in Figure 14.3.

Accompanying the sympathetic motor fibers that supply the viscera are sensory fibers of the afferent division of the peripheral nervous system returning from the viscera. These visceral sensory fibers pass, without synapsing, from the innervated structure, through the sympathetic chain ganglia and the white rami communicantes, to their cell bodies in the dorsal root ganglia. Therefore, the cell bodies of both visceral and somatic sensory neurons are located in the same ganglia.

Parasympathetic Division

The cell bodies of the preganglionic neurons of the **parasympathetic division** of the autonomic nervous system are located either within nuclei in the brain stem or within the lateral portions of the gray matter of the spinal cord in the second, third, and fourth sacral segments (Figure 14.1). Because of these origins, the parasympathetic division of the autonomic nervous system is also referred to as the **craniosacral division.** The distribution of the parasympathetic division differs from that of the sympathetic division in that parasympathetic fibers do not travel within the rami of the spinal nerves. Consequently, the sweat glands, arrector pili muscles, and cutaneous blood vessels do not receive parasympathetic innervation. In fact, with few exceptions, the parasympathetic division does not innervate blood vessels anywhere in the body.

Parasympathetic axons whose cell bodies are located in nuclei of the brain stem travel to the viscera of the head, thorax, and abdomen within the cranial nerves—specifically, the oculomotor, facial, glossopharyngeal, and vagus nerves. (The specific distribution of the parasympathetic axons in these cranial nerves is discussed on pages 431–437 and summarized in Table 13.1.) Pre-

◆ **TABLE 14.1 Distinguishing Anatomical Features of the Sympathetic and Parasympathetic Divisions of the Autonomic Nervous System**

FEATURE	SYMPATHETIC DIVISION	PARASYMPATHETIC DIVISION
Origin of preganglionic neuron	Thoracic and lumbar regions of spinal cord	Brain and sacral regions of spinal cord
Origin of postganglionic neuron (location of ganglion)	Sympathetic chain ganglia (near spinal cord) or collateral ganglia (about halfway between spinal cord and effector organs)	Terminal ganglia (in or near effector organs)
Lengths and types of neurons	Short cholinergic preganglionic neurons Long adrenergic postganglionic neurons (most) Long cholinergic postganglionic neurons (few)	Long cholinergic preganglionic neurons Short cholinergic postganglionic neurons
Effector organs innervated	Cardiac muscle, almost all smooth muscle, and exocrine glands	Cardiac muscle, most smooth muscle, and exocrine glands
Ratio of preganglionic to postganglionic neurons	Each preganglionic neuron synapses with many postganglionic neurons	Each preganglionic neuron synapses with comparatively few postganglionic neurons

ganglionic parasympathetic axons in the four cranial nerves synapse with postganglionic neurons in ganglia **(ciliary, pterygopalatine, otic, submandibular,** and **terminal ganglia)** that are located close to the structures innervated by the postganglionic neurons.

Preganglionic parasympathetic axons whose cell bodies are located in the sacral region of the spinal cord leave the cord in the ventral roots of the sacral spinal nerves. The parasympathetic axons then leave the ventral roots and join together to form the **pelvic splanchnic nerves,** which interconnect in the hypogastric plexus and supply the viscera of the pelvic cavity. The preganglionic axons of the sacral parasympathetic neurons synapse with postganglionic neurons in terminal ganglia located close to the organs they innervate.

Anatomical Differences between the Divisions

The sympathetic and parasympathetic divisions of the autonomic nervous system differ not only in the locations of the cell bodies of their preganglionic neurons but also in the lengths of their fibers and in the ratio between numbers of preganglionic and postganglionic neurons.

In the sympathetic division, most preganglionic axons are relatively short, synapsing in the chain ganglia close to the vertebral column. The postganglionic axons are long, extending from the chain ganglia to the structures they innervate. In contrast, the parasympathetic preganglionic axons are relatively long, passing uninterrupted from their origins in the central nervous system to terminal ganglia located on or close to the walls of the organs they supply. The postganglionic axons of the parasympathetic division are short, extending from the terminal ganglia to the organs innervated. Each sympathetic preganglionic neuron branches extensively and synapses with numerous postganglionic neurons. In contrast, parasympathetic preganglionic neurons synapse with comparatively few postganglionic neurons.

Table 14.1 summarizes the distinguishing anatomical features of the sympathetic and parasympathetic divisions of the autonomic nervous system.

Neurotransmitters of the Autonomic Nervous System

Parasympathetic preganglionic and postganglionic neurons, as well as sympathetic preganglionic neurons, release *acetylcholine*—the same neurotransmitter substance released by somatic efferent neurons supplying skeletal muscle (Figure 14.3). Therefore, these neurons are called **cholinergic neurons** *(ko-lin-ur´-jick).* In contrast, most sympathetic postganglionic neurons secrete *norepinephrine (noradrenaline).* Consequently, these neurons are called **adrenergic neurons** *(ad-ren-ur´-jick)* or, sometimes, noradrenergic neurons. Some postganglionic sympathetic neurons, however, secrete acetylcholine. For example, the sympathetic postganglionic neurons that innervate the sweat glands are cholinergic.

Norepinephrine (as well as the closely related substance epinephrine) is also secreted by the **adrenal medulla,** an endocrine gland. Like all endocrine glands, the adrenal medulla releases its secretions into the bloodstream. The adrenal medulla possesses physiological and biochemical properties similar to those of the sympathetic nervous system, and it is innervated by cholinergic preganglionic sympathetic neurons that do not synapse before reaching the gland. In fact, the adrenal medulla is essentially a modified sympathetic ganglion. Norepinephrine has the same effects, whether released into the bloodstream by the adrenal medulla or secreted directly onto an organ by a sympathetic postganglionic neuron. However, when norepinephrine is released into the bloodstream, it is carried to all parts of the body, and thus its effects may be more widespread.

Table 14.2 summarizes the sites where acetylcholine and norepinephrine are released in the peripheral nervous system.

◆ **TABLE 14.2 Sites of Release for Acetylcholine and Norepinephrine in the Peripheral Nervous System**

ACETYLCHOLINE	NOREPINEPHRINE
All preganglionic neurons of the autonomic nervous system	Most sympathetic postganglionic neurons
All parasympathetic postganglionic neurons	Adrenal medulla
Sympathetic postganglionic neurons at sweat glands	
Efferent neurons supplying skeletal muscle (motor neurons)	

Receptors for Autonomic Neurotransmitters

Once they are released, acetylcholine, norepinephrine, and epinephrine influence the activity of postganglionic neurons and effector cells by combining with receptors that are present on the plasma membranes of the neurons and effector cells. There are a number of different types of receptors (Table 14.3). Moreover, a given postganglionic neuron or effector cell can possess more than one receptor type.

Acetylcholine receptors are called *cholinergic receptors.* These receptors are described as being either *nicotinic* or *muscarinic* because the substance nicotine mimics the effects of acetylcholine at some cholinergic receptors, and the substance muscarine mimics the effects of acetylcholine at others.

Nicotinic receptors are present on postganglionic neurons within the ganglia of the autonomic nervous system (and also on skeletal muscle cells at neuromuscular junctions). These receptors are part of chemically gated ion channels that are permeable to small, positively charged ions (page 75). The effective combination of acetylcholine with nicotinic receptors on postganglionic neurons within autonomic ganglia can be blocked by the substance *tetraethylammonium.* The effective combination of acetylcholine with nicotinic receptors at neuromuscular junctions can be blocked by the substance *curare.* Thus, there appear to be at least two types of nicotinic receptors. (Both tetraethylammonium and curare compete with acetylcholine for binding sites on the receptors.)

Muscarinic receptors are present on many of the effectors supplied by parasympathetic postganglionic neurons. Muscarinic receptors are G-protein–linked receptors (page 70). For example, the binding of acetylcholine to muscarinic receptors on heart muscle cells activates G proteins tentatively called G_k. These G proteins, in turn, directly activate potassium ion channels in the plasma membranes of the cells. The opening of the potassium ion channels reduces the excitability of the cells and leads to a slowing of the heart. The effective combination of acetylcholine with muscarinic receptors can be blocked by the substance *atropine,* which competes with acetylcholine for binding sites on the receptors.

There are four principal receptors with which norepinephrine or epinephrine can combine. These receptors are called *adrenergic receptors,* and they are designated alpha$_1$, alpha$_2$, beta$_1$, and beta$_2$. All four receptors are G-protein–linked receptors (page 70). Norepinephrine combines effectively with alpha$_1$, alpha$_2$, and beta$_1$ receptors, but it generally combines either weakly or not at all with beta$_2$ receptors. Epinephrine combines effectively with all four receptor types.

Alpha$_1$ receptors are the most common alpha receptors on effector cells. The binding of norepinephrine or epinephrine to alpha$_1$ receptors activates G proteins tentatively called G_p. These G proteins, in turn, activate the

◆ **TABLE 14.3 Receptors for Acetylcholine, Norepinephrine, and Epinephrine**

RECEPTOR TYPE	MAJOR LOCATIONS OF RECEPTORS	NEUROTRANSMITTERS THAT COMBINE WITH RECEPTORS
CHOLINERGIC		
Nicotinic	Present on postganglionic neurons within autonomic ganglia and on skeletal muscle cells at neuromuscular junctions	Acetylcholine
Muscarinic	Present on many effector cells supplied by parasympathetic postganglionic neurons	Acetylcholine
ADRENERGIC		
$Alpha_1$	Most common alpha receptor present on effector cells	Norepinephrine or Epinephrine
$Alpha_2$	Less common alpha receptor; present on cells in uterus and brain	Norepinephrine or Epinephrine
$Beta_1$	Less common beta receptor; present on cells in heart and kidneys	Norepinephrine or Epinephrine
$Beta_2$	Most common beta receptor present on effector cells	Epinephrine (Norepinephrine generally combines either weakly or not at all)

enzyme phosphoinositide-specific phospholipase C and thereby initiate the phosphoinositide cascade of reactions (page 72).

$Alpha_2$ receptors are less common than $alpha_1$, receptors. However, some organs such as the uterus and brain appear to possess both $alpha_1$ and $alpha_2$ receptors. The binding of norepinephrine or epinephrine to $alpha_2$ receptors activates G proteins called inhibitory G proteins (G_i). These G proteins inhibit the enzyme adenylate cyclase and thereby reduce the formation of cyclic AMP within the cells (page 72).

The cells of some organs, such as the heart and the kidneys, possess $beta_1$ receptors. The binding of norepinephrine or epinephrine to $beta_1$ receptors activates G proteins called stimulatory G proteins (G_s). The stimulatory G proteins activate adenylate cyclase and thereby increase the formation of cyclic AMP within the cells (page 72).

$Beta_2$ receptors are the most common beta receptors on effector cells. The binding of epinephrine to $beta_2$ receptors activates stimulatory G proteins, which in turn activate adenylate cyclase.

The differences in the effectiveness of the combination of norepinephrine and epinephrine with the different adrenergic receptors are at least in part responsible for the fact that norepinephrine and epinephrine sometimes have different actions. For example, both norepinephrine and epinephrine stimulate the heart, which has mainly $beta_1$ receptors, and constrict certain blood vessels, which have $alpha_1$ receptors. Yet, only epinephrine dilates certain other blood vessels, which have $beta_2$ receptors.

Functions of the Autonomic Nervous System

Many body organs (for example, the heart, the intestines, and the lungs) are supplied by both the sympathetic and parasympathetic divisions of the autonomic nervous system. In such cases, the two divisions frequently (but not always) cause opposite responses. For example, if one division increases the activity of an organ, the other may decrease it. Although most organs are predominantly controlled by one division or the other, the dual innervation of an organ by both divisions of the autonomic nervous system contributes to the precise control of the organ's activity. The effects of sympathetic and parasympathetic stimulation on a number of organs are summarized in Table 14.4.

No generalization can indicate whether sympathetic or parasympathetic stimulation will excite or inhibit a

◆ **TABLE 14.4 Effects of the Autonomic Nervous System**

STRUCTURE	EFFECTS OF SYMPATHETIC STIMULATION	EFFECTS OF PARASYMPATHETIC STIMULATION
HEART	Increase rate	Decrease rate
LUNGS		
Bronchioles	Dilation	Constriction
Bronchial glands	Possible inhibition of secretion	Stimulation of secretion
SALIVARY GLANDS	Secretion of viscous fluid	Secretion of watery fluid
STOMACH		
Motility	Decreased	Increased
Secretion	Possible inhibition	Stimulation
INTESTINE		
Motility	Decreased peristalsis	Increased peristalsis
Secretion	Possible inhibition	Stimulation
PANCREAS (EXOCRINE PORTION)		Stimulation of secretion
LIVER	Increased release of glucose	
EYE		
Iris	Dilation of pupil (contraction of radial muscles)	Constriction of pupil (contraction of circular muscles)
Ciliary muscle	Slight relaxation	Contraction (accommodates for near vision)
SWEAT GLANDS	Stimulation of secretion (cholinergic)	
ADRENAL MEDULLA	Stimulation of secretion (cholinergic preganglionic neurons)	
URINARY BLADDER	Relaxation	Contraction
BLOOD VESSELS OF:		
Skin	Constriction	
Salivary glands	Constriction	
Abdominal viscera	Constriction	
External genitalia	Constriction	Dilation

particular organ. However, when viewed in broad terms, parasympathetic stimulation tends to produce responses that are primarily concerned with maintaining bodily functions under relatively quiet conditions. For example, parasympathetic stimulation decreases the heart rate and promotes digestive activities. Sympathetic stimulation, in contrast, tends to produce responses that prepare a person for strenuous physical activity, such as may be required in an emergency or in situations that lead to aggressive or defensive behavior. In fact, emotional states such as rage and fear are generally accompanied by a widespread activation of the sympathetic division of the autonomic nervous system. This broad sympathetic activity produces a group of responses—such as increased heart rate and dilation of the bronchii of the lungs—that increase the body's ability to perform vigorous muscular activity. These responses are particularly beneficial to a person who must defend against or flee from a physical threat or challenge; consequently, they are frequently called "fight-or-flight" responses.

A general comparison of the autonomic and somatic nervous systems is provided in Table 14.5.

ASPECTS OF EXERCISE PHYSIOLOGY

Exercise and Autonomic Effects on Heart Rate

Aerobic conditioning has a significant effect on a person's resting heart rate. Unconditioned individuals typically have resting heart rates between 70 and 80 beats per minute. Champion endurance athletes, on the other hand, commonly have resting heart rates of 40 beats per minute or less.

The mechanisms by which endurance training reduces resting heart rate are not completely clear. However, parasympathetic stimulation of the heart decreases heart rate, and sympathetic stimulation increases it. With training, parasympathetic tone (that is, tonic parasympathetic neural activity to the heart) appears to increase, and sympathetic activity appears to decrease. In addition, research has shown that conditioned athletes exhibit less cardiac reactivity to stress conditions than do sedentary individuals, and both resting and exercise levels of catecholamines are lower in endurance-trained athletes.

When exercise begins, a person's heart rate increases. It is believed that the initial increase is due to a reduction of parasympathetic tone, and further increases are due to increased sympathetic activity. The maximal heart rate during exercise is related to age and can be predicted by subtracting a person's age from 220.

The influence of the autonomic nervous system on heart rate during exercise is illustrated by a person who has received a heart transplant. In a person who has not had a transplant, the maximal heart rate during exercise is largely a consequence of increased sympathetic stimulation of the heart. A transplanted heart, however, is denervated. Therefore, in a person with a transplanted heart, the heart rate cannot be increased by direct sympathetic stimulation. When such a person exercises, the heart rate is slow to increase, and the individual cannot achieve the age-predicted maximal heart rate. The increase in heart rate that does occur during exercise is mediated by an increase in circulating catecholamines. Although the exercise tolerance of people with transplanted hearts is less than that of persons who have not had transplants, they can increase their aerobic conditioning, and some have even participated in marathons.

◆ **TABLE 14.5 Comparison of the Autonomic and the Somatic Nervous Systems**

FEATURE	AUTONOMIC NERVOUS SYSTEM	SOMATIC NERVOUS SYSTEM
Site of origin	Brain or lateral horn of spinal cord	Anterior horn of spinal cord
Number of neurons from origin in CNS to effector organ	Two-neuron chain (preganglionic and postganglionic)	Single neuron (motor neuron)
Organs innervated	Cardiac muscle, smooth muscle, exocrine glands	Skeletal muscle
Type of innervation	Most effector organs dually innervated by the two branches of this system (sympathetic and parasympathetic)	Effector organs innervated by motor neurons
Neurotransmitter at effector organs	May be acetylcholine (parasympathetic terminals) or norepinephrine (most sympathetic terminals)	Acetylcholine
Effects on effector organs	Either stimulation or inhibition (antagonistic actions of two branches)	Stimulation only (inhibition possible only centrally through inhibitory postsynaptic potentials [IPSPs] on cell body of motor neuron)
Types of control	Under involuntary control; may be voluntarily controlled with biofeedback techniques and training	Subject to voluntary control; much activity subconsciously coordinated

CONDITIONS OF CLINICAL SIGNIFICANCE

The Autonomic Nervous System

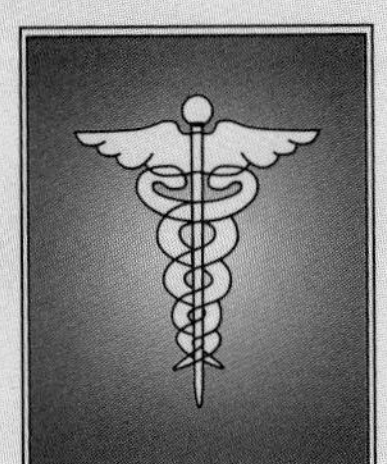

Biofeedback

The fact that the autonomic nervous system normally functions below the conscious level implies that a person has no control over the activities governed by this system. However, this is not entirely true.

Normally, a person receives only limited information at the conscious level about what is occurring within the body. For example, blood pressure may fluctuate or brain wave patterns change without the person becoming aware of the changes. Consequently, no conscious effort is made to react to or to control such changes. However, the technique of *biofeedback* provides a person with conscious information about body events that usually go unnoticed. Biofeedback utilizes electronic instruments to monitor some of the normally subconscious activities that occur within the body and raise them to the conscious level. The instruments provide information about such events as temperature changes, changes in heart rate, and variations in nerve impulse patterns. With such conscious knowledge of previously subconscious body activities, it has become possible in some cases for people to learn to control certain of these activities. For example, using biofeedback techniques, people have learned to lower their heart rate, lower their blood pressure, increase the circulation of blood through their limbs, relieve migraine headaches by reducing the blood pressure within the vessels of the head, and control epileptic seizures. Biofeedback, then, shows considerable promise as a self-administered therapeutic technique with broad applications.

Raynaud's Disease

Raynaud's disease is characterized by episodes of pallor or cyanosis of the extremities—especially the fingers and toes and, less frequently, the tip of the nose and the ears. It is thought to be the result of exaggerated vasomotor responses, both central and local, by the sympathetic division of the autonomic nervous system. These responses cause episodes of vasoconstriction of the blood vessels of the affected regions.

The episodes are generally first noticed in cold weather and may be infrequent. The course of Raynaud's disease is variable; it often remains as nothing more than a nuisance for years and in some cases subsides spontaneously. Occasionally, however, the condition becomes progressive and produces ulcerations and areas of gangrene on the fingertips.

Achalasia

Achalasia, or *cardiospasm,* is characterized by difficulty in swallowing accompanied by a feeling that food is sticking in the esophagus. It is the result of uncoordinated and ineffectual peristalsis of the esophagus and persistent contraction of the esophagus where it enters the cardiac region of the stomach. These conditions produce a functional obstruction of the esophagus. Achalasia is probably caused by a number of factors. Emotions and a hypersensitivity to the hormone gastrin are implicated, but there may also be structural or functional disorders of the portion of the parasympathetic nervous system that innervates the esophagus.

Hirschsprung's Disease

Hirschsprung's disease, or *megacolon,* is somewhat similar to achalasia, except that the functional obstruction occurs in the distal portion of the colon and the rectum. In response to this obstruction, the colon above the level of the obstruction dilates greatly (thus the name "megacolon"). This disorder is thought to be caused by a reduction in the parasympathetic innervation to the affected structures. The reduction of parasympathetic innervation allows the sympathetic neurons to inhibit peristalsis and maintain a chronic contraction in the affected region.

Study Outline

◆ AUTONOMIC NERVOUS SYSTEM p. 456

Visceral Motor (efferent) Neurons. Innervate and regulate cardiac muscle, smooth muscle, and glands; control involuntary body processes.

ENTIRELY MOTOR FUNCTIONS. Part of efferent division of peripheral nervous system.

◆ ANATOMY OF THE AUTONOMIC NERVOUS SYSTEM pp. 456–460

Efferent Pathways. Composed of two neurons.

PREGANGLIONIC (PRESYNAPTIC) NEURON. Cell body in CNS; axon synapses with postganglionic neurons.

POSTGANGLIONIC (POSTSYNAPTIC) NEURON. Located outside CNS; axons travel to effectors.

Sympathetic Division (Thoracolumbar Division). Preganglionic neuron cell bodies in lateral horns of spinal cord gray matter from T_1 through L_2. Neurons leave ventral rami and enter a series of sympathetic chain ganglia that form longitudinal pathways along both sides of vertebral column. *White rami communicantes* (14 pairs) are short pathways formed by myelinated preganglionic nerve fibers as they pass from ventral ramus to sympathetic chain ganglia. Preganglionic sympathetic neurons follow one of three courses upon entering sympathetic chain ganglia:

1. May synapse with postganglionic neurons in sympathetic chain ganglia at same level. Postganglionic axons return to the spinal nerve to innervate effectors located in skin. *Gray rami communicantes* are pathways formed by unmyelinated postganglionic axons as they pass from sympathetic chain ganglia to spinal nerves.
2. May travel up or down within the sympathetic trunks before synapsing with postganglionic neurons that supply effectors in skin, head, or thorax.
3. May pass through sympathetic chain ganglia without synapsing and synapse with postganglionic neurons in collateral ganglia; postganglionic neurons from collateral ganglia supply viscera of abdominopelvic cavity.

Parasympathetic Division (Craniosacral Division).

1. Preganglionic neuron cell bodies are located within nuclei in the brain or in the lateral portions of the gray matter of the spinal cord from S_2 through S_4.
2. Fibers do not travel through rami of spinal nerves.
3. Preganglionic parasympathetic axons in four cranial nerves (III, VII, IX, X) synapse with postganglionic neurons in ganglia close to structures innervated.
4. Sacral preganglionic parasympathetic axons leave ventral roots of spinal nerves and form a pelvic nerve that supplies viscera of the pelvic cavity.

Anatomical Differences between the Divisions.

LOCATION OF PREGANGLIONIC NEURON CELL BODIES. *Sympathetic.* Lateral horns of spinal cord gray matter from T_1 through L_2.
Parasympathetic. Brain stem and lateral horns of spinal cord gray matter from S_1 through S_4.

FIBER LENGTH. Short sympathetic preganglionic axons; long sympathetic postganglionic axons. Long parasympathetic preganglionic axons; short parasympathetic postganglionic axons.

◆ NEUROTRANSMITTERS OF THE AUTONOMIC NERVOUS SYSTEM pp. 460–461

Most postganglionic sympathetic neurons secrete norepinephrine; postganglionic parasympathetic neurons secrete acetylcholine.

◆ RECEPTORS FOR AUTONOMIC NEUROTRANSMITTERS pp. 461–462

Acetylcholine receptors are called cholinergic receptors. These receptors are described as being either nicotinic or muscarinic. There are four principal receptors with which norepinephrine or epinephrine can combine. These receptors are called adrenergic receptors, and they are designated alpha_1, alpha_2, beta_1, and beta_2.

◆ FUNCTIONS OF THE AUTONOMIC NERVOUS SYSTEM pp. 462–464

Many organs are innervated by both sympathetic and parasympathetic neurons, which generally cause opposite responses; parasympathetic stimulation tends to produce responses primarily concerned with maintaining bodily functions under relatively quiet conditions; sympathetic stimulation tends to produce responses that prepare a person for strenuous physical activity.

◆ CONDITIONS OF CLINICAL SIGNIFICANCE: THE AUTONOMIC NERVOUS SYSTEM p. 465

Biofeedback. Technique by which a person made aware of normally subconscious activities can exert some voluntary control over them.

Raynaud's Disease. Characterized by pallor or cyanosis of the extremities; thought to be caused by vasoconstriction of blood vessels in the area by the sympathetic nervous system.

Achalasia. Also called *cardiospasm;* characterized by difficulty in swallowing, with food remaining in esophagus; peristalsis of esophagus is uncoordinated and ineffectual, and esophagus is constricted where it enters stomach; structural or functional disorders of parasympathetic nervous system probably involved.

Hirschsprung's Disease. Also called *megacolon;* result of functional obstruction in distal portion of colon and rectum; probably caused by reduction in parasympathetic innervation to these structures; proximal portion of colon dilates greatly in response to obstruction.

Self-Quiz

1. The autonomic nervous system regulates the activity of: (a) cardiac muscle; (b) skeletal muscles; (c) smooth muscle; (d) both (a) and (c).
2. The autonomic nervous system is normally a voluntary system that functions at the conscious level. True or False?
3. The autonomic nervous system has both motor and sensory functions. True or False?
4. How many neurons compose the efferent pathways of the autonomic nervous system, which run from the central nervous system to the effectors? (a) 1; (b) 2; (c) 4.
5. The ventral ramus of each spinal nerve generally has associated with it: (a) a collateral ganglion; (b) a sympathetic chain ganglion; (c) a dorsal root ganglion.
6. Preganglionic sympathetic axons may synapse with the cell bodies of postganglionic neurons in the sympathetic chain ganglion located at the same level at which the preganglionic axons entered the ganglion. True or False?
7. Most postganglionic axons are myelinated. True or False?
8. Preganglionic sympathetic axons may pass through the sympathetic chain ganglia without synapsing. True or False?
9. Preganglionic sympathetic axons may travel up or down in the sympathetic trunks before synapsing with: (a) other preganglionic neurons; (b) postganglionic neurons; (c) sensory neurons.
10. The distribution of the sympathetic division differs from that of the parasympathetic division in that sympathetic neurons do not travel through the rami of the spinal nerves and, therefore, the cutaneous structures do not receive sympathetic innervation. True or False?
11. The parasympathetic division of the autonomic nervous system is also referred to as: (a) the thoracolumbar division; (b) the craniosacral division; (c) neither (a) nor (b).
12. The sympathetic and parasympathetic divisions of the autonomic nervous system differ in: (a) the location of the cell bodies of their preganglionic neurons; (b) the length of their preganglionic neurons; (c) both (a) and (b).
13. Most sympathetic postganglionic neurons secrete: (a) norepinephrine; (b) acetylcholine; (c) neither (a) nor (b).
14. The preganglionic neurons in the splanchnic nerves synapse within: (a) sympathetic chain ganglia; (b) terminal ganglia; (c) collateral ganglia.
15. Nicotinic acetylcholine receptors are G-protein–linked receptors. True or False?
16. Norepinephrine generally combines either weakly or not at all with: (a) alpha_1 receptors; (b) beta_2 receptors; (c) alpha_2 receptors.
17. The adrenal medulla causes reactions similar to those caused by the sympathetic nervous system. True or False?
18. The "fight-or-flight" responses are produced by widespread activation of the: (a) parasympathetic division; (b) somatic nervous system; (c) sympathetic division.
19. Parasympathetic stimulation tends to produce responses that are primarily concerned with maintaining bodily functions under relatively quiet conditions. True or False?
20. Under controlled conditions, it may be possible for a person to learn to control such responses of the autonomic nervous system as heart rate and blood pressure. True or False?

CHAPTER 15

Integrative Functions of the Nervous System

CHAPTER CONTENTS

LEARNING OBJECTIVES

After completing this chapter, you should be able to:

1. Describe the flexor reflex and the associated crossed extensor reflex.
2. Explain how general sensory input plays an important role in maintaining the arousal of the brain.
3. Distinguish between slow-wave sleep and paradoxical sleep.
4. Describe possible mechanisms involved in the sleep-wakefulness cycle.
5. Distinguish between short-term and long-term memory, and cite the possible physical-chemical mechanisms of each.
6. Define and cite an example of a conditioned response.
7. Describe and cite an example of instrumental learning.
8. Describe three neural mechanisms involved in supporting the body against the force of gravity.
9. Explain the role of the cerebellum in coordinating movements.
10. Describe the process by which skilled movements may be learned and performed.

The nervous system brings together information from many different sources and coordinates and regulates numerous interrelated activities that take place within the body. Consequently, the nervous system is an integrative system that is vital in maintaining homeostasis.

Integrative Spinal Reflexes

Even basic nervous system activities such as reflex responses exemplify the integrative function of the nervous system. This integrative function is evident in the **flexor,** or **withdrawal, reflex** and the associated **crossed extensor reflex.**

The flexor reflex results in the withdrawal of a limb, most commonly in response to some noxious stimulus. For example, if a person's foot is pricked by a pin, nerve impulses resulting from the stimulus (the prick) are transmitted to the spinal cord by afferent neurons (Figure 15.1). Within the spinal cord, the afferent neurons stimulate interneurons that, in turn, stimulate efferent neurons supplying the flexor muscles of the leg on the same side of the body as the site of the stimulation (the ipsilateral side). In addition, the afferent neurons stimulate interneurons that inhibit efferent neurons supplying the extensor muscles of the ipsilateral leg. Thus, the flexor muscles of the ipsilateral leg are activated, and the extensor muscles are inhibited. As a result, the leg is flexed and the foot is withdrawn.

The crossed extensor reflex, which is initiated by the same stimulus responsible for the flexor reflex, results in the extension of the corresponding limb on the side of the body opposite the site of stimulation (the contralateral side). For example, when a person's foot is pricked with a pin, the afferent neurons not only stimulate neurons involved in the flexor reflex, but they also stimulate commissural interneurons that cross to the opposite side of the spinal cord (Figure 15.1). There, the commissural interneurons stimulate interneurons that, in turn, stimulate efferent neurons supplying the extensor muscles of the contralateral leg. The commissural interneurons also stimulate interneurons that inhibit efferent neurons supplying the flexor muscles of the contralateral leg. Thus, the extensor muscles of the contralateral leg are activated and the flexor muscles are inhibited. As a result, the leg is extended.

Even though the flexor reflex and the associated crossed extensor reflex occur at the spinal cord level, they nevertheless result in a coordinated, adaptive integration of separate muscular activities. The withdrawal of the stimulated limb removes it from a potential

◆ **FIGURE 15.1 Diagrammatic representation of the flexor reflex and the associated crossed extensor reflex**

Stimulation is indicated by blue, and inhibition is indicated by black.

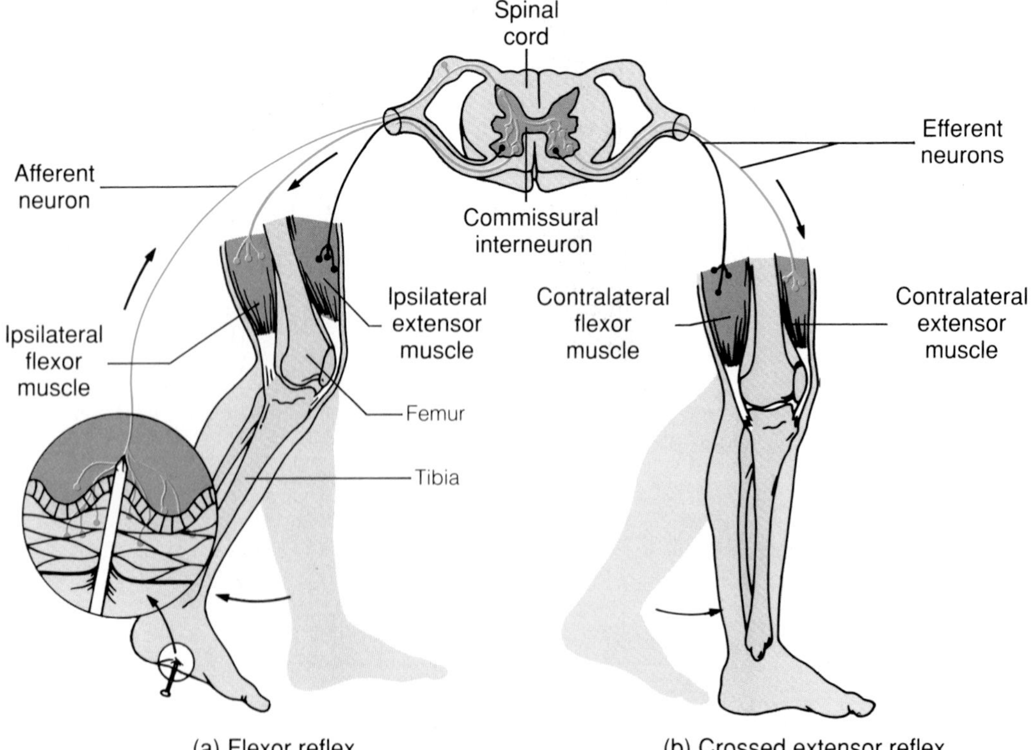

◆ **FIGURE 15.2 Recording the degree and pattern of electrical activity occurring within the brain by means of electrodes placed on the head**

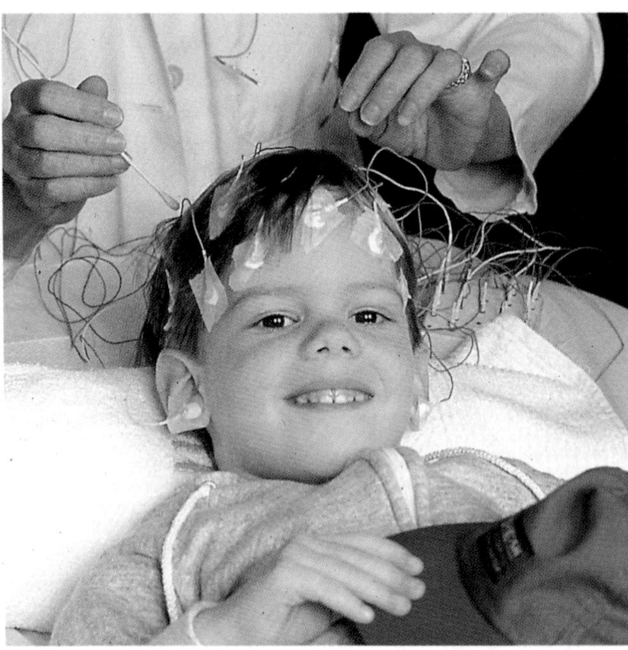

source of injury, and the extension of the corresponding contralateral limb can help the individual maintain balance.

Although some integrative activities occur at the spinal cord level, the brain is the major integrative center of the nervous system. For example, a stimulus such as a pin prick not only elicits reflex responses, but nerve impulses are transmitted to the brain where further integrative activities occur. Pain may be sensed, and conscious responses to the pin prick initiated. Ultimately, the event may be committed to memory, and learning may take place. This learning may take the form of avoiding conditions under which similar injury may occur in the future.

Mental Processes

The brain is an extremely complex organ. It has been studied intensively for many years, but researchers have so far achieved only enticing bits of understanding about its function. Bear in mind, then, that the following discussions of mental processes—particularly those dealing with complex phenomena such as memory and learning—are largely theoretical. As more information becomes available, some of these ideas may have to be abandoned in favor of new ones.

No area or structure of the brain acts entirely on its own. In many respects, the most complex of the so-called higher functions of the brain (such as memory and learning) can be viewed as whole-brain functions. That is, the performance of complex functions involves more than one brain area, each of which is probably involved in many functions. Even those functions that occur automatically below the conscious level generally entail input from several different sources.

The Electroencephalogram

The degree and pattern of electrical activity that occur within the brain can be detected by electrodes placed on the head (Figure 15.2). The record of the electrical waves produced during brain activity is called an **electroencephalogram (EEG).** The electroencephalogram is essentially a record of waves of different amplitudes (that is, heights) and frequencies that arise from the cerebral cortex, perhaps at times as a result of the influence of some subcortical center such as the thalamus. These "brain waves," as they are called, are always present, even during unconsciousness, indicating that the brain is continually active as long as it is alive.

There is some degree of correlation between brain waves and body activity (Figure 15.3). For example, when a person is awake, brain wave recordings show low-amplitude, high-frequency waves. When a person

◆ **FIGURE 15.3 Typical electroencephalograms**
(a) Awake and alert. (b) Relaxed with eyes closed. (c) Drowsy. (d) Asleep, slow-wave sleep. (e) Asleep, paradoxical sleep.

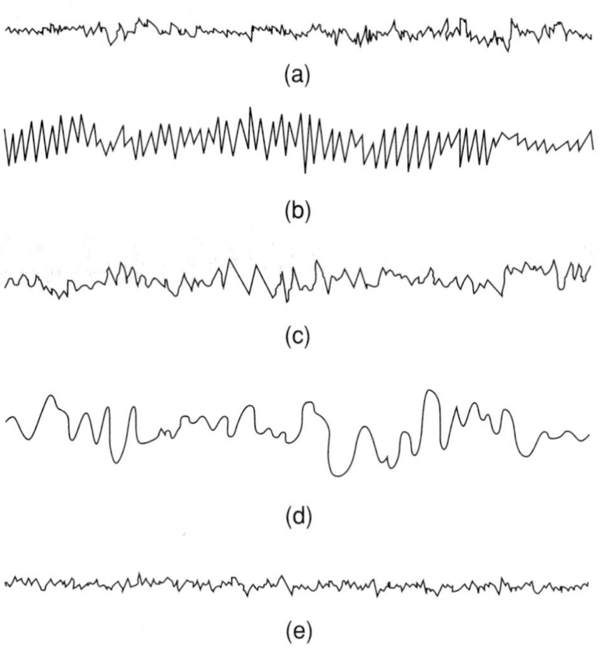

(a)

(b)

(c)

(d)

(e)

falls asleep, the frequency of the brain waves slows, but the amplitude increases. A person who is anesthetized also exhibits slow waves.

Consciousness

In order to perform most of its higher functions, the brain must be in a state of awareness, or **consciousness.** Consciousness, however, is difficult to define, and there are gradations of awareness between complete unconsciousness and alert wakefulness.

Consciousness is sometimes considered to be the state in which thoughts or instants of awareness occur. However, the occurrence of a thought probably requires the activity of several portions of the brain. That is, a thought is the result of a pattern of neural activities that occur simultaneously in several locations, such as the cerebral cortex, the thalamus, the limbic system, and the reticular formation. If the occurrence of a single thought requires the activity of several portions of the brain, the attainment of an overall state of consciousness would seem to require the activation or arousal of widespread brain areas.

The Aroused Brain

When the brain is aroused or awake, it is in a state of readiness and is able to react consciously to stimuli. The attainment of this state seems to depend in large part on nerve impulses sent throughout the brain by the reticular activating system, or RAS (see page 400). Stimulation of the RAS in a sleeping animal produces an EEG like that of an aroused brain and causes the animal to awaken. Conversely, destruction of the RAS produces a permanent coma, with an EEG characteristic of the sleeping state.

The activity of the RAS is influenced by nerve impulses from cutaneous, visual, auditory, muscular and visceral receptors. These impulses, which are called *arousal signals,* stimulate the RAS to send impulses throughout the brain to arouse it. Consequently, general sensory input plays an important role in maintaining the arousal of the brain.

Sleep

Sleep can be defined as a state of altered consciousness or partial unconsciousness from which a person can be aroused by appropriate stimuli. In contrast, a *coma* is a state of unconsciousness from which a person cannot be aroused.

As a person becomes drowsy and falls asleep, the EEG gradually shifts toward higher-amplitude, slower (lower-frequency) waves. The state of sleep characterized by high-amplitude, low-frequency EEG waves is called **slow-wave sleep.** During deep slow-wave sleep, the respiratory cycle is regular and deep. The respiratory rate, heart rate, and blood pressure are generally somewhat below waking levels.

Approximately once every 90 minutes during a normal night of sleep, the EEG of slow-wave sleep is interrupted by episodes (lasting 5–20 minutes) during which the EEG resembles that of an aroused brain. This state of sleep is called **paradoxical sleep.** Because the eyes move rapidly behind the closed eyelids during paradoxical sleep, it is also referred to as **rapid-eye-movement (REM)** sleep.

Paradoxical sleep occurs in conjunction with slow-wave sleep, and the first paradoxical sleep episode normally takes places 80 to 100 minutes after a person falls asleep. Although it is not certain what causes paradoxical sleep, a specific region of the pons called the locus caeruleus appears to induce it, and damage to this region prevents its occurrence. Neurons of the area contain an abundance of norepinephrine, leading some researchers to suggest that the release of norepinephrine leads to paradoxical sleep. However, other researchers report that injecting drugs that mimic acetylcholine into the pons can induce a paradoxical-like state, indicating that acetylcholine plays a role in the control of paradoxical sleep. Still other investigators have found that the injection of glutamic acid into the pons can bring on a paradoxical-like state.

During paradoxical sleep, muscle tone throughout the body is greatly depressed, although periodic twitching of the facial muscles and limbs occurs. In addition, respiration and heart rate are irregular, and blood pressure may rise or fall. A person who is awakened every time a period of paradoxical sleep begins, causing a paradoxical-sleep deficit to develop, spends a greater proportion of the total sleep time in paradoxical sleep when allowed to sleep undisturbed. Most dreaming occurs during paradoxical sleep. The characteristics of slow-wave sleep and paradoxical sleep are compared in Table 15.1.

Sleep-Wakefulness Cycle

The normal sleep-wakefulness cycle is one of the most obvious human rhythms, and both mental and physical benefits are derived from alternating periods of sleep and wakefulness. Although the processes that underlie this cycle are not fully understood, awakening appears to occur when signals from the RAS reach a sufficient intensity to arouse the brain. Conversely, sleep seems to result when the activity of the RAS declines to a level no longer adequate to maintain arousal.

Not only does the RAS transmit nerve impulses throughout the brain, but—in what is basically a positive-feedback response—brain areas such as the cerebral cortex also send stimulatory signals to the RAS

◆ **TABLE 15.1 Comparison of Slow-Wave and Paradoxical Sleep**

CHARACTERISTIC	SLOW-WAVE SLEEP	PARADOXICAL SLEEP
EEG	Displays slow waves	Similar to EEG of alert, awake person
Motor activity	Considerable muscle tone; frequent shifting	Muscle tone greatly depressed
Heart rate, respiratory rate, blood pressure	Minor reductions	Irregular
Dreaming	Rare (mental activity is extension of waking-time thoughts)	Common
Arousal	Sleeper easily awakened	Sleeper hard to arouse but apt to wake up spontaneously
Percentage of sleeping time	80%	20%
Other important characteristics	Sleeper must pass through this type of sleep first	Rapid eye movements

(Figure 15.4). In addition, a second positive-feedback response occurs in which the RAS not only receives nerve impulses from receptors in the peripheral musculature, but also transmits signals to the muscles. Thus, signals from the RAS that enhance the activity of the brain or the muscles lead to the transmission of stimulatory impulses from these structures to the RAS. These positive-feedback responses are believed to be part of a neural wakefulness circuit that operates as follows.

When the relatively dormant RAS of a sleeping individual is stimulated by arousal signals from various receptors, the RAS sends more intense signals to the brain and muscles. When this occurs, the positive-feedback responses just described further stimulate the RAS. The RAS, in turn, further stimulates the brain, and so on and so on. When the signals from the RAS to the brain become sufficiently intense, the brain is aroused and the individual awakens.

The wakefulness circuit contains numerous individual positive-feedback loops, and the degree of wakefulness at any moment depends on the number of active loops and their level of activity. If many loops are activated simultaneously in a sleeping person, the positive-feedback aspect of the wakefulness circuit produces a rapidly increasing response that could explain the often-sudden transition between the sleeping and the waking states. The ultimate level of activity of the positive-feedback loops is limited by the capacities of the neurons involved to transmit nerve impulses. Therefore, once particular loops are activated, their level of activity does not continue to increase indefinitely but finally stabilizes.

With the passage of time, the excitability of the wakefulness circuit is believed to diminish, and consequently the intensity of the signals from the RAS to the brain declines. As a result, the individual becomes drowsy and falls asleep. During sleep, the excitability of the wakefulness circuit increases, and arousal signals again set in motion the positive-feedback events that lead to awakening. Because of the cyclical variation in the excitability of the wakefulness circuit, arousal signals that are not strong enough to produce wakefulness early in the sleep period may be sufficient to awaken the individual later in the period. Moreover, the system exerts selectivity as to the types of signals that produce arousal. For example, a mother may quickly awaken if her child cries, but she may not be awakened by the sounds of traffic, trains, or airplanes.

Researchers have proposed a number of theories to explain the cyclical variation in the excitability of the

◆ **FIGURE 15.4 Positive-feedback loops from the reticular activating system to the brain and skeletal muscles**

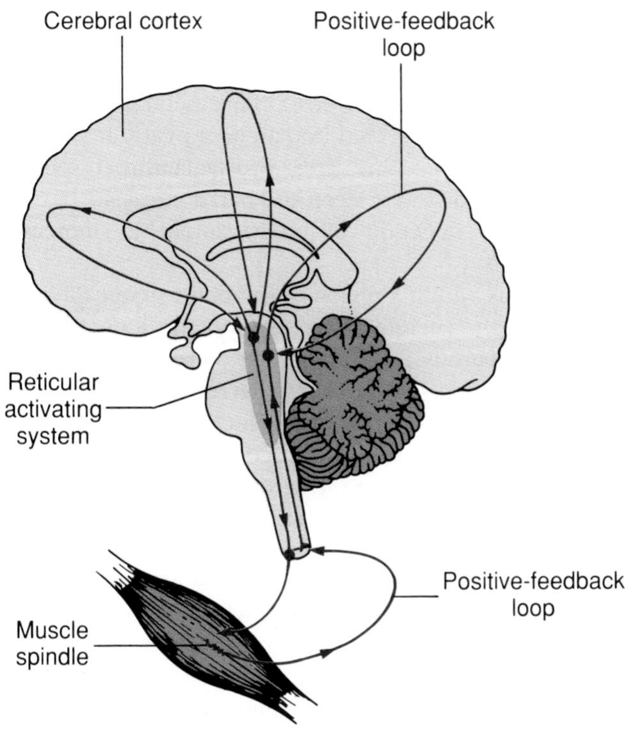

wakefulness circuit. One suggestion is that, with continued activity during the waking period, neurons of the positive-feedback loops become depleted of neurotransmitter substances. Consequently, the feedback diminishes, the activity of the RAS declines, and the person becomes drowsy and falls asleep. During the sleep period, the neurotransmitter substances are replenished, and the feedback loops recover their excitability.

Another suggestion is that the cyclical variation in the excitability of the wakefulness circuit is influenced by a complex interplay of various chemical substances. For example, inhibitory chemical substances could gradually build up during the waking period (and/or excitatory substances could diminish), leading to a decreased excitability of the wakefulness circuit and consequently to drowsiness and sleep. During the sleep period, the inhibitory substances could be removed (and/or excitatory substances could be produced), leading to an increased excitability of the wakefulness circuit and the arousal of the brain. Some evidence supports this possibility:

1. If dialyzed blood or whole cerebrospinal fluid from animals that have been kept awake for several days is injected into the brain ventricle systems of other animals, it can put them to sleep.
2. A group of intercellular messengers called **cytokines** are present in the brain, and a cytokine called **interleukin-1** is present in brain areas associated with sleep control, such as the anterior hypothalamus. When interleukin-1 is given to rabbits, the animals sleep 20% more than usual. Moreover, the amount of interleukin-1 in the cerebrospinal fluid fluctuates in parallel with the normal sleep-wakefulness cycle.
3. A group of hormonelike substances called **prostaglandins** *(pros-tuh-glan´-dinz)* have been injected into the brain. When injected into an area of the hypothalamus called the preoptic area, prostaglandin D_2 can increase both slow-wave sleep and paradoxical sleep. Conversely, the injection of prostaglandin E_2 increases wakefulness.
4. Neurons within a region of the brain stem called the median raphe secrete serotonin into the RAS. Damage to these neurons induces a state of relative sleeplessness, suggesting that serotonin secretion by these neurons contributes to the occurrence of sleep. In fact, some researchers have proposed that a feedback relationship exists between the serotonin-secreting neurons of the brain stem that may cause sleep and the neurons of the pons that appear to induce paradoxical sleep. According to this proposal, the brain-stem neurons facilitate the neurons of the pons, whose activity leads to paradoxical sleep. These neurons, in turn, stimulate the brain-stem neurons to take up serotonin that was previously secreted. This uptake decreases the extracellular serotonin concentration and lessens the inhibition of the RAS, permitting a return to the waking state. As a result, periods of paradoxical sleep would be important because they would remove an inhibitory chemical substance from the system.

Attention

The term **attention** refers to the ability of a person to maintain a selective awareness of some aspect of the environment as well as to the ability to respond selectively to a particular stimulus. A person does not always respond to every stimulus. Instead a selection process occurs so that the person pays attention to only certain stimuli. The person's previous experience is important in determining which stimuli warrant attention at any particular moment, and interest as well as actual biological urgency influence the priority given to particular stimuli.

A person's general level of attentiveness is influenced by the activity of the reticular activating system mechanisms involved in wakefulness and sleep. In addition, certain thalamic areas of the RAS can selectively activate specific areas of the cerebral cortex, thereby directing a person's attention to particular aspects of the environment. Moreover, other central nervous system mechanisms can selectively inhibit or enhance afferent signal input to the brain and thus focus a person's attention on a particular aspect of the environment. For example, the auditory portion of the cerebral cortex can inhibit or facilitate signals from the cochleas of the ears, and the visual cortex can influence the intensities of signals from the retinas of the eyes.

Emotions and Behavior

The integrative functions of the limbic system (see page 397) influence emotions such as fear and anxiety as well as emotional behavior patterns such as rage. In addition, the limbic system is important in determining whether a particular activity is pleasant and rewarding or painful and punishing. In animal experiments, the electrical stimulation of certain areas of the limbic system appears to please or satisfy the animal. Conversely, the stimulation of other areas appears to cause pain, fear, or other elements of punishment, and the animal may exhibit rage or escape behaviors. Similarly, the stimulation of certain limbic areas in humans during neurosurgery results in pleasurable sensations, and the stimulation of other areas leads to vague feelings of fear or anxiety.

The reward and punishment regions of the brain have a great influence on a person's behavior. In fact, much of what a person does depends on reward and punishment. If an activity is pleasurable or rewarding, a person generally will continue the activity. If an activity is unpleasant or punishing, the person generally will not continue the activity. Moreover, reward and punishment have much to do with learning (see page 479).

Neurotransmitter substances such as norepinephrine, dopamine, and serotonin appear to be involved in certain emotional and behavioral states and in learning. For example, drugs that enhance the effect of norepinephrine increase rage, and drugs that block its effect diminish rage behavior. Moreover, increased norepinephrine is associated with elevated mood, and decreased norepinephrine with depression. In this regard, some antidepressant drugs apparently act by increasing the brain concentration of norepinephrine.

◆ **TABLE 15.2 Characteristics of Pain**

FAST PAIN	SLOW PAIN
Carried by A-delta fibers	Carried by C fibers
Sharp, prickling sensation	Dull, aching, burning sensation
Easily localized	Poorly localized
Occurs first	Occurs second; persists for longer time, more unpleasant
Occurs upon stimulation of mechanical and thermal nociceptors	Occurs upon stimulation of polymodal nociceptors

Pain

The sensation of pain is a protective sensation that warns a person about harmful or potentially harmful occurrences. Stimuli strong enough to cause tissue damage commonly give rise to this sensation.

There are three categories of **nociceptors** (pain receptors), all of which are free nerve endings (page 349). Nociceptors that respond to mechanical damage such as cutting, crushing, or pinching are called *mechanical nociceptors.* Nociceptors that respond to temperature extremes—especially to high temperatures—are called *thermal nociceptors.* Nociceptors that respond equally well to a variety of noxious (harmful) stimuli are called *polymodal nociceptors.*

Polymodal nociceptors are apparently stimulated by the chemical substance **bradykinin** *(brad-ee-kigh´-nin).* Bradykinin is a peptide that contains nine amino acids, and it is one of the most potent pain-producing substances known. When tissues are injured or damaged, an enzyme called *kalikrein (kal-i-kree´-in)* is formed, and this enzyme cleaves bradykinin from large precursor molecules that are present in the blood and in a variety of tissues.

In addition to activating polymodal nociceptors, bradykinin triggers responses that promote the repair of injured tissues. It also causes the production of *prostaglandins* that greatly enhance the responses of all categories of nociceptors.

Nerve impulses resulting from nociceptor stimulation are transmitted to the central nervous system by one of two types of afferent nerve fibers (Table 15.2). Nerve impulses arising from mechanical and thermal nociceptors are transmitted by large-diameter, myelinated fibers called *A-delta fibers.* These fibers, which transmit nerve impulses rapidly (up to 80 meters/second), are called *fast pain fibers.* Nerve impulses arising from polymodal nociceptors are transmitted by small-diameter, unmyelinated nerve fibers called *C fibers.* These fibers, which transmit nerve impulses slowly (about 1 meter/second), are called *slow pain fibers.*

The pain sensations elicited by nerve impulses transmitted along fast pain fibers are typically described as sharp and prickling. The sensations are easily localized and relatively brief in duration. These sensations are generally the initial pain sensations to be perceived.

The pain sensations elicited by nerve impulses transmitted along slow pain fibers are typically described as dull, aching, and burning. The sensations are poorly localized, and they indicate that tissue damage has occurred. These sensations are more unpleasant, and they persist for a longer time than the sensations elicited by nerve impulses transmitted along fast pain fibers.

The pain signals originating at nociceptors are ultimately transmitted to a number of brain areas, including the reticular formation, the thalamus, and the somatic sensory areas of the cerebral cortex (Figure 15.5). Signals transmitted to the reticular formation increase the person's alertness. Signals transmitted to the thalamus appear to give rise to a general perception of pain, and signals transmitted to the somatic sensory areas of the cerebral cortex are apparently important in localizing the pain sensation. Signals transmitted from the thalamus and the reticular formation to the hypothalamus and limbic system lead to emotional reactions (crying, anxiety, or fear, for example) and behavioral responses (such as withdrawal or defensive responses), which accompany the occurrence of pain.

Pain Modulation

The body possesses a poorly understood analgesic (pain-suppressing) system that can inhibit the transmission of pain signals to the brain and thereby suppress pain perception. For example, one of the neurotransmitters released by afferent pain fibers at their synapses with ascending spinal cord neurons is substance P, which stimulates the ascending neurons. The periaqueductal gray matter, which surrounds the cerebral aqueduct, and the reticular formation within the brain stem apparently contain neurons that are part of a descending

◆ **FIGURE 15.5 Pain pathway**

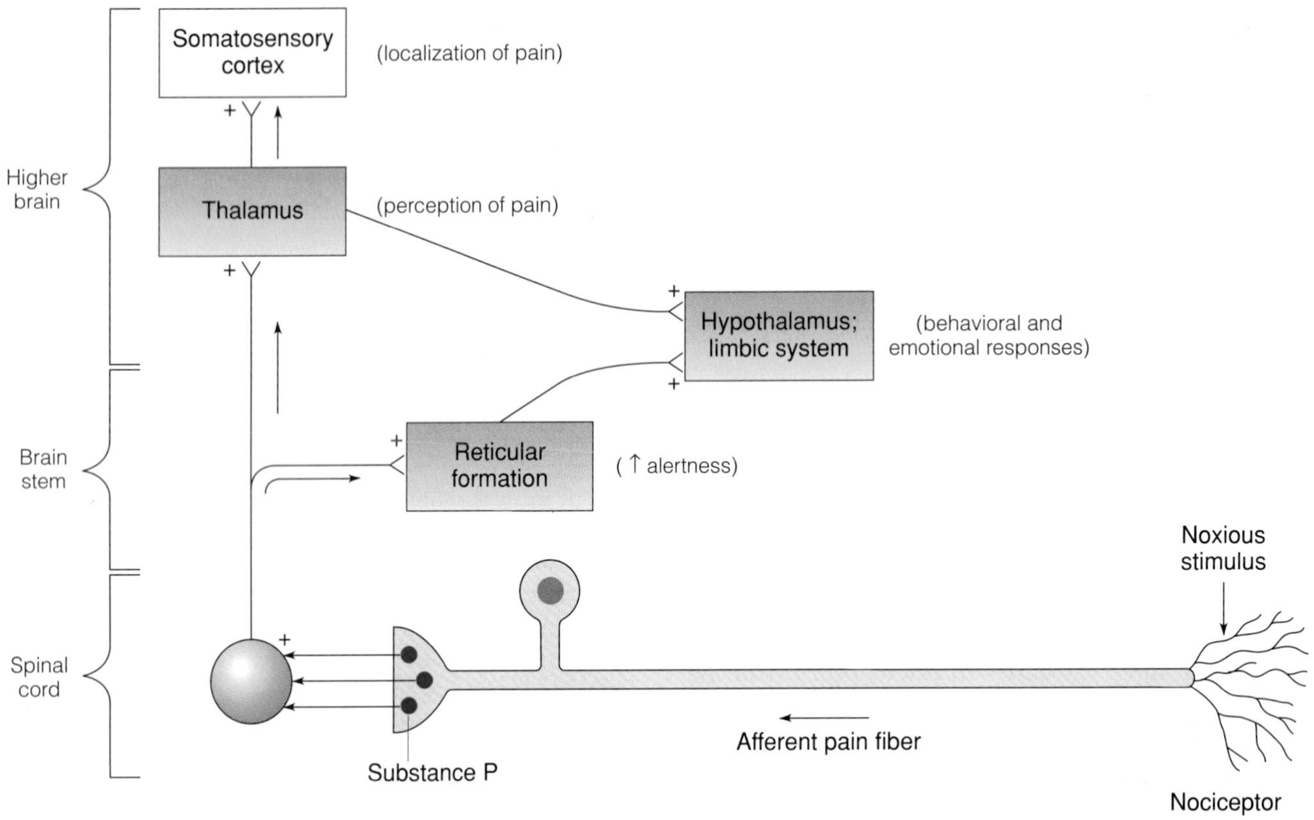

analgesic pathway (Figure 15.6). Neurons of this pathway are believed to inhibit the release of substance P from afferent pain fibers by a process of presynaptic inhibition (see page 373). In this process, descending neurons make synaptic contact with afferent pain fibers at synapses between the afferent fibers and ascending spinal cord neurons. When they are active, the descending neurons release neuropeptides called *enkephalins,*

◆ **FIGURE 15.6 Proposed analgesic pathway**

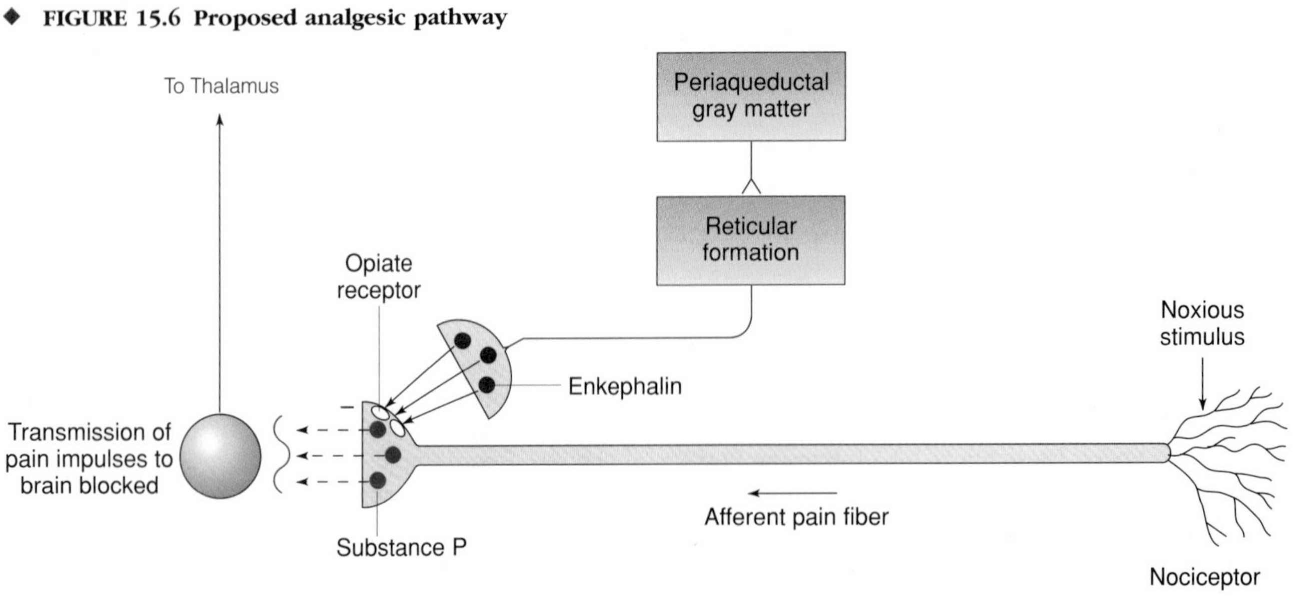

◆ **FIGURE 15.7 Cutaneous areas to which pain from certain viscera is referred**

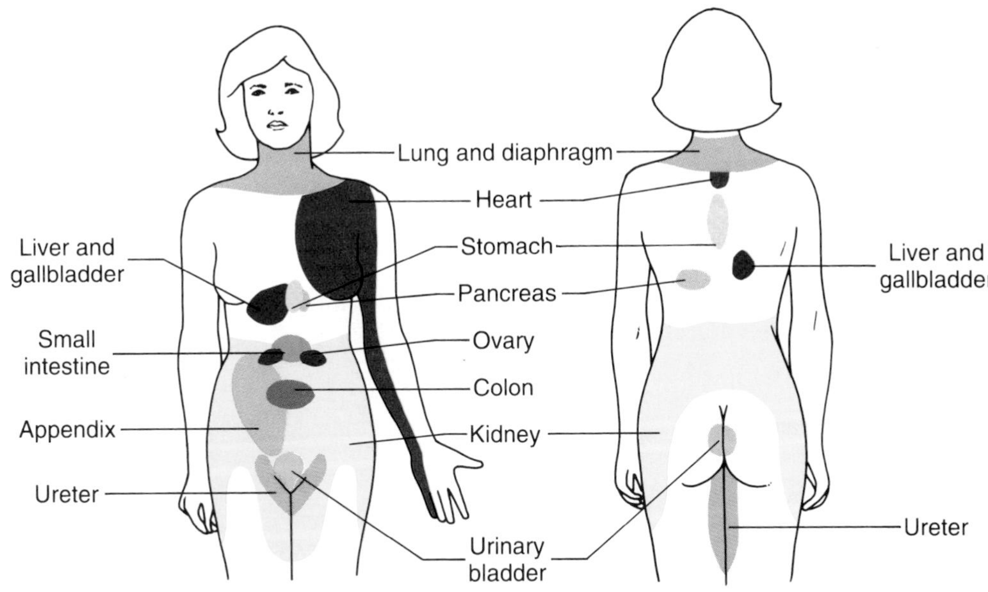

which bind to receptors known as *opiate receptors* on the afferent pain fibers. This binding inhibits the release of substance P from the afferent fibers.

Various drugs are commonly used to reduce a person's sensitivity to painful stimuli. Opiate analgesics such as morphine are believed to exert their effects by binding to opiate receptors in the brain and other areas. The analgesic effects of aspirin appear to be due at least in part to its ability to inhibit the synthesis of prostaglandins.

A number of local anesthetics (for example, procaine and tetracaine) exert their effects in circumscribed areas by decreasing the permeability of nerve cell membranes to sodium ions. This decreased sodium permeability reduces the excitability of the neurons so that they do not transmit nerve impulses. General anesthetics reduce pain perception by rendering a person unconscious. This result depends at least partially on the ability of the general anesthetic to inhibit or depress conduction in the reticular activating system.

Referred Pain

A person can often identify the location of a stimulus that produces pain because the brain usually projects the sensation of pain to the site of the stimulus. In some instances, however, the brain incorrectly interprets signals from nociceptors in internal organs as coming from areas quite distant from the actual sites of stimulation—particularly as coming from sites on the body surface (Figure 15.7). This phenomenon is called **referred pain.** The locations of some referred pains are so consistent that physicians use them in diagnosing visceral dysfunction. For example, a heart attack frequently causes referred pain in the skin over the heart, on the left shoulder, and down the medial surface of the left arm.

One attempt to explain the cause of referred pain suggests that afferent neurons that transmit pain signals from a particular area of the body surface and afferent neurons that transmit pain signals from an internal organ connect with the same ascending neurons within the spinal cord. Thus, these ascending neurons carry pain signals to the brain from both the particular body surface area and the internal organ. Because cutaneous pain is much more common than visceral pain, the brain interprets pain signals carried over the ascending neurons as having originated in the skin rather than in the viscera, and the pain sensation is projected to the skin site.

Phantom Pain

Phantom pain is the phenomenon whereby a person who has undergone an amputation continues to feel pain that he or she perceives as coming from the amputated body part. Phantom pain, like referred pain, is a case of the brain inaccurately projecting the pain sensation. The neurons that supplied the affected structure are, of course, severed as a result of the amputation. However, the remaining portions of the neurons may continue to send nerve impulses to the same area of the brain as previously. The brain continues for some time to interpret impulses from the severed neurons as originating from the same body region as they did before

◆ **FIGURE 15.8 Reverberating circuits**

(a) and (b) Simple reverberating circuits that include collateral branches from neurons that restimulate themselves. (c) A more complex reverberating circuit that includes both facilitatory and inhibitory fibers. (d) A complex reverberating circuit that includes many parallel fibers with collateral branches.

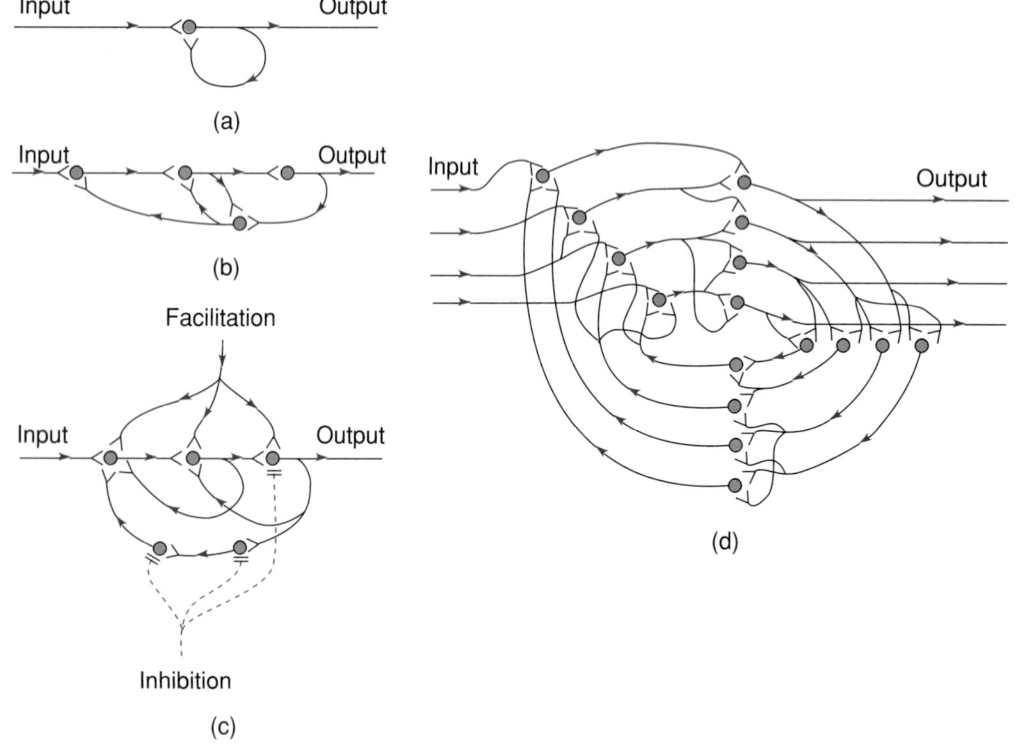

the amputation. As a result, the sensations evoked in the brain are projected to that region. In this manner, pain (and other sensations) may still be "felt," for example, in the toes, even after the foot has been amputated.

Memory

Memory refers to the ability to store experiences, thoughts, and sensations for later recall. Many questions about the processes that make memory possible are still unanswered, but it appears that there are at least two basic stages of memory: short term and long term.

Different physical-chemical mechanisms are believed to be responsible for the two stages of memory. In short-term memory, a bioelectrical process is thought to be involved, whereas in long-term memory, structural as well as biochemical changes in neurons or synapses are thought to occur.

Short-Term Memory

Short-term memory allows information to be recalled for only short periods following its initial presentation. Theorists believe that short-term memory decays in a matter of seconds unless the information within the short-term memory is rehearsed. Rehearsal, which is essentially the process of keeping one's attention on the information to be remembered, holds information within the short-term memory for an indefinite period of time. Rehearsal is also believed to be important in the transfer of information from short-term to long-term memory.

Some researchers suggest that short-term memory depends on the activation of *reverberating (oscillating) circuits* of neurons within the brain by signals from various receptors (Figure 15.8). The components of the reverberating circuits repeatedly reactivate each other for a period of time, thereby retaining information within the circuits. Without rehearsal, the activity of the reverberating circuits is believed to diminish, the short-term memory fades, and the person can no longer remember the information.

There is some evidence that a temporary facilitation or potentiation of nerve impulse transmission at the synapses of particular neuronal circuits contributes to short-term memory. In some cases, the process of presynaptic facilitation (page 373) may produce an increase in the release of neurotransmitters at certain synapses.

Long-Term Memory

The ability of the nervous system to recall information long after it was first presented is **long-term memory.** The consolidation of a bit of information within the long-term memory appears to take some time. During this time, various events can occur that interfere with the entrance of the information into the long-term memory. To some investigators, this possibility for interference suggests that information must remain within the short-term memory while the activities necessary to incorporate it into the long-term memory take place.

Theorists believe that long-term memory is the result of changes in the structure or biochemistry of neurons or synapses. Some investigators suggest that long-term memory involves alterations in the functional contacts between neurons in the brain (for example, changes in the number or area of synaptic junctions between neurons). In fact, there is evidence that mature neurons are constantly branching and establishing connections with new cells; at the same time, they are discontinuing connections with other cells. Such changes are thought to be part of the physiological basis of learning and memory.

Other investigators have suggested that the excitability of synapses is altered by a change in the ability of presynaptic neurons to secrete transmitter substances or by an alteration in the sensitivity of postsynaptic neurons. Moreover, some investigators have proposed that changes in protein synthesis—perhaps because of changes in gene expression—are important for the development of long-term memory. These changes are believed to influence the structures or activities of neurons.

Alterations in the structure or biochemistry of neurons or synapses, such as those just discussed, are thought to provide for memory by producing a long-term or permanent facilitation or potentiation of nerve impulse transmission at the synapses of particular neuronal circuits. This facilitation allows the circuits to be easily reexcited by incoming signals at later dates. Such hypothetical facilitated or potentiated neuronal circuits are called *memory traces (memory engrams).*

Learning

Learning, which is usually defined as a relatively permanent change in behavior as a result of experience, is a complex brain function that enables a person to adapt to a wide variety of circumstances and situations. Learning can occur very quickly—in the case of some adaptations in a second or less. Learning is also flexible—learned behaviors can be unlearned and new behaviors learned in their place. Although a number of behavioral responses such as the integrative spinal reflexes discussed earlier are not learned, most of a person's behaviors are learned.

A simple type of learning called **classical conditioning** leads to the development of conditioned responses. A *conditioned response* is a response to a stimulus (called a conditioned stimulus) that did not previously cause the response as a consequence of the pairing of this stimulus with another stimulus (called an unconditioned stimulus) that naturally produces the response.

A number of human behaviors can be influenced by classical conditioning, particularly those involving reflexes or emotional responses. For example, a puff of air blown into the eye is an unconditioned stimulus that naturally causes an eye-blink reflex. The sound of a bell is a conditioned stimulus that does not naturally cause this reflex. If a bell is rung just before a puff of air is blown into a person's eye, and the sequence is repeated several times, the person will eventually blink in response to the sound of the bell, even before the puff of air occurs. The blink in response to the sound of the bell is a conditioned response.

In a similar manner, fear, which is an emotional response to the threat of pain, is quite easily conditioned. If a person who is normally not afraid of dogs is bitten by a dog (an unconditioned stimulus), the person may, in the future, respond to the sight of the dog (a conditioned stimulus) with physical signs and subjective feelings of fear (a conditioned response). Moreover, the fear response may be generalized to all dogs, not just to the dog that bit the person.

Another type of learning is **instrumental learning,** or **operant conditioning.** In this process, the outcome or result of a particular behavior is important in determining whether the behavior will be strengthened (positively reinforced) or suppressed (negatively reinforced). If a particular behavior leads to a pleasant result or to a result that enables a person to escape from or avoid punishment, the behavior tends to be strengthened and becomes more likely to occur. If the behavior leads to the removal of a pleasant stimulus or to punishment, the behavior tends to be suppressed and becomes less likely to occur. Instrumental learning influences behavior in many situations, and parents often use instrumental learning techniques to teach children useful habits. For example, a parent may train a child to pick up his or her toys after play by smiling and praising the child each time the toys are picked up. In this situation, the picking-up behavior is positively reinforced by the pleasant result of parental approval. Consequently, the behavior becomes more likely to occur.

A more complex type of learning, called **cognitive learning,** also occurs in humans. The human brain is capable of forming new thought patterns by recalling and integrating previous experiences. The brain can correlate information—gained from previous experiences and stored in the memory—into new concepts by abstract reasoning.

Control of Body Movements

The nervous system coordinates and controls the contractions of the many skeletal muscles involved in body movements.

Support of the Body Against Gravity

Spinal reflexes such as the positive supportive reaction and muscle stretch reflexes, as well as higher centers within the brain, regulate the contractions of the skeletal muscles that support the body against the force of gravity and maintain an upright posture.

Reflexes

The **positive supportive reaction** is a complex reflex that is initiated when pressure is applied to the bottoms of the feet. Nerve impulses from pressure receptors are transmitted to the spinal cord, where the incoming signals are diverged through a complex pathway of interneurons, and stimulatory impulses are sent to the alpha motor neurons of the extensor muscles of the legs. The contractions of the extensor muscles stiffen the legs, helping to support the individual.

The skeletal muscles exist in a continual state of slight contraction known as **skeletal muscle tone (tonus)** that is important in supporting the body against gravity and in maintaining an upright posture. A basic neural mechanism involved in the maintenance of muscle tone is the stretch reflex (page 414). Consequently, the stretch reflex is important in maintaining an upright posture. For example, when the force of gravity causes the knees to buckle, the extensor muscles of the legs are stretched, initiating stretch reflexes that stimulate the alpha motor neurons to the extensor muscles. The contractions of the extensor muscles stiffen the legs, helping to support the individual.

Higher Centers

The excitability of the alpha motor neurons to skeletal muscles is influenced not only by nerve impulses from receptors involved in spinal reflexes, but also by nerve impulses from higher centers within the brain. Nerve impulses from these centers may be sent either to the alpha motor neurons themselves or to the intrafusal fibers of the muscle spindles. Impulses to the intrafusal fibers alter the sensitivity of the muscle spindles. This altered sensitivity, in turn, can result in an altered stimulation of the alpha motor neurons by way of the stretch reflex pathway.

The reticular formation and associated nuclei in the medulla-pons region (the vestibular nuclei) are intrinsically excitable and transmit stimulatory signals to motor areas of the spinal cord. Although the full activity of the reticular formation and vestibular nuclei is normally held in check by inhibitory signals from the basal nuclei, continuous impulses are sent to the spinal cord when a person is in a standing position. These impulses contribute to the excitation of the alpha motor neurons to the skeletal muscles, and they provide much of the intrinsic excitation required to maintain tone in the extensor muscles that support the body against gravity.

Locomotion

The alternating flexion of one limb and extension of the other that occur during locomotion (walking or running) depend on rhythmic patterns of activity occurring within the spinal cord. These patterns, it is thought, make use of reverberating circuits of interneurons to stimulate alternately the flexor and extensor muscles of each leg, while reciprocal inhibition mechanisms keep one leg extended when the other is flexed. These activity patterns are believed to be coordinated by the integrative actions of higher centers, including the cerebral cortex, the basal nuclei, the cerebellum, and the reticular formation. The higher centers control these activities according to the desires of the individual.

The Cerebellum and the Coordination of Movement

The cerebellum is an especially important controller of muscular activities, particularly very rapid activities such as typing or running. During movement, the cerebellum receives information about the signals being sent to the muscles from areas of the motor system such as the cerebral cortex, the basal nuclei, and the reticular formation. The cerebellum also receives information from muscle spindles, neurotendinous organs, and joint receptors (and also from the eyes and the vestibular apparatus of the ears, which are discussed in Chapter 16) about the position, rate of movement, inertia, momentum, and so forth of body parts. The cerebellum compares the information about the movement called for by the motor system with information about the actual status of the body parts. If necessary, it sends signals back to the motor system to bring the actual movement into line with the intended one.

The cerebellum appears to play an important role in providing the central nervous system with the ability to predict the future position of a body part in the next few hundredths of a second during a movement. Some time before a moving body part reaches its intended position, the cerebellum sends signals to slow the moving part and stop it at its intended point. This activity involves the excitation of antagonistic muscles near the end of a movement coupled with the inhibition of the muscles that started the movement.

The cerebellum is also important in maintaining body equilibrium. When the direction of movement changes, signals transmitted to the cerebellum from receptors such as the vestibular apparatus of the ears help a person anticipate an impending loss of equilibrium and take corrective action before the equilibrium loss actually occurs.

Skilled Movements

Almost all skilled movements (such as the hand movements involved in typing or playing a violin) are learned movements whose mastery requires repetition. Such movements are usually slow and awkward at first, but their speed and efficiency generally improve with practice. It has been proposed that as a skilled movement is learned, the pattern of neural activity necessary to perform the movement becomes stored within the brain, perhaps in the form of a facilitated or potentiated neuronal circuit (that is, as a memory trace). When the learned, preprogrammed movement is to be repeated, this stored pattern is called forth, and the muscle contractions necessary to perform the movement are repeated in the orderly sequence dictated by the pattern. The precise parts of the brain in which such patterns may be stored are not known. However, the motor cortex, the sensory cortex, and deeper centers such as the basal nuclei and even the cerebellum may be involved. Of particular interest is a cortical area in front of the primary motor cortex called the premotor area. Electrical stimulation of this area sometimes produces skilled patterns of movement such as hand movements, and this area is thought to be particularly involved in the occurrence of learned movements that can be rapidly performed.

As a skilled movement is being learned, it is usually performed slowly and deliberately. During this time, feedback from receptors in muscles and joints, from cutaneous receptors, and from the eyes is important in indicating the effectiveness and degree of success of the movement and in establishing a pattern of learned motor functions in the brain. Even after a precision movement has been mastered, feedback is important in monitoring and, if necessary, correcting the movement—if it is performed slowly. However, once they are learned, many skilled movements are performed so rapidly that there is no time for feedback control or correction to occur. Nevertheless, feedback is important in determining if the movement has been correctly performed and in correcting the skilled movement pattern if necessary so that the movement is executed properly the next time it is performed. The feedback signals are believed to be sent not only to the cerebellum but also to the sensory areas of the cerebral cortex, which, in turn, relay signals to the motor cortex.

The primary motor area of the cerebral cortex has been considered the area principally responsible for initiating voluntary motor activity. However, some studies have indicated that this is not the case. When monkeys were taught to perform simple hand movements in response to cues, it was found that neurons in the motor cortex, the cerebellum, and the basal nuclei all discharged impulses prior to the occurrence of the movement. Moreover, other studies indicated that cerebellar neurons become active before neurons of the motor cortex. These findings have led to the proposal that subcortical structures such as the cerebellum and basal nuclei are involved in initiating activity in motor cortex neurons. According to this view, signals related to the performance of learned, preprogrammed movements reach the motor cortex by way of the cerebellum and the basal nuclei. Moreover, it has been suggested that the cerebellum is particularly involved in the performance of rapid movements, the basal nuclei are important in the performance of slower movements, and the motor cortex is involved in the more precise integration of both rapid and slow movements.

◆ **FIGURE 15.9 Brain areas associated with language**

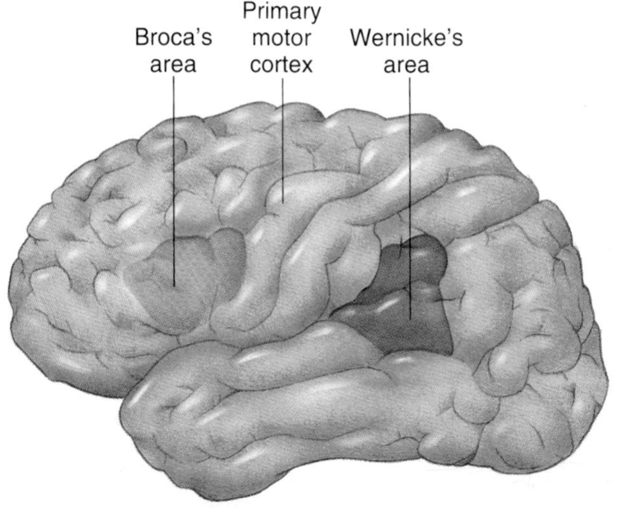

Language

Language is a complex form of communication in which spoken or written words represent objects or concepts. One person communicates with or transfers information to another person primarily by language.

The areas of the cerebral cortex concerned with language are usually located in only one cerebral hemisphere—the left in the vast majority of people (Figure 15.9). A portion of the temporal lobe known as *Wer-*

CLINICAL CORRELATION

Alzheimer's Disease

Case Report

THE PATIENT: A 61-year-old woman.

PRINCIPAL COMPLAINT: Loss of memory and disorientation.

HISTORY: The patient was reported by her sister to have shown a progressive decline in her mental capabilities over the past five years. Her family had attributed the mental changes to grief after the death of the patient's husband, but the increasing severity of the symptoms prompted them to seek medical evaluation. The family described the patient as very forgetful and reported that she became confused and disoriented in places other than her home and the small food market that she and her husband had owned for years. The patient was no longer capable of handling the bookkeeping for the market or of balancing her checkbook, although she had managed the market's finances for more than 30 years.

CLINICAL EXAMINATION: The patient was aware of her problems with memory and appeared concerned about the changes in her mental abilities. During an interview, she could not correctly state the current month, and although she correctly named the current president of the United States, she could not remember who preceded him. She had no difficulty with language, either receptive or expressive, no weakness of arms or legs, and no sensory deficits. Her gait was normal, and no abnormal movements were observed. Neuropsychological testing revealed a profound deficit in short-term memory, impaired visuospatial skills (manifested by the inability to duplicate three-dimensional constructions with blocks), and mild psychomotor slowing (decreased reaction time, decreased rate of finger tapping). A standard adult-intelligence test revealed that the patient's IQ was approximately 20 points lower than would be predicted by her performance in high school and her vocational history. Pertinent laboratory findings included a CT scan consistent with mild, diffuse cerebral atrophy, an electroencephalogram (EEG) diffusely slow, with no focal abnormality, normal thyroid function tests, and normal electrocardiogram (ECG) and blood pressure.

COMMENT: Clinically, *dementia* refers to a generalized diminution of mental abilities that consists mainly of memory loss, confusion and/or disorientation, personality changes, and generally slowed

nicke's area is concerned with language comprehension and plays a critical role in a person's ability to understand both spoken and written messages. Wernicke's area is also responsible for formulating coherent patterns of speech that express a person's thoughts in meaningful messages. A portion of the frontal lobe known as *Broca's area* is concerned with the motor processes of word formation.

Wernicke's area and Broca's area cooperate to produce spoken language. Wernicke's area formulates the appropriate speech patterns to express a person's thoughts and transmits signals to Broca's area. In Broca's area, the incoming signals are processed, and a detailed motor program for the muscular movements of vocalization is generated. This program is transmitted to the primary motor cortex, which, in turn, sends nerve impulses to the speech muscles.

Damage to Wernicke's area results in a condition called *sensory aphasia (a-fay´-zhuh)*. A person suffering from sensory aphasia is unable to comprehend spoken or written language. The person is able to articulate words, but the messages conveyed are generally inappropriate or meaningless. Damage to Broca's area results in a condition called *motor aphasia*. The speech of a person suffering from motor aphasia is slow and labored, and he or she has difficulty articulating words. The person can decide what he or she wants to say but cannot make the vocal system perform smoothly.

CLINICAL CORRELATION

"mental processing." The term dementia usually is reserved for an irreversible condition; similar cognitive changes caused by reversible processes (such as drug effects, toxins, or hypoxia) are referred to as *delirium.*

Dementia can be caused by brain tumors, vascular disease, nutritional deficits, or hydrocephalus and can accompany many specific neurological degenerative diseases. The most common cause, however, accounting for more than 40% of geriatric patients who have dementia, is Alzheimer's disease.

Patients in the early stages of Alzheimer's disease generally have impairments of recent memory. As the disease progresses, higher cognitive functions deteriorate, and the abilities to read, write, calculate, and use language appropriately are lost. Alzheimer's patients may also become irritable, display emotional lability, and have hallucinations. The diagnosis of Alzheimer's disease, except at autopsy, is made by excluding other possible causes of the patient's symptoms, especially in the early stages. Other causes, particularly those that are treatable, must be considered first.

A number of changes occur in the brains of Alzheimer's patients. These changes include the occurrence of *neuritic plaques,* which consist of abnormal axons (primarily axon terminals) associated with a core of extracellular amyloid (a complex protein that has a hyaline, structureless nature). Also present are *neurofibrillary tangles,* which are bundles of paired helical filaments such as cross-linked polypeptides that accumulate within the cell bodies of neurons. Neuritic plaques and neurofibrillary tangles are prominent in the cerebral cortex and hippocampus, and generally the more severe a patient's cognitive defects, the greater will be the density of neuritic plaques in the patient's cortex (as determined at autopsy). Alzheimer's patients have a particular deficiency of acetylcholine-releasing neural endings in the cerebral cortex and hippocampus, and it appears that this deficiency is due at least in part to the degeneration of certain neurons whose cell bodies are located in the basal forebrain. (The axons of these neurons normally extend into the cerebral cortex and the hippocampus.) In addition, there is evidence that a degeneration of other neural systems can also occur in Alzheimer's patients. Although the cause of Alzheimer's disease is unknown, there is evidence that the disease has a genetic component.

OUTCOME: Although there is no known effective treatment for Alzheimer's disease, the patient was given the tranquilizer diazepam (valium), 2 mg twice a day, to make her and her family more comfortable. Because of the progressive nature of the disorder, this patient probably will have to be placed in a special-care facility when the disease becomes more advanced.

Study Outline

◆ INTEGRATIVE SPINAL REFLEXES pp. 470–471
A noxious stimulus can cause flexion of limb on ipsilateral side of body; commissural interneuron connections lead to extension of corresponding limb on contralateral side of body.

◆ MENTAL PROCESSES pp. 471–479
No area or structure of the brain acts entirely on its own. The control of particular functions involves more than one area of the brain, and each brain area is probably involved in many functions.

The Electroencephalogram. Record of electrical waves produced during brain activity; spontaneous waves present, even during unconsciousness; some correlation between brain waves and body activity.

Consciousness. State of awareness; state in which thoughts or instants of awareness occur; result of activation or arousal of widespread areas of the brain.

THE AROUSED BRAIN. Able to consciously react to stimuli; reticular activating system important in arousing brain; general sensory input plays important role in maintaining brain arousal.

SLEEP. State of partial or complete unconsciousness from which a person can be aroused by appropriate stimuli.

Slow-wave sleep. High amplitude, low-frequency EEG.

Paradoxical sleep. Rapid-eye-movement (REM) sleep; dreams.

SLEEP-WAKEFULNESS CYCLE. Awakening appears to occur when impulses to the cerebral cortex from RAS reach sufficient intensity to arouse brain.

1. Positive-feedback responses involving brain areas such as cerebral cortex as well as peripheral musculature contribute to the activity of the RAS leading to wakefulness.
2. Over time, excitability of wakefulness circuit diminishes, intensity of signals from RAS declines, and individual becomes drowsy and falls asleep.
3. The depletion of chemical transmitter substances from positive-feedback loop neurons or the accumulation of inhibitory chemical substances may influence the excitability of the wakefulness circuit (for example, serotonin secretion may lead to sleep).
4. Brain-stem neurons that may contribute to slow-wave sleep by secreting serotonin may be involved in a feedback relationship with paradoxical sleep neurons of pons. The neurons of the pons may stimulate the serotonin-secreting neurons of the brain stem to take up serotonin, leading to wakefulness.

Attention.

1. Ability of an individual to maintain selective awareness of some aspect of the environment; ability to respond selectively to a particular stimulus.
2. Interest and biological urgency influence priority given to particular stimuli.
3. Influenced by specific thalamic areas of reticular activating system.
4. Other central nervous system mechanisms can selectively inhibit or enhance afferent signal input to brain.

Emotions and Behavior. Limbic system is important in emotions and emotional behavior patterns; it is also important in determining whether a particular activity is pleasant and rewarding or painful and punishing. Reward and punishment areas of brain have a great influence on a person's behavior. Norepinephrine, dopamine, and serotonin are involved in certain emotional and behavioral states and in learning.

Pain. Signals to brain from nociceptors elicit sensation of pain and also lead to such activities as emotional reactions and behavioral responses.

PAIN MODULATION. Body possesses analgesic system that can inhibit the transmission of pain signals to the brain and thereby suppress pain perception. Anesthetics reduce sensitivity to painful stimuli. Local anesthetics can decrease permeability of nerve cell membranes to sodium ions so neurons do not transmit impulses. General anesthetics render person unconscious, at least partially by inhibiting or depressing conduction in the reticular activating system.

REFERRED PAIN. Inaccurate projection of pain sensation to site other than stimulus site.

PHANTOM PAIN. Person feels pain from amputated body part.

Memory. Ability to store experiences, thoughts, and sensations for later recall.

SHORT-TERM MEMORY. Decays rapidly unless rehearsed; may depend on activation of reverberating circuits of neurons within the brain by signals from receptors.

LONG-TERM MEMORY. May involve actual alterations in structure or biochemistry of neurons or synapses; may involve changes in number or area of synaptic junctions between neurons; may involve facilitated neuronal circuits called memory traces (memory engrams).

Learning. Relatively permanent change in behavior as a result of experience.

1. *Classical conditioning.* Conditioned responses; produced by pairing an unconditioned stimulus with a conditioned stimulus.
2. *Instrumental learning (operant conditioning).* The outcome or result of a particular behavior is important in determining whether the behavior will be strengthened or suppressed.
3. *Cognitive learning.* Recall and integration of previous sensory experiences; concept formation; abstract reasoning.

◆ CONTROL OF BODY MOVEMENTS pp. 480–481

Nervous system coordinates and controls contractions of skeletal muscles.

Support of the Body Against Gravity. Involves reflexes and higher brain centers.

REFLEXES. Positive supportive reaction and muscle stretch reflexes are important in supporting the body against the force of gravity.

HIGHER CENTERS. Nerve impulses from higher centers (such as the reticular formation and vestibular nuclei) contribute to excitation of alpha motor neurons, helping to maintain tone in extensor muscles that support the body against gravity.

Locomotion.

1. Believed to depend on rhythmic patterns of activity occurring within spinal cord that involve reverberating neuronal circuits and reciprocal inhibition.
2. Integrative actions of higher centers (such as cerebral cortex, basal nuclei, cerebellum, reticular formation) coordinate rhythmic activity patterns to produce effective locomotion.
3. The higher centers control activities according to desires of the individual.

The Cerebellum and the Coordination of Movement.

1. During movement, the cerebellum receives signals from motor areas about intended movement and from periphery about actual movement.
2. Cerebellum compares information and, if necessary, sends signals to motor areas to bring actual movement into line with intended movement.
3. Cerebellum appears to play important role in enabling CNS to predict future position of a body part during movement.
4. Cerebellum is important in the maintenance of equilibrium.

Skilled Movements.

1. Almost all are learned movements that require repetition to master.
2. As a movement is learned, pattern of neural activity necessary to perform it may become laid down in brain.
3. When the movement is to be repeated, the pattern is called forth, and the necessary muscular contractions are repeated.
4. Premotor area of the cortex may be important in pattern storage.
5. Sensory feedback important in learning skilled movements and establishing patterns of learned motor function in brain; also involved in monitoring and perhaps correcting pattern of movement after movement is mastered.
6. Cerebellum and basal nuclei appear to be involved in initiating activity in neurons of motor cortex during learned, preprogrammed movements; cerebellum particularly involved in rapid movements, basal nuclei in slow movements.

◆ **LANGUAGE** pp. 481–482
Complex form of communication in which spoken or written words represent objects or concepts; *Wernicke's area* concerned with language comprehension and plays critical role in ability to understand spoken and written messages. Wernicke's area also responsible for formulating coherent patterns of speech that express thoughts in the form of meaningful messages. *Broca's area* concerned with motor processes of word formation.

Self-Quiz

1. During the flexor reflex and the associated crossed extensor reflex, the: (a) ipsilateral flexor muscles are stimulated; (b) contralateral extensor muscles are inhibited; (c) ipsilateral extensor muscles are stimulated.
2. Electroencephalogram records during deep sleep show waves of: (a) low amplitude; (b) low frequency; (c) no amplitude or frequency.
3. Electroencephalogram records of an alert or an awake brain show waves of: (a) low amplitude and high frequency; (b) high amplitude and low frequency; (c) high amplitude and high frequency.
4. Destruction of the reticular activating system produces: (a) temporary loss of memory; (b) permanent coma; (c) an inability to sleep soundly.
5. General sensory input may play an important role in maintaining the arousal of the brain. True or False?
6. REM sleep is the same as: (a) slow-wave sleep; (b) drowsy sleep; (c) paradoxical sleep.
7. Neurons within a region of the brain stem called the median raphe secrete which substance into the reticular activating system? (a) norepinephrine; (b) serotonin; (c) substance P.
8. The previous experience of an individual appears to be important in determining which sensory stimuli may warrant the individual's attention at any particular moment. True or False?
9. Brain structures that are particularly important in relation to emotions form the: (a) reticular activating system; (b) limbic system; (c) sensorimotor system.
10. The reward and punishment regions of the brain have a great influence on a person's behavior. True or False?
11. Local anesthetics are believed to exert their effect primarily by inhibiting the reticular activating system. True or False?
12. In order to keep information in the short-term memory for an indefinite period of time: (a) an alteration in the structure of synapses must occur; (b) an alteration in the structure of neurons must occur; (c) rehearsal is necessary.
13. Only relatively few of a person's behaviors are learned behaviors. True or False?
14. Learning in which the outcome of a particular behavior is important in determining whether the behavior will be strengthened or suppressed is called: (a) classical conditioning; (b) instrumental learning; (c) cognitive learning.
15. Impulses sent to motor areas of the spinal cord from the reticular formation and vestibular nuclei contribute to the excitation of the alpha motor neurons to the skeletal muscles. True or False?
16. The alternating flexion of one limb and extension of the other that occur during locomotion are believed to depend on rhythmic patterns of activity occurring within the: (a) spinal cord; (b) pons; (c) cerebral peduncles.
17. The cerebellum plays an important role in: (a) coordinating body movements; (b) emotion; (c) attention.
18. Almost all skilled movements are learned movements that require varying amounts of repetition to master. True or False?
19. Which brain structure may be particularly involved in the performance of learned, preprogrammed slow movements? (a) hypothalamus; (b) basal nuclei; (c) pons.
20. The brain area concerned with the motor processes of word formation is: (a) Wernicke's area; (b) Penfield's area; (c) Broca's area.

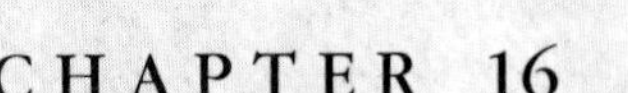

CHAPTER 16

The Special Senses

CHAPTER CONTENTS

LEARNING OBJECTIVES

After completing this chapter, you should be able to:

1. Name and describe the functions of the basic layers of the eye and their main component parts.
2. Describe the focusing of images on the retina.
3. Describe the structure of the retina.
4. Describe the responses of the rods to light.
5. Explain how colors are perceived.
6. Distinguish among presbyopia, astigmatism, and emmetropia.
7. Name the basic structural compartments of the ear and describe their main components.
8. Explain how a sound wave is converted into a perceived sound.
9. Describe mechanisms by which the source of a sound is localized.
10. Describe the function of the utricle.
11. Describe the function of the semicircular ducts.
12. On a neurophysiological level, explain how odors are detected.
13. On a neurophysiological level, explain how substances are tasted.

The body possesses specialized receptors for *vision* (sight), *hearing* (audition), *head position and movement* (the labyrinthine or vestibular sensations), *smell* (olfaction), and *taste* (gustation). These receptors are located in specific structures. The receptors for vision are located in the eyes; those for hearing and head position and movement are located in the ears; the receptors for smell are located in the nose; and the receptors for taste are found in the mouth and throat.

The Eye—Vision

The **eye** is the sense organ for vision. It is housed in a bony skull cavity called the *orbit,* which protects much of the eye from physical injury.

Embryonic Development of the Eye

By the fourth week of embryonic development, lateral outgrowths from either side of the diencephalon have formed structures called **optic vesicles.** As the optic vesicles enlarge, the distal portion of each invaginates, forming a double-layered **optic cup** (Figure 16.1). The optic cup is destined to become the retina of the eye. The internal layer of the cup thickens to form the nervous-tissue layers of the retina, which contain receptor cells and neural elements. The external layer develops into the pigmented layer of the retina. The proximal portion of each optic vesicle narrows into an **optic stalk,** which will become incorporated into the optic nerve.

As the optic cups approach the underside of the surface ectoderm of the embryo, the ectoderm thickens into a **lens placode** directly over each cup. As the cavity of each optic cup deepens, the lens placodes invaginate to form **lens vesicles,** each of which is surrounded by one of the optic cups. With further development, the lens vesicles separate from the surface

◆ **FIGURE 16.1 Embryonic formation of the eye**

(a) Section through diencephalon and optic vesicle. (The dotted line indicates the plane of the section.) The optic vesicle grows out from the diencephalon and contacts the surface ectoderm. The surface ectoderm then thickens to form a lens placode. (b) The lens placode invaginates, forming a hollow lens vesicle. The optic vesicle surrounds the lens vesicle, forming an optic cup that develops into both layers of the retina. (c) Saggital section of the optic cup and lens vesicle. The fibrous and vascular tunics of the eye are formed from mesodermal cells that migrate into the area and surround the optic cup. The optic stalk becomes part of the optic nerve.

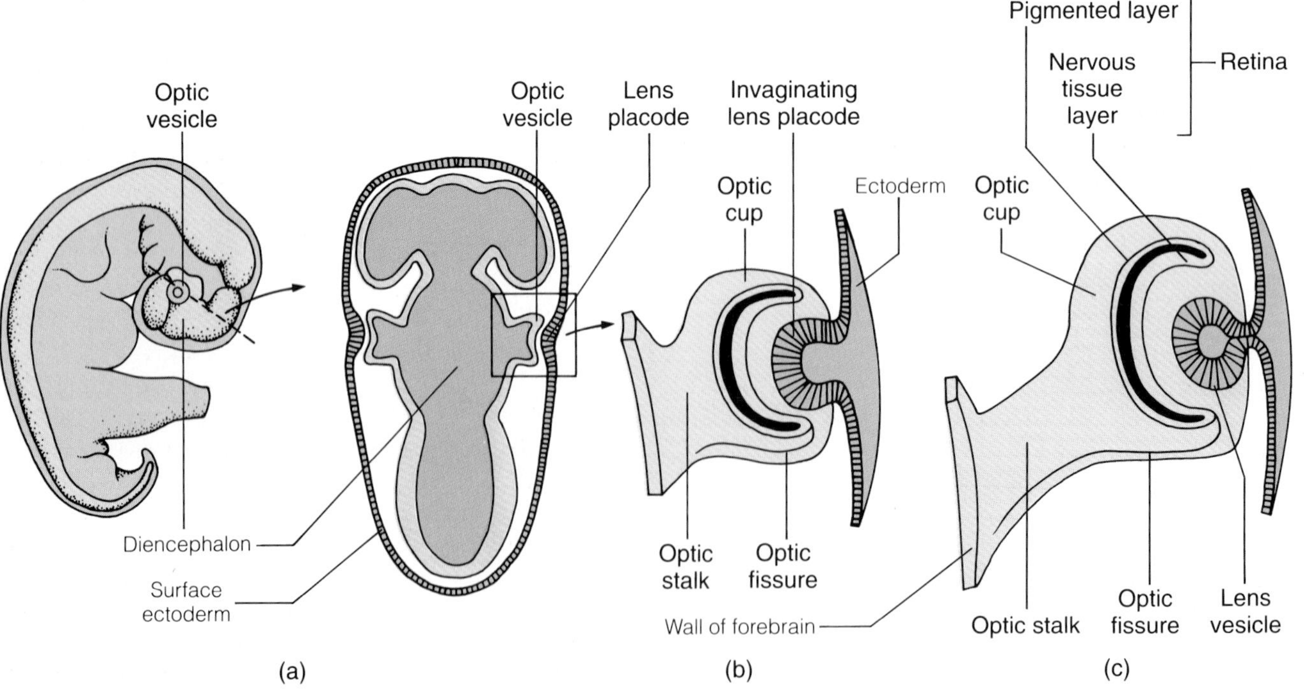

ectoderm and form a rounded body within the opening of each optic cup. The lens of each adult eye develops from these lens vesicles. Ectodermal cells superficial to the lens vesicles develop into the **cornea** of the eye.

On the undersurface of each optic cup and optic stalk is a groove called the **optic fissure.** The optic fissure serves as a path that directs the fibers of the optic nerve from the inner layer of the retina to the brain and that provides a means by which blood vessels are able to supply the interior of the eye. By the time of birth, the optic fissures have closed, completely encompassing the optic nerves and blood vessels.

Loose mesenchymal (mesoderm) cells accumulate around the outside of each optic cup. These cells differentiate into two connective-tissue layers of the eye: the fibrous tunic and the vascular tunic.

Structure of the Eye

The eye is essentially a spherical structure whose wall is composed of three basic layers or tunics: the *fibrous tunic,* the *vascular tunic,* and the internal tunic, or *retina* (Figure 16.2). The interior of the eye contains two cavities (anterior and posterior), which are filled with fluids called humors.

Fibrous Tunic

The outermost layer of the eye is called the **fibrous tunic.** The posterior five-sixths of the fibrous tunic, the **sclera** *(skleh´-rah),* is white and opaque. The sclera, composed of dense connective tissue, aids in protecting the inner structures of the eye and helps maintain the shape of the eye. The anterior one-sixth of the fibrous tunic is clear and is called the **cornea.** The cornea lacks blood vessels and is composed primarily of dense connective tissue, with an outer layer of stratified squamous epithelium. It has a greater curvature than the sclera, which causes it to protrude somewhat from the sclera. As light enters the eye, it passes through the cornea.

Vascular Tunic

Internal to the fibrous tunic is a layer called the **vascular tunic** (or **uvea**), which is composed of the *choroid,* the *ciliary body,* and the *iris.*

The **choroid** *(kō´-roid)* which lines most of the internal region of the sclera, is darkly pigmented and contains many blood vessels. Around the edge of the cornea, the choroid forms the **ciliary body,** a thickened ring of tissue that encircles the lens. The ciliary body contains smooth muscles called **ciliary muscles** and a series of folds called the **ciliary processes.** The ciliary muscles serve to control the shape of the lens, which is a clear structure that is part of the focusing system of the eye. A **suspensory ligament** extends from the ciliary body to the lens, encircling it. The ciliary processes secrete the fluid that fills the anterior cavity of the eye.

The anterior portion of the vascular tunic is the **iris.** The iris, which is largely a continuation of the choroid, is a thin, muscular diaphragm whose pigmentation is responsible for eye color. In the center of the iris is a rounded opening, the **pupil,** through which light enters the interior regions of the eye. Two sets of smooth muscles within the iris affect the diameter of the pupil, and thus how much light enters the eye. One set of muscles is circularly arranged, surrounding the pupil. Contraction of the circular muscles constricts the pupil and restricts the amount of light entering the eye. The other set of muscles is radially arranged, extending peripherally from the pupil. Contraction of the radial muscles dilates the pupil and allows more light to enter the eye. Contraction of the circular muscles is controlled by the parasympathetic nervous system, and contraction of the radial muscles is controlled by the sympathetic nervous system.

Retina

The innermost layer of the eye is the internal tunic, or **retina.** The retina consists of an outer **pigmented layer** and inner **nervous-tissue layers.** The outer pigmented layer is composed of a single layer of heavily pigmented cuboidal epithelial cells that lie in contact with, and are tightly adherent to, the choroid. Both the pigmented layer of the retina and the choroid contain the brown black pigment melanin. The dark pigmentation of these structures reduces the reflection of the light that enters the eye. The cells of the pigmented layer also act as phagocytes and store vitamin A needed by the photoreceptor cells of the nervous-tissue layers.

The nervous-tissue layers of the retina contain the actual receptors for light—photoreceptors called **rods** and **cones**—as well as numerous neural interconnections. After several synapses in the retina, nerve fibers converge and exit through the rear of the eye, slightly medial to the posterior pole of the eye, as the **optic nerve.** No photoreceptors are present at the point of exit, and this area is called the **blind spot,** or **optic disc** (Figure 16.3). The *central retinal* artery and vein that supply and drain the retina enter and leave the eye through the optic disc. These vessels can be examined with an ophthalmoscope, providing valuable information about a person's health. Lateral to the optic disc is a slightly yellow region known as the **macula.** The macula is located almost exactly at the posterior pole of the eye. In the center of the macula is a depression called the **fovea centralis** ("central depression"). The fovea, which is the portion of the retina where visual acuity is greatest, contains no rods but does contain very densely packed

◆ **FIGURE 16.2 The eye**

(a) The structures of the eye. (b) Photograph of a sagittal section of an eye.

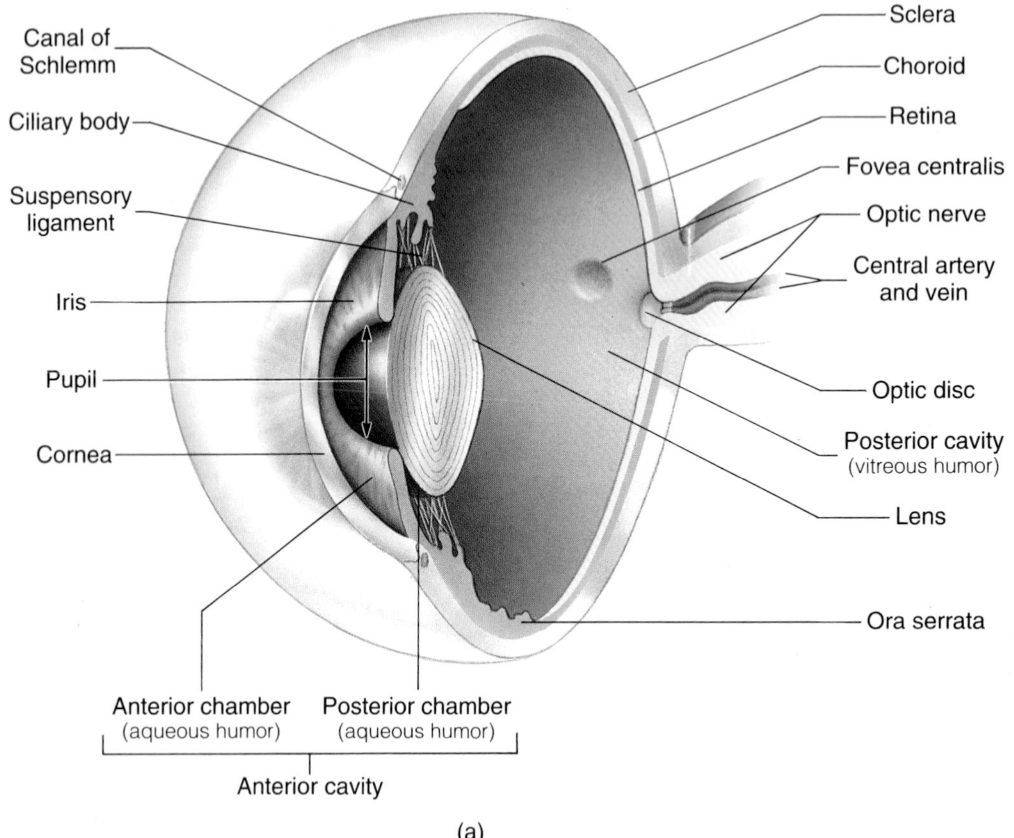

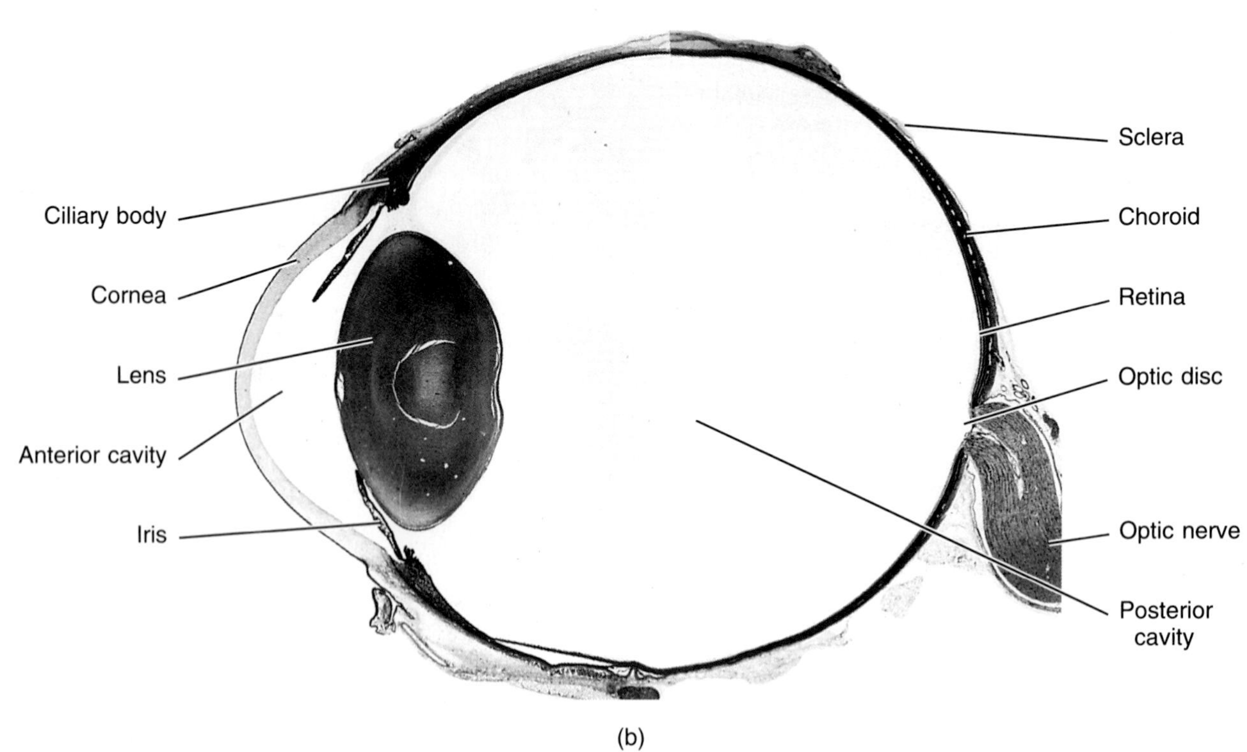

◆ **FIGURE 16.3 Demonstration of the blind spot**
Close your left eye and focus your right eye on the X. Move the page toward or away from you until the black spot disappears. When this occurs, the image of the spot has fallen on the blind spot (the optic disc) of the retina.

X ●

cones. When you look directly at an object, the image of the object generally is focused on the fovea.

The nervous-tissue layers appear to end anteriorally, near the ciliary body, in a scalloped margin called the *ora serrata.* However, they actually continue forward onto the inner surface of the ciliary body and the iris. The nervous-tissue layers are attached to the pigmented layer of the retina only around the optic nerve and at the ora serrata. Because the connection is loose, it is possible for the nervous-tissue layers of the retina to become detached from the pigmented layer. Such a separation can be repaired surgically, or the layers can be fused together using a laser.

Lens

Behind the pupil is a clear **lens,** which is held in position by the fibrous **suspensory ligament** that attaches to the ciliary body. The lens is a biconvex structure that is relatively elastic. By changing shape, the lens aids in properly focusing the light that enters the eye on the retina. The lens, like the cornea, has no blood supply. It receives its nourishment from fluids called the aqueous and vitreous humors within the cavities of the eye.

Cavities and Humors

The lens separates the interior of the eye into two cavities. In front of the lens is an **anterior cavity;** behind the lens is a **posterior cavity.** The anterior cavity can be subdivided into two chambers. The **anterior chamber** is located in front of the lens and iris and behind the cornea. The **posterior chamber** is located between the iris and the suspensory ligament. The anterior cavity contains a clear fluid, the **aqueous humor,** that has a composition similar to that of the blood plasma. The posterior cavity contains a transparent, semifluid, jelly-like substance called the **vitreous humor,** which consists of a viscous connective-tissue matrix containing a network of fine collagenous fibers.

The vitreous humor is replaced very slowly, if at all. In contrast, the aqueous humor is produced at a rate of about 5 to 6 ml per day by the ciliary processes, which project from the ciliary body. The aqueous humor drains into a canal (venous sinus) called the **canal of Schlemm** and eventually reaches the bloodstream. The canal of Schlemm is located near the junction of the cornea and the iris. The rates of production and removal of aqueous humor are such that a relatively constant pressure of about 15 mm of mercury (the *intraocular pressure*) is maintained in the eye.

Accessory Structures of the Eye

Several structures associated with the eye contribute in various ways to its functioning. These include the *eyelids,* the *conjunctiva,* the *lacrimal apparatus,* and the *extrinsic eye muscles.*

Eyelids

Skin-covered structures called upper and lower **eyelids,** or **palpebrae** *(pal´-pe-brah),* can be drawn over the anterior surface of the eye to protect the exposed portion (Figure 16.4). When closed, the eyelids prevent about 99% of the incident light from entering the eye. The movements of the eyelids are controlled by skeletal muscles. Fibers of the **orbicularis oculi** muscle extend into both eyelids and serve to close the lids. The **levator palpebrae superioris** *(lē-va´-tor pal-pe´-brah su-pe´-rē-oris)* muscle is inserted into the upper lid and functions to raise it (Figure 16.5). Much of the activity of these muscles occurs as the result of reflexes that produce an eye-blink response. This response may occur spontaneously about 25 times per minute when a person is awake, and it is particularly evident if a foreign object comes into contact with the surface of the eye or with the hairs **(eyelashes)** projecting from the borders of the eyelids. With each blink of the eyelids, fluid is spread over the front of the eye, protecting it from drying.

◆ **FIGURE 16.4 The accessory structures of the eye (anterior view)**
The lacrimal gland has been retracted to show the ducts.

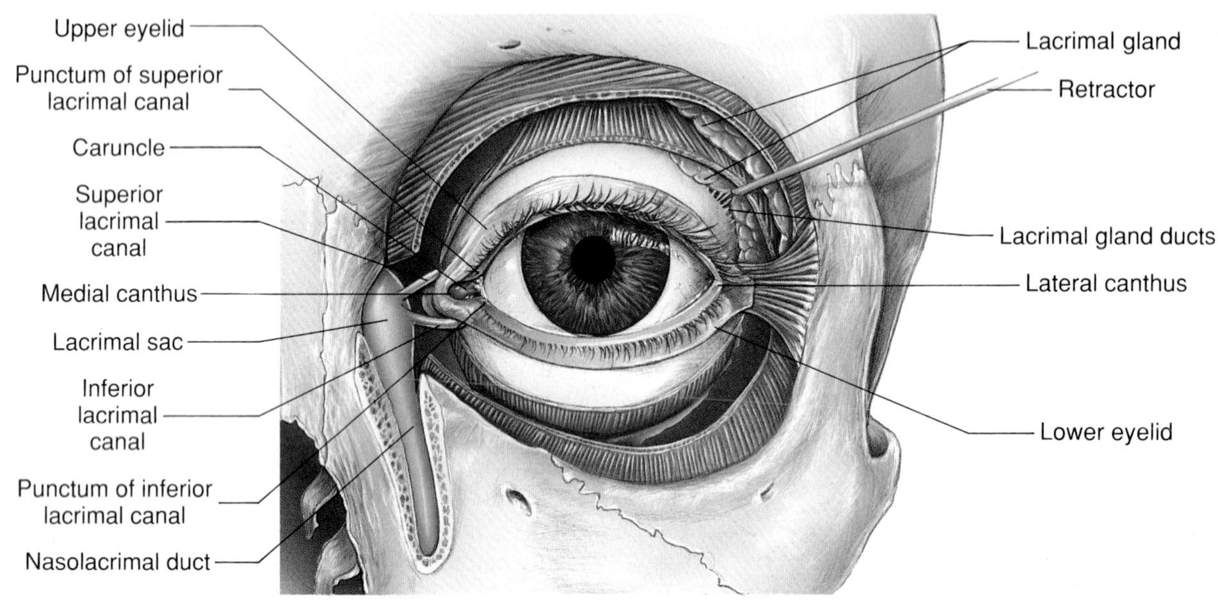

Within each eyelid is dense connective tissue that forms a structure called a **tarsal plate.** The tarsal plates are important in maintaining the shape of the eyelids. Sebaceous glands called **tarsal (meibomian) glands** are located close to the inner surfaces of the eyelids and are embedded in the tarsal plates. The ducts of the tarsal glands open onto the margins of the eyelids, and their secretions onto these margins help prevent tears from overflowing between the eyelids. An infection of the tarsal glands can produce a *cyst* on the eyelid.

The eyelids also possess modified sweat glands, called **ciliary glands,** as well as sebaceous glands located at the bases of the hair follicles of the eyelashes. Occasionally, these sebaceous glands may become infected, thereby producing a *sty.* Between the upper and lower eyelids of each eye is a gap through which the eye is exposed—the **palpebral fissure.** The angle formed by the lateral junction of the upper and lower eyelids is called the **lateral canthus;** the angle formed by the medial junction of the upper and lower eyelids is called the **medial canthus** (Figure 16.4). A small, reddish mound of tissue at the medial canthus of each eye, the **caruncle,** contains a few sebaceous glands.

Conjunctiva

Both the insides of the eyelids and the anterior surface of the eye itself are covered with a thin protective layer of epithelium that forms a mucous membrane called the **conjunctiva** *(kon″-junk-tī′-vah).* The conjunctiva produces mucus that lubricates and moistens the surface of the eye. The portion of the conjunctiva associated with the inner surface of the eyelids is called the **palpebral conjunctiva,** and the portion associated with the surface of the eye is called the **bulbar** *(ocular)* **conjunctiva** (Figure 16.5). The palpebral conjunctiva of the upper eyelid is reflected at its upper margin onto the anterior surface of the eye as the bulbar conjunctiva. The angle formed by the palpebral and bulbar portions of the conjunctiva at the line of reflection is called the **superior conjunctival fornix.** Similarly, the palpebral conjunctiva of the lower eyelid is reflected at its lower margin onto the anterior surface of the eye, and an **inferior conjunctival fornix** is located at the line of reflection. An inflammation of the conjunctiva, often causing dilation of the blood vessels beneath the conjunctival epithelium, is called **conjunctivitis.**

Lacrimal Apparatus

In the superior lateral region of the orbit is a gland called the **lacrimal** *(lak′-ra-mal)* **gland** (Figure 16.4). Six to 12 ducts arise from the lacrimal gland and carry its secretions onto the surface of the conjunctiva, primarily near the lateral portion of the superior conjunctival fornix. The lacrimal gland produces a watery secretion called **tears** that continually bathes the surface of the eye. Tears—which contain salts, some mucin, and a bac-

◆ **FIGURE 16.5 Accessory structures of the eye (sagittal section)**

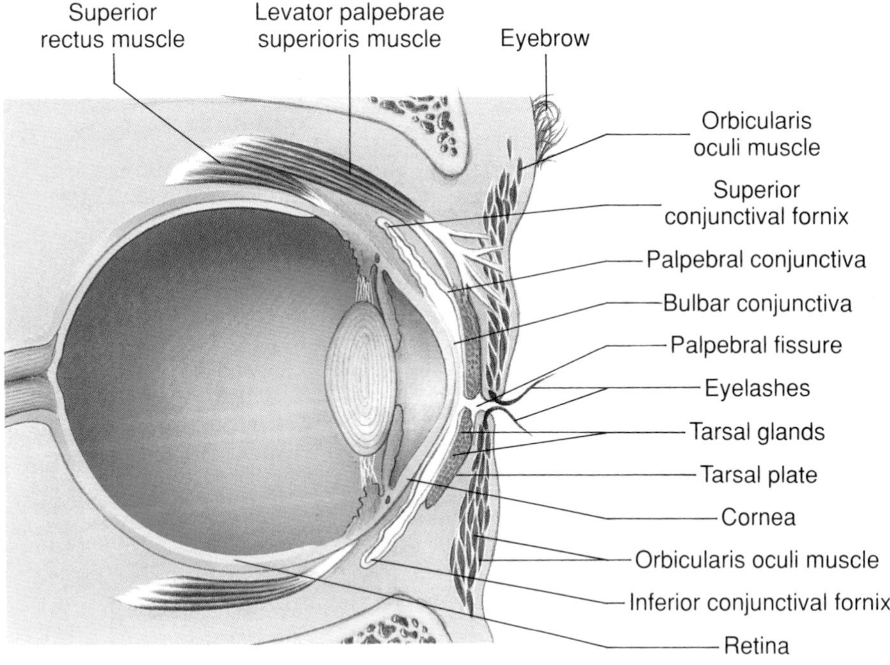

teriocidal enzyme called lysozyme—flow over the bulbar conjunctiva from their area of secretion to the medial canthus. Most tears simply evaporate. But excess fluid sometimes overflows from the eye, either through the palpebral fissure or by draining via ducts into the nasal cavity, which is why it is generally necessary to blow one's nose when crying.

Toward the medial end of each eyelid is a small projection called a **lacrimal papilla.** The lacrimal papilla contains the opening, or **punctum,** of a **lacrimal canal.** The lacrimal canals of the upper and lower eyelids converge medially from their openings and connect to a **lacrimal sac.** The inferior portion of the lacrimal sac opens into a **nasolacrimal duct,** which, in turn, opens into the nasal cavity beneath the inferior nasal concha (turbinate). Normally, the secretions of a lacrimal gland amount to less than 1 ml per day; but these secretions increase in amount during certain emotional states (such as crying) and when foreign objects or other irritants contact the eye.

Extrinsic Eye Muscles

Six straplike skeletal muscles that originate from the orbit and insert on the connective tissue of the eye are responsible for eye movement (Figure 16.6). Originating from the rear of the orbit and inserting on the lateral surface of the eye is the **lateral rectus** muscle, whose contraction principally turns the eye laterally (outward). The **medial rectus** muscle, which originates from the rear of the orbit and inserts on the medial surface of the eye, primarily turns the eye medially (inward). Originating from the rear of the orbit and inserting on the superior surface of the eye is the **superior rectus** muscle, which principally turns the eye upward and somewhat medially. Originating from the rear of the orbit and inserting on the inferior surface of the eye is the **inferior rectus** muscle, which principally turns the eye downward and somewhat medially. The **inferior oblique** muscle originates from the medial surface of the front of the orbit and inserts on the inferior surface of the eye; it primarily turns the eye upward and laterally. The **superior oblique** muscle originates from the posterior portion of the orbit, with the rectus muscles, and runs along the medial surface of the orbit. Anteriorly, its tendon passes through a pulleylike loop, the **trochlea,** and turns to insert on the superior surface of the eye. The superior oblique muscle principally turns the eye downward and laterally. As described in Chapter 13, the extrinsic eye muscles are innervated by the oculomotor (III), trochlear (IV), and abducens (VI) cranial nerves.

The extrinsic eye muscles are among the most rapidly acting and precisely controlled skeletal muscles in the body. The motor units of the extrinsic eye muscles each contain only a few muscle fibers, and relatively high frequencies of stimuli are required to produce tetanus (350 stimuli per second versus 100 stimuli per second for some other skeletal muscles).

◆ **FIGURE 16.6 The extrinsic muscles of the eye**
(a) Viewed from the side. (b) Viewed from above.

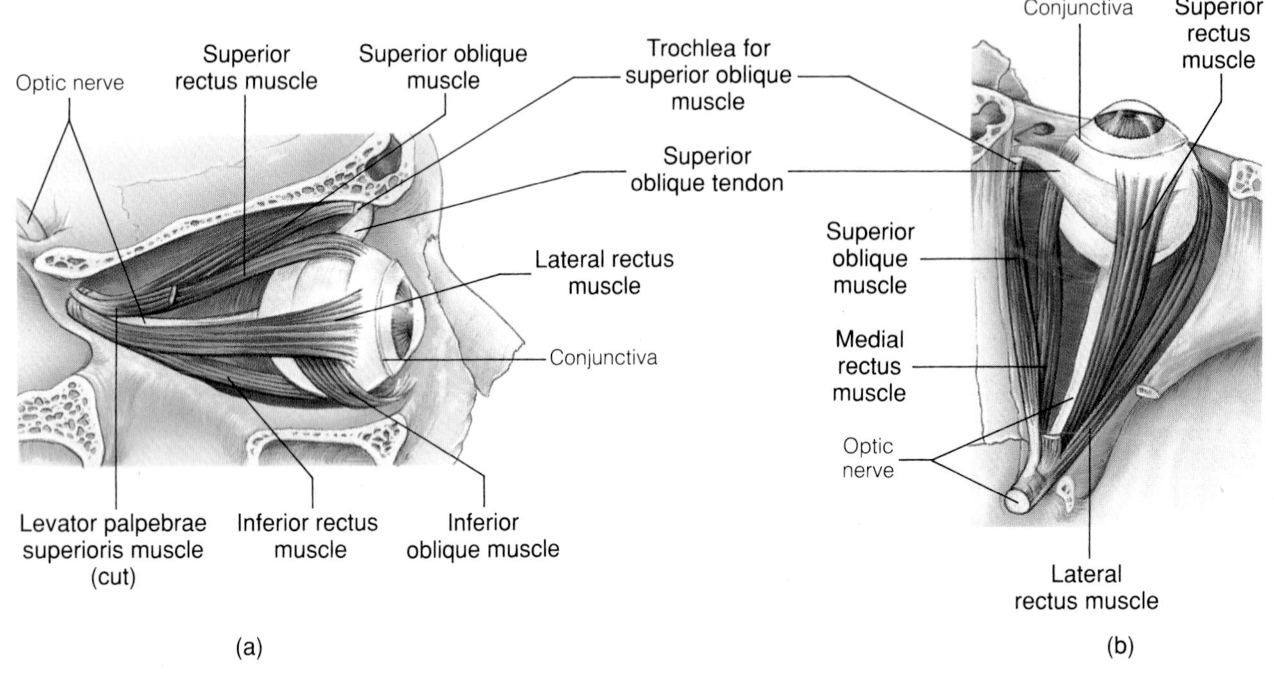

Light

Light is the small portion of the electromagnetic spectrum to which the receptors of the eye are sensitive (Figure 16.7). Light possesses both particlelike and wavelike properties. From a particlelike viewpoint, light is propagated in small, discrete packets of energy called photons. The travel of light, however, is commonly described in terms of wavelike properties such as wavelengths and frequencies (Figure 16.8).

The eye is stimulated by those wavelengths of electromagnetic radiation that range from about 400 to 700 nanometers (1 nanometer = one-billionth of a meter) (Figure 16.9). Within this range, the eye is able to distinguish an immense number of different wavelengths that are interpreted by the brain as different colors. The shorter wavelengths are interpreted as blues and violets, and the longer wavelengths are sensed as oranges and reds. A mixture of all the wavelengths of light is interpreted as white, and the absence of light is interpreted as black.

◆ **FIGURE 16.7 The electromagnetic spectrum**
The eye is sensitive to the limited range of wavelengths called light. Frequency is measured in cycles per second (Hz), and wavelength is given in meters (m).

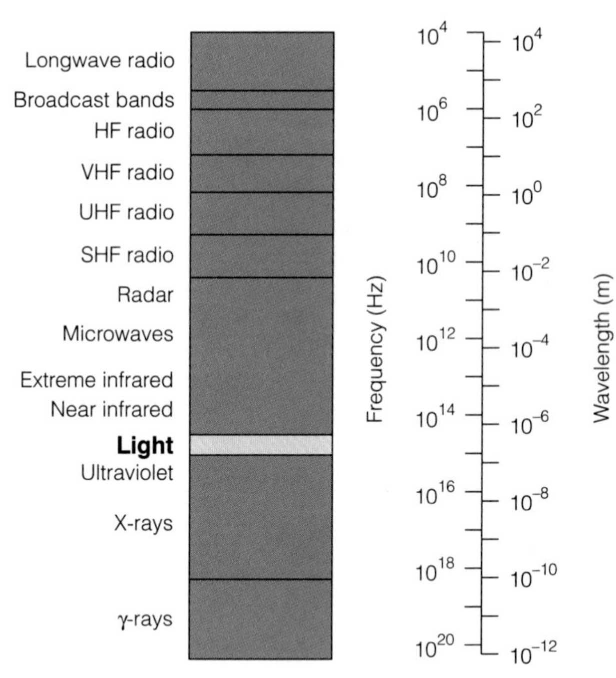

Optics

Light waves radiate outward in all directions from every point of a light source. The forward movement of a light wave in a particular direction is known as a *light ray*.

◆ **FIGURE 16.8 Wavelike properties**
The distance between successive peaks is the wavelength. The number of peaks that pass a fixed point in a given time—usually 1 second—is the frequency.

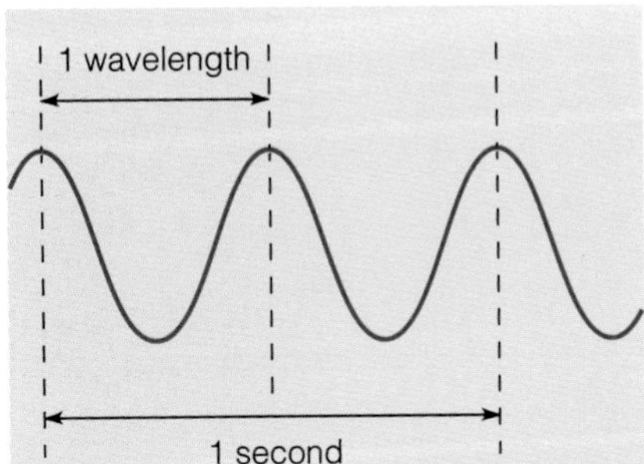

◆ **FIGURE 16.9 The visible portion of the electromagnetic spectrum**

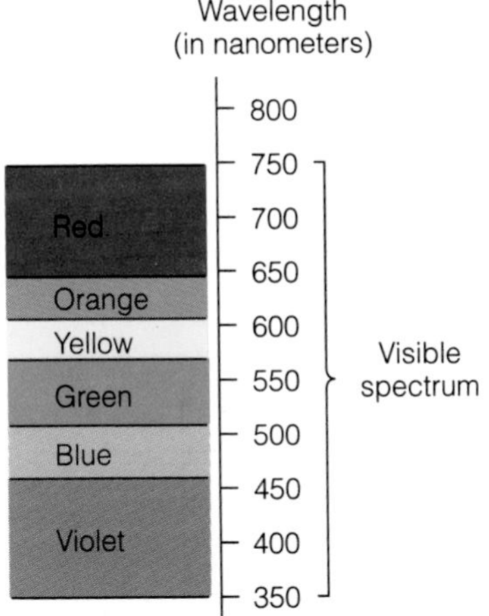

Light rays traveling in a straight line through a uniform medium of a given optical density (such as air) travel at a uniform speed. If the rays enter a second medium, with a different optical density (such as glass), their speed is altered. Moreover, if they strike the surface of the second medium at an angle other than perpendicular to the medium's surface, they are bent (refracted) (Figure 16.10). The greater the deviation from the perpendicular at which the rays strike the second medium, the greater the bending. If the optical density of the second medium is higher than that of the first (for example, when light rays traveling in air strike glass), the rays are bent toward the perpendicular. If the optical density of the second medium is lower than that of the first (for example, when light rays traveling through glass strike air), the rays are bent away from the perpendicular.

Suppose that light rays traveling through air from a point of light strike a spherical, biconvex glass lens (Figure 16.11). Rays from the point of light that strike the lens at angles other than the perpendicular are bent as they enter and leave the lens. Moreover, light rays striking the periphery of the lens are bent more than rays striking the central areas. This difference in bending occurs because rays striking the peripheral areas enter and leave the lens at angles more divergent from the perpendicular than rays striking the central areas. As a result of this differential refraction, the rays converge and come to a focus at a single point behind the lens called the *focal point.* The distance from the lens to the focal point increases as the curvature of the lens diminishes and also as the distance from the light source to the lens diminishes.

If the light source is an object instead of a point, an image of the object forms behind the lens. The image is inverted (upside down), and its right and left are reversed. Usually, the image is reduced in size, and the farther away the object, the smaller its image (Figure 16.12).

◆ **FIGURE 16.10 Bending of light rays as they travel from air into glass**
Note that in this case the light rays are bent toward the perpendicular because the optical density of the glass is greater than that of the air.

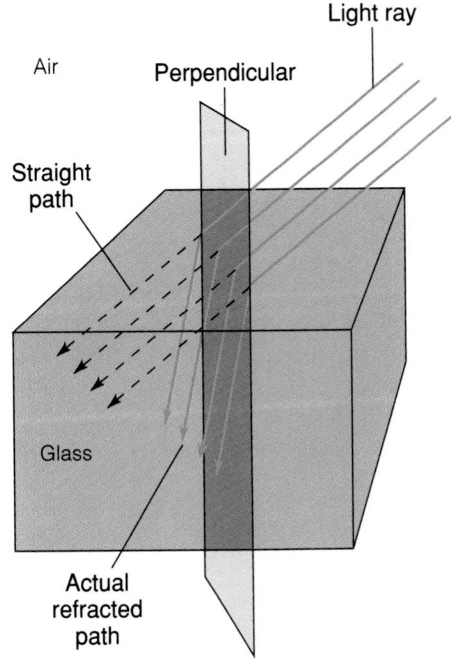

◆ **FIGURE 16.11 Bending of light rays from a point of light as they strike a bioconvex lens**

Line M–N is the perpendicular for ray A as it enters the lens. Line O–P is the perpendicular for ray A as it leaves the lens. Note that ray B, which strikes the very center of the lens perpendicular to the lens surface, is not bent.

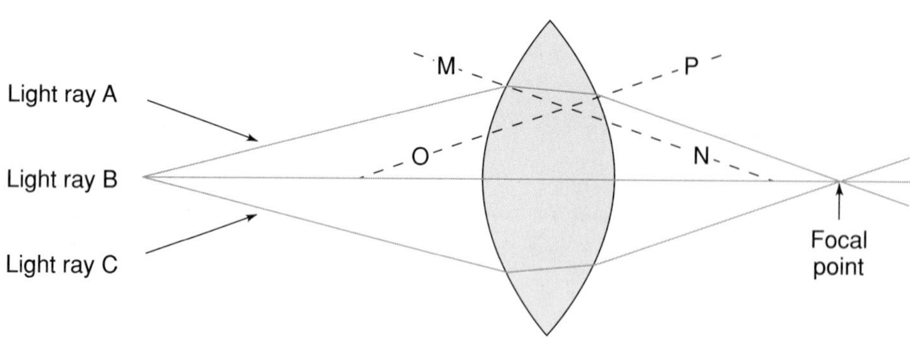

Focusing of Images on the Retina

When light rays strike the cornea, they are bent (refracted) in much the same way as light rays striking the lens in the previous example. In fact, when light rays enter the eye, the greatest bending occurs when the rays pass into the cornea. Additional bending occurs as the light rays pass from the cornea to the aqueous humor and also when the rays enter and leave the lens. The lens carries out the focusing adjustments required to ensure that images are formed sharply and clearly on the retina.

Emmetropia

The intraocular pressure exerted by the fluid within the eye forces the walls of the eye apart. Because the lens is attached to the walls by way of the suspensory ligament that connects to the ciliary body, the lens is under tension. Since the lens is elastic, the tension flattens it somewhat, giving it a less pronounced curvature than it might otherwise assume. In this condition, the normal, or **emmetropic** *(em˝-ĕ-trop´-ik),* eye focuses sharply onto the retina parallel light rays from distant points (light rays entering the eye from points farther away than 20 feet are considered parallel) (Figure 16.13a). Divergent light rays entering the eye from points closer than 20 feet are focused behind the retina. (Figure 16.13b).

Accommodation

The focusing system of the eye must adjust in order to focus onto the retina light rays from objects closer than 20 feet. The adjustment process is referred to as **accommodation.** It is accomplished by the ciliary muscles of the ciliary body, which contract in response to parasympathetic stimulation. The contraction of the ciliary muscles pulls the ciliary body slightly forward and inward, narrowing the ring of the ciliary body. This action lessens the tension on the suspensory ligament and permits the elastic lens to assume a more pronounced curvature (Figure 16.14). The greater curvature results in a greater bending of light rays that strike the lens at angles other

◆ **FIGURE 16.12 Retinal images are reversed (left to right) and upside down**

The farther away the object, the smaller is its image on the retina.

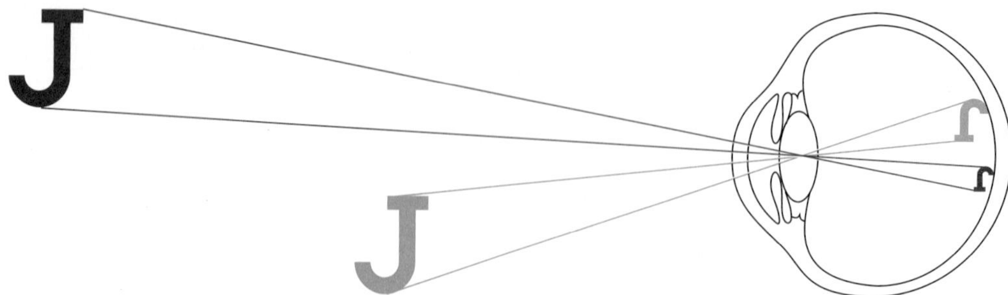

◆ **FIGURE 16.13 Focusing of light rays onto the retina**

Without accommodation, the tension exerted on the lens of the eye flattens the lens somewhat, giving it a less pronounced curvature than it might otherwise assume. In this condition, the normal eye (a) focuses sharply onto the retina parallel light rays from distant points (farther away than 20 feet) and (b) focuses behind the retina divergent light rays from points closer than 20 feet. When accommodation occurs, the lens assumes a more pronounced curvature. In this condition, the normal eye (c) focuses in front of the retina parallel light rays from distant points and (d) focuses sharply onto the retina divergent light rays from points closer than 20 feet.

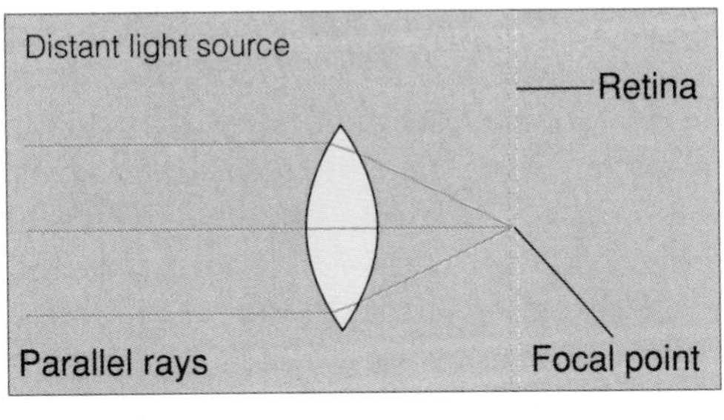

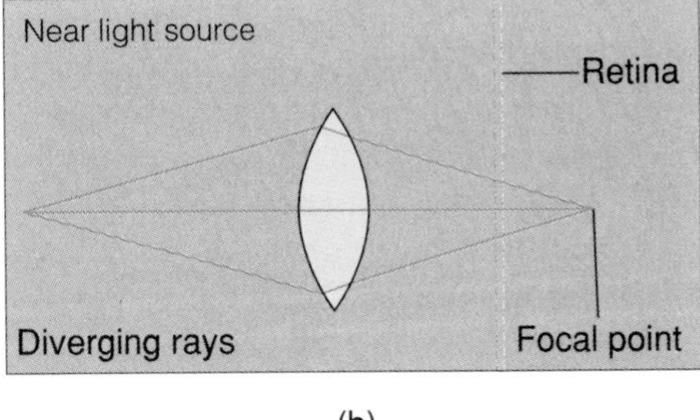

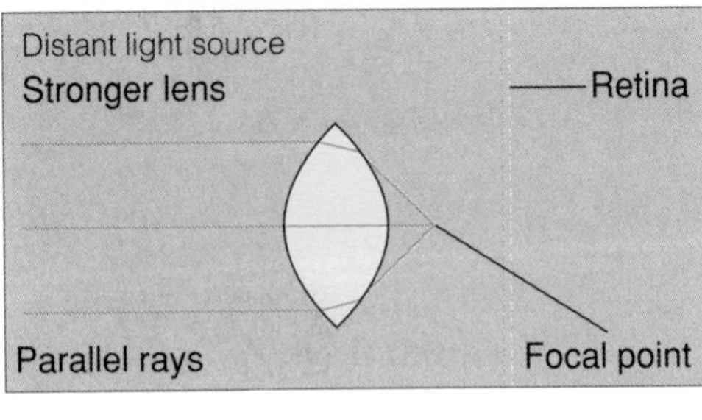

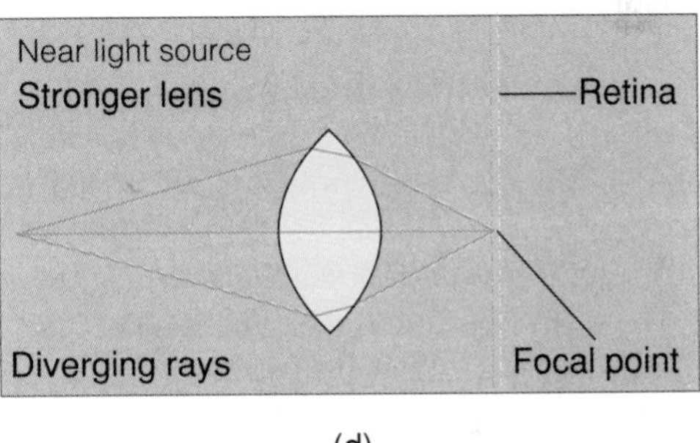

◆ **FIGURE 16.14 Accommodation**

(a) The suspensory ligament extends from the ciliary muscles to the lens. (b) Without accommodation, the lens is under tension, and it has a less pronounced curvature than it might otherwise assume. (c) When accommodation occurs, the ciliary muscles contract and the ring of the ciliary body narrows. This action lessens the tension on the suspensory ligament and permits the elastic lens to assume a more pronounced curvature.

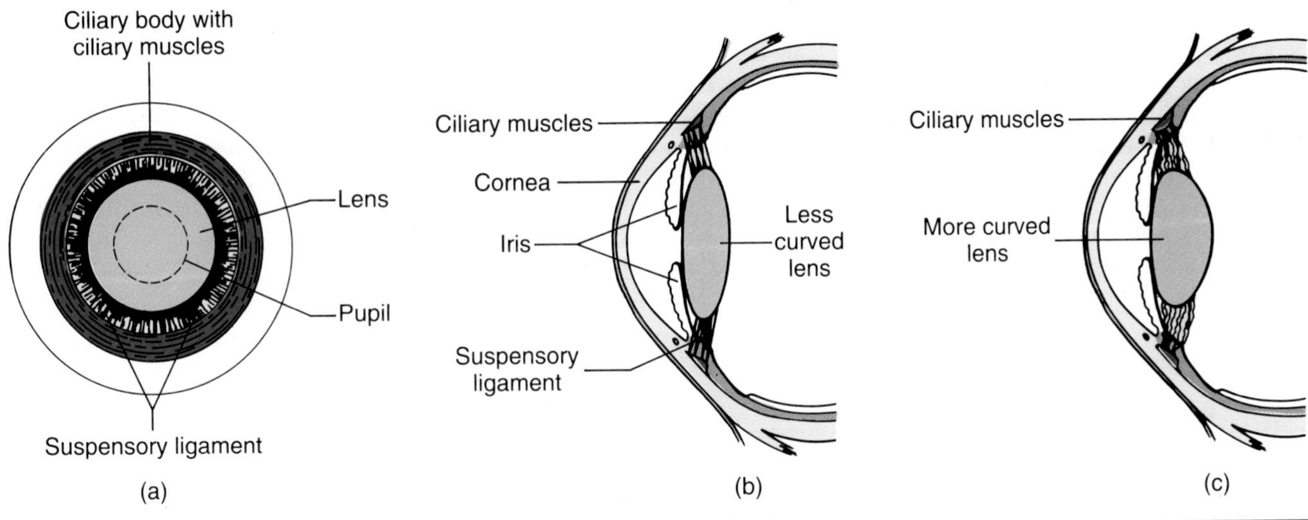

than the perpendicular, enabling the lens to focus onto the retina divergent light rays from points closer than 20 feet (Figure 16.13d). Under such conditions, light rays from distant points are focused in front of the retina (Figure 16.13c). The closer a point of light is to the eye, the greater must be the curvature of the lens—that is, the greater must be the degree of accommodation—if light rays from the point are to be focused onto the retina.

Information that reaches the visual association areas of the occipital cortex from the retina appears to be important in accommodation. When the image on the retina is out of focus, the visual association areas transmit signals to the ciliary muscles to alter the curvature of the lens and thus bring the image into focus. These focusing adjustments usually require less than a second.

During accommodation, the pupil constricts, eliminating the most divergent light rays that would otherwise pass through the most peripheral portions of the lens. The elimination of these rays is beneficial because even a good lens may not be perfect, and light rays that pass through the peripheral regions of the lens do not always focus at exactly the same point as those that pass through the central areas. Thus, the constriction of the pupil aids in the formation of a sharp image on the retina (and also reduces the amount of light entering the eye).

Another event associated with accommodation is the convergence of the eyes. That is, when viewing close objects, the eyes turn inward so the image falls on the fovea of each retina, which is the portion of the retina where visual acuity is greatest.

Near and Far Points of Vision

The closer an object is to the eye, the more divergent are the light rays from the object and the greater is the accommodation required to focus the light rays onto the retina. However, the ability of the eye to accommodate for viewing near objects is limited. The distance from the eye to the nearest point whose image can be focused on the retina is called the **near point of vision.** The near point of vision varies with age: it is close to the eye in youth and recedes farther and father from the eye as a person gets older.

The **far point of vision** is the distance from the eye to a point whose image can be focused on the retina without accommodation. The far point of vision for the normal emmetropic eye is infinity (any point beyond 20 feet).

Control of Eye Movements

The movements of the eyes, which are caused by the extrinsic eye muscles, are synchronized with one another so that both eyes move in a coordinated manner when viewing an object. When a person wants to look at something, he or she moves the eyes voluntarily to find the object and fix upon it. These *voluntary fixation movements* are controlled by a small area of the cerebral cortex located bilaterally in the premotor regions of the frontal lobes. Once the object is found, *involuntary fixation movements* keep the image of the object fixed on the foveal portions of the retinas. The involuntary fixation movements are controlled by the occipital region of the cortex, particularly the visual association areas. The visual fixation mechanisms depend on feedback from the retinas to keep the image of an object on the foveas. For example, each time the image drifts to the edge of the foveas, sudden flicking movements of the eyes occur. These movements, which are automatic reflex responses controlled by the involuntary fixation mechanisms, prevent the image from leaving the foveas.

Binocular Vision and Depth Perception

Humans have **binocular vision**—that is, each eye views a portion of the external world that overlaps considerably with that viewed by the other eye. However, the visual field of the left eye is not identical with that of the right eye (Figure 16.15). Moreover, the two eyes do not form exactly identical images of an object because they occupy slightly different locations. This *retinal disparity* contributes to a person's ability to judge relative distances when objects are nearby (that is, closer than about 70 meters). Depth perception with only one eye depends partly on the use of learned cues such as the fact that the sizes of objects in the environment appear to diminish with distance.

Diplopia (Double Vision)

Light rays coming from an object strike the retinas of both eyes. However, a person normally perceives only one object, not two. This is because the retinas possess *corresponding points* that, when stimulated, result in the peception of a single object. Normally, when an individual looks at an object, the eyes move in such a manner that light rays from the object fall on corresponding points of the foveas of the retinas. However, if the extrinsic eye muscles are weak or paralyzed, or if the muscles of the two eyes act in an unequal or uncoordinated fashion, light rays from an object may not fall on corresponding points of the retinas. As a result, the individual may develop **diplopia** (*dī-plō´-pe-ah;* double vision) and perceive two objects instead of one. Transient diplopia can develop in acute alcoholic intoxication due to the partial paralysis of the extrinsic eye muscles.

◆ **FIGURE 16.15 Visual fields of the eye**

The red arrows indicate the left portions of the visual field of each eye. The blue arrows indicate the right portions. Binocular vision occurs where these visual fields overlap.

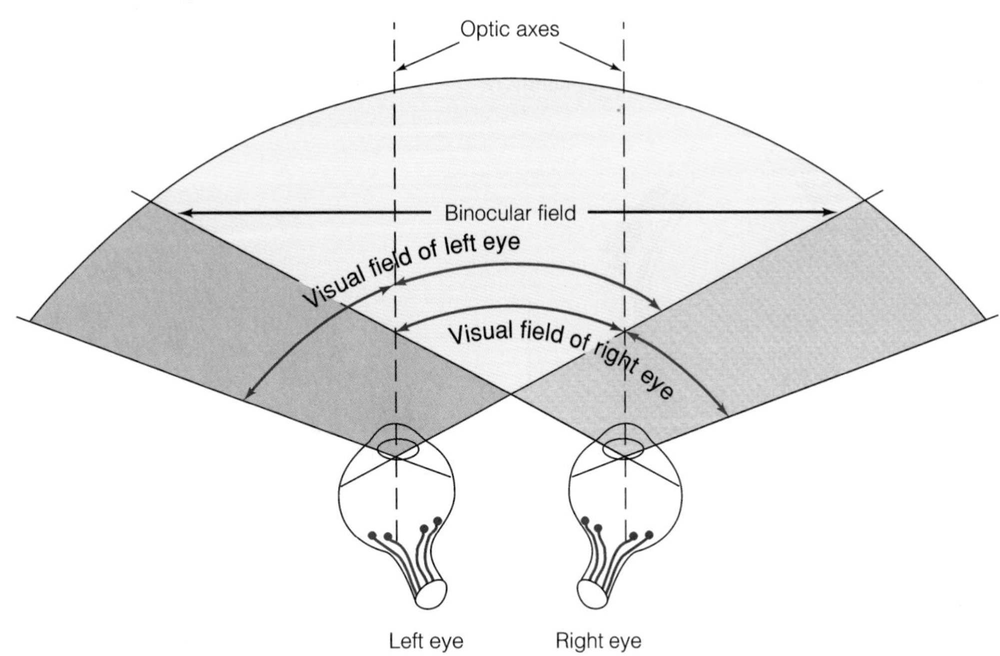

Strabismus

In the condition known as **strabismus** (*stra-biz´-mus;* squint, cross-eyedness), the movements of the two eyes are not properly coordinated with one another when a person looks at an object. In some cases of strabismus, the eyes alternate in fixing on an object. In others, the same eye is used all the time. In such cases, the image formed by the unused eye can eventually become suppressed, and if the condition is not corrected, the repressed eye may become functionally blind.

Photoreceptors of the Retina

The outermost portion of the nervous-tissue layers of the retina (the portion closest to the pigmented layer of the retina and the choroid) contains the light-sensitive photoreceptors: the **rods** and **cones** (Figure 16.16). Each rod cell or cone cell has four basic regions (Figure 16.17):

1. The *outer segment,* which faces the choroid, is rod-shaped in rods and cone-shaped in cones. The outer segment contains a series of stacked, flattened, membranous discs.
2. The *inner segment* contains rough endoplasmic reticulum and many mitochondria.
3. The *nuclear region* contains the nucleus of the cell.
4. The *synaptic region* makes synaptic contact with other cells.

Light-sensitive molecules called *photopigments* are embedded in the membranes of the flattened discs within the outer segments of the rods and cones. The configurations of the photopigments are altered when light strikes them and they absorb the light. These alterations lead to changes in the polarity of the rods and cones, ultimately resulting in the transmission of neural signals from the retina to the brain that are interpreted as visual events.

There are four different photopigments, each consisting of a protein called an *opsin,* to which a chromophore molecule called *retinal (retinene)* is attached. The opsins differ from pigment to pigment and confer specific light-sensitive properties on each photopigment. Retinal is produced from vitamin A_1, and retinal and vitamin A_1 can be interconverted. The pigmented layer of the retina contains stores of vitamin A_1.

Rods

The rods contain a photopigment called *rhodopsin (rō-dop´-sin).* Rhodopsin is composed of retinal and the opsin scotopsin. Vitamin A_1 absorbed by the rods is converted to retinal, which combines with scotopsin to form rhodopsin. When the eyes are not exposed to light, the concentration of rhodopsin can build up to a very high level.

When light of the proper wavelength strikes a rod and rhodopsin absorbs the light, the chemical configuration

◆ **FIGURE 16.16 The structure of the retina**

(a) Schematic view. (b) Scanning electron micrograph. (From *Tissues and Organs: A Text-Atlas of Scanning Electron Microscopy* by Richard C. Kessel and Randy H. Kardon. W. H. Freeman and Company. Copyright © 1979.) ×1422.

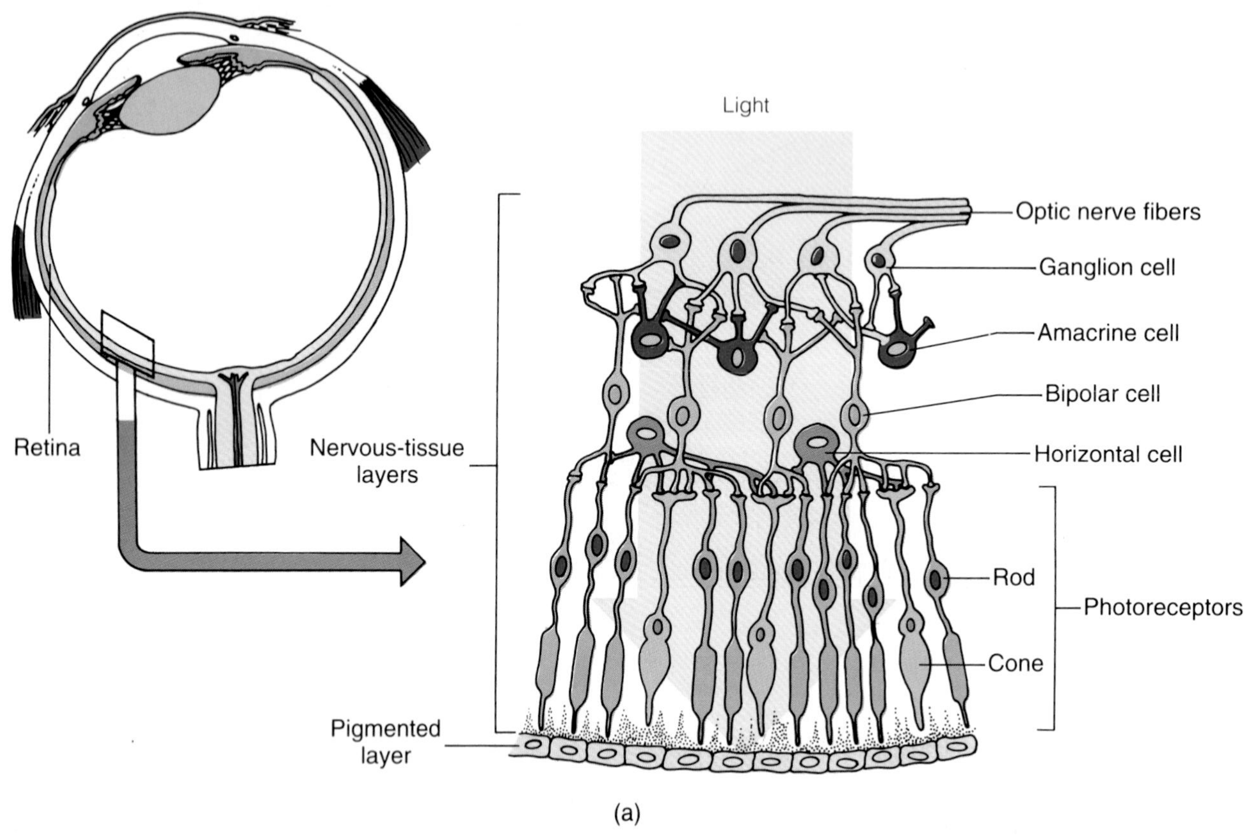

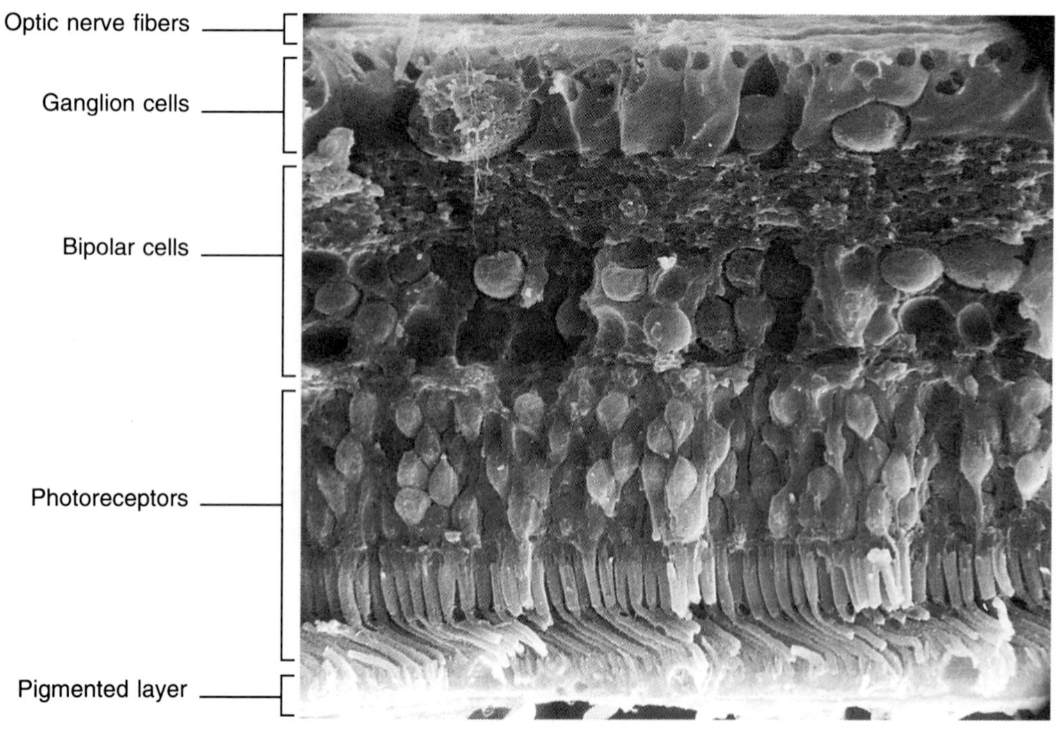

◆ **FIGURE 16.17 Photoreceptors**

(a) A rod and a cone. (b) Outer segments of photoreceptors showing membranous discs.

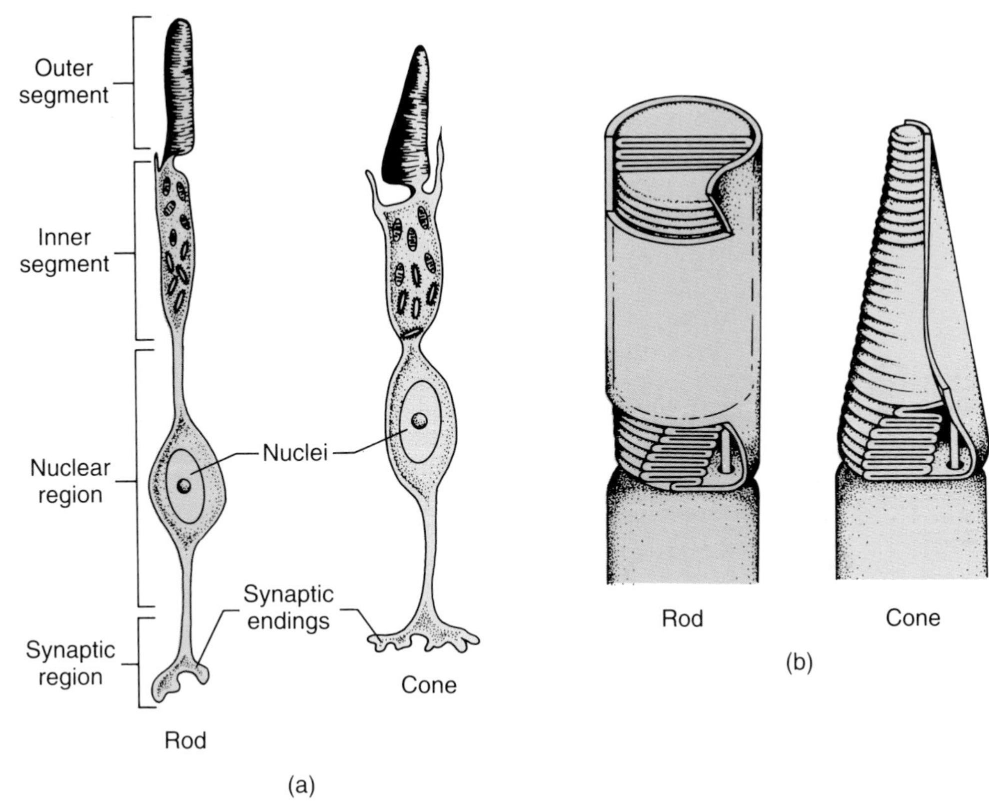

of the retinal is altered. The altered retinal begins to pull away from the scotopsin, forming a substance called bathorhodopsin (prelumirhodopsin) (Figure 16.18). Bathorhodopsin is very unstable, and it quickly decays into lumirhodopsin, which in turn rapidly decomposes into metarhodopsin I. Metarhodopsin I spontaneously forms metarhodopsin II, which in turn becomes pararhodopsin. Pararhodopsin then splits into altered retinal and scotopsin. Following these events, the altered retinal is rearranged and rejoined to scotopsin to form rhodopsin. The rearrangement of the altered retinal is enzymatically mediated and requires metabolic energy.

When rhodopsin absorbs light, the metarhodopsin II that is formed activates a G protein called *transducin.* Transducin, in turn, activates the enzyme *cyclic-GMP phosphodiesterase,* which breaks down the substance *cyclic guanosine monophosphate (cyclic GMP).* In a rod, cyclic GMP acts on ion channels in the plasma membrane of the outer segment to keep the channels open. When cyclic GMP is broken down, the channels close. The channels are particularly permeable to sodium ions, and their closure leads to a hyperpolarization of the cell. Consequently, a rod responds to an increase in the intensity of light of the proper wavelength by hyperpolarizing and reducing the release of neurotransmitter molecules from its synaptic region (Figure 16.19). It responds to a decrease in the intensity of the light by depolarizing and increasing the release of neurotransmitter molecules from its synaptic region.

The rods are very light sensitive and can respond to low levels of illumination, such as those present at night or in dimly lit areas. The responses of the rods indicate degrees of brightness. However, the rod system is characterized by a relative lack of color discrimination, and rod responses are interpreted principally as shades of gray.

Cones

The chemical events that occur when light strikes the cones are similar to those that take place when light strikes the rods. However, there are three different types of cones. Each type contains a different photopigment and is selectively sensitive to particular wavelengths of light (Figure 16.20). Red cones respond more intensely than the others to those wavelengths of light that the brain interprets as red. Green cones respond more intensely than the others to those wavelengths of light that the brain interprets as green. Blue cones respond most intensely to those wavelengths of light that the brain interprets as blue or violet.

◆ **FIGURE 16.18 The rhodopsin-retinal chemical cycle**
(a) Light energy alters the chemical configuration of the retinal of rhodopsin, leading to the formation of bathorhodopsin. Note that retinal can be formed from vitamin A_1. (b) Retinal, indicating the change in chemical configuration brought about by light. Enzymes and metabolic energy are required to convert altered retinal back to retinal.

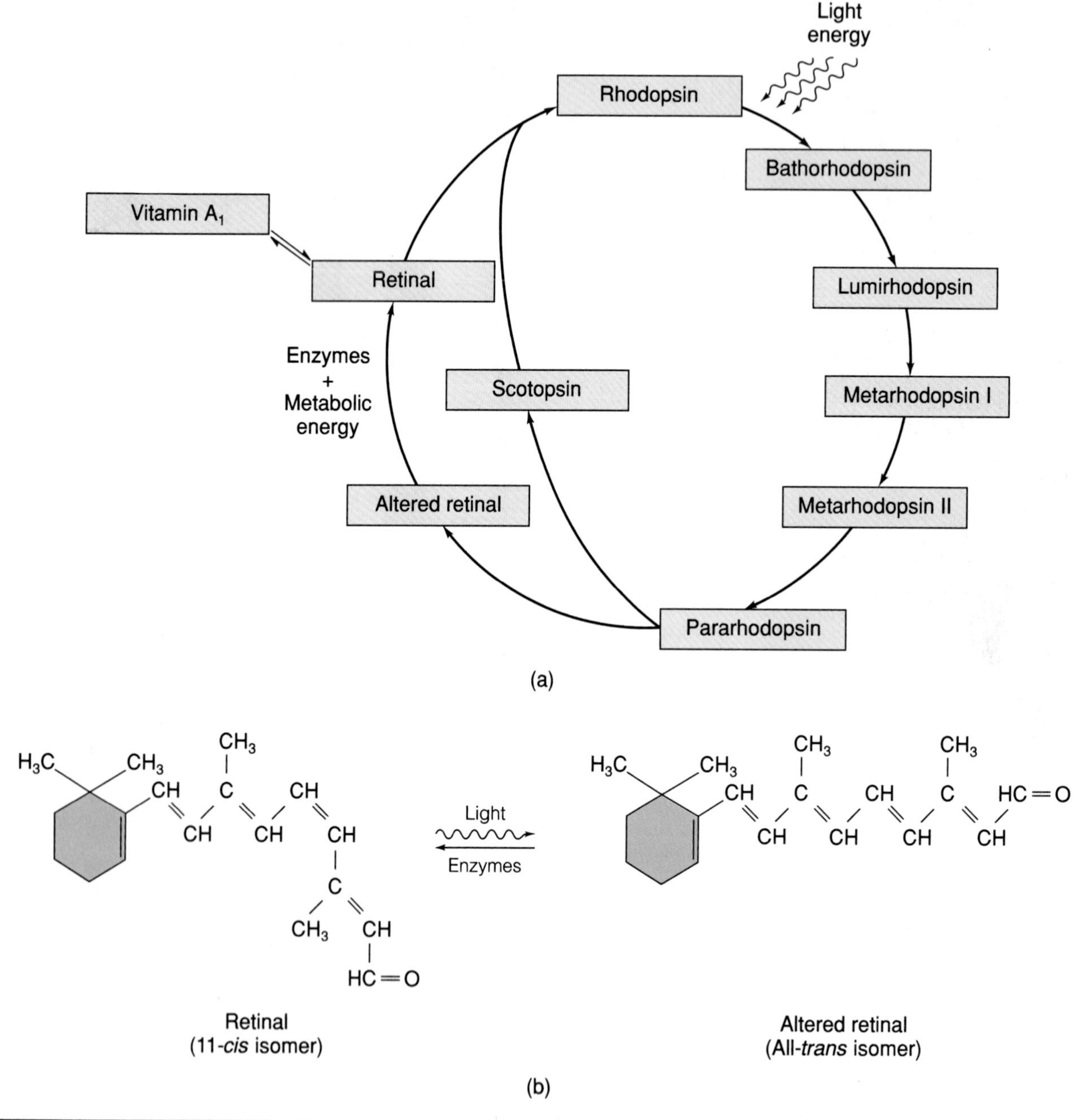

The cones are primarily responsible for color vision. The ability to perceive many different colors rather than only three colors (red, green, and blue) depends in part on the fact that different wavelengths of light striking the retina evoke different ratios of response from the three cone types. For example, when light with a wavelength of 580 nanometers strikes the retina, the red cones respond more intensely than the green cones, and the blue cones do not respond at all. Color perception also depends on the processing of cone responses by neural elements of the visual system. (The cone responses and neural processing that occur when light with a wavelength of 580 nanometers strikes the retina gives rise to the perception of the color yellow.)

◆ **FIGURE 16.19 Response of a rod cell to an increase in light intensity**

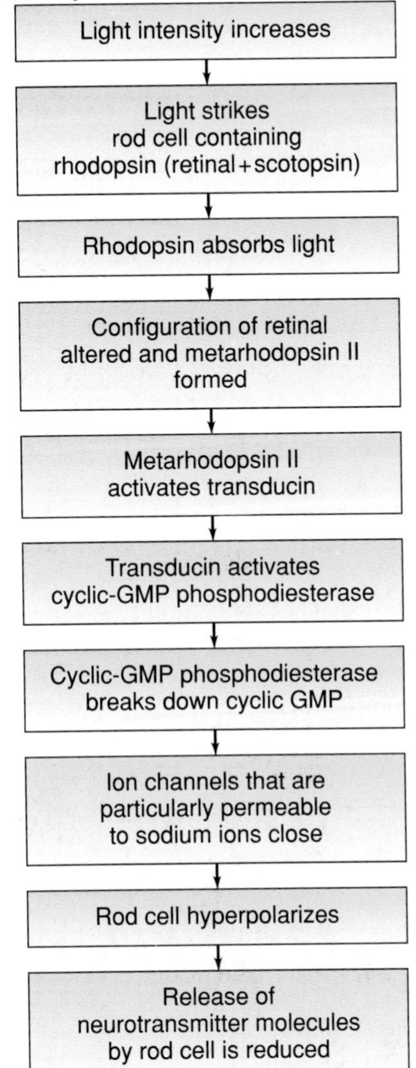

The cones operate only at relatively high levels of illumination, and they are the principal photoreceptors during daylight or in brightly lit areas. The cones are concentrated in the center of the retina, especially in the fovea, whereas the rods are more numerous in the peripheral retina (Table 16.1).

Neural Elements of the Retina

The portions of the nervous-tissue layers of the retina closest to the interior of the eye are composed of neural elements (Figure 16.16). The middle portion of the nervous-tissue layers contains **bipolar cells,** and the innermost portion contains **ganglion cells.** The rods and cones synapse with bipolar cells, which in turn synapse with ganglion cells.

◆ **FIGURE 16.20 Sensitivities of the three different types of cones to light of different wavelengths**

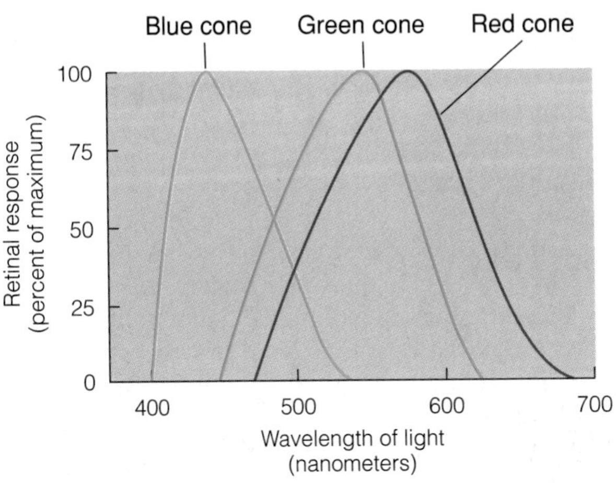

Ganglion cells produce and transmit action potentials. However, the rods and cones and almost all other neural elements of the retina produce only graded potentials.

In the dark, ganglion cells produce action potentials at low, steady frequencies. When rods and cones respond to an increase in light intensity by hyperpolarizing and decreasing their release of neurotransmitter molecules, ganglion cells can ultimately respond by either increasing or decreasing their frequencies of action potential production.

The neurotransmitter molecules released by rods and cones cause some bipolar cells to hyperpolarize. These bipolar cells depolarize when rods and cones decrease their release of neurotransmitter molecules. When the bipolar cells depolarize, they increase their release of neurotransmitter molecules, which increases the stimulation of the ganglion cells with which they synapse.

The neurotransmitter molecules released by rods and cones cause some bipolar cells to depolarize. These bipolar cells hyperpolarize when rods and cones decrease their release of neurotransmitter molecules. When the bipolar cells hyperpolarize, they decrease their release of neurotransmitter molecules, which decreases the stimulation of the ganglion cells with which they synapse.

In addition to bipolar cells and ganglion cells, the retina contains neural cells called **horizontal cells** and **amacrine cells** *(am´-ah-krin)* (Figure 16.16). These cells are important in lateral interactions that occur within the retina. For example, when the rods or cones of a given area of the retina respond to a change in light intensity, their responses influence horizontal cells as well as bipolar cells. If the bipolar cells directly affected by the rods and cones respond by hyperpolarizing, the horizontal cells act to cause adjacent bipolar cells to depolarize. If the bipolar cells directly affected by the rods and cones respond by depolarizing, the horizontal cells act to cause adjacent bipolar cells to hyperpolarize. Thus, horizontal cell activity tends to enhance the signal contrast between strongly stimulated regions of the retina and adjacent, more weakly stimulated regions.

Visual Pathways

The axons of the retinal ganglion cells come together and leave the eye as the **optic nerve.** (An optic nerve contains about one million nerve fibers.) The two optic nerves (one from each eye) meet at the **optic chiasma** *(kī-as´-mah),* located just anterior to the pituitary gland (Figure 16.21). Within the optic chiasma, ganglion cell axons from the medial half of each retina cross to the opposite side, and those from the lateral half of each retina remain on the same side. From the optic chiasma, the axons continue as the **optic tracts.** Thus, the left optic tract consists of ganglion cell axons from the lateral half of the retina of the left eye and the medial half of the retina of the right eye (the right portion of the visual field of each eye), and the right optic tract consists of axons from the lateral half of the retina of the right eye and the medial half of the retina of the left eye (the left portion of the visual field of each eye).

Most of the ganglion cell axons within the optic tracts travel to the **lateral geniculate bodies** of the thalamus. There they synapse with neurons that form pathways called **optic radiations,** which terminate in the visual cortex of the occipital lobes. Some of the axons within the optic tracts travel to midbrain nuclei called **pretectal nuclei,** where they are involved in a reflex called the pupillary light reflex (see page 506). Other optic tract axons pass to the **superior colliculi,** which serve as visual reflex centers controlling the extrinsic muscles of the eyes.

Processing of Visual Signals

The response of a visual pathway cell is commonly described in terms of its receptive field. The *receptive field* of a cell is the area of the retina that, when illuminated, influences the activity of the cell. In essence, it is an area of photoreceptors that provide input to the cell. This input is often provided indirectly by way of other cells, and it can be inhibitory as well as stimulatory.

The processing of visual signals begins in the retina, and the ganglion cells transmit to the brain a coded message rather than a simple mosaic representation of the image on the retina. The receptive fields of most ganglion cells are circular and are composed of essentially two parts: a small central area and a ringlike peripheral area (Figure 16.22). In the dark, a ganglion cell continually transmits nerve impulses to the brain. However, when the receptive field of a ganglion cell is illuminated, the frequency of nerve impulse transmission changes. In one type of ganglion cell, called an *on-center cell,* positioning a spot of light on the central area of a cell's receptive field increases the frequency of nerve impulse transmission. Conversely, illuminating the peripheral area decreases the frequency of nerve impulse transmission. In a second type of ganglion cell, called an *off-center cell,* positioning a spot of light on the central area of a cell's receptive field decreases the frequency of nerve impulse transmission, and illuminating the peripheral area increases the frequency of nerve impulse transmission.

◆ **TABLE 16.1 Properties of Rod and Cone Vision**

RODS	CONES
100 million per retina	3 million per retina
More numerous in periphery	Concentrated in fovea
Vision in shades of gray	Color vision
High sensitivity	Low sensitivity
Night vision	Day vision
Low acuity	High acuity

In either cell type, nerve impulse frequency varies with the degree of illumination of the central area of a cell's receptive field compared to the degree of illumination of the peripheral area. That is, when both areas are illuminated simultaneously, the activity of the ganglion cell is a reflection of both stimulatory and inhibitory influences. Consequently, the signals transmitted to the brain by a ganglion cell indicate such things as the light level in one small area of the visual scene compared to the average illumination of the immediately surrounding region.

The neurons of the lateral geniculate bodies of the thalamus that transmit nerve impulses to the visual cortex have receptive fields that are similar to those of ganglion cells—that is, their receptive fields are circular, with either an excitatory central area and an inhibitory peripheral region or vice versa.

In the visual cortex, there is a hierarchy of cells with receptive fields that vary widely in organization. Some cortical cells have circularly symmetrical receptive fields, and their responses resemble those of ganglion cells and lateral geniculate neurons. However, the receptive fields of other cortical cells are organized so the cells respond best not to spots of light but to specifically oriented line segments (such as narrow slits of light) positioned on the retina. For example, cells called *simple cells* respond only when the line is oriented at a

◆ **FIGURE 16.21 Neural pathways for vision**

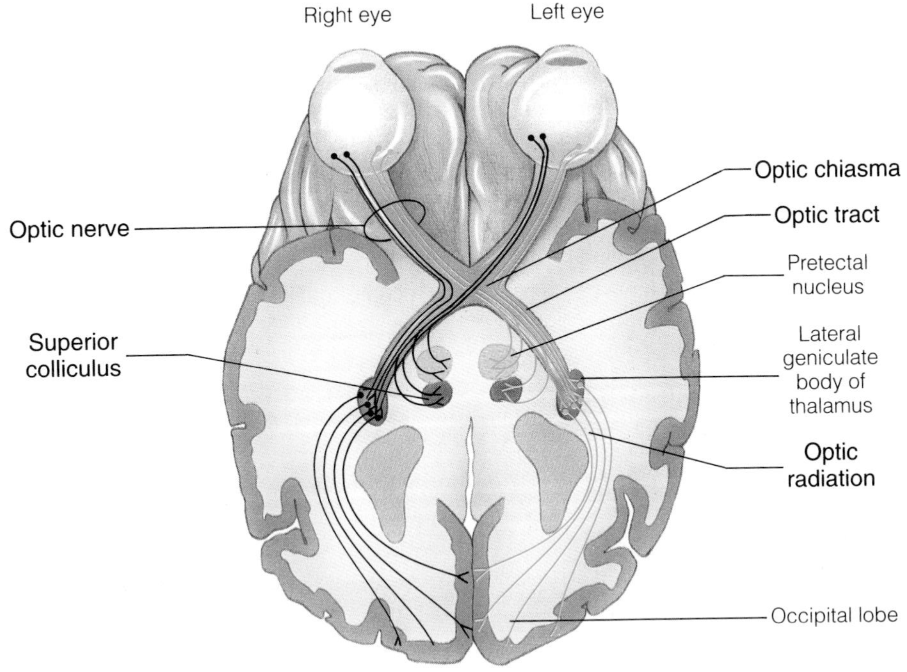

precise angle in a particular portion of a cell's receptive field (Figure 16.23a). The responses of these cells provide information about lines and borders. Other cells, called *complex cells,* respond when the line is oriented at a precise angle regardless of its position in a cell's receptive field (Figure 16.23b). The responses of these cells help indicate the movement of a visual stimulus.

Thus, the processing of visual signals that begins in the retina continues in other parts of the visual pathway. As a result, the components of the pathway transmit a coded message about certain aspects of the image on the retina rather than a mosaic pattern of the image. The message contains information about such characteristics of the image as contrast, movement, and color. The

◆ **FIGURE 16.22 Ganglion cell responses**

(a) When the central area of the receptive field of an on-center type of ganglion cell is illuminated by a spot of light, the frequency of nerve impulses transmitted by the cell increases. (b) When the central area of the receptive field of an off-center type of ganglion cell is illuminated, the frequency of nerve impulses transmitted by the cell decreases.

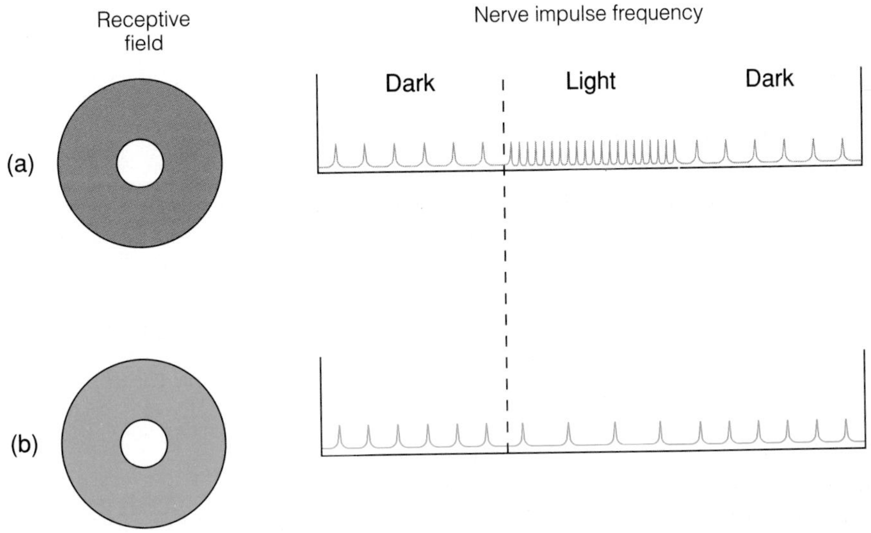

Receptive field

Nerve impulse frequency

Dark Light Dark

(a)

(b)

◆ **FIGURE 16.23 Responses of the cells of the visual cortex**

(a) A simple cell responds when a line is oriented at a precise angle in a particular portion of the cell's receptive field. (b) A complex cell responds when the line is oriented at a precise angle regardless of its position in the cell's receptive field. Thus, the cell continues to respond as a properly oriented line moves across its receptive field. (Some complex cells respond better when the line moves in one direction than in the other.)

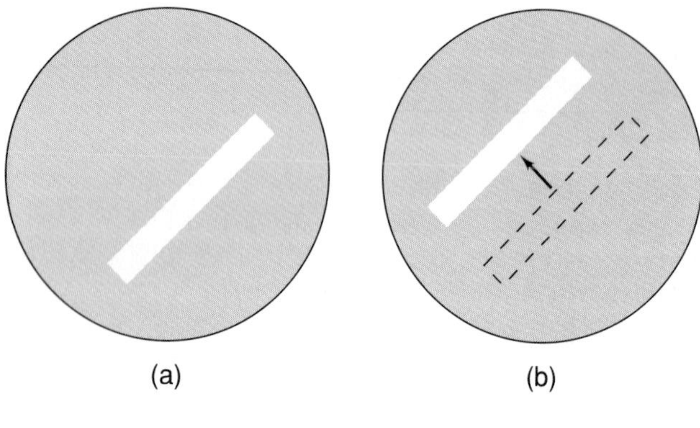

◆ **FIGURE 16.24 Resolving power of the eye**

In order for two point sources of light to be discrminated as separate, light rays from one point must stimulate a different receptor unit of the retina than light rays from the second point, with at least one unstimulated or only weakly stimulated unit between the two. The more closely the receptor units are packed, the more likely this will occur. In (a), two widely spaced points may be detected as separate. In (b), the two points will not be discriminated as separate. In (c), two closely spaced points may be detected as separate.

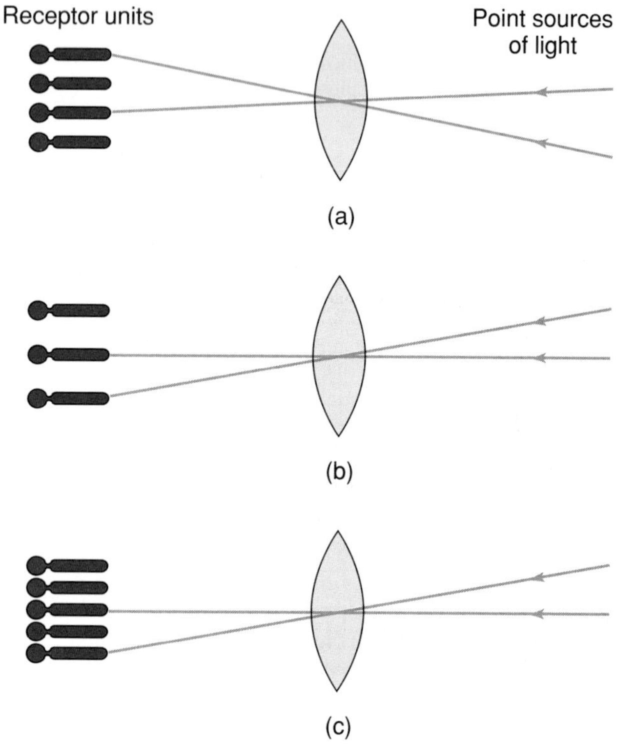

brain utilizes this information in developing a visual representation of the image.

Light and Dark Adaptation

When a person moves from a dimly lit area to a brightly lit area, the person's eyes often feel uncomfortable, and vision is poor for several minutes. With time, however, **light adaptation** occurs and vision improves. Similarly, when a person moves from a well-lit area to a dimly lit area, the person initially cannot see well. However, as time passes, **dark adaptation** occurs and vision gets better.

The rods and cones are able to adapt and maintain their responsiveness over a wide range of light intensities. However, the mechanisms of rod and cone adaptation are incompletely understood. The cone system apparently contributes to light adaptation by way of neural interconnections that inhibit rod function in bright-light conditions appropriate for cone function.

Pupillary Light Reflex

The **pupillary light reflex** constricts (reduces the diameter of) the pupils in bright light and helps regulate the amount of light entering the eyes. In this reflex, light striking the retina initiates neural signals that are sent to the pretectal nuclei of the midbrain. These signals ultimately lead to an increased stimulation of the circular smooth muscles of the iris by neurons of the parasympathetic division of the autonomic nervous system. The contraction of the smooth muscles constricts the pupils. In dim light, the stimulation of the circular smooth muscles decreases, and the pupils dilate (increase their diameter).

If bright light shines into only one eye, the pupils of both eyes will constrict. However, the constriction of the pupil of the eye into which the light shines will be greater than the constriction of the pupil of the other eye. The reflex leading to pupillary constriction in the eye into which the light shines is called the *direct light reflex*, and the reflex leading to pupillary constriction in the other eye is called the *consensual light reflex*.

Visual Acuity

Visual acuity is the ability of the eye to distinguish detail. It is related to resolving power, which is the eye's

CONDITIONS OF CLINICAL SIGNIFICANCE

The Eye

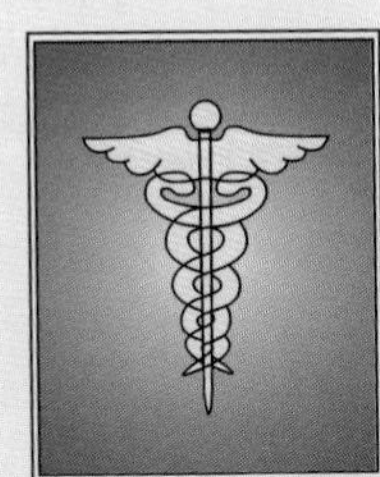

Many disorders can affect the eyes. Focusing problems are among the most common.

Myopia

Myopia (mī-ō´-pe-ah) is nearsightedness. In this condition, light rays from distant objects are focused in front of the retina, and only light rays from close objects can be focused accurately onto the retina. Myopia is caused by a focusing system that has too great a refractive power with respect to the position occupied by the retina. Most commonly this is due to an elongated eyeball, which can result from a weakness of the coats of the eye (Figure C16.1). In myopia, the near point of vision is closer than normal, and the far point of vision is closer than 20 feet. In all but the severest cases, myopia can be corrected by eyeglasses with concave lenses. Such lenses cause light rays to diverge slightly as they enter the eye, helping the refractive system of the eye focus them onto the retina.

Hypermetropia

Hypermetropia (hi´´-per-mē-tro´-pe-ah; hyperopia) is farsightedness. In this condition, light rays from distant objects are focused behind the retina when the eye is at rest, and accommodation is necessary in order to focus the rays onto the retina. Since the eyes can accommodate only so much, a hypermetropic individual has trouble viewing close objects. Moreover, accommodation is required to view any object, whether close or far. Thus, in hypermetropia there is no far point of vision, and the near point lies farther away than normal. Hypermetropia is most commonly due to a shortened eyeball (Figure C16.2). The condition can usually be corrected with eyeglasses having convex lenses. Such lenses cause light rays to converge as they enter the eye, and thus assist the eye's refractive system in focusing the rays onto the retina.

continued on next page

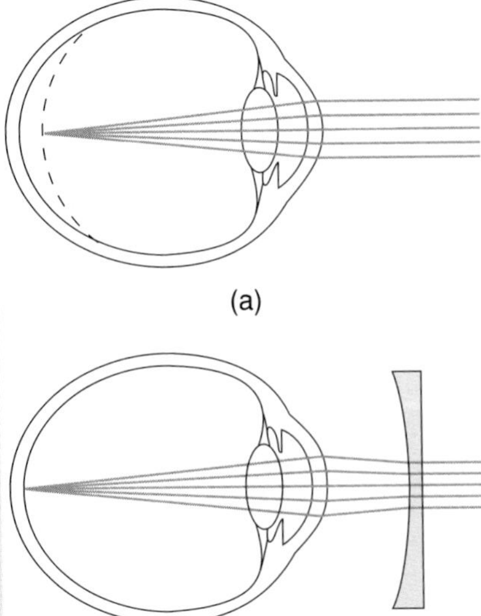

◆ **FIGURE C16.1 Myopia**
(a) Light rays from distant objects focus in front of the retina. (b) Most cases can be corrected by a concave lens, which causes light rays to diverge as they enter the eye.

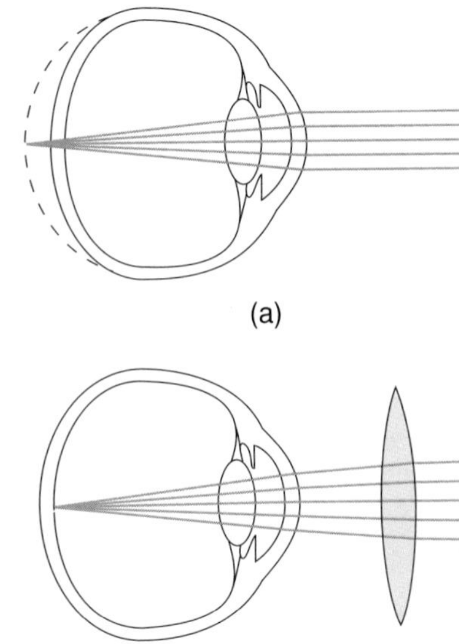

◆ **FIGURE C16.2 Hypermetropia**
(a) Light rays from distant objects focus behind the retina. (b) Most cases can be corrected by a convex lens, which causes light rays to converge as they enter the eye.

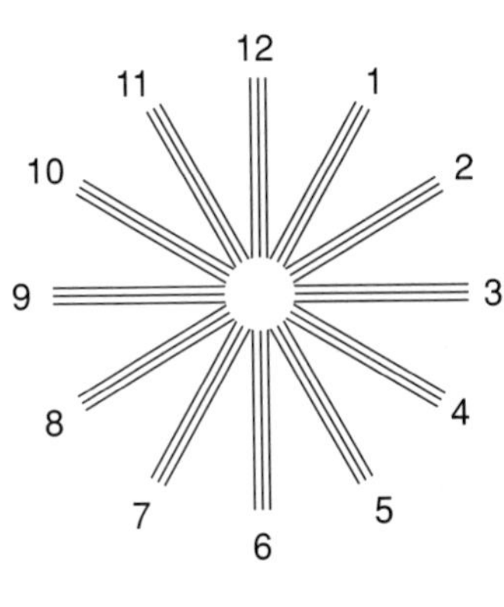

◆ **FIGURE C16.3 Detection of astigmatism**
Astigmatism may be detected by viewing a series of radiating lines. To astigmatic individuals, some lines appear more distinct than others.

CONDITIONS OF CLINICAL SIGNIFICANCE

Astigmatism

Previous considerations of the refractive system of the eye have assumed that all elements of the system possess uniformly curved surfaces (much like a marble). Often, however, the surface of one or more elements of the system is not uniformly curved in all planes. For example, the surface of the cornea may have a different curvature in the horizontal plane than in the vertical plane (much like a chicken's egg). As a result, light rays entering the eye in different planes are focused at different points, creating an out-of-focus image. The condition that results from unequal curvatures of portions of the eye's refractive system is called *astigmatism (ah-stig´-mah-tism).* If an astigmatic individual examines a pattern such as that in Figure C16.3, in which straight lines radiate from a central point, some lines are sharply focused on the retina and are seen clearly, whereas others are focused in front of or behind the retina and are seen indistinctly. All but very severe cases of astigmatism can be corrected by a lens (eyeglass) that has a greater curvature in one plane than another. The lens is oriented in front of the eye so that the differential bending of light rays passing through the lens compensates for the differential bending of light rays by the eye.

Cataract

In some cases, the lens of the eye or a portion of it becomes cloudy or opaque so that vision is impaired. This cloudiness or opacity is known as a *cataract (kat´-ah-rakt)* (Figure C16.4). Often, lenses with cataracts are removed surgically, and effective vision is restored by the use of lens implants, contact lenses, or special eyeglasses.

Glaucoma

Glaucoma (glaw-kō´-mah) is a condition in which the intraocular pressure is abnormally elevated, often resulting from deficient drainage of the aqueous humor. When severe, the high intraocular pressure of glaucoma squeezes shut blood vessels that supply the eye, leading to the degeneration of the retina and to blindness.

Color Blindness

Color blindness is a deficiency in color perception that ranges from an inability to distinguish certain shades of color to a complete lack of color perception. The condition is caused by a functional deficiency or absence of one or more of the different cone types involved in color vision. Difficulty in distinguishing reds and greens (red-green color blindness) is the most common form of color blindness. There is a strong hereditary component in color blindness, and about 8% of males but less than 1.5% of females are red-green color blind.

Effects of Aging

With aging, the lens of the eye becomes yellowed due to the effects of ultraviolet rays from such sources as sunlight. In addition, the lens is one of the few body structures that exhibits increased cellular growth with age, which may contribute to the development of cataracts.

As an individual ages, the pupil of the eye is no longer able to dilate fully, and the amount of light that reaches the photoreceptors of the retina by age 70 may be only 50% of the amount that reaches the retina during youth. The inability of the pupil to dilate fully can also contribute to poor drainage of the aqueous humor, resulting in increased intraocular pressure and a greater likelihood of glaucoma.

The continuous exposure of the rods and cones to light causes damage to their membranes and generates cellular debris. The debris can be removed by cells of the pigmented layer of the retina, and as long as these cells function properly, no problems occur. With aging, however, these cells can become congested, and debris from the rods and cones may accumulate, thus contributing to a loss of visual acuity. This loss is particularly striking if it occurs in the fovea.

With aging, the lens gradually loses its elasticity, and thereby loses some of its ability to change shape during accommodation for viewing near objects. As a result, the near point of vision recedes farther and farther from the eye. Although the loss of lens elasticity begins early in life, the greatest loss occurs after about the age of 40 (Figure C16.5). As the ability to accommodate for viewing near objects diminishes, it

CONDITIONS OF CLINICAL SIGNIFICANCE

becomes necessary to hold reading materials farther and farther from the eye in order to focus the printed letters onto the retina. Ultimately, books and newspapers may have to be held so far from the eye that the images of the letters on the retina are too small to be recognized. This condition, called *presbyopia,* can be corrected by the use of eyeglasses with convex lenses. The convex lens increases the convergence of the light rays so that the refractive system of the eye can focus them onto the retina when the printed matter is held reasonably close to the eye.

◆ FIGURE C16.4 Close-up of the eye, showing a cataract

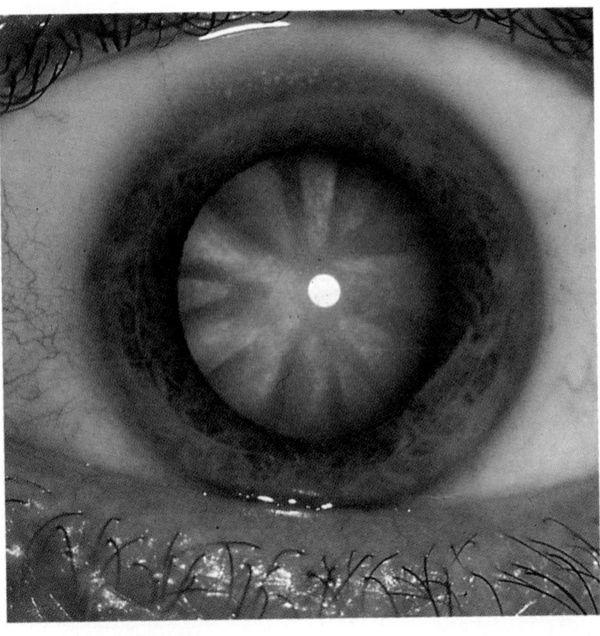

◆ FIGURE C16.5 Graph indicating changes in the near point of vision as lens elasticity decreases with age

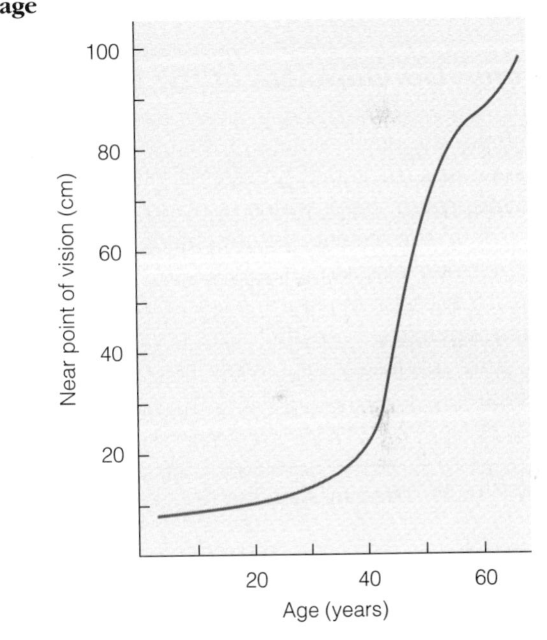

ability to distinguish two closely spaced points as separate. For two closely spaced points to be distinguished as separate, it is believed light rays entering the eye from one of the two points must stimulate a different receptor unit in the retina than light rays from the other point, and there must be at least one unstimulated or only weakly stimulated unit between the two receptor units (Figure 16.24). Whether or not this pattern of stimulation occurs depends on both the focusing system of the eye and the area of the retina struck by the light rays. Light rays entering the eye from two points have the greatest likelihood of stimulating different receptor units in the fovea, where the cones are packed closely together. Moreover, in the fovea the inner layers of the retina are generally displaced to one side, allowing light to pass relatively unimpeded to the cones. Therefore, visual acuity is greatest in the region of the fovea.

Visual acuity is often assessed with the aid of eye charts, which frequently consist of a series of letters of varying sizes. A person with normal visual acuity should be able to read the letters on the 20-foot line of the eye chart at a distance of 20 feet. If a person can read only the 40-foot line at 20 feet, the person is said to have 20/40 vision (meaning that the person's eyes cannot distinguish beyond 20 feet what normal eyes can distinguish at 40 feet).

The Ear—Hearing and Head Position and Movement

The ear contains receptors for hearing as well as receptors that detect head position and movement. The ear is divided into *external, middle,* and *inner* regions (Figure 16.25). The external ear is essentially a funnel-shaped structure used for collecting sound waves. The middle ear contains three small bones called ossicles, which transmit sound waves from the external ear to the inner ear. The inner ear is composed of a system of fluid-filled semicircular ducts and chambers that contain receptors for the detection of sound waves, as well as receptors concerned with head position and movement.

Embryonic Development of the Ear

In the embryo, the development of the ear begins with the formation of a thickened ectodermal plate called an **otic placode,** *(plak´-ode),* which is located on the side of the head in the vicinity of the hindbrain (Figure 16.26a). Each otic placode invaginates to form an **otic pit** (Figure 16.26b). By the fourth week of development, the otic pit separates from the surface ectoderm and develops into a closed sac called an **otic vesicle** (Figure 16.26c).

While the otic vesicles are forming, lateral pouches develop from the side of the pharynx. As these **pharyngeal pouches** expand, the surface ectoderm indents toward them, forming the **branchial grooves.**

With further development, the otic vesicle forms the *membranous labyrinth* of the inner ear. While this is occurring, fibers from the vestibulocochlear nerve (cranial nerve VIII), grow toward the otic vesicle and innervate it. The distal end of the first pharyngeal pouch forms the cavity of the middle ear. The proximal portion of the first pharyngeal pouch forms a tube called an *auditory (eustachian) tube,* which connects the middle-ear cavity with the pharynx (Figure 16.26d). The three ossicles form from condensations of mesenchymal cells in the middle-ear cavity. On the surface of the embryo, the branchial groove associated with the first pharyngeal pouch deepens and forms a canal called the *external auditory canal.* The *tympanic membrane (eardrum)* develops from membrane that separates the first pharyngeal pouch from the floor of the branchial groove. The *external ear (auricle)* forms from the coalescence of a series of elevations that develop around the external auditory meatus.

Structure of the Ear

The structures of the various regions of the ear are uniquely suited to collect sound waves, to convert the

◆ **FIGURE 16.25 The ear, showing the external, middle, and inner regions**

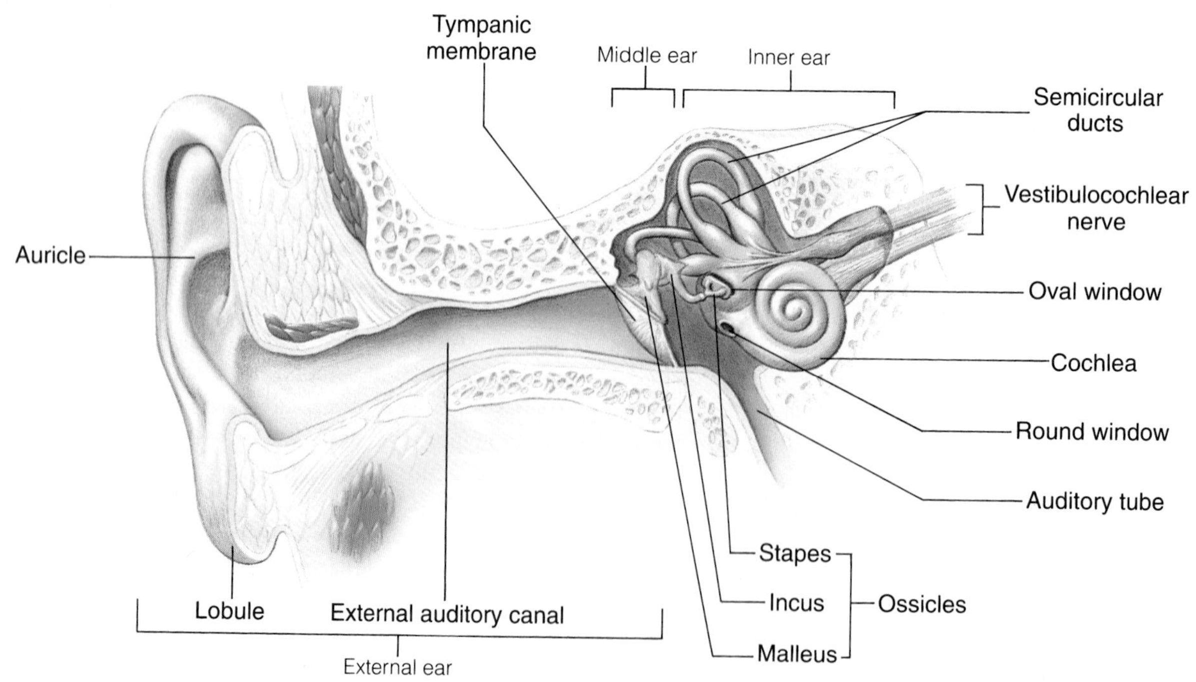

vibrations of sound waves traveling through air into other vibrations—first by small bones in the middle ear and then by fluid in the inner ear—and finally to convert the fluid vibrations into nerve impulses.

External Ear

The **auricle,** or **pinna,** is the most prominent portion of the external ear (Figure 16.25). It consists of an irregularly shaped framework of elastic cartilage covered with skin. The only part of the auricle that is not supported by cartilage is the **lobule,** a flap of skin-covered connective tissue that extends from the lower margin of the auricle. The auricle directs sound waves into the external auditory canal.

The **external auditory canal (meatus)** is a curved passageway approximately 2.5 cm long that extends from the auricle to the eardrum. The canal is lined with skin, and near its entrance are fine hairs and sebaceous glands. It also contains modified sweat glands called **ceruminous glands** *(sĕ-roo´-mĭ-nus)* that secrete *cerumen (earwax).* The hairs and the cerumen help to prevent small foreign objects from reaching the eardrum.

The external auditory canal serves as a resonator for the range of sound waves typical of human speech (2500 to 5000 cycles per second). Because of its resonating properties, the canal increases the sound pressure on the eardrum for sound waves in this frequency range.

Middle Ear

The middle ear is a small air-filled chamber in the temporal bone. It is separated from the external auditory canal by the **tympanic membrane (eardrum)** and

◆ **FIGURE 16.26 Embryonic development of the ear**
(a) Formation of the otic placodes and pharyngeal pouches. (b) Invagination of the otic placodes to form the otic pits, and beginning of the formation of the branchial grooves. (c) Otic vesicles form from the otic pits, and the branchial grooves deepen. (d) Inner-ear structures form from the otic vesicles. Each branchial groove develops into an external auditory canal, and each pharyngeal pouch becomes an auditory tube and a middle-ear cavity.

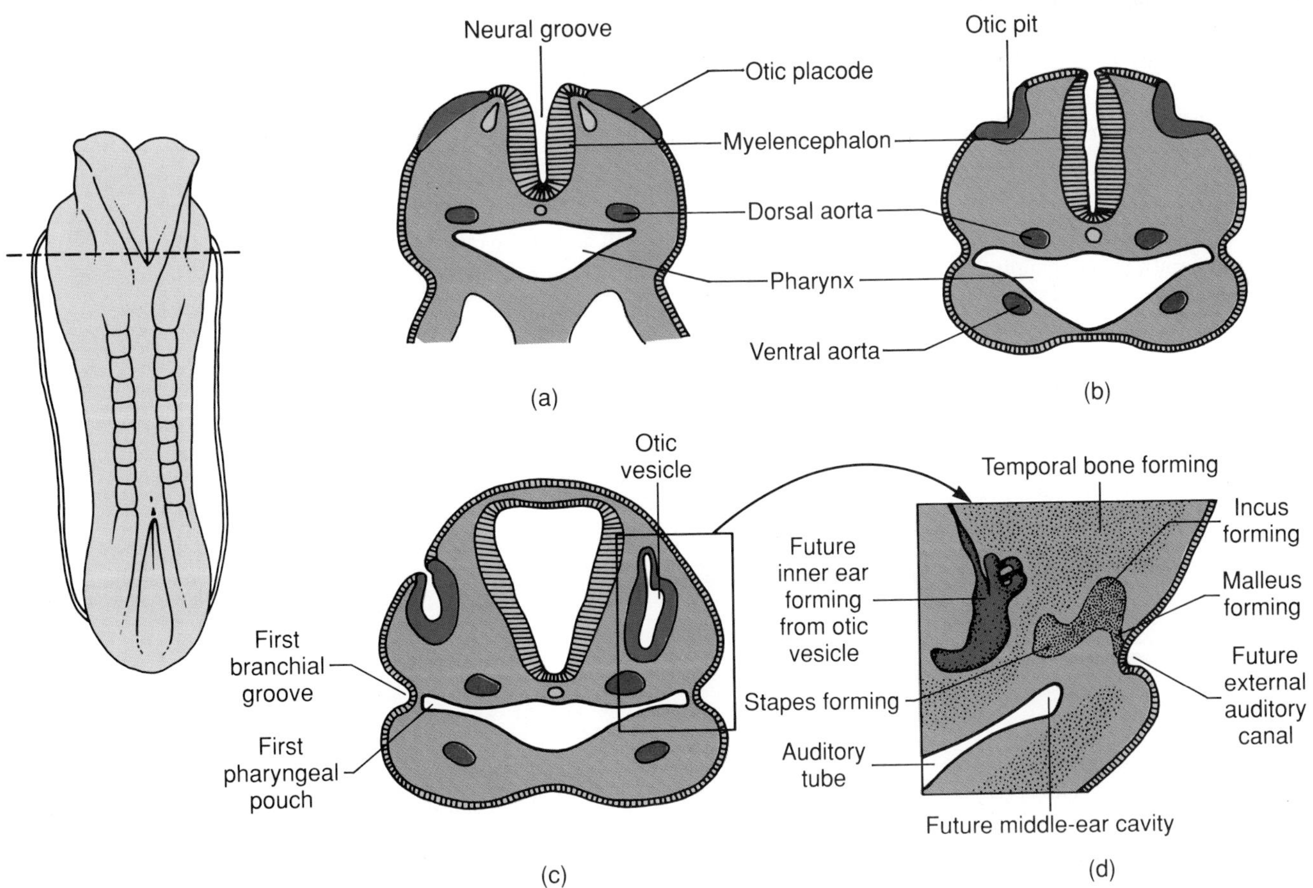

separated from the inner ear by a bony wall in which there are two small membrane-covered openings—the **oval window (fenestra vestibuli)** and the **round window (fenestra cochleae)** (Figure 16.25). An opening in the posterior wall of the middle ear leads to the mastoid sinuses in the mastoid portion of the temporal bone. Another opening connects the middle-ear chamber with the **auditory tube,** which leads to the nasopharynx.

The auditory tube provides a means by which the air pressure in the middle-ear chamber remains equalized with atmospheric pressure. When atmospheric pressure is reduced, as at higher altitudes, the tympanic membrane would bulge outward if the pressure in the middle-ear chamber were not correspondingly reduced. Not only would this be painful, but it would also impair hearing by interfering with the vibrations of the tympanic membrane. The auditory tube, which is closed most of the time in adults, can be opened by swallowing or yawning. This action allows the air pressure in the middle-ear chamber to equalize with the atmospheric pressure within the nasopharynx.

The mucous membranes that line the middle-ear chamber, the mastoid sinuses, and the auditory tubes are continuous with the mucous membrane of the throat. For this reason, infections of the throat can readily spread to the middle-ear chamber—particularly in young children, whose auditory tubes are straighter than in the adult and tend to remain open most of the time. Because of the opening between the middle-ear cavity and the mastoid sinuses, infections of the middle-ear cavity can spread to the mucous membrane that lines the mastoid sinuses, producing a condition called *mastoiditis.* The mastoid sinuses are separated from the brain only by thin, bony partitions. Therefore, it is also possible for infection to spread from the mastoid sinuses to the meninges of the brain.

The three ear **ossicles,** or middle-ear bones, form a flexible bridge across the middle-ear chamber (Figure 16.25).

The handle-shaped portion of the ossicle called the **malleus** *(hammer)* attaches to the inner surface of the tympanic membrane. The foot-plate portion of the ossicle called the **stapes** *(stirrup)* fits against the oval window on the medial wall of the middle-ear chamber. The third ossicle, the **incus** *(anvil),* lies between the malleus and the stapes and articulates with them. Thus, the three ossicles form a bridge between the tympanic membrane and the oval window. The articulations between the ear ossicles are freely movable synovial joints. The ossicles form a lever system that picks up vibrations of the tympanic membrane and transmits them to the oval window, which in turn leads into the inner ear. Two small muscles attach to the ear ossicles: the **stapedius muscle** attaches to the stapes, and the **tensor tympani muscle** attaches to the malleus. The functions of these muscles are discussed later in the chapter (page 521).

◆ **FIGURE 16.27 Structures of the inner ear**

The membranous labyrinth (smooth tan area) is filled with endolymph and is separated from the osseous labyrinth by perilymph (shown in purple).

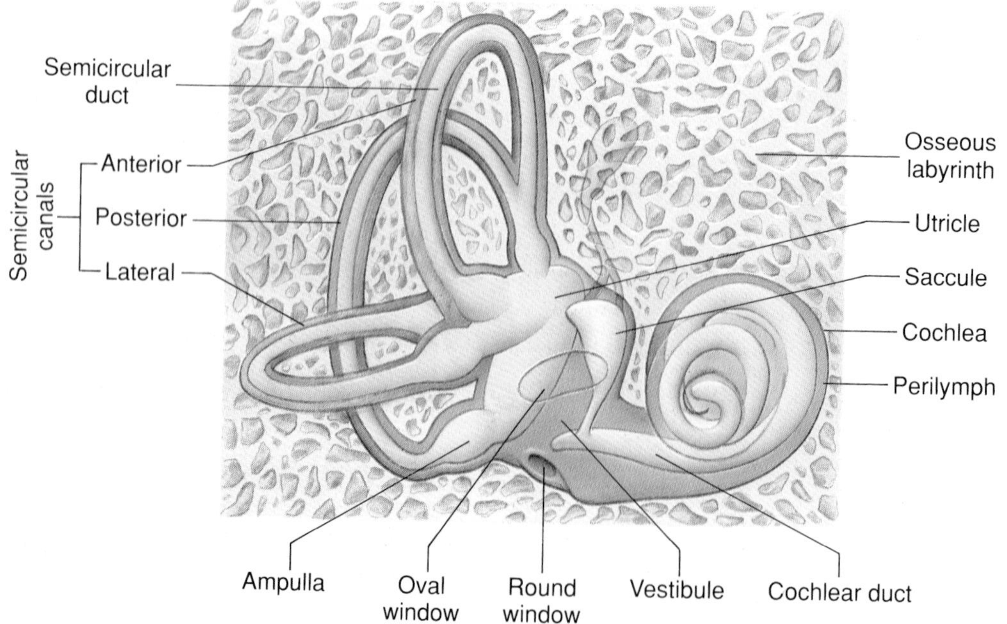

Inner Ear

The inner ear is located medial to the middle ear in the petrous portion of the temporal bone (Figure 16.27). It consists of a series of canals called the **osseous labyrinth,** which are hollowed out of the bone. Within the osseous labyrinth, and following its course, is a **membranous labyrinth.** The membranous labyrinth is filled with a fluid called **endolymph,** which has a high concentration of potassium ions and a low concentration of sodium ions. The membranous labyrinth is suspended in a fluid called **perilymph,** which has a low concentration of potassium ions and a high concentration of sodium ions. Perilymph therefore separates the walls of the membranous labyrinth from the osseous labyrinth. The osseous labyrinth is divided into three areas: the *vestibule,* the *semicircular canals,* and the *cochlea.*

Vestibule. The **vestibule** is a chamber just medial to the middle-ear chamber. Since the oval window forms a membranous partition between the middle-ear chamber and the vestibule of the inner ear, vibrations of the oval window induced by the stapes are transmitted to the perilymph of the vestibule. Within the vestibule are two enlargements of the membranous labyrinth: the **utricle** and the **saccule.** The utricle and the saccule contain receptor cells that detect the position and movement of the head, and information from these receptor cells contributes to the sense of balance. The utricle is connected to the portion of the membranous labyrinth that is located in the osseous semicircular canals; the saccule is connected with the portion of the membranous labyrinth located in the cochlea. The membranous labyrinth, therefore, is a continuous series of ducts that are filled with endolymph.

Semicircular Canals. Within the inner ear are three bony **semicircular canals,** which contain three membranous **semicircular ducts.** The semicircular canals are arranged at right angles to each other, forming anterior, lateral, and posterior canals. The anterior and posterior canals are vertical; the lateral canal is horizontal. Each membranous semicircular duct possesses an enlargement called an **ampulla,** which contains receptor cells that detect certain movements of the head and thereby provide information concerning equilibrium.

Cochlea. The portion of the inner ear associated with hearing is the **cochlea** *(kōk´-lē-ah).* It resembles a snail shell, spiraling 2½ turns around a central bony core called the **modiolus** *(mō-dī´-uh-lus)* (Figure 16.28a). A bony shelf called the **spiral lamina** extends into the cochlea from the modiolus (Figure 16.28b). Two membranes, the **vestibular membrane** and the **basilar membrane,** extend across the cochlea from the spiral lamina, dividing the cochlea into three longitudinal tunnels.

The central tunnel between the vestibular and basilar membranes is called the **cochlear duct,** or **scala media.** The cochlear duct is the membranous labyrinth of the cochlea and is filled with endolymph. One tunnel, the **scala vestibuli,** is separated from the cochlear duct by the vestibular membrane. The other tunnel, the **scala tympani,** is separated from the cochlear duct by the basilar membrane. The scala vestibuli and scala tympani both contain perilymph and are continuous with one another at the apex of the cochlea through an opening called the **helicotrema.** The scala tympani terminates at the round window, whereas the scala vestibuli ends at the oval window. The relationships of these structures are more easily understood if the cochlea is visualized as being uncoiled, as shown in Figure 16.29.

The **spiral organ (organ of Corti),** which contains the receptors for hearing, is located on the basilar membrane. It consists of a series of receptor cells and supporting cells (Figure 16.30). The receptor cells are called **hair cells** because they each have a tuft of large microvilli (page 110) projecting from their free surfaces. These microvilli are known as *stereocilia,* even though they are not true cilia.

The plasma membrane at the tips of the stereocilia of the hair cells apparently contains *mechanically gated ion channels* that open when the stereocilia are displaced in a particular direction. The channels are permeable to small, positively charged ions, and when they open, the hair cells depolarize.

The hair cells of the spiral organ are arranged in a single row of about 3500 *inner hair cells* and three or four rows of about 20,000 *outer hair cells* (Figure 16.31, page 516). The hair cells make synaptic contact with afferent neurons of the cochlear division of the vestibulocochlear nerve (cranial nerve VIII). These neurons pass through the modiolus and spiral lamina to reach the basilar membrane and the hair cells.

Overhanging the spiral organ is a flexible flap of fibrous and gelatinous tissue called the **tectorial membrane** *(tek-to´-rē-al).* Stereocilia of hair cells of the spiral organ are in contact with the tectorial membrane.

Mechanisms of Hearing

Hearing is the process of perceiving sound.

Sound Waves

Sound energy (a type of mechanical energy) travels through air in the form of pressure waves that are produced when air molecules are alternately compressed and rarefied. For example, when the arm of a vibrating tuning fork moves in one direction, air molecules ahead

◆ **FIGURE 16.28 The cochlea**

(a) Section through the cochlea. The scala tympani and scala vestibuli are continuous through the helicotrema. (b) Magnified cross section of one turn of the cochlea. The cochlear duct is formed by the membranous labyrinth and contains endolymph. The scala tympani and scala vestibuli contain perilymph. (c) Photomicrograph of the cochlea.

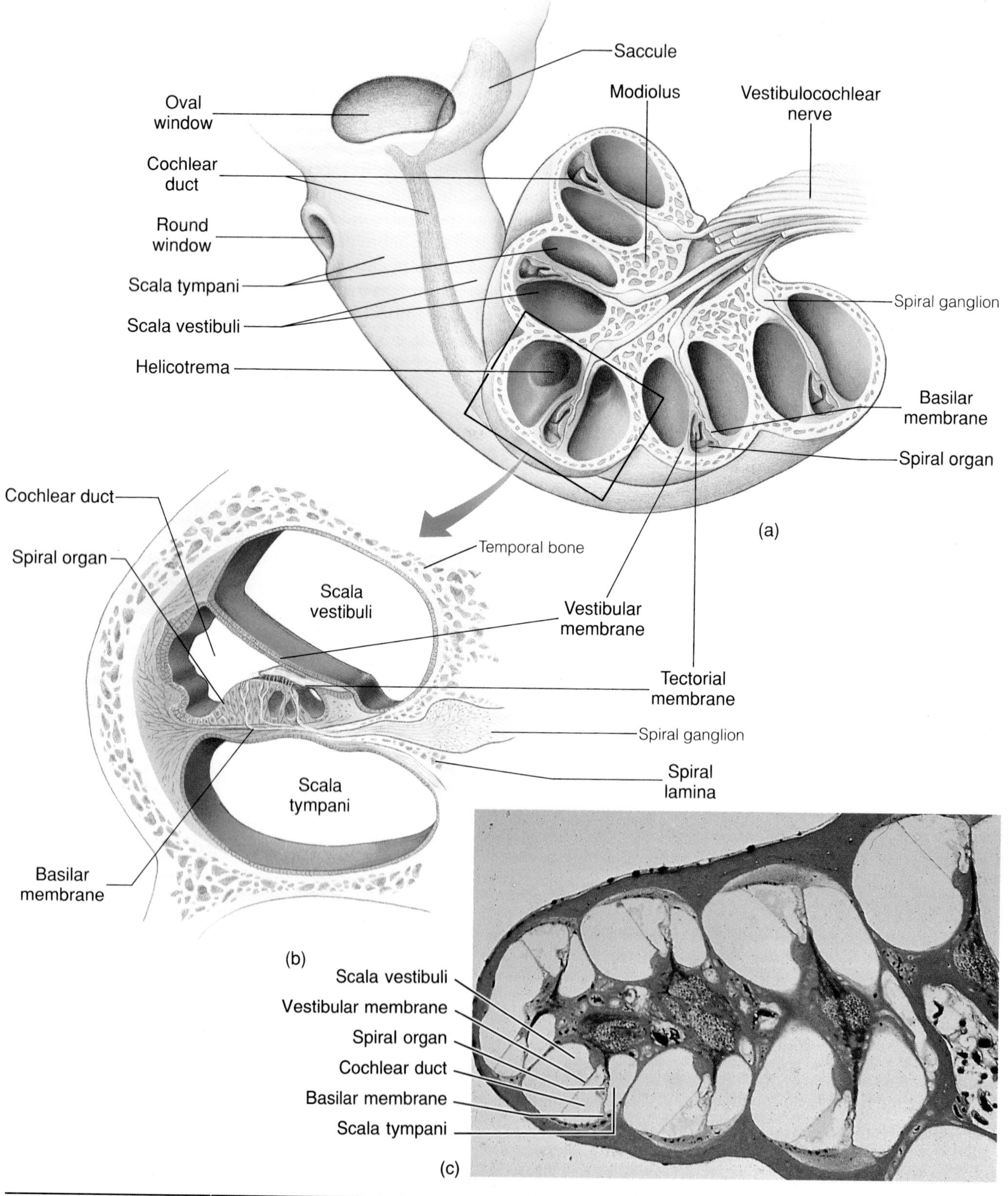

◆ **FIGURE 16.29 Diagrammatic representation of the structures of the inner ear**

The reddish areas are filled with endolymph and the gray areas contain perilymph.

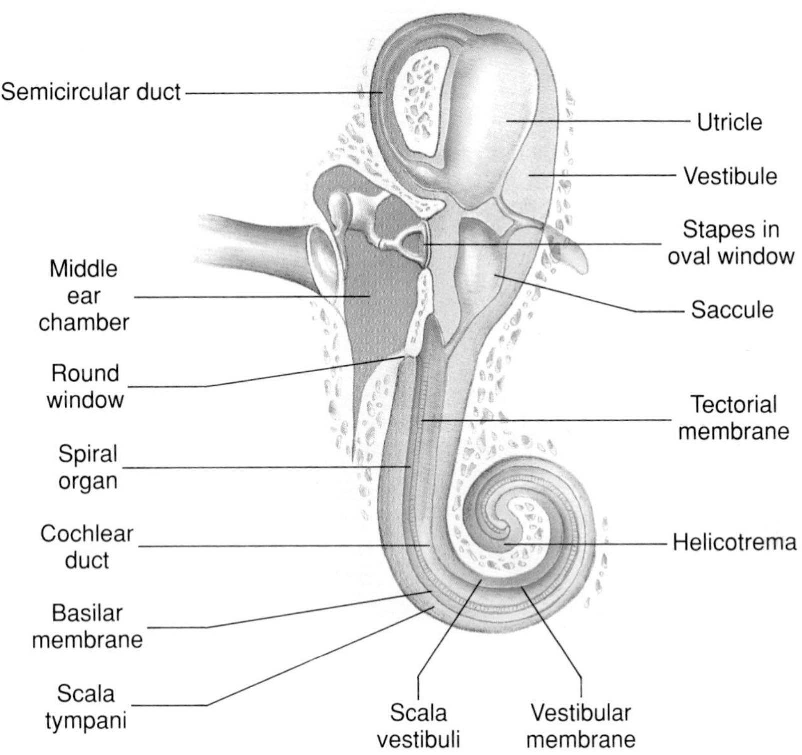

◆ **FIGURE 16.30 Section of the spiral organ**

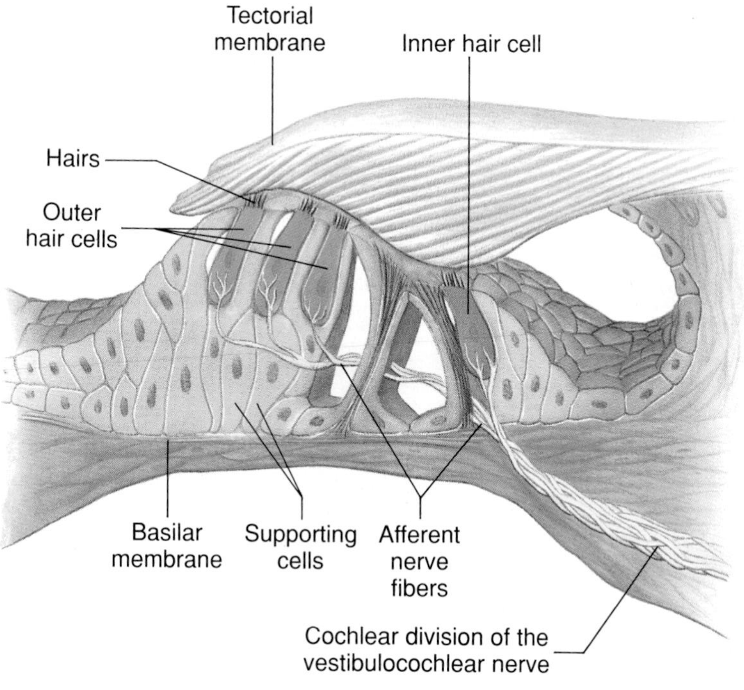

◆ **FIGURE 16.31 Scanning electron micrograph showing hair cells in part of the cochlea (×4000)**

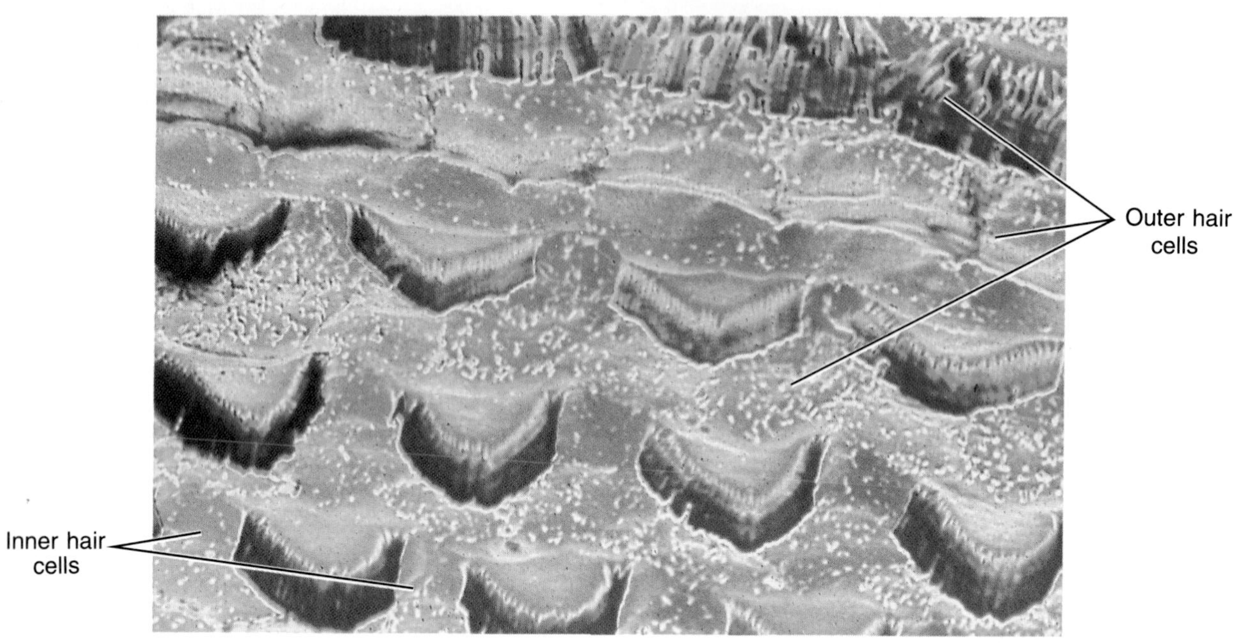

of the arm are pushed together, or compressed, and the pressure in the area increases (Figure 16.32). In turn, air molecules in the area of compression bump into other air molecules ahead of them, pushing these air molecules together and creating a new region of compression. Through many repetitions of this process, the initial compression of air molecules by the arm of the tuning fork can be transmitted a considerable distance, even though individual air molecules move only a short distance. Similarly, when the arm of the tuning fork moves back in the opposite direction, the compression of the air molecules and the pressure are decreased even below normal. These changes, too, are transmitted from air molecule to air molecule. Thus, a sound wave consists of regions of compression, in which the air molecules are close together and the pressure is relatively high, alternating with areas of rarefaction, where the molecules are farther apart and the pressure is lower.

◆ **FIGURE 16.32 Production of a sound wave**

As a sound source—here a tuning fork—vibrates, it creates alternating areas of compression (higher pressure), in which the air molecules are close together, and rarefaction (lower pressure), where the molecules are farther apart. This activity produces a pressure wave that radiates outward from the sound source. This pressure wave is called a sound wave.

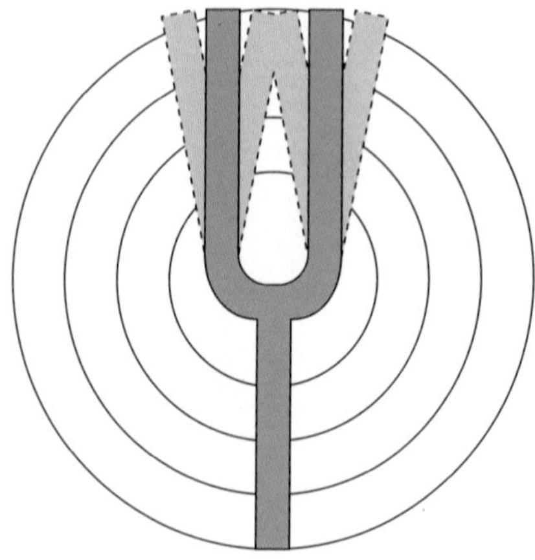

Sound Waves and the Loudness, Pitch, and Timbre of Sounds

A sound can be characterized according to its loudness, pitch, and timbre (quality).

1. The **loudness** of a sound is related to the intensity (or power) of the sound wave, which in turn, is related to the amplitude of the wave—that is, to the pressure difference between a zone of compression and a zone of rarefaction (Figure 16.33). In general, the greater the amplitude of a particular sound wave, the greater the intensity of the wave and the louder the sound.

2. The **pitch** of a sound (for example, whether the sound is a C note or an F note) is related to the frequency of the sound wave (Figure 16.33). The frequencies of sound waves are usually expressed in cycles per second, or Hertz (Hz). In general, the higher the frequency of a sound wave, the higher the pitch of the sound.

The human ear is capable of detecting sound waves with frequencies between about 20 and 20,000 cycles per second. However, the ear is most sensitive to sound waves with frequencies between 1000 and 4000 cycles per second. Sound waves with frequencies outside this range must have greater intensities than sound waves with frequencies within the range if they are to be detected.

3. The **timbre,** or **quality,** of a sound is related to the presence of additional sound-wave frequencies superimposed on a fundamental frequency (Figure 16.33). Although laboratory sound sources can produce relatively pure sound waves (that is, the sound waves all have the same fundamental frequency without any superimposed frequencies), most sound sources produce sound waves that have additional sound-wave frequencies superimposed on a fundamental frequency. Because these superimposed frequencies often differ from one sound source to another, the same fundamental frequency is often perceived differently when it is produced by different sound sources (that is, a C note played on a trumpet sounds different than a C note played on a piano).

Transmission of Sound Waves to the Inner Ear

If sounds are to be perceived—that is, if a person is to hear a sound—the vibratory movements of sound waves traveling in air must be transferred to the fluids within the inner ear. However, fluids have much greater inertia than air, and greater pressures are required to cause movements of fluids than are required to cause movements of air. The pressures necessary to move the fluids of the inner ear result in part from the arrangement of the ossicles of the middle ear and in part from the difference in area between the tympanic membrane and the oval window.

A sound wave enters the external auditory canal and strikes the tympanic membrane. The alternating higher and lower pressure regions of the sound wave cause the tympanic membrane to move forward and backward (vibrate) at the same frequency as the sound wave. The vibratory movements of the tympanic membrane are transmitted across the middle-ear cavity by the malleus, incus, and stapes to the oval window of the inner ear. Because of the arrangement of the lever system of the ossicles, the force exerted on the oval window by the stapes is about 1.3 times greater than the force exerted on the malleus at the tympanic membrane (although the distance of movement of the stapes at the oval window is only about 75% as great as the distance of movement of the malleus at the tympanic membrane). Moreover, the area of the tympanic membrane that the sound wave can strike (about 55 mm^2) is approximately 17 times greater than the area of the oval window (about 3.2 mm^2). Thus, the pressure applied to the perilymph within the cochlea by the movement of the stapes against the oval window is about 22 times (1.3×17) greater than the pressure that would be exerted if the sound wave were to strike the oval window directly. This higher pressure acts on the perilymph to cause its movement.

Function of the Cochlea

If the forward movement of the stapes at the oval window is very slow, the pressure exerted on the perilymph

◆ **FIGURE 16.33 Relationship between sound characteristics and sound-wave properties**

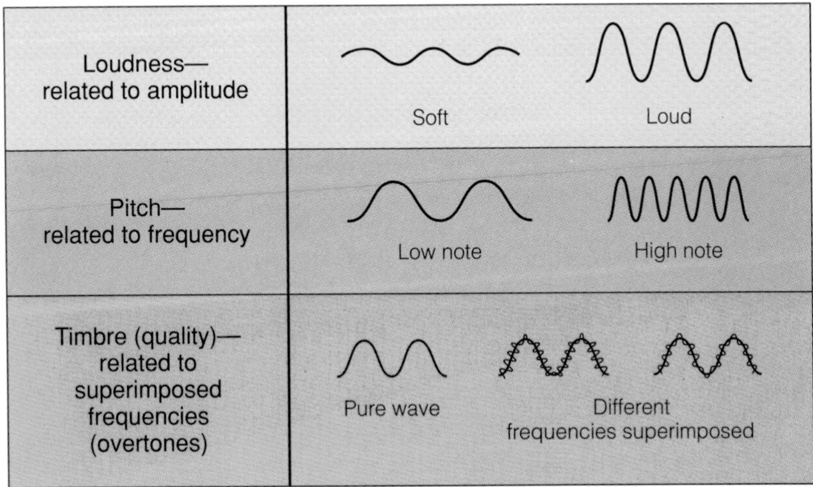

pushes the perilymph along the scala vestibuli, through the helicotrema, and into the scala tympani, causing the round window to bulge outward into the middle ear (Figure 16.34a). When the stapes moves backward and exerts less pressure, the perilymph moves back from the scala tympani into the scala vestibuli. Thus, very-low-frequency pressure waves have little effect on the basilar membrane. Because of the inertia of the fluids of the inner ear, however, higher-frequency pressure waves of the sort associated with sound perception do not follow this pathway. Instead, these pressure waves are transmitted from the perilymph of the scala vestibuli near the oval window, through the flexible vestibular membrane, to the endolymph of the cochlear duct. From the endolymph, the waves are transmitted through the basilar membrane to the perilymph of the scala tympani, and finally to the round window. The transmission of the pressure waves through the basilar membrane at the base of the cochlea near the oval and round windows causes this area of the membrane to vibrate (move downward and upward). The vibration of the basilar membrane near the base of the cochlea initiates a wave that travels along the basilar membrane toward the helicotrema.

The basilar membrane contains about 25,000 basilar fibers that project from the spiral lamina toward the outer wall of the cochlea. Although the basilar membrane extends completely across the cochlea from the spiral lamina, its fibers do not. Rather, the distal ends of the fibers are embedded in the basilar membrane.

The basilar fibers increase in length and decrease in thickness and rigidity from the base of the cochlea toward the helicotrema (Figure 16.34b). Because of these differences in the basilar fibers, the portion of the basilar membrane near the base of the cochlea tends to vibrate at high frequencies, and the portion near the helico-

◆ FIGURE 16.34 Transmission of pressure waves in the cochlea and the relative lengths of basilar fibers
(a) Schematic representation of the transmission of pressure waves in the cochlea. Blue arrows indicate the transmission pathway when the movement of the stapes is slow. Green arrows indicate the transmission pathway of higher-frequency pressure waves of the sort associated with sound perception. (b) Graphic representation of the basilar membrane showing that the end of the membrane toward the middle ear contains shorter basilar fibers than the end toward the helicotrema. Maximum vibrations occur in the region of the basilar membrane that has the same resonant frequency as the sound wave that causes the pressure wave. Low-frequency sound waves cause maximum vibration toward the apex of the basilar membrane. High-frequency waves have their maximal effect toward the base of the basilar membrane.

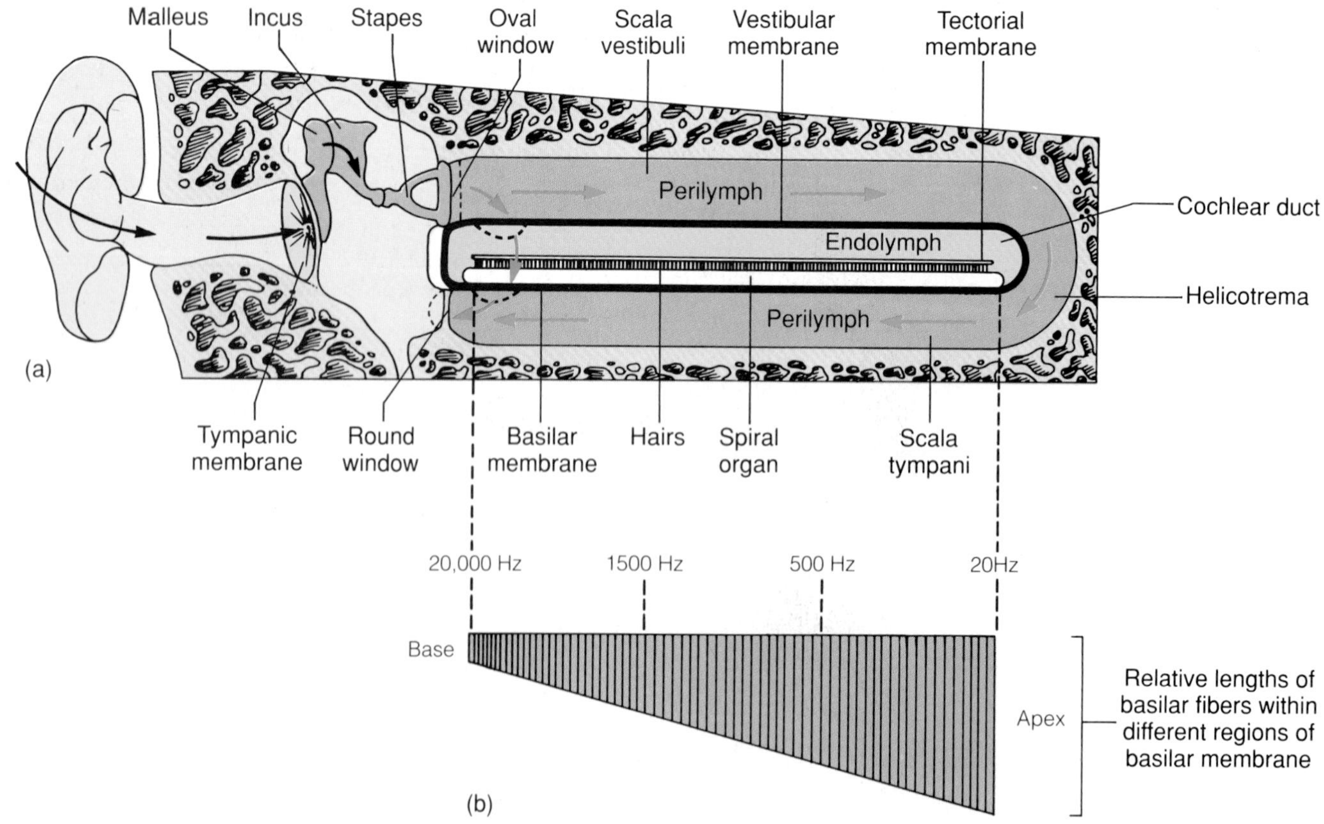

trema tends to vibrate at lower frequencies. Moreover, when a portion of the basilar membrane vibrates, the fluid between the vibrating portion of the membrane and the oval and round windows must also move. Since less fluid mass must move (that is, less inertia must be overcome) when a region of the membrane near the base of the cochlea vibrates than when a portion of the membrane near the helicotrema vibrates, the fluid loading of the basilar membrane also favors high-frequency vibration at the base of the cochlea near the oval and round windows and low-frequency vibration at the apex of the cochlea near the helicotrema. Thus, the sympathetic vibration, or resonance, of the basilar membrane in response to high-frequency sound waves is greatest near the base of the cochlea, and the sympathetic vibration, or resonance, of the basilar membrane in response to low-frequency sound waves is greatest near the apex of the cochlea.

As a wave of a particular frequency travels along the basilar membrane from its point of initiation near the oval and round windows, the vibration of the basilar membrane increases in amplitude and is greatest at that portion of the membrane that has the same natural resonant frequency as the frequency of the wave (Figure 16.35). At this point the energy of the wave is dissipated, and beyond this region the wave quickly dies away. Thus, high-frequency sound waves cause maximal vibration of the basilar membrane near the base of the cochlea near the oval and round windows, and low-frequency sound waves cause maximal vibration of the basilar membrane near the apex of the cochlea near the helicotrema.

The spiral organ, which contains receptor cells called hair cells, is located on the basilar membrane (page 513). When a region of the basilar membrane vibrates, the stereocilia of hair cells in the region are displaced, altering the polarity of the cells. When a region of the basilar membrane moves upward toward the scala vestibuli, hair cells in the region depolarize. The depolarization leads to an increased release of neurotransmitter molecules from the hair cells, which increases the stimulation of afferent neurons associated with them.

Auditory Pathways

Afferent neurons from the spiral organ pass within the vestibulocochlear nerve to the medulla, where they synapse in centers called *dorsal* and *ventral cochlear nuclei* (Figure 16.36). From these nuclei, other neurons extend to the *superior olivary nuclei,* also in the medulla. These neurons synapse in the olivary nuclei with neurons that form ascending tracts called the *lateral lemnisci.* Neurons forming the lemnisci synapse, in turn, in either the *inferior colliculi* of the midbrain or the *medial geniculate body* of the thalamus with neurons that pass to the auditory portions of the cerebral cortex. Some of the auditory pathway neurons from each ear cross over within the medulla. Therefore, each auditory cortex receives impulses from both ears.

Determination of Pitch

The ability to determine the pitch of most sounds is related to the fact that a sound wave of a specific fre-

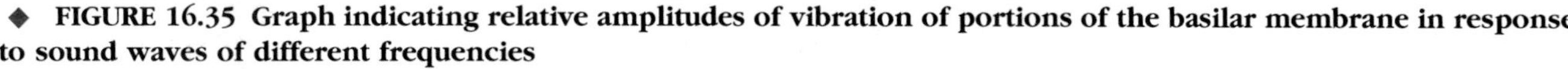

◆ **FIGURE 16.35 Graph indicating relative amplitudes of vibration of portions of the basilar membrane in response to sound waves of different frequencies**

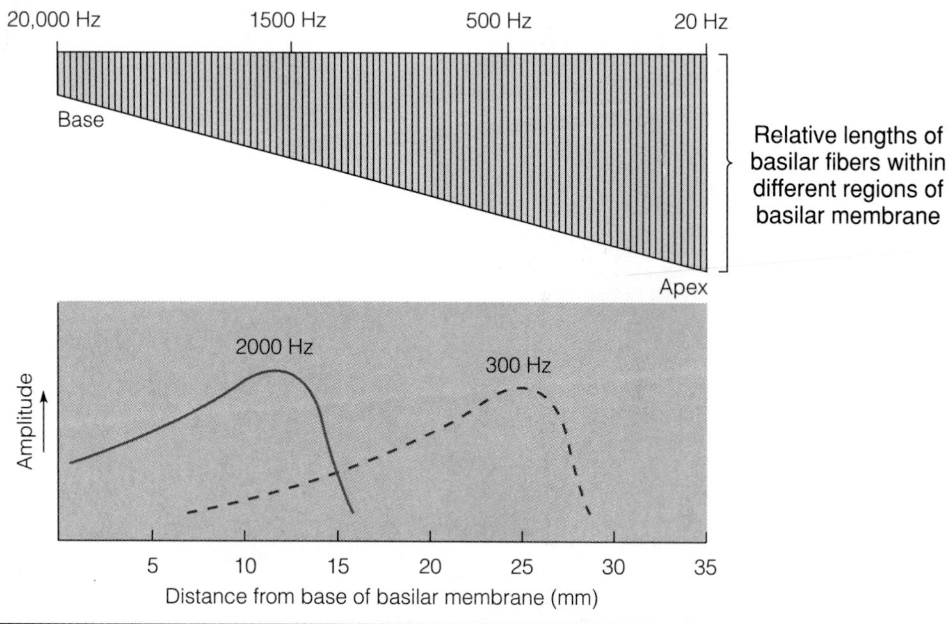

quency gives rise to a particular pattern of vibration of the basilar membrane in which a specific region of the membrane vibrates more intensely than other regions. This pattern of vibration, in turn, gives rise to a particular pattern of nerve impulses that are transmitted to the brain and interpreted as a sound of a particular pitch. When the maximal vibration of the basilar membrane is near the base of the cochlea, the sounds perceived are interpreted as high pitched; when the maximal vibration is in the middle of the cochlea, the sounds are interpreted as being of intermediate pitch; and when the maximal vibration is near the apex of the cochlea, the sounds are interpreted as low pitched.

Determination of Loudness

High-intensity (high-amplitude) sound waves cause a greater amplitude of vibration of the basilar membrane than low-intensity (low-amplitude) waves. As the amplitude of vibration of the basilar membrane increases, hair cells of the spiral organ are more strongly stimulated. In fact, certain hair cells apparently do not become strongly stimulated until the amplitude of vibration of the basilar membrane becomes quite high. The stronger stimulation of the hair cells activates more neurons and increases the frequency of transmission of afferent nerve impulses to the brain. The brain interprets this increased neural activity as an increase in the loudness of the sound.

Sound Localization

Several mechanisms are involved in localizing the source of a sound. The auditory system is able to detect differences in the time of arrival of a sound wave at each ear, and this ability provides one sound-localization mecha-

◆ **FIGURE 16.36 Auditory pathways from the spiral organ through the central nervous system to the cortex of the temporal lobes of the brain**

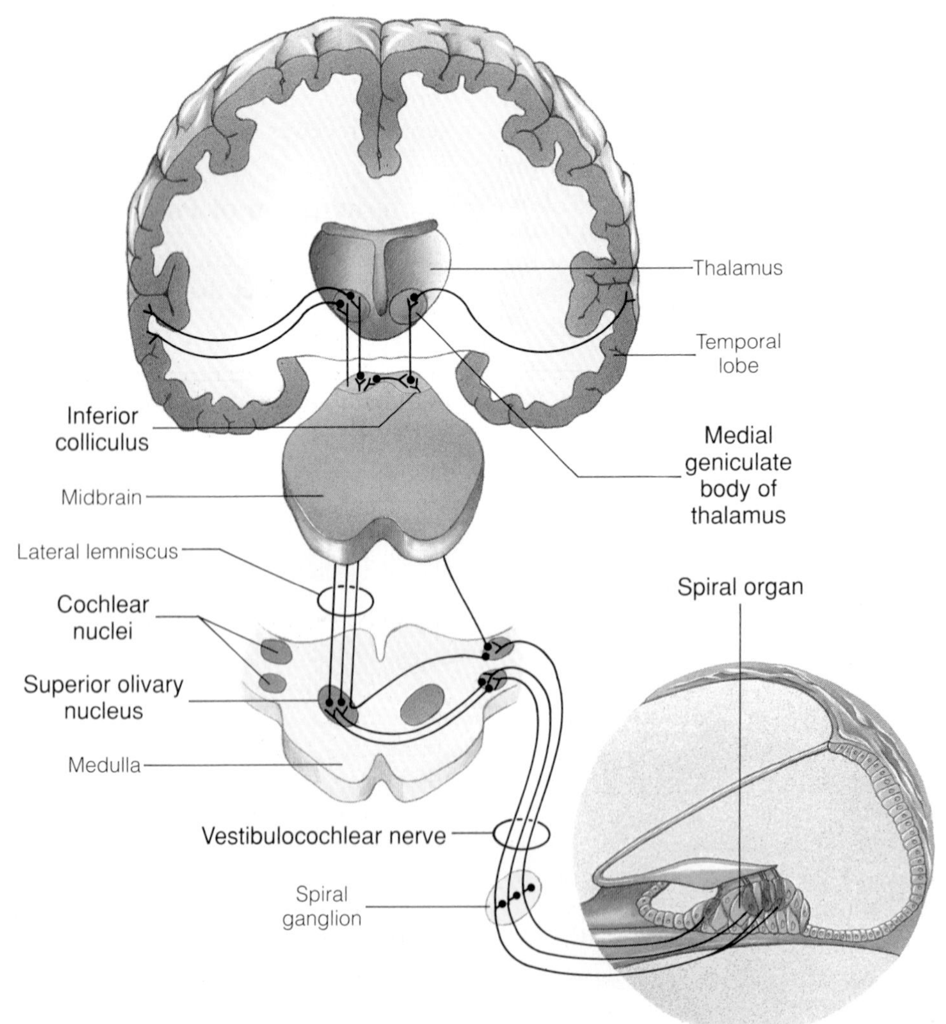

nism. For example, a sound wave coming from a point directly in front of a person reaches both ears at the same time, but a sound wave coming from a point to the person's right reaches the right ear slightly before it reaches the left ear. This aspect of sound localization is most effective at low frequencies (that is, frequencies less than 3000 cycles per second).

The auditory system is also able to detect differences in the intensity of a sound wave at one ear compared to the other, and this ability provides a second sound-localization mechanism. The head acts as a sound-wave barrier, particularly for higher-frequency sound waves. Thus, if a sound wave comes from a point directly in front of a person, both ears will receive the same intensity of stimulation. However, if the sound wave comes from a point to the person's right, the right ear will be more intensely stimulated than the left ear.

The Stapedius and Tensor Tympani Muscles

Two small muscles attach to the ossicles of the middle ear. The **stapedius** *(stuh-pē´-dē-us)* muscle attaches to the stapes, and the **tensor tympani** *(ten´-ser tim´-puh-nē)* muscle attaches to the malleus. An involuntary contraction of the stapedius muscle occurs in response to loud external sounds. Moreover, involuntary contractions of the stapedius muscle and probably also the tensor tympani muscle occur in association with a person's own vocalizations.

A reflex called the *acoustic stapedius reflex* occurs in response to loud external sounds. After a latent period of about 40 to 60 milliseconds following the occurrence of a loud external sound, there is a reflexive contraction of the stapedius muscle. This contraction dampens the vibrations of the ossicles (particularly the stapes) and decreases the transmission of sound waves through the middle ear to the cochlea. This reflex provides some protection to the auditory receptors of the cochlea from damage by excessively loud external sounds. Because of the latent period, however, the reflex provides little protection from sudden, explosive types of loud external sounds such as a gunshot. (Note that when a sufficiently loud sound is directed into only one ear, the stapedius muscles of both ears normally contract.)

The stapedius muscle, and probably also the tensor tympani muscle, contracts just before and during a person's own vocalizations. These contractions, which dampen the vibrations of the ossicles, are believed to provide some protection to the auditory receptors of the cochlea from fatigue, interference, and potential injury that could result from a person's own loud utterances.

The contractions of the stapedius and tensor tympani muscles are particularly effective at dampening low-frequency vibrations of the ossicles—that is, vibrations with frequencies below 1000 cycles per second. Because low-frequency vibrations reaching the auditory receptors of the cochlea tend to dominate and mask higher-frequency vibrations, the contractions of the middle-ear muscles help improve a person's ability to detect higher-frequency sound waves in an environment where low-frequency sound waves are also present. This ability is important because human speech contains many high-frequency components. In fact, individuals with nonfunctional stapedius muscles tend to have some difficulty distinguishing speech sounds when loud background noise is present. Moreover, a person's vocalizations produce intense low-frequency vibrations in the person's own ears. Consequently, the contractions of the stapedius and tensor tympani muscles that occur in association with vocalization help maintain the person's ability to hear soft sounds, such as other people talking, while the person is speaking.

Central Nervous System Influences on the Spiral Organ

Efferent neurons from the central nervous system make synaptic contact with hair cells of the spiral organ, particularly with the outer hair cells. Some researchers believe that the inner hair cells are the receptors primarily responsible for hearing, and there is some evidence that the outer hair cells are part of a feedback system that modulates the mechanical stimulus delivered to the inner hair cells. In any event, an increase in the activity of the efferent neurons can reduce the sensitivity of the spiral organ and make it less sharply tuned in particular portions of the audible frequency range. Such a central control of receptor sensitivity could help a person reduce the perception of background noise when concentrating on one part of the auditory range, such as another person's voice.

Head Position and Movement

In addition to its role in hearing, the inner ear provides information about the position and movement of the head. This information is utilized in the coordination of movements that maintain body equilibrium, or balance. Moreover, signals from the inner ear contribute to the maintenance of desired posture, to the perception of motion and orientation, and to the control of the extrinsic eye muscles so that the eyes can remain fixed on the same point despite changes in the position of the head. The portion of the inner ear involved in these activities is called the **vestibular apparatus.** Receptor cells of the vestibular apparatus are located within the utricle, the saccule, and the ampullae of the semicircular ducts.

CLINICAL CORRELATION

Hearing Loss Due to Aging

Case Report

THE PATIENT: A 76-year-old male.

PRINCIPAL COMPLAINT: Progressive hearing loss and difficulty understanding speech.

HISTORY: The patient had noticed gradually increasing difficulty with his hearing and a great change in his ability to understand conversations in a background of noise. He had no history of dizziness, middle-ear infections, head trauma, hypertension, or diabetes, and he was taking no medicines routinely.

CLINICAL EXAMINATION: Examination of the head and neck revealed no abnormality. No bruits (sounds caused by turbulent blood flow, due to vascular constriction or obstruction) were heard in the neck. Hearing tests showed a mild-to-moderate bilateral loss of the ability to hear high-frequency sounds. His understanding of monosyllabic, phonetically balanced speech was consistent with the degree of hearing loss. His understanding of nonsense sentences embedded in a background story was poorer than his understanding of single words.

COMMENT: The loss of hearing with aging, which is called *presbycusis,* is characterized by loss in the high-frequency range. The ability to understand simple spoken material is comparable to that of normal young people. When the pattern of speech is interrupted, the rate of speaking is changed, or background noise or unrelated speech interferes, older people do not comprehend as well as young people do. This difference of function may appear in persons as young as 40 to 50 years of age.

The pathophysiology of presbycusis is fourfold: peripheral, central, mechanical, and metabolic. Within the cochlea, hair cells, supporting cells, and the basilar membrane degenerate. Nerve fibers are lost, initially in the basal cochlea and later throughout the cochlea. Centrally, fewer neurons are found at each successive level within the auditory pathways. The elasticity of the basilar and tectorial membranes changes, which causes mechanical differences between a young system and an old one. Metabolically, the walls of certain capillaries in the cochlea thicken, and the nutrient supply to the cochlea is decreased. Moreover, the electrolyte balance of the cochlear fluids may also be altered by increasing age.

OUTCOME: A moderate-amplification, over-the-ear-style hearing aid was fitted to the patient, and his ability to hear was restored to normal.

Function of the Utricle

Receptor cells of the utricle provide information about the position of the head with respect to gravity and about the linear acceleration or deceleration of the head (such as would occur when a person starts to move forward in a straight line).

The receptor structure within the utricle is called a **macula,** and it is composed of hair cells and supporting cells (Figure 16.37). The hair cells are similar to the hair cells of the spiral organ (pages 513 and 519). However, each macula hair cell possesses a single true cilium called a *kinocilium* in addition to its stereocilia. (A hair cell of the spiral organ loses its kinocilium during development.)

The macula is covered by a gelatinous substance called the *otolithic membrane,* and the stereocilia of the hair cells project into this membrane. The otolithic membrane contains tiny particles of calcium carbonate

◆ **FIGURE 16.37 The macula and otolithic membrane of a utricle**

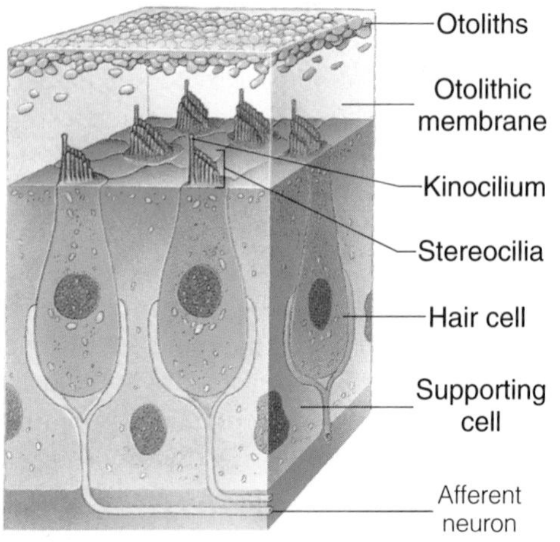

◆ **FIGURE 16.38 Displacement of the stereocilia of macula hair cells**
(a) Displacement of the stereocilia of macula hair cells from their original positions when the head is in a new position. (b) Displacement of the stereocilia of macula hair cells by linear acceleration (in this case, beginning to walk forward).

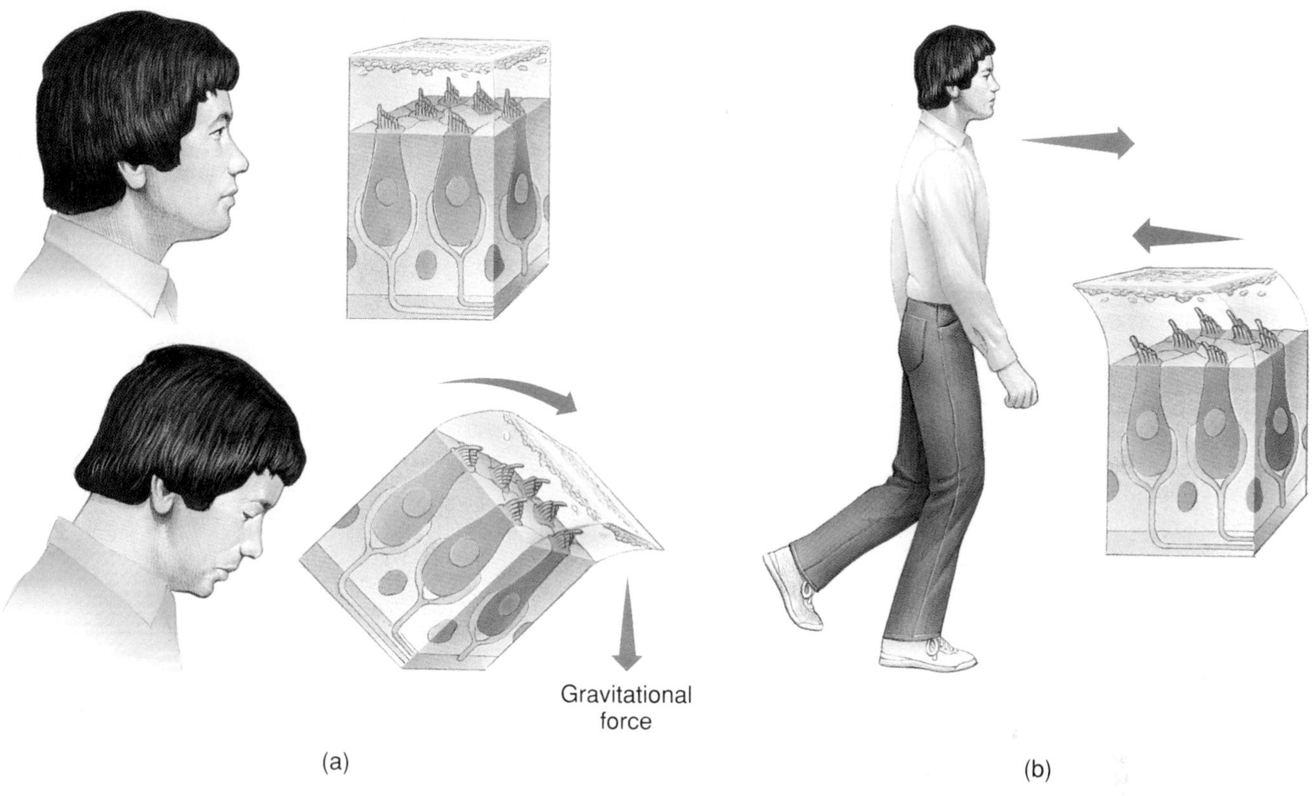

called *otoliths*. The otoliths make the otolithic membrane heavier than the endolymph within the utricle, and under the influence of gravity, the otolithic membrane exerts force on the hair cells.

Even when the head is upright, afferent neurons associated with the hair cells transmit a continuous series of nerve impulses to the brain by way of the vestibular division of the vestibulocochlear nerve. When the head position changes, the direction of force exerted by the otolithic membrane on the hair cells also changes, and the stereocilia of the hair cells are displaced from their original positions (Figure 16.38a). The displacement of the stereocilia alters the stimulation of the neurons associated with the hair cells, and a different pattern of nerve impulses is transmitted to the brain. For example, when the head is tilted to the left, the transmission of nerve impulses from the left utricle increases, and the transmission of impulses from the right utricle decreases. When the head is upright, it is possible to detect a change in head position of as little as one-half degree.

The hair cells of the macula also detect linear acceleration or deceleration. When the head undergoes linear acceleration, the inertia of the otolithic membrane causes it to lag behind the movement of the head. The lag of the membrane displaces the stereocilia of the hair cells, and an altered pattern of nerve impulses is transmitted to the brain. For example, the lag of the otolithic membrane that occurs when a person starts to move forward displaces the stereocilia of the hair cells to the rear (Figure 16.38b). This is similar to the displacement that would occur if the head was tilted backward so the face was oriented upward. If the linear movement of the head continues at a constant velocity, the inertia of the membrane is overcome, and it eventually moves at the same rate as the head is moving. During linear deceleration, the momentum of the otolithic membrane causes it to continue to move as the head slows and stops. The continued movement of the membrane also displaces the stereocilia of the hair cells and alters the pattern of nerve impulses transmitted to the brain. Thus, the hair cells of the macula of the utricle detect linear accleration or deceleration, but they provide little information about linear motion at a constant velocity.

◆ **FIGURE 16.39 The semicircular canals and ducts**
(a) The ampullae of the ducts have been cut open to show the locations of the receptors (cristae) within them. (b) Enlargement of a crista.

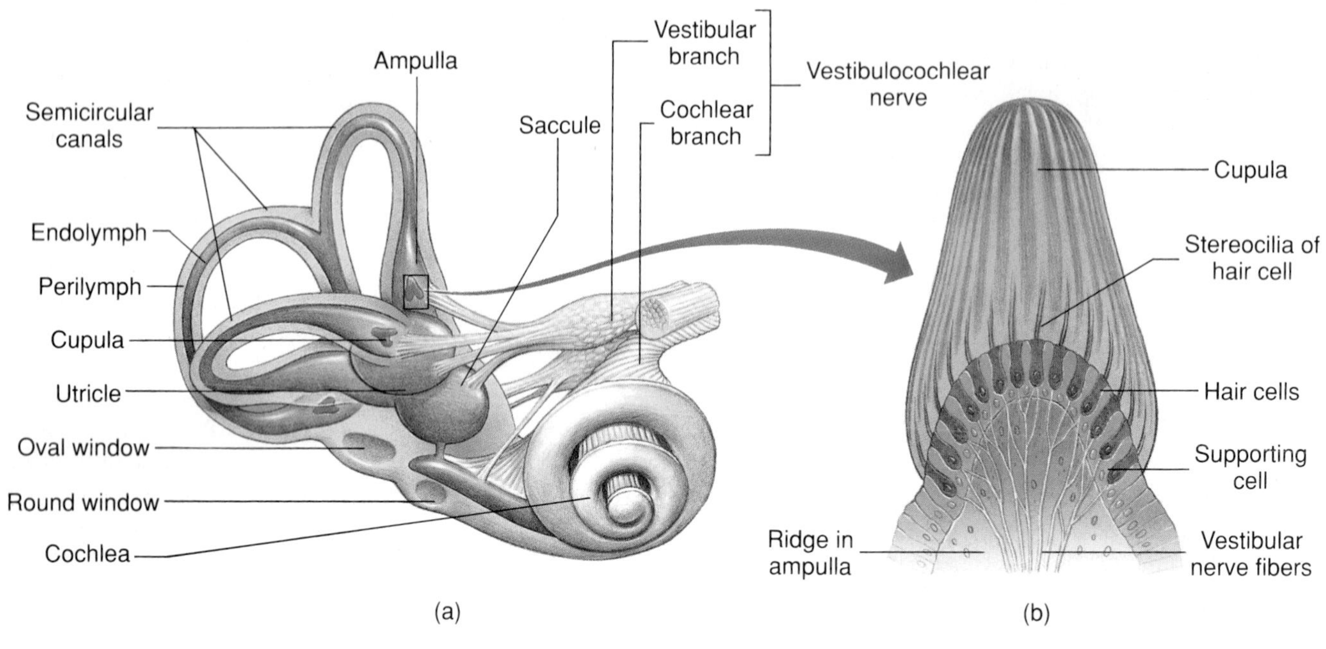

Function of the Saccule

The receptor cells of the saccule are organized into a macula like that of the utricle, and the saccule is generally considered to function in a similar manner as a component of the vestibular apparatus.

Function of the Semicircular Ducts

The semicircular ducts detect rotational (angular) acceleration or deceleration of the head (such as would occur when the head is suddenly turned to one side). The semicircular ducts are filled with endolymph, and the ampulla of each duct contains a receptor structure called a **crista ampullaris,** or **crista** (Figure 16.39). A crista is composed of hair cells, which resemble the hair cells of a macula, and supporting cells. The stereocilia of the hair cells project into a gelatinous mass called the **cupula.** Because each semicircular duct is oriented differently, the stereocilia of the hair cells within at least one semicircular duct are displaced by rotational acceleration or deceleration in any given plane.

When the head undergoes rotational acceleration, the inertia of the endolymph causes it to lag behind the movement of the semicircular duct. In effect, this response is equivalent to a flow of endolymph in a direction opposite to the movement of the duct—that is, opposite to the direction of the head's rotation. The lag of the endolymph alters the force exerted on the cupula within the ampulla of the duct and displaces the cupula and the stereocilia of the hair cells from their original position. The displacement of the stereocilia alters the stimulation of neurons associated with the hair cells, and a different pattern of nerve impulses is transmitted to the brain. If the rotation of the head continues at a constant velocity, the inertia of the endolymph is overcome, and it eventually moves at the same rate as the semicircular duct is moving. Under these conditions, the cupula returns to its original position, and the transmission of afferent nerve impulses returns to the original pattern within about 20 seconds.

During rotational deceleration, the momentum of the endolymph causes it to continue to move as the semicircular duct slows and stops. In effect, this response is equivalent to a flow of endolymph within the semicircular duct in the direction of rotation. The continued movement of the endolymph also alters the force exerted on the cupula within the ampulla of the duct and displaces the cupula and the stereocilia of the hair cells so that a different pattern of nerve impulses is transmitted to the brain. Thus, the hair cells of the cristae detect rotational acceleration or deceleration, but they provide little information during periods of rotation at a constant velocity.

Nystagmus

Nystagmus *(nis-tag´-mus)* is a characteristic movement of the eyes that is associated with their ability to remain fixed on the same point despite changes in the position of the head. A type of nystagmus called *vestibular nystagmus* occurs even when the eyes are closed. Vestibular nystagmus is produced when the cupulae and the stereocilia of the hair cells within the ampullae of the semicircular ducts are displaced. For example, if the head undergoes horizontal rotational acceleration (for example, the person suddenly spins around and around), the eyes move relatively slowly in a direction opposite to the direction in which the head is turning (as would occur if the eyes were fixed on some object in the environment). After the eyes have turned far to the side, they jump rapidly back to their forward position, and the movements are repeated. When the head undergoes horizontal deceleration (for example, the person suddenly stops spinning), nystagmus also occurs, but the directions of the eye movements are opposite to those that took place during acceleration (that is, the slow drift of the eyes occurs in the direction in which the rotation took place). These opposite directions of eye movements occur because, during deceleration, the cupulae and the stereocilia of the hair cells of the cristae are displaced in a direction opposite to that which took place during acceleration; therefore, a different pattern of nerve impulses is transmitted to the brain.

Another type of nystagmus, called *optokinetic nystagmus,* is produced when visual images move across the retinas of the eyes. For example, when the head moves horizontally in relation to the visual scene (such as when a person looks out of a window of a moving train), the eyes fix on an object and follow it, moving slowly until they have turned far to the side. The eyes then jump rapidly back to their forward position and fix upon a new object, and the movements are repeated. Optokinetic nystagmus does not occur when the eyes are closed.

Maintenance of Equilibrium

Neural signals from the vestibular apparatus of the ears are transmitted to the vestibular nuclei in the brain stem and to the cerebellum. The vestibular nuclei also receive signals from receptors in the eyes, skin, muscles, and joints. These signals contribute to the maintenance of equilibrium. They also contribute to the maintenance of desired posture, to the perception of motion and orientation, and to the control of eye movements so that the eyes remain fixed on the same point despite changes in the position of the head (Figure 16.40).

It is difficult to assign specific degrees of importance to the different types of information employed in maintaining equilibrium. Visual information is important, but so are other types of information. For example, receptors in the neck provide information about the position of the head relative to the body. This information is important because the vestibular apparatus of the ears provides information only about the position or movement of the head itself. Thus information from the neck receptors helps the system determine if signals from the vestibular apparatus are due to the movement of the head alone (such as bending the head to one side), in which case the equilibrium of the body may not be endangered, or to the movement of the whole body (such as tilting the entire body to one side), in which case the equilibrium of the body may be endangered and adjustments may be necessary to prevent a fall (Figure 16.41). Signals from the various receptors are integrated within the central nervous system to provide information about the position of the body and its parts in space, and this information helps determine the degree of activity in individual muscles that is required to maintain balance.

Smell (Olfaction)

The receptors for smell (olfaction) are specialized to respond to volatile chemicals that have dissolved in the mucous coating of the nasal cavity. The olfactory receptors are contained in a region of the nasal mucosa called the **olfactory epithelium** that is located in the upper portion of the nasal cavity on either side of the nasal septum (Figure 16.42, page 528). The olfactory epithelium contains three types of cells: (1) **olfactory receptor cells,** which are bipolar neurons; (2) **supporting cells,** which surround the receptor cells and secrete mucus that coats the epithelium; and (3) **basal cells,** which develop into receptor cells as the "old" receptor cells die. Olfactory receptor cells have a life span of about 60 days. They are the only neurons that are known to be replaced throughout adult life.

The receptive portion of an olfactory receptor cell terminates as a knob from which several long cilia called **olfactory hairs** protrude. The cilia are embedded in the mucous layer covering the olfactory epithelium. The afferent nerve fibers of the receptor cells pass through the cribriform plate of the ethmoid bone and constitute the olfactory nerve (cranial nerve I). These afferent nerve fibers synapse in the **olfactory bulbs** of the brain with neurons that form the **olfactory tracts** (Figure

◆ **FIGURE 16.40 Sources of neural signals that contribute to the maintenance of equilibrium**

The signals also contribute to the maintenance of desired posture, to the perception of motion and orientation, and to the control of eye movements.

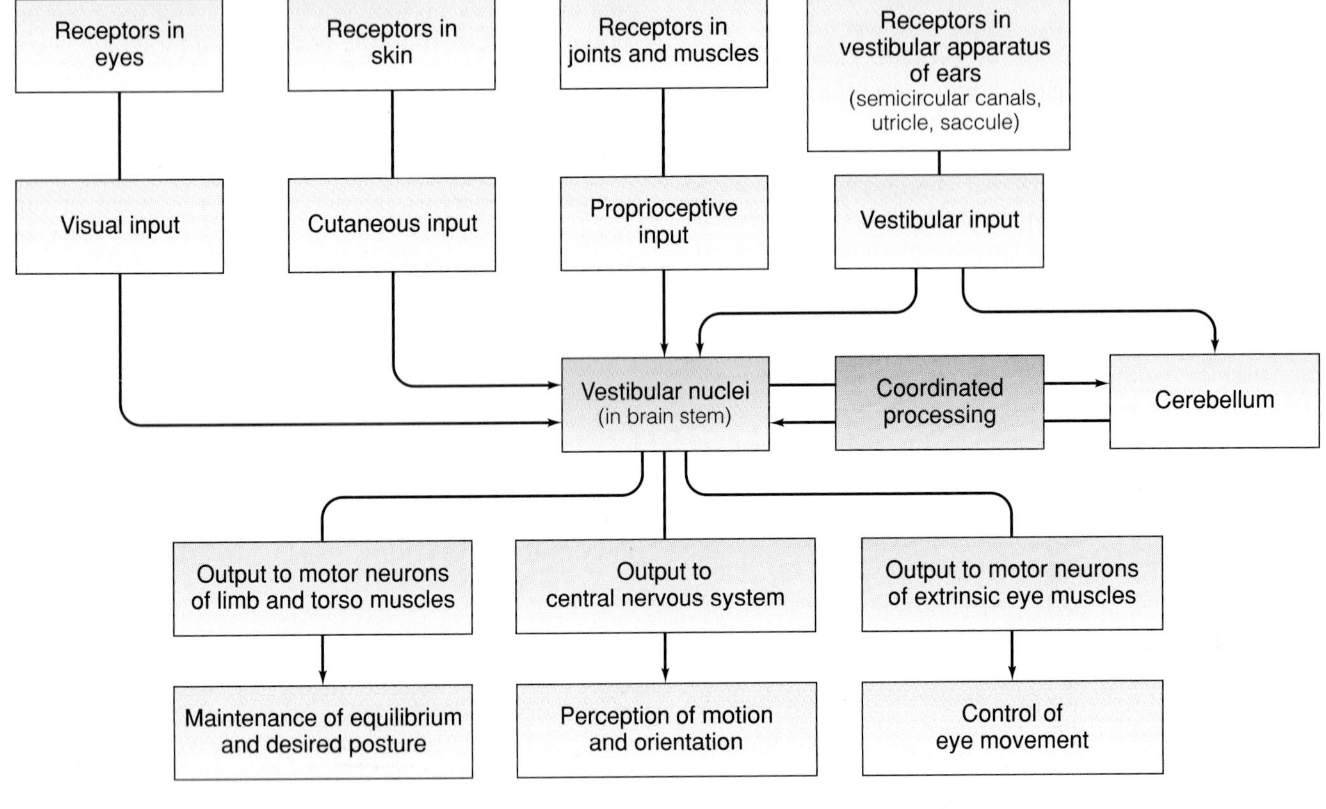

◆ **FIGURE 16.41 Signals from receptors in the neck and other body areas contribute to the maintenance of equilibrium**

Proprioceptive joint receptors in the neck (as well as receptors in other body areas) play an important role in maintaining equilibrium. They provide information to determine whether signals from receptors in the head such as the vestibular apparatus are due to the movement of the head alone or to the movement of the body as a whole.

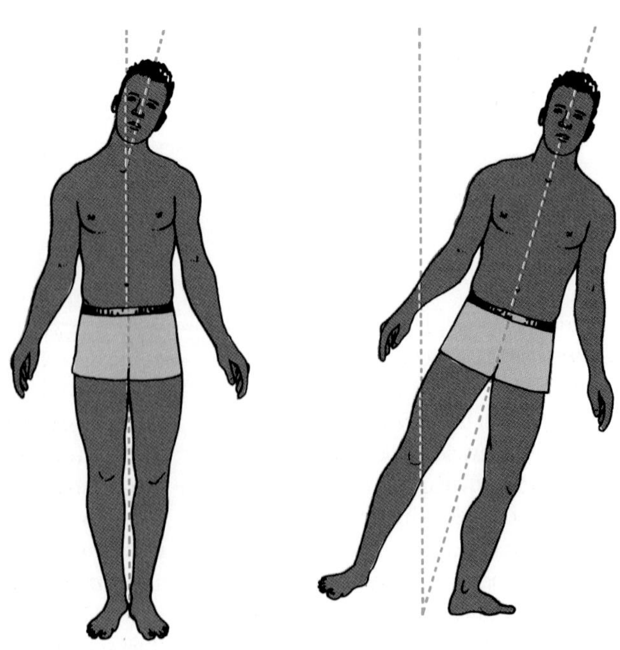

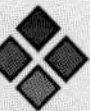

CONDITIONS OF CLINICAL SIGNIFICANCE

The Ear

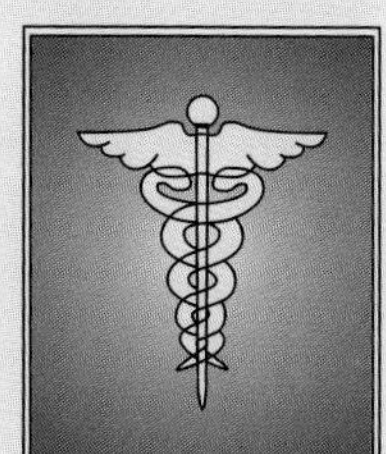

Middle-Ear Infections

Infections of the mucous membrane of the throat can travel through the auditory tube and cause an inflammation of the mucous membrane that lines the middle-ear cavity, including the inner surface of the tympanic membrane. The inflammation may cause fluid to collect within the middle ear, temporarily interfering with the ability to hear.

Deafness

Diseases of the ear or injuries to the ear can cause partial or total deafness. Such hearing losses can be classified as conductive deafness or as nerve deafness.

In *conductive deafness,* there is interference with the transmission of sound waves through the external or middle ear. The interference can be caused by a physical blockage of the external auditory canal by a foreign object or by cerumen (earwax), inflammation of the eardrum, adhesions between the ossicles, or thickening of the oval window. Neither the receptor cells of the spiral organ nor the nerve pathways are damaged in conductive deafness. Therefore, hearing aids that transmit sound waves to the inner ear through the bone of the skull rather than through the middle ear can aid a person suffering from this condition.

In *nerve deafness,* the loss of hearing results from disorders that affect the sound receptors in the inner ear, the vestibulocochlear nerve, or nerve pathways or centers within the central nervous system. Such disorders can be caused by a number of factors, including infections, tumors, and trauma. Hearing aids may not be helpful in cases of nerve deafness in which the hearing loss is due to destroyed receptor cells or nerve pathways.

Effects of Aging

Several age-related changes can contribute to the impairment of hearing. As the activity of sweat glands within the external auditory canal diminishes with age, the wax secreted by the ceruminous glands becomes drier and tends to accumulate. This buildup of earwax, by itself, is often a cause of hearing loss in older persons, especially in the low-frequency range.

Most hearing loss in older persons, however, is due to changes that occur in inner-ear structures. Due in part to a thickening of the walls of capillaries supplying the cochlea, the cells of the spiral organ and sensory ganglia often experience age-related degeneration. A decrease in the number of nerve fibers within both divisions of the vestibulocochlear nerve in persons over 45 years of age has also been reported. Thus, it is not uncommon for older persons to experience problems with balance and equilibrium as well as hearing.

16.43). Some of the neurons in the olfactory tracts pass to the thalamus and on to the olfactory regions (medial sides of the temporal lobes) of the cerebral cortex where the conscious perception and fine discrimination of smell occur. Other neurons in the olfactory tracts pass to the limbic system where they coordinate certain behavioral and emotional responses to particular smells. Thus, we spend money on perfumes and deodorants in an attempt to affect the responses of those around us.

For an odorous substance (an *odorant*) to be detected, it must reach the olfactory receptor cells in the upper portion of the nasal cavity. The movement of odorous substances into this region is aided by sniffing. Once in the region of the receptor cells, an odorous substance must dissolve in the mucous layer that covers the cells and interact with them. The interaction of an odorous substance with an olfactory receptor cell is believed to depolarize the cell, resulting in a generator potential that can initiate nerve impulses that are conveyed back to the brain, where they are interpreted as a particular smell.

Tens of thousands of different odors can be discriminated, but the basis for this discrimination is largely unknown. One theory is that the membrane of the cilia of a receptor cell possesses several different types of receptor sites where the molecules of various odorous

◆ **FIGURE 16.42 Location and structure of the olfactory receptors**

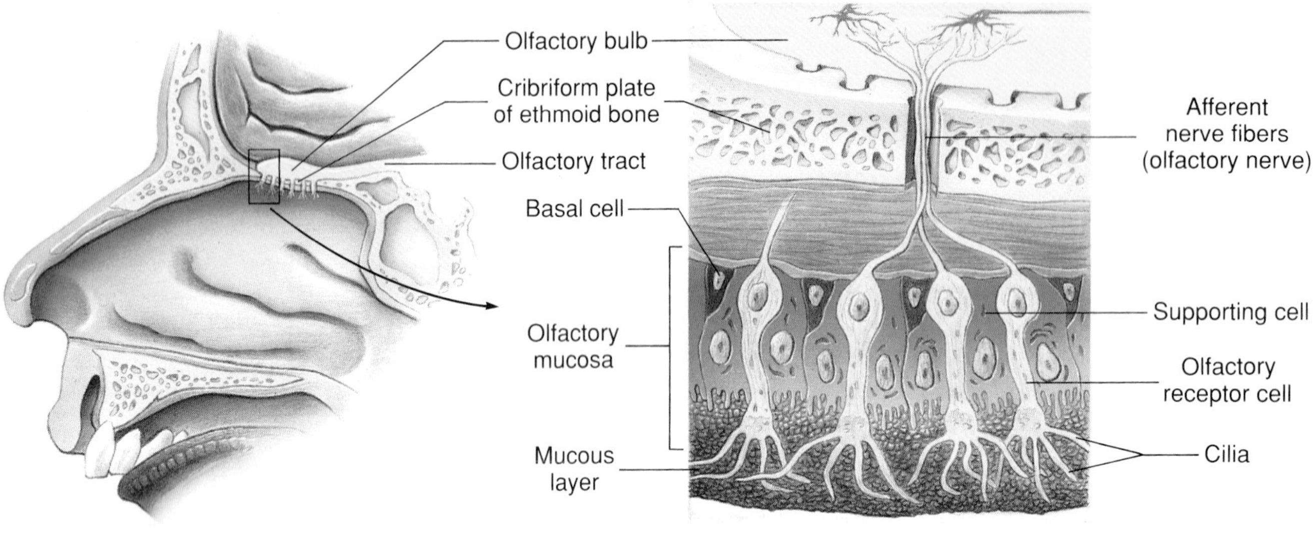

substances can interact and stimulate the depolarization of the cell. Although any one site is thought to be able to interact with a wide variety of odorous molecules, it responds best to molecules of particular sizes, polarities, and shapes. Thus, a particular molecule may interact well at some receptor sites and not as well at others. The better the interaction of the odorous molecule with a specific receptor site, the greater will be the depolarization resulting from the interaction. The depolarizations that occur at the different receptor sites of a single receptor cell are believed to be additive, resulting in a generator potential that determines the firing rate of the receptor cell. Different receptor cells are believed to have different combinations and concentrations of specific types of receptor sites. Hence, a given odorous substance may cause a simultaneous but differential stimulation of different receptor cells. The system is thus provided with the ability to discriminate many different odors.

The olfactory system adapts rapidly, and a person's sensitivity to a new odor quickly diminishes after a short period of exposure to it. The olfactory receptors them-

◆ **FIGURE 16.43 Olfactory pathways**

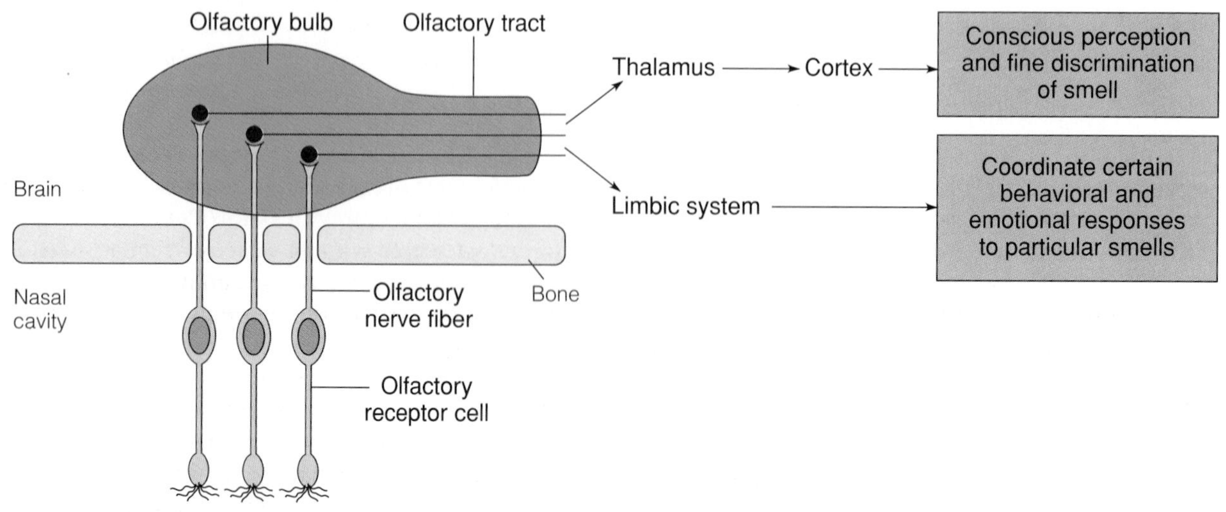

◆ **FIGURE 16.44 Taste buds**
(a) Structure of a taste bud. (b) Photomicrograph of a taste bud.

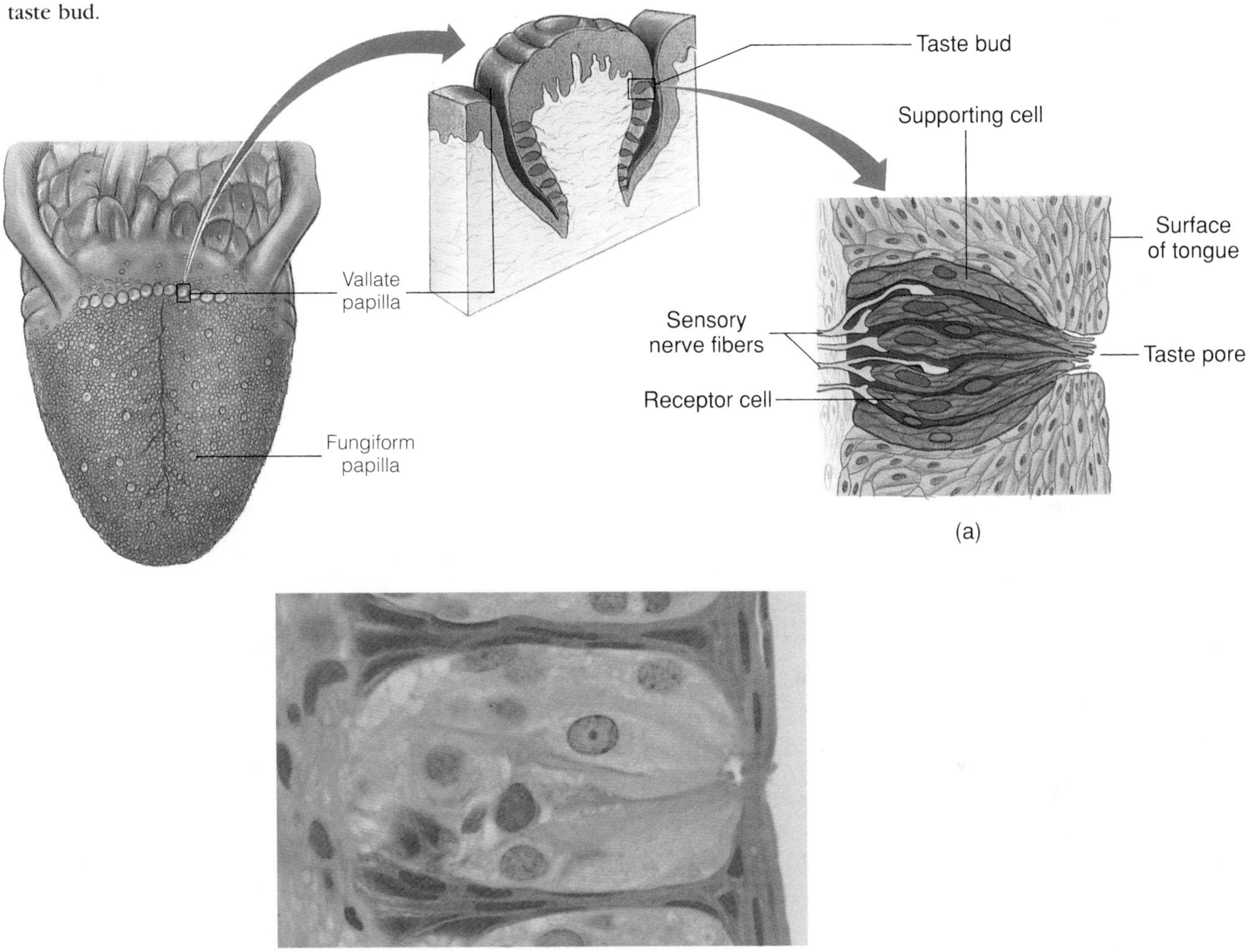

selves are slowly adapting, and the adaptation of the olfactory system appears to be due, at least in part, to inhibitory processes that occur in the central olfactory pathways.

Taste (Gustation)

The receptors for taste (gustation), like those for smell, are responsive to chemicals in solution. Taste receptors are specialized cells situated in structures called **taste buds.** Most of the approximately 10,000 taste buds are located on the surface and papillae of the tongue, but some are present on the roof of the mouth and in the pharynx and larynx. Each taste bud contains about 50 **receptor cells** surrounded by **supporting cells.** Although these names imply a specific functional role for each cell type, it is generally thought that the supporting cells are readily capable of differentiating into receptor cells. The receptor cells have a life span of only about one week, and new receptor cells must constantly be formed.

The free surfaces of the receptor cells contain microvilli that extend into a small opening, called a **taste pore,** at the surface of the taste bud (Figure 16.44). The microvilli greatly increase the surface area of the receptor cells, and their position in the taste pore allows them to be bathed by the fluids of the mouth.

Sensory nerve fibers synapse with the receptor cells of the taste buds. A single nerve fiber may synapse with several different receptor cells, and a single receptor cell may have several different nerve fibers synapsing with it. These fibers travel to the brain stem in the facial nerve (from the anterior two-thirds of the tongue), the glossopharyngeal nerve (from the posterior third of the tongue), and the vagus nerve (a few fibers from the pharyngeal region). The axons in these cranial nerves synapse in the brain stem with interneurons that travel through the medial lemnisci to the thalamus. In the thalamus the interneurons synapse with neurons that are conveyed to the tongue area of the somatosensory cortex in the postcentral gyrus of the parietal lobe.

If a substance is to evoke a taste sensation, it must dissolve in the fluids that bathe the tongue and interact with the receptor cells of the taste buds. Four primary taste sensations have traditionally been identified—sweet, salty, bitter, and sour, with each taste being detected best in specific regions of the tongue (Figure 16.45). It appears, however, that taste receptor cells do not have a corresponding specificity. A given taste receptor cell can interact with and respond to a variety of different substances that belong to more than one of the specific taste categories.

Taste receptor cells have been postulated to possess several different types of receptor sites on their microvilli that can form loose combinations with different types of molecules. The formation of these combinations gives rise to a depolarization of the receptor cell (that is, to a receptor potential), which leads to the generation of nerve impulses in the sensory nerve fibers that synapse with the receptor cell. Thus, the sensation of a specific taste is a complex phenomenon that involves the relative activities and firing patterns of a number of different sensory neurons from a variety of different taste receptor cells.

Much of what is commonly considered to be "taste" actually involves the stimulation of olfactory receptors. This fact is particularly evident when a person has a cold that produces congestion of the nasal mucosa that blocks olfactory stimulation. Under such conditions, the person often reports an inability to "taste" food, even though his or her actual taste system continues to function normally.

◆ **FIGURE 16.45 Areas of the tongue that are most sensitive to particular taste sensations**

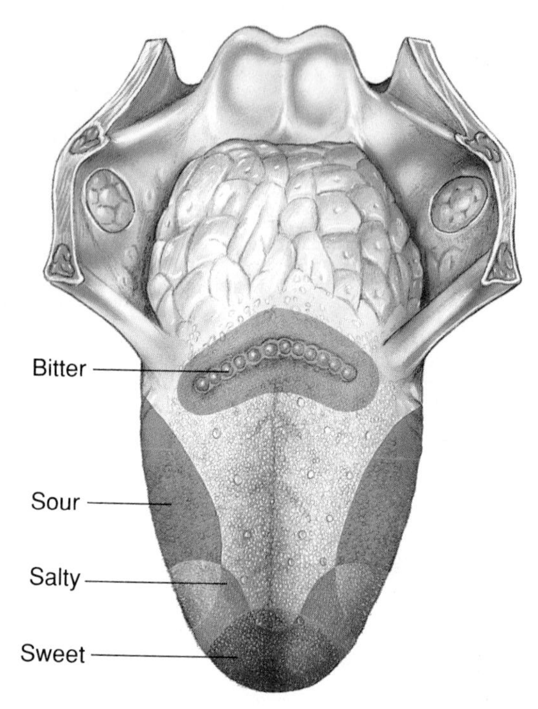

Study Outline

◆ **THE EYE—VISION** pp. 488-509

Embryonic Development of the Eye.

1. Outgrowths from lateral diencephalon form optic vesicles; each invaginates, forming optic cup, which becomes retina.
2. Optic stalk becomes incorporated into optic nerve.
3. Ectoderm over optic cup thickens into lens placode; placodes develop into lens vesicles that later separate from ectoderm, eventually forming lens of adult eye.
4. Cornea forms from cells superficial to lens vesicles.
5. Loose mesenchymal cells form fibrous tunic and vascular tunic of eye.

Structure of the Eye.

FIBROUS TUNIC. Outermost layer.

1. *Sclera.* Posterior; white; dense connective tissue; protects eye and maintains its shape.
2. *Cornea.* Anterior; clear; allows for light passage.

VASCULAR TUNIC. Beneath fibrous tunic.

1. *Choroid.* Posterior; dark; contains many blood vessels.
2. *Ciliary body.* Around edge of cornea; contains smooth ciliary muscles; ciliary processes produce aqueous humor.
3. *Iris.* Anterior; thin muscular diaphragm; pigmentation responsible for eye color.
4. *Pupil.* Opening in center of iris through which light passes.

RETINA. Innermost layer of eye (internal tunic) consists of outer pigmented layer and inner nervous-tissue layers.

1. *Pigmented layer.* Composed of epithelial cells in contact with choroid.
2. *Nervous-tissue layers.* Contain photoreceptors—rods and cones.
3. *Optic nerve.* Area of exit from eye is blind spot (optic disc).

4. *Macula.* Located at posterior pole of eye; at its center is fovea centralis with high concentration of cone photoreceptors.

LENS. Aids in focusing light onto retina; relatively elastic, biconvex structure.

CAVITIES AND HUMORS. Lens separates eye into two cavities: anterior cavity (contains aqueous humor) and posterior cavity (contains vitreous humor).

Accessory Structures of the Eye.

EYELIDS. Anterior, skin-covered structures; provide protection and light screening; controlled by skeletal muscles (orbicularis oculi and levator palpebrae superioris).

CONJUNCTIVA. Mucous membrane lining that covers inside of eyelids and anterior surface of eye.

LACRIMAL APPARATUS. Includes lacrimal gland and associated ducts; gland produces tears.

EXTRINSIC EYE MUSCLES. Lateral, medial, superior, and inferior rectus muscles; inferior and superior oblique muscles. Rapid acting; among the most precisely controlled skeletal muscles.

Light. Portion of electromagnetic spectrum to which eye receptors are sensitive.

Optics. Light rays are bent (refracted) when striking a biconvex lens; light rays converge at focal point behind lens.

Focusing of Images on the Retina.

EMMETROPIA. Lens under tension appears relatively flat; eye focuses parallel light rays from distant points (farther than 20 feet) sharply onto retina.

ACCOMMODATION. Focusing adjustments for viewing objects closer than 20 feet; ciliary muscles contract and permit greater lens curvature; pupil constriction eliminates divergent rays; convergence of eyes.

NEAR AND FAR POINTS OF VISION. Near point of vision varies with age—close to the eye when young, recedes with age; far point of vision: any object beyond 20 feet for normal eye can be focused onto retina without accommodation.

Control of Eye Movements. Movements of eyes are synchronized so both eyes move in a coordinated manner when viewing an object.

VOLUNTARY FIXATION MOVEMENTS. Controlled by small area of premotor regions of frontal lobes of cerebral cortex; move eyes voluntarily to find an object and fix on it.

INVOLUNTARY FIXATION MOVEMENTS. Controlled by occipital region of cortex; keep image of object focused on foveal portions of retinas.

Binocular Vision and Depth Perception. Retinal disparity enhances person's ability to judge relative distances when objects are nearby.

DIPLOPIA (DOUBLE VISION). May occur when light rays from an environmental object do not fall on corresponding points on retinas.

STRABISMUS. Improper coordination of eye movements. Eyes may alternate in fixing on environmental object, or one eye may be used all the time and other eye may become functionally blind.

Photoreceptors of the Retina. Rods and cones contain photopigments. A photopigment consists of a protein called an opsin to which a chromophore molecule called retinal is attached. Opsins differ from pigment to pigment. Light alters chemical configuration of retinal, and retinal breaks away from the opsin, leading to changes in polarity of receptor cells containing photopigment.

RODS. Responses indicate degrees of brightness, but rod system is characterized by relative lack of color discrimination; contain rhodopsin; numerous in peripheral retina.

CONES. Primarily responsible for color vision; three different cone types: red, green, blue; cones are concentrated in center of retina and fovea.

Neural Elements of the Retina.

1. Rods and cones synapse with bipolar cells, which synapse with ganglion cells.
2. Horizontal cells and amacrine cells are involved in lateral interactions that occur within retina.

Visual Pathways. Ganglion cell axons from medial half of each retina cross over at optic chiasma; optic tracts to thalamus; optic radiations to visual cortex.

Processing of Visual Signals. Begins in retina and continues throughout visual pathway. Components of pathway transmit coded message about certain aspects of image on retina.

Light and Dark Adaptation. Rods and cones are able to adapt and maintain their responsiveness over a wide range of light intensities.

Pupillary Light Reflex. Pupils constrict in bright light and dilate in dim light.

Visual Acuity. Ability of eye to distinguish detail—greatest in fovea; light rays stimulate different receptor units, with an unstimulated or weakly stimulated unit between them.

◆ CONDITIONS OF CLINICAL SIGNIFICANCE: THE EYE pp. 507–509

Myopia. Nearsightedness.

1. Light rays from distant objects are focused in front of retina.
2. Near point of vision closer than normal, far point of vision closer than 20 feet.
3. Corrected by concave lenses.

Hypermetropia. Farsightedness.

1. No far point of vision; near point farther away than normal.
2. Refraction increased with convex lenses.

Astigmatism. Unequal curvatures of portions of refraction system of eye. Light rays entering eye in one plane focus at a point different from that of light rays entering eye in another plane.

Cataract. Cloudiness or opaqueness of lens of eye or portion thereof.

Glaucoma. Abnormal elevation of intraocular pressure.

Color Blindness. Functional deficiency or absence of certain cones.

Effects of Aging.

1. Lens becomes yellow due to exposure to ultraviolet rays.
2. Potential for development of cataracts.
3. Pupil is less able to dilate, resulting in reduction of light reaching photoreceptors.
4. Poor drainage of aqueous humor increases likelihood of glaucoma.
5. Debris from rods and cones may cause loss of visual acuity.
6. Lens loses elasticity, leading to presbyopia.

◆ THE EAR—HEARING AND HEAD POSITION AND MOVEMENT pp. 510–525

Embryonic Development of the Ear.

1. Plate of thickened ectoderm called otic placode invaginates forming otic pit; otic pit separates from surface ectoderm to become otic vesicle; eventually forms membranous labyrinth of inner ear.
2. Part of first pharyngeal pouch forms eustachian tube.
3. Mesenchymal cells in middle-ear cavity form three ossicles.
4. Branchial groove external to first pharyngeal pouch forms external auditory canal.
5. Tympanic membrane develops from membrane separating first pharyngeal pouch and branchial groove.

Structure of the Ear.

EXTERNAL EAR.

1. *Auricle (pinna).* Directs sound waves.
2. *External auditory canal (meatus).* Contains hairs and cerumen; serves as resonator.

MIDDLE EAR. Air chamber in temporal bone; extends from tympanic membrane to oval window.

1. *Auditory tube.* Regulates pressure.
2. *Ossicles (malleus, incus, stapes).* Transmit and amplify vibrations of tympanic membrane to oval window.

INNER EAR. Series of fluid-filled (perilymph) canals called osseous labyrinth that contain membranous labyrinth, which is filled with endolymph.

1. *Vestibule.* Saccule and utricle; associated with sense of balance.
2. *Semicircular canals.* Equilibrium.
3. *Cochlea.* Portion of inner ear for hearing; spiral organ.

Mechanisms of Hearing.

SOUND WAVES. Sound energy travels through air in the form of pressure waves, consisting of regions of compression (air molecules are close together and pressure is relatively high) alternating with areas of rarefaction (molecules are farther apart and pressure is lower).

SOUND WAVES AND THE LOUDNESS, PITCH, AND TIMBRE OF SOUNDS.

1. Loudness of a sound is related to intensity (or power) of sound wave, which in turn, is related to amplitude of wave—that is, to pressure difference between a zone of compression and a zone of rarefaction.
2. Pitch of a sound is related to frequency of sound wave.
3. Timbre, or quality, of a sound is related to presence of additional sound-wave frequencies superimposed on a fundamental frequency.

TRANSMISSION OF SOUND WAVES TO THE INNER EAR. Sound wave enters external auditory canal and causes tympanic membrane to vibrate. Vibrations of tympanic membrane are transmitted across middle-ear cavity by malleus, incus, and stapes to oval window of inner ear.

FUNCTION OF THE COCHLEA. Pressure waves generated by movement of stapes against oval window cause basilar membrane to vibrate near base of cochlea, initiating a wave that travels along basilar membrane toward helicotrema. Vibration of basilar membrane alters polarity of hair cells of spiral organ (located on membrane), leading to generation of nerve impulses that are transmitted to the brain and interpreted as sounds. High-frequency sound waves cause maximal vibration of basilar membrane near base of cochlea; low-frequency sound waves cause maximal vibration near apex of cochlea.

AUDITORY PATHWAYS. Auditory signals from each ear are transmitted to both sides of the cortex, with synapses in the medulla and thalamus.

DETERMINATION OF PITCH. Ability to determine pitch of most sounds is related to fact that sound waves of a specific frequency give rise to a particular pattern of vibration of basilar membrane in which a specific region of the membrane vibrates more intensely than other regions. This pattern of vibration gives rise to particular pattern of nerve impulses that are transmitted to brain and interpreted as a sound of particular pitch.

DETERMINATION OF LOUDNESS. High-intensity sound waves cause greater vibration of basilar membrane than do low-intensity sound waves; the brain detects this difference and interprets greater vibration as louder sound.

SOUND LOCALIZATION.

1. Auditory system is able to detect differences in sound-wave arrival time at each ear.

2. Auditory system is able to detect differences in sound-wave intensity at each ear.

THE STAPEDIUS AND TENSOR TYMPANI MUSCLES. An involuntary contraction of the stapedius muscle occurs in response to loud external sounds. Involuntary contractions of the stapedius muscle and probably also the tensor tympani muscle occur in association with a person's own vocalizations.

CENTRAL NERVOUS SYSTEM INFLUENCES ON THE SPIRAL ORGAN. An increase in the activity of efferent neurons to the spiral organ can reduce its sensitivity and make it less sharply tuned in particular portions of the audible frequency range.

Head Position and Movement. Vestibular apparatus of inner ear provides information about head position and movement. Information is utilized in coordination of movements that maintain body equilibrium. Signals from inner ear also contribute to the maintenance of desired posture, to the perception of motion and orientation, and to the control of extrinsic eye muscles so eyes can remain fixed on the same point despite change in head position.

FUNCTION OF THE UTRICLE. Provides information about position of head with respect to gravity and about linear acceleration or deceleration of head.

MACULA. Receptor structure composed of hair cells and supporting cells. Hair-cell stereocilia project into gelatinous otolithic membrane containing calcium carbonate. Stereocilia are displaced and stimulation of neurons is altered by changing position of head or by linear acceleration or deceleration.

FUNCTION OF THE SACCULE. Considered to function like the utricle as component of vestibular apparatus.

FUNCTION OF THE SEMICIRCULAR DUCTS. Detect rotational acceleration or deceleration. Crista-receptor structure composed of hair cells and supporting cells. Hair-cell stereocilia project into gelatinous mass called the cupula. Stereocilia are displaced and stimulation of neurons is altered by rotational acceleration or deceleration.

NYSTAGMUS. Characteristic eye movement associated with ability of eyes to remain fixed on same point despite changes in head position.

Vestibular nystagmus. Produced when cupulae and stereocilia of hair cells within ampullae of semicircular ducts are displaced.
Optokinetic nystagmus. Produced when visual images move across retinas.

Maintenance of Equilibrium. Neural signals from various receptors, including the vestibular apparatus of the ears, provide information about position and movement of the body or its parts. Neural signals are integrated within CNS and contribute to maintenance of equilibrium.

◆ SMELL (OLFACTION) pp. 525–529

1. Receptors are neurons in epithelium of nasal mucosa in upper nasal cavity.
2. Odorous substances must dissolve in mucous layer that covers receptors, and must interact with receptors.
3. Different receptors may have different combinations and concentrations of specific types of receptor sites.

◆ CONDITIONS OF CLINICAL SIGNIFICANCE: THE EAR p. 527

Middle-Ear Infections. Inflammation of mucous membrane of middle ear.

Deafness.

CONDUCTIVE DEAFNESS. Interference with sound-wave transmission by physical blockage, inflammation of eardrum, ossicle adhesions, or oval window thickening.

NERVE DEAFNESS. Loss of hearing that involves damage to sensory cells or nerve pathways.

Effects of Aging.

1. Buildup of earwax.
2. Degeneration of cells in spiral organ and sensory ganglia.
3. Decrease in number of nerve fibers in both divisions of the vestibulocochlear nerve.

◆ TASTE (GUSTATION) pp. 529–530

1. Receptors are in taste buds on tongue, roof of mouth, pharynx, larynx.
2. Much of what is commonly considered to be "taste" actually involves stimulation of olfactory receptors.

Self-Quiz

1. The retina develops from the embryonic: (a) telencephalon; (b) diencephalon; (c) mesencephalon.
2. The iris belongs to the: (a) fibrous tunic; (b) internal tunic; (c) vascular tunic.
3. The anterior cavity of the eye is separated from the posterior cavity by the: (a) lens; (b) iris; (c) cornea.
4. The aqueous humor is produced by the: (a) cornea; (b) ciliary processes that project from the ciliary body; (c) vitreous body.
5. The extrinsic eye muscles are the least rapid acting but among the most precisely controlled skeletal muscles in the body. True or False?
6. The greater the deviation from the perpendicular at which light rays enter a block of glass: (a) the greater the wavelength of the light rays; (b) the less the bending of the light rays; (c) the greater the bending of the light rays.
7. During accommodation for viewing near objects: (a) the pupils constrict; (b) the ciliary muscles relax; (c) the eyes diverge.
8. Those eye receptors most responsible for vision in dim light are: (a) cones; (b) rods; (c) ossicles.
9. Visual acuity is greatest in the: (a) fovea centralis; (b) peripheral retina; (c) ciliary body.
10. The condition in which the intraocular pressure is abnormally elevated is called: (a) strabismus; (b) astigmatism; (c) glaucoma.
11. The loss of lens elasticity that occurs with aging is called: (a) presbyopia; (b) myopia; (c) hyperopia.
12. The middle ear contains the: (a) cochlea; (b) semicircular canals; (c) incus.
13. A membranous partition between the middle-ear chamber and the scala vestibuli of the inner ear is formed by the: (a) oval window; (b) round window; (c) fenestra cochleae.
14. The cochlea is the portion of the inner ear associated with hearing. True or False?
15. The spiral organ is located on the: (a) tympanic membrane; (b) basilar membrane; (c) vestibular membrane.
16. The loudness of a sound is related primarily to which aspect of a sound wave? (a) amplitude; (b) frequency; (c) pitch.
17. The endolymph within at least one semicircular duct is generally affected by rotational acceleration in any given plane. True or False?
18. Receptor cells of the semicircular ducts of the inner ear are best able to detect: (a) linear motion of the head; (b) rotational motion of the head at a constant velocity; (c) rotational acceleration.
19. A characteristic movement of the eyes that occurs while rotational acceleration of the body is taking place is termed: (a) dizziness; (b) nystagmus; (c) vertigo.
20. Sensory nerve fibers associated with taste travel to the brain in the: (a) glossopharyngeal nerve; (b) vagus nerve; (c) both.

CHAPTER 17

The Endocrine System

CHAPTER CONTENTS

LEARNING OBJECTIVES

After completing this chapter, you should be able to:

1. Describe two ways that hormones exert their effects at the cellular level.
2. List the hormones of the major endocrine glands, and describe their effects.
3. Explain how the release of pituitary hormones is influenced by the nervous system.
4. Explain the application of negative-feedback control to endocrine regulation.
5. Describe the synthesis, storage, and release of the thyroid hormones.
6. Describe how the adrenal medulla is controlled.
7. Describe four factors that influence the release of aldosterone.
8. Cite the symptoms of adrenal cortical hypofunction and hyperfunction, and relate them to the effects of the adrenal cortical hormones.
9. Distinguish between the roles of glucagon and insulin in controlling blood-glucose levels.
10. Cite the symptoms of diabetes mellitus, and relate them to the effects of the hormone insulin.

CHAPTER 17

There are two major controlling systems in the body: the nervous system and the endocrine system. The nervous system is a rapidly responding system that regulates the activities of muscle and secretory cells by means of nerve impulses and neurotransmitter molecules. The **endocrine system** is a slower responding system that influences the activities of virtually every cell in the body by means of chemical messengers called *hormones.* Endocrine-mediated responses are often more prolonged than neurally mediated responses, and some endocrine-mediated responses may last hours or even days. A number of the specific activities influenced by the endocrine system are listed in Table 17.1.

This chapter provides an overview of the activities of the endocrine system and the major endocrine glands. The activities of individual endocrine structures are considered further in those chapters dealing with the specific body processes they influence.

 Basic Endocrine Functions

In general, **endocrine glands** are ductless structures that contain epithelial cells specialized for the synthesis and secretion of specific hormones. The hormones are released from the cells into the surrounding extracellular spaces. From the extracellular spaces, the hormones enter the bloodstream, which transports them throughout the body. (In contrast, exocrine glands typically secrete their products by way of ducts onto body surfaces—either external surfaces like the skin, or internal surfaces like the lining of the digestive tract.) The major endocrine glands of the body are shown in Figure 17.1. Besides the clearly recognized endocrine glands such as the thyroid gland and the adrenal glands, other structures—for example, the gastrointestinal tract—also exhibit endocrine activity.

◆ **TABLE 17.1 Representative Activities Regulated or Influenced by the Endocrine System**

Intracellular and extracellular fluid volumes
Acid-base balance
Carbohydrate, protein, lipid, and nucleic acid metabolism
Food and fluid intake
Digestion, absorption, and nutrient distribution
Reproduction and lactation
Immune processes
Blood-cell formation
Blood pressure and blood flow
Resistance to stress
Adaptation to environmental change

◆ **FIGURE 17.1 The major endocrine glands**

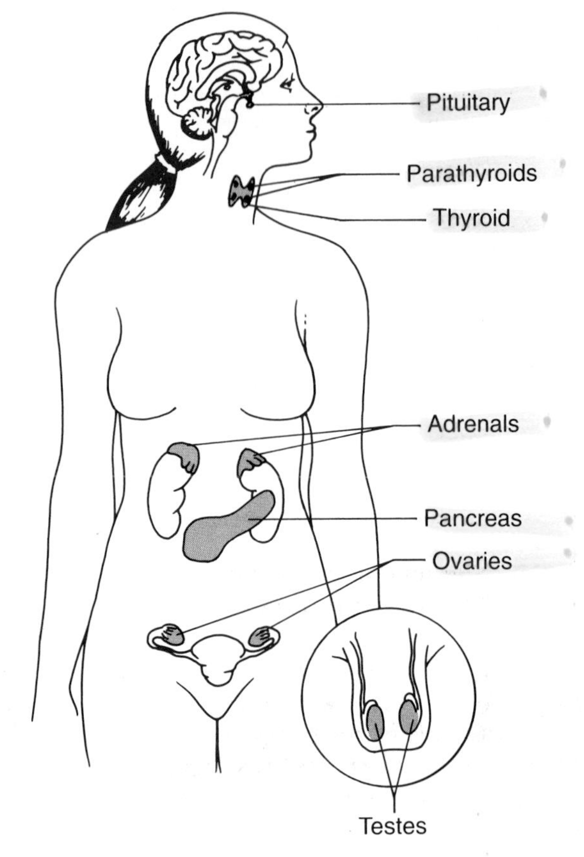

Hormones

Hormones are highly potent, specialized organic substances that function as intercellular signaling molecules. Many hormones are derivatives of amino acids. Included among these hormones are the ***protein hormones*** (for example, prolactin from the pituitary gland), the ***peptide hormones*** (for example, antidiuretic hormone from the pituitary gland), and the ***catecholamine hormones*** (for example, epinephrine and norepinephrine from the medulla of the adrenal gland). Other hormones are synthesized from cholesterol. Among these hormones are the ***steroid hormones*** (for example, testosterone from the testes, estrogens and progesterone from the ovaries, and cortisol and aldosterone from the cortex of the adrenal gland). Table 17.4, which is located at the end of the chapter, provides a summary of the major endocrine glands and the hormones they produce.

Transport of Hormones

Within the bloodstream, many hormones—particularly those that are not very water soluble, such as the steroid hormones and the thyroid hormones—bind to specific carrier proteins. This binding is reversible, and the free and bound forms of a hormone exist in equilibrium with one another.

The reversible association between a hormone and a carrier protein provides both a storage system and a buffer system for the hormone. Usually, only a small fraction of the hormone is present in the free form, and the protein-bound portion is a readily available reserve. The binding of a hormone to a carrier protein can protect the body against excessive concentrations of the hormone, and it can also delay the degradation of the hormone. In addition, the binding can protect certain small hormone molecules from being excreted by the kidneys.

Mechanisms of Hormone Action

The bloodstream transports hormones throughout the body. However, a particular hormone does not necessarily affect all the body's cells. Only certain cells, called **target cells,** respond to any given hormone. The target cells of a particular hormone possess receptors to which molecules of the hormone can bind.

The receptors for some hormones are part of the plasma membrane of a cell. The receptors for other hormones are located within a cell.

Binding of Hormones to Plasma-Membrane Receptors

A number of hormones—particularly those that are not very lipid soluble, such as the protein, peptide and catecholamine hormones—bind to receptors that are part of the plasma membrane of a cell. When a hormone molecule binds to a specific membrane receptor, the receptor undergoes a conformational change, and a series of events is triggered that influences such cellular activities as the rates of certain chemical reactions or the permeability of the plasma membrane, leading ultimately to the observed effects of the hormone on body processes. The specific events that take place when a hormone molecule (or some other signaling molecule) binds to a plasma-membrane receptor were discussed in Chapter 3 (page 70).

Binding of Hormones to Receptors within Cells

Some hormones—particularly those that are relatively lipid soluble, such as the steroid hormones and thyroid hormones—bind to receptors located within cells. In some cases, the receptors are located primarily within the cytoplasm, and in other cases they are located primarily within the nucleus. When a hormone molecule binds to one of these receptors, the conformation of the receptor is altered, and the receptor is activated (Figure 17.2). The activated receptor, in turn, binds to certain nucleotide sequences of the cell's DNA. In most cases, this binding activates particular genes, which leads to the synthesis of mRNA and ultimately to the production of proteins (for example, enzymes) that influence cellular reactions or processes. In some cases, however, the binding suppresses the activity of particular genes.

Hormonal Interrelationships

There are numerous interrelationships between different endocrine glands and hormones. In fact, almost every activity influenced by the endocrine system is affected by groups of hormones acting either together or in sequence. For example, hormones secreted by the pituitary gland, the thyroid gland, the adrenal glands, and the pancreas all influence carbohydrate metabolism.

In some cases, one hormone affects the activity of a second hormone. For example, the first hormone may

◆ **FIGURE 17.2 Hormones that bind to receptors located within cells activate (or, in some cases, suppress) particular genes within the cells**

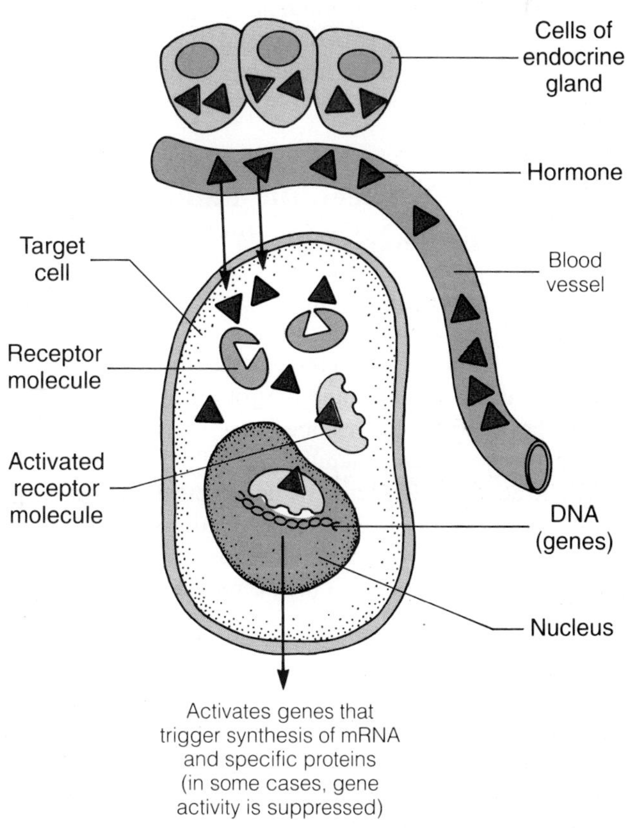

◆ **FIGURE 17.3 Stages in the embryonic development of the pituitary gland**
See text for detailed discussion.

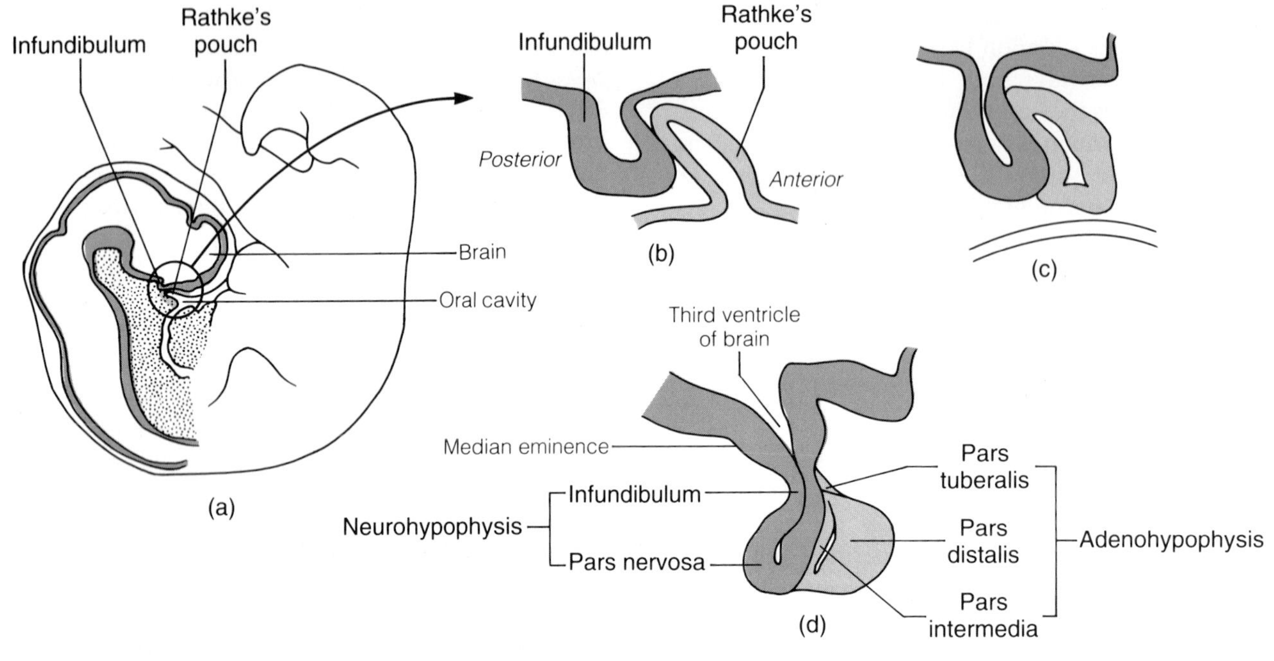

increase or decrease the number or affinity of the target-cell receptors for the second hormone, thereby increasing or decreasing the ability of the second hormone to influence its target cells.

A number of hormones influence the rates of production of other hormones. As discussed in subsequent sections, the pituitary gland produces several hormones that act in this fashion.

Relationship between the Endocrine System and the Nervous System

The regulatory role of the endocrine system is similar to that of the nervous system. In fact, the endocrine and nervous systems are very closely related. Hormones influence neural functions, and neurotransmitters contribute to the control of hormone secretion. Moreover, in certain cases a single chemical substance can act as a hormone at some sites and as a neurotransmitter or neuromodulator at others. The close relationship between the endocrine system and the nervous system is evident in the control and function of the pituitary gland.

Pituitary Gland

The hormones of the **pituitary gland,** or **hypophysis,** regulate several other endocrine glands and affect a number of diverse body activities.

Embryonic Development and Structure

The pituitary gland is located beneath the brain and is surrounded by the *sella turcica* of the sphenoid bone. The opening of the sella turcica is covered over by a ring-shaped fold of dura mater called the *diaphragm sellae.* The pituitary develops embryonically from two different ectodermal regions: the floor of the brain and the roof of the mouth (Figure 17.3).

Neurohypophysis

The portion of the pituitary known as the **neurohypophysis** *(noo-ro-hi-pof´-i-sis)* is an outgrowth of nervous tissue from the floor of the brain in the region of the hypothalamus. The fully developed neurohypophysis consists of a structure called the **pars nervosa (lobus nervosus),** which is connected to the brain by a stalklike **infundibulum.** Where the infundibulum connects the pars nervosa to the brain, there is a small elevation called the *median eminence* (Figures 17.3 and 17.4).

Adenohypophysis

The second portion of the pituitary arises from ectodermal tissue from the roof of the mouth (Figure 17.3b). An outpouching from the mouth called **Rathke's pouch** grows toward the developing neurohypophysis. (There is evidence that, during development, Rathke's

◆ **FIGURE 17.4 The pituitary gland**

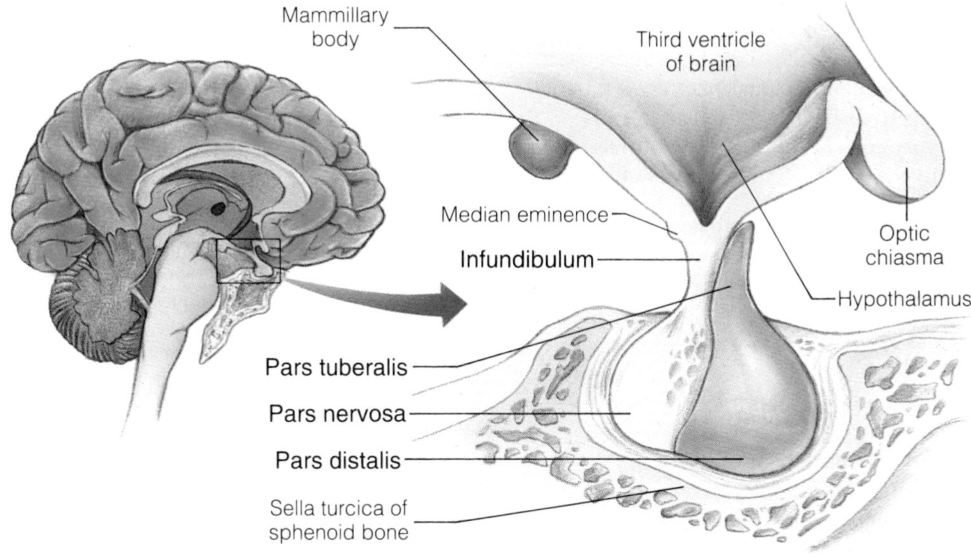

pouch is invaded by neural cells, which are believed to contribute to endocrine functions.) As it meets the neurohypophysis, Rathke's pouch loses its connection with the mouth (Figure 17.3c). The portion of the pituitary derived from Rathke's pouch becomes the **adenohypophysis** *(ad-e-no-hi-pof´-i-sis)* of the fully developed pituitary (Figure 17.3d). The adenohypophysis includes three regions called the **pars distalis, pars tuberalis,** and **pars intermedia.** The pars intermedia, however, is virtually nonexistent in the pituitary of the adult human, although it is present in most other vertebrates and in human fetuses.

Table 17.2 summarizes the terminology used to identify the divisions and subdivisions of the pituitary gland. Note that the pituitary gland may also be divided into an **anterior lobe,** which includes the pars distalis and pars tuberalis, and a **posterior lobe,** which includes the pars nervosa and, when present, the pars intermedia.

Relationship to the Brain

There is a close relationship between the pituitary and the nervous system. The neurohypophysis is connected to the brain by way of the infundibulum. The adenohypophysis is also closely associated with the brain. Within the median eminence of the lower hypothalamic region of the brain is a capillary bed known as the *primary capillary plexus* (Figure 17.5). This plexus receives blood from the internal carotid arteries and the cerebral arterial circle (see page 655). Blood in the primary capillary plexus flows downward along the stalk of the pituitary by way of *hypothalamic-hypophyseal portal veins* directly to a *secondary capillary plexus* surrounding the secretory cells of the adenohypophysis. Thus, the brain and the adenohypophysis are linked by the circulatory system.

Neurohypophyseal Hormones and Their Effects

The pars nervosa releases two peptide hormones: antidiuretic hormone (ADH) and oxytocin (Table 17.4).

Antidiuretic Hormone

Antidiuretic hormone (ADH), also called **vasopressin,** promotes the reabsorption of water from the urine-forming structures of the kidneys and thus helps retain fluid within the body. At moderate to high concentrations, ADH causes blood vessels called aterioles to constrict. ADH is a peptide hormone composed of nine amino acids.

◆ **TABLE 17.2 Terminology of the Pituitary Gland**

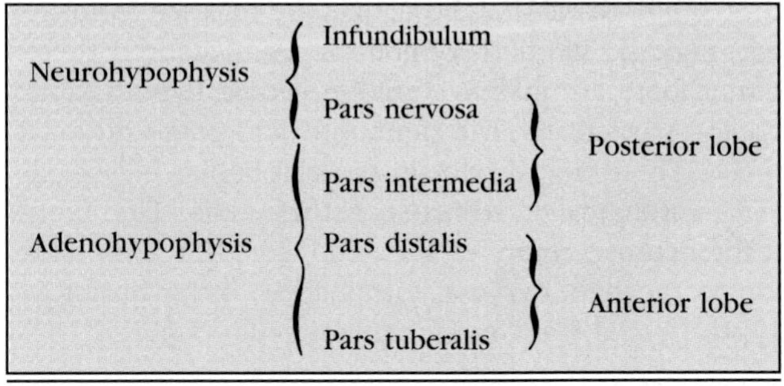

Division	Subdivision	Lobe
Neurohypophysis	Infundibulum	
	Pars nervosa	Posterior lobe
Adenohypophysis	Pars intermedia	Posterior lobe
	Pars distalis	Anterior lobe
	Pars tuberalis	Anterior lobe

◆ **FIGURE 17.5 Relationships between the brain and the pituitary gland**

A capillary bed known as the primary capillary plexus is located within the median eminence of the lower hypothalamic region of the brain. Blood in the primary capillary plexus flows by way of hypothalamic-hypophyseal portal veins to a secondary capillary plexus surrounding the secretory cells of the adenohypophysis. The axons of certain hypothalamic neurosecretory cells form the hypothalamic-hypophyseal tract as they extend along the infundibulum to the pars nervosa. The axons of other neurosecretory cells form the tuberoinfundibular tract as they pass through the tuber cinereum of the hypothalamus to the region of the primary capillary plexus within the median eminence.

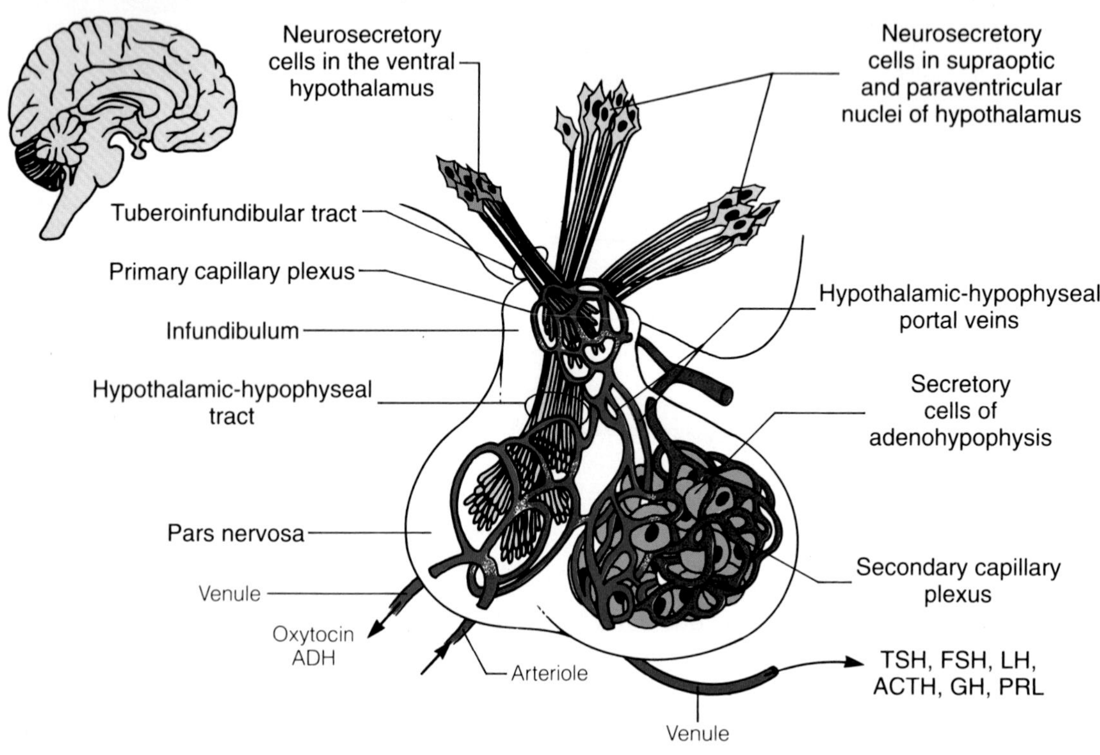

Oxytocin

Oxytocin stimulates the smooth muscle of the uterus. It also promotes the contraction of myoepithelial cells surrounding the saclike alveoli of the mammary glands, resulting in the ejection of milk during breast-feeding. The functions of oxytocin in males have not as yet been fully defined. Like ADH, oxytocin is a peptide hormone composed of nine amino acids.

Release of Neurohypophyseal Hormones

Although both ADH and oxytocin are released from the pars nervosa, neither hormone is manufactured there. Rather, both are manufactured in certain hypothalamic nuclei—particularly, the supraoptic and paraventricular nuclei. These nuclei contain the cell bodies of specialized neurons called **neurosecretory cells.** The axons of the neurosecretory cells extend along the infundibulum to the pars nervosa, forming the **hypothalamic-hypophyseal tract** in the infundibulum (Figure 17.5).

ADH and oxytocin are synthesized in the cell bodies of the neurosecretory cells within the hypothalamic nuclei. The hormones are packaged within membrane-bounded vesicles, which are transported along the axons of the neurosecretory cells to the axon terminals in the pars nervosa. ADH and oxytocin are released from the axon terminals by exocytosis and enter the bloodstream. An individual neurosecretory cell releases either ADH or oxytocin, but not both.

In addition to synthesizing ADH and oxytocin, the neurosecretory cells also conduct nerve impulses. The nerve impulses trigger the release of the neurohypophyseal hormones.

Because ADH and oxytocin are synthesized in the hypothalamus by neurosecretory cells, the synthesis and release of these hormones can be influenced by neural inputs to the brain. Other neurons make synaptic contact with the neurosecretory cells, and information from appropriate receptors is neurally relayed to the brain to alter the production and release of ADH or oxytocin in accordance with the regulatory requirements of the body.

ASPECTS OF EXERCISE PHYSIOLOGY

The Endocrine Response to the Challenge of Combined Heat and Exercise

When one exercises in a hot environment, maintenance of plasma volume becomes a critical homeostatic concern. Exercise in the heat results in the loss of large amounts of fluid through sweating. Simultaneously, blood is needed for shunting to the skin for cooling and for increased blood flow to nourish the working muscles. There must also be adequate venous return to maintain cardiac output. The hypothalamus-posterior pituitary neurosecretory system responds to these multiple, conflicting needs for fluid by releasing water-conserving antidiuretic hormone (ADH) in an attempt to reduce urinary fluid loss to preserve plasma volume.

Studies have generally shown that exercise in heat stimulates ADH release, which results in decreased urinary fluid loss. In one study conducted during an 18-mile road march in heat, the participants' average urine output dropped to 134 ml (the normal urine output during the same time period would be about twice that much), while sweat loss averaged four liters. Overhydration prior to exercise appears to decrease the intensity of this response, suggesting that increased ADH release is related to plasma osmolarity. If fluid loss is not adequately replaced, plasma osmolarity increases. This hypertonic condition is thought to be detected by hypothalamic receptors called osmoreceptors, which then promote increased secretion of ADH from the posterior pituitary. Some investigators believe, however, that increased ADH release results from other factors, such as changes in blood pressure or in renal blood flow. Regardless of the mechanism, ADH release is an important physiologic response to exercise in heat.

Adenohypophyseal Hormones and Their Effects

The adenohypophysis of the pituitary gland produces and releases several hormones (Table 17.4).

Gonadotropins

Two of the adenohypophyseal hormones are called **gonadotropins** because they particularly affect the gonads (ovaries and testes). These hormones are **follicle-stimulating hormone (FSH)** and **luteinizing hormone (LH).**

In females, ova (eggs) develop within structures known as follicles, which are located in the ovaries. FSH stimulates follicle development and induces the secretion of female sex hormones called *estrogens.* In this activity, FSH works in conjunction with LH. Surging levels of LH together with FSH lead to ovulation and the formation of a structure called the corpus luteum from an ovarian follicle. The corpus luteum produces estrogens and another female sex hormone called *progesterone.*

Both FSH and LH are also present in males. FSH promotes the maturation of cells within the testes called sustentacular cells (Sertoli cells), which are involved in the development and maturation of sperm. LH, which in males is sometimes called **interstitial-cell stimulating hormone (ICSH),** stimulates the interstitial cells of the testes to produce the male sex hormones (that is, androgens such as testosterone). Since the male sex hormones are themselves involved in sperm production, LH can be regarded as important in this process as well. Both FSH and LH are glycoproteins (proteins combined with carbohydrate molecules).

Thyrotropin

Thyrotropin, also called **thyroid-stimulating hormone (TSH),** stimulates the synthesis and release of the thyroid hormones from the thyroid gland. Like FSH and LH, thyrotropin is a glycoprotein.

Adrenocorticotropin

Adrenocorticotropin, or **adrenocorticotropic hormone (ACTH),** stimulates the release of hormones from the cortical regions of the adrenal glands, particularly the glucocorticoid hormone cortisol, which influences carbohydrate, lipid, and protein metabolism. ACTH is a polypeptide hormone composed of 39 amino acids.

Growth Hormone

Growth hormone (GH), also called **somatotropin,** promotes growth in general, and the growth of the skeletal system in particular. The specific actions of growth hormone are considered later in the chapter (page 545). Growth hormone is a protein hormone consisting of 191 amino acids.

Prolactin

In females, **prolactin (PRL)** is involved in the initiation and maintenance of milk production. In males, there is evidence that prolactin potentiates the stimulatory effects of LH on interstitial cells of the testes, perhaps by inducing the production of LH receptors. Prolactin may also potentiate the effects of testosterone on many of its target cells.

Endorphins

In addition to the hormones just discussed, the adenohypophysis contains several forms of **endorphins,** which are peptides that have opiatelike analgesic effects. In fact, the largest of these peptides—beta endorphin—is a much more potent analgesic than morphine. Although endorphins appear to be produced principally in the adenohypophysis, they occur in the medulla oblongata and hypothalamus of the brain and at other sites.

Pro-opiomelanocortin and the Production of Adenohypophyseal Hormones

A number of the peptide and protein hormones of the adenohypophysis, as well as other peptide and protein substances, are produced initially as parts of large precursor molecules that are subsequently split by enzymes into fragments of precise length and biological activity. For example, cells in the adenohypophysis and also in the hypothalamus of the brain synthesize a protein molecule called *pro-opiomelanocortin* that contains the amino acid sequences of an enkephalin, an endorphin, and ACTH, as well as the sequence of a substance called gamma melanocyte-stimulating hormone (gamma MSH), which may be used as a neurotransmitter by certain hypothalamic cells. Thus, pro-opiomelanocortin is a precursor molecule from which ACTH and other substances can be formed.

It should be noted that not all cells containing a precursor like pro-opiomelanocortin necessarily secrete the same hormones or other substances. In some cells, a precursor may be split to produce one substance, and in others it may be split to form another substance. Moreover, in some cases, the same cell produces more than one substance from a precursor molecule.

Release of Adenohypophyseal Hormones

The brain exercises a good deal of control over the synthesis and release of the adenohypophyseal hormones.

Releasing and Inhibiting Hormones

Within the hypothalamus of the brain, specialized neurosecretory cells manufacture a group of molecules known collectively as **releasing** and **inhibiting hormones.** The axons of these neurosecretory cells form the *tuberoinfundibular tract* as they pass through the tuber cinereum of the hypothalamus to the median eminence. Within the median eminence, the neurosecretory cells release their products in the region of the primary capillary plexus (Figure 17.5). The releasing or inhibiting hormones enter the blood and travel by way of the hypothalamic-hypophyseal portal veins to the adenohypophysis, where they stimulate or inhibit the release of adenohypophyseal hormones.

The major releasing or inhibiting hormones are:

- **Gonadotropin-releasing hormone (Gn-RH),** which stimulates the release of both FSH and LH.
- **Thyrotropin-releasing hormone (TRH),** which stimulates the release of TSH and also prolactin.
- **Corticotropin-releasing hormone (CRH),** which stimulates the release of ACTH.
- **Growth-hormone-releasing hormone (GH-RH),** which stimulates the release of growth hormone.
- **Prolactin-releasing hormone (PRH),** which stimulates the release of prolactin. (Unlike TRH, PRH affects only prolactin.)
- **Somatostatin,** which inhibits the release of growth hormone and TSH.
- **Prolactin-inhibiting hormone (PIH),** which inhibits the release of prolactin and is actually the substance *dopamine* (see page 370). (In addition to dopamine, there may be another hypothalamic substance that inhibits the release of prolactin.)

Nervous System Effects on Adenohypophyseal Hormone Release

Regardless of whether a particular releasing or inhibiting hormone promotes or inhibits the release of certain adenohypophyseal hormones, the adenohypophyseal hormones are under the influence of the nervous system. For example, other neurons make synaptic contact with the neurosecretory cells that produce releasing or inhibiting hormones, and information neurally relayed to the brain can alter the secretion of these hormones. An altered secretion of releasing or inhibiting hormones, in turn, influences the release of adenohypophyseal hormones. In addition, since many adenohypophyseal hormones stimulate the release of still other hormones (for example, those of the thyroid, adrenals, and gonads), a pathway exists by which neural activity can ultimately influence a wide variety of endocrine functions. Even brain activities of the sort that are responsible for emotions can affect hormonally controlled events. For instance, emotional trauma can upset the menstrual cycle or lactation.

Negative-Feedback Control of Endocrine Activity

Depending on the particular gland or hormone, endocrine activity may be stimulated or inhibited by a variety of factors, such as stress, exercise, or the levels of certain inorganic ions or other chemical substances in the body. In many cases, negative feedback (see page 17) contributes to the control of endocrine function.

In feedback terms, the level at which a hormone is to be maintained is indicated by the *set point* for the hormone (much as the temperature at which a room is to be maintained is indicated by the setting on a thermostat). A factor that stimulates or inhibits the production of a particular hormone (for example, stress, exercise, or the levels of certain inorganic ions or other chemical substances) can be viewed as raising or lowering the set point for the hormone.

Negative feedback operates to maintain the level of a hormone at its particular set-point level. For example, the hypothalamic releasing hormone CRH causes the pituitary gland to secrete ACTH (Figure 17.6). ACTH, in turn, causes the adrenal glands to secrete the hormone cortisol. In addition to exerting its other physiological effects, cortisol provides negative feedback to the hypothalamus to inhibit CRH secretion and to the pituitary to inhibit ACTH release. If, for some reason, the cortisol level rises above its set-point level, the higher cortisol level provides increased negative-feedback inhibition of CRH and ACTH release. The increased inhibition of CRH and ACTH release leads to a decreased secretion of cortisol, which tends to reduce the cortisol level to its set-point level. On the other hand, if the cortisol level falls below its set-point level, the lower cortisol level provides decreased negative-feedback inhibition of CRH and ACTH release. The decreased inhibition of CRH and ACTH release leads to an increased secretion of cortisol, which tends to raise the cortisol level to its set-point level.

◆ FIGURE 17.6 Negative-feedback control of cortisol release

See text for detailed discussion.

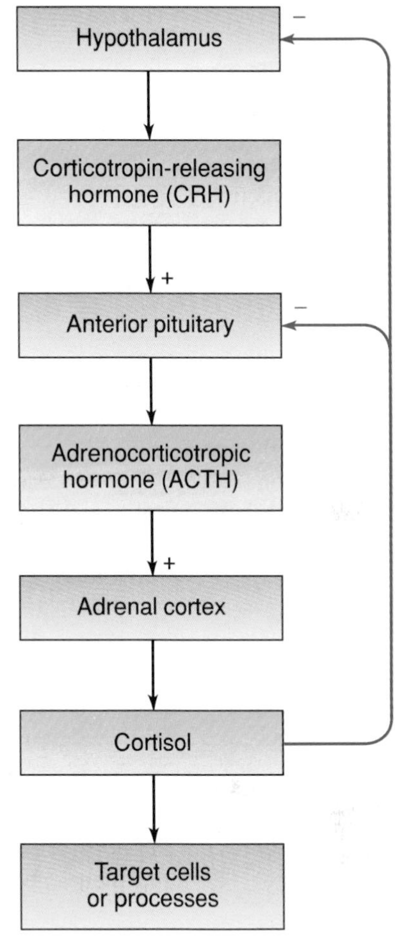

Actions of Growth Hormone

Ultimately, a person's maximum growth potential is genetically determined. However, the attainment of this potential depends on many factors, including proper nutrition, freedom from chronic disease, and a correct balance of hormones such as growth hormone, thyroid hormones, and sex hormones. Growth hormone is a particularly important stimulus for body growth, especially in young people.

A major growth-promoting effect of growth hormone is the stimulation of mitosis and cell division in many tissues. However, this effect is not due to the direct action of growth hormone itself. Rather, growth hormone stimulates the production by the liver and other tissues of a group of peptides called **somatomedins,** which include the substances *insulinlike growth factor I* and *insulinlike growth factor II.* The somatomedins (particularly, insulinlike growth factor I) stimulate mitosis and cell division.

CONDITIONS OF CLINICAL SIGNIFICANCE

Pituitary Gland

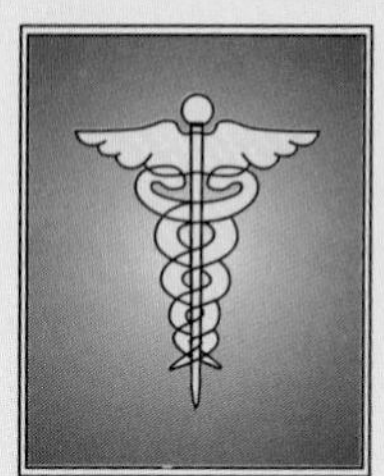

Disorders of the glands of the endocrine system generally result in either an underproduction or an overproduction of hormones, referred to as *hyposecretion* and *hypersecretion,* respectively. Pituitary malfunctions can be caused by disorders of the gland itself or by difficulties involving the releasing or inhibiting hormones from the brain. Pituitary disorders can involve the functioning of major segments of the gland, or they may be limited to disorders in the release of a single pituitary hormone.

Individual hormone disorders produce a number of observable conditions. Deficiencies of the gonadotropins upset normal gonadal function, and serious deficiencies can lead to gonadal inactivity in males and to the impairment or cessation of menstruation in females. TSH deficiency results in underactivity of the thyroid gland (hypothyroidism). Deficiencies of prolactin may result in a failure of lactation after a woman gives birth to a child, whereas prolactin overproduction can lead to lactation in a woman who has not recently given birth. ACTH deficiency leads to a deficient secretion of cortisol by the adrenal glands, whereas excessive ACTH results in an excess secretion of cortisol. A deficiency of growth hormone in the young can result in impaired growth and may produce pituitary dwarfs. An excess of growth hormone before the growth in length of the long bones is complete (that is, before the epiphyses have closed) results in increased height and gigantism. After this time, an excess secretion of growth hormone leads to acromegaly. In acromegaly, height does not increase as in gigantism, but bones thicken and soft tissues increase in size. Structures such as the jaw, hands, feet, and tongue enlarge, and there is a coarsening of the facial features. A deficiency of antidiuretic hormone results in diabetes insipidus, which is characterized by copious urination.

Growth hormone directly influences several aspects of protein, carbohydrate, and lipid metabolism:

1. Growth hormone enhances the entrance of amino acids into cells and stimulates their incorporation into protein. These actions, which favor growth, tend to decrease the concentrations of amino acids in the blood.
2. Growth hormone promotes glucose formation from liver glycogen. Although growth hormone may initially increase the rate of glucose uptake by adipose tissue, heart, and skeletal muscle cells, after some hours glucose uptake and utilization are reduced.
3. Growth hormone increases the release of fatty acids from adipose tissue into the blood, and it inhibits fat synthesis. The fatty acids can be taken up from the blood and used as energy sources by most cells.

Some researchers suggest that the influences of growth hormone on carbohydrate and lipid metabolism are important during periods of prolonged fasting. That is, the formation of glucose from liver glycogen and the reduction of glucose uptake and utilization by tissues such as skeletal muscle help keep glucose available for use by the brain, which requires glucose as an energy source. Moreover, the release of fatty acids from adipose tissue provides an alternative source of energy for tissues that have reduced their glucose utilization.

Although growth hormone produces its most dramatic effects during the period of body growth and development, it is secreted throughout life. In many instances, growth hormone acts synergistically with other hormones to enhance their effects. Other hormones also influence the activity of growth hormone. For example, thyroid hormones from the thyroid gland are necessary for the proper production and action of growth hormone.

Growth-hormone secretion exhibits a daily, or **circadian** *(sur-kay´-dee-un)* rhythm. When a person is awake, growth-hormone levels tend to be low and relatively constant. However, the secretion of growth hormone increases substantially about one hour after the onset of deep sleep and then falls over the next several hours.

The overall amount of growth hormone secreted is influenced by a number of factors (Figure 17.7). The release of growth hormone increases in response to declining concentrations of blood glucose and also in response to elevated blood levels of certain amino acids, particularly arginine. An increased secretion of growth hormone is associated with fasting, hypoglycemia (low blood-glucose levels), exercise, and certain kinds of stress.

In what are essentially negative-feedback effects, somatomedins and growth hormone itself apparently act

◆ **FIGURE 17.7 Control of growth-hormone secretion**

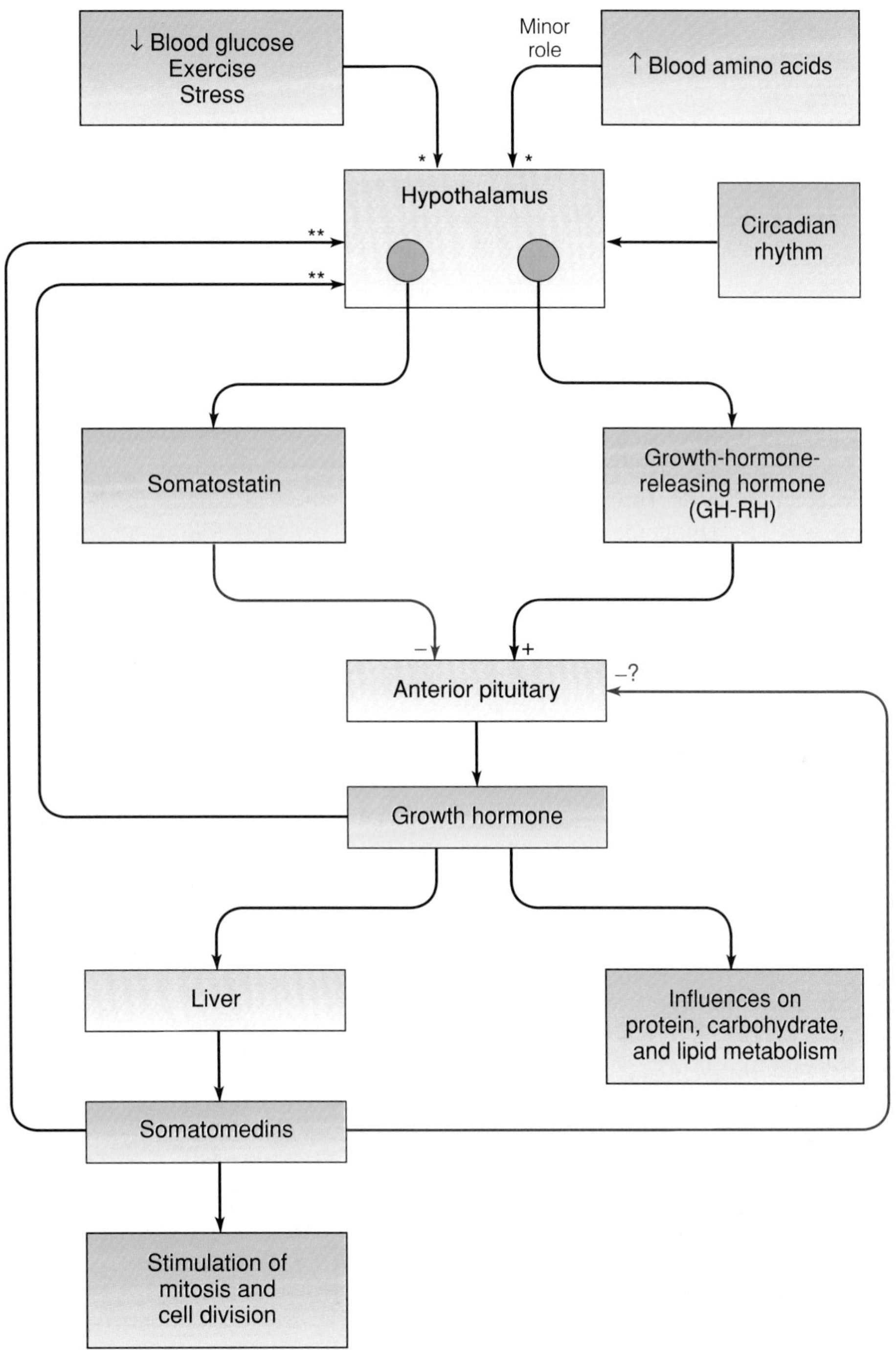

* These factors all increase growth-hormone secretion, but it is unclear whether they do so by stimulating GH-RH or inhibiting somatostatin, or both.

** These factors decrease growth-hormone secretion, but it is unclear whether they do so by inhibiting GH-RH or stimulating somatostatin, or both.

CLINICAL CORRELATION

Pituitary Gigantism

Case Report

THE PATIENT: A 19-year-old male.

PRINCIPAL COMPLAINT: Extreme growth (gigantism).

HISTORY: The patient has normal parents and siblings. He was normal himself until 10 years of age, when he began to experience abnormally rapid growth. The rate of growth continued at the 99th percentile for 2 years, and then remained between the 50th and 75th percentiles for 2 more years. At age 14 years, the concentration of growth hormone in the blood plasma was 113 nanograms (ng) per milliliter (normal: 10–20 ng/ml), and the level could not be increased by injection of insulin (which lowers the blood-glucose level) or decreased by injection of glucose. The fasting blood-glucose level was 110 milligrams (mg) per 100 ml (normal: 70–110 mg per 100 ml). The blood serum inorganic phosphorous was 8 mg per 100 ml (normal: 3–4.5 mg per 100 ml). In the initial period of rapid growth, the patient was strong and vigorous, but during the last 2 years he developed muscular weakness and atrophy. Specific weakness developed in his hands, making it difficult for him to hold a pencil and write or to use eating utensils.

CLINICAL EXAMINATION: At admission, all of the limb reflexes were markedly dulled. Cranial tomograms (sectional radiography) showed no significant abnormality of the sella turcica, and no signs of a tumor (such as impaired vision or increased intracranial pressure) were found. Nevertheless, on the basis of the increased levels of growth hormone in the blood plasma and its lack of susceptibility to stimulation or suppression, a pituitary tumor (microadenoma) was diagnosed.

TREATMENT: The pituitary gland was surgically removed, and tissue consistent with a tumor was also removed. Hormone-substitution therapy for pituitary insufficiency (due to removal of the gland) was initiated. Although the level of growth hormone in the blood plasma decreased considerably to 20–25 ng/ml, it was still slightly higher than normal. This level suggests that the tumor may not have been removed completely.

COMMENT: A pituitary tumor that produces one type of hormone often decreases production of other pituitary hormones by compressing pituitary tissue and destroying cells that produce other hormones. For example, the muscular weakness and atrophy that pituitary giants subsequently develop reflect adrenal cortical deficiency to some extent. Lack of secretion of pituitary gonadotropins delays puberty; hence the epiphyses do not close and bone growth continues. The menstrual cycles of adult women who develop such tumors often cease as a consequence of decreased FSH and LH. Neurological symptoms develop as abnormal growth of nerves causes compression in the channels through bones and joints. The weakness of the hands in this patient was caused by pressure on the median nerve in the space formed by the wrist bones and the transverse carpal ligament ("carpal tunnel syndrome").

OUTCOME: Although hormone-substitution therapy should bring about normal puberty and the subsequent limitation to linear growth, the continued high levels of growth hormone may produce some degree of acromegaly, and further surgery or radiation of the remaining tumor probably will be necessary.

on the hypothalamus to stimulate somatostatin release and/or inhibit GH-RH release. In either case, the effect is to reduce the release of growth hormone from the pituitary gland. Somatomedins probably also act directly on the pituitary to inhibit the stimulatory effects of GH-RH on growth-hormone secretion.

Thyroid Gland

The fully developed thyroid gland is among the largest of the body's endocrine organs. It is a butterfly-shaped structure located anterior to the upper part of the trachea, near its junction with the larynx.

◆ **FIGURE 17.8 Ventral view of the thyroid gland**

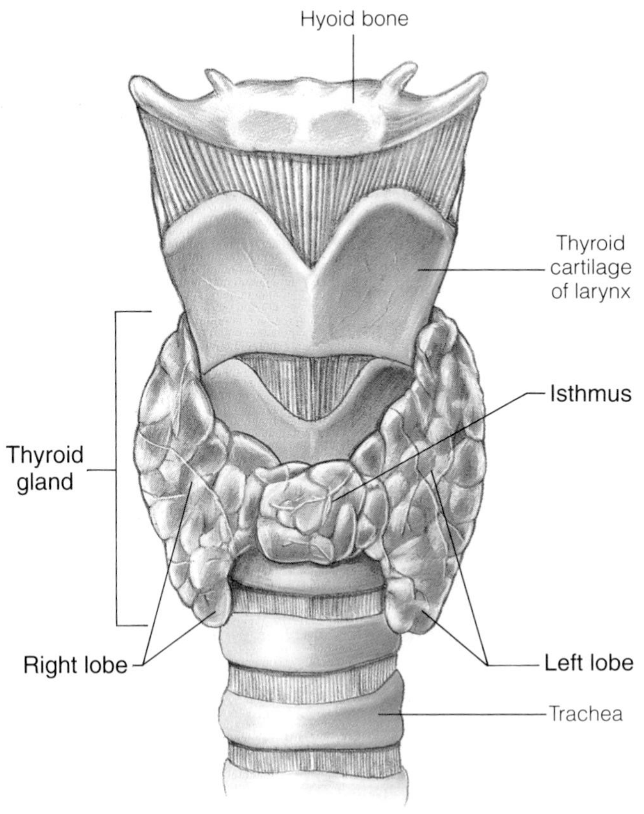

Embryonic Development and Structure

The **thyroid gland** originates as an epithelial thickening in the floor of the pharynx. The thickening grows outward from the pharynx, eventually loses its connection with the gastrointestinal tract, and comes to occupy a position around the trachea just below the larynx.

The thyroid gland is divided into right and left lobes that are joined across the trachea by a thin band called the **isthmus** (Figure 17.8). The gland is well vascularized, receiving blood from major arteries in the neck region.

The basic internal structure of the thyroid consists of hollow balls of cells called **follicles** that are bound together with connective tissue (Figure 17.9). The central region of each follicle contains a protein substance, the *colloid.* The colloid contains a stored form of thyroid hormones. The thyroid gland is different from all other endocrine glands in having extracellular storage sites—the colloid-containing regions of the follicles—for its hormones. The other endocrine glands store their hormones in the cells of the gland.

Production of the Thyroid Hormones: Triiodothyronine and Thyroxine

The thyroid hormones contain iodine and the amino acid tyrosine (Figure 17.10). Most of the body's iodine is obtained from the diet and appears in the blood as ionic iodide (I^-), which is actively taken up by thyroid follicle cells (Figure 17.11a). The iodide diffuses out of the cells into the colloid, and as it does, it is converted into an active form of iodine by enzymes located at the follicle cell–colloid border (Figure 17.11b).

The follicle cells synthesize a tyrosine-containing protein called **thyroglobulin,** which is packaged within membrane-bounded vesicles and secreted into the colloid by exocytosis (Figure 17.11c). As thyroglobulin is secreted, an iodine is attached to a tyrosine of thyroglobulin, forming monoiodotyrosine (MIT) (Figure 17.11d). A second iodine can also be attached to the tyrosine to form diiodotyrosine (DIT) (Figure 17.11e). Two diiodotyrosines are then coupled to form **tetraiodothyronine (T_4, or thyroxine)** (Figure 17.11f). Alternatively, a monoiodotryosine and a diiodotyrosine can combine to form **triiodothyronine (T_3)** (Figure 17.11g). (Coupling does not occur between two MITs.) The iodination and coupling processes produce thyro-

◆ **FIGURE 17.9 Photomicrograph of the thyroid gland**

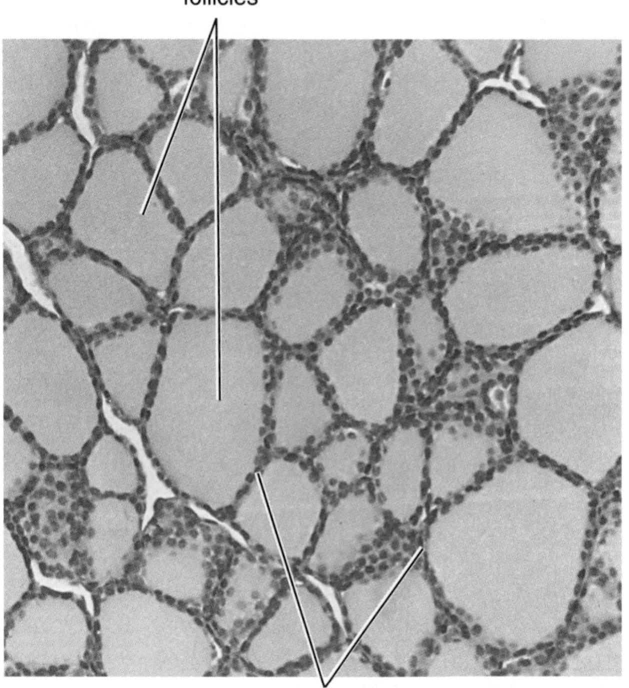

◆ **FIGURE 17.10 Structures of the thyroid hormones and substances important in their synthesis**

The thyroid hormones are triiodothyronine and tetraiodothyronine.

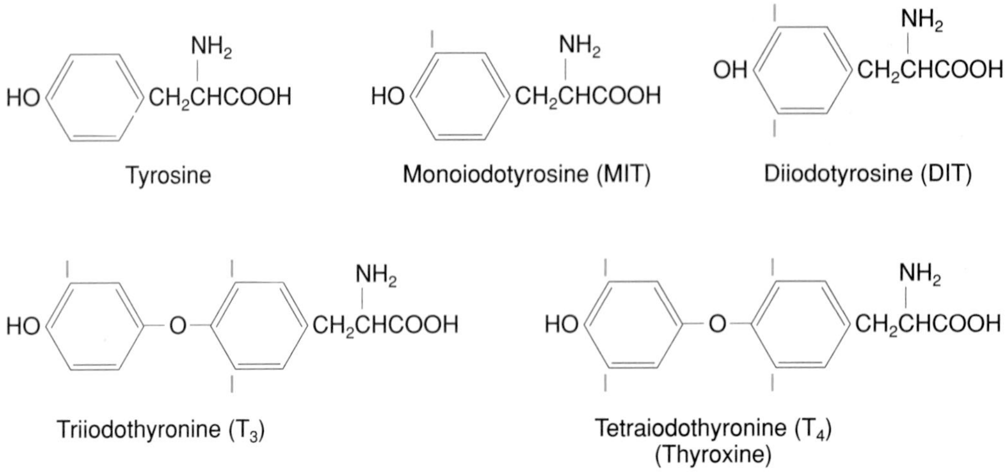

Tyrosine: HO–(benzene ring)–$CH_2CHCOOH$ with NH_2

Monoiodotyrosine (MIT): HO–(benzene ring, one I)–$CH_2CHCOOH$ with NH_2

Diiodotyrosine (DIT): OH–(benzene ring, two I)–$CH_2CHCOOH$ with NH_2

Triiodothyronine (T_3): HO–(benzene ring, one I)–O–(benzene ring, two I)–$CH_2CHCOOH$ with NH_2

Tetraiodothyronine (T_4) (Thyroxine): HO–(benzene ring, two I)–O–(benzene ring, two I)–$CH_2CHCOOH$ with NH_2

globulin molecules containing MIT, DIT, T_3, and T_4. The thyroglobulin molecules are stored in the colloid-containing regions of the follicles.

When the thyroid is actively secreting, thyroglobulin molecules are taken into the follicle cells by endocytosis (Figure 17.11h). Within the cells, lysosomes fuse with the endocytic vesicles, and lysosomal enzymes break down the thyroglobulin molecules, freeing MIT, DIT, T_3, and T_4 (Figure 17.11i). The MIT and DIT are deiodinated, and the iodine can be recycled for use in other iodinations (Figure 17.11j). The T_3 and T_4, which are collectively referred to as the thyroid hormones, pass out of the follicle cells and enter the bloodstream (Figure 17.11k). Within the bloodstream, almost all of the thyroid hormones are bound to plasma proteins such as *thyroid-binding globulin (TBG).*

The thyroid normally produces about 10% T_3 and 90% T_4. In the tissues, however, much of the T_4 is converted to T_3, which is the major active form of the thyroid hormones at the cellular level.

◆ **FIGURE 17.11 Synthesis, storage, and secretion of the thyroid hormones**

See text for detailed discussion.

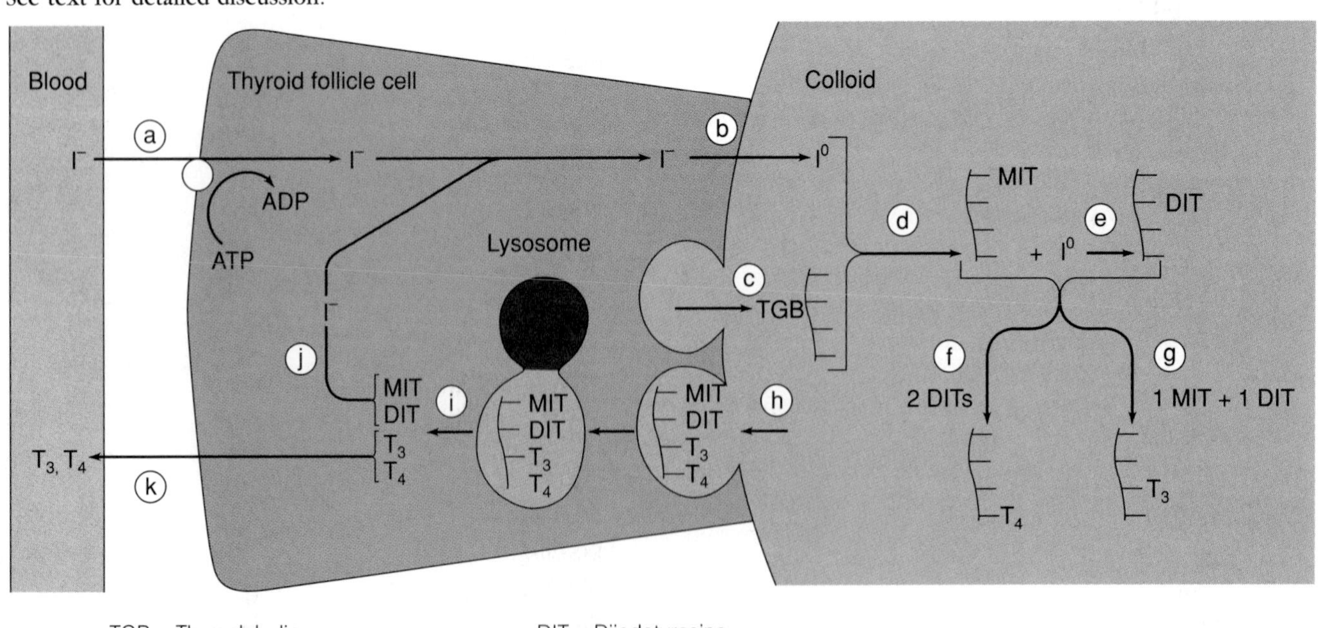

Thyroid-Hormone Actions

Among the most evident effects of the thyroid hormones are those associated with metabolism. The administration of thyroid hormones increases the body's oxygen consumption and heat production (the calorigenic effect). Although most body tissues are responsive to this influence, some tissues—including the spleen, brain, uterus, and testes—are not. When thyroid hormones are administered, the calorigenic effect normally does not become evident for some time, but it may last several days.

In physiological amounts, the thyroid hormones favor protein synthesis, but in excessive amounts they cause protein breakdown. Excessive amounts of the thyroid hormones can lead to muscle wasting and weakness, with the weakness being especially evident in the eye muscles and cardiac muscle.

The thyroid hormones affect almost all aspects of carbohydrate metabolism, and many of their influences depend on or are modified by other hormones, particularly the catecholamines and insulin. In small amounts, the thyroid hormones facilitate the conversion of glucose into glycogen (the storage form of glucose). In larger amounts, the thyroid hormones promote the breakdown of glycogen into glucose. In general, excessive amounts of the thyroid hormones increase the utilization of carbohydrates and accelerate the liberation of energy. Excessive amounts of the thyroid hormones deplete liver glycogen stores and may produce hyperglycemia (high blood-glucose levels).

The thyroid hormones stimulate many aspects of lipid metabolism, including synthesis, mobilization, and degradation. Generally, degradation effects predominate, and excess thyroid hormones cause a decrease in lipid stores.

The thyroid hormones promote the proliferation of beta adrenergic receptors for epinephrine and norepinephrine in many tissues, thereby increasing the responsiveness of the tissues to these substances. Consequently, the thyroid hormones facilitate the activity of the sympathetic nervous system.

The thyroid hormones are necessary for the proper production and action of growth hormone. Thus, they are required for normal body growth. The thyroid hormones are essential for the normal development of the central nervous system during fetal life and the first few months after birth. They are also necessary for normal nervous system function throughout life.

Release of the Thyroid Hormones

Thyrotropin (TSH) from the pars distalis of the pituitary stimulates the synthesis and release of the thyroid hormones, and TSH secretion is promoted by thyrotropin-releasing hormone (TRH) from the brain (Figure 17.12).

◆ FIGURE 17.12 Regulation of thyroid-hormone secretion

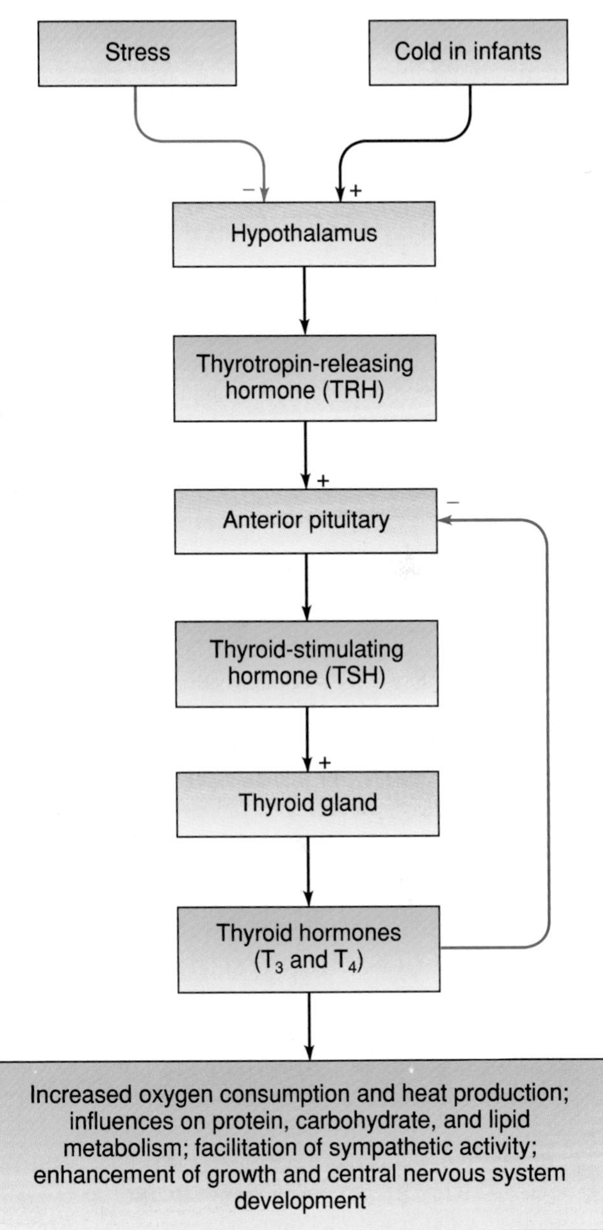

This system is influenced by a negative feedback of the thyroid hormones themselves. Most of the negative-feedback effect is manifested at the level of the pituitary gland by changing the sensitivity of the TSH-secreting cells to TRH.

In adults, the secretion rate of the thyroid hormones tends to be relatively steady. However, various types of stress can inhibit TSH secretion, presumably by way of neural influences that inhibit TRH release from the hypothalamus. In newborn infants, exposure to cold leads to an increase in TSH secretion and, consequently, to an increased secretion of the thyroid hormones. This re-

CONDITIONS OF CLINICAL SIGNIFICANCE

Thyroid Disorders

In young people, hypothyroidism (lowered thyroid function) retards growth and can lead to an abnormal development of bones, connective tissue, and reproductive structures. The condition caused by severe hypothyroidism during infancy is called *cretinism.* In this condition, the nervous system does not develop properly, leading to mental retardation. If the hypothyroid state is recognized early and thyroid-hormone replacement therapy is begun, the hypothyroid condition can be markedly improved. If treatment is delayed, the mental retardation becomes permanent.

Severe hypothyroidism in the adult is called *myxedema.* This condition, which is considerably more common in women than in men, is characterized by a puffiness of the face and eyelids and a swelling of the tongue and larynx. The skin becomes dry and rough, and the hair becomes scant. The individual has both a low basal-metabolic rate and a low body temperature. The sufferer also has poor muscle tone, lacks strength, and fatigues easily. His or her mental activity is generally sluggish. Myxedema can be alleviated by administering thyroid hormones.

An excess of thyroid hormones acting on the tissues is called *thyrotoxicosis.* One form of thyrotoxicosis is *hyperthyroidism* (thyroid overfunction). Hyperthyroidism is characterized by an elevated basal metabolism, elevated body temperature, and rapid heartbeat. Despite an increased appetite, there may be a large weight loss. The sufferer perspires freely and may be nervous, emotionally unstable, and unable to sleep. Hyperthyroidism can be treated surgically, with drugs that impair thyroid function, or with radioactive iodine that is taken up by the gland and destroys some of the thyroid cells.

The most common form of hyperthyroidism is known as *Graves' disease.* Graves' disease is an autoimmune disease in which the body produces an antibody whose target is the TSH receptors on thyroid cells. Like TSH, this antibody stimulates the secretion of thyroid hormones and the growth of the thyroid gland. However, the antibody is not subject to negative-feedback inhibition in response to increasing levels of the thyroid hormones. Thus, the thyroid gland continues to secrete and grow unchecked. In addition to the usual symptoms of hyperthyroidism, a person suffering from Graves' disease may experience a bulging of the eyes called *exophthalmos.* The eyes bulge because of mucopolysaccharide deposition and edema behind them. In severe cases, the person may not be able to completely close his or her eyelids, and the eyes may become dry and irritated.

A *goiter* is simply an enlargement of the thyroid gland. It may or may not be associated with hypothyroidism or hyperthyroidism. *Simple goiter* is a glandular enlargement without hypothyroidism or thyrotoxicosis. In Graves' disease, the thyroid usually exhibits a diffuse enlargement (goiter).

Endemic goiter is an enlargement of the thyroid gland caused by a lack of iodine in the diet. Without sufficient iodine, the thyroid gland is unable to produce adequate amounts of the thyroid hormones. Consequently, TSH levels remain high, and TSH continues to stimulate the thyroid. The thyroid gland enlarges, but may eventually atrophy as thyroid cells become exhausted. Endemic goiter is most common in regions of the world where people consume diets that are deficient in iodine (for example, in regions that have iodine-poor soil and lack an abundant supply of shellfish, which are rich in iodine). At one time, endemic goiter occurred in the midwestern United States, but the use of iodized salt has greatly reduced its incidence in this region.

sponse is believed to be mediated by an increased release of TRH from the hypothalamus.

Calcitonin

In addition to the thyroid hormones, the thyroid releases a hormone called **calcitonin,** which is produced by parafollicular cells (C cells) located between or adjacent to the thyroid follicles. Calcitonin lowers blood-calcium and -phosphate levels. It acts on bone cells to suppress bone resorption and enhance bone formation. Calcitonin is believed to be involved in the process of bone remodeling, and it may help prevent excessive bone resorption when there is a large calcium demand such as during pregnancy or breast-feeding.

The plasma-calcium level controls calcitonin secretion. When the concentration of calcium ions in the plasma rises, calcitonin secretion increases. Certain hormones released by the gastrointestinal tract during the digestion and absorption of food (for example, gastrin) also promote calcitonin secretion, and it has been proposed that a release of calcitonin associated with digestion may help the body conserve calcium obtained from the diet by preventing a rise in plasma-calcium levels that could lead to an increased urinary excretion of calcium. Calcitonin is a polypeptide consisting of 32 amino acids.

Parathyroid Glands

Embedded on the posterior surface of the lobes of the thyroid gland are four small **parathyroid glands** (Figure 17.13). The blood supply of these glands is the same as that of the thyroid gland, but the parathyroid and thyroid glands differ in their embryonic development, their structure, and their function.

Embryonic Development and Structure

The parathyroid glands develop from the dorsal halves of the third and fourth pairs of embryonic structures called pharyngeal pouches. With continued development they lose their attachments to the pouches and migrate to the neck, where they assume their adult positions on the posterior surfaces of the lateral lobes of the thyroid gland.

There are usually two masses of parathyroid tissue (*superior* and *inferior*) on each of the two thyroid lobes. This tissue is composed of densely packed masses or cords of cells.

Parathyroid Hormone and Its Effects

Parathyroid hormone (PTH, or **parathormone)** is a polypeptide, and it is currently believed that two or possibly three forms of the hormone may appear in the blood. Parathyroid hormone is a principal controller of calcium and phosphate metabolism, and it is involved in the remodeling of bone.

Parathyroid hormone increases the plasma-calcium concentration and decreases the plasma-phosphate concentration. It has three major modes of action:

1. Parathyroid hormone acts on bone cells to increase bone resorption and to release calcium and phosphate ions into the blood plasma.
2. Parathyroid hormone acts on the kidneys to decrease calcium excretion. However, a high plasma-calcium concentration due to the action of parathyroid hormone can often lead to an increased loss of calcium in the urine. Parathyroid hormone increases phosphate excretion in the urine. The increased phosphate excretion tends to decrease the plasma-phosphate concentration even though parathyroid hormone increases the release of phosphate ions into the plasma from bone.
3. Parathyroid hormone enhances a step in the metabolic transformation and activation of vitamin D_3. In the skin, the precursor substance provitamin D (7-dehydrocholesterol) is converted to vitamin D_3 (cholecalciferol) by exposure to ultraviolet radiation (sunlight). In the liver, vitamin D_3 subsequently is converted to 25-hydroxycholecalciferol. In the kidneys, 25-hydroxycholecalciferol is converted to 1,25-dihydroxycholecalciferol, and parathyroid hormone promotes this conversion. 1,25-Dihydroxycholecalciferol is a potent substance that enhances the absorption of calcium from the gastrointestinal tract. Consequently, parathyroid hormone indirectly promotes this absorptive activity.

The plasma-calcium level is the major controller of parathyroid-hormone secretion. When the plasma-calcium level falls, parathyroid-hormone secretion increases.

◆ **FIGURE 17.13 The parathyroid glands**

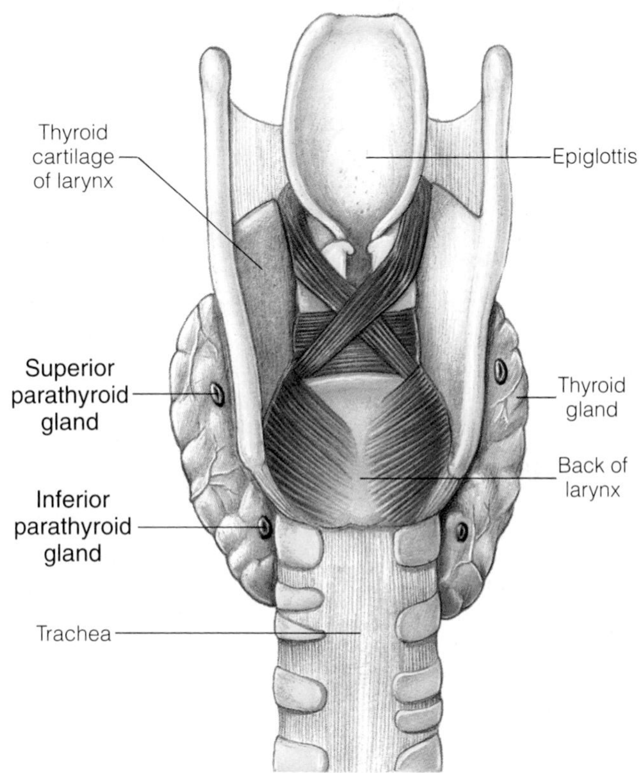

CONDITIONS OF CLINICAL SIGNIFICANCE

Parathyroid Disorders

Hyperparathyroidism, which results in an excess of parathyroid hormone, causes extensive bone decalcification and may lead to deformities and fractures. The plasma-calcium level rises, calcium-containing kidney stones may form, and the calcification of soft tissues may occur. A high plasma-calcium level reduces neural and muscular excitability, leading to abnormal reflexes and skeletal muscle weakness. Hyperparathyroidism is generally treated by the surgical removal of glandular tissue.

Hypoparathyroidism, which results in a deficiency of parathyroid hormone, leads to a lowered plasma-calcium level, which greatly increases neural and muscular excitability. A common symptom of a low plasma-calcium level is tingling of the extremities due at least in part to an increased activity of sensory nerve fibers that carry impulses associated with light touch. In addition, skeletal muscle twitching may occur. If the plasma-calcium level is low enough, skeletal muscle spasms (tetany) may occur. Hypoparathyroidism is generally treated by administering large doses of calcium salts and/or vitamin D.

Adrenal Glands

The two **adrenal glands,** or **suprarenal glands,** are pyramid-shaped organs located behind the peritoneum close to the superior border of each kidney (Figure 17.14). Each gland is surrounded by a connective-tissue capsule and embedded in fat. The adrenal glands are well supplied with blood vessels.

Each adrenal gland consists of two separate portions: an inner *medulla* and an outer *cortex* (Figure 17.15). The medulla and the cortex have different embryonic

◆ **FIGURE 17.14 Location of the adrenal glands**

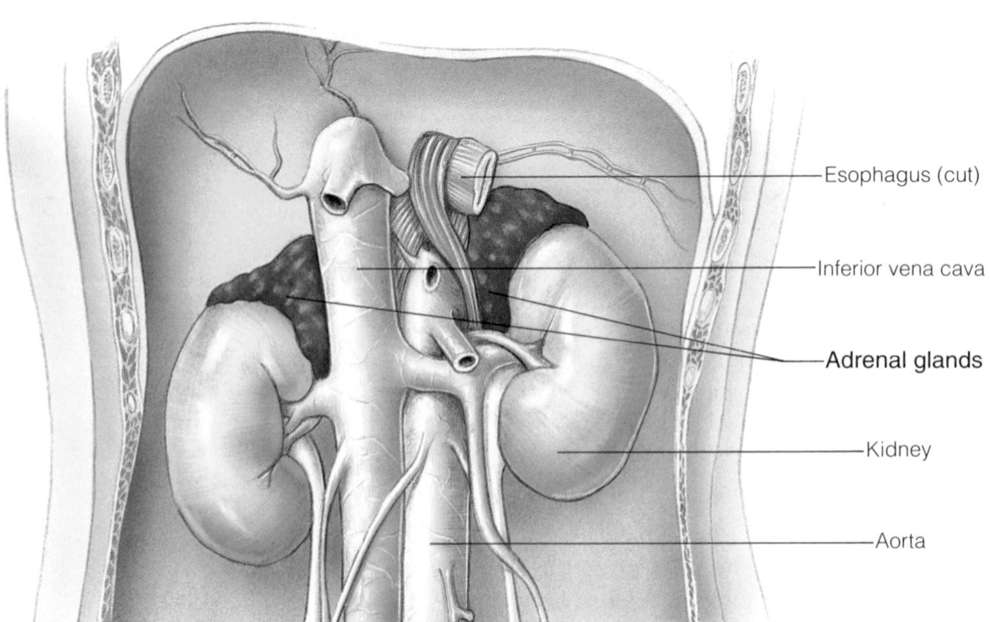

CLINICAL CORRELATION

Abnormal Secretion of Parathyroid Hormone

Case Report

THE PATIENT: A 34-year-old woman.

PRINCIPAL COMPLAINT: Persistent fatigue, fever, weight loss, and frequent headaches.

HISTORY: The symptoms had begun to develop about one year before admission, and the patient had first been seen one month before the present admission.

CLINICAL EXAMINATION: Because of the nonspecific nature of the symptoms, general laboratory studies were performed. The only significant abnormal findings were the plasma concentrations of calcium, 13.7 mg/100 ml (normal: 8.5–10.5 mg/100 ml); phosphorous, 1.2 mg/100 ml (normal: 3.4–5 mg/100 ml); and the enzyme alkaline phosphatase, 95 units (U) per liter (normal: 13–19 U/liter). These values are consistent with hyperparathyroidism. The concentration of parathyroid hormone in the plasma was determined to be 301 mU/ml (normal preovulatory level: 5–22 mU/ml).

TREATMENT: Two weeks after admission, the patient's neck was explored surgically. The four parathyroid glands appeared normal, and an examination of tissue samples from each revealed no abnormality. Since the evidence suggested a parathyroid adenoma (a benign tumor) in a supernumerary (extra) parathyroid gland, a search for the extra gland was begun. Injection of radiographic contrast material into the left internal mammary artery revealed an area of abnormally high blood supply, which is characteristic of parathyroid adenomas, in the anterior mediastinum. Subsequent surgical exploration of the area disclosed a large parathyroid nodule within the left lobe of the thymus gland.

OUTCOME: The abnormal tissue was removed, the patient's plasma-calcium concentration became normal, and the symptoms of hyperparathyroidism disappeared.

origins and different structures, and the actions of their hormones differ considerably (Table 17.4). Consequently, each adrenal gland is, in effect, actually two distinct endocrine organs.

Adrenal Medulla

The central portion of each adrenal gland, the **adrenal medulla,** is composed of cells arranged in groups or short cords surrounding blood capillaries and venules.

Embryonic Development and Structure

The adrenal medulla arises embryonically from neural crest cells. Since the neural-crest cells also give rise to postganglionic sympathetic neurons, the adrenal medulla can be regarded as a modified portion of the sympathetic division of the autonomic nervous system. In fact, the cells of the adrenal medulla function in a manner similar to postganglionic sympathetic cells. Because they stain with chromium salts, the cells of the adrenal medulla are called *chromaffin cells.*

Adrenal Medullary Hormones and Their Effects

The adrenal medulla produces two catecholamine hormones: epinephrine (adrenaline) and norepinephrine

◆ **FIGURE 17.15 An adrenal gland**

The gland is a dual structure that consists of a central medulla and a surrounding cortex enclosed in a fibrous capsule.

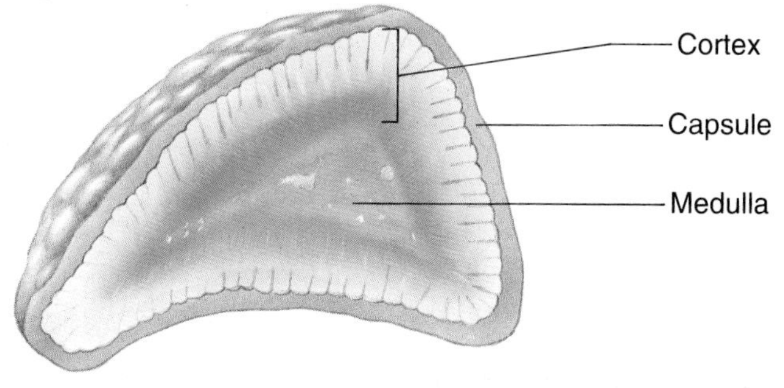

◆ **FIGURE 17.16 Structures of the adrenal medullary catecholamine hormones, epinephrine and norepinephrine**

Epinephrine

Norepinephrine

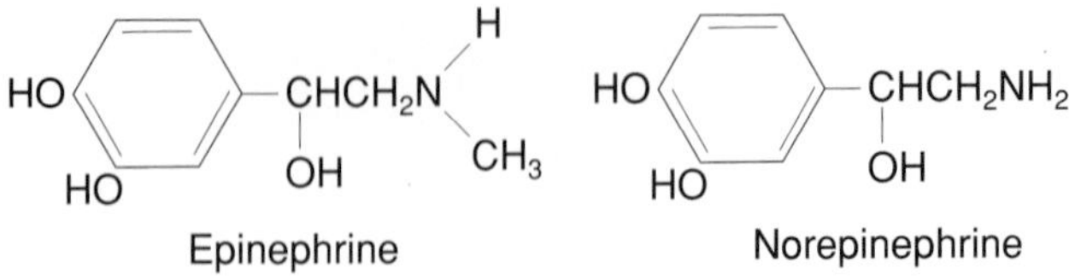

(noradrenaline) (Figure 17.16). (Recall that norepinephrine is also released from end terminals of postganglionic neurons of the sympathetic nervous system; thus, it may be present from sources other than the adrenal medulla.) Although epinephrine and norepinephrine are structurally similar molecules that exert a number of common effects, they are not identical in function.

Epinephrine. Epinephrine promotes the mobilization of stored carbohydrate and fat, and it elevates blood-glucose levels. Epinephrine acts on the liver to stimulate **glycogenolysis** (the breakdown of glycogen into glucose) and **gluconeogenesis** (the production of glucose from noncarbohydrate precursors such as amino acids). Epinephrine acts on muscle to reduce glucose uptake and to stimulate the production of pyruvic acid and lactic acid from glycogen. The pyruvic acid and lactic acid can be used by the liver to manufacture glucose. In adipose tissue, epinephrine promotes **lipolysis** (the breakdown of triglycerides into glycerol and fatty acids) and thereby increases the level of fatty acids in the blood.

Epinephrine increases the rate, force, and amplitude of the heartbeat. It causes blood vessels to constrict in a number of body areas, including the skin, mucous membranes, and kidneys, but it can induce vessels to dilate in some areas, such as skeletal muscles and the heart. Epinephrine also causes the dilation of respiratory passageways called bronchioles in the lungs by relaxing bronchiolar smooth muscle.

Norepinephrine. Norepinephrine increases the heart rate and the force of contraction of cardiac muscle. It also causes blood vessels to constrict in almost all areas of the body. In addition, norepinephrine promotes the breakdown of triglycerides in adipose tissue.

Release of the Adrenal Medullary Hormones

In general, the adrenal medulla releases a mixture of about 80% epinephrine and 20% norepinephrine, but these percentages vary considerably under different physiological conditions. The release of the catecholamine hormones of the adrenal medulla is controlled by preganglionic neurons to the medulla from the sympathetic division of the autonomic nervous system. A variety of conditions leads to the release of adrenal medullary hormones, including emotional excitement, injury, exercise, and low blood-glucose levels.

Because the adrenal medulla is controlled by sympathetic preganglionic neurons and because catecholamines are liberated by both the adrenal medulla and the sympathetic nervous system, it is often convenient to consider the two as a single sympatheticoadrenal system. Together, the divisions of this system maintain blood pressure and help regulate carbohydrate metabolism.

Adrenal Cortex

The outer portion of each adrenal gland, the *adrenal cortex,* makes up approximately 80% of the total weight of each fully developed gland.

Embryonic Development and Structure

The adrenal cortex is derived embryonically from the mesoderm of the region that gives rise to gonadal tissue. The outer portion of the adrenal cortex is surrounded by a connective-tissue capsule. Trabeculae (strands of connective tissue) extend from the capsule into the gland itself. The endocrine cells of the adrenal cortex are organized into three layers (Figure 17.17): the *zona glomerulosa,* the *zona fasciculata,* and the *zona reticularis.*

The **zona glomerulosa** is a relatively thin region in the cortex. It is located directly beneath the capsule and is composed of clusters of cells. Beneath the zona glomerulosa is a thick region called the **zona fasciculata.** The cells of the zona fasciculata are arranged in parallel columns that run at right angles to the surface of the gland. Occupying the deepest region of the adrenal cortex and lying adjacent to the adrenal medulla is the **zona reticularis,** in which the cells are arranged in a network of interconnecting cords.

Adrenal Cortical Hormones and Their Effects

Mineralocorticoids. The adrenal cortex produces hormones called **mineralocorticoids** *(min-ur-uh-lo-kor´-ti-koydz)* that regulate sodium and potassium metabolism. The principal mineralocorticoid is **aldosterone** *(al-dos´-te-rone),* which is produced by cells in the zona glomerulosa. Aldosterone promotes the reabsorption of sodium and the excretion of potassium by the urine-forming structures of the kidneys.

◆ **FIGURE 17.17 The layers of the adrenal cortex**
(a) Schematic representation. (b) Photomicrograph. (×136)

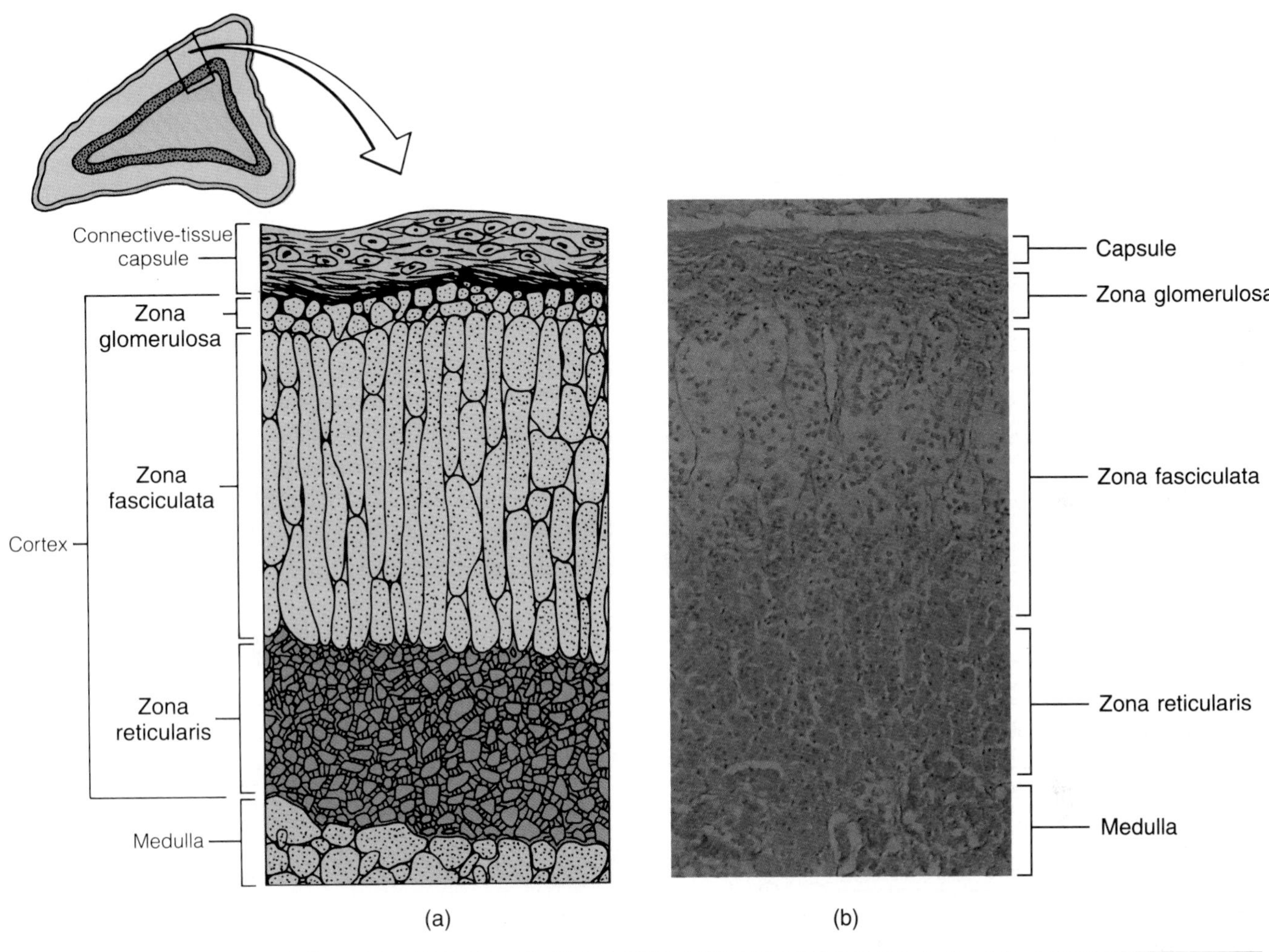

Glucocorticoids. The adrenal cortex also produces hormones called **glucocorticoids** *(gloo-ko-kor´-ti-koydz)* that affect carbohydrate, lipid, and protein metabolism. The principal glucocorticoid is **cortisol,** which is produced primarily by cells in the zona fasciculata and in lesser amounts by cells in the zona reticularis.

In general, cortisol favors gluconeogenesis and lipolysis, and it tends to increase the levels of glucose, amino acids, and fatty acids in the blood. In response to elevated levels of cortisol, glucose uptake and utilization in many peripheral tissues are inhibited, and muscle shifts from glucose to fatty acids for much of its metabolic energy. Elevated levels of cortisol accelerate the breakdown of proteins and inhibit amino acid uptake and protein synthesis by many tissues other than the liver. In the liver, however, amino acid uptake is enhanced as is the utilization of amino acids for protein and glucose synthesis. At pharmacological concentrations (that is, concentrations higher than those normally present in the body), cortisol has anti-inflammatory effects.

Estrogenic and Androgenic Substances. The adrenal cortex produces small quantities of estrogenic materials that resemble female sex hormones. It also produces some androgenic substances—particularly *dehydroepiandrosterone (DHEA)*—that resemble male sex hormones. DHEA is a much less potent androgen than testosterone, which is produced by the testes in males.

In males, testicular androgens are the dominant androgens, and adrenal androgens do not play a significant role. In females, adrenal androgens are largely responsible for the maintenance of sexual drive.

The adrenal cortical hormones and the male and female sex hormones are steroids that are synthesized from cholesterol (Figure 17.18).

CONDITIONS OF CLINICAL SIGNIFICANCE

Adrenal Disorders

Primary adrenal cortical hypofunction in humans is called *Addison's disease.* Inadequate amounts of aldosterone impair the body's ability to conserve sodium and excrete potassium. This impairment may lead to a decreased extracellular fluid volume, decreased plasma volume, low blood pressure, decreased cardiac size and output, general weakness, and shock. A deficiency of cortisol in Addison's disease may result in loss of appetite (anorexia), fasting hypoglycemia, apathy, weakness, and a diminished ability to withstand various types of physiological stress. Hormone administration can alleviate the symptoms of Addison's disease.

Adrenal cortical hyperfunction can result in a number of disorders. Hypercortisolism may produce *Cushing's syndrome,* which is characterized by increased blood-glucose levels, increased protein breakdown, a weakening of the skeleton, and muscle weakness. Hyperaldosteronism is characterized by potassium depletion and expansion of the extracellular fluid compartment, which may result in hypertension or edema (excessive accumulation of fluid in the tissue spaces). Certain adrenal tumors can secrete excessive quantities of androgenic substances, producing masculinizing effects that are particularly evident in females.

◆ **FIGURE 17.18 Pathways for the synthesis of the major steroid hormones**

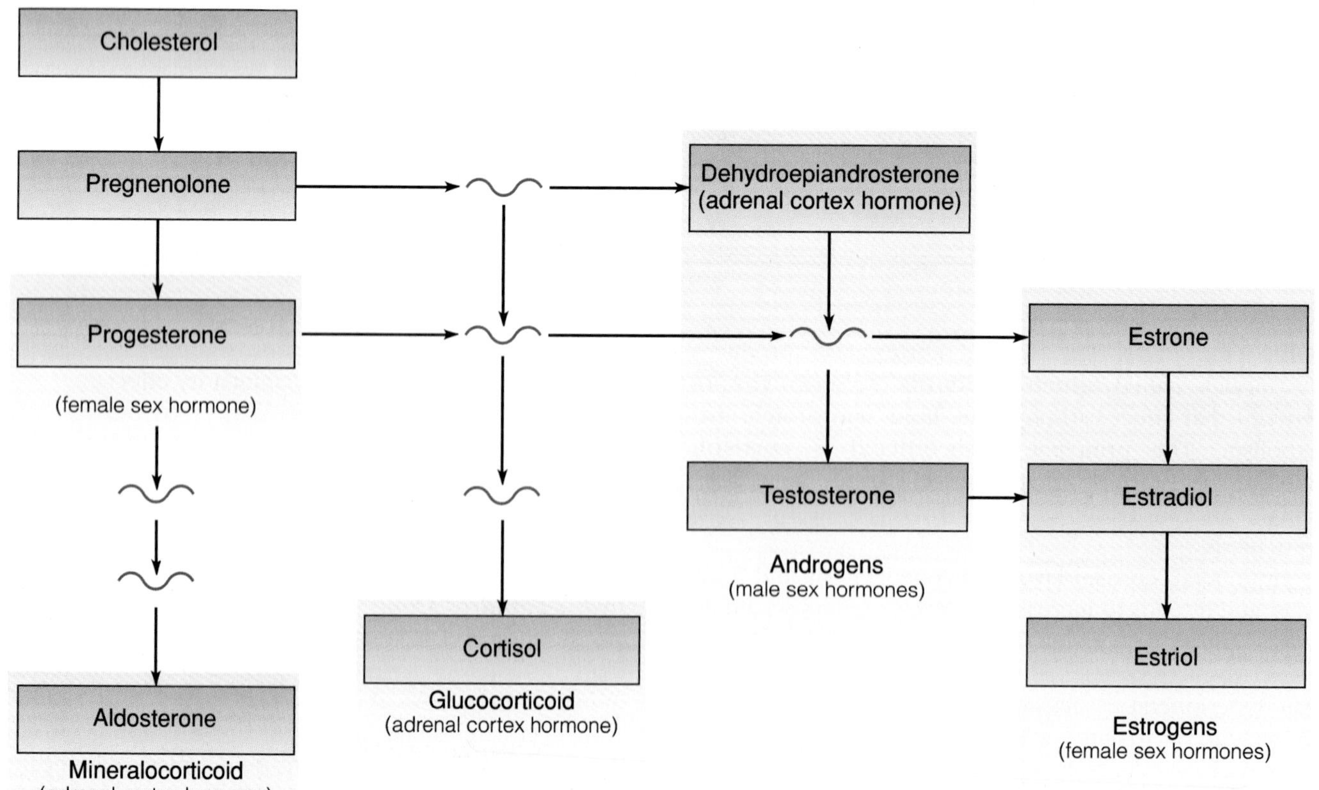

Release of the Adrenal Cortical Hormones

The release of aldosterone from the adrenal cortex is influenced by several factors. One of these is an enzyme called **renin** *(ree´-nin),* which is released by specialized cells of the kidneys. Renin acts on the precursor substance **angiotensinogen,** which is manufactured by the liver and is present in the blood. Renin converts angiotensinogen to **angiotensin I.** In turn, angiotensin I is converted into **angiotensin II** by a *converting enzyme* associated with the walls of blood capillaries, particularly in the lungs. Among the effects of angiotensin II is the stimulation of aldosterone release from the adrenal cortex. Renin secretion is itself controlled by a number of factors, including the activity of sympathetic nerves to the kidneys and the renal arterial blood pressure (decreased pressure increases renin secretion).

Aldosterone release is also stimulated by an elevation of the potassium concentration of the fluids bathing the adrenal glands, as well as by a decrease in the sodium content of the body.

The release of cortisol is controlled primarily by ACTH from the pituitary, and the release of ACTH, in turn, is influenced by cortiocotropin-releasing hormone (CRH) from the hypothalamus of the brain (Figure 17.19). Cortisol exerts an inhibitory influence over ACTH and CRH release by way of negative feedback.

The plasma-cortisol level exhibits a circadian rhythm, with the highest level occurring just before a person awakens. In the plasma, cortisol is bound to a glycoprotein called *transcortin* for transport through the circulation.

Various stressful situations (such as trauma or emotional stress) cause increased release of ACTH and, consequently, increased cortisol secretion. This effect is probably neurally mediated by way of CRH.

◆ **FIGURE 17.19 Control of cortisol secretion**

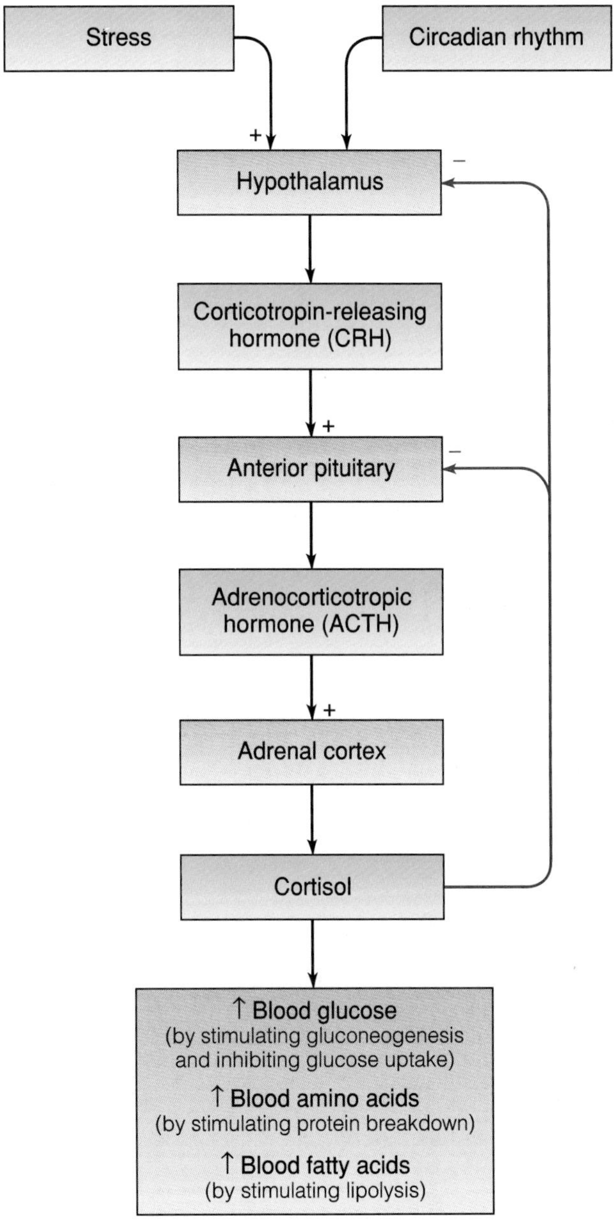

Pancreas

The **pancreas,** which is located behind the stomach, between the spleen and the duodenum of the small intestine, is both an endocrine gland that produces hormones and an exocrine gland that produces digestive enzymes. The endocrine portion of the pancreas produces hormones that have a role in regulating the metabolic activities of the body—particularly those associated with carbohydrate metabolism.

Embryonic Development and Structure

The pancreas arises embryonically through a fusion of two outgrowths from the duodenum of the small intestine. The exact origin of the cells that make up the endocrine portion of the pancreas remains uncertain. At least some of the endocrine cells are believed to bud off from the lining of the pancreatic ductules—either from endodermal cells that may have arisen at the site or from neuroectodermal cells that migrated into the intestinal mucosa at an earlier developmental stage.

The endocrine portion of the fully developed pancreas consists of aggregations of cells that are clustered into groups called **pancreatic islets (islets of Langerhans)** (Figure 17.20). The islet cells are arranged in irregular cords separated by a rich vascular system of capillary vessels or sinusoids. Nerves from both the

◆ **FIGURE 17.20 The pancreas**
The photomicrograph shows details of the pancreatic cellular structure.

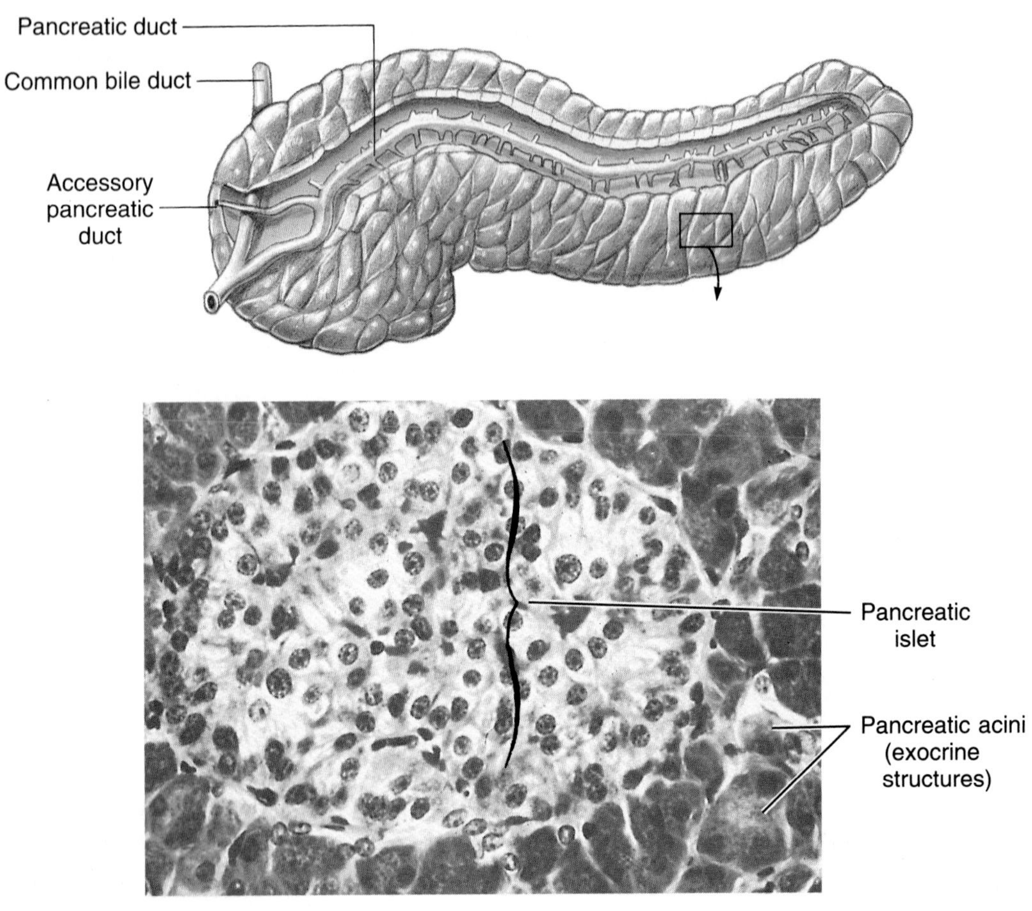

sympathetic and parasympathetic divisions of the autonomic nervous system supply the pancreas.

Pancreatic Hormones and Their Effects

The pancreatic islets contain a number of functionally different cell types:

◆ **A cells (alpha cells)** produce the hormone **glucagon.** Glucagon is a polypeptide composed of 29 amino acids.

◆ **B cells (beta cells)** produce the hormone **insulin.** Insulin is composed of two linked polypeptide chains. One chain consists of 21 amino acids, and the other consists of 30 amino acids.

◆ **D cells (delta cells)** produce **somatostatin**—the same substance that is produced in the brain and inhibits the release of growth hormone and TSH. There is some evidence that somatostatin, which is also secreted by the mucosa of the upper gastrointestinal tract and has been proposed as a possible neurotransmitter substance, inhibits the release of glucagon. Moreover, high concentrations of somatostatin appear to inhibit the release of insulin. Thus, somatostatin exemplifies the fact that a given chemical substance may be synthesized at different sites and may perform different functions.

◆ **F cells (PP cells)** produce a substance called **pancreatic polypeptide.** The precise function of pancreatic polypeptide is unclear, but it appears to inhibit gastrointestinal function.

Actions of Insulin

Insulin facilitates the uptake of glucose and the utilization of glucose as an energy source by many cells. It also promotes the synthesis of glycogen, triglycerides, and proteins. The actions of insulin tend to lower the levels of glucose, fatty acids, and amino acids in the blood.

Glucose enters most cells by means of the carrier-mediated transport process called facilitated diffusion (page 64). In many cells, particularly skeletal muscle and

◆ **FIGURE 17.21 Factors controlling insulin secretion**

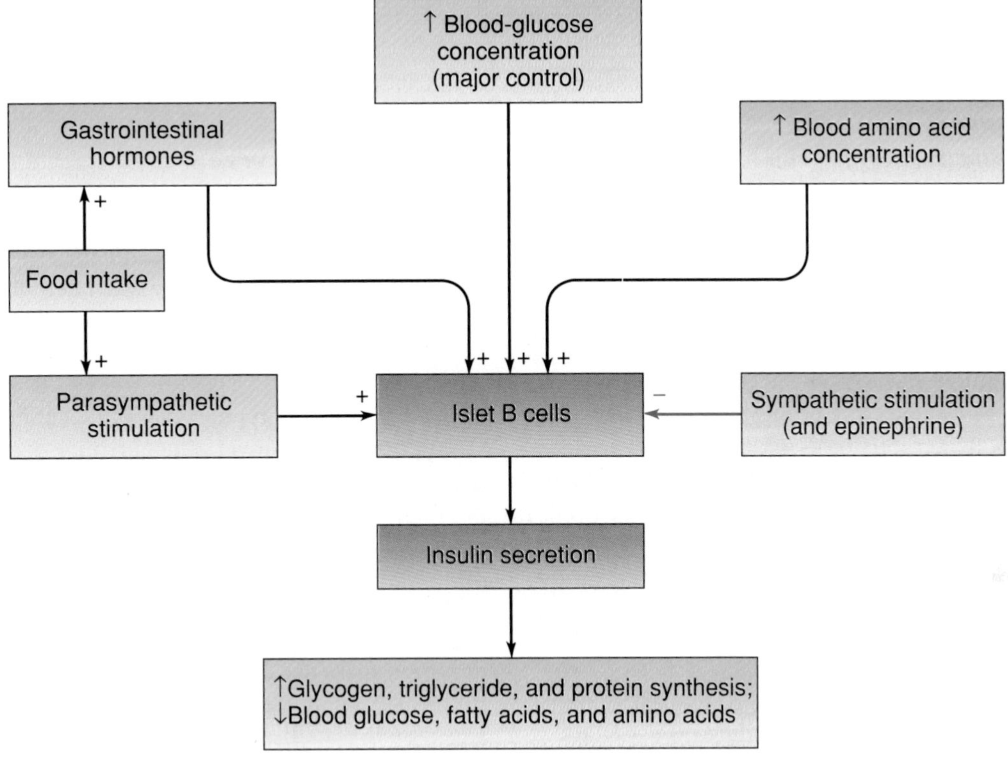

adipose tissue cells, insulin increases the facilitated diffusion of glucose into the cells by increasing the number of carrier molecules for glucose in the plasma membrane. (Insulin does not exert this effect in the liver or brain.)

In skeletal muscle cells, insulin promotes the formation of glycogen from glucose that is taken up but not used as an energy source, and it inhibits glycogen breakdown. In liver cells, insulin promotes the conversion of glucose into both glycogen and triglycerides (glucose is an important precursor of fatty acids and triglycerides). In addition, insulin inhibits glycogen breakdown and the production of glucose from noncarbohydrate precursors, thereby decreasing the release of glucose from liver cells into the blood.

In adipose tissue cells, insulin enhances triglyceride synthesis and inhibits the breakdown and mobilization of stored triglycerides. In muscle cells and other cells, insulin stimulates the active transport of amino acids into the cells and promotes the incorporation of amino acids into proteins. Insulin also inhibits protein breakdown.

Release of Insulin

The blood-glucose level is the major factor governing insulin release: the higher the blood-glucose level, the greater the insulin release (Figure 17.21). An elevated level of amino acids in the blood plasma also stimulates insulin release.

One or more hormones from the gastrointestinal tract (such as gastric inhibitory peptide, secretin, gastrin, and cholecystokinin) also promote insulin release. These gastrointestinal hormones are secreted when food is consumed.

Parasympathetic neural activity stimulates insulin release, and the parasympathetic nerves to the pancreas are particularly active when food is consumed. Sympathetic neural activity or an elevation of plasma epinephrine inhibits insulin in release.

Actions of Glucagon

The activities of glucagon are generally opposite to those of insulin, and glucagon tends to raise the blood-glucose level. Glucagon acts mainly on the liver, where it stimulates glycogen breakdown and inhibits glycogen synthesis. Glucagon also stimulates the formation of glucose in the liver from noncarbohydrate precursors.

Glucagon stimulates the breakdown of triglycerides in liver and adipose tissue. This breakdown increases the substances available for glucose production in the liver, and it also leads to the formation of substances called ketone bodies.

CONDITIONS OF CLINICAL SIGNIFICANCE

Pancreatic Disorders

The most common pancreatic endocrine disorder is the condition known as *diabetes mellitus.* This condition is caused by a relative insulin deficiency that can be due to a deficient pancreatic secretion of insulin or to a decreased sensitivity of target cells to insulin. There is also evidence that a relative or absolute excess of glucagon, as well as a deficiency of insulin, occurs in certain forms of diabetes. Diabetes mellitus has a strong hereditary or familial component, and two types of diabetes are recognized (Table C17.1). Type I (insulin-dependent) diabetes occurs most often in children or adolescents. Type II (insulin-independent) diabetes occurs most frequently in adults over 40 years of age.

Type I diabetes is due to an insulin deficiency resulting from B-cell destruction. The B-cell destruction is caused by an attack by the body's own defense mechanisms, which may be triggered by some type of infection.

Type I diabetes is characterized by increased levels of blood glucose coupled with a reduced entry of glucose into cells and an impairment of cellular ability to use glucose. Free fatty acids are mobilized and released from adipose tissue, and they provide energy sources for the cells. Such energy sources are necessary because of the deficient cellular uptake and utilization of glucose. The oxidation of fatty acids in the liver produces substances known as *ketone bodies* (for example, acetoacetic acid and beta-hydroxybutyric acid). Ketone bodies may be produced faster than other body tissues can metabolize them, and the level of these substances in the blood and other body fluids can increase considerably. In severe cases of type I diabetes, the body's buffering and excretory mechanisms are not able to cope with these acidic substances, and the pH of the body fluids falls. A sufficiently low pH of the body fluids can lead to altered respiration, nervous system depression, coma, and death.

Protein synthesis decreases and protein breakdown increases in type I diabetes. These effects contribute to weight loss, and they can impair the body's ability to combat infections and to repair injured tissues.

As the blood levels of glucose and ketone bodies rise in type I diabetes, these substances appear in the urine. Extra water is osmotically required to excrete them, and a large volume of urine is produced (polyuria). As a consequence of the substantial loss of water in the urine, strong sensations of thirst occur, and large volumes of fluid are consumed (polydipsia). If compensation for the water loss is inadequate, dehydration can occur, as can circulatory difficulties such as hypotension.

The destroyed B cells of a type I diabetic cannot secrete enough insulin to meet the person's needs. Consequently, the treatment of type I diabetes requires the person to take insulin by injection. (Insulin cannot be taken orally because it is a polypeptide that is broken down by digestive enzymes.)

Type II diabetes mainly strikes overweight people who are middle-aged or older. In fact, more than 80% of type II diabetics are or have been overweight. The B cells of type II diabetics can still produce insulin, and type II diabetics may have insulin levels that are normal or even above normal. However, in type II diabetes, insulin's target cells are relatively insensitive to the hormone. This insulin insensitivity, which is especially evident in skeletal muscle and liver cells, appears to have multiple causes. In some cases, various target cells for insulin apparently have a deficiency of insulin receptors.

The characteristics of type II diabetes are similar to those of type I diabetes. However, the symptoms of type II diabetes are usually slower in onset and less severe than the symptoms of type I diabetes. In general, type II diabetics have elevated blood-glucose levels, and glucose is present in the urine. However, most type II diabetics do not develop metabolic difficulties severe enough to produce the low pH of the body fluids that can cause coma and death.

In many type II diabetics, diet control, weight loss, and exercise can eliminate many of the manifestations of the disease. In some cases, orally administered drugs are used to stimulate additional insulin production.

Many diabetics develop long-term complications 20 or more years after the onset of the disease. Lesions can occur in blood vessels—particularly in the kidneys, where they can cause renal insufficiency, and in the retina, where they can impair vision and cause blindness. Atherosclerosis is also a frequent complication, and coronary artery disease is the most common cause of death among diabetics.

 CONDITIONS OF CLINICAL SIGNIFICANCE 

◆ **TABLE C17.1 Comparison of Type I and Type II Diabetes Mellitus**

CHARACTERISTIC	TYPE I DIABETES	TYPE II DIABETES
Level of insulin secretion	None or almost none	May be normal or exceed normal
Typical age of onset	Childhood	Adulthood
Percentage of diabetics	10–20%	80–90%
Basic defect	Destruction of B cells	Reduced sensitivity of insulin's target cells
Associated with obesity?	No	Usually
Genetic and environmental factors important in precipitating overt disease?	Yes	Yes
Speed of development of symptoms	Rapid	Slow
Development of ketosis	Common if untreated	Rare
Treatment	Insulin injections; dietary management	Dietary control and weight reduction; occasionally oral hypoglycemic drugs

Glucagon also has a stimulating effect on the breakdown of liver protein. The amino acids that result from protein breakdown can be converted to glucose.

Release of Glucagon

The blood-glucose level is the major regulator of glucagon release: the lower the blood-glucose level, the greater the glucagon release. In addition, an elevated level of amino acids in the blood plasma stimulates glucagon release. (Note that the influence of blood-glucose level on glucagon release is opposite to the influence of blood-glucose level on insulin release. However, an elevated amino acid level has the same influence on both insulin and glucagon release.) Sympathetic neural activity and epinephrine also promote glucagon release.

Gonads

The male gonads are the *testes,* which produce male gametes (sperm). They also produce male sex hormones—the **androgens,** particularly **testosterone.**

The female gonads are the *ovaries,* which produce female gametes (ova). They also produce female sex hormones—the **estrogens** and **progesterone.** The reproductive and endocrine roles of the gonads are discussed in Chapter 28.

Other Endocrine Tissues and Hormones

The digestive tract is the source of a number of hormones (for example, gastrin, secretin, and cholecystokinin). These hormones are discussed in Chapter 24. The placenta is also a source of hormones, as is discussed in Chapter 29.

The **thymus gland,** which is discussed in relation to immune responses in Chapter 22, produces a group of polypeptide hormones collectively called *thymosin.* Thymosin appears to be involved in the normal development of the immune response.

The **pineal body** of the brain is widely believed to be an endocrine gland. However, the precise function of the pineal body in humans is currently unclear. The pineal body produces a substance called *melatonin,* and some researchers suggest that this substance plays a role in the occurrence of puberty. Melatonin secretion is high during exposure to the dark and decreases during exposure to the light, and it may influence a variety of body rhythms.

Prostaglandins and Related Substances

The **prostaglandins** and a number of related substances—**prostacyclin, thromboxanes,** and **leukotrienes**—are chemical messengers, and one or another of them is present in almost every body tissue. In general, these substances are not believed to circulate widely in the blood, and thus they do not act as true hormones. Instead, they act primarily as local messengers that exert their effects in the tissues that synthesize them.

A fatty acid called *arachidonic acid* is the major compound from which prostaglandins, prostacyclin, thromboxanes, and leukotrienes are derived. Arachidonic acid is a part of phospholipids in the plasma membranes of cells. When a cell is appropriately stimulated (for example, by a hormone or neurotransmitter), a plasma-membrane enzyme called *phospholipase* A_2 is activated, and this enzyme splits arachidonic acid from the phospholipids. Arachidonic acid can be converted to prostaglandins, prostacyclin, or thromboxanes by way of metabolic pathways involving the enzyme *cyclooxygenase,* or it can be converted to leukotrienes by way of a pathway involving the enzyme *lipoxygenase.* Once prostaglandins, prostacyclin, thromboxanes, or leukotrienes are formed, they act locally within or near their sites of production and are rapidly inactivated by local enzymes.

The prostaglandins, prostacyclin, thromboxanes, and leukotrienes have a wide variety of actions, some of which are summarized in Table 17.3.

◆ **TABLE 17.3 Known or Suspected Actions of Prostaglandins and Related Substances**

BODY SYSTEM/ ACTIVITY	ACTIONS
Reproductive system	Promote sperm transport by action on smooth muscle in male and female reproductive tracts
	Important in menstruation
	Play role in ovulation
	Contribute to preparation of maternal portion of placenta
	Contribute to parturition
Respiratory system	Some promote bronchodilation, others bronchoconstriction
Urinary system	Increase renal blood flow
Digestive system	Inhibit hydrochloric acid secretion by stomach
	Stimulate intestinal motility
Nervous system	Modulate neurotransmitter release and action
	Act at hypothalamic "thermostat" to increase body temperature
Endocrine system	Enhance cortisol secretion
	Influence tissue responsiveness to hormones in many instances
Cardiovascular system	Modulate platelet aggregation
Fat metabolism	Inhibit fat breakdown
Inflammation	Promote many aspects of inflammation, including fever and development of pain

◆ **TABLE 17.4 Major Endocrine Glands and Their Hormones**

GLAND	HORMONES	REPRESENTATIVE EFFECTS	SELECTED DISORDERS
PITUITARY [*Fig. 17.4*]			
Neurohypophysis (hormones are actually manufactured in hypothalamus of the brain)	Antidiuretic hormone (ADH)	Promotes reabsorption of water from urine-forming structures of kidneys	Undersecretion leads to diabetes insipidus.

◆ **TABLE 17.4 Major Endocrine Glands and Their Hormones (continued)**

GLAND	HORMONES	REPRESENTATIVE EFFECTS	SELECTED DISORDERS
	Oxytocin	Stimulates contraction of uterine smooth muscle and myoepithelial cells around alveoli of mammary glands. As such, it is involved in birth processes and milk ejection during breast-feeding.	
Adenohypophysis	Follicle-stimulating hormone (FSH) and luteinizing hormone (LH)	Stimulate gonads to produce gametes and sex hormones	Undersecretion causes gonadal inactivity in males and impairment or cessation of menstruation in females.
	Thyrotropin (TSH)	Stimulates thyroid gland to secrete thyroid hormones	Undersecretion leads to symptoms of hypothyroidism.
	Adrenocorticotropin (ACTH)	Stimulates adrenal cortex to secrete glucocorticoids, particularly cortisol	Undersecretion leads to symptoms of adrenal cortical insufficiency. Oversecretion leads to symptoms of adrenal cortical hyperfunction.
	Growth hormone (GH)	Stimulates growth in general and growth of skeletal system in particular. It also affects metabolic functions.	Undersecretion produces pituitary dwarfs. Oversecretion causes gigantism or, in adults, acromegaly.
	Prolactin (PRL)	Involved in milk production in females. May potentiate stimulatory effects of LH on interstitial cells of the testes in males; may also potentiate the effects of testosterone on many of its target cells.	Undersecretion may cause failure of lactation after a woman gives birth to a child. Oversecretion may lead to lactation in a woman who has not recently given birth.
THYROID [*Fig. 17.8*]	Thyroxine (T_4 or tetraiodothyronine) and triiodothyronine (T_3)	Increase oxygen consumption and heat production (calorigenic effect). Important for normal growth and development. These hormones affect many metabolic processes.	Undersecretion leads to symptoms of hypothyroidism, possibly causing cretinism in children or myxedema in adults. Oversecretion leads to hyperthyroidism.
	Calcitonin	Lowers blood-calcium and -phosphate levels	
PARATHYROIDS [*Fig. 17.13*]	Parathyroid hormone	Affects calcium and phosphate metabolism to raise plasma-calcium levels and decrease plasma-phosphate levels	Undersecretion leads to increased neural and muscular excitability and possibly to tetanus. Oversecretion leads to bone decalcification, and calcification of soft tissues may occur.
ADRENALS [*Fig. 17.14*]			
Medulla	Epinephrine	Affects metabolic functions; generally tends to raise blood-glucose levels. Constricts vessels in skin, mucous membranes, and kidneys. Dilates respiratory passageways called bronchioles in the lungs	

continued on next page

◆ **TABLE 17.4 Major Endocrine Glands and Their Hormones (continued)**

GLAND	HORMONES	REPRESENTATIVE EFFECTS	SELECTED DISORDERS
ADRENALS [*Fig. 17.14*]			
Medulla	Norepinephrine	Increases heart rate and force of contraction of cardiac muscle; constricts blood vessels in almost all areas of the body	
Cortex	Mineralocorticoids, particularly aldosterone	Promote reabsorption of sodium and excretion of potassium from urine-forming structures of kidneys	Undersecretion may lead to decreased extracellular fluid volume and circulatory difficulties, and contribute to Addison's disease. Oversecretion may cause increased, extracellular fluid volume, edema, and hypertension.
	Glucocorticoids, particularly cortisol	Affect carbohydrate, lipid, and protein metabolism; generally tend to increase blood levels of glucose, amino acids, and fatty acids	Undersecretion contributes to Addison's disease. Oversecretion leads to Cushing's syndrome.
PANCREAS [*Fig. 17.20*]	Insulin	Affects many aspects of carbohydrate, lipid, and protein metabolism; generally tends to lower blood levels of glucose, fatty acids, and amino acids	Relative deficiency of insulin leads to hyperglycemia and diabetes mellitus
	Glucagon	Affects metabolism in fashion generally opposite of insulin; generally tends to raise blood-glucose levels	
GONADS			
Ovaries	Estrogens Progesterone	The gonadal hormones are involved in the processes of reproduction. Their functions are discussed in Chapters 28 and 29.	
Testes	Androgens, particularly testosterone		

Study Outline

◆ BASIC ENDOCRINE FUNCTIONS pp. 538–540

In general, endocrine glands are ductless structures that contain epithelial cells specialized for synthesis and secretion of specific hormones.

Hormones. Highly potent, specialized organic substances that function as intercellular signaling molecules.

Transport of Hormones. Many hormones, particularly those that are not very water soluble such as the steroid hormones and the thyroid hormones, bind to specific carrier proteins within bloodstream. Reversible association between a hormone and a carrier protein provides both storage system and buffer system for the hormone.

Mechanisms of Hormone Action. A particular hormone does not necessarily affect all cells—only its target cells. Target cells of a hormone possess receptors to which molecules of the hormone can attach.

BINDING OF HORMONES TO PLASMA-MEMBRANE RECEPTORS. Hormone molecule binds to receptor on plasma membrane; receptor undergoes conformational change; series of events is triggered that influences such cellular activities as rates of certain chemical reactions or permeability of the plasma membrane.

BINDING OF HORMONES TO RECEPTORS WITHIN CELLS. Hormone molecule enters cell and combines with receptor and activates receptor. Activated receptor interacts with DNA. In most cases, interaction activates certain genes; gene activation leads to mRNA synthesis, and ultimately to production of proteins (for example, enzymes) that influence cellular reactions or processes.

Hormonal Interrelationships. Almost every activity influenced by endocrine system is affected by groups of hormones acting together or in sequence. Some hormones influence other hormones' activities or rates of production.

Relationship between the Endocrine System and the Nervous System. Hormones influence neural functions; neurotransmitters contribute to the control of hormone secretion.

◆ PITUITARY GLAND pp. 540–548

Produces hormones that regulate several other endocrine glands and diverse body activities.

Embryonic Development and Structure.

NEUROHYPOPHYSIS. From ectodermal tissue of brain floor.

ADENOHYPOPHYSIS. From ectodermal tissue of roof of mouth.

Relationship to the Brain. Neurohypophysis connected to brain by way of infundibulum; brain and adenohypophysis linked by circulatory system.

Neurohypophyseal Hormones and Their Effects. Pars nervosa releases two peptide hormones.

ANTIDIURETIC HORMONE. Promotes reabsorption of water from urine-forming structures of kidneys; constricts arterioles.

OXYTOCIN. Stimulates smooth muscle of uterus; promotes contraction of myoepithelial cells that surround alveoli of mammary glands.

Release of Neurohypophyseal Hormones. Hormones synthesized in hypothalamic nuclei of brain by neurosecretory cells; transported to pars nervosa; neural inputs to brain influence release.

Adenohypophyseal Hormones and Their Effects. Adenohypophysis releases several hormones.

GONADOTROPINS. Follicle-stimulating hormone (FSH) and luteinizing hormone (LH); FSH and/or LH are involved in:

1. Ovarian follicle development and estrogen production.
2. Ovulation and formation of corpus luteum.
3. Development and maturation of sperm and androgen production.

THYROTROPIN. Stimulates synthesis and release of thyroid hormones.

ADRENOCORTICOTROPIN. Stimulates release of hormones from cortical region of adrenal gland, particularly the glucocorticoid hormone cortisol.

GROWTH HORMONE. Promotes growth in general and growth of skeletal system in particular.

PROLACTIN. Involved in initiation and maintenance of milk production in females. May potentiate stimulatory effects of LH on interstitial cells of testes in males. May also potentiate effects of testosterone on its target cells.

ENDORPHINS. Peptides in adenohypophysis that have opiatelike analgesic effects.

Pro-opiomelanocortin and the Production of Adenohypophyseal Hormones. Pro-opiomelanocortin is polypeptide molecule containing amino acid sequences of an enkephalin, an endorphin, and ACTH; is precursor molecule from which ACTH and other substances can be split.

Release of Adenohypophyseal Hormones. Brain exercises good deal of control.

RELEASING AND INHIBITING HORMONES. Manufactured in hypothalamus of brain by neurosecretory cells; travel through circulation by way of hypothalamic-hypophyseal portal veins to adenohypophysis, where they influence release of adenohypophyseal hormones.

NERVOUS SYSTEM EFFECTS ON ADENOHYPOPHYSEAL HORMONE RELEASE. Neural activity influences many endocrine functions.

Negative-Feedback Control of Endocrine Activity. Negative feedback contributes to control of endocrine function; long and short feedback loops.

Actions of Growth Hormone. Growth hormone stimulates mitosis and cell division in many tissues by way of somatomedins. It also influences several aspects of protein, carbohydrate, and lipid metabolism.

◆ CONDITIONS OF CLINICAL SIGNIFICANCE: PITUITARY GLAND p. 546

Can be caused by disorders of pituitary gland itself or by difficulties involving releasing or inhibiting substances from brain.

◆ THYROID GLAND pp. 548–553

Located anterior to upper part of trachea.

Embryonic Development and Structure. Originates as epithelial thickening in pharynx floor; right and left lobes joined across trachea by isthmus; consists of follicles capable of storing hormone.

Production of the Thyroid Hormones: Triiodothyronine and Thyroxine. Thyroglobulin contains iodine attached to tyrosine molecules.

Thyroid-Hormone Actions. Involved in wide range of metabolic activities; increase oxygen consumption and heat production. Excessive amounts of thyroid hormones increase utilization of carbohydrate and accelerate liberation of energy; stimulate many aspects of lipid metabolism.

Release of Thyroid Hormones. TSH from pituitary stimulates synthesis and release of thyroid hormones; thyrotropin-releasing hormone promotes TSH secretion.

Calcitonin. From parafollicular cells of thyroid; lowers blood-calcium and -phosphate levels.

◆ CONDITIONS OF CLINICAL SIGNIFICANCE: THYROID DISORDERS p. 552

1. *Cretinism.* Condition caused by severe hypothyroidism during infancy.
2. *Myxedema.* Severe hypothyroidism in the adult.
3. *Thyrotoxicosis.* Excess of thyroid hormones acting on tissues; hyperthyroidism, for example.
4. *Goiter.* Enlarged thyroid gland.

◆ PARATHYROID GLANDS p. 553

Embedded on posterior surface of thyroid gland.

Embryonic Development and Structure. From dorsal halves of third and fourth pairs of pharyngeal pouches; usually two masses of parathyroid tissue on posterior surface of each lateral lobe of thyroid.

Parathyroid Hormone and Its Effects. Two or three forms; controls calcium and phosphate metabolism.

◆ CONDITIONS OF CLINICAL SIGNIFICANCE: PARATHYROID DISORDERS p. 554

1. *Hyperparathyroidism.* Bone decalcification; calcification of soft tissues may occur.
2. *Hypoparathyroidism.* Lowered plasma-calcium level, increased neural and muscular excitability.

◆ ADRENAL GLANDS pp. 554–559

Located along superior border of each kidney; composed of inner medulla and outer cortex.

Adrenal Medulla.

EMBRYONIC DEVELOPMENT AND STRUCTURE. Arises from neural-crest cells; functions in manner similar to postganglionic sympathetic cells.

ADRENAL MEDULLARY HORMONES AND THEIR EFFECTS. Catecholamines released by adrenal medulla and sympathetic nervous system.

EPINEPHRINE. Tends to elevate blood-glucose levels; affects cardiovascular system.

NOREPINEPHRINE. Affects cardiovascular system; promotes breakdown of triglycerides in adipose tissue.

RELEASE OF THE ADRENAL MEDULLARY HORMONES. Controlled by preganglionic neurons from sympathetic division of autonomic nervous system; may be caused by emotional excitement, injury, exercise, low blood-glucose levels.

Adrenal Cortex.

EMBRYONIC DEVELOPMENT AND STRUCTURE. Derived from mesoderm of region that gives rise to gonadal tissue. Arranged in three layers: zona glomerulosa, zona fasciculata, and zona reticularis.

ADRENAL CORTICAL HORMONES AND THEIR EFFECTS.

MINERALOCORTICOIDS. Mainly aldosterone, which promotes reabsorption of sodium and excretion of potassium by urine-forming structures of kidneys.

GLUCOCORTICOIDS. Mainly cortisol, which favors gluconeogenesis, lipolysis, and elevation of glucose, amino acid, and fatty acid levels in the blood.

ESTROGENIC AND ANDROGENIC SUBSTANCES. Estrogenic materials that resemble female sex hormones. Androgenic substances—particularly dehydroepiandrosterone—that resemble male sex hormones.

RELEASE OF THE ADRENAL CORTICAL HORMONES. Renin-angiotensin system and concentration of potassium in fluids bathing adrenal glands influence release of aldosterone, as does sodium content of body; ACTH controls release of cortisol; CRH influences ACTH release.

◆ CONDITIONS OF CLINICAL SIGNIFICANCE: ADRENAL DISORDERS p. 558

1. *Addison's disease.* Primary hypofunction of adrenal cortex; inadequate aldosterone; impaired ability to conserve sodium and excrete potassium; inadequate glucocorticoids, fasting hypoglycemia, weakness.
2. *Cushing's syndrome.* Increased blood-glucose levels; increased protein breakdown.
3. *Hyperaldosteronism.* Potassium depletion; edema.
4. *Adrenal tumors:* May produce masculinizing effects, particularly evident in females.

◆ PANCREAS pp. 559–563

Both endocrine and exocrine gland.

Embryonic Development and Structure. Arises through fusion of two outgrowths from duodenum of small intestine; endocrine portion composed of cell clusters called pancreatic islets.

Pancreatic Hormones and Their Effects. A cells produce glucagon; B cells produce insulin; D cells produce somatostatin; F cells produce pancreatic polypeptide.

ACTIONS OF INSULIN. Facilitates uptake of glucose and utilization of glucose as an energy source by many cells; promotes synthesis of glycogen, triglycerides, and protein.

RELEASE OF INSULIN. The higher the blood-glucose level, the greater the insulin release. One or more hormones and autonomic neural activity also influence insulin release.

ACTIONS OF GLUCAGON. Activities are generally opposite to those of insulin; stimulates breakdown of liver glycogen.

RELEASE OF GLUCAGON. The lower the blood-glucose level, the greater the glucagon release. Sympathetic neural activity and epinephrine promote glucagon release.

◆ CONDITIONS OF CLINICAL SIGNIFICANCE: PANCREATIC DISORDERS pp. 562–563

Diabetes mellitus caused by relative insulin deficiency.

◆ **GONADS** p. 563
Testes and ovaries produce hormones and gametes.

◆ **OTHER ENDOCRINE TISSUES AND HORMONES** pp. 563–564
Digestive tract and placenta are sources of hormones; *thymus gland* produces thymosin; *pineal body* believed to be an endocrine gland.

◆ **PROSTAGLANDINS AND RELATED SUBSTANCES** p. 564
Prostaglandins, prostacyclin, thromboxanes, and leukotrienes are chemical messengers, and one or another of them is present in almost every body tissue. They act primarily as local messengers that exert their effects in the tissues that synthesize them.

Self-Quiz

1. Endocrine glands generally are ductless glands composed of epithelial cells that release their secretions directly into the target organ. True or False?
2. ADH and oxytocin are manufactured in the: (a) pars nervosa; (b) hypothalamus; (c) adenohypophysis.
3. Gonadotropins are: (a) produced by the ovaries and testes; (b) neurohypophyseal hormones; (c) adenohypophyseal hormones.
4. Surging levels of LH together with FSH are important in: (a) stimulation of the thyroid gland; (b) ovulation; (c) stimulation of the cortical region of the adrenal glands.
5. A hormone secreted by the adenohypophysis of the pituitary gland is: (a) thyrotropin; (b) antidiuretic hormone; (c) oxytocin.
6. Match the hormones with their appropriate lettered effects:

ADH	(a) Stimulates the release of cortisol from the adrenal glands
Oxytocin	(b) Stimulates the reabsorption of water from the urine-forming structures of the kidneys
LH	(c) Stimulates growth
TSH	(d) Stimulates the smooth muscle of the uterus
ACTH	(e) Stimulates the synthesis of the thyroid hormones
GH	(f) Acts on the interstitial cells of the testes to stimulate the production of the male sex hormones

7. Specialized neurosecretory cells within the brain manufacture a group of substances known collectively as: (a) gonadotropins; (b) luteinizing hormone; (c) releasing and inhibiting hormones.
8. Brain activities of the sort that are responsible for emotions are generally unable to influence hormonally controlled events. True or False?
9. Match the pituitary gland disorders with their possible lettered symptoms:

Lack of ACTH	(a) Underactivity of thyroid
Gonadotropin deficiency	(b) Menstrual impairment
TSH deficiency	(c) Pituitary dwarfs
Growth-hormone deficiency	(d) Acromegaly
Growth-hormone overproduction	(e) Deficient cortisol secretion
Antidiuretic-hormone deficiency	(f) Diabetes insipidus

10. Which of these glands produces hormones that contain iodine? (a) pituitary; (b) adrenal medulla; (c) thyroid.
11. The follicular storage form of the thyroid hormones is: (a) thyroglobulin; (b) thyroid-binding globulin; (c) free thyroxine.
12. The thyroid hormones generally travel freely in the circulation. True or False?
13. In the young, hypothyroidism may lead to: (a) elevated body temperature; (b) elevated basal metabolism; (c) abnormal bone development.
14. Enlargement of the thyroid gland is known as: (a) Addison's disease; (b) folliculitis; (c) goiter.
15. When the plasma-calcium level falls, parathyroid-hormone secretion: (a) decreases; (b) remains constant; (c) increases.
16. Hypoparathyroidism: (a) increases neural and muscular excitability; (b) raises plasma-calcium levels; (c) leads to calcification of soft tissues.
17. The release of epinephrine is controlled by preganglionic neurons to the adrenal medulla from the sympathetic division of the autonomic nervous system. True or False?
18. Epinephrine is able to: (a) elevate blood-glucose levels; (b) lower renal-potassium levels; (c) raise renal-sodium levels.
19. Steroid hormones: (a) are released only by the cortex of the adrenal glands; (b) may attach to carrier proteins within the bloodstream; (c) exert their effects solely by attaching to receptors at cell membranes.
20. Insulin release is stimulated by: (a) low blood-glucose levels; (b) the activity of sympathetic nerves to the liver; (c) gastrointestinal hormones.
21. A major effect of insulin is to facilitate the uptake of glucose by many cells. True or False?
22. Which hormone would most likely be released in increased amounts in response to elevated blood-glucose levels? (a) epinephrine; (b) glucagon; (c) insulin.

CHAPTER 18

The Cardiovascular System: The Blood

CHAPTER CONTENTS

LEARNING OBJECTIVES

After completing this chapter, you should be able to:

1. Describe four functions of the blood.
2. Describe three functions of the plasma proteins.
3. Explain the function of each of the formed elements of the blood.
4. Describe the formation and fate of erythrocytes.
5. Describe the structure and function of hemoglobin.
6. Discuss iron metabolism.
7. Explain the control of erythrocyte production.
8. Compare and contrast anemia and polycythemia.
9. Discuss the formation of a platelet plug as a hemostatic mechanism.
10. Explain the process of blood clotting, and distinguish between the extrinsic and intrinsic clotting pathways.
11. Explain why clots do not usually form within normal blood vessels, and describe the difficulties that can result from intravascular clotting.
12. Compare and contrast hemophilia and thrombocytopenia.

CHAPTER

18

The cells and tissues of the body are linked by the **cardiovascular system,** which includes the heart, the blood vessels, and the blood that flows within them. The blood transports materials throughout the body. It carries oxygen and food materials to and carbon dioxide and metabolic end products away from the cells. It also transports cellular products such as hormones between cells. In addition, the blood and its components help maintain homeostasis by buffering the tissues against extremes of pH, by aiding in the regulation of body temperature, and by protecting the tissues from toxic foreign materials or organisms.

In females, the general range of total blood volume is about 4 to 5 liters. In males, it is about 5 to 6 liters. The blood consists of both a liquid component, called *plasma,* and a nonliquid portion whose structures are collectively called *formed elements.*

Plasma

The liquid portion of the blood, the **plasma,** is approximately 91% water. The portion of the plasma that is not water consists of various dissolved or colloidal materials (Figure 18.1; Table 18.1). Hormones and other cellular products are present in the plasma, as are metabolic end products such as urea. The plasma contains plasma proteins, including various albumins, fibrinogen (which is involved in blood clotting), and globulins (some of which act as antibodies in immune responses, others of which serve as transport molecules). The plasma proteins act as buffers that help stabilize the pH of the internal environment, and they contribute importantly to the osmotic pressure and viscosity of the plasma. Various ions, including sodium (Na^+), chloride (Cl^-), and bicarbonate (HCO_3^-), are present within the plasma. These ions also contribute to plasma osmotic pressure, which averages about 300 milliosmoles per liter. The plasma contains food materials such as carbohydrates (for example, glucose), amino acids, and lipids, as well as gases such as oxygen, nitrogen, and carbon dioxide.

Formed Elements

The portion of the blood that is not plasma consists of **formed elements** (Figure 18.1; Table 18.2). The formed elements include *erythrocytes (red blood cells),* various types of *leukocytes (white blood cells),* and cell fragments called *platelets.* Among the different types of leukocytes are cells called *neutrophils, eosinophils, basophils, monocytes,* and *lymphocytes.*

Production of Formed Elements

The formed elements of the blood are produced in the **red bone marrow** *(myeloid tissue)* within bones by a process called **hemopoiesis.** (Prior to birth, formed elements are also produced in a number of other locations, including the liver and spleen.)

Red Bone Marrow

Red bone marrow contains precursor cells that give rise to the formed elements of the blood. The precursor cells, together with some fat cells, are packed within a fibrous meshwork of reticular connective tissue.

Red bone marrow is well supplied with blood by the nutrient artery of a bone. Capillaries from this artery form a system of thin-walled spaces called *sinusoids,* which permit the slow flow of blood throughout the marrow. Formed elements produced in the marrow enter the blood within the sinusoids.

In the fetus, almost all the bones contain red bone marrow. However, following birth, in most areas of bone marrow the number of precursor cells that give rise to the formed elements of the blood decreases, and the number of fat cells increases. Marrow that contains

◆ **TABLE 18.1 Principal Constituents of Plasma***

CONSTITUENT	APPROXIMATE AMOUNT
Water	91%
Plasma proteins	
Albumins	3.2 to 5.0 g/100 ml
Fibrinogen	0.20 to 0.45 g//100 ml
Globulins	
Alpha	0.40 to 0.98 g/100 ml
Beta	0.56 to 1.06 g/100 ml
Gamma	0.44 to 1.04 g/100 ml
Glucose	61 to 130 mg/100 ml
Cholesterol	128 to 347 mg/100 ml
Bilirubin *(total indirect reacting)*	0 to 1.1 mg/100 ml
Urea	13.8 to 39.8 mg/100 ml
Sodium	310 to 356 mg/100 ml
Potassium	12 to 21 mg/100 ml
Calcium *(total)*	8.2 to 11.6 mg/100 ml
Iron	0.04 to 0.21 mg/100 ml
Chloride	355 to 381 mg/100 ml

*Plasma also contains a wide variety of other substances too numerous to list here, including vitamins, hormones, enzymes, and metabolic end products.

◆ **FIGURE 18.1 Components of blood**
The values indicated are approximate, and in many cases there are individual variations and/or variations by sex.

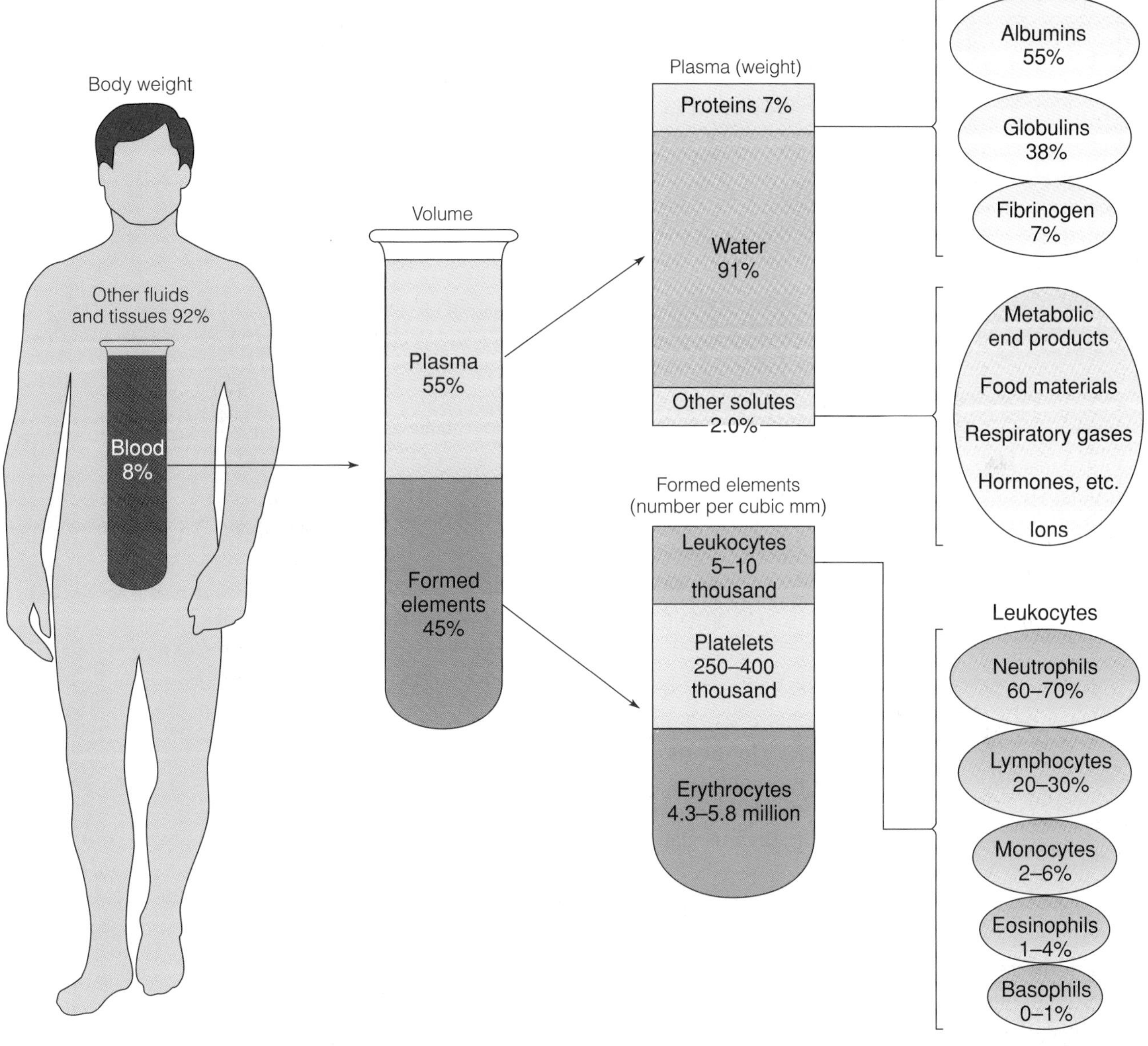

an abundance of fat cells appears yellow in color. Consequently, it is referred to as *yellow bone marrow.* Yellow bone marrow functions primarily as a site for fat storage and generally does not actively produce formed elements of the blood.

In the adult, most bones contain yellow bone marrow. Red bone marrow is present only in certain bones, such as the flat bones of the skull, the ribs, the sternum, the vertebrae, the pelvis, and the proximal epiphyses of the humerus and femur. The formed elements of the blood are normally produced only in these regions containing red bone marrow.

Hemopoiesis

Red bone marrow contains stem cells called **hemocytoblasts** *(pluripotent stem cells),* and all the formed elements of the blood are derived from these cells (Figure 18.2). The hemocytoblasts give rise to various **committed progenitor cells,** each of which develops into only one or a few types of formed elements. Some committed progenitor cells develop into erythrocytes. Others develop into large cells called *megakaryocytes,* which give rise to platelets. Still other committed progenitor cells develop into one or another of the different

◆ **TABLE 18.2 Formed Elements of the Blood**

FORMED ELEMENT	APPROXIMATE DIAMETER (microns)	APPROXIMATE ABUNDANCE (per mm³)	FUNCTION
ERTHROCYTES	7.5	Males: 5.1 to 5.8 million Females: 4.3 to 5.2 million	Transport oxygen and aid in the transport of carbon dioxide
PLATELETS	2.5	250,000 to 400,000	Involved in the processes of hemostasis and blood coagulation
LEUKOCYTES			
Granulocytes			
Neutrophils	12 to 14	3000 to 7000 (60 to 70% of total leukocytes)	Phagocytic cells that ingest bacteria and other foreign substances
Eosinophils	9	50 to 400 (1 to 4% of total leukocytes)	Particpate in the destruction of various types of parasites, such as parasitic worms; play a role in certain allergic inflammatory responses
Basophils	12	0 to 50 (0 to 1% of total leukocytes)	Believed to release chemicals such as histamine and heparin
Agranulocytes			
Monocytes	20 to 25	100 to 600 (2 to 6% of total leukocytes)	Develop into large phagocytic cells called macrophages within the tissues
Lymphocytes	9 (small) 12 to 14 (large)	1000 to 3000 (20 to 30% of total leukocytes)	Involved in specific immune responses, including antibody production

types of leukocytes, including *neutrophils, eosinophils, basophils, monocytes,* and *lymphocytes.* (In the tissues, monocytes develop into cells called *macrophages,* and some lymphocyte precursors leave the bone marrow and continue their development in the thymus gland.)

Factors That Influence Hemopoiesis

A number of glycoprotein substances promote the production of the formed elements of the blood, and in many cases they facilitate specific activities of mature leukocytes. These substances are usually referred to as **colony-stimulating factors (CSFs),** and they act by binding to plasma-membrane receptors.

A CSF called *interleukin-3* promotes the survival and proliferation of hemocytoblasts and most types of committed progenitor cells. Interleukin-3 and CSFs called *granulocyte/macrophage CSF, granulocyte CSF,* and *macrophage CSF* are believed to act in different combinations to regulate the selective production of neutrophils and macrophages. These four CSFs are produced by one or another of various cell types including lymphocytes, macrophages, fibroblasts, and endothelial cells. In response to acute bacterial infection, the blood levels of these CSFs rise, and large numbers of neutrophils are released from the bone marrow into the blood.

Researchers have described other factors that specifically promote the development of megakaryocytes, eosinophils, and basophils. However, these factors have not been clearly characterized.

A glycoprotein factor called *erythropoietin* acts on committed erythrocyte progenitors to stimulate erythrocyte production. Erythropoietin is discussed later in the chapter (page 578).

Erythrocytes

Erythrocytes, or **red blood cells (RBCs),** are small, circular, biconcave discs approximately 7.5 microns in diameter that have no nuclei (Figure 18.3; Table 18.2). Their principal function is to transport oxygen and carbon dioxide.

Erythrocytes are the most numerous of the formed elements contained in the blood, and they contribute importantly to total blood viscosity (which is normally 3.5 to 5.5 times that of water). Although their numbers are quite variable, a cubic millimeter of peripheral venous blood usually contains about 5.1 to 5.8 million erythrocytes in males and about 4.3 to 5.2 million erythrocytes in females.

The proportion of erythrocytes in a sample of blood is called the **hematocrit.** The hematocrit is obtained by centrifuging a blood sample in a hematocrit tube until

◆ **FIGURE 18.2 Production of blood cells from hemocytoblasts**

The existence of distinct lymphoid stem cells (which give rise to lymphocytes) and myeloid stem cells (which give rise to the other types of blood cells) is uncertain.

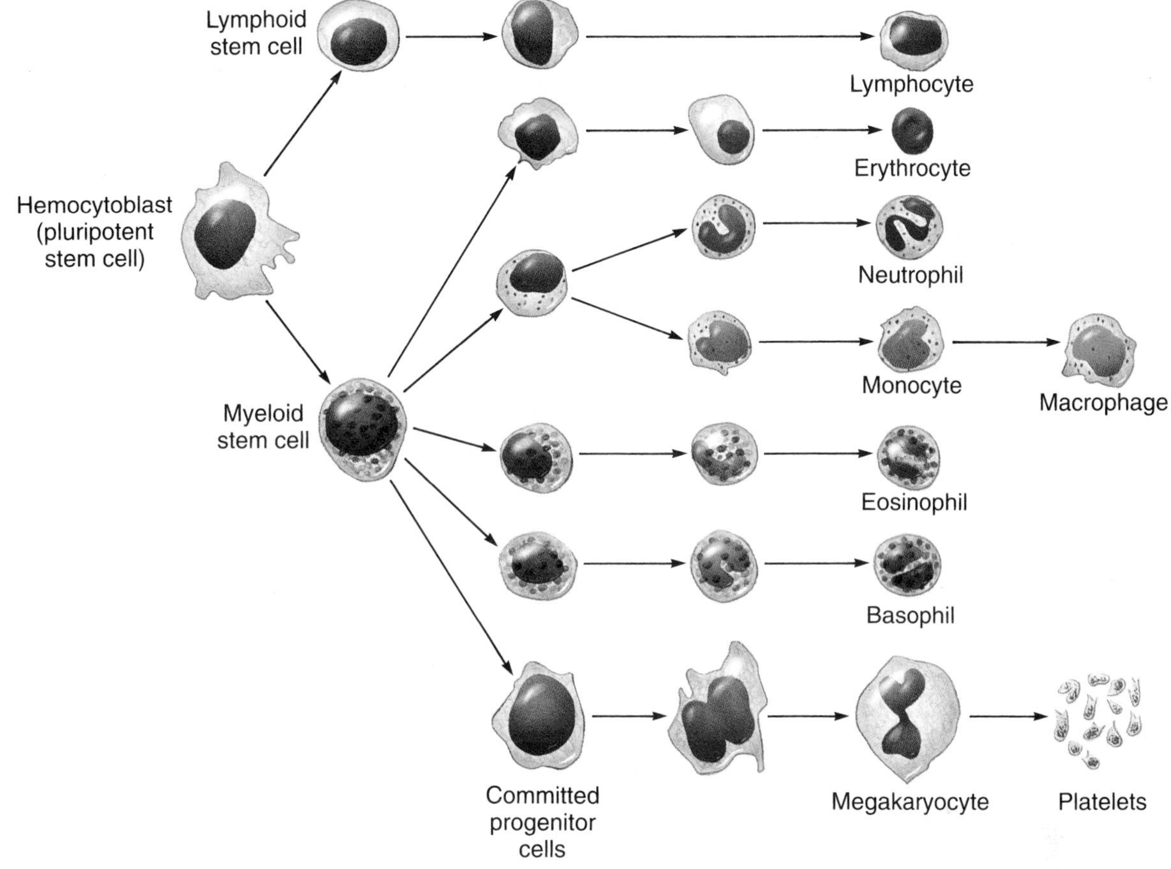

the erythrocytes are packed at the bottom of the tube with the plasma at the top of the tube. Between the erythrocytes and the plasma is the "buffy coat," a thin, whitish layer composed of leukocytes and platelets (Figure 18.4). The hematocrit is determined by measuring the ratio or percentage of the packed erythrocyte volume to the total sample volume. Although there is considerable variability, normal hematocrits of samples of peripheral venous blood average about 46.2% for males and 40.6% for females. The hematocrit of arterial blood is generally slightly lower than that of venous blood, and the hematocrit of blood in the very small vessels of the body is considerably lower than that of blood in the large arteries and veins.

Hemoglobin

A substance called **hemoglobin** is essential to the ability of erythrocytes to transport oxygen and carbon dioxide, and a single erythrocyte contains up to 300 million hemoglobin molecules. Males generally have more hemoglobin (14 to 18 grams per 100 ml of peripheral venous blood) than females (12 to 16 grams per 100 ml of peripheral venous blood).

A molecule of hemoglobin is composed of the protein *globin,* combined with four nonprotein groups called *hemes* (Figure 18.5). The globin protein consists of four polypeptide chains, each of which has a heme group

◆ **FIGURE 18.3 Erythrocytes**

A scanning electron micrograph showing the bioconcave disk shape of erythrocytes.

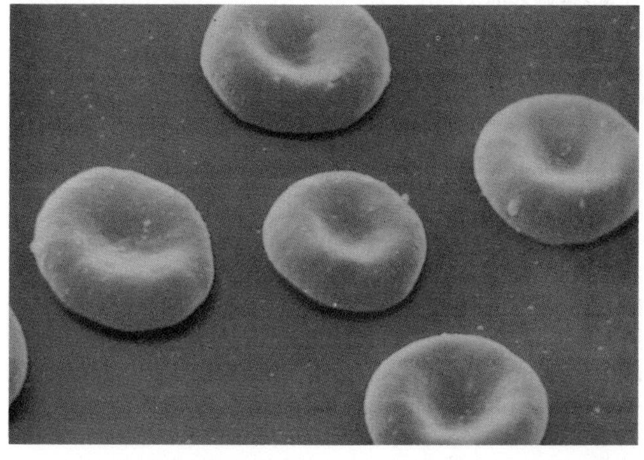

◆ **FIGURE 18.4 Hematocrit**

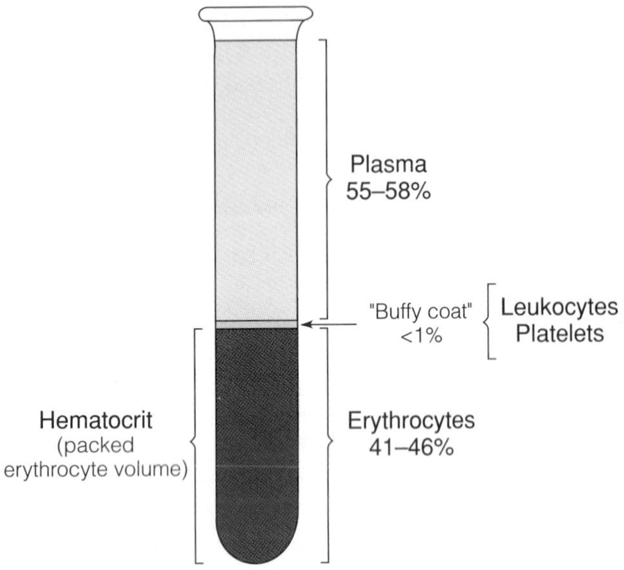

Plasma 55–58%

"Buffy coat" <1% — Leukocytes, Platelets

Erythrocytes 41–46%

Hematocrit (packed erythrocyte volume)

◆ **FIGURE 18.5 Hemoglobin**

(a) A hemoglobin molecule, showing its four polypeptide chains (2 α chains and 2 β chains). The structure in the center of each chain represents an iron-containing heme group. (b) Structure of a single iron-containing heme group.

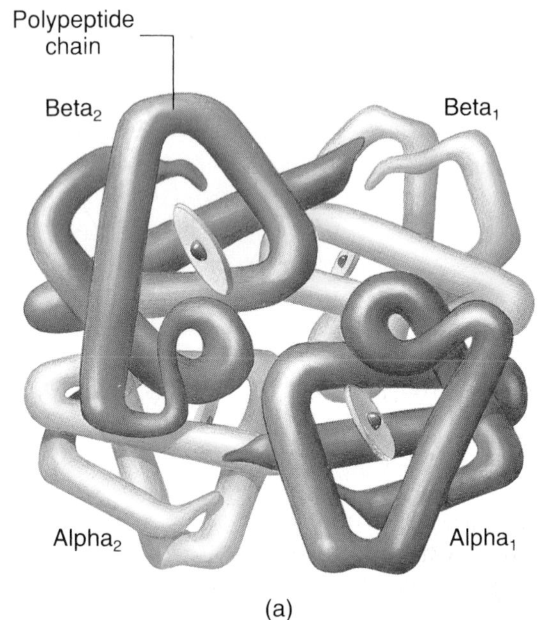

(a)

CH_3 CH_2CH_2COOH $CH_2{=}CH$ N Fe CH_2CH_2COOH CH_3 CH_3 $CH_2{=}CH$ CH_3

(b)

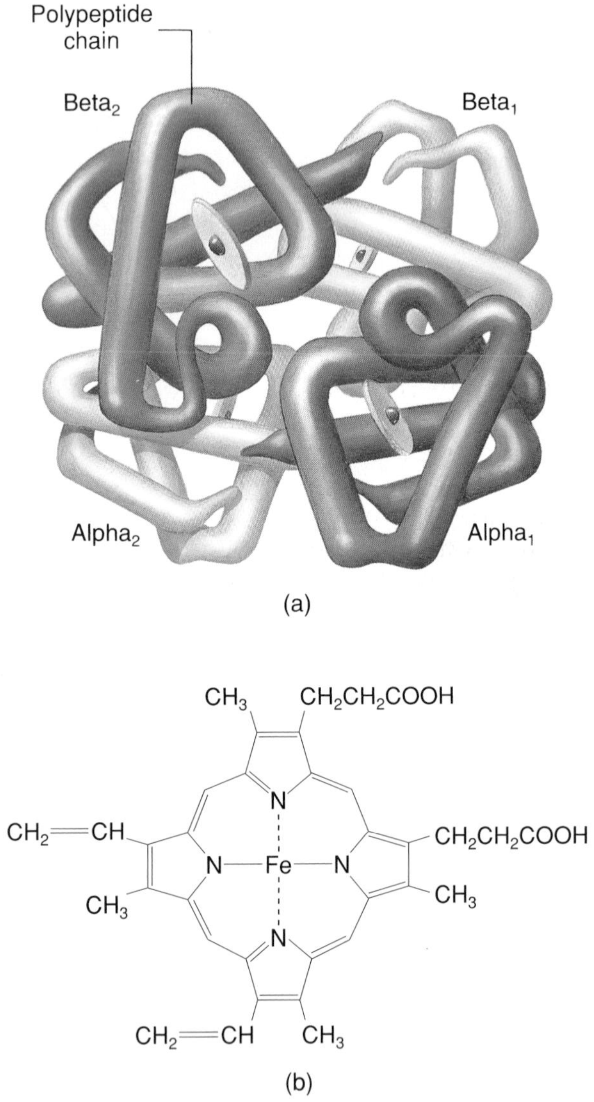

bound to it. Each heme group contains an iron atom (Fe^{+2}) that can combine reversibly with one molecule of oxygen. Therefore, a hemoglobin molecule can potentially associate with four oxygen molecules. When hemoglobin is combined with oxygen, it is called *oxyhemoglobin;* when it is not carrying oxygen, it is called *reduced hemoglobin.*

Under normal circumstances, a liter of blood traveling to the tissues contains approximately 198 ml of oxygen. Of this amount, about 195 ml are associated with hemoglobin molecules within erythrocytes. The remaining 3 ml are in physical solution dissolved in the plasma. Thus, erythrocytes and their hemoglobin account for almost all of the blood's ability to transport oxygen.

Hemoglobin can also combine with carbon dioxide and thus aid in the transport of this substance. However, in contrast to oxygen, which is carried in association with the iron of the heme groups of hemoglobin, carbon dioxide is carried in reversible association with the protein portion of the hemoglobin molecule. The transport of oxygen and carbon dioxide by hemoglobin is discussed further in Chapter 23 (beginning on page 734).

Iron

The production of hemoglobin requires iron, and the body normally contains about 4 grams of iron. Approximately 65% of the body's iron is found in hemoglobin, and about 15 to 30% is stored in the liver, spleen, bone marrow, and elsewhere, mainly in the form of intracellular iron-protein complexes called *ferritin* and *hemosiderin.* In the blood, iron forms a loose combination with a plasma protein called *transferrin,* and it is in this form that iron is transported in the plasma. Iron can be released from transferrin for use or storage.

Small amounts of iron are lost each day through the feces, urine, sweat, and cells sloughed from the skin. In women, additional iron is lost as a result of menstrual bleeding. The average daily loss of iron is about 0.9 mg in males and about 1.7 mg in females.

Iron losses must be replaced if iron homeostasis is to be maintained. Ingested iron is absorbed into the body by active processes, primarily in the upper portion of the small intestine, and total body iron is largely regulated by alterations in the intestinal absorption rate. When the body stores of iron are high, the intestinal absorption rate of iron is low; but when iron stores are depleted, the rate of iron absorption increases. In general, the rate of iron absorption is slow, and even at the maximum absorption rate only a few milligrams of iron can be absorbed per day.

Among the substances required for the production of erythrocytes is vitamin B_{12}. Vitamin B_{12} is required for DNA formation and, as a result, is necessary for nuclear maturation and division. Folic acid, another vitamin, is also required for DNA formation and thus for erythrocyte production.

Production of Erythrocytes

During erythrocyte production, some hemocytoblasts in the red bone marrow give rise to committed progenitor cells called *proerythroblasts* (Figure 18.6). The proerythroblasts are cells that have just begun to activate the biochemical machinery needed for the synthesis of hemoglobin. As hemoglobin synthesis occurs, the proerythroblasts differentiate into cells called *basophil erythroblasts.* As hemoglobin synthesis continues, the basophil erythroblasts give rise to cells called *polychromatophil erythroblasts,* which in turn differentiate into cells called *normoblasts (acidophil erythroblasts).*

When the normoblast cytoplasm has attained a hemoglobin concentration of about 34%, the nucleus is pinched off, enclosed in a portion of the plasma membrane and a thin layer of cytoplasm. The loss of the nucleus causes the cell to collapse inward, and it assumes a biconcave shape. This nonnucleated cell is called a *reticulocyte (polychromatophil erythrocyte).* Reticulocytes are essentially young erythrocytes, and it is these cells that are usually released from the bone marrow into the blood. Reticulocytes lose their endoplasmic reticulum, mitochondria, and ribosomes and become mature erythrocytes within one or two days after their release from the bone marrow. The normal proportion of reticulocytes in the blood is less than 0.5 to 1.0%.

The process of erythrocyte formation is called **erythropoiesis** *(e-rith-ro-poi-ee´-sis).* During erythropoiesis, the various cells continue to divide through the normoblast stage of development so that greater and greater numbers of cells are formed. Occasionally, when erythrocytes are being manufactured very rapidly, nucleated cells appear in the blood, but this is not a usual occurrence.

Control of Erythrocyte Production

A hormone called **erythropoietin** stimulates erythrocyte production. Erythropoietin is a glycoprotein produced in the kidneys and possibly elsewhere. The production of erythropoietin, and thus erythrocyte production, is regulated by a negative-feedback mechanism that is sensitive to the amount of oxygen delivered to the tissues. An inadequate supply of oxygen—particularly to the kidneys—leads to a greater production of erythropoietin and thereby to increased erythrocyte production. Conversely, an increased oxygen supply to the tissues leads to a decrease in erythropoietin levels and, as a result, to a decrease in erythropoiesis.

The male sex hormone testosterone enhances erythropoietin production, and the estrogenic female sex hormones tend to depress it. The influences of these hormones could account at least in part for the greater numbers of erythrocytes and higher levels of hemoglobin in males than in females.

Fate of Erythrocytes

Because a mature erythrocyte does not have a nucleus, endoplasmic reticulum, mitochondria, or ribosomes, it cannot synthesize proteins, and it cannot replace enzymes that are degraded. The loss of enzymes affects many cellular activities, including the ability to produce ATP, and a mature erythrocyte is unable to grow or divide. The average life of an erythrocyte is approximately 120 days in males and 109 days in females.

Aged, abnormal, or damaged erythrocytes are disposed of by phagocytic cells called *macrophages* (Figure 18.7). These cells are found in the spleen, liver,

◆ **FIGURE 18.6 Production of erythrocytes**

Hemocytoblasts proliferate and undergo sequential differentiation, forming a series of precursor cells.

Precursor cells called normoblasts extrude their nuclei, and the resulting reticulocytes are released into the blood.

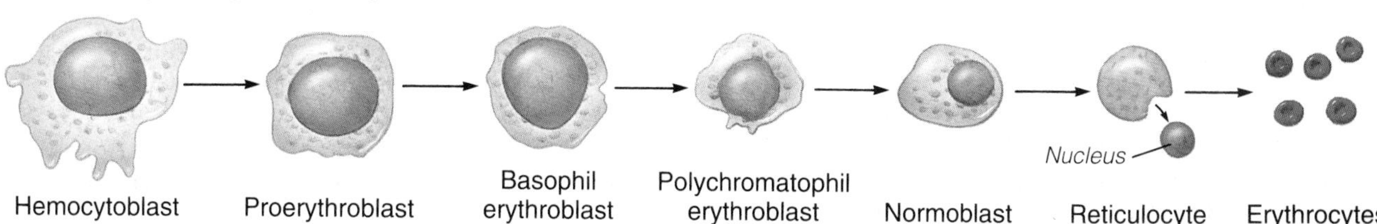

ASPECTS OF EXERCISE PHYSIOLOGY

Blood Doping: Is More of a Good Thing Better?

A continual supply of oxygen to exercising muscles is essential for generating energy to sustain endurance activities. Blood doping is a technique designed to temporarily increase the oxygen carrying capacity of the blood in an attempt to gain a competitive advantage. Blood doping involves removing blood from an athlete, then promptly reinfusing the plasma but freezing the red bloods cells for reinfusion one to seven days before a competitive event. One to four units of blood (one unit equals 450 ml) are usually withdrawn at three- to eight-week intervals before the competition. Increased erythropoietic activity restores the red blood cell count to a normal level in the intervening period between blood withdrawals.

Reinfusion of the stored red blood cells temporarily increases the red blood cell count and hemoglobin level above normal. Theoretically, blood doping would be beneficial to endurance athletes by improving the blood's oxygen carrying capacity. If too many red cells are infused, however, performance could suffer because the increased blood viscosity would decrease blood flow.

Early research into athletic performance after blood doping produced contradictory results. More recent research, however, indicates that, in a standard laboratory exercise test, athletes may realize a 5% to 13% increase in aerobic capacity; a reduction in heart rate during exercise compared to that during the same exercise in the absence of blood doping; improved performance; and reduced lactic acid levels in the blood. (Lactic acid is produced when muscles resort to less efficient anaerobic glycolysis for energy production—see page 256). Blood doping, although probably effective, is illegal in both collegiate athletics and Olympic competition for ethical and philosophical reasons.

bone marrow, and other tissues. The spleen is particularly important in the destruction of pathologic or defective erythrocytes. If the spleen is removed, the number of abnormal erythrocytes circulating in the blood increases considerably.

During erythrocyte destruction, macrophages degrade hemoglobin (Figure 18.7). The globin protein is broken down into its component amino acids, and the amino acids can be used in the synthesis of new proteins. The iron is liberated from the nonprotein heme, and the remainder of the heme is converted to *biliverdin (bil-i-ver´-din).* Biliverdin is in turn converted to *bilirubin (bil-i-roo´-bin),* which enters the blood and binds to albumin. Bilirubin is eventually taken up by the liver, conjugated to glucuronic acid, and secreted in the bile. The liberated iron also enters the blood, where it combines with transferrin. The iron can be used in the synthesis of new hemoglobin during erythropoiesis in the bone marrow, or it can be stored.

Blood Groups

An *antigen* is a substance that has the capacity to elicit a specific immune response, such as the production of a specialized protein called an *antibody.* An antibody can react with the specific antigen that stimulated its production. The surfaces of erythrocytes contain antigens that can react with particular antibodies. These reactions form the basis of classifying blood into various groups—most commonly, groups known as type A, type B, type AB, or type O. Because these classifications are based on antigen-antibody reactions, they are considered in detail with the specific immune responses in Chapter 22 (page 705).

Leukocytes

Leukocytes, or **white blood cells (WBCs),** are formed elements that defend the body against invasion by foreign organisms or chemicals and remove debris that results from dead or injured cells. Leukocytes are present within the blood in much smaller numbers than erythrocytes: they comprise about 1% of the total blood volume, and a cubic millimeter of blood usually contains between 5000 and 10,000 leukocytes.

Leukocytes are components of the immune system (Chapter 22), and they act primarily in the loose connective tissues and in lymphoid tissues, such as the lymph nodes. The leukocytes within the blood are mainly being transported by the circulation. Leukocytes leave the blood by a process called **diapedesis,** in which they pass between cells of the capillaries and enter the tissues.

◆ **FIGURE 18.7 Disposal of erythrocytes**

Aged, abnormal, or damaged erythrocytes are disposed of by macrophages. The iron that is released may be either stored or reused in the synthesis of new hemoglobin.

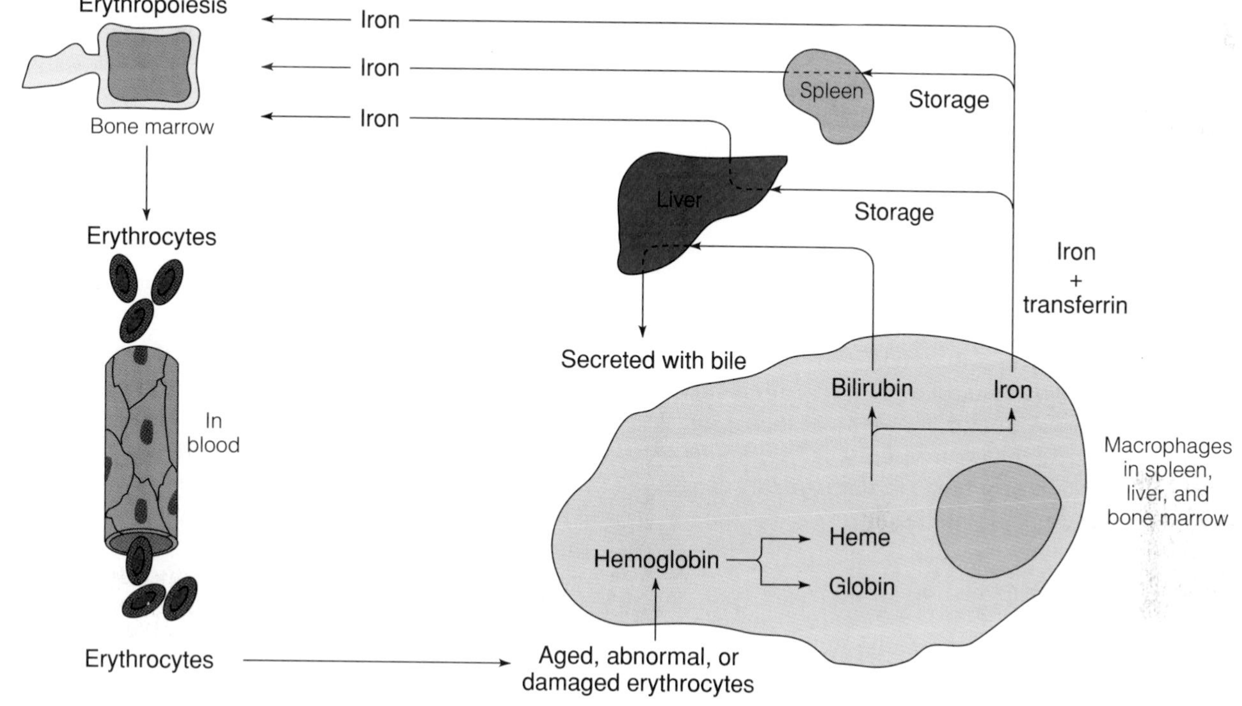

There are two major classes of leukocytes: granulocytes and agranulocytes (Figure 18.8; Table 18.2).

Granulocytes

Granulocytes are formed in the red bone marrow. They have lobed nuclei and clearly evident granules in their cytoplasm. Granulocytes circulate in the blood for only a few hours before leaving capillaries and entering the connective tissues, where they can survive for several days.

Three types of granulocytes are distinguished, based on their reactions to certain stains. They are neutrophils, eosinophils, and basophils.

Neutrophils. Neutrophils possess small cytoplasmic granules that appear light pink to blue-black when stained with Wright's stain. Many of these granules are lysosomes and secretory vesicles.

Neutrophils typically have a nucleus that varies in shape and consists of two or more lobes connected by

◆ **FIGURE 18.8 Types of leukocytes**

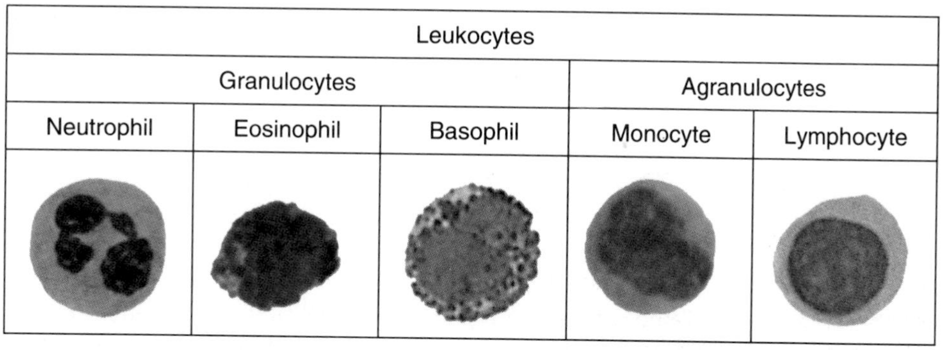

Leukocytes				
Granulocytes			Agranulocytes	
Neutrophil	Eosinophil	Basophil	Monocyte	Lymphocyte

CONDITIONS OF CLINICAL SIGNIFICANCE

Erythrocytes

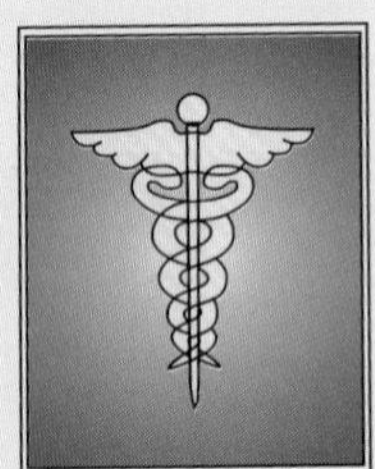

Anemia

Anemia is a condition characterized either by a decreased number of erythrocytes in the blood or by a decreased concentration of hemoglobin. Whatever the cause, anemia decreases the blood's ability to transport oxygen to the tissues. Because tissues cannot function at optimum levels without adequate supplies of oxygen, anemia is frequently associated with listlessness, fatigue, and a lack of energy. Anemia can result from a number of conditions, several of which are discussed in the following sections.

Hemorrhagic Anemia

Hemorrhagic anemia is the result of the loss of substantial quantities of blood. When blood is lost as the result of such occurrences as large wounds, stomach ulcers, or excessive menstrual bleeding, the lost volume is rather quickly replaced by fluid from the tissue spaces. The lost erythrocytes, however, are replaced by the slower processes of erythropoiesis. Until erythrocyte replacement is complete (which may take several weeks), there are fewer circulating erythrocytes than normal. Repeated or chronic hemorrhage can produce severe anemia, even if there is an increase in erythropoiesis.

Iron-Deficiency Anemia

Iron-deficiency anemia is due to an excessive loss, deficient intake, or poor absorption of iron. In iron-deficiency anemia, the erythrocytes produced are smaller than usual, and they contain less hemoglobin than normal.

Pernicious Anemia

Pernicious anemia is the result of an inability to absorb adequate amounts of vitamin B_{12} from the digestive tract. Vitamin B_{12}, which is required for the normal maturation of erythrocytes, is found in such foods as liver, kidney, milk, eggs, cheese, and meat. The absorption of vitamin B_{12} requires the presence of a glycoprotein substance called *intrinsic factor*, which is produced by cells in the epithelial lining of the stomach. Individuals who suffer from pernicious anemia lack sufficient quantities of functional intrinsic factor and cannot adequately absorb vitamin B_{12}. As a result, they produce fewer erythrocytes than normal, and those that are produced are larger and more fragile than normal. Pernicious anemia can be treated by injections of vitamin B_{12}, thus bypassing the absorption problem.

Sickle-Cell Anemia

Sickle-cell anemia is the result of the production of an abnormal hemoglobin (hemoglobin S; HbS) due to a genetic defect. In this condition, the globin portion of the molecule is abnormal. When the abnormal hemoglobin molecules are exposed to low concentrations of oxygen, they form fibrous precipitates within the erythrocytes, distorting them into the sickle shape characteristic of the disease. The misshapen erythrocytes are very fragile and often rupture

narrow strands. Consequently, neutrophils are also called *polymorphonuclear* ("many-shaped nucleus") *leukocytes.* However, this term is sometimes used to refer to all three types of granulocytes, since the shape of the nucleus varies somewhat in all of them.

Neutrophils are the most abundant type of leukocyte, comprising approximately 60–70% of the total leukocyte count. Neutrophils are phagocytic cells that are capable of ameboid movement. They are able to leave the blood and enter the tissues, where they protect the body by ingesting bacteria and other foreign substances.

Eosinophils. Eosinophils possess coarse cytoplasmic granules that appear reddish orange when stained with Wright's stain. Many of these granules are believed to be lysosomes.

Eosinophils participate in the destruction of various types of parasites, such as parasitic worms. They also play a role in certain allergic inflammatory responses.

Basophils. Basophils possess relatively large cytoplasmic granules that appear reddish purple to blue-

CONDITIONS OF CLINICAL SIGNIFICANCE

as they pass through the capillaries, and especially through the spleen. Sickled cells can also become trapped in small blood vessels, impeding blood flow. In its severest form sickle-cell anemia can be fatal.

Thalassemia

Like sickle-cell anemia, the various forms of *thalassemia* are the result of a genetic defect that causes the production of abnormal hemoglobin. In thalassemia, the erythrocytes are thin and fragile, and their numbers are severely depleted. Persons afflicted with thalassemia may experience extreme anemia and require frequent blood transfusions.

Aplastic Anemia

Aplastic anemia is the result of an inadequate production of erythrocytes due to the inhibition or destruction of the red bone marrow. Aplastic anemia can be caused by radiation, various toxins, and certain medications.

Polycythemia

Polycythemia is a condition in which there is a net increase in the total circulating erythrocyte mass in the body. There are several types of polycythemia.

Primary Polycythemia

Primary polycythemia (also called *polycythemia vera* or *erythremia*) occurs when excess erythrocytes are produced as a result of tumorous abnormalities of the tissues that produce blood cells. Often, excess white blood cells and platelets are also produced. In primary polycythemia, there may be 8 to 9 million and occasionally 11 million erythrocytes per cubic millimeter of blood, and the hematocrit may be as high as 70 to 80%. In addition, the total blood volume sometimes increases to as much as twice normal. The entire vascular system can become markedly engorged with blood, and circulation times for blood throughout the body can increase up to twice the normal value. The increased numbers of erythrocytes can increase the viscosity of the blood to as much as five times normal. Capillaries can become plugged by the very viscous blood, and the flow of blood through the vessels tends to be extremely sluggish.

Secondary Polycythemia

Secondary polycythemia is caused by either appropriate or inappropriate increases in the production of erythropoietin that result in an increased production of erythrocytes. In secondary polycythemia, there may be 6 to 8 million and occasionally 9 million erythrocytes per cubic millimeter of blood. A type of secondary polycythemia in which the production of erythropoietin increases appropriately is called *physiologic polycythemia.* Physiologic polycythemia occurs in individuals living at high altitudes (4275 to 5200 meters), where oxygen availability is less than at sea level. Such people may have 6 to 8 million erythrocytes per cubic millimeter of blood.

black when stained with Wright's stain. Many of these granules are secretory vesicles that contain a number of chemical substances, including *histamine* and *heparin.*

The exact function of basophils is still being investigated. They are believed to be functionally similar to cells called *mast cells,* which occur in connective tissue and in the extracellular spaces near blood vessels, particularly in the lungs. Mast cells have granules much like those in basophils, and they also contain histamine and heparin. Histamine causes vascular dilation and increases the permeability of blood vessels during inflammation. It also contributes to allergic responses. Heparin can prevent blood clotting and can also enhance the removal of fat particles from the blood after a fatty meal.

Agranulocytes

Certain leukocytes, called **agranulocytes** or *nongranular leukocytes,* do not have prominent granules in their cytoplasm. There are two types of agranulocytes: monocytes and lymphocytes.

CONDITIONS OF CLINICAL SIGNIFICANCE

Leukocytes

Leukemia

Leukemia is a cancerous condition in which an uncontrolled proliferation of leukocytes leads to a diffuse and almost total replacement of the red bone marrow with leukemic cells. If the leukemic cells are from the granulocyte or, in some cases, monocyte cell lines, the leukemia is referred to as *myelogenous leukemia.* If the leukemic cells are from the lymphocyte cell line, the leukemia is called *lymphocytic leukemia.* Leukemic cells often replace the cells that form erythrocytes, and anemia results. In addition, there is frequently a decrease in blood platelets, which are also formed in the red bone marrow. Since platelets are involved in blood clotting and other hemostatic processes, leukemia may be accompanied by bleeding and hemorrhage. In fact, one of the causes of death in leukemia is internal hemorrhage, especially cerebral hemorrhage. Many of the leukocytes that are formed in the red bone marrow in leukemia are immature or abnormal. These abnormal leukocytes are unable to defend the body adequately against invasion by foreign organisms, and leukemia victims can also die from infection. Leukemic cells can leave the bone marrow and infiltrate other tissues or organs in the body, where they can damage or interfere with the functions of the tissues or organs.

Treatment of leukemia may involve the use of radioactive isotopes or antileukemic drugs to destroy the rapidly dividing leukocytes. Although these therapies do not cure the disease, they may induce a remission that can last for several years. If a compatible donor is available, bone marrow transplants are sometimes used.

Infectious Mononucleosis

Infectious mononucleosis, which occurs mostly in children and young adults, is caused by a virus called the *Epstein-Barr virus.* The disease is characterized by an increase in the relative and absolute numbers of monocytes and lymphocytes in the blood, many of which are atypical. The symptoms of infectious mononucleosis include fatigue, sore throat, and slight fever. Affected individuals are usually watched carefully for complications during the course of the disease, which may last several weeks. Recovery is usually complete, without any ill effects.

Monocytes. Monocytes are formed in the red bone marrow. They are capable of ameboid movement and can leave the blood and enter the tissues where they develop into large cells called *macrophages.* Macrophages, which can survive in the tissues for months and perhaps even for years, are phagocytic cells that can ingest bacteria and other foreign substances.

Lymphocytes. Only a few of the total number of **lymphocytes** are present in the blood; most are lodged in the lymphoid tissues. Lymphocytes are classified on the basis of their size as small or large. Most of the lymphocytes in the blood are small. Large lymphocytes are located mainly in the lymphoid organs.

Lymphocytes are important in the body's specific immune responses, including antibody production. The development and specific functions of lymphocytes are discussed in Chapter 22.

Differential Count

It is important for diagnostic purposes to be able to estimate the relative abundance of each type of leukocyte in the blood. This can be done by a procedure called a **differential count.** A differential count involves staining a blood smear and then identifying and counting the leukocytes under a microscope. A differential count can also be obtained by placing a blood sample in a specialized machine that automatically identifies and tabulates the various types of leukocytes. A differential count provides a percentage figure for each type of leukocyte. The percentages of the different types of leukocytes present in the blood change in certain diseases, and a differential count can aid in their diagnosis.

CLINICAL CORRELATION

Monomyeloblastic Leukemia

Case Report

THE PATIENT: A 63-year-old woman.

PRINCIPAL COMPLAINTS: Weakness, fatigue, and dyspnea (difficult breathing).

HISTORY: The patient appeared to be chronically ill. During the last year, she had gradually lost 25 pounds in weight. The weakness and fatigue had become noticeable several months before she was seen and were progressing. She reported difficulty climbing even a single flight of stairs.

CLINICAL EXAMINATION: The jugular veins of the neck were visible (distended) almost to the angles of the mandible when the patient was sitting at 45° to the horizontal. The heart was moderately enlarged, the heart rate was 198/min (normal, <100/min), and the rate of respiration was 20/min (normal, <17/min). The liver was enlarged. The lower edge of the liver was felt 4–5 cm (normal, 1–2 cm) below the right costal margin (edge of the rib cage). The upper edge of the liver, determined by percussion, was at the normal level of the fifth rib. Pitting edema of the ankles was detected; that is, tissue in the ankle area was so swollen by fluid contained in the extracellular spaces (edema) that it could be indented by pressure with the fingers. The oral temperature was 37.9°C. The arterial blood pressure was 115/60 mm Hg (normal, 90–140/60–90 mm Hg). The hematocrit was 32% (normal, 37–48%). The white cell count was 66,000/mm^3 (normal, 4300–10,800/mm^3), with 1% neutrophils, 27% monocytes, 10% lymphocytes, 43% myeloblasts (immature granulocytes), and 26% promonocytes (immature monocytes) (normal: neutrophils, 60–70%; monocytes, 2.4–11.8%; lymphocytes, 20–53%; and immature forms, 0–25%). The blood levels of several enzymes were determined: serum glutamic oxaloacetic transaminase (SGOT) was 20 U/ml (normal, 10–40 U/ml); lactic dehydrogenase (LDH) was 295 U/ml (normal, 60–120 U/ml); and creatine phosphokinase (CPK) was 5 U/ml (normal, 5–35 U/ml). High blood levels of these enzymes are consistent with damage to heart muscle. An electrocardiogram (ECG) was obtained, and it displayed abnormalities that were consistent with acute myocardial infarction (an area of dead heart muscle cells like that resulting from a heart attack).

Digoxin, which increases cardiac contractility, was administered. Five days after admission, bone marrow was aspirated from the left posterior iliac crest. Pathological examination revealed large numbers of immature mononuclear cells, consistent with the diagnosis of acute monocyte leukemia. Six days after admission, the hematocrit was 27%, the white cell count was 167,000/mm^3, with 27% monocytes and 36% immature forms (monoblasts). The patient became increasingly weak and dyspneic and died nine days after admission.

COMMENT: Leukemia is generally characterized by an abnormally high concentration of leukocytes without any demonstrable stimulus. The high proportion of immature forms that occurs in many types of leukemia is believed to represent some cellular defect in maturation. The concentrations of normal blood cells usually are decreased. The abnormal leukocytes infiltrate the organs and systems of the body, although the leukemic cells are not invasive in the way that the cells of malignant carcinomas (malignant tumors of epithelial origin) are invasive. Enlarged spleen, liver, and lymph nodes are common. The central nervous system, the heart, the kidneys, and the gastrointestinal system can be affected adversely.

In this patient, infiltration of the heart caused heart failure and ECG abnormalities consistent with myocardial infarction. However, changes of serum enzymes characteristic of infarction (SGOT, LDH, and CPK) were not seen. Postmortem examination revealed leukemic infiltration of the heart, lungs, liver, and colon; the latter contained numerous leukemic ulcerations. Patients who have leukemia may die within a few days of diagnosis; or, with treatment, they may live for years. The anemia and organ degeneration of this patient precluded the usual chemotherapy and contributed to her early demise.

◆ **FIGURE 18.9 Megakaryocytes**

(a) Schematic drawing of a megakaryocyte among other cells in bone marrow. Processes extend from the megakaryocyte into an adjacent blood sinus. Adapted from Alberts, et al., *Molecular Biology of the Cell,* Garland Press. Figure 17-30, p. 977. (b) Scanning electron micrograph of the interior of a blood sinus in the bone marrow, showing the megakaryocyte processes (× 1225). (From *Tissues and Organs: A Text-Atlas of Scanning Electron Microscopy,* by Richard G. Kessel and Randy H. Kardon. W. H. Freeman and Company. Copyright © 1979. Reprinted with permission.)

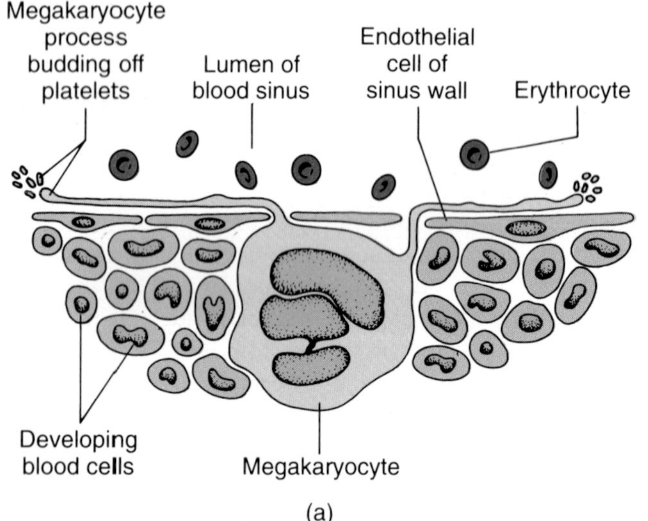

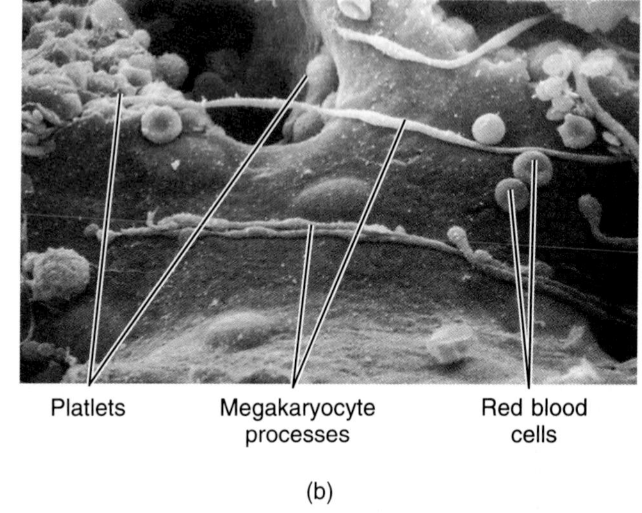

Platelets

Platelets are formed elements that are small cytoplasmic fragments. They are about 2.5 microns in diameter and contain many secretory vesicles (granules). There are about 250,000 to 400,000 platelets per cubic millimeter of blood.

Platelets are formed in the red bone marrow as pinched-off portions of large cells called **megakaryocytes.** Megakaryocytes are located close against the blood sinuses that drain the bone marrow, and they extend processes through holes in the endothelium lining the sinuses (Figure 18.9). Platelets pinch off from the processes, which are bathed by the blood within the sinuses. The platelets are carried away with the blood.

Platelets are involved in blood clotting. They also participate in other hemostatic processes.

Hemostasis

From time to time, blood vessels break and bleeding occurs. Occasionally, bleeding occurs from breaks in medium- or large-size blood vessels. During normal living, however, most bleeding occurs from breaks in small blood vessels.

The arrest of bleeding is called **hemostasis.** If a medium- or large-size artery breaks, the body is generally unable to achieve hemostasis without outside intervention, such as the application of pressure or the closing of the break with sutures. If a small blood vessel breaks, the body achieves hemostasis by mechanisms that include vascular spasm, the formation of a platelet plug, and blood clotting.

Vascular Spasm

When a blood vessel is cut or ruptured, one of the first noticeable responses is a constriction of the vessel. This constriction, which is called a **vascular spasm,** decreases blood flow and helps minimize blood loss. In addition, the constriction can bring opposing endothelial surfaces of the blood vessel together. The surfaces adhere to one another, further decreasing blood flow.

The mechanisms responsible for the vascular spasm include a contraction of vascular smooth muscle in direct response to injury and neurally mediated vascular constriction reflexes that occur in response to pain. In general, the greater the damage to a vessel, the greater the degree of constriction. Thus, a vessel ruptured by crushing usually bleeds less than a cleanly cut vessel.

Formation of a Platelet Plug

Within an undamaged vessel, platelets generally circulate freely and do not stick to the normal endothelial cells that line the vessel. However, in an injured vessel, where the deeper lying subendothelial tissues are exposed, platelets adhere to the underlying connective tissues—specifically collagen. (The adhesion of platelets to collagen is facilitated by a plasma protein called *von Willebrand factor,* which is secreted by endothelial cells.)

When platelets adhere to collagen, they release adenosine diphosphate (ADP) and other chemicals from their secretory vesicles. Many of the chemicals, including ADP, induce changes in platelet surfaces that cause the surfaces to become "sticky." As a result, additional platelets adhere to the original platelets. These additional platelets release chemicals from their secretory vesicles, and still more platelets adhere. Thus, by a process of **platelet aggregation,** a platelet plug or clump forms.

Platelet aggregation is enhanced by *thromboxane* A_2, which is produced in the platelet plasma membrane when platelets adhere to collagen. Thromboxane A_2 directly promotes platelet aggregation, and it triggers the release of chemicals from platelet secretory vesicles. (Thromboxane A_2 also stimulates vascular constriction.)

Platelet aggregation is inhibited by *prostacyclin,* which is produced by normal endothelial cells that line blood vessels. Prostacyclin plays an important role in preventing a platelet plug that develops in an area of vessel damage from spreading to normal vascular tissue.

The formation of a platelet plug may slow or completely stop bleeding from a damaged blood vessel that has a relatively small hole in it; but if the hole is large, blood clotting may be necessary to stop the bleeding. The ability of platelets to plug small vascular holes is important in closing minute tears that occur many times daily in the capillaries and other small vessels, and a person with an insufficient number of platelets often develops numerous small hemorrhagic areas under the skin and throughout the internal tissues.

Blood Clotting

Blood clotting, or **coagulation,** is a complex process that involves a number of different factors, many of which are present in the plasma (Table 18.3). The formation of a blood clot requires the conversion of a soluble protein called *fibrinogen* (normally present in the plasma) into an insoluble, threadlike polymer called *fibrin* (Figure 18.10). The insoluble fibrin threads form a network that entraps blood cells, platelets, and plasma to form the clot itself (Figure 18.11). The clot can adhere to surrounding tissue.

The conversion of fibrinogen to fibrin during clot formation normally requires the enzymatic action of a factor called *thrombin,* which is produced from the inactive plasma protein precursor *prothrombin.* Minute amounts of thrombin may be continually produced from prothrombin, but this thrombin is generally inactivated or destroyed relatively rapidly so that its concentration in the blood does not rise high enough to promote clotting. During cell, tissue, or platelet disruption, however, the formation of thrombin increases considerably, and its concentration may rise high enough to lead to clot formation.

The production of thrombin from prothrombin during clotting requires a *prothrombin activator,* which is formed either by way of an extrinsic pathway or by way of an intrinsic pathway. A tissue factor not normally present in the blood participates in the extrinsic pathway, but only factors present in the blood participate in the intrinsic pathway. When blood vessels are ruptured and tissues are damaged, both pathways are usually activated.

Extrinsic Pathway

The extrinsic pathway is activated when a substance called *tissue thromboplastin* (actually a complex mixture of lipoproteins that contains one or more phospholipid substances) is released from damaged tissues (Figure 18.10). Tissue thromboplastin, together with factor VII from the plasma and calcium ions, activates factor X. Activated factor X, together with the effective form of factor V and a phospholipid substance, forms prothrombin activator. Prothrombin activator acts enzymatically to catalyze the formation of thrombin from prothrombin. Thrombin, in turn, converts fibrinogen into fibrin monomers and fibrinopeptides. The fibrin monomers associate (polymerize) with one another, forming a loose network of fibrin threads. Thrombin also activates factor XIII, which augments the bonding between fibrin monomers, thereby strengthening and stabilizing the fibrin network.

Intrinsic Pathway

The intrinsic pathway is triggered when inactive factor XII in the plasma is activated by contact with a damaged vessel surface, particularly by contact with underlying collagen fibers. Activated factor XII activates factor XI, and activated factor XI in turn activates factor IX. Activated factor IX, together with the effective form of factor VIII, calcium ions, and a phospholipid substance, then activates factor X. From this point on, the sequence of events occurs in the same fashion as in the extrinsic pathway.

◆ **TABLE 18.3 Factors Involved in Blood Coagulation**

FACTOR NUMBER	NAME	TYPE OF FACTOR AND ORIGIN
I	Fibrinogen	A plasma protein produced in the liver
II	Prothrombin	A plasma protein produced in the liver
III	Tissue thromboplastin	A complex mixture of lipoproteins containing one or more phospholipid substances; released from damaged tissues
IV	Calcium ions	An ion in the plasma that is acquired from the diet and from bones
V	Proaccelerin (labile factor, accelerator globulin)	A plasma protein produced in the liver
VI	Not utilized	
VII	Serum prothrombin conversion accelerator (stable factor, proconvertin)	A plasma protein produced in the liver
VIII	Antihemophilic factor (antihemophilic globulin)	A plasma protein produced in the liver
IX	Plasma thromboplastin component (Christmas factor)	A plasma protein produced in the liver
X	Stuart factor (Stuart-Prower factor)	A plasma protein produced in the liver
XI	Plasma thromboplastin	A plasma protein produced in the liver
XII	Hageman factor	A plasma protein
XIII	Fibrin stabilizing factor	A protein present in plasma and in platelets

The intrinsic clotting pathway has more steps and is generally slower than the extrinsic pathway. In the intrinsic clotting sequence, the phospholipid substance required as a cofactor for some of the steps becomes exposed on the surfaces of platelets (as platelet factor III) during platelet aggregation, which is initiated when platelets contact collagen fibers.

Many of the reactions of the clotting process occur in what is called a "cascade" fashion. One factor becomes activated, and this factor in turn activates another that activates still another and so on. In this way, inactive factors in the blood are changed into active clotting factors.

Positive-Feedback Nature of Clot Formation

Once a blood clot starts to develop, thrombin acts in a positive-feedback fashion to promote the development of the clot. For example, thrombin activates some of the other clotting factors, and it enhances platelet aggregation.

Limiting Clot Growth

Despite the positive-feedback nature of the clotting process, clot formation normally occurs only locally at a site of damage. In general, once the various clotting factors are activated, they are rapidly inactivated or carried away by the blood so that their overall concentrations do not rise high enough to induce widespread clotting. For example, a plasma protein called *antithrombin III* inactivates thrombin. (Heparin promotes the activity of antithrombin III, and heparinlike molecules are present on the surfaces of endothelial cells.)

Clot Retraction

After the formation of a clot, a phenomenon known as *clot retraction* (syneresis) occurs. The fibrin meshwork shrinks and becomes denser and stronger. This activity helps pull the edges of a damaged vessel closer together. Clot retraction requires large numbers of platelets that, during clot formation, become trapped within the fibrin

◆ **FIGURE 18.10 Sequence of events that lead to the formation of a fibrin clot**

The factors indicated in blue are activated factors. See text for detailed discussion.

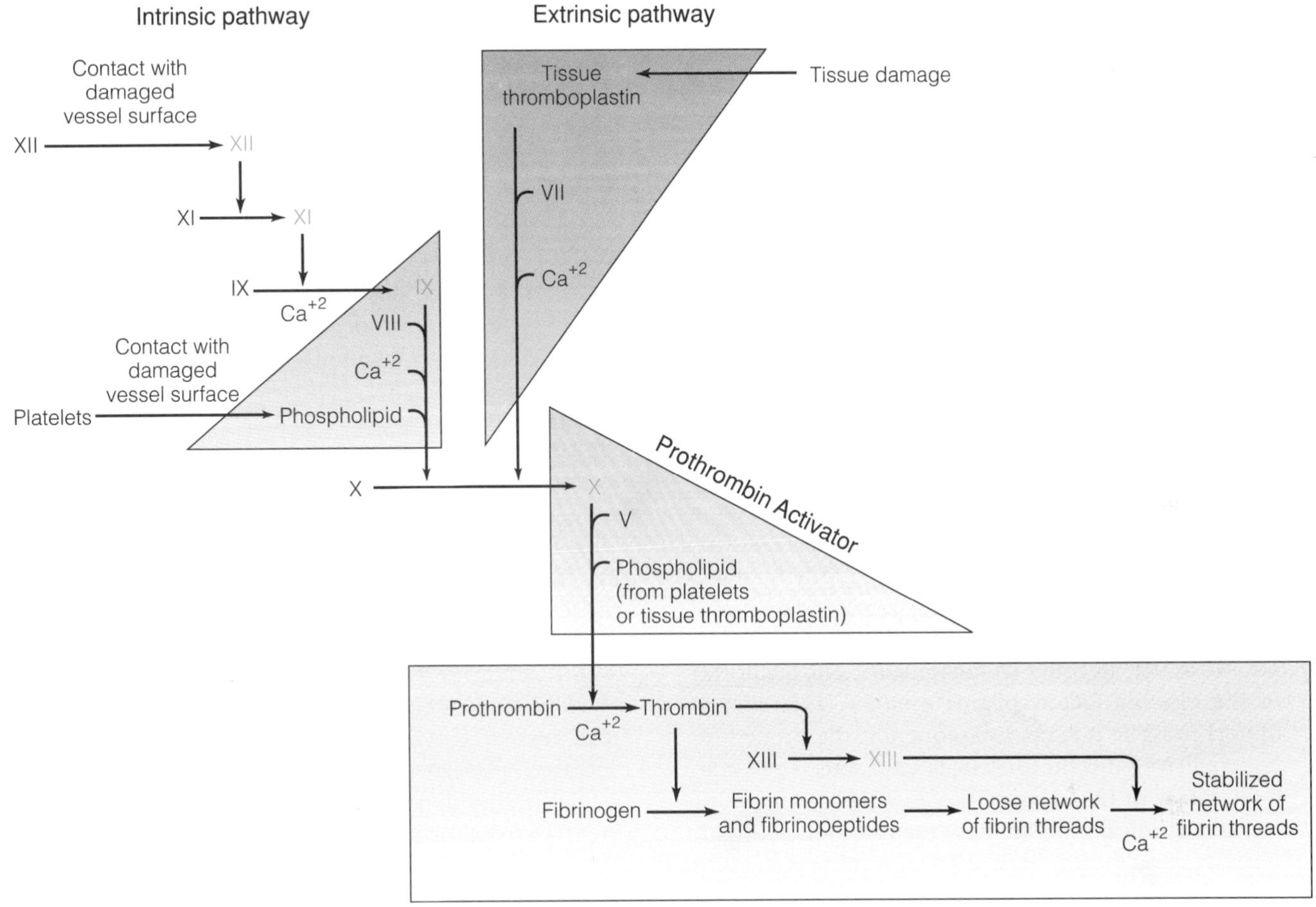

Intrinsic pathway
Contact with damaged vessel surface
XII → XII
XI → XI
IX → IX
Ca^{+2}
VIII
Ca^{+2}
Contact with damaged vessel surface
Platelets → Phospholipid
X → X

Extrinsic pathway
Tissue thromboplastin ← Tissue damage
VII
Ca^{+2}

Prothrombin Activator
V
Phospholipid (from platelets or tissue thromboplastin)

Prothrombin → Thrombin
Ca^{+2}
XIII → XIII
Fibrinogen → Fibrin monomers and fibrinopeptides → Loose network of fibrin threads → Stabilized network of fibrin threads
Ca^{+2}

meshwork and attached to the fibrin strands. The platelets contain an actomyosinlike contractile protein, and they contract and pull the fibrin threads closer together.

During clot retraction, a fluid known as *serum* is extruded from the fibrin meshwork. Serum is essentially plasma that lacks fibrinogen and some of the other clotting factors that are removed during the clotting process.

Clot Dissolution

An enzyme called **plasmin** can decompose fibrin and dissolve clots. Plasmin is normally present in the blood in an inactive form called *plasminogen,* and substances known as *plasminogen activators* can convert plasminogen to plasmin. Both activated factor XII and thrombin promote plasmin formation.

Normally, when clotting occurs, the plasminogen activators initiate a mechanism for the eventual dissolution of the clot. For example, vascular endothelial cells release a substance called *tissue plasminogen activator*

◆ **FIGURE 18.11 Portion of a blood clot showing fibrin network and entrapped erythrocytes**

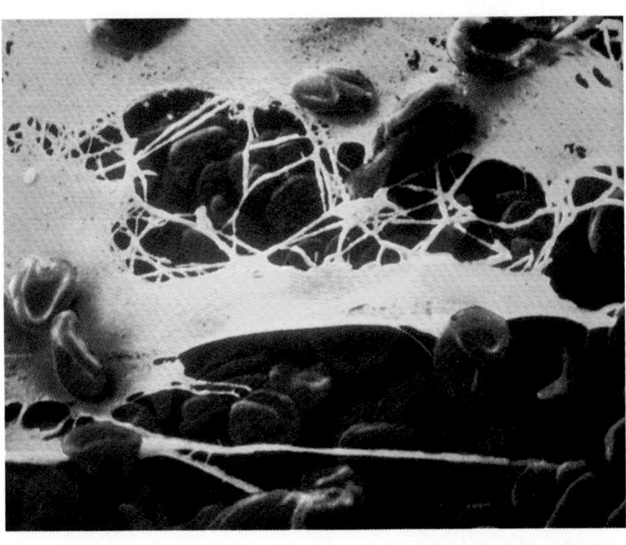

CONDITIONS OF CLINICAL SIGNIFICANCE

Excessive Bleeding

One of the factors required in the blood-clotting sequence is factor VIII (antihemophilic factor). An inherited genetic deficiency of an effective form of this factor results in a clotting defect called *hemophilia A.* A genetic deficiency of an effective form of another clotting factor, factor IX (plasma thromboplastin component), results in the less common defect *hemophilia B.* Both types of hemophilia are sex-linked conditions that are more common in males. The blood of individuals who suffer from hemophilia does not clot properly, and even minor damage to blood vessels can result in substantial bleeding. The bleeding may be controlled temporarily by injections of the appropriate clotting factor or by transfusions of blood plasma containing the clotting factor. However, in order to obtain sufficient quantities of the clotting factor, plasma samples from several donors must be pooled. Consequently, the procedure is expensive, and the patient is exposed to possible infection with blood-borne diseases such as hepatitis or AIDS. Clinical trials are now being conducted with factor VIII that has been manufactured by genetic engineering techniques. If the use of factor VIII obtained in this manner is successful, the danger of infection would be eliminated.

Nutritional deficiencies or liver disease can also lead to clotting problems and excessive bleeding. Vitamin K is required for the normal synthesis of prothrombin in the liver, and a deficiency of this vitamin interferes with prothrombin production. As a result, decreased levels of prothrombin may be present in the blood, leading to clotting difficulties. The liver is also involved in the production of a number of other clotting factors, and diseases that impair liver function can severely affect the clotting system.

Reduced numbers of platelets, a condition known as *thrombocytopenia,* causes bleeding states in which blood loss occurs through capillaries and other small vessels. Low platelet numbers can result from increased platelet destruction or from depressed platelet production due to pernicious anemia, certain drug therapies, or radiation.

(t-PA), which circulates in the blood. Circulating t-PA is relatively ineffective at converting plasminogen to plasmin. During clotting, however, both plasminogen and t-PA bind to fibrin. Binding to fibrin enhances the activity of the t-PA, and it converts the plasminogen to plasmin, thereby promoting the dissolution of the clot. In general, within a few days after a blood clot forms, the plasminogen activators cause the formation of enough plasmin to dissolve the clot.

Intravascular Clotting

Under most circumstances, blood does not clot as it flows along the blood vessels. The normal endothelial surfaces of undamaged blood vessels tend to prevent the activation of the intrinsic clotting mechanism, and plasmin quickly removes any small amounts of fibrin that form within the vessels.

Sometimes, however, clots do occur within a person's vascular system. If a clot blocks blood flow in a vessel that supplies a tissue critical to survival, the result can be very serious.

If a clot is fixed or adheres to a vessel wall, it is called a *thrombus.* A thrombus that forms in the coronary circulation may block the blood supply to heart tissue, leading to heart damage and ultimately to heart failure and death. The formation of a thrombus in a coronary artery is called a *coronary thrombosis.*

A number of studies have examined the causes of thrombus formation. Some investigations suggest that continuous, unreversed damage to a blood vessel wall that removes the inner endothelium and exposes deeper-lying tissues establishes conditions favorable to clot formation. Other studies have revealed alterations in blood-clotting characteristics in persons prone to thromboses that suggest a hyperreactivity of the clotting mechanism itself. Still others indicate that changes occur in a blood vessel wall as the result of the activity of viruses, and that the altered vessel wall may promote clot formation at the site. None of these occurrences is mutually exclusive, and all may play a role in clot formation.

If an intravascular clot is not fixed but floats within the blood, it is called an *embolus.* Emboli are potentially dangerous because they can become lodged in smaller vessels, thus blocking them and leading to tissue damage due to blood flow restriction. Emboli that lodge in the vessels of a lung can damage the lung itself and elicit cardiovascular reflexes leading to hypotension (low blood pressure) and death.

Study Outline

◆ **PLASMA** p. 572

Approximately 91% water; transports hormones and metabolic end products; contains proteins (such as albumin, fibrinogen, globulins), ions (such as sodium, bicarbonate, chloride), food materials, and gases.

◆ **FORMED ELEMENTS** pp. 572-584

Production of Formed Elements. Formed elements of blood produced in red bone marrow.

RED BONE MARROW. Contains precursor cells that give rise to formed elements.

HEMOPOIESIS. Hemocytoblasts give rise to committed progenitor cells.

FACTORS THAT INFLUENCE HEMOPOIESIS. Factors referred to as colony-stimulating factors promote the production of formed elements.

Erythrocytes. Small, circular, biconcave discs with no nuclei that function in transport of oxygen and carbon dioxide; *hematocrit* is proportion of erythrocytes in a sample of blood.

HEMOGLOBIN. Substance in erythrocytes that can bind with oxygen and carbon dioxide. Composed of protein called *globin* and four nonprotein groups called *hemes* that each contain an iron atom; oxygen binds reversibly with iron of heme groups; carbon dioxide binds reversibly with globin portion of hemoglobin, which is composed of four polypeptide chains.

IRON. Component of hemoglobin. Iron is transported in plasma in loose combination with a plasma protein called *transferrin.* Iron can be stored in the liver, spleen, and elsewhere in form of intracellular iron-protein complexes called *ferritin* and *hemosiderin.* Iron is absorbed into body in upper part of small intestine. Total body iron regulated largely by alterations in intestinal absorption rate.

PRODUCTION OF ERYTHROCYTES. Erythrocytes develop from hemocytoblasts, which give rise to proerythroblasts, which become basophil erythroblasts, which differentiate into polychromatophil erythroblasts, which become normoblasts. When hemoglobin content of normoblast cytoplasm reaches about 34%, nucleus is pinched off, and resulting reticulocytes lose their endoplasmic reticulum, mitochondria, and ribosomes as they mature into erythrocytes. Erythrocyte formation process (erythropoiesis) requires vitamin B_{12} and folic acid.

CONTROL OF ERYTHROCYTE PRODUCTION. Erythropoietin, which stimulates erythrocyte production, is produced in kidneys. Erythropoietin production, and thus erythrocyte production, is regulated by negative-feedback mechanism that is sensitive to amount of oxygen delivered to tissues; inadequate oxygen supply leads to increased erythropoiesis.

FATE OF ERYTHROCYTES. Aged, abnormal, or damaged erythrocytes are disposed of by phagocytic macrophages (especially in spleen, liver, and bone marrow); hemoglobin is degraded. Amino acids of globin can be used in synthesis of new protein. Iron is liberated from heme; remainder of heme is converted to biliverdin, which is in turn converted to bilirubin. Bilirubin is eventually taken up by liver, conjugated to glucuronic acid, and secreted with bile.

BLOOD GROUPS. A, B, AB, and O blood groups recognized, based on reactions between antigens on erythrocytes and particular antibodies.

◆ **CONDITIONS OF CLINICAL SIGNIFICANCE: ERYTHROCYTES** pp. 580-581

Anemia. Condition characterized by either decreased number of erythrocytes in blood or decreased concentration of hemoglobin.

HEMORRHAGIC ANEMIA. Result of substantial blood loss and relatively slow replacement of lost erythrocytes.

IRON-DEFICIENCY ANEMIA. Due to excessive loss, deficient intake, or poor absorption of iron; red blood cells are smaller than normal and contain lower-than-normal level of hemoglobin.

PERNICIOUS ANEMIA. Result of inability to absorb adequate amounts of vitamin B_{12} due to insufficient quantities of functional intrinsic factor. Erythrocytes are larger and more fragile than normal.

SICKLE-CELL ANEMIA. Result of production of abnormal hemoglobin due to genetic defect. Abnormal hemoglobin causes distortion of erythrocytes into sickle shape at low concentrations of oxygen; misshapen erythrocytes can rupture or block blood vessels.

THALASSEMIA. Genetic defect that produces abnormal hemoglobin. Red blood cells thin and fragile.

APLASTIC ANEMIA. Inadequate production of erythrocytes due to inhibition or destruction of red bone marrow.

Polycythemia. Condition in which there is a net increase in body's total circulating erythrocyte mass.

PRIMARY POLYCYTHEMIA. Production of excess erythrocytes due to tumorous abnormalities of tissues that produce blood cells.

SECONDARY POLYCYTHEMIA. Increased production of erythrocytes due to increased levels of erythropoietin formation; *physiologic polycythemia* is seen in people living at high altitudes.

Leukocytes. Formed elements of blood that act primarily in loose connective tissues and lymphoid tissues. Leukocytes in blood are mainly being transported by circulation. Important components of immune system.

GRANULOCYTES. Leukocytes with clearly evident granules in their cytoplasm; formed in red bone marrow.

NEUTROPHILS. Phagocytic cells; capable of ameboid movement; can leave blood and enter tissues, where they ingest bacteria and other foreign substances.

EOSINOPHILS. Participate in destruction of various types of parasites, such as parasitic worms; play a role in certain allergic inflammatory responses.

BASOPHILS. Contain histamine and heparin; may be functionally related to mast cells.

AGRANULOCYTES. Two types.

MONOCYTES. Ameboid cells formed in red bone marrow; can leave blood and enter tissues, where they develop into large, phagocytic cells called macrophages.

LYMPHOCYTES. Important in body's specific immune responses, including antibody production; many present in lymphoid tissues.

DIFFERENTIAL COUNT. Provides percentage figure for each type of leukocyte in blood.

◆ CONDITIONS OF CLINICAL SIGNIFICANCE: LEUKOCYTES p. 582

Leukemia. Cancerous condition involving excessive uncontrolled proliferation of leukocytes that leads to diffuse and almost total replacement of the red bone marrow with leukemic cells.

Infectious Mononucleosis. Caused by Epstein-Barr virus; characterized by increase in relative and absolute numbers of lymphocytes in blood.

Platelets. Small cytoplasmic fragments of megakaryocytes; involved in blood clotting and other hemostatic processes.

◆ HEMOSTASIS pp. 584–589

Vascular Spasm. Occurs when blood vessel is cut or ruptured; decreases blood flow and helps minimize blood loss.

Formation of a Platelet Plug. In damaged vessel, platelets adhere to exposed collagen and release ADP and other chemicals that lead to further platelet adhesion and the formation of platelet plug; plug may slow or stop bleeding if vessel damage is not too great.

Blood Clotting. Involves conversion of soluble plasma protein called *fibrinogen* into insoluble, threadlike polymer called *fibrin* by enzymatic action of thrombin; fibrin threads entrap blood cells, platelets, and plasma to form clot; thrombin can be formed from inactive plasma protein precursor by two pathways.

EXTRINSIC PATHWAY. Tissue thromboplastin released from damaged tissues acts with other clotting factors to form prothrombin activator; prothrombin activator catalyzes formation of thrombin from prothrombin.

INTRINSIC PATHWAY. Activation of factor XII and platelet aggregation may occur when inactive factor XII and platelets contact damaged vessel surface; phospholipid of platelets and active factor XII act with other clotting factors to form prothrombin activator.

POSITIVE-FEEDBACK NATURE OF CLOT FORMATION. Once clot formation begins, thrombin promotes clot development by activating some other clotting factors, and by enhancing platelet aggregation.

LIMITING CLOT GROWTH. Once the various clotting factors are activated, they are rapidly inactivated or carried away by the blood so their overall concentrations do not rise enough to induce widespread clotting.

CLOT RETRACTION. After clot formation, fibrin meshwork shrinks and becomes denser and stronger; requires large numbers of platelets that apparently contain actomyosinlike contractile protein.

CLOT DISSOLUTION. Enzyme *plasmin* can decompose fibrin and dissolve clots.

INTRAVASCULAR CLOTTING. Clots may occur within vascular system and block blood flow to tissues critical to survival; clot that is fixed or adheres to vessel wall is called *thrombus,* and clot that floats within blood is called *embolus.* Clots may form in blood vessels as result of continuous, unreversed damage to vessel walls, hyperreactivity of the clotting mechanisms, or viral activity.

◆ CONDITIONS OF CLINICAL SIGNIFICANCE: EXCESSIVE BLEEDING p. 588

Inherited genetic deficiencies of effective forms of factor VIII (anti-hemophilic factor) or factor IX (plasma thromboplastin component) may result in blood that does not clot properly and substantial bleeding; nutritional deficiencies and liver diseases can also cause clotting problems and excessive bleeding; thrombocytopenia (reduced numbers of platelets) may result in bleeding from capillaries.

Self-Quiz

1. In an adult, all bones contain red bone marrow, which is active in producing the formed elements of the blood. True or False?
2. The hematocrit of a sample of peripheral venous blood is usually higher in females than in males. True or False?
3. Most of the oxygen that is transported by the blood is in physical solution dissolved in the plasma. True or False?
4. Total body iron is regulated to a large extent by alterations in the intestinal absorption rate of iron. True or False?
5. Pernicious anemia is a condition that results from: (a) an inability to absorb adequate amounts of vitamin B_{12} from the digestive tract; (b) a deficiency of iron in the blood; (c) the destruction of the red bone marrow.
6. Leukocytes act primarily in the loose connective tissues and in lymphoid tissues, and those within the blood are mainly being transported by the circulation. True or False?
7. Match the following terms with their appropriate lettered descriptions:

Bilirubin	(a) A hormone that stimulates erythrocyte production
Heme	(b) A component of hemoglobin that contains iron
Erythropoietin	(c) A component of hemoglobin that can bind reversibly with carbon dioxide
Globin	(d) A storage form of iron
Ferritin	(e) A breakdown product of hemoglobin that is taken up by the liver, conjugated to glucuronic acid, and excreted with the bile

8. Which of the following leukocyte types leave the blood vessels and enter the tissues, where they protect the body by ingesting bacteria and other foreign substances? (a) neutrophils; (b) megakaryocytes; (c) lymphocytes.
9. Mast cells and basophils: (a) contain heparin; (b) produce antibodies; (c) secrete erythropoietin.
10. In the tissues, monocytes are converted to: (a) mast cells; (b) macrophages; (c) megakaryocytes.
11. The leukocyte type that is involved in specific immune responses, including antibody production, is the: (a) lymphocyte; (b) neutrophil; (c) megakaryocyte.
12. The Epstein-Barr virus is believed to cause: (a) sickle-cell anemia; (b) infectious mononucleosis; (c) leukemia.
13. Platelets are formed in the red bone marrow as pinched-off portions of cells called: (a) reticulocytes; (b) mast cells; (c) megakaryocytes.
14. Thromboxane A_2 facilitates the formation of a platelet plug. True or False?
15. During clot formation, fibrinogen is converted to fibrin by: (a) thrombin; (b) thromboplastin; (c) anti-thrombin III.
16. Blood clotting cannot occur unless tissue thromboplastin is released from damaged tissues. True or False?
17. Once it is formed, thrombin may contribute to clot development by acting enzymatically to convert prostacyclin to thromboplastin. True or False?
18. Clot retraction requires large amounts of: (a) anti-thrombin III; (b) plasmin; (c) platelets.
19. Fibrin can be decomposed and clots dissolved by: (a) thromboplastin; (b) plasmin; (c) fibrinogen.
20. A deficiency of vitamin K interferes with the normal production of prothrombin. True or False?

CHAPTER 19

The Cardiovascular System: The Heart

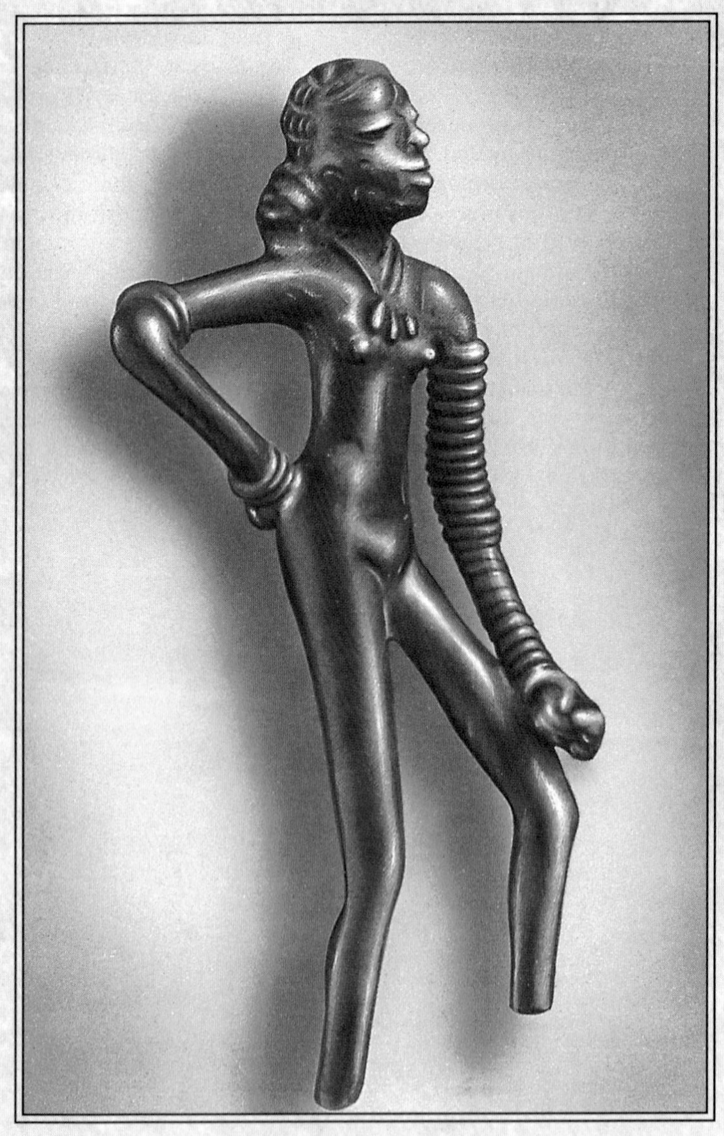

CHAPTER CONTENTS

LEARNING OBJECTIVES

After completing this chapter, you should be able to:

1. Describe the coverings of the heart.
2. Describe the structure of the heart, including its "skeleton."
3. Describe the flow of blood through the heart and the pumping action of the heart during a contractile cycle.
4. Describe the excitation process that occurs in the heart, and explain why the sinoatrial node acts as the pacemaker of the heart.
5. Describe the conduction of a stimulatory impulse through the heart.
6. Describe and explain the pressure changes, volume changes, and valve actions that occur within the left chambers of the heart during a cardiac cycle.
7. Define cardiac output, and discuss the interrelationship of heart rate and stroke volume with reference to cardiac output.
8. Describe the control of heart rate.
9. Describe the control of stroke volume.
10. Discuss the effects of exercise on cardiac function.
11. Describe the process of congestive heart failure.
12. Explain the causes of the normal heart sounds.
13. Describe a normal electrocardiogram, and explain its principal features.

CHAPTER 19

The **cardiovascular system** is a continuous closed system that includes the *heart,* which serves as a pump for the blood, and the *blood vessels,* which transport the blood throughout the body. Confined within the heart and the numerous vessels, the blood repeatedly travels through the heart, into arteries, then to capillaries, into veins, and back to the heart. Normally, blood does not leave this system, although some of the fluid part of the blood does pass through the walls of the capillaries to join the tissue fluid between the cells. However, even this fluid is returned to the cardiovascular system, either directly or by way of the *lymphatic system.* The heart is discussed in this chapter. The blood vessels are discussed in Chapter 20, and the lymphatic system is discussed in Chapter 21.

Embryonic Development of the Heart

Early in embryonic development, the future heart is a simple pulsating tube that receives blood from the veins at its posterior end and pumps it into the arterial system through its anterior end (Figure 19.1a). From this simple beginning, the heart must not only respond to the changing needs of the developing embryo but must also be capable of functioning under the vastly different conditions immediately following birth. To accomplish this, the tubular heart must develop into a four-chambered organ complete with valves and a midline partition. And the heart must undergo these alterations without interrupting its delivery of blood to the developing embryo.

By the fifth week of development it becomes apparent that the heart is beginning to undergo changes as it grows rapidly (Figure 19.1b) and evolves into an S-shaped structure (Figure 19.1c). With continued development, the anterior vessel, which carries blood away from the heart, divides into two vessels: the future *pulmonary trunk* which supplies blood to the lungs, and the *aorta,* which carries blood to the vessels that supply the rest of the body. At the same time, a midline septum (wall) is developing within the heart (Figure 19.1d). When completed, this wall will separate the blood flow through the heart into two channels; the blood on one side of the wall passes through the pulmonary trunk and to the lungs, and that on the other side is directed into the aorta and thus to the rest of the body (Figure 19.1e).

By the seventh week the embryonic heart has further divided into the four chambers it will retain in the adult—two atria and two ventricles. Through a series of changes, including the development of new segments in some veins and a degeneration of portions of other veins, the vessels that enter the posterior (venous) region of the heart develop into *superior* and *inferior venae cavae,* which return all the venous blood from the body to the heart. No other major changes occur in the heart until birth.

Position of the Heart

The adult heart is a cone-shaped organ about the size of a fist. It is located between the lungs, close to the midline of the body, in a region called the **mediastinum**

◆ **FIGURE 19.1 Successive stages in the embryonic development of the heart**
The arrows indicate the direction of blood flow.

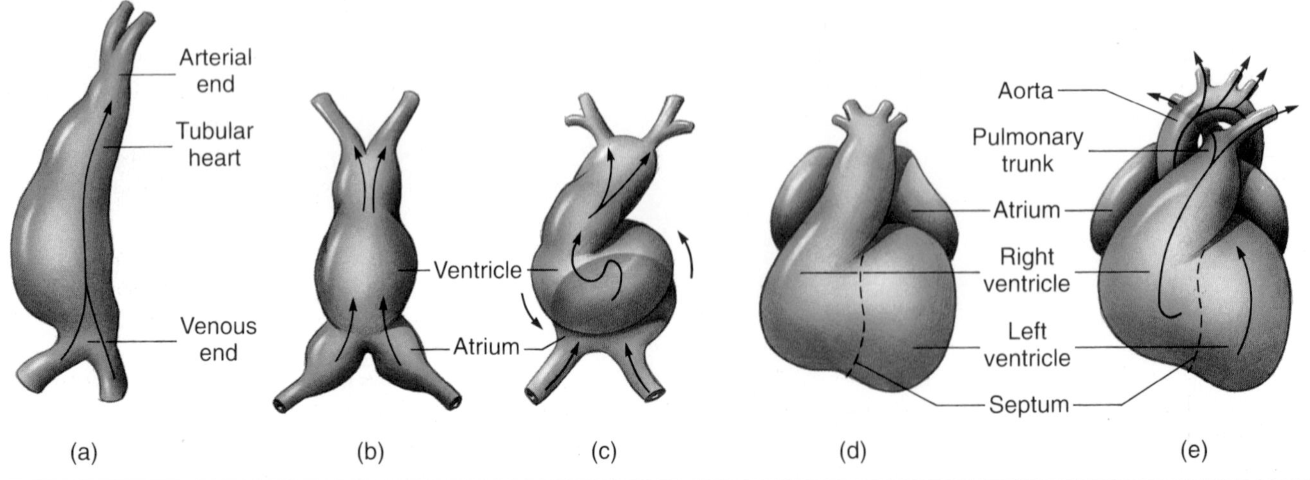

◆ **FIGURE 19.2 Anterior view of the thorax showing the position of the heart in the mediastinum**

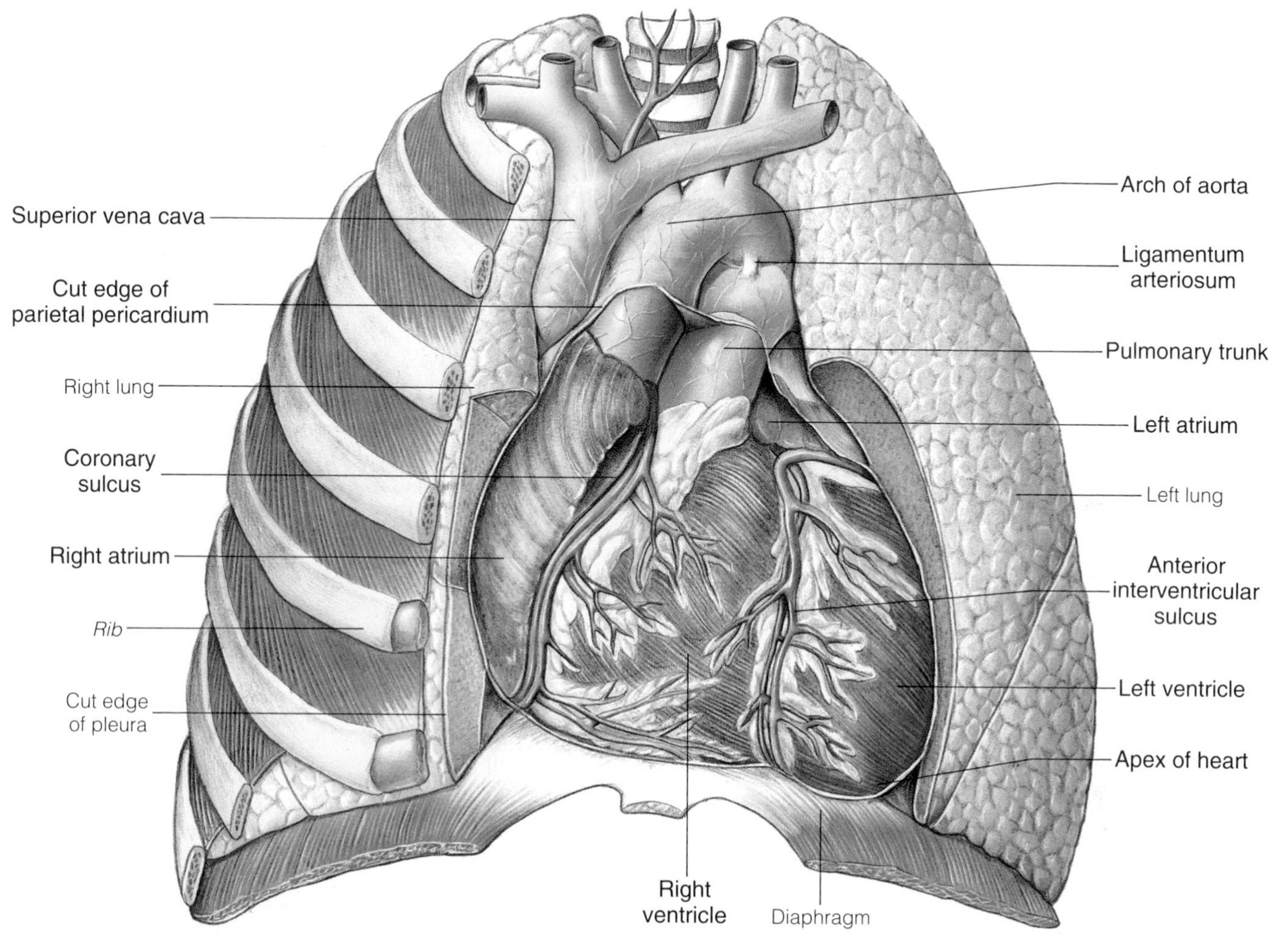

(Figure 19.2). The heart lies obliquely in the mediastinum, posterior to the sternum and anterior to the vertebral column. It is described as having a base and an apex, diaphragmatic and sternocostal surfaces, and four borders.

The *base* of the heart faces posteriorly and is located superiorly and to the right, behind the sternum, at about the level of the second and third ribs. It consists mainly of the left atrium, part of the right atrium, and the proximal portions of the large veins that enter the posterior wall of the heart. From the base the heart projects downward, anteriorly, and to the left, ending in a blunt *apex*. The apex reaches the fifth intercostal space, about 8 cm to the left of the midsternal line. When the heart beats, especially when it contracts forcefully, the apex can be felt thumping against the left chest wall. Because of this location of the discernible heart beats, there is a common misconception that the entire heart lies to the left of the midline of the body when, in fact, a portion of it extends to the right of the sternum. The *diaphragmatic surface* of the heart is that part between the base and the apex that rests upon the diaphragm. It involves the left and right ventricles. The anterior surface of the heart, which is formed mainly by the right ventricle and the right atrium, is referred to as the *sternocostal surface.*

The *superior border* of the heart is formed by both atria and is the region where the great vessels enter and leave the heart. It lies at about the level of the second intercostal space. The *inferior border* extends from behind the lower portion of the sternum to the left fifth intercostal space, where it ends at the apex; it is formed mostly by the right ventricle, plus a small portion of the left ventricle at the apex. The *right border* of the heart is formed by the right atrium and is located about 2.5 cm to the right of the sternum. The *left border* is formed mainly by the left ventricle, with the left atrium forming the upper portion. The left border extends to the apex of the heart, from the level of the junction of the left second rib with its costal cartilage.

◆ **FIGURE 19.3 The endocardium, myocardium, and pericardium**

(a) The serous layer of the parietal pericardium is reflected downward and forms the visceral pericardium on the surface of the myocardium. (b) The fluid-filled pericardial cavity is located between the visceral and parietal layers of the pericardium.

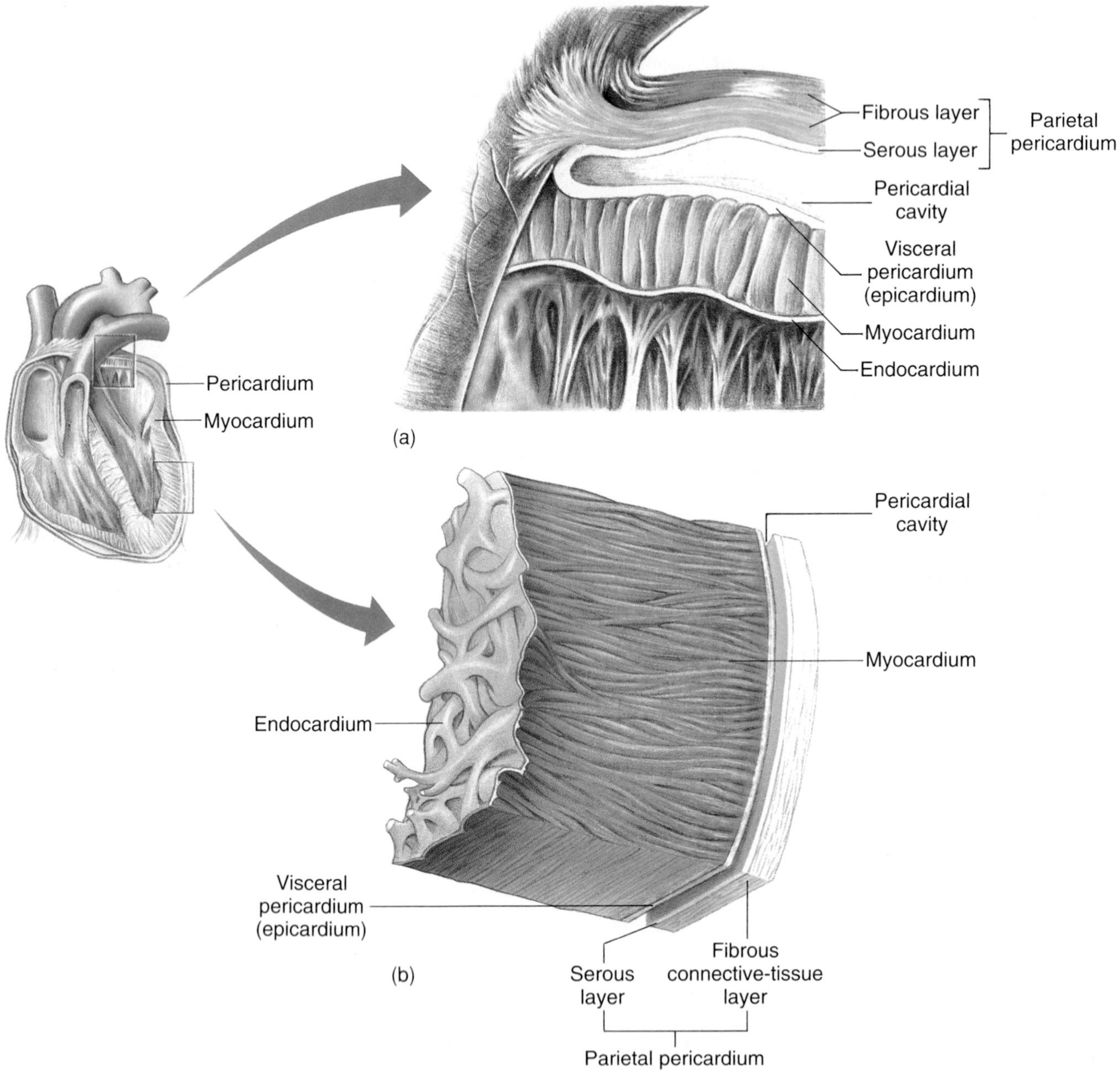

Coverings of the Heart

The heart is enclosed in a double-walled membranous sac called the **pericardium** *(per-i-kar´-dee-um)*. The outer wall of the sac is called the **parietal pericardium** and consists of two layers: an outer *fibrous* layer and an inner *serous* layer (Figure 19.3). The fibrous layer is quite strong, being composed of dense regular connective tissue, and serves to strengthen the pericardium and anchor it within the mediastinum. The serous layer is thin and smooth. It is composed of a layer of squamous epithelial cells underlain by a thin layer of loose connective tissue that adheres to the fibrous layer of the parietal pericardium. At the base of the heart, where the large vessels enter and leave the heart, the serous layer is reflected downward and forms a thin, tight, covering over the heart surface. This serous covering is called the **visceral pericardium** or **epicardium.**

Between the serous membranes of the visceral and parietal layers is a small space called the **pericardial cavity.** This cavity contains **pericardial fluid,** which is secreted by the cells of the serous membrane of the pericardium. The fluid lubricates the membranes, permitting them to slide over one another with a minimum of friction as the heart beats.

Inflammation of the pericardium, which is referred to as *pericarditis,* can result from a variety of causes. The amount and character of the pericardial fluid vary in the different forms of pericarditis. In some cases, the pericardial fluid is scanty; in others, it is abundant. In pericarditis, the serous layers of the pericardium become roughened, which causes pain as they move over one another and also interferes with the normal filling of the heart chambers. Some infections produce fibrin in the pericardial cavity and others produce pus.

Anatomy of the Heart

In order to function as a pump, the heart must have both receiving chambers and delivery chambers, valves to direct the flow of blood through the heart, a wall that is strongly contractile and thus provides the force to propel blood, and vessels to deliver blood to and from the heart.

Chambers of the Heart

The heart consists of four chambers: **right** and **left atria** (singular: atrium) and **right** and **left ventricles** (Figures 19.4–19.6). The atria are small and are located toward the superior region of the heart. The ventricles are larger and comprise the bulk of the heart. Located inferiorly, the ventricles form the apex of the heart. The right ventricle forms most of the heart's anterior surface; the left ventricle forms most of the inferior surface and left margin of the heart. The atria are separated from one another by a longitudinal partition called the **interatrial septum.** The ventricles are separated by an **interventricular septum.**

Vessels Associated with the Heart

Several large vessels enter or leave the base and the superior border of the heart (Figure 19.6, page 600):

1. Superior and **inferior venae cavae,** which return venous blood from the vessels of the body to the right atrium.

2. The **pulmonary trunk,** which divides into **right** and **left pulmonary arteries** and carries blood from the right ventricle to the lungs.

3. The four **pulmonary veins,** which carry blood from the lungs to the left atrium.

4. The **aorta,** which carries blood from the left ventricle into the vessels that supply the body.

Wall of the Heart

The heart is composed primarily of cardiac muscle anchored to a fibrous skeleton.

Epicardium, Myocardium, and Endocardium

The wall of the heart is formed of three layers: the epicardium, the myocardium, and the endocardium (Figure 19.3). The **epicardium (visceral pericardium)** is a thin serous membrane that adheres tightly to the outer surface of the heart. The thickest layer of the wall of the heart, the **myocardium** *(my-o-kar´-dee-um),* is composed of cardiac muscle. The myocardium is lined on the inside by the **endocardium** *(en-do-kar´-dee-um),* which is composed of a thin layer of connective tissue and a surface layer of squamous cells. Foldings of the endocardium form the valves that separate the atria from the ventricles—the *atrioventricular valves (a-tree-o-ven-trik´-yoo-lar)*—and the ventricles from the aorta and the pulmonary trunk—the *semilunar valves (sem-i-loo´-ner).* The endocardial lining of the heart is continuous with the endothelium that lines all the arteries, veins, and capillaries of the body.

The myocardium varies considerably in thickness from one heart chamber to another. Its thickness is related to the resistance encountered in pumping the blood from the different chambers. Since the muscles of the atria meet little resistance in pushing the blood into the ventricles, the walls of the atria are the thinnest part of the myocardium (Figure 19.6). In contrast, the ventricles must move the blood through the blood vessels of either the lungs or the rest of the body and back into a receiving chamber. Consequently, the myocardium of the ventricles is thicker than that of the atria. Moreover, the left ventricle, which propels blood through all parts of the body (other than the lungs) and back into the right atrium, has thicker walls than the right ventricle, which moves the blood only through the blood vessels of the lungs and back into the left atrium.

The inner surface of the myocardium of the ventricles is irregular, having folds and bridges—called **trabeculae carneae** *(trah-bek´-u-lē corn´-ē-uh)*—and cone-shaped **papillary muscles** that project into the lumen (Figure 19.6). Strong fibrous strands called **chordae tendineae** *(kor´-dah ten-din´-ē-uh)* run from the papillary muscles to the cusps (flaps) of the atrioventricular valves.

◆ **FIGURE 19.4 Anterior view of the heart**

(a) The vessels shown in red carry oxygenated blood. Those shown in blue carry deoxygenated blood. (b) Photograph of the anterior surface of the heart.

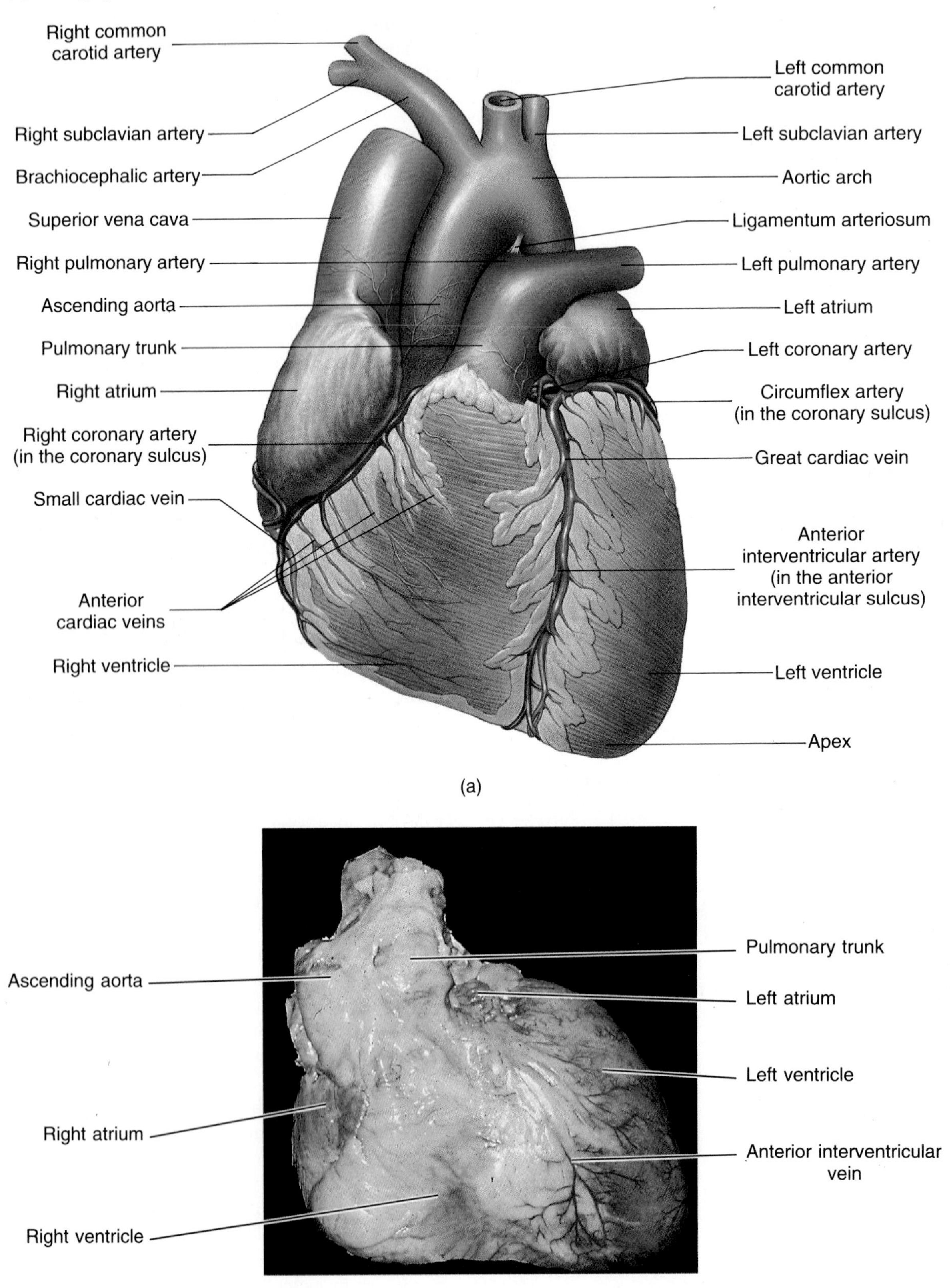

◆ **FIGURE 19.5 Posterior-inferior view of the heart**

(a) The vessels shown in red carry oxygenated blood. Those shown in blue carry deoxygenated blood. (b) Photograph of the posterior surface of the heart.

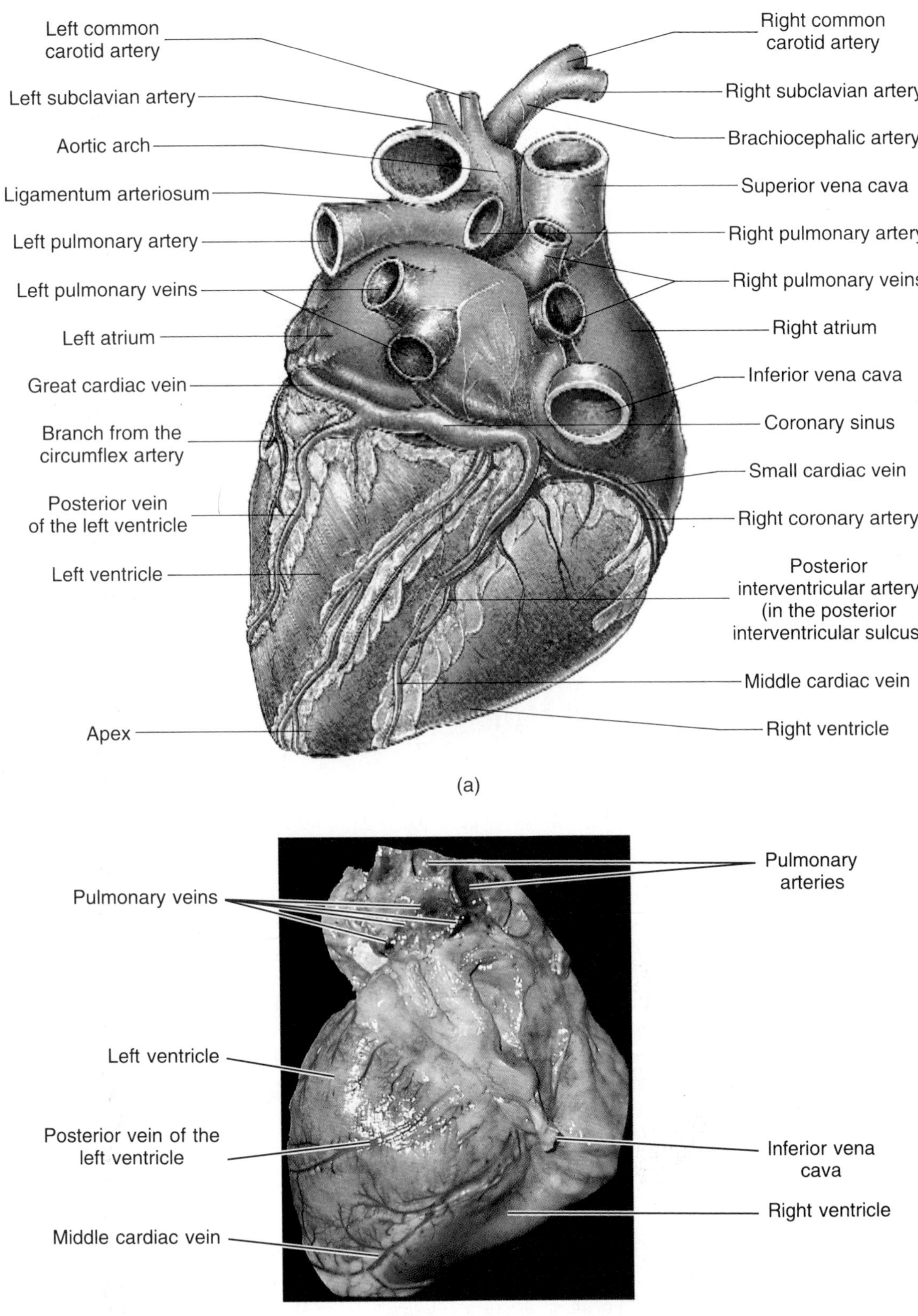

◆ **FIGURE 19.6 Frontal section of the heart**

The arrows indicate the path of the blood flow through the chambers, the valves, and the major vessels. Note that the branches of the right pulmonary vein pass behind the heart and enter the left atrium.

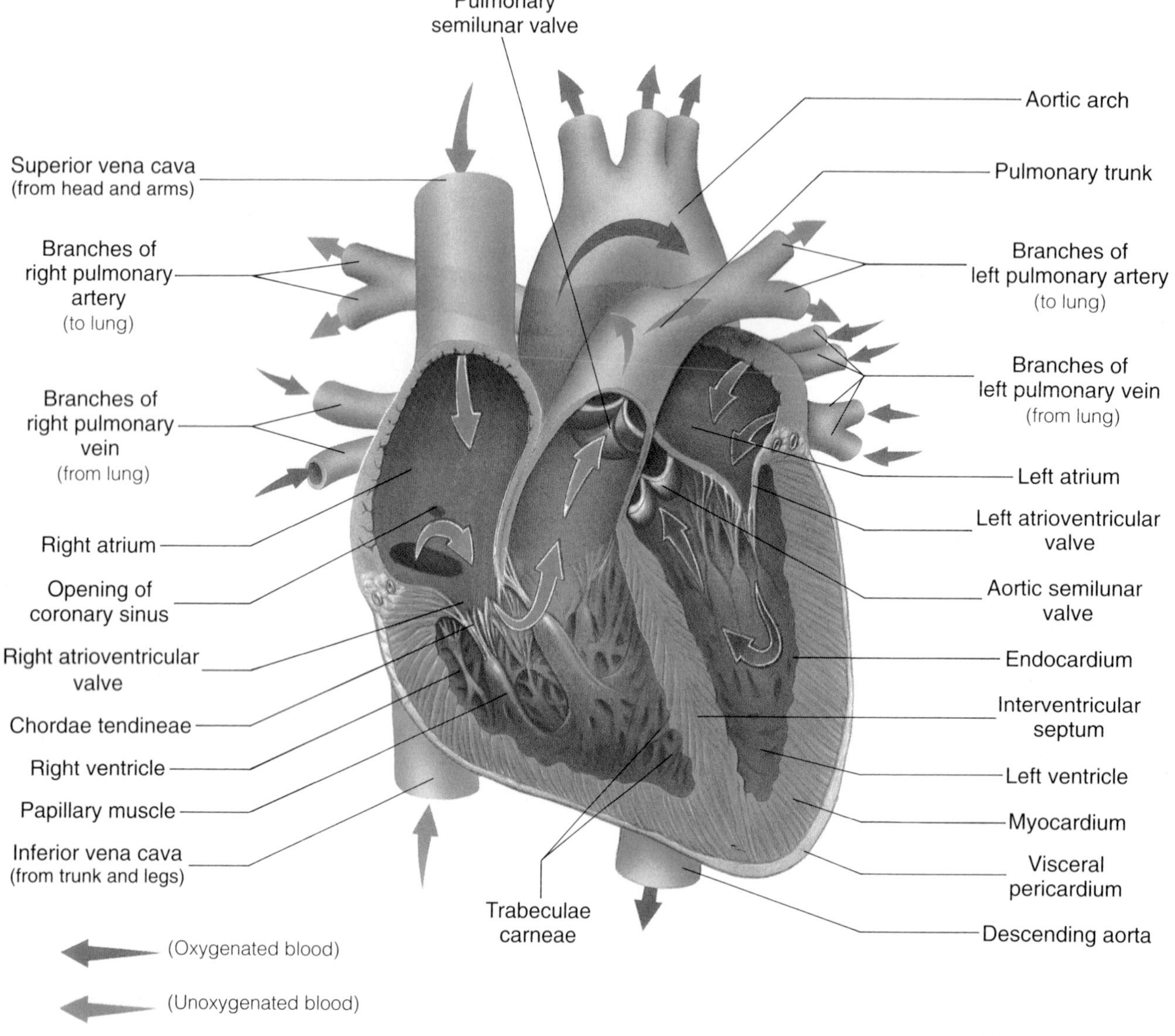

Several grooves visible on the surface of the heart indicate the margins of the chambers. The separation between the atria and the ventricles is marked by a deep horizontal groove called the **coronary sulcus.** The position of the interventricular septum is marked on the anterior surface of the heart by the **anterior interventricular sulcus,** and on the posterior surface by the **posterior interventricular sulcus.** These sulci contain varying amounts of fat and serve as pathways for the major blood vessels supplying the heart muscle.

Skeleton of the Heart

Horizontal **fibrous rings** surround the atrioventricular openings and the openings of the aorta and pulmonary trunk. The rings are fused together by additional fibrous tissue called **fibrous trigones** (Figure 19.7). Collectively, these fibrous supports are referred to as the **skeleton of the heart.** This fibrous skeleton not only serves as the attachment for the heart muscle and valves but also helps form the septa that separate the atria from the ventricles.

Vessels of the Myocardium

The myocardium receives an abundant blood supply through the **right** and **left coronary arteries** (Figures 19.4 and 19.5). These arteries arise from the aorta just beyond the point where it leaves the superior border of the heart (Figure 19.7). The coronary arteries obtain

◆ **FIGURE 19.7 Superior view of the heart with the atria removed showing the fibrous skeleton (yellow) that surrounds the valve openings**

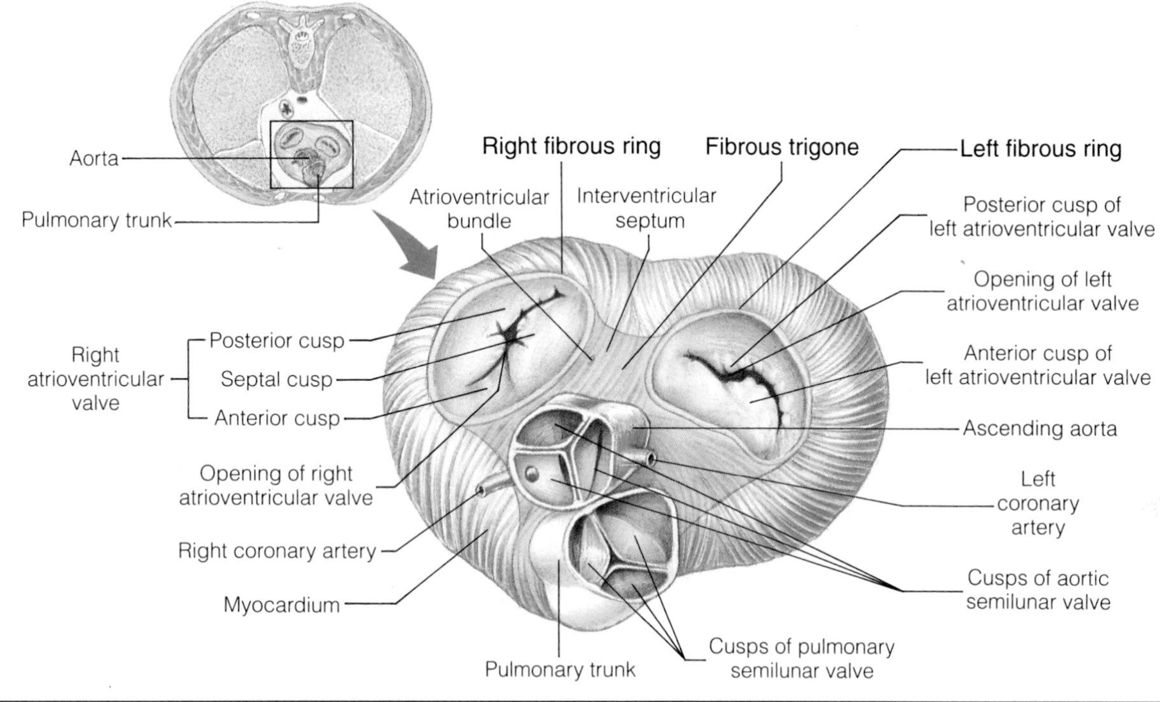

their blood from sinuses behind the cusps of the aortic semilunar valve (Figure 19.8).

The **right coronary artery** arises from the right anterior surface of the aorta and passes to the right margin of the heart in the coronary sulcus, which separates the atria from the ventricles (Figure 19.4). The right coronary artery extends around the margin of the heart to the posterior surface, sending branches to the atrium and the ventricle. On the posterior surface of the heart, the main branch of the artery turns downward toward the apex of the heart. This branch, the **posterior interventricular artery,** follows the posterior interventricular sulcus and supplies smaller branches to both ventricles (Figure 19.5).

The **left coronary artery** arises from the left anterior surface of the aorta, behind the pulmonary trunk. After

◆ **FIGURE 19.8 The origin of the coronary arteries**

(a) A closed aortic valve viewed from above. (b) An aortic orifice cut and opened to show the semilunar valves.

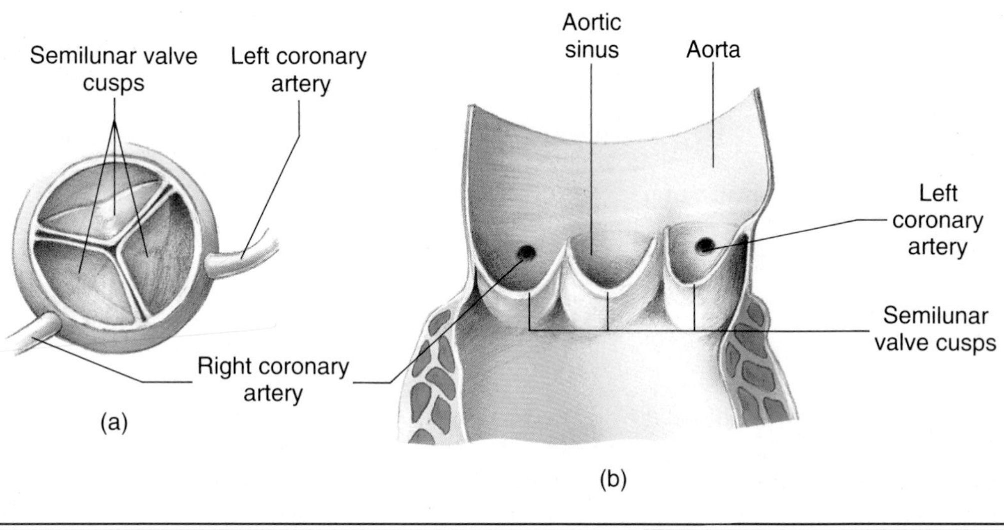

Semilunar valve cusps
Left coronary artery
Right coronary artery
(a)
Aortic sinus
Aorta
Left coronary artery
Semilunar valve cusps
(b)

passing a short distance in the coronary sulcus toward the left margin of the heart, it divides into anterior interventricular and circumflex branches. The **anterior interventricular artery** travels down the anterior interventricular sulcus toward the apex of the heart. It supplies branches to both ventricles. The **circumflex artery** travels in the coronary sulcus to the left margin of the heart. After passing around the left margin, the circumflex artery supplies branches to the posterior surfaces of the left atrium and left ventricle.

Any narrowing or blockage of the coronary arteries can restrict blood flow and interfere with the oxygen supply to the myocardium. If the restriction of blood flow is temporary, resulting in an inadequate oxygen supply to the myocardium for only a few seconds, a sharp pain in the chest results, often radiating down the left arm. This condition is called *angina pectoris.* If the restriction of blood flow results in an inadequate oxygen supply to the myocardium for an extended period of time, regions of the cardiac muscle may die. This condition, which can be disabling or fatal, is commonly called a *heart attack* (see page 622).

After passing through an extensive capillary network in the wall of the heart, the blood from the coronary arteries enters the **cardiac veins,** which travel alongside the arteries (Figures 19.4 and 19.5). The cardiac veins join together to form an enlarged vessel, the **coronary sinus,** which is located in the coronary sulcus on the posterior surface of the heart, between the atria and ventricles.

The anterior surface of the heart is drained primarily by the **great cardiac vein,** which passes upward alongside the anterior interventricular artery. The great cardiac vein begins at the apex of the heart and ascends to the base of the ventricles, where it becomes continuous with the coronary sinus. The largest veins on the posterior-inferior surface of the heart are the **posterior vein of the left ventricle,** which accompanies the circumflex artery, and the **middle cardiac vein,** which travels alongside the posterior interventricular artery. These veins, like the great cardiac vein, empty into the coronary sinus. The coronary sinus, in turn, empties into the right atrium. Also draining into the coronary sinus is a **small cardiac vein,** which travels along the right margin of the heart and enters the coronary sulcus on the posterior aspect of the heart. In addition to these veins, there are several small **anterior cardiac veins,** which drain the ventral surface of the right ventricle directly into the right atrium.

Valves of the Heart

Four sets of valves keep the blood flowing in the proper direction through the chambers of the heart—two sets of *atrioventricular valves* and two sets of *semilunar valves.*

Atrioventricular Valves

Located between the atria and the ventricles, the two **atrioventricular (AV) valves** are flaps of endocardium with an inner framework of fibrous connective tissue (Figures 19.6 and 19.7). The flaps are anchored to the papillary muscles of the ventricles by the chordae tendineae. The papillary muscles, which are extensions of the myocardium, exert tension on the valve cusps by way of the chordae tendineae, thus preventing them from being forced into the atria while the ventricles are contracting. The right atrioventricular valve, which separates the right atrium from the right ventricle, has three flaps, or *cusps,* and is therefore known as the *tricuspid valve.* The left atrioventricular valve is called the *bicuspid* or *mitral valve* because it has only two cusps. Both of the atrioventricular valves are usually open, but they are forced shut as the pressure in the ventricles increases, thus preventing the flow of blood back into the atria while the ventricles are contracting.

Semilunar Valves

The **semilunar valves** prevent blood from returning to the ventricles after the ventricles have completed their contractions (Figures 19.6 and 19.7). The **pulmonary semilunar valve** is located in the proximal end of the pulmonary trunk; the **aortic semilunar valve** is in the proximal end of the aorta. Both sets of semilunar valves have three cusps. Each cusp resembles a shallow cup that has been cut in half vertically, with the cut edges attached to the walls of the vessel (Figure 19.8). When the ventricles contract, the force of the blood pushes the cusps against the vessel wall. When the ventricles relax, the blood starts to flow back into them. As this occurs, blood fills the cusps and causes their free margins to meet in the middle of the vessel, thus preventing any further backflow.

The openings of the superior and inferior venae cavae, which empty into the right atrium, and those of the pulmonary veins, which empty into the left atrium, are not guarded by functional valves.

Surface Locations of the Valves

It is helpful to be able to locate the valves of the heart in relationship to the surface of the thorax (Figure 19.9). The semilunar valves are located toward the base of the heart. The aortic semilunar valve is behind the left side of the sternum, opposite the third intercostal space. The pulmonary semilunar valve lies slightly above and to the left of the aortic semilunar valve, behind the costal cartilage of the third rib. The atrioventricular valves are situated more centrally in the heart than are the semilunar valves. The right atrioventricular valve lies almost directly behind the sternum, extending between the lev-

◆ **FIGURE 19.9 Anterior view of the thorax showing the position of the heart and the heart valves in relationship to the ribs, sternum, and diaphragm**

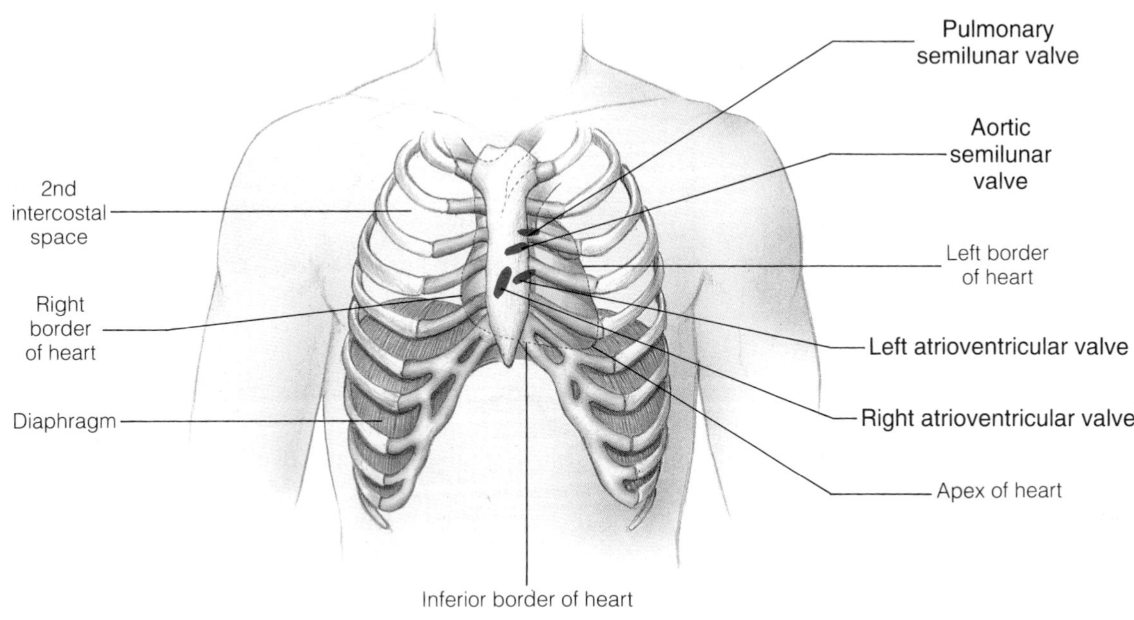

els where the fourth and fifth costal cartilages join with the sternum. The left atrioventricular valve is located at the level of the fourth costal cartilage, behind the left side of the sternum.

 Circulation through the Heart

Because the chambers on the right side of the heart are separated from those on the left by septa (the interatrial and interventricular septa), the heart functions as a double pump. Each pump has a receiving chamber (the atrium) and a propulsion chamber (the ventricle).

The right pump receives blood that has passed through the vessels of the body and sends it to the lungs, through the *pulmonary circuit* (Figure 19.10). Venous blood from the body enters the right atrium through several vessels:

1. The *superior vena cava,* which brings blood from the head, thorax, and upper limbs.
2. The *inferior vena cava,* which returns blood from the trunk, lower limbs, and abdominal viscera.
3. The *coronary sinus* and *anterior cardiac veins* that drain the myocardium.

From the right atrium, the blood enters the right ventricle, which pumps it through the pulmonary trunk and pulmonary arteries to the capillary network of the lungs. In the lungs, the blood gives up carbon dioxide and receives oxygen (see Chapter 23). Oxygenated blood from the lungs is returned to the left atrium by pulmonary veins.

The left pump receives the newly oxygenated blood from the lungs and sends it to the body through the *systemic circuit.* From the left atrium, the blood enters the left ventricle, which pumps it through the aorta to the body. Arteries that branch from the aorta transport the blood to specific body regions and organs. Thus, some of the blood from the left ventricle goes to the muscles, some goes to the digestive tract, some goes to the kidneys, some goes to the brain, and so on. Because of these direct arterial branches from the aorta, freshly oxygenated blood from the left ventricle is distributed to each body part and region, rather than passing from one tissue to another (Figure 19.11, page 606).

 Pumping Action of the Heart

The right and left pumps of the heart work in unison. When the heart beats, both atria contract simultaneously while the ventricles are relaxed; and then both ventricles contract simultaneously.

The period from the end of one heartbeat to the end of the next is called the *cardiac cycle.* During this cycle, blood from the superior and inferior venae cavae (as well as from the coronary sinus and anterior cardiac veins) moves through the right atrium, past the open

CLINICAL CORRELATION

Aortic Stenosis and Replacement of the Aortic Valve

Case Report

THE PATIENT: A 53-year-old man.

PRINCIPAL COMPLAINT: Fatigue and chest pain during exertion.

HISTORY: At the age of 31 years, a heart murmur (page 617) was detected during a routine physical examination for an insurance policy. Medical surveillance of the problem was not maintained. About a year before the present admission, the patient noted chest pain during exertion. The condition has progressed, with fatigue and chest pain during even slight exertion, and on one occasion the patient temporarily lost consciousness.

CLINICAL EXAMINATION: The patient appeared chronically ill. His arterial pressure was 95/80 mm Hg (normal: 90–140/60–90 mm Hg). His heart rate was 85 beats/min (normal: 60–100 beats/min), and his respiratory rate was 21 breaths/min (normal: $<$ 17 breaths/min). The electrocardiogram (ECG) suggested left ventricular hypertrophy, but no evidence of myocardial infarction was found. A chest radiograph confirmed that the heart was enlarged

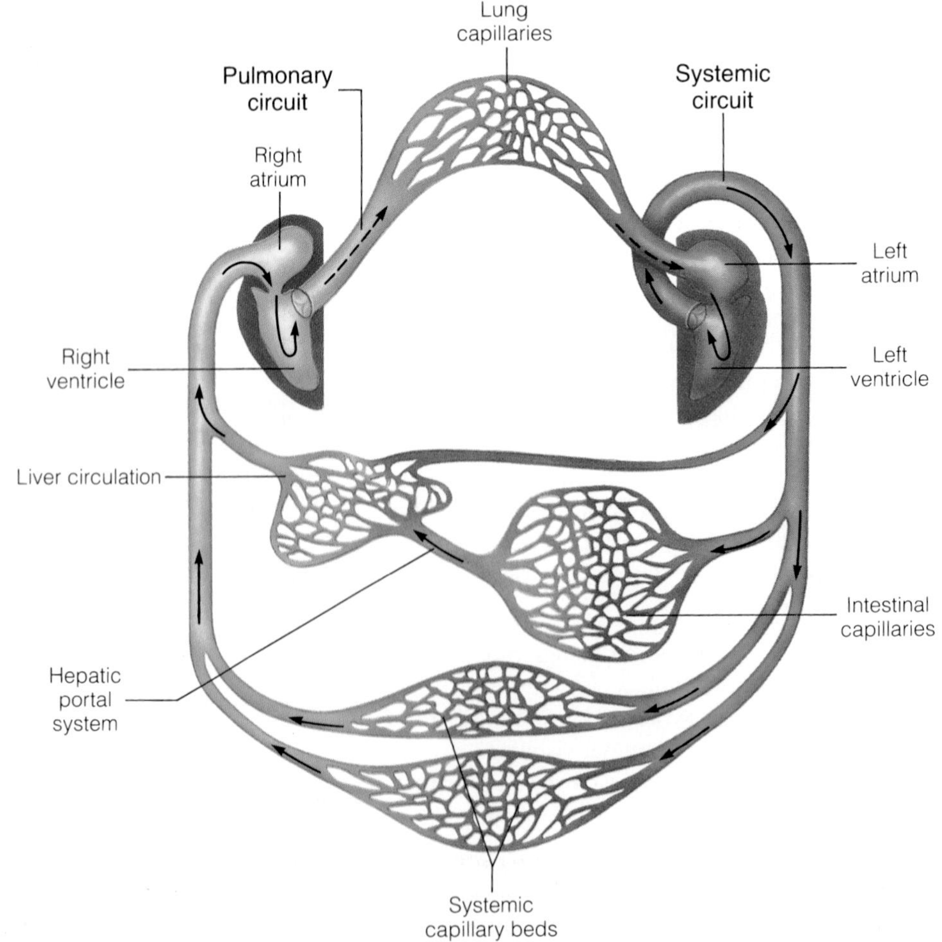

◆ **FIGURE 19.10 Schematic diagram of blood circulation**

Note that the right heart chambers propel blood into the pulmonary circuit (dashed arrows) and the left heart chambers propel blood into the systemic circuit (solid arrows).

CLINICAL CORRELATION

and indicated that the lungs were clear. Calcification was indicated in the region of the aortic semilunar valve. A harsh ejection murmur (a murmur occurring when the ventricles contract and eject blood) was heard, with maximal intensity at the apex of the heart.

CLINICAL COURSE: Digoxin (which increases cardiac contractility) was adminstered, and cardiac catheterization was performed. Coronary angiography (injection of contrast medium and observation under a fluoroscope) revealed no significant decrease in coronary blood flow. The aortic pressure was 90/50 mm Hg. The orifice of the aortic semilunar valve could not be penetrated. On the following day, the chest was opened, and the aortic semilunar valve was examined. It was heavily calcified, and only a pinhole-size orifice existed. The defective valve was excised, and a prosthetic aortic valve was installed.

OUTCOME: The patient recovered without event and was discharged three weeks after admission, on a regimen of digoxin and warfarin (an anticoagulant). At a follow-up examination six weeks after the operation, the patient felt well, the arterial pressure was 115/75 mm Hg, and no cardiac murmur was heard. The digoxin and warfarin were discontinued, and the patient was continued on aspirin (inhibits the aggregation of platelets) 150 mg twice a day.

right atrioventricular valve, and into the right ventricle (Figure 19.12). At the same time, blood from the pulmonary veins moves through the left atrium, past the open left atrioventricular valve, and into the left ventricle. The simultaneous contraction of both atria then squeezes more blood into the ventricles. Subsequently, the simultaneous contraction of both ventricles closes the atrioventricular valves and forces blood past the semilunar valves and into the pulmonary trunk (from the right ventricle) and the aorta (from the left ventricle).

Note from this discussion that the contraction of the atria is not essential for the movement of blood into the ventricles. In fact, even if the atria fail to function, the ventricles can still pump considerable quantities of blood.

Cardiac Muscle

The wall of the heart is composed almost entirely of cardiac muscle, and it is this muscle that is responsible for the pumping action of the heart.

Cellular Organization

Within the atria and the ventricles, the individual cells of cardiac muscle interconnect with one another to form branching networks. Where adjoining cells meet end to end, their junctions form structures called *intercalated discs (in-ter´-kah-lā-ted)* (Figure 19.13). Within the intercalated discs are two types of cell-to-cell membrane junctions: desmosomes, which attach one cell to another; and gap junctions, which allow electrical impulses to spread from cell to cell.

Cardiac muscle cells are cross-striated, and they possess thick myosin-containing filaments and thin actin-containing filaments that are arranged into regularly ordered sarcomeres and myofibrils. The basic contractile events that occur in cardiac muscle cells are believed to be similar to those that occur in skeletal muscle cells (Chapter 8). However, cardiac muscle cells have larger t tubules and a less-developed sarcoplasmic reticulum than skeletal muscle cells, and some of the calcium ions involved in contraction are obtained from the extracellular fluid.

Automatic Contraction

Like some smooth muscle cells, certain specialized cardiac muscle cells undergo spontaneous, rhythmical self-excitation. Moreover, a stimulatory impulse can spread from cell to cell through gap junctions, and an entire mass of interconnected cardiac muscle cells responds as a unit when specialized cells become excited. Thus, cardiac muscle is self-excitable, and it contracts automatically without external stimulation. However, the spontaneous activity of cardiac muscle is continually influenced by neurons of the autonomic nervous system that supply the heart and by certain chemicals and hormones that circulate within the blood.

◆ **FIGURE 19.11 Distribution of blood from the heart**

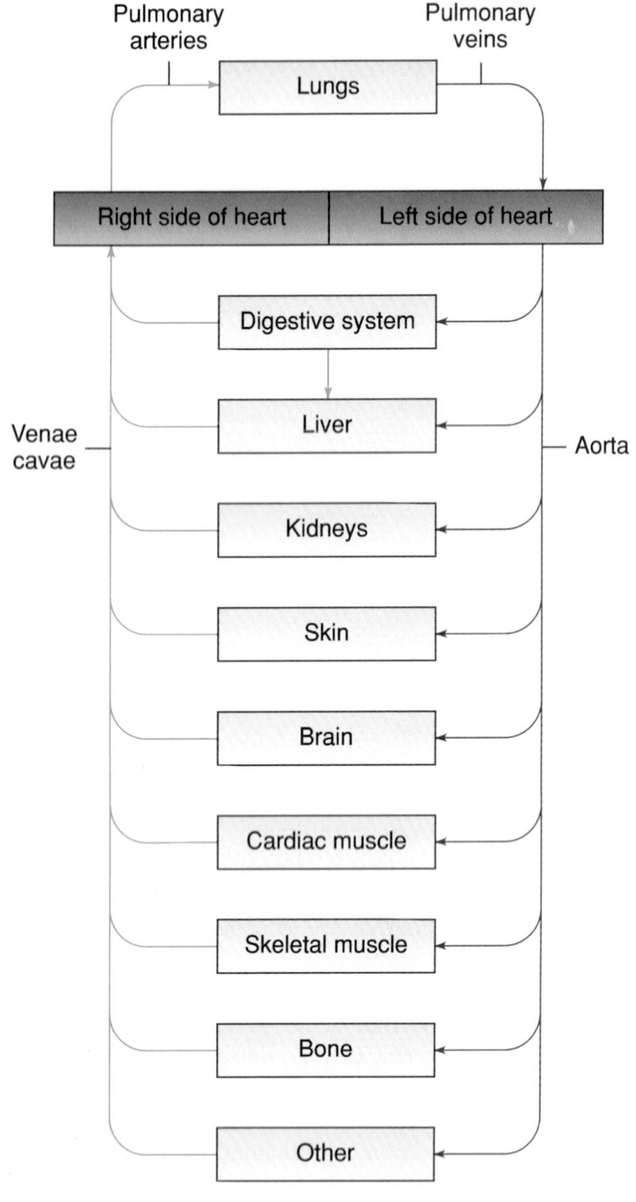

Strength of Contraction

When a skeletal muscle cell is stimulated to contract and calcium ions are released from the sarcoplasmic reticulum, the increase in the level of calcium ions in the cytosol is sufficient to completely activate the cell. However, when a cardiac muscle cell is stimulated to contract, the increase in the level of calcium ions in the cytosol generally is not sufficient to completely activate the cell. Consequently, factors that promote a further increase in the level of calcium ions in the cytosol of a cardiac muscle cell when the cell is stimulated can increase the strength of the contraction. (Norepinephrine released by sympathetic neurons supplying the heart is one factor that acts at least partially in this manner.)

Refractory Period

Following the excitation of a cardiac muscle cell, there is a relatively long period of time—called the *refractory period*—during which the cell cannot normally be reexcited. This period is particularly evident in ventricular cardiac muscle cells, for which the refractory period extends well into the relaxation phase of the contractile cycle (Figure 19.14). The relatively long refractory period normally prevents the heart from undergoing a prolonged tetanic contraction or a spasm, which would halt blood flow and therefore cause death.

Metabolism of Cardiac Muscle

The myocardium can obtain only insignificant amounts of energy—that is, ATP—from anaerobic metabolism. Therefore, cardiac muscle depends on aerobic metabolism for the continuous supply of energy required to support its contractile activity. For this reason, a substantial and continuous delivery of oxygen to the myocardium is essential, and the myocardium receives an abundant blood supply through the coronary arteries.

In a resting individual, the myocardium obtains most of its energy from the oxidation of fatty acids and small amounts from the oxidation of lactic acid and glucose. During increased activity—for example, during heavy physical exercise—the myocardium actively removes lactic acid from coronary blood and oxidizes it directly for the production of energy. In this regard, it should be noted that the anerobic metabolic processes utilized by skeletal muscles during vigorous activity produce lactic acid that is released into the blood.

Excitation and Conduction in the Heart

Approximately 99% of the cardiac muscle cells in the heart are contractile cells. These cells contract and produce force, but they normally do not undergo spontaneous self-excitation. The heart also contains specialized, noncontractile cardiac muscle cells that make up the conducting system of the heart (Figure 19.15). These cells initiate the impulses that cause the heart to beat, and they conduct the impulses throughout the myocardium. The conducting system coordinates the heartbeat, producing an efficient pumping action.

◆ **FIGURE 19.12 Blood flow through the heart during a single cardiac cycle**

(a) Blood fills both atria and enters both ventricles through the open AV valves. (b) The atria contract, which squeezes more blood into the ventricles. (c) The ventricles contract, which forces blood into the aorta and the pulmonary trunk and closes the AV valves.

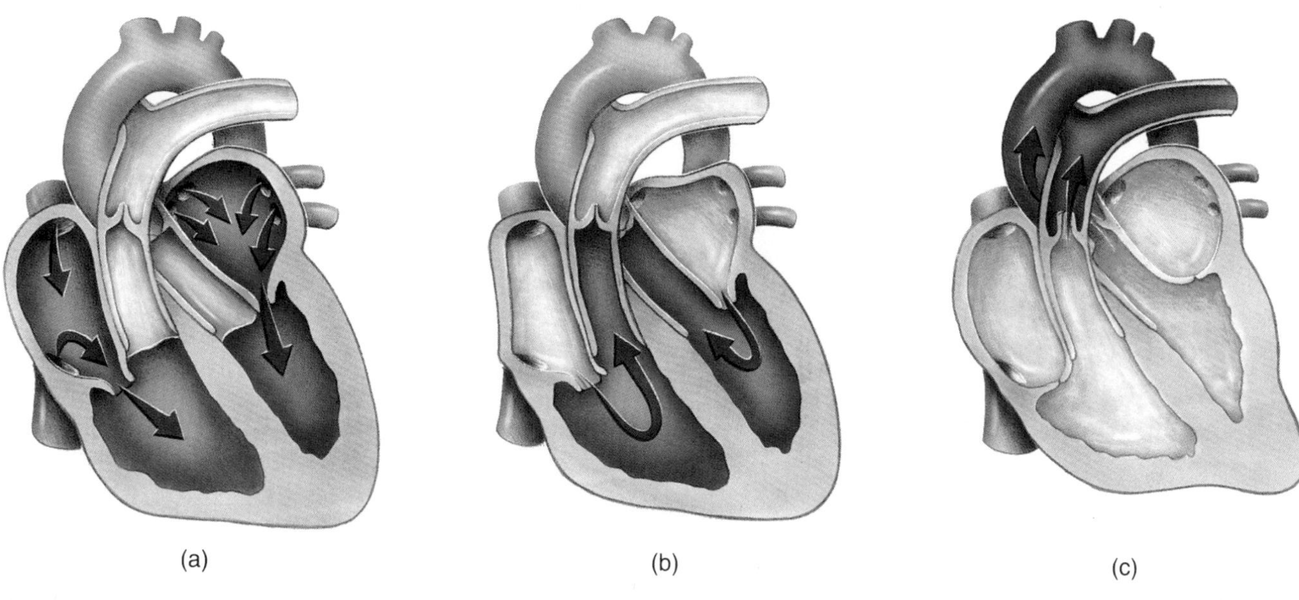

◆ **FIGURE 19.13 Microscopic appearance and cell-to-cell junctions of cardiac muscle**

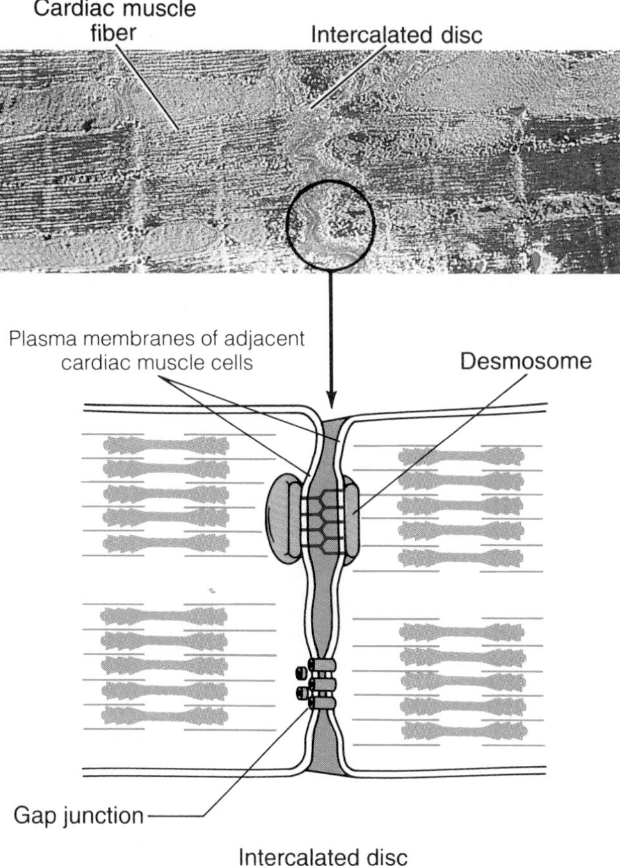

Excitation

Like other cells, cardiac muscle cells maintain an unequal distribution of ions on either side of their plasma membranes, and they are electrically polarized. If the plasma membrane of a cardiac muscle cell becomes depolarized to threshold, an action potential results and the cell is stimulated to contract.

In the wall of the right atrium, near the entrance of the superior vena cava, is a small mass of specialized

◆ **FIGURE 19.14 Contractile response of a ventricular cardiac muscle cell following its excitation at time 0**

The refractory period is indicated by the yellow area.

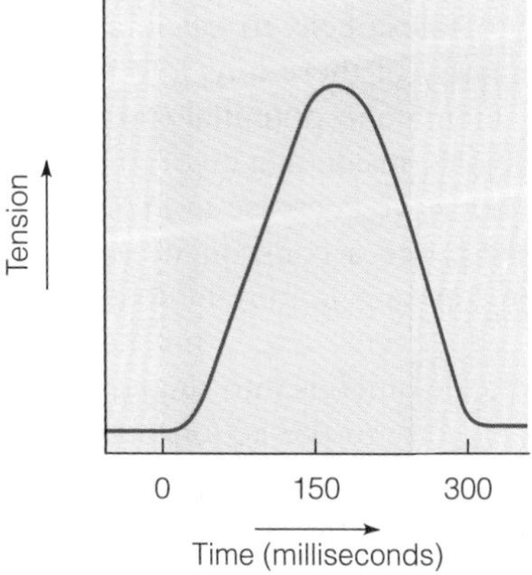

◆ **FIGURE 19.15 The conducting system of the heart**

The dashed arrows indicate the action potential from the sinoatrial node as it travels through the myocardium of both atria.

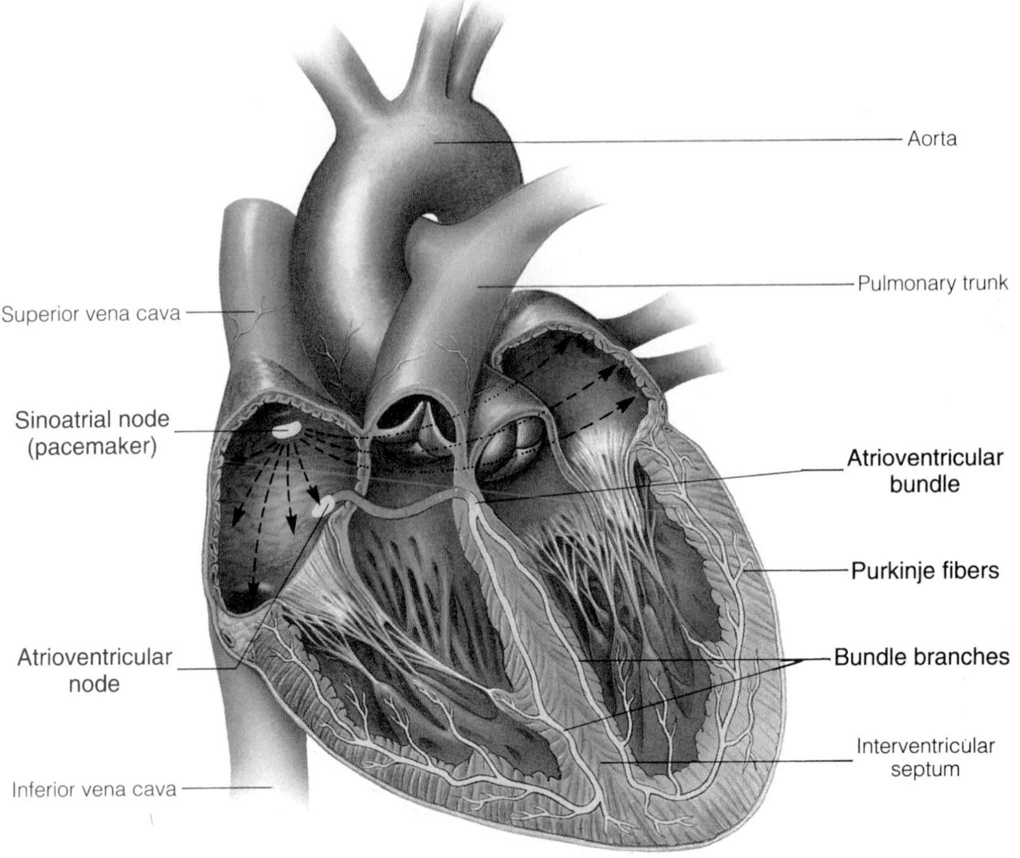

cardiac muscle cells called the **sinoatrial node** (*sī-no-a´-trē-al;* **SA node).** The cells of this node exhibit pacemaker activity (page 270), and in a resting adult, they spontaneously depolarize to threshold and generate an action potential approximately 70 to 80 times each minute (that is, about every 0.8 sec).

The reasons for the pacemaker activity of sinoatrial-node cells are not entirely clear. Some investigators suggest that a progressive increase occurs in the membrane permeability of these cells to calcium and sodium ions and that the entry of these ions into the cells gradually reduces the membrane potential (normally about −55 to −70 millivolts, inside negative) to threshold. Others believe a progressive decrease in membrane permeability to potassium and a consequent diminished outflow of potassium ions are important to the depolarization process. In any event, once threshold is reached, a movement of calcium ions into sinoatrial-node cells contributes significantly to the action potentials developed by the cells.

Specialized cardiac muscle cells in other portions of the conducting system of the heart can also undergo spontaneous depolarization to threshold and generate action potentials. However, they do so at slower rates than the cells of the sinoatrial node. As a result, impulses from the sinoatrial node reach these cells and stimulate them so frequently that they do not generate action potentials at their own inherent rates. Thus, the rate of discharge of the sinoatrial node sets the rhythm for the entire heart, and it is for this reason that the sinoatrial node is called the *pacemaker* of the heart.

Conduction

An impulse generated in the sinoatrial node spreads from cell to cell (and perhaps also along special atrial pathways) throughout the myocardium of the atria, stimulating the atria to contract. However, the fibrous skeleton of the heart that surrounds the openings between the atria and the ventricles as well as the openings of the aorta and the pulmonary trunk (Figure 19.7) cannot depolarize. Therefore, a stimulatory impulse transmitted from the sinoatrial node throughout the atria cannot pass directly to the myocardium of the ventricles. Instead, the impulse reaches the ventricles by way of the conducting system of the heart.

A group of specialized cardiac muscle cells called the **atrioventricular node** (*a-trē-o-ven-trik´-ū-lar;* **AV node)** is located within the interatrial septum just above the junction of the atria and the ventricles (Figure 19.15). From the atrioventricular node, a bundle of specialized muscle tissue called the **atrioventricular bundle** *(bundle of His)* passes to the ventricles. The atrioventricular bundle enters the interventricular septum, where it divides into **right** and **left bundle branches** that travel down the septum. Small groups of terminal conducting fibers called **Purkinje fibers** *(per-kin´-jee)* exit from the bundle branches and end on contractile cardiac muscle cells of the ventricles. At the apex of the heart, the Purkinje fibers pass to the outer wall of the ventricles and travel back toward the base of the heart.

As a stimulatory impulse from the sinoatrial node spreads throughout the atrial myocardium, it reaches the atrioventricular node. After a delay of approximately 0.1 sec, the impulse is transmitted from the antrioventricular node, along the atrioventricular bundle and the Purkinje fibers, to the cells of the ventricles. One of the reasons for the delay in the transmission of the impulse at the atrioventricular node is that the cells in the region are quite small and thus transmit impulses slowly. This delay is important to a coordinated heartbeat because it ensures that the atria will contract and squeeze blood into the ventricles before the ventricles contract.

In contrast to the delay in the conduction of the stimulatory impulse at the atrioventricular node, the rate of conduction of the impulse from the node to the ventricles along the specialized conducting fibers is quite rapid. Thus, the entire ventricular myocardium is stimulated almost simultaneously, producing a strong, effective pumping action.

The Cardiac Cycle

The main function of the heart is to receive blood from the veins at low pressure and pump it into the arteries at a pressure high enough to propel it through the vessels and back again to the heart.

In order to function as a pump, the heart repeats two alternating phases: (1) The chambers contract, forcing the blood within them out of the chambers. This phase is called **systole** *(sis´-ta-lee)*. (2) The chambers relax, allowing them to refill with blood. This phase is known as **diastole** *(dye-ass´-ta-lee)*. The cardiac cycle, therefore, consists of a period of contraction and emptying (systole) followed by a period of relaxation and filling (diastole). Although both the atria and the ventricles undergo systole and diastole, the terms usually refer to ventricular contraction or relaxation.

The pressure changes, volume changes, and valve actions that occur in the heart of a resting adult during the cardiac cycle are compared in Figure 19.16. Although this figure is concerned with the left chambers of the heart, the right chambers follow a similar pattern. However, the peak pressure within the right ventricle is considerably lower. The systolic pressure in the right ventricle—which only has to pump the blood through the pulmonary circuit—reaches about 30 millimeters of mercury (mm Hg), whereas, as the figure shows, the systolic pressure in the left ventricle reaches 120 mm Hg.

In a resting adult, each cardiac cycle lasts approximately 0.8 sec. Atrial systole requires about 0.1 sec, and ventricular systole about 0.3 sec. Thus, during each cardiac cycle, the atria are in diastole for 0.7 sec and the ventricles for 0.5 sec.

It is important to emphasize that during ventricular diastole, blood flows into the atria and through the open atrioventricular valves into the ventricles prior to the contraction of the atria. In fact, the ventricles are 70% filled before the atria contract. When the atria do contract, they force a small amount of additional blood (approximately 30% of the ventricular capacity) into the ventricles. However, since the ventricles are almost full prior to atrial contraction, defects that impair atrial pumping may not seriously impair cardiac function, at least in resting individuals. (During exercise, when the heart rate is elevated and the diastolic filling period is shortened, atrial contraction contributes more importantly to ventricular filling than it does in a resting individual.)

Pressure Curve of the Left Atrium

During most of ventricular diastole, the pressure within the left atrium is slightly higher than the pressure within the left ventricle. Moreover, the pressure within the left atrium gradually increases as blood flows into it (and continues into the left ventricle) from the pulmonary veins (Figure 19.16). When the left atrium contracts, there is a sudden further increase in the atrial pressure (curve a). Shortly thereafter, ventricular systole begins, and the pressure within the left ventricle rises above that in the left atrium, causing the left atrioventricular valve to close (line 1). The sudden increase in the intraventricular pressure is so great that it causes the left atrioventricular valve to bulge slightly into the left atrium. This activity contributes to a further increase in the atrial pressure (curve c) that lasts only until the aortic semilunar valve opens (line 2). The pressure within the left atrium, which is now in diastole, then quickly drops to almost 0 mm Hg. However, as blood from the pulmonary veins again fills the atrium, the atrial pressure shows a steady increase until the left atrioventricular valve reopens during the following ventricular diastole (line 4). The reopening of the left atrioventricular valve allows blood from the left atrium to enter the left ventricle. The flow of blood from the atrium into the ventricle causes the atrial pressure to drop again to almost 0 mm Hg. The atrium is now ready to begin another cycle.

Pressure Curve of the Left Ventricle

During most of ventricular diastole, the pressure within the left ventricle is slightly lower than the pressure within the left atrium (Figure 19.16). The contraction of the left atrium causes a slight increase in the pressure within the left ventricle (curve b) because atrial systole increases the volume of blood within the ventricle, as shown in the ventricular volume curve.

When ventricular systole begins, the pressure within the left ventricle rises rapidly (lines 1 to 2). When the pressure in the left ventricle exceeds the pressure in the left atrium, the left atrioventricular valve closes. When the pressure within the left ventricle becomes greater than the pressure within the aorta, the aortic semilunar valve is forced open (line 2), allowing blood to flow into the aorta. The pressure within the ventricle continues to rise to about 120 mm Hg and then begins to fall.

At the end of ventricular systole, the intraventricular pressure drops significantly below the pressure in the aorta. As a result, blood from the aorta begins to flow back toward the left ventricle, causing the aortic semilunar valve to close (line 3). The intraventricular pressure continues to fall precipitously until it becomes less than the atrial pressure (line 4), thus allowing blood in the left atrium to push open the left atrioventricular valve and flow into the left ventricle.

Pressure Curve of the Aorta

The aorta is an elastic artery. As blood is pumped into the aorta during ventricular systole, the vessel distends, or expands. Then, as the blood flow from the left ventricle into the aorta decreases and stops during late systole and diastole, the aorta recoils and squeezes down on the blood within it. This helps maintain a certain amount of pressure within the vessel (Figure 19.16). If the aorta and the other elastic arteries had rigid walls that were not distensible, the pressure within them would rise very abruptly during ventricular systole. If they were not capable of elastic recoil, the pressure within them would fall just as abruptly during ventricular diastole.

When the left ventricle is in diastole and no blood is being ejected into the aorta, the aortic pressure curve shows a steady but gradual decrease as the blood pumped into the aorta during the previous systole flows into other vessels of the body (Figure 19.16). The elastic recoil of the aortic wall maintains some pressure on the blood (the aortic blood pressure never drops much below 80 mm Hg), but it is not sufficient to keep the pressure constant. When the left ventricle contracts and the aortic semilunar valve opens (line 2), the aortic pressure increases rapidly as a large volume of blood is pumped into the aorta. The pressure peaks at about 120 mm Hg and then decreases as the flow of blood from the left ventricle into the aorta diminishes. However, the ventricular pressure falls more rapidly than the aortic pressure, and as it drops significantly below the aortic pressure, the aortic semilunar valve closes (line 3). The closure of the aortic semilunar valve produces a brief rise in the aortic pressure, which forms the incisura on the diagram.

Volume Curve of the Left Ventricle

The period of time between the closing of the left atrioventricular valve and the opening of the aortic semilunar valve is called the period of *isovolumetric ventricular contraction.* During this period, both valves are closed, and blood neither enters nor leaves the left ventricle. Consequently, the volume of blood within the ventricle remains constant (although the pressure within the ventricle rises).

The period of time when the aortic semilunar valve is open and blood is flowing from the left ventricle into the aorta is called the period of *ventricular ejection.* During this period, the volume of blood within the ventricle decreases.

The period of time between the closing of the aortic semilunar valve and the opening of the left atrioventricular valve is called the period of *isovolumetric ventricular relaxation.* During this period, both valves are closed, and blood neither enters nor leaves the left ventricle. Consequently, the volume of blood within the ventricle remains constant (although the pressure within the ventricle falls).

The period of time when the left atrioventricular valve is open and blood is flowing from the left atrium into the left ventricle is called the period of *ventricular filling.* During this period, the volume of blood within the ventricle increases. Most of the increase in volume occurs prior to atrial systole, and there is only a modest further increase during atrial systole.

Cardiac Output

Each cardiac cycle moves blood out of the ventricles into the pulmonary circuit and the systemic circuit. The term **cardiac output** is used to indicate the volume of

◆ **FIGURE 19.16 Comparisons of pressure changes, volume changes, and valve actions within the left chambers of the heart during a single cardiac cycle in a resting person**

Note that the duration of a complete cycle is 0.8 sec.

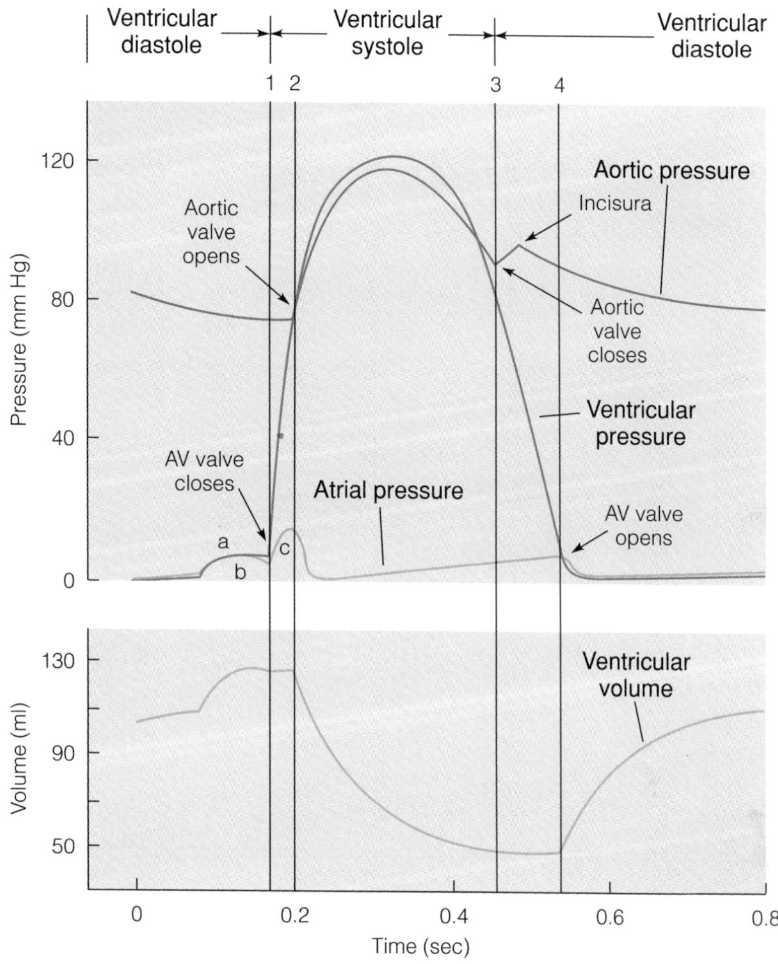

blood pumped per minute by either ventricle; for this reason, cardiac output is sometimes referred to as **cardiac minute output.**

Generally, cardiac output is discussed in terms of the left ventricle. However, over any period of time, the normal heart ejects the same amount of blood from each ventricle, although some variation may occur from beat to beat. If this were not the case and the left ventricle pumped a greater volume than the right, blood would accumulate in the systemic circuit. Similarly, if the right ventricle pumped more blood than the left, blood would accumulate in the pulmonary circuit. As we discuss later, unequal pumping by the two ventricles can be a problem in certain diseases of the heart.

Cardiac output is equal to the heart rate multiplied by the stroke volume. The **stroke volume** is the amount of blood pumped by one side of the heart per beat. The average stroke volume of a human heart is about 75 ml, and the heart rate of a resting individual is generally between 70 and 80 beats per minute. Therefore, the cardiac output under resting conditions is between 5 and 6 liters per minute (for example, 75 ml/beat × 70 beats/minute = 5250 ml/minute = 5.25 liters/minute). Under stressful conditions, the cardiac output can be greatly increased—up to 30 liters per minute in a trained athlete. The difference between the volume of blood moved per minute during rest and the volume that the heart is capable of moving per minute is called the **cardiac reserve.**

Changes in either the heart rate or the stroke volume can alter the cardiac output, and in general an increase in either factor increases the cardiac output. However, at very high heart rates—usually beginning between 170 and 250 beats per minute—cardiac output actually decreases. This decrease is due to the fact that stroke volume declines at very high heart rates, offsetting the effects of the increased rate. A major reason for the decline in stroke volume is that the length of diastole decreases as the heart rate increases, leaving a shorter time for the ventricles to fill with blood. Thus, at the

◆ **FIGURE 19.17 Some factors influencing cardiac output**

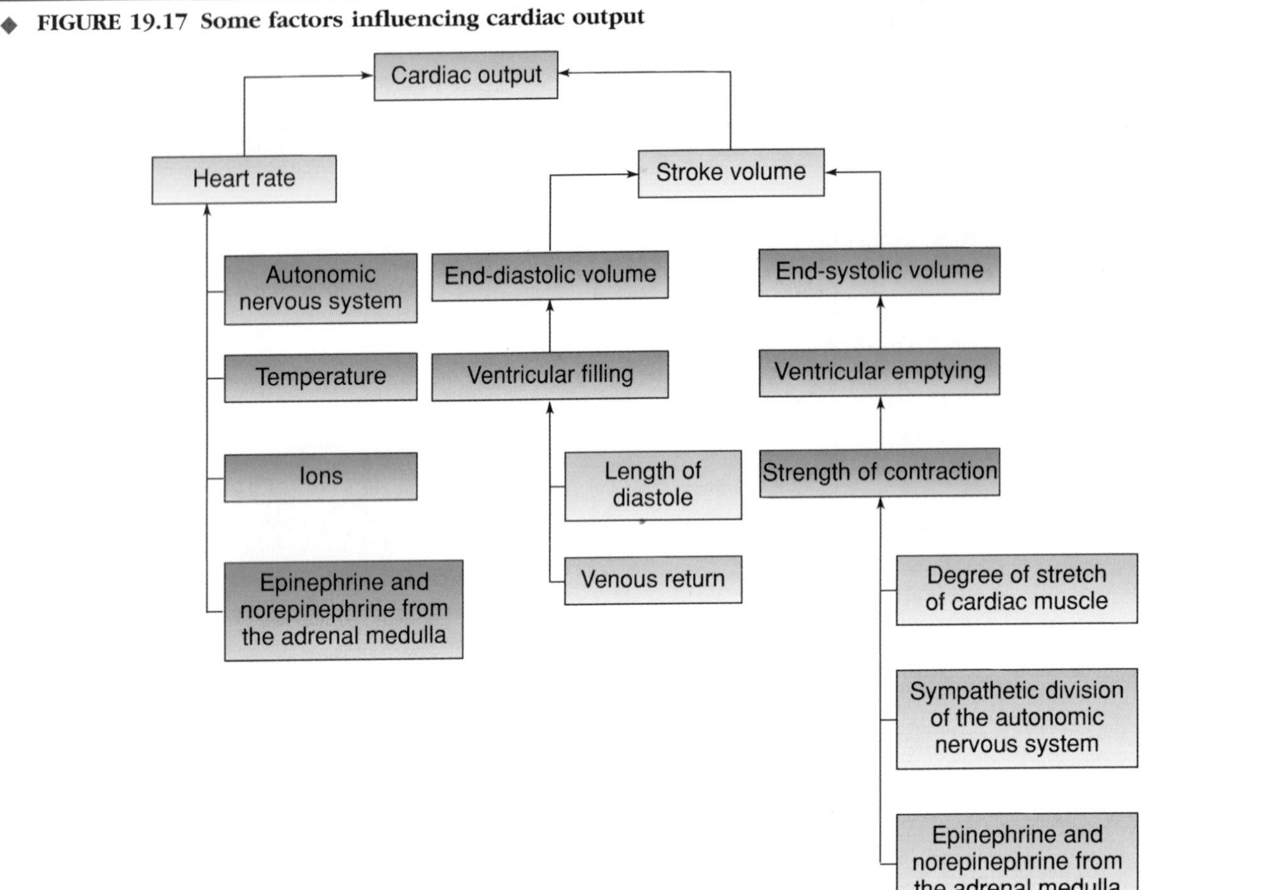

end of diastole, less blood is present within the ventricles to be pumped out during the following systole.

The cardiac output changes to meet the varying needs of the body, and since it is equal to the heart rate multiplied by the stroke volume, the mechanisms that control the heart rate and the stroke volume also control the cardiac output (Figure 19.17).

Control of Heart Rate

The most important factor in the control of heart rate is the effect of the autonomic nervous system. The sinoatrial node and the atrioventricular node are particularly well supplied by sympathetic and parasympathetic neurons, and the atria also receive both sympathetic and parasympathetic innervation. The ventricles are supplied mainly by sympathetic neurons, with far fewer parasympathetic neurons present.

Stimulation of the sympathetic neurons increases the heart rate, and maximal sympathetic stimulation can almost triple the heart rate. The sympathetic neurons release norepinephrine, which increases the rate of discharge of the sinoatrial node, decreases the conduction time through the atrioventricular node, and increases the excitability of all portions of the heart. Norepinephrine apparently exerts its effects by either decreasing membrane permeability to potassium ions or increasing membrane permeability to calcium ions and sodium ions, thereby enhancing the depolarization of excitable cells.

Parasympathetic fibers innervate the heart by way of the vagus nerves. The right vagus is distributed predominately to the sinoatrial node, and the left vagus innervates principally the atrioventricular node. Stimulation of the parasympathetic neurons to the heart decreases the heart rate. The parasympathetic neurons release acetylcholine, which decreases the rate of discharge of the sinoatrial node and increases the conduction time through the atrioventricular node. Acetylcholine increases membrane permeability to potassium ions, thereby causing hyperpolarization and making excitable tissue less excitable.

Intense parasympathetic stimulation can stop impulse generation in the sinoatrial node and block the transmission of impulses through the atrioventricular node to the ventricles. However, soon after the ventricles stop contracting, some other portion of the conducting system—often in the atrioventricular bundle—beings to discharge spontaneously and stimulate ventricular contraction. This phenomenon is known as *ventricular escape.*

The heart rate at any given moment is largely determined by the balance between the stimulatory effects of norepinephrine released from sympathetic neurons and the inhibitory effects of acetylcholine released from parasympathetic neurons. If the sympathetic nerves to the heart are blocked, the heart rate decreases (due to the unopposed parasympathetic activity), and if the parasympathetic nerves are blocked, the heart rate increases (due to the unopposed sympathetic activity). In a resting individual, parasympathetic activity is dominant, and if all autonomic innervation is blocked, the heart rate increases from the normal 70 to 80 beats per minute to about 100 beats per minute, which is the inherent rate of discharge of the sinoatrial node. During exercise, the activity of the sympathetic neurons increases, the activity of the parasympathetic neurons decreases, and the heart beats faster.

Control of Stroke Volume

The amount of blood that moves out of a ventricle with each beat—that is, the stroke volume—is equal to the difference between (1) the volume of blood within the ventricle at the end of diastole just as systole begins (the *end-diastolic volume*), and (2) the volume of blood remaining within the ventricle when the semilunar valves close at the end of systole (the *end-systolic volume*). A number of factors influence the end-diastolic and end-systolic volumes and, therefore, the stroke volume.

End-Diastolic Volume

The end-diastolic volume depends on the amount of blood that enters a ventricle during diastole. This amount, in turn, is determined by two main factors: the length of diastole and the venous return.

Length of Diastole. The longer the period of diastole, the more time is available for the ventricles to fill with blood. However, as the heart rate increases, the length of diastole becomes shorter, and less time is available for ventricular filling. In this regard, it should be noted that norepinephrine from the sympathetic nervous system not only increases the heart rate but also accelerates the rates at which cardiac muscle fibers shorten and relax. Thus, as norepinephrine increases the heart rate, the decreased time required for cardiac muscle fibers to shorten and relax provides somewhat more time for ventricular filling during diastole than would otherwise be the case.

Moreover, the rapid relaxation of the ventricles after contraction causes the intraventricular pressure to fall rapidly, which enhances the pressure gradient for the flow of blood from the atria into the ventricles. Nevertheless, in a very rapidly beating heart, the length of diastole may be so shortened that ventricular filling is significantly reduced.

Venous Return. The volume of blood that returns from the veins to one side of the heart in a given time has a major influence on the end-diastolic volume. (The term *venous return* usually refers to the volume of blood flowing from the systemic veins into the right side of the heart in a given time, but under normal circumstances the same volume of blood flows from the pulmonary veins into the left side of the heart.)

In general, if the length of diastole is constant, an increased venous return leads to increased ventricular filling and a relatively large end-diastolic volume that stretches the ventricle wall. If the length of diastole decreases (for example, when the heart rate increases), a ventricle may still fill completely if, as is often the case, the venous return increases.

As is discussed in Chapter 20 (page 642), conditions in the peripheral blood vessels have a major influence on the magnitude of the venous return. Moreover, the activity of the heart may facilitate the venous return and the movement of blood into the ventricles. For example, there is evidence that following ventricular systole, the ventricles tend to recoil and expand, creating a suction effect that aids ventricular filling.

End-Systolic Volume

The end-systolic volume depends on the amount of blood ejected from a ventricle during its contraction. A ventricle does not eject all of the blood it contains when it contracts, and the degree to which a ventricle is emptied during systole is influenced mainly by the strength of the ventricular contraction.

The Frank-Starling Law of the Heart. The Frank-Starling law of the heart states that, within limits, as cardiac muscle is stretched, its force of contraction increases. Thus, when ventricular filling produces a comparatively large end-diastolic volume that stretches the ventricle wall, the ventricle contracts more forcefully and ejects a larger volume of blood than when ventricular filling produces a smaller end-diastolic volume that stretches the ventricle wall a lesser amount. This phenomenon provides an inherent self-regulating mechanism by which the heart is able to adjust stroke volume to changing end-diastolic volumes. However, if a ventricle is overly stretched (that is, if the end-diastolic volume is too great), the effectiveness of the ventricular contraction diminishes.

The influence of stretch on the force of contraction of cardiac muscle helps ensure that over any period of time the normal heart ejects the same amount of blood from each ventricle, even though some variation may occur beat to beat. For example, suppose the stroke volume of the right ventricle increases while that of the left ventricle remains unchanged. The increased right ventricular stroke volume causes an increased volume of blood to enter the pulmonary circuit, which, in turn, leads to an increase in the volume of blood returned to the left ventricle. The increase in the volume of blood returned to the left ventricle produces a greater left ventricular end-diastolic volume, which stretches the ventricle wall. Consequently, the left ventricle contracts more forcefully, and its stroke volume increases, thereby restoring the balance in the amount of blood pumped by each ventricle.

Sympathetic Nervous System. Norepinephrine from the sympathetic nervous system is an important factor that influences the strength of ventricular contraction. Norepinephrine, which increases heart rate and accelerates the rates at which cardiac muscle fibers shorten and relax, also increases the force of contraction of cardiac muscle. At any particular end-diastolic volume (that is, degree of stretch), a more forceful contraction ejects a larger volume of blood from the ventricles, which in turn decreases the end-systolic volume and increases the stroke volume. Under normal conditions, sympathetic activity maintains the strength of ventricular contraction at a level about 20% greater than would be the case without stimulation, and maximal sympathetic stimulation can increase contractile strength to about 100% greater than normal. Norepinephrine increases contractile strength at least in part by increasing membrane permeability to calcium ions, which directly enhances the contractile activity of cardiac muscle cells.

In contrast to the effects of sympathetic activity, parasympathetic activity has comparatively little effect on the strength of ventricular contraction, and maximal parasympathetic stimulation decreases ventricular contractile strength by only about 30%.

Factors That Influence Cardiac Function

Cardiac function can be influenced by a number of factors.

Cardiac Center

Both the sympathetic and parasympathetic neurons that supply the heart are under the control of a center within the brain stem called the *cardiac center.* This center receives neural input from higher centers in the brain and from various receptors associated with the cardiovascular system. In general, when the sympathetic nerves to the heart are stimulated, the parasympathetic nerves are inhibited. Conversely, sympathetic inhibition and parasympathetic stimulation are usually elicited simultaneously. The role various inputs to the cardiac center play in the control of cardiovascular activity is considered in Chapter 20.

Exercise

Chronic endurance types of exercise generally cause cardiac muscle to hypertrophy and the ventricular chambers to enlarge, leading to an increase in stroke volume. For example, the hearts of marathon runners have been found to have stroke volumes that are approximately 40 to 50% larger than the stroke volumes of the hearts of untrained people. Compared to a heart with a smaller stroke volume, a heart with a larger stroke volume can pump the same amount of blood per minute with fewer beats, and it can pump more blood when the rates of the two hearts are the same. Thus, both at rest and during exercise, an individual in good physical condition can generally maintain the cardiac output required for a particular activity level with a lower heart rate than an individual in poor physical condition. In addition, the maximum cardiac output and cardiac reserve are greater in a person in good physical condition.

During exercise, cardiac output increases greatly (for example, from a resting value of about 5.25 liters/min to a maximum value of about 30 liters/min in a trained athlete). This increase in cardiac output is due to an increase in both stroke volume and heart rate. In an exercising marathon runner with a cardiac output of 30 liters/min, the stroke volume is about 50% greater than the resting value (162 ml compared to 105 ml), and the heart rate is about 270% greater (185 beats per minute compared to 50 beats per minute). As the cardiac output increases from the resting level to 30 liters/min, the stroke volume reaches its maximum (162 ml) at a cardiac output of approximately 15 liters/min. At this cardiac output, the heart rate is about 92 beats per minute. Increases in cardiac output above this level are achieved by further increases in heart rate.

Temperature

When the heart is warmed, the discharge rate of the sinoatrial node increases, and a rise in body temperature of 1°C increases the heart rate about 12 to 20 beats per minute. This may account for the rapid heart rate that

ASPECTS OF EXERCISE PHYSIOLOGY 

The What, Who, and When of Stress Testing

Stress tests, or graded exercise tests, are conducted primarily to aid in diagnosing or quantifying heart or lung disease and to evaluate the functional capacity of asymptomatic individuals. The tests are usually given on motorized treadmills or bicycle ergometers (stationary, variable-resistance bicycles). Workload intensity (how hard the subject is working) is adjusted by progressively increasing the speed and incline on the treadmill or by progressively increasing the pedaling frequency and resistance on the bicycle. The test starts at a low intensity and continues until a prespecified workload is achieved, physiologic symptoms occur, or the subject is too fatigued to continue.

During diagnostic testing, the patient is monitored with an electrocardiogram (ECG), and blood pressure is taken each minute. A cardiologist, or a pulmonary specialist in the case of lung disease, is present during testing. A test is considered to be positive if ECG abnormalities occur (such as ST segment depression, inverted T waves, or dangerous arrhythmias) or if physical symptoms such as angina pectoris develop. A test that is interpreted as being positive in a person who does not have heart disease is called a false positive test. In men, this occurs only about 10% to 20% of the time, so the diagnostic stress test for men has a *specificity* of 80% to 90%. Women have a greater frequency of false positive tests; the specificity for them is lower, at about 70%.

The *sensitivity* of a test means that those individuals with disease are correctly identified and there are few false negatives. The sensitivity of the stress test is reported to be 60% to 80%. This means that if 100 individuals with heart disease were tested, 60 to 80 would be correctly identified, but 20 to 40 would have a false negative test.

As more is learned about what causes false positive and false negative tests, the specificity and sensitivity of stress testing should improve. Although stress testing is now an important diagnostic tool, it is just one of several tests used to determine the presence of coronary artery disease.

Graded exercise tests are also conducted on individuals not suspected of having heart or lung disease to determine their present functional capacity. These functional tests are administered in the same way as diagnostic tests, but a physician need not be present. They are conducted by exercise physiologists and are used to establish safe exercise prescriptions, to aid athletes in establishing optimal training programs, and as research tools to evaluate the effectiveness of a particular training regimen. Functional stress testing is becoming more prevalent as more people are joining hospital- or community-based wellness programs for disease prevention.

accompanies fever. However, increased activity of the sympathetic nerves that supply the heart may also be involved. When the heart is cooled, the heart rate declines, and the heart ultimately stops. During some surgical procedures, the body temperature is artificially lowered, and the reduced heart rate that results makes it possible to perform operations that cannot be performed on a rapidly beating heart.

Ions

The levels of various inorganic ions in the blood and interstitial fluid can influence cardiac function. If the level of potassium ions is increased substantially, the heart rate drops and the heart becomes extremely dilated, flaccid, and weak. The elevated level of potassium ions is believed to decrease membrane potentials, which in turn decreases the intensities of action potentials, thereby weakening heart contractions. When the level of calcium ions is increased excessively, the heart contracts spastically, probably because of the effects of the calcium ions on the excitation and contraction processes of cardiac muscle cells. A substantial increase in the level of sodium ions slows the heart and depresses cardiac function, presumably by inhibiting the movement of calcium ions into cardiac muscle cells. The levels of calcium and sodium ions rarely rise sufficiently to alter cardiac function greatly.

Catecholamines

Epinephrine and norepinephrine from the adrenal medulla increase both the heart rate and the force of contraction, and they can provide a relatively small, slower-acting, but effective adjunct to the sympathetic innervation of the heart.

CONDITIONS OF CLINICAL SIGNIFICANCE

The Heart

Valvular Malfunctions

Valvular malfunctions can interfere with the normal movement of blood through the heart. They can decrease the amount of blood pumped out of a ventricle with each contraction, thereby making it necessary for the heart to work harder to maintain a given cardiac output.

Valvular Insufficiency

Valvular insufficiency is a condition in which the cusps of a valve do not form a tight seal when the valve is closed. As a result, blood leaks back, or regurgitates, into the chamber from which it came. If an atrioventricular valve does not close completely, blood flows back into the atrium when the ventricle contracts, and less than the normal amount of blood may be moved into the aorta or pulmonary trunk. Similarly, if a semilunar valve does not close completely, blood that moves into the aorta or pulmonary trunk during ventricular contraction flows back into the ventricle when it relaxes. Growths or scar tissue that form on a valve as a result of diseases such as rheumatic fever can prevent the valve from closing securely and cause valvular insufficiency.

Valvular Stenosis

Valvular stenosis is a condition in which the opening of a valve becomes so narrowed that it interferes with the flow of blood through it. If the opening of an atrioventricular valve is too narrow, the ventricle may not fill completely with blood, and a lower-than-normal amount of blood may be pumped out when the ventricle contracts. If the opening of a semilunar valve is too narrow, the ventricle may not eject a normal amount of blood into the aorta or pulmonary trunk when it contracts. Growths or scar tissue on a valve can cause valvular stenosis as well as valvular insufficiency. In many cases, valvular insufficiency and valvular stenosis occur in the same valve.

Congestive Heart Failure

Congestive heart failure is a condition in which the heart fails to pump enough blood to meet the body's needs. In congestive heart failure, an abnormal increase in blood volume and interstitial fluid occurs, and the heart is generally dilated with blood (as are the veins and capillaries). Congestive heart failure may be due to impaired contractile ability of cardiac muscle (for example, as a result of a heart attack) or to an increased workload placed on the heart (for example, as a result of valvular malfunction). In any event, the heart is unable to perform normally the work demanded of it.

In congestive heart failure, compensatory mechanisms may initially allow the cardiac output to be maintained—at least in resting individuals. In response to impaired cardiovascular function, the activity of sympathetic nerves to the heart increases, and the kidneys retain fluid in the body. The increased sympathetic activity increases the contractile strength of cardiac muscle, thereby helping to maintain cardiac output. The retention of fluid by the kidneys increases the blood volume (and the interstitial fluid volume). The increased blood volume leads to an increased venous return and end-diastolic volume that stretches the ventricular muscle. As described by the Frank-Starling law of the heart, the stretched muscle contracts more forcefully, increasing the stroke volume and thereby helping to maintain cardiac output. In addition, the myocardium may hypertrophy (particularly in cases of valvular malfunction), and although the contractile activity per unit weight of hypertrophied muscle may be below normal, the increased muscle mass permits an overall increase in work capacity.

If the heart continues to fail, the compensatory mechanisms may become detrimental. The continued retention of fluid can cause the end-diastolic volume to increase to the point where the cardiac muscle is stretched excessively and its contractile strength declines. In addition, hypertrophied cardiac muscle may not receive a sufficient blood supply to meet its needs. Thus, the stroke volume decreases and the cardiac output declines.

In congestive heart failure, only one side of the heart may fail initially. For example, if the left heart is failing, the right ventricle continues to pump blood normally into the pulmonary circuit, but blood returning from the lungs is not pumped efficiently into the systemic circuit by the left ventricle. As a con-

CONDITIONS OF CLINICAL SIGNIFICANCE

sequence, blood accumulates in the pulmonary circuit, and the pressure within the lung capillaries can rise to the point where fluid is forced out of the vessels. The accumulation of fluid in the lung tissues can result in potentially fatal pulmonary edema. Moreover, since the cardiovascular system is a closed circuit, the failure of the left side of the heart eventually produces an excessive strain on the right side of the heart that can result in total heart failure.

In addition to correcting the cause—for example, surgically repairing a defective valve—congestive heart failure is treated with drugs that increase the contractile strength of the failing cardiac muscle (for example, digitalis) and eliminate excess fluid (for example, diuretics).

Effects of Aging

In the absence of heart disease, the size of the heart does not change much with aging. However, because diseases of the heart and the blood vessels are so common, the heart is, in fact, often enlarged in older persons.

With aging, the strength of heart contraction and the cardiac output progressively decline. Several factors contribute to these declines:

1. The lower levels of physical activity in older people place reduced demands on the heart, and the heart rate may decrease.
2. Cardiac muscle cells are not capable of mitosis, and cardiac muscle cells that die are not replaced.
3. It is believed that the capillaries supplying the myocardium tend to undergo a progressive reduction.

The endocardium tends to become thicker with aging, due to the deposition of connective tissue, and the heart valves tend to thicken and become more rigid. Moreover, it is not uncommon for some calcification of the valves to occur after age 60.

An increase in fibrous connective tissue within the conducting system of the heart also occurs with aging. This increase may interfere with the initiation and transmission of the impulses that regulate heart contraction, leading to irregular beats or heart block.

Numerous studies have shown that older persons who follow a continuous exercise program can increase their cardiac output significantly. However, even if a person does not follow a regular program of exercise, the aging changes that occur in his or her heart generally do not hinder it from adequately supporting the reduced level of activity characteristic of most older persons. Only a disease condition reduces the capacity of the heart to support normal activities.

Heart Sounds

Normal Heart Sounds

Two principal sounds normally occur as blood moves through the heart during a cardiac cycle. These sounds are best described as "lub-dup." The first heart sound (the "lub") is associated with the closure of the atrioventricular valves at the beginning of ventricular contraction. It is largely due to vibrations of the taut atrioventricular valves immediately after closure and to the vibration of the walls of the heart and major veessels around the heart. The second sound (the "dup") is associated with the closure of the semilunar valves as the ventricles begin to relax following their contraction. This sound is due largely to vibrations of the taut, closed semilunar valves and to the vibration of the walls of the pulmonary artery, the aorta, and to some extent the ventricles. The areas of the chest where a stethoscope can best be placed to detect the sounds associated with the different valves are indicated in Figure C19.1

Abnormal Heart Sounds

By listening to the heart, a trained person can obtain considerable information about its condition. Abnormal sounds known as *heart murmurs* can be indicative of particular problems. These sounds, which are described as blowing or vibrating sounds, are caused by a turbulent flow of blood as it passes through the heart.

Both valvular insufficiency and valvular stenosis can cause heart murmurs. In valvular insufficiency, the backward movement of the regurgitated blood interferes with the normal pattern of blood flow through the heart, causing a detectable turbulence. In valvular stenosis, there is a rapid, turbulent flow of blood through the narrowed valvular opening. In addition, the walls around a narrowed valve are often roughened, further contributing to the turbulence.

Certain heart murmurs, called *functional murmurs,* are not pathological but are considered to be normal. For example, the rapid movement of blood

continued on next page

CONDITIONS OF CLINICAL SIGNIFICANCE

◆ **FIGURE C19.1 Areas of the chest where sounds associated with the different heart valves can best be detected**

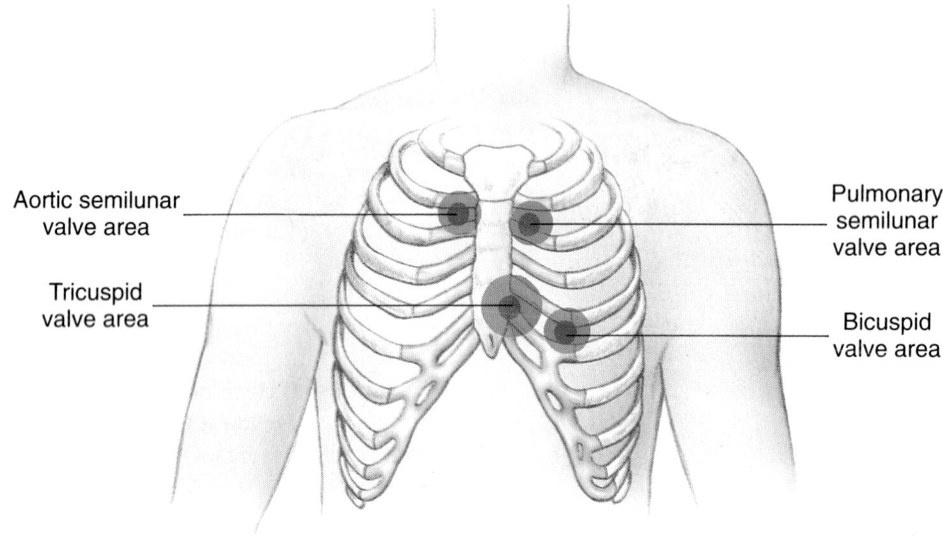

through the heart during heavy exercise may result in turbulence that produces a functional murmur. Functional murmurs are particularly common in young people.

Electrocardiography

The pattern of electrical activity associated with the contraction of cardiac muscle during a heartbeat can be recorded by **electrocardiography.** This procedure, which produces a recording called an **electrocardiogram (ECG),** is useful in detecting conditions that interfere with the normal conduction of impulses through the heart.

As an impulse generated by the sinoatrial node travels through the heart, it produces electrical currents that spread through the body fluids surrounding the heart and then continue onward to the body surface. By placing electrodes on the body surface, it is possible to detect and record the electrical potentials generated by the heart.

The electrodes used to obtain electrocardiograms are placed in various positions on the body surface. Commonly, electrocardiograms are obtained using three "standard" limb leads, or locations for electrode placements (Figure C19.2). Although all three leads give similar patterns of recordings, the amplitudes of the waves differ. These differences are often important in diagnosing various heart conditions.

Normal Electrocardiogram

A normal electrocardiogram of a single heartbeat consists of a regularly spaced series of waves designated **P, Q, R, S,** and **T** (Figure C19.3). (The letters simply indicate the order of appearance of the waves.) The P wave is caused by electrical currents that are produced as the atria depolarize prior to contraction. The QRS complex—actually three separate waves: a Q wave, an R wave, and an S wave—is caused by the depolarization of the ventricles. The T wave is caused by currents that are generated as the ventricles repolarize—that is, as the ventricles recover from being depolarized. The repolarization of the atria occurs during ventricular depolarization, and the atrial recovery wave is generally obscured by the QRS complex.

CONDITIONS OF CLINICAL SIGNIFICANCE

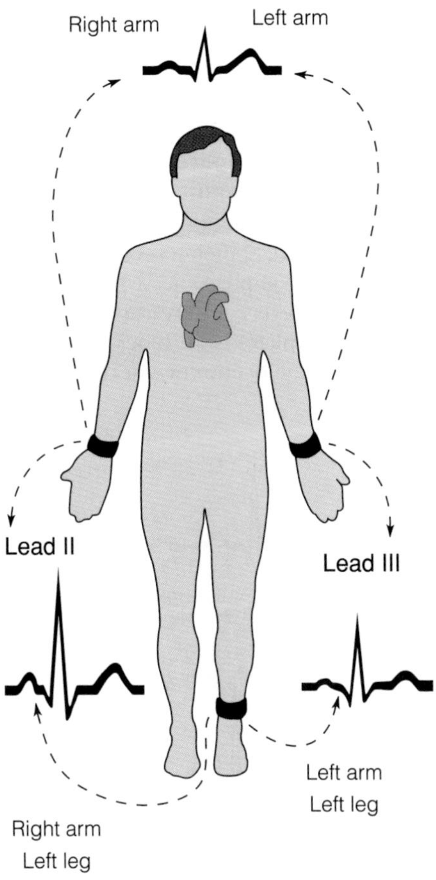

◆ **FIGURE C19.2 Locations of electrodes in the standard limb leads for an electrocardiogram**
The leads are designated I, II, and III.

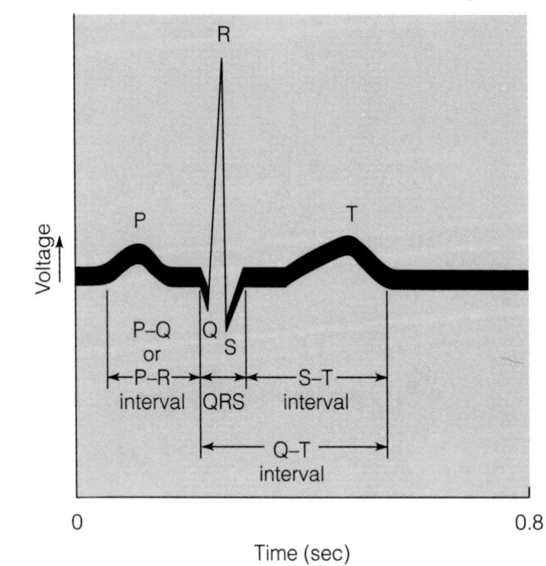

◆ **FIGURE C19.3 A normal electrocardiogram**

The time interval between the beginning of the P wave and the beginning of the QRS complex indicates the length of time between the beginning of the contraction of the atria and the beginning of the contraction of the ventricles. This period of time is referred to as either the P-Q interval or the P-R interval (because the Q wave is often absent). In a similar manner, the time between the beginning of the Q wave and the end of the T wave (the Q-T interval) provides a general indication of the duration of ventricular contraction.

Abnormal Electrocardiograms

The appearances of the different waves and the durations of the various intervals between the waves of the electrocardiogram are useful in diagnosing abnormalities that alter the conduction of impulses through the heart. The following are a few of the more common cardiac abnormalities detectable with electrocardiograms.

ABNORMAL HEART RATES. In a resting adult, the heart normally beats about 70 to 80 times a minute. When the rate drops below about 60 beats per minute, it is referred to as *bradycardia (brad″-ē-kar′-dē-ah)*. Bradycardia is generally not considered to be pathological. Much more serious and often associated with cardiovascular pathology is a resting heart rate of over 100 beats per minute, which is called *tachycardia (tak″-ē-kar′-dē-ah)*. When the heart rate is very fast, the ventricles do not have time to fill properly, and the movement of blood through the heart is impaired.

Somewhat coordinated atrial contractions that occur at a very rapid rate (often between 200 and 340 per minute) are called *atrial flutter* (Figure C19.4a).

continued on next page

CONDITIONS OF CLINICAL SIGNIFICANCE

◆ **FIGURE C19.4 Abnormal electrocardiograms**
(a) Atrial flutter. (b) Atrial fibrillation. (c) Ventricular fibrillation.

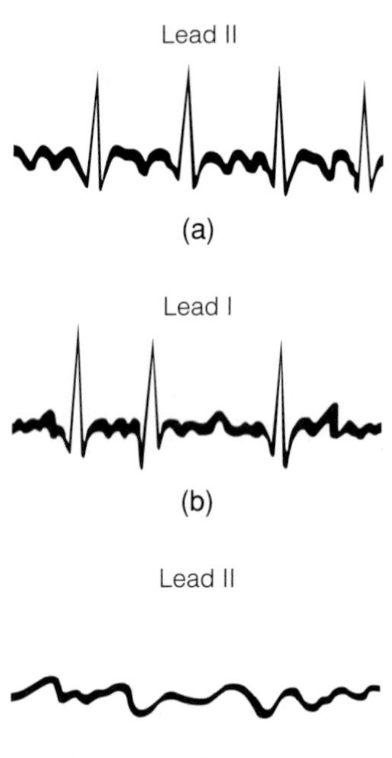

During atrial flutter, the atria pump almost no blood into the ventricles. Atrial flutter may occur in rheumatic and coronary heart disease. Extremely rapid, uncoordinated contractions of the atrial myocardium are called *atrial fibrillation* (Figure C19.4b). In atrial fibrillation, numerous impulses spread through the atria in all directions, and no P wave is evident on the electrocardiogram. The uncoordinated contractions of the atrial myocardium during atrial fibrillation are ineffective in pumping blood into the ventricles. Like atrial fibrillation, *ventricular fibrillation* is characterized by rapid, uncoordinated contractions of the ventricular myocardium. During ventricular fibrillation, a total irregularity of the QRS complex is evident on the electrocardiogram (Figure C19.4c). The uncoordinated contractions of the ventricular myocardium are ineffective in pumping blood from the ventricles, and ventricular fibrillation is generally fatal unless emergency measures (such as electrical defibrillation) are undertaken immediately.

HEART BLOCK. Occasionally, the conduction of a stimulatory impulse through the heart is blocked at some point.

Atrioventricular Block. Damage to or depression of the atrioventricular node or the atrioventricular bundle can impair the conduction of stimulatory impulses from the atria to the ventricles. In first-degree

◆ **FIGURE C19.5 Comparison of electrocardiograms (normal vs. atrioventricular heart blocks)**
(a) Normal electrocardiogram. (b) First-degree (incomplete) heart block showing prolonged P–R interval. (c) Second-degree (incomplete) heart block showing two atrial contractions for every ventricular contraction. (d) Third-degree (complete) heart block showing no correlation between the occurrence of the P wave and the QRS–T complex.

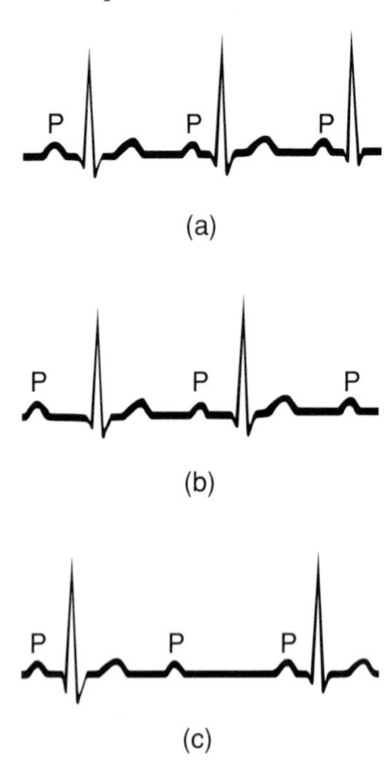

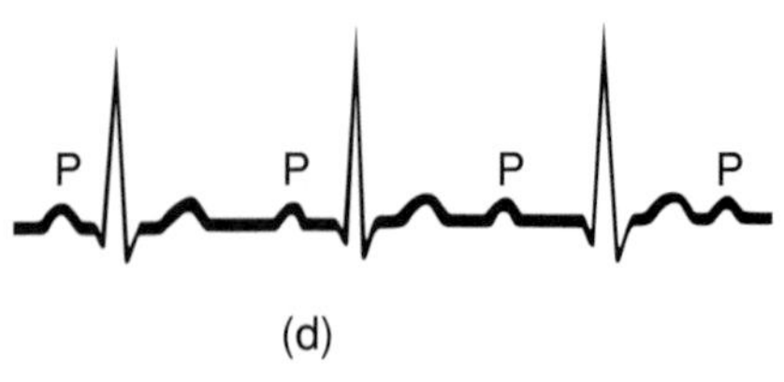

CONDITIONS OF CLINICAL SIGNIFICANCE

◆ **FIGURE C19.6 Comparison of electrocardiograms (normal vs. left bundle branch block)**

(a) Normal electrocardiograms. (b) Left bundle branch block. Notice that the QRS complex is greatly prolonged.

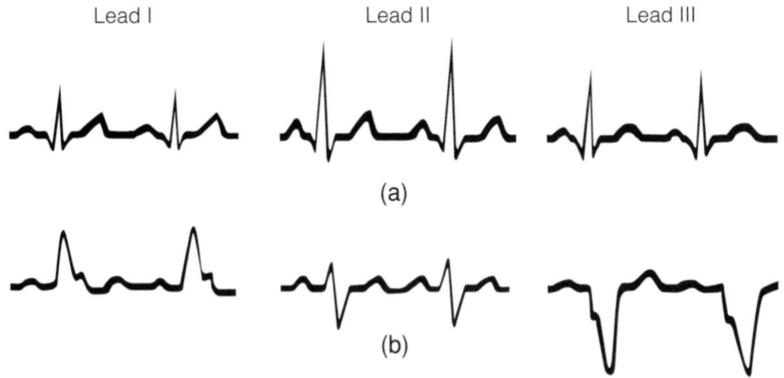

(incomplete) heart block, there is a longer-than-normal delay in the conduction of the impulse from the atria to the ventricles. In first-degree heart block, the electrocardiogram of a resting person's heart shows a P-R interval of greater than 0.20 sec (possibly up to between 0.35 and 0.45 sec) compared to a normal P-R interval of approximately 0.16 sec (Figure C19.5b). In second-degree (incomplete) heart block, some of the stimulatory impulses from the sinoatrial node fail to be transmitted to the ventricles. As a result, only every second (or third, or fourth, and so on) atrial contraction is followed by a ventricular contraction (Figure C19.5c). In this condition, the heart retains a definite, though altered, rhythm.

continued on next page

◆ **FIGURE C19.7 Comparison of electrocardiograms (normal vs. heart attacks)**

(a) Normal electrocardiograms. (b) Acute anterior wall infarction. (c) Acute posterior wall apical infarction.

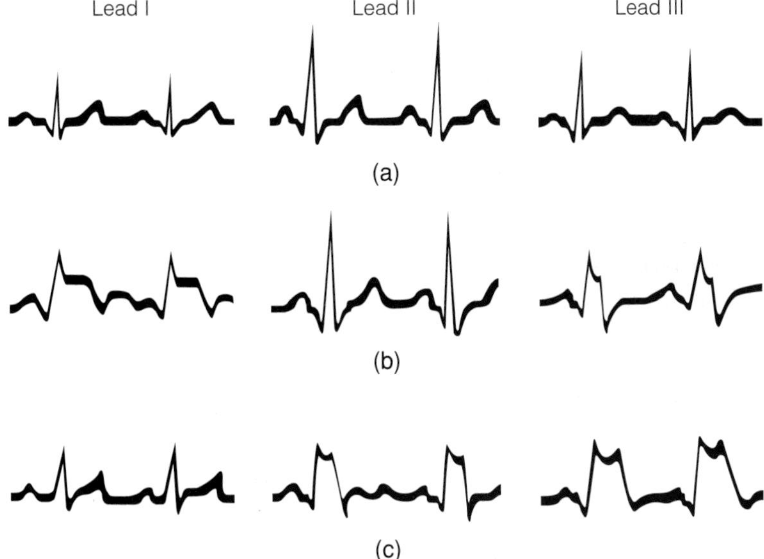

CONDITIONS OF CLINICAL SIGNIFICANCE

In third-degree (complete) heart block, the impairment of conduction is so severe that no stimulatory impulses are conducted from the atria to the ventricles. In this condition, the ventricles contract at a rate that is slower than and completely independent of the rate of the atria. Often, impulses that stimulate the ventricles originate spontaneously in the atrioventricular bundle (whose inherent rhythm is slower than that of the sinoatrial node). Thus, the atrioventricular bundle functions as an ectopic (that is, "out of place") pacemaker for the ventricles in place of the normal pacemaker (the sinoatrial node). In the electrocardiogram of a complete heart block, there is no correlation between the occurrences of the P wave and the QRS-T complex (Figure C19.5d).

Bundle Branch Block. If there is an impairment of conduction of the stimulatory impulse along one of the major branches of the atrioventricular bundle, the ventricle supplied by the impaired bundle branch depolarizes more slowly than the other ventricle. Such a block shows up on the electrocardiogram as a prolonged QRS complex (Figure C19.6). By varying the positions of the leads of the electrocardiograph, it is possible to determine whether the left or right bundle branch is blocked.

HEART ATTACKS. An insufficient flow of blood through the coronary arteries—due to an obstruction of a vessel, for example—can lead to myocardial damage and, if severe enough, to the deaths of cardiac muscle cells. An area of dead cardiac muscle cells resulting from inadequate blood flow is called a *myocardial infarct,* and a person who suffers a sudden obstruction of a coronary vessel (that is, a heart attack) generally incurs such damage. Because damaged or dead cardiac muscle cells do not conduct impulses normally, the presence of such areas produces altered ECG tracings (Figure C19.7).

Study Outline

◆ EMBRYONIC DEVELOPMENT OF THE HEART p. 594

1. Begins as pulsating tubule.
2. Tubule becomes S-shaped.
3. Anterior vessel becomes pulmonary trunk and aorta.
4. Posterior vessel becomes superior and inferior venae cavae.
5. Four chambers develop: two atria and two ventricles.

◆ POSITION OF THE HEART pp. 594–595

Cone-shaped organ in mediastinum; size of closed fist. Base behind sternum at level of second and third ribs; apex to left of midsternal line at level of fifth intercostal space.

◆ COVERINGS OF THE HEART pp. 596–597

1. Pericardium is a double-walled membranous sac (visceral and parietal layers).
2. Parietal pericardium has fibrous and serous layers.
3. Visceral pericardium has only a serous layer.
4. Pericardial cavity contains pericardial fluid, which minimizes friction as the heart beats.

◆ ANATOMY OF THE HEART pp. 597–603

Chambers of the Heart.

ATRIA (RIGHT AND LEFT). Small, located toward superior region of heart; separated by interatrial septum.

VENTRICLES (RIGHT AND LEFT). Larger than atria; located at apex of heart; separated by interventricular septum.

Vessels Associated with the Heart.

SUPERIOR AND INFERIOR VENAE CAVAE. Return venous blood from body to right atrium.

PULMONARY TRUNK. Delivers blood from right ventricle to the lungs.

PULMONARY VEINS. Carries blood from the lungs to the left atrium.

AORTA. Delivers blood from the left ventricle to the body.

Wall of the Heart.

EPICARDIUM. Serous membrane that adheres to outer surface of heart.

MYOCARDIUM. Cardiac muscle.

ENDOCARDIUM. Connective tissue, squamous cells; lines heart; folds to form valves.

SKELETON OF THE HEART. Fibrous rings separating atria from ventricles; serves for attachment for heart muscle and valves.

VESSELS OF THE MYOCARDIUM. Heart muscle supplied by coronary arteries; cardiac veins drain into coronary sinus.

Valves of the Heart.

ATRIOVENTRICULAR VALVES.

1. Right and left AV valves; tricuspid valve on right; bicuspid (mitral) valve on left.
2. Prevent backflow of blood into atria during ventricular contraction.

SEMILUNAR VALVES.

1. Pulmonary and aortic semilunar valves.
2. Prevent return of blood to ventricles from aorta and pulmonary trunk after contraction.

SURFACE LOCATIONS OF THE VALVES.

1. *Aortic semilunar valve.* Opposite left third intercostal space.
2. *Pulmonary semilunar valve.* Behind left third costal cartilage.
3. *Right AV valve.* Behind sternum, at level of fourth and fifth costal cartilages.
4. *Left AV valve.* At level of left fourth costal cartilage.

◆ CIRCULATION THROUGH THE HEART p. 603

1. Heart functions as a double pump: right pump receives blood from systemic circuit and pumps it into pulmonary circuit; left pump receives blood from pulmonary circuit and pumps it into systemic circuit.
2. Superior and inferior venae cavae, coronary sinus, and anterior cardiac veins return blood from body to right atrium. Blood then moves into right ventricle, which pumps it through pulmonary trunk and pulmonary arteries to capillary network of lungs. Blood from the lungs returns to left atrium by way of pulmonary veins, then moves into left ventricle, which pumps it through aorta to body.

◆ PUMPING ACTION OF THE HEART pp. 603-605

1. Both atria contract simultaneously, followed by simultaneous contraction of both ventricles.
2. Contraction of atria not essential for movement of blood into ventricles; even if atria fail to function, ventricles still pump considerable quantities of blood.

◆ CARDIAC MUSCLE pp. 605-606

Cellular Organization. Cardiac muscle cells form branching networks, and cells are connected end to end by intercalated discs. Cells coupled electrically by gap junctions. Contractile events believed similar to those in skeletal muscle cells.

Automatic Contraction. Cardiac muscle contracts automatically without external stimulation; spontaneous activity is continually influenced by neurons of autonomic nervous system and by certain chemicals and hormones.

Strength of Contraction. Factors that promote a further increase in the level of calcium ions in the cytosol of a cardiac muscle cell when the cell is stimulated can increase the strength of the contraction.

Refractory Period. Cardiac muscle cells have relatively long refractory periods that normally prevent heart from undergoing tetanic contraction or spasm.

Metabolism of Cardiac Muscle. Myocardium can obtain only insignificant amounts of energy from anaerobic metabolism, and cardiac muscle depends primarily on aerobic metabolism for continuous supply of energy required to support its contractile activity.

◆ EXCITATION AND CONDUCTION IN THE HEART pp. 606-609

Excitation. Sinoatrial node spontaneously generates stimulatory impulses approximately 70 to 80 times per minute in resting individual. Other portions of conducting system can also exhibit spontaneous activity, but sinoatrial node dominates them and is pacemaker of heart.

Conduction. Stimulatory impulse from sinoatrial node spreads throughout myocardium of atria, stimulating atrial contraction and reaching atrioventricular node. After delay of about 0.1 sec, impulse is transmitted from atrioventricular node along atrioventricular bundle, right and left bundle branches, and Purkinje fibers to cells of ventricles.

◆ THE CARDIAC CYCLE pp. 609-610

Heart repeats two alternating phases: systole (contraction) and diastole (relaxation).

Pressure Curve of the Left Atrium.

1. Pressure increases gradually as blood enters from pulmonary veins.
2. Pressure increases suddenly when atrium contracts.
3. Pressure increases when ventricular systole begins and atrioventricular valve bulges into atrium.
4. Pressure then drops quickly.

Pressure Curve of the Left Ventricle.

1. Slight pressure increase when atrium contracts.
2. Rapid rise in pressure during ventricular systole.
3. Pressure falls at end of ventricular systole to value below that in aorta.

Pressure Curve of the Aorta.

1. Pressure decreases gradually during ventricular diastole
2. Pressure increases rapidly when ventricle contracts and ejects blood into aorta.
3. Pressure then decreases, but rises briefly when aortic semilunar valve closes.

Volume Curve of the Left Ventricle.

1. *Period of isovolumetric ventricular contraction.* Volume of blood within the ventricle remains constant.
2. *Period of ventricular ejection.* Volume of blood within the ventricle decreases.
3. *Period of isovolumetric ventricular relaxation.* Volume of blood within the ventricle remains constant.
4. *Period of ventricular filling.* Volume of blood within the ventricle increases.

◆ CARDIAC OUTPUT pp. 610-614

Equal to heart rate multiplied by stroke volume. Stroke volume—amount of blood pumped by one side of heart per beat—decreases at very high heart rates. A major reason for this decrease is diminished time for ventricular filling due to shortened time of diastole.

Control of Heart Rate. Most important factor is autonomic nervous system: sympathetic stimulation increases heart rate and parasympathetic stimulation decreases heart rate.

Control of Stroke Volume. Stroke volume is equal to difference between end-diastolic volume and end-systolic volume.

END-DIASTOLIC VOLUME. Influenced mainly by:

LENGTH OF DIASTOLE.

VENOUS RETURN.

END-SYSTOLIC VOLUME. Influenced mainly by strength of ventricular contraction.

THE FRANK-STARLING LAW OF THE HEART. Stretching the ventricle wall increases strength of ventricular contraction.

SYMPATHETIC NERVOUS SYSTEM. Sympathetic stimulation increases strength of ventricular contraction.

◆ FACTORS THAT INFLUENCE CARDIAC FUNCTION pp. 614–615

Cardiac Center. Both sympathetic and parasympathetic neurons to heart are under control of a cardiac center in brain stem.

Exercise. Chronic endurance types of exercise generally cause cardiac muscle to hypertrophy and ventricular chambers to enlarge, leading to increase in stroke volume.

Temperature. Warming the heart increases heart rate; cooling the heart decreases heart rate.

Ions. When level of potassium ions is increased, heart rate drops and heart becomes extremely dilated, flaccid, and weak. When level of calcium ions is increased excessively, heart exhibits spastic contraction. Increase in level of sodium ions slows heart and depresses cardiac function.

Catecholamines. Epinephrine and norepinephrine from adrenal medullae can increase heart rate and force of contraction.

◆ CONDITIONS OF CLINICAL SIGNIFICANCE: THE HEART pp. 616–622

Valvular Malfunctions. Can make it necessary for heart to work harder to maintain a given cardiac output.

VALVULAR INSUFFICIENCY. Cusps of a valve do not form tight seal when valve is closed. As a result, blood leaks into chamber from which it came.

VALVULAR STENOSIS. Valve opening becomes so narrowed that it interferes with flow of blood through it.

Congestive Heart Failure. Condition in which heart fails to pump enough blood to meet body's needs. In congestive heart failure, an abnormal increase in blood volume and interstitial fluid occurs, and heart is generally dilated with blood. Kidneys retain fluid, activity of sympathetic nerves to heart increases, and myocardium may hypertrophy.

Effects of Aging. Although a progressive decline in cardiac output and strength of contraction occurs with age, a healthy heart is quite capable of meeting the reduced physical demands of older persons.

Heart Sounds.

NORMAL HEART SOUNDS. First heart sound ("lub") is associated with closure of atrioventricular valves at beginning of systole. Second heart sound ("dup") is associated with closure of semilunar valves as ventricles begin to relax following their contraction.

ABNORMAL HEART SOUNDS. Often due to turbulent flow and are called *murmurs.* Both valvular insufficiency and valvular stenosis can cause heart murmurs.

Electrocardiography. Electrodes placed on body surface detect electrical potentials generated by heart.

NORMAL ELECTRODARDIOGRAM. Consists of regularly spaced series of waves. P wave represents atrial depolarization. QRS complex represents ventricular depolarization. T wave represents ventricular repolarization. P–R interval represents length of time between beginning of contraction of atria and beginning of contraction of ventricles. Q–T interval provides general indication of duration of ventricular contraction.

ABNORMAL ELECTRODARDIOGRAMS.

ABNORMAL HEART RATES. Bradycardia: below 60 beats per minute; *tachycardia:* above 100 beats per minute. Somewhat coordinated atrial contractions occurring at very rapid rate are called *atrial flutter.* Extremely rapid, uncoordinated atrial contractions are called *atrial fibrillation* (no evident P wave on electrocardiogram). Extremely rapid, uncoordinated ventricular contractions are called *ventricular fibrillation* (total irregularity of QRS complex on electrocardiogram).

HEART BLOCK.

Atrioventricular Block. May delay or prevent impulse transmission from atria to ventricles. *First-degree block:* delayed impulse transmission (P–R interval exceeds 0.20 sec on electrocardiogram). *Second-degree block:* transmission of some impulses from sinoatrial node to ventricles is prevented (two, three, or more P waves for each QRS complex on electrocardiogram). *Third-degree block:* transmission of impulses from atria to ventricles is prevented (no correlation between P wave and QRS–T complex on electrocardiogram).

Bundle Branch Block. Impairment of conduction in one of major branches of atrioventricular bundle (prolonged QRS complex on electrocardiogram).

HEART ATTACKS. An area of dead cardiac muscle cells resulting from inadequate blood flow is called a *myocardial infarct;* areas of damaged or dead cardiac muscle cells result in altered electrocardiograms.

Self-Quiz

1. The heart is enclosed in a double-walled membranous sac called the: (a) mediastinum; (b) pericardium; (c) epicardium.
2. Match the various terms associated with the heart with the appropriate lettered description:

Pericardial cavity	(a) Separates the two atria
Atria	(b) Carries blood from the left ventricle into the systemic circuit
Ventricles	(c) The muscular layer of the wall of the heart
Interatrial septum	(d) Small chambers located toward the superior region of the heart
Inferior vena cava	(e) Folds and bridges of the inner surface of the myocardium
Aorta	(f) Conducts blood to the right atrium
Epicardium	(g) Space located between the visceral and parietal layers
Myocardium	(h) Foldings of this structure form the valves that separate the atria from the ventricles
Endocardium	(i) Large chambers that compose the bulk of the heart
Trabeculae carneae	(j) A thin serous membrane that adheres to the outer surface of the heart

3. Blood is prevented from returning to the ventricles after they have completed their contractions by the: (a) atrioventricular valves: (b) tricuspid valves; (c) semilunar valves.
4. Blood within the pulmonary veins returns to the: (a) right atrium; (b) right ventricle; (c) left atrium; (d) left ventricle.
5. The left atrium contracts at the same time as the: (a) right atrium; (b) right ventricle; (c) left ventricle.
6. The cardiac muscle of the heart: (a) commonly undergoes prolonged tetanic contractions; (b) does not contract unless stimulated by the nervous system; (c) obtains only an insignificant amount of energy from anaerobic metabolism.
7. The sinoatrial node is the only area of the myocardium that can undergo spontaneous depolarization to threshold and generate action potentials. True or False?
8. The stimulatory impulse from the sinoatrial node is normally delayed for a short time at the: (a) atrioventricular node; (b) atrioventricular bundle; (c) Purkinje fibers.
9. During left ventricular systole, which event occurs first? (a) the atrioventricular valve closes; (b) the semilunar valve opens; (c) the pressure within the ventricles peaks.
10. In a resting adult, during ventricular diastole, the ventricles are about 70% filled before the atria contract. True or False?
11. During most of ventricular diastole, the pressure within the left ventricle is slightly lower than the pressure within the left atrium. True or False?
12. Cardiac output is equal to the heart rate multiplied by the: (a) aortic pressure; (b) stroke volume; (c) ventricular end-systolic volume.
13. Stimulation of the parasympathetic neurons to the heart: (a) decreases the heart rate; (b) decreases the membrane permeability to potassium ions; (c) decreases the conduction time through the atrioventricular node.
14. Within limits, as cardiac muscle is stretched, its force of contraction increases. True or False?
15. Norepinephrine: (a) decreases the contractile strength of the ventricles; (b) decreases the heart rate; (c) accelerates the rates at which cardiac muscle fibers shorten and relax.
16. Chronic endurance types of exercise most likely will lead to: (a) an increased heart rate at rest; (b) an increased stroke volume; (c) a decreased cardiac reserve.
17. In general, when the level of potassium ions increases excessively, the heart rate increases and the heart exhibits spastic contraction. True or False?
18. Growths or scar tissue that form on a valve as a result of rheumatic fever can cause valvular insufficiency. True or False?
19. In congestive heart failure: (a) the blood volume decreases; (b) the kidneys excrete large volumes of fluid; (c) the activity of the sympathetic nerves to the heart increases.
20. The first heart sound is due largely to the: (a) vibrations of the taut atrioventricular valves immediately after closure; (b) vibrations of the taut, closed semilunar valves; (c) turbulent flow of blood past the open semilunar valves.
21. The P wave of a normal electrocardiogram indicates: (a) atrial depolarization; (b) ventricular depolarization; (c) atrial repolarization; (d) ventricular repolarization.
22. The Q–T interval of a normal electrocardiogram provides a general indication of: (a) the duration of atrial contraction; (b) the duration of ventricular contraction; (c) the length of time between the beginning of atrial contraction and the beginning of ventricular contraction; (d) the length of time that the stimulatory impulse is delayed at the atrioventricular node.
23. A P–R interval of 0.30 sec on an electrocardiogram would most likely indicate: (a) ventricular fibrillation; (b) an atrioventricular block; (c) atrial flutter.

CHAPTER 20

The Cardiovascular System: Blood Vessels

CHAPTER CONTENTS

LEARNING OBJECTIVES

After completing this chapter, you should be able to:

1. Describe and compare the structure of arteries, veins, and capillaries.
2. Describe how blood viscosity, vessel length, and vessel radius affect the resistance to blood flow.
3. Discuss the relationship among blood flow, pressure, and resistance.
4. Describe how baroreceptors in the aortic arch and carotid sinus influence arterial pressure.
5. Describe how local autoregulatory mechanisms affect blood flow.
6. Discuss the factors involved in the movement of fluid between the blood and the interstitial fluid.
7. Describe the factors that influence venous return.
8. Distinguish between the pulmonary and the systemic circuits.
9. Identify the principal pulmonary arteries and veins.
10. Identify the principal systemic arteries and veins.

CHAPTER 20

Upon leaving the heart, the blood enters the vascular system, which is composed of numerous **blood vessels.** The vessels transport the blood to all parts of the body; permit the exchange of nutrients, metabolic end products, hormones, and other substances between the blood and the interstitial fluid; and ultimately return the blood to the heart. Both the size of the vessels and the thickness of their walls vary, as does the blood pressure within them.

Types of Vessels

Large vessels called **arteries** carry the blood *away* from the heart. The major arteries divide into smaller arteries, then into still smaller **arterioles,** and finally into tiny **capillaries.** The capillaries converge into very small vessels called **venules,** which in turn join to form larger vessels called **veins.** The major veins *return* blood to the atria of the heart.

General Structure of Blood-Vessel Walls

Blood-vessel walls vary in thickness. This variation is due to the presence or absence of one or more of three layers of tissues and to the differences in their thicknesses (Figure 20.1). The **tunica intima** (or **tunica interna;** *tunic* = coat) is the innermost layer. It is formed of a layer of *simple squamous epithelium* called the **endothelium,** and a thin connective-tissue *basement membrane.* The endothelium of the tunica intima is the only layer present in vessels of all sizes. Moreover, it is continuous with the endocardium of the heart. Separating the tunica intima from the middle layer, and considered a part of the intima, there is often a thin layer of elastic fibers called the **internal elastic lamina.** The middle layer, the **tunica media,** is generally quite thick in arteries and is composed of *smooth muscle fibers* (mostly circularly arranged) mixed with *elastic fibers.* The outer border of the tunica media, where it contacts the outermost layer of the blood-vessel wall, is often in the form of a thin layer of elastic fibers called the **external elastic lamina.** The outermost layer is the **tunica adventitia** (*ad″-ven-tish′-ē-ah;* or **tunica externa**). This relatively thin layer of *connective tissue* contains elastic and collagenous fibers that run parallel to the long axis of the vessel. It serves to attach the vessel to the surrounding tissues.

The walls of the larger vessels are too thick to be nourished by diffusion from the blood in the vessel. Instead, they are supplied by their own small nutrient vessels—the **vasa vasorum** (*va′-sah va-sōr′-um;* "vessels of the vessels")—which are located in the tunica adventitia and arise either from the blood vessel itself or from other vessels located close by.

Structure of Arteries

The composition of the walls of the arteries differs, depending on the size of the vessels.

Elastic Arteries

The large arteries, such as the aorta and its major branches and the pulmonary trunk, are called **elastic arteries.** Because elastic arteries convey blood from the heart to medium-sized muscular arteries, they are sometimes referred to as **conducting arteries.** The walls of these arteries are composed of the three tunics just described. The tunica media of the large arteries is very thick, and it contains many elastic fibers in addition to smooth muscle fibers.

During ventricular systole, elastic arteries are stretched as blood is ejected from the heart. During diastole, the recoil of the elastic arteries helps maintain pressure within the vessels. Although these arteries are commonly considered to behave like passive elastic tubes, there is some evidence that this is not entirely the case. In animal experiments, it has been found that smooth muscle in the walls of large arteries may be stimulated by the cardiac pacemaker during the contraction of the heart, thus producing some resistance to stretch and decreasing the distensibility of the arteries. This response may prevent overdilation of the arteries when blood is pumped into them.

Muscular Arteries

The tunica media of the walls of most smaller arteries consists almost entirely of smooth muscle cells, with relatively few elastic fibers. Such arteries are called **muscular arteries** or—because they carry blood throughout the body—**distributing arteries.** Muscular arteries have well-defined internal and external elastic laminae. Because their walls contain few elastic fibers, muscular arteries are less distensible than elastic arteries.

Structure of Arterioles

When an arterial vessel has a diameter of less than 0.5 mm, it is referred to as an **arteriole.** Arterioles have

◆ **FIGURE 20.1 Comparison of the structure of blood vessels**

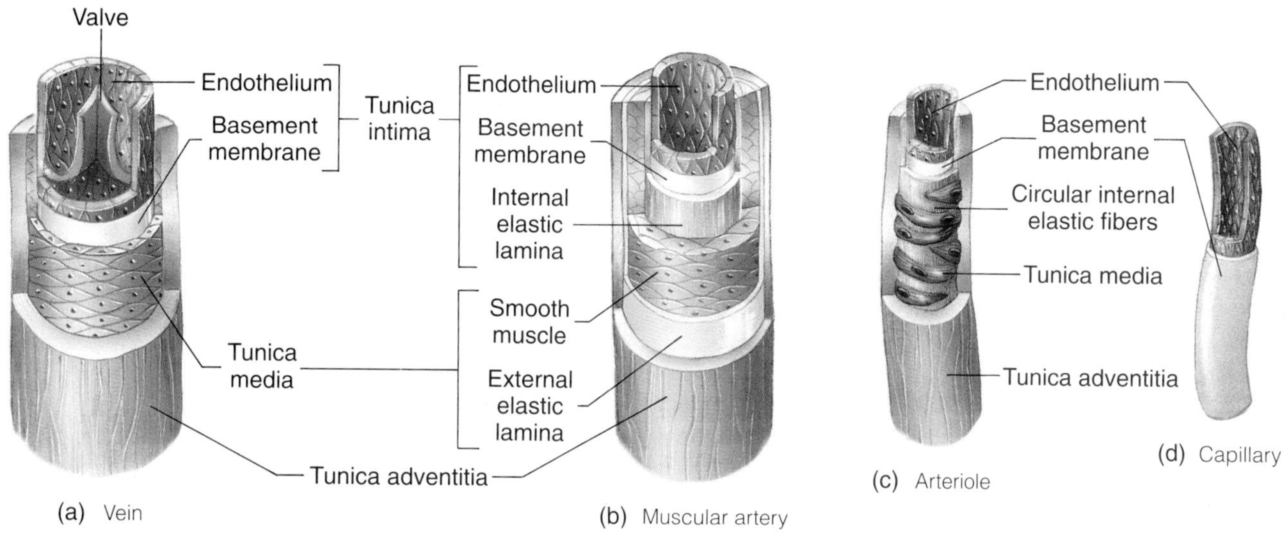

a small lumen and a relatively thick tunica media that is composed almost entirely of smooth muscle, with very little elastic tissue. In the smallest arterioles—that is, those closest to the capillaries—the external elastic membrane is lost and the tunica media is gradually reduced until it is composed of only a few scattered smooth muscle cells.

The smooth muscle of the tunica media has considerable **myogenic activity**—that is, an ability to contract without external stimulation. However, a number of factors that are discussed later in the chapter influence the degree of contraction of arteriole smooth muscle. An increase in the degree of contraction of arteriole smooth muscle leads to a narrowing of the internal cavities, or **lumens,** of the arterioles—that is, the vessels undergo **vasoconstriction** (Figure 20.2). A relaxation, or decrease in the degree of contraction, of the muscle leads to an enlargement of the lumens of the arterioles—that is, the vessels undergo **vasodilation.**

The arterioles play a major role in regulating the flow of blood into the capillaries. A constriction of the arterioles restricts the flow of blood into the capillaries, and

◆ **FIGURE 20.2 Cross sections of a dilated arteriole (left) and a constricted arteriole (right)**

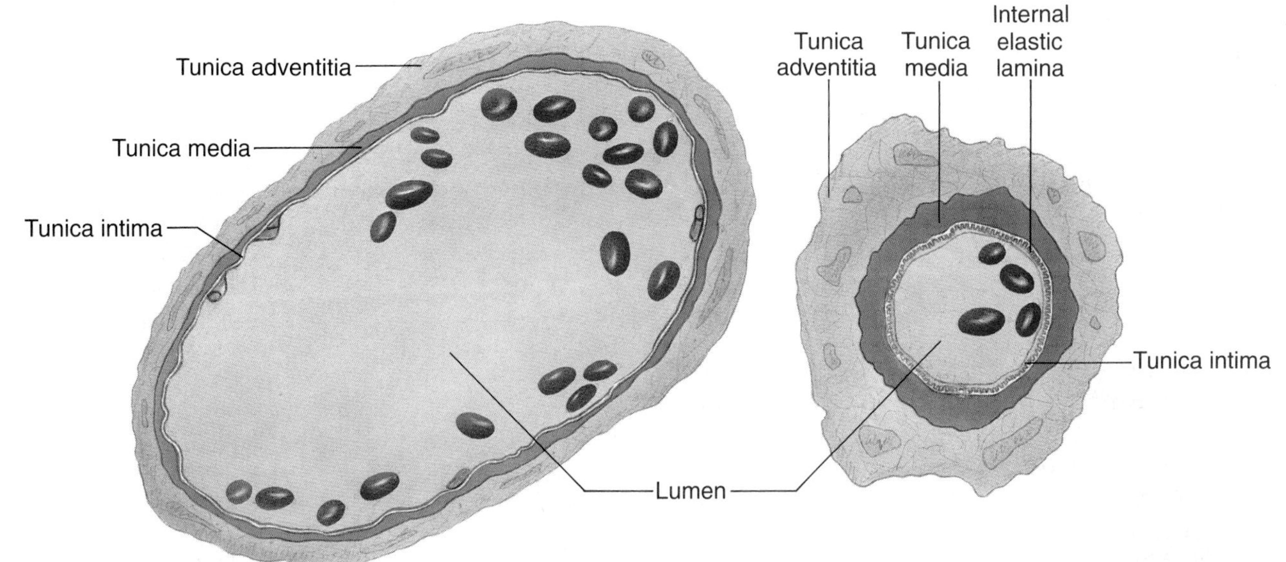

a dilation of the arterioles allows the blood to enter the capillaries more freely.

Structure of Capillaries

In most tissues, a capillary bed contains two types of vessels: (1) **thoroughfare channels (metarterioles),** which directly connect arterioles and venules, thereby largely bypassing the tissue cells, and (2) **true capillaries,** in which exchanges with tissue cells occur (Figure 20.3).

The proximal portions of the thoroughfare channels are surrounded by scattered smooth muscle cells. Contraction and relaxation of the muscle cells alter the internal diameters of the channels and thus help control blood flow through the capillary bed. The distal portions of the thoroughfare channels do not have any smooth muscle cells surrounding them. Thoroughfare channels empty directly into venules.

Most true capillaries branch from thoroughfare channels, but some branch directly from arterioles. A ring of smooth muscle called a **precapillary sphincter** usually surrounds each true capillary at the point where it arises from a thoroughfare channel or arteriole. Contraction and relaxation of the sphincters help regulate the flow of blood through the true capillaries. When the precapillary sphincters are relaxed, blood flow through the capillary bed is plentiful, and exchange between the blood and the tissue cells occurs freely. When the precapillary sphincters contract, blood flow through the capillary bed is reduced, most blood remains in the thoroughfare channels, and the tissue cells are largely bypassed (Figure 20.3b). For example, in inactive skeletal muscles, precapillary sphincters in the capillary beds are largely closed, and blood flow to the muscle cells is relatively low. During vigorous muscular activity, the precapillary sphincters in the capillary beds relax, and blood flow to the muscle cells increases greatly.

◆ **FIGURE 20.3 A capillary network**
(a) Thoroughfare channels directly connect arterioles and venules. True capillaries branch from and join with the thoroughfare channels. (b) When the precapillary sphincters contract, most blood flows through the thoroughfare channels, bypassing the true capillaries.

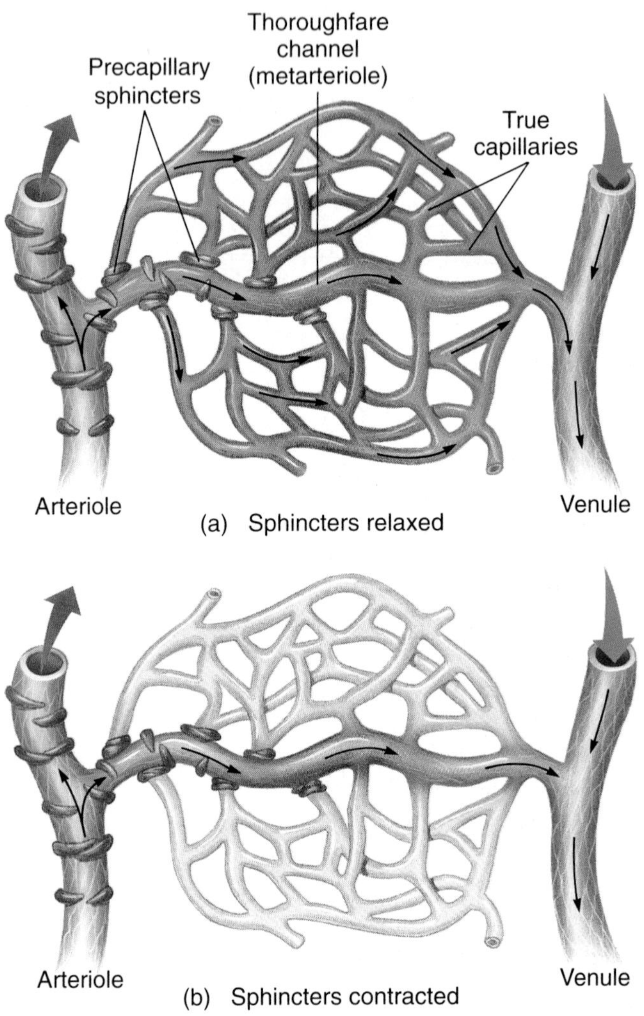

Capillaries have extremely thin walls. As a consequence, they are sites at which the exchange of materials between the blood and the interstitial fluid takes place. Capillary structure varies from one part of the body to another, but in general, a capillary consists of a single layer of endothelial cells surrounded by a thin basal lamina of the tunica intima. No tunica media or tunica adventitia is present. Endothelial cells are held to one another by tight junctions. Water-filled clefts occur between adjacent endothelial cells.

In the capillaries of muscle, connective tissues, and nervous tissue, a single endothelial cell may form the entire circumference of a capillary. Such capillaries are referred to as **continuous capillaries** (Figure 20.4).

In the capillaries of endocrine glands, the intestines, and the kidneys, the membranes of the endothelial cells contain numerous tiny pores called *fenestrations (fen″-es-trā′-shuns)*. These capillaries are referred to as **fenestrated capillaries.** In most fenestrated capillaries, the fenestrations are closed by a very thin diaphragm. However, the fenestrated capillaries of the kidneys lack a diaphragm and appear to be open. As a consequence, fluid moves across the walls of the fenestrated capillaries of the kidneys much faster than it moves across the walls of most unfenestrated capillaries.

Although a single capillary is only about 0.5–1 mm long and 0.01 mm in diameter, capillaries are so numerous that their total surface area within the body has been estimated to be more than 600 square meters. This provides a large surface across which the exchange of

◆ **FIGURE 20.4 Electron micrograph of a cross section of a continuous capillary**

Note that the entire circumference of the capillary is made up of a single endothelial cell.

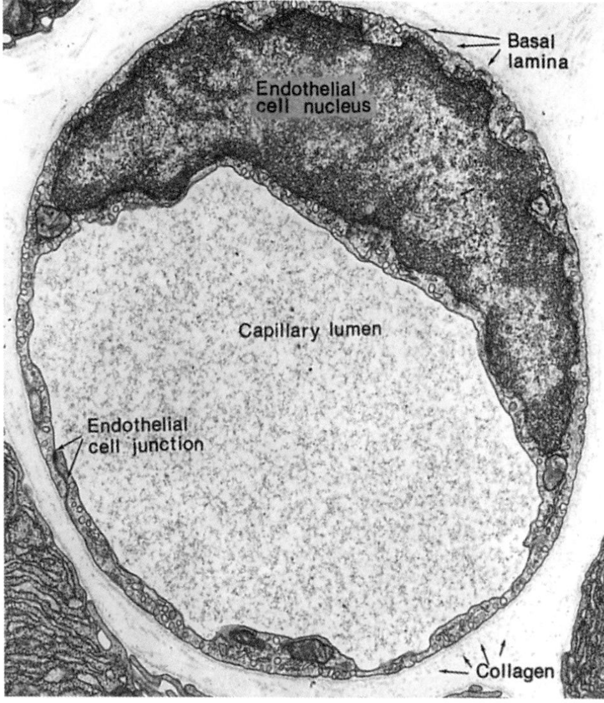

materials can occur. Substances may enter or leave capillaries by several routes: (1) through the water-filled clefts between endothelial cells; (2) across endothelial cells; and (3) in the case of capillaries with fenestrations, through the fenestrations.

The arterioles in certain structures, rather than connecting with capillaries, empty into relatively large vascular channels that have very thin walls. These channels are **sinusoids** *(sī´-nu-soids)*. Sinusoids are so thin-walled that they generally conform in shape to the space in which they are located rather than being cylindrical. In some sinusoids, the endothelium that lines them is discontinuous, with gaps present between the cells; in others the endothelium has small pores that are closed by thin diaphragms, as in fenestrated capillaries. Sinusoids are characteristic of the liver, spleen, bone marrow, and some endocrine glands.

Structure of Venules

The walls of the **venules,** which receive blood from the capillaries or the sinusoids, have an inner lining composed of the endothelium of the tunica intima, surrounded by a very thin tunica adventitia. The larger venules that are farther from the capillaries are encircled by a few smooth muscle fibers that form a thin tunica media.

Structure of Veins

The **veins,** which receive blood from the venules, have the same three coats as the arteries. In general, however, the tunica media of the veins is quite thin and has few muscle fibers. The tunica adventitia forms the greatest part of the wall, often being several times thicker than the media, although very little elastic tissue is present. Veins have no internal or external elastic laminae. They tend to have a larger lumen and thinner walls than the arteries they accompany (Figure 20.5). The walls of the large veins, like the walls of the large arteries, receive nourishment through tiny vasa vasorum. Some veins contain valves that allow blood to flow toward the heart but not away from it (Figure 20.6). These valves are folds of the tunica intima and are similar in form to the semilunar valves of the heart. When, in response to the pull of gravity, blood in the veins has a tendency to flow backward, away from the heart, the flaps of the valves fill with blood, thereby blocking the vessel. Valves are most common in the veins of the limbs.

The features of the various blood vessels are summarized in Table 20.1.

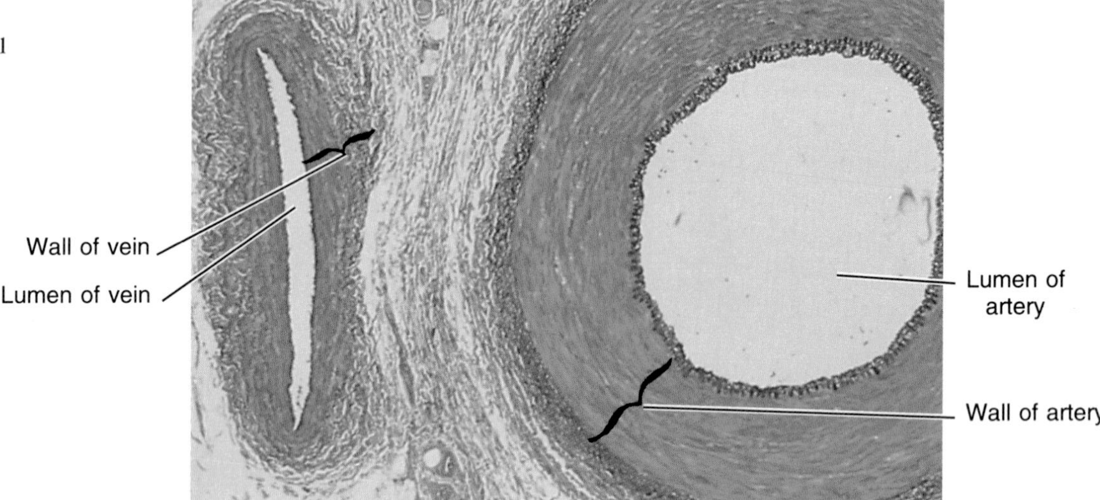

◆ **FIGURE 20.5 Photomicrograph of a portion of an artery and a vein**
Notice that the wall of the artery is much thicker than the wall of the vein.

◆ **FIGURE 20.6 Valves of a vein**
As the arrows indicate, the valves are forced open by pressure from below and shut by pressure from above. This arrangement of valves allows blood to move in only one direction—toward the heart.

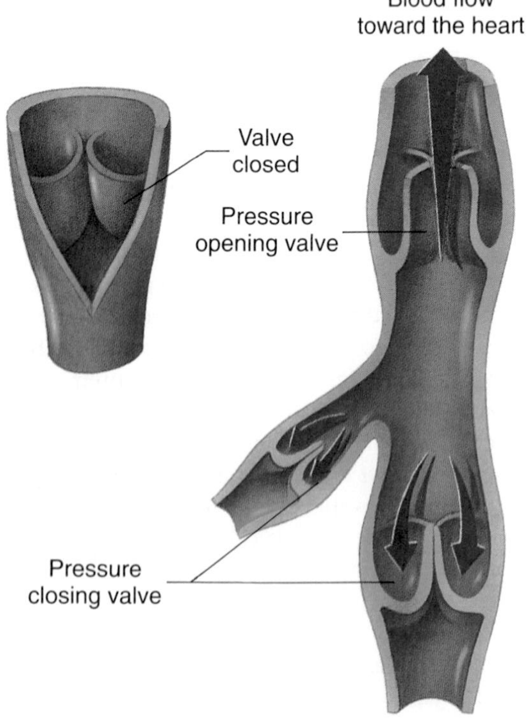

Principles of Circulation

Efficient circulation requires an adequate volume of circulating blood, blood vessels that are in good condition, and a heart that functions smoothly to pump the blood through the vessels.

Velocity of Blood Flow

In general, the velocity of blood flow in any segment of the cardiovascular system is inversely related to the total cross-sectional area of the vessels of the segment (Figure 20.7). In other words, the larger the total cross-sectional area, the lower the velocity of flow.

Although the lumen of an arteriole is smaller than the lumen of an artery, there are so many arterioles that their total cross-sectional area exceeds that of the arteries. Thus, the velocity of blood flow in the arterioles is lower than that in the arteries. Similarly, the smallest individual vessels, the capillaries, are so numerous that they have the greatest total cross-sectional area of the entire cardiovascular system. As a consequence, the velocity of blood flow in the capillaries is lower than in any other vessels. The slow blood flow in the capillaries allows adequate time for the exchange of nutrients, metabolic end products, and other substances between the blood and the interstitial fluid.

Pulse

When blood is ejected from the heart during ventricular systole, the pressure within the arteries rises and the arteries expand. During diastole, the arterial pressure

◆ **TABLE 20.1 Features of Blood Vessels**

FEATURE	VESSEL TYPE					
	Aorta	**Large Arteries**	**Arterioles**	**Capillaries**	**Large Veins**	**Venae Cavae**
Number	One	Several hundred	Half a million	Ten billion	Several hundred	Two
Wall thickness	2 mm (2000 μm)	1 mm (1000 μm)	20 μm	1 μm	0.5 mm (500 μm)	1.5 mm (1500 μm)
Internal radius	1.25 cm (12,500 μm)	0.2 cm (2000 μm)	30 μm	3.5 μm	0.5 cm (5000 μm)	3 cm (30,000 μm)
Total cross-sectional area	4.5 cm^2	20 cm^2	400 cm^2	6000 cm^2	40 cm^2	18 cm^2
Special features	Thick, distensible walls; large radii		Highly muscular, well-innervated walls; small radii	Thin walled; large total cross-sectional area	Relatively thin, flexible walls, large radii	
Functions	Passageway from heart to tissues; pressure reservoir		Primary resistance vessels; determine distribution of cardiac output	Site of exchange; determine distribution of extracellular fluid between plasma and interstitial fluid.	Passageway to heart from tissues; blood reservoir	

falls and the arteries recoil. The expansion and subsequent recoil of the arteries can be felt at various locations on the body surface. This is the **pulse.**

The pressure fluctuations within the arterioles are less extreme than those within the arteries. Within the capillaries and beyond, the pressures do not rise and fall greatly with the beating of the heart, but remain at relatively constant levels.

Blood Flow

The **blood flow** is the actual volume of blood that passes through a vessel in a given time (expressed, for example, in milliliters per minute). Blood flow is a function of the pressure forcing the blood through a vessel and the resistance of the vessel to the flow of blood through it.

Pressure

The pumping action of the heart imparts energy to the blood. This energy is evident as the **pressure** that drives the blood through the vessels.

Resistance

As the blood flows through the vessels, it encounters varying degrees of **resistance,** which is essentially a measure of friction. As a consequence, the energy imparted to the blood by the pumping action of the heart is ultimately dissipated as heat. This dissipation of energy is evident as a progressive drop in pressure as the blood moves through the vessels (Figure 20.8). Thus, the pressure is highest in the arteries, and it gradually drops as the blood flows through the arterioles, capillaries, venules, and veins. In general, the greater the resistance the blood encounters, the harder the heart must pump, and the more pressure it must generate in order to keep the blood circulating.

Factors That Affect Resistance

Several factors affect the magnitude of the resistance the blood encounters as it flows through the vessels.

Blood Viscosity. Viscosity *(vis-koss´-i-tee)* is the property of a fluid that makes it resist flow. Thus, the more

◆ FIGURE 20.7 The relationship between the velocity of blood flow and the total cross-sectional area of various segments of the cardiovascular system

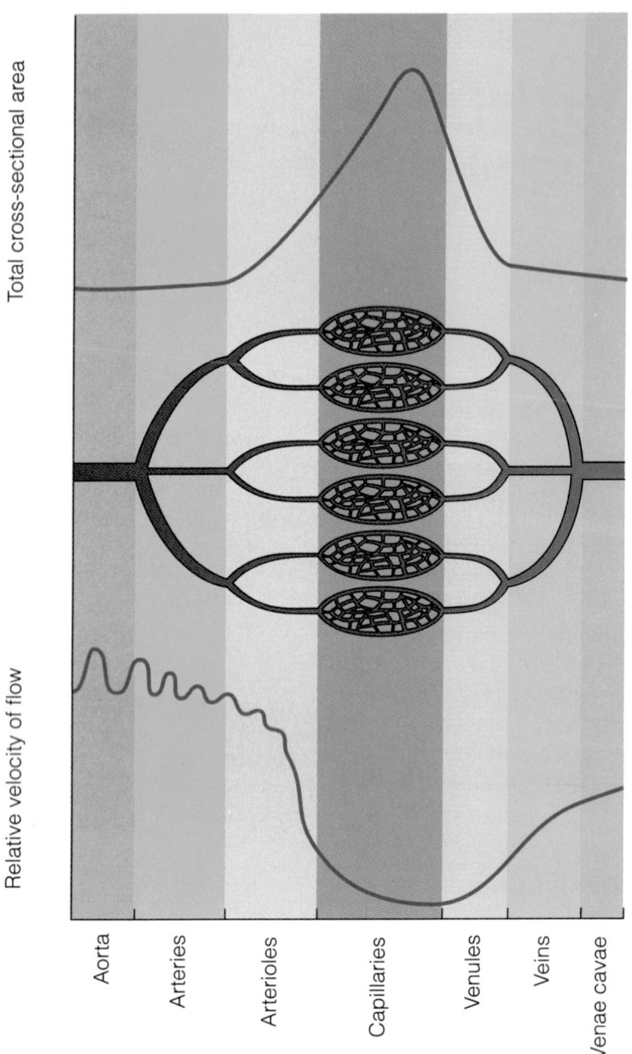

viscous the blood, the greater the resistance to its flow through any given vessel. The more protein and cells within the blood, the greater its viscosity.

Vessel Length. The longer a vessel, the greater the resistance encountered as blood flows through it.

Vessel Radius. The smaller the radius of a vessel, the greater its resistance to the flow of blood through it. The resistance of a vessel is inversely related to the fourth power of the vessel's radius:

$$\text{Resistance} \propto \frac{1}{\text{Radius}^4}$$

For example, if the radius of a vessel is decreased by one-half, its resistance increases 16 times.

Alterations in Resistance

The lengths of the blood vessels are constant, and under normal circumstances the viscosity of the blood does not vary greatly. However, the radii of blood vessels can change. For example, the radii of the muscular arterioles vary according to the degree of contraction or relaxation of the arteriole muscles, and the arterioles are of major importance in determining the amount of resistance the blood encounters as it flows through the vessels.

The resistance to the flow of blood offered by the entire systemic circulation is called the *total peripheral resistance.* The resistance to the flow of blood offered by the entire pulmonary circulation is called the *total pulmonary resistance.*

Relationship among Flow, Pressure, and Resistance

The relationship among blood flow, pressure, and resistance can be expressed as

$$\text{Flow } (F) = \frac{\text{Pressure}}{\text{Resistance}} \qquad (1)$$

The pressure term in this relationship is the pressure that drives the blood through a particular vessel or vessels. This pressure is represented by the pressure drop (ΔP) that occurs as the blood flows through the vessel, and it is equal to the pressure at the beginning of the vessel (P_1) minus the pressure at the end of the vessel (P_2). Thus, $\Delta P = P_1 - P_2$, and

$$F = \frac{\Delta P}{\text{Resistance}} \qquad (2)$$

The resistance term of this relationship includes the components of resistance discussed previously—blood viscosity, vessel length, and vessel radius. However, as previously pointed out, the component of resistance most likely to change and thus to affect the relationship is vessel radius, particularly arteriole radius.

The relationship among blood flow, pressure, and resistance expressed in equation 2 can be applied to the entire systemic or pulmonary circulation as well as to individual vessels or groups of vessels. For example, the volume of blood that flows through the systemic circulation in a given time is equal to the volume of blood ejected by the left ventricle during that time. If the time considered is one minute, then the flow term of the relationship is equal to the cardiac output (recall that

◆ **FIGURE 20.8 Pressures in various portions of the cardiovascular system**

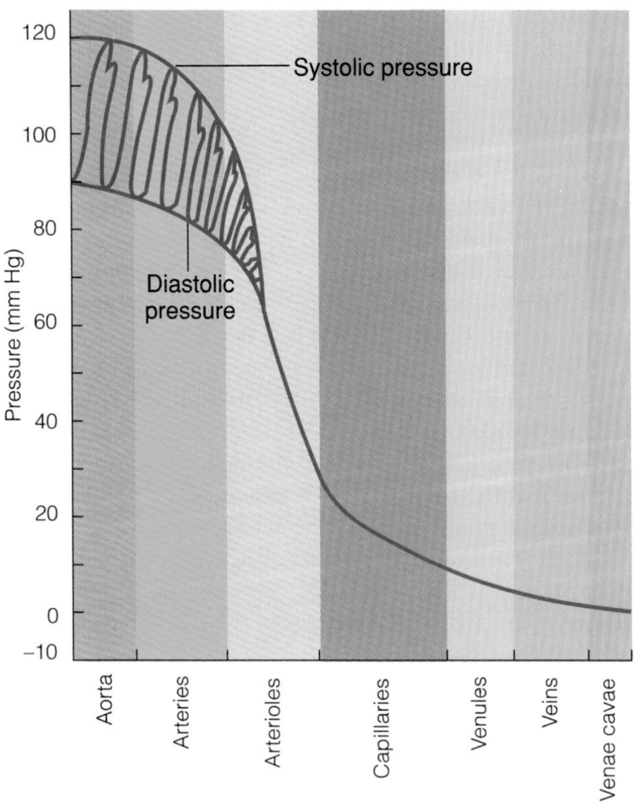

the cardiac output is the volume of blood pumped per minute by a ventricle). Thus,

$$\text{Cardiac Output} = \frac{\Delta P}{\text{Resistance}} \qquad (3)$$

The total resistance to the flow of blood through the entire systemic circulation is the total peripheral resistance. Thus, the resistance term of the relationship is equal to the total peripheral resistance, and

$$\text{Cardiac Output} = \frac{\Delta P}{\text{Total Peripheral Resistance}} \qquad (4)$$

The pressure at the end of the systemic circuit (P_2) is the pressure within the veins at the right atrium. Since this pressure is almost zero, the pressure drop (ΔP) that occurs as blood flows through the systemic circuit is essentially equal to the pressure at the beginning of the circuit. That is, $\Delta P = P_1 - P_2 = P_1 - 0 = P_1$. This pressure ($P_1$) is the pressure in the aorta at the left ventricle due to the pumping action of the heart. However, as discussed in Chapter 19, the pressure within the aorta rises and falls with the beating of the heart. As such, the mean aortic pressure, rather than the aortic pressure during ventricular systole or diastole, provides the best indication of the pressure at the beginning of the systemic circuit. Thus,

$$\text{Cardiac Output} = \frac{\text{Mean Aortic Pressure}}{\text{Total Peripheral Resistance}} \qquad (5)$$

The resistance to flow in the large systemic arteries is only slight, and the pressure changes very little as the blood flows through them. Therefore, the mean aortic pressure is essentially equal to the mean arterial pressure, which can be approximated as

$$\frac{\text{Systolic Pressure} + 2\ (\text{Diastolic Pressure})}{3}$$

and

$$\text{Cardiac Output} = \frac{\text{Mean Arterial Pressure}}{\text{Total Peripheral Resistance}} \qquad (6)$$

This relationship (equation 6) can be rearranged into the following form:

$$\text{Mean Arterial Pressure} = \text{Cardiac Output} \times \text{Total Peripheral Resistance} \qquad (7)$$

Thus, the mean arterial pressure is influenced by changes in either the cardiac output or the total peripheral resistance. Moreover, as discussed later, the body possesses receptors involved in sensing and regulating arterial pressure, and the activity of these receptors leads to changes in heart rate and contractile strength—which alter cardiac output—and to changes in vessel diameter—which alter total peripheral resistance.

Arterial Pressure

The pressure within the large systemic arteries—which is essentially equal to the pressure generated by the pumping action of the left ventricle of the heart—is the immediate driving force for blood flow through the body's organs and tissues. Therefore, this pressure must be carefully maintained in order to ensure adequate organ and tissue blood flows. A number of factors affect the arterial pressure through mechanisms that influence vessel resistance and cardiac function.

Neural Factors

Vasomotor Nerve Fibers

Vasomotor nerve fibers *(vā-zō-mō´-ter)* are efferent fibers that regulate blood vessel radii. With the exception of the capillaries, these fibers supply almost all types of vessels, particularly the arterioles. Changes in blood-vessel radii due to the activity of vasomotor nerve fibers alter the resistance of the vessels to blood flow, and this in turn can lead to a change in the mean arterial pressure (see equation 7).

Nerves of the sympathetic division of the autonomic nervous system contain *vasoconstrictor* vasomotor fibers. At their junctions with blood vessels, these fibers release norepinephrine, which stimulates the constriction of the vessels.

In some animals, *vasodilator* vasomotor fibers—as well as vasoconstrictor fibers—are present in the sympathetic nerves that innervate skeletal muscles. At their junctions with blood vessels, the vasodilator fibers release acetylcholine, which causes vessel dilation. However, it is not clear if such sympathetic vasodilator fibers are present in humans. In any event, the sympathetic vasoconstrictor fibers are the most widely distributed vasomotor fibers in the body, and most neurally mediated vessel dilation occurs as a result of diminished vasoconstrictor fiber activity.

Vasomotor Center

In the lower third of the pons and the medulla oblongata of the brain is an area known as the **vasomotor center.** The vasomotor center plays an important role in the regulation of blood-vessel resistance and thus in the regulation of arterial pressure (Figure 20.9). Nerve impulses from this center are ultimately transmitted by vasoconstrictor fibers to blood vessels, and the vasomotor center is continuously, or tonically, active in promoting some degree of vasoconstriction, which is called *vasomotor tone.*

The activity of the vasomotor center is influenced by nerve impulses that arrive at the center from receptors associated with the cardiovascular system, as well as by nerve impulses from sensory surfaces of the body and from higher brain centers. Hormones and other substances in the blood also affect the activity of the vasomotor center. For example, carbon dioxide strongly stimulates the vasoconstrictor activity of the vasomotor center. When carbon dioxide levels increase, greater vasoconstriction is noted, and this fact has been used to explain the rise in arterial pressure seen early in asphyxiation.

Cardiac Center

The cardiac center, described in Chapter 19, influences arterial pressure by virtue of its effects on cardiac function. Many of the inputs that affect the vasomotor center also influence the cardiac center.

Baroreceptors

In the arch of the aorta, in the slightly enlarged region where each common carotid artery divides into internal and external carotid arteries—that is, in the *carotid sinus*—and, to a lesser extent, in the walls of almost every large artery in the neck and thoracic regions are receptors that respond to the distension or stretch of the vessel walls. Since the degree of stretch of the vessel walls is directly related to the arterial pressure, these receptors function as pressure receptors, or **baroreceptors** (*ba-ro-ree-sep´-ters; baro* = pressure) (Figure 20.9). Nerve impulses from the baroreceptors ultimately inhibit the vasoconstrictor activity of the vasomotor center, and they also influence the cardiac center in such a manner that the activity of the parasympathetic nerves to the heart increases and the activity of the sympathetic nerves to the heart decreases.

When the arterial pressure rises, the rate of nerve impulse transmission from the baroreceptors increases (Figure 20.10a). As a consequence, blood vessels dilate, and the rate and contractile strength of the heart decrease, leading to a decline in cardiac output. These activities lower the arterial pressure.

When the arterial pressure falls, the rate of nerve impulse transmission from the baroreceptors diminishes, resulting in an increased vasoconstriction and an increased rate and contractile strength of the heart (Figure 20.10b). These activities raise the arterial pressure.

The arterial baroreceptors protect the cardiovascular system against relatively short-term changes in arterial pressure. However, they do little to protect the system from sustained long-term pressure changes. This situation is due to the fact that the baroreceptors display adaptation, and after a few days their rates of discharge return to normal levels regardless of continued high or low arterial pressures. The baroreceptors still function, but in essence they become "reset" to operate at a different pressure level.

Aortic and Carotid Bodies

The **aortic** and **carotid bodies,** which are located at the aortic arch and at the branchings of the common carotid arteries, contain chemoreceptors that are sensitive to arterial oxygen, carbon dioxide, and hydrogen ion concentrations. Like the baroreceptors, these receptors send impulses to the vasomotor center. Impulses sent in response to decreased arterial oxygen cause an increase in the arterial pressure. Changes in carbon dioxide and hydrogen ion concentrations can also alter the arterial pressure, but the effects of these substances by this pathway are relatively small.

Chemicals and Hormones

A number of chemicals and hormones influence the arterial pressure.

◆ **FIGURE 20.9 Diagrammatic representation of the neural pathways involved in the regulation of blood-vessel diameter**

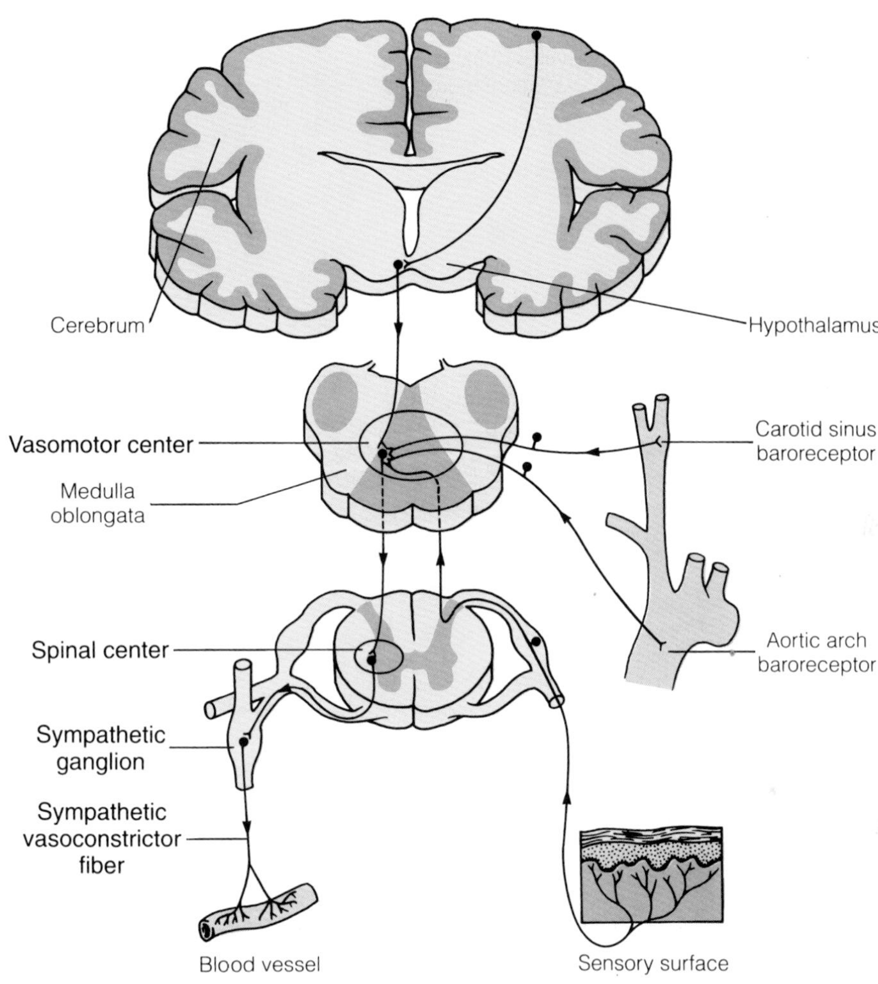

Angiotensin II

Angiotensin II (an-jee-o-ten´-sin) is a powerful vasoconstrictor that raises the arterial pressure. When the arterial pressure falls, the formation of angiotensin II, which is produced from precursors in the plasma by enzymatic action, increases.

Epinephrine

Epinephrine (ep-ĭ-nef´-rin) produced by the medullae of the adrenal glands causes a transitory increase in the systolic pressure within the arteries. Usually, when sympathetic nervous system activity causes widespread effects on blood vessels throughout the body, the adrenal medullae are stimulated, and epinephrine, as well as norepinephrine, is released.

Vasopressin

Vasopressin (vās-o-pres´-in), or antidiuretic hormone, raises the arterial pressure by stimulating arteriole constriction. When the arterial pressure falls, the release of vasopressin from the pars nervosa of the pituitary gland increases.

Blood Volume

The volume of blood within the circulatory system affects the arterial pressure. In general, an increase in blood volume tends to raise the arterial pressure, in large part by increasing the venous return and ventricular filling, which in turn leads to an increased cardiac output. Conversely, a decrease in blood volume tends

◆ **FIGURE 20.10 Baroreceptor reflexes**

(a) Baroreceptor reflex in response to a rise in arterial pressure. (b) Baroreceptor reflex in response to a fall in arterial pressure.

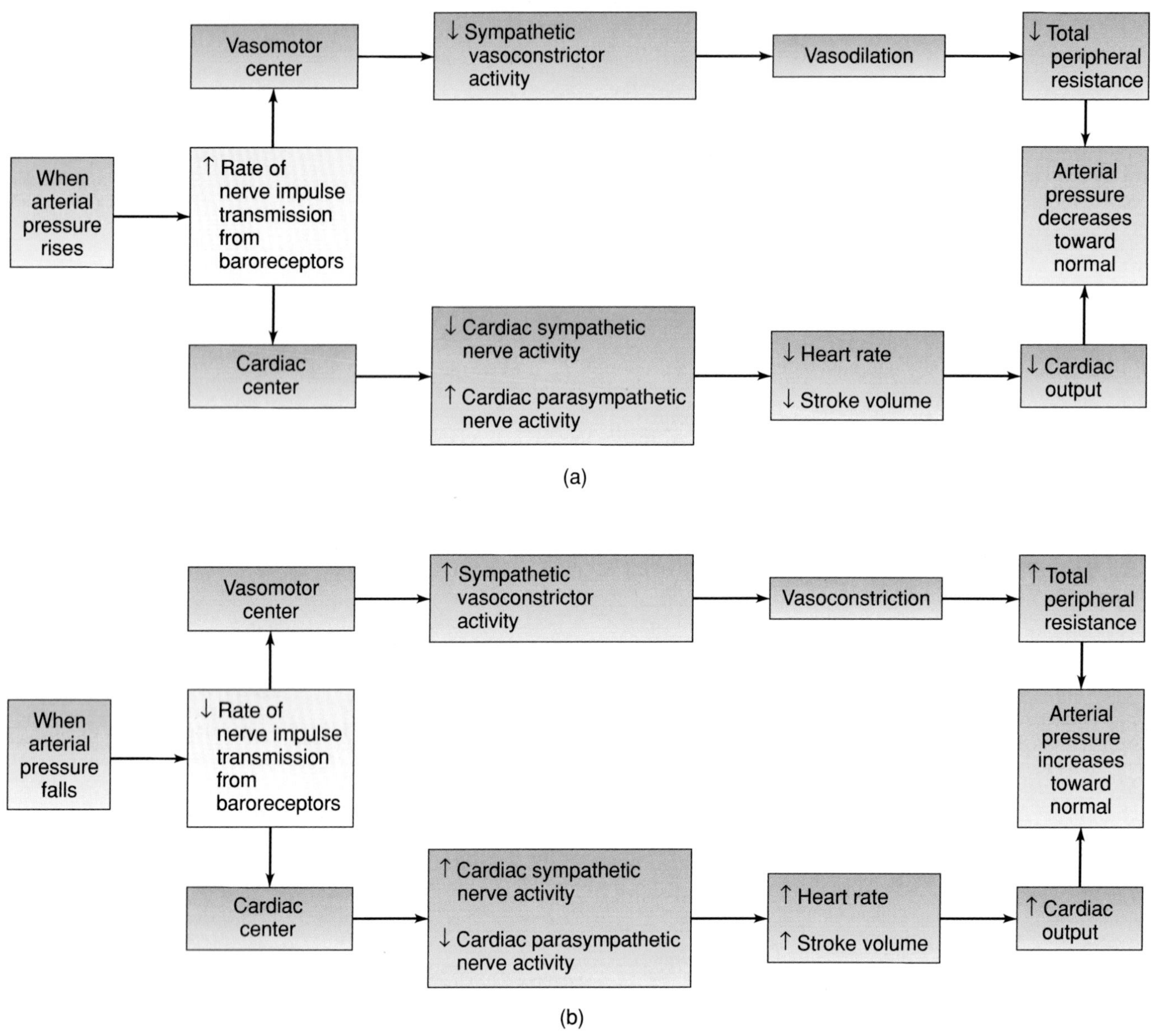

to lower the arterial pressure. Thus, mechanisms that alter blood volume can help control arterial pressure. For example, a rise in arterial pressure that leads to an elevation in capillary pressure favors the movement of fluid out of the capillaries and into the interstitial spaces. This movement of fluid decreases the blood volume and lowers the arterial pressure. Conversely, a fall in arterial pressure that leads to a decrease in capillary pressure favors the movement of fluid out of the interstitial spaces and into the capillaries. This movement of fluid tends to increase the blood volume and raise the arterial pressure.

The kidneys exert a significant effect on blood volume by virtue of their ability to regulate salt and water excretion, and they are particularly important in the long-term regulation of arterial pressure. A slight increase in the arterial pressure can cause a large increase in the formation of urine, which tends to decrease the blood volume and lower the arterial pressure. Conversely, a slight fall in the arterial pressure can lead to a substantial decrease in the formation of urine. As a consequence, fluids taken into the body tend to remain within the body, where they can increase the blood volume and thereby raise the arterial pressure.

◆ **FIGURE 20.11 Some factors that affect arterial pressure**

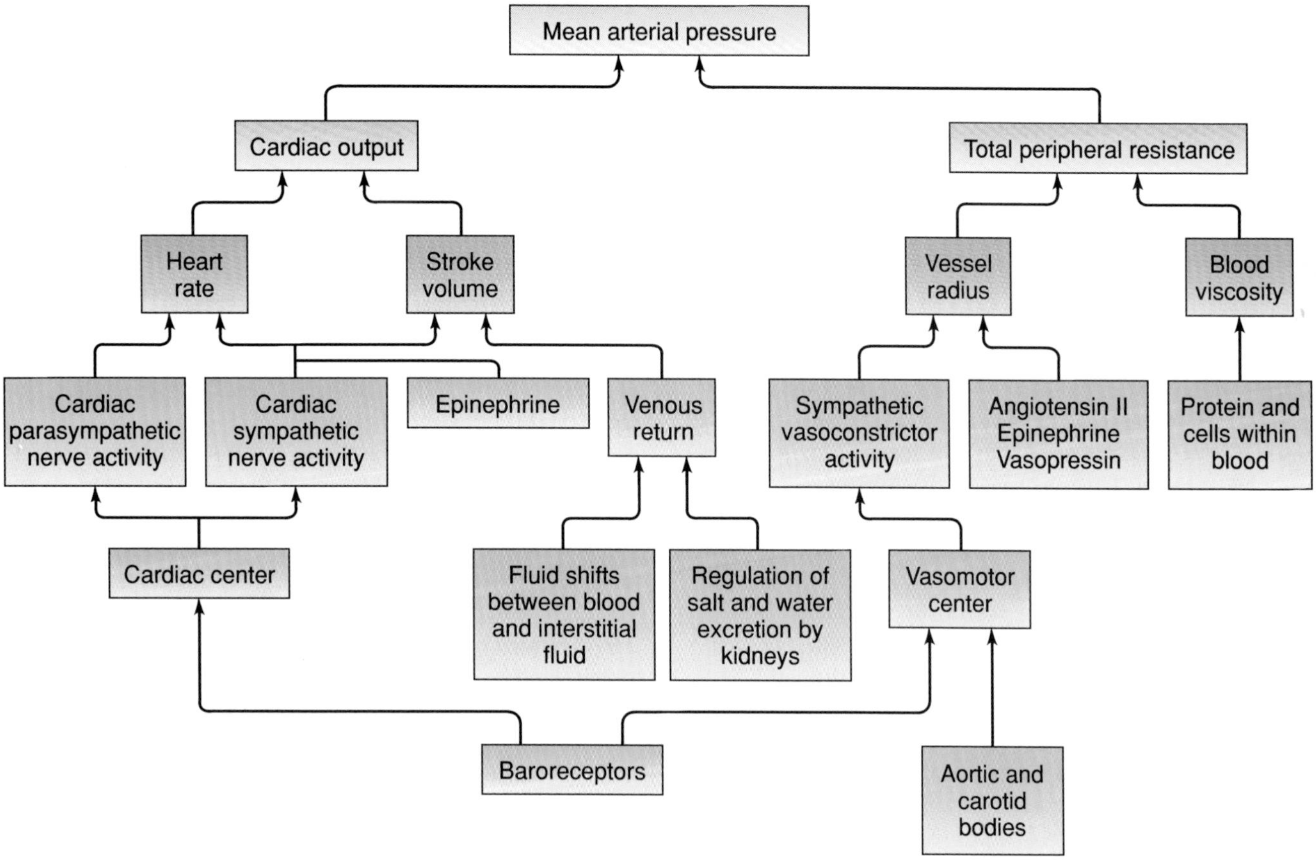

Mean arterial pressure

Cardiac output

Total peripheral resistance

Heart rate

Stroke volume

Vessel radius

Blood viscosity

Cardiac parasympathetic nerve activity

Cardiac sympathetic nerve activity

Epinephrine

Venous return

Sympathetic vasoconstrictor activity

Angiotensin II
Epinephrine
Vasopressin

Protein and cells within blood

Cardiac center

Fluid shifts between blood and interstitial fluid

Regulation of salt and water excretion by kidneys

Vasomotor center

Baroreceptors

Aortic and carotid bodies

Summary

From the previous discussion, it is evident that a number of different factors affect the arterial pressure (Figure 20.11). Consequently, the arterial pressure at any moment is usually the result of the combined influence of several factors. Moreover, different mechanisms work together to maintain the arterial pressure and, thus, to maintain a circulation that can meet the body's needs. For example, if the arterial pressure falls, neural mechanisms that involve baroreceptors, the cardiac and vasomotor centers, and vasomotor nerves, as well as chemical and hormonal mechanisms and mechanisms leading to changes in blood volume, may all act together to return the arterial pressure to normal.

Measurement of Arterial Pressure

The arterial pressure—which is commonly called the *blood pressure*—is usually measured indirectly (Figure 20.12). An inflatable cuff connected to a meter or mercury manometer (called a *sphygmomanometer*) is

◆ **FIGURE 20.12 Indirect measurement of blood pressure using a manometer**

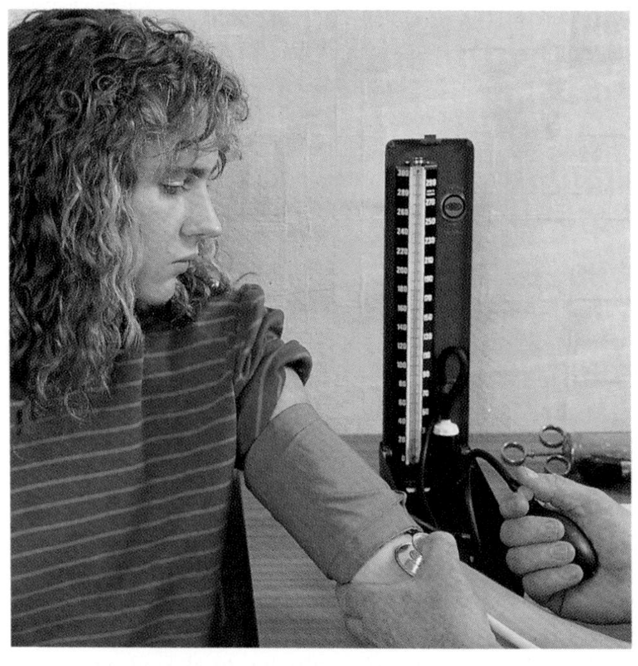

wrapped around the upper arm. The meter or manometer indicates the pressure within the cuff in millimeters of mercury (mm Hg). A pressure of 80 mm Hg, for example, is equal to the pressure exerted by a column of mercury 80 mm high. A stethoscope is placed below the cuff at the elbow in order to listen to the sounds in the brachial artery of the arm. The cuff is then inflated to a pressure sufficient to close the artery so that no blood flows past the cuff and no sound is heard with the stethoscope. The pressure is then slowly reduced until an intermittent thumping sound is heard. The pressure at which the sound is first heard is recorded as the *systolic pressure (sis-tol´ik)*. The sound is produced as follows: The contraction of the ventricles raises the pressure within the artery enough to overcome the resistance of the cuff as it presses the artery closed. Consequently, blood flows past the cuff into the lower arm in a turbulent fashion. It is the sound of this turbulence that is heard with the stethoscope. When the ventricles relax and the pressure in the artery drops, the cuff again closes the artery, no blood flows into the lower arm, and no sound is heard with the stethoscope.

Upon further reduction of the pressure in the cuff, the intermittent thumping sound becomes dull and muffled. The pressure at which this occurs is recorded as one diastolic *(dī-ah-stol´-ik)* end point *(diastolic pressure)*. This point corresponds to the point at which the pressure in the artery, even during the relaxation of the ventricles, is sufficient to overcome the resistance of the cuff, and thus the vessel never closes. However, the cuff still exerts enough pressure on the artery to produce a turbulent blood flow that can be heard with the stethoscope. If the pressure in the cuff is reduced further, the thumping sound eventually disappears completely. The pressure at which this occurs is recorded as a second diastolic end point. At this point, the flow of blood in the vessel is smooth, and no intermittent sound of turbulence can be heard with the stethoscope. Although the systolic pressure and both diastolic end points may be recorded—for example, 120/85 80—generally only the second diastolic end point is considered. Thus, arterial pressure is reported as a fraction such as 120/80, with the numerator being the systolic pressure and the denominator being the diastolic pressure. The difference between the systolic and diastolic pressures is called the *pulse pressure*. The pulse pressure tends to increase if the stroke volume of the ventricles increases or if the distensibility of the arteries decreases.

Blood Flow Through Tissues

Even when the mean arterial pressure is relatively constant, the blood flow through many body tissues varies from time to time. Changes in the blood flow through a particular tissue are primarily due to alterations in the resistance of the tissue's blood vessels. The resistance of the vessels is controlled by both local autoregulatory mechanisms and external factors.

Local Autoregulatory Control

Local autoregulatory mechanisms alter blood flow through a tissue according to the needs of the tissue. For example, if the blood flow through a tissue is inadequate, the oxygen level in the tissue tends to fall. In many tissues, a decrease in the oxygen level leads to a local dilation of arterioles and to an increased relaxation of precapillary sphincters. These responses result in an increased blood flow through the tissue, which increases the delivery of oxygen.

The metabolic processes of a tissue receiving an inadequate blood flow lead to the formation and accumulation of substances that dilate arterioles and precapillary sphincters in the tissue. This dilation decreases the resistance to blood flow, leading to an increased flow through the tissue. Among the substances that act as *local vasodilator substances* are carbon dioxide, adenosine, lactic acid, potassium ions, and hydrogen ions.

Some tissues have greater autoregulatory ability than others. This ability is particularly well developed in skeletal and cardiac muscle and in the gastrointestinal tract.

External Factors

The activity of external factors such as vasomotor nerve fibers and hormones can lead to the constriction or dilation of blood vessels in particular tissues and, therefore, to changes in blood flow through the tissues. However, external factors usually influence large segments of the vascular system and are generally more involved with mechanisms that function for the well-being of the entire body—such as those that maintain arterial blood pressure or body temperature—than they are with the control of tissue blood flow according to the needs of particular tissues. For example, in a warm environment, the activity of sympathetic vasoconstrictor fibers to skin vessels decreases, and the vessels dilate. This dilation results in an increased blood flow to the body surface and thereby increases the loss of body heat from the blood to the environment.

Capillary Exchange

Capillaries are sites at which the exchange of materials between the blood and the interstitial fluid takes place. Most capillaries are permeable to water and small particles such as glucose, inorganic ions, urea, amino acids, and lactic acid. These materials pass readily between the

 ASPECTS OF EXERCISE PHYSIOLOGY 

Muscle Blood Flow: Adaptations to Training

When a person exercises, active muscles receive an increased blood flow and, therefore, an increased delivery of nutrients and oxygen. However, even if additional oxygen is delivered to a muscle, the muscle must have sufficient oxidative capacity to use it.

Endurance training produces increases in the oxidative capacity of muscles. With training, the number and size of mitochondria within muscle cells increase as does the level of oxidative enzymes. An increase in oxidative capacity allows a muscle to use more of the oxygen available to it. For example, in an unconditioned person, if approximately 20 ml of oxygen are delivered to a muscle per 100 ml of blood, about 15 ml of oxygen remain in the blood after it passes through the muscle. In an endurance-trained person, only 5 or 10 ml of oxygen may remain in the blood.

Endurance training causes an increase in the capillary density within muscles (that is, an increase in the number of capillaries around muscle cells). This increase prolongs the transit times for blood flow through muscles and reduces diffusion distances from the blood to muscle cells, thereby enhancing the delivery of nutrients and oxygen to the cells and the removal of metabolic end products from them. Moderate training over several months can increase capillary density within muscles by 20 to 30%, and a prolonged, intensive training program can increase capillary density by about 40 to 50%.

Endurance training also produces an increase in the storage of energy reserves. (Glycogen storage may be increased about threefold by a combination of exercise and diet.)

Increases in the oxidative capacity of muscles, the capillary density within muscles, and the storage of energy reserves contribute to high levels of exercise tolerance. In fact, some highly conditioned athletes participate in ultra-endurance events in which they run 26 miles, swim 2 miles, and bicycle 50 miles.

blood and the interstitial fluid. Permeability to protein, however, is usually quite limited, although it varies from tissue to tissue. It is highest in the liver, less in muscle, and very limited in the central nervous system.

Capillary Blood Flow

As previously described, blood flows slowly through the capillaries, allowing adequate time for the exchange of materials between the blood and the interstitial fluid. Moreover, the precapillary sphincter muscles at the entrances of the capillaries undergo cycles of contraction and relaxation at frequencies of 2–10 cycles per minute. Because of this cyclical activity, which is called *vasomotion,* capillary blood flow is usually intermittent rather than continuous and steady.

Local autoregulatory factors, particularly oxygen levels, have a strong influence on vasomotion. For example, in many tissues, when the oxygen level in a particular region is low, the intermittent periods of blood flow through the capillaries of the region occur more often, and each period of flow lasts longer. Thus, the blood flow through the region increases, leading to a greater delivery of oxygen and nutrients.

Although the blood flow through any one capillary is intermittent, there are so many capillaries that their function on an organ level becomes averaged. Therefore, it is possible to consider capillary function in terms of average capillary pressures, rates of flow, and rates of transfer of substances between the blood and the interstitial fluid. This approach is used in the following sections.

Movement of Materials between the Blood and the Interstitial Fluid

Several processes contribute to the movement of materials across capillary walls between the blood and the interstitial fluid.

Diffusion

Diffusion is the most important means by which substances such as nutrients and metabolic end products pass between the blood and the interstitial fluid. If a capillary is permeable to a particular substance, and if there is a higher concentration of the substance within the capillary than outside it, the net diffusion of the substance will be outward. If the substance is in higher concentration within the tissue fluid than within the capillary, its net diffusion will be inward. Lipid-soluble substances, including oxygen and carbon dioxide, can penetrate plasma membranes and move across capillary endothelial cells. Water-soluble substances, such as glucose and amino acids, can pass through water-filled clefts between adjacent endothelial cells. However, the size of the cleft limits the size of the particles that can pass through.

Endocytosis and Exocytosis

Substances are taken into one side of capillary endothelial cells by endocytosis and released from the other side of the cells by exocytosis. However, there is disagreement about the importance of this activity in the exchange of materials between the blood and the interstitial fluid.

Fluid Movement

A general movement of fluid—that is, water and dissolved particles to which a capillary is permeable—takes place across the capillary wall. As previously indicated, under certain circumstances, the movement of fluid between the blood and the interstitial fluid leads to alterations in blood volume (and arterial pressure). However, under normal circumstances, this movement causes little change in the volume of either the blood or the interstitial fluid. Several factors are involved in the movement of fluid across the capillary wall, including fluid pressures and osmotic pressures.

Fluid Pressures. The pressure of the blood within the capillaries (that is, the *capillary pressure*) tends to force fluid out of the capillaries and into the tissue spaces by filtration through the capillary walls. Opposing this movement is the pressure of the interstitial fluid (that is, the *interstitial fluid pressure*), which tends to move fluid out of the tissue spaces and into the capillaries. The interstitial fluid pressure, however, is generally much less than the capillary pressure, and there is considerable evidence that it is even below atmospheric pressure.

Osmotic Pressures. The presence of nondiffusible proteins (particularly albumin) in the blood plasma leads to the development of an osmotic force called the *colloid osmotic pressure of the plasma,* or the *oncotic pressure (on-kot´-ik).* The oncotic pressure tends to draw water into the capillaries by osmosis. This activity is opposed by the *colloid osmotic pressure of the interstitial fluid,* which results from the presence of nondiffusible proteins in the interstitial fluid. The colloid osmotic pressure of the interstitial fluid tends to draw water out of the capillaries by osmosis.

When osmosis occurs across a barrier like the capillary wall, which contains pores or clefts that are relatively wide compared to the width of a water molecule, substantial quantities of water move rapidly across the barrier, dragging along solutes to which the barrier is freely permeable. Thus, the occurrence of osmosis across the capillary wall results in a movement of fluid similar to that due to the capillary and interstitial fluid pressures.

Direction of Fluid Movement. If the forces tending to move fluid out of a capillary (the capillary pressure and the colloid osmotic pressure of the interstitial fluid) are greater than the forces tending to move fluid into the capillary (the interstitial fluid pressure and the oncotic pressure), fluid leaves the capillary and enters the tissue spaces. If the inward forces are dominant, fluid enters the capillary from the tissue spaces.

The traditional view of fluid movement across the capillary wall is that under normal circumstances the capillary pressure at the arterial end of a systemic capillary is relatively high, and outward moving forces predominate (Figure 20.13). Thus, fluid moves out of the capillary into the tissues. As the blood flows through the capillary, the capillary pressure drops. Gradually, the balance of forces favoring movement outward is reversed, and fluid moves back into the capillary at the venous end.

An alternative view of fluid movement across the capillary wall is that when a precapillary sphincter muscle relaxes during the rhythmic cycling of vasomotion, the capillary pressure rises, and there is an outward movement of fluid all along the capillary. Then, when the precapillary sphincter constricts, the capillary pressure falls, and there is an inward movement of fluid all along the capillary.

In either case, not all of the fluid that moves out of the capillaries returns to the capillaries. Some of the fluid, including any protein that escapes from the capillaries, enters the lymphatic system, to be ultimately returned to the bloodstream with the lymph.

Venous Return

Blood that leaves the capillaries flows into venules and then into veins on its way back to the heart. The greater

◆ **FIGURE 20.13 Forces involved in the movement of fluid into and out of capillaries**

Values are general and may not apply to any particular capillary. They are expressed in mm Hg, with atmospheric pressure considered to be 0 mm Hg. Thus, negative values represent pressures below atmospheric pressure. CP = capillary pressure; OP = oncotic pressure; IFP = interstitial fluid pressure; IOP = colloid osmotic pressure of interstitial fluid.

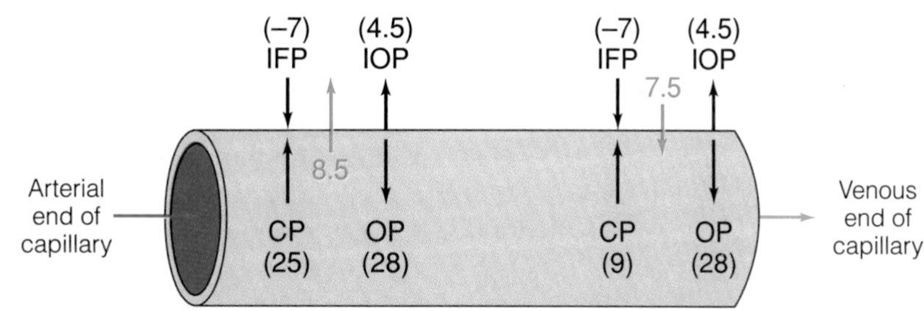

Out: CP + IOP: 25 + 4.5 = 29.5
In: IFP + OP: −7 + 28 = 21.0
8.5 Out

In: IFP + OP: −7 + 28 = 21.0
Out: CP + IOP: 9 + 4.5 = 13.5
7.5 In

the distance the blood travels, the greater the total resistance encountered and thus the greater the pressure drop. Blood in the venules and veins has traveled through the arteries, arterioles, and capillaries, and consequently the pressures within the venous circulation are quite low—for example, the pressures within the veins of the arms or legs average about 6–8 mm Hg.

The volume of blood that flows from the systemic veins into the right atrium of the heart in a given time is called the **venous return.** (Under normal circumstances, the same volume of blood flows from the pulmonary veins into the left atrium.) The magnitude of the venous return depends on the pressure driving blood through the veins and on the resistance of the veins to the flow of blood through them (recall that flow = ΔP/resistance). The pressure driving blood through the veins is represented by the difference between the pressure in the peripheral veins (the *peripheral venous pressure*) and the pressure at the right atrium of the heart. Consequently, factors that increase the peripheral venous pressure tend to increase the venous return, and factors that decrease the peripheral venous pressure tend to decrease it. For example, a decrease in the blood volume—as may result from hemorrhage—can lower the peripheral venous pressure and decrease the venous return, and an increase in the blood volume—as may result from a transfusion—can raise the peripheral venous pressure and increase the venous return. In a similar manner, factors that increase the right atrial pressure (which is normally close to 0 mm Hg) can also decrease the venous return. For example, a leaky right atrioventricular valve can cause an increase in the right atrial pressure and a decrease in the venous return.

The activities of skeletal muscles have an important influence on the venous return. The veins are quite flexible, and the movements of the skeletal muscles that surround them act in a pumping fashion to put pressure on the veins and compress them. This compression forces blood out of the veins and into the heart. The valves of the large veins contribute to the effectiveness of this pumping action by permitting blood leaving a compressed portion of a vein to flow only toward the heart and not in the opposite direction.

Breathing movements also influence the venous return. When a person inhales, the pressure within the thoracic cavity decreases, and the pressure within the abdominal cavity increases. These pressure changes facilitate the return of blood from the veins to the heart.

The veins have a large capacity, and they contain a substantial amount of the total blood volume. In fact, about 64% of the total circulating blood volume is normally in the systemic venous circulation, about 7% is in the heart, about 15% is in the systemic arterial vessels, about 5% is in the systemic capillaries, and about 9% is in the pulmonary vessels (Figure 20.14). Because of their large capacity, the veins serve as blood reservoirs. Although they are not heavily muscular, the veins can constrict and decrease their capacity, and even a slight venous constriction forces a good deal of blood into other portions of the cardiovascular system. Moreover, a constriction of the veins can decrease venous distensibility, raise the peripheral venous pressure, and increase the venous return. (In this regard it should be noted that the veins are relatively low-resistance vessels, and a moderate venous constriction does not greatly increase the overall resistance to blood flow.)

◆ **FIGURE 20.14 Percentage of the total blood volume in different portions of the cardiovascular system**

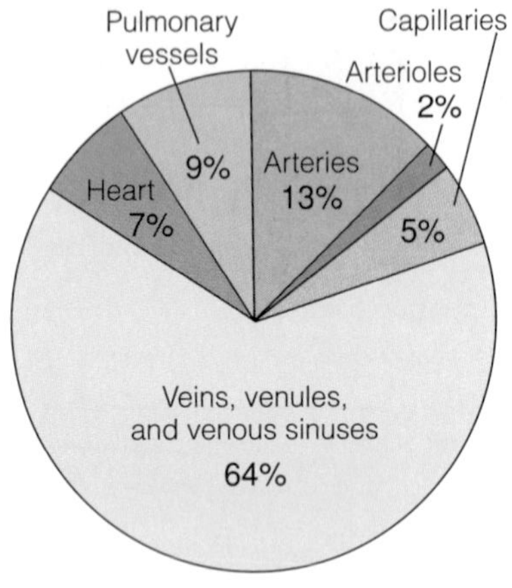

Effects of Gravity on the Cardiovascular System

Previous considerations of the pressures in various portions of the cardiovascular system have generally assumed that the body is in a horizontal position with the blood vessels at approximately the same level as the heart, and that pressures due to the force of gravity acting on the blood are negligible. However, when the body is in an upright position, the weight of the blood affects these pressures. For example, when a person moves from a supine position to a position in which he or she is standing upright and perfectly still, the weights of the columns of blood within the blood vessels raise the pressure within the vessels of the lower regions of the body (Figure 20.15a). The increased pressure distends the veins, increasing their capacity so they contain a greater amount of the total blood volume. Moreover, the increased pressure results in an increased filtration of fluid out of the capillaries and into the tissue spaces. The accumulation of blood in the veins and the increased filtration of fluid out of the capillaries reduces the effective circulating blood volume. If no adequate compensatory adjustments occur, the arterial pressure falls, the blood flow to the brain declines, and the person becomes dizzy or faints. Under most circumstances, however, several mechanisms compensate for the effects of gravity when a person is upright.

One mechanism is the baroreceptor mechanism described earlier. For example, when a person sits up or stands after lying down, the force of gravity causes the blood to accumulate in the veins of the lower limbs and abdomen, and the arterial pressure in the head and upper body falls. However, the falling pressure is sensed by the baroreceptors, which initiate a generalized vasoconstriction and increase in heart rate and contractile strength, thus preventing a severe pressure drop that could result in diminished blood flow to the brain and a loss of consciousness.

A second mechanism that compensates for the effects of gravity is the pumping action exerted on the veins by the skeletal muscles of the legs. The contractions of the muscles compress the veins and force blood toward the heart (Figure 20.15b). This activity enhances the venous return and reduces the pressure within the veins of the lower regions of the body. As a result, the upright position has less effect on venous distension and the accumulation of blood in the veins as well as on the capillary pressure and the filtration of fluid out of the cardiovascular system than would otherwise be the case. The importance of this mechanism is evident in the fact that when a person is standing perfectly still, the pressure within the veins of the feet can rise above 90 mm Hg, but the muscular activity of walking reduces the pressures within these veins to less than 25 mm Hg.

Circulation in Special Regions

Certain regions of the body have special needs or present special problems that are met by various circulatory adaptations. This section considers some of the adaptations that occur in these regions.

Pulmonary Circulation

The pulmonary arteries and arterioles generally have larger diameters and thinner walls than corresponding vessels of the systemic circulation, and the resistance to blood flow through the pulmonary circulation is less than that through the systemic circulation. As a result, less pressure is required to move blood through the pulmonary circuit than through the systemic circuit, and the pressure within the pulmonary arteries (about 22 mm Hg systolic pressure, 8 mm Hg diastolic pressure) is considerably lower than the pressure within the systemic arteries (about 120 mm Hg systolic pressure, 80 mm Hg diastolic pressure).

The blood flow through different local areas of the pulmonary circulation varies with the oxygen levels in the areas. A decreased oxygen level causes vasoconstriction, and an increased oxygen level causes vasodilation. Although the local effect of oxygen on pulmonary vessels is the opposite of its effect on many systemic arterioles, it is consistent with the role of the lungs in providing oxygen to the blood. For example, if a portion of the lungs is not functioning effectively, the level of oxygen in the region falls. The decline in the oxygen level causes the vessels supplying the area to constrict,

◆ **FIGURE 20.15 Influences of an upright position and movement on the pressure within blood vessels**
(a) When a person is standing upright and perfectly still, the weights of the columns of blood within the blood vessels raise the pressure within the vessels of the lower regions of the body. (b) During movement, muscular contractions compress the veins and force blood toward the heart. This activity reduces the pressure within the veins of the lower regions of the body, such as the veins of the feet.

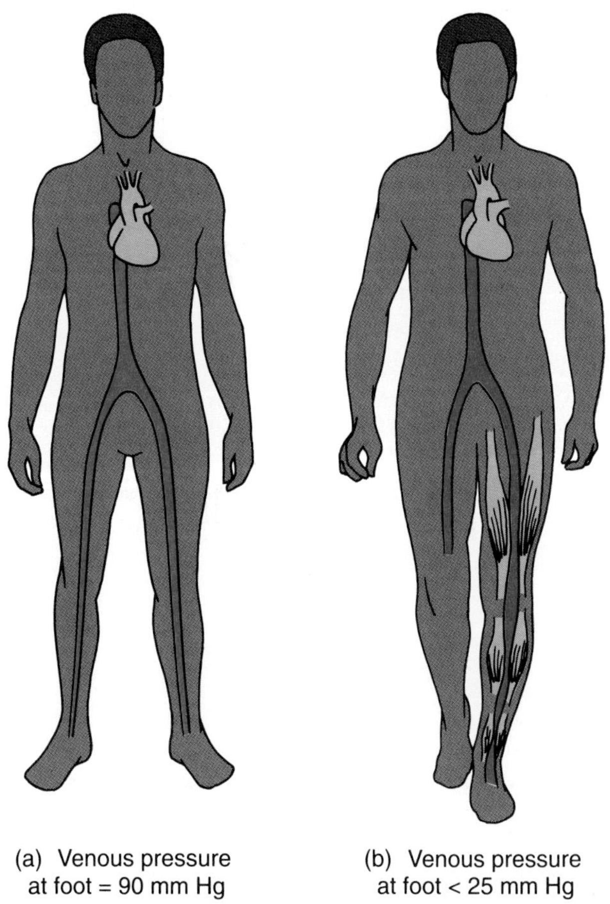

which results in a greater blood flow through the vessels in efficiently functioning areas of the lungs.

Coronary Circulation

The flow of blood through the coronary vessels is greatly influenced by the aortic pressure and by the pressures that build up in the walls of the heart chambers during systole—that is, the *intramural pressures.* The influence of these pressures is particularly evident in the left coronary artery (Figure 20.16). When the left ventricle contracts (line 1), the intramural pressure increases greatly, squeezing the coronary vessels and causing a significant reduction in the coronary flow (curve a). With the opening of the aortic semilunar valve (line 2), the effect of the intramural pressure of the ventricles is overcome by the increased aortic blood pressure. The increased aortic pressure causes an increase in blood flow through the left coronary artery (curve b). As ventricular systole progresses, the flow of blood into the aorta diminishes and the pressure within the aorta begins to fall (line 3). However, the left ventricle is still contracting at this time, and the intramural pressure remains high. As a result, there is a secondary slowing of the coronary flow (curve c). As ventricular systole ends (line 4), the intramural pressure drops, and the flow through the coronary vessels increases once again (curve d). During ventricular diastole, the flow through the coronary vessels is determined primarily by the aortic pressure, and the flow gradually decreases as the aortic pressure drops (curve e). Also, during this period, the atria contract and the pressure within their walls increases. This increase in the atrial intramural pressure also tends to decrease the flow of blood through the coronary vessels.

Under resting conditions, the blood flow through the coronary vessels is approximately 250 ml per minute, which is about 5% of the cardiac output. However, during strenuous exercise, the coronary vessels dilate, and the coronary blood flow can increase to four or five times the resting level—that is, up to 1250 ml per minute. The dilation of the coronary vessels during exercise is believed to be primarily due to the influence of local autoregulatory factors (such as low oxygen levels resulting from increased cardiac activity).

An increased blood flow to the heart during exercise is particularly important because, even in the resting condition, the heart removes about 65% of the oxygen in the arterial blood that flows to it. The removal of this much oxygen is very efficient compared to most other tissues, where as little as 25% of the oxygen is removed. However, it means that during activity, the heart cannot greatly increase the oxygen available to it by increasing the percentage of oxygen removed from the blood. Therefore, increasing the coronary blood flow—and thus the delivery of oxygen—is the principal way in which additional oxygen is made available to the myocardium.

Cerebral Circulation

The neurons of the brain are unable to store glycogen, and the brain must have a dependable blood flow in order to ensure that sources of metabolic energy such as glucose are available to it. In addition, brain tissue must have a steady, adequate supply of oxygen. Irreversible damage occurs if the brain's blood flow is stopped for over five minutes.

◆ FIGURE 20.16 Blood flow through the left coronary artery correlated with aortic blood pressure and ventricular systole and diastole

See text for detailed discussion.

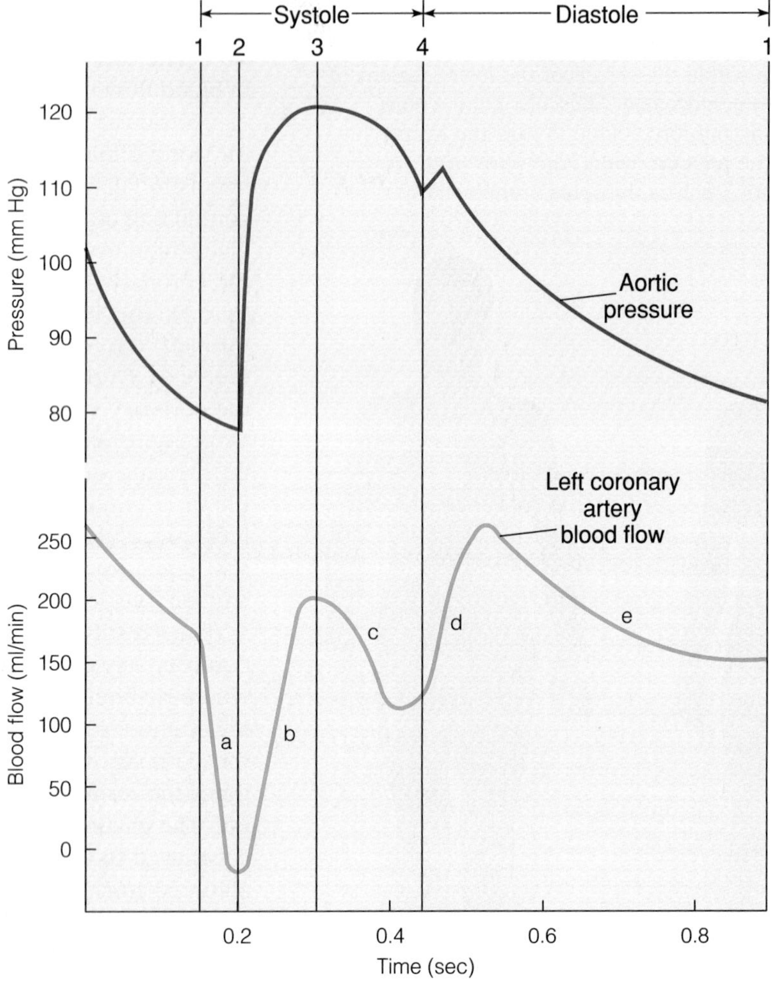

In adults, blood flow through the brain is about 50 to 55 milliliters per 100 grams of tissue per minute. The overall metabolic rate of the brain varies little under widely different physiological conditions—for example, intense mental activity, muscular activity, or sleep—and under most circumstances, cerebral blood flow does not vary greatly. However, cerebral blood flow can be altered by changes in the concentrations of carbon dioxide or oxygen in the blood. An elevated concentration of carbon dioxide in the blood or a decreased concentration of oxygen leads to a dilation of cerebral blood vessels, which increases cerebral blood flow.

Cerebral blood flow is relatively unaffected by changes in the systemic arterial pressure. Cerebral blood flow is remarkably constant over a very broad range of mean arterial pressures, and it does not decline significantly unless the mean arterial pressure is considerably below normal (Figure 20.17). This constancy of flow is believed to be due at least in part to regulatory mechanisms that cause cerebral vessels to constrict when the systemic arterial pressure rises and dilate when the systemic arterial pressure falls.

The blood flow to different local areas of the brain increases in response to increases in the activities of the areas. For example, clasping the right hand is accompanied by an increased blood flow in the motor cortex on the left side of the brain. Local autoregulatory mechanisms are believed to contribute to changes that occur in local blood flow. An elevated concentration of carbon dioxide in an active local area of the brain or a decreased concentration of oxygen is believed to cause a dilation of arterioles in the area, which increases blood flow to the area. (In general, the mechanisms that regulate brain blood flow seem to be more sensitive to carbon dioxide concentrations than they are to oxygen concentrations.)

If there is a temporary decrease in the cerebral blood flow, a person may faint, and factors that lower the ar-

◆ **FIGURE 20.17 Relationship of mean arterial pressure and cerebral blood flow**

The central area is the normal range of mean arterial pressure.

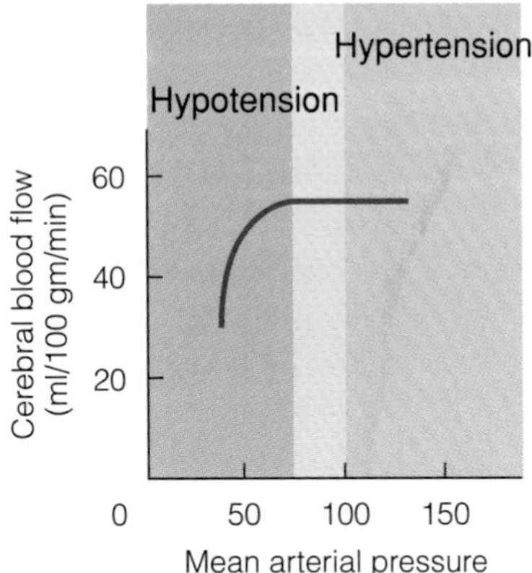

terial pressure below the level required to maintain an adequate cerebral blood flow can cause fainting. For example, the fainting that often accompanies severe emotional upset is due at least in part to a decreased sympathetic output to blood vessels, which leads to vasodilation and a decrease in total peripheral resistance. As a result, the arterial pressure declines, blood flow to the brain is impaired, and the person faints.

A more serious condition occurs if a cerebral vessel is obstructed or ruptured and the blood flow to the region of the brain supplied by the vessel is reduced. Such an occurrence is called a *cerebrovascular accident,* or *stroke.* The reduced blood flow results in an inadequate supply of oxygen to the cells of the affected area and may cause permanent brain damage. The neurological effects of a stroke vary, depending on the site of brain damage.

Skeletal Muscle Circulation

During exercise, the energy requirements of skeletal muscles increase so greatly that they can be met only by a tremendous increase in the blood flow to the muscles. During strenuous exercise, muscle blood flow can be more than 10 times greater than the flow at rest. The increased blood flow is made possible by the dilation of arterioles and the opening of previously closed capillaries within the muscles.

An important mechanism involved in increasing the blood flow through active skeletal muscles is the local autoregulatory response. During exercise, increased muscular activity leads to decreased levels of oxygen and/or to an accumulation of local vasodilator substances in active muscles. The decreased oxygen levels and/or the local vasodilator substances produce a vascular dilation that increases the blood flow through the muscles.

Cutaneous Circulation

The blood flow to the skin serves two functions: it provides nutrients to and carries metabolic end products away from the skin, and it brings heat from the internal structures of the body to the surface, where the heat can be lost from the body. The cutaneous blood flow is normally 20 to 30 times greater than is needed to meet the nutrient needs of the skin, and most of the flow is for purposes of temperature regulation.

The blood flow to the skin can vary tremendously. Under ordinary circumstances, cutaneous blood flow is approximately 400 ml per minute. However, under extreme conditions, the blood flow can increase to 2500 ml per minute. A unique arrangement of the blood vessels beneath the body surface makes such a large blood flow possible (Figure 20.18, page 650). In addition to the usual capillary beds, there are extensive subcutaneous venous plexuses (*plexus* = network) that can hold large volumes of blood. These plexuses are located close enough to the surface to allow heat to pass from the blood to the surface, and thus to be lost from the body.

The blood flow through the subcutaneous venous plexuses varies with the degree of constriction or dilation of the vessels that carry blood to the plexuses, and a temperature control center in the hypothalamus acts through sympathetic nerves to regulate the diameters of these vessels. In some locations—for example, the hands, feet, and ears—blood can bypass the capillary system and flow directly from the arteries into the venous plexuses by way of vascular shunts called *arteriovenous anastomoses (ah-nas-tō-mō´-ses).* The walls of these arteriovenous shunts are muscular, and their constriction decreases the flow of blood into the venous plexuses and reduces the loss of body heat.

Cardiovascular Adjustments During Exercise

During exercise, the dilation of blood vessels in active skeletal muscles due to local autoregulatory responses greatly increases blood flow to the muscles. At the same time, sympathetic vasoconstrictor activity causes a compensatory constriction of vessels elsewhere, particularly in the kidneys and gastrointestinal tract. The compensatory vasoconstriction does not fully offset the dilation of skeletal muscle vessels, however, and the total peripheral resistance decreases.

CONDITIONS OF CLINICAL SIGNIFICANCE

Blood Vessels

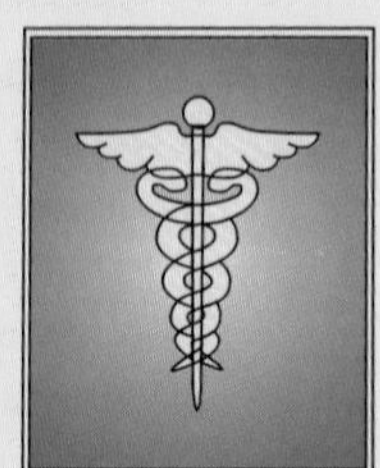

High Blood Pressure

High blood pressure, or *hypertension,* can lead to heart attack, heart failure, brain stroke, or kidney damage. Approximately 20% of the population can expect to have high blood pressure at some time during their lives. In general, arterial blood pressures above 140/90 mm Hg are considered to be hypertensive. The degree of elevation of both the systolic pressure and the diastolic pressure is important in hypertension, since both pressures affect the mean arterial pressure.

Hypertension due to an unknown cause is called *idiopathic* or *primary hypertension,* and approximately 95% of the cases of high blood pressure are classed as idiopathic. Many researchers believe that high blood pressure is frequently caused by an excessive dietary intake of sodium or an excessive retention of sodium by the body, which leads to an increase in blood volume and cardiac output. Researchers have also proposed that an increased peripheral resistance due to vascular constriction contributes to high blood pressure, and a number of studies suggest a sequence of initially high cardiac output followed by increased peripheral resistance.

High blood pressure is sometimes associated with kidney diseases or conditions that impair renal blood flow. In some cases, the formation of renin in the kidneys leads to the formation of angiotensin II and arteriolar constriction, although other factors are also likely to be involved. High blood pressure can also accompany the secretion of excessive amounts of norepinephrine and epinephrine by tumors of the adrenal medulla, and the secretion of excessive amounts of hormones by tumors of the adrenal cortex.

Atherosclerosis and Arteriosclerosis

As they grow older, a great number of people experience degenerative changes in the walls of their arteries that are known as *atherosclerosis (ath″-er-ō-skle-rō′-sis).*

Atherosclerosis is characterized by deposits (plaques) of abnormal smooth muscle cells, lipid materials, and connective tissue in arterial walls. The atherosclerotic plaques can gradually narrow or occlude blood vessels. In addition, the plaques can promote the formation of blood clots within the vessels. In any case, the flow of blood through the vessels is often restricted or blocked. Frequently, the first indication of atherosclerosis is the dysfunction of an organ supplied by an affected vessel. For example, a restricted or inadequate blood flow to the heart can cause pain during exertion or emotional stress—a condition known as *angina pectoris*—and a blocked coronary vessel can lead to the death of a portion of the myocardium—that is, to a *myocardial infarct.* Similarly, the occlusion of a vessel supplying the brain may kill brain tissue.

In very advanced cases, atherosclerotic plaques can become calcified, and there can be an extensive growth of fibrous tissue in arterial walls, giving rise to *arteriosclerosis (ar-te″-rē-ō-skle-rō′-sis),* or "hardening of the arteries." Arteriosclerosis can greatly reduce the elasticity of the vessel walls. As the walls become less elastic, the vessels cannot properly expand and recoil in response to the pressure changes produced by the beating heart. Consequently, in severe cases, the pressure within the vessels rises quite high during systole and falls unusually low during diastole.

Although atherosclerosis and arteriosclerosis can occur in any blood vessel, the aorta and the coronary arteries are most often affected. A number of methods are used to correct the circulatory problems caused by these conditions. One method is coronary bypass surgery, in which circulation to the heart muscle is improved by bypassing the clogged coronary arteries with vessels removed from the legs.

Several chemical agents are used to dissolve blood clots and atherosclerotic plaques within vessels. These substances help to quickly restore blood flow and significantly reduce heart muscle damage.

A procedure called *balloon angioplasty (an′-jē-ō-plas-tē)* is often used in conjunction with the clot dissolvers. In this procedure, a small catheter with a tiny balloon attached to its tip is threaded through blood vessels to the sites of obstruction. After the chemical clot dissolvers are administered, the balloon is inflated, compressing the plaque against the vessel wall and enlarging the lumen of the vessel. In yet another procedure, a catheter is used to direct laser beams to the obstruction in an attempt to destroy it.

CONDITIONS OF CLINICAL SIGNIFICANCE

Although these techniques have been successful in increasing blood flow through clogged vessels, they do not correct the conditions that caused the obstructions, and the vessels are often blocked again within a few months.

Edema

The accumulation of excess fluid in the tissues is called *edema.* Edema causes the tissues to swell, and it can impair tissue function because the excess fluid increases the distance that materials must diffuse between the capillaries and the cells. Many factors can contribute to edema. In general, any event that either enhances the movement of fluid out of the blood vessels and into the tissues or retards the return of fluid from the tissues back into the blood vessels can lead to edema. Thus, edema may result from (1) an increased capillary permeability, as may occur in certain allergic responses; (2) an elevated capillary pressure, due, for example, to blockage of a vein; (3) an increased interstitial fluid colloid osmotic pressure, caused, for example, by blocked lymphatic vessels; or (4) a decreased plasma colloid osmotic pressure, due, for example, to the loss of plasma proteins in the urine in certain kidney diseases.

Circulatory Shock

Circulatory shock is a condition in which the cardiac output is so reduced that the body tissues fail to receive an adequate blood supply. Any condition that decreases the cardiac output can lead to circulatory shock. Thus, circulatory shock can result from (1) diseases and weaknesses of the heart itself, caused, for example, by a heart attack; (2) reductions in blood volume, perhaps due to hemorrhage; or (3) vascular difficulties that reduce the venous return, due, for example, to extreme vasodilation and the accumulation of blood in the veins. In severe cases of shock, the inadequate blood supply to the tissues impairs tissue function so greatly that death results.

Aneurysms

An *aneurysm (an´-ū-rizm)* is a localized dilation of an artery due to a weakness of the artery wall. The most frequent site of aneurysm formation is the abdominal aorta. The greatest danger from an aneurysm is that the affected vessel will rupture. However, aneurysms are also common sites of thrombus formation, and a portion of a thrombus may break off and block a smaller vessel. In addition, aneurysms can obstruct or erode neighboring organs or tissues.

Phlebitis

The inflammation of a vein is called *phlebitis (fle-bī´-tis).* Phlebitis can result from a number of conditions. One type of phlebitis is caused by bacteria that invade a vein, perhaps from an abscess or where the vein passes through an area of inflammation. In phlebitis, the inflammation of a vein can lead to the formation of a thrombus, with all of its accompanying dangers. This condition is called thrombophlebitis.

Varicose Veins

Varicose veins are veins that are dilated, lengthened, and tortuous. Varicose veins are often brought about by a congenital and inherited weakness of the walls and valves of the veins. Such veins are unable to carry blood toward the heart efficiently, thus allowing the blood to accumulate in the lower limbs when the body is in an upright position. Other varicosities may be caused by pressure on veins, such as might occur during pregnancy, in obesity, or from an abdominal tumor. In severe cases, varicose veins can interfere with the return of blood to such an extent that muscle cramps and edema occur. Hemorrhage, phlebitis, and thrombosis are also possible complications.

Effects of Aging

One of the most common age-related structural changes in blood vessels, especially in the larger arteries, is a reduction in the elasticity of their walls. The diminished elasticity affects the capacity of arteries to resist stretch and to recoil after being stretched. Therefore the diminished elasticity may affect blood pressure. The reduced elasticity is due primarily to a progressive replacement of elastic fibers

continued on next page

CONDITIONS OF CLINICAL SIGNIFICANCE

in the walls of arteries with collagenous fibers. In addition, there is a tendency for calcium to bind to the elastic fibers in older persons, and this also reduces the elasticity of arterial walls. In a similar manner, the walls of veins may thicken with age because of an increase in connective tissue and calcium deposits. However, because the blood pressure is so low in veins, this is not thought to affect cardiovascular functioning significantly.

Cardiovascular problems cause more than half the deaths of the population over age 65. However, not all of these cardiovascular problems are necessarily due to aging alone. Evidence suggests that in many nonindustrialized societies there is little cardiovascular disease. By comparison, approximately 20% of the population of industrialized societies suffer from hypertension by the age of 65. Although age and genetic factors probably contribute to the changes in the walls of the blood vessels that lead to myocardial infarction or stroke, many investigators believe that in advanced societies environmental factors, especially diet and exercise, play a more significant role than age or genetic inheritance.

Various studies have concluded that (1) a substantial lowering of plasma cholesterol would greatly reduce the number of heart attacks; (2) maintaining diastolic blood pressure of under 75 mm Hg would result in a threefold reduction in the number of heart attacks; and (3) regular moderate exercise would reduce the incidence of fatal heart attacks by approximately half. In addition, nonsmokers have half the incidence of heart attacks as smokers. These studies suggest that aging may not be the major cause of cardiovascular disorders.

Also during exercise, the activity of the sympathetic nerves to the heart increases and the activity of the parasympathetic nerves decreases. In addition, venous return is enhanced by the increased pumping effects of the contracting skeletal muscles and by the constriction of the veins due to sympathetic stimulation. As a consequence of these activities, both the heart rate and the stroke volume increase, resulting in an increase in cardiac output. During most forms of exercise, the increase in cardiac output is somewhat greater than the decrease in total peripheral resistance, and the mean arterial pressure rises (recall that mean arterial pressure = cardiac output × total peripheral resistance).

It is believed that at the onset of exercise, or even in anticipation of exercise, discrete exercise centers in the brain induce many of the neurally mediated cardiac and

◆ **FIGURE 20.18 Diagrammatic representation of the blood vessels of the skin**

Note the direct arteriovenous anastomoses.

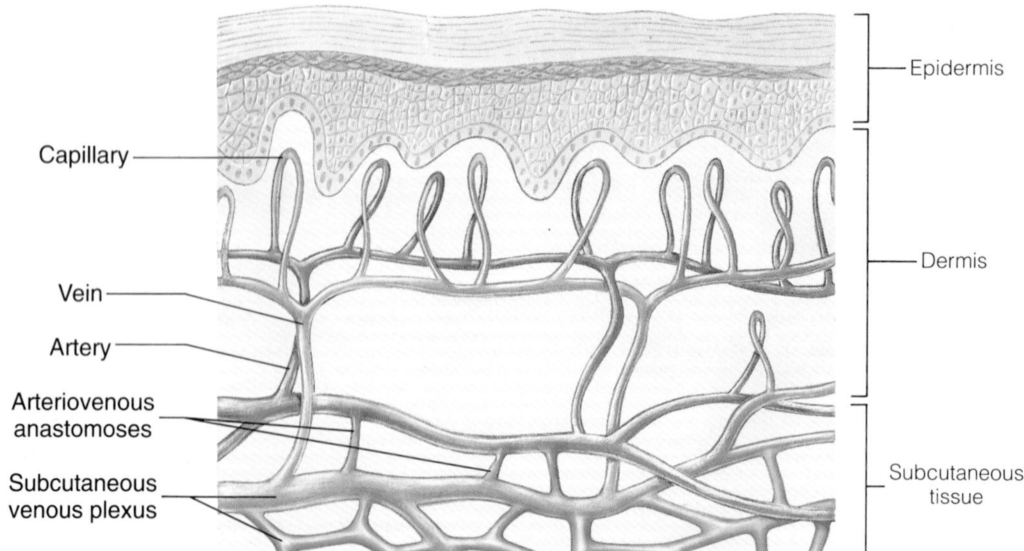

ASPECTS OF EXERCISE PHYSIOLOGY

The Ups and Downs of Hypertension and Exercise

When blood pressure is up, one way to bring it down is to increase the level of physical activity. An impressive epidemiologic study, which followed a group of 14,998 male graduates from Harvard for 16 to 50 years, found that participation in collegiate sports, climbing as many as 50 stairs a day, walking five blocks, or light sports activities were not protective against the development of primary hypertension. Participation in vigorous sports such as running, swimming, handball, tennis, and cross-country skiing, on the other hand, was protective against the development of hypertension, even if other risk factors were present. Other studies have supported the finding that participation in aerobic activities is protective against the development of hypertension.

A logical question to ask is whether exercise can be used as a therapy to reduce hypertension once it has already developed. Antihypertensive medication is available to lower blood pressure in severely hypertensive patients, but sometimes undesirable side effects occur. The side effects of diuretics include electrolyte imbalances, glucose intolerance, and increased blood-cholesterol levels. Side effects of drugs that manipulate total peripheral resistance, such as beta-blocker drugs, include increased blood triglyceride levels, weight gain, sexual dysfunction, and depression.

Patients with mild hypertension, arbitrarily defined as a diastolic blood pressure between 90 and 100 mm Hg and a systolic pressure of 140 mm Hg, pose a dilemma for physicians. The risks of taking the drugs may outweigh the benefits gained from lowering the blood pressure. Because of the drug therapy's possible side effects, nondrug treatment of mild hypertension may be most beneficial. The most common nondrug therapies are weight reduction, salt restriction, and exercise. Although losing weight will almost always reduce blood pressure, research has shown that weight reduction programs usually result in the loss of only 12 pounds, and the overall long-term success in keeping the weight off is only about 20%. Salt restriction is beneficial for most hypertensives, but adherence to a low-salt diet is difficult for many people because fast foods and foods prepared in restaurants usually contain high amounts of salt. Some recent studies employing exercise as a therapeutic tool have shown that blood pressure in cases of mild to moderate hypertension has been decreased whether or not salt was restricted or weight was lowered. The preponderance of evidence in the literature suggests that moderate aerobic exercise performed three times per week for 15 to 60 minutes is a beneficial therapy in mild to moderate hypertension. It is wise, therefore, to include a regular aerobic exercise program in conjunction with weight loss and salt reduction to optimally reduce high blood pressure without drug intervention.

vascular changes characteristic of exercise. Once exercise is underway, afferent inputs to the cardiac and vasomotor centers from receptors in active skeletal muscles lead to modifications of these changes as necessary. Moreover, once exercise begins, local autoregulatory responses produce a vascular dilation in active skeletal muscles that greatly increases blood flow to the muscles.

Anatomy of the Vascular System

The blood of the circulatory system is carried throughout the body in a complex network of vessels. Only the major blood vessels are named in this chapter, but you should bear in mind that there are many other smaller vessels that are not mentioned. The arteries branch into many small arterioles, which, in turn, branch into numerous microscopic capillaries where the exchange between tissues and blood occurs. From the capillaries, the blood enters tiny venules that join together to form the larger veins.

The vessels of the circulatory system can be divided into two separate circuits, each of which leaves and returns to the heart. The **pulmonary circuit** *(pull´-muh-na-ree)* carries blood from the right side of the heart to the lungs and back to the left side of the heart. The **systemic circuit** *(sis-tem´-ik)* carries blood that has just left the pulmonary circuit to the rest of the body and back to the right side of the heart.

◆ FIGURE 20.19 Pulmonary circulation showing the pulmonary trunk dividing into right and left pulmonary arteries, which in turn divide into lobar arteries
Blood from the lungs returns to the left atrium through the pulmonary veins. The arrows indicate the direction of blood flow.

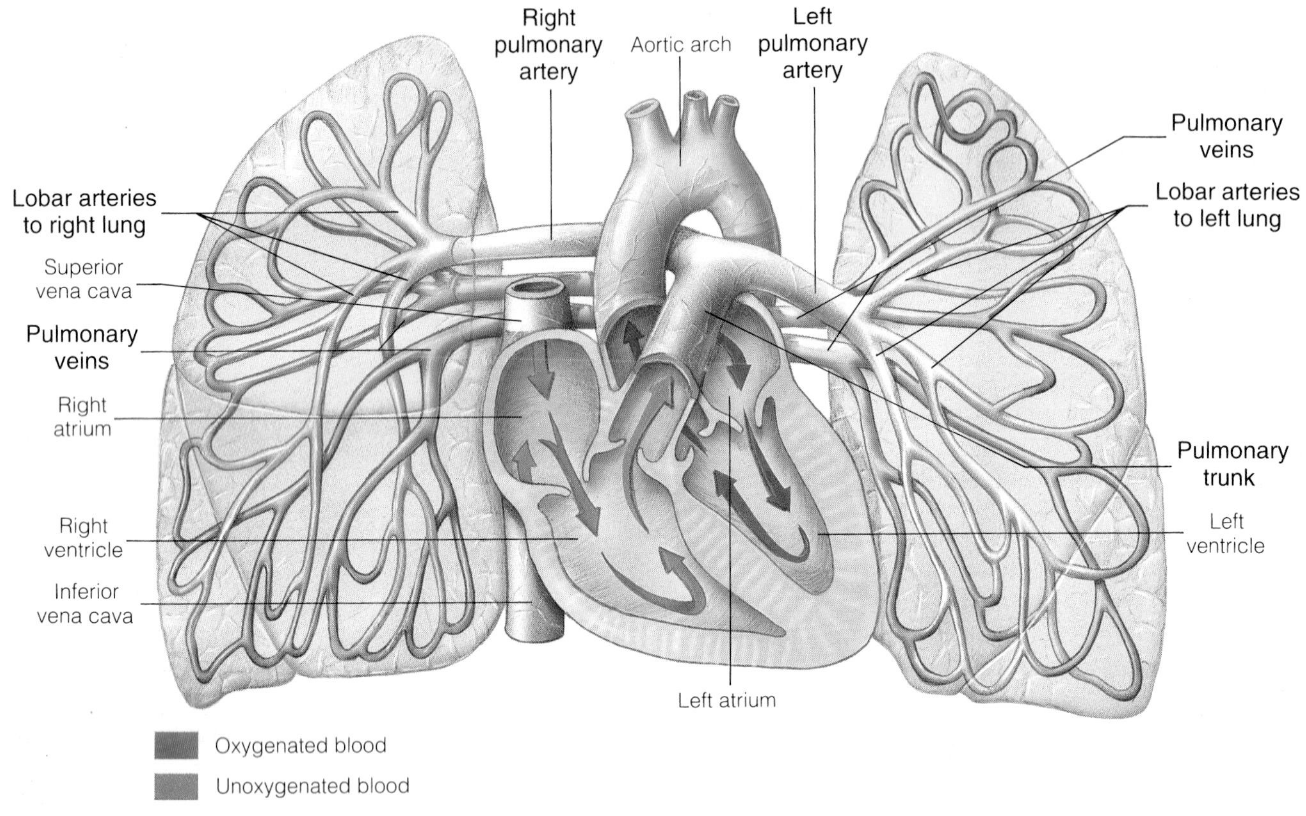

Pulmonary Circuit

Blood enters the pulmonary circuit from the right ventricle of the heart through the **pulmonary trunk** (Figure 20.19). The short pulmonary trunk divides into **left** and **right pulmonary arteries,** which pass directly to the left and right lungs. As they enter the lungs, the pulmonary arteries divide into several **lobar arteries,** one to each lobe of the lung. The right pulmonary artery divides into three lobar arteries; the left divides into two. (The right lung has three lobes and the left lung has two.) The lobar arteries divide into numerous orders of smaller arteries and arterioles and, eventually, into plexuses of **pulmonary capillaries** located in the walls of the tiny air sacs (alveoli) of the lungs. It is through these plexuses of pulmonary capillaries that gases are exchanged between the blood and the air.

From the pulmonary capillaries, the blood is collected into venules and then into progressively larger veins that lead into left and right **pulmonary veins,** generally two from each lung. The pulmonary veins empty into the left atrium. The pulmonary arteries carry blood that has a high carbon dioxide content and is low in oxygen. The blood in the pulmonary veins, however, has a high oxygen content and is low in carbon dioxide. This pattern is the exact opposite of that for the systemic arteries and veins. In other words, the vessels are named according to the direction of the blood flow within them, not the condition of the blood they carry. Thus, any vessel that carries blood away from the heart is an **artery,** and any vessel that returns blood to the heart is a **vein.**

The vessels of the pulmonary circuit supply only the alveoli. The metabolic needs of the remainder of the lung tissue are provided for by small *bronchial vessels* of the systemic circuit.

Systemic Circuit

Blood that is not in the pulmonary circuit is in the systemic circuit. The vessels of the systemic circuit transport blood to all tissues and organs of the body except the alveoli of the lungs.

◆ **FIGURE 20.20 Major systemic arteries of the body**

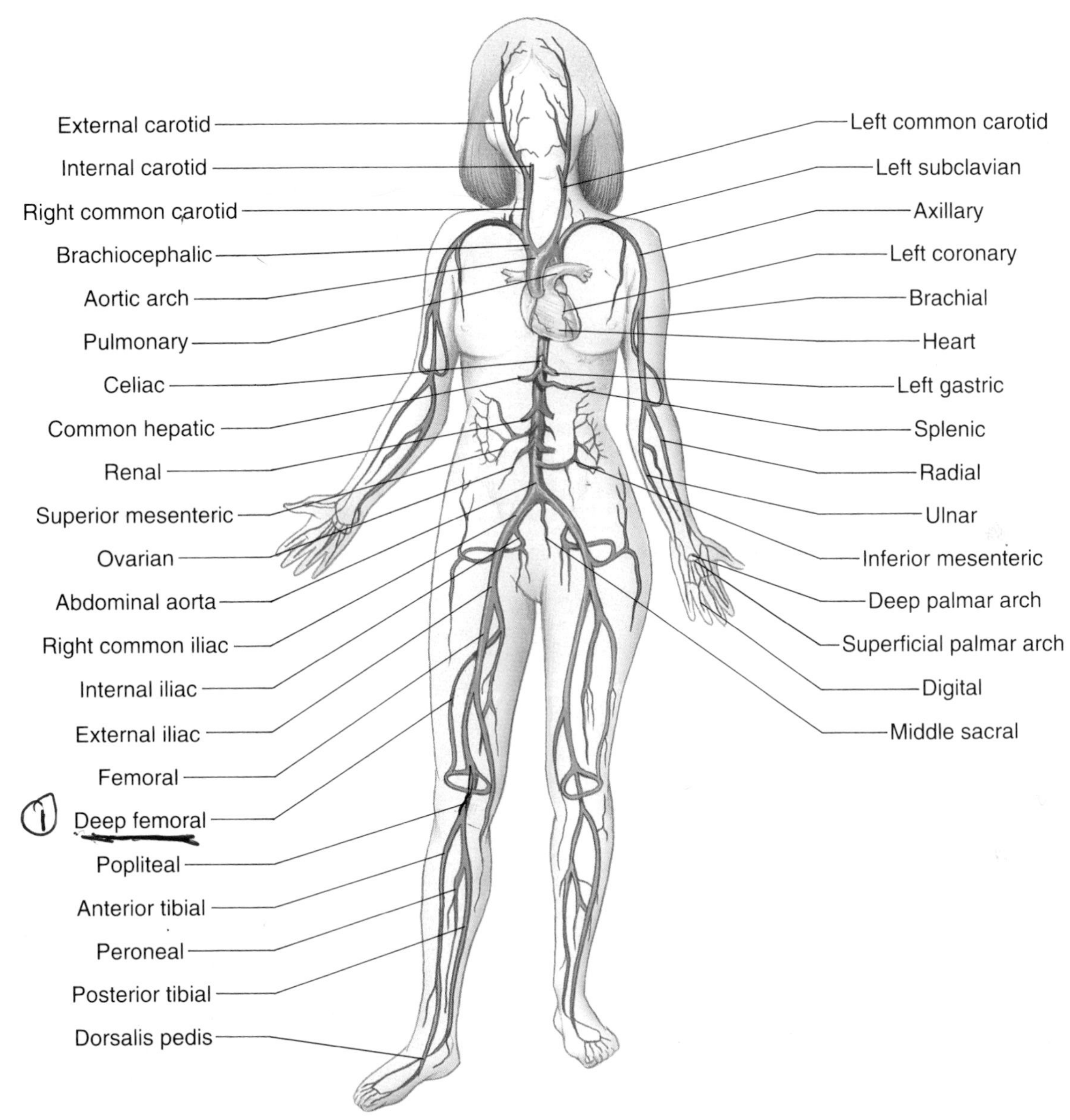

Systemic Arteries

Blood from the left ventricle enters the systemic circuit through the **aorta,** from which all the arteries of the systemic circuit branch (Figure 20.20). For descriptive purposes, it is convenient to divide the aorta into the **ascending aorta,** the **aortic arch,** and the **descending aorta,** which has **thoracic** and **abdominal** portions. The branches of these various portions are summarized in Table 20.2, page 662.

Branches of the Ascending Aorta. The ascending aorta is a short vessel that passes in front of the right pulmonary artery. Its only branches are the **right** and **left coronary arteries,** which arise from sinuses behind the cusps of the aortic valve and supply the heart muscle (Figures 19.7 and 19.8, page 601).

Branches of the Aortic Arch. As the aorta leaves the pericardial sac, it arches dorsally and to the left, forming the aortic arch. Three branches arise from the aortic arch: the **brachiocephalic artery** *(brăk-ē-ō-sĕ-fal´-ik),* the **left common carotid artery** *(kah-rot´-id),* and the **left subclavian artery** (Figure 20.20). These arteries furnish all the blood to the head, neck, and upper limbs.

◆ **FIGURE 20.21 Arteries of the right side of the head and neck**

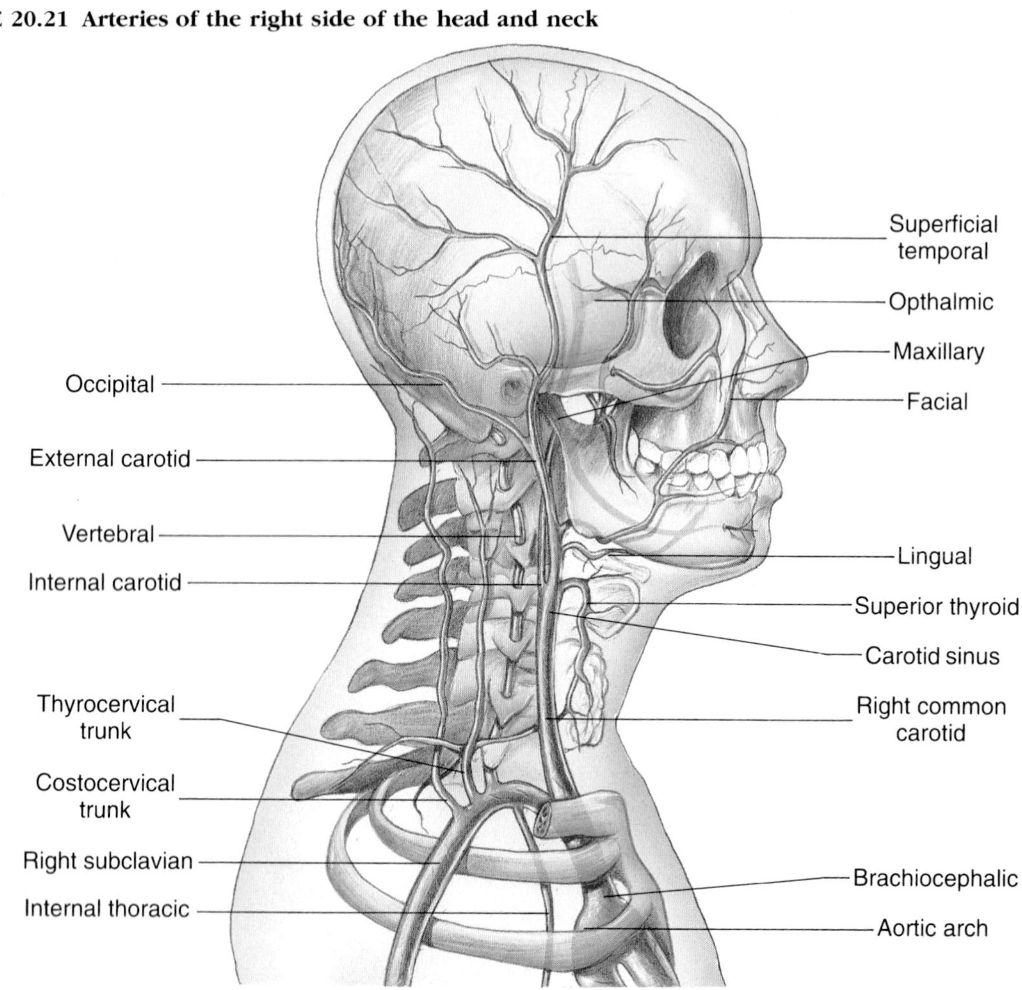

Arteries of the Head and Neck. Shortly after arising from the arch of the aorta, the brachiocephalic artery divides into a **right subclavian artery** and a **right common carotid artery** (Figure 20.21). The right common carotid artery and the **left common carotid artery** (which arises directly from the arch of the aorta) supply most of the blood to the head and neck. The common carotid arteries course upward alongside the trachea a short distance before each bifurcates (divides) into an **internal carotid artery** and an **external carotid artery.** At this point of bifurcation, the vessels enlarge slightly, forming the **carotid sinus.** The sinus contains baroreceptors that assist in regulating blood pressure. Near this same region is a small oval **carotid body** that contains chemoreceptors sensitive to changing levels of oxygen, carbon dioxide, and pH in the blood that travels toward the brain.

The **external carotid artery,** which supplies most of the head and neck except for the brain, lies more superficially than the internal carotid artery. It passes through the parotid salivary gland and terminates as the **maxillary** and **superficial temporal arteries** (Figure 20.21). Just before terminating, the external carotid artery gives off the **occipital artery** to the posterior region of the scalp. Through these branches, the external carotid artery supplies the muscles of mastication; the mucous membranes of the nose, the pharynx, and the palate; the teeth of the upper and lower jaws; some neck muscles; and the entire scalp region. While still in the neck, it sends branches to the thyroid gland and larynx **(superior thyroid artery),** the tongue **(lingual artery),** and the skin and the muscles of the face **(facial artery).**

The **internal carotid artery** enters the cranial cavity through the carotid canal of the temporal bone. It travels forward in the canal and soon leaves the bone to pass forward and upward along the side of the sella turcica. The internal carotid artery gives off an **ophthalmic artery** *(of-thal´-mik),* which follows the lower surface of the optic nerve and enters the orbit.

◆ **FIGURE 20.22 Arteries of the base of the brain, forming the cerebral arterial circle (circle of Willis) around the pituitary gland**

To provide an unobstructed view, part of the right temporal lobe and the right cerebellar hemisphere have been removed. Roman numerals indicate cranial nerves.

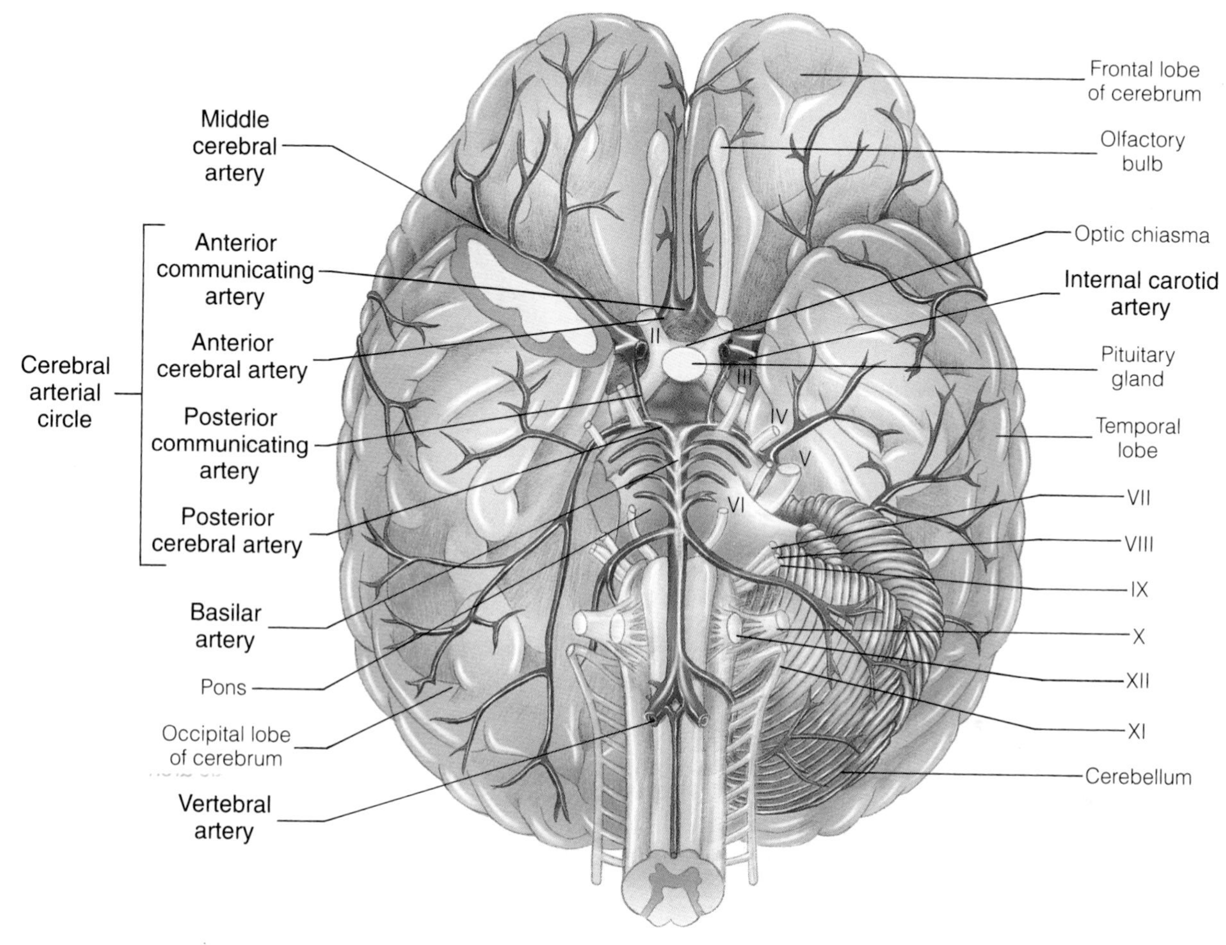

The internal carotid artery supplies the brain through its terminal branches: the **anterior cerebral artery** and the **middle cerebral artery.** A small **anterior communicating artery** connects the right and left anterior cerebral arteries. These vessels also help form the **cerebral arterial circle (circle of Willis)** (Figure 20.22), which underlies the hypothalamus and surrounds the infundibulum of the pituitary gland, the optic chiasma, the tuber cinereum, and the mammillary bodies.

The **vertebral artery,** which is the first branch of the subclavian artery, provides another major blood supply to the brain. The vertebral artery reaches the cranial cavity by passing through the transverse foramina of the cervical vertebrae and the foramen magnum. The right and left vertebral arteries join on the ventral surface of the pons of the brain, forming the **basilar artery** (Figure 20.22). The basilar artery continues forward and terminates as **right** and **left posterior cerebral arteries,** which supply the posterior regions of the cerebral hemispheres of the brain. The basilar artery also gives off branches that supply the pons and the cerebellum. The **posterior communicating arteries** from the internal carotid arteries join with the posterior cerebral arteries to complete the cerebral arterial circle around the infundibulum. Thus, all of the arterial blood to the brain is delivered through the internal carotid and vertebral arteries.

Arteries of the Upper Limbs. The upper limbs are supplied by the **subclavian arteries.** As noted earlier, the right subclavian artery is a branch of the brachiocephalic artery, and the left subclavian arises directly from the aortic arch. The first part of each subclavian artery

◆ **FIGURE 20.23 Arteries of the upper limb**

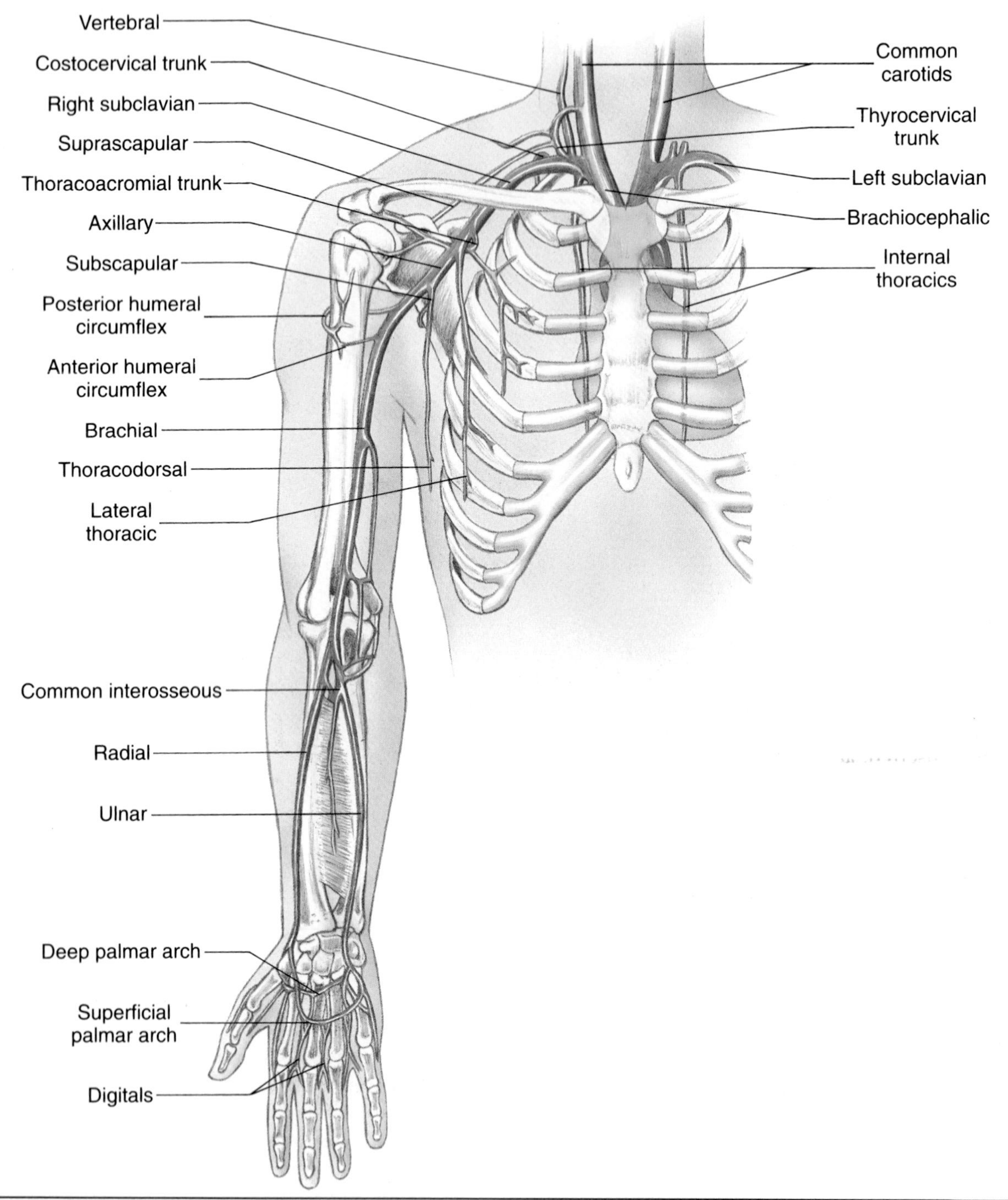

gives off branches that do not enter the upper limbs (Figure 20.23). Just lateral to the vertebral artery, the **thyrocervical trunk** and **costocervical trunk** send branches to the lower neck, the dorsal scapular, and the intercostal regions. The **internal thoracic (internal mammary) artery** arises from the subclavian artery and passes downward beneath the costal cartilage. It supplies branches to the thoracic and intercostal muscles, the mediastinum, and the diaphragm.

After passing beneath the clavicle, the subclavian artery becomes the **axillary artery.** The axillary artery gives off branches to the lateral wall of the thorax **(thoracoacromial trunk, lateral thoracic artery, subscapular artery,** and **thoracodorsal artery)** and to

CLINICAL CORRELATION

Traumatic Shock

Case Report

THE PATIENT: A 26-year-old woman.

REASON FOR ADMISSION: Multiple trauma.

CLINICAL COURSE: The patient had been well before she was involved in an automobile accident. She was transported to the emergency room by helicopter, dazed but conscious. Her arterial blood pressure was 55/30 mm Hg (normal: 90–140/60–90 mm Hg), and her heart rate was 96/min (normal: 60–100/min). No head injury was indicated, and the heart and lungs were normal. However, an X-ray film showed widening of the mediastinum (which suggests injury to the aorta) and three broken ribs. The abdomen was swollen and tender (which suggests intra-abdominal hemorrhage). The right tibia appeared to be broken, and the right patella was severely lacerated. The hematocrit was 34% (normal: 37–48%), and the platelet count was 20,000/mm^3 (normal: 250,000–400,000/mm^3). An intravenous line was inserted, and infusion of 10% glucose in saline was begun. Packed red cells, platelets, and fresh-frozen plasma were given. The urinary bladder was catheterized (for monitoring the production of urine).

Because of the evidence of internal bleeding, the patient was taken to the operating room, and an exploratory laparotomy (opening of the abdomen by way of an incision through the abdominal wall) was performed. One and one-half liters of blood and clots were removed from the peritoneal cavity, and several bleeding arteries were clamped and tied. The patient was given additional packed red cells, albumin, fresh-frozen plasma, and lactated Ringer's solution during the operation. Her arterial blood pressure stabilized in the range of 90/50 mm Hg. To assess the cause of the widened mediastinum, thoracic arteriography was done (fluoroscopy of contrast medium injected at the root of the aorta). A traumatic rupture was found just distal to the origin of the left subclavian artery. The vessel was repaired through a left thoracotomy (opening of the chest by way of an incision through the thoracic wall), using a Dacron graft. Additional cells and fluids were given, and the patient's arterial blood pressure increased to 115/70 mm Hg.

OUTCOME: The patient was fortunate to receive prompt and expert care. After the bleeding was stopped and the arterial blood pressure had stabilized, her external wounds were treated and her broken bones were set. Recovery was uneventful.

the region around the proximal end of the humerus **(anterior** and **posterior humeral circumflex arteries).** It then continues into the arm as the **brachial artery,** which travels down the medial side of the humerus, supplying the muscles of the arm.

The brachial artery crosses the anterior surface of the elbow and divides into **radial** and **ulnar arteries.** Both vessels travel down the anterior surface of the forearm—the radial artery alongside the radius, the ulnar artery along the ulna. The ulnar artery gives off a short **common interosseous artery,** which divides into anterior and posterior branches that supply the deep flexor and extensor muscles of the forearm. The radial artery is easily palpated on the lateral side of the wrist, where it is used to take the pulse. The radial and ulnar arteries join in the hand through interconnecting branches of **superficial** and **deep palmar arches. Digital arteries** from the palmar arches supply the fingers.

Branches of the Thoracic Aorta. The arteries that supply the thorax arise from the first part of the descending aorta, which lies to the left of the vertebral column (Figure 20.24). All the branches from this region are small. Most prominent are the **posterior intercostal arteries,** which travel between the ribs and anastomose with the **anterior intercostal arteries,** which arise from each internal thoracic artery. In addition, **pericardial, mediastinal, bronchial, esophageal, subcostal** (beneath the twelfth ribs), and **superior phrenic arteries** (which supply the diaphragm) also arise from the thoracic aorta.

Branches of the Abdominal Aorta. The descending aorta passes through the *aortic hiatus* of the diaphragm and enters the abdominopelvic cavity (Figure 20.25). It travels down the ventral surface of the vertebral column, supplying the posterior abdominal walls through

◆ **FIGURE 20.24 Branches of the thoracic aorta**
The anterior portion of the rib cage has been removed on the left side.

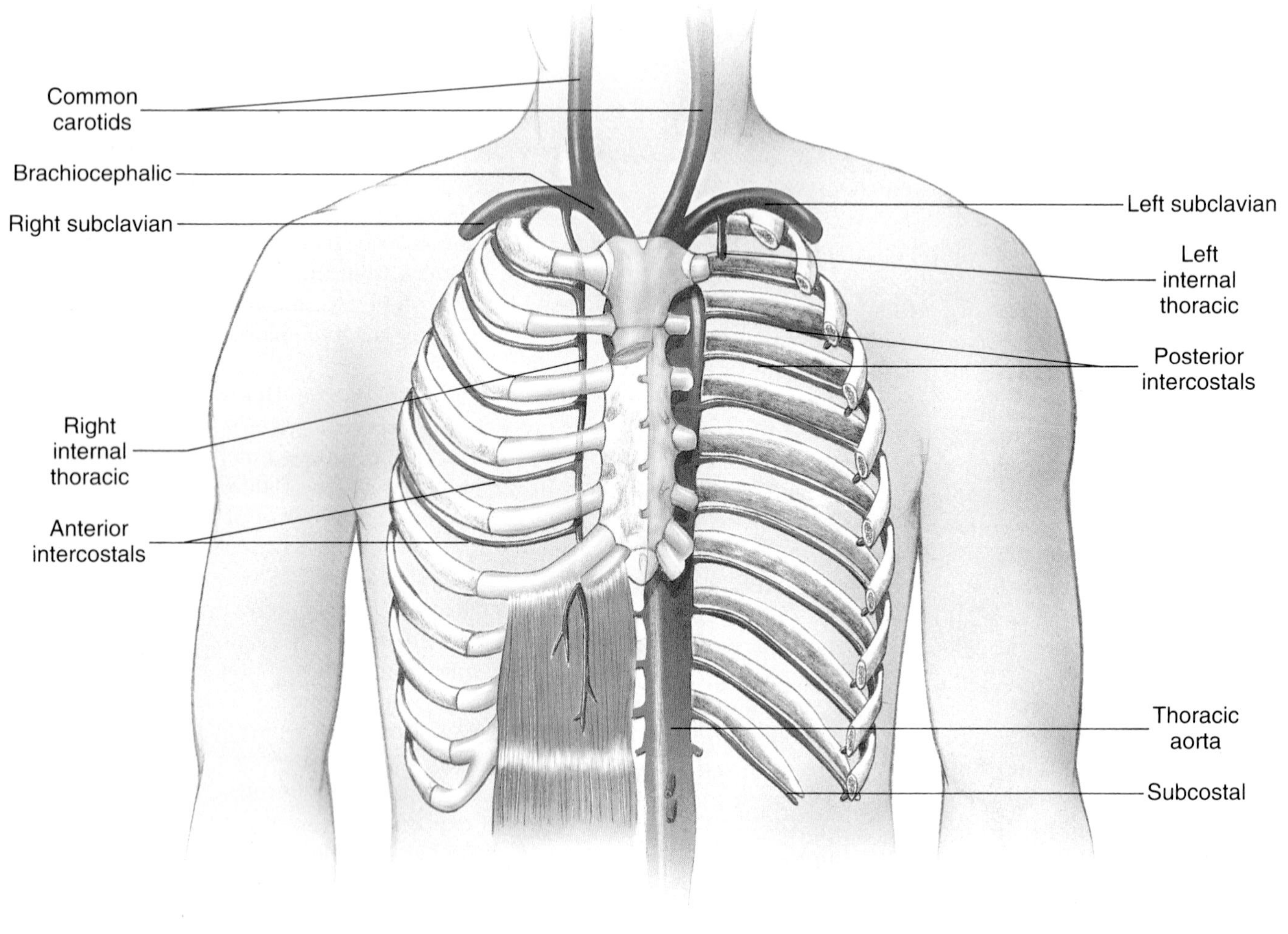

four pairs of **lumbar arteries.** The abdominal aorta terminates in front of the fourth lumbar vertebra by dividing into **right** and **left common iliac arteries** and a small **middle sacral artery.**

Immediately after entering the abdominopelvic cavity, the aorta gives off a pair of **inferior phrenic arteries** to the underside of the diaphragm. A single **celiac artery** arises from the aorta just below the inferior phrenic arteries, at about the level of the twelfth thoracic vertebra. The celiac artery is very short and almost immediately divides into *left gastric, splenic,* and *common hepatic arteries* (Figure 20.26). The **left gastric artery** follows the lesser curvature of the stomach, supplying the lower part of the esophagus and part of the stomach. The **splenic artery,** which supplies the spleen, also gives off small branches to the pancreas and stomach, as well as the large **left gastroepiploic artery,** which supplies the greater curvature of the stomach. As it travels to the liver, the **common hepatic artery** sends branches to the stomach, duodenum, and pancreas through **gastroduodenal** and **right gastric** branches. The right gastric artery supplies the lesser curvature of the stomach, and the **right gastroepiploic artery** *(gas″-trō-ep-i-plō′-ik)*—a branch of the gastroduodenal artery—supplies the greater curvature. After giving off these branches, the common hepatic artery divides into **right** and **left hepatic arteries,** which supply the liver and gallbladder.

The single **superior mesenteric artery** *(mes″-en-ter′-ik)* arises from the aorta just below the celiac artery. It travels within the mesentery of the intestines, where it gives off several branches that form anastomosing loops (Figure 20.27). The superior mesenteric artery supplies all of the small intestine, as well as the cecum and the ascending colon and most of the transverse colon of the large intestine.

Branching laterally from the aorta at the level of the superior mesenteric arteries are two **suprarenal arteries,** which supply the adrenal glands (Figure 20.25). These glands are also supplied by branches from the

◆ **FIGURE 20.25 Branches of the abdominal aorta**

Opening for inferior vena cava
Opening for esophagus
Aortic hiatus
Celiac artery
Kidney
Lumbar arteries
Abdominal aorta
Ureter
Middle sacral artery
External iliac artery
Diaphragm
Inferior phrenic artery
Suprarenal artery
Superior mesenteric artery
Renal artery
Left testicular or ovarian artery
Inferior mesenteric artery
Common iliac artery
Internal iliac artery

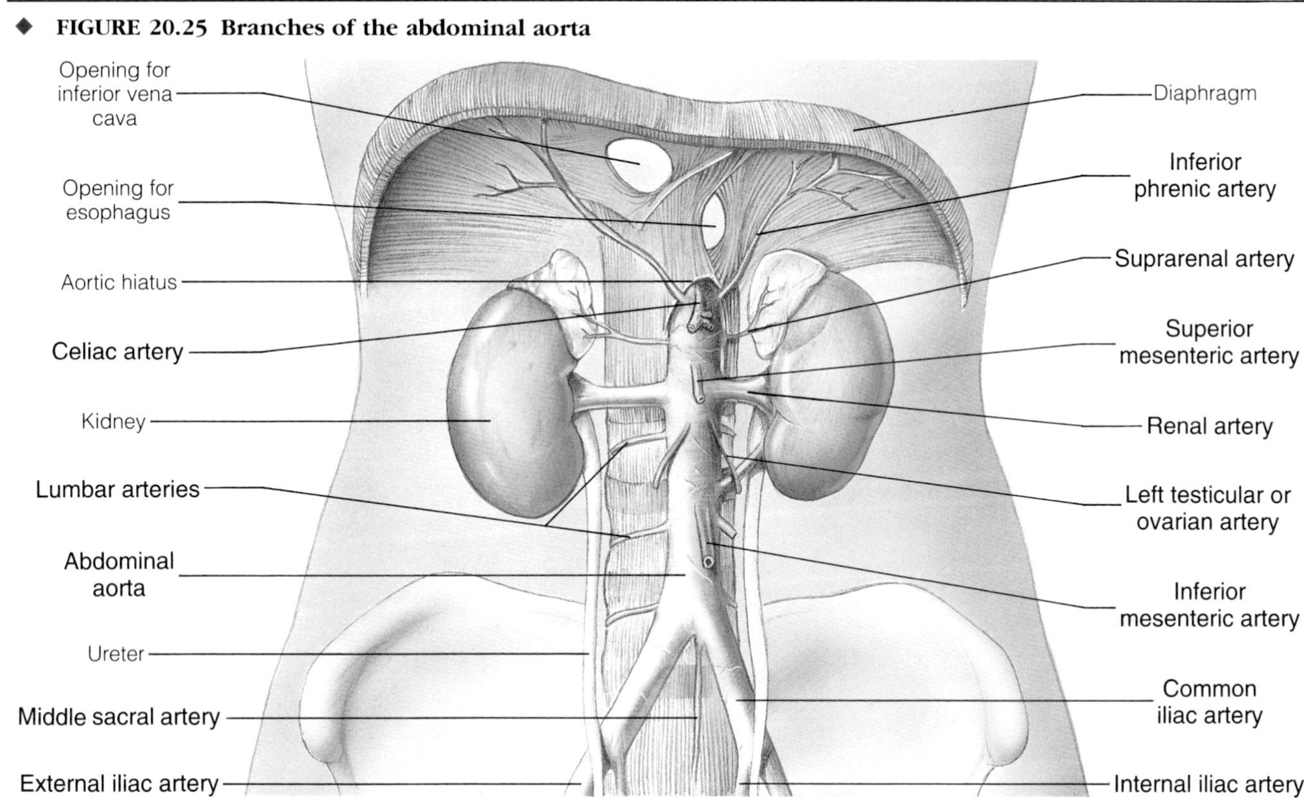

◆ **FIGURE 20.26 Branches of the celiac artery**

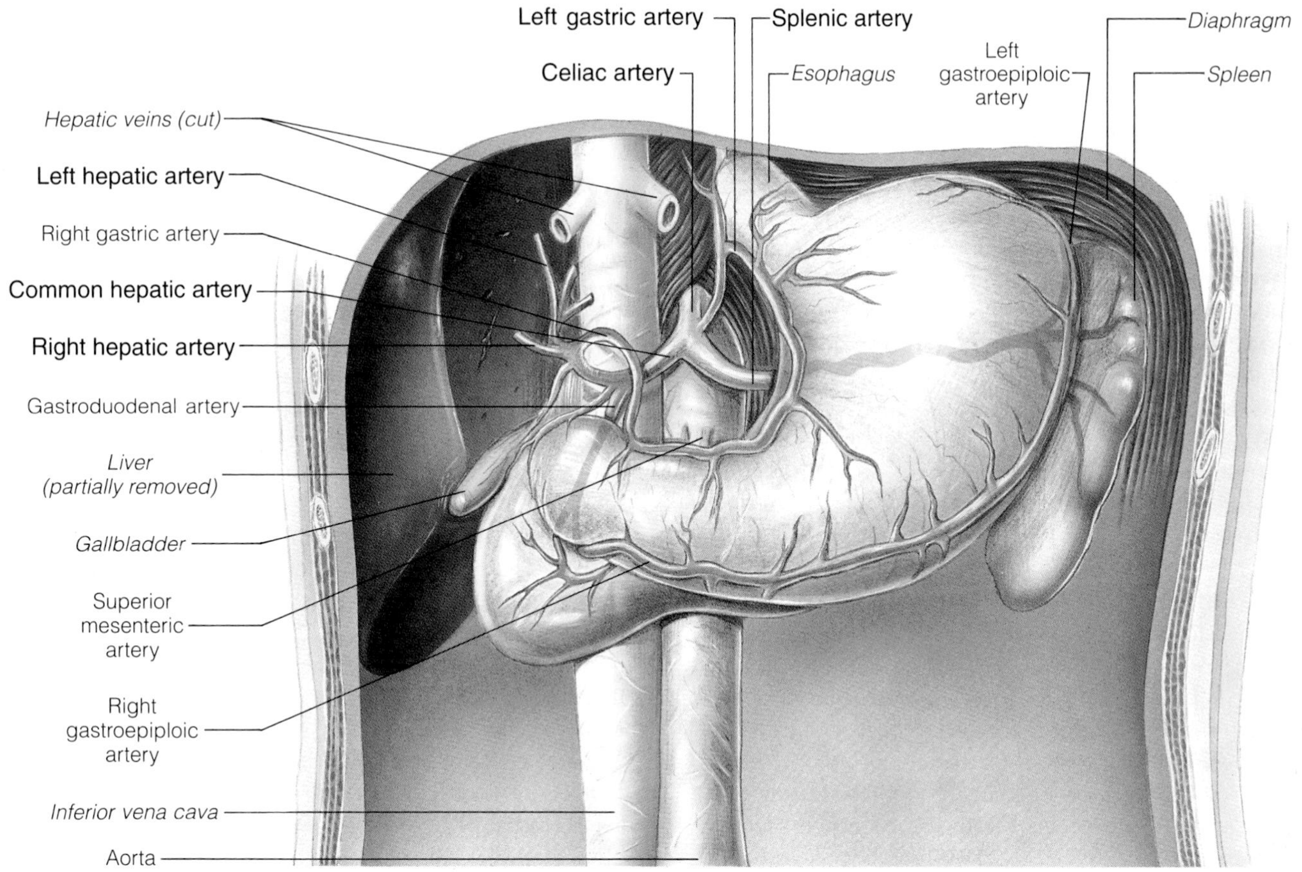

Left gastric artery
Splenic artery
Celiac artery
Esophagus
Left gastroepiploic artery
Diaphragm
Spleen
Hepatic veins (cut)
Left hepatic artery
Right gastric artery
Common hepatic artery
Right hepatic artery
Gastroduodenal artery
Liver (partially removed)
Gallbladder
Superior mesenteric artery
Right gastroepiploic artery
Inferior vena cava
Aorta

◆ **FIGURE 20.27 The superior and inferior mesenteric arteries**
The transverse colon has been lifted up from its normal position.

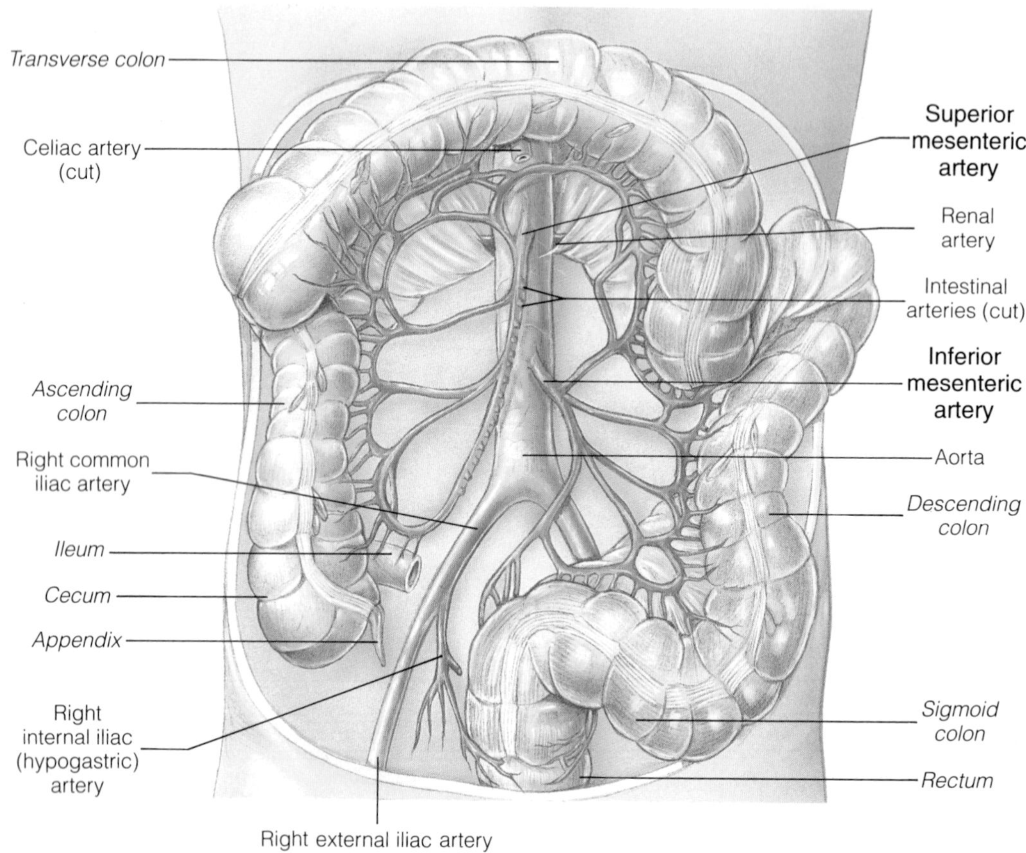

inferior phrenic and renal arteries. The paired **renal arteries** (Figure 20.25), which supply the kidneys, arise from the lateral margins of the aorta just inferior to the superior mesenteric artery.

The arteries that supply the gonads (ovaries and testes) arise from the ventral surface of the aorta a short distance below the renal arteries (Figure 20.25). In the female, the **ovarian arteries** pass downward and laterally into the pelvic cavity to supply the ovaries; in addition, they give off branches to the ureters and uterine tubes. The **testicular (internal spermatic) arteries** of the male are longer than the ovarian arteries, since they pass through the inguinal canal and enter the scrotum, where they supply the testes.

The **inferior mesenteric artery** is the final branch of the abdominal aorta (Figure 20.27). It is a single vessel that arises from the ventral surface of the aorta just above its bifurcation into the **common iliac arteries.** Branches of the inferior mesenteric artery supply part of the transverse colon, the descending colon, the sigmoid colon, and most of the rectum. This artery anastomoses with branches of the superior mesenteric artery.

Arteries of the Pelvic Region. Each of the common iliac arteries divides in front of the sacroiliac articulation into *internal* and *external iliac* arteries (Figure 20.28). The **internal iliac** ***(hypogastric)*** **arteries** enter the pelvic cavity and divide into branches that supply the pelvic viscera (urinary bladder, uterus, vagina, and rectum). In addition, branches from the internal iliac arteries supply the muscles of the gluteal and lumbar regions, the walls of the pelvis, the external genitalia, and the medial region of the thigh.

Arteries of the Lower Limbs. The **external iliac artery** is actually a continuation of the common iliac artery. It travels downward and laterally through the iliac fossa and enters the thigh as it passes beneath the inguinal ligament. While in the pelvic cavity, the external iliac artery supplies branches to the muscles and skin of the lower abdominal wall.

Upon entering the thigh, the external iliac artery becomes the **femoral artery** *(fem´-or-al)* (Figure 20.28). The femoral artery passes along the anterior medial region of the thigh. In the lower thigh, it passes to the

◆ **FIGURE 20.28 Arteries of the pelvis and thigh**

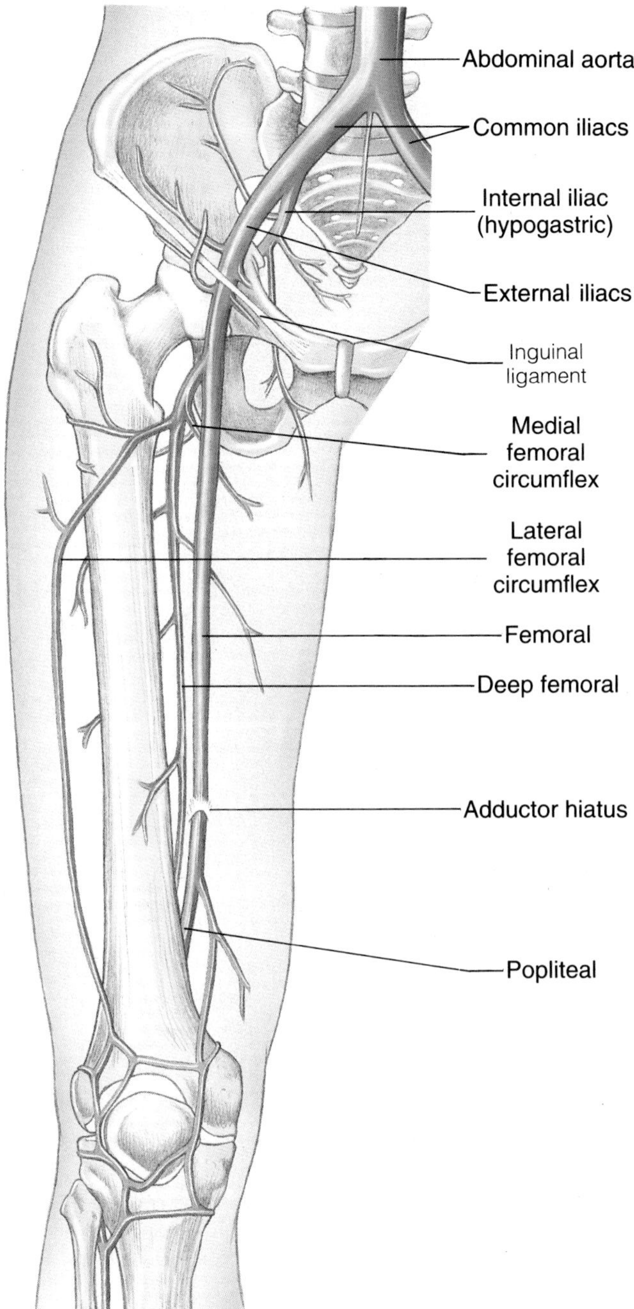

◆ **FIGURE 20.29 Arteries of the leg and foot**
(a) Anterior view. (b) Posterior view.

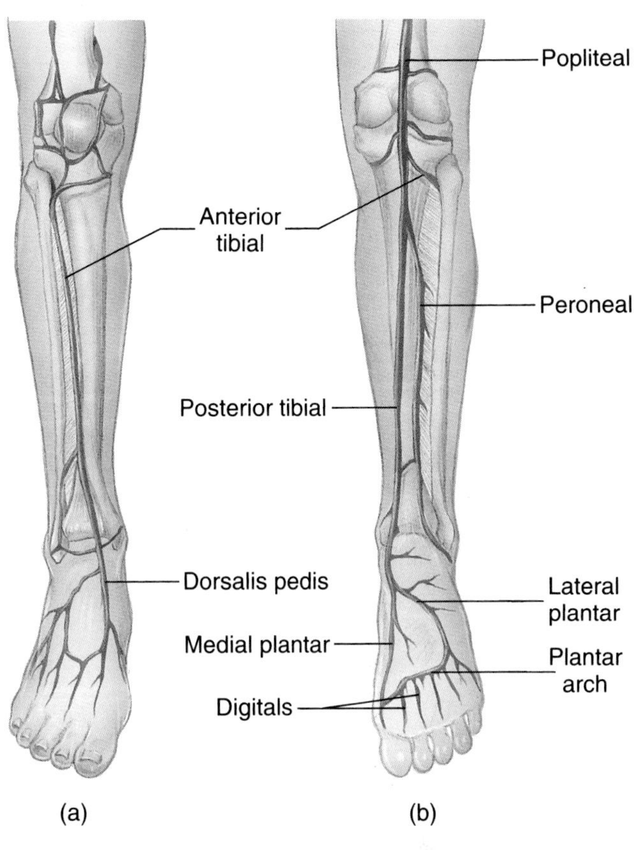

posterior surface of the knee through an opening *(adductor hiatus)* in the tendon of the adductor magnus muscle, and is then known as the **popliteal artery** *(pop-lit´-ē-ul)*. While in the thigh, the femoral artery sends several branches to the skin and muscles of the region. The **lateral** and **medial femoral circumflex arteries** supply the region around the proximal end of the femur, whereas the **deep femoral artery** passes posteriorly to supply muscles of the posterior compartment of the thigh.

The **popliteal artery** is the continuation of the femoral artery. It passes behind the knee (through the popliteal fossa), supplying the muscles and skin of the area, and then divides into an *anterior tibial artery* and a *posterior tibial artery* (Figure 20.29).

The **posterior tibial artery** continues downward behind the tibia, supplying the muscles of the posterior compartment of the leg. Behind the ankle, it divides into **medial** and **lateral plantar arteries,** which supply the sole of the foot and form the **plantar arch. Digital arteries** arise from the plantar arch. Near its origin, the posterior tibial artery gives rise to the **peroneal artery,** which supplies the peroneal muscles in the lateral compartment of the leg.

The **anterior tibial artery** passes through the interosseous membrane that connects the fibula to the tibia. It then travels down the anterior surface of the membrane, supplying the muscles of the anterior compartment of the leg. The anterior tibial artery passes in front of the ankle and ends on the dorsum of the foot as the **dorsalis pedis artery.** Branches of the dorsalis pedis artery supply the dorsum of the foot and anastomose with the plantar arch on the sole of the foot.

◆ **TABLE 20.2 Schematic Summary of the Systemic Arteries That Arise from Various Regions of the Aorta**

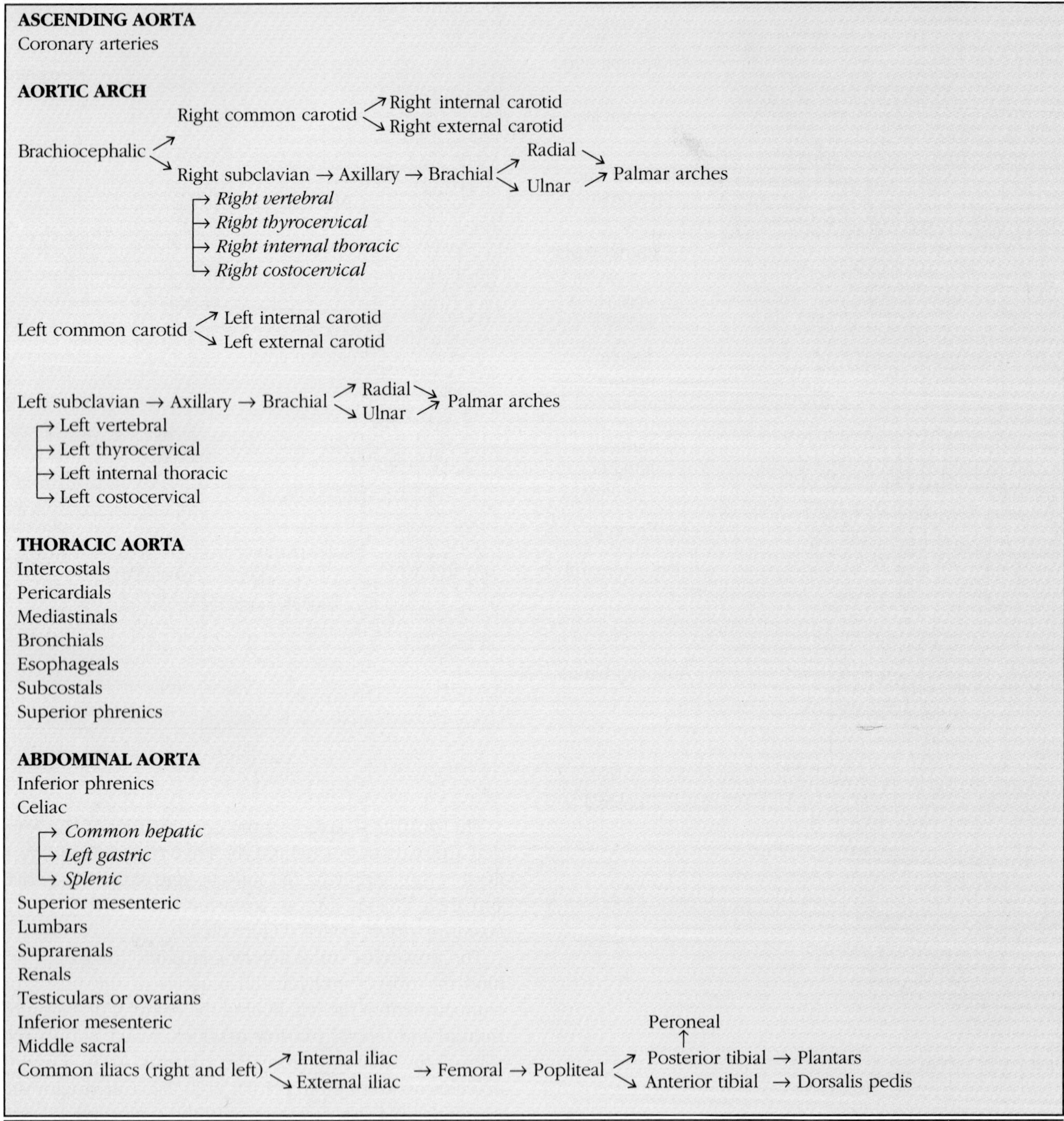

ASCENDING AORTA

Coronary arteries

AORTIC ARCH

Brachiocephalic ↗ Right common carotid ↗ Right internal carotid; ↘ Right external carotid

Brachiocephalic ↘ Right subclavian → Axillary → Brachial ↗ Radial ↘ / ↘ Ulnar ↗ Palmar arches

- → *Right vertebral*
- → *Right thyrocervical*
- → *Right internal thoracic*
- → *Right costocervical*

Left common carotid ↗ Left internal carotid; ↘ Left external carotid

Left subclavian → Axillary → Brachial ↗ Radial ↘ / ↘ Ulnar ↗ Palmar arches

- → Left vertebral
- → Left thyrocervical
- → Left internal thoracic
- → Left costocervical

THORACIC AORTA

Intercostals
Pericardials
Mediastinals
Bronchials
Esophageals
Subcostals
Superior phrenics

ABDOMINAL AORTA

Inferior phrenics
Celiac

- → *Common hepatic*
- → *Left gastric*
- → *Splenic*

Superior mesenteric
Lumbars
Suprarenals
Renals
Testiculars or ovarians
Inferior mesenteric
Middle sacral
Common iliacs (right and left) ↗ Internal iliac; ↘ External iliac → Femoral → Popliteal ↗ Posterior tibial → Plantars (Peroneal ↑); ↘ Anterior tibial → Dorsalis pedis

Systemic Veins

The blood that has been delivered throughout the body by the arteries is collected and returned to the heart through the veins. Since veins tend to be larger and more numerous than arteries, the capacity of the venous system is greater than that of the arterial system. The major veins, summarized in Table 20.3 (page 668), tend to travel alongside the arteries. Although the general pattern of veins (Figure 20.30) is quite similar to the pattern of arteries, there are some important differences, which are explained in this chapter. Systemic veins are classified as *deep veins, superficial veins,* or *venous sinuses.*

Most of the **deep veins** travel alongside an artery and have the same name as the artery. Certain deep veins of the head and vertebral column are exceptions to this pattern, however.

◆ **FIGURE 20.30 Major systemic veins of the body**

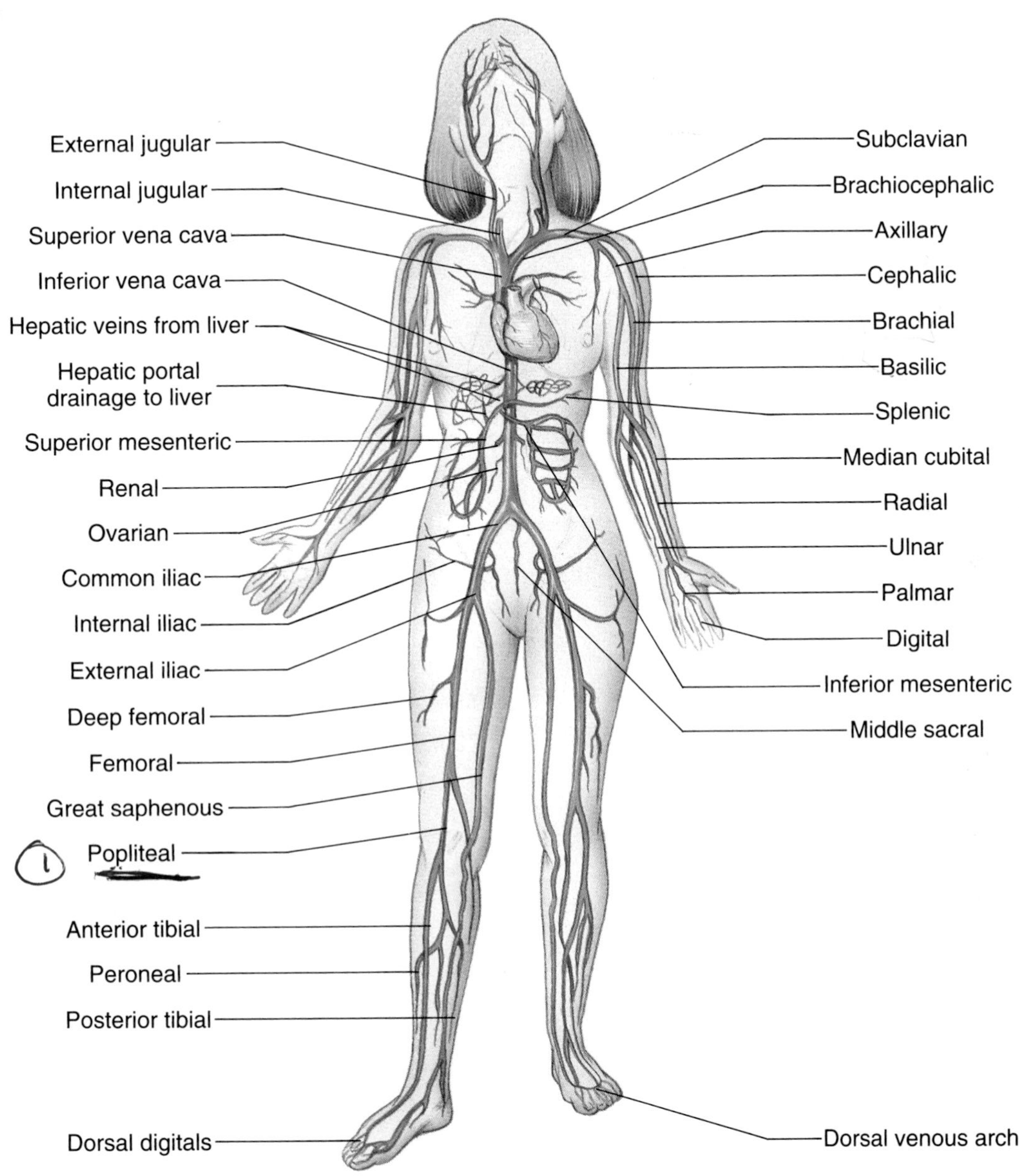

The **superficial veins,** which lie just beneath the skin (in the hypodermis), return blood from the skin and the subcutaneous regions to the deep veins. Since there are no large superficial arteries, the names of the superficial veins do not correspond to the names of arteries.

The **venous sinuses** are not actually vessels, but are spaces that collect blood in certain regions and return it to the veins. The walls of venous sinuses are composed of connective tissue, with no muscle present, and they are lined with endothelium that is continuous with the endothelium of the capillaries and veins. The coronary sinus is a venous sinus that collects blood from the cardiac veins of the heart and returns it to the right atrium. Still larger venous sinuses are located in the dura mater, the outer meningeal covering of the brain (Figure 20.31). Most of the blood from the brain travels through these **dural sinuses** before entering the veins that return blood to the heart.

Veins of the Head and Neck. Most of the venous blood from the head and neck regions returns to the heart through the *internal jugular veins,* the *external jugular veins,* and the *vertebral veins* (Figure 20.32).

The **internal jugular veins** are larger and deeper than the external jugular veins. Each internal jugular

◆ **FIGURE 20.31 Venous sinuses of the brain**

The arrows indicate the direction of blood flow. Note that blood from all of the dural sinuses enters the internal jugular veins.

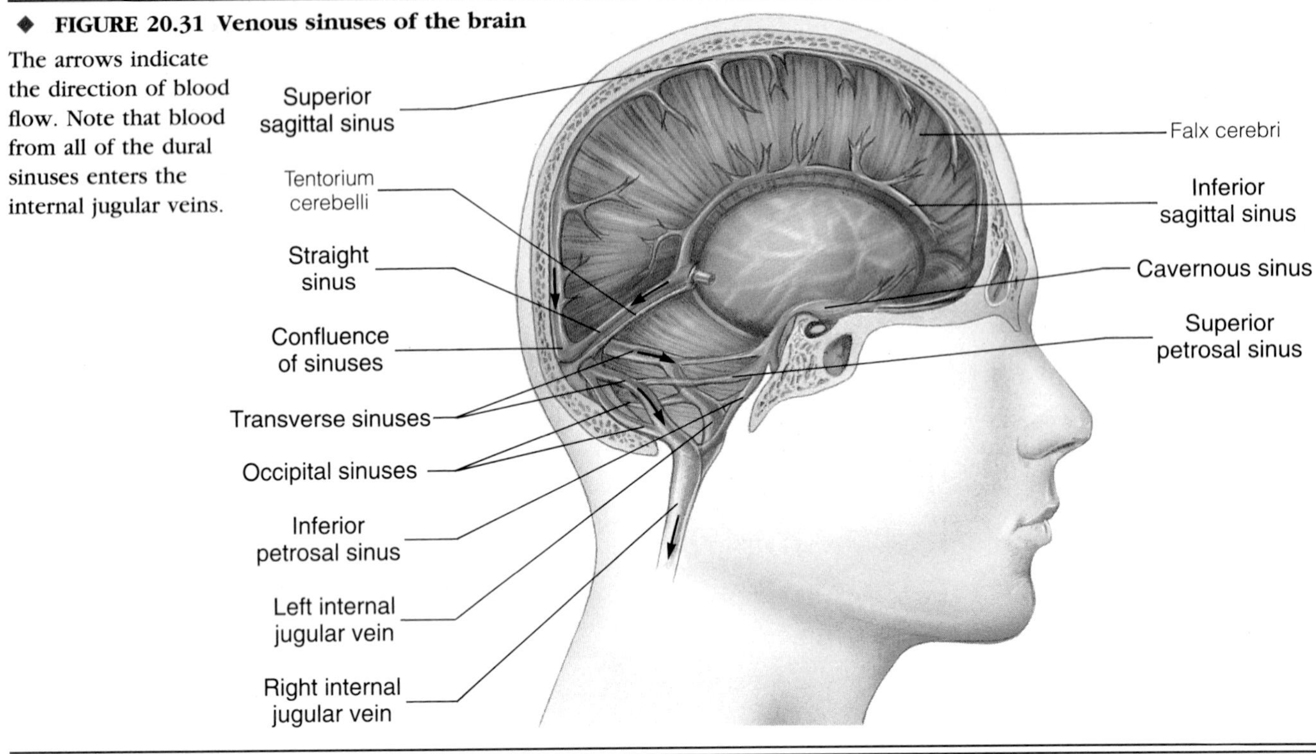

◆ **FIGURE 20.32 Veins of the head and neck**

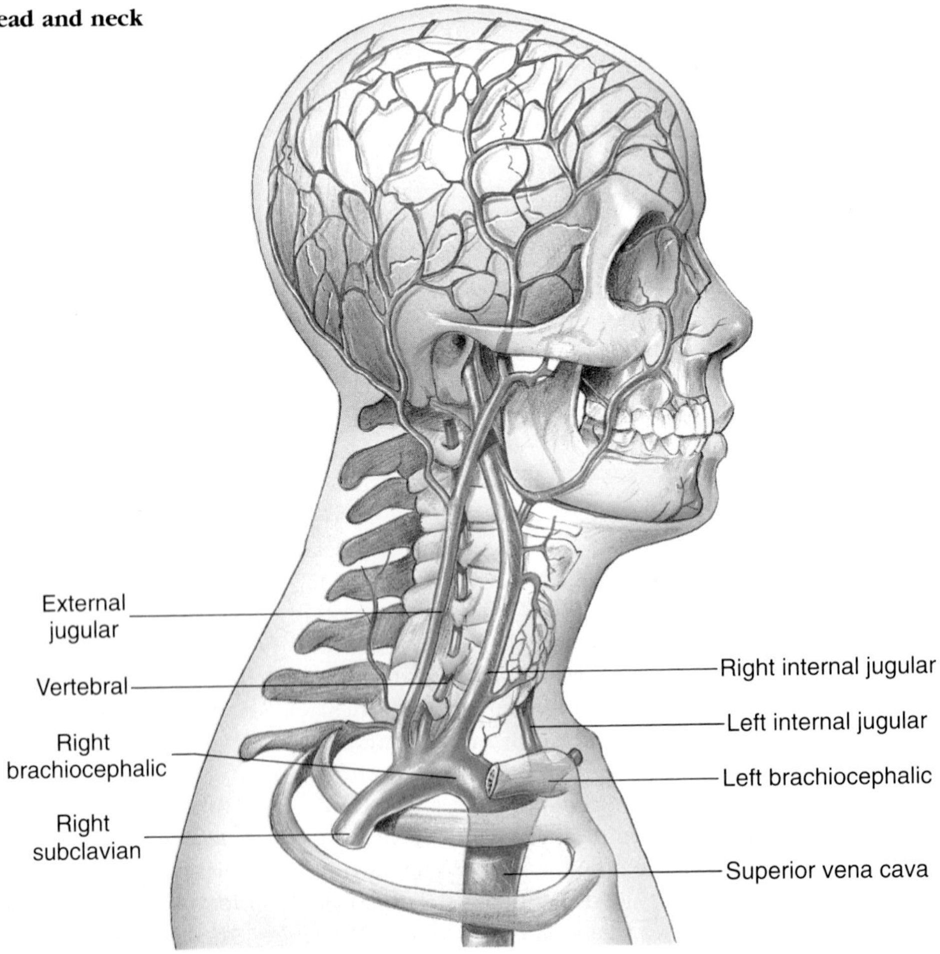

External jugular
Vertebral
Right brachiocephalic
Right subclavian
Right internal jugular
Left internal jugular
Left brachiocephalic
Superior vena cava

vein drains a **transverse sinus** that passes horizontally lateralward from the region of the external occipital protuberance after receiving blood from a **cavernous sinus,** the **superior sagittal sinus,** the **inferior sagittal sinus,** and the **straight sinus** (Figure 20.31). The internal jugular veins therefore serve as the major venous drainage of the brain. Each internal jugular vein leaves the skull through a jugular foramen, which is located in the junction between the petrous portion of the temporal bone and the occipital bone, and travels through the neck alongside a common carotid artery and a vagus nerve. The internal jugular vein joins with a **subclavian vein** to form a **brachiocephalic vein** (Figure 20.32).

The **vertebral veins** drain the posterior regions of the head (Figure 20.32). Each vertebral vein passes through the transverse foramina of the cervical vertebrae, alongside the vertebral artery, and joins a brachiocephalic vein.

The superficial regions of the head and neck, which are supplied by the external carotid arteries, are drained by the **external jugular veins** (Figure 20.32). Each external jugular vein passes superficially down the neck and empties into a **subclavian vein** lateral to the point at which the subclavian and internal jugular veins join to form a brachiocephalic vein. The right and left brachiocephalic veins join to form the **superior vena cava,** which empties into the right atrium of the heart.

◆ **FIGURE 20.33 Superficial veins of the upper limb**

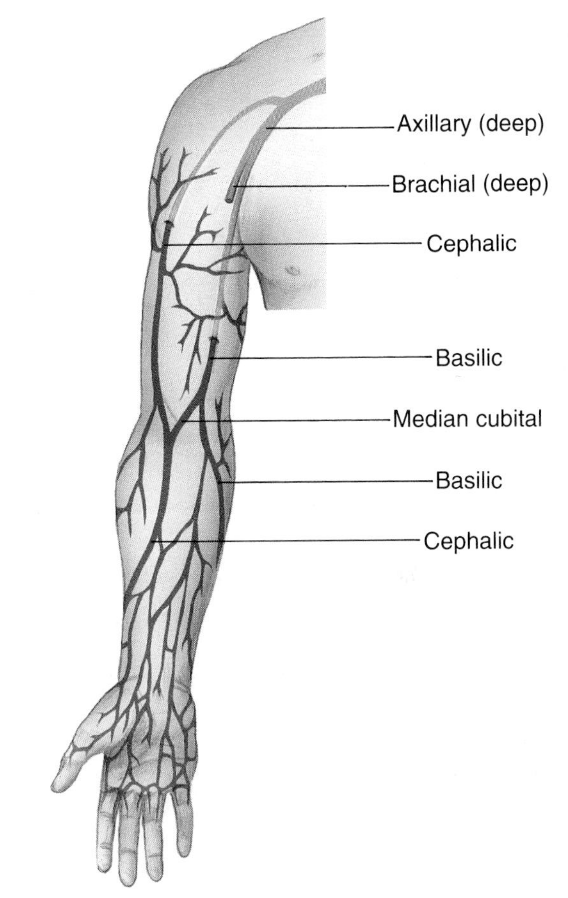

Veins of the Upper Limbs. The deep veins of the upper limbs follow the paths of the arteries and are given the same names: **axillary, brachial, radial,** and **ulnar veins** (Figure 20.30). The superficial veins (Figure 20.33) begin from venous networks that cover the dorsal and palmar surfaces of the hand. The **cephalic vein** arises from the lateral side of the dorsal veins of the hand, crosses to the ventral-lateral side of the forearm, and continues up the lateral side of the arm. At the shoulder, it goes deep and empties into the axillary vein. The **basilic vein** is the other major superficial vein of the upper limb. It arises from the medial side of the dorsal veins of the hand and ascends along the medial-posterior side of the forearm. Just below the elbow, the basilic vein travels to the front of the arm and goes deep above the elbow to join with the brachial vein, thus forming the axillary vein. There are several superficial branches between the basilic and cephalic veins. One of these, the **median cubital vein,** which connects the two vessels in front of the elbow, is commonly used to give or receive blood. All the venous blood from the head, neck, and upper limbs is returned to the heart through the superior vena cava.

Veins of the Thorax. The venous blood of the wall of the thorax also empties into the superior vena cava—in this case, by way of an *azygos system (az´-ĭ-gus)* of veins (Figure 20.34). The **azygos vein** begins in the upper right lumbar region and passes through the diaphragm via the *aortic hiatus* (along with the aorta and the thoracic duct of the lymphatic system) to enter the thorax. It continues up the posterior wall of the thorax, to the right of the vertebral column, and empties into the superior vena cava. The **hemiazygos vein** follows a similar course on the left side, passing in front of the vertebral column to empty into the azygos vein in the middle of the thorax. Most of the upper left region of the thorax is drained by an **accessory hemiazygos vein,** which also crosses the vertebral column to empty into the azygos vein. The upper three left intercostal veins drain directly into the left brachiocephalic vein through the left **superior intercostal vein.** The azygos system of veins also receives the **posterior intercostal, bronchial, esophageal,** and **pericardial veins.**

◆ **FIGURE 20.34 Veins of the posterior wall of the thorax**
The anterior wall of the thoracic cage has been removed.

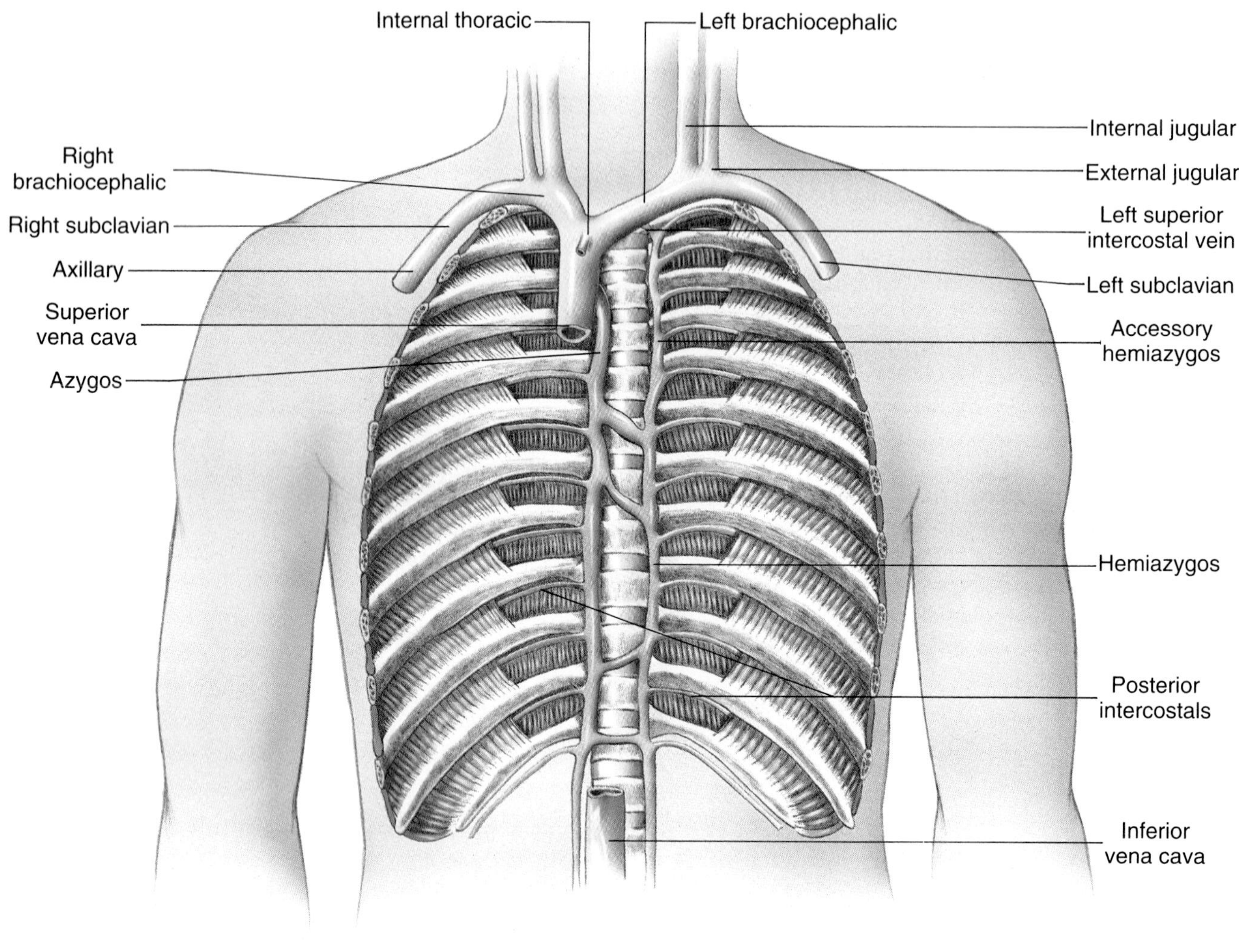

Veins of the Abdomen and Pelvis. Venous blood from the lower regions of the body is returned to the heart through the **inferior vena cava.** This large vessel is formed by the confluence of the **right** and **left common iliac veins** at the level of the fifth lumbar vertebra. It travels along the posterior body wall to enter the right atrium of the heart. As it passes through the abdomen, the inferior vena cava receives tributaries that correspond to most of the arteries that arise from the abdominal aorta—for example, **lumbar, renal, suprarenal, hepatic,** and **right testicular** or **ovarian veins.** (The **left testicular** or **ovarian veins** empty into the **left renal vein.)** The inferior vena cava does not receive blood directly from the digestive tract, pancreas, or spleen. The veins from these regions form the hepatic portal system, which is discussed in the next section.

The veins of the pelvic region follow the pattern of the arteries and are given the same names. Most of them empty into the **internal iliac veins,** which join with the **external iliac veins** from the lower limbs to form the **right** and **left common iliac veins.**

The Hepatic Portal System. Venous blood from the stomach, intestines, spleen, and pancreas is carried by small veins, most of which empty into three large vessels: the *splenic vein,* the *inferior mesenteric vein,* and the *superior mesenteric vein.* The **splenic vein** returns blood from the spleen. As it travels toward the midline of the body it receives tributaries from the stomach and

◆ **FIGURE 20.35 The veins of the hepatic portal system**
The arrows indicate the direction of blood flow.

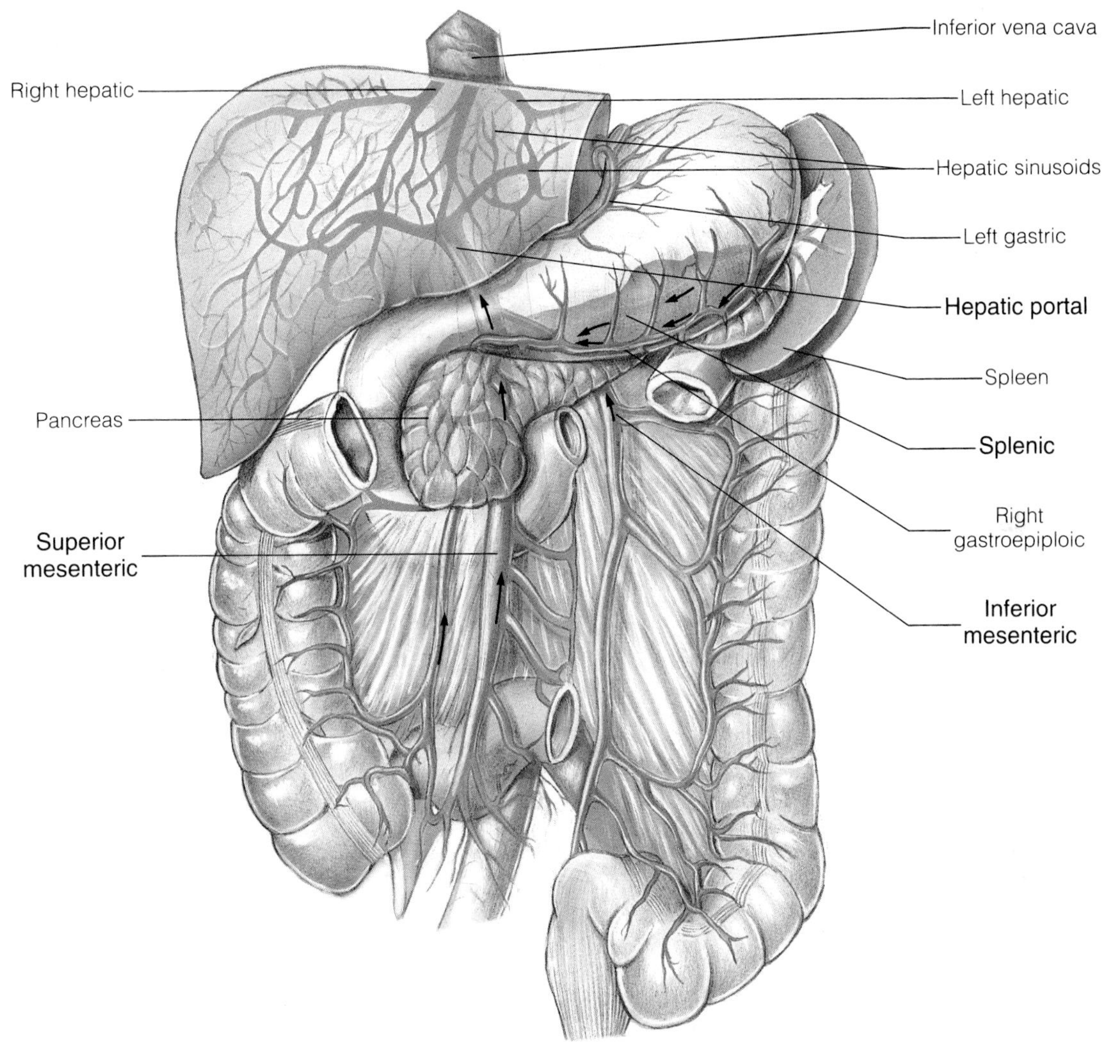

pancreas. The **inferior mesenteric vein** returns blood from the rectum and the descending limb of the large intestine. It ascends beneath the parietal peritoneum and joins the splenic vein behind the pancreas. The common vessel formed by the junction of the inferior mesenteric vein and the splenic vein joins the superior mesenteric vein behind the neck of the pancreas. The **superior mesenteric vein** returns blood from the small intestine, the cecum, and the ascending and transverse limbs of the large intestine. The junction of the superior mesenteric vein and the splenic vein forms the **hepatic portal vein** (Figure 20.35). The hepatic portal vein ascends in the right border of the lesser omentum and enters the inferior surface of the liver. Within the liver, blood from the hepatic portal vein empties into liver sinusoids and follows a unique pathway that is discussed in Chapter 24. After leaving the liver, blood travels through **hepatic veins** to enter the inferior vena cava.

A system of blood vessels where venous blood passes from veins to sinusoids (or capillaries) and then again into veins is referred to as a *portal system.* There are only a few portal systems in the body, and each serves specific functions. The **hepatic portal system,** in which blood from the digestive tract, pancreas, and spleen passes through the sinusoids of the liver before returning to the heart, makes it possible for the liver to perform its many functions. The venous blood in the hepatic portal vein contains nutrients and other substances that have been absorbed from the digestive tract.The hepatic portal system carries these substances directly to the liver, where they can be removed from the blood and stored, metabolized, or, in the case of harmful substances, detoxified.

Veins of the Lower Limbs. The deep veins of the lower limbs, like those of the upper limbs, travel along with the arteries and are given corresponding names: **external iliac, femoral, popliteal, anterior** and **posterior tibial,** and **peroneal veins** (Figure 20.30).

Two large superficial veins of the lower limbs arise from a **dorsal venous arch** on the top of the foot (Figure 20.36): the *great* and *small* saphenous veins. The **great saphenous vein** *(sub-fē´-nus)* is the longest vein in the body. It travels along the medial side of the foot, leg, and thigh, where it joins with the femoral vein just below the inguinal ligament. The **small saphenous vein** travels along the lateral side of the foot, crosses to the posterior surface of the leg, and joins the popliteal vein behind the knee. There are numerous connections between the great and small saphenous veins, as well as between them and the deep veins. Both the deep and the superficial veins of the lower limbs contain valves that assist in returning the blood to the heart against the force of gravity.

◆ **FIGURE 20.36 Superficial veins of the lower limb**
(a) Anterior view. (b) Posterior view.

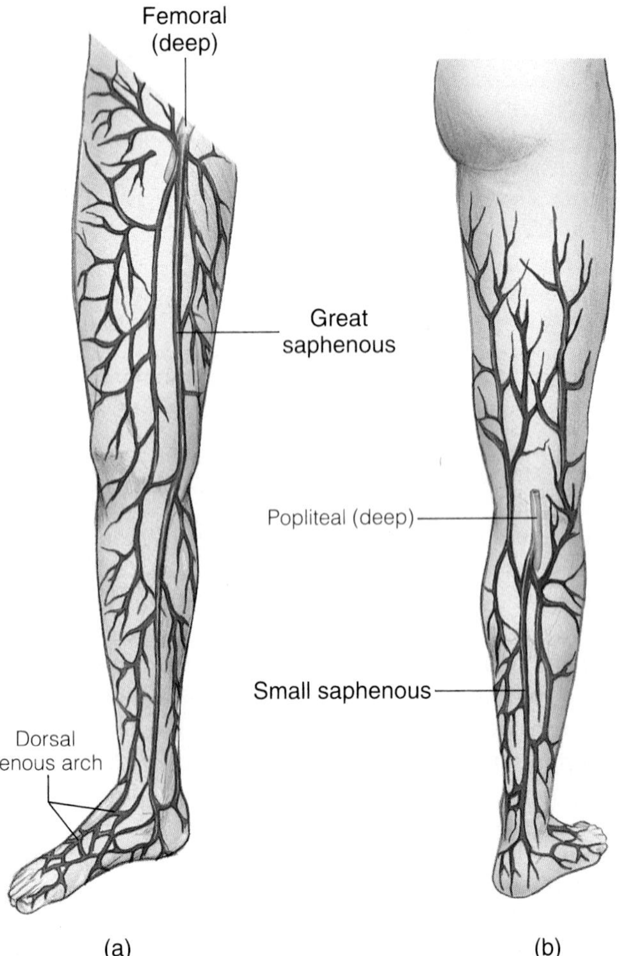

◆ **TABLE 20.3 Schematic Summary of the Major Veins That Empty into the Superior and Inferior Venae Cavae**

SUPERIOR VENA CAVA
AZYGOS
↑— *Hemiazygos*
└— *Accessory hemiazygos*

BRACHIOCEPHALIC
Internal jugular
Vertebral
Internal thoracic
Subclavian ← Axillary ← Brachial ← Ulnar
↑— External jugular; ↑ Cephalic (to Axillary); ↑ Basilic (to Brachial); ↖ Radial (to Brachial)

INFERIOR VENA CAVA
Inferior phrenics
Hepatics ← Liver ← Hepatic portal ← Inferior mesenteric / Splenic / Superior mesenteric
Right suprarenal
Renals
↑— *Left suprarenal*
└— *Left testicular or ovarian*
Right testicular or ovarian
Lumbars
Middle sacral
Common iliacs (right and left)
↑— *Internal iliac*
└— *External iliac* ← *Femoral* ← *Popliteal* ← *Posterior tibial* / *Anterior tibial*
Small saphenous (to *Popliteal*)
Great saphenous (to *Femoral*)

Study Outline

◆ TYPES OF VESSELS p. 628

Arteries. Large vessels that carry blood away from heart; *arterioles* and *capillaries* are progressively smaller vessels.

Venules, Veins. Vessels that return blood to heart.

◆ GENERAL STRUCTURE OF BLOOD-VESSEL WALLS p. 628

Tunica Intima. Present in all vessels; composed of endothelium, connective tissue, and basement membrane.

Tunica Media. Composed of elastic fibers and smooth muscle; often separated from tunica intima by an internal elastic lamina and from tunica adventitia by an external elastic lamina.

Tunica Adventitia. Composed of elastic and collagen fibers.

◆ STRUCTURE OF ARTERIES p. 628

Elastic Arteries (Conducting Arteries). Large arteries; thick tunica media with many elastic fibers.

Muscular Arteries (Distributing Arteries). Tunica media of smooth muscle cells; well-defined internal and external elastic laminae.

◆ STRUCTURE OF ARTERIOLES pp. 628-630

Diameter less than 0.5 mm; small lumen; thick tunica media of mostly muscle cells.

◆ STRUCTURE OF CAPILLARIES pp. 630-631

0.5-1 mm long; 0.01 mm diameter; walls have endothelial layer only. In most tissues, a capillary bed contains thoroughfare channels and true capillaries.

◆ STRUCTURE OF VENULES p. 631

Inner endothelium and some outer fibrous tissue; large venules encircled by a few smooth muscle fibers.

◆ STRUCTURE OF VEINS pp. 631-632

Thin tunica media with few muscle fibers; thick tunica adventitia with little elastic tissue; larger lumen, thinner walls than arteries; some have valves and vasa vasorum.

◆ PRINCIPLES OF CIRCULATION pp. 632-635

Efficient circulation requires adequate blood volume, blood vessels in good condition, and smoothly functioning heart.

Velocity of Blood Flow. In general, velocity of flow in any segment of cardiovascular system is inversely related to total cross-sectional area of vessels of the segment; velocity of flow is lowest in the capillaries.

Pulse. Elastic expansion and subsequent recoil of arteries that occurs in association with beating of heart; can be felt at various locations on body surface.

Blood Flow. Volume of blood that flows through a vessel or vessels in a given time.

Pressure. Pumping action of heart imparts energy to blood; this energy is evident as pressure that drives blood through vessels.

Resistance. Essentially a measure of friction.

FACTORS THAT AFFECT RESISTANCE.

BLOOD VISCOSITY, VESSEL LENGTH, VESSEL RADIUS.

ALTERATIONS IN RESISTANCE. Vessel length is constant and blood viscosity normally does not vary greatly. Most alterations in resistance are due to changes in vessel—particularly arteriole—radius.

Relation Among Flow, Pressure, and Resistance. Generally,

$$\text{flow} = \frac{\text{pressure}}{\text{resistance}}$$

and

$$\text{mean arterial pressure} = \text{cardiac output} \times \text{total peripheral resistance}$$

◆ ARTERIAL PRESSURE pp. 635-640

Affected by a number of factors through mechanisms that influence vessel resistance and cardiac function.

Neural Factors.

VASOMOTOR NERVE FIBERS. Affect vessel resistance by altering vessel diameters. Sympathetic vasoconstrictor fibers release norepinephrine and cause vasoconstriction.

VASOMOTOR CENTER. Tonically active brain-stem center that plays role in regulating blood-vessel resistance and, thus, in regulating arterial pressure; center influenced by inputs from higher brain centers, cardiovascular-system receptors, and receptors associated with sensory surfaces of body.

CARDIAC CENTER. Can influence arterial pressure through effects on cardiac function.

BARORECEPTORS. Influence vessel diameters, heart rate, and cardiac contractility, thereby affecting arterial pressure.

AORTIC AND CAROTID BODIES. Contain chemoreceptors that can influence arterial pressure in response to arterial oxygen, carbon dioxide, and hydrogen ion concentrations.

Chemicals and Hormones.

ANGIOTENSIN II. Extremely powerful vasoconstrictor that raises arterial pressure.

EPINEPHRINE. Causes transitory increase in systolic pressure within arteries.

VASOPRESSIN. Raises arterial pressure by stimulating arteriolar constriction.

Blood Volume. When increased, tends to raise arterial pressure; when decreased, tends to lower it. Kidney mechanisms that influence blood volume are particularly important in long-term regulation of arterial pressure.

Summary. A number of factors affect arterial pressure, and arterial pressure at any one moment is usually result of combined influence of several factors. Different mechanisms work together to maintain arterial pressure and thus to maintain a circulation that meets body's needs.

Measurement of Arterial Pressure. Usually done indirectly using inflatable cuff and manometer. Blood pressure is reported as systolic pressure/diastolic pressure—for example, 120/80. Pulse pressure is difference between systolic and diastolic pressures.

◆ BLOOD FLOW THROUGH TISSUES p. 640

Local Autoregulatory Control. Local autoregulatory mechanisms alter blood flow in accordance with tissue needs. Oxygen deficiency can lead to local dilation of arterioles and increased relaxation of precapillary sphincters. Carbon dioxide, adenosine, lactic acid, potassium ions, and hydrogen ions may act as local vasodilator substances.

External Factors. External factors such as vasomotor nerve fibers and hormones usually influence large segments of vascular system; generally more involved with mechanisms that function for well-being of entire body than with control of tissue blood flow according to needs of particular tissues.

◆ CAPILLARY EXCHANGE pp. 640–642

Most capillaries are permeable to water and small particles such as glucose, inorganic ions, urea, amino acids, and lactic acid; and these materials pass readily between blood and interstitial fluid. Permeability to protein generally quite limited.

Capillary Blood Flow. Blood flows slowly through capillaries, allowing time for exchange of materials between blood and interstitial fluid. Precapillary sphincters contract and relax cyclically, and capillary blood flow is usually intermittent rather than continuous and steady.

Movement of Materials between the Blood and the Interstitial Fluid. Several processes involved:

DIFFUSION. Most important means by which substances such as nutrients and metabolic end products pass between blood and interstitial fluid.

ENDOCYTOSIS AND EXOCYTOSIS. There is disagreement about the importance of this activity in exchange of materials between blood and interstitial fluid.

FLUID MOVEMENT. Normally causes little change in volume of blood or interstitial fluid.

FLUID PRESSURES. Capillary pressure tends to force fluid out of capillaries and into tissue space by filtration through capillary walls. Interstitial fluid pressure opposes this movement.

OSMOTIC PRESSURES. Colloid osmotic pressure of plasma tends to draw water into capillaries by osmosis; colloid osmotic pressure of interstitial fluid tends to draw water out of capillaries by osmosis.

DIRECTION OF FLUID MOVEMENT. If forces tending to move fluid out of a capillary are greater than forces tending to move fluid in, fluid will leave, and vice versa.

◆ VENOUS RETURN pp. 642–643

Influenced by blood volume, contraction of skeletal muscles, valves of veins, constriction of veins.

◆ EFFECTS OF GRAVITY ON THE CARDIOVASCULAR SYSTEM p. 644

When a person moves from supine position to standing upright and perfectly still, the weight of blood causes high pressures in blood vessels of lower regions of body. Baroreceptors and muscular contractions during movement help compensate for effects of gravity when a person is upright.

◆ CIRCULATION IN SPECIAL REGIONS pp. 644–647

Pulmonary Circulation. Pulmonary circuit is a low-pressure circuit. Low oxygen levels cause constriction of pulmonary vessels by local autoregulatory mechanisms.

Coronary Circulation. Greatly influenced by aortic pressure and by pressures that build up in walls of heart chambers during systole.

Cerebral Circulation. Total cerebral blood flow is remarkably constant over broad range of arterial pressures. Interference with cerebral blood flow can lead to fainting or stroke.

Skeletal Muscle Circulation. During exercise, blood flow to skeletal muscles increases greatly, primarily due to local autoregulatory mechanisms.

Cutaneous Circulation. Primarily for purposes of temperature regulation rather than skin nutrition. Blood flow can vary tremendously.

◆ CARDIOVASCULAR ADJUSTMENTS DURING EXERCISE pp. 647–651

Skeletal muscle blood flow increases greatly, due to dilation of skeletal muscle vessels. Other vessels constrict to compensate, but total peripheral resistance generally decreases. Heart rate and stroke volume increase, as does cardiac output. Mean arterial pressure usually rises.

◆ CONDITIONS OF CLINICAL SIGNIFICANCE: BLOOD VESSELS pp. 648–650

High Blood Pressure (Hypertension). Most cases due to unknown causes. Many cases may be due to excessive dietary intake of sodium or excessive retention of sodium by body, leading to increased blood volume and cardiac output. An increased peripheral resistance due to vascular constriction may also be involved.

Atherosclerosis and Arteriosclerosis. *Atherosclerosis* is characterized by deposits of abnormal smooth muscle cells, lipid materials, and connective tissue. Deposits can narrow or occlude blood vessels or favor clot formation. *Arteriosclerosis* can lead to loss of elasticity of vessel walls.

Edema. Accumulation of excess fluid in tissues. In general, can result from any event that enhances movement of fluid out of blood vessels and into tissues, or that retards return of fluid from tissues to bloodstream.

Circulatory Shock. Condition in which cardiac output is so reduced that body tissues fail to receive an adequate blood supply. Can be due to diseases or weaknesses of heart, a reduction in blood volume, or vascular difficulties that reduce venous return.

Aneurysms. Localized dilation of an artery due to weakness of artery wall.

Phlebitis. Inflammation of a vein that can lead to clotting.

Varicose Veins. Dilated, lengthened, and tortuous veins that interfere with venous return and lead to clotting.

Effects of Aging. Elasticity of arteries is reduced with age. Environment, lifestyle, and genetic inheritance may also contribute to cardiovascular problems associated with aging.

◆ ANATOMY OF THE VASCULAR SYSTEM pp. 651–668

Pulmonary Circuit.

1. From right ventricle to pulmonary trunk to pulmonary arteries to lobar arteries to plexuses of pulmonary capillaries for gas exchange to venules to pulmonary veins to left atrium.
2. Transports oxygenated venous blood and oxygen-poor arterial blood.

Systemic Circuit.

SYSTEMIC ARTERIES. Arise from aorta.

BRANCHES OF THE ASCENDING AORTA. Coronary arteries.

BRANCHES OF THE AORTIC ARCH. Brachiocephalic, left common carotid, and left subclavian arteries.

ARTERIES OF THE HEAD AND NECK. Common carotids, internal and external carotids, temporals, cerebral and vertebral arteries, and cerebral arterial circle.

ARTERIES OF THE UPPER LIMBS. Subclavian, axillary, brachial, radial, ulnar, and digital arteries.

BRANCHES OF THE THORACIC AORTA. Intercostal, pericardial, mediastinal, subcostal, bronchial, esophageal, and superior phrenic arteries.

BRANCHES OF THE ABDOMINAL AORTA. Lumbar, common iliac, middle sacral, inferior phrenic, celiac (gastric, splenic, common hepatic), suprarenal, renal, superior and inferior mesenteric, ovarian or testicular arteries.

ARTERIES OF THE PELVIC REGION. Internal and external iliac arteries.

ARTERIES OF THE LOWER LIMBS. Femoral, popliteal, anterior and posterior tibials, peroneal, plantar, digital, and dorsalis pedis arteries.

SYSTEMIC VEINS. Deep veins, superficial veins, venous sinuses.

VEINS OF THE HEAD AND NECK. Internal and external jugulars, subcalvian, brachiocephalic, superior vena cava, and vertebral veins.

VEINS OF THE UPPER LIMBS. *Deep:* axillary, brachial, radial, and ulnar veins. *Superficial:* cephalic, basilic, and median cubital veins.

VEINS OF THE THORAX. Azygos system: intercostal, bronchial, esophageal, and pericardial veins.

VEINS OF THE ABDOMEN AND PELVIS. Inferior vena cava; common iliac, lumbar, renal, suprarenal, hepatic, right testicular or ovarian veins.

THE HEPATIC PORTAL SYSTEM. Hepatic portal vein formed by joining of inferior mesenteric, superior mesenteric, and splenic veins; carries blood from stomach, intestines, spleen, and pancreas; enters liver for removal of nutrients.

VEINS OF THE LOWER LIMBS. Contain valves. *Deep:* external iliac, femoral, popliteal, anterior and posterior tibials, and peroneal veins. *Superficial:* great and small saphenous veins.

Self-Quiz

1. The tunica media and tunica adventitia are not present in: (a) veins; (b) capillaries; (c) arterioles.
2. The endothelium is part of the: (a) tunica adventitia; (b) tunica media; (c) tunica intima.
3. The walls of the larger arteries and veins are so thick that they are supplied by small nutrient vessels called vasa vasorum. True or False?
4. Valves are typically found in arteries as well as in the larger veins. True or False?
5. Which of the following blood-vessel layers is present in vessels of all sizes? (a) endothelium; (b) tunica media; (c) tunica adventitia.
6. Compared to corresponding arteries, veins have: (a) a better-developed tunica media; (b) a generally larger lumen; (c) generally thinner walls; (d) both (b) and (c); (e) all of these.
7. The velocity of blood flow is slowest in the: (a) arteries; (b) arterioles; (c) capillaries; (d) veins.
8. Blood pressure is generally lowest in the: (a) arteries; (b) capillaries; (c) veins.
9. In general, which of the following will cause an increased resistance to blood flow? (a) a decrease in vessel length; (b) a decrease in vessel diameter; (c) a decrease in blood viscosity.

10. The mean arterial pressure is equal to the cardiac output multiplied by the: (a) total peripheral resistance; (b) total systemic blood flow; (c) total venous return.

11. Which of the following is likely to lead to a fall in blood pressure? (a) decreased arterial oxygen concentration; (b) decreased activity of the carotid sinus baroreceptors; (c) stretching the wall of the aortic arch.

12. The kidneys exert a significant effect on blood volume by virtue of their ability to regulate salt and water excretion, and they are particularly important in the long-term regulation of arterial pressure. True or False?

13. When a person's blood pressure is reported to be 120/80, the 120 is the: (a) pulse pressure; (b) systolic pressure; (c) diastolic pressure.

14. Which of the following forces favors the movement of fluid into the capillaries from the tissue spaces? (a) colloid osmotic pressure of the plasma; (b) colloid osmotic pressure of the interstitial fluid; (c) capillary pressure.

15. Venous return is likely to be enhanced by a decreased peripheral venous pressure and an increased right atrial pressure. True or False?

16. The flow of blood through the coronary vessels is: (a) independent of the aortic pressure; (b) greatly influenced by the pressures that build up in the walls of the heart chambers during systole; (c) continuous and constant throughout the cardiac cycle.

17. Most of the cutaneous blood flow is for: (a) temperature regulation; (b) provision of nutrients to the skin; (c) removal of wastes from the skin.

18. During exercise: (a) the stroke volume of the heart generally decreases; (b) the total peripheral resistance generally increases; (c) the cardiac output generally increases.

19. The pulmonary arteries carry blood that has a low carbon dioxide content and a high oxygen content. True or False?

20. Match the terms associated with the systemic arteries with the appropriate lettered description:

Aorta
Coronary arteries
Internal carotid artery
Vertebral artery
Radial artery
Superior mesenteric artery
Inferior mesenteric artery
Common iliac arteries
Anterior tibial artery

(a) A major blood supplier to the brain; passes through the transverse foramina
(b) Supplies all of the small intestine
(c) Formed from bifurcation of abdominal aorta
(d) Vessel from which the pulse is usually taken
(e) Branches of ascending aorta
(f) Supplies the muscles of the anterior compartment of the leg
(g) Blood from left ventricle enters systemic circuit through this vessel
(h) Supplies the large intestine
(i) Supplies the brain through its terminal branches—the anterior and middle cerebral arteries

CHAPTER 21

The Lymphatic System

CHAPTER CONTENTS

LEARNING OBJECTIVES

After completing this chapter, you should be able to:

1. Describe the system of lymphatic vessels.
2. Discuss the functions of the lymphatic system.
3. Explain how excess interstitial fluid and plasma proteins are returned to the blood by the lymphatic vessels.
4. Describe the structure and function of the lymph nodes, spleen, thymus gland, tonsils, and Peyer's patches.

The **lymphatic system** consists of (1) an extensive network of *lymphatic capillaries* and larger *collecting vessels* that receive fluid from the loose connective tissues throughout the body and transport it to the cardiovascular system; (2) *lymph nodes,* which serve as filters of the fluid within the collecting vessels; and (3) the *lymphoid organs,* including lymphatic nodules, the spleen, the thymus gland, the tonsils, and Peyer's patches.

The lymphatic system is closely related both anatomically and functionally to the cardiovascular system. As indicated in Chapter 20, the volume of fluid that moves out of the capillaries and into the tissue spaces is slightly greater—by about 3 liters per day—than the volume of fluid that moves into the capillaries from the tissues. If this fluid were allowed to accumulate, the tissues would swell, producing edema. However, the lymphatic system returns this extra fluid (including any plasma protein that escapes from the capillaries) back to the blood.

Lymphatic Vessels

Interstitial fluid enters the lymphatic system by passing through the extremely thin walls of **lymphatic capillaries.** Once the fluid is within the vessels of the lymphatic system, it is called **lymph.** Since the lymphatic system is a one-way system—only *returning* fluid to the blood—the lymphatic capillaries are dead-end vessels (Figure 21.1). Their walls, like the walls of blood capillaries, are composed of a single layer of endothelium. However, lymphatic capillaries lack the surrounding basement membrane that ensheathes blood capillaries.

Another difference between lymphatic capillaries and blood capillaries is that the edges of adjacent endothelial cells in lymphatic capillaries are only loosely attached and overlap one another, forming inner and outer flaps. The outer flap is anchored to surrounding tissues by bundles of fine filaments (Figure 21.1, inset). This arrangement forms a functional one-way valve. Interstitial fluid outside lymphatic capillaries can enter the capillaries by pushing the inner flap of the overlapping endothelial cells inward while the outer flap remains anchored to the surrounding tissues. Fluid within the lymphatic capillaries is unable to reenter the intercellular spaces because when it attempts to do so, it pushes the inner flap of endothelial cells against the outer flap, closing the valve.

Because of this structural arrangement, lymphatic capillaries are more permeable than most blood capillaries, and virtually all the components of the interstitial fluid, including proteins and other large particles (such as disease-causing organisms), can enter these vessels and are transported throughout the body. Most tissues contain plexuses of lymphatic capillaries located among the vascular capillaries. A few tissues, including the central nervous system, bone, bone marrow, cartilage, and teeth do not. Special lymphatic capillaries called **lacteals** are located in the villi of the small intestine. The lacteals aid in the absorption of fat from the digestive tract and carry it to the blood as a milky fluid called *chyle.*

The lymphatic capillaries, which are widely distributed throughout the interstitial spaces of the body, join together to form progressively larger lymphatic vessels (Figure 21.2a). Generally, the larger lymphatic vessels, which are called *collecting vessels,* travel alongside the arteries and veins of the cardiovascular system and pass through one or more *lymph nodes* before emptying into either the *thoracic duct* or the *right lymphatic duct,* both of which return the lymph to the blood. The walls of the lymphatic collecting vessels are similar to the walls of veins, although they are thinner. Like veins, lymphatic collecting vessels contain valves that occur in pairs, each on opposite sides of the vessel, with their free edges pointing in the direction of lymph flow. The valves aid in the movement of lymph by preventing its backflow.

Lymph Nodes

Lymph nodes are small round or bean-shaped organs distributed along the course of the lymphatic vessels. There are groups of lymph nodes in the groin, axilla, and neck, as well as in numerous other deeper locations.

Each node is enclosed in a **fibrous capsule** (Figure 21.3). Strands of connective tissue called **trabeculae** *(tra-bek´-u-lē)* extend inward from the capsule and divide the node into several compartments (Figure 21.4). These compartments are further subdivided by a network of reticular fibers that extend between the trabeculae. The node consists of an outer *cortical* region and an inner *medullary* region. Within the cortex of each node are separate masses of lymphoid tissue called **germinal centers,** which serve as a source of lymphocytes. *Lymphoid tissue* is a modified type of loose connective tissue, with a prominent network of reticular fibers between which are many lymphocytes and macrophages. The cells of the medulla are arranged in strands called **medullary cords.**

Lymph enters the convex surface of the lymph node through several **afferent lymphatic vessels** and filters slowly through irregular channels in the node, called **sinuses.** The sinuses are spanned by networks of reticular fibers. Lymph from an afferent lymphatic vessel enters a *subcapsular sinus* located just beneath the capsule. From there it percolates through *cortical sinuses,* which penetrate the cortex, and enters *medullary sin-*

◆ **FIGURE 21.1 Lymphatic capillaries**

The right side of the figure is a schematic representation of the lymphatic capillaries showing their relationship to interstitial fluid, the blood vascular system, and tissue cells. The arrows indicate the directions of fluid movement. Note that lymphatic capillaries begin as dead-end vessels. The round inset on the left is an enlargement of a lymphatic capillary showing overlapping endothelial cells forming "valves."

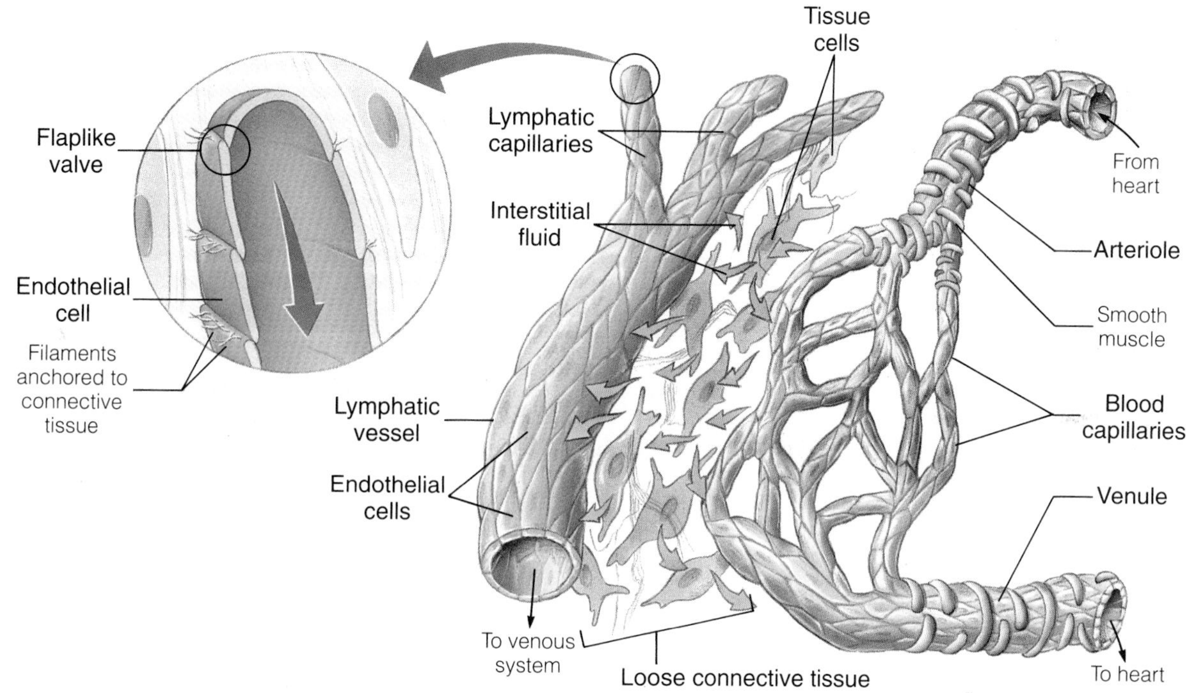

uses located between the medullary cords. From the medullary sinuses, lymph leaves by way of **efferent lymphatic vessels** at a small indentation called the *hilus.* Because there are fewer efferent vessels than afferent vessels, this has the effect of slowing the rate of lymph flow through the nodes, which allows phagocytic cells to remove foreign substances more effectively, thereby preventing their entry into the blood. Generally, all lymph passes through several lymph nodes before it is returned to the cardiovascular system.

Lymph that flows into the lymph nodes contains many foreign particles and microorganisms, some of which can cause diseases if they are not destroyed. As lymph percolates slowly through the sinuses within a node, large foreign particles such as bacteria become entrapped in the meshes of the reticular fibers that span the sinuses. The entrapped particles are soon attacked and destroyed by phagocytic cells, called **macrophages,** that line the sinuses. In addition, between the sinuses are masses of lymphoid tissue containing **lymphocytes** and **plasma cells** that produce antibodies for destroying certain foreign substances known as antigens.

Cancer cells sometimes enter lymphatic vessels and are carried to lymph nodes. Some cancer cells can survive and multiply within the nodes. Thus, the nodes serve as sites from which these cancer cells can spread throughout the body by way of the cardiovascular system. For this reason, swollen lymph nodes near cancer sites are often surgically removed.

The efferent collecting vessels from the lymph nodes of most regions of the body converge into larger vessels called **lymph trunks.** There are five major lymph trunks, four of them paired: (1) the unpaired *intestinal trunk,* which receives lymph from abdominal organs; (2) the *lumbar trunks,* which drain the lower limbs and some pelvic organs; (3) the *subclavian trunks,* which drain the arms and parts of the thorax and the back; (4) the *jugular trunks,* which drain the head and neck regions; and (5) the *bronchomediastinal trunks,* which drain the thorax. The lymph trunks, in turn, empty into either the thoracic duct or the right lymphatic duct.

Lymphatic Ducts

The **thoracic duct** arises from the **cisterna chyli** *(sis-ter´-na kī´-lē),* which is a saclike enlargement that lies in front of the second lumbar vertebra (Figure 21.5, page 680). The cisterna chyli receives lymph from the

◆ **FIGURE 21.2 The lymphatic system**

(a) Major lymphatic vessels and groups of lymph nodes. (b) Lymph from the pink area returns to the blood vascular system through the right lymphatic duct. Lymph from the remainder of the body travels through the thoracic duct.

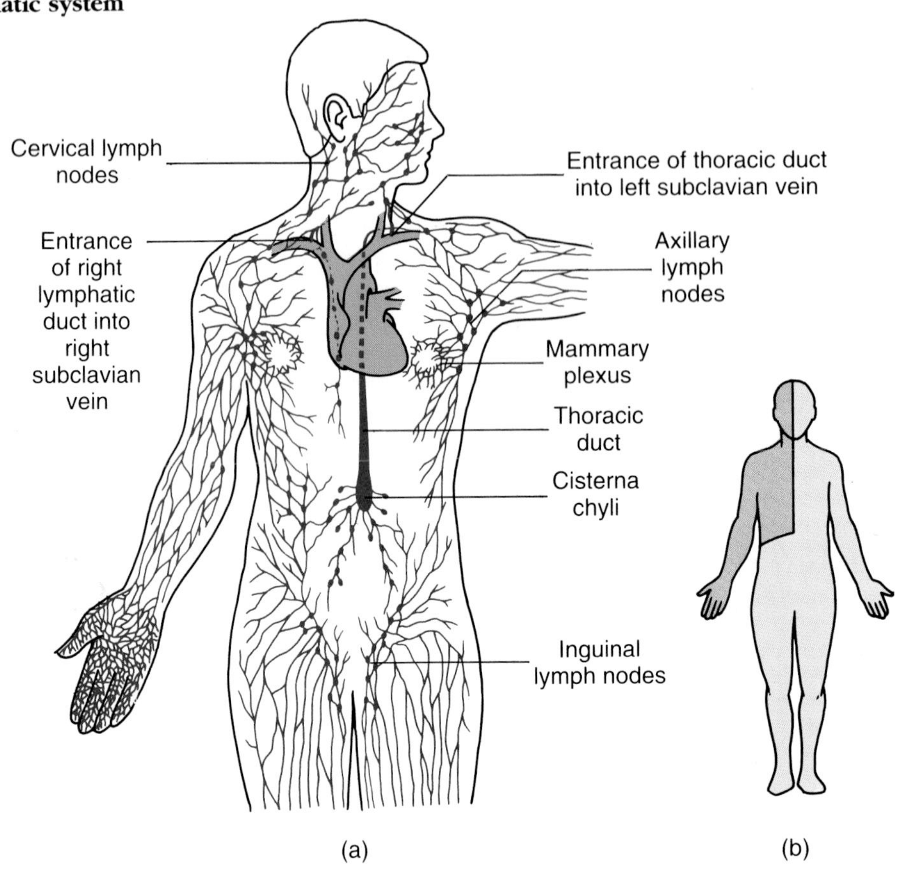

◆ **FIGURE 21.3 Photomicrograph of a portion of a lymph node**

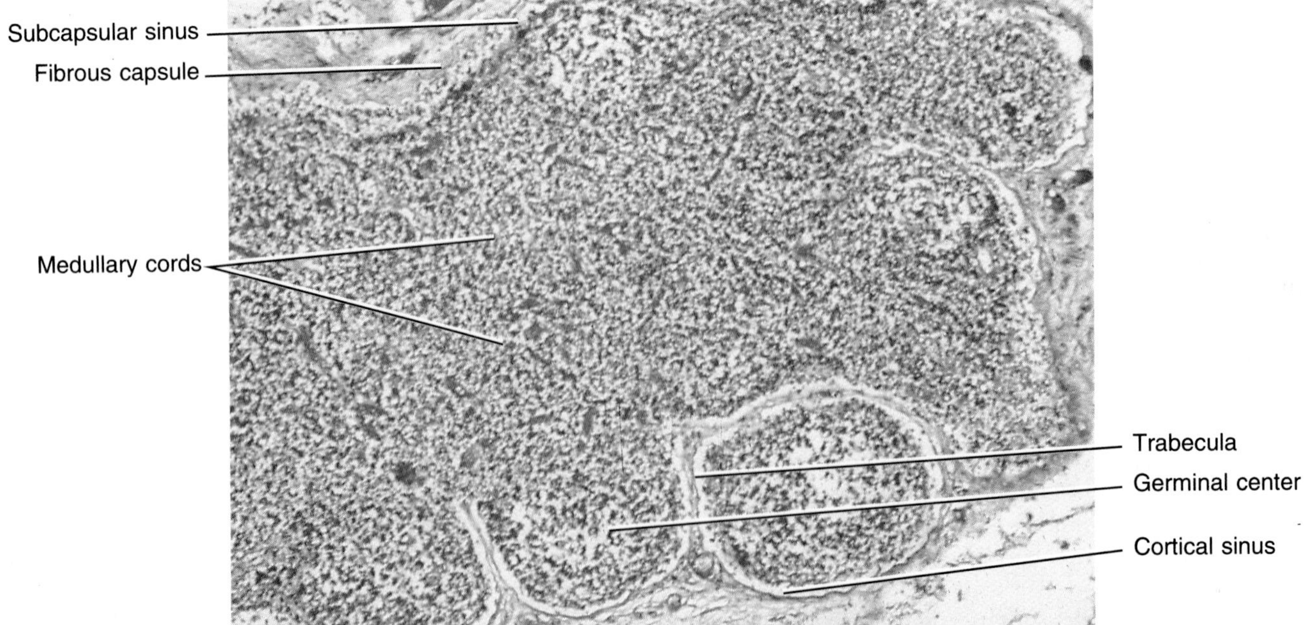

◆ **FIGURE 21.4 Cross section of a lymph node**

Note that there are more afferent vessels than efferent vessels, which has the effect of slowing the rate of lymph flow. The arrows indicate the direction of lymph flow.

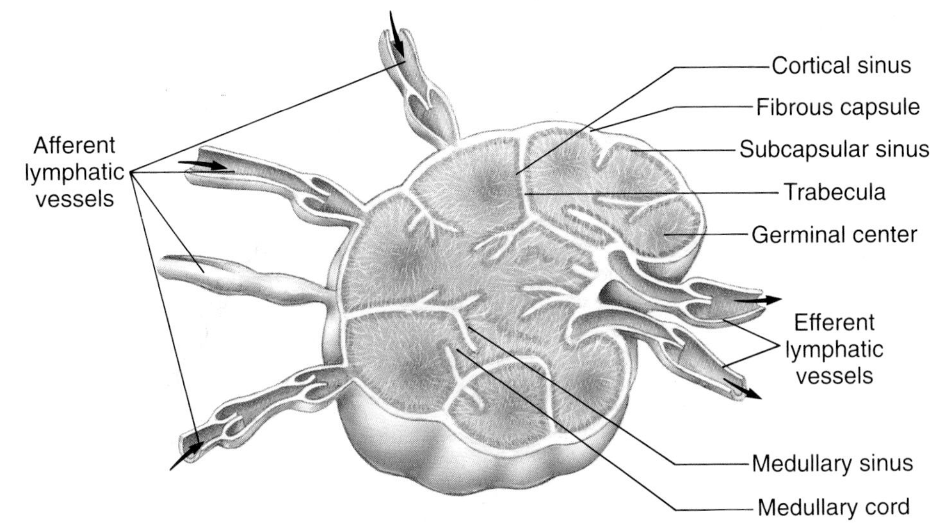

lumbar and intestinal trunks, which drain the lower limbs and the viscera of the abdominopelvic cavity. As the thoracic duct passes through the diaphragm (along with the aorta) and travels upward through the thoracic cavity in front of the vertebral column, it receives lymphatic vessels that drain the left side of the thorax. Behind the brachiocephalic vein, the thoracic duct curves to the left and usually receives the left subclavian and left jugular trunks from the upper limbs and the left side of the head and neck before emptying into the *left subclavian vein* nears its junction with the internal jugular vein.

As we have seen, the thoracic duct returns lymph to the blood from the entire body *except* the upper right limb and the right side of the thorax, neck, and head (Figure 21.2b). Lymph from these areas is returned to the *right subclavian vein* through a **right lymphatic duct.** The right lymphatic duct is a small vessel formed by the joining together of the right jugular, right subclavian, and right bronchomediastinal trunks, which drain lymph from these regions. The two lymphatic ducts thus convey all of the lymph that has been collected and filtered from throughout the body back into the cardiovascular system, and the lymph begins the circuit again as blood plasma.

Mechanisms of Lymph Flow

Lymph flows slowly; approximately 3 liters of lymph enter the blood every 24 hours. The flow is slow because, unlike the cardiovascular system, the lymphatic system does not have a pump like the heart to keep it moving. Rather, lymph flow depends on more subtle forces, such as contractions of skeletal muscles, which apply pressure on the lymph vessels and compress them. This action forces lymph along the vessels. In a similar manner, the pulsing of nearby arteries can compress lymph vessels and move the lymph along them. In addition, the walls of larger lymph vessels beyond the lymphatic capillaries contain some smooth muscle. When a section of one of these vessels is distended by lymph, the smooth muscle contracts slightly and thereby propels the lymph along the vessel. Changes in abdominal pressure and thoracic pressure that occur during breathing also contribute to the movement of lymph. The valves within lymph vessels permit lymph to flow only toward the blood vessels.

Functions of the Lymphatic System

The lymphatic system participates in several important activities, including the *destruction of bacteria,* the *removal of foreign particles from lymph,* the *specific immune responses,* and the *return of interstitial fluid to the blood.*

Destruction of Bacteria and Removal of Foreign Particles from Lymph

Bacteria and other foreign substances are removed from lymph by phagocytes—primarily macrophages—that are

◆ **FIGURE 21.5 Relationship of the thoracic duct and the right lymphatic duct to the blood vascular system**

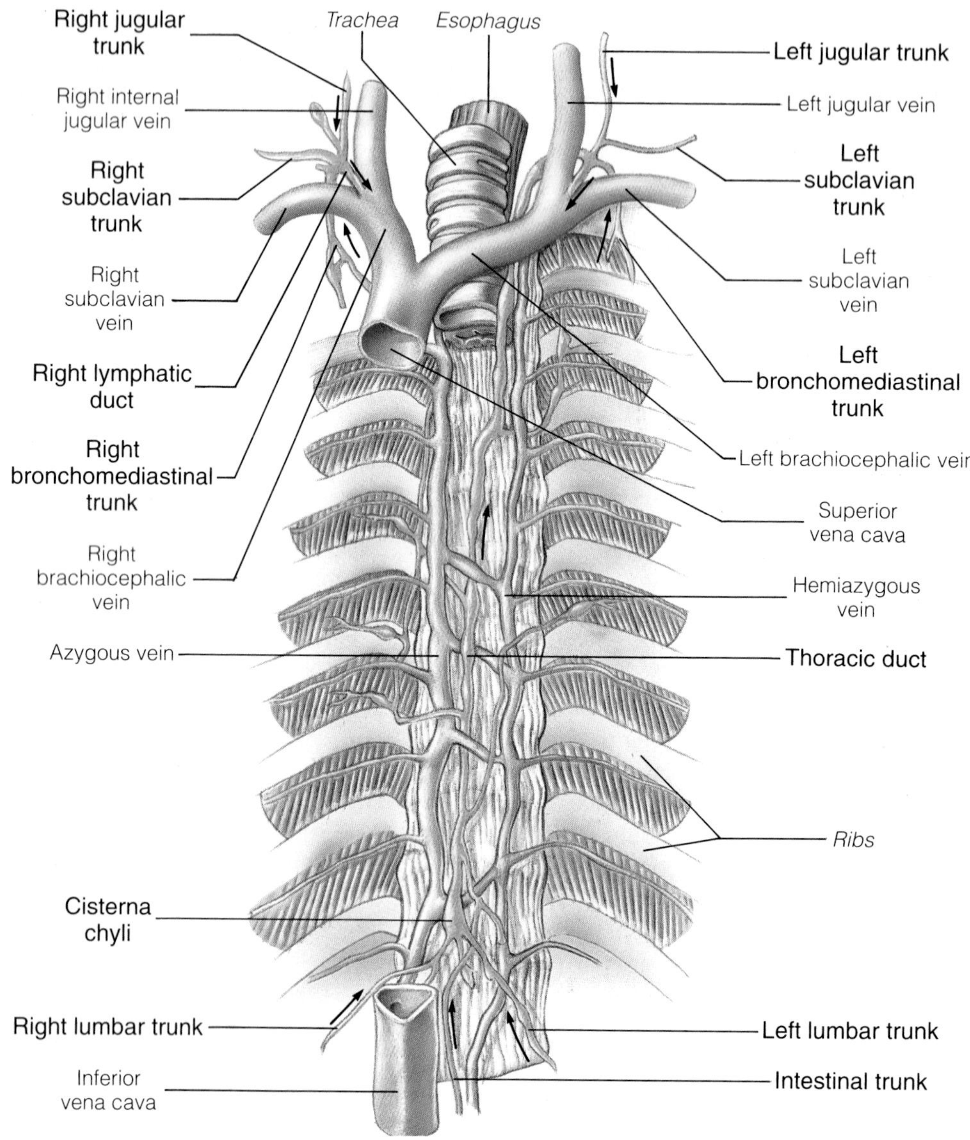

present in the lymph nodes. During infections, the rate of formation of macrophages within the nodes is so great that the nodes enlarge and become tender.

Specific Immune Responses

In response to the presence of bacteria or other foreign substances, certain cells in the lymph nodes participate in specific immune responses, such as the manufacture of antibodies, that result in the destruction of the foreign substances. These responses are described in Chapter 22.

Return of Interstitial Fluid to the Blood

As mentioned earlier, edema results if excess interstitial fluid is allowed to accumulate in the intercellular spaces. Moreover, the volume of blood is reduced, since it is the major source of interstitial fluid.

The small amount of protein that is lost from the blood capillaries into the interstitial fluid is normally returned to the blood by the lymphatic system. If this protein remained in the intercellular spaces, it would increase the osmotic pressure of the interstitial fluid and thereby affect the movement of fluid between the blood and interstitial fluid.

ASPECTS OF EXERCISE PHYSIOLOGY

Exercise: An Effective Preventative and Therapy for Edema

The accumulation of excess fluid in the tissues is called edema. In some cases, edema is due to an increased movement of fluid out of blood vessels and into tissues. In others, it is due to a decreased removal of fluid from tissues by the lymphatic system.

Edema of the feet and ankles can occur when a person stands upright and perfectly still for a long period of time, such as when a soldier stands at attention. Under these conditions, the pressure within the veins of the feet can rise above 90 mm Hg. The increased pressure distends the veins, increasing their capacity so they contain a greater amount of the total blood volume. Moreover, the increased pressure results in an increased movement of fluid out of the capillaries and into the tissues, producing some degree of edema.

The accumulation of blood in the veins and the increased movement of fluid out of the capillaries reduce the effective circulating blood volume, which can reduce the venous return. The reduced venous return can lead to a reduction in cardiac output sufficient to result in loss of consciousness. However, rhythmic contractions of the leg muscles can help prevent such an occurrence by exerting a pumping action on veins and lymph vessels. The contractions compress the vessels, forcing blood or lymph along them. This activity enhances the venous return and reduces the pressure within the veins of the lower regions of the body. In addition, the contractions increase the pressure of the interstitial fluid within the muscles. For example, contraction of the gastrocnemius muscle has been shown to increase the pressure of the interstitial fluid in the central region of the muscle to as high as 220 mm Hg. Such a high pressure would act as a deterrent to the movement of fluid out of the capillaries and the development of edema. Consequently, soldiers are instructed to contract their leg muscles isometrically while standing at attention.

In many cases, edema occurs following surgery that results in lymphatic obstruction or destruction. In women, edema commonly occurs in the arm after surgery for breast cancer, and in men, it occurs in the leg after surgery for prostate cancer. Edema of the hand, accompanied by inactivity, can lead to joint stiffness, pain, and dysfunction. Edema can also cause changes in the arches of both the hands and feet that cause pain and dysfunction.

Exercise, together with elevation of an affected limb, can help in reducing edema after surgery and in maintaining a reasonable range of motion. Once edema is reduced, elastic bandages may be used to deter its recurrence. A daily exercise regimen can help prevent edema in patients with damaged lymph systems.

Lymphoid Organs

In addition to the lymph nodes, several organs are lymphoid in nature; these include the *spleen,* the *thymus gland,* the *tonsils,* and *Peyer's patches.* These organs, which have no direct association with the lymphatic system of vessels or with lymph, are an integral part of the body's immune system.

Spleen

The **spleen** is the largest lymphoid organ. It lies in the left hypochondriac region, between the fundus of the stomach and the diaphragm. The spleen is usually about 12 cm in length; however, its size and weight vary from person to person, as well as in the same individual under different conditions. The *diaphragmatic surface* of the spleen is smooth and convex, conforming to the undersurface of the diaphragm, with which it is in contact. The *visceral surface* is divided into gastric, renal, and colic surfaces that conform to the portions of the stomach, the left kidney, and the left flexure of the colon, which lie adjacent to the spleen (Figure 21.6). Blood vessels enter and leave the spleen through a depression on the visceral surface called the *hilus (high´-lus).*

Like the lymph nodes, the spleen is covered by a strong fibrous capsule with trabeculae that extend into the organ and divide it into compartments. In the

◆ **FIGURE 21.6 The visceral surface of the spleen**

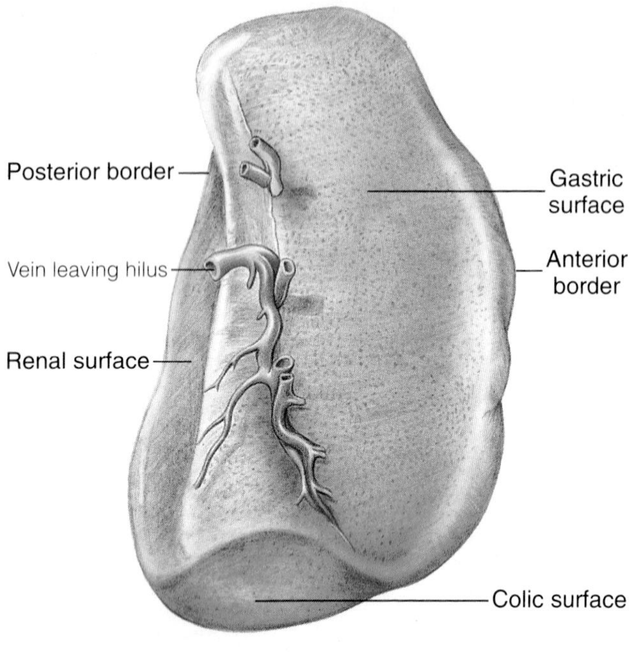

spleen, however, smooth muscle cells are also present in the capsule and the trabeculae. Two types of tissue, called *red pulp* and *white pulp,* are distinguishable within the compartments (Figure 21.7). **Red pulp** is the more abundant. It consists of branching venous sinuses separated from each other by columns of splenic tissue called *splenic cords.* Like other lymphoid tissues, the red pulp contains lymphocytes and macrophages; but it also contains red blood cells, which give the pulp its color. Scattered throughout the red pulp are round masses of **white pulp,** each surrounding an arteriole. White pulp, which does not contain red blood cells, is lymphoid tissue, and large numbers of lymphocytes are present in it. Blood enters the spleen at the hilus through the splenic artery. It may remain in the vessels that are surrounded by white pulp and enter the venous sinuses of the red pulp, or it may pass through the walls of the capillaries and percolate between the cells of the spleen before entering the venous sinuses (Figure 21.8). From the venous sinuses, blood leaves the spleen through the splenic vein and travels within the hepatic portal vein to the liver.

The spleen acts as a filter for the bloodstream much as the lymph nodes filter lymph. Like many other lymphoid tissues, the spleen is a site where specific immune responses (including antibody production) against foreign antigens are initiated. These activities are carried out primarily in the white pulp. In addition, the macrophages of the red pulp phagocytose old red blood cells as well as bacteria and other foreign particles.

The spleen is a storage site for blood platelets and also serves to a limited extent as a blood reservoir. About 200 ml of the blood contained within the venous sinuses of the red pulp may be forced from the spleen

◆ **FIGURE 21.7 Photomicrograph of a portion of the spleen**

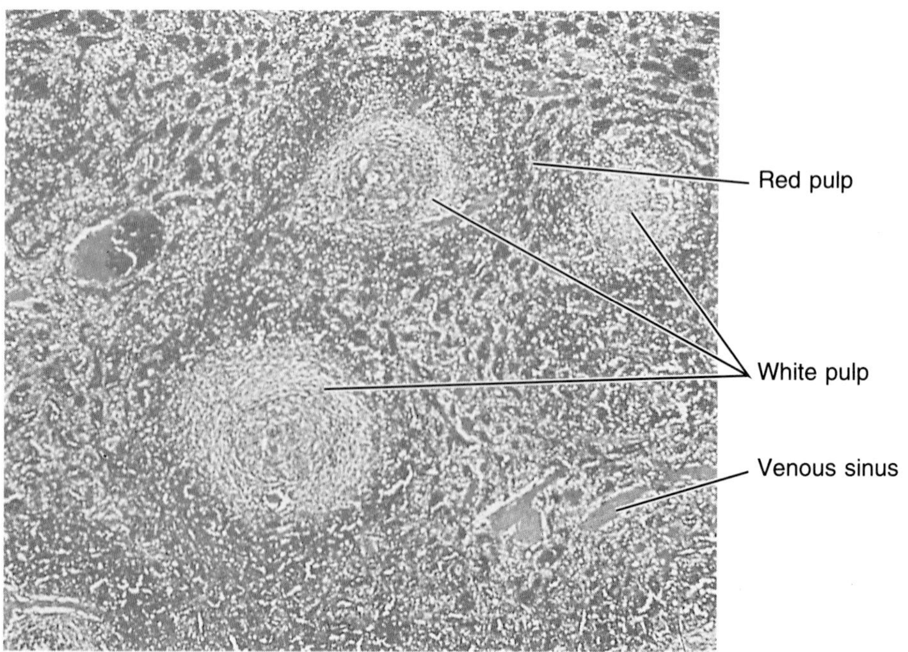

◆ **FIGURE 21.8 The path of blood through the spleen**
See text for explanation.

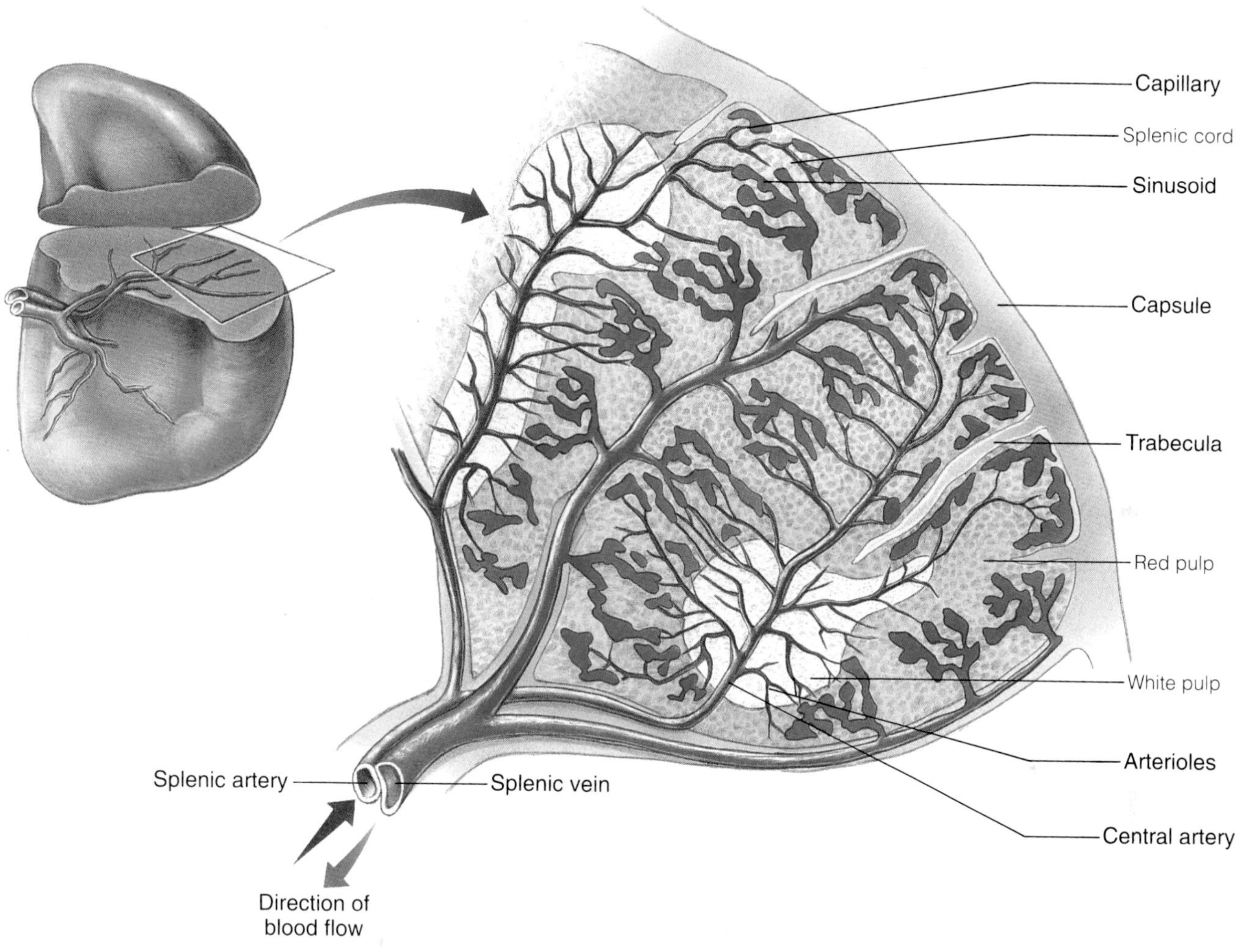

into the general vascular system by contraction of the smooth muscles of the capsule. This can help to offset blood lost through hemorrhage. In the developing embryo, the spleen forms red blood cells; but except in some abnormal conditions that require the replacement of large numbers of red blood cells, this capability is rarely utilized following birth.

Thymus Gland

The **thymus gland** is a bilobed mass of lymphoid tissue located deep to the sternum in the anterior region of the mediastinum. It increases in size during early childhood, then begins to atrophy slowly, and diminishes following puberty. In the adult, it may be entirely replaced by adipose tissue. The thymus gland is covered by a connective-tissue capsule that separates it into smaller lobules (Figure 21.9). Each lobule has an outer *cortex* and an inner *medulla.* The cortex is composed of densely packed lymphocytes. The medulla also contains lymphocytes, but in addition it has organized groups of cells referred to as **thymic corpuscles** (or *Hassall's corpuscles*). The significance of thymic corpuscles is not known.

The thymus confers on certain lymphocytes the ability to differentiate and mature into cells that can participate in specific immune responses (see Chapter 22). The thymus also releases a group of hormonal substances collectively called *thymosin,* which continue to influence the lymphocytes after they have left the gland.

Tonsils

The **tonsils** are small masses of lymphoid tissue embedded in the mucous membrane lining of the oral and

◆ **FIGURE 21.9 Photomicrograph of a portion of the thymus gland (×433)**

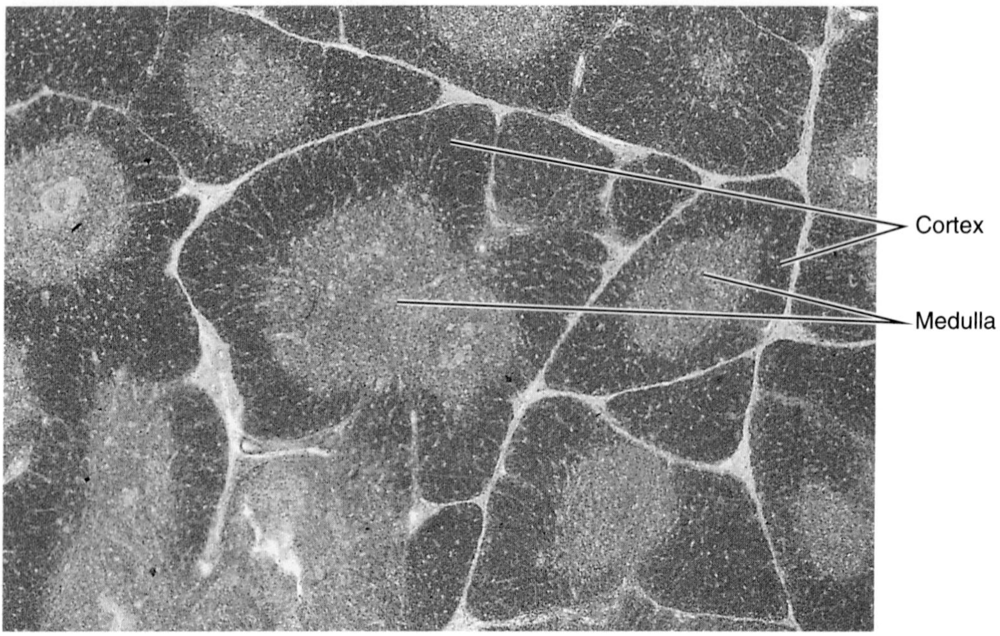

pharyngeal cavities. The **palatine tonsils** are located on the posteriolateral walls of the throat, one on each side. These are the tonsils that are noticeably enlarged when you suffer from a "sore throat." The **pharyngeal tonsils** are located on the posterior wall of the nasopharynx. They are at their peak of development during childhood, and when enlarged they are referred to as *adenoids.* On the dorsal surface of the tongue near its base is a group of **lingual tonsils.** Being composed of lymphoid tissue and surrounding the junction of the oral and nasal passageways, the tonsils serve as an additional defense against bacterial invasion of the body.

Peyer's Patches

Aggregations of lymphatic nodules located in the lining of the last portion (the ileum) of the small intestine are called **Peyer's patches.** Macrophages within Peyer's patches intercept and destroy microorganisms before they are absorbed by the digestive tract. Peyer's patches and the tonsils are referred to collectively as *gut-associated lymphatic tissue (GALT).* The lymphoid tissues included in GALT serve to limit invasion of the upper respiratory and digestive tracts by foreign substances.

Study Outline

◆ LYMPHATIC VESSELS p. 676

Lymphatic Capillaries. Allow interstitial fluid to enter lymphatic vessels.

◆ LYMPH NODES pp. 676–677

Lymphatic tissue enclosed in fibrous capsule; afferent lymphatic vessels transport lymph into nodes; efferent lymphatic vessels transport lymph out of nodes; efferent collecting vessels form into five major lymph trunks, which empty into thoracic duct or right lymphatic duct.

◆ LYMPHATIC DUCTS pp. 677–679

Thoracic Duct. From cisterna chyli; drains most of body; transports lymph to left subclavian vein.

Right Lymphatic Duct. Drains right upper portion of body; transports lymph to right subclavian vein.

◆ MECHANISMS OF LYMPH FLOW p. 679

Flows very slowly; depends on contractions of skeletal muscles squeezing on lymph vessels, the pulsing of nearby arteries, the presence of valves, and pressure changes during breathing.

◆ **FUNCTIONS OF THE LYMPHATIC SYSTEM** pp. 679–680

1. Destruction of bacteria and removal of foreign particles from lymph via phagocytes in nodes.
2. Specific immune responses, such as the manufacture of antibodies.
3. Return of interstitial fluid and plasma proteins to the blood, thus preventing edema.

◆ **LYMPHOID ORGANS** pp. 681–684
Have no direct association with lymphatic system; are an integral part of the immune system.

Spleen. Fibrous capsule with trabeculae; contains red pulp and white pulp. Functions include manufacture of antibodies and phagocytosis of old red blood cells and foreign particles; site of red-blood-cell formation in the embryo.

Thymus Gland. Gives certain lymphocytes ability to participate in specific immune responses.

Tonsils. Palatine, pharyngeal, and lingual; provide defense against bacteria and other foreign particles.

Peyer's Patches. Aggregations of lymphatic nodules in small intestine.

Self-Quiz

1. Normally, slightly more liquid tends to enter the capillaries of the cardiovascular system than leaves it. True or False?
2. The walls of the lymphatic capillaries are composed of: (a) columnar epithelium; (b) adventitia; (c) a single layer of endothelium.
3. Lymphatic capillaries in the villi of the small intestine are called: (a) lacteals; (b) cisterna chyli; (c) efferent lymphatic vessels; (d) trabeculae.
4. Rather large vessels that carry lymph to lymph nodes are called: (a) lymphatic trunks; (b) collecting vessels; (c) thoracic duct; (d) efferent lymphatic vessels.
5. Lymphoid tissue is a modified type of loose connective tissue, with a network of reticular fibers and containing many lymphocytes. True or False?
6. All of the major lymph trunks generally convey lymph into either the thoracic duct or the right lymphatic duct. True or False?
7. Which of the following assist in the flow of lymph within lymph vessels? (a) contraction of skeletal muscles; (b) pulsation of adjacent arteries; (c) valves within lymph vessels; (d) all of these.
8. Which is a function of the lymphatic system? (a) specific immune responses; (b) the return of tissue fluid to the blood; (c) the destruction of bacteria and removal of foreign particles; (d) all of these.
9. The red pulp of the spleen consists of venous sinuses separated from each other by columns of splenic tissue called splenic cords. True or False?
10. The direction of blood flow through the spleen is from arteries into the venous sinuses, through arteries surrounded by white pulp, and out the hilus in veins. True or False?

CHAPTER 22

Defense Mechanisms of the Body

CHAPTER CONTENTS

LEARNING OBJECTIVES

After completing this chapter, you should be able to:

1. Describe the role of body surfaces in preventing the entry of potentially damaging agents or organisms into the body.
2. Explain the changes in blood flow and vessel permeability that occur during acute, nonspecific inflammation.
3. Describe the role and activity of phagocytes during acute, nonspecific inflammation.
4. Describe the roles of histamine, kinins, the complement system, and prostaglandins in acute, nonspecific inflammation.
5. List and explain the local and systemic symptoms of inflammation.
6. Explain the activities of interferon in resisting viral invasions.
7. Describe the functions of antibodies.
8. Describe the events that occur during a humoral immune response.
9. Describe the events that occur during a cell-mediated immune response.
10. Distinguish between active and passive immunity.
11. Explain the anaphylactic hypersensitivity response.
12. Describe the A-B-O and Rh blood group systems, and explain the difficulties that can arise when a person is transfused with an incompatible blood type.

CHAPTER 22

The body is continually exposed to a wide variety of potentially harmful factors, including bacteria, viruses, and hazardous chemicals. Pathogenic bacteria that invade the body often release enzymes that break down cell membranes and organelles or give off toxins that disrupt the functions of organs and tissues. Viruses can enter cells and utilize cellular facilities to reproduce more virus particles. They can kill cells by depleting them of essential components or by causing them to produce toxic substances.

Fortunately, the body is able to resist many organisms and chemicals that can damage tissues. This ability is called **immunity.** Some of the body's immunity is provided by nonspecific defense mechanisms that do not require previous exposure to a particular foreign substance in order to react to it. These mechanisms include the barriers formed by body surfaces and certain inflammatory responses that occur when tissues are injured. Immunity is also provided by specific immune responses that depend on prior exposure to a specific foreign material, recognition of it upon subsequent exposure, and reaction to it. These responses are mediated by specialized lymphocytes and cells derived from them.

A wide variety of cells and chemical factors participate in the body's nonspecific defense mechanisms and specific immune responses. Table 22.6 at the end of the chapter lists many of the cells and chemical factors discussed in this chapter and summarizes their functions.

Body Surfaces

The skin and mucous membranes that form the body surfaces are the first barriers to the invasion of the body by potentially damaging factors. The intact skin prevents many microorganisms and chemicals from entering the body. The skin's sweat glands and sebaceous glands secrete substances that are toxic to many types of bacteria, and the outer portion of the epidermis contains a tough, water-insoluble protein called keratin that is resistant to weak acids and bases and many protein-digesting enzymes.

The mucous membranes lining body cavities that open to the exterior—for example, the digestive, respiratory, urinary, and reproductive tracts—secrete mucus, which entraps small particles that can then be swept away or engulfed by phagocytic cells. The mucous membrane of the upper respiratory tract contains ciliated cells that move mucus and its entrapped particles toward the throat (pharynx). The tears that bathe the mucous membranes of the eyes contain the enzyme *lysozyme,* which has antimicrobial activity, and the acidic gastric juice secreted by the stomach can destroy many bacteria and bacterial toxins.

Although the surface barriers are quite effective, from time to time potentially damaging agents or organisms do gain entrance to the body. It is then necessary for other defense mechanisms to neutralize or destroy them.

Inflammation

In its most basic form, **inflammation** is an acute, nonspecific, physiological response of the body to tissue injury caused by factors such as chemicals, heat, mechanical trauma, or bacterial invasion. Although some differences may be evident—depending on the causative agent, the site of injury, and the state of the body—the fundamental events of the *acute, nonspecific inflammatory response* are similar in virtually all cases (Table 22.1). These events include changes in blood flow and vessel permeability as well as the movement of leukocytes from the vessels into the tissues. The acute, nonspecific inflammatory response brings into the invaded or injured area plasma proteins and phagocytes that can inactivate or destroy the invaders, remove debris, and set the stage for tissue repair. The events of this response are mediated by a number of chemical substances that are released or generated in the injured area.

Changes in Blood Flow and Vessel Permeability

Although a transient vasoconstriction may occur immediately following an injury to the tissues, small vessels in the injured area soon dilate. This dilation leads to an increased blood flow to the area, which increases the delivery of plasma proteins and phagocytic leukocytes. In addition, the permeability of capillaries and venules in the area increases, and plasma fluid and solutes, including proteins and other large molecules, move out of the blood vessels and into the inflamed tissues. This movement is aided by the vasodilation that occurs, because the dilation of arterioles in the area leads to an increased pressure within capillaries and venules that favors the filtration of fluid and solutes out of the vessels. Moreover, the movement of proteins from the blood into the interstitial fluid due to the increased vessel permeability diminishes the difference in protein concentration between the plasma and the interstitial fluid, and thus favors the accumulation of fluid within the tissues.

As fluid and solutes move out of the blood vessels and into the tissues during inflammation, the viscosity of the

◆ **TABLE 22.1 Events of Inflammation**

EVENT	SIGNIFICANCE	CONTRIBUTING MEDIATORS OR MECHANISMS
Dilation of small blood vessels	Increases blood flow to injured area; increases delivery of plasma proteins and phagocytic leukocytes	Histamine, kinins, complement components, prostaglandins
Increased permeability of small vessels (particularly capillaries and venules)	Plasma fluid and solutes, including proteins and other large molecules, move out of the blood vessels and into inflamed tissues	Histamine, kinins, complement components, prostaglandins
Slowing of blood flow	Associated with margination and pavementing of leukocytes	Increased blood viscosity and clumping of erythrocytes due to loss of fluid and solutes
Walling off of injured area	Limits spread of toxic products or bacteria	Formation of fibrin in injured tissues
Margination and pavementing of leukocytes	Associated with emigration of leukocytes	Clumping of erythrocytes and slowing of blood flow, leukotrienes, endothelial leukocyte adhesion molecules
Emigration of leukocytes	Leukocytes leave the blood vessels and enter injured tissues	Amebalike activity of leukocytes
Aggregation of leukocytes and macrophages at site of injury	Provides large numbers of phagocytic cells to engulf invading agents or organisms	Amebalike activity of leukocytes and macrophages attracted by chemotactic factors (for example, kinins, complement components)
Phagocytosis by leukocytes and macrophages	Removes invading agents or organisms and debris	Certain surface characteristics of materials to be phagocytized; opsonins

blood increases and erythrocytes clump together. These occurrences increase the resistance to blood flow, and as inflammation progresses, blood flow through small vessels in the injured area slows and sometimes even stops.

Walling-Off Effect of Inflammation

One of the substances that moves from the blood into the tissue spaces during inflammation is fibrinogen. In the tissues, fibrinogen is converted into fibrin, forming a clot that walls off the injured area. This walling-off effect can delay or limit the spread of toxic products or bacteria.

Leukocyte Emigration

As erythrocytes clump together and the blood flow slows during inflammation, a process known as *margination* takes place. Leukocytes—particularly neutrophils and monocytes—are displaced to the periphery of the bloodstream, where they contact the endothelial linings of capillaries and venules in the inflamed area. At first the leukocytes slowly tumble or roll along the endothelial surfaces, but soon a phenomenon called *pavementing* occurs, and the leukocytes adhere to the vessel surfaces. The adhering leukocytes begin to exhibit an amebalike activity, and they squeeze between the endothelial cells of the capillaries and venules into the tissue spaces (Figure 22.1). The first leukocytes to arrive in the tissues are generally neutrophils, followed sometime later by monocytes. In the tissues, monocytes are transformed into macrophages, and some of the macrophages normally present in the tissues multiply and become motile.

Chemotaxis and Leukocyte Aggregation

Leukocytes and macrophages migrate through the tissues at varying rates of speed and aggregate at the site of injury or invasion. Their direction of movement is determined by chemical mediators released at the site of injury, and they are attracted to the highest concen-

◆ **FIGURE 22.1 Leukocytes leaving a blood vessel and entering the tissues**

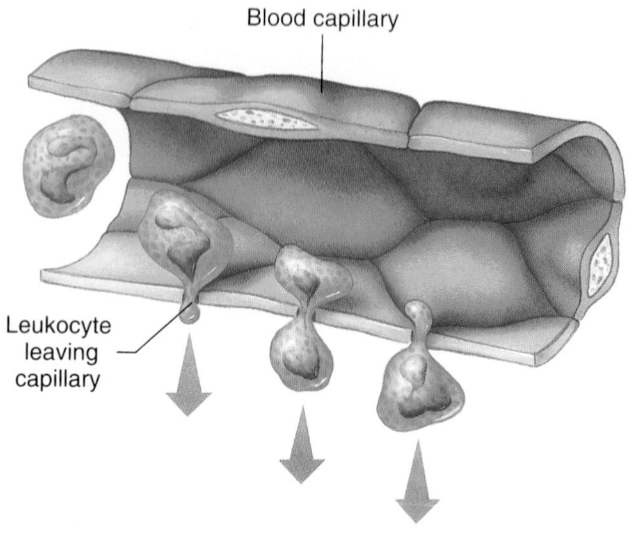

trations of certain chemical substances. Such a movement of cells in response to chemical factors is called *chemotaxis (kem-o-tak´-sis).* Positive chemotaxis is a movement toward a chemical substance, and negative chemotaxis is a movement away from a chemical substance.

In most acute, nonspecific inflammatory responses, leukocytes and macrophages aggregate at the site of injury or invasion in a fairly predictable sequence. Neutrophils usually predominate in the early stages of the response, and monocytes and macrophages predominate during the later stages. Neutrophils are present in greater numbers in the circulation and are more mobile than monocytes, and these facts probably contribute to their initial predominance. However, in the tissues, neutrophils die off faster than monocytes and macrophages. Moreover, neutrophils may produce or potentiate factors that facilitate monocyte emigration. These facts are believed to contribute to the predominance of monocytes and macrophages later in inflammation.

In contrast to the dominance of neutrophils, monocytes, and macrophages in most acute, nonspecific inflammatory responses, certain types of allergies and inflammatory responses to parasites are characterized by the presence of large numbers of eosinophils. In fact, a preponderance of eosinophils is often indicative of an allergic response.

Phagocytosis

A major benefit of inflammation is the phagocytosis of microorganisms, foreign materials, and debris by leukocytes—particularly neutrophils—and macrophages. The phagocytic cells identify the materials to be engulfed by "recognizing" certain surface characteristics of the materials.

The phagocytic activity of leukocytes and macrophages can be modified by many factors, including a number of proteins collectively called *opsonins.* Opsonins attach to specific foreign particles and thereby render them more susceptible to phagocytosis.

During phagocytosis, a phagocytic cell commonly attaches to a particle and engulfs it by endocytosis, forming a membrane-bounded phagocytic vesicle, or phagosome, that contains the engulfed particle within the phagocytic cell. Enzymes associated with the membranes of phagocytic vesicles catalyze the production of substances including hydrogen peroxide (H_2O_2), oxygen radicals (O_2^-), and halogenated oxides such as hypochlorite (OCl_2). The oxygen radicals and halogenated oxides are toxic substances that react with many organic molecules and inactivate many phagocytized microorganisms. In addition, lysosomes containing powerful digestive enzymes fuse with phagocytic vesicles. Since the lysosomes appear microscopically as granules, this process is called *degranulation.* The lysosomal enzymes break engulfed materials down into products of low molecular weight that can be utilized by the phagocytic cell. Macrophages, but apparently not neutrophils, can also extrude residual breakdown products from the cells.

A phagocytic cell continues to ingest and digest foreign particles until toxic substances from the foreign particles and from the activity of the cell itself accumulate and kill the cell. Macrophages are larger than neutrophils and less selective about what they engulf. Moreover, they generally phagocytize more particles before they die. As a result, macrophages are particularly important in clearing foreign material and tissue debris from an inflamed area.

In addition to their phagocytic activity, neutrophils can release a number of destructive chemicals, including lysosomal enzymes, directly into the interstitial fluid. These chemicals can destroy extracellular material without prior phagocytosis.

Abscess and Granuloma Formation

In most cases, the inflammatory response is able to overcome the injury or invasion of the body and clear the area for tissue repair. Occasionally, however, the inflammatory response is unable to overcome an invasion, and an abscess or a granuloma forms. An **abscess** is a sac of pus that consists of microbes, leukocytes, macrophages, and liquified debris walled off by fibroblasts and collagen. It does not diminish spontaneously and must be drained. A **granuloma** *(gran´´-yoo-lo´-muh)* can form when the invading agents are microbes that can survive

◆ **FIGURE 22.2 Role of histamine as a chemical mediator of the inflammatory response**

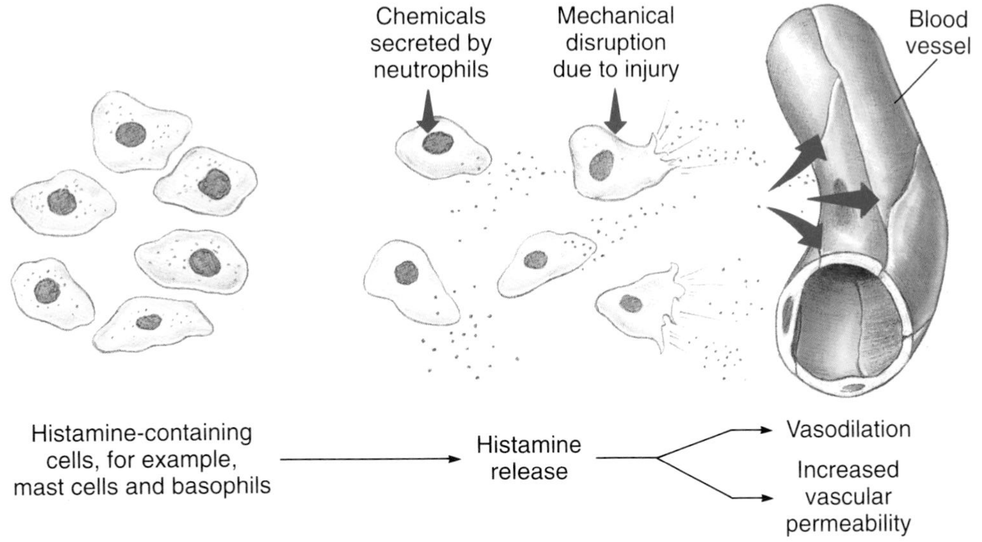

within the phagocytes or are materials that cannot be digested by the phagocytes. It consists of layers of phagocytic-type cells and is usually surrounded by a fibrous capsule. The central cells contain the offending agent. In this fashion, a person may harbor live bacteria such as tuberculosis-causing bacteria for many years without displaying any overt symptoms.

Chemical Mediators of Inflammation

Although tissue injury precipitates the acute, nonspecific inflammatory response, a variety of chemical substances mediate it. Some of these substances are released from tissue cells in the injured area, some are released from leukocytes, and some are generated by enzyme-catalyzed reactions.

Histamine

Histamine is present in many cells, particularly mast cells, basophils, and platelets (Figure 22.2). It causes vasodilation (particularly of arterioles) and increased vascular permeability (particularly in venules). Several factors cause histamine release, including mechanical disruption of cells due to injury and chemicals secreted by neutrophils attracted to the site.

Kinins

The **kinins** *(kī´-ninz)* are a group of polypeptides that dilate arterioles, increase vascular permeability, act as powerful chemotactic agents, and induce pain (Figure 22.3). They are formed from inactive precursors called kininogens that normally circulate in the plasma. The enzyme ***kallikrein,*** which is also present in the plasma in inactive form (prekallikrein), acts on kininogens to form kinins. Several factors can convert prekallikrein to kallikrein. Among them is active factor XII, which is also involved in blood clotting. Moreover, kallikrein is present in many body tissues, as well as in neutrophils, and once inflammation has begun, kallikrein from these sources contributes to kinin generation. In fact, the kinin-generating system displays a positive feedback in that kinins act chemotactically to attract neutrophils whose kallikrein can catalyze the production of still more kinins. In addition, tissue kallikrein cannot normally act on plasma kininogens because the kininogens are too large to pass out of the blood vessels. In inflammation, the kinins increase vascular permeability, allowing kininogens to enter the tissues and become activated by kallikrein.

Complement System

The **complement system** is particularly important in specific immune responses, but it can also contribute to acute, nonspecific inflammation. This system consists of a series of plasma proteins that normally circulate in the blood in inactive forms (Figure 22.4). The activation of the system's initial components initiates a cascading series of reactions that ultimately produce active complement molecules.

Active complement components can mediate virtually every event of the acute, nonspecific inflammatory response. Complement components enhance vasodilation

◆ **FIGURE 22.3 Role of kinins as chemical mediators of the inflammatory response**

The blue arrow indicates the positive-feedback effect of kinin generation on the formation of additional kinins. See text for details.

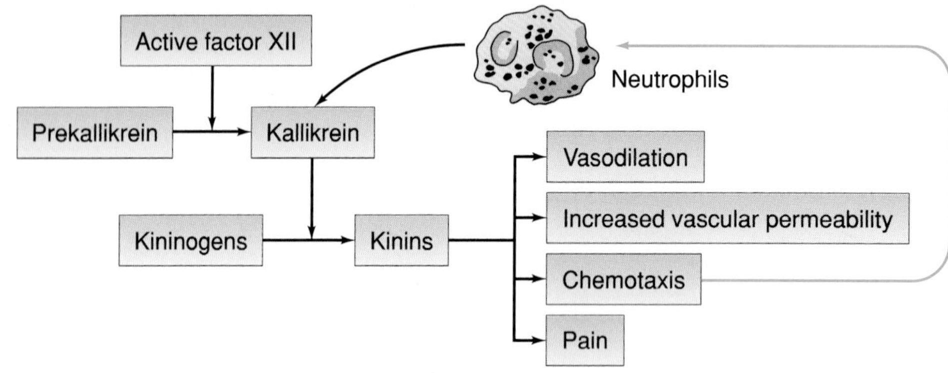

and vascular permeability by direct effects on blood vessels, by stimulating histamine release from mast cells and platelets, and by activating plasma kallikrein. They act as chemotactic agents in attracting neutrophils and also act as opsonins that enhance phagocytosis by attaching to microbes, where they serve as structures to which phagocytic cells can bind. In addition, complement components directly attack and kill cells such as invading microbes. The complement components involved in this microbe-killing activity become embedded in the surface of a microbe in groups called *membrane-attack complexes.* Membrane-attack complexes kill microbes by forming tunnels that make the microbes leaky.

As is discussed later in the chapter (page 696), structures called antigen-antibody complexes, which result from specific immune responses, are powerful activators

◆ **FIGURE 22.4 Role of complement components as chemical mediators of the inflammatory response**

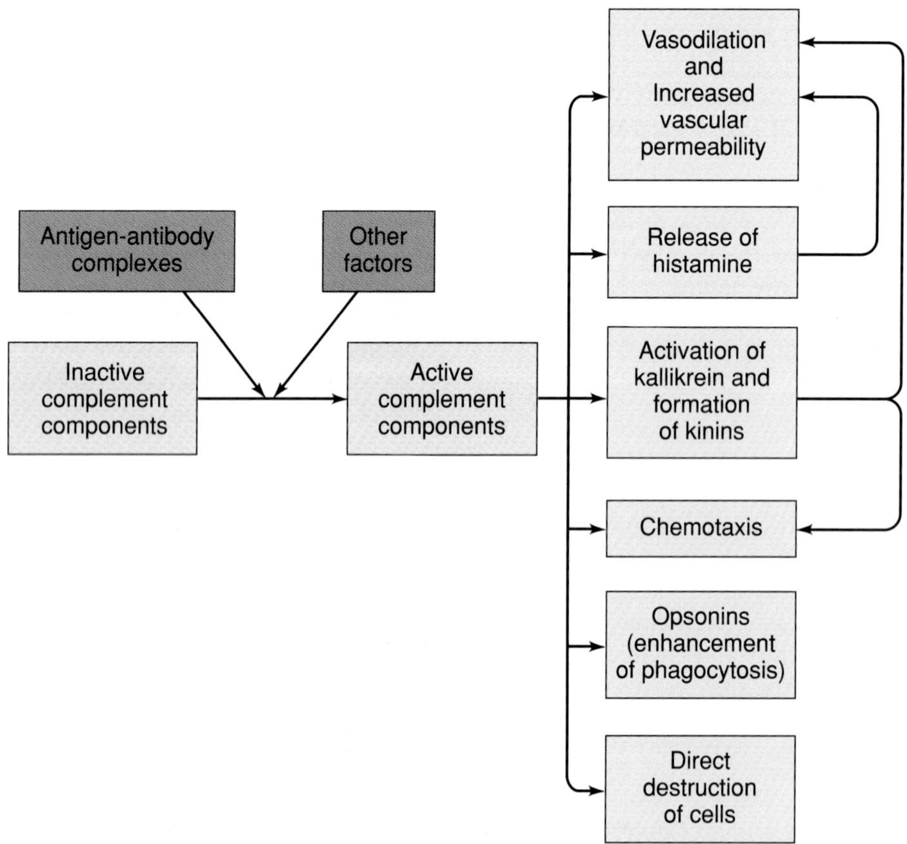

of the complement system. However, other factors—for example, polysaccharides in the cell walls of certain bacteria, protein A of staphylococcal microorganisms, and some lysosomal and bacterial enzymes—can also activate the complement system. Factors such as these may trigger the production of active complement components in acute, nonspecific inflammation.

Prostaglandins

Prostaglandins may also play a role in inflammatory responses. For example, prostaglandin PGE_1 induces vasodilation and increased vascular permeability, and it may stimulate leukocyte emigration through capillary walls. Prostaglandins are believed to contribute to the pain associated with inflammation, and in many cases they mediate the fever-producing effects of substances called endogenous pyrogens that are released from certain cells during inflammatory responses.

Leukotrienes

Leukotrienes are a group of substances produced by leukocytes. They are chemically related to the prostaglandins. Certain leukotrienes cause a relatively short-lived, transient constriction of arterioles that is followed by an increase in vascular permeability, particularly in venules. Some leukotrienes increase the adherence of leukocytes to endothelial cells of small venules, and some are chemotactic agents that attract neutrophils. Leukotrienes are very potent constrictors of peripheral air passageways in the lungs, and they stimulate mucus secretion in lung airways. Leukotrienes appear to play a significant role as mediators of asthmatic responses and certain allergic reactions.

Symptoms of Inflammation

Inflammation can produce both local and systemic symptoms.

Local Symptoms

The main local symptoms of inflammation are redness, heat, swelling, and pain (Table 22.2). The redness and heat (increased temperature) are due to the dilation of blood vessels and increased blood flow to the inflamed area. The swelling (edema) is caused by the increased vessel permeability and the movement of fluid and solutes out of the vessels and into the tissue spaces. The pain has been attributed to the increased pressure exerted on sensory nerve endings in the swollen tissues and to the effects of kinins and perhaps other chemical mediators of inflammatory responses on afferent nerve terminals.

◆ **TABLE 22.2 Local Symptoms of Inflammation**

SYMPTOM	CONTRIBUTING CAUSES
Redness	Dilation of vessels, increased blood flow to injured area
Heat (increased temperature)	Dilation of vessels, increased blood flow to injured area
Swelling (edema)	Increased vessel permeability, and movement of fluid and solutes out of the vessels and into the tissue spaces
Pain	Increased pressure on sensory nerve endings in swollen tissues; effects of some chemical mediators of inflammation (for example, kinins) on afferent nerve terminals

Systemic Symptoms

Fever is a common systemic symptom of inflammation, especially in inflammatory responses associated with the spread of organisms into the bloodstream. Macrophages involved in the inflammatory response release endogenous pyrogens, which include protein substances called *interleukin-1* and *tumor necrosis factor.* The endogenous pyrogens influence the body's temperature-regulating mechanisms and induce fever. Although an extremely high fever can be harmful, phagocytosis is favored by higher body temperatures. Thus, fever (as well as the local increase in temperature at the site of inflammation) may enhance this activity.

In addition to eliciting fever, endogenous pyrogens also depress the concentrations of iron and zinc in the blood. This action may retard the proliferation of bacteria that require a high concentration of iron to multiply.

Inflammation, particularly in response to many types of bacterial invasion, is often accompanied by an increased production and release of leukocytes. The leukocyte count in the peripheral blood may rise from about 7000 cells/mm^3 to 25,000 or more, and in most nonspecific inflammatory responses, the increase is due to an absolute as well as a relative rise in the number of circulating neutrophils. In contrast, infections caused by viruses and protozoa are often associated with decreased numbers of leukocytes in the blood.

Interferon

The glycoprotein **interferon** *(in″-tur-feer′-on)* provides some protection to the body against invasion by viruses, particularly until the more slowly reacting specific immune responses can take over. In general, when a virus invades the body, it enters a cell and reproduces more virus particles by utilizing cellular facilities to synthesize viral components, including nucleic acids and proteins. The new viruses are released to infect other, healthy cells and repeat the process. However, viruses (as well as other substances) also induce cells to produce interferon (Figure 22.5). One of the most effective inducers of interferon production is double-stranded RNA, and it has been suggested that viruses and other interferon inducers cause the production of double-stranded RNA within cells. The double-stranded RNA is believed to trigger the DNA of a cell to produce a messenger RNA that directs the production of interferon by the cell. The interferon leaves the cell and binds to receptors on the plasma membranes of other cells. This binding triggers the synthesis of enzymatic proteins within the cells that can prevent viral reproduction by breaking down messenger RNAs and inhibiting protein synthesis. The newly synthesized enzymatic proteins are inactive until the cells are infected by a virus or exposed to double-stranded RNA. This activation requirement may help protect the cells' mechanisms for synthesizing nucleic acids and proteins from being inhibited in the absence of viral infection.

It is thought that interferon inhibits the activity of some viruses by other mechanisms. Although interferon apparently cannot prevent the reproduction of certain viruses, the viruses are either not released from the cells or, if they are released, are not capable of infecting other cells. Moreover, there is evidence that, at least in some cases, interferon-stimulated cells can transfer their interferon-induced antiviral activity to adjacent cells by mechanisms requiring cell-to-cell contact.

All cells can manufacture interferon when they are appropriately stimulated, and at least three different

◆ **FIGURE 22.5 Mechanism by which interferon acts to prevent viral replication**

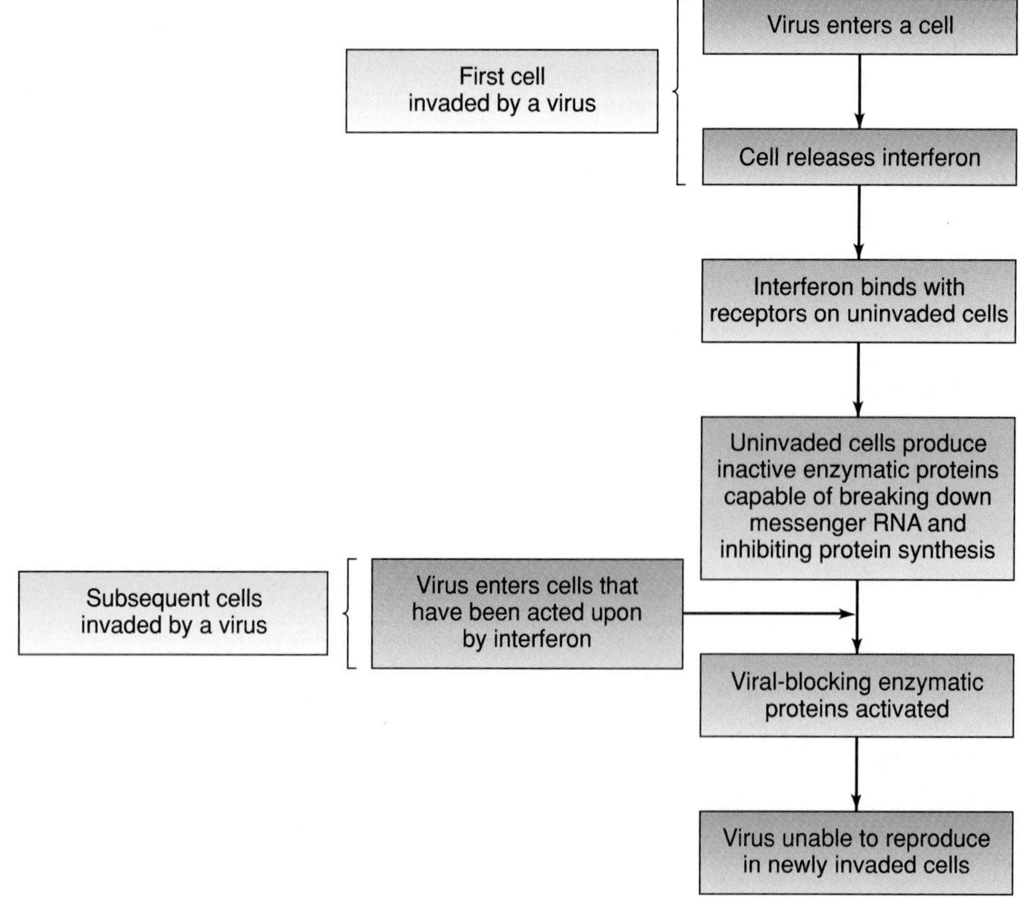

forms of interferon are produced. These different forms of interferon are named alpha interferon (also called leukocyte interferon), beta interferon (fibroepithelial interferon), and gamma interferon (immune interferon).

Interferon may play a role in protecting the body against some forms of cancer. Although viruses are suspected of causing some human cancers, interferon also seems to work against tumors thought to be caused by nonviral agents such as radiation and chemicals. In fact, interferon is increasingly viewed as a versatile agent whose effects on cells go beyond the inhibition of viral activity. For example, it slows cell division and inhibits the proliferation of both healthy and abnormal cells. It mobilizes and enhances the cell-killing properties of macrophages and stimulates the activity of a specialized group of cells called natural killer (NK) cells. Both macrophages and natural killer cells can attack tumor cells, and natural killer cells are believed to be important in *immune surveillance*—that is, in detecting and eliminating abnormal, potentially cancerous cells before they multiply and cause clinical cancer. In fact, it has been suggested that clinical cancer is due to ineffective immune surveillance mechanisms. Gamma interferon appears to be a more potent antitumor agent than alpha interferon or beta interferon.

Specific Immune Responses

The **specific immune responses** generally require previous exposure to a foreign agent or organism to be most effective. The specific immune responses are not always beneficial, however, and they are sometimes responsible for allergies and other harmful reactions.

Two major aspects of the specific immune responses are recognized. One aspect, known as **antibody-mediated immunity,** or **humoral immunity,** is mediated by specialized proteins called antibodies. Antibodies are produced by plasma cells that are progeny of lymphoid cells called B cells. The second aspect, known as **cell-mediated immunity,** is mediated by certain populations of lymphoid cells called T cells.

Antigens

An **antigen** *(an´-ti-jin)* is a substance that has the capacity to elicit a specific immune response. (An antigen that actually elicits a specific immune response when it is present in a person's body is called an *immunogen.*) Substances that act as antigens generally have molecular weights of 8000 to 10,000 or more, and proteins, polysaccharides, and nucleic acids commonly act as antigenic materials. Substances of lower molecular weight, collectively called *haptens,* can also act as antigens if they combine with larger molecules such as proteins.

A person ordinarily does not generate specific immune responses to the antigens of his or her own body—that is, to *self antigens.* However, a person usually does generate specific immune responses to other antigens—that is, to *foreign antigens (nonself antigens).*

A group of antigens called **major histocompatibility complex (MHC) antigens** are particularly important in specific immune responses. (The MHC antigens of humans are also referred to as *human-leukocyte-associated antigens,* or *HLA antigens.*) MHC antigens are displayed on the surfaces of body cells, and different individuals—except identical twins—possess different MHC antigens. Thus, MHC antigens serve as important markers that distinguish a person's own cells *(self cells)* from other cells *(foreign,* or *nonself, cells).* There are two principal classes of MHC antigens: class I MHC antigens, which are produced by virtually all nucleated cells of the body; and class II MHC antigens, which are produced by certain cells involved in the generation of specific immune responses.

Antibodies

An **antibody** is a specialized protein that is produced in response to the presence of an immunogenic antigen. An antibody can combine with the specific antigen that stimulated its production to form an *antigen-antibody complex.*

Antibodies belong to a family of proteins called globulins and are referred to as *immunoglobulins (Ig).* There are five immunoglobulin classes of antibodies: *IgA, IgD, IgE, IgG,* and *IgM.* Class IgG and class IgM antibodies are involved in specific immunity against bacteria and viruses. (Most of the antibodies in a person's blood are class IgG antibodies.) Class IgE antibodies are involved in specific immune responses to multicellular parasites and in certain allergic responses. Class IgA antibodies can be transported across certain epithelial cells. They are present predominantly in intestinal and respiratory secretions, in saliva, and in tears. Class IgA antibodies are particularly important in protecting epithelial surfaces from bacteria and other microorganisms. Class IgD antibodies are present in the body in relatively small amounts. Their precise function is unclear.

The basic structure of an antibody consists of two "heavy" and two "light" polypeptide chains, which are linked together to form a Y-shaped antibody *monomer* (Figure 22.6). One end of each chain contains a variable region whose amino acid sequence differs for each kind of antibody. The particular conformation of the variable regions enables an antibody to bind to a specific antigen in a manner that is analogous to the interaction between

◆ **FIGURE 22.6 Diagrammatic representation of the basic structure of an antibody**

The structure consists of four polypeptide chains, two heavy and two light. The composition of many regions of the chains is constant (C) among antibodies of a particular class. Only relatively small regions vary in composition (V) from antibody to antibody.

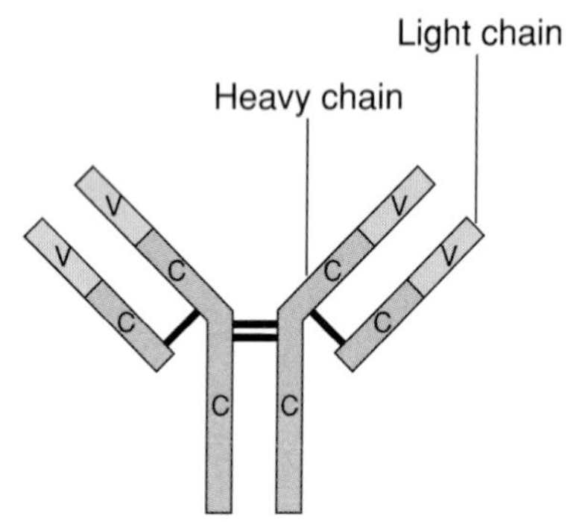

an enzyme and its substrate. The remainder of each polypeptide chain is called the constant region. The constant regions contain binding sites for molecules and cells that function as effectors of antibody-mediated activities. The constant regions of the heavy polypeptide chains differ from one class of antibody to another, and these differences provide the basis for distinguishing the various classes of antibodies.

Antibodies inactivate or destroy foreign substances in a number of ways. They can combine with bacterial toxins or destructive foreign enzymes and inactivate them by inhibiting their interactions with target cells or substrates. They can also bind to surface components of viruses, preventing the viruses from attaching to target cells and thereby keeping the viruses from entering the cells.

A major function of antibodies is to enhance the basic inflammatory response. Antigen-antibody complexes, particularly those involving antibodies of the IgG or IgM classes, are powerful activators of the complement system (Figure 22.4). As noted earlier, active complement components can mediate virtually every aspect of the acute, nonspecific inflammatory response. Moreover, if an antigen-antibody complex is on the surface of a foreign cell, active complement components attack the cell to cause lysis and cell death. Antibodies, particularly those of the IgG class, also act as opsonins that enhance phagocytosis by attaching to foreign substances, where they serve as structures to which phagocytic cells can bind.

B Cells

B cells are lymphocytes that are committed to differentiate into the antibody-producing plasma cells involved in antibody-mediated immunity. Like all lymphocytes, B cells are derived from stem-cell precursors in the red bone marrow (Figure 22.7). Stem cells that are destined to give rise to B cells proliferate and develop within the bone marrow (or within the liver and perhaps the spleen during fetal life). B cells continually circulate through the blood, tissues, and lymph, and at any one time large numbers of B cells are located in the lymph nodes, spleen, and other lymphoid tissues.

T Cells

T cells are a heterogeneous group of lymphocytes that include those cells committed to participating in cell-mediated immune responses. Lymphocyte stem cells from the red bone marrow that are destined to give rise to T cells proliferate and differentiate in the thymus gland (Figure 22.7). In addition, hormonal substances from the thymus—for example, thymosin—influence T cells that have left the gland. (Most T cells leave the thymus early in life, and the thymus gradually atrophies in adults.) Like B cells, T cells continually circulate through the blood, tissues, and lymph, and at any one time large numbers of T cells are located in the lymph nodes, spleen, and other lymphoid tissues.

There are two major classes of T cells: **helper T cells** and **cytotoxic T cells.** The functions of these cells are considered later in the chapter. The characteristics of B cells and T cells are compared in Table 22.3.

Accessory Cells

In addition to B cells and T cells, **accessory cells,** such as macrophages, play a significant role in specific immune responses. As is discussed in the next two sections, one of the major roles of accessory cells is to present foreign antigens, in combination with self MHC antigens, to certain other cells involved in generating specific immune responses.

Accessory cells present antigens by displaying combinations of foreign antigens and self MHC antigens on their surfaces. For example, in many cases, macrophages engulf and process a foreign protein antigen. During this processing, portions of the foreign antigen combine with self MHC antigens forming foreign-antigen–self-MHC-antigen complexes that become associated with the plasma membranes of the macrophages at the surfaces of the cells. (The portions of a foreign protein antigen—usually a few short peptide segments—that combine with self MHC antigens are said to be *immunodominant.*) Macrophages produce both class I and class II self MHC antigens, and they can display on their surfaces foreign antigens in combination with both class I and class II self MHC antigens.

◆ **FIGURE 22.7 Formation and differentiation of B cells and T cells**

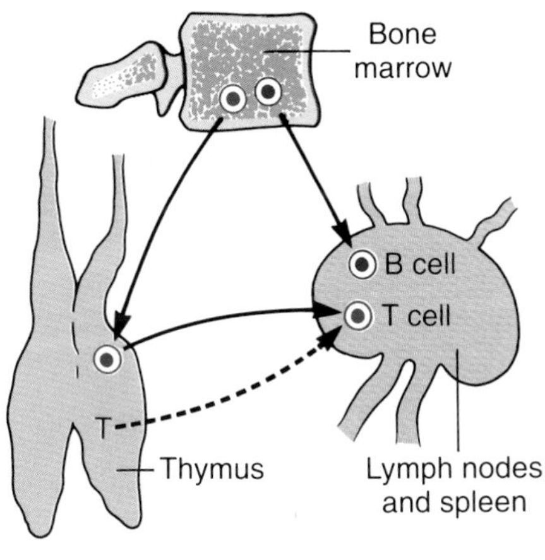

Several kinds of cells besides macrophages can also serve as accessory cells that present foreign antigens in combination with self MHC antigens. These antigen-presenting cells include *interdigitating cells,* which are located in the T-cell regions of lymphoid tissues; *dendritic reticular cells,* which are located in the B-cell regions of lymphoid tissues; and *Langerhans cells,* which are located in the skin and parts of the gastrointestinal tract. (A foreign antigen that penetrates the epithelial surfaces of the skin or gastrointestinal tract can be carried by Langerhans cells to lymph nodes.) Like macrophages, interdigitating cells, dendritic reticular cells, and Langerhans cells produce both class I and class II self MHC antigens.

The antigen-presenting activities of the various accessory cells are important because some cells involved in specific immune responses generally recognize and respond to foreign antigens only when the antigens are presented in combination with self MHC antigens. For example, helper T cells recognize foreign antigens presented in combination with class II self MHC antigens, and cytotoxic T cells recognize foreign antigens presented in combination with class I self MHC antigens.

Antibody-Mediated (Humoral) Immunity

Antibody-mediated immunity depends on B cells. It involves the production and release into the blood and lymph of antibodies to various antigens that the body recognizes as foreign. Antibody-mediated immune responses are important in providing specific immune resistance to bacterial toxins and to the extracellular phases of bacterial and viral infections.

B cells display antibody molecules on their surfaces (Figure 22.8). These cell-surface antibody molecules function as antigen receptors. Each B cell displays only a single kind of antibody molecule (in terms of antigen-binding specificity).

When a foreign antigen enters the body, the antigen binds to antibody molecules (receptors) specific for it displayed on the surfaces of particular B cells. The binding of antigen to antibody molecules on the surface of a B cell contributes to the *activation* of the B cell. The active B cell proliferates and differentiates. Most of the proliferating cells develop into *plasma cells* that manufacture the antibody displayed on the B cell's surface (other cells become *memory cells* as is discussed on page 702). The plasma cells release antibody molecules into the blood and lymph. The antibody molecules ultimately combine with the specific antigen that stimulated their production, and responses such as those described in the earlier section on antibodies occur (see page 695).

◆ **TABLE 22.3 Characteristics of B Cells and T Cells**

CHARACTERISTIC	B CELLS	T CELLS
Ancestral origin	Bone marrow	Bone marrow
Site of maturational processing	Bone marrow	Thymus
Receptors for antigens	Antibodies inserted in plasma membrane, which serve as surface receptors; highly specific	Surface receptors present but differing from antibodies; highly specific
Bind with	Extracellular antigens such as bacteria, free viruses, and other circulating foreign material	Foreign antigen in association with self MHC antigen
Types of active cells	Plasma cells	Active helper T cells, active cytotoxic T cells
Formation of memory cells	Yes	Yes

◆ **FIGURE 22.8 Diagrammatic representation of the basic events of an antibody-mediated immune response**
See text for details.

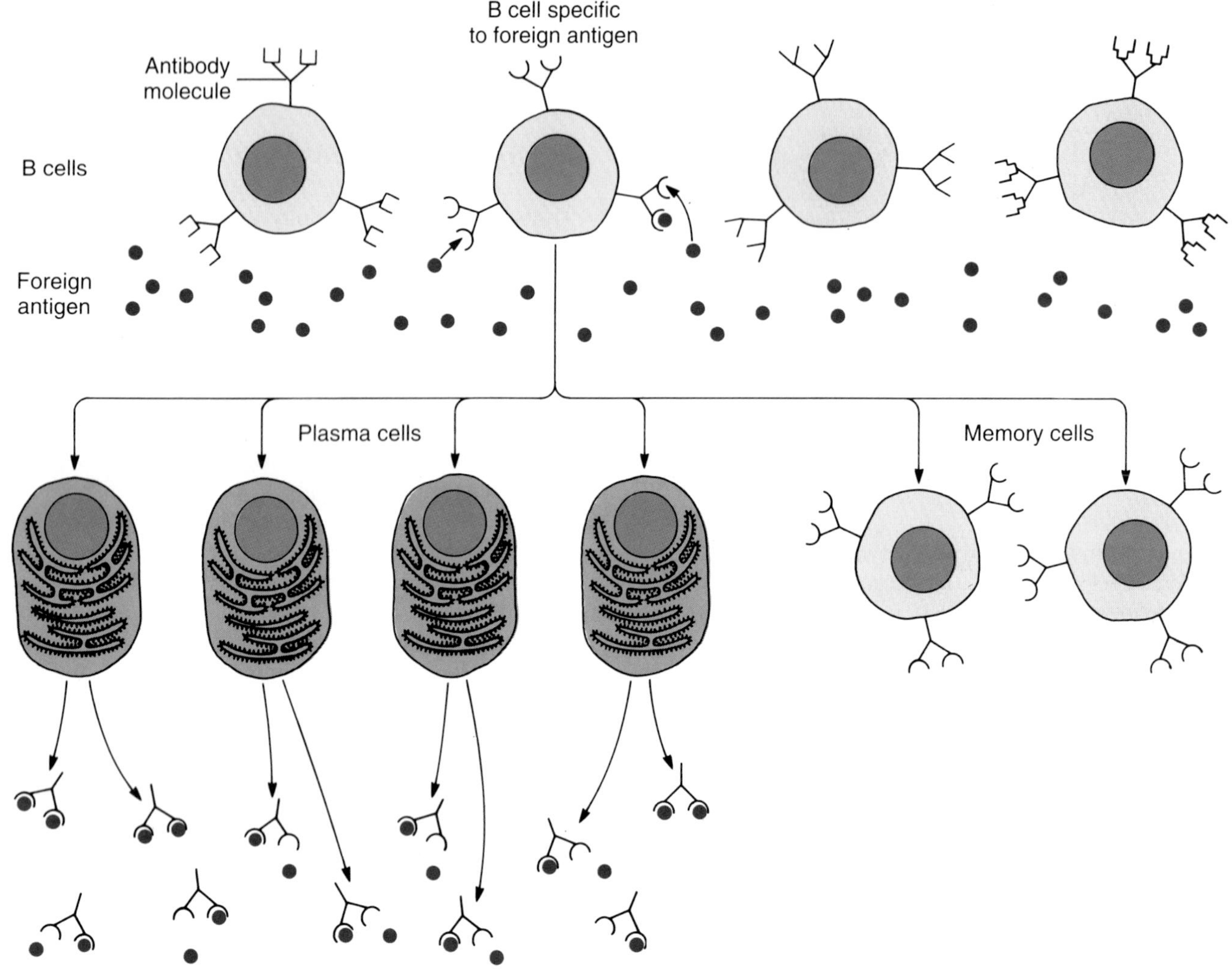

The body can produce enough different B cells to allow for the production of antibodies to essentially all the foreign antigens to which a person might ever be exposed.

In actual fact, B-cell activation and antibody production usually involve more than just the binding of a foreign antigen to antibody molecules (antigen receptors) on the surfaces of particular B cells. In most cases, accessory cells, such as macrophages, and helper T cells are also involved.

For example, in the case of a foreign protein antigen, accessory cells present the foreign antigen (actually a portion of the foreign antigen), in combination with class II self MHC antigens, to helper T cells (Figure 22.9a). Helper T cells possess specific cell-surface receptors, and the receptors of some helper T cells can bind to complexes composed of the foreign antigen and class II self MHC antigens displayed on the surfaces of the accessory cells. These helper T cells bind to the accessory cells and interact with them (Figure 22.9b). The interaction activates the helper T cells.

A number of factors are believed to be important for the activation of the helper T cells. One factor is antigen binding by the helper T cells (that is, the binding that occurs between the helper T cells' receptors and the complexes composed of the foreign antigen and class II self MHC antigens displayed on the surfaces of the accessory cells). Another factor is a chemical called *interleukin-1,* which is produced by the antigen-presenting accessory cells.

The active helper T cells produce a chemical called *interleukin-2 (T-cell growth factor).* They also express

◆ **FIGURE 22.9 Diagrammatic representation of the events of an antibody-mediated immune response**
See text for details.

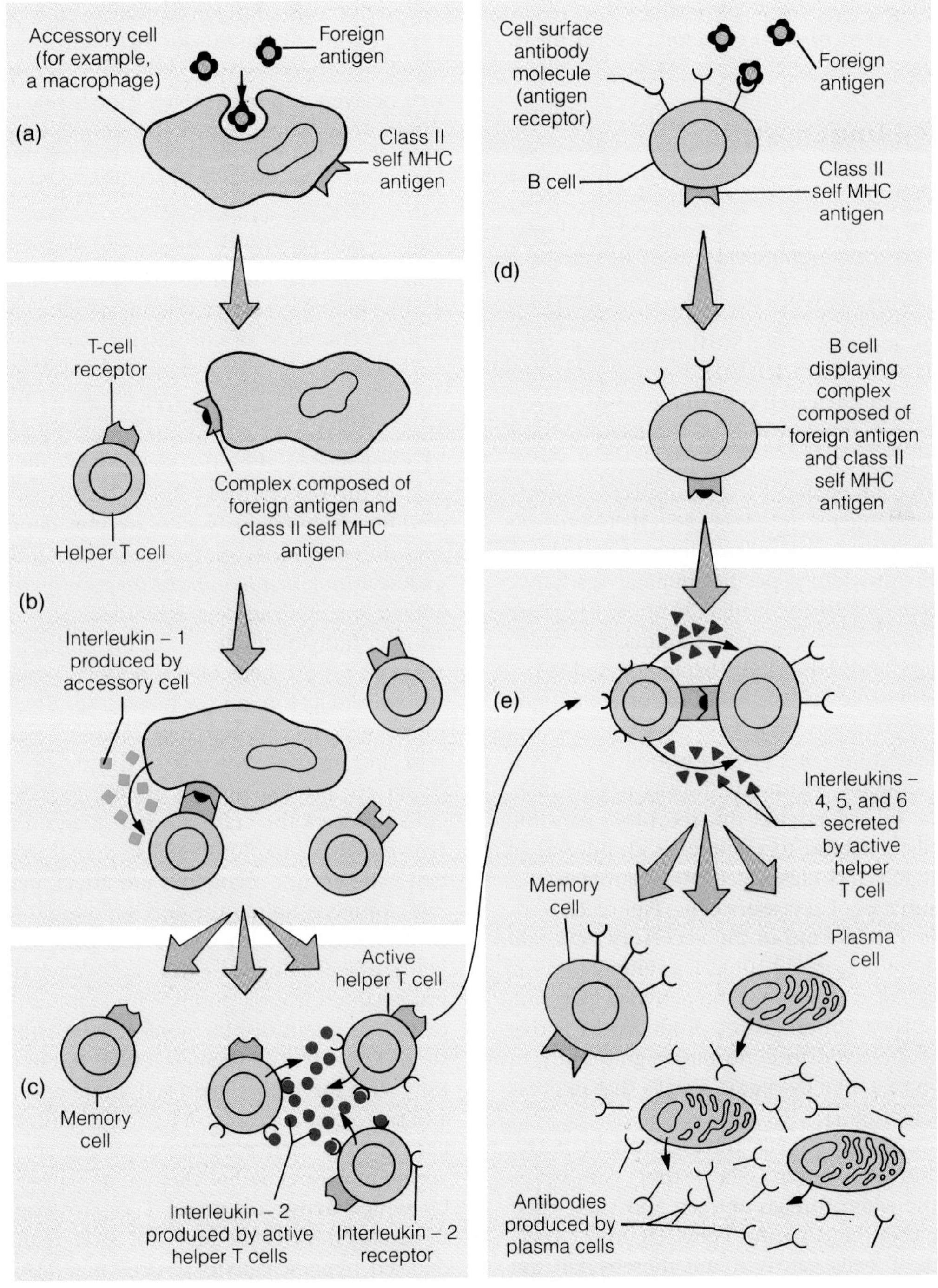

receptors for interleukin-2 on their surfaces. Interleukin-2 binds to the receptors and promotes the proliferation of the active helper T cells (Figure 22.9c).

A B cell not only binds a particular antigen, but it can also engulf the bound antigen, process it, and display portions of it in combination with class II self MHC antigens on its surface (Figure 22.9d). The active helper T cells bind with those B cells that display on their surfaces complexes composed of the same foreign antigen that initiated the activation of the helper T cells and class II self MHC antigens (Figure 22.9e). The active helper T cells promote the activation of the B cells and their proliferation and differentiation into antibody-secreting plasma cells, at least in part by secreting chemicals that include *interleukin-4 (B-cell stimulating factor-1), interleukin-5 (B-cell growth factor-2),* and *interleukin-6 (B-cell stimulating factor-2)* onto the B-cell surfaces.

Cell-Mediated Immunity

Cell-mediated immunity depends on T cells, and particularly on cytotoxic T cells. It is especially effective against cells that display combinations of foreign antigens and class I self MHC antigens on their surfaces. For example, most virus-infected cells display combinations of viral antigens and class I self MHC antigens on their surfaces. (Recall that class I self MHC antigens are produced by virtually all nucleated cells of the body.) Active cytotoxic T cells generated by the cell-mediated immune system in response to foreign viral antigens can attack and destroy virus-infected cells that display combinations of the viral antigens and class I self MHC antigens on their surfaces. Thus, cell-mediated immune responses are important in providing specific immune resistance to the intracellular phase of viral infections. They are also important in providing specific immune resistance to fungi, parasites, and intracellular bacteria. In addition, they play a major role in the rejection of solid-tissue transplants.

The cell-mediated immune system responds to a foreign antigen as follows. Cytotoxic T cells possess specific cell-surface receptors, and the receptors of some cytotoxic T cells can bind to complexes composed of the foreign antigen and class I self MHC antigens displayed on the surfaces of accessory cells (Figure 22.10a). These cytotoxic T cells bind to the accessory cells and interact with them (Figure 22.10b). The interaction activates the cytotoxic T cells, and the active cells proliferate (Figure 22.10c). (Interleukin-2 produced by active helper T cells is believed to contribute significantly to the proliferation of active cytotoxic T cells that express interleukin-2 receptors.)

The active cytotoxic T cells travel throughout the body. When they encounter cells bearing complexes composed of the same foreign antigen and class I self MHC antigens, they bind to the cells (Figure 22.10d). Active cytotoxic T cells can lyse and thereby kill the cells (Figure 22.11).

A group of helper-class T cells called **delayed hypersensitivity T cells** are also involved in cell-mediated immune responses. These cells are important mediators of long-term, chronic inflammatory responses. In general, delayed hypersensitivity T cells interact with accessory cells that display complexes composed of a foreign antigen and class II self MHC antigens on their surfaces. This interaction activates the delayed hypersensitivity T cells, and the active cells proliferate.

Active delayed hypersensitivity T cells travel throughout the body. When they encounter cells, such as Langerhans cells in the skin, that display complexes composed of the foreign antigen that initiated their production and class II self MHC antigens, the active delayed hypersensitivity T cells bind to the cells. The active delayed hypersensitivity T cells release a number of different chemical factors that generate inflammatory responses and enhance phagocytosis. Among these factors are a chemotactic factor that attracts macrophages to the area and a cytotoxic factor that can damage or kill many cells except lymphocytes. (In addition to active delayed hypersensitivity T cells, certain other groups of T cells may also release chemical factors that contribute to the generation of chronic inflammatory responses.)

Transplant Rejection

The cell-mediated immune responses play an important role in the rejection of solid-tissue transplants. The cells of different individuals who are not identical twins possess different MHC antigens (and also different antigens called *minor histocompatibility antigens*). When a tissue or organ from one individual is transplanted into another individual who is not the donor's identical twin, antigens on the cells of the donor's tissue or organ are recognized as foreign (or nonself) by the recipient's immune system, and active cytotoxic T cells that can attack and destroy the cells of the transplanted tissue or organ are produced. This ability of active cytotoxic T cells to attack the cells of a transplanted tissue or organ is somewhat puzzling because active cytotoxic T cells generally do not recognize and attack cells bearing foreign antigens unless the foreign antigens are displayed in combination with class I self MHC antigens on the cell surfaces. However, the cells of a tissue or organ transplanted from a donor who is not an identical twin of the recipient display nonself MHC antigens. Perhaps the nonself MHC antigens—either alone or in combination with foreign or even self antigens—are sufficiently similar to combinations of self MHC antigens and foreign antigens that active cytotoxic T cells produced by the recipient can recognize them in a manner similar to that by which active cytotoxic T cells recognize combinations of self MHC antigens and foreign antigens. (Active delayed hypersensitivity T cells may also be produced in response to the presence of foreign antigens from a transplanted tissue or organ, and the activities of these cells may contribute to the rejection of the transplant.)

In transplant surgery, the likelihood that a specific immune response will lead to destruction and rejection of a transplanted tissue or organ is lessened by trying to select a tissue or organ whose antigenic composition closely matches that of the recipient and by utilizing antilymphocyte serum or drugs that weaken or suppress specific immune responses. In many cases, however, the

◆ **FIGURE 22.10 Diagrammatic representation of a cell-mediated immune response involving cytotoxic T cells**
See text for details.

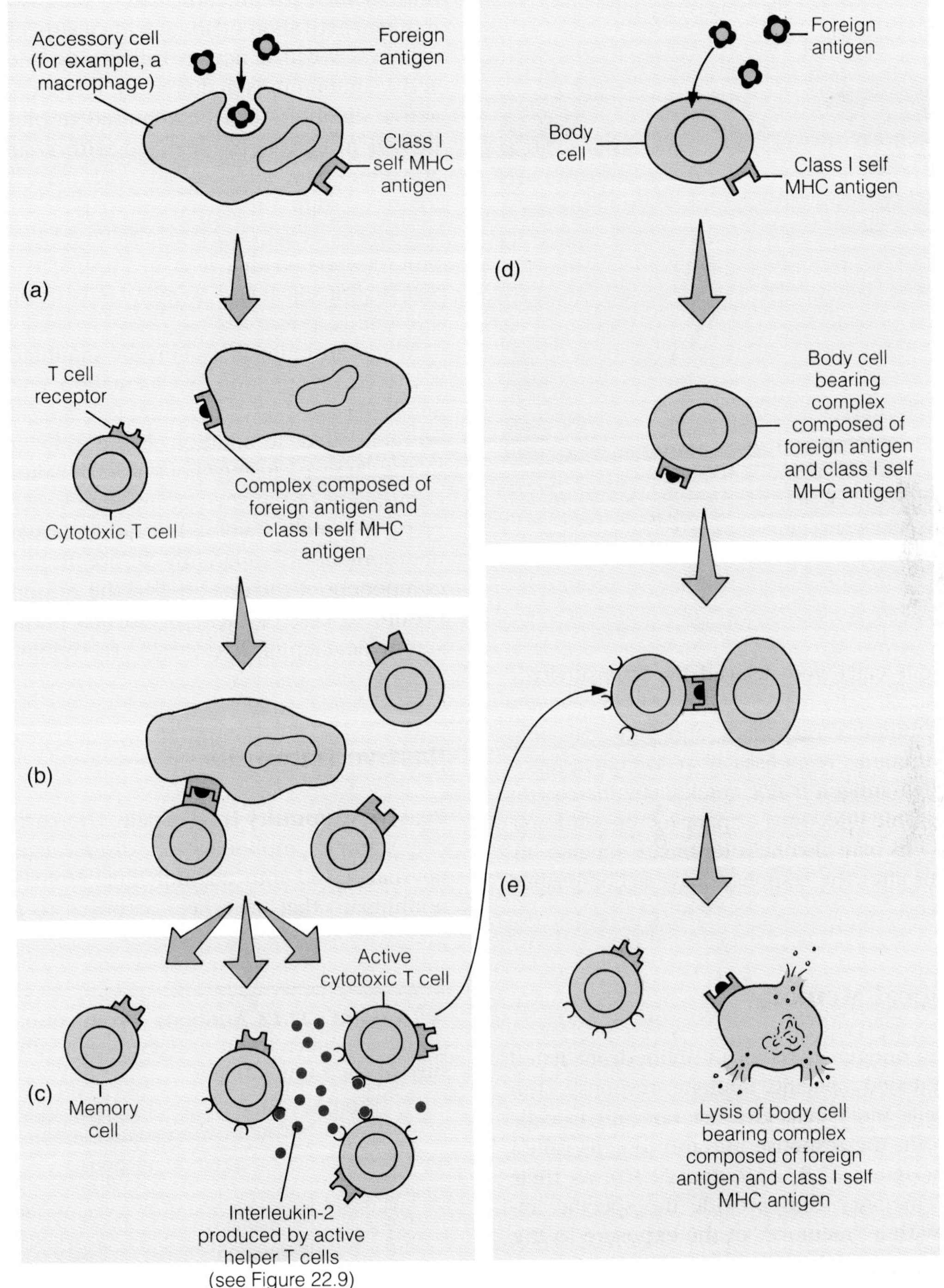

antilymphocyte serum or drugs not only weaken or suppress specific immune responses directed against the transplanted tissue or organ, but they also diminish the ability of the immune system to provide protection against other foreign antigens. Consequently, the transplant recipient becomes very susceptible to infection.

Suppressor T Cells

T cells referred to as **suppressor T cells** are believed to modulate or limit specific immune responses. In many cases, suppressor T cells are thought to inhibit the activities of helper T cells, and they may also inhibit B

◆ **FIGURE 22.11 Cell killing by an active cytotoxic T cell (×8450)**

An active cytotoxic T cell (smaller sphere) attaches to two cells recognized as foreign.

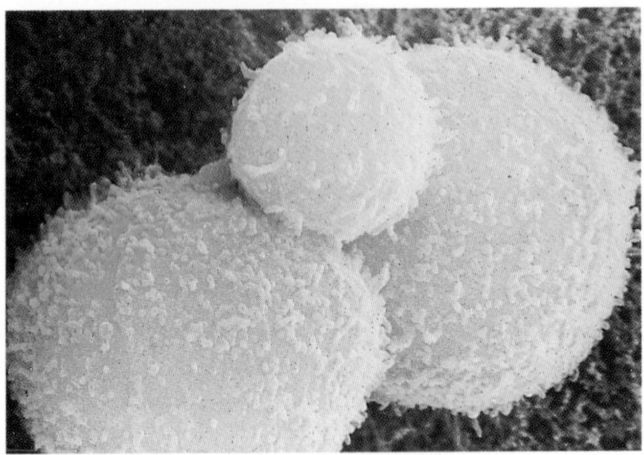

cells, cytotoxic T cells, and delayed hypersensitivity T cells. Suppressor T cells may help prevent excessively strong specific immune responses, and they may help limit specific immune responses to a foreign antigen when the foreign antigen is not quickly eliminated (for example, in chronic infections or parasitic infestations). Suppressor T cells may also help terminate specific immune responses after the responses have accomplished their purposes.

Immunological Memory

When a foreign antigen initiates an antibody-mediated and/or cell-mediated specific immune response, the stimulated B cells and T cells give rise not only to cells that participate in the response but also to cells called *memory cells* (Figures 22.8, 22.9, and 22.10). As their name implies, memory cells provide the specific immune system with a "memory" of the exposure to the antigen.

In general, the specific immune system responds rather slowly following an initial antigen exposure. Because of its memory component, however, the system responds quickly and vigorously to a subsequent exposure to the same antigen. For example, following an initial antigen exposure, it usually takes several days to build up substantial levels of antibodies (Figure 22.12). However, a subsequent exposure to the same antigen results in a much quicker and greater production of antibodies.

Active Immunity

A resistance to infection that results from an antigen-induced activation of an individual's specific immune responses is called **active immunity.** Most commonly, active immunity is acquired by infection. Alternatively, active immunity can be acquired through vaccination, in which a person is injected with a small amount of antigenic material. The material may be bacterial, a toxin, or some other substance. It is usually pretreated by drying, by exposure to ultraviolet light, or by some other means so that it is not strong enough to cause disease but still acts as an antigen that stimulates specific immune responses.

When an individual's specific immune responses are activated by infection or vaccination, antibodies and/or active T cells that can combat the disease-causing agent are generated. In addition, if the individual is exposed to the disease-causing agent at a later time, the memory component of the specific immune system enables it to respond rapidly. It should be noted, however, that some microorganisms apparently do not activate the memory component of the system, and the response to each exposure to one of these organisms is the same as the slowly developing response to the initial exposure.

Passive Immunity

Passive immunity is essentially "borrowed" immunity. For example, antibodies to a disease-causing agent can be transferred into an individual from sources (often nonhuman) that have been exposed to the agent. Pas-

◆ **FIGURE 22.12 Antibody production following an initial exposure to an antigen and a subsequent exposure to the same antigen**

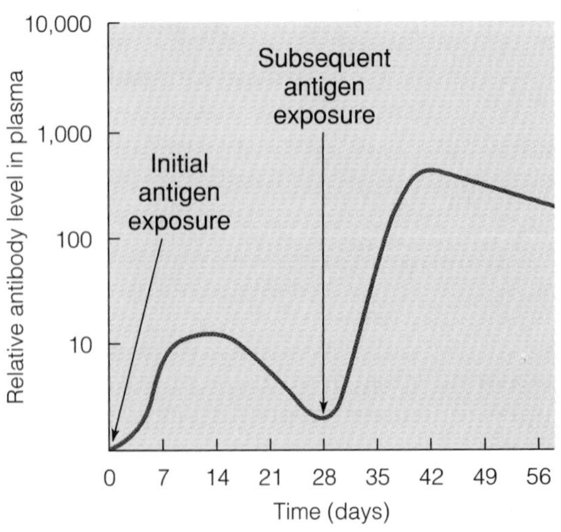

sive immunity is of relatively short duration—generally only several weeks. Moreover, the recipient's own specific immune system is not stimulated to produce antibodies or active T cells against the disease-causing agent.

A common instance of passive immunity occurs during pregnancy. Class IgG antibodies formed by the mother move across the placenta to the fetus. These antibodies protect the infant during the period just after birth, when the infant's own ability to produce antibodies is relatively poor.

Passive immunity can also be provided by injecting antibodies into an individual. For example, antibodies to the venom of a particular type of snake can be built up in a horse, recovered from the horse plasma, and injected into a person who has been bitten by that type of snake in order to provide immediate protection against the venom. A possible danger of such a procedure, however, is that injected antibodies may themselves be antigenic in the recipient's system—especially if they were obtained from nonhuman sources. Thus, the injected antibodies may stimulate the recipient to produce antibodies against them. This occurrence can lead to severe allergic responses.

The characteristics of active immunity and passive immunity are compared in Table 22.4.

Tolerance

Tolerance can be generally defined as the failure of the body to mount a specific immune response against a particular antigen. For example, the body is composed of proteins and other substances that are antigenic. Normally, however, the body exhibits self-tolerance and does not generate specific immune responses against its own antigens.

Tolerance is due primarily to the elimination or inactivation of specific B cells and T cells as a result of appropriate antigenic exposure. Thus, under proper conditions, exposure to an antigen can prevent rather than initiate a specific immune response. Two principal mechanisms—*clonal deletion* and *clonal anergy*—are believed to contribute to tolerance.

Clonal deletion results in the elimination of lymphocytes that could produce specific immune responses against self antigens. Clonal deletion appears to affect immature lymphocytes in particular. For example, as a T cell matures in the thymus, it apparently passes through a stage when appropriate contact with an antigen for which the T cell is specific leads to the death of the cell. Within the thymus, maturing T cells are exposed to a person's own antigens, and appropriate contact with these antigens eliminates maturing T cells that would develop into cells that could produce specific immune responses against them.

Clonal anergy results in the inactivation of lymphocytes that could produce specific immune responses against self antigens. Researchers have proposed that clonal anergy particularly affects mature lymphocytes. For example, depending on the conditions under which a mature B cell encounters an antigen, it appears that the B cell can either be activated or inactivated. Thus, mature B cells that could produce specific immune responses against self antigens are believed to encounter the self antigens in a manner that leads to their functional inactivation rather than to their activation (that is, the self-reactive B cells become anergic B cells).

◆ **TABLE 22.4 Characteristics of Active Immunity and Passive Immunity**

CHARACTERISTIC	ACTIVE IMMUNITY	PASSIVE IMMUNITY
Exposure to antigen required for immunity to develop?	Yes, either through natural exposure during infection or through artificial exposure during vaccination	No
Source of circulating antibodies	Antibodies self-generated on exposure to antigen	Antibodies "borrowed"—that is, produced from another source and transferred to the individual, either naturally, such as across the placenta, or artificially, such as through injection of antibodies harvested from another (often nonhuman) source
Injection required?	Injection of attenuated antigen required for artificial active immunity	Injection of borrowed antibodies required for artificial passive immunity
Duration of resistance	Long—may be lifelong	Short—less than a month

There is evidence that clonal deletion occurring within the thymus is the main tolerance mechanism for T cells with respect to at least some antigens. Clonal anergy appears to be a major tolerance mechanism for B cells. Suppressor T cells may play a role in tolerance by impeding the activation of self-reactive T cells and B cells that for some reason escape clonal deletion or clonal anergy.

Autoimmune Responses

Occasionally, the body's tolerance to its own antigens breaks down, and **autoimmune responses** occur—that is, the body generates specific immune responses against its own antigens. Autoimmune responses have several causes.

1. Drugs, environmental chemicals, viruses, or genetic mutations may cause the formation of new or altered antigens on cell surfaces. The body treats these antigens as foreign and generates specific immune responses against them.
2. Foreign antigens that are structurally very similar to some of the body's own antigens may stimulate the production of antibodies, active cytotoxic T cells, and/or active delayed hypersensitivity T cells that cross-react with body antigens. For example, certain streptococcal bacteria possess antigens that induce the formation of antibodies that cross-react with heart tissue, and severe, recurrent infections caused by streptococci sometimes lead to the development of rheumatic fever several weeks after the infection has subsided, suggesting an autoimmune response.
3. Under normal circumstances, certain body antigens may not be exposed to the specific immune system. Consequently, the system does not become tolerant to the antigens. If, at some time, tissue disruption following, for example, injury or infection exposes these antigens, the specific immune system treats them as foreign and generates antibody-mediated and/or cell-mediated responses against them. For example, antigens of the cornea of the eye do not seem to circulate in the body fluids. Should damage to the cornea expose these antigens, specific immune responses against them may cause corneal opacity.
4. Suppressor T cells may be involved in autoimmunity. If suppressor T cells contribute to tolerance, as was previously suggested, then an inhibition or deficiency of these cells could contribute to the development of autoimmune responses.

Hypersensitivity

Hypersensitivities are specific immune responses that can be harmful to the body.

Role of Complement Components in Hypersensitivity

In some types of hypersensitivity, active complement components generated during an antibody-mediated immune response damage body tissues and cells. In the type of hypersensitivity called **cytotoxic hypersensitivity,** antibodies are produced that can combine with antigens (often a person's own antigens) located on cell surfaces or on tissue components. The formation of antigen-antibody complexes activates the complement system, initiating inflammatory and cell-killing responses that damage cells or tissues. For example, in Goodpasture's syndrome, antibodies are produced that combine with basal lamina antigens in the lungs and kidneys. The inflammatory and cell-killing responses initiated by the formation of active complement components in this condition typically cause severe tissue injury in these organs.

In the type of hypersensitivity called **immune-complex hypersensitivity,** antigens combine with antibodies in the circulation, forming aggregates called immune complexes. These complexes can become deposited in blood-vessel walls, where the activation of the complement system can initiate an inflammatory response that damages the vessels. The deposition of immune complexes in the kidneys can lead to an inflammatory condition known as glomerulonephritis (see Chapter 26).

Role of Active Delayed Hypersensitivity T Cells and Active Cytotoxic T Cells in Hypersensitivity

In the type of hypersensitivity called **cell-mediated hypersensitivity,** or **delayed hypersensitivity,** the activities of active delayed hypersensitivity T cells and active cytotoxic T cells during a cell-mediated immune response cause damage to body tissues. For example, the toxin of poison ivy is a hapten that can bind to proteins in the skin. Consequently, exposure to poison ivy toxin can cause the production of active delayed hypersensitivity T cells and active cytotoxic T cells against the toxin, thus sensitizing the individual to the toxin. If such a sensitized person is again exposed to poison ivy toxin, the responses of active delayed hypersensitivity T cells and active cytotoxic T cells (including the release of chemical factors) induce a tissue-damaging inflammatory reaction at the site of exposure that is characterized by itching, swelling, and vesication. Hypersensitivity reactions involving active delayed hypersensitivity T cells and active cytotoxic T cells usually take from one to three days to reach their peak intensities.

Role of Class IgE Antibodies in Hypersensitivity

Class IgE antibodies are involved in a type of hypersensitivity called **anaphylactic hypersensitivity** *(an″-uh-fi-lak′-tik).* This type of hypersensitivity is the cause of many common allergic reactions. In these reactions, the exposure of a susceptible individual to a particular antigen—for example, dust, pollen, or a particular food—causes the production of class IgE antibodies from plasma cells. These antibodies bind to mast cells and basophils, thus sensitizing the individual to the antigen. When such a sensitized person is again exposed to the antigen, the antigen attaches to the antibodies bound to the mast cells and basophils. This attachment stimulates the release of substances that include histamine, leukotrienes, a chemotactic factor for eosinophils, and heparin. These substances induce inflammatory responses such as vasodilation and increased vascular permeability, as well as increased mucus secretion and the contractions of smooth muscles of the airways of the lungs.

Often, an anaphylactic-type allergic reaction is localized to a particular site in the body. For example, if the exposure to a particular antigen occurs in the nasal area, the result may be sneezing, runny nose, congestion, and other symptoms of hay fever due to the irritation, increased mucus secretion, increased blood flow, and increased protein leakage in the area.

Occasionally (particularly in response to injected antigens), a systemic rather than a localized anaphylactic allergic reaction occurs, and severe hypotension and airway constriction quickly develop. This reaction is called *anaphylactic shock,* and it can be fatal if countermeasures such as the injection of epinephrine are not taken promptly. (Epinephrine prevents the release of substances from mast cells and antagonizes the actions of histamine and leukotrienes on smooth muscles.) Some sensitized people develop anaphylactic shock in response to the antigen in a single insect sting or in response to drugs such as penicillin.

The administration of antihistamines often provides some relief from anaphylactic-type allergic reactions, but the relief is frequently incomplete because chemicals other than histamine (for example, leukotrienes) also participate in these reactions. When the particular antigen to which a person is sensitive has been identified, desensitization therapy may be utilized. Densensitization consists of injecting an individual with small but increasing quantities of the offending antigen. Some researchers think this treatment results in the production of class IgG rather than class IgE antibodies to the antigen. According to this view, when the person again encounters the particular antigen, the class IgG antibodies attach to it and prevent the antigen from attaching to class IgE antibodies bound to mast cells and basophils. Alternatively, it has been proposed that desensitization causes suppressor T cells to suppress the synthesis of class IgE antibodies directed against the antigen. In either case, the antigen does not attach to class IgE antibodies that are bound to mast cells and basophils, and the anaphylactic-type allergic reaction does not occur.

 Blood Groups and Transfusion

The surfaces of erythrocytes contain particular antigens (agglutinogens) that can react with appropriate antibodies (agglutinins). These reactions form the basis of various blood-typing classifications.

A-B-O System

The erythrocyte-surface antigens most often considered are those of the A-B-O system (Table 22.5). The antigens of the A-B-O system, which are inherited, are designated A and B, and the lack of both A and B antigens on the erythrocytes is designated O. A given individual may have either antigen A or antigen B, both antigens A and B (AB), or neither antigen A nor B (O).

Antibodies against antigens A or B begin to build up in the plasma shortly after birth. The antibody levels peak at about 8 to 10 years of age, and the antibodies remain present in declining amounts throughout the rest of life. The mechanism that stimulates the production of anti-A and anti-B antibodies is unclear, but it has been suggested that antibody development is initiated by small amounts of A- and B-type antigens that enter the body in the food, in bacteria, or by other means.

A person normally produces antibodies against those antigens that are not on his or her erythrocytes, but does not produce antibodies against those antigens that are present. Thus, a person with antigen A has anti-B antibodies; a person with antigen B has anti-A antibodies; a

◆ **TABLE 22.5 Summary of the A-B-O System**

BLOOD TYPE	ANTIGENS (AGGLUTINOGENS) ON ERYTHROCYTES	ANTIBODIES (AGGLUTININS) IN PLASMA
A	A	Anti-B
B	B	Anti-A
AB	Both A and B	Neither anti-A nor anti-B
O	Neither A nor B	Both anti-A and anti-B

 CONDITIONS OF CLINICAL SIGNIFICANCE 

AIDS

In addition to conditions such as autoimmune responses and hypersensitivities in which specific immune responses cause damage to the body, problems can also arise when specific immune responses are deficient or absent. For example, infection by a virus called HIV (human immunodeficiency virus) can damage the specific immune system and lead to a variety of disorders, including the condition known as AIDS (acquired immune deficiency syndrome).

HIV can invade many body cells, but among its primary targets are a set of T cells known as CD4 cells, which include the helper class of T cells. Following infection, the virus may remain dormant or inactive for long periods of time. However, once it becomes active and begins to reproduce new virus particles, it interferes with the normal function of and ultimately kills CD4 cells, which carry out activities necessary for effective specific immune responses. The depletion of functional CD4 cells produces a general immune suppression and deficiency that affects both antibody-mediated and cell-mediated specific immune responses and leaves the victim susceptible to a wide variety of infections and other problems that are ultimately fatal. At the present time, it is not certain what causes the dormant virus to become active and begin to reproduce.

HIV is spread from an infected person to an uninfected person by direct exposure of the uninfected person's bloodstream to virus-containing body fluids (such as blood or semen) from the infected individual. People infected by HIV generally produce antibodies against it, and the presence of these antibodies in a person's blood indicates that infection has occurred. However, the antibodies are not effective at eliminating the virus, and once infection occurs, it is believed to be permanent.

Some reports indicate that the initial infection of a person by HIV can be accompanied by symptoms that resemble those of acute mononucleosis (for example, fever, sore throat, pain in muscles and joints). However, it is not known if these symptoms always—or even usually—occur on initial infection. As previously stated, following infection the virus may remain dormant for long periods of time, and many infected people apparently do not develop any overt signs of illness, at least for a number of years. Others eventually develop what is called AIDS-related complex (ARC), which is characterized by symptoms such as a generalized swelling of the lymph nodes, persistent fever, diarrhea, and a rundown feeling. After a period of time, ARC may resolve, or it may develop into AIDS. (AIDS may also appear without being immediately preceded by ARC.)

AIDS is characterized by an inability to successfully resist infections by organisms that would normally be resisted and cause no harm. (Such infections are called opportunistic infections.) For example, AIDS victims frequently develop a type of pneumonia that results from infection by a protozoan called *Pneumocystis carinii,* and they may suffer from a fungal infection of the mouth called thrush or candidiasis. They also have a greater than normal tendency to develop Kaposi's sarcoma, which is a cancer of the lining of blood vessels. AIDS victims ultimately die from such opportunistic infections and malignancies.

person with neither antigen A nor B (O) has both anti-A and anti-B antibodies; and a person who possesses both antigens A and B has neither anti-A nor anti-B antibodies. The individual's blood type indicates the antigens he or she possesses, not the antibodies.

Transfusion

The mixing of incompatible blood types can cause erythrocyte destruction and other problems. For example, if a person with type A blood (antigen A on erythrocytes, anti-B antibodies in plasma) receives a transfusion of type B blood (antigen B on erythrocytes, anti-A antibodies in plasma), the recipient's anti-B antibodies will attack the incoming type B erythrocytes. The type B erythrocytes will be agglutinated (clumped), and hemoglobin will be released into the plasma—that is, the cells will undergo hemolysis. Incoming anti-A antibodies of the type B blood may also attack the type A erythrocytes of the recipient, with similar results. However, unless large amounts of blood are transfused, this problem is usually not as serious because the incoming antibodies are diluted in the recipient's plasma.

During a transfusion reaction resulting from the mixing of incompatible blood types, agglutinated erythrocytes can plug small blood vessels. When hemolysis is rapid, the released hemoglobin can precipitate in the urine-forming structures of the kidneys and contribute to kidney failure.

Since type AB individuals possess neither anti-A nor anti-B antibodies to attack incoming erythrocytes, they are often called universal recipients. On the other hand, since the erythrocytes of type O individuals will not be attacked by either anti-A or anti-B antibodies, these people are called universal donors. However, these terms are misleading because other erythrocyte antigens and plasma antibodies can cause transfusion problems. If possible, therefore blood for transfusion should be closely matched to the blood of the potential recipient.

Rh System

Another example of an erythrocyte antigen-antibody system is the Rh system (so named because it was initially studied in Rhesus monkeys). The Rh system consists of a group of surface antigens on erythrocytes, and certain Rh antigens are very likely to cause transfusion reactions. An individual who possesses these antigens is designated Rh positive, and an individual who lacks them is designated Rh negative. The antibody components of the system—the anti-Rh antibodies—are not normally present in the plasma but can be produced upon exposure and sensitization to Rh antigens. In general, an individual does not produce anti-Rh antibodies against Rh antigens that are present on his or her erythrocytes, but upon sensitization a person can produce anti-Rh antibodies against Rh antigens that are not on his or her erythrocytes.

Sensitization can occur if an Rh negative person receives a transfusion of Rh positive blood. It can also occur when an Rh negative mother carries a fetus who is Rh positive (due to the presence of Rh antigens inherited from the father). In this case, some of the fetal Rh antigens may enter the maternal circulation and sensitize the mother so that she begins to produce anti-Rh antibodies against the fetal antigens. Sensitization is most likely to occur near the time of birth, but because it takes some time for the mother to build up anti-Rh antibodies, the first Rh positive child carried by a previously unsensitized Rh negative mother is usually unaffected. However, if an Rh negative mother who has been sensitized by transfusion or by a previous Rh positive pregnancy subsequently carries an Rh positive fetus, maternal anti-Rh antibodies may enter the fetal circulation and cause the agglutination and hemolysis of fetal erythrocytes. This agglutination and hemolysis can result in an anemic condition known as hemolytic disease of the newborn (erythroblastosis fetalis). In this condition, the breakdown of the hemoglobin released during hemolysis results in the formation of sufficient bilirubin to give the infant's skin a yellow color—that is, jaundice. In some cases, bilirubin is deposited in nerve cells, causing brain damage. The usual treatment for severe hemolytic disease of the newborn resulting from Rh incompatibility is to remove the infant's Rh positive blood and replace it with Rh negative blood from an unsensitized donor. The Rh negative erythrocytes will not be attacked by the maternal anti-Rh antibodies in the infant's system, and the replacement of the infant's blood reduces the levels of these antibodies.

Since the sensitization of an Rh negative mother to the Rh antigens of an Rh positive fetus usually occurs near the time of birth, it is common to inject Rh negative mothers soon after the delivery of an Rh positive child with agents that prevent or limit sensitization. These agents contain anti-Rh antibodies that bind to any fetal Rh antigens that have entered the mother's system. This inhibits the antigens from sensitizing the mother and causing her to produce antibodies against them.

Metabolism of Foreign Chemicals

Many toxic foreign chemicals—for example, drugs and chemical pollutants in air, water, or food—can enter the body by way of the gastrointestinal tract, lungs, or skin. A number of these chemicals, particularly organic chemicals, are metabolized in the liver and also to some extent in the kidneys, skin and other organs.

Often, the metabolic transformation of a hazardous chemical decreases its toxicity and enhances its excretion. For example, a nonpolar, lipid-soluble chemical may be converted into a more polar, less lipid-soluble substance that can be eliminated from the body in the urine.

On the other hand, the metabolic processing of some foreign chemicals enhances their toxicities. For example, it is believed that many chemicals linked to the occurrence of cancer become carcinogenic only after metabolic transformation. Moreover, because the pathways involved in the metabolism of foreign chemicals generally process materials normally required by the body, the presence of foreign chemicals may upset the metabolism of necessary substances.

Resistance to Stress

A stressful event can be thought of as any event that leads to an increased release of glucocorticoids such as

ASPECTS OF EXERCISE PHYSIOLOGY

Exercise: A Help or Hindrance to Immune Defense?

Although many people who exercise claim that they have fewer infections when they are in good aerobic condition, few scientific studies have been conducted that can substantiate or refute this claim. On the basis of existing evidence, it cannot be stated conclusively that exercise influences resistance to disease. However, the studies that have been done suggest that exercise does have an impact on immune defense but that the intensity of exercise is an important factor in the effects on the immune system. The bulk of evidence indicates that moderate exercise may have beneficial effects on the immune system, whereas exhaustive exercise may be accompanied by reduced immune-system activity.

Animal studies have shown that high-intensity exercise after experimentally induced infection results in a more severe infection. Moderate exercise training done prior to infection or to tumor implantation, on the other hand, results in less severe infection and slower tumor growth in experimental animals.

In humans, moderate exercise training results in an increase in the circulating numbers of granulocytes and lymphocytes and an increase in the activity of cytotoxic T cells and natural killer cells. One study showed, on the other hand, that maximal exercise suppresses natural-killer-cell activity for one to two hours after exercise. It is possible that intense physical activity induces the stress response, which in turn suppresses immune functions.

Athletes with long, difficult training schedules and intense competition have increased incidence of respiratory infections. Studies have shown that these athletes have lower resting salivary IgA levels compared with control subjects and that their mucosal immunoglobulins are decreased after prolonged exhaustive exercise. These levels return to normal after 24 hours. It is possible that the athlete who is training with frequent bouts of exhaustive exercise may have a sustained reduction in mucosal immunity and therefore lower resistance to respiratory infection.

cortisol from the cortices of the adrenal glands, resulting in glucocorticoid levels in the blood plasma that are higher than those present at the same hour of the day in undisturbed, normal individuals on a similar activity (sleep-wake) cycle. Among the events that can have this effect are noxious or potentially noxious occurrences such as physical trauma, prolonged heavy exercise, and various infections. Many stressful events are believed to enhance glucocorticoid release by either directly or indirectly promoting an increased release of corticotropin-releasing hormone (CRH) from neurons of the hypothalamus of the brain. The CRH, in turn, enhances the release of adrenocorticotropin (ACTH) from the pituitary gland, and the ACTH increases the release of glucocorticoids from the adrenal cortices. Moreover, there is evidence that certain infections cause the production of ACTH or an ACTH-like substance by extrapituitary sources (possibly monocytes or lymphocytes), and this substance may be at least partly responsible for an increased release of glucocorticoids in such cases. In addition, monocytes produce substances (for example, interleukin-1 and hepatocyte-stimulating factor) whose effects apparently include the ability to stimulate ACTH release from the pituitary.

Many investigators have suggested that the glucocorticoids are particularly important to a person's ability to combat physical stress. Since a major response to overwhelming stress is vasodilation and circulatory failure, it has been proposed that the glucocorticoids combat stress by permitting norepinephrine to induce vasoconstriction while at the same time preventing an excessive vasoconstriction that can lead to tissue ischemia (deficiency of blood) by acting directly on vascular smooth muscle and stimulating heart muscle. It has also been suggested that the metabolic effects of the glucocorticoids—for example, raising blood-glucose levels—are particularly important in mobilizing the body's resources to resist physical stress.

A common response to stress is the activation of the sympathetic nervous system and the adrenal medullae. This response elevates the levels of epinephrine and norepinephrine, which leads to responses such as increased blood-glucose levels, elevated blood pressure, increased heart rate, and increased blood flow to

◆ **TABLE 22.6 Cells and Chemical Factors That Participate in the Body's Nonspecific Defense Mechanisms and Specific Immune Responses**

CELLS			
Cells	**Description**	**Representative Controlling Mechanisms**	**Representative Functions**
B lymphocytes or B cells (*see also* Plasma cells)	White blood cells	Activated by foreign antigen; helper T cells and interleukins usually participate	Give rise to plasma cells, which produce antibodies
Basophils	White blood cells	Binding of antigen to class IgE antibodies bound to the basophils	Release histamine; involved in anaphylactic hypersensitivity
Cytotoxic T cells	Class of T lymphocytes	Activated by foreign antigen in association with class I self MHC antigen; helper T cells and chemical factors (e.g., interleukins) participate	Destroy virus-infected cells or cancer cells
Delayed hypersensitivity T cells	Class of T lymphocytes	Activated by foreign antigen in association with class II self MHC antigen; accessory cells participate	Participate in cell-mediated specific immune responses; release chemical factors
Eosinophils	White blood cells	Stimulated by eosinophil chemotactic factor and presence of various types of parasites	Participate in destruction of various types of parasites, such as parasitic worms; play a role in certain allergic inflammatory responses
Helper T cells	Class of T lymphocytes	Activated by foreign antigen in association with class II self MHC antigen; accessory cells, such as macrophages, and chemical factors (e.g., interleukins) participate	Help activate B cells and other T cells; secrete chemical factors
Lymphocytes	White blood cells (two types: B and T lymphocytes)	Stimulated by products specific for each type of cell (*see* specific cells)	Involved in various aspects of specific immune responses
Macrophages	Large, phagocytic cells	Attracted by chemotactic agents; linked to target cells by opsonins; activity enhanced by interferon	Important in nonspecific defense; phagocytize foreign material and cellular debris; secrete chemical mediators that exert a variety of effects; clear way for tissue repair by removing debris; secret interferon; process and present antigen to B and T cells, thereby assisting in antibody production and T-cell activation
Memory cells	Cells produced when B and T cells are stimulated by antigen exposure but remaining dormant until subsequent exposure to same antigen	Stimulated by subsequent exposure to same antigen	Produce swifter, more powerful response on subsequent exposure

continued on next page

◆ **TABLE 22.6 Cells and Chemical Factors That Participate in the Body's Nonspecific Defense Mechanisms and Specific Immune Responses (continued)**

CELLS

Cells	Description	Representative Controlling Mechanisms	Representative Functions
Monocytes	White blood cells	Various factors stimulate their derivatives, the macrophages (*see* Macrophages)	Converted into macrophages after leaving the blood
Natural killer (NK) cells	Lymphocytelike cells	Stimulated by interferon	Nonspecifically lyse virus-infected cells and tumor cells on first exposure to them; secrete interferon
Neutrophils	White blood cells	Attracted by chemotactic agents; linked to target cells by opsonins	Important in nonspecific defense; important in inflammatory process; highly mobile phagocytic cells; release chemicals that exert a variety of effects
Plasma cells	Cells derived from B lymphocytes	A foreign antigen stimulates particular B cells to proliferate and differentiate, giving rise to antibody-secreting plasma cells; macrophages (or other accessory cells), helper T cells, and chemical factors (e.g., interleukins) usually participate	Secrete antibodies specific to the foreign antigen
Suppressor T cells	T lymphocytes		Modulate or limit-specific immune response
T lymphocytes or T cells (*see* Cytoxic T cells, Delayed hypersensitivity T cells, Helper T cells, and Suppressor T cells)	White blood cells	Activated by foreign antigen in association with self MHC antigen; chemical factors (e.g., interleukins) and accessory cells such as macrophages participate	Involved in various aspects of specific immune responses

CHEMICAL FACTORS

Factor	Description	Representative Controlling Mechanisms	Representative Functions
Antibodies	Specialized proteins produced in response to immunogenic antigens; same as immunoglobulins	Production and release stimulated by immunogenic antigens	Enhance nonspecific immune responses against the antigen that stimulated their production; in some instances, inactivate the antigen
Autoantibodies	Antibodies against self antigens	Breakdown of mechanisms for tolerance to self antigens	Attack one or more of body's own tissues
Chemotactic agents	Local chemical mediators	Released from a variety of sources at site of inflammation	Attract phagocytes to site of inflammation
Complement system	Group of inactive plasma proteins that can be sequentially activated	Activated by antigen-antibody complexes and by polysaccharides in cell walls of certain bacteria	Destroys foreign cells by forming tunnels in their plasma membranes; facilitates every step of inflammatory process (e.g., various components serve as opsonins, stimulate secretion of histamine, or activate kallikrein)

◆ **TABLE 22.6 Cells and Chemical Factors That Participate in the Body's Nonspecific Defense Mechanisms and Specific Immune Responses (continued)**

CHEMICAL FACTORS			
Factor	**Description**	**Representative Controlling Mechanisms**	**Representative Functions**
Endogenous pyrogens	Chemical mediators released by macrophages; include interleukin-1 and tumor necrosis factor	Released during inflammation	Cause development of fever
Eosinophil chemotactic factor	Chemical mediator in anaphylactic hypersensitivity	Released by mast cells or basophils	Attracts eosinophils to site of antigen contact
Fibrinogen (inactive); fibrin (active)	Final plasma protein in clotting cascade	Activated by clotting mechanisms	
Histamine	Chemical mediator released from mast cells, basophils, and platelets	Release stimulated by chemical mediators released from neutrophils, active complement components, and binding of antigen to class IgE antibodies bound to mast cells and basophils	Important in inflammation and anaphylactic hypersensitivity; causes vasodilation (particularly of arterioles) and increased vascular permeability (particularly in venules)
Immunoglobulins (*see* Antibodies)			
Interferon	Family of proteins	Released from cells invaded by virus	Stimulates noninvaded cells to produce substances that protect them from viral invasion; increases killing ability of macrophages and natural killer cells; slows cell division and suppresses tumor growth; exerts anticancer effect
Interleukin-1 (IL-1)	Chemical mediator released by antigen-presenting accessory cells such as macrophages	Release stimulated by helper T cell binding to antigen-presenting accessory cells	Helps activate helper T cells
Interleukin-2 (IL-2)	Chemical mediator secreted by helper T cells	Released by active helper T cells	Stimulates proliferation of active T cells
Interleukins 4, 5, and 6 (IL-4, IL-5, IL-6)	Chemical mediators secreted by helper T cells	Released by active helper T cells that bind to foreign antigen associated with class II self MHC antigen on B cells	Promote activation of B cells and their proliferation and differentiation into antibody-secreting plasma cells
Kallikrein	Chemical mediator present in inactive form in plasma and released from neutrophils	Activated by active factor XII and by a complement component; activation or release occurs during inflammation	Activates kinins
Kininogens (inactive); kinins (active)	Plasma proteins	Activation stimulated by kallikrein	Dilate arterioles and increase vascular permeability in inflammation; activate pain receptors; act as chemotactic agents

continued on next page

◆ **TABLE 22.6 Cells and Chemical Factors That Participate in the Body's Nonspecific Defense Mechanisms and Specific Immune Responses (continued)**

CHEMICAL FACTORS			
Factor	**Description**	**Representative Controlling Mechanisms**	**Representative Functions**
Leukotrienes	Local chemical mediators	Produced by certain cells during inflammatory responses	Mediators of some inflammatory activities; mediators of asthmatic responses and certain allergic reactions
Opsonins	Class IgG antibodies and one of activated proteins of complement system	Produced in response to factors that stimulate antibody secretion (*see* Antibodies) and complement activation (*see* Complement system)	Enhance phagocytosis by linking foreign cell to phagocytic cell
Prostaglandins	Local chemical mediators	Released from certain cells during inflammatory responses	Mediators of some inflammatory activities; raise set point of hypothalamic thermostat to produce fever
Thymosin	Collection of hormones produced by thymus		Enhances proliferation of T-cell colonies; enhances immune capabilities of existing T cells

skeletal muscles. These responses can also be useful in meeting physical stress.

The levels of other hormones are also frequently affected by stress. For example, the release of antidiuretic hormone (ADH) and aldosterone can increase during stress, resulting in a decreased urine output and in increased retention of water and sodium that leads to an increase in blood volume.

A number of psychological situations that elicit emotional responses such as fear, anger, or anxiety—for example, final exams for college students or awaiting a surgical operation—are also associated with an increased release of glucocorticoids and stress responses. In this regard it has been suggested that, although stress responses can be of benefit to an individual in meeting physical stress, they may not be as useful in resisting psychosocial stress. In fact, it has been proposed that stress responses elicited by chronic psychosocial stress—for example, high-pressure employment or anxiety-provoking economic situations—actually contribute to pathological conditions such as high blood pressure or heart disease.

Study Outline

◆ **BODY SURFACES** p. 688

Skin and mucous membranes are first barriers to the invasion of body by potentially damaging factors.

1. Skin contains keratin; sweat and sebaceous glands secrete chemicals toxic to many bacteria.

2. Mucus secreted by mucous membranes entraps small particles that may then be swept away or engulfed by phagocytic cells.

◆ **INFLAMMATION** pp. 688–693

Most basic form is acute, nonspecific physiological response of body to tissue injury caused by factors such as chemicals, heat, mechanical trauma, or bacterial invasion.

Changes in Blood Flow and Vessel Permeability.

1. Small vessels in injured area dilate, leading to increased blood flow that increases delivery of plasma proteins and phagocytic leukocytes.

2. Small-vessel permeability increases; plasma fluid and solutes (including proteins) move out of blood vessels and into inflamed tissues.

3. As inflammation progresses, blood flow through small vessels slows, sometimes stops.

Walling-Off Effect of Inflammation. In tissues, fibrinogen from plasma is converted to fibrin, forming a clot that walls off injured area; can delay or limit spread of toxic products or bacteria.

Leukocyte Emigration. As blood flow slows during inflammation, leukocytes marginate and pavement; leukocytes exhibit amebalike activity and squeeze between endothelial cells of capillaries and venules into tissue spaces. First leukocytes in tissues are generally neutrophils, followed later by monocytes, which become macrophages.

Chemotaxis and Leukocyte Aggregation. Chemical mediators determine direction of movement of leukocytes. In acute, nonspecific inflammation, neutrophils usually predominate initially; monocytes and macrophages predominate during later stages.

Phagocytosis. Leukocytes — particularly neutrophils — and macrophages phagocytize microorganisms, foreign materials, and debris; opsonins can attach to specific foreign particles and render them more susceptible to phagocytosis.

Abscess and Granuloma Formation. Occasionally, inflammatory response is unable to overcome invasion, and an abscess or granuloma forms.

1. *Abscess.* A sac of pus consisting of microbes, leukocytes, macrophages, and liquified debris walled off by fibroblasts or collagen.
2. *Granuloma.* Layers of phagocytic-type cells surrounded by fibrous capsule; central cells contain offending agent—for example, microbes that can survive within phagocytes or materials that cannot be digested by phagocytes.

Chemical Mediators of Inflammation.

HISTAMINE. Present in mast cells, basophils, and platelets; released in response to mechanical disruption of cells by injury and chemicals from neutrophils; causes vasodilation and increased vascular permeability.

KININS. Group of polypeptides formed by the action of kallikrein on kininogens; dilate arterioles, increase vascular permeability, act as chemotactic agents, and induce pain.

COMPLEMENT SYSTEM. Series of plasma proteins that normally circulate in inactive form. Active complement components can mediate virtually every event of acute, nonspecific inflammatory response; can directly attack and kill cells such as invading microbes.

PROSTAGLANDINS. May induce vasodilation and increased vascular permeability, stimulate leukocyte emigration through capillary walls, and contribute to pain and fever of inflammation.

LEUKOTRIENES. Produced by leukocytes. Certain leukotrienes apparently cause increased vascular permeability; some seem to be chemotactic agents that attract neutrophils.

Symptoms of Inflammation.

LOCAL SYMPTOMS. Redness and heat due to vessel dilation and increased blood flow; swelling due to increased vessel permeability and movement of fluid and solutes out of vessels; pain due to increased pressure on nerve endings and effects of some chemical mediators.

SYSTEMIC SYMPTOMS. Fever (macrophages release endogenous pyrogens); often, increased production and release of leukocytes.

◆ INTERFERON pp. 694–695

Provides some protection to body against viral invasion: Viruses induce cells to produce interferon, which then leaves cells to bind to receptors on plasma membranes of other cells; this binding triggers synthesis of enzymatic proteins within cells that can act to prevent viral reproduction. Interferon may prevent release of viruses from cells or make viruses incapable of infecting other cells; may also help protect body against some cancers; increasingly seen as versatile agent with effects that go beyond the inhibition of viral activity.

◆ SPECIFIC IMMUNE RESPONSES pp. 695–705

Two major aspects: antibody-mediated (humoral) immunity and cell-mediated immunity.

Antigens. Substances that can elicit specific immune responses; generally have molecular weights of 8000–10,000 or more; lower-molecular-weight substances (haptens) may elicit specific immune responses if they combine with larger molecules such as proteins.

Antibodies. An antibody is a specialized protein produced in response to presence of immunogenic antigen; can combine with specific antigen that stimulated its production, forming antigen-antibody complex. Antibodies inactivate or destroy foreign substances by:

1. Inhibiting interactions of bacterial toxins or destructive foreign enzymes with target cells or substrates.
2. Preventing attachment of viruses to target cells.
3. Enhancing basic inflammatory response.

B Cells. Lymphocytes committed to differentiate into antibody-producing plasma cells involved in humoral immunity.

T Cells. Heterogeneous group of lymphocytes; include cells committed to participating in cell-mediated immune responses.

Accessory Cells. Present foreign antigens, in combination with self MHC antigens, to certain other cells by displaying on their surfaces combinations of foreign antigens and self MHC antigens.

Antibody-Mediated (Humoral) Immunity. The binding of antigen to antibody molecules on the surface of a B cell contributes to the activation of the B cell. The active B cell proliferates and differentiates. Most of the proliferating cells develop into plasma cells that manufacture the antibody displayed on the B cell's surface. The antibody molecules combine with the antigen that stimulated their production to inactivate or destroy the antigen. In most cases, accessory cells, such as macrophages, and helper T cells are involved in B-cell activation and antibody production.

Cell-Mediated Immunity. Appropriate stimulation by an antigen activates certain cytotoxic T cells and the active cells proliferate. The active cytotoxic T cells travel throughout the body. When they encounter cells bearing complexes composed of the same antigen and class I self MHC antigens, they bind to the cells; they can lyse and thereby kill the cells. Active delayed hypersensitivity T cells can also participate in cell-mediated immune responses; they release chemical factors that generate inflammatory responses and enhance phagocytosis.

Transplant Rejection. Cell-mediated immune system contributes importantly to rejection of solid-tissue transplants.

Suppressor T Cells. Believed to modulate or limit specific immune responses.

Immunological Memory. When a foreign antigen initiates a specific immune response, stimulated T cells and B cells give rise not only to cells that participate in the response but also to memory cells. Memory cells provide the specific immune system with a "memory" of the exposure to the antigen, and the system responds quickly and vigorously to a subsequent exposure to the antigen.

Active Immunity. Resistance to infection resulting from antigen-induced activation of individual's specific immune responses; may be acquired by infection or by vaccination.

Passive Immunity. Essentially "borrowed" immunity. Examples: antibodies to disease-causing agent can be transferred to individual from sources (often nonhuman) that have been exposed to agent; maternal class IgG antibodies may pass to fetus during pregnancy.

Tolerance. Generally defined as failure of body to mount specific immune response against particular antigen. The body normally exhibits self-tolerance and does not generate specific immune responses against its own antigens.

Autoimmune Responses. Occasionally, body's tolerance to own antigens breaks down, and body generates specific immune responses against its own antigens.

Hypersensitivity. Specific immune response that can do harm.

ROLE OF COMPLEMENT COMPONENTS IN HYPERSENSITIVITY. Generation of active complement components during antibody-mediated immune response to an antigen can sometimes damage body tissues and cells.

ROLE OF ACTIVATED DELAYED HYPERSENSITIVITY T CELLS AND ACTIVATED CYTOTOXIC T CELLS IN HYPERSENSITIVITY. Activities of these cells during cell-mediated immune response can, in some cases, damage body tissues.

ROLE OF CLASS IgE ANTIBODIES IN HYPERSENSITIVITY. Exposure of susceptible individual to particular antigen can cause production of class IgE antibodies that bind to mast cells and basophils, thus sensitizing the individual to the antigen; on subsequent exposure to same antigen, antigen attaches to antibodies bound to mast cells and basophils, stimulating release of chemicals that induce inflammatory responses, increased mucus secretion, and contraction of smooth muscles of airways of lungs.

◆ CONDITIONS OF CLINICAL SIGNIFICANCE: AIDS p. 706

◆ BLOOD GROUPS AND TRANSFUSION pp. 705–707

Based on antigens (agglutinogens) on erythrocyte surfaces that react with appropriate antibodies (agglutinins).

A-B-O System. Based on inherited antigens on erythrocyte surfaces. Antibodies against antigens of this system that are not present on surfaces of a person's own erythrocytes begin to build up in plasma shortly after birth.

Transfusion. Mixing of incompatible blood types during transfusion can cause agglutination and hemolysis of erythrocytes.

Rh System. Based on inherited antigens on erythrocyte surfaces. Antibodies can be produced against Rh antigens not on a person's erythrocytes upon exposure and sensitization to the antigens.

◆ METABOLISM OF FOREIGN CHEMICALS p. 707

Many potentially toxic or poisonous foreign chemicals are metabolically transformed into less toxic, more easily excreted substances; metabolic transformations may sometimes enhance toxicity of foreign chemicals.

◆ RESISTANCE TO STRESS pp. 707–712

Stressful event can be thought of as one that leads to increased release of glucocorticoids such as cortisol from cortices of adrenal glands, resulting in higher plasma-glucocorticoid levels than those present at same hour in undisturbed, normal individuals on similar activity (sleep-wake) cycle; glucocorticoids may help combat physical stress. Stress can also activate sympathetic nervous system and adrenal medullae; their activities may also help combat physical stress. Some psychological situations elicit stress responses, but the responses may not help resist the stress.

◆ Self Quiz

1. The skin and mucous membranes serve as barriers to the invasion of the body by potentially damaging factors. True or False?
2. During an acute, nonspecific inflammatory response: (a) the permeability of small vessels decreases; (b) plasma proteins are unable to leave the circulatory system; (c) vessels of the microcirculation dilate.
3. Which cells usually predominate in the early stages of most acute, nonspecific inflammatory responses? (a) neutrophils; (b) monocytes; (c) eosinophils.
4. Histamine: (a) kills invading cells directly; (b) causes dilation of arterioles and venules; (c) decreases vessel permeability, particularly in venules.

5. Kininogens are converted to kinins by: (a) kallikrein; (b) opsonin; (c) chemotaxin.
6. Active complement components can mediate virtually every event of the acute, nonspecific inflammatory response. True or False?
7. The fever that often accompanies inflammation may be due to: (a) endogenous pyrogens; (b) kinins; (c) histamine.
8. One of the most effective inducers of interferon production is double-stranded DNA. True or False?
9. Haptens: (a) commonly elicit specific immune responses in their free, uncombined forms; (b) are substances of relatively low molecular weight that can elicit specific immune responses if they combine with larger molecules such as proteins; (c) generally have molecular weights greater than 10,000 in their free, uncombined forms.
10. An antibody is: (a) a substance that has the capacity to elicit a specific immune response; (b) a prostaglandin that circulates in the blood in an inactive form; (c) a specialized protein that is produced in response to an immunogenic antigen.
11. The constant regions of the polypeptide chains of an antibody molecule contain binding sites that allow the antibody to attach to a specific antigen. True or False?
12. B cells: (a) are committed to participating in cell-mediated immune responses; (b) are formed from macrophages; (c) can differentiate into plasma cells.
13. B cells and T cells: (a) are lymphocytes derived from stem-cell precursors in the red bone marrow; (b) complete their development in the thymus gland; (c) are present in large numbers in the lymph nodes but are not present in the spleen.
14. Antibodies are produced by: (a) active cytotoxic T cells; (b) mast cells; (c) plasma cells.
15. Active immunity can be acquired through (a) the transfer of antibodies from mother to fetus; (b) the injection of a pretreated bacterial toxin; (c) the injection of antibodies obtained from another person.
16. Anaphylactic hypersensitivity involves: (a) active complement components; (b) active cytotoxic T cells; (c) class IgE antibodies.
17. Some relief from an anaphylactic-type allergic reaction may be provided by the administration of: (a) opsonins; (b) antihistamines; (c) mast cells.
18. An adult with type A blood would normally have: (a) anti-A antibodies in the plasma; (b) anti-B antibodies in the plasma; (c) neither anti-A nor anti-B antibodies in the plasma.
19. During stress, which substance(s) is (are) released by the cortices of the adrenal glands? (a) glucocorticoids; (b) ACTH; (c) CRH.
20. Glucocorticoids may help a person combat physical stress. True or False?

CHAPTER 23

The Respiratory System

CHAPTER CONTENTS

LEARNING OBJECTIVES

After completing this chapter, you should be able to:

1. Distinguish between the regions of the pharynx, and describe their respiratory functions.
2. Describe the cartilages forming the framework of the larynx.
3. Describe the subdivisions of the bronchi.
4. Describe the gross anatomical structure of the lungs.
5. Describe the composition of the respiratory membrane.
6. Describe the factors responsible for the increased lung volume and the flow of air into the lungs during inspiration.
7. Describe the factors responsible for the reduced lung volume and the flow of air out of the lungs during expiration.
8. Discuss the factors that influence pulmonary airflow.
9. Explain the various lung volumes that are combined into the total lung capacity.
10. Describe the process by which gas exchange occurs between the lungs and the blood, and between the blood and the body tissues.
11. Explain how oxygen and carbon dioxide are transported in the blood.
12. Discuss the neural mechanisms involved in the control of respiration.
13. Describe the roles of oxygen, carbon dioxide, and hydrogen ions in the control of respiration.

CHAPTER 23

In order for the cells of the body to carry on their metabolic activities under aerobic conditions, they require a constant supply of oxygen and an efficient means of removing the carbon dioxide that their activities produce. Oxygen is supplied and carbon dioxide is removed by the **respiratory system,** with the assistance of the cardiovascular system. The respiratory system also makes vocalization possible. We are able to speak, sing, and laugh by varying the tension of the vocal folds as exhaled air passes over them.

Respiration is an integrated process, but for purposes of analysis it can be divided into five stages: (1) the movement of air into and out of the lungs; (2) the exchange of oxygen and carbon dioxide between the air in the lungs and the blood within the pulmonary capillaries; (3) the transport of oxygen and carbon dioxide throughout the body by the blood; (4) the exchange of oxygen and carbon dioxide between the blood and the interstitial fluid and cells; and (5) the utilization of oxygen and the production of carbon dioxide by metabolic processes within the cells. This chapter deals with the first four of these respiratory stages. The fifth is considered in Chapter 25.

The exchange of oxygen and carbon dioxide between the air and the blood occurs in the lungs. In order to reach the exchange sites in the lungs, air must flow through a series of conducting passageways that branch from one another much like the branches of a tree. Air that enters the nose or mouth passes into the **pharynx** and then into the **trachea.** The trachea branches into a **primary bronchus** to each lung. In each lung the primary bronchus divides into smaller bronchi that, in turn, branch into tiny tubules called **bronchioles.** The bronchioles themselves divide and eventually terminate in small air sacs called **alveoli.** Gaseous exchange between air and blood occurs in the alveoli.

Embryonic Development of the Respiratory System

The first indication of the development of the respiratory system in the embryo is the formation of an outpouch from the ventral surface of the endoderm of the digestive tract, just inferior to the pharynx (Figure 23.1). This outpouch, which appears in the four-week-old embryo, is called the **laryngotracheal bud.** As the bud elongates, the proximal portion develops into the trachea, and its distal end bifurcates, forming two buds that will develop into the bronchi. The bronchial buds continue to grow and rebranch, giving rise to many bronchioles. The closed terminal portions of the bronchiole buds become dilated as they develop into the alveoli of the lungs.

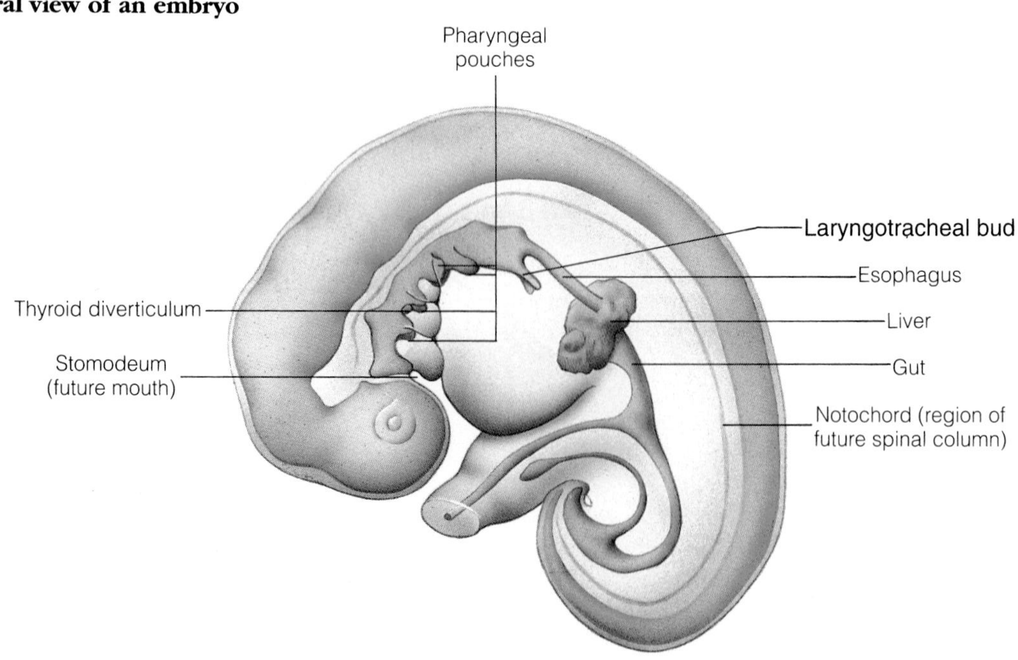

◆ **FIGURE 23.1 Lateral view of an embryo**
The pharyngeal pouches, the thyroid diverticulum, and the laryngotracheal bud develop as outpouches of the pharynx.

It is clear from the pattern of embryonic development that the epithelium lining the entire respiratory tract is derived from endoderm. In the adult this lining is called the **respiratory epithelium.** With the exception of the lining of the smallest bronchioles and the alveoli, the respiratory epithelium is composed of pseudostratified ciliated columnar cells and scattered mucus-secreting goblet cells. The cartilage, muscles, and connective tissues of the trachea and the connective tissues of the lungs develop from embryonic mesoderm that becomes massed around the laryngotracheal bud.

Anatomy of the Respiratory System

The respiratory system consists of the nose, nasal cavity, pharynx, larynx, trachea, bronchi, and lungs. The following sections describe the structures of these organs.

Nose and Nasal Cavity

Air enters the respiratory system through the **external nares (nostrils),** which lead to the **vestibule** of the nose. The lower part of the vestibule contains hairs that serve to trap the largest particles that might be drawn into the respiratory system during inspiration. The bridge of the nose is formed by the nasal bones. The rest of the framework of the nose consists of several plates of cartilage held together by fibrous connective tissue (Figure 23.2a). To form the **nasal septum,** one of the cartilages—the **septal cartilage**—joins with the nasal bones above, the vomer bone below, the perpendicular plate of the ethmoid posteriorly, and the maxillae inferiorly (Figure 23.2b). The septum divides the **nasal cavity** into right and left chambers. A deviation or deflection of the septum can interfere with the free flow of air through the nasal cavity, but this condition can be corrected surgically.

The bony roof of the nasal cavity is formed by the cribriform plate of the ethmoid bone (Figure 23.3). The lateral walls, which are irregular, are formed by the **superior** and **middle conchae** of the ethmoid bone and the **inferior conchae,** which are separate bones. Beneath the shelves formed by the conchae are recesses called the **superior, middle,** and **inferior meatuses.** The floor of the nasal cavity is formed by the bony **hard plate** (horizontal plates of the palatine bones and palatine processes of the maxillary bones) and the more posterior muscular **soft palate.** The palate separates the nasal cavity from the oral cavity. The nasal cavity opens posteriorly into the nasopharynx through the **internal nares** *(choanae)*.

Passageways from the **paranasal sinuses** drain into the nasal cavities. Most of them open into the meatuses formed by the conchae. The paranasal sinuses are air spaces located in the frontal, maxillary, ethmoid, and

◆ **FIGURE 23.2 The nose**

(a) Cartilage plates of the nose. (b) Midsagittal view of the skull showing the nasal septum.

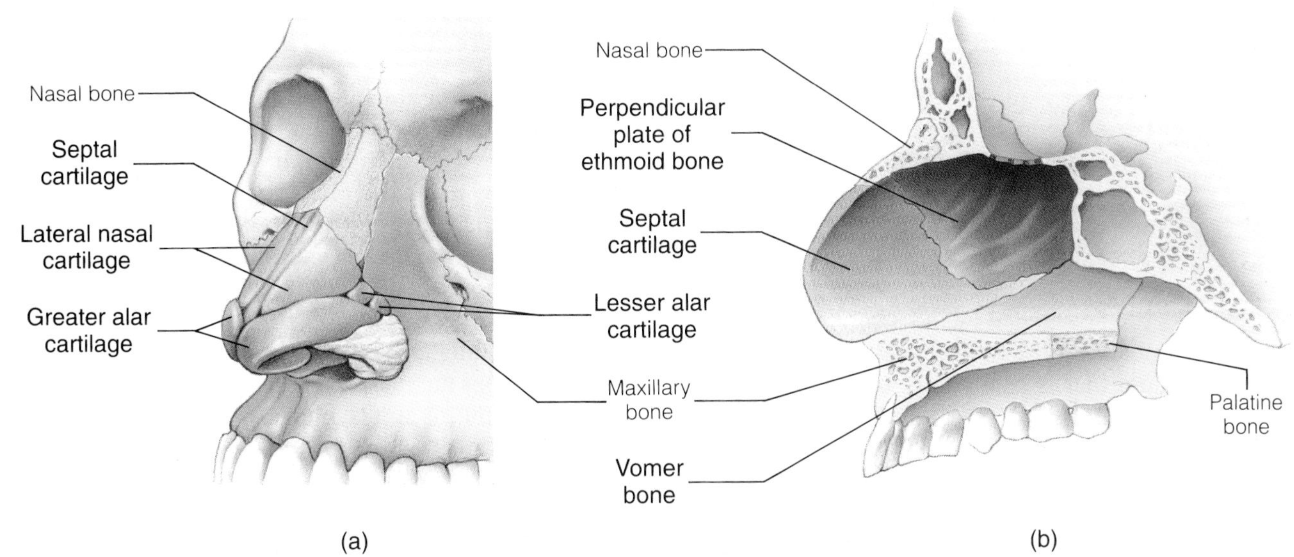

◆ **FIGURE 23.3 Sagittal view of the head and neck showing the mouth, the pharynx, and the lateral wall of the nasal cavity**

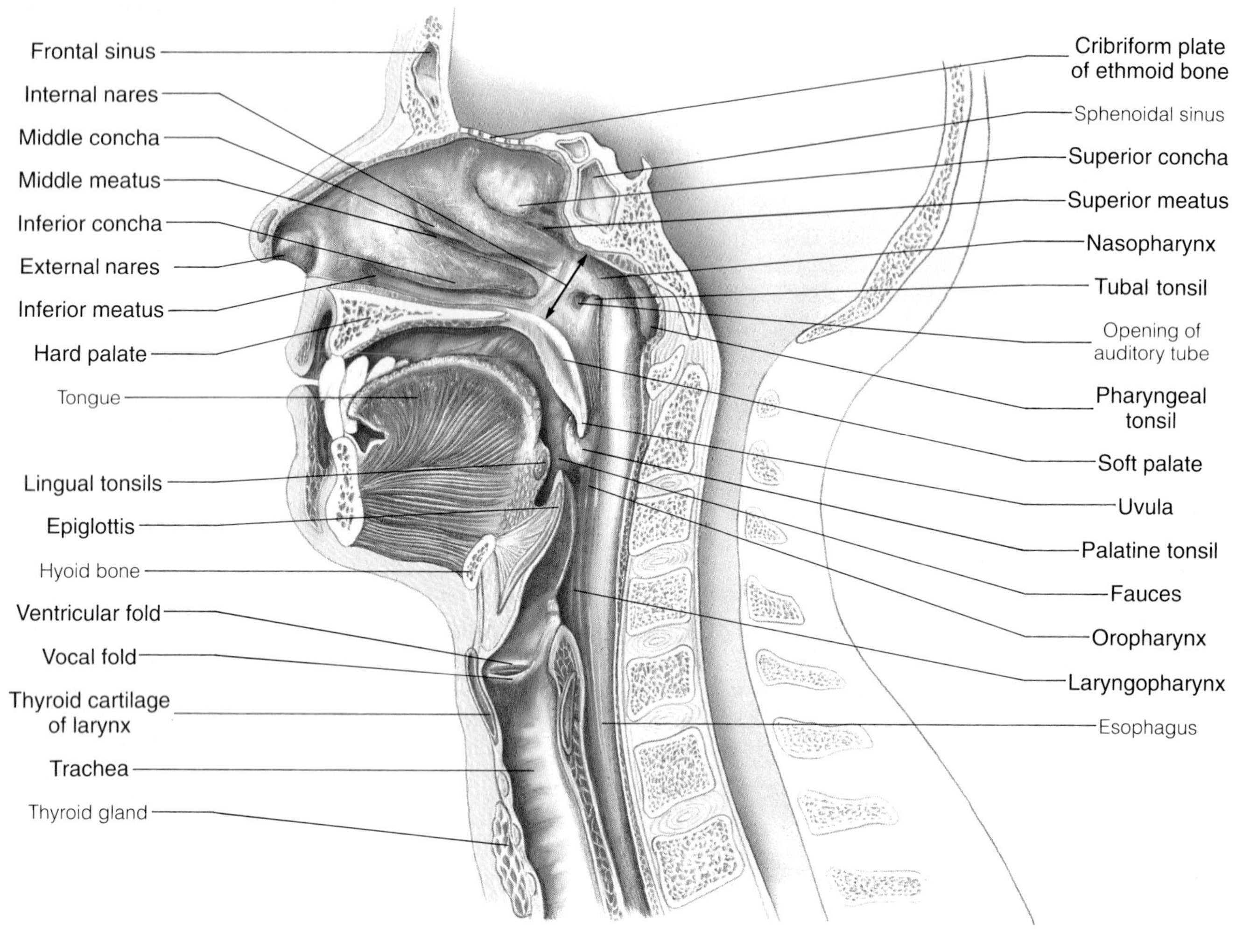

sphenoid bones (see Figure 6.25 on page 180). The **nasolacrimal ducts,** which drain fluid (tears) from the surface of the eyes, also empty into the nasal cavity.

Lining of the Nasal Cavity and the Paranasal Sinuses

The vestibule of the nose is lined with stratified squamous epithelium that is continuous with the skin. The rest of the nose, the nasal cavity, and the paranasal sinuses are lined with a continuous mucous membrane having a surface layer of pseudostratified ciliated columnar epithelium. The mucous membrane contains many mucus-secreting goblet cells. The part of the mucous membrane that is located at the top of the nasal cavity, just beneath the cribriform plate of the ethmoid bone, is a specialized epithelium called the **olfactory epithelium.** This region is supplied by the **olfactory nerves** (cranial nerve I), which pass through holes in the cribriform plate to reach the olfactory bulbs of the brain. Although the olfactory epithelium is sensitive to smells, it does not lie in the direct path of airflow. Consequently, in order to detect odors, it is helpful to sniff the air, causing it to be pulled upward into the region of the olfactory epithelium.

The mucous membrane has an extensive blood supply that warms the air as it is inhaled. Moreover, the mucous membrane saturates the air passing over it with water. The membranes of the nasal cavity and the delicate portions of the lungs are thus protected from becoming frozen or dried out. A sheet of mucus covering the mucous membrane further protects the respiratory system by trapping any small particles that get past the hairs guarding the external nares. The cilia of the membrane move in such a manner as to carry the particle-filled mucus toward the pharynx, where it can be removed by coughing or swallowing.

Infections of the Mucous Membranes

The mucous membranes of the nasal cavity may become inflamed because of infections (such as the common cold) or allergies. When inflamed, the blood vessels dilate, the membranes swell, and mucus secretion increases. The resulting congestion interferes with breathing and often causes a "runny nose." Such infection can spread into the mucous membranes of the paranasal sinuses, blocking their connections with the nasal cavity and causing them to fill with mucus. Since the sinuses act as resonance chambers, congestion changes the sound of the voice and can also cause a pressure increase in the sinuses such that a severe headache results.

Infections of the mucous membrane of the nasal cavity can also extend through the nasolacrimal duct to the covering (conjunctiva) of the eye. Thus, it is not uncommon to have red, watery eyes along with a cold. It is also possible for infection to travel from the mucous membrane of the nasal cavity into the pharynx, thus producing a "sore throat." From the pharynx, the infection can spread into the bronchi of the lungs, causing coughing and possibly bronchitis, or through the auditory (eustachian) tubes into the middle ear. The latter is particularly likely to occur in young children, in whom the lumina of the auditory tubes are relatively larger than in adults.

Pharynx

The **pharynx** *(fayr´-inks)* is a tube that is used by both the digestive system and the respiratory system. It communicates with the nasal cavity (through the internal nares), the oral cavity (through the fauces), the middle-ear cavity (through the auditory tubes), the larynx (through the glottis), and the esophagus. The pharynx is muscular and is lined with a mucous membrane that is continuous with the mucous membrane of the structures with which it communicates. For descriptive purposes the pharynx is divided into three regions: *nasopharynx, oropharynx,* and *laryngopharynx.*

Nasopharynx

The **nasopharynx** is located immediately behind the nasal cavity, above the soft palate. It is continuous with the nasal cavity through the internal nares (Figure 23.3). The surface layer of the mucous membrane of the nasopharynx, like that of the nasal cavity, is formed of pseudostratified columnar epithelium. On its lateral walls, the nasopharynx receives the **auditory (eustachian) tubes** that connect the nasopharynx with the cavity of the middle ear. Located near the openings of the auditory tubes are small masses of lymphoid tissue called **tubal tonsils.** On the posterior wall are the larger **pharyngeal tonsils.** When these tonsils become enlarged in response to an infection, they are called *adenoids.* Such enlargement can be chronic and can interfere with breathing through the nose, making it necessary to breathe through the mouth. The **soft palate** and **uvula** form the anterior wall of the nasopharynx.

Oropharynx

The **oropharynx** is a continuation of the nasopharynx, extending from the soft palate to the beginning of the laryngopharynx (Figure 23.3). It communicates with the oral cavity through the **fauces.** The oropharynx therefore receives food from the mouth and air from the nasopharynx. During exercise, air is also drawn into the oropharynx through the mouth, thus increasing the ventilation of the lungs. The free surface of the mucous membrane lining of the oropharynx is stratified squamous epithelium, which is the epithelium typically found in the mouth and the upper portion of the digestive tract. This type of epithelium serves to protect the region from the abrasiveness of swallowed food. On the side walls are two **palatine tonsils.** Embedded in the base of the tongue is an aggregate of **lingual tonsils** *(ling´-gwal).* The tonsils are formed of lymphoid tissues and are therefore part of the body's immune system.

Laryngopharynx

The **laryngopharynx** extends from the oropharynx above to the larynx and esophagus below (Figure 23.3). It communicates anteriorly with the larynx. Like the oropharynx, the laryngopharynx serves as a passageway for food and air and consequently is lined with a protective stratified squamous epithelium.

Larynx

The **larynx** connects the laryngopharynx with the trachea, which continues below the larynx. Air passes through the larynx on its way into and out of the lungs. Any solid substance that enters the larynx, such as food, is generally expelled by violent coughing. The larynx forms a **laryngeal prominence** *(Adam's apple)* on the anterior surface of the neck. This prominence is particularly noticeable in males following puberty, when the larynx becomes larger than in females and the anterior region of its framework forms a more acute angle.

Framework of the Larynx

The larynx is formed by a total of nine cartilages—three unpaired and six that are paired (Figure 23.4). These cartilages are held together, and attached to the hyoid bone above and the trachea below, by ligaments and muscles. The **thyroid cartilage** is the largest of the unpaired cartilages. It is formed by two broad plates that join anteriorly at an angle, producing the laryngeal prominence. The plates remain separated posteriorly, which leaves a wide opening in the laryngopharynx. Just below the thyroid cartilage is the ring-shaped **cricoid cartilage** *(krī´-koid)*, which is anchored to the thyroid cartilage above and the trachea below. The posterior region of the cricoid cartilage is longer than the anterior region. The third unpaired cartilage is the leaf-shaped **epiglottis** *(ep-e-glot´-tis)*. The epiglottis is attached by its narrow end to the inner surface of the anterior region of the thyroid cartilage. When air is flowing into or out of the larynx, the free upper portion of the epiglottis projects like a flap behind the base of the tongue, and the entrance to the trachea is exposed. During swallowing the larynx is pulled upward, tipping the epiglottis so that it tends to block the opening of the larynx and deflect solids and fluids into the esophagus.

The **arytenoid cartilages** *(ar″-ĕ-te´-noid)* are the most important of the paired cartilages. Each arytenoid cartilage is shaped like a small pyramid and rests on the superior-posterior border of the cricoid cartilage. The posterior ends of the vocal folds are attached to the arytenoid cartilages, and movement of the cartilages is responsible for varying the tension on the folds. The other paired cartilages, the **cuneiform** and **corniculate cartilages** *(kū-ne´-i-form; kor-nik´-ū-late)*, are small and lie just above the arytenoid cartilages (Figure 23.5).

All of the laryngeal cartilages except the epiglottis are composed of hyaline cartilage. The epiglottis is composed of elastic cartilage.

Mucous Membrane of the Larynx

The mucous membrane that covers the epiglottis and the upper parts of the larynx, where they are in direct

◆ **FIGURE 23.4 The larynx**
(a) Anterior view. (b) Sagittal section.

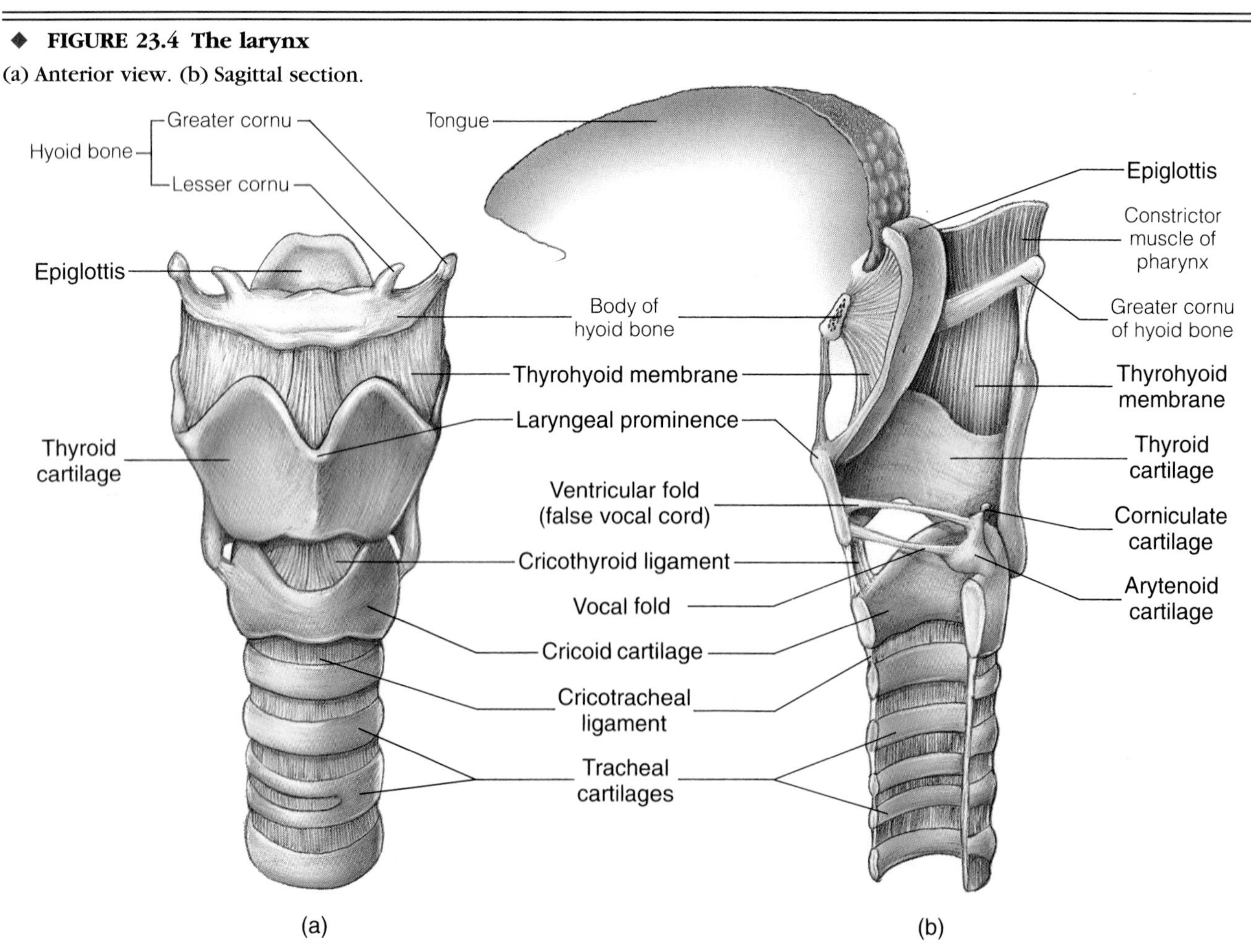

◆ **FIGURE 23.5 Superior view of the larynx**
(a) The glottis is closed. (b) The glottis is open.

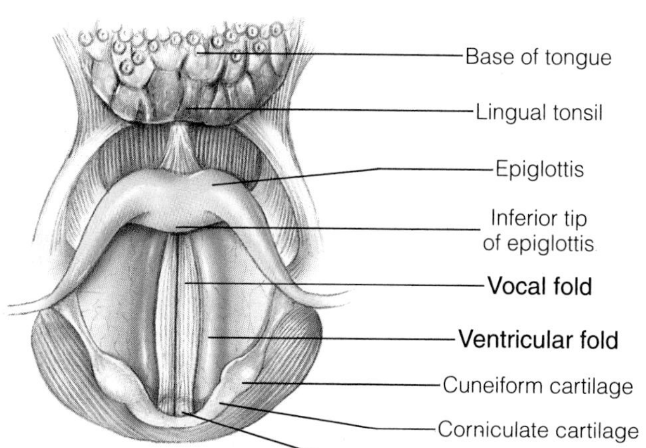

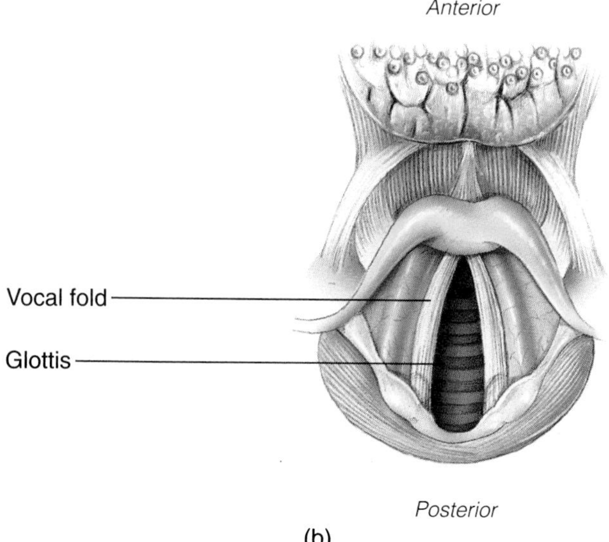

communication with the laryngopharynx, is lined with stratified squamous epithelium. The remainder of the larynx is lined with pseudostratified ciliated columnar epithelium. The mucous membrane near the entrance to the larynx forms two pairs of horizontal folds that extend posteriorly on each side from inside the angle of the thyroid cartilage to the arytenoid cartilages. The upper pair of folds are called the **ventricular folds** *(false vocal cords)*. The lower pair are the **vocal folds** *(true vocal cords)*. The opening between the vocal folds through which air enters the larynx is the **glottis** (Figure 23.5).

Within the vocal folds are bands of elastic ligaments that connect to the thyroid, cricoid, and arytenoid cartilages. The elastic vocal ligaments can be tightened or slackened by the actions of certain intrinsic muscles of the larynx. These muscles also rotate the arytenoid cartilages, thus further varying the amount of stretch on the vocal folds. As a result of the actions of the intrinsic muscles, the glottis can be narrowed or widened. Air passing through the glottis causes the vocal folds to vibrate and produce a sound. The frequency of the vibrations, and therefore the pitch of the sound produced, depends on the tension on the vocal folds.

The mucous membrane that covers the vocal folds becomes swollen when irritated or inflamed. This swelling interferes with the ability of the folds to vibrate, and hoarseness may result—a condition referred to as **laryngitis** *(lar-en-jigh´-tus)*.

Trachea

The **trachea** *(trā´-kē-uh)* or windpipe is a tube approximately 2.5 cm in diameter and 11 cm long. It extends from the larynx to the level of the sixth thoracic vertebra, where it divides into *left* and *right primary bronchi* (Figure 23.6a). The trachea lies against the anterior surface of the esophagus. The air passageway of the trachea is surrounded by a series of C-shaped hyaline cartilages that function to keep the trachea from collapsing, in the same manner as the wire rings in the hose of a vacuum cleaner. The cartilages are enclosed in an elastic fibrous membrane, and elastic fibers remain an important layer in the walls of all subsequent parts of the respiratory tract. Transverse smooth muscle fibers, called the **trachealis muscle,** and elastic connective tissue join the rings together posteriorly and close the opening of the C. When the trachealis muscle contracts, the diameter of the trachea is decreased, and expired air is expelled with greater force. The muscle contracts during coughing, which is particularly helpful in expelling mucus from the trachea. When the passage of air through the upper airways is impeded, a direct airway into the trachea is sometimes created surgically through the anterior surface of the neck, between the second and third cartilagenous rings. This procedure is called a *tracheotomy (trā´´-kē-ot´-ō-me)*.

The trachea is lined with a mucous membrane of pseudostratified ciliated columnar epithelium that contains numerous goblet cells. Since the cilia beat upward, they tend to carry foreign particles and excessive mucus away from the lungs to the pharynx, where they can be swallowed. Smoking diminishes ciliary activity and may eventually destroy the cilia. This effect is one reason smokers often have a persistent cough. Coughing is their only means of preventing mucus from accumulating in the lungs.

◆ **FIGURE 23.6 The trachea, bronchi, bronchioles, and alveoli**
(a) The trachea and bronchi. (b) The termination of bronchioles into alveoli.

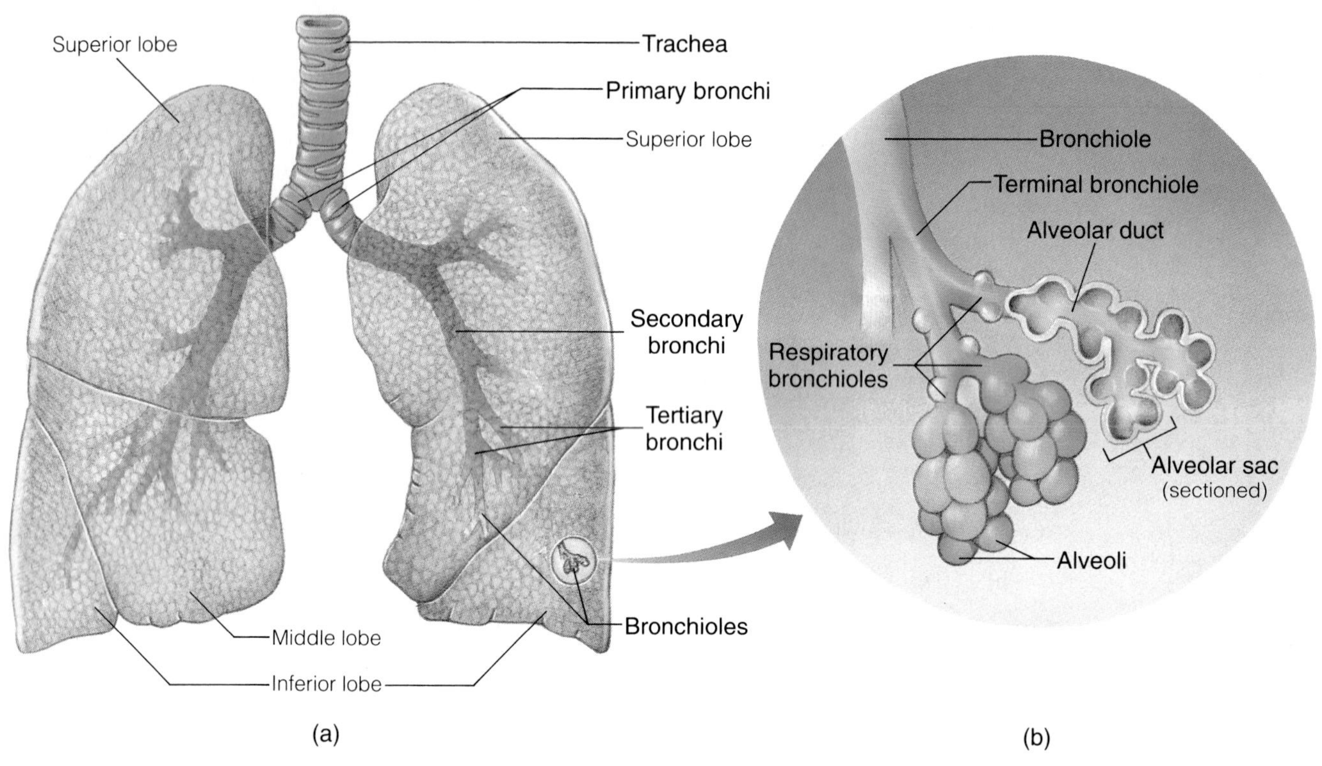

Bronchi, Bronchioles, and Alveoli

As the trachea passes behind the arch of the aorta, it divides into two smaller branches: the **left** and **right primary bronchi.** Each primary bronchus divides into still smaller **secondary (lobar) bronchi,** one for each lobe of the lung. The secondary bronchi, in turn, branch into many **tertiary (segmental) bronchi,** which further branch repeatedly, ultimately giving rise to tiny **bronchioles.** The bronchioles themselves subdivide many times, forming **terminal bronchioles,** each of which gives rise to several **respiratory bronchioles.** Each respiratory bronchiole subdivides into several **alveolar ducts,** which end in clusters of small, thin-walled air sacs called **alveoli** (Figure 23.6b). Often, several alveoli open into a common chamber called an **alveolar sac.** A few scattered alveoli are also present in the walls of the respiratory bronchioles.

The bronchial tree is lined with pseudostratified ciliated columnar epithelium. However, in the respiratory bronchioles the lining loses its cilia and changes to cuboidal and then to squamous epithelium as the bronchioles extend distally. The many tiny alveolar air sacs and the bronchial system of tubes through which the air travels to enter or leave the alveoli give the lungs a spongy texture (Figure 23.7).

The walls of the primary bronchi, like those of the trachea, are supported by incomplete cartilage rings. In the lungs, the rings are replaced by small plates of cartilage that completely encircle each bronchus. Smooth muscle also encircles the bronchi. With further branching, the cartilage plates gradually become smaller, forming incomplete rings, and the smooth muscle that surrounds the air passageways becomes more common. The walls of the bronchioles contain no cartilage and are completely surrounded by smooth muscle. The fact that bronchioles have walls of smooth muscle explains one of the most disturbing symptoms of an asthma attack. During an asthma attack, the muscles of the bronchioles undergo spasms, and since there is no supporting cartilage, the air passageways are squeezed shut, making breathing very difficult.

Lungs

Each lung is shaped somewhat like a cone, with the pointed **apex** fitting into the narrow space at the top of the thoracic cavity behind the clavicle (Figure 23.8). The **base** of each lung is broad and concave and rests on the convex surface of the diaphragm. A depression called the **hilus** is located on the mediastinal surface of

◆ **FIGURE 23.7 Photomicrograph of a lung section**
The numerous tiny openings are capillaries.

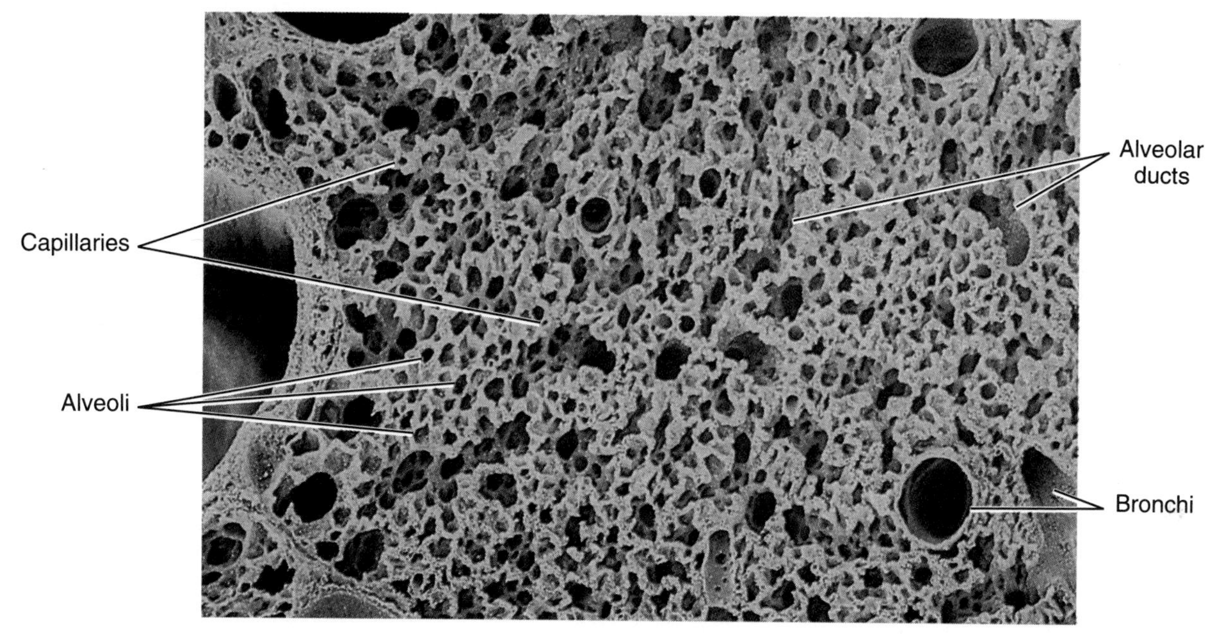

◆ **FIGURE 23.8 The lungs and the thoracic cavity**

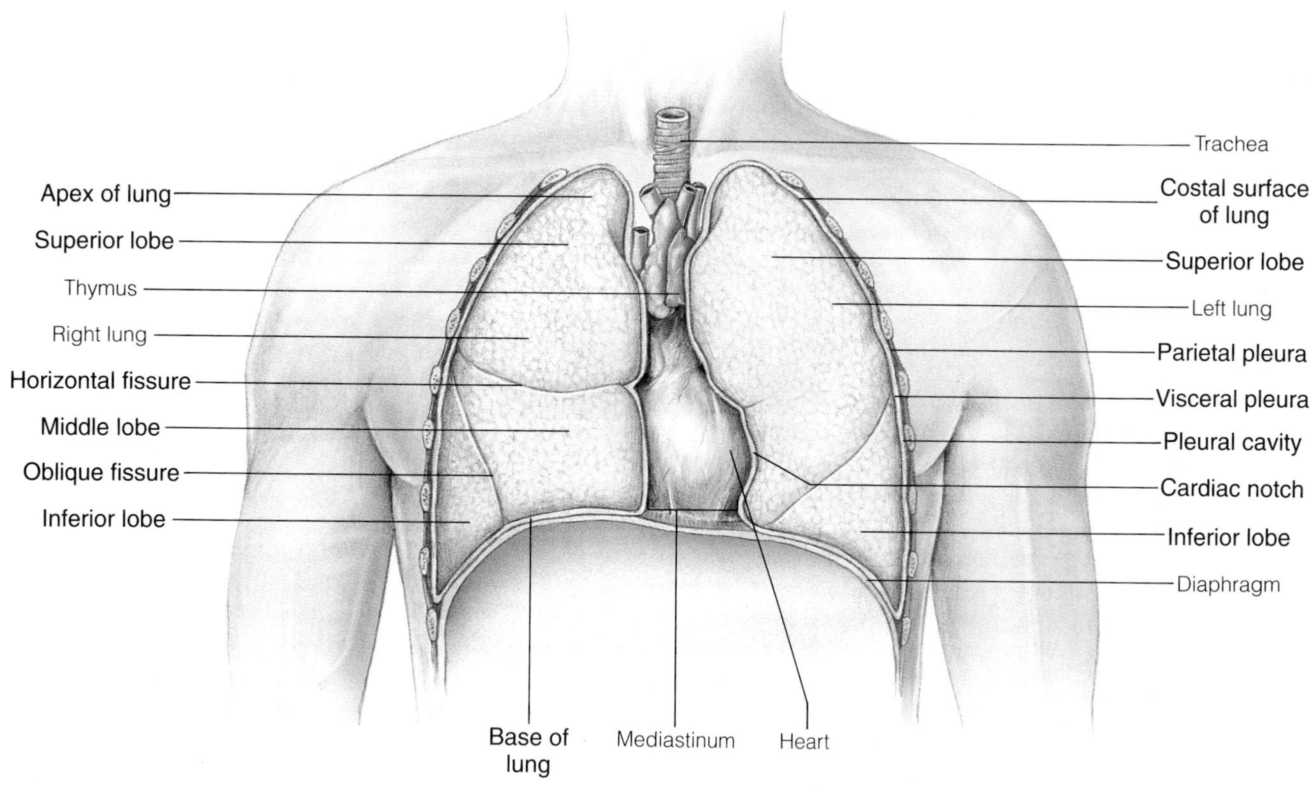

the lung. The hilus is the region where the structures that form the *root* of the lung—that is, the bronchus, blood vessels, lymphatics, and nerves—enter or leave the lung. The **costal surface,** which lies against the ribs, is rounded to match the curvature of the ribs. The left lung has a concavity for the heart, called the **cardiac notch,** on its medial surface.

Each lung is divided into **superior** and **inferior lobes** by an **oblique fissure.** The right lung is further divided by a **horizontal fissure,** which bounds a **middle lobe.** The right lung therefore has three lobes (and three secondary bronchi) whereas the left has only two lobes (and only two secondary bronchi).

In addition to being divided into lobes, which are visible externally, each lung is subdivided by connective tissue partitions into smaller units called **bronchopulmonary segments** *(brong″-kō-pul′-mō-ner″-ē)* (Figure 23.9). Each bronchopulmonary segment represents the portion of the lung that is supplied by a specific tertiary bronchus. The superior lobe of the right lung has three segments; the middle lobe has two segments; the inferior lobe has five segments. In the left lung, the superior lobe and the inferior lobe each have five segments, although some of them are not distinctly separated. The segments are important surgically because a diseased segment can be removed without having to remove an entire lobe or the entire lung. Also, the partitions that separate the segments impede the spread of disease, so pathology tends to be confined to one or several segments rather than spreading freely throughout the lung.

The two lungs are separated by a region called the **mediastinum.** Important structures are located in the mediastinum, including the heart, aorta, venae cavae, pulmonary vessels, esophagus, part of the trachea and bronchi, and the thymus gland.

The Pleura

Each lung is enclosed in a double-layered sac called the **pleura** *(ploor′-uh)* (Figure 23.8). Both layers of the pleura are composed of serous membrane. One layer, the **visceral pleura,** adheres firmly to the lung. The other layer, the **parietal pleura,** adheres firmly to the thoracic wall and diaphragm. The visceral and parietal layers are continuous at the hilus of the lung.

Between the two layers of the pleura is an extremely narrow **pleural cavity,** which is filled with **pleural fluid.** The pleural fluid is secreted by the pleura, and it acts as a lubricant to reduce the friction between the two layers during respiratory movements. The pleural fluid also couples the visceral pleura to the parietal pleura (and therefore the lungs to the thoracic wall and diaphragm). This coupling is analogous to the coupling of two sheets of glass by a thin film of water; it is extremely difficult to separate the sheets (the visceral pleura and the parietal pleura) by a force applied at right angles to their surfaces, even though the sheets can slide readily past one another. Because of this coupling, the movements of the lungs closely follow the movements of the thoracic wall and diaphragm.

Blood Supply of the Lungs

There is an important difference between the blood supplied to the alveoli and that supplied to the bronchi. The alveoli are supplied by branches of the pulmonary artery (which contains oxygen-poor blood), whereas small bronchial arteries from the descending aorta (which contains oxygen-rich blood) supply the bronchi. Thus, the pulmonary arteries carry blood that will be aerated within the alveoli, whereas the blood within the bronchial arteries provides nourishment for the rest of the lung tissue.

The Respiratory Membrane

The air in the alveoli is separated from the blood by a very thin **respiratory membrane,** formed by the alveolar epithelium and its basal lamina and the endothelium of the pulmonary capillaries and its basal lamina. In some places, there is a small amount of connective tissue, which contains reticular fibers and elastic fibers, between the two basal laminae (Figure 23.10). It is obviously advantageous to have a thin respiratory membrane—since oxygen and carbon dioxide diffuse across it during respiration—while still maintaining the integrity of the cardiovascular system. For the efficient diffusion of oxygen and carbon dioxide, the respiratory membrane must be moist, and the alveolar surfaces exposed to the air are coated with a thin layer of fluid.

The epithelium that lines the alveoli contains two types of cells. Most of the cells are of the squamous type through which gases diffuse readily. These cells are called *alveolar type I cells.* Scattered among the squamous cells are rounded or cuboidal cells—called *alveolar type II cells*—that may bulge into the lumens of the alveoli. The type II cells secrete a substance called *pulmonary surfactant,* which serves to lower the surface tension of the fluid that coats the avleoli, thus preventing the collapse of the alveoli and decreasing the muscular effort required to expand the lungs.

Phagocytic cells (macrophages) are commonly present within the respiratory membrane, and they may be found free in the spaces within the alveoli. These phagocytes are sometimes called *dust cells,* since they often contain carbon particles obtained from inhaled air.

◆ Mechanics of Breathing

In order to maintain concentrations of oxygen and carbon dioxide within the alveolar air that are favorable to

◆ **FIGURE 23.9 The bronchopulmonary segments of the lungs, and the branching of the bronchi into the segments**

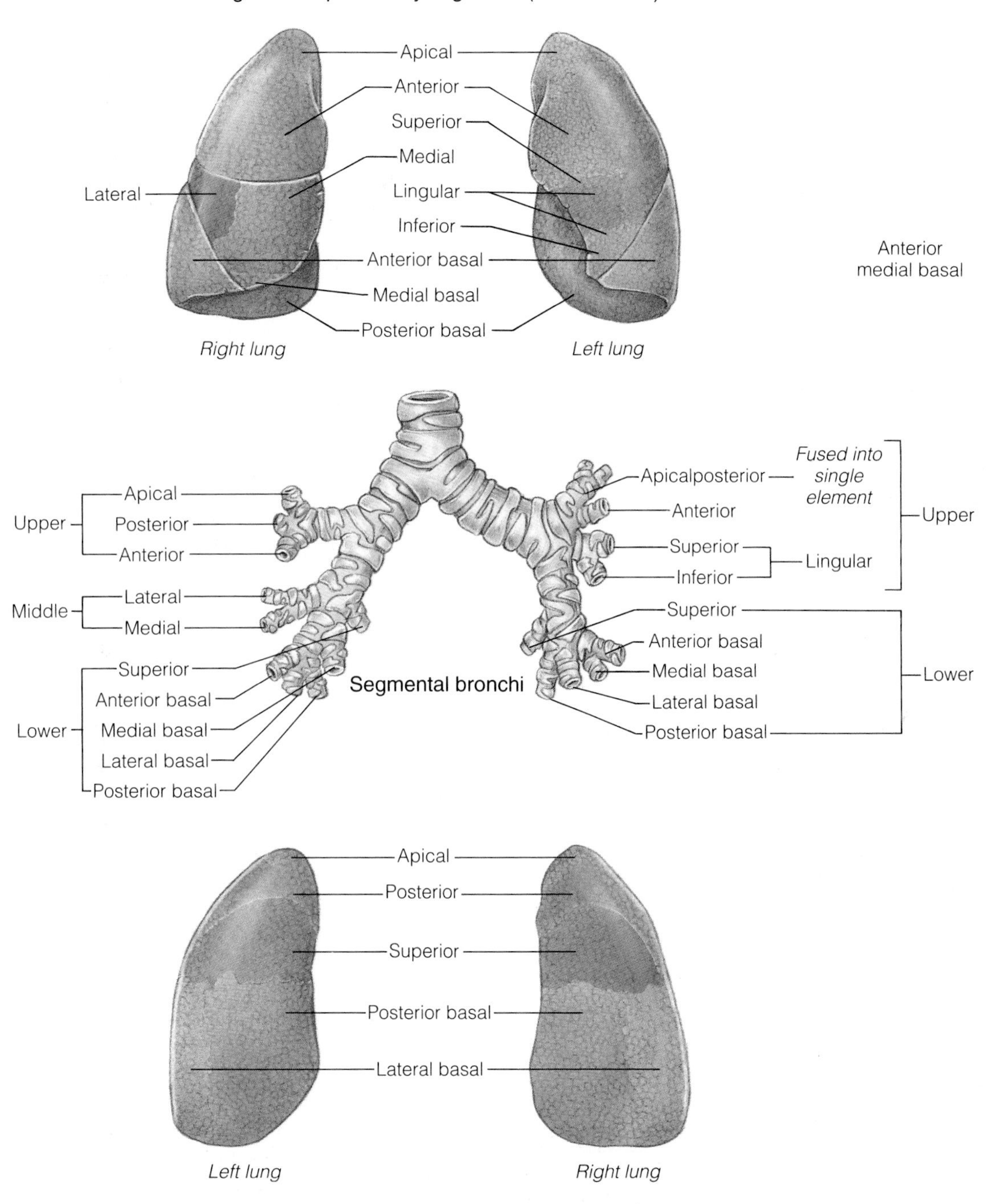

their diffusion across the respiratory membrane, it is necessary to constantly bring fresh air in and remove the air already within the lungs. The movement of air into and out of the lungs depends on pressure differences between the air in the atmosphere and the air in the lungs.

Atmospheric Pressure

At sea level, the gases of the atmosphere under the influence of gravity exert a pressure on the surface of the earth equivalent to that which would be exerted by a 760-millimeter-thick layer of mercury (Hg). Conse-

◆ **FIGURE 23.10 The respiratory membrane**
(a) An alveolus surrounded by capillaries, showing the respiratory membrane that separates the air in the alveolus from the blood in the capillaries. (b) The respiratory membrane of the lungs.

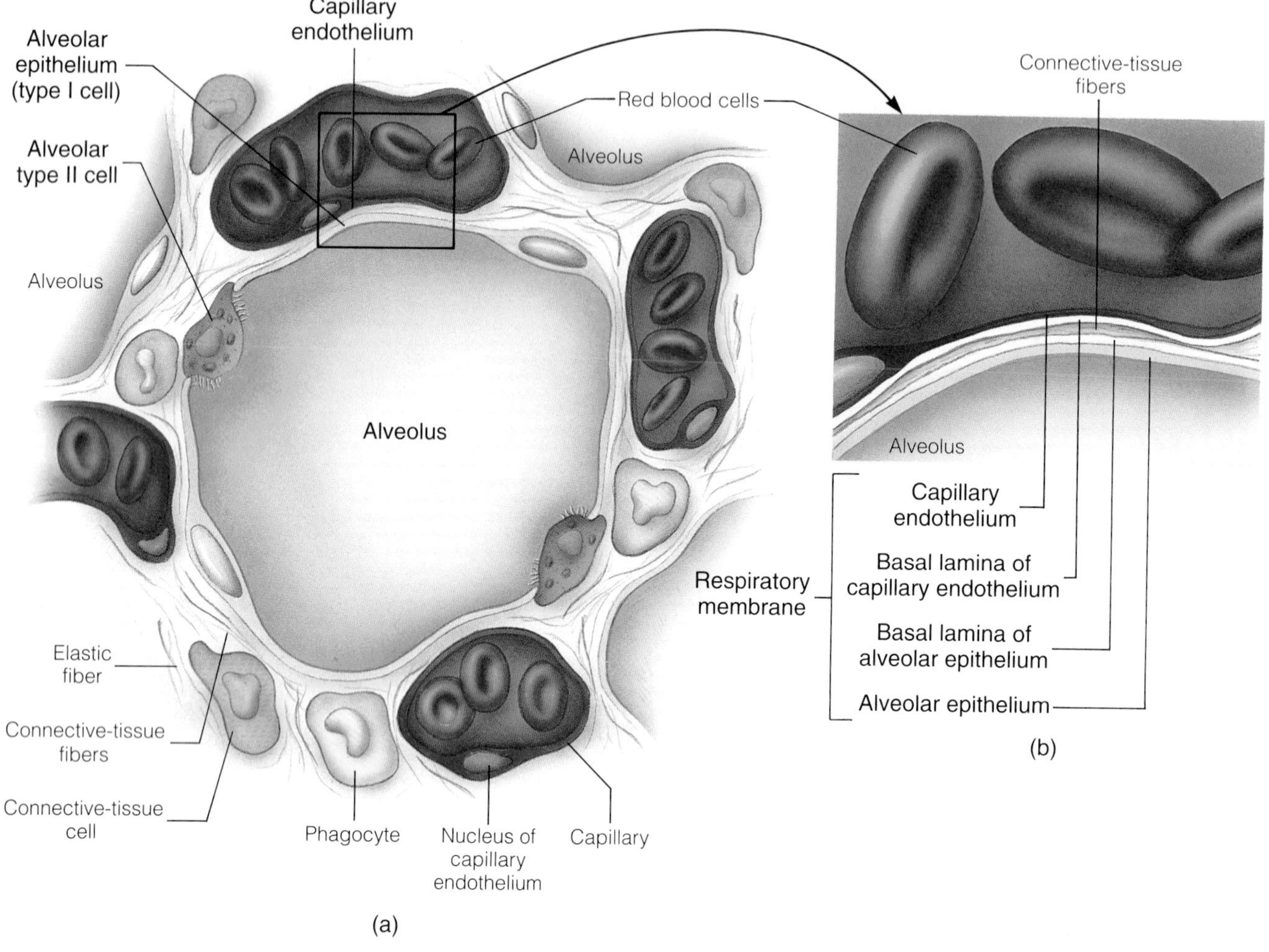

quently, the pressure exerted at sea level by the column of atmospheric gases above any point on the surface of the earth is the same as the pressure that would be exerted by a column of mercury 760 millimeters high. Thus, *atmospheric pressure* at sea level can be expressed as equal to 760 mm Hg.

Pressure Relationships in the Thoracic Cavity

The lungs are stretched within the thoracic cavity, and the visceral pleura (which adheres to each lung) is separated from the parietal pleura (which is attached to the thoracic wall and diaphragm) by only a thin layer of pleural fluid. This fluid couples the visceral pleura to the parietal pleura (and therefore the lungs to the thoracic wall and diaphragm).

The walls and partitions of the lungs contain elastic connective tissue whose recoil tends to reduce the size of the lungs. In addition, as is discussed later, the surface tension of the fluid that coats the exposed surfaces of the alveoli also favors a reduction in the size of the lungs. Thus, the lungs tend to pull away from the thoracic wall and diaphragm and collapse. When the lungs are in the resting position following a normal exhalation, the collapsing force of the lungs is equal to a pressure of about 4 mm Hg (Figure 23.11).

The pressure within the pleural cavity (the *intrapleural* or *intrathoracic pressure*) pushes against the outer walls of the lungs and also favors their collapse. However, the tendency of the lungs to pull away from the thoracic wall and diaphragm leads to a pressure within the pleural cavity that is somewhat less than the atmospheric pressure. When the lungs are in the resting position, the intrapleural pressure is approximately

◆ **FIGURE 23.11 Intrapulmonary and intrapleural pressures in the resting position**

All pressures are given in millimeters of mercury.

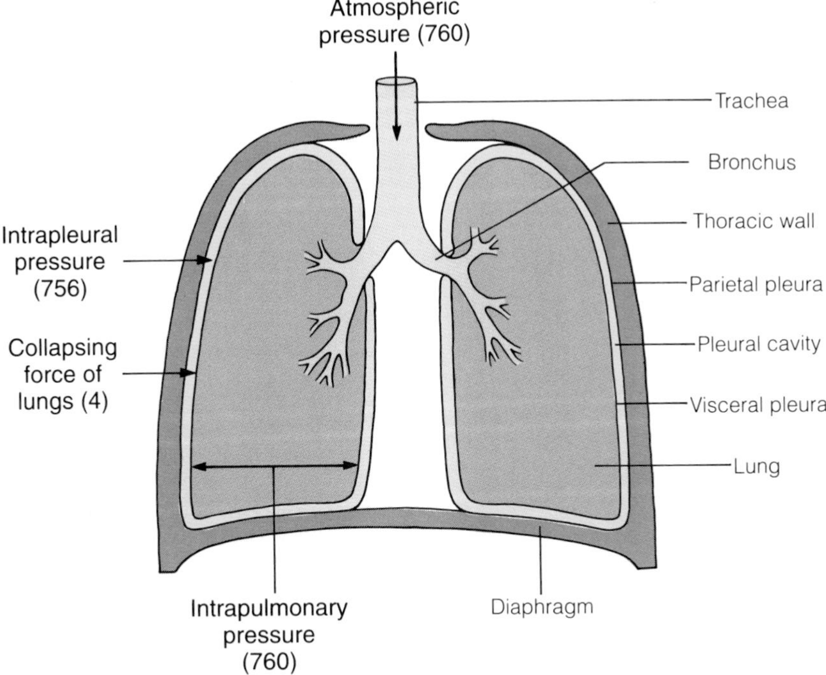

4 mm Hg less than the atmospheric pressure—that is, the intrapleural pressure is approximately 756 mm Hg at sea level.

The alveoli and passageways within the lungs communicate with the atmosphere through the trachea. Consequently, when the lungs are in the resting position, the pressure within them (the *intrapulmonary pressure*) is equivalent to the atmospheric pressure. Thus, at sea level, the resting intrapulmonary pressure is 760 mm Hg. This pressure pushes against the inner walls of the lungs and opposes their collapse.

Under normal circumstances, the forces favoring the collapse of the lungs (the 4-mm-Hg collapsing force of the lungs and the 756-mm-Hg intrapleural pressure) are balanced by a force that opposes their collapse (the 760-mm-Hg intrapulmonary pressure). As a consequence, the lungs normally remain expanded within the thoracic cavity and do not collapse. However, if a lung or the thoracic wall is punctured or torn so that the pleural cavity is exposed to the atmosphere, air enters the pleural cavity and the intrapleural pressure becomes equal to the atmospheric pressure. As a consequence, the forces tending to reduce the size of the lung (the 4-mm-Hg collapsing force of the lung and the now 760-mm-Hg intrapleural pressure) become greater than the force opposing a reduction in lung size (the 760-mm-Hg intrapulmonary pressure), and the lung collapses. The presence of air in the pleural cavity is known as *pneumothorax (noo-mo-thor´-aks)*, and the collapse of a lung is called *atelectasis (at´´-e-lek´-tah-sis)*.

Ventilation of the Lungs

Air flows from a region of higher pressure to a region of lower pressure. Thus, for air to flow into the lungs from the atmosphere, the pressure within the lungs—that is, the intrapulmonary pressure—must be less than the atmospheric pressure. Similarly, for air to flow out of the lungs, the intrapulmonary pressure must be greater than the atmospheric pressure.

As indicated by the relationship known as Boyle's law, the pressure exerted by a fixed number of gas molecules in a container is inversely proportional to the volume of the container (Figure 23.12). Thus, if the volume of the container increases, the pressure decreases, and if the volume of the container decreases, the pressure increases. Consequently, the pressure within the gas-filled spaces of the lungs can be altered by altering the volume of the lungs.

The volume of the lungs can be altered by changing the volume of the thoracic cavity (recall that under normal circumstances, the pleural fluid couples the visceral pleura to the parietal pleura, and the movements of the lungs closely follow the movements of the thoracic wall and diaphragm). Therefore, when the thoracic wall and diaphragm move in such a way that the volume of the thoracic cavity increases, so does the volume of the lungs. The increase in lung volume lowers the intrapulmonary pressure below the atmospheric pressure, and air flows into the lungs until the pressure within the lungs again equals the atmospheric pressure. Con-

◆ **FIGURE 23.12 Boyle's law**
Each of the containers has the same number of gas molecules within it. Given the random motion of gas molecules, the likelihood of a gas molecule striking the interior wall of a container and exerting pressure varies inversely with the volume of the container at any constant temperature. The gas in container (b) exerts more pressure than the same gas in larger container (c) but less pressure than the same gas in smaller container (a).

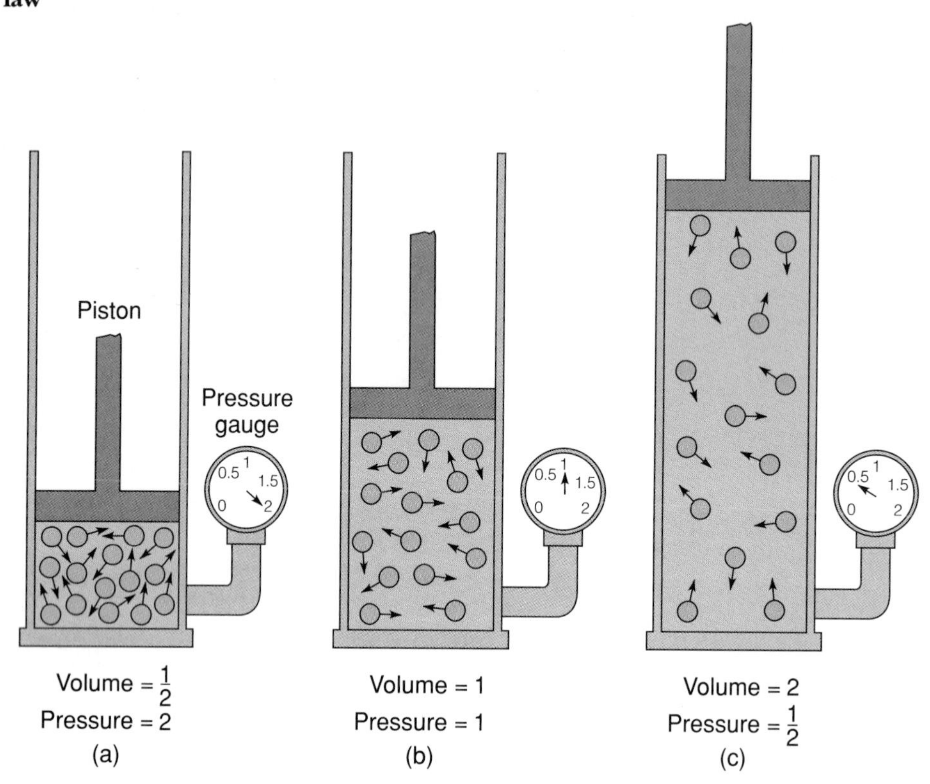

versely, when the thoracic wall and diaphragm move in such a way that the volume of the thoracic cavity decreases, so does the volume of the lungs. The decrease in lung volume compresses the air within the lungs, and the intrapulmonary pressure rises above the atmospheric pressure. As a consequence, air flows out of the lungs until the pressure within the lungs becomes equal to the atmospheric pressure.

Inspiration

Inspiration, or *inhalation,* refers to the movement of air into the lungs. As previously indicated, air moves into the lungs when the volume of the thoracic cavity—and thus the lungs—increases, and the intrapulmonary pressure falls. During quiet breathing, the intrapulmonary pressure during inspiration decreases about 1 mm Hg below the resting intrapulmonary pressure, which is the same as the atmospheric pressure.

There are two ways of increasing the volume of the thoracic cavity during inspiration. One way is by contracting the diaphragm. When it contracts, the diaphragm flattens, lowering its dome. This movement increases the longitudinal dimension of the thoracic cavity. The second way is by elevating the ribs. In the resting position, the ribs slant downward and forward from the vertebral column. The contraction of muscles such as the intercostal muscles, which are located between the ribs, pulls the ribs upward. This movement increases the anterior-posterior dimension of the thoracic cavity.

During normal, quiet inspiration, the contraction of the diaphragm is the dominant means of increasing the volume of the thoracic cavity and lowering the intrapulmonary pressure. The elevation of the ribs is most evident during forced inspiration.

Expiration

Expiration, or *exhalation,* refers to the movement of air out of the lungs. As previously indicated, air moves out of the lungs when the volume of the thoracic cavity—and thus the lungs—decreases, and the intrapulmonary pressure rises. During quiet breathing, the intrapulmonary pressure during expiration increases about 1 mm Hg above the resting intrapulmonary pressure.

During quiet breathing, the volume of the thoracic cavity is decreased by passive processes that do not involve muscular contractions. When the muscles involved in inspiration relax, the elastic recoil of the lungs, thoracic wall, and abdominal structures returns the ribs and diaphragm to their resting positions. This activity reduces the volume of the thoracic cavity and raises the intrapulmonary pressure.

During forced expiration, as occurs in exercise, muscles are involved in the further reduction of the volume of the thoracic cavity. The muscles of the anterior abdominal wall aid in forced expiration by exerting pressure on the abdominal viscera, thus forcing the diaphragm upward. The intercostals, the transversus thoracis, the quadratus lumborum, and the serratus posterior inferior muscles also assist in reducing the volume of the thoracic cavity by depressing the rib cage.

Traditionally, it has been thought that the roles of the intercostal muscles in respiratory movements could be distinguished, with the external intercostals assisting in forced inspiration and the internal intercostals assisting in forced expiration. However, some studies have found that both sets of intercostals are active in both forced inspiration and forced expiration, serving to control the distances between ribs.

Factors That Influence Pulmonary Airflow

The volume of air that flows in a given time between the atmosphere and the alveoli is influenced by the pressure that moves the air through the respiratory passageways and by the resistance the air encounters as it flows through the passageways. The relationship between these factors can be expressed as

$$\text{Flow} = \frac{\text{Pressure}}{\text{Resistance}}$$

The pressure that moves the air through the respiratory passageways is represented by the difference between the atmospheric pressure and the pressure within the alveoli. A major factor influencing resistance is passageway diameter: the smaller the diameter of a passageway, the greater its resistance. Smooth muscle associated with the respiratory passageways, particularly the bronchioles, can alter passageway diameters.

Under normal circumstances, the resistance of the respiratory passageways is slight, and a substantial airflow occurs in response to only a slight difference in atmospheric and alveolar pressures. During quiet inspiration, for example, the pressure difference between the atmosphere and the alveoli is only about 1 mm Hg; yet approximately 500 ml of air enters the lungs with each breath.

Under certain circumstances, the constriction of the passageways (particularly the bronchioles) or the accumulation of fluid or mucus within the passageways increases the resistance sufficiently to impede the flow of air between the atmosphere and the alveoli. Since the passageways tend to widen as the lung volume increases during inspiration and narrow as the lung volume decreases during expiration, an increased resistance can particularly affect the flow of air out of the lungs. If the resistance increases, the pressure difference between the atmosphere and the alveoli must increase in order to maintain the airflow. Within limits, the pressure difference can be increased by an increased contraction of the respiratory muscles.

Parasympathetic nervous stimulation, histamine, and leukotrienes cause smooth muscle of the respiratory passageways to contract and thereby increase passageway resistance. Parasympathetic nervous stimulation and histamine also increase mucus secretion. Although few, if any, sympathetic neurons supply the respiratory passageways, epinephrine from the adrenal medullae causes smooth muscle of the passageways to relax and thereby decreases passageway resistance.

Surface Tension and Pulmonary Surfactant

At an air-liquid interface, attractive forces between the molecules of a liquid such as water cause the molecules to be attracted more strongly into the liquid than into the air. This differential attraction gives rise to a tension at the surface of the liquid—the *surface tension*—that tends to reduce the liquid's surface area to the smallest possible value and to oppose any increase in the surface area. Thus, the surface tension of the fluid that coats the exposed surfaces of the alveoli tends to reduce the size of the alveoli and oppose the expansion of the lungs.

If the fluid that coats the exposed surfaces of the alveoli were pure water, which has a relatively high surface tension, the tendency of the alveoli to collapse would be so great that the expansion of the lungs during inspiration would require exhausting muscular effort. However, alveolar type II cells produce a phospholipoprotein complex called **pulmonary surfactant** *(sur-fak´-tant)*, which lowers the surface tension of the fluid that coats the alveoli, significantly decreasing the muscular effort required to expand the lungs. In fact, even during heavy exercise, the ventilation of the lungs requires only about 3–4% of the total energy expenditure.

Some children who are born prematurely do not produce sufficient quantities of pulmonary surfactant. The lack of adequate pulmonary surfactant production is believed to contribute to (or perhaps cause) an often fatal condition known as respiratory distress syndrome of the newborn, or hyaline membrane disease, in which breathing is difficult and labored.

Lung Volumes and Capacities

In young adult males, the lungs have a total capacity of approximately 5900 ml. However, a person cannot exhale all the air from the lungs, and about 1200 ml of air always remains no matter how forced the expiration. This remaining volume is called the **residual volume** (Figure 23.13).

◆ **FIGURE 23.13 Lung volumes and capacities**
The tidal volume indicated is the tidal volume during normal, quiet breathing.

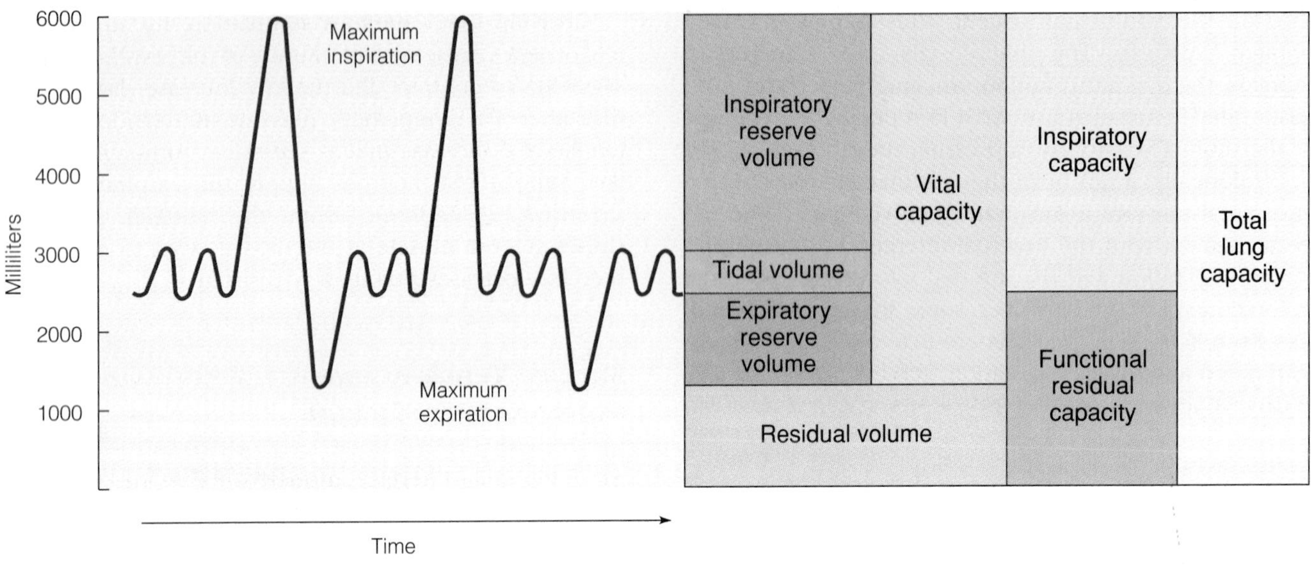

The volume of air that moves into and out of the lungs with each breath is called the **tidal volume.** During normal, quiet breathing, it is about 500 ml.

The **inspiratory reserve** is the volume of air (approximately 3000 ml) that can be inspired in addition to the normal, quiet tidal volume. The sum of the normal, quiet tidal volume and the inspiratory reserve volume is called the **inspiratory capacity.**

Following a normal passive expiration of the quiet tidal volume, additional air can be forced out. This **expiratory reserve** is about 1200 ml. The sum of the residual volume and the expiratory reserve volume is called the **functional residual capacity.**

The maximal amount of air that can be moved into and out of the lungs, from the deepest possible inspiration to the most forced expiration, is called the **vital capacity.** Therefore, the vital capacity represents the sum of the inspiratory reserve volume; the tidal volume during normal, quiet breathing; and the expiratory reserve volume. Note that it does not include the residual volume. In healthy young adult males, the vital capacity is about 4700 ml. It is somewhat less in women because they tend to have smaller thoracic cages and smaller lung capacities (Table 23.1).

Dead Space

The exchange of oxygen and carbon dioxide between air in the lungs and the blood occurs within the alveoli and, to some extent, within the small bronchioles and ducts that lead to the alveoli. However, for purposes of simplification, all these gas exchange areas will be considered together as the alveoli. No gas exchange takes place within the remaining respiratory passageways (nose, trachea, bronchi, and so on). These air-filled passageways are referred to, therefore, as *anatomical dead space.* For healthy young adults, the volume of the anatomical dead space can be estimated to be approximately 1 ml for each pound of a person's "ideal" body weight. Thus, for a normal 150-pound male, the volume of the anatomical dead space is approximately 150 ml.

In some cases, oxygen and carbon dioxide exchanges between the air within certain alveoli and the blood are unable to take place efficiently because no blood reaches certain alveoli or because the ventilation of the alveoli exceeds the blood flow to them. Such occurrences have the effect of increasing the volume of dead space within the lungs beyond that of the anatomical dead space alone.

◆ **TABLE 23.1 Average Lung Volumes (in ml) in Young Males and Females (Age 20–30 Years)**

	MALE	FEMALE
Total lung capacity	5900	4400
Vital capacity	4700	3400
Inspiratory reserve	3000	2100
Tidal volume (at rest)	500	500
Expiratory reserve	1200	800
Residual volume	1200	1000

The overall functional dead space within the lungs, including both the anatomical dead space and dead space attributable to the presence of alveoli that receive too little blood flow in relation to their airflow, is called the *physiological dead space.* In a healthy person, the volume of the physiological dead space and the volume of the anatomical dead space are the same or nearly the same. However, in certain disease conditions, the volume of the physiological dead space is significantly larger than that of the anatomical dead space.

Minute Respiratory Volume

The volume of air moved into the respiratory passageways in one minute is called the **minute respiratory volume.** It is equal to the respiratory rate (expressed as number of breaths per minute) multiplied by the volume of air that enters the respiratory passageways with each breath (the tidal volume). For example, during quiet breathing the respiratory rate is about 12 breaths per minute, and approximately 500 ml of air enters the respiratory passageways with each breath. Thus, the minute respiratory volume is 6000 ml/min (12 breaths/min × 500 ml/breath). The minute respiratory volume can be altered by altering either the respiratory rate or the amount of air entering the respiratory passageways with each breath.

Alveolar Ventilation

The exchange of gases between the lungs and the blood occurs within the alveoli and not within the respiratory passageways. Consequently, the volume of air that moves into the alveoli from the atmosphere is more important than the volume of air that enters the respiratory passageways.

With each inspiration, the first air to reach the alveoli is air that was left in the respiratory passageways—that is, in the anatomical dead space—from the previous expiration. Also, the last air to be inspired never reaches the alveoli. It remains within the respiratory passageways and is the first air to be exhaled. For a normal, 150-pound, young adult male, the volume of air within the respiratory passageways is approximately 150 ml. Therefore, of the approximately 500 ml of atmospheric air that enters the respiratory passsageways with each inspiration during quiet breathing, only 350 ml reaches the alveoli.

Note that although only 350 ml of atmospheric air enters the alveoli with each inspiration during quiet breathing, a total of 500 ml of air actually moves into the alveoli (the first 150 ml is air that remained in the respiratory passageways from the previous expiration). Note, too, that the air within the alveoli is not completely replaced with each breath. For example, during quiet breathing, an expiratory reserve volume of 800–1200 ml and a residual volume of 1000–1200 ml remain within the lungs at the end of expiration, and only about 500 ml of air is taken into the lungs with each breath.

The volume of *atmospheric air* that enters the alveoli (either per breath or in one minute) and that can participate in the exchange of gases between the alveoli and the blood is called the **alveolar ventilation.** Thus, the alveolar ventilation represents the physiologically effective ventilation.

The alveolar ventilation *per minute* can be estimated as follows:

$$\text{Alveolar Ventilation} = \text{Respiratory Rate} \times (\text{Tidal Volume} - \text{Physiological Dead Space Volume})$$

To express the alveolar ventilation in ml/min, the respiratory rate is expressed as breaths/minute, and the tidal volume and physiological dead space volume are expressed as ml/breath. Thus, for a normal, 150-pound, young adult male during quiet breathing at a rate of 12 breaths/min, the alveolar ventilation is 4200 ml/min [12 breaths/min × (500 ml/breath − 150 ml/breath)].

Assuming the physiological dead space remains essentially constant, the alveolar ventilation can be increased by increasing either the respiratory rate or the volume of air inspired with each breath (that is, the tidal volume). However, the alveolar ventilation is generally increased more effectively by increasing the volume of air inspired with each breath than by increasing the respiratory rate (Table 23.2). For example, if the respiratory rate doubles from its resting value (that is, doubles to 24 breaths/min) but the amount of air moved during each inspiration remains the same (that is, 500 ml/breath), the minute respiratory volume is 12,000 ml/min, and the alveolar ventilation is 8400 ml/min [24 breaths/min × (500 ml/breath − 150 ml/breath)]. If the respiratory rate remains the same (that is, 12 breaths/min) but the volume of air moved during each inspiration doubles (that is, to 1000 ml/breath), the minute respiratory volume is still 12,000 ml/min, but the alveolar ventilation is 10,200 ml/min [12 breaths/min × (1000 ml/breath − 150 ml/breath)]. During exercise and in most other situations where an increased supply of oxygen and an increased elimination of carbon dioxide are required, the increase in breathing depth—that is, the volume of air moved with each breath—is proportionally greater than the increase in respiratory rate.

Matching of Alveolar Airflow and Blood Flow

For efficient gas exchange, the airflow and blood flow to particular alveoli must be well matched. Poor gas exchange occurs when the alveolar airflow is either inadequate or excessive in relation to the blood flow.

◆ **TABLE 23.2 Effect of Different Breathing Patterns on Alveolar Ventilation**

BREATHING PATTERN	RESPIRATORY RATE*	TIDAL VOLUME†	PHYSIOLOGICAL DEAD SPACE VOLUME‡	PULMONARY VENTILATION§	ALVEOLAR VENTILATION‖
Normal, quiet breathing	12	500	150	6,000	4,200
More rapid breathing	24	500	150	12,000	8,400
Deeper breathing	12	1,000	150	12,000	10,200

* breaths/min † ml/min ‡ ml § ml/min = tidal volume × respiratory rate
‖ ml/min = respiratory rate × (tidal volume − physiological dead space volume)

Local autoregulatory mechanisms contribute to matching alveolar airflow and blood flow. One of these mechanisms was considered in Chapter 20, where it was pointed out that pulmonary vessels constrict and blood flow decreases in areas where the oxygen level is low (and pulmonary vessels dilate and blood flow increases in areas where the oxygen level is high). In addition, the respiratory passageways dilate and airflow to the alveoli increases in areas where the carbon dioxide level is high (and the respiratory passageways constrict and airflow to the alveoli decreases in areas where the carbon dioxide level is low).

These autoregulatory mechanisms act in the following manner to match alveolar airflow and blood flow. In an area of alveoli whose airflow is inadequate in relation to their blood flow, the oxygen level will be low and the carbon dioxide level will be high (Figure 23.14a). As a consequence, the pulmonary blood vessels in the area constrict and the blood flow decreases. In addition, the respiratory passageways dilate and the airflow increases. Thus, the alveolar airflow and blood flow become more closely matched. In an area of alveoli whose airflow is excessive in relation to their blood flow, the oxygen level will be high and the carbon dioxide level will be low (Figure 23.14b). As a consequence, the pulmonary blood vessels in the area dilate and the blood flow increases. In addition, the respiratory passageways constrict and the airflow decreases. Again, the alveolar airflow and blood flow become more closely matched.

Transport of Gases by the Blood

Partial Pressure

The atmosphere is a mixture of several gases, including nitrogen, oxygen, and carbon dioxide. The total atmospheric pressure at sea level is 760 mm Hg. According to Dalton's law of gases, the total pressure exerted by a mixture of gases is equal to the sum of the separate pressures exerted by each of the individual gases. Each individual gas in a mixture exerts a *partial pressure,* according to its percentage concentration in the mixture, that contributes to the total pressure exerted by the mixture. Thus, each atmospheric gas acts independently in generating the total atmospheric pressure, and each contributes partially to the total. For example, oxygen—which makes up approximately 21% of the atmospheric mixture of gases—contributes approximately 21% of the total pressure. Thus, the partial pressure of oxygen in the atmosphere at sea level is approximately 21%, or 160 mm Hg, of the total 760-mm-Hg atmospheric pressure at sea level.

Gases in Liquids

When a liquid is exposed to a gaseous atmosphere, gases enter the liquid and dissolve in it in proportion to their individual gas pressures. The number of molecules of a particular gas that enter a liquid, however, depends not only on the pressure of that gas in the atmosphere, but also on the solubility of the gas in the liquid. Nevertheless, by increasing the pressure of the gas, more molecules of it will dissolve in the liquid.

Under specific conditions, a gas that is dissolved in a liquid can be treated mathematically as though it were in a free state and able to exert an independent pressure or partial pressure. Thus, each of the gases of the atmosphere can, while in solution, be considered to exert a pressure or partial pressure.

Once a gas dissolves in a liquid, the gas molecules can diffuse, and the net diffusion is from regions of high pressure to regions of lower pressure. Moreover, if a liquid contains a particular gas at a higher pressure than the pressure of that gas in the surrounding atmosphere, molecules of the gas will leave the liquid and enter the atmosphere.

◆ **FIGURE 23.14 Local controls to match airflow and blood flow in an area of the lungs**

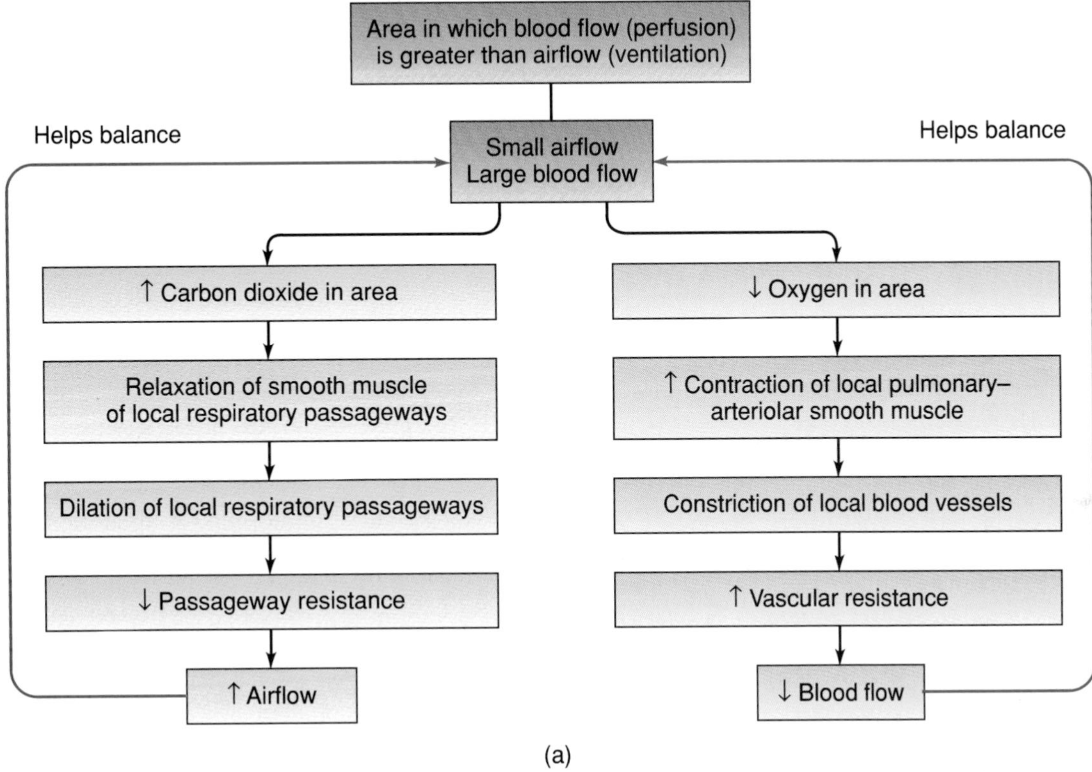

(a)

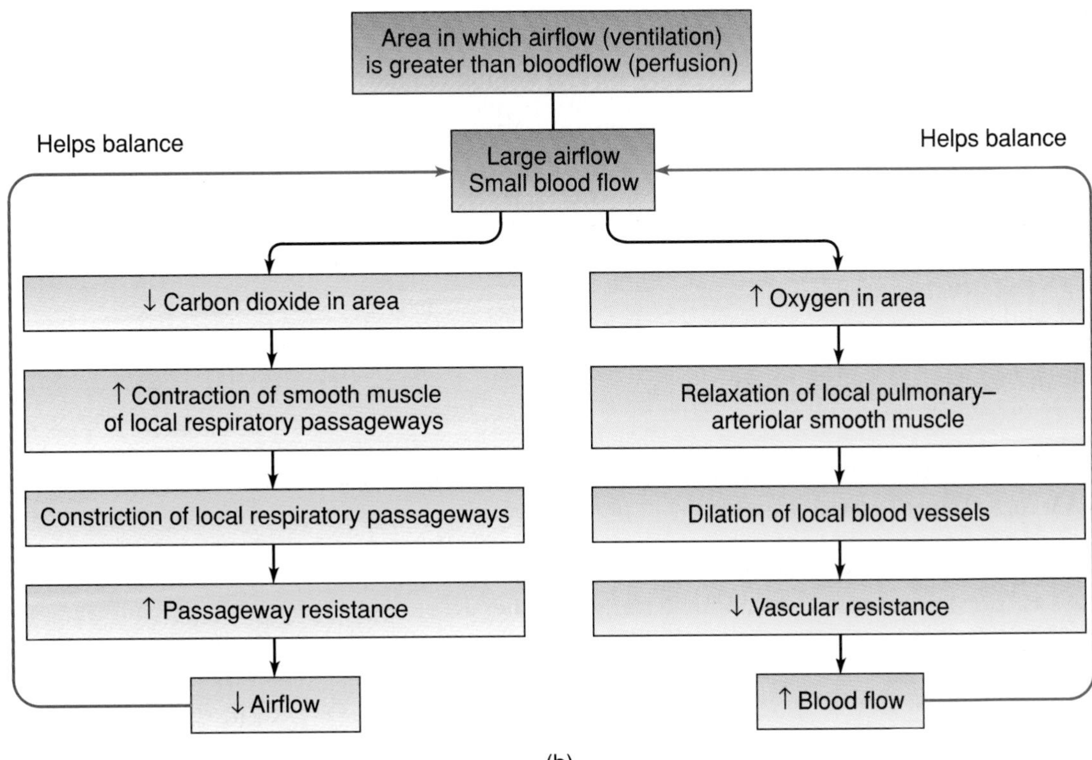

(b)

ASPECTS OF EXERCISE PHYSIOLOGY

How to Find Out How Much Work You're Capable of Doing

◆

The best single predictor of a person's work capacity is the determination of the maximum volume of oxygen the person is capable of using per minute to oxidize nutrient molecules for energy production. *Maximal oxygen consumption*, or *max* $\dot{V}O_2$, is measured by having the person engage in exercise, usually on a treadmill or bicycle ergometer (a stationary bicycle with variable resistance). The work load is incrementally increased until the person becomes exhausted. Expired air samples collected during the last minutes of exercise, when oxygen consumption is at a maximum because the person is working as hard as possible, are analyzed for the percentages of oxygen and carbon dioxide they contain. Furthermore, the volume of air expired is measured. Equations are then employed to determine the amount of oxygen consumed, taking into account the percentages of oxygen and carbon dioxide in the inspired air, the total volume of air expired, and the percentages of oxygen and carbon dioxide in the exhaled air.

Maximal oxygen consumption depends on three systems. The respiratory system is essential for ventilation and exchange of oxygen and carbon dioxide between the air and blood in the lungs. The cardiovascular system is required for delivery of oxygen to the working muscles. Finally, the muscles must have oxidative enzymes available to use the oxygen once it has been delivered.

Regular aerobic exercise can improve max $\dot{V}O_2$ by making the heart and respiratory system more efficient and therefore capable of delivering more oxygen to the working muscles. Exercised muscles also become better equipped to use oxygen once it is delivered. The number of functional capillaries is increased, as are the number and size of mitochondria, which contain the oxidative enzymes.

Max $\dot{V}O_2$ is measured in liters per minute and then converted into milliliters per kilogram of body weight per minute so that large and small people can be compared.

As would be expected, athletes have the highest values for max $\dot{V}O_2$. The max $\dot{V}O_2$ for male cross-country skiers has been recorded to be as high as 94 ml O_2/kg/min. Distance runners maximally consume between 65–85 ml O_2/kg/min, and football players have max $\dot{V}O_2$ values between 45–65 ml O_2/kg/min, depending on the position they play. Sedentary young men maximally consume between 25 and 45 ml O_2/kg/min. Female values for max $\dot{V}O_2$ are 20–25% lower than for males when expressed as ml/kg/min of total body weight. The difference in max $\dot{V}O_2$ between females and males is only 8–10% when expressed as ml/kg/min of lean body weight, however, because females generally have a higher percentage of body fat (the female sex hormone estrogen promotes fat deposition).

Available norms are used to classify people as being low, fair, average, good, or excellent in aerobic capacity for their age group. Exercise physiologists use determinations of max $\dot{V}O_2$ to prescribe or adjust training regimens to help people achieve their optimal level of aerobic conditioning.

Oxygen and Carbon Dioxide Exchange between Lungs, Blood, and Tissues

The partial pressure of oxygen (PO_2) in the atmosphere at sea level, as previously noted, is approximately 160 mm Hg. The atmospheric partial pressure of carbon dioxide (PCO_2) is only about 0.3 mm Hg. However, due to the humidification of the inhaled air and to the mixing of atmospheric air with air remaining in the lungs during each respiratory cycle, the gases in the lungs are not at the same partial pressures as in the atmosphere. The actual PO_2 and PCO_2 in the alveoli of the lungs, as well as the partial pressures of these gases in the blood entering and leaving the alveolar capillaries and in the body tissues of a resting person at sea level, are given in Figure 23.15. As can be seen from the figure, within the alveoli the PO_2 is about 104 mm Hg, and the PCO_2 is approximately 40 mm Hg. These pressures remain fairly constant throughout the average respiratory cycle.

As the figure also shows, blood entering the alveolar capillaries from the tissues contains oxygen at a partial pressure of only 40 mm Hg and carbon dioxide at a partial pressure of 45 mm Hg. As a result of the partial-pressure differences between the oxygen and carbon

◆ **FIGURE 23.15 Partial pressures of oxygen and carbon dioxide in the alveoli of the lungs, in the blood entering and leaving the alveolar capillaries, and in the body tissues of a resting person at sea level**

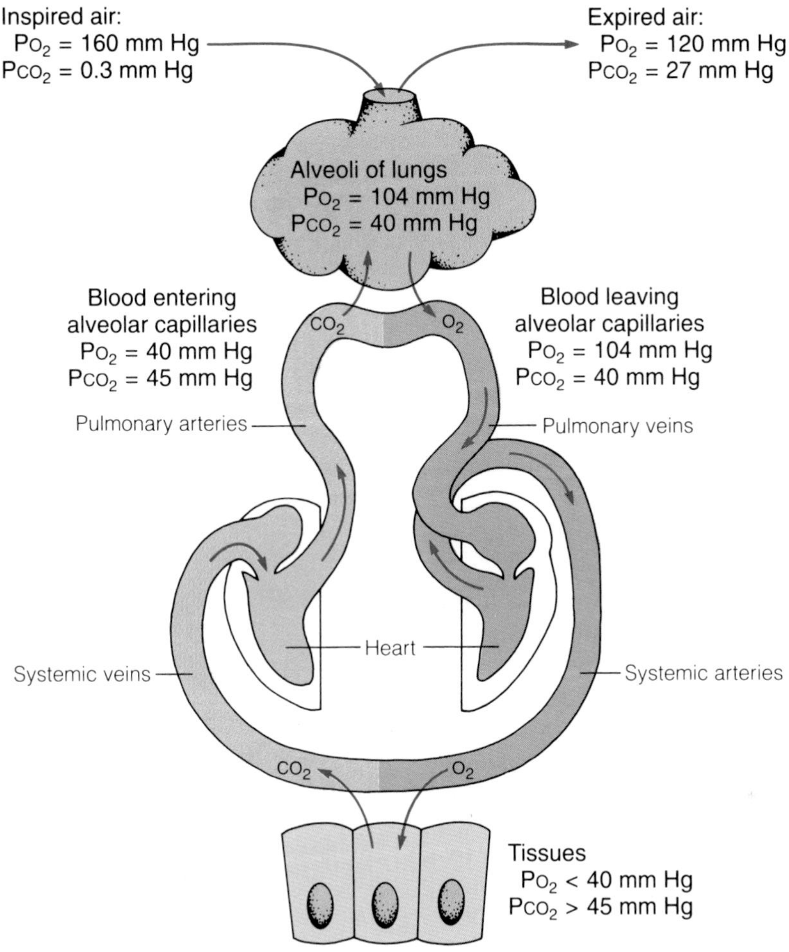

dioxide in the blood and the oxygen and carbon dioxide in the alveoli of the lungs, there is a net diffusion of oxygen into the blood from the alveoli and a net diffusion of carbon dioxide into the alveoli from the blood. As a consequence of these activities, the partial pressures of oxygen and carbon dioxide in the blood leaving the alveolar capillaries are essentially equal to the partial pressures of these gases in the alveoli (P_{O_2}: 104 mm Hg; P_{CO_2}: 40 mm Hg).

Within the tissues, metabolic reactions utilize oxygen and produce carbon dioxide. Thus, the P_{O_2} in the tissues is approximately 40 mm Hg, whereas the P_{CO_2} is about 45 mm Hg. When the blood from the lungs, which contains oxygen at a relatively high partial pressure, reaches the tissues, there is a net diffusion of oxygen from the blood to the tissues. In a similar fashion, there is a net diffusion of carbon dioxide from the tissues to the blood. The end result is that the venous blood leaving the tissues contains oxygen at a partial pressure of about 40 mm Hg and carbon dioxide at a partial pressure of about 45 mm Hg. These are the partial pressures that were previously described for blood entering the alveolar capillaries of the lungs.

Thus, the exchanges of oxygen and carbon dioxide that occur between the blood and alveoli on the one hand, and the blood and tissues on the other, take place by simple diffusion and depend on the pressure differences of the respective gases that exist between the blood and the alveoli, and the blood and the tissues.

Oxygen Transport in the Blood

Since oxygen is relatively insoluble in water, relatively little actually dissolves in the fluid of the blood. In fact, under normal circumstances, each liter of systemic arterial blood contains only about 3 ml of dissolved oxygen. However, the total oxygen content of a liter of systemic arterial blood is about 197 ml. The additional

◆ **FIGURE 23.16 Oxygen-hemoglobin dissociation curve**
The curve indicates the degree of hemoglobin saturation at different partial pressures of oxygen (Po_2) when the pH is 7.4 and the temperature is 38°C.

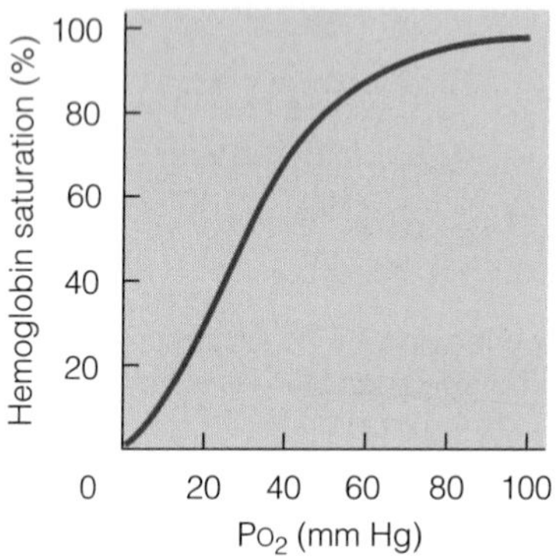

194 ml of oxygen is bound to the iron of the heme groups of hemoglobin molecules in erythrocytes.

The combination of hemoglobin with oxygen is reversible, and whether hemoglobin binds with or releases oxygen depends in large part on the oxygen partial pressure. The degree of saturation of hemoglobin with oxygen at any given Po_2 is indicated by the oxygen-hemoglobin dissociation curve (Figure 23.16). When the Po_2 is relatively high, hemoglobin binds with much oxygen and is essentially completely saturated. At lower oxygen partial pressures, hemoglobin binds with less oxygen and is only partially saturated.

The Po_2 is relatively high in the alveolar capillaries of the lungs. Here, oxygen from the alveoli diffuses into the plasma and then into the erythrocytes, where it binds with hemoglobin. Within the various body tissues, however, metabolic activities utilize oxygen, and there is a net diffusion of oxygen out of the blood. Thus, the Po_2 in the tissue capillaries is lower than that in the alveolar capillaries, and hemoglobin cannot bind with as much oxygen in the tissue capillaries as it can in the alveolar capillaries. As a result, the oxygen-rich hemoglobin that leaves the alveolar capillaries releases some of its oxygen when it reaches the tissue capillaries. This oxygen can then diffuse into the tissues for their use.

In addition to the Po_2, several other factors affect the binding of oxygen to hemoglobin. Under acidic conditions, the amount of oxygen that binds to hemoglobin at any given oxygen partial pressure is diminished. Thus, the higher the hydrogen ion concentration—that is, the lower the pH—the less oxygen is bound to hemoglobin at any given Po_2 (Figure 23.17). This effect is due to the fact that hydrogen ions bind with the hemoglobin molecules, altering their molecular structure and thereby reducing their affinity for oxygen.

The partial pressure of carbon dioxide has basically the same effect on the binding of oxygen and hemoglobin as the hydrogen ion concentration. That is, the higher the Pco_2, the less oxygen is bound to hemoglobin at any given Po_2 (Figure 23.18). This effect is due in part to the fact that carbon dioxide molecules bind with hemoglobin molecules, altering the molecular structure of the hemoglobin molecules and thereby reducing their affinity for oxygen. It is also due to the fact that carbon dioxide can influence pH in the following manner:

$$\underset{\text{Carbon Dioxide}}{CO_2} + \underset{\text{Water}}{H_2O} \rightleftharpoons \underset{\text{Carbonic Acid}}{H_2CO_3} \rightleftharpoons \underset{\text{Hydrogen Ion}}{H^+} + \underset{\text{Bicarbonate Ion}}{HCO_3^-}$$

◆ **FIGURE 23.17 The effects of acidity on the degree of hemoglobin saturation at different partial pressures of oxygen**

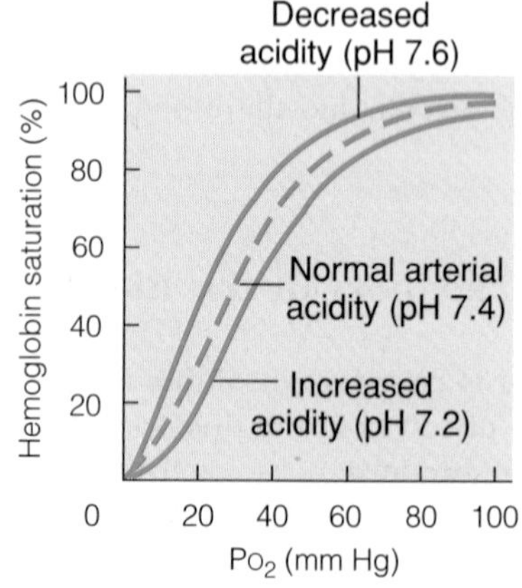

◆ **FIGURE 23.18 The effects of carbon dioxide on the degree of hemoglobin saturation at different partial pressures of oxygen**

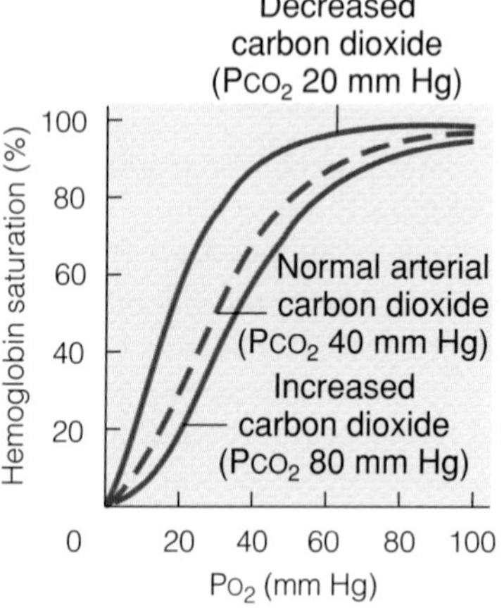

As these reactions show, carbon dioxide can combine with water to form carbonic acid. (Although not absolutely required for this reaction to occur, an enzyme within erythrocytes, *carbonic anhydrase,* greatly increases the rate of the reaction.) The carbonic acid can dissociate into hydrogen ions and bicarbonate ions, and the hydrogen ions contribute to the hydrogen ion concentration. Thus, an increase in the P_{CO_2} tends to increase the hydrogen ion concentration and therefore lower the pH. (The influence of carbon dioxide and hydrogen ions on the ability of hemoglobin to bind oxygen is known as the *Bohr effect.*)

Temperature also influences the binding of oxygen and hemoglobin. The higher the temperature, the less oxygen is bound to hemoglobin at any given P_{O_2} (Figure 23.19).

The influences of pH, P_{CO_2}, and temperature on the binding of oxygen by hemoglobin operate to ensure adequate deliveries of oxygen to active tissues that need it most. Active tissues tend to have higher hydrogen ion concentrations in their vicinities than do less active tissues (active skeletal muscles, for example, can produce lactic acid). They also tend to produce more carbon dioxide and to have higher temperatures as a result of their metabolic activity than do less active tissues. Consequently, hemoglobin binds less oxygen in active tissues (and thus provides more for delivery) than it does in less active tissues.

A substance called DPG (2,3-diphosphoglycerate) is produced by erythrocytes. DPG binds reversibly with hemoglobin molecules, altering their molecular structure and thereby reducing their affinity for oxygen.

DPG production increases when hemoglobin in the arterial blood is chronically undersaturated with oxygen. Some investigators believe this response helps maintain oxygen delivery to the tissues by promoting the release of oxygen from hemoglobin (an effect similar to that of increased acidity, P_{CO_2}, or temperature). However, other investigators point out that an increased level of DPG within erythrocytes not only favors the release of oxygen from hemoglobin in the tissues, but it can also make it more difficult for oxygen to combine with hemoglobin in the lung capillaries. Therefore, they point out that increased DPG production when hemoglobin in the arterial blood is chronically undersaturated with oxygen may not always be beneficial.

◆ FIGURE 23.19 The effects of temperature (°C) on the degree of hemoglobin saturation at different partial pressures of oxygen

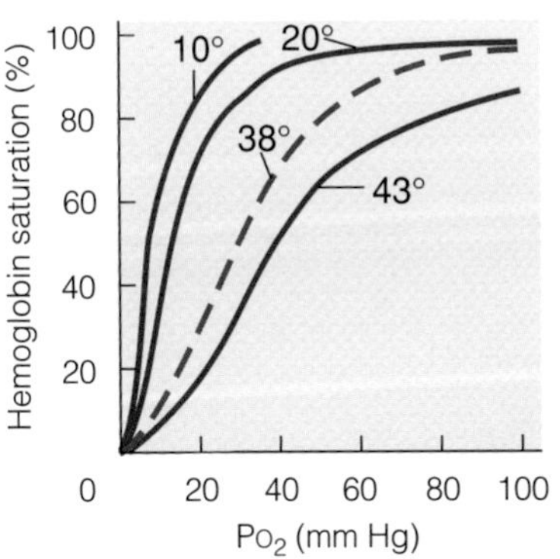

◆ TABLE 23.3 Methods of Gas Transport in the Blood

GAS	METHOD OF TRANSPORT IN BLOOD	PERCENTAGE CARRIED IN THIS FORM
O_2	Physically dissolved	1.5
	Bound to hemoglobin	98.5
CO_2	Physically dissolved	8
	As carbamino compounds	20
	As bicarbonate ions (HCO_3^-)	72

Carbon Dioxide Transport in the Blood

Carbon dioxide is transported by the blood in several different ways (Table 23.3).

Transport as Dissolved Carbon Dioxide

A small amount of carbon dioxide—perhaps 8%—is transported in physical solution dissolved in the plasma.

Transport as Carbamino Compounds

Some carbon dioxide—about 20%—is transported in reversible association with various blood proteins. These carbon dioxide-protein complexes are called *carbamino compounds (kar´-ba-min-ō).*

Carbon dioxide can combine with many proteins in the blood. However, the most abundant protein is the globin of hemoglobin, and many complexes form between carbon dioxide and the globin of hemoglobin. Thus, oxygen is carried by hemoglobin in association with the iron of the heme groups, and carbon dioxide is carried in reversible association with the globin protein portion of the hemoglobin molecules.

Hemoglobin that is not carrying oxygen *(reduced hemoglobin)* is able to form carbamino compounds more readily than hemoglobin that is carrying oxygen *(oxyhemoglobin).* Thus, in the tissues, the presence of hemoglobin molecules that have given up their oxygen favors the combination of carbon dioxide with hemoglobin. In the lungs, the binding of oxygen to reduced hemoglobin molecules decreases their affinity for carbon dioxide and favors the displacement of carbon dioxide from the molecules. The displaced carbon dioxide can diffuse into the alveoli for elimination from the body.

Transport as Bicarbonate Ions

Most of the carbon dioxide in the blood—about 72%—is transported as bicarbonate ions. Carbon dioxide produced in the tissues diffuses into the plasma and from there into erythrocytes (Figure 23.20a). Within the erythrocytes, the enzyme carbonic anhydrase facilitates the combination of carbon dioxide and water to form carbonic acid. The carbonic acid can dissociate into hydrogen ions and bicarbonate ions.

Most of the hydrogen ions associate with hemoglobin molecules. This association eliminates free hydrogen ions and thus helps prevent a substantial increase in the hydrogen ion concentration of the blood. In fact, because of the ability of hemoglobin molecules (and other substances in the blood) to combine with hydrogen ions, the pH of venous blood leaving the tissues (pH = 7.35) is only slightly lower than that of arterial blood flowing to the tissues (pH = 7.4).

The bicarbonate ions resulting from the dissociation of carbonic acid molecules diffuse out of the erythrocytes into the plasma, and in response to this movement of negatively charged bicarbonate ions, chloride ions from the plasma enter the erythrocytes. This exchange is known as the chloride shift. As a result of these events, carbon dioxide from the tissues is transported as bicarbonate ions in the blood.

At the lungs, the reverse of the preceding events occurs (Figure 23.20b). Carbon dioxide diffuses out of the blood into the alveoli, and the concentration of carbon dioxide within the plasma and the erythrocytes declines. Within the erythrocytes, hydrogen ions combine with bicarbonate ions to form carbonic acid, which is split by carbonic anhydrase into carbon dioxide and water. As the concentration of bicarbonate ions within the erythrocytes diminishes, bicarbonate ions from the plasma diffuse in and chloride ions leave. The carbon dioxide that is produced leaves the erythrocytes and diffuses into the alveoli for elimination from the body.

◆ **FIGURE 23.20 Transport of carbon dioxide in the blood as bicarbonate ions**

(a) Sequence of events that occurs in the tissues. (b) Sequence of events that occurs in the lungs. Hb is hemoglobin.

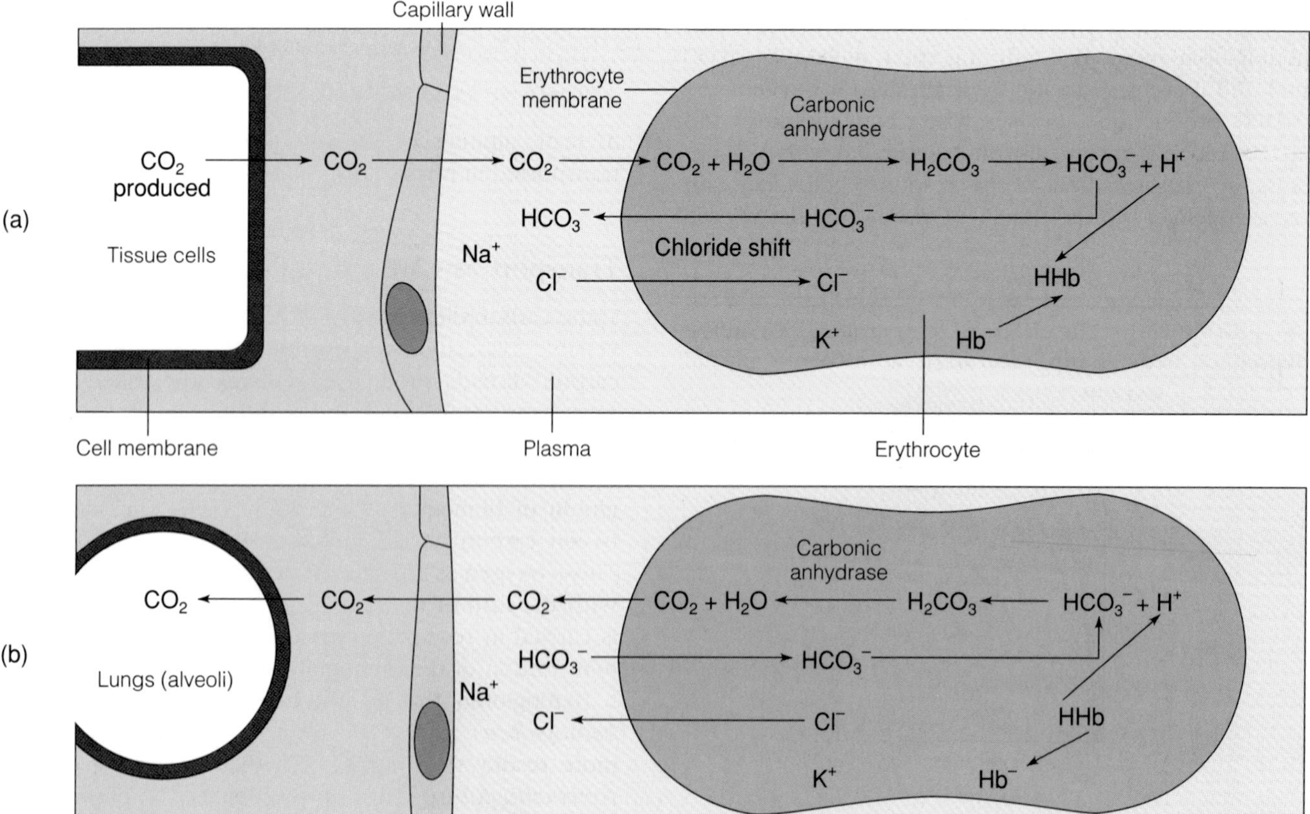

CONDITIONS OF CLINICAL SIGNIFICANCE

Problems of Gas Transport

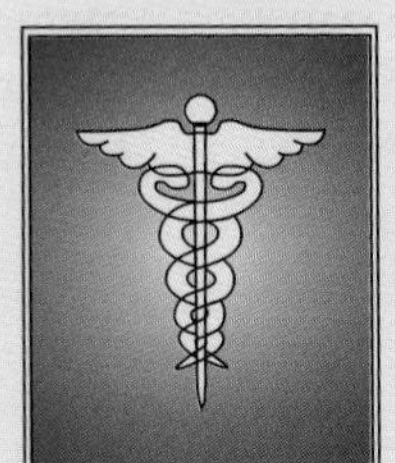

A considerable number of respiratory states are clinically important (Table C23.1), and the transport of oxygen and carbon dioxide by the blood can be affected adversely by a variety of conditions. These conditions include malfunctions of the cardiovascular system that interfere with the transport of gases by the blood, respiratory system problems that upset the gas exchange process, and deficiencies in the composition of the air breathed.

Cyanosis

Cyanosis (sī″-ah-no′-sis) is a bluish discoloration of the skin and mucous membranes. Reduced hemoglobin—that is, hemoglobin that is not combined with oxygen—appears darker than oxyhemoglobin, and cyanosis occurs when there is a significant concentration of reduced hemoglobin in the arterial blood. Cyanosis can result from inadequacies of either the respiratory or the cardiovascular system.

Hypoxia

Any state in which a physiologically inadequate amount of oxygen is available to or utilized by the tissues is called *hypoxia (hi-pok-sē-ah)*. Hypoxia can be caused by a number of factors.

In some cases, not enough oxygen reaches the blood, and the arterial Po_2 is below normal. Hypoxia due to such an occurrence is referred to as *hypoxemic hypoxia (arterial hypoxia)*. Hypoxemic hypoxia can result from insufficient oxygen in the air breathed or from respiratory problems—such as pneumonia, emphysema, or paralysis of the respiratory muscles—that prevent sufficient oxygen from reaching the blood within the lungs.

In other cases, enough oxygen reaches the blood, but there is not enough hemoglobin available to carry it to the tissues. Since the problem is often associated with anemia, hypoxia due to a deficiency of hemoglobin is referred to as *anemic hypoxia*.

In still other cases, the blood flow to the tissues is less than normal. The blood may be able to carry a normal amount of oxygen, but oxygen is not delivered to the tissues fast enough for the cells to maintain their normal metabolism. Hypoxia due to such an occurrence is called *stagnant hypoxia (hypokinetic hypoxia)*. Stagnant hypoxia can result from heart failure or the presence of emboli in the blood.

Carbon Monoxide Poisoning

Carbon monoxide (CO) is a colorless, odorless gas. It is present in the exhaust fumes of automobiles and

continued on next page

◆ **TABLE C23.1 Some Clinically Important Respiratory States**

Apnea	Transient cessation of breathing
Asphyxia	O_2 starvation of tissues, caused by either lack of O_2 in air, respiratory impairment, or inability of tissues to utilize O_2
Dyspnea	Difficult or labored breathing
Eupnea	Normal breathing
Hypercapnia	Excess CO_2 in arterial blood
Hyperpnea	Increased pulmonary ventilation that matches increased metabolic demands, as in exercise
Hyperventilation	Increased pulmonary ventilation in excess of metabolic requirements
Hypocapnia	Below-normal CO_2 in arterial blood
Hypoventilation	Underventilation in relation to metabolic requirements
Respiratory arrest	Permanent cessation of breathing (unless clinically corrected)
Suffocation	O_2 deprivation as a result of inability to breathe oxygenated air

CONDITIONS OF CLINICAL SIGNIFICANCE

is produced when carbon products such as wood and coal are burned. Carbon monoxide is hazardous because it competes with oxygen for the same hemoglobin binding sites. Hemoglobin has a much greater affinity for carbon monoxide than it does for oxygen, and when even small amounts of carbon monoxide are present in the air breathed, it preferentially occupies the oxygen-binding sites of the hemoglobin molecules. Moreover, the hemoglobin-carbon-monoxide bond is so strong that very little carbon monoxide is removed from the blood. Consequently, carbon monoxide causes drowsiness, coma, and ultimately death due to hypoxia.

The combination of hydrogen ions and bicarbonate ions to form carbonic acid and, ultimately, carbon dioxide and water is enhanced by the binding of oxygen to hemoglobin in the lung capillaries. When hemoglobin combines with oxygen, its ability to bind hydrogen ions decreases, and more hydrogen ions leave the hemoglobin molecules than would otherwise be the case. These hydrogen ions can combine with bicarbonate ions to form carbonic acid and, ultimately, carbon dioxide and water. (The influence of oxygen on the ability of hemoglobin to bind hydrogen ions and to bind carbon dioxide to form carbamino compounds is called the *Haldane effect.*)

As the preceding discussion indicates, carbonic anhydrase can catalyze the reaction:

$$CO_2 + H_2O \xrightleftharpoons{\text{Carbonic Anhydrase}} H_2CO_3 \rightleftharpoons H^+ + HCO_3^-$$

in either direction. The overall direction of the reaction depends on the partial pressure of carbon dioxide as well as on the pH and the bicarbonate ion concentration. In the tissues, where the P_{CO_2} is relatively high (and hydrogen ions are tied up by hemoglobin molecules), the formation of bicarbonate ions is favored. In the lungs, where the P_{CO_2} is lower (and hemoglobin molecules release hydrogen ions), the reaction moves from bicarbonate ions and carbonic acid toward carbon dioxide and water.

Control of Respiration

The muscles responsible for inspiration—for example, the diaphragm and the intercostals—are skeletal muscles that require neural stimulation to initiate their contractions. When nerve impulses activate the inspiratory muscles, the thorax expands and the lungs fill with air.

Following inspiration, nerve impulses to the inspiratory muscles diminish greatly, and the diaphragm and other muscles involved in inspiration relax. When the inspiratory muscles relax, expiration occurs as a result of the elastic recoil of structures such as the lungs and chest wall, which returns the thorax to its resting position. During forced expiration, nerve impulses activate expiratory muscles—such as the muscles of the anterior abdominal wall—that help depress the ribs and diminish the size of the thoracic cavity.

Generation of Rhythmical Breathing Movements

The rhythmic pattern of inspiration and expiration characteristic of normal breathing depends on the cyclical stimulation of the respiratory muscles. This cyclical stimulation is due primarily to the activity of neurons located in the medulla oblongata of the brain (Figure 23.21).

Dorsal-Respiratory-Group Inspiratory Neurons

The medulla contains a group of neurons called the **dorsal respiratory group (DRG),** which consists mostly of neurons known as *inspiratory neurons.* The DRG inspiratory neurons exhibit cyclical activity. Their discharges are synchronized with inspiration, and they cease discharging during expiration.

The DRG inspiratory neurons connect either directly or indirectly with motor neurons that supply the inspiratory muscles, and they provide cyclical stimulation to the motor neurons. Since expiration is basically a passive process that occurs when the muscles controlling inspiration relax, the cyclical activity of the DRG inspiratory neurons can account for alternating cycles of inspiration and expiration.

The mechanisms responsible for the cyclical activity of the DRG inspiratory neurons are not well understood. Some researchers suggest that these neurons (or neurons that provide input to them) are pacemaker neurons that possess an inherent automaticity and rhythmicity.

◆ **FIGURE 23.21 Diagrammatic representation of the pontine and medullary neural pathways involved in the control of respiration**

The dashed line indicates that the dorsal-respiratory-group inspiratory neurons apparently activate the ventral-respiratory-group expiratory neurons when the respiratory drive for increased pulmonary ventilation becomes greater than normal.

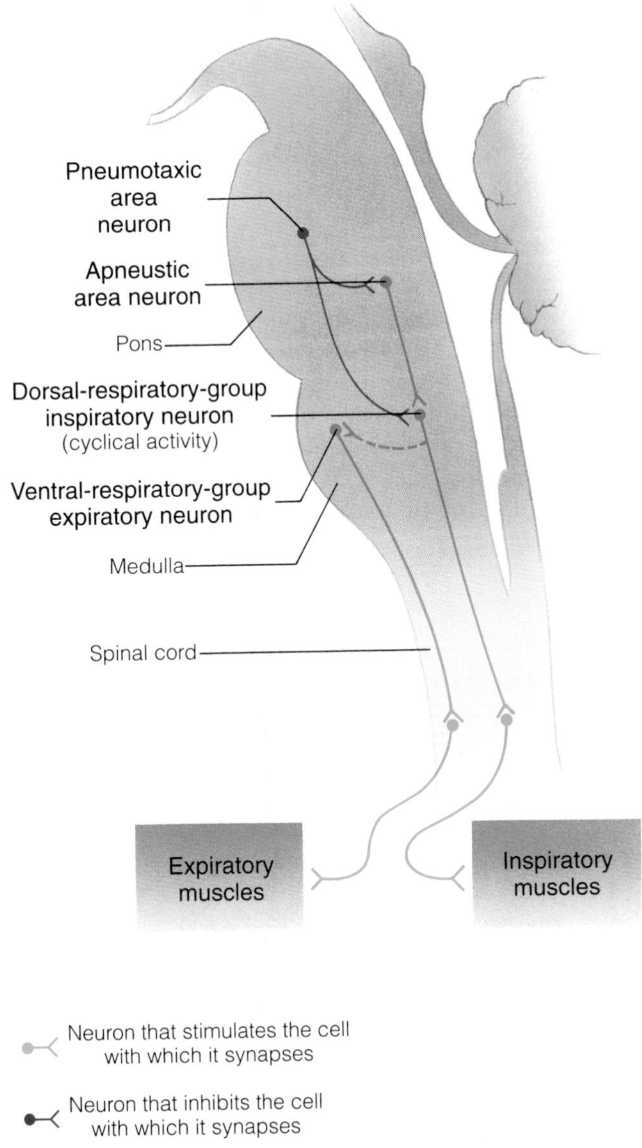

However, most researchers believe that, even if pacemaker neurons are involved, the generation of the cyclical activity of the DRG inspiratory neurons is a complex process that cannot be explained by the activity of pacemaker neurons alone.

Inputs from other neurons apparently modulate the activity of the DRG inspiratory neurons. Among these inputs are impulses from neurons of the pons and afferent impulses from stretch receptors in the lungs.

Neurons of the Pons

Within the pons are two areas called the *pneumotaxic area* and *apneustic area.* The pneumotaxic area continually sends inhibitory signals to the DRG inspiratory neurons that tend to limit the period of inspiration. When the signals from the pneumotaxic area are strong, the period of inspiration is short. When the signals are weak, the inspiratory period lasts longer. The apneustic area sends stimulatory signals to the DRG inspiratory neurons that tend to prolong inspiration. However, the pneumotaxic area normally overrides the apneustic area.

Lung Stretch Receptors

The lungs contain stretch receptors that initiate a reflex called the *Hering-Breuer reflex.* As the lungs expand during inspiration, the stretch receptors increase their activity, and afferent nerve impulses are transmitted to the brain more frequently. These impulses ultimately produce an inhibition of the DRG inspiratory neurons, which aids in terminating inspiration.

The inhibitory effect of the Hering-Breuer reflex is relatively weak unless the lungs are distended to a considerable extent. Consequently, this reflex does not appear to be of great importance in the control of respiration during normal breathing. Instead, the Hering-Breuer reflex is believed to be mainly a protective mechanism that prevents the lungs from overfilling.

Ventral-Respiratory-Group Expiratory Neurons

The medulla also contains a group of neurons called the **ventral respiratory group (VRG),** which includes neurons known as *expiratory neurons.* During most normal, quiet breathing, the VRG expiratory neurons are not active, and expiration is a passive process that occurs when the inspiratory muscles relax. However, when the respiratory drive for increased pulmonary ventilation becomes greater than normal, the expiratory neurons are activated—apparently by the DRG inspiratory neurons. The expiratory neurons send impulses to the expiratory muscles to facilitate expiration.

Control of Pulmonary Ventilation

Pulmonary ventilation can be varied by altering the respiratory rate and/or the tidal volume. Among the factors that influence pulmonary ventilation are (1) the partial pressure of oxygen in the arterial blood (the arterial P_{O_2}); (2) the partial pressure of carbon dioxide in the arterial blood (the arterial P_{CO_2}); and (3) the concentration of hydrogen ions in the arterial blood (the arterial pH). These factors influence pulmonary ventilation by way of receptors called central and peripheral chemoreceptors.

Central and Peripheral Chemoreceptors

The central and peripheral chemoreceptors provide input to the respiratory neurons in the brain. Neural signals originating at these receptors ultimately stimulate pulmonary ventilation.

The **central chemoreceptors** are a group of chemosensitive cells located within the medulla that are sensitive to hydrogen ions. When the pH of the cerebrospinal fluid and interstitial fluid of the brain falls (that is, when the hydrogen ion concentration of these fluids rises), the central chemoreceptors stimulate respiration.

The **peripheral chemoreceptors** are chemosensitive cells located at the bifurcations of the common carotid arteries, as well as at the arch of the aorta, in structures called the *carotid* and *aortic bodies.* The peripheral chemoreceptors are sensitive primarily to the arterial pH and P_{O_2}. When the arterial pH falls (that is, when the hydrogen ion concentration of the arterial blood rises), the peripheral chemoreceptors stimulate respiration. The peripheral chemoreceptors also stimulate respiration when the arterial P_{O_2} decreases significantly.

The peripheral chemoreceptors are somewhat sensitive to carbon dioxide, and carbon dioxide exerts a comparatively weak stimulatory effect on the receptors.

The arterial P_{O_2}, P_{CO_2}, and pH influence pulmonary ventilation by way of the central and peripheral chemoreceptors in the following manner.

Arterial P_{O_2}

If the arterial P_{O_2} is reduced below normal while the P_{CO_2} and pH are held constant, an increase in alveolar ventilation occurs (Figure 23.22). However, within the normal range of arterial partial pressures of oxygen (the normal arterial P_{O_2} is about 95 mm Hg), the effect of reduced oxygen on alveolar ventilation is relatively slight. It is not until low partial pressures of oxygen are present that a major stimulatory effect of reduced oxygen on alveolar ventilation is evident. This response is reasonable because hemoglobin remains almost completely saturated with oxygen until there is a considerable drop in the P_{O_2} (see Figure 23.16). Thus, the transport of oxygen to the tissues is not substantially diminished until the arterial P_{O_2} falls quite low.

Although the peripheral chemoreceptors mediate the respiratory effects of the arterial P_{O_2}, these receptors are not sensitive to the total oxygen content of the blood. Consequently, their activity is not altered under conditions in which the total oxygen content of the blood is moderately reduced but the arterial P_{O_2} does not change. For example, conditions of moderate anemia generally do not lead to alterations in ventilation even though the total oxygen content of the blood may be below normal. In anemia, the reduced oxygen content of the blood is due to a reduced number of functional hemoglobin molecules, and the arterial P_{O_2}, which is determined solely by the amount of dissolved oxygen, is not changed.

Arterial P_{CO_2}

The arterial P_{CO_2} is a major regulator of pulmonary ventilation. An elevation of only 5 mm Hg in the arterial P_{CO_2}, with the P_{O_2} and pH held constant, increases ventilation by 100% (Figure 23.22). Conversely, a lower-than-normal P_{CO_2} inhibits respiration.

To a small degree, carbon dioxide directly stimulates the peripheral chemoreceptors. However, hydrogen

◆ FIGURE 23.22 Effects of changing P_{O_2}, P_{CO_2}, and pH on alveolar ventilation in normal, healthy individuals when only one factor at a time changes and the others are held constant at normal levels

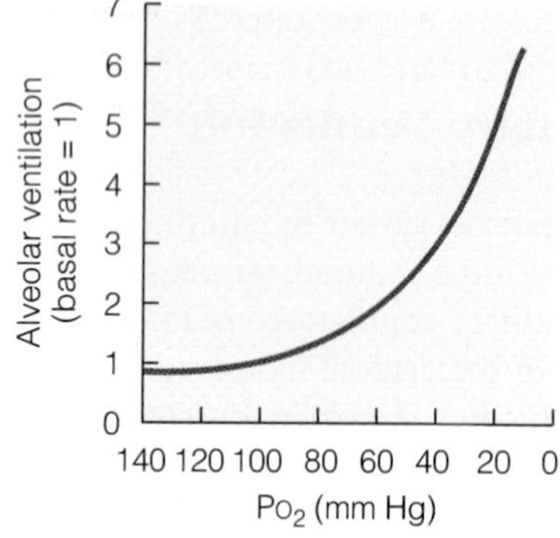

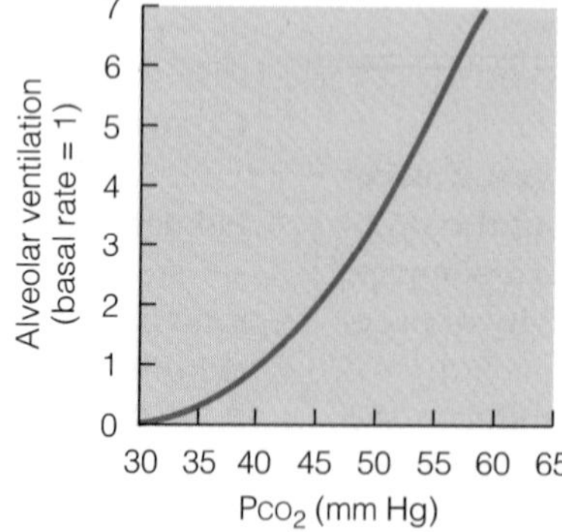

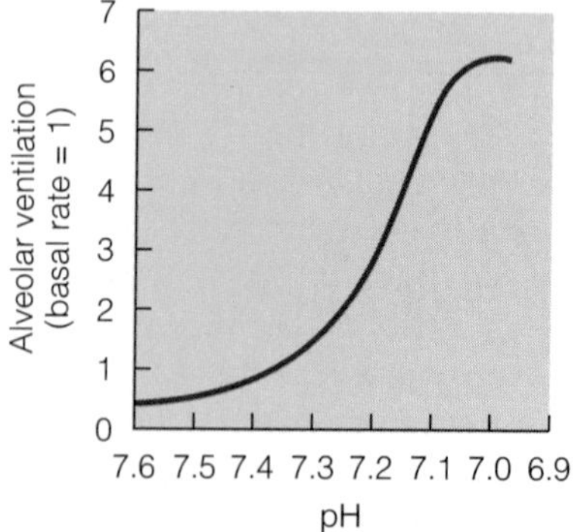

ions also stimulate the peripheral chemoreceptors, and carbon dioxide affects these receptors mainly by way of its influence on the hydrogen ion concentration of the blood. Recall that carbon dioxide can combine with water to form carbonic acid, which can dissociate into hydrogen ions and bicarbonate ions.

$$CO_2 + H_2O \rightleftharpoons H_2CO_3 \rightleftharpoons H^+ + HCO_3^-$$

Consequently, when the arterial P_{CO_2} rises, the hydrogen ion concentration of the arterial blood tends to rise, and when the arterial P_{CO_2} falls, the hydrogen ion concentration of the arterial blood tends to fall.

The arterial P_{CO_2} also influences the hydrogen ion concentration of the cerebrospinal fluid and interstitial fluid of the brain. Carbon dioxide can diffuse readily across the blood-brain barrier from the plasma to the cerebrospinal fluid and interstitial fluid of the brain. In the cerebrospinal fluid and interstitial fluid of the brain, carbon dioxide combines with water to form carbonic acid, which can dissociate into hydrogen ions and bicarbonate ions. Consequently, when the arterial P_{CO_2} rises, the concentration of hydrogen ions in the cerebrospinal fluid and interstitial fluid of the brain rises, and when the arterial P_{CO_2} falls, the concentration of hydrogen ions in the cerebrospinal fluid and interstitial fluid of the brain falls. The central chemoreceptors are sensitive to hydrogen ions, and when the hydrogen ion concentration of the cerebrospinal fluid and interstitial fluid of the brain rises, they stimulate respiration.

Overall, the arterial P_{CO_2} affects ventilation primarily by way of its influence on the hydrogen ion concentration of the cerebrospinal fluid and interstitial fluid of the brain, and the central chemoreceptors are considerably more important than the peripheral chemoreceptors in mediating the respiratory effects of the arterial P_{CO_2}.

Arterial pH

The arterial pH can be influenced by hydrogen ions derived from sources other than carbon dioxide—for example, by hydrogen ions derived from lactic acid produced by skeletal muscle metabolism.

If the arterial pH is decreased (that is, if the concentration of hydrogen ions in the arterial blood is increased) while the P_{O_2} and P_{CO_2} are held constant, alveolar ventilation increases substantially (Figure 23.22).

Hydrogen ions can stimulate both the central chemoreceptors and the peripheral chemoreceptors. However, the blood-brain barrier restricts the movement of hydrogen ions from the blood to the cerebrospinal fluid and interstitial fluid of the brain, and hydrogen ions penetrate the blood-brain barrier very slowly. Consequently, the peripheral chemoreceptors are the principal receptors mediating the respiratory effects of the arterial pH.

Interaction of Different Factors in the Control of Pulmonary Ventilation

The various factors that influence pulmonary ventilation interact with one another to regulate breathing. For example, when a low arterial P_{O_2} and a high arterial P_{CO_2} are present together, they act synergistically, and the combined influences of these factors on ventilation are considerably greater than the sum of their individual influences.

Effects of Exercise on the Respiratory System

The muscular activity associated with exercise rapidly consumes oxygen and produces carbon dioxide (Table 23.4). Thus, during exercise the body requires more oxygen and must eliminate more carbon dioxide than when at rest.

Increased Pulmonary Ventilation

During exercise both the rate and depth of breathing increase. During heavy exercise, the volume of air that reaches the alveoli may increase to as much as 20 times the resting volume.

The precise changes responsible for increased respiration during exercise are still not clearly understood, but several factors are thought to be involved. It is believed that when brain areas controlling movement transmit signals to exercising muscles, signals are also transmitted to the respiratory neurons to stimulate respiration. Consequently, increases in respiration occur even before muscular activity causes substantial changes in blood P_{O_2}, P_{CO_2}, or pH levels. Neural signals from joints and muscles involved in physical movements may also contribute to the stimulation of respiration during exercise.

If the neural mechanisms do not correctly adjust respiration to meet body requirements during exercise, the levels of chemical factors such as hydrogen ions may change, and these factors can then contribute to the overall control of respiration. The mechanisms that control respiration during exercise are quite effective, and in all but very strenuous exercise the arterial P_{O_2}, P_{CO_2}, and pH remain almost exactly normal.

Increased Delivery of Oxygen to the Tissues

Both at rest and during exercise, the blood that leaves the lungs through the pulmonary veins is approximately

CLINICAL CORRELATION

Pulmonary Thromboembolism

Case Report

THE PATIENT: A 45-year-old woman.

PRINCIPAL COMPLAINT: Sudden onset of severe dyspnea (difficult breathing).

HISTORY: The patient had been healthy until four days previously when she was admitted to the hospital for treatment of injuries suffered in an automobile accident. Her left femur and ulna were fractured, and she had extensive intramuscular hemorrhage and other less serious cuts and bruises. In addition to medication for pain, she was given low prophylactic doses of an oral anticoagulant agent. Her leg and arm were in casts, but otherwise she recovered uneventfully until she developed a slight fever during the third night. The next day she complained that she could not breathe.

CLINICAL EXAMINATION: The patient's blood pressure was 135/90 mm Hg (normal: 90–140/60–90 mm Hg). Her heart rate was 110 beats/min (normal: 60–100 beats/min). Some rales (abnormal rattling or scraping sounds) and localized wheezes were heard over the lower lung field. The arterial PO_2 was 53 mm Hg (normal: >75 mm Hg), the arterial PCO_2 was 35 mm Hg (normal: 35–45 mm Hg), and the arterial pH was 7.46 (normal: 7.35–7.45). Spirometric tests showed that her minute respiratory volume was 9100 ml/min, and her alveolar ventilation was 4100 ml/min, or 46% of the minute respiratory volume (predicted: 60–70%). Her physiological dead space was 5000 ml/min, or 54% of her minute respiratory volume (predicted 30–40%). Alveolar PO_2 was calculated to be 100 mm Hg (with the patient breathing room air), and the alveolar-arterial PO_2 difference (that is, the alveolar-arterial oxygen gradient) was 47 mm Hg. Pulmonary thromboembolism was suspected because it is often associated with fractures or other injuries of the lower extremities.

Because of the severity and potentially lethal nature of the suspected condition, further tests were ordered. The patient was transferred to the cardiac catheterization laboratory, where a catheter was placed into the pulmonary trunk. Cardiac ouptut was determined to be 3.8 liters/min (predicted 4.8 liters/min), and pulmonary arterial pressure was 60/45 mm Hg (normal: 12–28/3–13 mm Hg). The extent of vascular occlusion was determined by pulmonary angiography. A radiopaque material was injected into the pulmonary trunk, and the distribution of the material in the circulation of the lung was detected by X rays. This procedure showed decreased filling of two lobes of the right lung.

COMMENT: A major respiratory consequence of pulmonary emboli is an increase in the physiological dead space. The blood flow to alveoli in certain sections of the lungs may be completely or partially obstructed by clots, and ventilation of these alveoli is largely wasted. Because the patient frequently hyperventilates, arterial PCO_2 often is within normal limits, but alveolar PCO_2 is decreased (that is, closer to the PCO_2 of the inspired air). A decreased blood flow to a particular lung area also may decrease the availability of substrates for the synthesis of pulmonary surfactant, which causes alveolar instability and collapse.

The hemodynamic consequences of pulmonary emboli depend largely on the degree of vascular obstruction. Moderate obstruction may cause only an increased heart rate, and more serious problems may not arise if the patient remains quiet and the demand for oxygen does not increase. In this patient, the vascular obstruction was severe enough to produce arterial hypoxemia (a below-normal arterial oxygen tension), presumably due to a decreased functional area available for gas exchange at the lungs.

TREATMENT: Therapy was directed toward dissolving the blood clots and restoring the pulmonary circulation. Vigorous anticoagulant treatment was begun. Although anticoagulants do not cause the clots to dissolve, by preventing expansion of the clots, they enable the fibrinolytic system of the body to remove them. The patient initially received 10,000

CLINICAL CORRELATION

units of herapin intravenously and then continuous infusion of heparin at the rate of 1000 units every hour for seven days. She was switched to an oral anticoagulant several days before her discharge from the hospital, and this was continued for three months.

OUTCOME: The remainder of her recovery was uneventful.

97% saturated with oxygen. Therefore, the factor that seems to be most important in determining how much oxygen can be delivered to the tissues during exercise is not the pulmonary ventilation, but the amount of blood the heart is capable of pumping. If a person can increase his or her cardiac output through training, the pulmonary ventilation appears to be capable of increasing enough so that the extra volume of blood is maximally saturated with oxygen.

◆ **TABLE 23.4 Oxygen- and Carbon Dioxide–Related Variables During Exercise**

O_2- OR CO_2-RELATED VARIABLE	CHANGE	COMMENT
O_2 consumption	Marked increase	Active muscles are oxidizing nutrient molecules more rapidly to meet their increased energy needs.
CO_2 production	Marked increase	More actively metabolizing muscles produce more CO_2.
Alveolar ventilation	Marked increase	Alveolar ventilation generally keeps pace with or even slightly exceeds the increased metabolic demands during exercise.
Arterial P_{O_2}	Normal or slight increase	Despite marked increase in O_2 consumption and CO_2 production during exercise, alveolar ventilation generally keeps pace with or even slightly exceeds the stepped-up rate of O_2 consumption and CO_2 production.
Arterial P_{CO_2}	Normal or slight decrease	
O_2 delivery to muscles	Marked increase	Although arterial P_{O_2} remains normal, O_2 delivery to muscles is greatly increased by the increased blood flow to exercising muscles accomplished by increased cardiac output coupled with local vasodilation in active muscles.
O_2 extraction by muscles	Marked increase	Increased O_2 consumption lowers the P_{O_2} in active muscle tissue, which results in more O_2 dissociating from hemoglobin; this dissociation is enhanced by $\uparrow P_{CO_2}$, $\uparrow H^+$, and $\uparrow$ temperature.
CO_2 removal from muscles	Marked increase	The increased blood flow to exercising muscles removes the excess CO_2 produced by these actively metabolizing tissues.
Arterial H^+ concentration		
Mild to moderate exercise	Normal	Carbonic acid–generating CO_2 in arterial blood remains at essentially normal levels, and arterial H^+ concentration does not change.
Strenuous exercise	Modest increase	In strenuous exercise, when muscles resort to anaerobic metabolism, lactic acid is added to the blood.

CONDITIONS OF CLINICAL SIGNIFICANCE

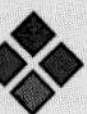

The Respiratory System

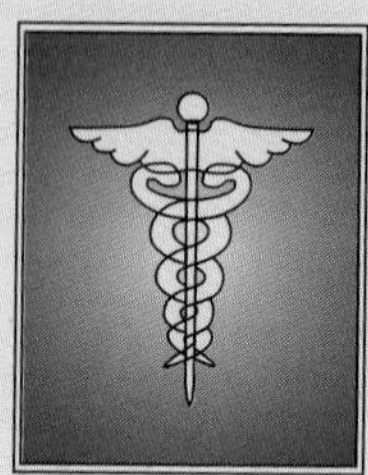

Common Cold

The *common cold,* or *acute coryza,* is an inflammation of the mucous membrane of the upper respiratory tract that is familiar to most people. The initial inflammation is caused by various viruses and is often followed by bacterial infections of the sinuses, ears, or bronchi. When inflamed, the mucous membrane becomes engorged and swollen, causing discomfort and difficulty in breathing. Later the mucous membrane discharges a watery fluid that makes its presence known in the form of a "runny nose." Such discharges from the mucous membranes that line the paranasal sinuses can irritate the larynx and trachea, producing a cough.

Bronchial Asthma

Bronchial asthma is an allergic (antigen-antibody) response to foreign substances that generally enter the body by way of either air breathed or food eaten. It is characterized by episodes of wheezing and difficult breathing. In response to the foreign substances, the mucous membranes of the respiratory system secrete excessive amounts of mucus, and the smooth muscle that surrounds the smaller bronchi and bronchioles goes into spasms. This narrows the passageways, making it difficult for air to move into or out of the alveoli.

Bronchitis

Bronchitis is an acute or chronic inflammation of the bronchial tree. It is caused by bacterial infection or by irritants in the inhaled air (such as smoke or chemicals). The mucous membranes of the respiratory system produce a sticky secretion that inhibits the normal protective function of the macrophages of the respiratory tract and hinders the self-cleaning actions of the cilia of the cells that line the bronchi. As the secretions accumulate within the bronchi, they are removed by coughing, which is annoying but serves the useful purpose of helping to keep the lungs clear.

Tuberculosis

Tuberculosis, an infection caused by the tubercle bacillus *(Mycobacterium tuberculosis),* can affect many parts of the body. However, because the bacterium most commonly enters the body by inhalation, tuberculosis of the lungs is the most prevalent form. Even if not inhaled, the bacilli can enter the lymphatic system or the bloodstream and thereby reach the lungs. When the bacilli reach the lungs, the lung tissue reacts by forming small clumps (tubercles) around the bacilli. Many of the bacilli are engulfed by phagocytes, and fibrous walls form around them. If the bacilli are not successfully walled off, the lung tissue is destroyed and the site of infection spreads. The process may continue until both lungs are extensively destroyed. Even if such a massive involvement of the lungs does not occur, the fibrosis in the affected portions interferes with the diffusion of gases and causes the lungs to lose their elasticity, thereby reducing the vital capacity.

Emphysema

Emphysema (em´´-fĭ-sē-mah) is a condition that develops slowly as a secondary response to other respiratory problems, such as chronic bronchitis and tuberculosis, or to environmental irritants, such as cigarette smoke and industrial pollutants. In persons suffering from emphysema, the alveoli become overdistended, and the walls of the alveoli break down and are often replaced by fibrous tissue. This greatly reduces the surface area across which gaseous exchange can occur. As a consequence, the partial pressure of oxygen in the blood is lowered, and even mild exercise increases the breathing rate. In addition, the elastic tissues of the overexpanded lungs are reduced, making expiration difficult. A reduced expiratory volume is an early symptom of emphysema.

Therefore, the victim of emphysema faces two basic problems: (1) the lungs are "fixed" in inspiration; and (2) the respiratory surfaces of the lungs have deteriorated so much that they are no longer

CONDITIONS OF CLINICAL SIGNIFICANCE

adequate to accomplish normal gas exchange. Unfortunately, the disease is progressive and irreversible.

Pneumonia

The inflammation of *pneumonia* causes a fibrinous exudate to be produced within the alveoli. The lung, or a part of it, becomes solid and airless, which makes it very difficult for gaseous exchange to occur within the alveoli. Most cases of pneumonia are probably caused by one of several viruses. Another common cause is the pneumococcus bacterium. However, pneumonia can also be caused by the inhalation of foods or other foreign bodies that obstruct a bronchus. This obstruction can lead to collapse of the lung, accumulation of fluids within the lung, and subsequent infection.

Pleurisy

Pleurisy (ploor´-i-se) is an inflammation of the pleural membranes that surround each lung. In the early stages, the inflamed membranes are "dry" and are covered with fibrous material. This condition causes pain during breathing. Adhesions between the layers of the pleura may develop as a result of pleurisy, and in severe cases, surgery may be necessary to remove them. In the later stages of pleurisy, there is often excessive secretion of pleural fluid into the pleural cavity. The pleura are most commonly infected with the pneumococci, the streptococci, or the tubercle bacilli.

Effects of Aging

Tissue changes that occur in the respiratory system with increased age cause the thoracic wall to become more rigid and the lungs to become less elastic. Thus, although the total lung volume does not change with age, the rigidity of the thoracic wall and the loss of elasticity in the lungs result in a diminished ventilating capacity. Because of these changes, the vital capacity decreases in males from about 4700 ml at age 20 to about 4000 ml at age 70. Accompanying these changes is a decrease in the arterial Po_2, which is quite pronounced when a person is lying in a supine position, where breathing is more difficult. For this reason elderly people tend to become hypoxic during sleep, and they are often more comfortable if supported by several pillows. Carbon dioxide diffuses through tissues much more rapidly than oxygen; therefore, in contrast to the Po_2, the Pco_2 of the arterial blood is not affected much by age.

With increasing age, there is a decrease in the phagocytic activity of macrophages and in the activity of the cilia of the epithelial linings of the respiratory tract. This decreased activity makes the cleaning of the respiratory tract lining less effective. The rigidity of the thoracic wall, the loss of elasticity in the lungs, and the diminished phagocytic activity and ciliary action all make elderly people more susceptible to pneumonia and other respiratory infections.

Study Outline

◆ EMBRYONIC DEVELOPMENT OF THE RESPIRATORY SYSTEM pp. 718–719

Laryngotracheal Bud. Diverticulum from ventral endoderm of digestive tract.

Trachea. Develops from proximal part of laryngotracheal bud.

Bronchi. Develop from bifurcation of distal end of laryngotracheal bud.

Bronchioles and Alveoli. Develop as closed tubes from bronchi.

Respiratory Epithelium. Derived from endoderm; lines entire respiratory tract.

◆ ANATOMY OF THE RESPIRATORY SYSTEM pp. 719–726

Respiratory system consists of nose, pharynx, larynx, trachea, bronchi, and lungs

Nose and Nasal Cavity. Air passes through external nares to vestibule of nose and to nasal cavity.

SEPTUM. Formed from septal cartilage, vomer bone, and perpendicular plate of ethmoid bone.

ROOF. Formed by cribriform plate of ethmoid.

LATERAL WALLS. Formed by superior, middle, and inferior conchae.

FLOOR. Formed by hard and soft palates.

VESTIBULE. Stratified squamous epithelial lining.

NASAL CAVITY, SINUSES. Mucous membrane of pseudostratified ciliated columnar epithelium. Blood supply warms air and saturates it with water; cilia trap small particles; internal nares connect nasal cavity with nasopharynx.

Pharynx. Muscular structure lined with mucous membrane.

NASOPHARYNX. Receives internal nares and auditory tubes; contains tubal and pharyngeal tonsils.

OROPHARYNX. Food and air passageway; contains palatine and lingual tonsils; communicates with oral cavity through fauces.

LARYNGOPHARYNX. Food and air passageway; between oropharynx and esophagus and larynx.

Larynx. Conducts air between laryngopharynx and lungs.

1. Framework of larynx consists of nine cartilages.
2. Mucous membranes of larynx are stratified squamous and pseudostratified ciliated columnar epithelium; form pairs of ventricular folds and vocal folds.
3. Glottis is opening between vocal folds.

Trachea (Windpipe).

1. Kept open by C-shaped hyaline cartilages.
2. Ciliated cells carry foreign particles away from lungs to pharynx.

Bronchi, Bronchioles, and Alveoli.

1. Primary, secondary, and tertiary bronchi→terminal and respiratory bronchioles→alveolar ducts→alveoli.
2. With branching, supportive cartilage gradually replaced by smooth muscle.

Lungs.

1. Three-lobed right lung; two-lobed left lung; each lobe further divided into bronchopulmonary segments.
2. Hilus—site where bronchi, blood vessels, lymphatics, and nerves pass into or out of lungs.
3. Mediastinum—space that separates lungs.

THE PLEURA.

1. Visceral and parietal pleura—double layered serous membrane sac that encloses each lung.
2. Pleural cavity between layers—filled with pleural fluid.

BLOOD SUPPLY OF THE LUNGS. Pulmonary artery (oxygen-poor blood) supplies alveoli; bronchial arteries from descending aorta (oxygen-rich blood) supply bronchi.

THE RESPIRATORY MEMBRANE. Thin membrane that separates alveolar air from blood—oxygen and carbon dioxide diffuse across membrane; consists of alveolar epithelium (and basal lamina) and capillary endothelium (and basal lamina); contains two types of cells: (1) *alveolar type I cells*—squamous cells that allow for gas exchange; (2) *alveolar type II cells*—rounded or cuboidal cells that secrete pulmonary surfactant.

◆ MECHANICS OF BREATHING pp. 726–734

Atmospheric Pressure. At sea level is 760 mm Hg.

Pressure Relationships in the Thoracic Cavity.

1. Elastic connective tissue of lungs and surface tension of fluid coating alveoli favor reduction in lung size, as does intrapleural pressure. These forces are balanced by intrapulmonary pressure, and lungs normally remain expanded and do not collapse.
2. If air enters pleural cavity, balance of forces maintaining lung expansion is upset, and lung collapses.

Ventilation of the Lungs. Air flows from high-pressure area to lower-pressure area; pressure within lungs is altered by altering volume of thoracic cavity.

INSPIRATION. Volume of thoracic cavity increased by contraction of diaphragm and elevation of ribs. This increases lung volume, and intrapulmonary pressure drops below atmospheric pressure. Air then flows into lungs.

EXPIRATION. When muscles involved in inspiration relax, volume of thoracic cavity—and lungs—decreases due to elastic recoil of respiratory structures such as lungs and thoracic wall. Decrease in lung volume raises intrapulmonary pressure above atmospheric pressure, and air flows out of lungs. During forced expiration, contraction of expiratory muscles further reduces volume of thoracic cavity.

Factors That Influence Pulmonary Airflow. Volume of airflow per given time between atmosphere and alveoli is influenced by pressure that moves air through respiratory passageways and by resistance air encounters as it flows through passageways. Passageway diameter affects resistance; parasympathetic stimulation, histamine, and leukotrienes cause smooth muscle of respiratory passageways to contract and thereby increase resistance; epinephrine causes smooth muscle of passageways to relax and thereby decreases resistance.

Surface Tension and Pulmonary Surfactant. Pulmonary surfactant produced by alveolar cells lowers surface tension of fluid that coats exposed surfaces of alveoli, allowing lungs to be expanded with reasonable muscular effort.

Lung Volumes and Capacities.

RESIDUAL VOLUME. About 1000–1200 ml of air that cannot be exhaled from lungs.

TIDAL VOLUME. Volume of air moving in and out of lungs with each breath (about 500 ml during normal, quiet breathing).

INSPIRATORY RESERVE. Extra volume of air that can be inspired in addition to normal, quiet tidal volume (about 2100–3000 ml).

INSPIRATORY CAPACITY. Sum of normal, quiet tidal volume and inspiratory reserve volume.

EXPIRATORY RESERVE. Extra volume of air that can be expired following normal, passive expiration of quiet tidal volume (about 800–1200 ml).

FUNCTIONAL RESIDUAL CAPACITY. Sum of residual volume and expiratory reserve volume.

VITAL CAPACITY. Sum of inspiratory reserve volume; normal, quiet tidal volume; and expiratory reserve volume (3400–4700 ml).

Dead Space. In some areas of respiratory system, no gas exchange takes place; volume of anatomical dead space is about 150 ml.

Minute Respiratory Volume. Volume of air moved into respiratory passageways in one minute; equal to respiratory rate multiplied by volume of air entering respiratory passageways with each breath (tidal volume).

Alveolar Ventilation. Volume of atmospheric air entering alveoli either per breath or in one minute that can participate in gas exchange between alveoli and blood.

Matching of Alveolar Airflow and Blood Flow. Local autoregulatory mechanisms are involved.

1. Low oxygen levels cause pulmonary vessels to constrict; high oxygen levels cause them to dilate.
2. Low carbon dioxide levels cause respiratory passsageways to constrict; high carbon dioxide levels cause them to dilate.

◆ TRANSPORT OF GASES BY THE BLOOD pp. 734–742

Partial Pressure. Oxygen, 21% of atmospheric gas mixture, contributes 21% of total atmospheric pressure.

Gases in Liquids. Gases enter a liquid and dissolve in proportion to their individual gas pressures.

Oxygen and Carbon Dioxide Exchange between Lungs, Blood, and Tissues.

LUNG. Partial pressure differences between O_2 and CO_2 in blood and in alveoli lead to net diffusion of O_2 into blood from alveoli and net diffusion of CO_2 into alveoli from blood.

TISSUES. Net diffusion of O_2 from blood to tissues; net diffusion of CO_2 from tissues to blood.

Oxygen Transport in the Blood. Most is transported bound to iron of hemoglobin in erythrocytes. Binding of oxygen to hemoglobin is affected by oxygen partial pressure, hydrogen ion concentration, carbon dioxide partial pressure, temperature.

Carbon Dioxide Transport in the Blood.

1. Dissolved in plasma.
2. Carbamino compounds (CO_2-protein complexes).
3. Bicarbonate ions.

◆ CONDITIONS OF CLINICAL SIGNIFICANCE: PROBLEMS OF GAS TRANSPORT pp. 741–742

Cyanosis. Bluish discoloration of skin and mucous membranes; occurs when there is significant concentration of reduced hemoglobin in arterial blood.

Hypoxia. Any state in which a physiologically inadequate amount of oxygen is available to or utilized by the tissues.

Carbon Monoxide Poisoning. Carbon monoxide occupies oxygen-binding sites of hemoglobin.

◆ CONTROL OF RESPIRATION pp. 742–745

Inspiration occurs when nerve impulses activate diaphragm and intercostals; thorax expands; lungs fill with air. Expiration occurs when nerve impulses to inspiratory muscles diminish and respiratory structures such as lungs and chest wall recoil. During forced expiration, nerve impulses activate expiratory muscles.

Generation of Rhythmical Breathing Movements.

DORSAL-RESPIRATORY-GROUP INSPIRATORY NEURONS. Exhibit cyclical activity.

NEURONS OF THE PONS.
Pneumotaxic area. Tends to limit period of inspiration.
Apneustic area. Tends to prolong inspiration; normally overridden by pneumotaxic area.

LUNG STRETCH RECEPTORS. May aid in terminating inspiration when lungs are considerably distended (Hering-Breuer reflex).

VENTRAL-RESPIRATORY GROUP EXPIRATORY NEURONS. When respiratory drive for increased pulmonary ventilation becomes greater than normal, these neurons send impulses to expiratory muscles to facilitate expiration.

Control of Pulmonary Ventilation.

CENTRAL AND PERIPHERAL CHEMORECEPTORS. Provide input to respiratory neurons in brain. Neural signals originating at these receptors ultimately stimulate pulmonary ventilation.

ARTERIAL P_{O_2}. Low P_{O_2} in arterial blood (with P_{CO_2} and pH constant) increases alveolar ventilation.

ARTERIAL P_{CO_2}. Elevated arterial P_{CO_2} (with P_{O_2} and pH constant) increases ventilation.

ARTERIAL pH. Increased arterial hydrogen ion concentration (with P_{O_2} and P_{CO_2} constant) increases ventilation.

INTERACTION OF DIFFERENT FACTORS IN THE CONTROL OF PULMONARY VENTILATION. The various factors that influence pulmonary ventilation interact with one another to regulate breathing.

◆ EFFECTS OF EXERCISE ON THE RESPIRATORY SYSTEM pp. 745–747

Increased Pulmonary Ventilation. During exercise, both rate and depth of breathing increase.

Increased Delivery of Oxygen to the Tissues. Both at rest and during exercise, blood leaving lungs is approximately 97% saturated with oxygen; most important factor in determining how much oxygen is delivered to tissues during exercise seems to be not pulmonary ventilation but amount of blood heart is capable of pumping.

◆ CONDITIONS OF CLINICAL SIGNIFICANCE: THE RESPIRATORY SYSTEM pp. 748–749

Common Cold. Viral inflammation and sometimes subsequent bacterial infection of mucous membranes of upper respiratory tract.

Bronchial Asthma. Allergic response characterized by wheezing caused by excessive mucus and spasms of smooth muscle of bronchioles.

Bronchitis. Acute or chronic inflammation of bronchial tree caused by bacterial infection or irritants.

Tuberculosis. Infection caused by *Mycobacterium tuberculosis,* which most commonly affects lungs.

Emphysema. Progressive condition secondary to other respiratory problems; overdistended alveoli and reduced elastic tissue of lungs reduce expiratory volume.

Pneumonia. Fibrinous exudate produced within alveoli, which causes part of lung to become solid and airless.

Pleurisy. Inflammation of pleural membranes.

Effects of Aging.

1. Thoracic wall rigidity.
2. Loss of lung elasticity.
3. Decreased ciliary activity of respiratory tract cells.
4. Decreased phagocytic activity, which increases susceptibility to infection.

Self-Quiz

1. In the embryo the respiratory epithelium develops from: (a) endoderm; (b) ectoderm; (c) mesoderm.
2. The nasal septum is formed from: (a) the vomer bone; (b) the ethmoid bone; (c) the septal cartilage; (d) all of these contribute to the nasal septum.
3. The floor of the nasal cavity is formed by: (a) the bony hard palate; (b) the membranous soft palate; (c) the inferior meatus; (d) both (a) and (b).
4. The paranasal sinuses, like the nasal cavity, are lined with stratified squamous epithelium. True or False?
5. The pharynx communicates with the nasal cavity via the: (a) fauces; (b) glottis; (c) middle meatus; (d) internal nares.
6. Which of the following is located in the oropharynx? (a) pharyngeal tonsils; (b) palatine tonsils; (c) lingual tonsils; (d) both (b) and (c); (e) all of these.
7. The largest of the unpaired cartilages of the larynx is the: (a) cricoid cartilage; (b) arytenoid cartilage; (c) thyroid cartilage; (d) cuneiform cartilage.
8. The vocal folds are attached to: (a) the thyroid cartilage; (b) the arytenoid cartilage; (c) the cricoid cartilage; (d) both (a) and (b).
9. The trachea divides directly into: (a) tertiary bronchi; (b) primary bronchi; (c) respiratory bronchi; (d) terminal bronchi; (e) alveolar ducts.
10. Alveoli are associated with the: (a) trachea; (b) pharynx; (c) lungs; (d) nose.
11. The region where the bronchi and blood vessels enter and leave the lung is called the: (a) costal surface; (b) oblique fissure; (c) hilus; (d) cardiac notch.
12. Match the terms associated with the lung with the appropriate lettered descriptions:

Costal surface	(a) The space that separates the two lungs
Cardiac notch	(b) A double-layered sac that encloses each lung
Oblique fissure	(c) The portion of the pleura that adheres firmly to the lungs
Mediastinum	(d) Lies against the ribs
Pleura	(e) Clusters of small, thin-walled air sacs
Visceral pleura	(f) The portion of the pleura that adheres firmly to the thoracic wall and diaphragm
Parietal pleura	(g) Divides each lung into a superior and inferior lobe
Alveoli	(h) A concavity in the left lung

13. When air is moving into the lungs during inspiration: (a) the intrapulmonary pressure is higher than the atmospheric pressure; (b) the atmospheric pressure is higher than the intrapulmonary pressure; (c) the intrapleural pressure is higher than the atmospheric pressure.
14. During expiration: (a) the volume of the thoracic cavity increases; (b) the diaphragm contracts; (c) the intrapulmonary pressure is higher than the atmospheric pressure.
15. The volume of gas that remains in the lungs even after the most forceful expiration is the: (a) residual volume; (b) tidal volume; (c) anatomical dead space volume.
16. Which has the greatest volume: (a) expiratory reserve volume; (b) vital capacity; (c) inspiratory reserve volume.
17. Smooth muscles of respiratory system passageways are relaxed and the resistance to the flow of air through the passageways is decreased by: (a) parasympathetic nervous stimulation; (b) histamine; (c) epinephrine.
18. The minute respiratory volume is equal to the volume of air that enters the respiratory passages with each breath, multiplied by the: (a) respiratory rate; (b) alveolar ventilation; (c) anatomical dead space volume.
19. In general, the binding of oxygen to hemoglobin will be increased by: (a) an increase in the oxygen partial pressure; (b) a decrease in pH; (c) an increase in temperature.
20. Which of the following is a true statement about the transport of respiratory gases? (a) Most of the CO_2 is carried as bicarbonate ions within the blood. (b) CO_2 is never carried as bicarbonate ions. (c) On the average, 27% of CO_2 is carried as bicarbonate ions within the blood.
21. The activity of the apneustic area tends to shorten the period of inspiration. True or False?
22. In moderate anemia: (a) the carotid and aortic bodies generally provide a greatly increased stimulus to the respiratory center; (b) the amount of dissolved oxygen in the blood is abnormally low; (c) the reduced oxygen content of the blood is due to a reduced number of functional hemoglobin molecules.
23. An elevation of 5 mm Hg of arterial P_{CO_2}, with P_{O_2} and pH held constant, will: (a) inhibit ventilation; (b) increase ventilation; (c) have no effect on ventilation.
24. The respiratory rate increases in response to: (a) increased arterial P_{O_2}; (b) lowered arterial pH; (c) decreased arterial P_{CO_2}.

The Digestive System

CHAPTER CONTENTS

LEARNING OBJECTIVES

After completing this chapter, you should be able to:

1. List the principal components of the digestive system, and cite the chief functions of each.
2. Describe the microscopic and gross anatomy of the components of the digestive system.
3. List the basic layers of the wall of the gastrointestinal tract.
4. Describe the modifications of the basic layers found in various regions of the gastrointestinal tract.
5. Describe the structure of the pancreas, liver, and gallbladder, and cite the functions of each.
6. Describe the factors that influence gastric motility.
7. Explain the movements that occur in the large intestine.
8. Explain the functions and control of the salivary secretions.
9. Discuss the phases of gastric secretion.
10. List the principal digestive fluids that are present in the small intestine, and describe the factors that stimulate the secretion of each.
11. Describe the processes involved in the digestion and absorption of carbohydrates.
12. Describe the processes involved in the digestion and absorption of proteins.
13. Describe the processes involved in the digestion and absorption of lipids.

Every cell in the body requires a constant source of energy in order to perform its particular functions, whether these functions be contraction, secretion, synthesis, or any other. Ingested food provides the basic materials from which this energy is obtained and new molecules are synthesized. Most food, however, cannot enter the bloodstream and be used by the cells of the body until it is broken down into simpler molecules. The **digestive system** alters the ingested food by mechanical and chemical processes so that it can ultimately cross the wall of the gastrointestinal tract and enter the cardiovascular and lymphatic systems for distribution to cells throughout the body. After entering the cells, digested food molecules may be used to provide energy to support body activity, or they may be incorporated into body structure.

The digestive system consists of a tube—called the **gastrointestinal tract,** or **alimentary canal**—extending from the mouth to the anus (Figure 24.1). As long as food remains in the gastrointestinal tract, it is technically still outside the body. To "enter" the body, it must cross the epithelium that lines the wall of the digestive tract. Emptying into the digestive tube are the secretions of the salivary glands, gastric glands, intestinal glands, liver, and pancreas, all of which assist in the digestion of food. Although the gastrointestinal tract is a continuous tube, it is divided into specialized regions that each perform specific functions in the digestion of food. These regions include the mouth, pharynx, esophagus, stomach, small intestine, and large intestine.

The activities of the digestive system can be divided into six basic processes:

1. Ingestion of food into the mouth
2. Movement of food along the digestive tract

◆ **FIGURE 24.1 Structures of the digestive system**

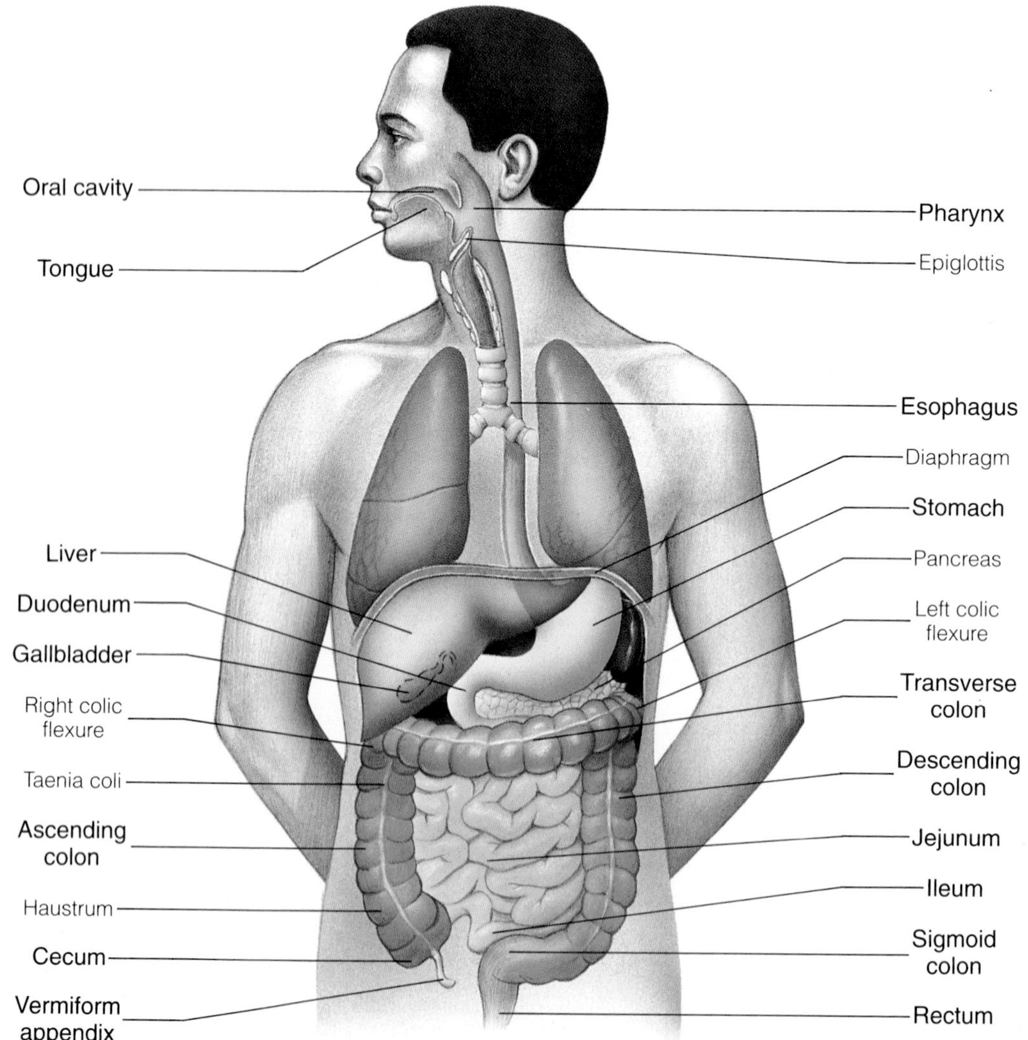

◆ **FIGURE 24.2 Sagittal section of an embryo (approximately 22 days old) showing the future foregut and hindgut forming**

Note that the midgut remains open to the yolk sac.

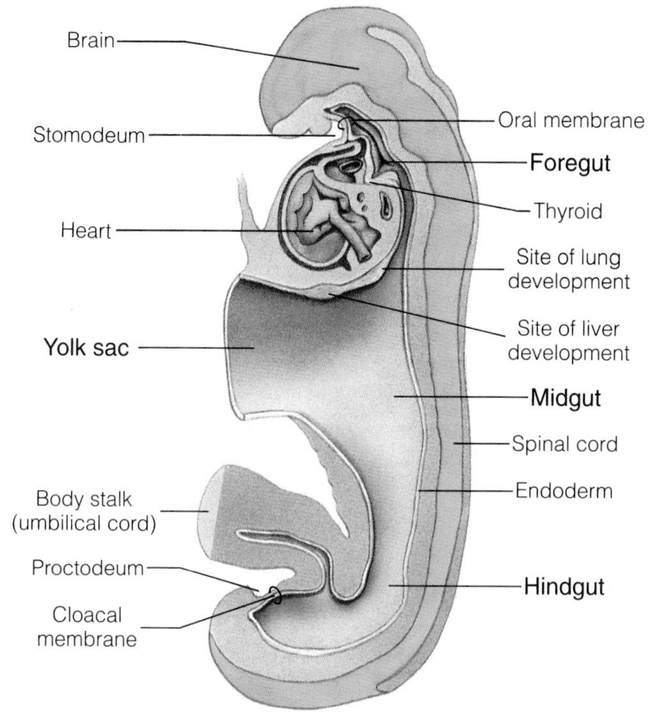

3. Mechanical preparation of food for digestion
4. Chemical digestion of food
5. Absorption of digested food into the blood and lymph for transport throughout the body
6. Elimination of indigestible substances and waste products from the body by defecation

Embryonic Development of the Digestive System

Early in development, the embryo is a hollow cylinder covered on the outside with ectoderm. Its internal cavity, which is lined with endoderm, is the developing digestive tract (Figure 24.2). The portion of the tract that extends anteriorly into the head region is the **foregut;** the portion that extends posteriorly is the **hindgut;** and the central region of the tract is the **midgut.** Until the fifth week of development, the midgut opens into a pouch called the yolk sac. After the fifth week, the attached portion of the yolk sac constricts, sealing the midgut.

Early in development, the anterior region of the foregut expands to form the pouches of the pharynx. With continued growth, the foregut contacts the surface ectoderm at the point where the ectoderm has formed a depression called the **stomodeum** (*sto-mo-dee´-um;* the future mouth). The **oral membrane,** which separates the foregut from the stomodeum, breaks through during the fourth week of development, and the foregut is then continuous with the outside of the embryo through the mouth. In a similar manner, the hindgut contacts the surface ectoderm at a depression called the **proctodeum** (*prok´´-tō-de´-um;* the future anus). With the rupture of the **cloacal membrane** (*klō-ā´-kal),* which separates the hindgut from the proctodeum, the digestive tract forms a continuous tube from the mouth to the anus.

With further development, hollow buds from the endoderm form at various places along the foregut. These buds will grow into the mesoderm that surrounds the digestive tract and give rise to the thyroid gland, parathyroid glands, salivary glands, liver, gallbladder, and pancreas. The thyroid and parathyroid glands later lose their connections with the digestive tract and function as endocrine glands. The salivary glands, liver, gallbladder, and pancreas, however, retain their connections with the digestive tube by means of ducts. Consequently, they serve as accessory glands to the digestive system.

Anatomy of the Digestive System

It is important to keep in mind that the entire digestive tract is lined with **mucous membrane.** This membrane protects the underlying tissues and, at the same time, allows for the absorption of digested food in the intestine.

In order to function in absorption, the membrane must be thin and moist. The secretion of *mucus* by cells of the mucous membrane not only keeps the membrane moist, but because it is viscous, the mucus also serves as a protective mechanism. Thus, the thin membrane that lines the absorptive regions of the digestive tract provides adequate protection as long as it is coated with mucus.

Mouth

The **mouth,** which is also referred to as the **oral cavity** or **buccal cavity** *(buk´-al)* is the first part of the digestive tract. It is bounded laterally by the cheeks, above by the hard and soft palates, and below by the tongue. The mouth extends from the lips to the oropharynx. The small space separating the lips and the cheeks from

◆ **FIGURE 24.3 Sagittal section of the oral and pharyngeal cavities**

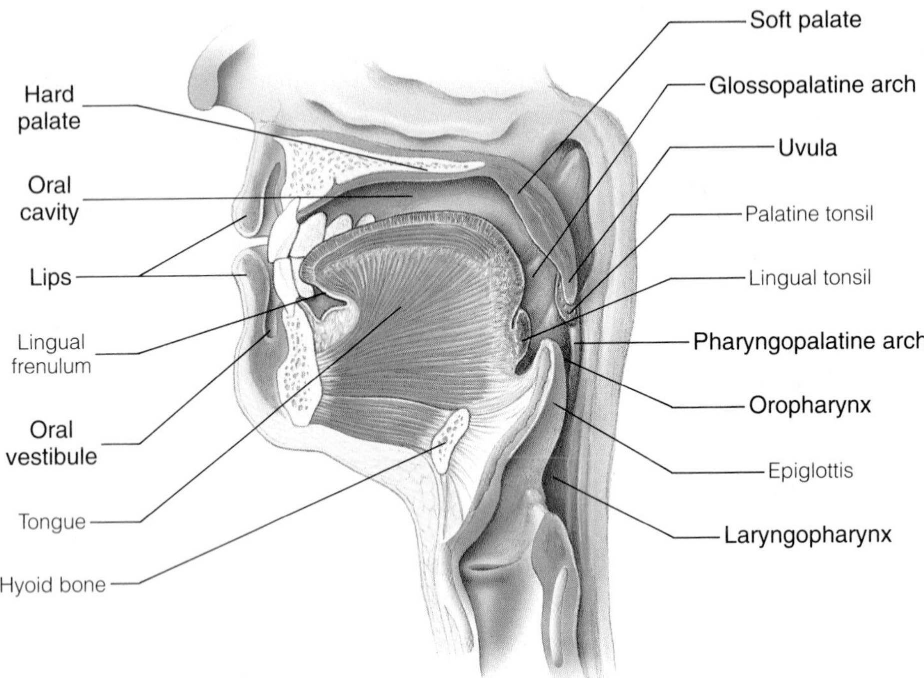

the teeth and gums is called the *oral vestibule* (Figure 24.3). The lips and cheeks aid in moving food between the upper and lower teeth during *mastication* (chewing) and also aid in speech.

The outer surfaces of the lips are covered with skin that has a relatively transparent surface layer of cells, allowing the underlying blood capillaries to show through. For this reason the lips appear red. Since the surface layer of the lips is not keratinized (horny), evaporation occurs from the lips. Consequently, the lips must be moistened frequently to prevent them from becoming dry and cracked. The inside surfaces of the lips and the rest of the mouth are lined with a mucous membrane that has a stratified squamous surface layer of nonkeratinized cells. No food is absorbed in the mouth, because the cells lining it are not capable of absorption; and the stratified surface layer affords an extra degree of protection from abrasive food particles.

The roof of the mouth is formed anteriorly by the **hard palate** and posteriorly by the **soft palate.** The hard palate is formed by the maxillary and palatine bones. The soft palate, which extends posteriorly from the hard palate, separates the oral cavity from the nasopharynx. It is composed primarily of muscles. The soft palate is pushed upward during swallowing, blocking the entrance from the pharynx to the nasal cavity and thus serving to prevent food and drink from entering the nasal cavity. The **uvula** *(ū´-vū-lah)* is a small muscular flap that hangs down from the posterior margin of the soft palate. It serves as a frictional pad to prevent the soft palate from being pushed into the nasal cavity during swallowing. The soft palate is attached laterally to the tongue by the **glossopalatine arches** *(glos´´-ō-pal´-ah-tine);* it is connected to the wall of the oropharynx by the **pharyngopalatine arches** *(fah-ring´´-go-pal´-ah-tine).* The palatine tonsils, which are composed mainly of lymphoid tissue, are located in the fossae between the two arches, a region called the **fauces** *(fau´-sēz).*

Tongue

The **tongue** forms the floor of the mouth (Figure 24.3). It is composed of interwoven bundles of skeletal muscles covered with mucous membrane. The **extrinsic muscles** of the tongue originate from the hyoid bone, the mandible, and the styloid processes of the temporal bones. These muscles protrude the tongue, retract it, and move it sideways. The **intrinsic muscles** originate and insert in the tongue. Their fibers run in various directions and modify the shape of the tongue in several different ways. Because of these two sets of muscles, the tongue is quite versatile and is used in manipulating food, swallowing, and speaking.

The mucous membrane covering the dorsum of the tongue is modified by the presence of numerous small projections called **papillae** (Figure 24.4). The papillae

◆ **FIGURE 24.4 Dorsal surface of the tongue**

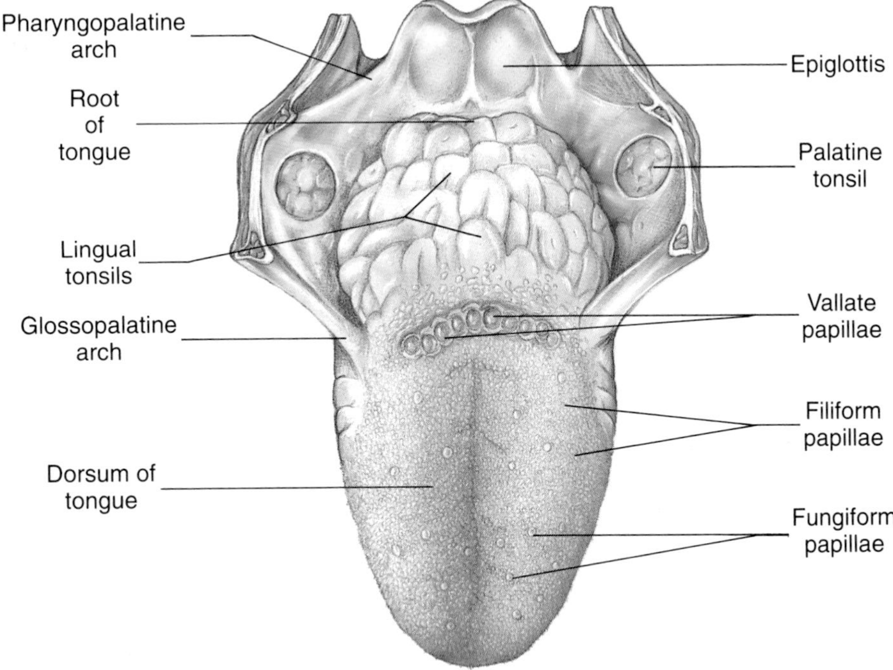

vary in shape; each type tends to be more common in certain regions of the tongue surface. The **filiform papillae,** which are small cones, are distributed in V-shaped rows over the entire dorsal surface of the tongue. Interspersed among the filiform papillae are flattened, mushroom-shaped **fungiform papillae** (Figure 24.5). About a dozen large **vallate papillae** form an inverted V toward the back of the tongue. **Taste buds** are found on the fungiform and vallate papillae. The filiform papillae make the surface of the tongue rough, allowing it to move food around during chewing. The posterior dorsal surface of the tongue contains an aggregate of small lymph nodules called the **lingual tonsil.**

The tongue is connected ventrally to the floor of the mouth by a fold of mucous membrane called the **lingual frenulum.** If the lingual frenulum is so short as to hinder the movement of the tongue, it interferes with speech. Such a condition, which can be corrected surgically, is called "tongue-tied."

Teeth

The teeth extend into the mouth from **alveoli** (sockets) located along the alveolar processes of the mandible and maxillary bones. **Gums** *(gingivae),* composed of stratified squamous epithelium and dense fibrous connective tissue, cover the alveolar processes. Each socket is lined with a fibrous membrane called the **periodontal ligament** that is attached to the root of the tooth, holding it firmly in the socket (Figure 24.6). Destruction of the

◆ **FIGURE 24.5 Filiform and fungiform papillae of the tongue (×938)**

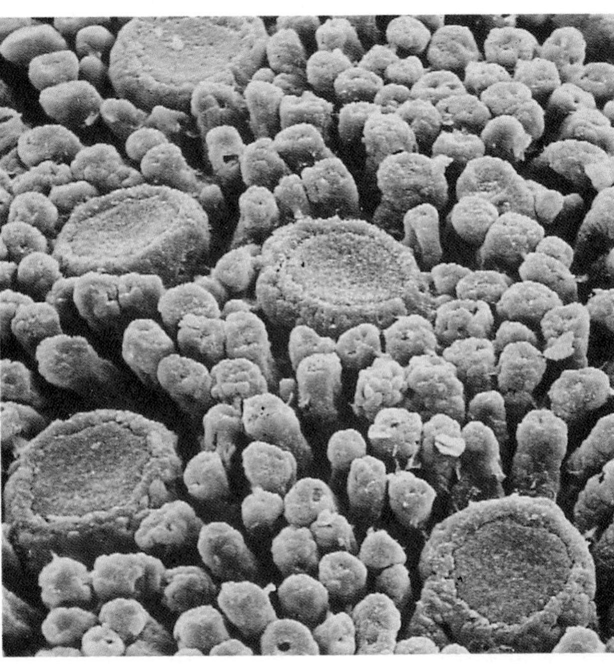

◆ **FIGURE 24.6 Longitudinal section of a molar tooth within an alveolus**

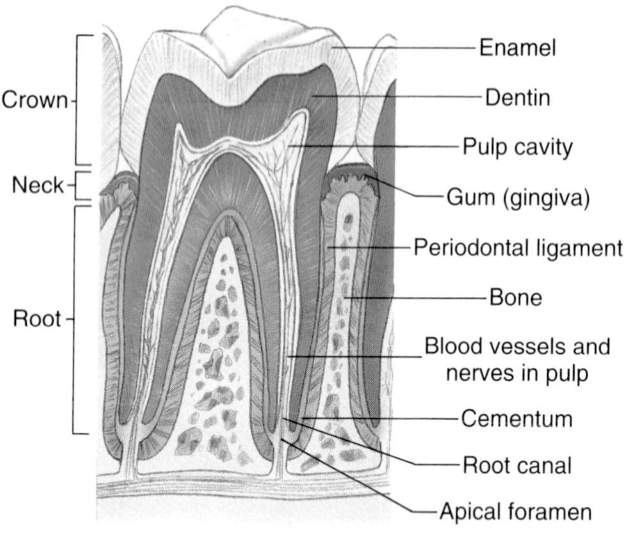

periodontal ligament by bacteria can lead to infection within a socket and loss of its tooth.

The portion of each tooth that extends from the gum into the mouth is called the **crown.** One or more **roots** anchor the tooth to the alveolus. Between the crown and the root is a slightly constricted **neck.**

Each tooth is composed mostly of a hard, calcified substance called **dentin.** The dentin of the crown is covered by **enamel,** which is even harder than dentin. A bonelike substance called **cementum** covers the dentin of the roots and anchors the tooth to the periodontal ligament lining the alveoli. The central region of the tooth contains the **pulp cavity,** in which are found blood vessels, nerves, and a connective tissue called **pulp.** The pulp cavity extends down into each root as a **root canal.** At the end of each root canal is an **apical foramen,** through which blood vessels and nerves enter the pulp cavity.

Certain bacteria found in the mouth can produce enzymes and acids capable of breaking down tooth enamel. The points of enamel destruction are called *dental caries,* or *cavities.* Once the enamel has been penetrated, the dentin may also be destroyed by the enzymes and acids. If the decay reaches the pulp, it can irritate nerves in the pulp cavity, causing a toothache. The application of fluoride, either directly to the teeth or in drinking water, seems to make the enamel more resistant to bacterial enzymes and acids.

There are four types of teeth, named according to function or shape (Figure 24.7). Each type of tooth performs a specific function in preparing food for digestion. The front chisel-shaped teeth, called **incisors,** are especially adapted for cutting. On each side of the incisors are conical **canines** *(cuspids),* which serve to tear food. The **premolars** *(bicuspids)* and **molars** have broad crowns with rounded cusps, which aid in crushing and grinding food.

Normally, two sets of teeth develop during a person's lifetime. There are 20 **deciduous (milk) teeth,** which erupt through the gums at regular intervals, beginning with the incisors at about 6 months of age. All the deciduous teeth are usually present by age 2½ years. There are 32 teeth in the **permanent set.** The permanent teeth begin to replace the deciduous teeth at about 6 years of age and continue to do so until about age 17,

◆ **FIGURE 24.7 The teeth**

Permanent teeth in (a) the upper jaw and (b) the lower jaw.

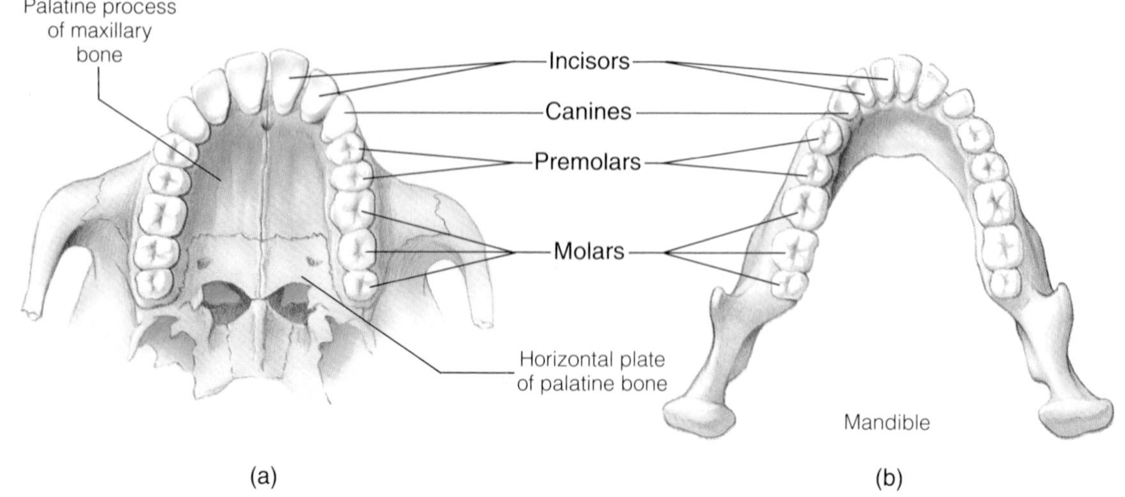

◆ **TABLE 24.1 Numbers of Specific Types of Teeth in Each Jaw**

	DECIDUOUS	PERMANENT
Incisors	4	4
Canines (cuspids)	2	2
Premolars (bicuspids)	0	4
Molars	4	6
TOTAL IN BOTH JAWS	20	32

when all of the temporary teeth generally have been replaced. The **third molars** *(wisdom teeth)* are the last to erupt, generally between the ages of 17 and 25. It is not unusual for the wisdom teeth to fail to erupt or to become wedged (impacted) in the jaw so that they are unable to erupt. Table 24.1 summarizes the numbers of each type of tooth present in each jaw, in both the deciduous and permanent sets.

In spite of extensive research, it is still uncertain precisely what causes teeth to emerge through the gums and to be replaced by other teeth in a fairly regular cycle. The development of a permanent tooth deep in the jaw puts pressure on the roots of the overlying deciduous tooth, causing the roots to disintegrate and the deciduous tooth to be shed. But researchers have shown that this action is not the complete answer. The presence of collagen and increased vascular pressure in the tissues surrounding the alveoli, as well as the regulation of hormonal levels by genes, have all been suggested as being involved in the cyclic development and replacement of teeth.

Salivary Glands

About 1000–1500 ml of saliva is secreted daily into the mouth. Some of this is produced by many small **buccal glands** *(buck´-kal)* located throughout the mucous membrane of the oral cavity. These glands secrete continuously and keep the mucous membranes moist. Most of the saliva, however, is secreted by three pairs of **salivary glands,** which are activated primarily by stimuli associated with food (Figure 24.8). These large salivary glands are of the compound tubuloalveolar type (see

◆ **FIGURE 24.8 The major salivary glands**

The right side of the mandible has been removed to expose the submandibular and sublingual glands.

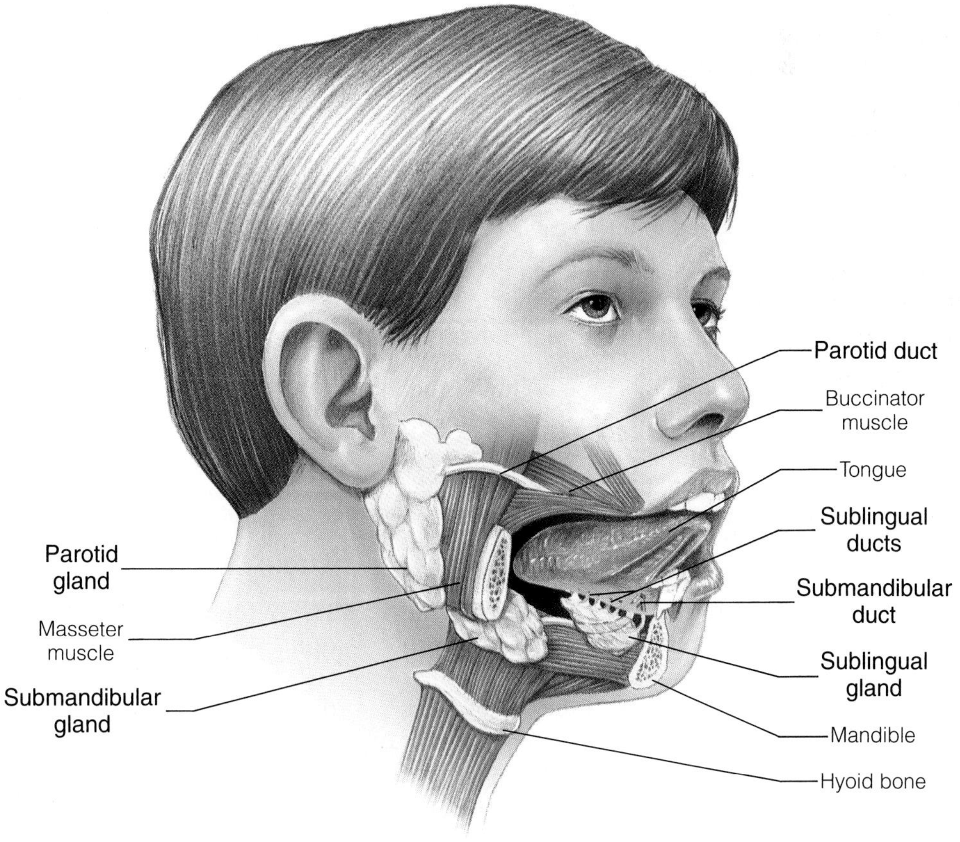

Chapter 4), being composed of blind-ended tubules and alveoli.

The largest of the paired salivary glands, the **parotid glands** *(pah-rot´-id),* are located below and anterior to the ear, on the posterior surface of the masseter muscle. Each parotid gland has a duct that crosses the masseter muscle, pierces the buccinator muscle, and empties into the vestibule of the mouth opposite the second upper molar tooth. The **submandibular glands** (sometimes called the submaxillary glands) lie medial to the angle of the mandible. The ducts of the submandibular glands travel anteriorly under the floor of the mouth to open at the base of the frenulum of the tongue. The **sublingual glands** are located on the floor of the mouth within a fold of mucous membrane. Each sublingual gland has several small ducts that open onto the floor of the mouth and, often, a larger duct that joins with the duct of the submandibular gland.

Pharynx

Food that is swallowed passes from the mouth into the oropharynx and then on into the laryngopharynx. These two portions of the **pharynx** serve as common passageways for the respiratory and digestive systems; they are described in more detail in Chapter 23. Upon leaving the laryngopharynx, food enters the esophagus.

Gastrointestinal Tract Wall

Before beginning a detailed study of the gastrointestinal tract wall, it would be helpful to review the membrane relationships described in Chapter 1 (pages 12–15). Recall that organs in the abdominopelvic cavity are covered by the *visceral peritoneum,* which is a continuation of the *parietal peritoneum* that lines the wall of the cavity. Also recall that most organs in the abdominopelvic cavity are suspended in the cavity by membranes referred to as *mesenteries.* The mesenteries are double-layered extensions of the parietal peritoneum. In the following sections, specific names will be used for the mesenteries that support particular organs or structures. Some of these mesenteries are illustrated in Figure 24.11.

Beginning in the esophagus and continuing all the way to the anus, the wall of the digestive tube has the same basic arrangement of four layers **(tunics),** with complex networks of nerves interconnecting the tunics (Figure 24.9). Although the structure of the wall is modified in various regions of the digestive tract, the four basic layers present are, from the lumen (cavity) of the gut outward, the *tunica mucosa,* the *tunica submucosa,* the *tunica muscularis,* and the *tunica serosa* or *adventitia.*

Tunica Mucosa

The **tunica mucosa** is the mucous membrane that lines the digestive tract. It consists of three layers:

1. An **epithelial layer** borders on the lumen. In the mouth, esophagus, and anus it is stratified squamous epithelium. In the remainder of the tract, it is simple columnar epithelium.
2. The **lamina propria** *(lam´-ĭ-nah pro´-pre-uh),* which is external to the epithelial layer, is composed of loose connective tissue to which the epithelial cells are attached. Blood vessels, lymph nodules, and small glands are generally located in this layer. In certain locations, such as the pharynx (tonsils), small intestine (Peyer's patches), and appendix, there are aggregations of lymph nodules within the lamina propria. The nodules serve as a means of defense against pathogens that might enter the body through the digestive system.
3. Outside the lamina propria is a thin layer of smooth muscle fibers called the **muscularis mucosae.**

Tunica Submucosa

The **tunica submucosa** is a thick layer of either dense or loose connective tissue located deep to the mucosa. It contains blood vessels, lymphatic vessels, nerves, and, in some regions, glands.

Tunica Muscularis

In most regions of the digestive tract, the **tunica muscularis** is a double layer of muscle tissue. The muscle fibers of the inner layer are arranged circularly around the tube, whereas the outer fibers are oriented longitudinally along its long axis. At several places along the tract, the fibers of the circular layer are thickened, forming sphincters that control the movement of food from one region of the digestive tract to another. In the upper part of the esophagus, the tunica muscularis is composed of skeletal (voluntary) muscle fibers. Throughout the rest of the tract, it is smooth (involuntary) muscle. Rhythmic contractions of these muscles mix the contents of the tract and move food toward the anus.

Tunica Serosa or Adventitia

The outermost tunic of the digestive tube is composed primarily of a layer of connective tissue. In the esophagus, this connective-tissue layer merges into the connective tissue of surrounding structures and is called the **adventitia.** Along the rest of the digestive tract, the connective tissue is covered with a serous membrane consisting of a single layer of squamous epithelial cells, forming the visceral peritoneum. In this case, the outer tunic is called the **serosa.**

◆ **FIGURE 24.9 Basic structure of the wall of the digestive tract**

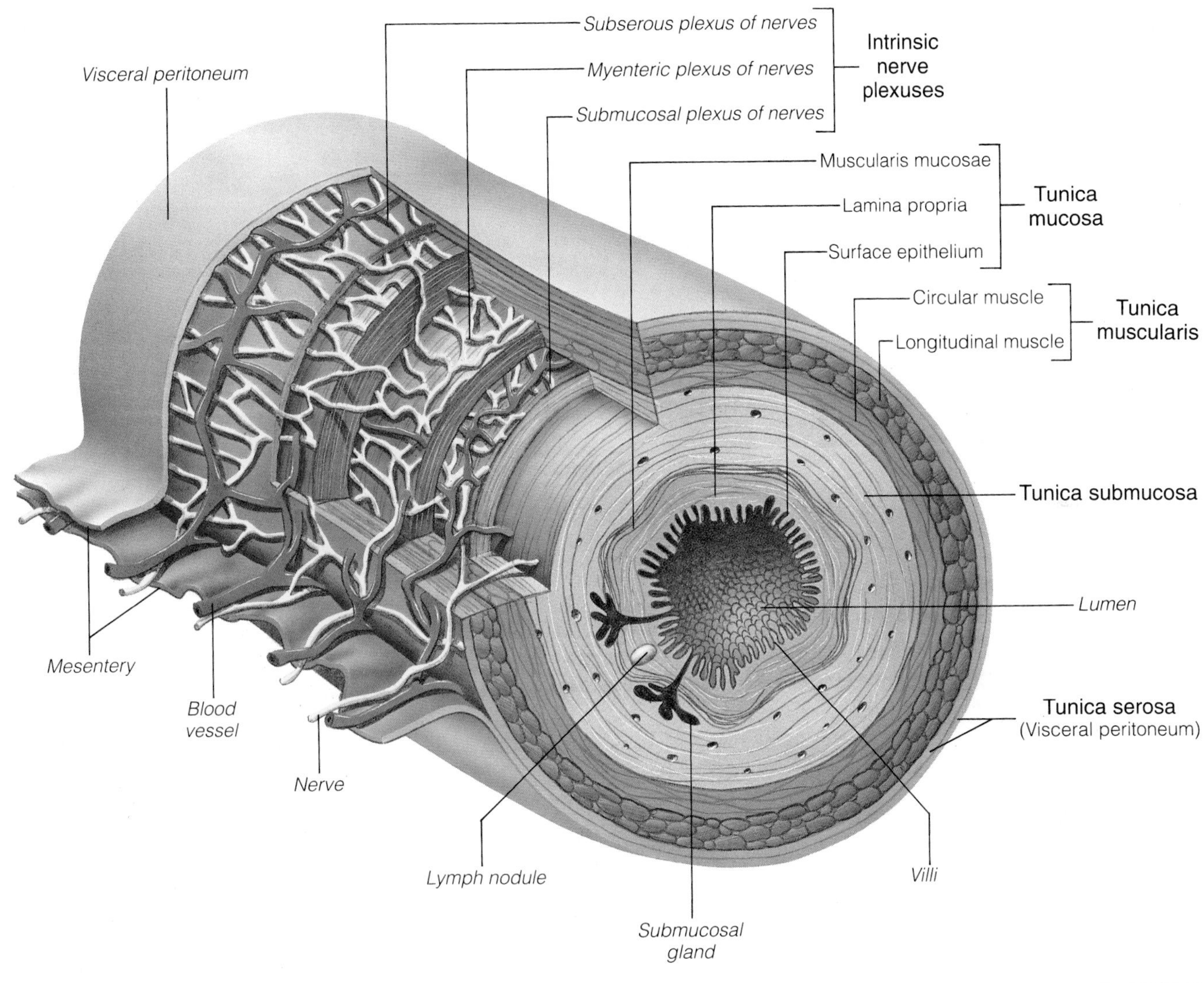

Intrinsic Nerve Plexuses and Reflex Pathways

The wall of the digestive tract contains complex interconnections of neurons that are organized into intrinsic nerve plexuses. The *submucosal plexus* is within the tunica submucosa, the *myenteric plexus* is between the circular and longitudinal layers of the tunica muscularis, and the *subserous plexus* is associated with the tunica serosa. The intrinsic nerve plexuses coordinate much of the activity of the digestive tract.

Two reflex pathways are particularly important in the control of digestive-tract activity. In **short reflexes,** neural signals originating at receptors in the wall of the digestive tract are transmitted by way of the intrinsic nerve plexuses to effector cells of the tract (for example, muscle or secretory cells). Thus, all the elements of the short-reflex pathway are located within the wall of the digestive tract.

In **long reflexes,** nerve impulses originating at receptors in the wall of the digestive tract are transmitted by afferent neurons to the central nervous system. From the central nervous system, nerve impulses are transmitted by autonomic neurons to the intrinsic nerve plexuses and effector cells of the digestive tract.

Both sympathetic and parasympathetic neurons supply the digestive tract, with parasympathetic innervation occurring primarily by way of the vagus nerves. In general, sympathetic neurons inhibit digestive-tract activity, and parasympathetic neurons stimulate it.

Esophagus

The **esophagus** *(e-sof´-ah-gus)* is a muscular tube connecting the pharynx with the stomach. It is located

◆ **FIGURE 24.10 Stomach and duodenum**
(a) Frontal section of the stomach and duodenum, showing the pancreatic and bile ducts. (b) X ray of the upper gastrointestinal tract following a barium (contrast medium) meal. The pyloric region of the stomach is contracted.

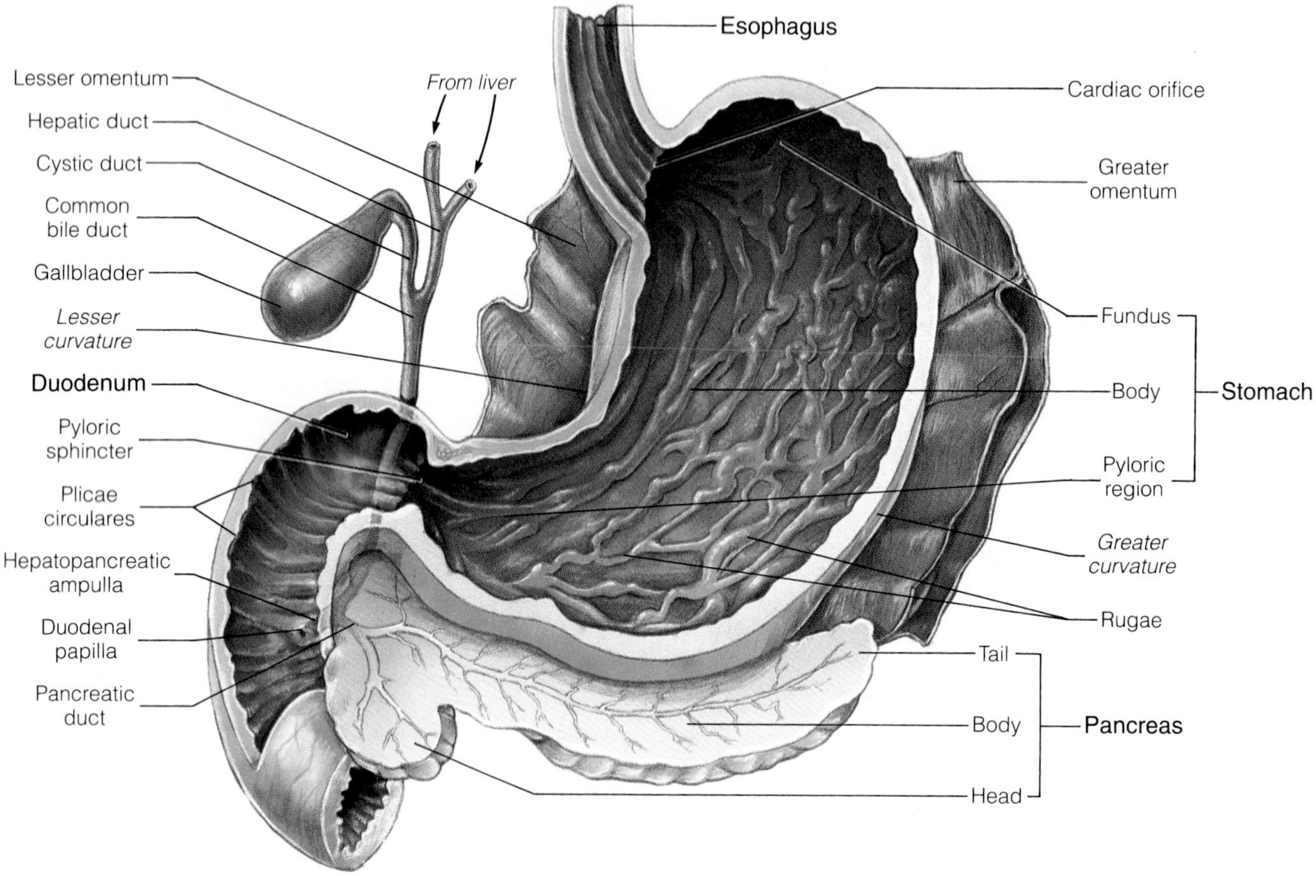

Esophagus
Lesser omentum
Hepatic duct
Cystic duct
Common bile duct
Gallbladder
Lesser curvature
Duodenum
Pyloric sphincter
Plicae circulares
Hepatopancreatic ampulla
Duodenal papilla
Pancreatic duct
From liver
Cardiac orifice
Greater omentum
Fundus
Body
Stomach
Pyloric region
Greater curvature
Rugae
Tail
Body
Pancreas
Head

(a)

Lesser curvature
Pyloric region
Greater curvature

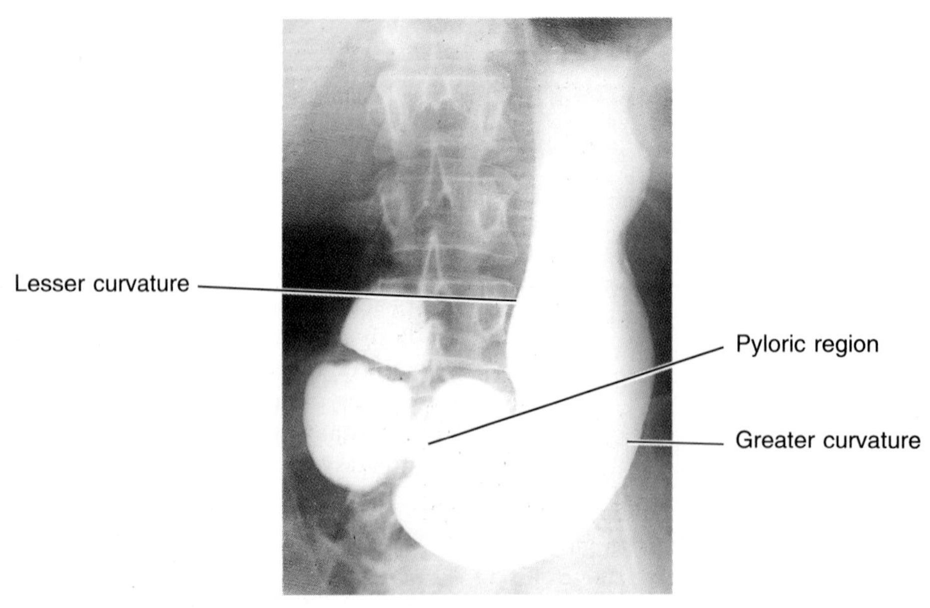

(b)

◆ **FIGURE 24.11 Sagittal section of the female abdominopelvic cavity, showing the peritoneum, the omenta, and the mesenteries**

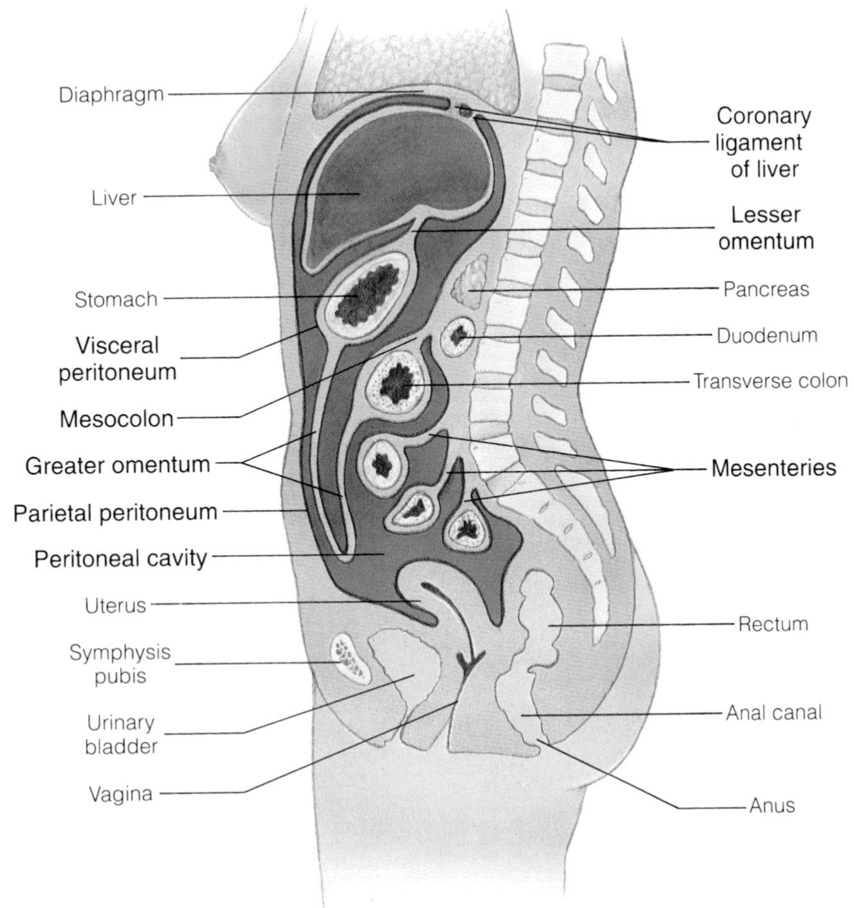

posterior to the trachea. The esophagus travels through the mediastinum of the thorax and passes through the diaphragm by means of an opening called the *esophageal hiatus (hi-a´-tus)*. Food is moved through the esophagus by waves of contractions of the muscles in its wall *(peristalsis)*. In the upper portion, near the pharynx, the tunica muscularis of the esophagus contains skeletal muscle. In the lower portions the wall of the esophagus contains smooth muscle.

Stomach

Shortly after passing through the diaphragm, the esophagus empties into the **stomach** (Figure 24.10). The stomach lies to the left of the midplane just beneath the diaphragm. The opening from the esophagus into the stomach is called the **cardiac orifice.** The exit from the stomach, where it joins with the small intestine, is restricted by the **pyloric sphincter.**

The right border of the stomach, which is concave, is called the **lesser curvature.** The convex left border is called the **greater curvature.** The lesser curvature is attached to the undersurface of the liver by a mesentery consisting of a double layer of visceral peritoneum. This mesentery is called the **lesser omentum** (Figure 24.11). The two layers of the lesser omentum separate at the lesser curvature and form the serosa on the anterior and posterior surfaces of the stomach. The layers of membrane rejoin at the greater curvature to form the **greater omentum.** The greater omentum is a folded mesentery forming an apron that covers the anterior surface of the transverse colon and the coils of the small intestine. It generally contains lymph nodules and deposits of fat between its layers.

The main portion of the stomach is called the **body.** The **fundus** is the portion that bulges above the entrance of the esophagus. The body of the stomach tapers inferiorly to form the **pyloric** region. The upper portion of the pyloric region is dilated and is called the

◆ **FIGURE 24.12 Structure of the gastric glands**

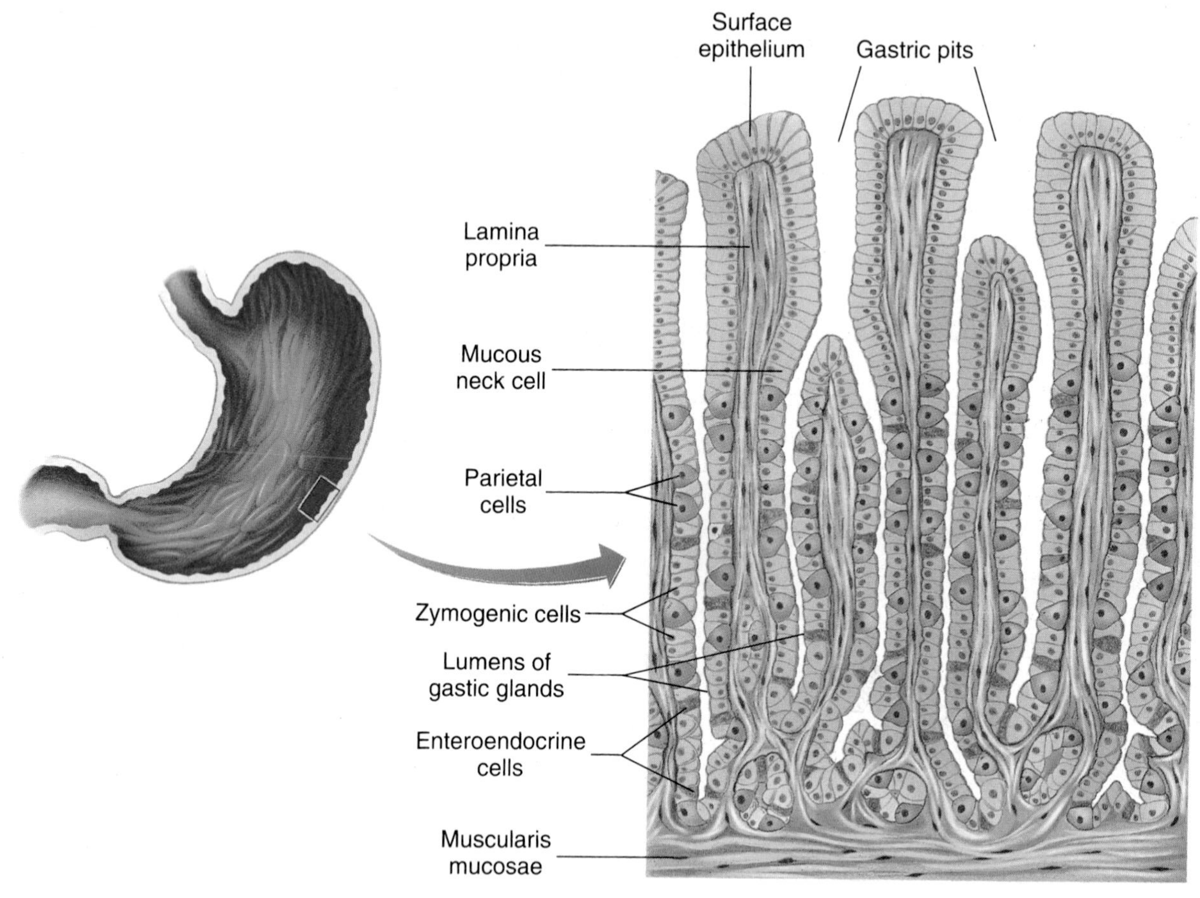

pyloric antrum. The antrum narrows into the **pyloric canal,** which joins the duodenum of the small intestine.

The wall of the stomach is composed of the four basic layers (tunics) that are typical of the digestive tract. In the stomach, however, the epithelial tissue of the mucosa is simple columnar rather than stratified squamous as in the esophagus. The epithelium of the mucosa remains simple columnar throughout the intestines. When the stomach is empty, the mucosa and submucosa form longitudinal folds called **rugae** *(roo´-gē).* These folds stretch and flatten as the stomach fills.

A modification of the mucosa in the stomach is the presence of many **gastric glands,** which occupy the lamina propria. These glands, which secrete gastric juice, empty onto the surface of the mucosa through small invaginations called **gastric pits** (Figure 24.12). The gastric glands located in the fundus and body of the stomach **(fundic glands** and **gastric glands proper)** contain several types of secretory cells:

1. **Mucous neck cells** are mucus-secreting cells located near the gastric pits.
2. **Parietal (oxyntic) cells** produce *hydrochloric acid.*
3. **Zymogenic (chief) cells** secrete *pepsinogen,* a precursor of the digestive enzyme *pepsin.*

The glands of the cardiac region **(cardiac glands)** and the pyloric region **(pyloric glands)** secrete mainly mucus. However, **enteroendocrine cells** that produce the hormone *gastrin* as well as *serotonin* and *histamine* are present throughout the stomach. Gastrin is of particular significance because it helps regulate stomach secretion and contraction. Its specific activities are discussed later in the chapter.

Another modification of the tunics of the stomach is found in the tunica muscularis. Apart from the circular and longitudinal coats, the stomach wall contains an oblique layer of muscle between the circular layer and the submucosa. The additional layer of muscle in the wall makes extra strong contractions possible in the stomach and aids in its major functions—mashing the food and mixing it with digestive juices.

Small Intestine

The stomach empties into the **small intestine**—the longest (approximately 6 m) and most convoluted portion of the digestive tract. The small intestine joins the large intestine at the ileocecal valve. The small intestine is lined with simple columnar epithelium that contains cells specialized to absorb nutrients, which is a major function of the small intestine. On the basis of differences in microscopic structure, the small intestine can be divided into three regions that are not otherwise distinct from each other:

1. The **duodenum** *(doo-o-dee´-num* or *doo-od´-denum)*, which is the first 25 cm or so of the small intestine, curves around the head of the pancreas (Figure 24.10a). The **common bile duct,** from the liver, and the **pancreatic duct,** from the pancreas, join together to form the **hepatopancreatic ampulla** *(ampulla of Vater),* which empties into the duodenum at the **duodenal papilla.** This opening is surrounded by a sphincter muscle called the **hepatopancreatic sphincter** *(sphincter of Oddi)*. The common bile duct carries bile; the pancreatic duct carries digestive enzymes. The duodenum is retroperitoneal—that is, situated behind the peritoneum—and is tightly attached to the posterior body wall (Figure 24.11).
2. The next 2.5 m or so of the small intestine is the **jejunum** *(je-joo´-num).* This portion is suspended in the abdominal cavity by a mesentery.
3. The **ileum** *(il´-ee-um)* is the remaining 3.5 m or so of the small intestine. The entrance of the ileum into the cecum of the large intestine is guarded by the **ileocecal valve** *(il´´-ē-ō-se´-kal),* which is composed of two folds of tissue. For 2 or 3 cm preceding the ileocecal valve, the muscle of the ileal wall is thickened, forming the **ileocecal sphincter.** The ileum, like the jejunum, is suspended from the posterior body wall by a mesentery. The mesentery allows the small intestine to move during the contractions of peristalsis, and it also provides support for blood vessels, lymphatic vessels, and nerves that supply the intestine.

The modifications in the wall of the small intestine are within the tunica mucosa and the submucosa. These layers form permanent circular shelflike folds **(plicae circulares;** *ply´-ka***)** that extend into the lumen of the small intestine (Figure 24.10). The plicae increase the surface area of the mucosa and help mix the food with intestinal fluid by causing it to move spirally through the lumen of the intestine. The surface area is increased further by **villi,** which are tiny leaflike, tonguelike, or fingerlike projections of the mucosa into the lumen (Figure 24.13). In many areas, the villi are so numerous and so close together that the mucosa of the small intestine has a velvety appearance. Each villus is covered with a single layer of epithelial cells and contains a lymphatic capillary called a *lacteal,* which is surrounded by a network of blood capillaries.

Three types of epithelial cells cover the villi: **goblet cells,** which secrete mucus; **absorptive cells,** which participate in the absorption (and also the digestion) of food materials; and, in the duodenum, **enteroendocrine cells,** which secrete cholecystokinin and secretin (see page 780), as well as other hormones or hormone-like substances. The free surfaces of the absorptive cells are folded into microscopic projections called **microvilli.** Microvilli greatly increase the total surface area of the intestinal mucosa. The increased surface area provided by the plicae circulares, villi, and microvilli aids in the absorption of digested foods, which must pass through the mucosa before entering blood capillaries or lacteals.

The mucosa of the small intestine contains many tubular **intestinal glands** *(crypts of Lieberkuhn)* located between the bases of the villi, and in the submucosa are aggregations of lymphatic nodules called **Peyer's patches** (see page 684). In the submucosa of the duodenum are mucous glands called **duodenal** *(Brunner's)* **glands,** or **submucosal glands.** These glands generally empty into the intestinal glands and provide a protective mucous barrier for the wall of the duodenum.

Large Intestine

The **large intestine,** which is about 1.5 m long, extends from the ileocecal valve to the anus (Figures 24.14, page 769 and 24.15, page 770). It is so named because its diameter in most regions is greater than that of the small intestine. Like the small intestine, the large intestine is lined with simple columnar epithelium with microvilli on the free surfaces of the cells. Absorptive cells and goblet (mucous) cells are present, with the latter being more abundant. Very few enzymes, if any, are produced by the epithelial cells of the large intestine.

The large intestine begins as a blind pouch called the **cecum** *(sē´-kum),* which receives the ileum of the small intestine. The **vermiform appendix** is a narrow blind tube that extends downward from the cecum. The wall of the appendix contains numerous lymphatic nodules. The large intestine extends upward from the cecum as the **ascending colon.** The ascending colon is not supported by a mesentery; instead, it lies tightly against the posterior wall of the abdomen. Just beneath the liver, the ascending colon bends sharply to the left, forming the **right colic** (or **hepatic**) **flexure,** and crosses the abdominal cavity as the **transverse colon.** This portion of the colon is suspended by a mesentery called the **mesocolon.** In the vicinity of the spleen, the transverse colon bends downward at the **left colic** (or **splenic**) **flexure** and forms the **descending colon.** The descending colon, like the ascending colon, is retroperitoneal. Where the descending colon reaches the left pelvic brim, it curves to the midplane via an S-shaped **sigmoid colon.**

The four tunics of the colon are unique in a number of ways. The tunica mucosa, for example, contains intestinal glands and a large number of mucous cells, but lacks villi. Moreover, while the longitudinal layer of the muscles of the tunica muscularis is continuous as a thin

◆ **FIGURE 24.13 Microscopic structure of the small intestine**
(a) Enlargement showing villi extending from a plica circularis. (b) Higher magnification of a section through two villi, showing blood vessels and lacteals. (c) Enlargement of two absorptive cells, showing microvilli on their free surfaces.

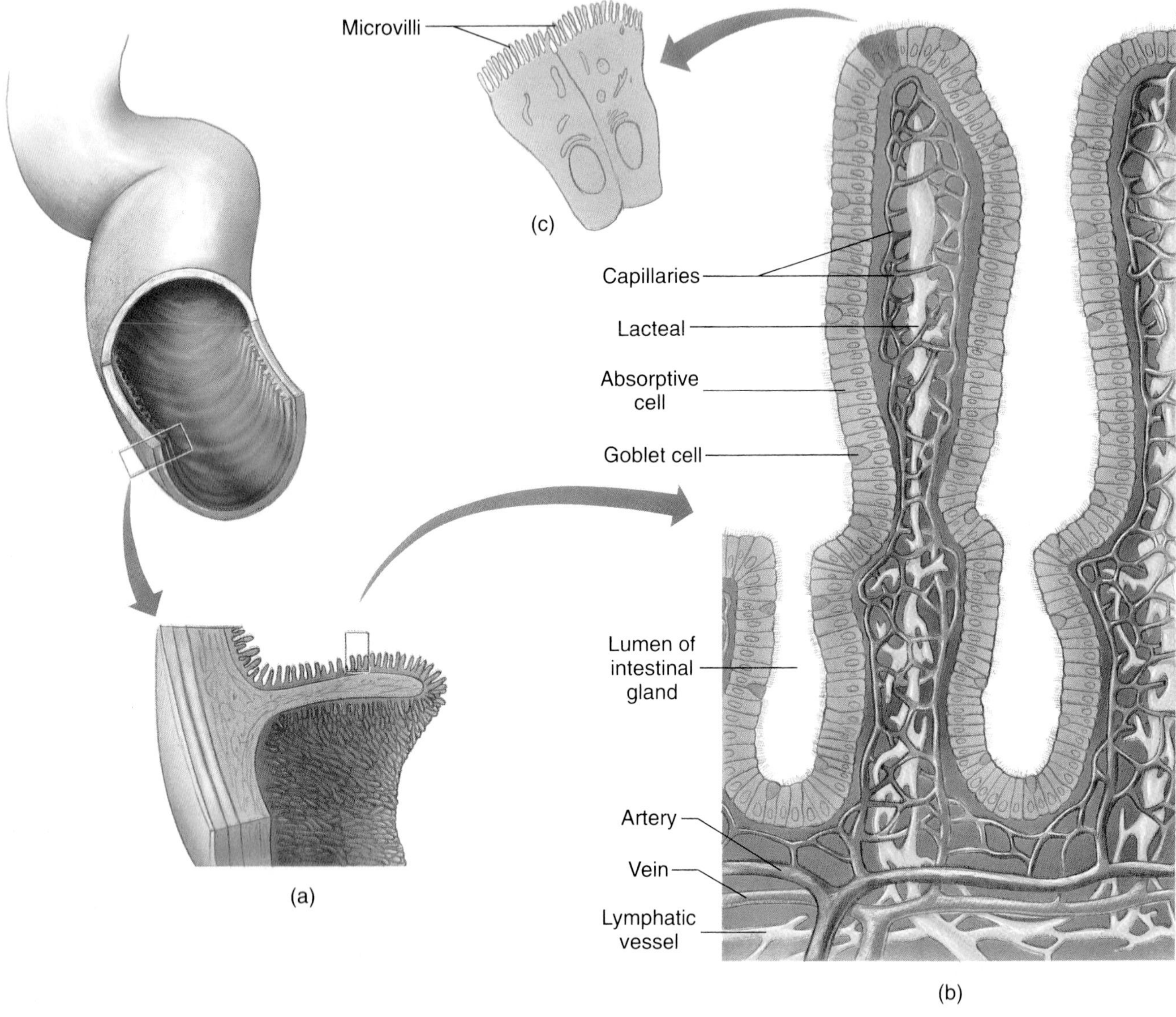

layer over the entire surface of the colon, most of the muscles of this layer are in the form of three bands—called **taeniae coli** *(tē´-nē-ah kō-li)*—that run the length of the colon. The taeniae are shorter than the colon and therefore cause the colon to form small pouches called **haustra** *(haw´-strah;* singular: *haustrum)*. The mucosa between the haustra is thrown into **plicae semilunares** (semilunar folds), which extend into the lumen of the gut. Unlike the plicae circulares of the small intestine, the folds of the colon extend only partway around the gut tube. Another unique characteristic of the colon is the presence on its external surface of small tabs called **epiploic appendages** *(ep˝-ĭ-plō-ik)*. These tabs are composed of folds of peritoneum filled with fat.

Rectum and Anal Canal

Beyond the sigmoid colon, the large intestine passes downward in front of the sacrum. This portion is straight and is called the **rectum** (Figure 24.14). The rectum has the same structure as the colon except that taeniae coli are not present, the longitudinal layer of muscles being again spread evenly over the entire surface.

◆ **FIGURE 24.14 The large intestine**

The cecum has been opened to expose the ileocecal valve.

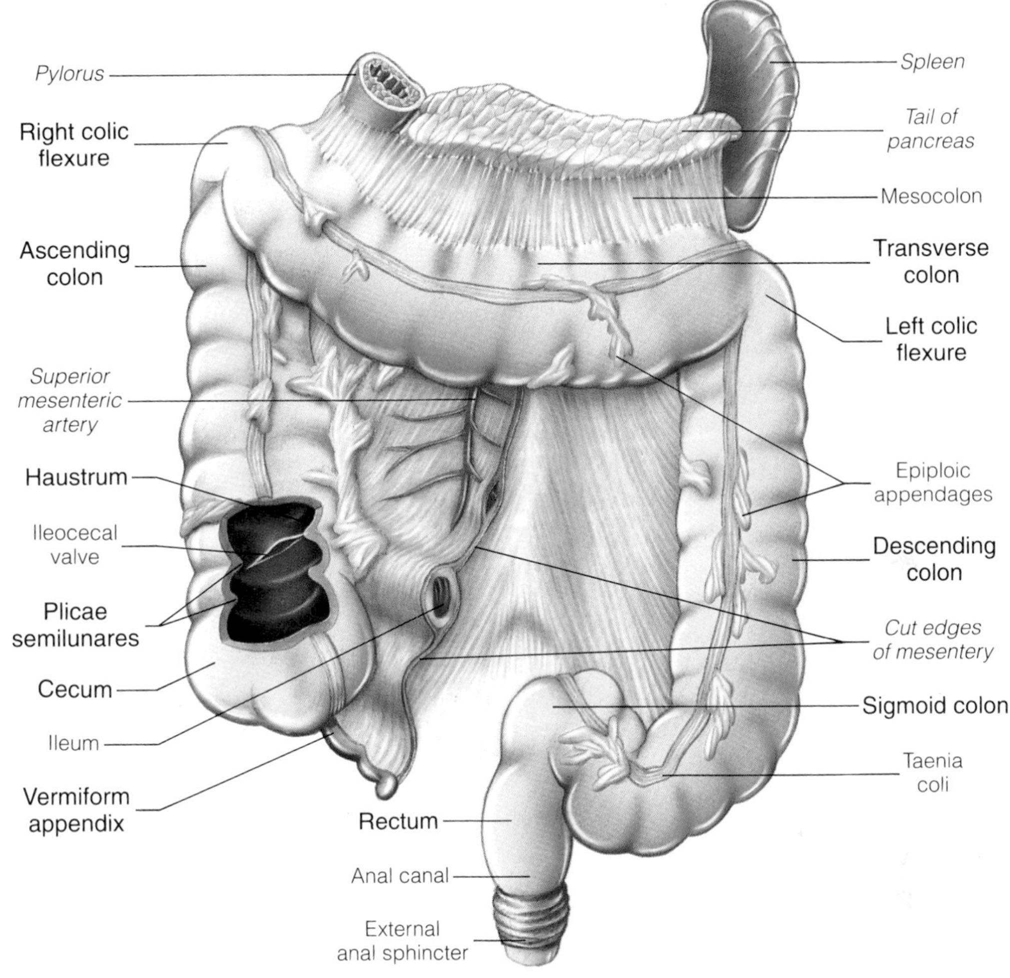

The terminal 3–4 cm of the large intestine is called the **anal canal** (Figure 24.16). This region is located below the pelvic diaphragm and thus is outside the pelvis.

The mucosa of the anal canal forms a series of longitudinal folds known as **anal columns.** The anal columns are separated from each other by furrows called **anal sinuses,** which end distally in membranous **anal valves.** The anal valves unite the lower ends of the anal columns. Within the anal canal, the epithelium changes from simple columnar, which is characteristic of the stomach and small and large intestines, to stratified squamous, which is typical of the mouth and esophagus.

The anal canal opens to the exterior through the **anus.** The anal canal is surrounded by **internal** and **external anal sphincter muscles.** The internal sphincter is formed from a thickening of the circular layer of the smooth muscles of the tunica muscularis and is therefore not under voluntary control. The external sphincter is formed by skeletal muscle and is under voluntary control. Normally, therefore, a person can voluntarily control bowel movements.

The circulation of blood through veins that travel the length of the anal columns is sometimes interfered with, causing them to become enlarged. This condition is called **hemorrhoids.** Hemorrhoids can be irritated by bowel movements, producing discomfort and bleeding, particularly if the feces are too dry, as in constipation.

Accessory Digestive Organs

In addition to the many small glands situated in the wall of the digestive tract, there are large glands located outside the tract. The secretions of these glands, which are very important in the digestion of food, are carried to the digestive tract by way of ducts. The ducts and the secretory portions of the accessory digestive glands are

◆ **FIGURE 24.15 X ray of the lower gastrointestinal tract following a barium (contrast medium) meal**

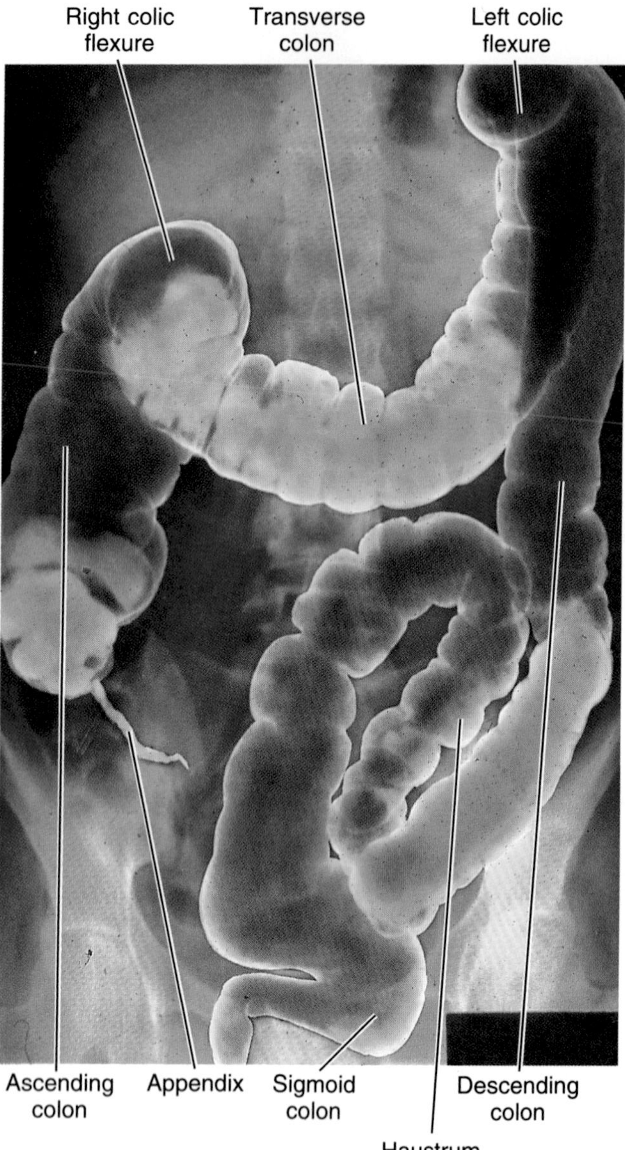

◆ **FIGURE 24.16 Longitudinal section of the anal canal**

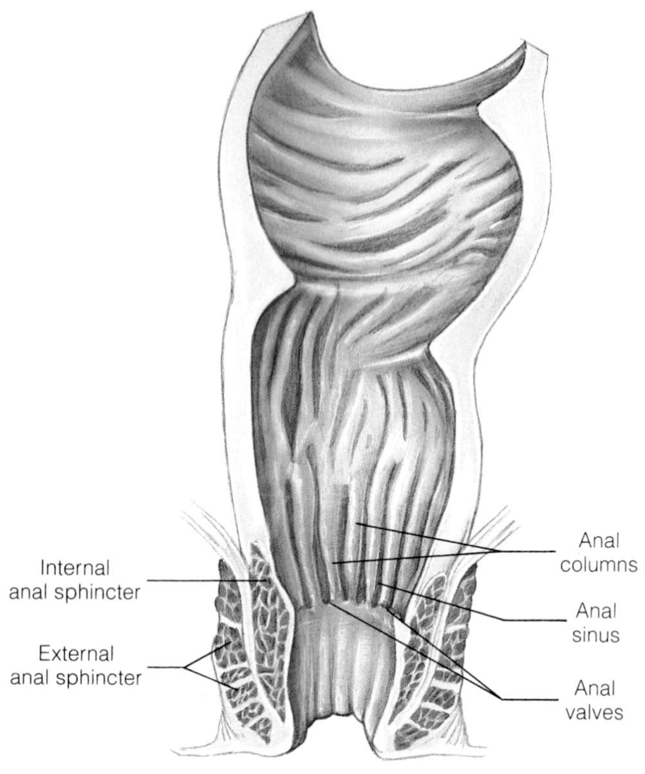

derived from the endodermal lining of the gut tube of the embryo. These glands include the *salivary glands* (discussed earlier in this chapter), whose ducts empty into the mouth, and the *pancreas* and *liver,* both of which empty their secretions into the duodenum of the small intestine.

Pancreas

The **pancreas,** which is both an exocrine gland and an endocrine gland, is located behind the peritoneum and beneath the stomach (Figure 24.10a). The **head** of the pancreas is nestled in the curve of the duodenum, with the **neck, body,** and **tail** regions extending to the left. The body and tail are retroperitoneal, and the tail reaches the vicinity of the spleen.

Microscopically, the exocrine portion of the pancreas resembles the salivary glands, containing many compound tubuloalveolar glands with their secretory cells arranged in short tubules or small sacs called **acini** (Figure 24.17). The acini consist of a single layer of pyramid-shaped epithelial cells, with their apices converging toward the lumen. In acinar cells that are actively secreting, vesicles called *zymogen granules* accumulate toward the apex of the cells.

The acinar cells secrete digestive enzymes. The digestive enzymes of the pancreas are thought to be synthesized in the cytoplasm at the base of the acinar cells, where they enter the endoplasmic reticulum and are transported to the Golgi apparatus in the apical region of the cell. The Golgi apparatus concentrates the enzymes into zymogen granules, which are released by exocytosis upon appropriate stimulation.

The exocrine pancreatic secretions, which are called **pancreatic juice,** are transported to the duodenum by the **pancreatic duct** *(duct of Wirsung).* The pancreatic duct usually joins with the common bile duct, and they enter the duodenum together. An **accessory pancreatic duct** *(duct of Santorini)* often branches from the pancreatic duct and empties into the duodenum independently.

The endocrine portion of the pancreas consists of clusters of cells called **pancreatic islets** *(islets of Lan-*

◆ **FIGURE 24.17 The pancreas**

One inset shows the form of its compound tubuloalveolar glands, and the other is a photomicrograph of acinar cells and an islet.

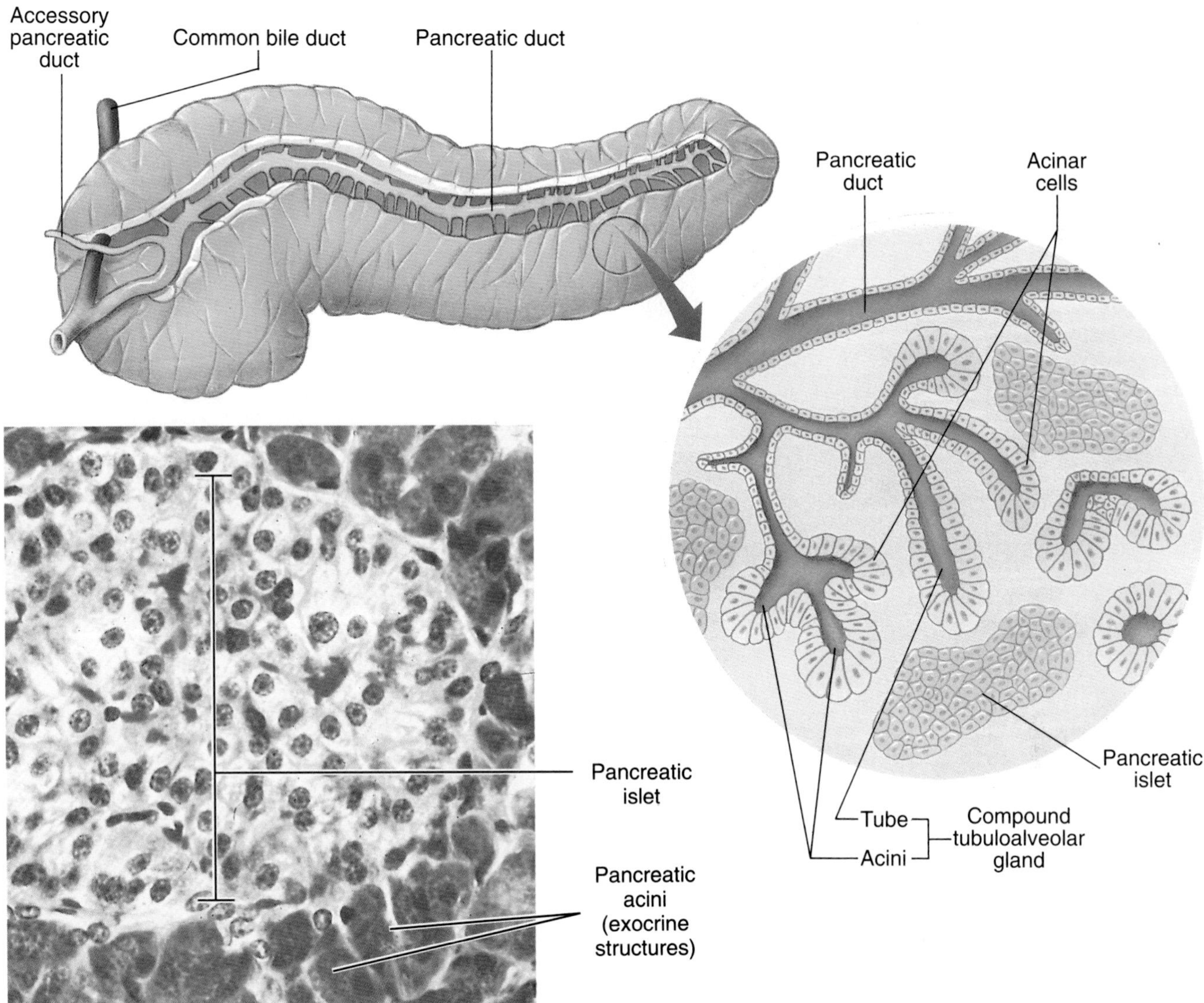

gerhans), which are scattered among the acini. The endocrine secretions of the pancreatic islets enter the blood and are carried throughout the body. The endocrine secretions of the pancreas are discussed with the endocrine glands in Chapter 17.

Liver

The **liver** is a large complex organ with numerous important and varied functions. Many of its functions are vital for metabolic and regulatory activities. These functions are discussed in Chapter 25. The liver produces a secretion called bile that is stored in the gallbladder and eventually transported to the duodenum. Bile is important in fat digestion, which is discussed later in this chapter (page 792).

The liver lies under the right side of the diaphragm. It is divided into two major regions: the **right** and **left lobes** (Figure 24.18a). On the inferior surfaces of the right lobe are smaller **caudate** and **quadrate lobes.** The right and left lobes are separated by a fold of parietal peritoneum called the **falciform ligament,** which attaches the liver to the anterior abdominal wall. In the free margin of the falciform ligament is a fibrous cord called the **ligamentum teres (round ligament).** The ligamentum teres is the remnant of the fetal umbilical vein that transported blood from the placenta to the liver. In postnatal life, it extends from the umbilicus to the undersurface of the liver. The falciform ligament is

◆ **FIGURE 24.18 The liver and its supporting ligaments**
(a) Anterior surface. (b) Superior (diaphragmatic) surface. (c) Posterior-inferior surface.

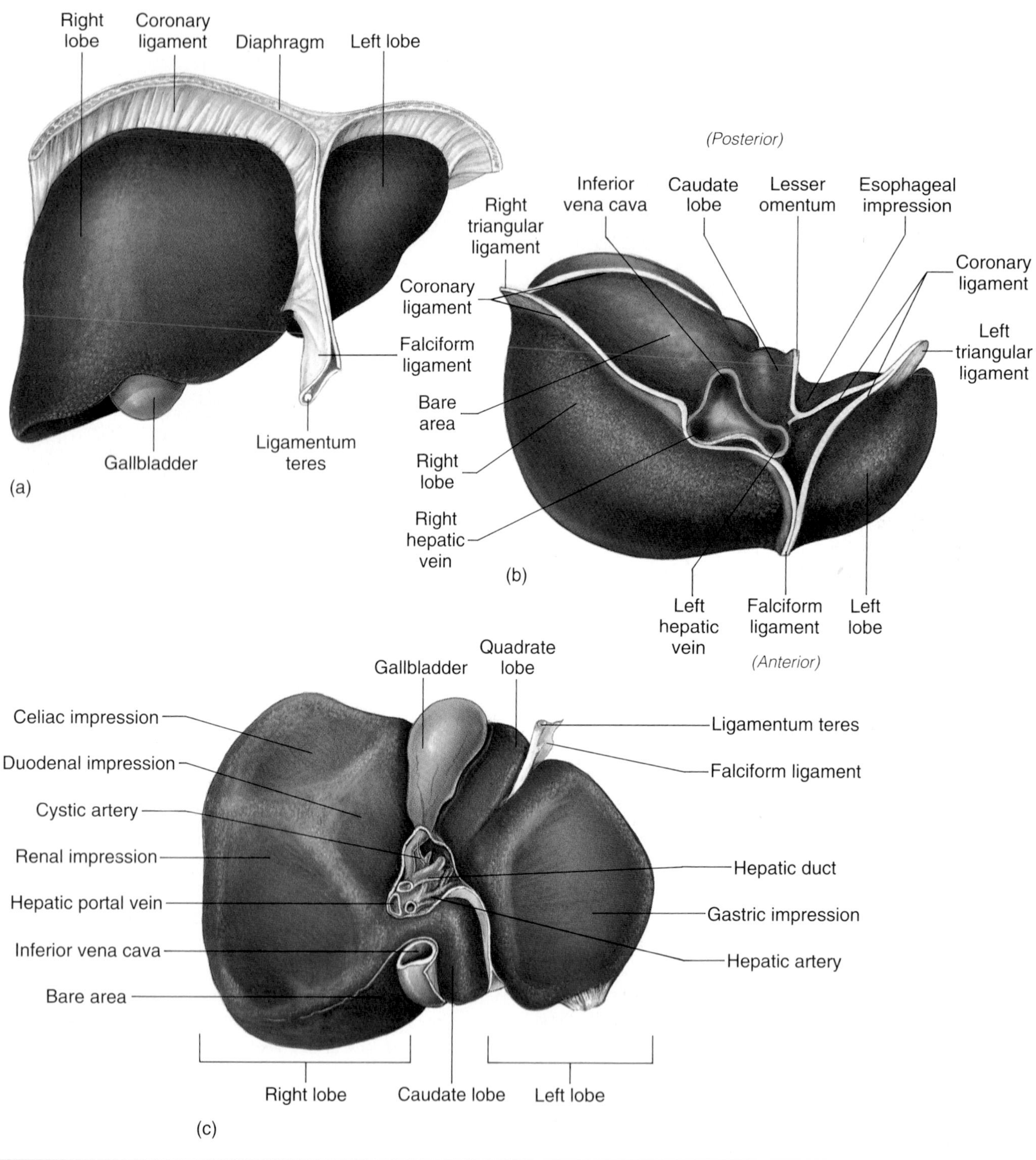

continuous on the superior surface of the liver with the **coronary ligament,** a fold of the parietal peritoneum that attaches the liver to the undersurface of the diaphragm (Figure 24.18b). The coronary ligament consists of anterior and posterior layers joined at their lateral margins by **right** and **left triangular ligaments.** Between the two layers of the coronary ligament is a region called the **bare area** of the liver. The bare area, which rests against the diaphragm, is the only portion of the liver not covered with visceral peritoneum. The undersurface of the liver is anchored to the lesser curvature of the stomach by the **lesser omentum,** through which travel the hepatic artery, the hepatic portal vein, and the common bile duct. The inferior vena cava is partially embedded on the posterior surface of the liver (Figure 24.18c).

The liver receives blood from two sources: the **hepatic artery,** which supplies it with oxygenated blood from the aorta, and the **hepatic portal vein,** which carries venous blood from the digestive tract, pancreas, and spleen. The oxygen saturation of the blood in the hepatic portal vein is relatively low, but the blood contains a high concentration of dissolved nutrients absorbed from the intestines as a result of digestion. These nutrients are acted on in various ways as the blood passes through the liver. Approximately 1500 ml of blood flows through the liver every minute, of which 1100 ml arrives from the hepatic portal vein and 400 ml from the hepatic artery.

The liver is composed of cells called **hepatocytes** *(heh-pa´-tō-sītz),* which are arranged in rows or sheets called **liver cords** or **plates.** The liver cords are organized into many tiny hexagonal compartments called **liver lobules,** with the corners of the compartments occupied by canals called **portal canals (porta hepatis).** Branches of the hepatic artery and the hepatic portal vein, as well as a **bile duct,** travel in the portal canals. These three structures in close approximation within a portal canal are referred to as a **hepatic triad** (Figure 24.19).

The most unusual aspect of hepatic circulation is that branches of both the hepatic artery and the hepatic portal vein empty into the same **sinusoids,** which therefore contain a mixture of arterial and venous blood (Figure 24.20). The sinusoids of each lobule pass between the hepatocytes and empty into a common **central vein** that passes through the center of each lobule, at right angles to the sinusoids. The central vein of each lobule drains into a **hepatic vein.** There are three hepatic veins, all emptying into the inferior vena cava, which transports the blood to the heart. The pathway of blood through the liver is summarized in Figure 24.21.

The liver sinusoids are lined with an endothelium that is highly permeable; it even allows proteins to diffuse out of the blood flowing through the sinusoids. Because of this permeability, substances absorbed into the blood from the intestines can freely leave the blood and enter the space **(perisinusoidal space** or **space of Disse)** that separates the endothelium lining the sinusoids from the hepatocytes. Microvilli of the hepatocytes extend into the space. The microvilli provide an increased surface area that greatly enhances the movement of substances from the blood into the hepatocytes, where the substances are metabolized or otherwise altered.

Interspersed between the endothelial cells that line the sinusoids are star-shaped cells called **stellate macrophages** *(Kupffer cells).* Projections of the macrophages extend into the lumen of the sinusoid. The macrophages are active phagocytes that remove bacteria and other foreign materials from the blood as it passes through the liver.

◆ **FIGURE 24.19 Microscopic anatomy of a liver lobule**
The arrows show the direction of the flow of blood and bile.

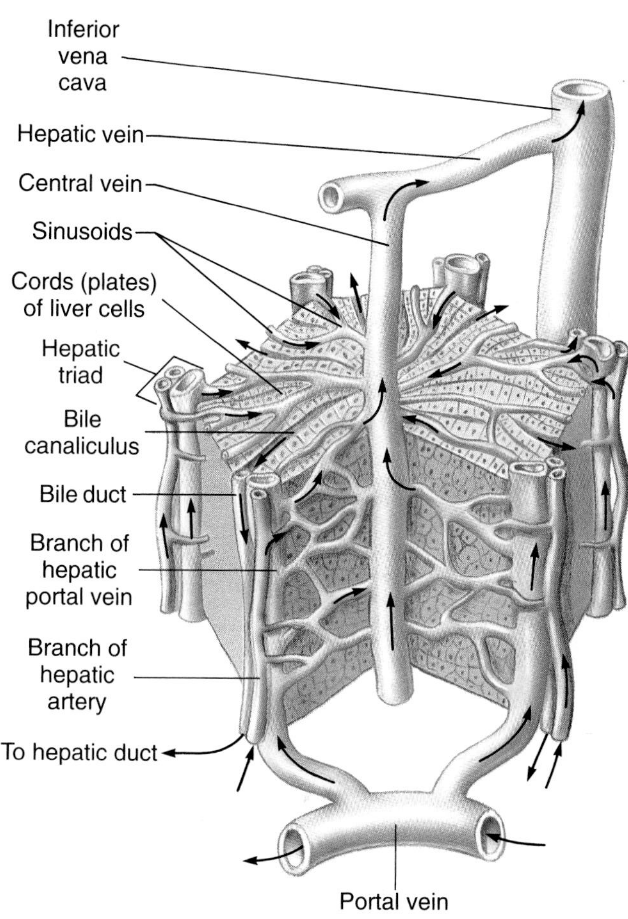

Tiny canals, called **bile canaliculi** *(kan˝-ah-lik´-ū-lī)* are located between the hepatocytes (Figure 24.22). The bile canaliculi carry bile to bile ducts at the periphery of each lobule. Within the canaliculi, the bile travels in the opposite direction from the blood flow in the sinusoids. The many bile ducts join together to form two main trunks that leave the liver and unite to form the **hepatic duct.**

Gallbladder and Bile Ducts

The **gallbladder,** a small sac on the inferior surface of the liver, is lined with columnar epithelium. It serves as a storage site for bile, which it receives from the liver. The gallbladder is drained by the **cystic duct,** which joins with the hepatic duct from the liver to form the

◆ **FIGURE 24.20 Microscopic structure of the liver**
(a) Enlargement showing parts of several lobules; two central veins are visible. (b) A higher magnification of two sinusoids, showing their relationship with the branches of the hepatic portal vein and the hepatic artery. The yellow arrows indicate the flow of blood through the sinusoids and into the central vein; the black arrows trace the flow of bile through the bile canaliculi and into the bile duct.

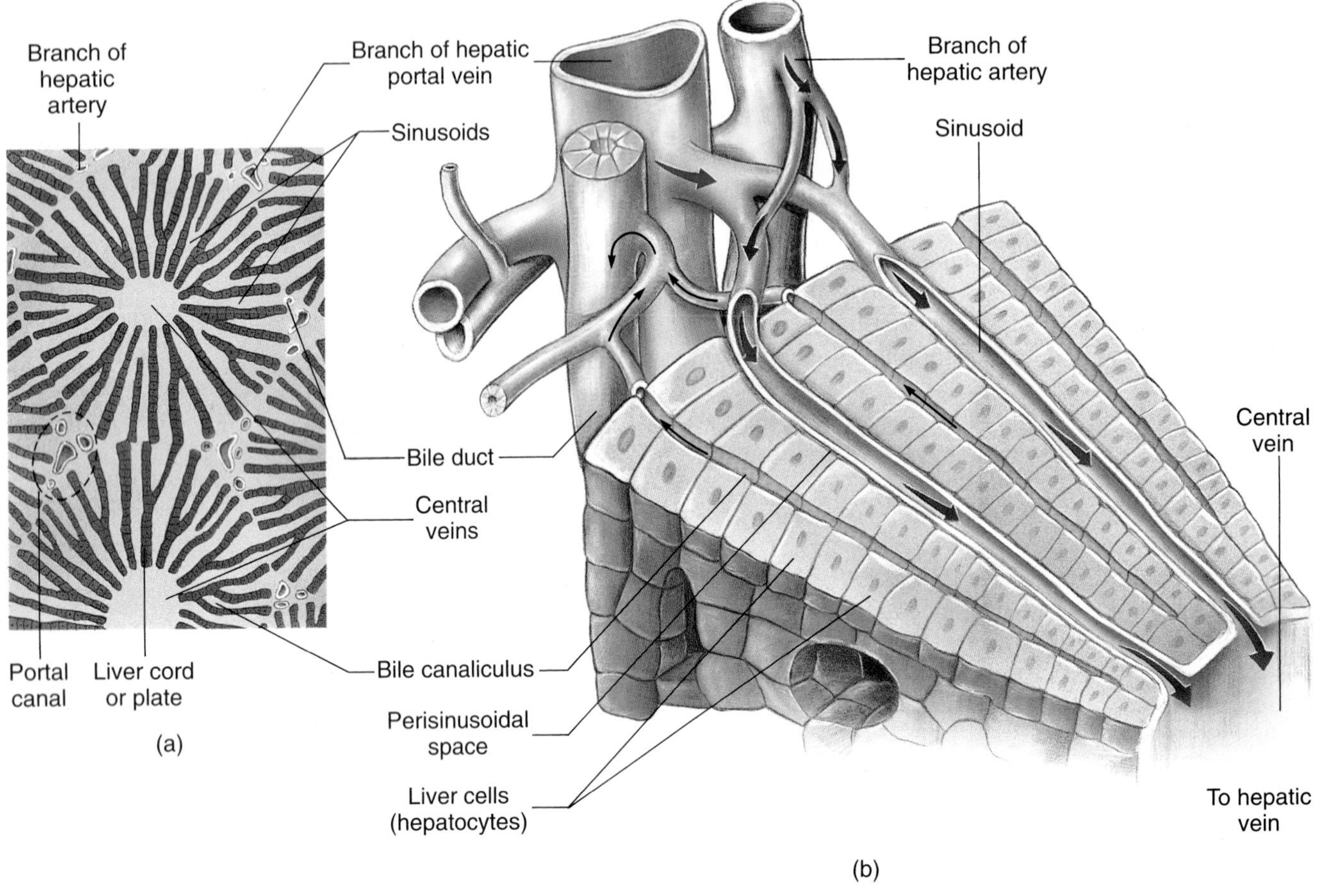

common bile duct (Figure 24.23). The pancreatic duct joins with the common bile duct, and the two ducts share a common entrance into the duodenum. This entrance is surrounded by the hepatopancreatic sphincter. When the sphincter relaxes and the smooth muscle in the wall of the gallbladder contracts, bile is propelled into the intestine. When the sphincter contracts, bile from the liver fills the common bile duct and then enters the gallbladder via the cystic duct.

Physiology of the Digestive System

The activities of the digestive system include **mechanical processes, secretion, chemical digestion,** and **absorption** (Table 24.2).

1. The mechanical processes occurring within the digestive system mix ingested food with digestive secretions and propel the contents of the gastrointestinal tract along the tract. These processes are due to muscular contractions.
2. The secretory activities of the digestive system release mucus, digestive enzymes, and other secretory products into the gastrointestinal tract.
3. Chemical digestion breaks down large food molecules into smaller molecules that can be absorbed from the gastrointestinal tract. For example, chemical digestion breaks down large starch molecules into smaller glucose molecules. Chemical digestion is accomplished by digestive enzymes.
4. The processes of absorption transfer the smaller molecules that result from chemical digestion—and also water, vitamins, and minerals—from the lumen of the gastrointestinal tract into the blood or lymph. Some substances are absorbed by active transport, and other substances are absorbed by passive processes.

◆ **FIGURE 24.21 The pathway of blood through the liver**

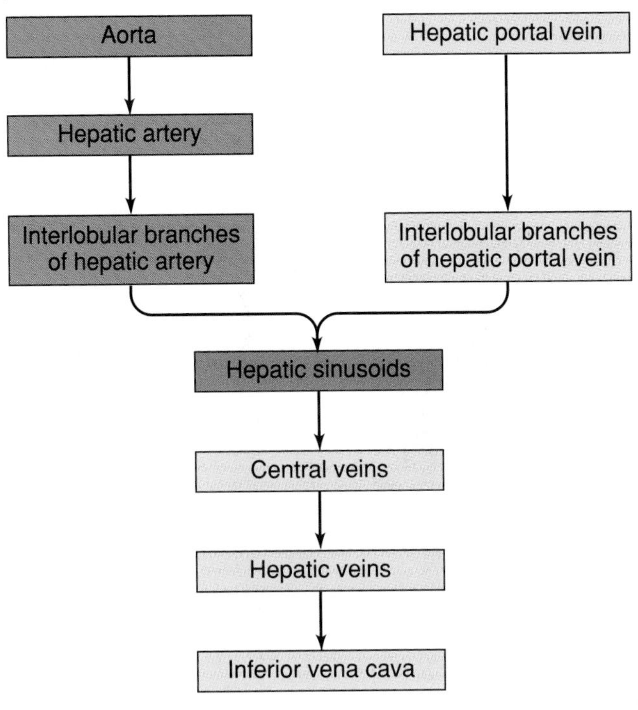

Mechanical Processes

Ingested food must be continually moved along the gastrointestinal tract so it can be acted upon by the digestive enzymes secreted into various regions of the tract. Moreover, the contents of the digestive tract must be constantly churned and mixed so all the food particles come into contact with the enzymes. This churning also brings the food into contact with the wall of the tract, which allows for the absorption of digested food.

The moving and mixing of food within the digestive tract is accomplished by the rhythmic contraction and relaxation of the muscles of the tract. The muscles of the mouth, pharynx, upper portion of the esophagus, and external anal sphincter are skeletal muscles that require nerve impulses to initiate their contractions. The muscles of the remainder of the gastrointestinal tract are smooth muscles whose contractions are strongly influenced by hormones and neural signals.

Two basic forms of movement—mixing and propulsion—occur within the gastrointestinal tract. A major mixing movement is **segmentation,** in which stationary muscular contractions that occur at intervals along a portion of the digestive tract divide the tract into constricted and unconstricted regions (Figure 24.24). As segmentation proceeds, constricted regions relax and

◆ **FIGURE 24.22 Microscopic anatomy of the liver (×1550)**

(From *Tissues and Organs: A Text-Atlas of Scanning Electron Microscopy* by Richard G. Kessel and Randy H. Kardon. W. H. Freeman and Company. Copyright © 1979. Reprinted with permission.)

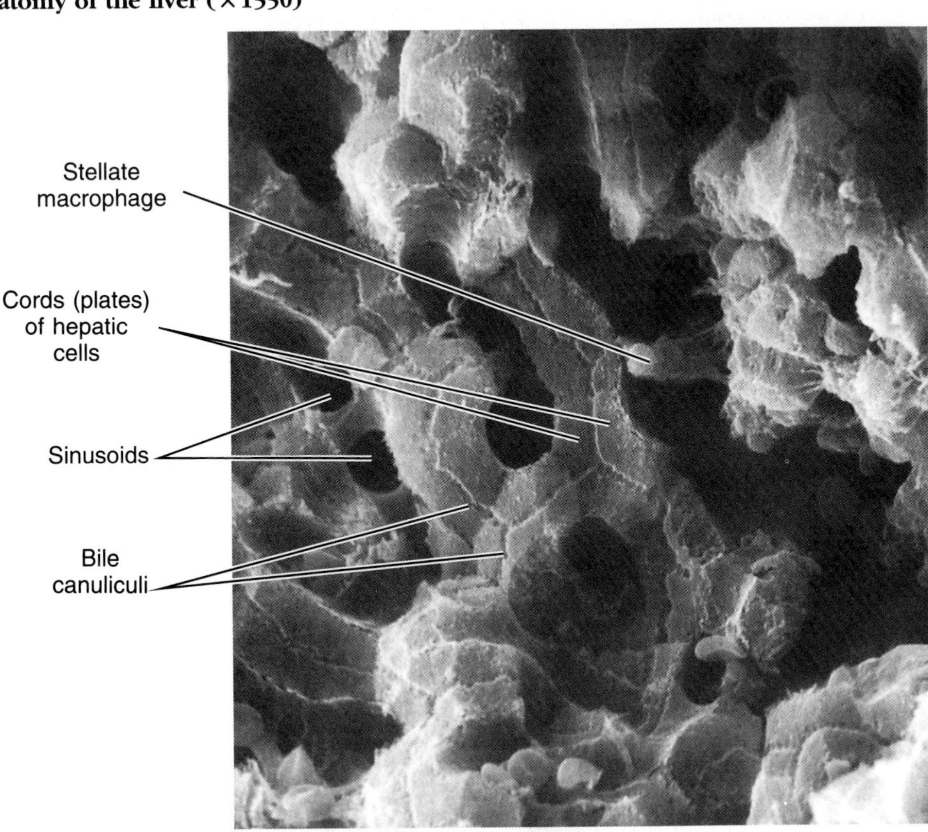

◆ **FIGURE 24.23 The gallbladder and bile ducts**

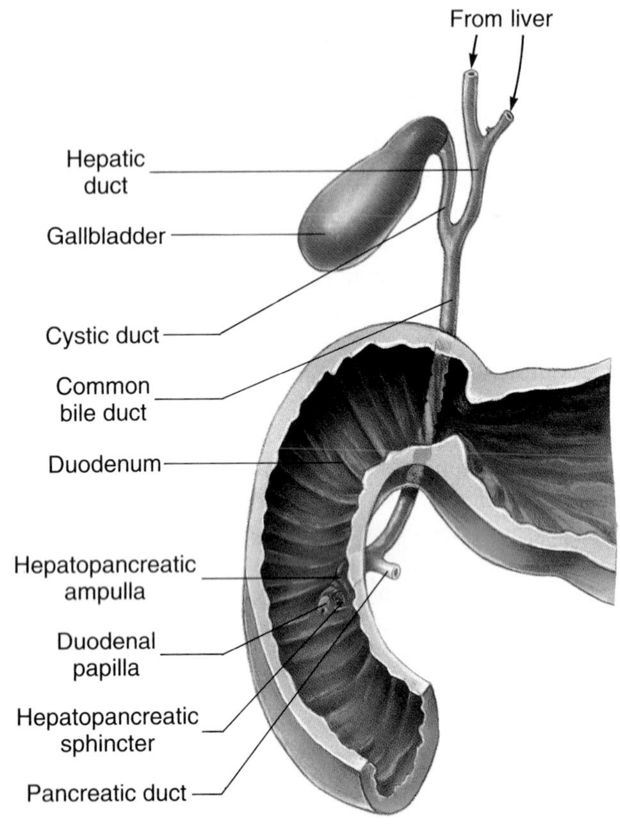

unconstricted regions constrict, thoroughly mixing the contents of the tract.

An important propulsive movement is **peristalsis,** in which the muscles surrounding a portion of the digestive tract undergo a wavelike contraction (Figure 24.25). This contraction produces a ring of constriction that moves along the tract, forcing materials within the tract to move ahead of it. Peristaltic waves normally travel toward the anus, and it has been suggested that this is due to the organization and influence of the intrinsic nerve plexuses.

◆ **FIGURE 24.24 The mixing action of segmentation in the small intestine**

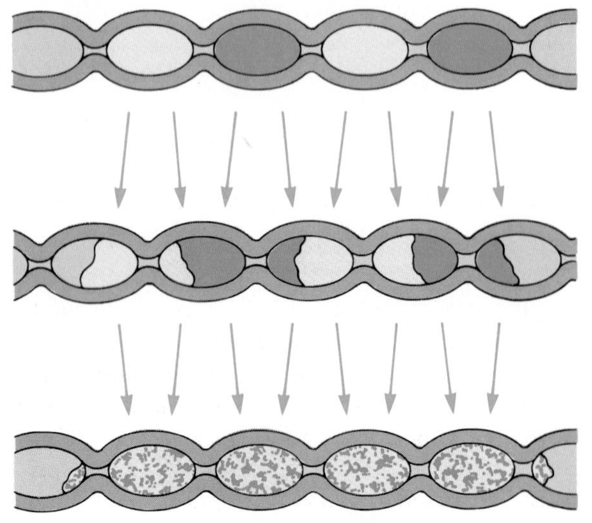

◆ **FIGURE 24.25 The movement of food by peristalsis, such as occurs within the esophagus and the small intestine**

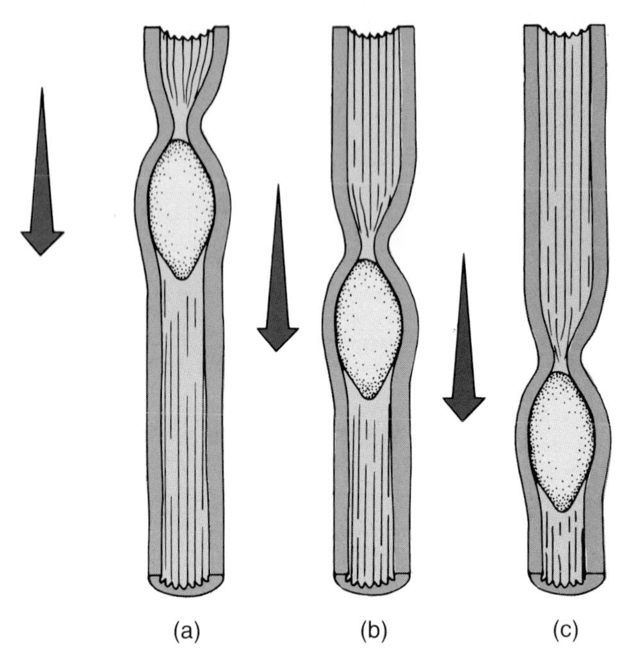

Chewing

Chewing *(mastication)* is accomplished by opening, closing, and lateral movements of the jaws, accompanied by the continual positioning of food between the teeth by the tongue and the muscles of the cheeks. Chewing is part voluntary and part reflex. The reflex aspect of chewing is initiated by tactile stimulation due to a mass of food (a *bolus*) pressing against the teeth, gums, and anterior portion of the roof of the mouth. This stimulation inhibits the muscles that raise the lower jaw. When the lower jaw drops, stretch reflexes are initiated, and the muscles contract and raise the jaw. This action again presses the bolus against the teeth, gums, and anterior portion of the roof of the mouth, and the cycle repeats itself. Chewing reduces ingested materials to smaller, more easily digested pieces, and thoroughly mixes them with the saliva secreted into the mouth by the salivary glands. The mixing of food with saliva moistens the food and enhances its formation into a manageable mass.

◆ **TABLE 24.2 Functions of Components of the Digestive System**

DIGESTIVE ORGAN	MOTILITY	SECRETION	DIGESTION	ABSORPTION
Mouth and salivary glands	Chewing	Saliva Amylase Mucus	Carbohydrate digestion begins	No foodstuffs; a few medications—for example, nitroglycerine
Pharynx and esophagus	Swallowing	Mucus	None	None
Stomach	Receptive relaxation; peristalsis	Gastric juice Hydrochloric acid Pepsin Mucus Intrinsic factor	Carbohydrate digestion continues in stomach; protein digestion begins	No foodstuffs; a few lipid-soluble substances, such as alcohol and aspirin
Exocrine pancreas	Not applicable	Pancreatic digestive enzymes, including Trypsin Chymotrypsin Carboxypeptidase Amylase Lipase Aqueous sodium bicarbonate secretion	Pancreatic enzymes accomplish much digestion in lumen of small intestine	Not applicable
Liver	Not applicable	Bile Bile salts Aqueous sodium bicarbonate secretion Bilirubin	Bile does not digest anything, but bile salts facilitate fat digestion and absorption in small intestine	Not applicable
Small intestine	Segmentation; migrating motility complex	Mucus Intestinal juice (Small intestine enzymes are not secreted but function at microvilli—for example, maltase, sucrase, and aminopeptidases)	In lumen, under influence of pancreatic enzymes and bile, carbohydrate and protein digestion continue and fat digestion occurs; carbohydrate and protein digestion completed at microvilli	All nutrients, most minerals, and water
Large intestine	Haustral contractions; mass movements	Mucus	None	Sodium chloride and water, converting contents to feces

Swallowing

Swallowing moves the bolus along the digestive tract from the mouth to the stomach. Swallowing is initiated by placing the tip of the tongue against the roof of the mouth and forcing the bolus into the pharynx (Figure 24.26). The bolus stimulates pressure receptors around the opening of the pharynx, and this stimulation elicits a complex swallowing reflex, which is coordinated by a swallowing center in the brain stem.

The swallowing reflex generates a peristaltic contraction of pharyngeal constrictor muscles that moves the bolus along the pharynx toward the esophagus. Muscles surrounding the upper portion of the esophagus act as a sphincter (the *upper esophageal sphincter*) that keeps the entrance to the esophagus closed. However, during

◆ **FIGURE 24.26 The movement of a bolus of food through the mouth, the pharynx, and the upper esophagus during swallowing**

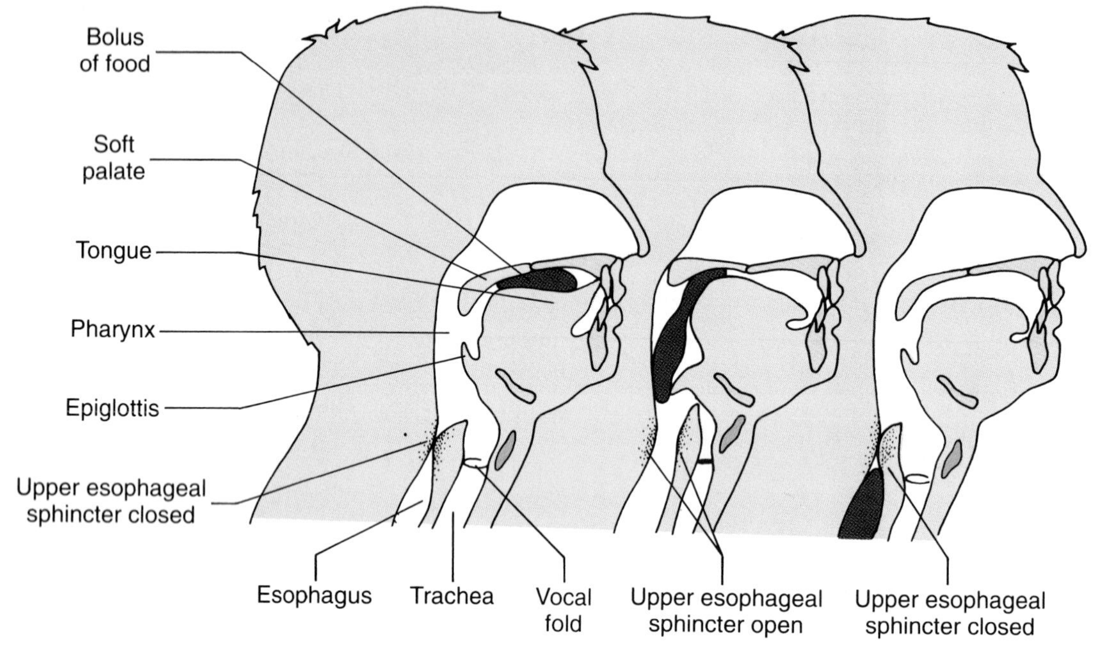

swallowing, the upper esophageal sphincter opens, allowing the bolus to enter the esophagus.

Instead of entering the esophagus, the bolus could go into the larynx or even upward into the nasal cavity. However, food is prevented from entering the larynx during swallowing by the approximation of the vocal folds, which blocks the laryngeal opening. In addition, the larynx is moved upward, and in this position the epiglottis tips backward and covers the laryngeal opening as the bolus passes by. The bolus is prevented from entering the nasal cavity by the elevation of the soft palate against the wall of the pharynx. Sometimes the coordination of these events by the swallowing center fails, and food does enter the larynx or nasal cavity. The usual response to food that enters the larynx is a violent cough, which expels the food back into the pharynx.

The bolus is moved along the esophagus by a primary peristaltic wave—that is, by a peristaltic contraction of the muscle of the tunica muscularis of the esophagus that is basically a continuation of the peristaltic wave that began in the pharynx. About 2–5 cm above the cardiac orifice, the smooth muscle of the esophagus is normally in a state of tonic contraction, and it acts as a physiological sphincter (the *lower esophageal sphincter*) that keeps the lower portion of the esophagus closed. During swallowing, the lower esophageal sphincter opens, allowing the bolus to enter the stomach. Food generally requires between 5 and 10 seconds to move from the mouth to the stomach.

The distension of the esophagus by materials within it stimulates receptors in the esophageal wall and elicits secondary peristaltic waves that are not directly associated with the act of swallowing. These waves, which begin in the region of the upper esophageal sphincter, are of value in continuing the movement of any food material that may not have been moved completely through the esophagus by the primary peristaltic wave.

Gastric Motility

The mechanical activities of the stomach include (1) storing ingested food until it can be utilized by the remainder of the gastrointestinal tract; (2) mixing the food with gastric secretions; and (3) moving the food into the duodenum of the small intestine at a rate consistent with efficient intestinal digestion and absorption.

The empty stomach has a volume of about 50 ml. When food is consumed, the stomach is able to expand to a volume of 1000–1500 ml without a great increase in the pressure within it. This ability is due in part to the ability of gastric smooth muscle to lengthen, when stretched, without a marked increase in tension. It is also due to reflex activity that causes gastric smooth muscle to relax as food is consumed. This activity, which is referred to as *receptive relaxation,* is mediated by the vagus nerves.

As food within the stomach is mixed with gastric juice, the mixture assumes a semifluid consistency and is referred to as *chyme.*

The opening between the stomach and duodenum is surrounded by the *pyloric sphincter,* which is generally

◆ **FIGURE 24.27 Influence of excitatory mechanical, neural, or hormonal inputs on the membrane potentials of gastric smooth muscle cells**

Excitatory mechanical, neural, or hormonal inputs can move the membrane potentials of gastric smooth muscle cells closer to threshold. If the membrane potentials of the cells are close enough to threshold, the basic electrical rhythm can depolarize them to threshold. When the cells are depolarized to threshold, they generate action potentials that trigger their contraction.

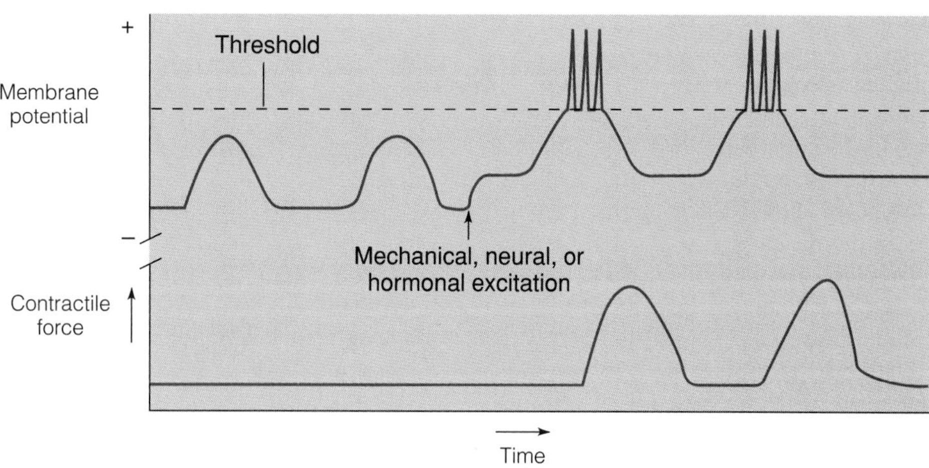

partially open and thus offers only limited resistance to the movement of the stomach contents into the duodenum. However, this resistance is usually sufficient to prevent chyme from entering the duodenum in the absence of gastric muscular activity.

The stomach undergoes peristaltic contractions that begin in the body of the stomach and proceed toward the duodenum. The contractions are relatively weak in the upper region of the stomach, but they increase in strength and speed of transmission in the lower region, where the muscle of the stomach wall is quite thick. A strong peristaltic contraction in the pyloric region of the stomach forces some of the chyme out of the stomach and into the duodenum. In addition, the arrival of a peristaltic wave at the terminal pyloric region constricts this area and forces much of the chyme back toward the body of the stomach. This activity provides an effective mixing action.

Gastric smooth muscle contains pacesetter cells that generate slow-wave potentials (page 270). These potentials spread from cell to cell throughout the muscle, providing it with a **basic electrical rhythm**—that is, with a spontaneous rhythm of depolarization and repolarization—that has a frequency of about three cycles per minute. If the basic electrical rhythm depolarizes gastric smooth muscle cells to threshold, the cells generate action potentials that trigger their contraction.

Mechanical, neural, and hormonal inputs influence the membrane potentials of gastric smooth muscle cells. Some inputs move the membrane potentials closer to threshold, and others move the membrane potentials farther from threshold. If the membrane potentials of the cells are close enough to threshold, the basic electrical rhythm can depolarize them to threshold (Figure 24.27). In general, the closer the membrane potentials of the cells are to threshold, the greater the number of action potentials generated when the basic electrical rhythm depolarizes them to threshold, and the more intense the contraction of the cells. Thus, mechanical, neural, and hormonal inputs are important in determining the ***intensity*** of gastric muscular contractions, and the basic electrical rhythm is important in determining the ***frequency*** of the contractions.

Gastric motility (particularly the intensity of gastric muscular contractions) and the rate of gastric emptying are under the combined influence of a variety of excitatory and inhibitory mechanisms. For example, the volume of material contained within the stomach exerts a moderate influence on gastric motility and the rate of gastric emptying. As the stomach fills and the volume of its contents increases, the stomach wall is distended and mechanoreceptors in the wall are stimulated. The stimulation of these receptors initiates reflexes that enhance gastric motility, cause some relaxation of the pyloric sphincter, and promote gastric emptying. (Both the short- and long-reflex pathways described on page 763 are involved in these reflexes.) In addition, the hormone **gastrin** is released by enteroendocrine cells in the pyloric region in response to distension (and also in response to the presence of partially digested protein and caffeine). Gastrin enters the blood and ultimately returns to the stomach, where it stimulates gastric motility, relaxes the pyloric sphincter, and enhances gastric emptying. In general, the more fluid the contents of the stomach, the more the rate of emptying increases with volume.

The volume and composition of the chyme that enters the duodenum exert a major influence on gastric motility and the rate of gastric emptying. This influence is exerted by way of both neural and hormonal pathways (Figure 24.28).

A reflex called the **enterogastric reflex** (*entero* = intestine) is initiated when the duodenum fills with chyme and its wall is distended. In this reflex, nerve impulses that inhibit gastric motility, produce a slight increase in the contraction of the pyloric sphincter, and slow down gastric emptying are transmitted from the duodenum to the stomach by way of both the short- and long-reflex pathways. Irritants in the duodenum, an excessive acidity of the duodenal chyme (that is, a pH below 3.5), or a relative hypertonicity or hypotonicity of the chyme also activates the enterogastric reflex and inhibits gastric motility and emptying.

The small intestine releases hormones collectively referred to as **enterogastrones,** which have inhibitory effects on stomach activity. These hormones are released in response to factors such as the acidity of the duodenal chyme and the presence of fat and certain amino acids and fatty acids in the chyme. The hormones enter the blood and are carried to the stomach, where they inhibit gastric motility and slow down gastric emptying. The precise identification of the enterogastrones is not yet complete, but they apparently include the intestinal hormone **secretin.** The intestinal hormones **cholecystokinin** and **gastric inhibitory peptide (glucose-dependent insulinotropic peptide)** may also function as enterogastrones.

The neural and hormonal duodenal mechanisms that inhibit stomach emptying tend to prevent additional chyme from entering the small intestine until the chyme already present has been effectively processed. Thus, the rate at which the stomach contents enter the small intestine is largely regulated by the small intestine itself.

Motility of the Small Intestine

Segmentation is a significant movement occurring in the small intestine. Although little or no segmentation occurs between meals, it is the most common movement of the small intestine during a meal and is quite vigorous immediately after a meal. Segmentation mixes the chyme thoroughly with the digestive juices and exposes different portions of the chyme to the intestinal mucosa.

Like its gastric counterpart, the smooth muscle of the small intestine exhibits a basic electrical rhythm that is important in determining the frequency of segmentation contractions. However, the frequency of this rhythm varies in different regions of the small intestine. The frequency is highest—about 11 or 12 cycles per minute—in the duodenum, and it decreases progressively along the length of the intestine to about 6 or 7 cycles per minute in the terminal ileum. Because of this progressive decrease in the frequency of the basic electrical rhythm, segmentation occurs with greater frequency in the upper part of the small intestine than in the lower part. Consequently, segmentation pushes more chyme toward the large intestine than it pushes back toward the stomach. Thus, in addition to mixing the chyme, segmentation slowly moves the chyme along the small intestine.

The intensity of segmentation contractions can be influenced by mechanical, neural, and hormonal inputs. For example, the distension of the intestine by chyme and parasympathetic neural activity increase the contractile force, and sympathetic neural activity decreases it.

After most of a meal has been absorbed, a pattern of intestinal contraction known as the **migrating motility complex** occurs. This complex consists of a series of weak peristaltic waves, each of which dies out after traveling only a short distance. The initial waves begin at the duodenum, and successive waves begin farther and farther along the small intestine. Over a period of 100–150 minutes, the site of origin of the peristaltic waves gradually migrates from the duodenum to the terminal ileum of the small intestine. When the terminal ileum is reached, peristaltic waves begin again at the duodenum, and the process repeats itself until the next meal when the migrating motility complex ceases and is replaced by segmentation. The peristaltic contractions of the migrating motility complex sweep the contents of the small intestine toward the large intestine.

The *ileocecal sphincter* is located at the end of the ileum near its junction with the cecum of the large intestine. The distension of the cecum initiates a reflex, by way of the intrinsic nerve plexuses, that stimulates the muscle of the ileocecal sphincter, and most of the time the sphincter is at least mildly constricted. The constriction of the sphincter facilitates absorption in the small intestine by slowing the movement of the chyme from the ileum into the large intestine.

When food enters the stomach, a reflex called the *gastroileal reflex* intensifies ileal contractions. In addition, the hormone gastrin, which is released by cells in the pyloric region of the stomach, increases ileal motility and relaxes the ileocecal sphincter. These events enhance the movement of the chyme from the ileum into the large intestine.

The *ileocecal valve,* which is composed of two folds of tissue, is located at the opening of the ileum into the cecum of the large intestine. The folds of tissue are pushed apart and the valve opens when the contents of the ileum move into the cecum, but the folds are forced together and the valve closes when the contents of the

◆ **FIGURE 24.28 Diagrammatic representation of the neural (enterogastric reflex) and hormonal pathways by which the volume and composition of the chyme that enters the duodenum can inhibit stomach motility**

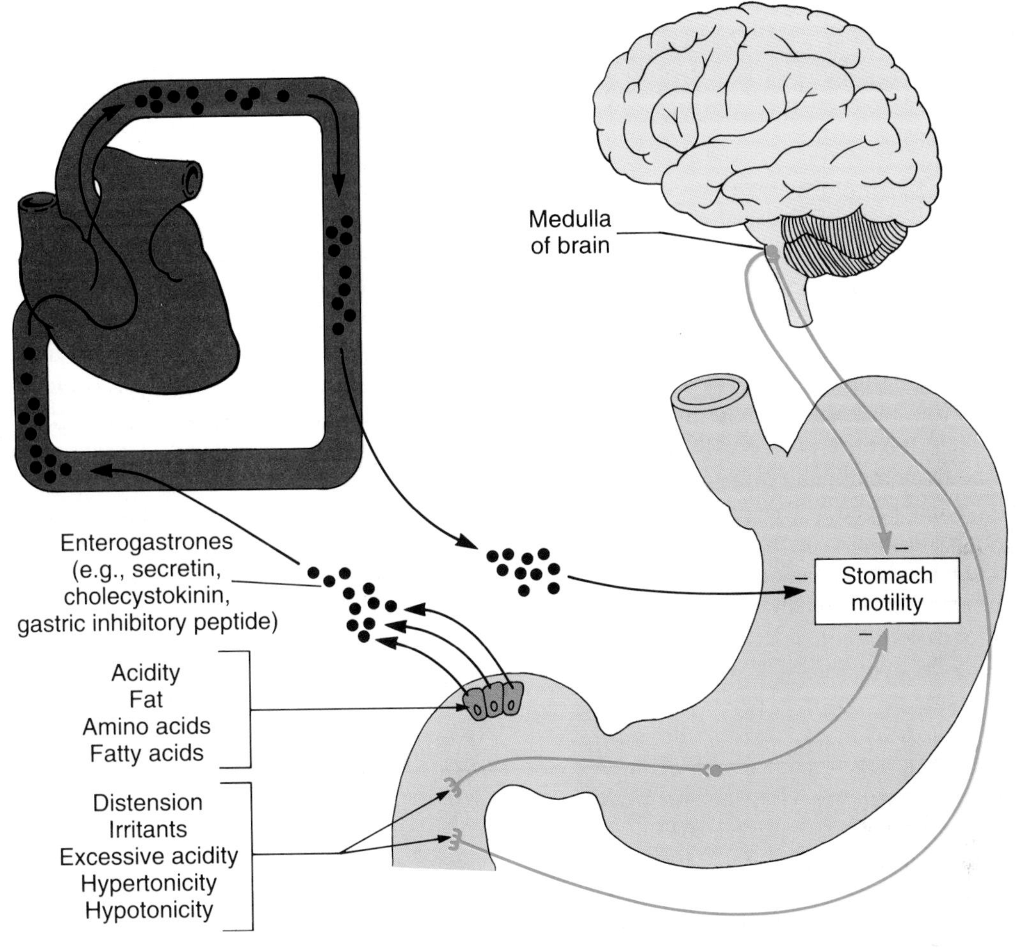

cecum attempt to move backward into the ileum. Thus, the ileocecal valve helps prevent regurgitation of the contents of the cecum back into the ileum.

Motility of the Large Intestine

The movements of the large intestine are generally quite sluggish, and it frequently takes 18–24 hours for material to pass through the colon. This slow movement allows time for bacterial growth, and many microorganisms inhabit the large intestine. However, these organisms normally cause no harm.

Mixing movements of a segmental type occur within the colon at a much lower frequency than segmentation within the small intestine. These movements, which are called *haustral contractions,* may occur only once every 30 minutes.

At infrequent intervals, perhaps three or four times per day, substantial segments of the colon undergo several strong contractions, called *mass movements,* that propel material along the colon for considerable distances. Mass movements often occur following a meal, indicating that the presence of food within the stomach and the duodenum activates *gastrocolic* and *duodenocolic* reflexes, which increase colon motility.

The distension of the wall of the rectum by a mass movement of fecal material initiates the *defecation reflex,* which tends to move material out of the lower colon and rectum. Nerve impulses associated with the defecation reflex are transmitted by way of the intrinsic nerve plexuses, and these impulses are strongly reinforced by impulses that travel to and from the sacral region of the spinal cord. The defecation reflex stimulates contractions of the sigmoid colon and rectum. It also causes the relaxation of the internal anal sphincter, which surrounds the anal opening. This combination of contraction and relaxation tends to propel the feces through the anus. However, a second sphincter, the external anal sphincter, also surrounds the anal opening.

CLINICAL CORRELATION

Spastic Colon

Case Report

THE PATIENT: A 43-year-old woman.

PRINCIPAL COMPLAINT: Abnormal bowel movements.

HISTORY: The patient related that her feces had been hard and pencil-thin and that she had experienced some crampy discomfort in the upper left quadrant of the abdomen. These conditions were not clearly related to meals and were partially relieved by bowel movements. The patient also expressed fear that she had cancer of the colon. She remembered her bowel movements as being normal until she entered high school, at which age she began to notice occasional constipation, and her mother shared a laxative with her. When the patient went to college, she began having episodes of explosive diarrhea, with bowel movements three or four times a day, but not severe enough at night to disturb her sleep. She married shortly after graduating from college and again developed constipation, passing feces in small, rock-hard segments. After the birth of her second child, the patient began to experience aching in the lower abdomen. At age 28 she had a hysterectomy (removal of the uterus). Her diarrhea and constipation had continued to alternate, with cycles of a few weeks to a few months. She developed dyspeptic symptoms (symptoms of disturbed digestion) in the upper right quadrant of the abdomen, particularly after fatty meals. Subsequently, at age 34, she had a cholecystectomy (removal of the gallbladder).

CLINICAL EXAMINATION: The patient appeared healthy, mildly obese, and slightly anxious. The results of the examination, which included proctoscopy (inspection of the anal canal and rectum) and barium enema (in which barium sulfate is introduced into the colon as a contrast medium for X-ray pictures), were normal. A diagnosis of irritable bowel syndrome, or "spastic colon," aggravated by psychological stress, was reached.

COMMENT: Spastic colon is an intermittent condition that may occur even in psychologically healthy persons at times of stress, such as school examinations, marriage, divorce, and bereavement, but the precise mechanisms are not known. With this patient, there was an abnormally close association between her emotions and the secretory and motor functions of her colon. Long-continued, excessive activity of the colonic muscle usually retards the passage of the contents, causing them to be excessively dry and hard. In diarrhea, the colon secretes more and contracts less; hence, the contents move faster.

OUTCOME: The patient was instructed to take a diet high in bran and other fiber, and her condition improved markedly. Unfortunately, not all patients who suffer from "spastic colon" are relieved by this kind of diet.

The external anal sphincter is composed of skeletal muscle and can be voluntarily controlled. If the external anal sphincter is voluntarily relaxed, defecation will occur. If the external anal sphincter is voluntarily contracted, defecation can be prevented, and the defecation reflex usually dies out after a few minutes. However, the reflex generally returns in several hours, when another mass movement propels additional fecal material into the rectum.

The defecation reflex is usually assisted through voluntary efforts. This assistance is accomplished by taking a deep inspiration and then closing the glottis and contracting the abdominal muscles. These actions raise the pressure within the abdomen, which aids defecation by increasing the pressure on the contents of the large intestine.

Secretion

In most regions, the lining of the gastrointestinal tract contains cells that secrete mucus. In addition, many of the digestive enzymes and other secretory products of the digestive system are produced and released by specialized epithelial cells that are organized into exocrine secretory glands (Figure 24.29). These exocrine glands release their products by way of ducts onto the surfaces of the gastrointestinal tract.

◆ **FIGURE 24.29 Diagrammatic representation of an exocrine gland**

Raw materials obtained from the blood are transformed by the gland cells into a primary secretion that is released into the glandular lumen. As the primary secretion moves along the glandular duct, active transport or diffusion across the duct walls may alter its composition.

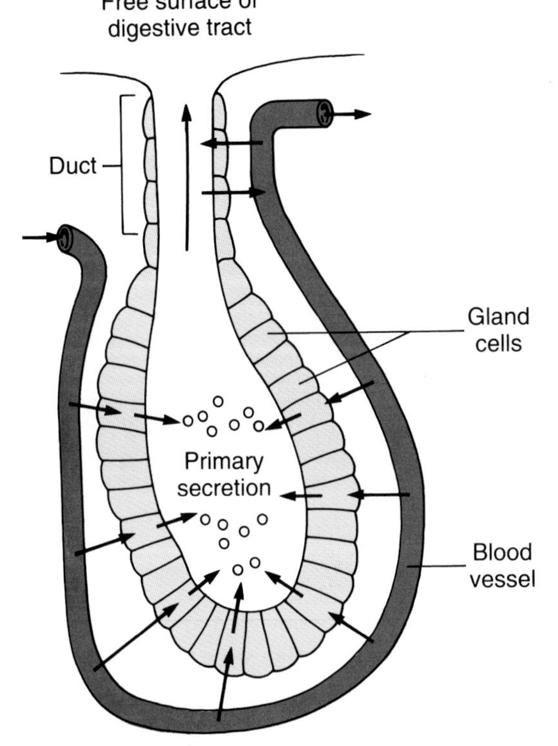

Mouth and Saliva

Three pairs of large salivary glands—the *parotid, sublingual,* and *submandibular (submaxillary) glands*—produce most of the saliva. The output of these glands is supplemented slightly by smaller buccal glands. Approximately two-thirds of the 1-1.5 liters of saliva produced per day by an average adult comes from the submandibular glands. Approximately one-quarter comes from the parotid glands, and the remainder is produced by the other glands.

Secretions of the Salivary Glands. The salivary glands produce two different secretions: (1) a mucous secretion that contains mucins; and (2) a serous secretion that contains the enzyme salivary amylase (ptyalin). The parotid glands produce serous secretions, the submandibular glands produce both mucous and serous secretions, and the sublingual and buccal glands primarily produce mucous secretions.

Mucins. The **mucins** are the major proteins of the saliva, and they have large polysaccharides attached to them. When mixed with water, the mucins form a highly viscous solution known as mucus, which lubricates the mouth and food.

Salivary Amylase. **Salivary amylase** splits starch molecules, which are composed of thousands of glucose units, into smaller fragments. Like other enzymes, salivary amylase has an optimal pH at which it functions best (6.9), but it is stable at pH values from 4 to 11. After food has been swallowed, the digestive action of salivary amylase continues in the stomach until the enzyme is inactivated by the acidic gastric juices.

Composition of Saliva. The exact composition of saliva depends on the glands from which it comes, on the secretion rate, and to some degree on the stimulus that evokes secretion. Generally, saliva is 97-99.5% water, and the pH of mixed saliva is usually from 6.0 to 7.0. Sodium, potassium, chloride, and bicarbonate ions are some of the main electrolytes of saliva. Saliva also contains kallikrein, as well as specific soluble blood-group substances of the sort responsible for blood type. Kallikrein acts enzymatically on plasma protein precursors to produce bradykinin, which is a vasodilator substance. The release of kallikrein from active salivary glands and the subsequent production of bradykinin are believed to increase blood flow to the glands by promoting local vasodilation.

Functions of Saliva. In addition to its lubricating and digestive functions, saliva assists in bolus formation and in swallowing by moistening food particles and holding them together. It also helps dissolve food so the food can be tasted. Saliva aids in speech by moistening the mouth and throat, and it has bacteriostatic properties.

Control of Salivary Secretion. Salivary secretion occurs in response to nerve impulses. Salivatory nuclei in the medulla-pons region of the brain receive impulses from the mouth and pharynx, as well as from higher brain centers. Nerve impulses travel to the salivary glands by way of the autonomic nervous system. Both parasympathetic and sympathetic stimulation cause salivary secretion, with parasympathetic stimulation having the greatest effect. The flow of saliva is enhanced by the presence of food in the mouth, as well as by the odor of food, the sight of food, the thought of food, or the presence of irritating food in the stomach and upper intestine. Chewing stimulates salivary secretion,

◆ **FIGURE 24.30 Potential mechanism for the secretion of hydrochloric acid (H^+Cl^-) by parietal cells**

Active transport is indicated by reddish colored circles.

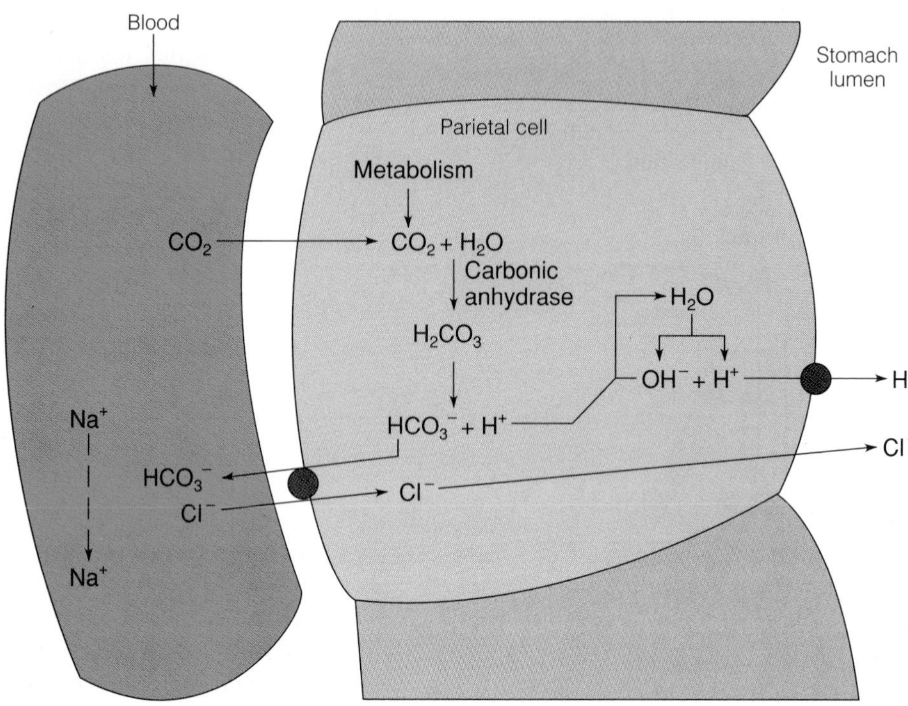

whereas intense mental effort, dehydration, fear, and anxiety tend to reduce it.

Stomach and Gastric Secretions

The gastric secretions include mucus, hydrochloric acid, and pepsinogen, which is a precursor of the protein-digesting enzyme pepsin.

Mucus. Cells that secrete a viscous, alkaline mucus line the surface of the stomach. The mucus, which adheres to the stomach walls in a layer 1–1.5 mm thick, lubricates the walls and protects the gastric mucosa. Even the slightest irritation of the mucosa directly stimulates these cells to secrete large amounts of mucus.

Glands that secrete mainly a thin mucus are present in the region of the cardiac orifice of the stomach (cardiac glands) and in the pyloric region (pyloric glands). Gastric glands in the body and fundus of the stomach produce some mucus as well as additional secretions.

Hydrochloric Acid. Cells in the gastric glands, called *parietal (oxyntic) cells,* produce hydrochloric acid, which is a strong acid that dissociates into hydrogen ions and chloride ions. The hydrochloric acid facilitates protein digestion in the stomach and also kills many of the bacteria that enter the digestive tract with the food. The parietal cells produce a hydrochloric acid solution that is about 0.16 molar (pH 0.8). Actual stomach acidity, however, depends on the rate of hydrochloric acid secretion as well as on the rate of neutralization and dilution of the acid by ingested food and by other secretions.

The precise details by which the parietal cells produce hydrochloric acid are not clear. One theory proposes that hydrogen ions from dissociated water molecules are actively transported from the interior of the parietal cells to the stomach lumen, leaving behind hydroxide ions (OH^-) (Figure 24.30). Carbon dioxide from the plasma diffuses into the cells, where it (and also carbon dioxide from cellular metabolism) combines with water to form carbonic acid. This reaction is facilitated by the enzyme carbonic anhydrase. Carbonic acid can dissociate into hydrogen and bicarbonate ions. Thus, hydrogen, bicarbonate, and hydroxide ions are present in the cells. The hydrogen ions and hydroxide ions combine to form water. An active-transport process moves the bicarbonate ions out of the cells into the blood in exchange for chloride ions. The chloride ions, in turn, leave the cells and enter the stomach lumen. As a result of the basic bicarbonate ions entering the blood, the pH of the venous blood leaving the actively secreting stomach is higher than that of the arterial blood flowing to the stomach.

Pepsinogen. Cells in the gastric glands of the stomach called *zymogenic (chief) cells* produce pepsinogen, which is a precursor of the active enzyme pepsin. In the acidic environment of the stomach, pepsinogen is spontaneously converted to pepsin. When the pH is below 5, pepsinogen undergoes a conformational change that exposes its active site. The exposed active site splits away a precursor segment from the pepsinogen to form pepsin.

Pepsin, which is optimally active at acid pHs, digests proteins by breaking peptide bonds that involve certain amino acids, such as tryptophan, phenylalanine, and tyrosine. This activity results in smaller *peptide chains* of amino acids. The production of peptide chains is about as far as protein digestion proceeds in the stomach.

Control of Gastric Secretion. The stomach glands produce as much as 2 to 3 liters of secretions (gastric juice) per day. Diet affects the amount of juice secreted, and a typical meal can result in the secretion of up to 700 ml of gastric juice.

Neural and hormonal mechanisms are involved in the control of gastric secretion. Neural mechanisms include parasympathetic impulses transmitted from the brain to the stomach by way of the vagus nerves. These impulses increase both pepsin and acid secretion and also cause some increase in mucus secretion. Hormonal mechanisms include the activity of the hormone gastrin, which stimulates gastric secretion, particularly the release of hydrochloric acid. The neural and hormonal mechanisms of control are interrelated because vagal activity also causes gastrin release.

The control of gastric secretion can be divided into three regulatory phases: the cephalic phase, the gastric phase, and the intestinal phase.

Cephalic Phase. The cephalic phase of gastric secretion is due to signals that originate in the head region. The sight, smell, or taste of food can stimulate gastric secretion. This response is due to nerve impulses that travel from sensory receptors to the central nervous system and then by way of the vagus nerves to the stomach.

The secretory response elicited by the sight or smell of food is basically a conditioned reflex that is not elicited when a person is afraid, depressed, or has no desire for food (no appetite). If the vagus nerves to the stomach are severed, secretion due to the cephalic phase ceases.

Gastric Phase. The gastric phase of gastric secretion is due to signals that originate in the stomach. The distension of the stomach by the presence of food stimulates the flow of gastric juice and also potentiates its acid and pepsin content (Figure 24.31). One way this response occurs is as a result of local short reflexes that stimulate gastric secretion. The distension of the stomach also results in a reflex in which impulses are sent to the brain stem and then back to the stomach by way of the vagus nerves to stimulate the flow of gastric juice (and the release of gastrin). In addition, local responses to the distension of the pyloric region of the stomach or the presence of substances called secretagogues—for example, partially digested protein and caffeine—lead to the release of gastrin.

Gastrin release is inhibited by high concentrations of hydrogen ions (acid) in the stomach. This activity provides a negative-feedback mechanism that helps prevent excessive stomach acidity and helps maintain an optimal pH for the function of the peptic enzymes. When the pH of the stomach contents reaches 2.0, the gastrin mechanism for the stimulation of gastric secretion becomes totally blocked.

Intestinal Phase. The intestinal phase of gastric secretion is due to signals that originate in the small intestine. The intestinal phase has two components: a weak excitatory component and a much stronger inhibitory component.

1. The excitatory component provides a slight stimulation of gastric secretion. This component may be evident when chyme from the stomach first begins to enter the small intestine.
2. The inhibitory component inhibits gastric secretion. This component becomes evident as chyme from the stomach continues to move into the small intestine. The inhibitory component is the major component of the intestinal phase, and it dominates the excitatory component.

The excitatory component is believed to be due mainly to the production of small amounts of a hormone called *intestinal gastrin* by the duodenal mucosa in response to the presence of partially digested protein in the chyme that first enters the small intestine. However, other hormones and reflexes are probably also involved.

The inhibitory component is due to both neural and hormonal mechanisms (Figure 24.32). The distension of the duodenum, irritants in the duodenum, an excessive acidity of the duodenal chyme, or a relative hypertonicity or hypotonicity of the chyme leads to a neural inhibition of gastric secretion by way of the enterogastric reflex. In this reflex, nerve impulses that inhibit gastric secretion are transmitted from the duodenum to the stomach by way of both short- and long-reflex pathways.

In addition, factors such as the acidity of the duodenal chyme and the presence of fat and certain amino acids and fatty acids in the chyme lead to the release of enterogastrones, which enter the blood and are carried to

◆ **FIGURE 24.31 Gastric-phase pathways that result in increased secretion by the stomach**

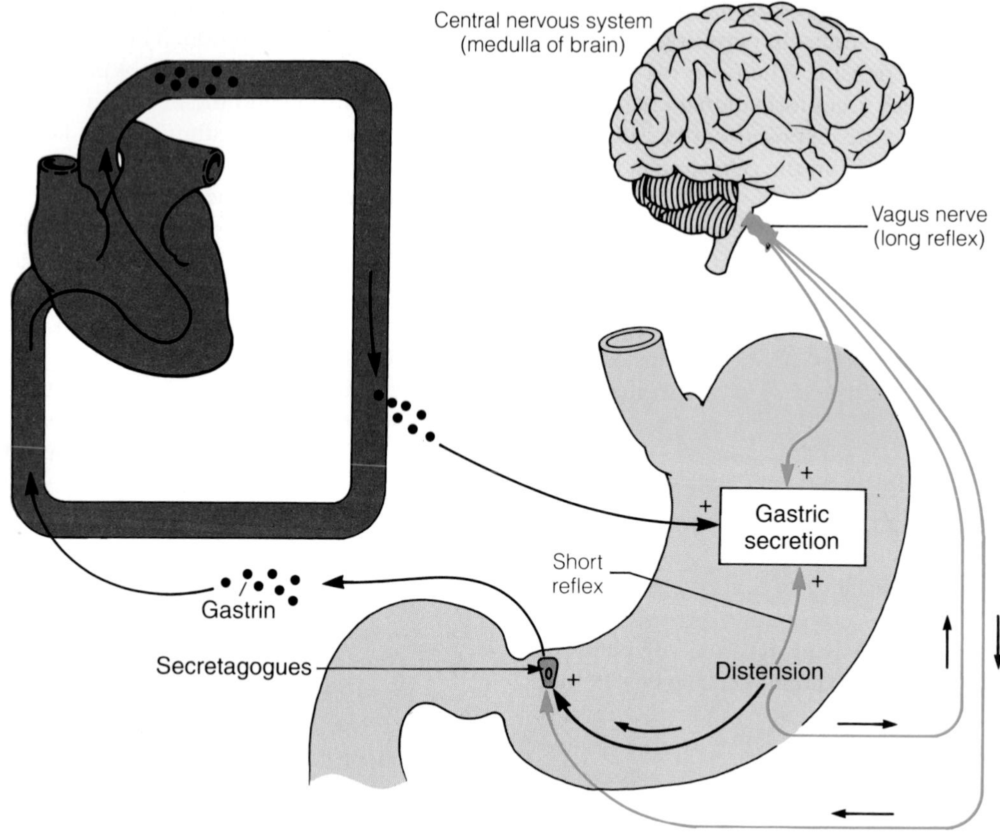

the stomach. The enterogastrones have an inhibitory effect on gastric secretion. (Recall that the enterogastric reflex and enterogastrones also have inhibitory effects on gastric motility [page 780].)

Protection of the Stomach. The mucosal surface of the stomach is composed of epithelial cells that are connected by tight junctions, and the stomach is believed to produce surface-active phospholipids that form an adsorbed hydrophobic layer between the gastric epithelium and the stomach contents. Moreover, the surface of the stomach is coated with a viscous, alkaline mucus. Together, these factors provide a protective barrier that prevents the hydrochloric acid and enzymes released into the stomach from digesting the stomach itself. In addition, damaged stomach cells are rapidly replaced, and the lining of the stomach is renewed about every three days.

Secretions Present in the Small Intestine

The secretions present in the small intestine include mucus, intestinal juice, pancreatic juice, and bile.

Mucus. The first part of the duodenum contains *duodenal glands (Brunner's glands),* which secrete mucus. The mucus provides a protective coat for the intestinal mucosa. The duodenal glands produce mucus in response to tactile or irritating stimulation of the mucosa, vagal stimulation, and intestinal hormones, particularly secretin. Cells called *goblet cells* that are located on the surface of the intestinal mucosa also produce mucus in response to tactile or chemical stimulation of the mucosa. The intestinal glands (crypts of Lieberkuhn) also contain mucus-producing goblet cells whose secretory activity is probably controlled by local nervous reflexes.

Intestinal Juice. Intestinal glands, which are located over essentially the entire surface of the small intestine, produce an intestinal juice that has a pH of 6.5–7.5 and is isotonic with the plasma. The main stimuli for the production of intestinal juice are local reflexes initiated by occurrences such as tactile or irritating stimulation of the intestinal mucosa. Hormones such as secretin and cholecystokinin also stimulate the secretion of intestinal juice, but these hormones are not considered to be major controlling factors.

◆ **FIGURE 24.32 Intestinal-phase factors that inhibit gastric secretion (particularly hydrochloric acid release)**

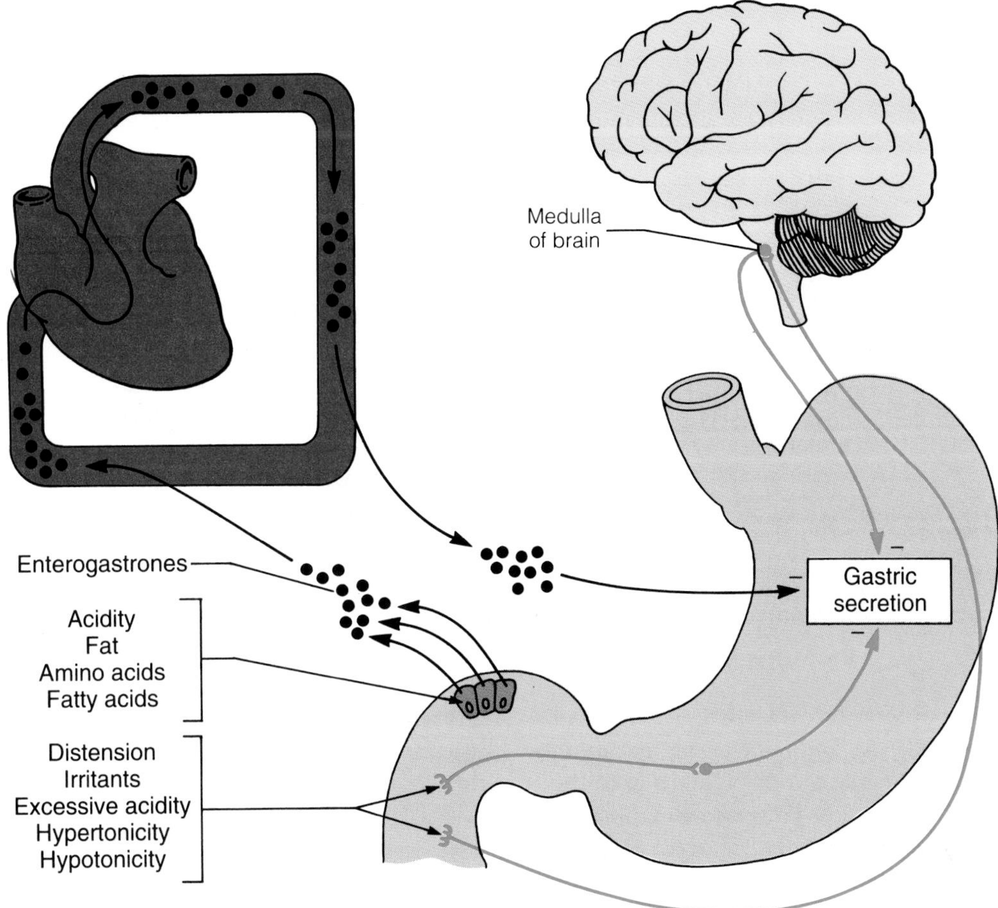

Epithelial cells of the small intestine synthesize digestive enzymes, but the enzymes are not secreted into the intestinal lumen. Instead, the enzymes are associated with the membranes of the microvilli of the epithelial cells. Among these enzymes are several different disaccharidases that participate in carbohydrate digestion and a number of peptidases that are involved in protein digestion.

Pancreatic Juice. The pancreas has a dual role as both an exocrine gland that produces digestive enzymes and an endocrine gland that produces hormones (Figure 24.33). The exocrine portion of the pancreas is composed of acinar cells and collecting ducts. The ducts from the pancreas join and open into the duodenal portion of the intestine, a short distance below the pyloric sphincter.

The exocrine portion of the pancreas produces an aqueous, isotonic fluid with a high bicarbonate ion concentration and a basic pH (about 8.0). In the intestine, this fluid, particularly the bicarbonate ions, helps neutralize the acidic chyme from the stomach. The exocrine pancreas also produces carbohydrate-, protein-, and lipid-digesting enzymes, as well as ribonuclease and deoxyribonuclease, which are enzymes that digest RNA and DNA, respectively. The combined secretions of the exocrine portion of the pancreas make up the pancreatic juice.

The exocrine secretory activities of the pancreas are controlled both hormonally and neurally (Figure 24.34). The intestinal hormone secretin, which is released primarily in response to the presence of acid, stimulates the release from the pancreas of a watery fluid that contains large amounts of bicarbonate ions. A second intestinal hormone, cholecystokinin, which is released principally in response to the presence of certain amino acids and fatty acids, mainly promotes the release of digestive enzymes from the pancreas.

Neurally, pancreatic secretion is stimulated by way of the vagus nerves, and the effect is mostly on enzymatic secretion. This response is most evident during the cephalic and gastric phases of stomach secretion.

◆ **FIGURE 24.33 Diagrammatic representation of the pancreas**

The glandular areas have been greatly enlarged.

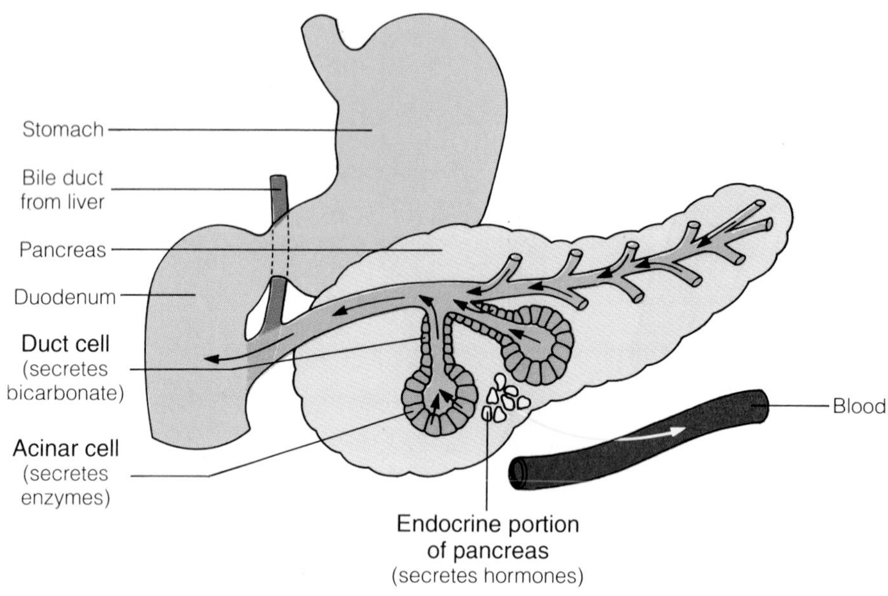

Bile. The secretory and excretory activities of the liver continually produce bile in amounts averaging about 600–1000 ml per day. The bile travels by ducts from the liver, where it is produced, and the gallbladder, where it is stored, to the duodenum. Bile is an aqueous solution of water and electrolytes such as sodium and bicarbonate that contains bile salts (of cholic and chenodeoxycholic acids), bile pigments (such as bilirubin from the breakdown of hemoglobin), cholesterol, neutral fats, and lecithin. Approximately 94% of the bile salts released into the duodenum are reabsorbed in the ileum. They are then returned to the liver by the blood and resecreted. This cycle is known as *enterohepatic circulation.*

The rate of bile secretion is chemically, hormonally, and neurally controlled (Figure 24.35). Chemically, bile salts present in the plasma as the result of enterohepatic circulation stimulate the secretion of additional bile salts. The hormone secretin increases the secretion of bile, but it enhances primarily water and sodium bicarbonate secretion and not bile-salt secretion. Neurally, parasympathetic impulses transmitted by way of the vagus nerves can increase the rate of bile secretion.

During periods when large quantities of bile are not required for digestion, bile is stored in the gallbladder, which has an approximate capacity of 40–70 ml. Within the gallbladder, water and electrolytes are rapidly absorbed, and the concentration of bile salts and pigments may increase five to ten times. Shortly after the ingestion of a meal, the gallbladder contracts, and bile is released into the duodenum. The contraction of the gallbladder is stimulated primarily by the hormone cholecystokinin. Parasympathetic impulses transmitted by way of the vagus nerves can also cause weak contractions of the gallbladder. A summary of the activities of the gastrointestinal hormones (gastrin, secretin, cholecystokinin, and gastric inhibitory peptide) is presented in Table 24.3.

Mucus Secretion in the Large Intestine

The mucosa of the large intestine possesses abundant mucus-secreting goblet cells. Moreover, intestinal glands within the large intestine consist almost entirely of goblet cells.

The rate of mucus production within the large intestine increases in response to direct tactile stimulation of the surface goblet cells and as a result of local reflexes that involve the goblet cells of the intestinal glands. Extrinsic innervation also influences mucus secretion.

Chemical Digestion and Absorption

Digestive enzymes catalyze the breakdown of large food molecules into smaller molecules that can be absorbed from the gastrointestinal tract. These enzymes function mainly by hydrolysis. That is, they split large molecules into smaller ones by introducing water into the molecular structures (Figure 24.36). In this fashion, large carbohydrate molecules such as starches are digested to monosaccharides such as glucose; proteins are degraded into their constituent amino acids; and triglycerides are broken down primarily into monoglycerides and free fatty acids. The smaller molecules that result from chemical digestion—and also water, vitamins, and minerals—

◆ **FIGURE 24.34 Schematic representation showing the regulation of pancreatic secretion**

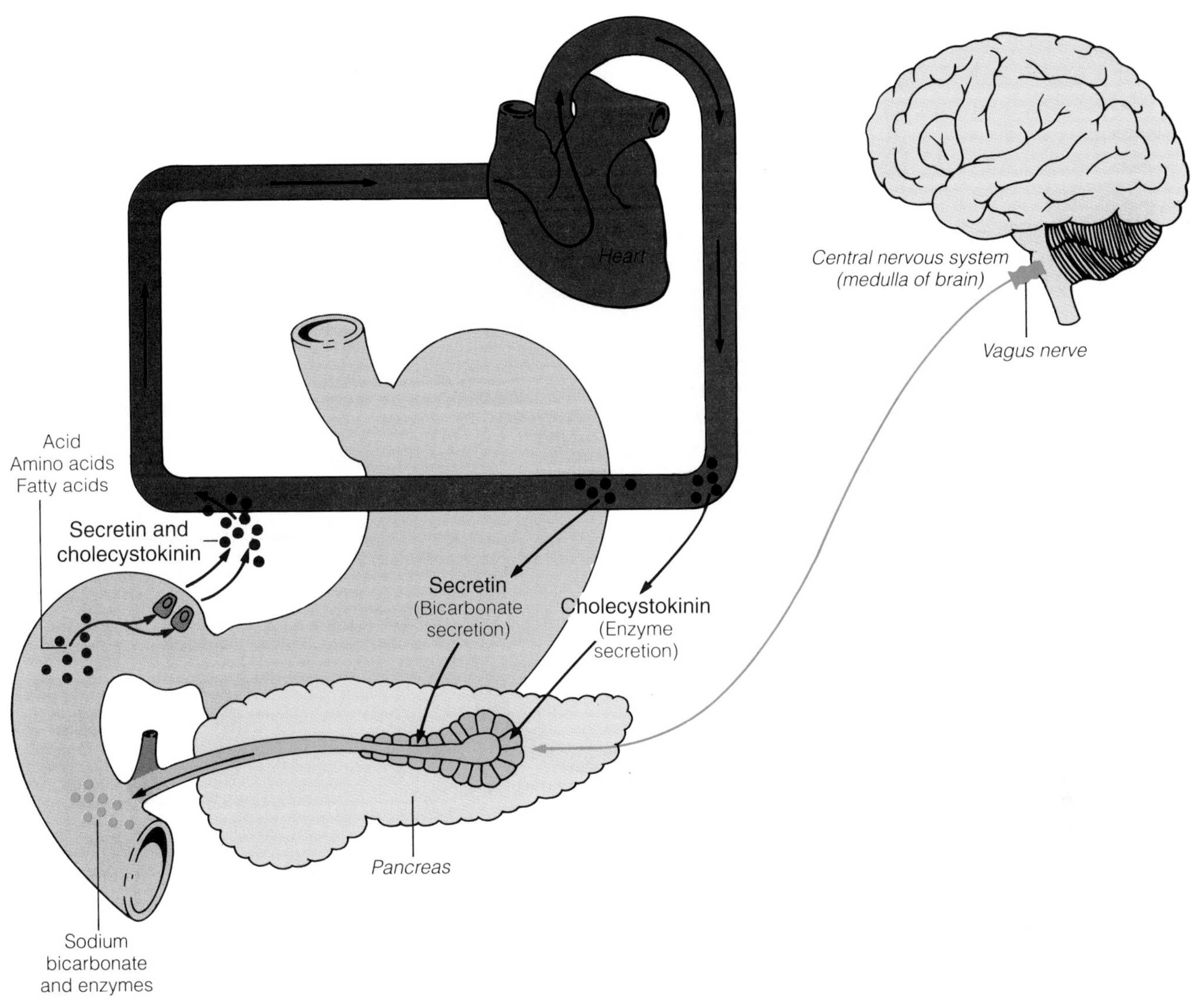

are absorbed from the lumen of the gastrointestinal tract into the blood or lymph for transport to the body's cells.

The most important portion of the digestive system as far as chemical digestion and absorption are concerned is the small intestine and its associated structures. Food that arrives at the small intestine has not been completely digested and is not yet prepared for absorption. Some carbohydrate digestion is begun by salivary amylase, and the protein-digesting enzymes of the stomach produce peptide chains of amino acids, but carbohydrate and protein digestion remain incomplete. Lipids that arrive at the small intestine are largely undigested. Thus, most digestive activity occurs within the small intestine, as does practically all the absorption.

Carbohydrates

Plant starches and sucrose are the major dietary carbohydrates. Small amounts of lactose (milk sugar) may also be present.

Amylases in both the saliva (salivary amylase) and the pancreatic juice (pancreatic amylase) break starch into alpha-dextrin, maltotriose, and maltose (Table 24.4, page 793).

Alpha-dextrin is composed of several monosaccharide glucose units. Maltotriose is composed of three glucose units, and maltose is made up of two glucose units. Sucrose is a disaccharide composed of the monosaccharides glucose and fructose. Lactose is a disaccharide made up of glucose and galactose. The final digestion of these substances to absorbable monosaccharide units is

◆ **FIGURE 24.35 Regulatory pathways for bile secretion and bile release from the gallbladder**

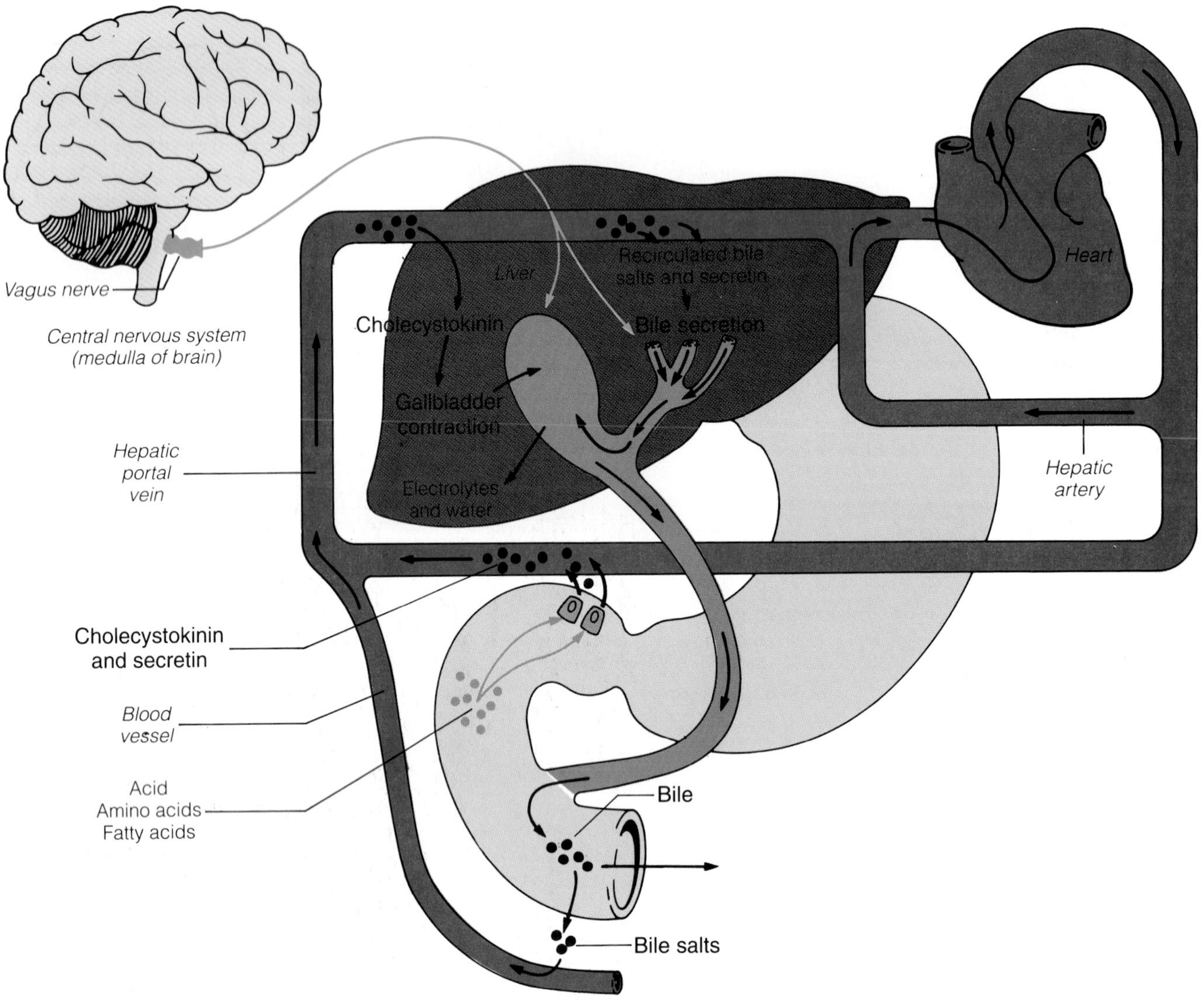

completed by enzymes of the microvilli of the small intestine. These enzymes include alpha-dextrinase, maltase, sucrase, and lactase.

Sugars are absorbed primarily in the duodenum and upper jejunum of the small intestine, and their absorption is practically complete by the time the chyme

◆ **FIGURE 24.36 Hydrolytic cleavage of an amino acid chain to liberate a single amino acid**
The peptide bond is broken by the introduction of water.

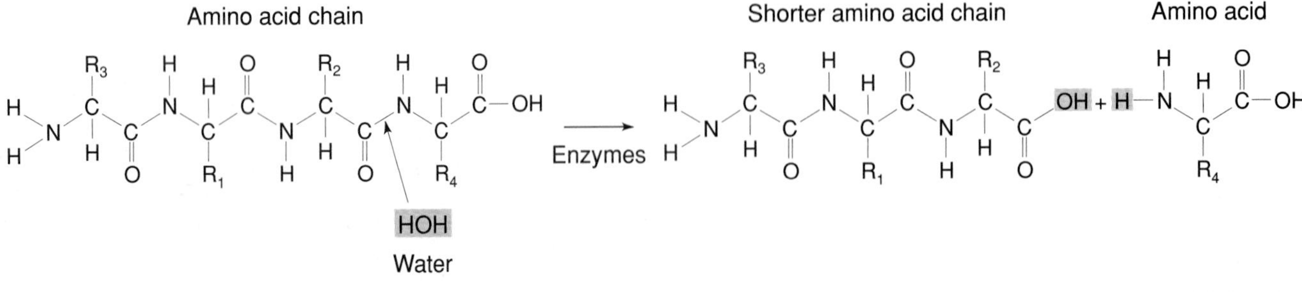

ASPECTS OF EXERCISE PHYSIOLOGY

Pregame Meal: What's In and What's Out?

◆

Many studies have been done to determine the effect of the pregame meal on athletic performance. Although substances such as caffeine have been shown to improve endurance in laboratory studies, no food substance that will greatly enhance performance has been identified. The athlete's prior training is the most important determinant of performance.

The greatest benefit of the pregame meal is to prevent hunger during competition. Because it can take from one to four hours for the stomach to empty, an athlete should eat at least three to four hours before competition begins. Excessive quantities of food should not be consumed before competition. Food that remains in the stomach during competition may cause nausea and possibly vomiting. This condition can be aggravated by nervousness, which slows digestion and delays gastric emptying by means of the sympathetic nervous system.

Foods that are slowly digested, such as high-fat fried foods, should be limited or avoided in the pregame meal. High-carbohydrate foods are recommended because they are removed from the stomach more easily than fat or protein is. Carbohydrates do not inhibit gastric emptying by means of cholecystokinin release, whereas fat and protein do.

Beverages or foods high in sugar should be avoided before competition because they trigger insulin release. Insulin is the hormone that enhances glucose entry into most body cells. Once the person begins exercising, insulin sensitivity increases (see page 258), which results in a decrease in the plasma-glucose level. A lowered plasma-glucose level induces feelings of fatigue and an increased use of muscle glycogen stores, which can limit performance in endurance events such as the marathon. It is best for the athlete to drink only plain water within an hour of competition.

reaches the ileum. Glucose is absorbed by active transport, and galactose is transported by the same carrier as glucose. Fructose is apparently absorbed by facilitated diffusion.

Proteins

The pancreatic juice contains the inactive protein-digesting enzymes trypsinogen, chymotrypsinogen, and procarboxypeptidase. In the intestine, trypsinogen is converted to active **trypsin** by an enzyme called *enterokinase* from the intestinal mucosa and by previously formed trypsin. In turn, trypsin converts chymotrypsinogen to active **chymotrypsin,** and procarboxypeptidase to active **carboxypeptidase.** These active enzymes further the protein digestion begun in the stomach by pepsin (Table 24.4). Trypsin breaks peptide bonds involving basic amino acids such as lysine and arginine, producing small peptide chains of amino acids. Chymotrypsin breaks peptide bonds involving aromatic amino acids such as tyrosine and phenylalanine, also producing small peptide chains. Carboxypeptidase frees the terminal amino acid from the carboxyl (acid) end of an amino acid chain. Together, the actions of these enzymes (as well as pepsin in the stomach) degrade proteins into an assortment of individual amino acids and small peptide chains.

A number of peptidase enzymes of the microvilli of the mucosal epithelium continue the process of protein digestion. Aminopeptidases liberate the terminal amino

◆ **TABLE 24.3 Representative Activities of the Gastrointestinal Hormones**

Hormone	Activities
Gastrin	Stimulates gastric secretion, particularly the secretion of hydrochloric acid; stimulates stomach motility Increases ileal motility Relaxes ileocecal sphincter
Secretin	Inhibits gastric acid secretion; inhibits stomach motility Stimulates the release from the pancreas of a watery fluid containing bicarbonate ions Stimulates the secretion of bile, particularly a watery fluid containing bicarbonate ions
Cholecystokinin	Stimulates the release of enzymes from the pancreas Stimulates the contraction of the gallbladder
Gastric inhibitory peptide	Inhibits gastric acid secretion; inhibits stomach motility Stimulates insulin secretion by the pancreas

acid from the amino end of a small peptide chain. Tetrapeptidases split the final amino acid from tetrapeptides (chains of four amino acids), producing tripeptides and free amino acids, and tripeptidases break tripeptides into dipeptides and free amino acids.

Much protein digestion occurs in the upper portion of the small intestine, and protein is 60–80% digested and absorbed by the time the chyme reaches the ileum. During absorption, amino acids, dipeptides, and even tripeptides are taken into absorptive intestinal epithelial cells by active transport. Within the cells, dipeptidase and tripeptidase enzymes break the dipeptides and tripeptides into their component amino acids, and individual amino acids leave the cells and enter the blood.

Lipids

Dietary fat, which consists mostly of triglycerides, is digested primarily in the small intestine. The first step in the intestinal digestion of fat is emulsification—that is, the dispersal of large water-insoluble fat droplets into a suspension of fine droplets that provide an increased surface area on which the water-soluble, fat-digesting enzymes can act (Figure 24.37). (Bile salts produced by the liver and released into the intestine with the bile are important in this process.) The small, emulsified fat particles are digested by lipase enzymes, primarily from the pancreas (Table 24.4). This digestion results mainly in the formation of monoglycerides and free fatty acids.

Bile-salt molecules are amphipathic, and they can aggregate with one another to form small, water-soluble structures called micelles (see page 34). The monoglycerides and free fatty acids formed during fat digestion associate with bile-salt micelles, and it is in this form that they reach the intestinal epithelium. During absorption, the monoglycerides and free fatty acids leave the micelles and enter absorptive epithelial cells by simple diffusion. The bile salts, which can be reused, are ultimately absorbed in the ileum as part of the enterohepatic circulation process.

During their entry into absorptive intestinal epithelial cells, many of the monoglycerides are digested to glycerol and fatty acids by an epithelial cell lipase. Within the endoplasmic reticulum of the epithelial cells, the free fatty acids combine with newly synthesized glycerol and small amounts of glycerol from the digested monoglycerides to form triglycerides. In addition, the cells synthesize phospholipids, cholesterol, and proteins. Small protein-coated globules of triglycerides, synthesized or absorbed phospholipids and cholesterol, and some free fatty acids are packaged within membrane-bounded vesicles. The globules leave the epithelial cells by exocytosis and enter the lacteals of the lymphatic system as minute droplets known as chylomicrons. Chylomicrons are, by weight, about 90% triglycerides, 5% phospholipids, 4% free fatty acids, 1% cholesterol, and

◆ **FIGURE 24.37 Processes of fat digestion and absorption, and chylomicron formation**

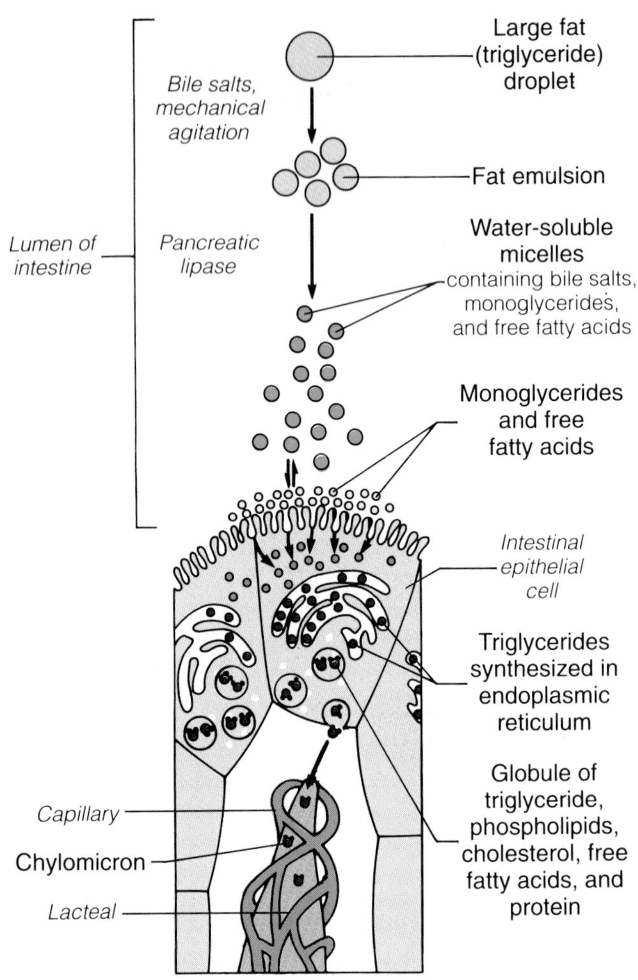

a small amount of protein. Fat absorption usually takes place in the duodenum and jejunum and is completed in the ileum.

Ingested cholesterol occurs both as free cholesterol and as cholesterol esters (cholesterol combined with a fatty acid). The cholesterol esters are digested to free cholesterol and free fatty acids by a pancreatic cholesterol esterase (Table 24.4). Free cholesterol can associate with micelles, and it is ultimately absorbed at the intestinal epithelium.

Vitamins

Fat-soluble vitamins such as A, D, E and K associate with micelles and are absorbed in conjunction with fat digestion. Water-soluble vitamins, such as vitamin C and the B vitamins (with the exception of vitamin B_{12}), are absorbed by passive transport in the proximal portion of the intestine. Vitamin B_{12} combines with a glycoprotein

◆ **TABLE 24.4 Principal Enzymes Involved in the Digestion of Carbohydrates, Proteins, and Lipids***

ENZYME	SOURCE	ACTION
CARBOHYDRATES		
Salivary amylase	Salivary glands	Breaks starch into alpha-dextrin, maltotriose, and maltose
Pancreatic amylase	Pancreas	Breaks starch into alpha-dextrin, maltotriase, and maltose
Alpha-Dextrinase	Intestinal epithelium	Breaks alpha-dextrin into glucose
Maltase	Intestinal epithelium	Breaks maltose and maltotriose into glucose
Sucrase	Intestinal epithelium	Breaks sucrose into glucose and fructose
Lactase	Intestinal epithelium	Breaks lactose into glucose and galactose
PROTEINS		
Pepsin (Pepsinogen)	Zymogenic cells of stomach	Breaks proteins into smaller peptide chains of amino acids
Trypsin (Trypsinogen)	Pancreas	Breaks proteins and peptides into smaller peptides
Chymotrypsin (Chymotrypsinogen)	Pancreas	Breaks proteins and peptides into smaller peptides
Carboxypeptidase (Procarboxypeptidase)	Pancreas	Frees terminal amino acid from the carboxyl (acid) end of an amino acid chain
Aminopeptidases	Intestinal epithelium	Free terminal amino acid from the amino end of an amino acid chain
Tetrapeptidases	Intestinal epithelium	Split final amino acid from tetrapeptides, producing tripeptides and free amino acids
Tripeptidases	Intestinal epithelium	Break tripeptides into dipeptides and free amino acids
Dipeptidases	Intestinal epithelium	Break dipeptides into free amino acids
LIPIDS		
Pancreatic lipase	Pancreas	Breaks triglycerides mainly into monoglycerides and free fatty acids
Epithelial cell lipase	Intestinal epithelium	Breaks monoglycerides into glycerol and fatty acids
Cholesterol esterase	Pancreas	Breaks cholesterol esters into free cholesterol and free fatty acids

*Substances in parentheses are precursors of active enzymes.

substance called intrinsic factor, which is produced by parietal cells in the stomach. Vitamin B_{12} is actively absorbed across the intestinal wall at specific sites in the terminal ileum.

Minerals

Sodium is actively absorbed from the small intestine. An active-transport mechanism for sodium is present within the epithelial cells of the mucosa, particularly in the jejunum. Potassium, magnesium, and phosphate can also be actively absorbed through the mucosa.

Chloride movement passively follows that of sodium in the upper part of the small intestine. However, chloride is actively transported in the ileum by a process in which the absorption of chloride from the small intestine is coupled with the secretion of bicarbonate ions into the intestine.

Calcium absorption occurs actively along the entire small intestine (particularly the duodenum) and requires vitamin D. The amount of calcium absorbed varies with body requirements and is controlled in part by parathyroid hormone (page 553).

Water

The small intestine can absorb about 200–400 ml of water per hour. Five to ten liters of water derived from food, drink, and digestive secretions enter the small intestine each day, but only about 0.5 liter enters the large intestine. The rest is absorbed, primarily through the upper small intestine. Water can move across the intes-

CLINICAL CORRELATION

Malabsorption

Case Report

THE PATIENT: A 12-year-old boy.

PRINCIPAL COMPLAINT: Failure to grow.

HISTORY: The patient had an older brother who died of "pneumonia" at age 3 years; three other siblings are alive and well. He had been suffering from "asthma" since approximately 6 or 7 years of age, and had two episodes of pneumonia over the past three years. Between these episodes he had an increase in coughing and production of purulent sputum (which contains pus). He grew normally until the age of 7 or 8 years, but at the time of admission was in a lower percentile for height and weight. His parents considered him a sickly child, and he had not been able to maintain a normal play schedule with his friends, particularly in the summer. He had a large appetite, with no particular food intolerance. He described his bowel habits as normal. However, his feces were large in mass, light colored, poorly formed, somewhat foul smelling, and floated in water.

CLINICAL EXAMINATION: Laboratory examination of the feces revealed a fat content that comprised about 30% of the amount ingested (normal: less than 6%). Examination of his serum revealed a lowered protein concentration and a prolonged prothrombin time (a test of blood clotting ability). All of these findings are consistent with cystic fibrosis.

COMMENT: Cystic fibrosis is an inherited disease in which the central problem is the secretion of an unduly viscous mucus. Infection of the air passages (pneumonia) occurs because the cilia do not adequately clear the mucus toward the trachea. Other obstructed or partly obstructed tubes (for example, ureters, auditory tubes) usually grow excess bacteria. The sweat contains high concentrations of sodium and chloride ions; hence, in hot weather salt depletion is a problem. The patient had failed to grow normally because of a decreased absorption of food, although this problem was partly offset by his large appetite. The large amount of feces signifies malabsorption, and the foul smell is caused by bacterial degradation of unabsorbed foodstuffs. The low concentration of serum protein is a sign of severe undernutrition, specifically of failure to absorb adequate protein (amino acids). The prolonged prothrombin time is caused by deficient absorption of vitamin K. (Prothrombin is formed in the liver by a process that requires vitamin K.) The malabsorption evident in cystic fibrosis is related to blockage of the pancreatic ducts by the viscid mucus and replacement of the pancreatic acini by fibrous tissue. The digestive enzymes, amylase (which digests carbohydrate), lipase (which digests fat), and the proteinases, are not secreted. Thus, digestion of all three major classes of foodstuff is inadequate, and these nutritive materials are poorly absorbed.

OUTCOME: The patient was given pancreatic enzymes by mouth, protected by antacids, and he began to grow. However, because of the effect of the viscous mucus on pulmonary function, his life span may be shorter than normal.

tinal wall in both directions, and the absorption of water from the intestine occurs osmotically in association with the absorption of particulate materials. In fact, the chyme within the duodenum becomes isosmotic with the blood plasma as a result of the osmotic removal or addition of water.

Absorption in the Large Intestine

Sodium is actively absorbed in the large intestine, and chloride absorption also occurs. Bicarbonate ions are actively secreted by the large intestine. In addition, the large intestine absorbs 300–400 ml of water per day as a consequence of sodium (and chloride) transport.

Bacteria within the large intestine synthesize some vitamins that can be absorbed (for example, vitamin K). However, under normal conditions, the body obtains only a small amount of the vitamins it requires in this manner.

The fecal material that ultimately leaves the gastrointestinal tract consists of water and solids such as undigested food residue, microorganisms, and sloughed-off epithelial cells.

CONDITIONS OF CLINICAL SIGNIFICANCE

The Digestive System

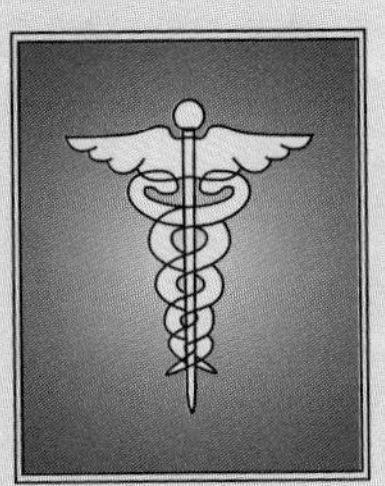

Vomiting

Vomiting, which ejects the contents of the stomach through the mouth, is a complex reflex coordinated by a vomiting center within the medulla of the brain. During vomiting, the lower esophageal sphincter relaxes, and the diaphragm and abdominal muscles contract. This increases the intraabdominal and intragastric pressures and forces the contents of the stomach through the esophagus and out of the mouth. The vomiting center receives input from many areas, both within and at the surface of the body. Thus, vomiting can result from dizziness, unpleasant odors or sights, and so forth, as well as from disturbances within the digestive tract. Excessive vomiting can lead to severe disturbances in the fluid, salt, and acid-base balances of the body.

Diarrhea

Diarrhea is characterized by watery stools, generally accompanied by frequent defecation. Diarrhea can be due to either an increased secretion of fluid into or a decreased absorption of fluid from the gastrointestinal tract. In either case, an increased volume of material within the large intestine can lead to more frequent defecation. Like vomiting, prolonged diarrhea can cause disturbances in the fluid, salt, and acid-base balances of the body.

Constipation

If the motility of the large intestine is decreased, digested materials remain within the colon and rectum for prolonged periods of time. The longer fecal materials remain within the colon, the more water is absorbed. As the feces lose water, they become drier and harder, making defecation more difficult and sometimes painful. This condition is referred to as *constipation.* Constipation can be caused by insufficient bulk in the diet, lack of exercise, or even by emotions.

Peptic Ulcer

A *peptic ulcer* is an erosion of the wall of the gastrointestinal tract in an area of the tract exposed to gastric juice containing acid and pepsin. Peptic ulcers are most commonly found in the stomach (gastric ulcers) and the duodenum (duodenal ulcers). They may be caused by an excessive acid-pepsin secretion, or they may result from an insufficient secretion of mucus, which normally protects the gastrointestinal mucosa from being digested by the acid-pepsin gastric juice.

A bacterium, *Helicobacter pylori,* has been found in almost all patients with duodenal ulcers and in more than 70 percent of patients with gastric ulcers. It is not certain, however, if the bacterium is a primary cause of the ulcers, and not everyone who harbors the bacterium gets ulcers.

The most common symptom of peptic ulcer is pain. However, in some cases, the pain can be temporarily relieved by the ingestion of food, apparently because the food provides some protection by coating the ulcer or by acting as a buffer.

If the erosion due to a gastric ulcer is sufficiently severe, blood vessels in the stomach wall are damaged, and bleeding occurs into the stomach itself (a bleeding ulcer). In extreme cases, a peptic ulcer can lead to a perforation—that is, to a hole entirely through the wall of the gastrointestinal tract. A perforation allows the contents of the gastrointestinal tract to pass into the abdominal cavity. Such an occurrence is very serious and requires immediate surgery.

Gastroenteritis

Gastroenteritis (gas″-trō-en-ter-ī′-tis) is an acute or chronic inflammation of the mucosa of the stomach and intestine. It is often caused by irritants such as excessive alcohol or cathartics (medicines that cause bowel movements), but it can have any of a wide variety of other causes, including viral infections, food allergies, overeating, and so forth.

continued on next page

CONDITIONS OF CLINICAL SIGNIFICANCE

Gallstones

Gallstones are particles—composed primarily of cholesterol or bile pigments—that sometimes form in the bile, particularly in the gallbladder. The stones can block the cystic duct from the gallbladder or the common bile duct to the duodenum. Gallstones are often painful, and those that block the cystic duct are especially so because the contractions of the gallbladder exert force on them.

In the case of cystic gallstones, bile is still able to reach the duodenum directly from the liver through the common bile duct. Blockage of the common bile duct, on the other hand, prevents bile from reaching the intestine and thus interferes with the proper absorption of fat. In addition, the bile pigments cannot reach the intestine to be excreted. These accumulate in the blood and are eventually deposited in the skin. This produces a yellow color in the skin, a condition that is called *jaundice (jawn´-dis).*

Pancreatitis

Pancreatitis is an inflammation of the pancreas that is often caused by the digestion of parts of the organ by pancreatic enzymes that are normally carried to the small intestine within the pancreatic ducts. In pancreatitis, the enzymes become activated within the ducts and may destroy the ducts and the pancreatic cells. Since pancreatic enzymes are very important in the digestion of carbohydrates, proteins, and fats, pancreatitis can produce severe nutritional problems.

Hepatitis

Hepatitis is an inflammation of the liver. Most commonly, it is due to a viral infection transmitted either by virus-infected blood or by contaminated food or water. The infected liver becomes enlarged, and its functioning is impaired, which can lead to jaundice. In some cases, liver function is depressed for a year or more.

Cirrhosis

Cirrhosis (sir-rō´-sis) is a chronic inflammation of the liver that is progressive and diffuse. In the affected areas of the liver, some cells are replaced by fibrous connective tissue, thereby interfering with liver function. Cirrhosis can cause a reduction in the production of bile, in the excretion of bile pigments, and in the production of blood-clotting factors and plasma albumin. Cirrhosis also reduces the liver's ability to detoxify the blood, and toxins accumulate.

Appendicitis

The appendix is a blind-ended pouch extending from the cecum. If it becomes obstructed—for example, with hardened fecal material—its venous circulation may be interfered with. Such interference reduces the oxygen supply to the area and permits bacteria to flourish. The appendix then becomes inflamed and filled with pus, a condition called *appendicitis.*

If an inflamed appendix is not removed soon enough, it can rupture and release its contents into the abdominal cavity. This occurrence can cause peritonitis, which is an inflammation of the lining of the abdominopelvic cavity.

Effects of Aging

With aging, the digestive secretions tend to be diminished somewhat, and the muscles in the walls of the tract become weaker, reducing the strength of the contractions.

A gradual atrophy of the salivary glands reduces the amount of saliva secreted and makes it more difficult to chew food efficiently. The loss of teeth may further interfere with chewing of food in older persons. Reduction in the volume of saliva not only causes the mouth to be dry, but also diminishes the ability to taste and provides less cleansing for the mouth and teeth. Swallowing difficulties are not uncommon in older persons and are often caused by incomplete opening of the lower esophageal sphincter.

Gradual atrophy of the gastric glands in the stomach reduces the secretion of mucus, hydrochloric

CONDITIONS OF CLINICAL SIGNIFICANCE

acid, and digestive enzymes. These reductions may particularly interfere with the digestion of proteins and cause a chronic inflammation of the stomach lining.

Atrophy of the cells lining the small intestine alters the structure of the villi, thereby reducing the surface area across which absorption may occur. The cells in the wall of the large intestine also undergo gradual atrophy. This atrophy may so weaken the wall that small saclike outpockets of the mucosa called *diverticula (dī″-ver-tik′-ū-luh)* protrude through the muscular wall of the large intestine. This condition is referred to as *diverticulosis.* If the diverticula become inflamed *(diverticulitis)* they may rupture, causing a perforation of the intestinal wall that allows fecal material to contaminate the abdominal cavity.

Study Outline

◆ EMBRYONIC DEVELOPMENT OF THE DIGESTIVE SYSTEM p. 757

1. Formed from endoderm.
2. Divided into foregut, midgut, hindgut.
3. Oral membrane breaks through to stomodeum, forming mouth.
4. Cloacal membrane ruptures through to proctodeum.
5. Endodermal buds along foregut, together with surrounding mesoderm, give rise to salivary glands, liver, gallbladder, pancreas.

◆ ANATOMY OF THE DIGESTIVE SYSTEM pp. 757–769

Mouth.

1. Lined with stratified squamous layer of nonkeratinized cells (except outer lips).
2. Lips and cheeks aid chewing and speech.
3. Roof formed anteriorly by hard palate; posteriorly by soft palate.

TONGUE. Forms floor of mouth.

1. Protruded, retracted, and moved sideways by extrinsic muscles.
2. Shape modified by intrinsic muscles.
3. Assists in food manipulation, swallowing, speech.
4. Papillae present on dorsal surface; taste buds present on some papillae.
5. Connected ventrally to mouth floor by frenulum, which can cause tongue-tied condition.

TEETH.

1. Enamel-covered crown anchored by roots into alveolus (socket).
2. Pulp cavity, root canal, and apical foramen enclose vessels and nerves.
3. Incisors, canines, premolars, molars.
4. Twenty deciduous teeth; 32 permanent teeth.

SALIVARY GLANDS.

Tubuloalveolar glands. Composed of small sacs (acini).
Parotid glands. Below and anterior to ear.
Submandibular glands. Medial to mandibular angle.
Sublingual glands. On floor of mouth.
Saliva. Moistens mucous membranes.

Pharynx. From oropharynx through laryngopharynx.

Gastrointestinal Tract Wall. Four tunics in wall of digestive tube; this basic structure modified in some regions of digestive tract.

TUNICA MUCOSA. Mucous membrane lining; consists of epithelial layer, lamina propria, and muscularis mucosae.

TUNICA SUBMUCOSA. Dense or loose connective tissue; contains blood vessels, lymphatic vessels, nerves, and glands.

TUNICA MUSCULARIS. Double layer of muscle tissue: inner layer circular, outer layer longitudinal.

TUNICA SEROSA (ADVENTITIA). Outer layer of connective tissue.

INTRINSIC NERVE PLEXUSES AND REFLEX PATHWAYS. Interconnections of neurons; coordinate much of the digestive-tract activity; make intratract reflexes possible.

Esophagus. Muscular tube connecting pharynx to stomach; upper portion contains skeletal muscles; lower portion has smooth muscles.

Stomach. From cardiac orifice to pyloric sphincter; body, fundus, and pyloric regions; greater and lesser curvatures.

TUNICA MUCOSA MODIFICATIONS.

Columnar epithelium

Rugae

Gastric glands

1. Mucous neck cells secrete mucus.
2. Parietal cells produce hydrochloric acid.
3. Zymogenic cells secrete pepsinogen.
4. Cardiac and pyloric glands secrete mainly mucus.

TUNICA MUSCULARIS MODIFICATION. Includes an oblique muscle layer, as well as circular and longitudinal layers.

Small Intestine.

1. Approximately 6 m long; site of most digestion and absorption.
2. Three regions:

DUODENUM. Retroperitoneal; receives common bile duct and pancreatic duct at duodenal papilla.

JEJUNUM. Suspended by mesentery.

ILEUM. Suspended by mesentery; entrance into large intestine surrounded by ileocecal valve; just preceding ileocecal valve is ileocecal sphincter.

3. Modifications in wall: plicae circulares, villi, and microvilli increase surface area of small intestine. Contains goblet cells, intestinal glands, duodenal glands.

Large Intestine.

1. Composed of cecum; ascending, transverse, descending, and sigmoid colons; rectum; anal canal.
2. Modifications in tunics—tunica mucosa: has many mucous cells, lacks villi; tunica muscularis: longitudinal muscles form bands (taeniae coli) that cause colon to form pouches (haustra); epiploic appendages on external surface.

RECTUM AND ANAL CANAL.

1. Rectum lies anterior to sacrum.
2. Anal canal mucosa forms longitudinal rectal columns.
3. Involuntary internal anal sphincter; voluntary external anal sphincter.

◆ ACCESSORY DIGESTIVE ORGANS pp. 769–774

Pancreas.

1. Cells in acini secrete digestive enzymes.
2. Endocrine secretions of pancreatic islets enter blood.

Liver.

1. Right and left lobes separated by falciform ligament.
2. Ligamentum teres is remnant of fetal umbilical vein.
3. Coronary ligament attaches liver to undersurface of diaphragm.
4. Lobules consist of rows of cuboidal cells radiating outward from a central vein that empties into a hepatic vein.
5. Bile ducts collect bile from canaliculi that transport bile in opposite direction from blood flow.
6. Hepatic circulation: Hepatic artery supplies oxygenated blood to liver from aorta. Hepatic portal vein carries venous blood from digestive tract, pancreas, and spleen. Liver sinusoids contain mixture of arterial and venous blood. Sinusoids lined with highly permeable endothelium with attached phagocytic stellate macrophages. Hepatic veins empty blood from sinusoids into inferior vena cava.

Gallbladder and Bile Ducts.

1. Gallbladder is small sac on inferior surface of liver; serves as bile storage site.
2. Cystic duct drains gallbladder, joins hepatic duct to form common bile duct, which joins with pancreatic duct to enter duodenum.

◆ PHYSIOLOGY OF THE DIGESTIVE SYSTEM pp. 774–794

Digestive system activities include mechanical processes, secretion, chemical digestion, and absorption.

Mechanical Processes.

1. *Segmentation.* Major mixing movement.
2. *Peristalsis.* Important propulsive movement.

CHEWING. Part voluntary, part reflex; tactile stimulation of food against teeth, gums, and anterior portion of roof of mouth leads to reflex relaxation of muscles that raise jaw; when jaw drops, stretch reflexes lead to muscle contractions that raise jaw.

SWALLOWING. Movement of bolus into pharynx elicits swallowing reflex.

1. Peristaltic contraction of pharyngeal constrictor muscles.
2. Opening of upper esophageal sphincter.
3. Approximation of vocal cords, upward movement of larynx, elevation of soft palate.
4. Primary peristaltic wave in esophagus.
5. Opening of lower esophageal sphincter.

GASTRIC MOTILITY.

1. Stomach can expand to accommodate food without great increase in intragastric pressure.
2. Stomach undergoes peristaltic contractions that move chyme into duodenum and provide effective mixing.
3. Basic electrical rhythm important in determining frequency of gastric contraction.
4. Gastric motility and emptying influenced by distension of stomach (via neural reflexes and gastrin) and by volume and composition of chyme in duodenum (via eneterogastric reflex and intestinal hormones).

MOTILITY OF THE SMALL INTESTINE. Includes segmentation and migrating motility complex. When food enters stomach, gastroileal reflex intensifies ileal contractions; gastrin relaxes ileocecal sphincter.

MOTILITY OF THE LARGE INTESTINE. Movements are generally sluggish; include haustral contractions and mass movements; defecation reflex leads to emptying of lower portion of large intestine.

Secretion. Includes mucus and digestive enzymes. Many digestive enzymes and other secretory products of the digestive system are produced and released by specialized epithelial cells that are organized into exocrine secretory glands.

MOUTH AND SALIVA. Saliva is produced by parotid, sublingual, submandibular, and buccal glands.

SECRETIONS OF THE SALIVARY GLANDS.

Mucins. Major proteins of saliva; form mucus with water; mucus lubricates mouth and food.

Salivary Amylase. An enzyme that breaks down large starch molecules.

COMPOSITION OF SALIVA. 97–99.5% water; pH 6.0 to 7.0; contains electrolytes (sodium, potassium, chloride, bicarbonate ions); kallikrein (leads to production of bradykinin, a vasodilator).

FUNCTIONS OF SALIVA. Lubrication; digestion; bolus formation; dissolves foods, aids taste; moistens mouth and throat; aids speech.

CONTROL OF SALIVARY SECRETION. Both parasympathetic and sympathetic stimulations cause secretion, with parasympathetic stimulation having greatest effect.

STOMACH AND GASTRIC SECRETIONS.

MUCUS. Lubricates stomach walls; protects gastric mucosa; produced by cardiac, pyloric, and gastric glands.

HYDROCHLORIC ACID. Produced by parietal (oxyntic) cells; facilitates protein digestion; kills bacteria.

PEPSINOGEN. Produced by zymogenic (chief) cells; digests proteins when activated to pepsin.

CONTROL OF GASTRIC SECRETION.

Cephalic Phase. Elicited by signals that originate in head region such as sight, smell, or taste of food.

1. Sensory receptors to CNS to vagus nerves to stomach for stimulation of gastric secretion.
2. Response elicited by sight and smell of food is basically a conditioned reflex.

Gastric Phase. Elicited by signals that originate in stomach such as distension; neural reflexes and gastrin involved.

Intestinal Phase. Elicited by signals that originate in small intestine; intestinal gastrin production involved in excitatory component; enterogastric reflex and enterogastrones involved in inhibitory component.

PROTECTION OF THE STOMACH. Tall, columnar epithelial cells; alkaline mucous secretions.

SECRETIONS PRESENT IN THE SMALL INTESTINE. Include mucus, intestinal juice, pancreatic juice, and bile.

MUCUS. Produced by duodenal glands (Brunner's glands), goblet cells, and intestinal glands; protects intestinal mucosa.

INTESTINAL JUICE. Produced by intestinal glands; isotonic with plasma; local reflex control. Digestive enzymes associated with microvilli of intestinal epithelial cells.

PANCREATIC JUICE.

1. Basic—helps neutralize acid chyme from stomach.
2. Carbohydrate-, protein-, and lipid-digesting enzymes.
3. Ribonuclease and deoxyribonuclease.
4. Hormonal control by secretin, cholecystokinin.
5. Neural control by vagus nerves.

BILE. Produced in liver; stored in gallbladder; travels to duodenum.

1. Contains bile salts, bile pigments, cholesterol, neutral fats, and lecithin.
2. Chemical control by bile salts.
3. Hormonal control by secretin.
4. Neural control by vagus nerves.
5. Gallbladder contraction stimulated by cholecystokinin.

MUCOUS SECRETION IN THE LARGE INTESTINE. Large intestine possesses abundant mucus-secreting goblet cells; intestinal glands within large intestine consist almost entirely of goblet cells.

Chemical Digestion and Absorption. Most digestive activity occurs within small intestine, as does practically all absorption.

CARBOHYDRATES.

1. Amylases break starch into alpha-dextrin, maltotriose, and maltose.
2. Enzymes of microvilli free monosaccharides.
3. Glucose and galactose absorbed by active transport. Fructose apparently absorbed by facilitated diffusion.

PROTEINS.

1. Enterokinase and previously formed trypsin convert trypsinogen to trypsin.
2. Trypsin activates chymotrypsinogen and procarboxypeptidase.
3. Trypsin, chymotrypsin, and carboxypeptidase produce assortment of individual amino acids and small peptide chains.
4. Peptidase enzymes of intestinal epithelium continue process of protein digestion.
5. Amino acids absorbed by active transport.

LIPIDS. (Digested primarily in small intestine.)

1. Emulsification; bile salts are important.
2. Lipase enzymes digest small, emulsified fat particles mainly to monoglycerides and free fatty acids, which associate with bile-salt micelles.
3. Absorption by simple diffusion.
4. Triglycerides resynthesized in intestinal epithelial cells.
5. Chylomicrons enter lymph.

VITAMINS.

1. Fat-soluble vitamins associate with micelles; absorbed in conjunction with fat digestion.
2. Water-soluble vitamins except B_{12} absorbed by passive transport in proximal intestine.
3. B_{12} combines with intrinsic factor; actively absorbed at specific sites in terminal ileum.

MINERALS. Sodium can be actively absorbed; potassium, magnesium, and phosphate can also be actively absorbed; calcium absorption requires vitamin D.

WATER. Absorbed mainly in small intestine.

ABSORPTION IN THE LARGE INTESTINE. Sodium, chloride, and small amount of water are absorbed. Fecal matter—composed of water, mucus, undigested food residue, microorganisms, and sloughed-off epithelium.

◆ CONDITIONS OF CLINICAL SIGNIFICANCE: THE DIGESTIVE SYSTEM pp. 795–797

Vomiting. Ejects contents of stomach through mouth; complex reflex leads to relaxation of lower esophageal sphincter and contraction of diaphragm and abdominal muscles.

Diarrhea. Watery stools generally accompanied by frequent defecation; can be due to either increased secretion of fluid into or decreased absorption of fluid from gastrointestinal tract.

Constipation. Decreased motility of large intestine leads to incrased water absorption and dry, hard feces.

Peptic Ulcer. Erosion of wall of gastrointestinal tract in area of tract exposed to gastric juice containing acid and pepsin.

Gastroenteritis. Acute or chronic inflammation of mucosa of stomach and intestine; caused by irritants, viral infection, or food allergy.

Gallstones. Particles containing cholesterol and bile salts in bile; can block cystic or common bile ducts.

Pancreatitis. Inflammation of pancreas; often caused by digestion of parts of organ by pancreatic enzymes.

Hepatitis. Inflammation of liver, usually due to virus; causes enlargement of liver and impaired liver function.

Cirrhosis. Chronic liver inflammation; may result in reduced bile production, reduced bile pigment excretion, reduced blood-clotting-factor production, and blood toxin accumulation.

Appendicitis. Inflammation of appendix; can be due to obstruction that interferes with venous circulation.

Effects of Aging. Digestive secretions diminished; peristaltic contractions weakened; reduced secretion of saliva; loss of teeth. Atrophy of mucous glands and digestive glands in stomach and small intestine. Wall of large intestine weakened.

Self-Quiz

1. The entire digestive tract is lined with: (a) cloacal membrane; (b) proctodeum; (c) mucous membrane.
2. The teeth specialized to tear food are the (a) canines; (b) incisors; (c) premolars.
3. Which of the four layers of the gastrointestinal tract is responsible for peristalsis? (a) tunica mucosa; (b) tunica submucosa; (c) tunica muscularis; (d) tunica serosa.
4. The point of exit from the stomach, where it joins with the small intestine, is guarded by a: (a) pyloric sphincter; (b) greater omentum; (c) lesser omentum.
5. The ducts from the pancreas and liver empty into the alimentary canal at the: (a) stomach; (b) duodenum; (c) ileum.
6. As the descending colon reaches the left pelvic brim, it curves to the midline via an S-shaped sigmoid colon. True or False?
7. The internal anal sphincter is formed by a thickening of the circular layer of the smooth muscles of the tunica muscularis and is therefore under voluntary control. True or False?
8. Bile is produced in the: (a) liver; (b) gallbladder; (c) pancreas.
9. Secondary peristaltic contractions in the esophagus: (a) move food from the stomach to the mouth during vomiting; (b) occur only during swallowing; (c) are generated in response to the distension of the esophagus.
10. The enterogastric reflex: (a) inhibits gastric motility; (b) stimulates the secretion of saliva; (c) causes defecation.
11. Mass movements characteristically move food through the stomach. True or False?
12. Salivary amylase is produced primarily by which glands? (a) buccal; (b) duodenal; (c) parotid.
13. The optimal pH at which salivary amylase splits large starch molecules into smaller fragments is: (a) 6.9; (b) 4.2; (c) 8.4; (d) 11.7.
14. Saliva contains: (a) pepsin; (b) vitamin K; (c) kallikrein.
15. The salivary substance that converts protein precursors to bradykinin is: (a) kallikrein; (b) salivary amylase; (c) mucin.
16. Parietal cells produce: (a) pepsinogen; (b) mucus; (c) hydrochloric acid.
17. The chloride ions of the hydrochloric acid of the stomach are believed to come from: (a) chloride ions derived from carbonic acid; (b) chloride ions from plasma that are transported across parietal cells to the stomach lumen; (c) water molecules that combine with bicarbonate ions.
18. Protein digestion in the stomach proceeds to the stage of the formation of: (a) monoglycerides; (b) chylomicrons; (c) peptide chains.

19. Most digestive activity and practically all absorption occur within the: (a) mouth; (b) small intestine; (c) large intestine.
20. The secretion of pancreatic enzymes occurs in response to: (a) cholecystokinin; (b) salivary amylase; (c) mucin.
21. Approximately 94% of the bile salts released into the duodenum are: (a) excreted with the feces; (b) reabsorbed in the ileum; (c) converted to bilirubin in the stomach by salivary amylase.
22. Match the following digestive agents with their appropriate lettered descriptions:

Pancreatic juice	(a) Produced by parietal cells in the stomach
Tetrapeptidases	(b) Aids in the emulsification of fats
Amylases	(c) Breaks certain peptide bonds that involve basic amino acids such as lysine and arginine
Bile salts	(d) Enzymes of the microvilli of the mucosal epithelium
Trypsin	(e) Contains the inactive protein-digesting enzymes trypsinogen, chymotrypsinogen, and procarboxypeptidase
Hydrochloric acid	(f) Contained in both the saliva and the pancreatic juice

23. Which element is actively absorbed along the entire small intestine (particulary the duodenum) and requires vitamin D for absorption? (a) potassium; (b) calcium; (c) sodium.

CHAPTER 25

Metabolism, Nutrition, and Temperature Regulation

CHAPTER CONTENTS

LEARNING OBJECTIVES

After completing this chapter, you should be able to:

1. Outline the events of glycolysis and the Krebs cycle.
2. Describe the movement of electrons through the electron transport system.
3. Discuss the interconversion of carbohydrates, proteins, and triglycerides.
4. Explain the utilization of carbohydrates, proteins, and triglycerides during the absorptive and postabsorptive metabolic states.
5. Distinguish between the two major classes of vitamins.
6. Describe the roles of vitamins and minerals in the body.
7. Describe the concept of and the determination of the basal metabolic rate.
8. Explain three theories that deal with the nature of the signals that influence food intake.
9. Explain three ways in which heat is transferred between the body and the environment.
10. Distinguish between heat stroke and heat exhaustion.

CHAPTER 25

The chemical reactions that continually occur within the body are grouped together under the category of **metabolism.** The metabolic reactions produce heat, and this heat is important in body-temperature regulation. Many metabolic reactions process absorbed nutrients such as glucose, amino acids, and triglycerides for use by the body. Portions of these substances are utilized for the synthesis of new structural materials, but the majority are used for the production of adenosine triphosphate (ATP), which provides energy to support the body's activities.

Utilization of Glucose as an Energy Source

One of the principal food materials utilized in the formation of ATP is the six-carbon sugar glucose. The breaking of a glucose molecule's chemical bonds releases energy that can be trapped in the form of ATP.

Overview

In the body, glucose is broken down in a precisely controlled series of chemical reactions that releases energy in a stepwise fashion. Each reaction is catalyzed by specific enzymes, often with the help of additional agents such as coenzymes or inorganic ions. The individual reactions are linked together into metabolic pathways in which the product of one reaction is a substrate for the next.

A metabolic pathway known as **glycolysis** breaks a glucose molecule into two molecules of pyruvic acid, and energy released in this process is used to produce a small amount of ATP (Figure 25.1). In the absence of oxygen, the pyruvic acid molecules are converted to lactic acid. However, when oxygen is available, as is usually the case, pyruvic acid is broken down, and carbon dioxide and water are produced. This process, which is accomplished by a series of reactions called the *Krebs cycle* and a system known as the **electron transport system,** releases a considerable amount of energy, much of which is trapped in the form of ATP.

Glycolysis

The enzymes for the glycolytic reactions are located within the cytoplasm of the cells. These reactions are **anaerobic**—that is, they require no molecular oxygen.

The chemical reactions that occur during the breakdown of a glucose molecule into two molecules of pyruvic acid by glycolysis are outlined in Figure 25.2. Two key points may be observed about these reactions:

1. Two molecules of ATP are converted to ADP, and four molecules of ADP are converted to ATP, directly yielding a net profit of two ATP.
2. Two molecules of a coenzyme called nicotinamide adenine dinucleotide (NAD^+) each acquire two electrons and a proton. (A hydrogen ion, H^+, is a proton.) In chemical terms, the gain of electrons by a molecule is called **reduction,** and the loss of electrons is called **oxidation.** Thus, during glycolysis, two molecules of NAD^+ are *reduced* to two molecules of NADH.

◆ **FIGURE 25.1 Utilization of glucose as an energy source**
A metabolic pathway called glycolysis breaks glucose down to pyruvic acid, and the energy released in this process is used to produce a small amount of ATP. When oxygen is unavailable, pyruvic acid is converted to lactic acid. When oxygen is available, pyruvic acid is broken down, and carbon dioxide and water are produced. This process, which is accomplished by the Krebs cycle and the electron transport system, releases a considerable amount of energy, which is used to produce a large amount of ATP.

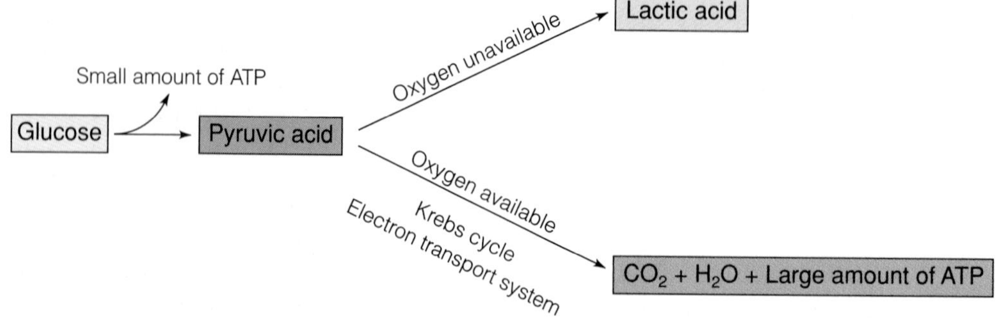

◆ **FIGURE 25.2 The reactions of glycolysis by which a molecule of glucose is transformed into two molecules of pyruvic acid**
The number of carbons in each reactant is indicated in parentheses.

Reactions 1-4. These reactions transform a six-carbon glucose molecule into two three-carbon molecules (dihydroxyacetone phosphate and glyceraldehyde-3-phosphate). The first and third reactions each require a molecule of ATP (adenosine triphosphate), which provides a phosphate group and leaves behind a molecule of ADP (adenosine diphosphate). Thus, the initial reactions of glycolysis do not produce energy, but actually use energy in the form of ATP.

Reaction 5. This reaction transforms dihydroxyacetone phosphate into glyceraldehyde-3-phosphate. Thus, each original glucose molecule forms two molecules of glyceraldehyde-3-phosphate, and the remaining reactions of glycolysis occur twice for every glucose molecule used. For simplicity's sake, however, each of the remaining reactions is shown only once in this figure.

Reaction 6. During this reaction, two electrons and a proton (a hydrogen ion, H^+, is a proton) are transferred to the coenzyme NAD^+ (nicotinamide adenine dinucleotide), forming NADH. (The symbol P_i represents an inorganic phosphate group.)

Reaction 7. In this reaction, a molecule of ATP is produced.

Reactions 8-10. These reactions conclude glycolysis with the production of pyruvic acid. During reaction 10, a molecule of ATP is produced.

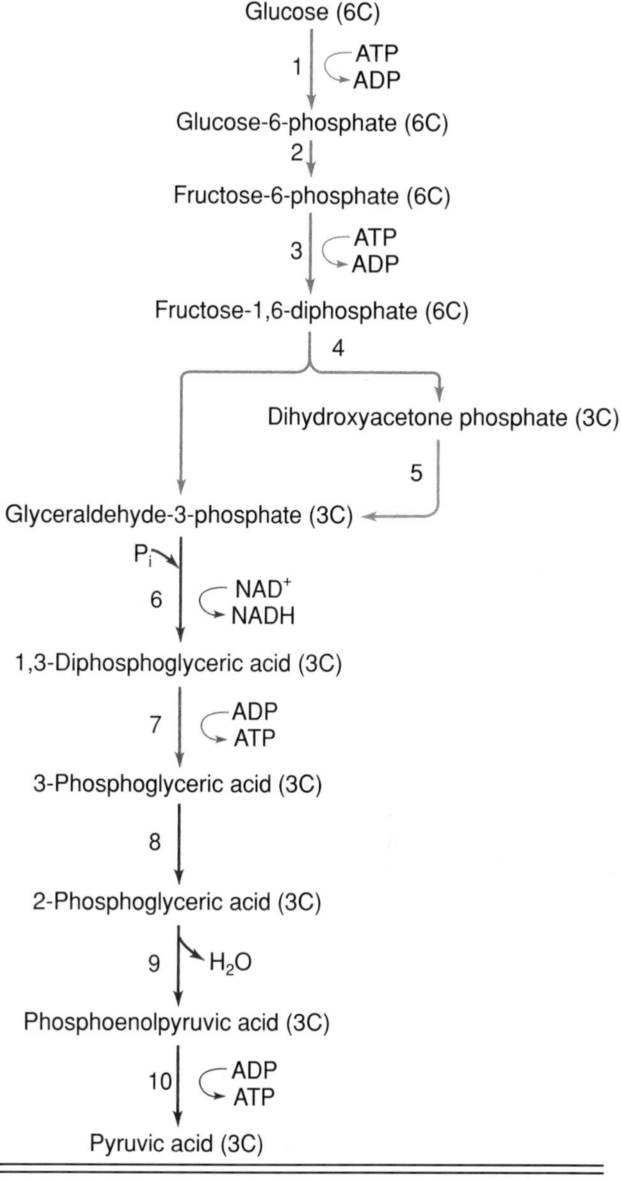

Fate of Pyruvic Acid

NAD^+ and other coenzymes are present in cells in only small amounts. Consequently, the NAD^+ that is reduced to NADH during glycolysis must be regenerated if glycolysis is to continue processing glucose molecules. In order to regenerate NAD^+, NADH must lose two electrons (and the accompanying proton) by transferring them to some acceptor molecule.

When oxygen is unavailable, pyruvic acid acts as an electron acceptor, and it is converted to lactic acid in a reaction that regenerates NAD^+ (Figure 25.3). Like the reactions of glycolysis, this reaction is anaerobic. Consequently, the breakdown of a glucose molecule to two lactic acid molecules is an anaerobic process. However, much of the energy contained within the glucose molecule remains within the lactic acid molecules, and the anaerobic breakdown of a glucose molecule to two lactic acid molecules yields a net profit of only two ATP (that is, the ATP profit from glycolysis).

Glycolysis alone cannot produce enough ATP to support the body's activities. However, it does allow cells to produce some ATP under anaerobic conditions. This ability is particularly important in skeletal muscle cells during heavy exercise.

To produce enough ATP to support the body's activities, **aerobic** processes—that is, processes that require molecular oxygen—are necessary. When oxygen is avail-

◆ **FIGURE 25.3 The conversion of pyruvic acid to lactic acid**

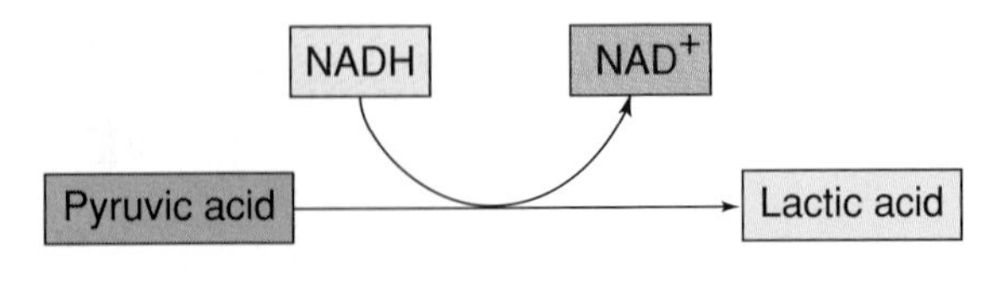

able, as is usually the case, it acts as an electron acceptor by way of the electron transport system, and the pyruvic acid produced by glycolysis is not converted to lactic acid. Instead, it enters the Krebs cycle and is broken down. These aerobic events release a substantial amount of energy, which is used to produce a large amount of ATP.

Krebs Cycle

The enzymes of the Krebs cycle (or citric acid cycle, as it is also called) are contained within the mitochondria of the cells. In order to enter the Krebs cycle, a molecule of pyruvic acid enters a mitochondrion and combines with a molecule of coenzyme A to form acetyl-coenzyme A (Figure 25.4). During the reaction, one of the three carbons of pyruvic acid is lost as carbon dioxide, and an NAD^+ is reduced to NADH. The acetyl-coenzyme A molecule enters the Krebs cycle by combining with a molecule of oxaloacetic acid to form a molecule of citric acid. The Krebs cycle itself involves the stepwise rearrangement of citric acid to reform oxaloacetic acid. Along the way, two carbons are lost as carbon dioxide. At the conclusion of the Krebs cycle, three additional NAD^+ have been reduced to NADH, and electrons and protons have been transferred to the coenzyme flavin adenine dinucleotide (FAD), forming $FADH_2$. In addition, a molecule of ATP has been formed from a molecule of guanosine triphosphate (GTP), which was itself formed from guanosine diphosphate (GDP).

Since glycolysis produces two pyruvic acid molecules from one glucose molecule, the results when two pyruvic acid molecules each form acetyl-coenzyme A and enter the Krebs cycle are six carbon dioxide molecules, two ATP, eight NADH, and two $FADH_2$. As is the case for glycolysis, the coenzymes NAD^+ and FAD must be regenerated from NADH and $FADH_2$ if the Krebs cycle is to continue. This regeneration is accomplished by the oxygen-requiring processes of the electron transport system.

Electron Transport System

The final steps in the aerobic generation of ATP involve the transfer of electrons from the reduced coenzymes NADH and $FADH_2$ to oxygen, and the combination of protons (H^+) with the oxygen to form water and regenerate NAD^+ and FAD. These activities are accomplished by the electron transport system, which is associated with the inner membrane of the mitochondria (Figure 25.5a).

The electron transport system consists of a series of electron carriers, each of which can accept and donate electrons. Some of the carriers are **coenzymes** that incorporate vitamins such as thiamine (vitamin B_1) or riboflavin (vitamin B_2). Others are proteins called **cytochromes,** which contain a heme group whose iron atom acts as an electron carrier.

The initial component of the electron transport system is a large protein complex called the **NADH dehydrogenase complex** (Figure 25.5b). Electrons that pass through this complex are shuttled by **coenzyme Q** to a second complex (the **cytochrome reductase complex**). From this second complex, **cytochrome C** transfers electrons to a third complex (the **cytochrome oxidase complex**). Finally, electrons are passed to oxygen, which combines with protons to form water.

As electrons pass through each of the three complexes, they fall to a lower energy state and thus release energy. The energy is used to actively transport protons across the inner mitochondrial membrane from the matrix to the intermembrane space (Figure 25.5b). This proton transport produces a higher concentration of protons in the intermembrane space than in the matrix. Moreover, because protons (H^+) are positively charged, the transport generates a voltage across the inner mitochondrial membrane (negative on the matrix side and positive on the intermembrane-space side). Thus, the proton transport creates an *electrochemical proton gradient* across the inner membrane.

Protons move down this electrochemical gradient from the intermembrane space to the matrix by passing through an enzyme complex called **ATP synthetase,** which is associated with the inner membrane (Figure 25.5b). As the protons move, they provide energy that drives the production of ATP by ATP synthetase. Thus, the energy released when electrons pass through the electron transport system is initially used to actively transport protons but is ultimately used to synthesize ATP. The energy released when a pair of electrons pass through one of the complexes of the electron transport system is sufficient to drive the synthesis of one molecule of ATP.

Each NADH generated within the mitochondria during the breakdown of pyruvic acid donates a pair of electrons to the NADH dehydrogenase complex (Figure 25.5b). These electrons pass through all three com-

◆ **FIGURE 25.4 The reactions of the Krebs cycle**
P_i is inorganic phosphate.

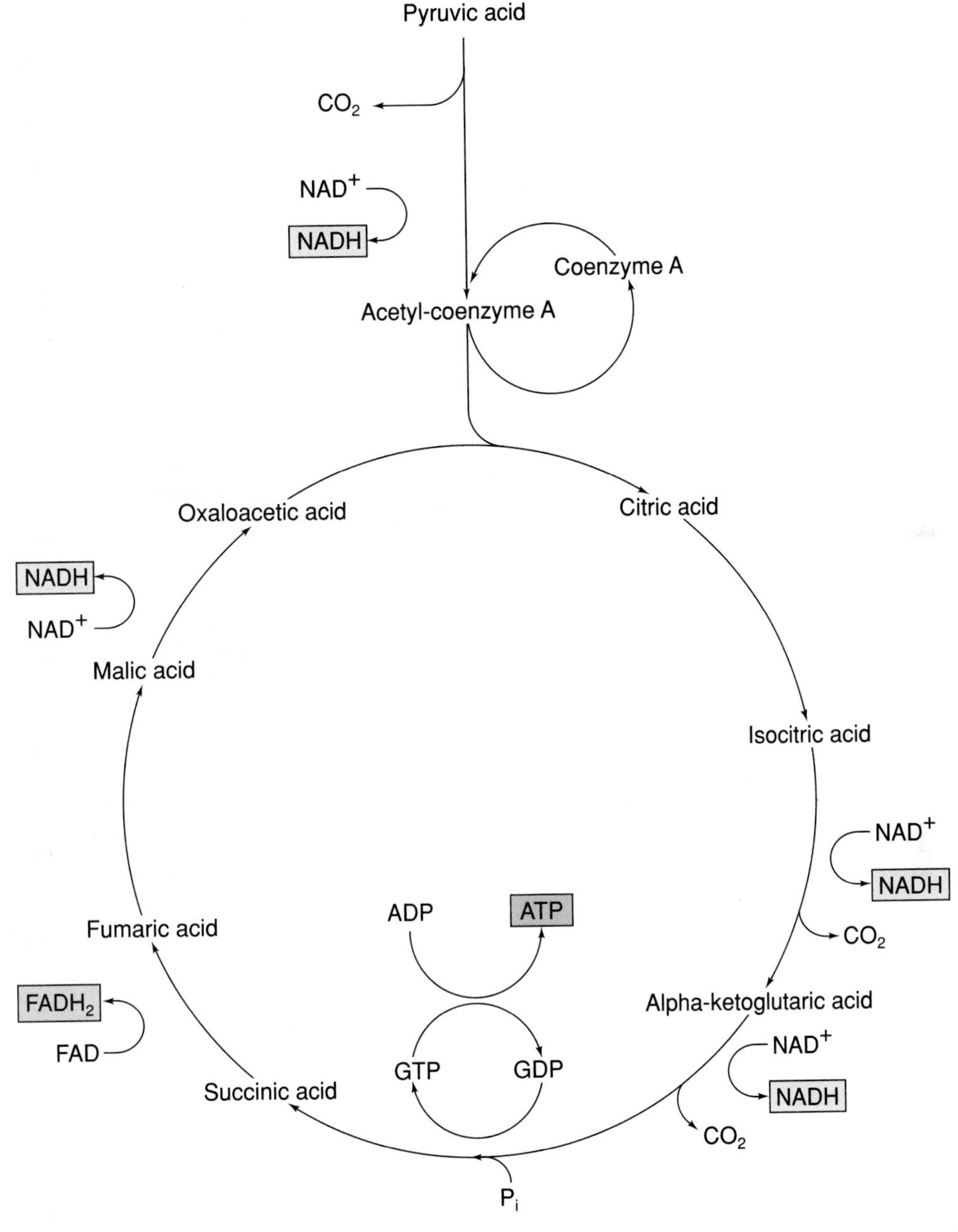

plexes of the electron transport system. Consequently, three molecules of ATP are produced for each of these NADH.

Each $FADH_2$ generated during the Krebs cycle donates a pair of electrons to coenzyme Q. These electrons do not pass through the NADH dehydrogenase complex, and only two molecules of ATP are produced for each $FADH_2$.

The NADH generated in the cytoplasm during glycolysis cannot enter the mitochondria to donate electrons directly to the electron transport system. Instead, they donate their electrons to shuttle molecules that can penetrate the mitochondrial membrane. The electrons carried by one shuttle molecule, *malic acid,* are donated to the NADH dehydrogenase complex and pass through all three complexes of the electron transport system. The electrons carried by another shuttle molecule, *glycerol phosphate,* are donated to coenzyme Q and do not pass through the NADH dehydrogenase complex. Consequently, sometimes three and sometimes only two

◆ **FIGURE 25.5 A mitochondrion, the components of the electron transport system, and ATP synthetase**
(a) The basic structure of a mitochondrion. (b) Diagrammatic representation of the inner membrane of a mitochondrion showing the components of the electron transport system and ATP synthetase. A pair of electrons is indicated by e^-.

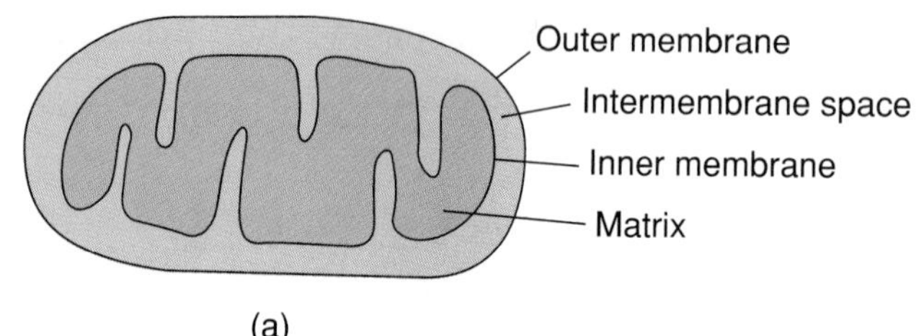

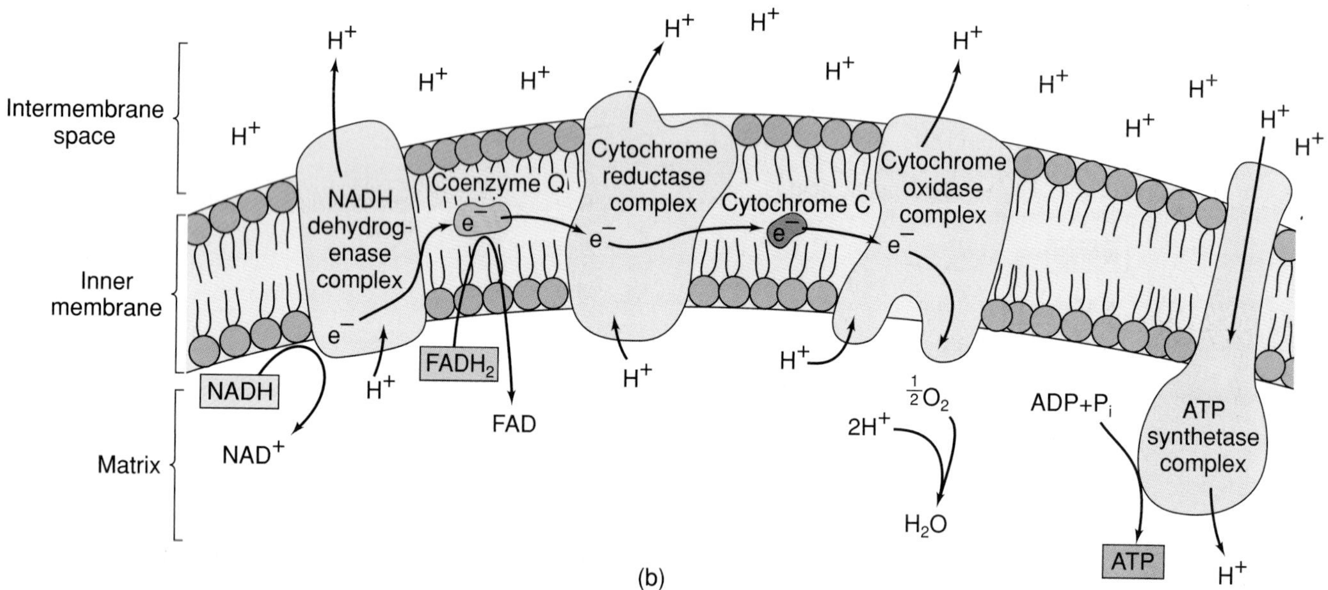

ATP are produced for each NADH generated during glycolysis. (The malate shuttle is particularly evident in the heart and liver.)

ATP and Energy Yield from the Aerobic Breakdown of Glucose

Overall, the aerobic breakdown of a single molecule of glucose produces a net profit of 36–38 ATP (Figure 25.6). Two ATP are produced from glycolysis and 2 from the Krebs cycle. The eight NADH generated within the mitochondria yield 24 ATP when they donate electrons to the electron transport system, and the two $FADH_2$ yield 4 ATP. Depending on the shuttle molecules involved, the two NADH from glycolysis yield 4–6 ATP. The process of forming ATP during glycolysis and the Krebs cycle is called **substrate-level phosphorylation,** and the formation of ATP as the result of electron transport is called **oxidative phosphorylation.**

The uncontrolled combustion of glucose in the presence of oxygen produces carbon dioxide and water and releases about 686,000 calories of energy per mole (1 calorie is the amount of heat energy required to raise the temperature of 1 gram of water from 15°C to 16°C). Although dependent on pH, the production of a mole of ATP from ADP traps about 7300 calories of energy. Therefore, the net profit to the cell from the aerobic breakdown of a mole of glucose is 277,400 calories of energy ($38 \times 7300 = 277{,}400$). Thus, by breaking down glucose in a controlled series of reactions, the aerobic processes can trap in the form of ATP about 40% of the energy released. The other 60% appears as heat. Remaining at the end of the reactions are carbon dioxide and water. In actual fact, however, the ATP yield in living cells is often lower than the maximum (38) used in these calculations.

Utilization of Amino Acids and Triglycerides as Energy Sources

In addition to glucose (which is a carbohydrate), molecules of other substances—most importantly amino

ASPECTS OF EXERCISE PHYSIOLOGY

Aerobic Exercise: What For and How Much?

◆

Aerobic ("with oxygen") exercise involves large muscle groups and is performed at a low-enough intensity and for a long-enough period of time that fuel sources can be converted to ATP by using the Krebs cycle and the electron transport system as the predominant metabolic pathways. Aerobic exercise can be sustained for from 15 to 20 minutes to several hours at a time. Short-duration, high-intensity activities, such as weight training and the 100-meter dash, which last for a matter of seconds and rely solely on energy stored in the muscles and on glycolysis, are forms of *anaerobic* ("without oxygen") exercise.

Inactivity is associated with increased risk of developing both hypertension (high blood pressure) and coronary artery disease (which can narrow arteries that supply the heart). To reduce the risk of hypertension and coronary artery disease and to improve physical work capacity, the American College of Sports Medicine recommends that an individual participate in aerobic exercise a minimum of three times per week for 20 to 60 minutes. The intensity of the exercise should be based on a percentage of the individual's maximal capacity to work. The easiest way to establish the proper intensity of exercise and to monitor intensity levels is by checking heart rate. The estimated maximal heart rate is determined by subtracting the person's age from 220. Significant benefits can be derived from aerobic exercise performed between 70% and 80% of maximal heart rate. For example, the estimated maximal heart rate for a 20 year old is 200 beats per minute. If this person exercised three times per week for 20 to 60 minutes at an intensity that increased the heart rate to 140 to 160 beats per minute, the participant should significantly improve his or her aerobic work capacity and reduce the risk of cardiovascular disease.

◆ **FIGURE 25.6 ATP yield from the aerobic breakdown of glucose**

The aerobic breakdown of a single molecule of glucose produces a net profit of 36–38 ATP. A net profit of 4 ATP is produced by substrate-level phosphorylation during glycolysis and the Krebs cycle, and 32–34 ATP are produced by oxidative phosphorylation during electron transport. The glycolytic reactions occur in the cytoplasm of the cell. The other reactions take place within the mitochondria.

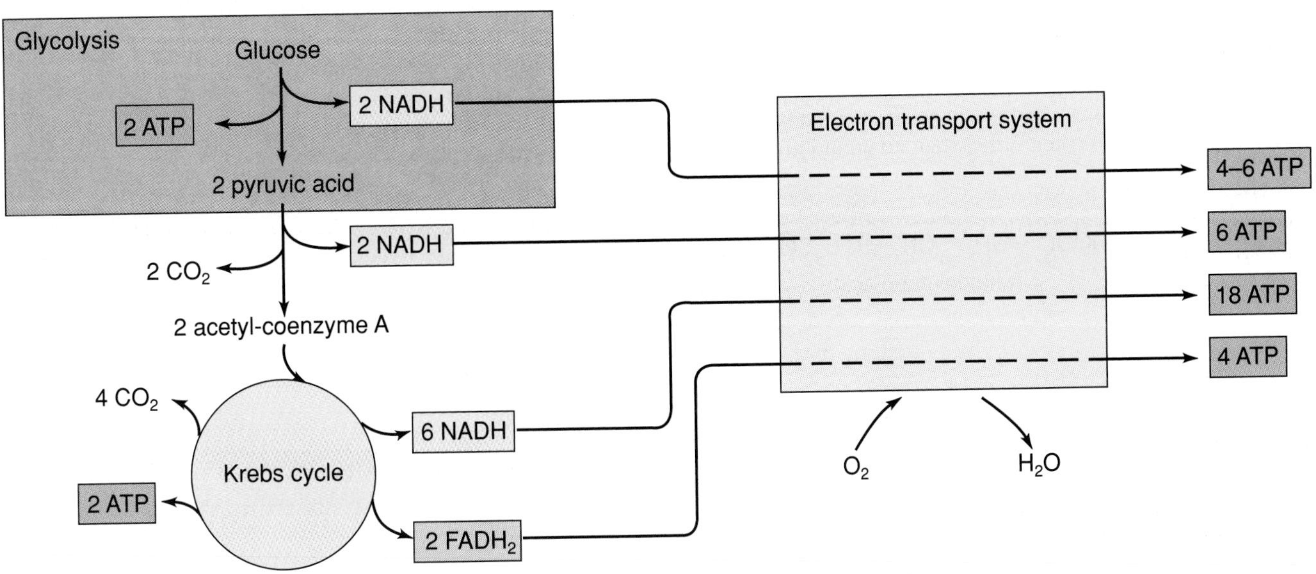

acids and triglycerides—can undergo metabolic conversion to molecules that can enter the glycolytic and Krebs cycle pathways and thereby provide energy for body activities.

Amino Acids

Amino acids can lose their nitrogen-containing amino groups and be converted to **α-keto acids (alpha-keto acids)** that can ultimately enter the Krebs cycle—for example, by way of pyruvic acid or the Krebs cycle component oxaloacetic acid, both of which are α-keto acids. (An α-keto acid is similar to an amino acid except that it has oxygen rather than an amino group bonded to its alpha carbon.)

The process of converting an amino acid to an α-keto acid generally involves a reaction known as transamination. In this reaction, the amino group of an amino acid is exchanged with the oxygen of an α-keto acid, most commonly α-ketoglutaric acid (Figure 25.7a). The acquisition of an amino group by α-ketoglutaric acid in exchange for its oxygen produces a molecule of glutamic acid (an amino acid), and the amino acid that provides the amino group is converted to an α-keto acid. Glutamic acid can then undergo a reaction called oxidative deamination. In this reaction, the amino group of glutamic acid is removed, with the resultant production of ammonia, and α-ketoglutaric acid is reformed (Figure 25.7b).

The overall result of the transamination reaction involving α-ketoglutaric acid and the subsequent deamination of glutamic acid is to transform an amino acid into an α-keto acid, with the resultant production of ammonia. As explained previously, the α-keto acids produced in this process can ultimately enter the Krebs cycle. The ammonia is converted to urea by the liver, and the urea enters the bloodstream and is excreted by the kidneys.

Triglycerides

Triglycerides can be broken down into their components, glycerol and fatty acids. The glycerol can be converted to glyceraldehyde-3-phosphate, which is one of the substances of the glycolytic pathway. The fatty acids can undergo a stepwise breakdown into molecules of acetyl-coenzyme A by the process of beta oxidation (Figure 25.8). In this process, a fatty acid molecule is activated by combination with a molecule of coenzyme A. After a series of reactions, two carbons of the fatty acid carbon chain are broken away as a molecule of acetyl-coenzyme A, which can enter the Krebs cycle. The remainder of the fatty acid carbon chain can be recycled through the reactions and another two carbons can be split off as acetyl-coenzyme A. This process can be repeated until the entire carbon chain of the fatty acid is utilized.

Other Uses of Food Molecules

In addition to being broken down and utilized as sources of energy, food molecules are also built into larger structures. For example, amino acids can be as-

◆ **FIGURE 25.7 Transamination and oxidative deamination reactions**

(a) Transamination reaction involving α-ketoglutaric acid and an amino acid. In this reaction, the amino group of the amino acid is exchanged with the oxygen of α-ketoglutaric acid. This exchange converts α-ketoglutaric acid to glutamic acid, and the amino acid to an α-keto acid. (Note that an α-keto acid is similar to an amino acid, except that it has an oxygen rather than an amino group bonded to its alpha carbon.) (b) Oxidative deamination of glutamic acid. This reaction forms α-ketoglutaric acid and ammonia. The overall result of the transamination reaction involving α-ketoglutaric acid and the subsequent oxidative deamination of the glutamic acid that is produced is to transform an amino acid into an α-keto acid, with the resultant production of ammonia.

$$\underset{\text{Amino acid}}{R_1-\overset{NH_2}{\overset{|}{CH}}-COOH} + \underset{\alpha\text{-ketoglutaric acid}}{HOOC-CH_2-CH_2-\overset{O}{\overset{\|}{C}}-COOH} \xrightleftharpoons{\text{Enzymes}} \underset{\alpha\text{-keto acid}}{R_1-\overset{O}{\overset{\|}{C}}-COOH} + \underset{\text{Glutamic acid}}{HOOC-CH_2-CH_2-\overset{NH_2}{\overset{|}{CH}}-COOH}$$

(a)

$$\underset{\text{Glutamic acid}}{HOOC-CH_2-CH_2-\overset{NH_2}{\overset{|}{CH}}-COOH} + H_2O \xrightarrow[NAD^+ \rightarrow NADH]{\text{Enzyme};\ H^+} \underset{\alpha\text{-ketoglutaric acid}}{HOOC-CH_2-CH_2-\overset{O}{\overset{\|}{C}}-COOH} + \underset{\text{Ammonia}}{NH_3}$$

(b)

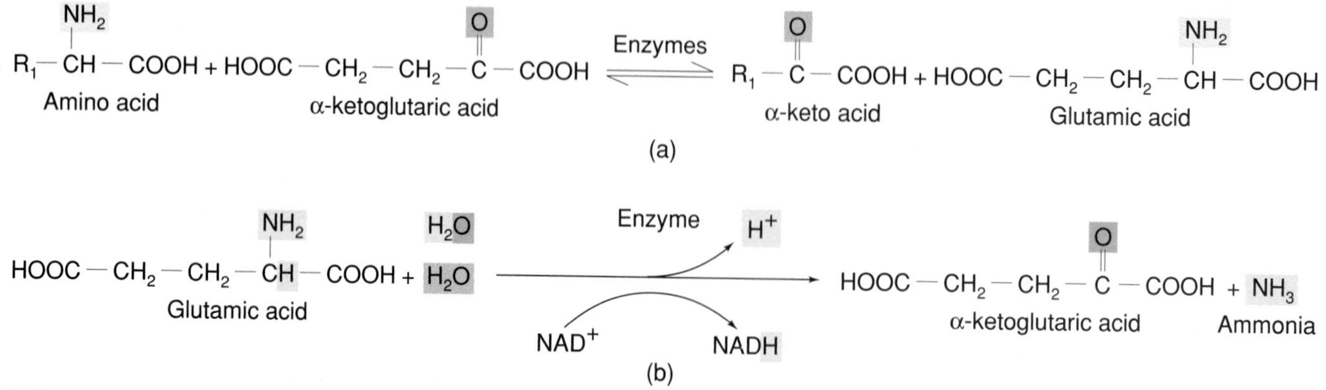

◆ **FIGURE 25.8 Breakdown of a molecule of fatty acid by beta oxidation**
CoA-SH is coenzyme A.

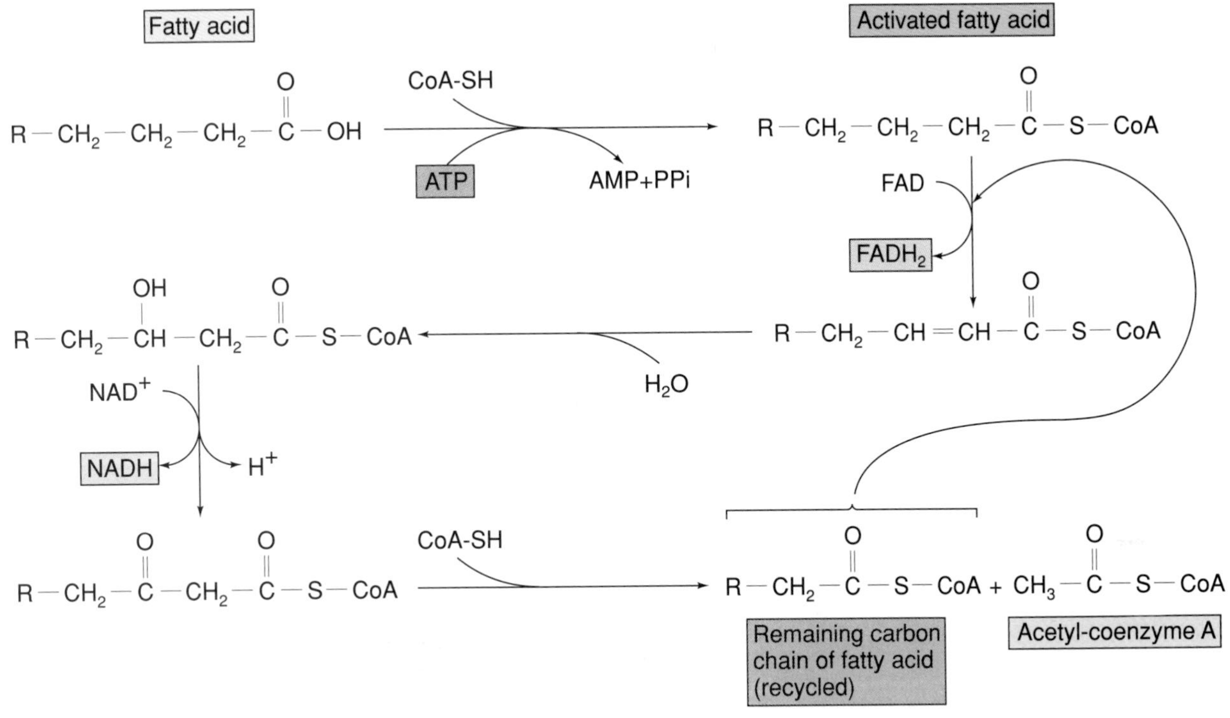

sembled into proteins that function as enzymes or structural cellular components, and glucose molecules can be linked together to form large glycogen molecules, which are a storage form of carbohydrate. Moreover, when considering the uses of the various food molecules, it is important to realize that the metabolic pathways followed by carbohydrates (such as glucose), proteins (amino acids), and triglycerides are interrelated (Figure 25.9). Thus, pathways exist by which amino acids or glucose, for example, can be converted to triglycerides, or by which amino acids can form glucose. The liver is especially active in such interconversions; so is adipose tissue.

The ability to perform interconversions enables the body to synthesize many of the substances it needs. But some essential substances cannot be manufactured by the body, and these must be obtained from external sources. For example, the amino acids of proteins are essential sources of nitrogen. Although the body can form some amino acids, it cannot synthesize all of the amino acids it needs. Therefore, certain amino acids, called essential amino acids, must be obtained in the diet. In addition, some fatty acids cannot be synthesized, and these must also come from dietary sources. The same is largely true of vitamins, many of which serve mainly as coenzymes in chemical reactions.

Nutrient Pools

In any consideration of metabolism, it must be remembered that the human body is not a static entity but rather exists in a dynamic steady state. The processes involved in the synthesis of protoplasm from simpler food molecules **(anabolism)** occur simultaneously with the breakdown of molecules to provide energy **(catabolism).** Thus, almost all the body's atoms and molecules are in a continual state of flux; even the atoms and molecules of bone are continually being exchanged.

In view of the continual turnover of the body's atoms and molecules, it is useful to consider the existence of nutrient pools within the body that can be drawn upon to meet its needs. For example, ingested protein (as well as the body's own protein) can be broken down to amino acids that contribute to an amino acid pool (Figure 25.10). Amino acids from this pool can be used for the synthesis of proteins and other nitrogen-containing structures required by the body. Amino acids can also have their nitrogen removed and be used as energy sources or be converted to carbohydrate or triglycerides.

Since carbohydrates are readily converted to triglycerides, carbohydrate and triglyceride pools are often

◆ **FIGURE 25.9 The interrelation of the metabolic pathways followed by carbohydrates, proteins, and triglycerides**

By means of such interrelated pathways, different molecular classes may often be interconverted.

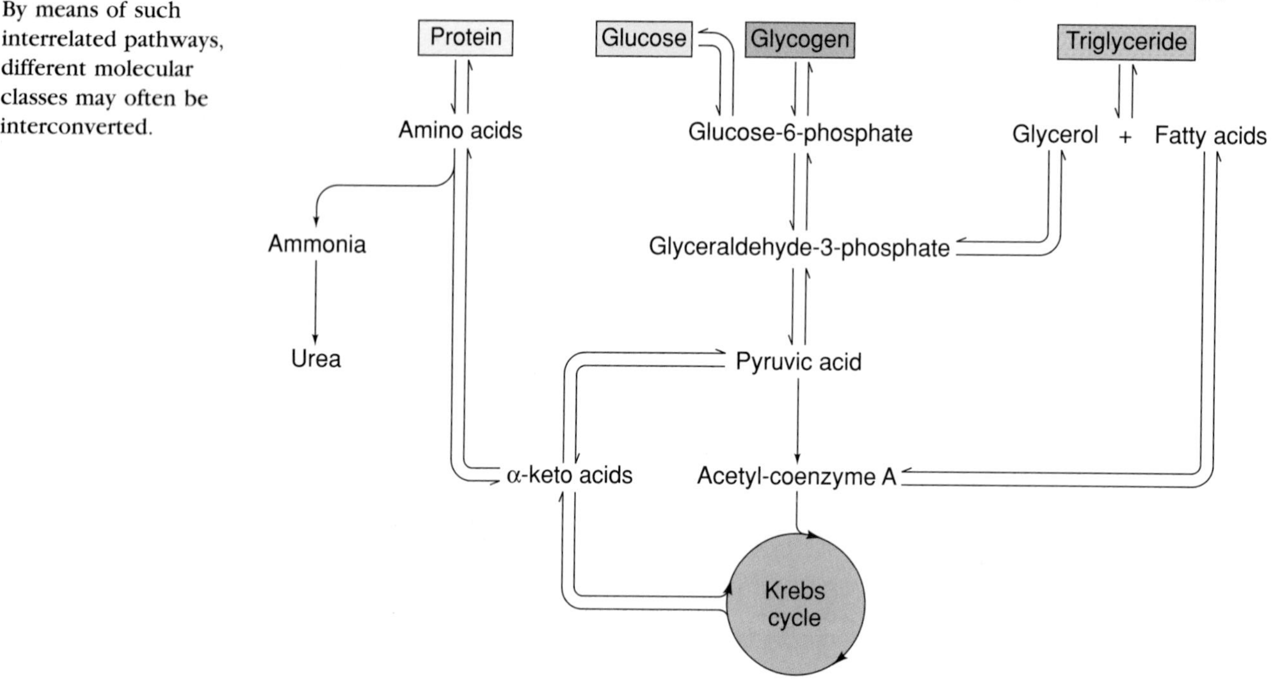

considered together (Figure 25.11). Carbohydrates and triglycerides are major energy sources for the body. The body is able to store excess triglyceride as well as some excess carbohydrate. Triglyceride stores, however, provide the major energy reserves for the body, and only comparatively small amounts of energy are stored as carbohydrate (glycogen).

Metabolic States

The human body actually exists in two different metabolic states. After eating, the body is in an *absorptive state* in which food substances are being absorbed from the gastrointestinal tract into the blood and lymph.

◆ **FIGURE 25.10 Amino acid pool, showing contributors to the pool and the uses of pool amino acids**

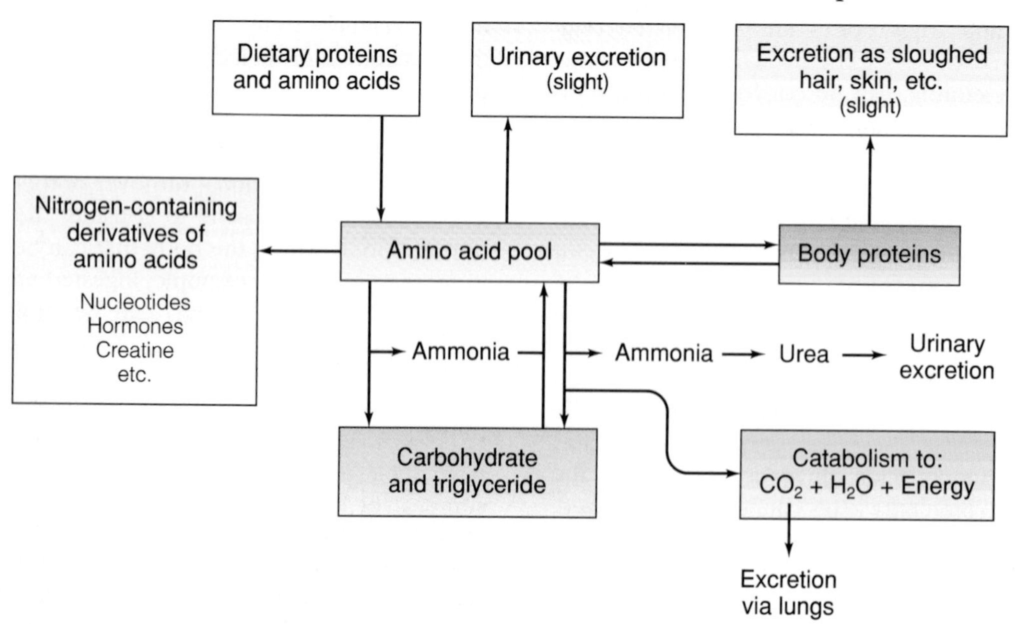

◆ **FIGURE 25.11 Carbohydrate and triglyceride pools, showing contributors to the pools and the uses of pool carbohydrates and triglycerides**

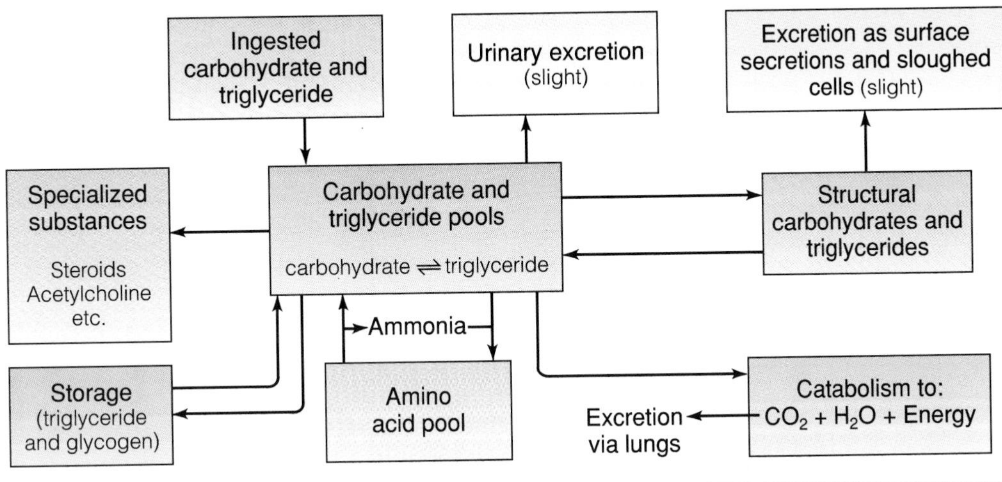

When absorption is complete, the body is in a *post-absorptive*, or fasting, *state*. In the postabsorptive state, a person's needs must be met by the use of materials already present within the body.

Absorptive State

Carbohydrate

During the absorption of a typical meal (65% carbohydrate, 25% protein, and 10% triglyceride), the body usually utilizes carbohydrate (glucose) as its major energy source, although some energy is derived from absorbed amino acids and triglycerides (Figure 25.12). During the absorptive state, the body stores as glycogen some carbohydrate not used for energy, and it converts much to adipose-tissue triglyceride.

Carbohydrate is absorbed from the digestive tract largely in the form of the monosaccharides glucose, galactose, and fructose. However, glucose is the principal monosaccharide, and most of the galactose and fructose are either converted to glucose or enter essentially the same metabolic pathways as glucose. Therefore, the absorbed monosaccharides can be considered simply as glucose.

Glucose absorbed from the digestive tract enters the blood and is carried to the liver, where a large portion of it enters liver cells. The liver itself uses little glucose for energy during the absorptive state, and it converts much into triglyceride or glycogen. Some of the triglyceride is stored in the liver, but most is packaged into particles called **very-low-density lipoproteins (VLDLs),** which are released into the blood. (In addition to triglyceride, VLDLs contain cholesterol, phospholipid, and protein.) As is discussed later, many of the fatty acids of the VLDL triglycerides are utilized in the formation of adipose-tissue triglyceride. Moreover, adipose-tissue cells can take up and convert to triglyceride absorbed glucose in the blood that is not taken up by the liver. Skeletal muscle and certain other tissues can also take up some glucose and store it as glycogen.

Protein

During the absorptive state, the body uses amino acids for protein synthesis, and amino acids are also converted into adipose-tissue triglyceride (Figure 25.12). Some of the amino acids from digested protein are taken up by the liver, where their nitrogen is removed and they are converted to α-keto acids. The α-keto acids can ultimately be metabolized by the Krebs cycle to provide energy for the liver cells. They can also be converted to fatty acids, which can be synthesized into triglyceride, much of which is packaged into very-low-density lipoproteins and released into the blood. The liver also uses amino acids to synthesize proteins such as plasma proteins. Skeletal muscle cells and other cells take up many amino acids and synthesize proteins from them.

Triglyceride

During the absorptive state, the body uses some triglyceride for synthesis, and some is used as an energy source. However, much triglyceride is stored in adipose tissue (Figure 25.12). Triglycerides from the digestive system enter the lacteals of the lymphatic system as a major component of chylomicrons. The lymphatic system delivers the chylomicrons to the bloodstream. The enzyme lipoprotein lipase, which is associated with the

◆ **FIGURE 25.12 Absorptive-state metabolic pathways**
Although the diagram shows only muscle and liver tissue utilizing amino acids for protein synthesis, other tissues can also take up amino acids and synthesize protein from them.

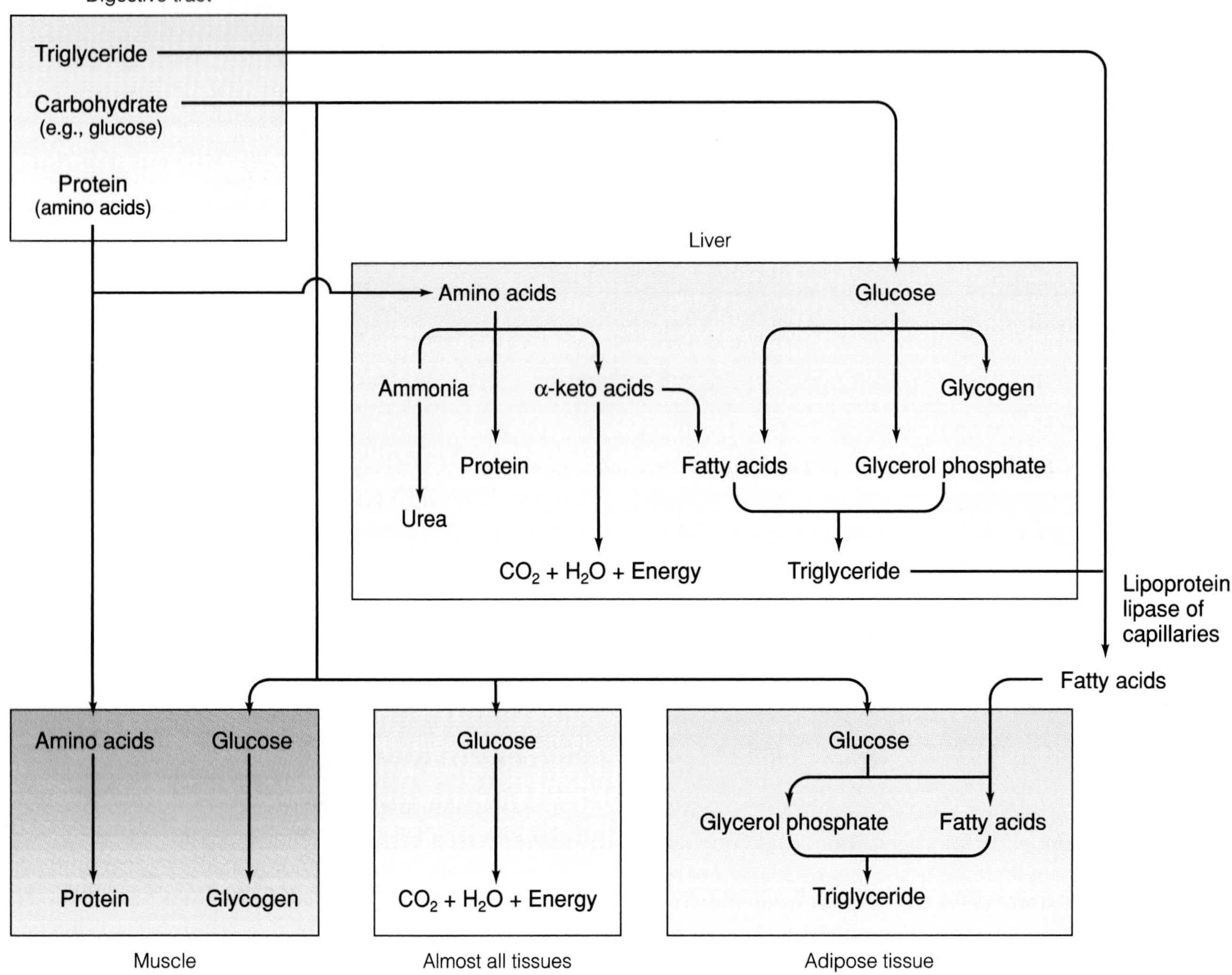

endothelium of capillaries, particularly in skeletal muscle and adipose tissue, breaks down the triglycerides of the chylomicrons (and also the triglycerides of the very-low-density lipoproteins from the liver), releasing free fatty acids. Some of the free fatty acids are taken up and utilized as an energy source by the muscle cells (not shown in Figure 25.12), and many are taken up and resynthesized into triglycerides by the adipose-tissue cells. The glycerol portion of the resynthesized triglycerides is derived from glucose.

When the absorptive-state uses of carbohydrates and amino acids as well as triglycerides are considered, it is evident that the components of adipose-tissue triglycerides are derived primarily from (1) chylomicron triglycerides, (2) triglycerides produced as a result of the synthetic activities of the liver, and (3) glucose.

Postabsorptive State

During the postabsorptive state, the blood-glucose level must be maintained, especially for the support of the nervous system, even though no glucose is being absorbed. The level is maintained by using body-glucose sources and glucose-sparing, fat-utilizing metabolic pathways (Figure 25.13).

Glucose Sources

The liver contains stores of glycogen that can be broken down to glucose. These glycogen stores, however, can support the body's activities for only about four hours. Other tissues, including skeletal muscle, also contain gly-

◆ **FIGURE 25.13 Postabsorptive-state metabolic pathways**

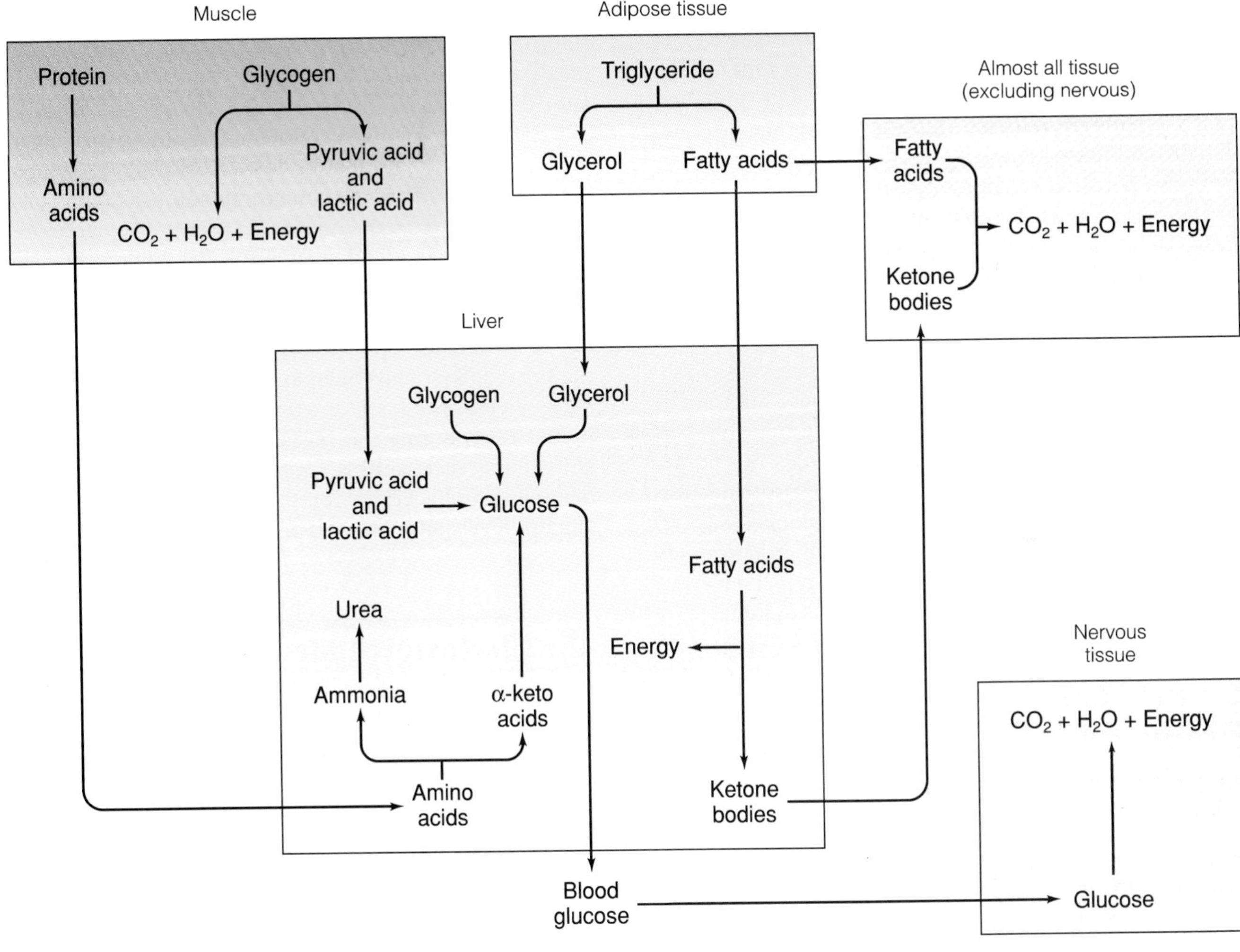

cogen reserves, but these too are limited. Moreover, skeletal muscle lacks the enzyme required to form free glucose from glycogen, and it cannot directly release glucose to maintain the blood-glucose level. Instead, skeletal muscle breaks glycogen down to glucose-6-phosphate, which it can catabolize for energy by way of glycolysis and the Krebs cycle. In addition, the pyruvic acid and, under anaerobic conditions, lactic acid that are produced by muscle glycolysis can enter the blood and travel to the liver, where they can be converted to glucose. As a result, muscle glycogen can serve as an indirect source of blood glucose.

The catabolism of triglycerides to glycerol and fatty acids, particularly in adipose tissue, can also provide some glucose, because glycerol can be converted to glucose by the liver.

Protein, however, is the major blood-glucose source during prolonged fasting. Large quantities of muscle protein that are not essential to cellular function (and to a lesser extent similar protein in other tissues) can be catabolized to amino acids, which in turn can be converted to glucose by the liver. During prolonged fasting, the kidneys as well as the liver produce substantial amounts of glucose for the body.

Glucose Sparing and Triglyceride Utilization

During the postabsorptive state, almost all the organs and tissues of the body—the nervous system is the major exception—diminish their use of glucose for energy and depend primarily on triglyceride. This adjustment spares glucose from the liver for use by the nervous system, which normally utilizes glucose as its principal energy source. During the postabsorptive state, adipose-tissue triglyceride is broken down and fatty acids are released into the blood. Almost all tissues except the nervous system can take up the fatty acids, which ultimately enter the Krebs cycle to provide energy for body activities.

During the postabsorptive state, the liver not only utilizes fatty acids as an energy source, but it also processes them into a group of substances called ketone bodies (acetoacetic acid, beta-hydroxybutyric acid, and acetone), which are released into the blood. Many tissues can take up these ketone bodies (specifically acetoacetic acid and beta-hydroxybutyric acid) and, by way of the Krebs cycle, utilize them as an important source of energy. Thus, the breakdown of adipose-tissue triglyceride during fasting ultimately provides energy for the body and spares glucose for the nervous system.

With prolonged fasting, a noteworthy change in brain metabolism occurs. Many areas of the brain can metabolize ketone bodies. As the levels of these substances increase after several days of fasting, the brain utilizes both ketone bodies and glucose as sources of energy, thereby reducing its use of glucose. Since protein is the major source of blood glucose during prolonged fasting, the utilization of ketone bodies by the brain conserves body protein that would otherwise have to be converted to glucose for the support of the nervous system.

Control of Absorptive- and Postabsorptive-State Events

Several hormonal and neural mechanisms control and regulate metabolism in both the absorptive and postabsorptive states. For example, when food is being absorbed following a meal—that is, during the absorptive state—the blood-glucose level may rise from about 85 mg per 100 ml to about 120 mg per 100 ml. The rise in the blood-glucose level (as well as one or more gastrointestinal hormones and increased parasympathetic neural activity to the pancreas) causes an increased release of the hormone insulin. With the notable exception of most areas of the brain, insulin facilitates the entry of glucose into many cells, particularly skeletal muscle and adipose-tissue cells. It favors glycolysis and the formation of glycogen within cells, and it inhibits glycogen breakdown. Insulin stimulates the active transport of amino acids into most cells, and it promotes protein synthesis and inhibits protein breakdown. Insulin also promotes the synthesis and inhibits the breakdown of triglycerides in adipose-tissue cells. Thus, in general, insulin promotes metabolic activities of the sort that occur during the absorptive state.

When absorption is completed and blood-glucose levels begin to fall—that is, during the postabsorptive state—insulin release diminishes. A diminished release of insulin favors the breakdown of glycogen, proteins, and triglycerides and the release of glucose, amino acids, glycerol, and fatty acids into the blood.

A declining blood-glucose level also promotes an increased release of the hormone glucagon. Glucagon stimulates the breakdown of glycogen and the production of carbohydrate from noncarbohydrate precursors (such as amino acids) in the liver. Glucagon also stimulates the breakdown of triglycerides in adipose tissue.

During acute hypoglycemia (but not during prolonged fasting or the ingestion of low-calorie diets), sympathetic neural activity to the liver, to adipose-tissue cells, and to the adrenal medullae increases. The increased sympathetic activity stimulates the breakdown of liver glycogen and adipose-tissue triglycerides. It also stimulates the release of the adrenal medullary hormones. These hormones—particularly epinephrine—stimulate the breakdown of glycogen in the liver and skeletal muscle as well as the breakdown of triglycerides in adipose tissue. Thus, during the postabsorptive state, hormonal and neural mechanisms mobilize carbohydrates, proteins, and triglycerides for the maintenance of blood-glucose levels and the support of body activities.

Cholesterol Metabolism

Cholesterol is both obtained from the diet and synthesized by cells, particularly by liver cells and to a lesser extent by intestinal cells. It is a component of cell membranes, and steroid hormones and bile salts are synthesized from it. However, a high level of cholesterol in the blood plasma is a significant risk factor for the development of atherosclerosis, which can lead to heart attacks and strokes.

Lipoproteins

Cholesterol is present in the plasma as a component of lipid-protein complexes called **lipoproteins.** There are several categories of lipoproteins, including very-low-density lipoproteins **(VLDLs),** low-density lipoproteins **(LDLs),** intermediate-density lipoproteins **(IDLs),** and high-density lipoproteins **(HDLs).** (In general, the greater the lipid content of a lipoprotein, the lower its density.) Chylomicrons present in the blood as a result of digestive activities are a separate category of lipoprotein.

Relationships among Lipoproteins

Lipoprotein lipase associated with capillary endothelial cells breaks down chylomicron triglycerides (page 813). The chylomicron remnants, which include dietary cholesterol, are taken up by the liver.

The liver produces cholesterol-containing VLDLs, which are released into the blood. Lipoprotein lipase breaks down VLDL triglycerides, and the VLDL remnants become IDLs. Some of the IDLs are rapidly taken up by the liver. Others remain in the blood and are converted to LDLs. LDLs are the major cholesterol carrier in the blood.

HDLs pick up cholesterol released from cells. HDL cholesterol is transferred to IDLs.

Regulation of Cellular Cholesterol

In addition to liver cells and intestinal cells, many other cells can synthesize cholesterol. However, these cells normally obtain most of the cholesterol they require by taking up LDLs. In this process, LDLs bind to LDL receptors on plasma membranes and are taken into the cells by receptor-mediated endocytosis. When the cholesterol content of the cells is high, the synthesis of cholesterol by the cells is suppressed. In addition, the synthesis of LDL receptors is suppressed, thereby reducing the uptake of LDLs.

HDLs also bind to receptors on the plasma membranes of cells. This binding apparently signals the cells to release cholesterol, which can be picked up by HDLs. Experiments indicate that when the cholesterol content of the cells is low, the synthesis of HDL receptors is suppressed. Conversely, when the cholesterol content of the cells is high, more HDL receptors are synthesized.

Role of the Liver

The liver has a central role in cholesterol metabolism and carries out the following activities:

1. The liver is the major site of cholesterol synthesis in the body.
2. The liver releases cholesterol-containing VLDLs.
3. The liver synthesizes bile salts from cholesterol. (Although most of the bile salts secreted with the bile are reabsorbed from the intestine, some are lost in the feces. This loss is a major route by which cholesterol is eliminated from the body.)
4. The liver takes up chylomicron remnants, which include dietary cholesterol. (When the levels of dietary cholesterol are high, cholesterol synthesis by the liver is suppressed.)
5. The liver takes up IDLs.
6. The liver takes up LDLs. (The liver is the major site of expression of LDL receptors, and it is believed that when excess dietary cholesterol accumulates in the liver, the production of liver LDL receptors is suppressed.)

Cholesterol and Atherosclerosis

Several studies have indicated that people with high plasma levels of cholesterol—particularly LDL cholesterol—have an increased risk of developing atherosclerosis. On the other hand, some researchers suggest that high plasma levels of HDL cholesterol reduce the risk of developing atherosclerosis. Thus, LDL cholesterol is sometimes viewed as "bad" cholesterol, and HDL cholesterol is thought of as "good" cholesterol.

Cholesterol is synthesized by body cells, and it is difficult to substantially lower the plasma cholesterol level solely by reducing the dietary intake of cholesterol. Moreover, when people habitually consume diets low in animal fats, their plasma LDL-cholesterol levels tend to remain low, but when even moderate amounts of animal fat are introduced into the diet, the plasma cholesterol level rises. Therefore, many nutritionists recommend the consumption of a low-cholesterol, low-fat diet that contains only limited amounts of animal fats.

Vitamins

Normal metabolism and growth require not only adequate supplies of carbohydrates, proteins, and fats, but also minute amounts of other organic nutrients called **vitamins.** Vitamins do not provide energy or serve as structural components. Rather, most vitamins function either directly or after chemical alteration as coenzymes or components of coenzymes involved in various metabolic reactions.

In general, vitamins cannot be synthesized by the body; they must be obtained from external sources. The principal source of vitamins is ingested food, although some vitamins, such as vitamin K, can be produced by bacteria in the large intestine. If raw materials, called provitamins, are provided, the body can assemble some vitamins. For example, the provitamin carotene found in foods such as spinach, carrots, liver, and milk can be used by the body in the production of vitamin A.

Vitamins are divided into two major groups: water-soluble vitamins and fat-soluble vitamins. Water-soluble vitamins, which include the B vitamins and vitamin C, are generally absorbed from the gastrointestinal tract by diffusion or mediated transport. (Vitamin B_{12} absorption requires intrinsic factor.) Fat-soluble vitamins, which include vitamins A, D, E, and K, are absorbed with digested dietary triglycerides.

As a general rule, the storage of the water-soluble vitamins by the body is slight. The fat-soluble vitamins, with the exception of vitamin K, are usually stored to a greater extent. Vitamins A and D, in particular, are

CLINICAL CORRELATION

Vitamin B_{12} Deficiency on a Vegan Diet

Case Report

THE PATIENT: A 46-year-old woman.

PRINCIPAL COMPLAINTS: Loss of appetite, weakness, sore tongue.

HISTORY: The patient had been well until about six months before admission, when she began to experience the symptoms that brought her to admission. She had experienced a psychological change five years earlier and become a "vegan" (one who eats no food of animal origin).

CLINICAL EXAMINATION: The patient was pale, with a moderate yellowish tinge to the skin, but she did not appear acutely ill. The tongue was red and smooth, with a glazed appearance. The respiratory rate was 16/min (normal: 12–17/min). The chest was clear, no masses were felt in the abdomen, and kidney function was normal. The heart rate was 82 beats/min (normal: 60–100 beats/min), and the arterial blood pressure was 125/75 mm Hg (normal 90–140/60–90 mm Hg). The hematocrit was 34% (normal: 37–48%), and the hemoglobin content was 11 grams/100 ml (normal: 12–16 grams/100 ml). The mean corpuscular volume (the mean volume of a red blood cell determined by dividing the hematocrit by the red cell count) was 101 μm^3 (normal: 83–97 μm^3). The leukocyte count was 4500 cells/mm^3 (normal: 4300–10,800 cells/mm^3), with a normal differential (distribution of cell types).

The microscopic examination of the blood revealed macrocytosis (abnormally large erythrocytes) and hypersegmented neutrophils. Since these findings suggested pernicious anemia, the serum level of vitamin B_{12} was measured and found to be 48 picograms/ml (normal: 200–600 picograms/ml). Folic acid levels were normal. An examination of the bone marrow revealed large, immature red cells (megaloblasts), large leukocytes with bizarrely shaped nuclei, and decreased stores of iron. The concentration of iron in the serum was 89 micrograms/100 ml (normal: 50–150 micrograms/100 ml), and the total iron-binding capacity was 302 micrograms/100 ml (normal: 250–410 micrograms/100 ml). (Normally, only about 25% of the binding sites on transferrin are occupied by iron, and the remain-

stored by the liver. Specific vitamins, examples of their functions in or importance to the body, and some of the results of their deficiencies are listed in Table 25.1.

Minerals

The body needs not only organic compounds but also certain inorganic elements, or *minerals.* Minerals make up 4–5% of the body by weight, and many of them are essential for life. Minerals are involved in numerous body functions, and the activity of many enzymes depends on their presence. They are frequently found in chemical unions with organic compounds—for example, the iron of the hemoglobin molecule. The minerals used by the body are generally present in ionized form. Specific minerals and examples of their functions in or importance to the body are listed in Table 25.2.

Metabolic Rate

The **metabolic rate** is the total energy expended by the body per unit time. This energy is derived from the breakdown of carbohydrates, proteins, triglycerides, and other organic molecules that are originally obtained in the diet. The energy is used for biological work such as the muscle contractions that move the body from one place to another or allow a person to manipulate objects in the external environment. It also appears as heat. Actually, only a relatively small amount of the energy is used for work; most appears as heat. However, the heat can be valuable in body-temperature regulation.

A reasonable estimate of the metabolic rate can be obtained by measuring the body's rate of oxygen consumption (Figure 25.14). For this purpose it is generally assumed that for every liter of oxygen consumed, the body produces 4.825 kilocalories of energy (1 kilocalorie = 1000 calories).

CLINICAL CORRELATION

der represent the latent iron-binding capacity of the plasma. If iron is deficient, fewer transferrin sites are occupied and the iron-binding capacity is increased.)

Because the principal cause of pernicious anemia is failure to absorb vitamin B_{12} due to loss of functional intrinsic factor associated with gastric atrophy, gastric secretory function was tested. Stimulation of the gastric parietal cells (which release both gastric acid and intrinsic factor) by gastrin revealed normal gastric secretion, and no antibodies to parietal cells or intrinsic factor were found. The absorption of B_{12} (as measured by the Schilling test) was found to be normal.

COMMENT: The normal gastric secretory function and uptake of vitamin B_{12} indicated that the low concentration of vitamin B_{12} in this patient's plasma (which indicates inadequate stores of vitamin B_{12} in the body) was related to her diet. The body normally stores about 5000 micrograms of vitamin B_{12}, mostly in the liver, and several years may be required for its depletion, even when no vitamin B_{12} is absorbed. The patient had been on her exclusively vegetable (vegan) diet for about five years. The best sources of vitamin B_{12} in the diet are fresh meat and dairy products. Vegetable tissues contain no vitamin B_{12}.

Deficiency of vitamin B_{12} alters DNA synthesis in many tissues of the body, but its effects are especially evident in the hematopoietic (blood-forming) system. Impaired synthesis of DNA retards mitosis, and this retardation causes the production of large blood cells, both erythrocytes and leukocytes. Mitosis of epithelial cells also is affected, which probably accounts for the abnormalities of the mouth and the gastrointestinal system.

The patient's diet probably also accounts for the mild deficiency of iron that was observed, because iron from vegetable sources is not absorbed well.

TREATMENT: The patient was given vitamin B_{12} supplement, 5 micrograms/day, and her condition improved dramatically.

OUTCOME: Six weeks after the beginning of therapy all of the symptoms relating to deficiency of vitamin B_{12} were gone, and the patient's hematologic profile was normal. Since she is committed to her vegan diet, she will have to continue to take supplementary vitamin B_{12}.

Basal Metabolic Rate

Many factors—including eating, exercise, and exposure to extreme environmental temperatures—significantly influence a person's metabolic rate. Therefore, for purposes of comparison between individuals, the metabolic rate is often determined under specific, standard conditions, which eliminate many of these variables. The metabolic rate determined under these conditions is called the **basal metabolic rate,** and it is usually expressed in terms of kilocalories of energy expended per square meter of body surface area per hour.

The conditions under which the basal metabolic rate is determined include having the person awake but in a supine position at mental and physical rest in a room with a temperature between 20°C and 25°C (68°F to 77°F). The person should also be in a postabsorptive state and have had nothing to eat for 12–18 hours. Under these conditions, no energy is expended in digesting, absorbing, and processing ingested nutrients, and none is expended by muscular contractions that do external work, such as moving objects in the environment. Moreover, the person is in moderate thermal surroundings.

The basal metabolic rate is not the body's lowest metabolic rate. The metabolic rate during sleep is 10–15% lower than basal, and this lower rate is presumably due, at least in part, to a more complete muscular relaxation during sleep.

Factors That Affect Metabolic Rate

Both sex and age influence metabolic rates. Basal metabolic rates tend to be lower in females than in males, and they tend to increase with age to approximately two or three years and decrease with age thereafter.

The ingestion of a typical meal containing carbohydrate, protein, and fat increases the metabolic rate 10–20% over basal. The effect of food consumption on me-

◆ **TABLE 25.1 Vitamins**

VITAMIN	EXAMPLES OF FUNCTIONS OR IMPORTANCE	SOME RESULTS OF DEFICIENCY
WATER-SOLUBLE VITAMINS		
Vitamin B_1 (Thiamine)	Involved in the metabolism of carbohydrates and many amino acids	Deficiency leads to beriberi, which is characterized in some cases by cardiac muscle weakness and heart failure, and in other cases by peripheral neuritis, with muscle atrophy and occasionally paralysis.
Vitamin B_2 (Riboflavin)	Component of FAD, which is involved in Krebs cycle reactions	Deficiency leads to cracking at the corners of the mouth.
Niacin (Nicotinic acid)	Component of NAD^+, which is involved in glycolytic, Krebs cycle, and other reactions	Deficiency gives to pellagra, which is characterized by diarrhea, dermatitis, and mental disturbances.
Vitamin B_6 (Pyridoxine)	Involved in amino acid and protein metabolism	Deficiency leads to dermatitis and perhaps convulsions and gastrointestinal disturbances in children.
Vitamin B_{12} (Cyanocobalamin)	Involved in erythrocyte maturation and nucleic acid metabolism	Deficiency results in macrocytic anemia, loss of peripheral sensation, and, in severe cases, paralysis.
Pantothenic acid	Component of coenzyme A, which functions in the entry of pyruvic acid into the Krebs cycle and in the degradation of fatty acids	
Folic acid	Involved in the synthesis of purines and thymine, which are required for DNA formation; important in erythrocyte maturation	Deficiency leads to macrocytic anemia, which is characterized by the production of abnormally large erythrocytes.
Biotin	Involved in amino acid and protein metabolism	Deficiency leads to muscle pain, fatigue, and poor appetite.
Vitamin C (Ascorbic acid)	Important in the formation of collagen and in the maintenance of normal intercellular substances throughout the body	Deficiency leads to scurvy, which is characterized by poor wound healing, defective bone formation and maintenance, and fragile blood-vessel walls.
FAT-SOLUBLE VITAMINS		
Vitamin A	Important in the formation of photopigments; apparently needed for the maintenance of certain epithelial surfaces, such as the mucous membranes of the eyes	Deficiency results in night blindness and corneal opacity.
Vitamin D	Needed for the proper absorption of calcium from the gastrointestinal tract	Deficiency leads to rickets in children and osteomalacia in adults.
Vitamin E	Inhibits the breakdown of certain fatty acids	
Vitamin K	Required for the synthesis of prothrombin and certain other clotting factors in the liver	Deficiency leads to retarded blood clotting and excessive bleeding.

tabolism is called the *specific dynamic action* (or *food-induced thermogenesis*). Protein ingestion has the greatest effect, and the ingestion of protein alone increases metabolism as much as 30%. The ingestion of carbohydrate alone may increase metabolism only 5%, and the ingestion of lipid alone increases metabolism approximately 8%. The energy expended in the digestion and absorption of food accounts for only a small portion of

◆ **TABLE 25.2 Minerals**

MINERAL	EXAMPLES OF FUNCTIONS OR IMPORTANCE
Calcium	Needed for the formation of bones and teeth; required for blood clotting; necessary for normal skeletal and cardiac muscle activity; important in normal nerve function
Sodium	Exerts a major influence on the osmotic pressure of the extracellular fluid; important in muscle and nerve function; major cation (positively charged ion) of the extracellular fluid
Potassium	Important in normal muscle and nerve function; major cation of the intracellular fluid
Phosphorus	Needed for the formation of bones and teeth; a component of energy compounds (such as ATP) that are involved in energy storage and transfer; a component of nucleic acids (DNA, RNA)
Magnesium	Important in normal muscle and nerve function; activates a number of enzymes
Iron	A component of hemoglobin, which is important in oxygen transport; a component of myoglobin; involved in the formation of ATP as a component of cytochromes of the electron transport system
Iodine	Essential for the formation of the thyroid hormones thyroxine and triiodothyronine, which help govern metabolism
Copper	Required in the manufacture of hemoglobin and the pigment melanin
Zinc	A component of several enzymes, such as carbonic anhydrase and carboxypeptidase, and thus important for reactions catalyzed by these enzymes
Fluorine	Normally present in teeth and bones; appears to provide protection against dental caries (cavities)
Manganese	Needed for the formation of urea; activates some enzymes
Cobalt	A component of vitamin B_{12}, which is necessary for the normal maturation of erythrocytes

◆ **FIGURE 25.14 Measurement of the metabolic rate by measuring oxygen consumption**
The utilization of 1 liter of oxygen is approximately equal to the liberation of 4.825 kilocalories (kcal) of energy.

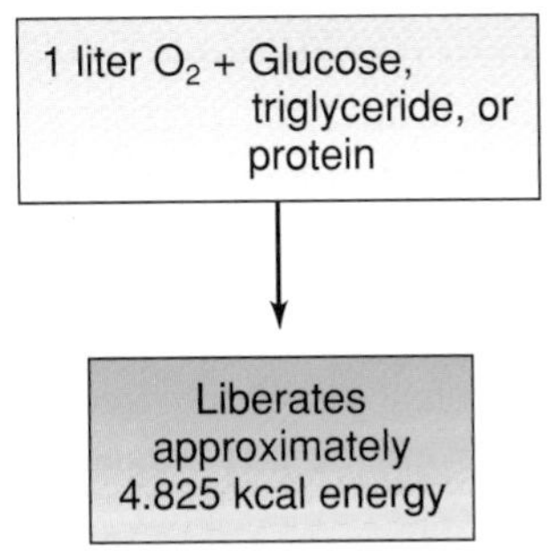

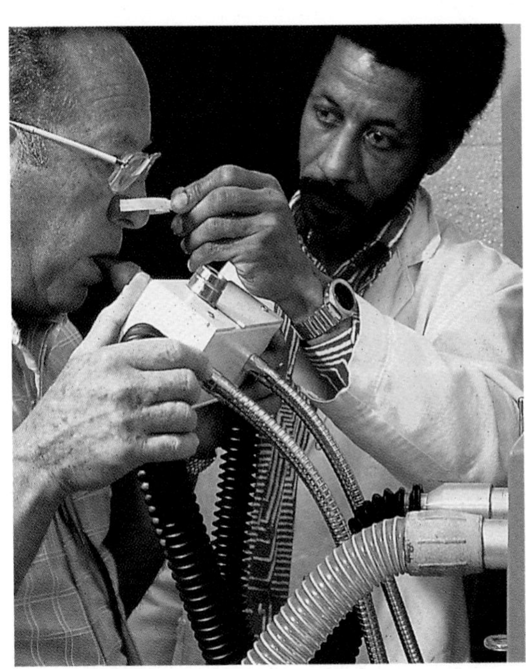

ASPECTS OF EXERCISE PHYSIOLOGY

All Fat Is Not Created Equal

There are different ways to be fat, and one way is more dangerous than the other. Obese patients can be classified into two categories—*android,* a male-type of adipose-tissue distribution, and *gynoid,* a female-type distribution—based on the anatomical distribution of adipose tissue measured as the ratio of waist circumference to hip circumference. Android obesity is characterized by abdominal fat distribution or upper body obesity, whereas gynoid obesity is characterized by fat distribution in the hips and thighs. Females can have android obesity as well as males, and males can exhibit gynoid obesity.

Android obesity is associated with a number of disorders, including insulin resistance, excessive insulin secretion, Type II diabetes mellitus, excess blood-lipid levels, hypertension, coronary heart disease, and stroke. Gynoid obesity is not associated with high risk for these diseases.

Because android obesity is associated with increased risk for disease, it is most important for overweight individuals with this type of adipose distribution to reduce their fat stores. Research on the success of weight-reduction programs indicates that it is very difficult for people to lose weight, but when weight loss occurs, it is from the area of increased stores.

Research has shed some light on the problems of obesity. Studies indicate that the resting metabolic rate might be an inherited trait. In one study, at three months of age, babies of obese parents showed 20% less energy expenditure than babies of lean parents. Also, specific dynamic action (SDA), which is the increase in metabolism that follows consumption of food, has been shown to be lower in those who have been obese since childhood. In other words, people who have been obese since childhood are very efficient at storing the excess calories they ingest. Lean people may metabolize more of the calories they ingest, with the energy being given off as heat.

Even after obese people reduce their weight to normal levels, their SDA remains lower than the SDA of someone of the same weight who has always been lean. This would be an admirable physiologic trait in times of food deprivation, but in times of food abundance, low SDA and a low metabolic rate can predispose an individual to obesity. A person with these traits has to eat less than his or her lean counterpart to maintain normal weight.

Because very-low-calorie diets are difficult to maintain, an alternative to severely cutting caloric intake to lose weight is to increase energy expenditure through physical exercise. An aerobic-exercise program further helps reduce the risk of the disorders associated with android obesity and helps to reduce fat stores.

the increase. Most of the increase appears to be due to the processing of exogenous nutrients by the liver.

The greatest changes in metabolic rates are produced by muscular activity. Strenuous exercise can increase metabolism as much as 15 times.

Regulation of Total Body Energy Balance

The concept of energy balance implies a relationship between energy intake and energy expenditure, or outflow. If an energy balance is to be maintained, the energy acquired through food must equal the energy expended by the body. If the energy taken in exceeds the body's needs, some is stored—for example, as adipose-tissue triglyceride—and the person gains weight. If the energy taken in is less than the body needs, stored forms of energy are utilized, and the person loses weight. If the body's energy requirements and energy intake are balanced, body weight remains stable.

Since the body weights of most adults do remain relatively stable, there must be some sort of regulation that maintains a reasonable balance between energy intake and energy expenditure. In this regard, it appears that controlling food (energy) intake in accordance with the energy expenditure requirements of the body is one of the major means of maintaining an energy balance.

Control of Food Intake

A number of brain areas, including certain areas of the hypothalamus, have been implicated in the control of food intake, and there are several theories about the

◆ **FIGURE 25.15 Possible inputs controlling food intake**

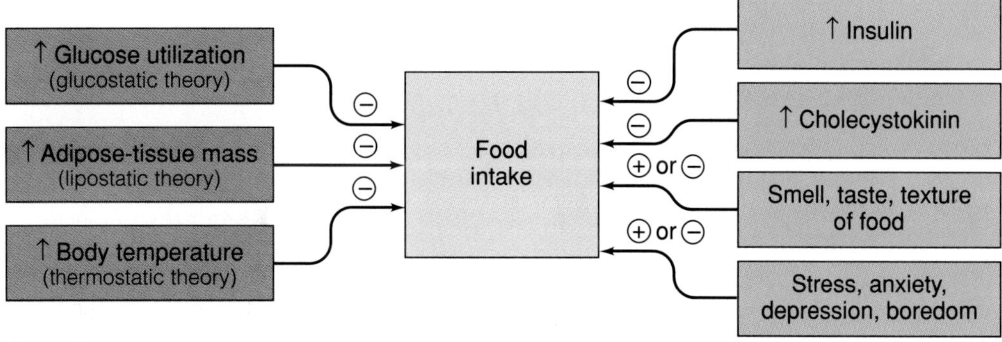

nature of the signals that influence these areas and thus alter feeding activity (Figure 25.15).

Glucostatic Theory

According to the glucostatic theory of hunger and feeding regulation, the brain contains glucose receptor cells (neurons) that are sensitive to their own rate of glucose utilization. It is believed that when their rate of glucose utilization is high (as may occur when blood-glucose levels increase, perhaps as a result of eating), the glucose receptor cells increase their activity, and feeding is depressed. Conversely, when their rate of glucose utilization is low (as may occur when blood-glucose levels decrease, perhaps as a result of fasting), the glucose receptor cells decrease their activity, and feeding is enhanced. In support of this theory is the fact that a decrease in the concentration of blood glucose can often be associated with the development of hunger. In addition, an increase in the blood-glucose level increases the electrical activity of certain regions of the hypothalamus.

Increased levels of amino acids in the blood also tend to depress feeding, whereas decreased amino acid levels tend to enhance it. Although this effect is not nearly as great as that involving blood glucose, it may contribute to food intake regulation in a manner similar to that seen for blood glucose.

Lipostatic Theory

According to the lipostatic theory of hunger and feeding regulation, some substance or substances (perhaps fatty acids and/or glycerol) are released from fat stores in adipose tissue in direct proportion to the total adipose-tissue mass. These substances act in a manner similar to that of glucose to depress feeding. Thus, the greater the total adipose-tissue mass, the less food is consumed. It is believed that such a mechanism, if it exists, would provide a reasonable form of long-term feeding regulation. In support of the lipostatic theory is the observation that increases in the quantity of adipose tissue in the body decrease the overall degree of feeding. In addition, the basal rate of glycerol release from adipose tissue appears to be directly related to the size of the adipose-tissue cells, and the long-term average concentration of free fatty acids in the blood is directly proportional to the amount of adipose tissue in the body.

Thermostatic Theory

According to the thermostatic theory of hunger and feeding regulation, brain areas involved in the control of body temperature interact with areas that control food intake. As a result of this interaction, higher body temperatures (as may be associated with exposure to warm environments or the specific dynamic action of food) tend to depress feeding. Conversely, lower body temperatures tend to enhance feeding. The thermostatic theory is supported by the observation that animals exposed to heat tend to undereat, whereas animals exposed to cold tend to overeat.

Hormonal Influences

The plasma concentrations of various hormones that affect the metabolism of glucose, amino acids, and triglycerides may influence brain areas that control food intake. For example, insulin, which is secreted in increased amounts during food absorption, tends to suppress hunger.

The gastrointestinal hormone cholecystokinin, which is secreted in association with the digestion of food, is believed to depress feeding. (Cholecystokinin is apparently also released in the brain during a meal, and it may depress feeding mainly by binding to receptors for cholecystokinin in the brain.)

CONDITIONS OF CLINICAL SIGNIFICANCE

Obesity

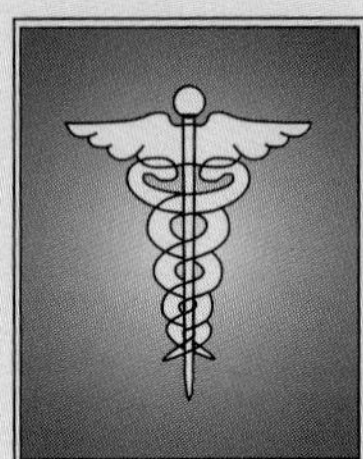

Obesity (the condition of being overweight) is a serious problem that often accompanies and contributes to a number of degenerative diseases, including high blood pressure, atherosclerosis, and diabetes. According to one view, obesity is the result of consuming too much energy (food) in relation to energy expenditure (physical activity). Indeed, there is evidence that many sedentary people eat more in relation to their degree of physical activity than active people, and thus tend to become obese. Most cases of obesity, however, can be only partly explained by lack of exercise. It is only during the time that obesity is developing that energy intake exceeds energy expenditure, and in many obese people, body weight eventually reaches some higher-than-normal value at which it remains relatively stable. Thus, the mechanisms that control energy intake in relation to energy expenditure appear to be operative, but they maintain body weight at a higher-than-normal level. Perhaps the brain areas that control food intake malfunction or the incoming signals are somehow upset. Some researchers think habits and social customs as well as genetic factors contribute to the problem.

Psychological Factors

Caloric balance may be influenced by such reinforcing factors as the smell, taste, and texture of food. Moreover, psychological factors such as stress, anxiety, depression, and boredom may affect food consumption.

Regulation of Body Temperature

Humans maintain a relatively constant body temperature, which is an important aspect of homeostasis because it allows the heat-sensitive chemical reactions of the body to proceed in a stable fashion. Under most conditions, the body temperature is higher than the environmental temperature. Thus, the heat produced by metabolism is of value in maintaining body temperature.

When determined in the morning under carefully controlled conditions, the oral temperature is approximately 36.7°C (98.1°F). However, not all parts of the body are the same temperature. The core or internal body temperature is usually considerably higher than the temperatures of the skin and external body parts. Moreover, the body temperature varies somewhat with activity, and there is even a daily or circadian *(sir-kay´-dee-an)* body-temperature cycle, with the lowest temperature generally occurring in the early morning after a night's rest, and the highest temperature occurring in the evening.

Despite these variations, however, body temperature does remain relatively constant. For this constancy to occur, heat gain must equal heat loss, and both the production of heat by the body and the loss of heat from the body are subject to some degree of control.

Muscle Activity and Heat Production

When a person is exposed to cold, muscle tone increases. This increase in muscle tone ultimately leads to oscillating, rhythmic muscle tremors (10 to 20 per second) known as shivering, which can increase body-heat production several times.

When a person is exposed to a warm environment, muscle tone decreases. This decrease in tone may not cause much of a decline in body-heat production, however, because muscle tone is normally quite low. Moreover, high environmental temperatures tend to raise body temperature, which increases the rates of the body's chemical reactions and consequently results in the production of additional heat.

Chronic exposure to cold causes some organisms to increase their heat production by means other than increased muscular activity. This phenomenon, called nonshivering (chemical) thermogenesis, is believed to be mediated mainly by an increased secretion of epinephrine, which substantially increases metabolism in certain tissues. (Thyroid hormones may also be involved.) Nonshivering thermogenesis occurs to some degree in newborn humans, but it does not appear to be a significant phenomenon in adults.

Heat Transfer Mechanisms

Heat is transferred between the body and the environment by three principal means: radiation, conduction and convection, and evaporation.

Radiation

All dense objects, including the body, continually emit heat in the form of infrared rays, which are a type of electromagnetic radiation. Thus, the body is continually exchanging heat by **radiation** with objects in the environment. The net direction of the heat exchange between objects by radiation depends on the surface temperatures of the objects, with the net direction being from objects with warmer surfaces to objects with cooler ones.

The surface temperature of the body is usually higher than the surface temperatures of most objects in the environment. Consequently, the net movement of heat by radiation is usually away from the body. However, the reverse may be true in a very warm environment or in the case of an exchange involving a warm individual object in the environment.

Conduction and Convection

Conduction is the transfer of thermal energy from atom to atom or molecule to molecule as the result of direct contact between two objects. In this transfer, heat moves from the object with the higher surface temperature to the object with the lower one. Most commonly, body surfaces are in direct contact with air, and conductive heat exchanges occur between air molecules and the body. Usually, the air temperature is lower than the temperature of the body surface, and the body loses heat to the air by conduction.

When heat is conducted from the body surface to air molecules, the air molecules are warmed. Since warm air is less dense than cool air, the warm air molecules rise away from the body surface and are replaced by cooler air molecules. This process, which is called **convection,** greatly enhances the conductive exchange of heat by constantly bringing cool air molecules into contact with the body surface. In fact, without convection, the conductive exchange of heat between the body surface and the air would be negligible. Convection, and thus the loss of heat from the body to the air by conduction, is enhanced by wind.

Evaporation

The **evaporation** of water requires the input of heat (heat of vaporization). Thus, when water evaporates from body surfaces, it carries heat with it, and the evaporation of 1 gram of water removes about 580 calories of heat from the body. Water is lost from body surfaces in two ways: by insensible water loss, which is not controlled for purposes of temperature regulation, and by sweating, which is an important thermoregulatory activity.

Insensible Water Loss. The skin is not perfectly impermeable to water, and some evaporative water loss continually occurs across external body surfaces. Moreover, respiration continually passes air in and out of the moist respiratory passages and the lungs. The inhaled air becomes humidified, and as a result, water is lost from the body with each expiration. Insensible water loss, which is so named because a person is not usually aware of its occurrence, can be as great as 600 ml per day.

Sweating. Sweating requires active fluid secretion by sweat glands in the skin. Sweat is a dilute solution of sodium chloride that also contains urea, lactic acid, and potassium. In hot environments, sweat can be produced at rates exceeding 1.5 liters per hour. The sweat covers the body surface and evaporates, carrying heat with it.

Sweat must evaporate to produce its cooling effects. If it simply drips from the body, little benefit occurs. If the humidity (water-vapor concentration of the air) is high and the air is already saturated with water, no additional water can evaporate into the air. Thus, sweating is a more effective means of cooling in low-humidity environments than in high-humidity environments.

Blood Flow and Heat Exchange

The rate of heat exchange between the body and the environment by radiation and conduction depends in large measure on the temperature gradient (difference) between the body surface and the environment. This gradient is influenced by the blood flow to the body surface (Figure 25.16).

In warm environments, where the body must lose much heat in order to maintain a constant temperature, vessels to the skin dilate, and a large amount of warm blood flows from the core of the body to the surface for heat exchange with the environment. This large blood flow increases the temperature of the body surface and creates a more favorable gradient for heat loss.

In cool environments, where heat conservation is important, blood vessels to the skin constrict, reducing blood flow to the body surface. This reduction in blood

◆ **FIGURE 25.16 How the adjustment of blood flow to the body surface affects the thermal gradient for heat exchange by radiation and conduction**

(a) Vasodilation when exposed to heat. (b) Vasoconstriction when exposed to cold. All temperatures are in °C.

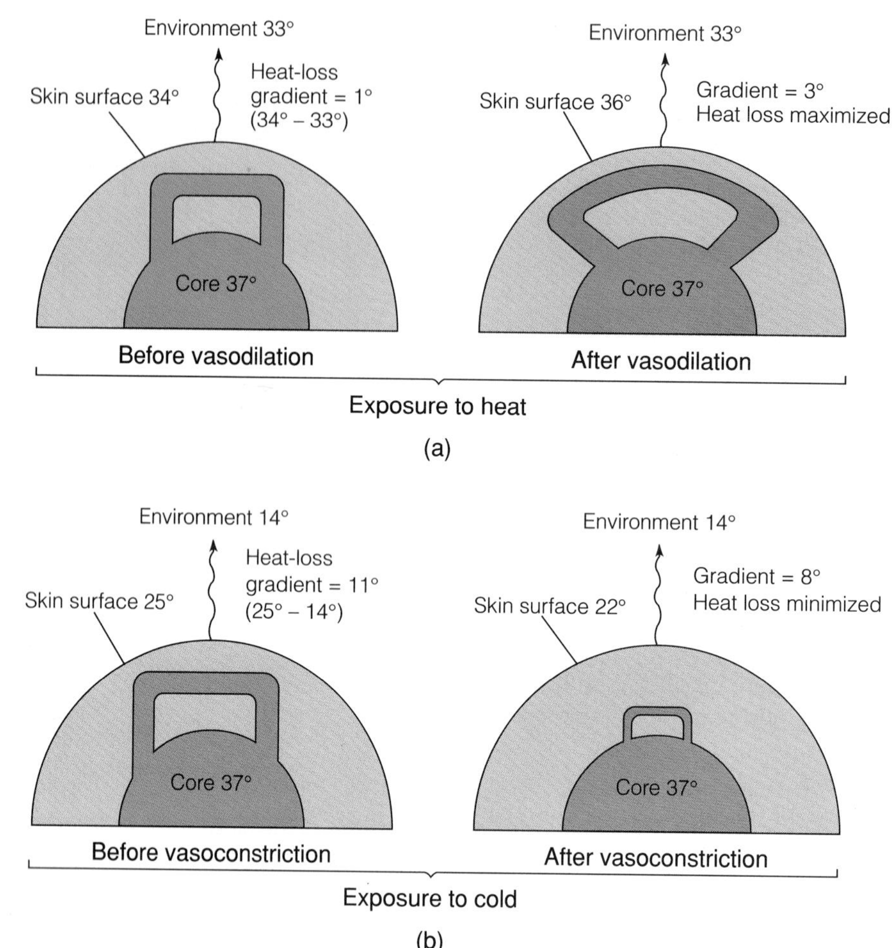

flow retains needed heat within the core of the body and reduces the temperature of the body surface, which decreases the gradient for heat loss to the environment.

Adjusting blood flow to the body surface is a generally effective means of balancing heat production and heat loss within the moderate temperature range of 20–28°C (68–82°F). At temperatures lower than about 20°C, large heat losses require increased heat production—for example, by shivering—for compensation. At temperatures above about 28°C, additional heat loss by the body as a result of the evaporation of water—that is, sweating—becomes necessary. In fact, when the environmental temperature is higher than the body temperature, the body gains heat by radiation and conduction, and evaporation is the only significant avenue of heat loss.

Humans do, of course, utilize other means of adjusting to high and low environmental temperatures. A person can, for example, reduce the body surface area over which heat loss occurs by curling into a ball. In addition, clothing helps to insulate the body from both heat and cold, as does shelter.

Brain Areas Involved in Thermoregulation

A number of brain areas, including certain areas of the hypothalamus, are involved in the integration of thermoregulatory activities (Figure 25.17). These areas receive input from temperature receptors in the skin and certain mucous membranes (peripheral thermoreceptors), as well as from receptors within the hypothalamus itself, the spinal cord, abdominal organs, and other internal structures (central thermoreceptors). The output of the integrating areas controls thermoregulatory activities such as adjustments in blood flow to the body surface, shivering, and sweating.

The operation of the body's temperature-control system appears to be analogous to that of a thermostat in which a particular temperature is set or called for (see page 17). If inputs to the integrating areas from various thermoreceptors indicate that the actual body temperature differs from the called-for temperature, heat-gain or heat-loss adjustments are instituted as necessary to bring

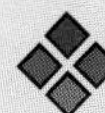

CONDITIONS OF CLINICAL SIGNIFICANCE

Thermoregulation

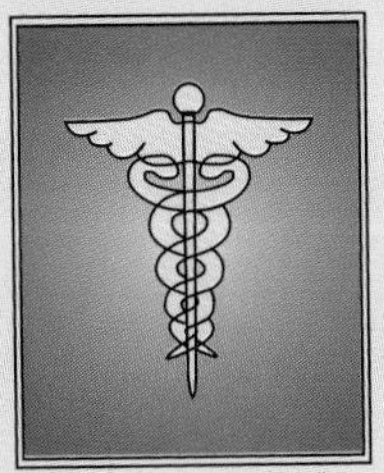

Fever

The discussed view of temperature regulation envisions fever as a resetting of the body-temperature thermostat to a new, higher level. This resetting can be accomplished by substances called endogenous pyrogens, which are released during conditions such as infection or inflammation (page 693). The effects of the endogenous pyrogens are believed to be mediated by prostaglandins. (The synthesis of prostaglandins is inhibited by aspirin, which may explain why aspirin reduces fever.)

When the body-temperature thermostat is set to a higher level in fever, the actual body temperature is initially below that level, and shivering, vasoconstriction, and chills can occur as the body attempts to raise its temperature to the new, higher setting. By the same token, when a fever breaks, the thermostat is reset to its original level, and the body temperature is now above that level. Consequently, sweating and vasodilation can occur as the body brings its temperature back to the lower setting.

Heat Stroke and Heat Exhaustion

Occasionally, under heat stress, the body's temperature-control system breaks down. In such a case, the body temperature rises rapidly, accompanied by a dry skin and the absence of sweating. This condition is called *heat stroke.*

A more common condition experienced by individuals exposed to heat is *heat exhaustion.* In this condition, the body's temperature-control system remains functional, but as a result of extreme sweating (fluid loss) and vasodilation to lose heat, the individual collapses and has low blood pressure (hypotension) and a cool, clammy skin, with little rise in body temperature. Heat exhaustion is usually much less dangerous than heat stroke.

the actual temperature to the called-for temperature. Thus, if the body temperature is lower than called for, heat-conserving and heat-producing activities such as vasoconstriction and shivering are stimulated. If the temperature is higher than called for, heat-loss activities such as vasodilation and sweating are initiated.

◆ **FIGURE 25.17 Thermoregulatory pathways and mechanisms**

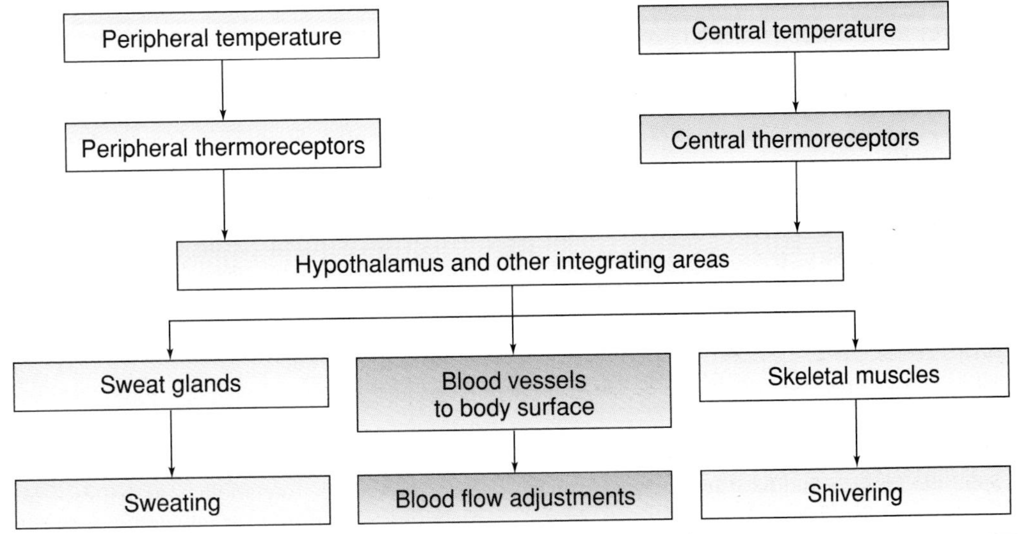

Researchers have suggested that inputs from the peripheral thermoreceptors enable the body's temperature-control system to make thermoregulatory adjustments before there is a change in the central or core body temperature. For example, if a person is exposed to cold, inputs from the peripheral thermoreceptors are believed to cause vasoconstriction and shivering before there is a decrease in the core body temperature. This type of response would be particularly useful in maintaining a stable body temperature in environments where the temperature fluctuates.

Study Outline

◆ UTILIZATION OF GLUCOSE AS AN ENERGY SOURCE pp. 804–808

Glucose is one important food material utilized in ATP formation.

Overview. In the body, glucose is broken down in a precisely controlled series of chemical reactions that release energy in a stepwise fashion.

Glycolysis.

1. Enzymes in cytoplasm; no molecular O_2 required.
2. Glucose molecule broken into two molecules of pyruvic acid; two ATP spent, four ATP generated, two NADH produced.

Fate of Pyruvic Acid. When O_2 is unavailable, pyruvic acid can be converted to lactic acid; uses two NADH, regenerating NAD^+. When O_2 is available, pyruvic acid enters the Krebs cycle.

Krebs Cycle.

1. Enzymes in mitochondria.
2. Two molecules of pyruvic acid each form acetyl-coenzyme A and enter Krebs cycle; produce six CO_2, two ATP, eight NADH, and two $FADH_2$.

Electron Transport System.

1. Transfer of electrons from NADH and $FADH_2$ to O_2, and combination of protons with O_2 to form water and regenerate NAD^+ and FAD; energy from electrons used to produce ATP.
2. Molecular oxygen required.

ATP and Energy Yield from the Aerobic Breakdown of Glucose. Glycolysis, Krebs cycle, electron transport together yield net profit of 36–38 ATP per glucose molecule.

◆ UTILIZATION OF AMINO ACIDS AND TRIGLYCERIDES AS ENERGY SOURCES pp. 808–810

Amino acids and triglycerides can undergo metabolic conversions to molecules that can enter glycolytic and Krebs cycle pathways, thereby providing energy for body activities.

Amino Acids. Lose amino groups and form α-keto acids that can ultimately enter glycolytic or Krebs cycle pathways.

Triglycerides. Split into glycerol and fatty acids; glycerol enters glycolytic pathway as glyceraldehyde-3-phosphate; fatty acids broken by beta oxidation to acetyl-coenzyme A, which can enter Krebs cycle.

◆ OTHER USES OF FOOD MOLECULES pp. 810–811

Built into larger structures; interconverted—for example, amino acids or glucose form triglycerides; essential amino acids, some fatty acids, and most vitamins must be obtained in diet.

NUTRIENT POOLS pp. 811–812

Amino Acid Pool. Protein synthesis; energy sources; conversion to carbohydrate or triglyceride.

Carbohydrate and Triglyceride Pools. Energy sources; storage.

◆ METABOLIC STATES pp. 812–816

Absorptive State.

CARBOHYDRATE. Usually major energy source (glucose); converted to adipose-tissue triglyceride; some stored as glycogen.

PROTEIN. Some protein synthesis; converted to adipose-tissue triglyceride.

TRIGLYCERIDE. Some used for synthesis; energy; most stored as adipose-tissue triglyceride.

Postabsorptive State.

GLUCOSE SOURCES. Liver glycogen supply; skeletal muscle glycogen supply; catabolism of triglycerides; protein provides major blood-glucose source.

GLUCOSE SPARING AND TRIGLYCERIDE UTILIZATION. Glucose reserved for nervous system; body energy largely from triglycerides.

Control of Absorptive- and Postabsorptive-State Events. Hormonal and neural mechanisms involved; insulin promotes absorptive-state metabolic activities; glucagon favors mobilization of stored carbohydrates and triglycerides.

◆ **CHOLESTEROL METABOLISM** pp. 816–817
Cholesterol is a component of cell membranes; steroid hormones and bile salts synthesized from it.

Lipoproteins. Lipid-protein complexes; cholesterol is present in plasma as component of lipoproteins. Several lipoprotein categories, including VLDLs, LDLs, IDLs, HDLs, chylomicrons.

Relationships among Lipoproteins. Lipoprotein lipase breaks down chylomicron triglycerides; chylomicron remnants taken up by liver. Liver releases VLDLs. Lipoprotein lipase breaks down VLDL triglycerides; VLDL remnants become IDLs, which are converted to LDLs. HDLs transfer cholesterol to IDLs.

Regulation of Cellular Cholesterol. Many cells obtain most of the cholesterol they require by taking up LDLs.

Role of the Liver. Liver synthesizes cholesterol; releases VLDLs; takes up chylomicron remnants, IDLs, LDLs; synthesizes bile salts.

Cholesterol and Atherosclerosis. Several studies indicate that people with high plasma levels of cholesterol—particularly LDL cholesterol—have increased risk of developing atherosclerosis. Some researchers suggest that high plasma levels of HDL cholesterol reduce risk of developing atherosclerosis.

◆ **VITAMINS** pp. 817–818
Most vitamins function either directly or after chemical alteration as coenzymes or components of coenzymes.

Water-Soluble Vitamins. Generally absorbed by diffusion or mediated transport; examples: most B vitamins, vitamin C; vitamin B_{12} absorption requires intrinsic factor.

Fat-Soluble Vitamins. Absorbed with digested dietary triglycerides; examples: vitamins A, D, E, K.

◆ **MINERALS** p. 818
Involved in numerous body functions; enzyme activity.

◆ **METABOLIC RATE** pp. 818–822
Total energy liberated by body per unit time; measured by determining rate of oxygen consumption.

Basal Metabolic Rate. Metabolic rate determined on awake, supine person; 20–25°C environment; postabsorptive state (12–18-hour fast period).

Factors That Affect Metabolic Rate. Age, sex, specific dynamic action of food, muscle activity.

◆ **REGULATION OF TOTAL BODY ENERGY BALANCE** pp. 822–824
If energy requirements and energy intake are balanced, body weight remains stable.

Control of Food Intake. A number of brain areas implicated; several theories about signals that influence these areas. Theories include the glucostatic theory, the lipostatic theory, and the thermostatic theory.

HORMONAL INFLUENCES. Plasma concentrations of hormones that affect metabolism of glucose, amino acids, and triglycerides (for example, insulin) may influence brain areas that control food intake. Cholecystokinin may depress feeding.

PSYCHOLOGICAL FACTORS. Caloric balance may be influenced by smell, taste, texture of food. Stress, anxiety, depression, and boredom may affect food consumption.

◆ **CONDITIONS OF CLINICAL SIGNIFICANCE: OBESITY** p. 824
Serious problem; often accompanies and contributes to a number of degenerative diseases.

◆ **REGULATION OF BODY TEMPERATURE** pp. 824–828
Human body temperature generally relatively constant; thus, heat gain equals heat loss.

Muscle Activity and Heat Production. In cold, shivering can increase body-heat production several times.

Heat Transfer Mechanisms.

RADIATION. Emission of heat in form of infrared rays; net direction of heat exchange is from objects with warmer surfaces to objects with cooler ones.

CONDUCTION AND CONVECTION. *Conduction:* transfer of thermal energy as result of direct contact between two objects; aided by *convection:* air molecules exchange heat with body surface and move away to be replaced by other molecules.

EVAPORATION. Heat is lost when water evaporates from body surfaces.

INSENSIBLE WATER LOSS. Occurs by way of exhaled air and through skin.

SWEATING. Active fluid secretion; cools body by evaporation.

Blood Flow and Heat Exchange. Adjusting blood flow to body surface is generally effective in balancing heat production and heat loss within moderate temperature range (20–28°C).

Brain Areas Involved in Thermoregulation.

1. Receive input from peripheral and central thermoreceptors.
2. If inputs to integrating centers indicate an actual body temperature different from called-for temperature, heat-gain or heat-loss adjustments are instituted as necessary.

◆ **CONDITIONS OF CLINICAL SIGNIFICANCE: THERMOREGULATION** p. 827

Fever. Body-temperature thermostat set higher; aspirin may inhibit synthesis of prostaglandins.

Heat Stroke and Heat Exhaustion.

HEAT STROKE. Failure of body-temperature-control system results in rapid rise of body temperature, dry skin, and no sweating.

HEAT EXHAUSTION. Extreme sweating and vasodilation lead to collapse, with low blood pressure, cool skin, and little rise in body temperature.

Self-Quiz

1. The breakdown of glucose by glycolysis: (a) occurs within mitochondria; (b) does not liberate energy in the form of ATP; (c) requires no molecular oxygen.
2. When oxygen is not available, which substance serves as an electron acceptor to regenerate NAD^+ from NADH? (a) glucose; (b) pyruvic acid; (c) lactic acid.
3. Lactic acid molecules are degraded to carbon dioxide and water during glycolysis. True or False?
4. Unlike glucose, α-keto acids are unable to enter the Krebs cycle. True or False?
5. The fatty acids of triglycerides can be split by beta oxidation into molecules of: (a) glycerol; (b) acetyl-coenzyme A; (c) glyceraldehyde-3-phosphate.
6. During the absorptive state, the major energy source of the body is usually: (a) carbohydrate; (b) amino acids; (c) triglycerides.
7. Lipoprotein lipase breaks down chylomicron triglycerides, releasing free fatty acids. True or False?
8. During the postabsorptive state, glycogen in skeletal muscle cells is broken down to glucose, which is released from the cells into the blood. True or False?
9. Pyruvic acid and lactic acid produced by muscle glycolysis can be transported by the blood to the liver, where they can be converted to glucose. True or False?
10. During periods of fasting, which tissue has the greatest need for glucose to supply itself with energy? (a) nervous system; (b) muscular system; (c) liver.
11. Which hormone tends to promote metabolic activities of the sort that occur during the absorptive state? (a) epinephrine; (b) glucagon; (c) insulin.
12. Vitamins function primarily as: (a) energy sources; (b) structural components of the body; (c) coenzymes.
13. For a determination of the basal metabolic rate, a person must be: (a) in a postabsorptive state; (b) asleep; (c) in a room with a temperature of 5°C.
14. The greatest change in metabolic rate occurs as the result of: (a) protein ingestion; (b) muscular activity; (c) carbohydrate intake.
15. The basal metabolic rate is determined when a person is asleep and in a postabsorptive state. True or False?
16. In general, if the body's energy requirements and energy intake are balanced, body weight remains stable. True or False?
17. The transfer of thermal energy from atom to atom or molecule to molecule that occurs as the result of direct contact between two objects is called: (a) conduction; (b) radiation; (c) convection.
18. In warm environments, vessels to the skin: (a) constrict; (b) dilate; (c) remain essentially the same as when in a cold environment.
19. A resetting of the body's thermostat to a new, higher level may be accomplished by: (a) aspirin; (b) endogenous pyrogens; (c) thermostatin.
20. A person suffering from heat exhaustion is likely to exhibit: (a) an absence of sweating; (b) low blood pressure; (c) an extremely high body temperature.

CHAPTER 26

The Urinary System

CHAPTER CONTENTS

LEARNING OBJECTIVES

After completing this chapter, you should be able to:

1. State the general functions of the kidneys.
2. Name the components of the urinary system.
3. Describe the embryonic development of the kidneys.
4. Describe the gross anatomical structure of the kidneys.
5. Describe the microscopic anatomy of a nephron.
6. Describe the filtration barrier that separates the blood in the glomerulus from the capsular space.
7. Explain the process of glomerular filtration.
8. Explain the processes of tubular reabsorption and tubular secretion.
9. Explain how the kidneys maintain a concentrated interstitial fluid in their medullary regions.
10. Explain how the kidneys produce different concentrations and volumes of urine.
11. List the physiological and neurophysiological chain of events leading to micturition.

CHAPTER 26

If the cells of the body are to survive and carry out their functions effectively, they must be surrounded by a stable environment. As noted in Chapter 1 (page 15), the maintenance of a relatively constant internal environment in the body is called *homeostasis.* To maintain homeostasis the concentrations of such substances as water, sodium, potassium, calcium, and hydrogen ions must remain relatively constant, as must the concentrations of a wide variety of cellular nutrients and products. Cellular metabolism constantly tends to upset the body's internal environmental balance by consuming some substances (such as oxygen and glucose) and producing others (such as carbon dioxide and urea). Also, substances may be added to the internal environment as a result of ingestion and removed from the internal environment by excretion.

Most body systems are involved in maintaining homeostasis. The digestive system, for example, supplies nutrients and also serves as a means of excreting some waste products. The lungs supply oxygen to the body and eliminate carbon dioxide. The skin also plays a minor role in excretion—sweat, for example, contains small amounts of urea and ammonia. The **kidneys,** however, are the main excretory organs, and they are critically important in maintaining the balance of substances required for internal constancy.

The kidneys eliminate from the body a variety of metabolic products, such as urea, uric acid, and creatinine. Further, the kidneys conserve or excrete water and electrolytes as required so that the internal balance of these substances is maintained. In fact, kidney malfunction can cause severe and even fatal problems as a result of upsets in fluid and electrolyte balance. The kidneys also act as endocrine structures. For example, they produce a hormone called *erythropoietin,* which stimulates the production of red blood cells (see page 578).

Because they are selective in what they excrete, the kidneys are able to maintain the internal environment within a range that is optimal for the survival of the body's cells. As a result of this selectivity, some substances that are vital to the cells may not be excreted at all, and other substances are excreted in varying amounts that depend largely on the needs of the body. Therefore, the kidneys have a major role in regulating the composition and pH of the interstitial fluid. Similarly, the amount of water removed by the kidneys from the blood, and thus from the interstitial fluid, varies with the needs of the body.

If the kidneys fail, there is no way for many of the substances they normally excrete to be removed from the blood. As a consequence, these substances accumulate in the blood and eventually in the interstitial fluid. Within a matter of days following kidney failure, the internal environment of the body can change so much that the body's cells no longer function. To prevent death, it is necessary to replace the nonfunctioning kidneys with healthy ones by means of a kidney transplant or to remove potentially harmful substances regularly from the blood with an artificial kidney machine.

Components of the Urinary System

The urinary system consists of the **kidneys,** which produce urine; the **ureters** *(ū-rē´-ters),* which carry urine to the **urinary bladder,** where it is temporarily stored; and the **urethra** *(ū-rē´-thra),* which transports urine to the outside of the body (Figure 26.1).

Embryonic Development of the Kidneys

The kidneys develop in the embryo from columns of mesodermal cells called *intermediate mesoderm* that form just lateral to the somites. The somites give rise to the vertebrae that will form the vertebral column and to the trunk muscles. The columns of intermediate mesoderm begin to form in the superior trunk region of the three-week-old human embryo. As the more inferior regions of the intermediate mesoderm develop into kidneys, the older, more superior regions degenerate.

During embryonic development, three pairs of kidneys form in the intermediate mesoderm (Figure 26.2). The first and most superiorly located kidney to form is called the **pronephros** *(pro-nef´-ros).* Even though the pronephros is never functional in humans, a **pronephric duct** develops and connects each pronephros with the exterior of the body through the embryonic cloaca. The pronephros begins to degenerate in the fourth week of embryonic development and is completely gone by about the sixth week. The pronephric ducts remain, however, and are utilized by the second pair of kidneys, the **mesonephros** *(mes-ō-nef´-ros).*

As the pronephros degenerates, the mesonephros develops from the columns of intermediate mesoderm inferior to the region of the pronephros. Tubules from the mesonephros join with the pronephric ducts, which are then called **mesonephric ducts.** By the sixth week, when development of the mesonephros has reached its inferior limit, its superior portions begin to degenerate. By about the eighth week, only the most inferior regions of the mesonephros remain. Although there is no direct evidence that the mesonephros functions in the human embryo, it appears to be structurally capable of producing urine.

◆ **FIGURE 26.1 Organs of the urinary system**
The anterior abdominal wall and most of the abdominal organs have been removed.

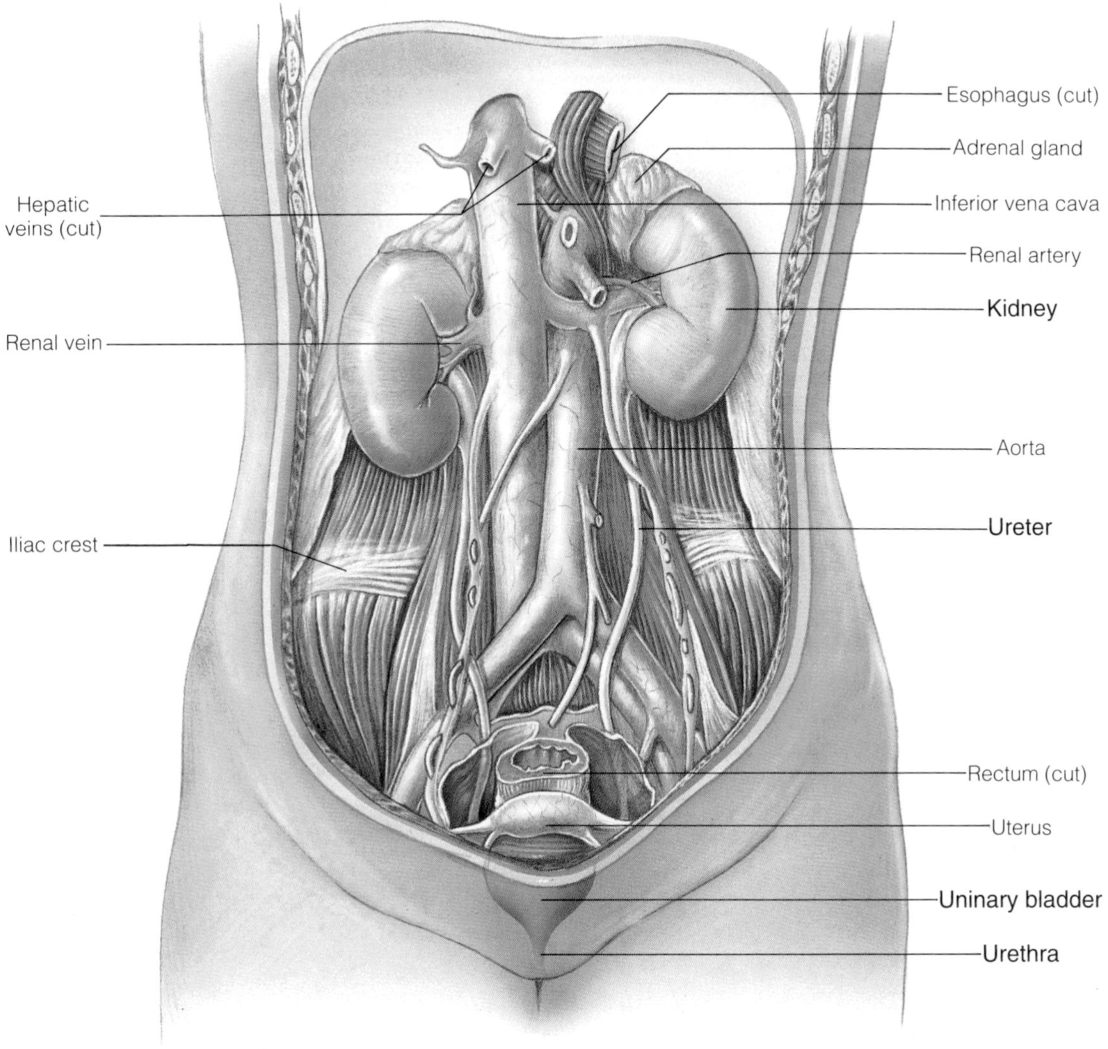

The **metanephros** *(met-a-nef´-ros)* is the third pair of kidneys to develop from the embryonic intermediate mesoderm. This pair develops into the adult kidneys. In about the fifth week of embryonic development, a hollow outgrowth called a **ureteric bud** *(u˝-rĕ-ter´-ik)* arises from the distal end of each mesonephric duct, near its junction with the embryonic cloaca. The ureteric buds grow dorsally and cranially into the most inferior regions of the intermediate mesoderm. The upper ends of the ureteric buds, which come into contact with the intermediate mesoderm, enlarge. Each ureteric bud then elongates and carries its cap of mesoderm cranially, eventually reaching what will be the permanent positions of the kidneys in the upper lumbar region. With further development, the *nephrons,* which are actively involved in the production of urine, form from the cap of intermediate mesoderm that covers each ureteric bud. The ureteric buds themselves develop into the *collecting tubules,* the *calyces (kay´-li-sees),* the *renal pelvises,* and the *ureters.* The collecting tubules are also actively involved in urine production, and the calyces, the renal pelvises, and the ureters transport to the urinary bladder the urine that is formed in the nephrons and collecting tubules. These structures remain in the adult; they are described in the next section.

 Anatomy of the Kidneys

The kidneys are paired reddish brown organs situated on the posterior wall of the abdominal cavity, one on each side of the verterbral column (Figure 26.1). Each

◆ **FIGURE 26.2 Embryonic development of the kidneys**
The illustration shows the sequential development of the pronephros, the mesonephros, and the metanephros on one side of the embryo.

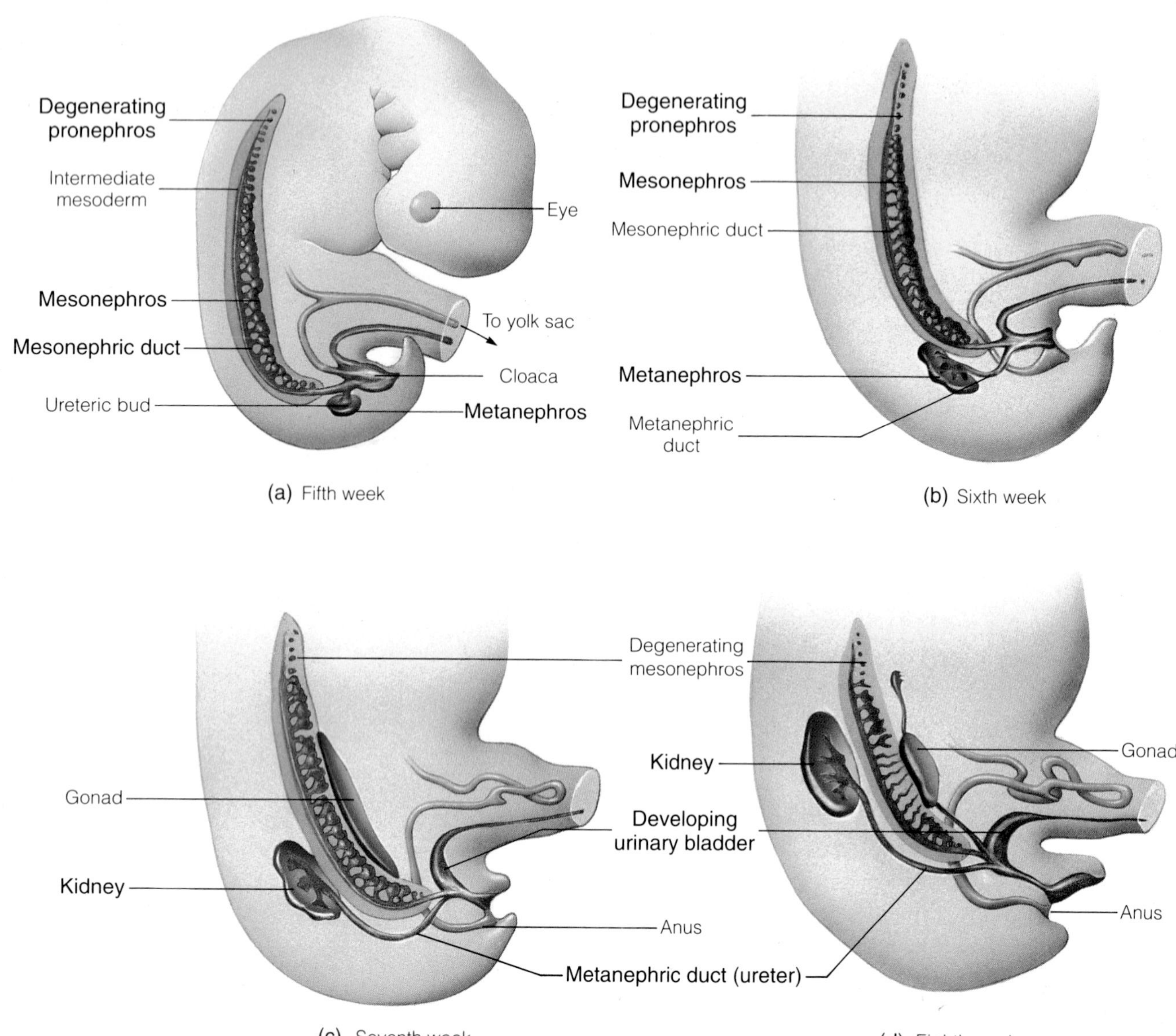

kidney is capped by an endocrine gland called the **adrenal gland.** The kidneys are approximately 11 cm long, extending from the level of the eleventh or twelfth thoracic vertebra to the third lumbar vertebra. Because of the presence of the liver, the right kidney is generally slightly lower than the left. The kidneys are located between the muscles of the back and the peritoneal cavity (Figure 26.3). This retroperitoneal location makes it possible for the kidneys to be exposed surgically through the posterior body wall, without opening the peritoneal cavity.

Tissue Layers Surrounding the Kidneys

Each kidney is surrounded by three layers of tissues (Figure 26.3). The innermost layer, which covers the surface of the kidney, is the fibrous **renal capsule.** Surrounding the renal capsule is a mass of perirenal fat called the **adipose capsule.** The third tissue layer that covers the kidney is a double layer of fascia called the **renal fascia.** The renal fascia surrounds the kidney and the adipose capsule, completely enclosing them and anchoring the kidney to the posterior abdominal wall. There is an ad-

◆ **FIGURE 26.3 Transverse section of the body trunk showing the retroperitoneal location of the kidneys and the renal fascia that surrounds them**

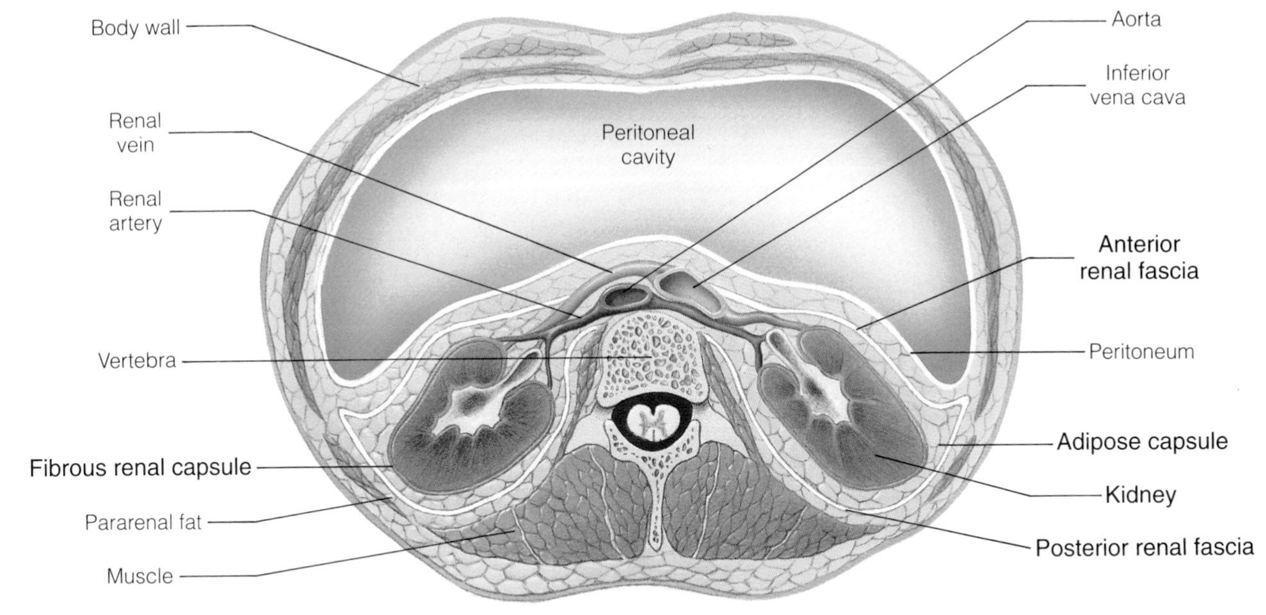

ditional accumulation of fat *(pararenal fat)* outside of the renal fascia.

External Structure of the Kidneys

The kidneys are bean-shaped, with convex lateral borders and concave medial borders. The medial border of each kidney contains an indentation, called the **renal hilus,** through which the renal arteries enter the kidney and the renal veins and the ureter leave. The hilus opens into a space, called the **renal sinus,** in which the renal vessels and the renal pelvis are located.

Internal Structure of the Kidneys

Three general regions can be distinguished in each kidney: the *cortex,* the *medulla,* and the *pelvis* (Figure 26.4).

The **cortex** is the outer layer of the kidney, just deep to the renal capsule. Extensions of the cortex, called **renal columns,** pass into the medulla of the kidney.

The **medulla** is located deep to the cortex and consists of several (up to 18) triangular **renal pyramids.** The pyramids are oriented so that their broad bases are covered by the cortex and their tips **(papillae;** *pa-pil´-lie***)** project toward the renal pelvis. The pyramids are separated from one another by the cortical renal columns. Blood vessels that supply the cortex and medulla pass through the renal columns.

The papilla of each pyramid projects into a funnel-shaped chamber called a **minor calyx;** however, one minor calyx may receive two or more papillae. Several minor calyces join together to form a **major calyx.** There are generally 2 or 3 major calyces and up to 13 minor calyces in each kidney. The major calyces join with one another to form the **renal pelvis,** which is the expanded upper end of the ureter. Urine passes as droplets from tiny pores in the papillae into the minor calyces. From there it travels into the major calyces, the renal pelvis, and finally into the ureter, which carries it to the urinary bladder.

Nephrons

The functional units of the kidneys where urine is formed are called **nephrons.** There are estimated to be over 1 million nephrons in each kidney. Some nephrons—called **juxtamedullary nephrons** *(jux´-ta-med´-u-lair-ee)*—have portions that extend deep into the medulla. Other nephrons—called **cortical nephrons**—do not penetrate as deeply into the medulla (Figure 26.5). Each nephron consists of two parts: (1) a network of parallel capillaries called a **glomerulus** *(glo-mer´-yoo-lus)* and (2) a **renal tubule.** Various regions of the renal tubule differ from one another ana-

◆ **FIGURE 26.4 Longitudinal section of a kidney illustrating its internal structure**

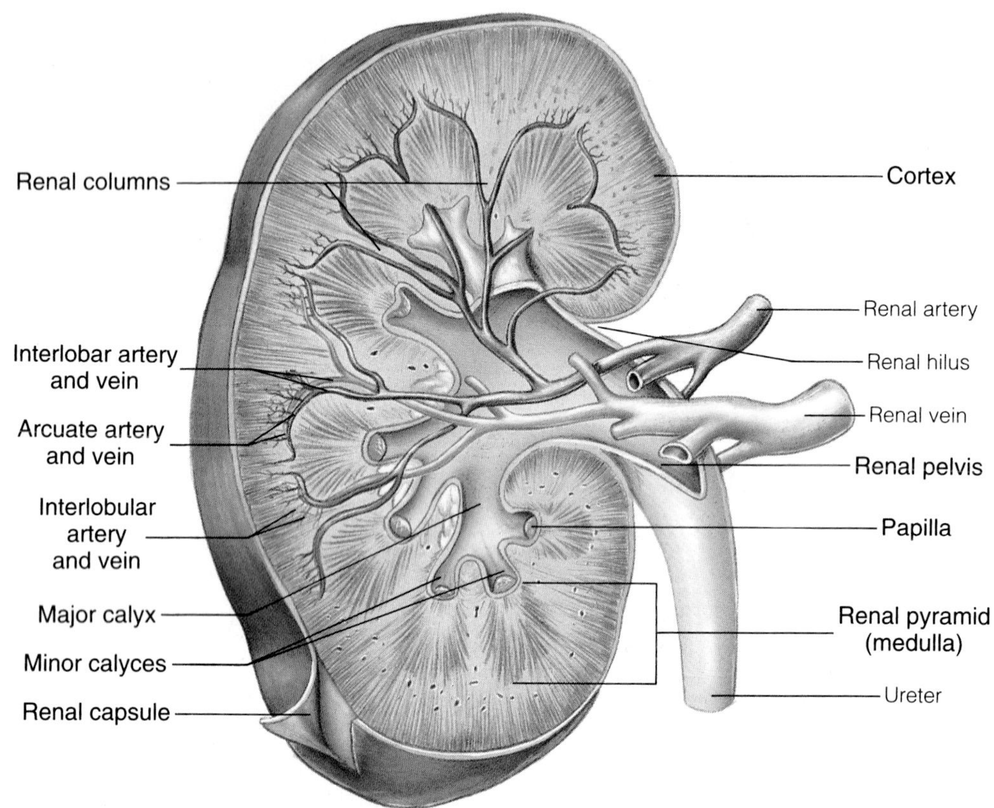

◆ **FIGURE 26.5 Longitudinal section of a kidney showing the location of cortical and juxtamedullary nephrons**

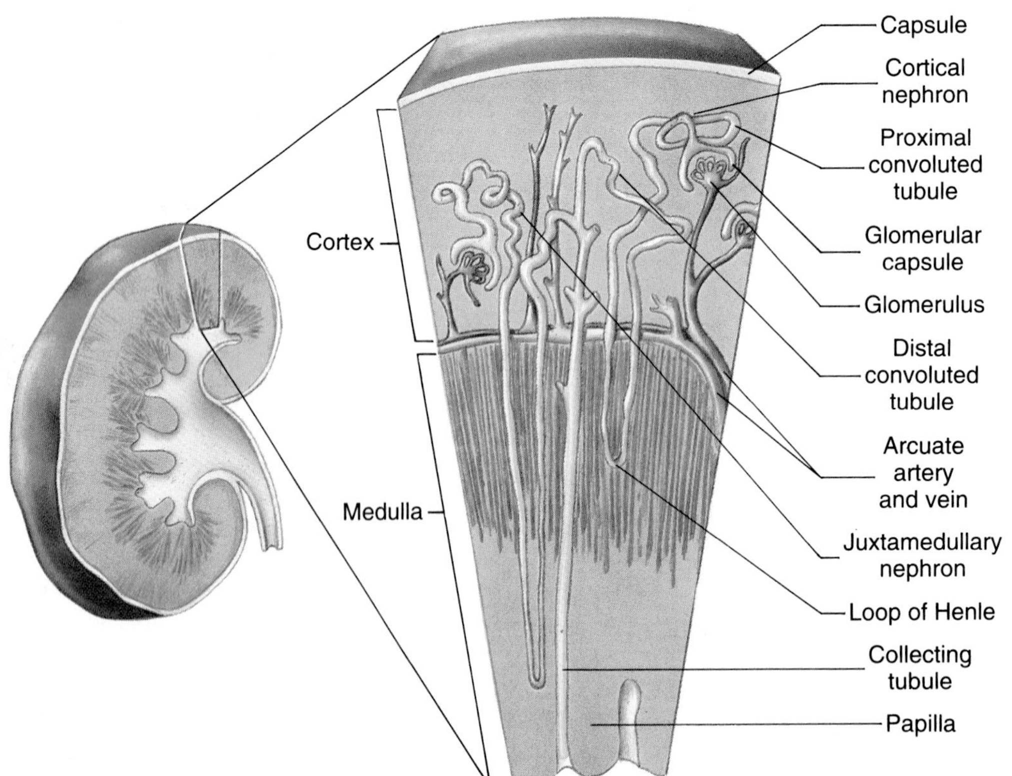

◆ **FIGURE 26.6 Longitudinal section of a renal corpuscle**
(a) Drawing. (b) Photomicrograph.

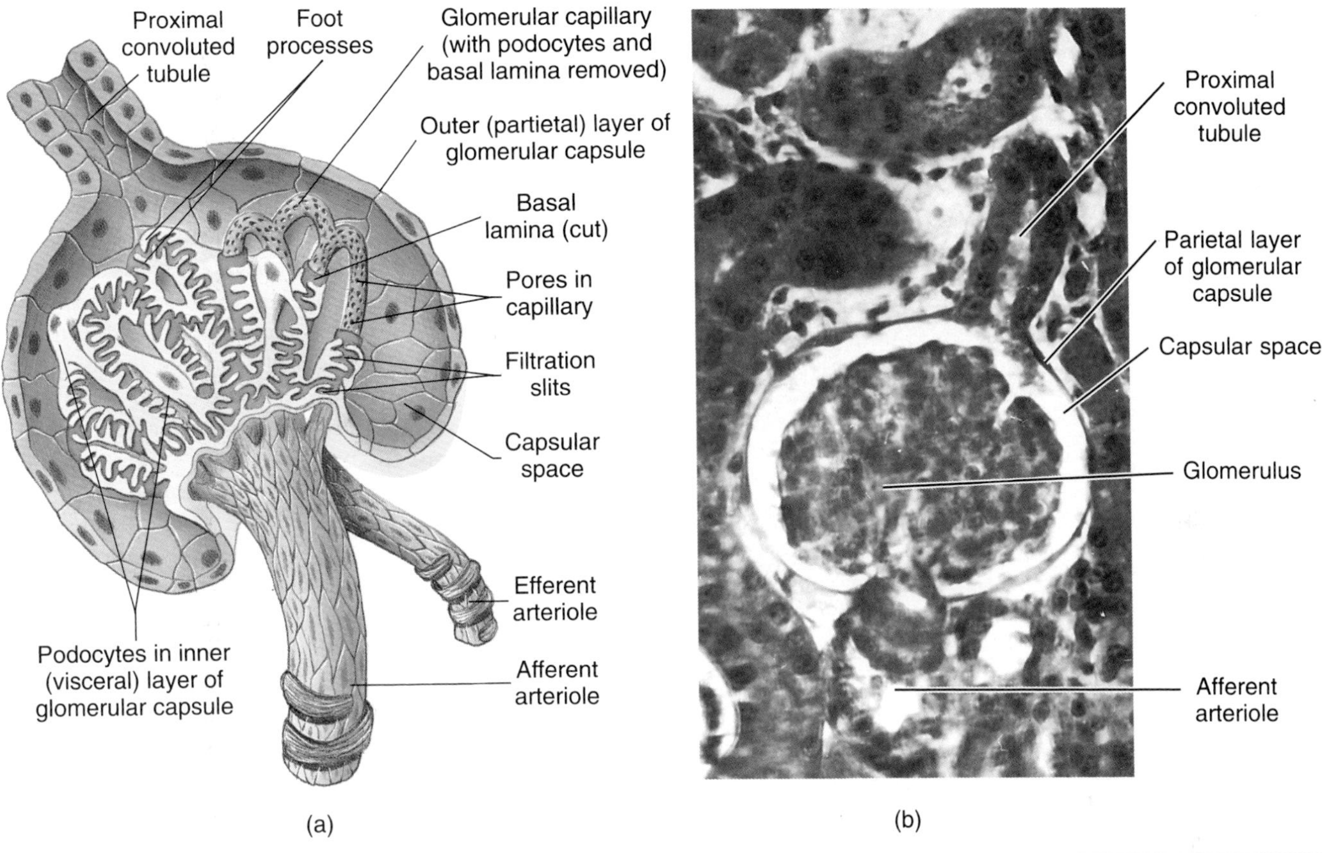

tomically. Epithelial variations along the length of the tubule are related to variations in function.

The proximal end of the renal tubule forms a double-walled cup known as the **glomerular capsule** (or *Bowman's capsule*), which surrounds the glomerulus. The capsule and the glomerulus together are called the **renal corpuscle** (Figure 26.6). The renal corpuscles are located in the cortical regions of the kidneys, and a process known as **glomerular filtration** occurs within them. In this process, which is an important aspect of urine production, some of the blood plasma (except for most proteins) passes out of the glomerular capillaries and into the space between the inner and outer layers of the glomerular capsule, forming a fluid called the **glomerula filtrate.**

The outer (parietal) layer of the glomerular capsule is composed of simple squamous epithelium that rests on a thin basal lamina. The inner (visceral) layer of the capsule is composed of specialized cells called **podocytes** *(pod´-o-sites).* The podocytes have several processes that radiate from a central cell body and adhere to the basal lamina covering the squamous cells of the endothelium of the capillaries that form the glomerulus (Figure 26.7). These processes, in turn, branch into secondary and tertiary processes. The tertiary processes of the podocytes are called **foot processes,** or **pedicels.** The foot processes of one podocyte interdigitate with those of an adjacent podocyte, leaving an elaborate network of small clefts between them. These intercellular clefts are called **filtration slits,** or **slit pores.** A very thin **slit membrane** extends between the foot processes of adjacent cells, forming a diaphragm-like barrier that restricts the passage of molecules through the filtration slits. The capillaries that form the glomeruli are similar to other capillaries in that they consist of endothelium formed of a single layer of squamous cells; but they are a bit different due to the presence of many small, open pores that perforate the endothelial cells. Because of these pores, the endothelium of the glomeruli is referred to as **fenestrated endothelium** *(fen´-es-tray-ted; fenestra* = window).

The **filtration barrier** that separates the blood in the glomerular capillaries from the space in a glomerular capsule **(capsular space)** consists of only (1) the fenestrated endothelium, (2) the basal lamina, and (3) the slit membranes that cover the filtration slits (Figure 26.8). Consequently, many substances are able to pass through this barrier during glomerular filtration. Not all molecules are able to pass through the filtration barrier, however. The endothelial pores restrict the movement of

◆ **FIGURE 26.7 Scanning electron micrograph of podocytes surrounding glomerular capillaries (×4940)**

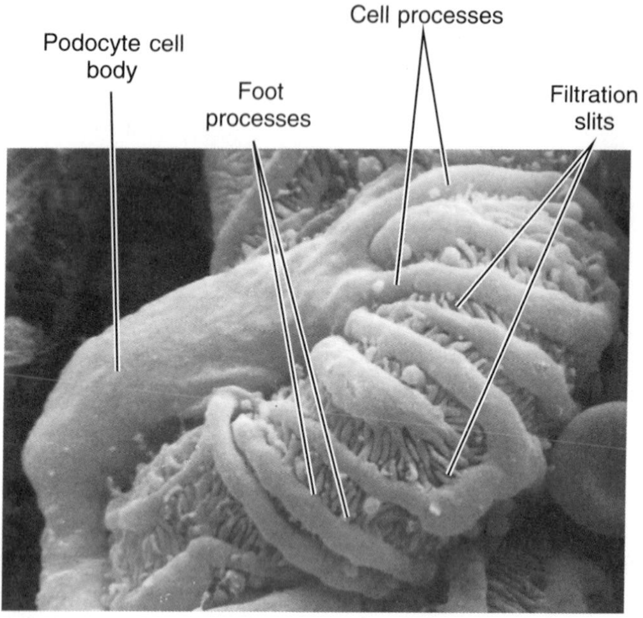

blood cells and molecules larger than 16 nanometers in diameter, and the basal lamina and the slit membranes act as barriers to smaller molecules, allowing only those smaller than about 7 nanometers in diameter to pass through. As a result, the glomerular filtrate that enters a glomerular capsule includes most of the substances present in the blood plasma except most plasma proteins (Figure 26.9).

Negatively charged glycoproteins are associated with the filtration barrier and influence the movement of molecules across the barrier. Consequently, positively charged molecules pass through the barrier faster than neutral molecules of the same size, and negatively charged molecules penetrate more slowly.

Beyond its glomerular capsule, each nephron forms a tightly looping tubule whose lumen is continuous with the capsular space. This region of the nephron is referred to as the **proximal convoluted tubule** because it is located in the cortex, close to the capsule, and is twisted (Figure 26.10). The wall of each proximal convoluted tubule consists of a single layer of thick cuboidal or pyramidal cells. The free surfaces of these cells, facing the lumen of the tubule, have many microvilli.

Beyond its proximal convoluted tubule, each nephron has a straight portion, the **proximal straight tubule,** and it then forms a **loop of Henle (ansa nephroni),** which passes into a pyramid of the medulla of the kidney (Figure 26.10). The loops of Henle of juxtamedullary nephrons are longer than those of cortical nephrons. As a consequence, the loops of Henle of juxtamedullary nephrons extend deeper into the medulla than do the loops of Henle of cortical nephrons. The portion of each loop of Henle that descends into the medulla is called the **descending limb** of the loop of Henle. Because the epithelium of the tubule wall changes to thin squamous cells in the descending limb, this region is also referred to as the **thin segment** of the loop of Henle. Within the medulla, the tubule makes a hairpin turn and ascends toward the cortex as the **ascending limb** of the loop of Henle, which passes out of the medulla and back into the cortex. The wall of the ascending limb is composed primarily of cuboidal cells, and this portion of the nephron is therefore referred to as the **thick segment** of the loop of Henle. The thick segment of the loop of Henle is also known as the **distal straight tubule.** In juxtamedullary nephrons, the thin segment often extends around the hairpin turn and into the ascending limb. In these nephrons, therefore, the thick segment does not begin until the upper portion of the ascending limb.

Beyond its distal straight tubule, each nephron again becomes highly coiled. This portion is called the **distal convoluted tubule.** The wall of each distal convoluted tubule is composed of a single layer of cuboidal cells that have very few microvilli.

The distal convoluted tubules of several nephrons empty into a common **collecting tubule** that passes

◆ **FIGURE 26.8 Transmission electron micrograph of a cross section of a glomerular capillary**

The filtration barrier is composed of fenestrated endothelium, basal lamina, and slit membranes.

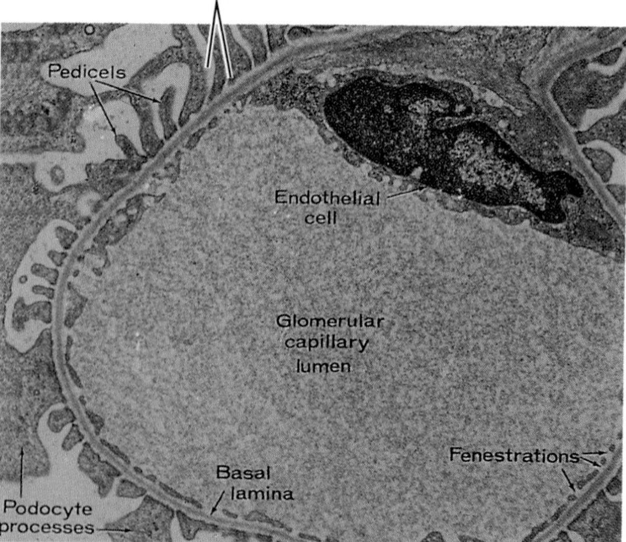

◆ **FIGURE 26.9 The filtration barrier**
(a) A glomerular capillary surrounded by basal lamina and a podocyte.
(b) Restrictions imposed by the various layers of the filtration barrier.

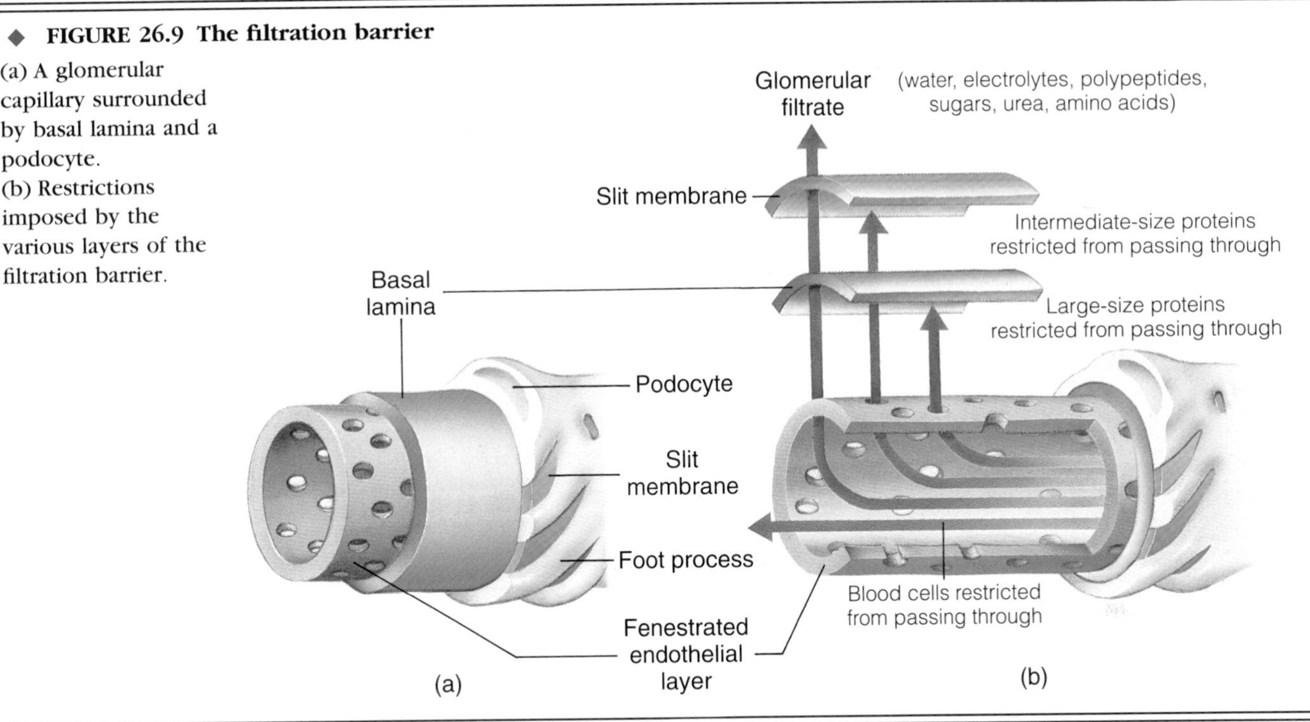

through a renal pyramid of the medulla. Near the tip of the pyramid, adjacent collecting tubules join together and form larger ducts that open on the papilla, where they empty into a minor calyx. The cells that make up the walls of the collecting tubules vary in shape from cuboidal to columnar.

Juxtaglomerular Complex

Within the cortex of the kidney, the distal straight tubule contacts the blood vessel (the *afferent arteriole*) that carries blood to the glomerulus (Figure 26.10). At this point of contact, the cells of both the blood vessel and the tubule are modified. Among the smooth muscle cells of the tunica media of the afferent arteriole are cells that contain prominent secretory granules in their cytoplasm. These cells are called **juxtaglomerular cells** *(juks″-tah-glō-mer′-u-lar).* Where the distal straight tubule contacts the juxtaglomerular cells, there is a concentration of nuclei, and the cells appear higher than in other parts of the distal tubule. This region is the **macula densa.**

The juxtaglomerular cells plus the macula densa form a structure called the **juxtaglomerular complex** (or **juxtaglomerular apparatus**) (Figure 26.11). The juxtaglomerular complex is involved in the regulation of blood pressure, sodium balance, and certain aspects of renal function. These activities will be described later in this chapter and in Chapter 27.

Blood Vessels of the Kidneys

The importance of the relationship between the kidneys and the blood vascular system becomes apparent when one considers the large size of the **renal arteries,** which supply the organs. It has been estimated that, at rest, these vessels carry to the kidneys about 20% of the total cardiac output. In young adults, approximately 1100 ml of blood passes through the two kidneys every minute. Very little of this blood supplies the kidneys' nutritive needs. Rather, the large blood flow is related to the fact that the kidneys can maintain the homeostasis of the blood only if a considerable amount of blood passes through them.

Shortly after entering the hilus of the kidney, the renal artery divides into ventral and dorsal sets of vessels that pass on either side of the renal pelvis (Figure 26.4). These vessels travel between the pyramids and through the renal columns as the **interlobar arteries.** At the bases of the pyramids, which represent the junction of the medulla and the cortex, the interlobar arteries form arching branches that run parallel to the surface of the kidney. These are the **arcuate arteries.** At intervals, the arcuate arteries give off small **interlobular arteries,** which travel through the cortex toward the surface of the kidney. The interlobular arteries divide into several **afferent arterioles,** each of which supplies a renal corpuscle and forms a capillary network called the **glomerulus** (Figure 26.12). As mentioned earlier, it is at the glomerulus that blood comes in close contact

◆ **FIGURE 26.10 A juxtamedullary nephron and its blood supply**
(a) Anatomical illustration. (b) Schematic representation.

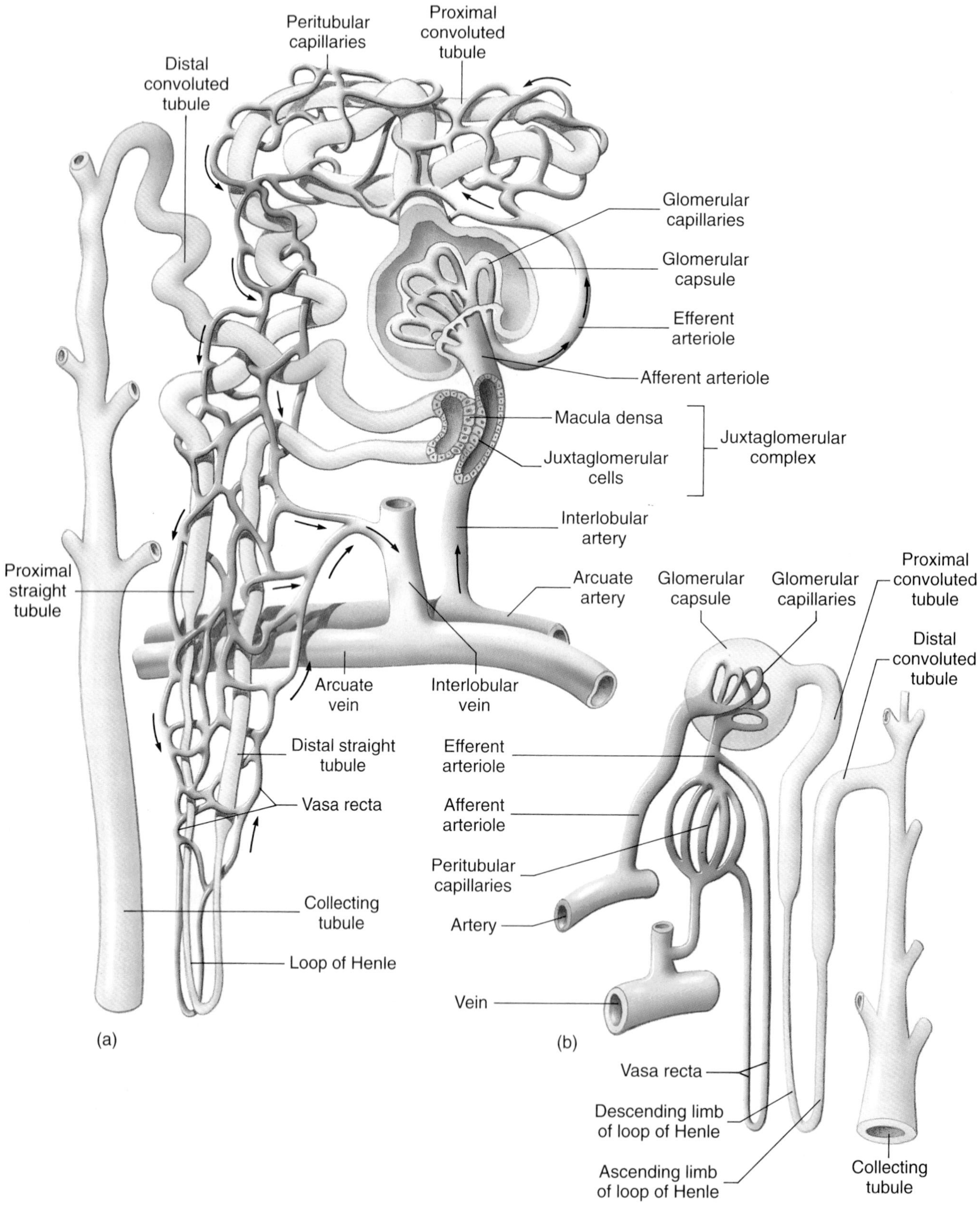

◆ **FIGURE 26.11 Juxtaglomerular complex**

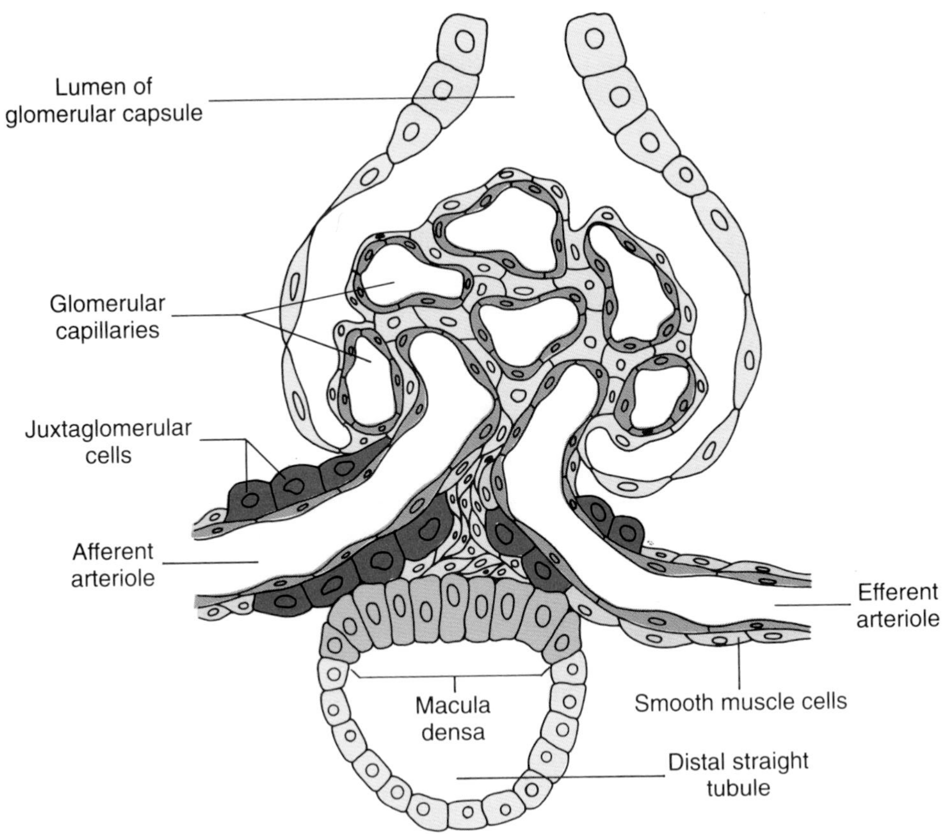

◆ **FIGURE 26.12 Scanning electron micrograph of the blood vessels associated with glomeruli (×299)**

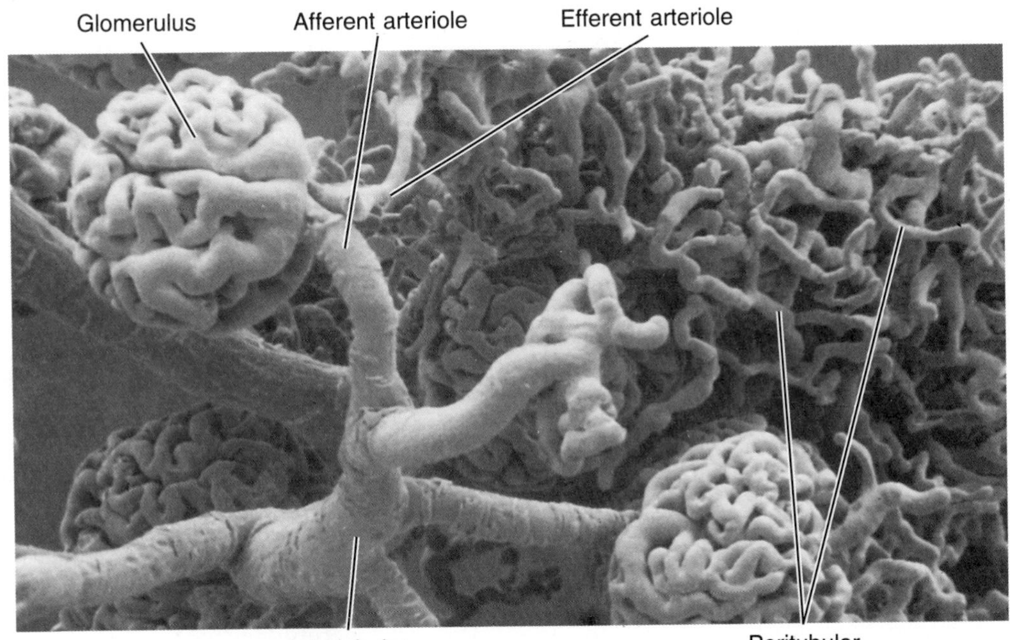

◆ **FIGURE 26.13 Summary of the pathway of blood through the kidney**

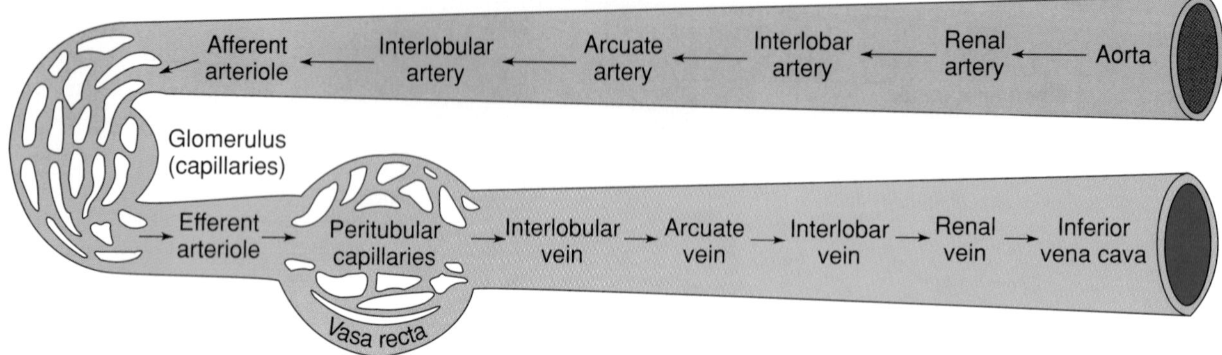

with the cells of a glomerular capsule and the glomerular filtrate is formed.

Blood leaves each glomerulus through an **efferent arteriole** that passes out of the renal corpuscle. The efferent arterioles divide into networks of capillaries that surround the proximal and distal convoluted tubules of the nephron as well as the cortical portions of the proximal and distal straight tubules and the collecting tubules. These capillaries are called the **peritubular capillaries** (Figure 26.10). Thin-walled vessels called **vasa recta** *(va´-sah rek´-tah)* extend from the efferent arterioles of the juxtamedullary nephrons to supply the loops of Henle and the collecting tubules in the medulla of the kidney, particularly in the deep regions of the medulla. The vasa recta play a very important role in the formation of concentrated urine. The peritubular capillaries and vasa recta empty into **interlobular veins,** which in turn empty into **arcuate veins.** The arcuate veins lead into **interlobar veins,** which join with the **renal veins** that leave the kidney. The veins follow the same general course as the arteries of the same names. The pathway of blood through the kidney is summarized in Figure 26.13.

Blood flow through the kidney exhibits several unusual and important features. As blood moves through the kidney, it flows through two sequential series of capillary beds—the glomerulus and the peritubular capillaries. The kidney is one of the few places in the body where this arrangement occurs. Generally, when blood leaves capillaries, it enters venules. In the kidney, however, efferent arterioles transport the blood from the first capillary bed (glomerulus) to the second capillary bed (peritubular capillaries). The presence of arterioles leading to and from the glomeruli has great functional significance. Like all arterioles, these vessels have smooth muscle in their walls that permits them either to constrict or to dilate in response to various types of stimulation. Thus, it is possible for a fairly constant blood pressure to be maintained in the gomeruli, even if the systemic pressure fluctuates. This constant blood pressure contributes to the efficient functioning of the kidney in spite of varying systemic conditions.

 Physiology of the Kidneys

As noted at the beginning of the chapter (page 834), the excretory and regulatory activities of the kidneys are extremely important for the maintenance of homeostasis. These activities are carried out in large measure by the nephrons and collecting tubules.

Basic Renal Processes

Three basic processes—*glomerular filtration, tubular secretion,* and *tubular reabsorption*—occur within the kidneys (Figure 26.14).

1. As blood flows through the kidneys, some of the plasma is filtered out of the glomerular capillaries and into the glomerular capsules of the renal tubules. This filtration process is called **glomerular filtration.**

Glomerular filtration is primarily due to the pressure of the blood within the glomerular capillaries. It is not very selective, and water and almost all other plasma components—with the exception of large protein molecules—can enter the glomerular capsules.

The water and other plasma components that enter the glomerular capsules during glomerular filtration are called the **glomerular filtrate.** From the glomerular capsules, the filtrate flows along the other portions of the renal tubules and the collecting tubules as the **tubular fluid.**

2. Not all the plasma enters the glomerular capsules during glomerular filtration. However, in some cases,

◆ **FIGURE 26.14 The processes of glomerular filtration, tubular secretion, and tubular reabsorption**

(a) Glomerular filtration. As blood flows through the kidneys, some of the plasma (except most proteins) is filtered out of the glomerular capillaries and into the glomerular capsules of the renal tubules. (b) Tubular secretion. In some cases, materials within the remaining plasma enter the fluid within the tubules. (c) Tubular reabsorption. Water, inorganic ions, glucose, and other substances required by the body are reabsorbed from the fluid within the tubules and returned to the blood.

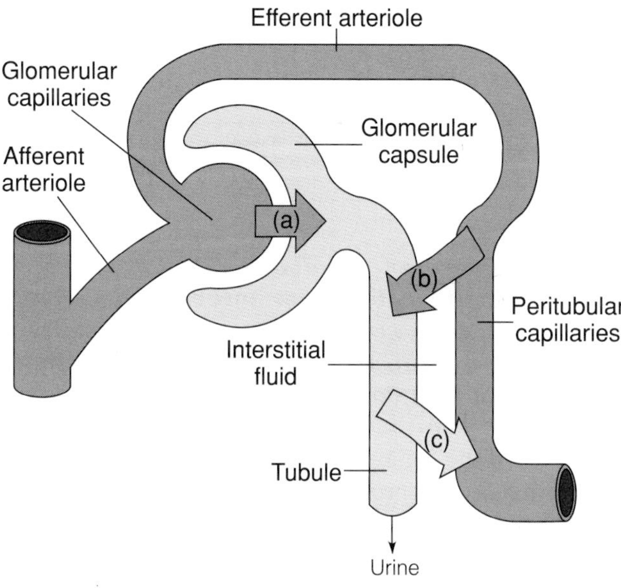

materials within the remaining plasma can enter the tubular fluid by processes collectively called **tubular secretion.** Some tubular secretion occurs by active transport and some by passive mechanisms.

3. Water, inorganic ions, glucose, and other substances required by the body can be removed from the tubular fluid and returned to the blood by processes collectively called **tubular reabsorption.** As is the case for tubular secretion, some tubular reabsorption occurs by active transport and some by passive mechanisms.

Many tubular secretion and tubular reabsorption processes are under hormonal control. This control enables the body to regulate the amounts of many materials—particularly water and inorganic ions—that are secreted into the tubular fluid or reabsorbed from the fluid and returned to the blood.

When the tubular secretion and tubular reabsorption processes are completed, the fluid remaining within the tubules is transported to other components of the urinary system to be excreted as urine. Thus, **urine** consists of water and other materials that were filtered or secreted into the tubules but not reabsorbed.

Glomerular Filtration

Approximately 16–20% of the blood plasma that enters the kidneys is filtered from the glomerular capillaries into the glomerular capsules of the renal tubules as the glomerular filtrate. The remaining plasma continues into the efferent arterioles and on into the peritubular capillaries and vasa recta.

Under normal circumstances, the kidneys produce approximately 180 liters (about 45 gallons) of glomerular filtrate per day. However, approximately 99% of the fluid volume of the filtrate is reabsorbed from the renal tubules and collecting tubules, and only about 1% (1–2 liters per day) is excreted as urine.

The filtration barrier between the plasma and the internal portion of a glomerular capsule is from 100 to 1000 times more permeable than a typical capillary, and water and solutes of small molecular dimension pass freely from the plasma into glomerular capsules. On the other hand, this barrier is relatively impermeable to large molecules, such as plasma proteins. As a result, the glomerular filtrate contains most plasma components at essentially the same concentrations as they are found within the plasma, but it is basically protein free. (Actually, a small amount of protein does appear in the glomerular filtrate, but this is normally reabsorbed by pinocytosis in the proximal convoluted tubules so that essentially no protein is present in the urine.)

Net Filtration Pressure

The pressure of the blood within the glomerular capillaries (that is, the glomerular capillary pressure) is a major factor in glomerular filtration. This pressure—about 60 mm of mercury (Hg) or more—is higher than the pressure in most capillaries, and it favors filtration out of the glomerular capillaries and into the glomerular capsules. The glomerular capillary pressure is opposed by the pressure of the fluid within glomerular capsules (about 18 mm Hg) and by the osmotic force exerted by unfiltered plasma proteins that remain within the glomerular capillaries (about 32 mm Hg). Thus, a *net filtration pressure* of about 10 mm Hg ($60 - 18 - 32 = 10$) favors the movement of materials out of the glomeruli and into the glomerular capsules (Figure 26.15). This net filtration pressure is responsible for glomerular filtration, which occurs as a bulk flow of water and small dissolved particles from the glomerular capillaries into glomerular capsules.

Factors That Influence Glomerular Filtration

The rate of glomerular filtration at any moment depends on the relationship between the blood pressure within the glomerular capillaries that favors filtration and the

◆ **FIGURE 26.15 Net filtration pressure**

The pressure of the blood within the glomerular capillaries (GCP) favors the filtration of materials into glomerular capsules. The pressure of the fluid within glomerular capsules (CP) and the osmotic pressure exerted by unfiltered plasma proteins that remain within the glomerular capillaries (COP) oppose filtration.

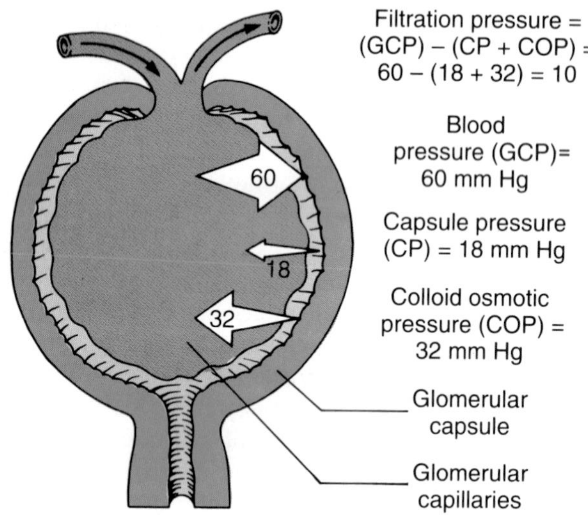

forces that oppose filtration, which include the pressure of the fluid within the glomerular capsules and the osmotic force exerted by plasma proteins within the glomerular capillaries. Variations in any of these factors can alter the glomerular filtration rate.

Influence of Arterial Blood Pressure and Autoregulatory Mechanisms. It might be expected that changes in the arterial blood pressure would lead to comparable changes in the pressure within the glomerular capillaries and, consequently, to alterations in the glomerular filtration rate. However, the influence of arterial blood pressure on the glomerular filtration rate is not as great as expected.

The kidneys possess intrinsic autoregulatory mechanisms that maintain a fairly stable glomerular filtration rate over a rather wide range of systemic arterial blood pressures. Thus, even when the arterial pressure increases from 100 to 160 mm Hg, the glomerular filtration rate increases by only a few percent (Figure 26.16).

The intrinsic autoregulatory mechanisms of the kidneys are not completely understood. However, they are believed to include both a myogenic mechanism and a tubulo-glomerular feedback mechanism.

Myogenic Mechanism. The **myogenic mechanism** is based on an inherent response of arteriole smooth muscle: When arteriole smooth muscle is stretched, it tends to contract and thereby constrict the arterioles. This response contributes to autoregulation in the following manner.

When a rise in arterial pressure leads to an increase in pressure within the afferent arterioles that stretches their walls, the afferent arterioles automatically constrict. The constriction limits blood flow into the glomeruli and prevents a rise in glomerular capillary pressure and glomerular filtration rate despite the rise in arterial pressure.

When a fall in arterial pressure leads to a decrease in pressure within the afferent arterioles that reduces the stretch of their walls, the smooth muscles relax, and the afferent arterioles dilate. The dilation permits an increased blood flow into the glomeruli and prevents a fall in glomerular capillary pressure and glomerular filtration rate despite the fall in arterial pressure.

Tubulo-Glomerular Feedback Mechanism. The **tubulo-glomerular feedback mechanism** involves the juxtaglomerular complex (page 841). Macula-densa cells of this structure are believed to detect the rate at which tubular fluid flows past them (and/or the osmolarity of the fluid). If the tubular fluid is flowing too rapidly (or if the osmolarity of the fluid is too high), the macula-densa cells promote the constriction of the afferent arteriole. Conversely, if the tubular fluid is flowing too slowly (or if the osmolarity of the fluid is too low), the macula-densa cells promote vasodilation of the afferent arteriole. These responses contribute to autoregulation in the following manner.

If the glomerular filtration rate increases (perhaps due to a rise in arterial pressure), fluid flows rapidly past the macula-densa cells, and the cells promote constriction of the afferent arteriole (Figure 26.17). This constriction reduces blood flow into the glomerulus and lowers the glomerular capillary pressure, thereby lowering the glomerular filtration rate.

◆ **FIGURE 26.16 The effects of changes in arterial blood pressure on the glomerular filtration rate**

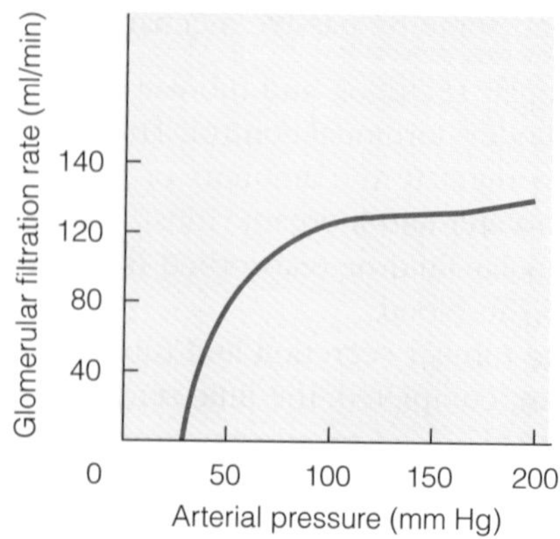

◆ **FIGURE 26.17 Tubulo-glomerular feedback mechanism of autoregulation**

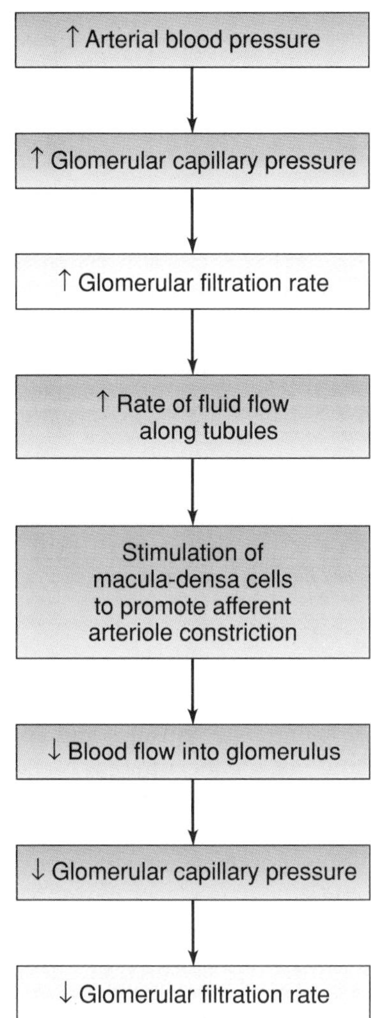

If the glomerular filtration rate decreases (perhaps due to a fall in arterial pressure), fluid flows slowly past the macula-densa cells, and the cells promote dilation of the afferent arteriole. This dilation permits an increased blood flow into the glomerulus and a rise in glomerular capillary pressure, thereby increasing the glomerular filtration rate.

Influence of Sympathetic Neural Stimulation. The activity of the sympathetic neurons that supply the kidneys can cause changes in the glomerular filtration rate. For example, an increase in sympathetic neural stimulation generally results in a preferential constriction of the afferent arterioles. This constriction reduces blood flow into the glomeruli and lowers the glomerular capillary pressure, thereby lowering the glomerular filtration rate.

Under appropriate circumstances, sympathetic neural activity can override the kidneys' autoregulatory mechanisms. As is discussed in Chapter 27, mechanisms involved in blood-pressure regulation can alter the activity of sympathetic neurons supplying the kidneys and thereby alter the glomerular filtration rate.

General Mechanisms of Tubular Secretion and Tubular Reabsorption

Tubular secretion and tubular reabsorption are similar activities that differ in direction. Substances enter the tubular fluid by tubular secretion, and substances leave the tubular fluid by tubular reabsorption (see Figure 26.14).

Tubular secretion and tubular reabsorption generally occur by diffusion or by means of discrete transport mechanisms. Carrier molecules are often involved, and a single transport mechanism is sometimes responsible for the movement of several different substances with similar structures.

In many cases, tubular secretion and tubular reabsorption are active processes that require energy to transport substances across the tubule wall. For example, in the proximal portion of the renal tubule, sodium ions are actively reabsorbed from the tubular fluid. In this process, sodium ions enter tubular epithelial cells by diffusing passively across the luminal membranes of the cells (Figure 26.18). Then, energy-requiring active-transport mechanisms move sodium ions across the basolateral membranes of the cells into the interstitial fluid. From the interstitial fluid, sodium ions enter the peritubular capillaries.

Note that, in this example, the movement of sodium ions across the tuble wall requires the ions to cross more than one membrane. Sodium ions enter tubular epithelial cells across the cells' luminal membranes, and they exit across the basolateral membranes. Although only the exit of sodium ions from the cells is an active energy-requiring process, if any one step is active, the whole process is considered to be active. Thus, sodium ions are actively reabsorbed from the proximal portions of the renal tubules.

In subsequent sections of this chapter, the processes of tubular secretion and tubular reabsorption are considered as if only a single membrane is involved, and if any transport step is active, the whole process is considered to be active.

Tubular Secretion

A number of substances can enter the tubular fluid by tubular secretion. Among these substances are potassium ions and hydrogen ions, which are secreted into the tubules by active-transport mechanisms.

◆ **FIGURE 26.18 Reabsorption of sodium ions from the proximal tubule**

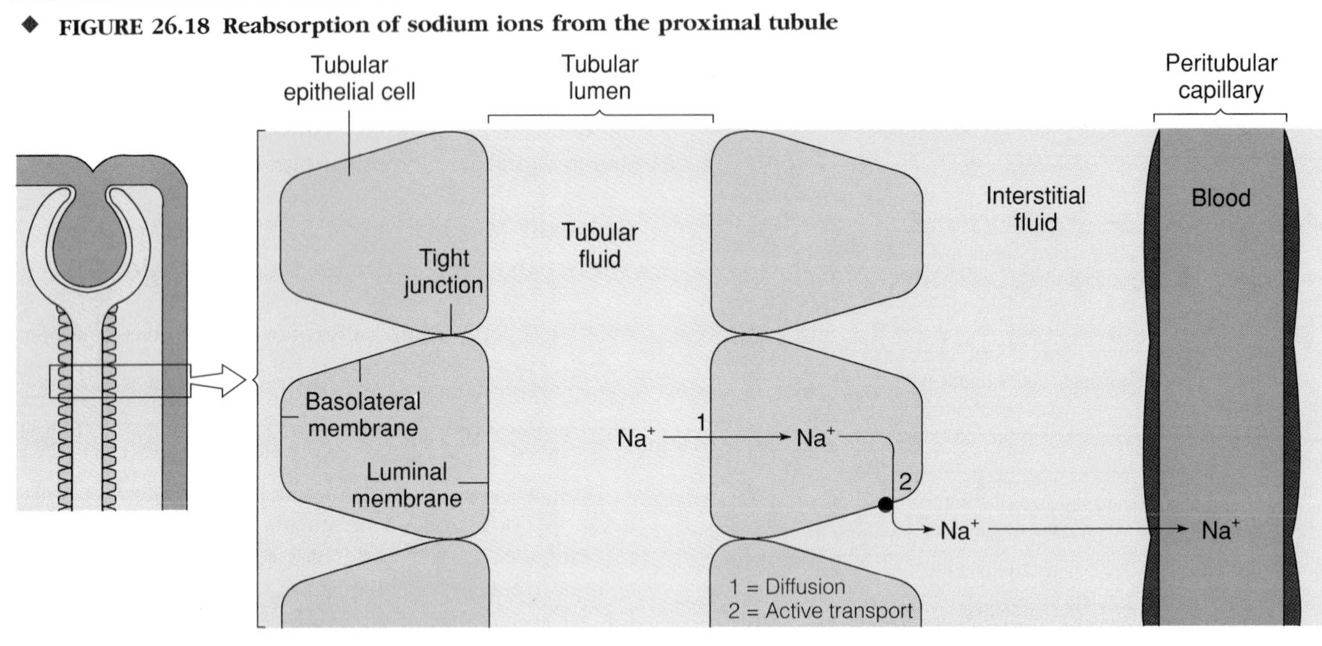

A variety of organic ions can also enter the tubular fluid by tubular secretion. The proximal tubule contains two distinct types of secretory carriers for organic ions—one type for positvely charged ions and one type for negatively charged ions. Both types of carriers are relatively nonspecific, and the organic ions secreted include foreign organic chemicals such as food additives, environmental pollutants (for example, certain pesticides), and drugs (for example, penicillin).

Many foreign organic chemicals are not ionic in their original forms. Consequently, they cannot be secreted into the tubular fluid by the secretory carriers for organic ions. However, the liver converts many of these chemicals into ionic forms. This conversion facilitates their secretion by the secretory carriers, and thus accelerates their elimination from the body.

Tubular Reabsorption

A number of substances, including glucose, amino acids, and vitamins, as well as sodium, calcium, and chloride ions, are reabsorbed from the tubular fluid by active-transport mechanisms. Most of these mechanisms utilize carrier molecules that can become saturated (see page 63). Consequently, almost every substance that is actively reabsorbed has a **tubular transport maximum (T_m)**—that is, a maximum amount of the substance that can be reabsorbed per unit time. If the concentration of such a substance (for example, glucose) increases substantially in the plasma and subsequently in the glomerular filtrate, not all of the substance can be reabsorbed, and some appears in the urine.

Some substances are reabsorbed by passive processes. For example, urea is freely filtered from the glomerular capillaries into glomerular capsules and appears in the glomerular filtrate at a concentration that is essentially equal to that of urea in the plasma. As the filtrate moves along the tubules, however, water is reabsorbed. The reabsorption of water increases the concentration of urea in the tubular fluid to a level above that of the urea concentration in the interstitial fluid. As a consequence, urea diffuses from the tubles to the interstitial fluid. Thus, urea reabsorption is passive and depends on water reabsorption and also on the permeability of the tubules to urea. Since the tubules are generally not as permeable to urea as they are to water, only about 40–60% of the filtered urea is normally reabsorbed, even though about 99% of the filtered water is reabsorbed.

Some of the major secretory and reabsorptive activities occurring across different tubular segments are summarized in Table 26.1.

Excretion of Water and Its Effects on Urine Concentration

A person who has just consumed a large quantity of water must eliminate a considerable volume of fluid without losing excessive amounts of ions and other vital substances if homeostasis is to be maintained. Under such conditions it is beneficial to produce a large volume of dilute urine. Conversely, a dehydrated individual who has had nothing to drink for some time must still produce urine for the elimination of wastes. In this case, it is advantageous to produce a small volume of concen-

ASPECTS OF EXERCISE PHYSIOLOGY

When Protein in the Urine Does Not Mean Kidney Disease

Usually, urinary loss of proteins signifies kidney disease (nephritis). However, a urinary protein loss similar to that of nephritis often occurs following exercise, but the condition is harmless, transient, and reversible. The term *athletic pseudonephritis* is used to describe this postexercise (after exercise) proteinuria (protein in the urine). Studies indicate that 70–80% of athletes have proteinuria after very strenuous exercise. This occurs in participants in both noncontact and contact sports, so the condition does not arise from physical trauma to the kidneys. In one study, subjects who engaged in maximal short-term running excreted more protein than when they were bicycling, rowing, or swimming at the same work intensity. The reason for this difference is unknown.

Usually, only a very small fraction of the plasma proteins that enter the glomerulus are filtered; those that are filtered are reabsorbed in the tubules, so no plasma proteins normally appear in the urine. Two basic mechanisms can cause proteinuria: (1) increased glomerular permeability with no change in tubular reabsorption or (2) impairment of tubular reabsorption. Research has shown that during mild to moderate exercise, the proteinuria that occurs results from changes in glomerular permeability, whereas during short-term exhaustive exercise, the proteinuria seems to be caused by both increased glomerular permeability and tubular dysfunction.

This reversible kidney dysfunction is believed to result from circulatory and hormonal changes that occur with exercise. Several studies have shown that renal blood flow is reduced during exercise as the renal vessels are constricted and blood is diverted to the exercising muscles. This reduction is positively correlated with the intensity of the exercise. With intense exercise, the renal blood flow may be reduced to 20% of normal. The glomerular filtration rate is also reduced, but not to the same extent as renal blood flow, presumably because of autoregulatory mechanisms. Some investigators propose that decreased glomerular blood flow enhances movement of proteins into the tubular lumen because, as the more slowly flowing blood spends more time in the glomerulus, a greater proportion of the plasma proteins have time to escape through the filtration barrier. Hormonal changes that occur with exercise may also affect glomerular permeability. For example, renin injection is a well-recognized way to experimentally induce proteinuria. Plasma renin activity is increased during strenuous exercise and may contribute to postexercise proteinuria. It is also hypothesized that maximal tubular reabsorption is reached during severe exercise, which could result in impaired protein reabsorption.

trated urine, which requires the excretion of only small amounts of water. Thus, if the kidneys are to be effective as regulatory organs, they must be able to produce a range of urine volumes and concentrations according to the body's needs. In fact, the kidneys can produce urine as dilute as 60–75 milliosmoles per liter or as concentrated as 1200 milliosmoles per liter compared to the concentration of the blood plasma, which is about 300 milliosmoles per liter.

The ability to produce a low-volume, concentrated urine depends on the presence of a highly concentrated interstitial fluid within the medullary region of each kidney. Therefore, the mechanisms responsible for maintaining this high interstitial fluid concentration will be considered first, and then the processes involved in producing different volumes and concentrations of urine will be examined.

Maintenance of a Concentrated Medullary Interstitial Fluid

Within the medullary region of a kidney, the interstitial fluid concentration increases from about 300 milliosmoles per liter at the cortex to as much as 1200 milliosmoles per liter at the tips of the pyramids. The high medullary interstitial fluid concentration is due to the activities and anatomical arrangement of the loops of Henle of the nephrons (particularly those of the juxtamedullary nephrons), the collecting tubules, and the vasa recta.

Activities of the Loops of Henle and the Collecting Tubules. The loops of Henle of juxtamedullary nephrons extend deep into the medulla, and the collecting

◆ **TABLE 26.1 Examples of Transport across Different Tubular Segments**

PROXIMAL TUBULE	DESCENDING LIMB OF LOOP OF HENLE	ASCENDING LIMB OF LOOP OF HENLE	DISTAL CONVOLUTED TUBULE	COLLECTING TUBULE
Variable H^+ secretion; dependent on acid-base status of body Organic ion secretion; not subject to control All filtered glucose and amino acids reabsorbed by active transport; not subject to control 67% of filtered Na^+ actively reabsorbed; not subject to control; Cl^- follows passively All filtered K^+ reabsorbed; not subject to control 65–70% of filtered water osmotically reabsorbed; not subject to control	10–15% of filtered water osmotically reabsorbed into medullary interstitial fluid; not controlled	25% of filtered NaCl actively reabsorbed into medullary interstitial fluid; not subject to control; region essentially impermeable to water	Variable H^+ secretion; dependent on acid-base status of body Variable K^+ secretion; controlled by aldosterone Variable Na^+ reabsorption; controlled by aldosterone; Cl^- follows passively	Variable H^+ secretion; dependent on acid-base status of body Variable water reabsorption; controlled by antidiuretic hormone (ADH)

tubules pass through the medulla to the tips of the pyramids. As fluid flows along the loops of Henle and the collecting tubules, the reabsorption of solutes—particularly sodium ions, chloride ions, and urea—raises the concentration of the medullary interstitial fluid.

Sodium and Chloride Ions. The reabsorption of sodium and chloride ions from the thick segments of the ascending limbs of the loops of Henle as well as from the collecting tubules and the thin segments of the ascending limbs contributes importantly to the high concentration of the medullary interstitial fluid (Figure 26.19).

1. Sodium and chloride ions are actively transported out of the fluid in the thick segments of the ascending limbs of the loops of Henle into the interstitial fluid of the outer medulla. This transport not only raises the concentration of the outer medullary interstitial fluid, but as is discussed later, sodium and chloride ions are carried downward into the inner medulla by blood flowing along the vasa recta.

2. Sodium ions are actively transported out of the fluid in the collecting tubules. Because of electrical considerations, chloride ions (which are negatively charged) passively follow the sodium ions (which are positively charged).

3. Sodium and chloride ions can move out of the thin segments of the ascending limbs of the loops of Henle into the interstitial fluid of the inner medulla. This movement apparently occurs by passive processes.

Urea. Water can be reabsorbed from the fluid within the collecting tubules, and as discussed later, the amount of water reabsorbed is controlled hormonally by antidiuretic hormone. When a large amount of water is reabsorbed from the collecting tubules, a substantial amount of urea can diffuse out of the final portions of the tubules into the interstitial fluid of the inner medulla and thus contribute to the high concentration of the medullary interstitial fluid. This movement of urea occurs as follows.

The first portions of the collecting tubules are quite impermeable to urea. Consequently, when a large amount of water is reabsorbed, the concentration of urea in the tubular fluid increases considerably, and much of it diffuses out of the final portions of the collecting tubules into the interstitial fluid of the inner me-

◆ **FIGURE 26.19 Mechanisms involved in maintaining a concentrated medullary interstitial fluid and producing a low-volume, concentrated urine**

See text for detailed discussion. The concentrations indicated are in milliosmoles per liter. They are the total concentrations of all solutes present, not the concentrations of individual substances.

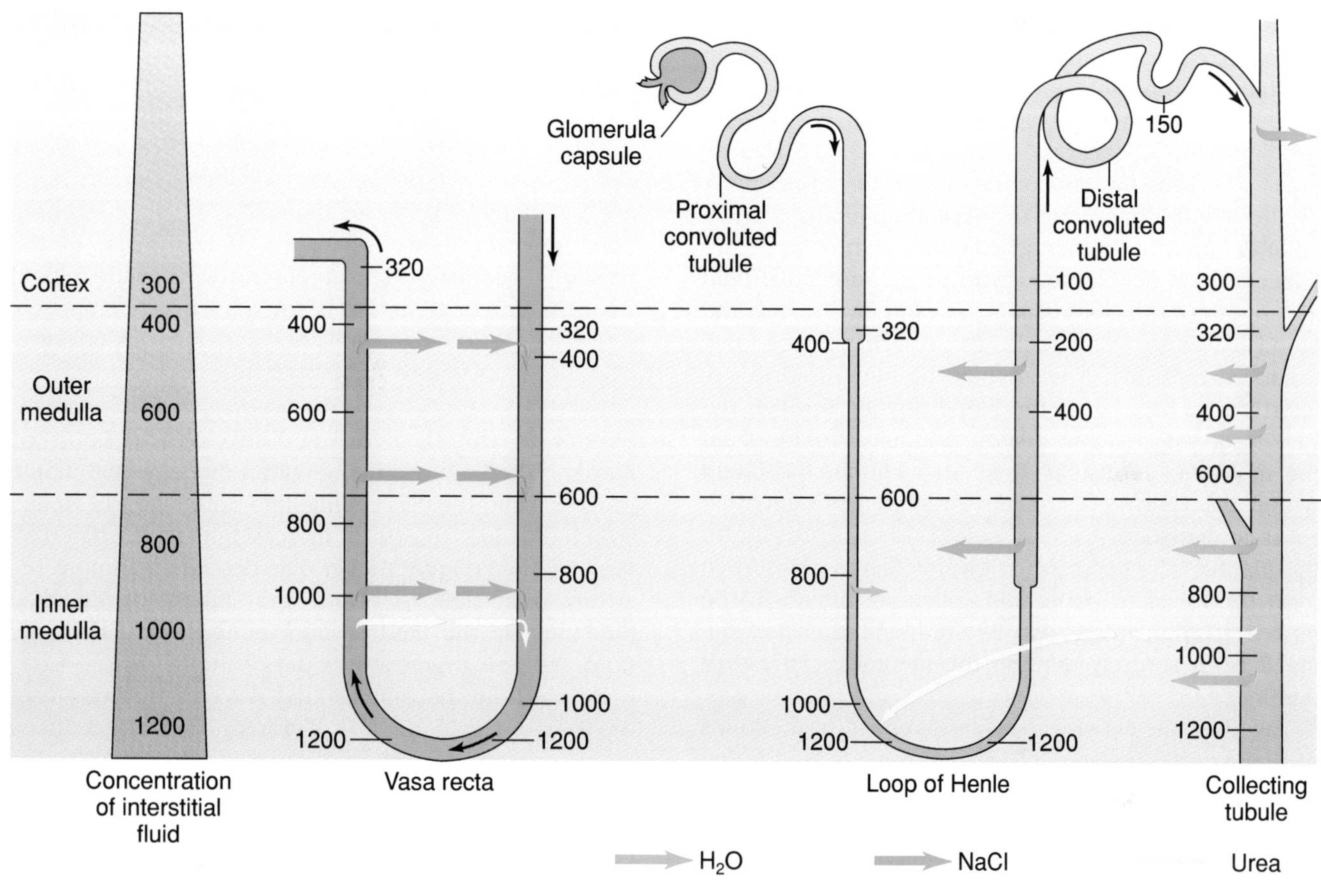

dulla (Figure 26.19). This activity raises the urea concentration of the interstitial fluid above that of the tubular fluid in the bottoms of the loops of Henle. As a result, some of the urea diffuses into the tubular fluid at the bottoms of the loops, raising its concentration.

The tubular fluid is then carried through the ascending limbs of the loops of Henle and the distal convoluted tubules (both of which are essentially impermeable to urea) to the collecting tubules, where the reabsorption of water can further increase the fluid's urea concentration. Even more urea then diffuses out of the final portions of the collecting tubules, and the cycle is repeated. Thus, when much water is reabsorbed from the collecting tubules, this cycling process can multiply the urea concentration of the interstitial fluid of the inner medulla (and the urea concentration of the fluid within the final portions of the collecting tubules) to quite high levels.

Together, the mechanisms that raise the concentration of the medullary interstitial fluid are frequently referred to as the **countercurrent mechanism** of the loop of Henle because some aspects of the different mechanisms—for example, the cycling process involved in urea transport—depend on the fact that fluid flows in opposite directions in the essentially parallel ascending and descending limbs of the loops of Henle and the collecting tubules.

Effect of Vasa Recta Blood Flow. If a high medullary interstitial fluid concentration is to be maintained, the reabsorbed solutes (for example, sodium, chloride, and urea) that raise the concentration of the fluid must not be removed by the blood. The removal of the solutes is minimized by the pattern of blood flow in the vasa recta, which extend into the medulla.

The vasa recta form a countercurrent system. In this system, blood flows in opposite directions within the descending and ascending limbs of the vasa recta, and most solutes can pass readily between the blood and the interstitial fluid. The system operates essentially as follows.

Blood with a concentration of about 300 milliosmoles per liter flows from the cortex into the medulla along the descending limbs of the vasa recta, which extend into regions of increasingly concentrated interstitial fluid (Figure 26.19). As the blood flows along the descending limbs, sodium, chloride, urea, and other solutes that are in higher concentration in the interstitial fluid than in the blood diffuse into the blood, raising its concentration. The concentrated blood then flows into the ascending limbs of the vasa recta, which extend into regions of decreasing concentratioins of interstitial fluid. As the blood flows along the ascending limbs, sodium, chloride, urea, and other solutes that are now in higher concentration in the blood than in the interstitial fluid diffuse out of the vasa recta. As a result, by the time the blood leaves the vasa recta, its concentration is only slightly higher than it was when the blood initially entered the vessels. Thus, the blood flowing through the vasa recta removes only a small amount of solutes from the medullary interstitial fluid. In addition, the blood within the vasa recta contains unfiltered proteins that exert an osmotic force, the colloid osmotic pressure of the plasma. This force causes water that is reabsorbed from the loops of Henle and collecting tubules to be picked up and carried away by the blood, thus preventing it from severely diluting the medullary interstitial fluid.

Together, the activities and arrangement of the loops of Henle of the juxtamedullary nephrons, the collecting tubules, and the vasa recta maintain a high medullary interstitial fluid concentration that increases progressively from about 300 milliosmoles per liter at the cortex to as much as 1200 milliosmoles per liter at the tips of the pyramids. The kidneys utilize this highly concentrated interstitial fluid to produce a low-volume, concentrated urine.

Production of Different Urine Volumes and Concentrations

As previously indicated, the vast majority of the approximately 180 liters of glomerular filtrate formed per day is reabsorbed from the tubules and returned to the body. Moreover, the hormonal mechanism involving antidiuretic hormone that controls water reabsorption in the collecting tubules strongly influences the actual volume of filtrate reabsorbed. Consequently, this mechanism also influences the volume and concentration of the urine that is produced.

Events in the Proximal Tubule. The concentration of the glomerular filtrate within a glomerular capsule is essentially the same as that of the plasma—that is, about 300 milliosmoles per liter. As the filtrate flows along the proximal portion of a renal tubule, sodium ions are actively transported out of the tubule and into the interstitial fluid, with chloride ions following passively (Figure 26.20). In addition, a number of other substances, such as glucose and amino acids, are reabsorbed by various transport mechanisms. The proximal tubule is permeable to water, and water leaves osmotically, since the loss of sodium, chloride, and other substances transiently lowers the concentration of the tubular fluid and increases the concentration of the interstitial fluid.

The net result of the removal of solutes and water from the proximal tubule is to reduce the volume of the original glomerular filtrate about 65–70%. Its concentration, however, remains at approximately that of the plasma. The materials that enter the interstitial fluid from the proximal portion of the renal tubule ultimately return to peritubular capillaries and are carried away from the kidney with the blood.

Events in the Descending Limb of the Loop of Henle. The fluid remaining within the proximal tubule proceeds into the descending limb of the loop of Henle, which dips into regions of increasingly concentrated interstitial fluid (Figure 26.19). The descending limb is relatively permeable to water, and since the interstitial fluid outside the limb is more concentrated than the fluid within it, water moves out of the tubule.

By the time the tubular fluid reaches the bottom of the loop of Henle, the loss of water has further reduced its volume and also raised its concentration above that of the plasma. In fact, at the bottom of the loop of Henle of a juxtamedullary nephron, the loss of water—together with the previously discussed entry of urea into the tubule—can raise the concentration of the tubular fluid to approximately 1200 milliosmoles per liter.

Events in the Ascending Limb of the Loop of Henle. As already noted, within the ascending limb of the loop of Henle, sodium and chloride ions move out of the tubule and into the interstitial fluid (Figure 26.19). In the thin segment of the ascending limb, the outward movement of sodium and chloride ions apparently occurs passively by diffusion as the concentrated tubular fluid from the bottom of the loop of Henle flows up the ascending limb, which extends into regions of decreasing concentrations of interstitial fluid. In the thick segment of the ascending limb, sodium and chloride ions are actively transported out of the tubule.

The ascending limb of the loop of Henle—particularly the thick segment—is essentially impermeable to water. Consequently, as sodium and chloride ions move out of the tubular fluid within the limb, the concentration of the fluid drops, but its volume is not greatly altered. By the time the fluid completes its passage through the ascending limb and enters the distal convoluted tubule, the reabsorption of sodium and chloride ions has lowered its concentration below that of the plasma.

◆ **FIGURE 26.20 Sodium, chloride, and water movements in the proximal tubule**

(a) Sodium ions are actively transported out of the tubule. (b) Because of electrical considerations, negatively charged chloride ions passively follow the positively charged sodium ions. (c) The reabsorption of sodium and chloride establishes an osmotic gradient that causes water to be reabsorbed from the proximal tubule. (d) Summary of the movements of sodium, chloride, and water in the proximal tubule.

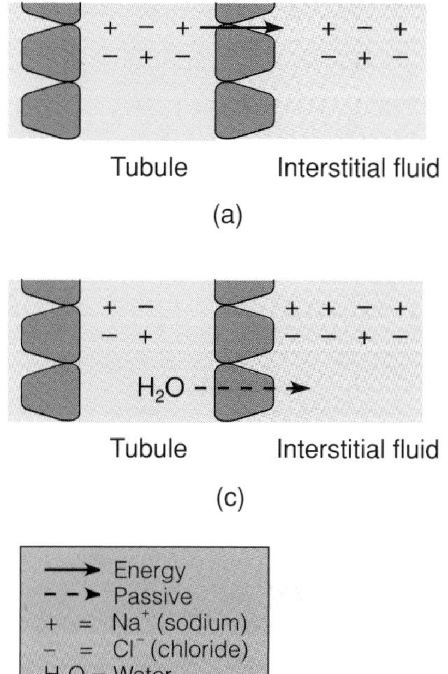

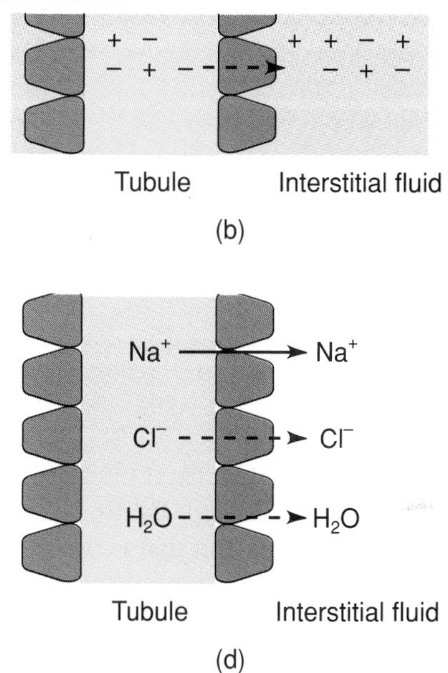

Events in the Distal Convoluted Tubule. The processes occurring within the distal convoluted tubule are complex. Certain ions are reabsorbed from the fluid within the tubule, whereas others are secreted into the fluid, and much of this activity is influenced by physiological regulatory mechanisms. In monkeys and presumably in other primates, such as humans, the distal convoluted tubule is relatively impermeable to water. In general, the concentration of the fluid within the distal convoluted tubule remains below that of the plasma (and it can even fall below that of the fluid leaving the ascending limb of the loop of Henle), but its volume does not change greatly.

Up to this point, much of the fluid volume of the glomerular filtrate has been reabsorbed—primarily in the proximal tubule. Moreover, many nutrients required by the body, such as glucose and amino acids, are almost completely reabsorbed in the proximal tubule. In the case of a juxtamedullary nephron, the sodium and chloride ions that move out of the ascending limb of the loop of Henle contribute to the high concentration of the medullary interstitial fluid, which is required for the production of a low-volume, concentrated urine.

Events in the Collecting Tubule. From the distal convoluted tubule, the fluid enters a collecting tubule, which extends from the cortex into the medullary region of the kidney where there are increasing concentrations of interstitial fluid. The permeability of the collecting tubule to water is controlled by antidiuretic hormone (ADH), which is manufactured in the hypothalamus of the brain and released from the pituitary gland.

When the ADH concentration of the plasma is high, the collecting tubule is very permeable to water. As a result, water leaves the tubular fluid and enters the concentrated interstitial fluid. (Note that even though the permeability of the collecting tubule is high, the movement of water out of the tubule depends on the presence of the concentrated medullary interstitial fluid to establish the proper osmotic gradient.) The outward movement of water reduces the volume of the fluid within the collecting tubule and increases its concentration. This low-volume, concentrated fluid then leaves the tubule at the tip of a pyramid to be excreted as urine.

When the plasma-ADH concentration is low, the collecting tubule is not very permeable to water, and little water is reabsorbed from the tubular fluid. Consequently, the dilute fluid that flows along the distal convoluted tubule into a collecting tubule remains dilute, and its volume is not significantly reduced. (Moreover, ions can be actively reabsorbed from the collecting tubule, and this activity can lower the concentration of the tubular fluid even further.) In this case, a high-volume, low-concentration fluid leaves the collecting tubules as urine.

Mechanism of Water Reabsorption

As is evident from the preceding discussion, the tubular reabsorption of water occurs by passive transport (osmosis) and depends on the reabsorption of solutes such as sodium and chloride ions to establish the proper osmotic gradient. However, once a gradient is established, the volume of water reabsorbed from the collecting tubules varies with their permeability to water, which is controlled by ADH. Consequently, ADH is important in regulating the volume of fluid excretion and in providing the kidneys with the ability to produce a range of urine volumes and concentrations according to the body's needs. Factors influencing ADH release are discussed in Chapter 27.

Water reabsorption that is regulated by ADH according to the body's needs is called **facultative water reabsorption.** Water reabsorption that is not regulated and occurs regardless of the body's needs is called **obligatory water reabsorption.** Approximately 80% of the water that enters glomerular capsules during glomerular filtration is obligatorily reabsorbed in the proximal tubules and loops of Henle (see Table 26.1). More or less of the remaining 20% can be facultatively reabsorbed.

Minimum Urine Volumes

The body must excrete about 600 milliosmoles of waste products such as urea, sulfates, phosphates, and the like each day. The elimination of these materials, even at a urine concentration of 1200 milliosmoles per liter, requires a certain degree of water loss:

$$\frac{600 \text{ Milliosmoles/day}}{1200 \text{ Milliosmoles/liter}} = 0.5 \text{ liter/day}$$

Regardless of water intake or availability, this volume of fluid must be excreted simply to eliminate wastes. Beyond this requirement, however, physiological mechanisms come into play to vary the actual volume, concentration, and composition of the urine produced in order to compensate for altered intakes of fluids, inorganic ions, and other substances, as well as to compensate for varying losses of these materials by other pathways, including the lungs and sweat glands. The composition of a representative sample of urine is given in Table 26.2.

Plasma Clearance

The **plasma clearance** of a substance is the rate at which the substance is eliminated or cleared from the plasma by the kidneys. By determining the plasma clearance of certain specific substances, the glomerular filtration rate and the rate of plasma flow through the kidneys can be determined. These types of information are useful in assessing kidney function.

The measurement of the plasma clearance of a substance requires the determination of (1) the rate of urine formation, (2) the concentration of the substance in the arterial plasma flowing to the kidneys, and (3) the concentration of the substance in the urine. With these values, the plasma clearance of the substance can be determined according to the following formula:

$$\text{Plasma Clearance of } A \text{ (ml/min)} = \frac{\text{Rate of Urine Formation (ml/min)} \times \text{Concentration of } A \text{ in Urine}}{\text{Concentration of } A \text{ in Plasma}}$$

For example, suppose the concentration of urea in the plasma is 0.30 mg/ml. Suppose further than in one hour 66 ml of urine is formed (a formation rate that equates to 1.1 ml/min), and that this urine has a urea concentration of 17 mg/ml. The plasma clearance of urea would then be:

$$\text{Plasma Clearance of Urea (ml/min)} = \frac{(1.1 \text{ ml/min}) \times (17 \text{ mg/ml})}{0.30 \text{ mg/ml}} = 62 \text{ ml/min}$$

This plasma clearance value means that in one minute an amount of urea is removed from the plasma that is equivalent to the amount of urea in 62 ml of plasma. (It does not mean that any single milliliter of plasma has all of the urea removed from it during one transit through the renal circulation.)

◆ **TABLE 26.2 Representative Components and Characteristics of a 24-Hour Sample of Urine**

COMPONENT	VALUE
Calcium	0.01 to 0.30 g
Chloride	6.0 to 9.0 g
Creatinine	1.0 to 1.5 g
Potassium	2.5 to 3.5 g
Sodium	4.0 to 6.0 g
Urea	20.0 to 35.0 g
CHARACTERISTIC	**VALUE**
pH	4.8 to 7.4
Specific gravity	1.015 to 1.022
Volume	1000 to 1600 ml

CLINICAL CORRELATION

Inappropriate Secretion of ADH

Case Report

THE PATIENT: A 43-year-old man.

PRINCIPAL COMPLAINTS: Headache, nausea, weakness, thirst, and abrupt loss of vision in one eye.

HISTORY: At the age of 33 (ten years before this report), the patient was diagnosed as having multiple sclerosis (MS). He had developed numbness and weakness on the right side of the body that had progressed to quadriplegia (paralysis affecting the four extremities of the body). He was treated with adrenocorticotropic hormone (ACTH) and dismissed in improved condition. During the intervening years, he experienced episodes of hemiplegia (paralysis of one side of the body) or paraplegia (paralysis of the lower limbs) that were ameliorated by ACTH.

CLINICAL EXAMINATION: Loss of vision was complete in the right eye, and the pupil was fixed. Deep tendon reflexes in the right leg were absent, and deep sensation was decreased in both legs. Sodium concentration in the serum was 110 mEq/liter (normal: 135–145 mEq/liter), potassium concentration was 4 mEq/liter (normal: 3.5–5 mEq/liter), and chloride concentration was 75 mEq/liter (normal: 100–106 mEq/liter). Blood urea nitrogen (BUN; nitrogen in the form of urea) was 7 mg/100 ml (normal: 8–25 mg/100 ml), and creatinine (a metabolic end product derived from creatine phosphate) was 0.3 mg/100 ml (normal: 0.6–1.5 mg/100 ml). The hematocrit was 33% (normal: 45–52%), and serum osmolality was 235 milliosmoles/kg water (normal: 285–295 milliosmoles/kg water). The patient weighed 60.4 kg (usual weight: 57.4 kg). The sodium concentration of the urine was 125 mEq/liter (normal: 50–130 mEq/liter), the potassium concentration was 45 mEq/liter (normal: 20–70 mEq/liter), the specific gravity was 1.020 (normal: 1.015–1.022), and the osmolality was 615 milliosmoles/kg water (normal: 500–800 milliosmoles/kg water).

COMMENT: The findings of normal osmolality and sodium concentration of the urine despite hypotonic plasma indicate inappropriate secretion of antidiuretic hormone (ADH). The concentrations of several constituents of the plasma—sodium, chloride, BUN, creatinine, and red cells (hematocrit)—were low, and the osmolality of the blood was low. Under these conditions, a dilute urine should have been excreted to increase the osmolality of the blood. For this to occur, the secretion of ADH should be low. Inappropriate secretion of ADH in this patient probably is related to demyelinating lesions of the hypothalamus (particularly in areas involved in the control and secretion of ADH) as one of the manifestations of MS. (Other effects of hypothalamic damage also have been seen in patients who have MS.) Consistent with this hypothesis, the development of hypotonic plasma in this patient occurred during worsening of the MS characterized by focal disorders of the central nervous system. The symptoms of headache, nausea, and weakness probably are related to the patient's low blood-sodium level.

TREATMENT: The patient was given hypertonic (5%) sodium chloride solution intravenously, and his fluid intake was restricted to establish normal osmolality of the plasma. He was given both democlocycline—which antagonizes the action of ADH—and ACTH. He also was given cyclophosphamide, which suppresses the immune system of the body. Evidence indicates that MS is an autoimmune disease associated in some way with viral infection.

OUTCOME: The patient improved rapidly, and the vision in his affected eye began to return after two days. He produced large amounts of urine, and his weight returned to normal in three days. He was discharged after five days, with his blood chemistry normal and the MS in remission. However, MS is a progressive illness, and he will no doubt suffer relapse again.

The clearance concept can be used to determine the glomerular filtration rate by measuring the plasma clearance of the carbohydrate inulin. Inulin is injected into the plasma, and its plasma concentration is determined. Inulin is freely filtered from the glomerular capillaries into glomerular capsules with the glomerular filtrate, but it does not undergo tubular secretion or tubular reabsorption. As a result, the amount of inulin in a timed sample of urine is essentially equal to the amount of inulin filtered from the plasma into glomerular capsules in the same amount of time. For example, suppose that the one-hour, 66-ml urine sample of the previous example contained 1500 mg of inulin (22.7 mg/ml) and that the plasma inulin concentration was 0.2 mg/ml. Since inulin does not undergo tubular secretion or tubular reabsorption, all of the inulin in this one-hour sample of urine must have been filtered into glomerular capsules from the plasma with the glomerular filtrate in the same hour. Moreover, because the plasma inulin concentration is only 0.2 mg/ml, in order to get 1500 mg of inulin into the urine in one hour, 7500 ml of plasma must have been filtered at the glomeruli in that hour. This activity equates to a glomerular filtration rate of 125 ml/min [(7500 ml per hour)/(60 min per hr)]. The glomerular filtration rate could also have been obtained using the plasma clearance formula, as follows:

$$\text{Plasma Clearance of Inulin (ml/min)} = \frac{(1.1 \text{ ml/min}) \times (22.7 \text{ mg/ml})}{0.2 \text{ mg/ml}} = 125 \text{ ml/min}$$

In the case of inulin, the plasma clearance (of inulin) and the glomerular filtration rate are essentially equal.

The concept of plasma clearance can also be employed to determine the rate of plasma flow through the kidneys. The substance utilized for this purpose is para-aminohippuric acid (PAH). PAH is freely filtered at the glomerulus, and it undergoes tubular secretion. However, it does not undergo tubular reabsorption. As a result, the plasma that passes through the kidneys is almost completely (actually about 91%) cleared of PAH.

Suppose, to continue the previous examples, that sufficient PAH is injected into the plasma to achieve a plasma concentration of 0.01 mg/ml. Suppose also that the 1.1 ml of urine produced in one minute contains 6.25 mg of PAH (5.68 mg/ml). This means that a minimum of 625 ml of plasma [(6.25 mg per min)/(0.01 mg per ml)] must have passed into the kidneys in one minute to provide the amount of PAH that appears in the urine in one minute, even if all of the PAH is removed from the plasma. The same conclusion can also be reached by using the plasma clearance formula:

$$\text{Plasma Clearance of PAH (ml/min)} = \frac{(1.1 \text{ ml/min}) \times (5.68 \text{ mg/ml})}{0.01 \text{ mg/ml}} = 625 \text{ ml/min}$$

This value is called the effective renal plasma flow (ERPF). In fact, since only about 91% of the para-aminohippuric acid is in reality cleared from the plasma, the actual total renal plasma flow is about 687 ml/min [(625 ml per min)/0.91], or 989 liters per day. If the hematocrit is known (for example, 40%), then the total renal blood flow (1648 liters per day) can also be determined.

◆ FIGURE 26.21 Colorized X-ray of the urinary tract by intravenous pyelogram

The kidneys are shown in green. The ureters and urinary bladder in red.

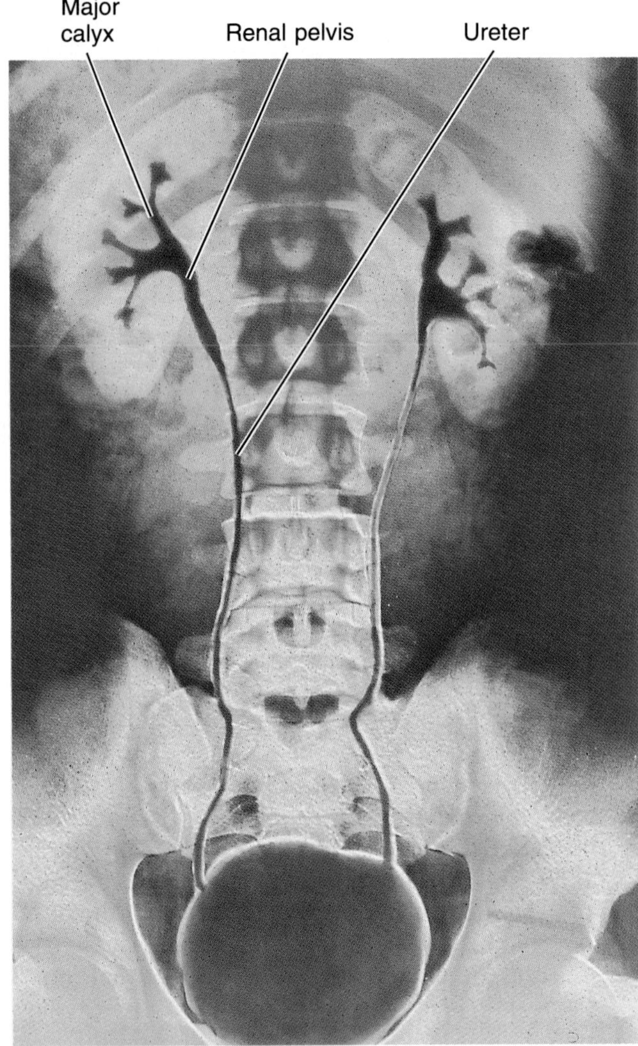

Ureters

The urine drips from the collecting tubules at the tips of the papillae and enters the minor calyces. The minor calyces join with the major calyces, which in turn join with the renal pelvis. From the renal pelvis, urine is transported to the urinary bladder by **ureters,** one from each kidney (Figure 26.21).

The ureters descend between the parietal peritoneum and the body wall to the pelvic cavity, where they turn medially and enter the urinary bladder on its posterior lateral surfaces. Before opening into the bladder, the ureters travel obliquely through the bladder wall. As a result, contraction of the muscles of the bladder wall can compress the ureters and help prevent urine from flowing back into the ureters from the bladder, especially during bladder emptying *(micturition).* In effect, the muscles of the bladder wall act as sphincters on the ureters. Valvelike folds of the mucous membrane lining the bladder cover the orifices of the ureters and assist in preventing urine from flowing back into the ureters during micturition (urination).

The walls of the ureters are composed of three layers: an inner *muscosal layer,* a middle *muscular layer,* and an outer *fibrous layer.* The mucosa has a surface layer of transitional epithelium, which is also typical of the urinary bladder and urethra. The muscular layer of the ureter is capable of undergoing peristaltic contractions, thus propelling urine to the bladder.

Urinary Bladder

The **urinary bladder** is a hollow muscular organ used to store urine (Figure 26.22). The bladder rests on the floor of the pelvic cavity, and like other urinary structures, it is retroperitoneal. The anterior surface of the bladder lies just behind the pubic symphysis. In males, the bladder is in front of the rectum. In females, it lies just anterior to the uterus and the superior portion of the vagina. When full, the bladder is spherical; but when it is empty, it resembles an inverted pyramid.

The urinary bladder, like the ureters, is lined with a mucous membrane of transitional epithelium. Covering the transitional epithelium is a tunic of muscle called the **detrusor muscle** *(de-trū´-sor)* which consists of three layers of smooth muscle—inner and outer longitudinal layers on either side of a prominent circular layer. Throughout most of the bladder, the mucous membrane is loosely attached to the underlying muscular coat, and it appears wrinkled when the bladder is contracted. However, the internal opening of the urethra anteriorly and the openings of the two ureters laterally mark the corners of a triangular area in which the mucous membrane is firmly attached to the muscular coat and where it is therefore always smooth. This smooth, triangular region is called the **trigone** *(trī´-gōn)* of the bladder.

As the bladder fills with urine, the pressure within it increases somewhat initially and then remains fairly constant up to a volume of about 300–400 ml (Figure 26.23). Beyond this point, the pressure rises rapidly. The bladder can hold 600–800 ml of urine, but it is generally emptied before it reaches this capacity.

Urethra

The **urethra** is a muscular tube, lined with mucous membrane, that exits from the inferior surface of the urinary bladder. It carries urine from the bladder to the exterior of the body.

At the junction of the urethra and bladder, the smooth muscle of the bladder surrounds the urethra and acts as a sphincter (the **internal urethral sphincter**) that tends to keep the urethra closed. During bladder emptying, the contraction of the bladder and the resulting change in its shape pull the sphincter open. Thus, no special mechanism is required to relax this sphincter.

As the urethra passes through the pelvic floor **(urogenital diaphragm),** it is surrounded by skeletal muscles that form the **external urethral sphincter.** When constricted, this sphincter—which is under voluntary control—is able to hold the urethra closed against strong bladder contractions.

In females, the urethra is short (approximately 4 cm) and runs along the anterior surface of the vagina (Figure 26.22a). It opens to the exterior at the **external urethral orifice,** which is located between the clitoris and the vaginal orifice.

In males, the urethra is about 20 cm long and extends to the external urethral orifice at the tip of the penis (Figure 26.22b). It is divisible into three parts—the *prostatic, membranous,* and *spongy urethrae,* which are named according to the regions through which the urethra passes.

The **prostatic urethra** passes through the prostate gland and receives the ejaculatory ducts of the reproductive system on its posterior walls. Beyond the point of junction with the ejaculatory ducts, the urethra in males is used for reproduction as well as urine transport.

The short portion of the urethra that passes through the urogenital diaphragm is called the **membranous urethra.**

The **spongy** *(cavernous)* **urethra** is the longest portion, extending from just below the urogenital diaphragm to the external urethral orifice on the glans penis. Within the glans penis, the urethra dilates into a small chamber called the **fossa navicularis.** A short distance below the urogenital diaphragm, the spongy urethra receives the ducts of the bulbourethral glands (Cowper's glands) of the reproductive system. In the penis, the urethra passes through the **corpus spongiosum,** one of three columns of erectile tissue.

Micturition

Emptying the bladder is called **micturition** *(mik´´-tū-rish´-un), urination,* or *voiding.* Micturition is basically a reflex act. As the bladder fills with urine, its walls are stretched, and mechanoreceptors within the walls initiate increasing numbers of nerve impulses, which are transmitted to the sacral region of the spinal cord.

Impulses initiated by the mechanoreceptors ultimately stimulate parasympathetic neurons that pass

◆ **FIGURE 26.22 Sagittal section of the pelvis showing the urinary bladder and urethra**
(a) A female. (b) A male.

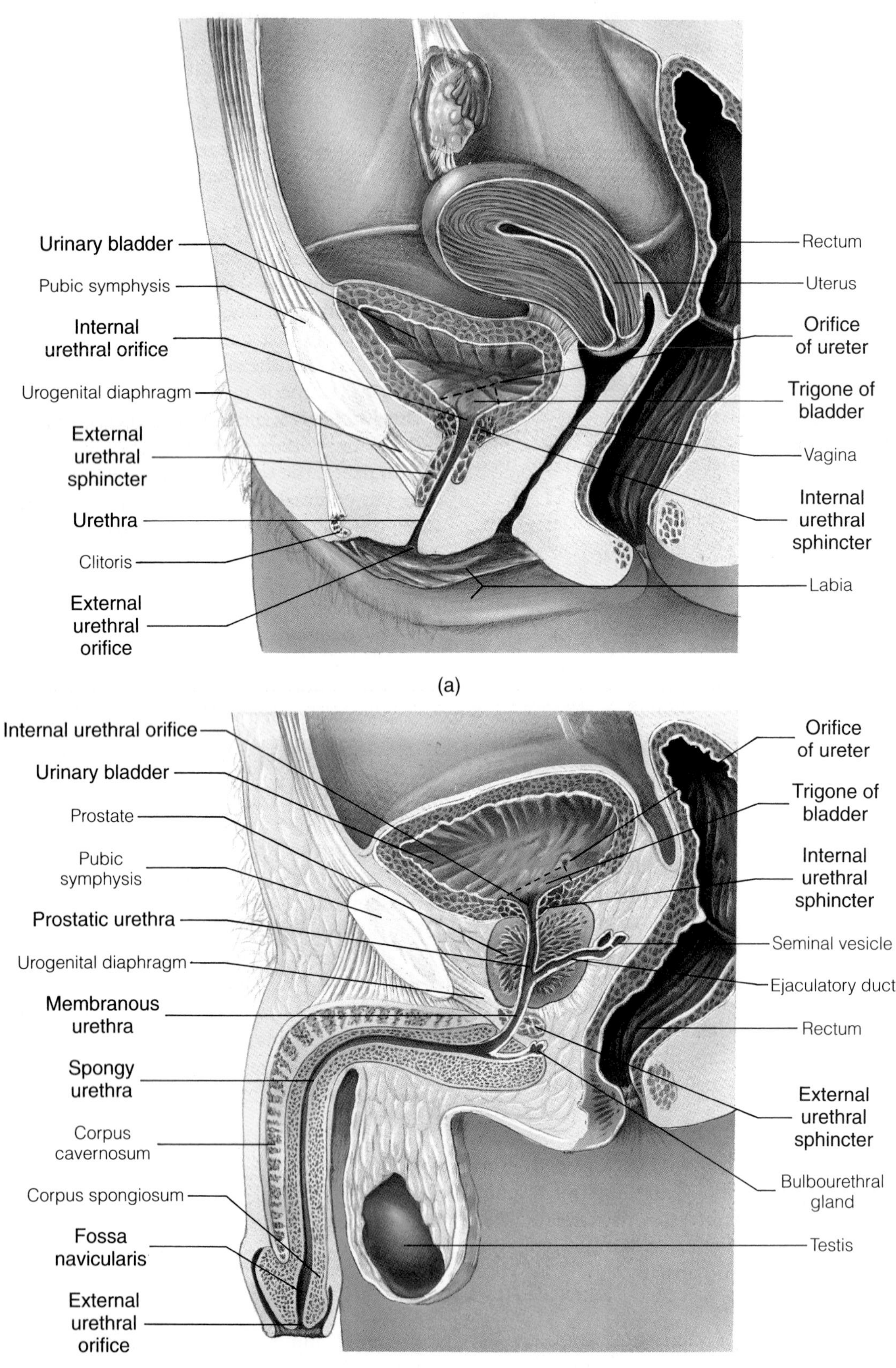

from the spinal cord to the bladder through pelvic nerves. Impulses transmitted by these parasympathetic neurons cause the smooth muscle forming the detrusor muscle of the bladder wall to contract. When the muscle contracts, the internal urethral sphincter opens.

At the same time, impulses initiated by the mechanoreceptors inhibit somatic motor neurons that supply skeletal muscles of the external urethral sphincter and the muscles of the urogenital diaphragm that surround it. Consequently, these muscles relax. Therefore, when approximately 300 ml of urine has accumulated within the bladder, a reflex response is triggered that causes the bladder to contract and the sphincters to open, and micturition occurs.

Although micturition as just described is essentially the result of a local spinal reflex, it is also influenced by higher brain centers, particularly centers in the brain stem and cerebral cortex. In addition to being transmitted to the spinal cord, nerve impulses initiated by mechanoreceptors within the bladder walls are sent to higher brain centers. These impulses can lead to the sensation of a full bladder and to a feeling of a need to urinate. Moreover, impulses sent from the brain can either facilitate or inhibit the reflex emptying of the bladder, largely by consciously stimulating or inhibiting the external urethral sphincter and the muscles of the urogenital diaphragm. With training it is possible to gain a high degree of voluntary control over micturition. As a result, urination can be either voluntarily induced or postponed until an opportune time. However, until the central nervous system develops sufficiently so that voluntary control over the external urethral sphincter can be established and training is possible, the reflex response is the dominant factor governing bladder emptying. Therefore, young babies, in which the central nervous system tracts are not yet completely developed, urinate whenever the bladder is sufficiently full to activate the spinal reflex.

◆ FIGURE 26.23 Pressure changes within the urinary bladder as the bladder fills with urine

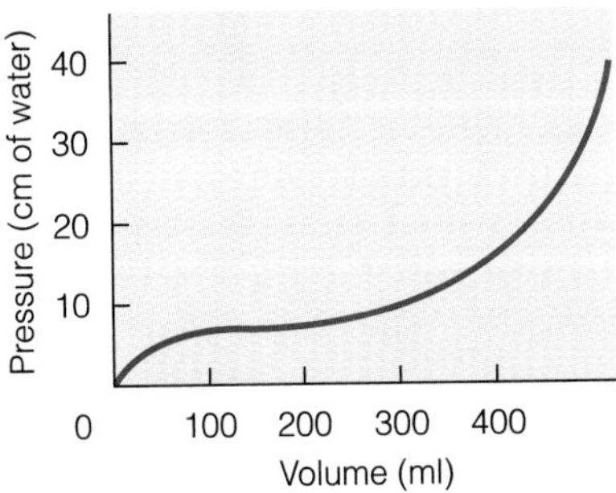

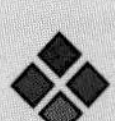

CONDITIONS OF CLINICAL SIGNIFICANCE

The Urinary System

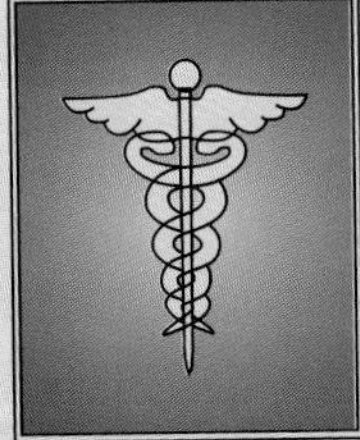

Pressure-Related Pathologies

Factors that upset the pressure relationships determining glomerular filtration rates can interfere with normal kidney function. (Recall that the glomerular filtration rate depends on the net filtration pressure, which equals glomerular capillary pressure minus capsular pressure minus colloid osmotic pressure of the plasma.) The two most common pressure-related pathologies of the urinary system are prostate hypertrophy and low arterial blood pressure.

1. *Prostate hypertrophy:* In males, it is not unusual for the prostate gland, which surrounds the urethra just below the bladder, to hypertrophy (enlarge) and compress the urethra so it becomes difficult for urine to leave the bladder. As urine accumulates, the pressure within the bladder rises. If the pressure increases enough, it causes urine to back up in the ureters, which produces a dilation of the renal pelvis and calyces and leads to an increased pressure within the glomerular capsules. The increased capsular pressure reduces the glomerular filtration rate and thus interferes with the kidneys' regulation of body-fluid composition.

2. *Low arterial blood pressure:* Blood pressure changes associated with heart failure, hemorrhage, or shock can cause renal failure. In all of these conditions, there is a drop in arterial blood pressure, which reduces the ability of the kidneys to form glomerular

continued on next page

CONDITIONS OF CLINICAL SIGNIFICANCE

filtrate. In addition, a substantial drop in blood pressure can activate reflexes involved in maintaining normal pressure within the major arteries of the body. These reflexes cause the afferent arterioles of the kidneys to constrict. The constriction of the afferent arterioles helps elevate the general body blood pressure, but it diminishes the blood pressure within the glomeruli and further reduces the formation of glomerular filtrate. The reduction in the formation of glomerular filtrate diminishes the kidneys' ability to excrete wastes and regulate body-fluid composition.

Acute Glomerulonephritis

Acute glomerulonephritis (glō-mer″-u-lō-nĕ-frī´-tis) is an inflammatory condition of the glomeruli that is the result of a hypersensitivity response in which immune complexes of antigen and antibody pass between the endothelial cells of the glomerular capillaries and deposit along the basal lamina where they activate the complement system. It is frequently associated with streptococcal infections, such as throat infections. In acute glomerulonephritis, many of the glomeruli become blocked by the inflammatory response, and others become so permeable that they allow erythrocytes and large amounts of plasma proteins to pass into the glomerular capsules with the glomerular filtrate. In severe cases, total or almost total renal shutdown can occur. However, the vast majority of individuals afflicted with acute glomerulonephritis return to normal renal function within a few months.

Pyelonephritis

Pyelonephritis (pī″-ĕ-lō-nĕ-frī´-tis) is a bacterial infection of the renal pelvis and the surrounding tissues of the kidney. The bacteria reach the kidneys from other sites of infection by way of the bloodstream or lymphatics, or the infection may spread up the ureters from the bladder. In pyelonephritis, the kidney may become swollen and congested, and the pelvis may become inflamed and filled with pus. Abscesses often develop on the surface of the kidney. Pyelonephritis generally responds well to treatment with antibiotics. In chronic cases, however, extensive scar tissue is formed in the kidney, and renal failure becomes a possibility.

Proteinuria

In a number of renal diseases, the permeability of the glomerular capillaries may be increased to such an extent that large amounts of plasma proteins (mostly albumin) pass into the glomerular filtrate and are excreted in the urine. This condition is called *proteinuria,* or *albuminuria.* In severe cases, the loss of plasma proteins can be so great that the plasma osmotic pressure decreases substantially. As a result of the decrease in plasma osmotic pressure, there is an increased tendency for fluid to leave the systemic blood vessels and enter the tissue spaces—a condition that produces generalized edema (swelling) of the body.

Uremia

If the metabolic products (such as urea) that are derived from the breakdown of proteins are not excreted properly, they accumulate in the blood and produce a condition called *uremia (ū-rē´-me-ah).* Uremia affects several body systems, including the nervous system (convulsions and coma), digestive system (vomiting and diarrhea), and respiratory system (dyspnea, or labored breathing).

Kidney Stones

Kidney stones (renal calculi) sometimes form within the renal pelvis. The stones generally consist of uric acid, calcium oxalate, calcium phosphate, and certain other substances. What causes the stones to form is not known. However, several factors may contribute to their formation. These factors include kidney infections, irregularities in kidney structure, certain metabolic problems, vitamin A deficiency, low fluid intake, a diet too high in milk and other calcium-rich foods, and in rare cases, hyperparathyroidism caused by a tumor.

A stone formed in the renal pelvis may remain there, or it may enter the ureter and pass to the bladder. The stone often causes severe, painful contractions of the ureter as it travels through it. A more serious condition results if a stone becomes lodged within the ureter, obstructing the flow of urine to the bladder. In addition, to the retention of urine, kidney stones can cause ulcerations in the lining of the uri-

CONDITIONS OF CLINICAL SIGNIFICANCE

nary tract, which makes the tract more prone to infections.

Cystitis

Cystitis (sis-tī´-tis) is an inflammation of the urinary bladder accompanied by frequent and burning urination and by blood in the urine. In acute cystitis, the mucous membrane lining the bladder becomes swollen, and some bleeding occurs. In the chronic condition, the wall of the bladder can become thickened and its capacity reduced.

The bladder is generally quite resistant to bacterial infections, but under certain conditions bacteria become established in the bladder lining, thus producing cystitis. Cystitis can also be caused by chemicals or by mechanical irritation, such as catheterization. Women have a higher incidence of cystitis than men, probably due to their short urethra, which makes it easier for bacteria to reach the bladder from outside the body. It is not uncommon for *E. coli* bacteria from the woman's anal region to infect the urethra as a result of improper cleansing of the area.

Artificial Kidney

Effective kidney function is necessary for survival, and severe kidney disease can be fatal. The development of the artificial kidney machine, however, has greatly alleviated a number of kidney disease problems. The artificial kidney employs the principal of dialysis (page 52) to remove waste materials from the blood.

In the artificial kidney, the patient's blood is passed through a dialysis tubing that allows urea, electrolytes, and other small molecules to move freely across its wall, but does not allow the movement of large protein molecules. The dialysis tubing is immersed in a bath that contains various substances. If urea or some other small molecule is present in the blood but not in the bath, it diffuses out of the dialysis tubing and into the bath and thus is removed from the blood. If a specific low-molecular-weight substance is present in the bath at the same concentration as within the blood, then it diffuses into the tubing as fast as it diffuses out, and there is no net loss of the material from the blood.

Thus, by regulating the composition and concentration of materials within the bath, the types and amounts of materials that leave the blood by net diffusion out of the dialysis tubing can be regulated. In this way, waste products are removed from the blood while needed constituents are retained.

Effects of Aging

Kidney function declines progressively with age. At age 70, the glomerular filtration rate is only about 50% of the rate at age 40. Renal blood flow decreases from approximately 1100 ml per minute at ages 20–45 to only about 475 ml per minute at 80-89 years of age. There is a corresponding decrease in the function of the renal tubules and in their ability to concentrate the tubular fluid. The decrease in tubular functioning may slow the removal from the body of drugs administered for medical treatments, causing the medications to accumulate in the blood at higher than optimal levels. For this reason, the dosages of certain medications may need to be adjusted as a person becomes older. However, the kidneys do retain their ability to regulate the acid-base balance of the body, although they respond less quickly to a sudden, large acid load.

Among the more common pathological changes that occur in aging kidneys are the degeneration and sclerosis (hardening) of glomeruli. As glomeruli undergo sclerosis, the passage of substances from the blood into the glomerular capsules is restricted, and the glomeruli are no longer able to function as filters of the blood as efficiently as they did before they hardened.

Study Outline

◆ **COMPONENTS OF THE URINARY SYSTEM** p. 834
Kidneys, ureters, urinary bladder, and urethra.

◆ **EMBRYONIC DEVELOPMENT OF THE KIDNEYS** pp. 834–835

From Intermediate Mesoderm. Three successive pairs of embryonic kidneys develop:
Pronephros (with pronephric duct).
Mesonephros (with mesonephric duct).
Metanephros (becomes adult kidney).

Ureteric Buds. Develop into collecting tubules, calyces, renal pelvis, and ureters.

Nephrons. Develop from cap of intermediate mesoderm that covers ureteric bud.

◆ **ANATOMY OF THE KIDNEYS** pp. 835–844
Paired reddish brown, bean-shaped organs on posterior abdominal wall; retroperitoneal.

Tissue Layers Surrounding the Kidneys. Three layers: *renal capsule* (fibrous), *adipose capsule* (perirenal fat), *renal fascia* (double-layered).

External Structure of the Kidneys. Renal hilus is medial indentation; passageway for blood vessels and ureters. Renal pelvis located in renal sinus.

Internal Structure of the Kidneys.

1. *Cortex.* Outer layer; also forms renal columns that pass into medulla.
2. *Medulla.* Consists of renal pyramids.
3. *Renal pelvis.* Expanded upper end of ureter; formed from joining of major calyces.

Nephrons. Functional units of kidneys. Each nephron consists of a glomerulus (network of parallel capillaries) and a renal tubule (proximal end forms glomerular capsule).

RENAL CORPUSCLE. Capsule and glomerulus; site of transfer between blood and nephron.

1. Visceral layer of glomerular capsule composed of podocytes; foot processes separated by filtration slits, which are covered by slit membrane.
2. Glomeruli have fenestrated endothelium.

FILTRATION BARRIER. Formed of fenestrated endothelium, basal lamina, and slit membrane.

PROXIMAL CONVOLUTED TUBULE. Single layer of cuboidal cells with microvilli.

LOOP OF HENLE. Descending and ascending limbs.

DISTAL CONVOLUTED TUBULE. Empties into collecting tubule.

Juxtaglomerular Complex. Distal straight tubule contacts afferent arteriole; composed of juxtaglomerular cells of afferent arteriole and macula densa of tubule.

Blood Vessels of the Kidneys.

RENAL ARTERY. Provides substantial blood flow, which allows kidneys to maintain blood homeostasis. Interlobar artery → arcuate artery → interlobular artery → afferent arteriole → glomerulus → efferent arteriole → peritubular capillaries → interlobular vein → arcuate vein → interlobar vein → renal vein.

TWO CAPILLARY BEDS. Glomerulus and peritubular capillaries.

AFFERENT AND EFFERENT ARTERIOLES. Maintain constant blood pressure in glomerulus.

◆ **PHYSIOLOGY OF THE KIDNEYS** pp. 844–856

Basic Renal Processes. Glomerular filtration, tubular secretion, tubular reabsorption.

Glomerular Filtration.

1. 16–20% of blood plasma entering kidneys is filtered from glomerular capillaries and into glomerular capsules.
2. Glomerular filtrate contains most plasma components at essentially same concentrations as plasma, but is basically protein free.

NET FILTRATION PRESSURE. Filtration is favored by pressure of blood within glomerular capillaries and opposed by pressure of fluid within glomerular capsules and by osmotic force exerted by unfiltered plasma proteins in glomerular capillaries.

FACTORS THAT INFLUENCE GLOMERULAR FILTRATION.

INFLUENCE OF ARTERIAL BLOOD PRESSURE AND AUTOREGULATORY MECHANISMS. Arterial blood pressure influences glomerular filtration rate, but effect is not as great as expected due to intrinsic autoregulatory mechanisms that maintain a fairly stable glomerular filtration rate over wide range of systemic blood pressures.

Myogenic Mechanism. Afferent arterioles constrict when pressure rises and dilate when pressure falls.

Tubulo-Glomerular Feedback Mechanism. Involves juxtaglomerular complex.

INFLUENCE OF SYMPATHETIC NEURAL STIMULATION. Increased sympathetic stimulation preferentially constricts afferent arterioles, lowering glomerular capillary pressure and glomerular filtration rate.

General Mechanisms of Tubular Secretion and Tubular Reabsorption. Generally occur by diffusion or by means of discrete transport mechanisms. In many cases, they are active processes.

Tubular Secretion. Means by which some substances enter tubular fluid; may be active or passive; potassium ions, hydrogen ions, and organic ions undergo tubular secretion.

Tubular Reabsorption. Removes materials from tubular fluid.

ACTIVE TUBULAR REABSORPTION. Energy-requiring; glucose, amino acids, and vitamins, as well as sodium, calcium, and chloride ions are actively reabsorbed.

PASSIVE TUBULAR REABSORPTION. Does not require energy; urea is passively reabsorbed.

Excretion of Water and Its Effect on Urine Concentration. Kidneys can produce urine as dilute as 65–70 milliosmoles per liter or as concentrated as 1200 milliosmoles per liter.

MAINTENANCE OF A CONCENTRATED MEDULLARY INTERSTITIAL FLUID. Activities and anatomical arrangement of loops of Henle of nephrons (particularly of juxtamedullary nephrons), collecting tubules, and vasa recta are important.

ACTIVITIES OF THE LOOPS OF HENLE AND THE COLLECTING TUBULES. As fluid flows along loops of Henle and collecting tubules, reabsorption of solutes raises concentration of medullary interstitial fluid.

EFFECT OF VASA RECTA BLOOD FLOW. Pattern of blood flow along vasa recta minimizes solute removal from medullary interstitial fluid.

PRODUCTION OF DIFFERENT URINE VOLUMES AND CONCENTRATIONS.

1. Sodium ions actively transported out of proximal tubule; chloride ions follow passively. Substances such as glucose and amino acids are also reabsorbed. Water leaves osmotically. Volume of glomerular filtrate reduced 65–70%; concentration remains essentially that of plasma.
2. Water moves out of descending limb of loop of Henle. Tubular fluid is reduced in volume; its concentration is increased.
3. Sodium and chloride ions move out of ascending limb of loop of Henle, which is relatively impermeable to water. Concentration of tubular fluid is diminished; volume is not greatly altered.
4. Certain ions are reabsorbed from the fluid within distal convoluted tubule; others are secreted into the fluid. In monkeys and presumably other primates (such as humans), distal convoluted tubule is relatively impermeable to water. In general, concentration of fluid within distal convoluted tubule remains below that of plasma; its volume does not change greatly.
5. Permeability of collecting tubule to water controlled by ADH. If much ADH is present, permeability to water is high; water leaves and tubular fluid becomes concentrated and reduced in volume.

MECHANISM OF WATER REABSORPTION. Occurs by passive transport (osmosis); influenced by ADH.

MINIMUM URINE VOLUMES. Need to eliminate waste products such as urea, sulfates, phosphates requires daily excretion of about 600 milliosmoles of these materials.

Plasma Clearance. Rate at which a particular substance is eliminated or cleared from plasma by kidneys. Plasma clearance of inulin indicates glomerular filtration rate.

◆ URETERS pp. 856–857

Urine Transport. From renal pelvis to urinary bladder; retroperitoneal.

◆ URINARY BLADDER p. 857

Hollow Muscular Organ. Urine storage.

1. Retroperitoneal.
2. Lined with transitional epithelium.
3. Three layers of smooth muscle.

◆ URETHRA p. 857

Muscular tube lined with mucous membrane; carries urine from bladder to exterior of body; surrounded by internal urethral sphincter where it leaves bladder; surrounded by external urethral sphincter where it passes through urogenital diaphragm.

Female. Short (4 cm); runs along anterior surface of vagina.

Male. Urine passage and reproduction; long (20 cm).

◆ MICTURITION pp. 857–859

1. As bladder fills, local spinal reflex causes bladder to contract and internal and external urethral sphincters to open, resulting in micturition.
2. Impulses from brain can facilitate or inhibit reflex emptying of bladder; with training, high degree of voluntary control can be exercised.

◆ CONDITIONS OF CLINICAL SIGNIFICANCE: THE URINARY SYSTEM pp. 859–861

Pressure-Related Pathologies. Prostrate hypertrophy, low arterial blood pressure.

Acute Glomerulonephritis. Inflammatory condition that affects glomeruli. Frequently associated with streptococci bacterial infection in other body parts.

Pyelonephritis. Bacterial infection of renal pelvis and surrounding tissue.

Proteinuria. Plasma proteins pass into urine.

Uremia. Metabolic products, including urea, accumulate in blood due to improper excretion.

Kidney Stones (Renal Calculi). Formed in renal pelvis or urinary bladder; generally consist of uric acid, calcium oxalate, and calcium phosphate; may cause urine retention, pain, and infection due to blockage of ureters and ulceration of urinary tract lining.

Cystitis. Urinary bladder inflammation that produces frequent, burning urination and blood in urine. More common in women, probably because urethra is short.

Artificial Kidney. Selective removal of waste materials from the blood by dialysis.

Effects of Aging.

1. Progressive decline in function: decreased glomerular filtration rate; decreased renal blood flow; decreased renal tubule function.
2. Acid-base regulation ability retained by aging kidneys.
3. Pathological changes: degeneration and sclerosis of glomeruli.

 Self-Quiz

1. The medulla of the kidney contains: (a) the adipose capsule; (b) glomeruli; (c) pyramids.
2. The expanded, funnel-shaped upper end of the ureter within the kidney is the: (a) renal pelvis; (b) renal hilus; (c) urethra; (d) nephron.
3. The capillary found in the glomerular capsule of a nephron is the: (a) glomerulus; (b) proximal convoluted tubule; (c) adipose capsule; (d) juxtaglomerular complex.
4. The space in each kidney that contains the renal vessels and the renal pelvis is called the renal: (a) capsule; (b) sinus; (c) cortex.
5. From the collecting tubules, urine enters the: (a) renal pelvis; (b) major calyces; (c) minor calyces; (d) distal tubules.
6. Podocytes are responsible for the formation of: (a) filtration slits; (b) fenestrated endothelium; (c) capsular space; (d) pronephros.
7. Renal corpuscles are located in the cortical region of the kidney. True or False?
8. The juxtaglomerular complex is formed from modification of the cells of: (a) the distal convoluted tubule; (b) the afferent arteriole; (c) both (a) and (b).
9. Thin-walled vessels that supply the loops of Henle and the collecting ducts are called: (a) glomeruli; (b) vasa recta; (c) macula densa; (d) arcuate veins.
10. Essentially 100% of the plasma entering the kidneys is filtered from the glomerular capillaries and into the glomerular capsules as the glomerular filtrate. True or False?
11. Which force favors the movement of fluid out of the glomerular capillaries and into the glomerular capsules? (a) the osmotic force exerted by unfiltered plasma proteins within the glomerular capillaries; (b) the pressure of the fluid within the glomerular capsules; (c) the glomerular capillary pressure.
12. In general, an increase in the activity of the sympathetic neurons that supply the kidneys leads to a dilation of the afferent arterioles. True or False?
13. A rise in the arterial blood pressure: (a) reduces the glomerular filtration rate; (b) increases the output of urine; (c) has no effect on kidney function because of the influence of the kidney's intrinsic autoregulatory mechanisms.
14. Urea reabsorption from the tubular fluid is passive and depends on water reabsorption. True or False?
15. At the proximal convoluted tubule of a nephron: (a) sodium ions are actively transported out of the tubule; (b) there is a net movement of water into the tubule; (c) there is a net movement of chloride ions into the tubule.
16. As fluid flows along the descending limb of the loop of Henle of a juxtamedullary nephron, there is a net movement of water out of the tubule. True or False?
17. Fluid at the bottom of the loop of Henle of a juxtamedullary nephron is: (a) less concentrated than the glomerular filtrate; (b) less concentrated than fluid within the distal convoluted tubule; (c) more concentrated than fluid within the proximal convoluted tubule.
18. With no antidiuretic hormone: (a) large volumes of urine are produced; (b) large amounts of protein appear in the urine; (c) glucose is not reabsorbed from the kidney tubules.
19. The urethra of the female is divisible into three parts—the prostatic, membranous, and spongy urethra, which are named according to the regions through which the urethra passes. True or False?
20. Which of the following is least affected by aging? (a) glomerular filtration rate; (b) renal blood flow; (c) renal tubule function; (d) renal regulation of body pH.

CHAPTER 27

Fluid and Electrolyte Regulation and Acid-Base Balance

CHAPTER CONTENTS

LEARNING OBJECTIVES

After completing this chapter, you should be able to:

1. List similarities and differences in the composition of the plasma, the interstitial fluid, and the intracellular fluid.
2. Describe the factors that influence the renal excretion of sodium.
3. Cite possible causes for a feeling of thirst.
4. Describe the factors that influence the renal excretion of water.
5. Explain how body potassium is regulated, and cite the regulatory factors involved.
6. Explain the actions of hormones involved in calcium regulation.
7. State what distinguishes a strong acid from a weak acid.
8. Explain the function of buffer systems in the body.
9. Explain how the respiratory system can have an important influence on the pH of the internal environment.
10. Explain how the kidneys act as essential regulators of acid-base balance.

A substance that can conduct an electric current when it is dissolved in water is called an **electrolyte.** In general, substances that exist as ions in aqueous solution are electrolytes. For example, sodium chloride, which exists as sodium ions and chloride ions when dissolved in water, is an electrolyte.

In order to maintain homeostasis, the body must regulate the concentrations of water and electrolytes, including acids and bases, in its internal environment. The kidneys are particularly important in the regulation of these substances, as are the lungs and the gastrointestinal tract.

Body Fluids

There are two general categories of body fluid: (1) **intracellular fluid,** or fluid within cells; and (2) **extracellular fluid,** or fluid outside cells (Figure 27.1; Table 27.1). The extracellular fluid forms the body's internal environment. It is composed primarily of **plasma** (the fluid within the blood vessels) and **interstitial fluid** (the fluid that surrounds the cells).

Fluids such as lymph, cerebrospinal fluid, intraocular fluid, synovial fluid, pericardial fluid, pleural fluid, and peritoneal fluid are generally included with the extracellular fluid. However, the volume of these fluids is very small compared to the volume of the plasma and interstitial fluid.

◆ **FIGURE 27.1 Distribution of body fluids**

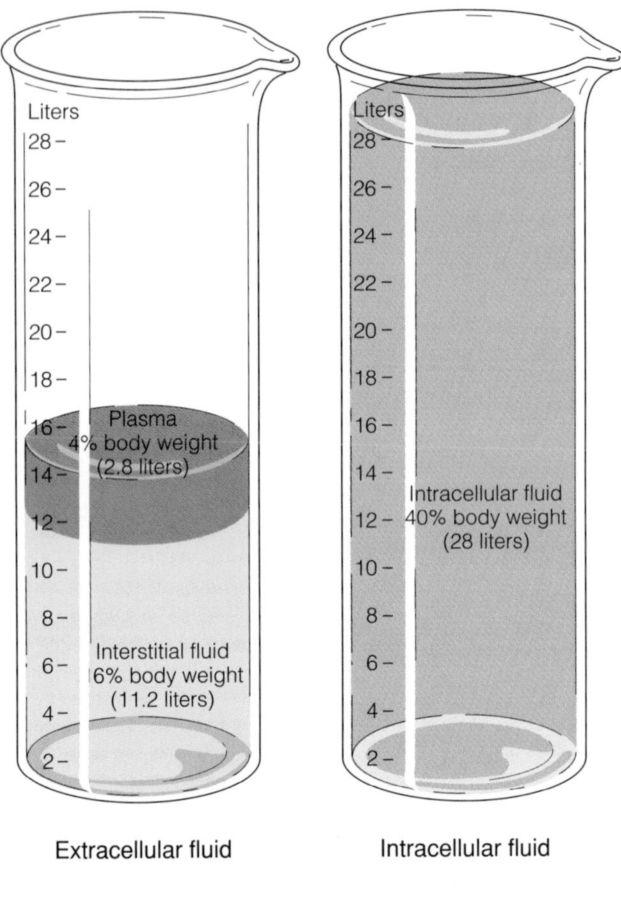

Composition of Plasma and Interstitial Fluid

The different body fluids are not isolated from one another, and a continuous exchange of materials occurs between them. As can be seen from Figure 27.2, the compositions of plasma and interstitial fluid are quite similar. These fluids are separated from one another by the walls of the capillaries. A relatively free exchange of low-molecular-weight solutes occurs across capillary walls, and this exchange accounts for the similarity between the two fluids. The major difference between plasma and interstitial fluid is that plasma contains more proteins. This difference is due to the fact that plasma proteins tend to remain within the blood vessels, since capillaries are generally not very permeable to them. Under physiological conditions, these proteins normally exist as anions—that is, as negatively charged ions. For this reason, there is somewhat more sodium (a positively charged ion, or cation) and somewhat less chloride (an anion) within the plasma than within the interstitial fluid.

Since a relatively free exchange of materials occurs between plasma and interstitial fluid, the regulation of the fluid, electrolyte, and acid-base composition of the plasma by the kidneys (and to some degree by the lungs and gastrointestinal tract) also serves to regulate the fluid, electrolyte, and acid-base composition of the interstitial fluid.

Composition of Intracellular Fluid

The composition of intracellular fluid is considerably different from that of plasma and interstitial fluid. Intracellular fluid is separated from interstitial fluid by the membranes of the cells, which are relatively impermeable to proteins. Consequently, cellular proteins tend to remain within the cells. Moreover, the action of the cell membrane's sodium-potassium pump moves sodium ions out of the cells and accumulates potassium ions within them.

◆ **TABLE 27.1 Classification of Body Fluid**

	VOLUME OF FLUID (in liters)	PERCENTAGE OF BODY FLUID	PERCENTAGE OF BODY WEIGHT
Total body fluid	42	100	60
Intracellular fluid (ICF)	28	67	40
Extracellular fluid (ECF)	14	33	20
Plasma	2.8	6.6 (20% of ECF)	4
Interstitial fluid	11.2	26.4 (80% of ECF)	16
Lymph	Negligible	Negligible	Negligible
Other (cerebrospinal fluid, intraocular fluid, synovial fluid, pericardial fluid, pleural fluid, peritoneal fluid)	Negligible	Negligible	Negligible

Regulation of the Body's Internal Environment

In order to maintain homeostasis, the addition of substances such as water and electrolytes to the body's internal environment must be balanced by the removal of equal amounts of these materials. Ingestion, of course, adds materials to the internal environment, whereas excretion by organs such as the kidneys removes materials. In fact, the kidneys' role in controlling the excretion of water and electrolytes is one of their most important regulatory functions.

Sodium Regulation

Sodium is the major extracellular cation. Because of sodium's osmotic effects, variations in body-sodium content can cause changes in extracellular fluid volume, including plasma volume. Changes in plasma volume can lead to changes in blood pressure, and mechanisms involved in plasma-volume and blood-pressure regulation are important in controlling body-sodium content.

Sodium Ingestion

Animal studies indicate that two components are involved in determining the amount of sodium ingested (in the form of salt). One is the regulatory component, which governs the intake of sodium in such a way that a balance between sodium intake and outflow is maintained. The other is the hedonistic component, by which an animal demonstrates a strong preference for salt regardless of regulatory requirements.

The degree to which regulatory and hedonistic components operate in humans is not entirely clear. People severely depleted of sodium chloride often develop a desire for salt. However, people also seem to have a strong hedonistic salt appetite, and they consume large amounts of it whenever it is inexpensive and easily available. For example, in the United States, an average person consumes 10–15 g of salt per day even though less than 0.5 g per day is normally needed to meet regulatory requirements. Consequently, for humans it appears that the regulation of the body's sodium content is achieved mainly by controlling sodium excretion by the kidneys rather than by controlling sodium ingestion.

Renal Excretion of Sodium

The relationship between the glomerular filtration rate of sodium and the tubular reabsorption rate of sodium is very important in determining the amount of sodium excreted (Figure 27.3). For example, if much sodium is filtered into the glomerular capsules but only a little is reabsorbed, then a considerable amount of sodium will be excreted.

Since sodium is freely filtered from the plasma into the glomerular capsules, any factors that alter the general rate of glomerular filtration will also alter the glomerular filtration rate of sodium. Among the factors that influence the glomerular filtration rate are (1) the arte-

◆ **FIGURE 27.2 Composition of body fluids**

Numbers on the left of the scale indicate amounts of cations or anions. Numbers on the right indicate the sum of cations and anions.

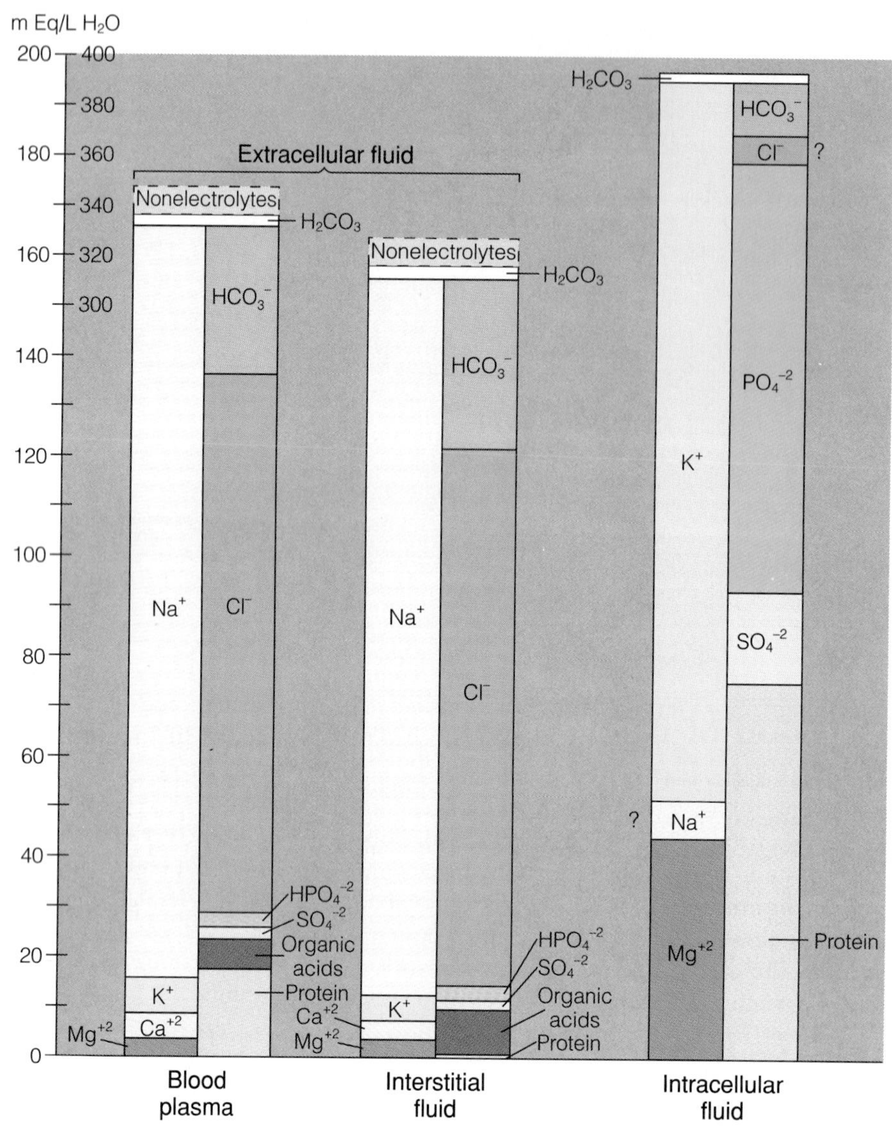

rial blood pressure, and (2) the activity of sympathetic nerves to the kidneys, which alters the diameters of the afferent arterioles.

The tubular reabsorption of sodium is subject to physiological control, and the control of sodium reabsorption is believed to be particularly important in the long-term regulation of sodium excretion. A major factor in this control is the adrenal cortical hormone aldosterone. Aldosterone stimulates sodium reabsorption, particularly from the last portions of the distal convoluted tubules and the collecting tubules. When much aldosterone is present, almost all the sodium that enters the tubules is reabsorbed, and as little as as 0.1 g of sodium is excreted per day. When little aldosterone is present, as much as 30–40 g of sodium can be excreted per day.

The release of aldosterone is enhanced indirectly by the substance renin, which is secreted into the blood by specialized cells (juxtaglomerular cells) of kidney arterioles in response to a number of factors. Although it is currently impossible to assign a precise quantitative role to each of these factors, a decrease in the renal arterial pressure and/or an increase in the activity of sympathetic nerves to the kidneys apparently stimulates renin secretion.

Renin converts the precursor substance angiotensinogen, which is manufactured by the liver and is present in the blood, into angiotensin I. In turn, angiotensin I is converted into angiotensin II by a converting enzyme associated with the walls of blood capillaries, particularly in the lungs. One of the effects of angiotensin II is

◆ **FIGURE 27.3 Sodium excretion by the kidneys**
The amount of sodium excreted by the kidneys may be altered by varying either the glomerular filtration rate (filtered sodium) or the tubular reabsorption rate of sodium.

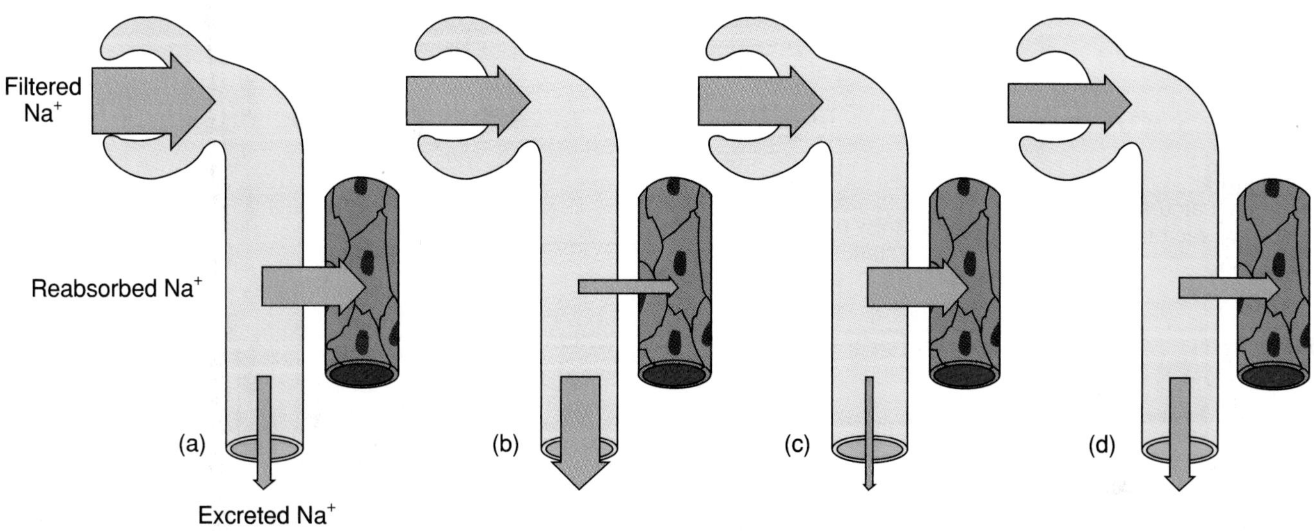

the stimulation of aldosterone release from the adrenal glands. Thus, an increased release of renin ultimately leads to an increased release of aldosterone. The aldosterone then stimulates sodium reabsorption from the last portions of the distal convoluted tubules and the collecting tubules.

Factors That Influence Sodium Excretion

As previously indicated, a change in body-sodium content can lead to a change in both plasma volume and blood pressure, and mechanisms involved in plasma-volume and blood-pressure regulation are important in controlling body-sodium content. For example, an increase in systemic blood pressure, due perhaps to an increase in plasma volume, can influence sodium excretion, and thus body-sodium content, in several ways (Figure 27.4).

1. The kidneys' ability to autoregulate the glomerular filtration rate is not perfect, and an increase in systemic blood pressure acts directly on the kidneys to cause some increase in the glomerular filtration rate.
2. An increase in systemic blood pressure causes an increase in the rate at which nerve impulses are transmitted to the central nervous system from circulatory pressure receptors such as the aortic arch and carotid sinus baroreceptors and receptors in the atria of the heart. The increased rate of nerve-impulse transmission from these receptors leads to a decrease in the activity of sympathetic nerves to the kidneys. The decrease in sympathetic activity, in turn, leads to a dilation of the afferent arterioles and, consequently, to an increase in blood pressure within the glomerular capillaries, which increases the glomerular filtration rate.
3. The increase in systemic blood pressure (including the renal arterial pressure) and the decrease in the activity of sympathetic nerves to the kidneys cause a decreased release of renin. The decreased release of renin ultimately leads to a decrease in the level of aldosterone and, consequently, to a diminished tubular reabsorption of sodium.
4. An increase in plasma volume can lead to an increased filling of the atria of the heart, which stretches the atrial walls. In response to stretch, atrial cardiac muscle cells release a polypeptide hormone called **atrial natriuretic factor** *(nay″-tree-yoo-ret′-ick).* Atrial natriuretic factor modifies the activity of the renin-angiotensin system by inhibiting the secretion of renin and also by directly inhibiting the adrenal secretion of aldosterone. It also acts directly at various sites in the kidneys to inhibit sodium reabsorption.

Together, the foregoing responses increase the amount of sodium filtered into the glomerular capsules and decrease the amount of sodium reabsorbed from the tubules. As a result, the urinary excretion of sodium increases, and the total body mass of sodium declines. Because sodium, by virtue of its osmotic effects, has a direct influence on extracellular volume, the loss of sodium leads to a decrease in plasma volume, which tends to return the blood pressure to normal.

◆ **FIGURE 27.4 Effect of an increase in plasma volume on sodium excretion by the kidneys**

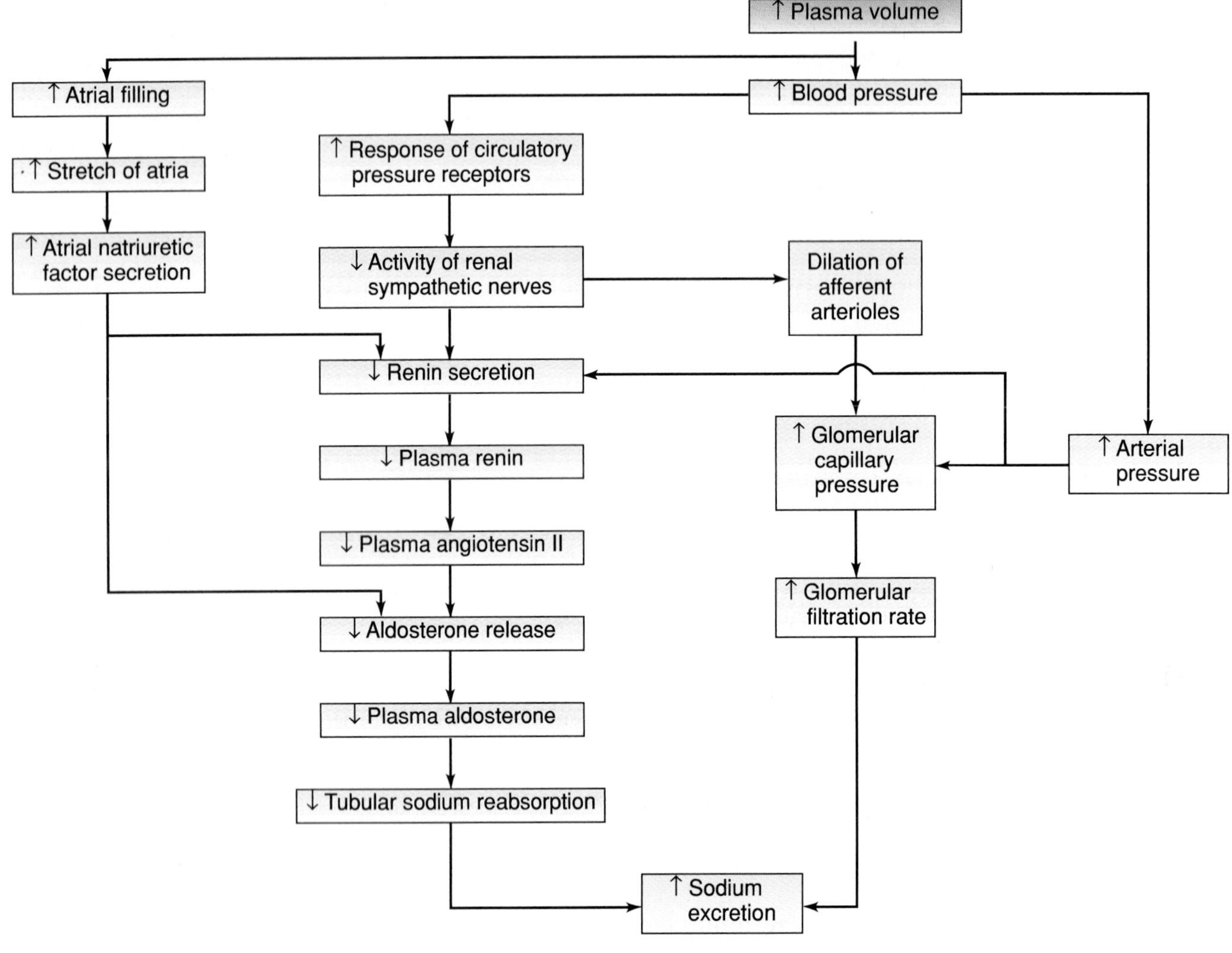

Sodium Content and Sodium Concentration

The loss or gain of sodium by the body changes the total sodium *content* of the body, but it does not necessarily have a great effect on the *concentration* of sodium in the internal environment. As discussed in the next section, the body possesses receptors called **osmoreceptors** *(oz˝-mo-re-sep´-turz)*, which are sensitive to the osmotic pressure of the extracellular fluid. Regulatory mechanisms brought into play by these receptors compensate for changes in the osmotic concentration of the extracellular fluid (which is largely due to sodium) by stimulating or retarding water ingestion and water excretion in order to maintain the proper osmotic concentration. As a result, these receptors are of considerable importance in the regulation of the actual sodium concentration of the extracellular fluid.

Water Regulation

The water content of the body must be regulated because of its effect on both the osmotic pressure and the fluid volume of the internal environment.

Water Ingestion and Thirst

Although fluid intake is often influenced more by habit and social factors than by the need to regulate body water, water ingestion does seem to depend at least in part on the regulatory needs of the body. A feeling of thirst results from an increased osmotic concentration of the extracellular fluid and also from a substantial reduction in the plasma volume. It is thought that osmoreceptors in a hypothalamic thirst center in the brain, and perhaps also circulatory pressure receptors, stimulate thirst. Moreover, angiotensin II stimulates thirst by a direct effect on the brain.

Often when a dehydrated person drinks water, the amount ingested approximates the amount required to bring the extracellular fluid back to the proper osmotic concentration. This fact is interesting because the person usually stops drinking before the ingested water has had time to be absorbed from the gastrointestinal tract and actually return the extracellular osmolarity to normal. Thus, the gastrointestinal tract may possess some sort of water-metering system that is able to regulate water intake in accordance with the needs of the body. Moreover, the control of thirst may have a learned or anticipatory component that enables a person to anticipate his or her fluid needs and therefore to drink sufficient water to prevent dehydration. For example, the amount of fluid consumed in association with eating appears to be a learned activity that may prevent dehydration.

Renal Excretion of Water

Body water is also controlled by the excretory pathways of the kidneys, and the relationship between the glomerular filtration rate of water and the tubular reabsorption rate of water is of major importance in determining the amount of water excreted. For example, if much water is filtered into the glomerular capsules but only a little is reabsorbed, then a considerable amount of water will be excreted.

Since water is freely filtered from the plasma into the glomerular capsules, any factors that alter the general rate of glomerular filtration will also alter the glomerular filtration rate of water. Some of these factors—particularly some associated with blood-pressure regulation—were discussed in the previous section on sodium regulation. For example, an increase in systemic blood pressure, due perhaps to an increase in plasma volume, leads to an increase in the glomerular filtration rate. (Conversely, a decrease in systemic blood pressure leads to a decrease in the glomerular filtration rate.)

As explained in Chapter 26 (page 853), the amount of water reabsorbed from the tubules depends in large part on the collecting tubules' permeability to water, which is controlled by antidiuretic hormone (ADH). When the plasma-ADH level is low, the collecting tubules' permeability to water is low, and relatively little water is reabsorbed. Consequently, a large volume of urine is produced. Conversely, when the plasma-ADH level is high, the collecting tubules' permeability to water is high, much water is reabsorbed, and only a relatively small volume of urine is produced.

Inputs to neural centers that influence ADH release come both from osmoreceptors located in the hypothalamus and from circulatory pressure receptors, particularly those in the left atrium of the heart. The osmoreceptors, which are stimulated by increased extracellular osmolarity, stimulate the cells that secrete ADH. The pressure receptors, which are stimulated by increased atrial blood pressure (possibly resulting from increased plasma volume), inhibit the ADH-producing cells. Thus, considerable ADH release and substantial water reabsorption result from increased extracellular osmolarity and decreased blood pressure (plasma volume). Conversely, strong ADH inhibition and substantial water excretion result from decreased extracellular osmolarity and increased blood pressure. These responses, especially when coupled with the previously discussed changes in glomerular filtration rate that occur in response to changes in blood pressure, tend to return the extracellular osmolarity and the blood pressure (plasma volume) to normal when they either increase or decrease.

Factors That Influence Water Excretion

From the preceding discussion it is evident that the osmotic pressure of the extracellular fluid and the blood pressure (plasma volume) are major factors influencing water excretion. Moreover, the factors that influence water excretion are very closely related to the factors that influence sodium excretion. Both groups of factors are involved in the regulation of blood pressure as well as in the maintenance of the proper volume and concentration of the body's internal environment.

Indeed, the activity of the mechanisms that influence water excretion and the mechanisms that influence sodium excretion can rarely be separated completely. For example, when an increase in systemic blood pressure causes an increase in the glomerular filtration rate, both sodium and water are affected. Similarly, an increase in the plasma volume that increases atrial and systemic blood pressure can decrease the reabsorption of sodium by way of the renin-angiotensin mechanism and reduce the reabsorption of water by way of the ADH mechanism. Moreover, a change in body sodium content due to a change in the excretion of sodium causes a change in the osmotic pressure of the extracellular fluid. The change in osmotic pressure is sensed by osmoreceptors that, in turn, cause the ADH mechanism to increase or decrease the reabsorption of water as necessary to reestablish the proper osmotic concentration of the extracellular fluid.

Potassium Regulation

Potassium is important in the excitability of nerve and muscle, and the potassium concentration of the extracellular fluid is closely regulated. Humans normally excrete potassium in amounts equal to ingested amounts and thus maintain potassium balance.

Potassium is freely filtered from the plasma into the glomerular capsules, and it can be both reabsorbed and secreted by the tubules. Normally, almost all the filtered potassium is actively reabsorbed, and adjustments in the amount of potassium excreted depend mainly on how much is secreted into the tubules.

The amount of potassium secreted into the tubules is related to the potassium concentration in the tubular cells. When the potassium concentration of the extracellular fluid is high, more potassium will be present in the tubular cells, and as a consequence, more potassium will be secreted than when the potassium concentration of the extracellular fluid is low.

A second factor that influences the elimination of potassium is aldosterone. Besides promoting the tubular reabsorption of sodium, this hormone enhances the tubular secretion of potassium. In fact, the potassium concentration of the extracellular fluid that bathes the adrenal glands is a major stimulus for the release of aldosterone from the adrenal cortex. When the potassium concentration of the extracellular fluid rises, so does the release of aldosterone. When the potassium concentration falls, the release of aldosterone decreases. Once aldosterone is released, however, it exerts both of its effects and enhances sodium reabsorption as well as potassium secretion.

Calcium Regulation

The calcium level of the extracellular fluid is closely regulated, and low levels of calcium increase nerve and muscle excitability. Calcium regulation involves bone, which contains 99% of the body's calcium, as well as the kidneys and the gastrointestinal tract. The calcium-regulating activities of all three sites are influenced either directly or indirectly by parathyroid hormone (parathormone, or PTH) from the parathyroid glands.

Parathyroid hormone increases the plasma-calcium concentration (and decreases the plasma-phosphate concentration). It increases the movement of calcium and phosphate from bone into the extracellular fluid. This activity is believed to be due at least in part to the ability of parathyroid hormone to stimulate the activity of cells called osteoclasts, which break down bone. In the kidneys, parathyroid hormone decreases the urinary excretion of calcium and increases the excretion of phosphate. Parathyroid hormone also enhances a step in the metabolic transformation of vitamin D_3 to a substance called 1,25-dihydroxycholecalciferol. This substance stimulates active calcium absorption from the intestine.

The release of parathyroid hormone is controlled by the calcium concentration of the extracellular fluid that bathes the parathyroid glands. When the calcium concentration is low, parathyroid hormone release increases, and this response tends to raise the extracellular fluid calcium concentration toward normal. Conversely, when the calcium concentration is high, parathyroid hormone release decreases, and this response tends to lower the extracellular fluid calcium concentration toward normal.

Calcitonin, a hormone from the thyroid gland, lowers the plasma-calcium level, primarily by inhibiting the removal of calcium from bone. Thus, the influence of calcitonin on the plasma-calcium level is opposite to that of parathyroid hormone. The secretion of calcitonin is controlled by the calcium concentration of the fluid that bathes the thyroid gland; high calcium concentrations enhance secretion. Calcitonin, however, appears to play a less important role in calcium regulation than parathyroid hormone does.

Hydrogen Ion and Acid-Base Regulation

Because hydrogen ions affect enzyme action, most metabolic reactions are sensitive to the hydrogen ion concentration of the fluid in which they occur. Consequently, the proper regulation of hydrogen ion concentration is essential for effective cellular function.

Acids and Bases

As explained in Chapter 2 (page 48), acids are substances that liberate hydrogen ions, and bases are substances that accept them. The concentration of free hydrogen ions (H^+) determines the acidity of a solution. A solution's acidity is measured on the pH scale, with lower pH values indicating higher hydrogen ion concentrations (acidity).

Acids can be grouped as either strong acids or weak acids. Strong acids are those that dissociate virtually completely in solution, providing large numbers of free hydrogen ions. Hydrochloric acid (HCl), which in solution dissociates into hydrogen ions and chloride ions, is an example of a strong acid. Weak acids are those that do not dissociate completely and do not provide as many free hydrogen ions when they are placed in solution. For example, when dissolved in water, some of the molecules of the weak acid carbonic acid dissociate into hydrogen ions and bicarbonate ions. Substantial numbers of carbonic acid molecules, however, do not dissociate into hydrogen ions and bicarbonate ions.

Weak acids such as carbonic acid generally dissociate in a predictable fashion. A certain proportion of the molecules dissociate to provide free hydrogen ions, and a certain proportion do not undergo this dissociation.

pH of the Blood and Interstitial Fluid

The pH of arterial blood is normally 7.4, whereas the pH of venous blood and interstitial fluid is 7.35. However, the body's metabolic activities generate acidic products that tend to raise the hydrogen ion concentration (and decrease the pH) of the internal environment. The phosphorous and sulfur of proteins, for example, are potential sources of phosphoric and sulfuric acids, and metabolic reactions produce organic acids such as fatty acids and lactic acid. Thus, under normal circumstances, the body must deal with a continuing production of hydrogen ions in order to maintain the proper pH of the internal environment.

The proper pH of the internal environment is maintained by (1) buffer systems, (2) the respiratory system, and (3) the kidneys. Buffer systems act to resist changes in the hydrogen ion concentration of the internal environment. If a change does occur, the respiratory system responds within minutes to compensate for the change. The kidneys usually require hours or days to compensate for a change in the hydrogen ion concentration of the internal environment, but overall they are the most potent regulators of acid-base balance.

Buffer Systems

The body contains buffer systems that help stabilize the pH of the body fluids. A buffer system consists of a solution containing two or more chemical substances that can prevent extreme changes in the hydrogen ion concentration (pH) of the solution when either an acid or a base is added to it. A buffer system works by chemically combining with hydrogen ions as their concentration starts to rise and releasing them as their concentration starts to fall.

A solution of carbonic acid and sodium bicarbonate, for example, acts as a buffer system. When carbonic acid and sodium bicarbonate are placed together in solution, the sodium bicarbonate dissociates into sodium ions and bicarbonate ions. The carbonic acid, however, is a weak acid, and it does not dissociate fully into hydrogen ions and bicarbonate ions (Figure 27.5). Thus the carbonic acid–sodium bicarbonate buffer system contains undissociated carbonic acid, sodium ions, hydrogen ions, and bicarbonate ions. Moreover, a certain proportion of the carbonic acid is always dissociated into hydrogen ions and bicarbonate ions, and a certain proportion does not undergo this dissociation.

If hydrogen ions (for example, in the form of hydrochloric acid) are added to a buffer system such as the carbonic acid–sodium bicarbonate buffer system, many of the hydrogen ions do not remain in the free state to affect pH as they would if they were added to pure water (Figure 27.6). In the carbonic acid–sodium bicarbonate buffer system, the addition of hydrogen ions (from hydrochloric acid) results in the presence of many hydrogen ions and many bicarbonate ions (from sodium bicarbonate) in the solution. These ions are the equivalent of carbonic acid molecules that have dissociated into hydrogen ions and bicarbonate ions. Consequently,

◆ FIGURE 27.5 Schematic representation of the carbonic acid–sodium bicarbonate buffer system

In solution, the sodium bicarbonate dissociates into sodium ions (Na^+) and bicarbonate ions (HCO_3^-). Because it is a weak acid, only a relatively small amount of the carbonic acid dissociates into hydrogen ions (H^+) and bicarbonate ions.

the presence of these ions upsets the normal proportion that must be maintained between carbonic acid molecules that dissociate into hydrogen ions and bicarbonate ions, and carbonic acid molecules that do not undergo this dissociation. In order to reestablish the normal proportion, some of the hydrogen ions and bicarbonate ions combine into undissociated carbonic acid molecules. This combination removes from the solution some of the hydrogen ions resulting from the addition of the hydrochloric acid. These hydrogen ions would otherwise have been free to increase the hydrogen ion concentration of the solution and lower the pH. Instead, many of the hydrogen ions do not remain free, and the effect of the hydrogen ions from the hydrochloric acid on the hydrogen ion concentration (pH) of the solution is minimized.

Conversely, the removal of hydrogen ions from the carbonic acid–sodium bicarbonate buffer system—due, for example, to the addition of a base—also upsets the normal proportion that must be maintained between carbonic acid molecules that dissociate into hydrogen ions and bicarbonate ions, and carbonic acid molecules that do not undergo this dissociation. In this case, some of the carbonic acid molecules dissociate into hydrogen ions and bicarbonate ions in order to restore the normal proportion. This dissociation minimizes the effect of removing hydrogen ions on the pH of the solution.

In the body, the carbonic acid–sodium bicarbonate buffer system acts in the manner just described to sta-

◆ **FIGURE 27.6 Effect of adding strong acid to the carbonic acid–sodium bicarbonate buffer system**

(a) Hydrochloric acid (HCl) added to the solution dissociates into hydrogen ions (H^+) and chloride ions (Cl^-). This dissociation upsets the proportion that must be maintained between carbonic acid molecules that are dissociated into hydrogen ions and bicarbonate ions, and carbonic acid molecules that have not undergone that dissociation. (b) As a result, some of the hydrogen ions and bicarbonate ions combine into undissociated carbonic acid molecules. (c) This combination ties up hydrogen ions from hydrochloric acid that would otherwise have been free to lower the pH of the solution.

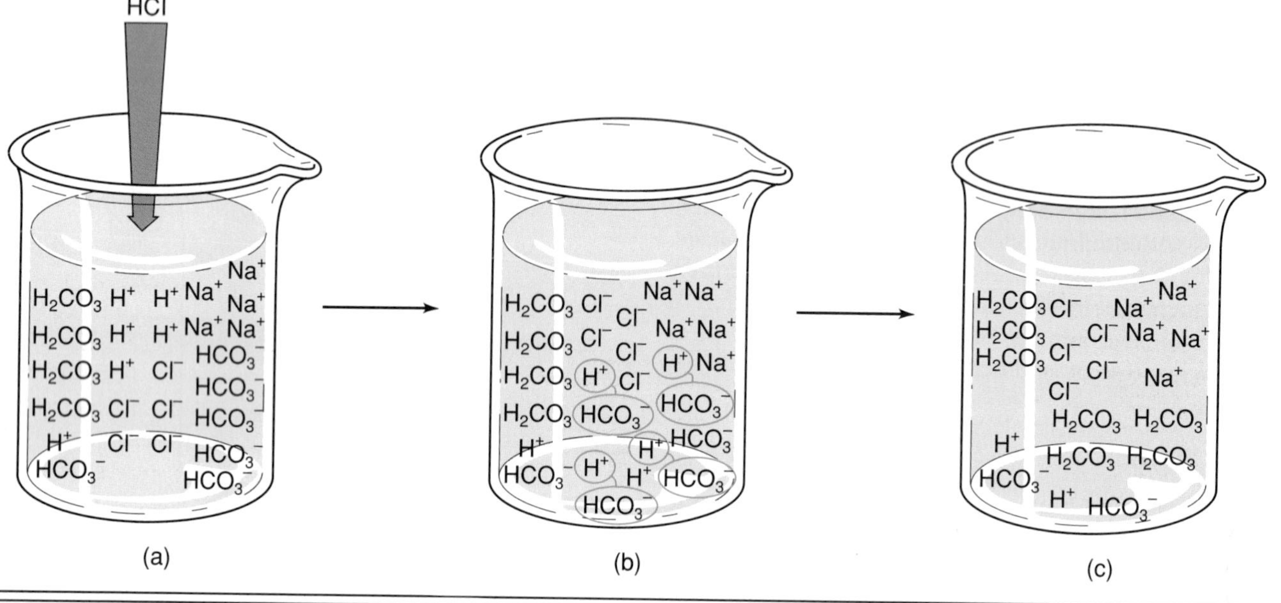

bilize the pH of the body fluids. Moreover, other body substances also act as buffers to resist changes in pH. Among these are large anions, such as plasma proteins, intracellular phosphate complexes, and hemoglobin molecules (reduced hemoglobin has a much greater affinity for hydrogen ions than oxyhemoglobin does).

Although the body's buffer systems can tie up some hydrogen ions and resist changes in pH to some degree, the continued production of acidic products by the body's metabolic reactions would eventually overcome their buffering capacity. Consequently, the activities of the respiratory system and the kidneys are important in maintaining the normal pH of the internal environment.

Respiratory Regulation of Acid-Base Balance

The respiratory regulation of acid-base balance makes use of the fact that in solution carbonic acid exists in reversible equilibrium with carbon dioxide and water (dissolved carbon dioxide), as follows:

$$CO_2 + H_2O \rightleftharpoons H_2CO_3 \rightleftharpoons H^+ + HCO_3^-$$

Thus, the higher the concentration of carbon dioxide in the internal environment of the body, the more carbonic acid is formed. As more carbonic acid is formed, the normal proportion that must be maintained between carbonic acid that dissociates into hydrogen ions and bicarbonate ions, and carbonic acid that does not undergo this dissociation is upset, and some of the carbonic acid dissociates into hydrogen ions and bicarbonate ions. This dissociation tends to increase the acidity (lower the pH) of the internal environment.

Conversely, the lower the concentration of carbon dioxide in the internal environment, the more carbonic acid forms carbon dioxide and water. This activity also upsets the normal proportion that must be maintained between carbonic acid molecules that dissociate into hydrogen ions and bicarbonate ions, and carbonic acid molecules that do not undergo this dissociation. As a result, some of the dissociated hydrogen ions and bicarbonate ions of carbonic acid combine into undissociated carbonic acid molecules. This combination removes hydrogen ions and tends to decrease the acidity (raise the pH) of the internal environment.

As just discussed, the carbon dioxide concentration of the body's internal environment influences the pH of this environment. Moreover, carbon dioxide is eliminated from the body at the lungs. Consequently, the activity of the respiratory system influences the pH of the internal environment. If much carbon dioxide is eliminated at the lungs, the carbon dioxide concentration in the internal environment decreases. This decrease results in a less acidic, more basic internal environment, with a higher pH. If only a small amount of carbon dioxide is eliminated at the lungs, its concentration in the internal environment increases. This increase results in a more acidic internal environment, with a lower pH.

By controlling the carbon dioxide concentration of the internal environment, the respiratory system plays an important role in maintaining the pH of this environment. For example, if the pH of the internal environment falls below normal values, an increased elimination of carbon dioxide at the lungs tends to raise the pH. Conversely, if the pH of the internal environment rises above normal levels, a decreased elimination of carbon dioxide at the lungs tends to decrease the pH.

In view of the ability of the respiratory system to influence acid-base balance, it is not surprising that respiration is itself sensitive to such factors as carbon dioxide concentration and pH. The influence of these factors on respiration was discussed in Chapter 23 (page 744).

Renal Regulation of Acid-Base Balance

The kidneys contribute to the acid-base balance of the body by (1) secreting hydrogen ions into the tubules and (2) reabsorbing bicarbonate ions from the tubules.

Secretion of Hydrogen Ions. Hydrogen ions are secreted into the tubules by tubular cells. The secretion mechanism derives hydrogen ions from carbonic acid (Figure 27.7). The enzyme carbonic anhydrase is present within tubular cells, and it catalyzes the formation of carbonic acid from carbon dioxide and water. Some of the carbonic acid dissociates into hydrogen ions and bicarbonate ions. The hydrogen ions are actively transported into the lumen of the tubule. The bicarbonate ions pass out of the cells and into the blood. Thus, for each hydrogen ion secreted, a bicarbonate ion enters the blood.

Reabsorption of Bicarbonate Ions. The secretion of hydrogen ions by the mechanism described in the preceding section provides a way of recovering bicarbonate ions that are filtered from the plasma into the glomerular capsule (Figure 27.7). Bicarbonate ions within the tubular lumen combine with secreted hydrogen ions to form carbonic acid molecules. The carbonic acid molecules then dissociate into carbon dioxide and water. (This dissociation is facilitated by carbonic anhydrase associated with the membranes of tubular cells in the proximal convoluted tubule.) The carbon dioxide enters the tubular cells, where carbonic anhydrase catalyzes its combination with water to form carbonic acid. Some of the carbonic acid dissociates into hydrogen ions and bicarbonate ions. The hydrogen ions are actively transported into the lumen of the tubule. The bicarbonate ions pass out of the tubular cells and into the blood.

Control of Hydrogen Ion Secretion and Bicarbonate Ion Reabsorption. As can be seen from the preceding discussion, the tubular reabsorption of bicarbonate ions that are filtered from the plasma into the

◆ **FIGURE 27.7 Secretion of hydrogen ions by cells of the proximal tubule**

Also shown is the pathway by which hydrogen ions in the tubular lumen participate in the reabsorption of bicarbonate ions from the tubule.

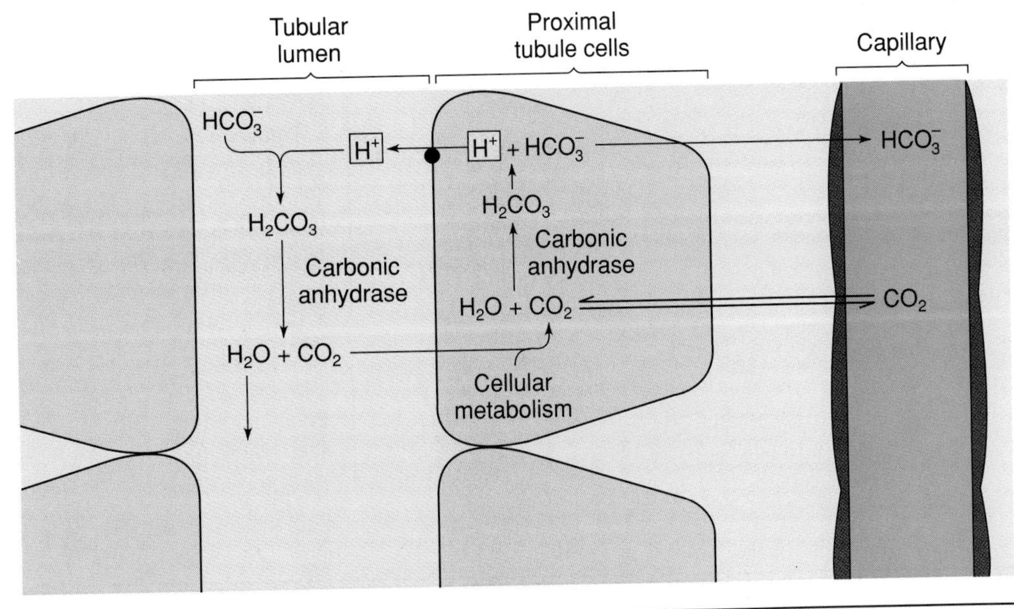

glomerular capsules depends on the presence of hydrogen ions within the tubules. Moreover, the rate of secretion of hydrogen ions into the tubules varies directly with the acidity of the internal environment.

Normally, the balance between the rate of filtration of bicarbonate ions into the glomerular capsules and the rate of secretion of hydrogen ions into the tubules is such that there are slightly more hydrogen ions secreted than there are bicarbonate ions filtered. As a result, essentially all the bicarbonate ions are reabsorbed, and the slight excess of hydrogen ions that remain after bicarbonate ion reabsorption is excreted in the urine.

If the concentration of hydrogen ions in the internal environment falls, the rate of secretion of hydrogen ions into the tubules diminishes relative to the filtration rate of bicarbonate ions into the glomerular capsules. As a consequence, not all of the bicarbonate ions are reabsorbed, and bicarbonate ions are excreted in the urine. The urinary excretion of bicarbonate ions diminishes the concentration of bicarbonate ions in the internal environment. The diminished concentration of bicarbonate ions leads to a decrease in the number of hydrogen ions tied up by the carbonic acid–sodium bicarbonate buffer system (as well as by other buffer systems), and the hydrogen ion concentration of the internal environment tends to rise toward normal.

If the concentration of hydrogen ions in the internal environment rises, the rate of secretion of hydrogen ions into the tubules increases considerably in comparison to the rate of filtration of bicarbonate ions into the glomerular capsules. Many of the secreted hydrogen ions are not needed for the reabsorption of bicarbonate ions, and they are excreted in the urine. Moreover, as previously described, the process of secreting hydrogen ions into the tubules produces bicarbonate ions that enter the blood. Consequently, when the rate of secretion of hydrogen ions increases, more bicarbonate ions enter the blood, thereby increasing the concentration of bicarbonate ions in the internal environment. The increased concentration of bicarbonate ions in the internal environment leads to an increase in the number of hydrogen ions tied up by the carbonic acid–sodium bicarbonate buffer system (as well as by other buffer systems). This response tends to decrease the hydrogen ion concentration of the internal environment toward normal.

◆ **FIGURE 27.8 Secretion of hydrogen ions into the tubular lumen by cells of the distal tubule**

The hydrogen ions are excreted in combination with phosphate compounds.

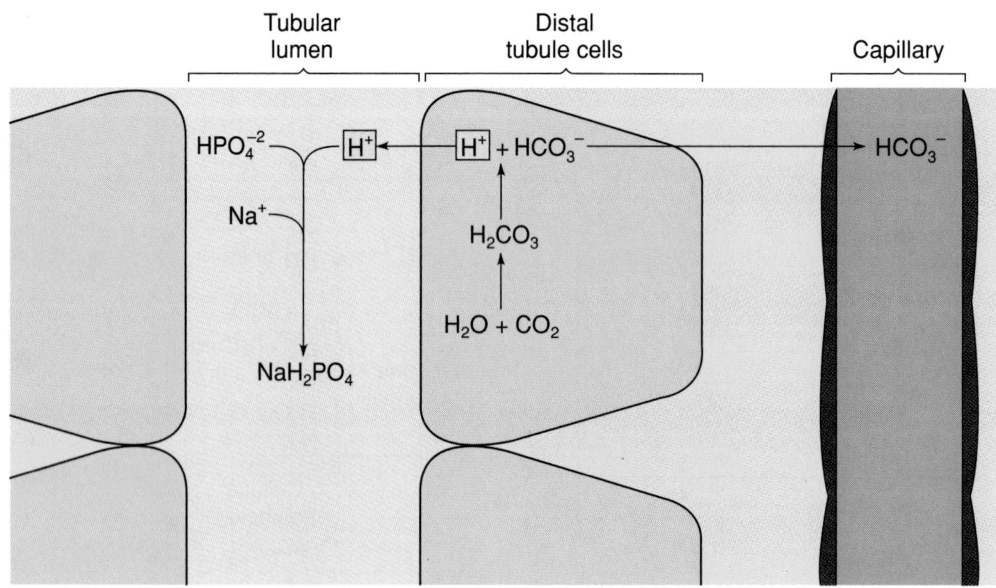

Buffering of Secreted Hydrogen Ions. The mechanism that secretes hydrogen ions into the tubules can achieve a maximum concentration of free hydrogen ions in the tubular fluid that is equivalent to a pH of 4.5. At normal rates of urine flow, this concentration is insufficient to excrete all the hydrogen ions that must be eliminated from the body. Consequently, buffers must be present within the tubular fluid to tie up secreted hydrogen ions so that the pH of the tubular fluid remains above 4.5, thereby allowing additional hydrogen ions to be secreted.

Phosphate compounds (HPO_4^{-2}) and ammonia (NH_3) act as buffers to tie up hydrogen ions in the tubular fluid. The phosphate compounds are filtered at the glomeruli and are only poorly reabsorbed (Figure 27.8). They combine with hydrogen ions ($H^+ + HPO_4^{-2} \rightarrow H_2PO_4^-$) and are excreted in combination with a cation such as sodium (Na^+) in the form of a weakly acidic substance ($Na^+ + H_2PO_4^- \rightarrow NaH_2PO_4$). This same type of excretion can occur with other anions, such as lactate.

Ammonia is formed in the tubular cells by the deamination of certain amino acids, particularly glutamic acid (Figure 27.9). The ammonia diffuses from the tubular cells into the tubules, where it combines with hydrogen ions to form ammonium ions ($H^+ + NH_3 \rightarrow NH_4^+$). This combination removes ammonia (as well as hydrogen ions) from the tubular fluid and allows the diffusion of still more ammonia from the tubular cells into the tubular fluid. The ammonium ions are excreted in combination with anions such as chloride (Cl^-) in the form of very weakly acidic or neutral substances ($NH_4^+ + Cl^- \rightarrow NH_4Cl$) that do not significantly lower the pH of the urine.

It is interesting to note that the production rate of ammonia increases when the hydrogen ion concentration has been elevated for two or three days. This increase in ammonia production provides additional buffering capacity for hydrogen ion excretion, and it allows extra hydrogen ions to be excreted without exceeding the maximum allowable acidity of the tubular fluid.

◆ **FIGURE 27.9 Secretion of hydrogen ions and ammonia into the tubular lumen by cells of the distal tubule**

The hydrogen ions are excreted in combination with ammonia in the form of ammonium salts.

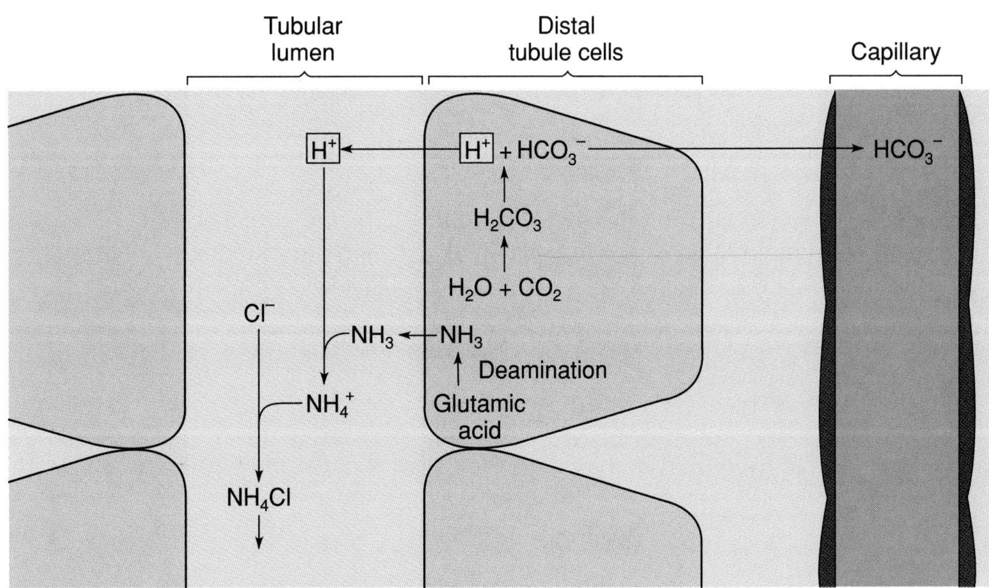

CONDITIONS OF CLINICAL SIGNIFICANCE

Disorders of Acid-Base Balance

Acidosis and Alkalosis

As previously indicated, the pH of arterial blood is normally 7.4, whereas the pH of venous blood and interstitial fluid is 7.35. However, certain conditions tend to alter the pH of the blood.

Acidosis is an abnormal condition in which the hydrogen ion concentration of the blood tends to increase and the pH of the blood tends to decrease. There are two categories of acidosis: respiratory acidosis and metabolic acidosis.

Respiratory acidosis is acidosis due to an increase in the carbon dioxide concentration, and consequently the carbonic acid concentration, of the body fluids. It frequently occurs as a result of asphyxia or as a result of diseases that interfere with the elimination of carbon dioxide by the lungs.

Metabolic acidosis is acidosis due to all other factors besides an excess of carbon dioxide in the body fluids. It can occur as a result of severe diarrhea or as a result of vomiting the lower intestinal contents. Either event causes the loss of bicarbonate-rich fluids from the gastrointestinal tract. This loss has the same effect as the excretion of bicarbonate ions by the kidneys—that is, a tendency to decrease the pH of the internal environment. Metabolic acidosis can also occur as a result of the production of an excessive amount of ketone bodies (for example, acetoacetic acid) during diabetes mellitus or as a result of the ingestion of acidic drugs.

Alkalosis is an abnormal condition in which the hydrogen ion concentration of the blood tends to decrease and the pH tends to increase. As was the case for acidosis, there are two categories of alkalosis: respiratory alkalosis and metabolic alkalosis.

Respiratory alkalosis is alkalosis due to a decrease in the carbon dioxide concentration, and consequently the carbonic acid concentration, of the body fluids. It can occur as a result of intense overventilation, which eliminates an excessive amount of carbon dioxide from the body. However, respiratory alkalosis is generally much less common than respiratory acidosis.

Metabolic alkalosis is alkalosis due to all other factors besides a deficiency of carbon dioxide in the body fluids. It can occur as a result of vomiting the acidic gastric contents alone, which rarely happens, or as a result of the ingestion of alkaline drugs.

Acidemia and Alkalemia

If the pH of the arterial blood falls below 7.35, a state of **acidemia** exists; and if the pH of the arterial blood rises above 7.45, a state of **alkalemia** exists. The occurrence of either acidemia or alkalemia is a serious problem, and a person generally cannot live more than a few hours if the pH of the arterial blood falls to about 7.0 or rises to about 8.0. Acidemia depresses the central nervous system, and a person suffering from severe acidemia usually enters a coma. Conversely, severe alkalemia produces an overexcitability of the nervous system that can lead to muscular tetany, to extreme nervousness, and in certain individuals, to convulsions.

CLINICAL CORRELATION

Metabolic Acidosis

Case Report

THE PATIENT: A 63-year-old man.

PRINCIPAL COMPLAINTS: Lethargy and confusion.

HISTORY: The patient had a long history of hypertension and congestive heart failure, for which he took digitalis (a cardiac stimulant) and a diuretic agent (an agent that increases the volume of urine). He also had moderate renal disease. The patient occasionally followed folk medicine, and on the advice of a friend, had begun to consume sulfur (Sublimed Sulfur, U.S.P., a fine powder of pure sulfur) for his malaise (general body weakness) and dyspnea (difficult breathing). He had been taking sulfur for a week before current admission, and it was estimated that he may have consumed as much as 75 grams/day during each of the last two days before admission. He reported no bowel movements for at least three days before admission. Because of his apparently deteriorated condition, his family brought him to the hospital.

CLINICAL EXAMINATION: At admission, the patient's body temperature was normal, and the heart rate was 95/min (normal: 60–100/min). The arterial blood pressure was 165/95 mm Hg (normal: 90–140/60–90 mm Hg), and the respiratory rate was 22/min (normal: 12–17/min). The pH of the arterial blood was 7.18 (normal: 7.35–7.45). The serum-sodium concentration was 125 mEq/liter (normal: 135–145 mEq/liter), the serum-potassium concentration was 8.7 mEq/liter (normal: 3.5–5 mEq/liter), and the serum-chloride concentration was 103 mEq/liter (normal: 100–105 mEq/liter). The arterial partial pressure of carbon dioxide was 28 mm Hg (normal: 35–45 mm Hg), the bicarbonate concentration was 10 mEq/liter (normal: 21–28 mEq/liter), and the arterial partial pressure of oxygen was 90 mm Hg (normal: 75–100 mm Hg). The pH of the urine was 4.9, the concentration of protein in the urine was 0.3 g/100 ml (normal: undetectable), and the numbers of red blood cells and white blood cells were excessive.

TREATMENT: Sodium bicarbonate solution was infused, and enemas were given to remove sulfur from the colon. After four days, the blood gases, plasma electrolytes, and pH of the arterial blood were normal, and the patient was discharged.

COMMENT: Elemental sulfur is a relatively nontoxic substance, but it is converted to sulfide and then to sulfate by bacteria in the colon. Sulfur often produces diarrhea because of irritation produced by its metabolites, but such did not occur in this patient. Thus, acidic products of sulfur were absorbed, producing severe metabolic acidosis. The patient's renal disease also contributed to the condition, because the excretion of hydrogen ions was slow. The decreased arterial partial pressure of carbon dioxide reflects respiratory compensation for the metabolic acidosis, and the decreased bicarbonate concentration reflects the shift of the following reaction toward carbon dioxide and water.

$$CO_2 + H_2O \rightleftharpoons H_2CO_3 \rightleftharpoons H^+ + HCO_3^-$$

Complete correction of the condition requires the excretion of hydrogen ions by the kidneys.

Study Outline

◆ **BODY FLUIDS** p. 868
Intracellular fluid and extracellular fluid; extracellular fluid is composed primarily of plasma and interstitial fluid.

Composition of Plasma and Interstitial Fluid. Plasma and interstitial fluid are separated by capillary walls; are of similar composition.

Composition of Intracellular Fluid. Different from that of plasma and interstitial fluid.

◆ **REGULATION OF THE BODY'S INTERNAL ENVIRONMENT** pp. 869–879
To maintain homeostasis, any addition of substances such as water and electrolytes to body's internal environment must be balanced by removal of equal amounts of these materials.

Sodium Regulation. Mechanisms involved in plasma-volume and blood-pressure regulation are important.

SODIUM INGESTION. Two components—regulatory and hedonistic; control of sodium ingestion does not appear to be major means of regulating body-sodium content.

RENAL EXCRETION OF SODIUM. Relation between glomerular filtration rate and tubular reabsorption rate of sodium is important.

FACTORS THAT INFLUENCE SODIUM EXCRETION. Mechanisms involved in regulation of plasma volume and blood pressure are important factors.

SODIUM CONTENT AND SODIUM CONCENTRATION. Sodium loss or gain changes overall body-sodium content but not necessarily the actual concentration of sodium in internal environment.

Water Regulation. Essential because of water's effects on osmotic pressure and fluid volume of internal environment.

WATER INGESTION AND THIRST. Feeling of thirst results from increased osmotic concentration of extracellular fluid or substantial reduction in plasma volume; osmoreceptors and perhaps circulatory pressure receptors stimulate thirst.

RENAL EXCRETION OF WATER. Relation between glomerular filtration rate and tubular reabsorption rate of water is important.

FACTORS THAT INFLUENCE WATER EXCRETION. Osmotic pressure of extracellular fluid and blood pressure (plasma volume) are major factors; factors that influence sodium excretion are very closely related to factors that influence water excretion.

Potassium Regulation. Ingestion and excretion are normally in balance; most filtered potassium is actively reabsorbed; amount excreted depends mainly on secretion into renal tubules. Amount of potassium secreted is related to potassium concentration of renal tubular cells; influenced by aldosterone.

Calcium Regulation. Involves bone, kidneys, and gastrointestinal tract; influenced either directly or indirectly by parathyroid hormone. Calcitonin inhibits removal of calcium from bone.

Hydrogen Ion and Acid-Base Regulation.

ACIDS AND BASES. Acids liberate hydrogen ions, bases accept them; concentration of free hydrogen ions determines solution's acidity.

pH OF THE BLOOD AND INTERSTITIAL FLUID. pH of arterial blood normally 7.4; pH of venous blood and interstitial fluid normally 7.35.

BUFFER SYSTEMS.

1. Solution containing two or more chemical substances that can prevent extreme changes in solution's pH when acid or base is added.
2. Body substances that act as buffers include carbonic acid–sodium bicarbonate buffer system, plasma proteins, intracellular phosphate complexes, and hemoglobin.

RESPIRATORY REGULATION OF ACID-BASE BALANCE. Makes use of fact that in solution carbonic acid exists in reversible equilibrium with carbon dioxide and water. Large carbon dioxide elimination at lungs tends to reduce acidity (raise pH) of internal environment and vice versa. Respiration is sensitive to carbon dioxide concentration and pH.

RENAL REGULATION OF ACID-BASE BALANCE. Hydrogen ions secreted into tubules; bicarbonate ions filtered into glomerular capsules can be reabsorbed by tubular cells; phosphate compounds and ammonia buffer hydrogen ions in tubular fluid.

◆ **CONDITIONS OF CLINICAL SIGNIFICANCE: DISORDERS OF ACID-BASE BALANCE** p. 880

Self-Quiz

1. There are only slight differences in composition between the: (a) plasma and interstitial fluid; (b) intracellular fluid and plasma; (c) intracellular fluid and interstitial fluid.
2. It appears that humans normally regulate their body sodium content by carefully controlling the amount of sodium they ingest. True or False?
3. The arterial blood pressure can influence the glomerular filtration rate. True or False?
4. Angiotensin II is a factor that stimulates the release of: (a) renin; (b) sodium; (c) aldosterone.
5. The loss or gain of sodium by the body does not necessarily have a great effect on the actual concentration of sodium in the internal environment. True or False?
6. A feeling of thirst results from: (a) an increased osmotic concentration of the extracellular fluid; (b) a substantial increase in plasma volume; (c) an increased volume of extracellular fluid.
7. The permeability of the kidneys' collecting tubules to water is controlled hormonally by: (a) aldosterone; (b) ADH; (c) angiotensin II.
8. ADH release is likely to increase in response to an increased: (a) blood pressure; (b) plasma volume; (c) extracellular osmolarity.
9. In most cases, an increase in the glomerular filtration rate of water is accompanied by a simultaneous decrease in the glomerular filtration rate of sodium. True or False?
10. Adjustments in the amount of potassium excreted depend mainly on how much is: (a) filtered into the glomerular capsules; (b) reabsorbed from the tubules; (c) secreted into the tubules.
11. Which hormone enhances the tubular secretion of potassium? (a) aldosterone; (b) ADH; (c) parathyroid hormone.
12. Parathyroid hormone decreases the: (a) movement of calcium from bone into the extracellular fluid; (b) movement of phosphate from bone into the extracellular fluid; (c) urinary excretion of calcium.
13. The normal pH of arterial blood is about: (a) 7.0; (b) 7.7; (c) 7.4.
14. In general, the body's metabolic activities generate basic products that tend to lower the hydrogen ion concentration (and increase the pH) of the internal environment. True or False?
15. When the carbon dioxide concentration of the internal environment rises, the pH of the internal environment also tends to rise. True or False?
16. If a large amount of carbon dioxide is eliminated at the lungs, the carbon dioxide concentration in the internal environment decreases, leading to a more acidic, less basic internal environment, with a higher pH. True or False?
17. When the pH of the internal environment of the body rises, the rate of secretion of hydrogen ions into the tubules: (a) diminishes in comparison to the rate of filtration of bicarbonate ions into the glomerular capsules; (b) increases in comparison to the rate of filtration of bicarbonate ions into the glomerular capsules.
18. Within the tubules of the kidneys, hydrogen ions are buffered by: (a) chloride; (b) ammonia; (c) sodium.
19. Metabolic acidosis can be caused by: (a) asphyxia; (b) severe diarrhea; (c) vomiting the acidic gastric contents alone.
20. Severe acidemia often results in: (a) muscular tetany; (b) convulsions; (c) coma.

CHAPTER 28

The Reproductive System

CHAPTER CONTENTS

LEARNING OBJECTIVES

After completing chapter, you should be able to:

1. Describe the embryonic development of the internal and external reproductive structures.
2. Describe the structure and function of each part of the male internal reproductive structures and external genitalia.
3. Describe the process of spermatogenesis.
4. Discuss the hormonal control of testicular function.
5. Describe the structure and function of each part of the female internal reproductive structures and external genitalia.
6. Describe the process of oogenesis.
7. Describe the events of the ovarian cycle.
8. Discuss the hormonal control of ovarian function.
9. Describe the events of the menstrual cycle.
10. Discuss the hormonal control of the menstrual cycle.

CHAPTER 28

The organs of the male and female reproductive systems ensure the continuance of the human species. They do this by producing **gametes** (or *germ cells*) and by providing a method by which the gametes of the male **(spermatozoa,** or **sperm)** can be introduced into the body of the female, where one of the male gametes joins with a gamete **(ovum)** of the female. This union of an ovum and a spermatozoon is called *fertilization.* The organs of the female reproductive system provide a suitable environment in which the fertilized ovum **(zygote)** can develop to a stage at which the baby is capable of surviving outside the mother's body.

The organs that produce the gametes are referred to as the **primary sex organs,** or **essential sex organs.** These are the **gonads**—the **testes** in the male and the **ovaries** in the female. In addition to producing gametes, the primary sex organs also produce hormones. In the male, specialized interstitial cells in the testes produce a group of hormones called *androgens.* The principal androgen is *testosterone.* In the female, the ovaries produce *estrogens* and *progesterone.*

The structures that transport, protect, and nourish the gametes after they leave the gonads are called **accessory sex organs.** In the male, the accessory sex organs include the *epididymis, ductus deferens, seminal vesicles, prostate gland, bulbourethral glands, scrotum,* and *penis.* Female accessory sex organs include the *uterine tubes, uterus, vagina,* and *vulva.*

Embryonic Development of the Reproductive System

The gonads are formed within retroperitoneal elevations called **genital ridges,** which protrude into the coelom (body cavity) just medial to the developing kidneys (mesonephros). Although the sex of the embryo is determined at the time of fertilization, the genital ridges do not begin to differentiate into testes until the seventh week or ovaries until the end of the eighth week of embryonic development. Cells that will become gametes originate in the yolk sac and migrate to the genital ridges in the sixth week of development, prior to the differentiation of the ridges. During the undifferentiated period, the gonads develop in close approximation to the **mesonephric ducts,** which drain the embryonic kidneys. A second pair of ducts—the **paramesonephric ducts**—also develops. The paramesonephric ducts run parallel to the mesonephric ducts and empty into the urogenital sinus at the posterior end of the embryo (Figure 28.1a). Males retain the mesonephric ducts for transport of sperm, while females retain the paramesonephric ducts for the transport of ova and the nourishment of the fetus.

Development of the Internal Reproductive Structures

Following the undifferentiated period, during which all embryos have the same structure, male and female embryos take different developmental paths. As a result, unique internal reproductive structures are formed in each sex.

Male Embryo

In the male embryo, the inner portion **(medulla)** of the undifferentiated gonads develops tubules that join with the mesonephric duct. These tubules become the **seminiferous tubules** *(sem-i-nif´-er-us)* of each testis, in which spermatozoa eventually are produced. In the male, the mesonephric duct is utilized for the transport of sperm from the testes to the exterior of the body. With further development, each mesonephric duct forms **efferent ductules,** an **epididymis** *(ep-i-did´-i-mis),* and a **ductus deferens** of the adult (Figure 28.1b). Shortly after the gonads begin to differentiate into testes, the paramesonephric ducts degenerate.

Female Embryo

In the female embryo, the outer portion **(cortex)** of the undifferentiated gonads undergoes the greater development and forms follicles, in which ova develop. Coinciding with the differentiation of the gonads into ovaries, the distal ends of the paramesonephric ducts join together to form the **uterus** and **vagina.** The portion of the paramesonephric ducts between the ovaries and the uterus forms the **uterine tubes** (*fallopian tubes,* or *oviducts*), through which ova travel to the uterus (Figure 28.1c). In the female, the mesonephric ducts degenerate without contributing any functional structures to the reproductive system.

Effects of Hormones on Genital Duct Development

The development of the genital ducts into either male or female structures appears to depend upon the presence or absence of hormones produced by the testes. In the presence of the male sex hormone **testosterone,** which is produced after the development of the genital ridges into testes, the mesonephric ducts remain and develop further. In addition, the embryonic testes produce a peptide that inhibits the paramesonephric ducts and causes them to degenerate.

◆ **FIGURE 28.1 Embryonic development of male and female internal reproductive structures**
Notice that the testes descend out of the abdominopelvic cavity into the scrotum, but the ovaries, which also descend, remain within the pelvic cavity.

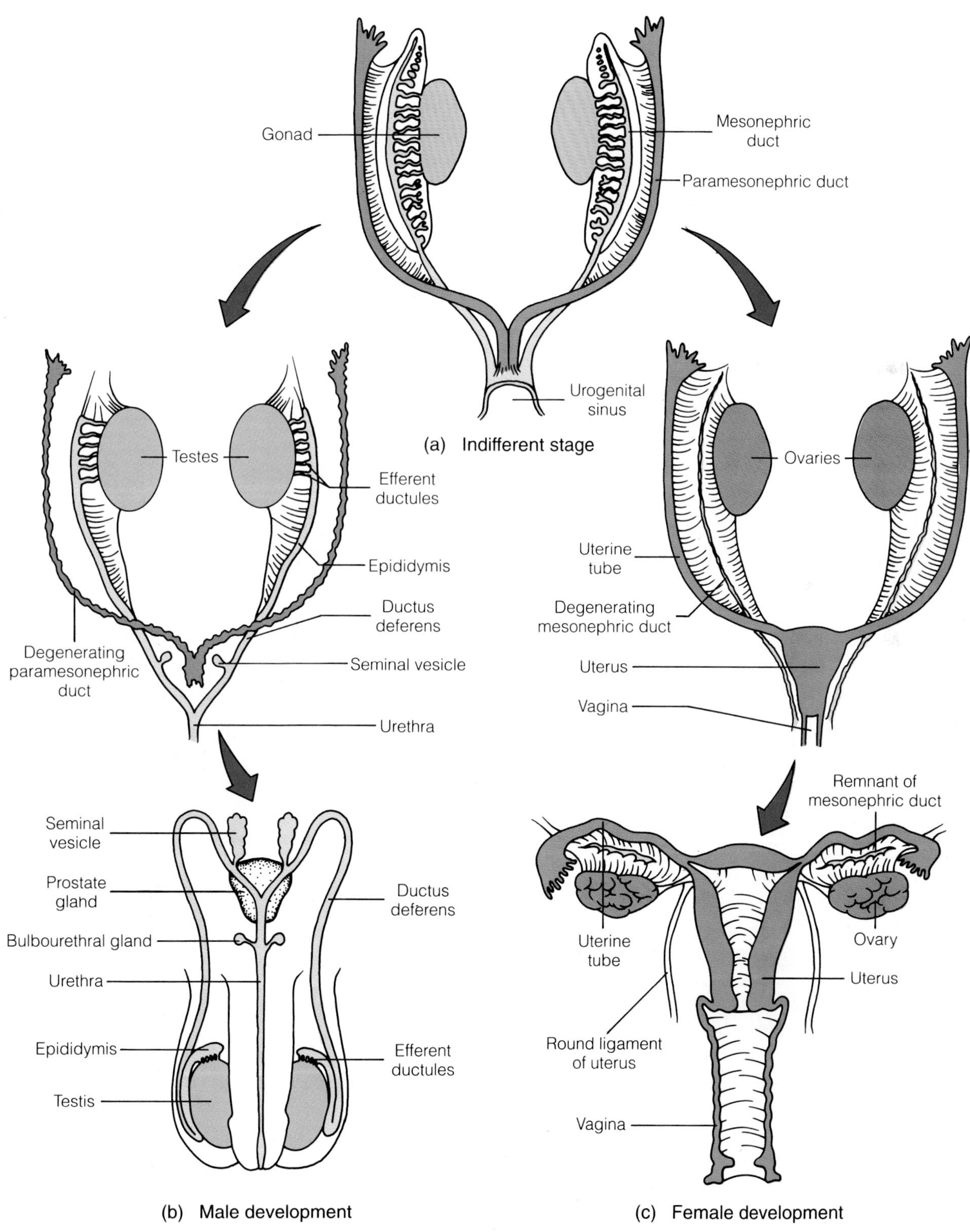

◆ **FIGURE 28.2 Embryonic development of male and female external reproductive structures**

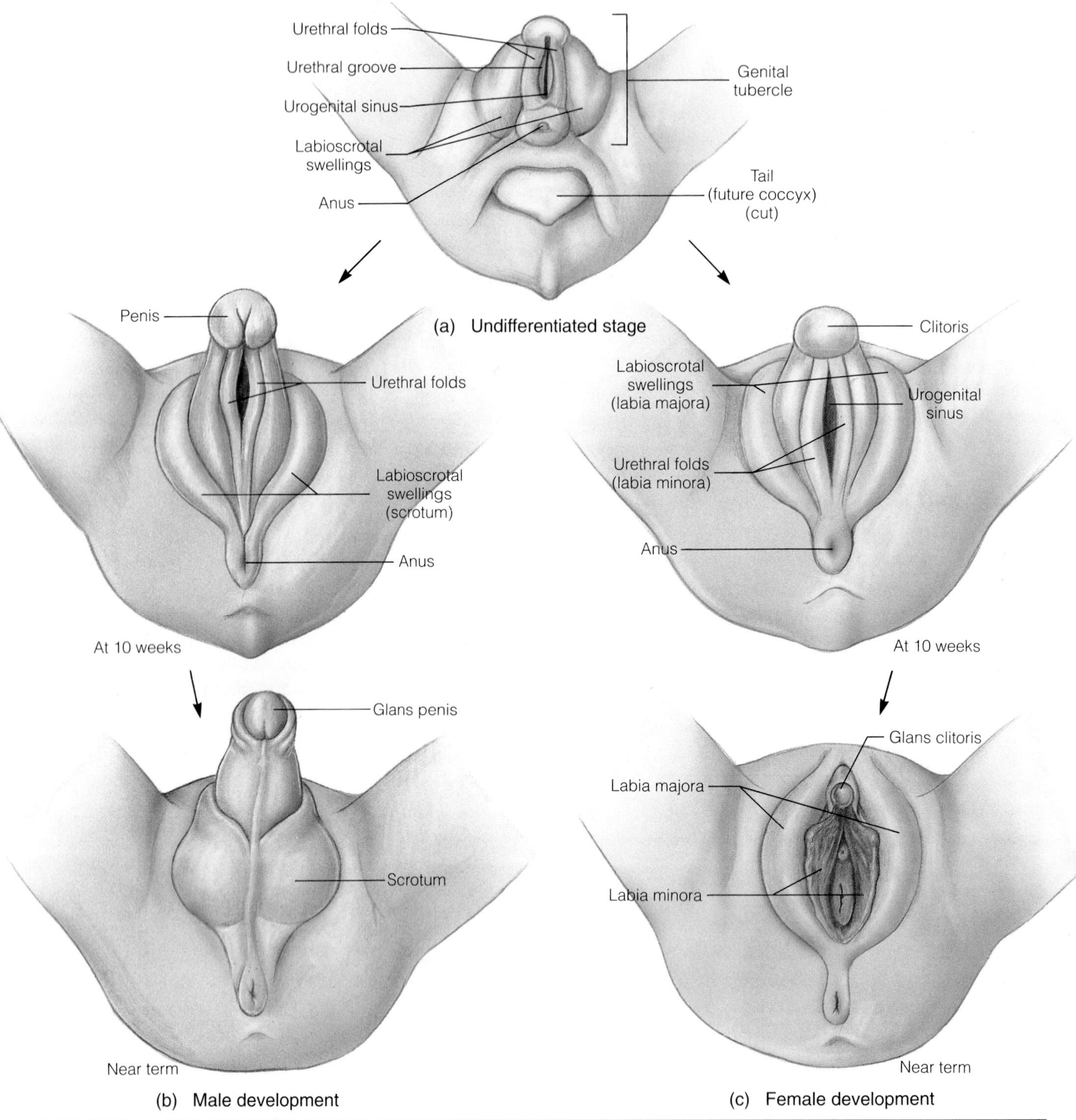

If the testes are removed from a genetically male embryo (XY), the embryo develops the reproductive structures of a female. That is, the paramesonephric ducts undergo further development, and the mesonephric ducts degenerate. Similarly, the injection of testosterone into a genetic female (XX) causes the embryo to develop according to the male pattern. These results indicate that in the absence of testosterone, all embryos, regardless of their genetic makeup, would develop into females. Under normal conditions, of course, embryos that are genetically male produce testosterone and therefore develop male reproductive structures.

Development of the External Reproductive Structures

The embryonic development of the external genitalia is also controlled by hormones produced by the gonads. The external genitalia, like the internal reproductive organs, remain in an undifferentiated state until about the eighth week. Before the production of hormones begins, all embryos develop a conical elevation—called the **genital tubercle**—at the point where the mesonephric and paramesonephric ducts open to the exterior of the body (Figure 28.2a). On the inferior surface of this tubercle is a shallow depression called the **urethral groove,** which opens into the urogenital sinus of the embryo. On either side of the urethral groove are slight elevations called the **urethral folds.** Bounding the folds laterally are rounded elevations called the **labioscrotal swellings.**

Male Embryo

If the embryonic gonads are producing testosterone, as occurs when the individual is a male, the genital tubercle elongates to form the **penis** (Figure 28.2b). The urethral folds fuse, leaving an opening only at the distal end of the penis. The tube thus formed becomes the spongy portion of the urethra. The labioscrotal swellings develop into a pouch called the **scrotum,** which eventually receives the testes.

Descent of the Testes

As discussed earlier, the testes begin their development as retroperitoneal structures in the abdominopelvic cavity, just below the kidneys. As development proceeds, the testes move caudally toward the labioscrotal swellings, which develop into the scrotum (Figure 28.3). As the testes move toward the developing scrotum, two outpouchings of the peritoneum called the **vaginal processes** protrude over the superior ramus of each pubic bone and extend through an **inguinal canal** *(ing´-gwi-nal)* into the scrotum. The testes, which remain behind the peritoneum, follow the vaginal processes out of the abdominopelvic cavity and through the inguinal canals into the scrotum.

As the testes descend into the scrotum from their original position in the abdominopelvic cavity, their blood supply travels with them. Therefore, the testicular arteries leave the aorta—and the testicular veins join the inferior vena cava (the left testicular vein via the left renal vein)—in the region of the kidneys, near the original site of the testes, and follow their paths of descent into the scrotum.

After the testes have entered the scrotum, the inguinal canals narrow, constricting the upper portion of the vaginal processes. The lower portion of each vaginal process forms a doubled-layered sac that covers a testis. This portion is called the **tunica vaginalis.** If the upper portion of a vaginal process does not close off completely, it is possible for small loops of intestine to protrude into an inguinal canal. This condition is known as

◆ **FIGURE 28.3 Descent of a testis**
(a) Formation of a vaginal process within the scrotal swellings. (b) Descent almost complete. (c) Upper portion of the vaginal process is obliterated, while the lower portion remains as the double-layered tunica vaginalis.

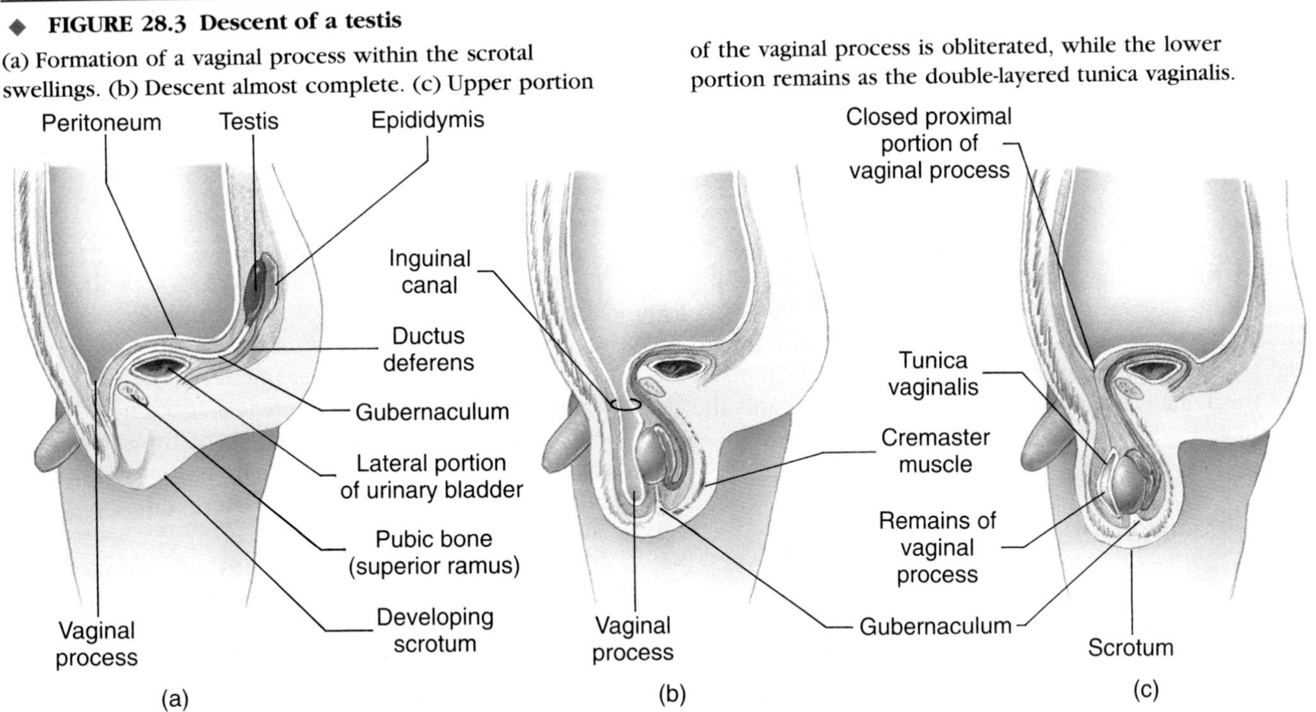

an *inguinal hernia.* Even if the vaginal process closes completely, the inguinal canal is an area of weakness in males, and thus a potential site of hernia. Inguinal canals also are present in females, but the gonads do not pass through the body wall and thus do not stretch and weaken the abdominal muscles surrounding the canals. Consequently, inguinal hernias are less common in females than in males.

The actual cause of the descent of the testes into the scrotum is not known. However, it seems to be initiated by testosterone from the testes and certain hormones from the pituitary gland. A fibromuscular band called the **gubernaculum** *(gu″-ber-nak′-ū-lum)* is thought to assist in the descent, but the precise manner in which it does so is not certain. The gubernaculum extends from the caudal surface of each testis, passes through the body wall, and is attached to the floor of the scrotum. As the embryo grows, the gubernaculum shortens, but it is questionable whether it is strong enough actually to pull each testis toward the scrotum.

Occasionally, one or both testes do not descend out of the abdominopelvic cavity. This condition is called *cryptorchidism (krip-tor′-kĭ-dizm).* Sperm production does not occur normally in an undescended testis, and the testis eventually atrophies. If both testes fail to descend, the person will be sterile.

Female Embryo

In female embryos, in the absence of testosterone, the genital tubercle does not elongate as much as in the male, and it becomes the **clitoris** (Figure 28.2c). Nor do the urethral folds fuse in the female. Rather, they remain as the **labia minora,** surrounding the entrance into the vagina. The labioscrotal swellings are not destined to receive gonads in the female. They remain as elevations called the **labia majora,** which flank the labia minora.

Homologues in the External Genitalia

Adult structures that develop from the same embryonic tissues are said to be *homologous (hō-mah′-luh-gus).* Thus, the male penis and female clitoris are homologues, as are the spongy portion of the urethra in males and the female labia minora, and the scrotum and the labia majora (Table 28.1).

Sexual Malformations during Embryonic Development

Understanding the patterns of development of the reproductive structures makes it apparent how even slight malformations during embryonic development can cause any number of abnormalities in the adult.

◆ **TABLE 28.1 Homologous Structures in the External Genitalia**

ADULT STRUCTURE IN MALES	EMBRYONIC STRUCTURE	ADULT STRUCTURE IN FEMALES
Penis ←	Genital tubercle	→ Clitoris
Spongy portion of urethra ←	Urethral folds	→ Labia minora
Scrotum ←	Labioscrotal swellings	→ Labia majora

It is estimated that some degree of sexual abnormality occurs in one out of every thousand persons. In a male, for example, the urethral folds may not join completely, leaving openings into the urethra on the undersurface of the penis. In a female with normal external genitalia, a ductus deferens may develop in addition to female internal sexual organs. Many other sexual abnormalities occur, but they are beyond the scope of this book.

Anatomy of the Male Reproductive System

The male reproductive system includes the *testes,* which produce spermatozoa; a number of *ducts* that store, transport, and nourish the spermatozoa; several accessory *glands* that contribute to the formation of semen; and the *penis,* through which semen is conveyed to the exterior of the body.

The Male Perineum

The **perineum** *(per″-i-nē′-um)* includes all the structures that are located between the pubic symphysis anteriorly, the coccyx posteriorly, and the ischiopubic rami and the sacrotuberous ligaments laterally. The portion of the sacrotuberous ligament that bounds the perineum runs from the lateral margin of the sacrum and coccyx to the ischial tuberosity. The muscles of the perineum are described on page 294. Here only the surface anatomy of the region in males is considered.

The male perineum can be divided into two triangles by a transverse line that passes between the ischial tuberosities (Figure 28.4). The anterior triangle, which contains the external genitalia, is called the **urogenital triangle.** The posterior triangle is called the **anal triangle** because it contains the anus.

◆ **FIGURE 28.4 Divisions of the male perineum**

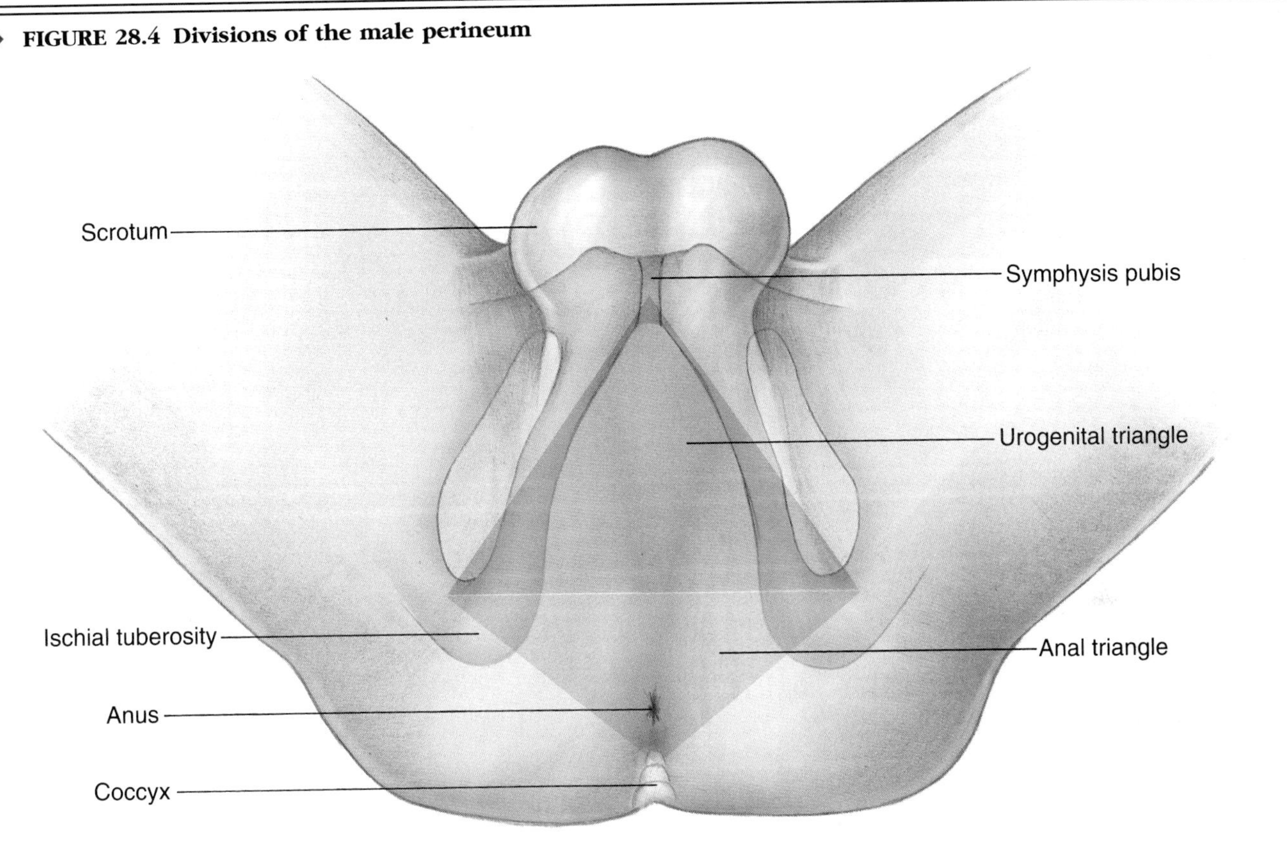

Testes and Scrotum

The **testes** are the organs in which spermatozoa are produced. The testes are contained within a pouch called the **scrotum,** which is located just behind the penis in the anterior portion of the urogenital triangle of the perineum. The scrotum consists of an outer layer of skin, which covers a thin layer of smooth muscle called the **dartos tunic** *(dar´-tōs).* The contraction of this muscle gives the scrotum a wrinkled appearance.

The location of the testes in the scrotum allows them to be maintained at a temperature lower than that of the abdominopelvic cavity. This lower temperature is necessary because the temperature of the abdominopelvic cavity is too high to permit normal sperm production.

The temperature of the testes can be regulated to some extent by the activity of a muscle called the **cremaster muscle** *(krē-mas´-ter).* This muscle, which is a continuation of the internal abdominal oblique muscle, extends down over the testes as a series of loops. When the environmental temperature is low, the cremaster muscle contracts, moving the testes close to the warmth of the body. At higher environmental temperatures, the cremaster muscle relaxes, allowing the testes to move away from the warmth of the body.

Each testis is an oval organ surrounded by a connective-tissue capsule called the **tunica albuginea** *(al˝-bu-jin´-ē-uh).* Inward extensions **(septa)** of the tunica divide the testis into compartments (Figure 28.5). Each compartment contains several highly convoluted **seminiferous tubules** *(se˝-mĭ-nif´-er-us)* and each tubule is bounded by a basement membrane and a layer of cells resembling smooth muscle.

The seminiferous tubules contain cells that give rise to spermatozoa. They also contain cells called **sustentacular** (*sus˝-ten-tak´-ū-lar;* Sertoli) **cells,** which are involved in the process of sperm formation. The process of sperm formation, which is called **spermatogenesis** *(sper˝-mah-toe-jen´-ĕ-sis),* will be discussed later in the chapter (page 896).

The seminiferous tubules in each compartment of a testis join together to form a short, straight tube called the **tubulus rectus** (*rectus* = straight). Toward the posterior portion of the testis, the tubuli recti from all the compartments form a network of tubes termed the **rete testis.** The tubes of the rete testis, in turn, empty into the **efferent ductules,** which leave the testis and enter the epididymis.

Clusters of cells called **interstitial endocrinocytes** *(Leydig cells)* are located in the loose connective tissue

◆ **FIGURE 28.5 A testis**
(a) Sagittal section of the ductus deferens, the epididymis, and the testis. (b) Seminiferous tubule within a single compartment of the testis.

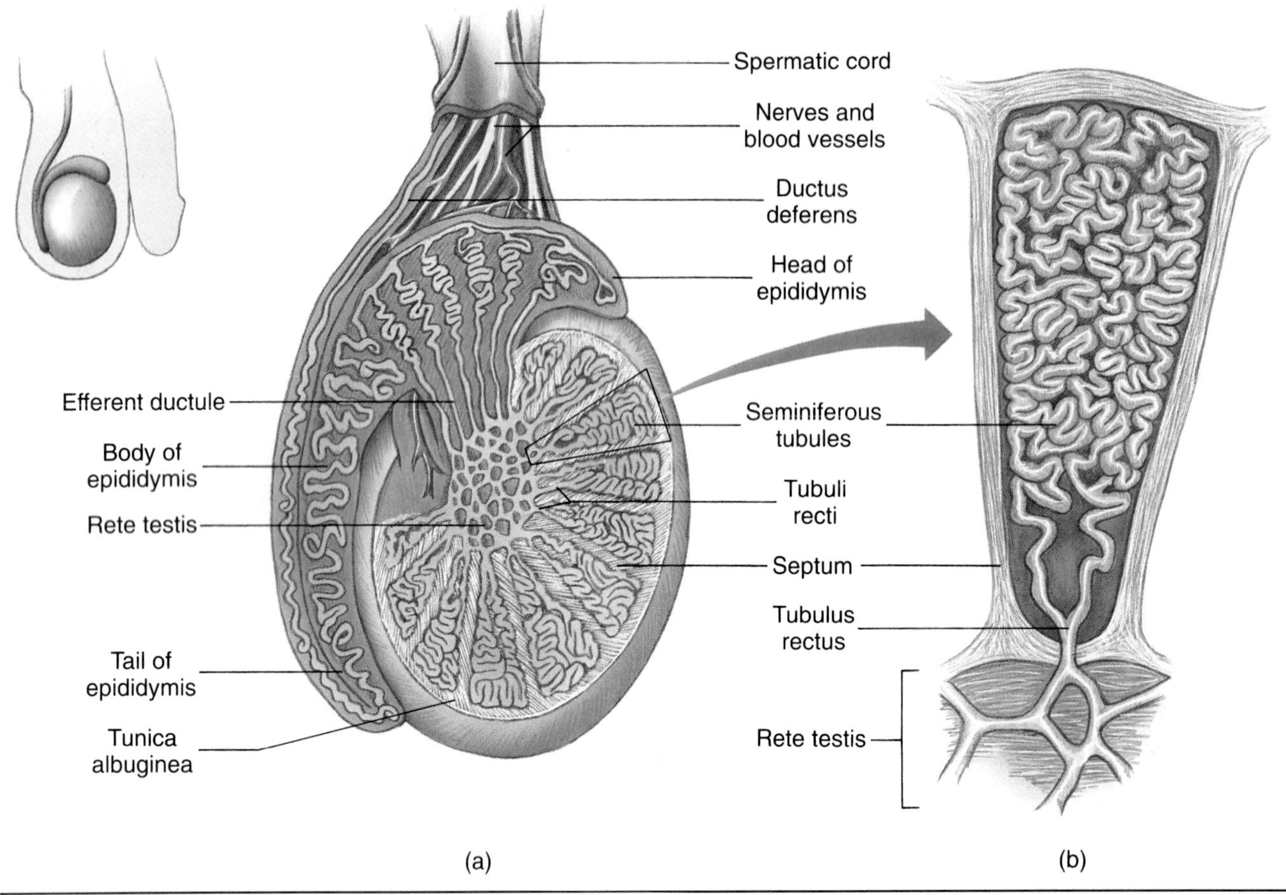

between the seminiferous tubules. These interstitial cells secrete the male sex hormone **testosterone.**

Epididymis

The **epididymis** *(ep″-ĭ-did′-ĭ-mis),* which is located in the scrotum, is the first portion of the duct system that transports sperm from the testes to the exterior of the body (Figure 28.5). Each epididymis is an elongated structure that fits tightly against the posterior surface of the testis. It consists of a tightly coiled tube that receives the sperm from the testis through the efferent ductules. The tube is lined with pseudostratified columnar epithelium. The free surfaces of some of the columnar cells have long, nonmotile microvilli called *stereocilia.* The stereocilia serve to facilitate the passage of nutrients from the epithelium to the sperm.

If uncoiled, the epididymal tube would be 4-6 m in length, and it is generally considered to serve as a storage site for sperm. The efferent ductules enter the upper region *(head)* of the epididymis. In its lower region (the *tail*), the tube becomes continuous with the ductus deferens. Smooth muscles in the wall of the epididymal tube contract during ejaculation. These contractions move sperm into the ductus deferens. During their slow passage through the epididymis, which takes about 12 days, the sperm undergo a maturation process without which they would be nonmotile and infertile when they enter the female reproductive tract.

Ductus Deferens

The **ductus (vas) deferens** is a continuation of the duct of the epididymis (Figure 28.6). Each ductus deferens is a straight tube that passes along the posterior surface of the testis, medial to the epididymis, and ascends through the scrotum. The ductus deferens passes through the body wall by way of the inguinal canal and enters the abdominopelvic cavity. As it passes from the epididymis to the point where it enters the abdominopelvic cavity, the ductus deferens lies next to the vessels and nerves supplying the testis. All of these structures are enclosed in a sheath of fascia. The fascia and the enclosed structures are collectively referred to as the

◆ **FIGURE 28.6 Sagittal section of the male pelvis, with a portion of the left pubic bone attached to illustrate the path of the left ductus deferens**

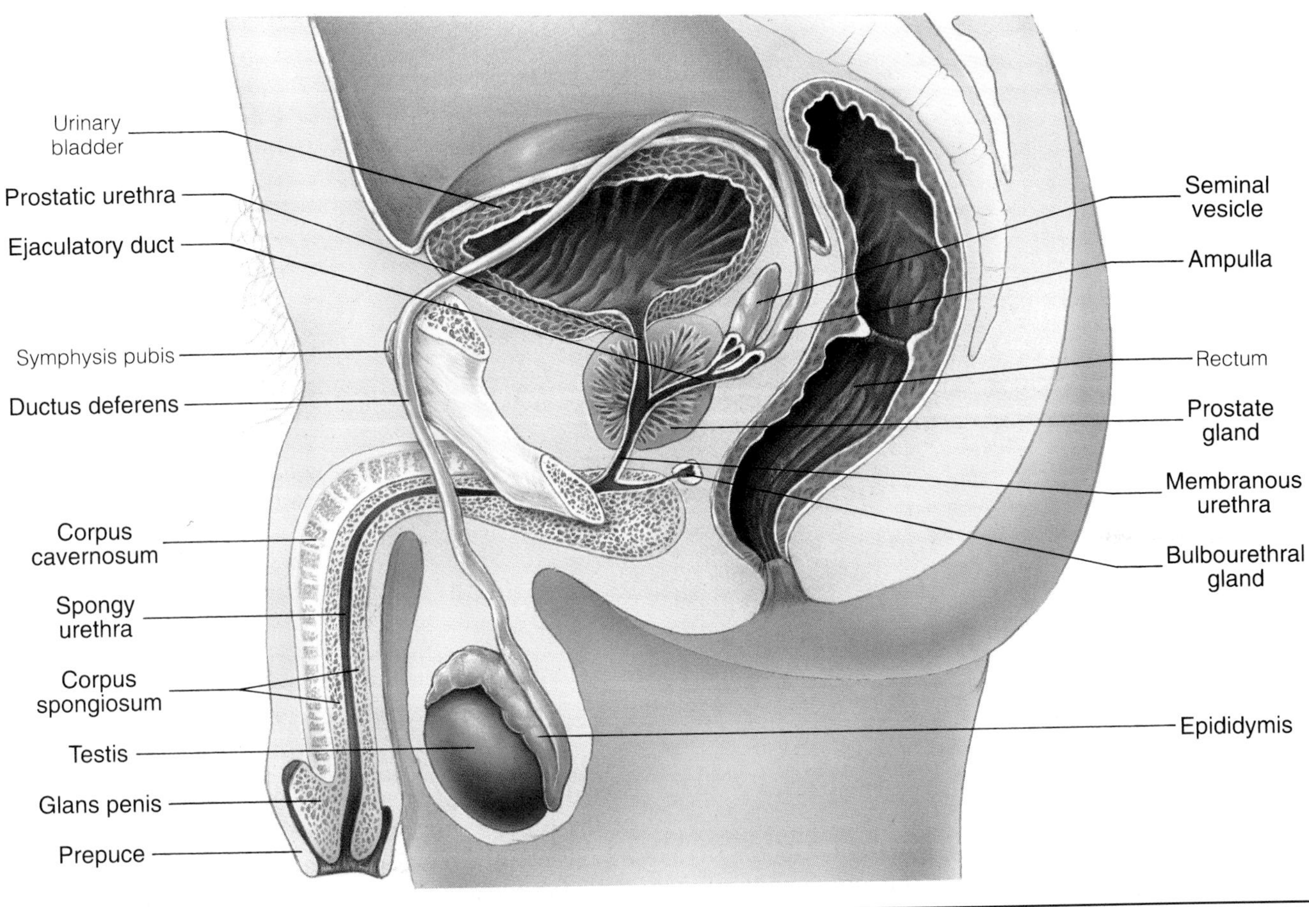

spermatic cord. Included in the spermatic cord along with the ductus deferens are the testicular artery, testicular vein, lymphatic vessels, and nerves. The veins returning from the testes form a network of connecting branches—the *pampiniform plexus*—around the testicular artery. This plexus is thought to absorb heat from the blood in the testicular artery, thereby lowering the temperature of the arterial blood before it reaches the testes and thus helping to maintain the temperature of the testes about 3°C below body-core temperature. This lower temperature is essential for normal spermatogenesis.

Inside the abdominopelvic cavity, the ductus deferens travels beneath the peritoneum along the lateral wall of the cavity, crosses over the top of the ureter, and then descends along the posterior surface of the urinary bladder, where it enlarges to form an **ampulla** (Figure 28.7). On reaching the inferior surface of the urinary bladder, each ductus deferens is joined by the duct of a seminal vesicle to form a short **ejaculatory duct** that passes through the prostate gland to join the prostatic portion of the urethra. The ductus deferens, like the epididymis, is lined with pseudostratified columnar epithelium that has stereocilia on its free surface. The muscular coat of the wall of the ductus deferens is quite thick, consisting of three layers of smooth muscle (Figure 28.8). The muscle of the wall of the duct undergoes peristaltic contractions during ejaculation, propelling sperm into the ejaculatory duct.

Seminal Vesicles

The **seminal vesicles** are two membranous pouches located lateral to the ductus deferens on the posterior surface of the urinary bladder (Figures 28.7 and 28.9). The duct of each seminal vesicle joins with a ductus deferens to form an **ejaculatory duct.** The ejaculatory ducts penetrate the prostate gland and enter the urethra just below its point of exit from the urinary bladder. Contraction of the ejaculatory ducts propels spermatozoa from the ductus deferens, along with secretions from the seminal vesicles, into the urethra. The seminal vesicles secrete a viscous fluid that contributes to the formation of semen.

◆ **FIGURE 28.7 Posterior view of the male urinary bladder, illustrating its relationship to the ductus deferens, the ureters, the seminal vesicles, and the prostate gland**

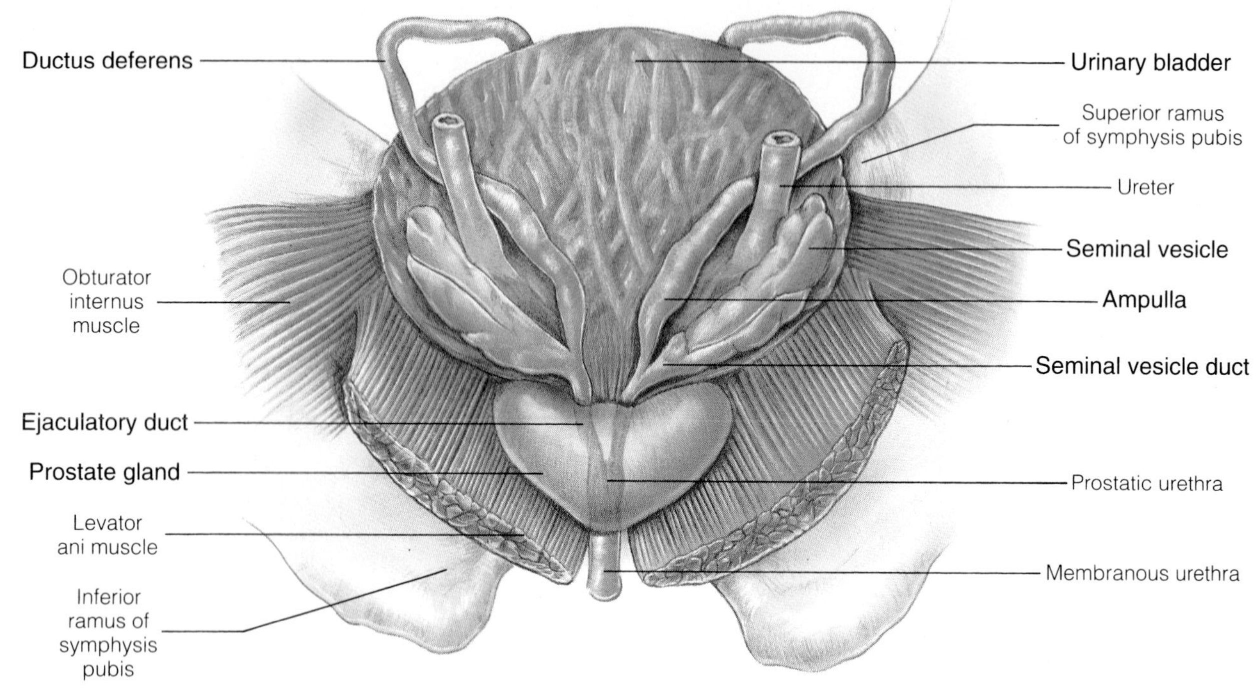

Prostate Gland

The **prostate gland** *(pros´-tāt)* (Figures 28.6 and 28.9) is a single organ that encompasses the urethra just below the bladder. Because it is located directly in front of the rectum, the prostate can be manually examined by means of a rectal examination in which a finger is placed in the anal canal and the gland palpated through

◆ **FIGURE 28.8 Photomicrograph of a cross section of a ductus deferens**

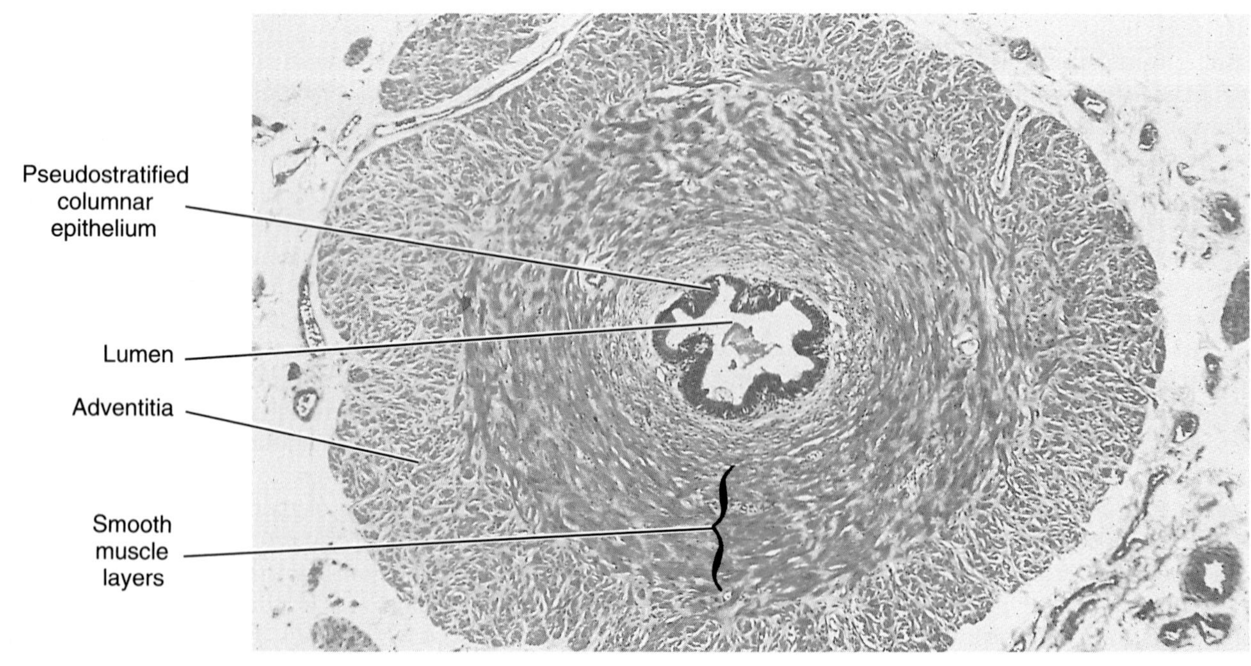

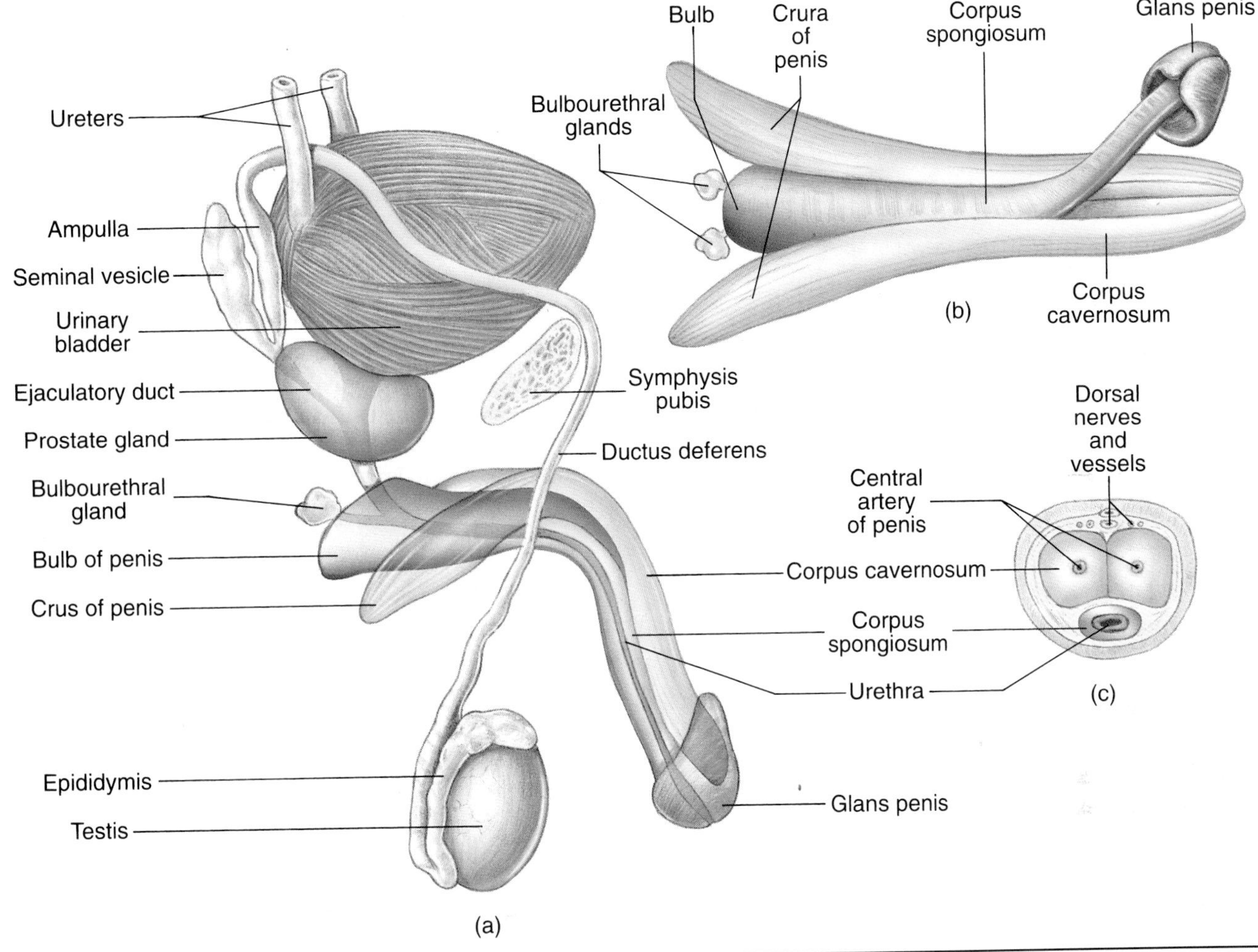

◆ **FIGURE 28.9 The male reproductive system** (a) Schematic representation of the male reproductive system. (b) Inferior view of a dissected penis, with the distal portion of the corpus spongiosum displaced to one side. (c) Cross section of a penis.

the anterior wall of the rectum. The gland is composed of about 30 small tubuloalveolar glands from which a number of excretory ducts open independently into the urethra. It is enclosed in a capsule of fibrous connective tissue and smooth muscle fibers that extend into the gland and divide it into indistinct lobes. The prostate secretes a thin, milky alkaline fluid that contributes about one-third of the volume of semen. The prostate tends to enlarge in older men and may cause difficulty in urination by compressing the prostatic portion of the urethra.

Bulbourethral Glands

The **bulbourethral** *(Cowper's)* **glands** (Figure 28.9) are a pair of small glands located below the prostate on either side of the membranous portion of the urethra. They secrete a thick alkaline mucus, which is carried to the urethra by a duct from each gland. The secretion of the bulbourethral glands is released early in sexual stimulation, prior to ejaculation. It serves as a lubricant for sexual intercourse and is believed to protect the sperm by neutralizing any acidic urine that may remain in the urethra.

Penis

The **penis** is the copulatory organ by which spermatozoa are placed in the female reproductive tract. It consists of a shaft covered by loosely attached skin and an expanded tip, the **glans penis** (Figure 28.9). The skin continues over the glans as the **prepuce** (*prē´-puce;* **foreskin).** In order to make it easier to keep the glans clean, the prepuce is often surgically removed shortly after birth in a procedure called *circumcision.*

The penis contains three cylindrical bodies, each of which is surrounded by a fibrous capsule. The three cylindrical bodies are held together by a connective-tissue sheath that is covered by skin. Each cylindrical body is composed of highly vascular connective tissue called *erectile tissue,* which contains numerous spongelike spaces that fill with blood during sexual arousal, causing the penis to enlarge and become firm. This phenomenon is referred to as an *erection.* The two dorsal cylindrical bodies are called **corpora cavernosa penis.** The single ventral body is the **corpus spongiosum penis.** The urethra passes through the corpus spongiosum. The expanded distal end of the corpus spongiosum forms the glans penis. The proximal end of the corpus spongiosum is slightly enlarged, forming the **bulb** of the penis. The bulb is attached to the urogenital diaphragm, which forms the floor of the pelvic cavity. The two corpora cavernosa separate at their proximal ends and form the **crura** *(kroo´-rah)* of the penis, which are anchored to the rami of the ischial and pubic bones.

Semen

Semen is a mixture of spermatozoa from the testes and fluids from the seminal vesicles, prostate gland, and bulbourethral glands. The secretion of the seminal vesicles, which contains fructose and prostaglandins, contributes about 60% of the bulk of the semen. Fructose is a major energy source for ejaculated sperm, which contain very little cytoplasm and therefore have only limited intracellular glycogen available from which to derive energy. The prostaglandins are thought to facilitate the movement of sperm along the female reproductive tract by helping to stimulate (1) contractions of uterine smooth muscle that disperse the sperm throughout the uterine cavity and (2) reverse peristaltic contractions of uterine tube smooth muscle.

The alkaline fluid secreted by the prostate gland counteracts the mild acidity of other portions of the semen so that the semen itself is slightly alkaline, with a pH of about 7.5. The alkaline nature of the semen helps counteract the acid secretions of the female vagina (pH 3.5–4.0). This action is believed to be important because sperm do not become optimally active until the pH of their environment rises to about 6.0–6.5.

Semen also contains a substance called *seminalplasmin,* which is capable of destroying bacteria.

Table 28.2 summarizes the functions of the components of the male reproductive system.

Puberty in the Male

The time of life at which reproduction becomes possible is called **puberty.** Its age of onset varies considerably among individuals, but it most commonly occurs between 10 and 15 years of age.

In males, puberty is marked by the first ejaculate of semen that contains mature spermatozoa, as well as by more gradual and subtle changes in personality and body form. The changes that occur in body form are called **secondary sex characteristics.** The penis, scrotum, and testes enlarge, and hair grows on the face, axilla, and pubic area. The voice deepens, and there is a tendency toward increased development of skeletal muscles, accompanied by a decrease in the amount of body fat.

The changes that occur at puberty are the result of an increased production of testosterone by the testes, and a continued production of adequate levels of testosterone is required to sustain them.

During fetal development, the testes secrete testosterone, which leads to the development of the male reproductive system. However, a few months after birth, testosterone levels decrease, and they remain very low throughout childhood. During childhood, the hypothalamus of the brain secretes only small amounts of gonadotropin-releasing hormone (Gn-RH). Consequently, there is little release of the gonadotropins—follicle-stimulating hormone (FSH) and luteinizing hormone (LH)—from the pituitary, and the testes are not stimulated to produce large amounts of testosterone. At puberty, an alteration in brain function occurs that leads to an increased release of Gn-RH, which ultimately results in an increased production of testosterone.

Some researchers believe that, during childhood, the hypothalamus is extremely sensitive to the negative-feedback effects of testosterone. Thus, even though only very low levels of testosterone are present, Gn-RH release is suppressed. But at puberty, the hypothalamus decreases in sensitivity to the negative-feedback effects of testosterone, and Gn-RH is released in greater quantities. Alternatively, other researchers propose that, during childhood, the hypothalamic region that releases Gn-RH is inhibited by neural signals from the anterior hypothalamus, and Gn-RH release is suppressed. At puberty, this inhibition is relaxed, and Gn-RH is released in greater quantities.

Spermatogenesis

The production of spermatozoa within the testes is called **spermatogenesis.** During spermatogenesis, a unique type of cell division called meiosis takes place. (Meiosis is described on page 97, and the specific events of this process should be reviewed at this time.) The purpose of meiosis, which is aptly called a reduction division, is to reduce the number of chromosomes in each gamete by half.

◆ **TABLE 28.2 Representative Functions of Components of the Male Reproductive System**

TESTES
Produce sperm
Secrete testosterone
EPIDIDYMIS AND DUCTUS DEFERENS
Serve as sperm exit route from testes
Serve as site for maturation of sperm for motility and fertility
Concentrate and store sperm
SEMINAL VESICLES
Supply fructose to nourish ejaculated sperm
Secrete prostaglandins that stimulate motility within female reproductive tract to help transport sperm
Provide bulk of semen
Provide precursors for clotting of semen
PROSTATE GLAND
Secretes alkaline fluid that neutralizes acidic vaginal secretions
Triggers clotting of semen to keep sperm in vagina during penis withdrawal
BULBOURETHRAL GLANDS
Secrete mucus for lubrication

The cells of the human body are *diploid;* that is, they contain 23 chromosomes from each parent, for a total of 46. Cells undergoing meiosis lose half of their total number of chromosomes, retaining 23 chromosomes. Such cells are said to be *haploid.* The formation of haploid spermatozoa (as well as haploid ova in the female) ensures that, following fertilization, when a male gamete and a female gamete unite, the resulting cell will be a diploid cell that has 46 chromosomes.

Spermatogenesis occurs within the seminiferous tubules of the testes. Located within each tubule are cells called **spermatogonia** *(sper″-mah-tō-gō′-ne-ah),* which have the diploid number of chromosomes (44 autosomes plus one X and one Y sex chromosome). Spermatogonia divide mitotically (page 95). Some of the daughter cells remain spermatogonia. These cells provide a continuous source of new cells that are used for the production of spermatozoa. Other daughter cells give rise to diploid cells called **primary spermatocytes** (Figure 28.10).

The primary spermatocytes increase considerably in size and undergo the first meiotic division. This division results in the formation of two **secondary spermatocytes** from each primary spermatocyte. One of the secondary spermatocytes contains 22 autosomes and the X sex chromosome; the other contains 22 autosomes and the Y sex chromosome.

The secondary spermatocytes then undergo the second meiotic division. This division results in the formation of two **spermatids** from each secondary spermatocyte. Thus, meiosis results in the formation of four haploid spermatids from each diploid primary spermatocyte.

Each spermatid undergoes a process called **spermiogenesis** by which it develops into a **spermatozoon.** During this process, much of the cytoplasm of the spermatid is discarded, and a tail (flagellum) forms.

During sperm production, the developing gametes are closely associated with **sustentacular cells** (Figure 28.11). These cells extend from the basement membrane of a seminiferous tubule to the lumen. The sustentacular cells are connected to one another by tight junctions (page 111).

The interconnected sustentacular cells form a continuous layer within the seminiferous tubule. This layer acts as a **blood-testes barrier** that keeps many substances from moving between the blood and the lumen of the tubule. The blood-testes barrier helps provide a stable microenvironment within the tubule that is favorable for the development of sperm. Moreover, it is believed to prevent sperm antigens from entering the blood. Sperm are not produced during the first 12 years or so of life, and the immune system does not become tolerant to sperm antigens (see page 703). If sperm antigens entered the blood, they would be treated as foreign, and specific immune responses would be generated against them.

The tight junctions that connect the sustentacular cells divide the seminiferous tubule into (1) a *basal compartment,* which extends from the basement membrane to the junctions, and (2) a *central (adluminal) compartment,* which extends from the junctions into the lumen.

Spermatogonia are located in the basal compartment between the sustentacular cells. During spermatogenesis, they undergo mitosis within this compartment. The tight junctions open, and primary spermatocytes that are formed move into the central compartment. New tight junctions form behind the primary spermatocytes, and the remaining events of spermatogenesis take place in the central compartment.

The sustentacular cells nourish the developing gametes, and certain hormones that influence spermatogenesis exert their effects by way of these cells. The sustentacular cells also secrete fluid into the tubular lumen. The fluid contains **androgen-binding protein,** which binds testosterone and thereby maintains a high level of this hormone within the male reproductive tract.

In humans, the process of sperm production, from the formation of primary spermatocytes to spermatozoa, requires about 64 days. Spermatogenesis occurs continuously beginning at puberty. In a normal adult male, approximately 30 million sperm are produced each day.

◆ **FIGURE 28.10 Diagrammatic representation of the process of spermatogenesis**
Note that the secondary spermatocytes and spermatids arising from a single primary spermatocyte remain linked by cytoplasmic bridges.

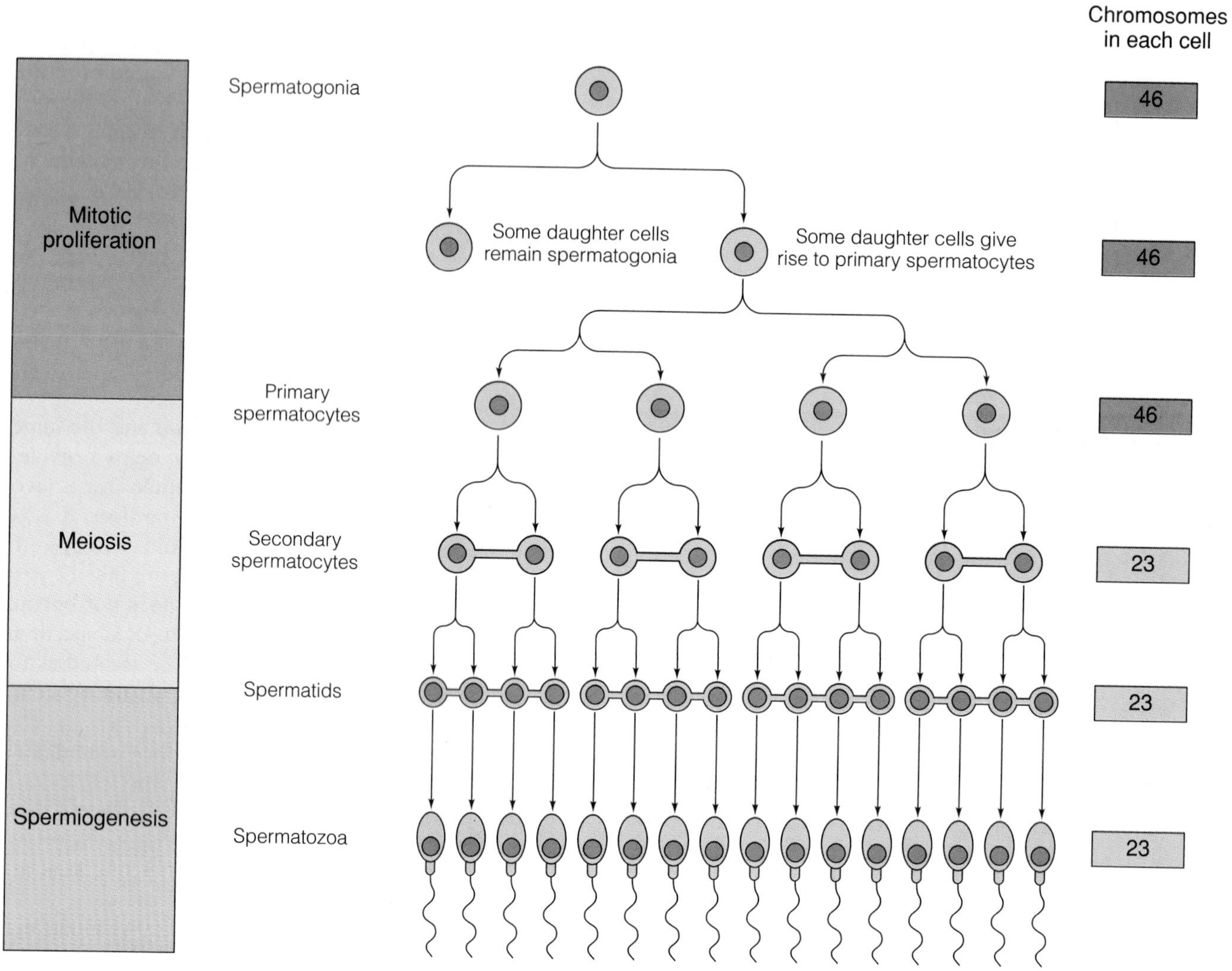

Chromosomes in each cell

Mitotic proliferation

Spermatogonia — 46

Some daughter cells remain spermatogonia

Some daughter cells give rise to primary spermatocytes — 46

Meiosis

Primary spermatocytes — 46

Secondary spermatocytes — 23

Spermiogenesis

Spermatids — 23

Spermatozoa — 23

The Spermatozoon

The **spermatozoon** *(sper″-mah-tō-zō´-on),* or **sperm,** is a cell that is divisible into **head, neck, middle piece,** and **tail** regions (Figure 28.12). The head consists almost entirely of the condensed nucleus, capped with an enzyme-containing vesicle called the **acrosome** *(ak´-rō-sōm)* at its tip. The acrosome develops from the Golgi apparatus of the spermatid. Acrosomal enzymes play an important role in the process by which a sperm fertilizes an ovum.

Located in the neck region are two centrioles at right angles to each other. The distal centriole is oriented along the long axis of the sperm. Microtubules extend from it into the tail (flagellum), which gives the sperm motility.

The middle piece contains mitochondria that are oriented end to end in a spiral pattern. Most of the small amount of cytoplasm present in a mature sperm is located around the mitochondria.

Hormonal Control of Testicular Function

Gonadotropin-releasing hormone (Gn-RH) from the hypothalamus stimulates the release of the pituitary gonadotropins—follicle-stimulating hormone (FSH) and luteinizing hormone (LH) (Figure 28.13). In a normal adult male, Gn-RH release takes place in bursts that occur approximately every two or three hours. Virtually no

◆ **FIGURE 28.11 Section of testis showing spermatogenesis within the seminiferous tubules**

(a) Sagittal section of a testis. (b) Cross section of several seminiferous tubules. (c) Flow chart of the events of spermatogenesis. 2n refers to diploid cells with 46 chromosomes; n refers to haploid cells with 23 chromosomes. (d) Magnified view of a portion of a seminiferous tubule showing the relative positions of various spermatogenic cells.

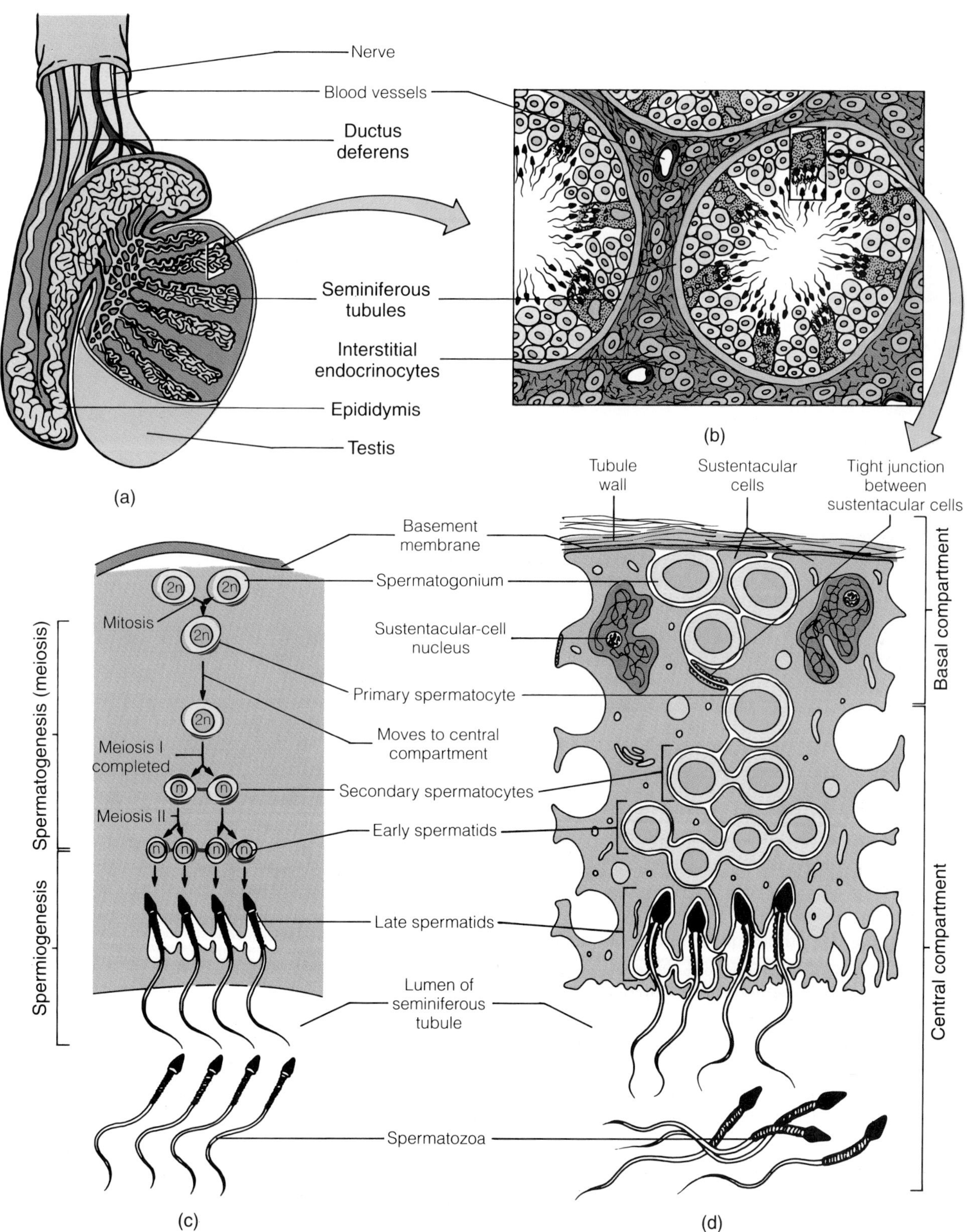

◆ **FIGURE 28.12 Spermatozoa**

(a) Phase contrast micrograph of a group of human sperm. (b) Schematic drawing of the structure of a spermatozoon.

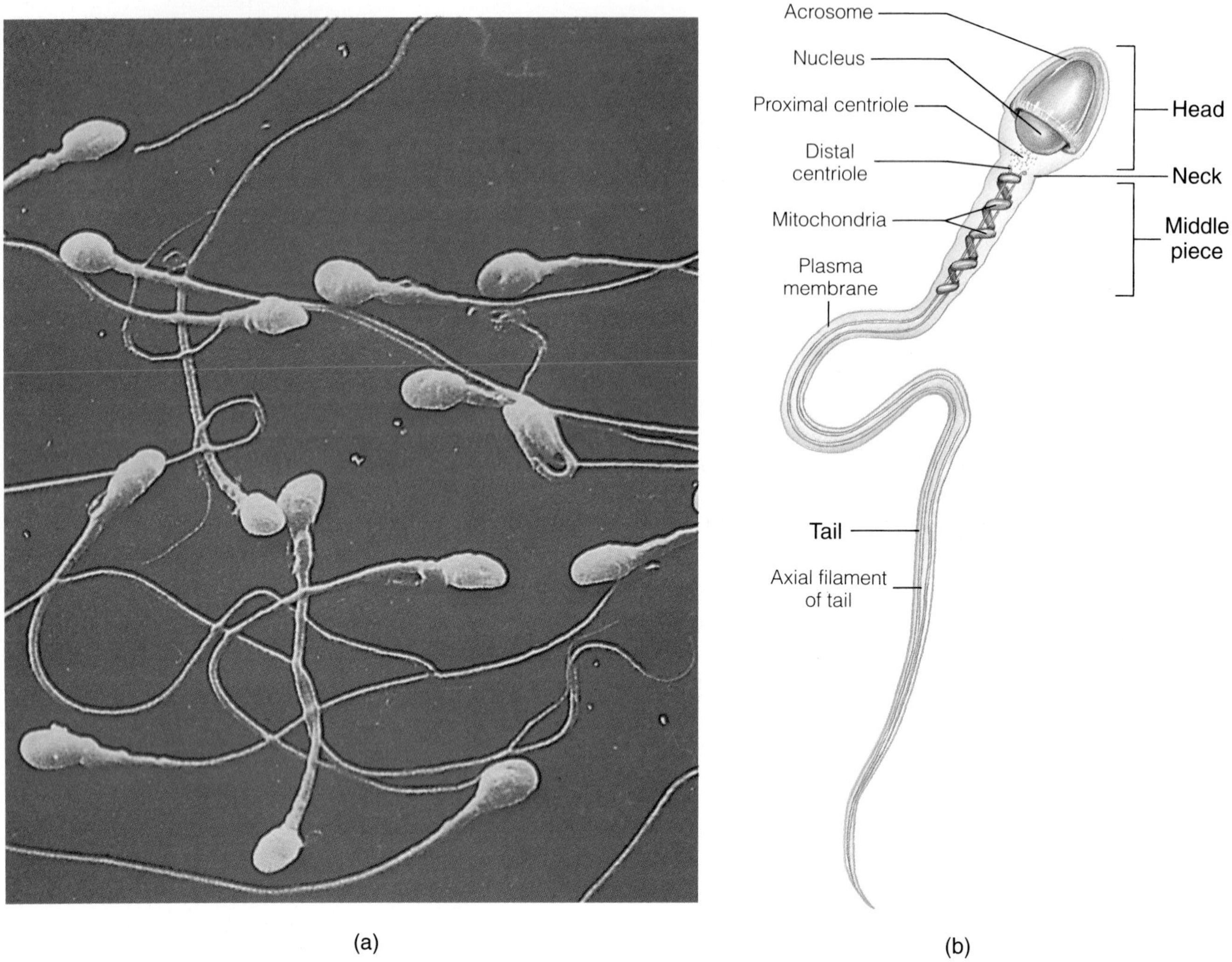

Gn-RH secretion occurs between bursts. Since Gn-RH is released in bursts, FSH and LH release occurs in similar episodic bursts. Once released, LH is removed from the blood more rapidly than FSH.

LH acts on the interstitial endocrinocytes of the testes to stimulate testosterone secretion. FSH and testosterone act on the sustentacular cells, which are in close contact with the spermatogenic cells at all stages of development. Both FSH and testosterone are required for the initiation of spermatogenesis at puberty. Testosterone is required to maintain spermatogenesis beyond puberty, but it is not clear if FSH is necessary.

In addition to stimulating spermatogenesis, testosterone acts in negative-feedback fashion on the hypothalamus to decrease the episodes of Gn-RH secretion. It also acts in negative-feedback fashion on the pituitary to reduce the responsiveness of LH-secreting cells to Gn-RH. Consequently, testosterone exerts a greater inhibitory effect on LH secretion than on FSH secretion.

A glycoprotein hormone called **inhibin** is produced by the sustentacular cells of the seminiferous tubules. There is strong evidence that inhibin acts directly on the pituitary to inhibit FSH release.

Male Sexual Responses

Erection

Sexual intercourse (coitus) is the usual method by which male sperm are deposited within the vagina of the female. In order to accomplish sexual intercourse,

◆ **FIGURE 28.13 Hormonal control of testicular function**

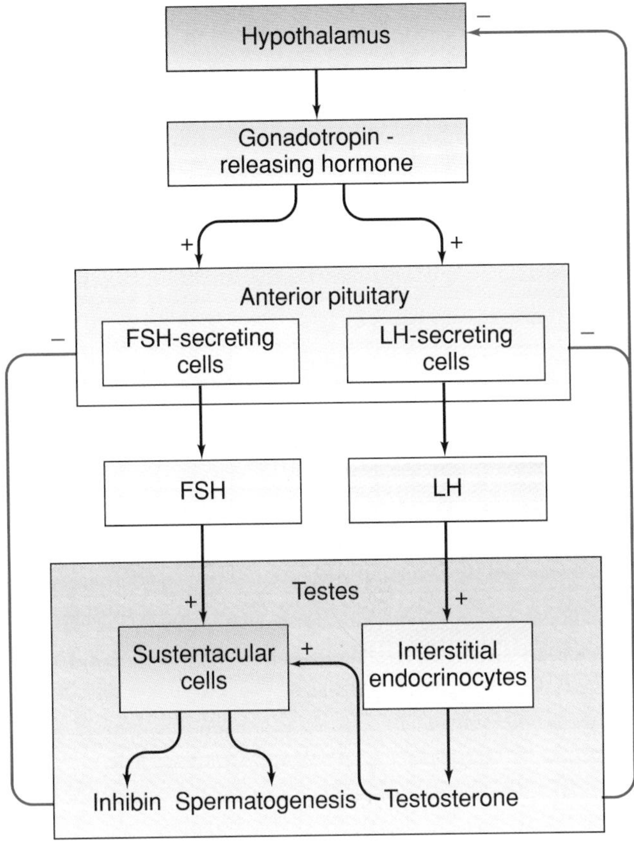

the male penis must become enlarged and firm, a process referred to as **erection.**

Erection is a vascular phenomenon in which the spaces within the corpora cavernosa and corpus spongiosum of the penis become engorged with blood. In the absence of sexual arousal, the arterioles that supply the corpora of the penis are partially constricted, which allows only limited blood flow into the penis. Under these conditions the penis is flaccid. During sexual excitation, increased parasympathetic activity and decreased sympathetic activity from the sacral region of the spinal cord to the arterioles of the penis cause the arterioles to dilate (Figure 28.14). (This is one of the few instances in which there is parasympathetic control over changes in blood-vessel diameter, which is typically under sympathetic control.) Dilation of the arterioles allows more blood to enter the vascular spaces of the erectile tissues.

As the erectile tissues of the penis expand due to the increased blood volume, the veins that drain the penis are compressed against the coverings of the penis. The compression of the veins reduces the flow of blood out of the penis, which further increases the engorgement of the erectile tissues. As a result, the penis becomes enlarged and firm. (Parasympathetic neural activity also promotes the secretion of mucus by the bulbourethral glands, providing lubrication for the glans penis.)

An erection can result from visual or psychic stimuli, including thoughts or emotions that originate in the brain, or from the physical stimulation of touch receptors on the penis and various other regions of the body. Moreover, psychic factors can greatly influence the effectiveness of physical stimuli. Indeed under adverse psychological conditions, the physical stimulation of the penis frequently will not cause an erection.

Ejaculation

In addition to causing the penis to become erect, continued tactile stimulation of the penis ultimately results in a forceful expulsion, or **ejaculation,** of semen. Ejaculation can be divided into two phases: an emission phase and an expulsion phase.

1. When sexual stimulation becomes sufficiently intense, sympathetic nerve impulses from the lumbar region of the spinal cord cause rhythmic contractions of the epididymis, ductus deferens, ejaculatory ducts, seminal vesicles, and prostate gland. These contractions propel the contents of the ducts and glands into the

◆ **FIGURE 28.14 Erection of the penis**

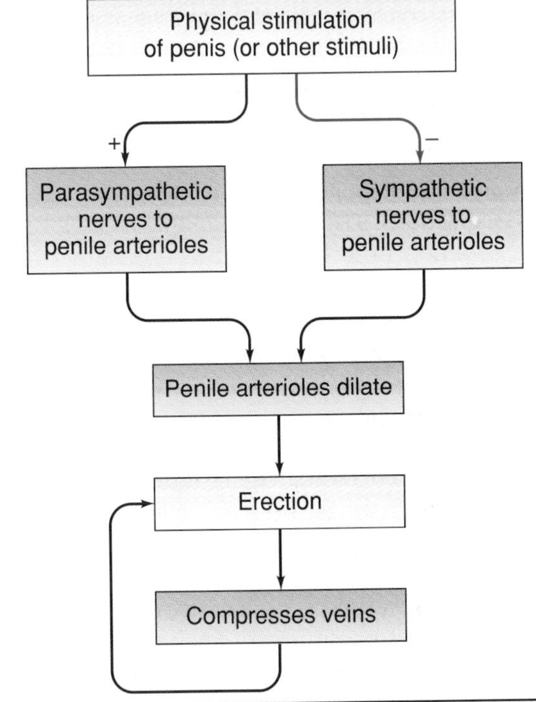

◆ **FIGURE 28.15 Structures of the female reproductive system as seen from behind**
The posterior walls of the vagina, the left side of the uterus, and the left uterine tube have been removed, as has the entire left broad ligament.

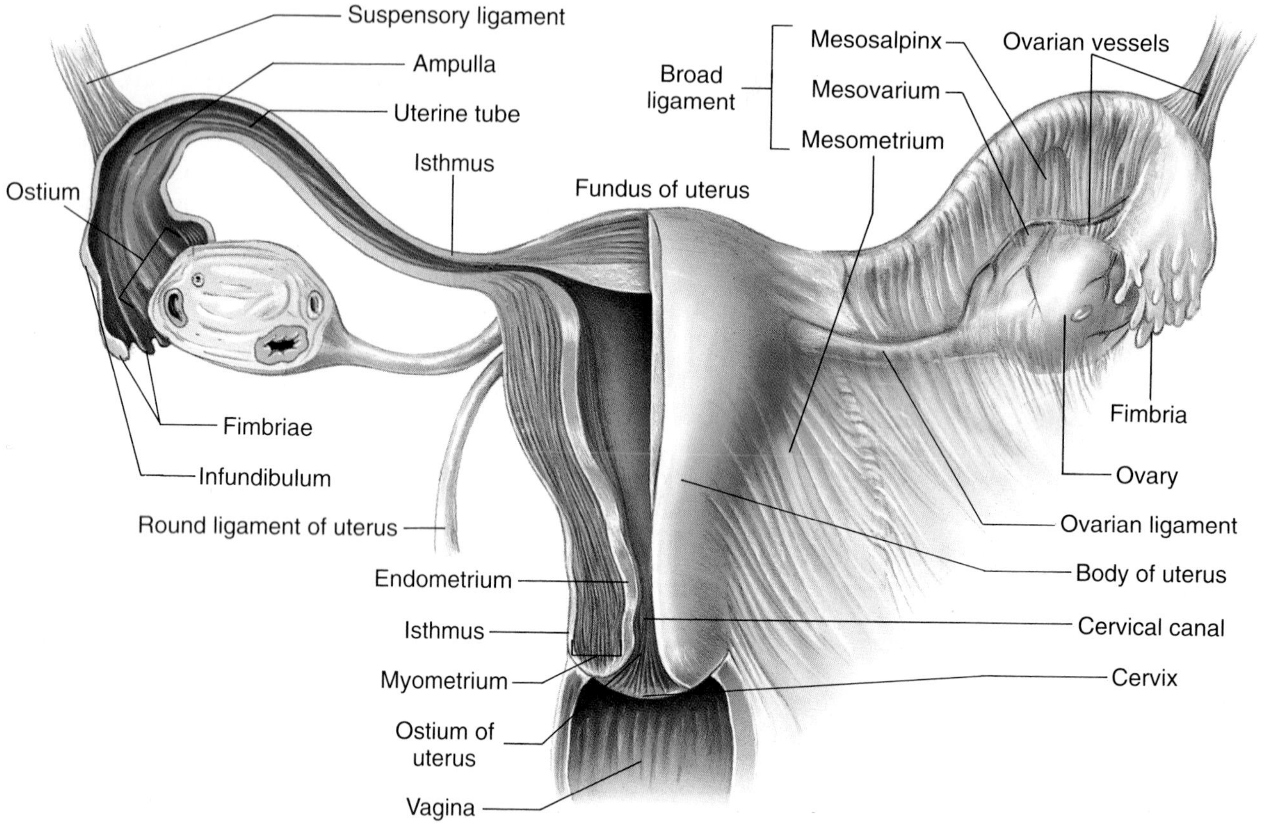

urethra, where they form semen. This phase of ejaculation is referred to as **emission.**
2. Following emission, skeletal muscles at the base of the penis undergo rhythmic contractions that expel the semen from the urethra. This phase of ejaculation is referred to as **expulsion.**

The volume and sperm content of each ejaculation vary somewhat, depending on the length of time between ejaculations. The average volume is 3 ml, but following a period of abstinence from sexual activity, it may be as much as 6 ml. An average ejaculate contains 300–400 million sperm.

The rhythmic contractions that produce an ejaculation are accompanied by intensely pleasurable sensations and widespread muscular contractions. This reaction is called an **orgasm.**

Semen ejaculated into a vagina rapidly forms into a clot as enzymes from the prostate gland act on fibrinogen that is contributed to semen by the seminal vesicles. The clotting enzymes change fibrinogen into fibrin, which forms the core of the semen clot. The clot is believed to serve the purpose of keeping the ejaculated sperm in the female reproductive tract as the penis is withdrawn. The clot liquefies and releases motile sperm within the female tract within a few minutes due to the presence of a fibrin-degrading enzyme also released by the prostate.

Following ejaculation, the arterioles that supply the penis return to their usual constricted condition, which decreases blood flow into the penis. This decrease in blood flow relieves the vascular congestion within the corpora of the penis and allows blood to return freely through the veins to the general circulation. As the accumulated blood leaves the corpora, the penis returns to a flaccid state.

Anatomy of the Female Reproductive System

The female reproductive system includes the *ovaries,* which produce ova; the *uterine tubes,* which transport, protect, and nourish an ovum that has been released from an ovary; the *uterus,* which provides a suitable environment for the development of an embryo; and the *vagina,* which serves as a receptacle for sperm.

◆ **FIGURE 28.16 Median section of the female pelvis**

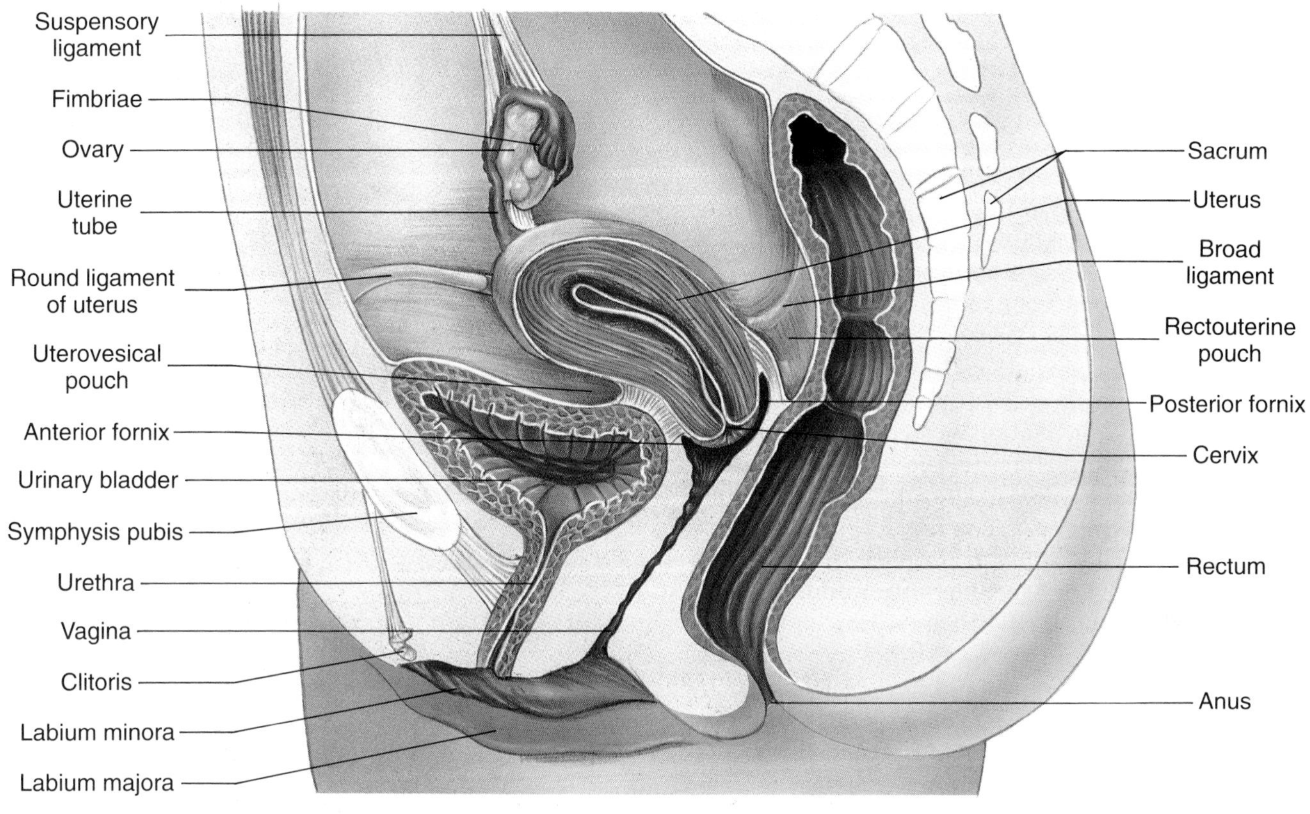

Ovaries

The female gonads, or primary sex organs, are the **ovaries,** in which the female gametes **(ova)** are produced. Moreover, *estrogens* and *progesterone*—hormones that influence the development of the female accessory sex organs and secondary sex characteristics—are secreted by the ovaries.

Each of the paired ovaries is oval and slightly smaller than a testis. Like all the organs of both the female and the male reproductive systems, the ovaries are retroperitoneal. The ovaries do not migrate as extensively during embryonic development as do the testes. Rather, they descend only as far as the pelvis and lie against the lateral wall on either side of the uterus (Figures 28.15 and 28.16). The ovaries are held in this position by several ligaments. The largest is the **broad ligament,** which also supports the uterine tubes, the uterus, and the vagina. The ovaries are suspended from the posterior surface of the broad ligament by a short fold of peritoneum called the **mesovarium.** A fibrous band called the **ovarian ligament** lies within the broad ligament and extends from the superior-lateral margins of the uterus to the ovary. The lateral margins of the broad ligament form a fold that attaches the ovary to the pelvic wall. This fold, through which the ovarian vessels reach the ovary, is known as the **suspensory ligament.**

The peritoneum that covers the surface of the ovary is composed of relatively short, simple cuboidal cells rather than squamous cells as is typical of the rest of the peritoneum. This outer epithelial covering of the ovary is called the **germinal epithelium.** In spite of what the name implies, the germinal epithelium does not give rise to germ cells. Beneath the germinal epithelium is a fibrous connective-tissue layer, the **tunica albuginea.** The ovary itself can be divided into an outer **cortex** surrounding a central **medulla.** The medulla is composed of connective tissue and contains blood vessels, lymphatic vessels, and nerves that enter or leave the ovary where it attaches to the mesovarium.

At birth, the cortices of the two ovaries contain approximately 700,000 immature structures called **primordial follicles.** Each primordial follicle consists of an undeveloped ovum (called an **oocyte;** *ō´-ō-sīt*) located within a small sphere that is composed of a single layer of flattened cells, called *follicle cells.* Beginning at the time of puberty and continuing throughout the 30–40 years during which a female is capable of reproduction, one follicle usually matures fully and ruptures each month (approximately every 28 days), releasing an ovum that is then available for fertilization. This being the case, only about 400 of the total number of follicles reach full maturity during a woman's lifetime. The rest reach various stages of development and then regress and degenerate.

Uterine Tubes

An ovum released at ovulation is carried to the uterus by a **uterine** *(fallopian)* **tube** (Figure 28.15). Each uterine tube, which extends from the vicinity of the ovary to the superior lateral angle of the uterus, lies between the layers of the broad ligament. The portion of the broad ligament that anchors each uterine tube is called the **mesosalpinx** *(mes″-ō-sal′-pinks)*. The medial constricted portion of the uterine tube is the **isthmus** *(is′-mus)*. It opens into the uterus. The uterine tube is expanded somewhat in the region where it curves around the ovary. This region is called the **ampulla.** Fertilization generally occurs within the ampulla. The distal end of each uterine tube is called the **infundibulum** *(in″-fun-dib′-ū-lum)*. The infundibulum opens into the abdominopelvic cavity very close to an ovary. This opening, called the **ostium** *(os′-tē-um)*, is surrounded by small, ciliated, fingerlike projections called **fimbriae** *(fim′-brē-ē)*. One of the fimbriae is generally attached to the ovary. The movements of the fimbriae and their cilia are thought to cause currents of peritoneal fluid to enter the uterine tube and thus to carry the ovum that has been released from its follicle into the tube. Since the reproductive tract of the female opens into the peritoneal cavity through the two uterine tubes (one tube to each ovary) and to the exterior of the body through the uterus and vagina, infections of the lower reproductive tract can spread into the body cavity and cause peritonitis. (In males, the reproductive tract has no opening into the body cavity.)

The wall of the uterine tube is covered by the peritoneum. The *muscular layer* of the uterine tube, which is deep to the peritoneum, contains both circular and longitudinal layers of smooth muscles. The inner *mucosa* is thick and its epithelium is composed of simple columnar cells. The mucosa forms numerous branched folds that extend into the lumen of the uterine tube (Figure 28.17). Two types of cells are present in the epithelium of the mucosa: those with cilia and those without. Most of the cells have cilia that beat rhythmically toward the uterus. Thus, once within the uterine tube, the ovum is carried to the uterus by a weak fluid current caused by the beating of the cilia, possibly aided by the peristaltic contractions of the smooth muscles in the walls of the uterine tube. Interspersed among the ciliated cells are cells that lack cilia. These cells are thought to be secretory cells that maintain a moist environment in the uterine tube and may serve as a source of nourishment for the ovum.

◆ **FIGURE 28.17 Photomicrograph of a uterine tube**

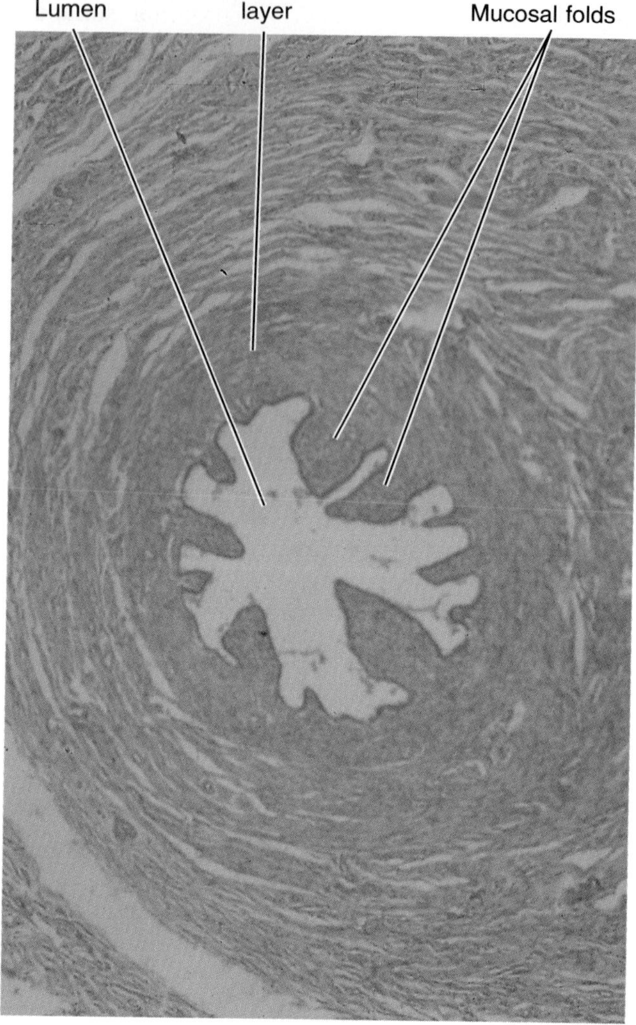

Uterus

The **uterus** is a single, hollow, pear-shaped organ that receives the uterine tubes at its upper lateral angles and is continuous below with the vagina (Figure 28.15). The upper portion of the uterus is called the **body.** Below its body, the uterus narrows into the **isthmus,** and where it joins with the vagina, it forms the cylindrical **cervix.** The opening of the uterus into the vagina is the **ostium** of the uterus. The dome-shaped region of the uterine body above and between the points of entrance of the uterine tubes is the **fundus.**

The uterus lies in the pelvis, behind the urinary bladder and in front of the sigmoid colon and rectum. The peritoneum that covers the superior and posterior surfaces of the urinary bladder is folded back up the anterior surface of the uterus from the floor of the pelvic cavity. The space thus formed is called the **uterovesical**

pouch *(ū″-ter-ō-ves′-ĭ-kal)* (Figure 28.16). In a similar manner, the **rectouterine pouch** *(rek″-tō-ū′-ter-in)* is formed where the peritoneum that continues over the top of the uterus and down its posterior surface is folded back up the anterior surface of the rectum from the pelvic cavity floor.

The layers of peritoneum that cover the uterus anteriorly and posteriorly fuse along its lateral margins and extend to the lateral walls and floor of the pelvic cavity as the **broad ligament** (Figure 28.15). The portion of the broad ligament that is below the mesovarium and that anchors the uterus is called the **mesometrium** *(mes″-ō-mē′-tre-um).* Blood vessels and nerves reach the uterus and uterine tubes by traveling between the peritoneal layers of the broad ligament. The *uterine arteries,* which are branches of the internal iliac arteries, travel within the broad ligament to reach the cervical portion of the uterus. They then follow its lateral margin up to the isthmus of the uterine tube, where they give off tubal and ovarian branches.

Assisting the broad ligament in holding the uterus in place are three pairs of suspensory ligaments: the round ligaments, the uterosacral ligaments, and the cardinal ligaments. The **round ligaments** are fibrous bands that run within the broad ligaments from the lateral margins of the uterus, near the junctions with the uterine tubes, through the inguinal canals and into the labia majora. The ovarian and round ligaments form a continuum that is *homologous* to—that is, formed from the same embryonic tissue as—the gubernaculum of the male. Recall that the labia majora and the scrotum are also homologous structures. The **uterosacral ligaments** pass from the lateral surfaces of the uterus to the anterior surface of the sacrum. The **cardinal (lateral cervical) ligaments** pass from the cervix of the uterus and the upper region of the vagina to the lateral walls of the pelvis. The uterosacral and cardinal ligaments tend to prevent the uterus from moving downward.

The ligaments do not hold the uterus tightly in place; rather, they allow it to have limited movement. Normally, the uterus is tipped forward, its longitudinal axis forming an angle of about 90° with the longitudinal axis of the vagina. This angle varies, however, depending on how full the urinary bladder or the rectum is. The ligaments do not in fact provide much support for the uterus. Its chief support is provided from below by the muscles and membranes that form the floor of the pelvic cavity—the region known as the **pelvic diaphragm** and the **urogenital diaphragm** (see pages 294–295). These muscles all attach to a small, circular tendon located just posterior to the vaginal opening. This tendon is called the *central tendon* of the perineum. If the central tendon is weakened, which sometimes occurs as a result of childbirth, the uterus can move downward into the vagina. In this condition, the uterus is said to be *prolapsed.*

The wall of the uterus is covered by the **perimetrium (serosa)** which is formed by the peritoneum that is folded back over the uterus from the broad ligament. Beneath the perimetrium is a thick, muscular layer called the **myometrium** *(mī-ō-mē′-trē-um)* (Figure 28.15). The myometrium, which makes up most of the thickness of the wall of the uterus, is composed of bundles of smooth muscles running in various directions. The cavity in the uterus is lined with an epithelium of ciliated columnar cells. This surface epithelium plus an underlying connective-tissue lamina propria called the *endometrial stroma* constitute the **endometrium** *(en-dō-mē′-trē-um).* The surface epithelium invaginates to form numerous tubular *uterine glands,* which extend down into the stroma. The endometrium consists of a thick superficial layer called the **functional layer** and a thin, deep **basal layer.** The functional layer undergoes marked developmental changes during the menstrual cycle. Each month it thickens and becomes engorged with blood in preparation for receiving a fertilized ovum. If the ovum is not fertilized, the functional layer of the endometrium is sloughed off as the menstrual flow.

Vagina

The **vagina** is the canal that leads from the cervix of the uterus to the exterior of the body (Figure 28.16). The smooth muscle tunic in the wall of the vagina is much thinner than the muscular tunic of the uterus. The mucosa lining the vagina has a protective surface layer of stratified squamous epithelium, as is typical of all the canals that open onto the body surface. The vaginal mucosa proliferates during the menstrual cycle in a manner similar to the endometrial changes of the uterus, but there is considerably less development in the mucosa of the vagina than in the mucosa of the uterus. The mucosa of the vagina contains numerous transverse ridges, or rugae.

Near the entrance to the vagina, the mucosa usually forms a vascular fold called the *hymen.* Generally, the hymen only partially blocks the vaginal opening, but in some cases it completely closes the orifice. The hymen, which is stretched and often torn during sexual intercourse, may also be torn by other means. Since the hymen may persist after sexual intercourse, its presence or absence is not a reliable method by which to prove or disprove virginity.

The lumen of the vagina is generally quite small, and the walls that surround it are usually in contact with each other. The canal is capable of considerable stretching, however, as when the vagina receives the male penis during sexual intercourse or when it serves as the birth canal. The upper end of the vaginal canal surrounds the cervix of the uterus, forming a recess (the

◆ **FIGURE 28.18 External genital organs of the female, showing divisions of the perineum**

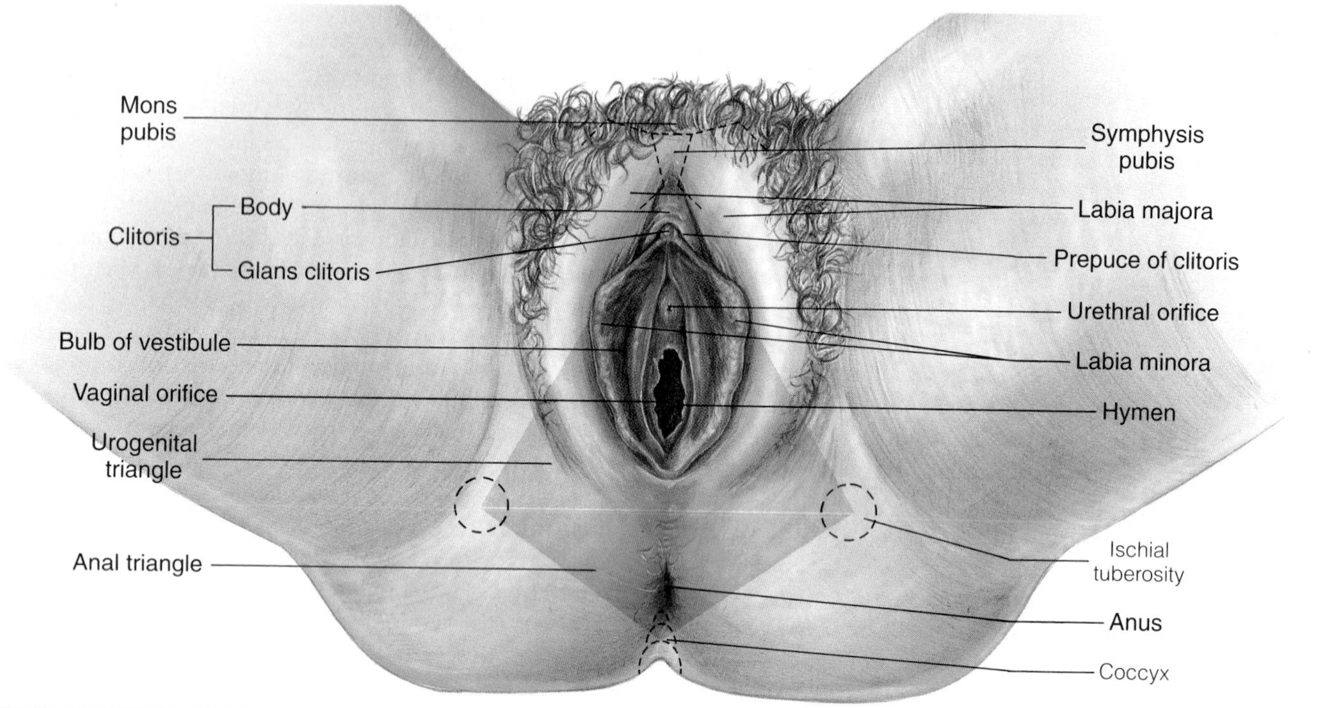

vaginal fornix) around the cervix (Figure 28.16). The largest portion of the recess lies dorsal to the cervix and is called the **posterior fornix.** The **anterior fornix** and the two **lateral fornices** are not as deep.

The mucosal cells lining the vagina contain large quantities of glycogen, which is metabolized to lactic acid by bacteria that are normally present in the vagina. Sufficient lactic acid is produced to cause the vagina to be quite acidic, with a pH of 3.5 to 4. This acidity is beneficial to the vagina because it retards infection, but it is injurious to sperm cells. Thus, it is necessary for semen to neutralize the acidity of the vagina if sperm are to survive.

Female External Genital Organs

When considered collectively, the external genital organs of the female are known as the **vulva,** or **pudendum** (Figure 28.18).

Under the influence of estrogens, there is a tendency in the female for fatty tissue to be deposited over the pubic symphysis. This deposition produces a mound called the **mons pubis.** (Because of the lack of estrogens, this fat deposit is lacking in males.) The skin over the mons becomes covered with hair after puberty.

Two rounded folds—the **labia majora**—extend backward from the mons. The outer surfaces of the labia majora are pigmented and covered with hair. Their inner surfaces are smooth, lack hair, and are moist because of the presence of large sebaceous glands.

The **labia minora** are two smaller folds located medial to the labia majora. Anteriorly, they surround the clitoris. The labia minora are highly vascular and lack hair. They surround a space (the **vestibule**) into which the vagina and urethra open. The vaginal opening into the vestibule may be partially blocked by the hymen. A number of glands open into the vestibule and keep its walls moist. The ducts of the **greater vestibular glands,** which are homologous to the bulbourethral glands of the male, open into the vestibule on either side of the vagina. On each side of the external urethral orifice are the openings for the **lesser vestibular glands.** The vestibular glands lubricate the vestibule and thus facilitate sexual intercourse.

The **clitoris** *(klit´-ō-ris)* is a small elongated structure located at the anterior junction of the labia minora. Most of the body of the clitoris is enclosed by a **prepuce** (foreskin) formed by the labia minora. The free exposed portion of the clitoris is called the **glans clitoris.** The clitoris is homologous to the dorsal portion of the male penis and, like the penis, is composed of erectile tissue. The clitoris contains two **corpora cavernosa clitoris,** but except for the glans, it does not contain a corpus spongiosum (which, in the male, surrounds the urethra). The female's urethra is not within the clitoris. Rather, it opens separately, posterior to the clitoris. The crura of the corpora cavernosa attach the clitoris to the rami of the pubic and ischial bones. The clitoris, which is very sensitive to touch, becomes engorged with blood when stimulated, contributing to the sexual arousal of the woman.

Just deep to the labia are two elongated masses of erectile tissue called the **bulb of the vestibule.** The bulb, which is homologous to the corpus spongiosum penis and bulb of the penis, passes on either side of the vaginal orifice. During sexual arousal, the bulb becomes engorged with blood, thus narrowing the opening into the vagina and squeezing against the male penis during sexual intercourse.

The Female Perineum

As in the male, the female perineum can be divided by a transverse line between the ischial tuberosities into an anterior **urogenital triangle,** which contains the external genitalia, and a posterior **anal triangle,** which contains the anus (Figure 28.18). The region between the vagina and the anus is referred to as the *clinical perineum* because this area is sometimes torn by the stretching that occurs during childbirth. Sometimes the tears even damage the anal sphincter. To prevent tearing, an incision called an *episiotomy (ĕ-pēz´´-ē-ot´-ō-mē)* is often made in the perineum during delivery. This operation enlarges the vestibule and makes delivery easier. Moreover, a surgical incision is easier to repair than a tear.

Mammary Glands

Even though **mammary glands (breasts)** are also present in males, we discuss their anatomy under the female reproductive system because it is only in the female that the breasts are at all involved in reproduction. Even in the female, they are not involved in the actual reproductive process, however, but serve to produce milk for the nourishment of the infant.

Each mammary gland is a skin-covered hemispherical elevation located superficial to the pectoral muscles (Figure 28.19). Just below the center of each mammary gland is a protruding **nipple** that is surrounded by a circular **areola** *(uh-rē´-ō-luh).* The areola has many small elevations due to the presence of numerous large sebaceous glands called **areolar glands.** The areolar glands produce a waxy secretion that prevents chafing of the nipple during nursing. Both the nipple and the areola are pigmented and have capillary beds located close to their surfaces. The pigmentation becomes darker during pregnancy and diminishes somewhat afterward. Smooth muscles in the areola and nipple cause the nipple to become erect as a consequence of stimulation.

Internally, the periphery of each mammary gland is composed of adipose tissue held in place by a connective-tissue stroma. Bands of connective tissue extend from the anterior aspect of the stroma and attach into the dermis. These connective-tissue septa are called the **suspensory ligaments** of the breast. The septa subdivide the fat that lies superficial to the glandular tissue, and they give a smooth contour to the breast. Centrally, there are about 15–25 lobes of glandular tissue arranged radially around the nipple, each consisting of a separate compound tubuloalveolar gland. Each lobe is drained by a single **lactiferous duct** *(lak-tif´-er-us)* which opens onto the nipple. The nipple, therefore, is perforated by numerous openings. Just before reaching the nipple, each lactiferous duct expands into a small milk reservoir called an **ampulla.** Glandular tissue also extends beyond the breast, into the axilla. Milk is released from the glands by a modified form of apocrine secretion.

Puberty in the Female

Unlike the situation in males, where testosterone is secreted by the testes prior to birth, the female sex hormones are not produced in significant amounts until puberty. The female reproductive tract automatically develops in the absence of testosterone and does not require the presence of estrogens.

On the average, females reach puberty one or two years earlier than males. In females, puberty is marked by the first episode of menstrual bleeding, which is called the **menarche** *(meh-nar´-kē).* At puberty, the vagina, uterus, and uterine tubes enlarge, and the deposition of fat in the breasts and hips causes them to become more prominent. In addition, the glandular portion of the breasts begins to develop. Axillary and pubic hair grows, and females experience a growth spurt that causes them to grow more quickly than males between the ages of 12 and 13. However, the epiphyseal plates of females close earlier than those of males, and females reach their adult height between 15 and 17 years of age. In contrast, males continue to grow in height until about age 21.

Although the growth of axillary and pubic hair is caused by increased secretion of androgens by the adrenal glands, most of the changes associated with puberty in females are largely the result of an increased production of estrogens by the ovaries. Prior to puberty, estrogens are secreted at very low levels, probably due mainly to the secretion of only small amounts of Gn-RH by the hypothalamus. At puberty, an alteration in brain function leads to increased Gn-RH release. The increased Gn-RH release stimulates the secretion of FSH and LH by the pituitary gland, which ultimately leads to an increased production of estrogens by the ovaries.

◆ **FIGURE 28.19 The mammary gland**
(a) Anterior view of a partially dissected left breast. (b) Sagittal section.

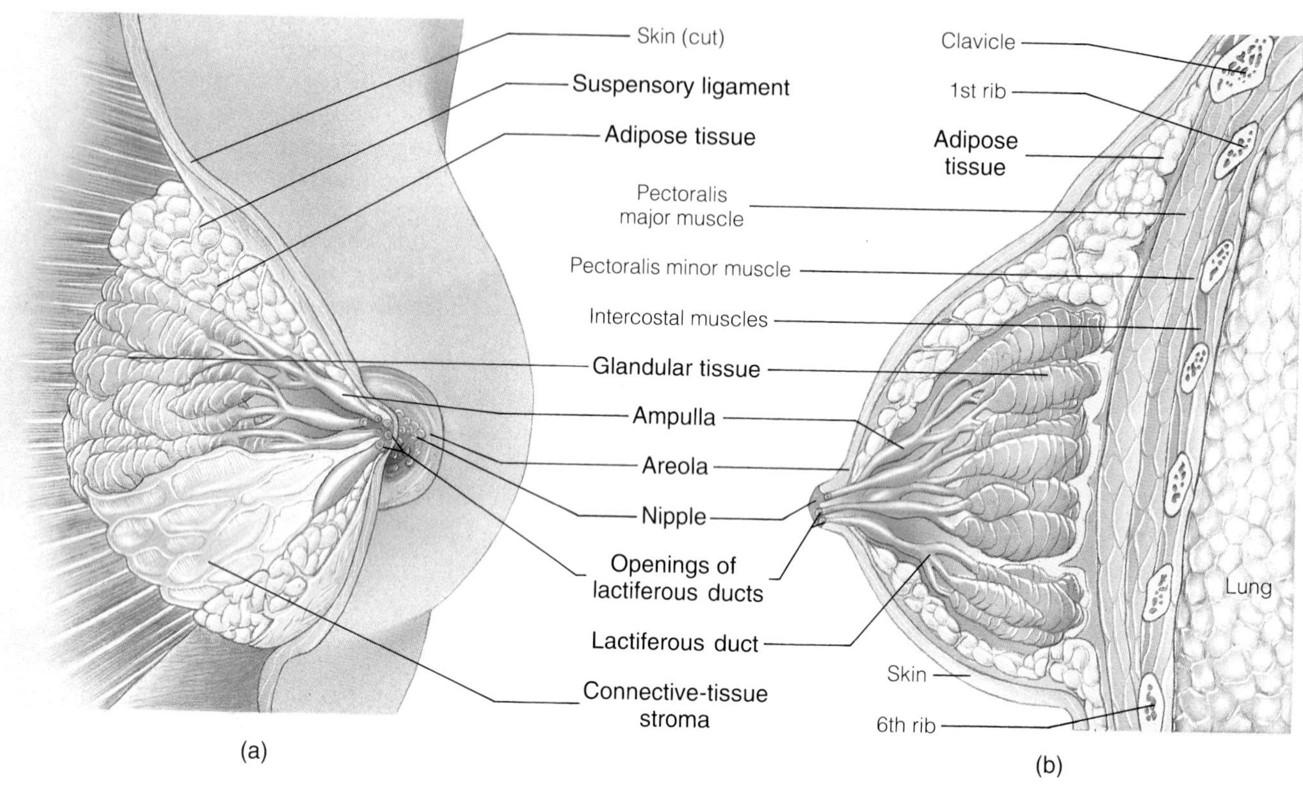

Oogenesis

The production of ova within the ovaries is called **oogenesis** *(ō-ō-jen´-e-sis)*. As in the case of spermatogenesis, meiosis occurs during this process (Figure 28.20).

During the embryonic and fetal development of a female, precursor cells called **oogonia** divide mitotically and thereby provide the cells that will develop into ova. Each oogonium is a diploid cell containing 44 autosomes and two X sex chromosomes.

By the time fetal development is complete, several hundred thousand oogonia have undergone a growth phase and have become **primary oocytes.** The primary oocytes enter prophase of the first meiotic division, but do not complete it and are in first prophase at the time a female is born. Each primary oocyte is located within a small sphere composed of a single layer of flattened cells, thus forming a **primordial follicle.**

It is not until puberty, when full follicle maturation begins to occur, that a primary oocyte develops further. As a follicle matures, the primary oocyte within it enlarges and completes the first meiotic division, producing two haploid cells that are of unequal size. The larger cell, called a **secondary oocyte,** receives half of the chromosomes and almost all of the cytoplasm. The smaller cell, called the **first polar body,** receives the other half of the chromosomes but very little of the cytoplasm. Thus, the secondary oocyte and the first polar body each contain 23 autosomes and one X sex chromosome. The secondary oocyte immediately begins the second meiotic division; however, it halts at metaphase and is released from the ovary in this stage when the follicle ruptures at ovulation.

The secondary oocyte does not complete the second meiotic division unless it is entered by a sperm. However, if a sperm does enter, meiosis is quickly completed. As was the case in the first meiotic division, the completion of the second meiotic division produces one large cell, the mature **ovum,** and one small cell, a **second polar body.** The first polar body also often undergoes the second meiotic division, producing two additional polar bodies. Therefore, during oogenesis, one ovum and as many as three polar bodies are produced from an oogonium. Beginning at puberty, one follicle generally matures fully, and consequently one secondary oocyte is released at ovulation each month (approximately every 28 days).

◆ **FIGURE 28.20 Diagrammatic representation of the process of oogenesis**

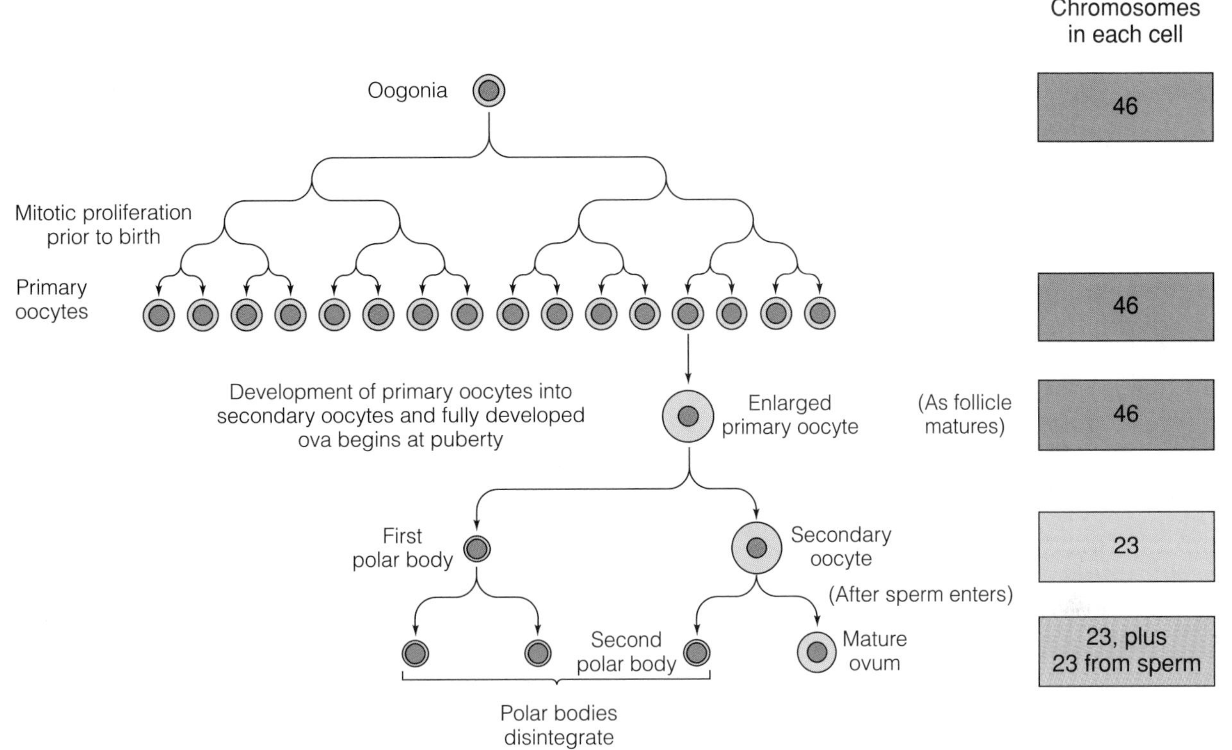

The Ovarian Cycle

The **ovarian cycle** consists of a series of events that occur within an ovary, including the full maturation of a follicle, the release of a secondary oocyte from the mature follicle at ovulation, and the formation of a structure called the corpus luteum. The length of the ovarian cycle ranges from about 20 to 40 days, with variability both between women and in the same woman during different cycles. Usually, the ovarian cycle is considered to be a 28-day cycle.

The ovarian cycle is commonly divided into two phases, which are separated by ovulation. The phase preceding ovulation is called the **follicular phase.** During this phase, a follicle reaches full maturity. The phase following ovulation is called the **luteal phase.** During this phase, the corpus luteum forms and then degenerates. In a 28-day cycle, both the follicular phase and the luteal phase are approximately 14 days long.

Follicle Maturation

Follicle maturation begins when the flattened cells that form the wall of a primordial follicle become cuboidal in shape (Figure 28.21). At this point, the follicle is called a **primary follicle,** and it consists of a primary oocyte surrounded by a single layer of cuboidal cells.

As follicle maturation continues, the primary oocyte increases in size. In addition, the single layer of cells that forms the wall of the follicle proliferates, creating a stratified layer of cells, now called **granulosa cells** *(gran″-ū-lō′-sah).* At this point, the follicle is called a **growing** (or **secondary**) **follicle.**

Some cells in the connective tissue outside the follicle condense into a layer around the follicle. The layer is called the **theca.** Thecal cells join with the granulosa cells to become part of the follicle. Collectively, thecal cells and granulosa cells are referred to as **follicle cells.**

Granulosa cells and the primary oocyte secrete a gel-like glycoprotein material that forms a thick, transparent membrane around the oocyte. This membrane is called the **zona pellucida** *(peh-lū′-sih-duh).* The zona pellucida appears to separate the primary oocyte from the surrounding granulosa cells. However, cytoplasmic processes from the inner layer of granulosa cells extend through the zona pellucida and form gap junctions with the oocyte. Chemical messengers and nutrients can reach the oocyte through these junctions.

As the solid, growing follicle becomes larger, a fluid-filled cavity called the **antrum** forms in the midst of the granulosa cells. The fluid, which is called *antral* (or *follicular*) *fluid,* consists at least in part of granulosa-cell secretions.

◆ **FIGURE 28.21 Follicle maturation**

(a) Stages in follicle maturation from primordial follicle through mature follicle. (b) Ovary showing presence of mature follicle.

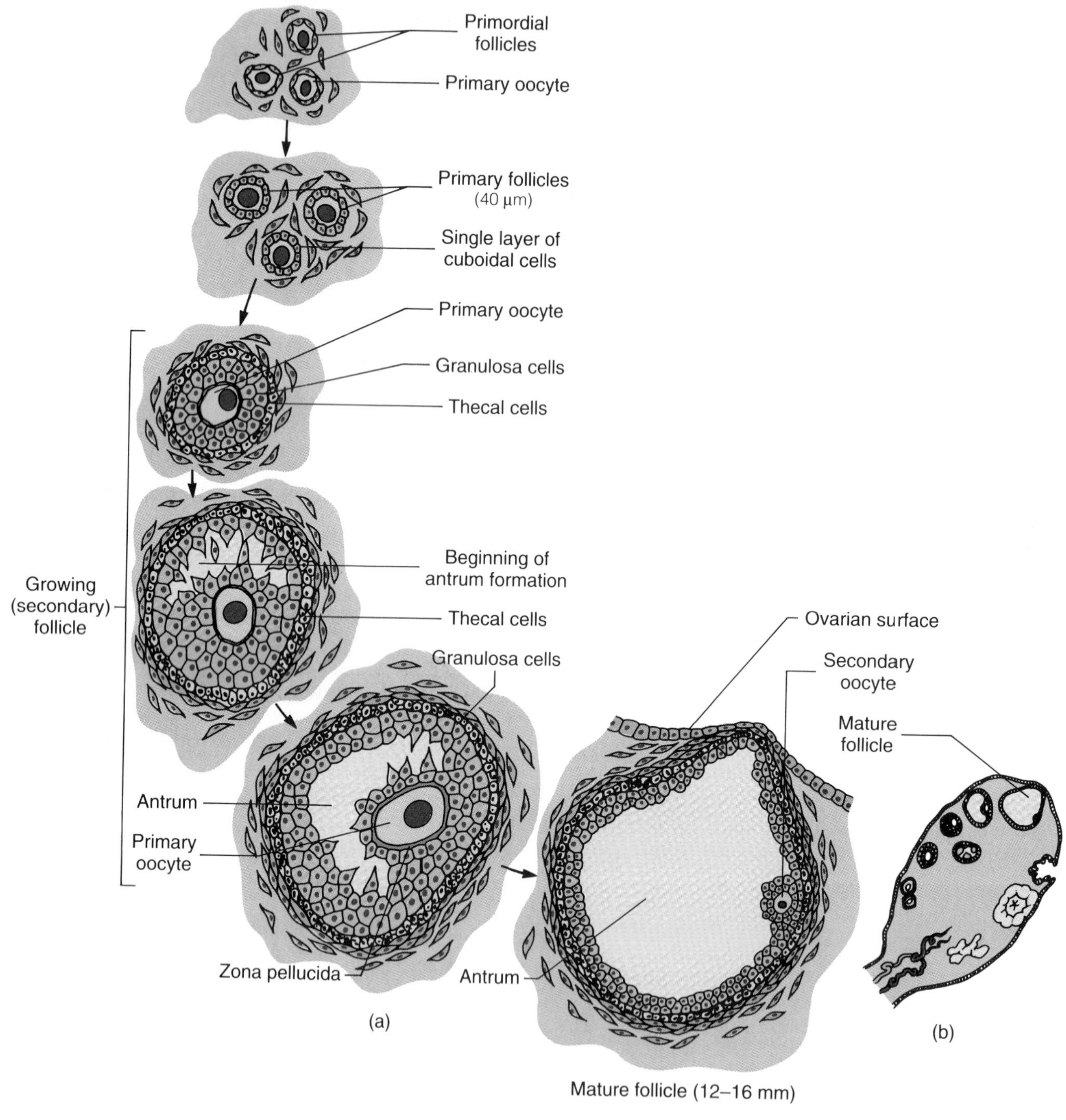

The primary oocyte and a few layers of granulosa cells that surround it are displaced to one side of the follicle by the antrum. As more fluid accumulates within the antrum, the follicle continues to enlarge and moves toward the surface of the ovary, where it produces a bulge. The follicle is then referred to as a **mature** *(Graafian)* **follicle** (Figure 28.22).

The primary oocyte reaches full size by the time the antrum begins to form. Just prior to ovulation, it completes the first meiotic division, giving rise to a second-

◆ **FIGURE 28.22 Photomicrograph of a mature follicle (×472)**

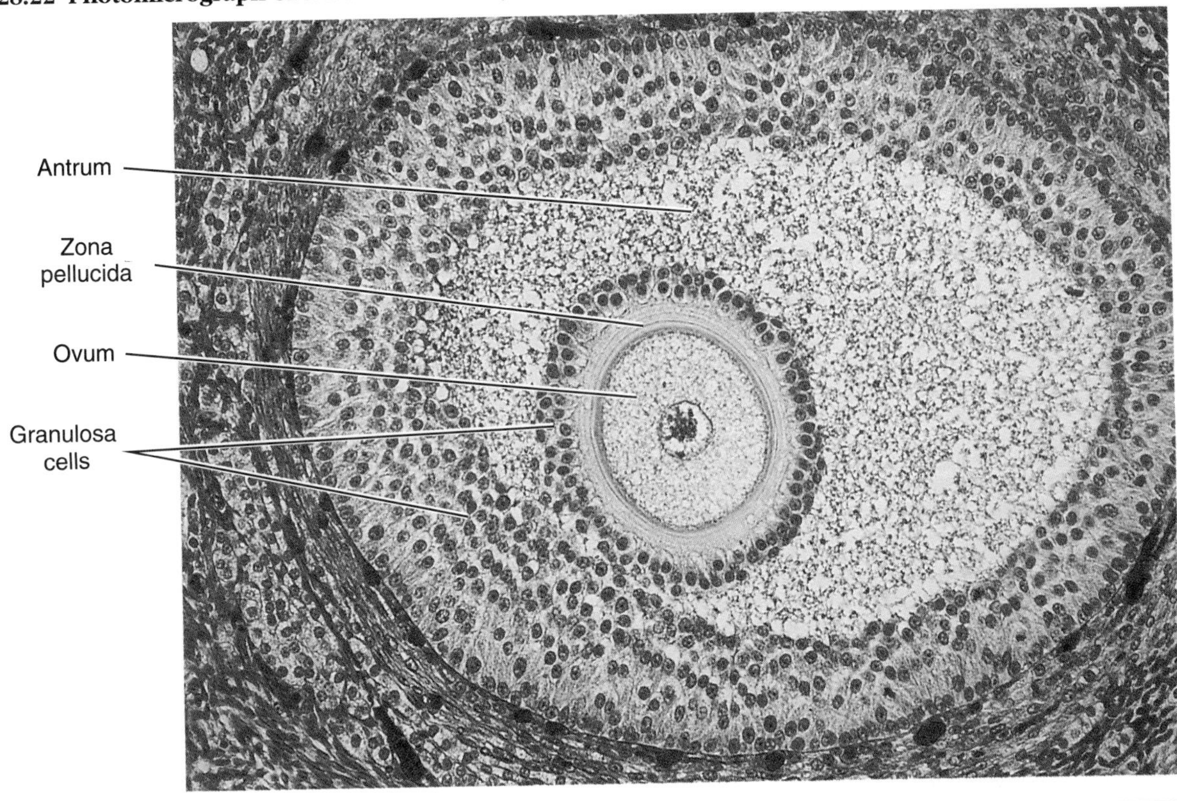

ary oocyte. The secondary oocyte immediately begins the second meiotic division. However, it halts at metaphase and is at this stage at ovulation.

In the ovaries of a sexually mature female, follicles continually develop through the preantral and early antral stages. Thus, at any moment some preantral follicles and a few small-antral follicles are present. At the beginning of each ovarian cycle, several of these follicles begin to develop into larger antral follicles. However, only one usually reaches the mature stage. The rest degenerate. The process of degeneration of a follicle—and of the oocyte within it—is called *atresia (ah-tre´-ze-ah),* and follicles undergoing atresia are referred to as *atretic follicles.*

Ovulation

Ovulation takes place when the mature follicle ruptures and releases its secondary oocyte (commonly referred to as an ovum) into the abdominopelvic cavity (Figure 28.23a). During ovulation, those follicle cells that directly surround the secondary oocyte remain with it. Therefore, the ovulated secondary oocyte is surrounded by a zona pellucida and a sphere of follicle cells that is now called the **corona radiata.**

The secondary oocyte, with its surrounding zona pellucida and corona radiata, is swept into a uterine tube by a flow of fluid from the abdominopelvic cavity into the tube. This flow is produced by movements of the fimbriae at the opening of the tube and by the beating of cilia that line the fimbriae.

Formation of the Corpus Luteum

Following ovulation, the ruptured mature follicle collapses. Shortly thereafter, follicle cells increase in size and take on a yellowish color, due in part to the accumulation of lipid droplets (Figure 28.23b). The structure that results is called a **corpus luteum** ("yellow body").

The future of the corpus luteum depends on the fate of the ovulated secondary oocyte. If the oocyte is not fertilized and pregnancy does not occur, the corpus luteum reaches its maximum development about ten days after ovulation and then begins to degenerate (Figure 28.23c). Eventually, all that remains is a white connective-tissue scar called the *corpus albicans* ("white body"). However, if the oocyte is fertilized and pregnancy does occur, the corpus luteum continues to develop and remains until near the end of the pregnancy.

◆ **FIGURE 28.23 Ovulation and development and degeneration of corpus luteum**

(a) Rupture of a mature follicle and release of secondary oocyte at ovulation. (b) Formation of corpus luteum from remaining follicle cells following ovulation. (c) Degeneration of corpus luteum if released secondary oocyte is not fertilized.

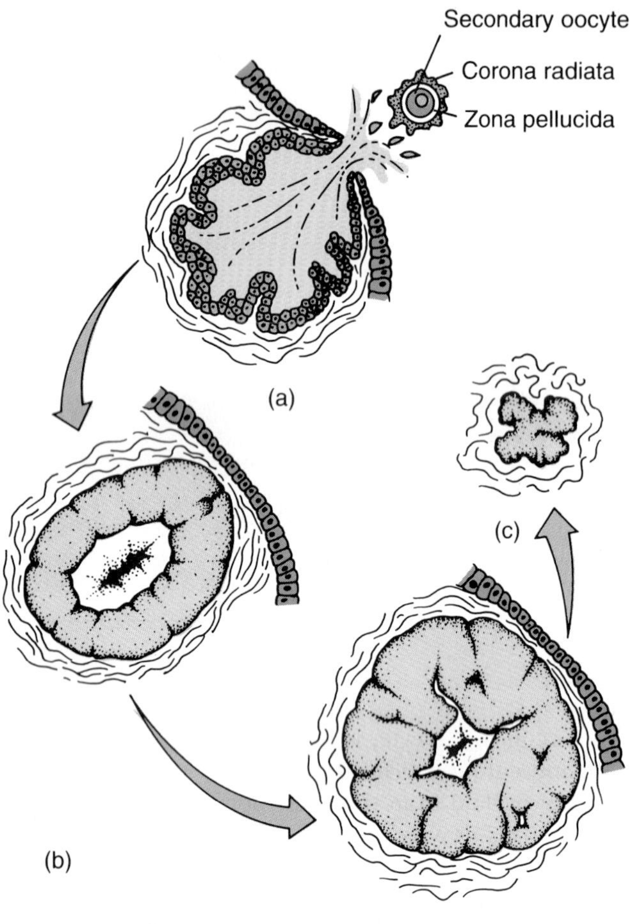

Hormonal Control of Ovarian Function

Gonadotropin-releasing hormone (Gn-RH) from the hypothalamus and follicle-stimulating hormone (FSH) and luteinizing hormone (LH) from the pituitary are involved in the control of ovarian function. Gn-RH is released in pulses, and its release is influenced by the female sex hormones (and probably by other brain areas). Gn-RH, in turn, stimulates the release of both FSH and LH.

The female sex hormones—the estrogens and progesterone—are produced by the ovaries. During the follicular phase of the ovarian cycle, follicle cells are the primary source of estrogens. Both thecal cells and granulosa cells are involved in estrogen production. Thecal cells produce androgens, which diffuse to granulosa cells where they are converted to estrogens.

During the luteal phase of the ovarian cycle, cells of the corpus luteum produce estrogens. The corpus luteum is also the primary source of progesterone.

As previously indicated, in the ovaries of a sexually mature female, follicles continually develop through the preantral and early antral stages. Thus, at any moment some preantral follicles and a few small-antral follicles are present. The development of larger antral follicles requires specific hormonal stimulation, which occurs during the ovarian cycle.

The plasma levels of FSH, LH, estrogens, and progesterone during a typical ovarian cycle are illustrated in Figure 28.24. These hormones interact with one another during the cycle and influence the events of the cycle.

1. At the beginning of an ovarian cycle, an increased release of Gn-RH leads to an increased secretion of FSH and LH. FSH and LH, in turn, stimulate some follicles to begin development into larger antral follicles.
2. During the first week or so of the cycle, the follicles enlarge, and the secretion of estrogens gradually increases (recall that follicle cells are a primary source of estrogens). Both LH and FSH contribute to follicle growth and estrogen production. LH stimulates thecal cells to proliferate and produce androgens. FSH stimulates granulosa cells to proliferate and convert the androgens into estrogens. (Estrogens themselves may influence granulosa cells during follicle development.)
3. At relatively low levels such as are present during the first 11 days or so of the ovarian cycle, estrogens act in negative-feedback fashion on the pituitary to limit the release of FSH and LH in response to Gn-RH. Estrogens may also act on the hypothalamus to limit the release of Gn-RH. Moreover, the hormone inhibin, which inhibits the release of FSH, is produced by granulosa cells. This hormone contributes to the decline in the level of FSH that occurs during the second week of the cycle.
4. As the level of FSH declines, all but one of the enlarging follicles regress. The one follicle that continues to develop produces large amounts of estrogens, and toward the end of the second week of the cycle, the level of estrogens reaches its peak.
5. The high level of estrogens present near the end of the second week of the cycle has a different effect on the pituitary and hypothalamus than the relatively low levels present earlier in the cycle. The high level of estrogens enhances the sensitivity of the LH-releasing cells of the pituitary to Gn-RH. It may also stimulate the release of Gn-RH from the hypothalamus. As a consequence, there is a sharp increase in LH secretion called the LH surge (and a lesser increase in FSH secretion).
6. The LH surge leads to the final maturation of the follicle, the rupture of the follicle, and ovulation. Under the influence of LH, the primary oocyte completes the first meiotic division, the follicular antrum increases in size, and blood flow to the follicle increases. Granulosa

◆ **FIGURE 28.24 The plasma levels of FSH, LH, estrogens, and progesterone during a typical ovarian cycle**

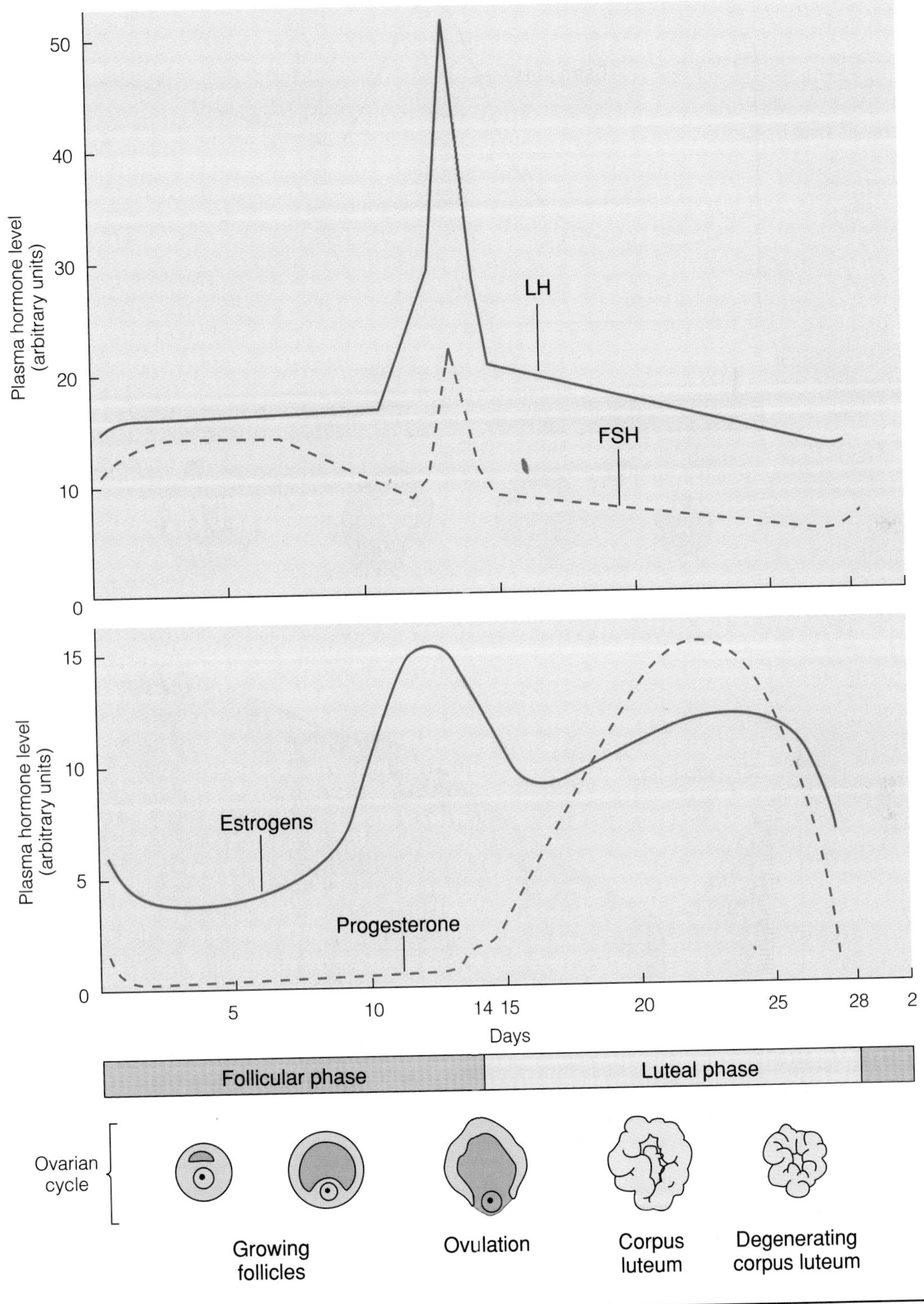

cells reduce their secretion of estrogens and begin secreting a small amount of progesterone. Enzymes and prostaglandins are produced, and these substances contribute to the rupture of the follicle wall and ovulation.

7. The LH surge also initiates the series of events that result in the formation of the corpus luteum following ovulation. The corpus luteum secretes substantial quantities of estrogens and progesterone. In the presence of

◆ **FIGURE 28.25 The menstrual cycle, illustrating cyclic changes in the endometrium of the uterus**

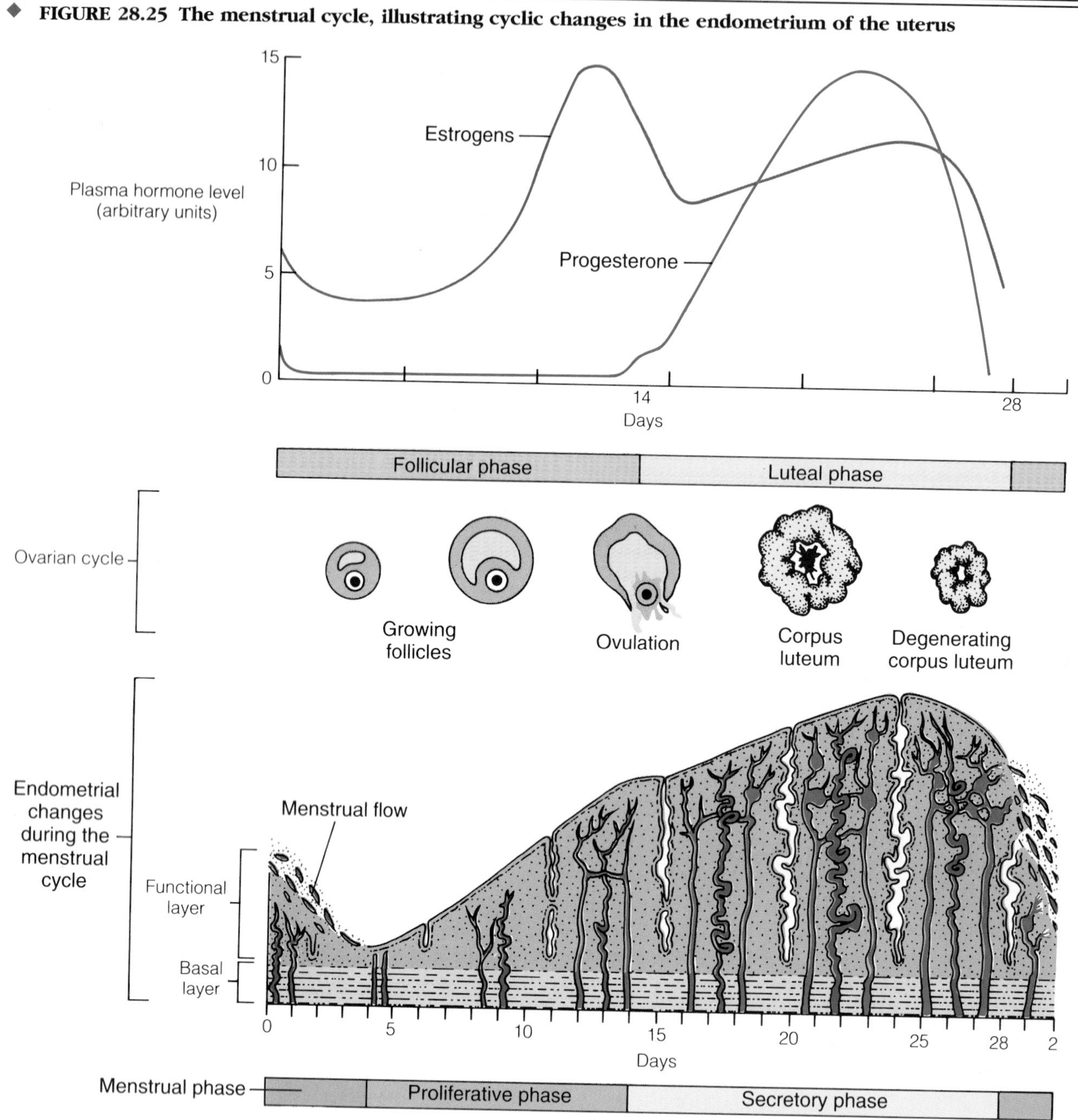

estrogens, the high level of progesterone acts on the hypothalamus in a negative-feedback fashion to inhibit the release of Gn-RH. As a result, FSH and LH decline to very low levels during the luteal phase of the ovarian cycle. Nevertheless, LH continues to stimulate the corpus luteum during this phase.

8. If the secondary oocyte released at ovulation is not fertilized and pregnancy does not occur, the corpus luteum begins to degenerate about ten days after ovulation. As it degenerates, the levels of estrogens and progesterone fall, removing the inhibition of Gn-RH secretion. As more Gn-RH is released, the levels of FSH and LH increase, other follicles are stimulated to begin development into larger antral follicles, and another ovarian cycle begins.

The Menstrual Cycle

The **menstrual cycle** *(uterine cycle)* consists of a series of changes that occur in the uterus (and to a lesser degree in the vagina). It is closely associated with the ovarian cycle, and estrogens and progesterone produced during the ovarian cycle control the events of the menstrual cycle (Figure 28.25).

ASPECTS OF EXERCISE PHYSIOLOGY

Menstrual Cycle Irregularity in Athletes

Many women experience changes in their reproductive cycles as a result of athletic participation. These changes, which are referred to as *athletic menstrual cycle irregularity (AMI),* can vary in severity from amenorrhea (absence of menstrual periods) to oligomenorrhea (periods at infrequent intervals) to reproductive cycles that are normal in length but are anovulatory (no ovulation) or that have a short or inadequate luteal phase.

In early research studies using surveys and questionnaires to determine the prevalence of the problem, the frequency of sport-related menstrual cycle disorders varied from 2% to 51%. In contrast, the rate of occurrence of menstrual cycle disorders in females of reproductive age in the general population is 2% to 5%. A major problem in determining the frequency of menstrual cycle irregularity by survey is the questionable accuracy of recall of menstrual periods. Furthermore, without blood tests to determine hormone levels throughout the reproductive cycle, a woman would not know whether she was anovulatory or had a shortened luteal phase. Studies in which hormone levels have been determined throughout the cycle have demonstrated that seemingly normal cycles in athletes frequently have a short luteal phase (less than ten days long with low progesterone levels).

In a study conducted to determine whether strenuous exercise spanning two menstrual cycles would induce disorders, 28 initially untrained college women with documented ovulation and luteal adequacy served as subjects. The women participated in an eight-week exercise program in which they initially ran four miles per day and progressed to ten miles per day by the fifth week. They were expected to participate daily in three and one-half hours of moderate-intensity sports. Only four women had normal cycles during the training. Abnormalities that occurred as a result of training included abnormal bleeding, delayed menstrual periods, abnormal luteal function, and loss of LH surge. All women returned to normal cycles within six months after training. The results of this study suggest that the frequency of AMI with strenuous exercise may be much greater than indicated by questionnaires alone. In other studies using low-intensity exercise regimens, AMI was much less frequent.

The mechanisms of AMI are unknown at present, although studies

continued on next page

During the follicular phase of the ovarian cycle, the level of estrogens increases. The estrogens stimulate the functional layer of the endometrium of the uterus, causing it to proliferate. Straight tubular glands form, and blood vessels invade the new endometrial epithelium. Thus, the thickness and vascularity of the functional layer of the endometrium increase. The estrogens also promote growth of the myometrium of the uterus, and they induce endometrial cells to synthesize receptors for progesterone.

Following ovulation and the formation of the corpus luteum, both estrogens and progesterone are produced. Progesterone acts on the estrogen-primed endometrium to stimulate its continued development. Under the influence of progesterone, the endometrial glands continue to grow and begin to secrete a glycogen-rich fluid. In addition, arteries within the endometrium become enlarged and spiraled. Progesterone also inhibits contractions of the uterine muscle. In this condition, the uterus is prepared to receive the embryo should fertilization occur.

If fertilization and pregnancy do not occur, the corpus luteum degenerates, and the levels of estrogens and progesterone decline. As the levels of these hormones decrease, their influence on the uterus diminishes, and prostaglandins are released from the endometrium. The prostaglandins cause blood vessels of the endometrium to undergo prolonged constriction, which reduces blood flow to the area of endometrium supplied by the vessels. The resulting lack of blood causes tissues of the affected region to degenerate. The prostaglandins also stimulate mild, rhythmic contractions of the uterine muscle.

After some time, the blood vessels dilate, allowing blood to flow through them again. However, capillaries in the area have become so weakened that blood leaks through them. This blood and the deteriorating endometrial tissue are discharged from the uterus as the menstrual flow. As a new ovarian cycle begins and the levels of estrogens rise, the functional layer of the endo-

ASPECTS OF EXERCISE PHYSIOLOGY

have implicated rapid weight loss, decreased percentage of body fat, dietary insufficiencies, prior menstrual dysfunction, stress, age at onset of training, and the intensity of training as factors that play a role.

Epidemiologists have shown that menarche (the first menstrual period) is delayed in girls who participate in vigorous sports before menarche. On the average, athletes have their first period when they are about three years older than their nonathletic counterparts. Furthermore, females who participate in sports before menarche seem to have a higher frequency of AMI throughout their athletic careers than those who begin to train after menarche.

Hormonal changes that have been found in female athletes include (1) severely depressed FSH levels, (2) elevated LH levels, (3) low progesterone during the luteal phase, (4) low estrogen levels in the follicular phase, and (5) an FSH/LH environment totally unbalanced as compared to age-matched nonathletic women. The preponderance of evidence indicates that cycles return to normal once vigorous training is stopped.

The major problem associated with athletic amenorrhea is a reduction in bone-mineral density. Studies have shown that the mineral density in the vertebrae of the lower spine of those with athletic amenorrhea is lower than in athletes with normal menstrual cycles and lower than in age-matched nonathletes. However, amenorrheic runners have higher bone-mineral density than amenorrheic nonathletes, presumably because the mechanical stimulus of exercise helps to retard bone loss. Studies have shown that amenorrheic athletes are at higher risk for stress-related fractures than athletes with normal menstrual cycles. One study, for example, found stress fractures in 6 of 11 amenorrheic runners but in only 1 of 6 runners with normal menstrual cycles. The mechanism for bone loss is probably the same as is found in postmenopausal osteoporosis—lack of estrogens (see p. 195). The problem is serious enough that an amenorrheic athlete should discuss the possibility of estrogen replacement therapy with her physician.

On the positive side, an epidemiologic study to determine if the long-term reproductive and general health of women who had been college athletes differed from that of college nonathletes showed that former athletes had less than half the lifetime occurrence rate of cancers of the reproductive system and half the breast cancer occurrence compared to nonathletes. Because these are hormone-sensitive cancers, the delayed menarche and lower estrogen levels found in women athletes may play a key role in decreasing the risk of cancer of the reproductive system and breast.

metrium undergoes repair and once again begins to proliferate.

The menstrual cycle is commonly divided into three phases: (1) the **menstrual phase,** (2) the **proliferative phase,** and (3) the **secretory phase.** The menstrual phase is the period during which menstrual bleeding occurs. It lasts about 3 to 5 days in a typical 28-day cycle, and the first day of menstrual bleeding is considered to be the first day of the cycle. The proliferative phase follows the menstrual phase and lasts about 10 days until ovulation occurs. The secretory phase follows ovulation and lasts about 14 days.

The menstrual phase basically corresponds with the initial portion of the follicular phase of the ovarian cycle when the level of estrogens is low. The proliferative phase corresponds with the remaining portion of the follicular phase when the level of estrogens progressively increases. The secretory phase essentially corresponds with the luteal phase of the ovarian cycle when both estrogens and progesterone are produced.

Female Sexual Responses

The responses of the female to sexual stimulation, like those of the male, are governed both by psychological stimuli and by tactile stimulation of the genital organs and other areas of the body. The neural pathways over which sexual stimuli are carried in the female are basically the same as in the male. The motor impulses causing the various responses to sexual stimulation are carried over autonomic (that is, sympathetic and parasympathetic) nerves.

During sexual stimulation, the clitoris becomes engorged with blood in a manner similar to that of the penis, while eliciting widespread sexual sensations. Erectile tissue in the area of the vaginal opening also becomes engorged with blood. This narrows the vaginal orifice and facilitates the stimulation of the penis during intercourse. In addition, the epithelial lining of the vagina becomes lubricated through the copious secretion of mucus by glands located around the cervix of the

CONDITIONS OF CLINICAL SIGNIFICANCE

The Reproductive System

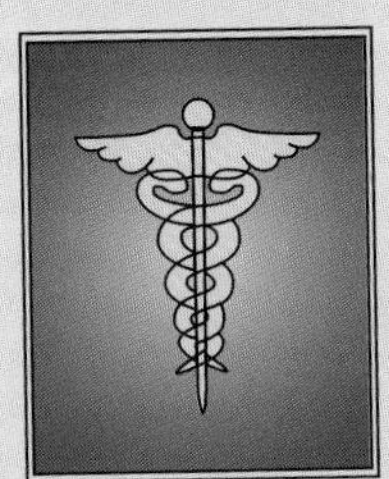

Sexually Transmitted Diseases

The most prevalent diseases of the reproductive systems of both males and females are those that are spread through sexual contact. In the past, these have been referred to as venereal diseases, but they are now more commonly called *sexually transmitted diseases.*

Gonorrhea

Gonorrhea (gon″-ō-rē′-ah) is an inflammation of the mucous membranes of the urogenital tract or rectum caused by the bacterium *Neisseria gonorrhoeae.* This inflammation usually results in a discharge of pus from the urethra and painful urination. In females, the cervix of the uterus and the uterine tubes can also become infected and inflamed, and *N. gonorrhoeae* is a frequent cause of pelvic inflammatory disease (see page 919). However, it is not unusual for a woman to have gonorrhea and transmit it through sexual contacts, yet have no apparent symptoms herself. In males, the spread of the gonococcus within the reproductive tract can lead to inflammation of the prostate, the seminal vesicles, and the epididymis.

Syphilis

Syphilis (sif′-ĭ-lis) is a sexually transmitted disease caused by the bacterium *Treponema pallidum.* At the site where it enters the body, the bacterium usually produces a lesion called a *chancre (shang′-ker).* Most commonly, the chancre is on the penis or in the vagina. The chancre soon heals, and there may be no other symptoms for several weeks. However, during that time the infection spreads throughout the body by way of the bloodstream. After about six weeks, a skin rash accompanied by fever and aching joints develops. These secondary symptoms then disappear, and the disease enters a latent (inactive) period that can last for many years. During this latent period, the body may develop an immunity to the bacterium and destroy it, or the bacterium may spread to many different sites, including the nervous and vascular systems as well as various organs, causing damage to these structures. This latter occurrence produces severe and varied symptoms, depending on the structures affected. Syphilis can be detected by several different blood tests, one of which is called the Wassermann test.

Genital Herpes

Genital herpes (her′-pēz) is an increasingly common sexually transmitted disease. It is caused by the herpes simplex virus Type 2, which produces painful blisters on the reproductive organs. After some time, the lesions heal, but they can recur periodically for years. The virus can also cause serious malformations of children born to infected mothers, and many researchers believe it is implicated in the occurrence of cervical cancer.

Chlamydia

Chlamydia (klah-mid′-ē-ah) is a sexually transmitted disease that has been recognized only since 1974. Prior to this time, any inflammation of the urethra that was not caused by gonorrhea was called nongonococcal urethritis. In 1974 it was determined that

continued on next page

uterus and the area of the vaginal opening. The lubrication provides for easy entrance of the penis into the vagina and facilitates rhythmic massaging stimulation of the female external genital organs as well as of the penis.

When sexual stimulation reaches sufficient intensity, the female undergoes an orgasm. The physiological changes that occur during an orgasm in the female are similar to those of the male. However, there is no ejaculation in the female. The question as to whether a female orgasm in some way facilitates fertilization is not yet adequately answered. Certainly, it is not necessary for a female to experience an orgasm for fertilization to occur.

CONDITIONS OF CLINICAL SIGNIFICANCE

most of these inflammations were caused by the bacterium *Chlamydia trachomatis,* and the condition is now more commonly referred to simply as chlamydia. Despite the fact that chlamydia responds well to antibiotic therapy, it is now the most common sexually transmitted disease in the United States.

In males the symptoms of chlamydia are often similar to those of gonorrhea, including discharge from the penis and painful and frequent urination. Women may show no symptoms for extended periods of time, during which they can pass the infection to sexual partners. When symptoms are present, they include vaginal discharge, urethritis, and painful intercourse. The infection may spread from the urethra throughout the woman's reproductive tract. If a woman gives birth while infected with chlamydia, the child may have a chronic infection of the eyes.

Male Disorders

Prostate Conditions

Although disorders of the male reproductive system are not restricted to the prostate gland, this structure is frequently affected. *Prostatitis* is an inflammation of the prostate that is often due to a bacterial infection. In this condition, the gland is swollen and tender, and in severe cases, abscesses can form. The prostate gland is also a common site of cancerous tumors, and an enlargement of the gland can compress the urethra, making urination difficult.

Impotence

Impotence is a fairly common male disorder in which a man is unable to attain an erection of the penis or to retain an erection long enough to complete sexual intercourse. Impotence can result from physical disorders of the vascular or nervous systems, and it is often a consequence of psychological or emotional problems.

Infertility

Male *infertility* is the inability of the male to fertilize the ovum. It is caused by either inadequate production of spermatozoa, the production of abnormal or nonmotile spermatozoa, or an obstruction that prevents the delivery of spermatozoa from the testes to the female vagina. The production of normal sperm can be interfered with by exposure to X rays, malnutrition, and certain diseases—including mumps.

Female Disorders

Abnormal Menstruation

Abnormalities of the menstrual cycle, which are among the most common disorders of the female reproductive system, can result from infections of the reproductive organs or malfunctions of the ovaries or the pituitary gland. Moreover, emotional or psychological factors are often involved.

Amenorrhea—the complete absence of menstrual periods—can be caused by disorders of the ovaries, the pituitary, or the hypothalamus. *Dysmenorrhea,* or painful menstruation, is usually associated with the occurrence of uterine cramps (that is, strong contractions of the uterine muscle). The cramps are apparently due to an overproduction of prostaglandins by the uterine endometrium. Antiprostaglandin drugs are often effective in treating dysmenorrhea, and the condition frequently disappears following the first pregnancy.

Endometriosis

Some cases of severe dysmenorrhea and pelvic pain are caused by a condition called *endometriosis (en″-dō-mē-tre-ō′-sis).* In this condition endometrial tissue develops in abnormal locations, often outside the uterus. The most common site for endometriosis is in the ovaries, but aberrant endometrial tissue has also been reported in the uterine ligaments, the pelvic peritoneum, and occasionally in various other locations. The tissue is thought either to enter the pelvic cavity from the uterus by way of the uterine tubes during menstruation or else to originate in the pelvic cavity as the result of abnormal embryonic differentiation of the epithelial cells that line the cavity.

Endometriosis becomes clinically important during menstruation, when the aberrant endometrial tissue, like the endometrium of the uterus, apparently reacts to hormonal levels in the body and undergoes cyclic bleeding. Unlike the endometrial tissue in the uterus, blood from the aberrant endometrial tissue generally has no means by which it can drain to the outside of the body. As a consequence, the blood collects in the

CONDITIONS OF CLINICAL SIGNIFICANCE

aberrant tissue, causing pain and, in some cases, more serious complications. The condition is most common during a woman's reproductive life and declines in frequency after the age of 40.

Pelvic Inflammatory Disease

Pelvic inflammatory disease (PID) is an infection of the pelvic organs of the female reproductive tract, including the uterus and uterine tube. It is caused by bacteria—usually gonococci, streptococci, or staphylococci—that generally reach the pelvic structures by passing through the vagina to the uterus and the uterine tubes. In some cases, bacteria from distant sites of infection reach the pelvic structures through the blood.

Tumors

Tumors that develop within the female reproductive system can be either malignant (cancerous) or benign (noncancerous), and malignant tumors can metatasize—that is, spread to other locations within the body.

Ovarian tumors can be solid, or they can occur as cysts (hollow sacs) that frequently contain fluid. Ovarian cysts are usually noncancerous, but a high percentage of solid ovarian tumors are malignant.

Tumors of the uterus and particularly the cervix are quite common. Fortunately, these tumors can be diagnosed early in their development by means of a simple test called a Pap smear. This test involves the removal (by a swab) of cells from the cervix and the surrounding area. The cells are then examined microscopically for signs of malignancy. With this procedure, cancerous cells can often be detected before any symptoms have appeared. If identified early, uterine tumors can be removed surgically or destroyed with radiation treatments. Since many women now have regular gynecological examinations that include Pap smears, the number of deaths due to uterine cancer has declined steadily.

Tumors are also common in the breasts, particularly after age 30. Breast tumors, whether malignant or benign, can be best detected by regular manual self-examination, plus periodic x rays of the breasts *(mammography)*. Breast tumors are often removed surgically, and they are also treated by radiotherapy and chemotherapy.

Effects of Aging

Males

In males, the secretion of testosterone undergoes a slight, but gradual, age-related decline, which may cause a reduction in muscle strength, contribute to decreased sexual desire, and affect sperm production. However, even in quite elderly men, abundant viable sperm may be produced. The prostate gland begins to atrophy by about 50 years of age, and the volume of its secretion contributed to semen and the force behind its contraction during orgasm are both reduced. The penis undergoes some atrophy with age, including changes in the walls of the penile blood vessels that may affect an older man's ability to attain an erection.

Females

In the reproductive tracts of women, age-related changes occur first in the ovaries and result in reduced secretion of estrogens and progesterone by them. The declining levels of these hormones, in turn, cause degenerative changes in the uterus, vagina, and external genitalia.

At about age 50, the ovarian and menstrual cycles gradually become irregular. Ovulation fails to occur during many of the irregular cycles, and in most women the cycles cease altogether over the next several months or, at most, within a few years. The cessation of the menstrual cycle is referred to as **menopause,** and the entire period of decline is called the **female climacteric** *(klī-mak´-ter-ik).*

The female climacteric is thought to be caused by an inability of the ovaries to respond to hormonal signals, most probably due to a shortage of follicles that results from their ovulation or degeneration during the reproductive years. As a consequence, the production of estrogens and progesterone by the ovaries is quite low, and estrogen-dependent tissues such as the genital organs and breasts gradually atrophy.

During the climacteric, the production of FSH and LH by the pituitary is quite high, apparently due to a partial release of the Gn-RH/FSH/LH system from the negative-feedback influences of ovarian hormones.

continued on next page

CONDITIONS OF CLINICAL SIGNIFICANCE

Because of changes in hormonal levels, some women experience such symptoms as sensations of warmth accompanied by sweating ("hot flashes"), irritability, fatigue, and anxiety.

After menopause, glands that normally lubricate the vagina gradually atrophy. As a consequence, the vagina becomes dry, and sexual intercourse may be painful.

Study Outline

◆ EMBRYONIC DEVELOPMENT OF THE REPRODUCTIVE SYSTEM pp. 886–890

Early Development. Genital ridges, paramesonephric ducts, and mesonephric ducts.

Development of the Internal Reproductive Structures.

MALE EMBRYO.

1. Medulla of gonad gives rise to seminiferous tubules.
2. Mesonephric ducts form efferent ductules, epididymis, and ductus deferens; paramesonephric ducts degenerate.

DESCENT OF THE TESTES. Testes follow the vaginal processes through the inguinal canals into the scrotum.

FEMALE EMBRYO.

1. Cortex of gonad becomes site of ova production.
2. Mesonephric ducts degenerate; paramesonephric ducts form uterus, vagina, and uterine tubes.

Development of the External Reproductive Structures.

1. Genital tubercle forms penis or clitoris.
2. Urethral folds fuse in male but not in female.
3. Labioscrotal folds form scrotum in male, labia majora in female.

◆ ANATOMY OF THE MALE REPRODUCTIVE SYSTEM pp. 890–896

The Male Perineum. Divided into anterior urogenital triangle and posterior anal triangle.

Testes and Scrotum.

SEMINIFEROUS TUBULES. Contain cells that give rise to spermatozoa.

INTERSTITIAL ENDOCRINOCYTES. Secrete testosterone.

Epididymis. First portion of duct system; storage of mature nonmotile sperm.

Ductus Deferens. Continuation of duct of epididymis; heavy wall of smooth muscle; located in spermatic cord.

Seminal Vesicles. Two membranous pouches lateral to ductus deferens; secrete viscid alkaline fluid that contributes to semen.

Prostate Gland. Encompasses urethra below bladder; secretes fluid that contributes to semen.

Bulbourethral Glands. Pair of glands below prostate; secretion is a thick, alkaline mucus.

Penis. Copulatory organ; contains three cylindrical cavernous bodies; deposits spermatozoa in female reproductive tract.

Semen. Mixture of spermatozoa from testes and fluids from seminal vesicles, prostate, and bulbourethral glands; alkaline; fructose provides energy source for ejaculated spermatozoa.

◆ PUBERTY IN THE MALE p. 896

1. The time of life at which reproduction becomes possible.
2. Events of puberty (for example, enlargement of penis, scrotum, and testes; characteristic hair growth, voice changes) result from increased testosterone production by testes.
3. At puberty, alteration in brain function leads to increased release of Gn-RH, which ultimately results in increased testosterone production.

◆ SPERMATOGENESIS pp. 896–897

1. Occurs within seminiferous tubules of testes.
2. Some spermatogonia (diploid) give rise to primary spermatocytes (diploid).
3. Primary spermatocytes undergo first meiotic division, forming secondary spermatocytes (haploid).
4. Secondary spermatocytes undergo second meiotic division, producing spermatids.
5. Spermatids develop into spermatozoa by a process called spermiogenesis.

◆ THE SPERMATOZOON p. 898

Flagellated; contains acrosome, mitochondria, centrioles, and little cytoplasm.

◆ HORMONAL CONTROL OF TESTICULAR FUNCTION pp. 898–900

1. FSH and LH release from pituitary is stimulated by Gn-RH from hypothalamus.
2. Testosterone inhibits Gn-RH release and reduces responsiveness of LH-secreting cells to Gn-RH.
3. FSH release believed to be inhibited by inhibin from sustentacular cells.

◆ MALE SEXUAL RESPONSES pp. 900–902

Erection. Penis becomes enlarged and firm.

1. Results from vascular engorgement of spaces within corpora cavernosa and corpus spongiosum of penis.
2. Can be caused by physical stimulation or by visual or psychic stimuli.

Ejaculation. Results in forceful expulsion of semen from penis. Semen temporarily clots in vagina.

◆ ANATOMY OF THE FEMALE REPRODUCTIVE SYSTEM pp. 902–907

Ovaries. Production of ova; estrogen and progesterone production; cortex contains primordial follicles.

Uterine Tubes. Carry ovum to uterus.

1. Fimbriae surround opening.
2. Thick mucosa of simple columnar cells; some ciliated.
3. Cilia and peristaltic contractions carry ovum to uterus.

Uterus. Composed of body, isthmus, and cervix.

1. Supported by central tendon of pelvic and urogenital diaphragms.
2. Wall composed of perimetrium, myometrium, and endometrium.

Vagina. Canal from cervix to exterior of body; mucosa of stratified squamous epithelium; entrance partially covered by hymen.

Female External Genital Organs. (Vulva, or pudendum)

1. Labia majora and minora.
2. Greater and lesser vestibular glands lubricate vestibule.
3. Clitoris homologous to male penis.

The Female Perineum. Divided into anterior urogenital triangle and posterior anal triangle. *Clinical perineum* is located between vagina and anus.

Mammary Glands.

1. Nipple and areola pigmented.
2. Composed of adipose tissue and lobes of compound tubuloalveolar glands.
3. Ampullae serve as milk reservoirs.

◆ PUBERTY IN THE FEMALE p. 907

1. Events of puberty (such as enlargement of vagina, uterus, and uterine tubes; deposition of fat in breasts and hips) largely a result of increased production of estrogens by ovaries.
2. At puberty, alteration in brain function leads to increased release of Gn-RH, which ultimately results in increased production of estrogens.

◆ OOGENESIS p. 908

1. Oogonia (diploid) develop into primary oocytes (diploid).
2. Primary oocytes undergo first meiotic division, forming secondary oocytes (haploid) and polar bodies.
3. Secondary oocyte begins second meiotic division and completes it, if fertilized, to become a mature ovum.
4. Beginning at puberty, one secondary oocyte is usually released at ovulation each month.

◆ THE OVARIAN CYCLE pp. 909–911

Follicle Maturation.

1. Some primordial follicles become primary follicles, which become growing follicles that undergo further development.
2. A zona pellucida forms within growing follicle, as does a fluid-filled cavity called the antrum.

Ovulation.

1. Mature follicle ruptures and releases its secondary oocyte.
2. Ovulated secondary oocyte is surrounded by zona pellucida and sphere of follicle cells called corona radiata.

Formation of the Corpus Luteum.

1. Cells of ruptured mature follicle increase in size and take on yellowish color, forming corpus luteum.
2. If secondary oocyte is not fertilized and pregnancy does not occur, corpus luteum begins to degenerate about 10 days after ovulation.
3. If secondary oocyte is fertilized and pregnancy does occur, corpus luteum does not degenerate but remains until near end of pregnancy.

◆ HORMONAL CONTROL OF OVARIAN FUNCTION pp. 912–914

1. At beginning of ovarian cycle, increased secretion of FSH and LH stimulates some follicles to begin development into larger antral follicles.
2. As follicle development proceeds, secretion of estrogens gradually increases.
3. At relatively low levels, estrogens act on pituitary to limit FSH and LH release and may act on hypothalamus to inhibit release of Gn-RH.
4. High levels of estrogens enhance sensitivity of LH-releasing cells of the pituitary to Gn-RH and may stimulate Gn-RH release. Result is LH surge, which leads to final maturation of follicle, rupture of follicle, and ovulation.
5. In presence of estrogens, high level of progesterone secreted by corpus luteum inhibits Gn-RH release.

6. As corpus luteum degenerates, levels of estrogens and progesterone fall and levels of FSH and LH increase, stimulating other follicles to begin development into larger antral follicles, and another cycle begins.

◆ THE MENSTRUAL CYCLE pp. 914–916

1. Estrogens produced during the follicular phase of ovarian cycle stimulate functional layer of endometrium of uterus to proliferate. Thickness and vascularity of the layer increase.
2. Following ovulation, progesterone from corpus luteum acts on estrogen-primed endometrium to stimulate its continued development.
3. As corpus luteum degenerates and levels of estrogens and progesterone fall, the functional layer of endometrium degenerates and menstruation occurs.

◆ FEMALE SEXUAL RESPONSES pp. 916–917

1. Clitoris becomes engorged with blood.
2. Epithelial lining of vagina becomes lubricated.
3. Erectile tissue in the area of vaginal opening becomes engorged with blood.

◆ CONDITIONS OF CLINICAL SIGNIFICANCE: THE REPRODUCTIVE SYSTEM pp. 917–920

Sexually Transmitted Diseases.

GONORRHEA. Inflammation of mucous membranes of genital tract or rectum caused by bacterium *Neisseria gonorrhoeae.*

SYPHILIS. Infection caused by bacterium *Treponema pallidum.*

GENITAL HERPES. Infection caused by herpes simplex virus Type 2.

CHLAMYDIA. Urethritis caused by bacterium *Chlamydia trachomatis.* Infection may spread throughout a woman's reproductive tract.

Male Disorders.

PROSTATE CONDITIONS. Such as bacterial inflammation (prostatitis) and tumors.

IMPOTENCE. Inability to attain erection or maintain it long enough to accomplish sexual intercourse.

INFERTILITY. Inability to fertilize ovum.

Female Disorders.

ABNORMAL MENSTRUATION. May be due to infections or glandular malfunctions, as well as emotional and psychological factors.

ENDOMETRIOSIS. Endometrial tissue in aberrant locations.

PELVIC INFLAMMATORY DISEASE. Inflammation that can involve uterine tubes, ovaries, or peritoneum of abdominopelvic cavity.

TUMORS. Can occur in ovaries, uterus, or breasts.

Effects of Aging.

MALES. Decline in secretion of testosterone; atrophy of prostate gland and penis.

FEMALES. Reduced secretion of estrogens and progesterone by ovaries causes degenerative changes in uterus, vagina, and external genitalia. Cessation of menstrual cycles referred to as menopause.

Self-Quiz

1. In the male embryo, which of these structures degenerate shortly after the gonads begin to differentiate into testes? (a) the paramesonephric ducts; (b) the seminiferous tubules; (c) the mesonephric ducts.
2. In the absence of the male hormones, all embryos, regardless of their genetic makeup, will develop into females. True or False?
3. Spermatogenesis occurs in the: (a) tunica albuginea; (b) seminiferous tubules; (c) sustentacular cells; (c) prostate gland.
4. The interstitial endocrinocyte cells produce: (a) sperm; (b) testosterone; (c) luteinizing hormone.
5. Occasionally, one or both of the testes fail to descend out of the abdominopelvic cavity, a condition called cryptorchidism. True or False?
6. Which structure helps regulate the temperature of the testes? (a) gubernaculum; (b) cremaster muscle; (c) epididymis.

7. Match the structures of the male reproductive system with the appropriate lettered description:

Epididymis	(a) Structures that penetrate the prostate gland and enter the urethra just below its exit from the bladder
Ductus deferens	(b) Encompasses the urethra just below the bladder
Seminal vesicles	(c) Structure that transports sperm from the epididymis to the urethra
Ejaculatory ducts	(d) The first portion of the duct system that transports sperm from the testes to the outside of the body
Prostate gland	(e) Located below the prostate on either side of the membranous portion of the urethra
Bulbourethral glands	(f) Structures that join with the ductus deferens to form the ejaculatory ducts

8. The acidity of the semen counteracts the alkaline pH of the vagina. True or False?
9. In the male, prior to puberty: (a) the testes continually produce large amounts of testosterone; (b) there is little release of follicle-stimulating hormone and luteinizing hormone from the pituitary; (c) the hypothalamus secretes large amounts of gonadotropin-releasing hormone.
10. Spermatids: (a) are haploid cells; (b) develop within the bulbourethral glands; (c) divide mitotically, giving rise to spermatogonia.
11. In the male, testosterone: (a) strongly stimulates FSH release; (b) is required to maintain spermatogenesis beyond puberty; (c) is not involved in sperm production.
12. The ligament that supports the ovary from the posterior surface of the broad ligament is: (a) mesosalpinx; (b) mesovarium; (c) round ligament; (d) ovarian ligament.
13. Oocytes develop in the: (a) uterus; (b) corpus luteum; (c) follicles; (d) cervix.
14. The vagina is lined with stratified squamous epithelium. True or False?
15. Collectively, the external genital organs of the female are known as: (a) the mons pubis; (b) the labia majora; (c) the vulva; (d) the pudendum; (e) both (c) and (d).
16. The structure in the female that is homologous to the dorsal portion of the male penis is the: (a) hymen; (b) mons pubis; (c) clitoris; (d) labia majora.
17. The corpus luteum secretes: (a) luteinizing hormone; (b) follicle-stimulating hormone; (c) progesterone.
18. During the ovarian cycle, a high level of luteinizing hormone leads to the final maturation of the follicle, the rupture of the follicle, and ovulation. True or False?
19. During the first portion of the ovarian cycle, the functional layer of the endometrium of the uterus is stimulated to proliferate by: (a) progesterone; (b) estrogens; (c) endometriotropin.
20. During the female climacteric, the level of: (a) estrogens is relatively high; (b) FSH is relatively high; (c) progesterone is relatively high.

CHAPTER 29

Pregnancy, Embryonic Development, and Inheritance

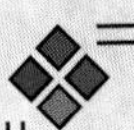

CHAPTER CONTENTS

LEARNING OBJECTIVES

After completing this chapter, you should be able to:

1. Describe the process of fertilization.
2. Describe the implantation of the blastocyst.
3. Describe the formation and function of the placenta, and indicate the relationship between maternal and fetal blood in the placenta.
4. Describe how each embryonic membrane is formed and state its function.
5. Describe the process of parturition.
6. Discuss the development of the breasts.
7. Discuss the hormonal activities involved in lactation.
8. Describe the unique features of fetal circulation.
9. Describe the circulatory changes that occur at birth.
10. Discuss how inherited traits are passed from one generation to the next.

CHAPTER 29

Two of the main functions of the reproductive system are to provide a means by which fertilization can occur and to furnish an environment that is conducive to the development of the fertilized ovum *(zygote)* to a stage at which the baby is capable of surviving on its own outside the mother's body. During the time that the baby is developing within a woman, the woman is said to be **pregnant.** This chapter describes how fertilization occurs and discusses the major changes that occur in a woman during pregnancy. It also examines embryonic development and basic aspects of human genetics.

Pregnancy

For pregnancy to occur, the secondary oocyte released at ovulation must be fertilized by a spermatozoon. The secondary oocyte remains capable of being fertilized for about 12–24 hours following ovulation. During this time, it is usually moving along the upper portion of a uterine tube. Consequently, fertilization normally takes place in this location. If the secondary oocyte is not fertilized, it begins to disintegrate and is phagocytized by cells that line the tube.

Sperm remain viable within the female reproductive tract for about 24–72 hours after ejaculation. Therefore, for fertilization to occur, sperm should be deposited within the female tract no earlier than about 72 hours before ovulation and no later than about 24 hours after ovulation.

Transport of the Secondary Oocyte

The secondary oocyte released at ovulation is swept into a uterine tube by movements of the fimbriae at the opening of the tube and by the beating of cilia that line the fimbriae. Initially, contractions of smooth muscle of the tube and ciliary action propel the secondary oocyte rapidly along the tube. The muscle contractions soon diminish, but ciliary action continues to move the secondary oocyte slowly along the tube.

Transport of Sperm

The opening of the uterus into the vagina contains mucus secreted by the cervix. During most of the ovarian cycle, the cervical mucus is quite viscous and impenetrable by sperm. However, under the influence of the high level of estrogens present near the time of ovulation, the mucus becomes thin and watery, and sperm can penetrate it. During this period, sperm deposited within the vagina move into the uterus and reach the upper portion of the uterine tubes rather quickly, often requiring only minutes to do so.

The movement of sperm to the upper portion of the uterine tubes is too rapid to be due solely to the motility of the sperm. (On their own, sperm move at a rate of about 1–4 mm/min.) Therefore, researchers have suggested that the high level of estrogens present near the time of ovulation and prostaglandins present in semen assist sperm transport by stimulating contractions of smooth muscles of the uterus and uterine tubes.

The semen deposited in the vagina during sexual intercourse usually contains several hundred million sperm. However, sperm mortality is high, and only a few hundred reach the upper portion of a uterine tube. The acidity of the vagina causes many sperm to die very soon after being deposited in the vagina, and once in the uterus, many sperm are destroyed by phagocytic leukocytes. The high mortality rate of sperm is one reason why large numbers of sperm are required for pregnancy to occur. In fact, a male is considered to be clinically infertile if his semen contains fewer than 20 million sperm per milliliter.

Capacitation

Following ejaculation, sperm must remain within the female reproductive tract for several hours before they become capable of fertilizing the secondary oocyte. During this period, they undergo a process called **capacitation** *(kah-pas″-ĭ-tā′-shun).*

The mechanism of capacitation is not well understood. However, it apparently involves an alteration in the glycoprotein and lipid composition of the sperm plasma membrane and an increase in sperm metabolism and motility.

Fertilization

The secondary oocyte released at ovulation is surrounded by the corona radiata and zona pellucida. Capacitated sperm penetrate the corona radiata and bind to specific receptor sites at the outer surface of the zona pellucida.

The sperm undergo an **acrosomal reaction** during which they release enzymes from their acrosomes. The enzymes—for example, hyaluronidase and proteases—are essential for penetration of the zona pellucida.

The first sperm to penetrate the zona pellucida contacts the secondary oocyte. The membranes of the sperm and oocyte fuse, and the head of the sperm is drawn into the oocyte cytoplasm (Figure 29.1). The tail of the sperm is frequently lost in this process. The

◆ **FIGURE 29.1 Process of fertilization**

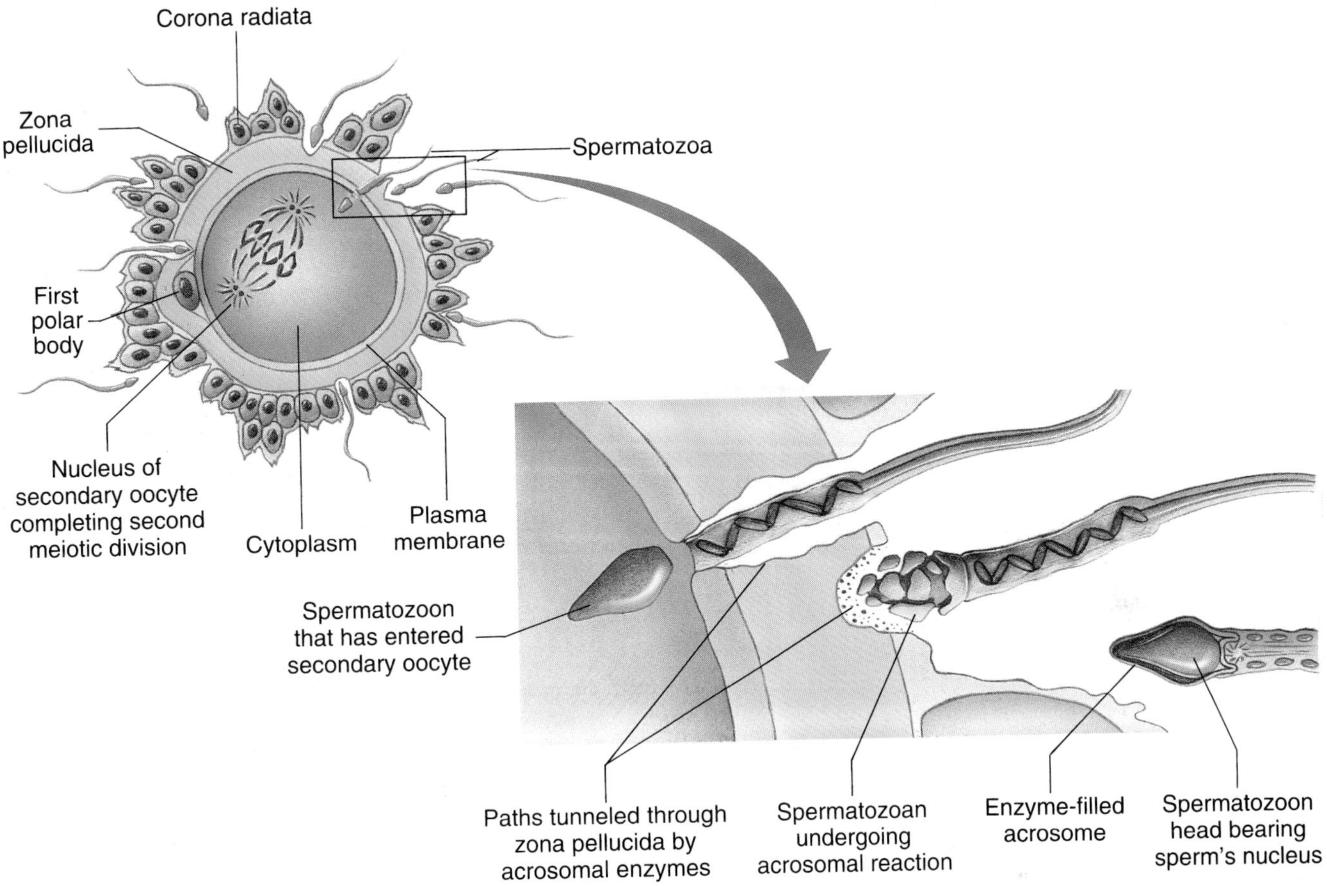

Corona radiata
Zona pellucida
Spermatozoa
First polar body
Nucleus of secondary oocyte completing second meiotic division
Cytoplasm
Plasma membrane
Spermatozoon that has entered secondary oocyte
Paths tunneled through zona pellucida by acrosomal enzymes
Spermatozoan undergoing acrosomal reaction
Enzyme-filled acrosome
Spermatozoon head bearing sperm's nucleus

secondary oocyte then completes the second meiotic division, giving rise to a mature ovum. The sperm nucleus and the ovum nucleus swell and, when swollen, are referred to as *pronuclei.* The membranes of the pronuclei rupture, and their chromosomes come together. Thus, fertilization is accomplished, and the fertilized ovum, or **zygote,** contains the full complement of 46 chromosomes.

Block to Polyspermy

The fusion of the membrane of the sperm that enters the secondary oocyte with the membrane of the oocyte brings about changes that prevent the entry of additional sperm—that is, that prevent polyspermy.

Secretory vesicles called *cortical granules* are located in the cytoplasm of the secondary oocyte, just beneath the plasma membrane. The fusion of the sperm and oocyte membranes triggers a *cortical reaction* in which cortical granules fuse with the oocyte membrane and release their contents by exocytosis. Substances released from the cortical granules prevent the entry of additional sperm in one or more of the following ways:

1. By altering the zona pellucida so that sperm cannot bind to it.
2. By altering the zona pellucida so that sperm cannot penetrate it.
3. By altering the secondary-oocyte membrane so that sperm cannot fuse with it.

Sex Determination

The genetic sex of an individual is determined at the time of fertilization, even though the individual's sex does not become apparent until about the eighth week of development. If a secondary oocyte is fertilized by a sperm containing an X sex chromosome, a female (XX) normally develops. If it is fertilized by a sperm containing a Y sex chromosome, a male (XY) normally develops.

Development and Implantation of the Blastocyst

The zygote, which immediately begins to divide, or undergo **cleavage** (Figures 29.2b and c), passes along the uterine tube to the uterus—a journey that requires about three to four days. Indeed, if the zygote were to reach the lower end of the uterine tube much sooner, it could not enter the uterus. Smooth muscle of this region of the uterine tube remains spastically contracted, constricting the tube, for about three days following ovulation. Then, when progesterone levels become high enough, the muscle relaxes and the uterine tube is no longer constricted.

By the time the uterus is reached, or shortly thereafter, enough cell divisions have occurred for the zygote to have developed into a fluid-filled sphere of cells called a **blastocyst** (Figure 29.2d). The outer layer of cells of the blastocyst is called the **trophoblast** *(trophectoderm)*. This layer is instrumental in enabling the blastocyst to become implanted in the endometrium. Once implantation occurs, the trophoblast will contribute to the formation of the placenta. The blastocyst contains an accumulation of cells called the **inner cell mass.** The embryo develops from the inner cell mass.

The blastocyst remains free within the lumen of the uterus for several days before it implants itself in the endometrium. During this period, it obtains nourishment from the uterine fluids and continues to undergo cell division. By this time, the zona pellucida has disintegrated, allowing the trophoblast to expand.

About seven days after ovulation, the blastocyst undergoes implantation in the endometrium (Figure 29.2e). By this time, the endometrium of the uterus has been prepared for implantation by estrogens and progesterone. The portion of the trophoblast near the inner cell mass contacts the endometrium and adheres to it. Trophoblast cells release enzymes that digest cells of the endometrium, thus eroding the endometrial surface. This erosion makes additional fluids and nutrients avail-

◆ **FIGURE 29.2 Fertilization, cleavage, and blastocyst formation and implantation**

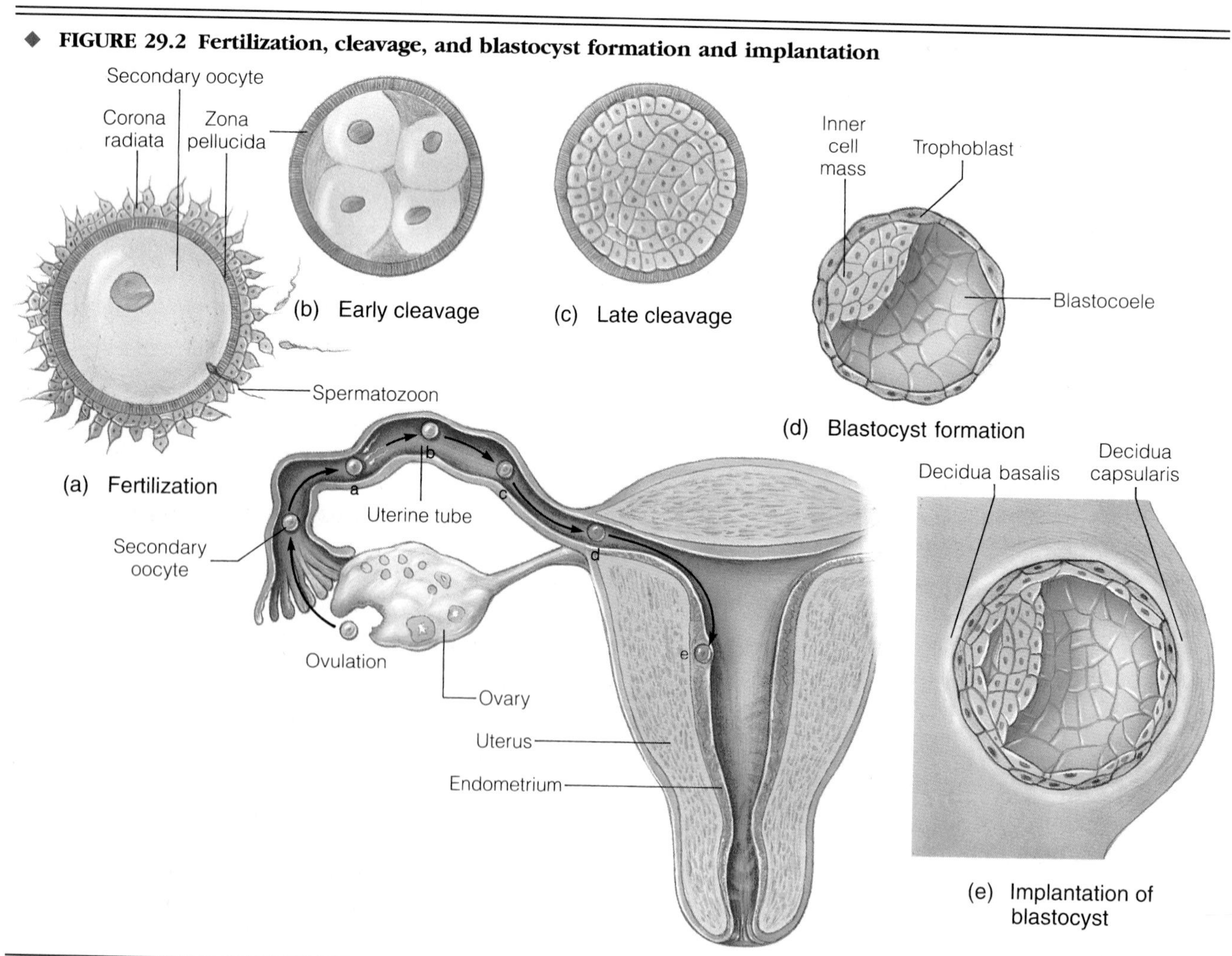

able to the blastocyst and allows the blastocyst to burrow into the endometrium. As the blastocyst erodes deeper, the endometrium grows over it, and within a few days the blastocyst is completely implanted within the endometrium.

For about seven days following implantation, the embryo obtains all of its nutrients from the destruction of endometrial cells by the trophoblast. Eventually, nutrients are obtained through the placenta, but erosion by the trophoblast continues to supply a significant fraction of the embryo's nutrients for the first two months of development, while the placenta is gradually enlarging and becoming increasingly functional.

Maintenance of the Endometrium

Because of the importance of the endometrium for the nourishment of the embryo, it is vital that the endometrium be maintained in a highly developed state for pregnancy to continue. Thus, the menstrual cycle must be interrupted, and menses must be prevented from occurring.

The endometrium is maintained in a highly developed state and menses is prevented from occurring by high levels of estrogens and progesterone that are present during pregnancy (Figure 29.3). The high levels of estrogens and progesterone also inhibit the secretion of Gn-RH, thus interrupting the ovarian cycle and preventing the further development of any follicles.

Recall that estrogens and progesterone are produced by the corpus luteum (page 912). In the absence of pregnancy, the corpus luteum begins to degenerate about ten days after ovulation. However, when pregnancy occurs, the corpus luteum remains and continues to secrete estrogens and progesterone.

A major factor that causes the corpus luteum to remain and to continue to secrete its hormones is **human chorionic gonadotropin** *(ko″-rē-on′-ik go-nad-o-tro′-pin)*, or **hCG.** This hormone is produced by the **chorion** *(kor′-ee-on)*, a fetal membrane that develops from the cells of the trophoblast and later becomes involved in the formation of the placenta. hCG has properties very similar to those of LH. It is secreted as early as the second week of pregnancy, and its level reaches a peak during the third month and then sharply declines, remaining low throughout the remainder of pregnancy. Because hCG is excreted in the urine, where it is detectable as early as the first month of pregnancy, its presence or absence is used as a basis for pregnancy tests.

Another source of estrogens and progesterone during pregnancy is the placenta. By about the end of the second month, the placenta has developed considerably and is secreting significant amounts of estrogens and progesterone. Moreover, the substantial increases in the levels of estrogens and progesterone that occur as pregnancy progresses are due almost entirely to the secretion of these hormones by the placenta. Thus, beyond the second month, the continuation of the pregnancy no longer depends on the secretion of estrogens and progesterone by the corpus luteum.

◆ **FIGURE 29.3 Relative amounts of human chorionic gonadotropin, estrogens, and progesterone in the maternal blood during pregnancy**

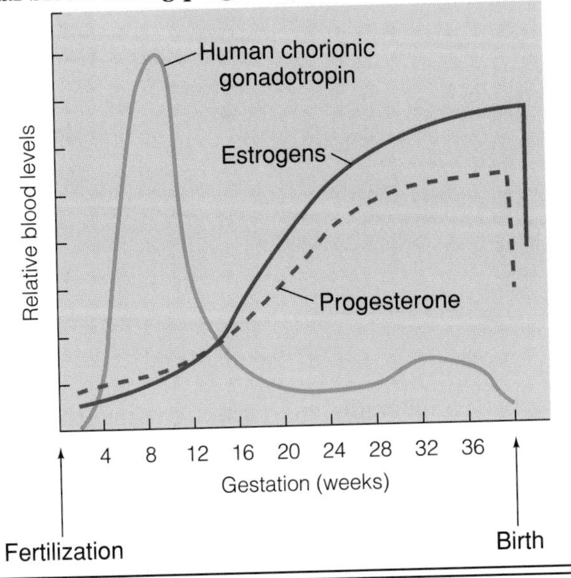

Development of the Placenta

The **placenta** is an organ that, from the third month to the time of birth some six months later, serves to supply nutrients to, and remove wastes from, the embryo—which after the second month is called a **fetus.** Moreover, the placenta serves as an endocrine gland that produces a number of hormones, including estrogens and progesterone.

Because the superficial layer of the thickened endometrium is discarded by the body, either with the menses or at birth, it is called the **decidua** *(dē-sid′-ū-ah;* "falling off"). The portion of the decidua that lies deep to the blastocyst is called the **decidua basalis** (Figure 29.2e). The decidua basalis is involved in the formation of the maternal portion of the placenta. The portion of the decidua superficial to the blastocyst and extending into the lumen of the uterus is the **decidua capsularis.**

As the blastocyst becomes implanted in the endometrium, the trophoblast separates into two layers: an outer **syncytiotrophoblast** *(sin-sit″-ē-ō-trof′-ō-blast)*, in which the cell boundaries disappear, and an inner **cytotrophoblast** *(si″-tō-trof′-ō-blast)*, which is composed of distinctly separate cells (Figure 29.4b). At this

◆ **FIGURE 29.4 Successive stages of embryonic development, showing proliferation of the trophoblast, and early formation of the amnion, the chorion, the yolk sac, and the three primary germ layers**

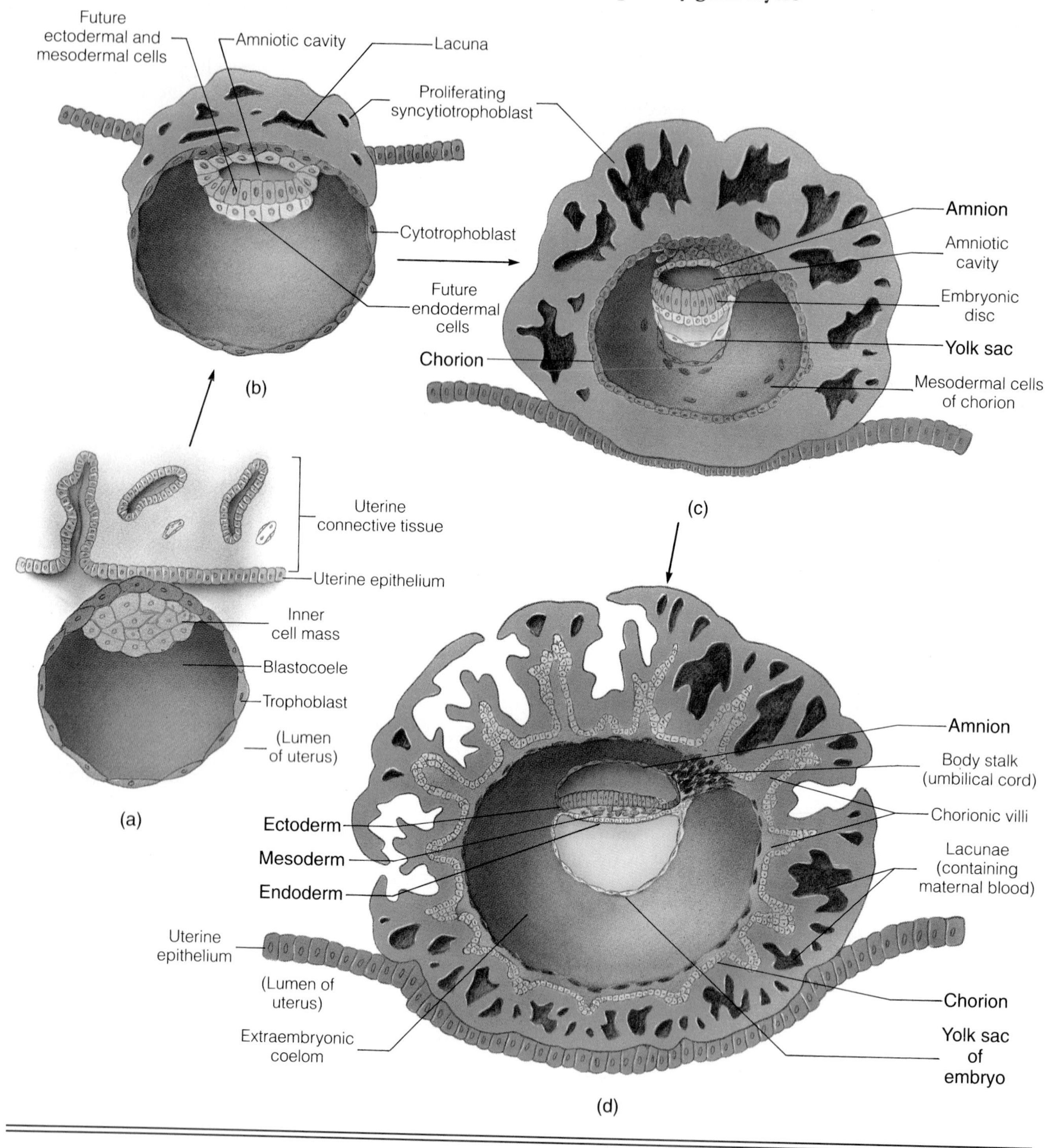

time, a layer of mesodermal cells develops on the inside of the trophoblast. This layer of cells combines with the trophoblast to form a membrane called the **chorion** (Figure 29.4c). As mentioned earlier, the chorion is the source of human chorionic gonadotropin. But it also forms an important part of the placenta. The syncytiotrophoblast proliferates and releases enzymes that erode the endometrium. At the same time, the chorion develops many fingerlike projections called **chorionic villi** (Figure 29.4d). Branches from the umbilical artery and vein of the fetus grow into the villi and form capillary beds. With continued development, the villi become surrounded by pools of the mother's blood, which has collected in sinuses, called *lacunae,* within the endometrium. These blood pools result from the syncytiotrophoblast's erosion of the decidua basalis region of the endometrium and the capillaries in the endometrium. Thus, while maternal and fetal blood are separated by only a few layers of cells (chorion and endothelium of the fetal capillaries), normally no actual mixing of the two blood supplies occurs.

Oxygen, carbon dioxide, glucose, amino acids, fatty acids, various ions, and other substances can move in one direction or another between the maternal blood and the fetal blood either by diffusion or by active transport. Unfortunately, many potentially harmful substances can also cross the placenta and enter the fetal blood. A pregnant woman should be aware that alcohol, nicotine, and many drugs (for example, cocaine) can cross the placenta and affect the fetus. Certain maternal infections such as German measles and AIDS can also cross the placenta.

The presence of a large number of villi provides an extensive surface area across which substances can pass. It is estimated that a fully developed placenta has a surface area of about 16 square meters.

It is clear, from the manner in which it is formed, that the placenta is a combination of maternal tissue (decidua basalis) and fetal tissue (chorion) (Figure 29.5). The fetal blood, which remains in the fetal vessels, enters the placenta through the **umbilical arteries** and passes through capillaries in the villi (where the ex-

◆ **FIGURE 29.5 The vascular arrangement of the placenta**
The arrows indicate the direction of blood flow.

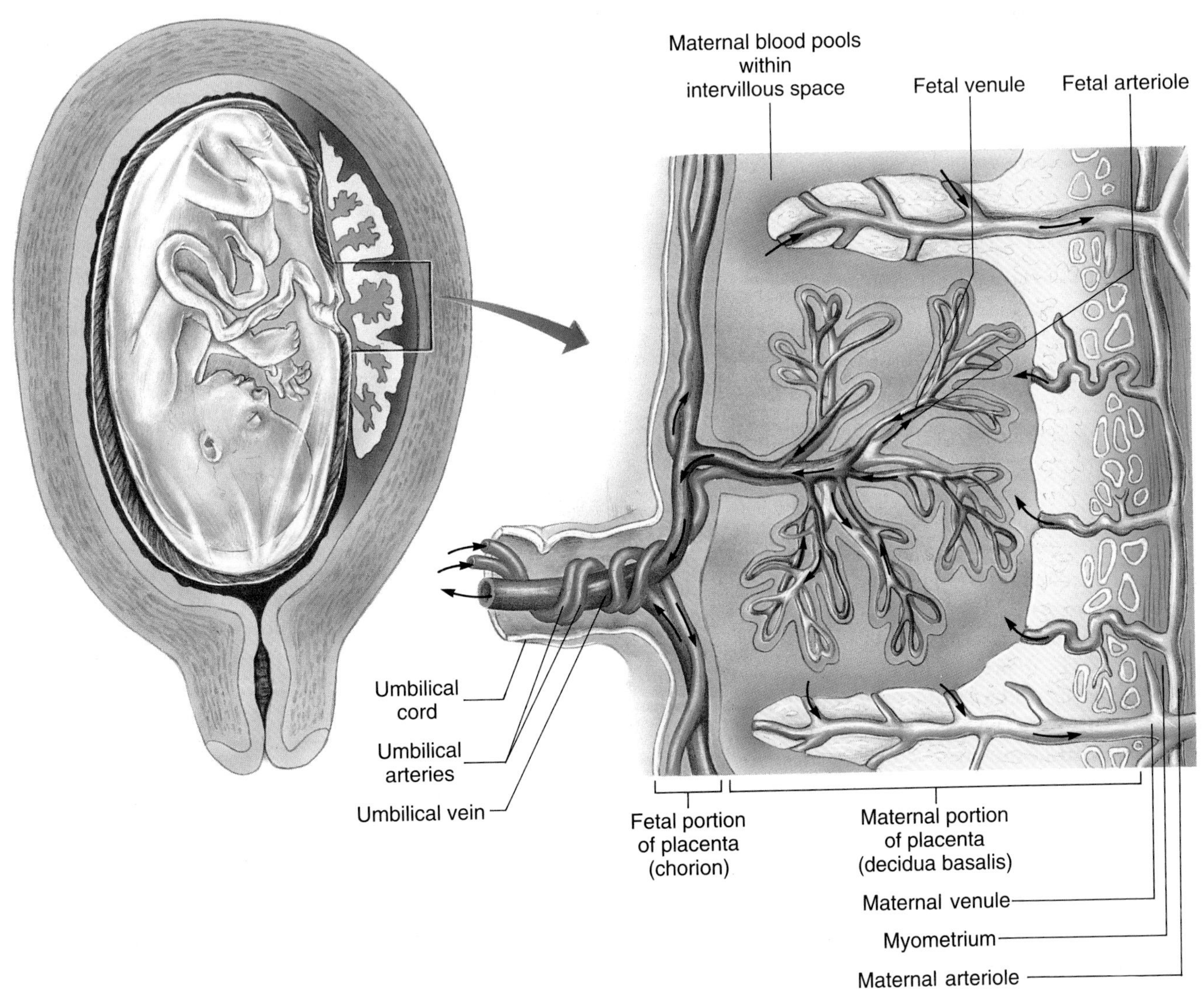

change of substances occurs). The fetal blood then leaves the placenta—and returns to the fetus—through the **umbilical vein.** The umbilical vessels travel between the placenta and the fetus in the **umbilical cord.** The mother's blood enters the placenta through the **uterine arteries** (branches of the internal iliac arteries), flows through the blood pools of the endometrium (where the exchange of substances occurs), and leaves the placenta through the **uterine veins.**

Formation of Embryonic Membranes and Germ Layers

The placenta makes continued development of the embryo possible. Early in development, four embryonic membranes form from the embryo—the *amnion, yolk sac, allantois,* and *chorion.* These membranes assist in the embryo's development in various ways, although they contribute very little to the body of the embryo itself.

While the embryonic membranes are forming, the cells of the inner cell mass undergo movements that result in the formation of three germ layers: the *endoderm, ectoderm,* and *mesoderm.* The embryo—including all of its organs and structures—develops from these germ layers.

Embryonic Membranes

Amnion. Following implantation, while the placenta is developing, the inner cell mass proliferates, and a cavity forms that separates it from the trophoblast. The membrane that surrounds this cavity is the **amnion** *(am´-nē-on).* The amnion is a fluid-filled structure that eventually surrounds the fetus (Figures 29.4 and 29.6). The amniotic fluid protects the embryo against physical injury, helps to maintain the constancy of its temperature, and allows it to move freely. Those cells of the inner cell mass that form the floor of the amniotic cavity make up the **embryonic disc.** The cells of the embryonic disc, from which the embryo will be formed, separate into two layers: a layer of *ectodermal cells* that face the amniotic cavity and a layer of *endodermal cells* on the side toward the space *(blastocoele)* within the blastocyst. Included in the layer of ectodermal cells are cells that will contribute to the formation of mesoderm.

Amniotic fluid is originally derived entirely from maternal blood, but as the fetus develops and its kidneys begin to function, their urine is excreted into the amniotic fluid and increases the total volume of the fluid. Amniotic fluid is continually being added to and eliminated, and it is thought to be completely exchanged several times each day. Cells that have been sloughed off from the fetus are present in the amniotic fluid.

Yolk Sac. The endoderm cells of the embryonic disc undergo rapid mitosis and form the second embryonic membrane: the **yolk sac.** The yolk sac develops as a small cavity on the undersurface of the embryonic disc and eventually extends into the umbilical cord. In humans, where very little yolk is present in the embryo, the yolk sac never becomes very large and serves no nutritive function. The yolk sac is, however, the earliest site of development of blood cells and serves as the source of the primordial germ cells that migrate to the gonads. With continued development, the endodermal cells on the undersurface of the embryonic mass that forms the upper portion of the yolk sac fold together, forming the gut tube of the embryo. The midgut region remains open to the yolk sac.

Allantois. Another diverticulum develops from the endoderm of the hindgut. This pouch is the **allantois,** the third embryonic membrane. In humans, the allantois becomes part of the urinary bladder. Like the yolk sac, the allantois is in the umbilical cord. The blood vessels that carry the fetal blood to and from the placenta (umbilical arteries and the left umbilical vein) also supply the allantois (Figure 29.7).

Chorion. The fourth embryonic membrane, the **chorion,** is formed from the trophoblast of the blastocyst plus the mesoderm that lines the inside of the trophoblast. The chorion surrounds the entire embryo and the other three embryonic membranes. As the amnion enlarges, it fuses with the inner layer (mesoderm) of the chorion. In combination with the umbilical blood vessels, the chorion forms the fetal portion of the placenta.

Germ Layers

In the second week of development, some of the surface cells of the embryonic disc settle deeper in the disc and spread between the **ectoderm** and **endoderm** layers that were formed by an earlier separation of the embryonic disc (Figure 29.8). These cells will form the **mesoderm.** As the sheet of mesoderm moves between the ectoderm and endoderm, it splits into two layers: *parietal* and *visceral.* The space between the layers becomes the **coelom,** or body space. Table 29.1 summarizes the main structures formed by the three primary germ layers.

Gestation

The **gestation** (pregnancy) period usually lasts about 280 days from the last menstrual flow. During this time, the developing individual increases in size, and all the

◆ **FIGURE 29.6 Successive stages of development of the embryo and the embryonic membranes, showing their relationships with the uterus**

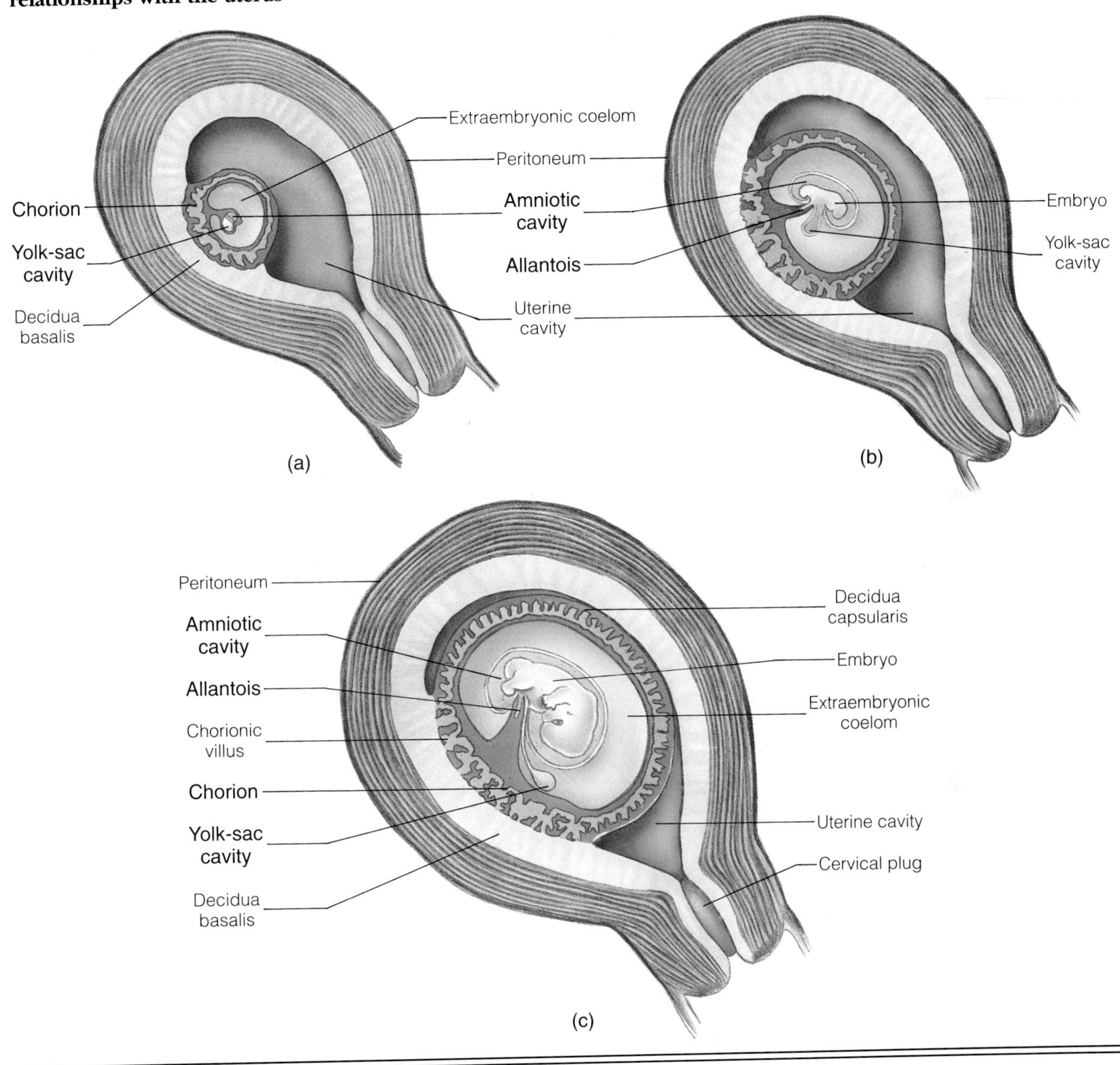

organs and structures of the adult are formed. During gestation, the total mass of the uterus increases to accommodate the growing fetus. In the later stages of pregnancy, the uterus occupies such a large portion of the abdominopelvic cavity that it can exert pressure on the rectum and the urinary bladder, causing constipation and frequent urination (Figure 29.9).

The presence of a growing fetus places extra demands on a woman's body systems. For example, a pregnant woman generally produces somewhat more urine than she did before pregnancy because her kidneys must process the excretory products of the fetus as well as her own excretory products. Diet is especially important during pregnancy, and an expectant mother's diet must include sufficient amounts of vitamins, minerals, proteins, and other substances to supply both her own needs and those of the fetus. The blood volume of a pregnant woman increases as much as 30% as the placenta develops, and her respiratory system increases its activity in order to supply the additional oxygen required by the fetus and to eliminate the carbon dioxide produced by it.

During pregnancy, the placenta produces a hormone called **human chorionic somatomammotropin**

◆ **FIGURE 29.7 Early blood vessels of the embryo and the fetal portion of the placenta**
The amount of color in a vessel indicates the relative proportion of oxygenated blood carried by the vessel. Red indicates highy oxygenated blood; blue indicates deoxygenated blood; purple indicates a mixture of oxygenated and deoxygenated blood.

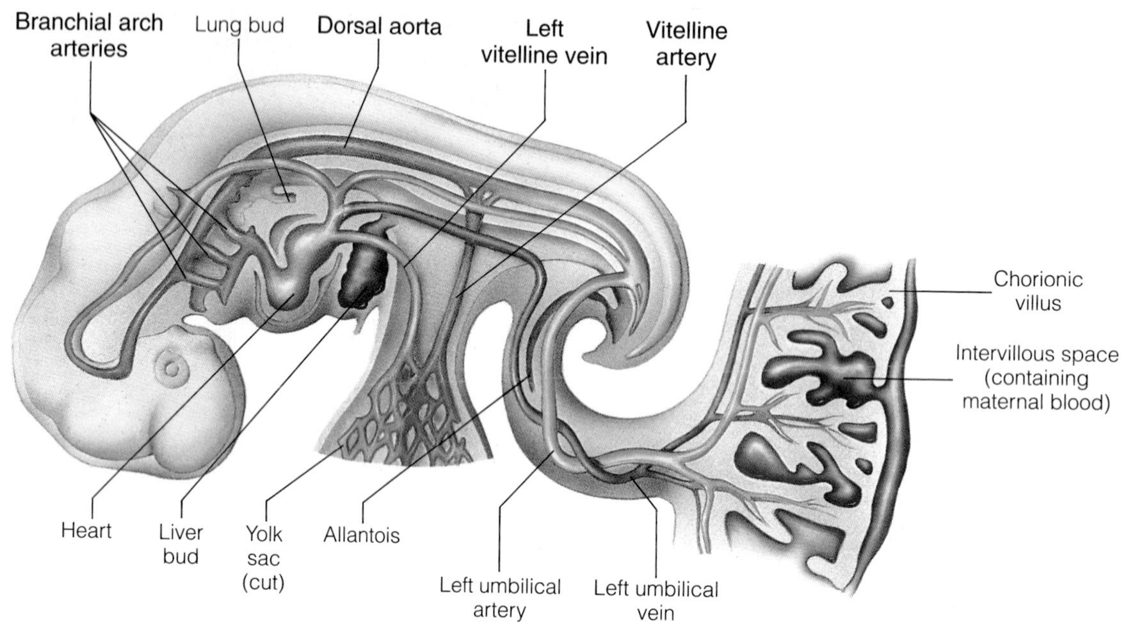

◆ **FIGURE 29.8 Schematic cross section of an embryo showing the development of the germ layers**
(a) Early development: a sheet of mesoderm is moving between the endoderm of the gut and the ectoderm of the body surface.
(b) Later development: the coelom has formed between the parietal and visceral layers of the mesoderm.

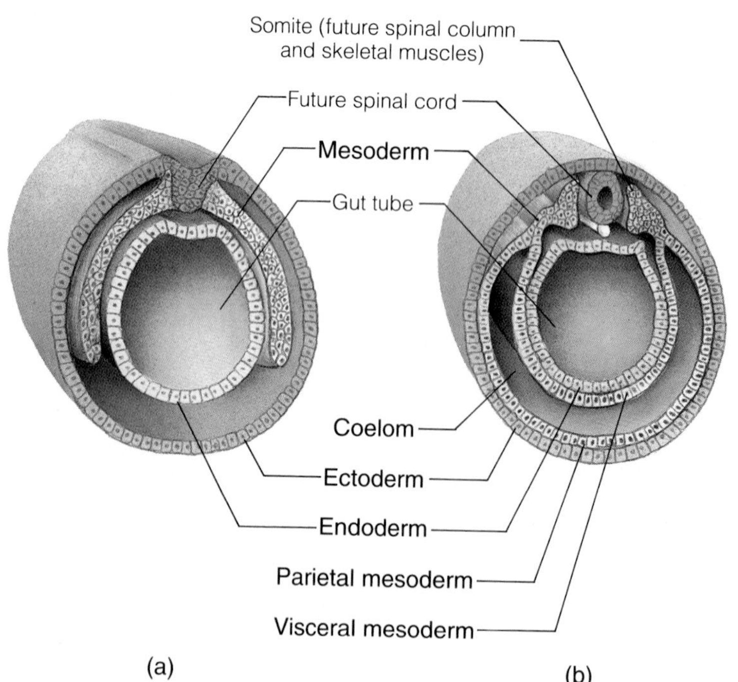

◆ **TABLE 29.1 Structures Formed by the Three Primary Germ Layers**

ENDODERM
Epithelium of the digestive tract and its glands (for example, the liver and pancreas)
Epithelium of the urinary bladder and urethra
Epithelium of the pharynx, auditory tubes, larynx, trachea, bronchi, and lungs
Epithelium of the tonsils and the thyroid, parathyroid, and thymus glands
MESODERM
Skeletal, smooth, and cardiac muscle
Cartilage, bone, and other connective tissues
Blood, bone marrow, and lymphoid tissue
Epithelium of blood vessels and lymphatics
Epithelium of the coelom and joint cavities
Epithelium of the kidneys and ureters
Epithelium of the gonads and reproductive ducts
Epithelium of the adrenal cortex
Dermis of the skin
ECTODERM
Epidermis of the skin
Hair, nails, and glands of the skin
Lens of the eye
Receptor cells of the sense organs
Epithelium of the mouth, nostrils, sinuses, and anal canal
Enamel of the teeth
All nervous tissue
Adrenal medulla

◆ **FIGURE 29.9 Sagittal section of a woman in a late stage of pregnancy**

Notice how the rectum and the urinary bladder are compressed by the enlarged uterus.

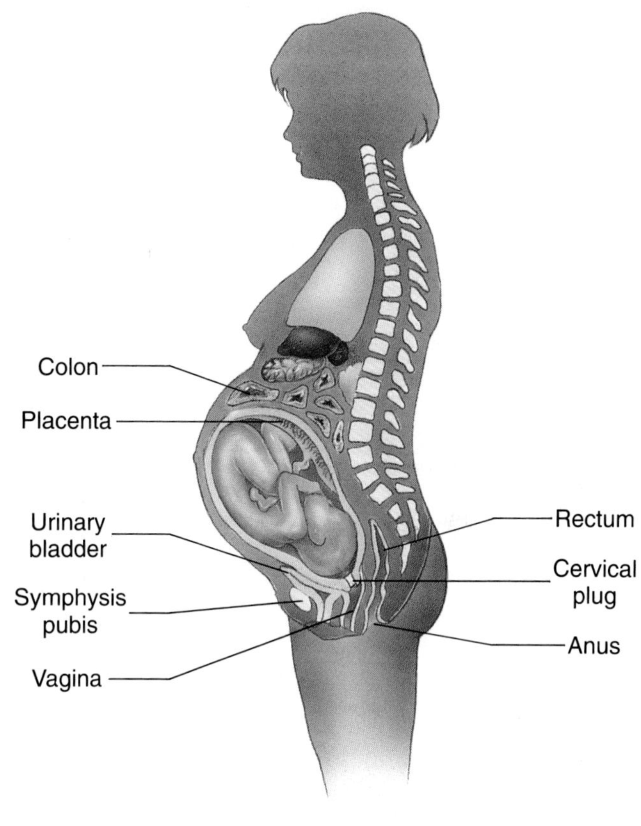

(human placental lactogen), and researchers believe this hormone affects the metabolism of the mother. Human chorionic somatomammotropin begins to be produced by about the fifth week of pregnancy, and its level increases throughout pregnancy. Human chorionic somatomammotropin decreases the mother's use of glucose; thus, more glucose becomes available for use by the fetus. This effect is believed to be useful in promoting the development of the fetus because the fetus uses glucose as a major source of energy. Human chorionic somatomammotropin also increases the mobilization of fatty acids from maternal adipose tissue. The fatty acids can be used in place of glucose as an energy source by the mother.

At five weeks, the developing human embryo has short limb buds, which eventually develop into arms, legs, fingers, and toes (Figure 29.10a). A rudimentary tail is present as well as primitive branchial arches. At this stage, the brain is covered by only a thin layer of tissue, the eyes are just beginning to form, and the heart is larger in proportion to the body than at any other time in development.

By seven weeks, the tail has degenerated, and the branchial arches have developed into jaws and other structures (Figure 29.10b). Gender is now determinable, and most of the major organs are present, including the liver, pancreas, and heart. By this stage, the fetus has acquired reflexes and is thus aware of touch, although it is unaware of sound (except for the mother's heartbeat).

During the third month, the fetus grows rapidly and attains definite human characteristics, with the trunk region becoming relatively more slim and the limbs well defined (Figure 29.10c).

By 20 weeks, eyelids are able to open and close, fingernails are present, and respiratory, circulatory, and nervous systems are established (Figure 29.10d).

Twins

Twins occur in about one out of every 85 births. There are two types of twins: identical and fraternal. Each results from a different series of events.

◆ **FIGURE 29.10 Selected stages of gestation**
(a) 5–6 weeks; (b) 7 weeks; (c) 10–13 weeks; (d) 16–20 weeks.

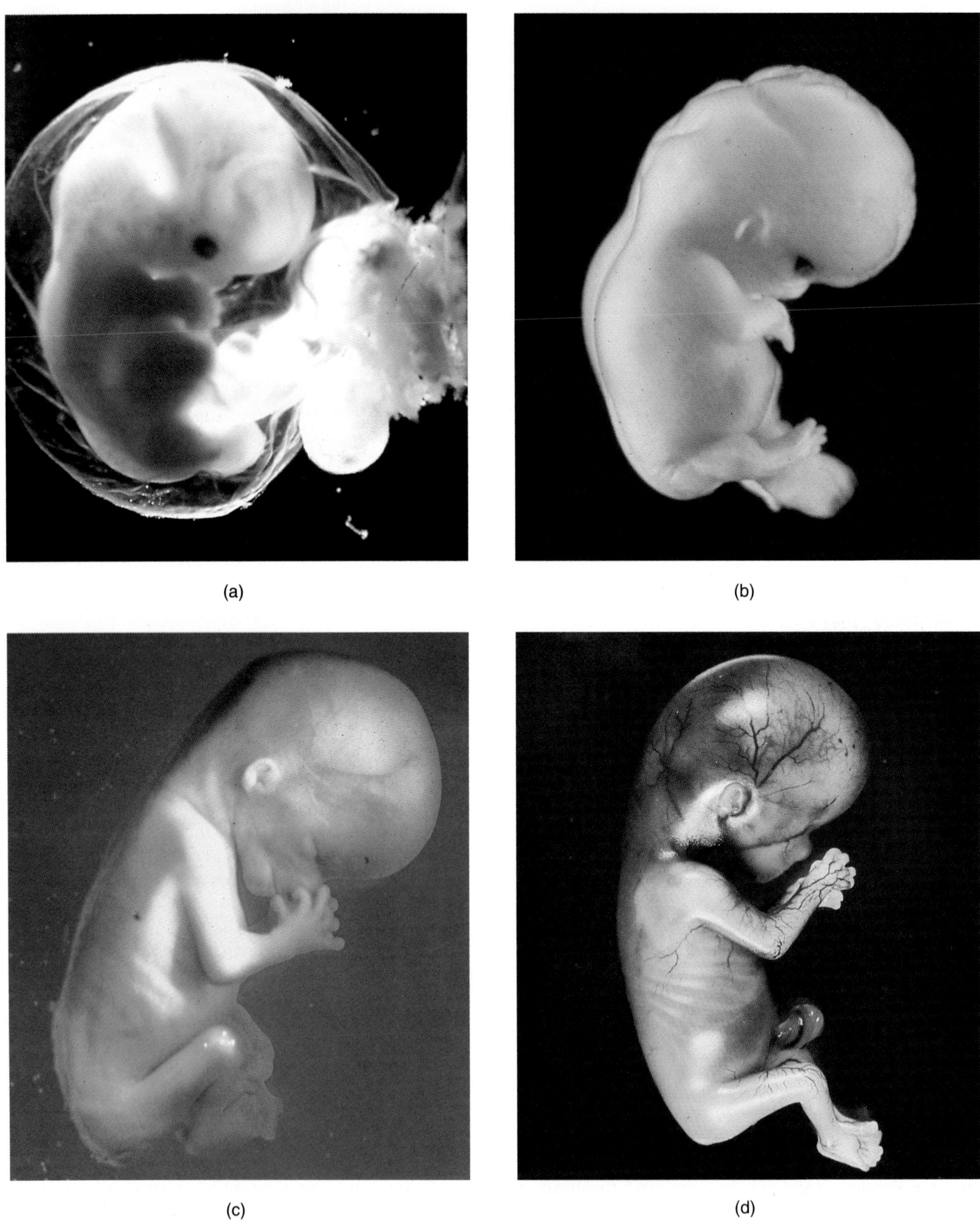

Identical, or *monozygotic, twins* occur when an inner cell mass that develops from a single fertilized ovum separates into two masses, each of which gives rise to a complete embryo. Because identical twins are the result of the fertilization of a single ovum by a single spermatozoon, they always are the same sex and have identical genes. However, they may have separate placentae, depending on the stage of development at which the inner cell mass separates.

Fraternal, or *dizygotic, twins* occur when two ova are released during a single ovarian cycle and both are subsequently fertilized by separate spermatozoa. Fraternal twins, therefore, can be of the opposite sex and can differ just as much as any siblings. Fraternal twins always have separate placentae.

Parturition

Parturition *(par-tū-rish´-un)* is the act of giving birth to a baby at the end of pregnancy. The series of events by which parturition is accomplished is referred to as **labor.** During the last portion of pregnancy, several activities take place in preparation for parturition.

Near the end of pregnancy, collagen fibers in the cervix of the uterus dissociate, and the cervix softens (or "ripens"). The softening is due at least in part to the influence of a hormone called **relaxin,** which is secreted by the corpus luteum and placenta.

During the last three months of pregnancy, the uterus undergoes weak, irregular contractions called **Braxton-Hicks contractions.** During the final weeks of pregnancy, the contractions gradually become stronger and more frequent. Then, within a few hours, the contractions become quite strong and occur about every 15 to 30 minutes. The occurrence of these relatively strong, regular contractions marks the beginning of labor.

As uterine contractions become stronger, they exert increasing pressure on the amniotic sac. When the pressure becomes great enough, the amnion bursts, releasing the amniotic fluid, which escapes through the vagina. The amnion frequently bursts at the onset of labor or early in labor.

Labor

During labor, uterine contractions travel in peristaltic waves from the upper portion of the uterus downward toward the cervix. As labor progresses, the contractions become stronger and more frequent, eventually occurring only two to three minutes apart and lasting about one minute.

Labor is commonly divided into three stages: (1) cervical dilation, (2) delivery of the baby, and (3) delivery of the placenta (Figure 29.11).

First Stage: Cervical Dilation

During the first stage of labor, uterine contractions push the fetus against the cervix, forcing the cervix to dilate. In most births, the head of the fetus is positioned downward against the cervix, and the cervix dilates sufficiently to accommodate it (usually a maximum of 10 cm).

The first stage of labor is generally the longest. It may last 6 to 12 hours or more, particularly in a woman who has not previously given birth.

Second Stage: Delivery of the Baby

When the cervix is sufficiently dilated, uterine contractions (assisted by contractions of abdominal muscles) move the fetus along the cervical and vaginal canals to the exterior. In most births, the head of the fetus is delivered first, allowing the fetus to breathe safely even before it is completely free of the birth canal.

The second stage of labor usually lasts between 30 and 90 minutes.

Third Stage: Delivery of the Placenta

Within a few minutes after delivery of the baby, uterine contractions separate the placenta from the wall of the uterus and expel the placenta as the *afterbirth.* Bleeding is generally minimal during this process because the placental blood vessels constrict, and the contraction of the uterus squeezes closed the uterine vessels that supply blood to the placenta.

The third stage of labor is typically the shortest. It is usually completed 15 to 30 minutes after the delivery of the baby.

If all placental fragments are not expelled from the uterus, the woman may continue to experience uterine bleeding following birth *(postpartum bleeding).* Moreover, the absence of one of the umbilical arteries supplying the placenta is often associated with certain disorders in the child. For these reasons, the placenta and the umbilical cord are closely examined following their delivery.

Within about five weeks following birth, the uterus returns to its normal size, and the endometrial surface heals. If the mother is not nursing her child, menstrual cycles generally resume by about the end of the sixth week.

◆ **FIGURE 29.11 Stages of labor**
(a) Position of fetus near the end of pregnancy. (b) First stage of labor: cervical dilation. (c) Second stage of labor: delivery of the baby. (d) Third stage of labor: delivery of the placenta.

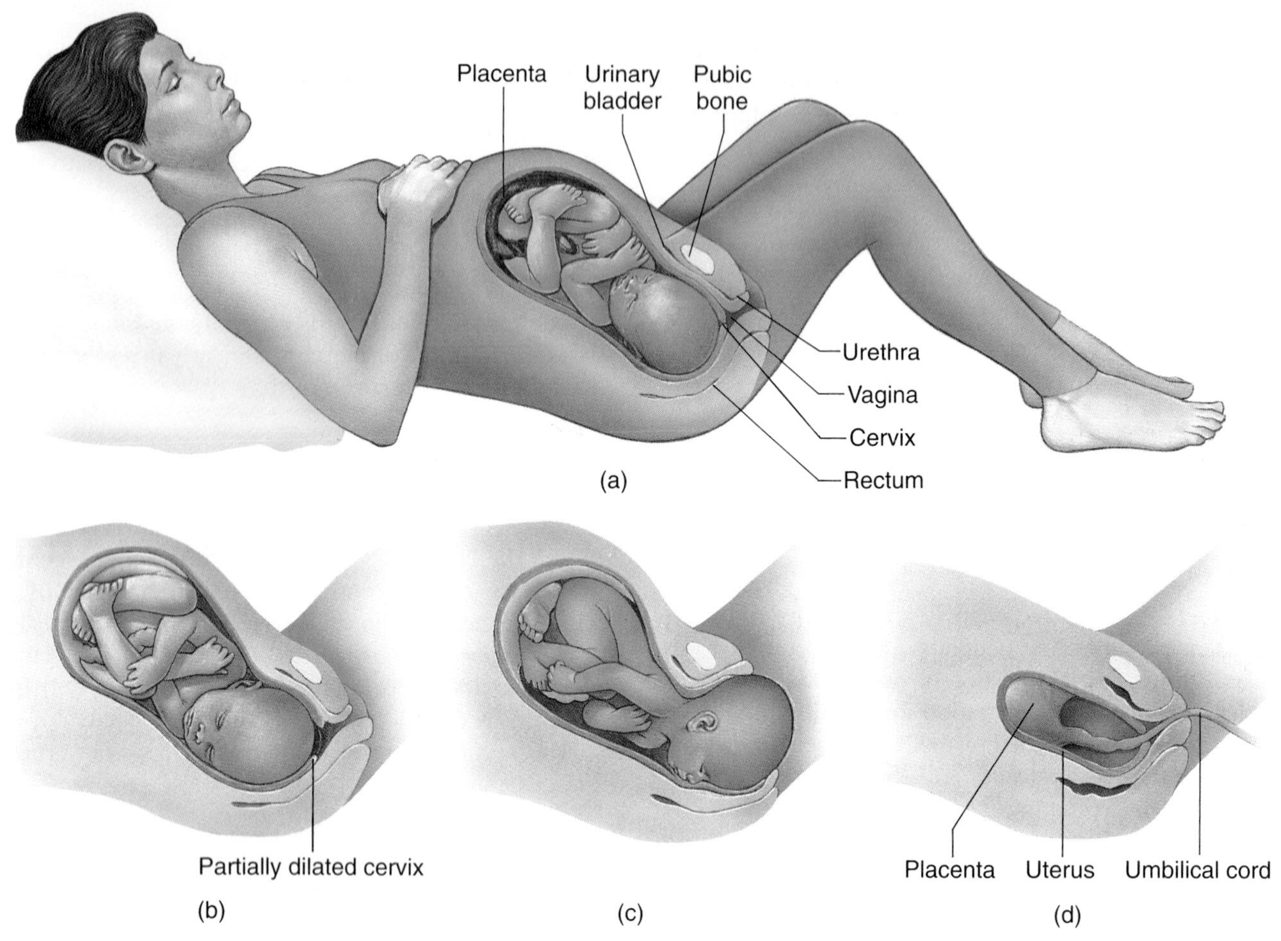

Factors Associated with Parturition

It is still not certain what initiates parturition. However, a number of different factors, including the following, are believed to be involved:

1. The hormone oxytocin, which is released from the posterior lobe of the pituitary gland, is particularly effective at stimulating uterine contractions late in pregnancy, although it is less effective early in pregnancy. Before the onset of labor, oxytocin levels are relatively low in both maternal and fetal (umbilical artery) blood. In very early labor, the oxytocin level in maternal blood remains low, but the level in fetal blood rises. In advanced labor, the oxytocin levels in both maternal and fetal blood are elevated.

2. Prostaglandins that can stimulate uterine smooth muscle are synthesized by the placenta, and their release increases during labor. Moreover, there is evidence that oxytocin stimulates prostaglandin production by the placenta.

3. The concentration of oxytocin receptors in the uterus and the placenta increases during pregnancy and is high at parturition. In experimental animals, it has been found that estrogens and uterine distension can increase the concentration of oxytocin receptors. During pregnancy, the level of estrogens gradually increases (see Figure 29.3), and near the time of birth, the fetus grows rapidly, accelerating uterine distension.

These observations have resulted in the following theory to explain the initiation of parturition.

During pregnancy, the concentration of oxytocin receptors in the uterus and the placenta increases greatly, probably under the influence of rising estrogen levels and uterine distension. Although there is no substantial elevation in circulating oxytocin at the onset of labor, the rising concentration of receptors increases the oxytocin sensitivity of the uterus to the point where relatively strong uterine contractions are stimulated. In addition, oxytocin (much of which may come from the

CONDITIONS OF CLINICAL SIGNIFICANCE

Pregnancy

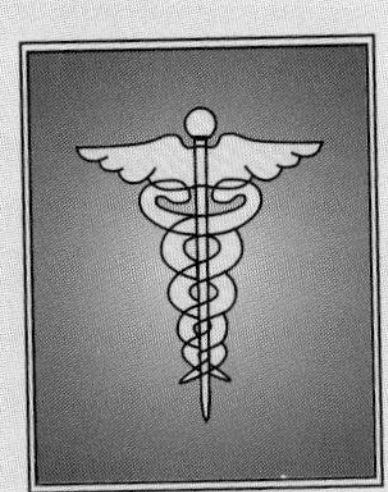

Disorders of Pregnancy

Ectopic Pregnancy

The blastocyst normally becomes implanted in the upper portion of the uterine endometrium. However, in some cases, implantation occurs at a site other than the uterus. Such an occurrence is called an *ectopic pregnancy (ek-top´-ik).* The most common ectopic pregnancy is a *tubal pregnancy,* which occurs within a uterine tube.

Ectopic pregnancies can endanger the life of the mother. They frequently cause internal hemorrhaging, since the trophoblast of the blastocyst destroys tissue at the site of implantation. The endometrium, which is the normal site of implantation, can withstand the destructive effects of the trophoblast, but most tissues simply break down. The lack of room for expansion is another danger in ectopic pregnancies. In the case of a tubal pregnancy, for example, the uterine tube may rupture as the embryo grows.

Placenta Previa

Occasionally, when implantation occurs in the lower regions of the uterus, the placenta covers the inner opening of the cervical canal. This condition is called *placenta previa*—that is, placenta leading the way. In this position, the placenta can irritate the cervix, which causes the uterus to contract and can result in a spontaneous abortion. If the pregnancy goes to term, the placenta is expelled before the baby is born, which causes the mother to hemorrhage severely. The expulsion of the placenta prior to the expulsion of the fetus can also stimulate the fetus to begin breathing while still within the birth canal.

Toxemia of Pregnancy

Toxemia (poison blood) associated with pregnancy continues to be a serious problem. Despite intensive research, the cause of toxemia is not known. It is probable that numerous factors produce toxemia—including malnutrition, metabolic disorders, endocrine imbalances, production of toxins, and a decreased blood supply to the uterus. As a prelude to toxemia, the woman often shows a strong tendency to reabsorb salt from the kidney tubules, and with it, water. This excessive reabsorption of salt and water can usually be controlled with a low-salt diet or by the use of diuretic drugs to increase water loss.

If toxemia is unchecked, among other symptoms there may be excessive weight gain; edema of the lungs, kidneys, and brain; visual disturbances; and, eventually, convulsions. If the convulsive stage is reached, the condition is called *eclampsia (e-klamp´-sē-ah).*

Birth Control

Over the years much effort has been expended in attempting to develop a convenient, safe, and effective method of birth control, and a variety of birth control methods are now in widespread use.

Birth Control Methods Available to the Male

COITUS INTERRUPTUS. *Coitus interruptus* is the withdrawal of the penis just before the male orgasm so that ejaculation does not occur within the female tract. It is not a reliable method because it requires an unnatural response and perfect timing by the male.

CONDOM. A *condom* is a very thin sheath of rubber, plastic, or animal membranes that is pulled onto the erect penis just before intercourse. Condoms are an effective method of preventing fertilization as long as they do not tear or slip off after orgasm and allow semen to enter the vagina. They have the added advantage of providing protection against sexually transmitted diseases and AIDS.

VASECTOMY. *Vasectomy (vah-sek´-tō-mē)* is a form of surgical sterilization in which each ductus deferens is tied and cut. This procedure does not interfere with normal ejaculation, but spermatozoa are prevented from entering the semen.

continued on next page

CONDITIONS OF CLINICAL SIGNIFICANCE

Birth Control Methods Available to the Female

DETECTION OF OVULATION. Several methods of birth control rely on a woman's ability to determine the time of ovulation and to abstain from sexual intercourse for several days preceding and following ovulation.

In one method, called the *rhythm method,* a woman keeps records of the lengths of her menstrual cycles and from these records predicts the length and likely time of ovulation for her current cycle. The rhythm method may be an effective method of birth control if the woman's menstrual cycles are regular and if the period of abstinence from sexual intercourse is long enough. However, menstrual cycles can vary in length, and the timing of ovulation can vary even when menstrual cycles are regular. Consequently, the likely time of ovulation during a current cycle can be difficult to predict accurately from records of past cycles.

A second method used to determine the time of ovulation involves the measurement of body temperature. In this method, a woman determines her basal body temperature with a thermometer each morning upon awakening and prior to getting out of bed. Generally, a woman's basal body temperature will be somewhat lower prior to ovulation than it will be following ovulation.

A third method for determining the time of ovulation is observation of the mucus secretions of the uterine cervix that are discharged at the vagina. Prior to ovulation, estrogens increase the quantity of alkaline cervical mucus that is secreted and decrease the viscosity and the cellularity of the mucus. These occurrences tend to favor the survival and transport of sperm. Following ovulation, progesterone causes the cervical mucus to be thick, tenacious, and cellular. This mucus contributes to the occurrence of conditions unfavorable for the penetration and survival of spermatozoa. For several days following menstrual bleeding, no mucus discharge is evident at the vagina. These "dry days" are followed by the onset of "mucous symptoms" characterized by the appearance of increasing quantities of cloudy or sticky secretions that, in turn, are followed by the appearance of a clear, slippery, lubricative mucus. The clear, slippery, lubricative mucus is then followed by the occurrence of thick, tacky, opaque mucus. Generally, the period from the beginning of the mucous symptoms until the fourth day after the last day of appearance of clear, slippery, lubricative mucus is considered to be a woman's fertile period. (With this method, the period of menstrual bleeding is also considered to be a fertile period).

DIAPHRAGM. A *diaphragm* is a thin hemispherical dome of rubber or plastic with a spring margin. It covers the cervix, thereby preventing sperm from entering the uterus. A diaphragm, which is initially fitted by a physician, is generally used in combination with a spermicidal cream or jelly. Although a properly fitted diaphragm is quite effective, its use requires the woman to predict the occurrence of sexual intercourse so the diaphragm will be in place when needed. Otherwise she must interrupt the prelude to the act while she inserts it.

INTRAUTERINE DEVICE. An *intrauterine device (IUD)* consists of a small spiral, ring, or loop made of plastic stainless steel, or copper that is inserted into the uterine cavity by a physician. Once in place, the IUD can remain within the uterus for a long time. An IUD does not affect the menstrual cycle, but it does prevent the implantation of the blastocyst into the endometrium. Some women's bodies cannot tolerate an IUD and expel it from the uterus. There have been so many health problems with certain forms of IUDs that they have been removed from public use.

ORAL CONTRACEPTIVES. *Oral contraceptives,* better known as "the pill," are extremely effective methods of preventing pregnancy. Several types of pills are available, most of which contain a combination of synthetic estrogenlike and progesteronelike substances. When taken daily for the first 20 or 21 days of the menstrual cycle, these substances apparently prevent the rise in luteinizing hormone that leads to ovulation. Even if ovulation should occur (as it apparently does in some cases), oral contraceptives can still prevent pregnancy by thickening and chemically altering the cervical mucus so that it is more hostile to sperm and by making the uterine endometrium unreceptive to implantation. In some women, oral contraceptives have undesirable side effects, the most serious of which is a tendency to form blood clots.

CONDITIONS OF CLINICAL SIGNIFICANCE

CHEMICAL CONTRACEPTION. Various types of douches, foams, suppositories, creams, and jellies that are introduced into the vagina either just before or after intercourse are available. Most of these prevent fertilization by acting as physical barriers to sperm and by serving as spermicides. They are most effective when used in combination with a diaphragm.

TUBAL LIGATION. *Tubal ligation* is a form of surgical sterilization in which the uterine tubes are tied and severed. This procedure prevents spermatozoa from reaching the ovum to fertilize it and prevents the developing embryo from reaching the uterus.

SUBCUTANEOUS IMPLANTS. One of the newest methods of birth control available to women involves the insertion of several small capsules beneath the skin of the upper arm. The capsules contain a synthetic progesteronelike substance, which prevents pregnancy by inhibiting ovulation. It also causes the cervical mucus to become thicker, helping to prevent sperm from entering the uterus.

This method of birth control is long-lasting and very effective, providing continuous protection against pregnancy for about five years. Because the implants do not contain estrogenlike substances, they have fewer side effects than birth control pills. (Estrogenlike substances cause many of the side effects of birth control pills, such as fluid retention, breast tenderness, and nausea.) However, irregular or prolonged menstrual bleeding may occur in some women.

Because the implants must be inserted and removed by trained medical personnel, one disadvantage of this method of birth control may be its initial cost. Overall, however, it may be less expensive than a five-year supply of birth control pills.

VAGINAL POUCHES. Another new method of birth control available to women is the *vaginal pouch*. A vaginal pouch is shaped much like a condom, but it is inserted into the vagina rather than being pulled over the penis.

Vaginal pouches are made of thicker latex than are condoms, and thus are less likely to leak or tear. Because the diameter of the pouches is greater than the diameter of condoms, they fit more loosely around the penis during sexual intercourse. As a consequence, men have reported that the pouches do not reduce sensation as much as do condoms.

Vaginal pouches not only reduce the number of unwanted pregnancies, but they also provide protection against sexually transmitted diseases.

INDUCED ABORTION. In recent years, *induced abortion* of the embryo or fetus has become a more common method of birth control, although the moral issues involved continue to be debated. Induced abortion involves using various means to separate the implanted embryo or fetus from the wall of the uterus. The detachment of the embryo or fetus is generally accomplished either by scraping, by saline solution rinses, or by vacuum aspiration (suction).

fetus) binds to placental receptors, stimulating prostaglandin production, and other substances released by the fetus at or near the time of birth may enhance prostaglandin production. Prostaglandins diffuse to the uterine muscle, where they act with oxytocin to produce even stronger uterine contractions. (In fact, some researchers believe that prostaglandins provide the major stimulus for uterine contractions at parturition.)

As uterine contractions move the fetus downward toward the birth canal, the fetus stimulates the cervix of the uterus. This stimulation triggers a neuroendocrine reflex that further strengthens the contractions. Nerve impulses transmitted from the cervix to the hypothalamus cause the posterior lobe of the maternal pituitary to release additional oxytocin, which enhances the contractions of the uterine muscle (Figure 29.12). Moreover, the stimulation of the cervix also seems to enhance uterine contractions by way of signals transmitted either neurally or along the uterine muscle to the body of the uterus. The enhanced uterine contractions push the fetus even more forcefully against the cervix, leading to the release of more oxytocin and still stronger uterine contractions. Thus, as described on page 19, these activities act in a positive-feedback fashion to produce progressively stronger uterine contractions that expel the fetus from the uterus.

◆ **FIGURE 29.12 Neuroendocrine reflex by which oxytocin from the maternal pituitary enhances uterine contractions during parturition**

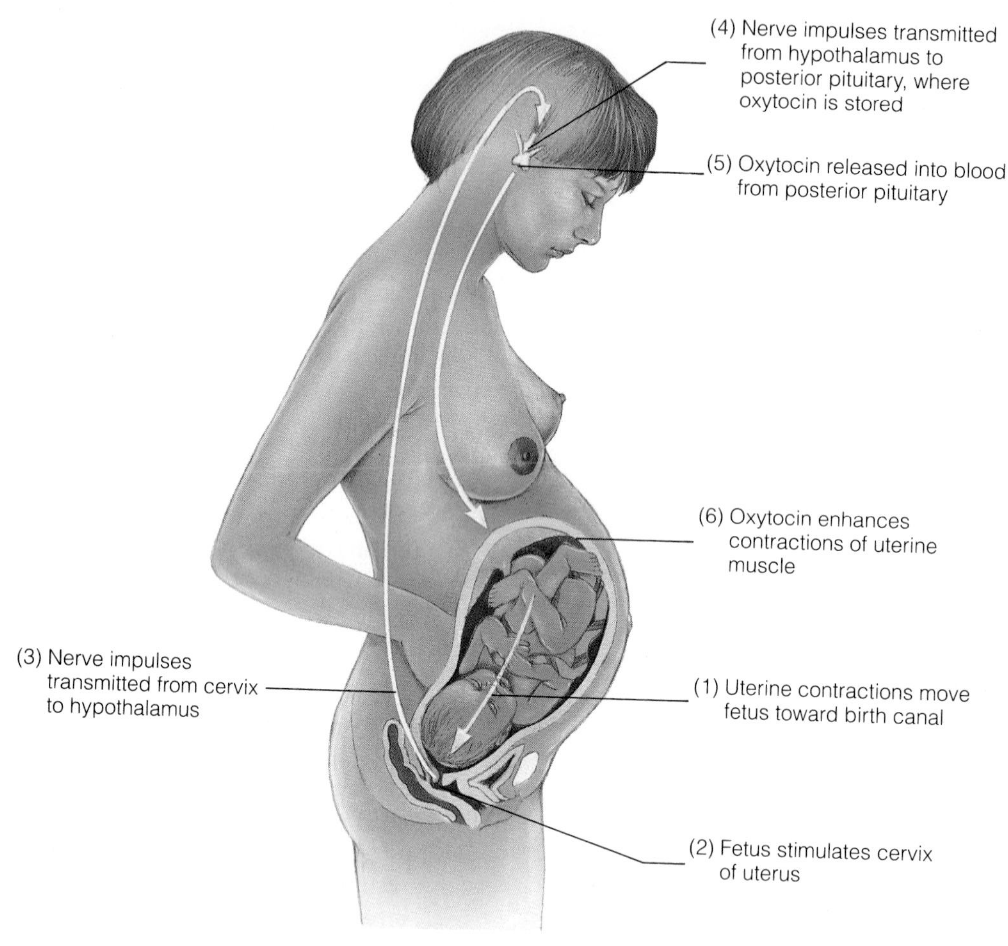

Lactation

Lactation, or the production of milk by the breasts (mammary glands), requires the interaction of several hormones.

A fully developed breast is composed of glandular tissue surrounded by adipose tissue (Figure 29.13). The glandular tissue consists of little sacs of secretory cells called *alveoli* that secrete milk into epithelium-lined ducts. The ducts carry the milk to the surface of the breast at the nipple. The alveoli and the initial portions of the ducts are surrounded by specialized contractile cells called *myoepithelial cells.*

Preparation of the Breasts for Lactation

Prior to puberty, the breasts are largely undeveloped. At puberty, the increased level of estrogens stimulates the growth and branching of the duct system within the breasts, and there is considerable breast enlargement due to fat deposition. The increased level of progesterone at puberty also contributes to breast growth. However, the alveoli do not develop completely, and the changes that occur at puberty do not result in milk production.

During pregnancy, the breasts enlarge still more, and the alveoli develop fully. These changes are due to the stimulatory effects of the high concentrations of estrogens and progesterone during pregnancy, as well as to the influence of the pituitary hormone **prolactin.** Prior to puberty, the prolactin level is low. However, estrogens stimulate prolactin secretion, and the prolactin level increases at puberty and rises still higher during pregnancy. (It is not known if estrogens influence prolactin secretion by way of the hypothalamus or by direct action on the pituitary.)

Milk and Colostrum

Human milk contains four principal components: water, protein, fat, and lactose (milk sugar). Following parturition, there is a delay of several days before true milk secretion begins. During this period, the breasts secrete

◆ **FIGURE 29.13 Anterior view of a partially dissected left breast**

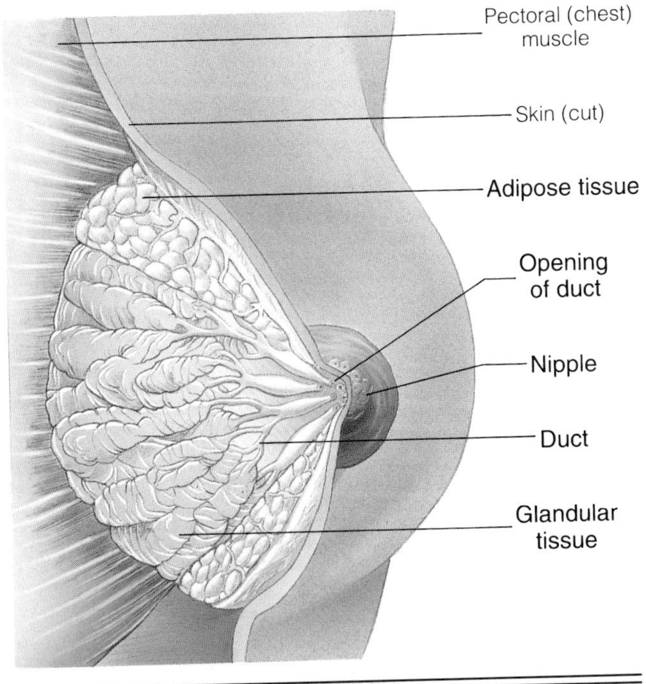

a yellowish fluid called **colostrum** *(ko-los´-trum).* Colostrum contains high amounts of proteins, minerals, vitamin A, and antibodies, but is low in lactose and fat. Some evidence suggests that the immune systems of newborns who nurse become active earlier than do the immune systems of babies who are bottle-fed on a formula derived from cow's milk.

Production of Milk

Prolactin is the major hormone responsible for milk production. However, despite its relatively high levels (and the full development of the breasts) during pregnancy, significant milk production does not occur. This lack of milk production is due to the fact that the high levels of estrogens and progesterone present during pregnancy inhibit the milk-producing action of prolactin on the breasts (even though estrogens enhance prolactin secretion and act with it in promoting breast growth and development). At parturition, the loss of the placenta leads to a decline in the levels of estrogens and progesterone, and prolactin is able to stimulate milk production.

Following parturition, a reflex response is important in stimulating prolactin secretion. Mechanical stimulation of the nipples by the suckling infant initiates nerve impulses that are transmitted to the hypothalamus. The nerve impulses ultimately lead to prolactin release by inhibiting the release of prolactin-inhibiting hormone or stimulating the release of prolactin-releasing hormone or both. Thus, even though the basal level of prolactin secretion returns to the prepregnancy level several weeks after parturition, when suckling is vigorous and frequent, a substantial surge in prolactin secretion occurs each time the mother nurses her infant. The prolactin acts on the breasts to continue milk production.

If suckling is less vigorous and frequent (as may occur after several months of nursing as the infant begins to eat significant amounts of other food), prolactin levels decline, and within about the three months there is relatively little prolactin secretion in response to suckling. (In this regard, it is interesting to note that the rate of milk production by many nursing mothers declines substantially within seven to nine months after parturition.) If nursing is stopped (or if it never began), prolactin release in response to suckling does not occur and lactation ceases.

Milk Ejection Reflex

Milk produced by the mammary glands tends to remain within the alveoli unless the specialized myoepithelial cells that surround the alveoli contract and force it into the ducts. Within a minute after suckling begins, the ***milk ejection reflex (milk let-down reflex)*** leads to the contraction of the myoepithelial cells and the ejection of milk into the ducts of both breasts—not just the one being nursed. In this reflex, the stimulation of the nipple by suckling initiates nerve impulses that are sent to the hypothalamus (Figure 29.14). The impulses ultimately lead to the release of oxytocin from the posterior lobe of the pituitary gland. Oxytocin causes the myoepithelial cells to contract, and this contraction ejects milk into the ducts, where it can be easily obtained by the nursing infant.

Higher brain centers also influence milk ejection. For example, it is not unusual for a nursing mother to leak milk in response to the sound of a crying infant, even prior to physical contact with the child.

Fetal Circulation

The circulatory system of the fetus contains some unique features that are well adapted to development in the uterus but must change dramatically following birth if the child is to survive.

Fetal Blood Pathways

Fetal blood travels to the placenta in the **umbilical arteries,** which are branches of the internal iliac arteries (Figure 29.15). After circulating through the capillaries of the chorionic villi, where exchanges of gases, nutrients, and wastes occur, the blood leaves the placenta

◆ **FIGURE 29.14 The milk-ejection reflex**

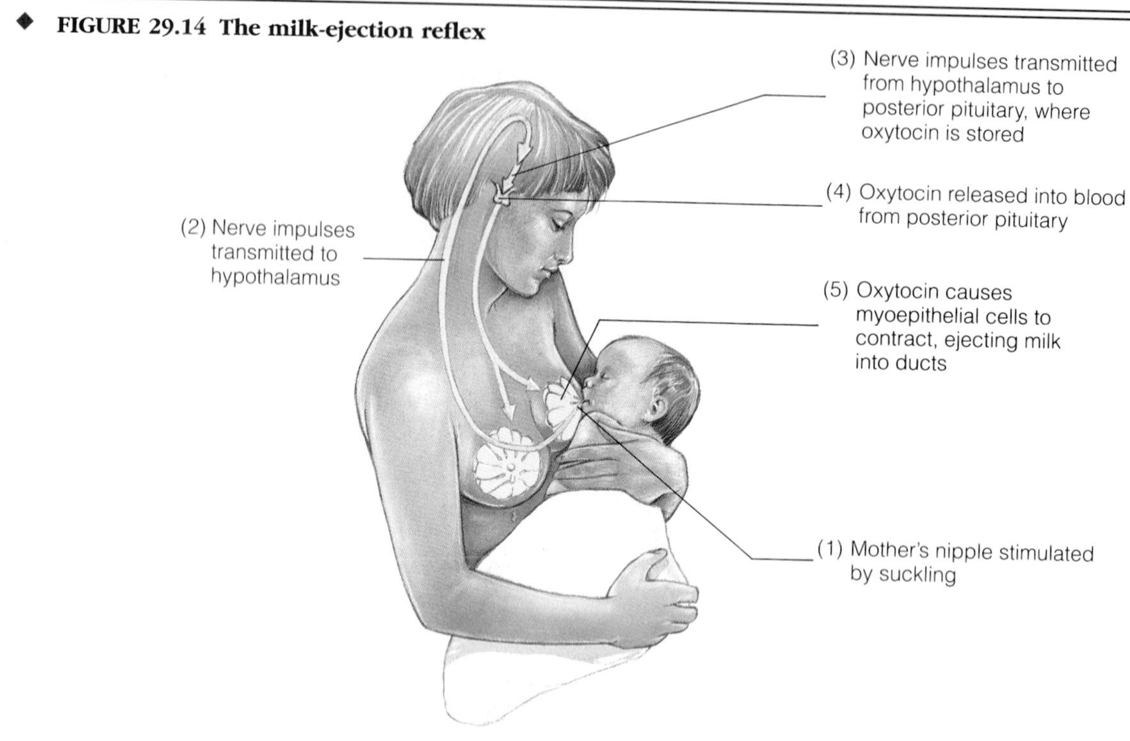

in a single left **umbilical vein,** the right vein having atrophied. The umbilical vein, therefore, carries blood that is rich in nutrients and has a high oxygen content. It is the only fetal vessel to do so. All other vessels of the fetus carry mixtures of arterial and venous blood. Between the placenta and the **umbilicus (navel)** of the fetus, the umbilical arteries and the umbilical vein travel within the umbilical cord. Part of the blood in the umbilical vein enters the liver and follows the usual path of blood through that organ and into the inferior vena cava. The remainder joins with blood from the hepatic portal veins and enters a bypass around the liver called the **ductus venosus.** The ductus venosus directly enters the inferior vena cava. This bypass of the fetal liver by much of the fetal blood causes no functional problems, since the mother's liver serves the needs of the fetus as the result of the exchange of materials across the placenta.

In the fetus, blood from the inferior vena cava enters the right atrium, from which it can follow several pathways. It can follow the usual route, for example, entering the right ventricle and traveling to the lungs within the pulmonary arteries. Since the lungs are collapsed and nonfunctioning, however, they offer considerable resistance to blood flow. Therefore, most of the blood that leaves the right ventricle never reaches the lungs. Instead, it flows through a low-resistance shunt called the **ductus arteriosus,** which connects the pulmonary trunk to the arch of the aorta.

In another pathway, the blood in the right atrium of the fetus enters the left atrium directly by means of a flap-covered opening, called the **foramen ovale,** in the interatrial septum. The blood then passes into the left ventricle and the aorta. The foramen ovale is located in such a position that most of the blood from the inferior vena cava (which is highly oxygenated) is directed through it. Since the brain is supplied by arteries that branch off the aortic arch, the brain thus receives a good portion of this highly oxygenated blood immediately after it leaves the left ventricle. In contrast, much of the blood that enters the heart from the superior vena cava (venous blood) is directed into the right ventricle and pulmonary trunk. As we have noted, most of this blood passes through the ductus arteriosus, which joins the aorta beyond the arch. This blood is destined for the trunk and lower limbs of the fetus, where the oxygen demand is not as vital as that of the brain tissues.

Circulatory Changes at Birth

Separation of the infant from the placenta, which occurs when the umbilical cord is cut following birth, suddenly deprives the newborn of its oxygen source and its means of eliminating carbon dioxide. The resultant oxygen shortage and buildup of carbon dioxide cause the infant to take its first breath. The expansion of the lungs immediately lowers their resistance to blood flow,

◆ **FIGURE 29.15 Fetal circulation**
Blood that is highly oxygenated and rich in nutrients leaves the placenta and enters the fetus through the umbilical vein. After circulating through the fetus, the blood returns to the placenta through the umbilical arteries. The ductus venosus, the foramen ovale, and the ductus arteriosus allow the blood to bypass the fetal liver and lungs. The amount of color in a vessel indicates the relative proportion of oxygenated blood carried by the vessel. Red indicates highly oxygenated blood; blue indicates deoxygenated blood; purple indicates a mixture of oxygenated and deoxygenated blood.

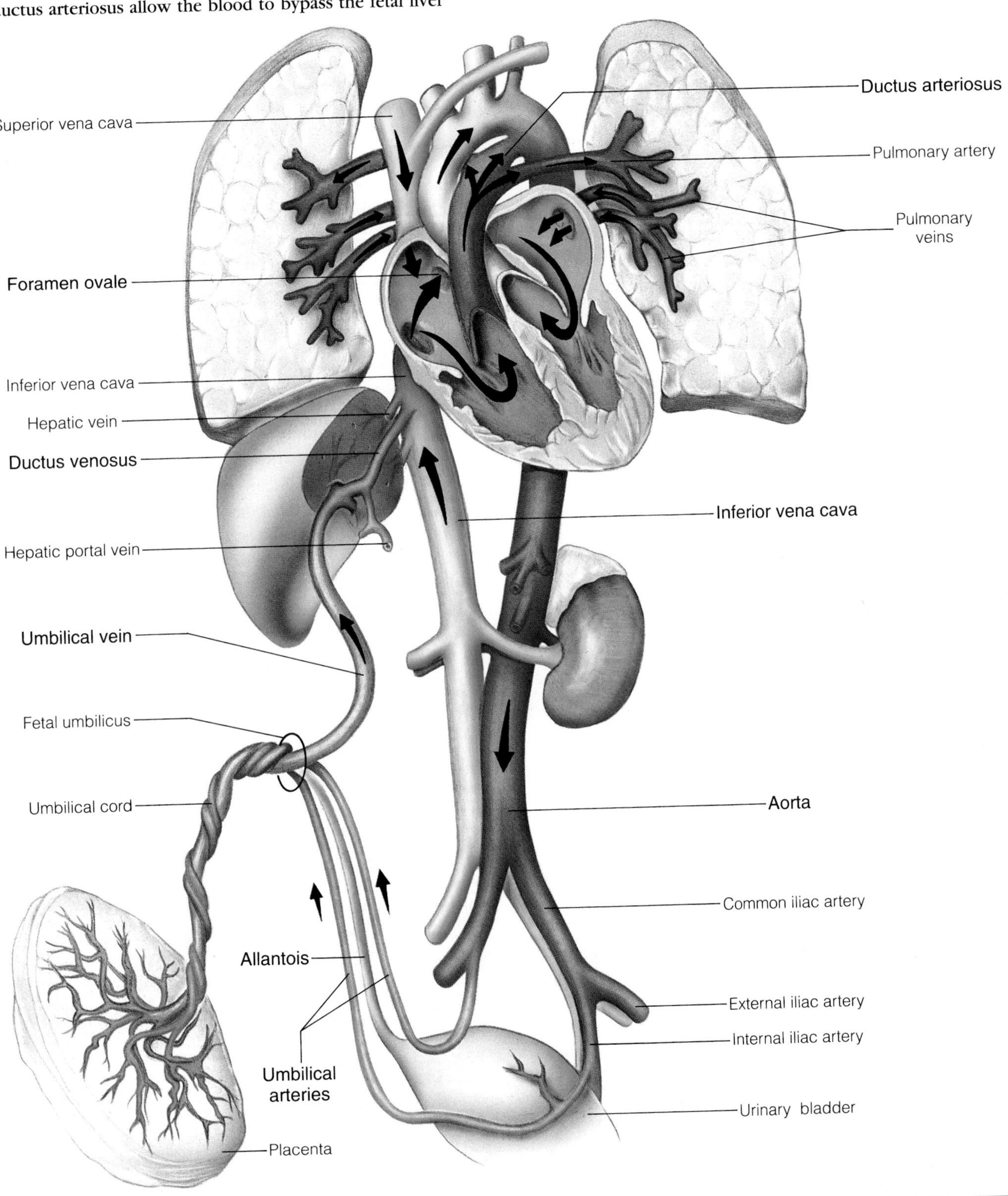

which allows as much as five times more blood to pass through the lungs after birth than before. This action also increases the return of blood from the lungs to the heart, which in turn increases the pressure in the left atrium. The increased pressure in the left atrium forces the flap of the foramen ovale against the interatrial septum, which immediately blocks the opening. All blood that enters the right atrium must then travel to the right ventricle and the pulmonary trunk. Eventually, the foramen ovale becomes permanently closed with fibrous connective tissue, leaving an indentation, called the **fossa ovalis,** on the interatrial septum (Figure 29.16).

If the infant's foramen ovale fails to close, the higher pressure in the left atrium forces some of the blood that has returned from the lungs into the right atrium, thus recirculating the blood back to the lungs again. If the defect is large enough, the volume of blood flowing from the left atrium into the right atrium can cause pulmonary congestion and failure of the right ventricle. A severe defect in the interatrial septum must be surgically corrected.

With the lowered resistance in the lungs following birth, the pressure in the pulmonary arteries and the chambers of the right side of the heart is also reduced. As a result, blood is no longer forced through the ductus arteriosus into the aorta, where the pressure is now greater. In fact, for a short time following birth, blood may flow backward through the duct, from the aorta to the pulmonary artery. Within a few hours, however, the ductus arteriosus constricts, which prevents the reverse flow of blood from the aorta to the pulmonary trunk. Within a few weeks, the ductus arteriosus is generally completely obliterated by fibrous connective tissue and is then called the **ligamentum arteriosum.**

Once the umbilical cord is cut, blood no longer flows through the umbilical vessels or the ductus venosus. The proximal portions of the umbilical arteries continue to supply the urinary bladder—which has developed from the proximal portion of the allantois of the fetus—as **superior vesical arteries.** Beyond the bladder, the umbilical arteries fill in with connective tissue and travel on the inner wall of the abdomen to the umbilicus as the **umbilical ligaments.** In a similar manner, the umbilical vein becomes obliterated with connective tissue and is then called the **ligamentum teres** *(round ligament of the liver).* Because blood is no longer carried in the umbilical vein, the ductus venosus gradually constricts, shunting more and more of the newborn's blood from the hepatic portal vein into the liver. Within a few weeks, the ductus venosus is permanently obliterated, remaining as the **ligamentum venosum.** With the closure of the ductus venosus, the bypass around the liver is lost, so all the blood from the hepatic portal vein must then pass through the liver. This change is of vital importance because the hepatic portal vein drains the digestive system and the infant no longer has the mother's liver available to act on the various substances carried in the blood.

Human Genetics

The study of heredity in humans is called **human genetics.** A person's hereditary material is the DNA of the cells. The DNA provides a program, or set of instructions, that determines personal characters such as eye color or blood type. The DNA program is subdivided into informational units called **genes,** which control specific characters by presiding over the synthesis of chemical components of a person's cells. Each gene is a segment of a DNA chain, and most genes operate by specifying the sequence in which amino acids are joined to form polypeptides or proteins (see Chapter 3). Once formed, the polypeptides or proteins act as enzymes, as hormones, and in other ways to influence particular characters (the actual form of a character that is expressed in a person—for example, brown eyes or blue eyes—is referred to as a *trait*).

Human Chromosomes

Each human somatic cell (all cells except the reproductive cells) is a diploid cell that contains 46 chromosomes. Two of the chromosomes are **sex chromosomes** (two X chromosomes in females and one X and one Y chromosome in males). The remaining 44 chromosomes are called **autosomes.** The 44 autosomes can be grouped into 22 pairs of similar-appearing chromosomes (see Figure 3.29, page 98). One member of each pair contains genes derived from the person's father, and the other member of each pair contains genes derived from the person's mother. Each pair comprises a set of **homologous chromosomes.** The two sex chromosomes of the female (XX) are also homologous, but the two sex chromosomes of the male (XY) are not.

Genes are arranged linearly along the chromosomes, and each gene occupies a specific position, or **locus,** on a particular chromosome. Homologous chromosomes each possess genes that control the same characters.

Genes that control the same character and occupy the same locus on homologous chromosomes are called **alleles** *(ah-lē´-uls).* If a gene (an allele) for a particular character on one chromosome of a homologous pair is identical to the gene (the allele) for that character on the other chromosome of the pair, the person is **homozygous** *(ho-mo-zī´-gus)* for the character. If the two genes differ from one another, the person is **heterozygous** *(het″-er-o-zī´-gus)* for the character.

Dominance

In some cases, one allele for a particular character can mask, or suppress, the effect of another allele for the character. In such a case, the first allele is said to be

◆ **FIGURE 29.16 Changes in circulatory pathways following birth**

These changes result from the closure of the foramen ovale and the transformation of the ductus venosus, the ductus arteriosus, the umbilical vein, and the distal portions of the umbilical arteries into ligaments. Red vessels carry oxygenated blood; blue vessels carry deoxygenated blood.

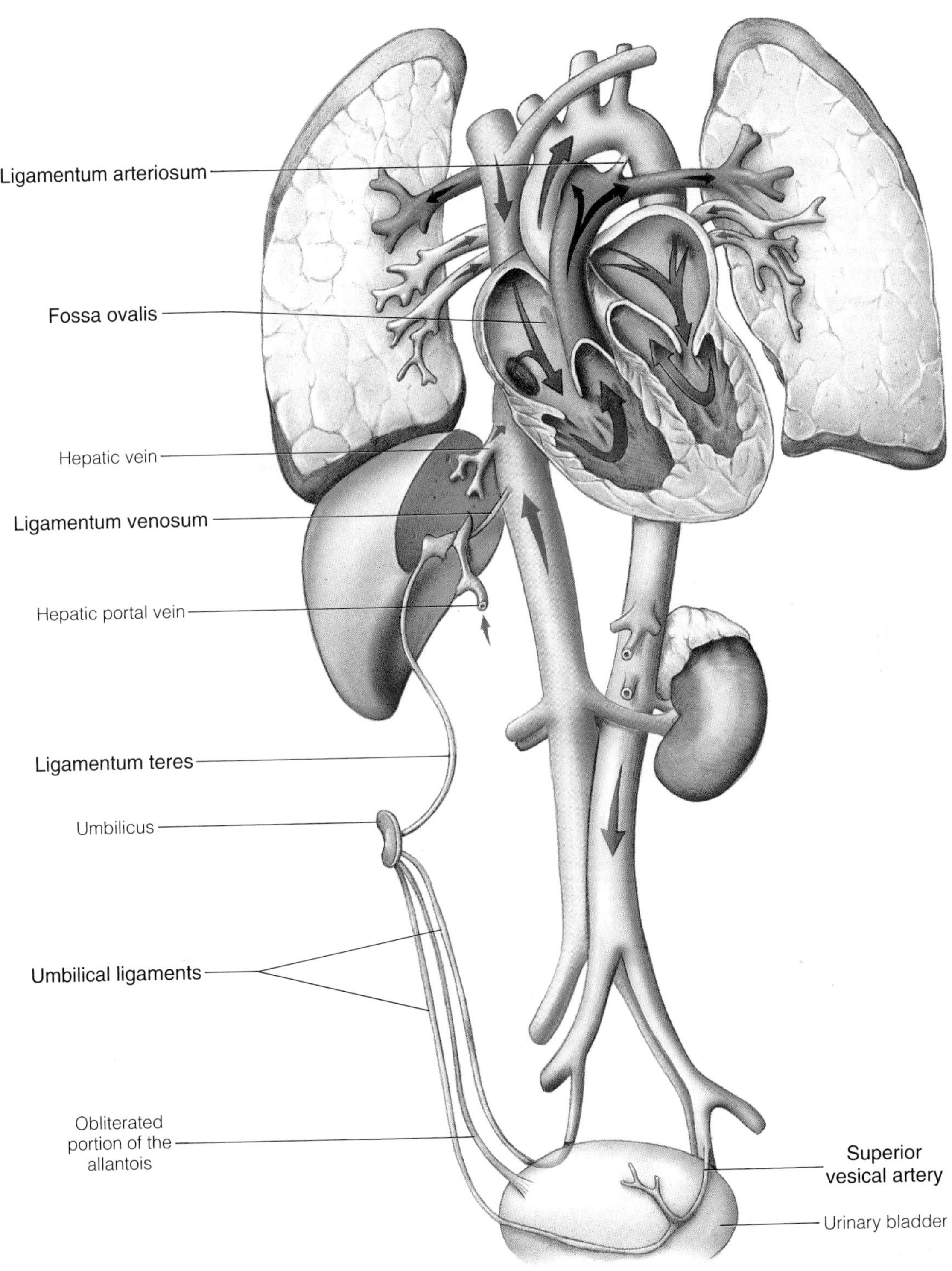

dominant, and the masked allele is referred to as **recessive.** In a homozygous person who possesses two dominant alleles for a particular character (a homozygous dominant), the effect of the dominant alleles is expressed, and the person is said to display a dominant trait with regard to the character. In a homozygous person who possesses two recessive alleles for the character (a homozygous recessive), the effect of these alleles is expressed, and the person displays a recessive trait. In a heterozygous person who possesses both dominant and recessive alleles for the character, the effect of the dominant allele is expressed, and the effect of the recessive allele is masked. Consequently, the person displays the dominant trait.

It is not always the case that one allele for a particular character masks the effects of a second allele for the character—that is, sometimes neither allele for a character on homologous chromosomes is completely dominant or completely recessive. In such a situation, the effects of both alleles are often expressed to some degree, and the person displays a trait that is intermediate between the traits that would be evident if one allele or the other were the only form of the gene present.

Genotype and Phenotype

A person's actual genetic makeup is called his or her **genotype.** The outward expression of the genotype—that is, the expressed traits—is the person's **phenotype.** In the case of dominant and recessive alleles, for example, homozygous dominant, homozygous recessive, and heterozygous individuals have different genotypes, but homozygous dominant and heterozygous individuals have the same phenotype.

Inheritance Patterns

Genetic traits can be passed from one generation to the next. For example, the ability to roll one's tongue is inherited in the manner expected for a dominant allele located on an autosomal chromosome. (The dominant allele for tongue rolling will be indicated by T and the recessive allele by t.) Thus, a person who is either homozygous dominant (TT) or heterozygous (Tt) will be able to roll his or her tongue, but a person who is homozygous recessive (tt) will not.

When a heterozygous individual who possesses both dominant and recessive alleles for tongue rolling forms haploid reproductive cells, the events of meiosis separate the two alleles in such a manner that half the gametes produced contain the dominant allele (T), and half contain the recessive allele (t). If a heterozygous male mates with a heterozygous female, there is a 50% probability that the sperm that fertilizes the ovum will contain the dominant allele (T) and a 50% probability it will contain the recessive allele (t). Similarly, there is a 50% probability that the ovum will contain the dominant allele and a 50% probability it will contain the recessive allele. This relationship and an indication of the possible genotypes of the fertilized ovum can be demonstrated in a *Punnett square,* as in Figure 29.17. From this Punnett square, it is evident that there is a 25% probability that the person developing from the fertilized ovum will be homozygous dominant (TT) for tongue rolling, a 50% probability that the person will be heterozygous (Tt), and a 25% probability that the person will be homozygous recessive (tt). Phenotypically, there is a 75% probability that the person will display tongue-rolling ability (recall that a person who is either homozygous dominant or heterozygous for tongue rolling will be able to roll his or her tongue), and a 25% probability that the person will not.

◆ FIGURE 29.17 Use of a Punnett square to indicate possible genotypes of a fertilized ovum when a male and female who are both heterozygous for tongue rolling mate with one another

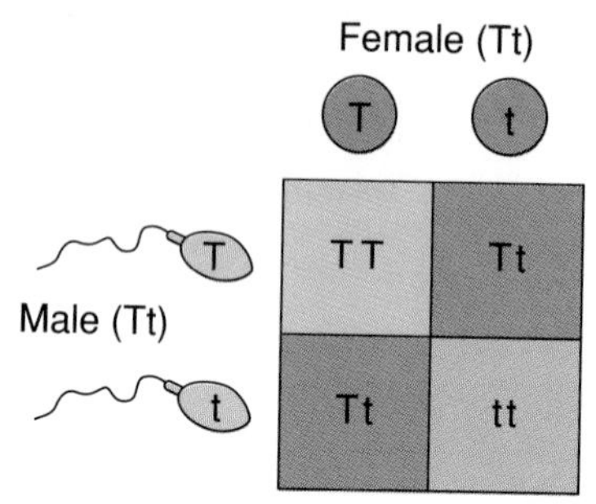

The pattern of inheritance of a genetic trait in the members of different generations of a family can be recorded by constructing a *pedigree pattern.* Such patterns are used in human genetics to examine the manner in which particular traits are inherited. For example, the pedigree pattern for tongue-rolling ability illustrated in Figure 29.18 indicates that this ability is inherited in the manner expected for a dominant allele.

Family pedigrees have proven to be useful for *genetic counseling,* during which prospective parents are advised concerning the chances of their children having a given abnormality. For some conditions, the counseling can rely heavily on pedigree patterns. For example, if one parent is heterozygous for the sickle-cell anemia gene (see Chapters 3 and 18)—which causes abnormal hemoglobin in red blood cells—each child has a 50% chance of inheriting the gene.

Most situations that require genetic counseling are too complicated to rely entirely on pedigree patterns, since they involve rare recessive alleles that remain masked in heterozygous individuals for many generations and are

◆ **FIGURE 29.18 Pedigree of a hypothetical family illustrating the inheritance of tongue-rolling ability**

The facts that parents with tongue-rolling ability can have children without this ability and parents without this ability have only non-tongue-rolling children indicates that tongue-rolling ability is inherited in the manner expected for a dominant allele. Moreover, the fact that parents who both have tongue-rolling ability have a child without this ability shows that the parents are heterozygous for this ability.

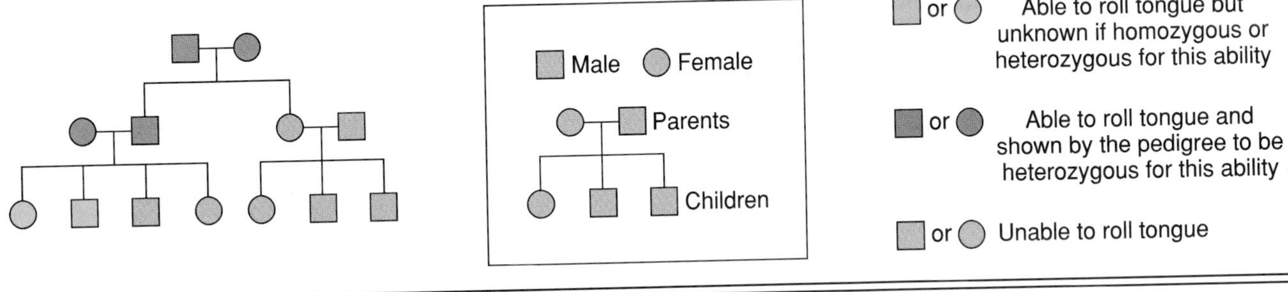

harmful only when present in the homozygous form. To detect such conditions, it is often useful to perform biochemical analyses of cells and blood taken from the parents in an attempt to identify products of abnormal recessive alleles that may be present.

X-Linked Traits

Only a single X chromosome occurs in males. Consequently, the effect of a recessive allele located on the X chromosome of a male is generally expressed because there is no dominant allele to mask it. (The Y chromosome apparently has few or none of the same genes as the X chromosome, and it therefore does not mask recessive alleles on the X chromosome.) As a result, the effects of recessive alleles located on an X chromosome are more frequently expressed in males than in females. This situation is evident in the case of the blood-clotting abnormality hemophilia A (see Chapter 18), which is inherited in the manner expected for a recessive allele located on the X chromosome. To display this abnormality, females must have the recessive allele on both X chromosomes, but males must have the allele only on their single X chromosome. If a heterozygous female who has the recessive allele for hemophilia A on one of her X chromosomes mates with a normal male, none of their female children will have hemophilia A. However, there is a 50% probability that their male children will have this disorder (Figure 29.19). In this case, neither parent has hemophilia A, but the mother is termed a *carrier* because she possesses a recessive allele for hemophilia A that can be passed on to her children.

◆ **FIGURE 29.19 Inheritance of hemophilia A**

Use of a Punnett square to indicate possible genotypes of a fertilized ovum when a normal male who does not have hemophilia A mates with a female who does not have hemophilia A but does have the recessive allele for hemophilia A on one of her X chromosomes (indicated by X^h).

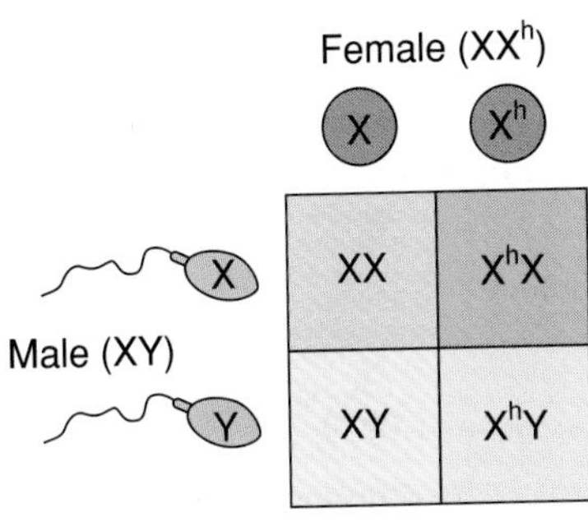

Expression of Genetic Potential

In previous sections it was assumed that the effect of a dominant allele for a particular character is always expressed and, consequently, that a person who possesses such an allele will display the dominant trait with regard to the character. Although this assumption is generally true, cases are known in which a person possessing a dominant allele for a character does not display the dominant trait. These cases serve to emphasize the fact that individual genes operate as part of a person's overall genetic makeup and that, at times, the activities of genes for other characters may modify or otherwise influence the expression of the effect of a particular gene.

In addition, there is evidence that a phenomenon called *genomic imprinting* takes place in parental cells. This phenomenon causes some genes to behave differently when they are inherited from a person's mother than when they are inherited from the person's father. Genomic imprinting, for example, may lead to the expression of the effect of a particular gene when the gene is inherited from a person's mother but not when it is inherited from the person's father (or vice versa).

Environmental factors can also influence the expression of the effects of particular genes. For example, a person may have the genetic capacity to attain great height, but the actual attainment of this height depends on the availability of the proper amounts and types of food. Thus, particular genes represent potentialities whose full effects and complete expression may be influenced by the activities of other genes, by genomic imprinting, and by environmental factors.

Fetal Testing

When it is suspected that a child may be born with a genetic abnormality, it is useful to determine the condition of the fetus before birth. Fetal testing is often accomplished by collecting samples of the amniotic fluid through a procedure called *amniocentesis (am″-ne-o-sen-tē′-sis)* (Figure 29.20).

In this procedure, the locations of the fetus and placenta are first determined by recording the echoes of ultrasound waves directed at the uterus through the pregnant woman's abdominal wall. A hollow needle attached to a syringe is then inserted through the abdominal wall and uterus into the amniotic fluid surrounding the fetus. Some of the amniotic fluid and fetal cells present in the fluid are withdrawn through the needle and collected in the syringe.

The amniotic fluid can be tested to determine the presence of certain chemicals that indicate specific diseases of the fetus. The cells collected during amniocentesis are cultured in laboratory dishes for several weeks and then subjected to microscopic examination and biochemical testing that can detect chromosomal abnormalities or DNA markers of genetic disease.

Because of the danger of injury to the fetus during amniocentesis if sufficient amniotic fluid is not present, this procedure is not usually done until the fourteenth week of pregnancy.

Another method of fetal testing is *chorionic villi sampling.* In this procedure, a small tube is inserted into the

◆ **FIGURE 29.20 Amniocentesis**

A sample of amniotic fluid containing loose fetal cells is collected in a syringe and transferred to a test tube. The sample is centrifuged, causing the fetal cells to collect at the bottom of the tube. The cells are then cultured and subjected to various biochemical tests and microscopic examination.

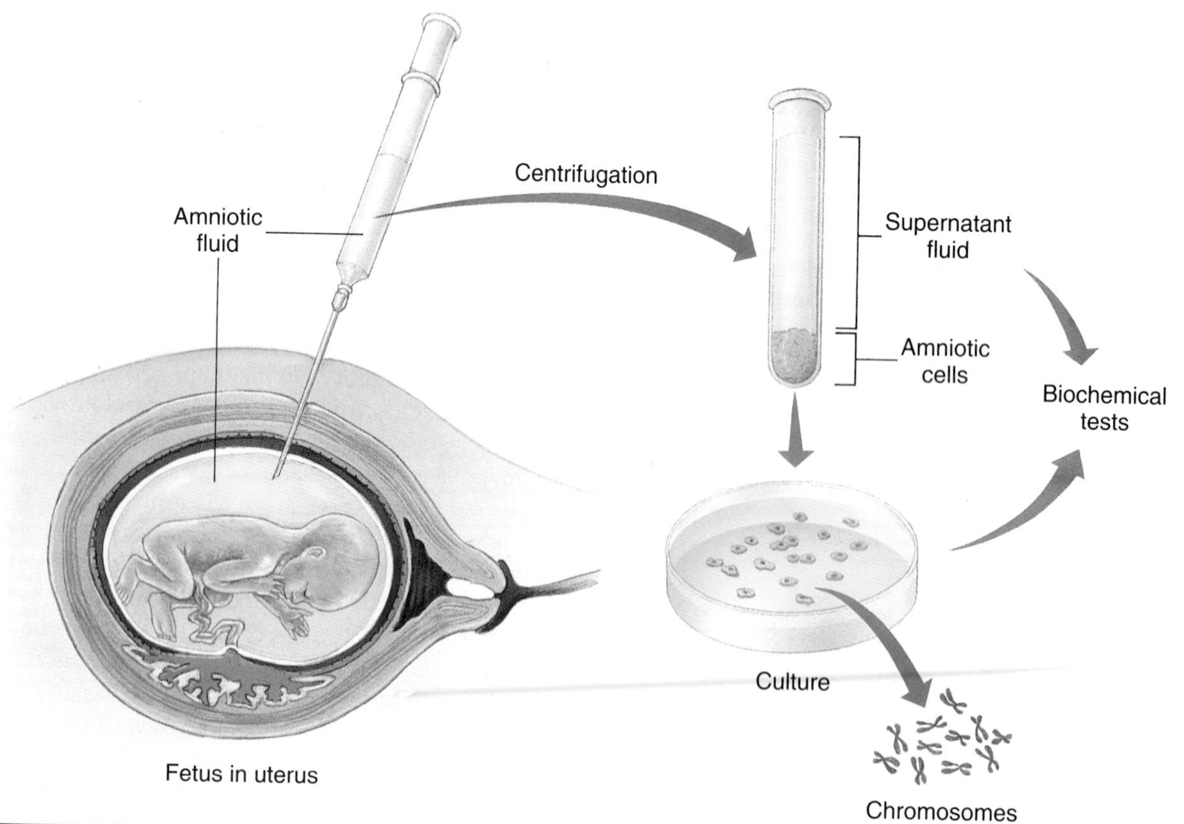

CONDITIONS OF CLINICAL SIGNIFICANCE

Galactosemia

Galactosemia *(gah-lak″-to-sē′-me-ah)* is a genetic disorder that is inherited in the manner expected for a recessive allele located on an autosomal chromosome. Babies afflicted with this condition cannot tolerate the milk sugar lactose. If they drink milk, they develop cataracts of the eyes and show symptoms of damage to the brain and liver. However, if newborn galactosemic babies are taken off milk soon after birth and fed special lactose-free formulas, they can develop normally.

Study Outline

◆ **PREGNANCY** pp. 926–937

For pregnancy to occur, the secondary oocyte released at ovulation must be fertilized by a spermatozoon.

Transport of the Secondary Oocyte. The secondary oocyte released at ovulation is swept into a uterine tube and moved along the tube.

Transport of Sperm. When sperm can penetrate the cervical mucus, they move into the uterus and reach the upper portion of the uterine tubes. Mechanisms of transport other than the motility of the sperm are apparently involved.

Capacitation. Following ejaculation, sperm must remain within the female reproductive tract for several hours before they become capable of fertilizing the secondary oocyte. During this period, they undergo capacitation.

Fertilization. The first sperm to penetrate the zona pellucida contacts the secondary oocyte. Membranes of the sperm and oocyte fuse, and the sperm's head is drawn into oocyte cytoplasm. The secondary oocyte completes second meiotic division, giving rise to mature ovum. Membranes of sperm and ovum pronuclei rupture, and their chromosomes come together.

Block to Polyspermy. Fusion of the membrane of the sperm that enters the secondary oocyte with the membrane of the oocyte brings about changes that prevent the entry of additional sperm.

Sex Determination.

1. If secondary oocyte is fertilized by a sperm containing X sex chromosome, a female normally develops.
2. If secondary oocyte is fertilized by a sperm containing Y sex chromosome, a male normally develops.

Development and Implantation of the Blastocyst.

1. Blastocyst is fluid-filled sphere of cells. Embryo develops from inner cell mass of blastocyst; outer layer of cells is *trophoblast.*
2. After several days within lumen of uterus, trophoblast cells release enzymes that digest cells of the endometrium, and blastocyst implants in uterine wall.

Maintenance of the Endometrium.

1. Endometrium is maintained and menses is prevented by high levels of estrogens and progesterone.
2. During pregnancy, estrogens and progesterone are produced by corpus luteum and placenta.
3. Human chorionic gonadotropin is major factor causing corpus luteum to remain when pregnancy occurs.

Development of the Placenta. Provides fetal nutrient supply and waste removal; chorionic villi cover extensive surface area; connected to fetus by blood vessels in umbilical cord.

DECIDUA. Portion of endometrium that is discarded at birth or with menses.

DECIDUA BASALIS AND FETAL CHORION. Combine to form placenta.

Formation of Embryonic Membranes and Germ Layers.

EMBRYONIC MEMBRANES.

AMNION. Fluid filled, surrounds fetus; provides protection; permits movement; maintains temperature.

YOLK SAC. Develops from endoderm of embryonic disc; source of early blood cells and germ cells.

ALLANTOIS. Develops from endoderm of embryonic disc; allows for waste storage; allantoic blood vessels carry fetal blood to and from placenta.

CHORION. Develops from trophoblast plus mesoderm; forms fetal portion of placenta.

GERM LAYERS. Ectoderm and endoderm formed by separation of embryonic disc into two layers; mesoderm formed from cells of embryonic disc that spread between ectoderm and endoderm. *Coelom* formed when mesoderm sheet splits into parietal and visceral layers.

Gestation. Normally lasts about 280 days from the last menstrual flow.

Twins.

IDENTICAL TWINS. An inner cell mass that develops from single fertilized ovum separates.

FRATERNAL TWINS. Two ova fertilized by separate spermatozoa.

◆ **PARTURITION** pp. 937–941

Act of giving birth to a baby at the end of pregnancy.

Labor. Series of events by which parturition is accomplished.

FIRST STAGE: CERVICAL DILATION. Uterine contractions push fetus against cervix, forcing cervix to dilate.

SECOND STAGE: DELIVERY OF THE BABY. Uterine contractions (assisted by contractions of abdominal muscles) move fetus along cervical and vaginal canals to exterior.

THIRD STAGE: DELIVERY OF THE PLACENTA. Uterine contractions separate placenta from wall of uterus and expel placenta as afterbirth.

Factors Associated with Parturition. Though what initiates parturition is not certain, estrogens and prostaglandins are thought to be involved; increasing uterine sensitivity to oxytocin as pregnancy progresses may also be important.

◆ **CONDITIONS OF CLINICAL SIGNIFICANCE: PREGNANCY** pp. 939–941

Disorders of Pregnancy.

ECTOPIC PREGNANCY. Implantation occurs at a site other than uterus.

PLACENTA PREVIA. Placenta covers inner opening of cervical canal and is expelled before baby is born, causing mother to hemorrhage.

TOXEMIA OF PREGNANCY. May be caused by numerous factors; often shows tendency for salt and fluid retention; results in lung, kidney, and brain edema.

Birth Control.

BIRTH CONTROL METHODS AVAILABLE TO THE MALE.

COITUS INTERRUPTUS. Withdrawal of penis before ejaculation.

CONDOM. Cover pulled onto penis.

VASECTOMY. Ductus deferens is tied and cut.

BIRTH CONTROL METHODS AVAILABLE TO THE FEMALE.

DETECTION OF OVULATION. Requires abstinence from sexual intercourse during fertile period.

DIAPHRAGM. Cover for cervix to prevent entrance of sperm into uterus.

INTRAUTERINE DEVICE. Device inserted into uterus that prevents implantation of blastocyst.

ORAL CONTRACEPTIVES. Pills containing synthetic estrogenlike and progesteronelike substances that inhibit ovulation, fertilization, or implantation.

CHEMICAL CONTRACEPTION. Chemicals introduced into vagina that act as physical barriers to sperm and as spermicides.

TUBAL LIGATION. Tying and severing uterine tubes.

SUBCUTANEOUS IMPLANTS. Capsules of synthetic progesteronelike substance implanted under skin; inhibit ovulation for about 5 years.

VAGINAL POUCHES Condomlike pouches inserted into the vagina.

INDUCED ABORTION. Removal of implanted embryo or fetus from uterus.

◆ **LACTATION** pp. 942–943

Preparation of the Breasts for Lactation.

1. At puberty, estrogens stimulate growth and branching of duct system within breasts, and fat deposition leads to considerable breast enlargement. Increased level of progesterone at puberty also contributes to breast growth.
2. During pregnancy, breasts become fully developed due to influences of high levels of estrogens and progesterone as well as prolactin.

Milk and Colostrum.

1. There is a delay of several days following parturition before true milk is secreted.
2. During this period, the breasts secrete a fluid called colostrum.

Production of Milk.

1. Prolactin is major hormone responsible.
2. Mechanical stimulation of nipples by suckling initiates reflex that stimulates prolactin secretion for some time after childbirth.

Milk Ejection Reflex. Suckling reflexively causes oxytocin release, and oxytocin causes myoepithelial cells to contract, ejecting milk into ducts of breasts.

◆ **FETAL CIRCULATION** pp. 943–946

Fetal Blood Pathways.

BLOOD TO PLACENTA. Carried by umbilical arteries.

UMBILICAL VEIN. Carries high-nutrient, high-oxygenated blood from placenta.

FETAL VESSELS. Most carry mixed arterial and venous blood.

FETAL CIRCUIT.

Lungs. Collapsed; therefore provide much resistance to blood flow.

Ductus venosus. Serves as liver bypass.

Ductus arteriosus. Carries blood from pulmonary trunk to aorta.

Foramen ovale. Opening in interatrial septum.

Circulatory Changes at Birth.

1. Pulmonary circuit becomes functional.
2. Foramen ovale closes; forms fossa ovalis.
3. Ductus arteriosus is obliterated; forms ligamentum arteriosum.
4. Umbilical arteries become obliterated; form umbilical ligaments.
5. Ductus venosus is obliterated; forms ligamentum venosum.

◆ **HUMAN GENETICS** pp. 946–951
The study of heredity in humans.

Human Chromosomes. Each human somatic cell is diploid cell containing 46 chromosomes: 44 autosomes and 2 sex chromosomes. Genes controlling same character and occupying same locus on homologous chromosomes are called *alleles.*

Dominance. In some cases, one allele for a particular character can mask, or suppress, effect of another allele for that character. First allele is *dominant;* masked allele is *recessive.*

Genotype and Phenotype. *Genotype:* a person's actual genetic makeup. *Phenotype:* outward expression of genotype—that is, person's expressed traits.

Inheritance Patterns. Genetic traits can be passed from one generation to next, for example, inheritance of tongue-rolling ability.

X-Linked Traits. Only a single X chromosome occurs in males; effects of recessive alleles located on X chromosome more frequently expressed in males than females.

Expression of Genetic Potential. Particular genes represent potentialities whose full effects and complete expression may be influenced by activities of other genes and by environmental factors.

Fetal Testing. Chromosomal abnormalities and genetic disease can be detected by amniocentesis and chorionic villi sampling.

◆ **CONDITIONS OF CLINICAL SIGNIFICANCE: INHERITANCE** pp. 952–953
Many diseases and other abnormalities have genetic origins.

Nondisjunction. A major cause of abnormal numbers of chromosomes in cells. Occurs during meiosis.

Down's Syndrome. Most common and extreme form is due to three copies of autosome 21. Characterized by consistent pattern of physical features, almost always by mental retardation.

Abnormal Numbers of Sex Chromosomes. Cause a number of different abnormalities.

Galactosemia. Genetically based inability to tolerate the milk sugar lactose.

Self-Quiz

1. Fertilization normally occurs within the: (a) ovary; (b) uterine tube; (c) uterus.
2. The fusion of the membrane of the sperm that enters the secondary oocyte with the membrane of the oocyte brings about changes that prevent the entry of additional sperm. True or False?
3. When fertilization involves a spermatozoon containing a Y chromosome the zygote normally: (a) develops into a male; (b) develops into a female; (c) has an equal probability of developing into either a male or a female.
4. During implantation, the cells of the inner cell mass release enzymes that digest the cells of the trophoblast. True or False?
5. During pregnancy, the corpus luteum secretes: (a) progesterone; (b) human chorionic somatomammotropin; (c) human chorionic gonadotropin.
6. During pregnancy, the corpus luteum degenerates about ten days after implantation. True or False?
7. The superficial layer of the endometrium is called the: (a) chorion; (b) decidua; (c) embryonic disc; (d) zona pellucida.
8. The placenta is formed from: (a) the decidua basalis; (b) the chorion; (c) (a) and (b) combined.
9. Maternal blood normally mixes with fetal blood within the placenta in order to exchange food materials and oxygen. True or False?
10. The placenta produces: (a) estrogens; (b) follicle-stimulating hormone; (c) luteinizing hormone.
11. The membrane surrounding the cavity that forms in the inner cell mass is the: (a) amnion; (b) yolk sac; (c) allantois; (d) chorion.

12. The layer of cells on the side of the embryonic disc toward the yolk sac is composed of endodermal cells. True or False?
13. The chorion forms when mesoderm lines the inside of the trophoblast. True or False?
14. Which of the following is supplied by the blood vessels that carry fetal blood to and from the placenta? (a) yolk sac; (b) amnion; (c) allantois; (d) chorion.
15. Which of the following are *not* formed from mesoderm? (a) muscles; (b) bone; (c) blood; (d) all of these are formed from mesoderm.
16. Oxytocin is particularly effective at stimulating uterine contractions early in pregnancy, but it is unable to do so late in pregnancy. True or False?
17. At puberty, the alveoli of the breasts develop fully, and milk secretion begins. True or False?
18. Prolactin production: (a) occurs at a relatively high level prior to puberty; (b) ceases when the placenta is lost at parturition; (c) can be stimulated for some time following parturition by vigorous, frequent suckling.
19. The umbilical arteries of the fetus carry blood that is rich in nutrients and has a high oxygen content. True or False?
20. Most of the blood that leaves the right ventricle of the fetus never reaches the lungs. Instead, it enters the aorta through a shunt called the: (a) ductus venosus; (b) foramen ovale; (c) ductus arteriosus; (d) umbilicus.
21. Following birth the umbilical vein becomes obliterated with connective tissue and is called the: (a) fossa ovalis; (b) ligamentum teres; (c) ligamentum venosum; (d) umbilical ligament.
22. A person's actual genetic makeup is called his or her: (a) phenotype; (b) genotype; (c) heterotype.
23. In a male, the effects of recessive alleles located on the X chromosome are always masked by dominant alleles located on the Y chromosome. True or False?

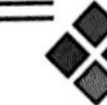

APPENDIX 1

Regional Anatomy

The best way to learn human anatomy is to dissect human cadavers. However, for a number of reasons, most undergraduate anatomy courses dissect some animal other than a human—generally a cat or fetal pig. Because figures in textbooks are of humans, what is observed in the laboratory animals must be extrapolated to the text illustrations.

One problem with relying so heavily on drawings is that they tend to be somewhat idealized. It is a simple matter to alter a drawing to include or exclude particular structures. Therefore, the drawings often do not exactly represent the spatial relationships that exist in the body. Students who rely entirely on drawings may assume that the body is just as neatly organized, with the various structures as clearly separated, as are the drawings.

The following photographs of actual human dissections clarify the relationships between the structures. All of the labeled structures in the photographs are also illustrated and discussed in the text. The photographs should be used to reinforce the anatomical relationships seen in the drawings and the laboratory.

◆ **FIGURE A.1 Superficial structures of the right anterior neck**

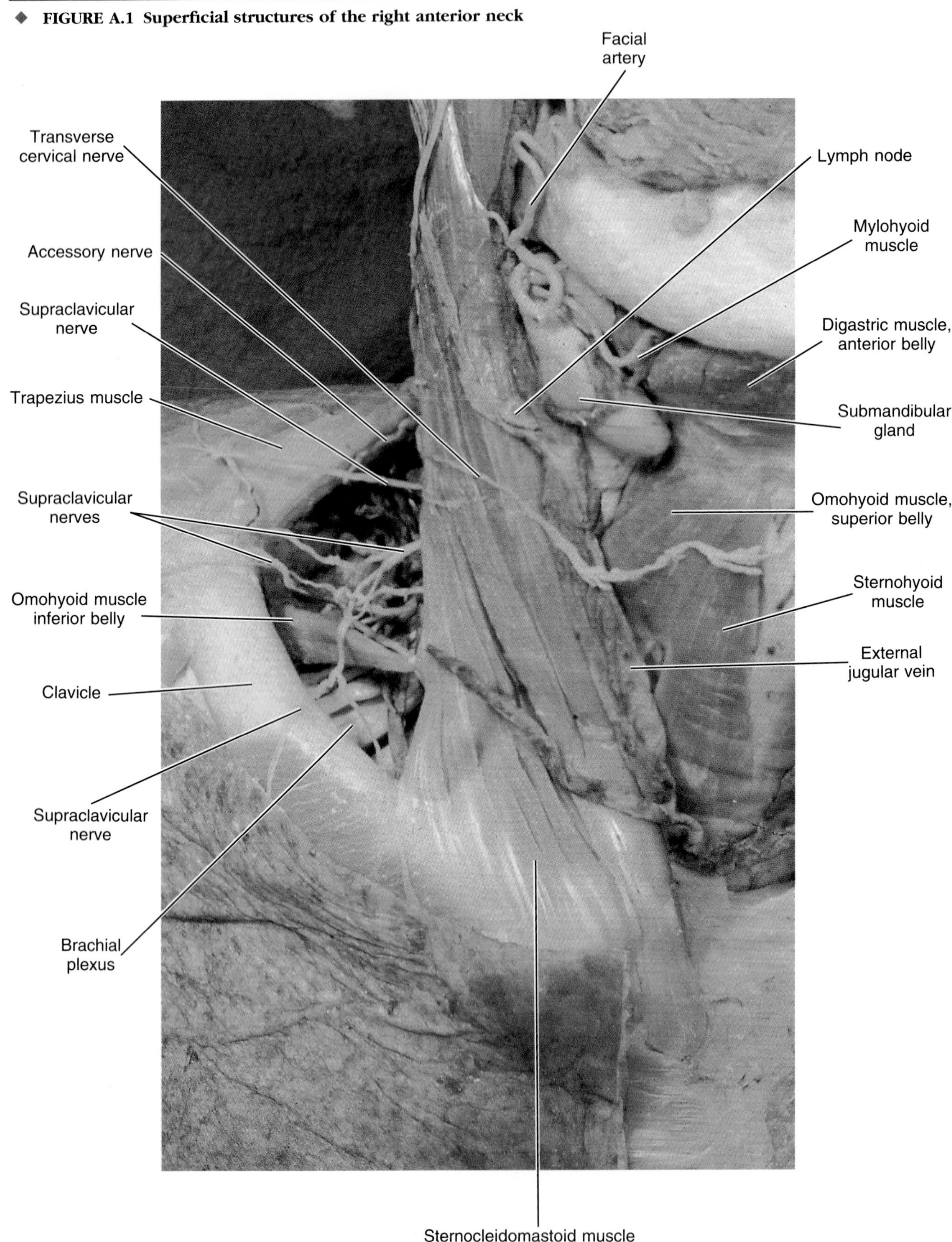

◆ **FIGURE A.2 Deeper structures of the right anterior neck**

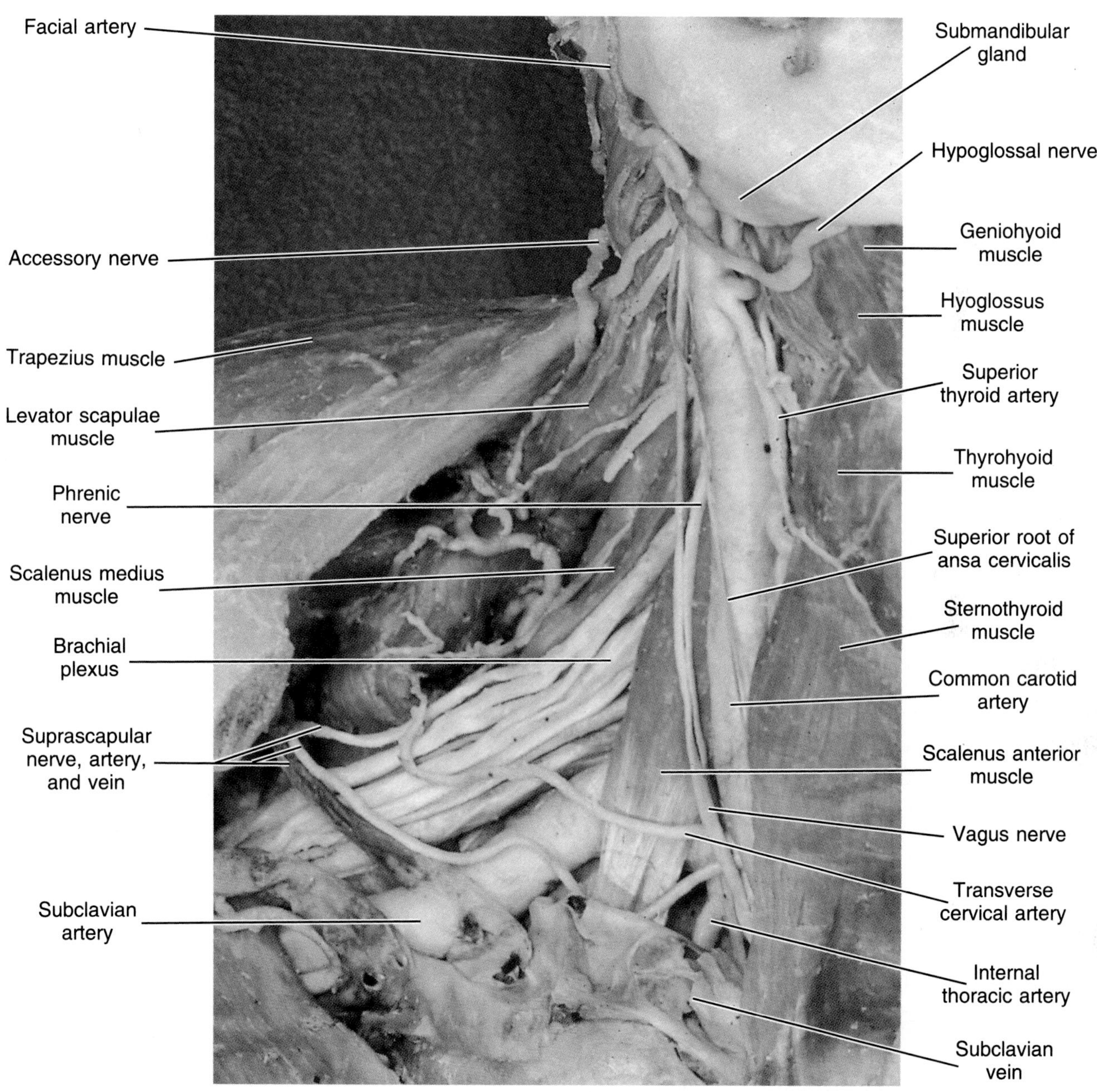

◆ **FIGURE A.3 Superficial muscles of the right anterior chest and arm**

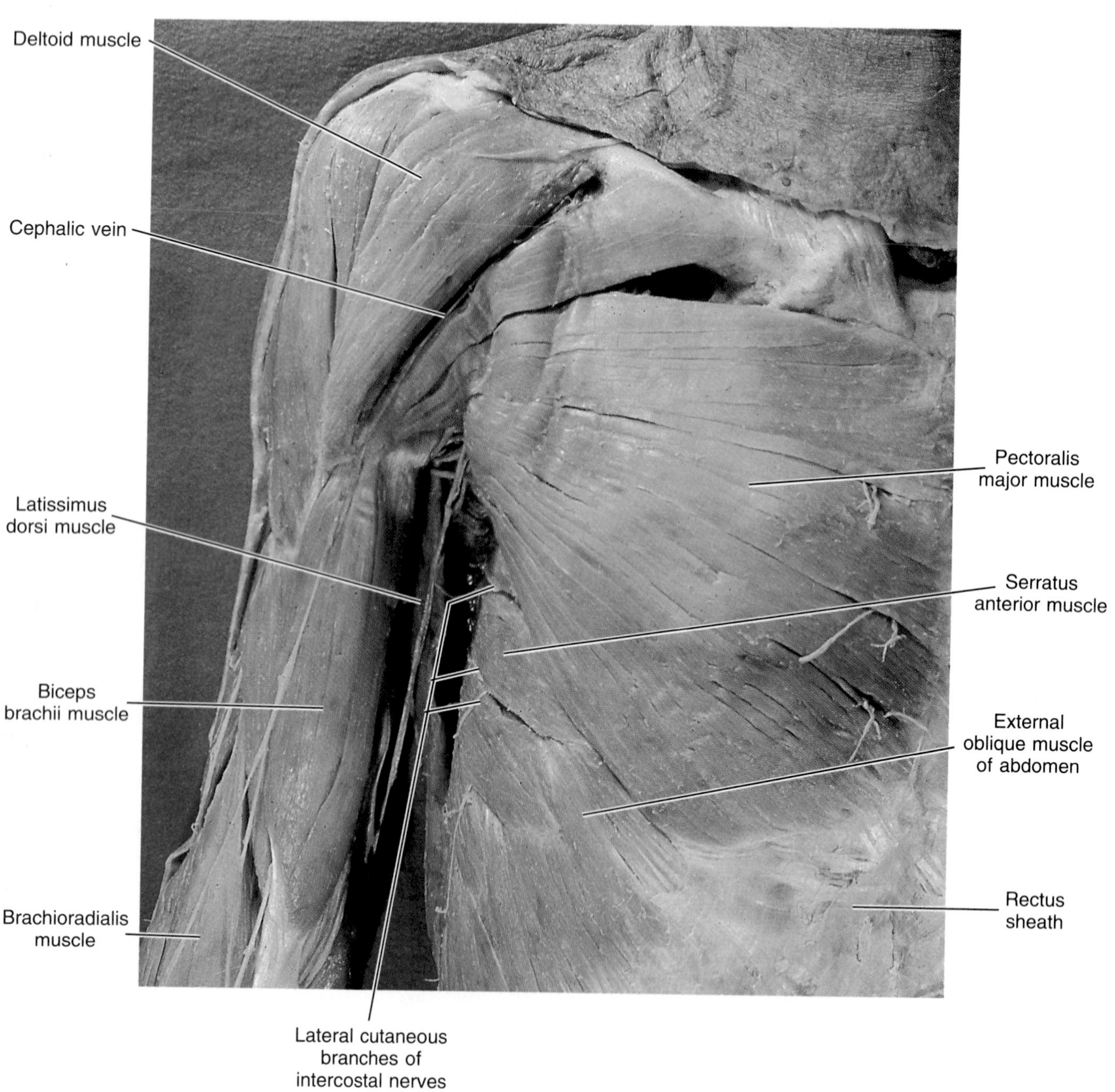

◆ **FIGURE A.4 Deeper muscles of the right anterior chest and arm**

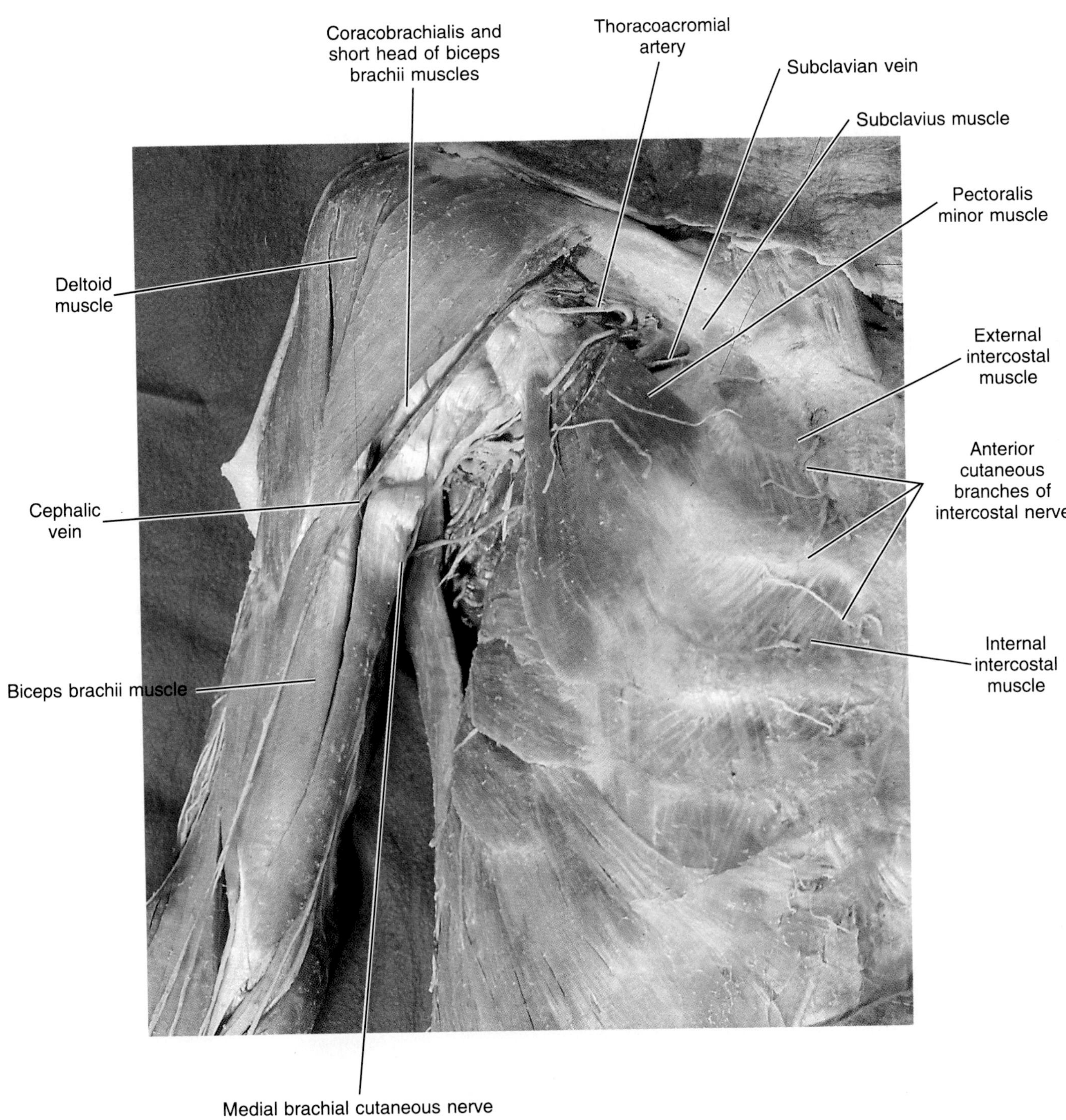

◆ **FIGURE A.5 Anterior view of the contents of the thorax and upper abdomen**

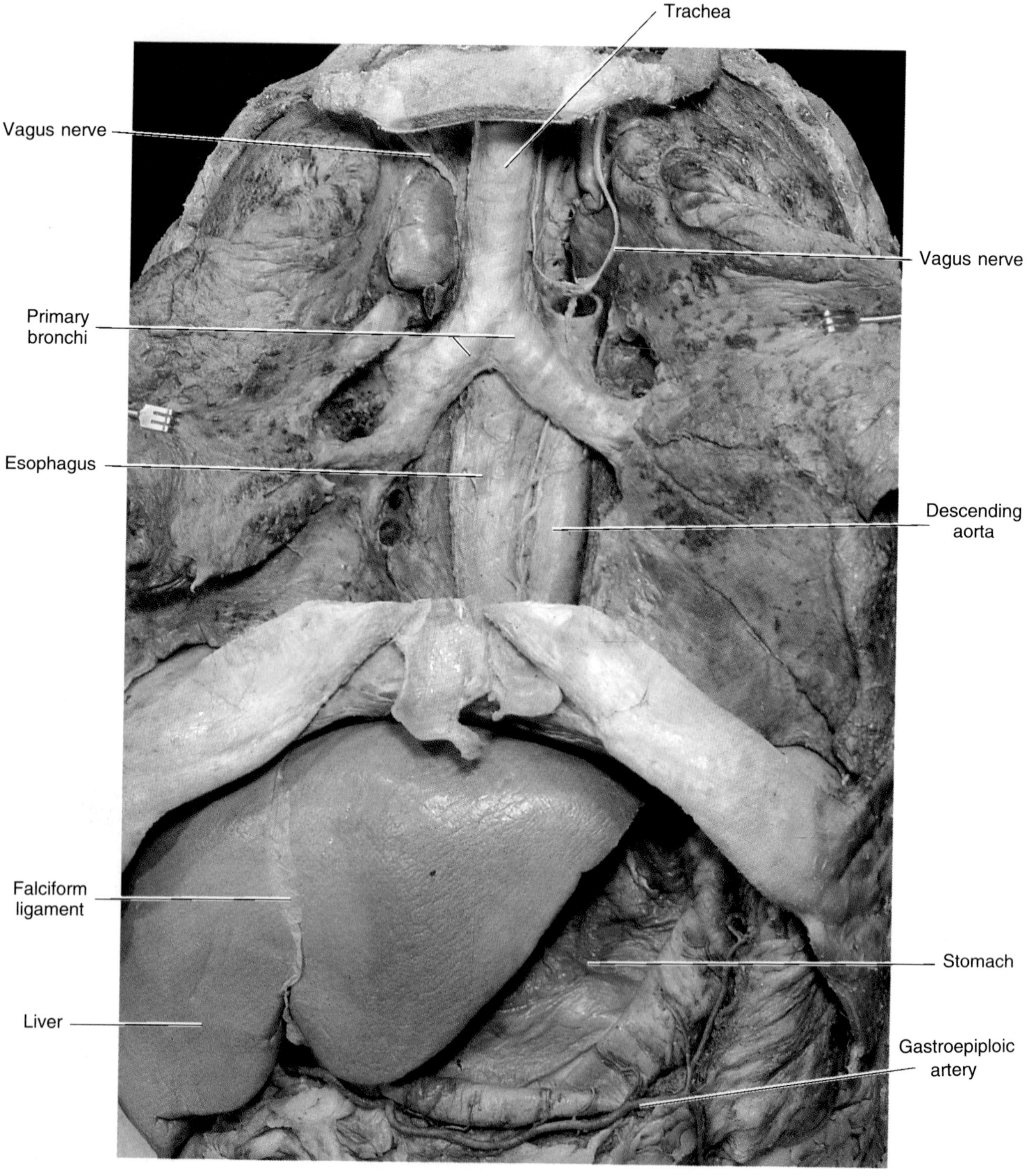

◆ **FIGURE A.6 Anterior view of the contents of the abdominal cavity**

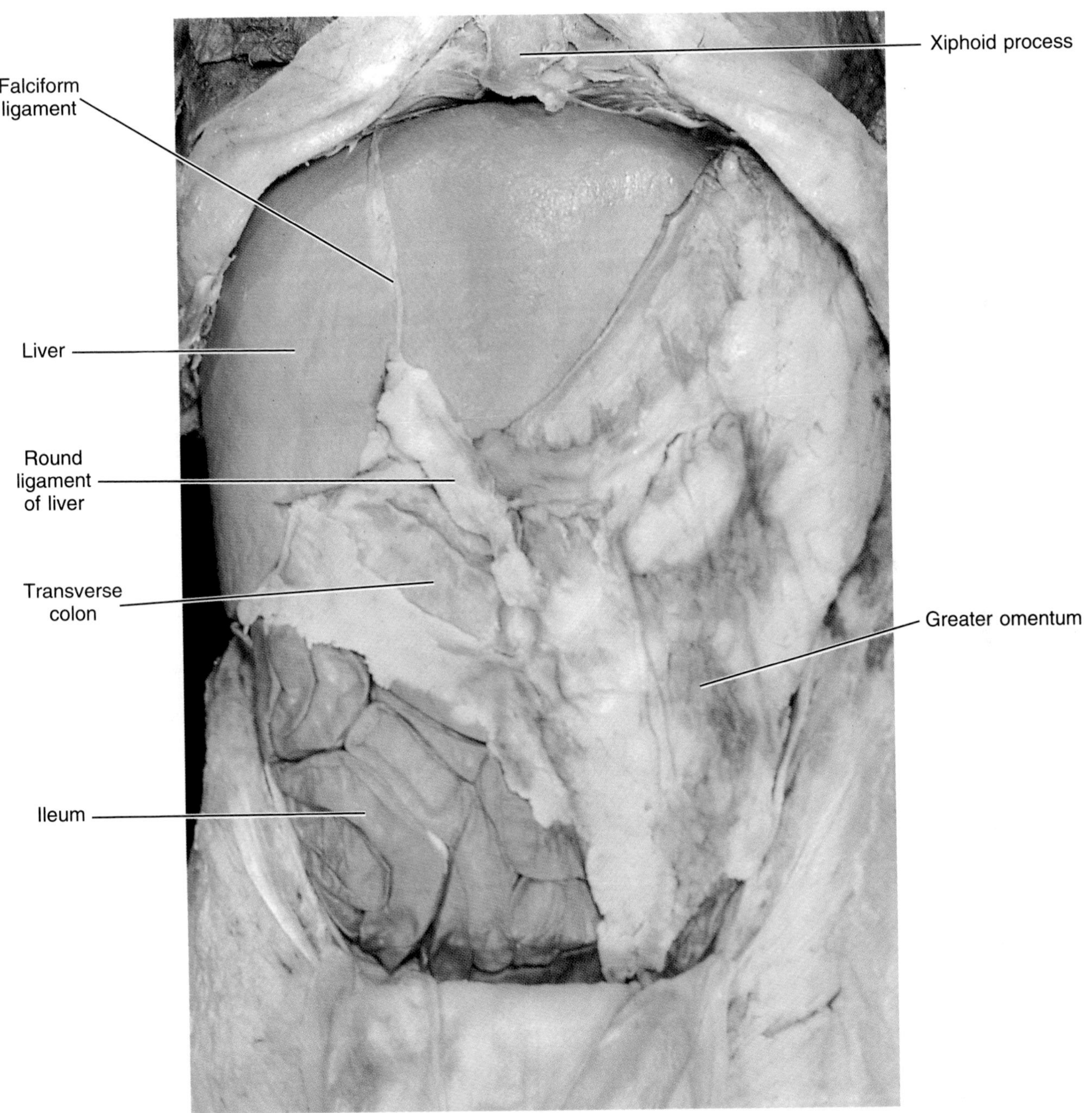

◆ **FIGURE A.7 Superficial posterior muscles of the upper trunk and arm**

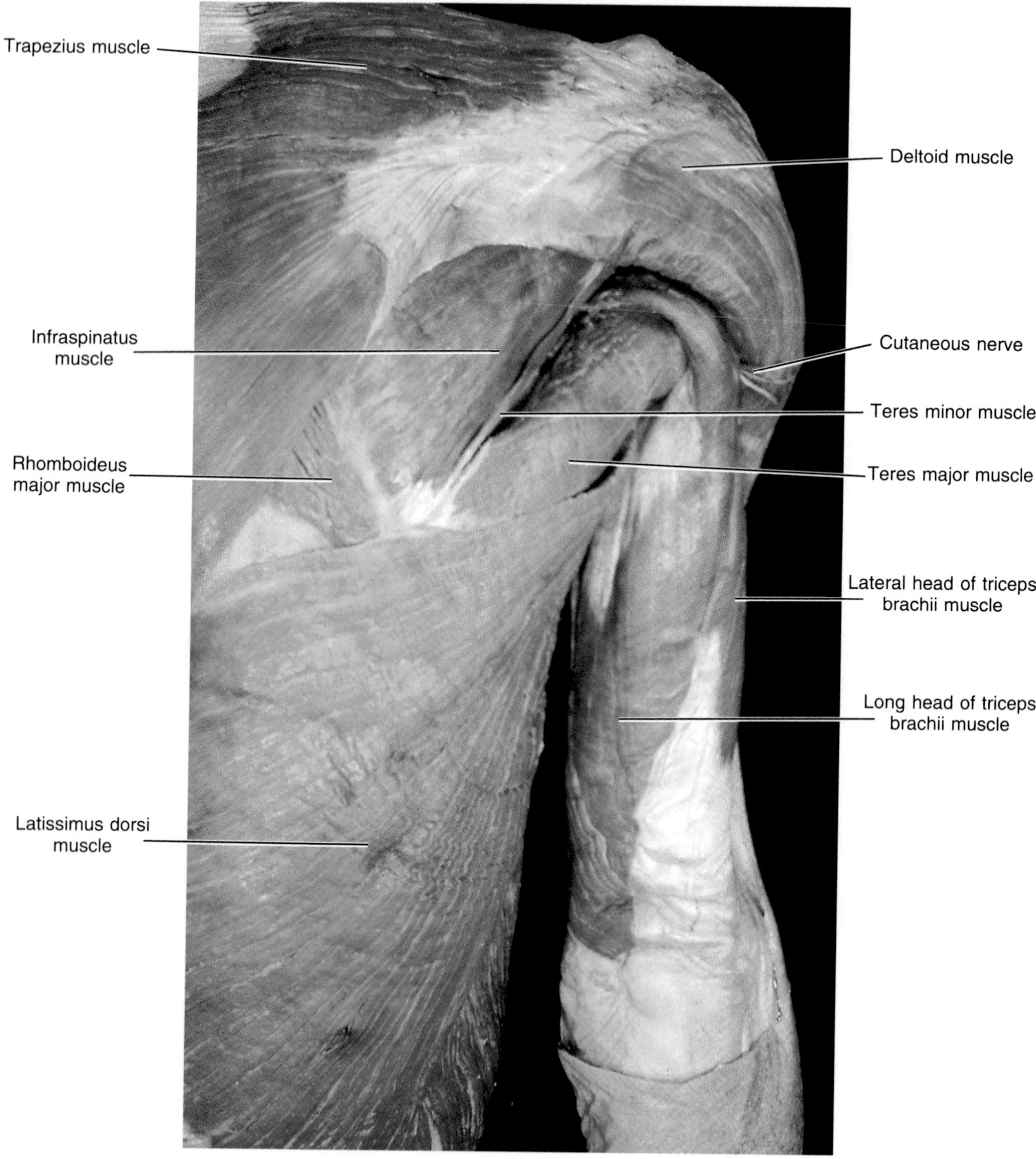

◆ **FIGURE A.8 Deeper posterior muscles of the upper trunk and arm**

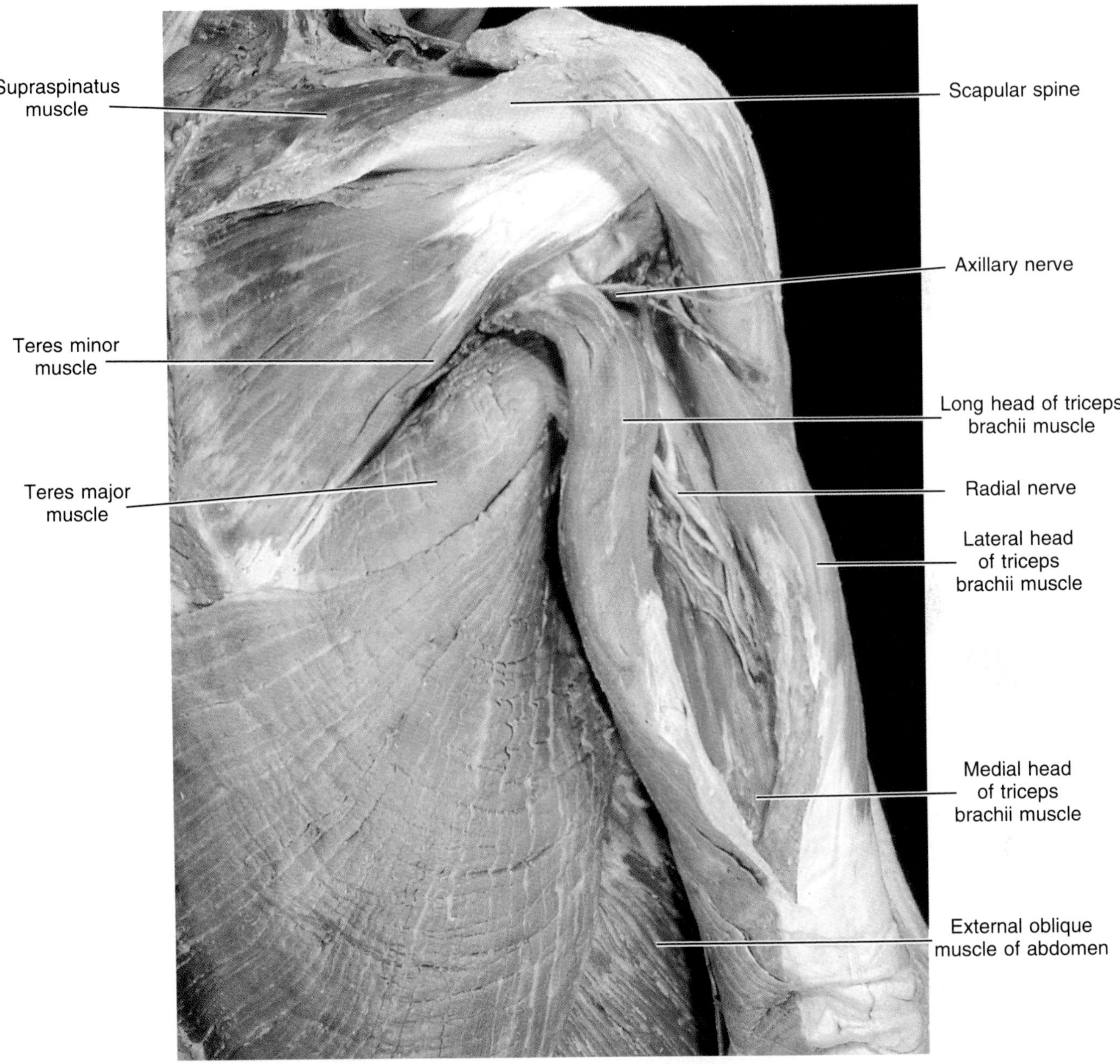

◆ **FIGURE A.9 Superficial muscles of the right lower back and buttocks**

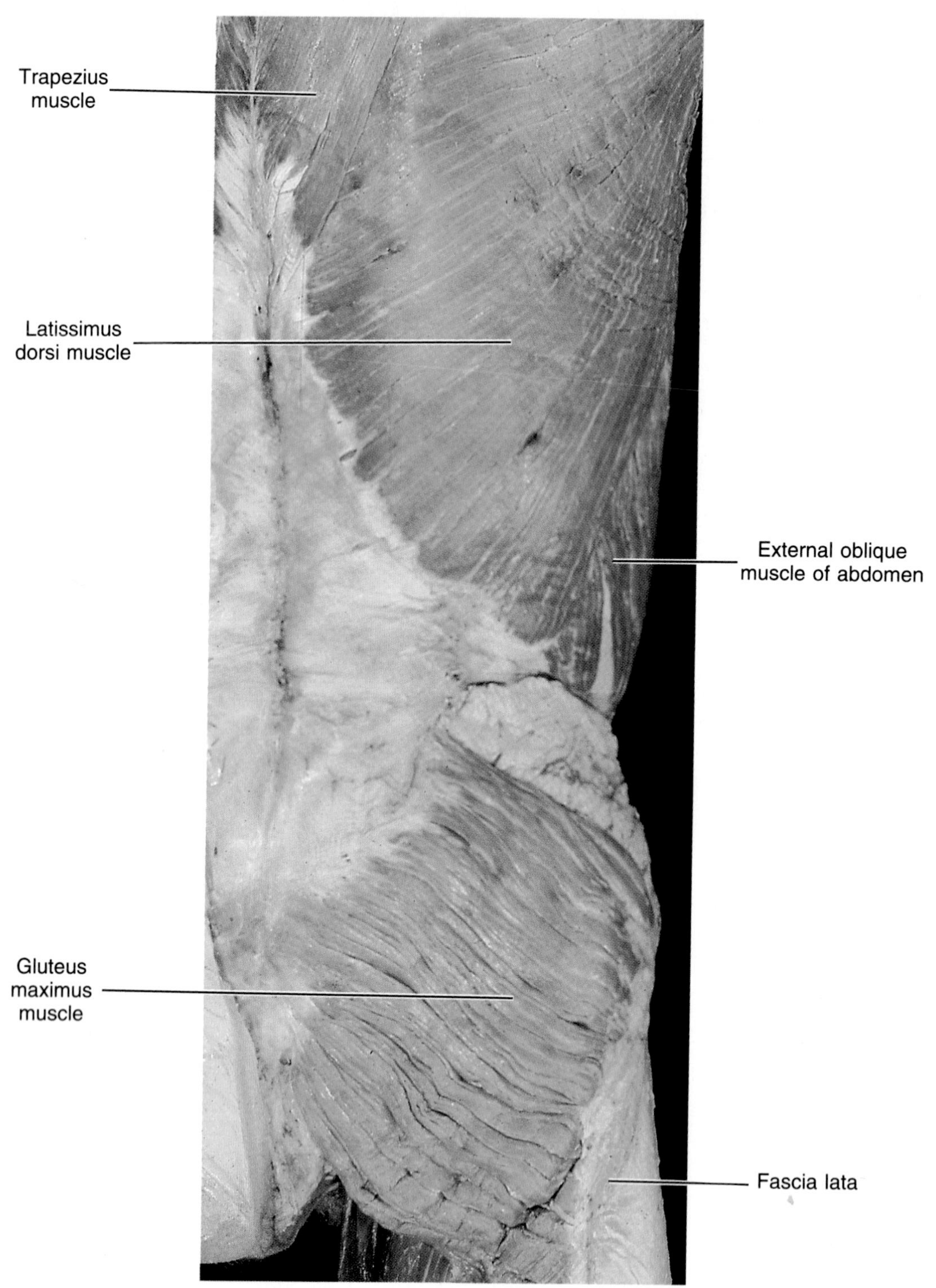

◆ **FIGURE A.10 Deeper muscles of the right lower back and buttocks**

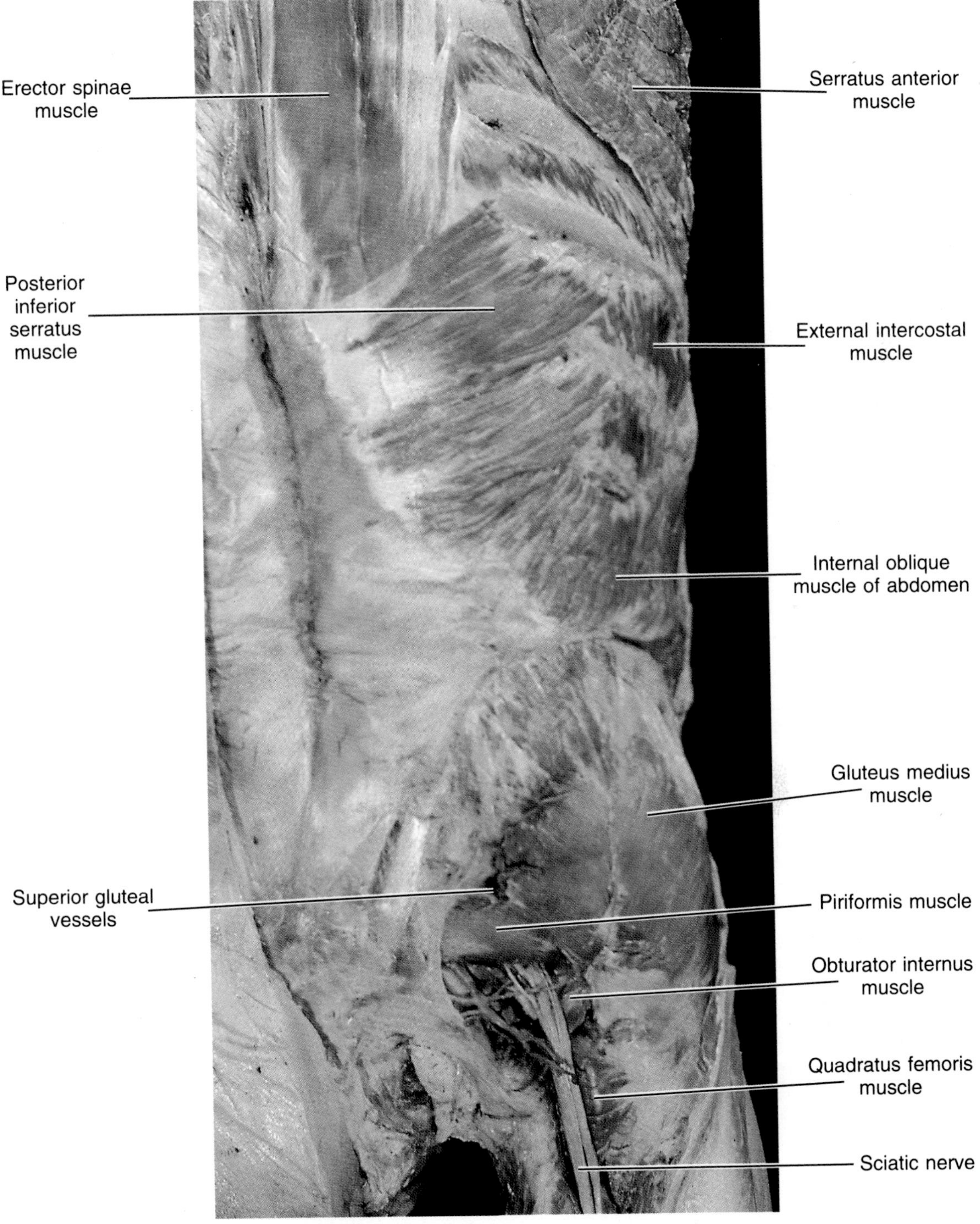

◆ **FIGURE A.11 Superficial muscles of the right anterior forearm**

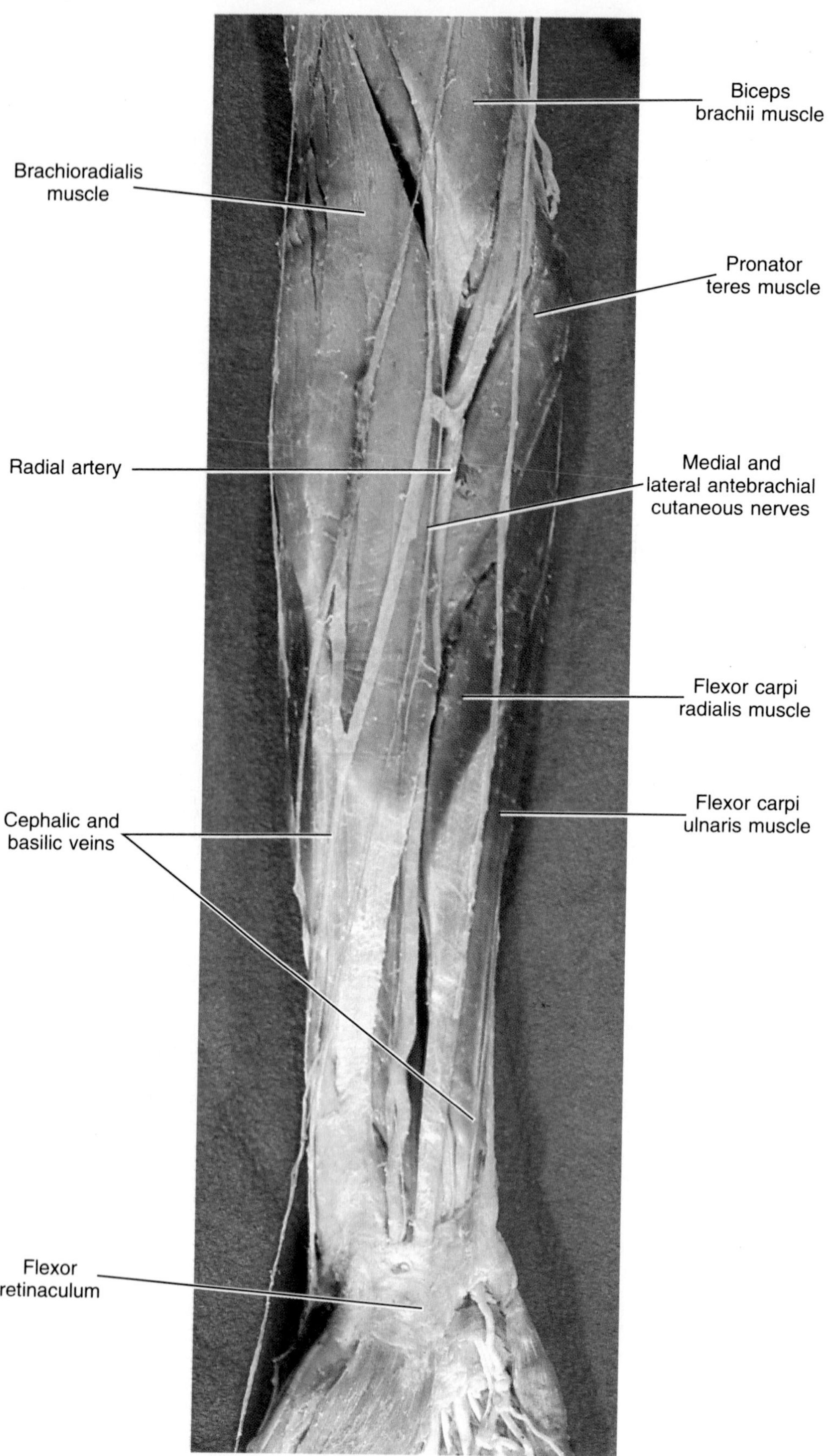

◆ **FIGURE A.12 Deeper muscles of the right anterior forearm**

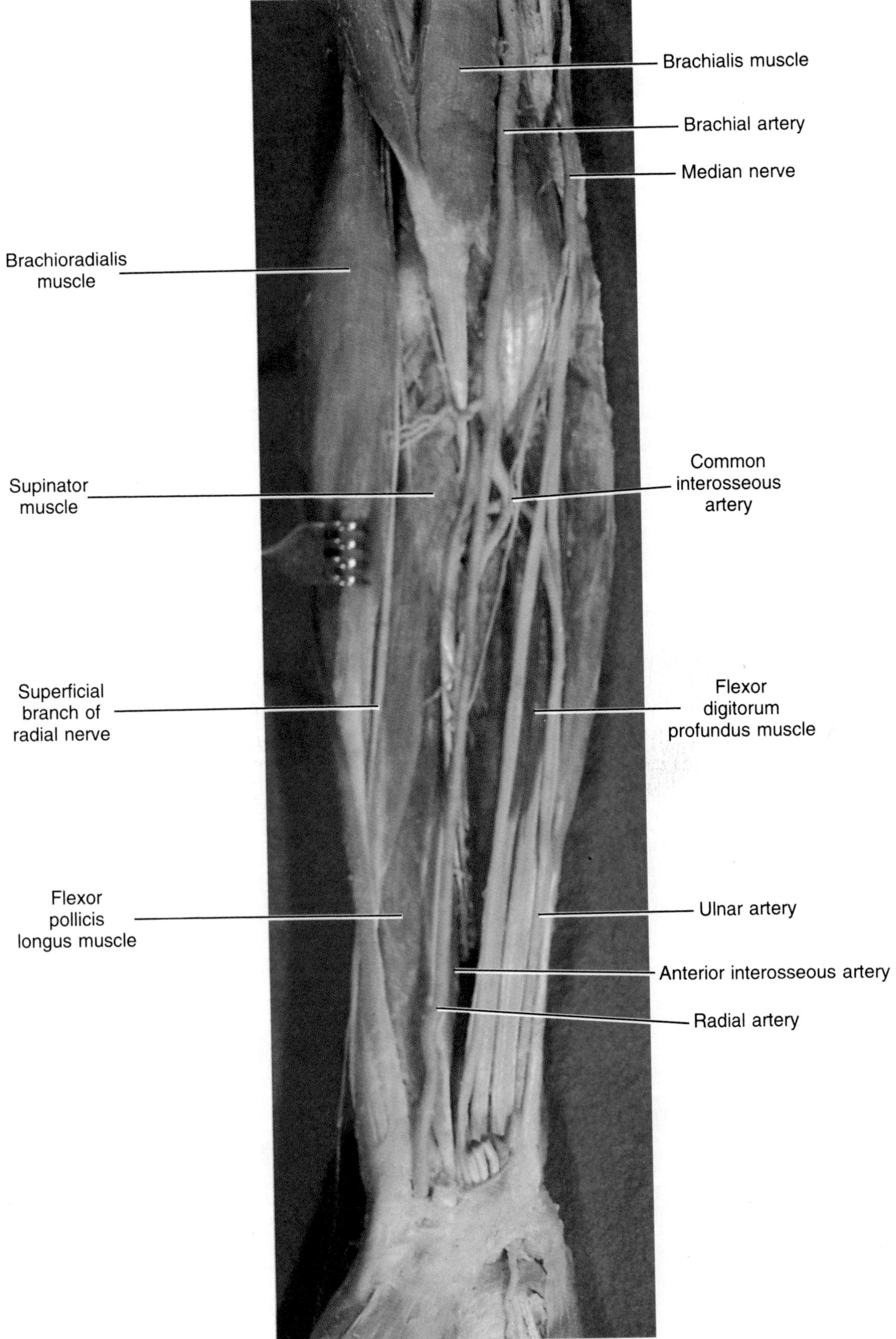

◆ **FIGURE A.13 Deepest muscles of the right anterior forearm**

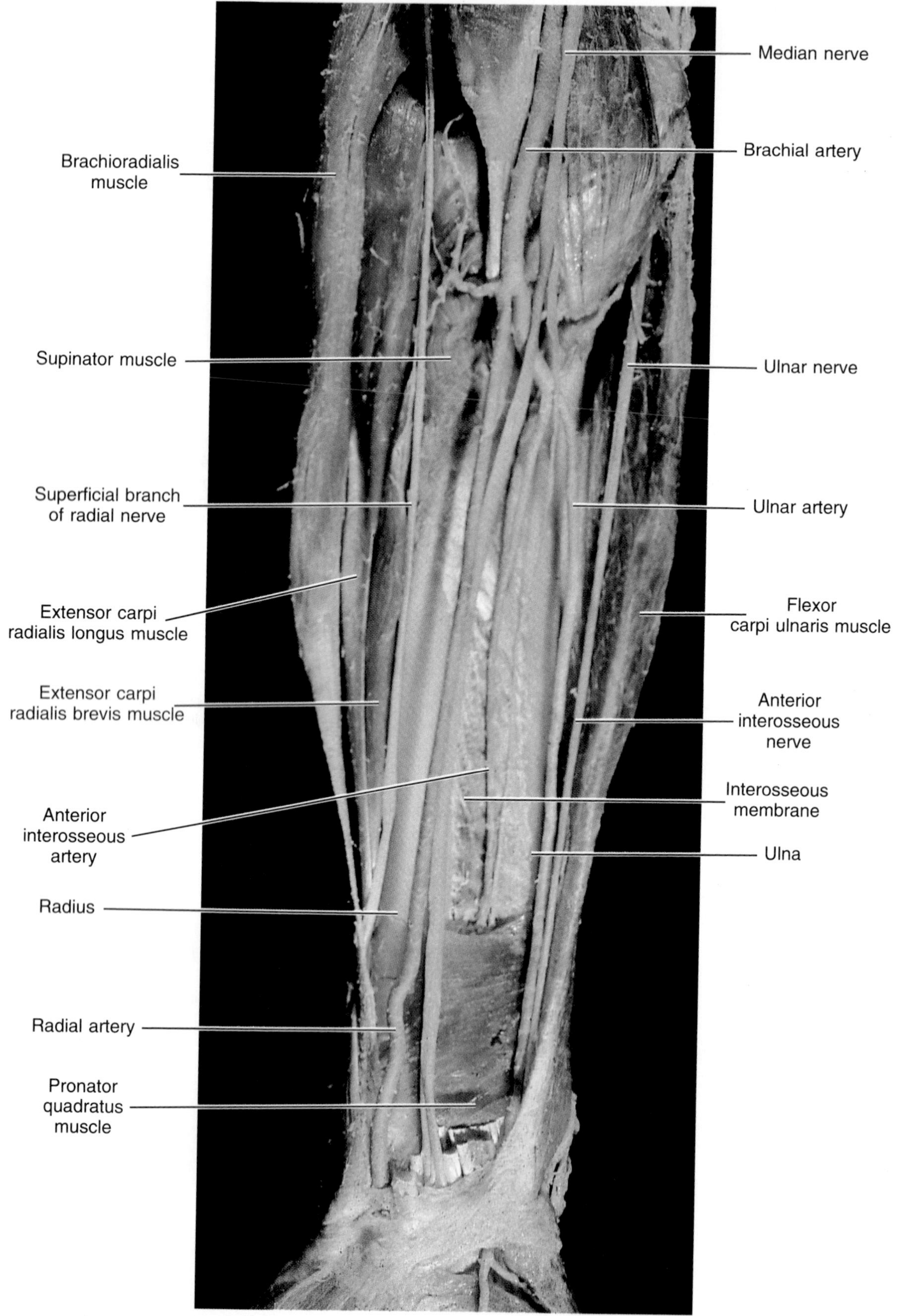

◆ **FIGURE A.14 Superficial muscles of the right posterior forearm and hand**

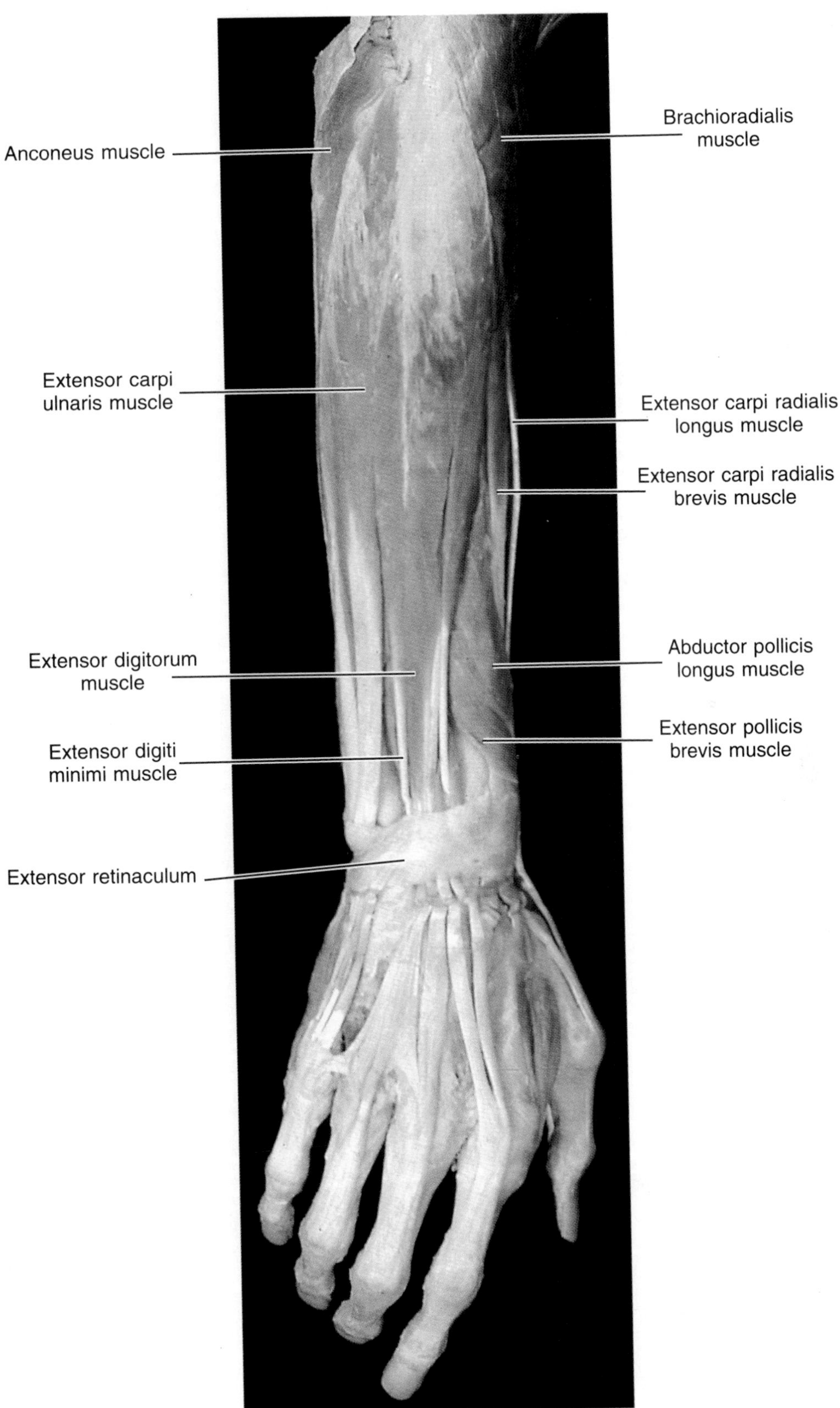

◆ **FIGURE A.15 Deeper muscles of the right posterior forearm and hand**

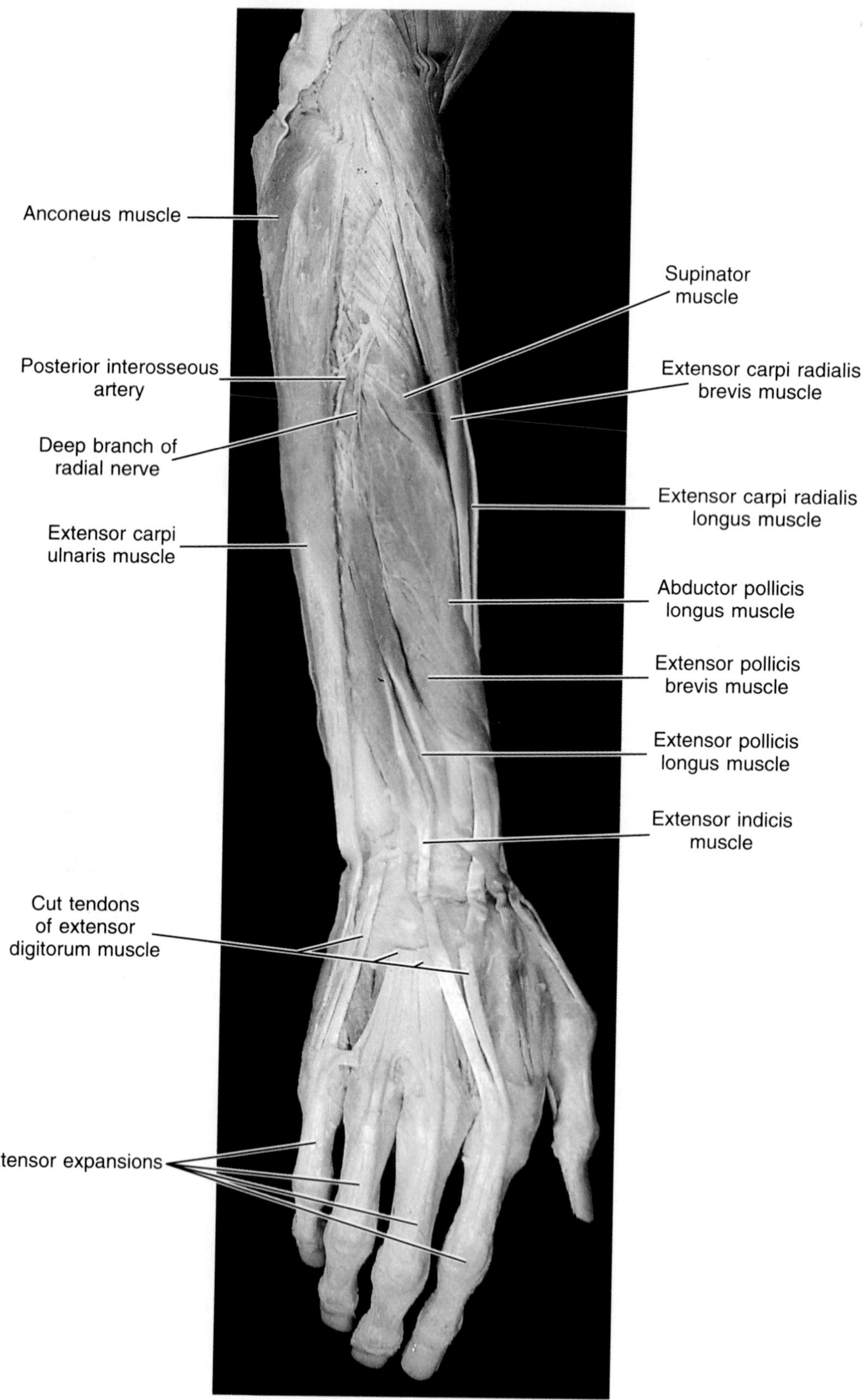

◆ **FIGURE A.16 Superficial muscles of the anterior and medial compartments of the right thigh**

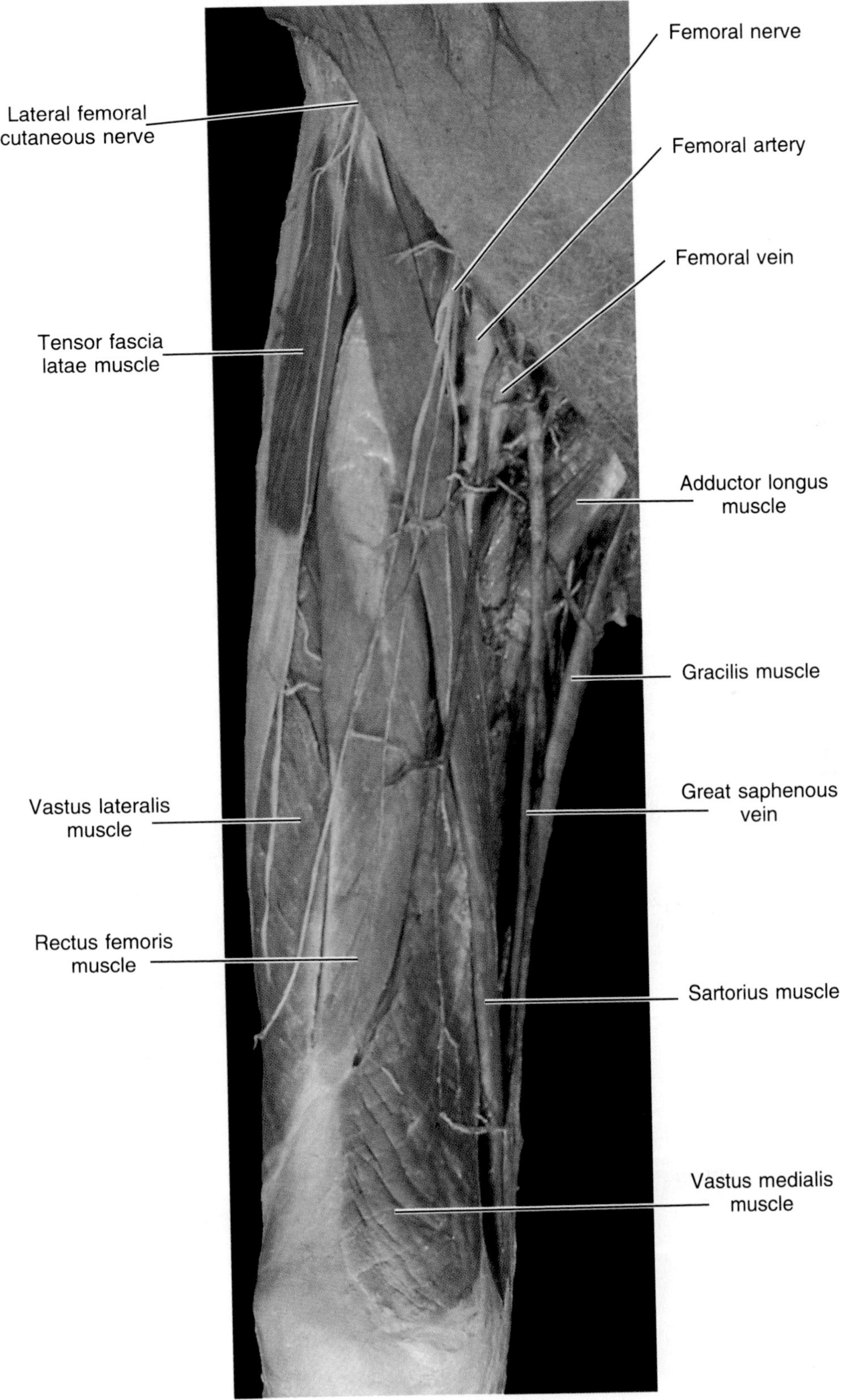

◆ **FIGURE A.17 Superficial muscles of the posterior compartment of the right thigh**

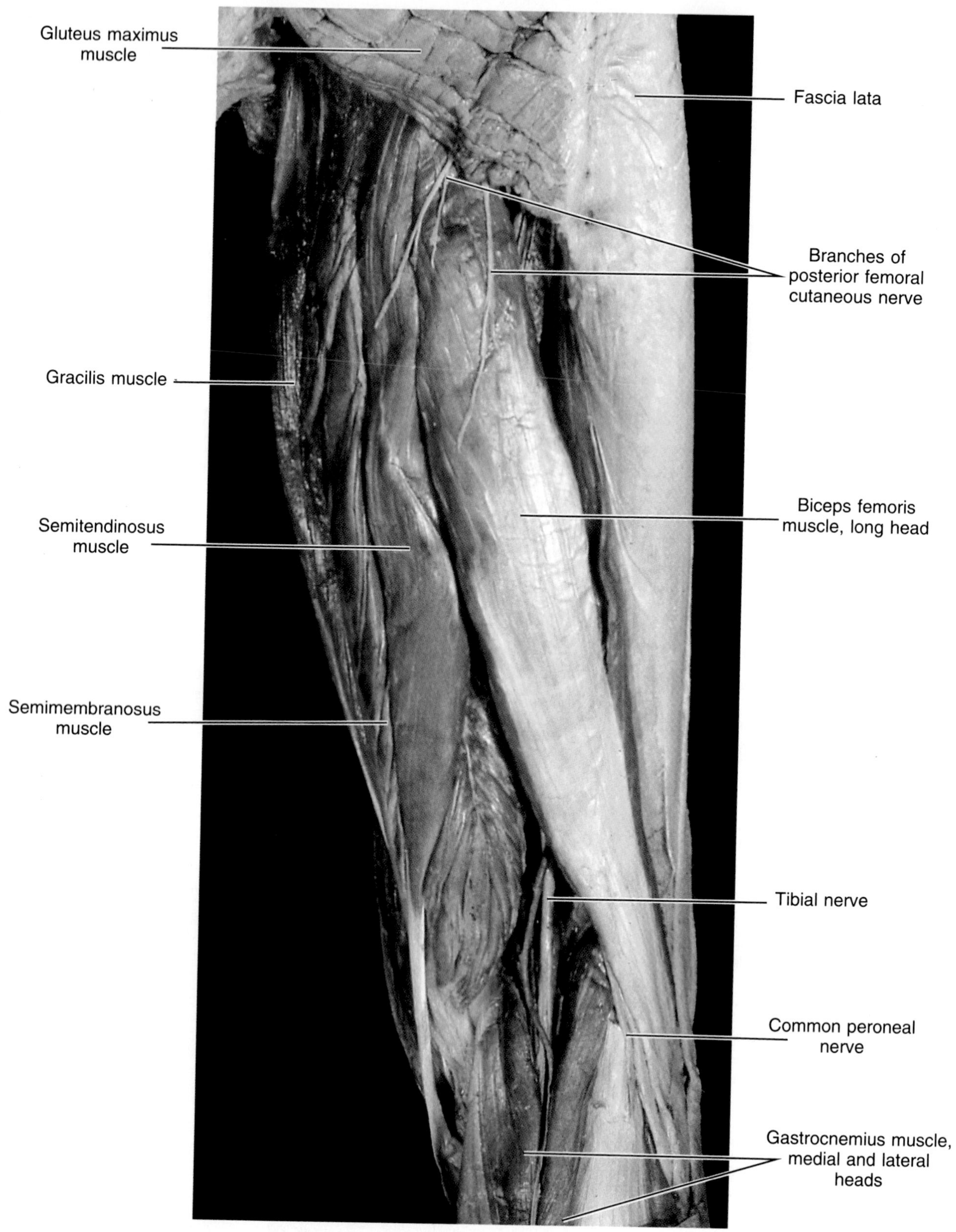

◆ **FIGURE A.18 Superficial muscles of the anterior and medial compartments of the right leg and the dorsal aspect of the foot**

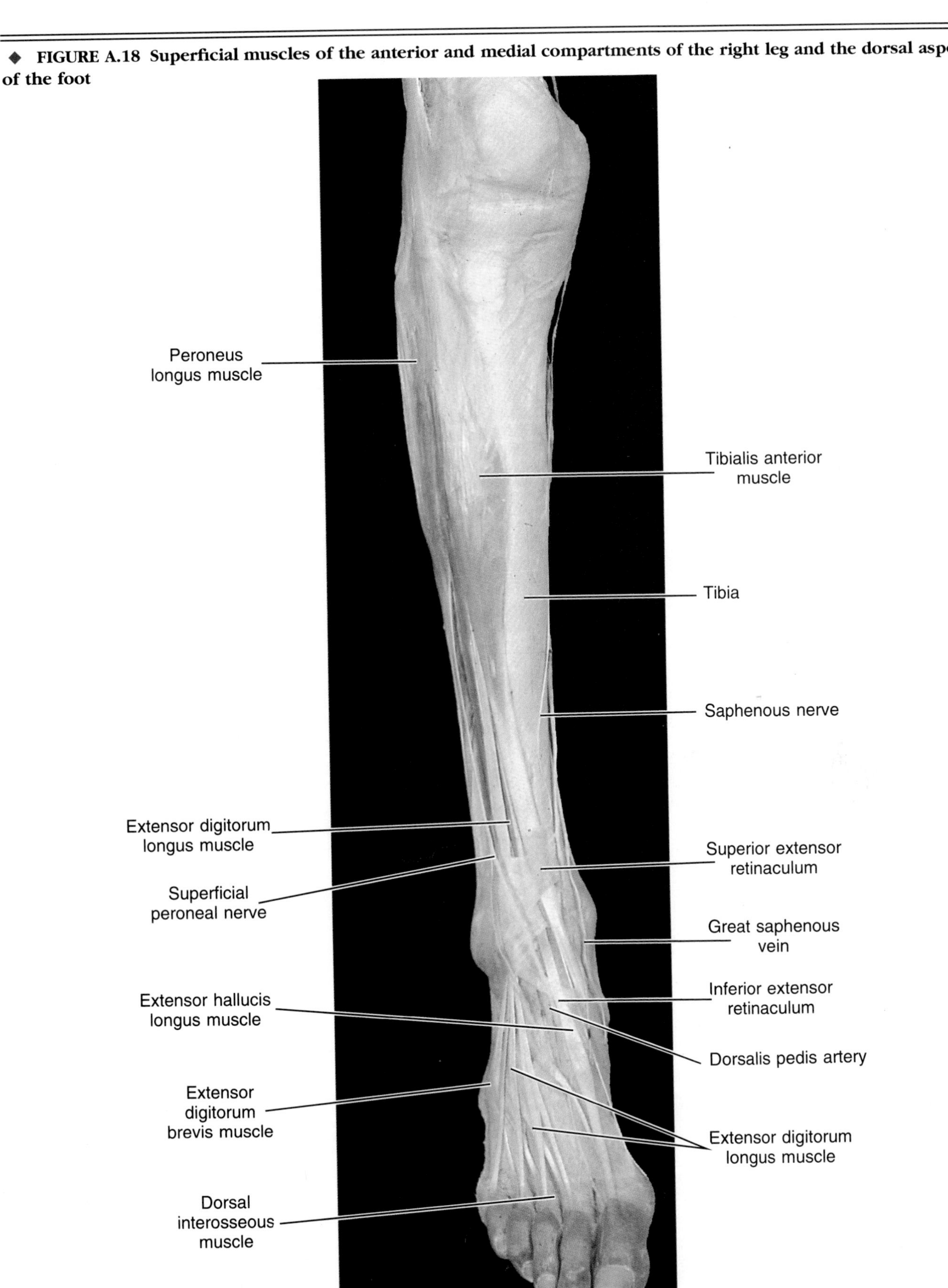

◆ **FIGURE A.19 Superficial muscles of the posterior compartment of the right leg**

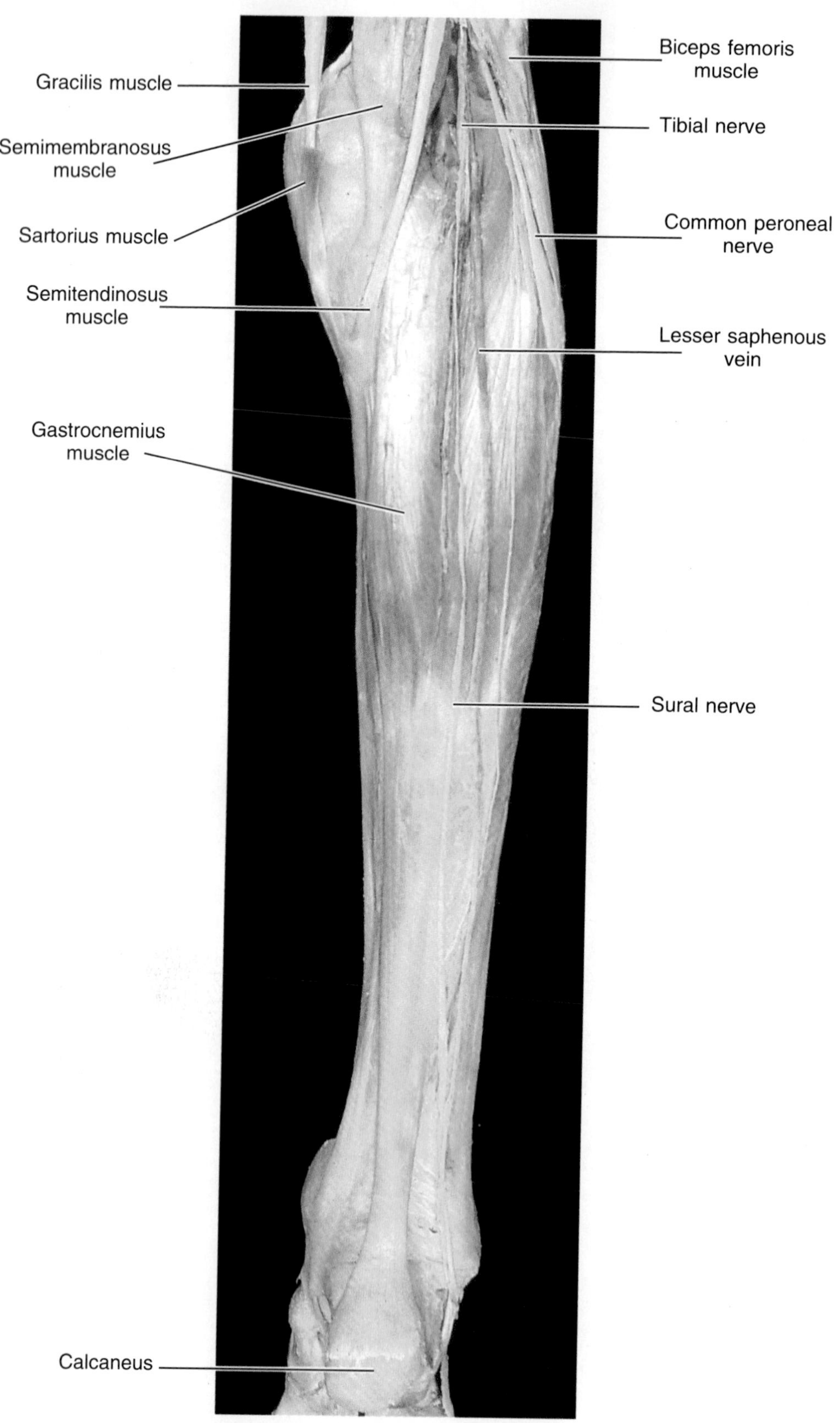

◆ **FIGURE A.20 Midsagittal section of a male pelvis**

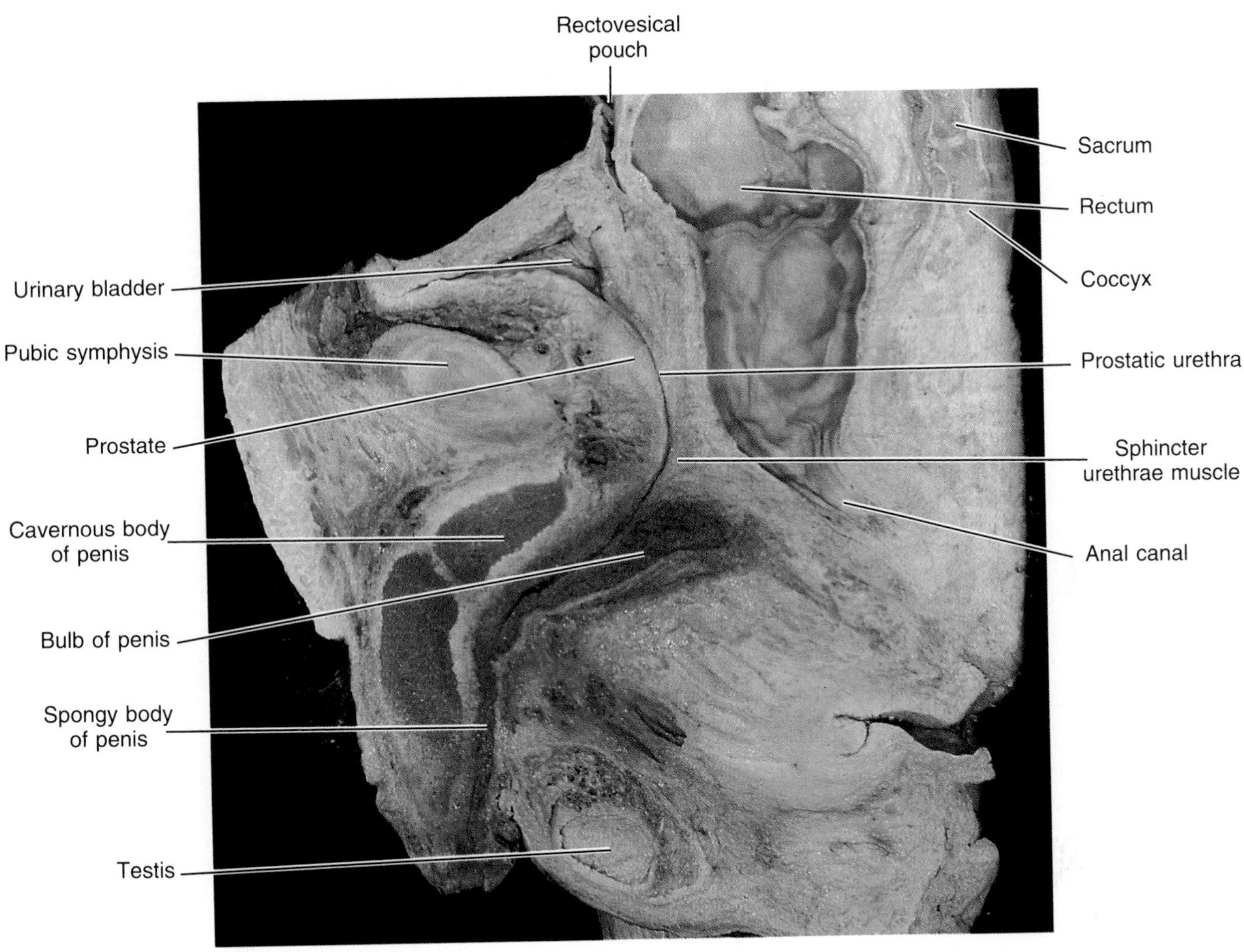

◆ **FIGURE A.21 Midsagittal section of a female pelvis**

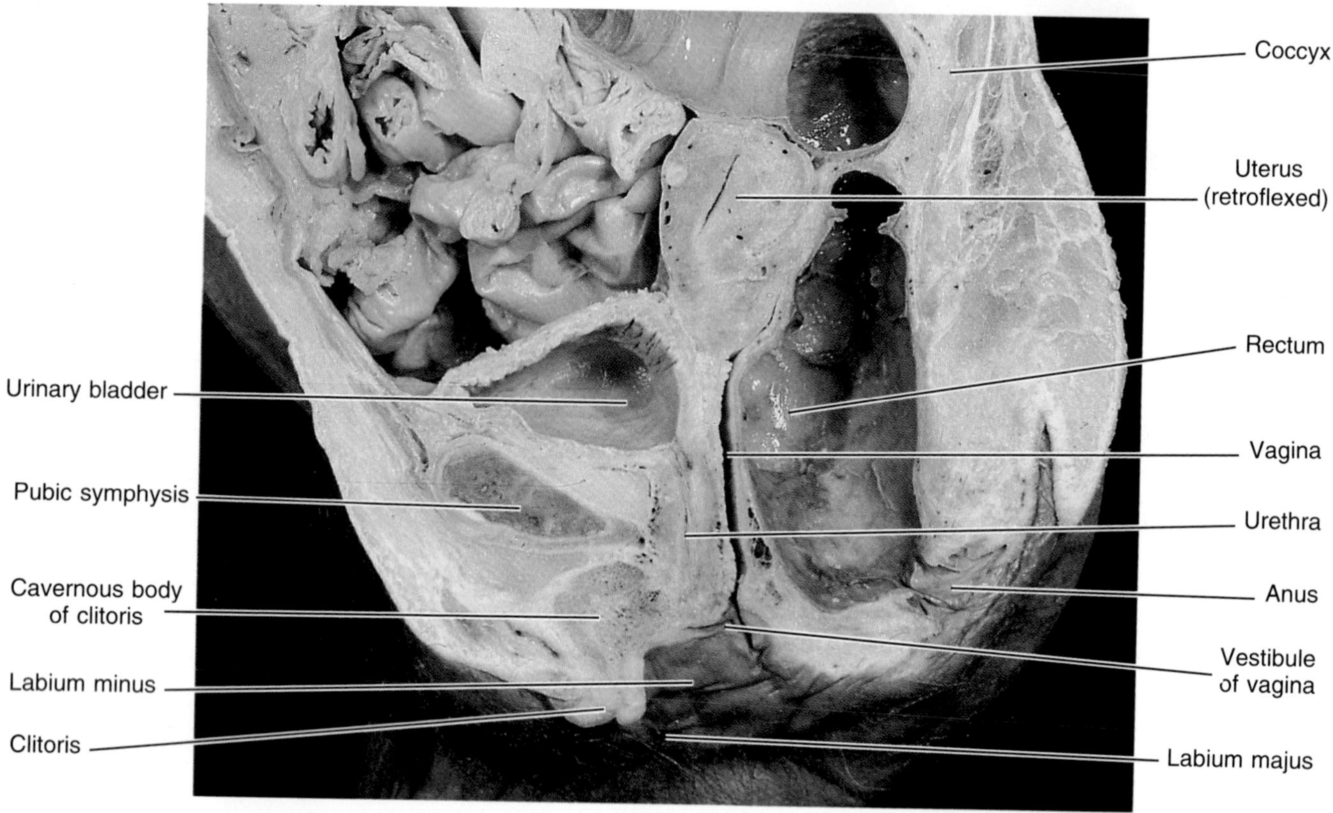

◆ **FIGURE A.22 Lateral aspect of a nasal cavity**

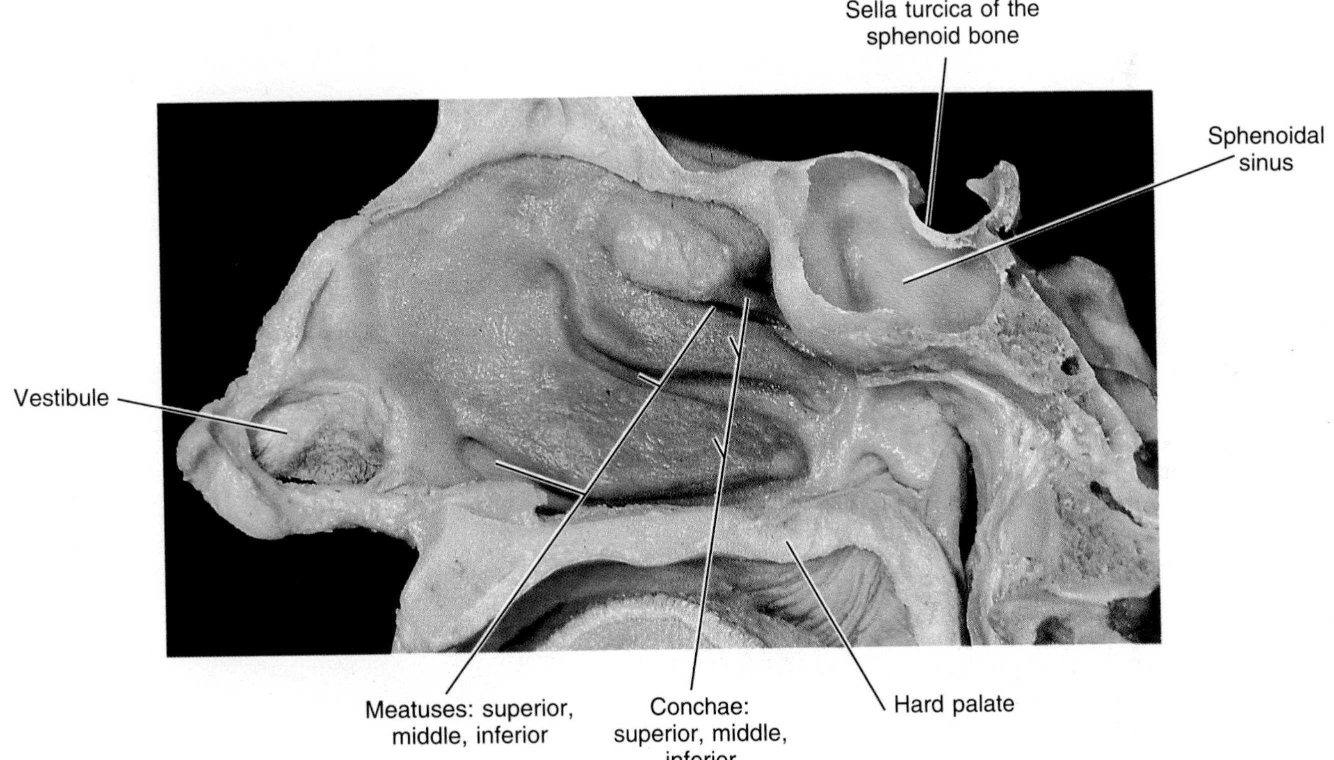

◆ **FIGURE A.23 Right lateral aspect of the falx cerebri**

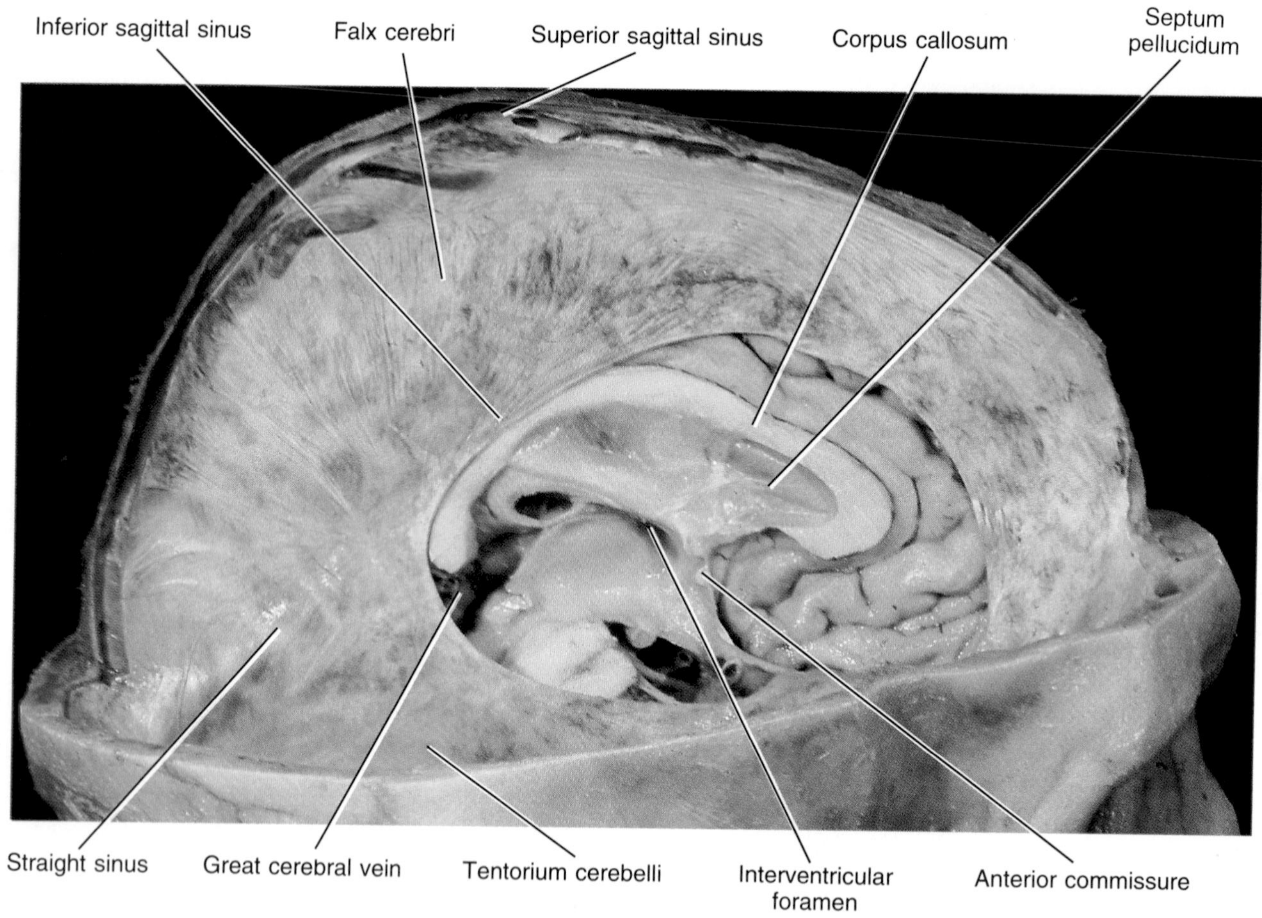

◆ **FIGURE A.24 Posterior aspect of the cervical region of the spinal cord (right half)**

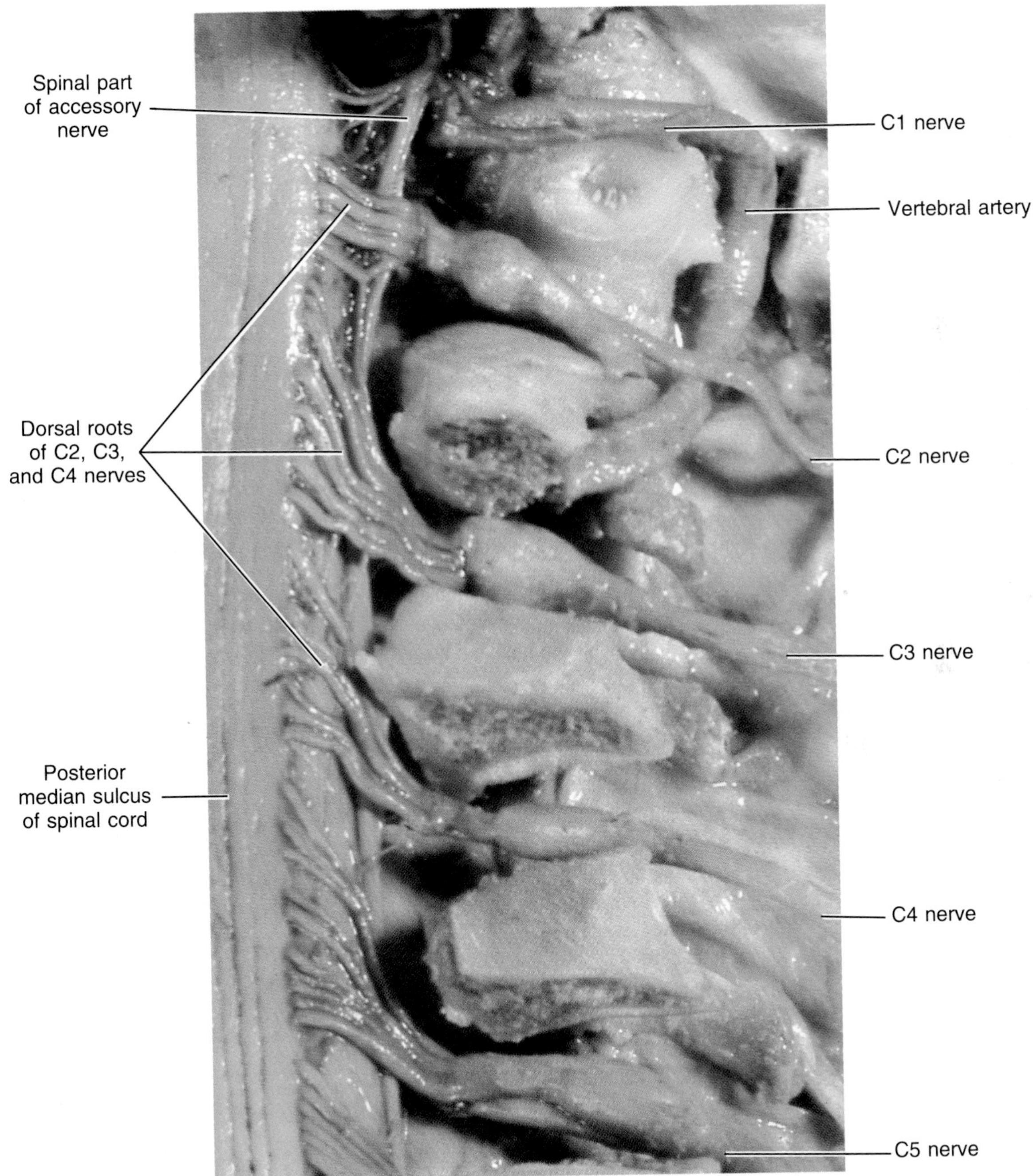

APPENDIX 2

Self-Quiz Answers

CHAPTER 1

1. c
2. b
3. a
4. b
5. c
6. Epithelial tisues: c, e
   Muscular tissues: a, g
   Nervous tissues: d, f
   Connective tissues: b, h
7. Prone position: k
   Anterior (ventral): g
   Posterior (dorsal): l
   Anatomical position: h
   Superior (cranial): a
   Medial: n
   Supine position: c
   Inferior (caudal): d
   Lateral: b
   Proximal: e
   Cervical: m
   Plantar: f
   Distal: i
   Lumbar: j
   Palmar: o
8. a
9. T
10. c
11. b
12. T
13. a
14. a
15. c
16. b
17. F
18. c
19. T
20. b

CHAPTER 2

1. T
2. c
3. a
4. b
5. F
6. d
7. c
8. b
9. Proteins: c
   Amino acids: g
   Peptide bond: b
   Dipeptide: e
   Polypeptide: i
   Primary structure: a
   Secondary structure: f
   Tertiary structure: h
   Quaternary structure: d
10. F
11. T
12. F
13. b
14. Solvent: c
    Solution: d
    Solute: a
    Suspension: b
    Colloid: e
15. c
16. a
17. F
18. b
19. a
20. c

CHAPTER 3

1. c
2. Specificity: b
   Saturation: d
   Competition: e
   Facilitated diffusion: a
   Active transport: c
3. b
4. b
5. a
6. T
7. F
8. b
9. c
10. a
11. c
12. b
13. a
14. c
15. b
16. centrosome: d
    Golgi apparatus: e
    Peroxisomes: a
    Basal bodies: b
    Microfilaments: c
17. F
18. c
19. T
20. Autosomes: h
    XY: g
    Homologous chromosomes: a
    46: f
    Diploid: c
    23: i
    Haploid: e
    Interkinesis: b
    Crossing over: d

CHAPTER 4

1. b
2. a
3. T
4. a
5. Squamous cells: c
   Stratified epithelium: f
   Simple columnar: e
   Columnar cells: a
   Cuboidal cells: d
   Simple squamous epithelium: b
6. b
7. c
8. T
9. Mucous membranes: d
   Lamina propria: c
   Serous membranes: f
   Mucus: g
   Mesothelium: b
   Endothelium: a
   Goblet cells: e
10. b
11. Fibroblasts: g
    Macrophages: f
    Ground substance: d
    Fibrocytes: a
    Fibers: c
    Collagenous: h
    Elastic: e
    Reticular: b

12. F
13. c
14. a
15. c
16. c
17. T
18. F
19. T
20. c

CHAPTER 5

1. b
2. T
3. stratum basale: e
Keratohyalin: g
Melanin: i
Stratum corneum: a
Eleidin: b
Stratum granulosum: h
Carotene: c
Stratum lucidum: d
Keratin: f
4. b
5. T
6. a
7. F
8. c
9. F
10. a
11. T
12. b.
13. a
14. T
15. c
16. c
17. Acne: d
Warts: g
Psoriasis: f
Impetigo: h
Moles: c
Herpes simplex: a
Shingles: i
Cancers: e
Eczema: b
18. a
19. b
20. T

CHAPTER 6

1. c
2. b
3. a
4. c
5. F
6. c
7. b
8. c
9. b
10. F
11. a
12. Process: i
Trochanter: f
Tubercle: j
Condyle: d
Sulcus: k
Crest: h
Facet: a
Fossa: g
Fovea: b
Foramen: c
Meatus: e
13. Fontanels: f
Frontal bone: j
Sinuses: d
Parietal bones: a
Occipital bone: h
Ethmoid bone: b
Temporal bones: l
Maxillary bones: c
Zygomatic bones: i
Lacrimal bones: k
Mandible: e
Vomer bone: g
An auditory ossicle: p
Hyoid bone: n
Cervical vertebrae: m
Thoracic vertebrae: q
Sacral vertebrae: o
14. a
15. c
16. a
17. d
18. b
19. c
20. a
21. c
22. c
23. Ilium: h
Ischium: k
Pubis: a
Acetabulum: m
False pelvis: b
Pelvic outlet: j
Femur: e
Greater trochanter: i
Condyles: c
Adductor tubercle: g
Tibia: d
Anterior crest: f
Fibula: l

CHAPTER 7

1. F
2. b
3. c
4. a
5. T
6. Synarthroses: d
Sutures: g
Syndesmoses: b
Synostosis: a
Amphiarthroses: f
Synchondroses: h
Symphyses: c
Synovial: e
7. b
8. F
9. F
10. a
11. Flexion: d
Extension: f, b
Abduction: c
Adduction: a
Plantar flexion: e
Dorsiflexion: g
12. F
13. c
14. b
15. b
16. Nonaxial: d, g
Gliding: g
Uniaxial: f, i
Hinge: i
Pivot: e
Biaxial: a, b, h
Condyloid: h
Saddle: b
Triaxial: c
Ball and Socket: c
17. a
18. a
19. b
20. Distal tibia/fibula: a, g
Pubic/pubic: b
Ulna/humerus: f
Sternum/clavicle: d, e
Radius/carpals: c
Tarsal/tarsal: d, e
Occipital bone/atlas: c
21. T
22. c

CHAPTER 8

1. T
2. a
3. Epimysium: c
Perimysium: h
Endomysium: f
Aponeuroses: d
Origin: a
Insertion: g
Unipennate: b
Bipennate: i
Multipennate: e
4. c
5. b
6. a
7. b
8. T
9. b
10. F
11. c
12. c
13. b
14. c
15. b
16. c
17. b
18. T
19. c
20. b
21. a

CHAPTER 9

1. Intercostalis: d
Rhomboideus: a
Pronator: f
Coracobrachialis: c
Gluteus maximus: b
Trapezius: a
Rectus abdominis: d, f
Pectoralis minor: b
Flexor carpi radialis: f
Biceps brachii: d, e
Quadriceps femoris: d, e
2. b
3. Depressor labii inferioris: d
Epicranius occipitalis: c
Corrugator: b
Orbicularis oculi: f
Orbicularis oris: g
Procerus: e
Risorius: a
4. a
5. T
6. b
7. T
8. Multifidus: c
Interspinales: d
Splenius capitis and cervicis: b
Spinalis thoracis and cervicis: a
9. c
10. T
11. F
12. a
13. b
14. Subclavius: b
Rhomboideus major: c
Pectoralis minor: a
15. Deltoid: a, b, d, e, f
Latissimus dorsi: b, c, e
Teres minor: f
Teres major: b, c, e
Coracobrachialis: a, c
Infraspinatus: c, f
16. a, b, e

17. Flexor carpi radialis: h
Extensor indicis: d
Extensor carpi ulnaris: e
Extensor pollicis longus: f
Extensor carpi radialis longus: b
Extensor digiti minimi: c
Flexor carpi ulnaris: g
Flexor digitorum superficialis: a
18. b, e
19. Gluteus maximus: d
Adductor magnus: e
Pectineus: b
Adductor longus: c
Gluteus medius: a
20. c
21. b
22. Tibialis anterior: c
Peroneous longus: b
Extensor digitorum longus: e
Flexor hallucis longus: a
Popliteus: d

CHAPTER 10

1. T
2. c
3. a
4. a
5. F
6. c
7. a
8. F
9. c
10. Nuclei: d
Soma: f
Center: h
Ganglia: a
Dendrite: b
Axon: i
Nerve fiber: c
Sheath of Schwann: e
Myelin: g
11. c
12. T
13. c
14. a
15. T
16. Motor neuron endings: e
Sensory neuron endings: a
Pacinian corpuscles: b
End-bulbs of Krause: f
Ruffini's corpuscles: d
Muscle spindles: h
Neurotendinous organs: i
Free nerve endings: c
Meissner's corpuscles: g
17. b
18. a
19. b
20. T
21. b
22. c
23. a

CHAPTER 11

1. a
2. b
3. c
4. b
5. b
6. c
7. T
8. a
9. a
10. F
11. b
12. a
13. F
14. b
15. c
16. T
17. a
18. c
19. F
20. T

CHAPTER 12

1. c
2. a
3. a
4. F
5. b
6. b
7. c
8. b
9. T
10. c
11. b
12. T
13. b
14. a
15. b
16. Cerebrum: i
Projection tracts: e
Basal nuclei: m
Cerebral hemisphere: a
Insula: k
Primary motor area: p
Visual area: b
Olfactory area: l
Thalamus: c
Cerebral peduncles: n
Metencephalon: o
Pons: d
Hypothalamus: g
Ventricles: h
Telencephalon: f
Reticular formation: j
17. T
18. a
19. Cauda equina: g
Meninges: d
White matter: a
Gray matter: j
Ventral roots: c
Fasciculus gracilis: i
Spinothalamic tracts: b
Spinocerebellar tracts: h
Pyramidal tracts: f
Extrapyramidal tracts: e
20. b
21. a

CHAPTER 13

1. c
2. a
3. c
4. a
5. a
6. b
7. a
8. c
9. c
10. a
11. b
12. a
13. d
14. c
15. T
16. T
17. F
18. c
19. a
20. c
21. T
22. T

CHAPTER 14

1. d
2. F
3. F
4. b
5. b
6. T
7. F
8. T
9. b
10. F
11. b
12. c
13. a
14. c
15. F
16. b
17. T
18. c
19. T
20. T

CHAPTER 15

1. a
2. b
3. a
4. b
5. T
6. c
7. b
8. T
9. b
10. T
11. F
12. c
13. F
14. b
15. T
16. a
17. a
18. T
19. b
20. c

CHAPTER 16

1. b
2. c
3. a
4. b
5. F
6. c
7. a
8. b
9. a
10. c
11. a
12. c
13. a
14. T
15. b
16. a
17. T
18. c
19. b
20. c

CHAPTER 17

1. F
2. b
3. c
4. b
5. a
6. ADH: b
Oxytocin: d

LH: f
TSH: e
ACTH: a
GH: c
7. c
8. F
9. Lack of ACTH: e
Gonadotropin deficiency: b
TSH deficiency: a
Growth-hormone deficiency: c
Growth-hormone overproduction: d
Antidiuretic-hormone deficiency: f
10. c
11. a
12. F
13. c
14. c
15. c
16. a
17. T
18. a
19. b
20. c
21. T
22. c

CHAPTER 18

1. F
2. F
3. F
4. T
5. a
6. T
7. Bilirubin: e
Heme: b
Erythropoietin: a
Globin: c
Ferritin: d
8. a
9. a
10. b
11. a
12. b
13. c
14. T
15. a
16. F
17. F
18. c
19. b
20. T

CHAPTER 19

1. b
2. Pericardial cavity: g
Atria: d
Ventricles: i
Interatrial spetum: a
Inferior vena cava: f
Aorta: b
Epicardium: j
Myocardium: c
Endocardium: h
Trabeculae carneae: e
3. c
4. c
5. a
6. c
7. F
8. a
9. a
10. T
11. T
12. b
13. a
14. T
15. c
16. b
17. F
18. T
19. c
20. a
21. a
22. b
23. b

CHAPTER 20

1. b
2. c
3. T
4. F
5. a
6. d
7. c
8. c
9. b
10. a
11. c
12. T
13. b
14. a
15. F
16. b
17. a
18. c
19. F
20. Aorta: g
Coronary arteries: e
Internal carotid artery: i
Vertebral artery: a
Radial artery: d
Superior mesenteric artery: b
Inferior mesenteric artery: h
Common iliac arteries: c
Anterior tibial artery: f

CHAPTER 21

1. F
2. c
3. a
4. b
5. T
6. T
7. d
8. d
9. T
10. F

CHAPTER 22

1. T
2. c
3. a
4. b
5. a
6. T
7. a
8. F
9. b
10. c
11. F
12. c
13. a
14. c
15. b
16. c
17. b
18. b
19. a
20. T

CHAPTER 23

1. a
2. d
3. d
4. F
5. d
6. d
7. c
8. d
9. b
10. c
11. c
12. Costal surface: d
Cardiac notch: h
Oblique fissure: g
Mediastinum: a
Pleura: b
Visceral pleura: c
Parietal pleura: f
Alveoli: e
13. b
14. c
15. a
16. b
17. c
18. a
19. a
20. a
21. F
22. c
23. b
24. b

CHAPTER 24

1. c
2. a
3. c
4. a
5. b
6. T
7. F
8. a
9. c
10. a
11. F
12. c
13. a
14. c
15. a
16. c
17. b
18. c
19. b
20. a
21. b
22. Pancreatic juice: e
Tetrapeptidases: d
Amylases: f
Bile salts: b
Cholecystokinin: a
Trypsin: c
23. b

CHAPTER 25

1. c
2. b
3. F
4. F
5. b
6. a
7. T
8. F
9. T
10. a
11. c
12. c
13. a
14. b

15. F
16. T
17. a
18. b
19. b
20. b

CHAPTER 26

1. c
2. a
3. a
4. b
5. c
6. a
7. T
8. c
9. b
10. F
11. c
12. F
13. b
14. T
15. a
16. T
17. c
18. a
19. F
20. d

CHAPTER 27

1. a
2. F
3. T
4. c
5. T
6. a
7. b
8. c
9. F
10. c
11. a
12. c
13. c
14. F
15. F
16. F
17. a
18. b
19. b
20. c

CHAPTER 28

1. a
2. T
3. b
4. b
5. T
6. b
7. Epididymis: d
   Ductus deferens: c
   Seminal vesicles: f
   Ejaculatory ducts: a
   Prostrate gland: b
   Bulbourethral glands: e
8. F
9. b
10. a
11. b
12. b
13. c
14. T
15. e
16. c
17. c
18. T
19. b
20. b

CHAPTER 29

1. b
2. T
3. a
4. F
5. a
6. F
7. b
8. c
9. F
10. a
11. a
12. T
13. T
14. c
15. d
16. F
17. F
18. c
19. F
20. c
21. b
22. b
23. F

APPENDIX 3

Word Roots, Prefixes, Suffixes, and Combining Forms

Prefixes and Combining Forms

a-, an- *absence or lack* acardia, lack of a heart; anaerobic, in the absence of oxygen

ab- *departing from; away from* abnormal, departing from normal

acou- *hearing* acoustics, the science of sound

acr-, acro- *extreme or extremity; peak* acrodermatitis, inflammation of the skin of the extremities

ad- *to or toward* adorbital, toward the orbit

aden-, adeno- *gland* adeniform, resembling a gland in shape

amphi- *on both sides; of both kinds* amphibian, an organism capable of living in water and on land

angi- *vessel* angiitis, inflammation of a lymph vessel or blood vessel

ant-, anti- *opposed to; preventing or inhibiting* anticoagulant, a substance that prevents blood coagulation

ante- *preceding; before* antecubital, in front of the elbow

arthr-, arthro- *joint* arthropathy, any joint disease

aut-, auto- *self* autogenous, self-generated

bi- *two* bicuspid, having two cusps

bio- *life* biology, the study of life and living organisms

blast- *bud or germ* blastocyte, undifferentiated embryonic cell

broncho- *bronchus* bronchospasm, spasmodic contraction of bronchial muscle

bucco- *cheek* buccolabial, pertaining to the cheek and lip

caput- *head* decapitate, remove the head

carcin- *cancer* carcinogen, a cancer-causing agent

cardi-, cardio- *heart* cardiotoxic, harmful to the heart

cephal- *head* cephalometer, an instrument for measuring the head

cerebro- *brain, especially the cerebrum* cerebrospinal, pertaining to the brain and spinal cord

chondr- *cartilage* chondrogenic, giving rise to cartilage

circum- *around* circumnuclear, surrounding the nucleus

co-, con- *together* concentric, common center, together in the center

contra- *against* contraceptive, agent preventing conception

cost- *rib* intercostal, between the ribs

crani- *skull* craniotomy, a skull operation

crypt- *hidden* cryptomenorrhea, a condition in which menstrual symptoms are experienced but no external loss of blood occurs

cyt- *cell* cytology, the study of cells

de- *undoing, reversal, loss, removal* deactivation, becoming inactive

di- *twice, double* dimorphism, having two forms

dia- *through, between* diaphragm, the wall through or between two areas

dys- *difficult, faulty, painful* dyspepsia, disturbed digestion

ec, ex, ecto- *out, outside, away from* excrete, to remove materials from the body

en-, em- *in, inside* encysted, enclosed in a cyst or capsule

entero- *intestine* enterologist, one who specializes in the study of intestinal disorders

epi- *over, above* epidermis, outer layer of skin

eu- *well* euesthesia, a normal state of the senses

exo- *outside, outer layer* exophthalmos, an abnormal protrusion of the eye from the orbit

extra- *outside, beyond* extracellular, outside the body cells of an organism

gastr- *stomach* gastrin, a hormone that influences gastric acid secretion

glosso- *tongue* glossopathy, any disease of the tongue

hema-, hemato-, hemo- *blood* hematocyst, a cyst containing blood

hemi- *half* hemiglossal, pertaining to one half of the tongue

hepat- *liver* hepatitis, inflammation of the liver

hetero- *different or other* heterosexuality, sexual desire for a person of the opposite sex

hist- *tissue* histology, the study of tissues
hom-, homo- *same* homeoplasia, formation of tissue similar to normal tissue; homocentric, having the same center
hydr-, hydro- *water* dehydration, loss of body water
hyper- *excess* hypertension, excessive tension
hypno- *sleep* hypnosis, a sleeplike state
hypo- *below, deficient* hypodermic, beneath the skin; hypokalemia, deficiency of potassium
hyster-, hystero- *uterus or womb* hysterectomy, removal of the uterus; hysterodynia, pain in the womb
im- *not* impermeable, not permitting passage, not permeable
inter- *between* intercellular, between the cells
intra- *within, inside* intracellular, inside the cell
iso- *equal, same* isothermal, equal, or same, temperature
leuko- *white* leukocyte, white blood cell
lip-, lipo *fat, lipid* lipophage, a cell that has taken up fat in its cytoplasm
macro- *large* macromolecule, large molecule
mal- *bad, abnormal* malfunction, abnormal functioning of an organ
mamm- *breast* mammary gland, breast
mast- *breast* mastectomy, removal of a mammary gland
meningo- *membrane* meningitis, inflammation of the membranes of the brain
meso- *middle* mesoderm, middle germ layer
meta- *beyond, between, transition* metatarsus, the part of the foot between the tarsus and the phalanges
metro- *uterus* metroscope, instrument for examining the uterus
micro- *small* microscope, an instrument used to make small objects appear larger
mito- *thread, filament* mitochondria, small, filamentlike structures located in cells
mono- *single* monospasm, spasm of a single limb
morpho- *form* morphology, the study of form and structure of organisms
multi- *many* multinuclear, having several nuclei
myelo- *spinal cord, marrow* myeloblasts, cells of the bone marrow
myo- *muscle* myocardium, heart muscle
narco- *numbness* narcotic, a drug producing stupor or numbed sensations
nephro- *kidney* nephritis, inflammation of the kidney
neuro- *nerve* neurophysiology, the physiology of the nervous system
ob- *before, against* obstruction, impeding or blocking up
oculo- *eye* monocular, pertaining to one eye
odonto- *teeth* orthodontist, one who specializes in proper positioning of the teeth in relation to each other
ophthalmo- *eye* ophthalmology, the study of the eyes and related disease
ortho- *straight, direct* orthopedic, correction of deformities of the musculoskeletal system
osteo- *bone* osteodermia, bony formations in the skin
oto- *ear* otoscope, a device for examining the ear
oxy- *oxygen* oxygenation, the saturation of a substance with oxygen
pan- *all, universal* panacea, a cure-all
para- *beside, near* paraphrenitis, inflammation of tissues adjacent to the diaphragm
peri- *around* perianal, situated around the anus
phago- *eat* phagocyte, a cell that engulfs and digests particles or cells
phleb- *vein* phlebitis, inflammation of the veins
pod- *foot* podiatry, the treatment of foot disorders
poly- *multiple* polymorphism, multiple forms
post- *after, behind* posterior, places behind (a specific) part
pre-, pro- *before, ahead of* prenatal, before birth
procto- *rectum, anus* proctoscope, an instrument for examining the rectum
pseudo- *false* pseudotumor, a false tumor
psycho- *mind, psyche* psychogram, a chart of personality traits
pyo- *pus* pyocyst, a cyst that contains pus
retro- *backward, behind* retrogression, to move backward in development
sclero- *hard* sclerodermatitis, inflammatory thickening and hardening of the skin
semi- *half* semicircular, having the form of half a circle
steno- *narrow* stenocoriasis, narrowing of the pupil
sub- *beneath, under* sublingual, beneath the tongue
super- *above, upon* superior, quality or state of being above others or a part
supra- *above, upon* supracondylar, above a condyle
sym-, sny- *together, with* synapse, the region of communication between two neurons
tachy- *rapid* tachycardia, abnormally rapid heartbeat
therm- *heat* thermometer, an instrument used to measure heat
tox- *poison* antitoxic, effective against poison
trans- *across, through* transpleural, through the pleura
tri- *three* trifurcation, division into three branches
viscero- *organ, viscera* visceroinhibitory, inhibiting the movements of the viscera

Suffixes

-able *able to, capable of* viable, ability to live or exist
-ac *referring to* cardiac, referring to the heart
-algia *pain in a certain part* neuralgia, pain along the course of a nerve
-ary *associated with, relating to* coronary, associated with the heart
-atresia *imperforate* proctatresia, an imperforate condition of the rectum or anus
-cide *destroy or kill* germicide, an agent that kills germs
-ectomy *cutting out, surgical removal* appendectomy, cutting out of the appendix
-emia *condition of the blood* anemia, deficiency of red blood cells
-ferent *carry* efferent nerves, nerves carrying impulses away from the CNS
-fuge *driving out* vermifuge, a substance that expels worms of the intestine
-gen *an agent that initiates* pathogen, any agent that produces disease
-gram *data that are systematically recorded, a record* electrocardiogram, a recording showing action of the heart
-graph *an instrument use for recording data or writing* electrocardiograph, an instrument used to make an electrocardiogram

-ia *condition* insomnia, condition of not being able to sleep
-iatrics *medical specialty* geriatrics, the branch of medicine dealing with disease associated with old age
-itis *inflammation* gastritis, inflammation of the stomach
-logy *the study of* pathology, the study of changes in structure and function brought on by disease
-lysis *loosening or breaking down* hydrolysis, chemical decomposition of a compound into other compounds as a result of taking up water
-malacia *soft* osteomalacia, a process leading to bone softening
-mania *obsession, compulsion* erotomania, exaggeration of the sexual passions
-odyn *pain* coccygodynia, pain in the region of the coccyx
-oid *like, resembling* cuboid, shaped as a cube
-oma *tumor* lymphoma, a tumor of the lymphatic tissues
-opia *defect of the eye* myopia, nearsightedness
-ory *referring to, of* auditory, referring to hearing
-pathy *disease* psychopathy, any disease of the mind
-phobia *fear* acrophobia, fear of heights
-plasty *reconstruction of a part, plastic surgery* rhinoplasty, reconstruction of the nose through surgery
-plegia *paralysis* paraplegia, paralysis of the lower half of the body or limbs
-rrhagia *abnormal or excessive discharge* metrorrhagia, uterine hemorrhage
-rrhea *flow or discharge* diarrhea, abnormal emptying of the bowels
-scope *instrument used for examination* stethoscope, instrument used to listen to sounds of various parts of the body
-stasis *arrest, fixation* hemostasis, arrest of bleeding
-stomy *establishment of an artificial opening* enterostomy, the formation of an artificial opening into the intestine through the abdominal wall
-tomy *to cut* appendectomy, surgical removal of the appendix
-ty *condition of, state* immunity, condition of being resistant to infection or disease
-uria *urine* polyuria, passage of an excessive amount of urine

APPENDIX 4

Units of the Metric System

UNIT	METRIC EQUIVALENT	SYMBOL	ENGLISH EQUIVALENT
LINEAR MEASURE			
1 kilometer	= 1000 meters	km	0.62137 mile
1 meter	= 10 decimeters	m	39.37 inches
1 decimeter	= 10 centimeters	dm	3.937 inches
1 centimeter	= 10 millimeters	cm	0.3937 inch
1 millimeter	= 1000 microns	mm	no English equivalents
1 micron	= 1/1000 millimeter or 1000 millimicrons	μ	no English equivalents
1 millimicron	= 1 nanometer = 10 angstrom units	nm, mμ	no English equivalents
1 angstrom unit	= 1/100,000,000 centimeter	Å	no English equivalents
MEASURES OF CAPACITY			
1 kiloliter	= 1000 liters	kl	35.15 cubic feet or 264.16 gallons
1 liter	= 10 deciliters	l	1.0567 U.S. liquid quarts
1 deciliter	= 1000 milliliters	dl	0.03 fluid ounce
1 milliliter	volume of 1 g of water at standard temperature and pressure (STP)	ml	
MEASURES OF MASS			
1 kilogram	= 1000 grams	kg	2.2046 pounds
1 gram	= 100 centigrams	g	15.432 grains
1 centigram	= 10 milligrams	cg	0.1543 grain
1 milligram	= 1/1000 gram	mg	0.01 grain (about)
MEASURES OF VOLUME			
1 cubic meter	= 1000 cubic decimeters	m^3	
1 cubic decimeter	= 1000 cubic centimeters	dm^3	
1 cubic centimeter	= 1000 cubic millimeters	cm^3	
1000 cubic millimeters	= 1 milliliter (ml)	mm^3	

GLOSSARY

A

Abdomen *ab´-do-men* the portion of the body between the diaphragm and the pelvis.

Abduct *ab-duct´* to move away from the midline of the body.

Abortion *a-bor´-shun* termination of a pregnancy before the embryo or fetus is viable outside the uterus.

Abscess *ab´-ses* a localized accumulation of pus and disintegrating tissue.

Absolute refractory period *ree-frac´-to-ree* period following stimulation during which no additional action potential can be evoked.

Absorption *ab-sorp´-shun* passage of a substance into or across a membrane or blood vessel.

Accommodation (1) adaptation or adjustment by an organ or organism in response to differences or needs; (2) adjustment of the eye for seeing objects at various distances.

Acetabulum *as-e-tab´-u-lum* the cuplike cavity where the femur fits on the lateral surface of the pelvic bone.

Acetylcholine *a-see-til-ko´-leen* a chemical transmitter substance released by nerve endings.

Acetylcholinesterase *es´ter-ase* the enzyme that inactivates acetylcholine; present in muscle and nervous tissue.

Achilles tendon *a-kil´-leze ten´-don* the tendon at the back of the heel (calcaneus) that attaches the calf muscles.

Achondroplasia *a-kon-dro-play´-zee-ah* a condition, sometimes influenced by hereditary factors as well as hormonal levels, that produces a dwarf with short arms and legs but normal trunk and head.

Acid a substance that dissociates into hydrogen ions and anions when in an aqueous solution (compare with *base*).

Acidosis *ass-i-do´-sis* a condition in which the hydrogen ion concentration of the blood tends to increase, and the pH of the blood tends to decrease.

Acne inflammatory disease of the skin.

Acromegaly *ak-ro-meg´-a-lee* an abnormal pattern of bone and connective-tissue growth characterized by enlarged hands, face, and feet and associated with excessive pituitary growth hormone that is secreted after the epiphyseal cartilages have been replaced.

Acromion *ak-ro´-mee-un* the outer projection of the spine of the scapula; forms the highest point of the shoulder.

Acrosome *ak-ro-some* crescent-shaped structure molded to the nucleus and forming the anterior sperm head.

Actin *ak´-tin* a contractile protein of muscle fiber; myosin is another protein found in muscle.

Action potential an event occurring when a stimulus of sufficient intensity is applied to a neuron, allowing sodium ions to move into the cell and reverse the polarity.

Active state the development of tension by the contractile proteins of muscle.

Active transport net movement of a substance across a membrane against a concentration gradient; requires release and use of energy.

Adaptation (1) any change in structure, form, or habits to suit a new environment; (2) decline in the frequency of sensory nerve excitation when a receptor is stimulated continuously.

Addison's disease condition resulting from abnormal, deficient secretion of adrenal cortical hormones.

Adduct *ad-duct´* to move toward the midline of the body.

Adenohypophysis *ad-e-no-high-pof´-i-sis* the part of the pituitary gland that develops from Rathke's pouch.

Adenoids *ad´-e-noidz* enlargement of the pharyngeal tonsils in children.

Adenosine diphosphate (ADP) *a-den´-o-sene di-fos´-fate* the substance formed when ATP is split apart and releases energy.

Adenosine triphosphate (ATP) *a-den´-o-sene tri-fos´-fate* the compound that is an important intracellular energy source.

Adipose *ad´-i-pos* fatty.

Adrenal glands *ad-reen´-al* hormone-producing glands located superior to the kidneys; each consists of medulla and cortex areas.

Adrenalin *ad-ren´-a-lin* trademark name for epinephrine.

Adrenergic fibers *ad-ren-ur´-jick* nerve fibers whose terminals release norepinephrine or epinephrine upon stimulation.

Adrenocorticotropic hormone (ACTH) *a-dree´-no-kort-i-ko-tro´-pik* a hormone that influences the activity of the adrenal cortex and is released by the anterior portion of the pituitary.

Adventitia *ad-ven-tish´-yah* the outermost layer or covering of an organ.

Aerobic *ay-er-o´-bick* requiring oxygen to live or grow.

Afferent *af´-er-ent* carrying to or toward a center.

Afferent neuron *noo´-ron* nerve cell that carries impulses toward the central nervous system.

Afterbirth *af´-ter-berth* the expelled placenta after a baby has been born.

Agglutinin *a-gloo´-tin-in* an antibody in blood plasma that causes clumping of corpuscles or bacteria.

Agglutinogen *a-gloo´-tin-a-jen* (1) an antigen that stimulates the formation of a specific agglutinin; (2) an antigen found on red blood cells that is responsible for determining the ABO blood group classification.

Agonist *ag´-o-nist* a muscle whose contraction opposes the action of another muscle, its antagonist, which at the same time relaxes.

Albumin *al-bu´-min* a protein found in virtually all animal tissue and fluid.

Albuminuria *al-bu-min-oor´-ee-ah* the presence of albumin in the urine.

Aldosterone *al-dos´-ter-own* a hormone produced by the adrenal cortex that is important in sodium retention and reabsorption.

Alimentary *al-im-en´-ta-ree* pertaining to nourishment or the digestive organs.

Alkalosis *al-ka-lo´-sis* a condition in which the hydrogen ion concentration of the blood tends to decrease, and the pH of the blood tends to increase.

Allantois *a-lan´-to-is* an embryonic membrane; its blood vessels develop into blood vessels of the umbilical cord.

Allele *al´-eel* one of two or more alternate forms of a gene.

Allergy *al´-ler-jee* a condition in which there is hypersensitive response to a particular substance or environmental element.

Alveolus *al-ve´-o-lus* (1) a general anatomic term referring to a small cavity or depression; (2) an air sac in the lungs.

Amenorrhea *a-men-or-ree´-ah* absence of menstruation.

Amino acid *a-mee´-no* an organic compound containing nitrogen, carbon, hydrogen, and oxygen; the building block of protein.

Amnion *am´-nee-on* the innermost fetal membrane; it forms a fluid-filled sac for the embryo.

Amniocentesis *am´-nee-o-sen-tee´-sis* a procedure for obtaining amniotic fluid. Amniotic fluid is removed by inserting a needle through a pregnant woman's abdomen into the sac containing the fetus. Fetal cells in the fluid can be grown in a laboratory and subjected to chromosomal and/or biochemical analysis.

Amphiarthroses *am-fee-ar-throw´-seez* a type of articulation in which little motion occurs because of fibrocartilaginous connection of the articulating bony surfaces.

Amplitude *am´-pli-tude* largeness; fullness or wideness of extent or range.

Ampulla *am-pul´-lah* a saclike dilation of a duct or tube.

Anabolism *an-ab´-o-li-zem* the energy-requiring building-up phase of metabolism in which simpler substances are synthesized into more complex substances.

Anaerobic *an-ay-er-o´-bick* requiring no oxygen to live or grow.

Anastomosis *a-nas-to-mo´-sis* a union or joining of blood vessels or other tubular structures.

Anatomy the science of the structure of organs of organic beings.

Androgen *an´-dro-jen* a hormone or other substance that controls male secondary sex characteristics.

Anemia *a-nee´-mee-ah* a decreased number of erythrocytes or decreased percentage of hemoglobin in the blood.

Aneurysm *an´-yu-riz-em* a localized blood-filled sac in an artery wall caused by dilation or weakening of the wall.

Angina pectoris *an-jee´-nah pek´-tor-is* a severe, suffocating chest pain caused by brief lack of oxygen to heart muscle.

Angiotensin *an-jee-o-ten´-sin* a vasoconstrictor substance found in the blood.

Anion *an´-eye-on* an ion carrying one or more negative charges and therefore attracted to a positive pole.

Annulus fibrosus *an´-u-lus fi-bro´-sus* firm, ring-shaped outer portion of the intervertebral disc.

Anorexia *an-or-ek´-see-ah* loss of appetite or desire for food.

Anorexia nervosa *ner-vo´-sah* a nervous condition in which an extreme loss of appetite leads to emanciation, malnutrition, and possible worse consequences.

Anoxia *an-ok´-see-uh* a deficiency of oxygen.

Antagonist *an-tag´-o-nist* a muscle that acts in opposition to an agonist or prime mover.

Anterior the front of an organ or part; the ventral surface.

Antibody a specialized substance produced by the body that can provide immunity against a specific antigen.

Antidiuretic hormone (ADH) *an-tee-dye-u-re´-tik* a pituitary gland hormone that controls the reabsorption of water by the kidney.

Antigen *an´-ti-jen* any substance—including toxins foreign proteins, or bacteria—which, when introduced to the body, causes antibody formation.

Antithrombin *an-ti-throm´-bin* a substance in blood plasma that neutralizes thrombin, thereby inhibiting coagulation.

Antrum *an´-trum* an open space or chamber; a cavity, especially within a bone.

Anus *ay´-nus* the distal end of the digestive tract and the outlet of the rectum.

Aorta *ay-or´-tah* the major systemic artery; arises from the left ventricle of the heart.

Aortic arch *ay-or´-tik* the curved and most superior portion of the aorta.

Aortic body a receptor in the aortic arch sensitive to changing oxygen, carbon dioxide, and pH levels of the blood.

Aortic hiatus *high-ay´-tus* an opening behind the diaphragm giving passage to the aorta.

Aphasia *a-fay´-zhee-ah* a loss of the power of speech or a defect in speech.

Aplastic anemia *a-plas´-tic* anemia caused by inadequate production of erythrocytes resulting from inhibition or destruction of the red bone marrow.

Apocrine gland *ap´-o-krin* a gland in which the secretions accumulate toward the outer ends of the secreting cells; the cell ends pinch off and are released with the secretions.

Aponeurosis *ap-o-noo-ro´-seez* the fibrous or membranous sheet serving as a connection between muscle and the part it moves.

Appendix *a-pen´-dicks* a wormlike sac attached to the large intestine.

Aqueous humor *ay´-kwee-us hyu´-mer* the watery fluid of the anterior and posterior chambers of the eye.

Arachnoid *a-rak´-noid* weblike; specifically, the weblike, middle layer of the three meninges.

Areola *ah-ree´-o-lah* (1) any minute opening or space in a tissue; (2) the circular, pigmented area surrounding the nipple.

Arrector pili *ar-rek´-tor- pih´lee* tiny, smooth muscles of the skin that, upon contraction, pull the hair follicle, causing the hair to stand up.

Arteriole *ar-te´-ree-ole* a minute artery.

Arteriosclerosis *ar-te-ree-o-skle-ro´-sis* any of a number of proliferative and degenerative changes in the arteries.

Artery a vessel that carries blood away from the heart.

Arthritis *ar-thright´-us* inflammation of the joints.

Articulate *ar-tik´-u-late* to join together in such a way as to allow motion between the parts.

Ascites *as-sign´teez* accumulation of serous fluid in the abdominal cavity.

Asphyxia *as-fik´-see-ah* a loss of consciousness resulting from a deficiency in the oxygen supply.

Asthma *az´-ma* a disease or allergic response characterized by bronchial spasms and difficult breathing.

Astigmatism *a-stig´-ma-tiz-em* a visual defect resulting from irregularity in the lens or cornea of the eye causing the image to be out of focus.

Ataxia *a-tak´-see-a* lack of muscular coordination.

Atelectasis *at-e-lek´-ta-sis* lung collapse or incomplete expansion of a lung.

Atherosclerosis *ath-er-o-skle-ro´-sis* changes in the walls of large arteries consisting of lipid deposits on the artery walls.

Atlas the first cervical vertebrae; articulates with the occipital bone of the skull and the second cervical vertebra (axis).

Atom the smallest particle of an element; composed of electrons, protons, and neutrons; capable of existing individually or in combination with atoms of the same or another element.

Atresia *a-tree´-zhee-ah* (1) the abnormal closure of a body canal or opening; (2) degeneration of the ovarian follicle and the ovum within it.

Atrioventricular bundle (AV bundle) *a-tree-o-ven-trik´-yoo-lar* the bundle of specialized fibers serving to conduct impulses from the AV node to the right and left ventricles; failure of the AV bundle results in heart block; also called bundle of His.

Atrioventricular node (AV node) a specialized compact mass of conducting cells located at the atrioventricular junction in the heart.

Atrium *ay´-tree-um* a chamber of the heart receiving blood from the veins.

Atrophy *a´tro-fee* a reduction in size or wasting away of an organ or cell resulting from disease or lack of use.

Auditory *aw´-di-to-ree* pertaining to the sense of hearing.

Auditory ossicles *aw´-di-to-ree oss´-i-kuls* the three tiny bones serving as transmitters of vibrations and located within the middle ear; the malleus, incus, and stapes.

Auditory (eustachian) tube *yoo-stay´-shee-un* the connection between the middle ear and the pharynx.

Auricle *aw´-ri-kel* the external ear.

Auscultation *aws-kul-tay´-shun* the act of examination by listening to body sounds.

Autoimmune response *aw-to-im-myoon´* the production of antibodies or effector T cells that attack a person's own tissue.

Automaticity *aw-to-ma-ti´-si-tee* the ability of a structure, organ, or system to initiate its own activity.

Autonomic *aw-to-nom´-ik* self-directed; self-regulating; regulating; independent.

Autonomic nervous system the division of the nervous system that functions involuntarily and is responsible for innervating cardiac muscle, smooth muscle, and glands.

Autosome *aw´-to-some* a chromosome that is not a sex chromosome.

Axial skeleton *aks´-ee-al* the bones of the head and trunk: the skull, vertebral column, thorax, and sternum.

Axilla *aks-il´-ah* the armpit.

Axis (1) the second cervical vertebra; has a vertical projection called the odontoid process around which the atlas rotates; (2) the imaginary line about which a joint or structure revolves.

Axon *aks´-on* the process of a nerve cell by which impulses are carried away from the cell; efferent process; the conducting portion of a nerve cell.

B

Bacteria any of a large group of microorganisms, usually non-spore-producing and generally one celled; found in humans and other animals, plants, soil, air, and water; have a broad range of functions.

Baroreceptor *ba-ro-ree-sep´-ter* a receptor that is stimulated by pressure changes.

Basal layer columnar cells in which mitosis takes place; also called stratum basale of the epidermis.

Basal metabolic rate *met-ah-bol´-ik* the rate at which energy is expended (heat produced) by the body per unit time under controlled (basal) conditions: 12 hours after a meal, at rest.

Basal nuclei *bay´-zel noo´-klee-i* gray matter structures deep inside each of the cerebral hemispheres.

Base a substance that accepts hydrogen ions; capable of uniting with an acid to form a salt.

Basement membrane a thin layer of substance to which epithelial cells are attached in mucous surfaces.

Basophil *bay´-so-fil* white blood cells that readily take up basic dye; have a relatively pale nucleus and granular-appearing cytoplasm.

B cells lymphocytes that are committed to differentiate into the antibody producing plasma cells involved in humoral immunity.

Benign *bee-nine´* not malignant.

Biceps *bigh´-seps* two-headed, especially applied to certain muscles.

Bicuspid *bigh-kus´-pid* having two points or cusps.

Bifurcation *bi-fur-ka´-shun* division into two branches.

Bile a greenish yellow or brownish fluid produced in and secreted by the liver, stored in the gallbladder, and released into the small intestine.

Bilirubin *bil-i-roo´-bin* the red pigment of bile.

Biliverdin *bil-i-ver´-din* the green pigment of bile.

Biofeedback an area of research in which subconscious feedback is raised to the conscious level.

Biopsy *bigh´-op-see* the removal and examination of live tissue

Bipennate *bigh-pen´-nate* muscles in which the fibers are attached obliquely on both sides of the tendon.

Blastocyst *blas´-to-sist* a stage of early embryonic development.

Blood-brain barrier a mechanism that inhibits passage of materials from the blood into brain tissues and cerebrospinal fluid.

Bolus *bo´-lus* (1) a rounded mass of food prepared by the mouth for swallowing; (2) a concentrated mass of a pharmaceutical preparation, usually given intravenously.

Bowman's capsule *bo´-manz* the double-walled cup at the end of a nephron.

Brachial *bray´-kee-al* pertaining to the arm.

Bradycardia *brad-i-kar´-dee-ah* slowness of the heart rate; below 60 beats per minute.

Brain stem the portion of the brain consisting of the medulla, pons, and midbrain.

Bronchitis *brong-kigh´-tis* inflammation of the bronchi.

Bronchus *brong´-kus* one of the two large branches of the trachea leading to the lungs.

Buccal *buck´-kal* pertaining to the cheek.

Buffer a substance or substances that tend to stabilize the pH of a solution.

Bundle branch block a blocking of heart action resulting from a local lesion of the bundle of His; delayed contraction of one ventricle.

Bursa *bur´-sah* a small sac or cavity filled with fluid and located at friction points, especially joints.

C

Calcitonin *kal-si-to´-nin* a hormone released by the thyroid that lowers calcium and phosphate levels of the blood.

Calculus *kal´-ku-lus* a stone formed within various body parts.

Callus *kal´-us* (1) a bonelike material that protrudes between ends of a fractured bone; (2) a thickening of the skin caused by rubbing or friction.

Calyx *kal´-liks* a cuplike division of pelvis of the kidney.

Calorie *kal´-or-ee* a unit of heat; the amount of heat required to raise the temperature of 1 g of water 1° C is the small calorie; the large calorie is the amount of heat required to raise 1 kg of water 1° C—used in metabolic and nutrition studies.

Canal a duct or passageway; a tubular structure.

Canaliculi *kan-al-ik´-u-lee* extremely small tubular passages or channels.

Cancer a malignant, invasive cellular tumor that has the capability of spreading throughout the body or body parts.

Capacitation *ka-pass-i-tay´-shun* the process in which sperm undergo changes in the female reproductive tract making them capable of fertilization.

Capillary *kap´-a-lar-ee* a minute blood vessel connecting arterioles with venules.

Carbohydrates *kar-bo-high´-drates* organic compounds composed of carbon, hydrogen, and oxygen with the hydrogen and oxygen present in a 2:1 ratio; include starches, sugars, cellulose.

Carbonic anhydrase *kar-bon´-ik an-high´-drase* an enzyme that facilitates the combination of carbon dioxide with water to form carbonic acid.

Carcinogen *kar-sin´-o-jen* cancer-causing agent.

Carcinoma *kar-sin-o´-ma* cancer; a malignant growth.

Cardiac *kar´-dee-ak* pertaining to the heart.

Cardiac muscle specialized muscle of the heart.

Cardiac output the blood volume (in liters) ejected per minute by the left ventricle.

Carotene *kar´-o-teen* a yellow pigment; influences skin color.

Carotid *ka-rot´-id* the main artery in the neck.

Carotid body a receptor at the bifurcations of the common carotid arteries sensitive to changing oxygen, carbon dioxide, and pH levels of the blood.

Carotid sinus *sigh´-nus* a dilation of a common carotid artery at its bifurcation; involved in regulation of systemic blood pressure.

Carpals *kar´-puls* the eight bones of the wrist.

Cartilage *kar´-ti-lej* white, semiopaque, fibrous connective tissue.

Caruncle *kar´-un-kul* a small fleshy protuberance.

Catabolism *ka-tab´-a-liz-em* the process in which living cells break down more complex substances into simpler substances; destructive metabolism.

Cataract *kat´-a-rakt* partial or complete loss of transparency of the crystalline lens of the eye or its capsule.

Catheter *ka´-the-ter* a narrow, hollow tube that can be inserted into a body cavity for withdrawal of fluids.

Cation *kat´-eye-on* an ion with a positive charge and therefore attracted toward a negative pole (cathode) in electrolytic cells.

Caudal *kaw´dal* in humans, the inferior portion of the anatomy.

Cecum *see´kum* the blind-end pouch at the beginning of the large intestine.

Cell the basic biological unit of living organisms (except viruses), containing a nucleus and a variety of organelles; usually enclosed in the membrane.

Cell-mediated immunity immunity mediated by certain populations of lymphoid cells called T cells.

Cell membrane the selectively permeable membrane forming the outer layer of most animal cells; also called plasma membrane or unit membrane.

Cellulose *sel´-u-lose* a fibrous carbohydrate that is the main structural component of plant tissues.

Cementum *se-men´-tum* the bony connective tissue that covers the root of a tooth.

Central nervous system (CNS) the brain and the spinal cord.

Centriole *sen´-tree-ole* a minute body found in the nucleus of the cell; active in cell division.

Cerebellum *ser-e-bel´-lum* part of the hindbrain; controls movement coordination; consists of two hemispheres and a central portion (vermis).

Cerebral aqueduct *ser-ee´-bral a´-kwe-duct* the elongated, slender cavity of the midbrain that connects the third and fourth ventricles; also called the aqueduct of Sylvius.

Cerebrospinal fluid *ser-ee-bro-spy´-nul* the fluid produced in the cerebral ventricles; fills the ventricles and surrounds the central nervous system.

Cerebrum *ser´-i-brum* or *se-ree´-brum* the largest part of the brain; consists of right and left cerebral hemispheres.

Cerumen *se-roo´-men* earwax.

Cervical *ser´-vi-kal* refers to the neck or the necklike portion of an organ or structure.

Cervix *ser´-vix* (1) the cylindrical, inferior portion of the uterus leading to the vagina; (2) any necklike structure.

Chemoreceptor *kem-o-ree-sep´-tor* receptors sensitive to various chemical stimulations and changes.

Chemotaxis *kem-o-tak´-sis* movement of a cell, organism, or part of an organism toward or away from a chemical substance.

Chiasma *kee-az´-ma* a crossing or intersection of two structures, such as the optic nerves.

Cholecystectomy *kol-e-sis-tek´-tom-ee* removal of the gallbladder.

Cholecystokinin *kol-e-sis-toe´-kine-in* an intestinal hormone that stimulates gallbladder contraction and pancreatic enzyme release, but inhibits gastric acid secretion when the stomach is actively secreting.

Cholesterol *ko-les´-ter-ol* an organic alcohol found in animal fats and oil as well as in most body tissues, especially bile.

Cholinergic fibers *ko-lin-ur´-jick* nerve endings that, upon stimulation, release acetylcholine at their terminations.

Chondroblast *kon´-dro-blast* a cell that forms the fibers and matrix of cartilage.

Chondrocyte *kon´-dro-site* a mature cartilage cell.

Chorion *kor´-ee-on* the outermost fetal membrane; forms the placenta.

Choroid *ko´-roid* (1) skinlike; (2) the pigmented covering of the eye.

Chromosome *kro´-mo-some* the structures in the nucleus that carry the hereditary factors (genes).

Chyle *kile* a milky fluid consisting of fat globules in lymph formed in the small intestine during digestion.

Chyme *kime* the semifluid contents of the stomach consisting of partially digested food and gastric secretions.

Cilia *sil´-lee-ah* tiny, hairlike projections on cell surfaces that move in a wavelike manner.

Circadian *sir-kay´-dee-an* daily; on a daily cycle.

Circle of Willis a union of arteries at the base of the brain.

Circumcision *sur-kum-si´-zhun* removal of the foreskin of the penis.

Circumduction *sur-kum-duk´-shun* circular movement of a body part.

Cirrhosis *sir-o´-sis* a chronic disease, particularly of the liver, characterized by an overgrowth of connective tissue.

Clitoris *kli´-to-ris* a small, erectile structure in the female, homologous to the penis in the male.

Clone one or more offspring of like genetic constitution produced by asexual reproduction.

Cochlea *koak´-lee-ah* a cavity of the inner ear resembling a snail shell.

Coenzyme *ko-en´-zime* a nonprotein substance associated with and activating an enzyme.

Coitus *ko´-i-tus* sexual intercourse.

Colloid *kol´-loid* solute particles dispersed in a medium; particles do not settle out readily and do not pass through natural membranes.

Colloidal osmotic pressure (COP) *kol-loi´-dal os-mot´-ick* the pressure exerted on a membrane by the particles in a colloid; usually refers to the osmotic pressure of blood plasma and body fluids resulting from the presence of protein.

Colostrum *kol-os´-trum* the first milk secreted after parturition.

Coma *ko´-ma* unconsciousness from which the person cannot be aroused.

Complement system a system consisting of a series of plasma proteins that normally circulate in the blood in inactive forms. The activation of the system's initial components initiates a cascading series of reactions that ultimately produce active complement molecules, which can act as chemotactic agents, enhance phagocytosis, and exert other effects.

Compound a substance composed of two or more different elements.

Concave *kon´-kave* having a curved, depressed surface.

Concha *kon´-kah* a shell-shaped structure.

Condom *kon´-dom* a sheath worn over the penis during sexual intercourse to prevent conception and/or infection.

Conduction *kon-duk´-shun* the transfer of energy from molecule to molecule by direct contact.

Condyle *kon´-dile* a rounded projection at the end of a bone that articulates with another bone.

Cones one of the two types of photosensitive cells in the retina of the eye.

Congenital *kon-jen´-i-tal* existing at birth.

Conjunctiva *kon-junk-tigh´-vah* the thin protective membrane on the insides of the eyelids and the anterior surface of the eye itself.

Conjunctivitis *kon-junk-ti-vigh´-tis* an inflammation of the conjunctiva of the eye.

Connective tissue a primary type of tissue; form varies extensively, as does function, which includes support and storage.

Contraception *kon-tra-sep´-shun* the prevention of conception; birth control.

Contractility *kon-trak-til´-i-tee* a substance's ability to shorten or develop tension upon the application of a stimulus.

Contralateral *kon-tra-lat´-er-al* opposite; acting in unison with a similar part on the opposite side of the body.

Convection *kon-vek´-shun* the constant transfer of heat by movement of heated particles.

Convergence *kon-verj´-ence* turning toward or approaching a common point from different directions.

Convoluted *kon´-vo-lu-ted* rolled, coiled, or twisted.

Copulation *kop-u-lay´-shun* sexual intercourse.

Cornea *kor´-nee-ah* the transparent anterior portion of the eyeball.

Corona radiata *ko-ro´-nah ray-dee-aw´-tah* (1) the arrangement of elongated follicle cells around a mature ovum; (2) the crownlike arrangement of nerve fibers radiating from the inner capsule of the brain to every part of the cerebral cortex.

Corpus *kor´-pus* body; the major portion of an organ.

Cortex *kor´-teks* the outer surface layer of an organ.

Corticoid *kor´-ti-koid* a substance whose function and properties are similar to those of corticosteriods.

Corticosteroids *kor´-ti-ko-ste´-roidz* the steriod hormones released by the adrenal cortex.

Cortisol *kor´-ti-sol* a glucocorticoid produced by the adrenal cortex.

Costal *kos´-tal* pertaining to the ribs.

Cramp a painful, involuntary contraction of a muscle.

Cranial *kray´-nee-al* pertaining to the skull.

Creatine phosphate *kree´-uh-tin fos´-fate* a compound that serves as an alternative energy source for muscle tissue.

Crenation *kre-nay´-shun* the shriveling of an erythrocyte resulting from withdrawal of water.

Cretinism *kree´-tin-izm* a severe thyroid deficiency in the young that leads to stunted physical and mental growth.

Crista *kirs´-ta* a crest or ridge.

Cryptorchidism *kript-or´-kid-izm* a developmental defect in which the testes fail to descent into the scrotum.

Crystalloid *kris´-tal-oid* a substance having a crystalline structure as opposed to a colloidal composition; particles can pass through a natural membrane.

Cubital *ku´-bi-tal* pertaining to the forearm.

Cupula *ku´-pu-la* a domelike structure.

Cushing's syndrome *ku´-shingz sin´-drome* a disease produced by excess secretion of adrenocortical hormone; characterized by adipose-tissue accumulation, weight gain, and osteoporosis, for example.

Cutaneous *ku-tay´-nee-us* pertaining to the skin.

Cyanosis *sigh-a-no´-sis* a bluish coloration of the mucous membranes and skin caused by deficient oxygenation of the blood.

Cyst *sist* a sac with a distinct wall, containing fluid or other material; may be pathological or normal.

Cystitis *sis-tigh´-tis* inflammation of the urinary bladder.

Cytokinesis *sigh-to-ki-nee´-sis* the changes in the cytoplasm during cell division.

Cytology *sigh-tol´-o-jee* the science concerned with the study of cells.

Cytoplasm *sigh´-to-plaz-um* the protoplasm of a cell other than that of the nucleus.

D

Deamination *dee-am-i-nay´-shun* the removal of an amino group from an organic compound by reduction, hydrolysis, or oxidation.

Deciduous *dee-sid´-u-us* temporary.

Deciduous (milk) teeth the 10 temporary teeth replaced by permanent teeth; "baby" teeth.

Defecation *def-e-kay´-shun* the elimination of the contents of the bowels (feces).

Deglutition *dee-gloo-tish´-un* the act of swallowing.

Dehydration *dee-high-dray´-shun* a condition resulting from excessive loss of water.

Dendrite *den´-drite* branching; the branching neuron process that transmits the nerve impulse to the cell body; the receptive portion of a nerve cell.

Dental caries *den´-tal kar´-eez* tooth cavity.

Dentin *den´-tin* the calcified tissue beneath the enamel forming the major part of a tooth.

Deoxyribonucleic acid (DNA) *dee-ox-i-rye-bo-nu-klee´-ik* a nucleic acid found in all living cells: carries the organism's hereditary information.

Depolarization *dee-po-lar-i-zay´-shun* the neutralization to a state of nonpolarity; the loss of a negative charge inside the cell.

Depressor *dee-pres´sor* any substance that causes slowing, reduction of activity, or inhibition of another structure, organ, or substance.

Dermatitis *der-ma-tigh´-tis* an inflammation of the skin; nonspecific skin allergies.

Dermis *der´-mis* the deep layer of dense, irregular connective tissue of the skin; also called corium.

Desmosome *des´-mo-some* small, apposed, ellipsoidal plates in membranes of adjacent cells; also called macula adherens.

Diabetes insipidus *dye-uh-bee´-teez in-sip´-i-dus* a disease characterized by passage of a large quantity of urine of low specific gravity plus intense thirst and dehydration; a hypothalamic disorder is the cause.

Diabetes mellitus *mel´-li-tus* a disease caused by deficient insulin release, leading to failure of the body tissue to oxidize carbohydrates at a normal rate.

Dialysis *dye-al´-i-sis* the separation of substances from one another in a solution by taking advantage of their differing rate of diffusion through a semipermeable membrane.

Diapedesis *dye-a-pe-dee´-sis* the passage of blood cells through unruptured vessel walls into the tissues.

Diaphragm *dye´-a-fram* (1) any partition or wall separating one area from another; (2) a muscle that separates the upper thoracic cavity from the lower abdominopelvic cavity.

Diarthrosis *dye-ar-throw´-sis* a freely movable joint; a synovial joint.

Diastole *dye-ass´-ta-lee* a period (between contractions) of relaxation and dilation of the heart during which it fills with blood.

Diencephalon *dye-en-sef´-uh-lon* that part of the forebrain between the telencephalon and the mesencephalon including the thalami and most of the third ventricle.

Diffusion *di-fu´-zhun* the spreading of particles in a gas or solution with a movement toward uniform distribution of particles.

Digestion *di-jest´-yun* the bodily process of breaking down foods chemically and mechanically into compounds capable of being absorbed by body cells.

Dilate *dye´-late* to stretch; to open; to expand.

Distal *diss´tal* farthest from the center or midpoint of a limb structure.

Diverticulum *dye-ver-tik´-u-lum* a pouch or sac in the walls of a hollow organ or structure.

Dominant an allele, or the corresponding trait, that is expressed in all heterozygous cells or organisms.

Dorsal *dor´-sal* pertaining to the back; posterior.

Downs syndrome *dounz sin´-drome* an abnormality resulting from the presence of an extra copy of the genetic material contained in chromosome number 21. Characteristics include mental retardation and altered physical appearance.

Duct a canal or passageway.

Duodenum *doo-o-dee´-num* the first part of the small intestine.

Dura mater *doo´-rah may´-ter* the outermost and toughest of the three membranes (meninges) covering the brain and spinal cord.

Dysfunction *dis-funk´-shun* lack of normal function; disorder.

Dysmenorrhea *dis-men-or-ree´-ah* difficult, painful menstruation.

Dyspnea *disp´-nee-ah* labored, difficult breathing.

Dystrophy *dis´-tro-fee* a disorder caused by a defect or dysfunction of nutrition.

E

Ectoderm *ek´-to-derm* tissue forming the outer covering of the body and nervous tissues.

Ectopic *ek-to´-pik* not in the normal place; for example, in an ectopic pregnancy the egg is implanted at a place other than the uterus.

Edema *e-dee´-mah* an abnormal accumulation of fluid in body parts or tissues; causes swelling.

Effector *ef-fek´-tor* a motor or sensory nerve ending in an organ, gland, or muscle.

Efferent *ef´-er-ent* carrying away or away from, especially a nerve fiber that carries impulses away from the central nervous system.

Ejaculation *e-jak-u-lay´-shun* the sudden ejection of a fluid from a duct, especially semen from the penis.

Elastin *e-las´-tin* the main protein in elastic fibers of connective tissues.

Electrocardiogram (ECG) *ee-kel-tro-kar´-dee-o-gram* a graphic record of the electric current associated with heartbeats.

Electroencephalogram (EEG) *ee-lek-tro-en-sef´-a-lo-gram* a graphic record of the activity of nerve cells in the brain.

Electrolyte *ee-lek´-tro-lite* a substance that breaks down into ions when in solution and is capable of conducting an electric current.

Electron *ee-lek´-tron* a negatively charged particle in motion around the nucleus of an atom.

Embolism *em´-bo-liz-em* the obstruction of a blood vessel by a clot floating in the blood; may also be a bubble of air in the vessel (air embolism).

Embryo *em´-bree-oh* an organism in its early stages of development; in humans, the first two months after conception.

Emesis *em´-e-sis* vomiting.

Emphysema *em-fi-see´-muh* a condition caused by overdistension of the pulmonary alveoli or abnormal presence of air or gas in body tissues.

Enamel the hard, calcified substance that covers the crown of a tooth.

Encephalitis *en-sef-a-ligh´-tis* an inflammation of the brain.

Endocardial *en-do-kar´-di-al* pertaining to the inner lining of the heart.

Endocarditis *en-do-kar-di´-tis* an inflammation of the inner lining of the heart.

Endocardium *en-do-kar´-dee-um* the membrane lining the interior of the heart; endothelium and connective tissue.

Endochondral *en-do-kon´-dral* pertaining to the development of structures in cartilage.

Endocrine glands *en´-do-krin* ductless glands that empty their secretions directly into the blood.

Endoderm *en´-do-derm* tissue forming the digestive tube and its associated structures.

Endometrium *en-do-me´-tree-um* the mucous membrane lining of the uterus.

Endomysium *en-doo-mis´-ee-um* the thin connective tissue between the fibers of a muscle bundle.

Endoplasmic reticulum *en-do-plaz´-mik re-tik´-u-lum* a membranous network of tubular or saclike channels through the cytoplasm of a cell.

Endothelium *en-do-thee´-lee-um* the single layer of simple squamous cells that line the walls of the heart and the vessels that carry blood and lymph.

Energy the capacity to do work.

Enzyme *en´-zime* a substance formed by living cells that acts as a catalyst in bodily chemical reactions.

Eosinophil *ee-o-sin´-o-fil* a granular white blood cell whose granules readily take up a stain called eosin.

Epidermis *e-pi-der´-mis* the outer layer of cells of the skin.

Epididymis *ep-i-did´-i-mis* that portion of the seminal duct in which sperm mature and are transported from testes to body exterior.

Epiglottis *ep-e-glot´-tis* the elastic membrane-covered cartilage at the back of the throat; guards the glottis during swallowing.

Epimysium *e-pi-mis´-ee-um* the sheath of connective tissue surrounding a muscle.

Epinephrine *ep-i-nef´-rin* the chief hormone of the adrenal medulla.

Epiphysis *e-pif´-i-sis* the extremities of a long bone.

Epithelium *e-pi-thee´-lee-um* one of the primary tissues; covers the surface of the body and lines the body cavities, ducts, and vessels.

Eponychium *ep-o-neech´-ee-um* the fold of skin overlying the root of the nail.

Equilibrium *ee-kwi-lib´-ri-um* balance; a state when opposite reactions or forces counteract each other exactly.

Erythrocyte *e-rith´-ro-site* red blood cell.

Erythropoiesis *e-rith-ro-poi-ee´-sis* the formation process of erythrocytes.

Estrogen *es´-tro-jen* any substance that stimulates female secondary sex characteristics, female sex hormones.

Eupnea *yoop-nee´-ah* easy, normal breathing.

Excretion *eks-kree´-shun* the elimination of waste products from the body.

Exocrine glands *eks´-o-krin* glands that have ducts through which their secretions are carried to a particular site.

Exogenous *eks-og´-en-us* developing or originating outside the organ or part.

Expiration *ex-pi-ray´-shun* the act of expelling air from the lungs.

Exteroceptor *eks´-ter-o-sep-tor* an end organ that responds to stimuli from the external world.

Extracellular *eks-tra-sel´-u-lar* outside a cell.

Extracellular fluid fluid within the body but outside the cells.

Extrinsic *eks-trin´-zik* originating from outside an organ or part.

Exudate *eks´-yu-date* material including fluid, pus, or cells that has escaped from blood vessels and has been deposited in tissues.

F

Facet *fas´-et* a smooth, nearly flat surface on a bone for articulation.

Fallopian tube *fal-low´-pee-an* the oviduct or uterine tube; the tube through which the ovum is transported to the uterus.

Fascia *fash´-ee-ah* the layers of fibrous tissue under the skin or covering and separating muscles.

Fasciculus *fa-sik´-yoo-lus* a bundle of nerve, muscle, or tendon fibers separated by connective tissues.

Feces *fee´-seez* material discharged from the bowel composed of food residue, secretions, and bacteria.

Fenestrated *fen´-es-tray-ted* pierced with one or more small openings.

Fetus *fee´-tus* the unborn young; in humans, from the third month in the uterus until birth.

Fibrillation *fib-ri-lay´-shun* irregular, uncoordinated contraction of muscle fibers.

Fibrin *figh´-brin* the fibrous insoluble protein formed during the clotting of blood.

Fibrinogen *figh-brin´-o-jen* a protein that is converted to fibrin during blood clotting.

Filtration *fil-tray´-shun* the passage or straining of a solvent and dissolved substances through a membrane or filter.

Fissure *fis´-sure* (1) a groove or cleft; (2) the deepest linear depressions on the brain.

Fistula *fis´-tu-lah* an abnormal passage between organs or between a body cavity and the outside.

Fixator *fiks´-ay-tor* a muscle acting to immobilize a joint or a bone; fixes the origin of prime movers so that muscle action can be exerted at the insertion.

Flaccid *flak´-sid* soft; flabby; relaxed.

Flagella *fla-jel´-ah* long whiplike extensions of the cell membrane of some bacteria; serve as agents for locomotion.

Flexion *flek´-shun* bending; the movement that decreases the angle between bones.

Focal length *fo´-kal* the distance from the lens to the focal point.

Focal point the point at which light rays converge behind the lens.

Follicle *fol´-i-kal* a small sac or gland.

Follicle-stimulating hormone (FSH) a hormone produced by the anterior pituitary that stimulates ovarian follicle production in females and sperm production in males.

Fontanels *fon-tan-els´* the fibrous membranes at the body area where bones have not yet formed; babies' "soft spots."

Foramen *fo-ra´-men* a hole or opening in a bone or between body cavities.

Forebrain the anterior portion of the brain consisting of the telencephalon and the diencephalon.

Fossa *fos´-ah* a depression; often used as an articular surface.

Fovea *fo´-vee-ah* a pit, generally used for attachment rather than for articulation.

Frontal (coronal) plane a longitudinal section that runs at right angles to sagittal planes dividing the body into anterior and posterior parts.

Fulcrum *ful´-krum* the pivot point of a lever.

Fundus *fun´-dus* the base of an organ; that part farthest from the opening of the organ.

Funiculi *fu-nik´-ye-lee* (1) cordlike structures; (2) anterior and lateral divisions of white matter in the spinal cord.

G

Gallbladder the sac beneath the right lobe of the liver used for bile storage.

Gallstones particles of cholesterol or calcium carbonate that are occasionally formed in gallbladder and bile ducts.

Gamete *gam´-eet* male or female reproductive cell.

Gametogenesis *gam-e-to-jen´-e-sis* the origin and formation of gametes.

Ganglion *gan´-glee-on* a group of nerve-cell bodies, usually located in the peripheral nervous system.

Gap junction intercellular specialization with the cell membranes of adjacent cells only 20 Å apart; also called nexus.

Gastrin a hormone that stimulates gastric secretion, especially hydrochloric acid release.

Gastroenteritis *gas-tro-en-ter-i´-tis* an inflammation of mucosa of stomach and intestine.

Gastroesophageal sphincter *gas-tro-e-soff-a-jee´-al sfink´-ter* narrowing between the esophagus and the stomach.

Gene *jeen* one of the biological units of heredity located on chromosomes; transmits hereditary message.

Genetic counseling *jen-e´-tik* the counseling of parents about the chances of their children having particular genetic abnormalities.

Genetics *jen-e´-tiks* the science of heredity.

Genitalia *jen-i-tay´-lee-ah* the external sex organs.

Genotype *jen´-o-tipe* the genetic composition of an individual.

Gestation *jes-tay´-shun* the period of pregnancy; about 280 days for humans.

Gingiva *jin-jigh´-va* or *jin´-ji-vah* the gums.

Gland an organ specialized to secrete or excrete substances for further use in the body or for elimination.

Glans a small glandlike mass of erectile tissue at the tip of the penis and the clitoris.

Glaucoma *glaw-ko´-mah* an abnormal elevation of the pressure within the eye.

Glia *glee´-a* see *neuroglia.*

Globin *glow´-bin* the protein component of hemoglobin.

Glomerulus *glo-mer´-yoo-lus* a knot of coiled capillaries in the kidney.

Glottis *glot´-tis* the opening between the vocal cords; entrance to the larynx.

Glucagon *gloo´-ka-gon* a hormone formed by islets of Langerhands in the pancreas; raises the glucose level of blood.

Glucocorticoids *gloo-ko-kor´-ti-koidz* the adrenal cortex hormones that affect metabolism of fats and carbohydrates.

Glucose *gloo´-kose* the principal sugar in the blood.

Glycerol *gliss-e-rol* an important alcoholic component of fat.

Glycogen *gligh-ko-jen* an animal starch; the main carbohydrate stored in animal cells.

Glycogenesis *gleye-ko-jen´-e-sis* the body's formation of glycogen from other carbohydrates.

Glycogenolysis *gleye-ko-jen-ol´-i-sis* the body's breakdown of glycogen to glucose.

Glycolysis *gligh-kol´-a-sis* the body's breakdown of glucose into simpler compounds, especially lactic acid.

Goblet cells the individual cells of the respiratory and digestive tracts that function as glands.

Goiter *goi´-ter* an enlargement of the thyroid gland.

Gonad *go´-nad* a gland or organ producing gametes; an ovary or testis.

Gonadotropins *go-nad-o-tro´-pinz* the gonad-stimulating hormones produced by the anterior pituitary.

Graafian follicle *graf´-ee-an fol´-i-kal* a mature ovarian follicle.

Graded response a response whose magnitude varies directly with the strength of the stimulus.

Gray matter the gray area of the central nervous system; contains neurons.

Groin the junction of the thigh and the trunk; the inguinal area.

Growth hormone a hormone that stimulates growth in general; produced in the anterior pituitary; also called somatotropin.

Gubernaculum *goo-ber-nak´-yoo-lum* a guiding, connecting cord structure; the cord between the testis and the scrotal sac, for example.

Gustation *gus-tay´-shun* taste.

Gyrus *jigh´-rus* a convolution on the surface of the cerebral cortex.

H

Hamstring muscles the posterior thigh muscles; the biceps femoris, semimembranosus, and semitendinosus.

Haustra *haw´-strah* pouches of the colon.

Haversian system or osteon *ha-ver´-zee-an, os´-tee-on* an organized system of interconnecting canals in the microscopic structure of adult compact bone.

Hay fever an acute allergic reaction of conjunctiva and upper air passages due to pollen sensitivity.

Heart block a defective transmission of impulses from atrium to ventricle.

Heart murmur an abnormal heart sound.

Heat exhaustion the collapse of an individual after heat exposure without failure of the body's heat-regulating mechanism.

Heat stroke the failure of the heat-regulating ability of an individual under heat stress.

Hematocrit *hem-at´-o-krit* the percentage of erythrocytes to total blood volume.

Heme *heem* the iron-containing pigment that is essential to oxygen transport by hemoglobin.

Hemiplegia *hem-i-plee´-jee-ah* paralysis of one side of the body.

Hemocytoblasts *hee-mo-sigh´-to-blasts* stem cells that give rise to all the formed elements of the blood.

Hemoglobin *hee-mo-glo´-bin* the oxygen-transporting component of erythrocytes composed of heme and globin.

Hemolysis *hee-mol´-i-sis* the destruction of erythrocytes.

Hemophilia *hee-mo-phil´-i-a* a clotting defect caused by an inherited genetic absence of a blood-clotting factor.

Hemopoiesis *hee-mo-poi-ee´-sis* the formation of blood.

Hemorrhage *hem´-o-ridj* the escape of blood from the vessels by flow through ruptured walls; bleeding.

Heparin *hep´-a-rin* a substance that prevents clotting found in many tissues, especially the liver.

Hepatic portal system *he-pat´-ik* the liver circulatory arrangement where the hepatic portal vein carries dissolved nutrients to the liver tissues.

Hepatitis *hep-at-eye´-tis* an inflammation and/or infection of the liver.

Hernia *her´-nee-ah* the abnormal protrusion of an organ or a body part through the containing wall of its cavity.

Herpes simplex *her´-peez sim´-pleks* a fever blister or cold sore; a virus-caused condition.

Herpes zoster *zos´-ter* an infection of the dorsal root ganglia of spinal nerves by a virus, causing pain and fluid-filled vesicles on the skin; also called shingles.

Heterosexuality *he-ter-o-seks-u-al´-i-tee* sexual interest in or desire for persons of the opposite sex.

Heterozygous *het-er-o-zigh´-gus* a situation in which two different alleles occur at a given locus on homologous chromosomes.

Hilum, hilus *high´-lum, high´-lus* the notched or depressed area where vessels enter and leave an organ.

Histamine *his´-ta-meen* a substance present in many cells which causes vasodilation and increased vascular permeability.

Histology *his-tol´-o-jee* the branch of anatomy dealing with the microscopic structure of tissues.

Holocrine glands *hol´-o-krin* glands that accumulate their secretions within their cells; secretions are discharged only upon rupture and death of the cell.

Homeostasis *hom-ee-o-stas´-sis* the state when the body organs function together to maintain a stable internal environment for the general well-being of the entire body; a state of body equilibrium.

Homeotherm *ho´-me-o-therm* an organism that produces its own heat and maintains a constant body temperature.

Homologous *ho-mol´-o-gus* parts or organs corresponding in structure but not necessarily in function.

Homosexuality *ho-mo-seks-u-al´-i-tee* sexual interest in or desire for persons of the same sex.

Homozygous *ho-mo-zigh´-gus* a situation in which the same allele occurs at a given locus on homologous chromosomes.

Hormones *hor´-mones* the secretions of endocrine glands; responsible for specific regulatory effects on certain parts or organs.

Humoral immunity *hyoo´-mer-al im-myoo´-ni-tee* immunity mediated by specialized proteins called antibodies.

Hyaline *high´-a-lin* glassy; transparent.

Hydrocarbon *high-dro-kar´-bon* a molecule composed of only carbon and hydrogen.

Hydrochloric acid *high-dro-klo´-rik* HCl; facilitates protein digestion in the stomach; produced by parietal cells.

Hydrolysis *high-drol´-i-sis* the process in which a chemical compound unites with water and is then split into smaller molecules.

Hydrostatic pressure *high-dro-stat´-ik* the pressure of fluid in a system.

Hyperopia *high-per-o´-pi-a* farsightedness.

Hypertension *high-per-ten´-shun* high blood pressure.

Hypertonic *high-per-ton´-ik* excessive, above normal, tone or tension.

Hypertrophy *high-per´-tro-fee* an increase in the size of a tissue or organ independent of the body's general growth.

Hypodermis *high-po-der´-mis* the subcutaneous connective tissue; also called superficial fascia.

Hyponychium *high-po-neech´-ee-um* the thickened horny zone of the epidermis beneath the free border of the nail.

Hypothalamus *high-po-thal´-a-mus* the region of the diencephalon forming the floor of the third ventricle of the brain.

Hypothermia *high-po-ther´-mi-ah* subnormal body temperature.

Hypotonic *high-po-ton´-ik* below normal tone or tension.

Hypoxemic hypoxia *high-pok-see´-mik high-pock´-see-uh* a condition in which decreased oxygen is available to tissues because of decreased Po_2 in arterial blood.

Hypoxia *hi-pox´-ee-a* a condition in which a physiologically inadequate amount of oxygen is available to tissues.

I

Ileum *il´-ee-um* the lower part of the small intestine between the jejunum and the cecum of the large intestine.

Immune surveillance *im-myoon´ sir-vail´-lantz* the body's immune response to cancer.

Immunity *im-myoon´-i-tee* the body's ability to resist many organisms and chemicals that can damage tissues.

Immunoglobulin *im-myoo-no-glob´-you-lin* an antibody.

Impetigo *im-pe-tee´-go* a highly contagious skin infection common in children.

Impotence *im´-po-tense* (1) lack of power, inability; (2) a male's inability to have sexual intercourse or maintain an erection.

In vitro *in vee´-tro* in a test tube, glass, or artificial environment.

In vivo *in vee´-vo* in the living body.

Infarct *in-farkt´* a region of dead, deteriorating tissue resulting from blood flow interference.

Inferior (caudal) pertaining to a position near the tail end of the long axis of the body.

Inflammation *in-flam-may´-shun* a physiological response of the body to tissue injury; includes dilation of blood vessels and an increase in vessel permeability.

Infundibulum *in-fun-dib´-yoo-lum* a funnel-shaped body part or passageway.

Inguinal *ing´-gwi-nal* pertaining to the groin region.

Inner cell mass an accumulation of cells in the blastocyst from which the embryo develops.

Innervation *in-ner-vay´-shun* the supply or distribution of nerves or nerve stimuli to a part.

Insertion *in-ser´-shun* the place or mode of attachment of a muscle; the movable part of a muscle as opposed to origin.

Inspiration *in-spi-ray´-shun* the drawing of air into the lungs.

Insulin *in´-su-lin* the hormone produced in the pancreas affecting carbohydrate and fat metabolism, blood-glucose levels, and other systemic processes.

Integumentary system *in-teg-u-men´-tar-ee* the skin and its accessory structures.

Intercellular *in-ter-sel´-u-lar* between the cells of the body or part.

Intercellular matrix *may´-triks* the material between adjoining cells.

Interferon *in-ter-fer´-on* a chemical that is able to provide some protection against virus invasion of the body; inhibits viral growth.

Internal capsule the band of white matter between the basal nuclei and the thalamus.

Internal environment the environment within the body.

Internal respiration the exchange of gases between blood and tissue fluid and between tissue fluid and cells.

Interoceptor *in´-ter-o-sep-tor* a nerve ending situated in the viscera sensitive to changes and stimuli within the body's internal environment; also called visceroceptor.

Interstitial fluid *in-ter-stish´-al* the fluid between the cells or body parts.

Intervertebral discs *in-ter-ver´-te-bral* the discs of fibrocartilage between bodies of vertebrae.

Intervertebral foramina *fo-rah´-mi-nah* the openings between the pedicles of adjacent vertebrae through which the spinal nerves pass.

Intracellular *in-tra-sel´-u-lar* within a cell.

Intracellular fluid fluid within a cell.

Intramural pressure *in-tra-mu´-ral* the pressure built up in the walls of an organ.

Intrinsic factor *in-trin´-sik* a glycoprotein substance required for vitamin B_{12} absorption.

Intrinsic muscle a muscle that has both its origin and its insertion in an organ.

Invert to turn inward.

Involuntary muscle a muscle not under control of the will; independent muscle.

Ion *eye´-on* an atom with a positive or negative electric charge.

Ionic bond *eye-on´-ik* a bond in which oppositely charged ions are held together by the attraction between opposite charges.

Ipsilateral *ip-si-lat´-er-al* situated on the same side.

Iris *eye´-ris* the pigmented, circular diaphragm in front of the eye's lens.

Ischemia *is-kee´-mee-uh* a local decrease in blood supply resulting from obstruction of arterial inflow.

Isometric *i-so-met´-rik* of the same length.

Isotonic *i-so-ton´-ik* having uniform tension under pressure or stimulation.

Isotope *i´so-tope* a different form of a given element; isotopes have the same atomic number but different mass numbers.

J

Jaundice *jawn´-dis* an accumulation of bile pigments in the blood producing a yellow color of the skin.

Jejunum *je-joo´-num* the part of the small intestine between the duodenum and the ileum.

Joint the junction of two or more bones; an articulation.

Junctional complex the junction between cells in columnar and some cuboidal epithelium.

Juxtamedullary *jux´-ta-med´-u-lair-ee* referring to the inner portion of the cortex of the kidney, adjacent to the medulla.

K

Karyokinesis *kar-ee-o-ken-ee´-sis* see *mitosis.*

Keratin *ker´-a-tin* a fibrous insoluble protein found in tissues such as hair or nails.

Ketosis *kee-to´-sis* an abnormal condition during which an excess of ketone bodies is produced.

Kinesthesia *ki-nez-thee´-zhah* the ability to perceive muscle movement.

Kinetic energy *ki-net´-ik* the energy of motion.

Kinins *kigh´-ninz* group of polypeptides that dilate arterioles, increase vascular permeability, act as powerful chemotactic agents, and induce pain.

Krebs cycle the citric acid cycle; the series of reactions during which energy is liberated from metabolism of carbohydrates, fats, and amino acids.

L

Labia *lay´-bee-ah* lips.

Labor *la´-bor* the period characterized by strong, rhythmic uterine contractions that precedes the birth of a baby.

Lacrimal *lak´-ra-mal* pertaining to tears.

Lactation *lak-tay´-shun* the production and secretion of milk.

Lacteal *lak´-tee-al* the special lymphatic capillaries of the small intestine that take up chyle.

Lactic acid *lak´-tik* the product of anaerobic glycolysis, especially in muscle.

Lacuna *la-ku´-nah* a little depression or space; in bone or cartilage, lacunae are occupied by cells.

Lamina *lam´-i-nah* (1) a thin layer or flat plate; (2) the portion of a vertebra between the transverse process and the spinous process.

Laryngeal prominence *la-rin´-jul prom´-i-nense* the tubercle of the thyroid cartilage; Adam's apple.

Laryngitis *lar-en-jigh´-tis* an inflammation of the larynx.

Larynx *lar´-inks* the organ of the voice; located between the trachea and the base of the tongue.

Lateral *lat´-ur-al* away from the midline of the body.

Lateral sacs the reticular sites in muscle from which calcium is released.

Lens the elastic, doubly convex structure behind the pupil of the eye; focuses the light entering the eye on the retina.

Lesion *lee´-zhun* a tissue injury or wound.

Leukemia *loo-kee´-mee-ah* a cancerous condition in which there is an excessive production of leukocytes.

Leukocyte *loo´-ko-site* a white blood cell.

Ligament *lig´-a-ment* a band or sheetlike fibrous tissue connecting bones or parts.

Lingual *ling´-gwal* pertaining to the tongue.

Lipid *lip´-id* a substance that is almost insoluble in water but soluble in fat solvents; fatty acids and fats.

Lobe a curved, rounded structure or projection.

Locus *low´-kus* the position that a gene occupies on a chromosome.

Lordosis *lor-do´-siss* an excessive curve in the anterior lumbar spine; otherwise known as swayback.

Lumbar *lum´-bar* the portion of the back between the thorax and the pelvis.

Lumbar puncture a procedure involving insertion of a needle between the third and fourth lumbar vertebrae and into the subarachnoid space to sample cerebrospinal fluid.

Lumbosacral trunk *lum-bo-sak´-ral* a group of nerves that connects the lumbar plexus and the sacral plexus.

Lumen *loo´-men* the space inside a tube, blood vessel, or intestine.

Luteinizing hormone *lu´-tee-in-eye-zing* an anterior pituitary hormone that stimulates maturation of cells in the ovary and acts on interstitial cells of the male testis.

Lymph *limf* the watery fluid in the lymph vessels collected from the tissue fluids.

Lymph node a mass of lymphatic tissue.

Lymphatic system *lim-fat´-ik* a system of vessels carrying lymph closely related anatomically and functionally to the circulatory system.

Lymphocyte *lim´-fo-site* a granular white blood cell formed in the lymphoid tissue.

Lymphokines *lim´-fo-kines* substances involved in cell-mediated immune responses that enhance the basic inflammatory response and subsequent phagocytosis.

Lysosomes *ligh´-so-somes* tiny organelles that originate from the Golgi apparatus and contain strong digestive enzymes.

Lysozyme *ligh´-so-zime* an enzyme capable of destroying certain kinds of bacteria.

M

Macula *mak´-yoo-la* the slightly yellow region lateral to the optic disc.

Macula adherens *mak´-u-la ad-hear´-uns* see *desmosome.*

Malignant *ma-lig´-nant* life threatening; pertains to diseases that spread and lead to death, such as cancer.

Mammary glands *mam´-mar-ee* milk-producing glands of the breasts.

Margination *mar-jin-ay´-shun* the adhesion of white blood cells to the walls of capillaries in the early stage of inflammation.

Mass number the combined number of protons and neutrons in the nucleus of an atom.

Mastication *mas-ti-kay´-shun* the act of chewing.

Matrix *may´-trix* the homogeneous intercellular substance of any tissue.

Meatus *mee-ay´-tus* the external opening of a canal.

Mechanoreceptor *mek-an-o-ree-sep´-tor* a receptor sensitive to mechanical pressures such as touch, sound, or contractions.

Medial *mee´-dee-al* toward the midline of the body.

Mediastinum *mee-dee-a-stigh´-num* the portion of the thoracid cavity between the lungs.

Mediated transport *mee´-dee-ay-ted* transport involving protein carrier molecules within the cell.

Medulla *ma-dul´-ah* the central portion of certain organs.

Meiosis *my-o´-sis* the last two cell divisions in gamete formation producing nuclei with half the full number of chromosomes (haploid).

Melanin *mel´-a-nin* the dark pigment responsible for skin color.

Melanocyte *me-lan´-o-site* a cell that produces melanin.

Menarche *me´-nar-kee* the onset of menstruation.

Meninges *men-in´-jeez* the membranes that cover the brain and spinal cord.

Meningitis *men-in-ji´-tis* an inflammation of the meninges covering the brain and spinal cord.

Menopause *men´-o-pawz* the physiological termination of menstrual cycles.

Menses *men´-seez* the recurrent monthly discharge of menstruation.

Menstruation *men-stroo-ay´-shun* the periodic, cyclic discharge of blood, secretions, tissue, and mucus from the mature female genital canal in the absence of pregnancy.

Merocrine glands *mer´-o-krin* glands that produce secretions intermittently; secretions do not accumulate in the gland.

Mesencephalon *mes-en-sef´-uh-lon* the midbrain.

Mesenteries *mes´-en-ter-eez* the double-layered membranes of the peritoneum that support most of the organs in the abdominal cavity.

Mesoderm *mes´-o-derm* tissue that forms the skeleton and muscles of the body.

Metabolic rate *met-a-bol´-ik* the energy expended by the body per unit time.

Metabolism *me-tab´-o-liz-em* the sum total of the chemical reactions that occur in the body.

Metabolize *me-tab´-o-lize* to transform substances into energy or materials the body can use or store by means of anabolism or catabolism.

Metacarpals *met-a-kar´-puls* the five bones of the palm of the hand.

Metastasize *me-tas´-ta-size* the spread of disease from one body part or organ into another not directly connected to it.

Metatarsals *met-a-tar´-suls* the five bones between the instep and the phalanges.

Metencephalon *met-en-sef´un-lon* the anterior part of the hindbrain composed of the cerebellum and pons.

Microbodies *my-kro-bod´-eez* membrane-bound cytoplasmic structures containing oxidative enzymes; also called peroxisomes.

Microfilament *my-kro-fil´-a-ment* filaments associated with contractile activities of the cell and developmental modifications of cell and organ shape.

Microtubule *my-kro-too´-bule* tiny cylindrical tubes in the cytoplasm of many cells.

Microvilli *my-kro-vil´-lee* the tiny protoplasmic projections formed on the free surfaces of some epithelial cells; appear brushlike when viewed through a microscope.

Mineralocorticoid *min-er-al-o-kor´-ti-koid* an adrenal cortical steroid hormone that regulates mineral metabolism and fluid balance.

Minerals the inorganic chemical compounds found in nature.

Mitochondria *my-to-kin´-dree-ah* the cytoplasmic organelles in the form of granules, rods, and filaments responsible for generation of metabolic energy for cellular activities.

Mitosis *my-to´-sis* the division of the cell nucleus; often followed by division of the cytoplasm of a cell; also called karyokinesis.

Mixed nerves the nerves containing the processes of motor and sensory neurons; their impulses travel to and from the central nervous system.

Molar *mo´-lar* a solution concentration determined by mass of solute—one liter of solution contains an amount of solute equal to its molecular weight in grams.

Molecule *mol´-e-kewl* a very small mass of matter composed of atoms held together as a unit.

Moles elevations of the skin that are pigmented.

Monocyte *mon´-o-site* a large, single-nucleus white blood cell.

Monosomy *mo-no´-so-me´* the condition in which one of two homologous chromosomes is missing.

Mons pubis *monz pu-bus* the fatty eminence over the pubic symphysis in the female.

Motor nerve cells the nerves that carry impulses leaving the brain and spinal cord.

Motor unit a neuron and the muscle cells it supplies.

Mucus *mew´-kus* a sticky, thick fluid secreted by mucous glands and mucous membranes that keeps the free surface of membranes moist.

Mucous membranes the membranes that form the linings of the digestive, respiratory, urinary, and reproductive tracts.

Multipennate *mul-ti-pen´-nate* the muscles in which the fibers have a complex arrangement involving converging of tendons.

Multiple sclerosis *skler-o´-sis* a chronic condition characterized by destruction of the myelin sheaths of neurons in the spinal cord and the brain.

Multipolar neurons *mul-ti-pol´-ar* neurons that have one long axon and numerous dendrites.

Muscle fibers muscle cells.

Muscle spindles the complex capsules found in skeletal muscles that are sensitive to stretching.

Muscle twitch a single rapid contraction of a muscle followed by relaxation.

Muscular dystrophy *dis´-tro-fee* a progressive disorder marked by atrophy and stiffness of the muscles.

Mutation *mu-tay´-shun* an alteration in the genetic material.

Myelencephalon *mile-len-sef´-uh-lon* the lower part of the hindbrain, especially the medulla oblongata.

Myelin *my´-uh-lin* the white, fatty lipid substance forming a sheath around some nerves.

Myelinated fibers *my´-uh-li-nay-ted* axons (projections of a nerve cell) covered with myelin.

Myelitis *my-a-light´-us* an inflammation of the spinal cord.

Myocardial infarction *my-o-kar´-dee-al in-fark´-shun* a condition characterized by dead tissue areas in the myocardium caused by interruption of blood supply to the area.

Myocardium *my-o-kar´-dee-um* the cardiac muscle layer of the wall of the heart.

Myofibril *my-o-figh´-bril* a fibril found in the cytoplasm of muscle.

Myofilament *my-o-fil´-a-ment* the filamentous structures making up a sarcomere consisting of thick and thin types.

Myogenic *my-o-jen´-ik* having the potential to contract automatically without nervous system stimulation.

Myoglobin *my-o-glo´-bin* muscle hemoglobin.

Myometrium *my-o-me´-tree-um* the thick uterine musculature.

Myopia *my-o´-pee-ah* nearsightedness.

Myosin *my´-o-sin* one of the principal proteins found in muscle.

Myotome *my´-a-tome* that part of a somite that differentiates into skeletal muscle.

N

Nares *nar´-ezz* the nostrils.

Necrosis *ne-kro´-sis* the death or disintegration of a cell or tissues caused by disease or injury.

Negative feedback feedback that tends to cause the levels of a variable to change in a direction opposite to that of an initial change.

Nephron *nef´-ron* the functional part of the kidney.

Nerve fiber axon (nerve cell projection) together with certain sheaths or coverings.

Neuralgia *noo-ral´-jee-ah* a severe paroxysmal (spasm-producing) pain along the course of a nerve.

Neuritis *noo-righ´-tis* an inflammatory or degenerative condition of the nerves.

Neurofibril *noo-ro-fi´-bril* the fibril of a nerve cell, usually extending from the processes and traversing the cell body.

Neuroglia *noo-rog´-lee-uh* the nonneuronal tissue of the central nervous system that performs supportive and other functions; also called glia.

Neurohypophysis *noo-ro-high-pof´-i-sis* the portion of the pituitary gland derived from the brain.

Neuromuscular junction *noo-ro-mus´-kyoo-lar* the region where a motor neuron approaches skeletal muscle sarcolemma.

Neurons *noo´-ronz* the nerve cells that transmit messages throughout the body.

Neutron *noo´-tron* an uncharged particle located in the nucleus of an atom.

Neutrophil *noo´-tro-fil* the most abundant of the white blood cells.

Nexus *nex´-us* see *gap junction.*

Nitrogen balance the state in a normal adult in which the nitrogen excreted equals the nitrogen intake in the form of food.

Nondisjunction *non-dis-jungk´-shun* failure of paired chromosomes to separate during cell division.

Nuchal lines *nu´-kal* the three slight ridges on the external surface of the occipital bone.

Nucleoli *noo-klee´-o-lee* the small spherical bodies in the cell nucleus.

Nucleotide *noo´-klee-o-tide* a component of DNA and RNA consisting of a sugar, a base, and a phosphate group.

Nucleus *noo´-klee-us* (1) a dense central body in most cells containing the genetic apparatus of the cell; (2) the core of an atom.

Nucleus pulposus *noo´-klee-us pul-po´-sus* the central gelatinous part of the intervetebral disc.

Nystagmus *niss-tag´-mus* an oscillatory movement of the eyeballs.

O

Obesity *o-bee´-sit-ee* a condition of a person being overweight; often leads to other complications.

Occipital *ok-sip´-i-tal* pertaining to the back of the head.

Occlusion *o-kloo´-zhun* closure or obstruction.

Olfaction *ol-fak´-shun* smell.

Oncotic pressure *on-kot´-ik* the osmotic force exerted by plasma proteins.

Oogenesis *oh-oh-jen´-e-sis* the process of origin, growth, and formation of the ovum.

Ophthalmic *of-thal´-mik* pertaining to the eye.

Opsonins *op´-se-ninz* proteins that coat specific foreign particles and thereby render them more susceptible to phagocytosis.

Optic *op´-tik* pertaining to the eye.

Optic chiasma *op´-tik kigh-az´-mah* the meeting of the optic nerves after entering the cranium.

Oral relating to the mouth.

Ora serrata *o´-ra ser-rat´-a* the serrated margin of the retina.

Organ a part of the body combining two or more tissues to perform a specialized function.

Organelle *or-gan-el´* a specialized structure or part of a cell having a definite function to perform.

Organic compound any hydrocarbon or hydrocarbon derivative.

Orgasm *or´-gaz-um* the intense emotional and physical climax associated with sexual stimulation.

Origin the end of attachment of a muscle that remains relatively fixed during muscular contraction.

Osmoreceptor *os-mo-ree-sep´-tor* a structure sensitive to osmotic pressure.

Osmosis *os-mo´-sis* the passage (diffusion) of a solvent through a membrane from a dilute solution into a more concentrated one.

Ossicles *os´-si-kalz* the three bones of the middle ear: malleus, stapes, and incus.

Osteoblasts *os´-tee-o-blasts* the bone-forming cells.

Osteoclasts *os´-tee-o-klasts* the large cells that reabsorb or erode bone substance.

Osteocyte *os´-tee-o-site* a mature bone cell found in each lacuna.

Osteomalacia *os-tee-o-ma-lay´-she-ah* the softening of bone resulting from vitamin D deficiency in the adult.

Osteomyelitis *os-tee-o-my-a-light´-us* a disease in which the periosteum, the contents of the marrow cavity, and the bone tissue become inflamed.

Osteoporosis *os-tee-o-por´-o-sis* an increased softening of the bone resulting from a gradual reduction in the rate of bone formation while the rate of bone absorption remains normal; a common condition in older people.

Ostium *os´-tium* a small opening into a hollow structure.

Otic *o´-tik* pertaining to the ear.

Otitis media *o-tigh´-tis mee´-dee-ah* middle-ear infection.

Otolith *oh´-to-lith* one of the small calcareous masses in the utricle and saccule of the inner ear.

Ovarian cycle *o-va´-ree-an* the monthly cycle of follicle development, ovulation, and corpus luteum formation in an ovary.

Ovary *o´-va-ree* the female sex organ in which ova (eggs) are produced.

Ovulation *ov-u-lay´-shun* the maturation and release of an ovum.

Ovum *o´-vum* the female gamete (germ cell); an egg cell.

Oxidation *oks-i-day´-shun* the loss of electrons by a molecule; the process of substances combining with oxygen.

Oxygenated *oks´-i-je-nay-ted* the condition in which a substance is saturated with oxygen.

Oxygen debt *oks´-i-jen* the volume of oxygen required after exercise in excess of oxygen consumption in the resting state for an equivalent length of time.

Oxyhemoglobin *oks-i-hee´-mo-glo-bin* oxidized hemoglobin.

Oxytocin *oks-i-to´-sin* a hormone released by the posterior pituitary that stimulates contractility of smooth muscles of the uterus during labor and myoepithelial cells surrounding the alveoli of the mammary glands during lactation.

P

Palate *pal´-et* the roof of the mouth.

Palmar *pal´-mar* the anterior surface of the hands.

Palpation *pal-pay´-shun* examination by touch.

Palpebral fissure *pal´-pe-bral* the gap between the upper and lower eyelids.

Pancreas *pan´-kree-as* the gland located behind the stomach, between the spleen and the duodenum, producing both endocrine and exocrine secretions.

Pancreatic juice *pan-kree-at´-ik* a secretion of the pancreas containing enzymes for digestion of carbohydrates.

Pancreatitis *pan-kree-a-tigh´-tis* an inflammation of the pancreas.

Papilla *pa-pil´-lah* a small elevation, nipple-shaped or cone-shaped.

Papillary muscles *pa´-pil-lar-ee* cone-shaped muscles such as those that project into the cardiac ventricular lumen.

Paralysis *pa-ral´-i-sis* the loss of muscle function or of sensation.

Paraplegia *par-a-plee´-gee-ah* paralysis of the lower limbs.

Parasympathetic division *par-a-sim-pa-thet´-ik* a division of the autonomic nervous system; also referred to as the craniosacral division.

Parathyroid glands *par-a-thigh´-roid* the several small endocrine glands posterior to the capsule of the thyroid gland.

Parathyroid hormone the hormone released by the parathyroid glands that regulate blood-calcium level.

Parenchyma *par-en´-ki-mah* the functional components of an organ.

Parenteral *par-en´-ter-al* occurring through some route other than the alimentary canal, such as intravenous.

Parietal *par-eye´-i-tal* pertaining to the walls of a cavity.

Parotid *pa-rot´-id* located near the ear.

Parturition *par-toor-ish´-un* the act of giving birth.

Patella *pa-tel´-ah* the kneecap.

Pathogenesis *path-o-jen´-e-sis* the development of a disease.

Pectoral *pek´-to-ral* pertaining to the chest.

Pedicle *ped´-i-kul* the portion of the neural arch between the centrum and the transverse process.

Peduncle *pe-dun´-kal* a stalk of fibers, especially that connecting the cerebellum to the pons, mesencephalon, and medulla oblongata.

Pelvis a basin-shaped structure, especially the lower portion of the body's trunk.

Penis *pee´-nis* the male organ of copulation and urinary excretion.

Pepsin an enzyme capable of digesting proteins in an acid pH.

Peptide bond *pep´-tide* a bond joining the amino group of one amino acid to the acid carboxyl group of a second amino acid with the loss of a water molecule.

Pericardium *per-i-kar´-dee-um* the closed membranous sac enveloping the heart.

Perichondrium *per-i-kon´-dree-um* a fibrous, connective-tissue membrane covering the external surface of cartilaginous structures.

Perimysium *per-i-mis´-ee-um* the connective tissue enveloping bundles of muscle fibers.

Perineum *per-i-nee´-um* that region of the body extending from the anus to the scrotum in males and from the anus to the vulva in females.

Periosteum *per-ee-os´-tee-um* a double-layered connective tissue that covers, invests, and nourishes the bone.

Peripheral nervous system (PNS) *per-if´-er-al* a system of nerves that connect the outlying parts of the body and their receptors with the central nervous system.

Peripheral resistance the impedance to blood flow offered by the systemic blood vessels.

Peristalsis *per-is-tal´-sis* the progressive wave of contraction seen in tubes.

Peritoneum *per-i-ton-ee´-um* the membrane lining of the interior of the abdominal cavity.

Peritonitis *per-i-ton-eye´-tis* an inflammation of the peritoneum.

Permeability *per-mee-a-bil´-i-tee* the property of membranes that permits transit of molecules and ions.

Peroneal *per-o-ne´-al* pertaining to the outer side of the leg.

Peroxisome *per-ox´-i-some* a membrane-bounded organelle in cells that contains oxidase and peroxidase; also called microbody.

pH the symbol for hydrogen ion concentration; a measure of the relative acid or base level of a substance or solution.

Phagocyte *fag´-o-site* a cell capable of engulfing and digesting particles or cells harmful to the body.

Phagocytosis *fag-o-sigh-to´-sis* the ingestion of foreign solids by cells.

Phalanges *fay-lan´-jeez* the bones of the finger or toe.

Phantom pain a phenomenon whereby a person who has undergone amputation continues to feel pain from the amputated body part.

Pharynx *far´-inks* the muscular, membranous tube extending from the oral cavity to the esophagus.

Phenotype *fee´-no-tipe* the observable characteristics of an individual as determined both genetically and environmentally.

Phlebitis *fle-by´-tis* an inflammation of a vein.

Photoreceptor *fo-to-ree-sep´-tor* the specialized receptor cells that can convert light energy into a nerve impulse.

Physiology *fiz-ee-ol´-o-jee* the science of the functions of organic beings.

Pinna *pin´-nah* the irregularly shaped elastic cartilage covered with skin forming the most prominent portion of the outer ear.

Pinocytosis *pi-no-sigh-to´-sis* the engulfing of liquid by cells.

Pituitary gland the gland located beneath the brain that serves a variety of functions including regulation of the gonads, thyroid, adrenal cortex, and other endocrine glands.

Placenta *pla-sen´-tah* the organ to which the embryo attaches by the umbilical cord for nourishment and waste removal; has an endocrine function as well.

Placode *plak´-ode* an ectodermal thickening in the early embryo from which a sense organ or structure develops.

Plantar *plan´-tar* pertaining to the sole of the foot.

Plasma the fluid portion of the blood or lymph.

Plasma cells *plaz´-ma* cells that produce antibodies.

Platelet *plate´-let* one of the components of blood; involved in clotting.

Pleura *ploor´-ah* the membrane covering the lungs.

Pleurisy *plur´-i-see* prolonged inflammation of the visceral and parietal pleuras, making breathing painful.

Plexus *plek´-sus* a network of interlacing nerves or anastomosing blood vessels or lymphatics.

Plica *ply´-ka* a fold.

Pneumothorax *noo-mo-thor´-aks* the presence of air or gas in a pleural cavity.

Podocyte *pod´-o-site* an epithelial cell located on the basement membrane of the glomerulus, spreading thin cytoplasmic projections over the membrane.

Polar body a minute cell given off by the ovum during maturation divisions.

Polarized *po´-lar-ized* the state of an unstimulated neuron in which the inside of the cell is relatively negative in comparison to the outside.

Poliomyelitis *po´-lee-o-my-a-light´-us* the viral destruction of nerve cell bodies within the anterior horns of the spinal cord.

Polycythemia *pol-ee-sigh-theem´-ee-a* the presence of an abnormally large number of erythrocytes in the blood.

Polypeptide *pol-ee-pep´-tide* a small chain of amino acids.

Polyribosome *pol-ee-ribe´-o-some* a multiple structure composed of several ribosomes held together by a molecule of messenger RNA; also called polysome.

Polysome *pol´-ee-some* see *polyribosome.*

Pons (1) any bridgelike structure or part; (2) the structure connecting the cerebellum with the brain stem providing linkage between upper and lower levels of the central nervous system.

Positive feedback feedback that tends to cause the level of a variable to change in the same direction as an initial change.

Postganglionic (postsynaptic) neuron *post-gang-lee-on´-ik* a neuron of the autonomic nervous system having its cell body in a ganglion with the axon extending to an organ or tissue.

Preganglionic (presynaptic) neuron *pree-gang-lee-on´-ik* a neuron of the autonomic nervous system having its cell body in the brain or spinal cord and its axon terminating in a ganglion.

Prepuce *pree´-puce* the loose fold of skin that covers the glans penis or glans clitoris.

Pressoreceptor *pre-so-ree-sep´-tor* a nerve ending in the wall of the carotid sinus and aortic arch sensitive to vessel stretching.

Prime movers those muscles whose contractions are primarily responsible for a particular movement.

Process (1) a prominence or projection; (2) a series of actions or method of action for a specific purpose.

Proctologic *prok-to-loj´-ik* pertaining to the rectum or anus.

Progesterone *pro-jes´-ter-own* a hormone responsible for preparing the uterus for the fertilized ovum.

Pronation *pro-nay´-shun* the inward rotation of the forearm causing the radius to cross diagonally over the ulna—palms face posteriorly.

Prone refers to a body lying horizontally with the face downward.

Pronucleus *pro-noo´-clee-us* one of two nuclear bodies (one male and one female) of a newly fertilized ovum, the fusion of which results in formation of a cleavage nucleus.

Proprioceptor *pro-pree-o-sep´-tor* a receptor located in a muscle, tendon, joint, or vestibular apparatus of the internal ear; concerned with locomotion and posture.

Prostaglandin *pros-ta-glan´-din* a substance included in seminal vesicle secretion thought to facilitate fertilization; causes uterine contractions.

Protein *pro´-teen* a complex nitrogenous substance found in various forms in animals and plants as the principal component of protoplasm.

Proteinuria *pro-te-in-oo´-ree-ah* the passage of albumin or other protein in the urine.

Proton *pro´-ton* a particle carrying a positive charge; located in the nucleus of an atom.

Protrude *pro-trood´* to project or assume an abnormally prominent position.

Proximal *proks´-i-mal* toward the attached end of a limb or the origin of a structure.

Psoriasis *so-rye´-uh-sis* a chronic inflammatory skin disease characterized by development of red patches covered with silvery, overlapping scales.

Ptyalin *tie´-a-lin* a starch-splitting enzyme contained in saliva.

Puberty *pu´-ber-tee* the period at which reproductive organs become functional.

Pulmonary *pull´-muh-na-ree* pertaining to the lungs.

Pulmonary circuit the circulatory vessels of the lungs.

Pulmonary edema *e-dee´-mah* an effusion of fluid into the air sacs and interstitial tissue of the lungs.

Pulse the rhythmic expansion in arteries resulting from heart contraction; can be felt from the outside of the body.

Pupil an opening in the center of the iris through which light enters the eye.

Purkinje fibers *per-kin´-jee* the modified cardiac muscle fibers of the conduction system of the heart.

Pus the fluid product of inflammation composed of white blood cells, the debris of dead cells, and a thin fluid.

Pyelonephritis *pie-el-o-nef-rye´-tis* an inflammation of the kidney pelvis and surrounding kidney tissues.

Pyloric glands *pie-lor´-ik* the glands that secrete thin mucus; located in the region of the pylorus of the stomach.

Pyloric region the final portion of the stomach preceding the duodenum.

Pyramid any conical eminence of an organ, especially a body of longitudinal fibers on each side of the anterior median fissure of the medulla oblongata.

Pyrogen *pie´-ro-jen* an agent that induces fever.

Q

Quadriplegia *quad-ri-plee´-jee-ah* the paralysis of all four limbs.

R

Radiate *ray´-dee-ate* diverging from a central point.

Ramus *ray´-mus* a branch of a nerve, artery, vein, or bone, especially a primary division.

Receptor *ree-sep´-tor* a peripheral nerve ending in the skin; specialized for response to particular types of stimuli.

Recessive *ree-seh´-siv* an allele, or the corresponding trait, that is expressed only in homozygous cells or organisms.

Reduction the gain of electrons by a molecule.

Reflex automatic, stereotyped reactions to stimuli.

Refracted bent.

Refracting media substances that bend light.

Refractory period *ree-frak´-to-ree* the period of resistance to stimulation immediately after responding.

Relative refractory period the period following stimulation during which only a stronger than usual stimulation can evoke an action potential.

Renal *ree´-nal* pertaining to the kidney.

Renal calculus *ree´-nal kal´-ku-lus* a kidney stone formed in the renal pelvis or urinary bladder.

Renin *ren´-in* a substance released by the juxtaglomerular complex of the kidneys; involved with raising blood pressure.

Rete *ree´-tee* a network; often composed of nerve fibers or blood vessels.

Reticulocyte *re-tik´-u-lo-site* a nonnucleated, young erythrocyte.

Reticulum *re-tik´-u-lum* a fine network.

Retract to draw back, shorten, contract.

Rhinencephalon *rye-nen-sef´-a-lon* that portion of the cerebrum concerned with reception and integration of olfactory impulses.

Rhodopsin *ro-dop´-sin* the photopigment contained in the rods of the retina.

Ribonucleic acid (RNA) *rye-bo-nu-kle´-ik* the nucleic acid that contains ribose.

Ribosomes *rye´-bo-somes* the cytoplasmic structures at which proteins are synthesized.

Rickets *rik´-ets* a disease occurring in infants and young children characterized by softening of the bone caused by demineralization from malnutrition.

Rods one of the two types of photosensitive cells in the retina.

Roentgenogram *rent-gen´-o-gram* an x-ray film.

Rotate to turn about an axis.

Rugae *ru´-je* elevations or ridges, as in the mucosa of the stomach, uterus, and palate.

S

Sacral *sa´-kral* the lower portion of the back, just superior to the buttocks.

Sagittal plane *saj´-i-tal* a longitudinal section that divides the body or any of its parts into right and left portions.

Saliva *sa-ligh´-vah* the combined secretions of salivary and mucous glands of the mouth.

Salt a compound that, when dissolved in water, dissociates into cations other than hydrogen ions and anions other than hydroxide ions.

Sarcoplasmic reticulum *sar-ko-plas´-mik re-tik´-u-lum* the membranous network running through skeletal muscle cells.

Sclera *skleh´-rah* the firm, fibrous outer layer of the eyeball; functions for protection and maintenance of eyeball shape.

Scoliosis *sko-lee-o´-siss* a lateral curve in the vertebral column.

Scrotum *skro´-tum* the two-layered sac enclosing the testes.

Sebaceous glands *se-bay´-shus* glands that develop from and empty their sebum secretion into hair follicles.

Sebum *see´-bum* the secretion of sebaceous glands; oily substance rich in lipids.

Secretion *see-kree´-shun* the passage of material formed by a cell from the inside to the outside of its plasma membrane.

Segmentation *seg-men-tay´-shun* the process of cleavage or splitting; the division of an organism into somites.

Semen *see´-men* the fluid produced by male reproductive structures; contains sperm.

Semilunar valves *sem-i-loo´-ner* valves that prevent blood return to the ventricle after contraction.

Seminiferous tubules *sem-i-nif´-er-us* highly convoluted tubes within the testes containing spermatocytes.

Sensory nerve a nerve that contains only the processes of sensory neurons and carries nerve impulses to the central nervous system.

Sensory nerve cell an initiator of nerve impulses following receptor activity.

Serous fluid *ser´-us* a clear, watery fluid secreted by the cells of the mesothelium.

Serum *se´-rum* the amber-colored fluid that exudes from clotted blood as the clot shrinks.

Sesamoid *ses´-a-moid* denoting a small bone that is embedded in a tendon or in a joint capsule.

Sex chromosome *kro´-mo-some* chromosome that determines sex of the fertilized egg; X and Y chromosomes.

Sinoatrial node *sigh-noo-ay´-tree-al* the mass of specialized myocardial cells in the wall of the right atrium.

Sinus *sigh´-nus* (1) a mucous-membrane-lined, air-filled cavity in certain cranial bones; (2) a dilated channel for the passage of blood or lymph, which lacks the coats of an ordinary vessel.

Smooth (nonstriated) muscle muscle consisting of spindle-shaped, unstriped (nonstriated) muscle cells; involuntary muscles.

Solute *sol´-yoot* the dissolved substance in a solution.

Solution a homogenous mixture of two or more components.

Somatic nervous system *so-mat´-ik* a division of the peripheral nervous system; also called the voluntary nervous system.

Somite *so´-mite* a segment of the body of an embryo.

Sperm that mature male germ cell, a spermatozoon.

Spermatogenesis *sper-mat-o-jen´-e-sis* the process of meiosis (cell division) in the male to produce mature male germ cells.

Sphenoid *sfen´-oid* wedgelike.

Sphincter *sfink´-ter* a muscle surrounding and enclosing an orifice.

Sprain the wrenching of a joint producing stretching or laceration of the ligaments.

Squamous *skway´-mus* pertaining to flat and thin cells that form the free surface of epithelial tissue.

Stagnant hypoxia *stag´-nant high-pock´-see-uh* a condition marked by reduced available oxygen caused by slowed blood circulation.

Stasis *stay´-sis* (1) a decrease or stoppage of flow; (2) a state of equilibrium.

Static balance *stat´-ic* balance concerned with changes in the position of the head.

Stenosis *sten-o´-sis* constriction or narrowing.

Steroids *ster´-oidz* a specific group of chemical substances including certain endocrine secretions and cholesterol.

Stimulus *stim´-u-lus* an excitant or irritant; an alteration in the environment of a living thing producing a response.

Stomodeum *sto-mo-dee´-um* the primitive oral cavity of the embryo.

Strabismus *stra-biz´-mus* a squint; that abnormality of the eyes in which the visual axes do not meet at the desired objective point.

Stratum *stray´-tum* a layer.

Stress any stimulus that directly or indirectly causes neurons of the hypothalamus of the brain to release corticotropin releasing hormone (CRH).

Straited muscle *stry´-ay-ted* muscle consisting of cross-straited (cross-striped) muscle fibers.

Stricture *strik´-chur* a contraction or inward pinching of a canal or duct.

Stroke a condition in which a cerebral blood vessel is blocked.

Stroke volume a volume of blood ejected by the left ventricle during a systole.

Stroma *stro´-mah* the supporting framework of an organ including connective tissue, vessels, and nerves.

Sty an inflammation of the connective tissue of the eyelid, near a hair follicle.

Subcutaneous *sub-ku-tay´-nee-us* beneath the skin.

Sublingual *sub-ling´-gwal* located beneath the tongue.

Sulcus *sul´-kus* a furrow or linear groove; on the brain, a less deep depression than a fissure.

Summation *sum-may´-shun* the accumulation of effects, especially those of muscular, sensory, or mental stimuli.

Superficial (external) located close to or on the body surface.

Superior refers to the head or upper; higher.

Supination *soo-pa-nay´-shun* the outward rotation of the forearm causing palms to face anteriorly.

Supine refers to a body lying with the face upward.

Surface tension the contractile surface of a liquid or structure by which it tends to present the least possible surface; liquid meniscus formation.

Surfactant *sur-fact´-ant* a substance on pulmonary alveoli walls which reduces surface tension thus preventing collapse.

Suspension *sus-pen´-shun* a dispersing of particles throughout a body of liquid.

Suspensory ligament of an eye *sus-pen´-so-ree lig´-a-ment* fibrous ligament that holds the lens in place in the eye.

Sutures *soo´-churz* the immovable joints that connect the bones of the adult skull.

Sweat glands the glands that secrete a watery solution of sodium chloride (salt water).

Sympathetic division a division of the autonomic nervous system; opposes parasympathetic functions.

Synapse *sin´-apse* the region of communication between neurons.

Synaptic cleft *sin-ap´-tik* the space at a synapse between neurons.

Synaptic delay *sin-ap´-tik* the time required for an impulse to cross a synapse between two neurons.

Synarthrosis *sin-ar-throw´-sis* a fibrous joint; two types: sutures and syndesmoses.

Synergist *sin´-er-jist* a muscle cooperating with another to produce a movement neither alone can produce.

Synostosis *sin-os-tow´-sis* a union of originally separate bones by osseous material.

Synovial fluid *sa-no´-vee-al* a fluid secreted by the synovial membrane; lubricates joint surfaces and nourishes articular cartilages.

System a group of organs that function cooperatively to accomplish a common purpose; there are eleven major systems in the human body.

Systemic *sis-tem´-ik* general, pertaining to the whole body.

Systemic circuit the circulatory vessels of the body.

Systemic edema *e-dee´-ma* an accumulation of fluid in body organs or tissues.

Systole *sis´-ta-lee* the contraction phase of a cardiac cycle.

Systolic pressure *sis-tol´-ik* the pressure generated by the left ventricle during systole.

T

T cells a heterogeneous group of lymphocytes that include those cells committed to participating in cell-mediated immune responses.

Tachycardia *tak-ee-kar´-dee-ah* the abnormal, excessive rapidity of heart action; over 100 beats per minute.

Tarsals *tar´-suls* the seven bones that form the ankle and heel.

Taste buds the receptors for taste on the tongue, roof of mouth, pharynx, and larynx.

Telencephalon *tel-en-sef´-uh-lon* the anterior subdivision of the primary forebrain that develops into olfactory lobes, cerebral cortex, and corpora striata.

Temporal summation *tem´-po-ral sum´-may-shun* the arrival of many nerve impulses at a synapse within a short period causing release of sufficient transmitters to initiate a nerve impulse in the postsynaptic cell.

Tendon *ten´-don* a band of dense fibrous tissue forming the termination of a muscle and attaching the muscle to a bone.

Testis *tes´-tis* the male primary sex organ that produces sperm.

Tetanus *tet´-a-nus* (1) the tense, contracted state of a muscle; (2) an infectious disease.

Thalamus *thal´-a-muss* the mass of gray matter at the base of the brain.

Theca *thee´-ka* a sheath.

Thermoreceptor *ther-mo-ree-sep´-tor* a receptor sensitive to temperature changes.

Thoracic *tho-ras´-ik* refers to the chest.

Thorax *tho´-raks* that portion of the trunk above the diaphragm and below the neck.

Threshold the lower limit of stimulus capable of producing an impression on consciousness or evoking a response in an irritable tissue.

Thrombin *throm´-bin* an enzyme that induces clotting by converting fibrinogen to fibrin.

Thrombocyte *throm´-bo-site* a blood platelet thought to be part of the blood-clotting mechanism.

Thrombocytopenia *throm-bo-sigh-to-pee´-nee-ah* a condition in which there is a decrease in the number of blood platelets below normal.

Thrombophlebitis *throm-bo-fle-by´-tis* an inflammation of a vein associated with blood-clot formation.

Thrombus *throm´-bus* a clot that is fixed or stuck to a vessel wall.

Thymus gland *thigh´-mus* a potential source of hormonal material; active in immune response.

Thyroid gland *thigh´-roid* one of the largest of the body's endocrine glands.

Tissue a group of similar cells and fibers forming a distinct structure.

Tolerance *tol´-er-antz* the failure of the body to mount a specific immune response against a particular antigen.

Tomography *toe-mog´-ra-fee* x-ray photography of a specific structure in a certain layer of tissue in the body, in which images of structures in other layers are eliminated.

Tonic *ton´-ik* refers to the state of continuous muscular or neuron activity.

Tonofibril *tone-o-figh´-bril* one of the fine fibrils in the cytoplasm of epithelial cells.

Toxemia *tok-see´-mee-ah* a condition in which blood contains poisonous products.

Toxic *tok´-sik* poisonous.

Trabecula *tra-bek´-u-lah* any one of the fibrous bands extending from the capsule into the interior of an organ.

Trachea *tray´-kee-uh* the windpipe; the cartilaginous and membranous tube extending from larynx to bronchi.

Tract a collection of nerve fibers having the same origin, termination, and function.

Trait *trate* a characteristic or quality of an individual.

Transducer *trans-doo´-ser* an agent that converts energy forms; for example, receptors of the nervous system transfer light into a nerve impulse.

Transmutation *trans-mu-tay´-shun* the change of one element into another.

Transverse processes the projections that extend laterally from each neural arch.

Trauma *traw´-mah* an injury, wound, or shock; usually produced by external forces.

Trisomy *trigh´-so-mee* the condition in which one chromosome is represented three times rather than the normal two.

Trochanter *tro-kan´-tur* a large, somewhat blunt process.

Trophic *tro´-fik* pertains to nutrition.

Trophoblast *trof´-o-blast* outer sphere of cells of the blastocyst.

Trypsin *trip´-sin* an active enzyme that splits proteins.

Tubal ligation *too´-bal li-gay´-shun* a form of surgical sterilization in which the uterine tubes are tied and severed.

Tubal pregnancy an ecotopic pregnancy that occurs within a uterine tube.

Tubercle *too´-bur-kil* a nodule or small rounded process.

Tuberosity *too-bur-os´-i-tee* a broad process, larger than a tubercle.

Tumor an abnormal growth of cells; a swelling; cancerous at times.

Tunica *too´-ni-kah* a covering or tissue coat; membrane layer.

Twitch a brief contraction of muscle in response to a stimulus.

Tympanic membrane *tim-pan´-ik* the eardrum.

U

Ulcer *ul´-ser* a lesion or erosion of the mucous membrane, such as gastric ulcer of stomach.

Umbilical cord *um-bil´-i-kal* a structure bearing arteries and veins connecting the placenta and the fetus.

Umbilicus *um-bil´-i-kus* the navel; marks site that gave passage to umbilical vessels in the fetal stage.

Unipennate *yoo-na-pen´-it* muscles in which all fasciculi insert onto one side of the tendon.

Unipolar neurons *yoon-i-pol´-ar* neurons in which embryological fusion of the two processes leaves only one process extending from the cell body.

Unitary smooth muscle muscle in which the cells contact one another at gap junctions and the impulses spread from cell to cell.

Unmyelinated fibers *un-my´-e-li-nay-ted* nerve axons that are not covered with myelin.

Urea *yoo-ree´-ah* the main nitrogen-containing waste excreted in the urine.

Uremia *yoo-ree´-mee-ah* a toxic accumulation in the blood of substances normally excreted by the kidneys.

Ureter *yoo-ree´-ter* the tube that carries urine from kidney to bladder.

Urethra *yoo-ree´-thrah* the canal through which urine passes from the bladder to the outside of the body.

Uvula *yoo´-vu-la* conical appendix hanging from soft palate.

V

Vacuole *vak´-yoo-ole* a clear space in a cell.

Valvular insufficiency *val´-vu-lar* a condition in which the cusps of the cardiac valves do not close tightly.

Varicose vein *var-uh-kos´* a dilated, knotted, tortuous blood vessel.

Vas a duct; vessel.

Vasa recta *va´-sa rek´-ta* capillary branches that supply loops of Henle and collecting ducts.

Vascular *vas´-ku-lar* pertaining to channels or vessels.

Vasoconstriction *vaz-o-kon-strik´-shun* the narrowing of blood vessels.

Vasodilatation *vaz-o-die-lay-tay´-shun* the relaxation of the smooth muscles of the vascular system producing dilated vessels.

Vasomotion *vaz-o-mo´-shun* an increase or decrease in caliber of a blood vessel.

Vasomotor center *vaz-o-mo´-tor* an area of brain concerned with regulation of blood-vessel resistance.

Vasomotor nerve fibers *vaz-o-mo´-tor* the motor nerve fibers that regulate the constriction or dilation of blood vessels.

Vasopressin *vaz-o-pres´-sin* another name for antidiuretic hormone.

Vein a vessel carrying blood away from the tissues toward the heart.

Ventral *ven´-tral* anterior or front.

Ventricle *ven´-tri-kal* (1) a small cavity or pouch; (2) a blood propulsion chamber of the heart.

Ventricles of the brain *ven´-tri-kalz* the cavities in the interior of the brain.

Venule *ven´-yool* a small vein.

Vertigo *ver´-tee-go* dizziness; the feeling of movement such as a sensation that the external environment is revolving.

Vesicle *ves´-i-kal* a small liquid-filled sac or bladder.

Viscera *vis´-ser-ah* the internal organs.

Visceral *vis´-ser-al* pertaining to the internal part of a structure or the internal organs.

Viscosity *vis-koss´-i-tee* the state of being sticky or thick.

Visual acuity *a-kyoo´-i-ty* the ability of the eye to distinguish detail.

Vital capacity the volume of air that can be expelled from the lungs by forcible expiration after the deepest inspiration.

Vitamins the organic compounds required by the body in minute amounts for physiological maintenance and growth.

Voluntary muscle muscle under control of the will; skeletal muscle.

Vulva *vul´-va* female external genitalia.

W

White matter the white substance of the central nervous system; composed of myelinated nerve fibers.

Y

Yolk sac *yoke* endodermal sac that serves as the source for primordial germ cells.

Z

Zonula *zon´-u-la* a small zone, usually beltlike.

Zonula adherens that part of a junctional complex of epithelial cells where the cell membranes are not modified and are separated by a gap of 200Å.

Zonula occludens that part of a junctional complex of epithelial cells where the outer layer of cell membranes of adjacent tissues fuse.

Zygote *zy´-goat* the fertilized ovum before splitting (cleavage); produced by union of two gametes.

INDEX

A

B

D

E

F

G

H

I

J

K

L

M

O

Q

R

S

T

U

V

W

X

Y

Z

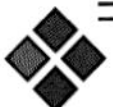

CREDITS

Chapter Opener Photographs

Chapter 1, page 1: Matisse, Henri. *Grand Nu Assis.* Bronze. 1925.
The Minneapolis Institute of Arts.

Chapter 2, page 22: Moore, Henry. *Family Group.* 1949.
The Granger Collection, New York.

Chapter 3, page 58: African art. Bena Lulua, Congo. Wooden statue of a man.
The Granger Collection, New York.

Chapter 4, page 108: Nigeria. Benin art. Female statue.
Giraudon/Art Resource, New York.

Chapter 5, page 132: Mexico. Nayarit. *Seated Ball Player.* Polychromed ceramic. 200 B.C.-A.D. 300.
The Minneapolis Institute of Arts.

Chapter 6, page 152: Rinehart, William H. *Harriet Newcomer.* 1868.
Nefsky/Art Resource, New York.

Chapter 7, page 216: Trova, Ernest. *Study: Falling Man (Wheel Man).* Silicon bronze. 1965.
Collection, Walker Art Center Minneapolis, Gift of the T. B. Walker Foundation, 1965.

Chapter 8, page 242: Degas, Edgar. *Dancer Putting on Her Stocking.* Bronze. ca. 1900.
The Minneapolis Institute of Arts.

Chapter 9, page 278: Ivory Coast, Africa. Primordial couple. Probably from the Senufo culture. Wood.
Werner Forman Archive.

Chapter 10, page 334: Houdon. *Diana as Huntress.* 1790.
Giraudon/Art Resource, New York.

Chapter 11, page 356: Egyptian painted limestone statuette of a seated scribe. Vth Dynasty. ca. 2500 B.C. From Sakkara.
The Granger Collection, New York.

Chapter 12, page 382: Moore, Henry. *King and Queen.* 1952-53.
Scala/Art Resource, New York.

Chapter 13, page 426: Mexico. Nayarit Standing Figure. Polychromed ceramic. 200 B.C.-A.D. 300.
The Minneapolis Institute of Arts.

Chapter 14, page 454: India. Vijayanagara Parvati. Bronze.
The Minneapolis Institute of Arts.

Chapter 15, page 468: Rodin. *The Thinker.*
Vanni/Art Resource, New York.

Chapter 16, page 486: Egypt. 18th Dynasty. Young woman carrying an amphora. Wood.
Giraudon/Art Resource, New York.

Chapter 17, page 536: Guinea. Baga people. Male and female figures.
Aldo Tutino/Art Resource, New York.

Chapter 18, page 570: Chinese stoneware with colored glazes. Liao Dynasty. A.D. 907-1125.
The Granger Collection, New York.

Chapter 19, page 592: Indian art. Dancing girl, from Mohenjo-Daro. Bronze. ca. 2500 B.C.
The Granger Collection, New York.

Chapter 20, page 626: Egyptian limestone statuette of the chief scribe Raherka and his wife Mersankh. Vth Dynasty.
The Granger Collection, New York.

Chapter 21, page 674: Sumaria. 2150 B.C.
Scala/Art Resource, New York.

Chapter 22, page 686: Austria. The Venus of Willendorf. Late Paleolithic stone statuette.
The Granger Collection, New York.

Chapter 23, page 716: Ivory Coast, Africa. Probably an ancestor figure. Baule culture.
Werner Forman Archive.

Chapter 24, page 754: Mayoide style. Nude female figure.
Scala/Art Resource, New York.

Chapter 25, page 802: Moore, Henry. *Recumbent Figure.* 1938.
Tate Gallery, London/Art Resource, New York.

Chapter 26, page 832: Japan. Kamakura Period. Nio Guardian Figure, Naraen Kongo. Hinoki wood. ca. 1360.
The Minneapolis Institute of Arts.

Chapter 27, page 866: Reggio Calabria. Bronze.
Scala/Art Resource, New York.

Chapter 28, page 884: Nippur, Bagdad. Statue of man and woman.
Scala/Art Resource, New York.

Chapter 29, page 924: Harati Gandhara. Third century.
Edward Owen/Art Resource, New York.

Interior Photographs

Figure 1.1a, page 3: Simon Fraser/Science Photo Library/Photo Researchers, Inc.

Figure 1.1b, page 3: Simon Fraser/Science Photo Library/Photo Researchers, Inc.

Figure 1.2a, page 4: Reproduction with permission from the JIRA.

Figure 1.2b, page 4: Reproduced with permission from S. M. Jorgensen et al.: The dynamic spatial reconstructor: a high speed, stop action, 3-D, digital radiographic image of moving internal organs and blood. In: Shaw L. L., Jaanimagi, P. A., Neyer, B. T. (Eds.); Ultrahigh- and High-Speed Photography, Videograph, Photonics, and Velocimetry '90. Proceedings of SPIE—The International Society for Optical Engineering 1346: 180–191, 1990.

Figure 1.3, page 5: St. Bartholomew's Hospital/Science Photo Library/Photo Researchers, Inc.

Figure 1.4, page 5: NIH/Science Source/ Photo Researchers, Inc.

Figure 1.6, page 9: © David Young-Wolff/PhotoEdit.

Figure 1.8, page 12: © David Young-Wolff/PhotoEdit.

Figure 3.10a, page 78: © D. W. Fawcett/Photo Researchers, Inc.

Figure 3.15b, page 85: © K. G. Murti/Visuals Unlimited.

Figure 3.16b, page 86: © David M. Phillips/Visuals Unlimited.

Figure 3.17a, page 87: © David M. Phillips/Visuals Unlimited.

Figure 3.19b, page 89: © G. Musil/Visuals Unlimited.

Figure 3.20, page 90: © M. Schliwa/Visuals Unlimited.

Figure 3.21b, page 91: © K. G. Murti/Visuals Unlimited.

Figure 3.22, page 92: © M. Schliwa/Visuals Unlimited.

Figure 3.23a1, page 93: © David M. Phillips/Visuals Unlimited.

Figure 3.23b1, page 93: ©Fawcett/de Harven/Kalnins/Photo Researchers, Inc.

Figure 3.29a, page 98: © Dr. R. Verma/Phototake.

Figure 3.29b, page 98: © CNRI/Phototake.

Figure 4.1a, page 111: © J. F. Gennaro and L. R. Grillone, Photo Researchers, Inc.

Figure 4.1b, page 111: © David M. Phillips/The Population Council/Science Source/Photo Researchers, Inc.

Figure 4.2a, page 113: © Ed Reschke/Peter Arnold, Inc.

Figure 4.3a, page 113: © Fred Hossler/Visuals Unlimited.

Figure 4.4a, page 114: © Ed Reschke/Peter Arnold, Inc.

Figure 4.5a, page 115: © Biophoto Associates/Photo Researchers, Inc.

Figure 4.6a, page 115: © Biophoto Associates/Photo Researchers, Inc.

Figure 4.7a, page 116: © Manfred Kage/Peter Arnold, Inc.

Figure 4.8a, page 116: © Dr. Mary Notter/Phototake.

Figure 4.12a, page 120: © Biophoto Associates/Science Source.

Figure 4.13a, page 121: © Carolina Biological/Visuals Unlimited.

Figure 4.14a, page 122: © Cabisco/Visuals Unlimited.

Figure 4.15a, page 123: © Ed Reschke/Peter Arnold, Inc.

Figure 4.16a, page 123: © Ed Reschke/Peter Arnold, Inc.

Figure 4.17a, page 124: © Biophoto Associates/Science Source.

Figure 4.18a, page 124: © Ed Reschke/Peter Arnold, Inc.

Figure 4.19a, page 125: © Biology Media/Photo Researchers, Inc.

Figure 4.20a, page 126: © Ed Reschke/Peter Arnold, Inc.

Figure 4.21a, page 126: © John D. Cunningham/Visuals Unlimited.

Figure 4.22a, page 127: © Dr. R. Kessel/Peter Arnold, Inc.

Figure 4.23a, page 128: © John Cunningham/Visuals Unlimited.

Figure 4.24a, page 128: © Ed Reschke/Peter Arnold, Inc.

Figure 5.1, page 134: © Ed Reschke/Peter Arnold, Inc.

Figure 5.4, page 140: © David Scharf/Peter Arnold, Inc.

Figure C5.1, page 145: © Leonard Kamsler/Medichrome.

Figure 6.2b, page 156: L. V. Bergman & Associates, Inc.

Figure 6.4, page 158: © Ed Reschke/Peter Arnold, Inc.

Figure 6.5, page 158: © Biophoto Associates/Photo Researchers, Inc

Figure 6.7a, page 160: © CNRI/Phototake.

Figure 6.10b, page 163: © George J. Wilder/Visuals Unlimited.

Figures C6.2a, C6.2b, C6.2c, page 166: Courtesy of Dr. Henry H. Jones, MD, Stanford University Medical Center.

Figure 6.17b, page 175: © John Watney/Photo Researchers, Inc.

Figure 6.25, page 180: Custom Medical Stock Photo: Bates.

Figure 6.28b, page 185: © SIU/Custom Medical Stock Photo.

Figure 6.38c, page 196: James Stevenson/Science Photo Library/Photo Researchers, Inc.

Figure 6.39b, page 197: © Lester V. Bergman & Associates, Inc.

Figure 6.40b, page 200: © Biophoto Associates/Photo Researchers, Inc.

Figure 6.43c, page 204: Lester V. Bergman & Associates, Inc.

Figure 6.44c1, page 205: Lester V. Bergman & Associates, Inc.

Figure 6.44c2, page 205: Lester V. Bergman & Associates, Inc.

Figure 6.45, page 206: © Dr. Howells/Science Photo Library/Photo Researchers, Inc.

Figure 6.46c, page 207: © Lester V. Bergman & Associates, Inc.

Figure 6.47c, page 208: © Susan Leavines/Photo Researchers, Inc.

Figure 7.8a, page 224: © David Young-Wolff/PhotoEdit.

Figure 7.14b, page 232: Courtesy of Dr. H. Griffiths.

Figure C7.2, page 236: Princess Margaret Rose Orthopaedic Hospital/Science Photo Library/Photo Researchers, Inc.

Figure C7.3, page 237: David York/Medichrome.

Figure 8.4a, page 248: Clara Franzini-Armstrong/University of Pennsylvania.

Figure 10.9b, page 343: © Dennis Kunkel/CNRI/Phototake.

Figure 10.13, page 345: © David M. Phillips/Visuals Unlimited.

Figure 10.15b, page 347: © Manfred Kage/Peter Arnold, Inc.

Figure 10.17a1, page 352: © Biophoto Associates/Science Source/Photo Researchers, Inc.

Figure 10.17b1, page 352: John D. Cunningham/Visuals Unlimited.

Figure 10.17c1, page 352: John D. Cunningham/Visuals Unlimited.

Figure 12.1b, page 385: Lester V. Bergman & Associates, Inc.

Figure 12.2b, page 388: © Manfred Kage/Peter Arnold, Inc.

Figure 12.7c, page 392: © Manfred Kage/Peter Arnold, Inc.

Figure 12.9b, page 394: Lester V. Bergman & Associates, Inc.

Figure 12.10b, page 395: CNRI/Phototake.

Figure 12.19c, page 405: Harry J. Przekop, Jr./Medichrome.

Figure 12.24, page 409: Manfred Kage/Peter Arnold, Inc.

Figure 15.2, page 471: Larry Mulvehill/Science Source/Photo Researchers, Inc.

Figure 16.2b, page 490: Dr. D. W. Fawcett/Mizoguti/Photo Researchers, Inc.

Figure 16.16b, page 500: From *Tissues and Organs: A Text-Atlas of Scanning Electron Microscopy.* By Richard Kessel and Randy H. Kardon. Copyright © 1979 by W. H. Freeman and Company. Reprinted with permission.

Figure C16.4, page 509: Science Photo Library/Photo Researchers, Inc.

Figure 16.28c, page 515: © Biophoto Associates/Photo Researchers, Inc.

Figure 16.31, page 517: CNRI/Science Photo Library/Photo Researchers, Inc.

Figure 16.44b, page 529: © Carolina Biological/Visuals Unlimited.

Figure 17.9, page 549: © Carolina Biological/Visuals Unlimited.

Figure 17.17b, page 557: © John D. Cunningham/Visuals Unlimited.

Figure 17.20b, page 560: © Biophoto Associates/Photo Researchers, Inc.

Figure 18.3, page 575: © Stanley Flegler/Visuals Unlimited.

Figure 18.8a, page 579: John D. Cunningham/Visuals Unlimited.

Figure 18.9b, page 584: From *Tissues and Organs: A Text-Atlas of Scanning Electron Microscopy.* By Richard Kessel and Randy H. Kardon. Copyright © 1979 by W. H. Freeman and Company. Reprinted by permission.

Figure 18.11, page 587: © Boehringer Ingelheim International GmbH, photo Lennart Nilsson.

Figure 19.4b, page 598: Lester V. Bergman & Associates, Inc.

Figure 19.5b, page 599: Lester V. Bergman & Associates, Inc.

Figure 19.13a, page 607: CNRI/Science Photo Library/Photo Researchers, Inc.

Figure 20.4, page 631: © R. J. Bolander—D. W. Fawcett/Photo Researchers, Inc.

Figure 20.5, page 632: © Carolina Biological/Visuals Unlimited.

Figure 20.12, page 639: Sheila Terry/Science Photo Library/Photo Researchers, Inc.

Figure 21.3, page 678: © Ida Wyman/PhotoTake.

Figure 21.7, page 682: © Biophoto Associates/Photo Researchers, Inc.

Figure 21.9, page 684: © John D. Cunningham/Visuals Unlimited.

Figure 22.11, page 702: Dr. A. Liepins/Science Photo Library/Photo Researchers, Inc.

Figure 23.7, page 725: © David M. Phillips/Visuals Unlimited.

Figure 24.5, page 759: © Omikron/Science Source/Photo Researchers, Inc.

Figure 24.10b, page 764: © Biophoto Associates/Science Source/Photo Researchers, Inc.

Figure 24.15, page 770: CNRI/Science Photo Library/Photo Researchers, Inc.

Figure 24.17c, page 771: © Biophoto Associates/Photo Researchers, Inc.

Figure 24.22, page 775: From *Tissues and Organs: A Text-Atlas of Scanning Electron Microscopy.* By Richard Kessel and Randy H. Kardon. Copyright © 1979 by W. H. Freeman and Company. Reprinted with permission.

Figure 25.14b, page 821: NIH/Science Source/Photo Researchers, Inc.

Figure 26.6b, page 839: Lester V. Bergman & Associates, Inc.

Figure 26.7, page 840: © G. Shih and R. Kessel/Visuals Unlimited.

Figure 26.8, page 840: © G. Tyron, R. Bulger, D. W. Fawcett/Visuals Unlimited.

Figure 26.12, page 843: © G. Shih and R. Kessel/Visuals Unlimited.

Figure 26.21, page 856: CNRI/Science Photo Library/Photo Researchers, Inc.

Figure 28.8, page 894: © Biophoto Associates/Photo Researchers, Inc.

Figure 28.12a, page 900: David M. Phillips/Visuals Unlimited.

Figure 28.17, page 904: © Biophoto Associates/Photo Researchers, Inc.

Figure 28.22, page 911: © Manfred Kage/Peter Arnold, Inc.

Figure 29.10a, page 936: © Petit Format/Nestle/Photo Researchers, Inc.

Figure 29.10b, page 936: © Garbis Kerimian/Peter Arnold, Inc.

Figure 29.10c, page 936: C. Meitchik/Custom Medical Stock Photo.

Figure 29.10d, page 936: Custom Medical Stock Photo.

Figures A.1–A.24, pages A2–A25: Reproduced by permission from *Photographic Atlas of the Human Body* by Drs. B. Vidić and F. R. Suarez. Copyright © 1984 by The C. V. Mosby Company.

Illustrations

The following illustrations are all reprinted by permission from *Human Physiology: From Cells to Systems.* Copyright © 1989 by West Publishing Company. All rights reserved.

Figure 3.3, page 63. Figure C3.1, page 67.
Figure 3.14, page 82. Figure C3.4, page 84.
Figure 3.15a, page 85. Figure 3.16a, page 86.
Figure 3.19a, page 89. Figure 3.21a, page 91.
Figure 3.24, page 94. Figure 3.28, page 97.
Figure 6.9, page 162. Figure 8.20, page 262.
Figure 8.24, page 270. Figure 10.14, page 346.
Figure 11.3, page 361. Figure 11.5, page 362.
Figure 12.17, page 403. Figure 15.5, page 476.
Figure 15.6, page 476. Figure 16.13, page 497.
Figure 16.14, page 497. Figure 16.17b, page 501.
Figure 16.19, page 503. Figure 16.33, page 517.
Figure 16.37, page 522. Figure 16.38, page 523.
Figure 16.39, page 524. Figure 16.40, page 526.
Figure 16.43, page 528. Figure 17.6, page 545.
Figure 17.7, page 547. Figure 17.11, page 550.
Figure 17.12, page 551. Figure 17.18, page 558.
Figure 17.19, page 559. Figure 17.21, page 561.
Figure 18.4, page 576. Figure 18.8, page 579.
Figure 19.1, page 594. Figure 19.11, page 606.
Figure 19.13, page 607. Figure 20.7, page 634.
Figure 20.10, page 638. Figure 20.11, page 639.
Figure 22.1, page 690. Figure 22.5, page 694.
Figure 22.8, page 698. Figure 23.12, page 730.
Figure 23.14, page 735. Figure 25.15, page 823.
Figure 26.11, page 843. Figure 26.17, page 847.
Figure 26.18, page 848. Figure 28.10, page 898.
Figure 28.11a, page 899. Figure 28.13, page 901.
Figure 28.14, page 901. Figure 28.20, page 909.
Figure 28.21, page 910. Figure 28.23, page 912.
Figure 29.1, page 927. Figure 29.11, page 938.

Tables

The following tables are all reprinted by permission from *Human Physiology: From Cells to Systems* by Lauralee Sherwood. Copyright © 1989 by West Publishing Company. All rights reserved.

Table 3.1, page 61. Table 3.2, page 71.
Table 3.5, page 90. Table 8.2, page 266.
Table 8.3, pages 272–273. Table 11.2, page 363.
Table 12.2, pages 386–387. Table 14.1, page 460.
Table 14.2, page 461. Table 14.5, page 464.
Table 15.1, page 473. Table 15.2, page 475.
Table 16.1, page 504. Table C17.1, page 563.
Table 17.3, page 564. Table 20.1, page 633.
Table 22.3, page 697. Table 22.4, page 703.
Table 22.6, pages 709–712. Table 23.2, page 734.
Table 23.3, page 739. Table C23.1, page 741.
Table 23.4, page 747. Table 24.2, page 777.
Table 26.1, page 850. Table 27.1, page 869.
Table 28.2, page 897.

Aspects of Exercise Physiology

The following Aspects of Exercise Physiology features are all reprinted by permission from *Human Physiology: From Cells to Systems.* Copyright © 1989 by West Publishing Company. All rights reserved.

What Is Exercise Physiology?, page 17.

What the Scales Don't Tell You, page 138.

Osteoporosis: The Bane of Brittle Bones, page 195.

Exercising Muscles Increases Their Uptake of Glucose, page 258.

Adverse Effects of Steroid Use, page 329.

Back Swings and Pre-Jump Crouches: What Do They Share in Common?, page 350.

Relation between Excitatory Postsynaptic Potentials and Muscle Strength, page 373.

Swan Dive or Belly Flop: It's a Matter of CNS Control, page 417.

Loss of Muscle Mass: A Problem of Space Travel, page 443.

The Endocrine Response to the Challenge of Combined Heat and Exercise, page 543.

Blood Doping: Is More of a Good Thing Better?, page 577.

The What, Who, and When of Stress Testing, page 615.

The Ups and Downs of Hypertension and Exercise, page 651.

Exercise: A Help or Hindrance to Immune Defense?, page 708.

How to Find Out How Much Work You're Capable of Doing, page 736.

Pregame Meal: What's In and What's Out?, page 791.

All Fat Is Not Created Equal, page 822.

Aerobic Exercise: What For and How Much?, page 809.

When Protein in the Urine Does Not Mean Kidney Disease, page 849.

Menstrual Cycle Irregularity in Athletes, page 915.